Prealgebra & Introductory Algebra

K. Elayn Martin-Gay

University of New Orleans

Upper Saddle River, New Jersey 07458

Library of Congress Cataloging-in-Publication Data
Martin-Gay, K. Elayn

p. cm.
Includes index.
ISBN 0-13-144972-9

1. Arithmetic. I. Title

CIP data available.

Editor in Chief: *Christine Hoag*
Project Manager: *Mary Beckwith*
Production Editor: *Lynn Savino Wendel*
Vice President/Director of Production and Manufacturing: *David W. Riccardi*
Senior Managing Editor: *Linda Mihatov Behrens*
Executive Managing Editor: *Kathleen Schiaparelli*
Assistant Manufacturing Manager/Buyer: *Michael Bell*
Manufacturing Manager: *Trudy Pisciotti*
Executive Marketing Manager: *Eilish Collins Main*
Marketing Assistant: *Annett Uebel*
Marketing Project Manager: *Barbara Herbst*
Development Editor: *Elka Block*
Editor in Chief, Development: *Carol Trueheart*
Media Project Manager, Developmental Math: *Audra J. Walsh*
Assistant Managing Editor, Math Media Production: *John Matthews*
Assistant Editor: *Christina Simoneau*
Art Directors: *Geoffrey Cassar/Maureen Eide*
Interior Designer: *Circa 86*
Cover Designer: *Jack Robol; Kristine Carney*
Art Editor: *Tom Benfatti*
Creative Director: *Carole Anson*
Director of Creative Services: *Paul Belfanti*
Director, Image Resource Center: *Melinda Reo*
Manager, Rights and Permissions: *Zina Arabia*
Interior Image Specialist: *Beth Brenzel*
Image Permission Coordinator: *Charles Morris*
Photo Researcher: *Elaine Soares*
Cover Art: ©Dale Chihuly, Teal & Orange Persian with Process Lip Wrap. 1992 13 x 23 x 18".
Photo: Claire Garoutte 92/92.738.p1
Art Studio: Artworks
Managing Editor, AV Management & Production: *Patty Burns*
Production Manager: *Ronda Whitson*
Production Technologies Manager: *Matthew Haas*
Illustrators: *Dan Knopsnyder, Scott Wieber, Stacy Smith, Nate Storck, Mark Landis, Ryan Currier, Audrey Simonetti*
Quality Assurance: *Pamela Taylor, Ken Mooney, Tim Nguyen*
Formatting Manager: *Jim Sullivan*
Electronic Production Specialists: *Karen Noferi, Joanne Del Ben, Karen Stephens, Jackie Ambrosius, Vicki Croghan, Julita Nazario*

Pearson Prentice Hall
Pearson Education, Inc., Upper Saddle River, New Jersey 07458

Printed in the United States of America

10 9 8 7 6 5 4 3 2 1

Photo Credits appear on page I-11, which constitutes a continuation of the copyright page.
ISBN 0-13-144972-9

Pearson Education Ltd., London
Pearson Education Australia Pty. Limited, Sydney
Pearson EducationSingapore Pte. Ltd.
Pearson Education North Asia, Ltd, Hong Kong
Pearson Education Canada, Ltd., Toronto
Pearson Educacion de Mexico, S.A.,de C.V.
Pearson Education, Japan, Tokyo
Pearson Education Malaysia, Pte. Ltd.

CONTENTS

CHAPTER 4 Fractions 231

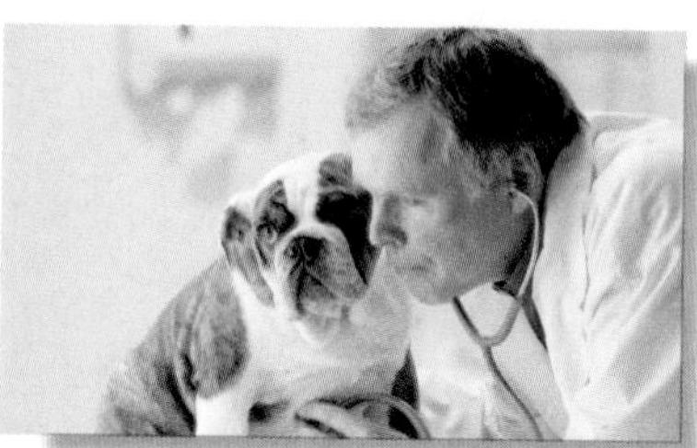

CHAPTER 5 Decimals 359

CHAPTER 6 Percent 457

CHAPTER 7 Graphing and Introduction to Statistics 543

CHAPTER 8 Geometry and Measurement 603

CHAPTER 9 Equations, Inequalities, and Problem Solving 729

CHAPTER 10 Exponents and Polynomials 805

CHAPTER 11 Factoring Polynomials 881

CHAPTER 12 Rational Expressions 951

CHAPTER 13 Graphing Equations and Inequalities 1029

CHAPTER 14 Systems of Equations 1145

CHAPTER 15 Roots and Radicals 1201

CHAPTER 16 Quadratic Equations 1261

PREFACE

About the Book

Prealgebra & Introductory Algebra was written in response to the needs of those teaching combined courses. My goals were to help students make the transition from arithmetic to algebra and to provide a solid foundation in algebra. To achieve these goals, I introduce algebraic concepts early and repeat them often, as traditional arithmetic topics are discussed, thus laying the groundwork for algebra. Specific care was taken to ensure that all core topics of an introductory algebra course are covered and that students have the most up to date relevant text preparation for future courses that require an understanding of algebraic fundamentals.

In preparing this edition, I considered the comments and suggestions of colleagues throughout the country. This edition emphasizes study and test preparation skills, attention to geometric concepts, and real-life applications. I have carefully chosen pedagogical features to help students understand and retain concepts. As suggested by the AMATYC Crossroads Document and the NCTM Standards (plus Addenda), real-life and real-data applications, data interpretation, conceptual understanding, problem solving, writing, cooperative learning, appropriate use of technology, mental mathematics, number sense, estimation, critical thinking, and geometric concepts are emphasized and integrated throughout the book. In addition, *Prealgebra & Introductory Algebra* is accompanied by a new resource—the Chapter Test Prep Video CDs. With these CDs/videos, students have instant access to video solutions for each of the chapter test questions contained in the text. It's designed to help them study efficiently.

Prealgebra & Introductory Algebra is part of a series of texts that can include *Basic Mathematics*, Second Edition; *Intermediate Algebra*, Second Edition; and a combined text, *Algebra A Combined Approach*, Second Edition. Throughout the series pedagogical features are designed to develop student proficiency in algebra and problem solving and to prepare students for future courses. Key pedagogical features and resources are described on the following pages.

Key Pedagogical Features and Content

Readability and Connections I have tried to make the writing style as clear as possible while retaining the mathematical integrity of the content. When a new topic is presented, an effort has been made to relate the new ideas to those that students may already know. Constant reinforcement and connections within problem-solving strategies, data interpretation, geometry, patterns, graphs, and situations from everyday life can help students gradually master both new and old information. In addition, each section begins with a list of objectives covered in the section. Clear organization of section material based on objectives further enhances readability.

Problem-Solving Process This is formally introduced in Chapter 3 with a four-step process that is integrated throughout the text. The four steps are **Understand, Translate, Solve,** and **Interpret**. The repeated use of these steps in a variety of examples shows their wide applicability. Reinforcing the steps can increase students' comfort level and confidence in tackling problems.

Applications and Connections Every effort was made to include as many interesting and relevant real-life applications as possible throughout the text in both worked-out examples and exercise sets. The applications help to motivate students and strengthen their understanding of mathematics in the real world. They show connections to a wide range of fields including agriculture, allied health, anthropology, art, astronomy, biology, business, chemistry, construction, consumer affairs, earth science, education, entertainment, environmental issues, finance, geography, government, history, medicine, music, nutrition, physics, sports, travel, and weather. Many of the applications are based on recent real data. Sources for data include newspapers, magazines, publicly held companies, government agencies, special-interest groups, research organizations, and reference books. Opportunities for obtaining your own real data are also included. See the Applications Index on page xviii.

Practice Problems Throughout the text, each worked-out example has a parallel Practice Problem placed next to the example in the margin. These invite students to be actively involved in the learning process before beginning the end-of-section exercise set. Practice Problems immediately reinforce a skill after it is developed. Answers appear at the bottom of the page for quick reference.

Concept Checks These margin exercises are appropriately placed throughout the text. They allow students to gauge their grasp of an idea as it is being explained in the text. Concept Checks stress conceptual understanding at the point of use and help suppress misconceived notions before they start. Answers appear at the bottom of the page.

△ **Integration of Geometry Concepts** In addition to the traditional topics in prealgebra and introduction to algebra courses, this text contains a strong emphasis on problem solving and geometric concepts, which are integrated throughout. The geometry concepts presented are those most important to a student's understanding, and I have included many applications and exercises devoted to this topic. These are marked with the the geometry icon △. Also, Chapter 8, Geometry and Measurement, provides a focused treatment of the topics. There are also dedicated appendices.

Helpful Hints Helpful Hints contain practical advice on applying mathematical concepts. These are found throughout the text and strategically placed where students are most likely to need immediate reinforcement. Helpful Hints are highlighted for quick reference.

Visual Reinforcement of Concepts This edition contains a wealth of graphics, models, photographs, and illustrations to visually clarify and reinforce concepts. These include bar graphs, line graphs, calculator screens, application illustrations, and geometric figures.

Calculator Explorations These optional explorations offer point-of-use instruction, through examples and exercises, on the proper use of scientific and graphing calculators as tools in the mathematical problem-solving process. Placed appropriately throughout the text, Calculator Explorations also reinforce concepts learned in the corresponding section.

Additional exercises building on the skill developed in the Explorations may be found in exercise sets throughout the text. Exercises requiring a calculator are marked with the 🖩 icon.

Study Skills Reminders Study Skills Reminder boxes are integrated throughout the text. They are strategically placed to constantly remind and encourage students as they hone their study skills. **Section 1.1**, Tips on Success in Mathematics, provides an overview of the Study Skills needed to succeed in math. These are reinforced by the Study Skills Reminder boxes throughout the text.

Focus On Appropriately placed throughout each chapter, these are divided into Focus on Mathematical Connections, Focus on Business and Career, Focus on the Real World, and Focus on History. They are written to help students develop effective habits for engaging in investigations of other branches of mathematics, understanding the importance of mathematics in various careers and in the world of business, and seeing the relevance of mathematics in both the present and past through critical thinking exercises and group activities.

Chapter Highlights Found at the end of each chapter, these contain key definitions, concepts, and examples to help students understand and retain what they have learned and help them organize their notes and study for tests.

Chapter Activity This feature occurs once per chapter at the end of the chapter, often serving as a chapter wrap-up. For individual or group completion, the Chapter Activity, usually hands-on or data-based, complements and extends to concepts of the chapter, allowing students to make decisions and interpretations and to think and write about algebra.

Integrated Reviews These "midchapter reviews" are appropriately placed once per chapter. Integrated Reviews allow students to review and assimilate the many different skills learned separately over several sections before moving on to related material in the chapter.

Pretests Each chapter begins with a Pretest that is designed to help students identify areas where they need to pay special attention in the upcoming chapter.

Chapter Review and Test The end of each chapter contains a review of topics introduced in the chapter. The Chapter Review offers exercises that are keyed to sections of the chapter. The Chapter Test is a practice test and is not keyed to sections of the chapter. This text is accompanied by the Chapter Test Prep Video CDs, which give students instant access to a step-by-step video solution to each chapter test question.

Cumulative Review These features are found at the end of each chapter (except Chapter 1). Each odd problem contained in the Cumulative Review is an earlier worked example in the text that is referenced in the back of the book along with the answer. Students who need to see a complete worked-out solution, with explanation, can do so by turning to the appropriate example in the text. The evens are not keyed to examples.

Functional Use of Color and New Design Elements of this text are highlighted with color or design to make it easier for students to read and study. Special care has been taken to use color within solutions to examples or in the art to **help clarify, distinguish, or connect concepts**.

Exercise Sets Each text section ends with an Exercise Set. Each exercise in the set, except those found in parts labeled Review and Preview or Combining Concepts, is keyed to one of the objectives of the section. Wherever possible, a specific example is also referenced. In addition, exercises may also be found in the Pretests, Integrated Reviews, Chapter Reviews, Chapter Tests, and Cumulative Reviews.

Exercises and examples marked with a video icon have been worked out step-by-step by the author in the lecture videos that accompany this text.

Throughout the exercises in the text there is an emphasis on data and graphical interpretation via tables, charts, and graphs. The ability to interpret data and read and create a variety of types of graphs is developed gradually so students become comfortable with it. Geometric concepts—such as perimeter and area—are integrated throughout the text. Exercises and examples marked with a geometry icon have been identified for convenience. In addition, Chapter 8 provides a focused treatment of the topic.

Each exercise set contains one or more of the following features.

Mental Math Found at the beginning of an exercise set, these mental warmups reinforce concepts found in the accompanying section and increase students' confidence before they tackle an exercise set. By relying on their own mental skills, students increase not only their confidence in themselves but also their number sense and estimation ability.

Review and Preview These exercises occur in each exercise set (except for those in Chapter 1) after the exercises keyed to the objectives of the section. Review and Preview problems are keyed to earlier sections and review concepts learned earlier in the text that are needed in the next section or in the next chapter. These exercises show the links between earlier topics and later material.

Combining Concepts These exercises are found at the end of each exercise set after the Review and Preview exercises. Combining Concepts exercises require students to combine several concepts from that section or to take the concepts of the section a step further by combining them with concepts learned in previous sections. For instance, sometimes students are required to combine the concepts of the section with the problem-solving process they learned in Chapter 3 to try their hand at solving an application problem.

Writing Exercises These exercises occur in almost every exercise set and are marked with an icon. They require students to assimilate information and provide a written response to explain concepts or justify their thinking. Guidelines recommended by the American Mathematical Association of Two Year Colleges (AMATYC) and other professional groups recommend incorporating writing in mathematics courses to reinforce concepts.

Vocabulary Checks Vocabulary Checks provide an opportunity for students to become more familiar with the use of mathematical terms as they strengthen their verbal skills. These appear at the end of the chapter before the Chapter Highlights.

Data and Graphical Interpretation There is an emphasis on data interpretation in exercises via tables and graphs. The ability to interpret data and read and create a variety of types of graphs is developed gradually so students become comfortable with it.

Resources for the Instructor

Printed Supplements

Annotated Instructor's Edition (0-13-147322-0)

- Answers to all exercises printed on the same text page.
- Teaching Tips throughout the text placed at key points in the margin.

Instructor's Solution Manual (0-13-117629-3)

- Solutions to even-numbered section exercises.
- Solutions to every (even and odd) Mental Math exercise.

- Solutions to every (even and odd) Practice Problem (margin exercise).
- Solutions to every (even and odd) exercise found in the Pretests, Integrated Reviews (mid-chapter reviews), Chapter Reviews, Chapter Tests, and Cumulative Reviews.

Instructor's Resource Manual with Tests (0-13-117626-9)

- Notes to the Instructor including suggested homework assignments for each section, guidelines for course pacing, and much more.
- Two free-response Pretests per chapter.
- Eight Chapter Tests per chapter (3 multiple-choice, 5 free-response).
- Two Cumulative Review Tests (one multiple-choice, one free-response) every two chapters (after chapters 2, 4, 6, 8, 10).
- Eight Final Exams (4 multiple-choice, 4 free-response).
- Twenty additional exercises per section for added test exercises if needed.
- Additional examples for each section for use in the classroom.
- Group Activities (an average of two per chapter; providing short group activities in a convenient, ready-to-use format).
- Answers to all items.

Media Supplements

TestGen with QuizMaster TestGen enables instructors to build, edit, print, and administer tests using a computerized bank of questions developed to cover all the objectives of the text. Instructors can modify test bank questions or add new questions by using the built-in question editor, which allows users to create graphs, import graphics, and insert math notation, variable numbers, or text. Tests can be printed or administered online via the Internet or another network. TestGen comes packaged with QuizMaster, which allows students to take tests on a local area network. The software is available on a dual-platform Windows/Macintosh CD-ROM.

MyMathLab® (instructor) MyMathLab® is a series of text-specific, easily customizable online courses for Prentice Hall mathematics textbooks. MyMathLab® is powered by CourseCompass™—Pearson Education's online teaching and learning environment—and by MathXL®—our online homework, tutorial, and assessment system. MyMathLab gives you the tools you need to deliver all or a portion of your course online, whether your students are in a lab setting or working from home.

MyMathLab® provides a rich and flexible set of course materials, featuring free-response exercises that are algorithmically generated for unlimited practice and mastery. Students can also use online tools such as video lectures, animations, and a multimedia textbook to independently improve their understanding and performance. Instructors can use MyMathLab's® homework and test managers to select and assign online exercises correlated directly to the textbook, and they can also import TestGen tests into MyMathLab for added flexibility. MyMathLab's online gradebook—designed specifically for mathematics—automatically tracks students' homework and test results and gives the instructor control over how to calculate final grades.

MyMathLab® is available to qualified adopters. For more information, visit our website at *www.mymathlab.com* or contact your Prentice Hall sales representative for a product demonstration.

MathXL® MathXL® is a powerful online homework, tutorial, and assessment system that accompanies your Prentice Hall mathematics textbook. With MathXL, instructors can create, edit, and assign online homework and tests using algorithmically–generated exercises correlated at the objective level to your textbook. All student work is tracked in MathXL's online gradebook. Students can take chapter tests in MathXL and receive personalized study plans based on their test results. The study plan diagnoses weaknesses and links students directly to tutorial exercises for the objectives they need to study and retest. Students can also access supplemental animations and video clips directly from selected exercises. MathXL is available to qualified adopters. For more information, visit our website at *www.mathxl.com*, or contact your Prentice Hall sales representative for a product demonstration.

MathXL® Tutorials on CD This interactive tutorial CD-ROM provides algorithmically–generated practice exercises that are correlated at the objective level to the exercises in the textbook. Every practice exercise is accompanied by an example and a guided solution designed to involve students in the solution process. Selected exercises may also include a video clip to help students visualize concepts. The software tracks student activity and scores and can generate printed summaries of students' progress.

Resources for the Student

Printed Supplements

Student's Solution Manual

- Solutions to odd-numbered section exercises.
- Solutions to every (even and odd) Mental Math exercise.
- Solutions to every (even and odd) Practice Problem (margin exercise).
- Solutions to every (even and odd) exercise found in the Pretests, Integrated Reviews (mid-chapter reviews), Chapter Reviews, Chapter Tests, and Cumulative Reviews.

Media Supplements

- New **Chapter Test Prep Video CDs** provide students with instant access to video solutions for each of the chapter test questions contained in the text. The Video CDs are packaged in each new student text. They are also available for purchase through your campus bookstore.
- The Video CD's easy navigation is designed to help students study efficiently. Students select the exact chapter test questions they missed or need help with and instantly view the solution worked by K. Elayn Martin-Gay.
- Start making the most of your study time today. Turn to the back of the text to access the Video CDs. You'll also find a Note to the Students with steps to success for studying for tests and using the video.

MyMathLab® (student) MyMathLab® is a complete online course designed to help you succeed in learning and understanding mathematics. MyMathLab contains an online version of your textbook with links to multimedia resources—such as video clips, practice exercises, and animations—that are correlated to the examples and exercises in the text. MyMathLab also provides you with online homework and tests and generates a personalized study plan based on your test results. Your study plan links directly to unlimited tutorial exercises for the areas you need to study and re-test, so you can practice until you have mastered the skills and concepts in your textbook. All of the online homework, tests, and tutorial work you do is tracked in your MyMathLab gradebook.

MathXL® MathXL® is a powerful online homework, tutorial, and assessment system that accompanies your Prentice Hall mathematics textbook. With MathXL, instructors can create, edit, and assign online homework and tests using algorithmically generated exercises correlated at the objective level to your textbook. All student work is tracked in MathXL's online gradebook. Students can take chapter tests in MathXL and receive personalized study plans based on their test results. The study plan diagnoses weaknesses and links students directly to tutorial exercises for the objectives they need to study and retest. Students can also access supplemental animations and video clips directly from selected exercises. MathXL is available to qualified adopters. For more information, visit our website at *www.mathxl.com*, or contact your Prentice Hall sales representative for a product demonstration.

MathXL® Tutorials on CD This interactive tutorial CD-ROM provides algorithmically–generated practice exercises that are correlated at the objective level to the exercises in the textbook. Every practice exercise is accompanied by an example and a guided solution designed to involve students in the solution process. Selected exercises may also include a video clip to help students visualize concepts. The software tracks student activity and scores and can generate printed summaries of students' progress.

Videotape Series

- Written and presented by K. Elayn Martin-Gay.
- Keyed to each section of the text.
- Step-by-step solutions to exercises from each section of the text. Exercises that are worked in the videos are marked with a video icon.

Digitized Lecture Videos on CD-ROM (0-13-117633-1)

- The entire set of *Prealgebra & Introductory Algebra* lecture videotapes in digital form.
- Convenient access anytime to video tutorial support from a computer at home or on campus.
- Available shrinkwrapped with the text or stand-alone.

Prentice Hall Tutor Center

SSM
TUTOR CENTER

- Staffed with qualified math instructors and open 5 days a week, 7 hours per day.
- Obtain help for examples and exercises in Martin-Gay, *Prealgebra & Introductory Algebra*, via toll-free telephone, fax, or e-mail.
- The Prentice Hall Tutor Center is accessed through a registration number that may be bundled with a new text or purchased separately with a used book. Visit http://www.prenhall.com/tutorcenter to learn more.

Acknowledgments

First, as usual, I would like to thank my husband, Clayton, for his constant encouragement. I would also like to thank my children, Eric and Bryan, for their sense of humor and their help finding data for this text. I now buy precooked bacon that goes in the microwave.

I would also like to thank my extended family for their invaluable help and also their sense of humor. Their contributions are too numerous to list. They are Rod and Karen, Pasch; Michael, Christopher, Matthew, and Jessica Callac; Stuart Martin, Josh, Mandy, Bailey, Ethen and Avery Barnes; Mark, Sabrina, and Madison Martin; Leo and Barbara Miller; and Jewett Gay.

I would like to thank the following reviewers who commented on portions of this book for their input and suggestions:

Reviewers

Bob Angus, Northeastern University, Nanci Barker, Caroll Community College, Rob Dengler, Southeastern Community College, Kathryn Gunderson, Three Rivers Community College, Kelly Houlton, University of Alaska–Fairbanks, Pat Kline, Southwestern Community College, Rose Turn, Westchester Community College, Aretha Vernette, Indian River Community College, Kinley Alston, Trident Technical College, John Close, Salt Lake Community College, Kay Haralson, Austin Peay State Community College, Barbara Hughes, San Jacinto Community College, Karen Pagel, Dona Ana Branch, Jonathan Weissman, Essex Community College, Susan F. Akers, Northeast State Technical Community College, Matthew Cardner, North Hennepin Community College, Betty Dennison, Roane State Community College, Joye Elaine Gowan, Roane State Community College, Elizabeth Hill, University of North Carolina-Charlotte, Pat Corey Horacek, Pensacola Junior College, Marilyn Platt, Gaston College, Lee Ann Spahr, Durham Technical Community College, Kevin Yokoyama, College of the Redlands.

There were many people who helped me develop this text and I will attempt to thank some of them here. Elka Block, Chris Callac, and Miriam Daunis were invaluable for their many suggestions and contributions during the development and writing of this edition. Lynn Savino Wendel provided guidance throughout the production process. I thank Richard Semmler and Carrie Green for all their work on the solutions, text, and accuracy.

Sadly, executive editor of this project, Karin Wagner, passed away this year. She will be dearly remembered by me and the rest of the staff at Prentice Hall for her integrity, wisdom, commitment, and especially her sense of humor and infectious laugh.

A very special thank you to my project manager, Mary Beckwith, for taking over during a difficult period for all of us.

Lastly, my thanks to the staff at Prentice Hall for all their support: Linda Behrens, Mike Bell, Patty Burns, Tom Benfatti, Paul Belfanti, Jim Sullivan, Karen Noferi, Joanne Del Ben, Karen Stephens, Beth Gschwind, Jacqueline Ambrosius, Julita Nazario, Geoffrey Cassar, Eilish Main, Patrice Jones, John Tweeddale, Chris Hoag, Paul Corey, and Tim Bozik.

K. Elayn Martin-Gay

About the Author

K. Elayn Martin-Gay has taught mathematics at the University of New Orleans for more than 20 years and has received numerous teaching awards including the local University Alumni Association's Award for Excellence in Teaching.

Over the years, Elayn has developed a videotaped lecture series to help her students understand algebra better. This highly successful video material is the basis for her books: *Basic College Mathematics*, Second Edition; *Prealgebra*, Fourth Edition: *Introductory Algebra*, Second Edition; *Intermediate Algebra*, Second Edition; *Algebra A Combined Approach*, Second Edition; and her hardback series: *Beginning Algebra*, Fourth Edition; *Intermediate Algebra*, Fourth Edition; *Beginning and Intermediate Algebra*, Third Edition; and *Intermediate Algebra: A Graphing Approach*, Third Edition.

APPLICATIONS INDEX

HIGHLIGHTS OF PREALGEBRA & INTRODUCTORY ALGEBRA

Prealgebra & Introductory Algebra is the primary learning tool in a fully integrated learning package to help you succeed in this course. Author K. Elayn Martin-Gay focuses on enhancing the traditional emphasis of mastering the basics with innovative pedagogy and a meaningful learning program. There are three goals that drive her authorship:

- ▲ **Master and apply skills and concepts**
- ▲ **Build confidence**
- ▲ **Increase motivation**

Take a few moments now to examine some of the features that have been incorporated into *Prealgebra & Introductory Algebra* to help students excel.

Solving Equations and Problem Solving

CHAPTER

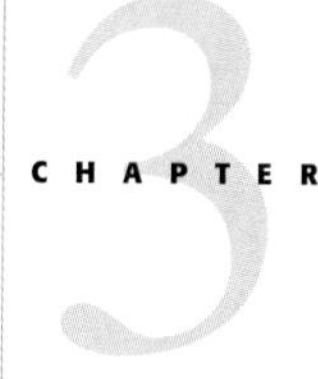

Throughout this text, we have been making the transition from arithmetic to algebra. We have said that in algebra letters called variables represent numbers. Using variables is a very powerful method for solving problems that cannot be solved with arithmetic alone. This chapter introduces operations on algebraic expressions and solving variable equations.

Americans are in love with traveling the open roads. Throughout the country, roads and highways lead travelers on an exciting journey through every period of U.S. history. The *Santa Fe Trail Scenic and Historic Byway* in Colorado retraces the famous road taken by wagon trains of the Old West. The *Seward Highway* in Alaska is named for the U.S. Secretary of State who arranged the purchase of Alaska from Russia in 1867. The *Selma to Montgomery March Byway* in Alabama traces the historic route taken by Dr. Martin Luther King, Jr. in 1965. Whether by scenic byways or interstate highways, touring the country by automobile remains a favorite pastime of vacationers today. Before you venture out on the open road, check out Exercises 107–110 on page 198 and see how we can use equations to relate the distance, rate, and time of an automobile trip.

169

Page 169

◀ Real-World Applications

Chapter-opening real-world applications introduce you to everyday situations that are applicable to the mathematics you will learn in the upcoming chapter, showing the relevance of mathematics in daily life.

Become a Confident Problem Solver

A goal of this text is to help you develop problem-solving abilities.

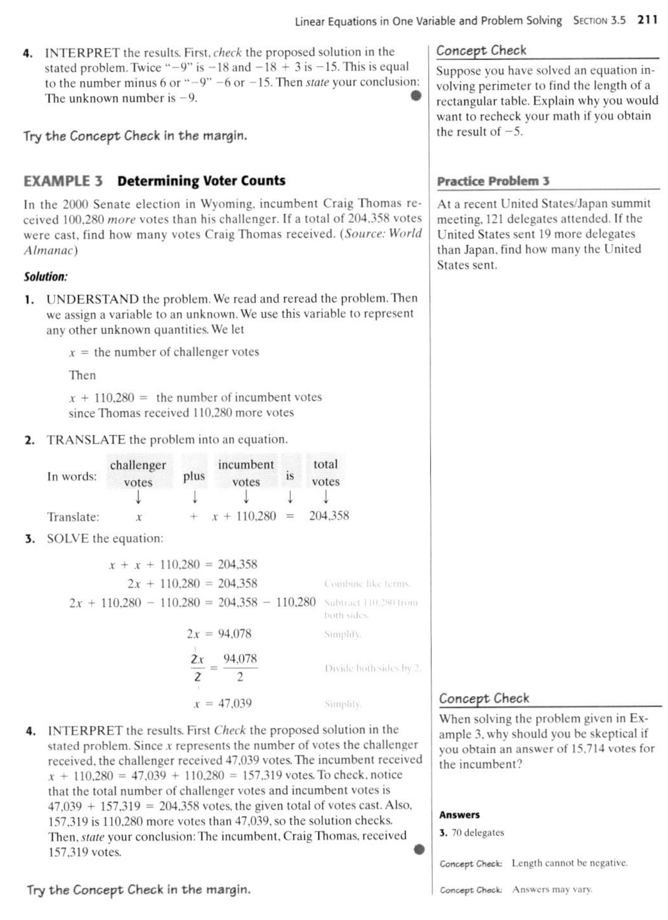

4. INTERPRET the results. First, *check* the proposed solution in the stated problem. Twice "−9" is −18 and −18 + 3 is −15. This is equal to the number minus 6 or "−9" −6 or −15. Then *state* your conclusion: The unknown number is −9.

Try the Concept Check in the margin.

Concept Check

Suppose you have solved an equation involving perimeter to find the length of a rectangular table. Explain why you would want to recheck your math if you obtain the result of −5.

EXAMPLE 3 Determining Voter Counts

In the 2000 Senate election in Wyoming, incumbent Craig Thomas received 100,280 *more* votes than his challenger. If a total of 204,358 votes were cast, find how many votes Craig Thomas received. (*Source: World Almanac*)

Practice Problem 3

At a recent United States/Japan summit meeting, 121 delegates attended. If the United States sent 19 more delegates than Japan, find how many the United States sent.

Solution:

1. UNDERSTAND the problem. We read and reread the problem. Then we assign a variable to an unknown. We use this variable to represent any other unknown quantities. We let

x = the number of challenger votes

Then

$x + 110{,}280$ = the number of incumbent votes since Thomas received 110,280 more votes

2. TRANSLATE the problem into an equation.

In words:	challenger votes	plus	incumbent votes	is	total votes
	↓	↓	↓	↓	↓
Translate:	x	+	$x + 110{,}280$	=	204,358

3. SOLVE the equation:

$$x + x + 110{,}280 = 204{,}358$$

$$2x + 110{,}280 = 204{,}358 \quad \text{Combine like terms.}$$

$$2x + 110{,}280 - 110{,}280 = 204{,}358 - 110{,}280 \quad \text{Subtract 110,280 from both sides.}$$

$$2x = 94{,}078 \quad \text{Simplify.}$$

$$\frac{2x}{2} = \frac{94{,}078}{2} \quad \text{Divide both sides by 2.}$$

$$x = 47{,}039 \quad \text{Simplify.}$$

4. INTERPRET the results. First *Check* the proposed solution in the stated problem. Since x represents the number of votes the challenger received, the challenger received 47,039 votes. The incumbent received $x + 110{,}280 = 47{,}039 + 110{,}280 = 157{,}319$ votes. To check, notice that the total number of challenger votes and incumbent votes is $47{,}039 + 157{,}319 = 204{,}358$ votes, the given total of votes cast. Also, 157,319 is 110,280 more votes than 47,039, so the solution checks. Then, *state* your conclusion: The incumbent, Craig Thomas, received 157,319 votes.

Try the Concept Check in the margin.

Concept Check

When solving the problem given in Example 3, why should you be skeptical if you obtain an answer of 15,714 votes for the incumbent?

Answers

3. 70 delegates

Concept Check: Length cannot be negative.

Concept Check: Answers may vary.

Page 211

◀ General Strategy for Problem-Solving

Save time by having a plan. This text's organization can help you. Note the outlined problem-solving steps, *Understand, Translate, Solve,* and *Interpret.*

Problem solving is introduced early, emphasized and integrated throughout the book. The author provides patient explanations and illustrates how to apply the problem-solving procedure to the in-text examples.

Geometry ▶

Geometric concepts are integrated throughout the text. Examples and exercises involving geometric concepts are identified with a triangle icon △. This text includes chapter 8 on Geometry and Measurement as well as dedicated Appendices.

△ **EXAMPLE 15** Find the area of the rectangular deck.

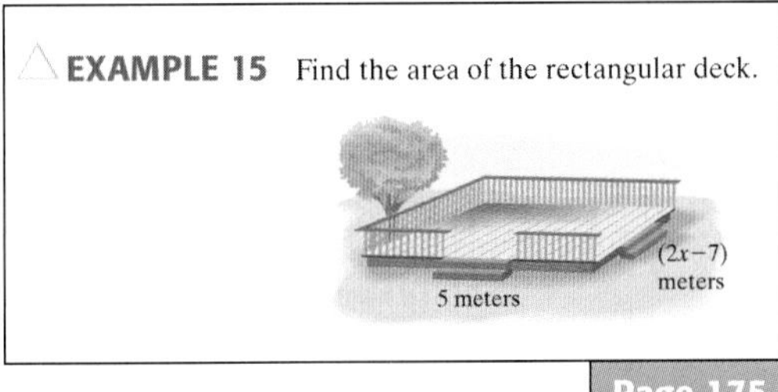

Page 175

Master and Apply Basic Skills and Concepts

K. Elayn Martin-Gay provides thorough explanations of key concepts and enlivens the content by integrating successful and innovative pedagogy. *Prealgebra & Introductory Algebra* integrates skill building throughout the text and provides problem-solving strategies and hints along the way. These features have been included to enhance your understanding of algebraic concepts.

Concept Check

What number should be added to or subtracted from both sides of the equation in order to solve the equation $-3 = y + 2$?

Page 184

◀ Concept Checks

Concept Checks are special margin exercises found in most sections. Work these to help gauge your grasp of the concept being developed in the text.

Combining Concepts ▶

Combining Concepts exercises are found at the end of each exercise set. Solving these exercises will expose you to the way mathematical ideas build upon each other.

Page 78

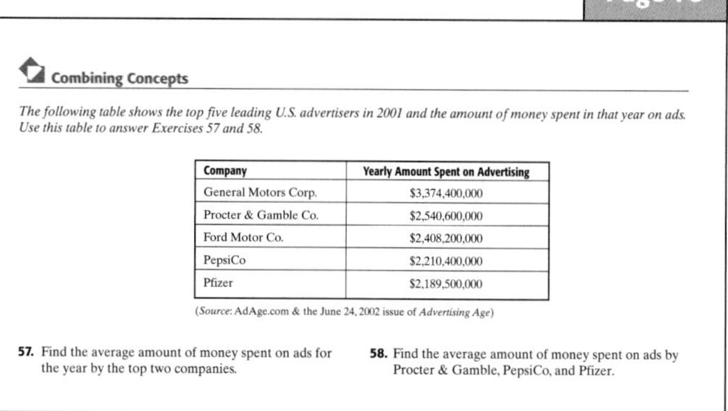

Combining Concepts

The following table shows the top five leading U.S. advertisers in 2001 and the amount of money spent in that year on ads. Use this table to answer Exercises 57 and 58.

Company	Yearly Amount Spent on Advertising
General Motors Corp.	\$3,374,400,000
Procter & Gamble Co.	\$2,540,600,000
Ford Motor Co.	\$2,408,200,000
PepsiCo	\$2,210,400,000
Pfizer	\$2,189,500,000

(*Source:* AdAge.com & the June 24, 2002 issue of *Advertising Age*)

57. Find the average amount of money spent on ads for the year by the top two companies.

58. Find the average amount of money spent on ads by Procter & Gamble, PepsiCo, and Pfizer.

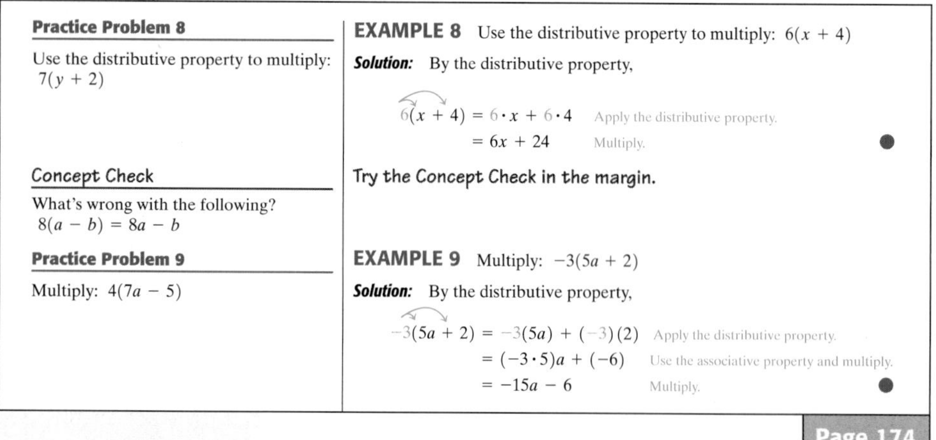

Practice Problem 8

Use the distributive property to multiply: $7(y + 2)$

Concept Check

What's wrong with the following? $8(a - b) = 8a - b$

Practice Problem 9

Multiply: $4(7a - 5)$

EXAMPLE 8 Use the distributive property to multiply: $6(x + 4)$

Solution: By the distributive property,

$$6(x + 4) = 6 \cdot x + 6 \cdot 4 \quad \text{Apply the distributive property.}$$
$$= 6x + 24 \quad \text{Multiply.}$$

Try the Concept Check in the margin.

EXAMPLE 9 Multiply: $-3(5a + 2)$

Solution: By the distributive property,

$$-3(5a + 2) = -3(5a) + (-3)(2) \quad \text{Apply the distributive property.}$$
$$= (-3 \cdot 5)a + (-6) \quad \text{Use the associative property and multiply.}$$
$$= -15a - 6 \quad \text{Multiply.}$$

Page 174

◀ Practice Problems

Practice Problems occur in the margins next to every Example. Work these problems after an example to immediately reinforce your understanding.

Writing Exercises ▶

Writing Exercises, marked by an icon, are found in most practice sets.

Page 207

85. A classmate tries to solve $3x = 39$ by subtracting 3 from both sides of the equation. Will this step solve the equation for x? Why or why not?

Test Yourself and Check Your Understanding

Good exercise sets and an abundance of worked-out examples are essential for building student confidence. The exercises you will find in this worktext are intended to help you build skills and understand concepts as well as motivate and challenge you. In addition, features like Chapter Highlights, Chapter Reviews, Chapter Tests, and Cumulative Reviews are found at the end of each chapter to help you study and organize your notes.

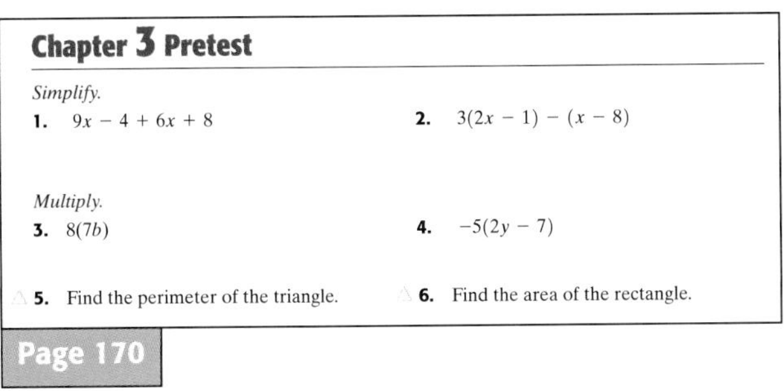

Chapter 3 Pretest

Simplify.

1. $9x - 4 + 6x + 8$ **2.** $3(2x - 1) - (x - 8)$

Multiply.

3. $8(7b)$ **4.** $-5(2y - 7)$

5. Find the perimeter of the triangle. **6.** Find the area of the rectangle.

Page 170

◀ Pretests

Pretests open each chapter. Take a Pretest to evaluate where you need the most help before beginning a new chapter.

Page 199

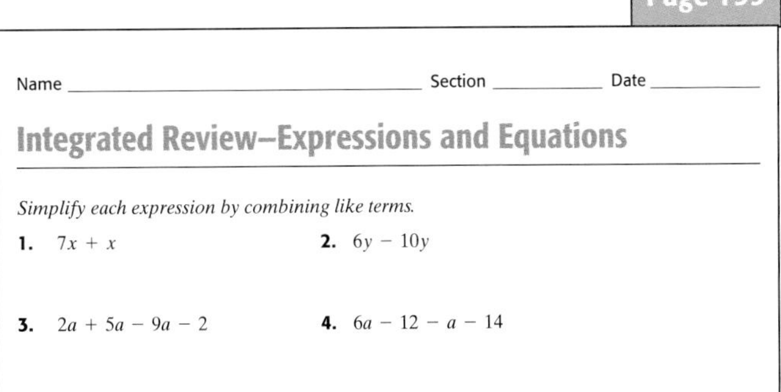

Name ______________ Section ________ Date ________

Integrated Review—Expressions and Equations

Simplify each expression by combining like terms.

1. $7x + x$ **2.** $6y - 10y$

3. $2a + 5a - 9a - 2$ **4.** $6a - 12 - a - 14$

Integrated Reviews ▶

Integrated Reviews serve as mid-chapter reviews and help you to assimilate the new skills you have learned separately over several sections.

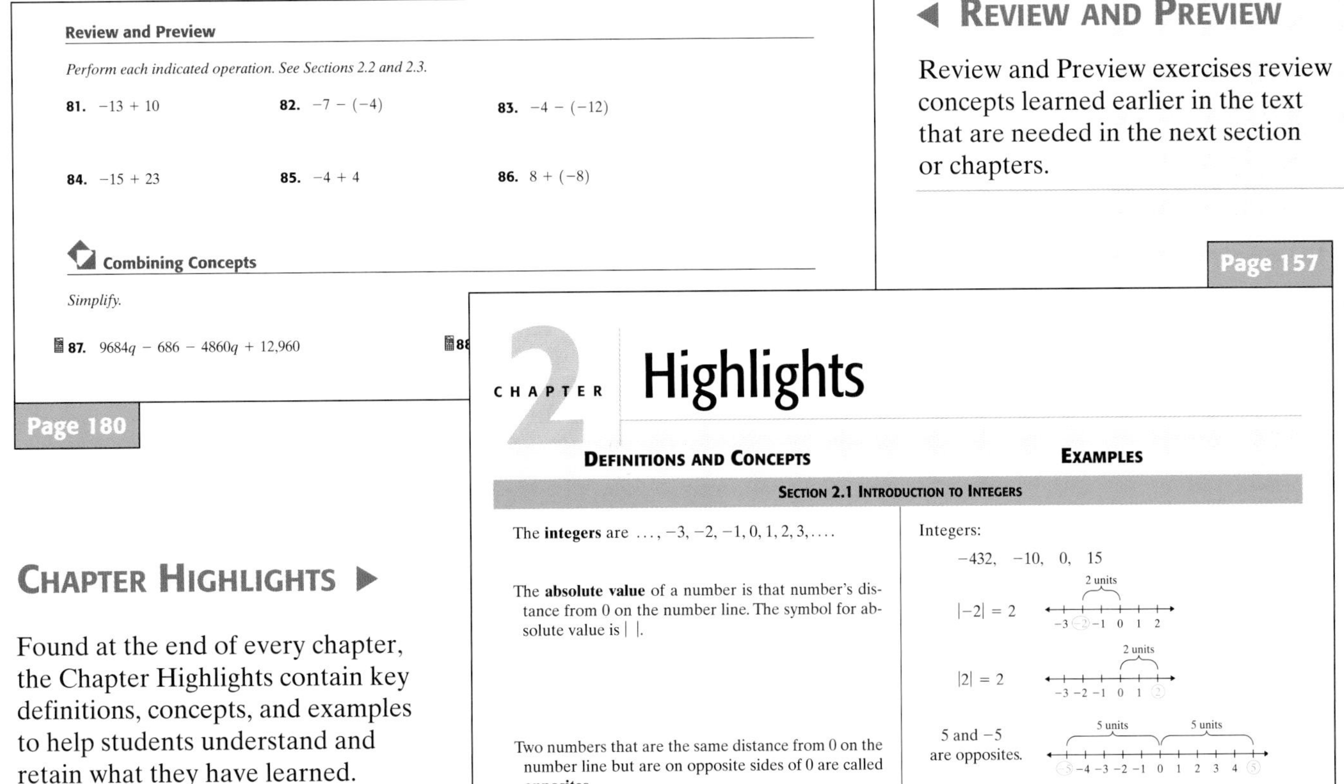

Review and Preview

Perform each indicated operation. See Sections 2.2 and 2.3.

81. $-13 + 10$ **82.** $-7 - (-4)$ **83.** $-4 - (-12)$

84. $-15 + 23$ **85.** $-4 + 4$ **86.** $8 + (-8)$

Combining Concepts

Simplify.

87. $9684q - 686 - 4860q + 12{,}960$

Page 180

Page 157

CHAPTER 2 **Highlights**

Definitions and Concepts	**Examples**
Section 2.1 Introduction to Integers	
The **integers** are …, −3, −2, −1, 0, 1, 2, 3, ….	Integers: −432, −10, 0, 15
The **absolute value** of a number is that number's distance from 0 on the number line. The symbol for absolute value is | |.	$\|-2\| = 2$ (2 units); $\|2\| = 2$ (2 units)
Two numbers that are the same distance from 0 on the number line but are on opposite sides of 0 are called **opposites**.	5 and −5 are opposites. (5 units, 5 units)

◀ Review and Preview

Review and Preview exercises review concepts learned earlier in the text that are needed in the next section or chapters.

Chapter Highlights ▶

Found at the end of every chapter, the Chapter Highlights contain key definitions, concepts, and examples to help students understand and retain what they have learned.

Increase Motivation

Throughout *Prealgebra & Introductory Algebra*, K. Elayn Martin-Gay provides interesting real-world applications to strengthen your understanding of the relevance of math in everyday life. When a new topic is presented, an effort has been made to relate the new ideas to those that students may already know. This edition emphasizes visualization to clarify and reinforce key concepts.

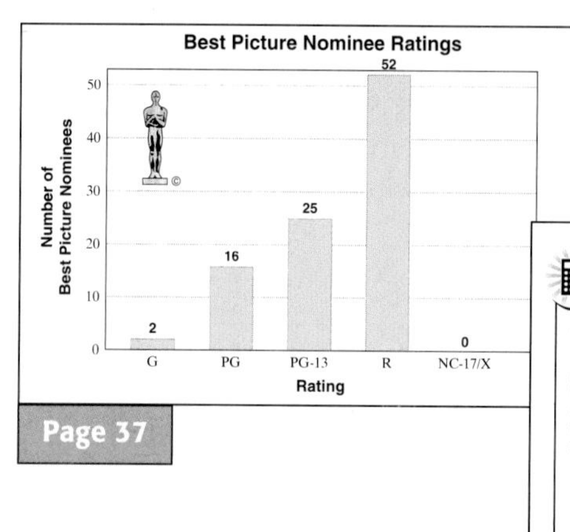

◀ Real data is integrated throughout the worktext, drawn from current and familiar sources.

CALCULATOR EXPLORATIONS

Checking Equations

A calculator can be used to check possible solutions of equations. To do this, replace the variable by the possible solution and evaluate each side of the equation separately. For example, to see whether 7 is a solution of the equation $52x = 15x + 259$, replace x with 7 and use your calculator to evaluate each side separately.

Equation: $52x = 15x + 259$

$52 \cdot 7 \stackrel{?}{=} 15 \cdot 7 + 259$ Replace x with 7.

Evaluate left side: [52] [×] [7] [=] or [ENTER]. Display: [364].

Evaluate right side: [15] [×] [7] [+] [259] [=] or [ENTER]. Display: [364].

Since the left side equals the right side, 7 is a solution of the equation $52x = 15x + 259$.

Use a calculator to determine whether the numbers given are solutions of each equation.

1. $76(x - 25) = -988$; 12
2. $-47x + 862 = -783$; 35
3. $x + 562 = 3x + 900$; -170
4. $55(x + 10) = 75x + 910$; -18
5. $29x - 1034 = 61x - 362$; -21
6. $-38x + 205 = 25x + 120$; 25

Page 204

▲ **CALCULATOR EXPLORATIONS**

Optional Calculator and Graphing Calculator Explorations and exercises appear in appropriate sections.

CHAPTER 3 ACTIVITY Using Equations

This activity may be completed by working in groups or individually.

A hospital nurse working second shift has been keeping track of the fluid intake and output on the chart of one of her patients, Mr. Ramirez. At the end of her shift, she totals the intakes and outputs and returns Mr. Ramirez's chart to the floor nurses' station.

Suppose you are one of the night nurses for this floor and have been assigned to Mr. Ramirez. Shortly after you come on duty, someone at the nurses' station spills coffee over several patients' charts, including the one for Mr. Ramirez. After blotting up the coffee on the chart, you notice that several entries on his chart have been smeared by the coffee spill.

MEDICAL CHART

Patient: Juan Ramirez Room: 314

Shift: Second Nurse's Initials: SRT Date: 3/27

Fluid Intake (cubic centimeters):					Totals
Blood	500				500
Intravenous	250		100		500
Oral Fluid	300	150		100	900
Oral Meds	4 doses of		cc each		200
TOTAL					2100

Fluid Output (cubic centimeters):				Totals
Urine	240	310		550
Emesis	120			120
Irrigation	50	25		75
TOTAL				815

Page 218

▲ Graphics, models, and illustrations provide visual reinforcement.

Focus On Boxes ▶

Focus On boxes found throughout each chapter help you see the relevance of math through critical-thinking exercises and group activities. Try these on your own or with a classmate. Focus On covers the areas of: History, Mathematical Connections, Real World, and Business and Career.

Page 148

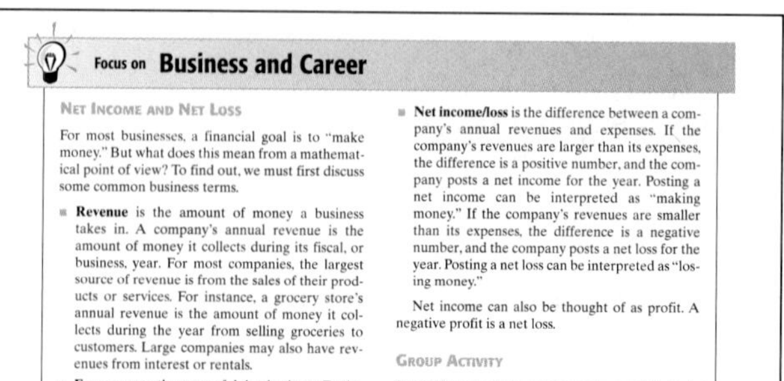

Focus on Business and Career

NET INCOME AND NET LOSS

For most businesses, a financial goal is to "make money." But what does this mean from a mathematical point of view? To find out, we must first discuss some common business terms.

- **Revenue** is the amount of money a business takes in. A company's annual revenue is the amount of money it collects during its fiscal, or business, year. For most companies, the largest source of revenue is from the sales of their products or services. For instance, a grocery store's annual revenue is the amount of money it collects during the year from selling groceries to customers. Large companies may also have revenues from interest or rentals.
- **Expenses** are the costs of doing business. For instance, a large part of a grocery store's expenses
- **Net income/loss** is the difference between a company's annual revenues and expenses. If the company's revenues are larger than its expenses, the difference is a positive number, and the company posts a net income for the year. Posting a net income can be interpreted as "making money." If the company's revenues are smaller than its expenses, the difference is a negative number, and the company posts a net loss for the year. Posting a net loss can be interpreted as "losing money."

Net income can also be thought of as profit. A negative profit is a net loss.

GROUP ACTIVITY

Search for corporate annual reports or articles in financial newspapers and magazines that report a

Build Confidence

Several features of this text can be helpful in building your confidence and mathematical competence. As you study, also notice the connections the author makes to relate new material to ideas that you may already know.

1.1 Tips for Success in Mathematics

Before reading this section, remember that your instructor is your best source for information. Please see your instructor for any additional help or information.

Getting Ready for This Course

Now that you have decided to take this course, remember that a *positive attitude* will make all the difference in the world. Your belief that yo
ceed is just as important as your commitment to this course. Make

Page 3

◀ TIPS FOR SUCCESS

Coverage of study skills in Section 1.1 reinforces this important component to success in this course.

Page 148

STUDY SKILLS REMINDER

How are your homework assignments going?

It is so important in mathematics to keep up with homework. Why? Many concepts build on each other. Often, your understanding of a day's lecture in mathematics depends on an understanding of the previous day's material.

Remember that completing your homework assignment involves a lot more than attempting a few of the problems assigned.

To complete a homework assignment, remember these four things:

1. Attempt all of it.
2. Check it.
3. Correct it.
4. If needed, ask questions about it.

STUDY SKILLS REMINDERS ▶

Study Skills Reminders are integrated throughout the book to reinforce section 1.1 and encourage the development of strong study skills.

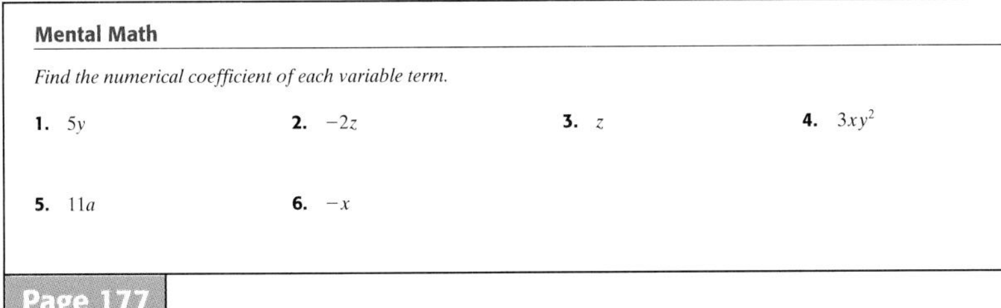

Mental Math

Find the numerical coefficient of each variable term.

1. $5y$
2. $-2z$
3. z
4. $3xy^2$
5. $11a$
6. $-x$

Page 177

◀ MENTAL MATH

Mental Math warm-up exercises reinforce concepts found in the accompanying section and can increase your confidence before beginning an exercise set.

HELPFUL HINTS ▶

Found throughout the text, these contain practical advice on applying mathematical concepts. They are strategically placed where you are most likely to need immediate reinforcement.

Page 113

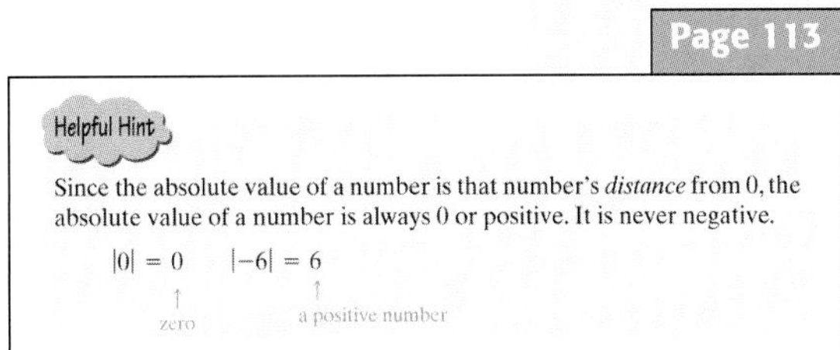

Helpful Hint

Since the absolute value of a number is that number's *distance* from 0, the absolute value of a number is always 0 or positive. It is never negative.

$|0| = 0$ $\quad |-6| = 6$

↑ zero $\quad$ ↑ a positive number

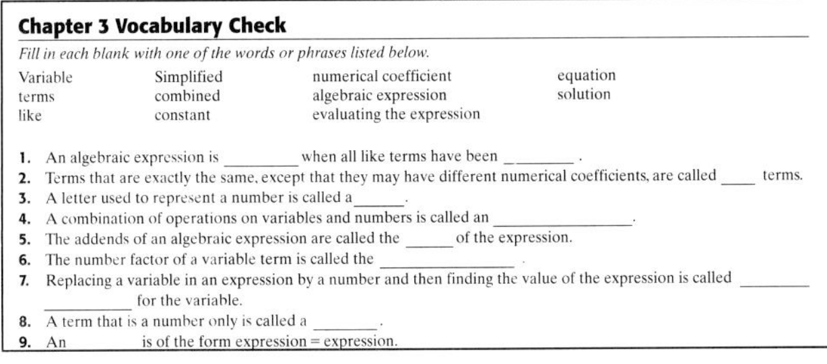

Chapter 3 Vocabulary Check

Fill in each blank with one of the words or phrases listed below.

Variable	Simplified	numerical coefficient	equation
terms	combined	algebraic expression	solution
like	constant	evaluating the expression	

1. An algebraic expression is ________ when all like terms have been ________ .
2. Terms that are exactly the same, except that they may have different numerical coefficients, are called ____ terms.
3. A letter used to represent a number is called a______.
4. A combination of operations on variables and numbers is called an ______________.
5. The addends of an algebraic expression are called the _____ of the expression.
6. The number factor of a variable term is called the ______________ .
7. Replacing a variable in an expression by a number and then finding the value of the expression is called ________ ________ for the variable.
8. A term that is a number only is called a _______ .
9. An ________ is of the form expression = expression.

Page 219

◀ VOCABULARY CHECKS

Vocabulary Checks allow you to write your answers to questions about chapter content and strengthen verbal skills.

Enrich Your Learning

Good study skills are essential for success in mathematics. Use the student resources that accompany *Prealgebra & Introductory Algebra* to make the most of your valuable study time.

The Chapter Test Prep Video CDs provide students with instant access to video solutions for each of the chapter test questions contained in the text.

Step-by-step solutions, presented by the textbook author, are included for every problem on the Chapter Tests in *Prealgebra & Introductory Algebra.* To make the most of this resource, when studying for a test, follow these three steps:

1. Take the Chapter Test found at the end of the chapter.
2. Check your answers in the back of the text.
3. Use the videos to review *every step* of the worked out solution for those specific questions you answered in correctly or need help with

The Video CDs' easy navigation is designed to help students study efficiently. Students select the exact chapter test questions they missed or need help with and instantly view the solution worked by K. Elayn Martin-Gay.

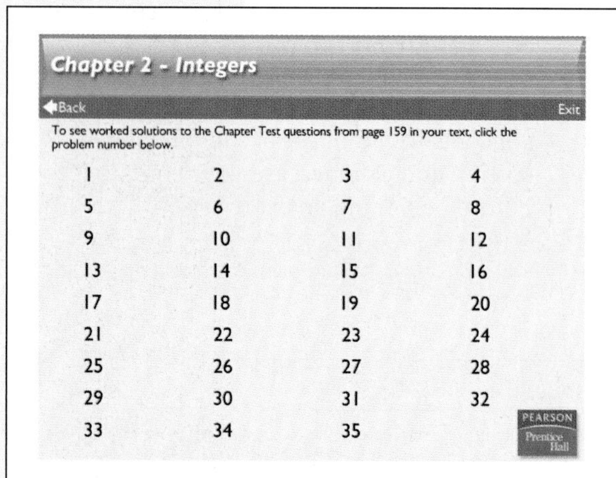

Chapter test menu allows students to select just those test questions they wish to view.

Previous and Next buttons allow for easy navigation within tests.

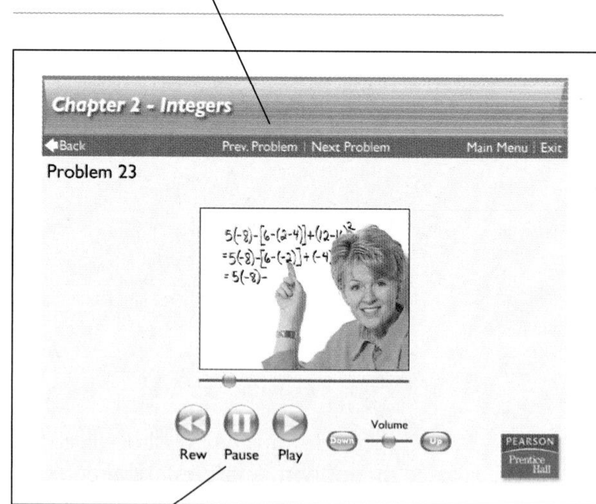

Rewind, Pause, and Play buttons let students view the solutions at their own pace.

The Video CDs include an Introduction with instructions on how to use the Chapter Test Prep Video CDs as well as test taking study skills tips. To begin preparing for your test today, see the Video CDs and Note to Students that accompany this text.

Also Available

- Online and CD based tutorial programs. For descriptions see the Preface listings for Instructor and Student Resources.
- Student Solutions Manual
- CD Lecture Series – For complete instruction, see the text specific videos available on CD. They are hosted by K. Elayn Martin-Gay and cover each objective in every chapter section of your text.

Prentice Hall Tutor Center

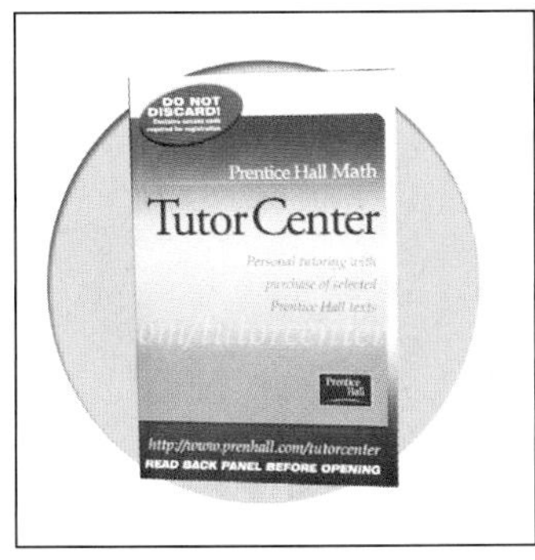

Prentice Hall Tutor Center provides text-specific tutoring via phone, fax, and e-mail. Visit http://prenhall.com/tutorcenter for details.

To Jewett B. Gay, and to the memory of her husband, Jack Gay

Whole Numbers and Introduction to Algebra

CHAPTER 1

Mathematics is an important tool for everyday life. Knowing basic mathematical skills can help simplify tasks such as creating a monthly budget. Whole numbers are the basic building blocks of mathematics. The whole numbers answer the question "How many?"

This chapter covers basic operations on whole numbers. Knowledge of these operations provides a good foundation on which to build further mathematical skills.

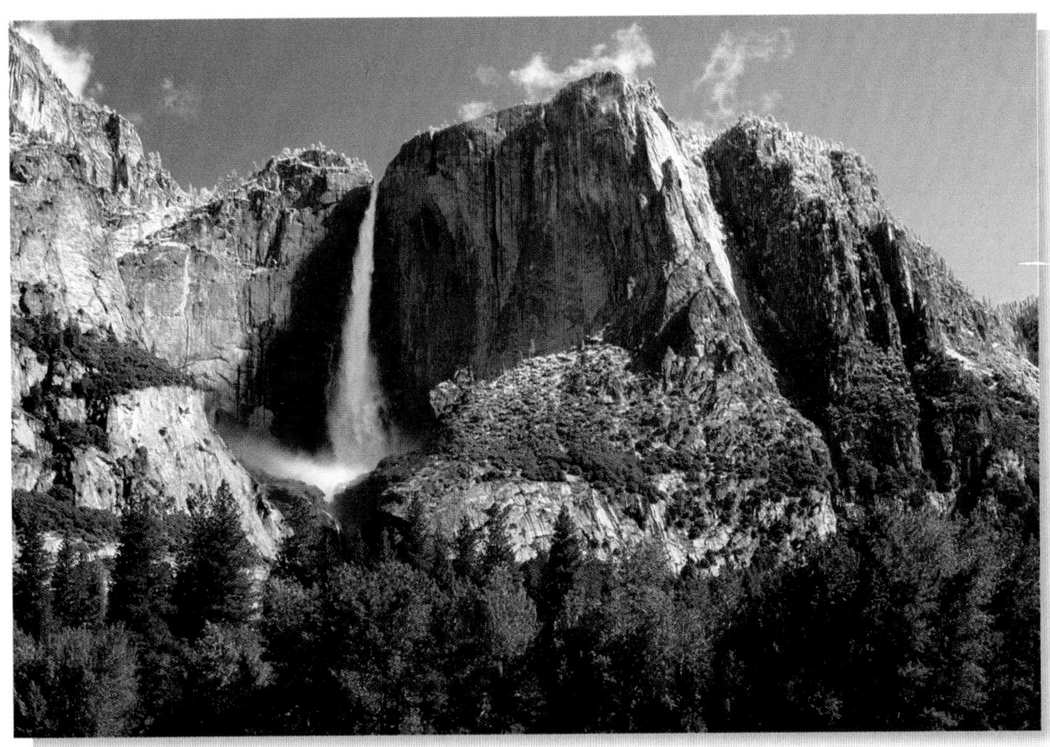

Yosemite National Park was established on October 1, 1890. With over 4 million visitors annually, it is a favorite tourist destination in the Sierra Nevada Mountains in central California. Its nearly 750,000 acres are home to many of nature's most beautiful sites, including rock formations, giant sequoias, and waterfalls. In Exercise 59 on page 30, we will see how whole numbers can be used to measure the height of Yosemite Falls, the highest waterfall in the United States.

Answers

1.

2.

3.

4.

5.

6.

7.

8.

9.

10.

11.

12.

13.

14.

15.

16.

17.

18.

19.

20.

21.

22.

Name ______________ Section __________ Date__________

Chapter 1 Pretest

1. Determine the place value of the digit 7 in the whole number 5732.

2. Write the whole number 23,490 in words.

3. Add: 58 + 29

4. Multiply: $\begin{array}{r} 413 \\ \times\ 9 \\ \hline \end{array}$

Subtract. Check by adding.

5. $\begin{array}{r} 857 \\ -\ 231 \\ \hline \end{array}$

6. $\begin{array}{r} 51 \\ -\ 19 \\ \hline \end{array}$

Solve.

7. Karen Lewis is reading a 329-page novel. If she has just finished reading page 193, how many more pages must she read to finish the novel?

8. Round 9045 to the nearest ten.

9. Round each number to the nearest hundred to find an estimated sum.

$\begin{array}{r} 382 \\ 436 \\ 2084 \\ +\ \ 176 \\ \hline \end{array}$

10. Use the distributive property to rewrite the following expression: 9(3 + 11)

△ **11.** Find the perimeter of the following figure:

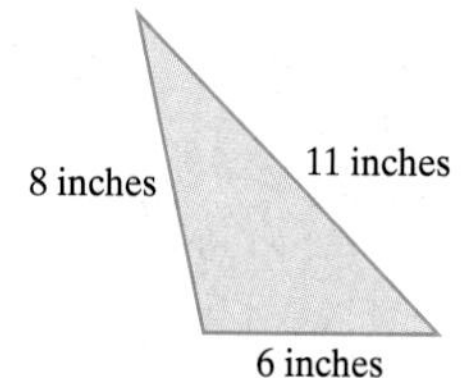

△ **12.** Find the area of the following rectangle:

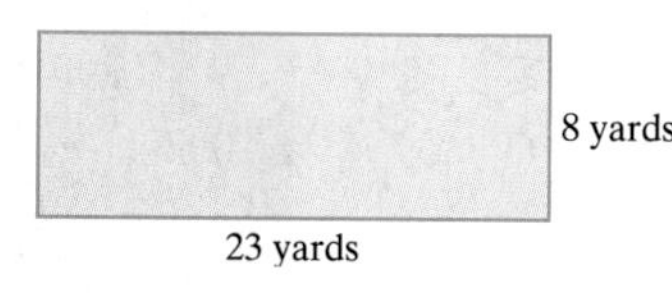

13. The seats in the history lecture hall are arranged in 32 rows with 18 seats in each row. Find how many seats are in this room.

Divide and then check by multiplying.

14. 2187 ÷ 9

15. $\frac{5361}{12}$

Solve.

16. Find the average of the following list of numbers: 29, 36, 84, 41, 6, 12, 65

17. Paul Crandall left $29,640 in his will to go to his three favorite nephews. If each boy was to receive the same amount of money, how much did each boy receive?

18. Write $9 \cdot 9 \cdot 9 \cdot 9 \cdot 9 \cdot 9 \cdot 9 \cdot 9$ using exponential notation.

19. Evaluate: 7^4.

20. Simplify: 36 + 18 ÷ 6

21. Evaluate $\frac{2x^2 + 3}{y}$ for $x = 3$ and $y = 7$.

22. Write the phrase as a variable expression. Use x to represent "a number." The sum of twice a number and 6

1.1 Tips for Success in Mathematics

OBJECTIVES

- A Get ready for this course.
- B Understand some general tips for success.
- C Understand how to use this text.
- D Get help as soon as you need it.
- E Learn how to prepare for and take a test.
- F Develop good time management.

Before reading this section, remember that your instructor is your best source for information. Please see your instructor for any additional help or information.

A Getting Ready for This Course

Now that you have decided to take this course, remember that a *positive attitude* will make all the difference in the world. Your belief that you can succeed is just as important as your commitment to this course. Make sure that you are ready for this course by having the time and positive attitude that it takes to succeed.

Next, make sure that you have scheduled your math course at a time that will give you the best chance for success. For example, if you are also working, you may want to check with your employer to make sure that your work hours will not conflict with your course schedule. Also, schedule your class during a time of day when you are more attentive and do your best work.

On the day of your first class period, double-check your schedule and allow yourself extra time to arrive in case of traffic problems or difficulty locating your classroom. Make sure that you bring at least your textbook, paper, and a writing instrument. Are you required to have a lab manual, graph paper, calculator, or other supplies besides this text? If so, also bring this material with you.

B General Tips for Success

Below are some general tips that will increase your chance for success in a mathematics class. Many of these tips will also help you in other courses you may be taking.

Exchange names and phone numbers with at least one other person in class. This contact person can be a great help if you miss an assignment or want to discuss math concepts or exercises that you find difficult.

Choose to attend all class periods and be on time. If possible, sit near the front of the classroom. This way, you will see and hear the presentation better. It may also be easier for you to participate in classroom activities.

Do your homework. You've probably heard the phrase "practice makes perfect" in relation to music and sports. It also applies to mathematics. You will find that the more time you spend solving mathematics problems, the easier the process becomes. Be sure to schedule enough time to complete your assignments before the next class period.

Check your work. Review the steps you made while working a problem. Learn to check your answers in the original problems. You may also compare your answers with the answers to selected exercises section in the back of the book. If you have made a mistake, try to figure out what went wrong. Then correct your mistake. If you can't find what went wrong don't erase your work or throw it away. Bring your work to your instructor, a tutor in a math lab, or a classmate. It is easier for someone to find where you had trouble if they look at your original work.

Learn from your mistakes and be patient with yourself. Everyone, even your instructor, makes mistakes. Use your errors to learn and to become a better math student. The key is finding and understanding your errors.

Was your mistake a careless one, or did you make it because you can't read your own math writing? If so, try to work more slowly or write more neatly and make a conscious effort to carefully check your work.

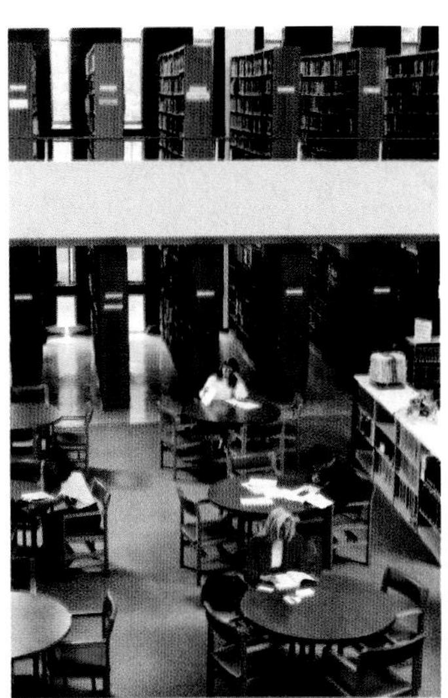

Did you make a mistake because you don't understand a concept? Take the time to review the concept or ask questions to better understand it.

Did you skip too many steps? Skipping steps or trying to do too many steps mentally may lead to preventable mistakes.

Know how to get help if you need it. It's all right to ask for help. In fact, it's a good idea to ask for help whenever there is something that you don't understand. Make sure you know when your instructor has office hours and how to find his or her office. Find out whether math tutoring services are available on your campus. Check out the hours, location, and requirements of the tutoring service. Videotapes and software are available with this text. Learn how to access these resources.

Organize your class materials, including homework assignments, graded quizzes and tests, and notes from your class or lab. All of these items will make valuable references throughout your course especially when studying for upcoming tests and the final exam. Make sure that you can locate these materials when you need them.

Read your textbook before class. Reading a mathematics textbook is unlike leisure reading such as reading a book or newspaper. Your pace will be much slower. It is helpful to have paper and a pencil with you when you read. Try to work out examples on your own as you encounter them in your text. You should also write down any questions that you want to ask in class. When you read a mathematics textbook, some of the information in a section may be unclear. But after you hear a lecture or watch a videotape on that section, you will understand it much more easily than if you had not read your text beforehand.

Don't be afraid to ask questions. Instructors are not mind readers. Many times we do not know a concept is unclear until a student asks a question. You are not the only person in class with questions. Other students are normally grateful that someone has spoken up.

Hand in assignments on time. This way you can be sure that you will not lose points for being late. Show every step of a problem and be neat and organized. Also be sure that you understand which problems are assigned for homework. You can always double-check the assignment with another student in your class.

C Using This Text

There are many helpful resources that are available to you in this text. It is important that you become familiar with and use these resources. They should increase your chances for success in this course.

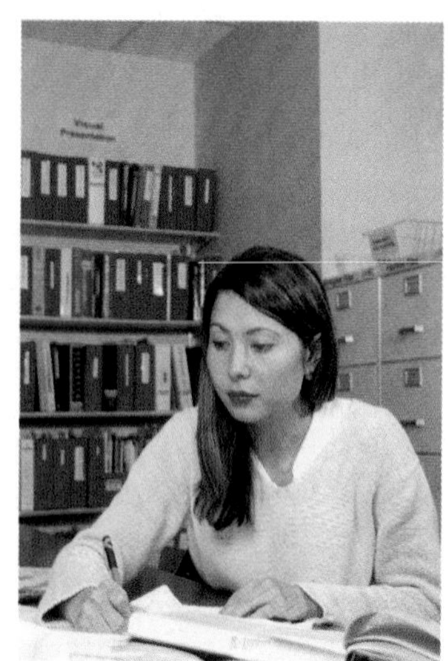

- Each example in every section has a parallel Practice Problem. As you read a section, try each Practice Problem after you've finished the corresponding example. This "learn-by-doing" approach will help you grasp ideas before you move on to other concepts.
- The main section of exercises in each exercise set is referenced by an objective, such as A or B, and also via examples. Use this referencing if you have trouble completing an assignment from the exercise set.
- If you need extra help in a particular section, look at the beginning of the section to see what videotapes and software are available.
- Make sure that you understand the meaning of the icons that are beside many exercises. The video icon tells you that the corresponding exercise may be viewed on the videotape that corresponds to that section. The pencil icon tells you that this exercise is a writing exercise in which you should answer in complete sentences. The △ icon tells you that the exercise involves geometry.

- Integrated Reviews in each chapter offer you a chance to practice—in one place—the many concepts that you have learned separately over several sections.
- There are many opportunities at the end of each chapter to help you understand the concepts of the chapter.

  Chapter Highlights contain chapter summaries and examples.
  Chapter Reviews contain review problems organized by section.
  Chapter Tests are sample tests to help you prepare for an exam.
  Cumulative Reviews are reviews consisting of material from the beginning of the book to the end of that particular chapter.

See the preface at the beginning of this text for a more thorough explanation of the features of this text.

D Getting Help

If you have trouble completing assignments or understanding the mathematics, get help as soon as you need it! This tip is presented as an objective on its own because it is so important. In mathematics, usually the material presented in one section builds on your understanding of the previous section. This means that if you don't understand the concepts covered during a class period, there is a good chance that you will not understand the concepts covered during the next class period. If this happens to you, get help as soon as you can.

Where can you get help? Many suggestions have been made in the section on where to get help, and now it is up to you to do it. Try your instructor, a tutoring center, or a math lab, or you may want to form a study group with fellow classmates. If you do decide to see your instructor or go to a tutoring center, make sure that you have a neat notebook and are ready with your questions.

E Preparing for and Taking a Test

Make sure that you allow yourself plenty of time to prepare for a test. If you think that you are a little "math anxious," it may be that you are not preparing for a test in a way that will ensure success. The way that you prepare for a test in mathematics is important. To prepare for a test,

1. Review your previous homework assignments.
2. Review any notes from class and section-level quizzes you have taken. (If this is a final exam, also review chapter tests you have taken.)
3. Review concepts and definitions by reading the Highlights at the end of each chapter.
4. Practice working out exercises by completing the Chapter Review found at the end of each chapter. (If this is a final exam, go through a Cumulative Review. There is one found at the end of each chapter except Chapter 1. Choose the review found at the end of the latest chapter that you have covered in your course.) *Don't stop here!*
5. It is important that you place yourself in conditions similar to test conditions to find out how you will perform. In other words, as soon as you feel that you know the material, get a few blank sheets of paper and take a sample test. There is a Chapter Test available at the end of each chapter. During this sample test, do not use your notes or your textbook. Once you complete the Chapter Test, check your answers in the back of the book. If any answer is incorrect, there is a CD available with each exercise of each chapter test worked. Use this CD or your instructor to correct your sample test. Your instructor may also provide

you with a review sheet. If you are not satisfied with the results, study the areas that you are weak in and try again.

6. Get a good night's sleep before the exam.
7. On the day of the actual test, allow yourself plenty of time to arrive at where you will be taking your test.

When taking your test,

1. Read the directions on the test carefully.
2. Read each problem carefully as you take the test. Make sure that you answer the question asked.
3. Watch your time and pace yourself so that you can attempt each problem on your test.
4. If you have time, check your work and answers.
5. Do not turn your test in early. If you have extra time, spend it double-checking your work.

F Managing Your Time

As a college student, you know the demands that classes, homework, work, and family place on your time. Some days you probably wonder how you'll ever get everything done. One key to managing your time is developing a schedule. Here are some hints for making a schedule:

1. Make a list of all of your weekly commitments for the term. Include classes, work, regular meetings, extracurricular activities, etc. You may also find it helpful to list such things as laundry, regular workouts, grocery shopping, etc.
2. Next, estimate the time needed for each item on the list. Also make a note of how often you will need to do each item. Don't forget to include time estimates for the reading, studying, and homework you do outside of your classes. You may want to ask your instructor for help estimating the time needed.
3. In the following Exercise Set, you are asked to block out a typical week on the schedule grid given. Start with items with fixed time slots like classes and work.
4. Next, include the items on your list with flexible time slots. Think carefully about how best to schedule some items such as study time.
5. Don't fill up every time slot on the schedule. Remember that you need to allow time for eating, sleeping, and relaxing! You should also allow a little extra time in case some items take longer than planned.
6. If you find that your weekly schedule is too full for you to handle, you may need to make some changes in your workload, classload, or in other areas of your life. You may want to talk to your advisor, manager or supervisor at work, or someone in your college's academic counseling center for help with such decisions.

Note: In this chapter, we begin a feature called Study Skills Reminder. The purpose of this feature is to remind you of some of the information given in this section and to further expand on some topics in this section.

Name ______________________ Section ____________ Date ____________

EXERCISE SET 1.1

1. What is your instructor's name?

2. What are your instructor's office location and office hours?

3. What is the best way to contact your instructor?

4. What does the icon mean?

5. What does the icon mean?

6. What does the icon mean?

7. Where are answers located in this text?

8. What Exercise Set answers are available to you in the answers section?

9. What Chapter Review, Chapter Text, and Cumulative Text answers are available to you in the answer section?

10. Are there worked-out solutions to exercises in this text?

11. If the answer to Exercise 10 is yes, what worked-out solutions are available to you in this text?

12. Go to the Highlights Section at the end of this chapter. Describe how this section may be helpful to you when preparing for a test.

13. Do you have the name and contact information of at least one other student in class?

14. Will your instructor allow you to use a calculator in this class?

15. Are videotapes, CDs, and/or tutorial software available to you? If so, where?

16. Is there a tutoring service available? If so, what are its hours?

17. Have you attempted this course before? If so, write down ways that you might improve your chances of success during this second attempt.

18. List some steps that you can take if you begin having trouble understanding the material or completing an assignment.

19. Read or reread objective F and fill out the schedule grid below.

	Monday	Tuesday	Wednesday	Thursday	Friday	Saturday	Sunday
7:00 a.m.							
8:00 a.m.							
9:00 a.m.							
10:00 a.m.							
11:00 a.m.							
12:00 a.m.							
1:00 p.m.							
2:00 p.m.							
3:00 p.m.							
4:00 p.m.							
5:00 p.m.							
6:00 p.m.							
7:00 p.m.							
8:00 p.m.							
9:00 p.m.							

20. Study your filled-out grid from Exercise 19. Decide whether you have the time necessary to successfully complete this course and any other courses you may be registered for.

1.2 Place Value and Names for Numbers

OBJECTIVES

- **A** Find the place value of a digit in a whole number.
- **B** Write a whole number in words and in standard form.
- **C** Write a whole number in expanded form.
- **D** Compare two whole numbers.
- **E** Read tables.

The **digits** 0, 1, 2, 3, 4, 5, 6, 7, 8, and 9 can be used to write numbers. For example, the **natural numbers** are

1, 2, 3, 4, 5, 6, 7, 8, 9, 10, 11, ...

The natural numbers are also called the counting numbers. The three dots (...) after the 11 mean that this list continues indefinitely. That is, there is no largest natural number. The smallest natural number is 1.

If 0 is included with the natural numbers, we have the whole numbers. The whole numbers are

0, 1, 2, 3, 4, 5, 6, 7, 8, 9, 10, 11, 12, ...

There is no largest whole number. The smallest whole number is 0.

A Finding the Place Value of a Digit in a Whole Number

The position of each digit in a number determines its **place value**. A place-value chart is shown next for the whole number 48,337,000.

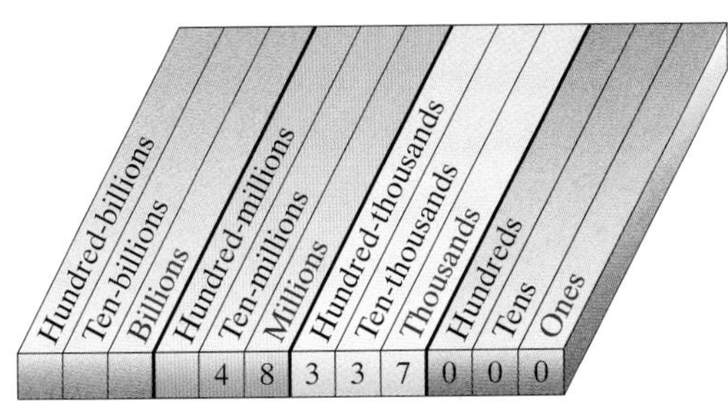

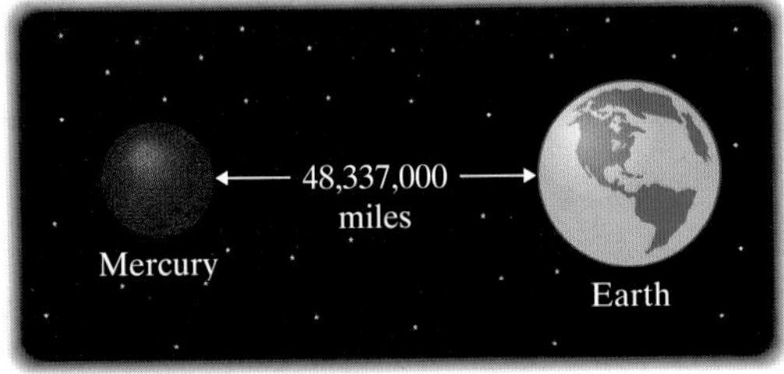

The two 3s in 48,337,000 represent different amounts because of their different placements. The place value of the 3 on the left is hundred-thousands. The place value of the 3 on the right is ten-thousands.

EXAMPLES Find the place value of the digit 4 in each whole number.

1. 48,761
↑
ten-thousands

2. 249
↑
tens

3. 524,007,656
↑
millions

Practice Problems 1–3

Find the place value of the digit 7 in each whole number.

1. 72,589,620
2. 67,890
3. 50,722

Answers

1. ten-millions **2.** thousands **3.** hundreds

B Writing a Whole Number in Words and in Standard Form

A whole number such as 1,083,664,500 is written in **standard form**. Notice that commas separate the digits into groups of three, starting from the right. Each group of three digits is called a **period**. The names of the first four periods are shown in red and above the chart.

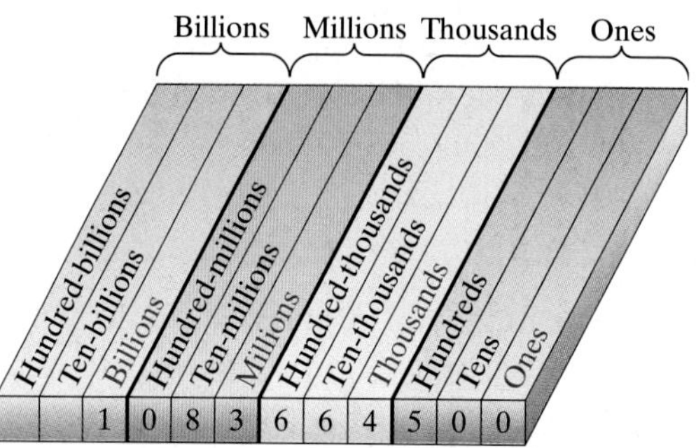

Writing a Whole Number in Words

To write a whole number in words, write the number in each period followed by the name of the period. (The ones period is usually not written.) This same procedure can be used to read a whole number.

For example, we write 1,083,664,500 as

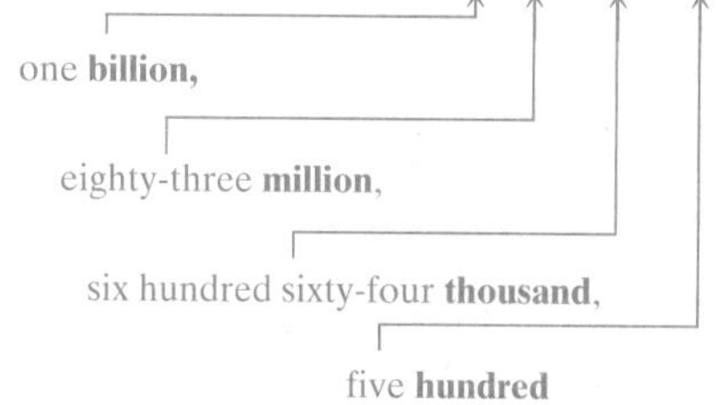

Helpful Hint

The name of the ones period is not used when reading and writing whole numbers. For example,

9,265

is read as

"nine **thousand**, two **hundred** sixty-five."

EXAMPLES Write each number in words.

4. 126 one hundred twenty-six
5. 3005 three thousand five

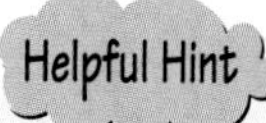

The word "and" is *not* used when reading and writing whole numbers. It is used when reading and writing mixed numbers and some decimal values, as shown later in this text.

Practice Problems 4–5

Write each number in words.

4. 395
5. 2807

Answers

4. three hundred ninety-five
5. two thousand eight hundred seven

EXAMPLE 6 Write 106,052,447 in words.

Solution: 106,052,447 is written as

one hundred six million, fifty-two thousand, four hundred forty-seven

Try the Concept Check in the margin.

Writing a Whole Number in Standard Form

To write a whole number in standard form, write the number in each period, followed by a comma.

EXAMPLES Write each number in standard form.

7. sixty-one 61

8. eight hundred five 805

9. two million, five hundred sixty-four thousand, three hundred fifty

2,564,350

10. nine thousand, three hundred eighty-six

9,386 or 9386

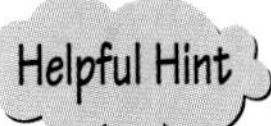

A comma may or may not be inserted in a four-digit number. For example, both

9,386 and 9386

are acceptable ways of writing nine thousand, three hundred eighty-six.

C Writing a Whole Number in Expanded Form

The place value of a digit can be used to write a number in expanded form. The **expanded form** of a number shows each digit of the number with its place value. For example, 5672 is written in expanded form as

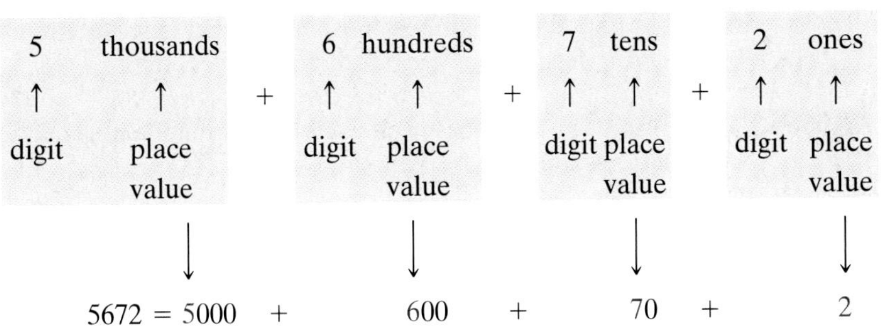

Practice Problem 6

Write 321,670,200 in words.

Concept Check

True or false? When writing a check for $2600, the word name we write for the dollar amount of the check would be "two thousand sixty." Explain your answer.

Practice Problems 7–10

Write each number in standard form.

7. twenty-nine
8. seven hundred ten
9. twenty-six thousand seventy-one
10. six thousand, five hundred seven

Concept Check: false

Answers

6. three hundred twenty-one million, six hundred seventy thousand, two hundred

7. 29 **8.** 710 **9.** 26,071 **10.** 6507

Practice Problem 11

Write 1,047,608 in expanded form.

EXAMPLE 11 Write 706,449 in expanded form.

Solution: 700,000 + 6000 + 400 + 40 + 9

STUDY SKILLS REMINDER

Many of the terms used in this text may be new to you. It will be helpful to make a list of new mathematical terms and symbols as you encounter them and to review them frequently. Placing these new terms (including page references) on 3 × 5 index cards might help you later when you're preparing for a quiz.

D Comparing Whole Numbers

We can picture whole numbers as equally spaced points on a line called the **number line**. The whole numbers are written in "counting" order on the number line beginning with 0. An arrow at the right end of the number line means that the whole numbers continue indefinitely.

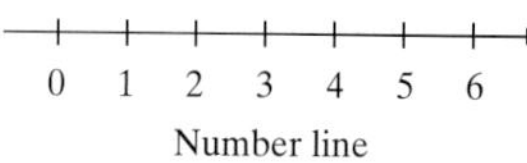

Number line

A whole number is **graphed** by placing a dot on the number line at the point corresponding to the number. For example, 5 is graphed on the number line as shown.

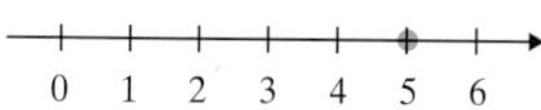

For any two numbers graphed on a number line, the number to the *right* is the **greater number**, and the number to the *left* is the **smaller number**. For example, 3 is *to the left* of 6, so 3 **is less than** 6. Also, 6 is *to the right* of 3, so 6 **is greater than** 3.

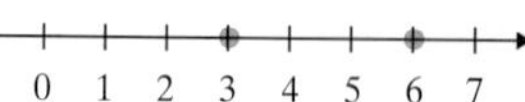

We use the symbol $<$ to mean "is less than" and the symbol $>$ to mean "is greater than." The symbols $<$ and $>$ are called **inequality** symbols. For example, the true statement "*3 is less than 6*" can be written in symbols as

$3 < 6$

Also, "*6 is greater than 3*" is written as

$6 > 3$

The statements $3 < 6$ and $6 > 3$ are both *true statements*. The statement $5 < 2$ is a *false statement* because 5 is greater than 2, not less than.

Practice Problem 12

Insert $<$ or $>$ between each pair of numbers to make a true statement.

a. 0 19 b. 18 32 c. 107 103

EXAMPLE 12 Insert $<$ or $>$ between each pair of numbers to make a true statement.

a. 5 50 **b.** 101 0 **c.** 29 27

Solution:

a. $5 < 50$ **b.** $101 > 0$ **c.** $29 > 27$

Answers

11. 1,000,000 + 40,000 + 7000 + 600 + 8
12. a. $<$ **b.** $<$ **c.** $>$

One way to remember the meaning of the inequality symbols $<$ and $>$ is to think of them as arrowheads "pointing" toward the smaller number. For example, $5 < 11$ and $11 > 5$ are both true statements.

E Reading Tables

Now that we know about place value and names for whole numbers, we introduce one way that whole number data may be presented. **Tables** are often used to organize and display facts that contain numbers. The table below shows the countries that have won the most medals during the Olympic winter games for the years 1924 through 2002. (Although the medals are truly won by athletes from the various countries, for simplicity we will state that countries have won the medals.)

Most Medals Olympic Winter (1924 – 2002) Games

	Gold	Silver	Bronze	Total
Germany[1]	107	104	86	297
Russia	113	83	78	274
Norway	94	92	74	260
United States	69	71	51	191
Austria	41	57	64	162
Finland	41	51	49	141
Sweden	39	30	39	108
Switzerland	32	33	38	103
Canada	30	28	36	94

(*Source:* The Sydney Morning Herald)
[1]Includes West Germany 1952, 1968–1988; East Germany 1968–1988

For example, by reading from left to right along the row marked "U.S." we find that the United States won 69 gold, 71 silver, and 51 bronze medals during the Olympic Winter Games 1924–2002.

Practice Problem 13

Use the winter Games table to answer the following questions:

a. How many gold medals did Russia win during the winter Games of the Olympics?

b. Which countries shown have won more than 100 gold medals?

Answers

13. a. 113 **b.** Germany and Russia

EXAMPLE 13 Use the Winter Games table to answer each question.

a. How many silver medals did Sweden win during the Winter Games of the Olympics?

b. Which countries shown have won fewer bronze medals than Austria?

Solution:

a. Find "Sweden" in the left column. Then read from left to right until the "Silver" column is reached. We find that Sweden has won 30 silver medals.

b. Austria has won 64 bronze medals. United States, Finland, Sweden, Switzerland, and Canada have each won fewer than 64 bronze medals.

Name ______________________ Section __________ Date __________

EXERCISE SET 1.2

A *Determine the place value of the digit 5 in each whole number. See Examples 1 through 3.*

1. 352

2. 905

3. 5890

4. 6527

5. 62,500,000

6. 79,050,000

7. 5,070,099

8. 51,682,700

B *Write each whole number in words. See Examples 4 through 6.*

9. 5420

10. 3165

11. 26,990

12. 42,009

13. 1,620,000

14. 3,204,000

15. 53,520,170

16. 47,033,107

Write each number in the sentence in words. See Examples 4 through 6.

17. Bermuda is a British dependency consisting of many small islands located in the Atlantic Ocean east of North Carolina. At this writing, the population of Bermuda is 63,960. (*Source*: The World Almanac, 2003)

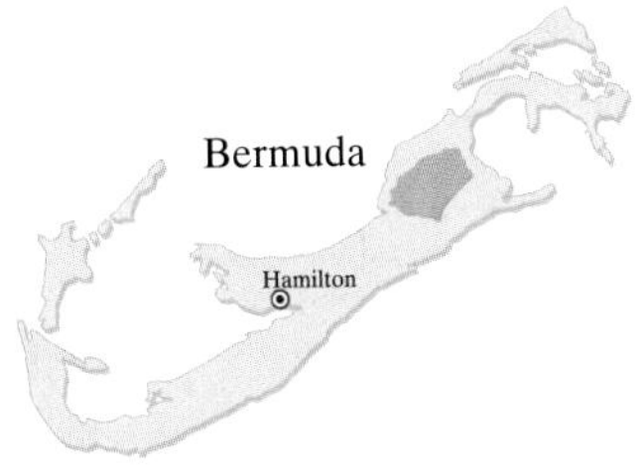

18. The Goodyear blimp *Eagle* holds 202,700 cubic feet of helium. (*Source*: The Goodyear Tire & Rubber Company)

19. The world's tallest building, the PETRONAS Twin Towers in Kuala Lumpur, Malaysia, is 1483 feet tall. (*Source*: Council on Tall Buildings and Urban Habitat)

20. Liz Harold has the number 16,820,409 showing on her calculator display.

21. The highest point in Idaho is at Granite Peak, at an elevation of 12,662 feet. (*Source*: U.S. Geological Survey)

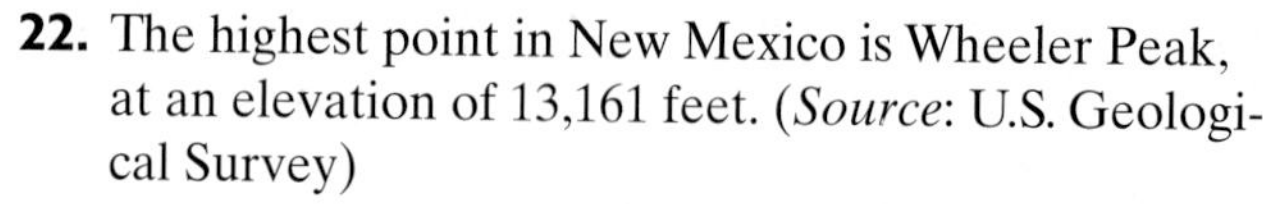

22. The highest point in New Mexico is Wheeler Peak, at an elevation of 13,161 feet. (*Source*: U.S. Geological Survey)

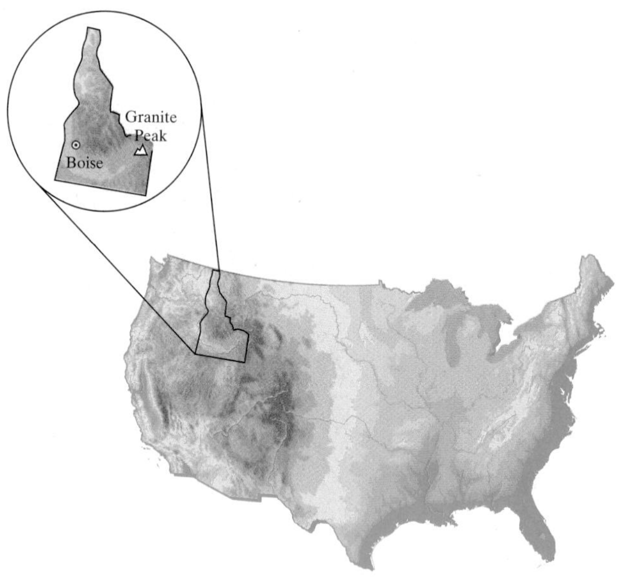

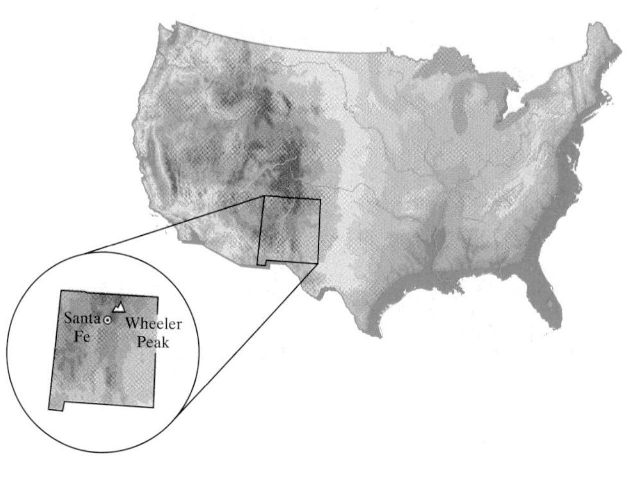

23. Each day, UPS delivers 13,600,000 packages and documents worldwide. (*Source*: United Parcel Service of America, Inc.)

24. In a recent year, zinc mines in the United States mined 799,000 metric tons of zinc. (*Source*: U.S. Dept. of Interior)

25. In a recent year, there were 3895 patients in the United States waiting for a heart transplant. (*Source*: United Network for Organ Sharing)

26. Each Home Depot store in the United States and Canada stocks at least 40,000 different kinds of building materials, home improvement supplies, and lawn and garden products. (*Source*: The Home Depot, Inc.)

Write each whole number in standard form. See Examples 7 through 10.

27. Six thousand, five hundred eight

28. Three thousand, three hundred seventy

29. Twenty-nine thousand, nine hundred

30. Forty-two thousand, six

31. Six million, five hundred four thousand, nineteen

32. Ten million, thirty-seven thousand, sixteen

33. Three million, fourteen

34. Seven million, twelve

Write the whole number in each sentence in standard form. See Examples 7 through 10.

35. The International Space station orbits above Earth at an altitude of two hundred twenty miles. (NASA)

36. The average distance between the surfaces of the Earth and the Moon is about 234 thousand miles.

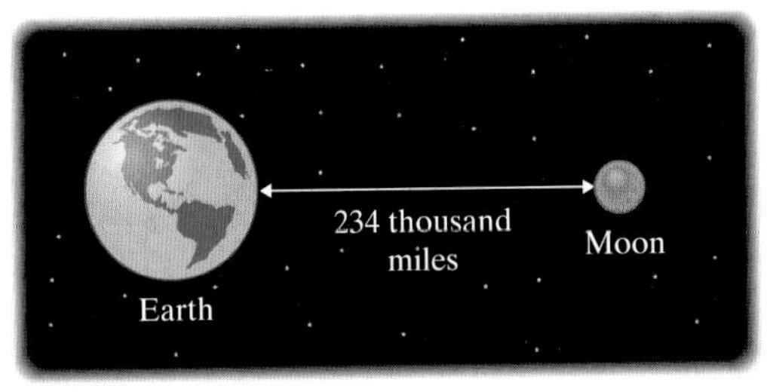

37. Hank Aaron holds the career record for home runs in Major League baseball, with a total of seven hundred fifty-five home runs. (*Source*: Major League Baseball)

38. The average annual salary for an NHL hockey player for the 2000–2001 season was one million, four hundred thousand dollars. (*Source*: NHL Players Association)

39. In 2002, there were seventy-three million, five hundred thousand U.S. households that have cable TV. (*Source*: Nielsen Media Research)

40. There were thirty thousand slices of pizza consumed at a recent Daytona 500 race. (*Source*: Americrown Service Corporation)

41. The world's tallest self-supporting structure is the CN Tower in Toronto, Canada. It is one thousand, eight hundred fifteen feet tall. (*Source*: The World Almanac, 2003)

42. At this writing, there are about 3 billion Web pages.

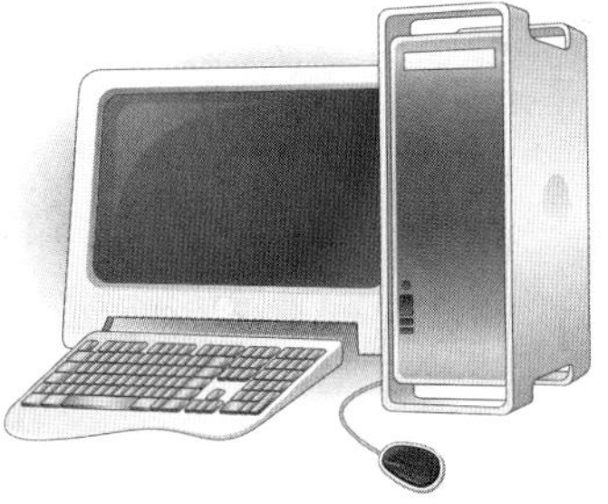

43. As of 2003, there were one thousand, two hundred sixty-two species classified as either threatened or endangered in the United States. (*Source*: U.S. Fish & Wildlife Service)

44. Grain storage facilities in Toledo, Ohio, have a capacity of sixty-three million, one hundred thousand bushels. (*Source*: Chicago Board of Trade Market Information Department)

45. FedEx employees retrieve packages from over forty-five thousand FedEx drop boxes around the world each business day. (*Source*: Federal Express Corporation)

46. One hundred million pounds of nonhazardous waste is recycled at Kodak Park each year. (*Source*: Eastman Kodak Company)

C *Write each whole number in expanded form. See Example 11.*

47. 406 **48.** 789 **49.** 5290 **50.** 6040

51. 62,407 **52.** 20,215 **53.** 30,680 **54.** 99,032

55. 39,680,000 **56.** 47,703,029 **57.** 1006 **58.** 20,304

D *Insert* $<$ or $>$ *to make a true statement. See Example 12.*

59. 3 8 **60.** 7 10 **61.** 9 0 **62.** 0 12

63. 6 2 **64.** 15 14 **65.** 22 0 **66.** 39 47

E *The table shows the five longest rivers in the world. Use this table to answer Exercises 67 through 72. See Example 13.*

River	Miles
Chang jiang-Yangtze (China)	3964
Amazon (Brazil)	4000
Tenisei-Angara (Russia)	3442
Mississippi-Missouri (U.S.)	3740
Nile (Egypt)	4145

67. Write the length of the Amazon River in words.

68. Write the length of the Tenisei-Angara River in words.

69. Write the length of the Nile River in expanded form.

70. Write the length of the Mississippi-Missouri River in expanded form.

71. Which river is the longest in the world?

72. Which river is the second longest in the world?

The table shows the top ten breeds of dogs in 2001 according to the American Kennel Club. Use this table to answer Exercises 73 through 76. See Example 13.

Top Ten American Kennel Club Registrations in 2001	
Breed	**Number of Registered Dogs**
Beagles	50,419
Boxers	37,035
Chihuahuas	36,627
Dachshunds	50,478
German Shepherd Dogs	51,625
Golden Retrievers	62,497
Labrador Retrievers	165,970
Poodles	40,550
Shih Tzu	33,240
Yorkshire Terriers	42,025

73. Which breed has the most American Kennel Club registrations? Write the number of registrations for this breed in words.

74. Which of the listed breeds has the fewest registrations? Write the number of registered dogs for this breed in words.

75. Which breed has more dogs registered, Chihuahua or Golden retriever?

76. Which breed has fewer dogs registered, Beagle or Yorkshire terrier?

Combining Concepts

77. Write the largest four-digit number that can be made from the digits 3, 6, 7, and 2 if each digit must be used once. ____ ____ ____ ____

78. Write the largest five-digit number that can be made using the digits 4, 5, and 3 if each digit must be used at least once. ____ ____, ____ ____ ____

79. If a number is given in words, describe the process used to write this number in standard form.

80. If a number is written in standard form, describe the process used to write this number in expanded form.

81. The Pro-Football Hall of Fame was established on September 7, 1963, in this town. Use the information and the accompanying diagram to find the name of the town.

- Alliance is east of Massillon.
- Dover is between Canton and New Philadelphia.
- Massillon is not next to Alliance.
- Canton is north of Dover.

1.3 Adding Whole Numbers and Perimeter

OBJECTIVES

- **A** Add whole numbers.
- **B** Find the perimeter of a polygon.
- **C** Solve problems by adding whole numbers.

SSM TUTOR CENTER SG CD & VIDEO MATH PRO WEB

A Adding Whole Numbers

If one computer in an office has a 2-megabyte memory and a second computer has a 4-megabyte memory, the total memory in the two computers can be found by adding 2 and 4.

2 megabytes + 4 megabytes = 6 megabytes

The **sum** is 6 megabytes of memory. Each of the numbers 2 and 4 is called an **addend**.

$$\underset{\text{addend}}{\underset{\uparrow}{2}} + \underset{\text{addend}}{\underset{\uparrow}{4}} = \underset{\text{sum}}{\underset{\uparrow}{6}}$$

To add whole numbers, we add the digits in the ones place, then the tens place, then the hundreds place, and so on. For example, let's add 2236 + 160.

$$\begin{array}{r} 2236 \\ +\ 160 \\ \hline 2396 \end{array}$$

Line up numbers vertically so that the place values correspond. Then add digits in corresponding place values, starting with the ones place.

In 2396: 6 — sum of ones; 9 — sum of tens; 3 — sum of hundreds; 2 — sum of thousands

EXAMPLE 1 Add: 23 + 136

Solution:

$$\begin{array}{r} 23 \\ +136 \\ \hline 159 \end{array}$$

Practice Problem 1

Add: 7235 + 542

When the sum of digits in corresponding place values is more than 9, "carrying" is necessary. For example, to add 365 + 89, add the ones-place digits first.

$$\begin{array}{r} \overset{1}{3}65 \\ +\ 89 \\ \hline 4 \end{array}$$

5 ones + 9 ones = **14 ones** or **1 ten** + **4 ones**

Write the 4 ones in the ones place and carry the 1 ten to the tens place.

Next, add the tens-place digits.

Answer

1. 7777

$$\begin{array}{r} {}^{1\,1} \\ 365 \\ +\ 89 \\ \hline 54 \end{array}$$

1 ten + 6 tens + 8 tens = **15 tens** or **1 hundred** + **5 tens**
Write the 5 tens in the tens place and carry the 1 hundred to the hundreds place.

Next, add the hundreds-place digits.

$$\begin{array}{r} {}^{1\,1} \\ 365 \\ +\ 89 \\ \hline 454 \end{array}$$

1 hundred + 3 hundreds = 4 hundreds

Write the 4 hundreds in the hundreds place.

Practice Problem 2

Add: 27,364 + 92,977

EXAMPLE 2 Add: 34,285 + 149,761

Solution:

$$\begin{array}{r} {}^{1\,1\ \ 1} \\ 34{,}285 \\ +\ 149{,}761 \\ \hline 184{,}046 \end{array}$$

Concept Check

What is wrong with the following computation?

$$\begin{array}{r} 394 \\ +\ 283 \\ \hline 577 \end{array}$$

Try the Concept Check in the margin.

Before we continue adding whole numbers, let's review some properties of addition that you may have already discovered. The first property that we will review is the **addition property of 0.** This property reminds us that the sum of 0 and any number is that same number.

Addition Property of 0

The sum of 0 and any number is that number. For example,

$7 + 0 = 7$

$0 + 7 = 7$

Next, notice that we can add any two whole numbers in any order and the sum is the same. For example,

$4 + 5 = 9$ and $5 + 4 = 9$

We call this special property of addition the **commutative property of addition.**

Commutative Property of Addition

Changing the **order** of two addends does not change their sum. For example,

$2 + 3 = 5$ and $3 + 2 = 5$

Another property that can help us when adding numbers is the **associative property of addition.** This property states that when adding numbers, the grouping of the numbers can be changed without changing the sum. We use parentheses to group numbers. They indicate what numbers to add first. For example, let's use two different groupings to find the sum of 2 + 1 + 5.

$2 + (1 + 5) = 2 + 6 = 8$

Also,

$(2 + 1) + 5 = 3 + 5 = 8$

Both groupings give a sum of 8.

Concept Check: forgot to carry 1 hundred to the hundreds place

Answer

2. 120,341

Associative Property of Addition

Changing the grouping of addends does not change their sum. For example,

$$3 + (5 + 7) = 3 + 12 = 15 \quad \text{and} \quad (3 + 5) + 7 = 8 + 7 = 15$$

The commutative and associative properties tell us that we can add whole numbers using any order and grouping that we want.

When adding several numbers, it is often helpful to look for two or three numbers whose sum is 10, 20, and so on.

EXAMPLE 3 Add: 13 + 2 + 7 + 8 + 9

Solution: 13 + 2 + 7 + 8 + 9 = 39

20 + 10 + 9

39

Practice Problem 3

Add: 11 + 7 + 8 + 9 + 13

Remember these properties as we continue practicing addition of whole numbers.

EXAMPLE 4 Add: 1647 + 246 + 32 + 85

Solution:

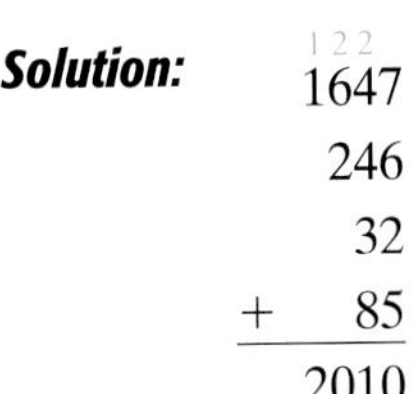

```
  122
  1647
   246
    32
+   85
------
  2010
```

Practice Problem 4

Add: 19 + 5042 + 638 + 526

B Finding the Perimeter of a Polygon

A special application of addition is finding the perimeter of a polygon. A **polygon** can be described as a flat figure formed by line segments connected at their ends. (For more review, see Appendix D.) Geometric figures such as triangles, squares, and rectangles are called polygons.

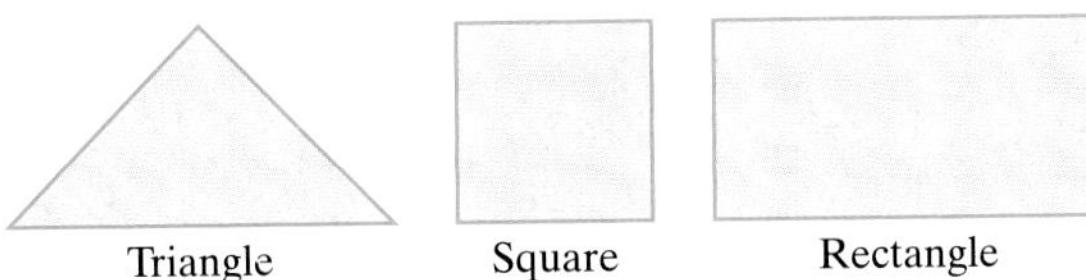

The **perimeter** of a polygon is the *distance around* the polygon. This means that the perimeter of a polygon is the sum of the lengths of its sides.

EXAMPLE 5 Find the perimeter of the polygon shown.

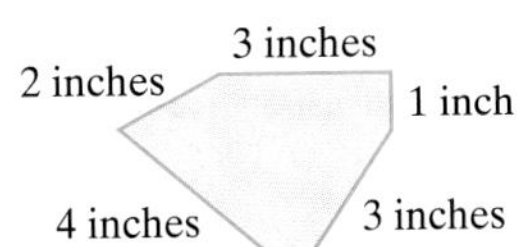

Solution: To find the perimeter (distance around), we add the lengths of the sides.

2 in. + 3 in. + 1 in. + 3 in. + 4 in. = 13 in.

The perimeter is 13 inches.

Practice Problem 5

Find the perimeter of the polygon shown. (A centimeter is a unit of length in the metric system.)

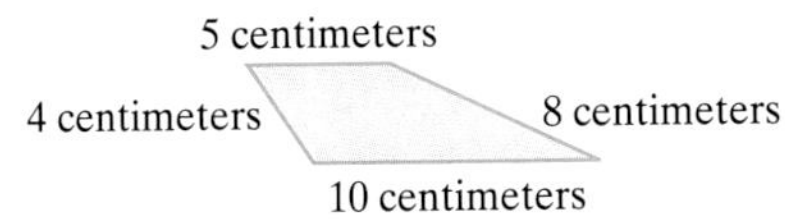

Answers

3. 48 **4.** 6225 **5.** 27 cm

Practice Problem 6

A new shopping mall has a floor plan in the shape of a triangle. Each of the mall's three sides is 532 feet. Find the perimeter of the building.

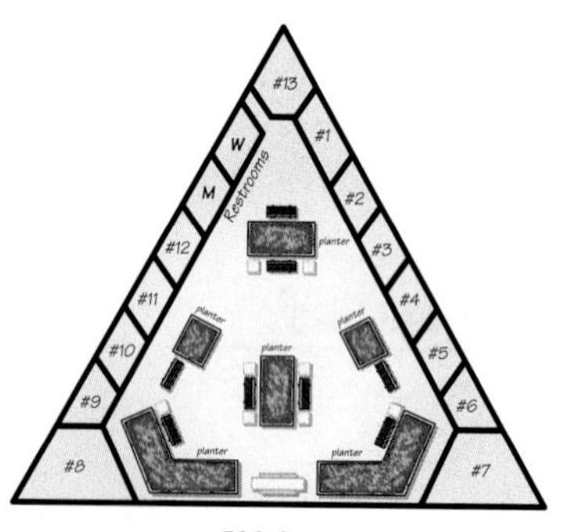

Answer

6. 1596 ft

EXAMPLE 6 Calculating the Perimeter of a Building

The largest commercial building in the world under one roof is the flower auction building of the cooperative VBA in Aalsmeer, Netherlands. The floor plan is a rectangle that measures 776 meters by 639 meters. Find the perimeter of this building. (A meter is a unit of length in the metric system.) (*Source: The Handy Science Answer Book*, Visible Ink Press)

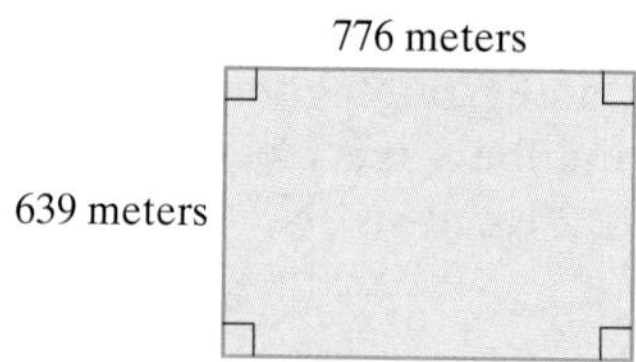

Solution: Recall that opposite sides of a rectangle have the same lengths. To find the perimeter of this building, we add the lengths of the sides. The sum of the lengths of its sides is

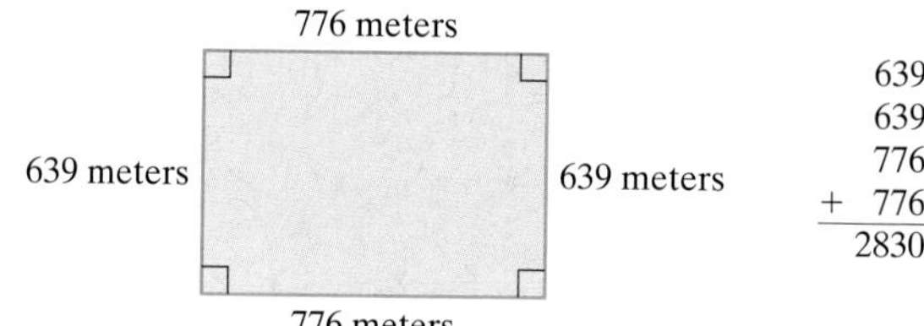

$$\begin{array}{r} 639 \\ 639 \\ 776 \\ +\ 776 \\ \hline 2830 \end{array}$$

The perimeter of the building is 2830 meters.

C Solving Problems by Adding

Often, real-life problems occur that can be solved by writing an addition statement. The first step in solving any word problem is to *understand* the problem by reading it carefully. Descriptions of problems solved through addition *may* include any of these key words or phrases:

Key Words or Phrases	Example	Symbols
added to	5 added to 7	$7 + 5$
plus	0 plus 78	$0 + 78$
increased by	12 increased by 6	$12 + 6$
more than	11 more than 25	$25 + 11$
total	the total of 8 and 1	$8 + 1$
sum	the sum of 4 and 133	$4 + 133$

To solve a word problem that involves addition, we first use the facts described to write an addition statement. Then we write the corresponding solution of the real-life problem. It is sometimes helpful to write the statement in words (brief phrases) and then translate to numbers.

EXAMPLE 7 Finding a Salary

The governor's salary in the state of Florida was recently increased by $3004. If the old salary was $120,171, find the new salary.
(*Source: The World Almanac and Book of Facts*, 2003)

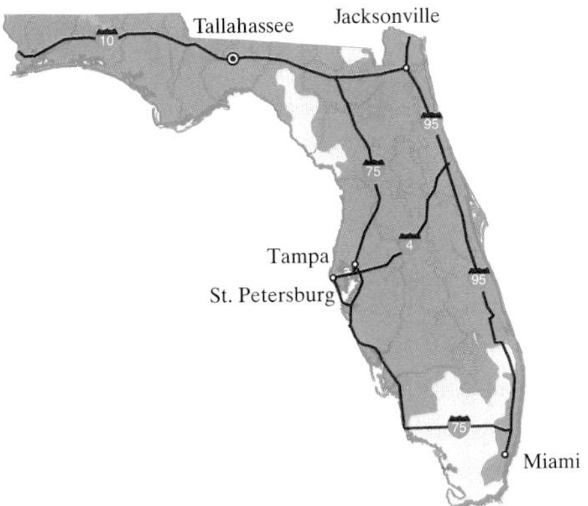

Solution: The key phrase here is "increased by," which suggests that we add. To find the new salary, we add the increase, $3004, to the old salary.

In Words		**Translate to Numbers**
old salary	→	120,171
increased by 3004	→	+ 3004
new salary	→	123,175

The Florida governor's salary is now $123,175.

EXAMPLE 8 Determining the Number of Baseball Cards in a Collection

Alan Mayfield collects baseball cards. He has 109 cards for the New York Yankees, 96 for the Chicago White Sox, 79 for the Kansas City Royals, 42 for the Seattle Mariners, 67 for the Oakland Athletics, and 52 for the California Angels. How many cards does he have in total?

Solution: The key word here is "total." To find the total number of Alan's baseball cards, we find the sum of the quantities for each team.

In Words		**Translate to Numbers**
		3 3
New York Yankees cards	→	109
Chicago White Sox cards	→	96
Kansas City Royals cards	→	79
Seattle Mariners cards	→	42
Oakland Athletics cards	→	67
+ California Angels cards	→	+ 52
Total cards	→	445

Alan has a total of 445 baseball cards.

Practice Problem 7

Texas produces 90 million pounds of pecans per year. Georgia is the world's top pecan producer and produces 15 million pounds more pecans than Texas. How much does Georgia produce? (*Source:* Absolute Trivia.com)

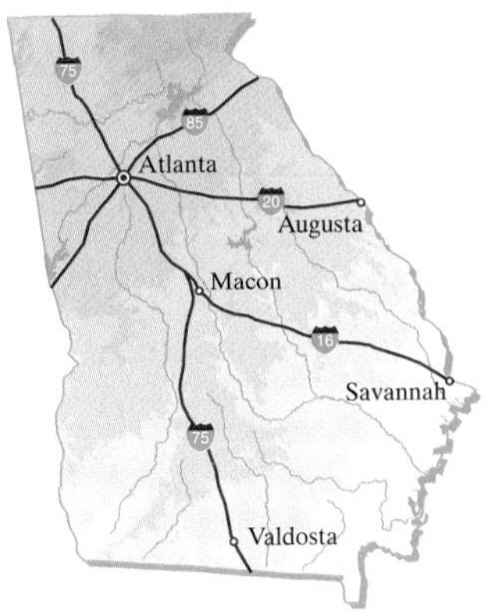

Practice Problem 8

Elham Abo-Zahrah collects thimbles. She has 42 glass thimbles, 17 steel thimbles, 37 porcelain thimbles, 9 silver thimbles, and 15 plastic thimbles. How many thimbles are in her collection?

Answers

7. 105 million lb **8.** 120 thimbles

CALCULATOR EXPLORATIONS

Adding Numbers

To add numbers on a calculator, find the keys marked [+] and [=] or [ENTER].

For example, to add 5 and 7 on a calculator, press the keys [5] [+] [7] [=] or [ENTER].
The display will read [12].
Thus, 5 + 7 = 12.
To add 687 and 981 on a calculator, press the keys [687] [+] [981] [=] or [ENTER].
The display will read [1668].
Thus, 687 + 981 = 1668. (Although entering 687, for example, requires pressing more than one key, here numbers are grouped together for easier reading.)

Use a calculator to add.

1. 89 + 45

2. 76 + 97

3. 285 + 55

4. 8773 + 652

5.
$$\begin{array}{r} 985 \\ 1210 \\ 562 \\ +\ \ 77 \\ \hline \end{array}$$

6.
$$\begin{array}{r} 465 \\ 9888 \\ 620 \\ +\ 1550 \\ \hline \end{array}$$

Name ______________________ Section __________ Date __________

Mental Math

Find each sum.

1. 5 + 7 **2.** 20 + 30 **3.** 5000 + 4000 **4.** 4300 + 26 **5.** 1620 + 0 **6.** 6 + 126 + 4

EXERCISE SET 1.3

 *Add. See Examples 1 through 4.*

1. $\begin{array}{r} 14 \\ +\ 22 \\ \hline \end{array}$ **2.** $\begin{array}{r} 27 \\ +\ 31 \\ \hline \end{array}$ **3.** $\begin{array}{r} 62 \\ +\ 30 \\ \hline \end{array}$ **4.** $\begin{array}{r} 37 \\ +\ 42 \\ \hline \end{array}$ **5.** $\begin{array}{r} 12 \\ 13 \\ +\ 24 \\ \hline \end{array}$ **6.** $\begin{array}{r} 23 \\ 45 \\ +\ 30 \\ \hline \end{array}$

 **7.** $\begin{array}{r} 5267 \\ +\ 132 \\ \hline \end{array}$ **8.** $\begin{array}{r} 236 \\ +\ 6243 \\ \hline \end{array}$ **9.** 53 + 64 **10.** 41 + 74 **11.** 22 + 49 **12.** 35 + 47

13. 38 + 79 **14.** 92 + 37 **15.** $\begin{array}{r} 8 \\ 9 \\ 2 \\ 5 \\ +\ 1 \\ \hline \end{array}$ **16.** $\begin{array}{r} 3 \\ 5 \\ 8 \\ 5 \\ +\ 7 \\ \hline \end{array}$ **17.** $\begin{array}{r} 6 \\ 21 \\ 14 \\ 9 \\ +\ 12 \\ \hline \end{array}$ **18.** $\begin{array}{r} 12 \\ 4 \\ 8 \\ 26 \\ +\ 10 \\ \hline \end{array}$

19. $\begin{array}{r} 81 \\ 17 \\ 23 \\ 79 \\ +\ 12 \\ \hline \end{array}$ **20.** $\begin{array}{r} 64 \\ 28 \\ 56 \\ 25 \\ +\ 32 \\ \hline \end{array}$ **21.** 62 + 18 + 14 **22.** 23 + 49 + 18

23. 40 + 800 + 70 **24.** 30 + 900 + 20 **25.** 7542 + 49 + 682 **26.** 1624 + 1832 + 1976

27. 24 + 9006 + 489 + 2407 **28.** 16 + 748 + 1056 + 770 **29.** $\begin{array}{r} 627 \\ 628 \\ +\ 629 \\ \hline \end{array}$ **30.** $\begin{array}{r} 427 \\ 383 \\ +\ 229 \\ \hline \end{array}$

31. $\begin{array}{r} 6820 \\ 4271 \\ +\ 5626 \\ \hline \end{array}$ **32.** $\begin{array}{r} 6789 \\ 4321 \\ +\ 5555 \\ \hline \end{array}$ **33.** $\begin{array}{r} 507 \\ 593 \\ +\ \ 10 \\ \hline \end{array}$ **34.** $\begin{array}{r} 864 \\ 733 \\ +\ 356 \\ \hline \end{array}$ **35.** $\begin{array}{r} 4200 \\ 2107 \\ +\ 2692 \\ \hline \end{array}$

36. $\begin{array}{r} 5000 \\ 400 \\ +\ 3021 \\ \hline \end{array}$

37. $\begin{array}{r} 49 \\ 628 \\ 5762 \\ +\ 29,462 \\ \hline \end{array}$

38. $\begin{array}{r} 26 \\ 582 \\ 4763 \\ +\ 62,511 \\ \hline \end{array}$

39. $\begin{array}{r} 121,742 \\ 57,279 \\ 6586 \\ +\ 426,782 \\ \hline \end{array}$

40. $\begin{array}{r} 504,218 \\ 321,920 \\ 38,507 \\ +\ 594,687 \\ \hline \end{array}$

B *Find the perimeter of each figure. See Examples 5 and 6.*

△ **41.**

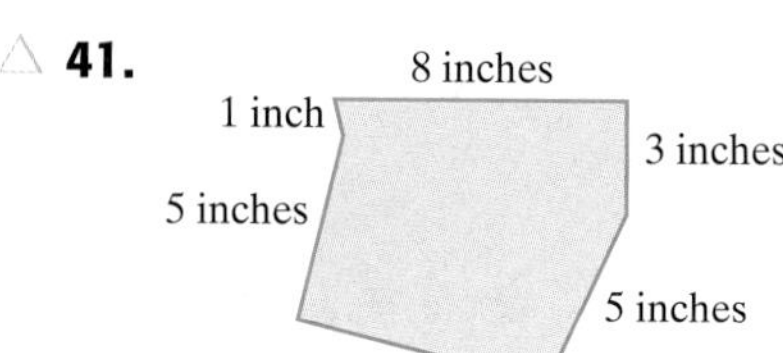

△ **42.**

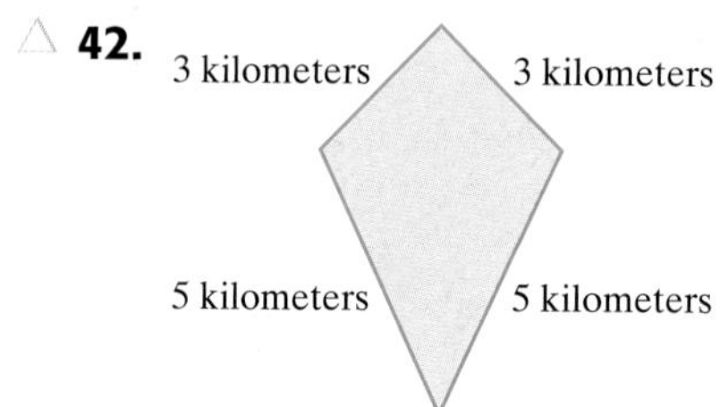

43.

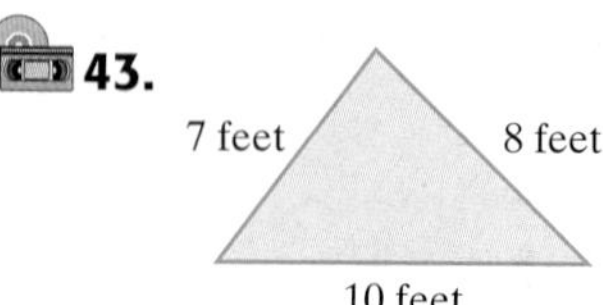

△ **44.**

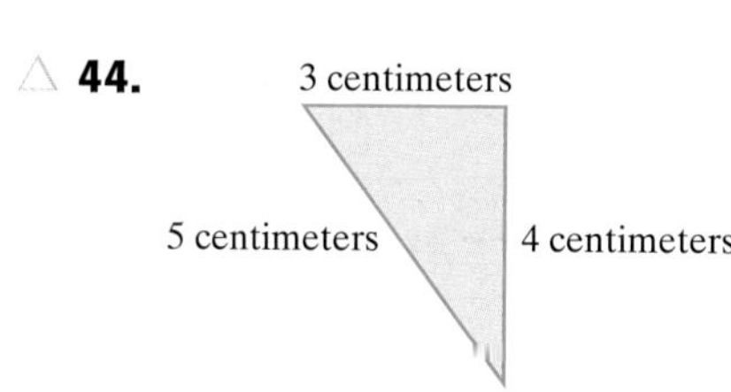

△ **45.**

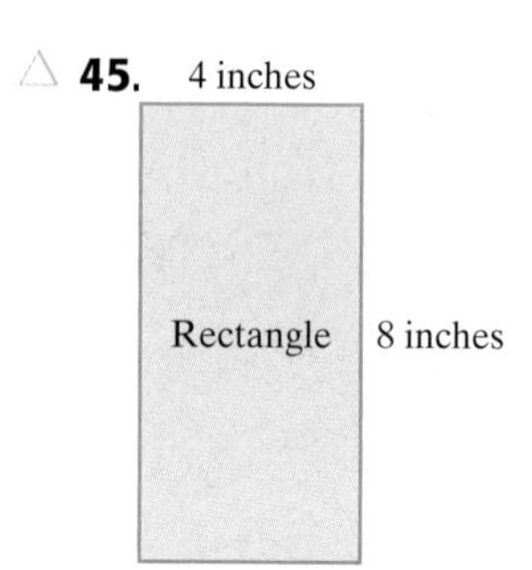

△ **46.**

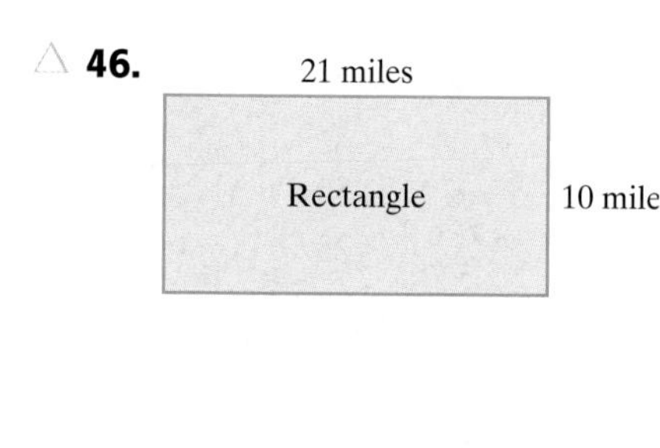

△ **47.**

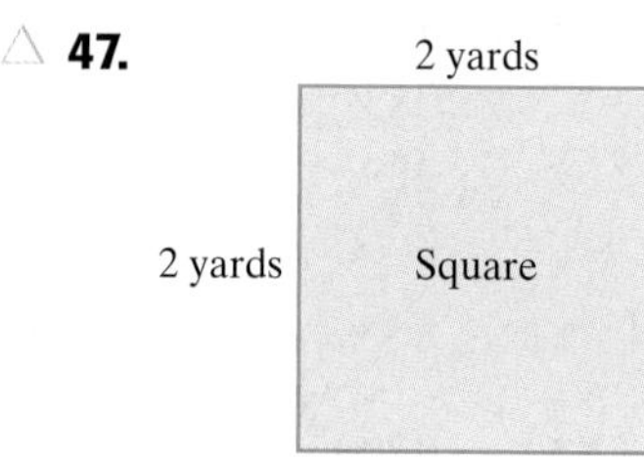

△ **48.**

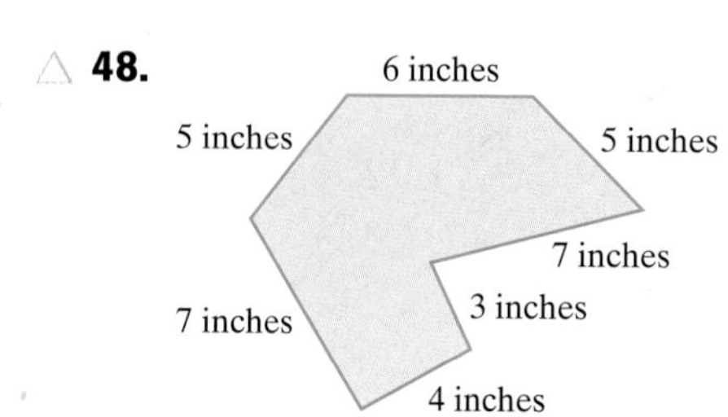

C *Solve. See Examples 7 and 8.*

49. The highest point in South Carolina is Sassafras Mountain at 3560 feet above sea level. The highest point in North Carolina is Mt. Mitchell, whose peak is 3124 feet higher than Sassafras Mountain. Find how high Mt. Mitchell is. (*Source:* U.S. Geological Survey)

50. The distance from Kansas City, Kansas, to Hays, Kansas, is 285 miles. Colby, Kansas, is 98 miles farther from Kansas City than Hays. Find how far it is from Kansas City to Colby.

51. Leo Callier is installing an invisible fence in his backyard. How many feet of wiring are needed to enclose the yard below?

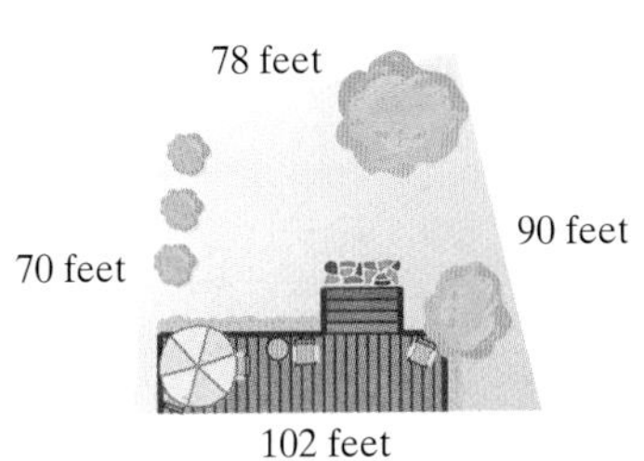

52. A homeowner is considering installing gutters around her home. To help her estimate the cost of gutters find the perimeter of the edge of her roof.

53. In 2001, Harley-Davidson shipped 234,500 motorcycles worldwide. In 2002, it shipped 29,200 more motorcycles. What was the total number of Harley-Davidson motorcycles shipped in 2002? (*Source:* Harley-Davidson, Inc.)

54. Dan Marino holds the NFL career record for most passes completed. He completed 2305 passes from the beginning of his NFL career in 1983 through 1989. He completed another 2662 passes from 1990 through 1999, his last season before retiring from professional football. How many passes did he complete during his NFL career? (*Source:* National Football League)

55. During the spring of 2001, Kellogg Company acquired Keebler Foods Company. Before the merger, Kellogg employed 15,000 people and Keebler employed 13,000 people. Assuming there were no layoffs after the merger, how many employees did the newly combined company have? (*Source:* Kellogg Company)

56. The DVD video format was introduced in March 1997. From 1997 through 1999, a total of 5,423,786 DVD players were sold in the United States. A total of 8,498,545 DVD players were sold in the United States in 2000. How many DVD players in all were sold from 1997 through the end of 2000? (*Source:* Consumer Electronics Association)

57. In 2001, there were 7981 Blockbuster video rental stores worldwide. In 2002, the Blockbuster store tally had increased by 564 stores. How many Blockbuster stores were there in 2002? (*Source:* Blockbuster Inc.)

58. Wilma Rudolph, who won three gold medals in track and field events in the 1960 Summer Olympics, was born in 1940. Marion Jones, who also won three gold medals in track and field events but in the 2000 Summer Olympics, was born 35 years later. In what year was Marion Jones born?

59. The highest waterfall in the United States is Yosemite Falls in Yosemite National Park in California. Yosemite Falls is made up of three sections, as shown in the graph. What is the total height of Yosemite Falls? (*Source:* U.S. Department of the Interior)

60. Jordan White, a nurse at Mercy Hospital, is recording fluid intake in a patient's medical chart. During his shift, the patient had the following types and amounts of intake measured in cubic centimeters (cc). What amount should Jordan record as the total fluid intake for this patient?

Oral	Intravenous	Blood
240	500	500
100	200	
355		

61. The State of Alaska has 1795 miles of urban highways and 11,460 miles of rural highways. Find the total highway mileage in Alaska. (*Source:* U.S. Federal Highway Administration)

62. The state of Hawaii has 1851 miles of urban highways and 2291 miles of rural highways. Find the total highway mileage in Hawaii. (*Source:* U.S. Federal Highway Administration)

63. As of January 3, 2003, the Broadway show *Cats* had staged 953 more performances than the Broadway show *Les Miserables*. At that time, *Les Miserables* had staged 6532 performances. How many performances had *Cats* staged by January 3, 2003? (*Source:* The League of American Theatres and Producers)

64. Charles Schulz, the creator of the Peanuts comic strip, was born in 1922. Scott Adams, the creator of the Dilbert comic strip, was born 35 years later. In what year was Scott Adams born? (*Source: The World Almanac*)

65. There were 9821 kidney transplants and 592 kidney-pancreas transplants performed in the United States in 2002. What was the total number of organ transplants performed in 2002 that involved a kidney? (*Source:* Organ Procurement and Transplantation Network)

66. There were 1450 heart transplants and 23 heart-lung transplants performed in the United States in 2002. What was the total number of organ transplants performed in 2002 that involved a heart? (*Source:* Organ Procurement and Transplantation Network)

Combining Concepts

The table shows the number of Wal-Mart stores in ten states. Use this table to answer Exercises 67 through 72.

67. Which state listed in the table has the most Wal-Mart stores?

68. Which state listed in the table has the fewest Wal-Mart stores?

The Top States for Wal-Mart Stores in 2002

State	Number of Stores
Pennsylvania	102
California	154
Florida	182
Georgia	115
Illinois	140
North Carolina	112
Tennessee	104
Missouri	128
Ohio	116
Texas	342

(*Source:* Wal-Mart Corporation)

69. What is the total number of Wal-Mart stores located in the three states with the most Wal-Mart stores?

70. Which pair of neighboring states have more Wal-Mart stores, Pennsylvania and Ohio or Florida and Georgia?

71. How many Wal-Mart stores are located in the ten states listed in the table? Use a calculator to check your total.

72. Wal-Mart operates stores in all 50 states. There are 1287 Wal-Mart stores located in the states not listed in the table. What is the total number of Wal-Mart stores in the United States?

73. In your own words, explain the commutative property of addition.

74. In your own words, explain the associative property of addition.

75. Add: 78,962 + 129,968,350 + 36,462,880

76. Add: 56,468,980 + 1,236,785 + 986,768,000

77. A student shows you the following computation and asks you to check it for him. What is your response and why?

$$\begin{array}{r} 1288 \\ +\ \ 377 \\ \hline 1555 \end{array}$$

1.4 Subtracting Whole Numbers

OBJECTIVES

- A Subtract whole numbers.
- B Subtract whole numbers when borrowing is necessary.
- C Solve problems by subtracting whole numbers.

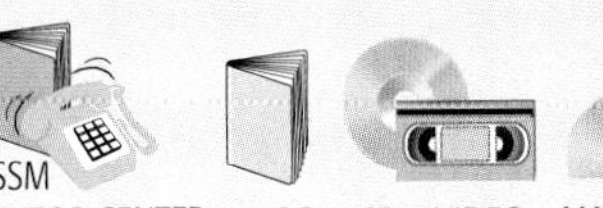

A Subtracting Whole Numbers

If you have \$5 and someone gives you \$3, you have a total of \$8, since $5 + 3 = 8$. Similarly, if you have \$8 and then someone borrows \$3, you have \$5 left. **Subtraction** is finding the **difference** of two numbers.

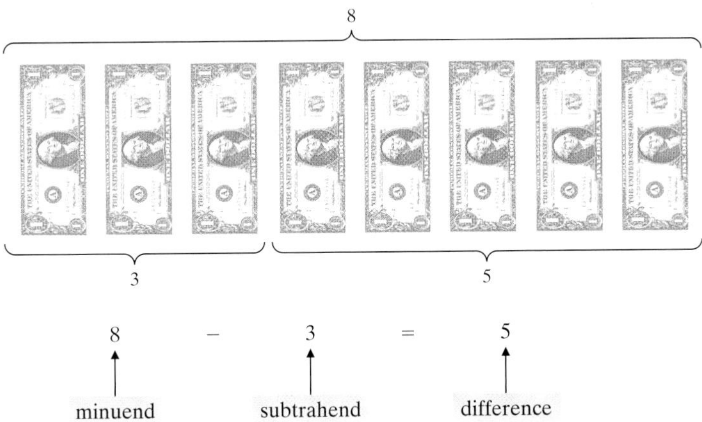

$$\underset{\text{minuend}}{8} - \underset{\text{subtrahend}}{3} = \underset{\text{difference}}{5}$$

Notice that addition and subtraction are very closely related. In fact, subtraction is defined in terms of addition.

$8 - 3 = 5$ because $5 + 3 = 8$

This means that subtraction can be *checked* by addition, and we say that addition and subtraction are reverse operations.

EXAMPLE 1 Subtract. Check each answer by adding.

a. $12 - 9$ **b.** $11 - 6$ **c.** $5 - 5$ **d.** $7 - 0$

Solution:

a. $12 - 9 = 3$ because $3 + 9 = 12$
b. $11 - 6 = 5$ because $5 + 6 = 11$
c. $5 - 5 = 0$ because $0 + 5 = 5$
d. $7 - 0 = 7$ because $7 + 0 = 7$

Practice Problem 1

Subtract. Check each answer by adding.

a. $14 - 9$
b. $9 - 9$
c. $4 - 0$

Look again at Examples 1(c) and 1(d).

1(c) $5 - 5 = 0$ — 5 and 5: same number; 0: difference is 0

1(d) $7 - 0 = 7$ — $7 - 0$: a number minus 0; 7: difference is the same number

These two examples illustrate the subtraction properties of 0.

Subtraction Properties of 0

The difference of any number and that same number is 0. For example,

$11 - 11 = 0$

The difference of any number and 0 is that same number. For example,

$45 - 0 = 45$

Answers

1. a. 5 **b.** 0 **c.** 4

When subtraction involves numbers of two or more digits, it is more convenient to subtract vertically. For example, to subtract 893 − 52,

```
  8 9 3   ← minuend
−   5 2   ← subtrahend
  8 4 1   ← difference
  ↑ ↑ ↑
  | | 3 − 2
  | 9 − 5
  8 − 0
```

Line up the numbers vertically so that the minuend is on top and the place values correspond. Subtract digits that are in corresponding places, starting with the ones place.

To check, add.

```
  difference    or    841
+ subtrahend        +  52
  minuend             893  ← Since this is the original minuend, the problem checks.
```

Practice Problem 2

Subtract. Check by adding.

a. 4689 − 253

b. 981 − 630

EXAMPLE 2 Subtract: 7826 − 505. Check by adding.

Solution:

```
  7826
−  505
  7321
```

Check:

```
  7321
+  505
  7826
```

B Subtracting with Borrowing

When a digit in the second number (subtrahend) is larger than the corresponding digit in the first number (minuend), **borrowing** is necessary. For example, consider

```
  81
− 63   ← We cannot take away 3 ones from 1 one, so we borrow.
```

Since the 3 in the ones place of 63 is larger than the 1 in the ones place of 81, borrowing is necessary. We borrow 1 ten from the tens place and add it to the ones place.

Borrowing

```
8 tens − 1 ten = 7 tens  →    7 11   ← 1 ten + 1 one = 11 ones
                              8 1
                            − 6 3
```

Now we subtract the ones-place digits and then the tens-place digits.

```
  7 11
  8 1
− 6 3
  1 8 ← 11 − 3 = 8
  ↑
  └── 7 − 6 = 1
```

Check:

```
  18
+ 63
  81   The original minuend
```

Answers

2. a. 4436 b. 351

EXAMPLE 3 Subtract: 43 − 29. Check by adding.

Solution:

$$\begin{array}{r} \scriptstyle 3\ 13 \\ \not{4}\ \not{3} \\ -2\ 9 \\ \hline 1\ 4 \end{array}$$

Check:

$$\begin{array}{r} 14 \\ +\ 29 \\ \hline 43 \end{array}$$

Sometimes we may have to borrow from more than one place value. For example, to subtract 7631 − 152, we first borrow from the tens place.

$$\begin{array}{r} \scriptstyle 2\ 11 \\ 7\ 6\ \not{3}\ \not{1} \\ -\ 1\ 5\ 2 \\ \hline 9 \end{array} \leftarrow 11 - 2 = 9$$

In the tens place, 5 is greater than 2, so we borrow again. This time we borrow from the hundreds place

$$\begin{array}{r} \scriptstyle 12 \\ \scriptstyle 5\ \not{2}\ 11 \\ 7\ \not{6}\ \not{3}\ \not{1} \\ -\ 1\ 5\ 2 \\ \hline 7\ 4\ 7\ 9 \end{array}$$

6 hundreds − 1 hundred = 5 hundreds

1 hundred + 2 tens or
10 tens + 2 tens = 12 tens

Check:

$$\begin{array}{r} 7479 \\ +\ \ 152 \\ \hline 7631 \end{array}$$ The original minuend

EXAMPLE 4 Subtract: 900 − 174. Check by adding.

Solution: In the ones place, 4 is larger than 0, so we borrow from the tens place. But the tens place of 900 is 0, so to borrow from the tens place we must first borrow from the hundreds place.

$$\begin{array}{r} \scriptstyle 8\ 10\ \ \\ \not{9}\ \not{0}\ 0 \\ -\ 1\ 7\ 4 \\ \hline \end{array}$$

Now borrow from the tens place.

$$\begin{array}{r} \scriptstyle 9\ \ \\ \scriptstyle 8\ \not{10}\ 10 \\ \not{9}\ \not{0}\ \not{0} \\ -1\ 7\ 4 \\ \hline 7\ 2\ 6 \end{array}$$

Check:

$$\begin{array}{r} \scriptstyle 1\,1\ \\ 726 \\ -\ 174 \\ \hline 900 \end{array}$$

Ⓒ Solving Problems by Subtracting

Descriptions of real-life problems that suggest solving by subtraction include these key words or phrases:

Practice Problem 3

Subtract. Check by adding.

a. $\begin{array}{r} 227 \\ -\ 175 \\ \hline \end{array}$

b. $\begin{array}{r} 1136 \\ -\ 914 \\ \hline \end{array}$

c. $\begin{array}{r} 8627 \\ -\ 4119 \\ \hline \end{array}$

Practice Problem 4

Subtract. Check by adding.

a. $\begin{array}{r} 400 \\ -\ 164 \\ \hline \end{array}$

b. $\begin{array}{r} 200 \\ -\ 45 \\ \hline \end{array}$

c. $\begin{array}{r} 1000 \\ -\ 762 \\ \hline \end{array}$

Answers

3. a. 52 **b.** 222 **c.** 4508

4. a. 236 **b.** 155 **c.** 238

Key Words or Phrases	Examples	Symbols
subtract	subtract 5 from 8	$8 - 5$
difference	the difference of 10 and 2	$10 - 2$
less	17 less 3	$17 - 3$
take away	14 take away 9	$14 - 9$
decreased by	7 decreased by 5	$7 - 5$
subtracted from	9 subtracted from 12	$12 - 9$

Helpful Hint

Be careful when solving applications that suggest subtraction. Although order *does not* matter when adding, order *does* matter when subtracting. For example, $10 - 3$ and $3 - 10$ do *not* simplify to the same number.

Concept Check

In each of the following problems, identify which number is the minuend and which number is the subtrahend.

a. What is the result when 9 is subtracted from 20?

b. What is the difference of 15 and 8?

c. Find a number that is 15 fewer than 23.

Try the Concept Check in the margin.

Practice Problem 5

The radius of Earth is 6378 kilometers. The radius of Mars is 2981 kilometers less than the radius of Earth. What is the radius of Mars? (*Source:* National Space Science Data Center)

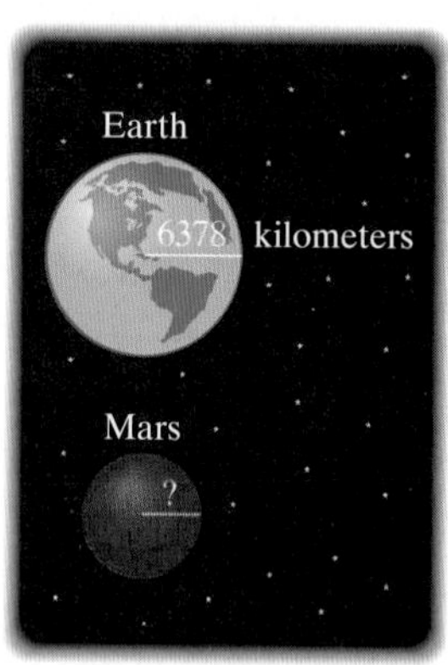

EXAMPLE 5 Finding the Radius of a Planet

The radius of Venus is 6052 kilometers. The radius of Mercury is 3612 kilometers less than the radius of Venus. Find the radius of Mercury. (*Source:* National Space Science Data Center)

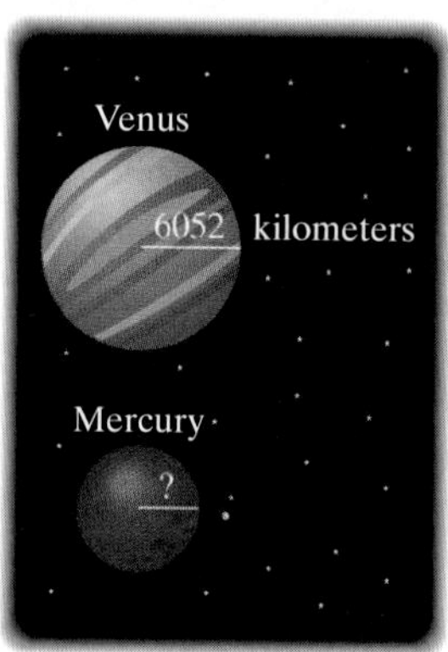

Solution:

In Words		Translate to Numbers
radius of Venus	→	6052
less 3612	→	−3612
radius of Mercury	→	2440

The radius of Mercury is 2440 kilometers.

Practice Problem 6

A new suit originally priced at \$92 is now on sale for \$47. How much money was taken off the original price?

EXAMPLE 6 Calculating Miles Per Gallon

A subcompact car gets 42 miles per gallon of gas. A full-size car gets 17 miles per gallon of gas. How many more miles per gallon does the subcompact car get than the full-size car?

Solution:

In Words		Translate to Numbers
subcompact miles per gallon	→	$\overset{3}{\cancel{4}}\,\overset{12}{\cancel{2}}$
− full-size miles per gallon	→	− 1 7
more miles per gallon		2 5

The subcompact car gets 25 more miles per gallon than the full-size car.

Answers

5. 3397 km **6.** \$45

Concept Check: **a.** minuend: 20; subtrahend: 9
b. minuend: 15; subtrahend: 8
c. minuend: 23; subtrahend: 15

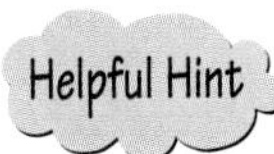

Since subtraction and addition are reverse operations, don't forget that a subtraction problem can be checked by adding.

Graphs can be used to visualize data. The graph shown next is called a **bar graph**. Notice that the height of each bar and each number above it correspond to the number's graph on the vertical number line, or axis.

EXAMPLE 7 Reading a Bar Graph

The graph below shows the ratings of Best Picture nominees since PG-13 was introduced in 1984. In this graph, each bar represents a different rating, and the height of each bar represents the number of Best Picture nominees for that rating.

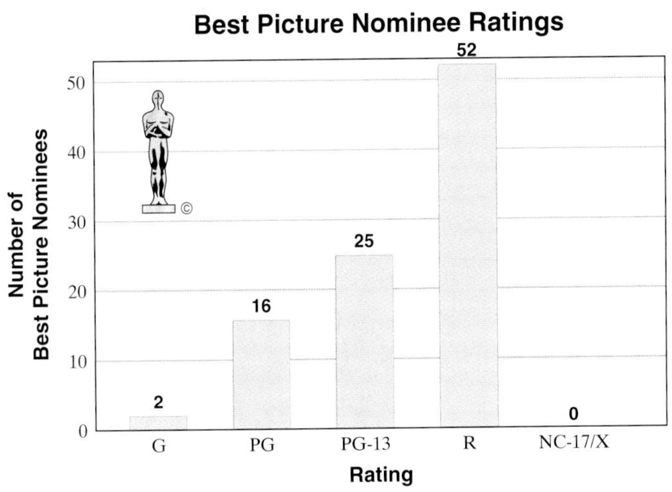

Source: The internet Movie Database and Cuadra Associates Movie Star Database, 2003.

a. Which rating did most Best Picture nominees have?

b. How many more Best Picture nominees were rated PG-13 than PG?

Solution:

a. The rating for most Best Picture nominees is the one corresponding to the highest bar, which is an R rating.

b. The number of Best Picture nominees rated PG-13 is 25. The number of Best Picture nominees rated PG is 16. To find how many more pictures were rated PG-13, we subtract.

$$25 - 16 = 9$$

Nine more Best Picture nominees were rated PG-13 than PG.

Practice Problem 7

Use the graph in Example 7 to answer the following:

a. Which rating had the least number of Best Picture nominees?

b. How many more Best Picture nominees were rated R than G?

Answers

7. a. NC-17/X **b.** 50

CALCULATOR EXPLORATIONS

Subtracting Numbers

To subtract numbers on a calculator, find the keys marked [−] and [=] or [ENTER].

For example, to find 83 − 49 on a calculator, press the keys [83] [−] [49] [=] or [ENTER].

The display will read [34]. Thus, 83 − 49 = 34.

Use a calculator to subtract.

1. 865 − 95

2. 76 − 27

3. 147 − 38

4. 366 − 87

5. 9625 − 647

6. 10,711 − 8925

FOCUS ON Mathematical Connections

MODELING SUBTRACTION OF WHOLE NUMBERS

A mathematical concept can be represented or modeled in many different ways. For instance, subtraction can be represented by the following symbolic model:

11 − 4

The following verbal models can also represent subtraction of these same quantities:

"Four subtracted from eleven" or
"Eleven take away four"

Physical models can also represent mathematical concepts. In these models, a number is represented by that many objects. For example, the number 5 can be represented by five pennies, squares, paper clips, tiles, or bottle caps.

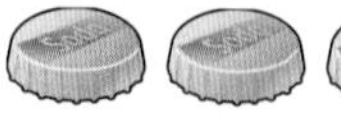
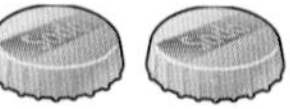

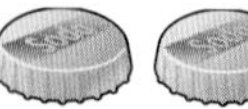

A physical representation of the number 5

Take-Away Model for Subtraction: 11 − 4

- Start with 11 objects.
- Take 4 objects away.
- How many objects remain?

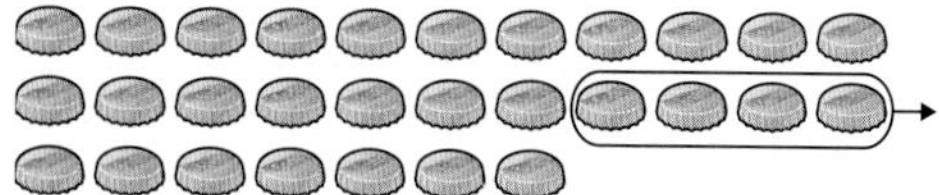

Comparison Model for Subtraction: 11 − 4

- Start with a set of 11 of one type of object and a set of 4 of another type of object.
- Make as many pairs that include one object of each type as possible.
- How many more objects are in the larger set?

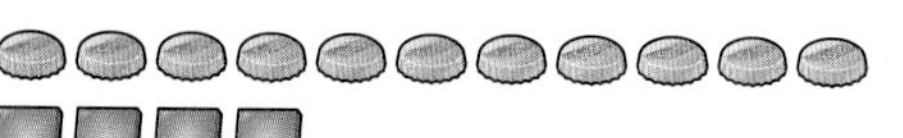

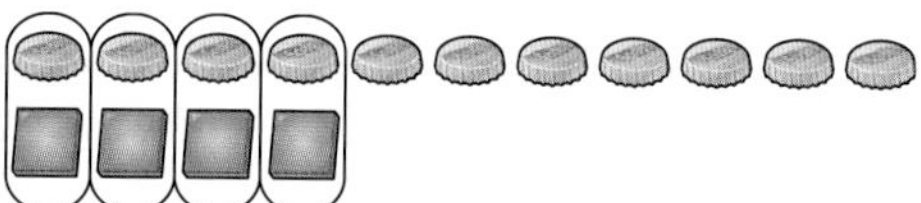

Missing Addend Model for Subtraction: 11 − 4

- Start with 4 objects.
- Continue adding objects until a total of 11 is reached.
- How many more objects were needed to give a total of 11?

CRITICAL THINKING

Use an appropriate physical model for subtraction to solve each of the following problems. Explain your reasoning for choosing each model.

1. Sneha has assembled 12 computer components so far this shift. If his quota is 20 components, how many more components must he assemble to reach his quota?

2. Yuko wants to plant 14 daffodil bulbs in her yard. She planted 5 bulbs in the front yard. How many bulbs does she have left for planting in the backyard?

3. Todd is 19 years old and his sister Tanya is 13 years old. How much older is Todd than Tanya?

Name ______________________ Section ____________ Date ____________

Mental Math

Find each difference.

1. 9 − 2 **2.** 6 − 6 **3.** 5 − 0 **4.** 44 − 22 **5.** 93 − 93

6. 700 − 400 **7.** 700 − 300 **8.** 700 − 700 **9.** 600 − 100 **10.** 600 − 0

EXERCISE SET 1.4

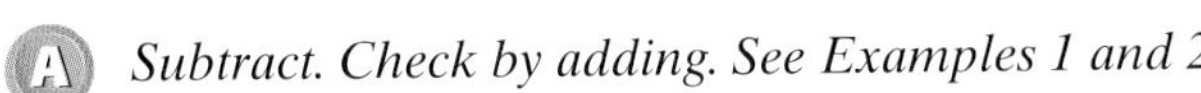

A *Subtract. Check by adding. See Examples 1 and 2.*

1. 67 − 23 **2.** 72 − 41 **3.** 82 − 22 **4.** 27 − 10 **5.** 389 − 124 **6.** 572 − 321

7. 677 − 423 **8.** 766 − 324 **9.** 998 − 453 **10.** 912 − 610 **11.** 749 − 149 **12.** 257 − 257

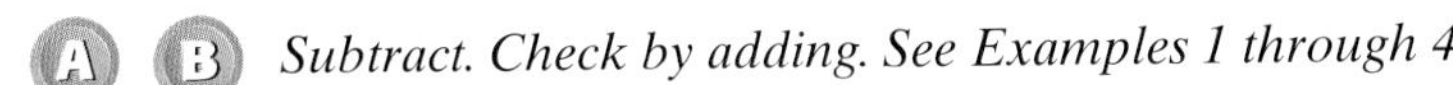

A B *Subtract. Check by adding. See Examples 1 through 4.*

13. 62 − 37 **14.** 55 − 29 **15.** 70 − 25 **16.** 80 − 37 **17.** 938 − 792 **18.** 436 − 275

19. 922 − 634 **20.** 674 − 299 **21.** 600 − 432 **22.** 300 − 149 **23.** 42 − 36 **24.** 73 − 29

25. 923 − 476 **26.** 813 − 227 **27.** 6283 − 560 **28.** 5349 − 720 **29.** 533 − 29 **30.** 724 − 16

31. 200 − 111 **32.** 300 − 211 **33.** 1983 − 1904 **34.** 1983 − 1914 **35.** 56,422 − 16,508 **36.** 76,652 − 29,498

37. 50,000 − 17,289 **38.** 40,000 − 23,582 **39.** 7020 − 1979 **40.** 6050 − 1878

41. 51,111 − 19,898 **42.** 62,222 − 39,898 **43.** Subtract 5 from 9. **44.** Subtract 9 from 21.

45. Find the difference of 41 and 21.

46. Find the difference of 16 and 5.

47. Subtract 56 from 63.

48. Subtract 41 from 59.

49. Find 108 less 36.

50. Find 25 less 12.

51. Find 12 subtracted from 100.

52. Find 86 subtracted from 90.

Solve. See Examples 5 through 7.

53. Dyllis King is reading a 503-page book. If she has just finished reading page 239, how many more pages must she read to finish the book?

54. When Lou and Judy Zawislak began a trip, the odometer read 55,492. When the trip was over, the odometer read 59,320. How many miles did they drive on their trip?

55. In 1997, the hole in the Earth's ozone layer over Antartica was about 21 million square kilometers in size. In 2001, the hole had grown to 25 million square kilometers. By how much has the hole grown from 1997 to 2001? (*Source*: U.S. Environmental Protection Agency EPA: *http://www.epa.gov/ozone/science/hole/sizedata.html#areatime*)

56. Bamboo can grow to 98 feet while Pacific giant kelp (a type of seaweed) can grow to 197 feet. How much taller is the kelp than the bamboo?

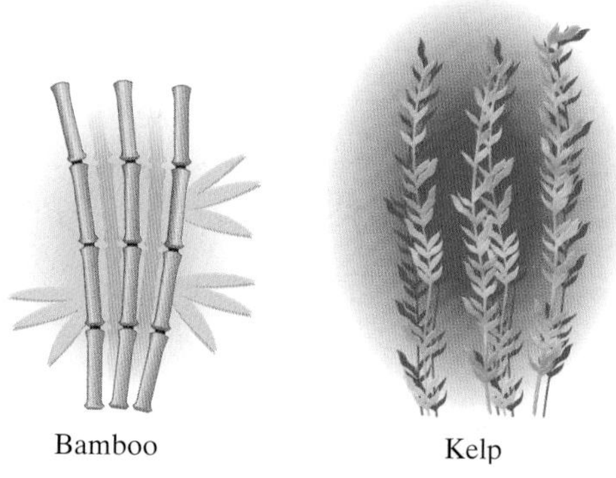

57. The peak of Mt. McKinley in Alaska is 20,320 feet above sea level. The peak of Long's Peak in Colorado is 14,255 feet above sea level. How much higher is the peak of Mt. McKinley than Long's Peak? (*Source:* U.S. Geological Survey)

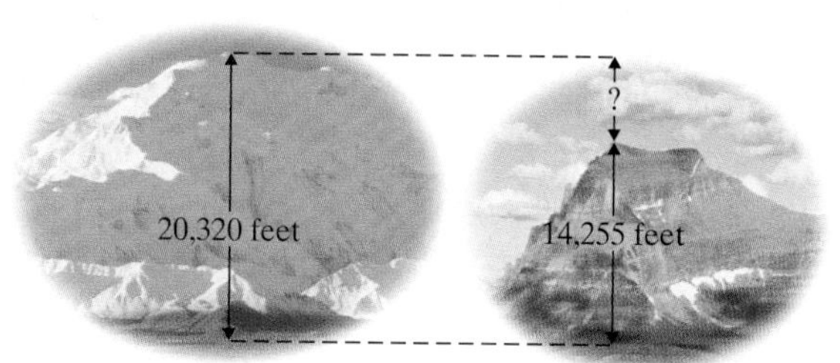

58. On one day in May the temperature in Paddin, Indiana, dropped 27 degrees from 2 p.m. to 4 p.m. If the temperature at 2 p.m. was 73° Fahrenheit, what was the temperature at 4 p.m.?

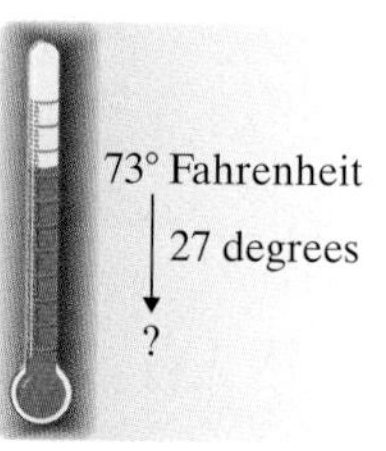

59. Buhler Gomez has a total of $539 in his checking account. If he writes a check for each of the items below, how much money will be left in his account?

Bell South	$27
Cleco	$101
Mellon Finance	$236

60. Pat Salanki's blood cholesterol level is 243. The doctor tells him it should be decreased to 185. How much of a decrease is this?

61. The distance from Kansas City to Denver is 645 miles. Hays, Kansas, lies on the road between the two and is 287 miles from Kansas City. What is the distance between Hays and Denver?

62. Nhoc Tran is trading his car in on a new car. The new car costs $ 15,425. His car is worth $7998. How much more money does he need to buy the new car?

63. A new home theater system with DVD player costs $732. Pat Gomez has $971 in his checking account. How much will he have left in his checking account after he buys the home theater?

64. A stereo that regularly sells for $547 is discounted by $99 in a sale. What is the sale price?

65. During the 2000–2001 regular season, Jerry Stackhouse of the Detroit Pistons led the NBA in total points scored with 2380. The Philadelphia 76ers' Allen Iverson placed second for total points scored with 2207. How many more points did Stackhouse score than Iverson during the 2000–2001 regular season? (*Source:* National Basketball Association)

66. In 1999, Americans bought 233,125 Ford Expeditions. In 2000, 19,642 fewer Expeditions were sold in the United States. How many Expeditions were sold in the United States in 2000? (*Source:* Ford Motor Company)

67. In 2000, there were 38,803 boxers registered with the American Kennel Club. In 2001, there were 1768 fewer boxers registered. How many boxers were registered with the AKC in 2001? (*Source:* American Kennel Club)

68. In the United States, there were 41,589 tornadoes from 1950 through 2000. In all, 13,205 of these tornadoes occurred from 1990 through 2000. How many tornadoes occurred during the period prior to 1990? (*Source:* Storm Prediction Center, National Weather Service)

69. Jo Keen and Trudy Waterbury were candidates for student government president. Who won the election if the votes were cast as follows? By how many votes did the winner win?

Class	Candidate	
	Jo	Trudy
Freshman	276	295
Sophomore	362	122
Junior	201	312
Senior	179	18

70. Two students submitted advertising budgets for a student government fund-raiser. If $1200 is available for advertising, how much excess would each budget have?

	Student A	Student B
Radio ads	$600	$300
Newspaper ads	$200	$400
Posters	$150	$240
Handbills	$120	$170

71. The population of Florida grew from 12,937,926 in 1990 to 15,982,378 in 2000. What was Florida's population increase over this time period? (*Source:* U.S. Census Bureau)

72. The population of El Paso, Texas, was 515,342 in 1990 and 563,662 in 2000. By how much did the population of El Paso grow from 1990 to 2000? (*Source:* U.S. Census Bureau)

73. Until recently, the world's largest permanent maze was located in Ruurlo, Netherlands. This maze of beech hedges covers 94,080 square feet. A new hedge maze using hibiscus bushes at the Dole Plantation in Wahiawa, Hawaii, covers 100,000 square feet. How much larger is the Dole Plantation maze than the Ruurlo maze? (*Source*: The Guinness Book of Records)

74. There were only 27 California condors in the entire world in 1987. By 2002, the number of California condors had increased to 200. How much of an increase was this? (*Source:* California Department of Fish and Game)

75. The Mackinac Bridge is a suspension bridge that connects the lower and upper peninsulas of Michigan across the Straits of Mackinac. Its total length is 26,372 feet. The Lake Pontchartrain Bridge is a twin concrete trestle bridge in Slidell, Louisiana. Its total length is 28,547 feet. Which bridge is longer and by how much? (*Sources:* Mackinac Bridge Authority and Federal Highway Administration, Bridge Division)

76. Papa John's is the third largest pizza chain in the United States. In 1999, there were 2486 Papa John's restaurants worldwide. By 2000, the number of Papa John's restaurants had grown to 2817. How many new Papa John's restaurants were added during 2000? (*Source:* Papa John's International Inc.)

The bar graph shows the number of passenger arrivals and departures in 2001 (in thousands) for the top busiest airports in the U.S. Use this graph to answer Exercises 77 through 80. See Example 7.

77. Which airport is the busiest?

78. Which airports have fewer than 60 million passenger arrivals and departures per year?

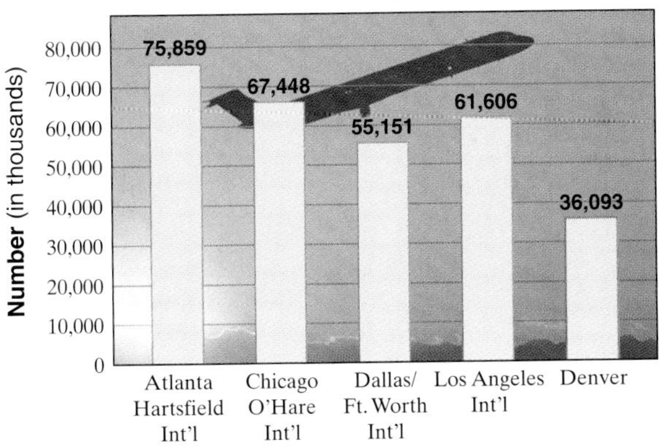

Source: Airports Council International

79. How many more passenger arrivals and departures per year does the Chicago O'Hare International Airport have than the Los Angeles International Airport?

80. How many more passenger arrivals and departures per year does the Atlanta Hartsfield International Airport have than the Dallas/Ft. Worth International Airport?

Combining Concepts

The table shows the top ten leading advertisers in the United States in 2001 and the amount of money spent in that year on advertising. Use this table to answer Exercises 81 through 85.

Company	Yearly Amount Spent on Advertising
General Motors Corp.	$3,374,400,000
Procter & Gamble Co.	$2,540,600,000
Ford Motor Co.	$2,408,200,000
PepsiCo	$2,210,400,000
Pfizer	$2,189,500,000
DaimlerChrysler	$1,985,300,000
AOL Time Warner.	$1,885,300,000
Phillip Morris Cos	$1,815,700,000
Walt Disney Co.	$1,757,300,000
Johnson & Johnson	$1,618,100,000

(*Source:* AdAge.com & the June 24, 2002 issue of *Advertising Age*)

81. Which company spent more than $3 billion on advertising?

82. Which companies spent fewer than $2 billion on advertising?

83. How much less money did DaimlerChrysler spend on advertising than General Motors Corp.?

84. How much more money did PepsiCo spend on advertising than Pfizer?

85. Find the total amount of money spent by these ten companies on advertising.

86. The local college library is having a Million Pages of Reading promotion. The freshmen have read a total of 289,462 pages; the sophomores have read a total of 369,477 pages; the juniors have read a total of 218,287 pages; and the seniors have read a total of 121,685 pages. Have they reached a goal of one million pages? If not, how many more pages need to be read?

Fill in the missing digits in each problem.

87.
```
  526_
- 2_85
  28_4
```

88.
```
  10,_4_
-  85_4
  _710
```

89. Is there a commutative property of subtraction? In other words, does order matter when subtracting? Why or why not?

90. The Gap, Inc., had net income of $13,847,873,000 in 2001. In 1997 The Gap's net income was only $6,507,825,000. How much did The Gap's net income increase from 1997 to 2001? (*Source:* The Gap, Inc.)

FOCUS ON Business and Career

THE EARNING POWER OF A COLLEGE DEGREE

According to data from the U.S. Bureau of the Census, average annual wages tend to increase with additional education. The following table shows the average annual earnings for both men and women in 1996 and 2001 for workers with only a high school diploma and for those who have earned a bachelor's degree.

		Average Annual Wages	
Gender	**Year**	**High School Diploma**	**Bachelor's Degree**
Men	1996	$24,814	$39,624
	2001	$28,343	$49,985
Women	1996	$12,702	$25,192
	2001	$15,665	$30,973

CRITICAL THINKING

Use the table to answer each question.

1. First, analyze men's earnings.
   a. Did the average annual wages for men with a high school diploma increase or decrease from 1996 to 2001? By how much?
   b. Did the average annual wages for men with a bachelor's degree increase or decrease from 1996 to 2001? By how much?
   c. How much more could a man with a bachelor's degree earn than a man with a high school diploma in 1996? In 2001?
2. Now analyze women's earnings.
   a. Did the average annual wages for women with a high school diploma increase or decrease from 1996 to 2001? By how much?
   b. Did the average annual wages for women with a bachelor's degree increase or decrease from 1996 to 2001? By how much?
   c. How much more could a woman with a bachelor's degree earn than a woman with a high school diploma in 1996? In 2001?
3. Now compare men's and women's earnings.
   a. Find the difference between average annual wages (the "wage gap") for men and women with high school diplomas in 1996.
   b. Find the wage gap for men and women with high school diplomas in 2001.
   c. Find the wage gap for men and women with bachelor's degrees in 1996.
   d. Find the wage gap for men and women with bachelor's degrees in 2001.
4. Write a paragraph summarizing the conclusions that can be drawn from this table of data. Identify any apparent trends.

1.5 Rounding and Estimating

OBJECTIVES

- (A) Round whole numbers.
- (B) Use rounding to estimate sums and differences.
- (C) Solve problems by estimating.

(A) Rounding Whole Numbers

Rounding a whole number means approximating it. A rounded whole number is often easier to use, understand, and remember than the precise whole number. For example, instead of trying to remember the Iowa state population as 2,851,792, it is much easier to remember it rounded to the nearest million: 3 million people.

To understand rounding, let's look at the following illustrations. The whole number 36 is closer to 40 than 30, so 36 rounded to the nearest ten is 40.

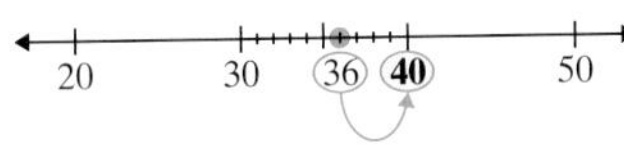

The whole number 52 rounded to the nearest ten is 50 because 52 is closer to 50 than to 60.

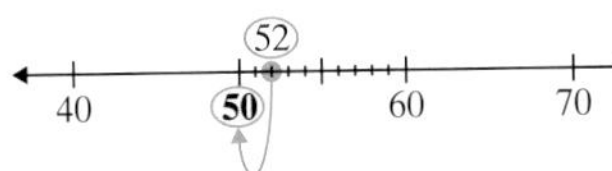

In trying to round 25 to the nearest ten, we see that 25 is halfway between 20 and 30. It is not closer to either number. In such a case, we round to the larger ten, that is, to 30.

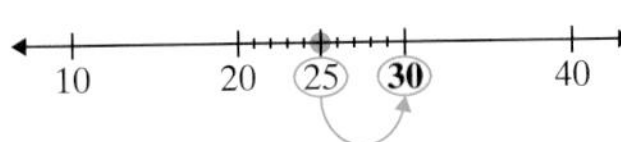

To round a whole number without using a number line, follow these steps:

Rounding Whole Numbers to a Given Place Value

Step 1. Locate the digit to the right of the given place value to be rounded.

Step 2. If this digit is 5 or greater, add 1 to the digit in the given place value and replace each digit to its right by 0.

Step 3. If this digit is less than 5, keep the digit in the given place value and replace each digit to its right by 0.

EXAMPLE 1 Round 568 to the nearest ten.

Solution: 5 6 (8)

↑ tens place — The digit to the right of the tens place is the ones place, which is circled.

5 6 (8)

↑ Add 1. Replace with 0. — Since the circled digit is 5 or greater, add 1 to the 6 in the tens place and replace the digit to the right by 0.

We find that 568 rounded to the nearest ten is 570.

Practice Problem 1

Round to the nearest ten.

a. 46
b. 731
c. 125

Answers

1. a. 50 **b.** 730 **c.** 130

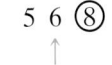

Practice Problem 2

Round to the nearest thousand.

a. 56,702
b. 7444
c. 291,500

EXAMPLE 2 Round 278,362 to the nearest thousand.

Solution:

Thousands place ↓, 3 is less than 5.

278,③62

Do not add 1. Replace with zeros.

The number 278,362 rounded to the nearest thousand is 278,000.

Practice Problem 3

Round to the nearest hundred.

a. 2777
b. 38,152
c. 762,955

EXAMPLE 3 Round 248,982 to the nearest hundred.

Solution:

Hundreds place ↓, 8 is greater than or equal to 5

248,9⑧2

Add 1. 9 + 1 = 10, so replace the digit 9 by 0 and carry 1 to the place value to the left.

2 4 8, 9 8 2 (8+1, 0)

Add 1. Replace with zeros.

The number 248,982 rounded to the nearest hundred is 249,000.

Concept Check

Round each of the following numbers to the nearest *hundred*. Explain your reasoning.

a. 79
b. 33

Try the Concept Check in the margin.

B Estimating Sums and Differences

By rounding addends, we can estimate sums. An estimated sum is appropriate when an exact sum is not necessary. To estimate the sum shown, round each number to the nearest hundred and then add.

768	rounds to	800
1952	rounds to	2000
225	rounds to	200
+ 149	rounds to	+ 100
		3100

The estimated sum is 3100, which is close to the exact sum of 3094.

Practice Problem 4

Round each number to the nearest ten to find an estimated sum.

$$\begin{array}{r} 79 \\ 35 \\ 42 \\ 21 \\ +\ 98 \\ \hline \end{array}$$

EXAMPLE 4 Round each number to the nearest hundred to find an estimated sum.

$$\begin{array}{r} 294 \\ 625 \\ 1071 \\ +\ 349 \\ \hline \end{array}$$

Solution:

294	rounds to	300
625	rounds to	600
1071	rounds to	1100
+ 349	rounds to	+ 300
		2300

The estimated sum is 2300. (The exact sum is 2339.)

Answers

2. a. 57,000 **b.** 7000 **c.** 292,000 **3. a.** 2800 **b.** 38,200 **c.** 763,000 **4.** 280

Concept Check:
a. 100 **b.** 0

EXAMPLE 5 Round each number to the nearest hundred to find an estimated difference.

$$\begin{array}{r} 4725 \\ -\ 2879 \\ \hline \end{array}$$

Solution:

4725	rounds to	4700
− 2879	rounds to	− 2900
		1800

The estimated difference is 1800. (The exact difference is 1846.)

Practice Problem 5

Round each number to the nearest thousand to find an estimated difference.

$$\begin{array}{r} 4725 \\ -\ 2879 \\ \hline \end{array}$$

C Solving Problems by Estimating

Making estimates is often the quickest way to solve real-life problems when their solutions do not need to be exact.

EXAMPLE 6 Estimating Distances

Jose Guillermo is trying to estimate quickly the distance from Temple, Texas, to Brenham, Texas. Round each distance given on the map to the nearest ten to estimate the total distance.

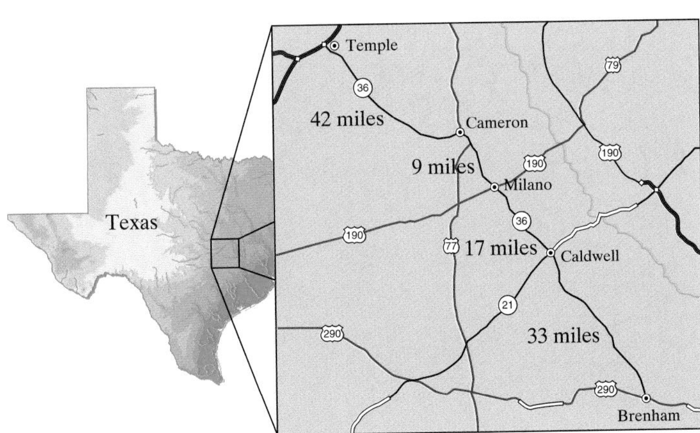

Solution:

Distance		Estimation
42	rounds to	40
9	rounds to	10
17	rounds to	20
+ 33	rounds to	+ 30
		100

It is approximately 100 miles from Temple to Brenham. (The exact distance is 101 miles.)

Practice Problem 6

Tasha Kilbey is trying to estimate how far it is from Grove, Kansas, to Hays, Kansas. Round each given distance on the map to the nearest ten to estimate the total distance.

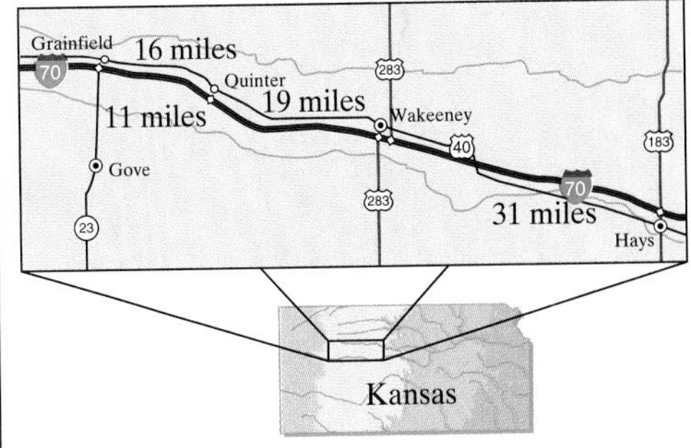

Answers

5. 2000 **6.** 80 mi

Practice Problem 7

In a recent year, there were 120,624 reported cases of chicken pox, 22,866 reported cases of tuberculosis, and 45,970 reported cases of salmonellosis in the United States. Round each number to the nearest ten-thousand to estimate the total number of cases reported for these diseases. (*Source*: Centers for Disease Control and Prevention)

Answer

7. 190,000

EXAMPLE 7 Estimating Data

In three recent years the numbers of reported cases of mumps in the United States were 906, 1537, and 1692. Round each number to the nearest hundred to estimate the total number of cases reported over this period. (*Source*: Centers for Disease Control and Prevention)

Solution:

Number of Cases		Estimation
906	rounds to	900
1537	rounds to	1500
+ 1692	rounds to	+ 1700
		4100

The approximate number of cases reported over this period is 4100. ●

Name ______________________ Section ____________ Date ____________

EXERCISE SET 1.5

A *Round each whole number to the given place. See Examples 1 through 3.*

1. 632 to the nearest ten

2. 273 to the nearest ten

3. 635 to the nearest ten

4. 275 to the nearest ten

5. 792 to the nearest ten

6. 394 to the nearest ten

7. 395 to the nearest ten

8. 582 to the nearest ten

9. 1096 to the nearest ten

10. 2198 to the nearest ten

11. 42,682 to the nearest thousand

12. 42,682 to the nearest ten-thousand

13. 248,695 to the nearest hundred

14. 179,406 to the nearest hundred

15. 36,499 to the nearest thousand

16. 96,501 to the nearest thousand

17. 99,995 to the nearest ten

18. 39,994 to the nearest ten

19. 59,725,642 to the nearest ten-million

20. 39,523,698 to the nearest million

Complete the table by rounding the given number to the given place value.

		Tens	Hundreds	Thousands
21.	5281			
22.	7619			
23.	9444			
24.	7777			
25.	14,876			
26.	85,049			

27. Round to the nearest thousand the 2001–2002 enrollment of East Tennessee State University: 11,331. (*Source:* Peterson's)

28. Round to the nearest hundred the hourly cost of operating a B747-400 aircraft: $6592. (*Source:* Air Transport Association of America)

Round each number to the indicated place.

29. Kareem Abdul-Jabbar holds the NBA record for points scored, a total of 38,387 over his NBA career. Round this number to the nearest thousand. (*Source:* National Basketball Association)

30. It takes 10,759 days for Saturn to make a complete orbit around the Sun. Round this number to the nearest hundred. (*Source:* National Space Science Data Center)

31. In 2001, U.S. farms produced 116,856,000 bushels of oats. Round the oat production figure to the nearest ten-million. (*Source:* U.S. Department of Agriculture)

32. In 2002, there were 485,536 U.S. Army personnel on active duty. Round this personnel figure to the nearest ten-thousand.

33. A total of 18,188 women participate in college soccer in the United States. Round this number to the nearest thousand. (*Source:* NCAA Sports)

34. The United States currently has 110,547,000 cellular mobile phone users (about 40% of population) while Austria has 6,150,000 users (about 75% of population). Round each of the user numbers to the nearest million. (*Note*: We will study percents in a later chapter.) (*Source*: Siemens AG, International Telecom Statistics, 2001)

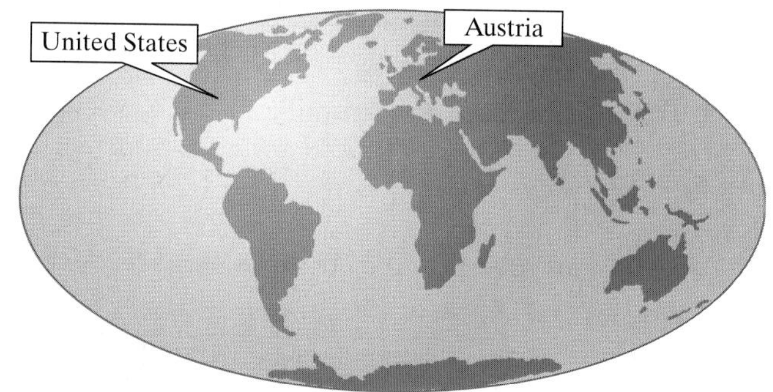

B *Estimate the sum or difference by rounding each number to the nearest ten. See Examples 4 and 5.*

35. $\begin{array}{r} 29 \\ 35 \\ 42 \\ +\ 16 \\ \hline \end{array}$

36. $\begin{array}{r} 62 \\ 72 \\ 15 \\ +\ 19 \\ \hline \end{array}$

37. $\begin{array}{r} 649 \\ -\ 272 \\ \hline \end{array}$

38. $\begin{array}{r} 555 \\ -\ 235 \\ \hline \end{array}$

Estimate the sum or difference by rounding each number to the nearest hundred. See Examples 4 and 5.

39. $\begin{array}{r} 1812 \\ 1776 \\ +\ 1945 \\ \hline \end{array}$

40. $\begin{array}{r} 2010 \\ 2001 \\ +\ 1984 \\ \hline \end{array}$

41. $\begin{array}{r} 1774 \\ -\ 1492 \\ \hline \end{array}$

42. $\begin{array}{r} 1989 \\ -\ 1870 \\ \hline \end{array}$

43. $\begin{array}{r} 2995 \\ 1649 \\ +\ 3940 \\ \hline \end{array}$

44. $\begin{array}{r} 799 \\ 1655 \\ +\ 271 \\ \hline \end{array}$

Estimation is useful to check for incorrect answers when using a calculator. For example, pressing a key too hard may result in a double digit, while pressing a key too softly may result in the number not appearing in the display.

Two of the given calculator answers below are incorrect. Find them by estimating each sum.

45. 362 + 419   781

46. 522 + 785   1307

47. 432 + 679 + 198   1139

48. 229 + 443 + 606   1278

49. 7806 + 5150   12,956

50. 5233 + 4988   9011

51. 31,439 + 18,781   50,220

52. 68,721 + 52,335   121,056

Solve each problem by estimating. See Examples 6 and 7.

53. Campo Appliance Store advertises three refrigerators on sale at $799, $1299, and $999. Round each cost to the nearest hundred to estimate the total cost.

54. Jared Nuss scored 89, 92, 100, 67, 75, and 79 on his calculus tests. Round each score to the nearest ten to estimate his total score.

55. Arlene Neville wants to estimate quickly the distance from Stockton to LaCrosse. Round each distance given on the map to the nearest ten miles to estimate the total distance.

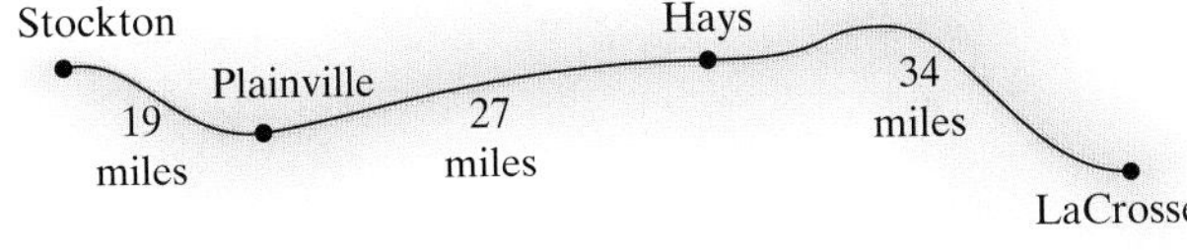

56. Carmelita Watkins is pricing new stereo systems. One system sells for $1895 and another system sells for $1524. Round each price to the nearest hundred dollars to estimate the difference in price of these systems.

57. The peak of Mt. McKinley, in Alaska, is 20,320 feet above sea level. The top of Mt. Rainier, in Washington, is 14,410 feet above sea level. Round each height to the nearest thousand to estimate the difference in elevation of these two peaks. (*Source:* U.S. Geological Survey)

58. The Gonzales family took a trip and traveled 458, 489, 377, 243, 69, and 702 miles on six consecutive days. Round each distance to the nearest hundred to estimate the distance they traveled.

59. In 2000 the population of Chicago was 2,896,016, and the population of Philadelphia was 1,517,550. Round each population to the nearest hundred-thousand to estimate how much larger Chicago was than Philadelphia. (*Source:* U.S. Census Bureau, 2000 census)

60. The distance from Kansas City to Boston is 1429 miles and from Kansas City to Chicago, 530 miles. Round each distance to the nearest hundred to estimate how much farther Boston is from Kansas City than Chicago is.

61. In the 1964 presidential election, Lyndon Johnson received 41,126,233 votes and Barry Goldwater received 27,174,898 votes. Round each number of votes to the nearest million to estimate the number of votes by which Johnson won the election.

62. Enrollment figures at Normal State University showed an increase from 49,713 credit hours in 1988 to 51,746 credit hours in 1989. Round each number to the nearest thousand to estimate the increase.

63. Head Start is a national program that provides developmental and social services for America's low-income preschool children ages three to five. Enrollment figures in Head Start programs showed an increase from 750,696 children in 1995 to 905,235 children in 2000. Round each number of children to the nearest thousand to estimate this increase. (*Source:* Head Start Bureau)

64. In 2000, General Motors produced 271,800 Saturn cars. Similarly, in 1999 only 232,570 Saturns were produced. Round each number of cars to the nearest thousand to estimate the increase in Saturn production from 1999 to 2000. (*Source:* General Motors Corporation)

65. Jupiter has the largest planetary moon. It is called Ganymede and has a diameter of 3274 miles. In contrast, the diameter of our moon is 2159 miles. Round each of these numbers to the nearest hundred to estimate the difference in the diameters.

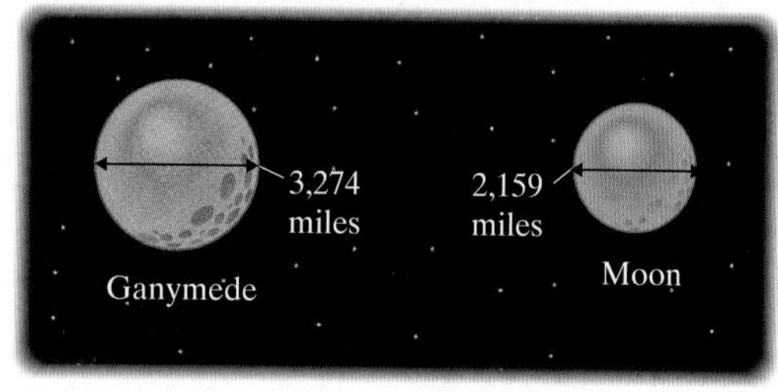

66. AOL Time Warner has the most visited website with 39,458,234 unique visitors per week. In second place is Yahoo! with 32,514,019 visitors per week. Round each number to the nearest million to estimate how many more visitors AOL Time Warner has than Yahoo!. (*Source*: Nielsen/Net Ratings Audience Measurement Service)

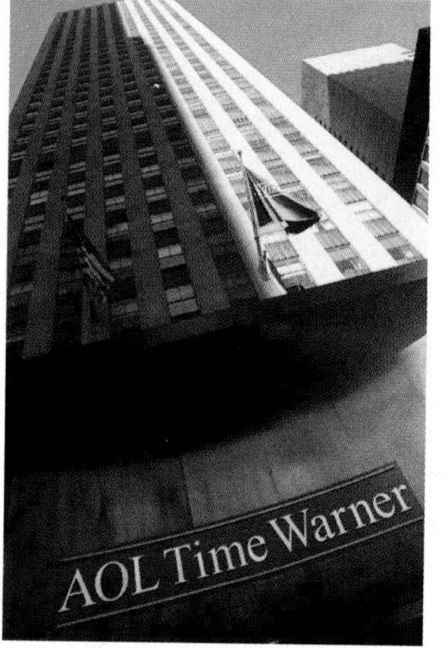

The following table (from Section 1.4) shows the top ten leading advertisers in the United States for 2001 and the amount of money spent in that year on advertising. Use this table to answer Exercises 67 through 70.

Company	Yearly Amount Spent on Advertising
General Motors Corp.	$3,374,400,000
Procter & Gamble Co.	$2,540,600,000
Ford Motor Co.	$2,408,200,000
PepsiCo	$2,210,400,000
Pfizer	$2,189,500,000
DaimlerChrysler	$1,985,300,000
AOL Time Warner.	$1,885,300,000
Phillip Morris Cos	$1,815,700,000
Walt Disney Co.	$1,757,300,000
Johnson & Johnson	$1,618,100,000

(*Source:* Ad Age.com & the June 24, 2002 issue of *Advertising*)

67. Approximate the amount of money spent on advertising by General Motors Corp. to the nearest hundred-million.

68. Approximate the amount of money spent on advertising by Johnson & Johnson Ltd. to the nearest hundred-million.

69. Approximate the amount of money spent on advertising by PepsiCo to the nearest billion.

70. Approximate the amount of money spent on advertising by Procter & Gamble Co. to the nearest billion.

Combining Concepts

71. A number rounded to the nearest hundred is 8600. Determine the smallest possible number.

72. Determine the largest possible number.

73. On August 23, 1989, it was estimated that 1,500,000 people joined hands in a human chain stretching 370 miles to protest the fiftieth anniversary of the pact that allowed what was then the Soviet Union to annex the Baltic nations in 1939. If the estimate of the number of people is to the nearest hundred-thousand, determine the largest possible number of people in the chain.

74. In your own words, explain how to round a number to the nearest thousand.

75. Estimate the perimeter by first rounding each length to the nearest hundred.

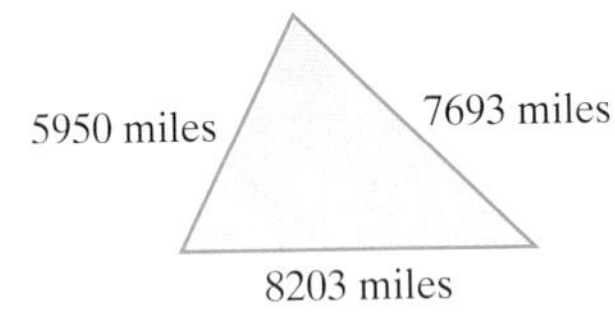

Internet Excursions

Go To: 

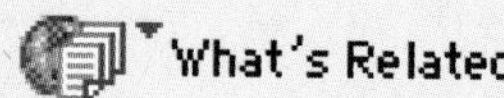

This World Wide Web site provides access to a site that allows the user to find the distance, as the crow flies, between two cities. It also gives population and elevation figures for each city. Visit this site and answer the following questions:

76. Choose three cities in your state: (A) _____, (B) _____, and (C) _____.

77. Find the distance between city A and city B. Then find the distance from city B to city C. If a crow were to fly directly from city A to city B and then continue on to city C, how far would it fly?

78. List the population for each city. Which city has the greatest population? How many more people does that city have than the city with the least population? Check your results by estimating.

1.6 Multiplying Whole Numbers and Area

OBJECTIVES

- **A** Use the properties of multiplication.
- **B** Multiply whole numbers.
- **C** Find the area of a rectangle.
- **D** Solve problems by multiplying whole numbers.

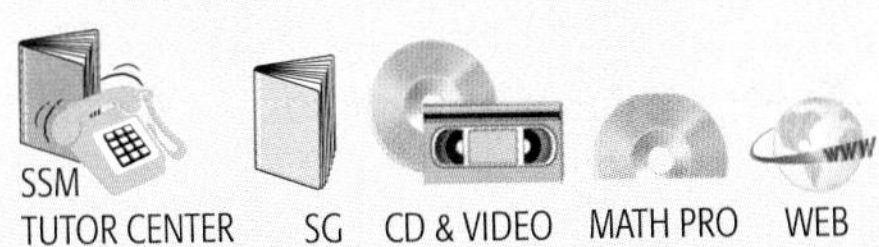

Suppose that we wish to count the number of desks in a classroom. The desks are arranged in 5 rows, and each row has 6 desks.

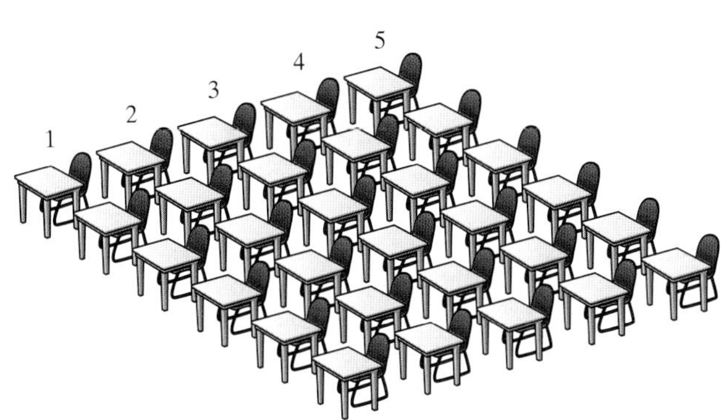

Adding 5 sixes gives the total number of desks: $6 + 6 + 6 + 6 + 6 = 30$ desks. When each addend is the same, we refer to this as **repeated addition**. **Multiplication** is repeated addition but with different notation.

$$\underbrace{6 + 6 + 6 + 6 + 6}_{\text{5 sixes}} = \underset{\text{factor}}{\underset{\uparrow}{5}} \times \underset{\text{factor}}{\underset{\uparrow}{6}} = \underset{\text{product}}{\underset{\uparrow}{30}}$$

The $\times$ is called a **multiplication sign**. The numbers 5 and 6 are called **factors**. The number 30 is called the **product**. The notation 5×6 is read as "five times six." The symbols $\cdot$ and () can also be used to indicate multiplication.

$$5 \times 6 = 30, \quad 5 \cdot 6 = 30, \quad (5)(6) = 30, \quad \text{and} \quad 5(6) = 30$$

Try the Concept Check in the margin.

Concept Check

a. Rewrite $4 + 4 + 4 + 4 + 4 + 4 + 4$ using multiplication.

b. Rewrite 3×16 as repeated addition. Is there more than one way to do this? If so, show all ways.

A Using the Properties of Multiplication

As with addition, we memorize products of one-digit whole numbers and then use certain properties of multiplication to multiply larger numbers. Notice in the appendix that when any number is multiplied by 0, the result is always 0. This is called the **multiplication property of 0**.

Multiplication Property of 0

The product of 0 and any number is 0. For example,

$$5 \cdot 0 = 0$$
$$0 \cdot 8 = 0$$

Also notice in the appendix that when any number is multiplied by 1, the result is always the original number. We call this result the **multiplication property of 1**.

Multiplication Property of 1

The product of 1 and any number is that same number. For example,

$$1 \cdot 9 = 9$$
$$6 \cdot 1 = 6$$

Answers

Concept Check:

a. $7 \times 4 = 28$
b. $16 + 16 + 16 = 48$; yes
$3 + 3 + 3 + 3 + 3 + 3 + 3 + 3 + 3 + 3 + 3 + 3 + 3 + 3 + 3 + 3 = 48$

Practice Problem 1

Multiply.

a. 3×0
b. $4(1)$
c. $(0)(34)$
d. $1 \cdot 76$

EXAMPLE 1 Multiply.

a. 6×1 **b.** $0(8)$ **c.** $1 \cdot 45$ **d.** $(75)(0)$

Solution:

a. $6 \times 1 = 6$ **b.** $0(8) = 0$

c. $1 \cdot 45 = 45$ **d.** $(75)(0) = 0$

Like addition, multiplication is commutative and associative. Notice that when multiplying two numbers, the order of these numbers can be changed without changing the product. For example,

$$3 \cdot 5 = 15 \quad \text{and} \quad 5 \cdot 3 = 15$$

This property is the **commutative property of multiplication**.

Commutative Property of Multiplication

Changing the **order** of two factors does not change their product. For example,

$$9 \cdot 2 = 18 \quad \text{and} \quad 2 \cdot 9 = 18$$

Another property that can help us when multiplying is the **associative property of multiplication**. This property states that when multiplying numbers, the grouping of the numbers can be changed without changing the product. For example,

$$2 \cdot (3 \cdot 4) = 2 \cdot 12 = 24$$

Also,

$$(2 \cdot 3) \cdot 4 = 6 \cdot 4 = 24$$

Both groupings give a product of 24.

Associative Property of Multiplication

Changing the **grouping** of factors does not change their product. For example,

$$5 \cdot (3 \cdot 2) = (5 \cdot 3) \cdot 2$$

$$5 \cdot (3 \cdot 2) = 5 \cdot 6 = 30 \quad \text{and} \quad (5 \cdot 3) \cdot 2 = 15 \cdot 2 = 30$$

With these properties, along with the **distributive property**, we can find the product of any whole numbers. The distributive property says that multiplication **distributes** over addition. For example, notice that $3(2 + 5)$ is the same as $3 \cdot 2 + 3 \cdot 5$.

$$3(2 + 5) = 3(7) = 21$$

$$3 \cdot 2 + 3 \cdot 5 = 6 + 15 = 21$$

Notice in $3(2 + 5) = 3 \cdot 2 + 3 \cdot 5$ that each number inside the parentheses is multiplied by 3.

Answers

1. a. 0 **b.** 4 **c.** 0 **d.** 76

Distributive Property

Multiplication distributes over addition. For example,

$$2(3 + 4) = 2 \cdot 3 + 2 \cdot 4$$

EXAMPLE 2 Rewrite each using the distributive property.

a. $3(4 + 5)$ **b.** $10(6 + 8)$ **c.** $2(7 + 3)$

Solution: Using the distributive property, we have

a. $3(4 + 5) = 3 \cdot 4 + 3 \cdot 5$
b. $10(6 + 8) = 10 \cdot 6 + 10 \cdot 8$
c. $2(7 + 3) = 2 \cdot 7 + 2 \cdot 3$

Practice Problem 2

Rewrite each using the distributive property.

a. $5(2 + 3)$
b. $9(8 + 7)$
c. $3(6 + 1)$

B Multiplying Whole Numbers

Let's use the distributive property to multiply 7(48). To do so, we begin by writing the expanded form of 48 and then applying the distributive property.

$$\begin{aligned} 7(48) &= 7(40 + 8) \\ &= 7 \cdot 40 + 7 \cdot 8 && \text{Apply the distributive property.} \\ &= 280 + 56 && \text{Multiply.} \\ &= 336 && \text{Add.} \end{aligned}$$

This is how we multiply whole numbers. When multiplying whole numbers, we will use the following notation.

$$\begin{array}{r} \scriptstyle 5 \\ 48 \\ \times \quad 7 \\ \hline 336 \end{array}$$

$336 \leftarrow 7 \cdot 8 = 56$ Write 6 in the ones place and carry 5 to the tens place.

$7 \cdot 4 = 28$ and $28 + 5 = 33$

EXAMPLE 3 Multiply: $\begin{array}{r} 25 \\ \times \quad 8 \\ \hline \end{array}$

Solution:

$$\begin{array}{r} \scriptstyle 4 \\ 25 \\ \times \quad 8 \\ \hline 200 \end{array}$$

Practice Problem 3

Multiply.

a. $\begin{array}{r} 36 \\ \times \quad 4 \\ \hline \end{array}$ b. $\begin{array}{r} 92 \\ \times \quad 9 \\ \hline \end{array}$

To multiply larger whole numbers, use the following similar notation. Multiply 89×52.

Step 1

$$\begin{array}{r} \scriptstyle 1 \\ 89 \\ \times \quad 52 \\ \hline 178 \end{array}$$

$178 \leftarrow$ Multiply 89×2.

Step 2

$$\begin{array}{r} \scriptstyle 4 \\ 89 \\ \times \quad 52 \\ \hline 178 \\ 4450 \\ \hline \end{array}$$

$4450 \leftarrow$ Multiply 89×50.

Step 3

$$\begin{array}{r} 89 \\ \times \quad 52 \\ \hline 178 \\ 4450 \\ \hline 4628 \end{array}$$

4628 Add.

The numbers 178 and 4450 are called **partial products**. The sum of the partial products, 4628, is the product of 89 and 52.

Answers

2. **a.** $5(2 + 3) = 5 \cdot 2 + 5 \cdot 3$
b. $9(8 + 7) = 9 \cdot 8 + 9 \cdot 7$
c. $3(6 + 1) = 3 \cdot 6 + 3 \cdot 1$
3. **a.** 144 **b.** 828

Practice Problem 4

Multiply.

a. $\begin{array}{r} 594 \\ \times \ 72 \\ \hline \end{array}$ b. $\begin{array}{r} 306 \\ \times \ 81 \\ \hline \end{array}$

Practice Problem 5

Multiply.

a. $\begin{array}{r} 726 \\ \times 142 \\ \hline \end{array}$ b. $\begin{array}{r} 4 \\ \times 288 \\ \hline \end{array}$

Concept Check

Find and explain the error in the following multiplication problem:

$$\begin{array}{r} 102 \\ \times \ 33 \\ \hline 306 \\ 306 \\ \hline 612 \end{array}$$

Answers

4. a. 42,768 **b.** 24,786
5. a. 103,092 **b.** 1152

Concept Check: $\begin{array}{r} 102 \\ \times \ 33 \\ \hline 306 \\ 3060 \\ \hline 3366 \end{array}$

EXAMPLE 4 Multiply: 236 × 86

Solution:

$$\begin{array}{rl} 236 & \\ \times \ 86 & \\ \hline 1416 & \leftarrow 6(236) \\ 18{,}880 & \leftarrow 80(236) \\ \hline 20{,}296 & \text{Add.} \end{array}$$

EXAMPLE 5 Multiply: 631 × 125

Solution:

$$\begin{array}{rl} 631 & \\ \times \ 125 & \\ \hline 3155 & \leftarrow 5(631) \\ 12{,}620 & \leftarrow 20(631) \\ 63{,}100 & \leftarrow 100(631) \\ \hline 78{,}875 & \text{Add.} \end{array}$$

Try the Concept Check in the margin.

A Finding the Area of a Rectangle

A special application of multiplication is finding the area of a region. Area measures the amount of surface of a region. For example, we measure a plot of land or the living space of a home by area. The figures show two examples of units of area measure. (A centimeter is a unit of length in the metric system.)

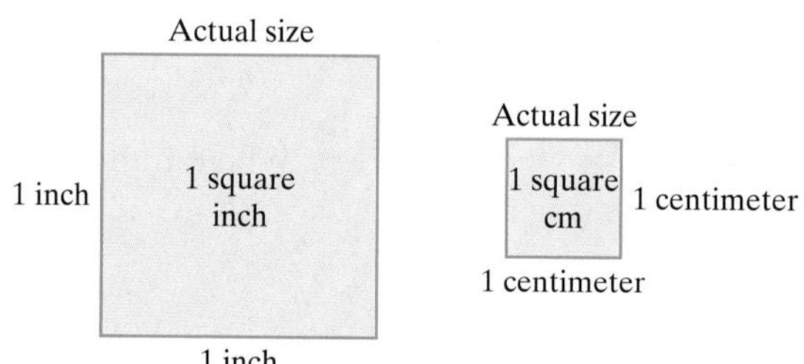

To measure the area of a geometric figure such as the rectangle shown, count the number of square units that cover the region.

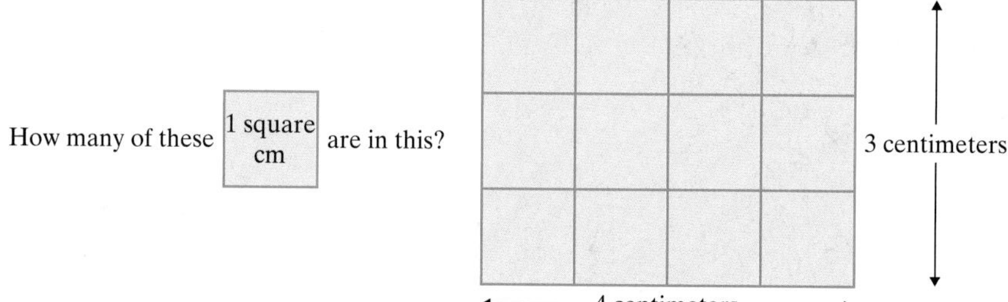

This rectangular region contains 12 square units, each 1 square centimeter. Thus, the area is 12 square centimeters. This total number of squares can be found by counting or by multiplying **4 · 3** (length · width).

$$\begin{aligned} \text{Area of a rectangle} &= \text{length} \cdot \text{width} \\ &= (4 \text{ centimeters})(3 \text{ centimeters}) \\ &= 12 \text{ square centimeters} \end{aligned}$$

This example above shows why area is always measured in square units. In this section, we find the areas of rectangles only. In later sections, we find the areas of other geometric regions.

Remember that *perimeter* (distance around a plane figure) is measured in units. *Area* (space enclosed by a plane figure) is measured in square units.

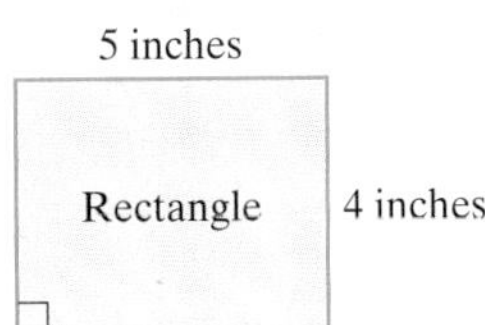

$$\text{Perimeter} = 5 \text{ inches} + 4 \text{ inches} + 5 \text{ inches} + 4 \text{ inches} = 18 \text{ inches}$$

$$\text{Area} = (5 \text{ inches})(4 \text{ inches}) = 20 \text{ square inches}$$

EXAMPLE 6 Finding the Area of a State

The state of Colorado is in the shape of a rectangle whose length is about 380 miles and whose width is about 280 miles. Find its area.

Solution: The area of a rectangle is the product of its length and its width.

$$\begin{aligned} \text{Area} &= \text{length} \cdot \text{width} \\ &= (380 \text{ miles})(280 \text{ miles}) \\ &= 106{,}400 \text{ square miles} \end{aligned}$$

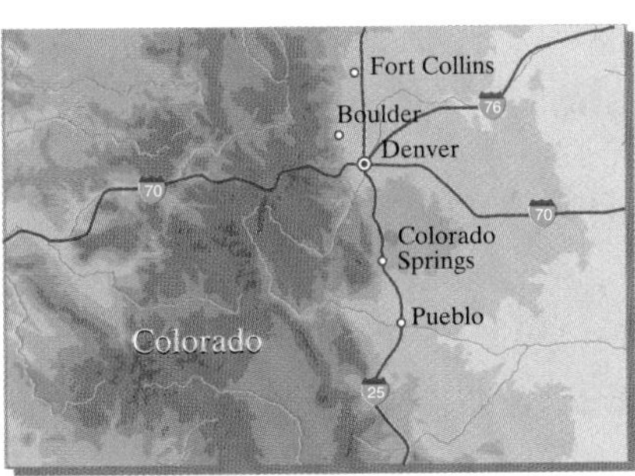

The area of Colorado is approximately 106,400 square miles.

[Note: The abbreviations sq in. and in.^{2} both mean square inches. In this text, we will use the notation sq in.]

Practice Problem 6

The state of Wyoming is in the shape of a rectangle whose length is 360 miles and whose width is 280 miles. Find its area.

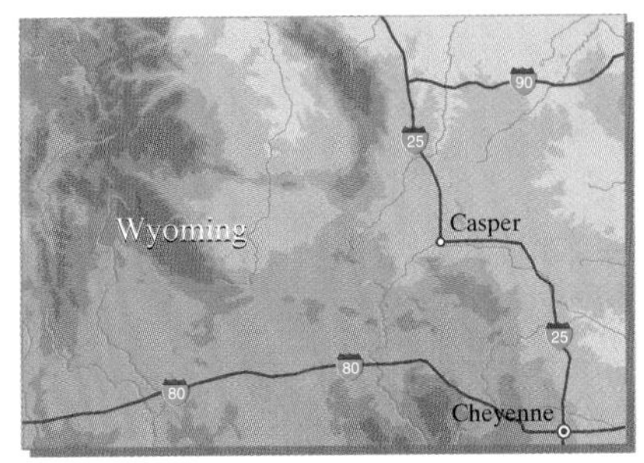

D Solving Problems by Multiplying

There are several words or phrases that indicate the operation of multiplication. Some of these are as follows:

Key Words or Phrases	Example	Symbols
multiply	multiply 5 by 7	$5 \cdot 7$
product	the product of 3 and 2	$3 \cdot 2$
times	10 times 13	$10 \cdot 13$

Many key words or phrases describing real-life problems that suggest addition might be better solved by multiplication instead. For example, to find the **total** cost of 8 shirts, each selling for \$27, we can either add $27 + 27 + 27 + 27 + 27 + 27 + 27 + 27$, or we can multiply 8(27).

Answer

6. 100,800 square mile

Practice Problem 7

A computer printer can print 240 characters per second in draft mode. How many total characters can it print in 15 seconds?

EXAMPLE 7 Finding Disk Space

A certain computer disk can hold about 1510 thousand bytes of information. How many total bytes can 42 such disks hold?

Solution: Forty-two disks will hold 42 × 1510 thousand bytes.

In Words		Translate to Numbers
bytes per disk	→	1510
× disks	→	× 42
		3020
		60,400
total bytes		63,420

Forty-two disks will hold 63,420 thousand bytes.

Practice Problem 8

Softball T-shirts come in two styles: plain at $6 each and striped at $7 each. The team orders 4 plain shirts and 5 striped shirts. Find the total cost of the order.

EXAMPLE 8 Budgeting Money

Earline Martin agrees to take her children and their cousins to the San Antonio Zoo. The ticket price for each child is $4 and for each adult, $6. If 8 children and 1 adult plan to go, how much money is needed for admission?

Solution: If the price of one child's ticket is $4, the price for 8 children is $8 \cdot 4 =$ $32. The price of one adult ticket is $6, so the total cost is

In Words		Translate to Numbers
price of 8 children	→	32
+ price of adult	→	+ 6
total cost		38

The total cost is $38.

Practice Problem 9

If an average page in a book contains 259 words, estimate, rounding to the nearest hundred, the total number of words contained on 195 pages.

EXAMPLE 9 Estimating Word Count

The average page of a book contains 259 words. Estimate, rounding to the nearest ten, the total number of words contained on 22 pages.

Solution: The exact number of words is 259 × 22. Estimate this product by rounding each factor to the nearest ten.

259	rounds to	260 (carry 1)
× 22	rounds to	× 20
		5200

There are approximately 5200 words contained on 22 pages.

Answers

7. 3600 characters **8.** $59 **9.** 60,000 words

CALCULATOR EXPLORATIONS

Multiplying Numbers

To multiply numbers on a calculator, find the keys marked [×] and [=] or [ENTER]. For example, to find $31 \cdot 66$ on a calculator, press the keys [31] [×] [66] [=] or [ENTER]. The display will read [2046]. Thus, $31 \cdot 66 = 2046$.

Use a calculator to multiply.

1. 72 × 48

2. 81 × 92

3. $163 \cdot 94$

4. $285 \cdot 144$

5. 983(277)

6. 1562(843)

Name ______________________ Section ____________ Date ____________

Mental Math

Multiply. See Example 1.

1. $1 \cdot 24$ **2.** $55 \cdot 1$ **3.** $0 \cdot 19$ **4.** $27 \cdot 0$

5. $8 \cdot 0 \cdot 9$ **6.** $7 \cdot 6 \cdot 0$ **7.** $87 \cdot 1$ **8.** $1 \cdot 41$

EXERCISE SET 1.6

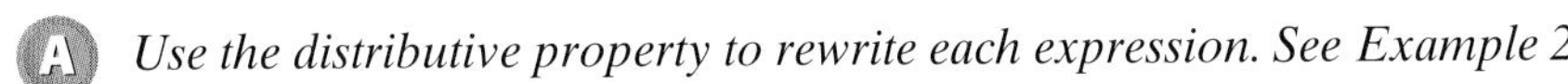

Use the distributive property to rewrite each expression. See Example 2.

1. $4(3 + 9)$ **2.** $5(8 + 2)$ **3.** $2(4 + 6)$ **4.** $6(1 + 4)$ **5.** $10(11 + 7)$ **6.** $12(12 + 3)$

B *Multiply. See Example 3.*

7. 42×6 **8.** 79×3 **9.** 624×3 **10.** 638×5

11. 227×6 **12.** 882×2 **13.** 1062×5 **14.** 9021×3

Multiply. See Examples 4 and 5.

15. 298×14 **16.** 591×72 **17.** 231×47 **18.** 526×23 **19.** 809×14 **20.** 307×16

21. (620)(40) **22.** (720)(80) **23.** (998)(12)(0) **24.** (593)(47)(0) **25.** (590)(1)(10)

26. (240)(1)(20) **27.** 1234×48 **28.** 1357×79 **29.** 609×234 **30.** 505×127

31. 5621×324 **32.** 1234×567 **33.** 1941×235 **34.** 1876×437

35. 589×110 **36.** 426×110 **37.** 964×207 **38.** 462×305

Estimate the products by rounding each factor to the nearest hundred. See Example 9.

39. 576×354 **40.** 982×650 **41.** 604×451 **42.** 111×999

Estimate the products by rounding each factor to the nearest ten. See Example 9.

43. 872×27 **44.** 126×41 **45.** 36×87 **46.** 57×77

C *Find the area of each rectangle. See Example 6.*

47.

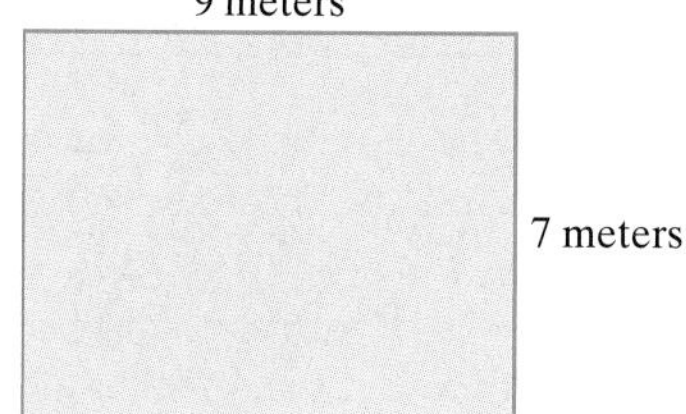

48.

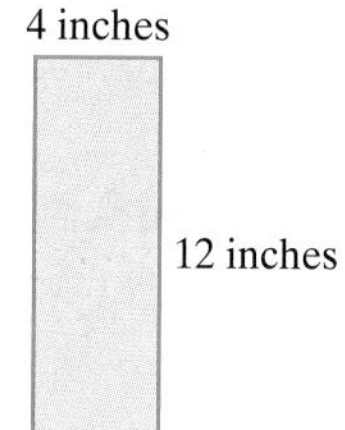

49.

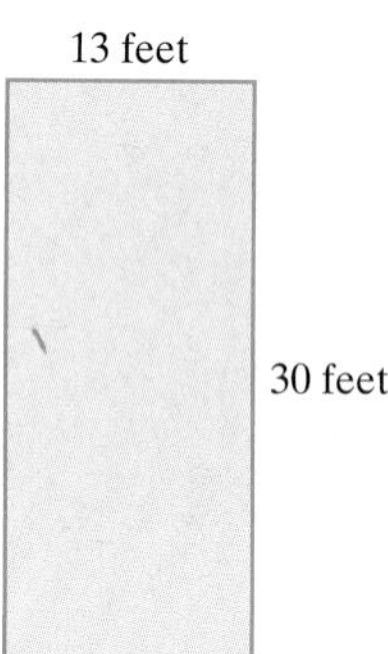

50.

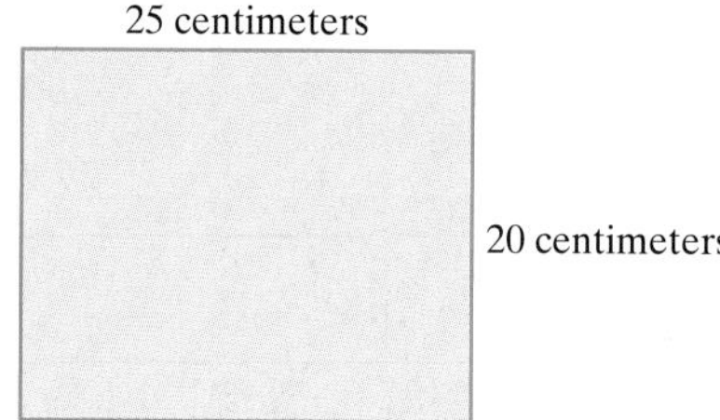

C **D** *Solve. See Examples 6 through 9.*

51. One tablespoon of olive oil contains 125 calories. How many calories are in 3 tablespoons of olive oil? (*Source: Home and Garden Bulletin No. 72*, U.S. Department of Agriculture).

52. One ounce of hulled sunflower seeds contains 14 grams of fat. How many grams of fat are in 6 ounces of hulled sunflower seeds? (*Source: Home and Garden Bulletin No. 72*, U.S. Department of Agriculture).

53. The textbook for a course in Civil War history costs \$54. There are 35 students in the class. Find the total cost of the history books for the class.

54. The seats in the mathematics lecture hall are arranged in 12 rows with 6 seats in each row. Find how many seats are in this room.

55. A case of canned peas has *two layers* of cans. In each layer are 8 rows with 12 cans in each row. Find how many cans are in a case.

56. An apartment building has *three floors*. Each floor has five rows of apartments with four apartments in each row. Find how many apartments there are.

57. A plot of land measures 90 feet by 110 feet. Find its area.

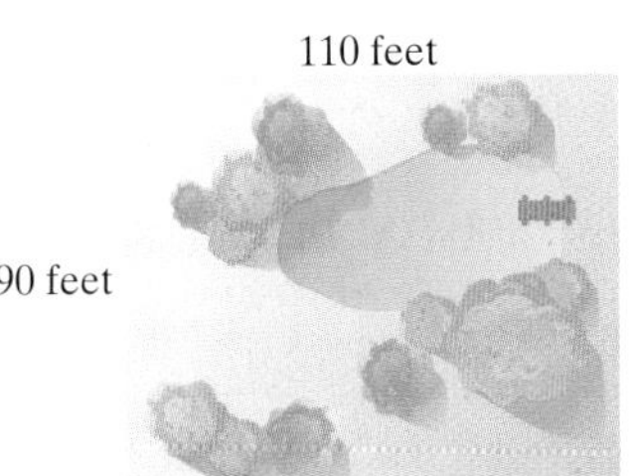

58. A house measures 45 feet by 60 feet. Find the floor area of the house.

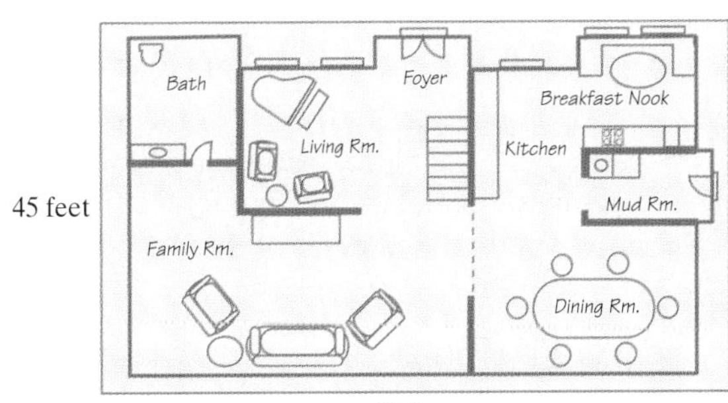

59. The largest lobby can be found at the Hyatt Regency in San Francisco, CA. It is in the shape of a rectangle that measures 350 feet by 160 feet. Find its area.

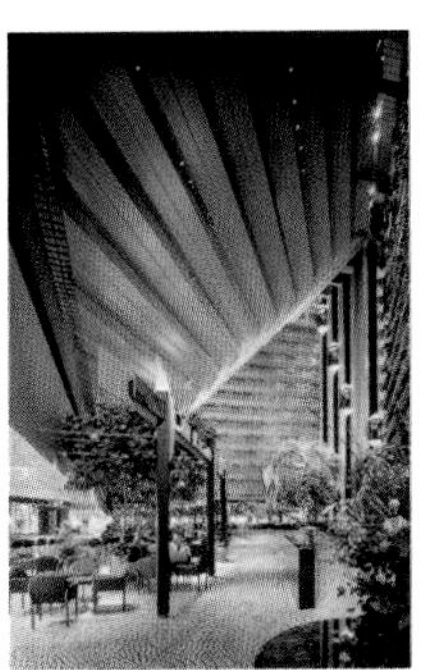

60. Recall from an earlier section that the largest commercial building in the world under one roof is the flower auction building of the cooperative VBA in Aalsmeer, Netherlands. The floor plan is a rectangle that measures 776 meters by 639 meters. Find the area of this building. (A meter is a unit of length in the metric system.) (*Source: The Handy Science Answer Book*, Visible Ink Press)

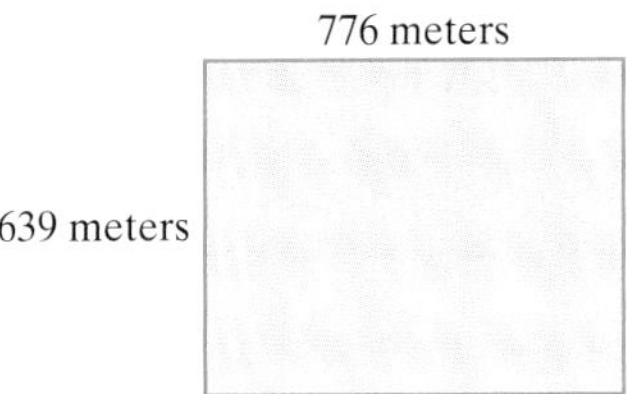

61. A pixel is a rectangular dot on a graphing calculator screen. If a graphing calculator screen contains 62 pixels in a row and 94 pixels in a column, find the total number of pixels on a screen.

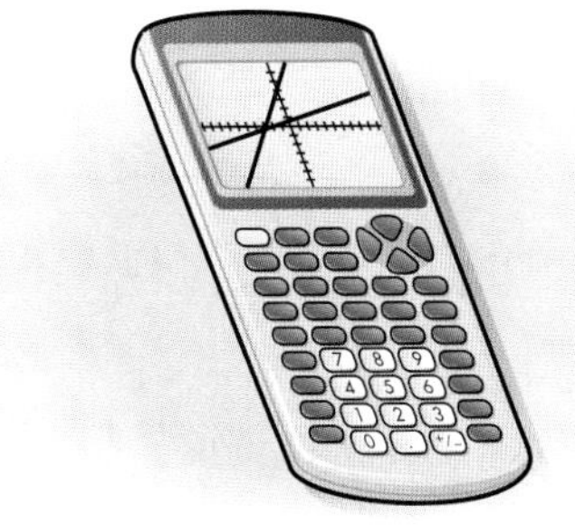

62. A CD (compact disk) can hold 650 megabytes (MB) of information. How many MBs can 17 disks hold?

63. A line of print on a computer contains 80 characters (letters, spaces, punctuation marks). Find how many characters there are in 25 lines.

64. An average cow eats 3 pounds of grain per day. Find how much grain a cow eats in a year. (Assume 365 days in 1 year.)

65. One ounce of Planters® Dry Roasted Peanuts has 160 calories. How many calories are in 8 ounces? (*Source:* RJR Nabisco, Inc.)

66. One ounce of Planters® Dry Roasted Peanuts has 13 grams of fat. How many grams of fat are in 8 ounces? (*Source:* RJR Nabisco, Inc.)

67. The diameter of the planet Saturn is 9 times as great as the diameter of Earth. The diameter of Earth is 7927 miles. Find the diameter of Saturn.

68. The planet Uranus orbits the Sun every 84 Earth years. Find how many Earth days two orbits take. (Assume 365 days in 1 year.)

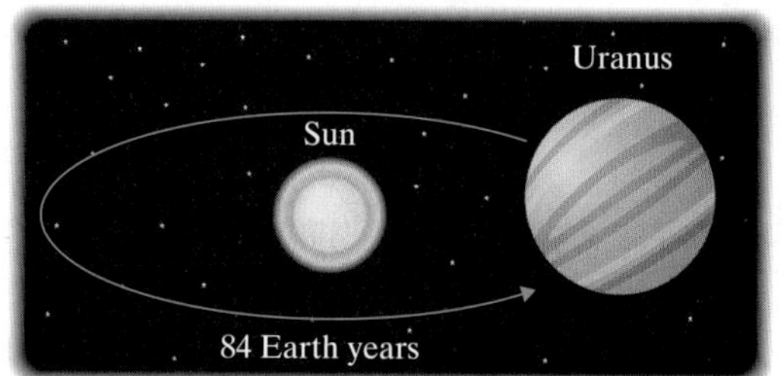

69. A window washer in New York City is bidding for a contract to wash the windows of a 23-story building. To write a bid, the number of windows in the building is needed. If there are 7 windows in each row of windows on 2 sides of the building and 4 windows per row on the other 2 sides of the building, find the total number of windows.

70. In North America, the average toy expenditure per child is $328 per year. On average, how much is spent on toys for a child by the time he or she reaches age 18? (*Source:* The NPD Group Worldwide)

71. Hershey's main chocolate factory in Hershey, Pennsylvania, uses 700,000 quarts of milk each day. How many quarts of milk would be used during the month of March, assuming that chocolate is made at the factory every day of the month? (*Source:* Hershey Foods Corp.)

72. Among older Americans (age 65 years and older), there are 4 times as many widows as widowers. There were 1,994,000 widowers in 2000. How many widows were there in 2000? (*Sources:* Administration on Aging, U.S. Census Bureau)

73. In 2001, the average cost of the Head Start program was $6,633 per child. That year, 905,235 children participated in the program. Round each number to the nearest thousand and estimate the total cost of the Head Start program in 2001. (*Source:* Head Start Bureau)

74. American households currently spend an average of $4810 on food each year. A small community encompasses 1643 households. Round each number to the nearest hundred and estimate the total annual food expenditure for the residents of this community. (*Source:* U.S. Bureau of Labor Statistics)

Combining Concepts

In a survey, college students were asked to name their favorite fruit. The results are shown in the picture graph. Use this graph to answer Exercises 75 through 78.

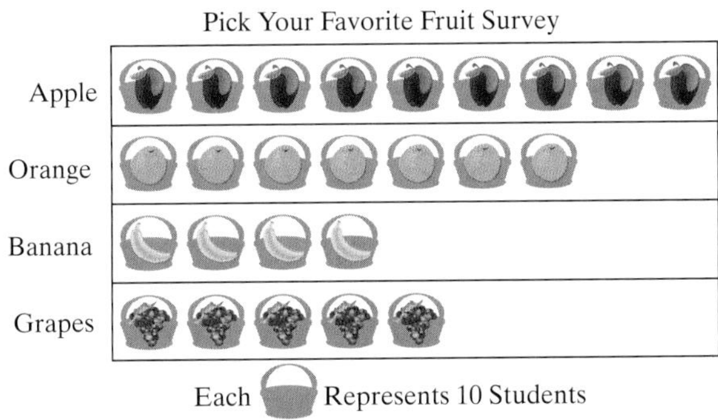

75. How many students chose grapes as their favorite fruit?

76. How many students chose bananas as their favorite fruit?

77. Which two fruits were the most popular?

78. Which two fruits were chosen by a total of 110 students?

Fill in the missing digits in each problem.

79.

```
    4_
 ×  _3
  ----
   126
  3780
  ----
  3906
```

80.

```
    _7
 ×  6_
  ----
   171
  3420
  ----
  3591
```

81. Explain how to multiply two 2-digit numbers using partial products.

82. A slice of enriched white bread has 65 calories. One tablespoon of jam has 55 calories, and one tablespoon of peanut butter has 95 calories. Suppose a peanut butter and jelly sandwich is made with two slices of white bread, one tablespoon of jam, and one tablespoon of peanut butter. How many calories are in two such peanut butter and jelly sandwiches? (*Source: Home and Garden Bulletin No. 72,* U.S. Department of Agriculture)

83. During the NBA's 2000–2001 season, Kobe Bryant of the Los Angeles Lakers scored 61 three-point field goals, 640 two-point field goals, and 475 free throws (worth one point each). How many points did Kobe Bryant score during the 2000–2001 season? (*Source:* National Basketball Association)

FOCUS ON The Real World

JUDGING DISTANCES

Do you know how to estimate a distance without using a tape measure? One easy way to do this is to use the length of your own stride. First, measure the length of your stride in inches. You can do this with the following steps:

- Lay a yardstick on the floor.
- Stand next to the yardstick with feet together so that both heels line up with the 0-mark on the yardstick.
- Take a normal-sized step forward.
- Find the whole-inch mark nearest the toe of the foot that is farthest from the 0-mark. This is roughly the length of your stride.

To judge a distance, simply pace it off using normal-sized strides. Multiply the number of strides by the length of your stride to get a rough estimate of the distance.

Suppose you need to measure a distance that can't easily be paced, such as a pond or a busy street. You can easily "transfer" the distance to a more easily paced area by using a baseball cap.

- Stand at the edge of the pond or street while wearing a baseball cap.
- Bend your head until your chin rests on your chest.
- Pull the bill of the cap up or down until it appears to touch the other side of the pond or street.
- Without moving your head or the cap, pivot your body to the right until you have a clear path straight ahead for walking.
- Notice where the bill seems to be touching the ground now. The distance to this point is the same as the distance across the pond or street you are measuring.
- Pace off the distance to this point and find an estimate of the distance as before.

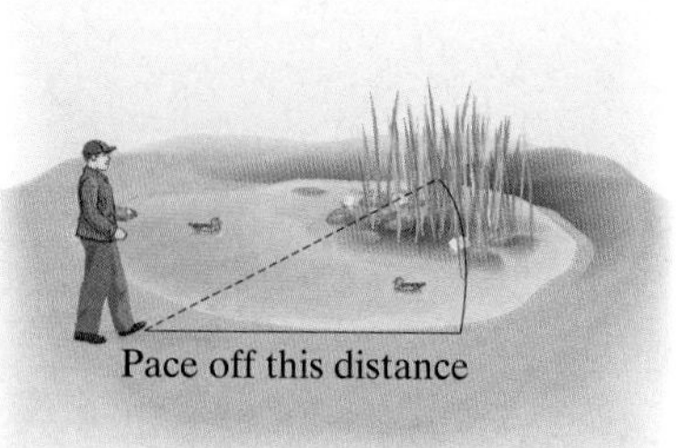

Suppose a distance estimate using this method is 1302 inches. To write the distance in terms of feet, divide the estimate by 12. The quotient is the number of whole feet, and the remainder is the number of inches. A distance of 1302 inches is the same as 108 feet 6 inches.

$$
\begin{array}{r}
108 \text{ R } 6 \\
12\overline{)1302} \\
\underline{12} \\
10 \\
\underline{0} \\
102 \\
\underline{96} \\
6
\end{array}
$$

GROUP ACTIVITY

Materials: yardstick, baseball cap

Use the baseball cap procedure to estimate the distance across a river, stream, pond, or busy road on or near your campus. Have each person in your group estimate the same distance using the length of his or her own stride. Compare your results. Write a brief report summarizing your findings. Be sure to include what distance your group estimated, the length of each member's stride, the number of strides each member paced off, and each member's distance estimate. Conclude by discussing reasons for any differences in estimates.

1.7 Dividing Whole Numbers

Suppose three people pooled their money and bought a raffle ticket at a local fund-raiser. Their ticket was the winner and they won a $60 cash prize. They then divided the prize into three equal parts so that each person received $20.

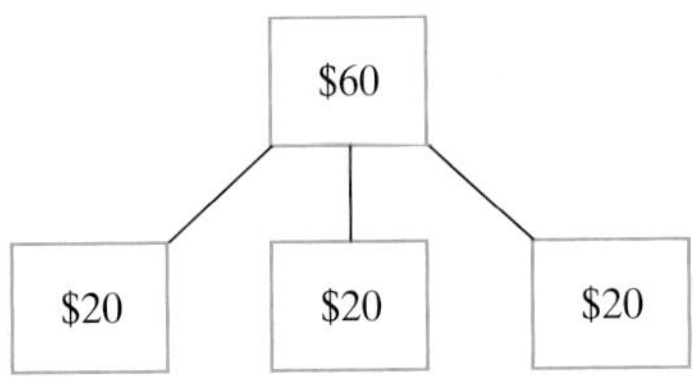

OBJECTIVES

A Divide whole numbers.

B Perform long division.

C Solve problems that require dividing by whole numbers.

D Find the average of a list of numbers.

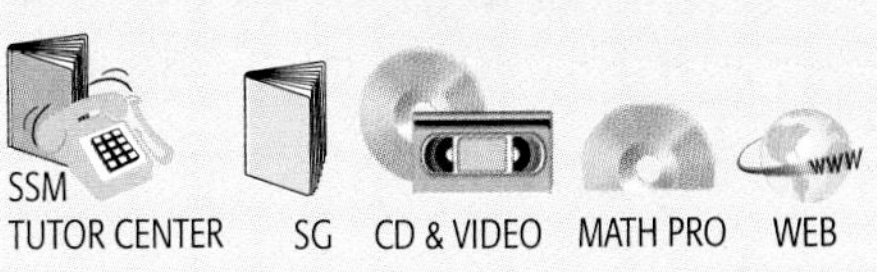

A Dividing Whole Numbers

The process of separating a quantity into equal parts is called **division**. Division can be symbolized by several notations.

$3\overline{)60}$ with quotient 20 (labels: quotient, dividend, divisor)

$\frac{60}{3} = 20$ (labels: dividend, divisor, quotient)

$60 \div 3 = 20$ (labels: dividend, divisor, quotient)

(In the notation $\frac{60}{3}$, the bar separating 60 and 3 is called a **fraction bar**.) Just as subtraction is the reverse of addition, division is the reverse of multiplication. This means that division can be checked by multiplication.

$$\begin{array}{r}20\\3\overline{)60}\end{array} \text{ because } 20 \cdot 3 = 60$$

Since multiplication and division are related in this way, you can use the multiplication table in Appendix B to review quotients of one-digit divisors if necessary.

EXAMPLE 1 Find each quotient. Check by multiplying.

a. $42 \div 7$ **b.** $\frac{81}{9}$ **c.** $4\overline{)24}$

Solution:

a. $42 \div 7 = 6$ because $6 \cdot 7 = 42$

b. $\frac{81}{9} = 9$ because $9 \cdot 9 = 81$

c. $\begin{array}{r}6\\4\overline{)24}\end{array}$ because $6 \cdot 4 = 24$

EXAMPLE 2 Find each quotient. Check by multiplying.

a. $1\overline{)8}$ **b.** $11 \div 1$ **c.** $\frac{9}{9}$ **d.** $7 \div 7$ **e.** $\frac{10}{1}$ **f.** $6\overline{)6}$

Solution:

a. $\begin{array}{r}8\\1\overline{)8}\end{array}$ because $8 \cdot 1 = 8$

b. $11 \div 1 = 11$ because $11 \cdot 1 = 11$

Practice Problem 1

Find each quotient. Check by multiplying.

a. $8\overline{)48}$

b. $35 \div 5$

c. $\frac{49}{7}$

Practice Problem 2

Find each quotient. Check by multiplying.

a. $\frac{8}{8}$ b. $3 \div 1$ c. $1\overline{)12}$

d. $2 \div 1$ e. $\frac{5}{1}$

Answers

1. a. 6 **b.** 7 **c.** 7 **2. a.** 1 **b.** 3 **c.** 12 **d.** 2 **e.** 5

c. $\frac{9}{9} = 1$ because $1 \cdot 9 = 9$

d. $7 \div 7 = 1$ because $1 \cdot 7 = 7$

e. $\frac{10}{1} = 10$ because $10 \cdot 1 = 10$

f. $6\overline{)6}$ with quotient 1 because $1 \cdot 6 = 6$

Example 2 illustrates important properties of division as described next:

Division Properties of 1

The quotient of any number and that same number is 1. For example,

$8 \div 8 = 1 \qquad \frac{7}{7} = 1 \qquad 4\overline{)4}$ with quotient 1

The quotient of any number and 1 is that same number. For example,

$9 \div 1 = 9 \qquad \frac{6}{1} = 6 \qquad 1\overline{)3}$ with quotient 3 $\qquad \frac{0}{1} = 0$

Practice Problem 3

Find each quotient. Check by multiplying.

a. $\frac{0}{7}$ b. $5\overline{)0}$ c. $9 \div 0$

d. $0 \div 6$

EXAMPLE 3 Find each quotient. Check by multiplying.

a. $9\overline{)0}$ **b.** $0 \div 12$ **c.** $\frac{0}{5}$ **d.** $\frac{3}{0}$

Solution:

a. $9\overline{)0}$ with quotient 0 because $0 \cdot 9 = 0$

b. $0 \div 12 = 0$ because $0 \cdot 12 = 0$

c. $\frac{0}{5} = 0$ because $0 \cdot 5 = 0$

d. If $\frac{3}{0}$ = a **number**, then the **number** times $0 = 3$. Recall that any number multiplied by 0 is 0 and not 3. We say, then, that $\frac{3}{0}$ is **undefined**.

Example 3 illustrates important division properties of 0.

Division Properties of 0

The quotient of 0 and any number (except 0) is 0. For example,

$0 \div 9 = 0 \qquad \frac{0}{5} = 0 \qquad 14\overline{)0}$ with quotient 0

The quotient of any number and 0 is not a number. We say that

$\frac{3}{0}, \qquad 0\overline{)3}, \qquad$ and $\qquad 3 \div 0$

are **undefined**.

Answers

1. a. 0 **b.** 0 **c.** undefined **d.** 0

B Performing Long Division

When dividends are larger, the quotient can be found by a process called **long division**. For example, let's divide 2541 by 3.

$$3\overline{)2541}$$

We can't divide 3 into 2, so we try dividing 3 into the first two digits.

$$\begin{array}{r} 8 \\ 3\overline{)2541} \end{array}$$

$25 \div 3 = 8$ with 1 left, so our best estimate is 8. We place 8 over the 5 in 25.

Next, multiply 8 and 3 and subtract this product from 25. Make sure that this difference is less than the divisor.

$$\begin{array}{r} 8 \\ 3\overline{)2541} \\ -\ 24 \\ \hline 1 \end{array}$$

$8(3) = 24$

$25 - 24 = 1$, and 1 is less than the divisor 3.

Bring down the next digit and go through the process again.

$$\begin{array}{r} 84 \\ 3\overline{)2541} \\ -\ 24\downarrow \\ \hline 14 \\ -\ 12 \\ \hline 2 \end{array}$$

$14 \div 3 = 4$ with 2 left

$4(3) = 12$

$14 - 12 = 2$

Once more, bring down the next digit and go through the process.

$$\begin{array}{r} 847 \\ 3\overline{)2541} \\ -\ 24 \\ \hline 14 \\ -\ 12\downarrow \\ \hline 21 \\ -\ 21 \\ \hline 0 \end{array}$$

$21 \div 3 = 7$

$7(3) = 21$

$21 - 21 = 0$

The quotient is 847. To check, see that $847 \times 3 = 2541$.

EXAMPLE 4 Divide: $3705 \div 5$. Check by multiplying.

Solution:

$$\begin{array}{r} 7 \\ 5\overline{)3705} \\ -\ 35\downarrow \\ \hline 20 \end{array}$$

$37 \div 5 = 7$ with 2 left. Place this estimate, 7, over the 7 in 37.

$7(5) = 35$

$37 - 35 = 2$, and 2 is less than the divisor 5.

Bring down the 0.

$$\begin{array}{r} 74 \\ 5\overline{)3705} \\ -\ 35 \\ \hline 20 \\ -\ 20\downarrow \\ \hline 05 \end{array}$$

$20 \div 5 = 4$

$4(5) = 20$

$20 - 20 = 0$, and 0 is less than the divisor 5.

Bring down the 5.

Practice Problem 4

Divide. Check by multiplying.

a. $6\overline{)5382}$

b. $4\overline{)2212}$

Answers

4. a. 897 **b.** 553

$$\begin{array}{r} 741 \\ 5\overline{)3705} \\ -\ 35 \\ \hline 20 \\ -\ 20 \\ \hline 5 \\ -\ 5 \\ \hline 0 \end{array} \quad \begin{array}{l} 5 \div 5 = 1 \\ \\ \\ \\ \\ \\ \\ 1(5) = 5 \\ 5 - 5 = 0 \end{array}$$

Check:

$$\begin{array}{r} 741 \\ \times\ \ 5 \\ \hline 3705 \end{array}$$

Helpful Hint

Since division and multiplication are reverse operations, don't forget that a division problem can be checked by multiplying.

Practice Problem 5

Divide and check.

a. $3\overline{)2397}$

b. $7\overline{)2520}$

EXAMPLE 5 Divide and check: $1872 \div 9$

Solution:

$$\begin{array}{r} 208 \\ 9\overline{)1872} \\ -\ 18 \\ \hline 07 \\ -\ 0 \\ \hline 72 \\ -\ 72 \\ \hline 0 \end{array} \quad \begin{array}{l} \\ \\ 2(9) = 18 \\ 18 - 18 = 0\text{; bring down the 7.} \\ 0(9) = 0 \\ 7 - 0 = 7\text{; bring down the 2.} \\ 8(9) = 72 \\ 72 - 72 = 0 \end{array}$$

Check: $208 \cdot 9 = 1872$

Naturally, quotients don't always "come out even." Making 4 rows out of 26 chairs, for example, isn't possible if each row is supposed to have exactly the same number of chairs. Each of 4 rows can have 6 chairs, but 2 chairs are still left over.

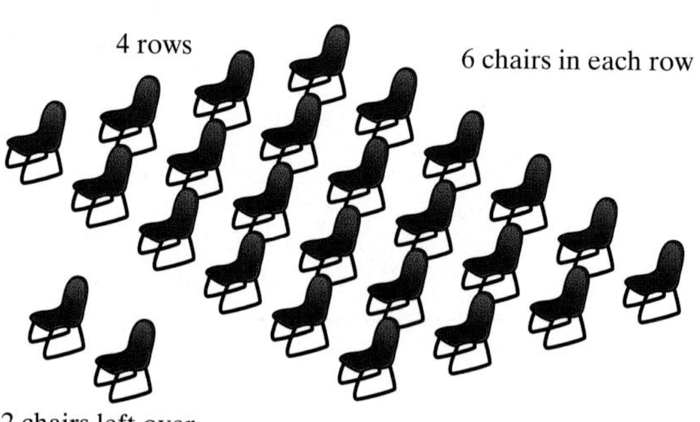

We signify "leftovers" or **remainders** in this way:

$$\begin{array}{r} 6\ \text{R}\,2 \\ 4\overline{)26}\phantom{\ \text{R}\,2} \\ -24\phantom{\ \text{R}\,2} \\ \hline 2\phantom{\ \text{R}\,2} \end{array}$$

Answers

5. a. 799 **b.** 360

The **whole number part of the quotient** is 6; the **remainder part of the quotient** is 2. Checking by multiplying,

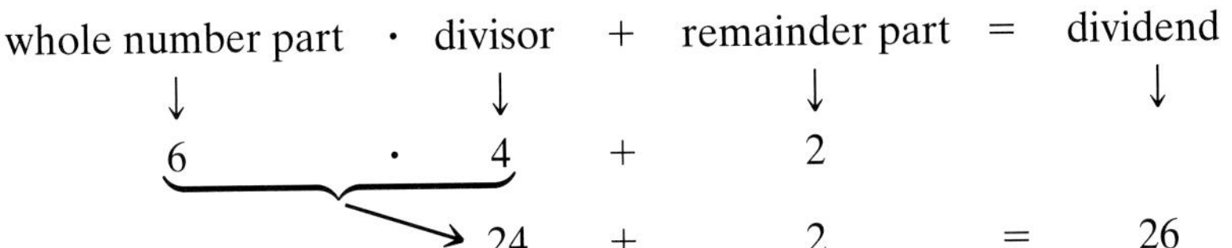

EXAMPLE 6 Divide and check: 2557 ÷ 7

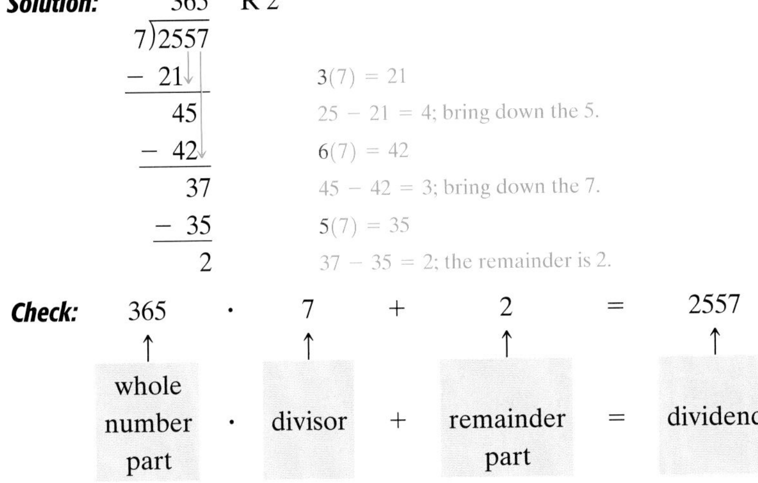

Solution:

$$\begin{array}{r} 365 \text{ R } 2 \\ 7\overline{)2557} \\ -21 \\ \hline 45 \\ -42 \\ \hline 37 \\ -35 \\ \hline 2 \end{array}$$

3(7) = 21
25 − 21 = 4; bring down the 5.
6(7) = 42
45 − 42 = 3; bring down the 7.
5(7) = 35
37 − 35 = 2; the remainder is 2.

Check: 365 · 7 + 2 = 2557

whole number part · divisor + remainder part = dividend

Practice Problem 6

Divide and check.

a. $5\overline{)949}$

b. $6\overline{)4399}$

EXAMPLE 7 Divide and check: 56,717 ÷ 8

Solution:

$$\begin{array}{r} 7089 \text{ R } 5 \\ 8\overline{)56717} \\ -56 \\ \hline 07 \\ -0 \\ \hline 71 \\ -64 \\ \hline 77 \\ -72 \\ \hline 5 \end{array}$$

7(8) = 56
Subtract and bring down the 7.
0(8) = 0
Subtract and bring down the 1.
8(8) = 64
Subtract and bring down the 7.
9(8) = 72
Subtract. The remainder is 5.

Check: 7089 · 8 + 5 = 56,717

whole number part · divisor + remainder part = dividend

Practice Problem 7

Divide and check.

a. $5\overline{)40,841}$

b. $7\overline{)22,430}$

When the divisor has more than one digit, the same pattern applies. For example, let's find 1358 ÷ 23.

$$\begin{array}{r} 5 \\ 23\overline{)1358} \\ 115 \\ \hline 208 \end{array}$$

135 ÷ 23 = 5 with 20 left over. Our estimate is 5.
5(23) = 115
135 − 115 = 20. Bring down the 8.

Now we continue estimating.

Answers

6. a. 189 R 4 **b.** 733 R 1 **7. a.** 8168 R 1 **b.** 3204 R 2

```
      59  R 1      208 ÷ 23 = 9 with 1 left over.
 23)1358
  - 115
    208
  - 207            9(23) = 207
      1            208 - 207 = 1. The remainder is 1.
```

To check, see that $59 \cdot 23 + 1 = 1358$.

Practice Problem 8

Divide: 5740 ÷ 19

EXAMPLE 8 Divide: 6819 ÷ 17

Solution:

```
      401  R 2
  17)6819
   - 68↓          4(17) = 68
      01          Subtract and bring down the 1.
    -  0↓         0(17) = 0
      19          Subtract and bring down the 9.
    - 17          1(17) = 17
       2          Subtract. The remainder is 2.
```

To check, see that $401 \cdot 17 + 2 = 6819$.

Practice Problem 9

Divide: 16,589 ÷ 247

EXAMPLE 9 Divide: 51,600 ÷ 403

Solution:

```
        128  R 16
  403)51600
    - 403↓         1(403) = 403
      1130         Subtract and bring down the 0.
     - 806↓        2(403) = 806
      3240         Subtract and bring down the 0.
    - 3224         8(403) = 3224
        16         Subtract. The remainder is 16.
```

To check, see that $128 \cdot 403 + 16 = 51{,}600$.

Concept Check

Which of the following is the correct way to represent "the quotient of 20 and 5"? Or are both correct? Explain your answer.

a. 5 ÷ 20

b. 20 ÷ 5

Try the Concept Check in the margin.

C Solving Problems by Dividing

Below are some key words and phrases that indicate the operation of division:

Key Words or Phrases	Examples	Symbols
divide	divide 10 by 5	$10 \div 5$ or $\frac{10}{5}$
quotient	the quotient of 64 and 4	$64 \div 4$ or $\frac{64}{4}$
divided by	9 divided by 3	$9 \div 3$ or $\frac{9}{3}$
divided or **shared equally among**	$100 divided equally among five people	$100 \div 5$ or $\frac{100}{5}$

Answers

8. 302 R 2

9. 67 R 40

Concept Check: **a.** incorrect **b.** correct

EXAMPLE 10 Finding Shared Earnings

Zachary, Tyler, and Stephanie McMillan share a paper route to earn money for college expenses. The total in their fund after expenses was $2895. How much is each person's equal share? Use estimation to check your result.

Solution: Each person's equal share is (total) ÷ (number of people) or
2895 ÷ 3

Then

```
   965
 3)2895
 - 27
   19
  - 18
    15
   - 15
     0
```

Each person's share is $965.

Check: To check, we can round $2895 to $3000 and mentally divide $3000 by 3. The result is $1000, which is close to our actual answer of $965. ●

Practice Problem 10

Marina, Manual, and Min bought 10 dozen high-density computer diskettes to share equally. How many diskettes did each person get?

EXAMPLE 11 Calculating Shipping Needs

How many boxes are needed to ship 56 pairs of Nikes to a shoe store in Texarkana if 9 pairs of shoes will fit in each shipping box?

Solution:

number of boxes	=	total pairs of shoes	÷	how many pairs in a box
↓		↓		↓
number of boxes	=	56	÷	9

```
   6 R2
 9)56
 - 54
    2
```

There are 6 full boxes with 2 pairs of shoes left over, so 7 boxes will be needed. ●

Practice Problem 11

Peanut butter and cheese cracker sandwiches come in 6 sandwiches to a package. How many full packages are formed with 195 sandwiches?

EXAMPLE 12 Dividing Holiday Favors among Students

Mary Schultz has 48 kindergarten students. She buys 260 stickers as Thanksgiving Day favors for her students. Can she divide the stickers up equally among her students? If not, how many stickers will be left over?

Solution:

number of stickers	÷	number of students
↓		↓
260	÷	48

```
     5 R20
 48)260
  - 240
     20
```

No, the stickers cannot be divided equally among her students since there is a nonzero remainder. There will be 20 stickers left over. ●

Practice Problem 12

Calculators can be packed 24 to a box. If 497 calculators are to be packed but only full boxes are shipped, how many full boxes will be shipped? How many calculators are left over and not shipped?

Answers

10. 40 diskettes **11.** 32 full packages
12. 20 full boxes; 17 calculators left over

D Finding Averages

A sp cial application of division (and addition) is finding the average of a list f numbers. The **average** of a list of numbers is the sum of the numbers divided by the number of numbers.

$$\text{average} = \frac{\text{sum of numbers}}{\textit{number} \text{ of numbers}}$$

Practice Problem 13

To compute a safe time to wait for reactions to occur after allergy shots are administered, a lab technician is given a list of elapsed times between administered shots and reactions. Find the average of the times 5 minutes, 7 minutes, 20 minutes, 6 minutes, 9 minutes, 3 minutes, and 48 minutes.

Answer

13. 14 min

EXAMPLE 13 Averaging Scores

Liam Reilly's scores in his mathematics class so far are 93, 86, 71, and 82. Find his average score.

Solution: To find his average score, we find the sum of his scores and divide by 4, the number of scores.

$$\begin{array}{r} 93 \\ 86 \\ 71 \\ +\ 82 \\ \hline 332 \end{array} \text{ sum}$$

$$\text{average} = \frac{332}{4} = 83$$

$$\begin{array}{r} 83 \\ 4\overline{)332} \\ -\ 32 \\ \hline 12 \\ -\ 12 \\ \hline 0 \end{array}$$

His average score is 83.

CALCULATOR EXPLORATIONS

Dividing Numbers

To divide numbers on a calculator, find the keys marked [÷] and [=] or [ENTER]. For example, to find 435 ÷ 5 on a calculator, press the keys [435] [÷] [5] [=] or [ENTER]. The display will read [87]. Thus, 435 ÷ 5 = 87.

Use a calculator to divide.

1. 848 ÷ 16

2. 564 ÷ 12

3. $95\overline{)5890}$

4. $27\overline{)1053}$

5. $\frac{32{,}886}{126}$

6. $\frac{143{,}088}{264}$

7. 0 ÷ 315

8. 315 ÷ 0

Name ______________________ Section ____________ Date ____________

Mental Math

A *Find each quotient. See Examples 1 through 3.*

1. $40 \div 8$ **2.** $72 \div 9$ **3.** $45 \div 5$ **4.** $24 \div 3$ **5.** $0 \div 5$

6. $0 \div 8$ **7.** $9 \div 1$ **8.** $12 \div 1$ **9.** $\frac{16}{16}$ **10.** $\frac{49}{49}$

11. $\frac{25}{5}$ **12.** $\frac{45}{9}$ **13.** $6 \div 0$ **14.** $\frac{12}{0}$ **15.** $7 \div 1$

16. $6 \div 6$ **17.** $0 \div 4$ **18.** $7 \div 0$ **19.** $16 \div 2$ **20.** $18 \div 3$

EXERCISE SET 1.7

B *Divide and then check by multiplying. See Examples 4 and 5.*

1. $9\overline{)108}$ **2.** $5\overline{)85}$ **3.** $6\overline{)222}$ **4.** $8\overline{)640}$ **5.** $3\overline{)1014}$ **6.** $4\overline{)504}$

Divide and then check by multiplying. See Examples 6 and 7.

7. $6\overline{)98}$ **8.** $7\overline{)422}$ **9.** $2\overline{)1127}$ **10.** $3\overline{)1240}$

11. $186 \div 5$ **12.** $167 \div 3$ **13.** $2121 \div 8$ **14.** $333 \div 4$

Divide and then check by multiplying. See Examples 8 and 9.

15. $23\overline{)1127}$ **16.** $42\overline{)2016}$ **17.** $55\overline{)715}$ **18.** $32\overline{)1856}$ **19.** $97\overline{)9449}$

20. $1938 \div 44$ **21.** $3708 \div 18$ **22.** $7224 \div 12$ **23.** $6578 \div 13$ **24.** $5670 \div 14$

25. $9299 \div 46$ **26.** $2539 \div 64$ **27.** $\frac{10{,}620}{236}$ **28.** $\frac{5781}{123}$ **29.** $\frac{10{,}194}{103}$

30. $\frac{23,048}{240}$

31. 20,619 ÷ 102

32. 40,803 ÷ 203

33. 45,046 ÷ 223

34. 164,592 ÷ 543

C *Solve. See Examples 10 through 12.*

35. Kathy Gomez teaches Spanish lessons for $85 per student for a 5-week session. From one group of students, she collects $4930. Find how many students are in the group.

36. Martin Thieme teaches American Sign Language classes for $55 per student for a 7-week session. He collects $1430 from the group of students. Find how many students are in the group.

37. Twenty-one people pooled their money and bought lottery tickets. One ticket won a prize of $5,292,000. Find how many dollars each person received.

38. The gravity of Jupiter is 318 times as strong as the gravity of Earth, so objects on Jupiter weigh 318 times as much as they weigh on Earth. If a person would weigh 52,470 pounds on Jupiter, find how much the person weighs on Earth.

39. A truck hauls wheat to a storage granary. It carries a total of 5810 bushels of wheat in 14 trips. How much does the truck haul each trip if each trip it hauls the same amount?

40. The white stripes dividing the lanes on a highway are 25 feet long, and the spaces between them are 25 feet long. Find how many whole stripes there are in 1 mile of highway. (A mile is 5280 feet.)

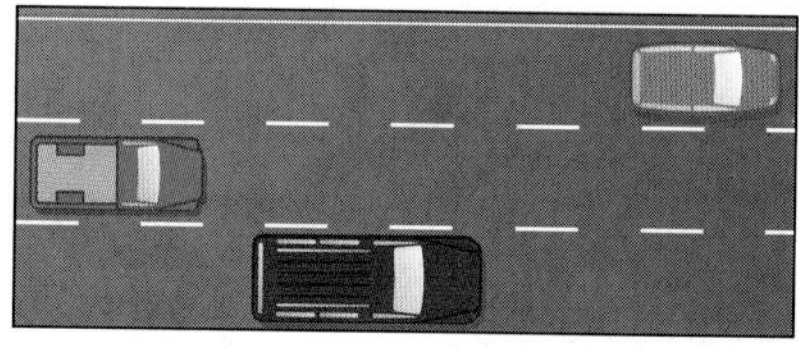

41. There is a bridge over highway I-35 every three miles. The first bridge is at the beginning of the 265-mile stretch of highway. Find how many bridges there are over 265 miles of I-35.

42. An 18-hole golf course is 5580 yards long. If the distance to each hole is the same, find the distance between holes.

43. Wendy Holladay has a piece of rope 185 feet long that she wants to cut into pieces for an experiment in her second-grade class. Each piece of rope is to be 8 feet long. Determine whether she has enough rope for her 22-student class. Determine the amount extra or the amount short.

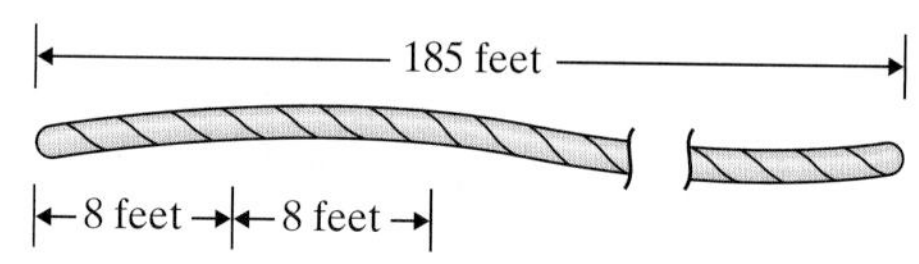

44. Jesse White is in the requisitions department of Central Electric Lighting Company. Light poles along a highway are placed 492 feet apart. The first light pole is at the beginning of the 1-mile strip. Find how many poles he should order for a 1-mile strip of highway. (A mile is 5280 feet.)

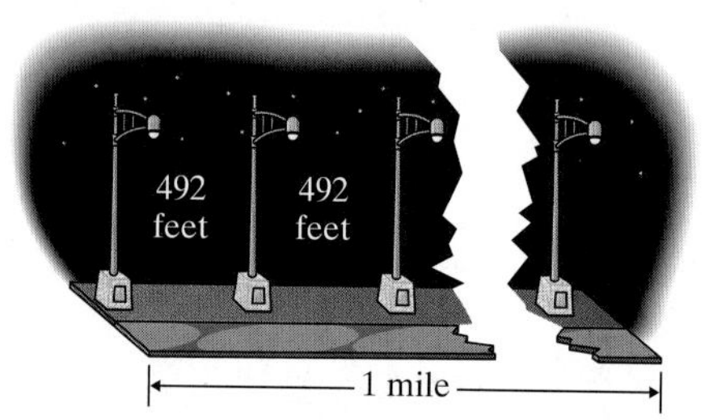

45. Priest Holmes led the NFL in touchdowns during the 2002 football season, scoring a total of 144 points as touchdowns. If a touchdown is worth 6 points, how many touchdowns did Priest make during 2002? (*Source:* National Football League)

46. Broad Peak in Pakistan is the twelfth-tallest mountain in the world. Its elevation is 26,400 feet. A mile is 5280 feet. How many miles high is Broad Peak? (*Source:* National Geographic Society)

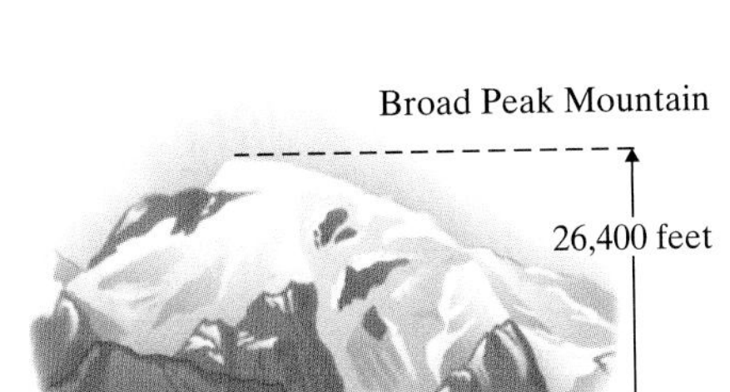

47. Find how many yards are in 1 mile. (A mile is 5280 feet; a yard is 3 feet.)

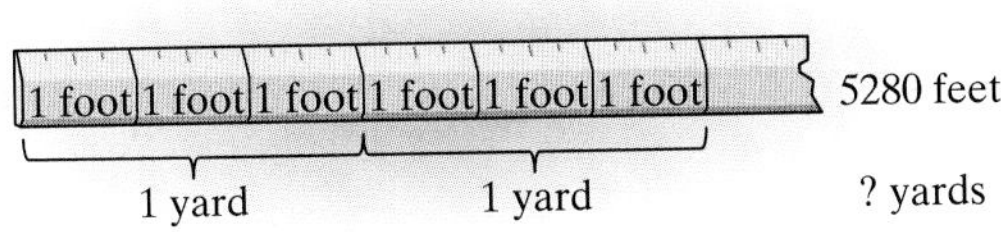

48. Find how many whole feet are in 1 rod. (A mile is 5280 feet; 1 mile is 320 rods.)

D *Find the average of each list of numbers. See Example 13.*

49. 14, 22, 45, 18, 30, 27

50. 37, 26, 15, 29, 51, 22

51. 204, 968, 552, 268

52. 121, 200, 185, 176, 163

53. 86, 79, 81, 69, 80

54. 92, 96, 90, 85, 92, 79

The normal monthly temperature in degrees Fahrenheit for Minneapolis, Minnesota, is given in the graph. Use this graph to answer Exercises 55 and 56. (Source: National Climatic Data Center)

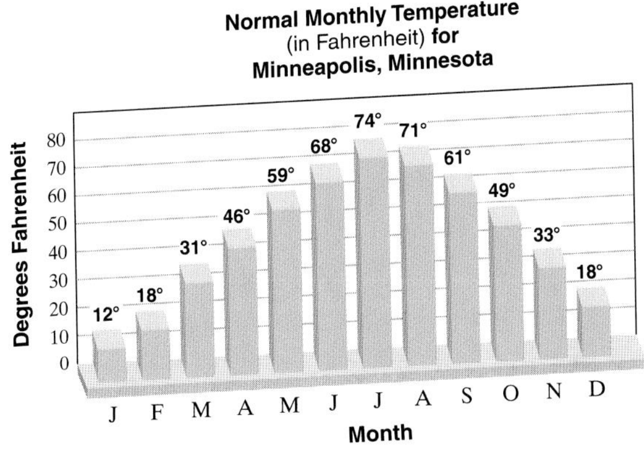

55. Find the average temperature for December, January, and February.

56. Find the average temperature for the entire year.

Combining Concepts

The following table shows the top five leading U.S. advertisers in 2001 and the amount of money spent in that year on ads. Use this table to answer Exercises 57 and 58.

Company	Yearly Amount Spent on Advertising
General Motors Corp.	\$3,374,400,000
Procter & Gamble Co.	\$2,540,600,000
Ford Motor Co.	\$2,408,200,000
PepsiCo	\$2,210,400,000
Pfizer	\$2,189,500,000

(*Source*: AdAge.com & the June 24, 2002 issue of *Advertising Age*)

57. Find the average amount of money spent on ads for the year by the top two companies.

58. Find the average amount of money spent on ads by Procter & Gamble, PepsiCo, and Pfizer.

In Example 13 in this section, we found that the average of 93, 86, 71, and 82 is 83. Use this information to answer Exercises 59 and 60.

59. If the number 71 is removed from the list of numbers, does the average increase or decrease? Explain why.

60. If the number 93 is removed from the list of numbers, does the average increase or decrease? Explain why.

61. Without computing it, tell whether the average of 126, 135, 198, 113 is 86. Explain why it is or why it is not.

62. If the area of a rectangle is 30 square feet and its width is 3 feet, what is its length?

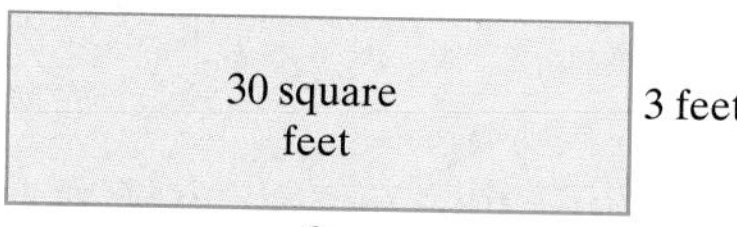

Name ______________________ Section __________ Date __________

Integrated Review—Operations on Whole Numbers

Perform each indicated operation. Use estimation to see if your results are reasonable.

1. $\begin{array}{r} 23 \\ 46 \\ +\ 79 \\ \hline \end{array}$

2. $\begin{array}{r} 7006 \\ -\ 451 \\ \hline \end{array}$

3. $\begin{array}{r} 36 \\ \times\ 45 \\ \hline \end{array}$

4. $8\overline{)4496}$

5. $1 \cdot 79$

6. $\frac{36}{0}$

7. $9 \div 1$

8. $9 \div 9$

9. $0 \cdot 13$

10. $7 \cdot 0 \cdot 8$

11. $0 \div 2$

12. $12 \div 4$

13. $\begin{array}{r} 4219 \\ -\ 1786 \\ \hline \end{array}$

14. $\begin{array}{r} 1861 \\ 7965 \\ 199 \\ +\ 2870 \\ \hline \end{array}$

15. $5\overline{)1068}$

16. $\begin{array}{r} 1259 \\ \times\ 63 \\ \hline \end{array}$

17. $3 \cdot 9$

18. $45 \div 5$

19. $\begin{array}{r} 207 \\ -\ 69 \\ \hline \end{array}$

20. $\begin{array}{r} 207 \\ +\ 69 \\ \hline \end{array}$

21. $7\overline{)7695}$

22. $9\overline{)1000}$

23. $32\overline{)21,222}$

24. $65\overline{)70,000}$

25. $4000 - 2976$

26. $10,000 - 101$

27. $\begin{array}{r} 303 \\ \times\ 101 \\ \hline \end{array}$

28. $(475)(100)$

29. $7\overline{)0}$

30. $\frac{14}{0}$

31. $\frac{0}{6}$

32. $0 \div 105$

33. Subtract 14 from 100.

34. Find the difference of 43 and 21.

Answers

1. ______
2. ______
3. ______
4. ______
5. ______
6. ______
7. ______
8. ______
9. ______
10. ______
11. ______
12. ______
13. ______
14. ______
15. ______
16. ______
17. ______
18. ______
19. ______
20. ______
21. ______
22. ______
23. ______
24. ______
25. ______
26. ______
27. ______
28. ______
29. ______
30. ______
31. ______
32. ______
33. ______
34. ______

39. ____

40. ____

41. ____

42. ____

43. ____

44. ____

45. ____

46. ____

Complete the table by rounding the given number to the given place value.

		Tens	Hundreds	Thousands
35.	8625			
36.	1553			
37.	10,901			
38.	432,198			

Find the perimeter and area of each figure.

△ **39.**

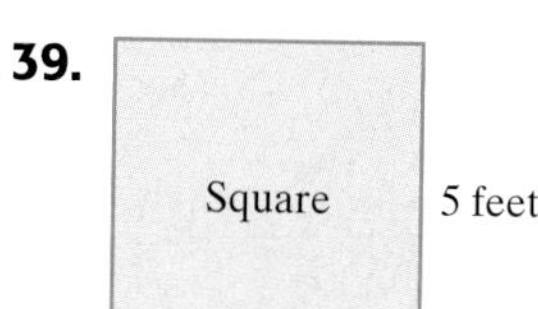

△ **40.**

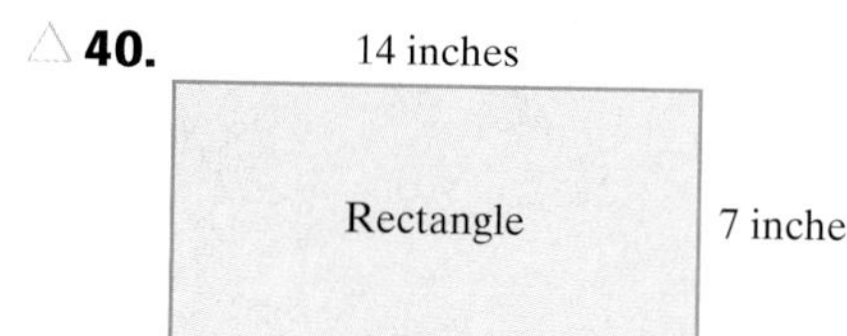

Find the perimeter of each figure.

△ **41.**

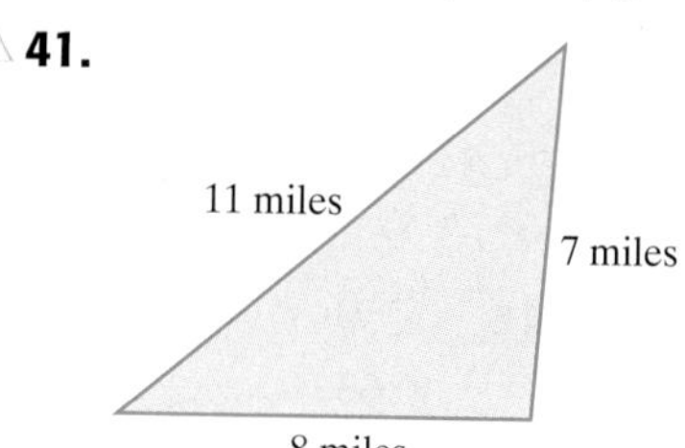

△ **42.**

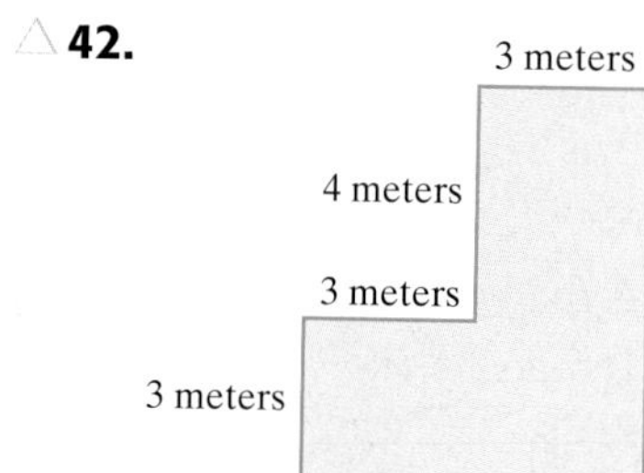

43. The length of the southern boundary of the conterminous United States is 1933 miles. The length of the northern boundary of the conterminous United States is 2054 miles longer than this. What is the length of the northern boundary? (*Source:* U.S. Geological Survey)

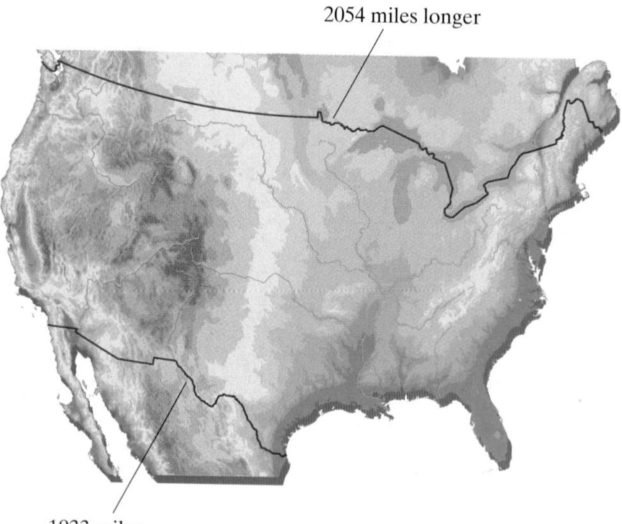

44. In humans, 14 muscles are required to smile. It takes 29 more muscles to frown. How many muscles does it take to frown?

45. How many cases can be filled with 9900 cans of jalapeños if each case holds 48 cans? How many cans will be left over? Will there be enough cases to fill an order for 200 cases?

46. The director of a learning lab at a local community college is working on next year's budget. Thirty-three new video players are needed at a cost of $540 each. What is the total cost of these video players?

1.8 Exponents and Order of Operations

OBJECTIVES

A Write repeated factors using exponential notation.

B Evaluate expressions containing exponents.

C Use the order of operations.

D Find the area of a square.

SSM TUTOR CENTER SG CD & VIDEO MATH PRO WEB

A Using Exponential Notation

An **exponent** is a shorthand notation for repeated multiplication. When the same number is a factor several times, an exponent may be used. In the product

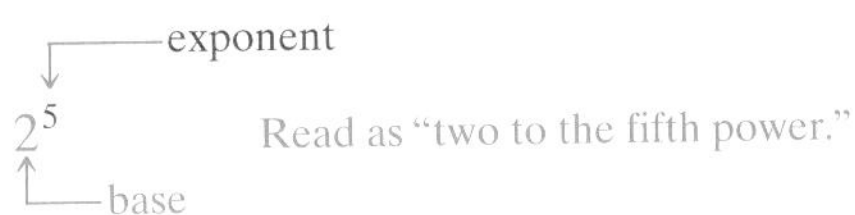

Using an exponent, this product can be written as

2^5 (exponent: 5; base: 2) Read as "two to the fifth power."

Thus,

$$2 \cdot 2 \cdot 2 \cdot 2 \cdot 2 = 2^5$$

This is called **exponential notation**. The **exponent**, 5, indicates how many times the **base**, 2, is a factor.

Certain expressions are used when reading exponential notation.

$5 = 5^1$ is read as "five to the first power."

$5 \cdot 5 = 5^2$ is read as "five to the second power" or "five squared."

$5 \cdot 5 \cdot 5 = 5^3$ is read as "five to the third power" or "five cubed."

$5 \cdot 5 \cdot 5 \cdot 5 = 5^4$ is read as "five to the fourth power."

Helpful Hint

Usually, an exponent of 1 is not written, so when no exponent appears, we assume that the exponent is 1. For example, $2 = 2^1$ and $7 = 7^1$.

EXAMPLES Write using exponential notation.

1. $4 \cdot 4 \cdot 4 = 4^3$
2. $7 \cdot 7 = 7^2$
3. $5 \cdot 5 \cdot 5 \cdot 5 = 5^4$
4. $6 \cdot 6 \cdot 6 \cdot 8 \cdot 8 \cdot 8 \cdot 8 \cdot 8 = 6^3 \cdot 8^5$

Practice Problems 1–4

Write using exponential notation.

1. $2 \cdot 2 \cdot 2$
2. $3 \cdot 3$
3. $10 \cdot 10 \cdot 10 \cdot 10 \cdot 10 \cdot 10$
4. $5 \cdot 5 \cdot 4 \cdot 4 \cdot 4$

B Evaluating Exponential Expressions

To **evaluate** an exponential expression, we write the expression as a product and then find the value of the product.

EXAMPLES Evaluate.

5. $8^2 = 8 \cdot 8 = 64$
6. $7^1 = 7$
7. $2^5 = 2 \cdot 2 \cdot 2 \cdot 2 \cdot 2 = 32$
8. $5 \cdot 6^2 = 5 \cdot 6 \cdot 6 = 180$

Example 8 illustrates an important property: An exponent applies only to its base. The exponent 2, in $5 \cdot 6^2$, applies only to its base, 6.

Practice Problems 5–8

Evaluate.

5. 2^3
6. 5^2
7. 10^1
8. $4 \cdot 5^2$

Answers

1. 2^3 **2.** 3^2 **3.** 10^6 **4.** $5^2 \cdot 4^3$ **5.** 8 **6.** 25 **7.** 10 **8.** 100

Helpful Hint

An exponent applies only to its base. For example, $4 \cdot 2^3$ means $4 \cdot 2 \cdot 2 \cdot 2$.

Helpful Hint

Don't forget that 2^4, for example, is *not* $2 \cdot 4$. 2^4 means repeated multiplication of the same factor.

$$2^4 = 2 \cdot 2 \cdot 2 \cdot 2 = 16, \text{ whereas } 2 \cdot 4 = 8$$

Concept Check

Which of the following statement(s) is correct?

a. 3^6 is the same as $6 \cdot 6 \cdot 6$.

b. "Eight to the fourth power" is the same as 8^4.

c. "Ten squared" is the same as 10^3.

d. 11^2 is the same as $11 \cdot 2$.

e. $5^2 = 10$

Try the Concept Check in the margin.

C Using the Order of Operations

Suppose that you are in charge of taking inventory at a local bookstore. An employee has given you the number of a certain book in stock as the expression

$$3 + 2 \cdot 10$$

To calculate the value of this expression, do you add first or multiply first? If you add first, the answer is 50. If you multiply first, the answer is 23.

Mathematical symbols wouldn't be very useful if two values were possible for one expression. Thus, mathematicians have agreed that, given a choice, we multiply first.

$$3 + 2 \cdot 10 = 3 + 20 \quad \text{Multiply.}$$
$$= 23 \quad \text{Add.}$$

This agreement is one of several **order of operations** agreements.

Order of Operations

1. Perform all operations within grouping symbols such as parentheses or brackets.
2. Evaluate any expressions with exponents.
3. Multiply or divide in order from left to right.
4. Add or subtract in order from left to right.

For example, using the order of operations, let's evaluate $2^3 \cdot 4 - (10 \div 5)$.

$$2^3 \cdot 4 - (10 \div 5) = 2^3 \cdot 4 - 2 \quad \text{Simplify inside parentheses.}$$
$$= 8 \cdot 4 - 2 \quad \text{Write } 2^3 \text{ as 8.}$$
$$= 32 - 2 \quad \text{Multiply } 8 \cdot 4.$$
$$= 30 \quad \text{Subtract.}$$

ANSWER

Concept Check: b

EXAMPLE 9 Simplify: $2 \cdot 4 - 3 \div 3$

Solution: There are no parentheses and no exponents, so we start by multiplying and dividing, from left to right.

$$\begin{aligned} 2 \cdot 4 - 3 \div 3 &= 8 - 3 \div 3 && \text{Multiply.} \\ &= 8 - 1 && \text{Divide.} \\ &= 7 && \text{Subtract.} \end{aligned}$$

Practice Problem 9

Simplify: $16 \div 4 - 2$

EXAMPLE 10 Simplify: $4^2 \div 2 \cdot 4$

Solution: We start by evaluating 4^2.

$$4^2 \div 2 \cdot 4 = 16 \div 2 \cdot 4 \quad \text{Write } 4^2 \text{ as 16.}$$

Next we multiply or divide *in order* from left to right. Since division appears before multiplication from left to right, we divide first, then multiply.

$$\begin{aligned} 16 \div 2 \cdot 4 &= 8 \cdot 4 && \text{Divide.} \\ &= 32 && \text{Multiply.} \end{aligned}$$

Practice Problem 10

Simplify: $18 \div 3^2 \cdot 2^2$

EXAMPLE 11 Simplify: $(8 - 6)^2 + 2^3 \cdot 3$

Solution:

$$\begin{aligned} (8 - 6)^2 + 2^3 \cdot 3 &= 2^2 + 2^3 \cdot 3 && \text{Simplify inside parentheses.} \\ &= 4 + 8 \cdot 3 && \text{Write } 2^2 \text{ as 4 and } 2^3 \text{ as 8.} \\ &= 4 + 24 && \text{Multiply.} \\ &= 28 && \text{Add.} \end{aligned}$$

Practice Problem 11

Simplify: $(9 - 8)^3 + 3 \cdot 2^4$

EXAMPLE 12 Simplify: $4^3 + [3^2 - (10 \div 2)] - 7 \cdot 3$

Solution: Here we begin with the innermost set of parentheses.

$$\begin{aligned} 4^3 + [3^2 - (10 \div 2)] - 7 \cdot 3 &= 4^3 + [3^2 - 5] - 7 \cdot 3 && \text{Simplify inside parentheses.} \\ &= 4^3 + [9 - 5] - 7 \cdot 3 && \text{Write } 3^2 \text{ as 9.} \\ &= 4^3 + 4 - 7 \cdot 3 && \text{Simplify inside brackets.} \\ &= 64 + 4 - 7 \cdot 3 && \text{Write } 4^3 \text{ as 64.} \\ &= 64 + 4 - 21 && \text{Multiply.} \\ &= 47 && \text{Add and subtract from left to right.} \end{aligned}$$

Practice Problem 12

Simplify: $24 \div [20 - (3 \cdot 4)] + 2^3 - 5$

EXAMPLE 13 Simplify: $\dfrac{7 - 2 \cdot 3 + 3^2}{2^2 + 1}$

Solution: Here, the fraction bar is like a grouping symbol. We simplify above and below the fraction bar separately.

$$\begin{aligned} \frac{7 - 2 \cdot 3 + 3^2}{2^2 + 1} &= \frac{7 - 2 \cdot 3 + 9}{4 + 1} && \text{Write } 3^2 \text{ as 9 and } 2^2 \text{ as 4.} \\ &= \frac{7 - 6 + 9}{5} && \text{Multiply } 2 \cdot 3 \text{ in the numerator and add 4 and 1 in the denominator.} \\ &= \frac{10}{5} && \text{Add and subtract from left to right.} \\ &= 2 && \text{Divide.} \end{aligned}$$

Practice Problem 13

Simplify: $\dfrac{60 - 5^2 + 1}{3(1 + 1)}$

Answers

9. 2 **10.** 8 **11.** 49 **12.** 6 **13.** 6

D Finding the Area of a Square

Since a square is a special rectangle, we can find its area by finding the product of its length and its width.

Area of a rectangle = length · width

By recalling that each side of a square has the same measurement, we can use the following procedure to find its area:

$$\begin{aligned}\text{Area of a square} &= \text{length} \cdot \text{width}\\ &= \text{side} \cdot \text{side}\\ &= (\text{side})^2\end{aligned}$$

Square Side

Side

Practice Problem 14

Find the area of a square whose side measures 11 centimeters.

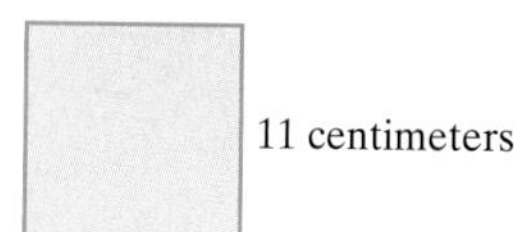

Answer

14. 121 sq cm

EXAMPLE 14 Find the area of a square whose side measures 5 inches.

Solution:

$$\begin{aligned}\text{Area of a square} &= (\text{side})^2\\ &= (5 \text{ inches})^2\\ &= 25 \text{ square inches}\end{aligned}$$

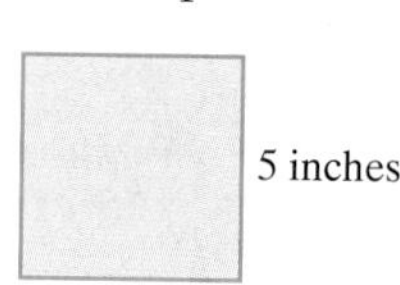

The area of the square is 25 square inches.

CALCULATOR EXPLORATIONS

Exponents

To evaluate an exponent such as 4^7 on a calculator, find the keys marked [y^x] and [=] or [ENTER]. To evaluate 4^7 press the keys [4] [y^x] [7] [=] or [ENTER]. The display will read [16384]. Thus, $4^7 = 16{,}384$.

Use a calculator to evaluate.

1. 3^6

2. 5^6

3. 4^5

4. 7^6

5. 2^{11}

6. 6^8

Order of Operations

To see whether your calculator has the order of operations built in, evaluate $5 + 2 \cdot 3$ by pressing the keys [5] [+] [2] [×] [3] [=] or [ENTER]. If the display reads [11], your calculator does have the order of operations built in. This means that most of the time you can key in a problem exactly as it is written and the calculator will perform operations in the proper order. When evaluating an expression containing parentheses, key in the parentheses. (If an expression contains brackets, key in parentheses.) For example, to evaluate $2[25 - (8 + 4)] - 11$, press the keys [2] [×] [(] [25] [−] [(] [8] [+] [4] [)] [)] [−] [11] [=] or [ENTER].

The display will read [15].

Use a calculator to evaluate.

7. $7^4 + 5^3$

8. $12^4 - 8^4$

9. $63 \cdot 75 - 43 \cdot 10$

10. $8 \cdot 22 + 7 \cdot 16$

11. $4(15 \div 3 + 2) - 10 \cdot 2$

12. $155 - 2(17 + 3) + 185$

Name ______________ Section ________ Date ________

EXERCISE SET 1.8

A *Write using exponential notation. See Examples 1 through 4.*

1. $3 \cdot 3 \cdot 3 \cdot 3$

2. $5 \cdot 5 \cdot 5$

3. $7 \cdot 7 \cdot 7 \cdot 7 \cdot 7 \cdot 7 \cdot 7 \cdot 7$

4. $6 \cdot 6 \cdot 6 \cdot 6 \cdot 6$

5. $12 \cdot 12 \cdot 12$

6. $10 \cdot 10$

7. $6 \cdot 6 \cdot 5 \cdot 5 \cdot 5$

8. $4 \cdot 4 \cdot 4 \cdot 3 \cdot 3$

9. $9 \cdot 9 \cdot 9 \cdot 8$

10. $7 \cdot 7 \cdot 7 \cdot 4$

11. $3 \cdot 2 \cdot 2 \cdot 2 \cdot 2 \cdot 2$

12. $4 \cdot 6 \cdot 6 \cdot 6 \cdot 6$

13. $3 \cdot 2 \cdot 2 \cdot 5 \cdot 5 \cdot 5$

14. $6 \cdot 6 \cdot 2 \cdot 9 \cdot 9 \cdot 9 \cdot 9$

B *Evaluate. See Examples 5 through 8.*

15. 5^2

16. 6^2

17. 5^3

18. 6^3

19. 2^6

20. 2^7

21. 2^{10}

22. 1^{12}

23. 7^1

24. 8^1

25. 3^5

26. 5^4

27. 2^8

28. 3^3

29. 4^3

30. 4^4

31. 9^2

32. 8^2

33. 9^3

34. 8^3

35. 10^2

36. 10^3

37. 10^4

38. 10^5

39. 10^1

40. 14^1

41. 1920^1

42. 6849^1

43. 3^6

44. 4^5

C *Simplify. See Examples 9 through 13.*

45. $15 + 3 \cdot 2$

46. $24 + 6 \cdot 3$

47. $20 - 4 \cdot 3$

48. $17 - 2 \cdot 6$

49. $5 \cdot 9 - 16$

50. $8 \cdot 4 - 10$

51. $28 \div 4 - 3$

52. $42 \div 7 - 6$

53. $14 + \frac{24}{8}$

54. $32 + \frac{8}{2}$

55. $6 \cdot 5 + 8 \cdot 2$

56. $3 \cdot 4 + 9 \cdot 1$

57. $0 \div 6 + 4 \cdot 7$

58. $0 \div 8 + 7 \cdot 6$

59. $6 + 8 \div 2$

60. $6 + 9 \div 3$

61. $(6 + 8) \div 2$

62. $(6 + 9) \div 3$

63. $(6^2 - 4) \div 8$

64. $(7^2 - 7) \div 7$

65. $(3 + 5^2) \div 2$

66. $(13 + 6^2) \div 7$

67. $6^2 \cdot (10 - 8)$

68. $5^3 \div (10 + 15)$

69. $\frac{18 + 6}{2^4 - 4}$

70. $\frac{15 + 17}{5^2 - 3^2}$

71. $(2 + 5) \cdot (8 - 3)$

72. $(9 - 7) \cdot (12 + 18)$

73. $\frac{7(9 - 6) + 3}{3^2 - 3}$

74. $\frac{5(12 - 7) - 4}{5^2 - 2^3 - 10}$

75. $5 \div 0 + 24$

76. $18 - 7 \div 0$

77. $3^4 - [35 - (12 - 6)]$

78. $[40 - (8 - 2)] - 2^5$

79. $(7 \cdot 5) + [9 \div (3 \div 3)]$

80. $(18 \div 6) + [(3 + 5) \cdot 2]$

81. $8 \cdot [4 + (6 - 1) \cdot 2] - 50 \cdot 2$

82. $35 \div [3^2 + (9 - 7) - 2^2] + 10 \cdot 3$

83. $7^2 - \{18 - [40 \div (4 \cdot 2) + 2] + 5^2\}$

84. $29 - \{5 + 3[8 \cdot (10 - 8)] - 50\}$

D *Find the area of each square. See Example 14.*

△ 85.

△ 86.

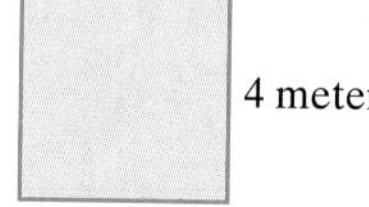

△ 87.

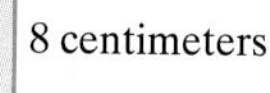

△ 88.

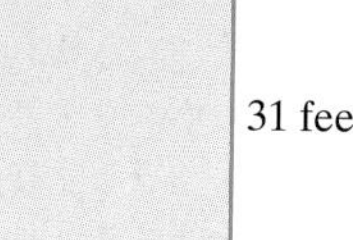

89. The Eiffel Tower stands on a square base measuring 100 meters on each side. Find the area of the base.

90. A square lawn that measures 72 feet on each side is to be fertilized. If 5 bags of fertilizer are available and each bag can fertilize 1000 square feet, is there enough fertilizer to cover the lawn?

Combining Concepts

Insert grouping symbols (parentheses) so that each given expression evaluates to the given number.

91. $2 + 3 \cdot 6 - 2$; evaluate to 28

92. $2 + 3 \cdot 6 - 2$; evaluate to 20

93. $24 \div 3 \cdot 2 + 2 \cdot 5$; evaluate to 14

94. $24 \div 3 \cdot 2 + 2 \cdot 5$; evaluate to 15

95. A building contractor is bidding on a contract to install gutters on seven homes in a retirement community, all in the shape shown. To estimate her cost of materials, she needs to know the total perimeter of all seven homes. Find the total perimeter.

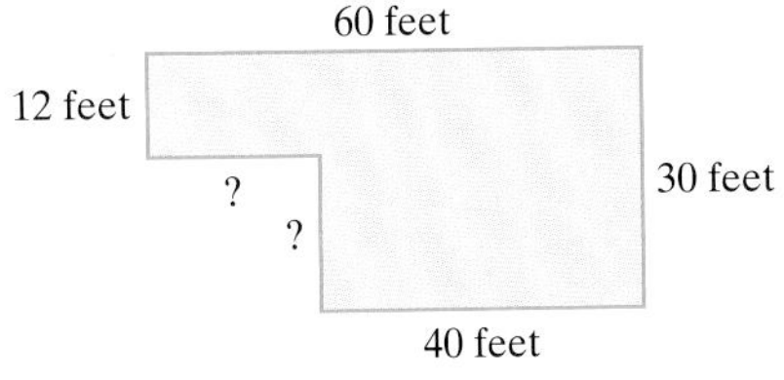

Simplify.

96. $25^3 \cdot (45 - 7 \cdot 5) \cdot 5$

97. $(7 + 2^4)^5 - (3^5 - 2^4)^2$

98. Explain why $2 \cdot 3^2$ is not the same as $(2 \cdot 3)^2$.

FOCUS ON Mathematical Connections

Investigating Whole Numbers and Order of Operations

This activity may be completed by working in groups or individually.

Try the following activity. You will need 30 index cards.

1. Label each index card with a number from 1 to 30, using each number once.
2. Shuffle the cards. Deal 5 cards face up to each player or team. Then deal a single card face down to each player or team. This single card is the goal card.
3. Play begins when the goal cards are turned over simultaneously. The object is for each player or team to use all 5 number cards (each only once) along with any of the operations $+$, $-$, $\times$, or $\div$ to obtain the number on the goal card. The winner is the first team to correctly write out the sequence of numbers and operations equaling the number on the goal card, according to the order of operations. Parentheses or other grouping symbols may be used as needed.
4. If no player/team is able to obtain the number on the goal card, the player/team that is able to come closest to their goal wins.

Example:

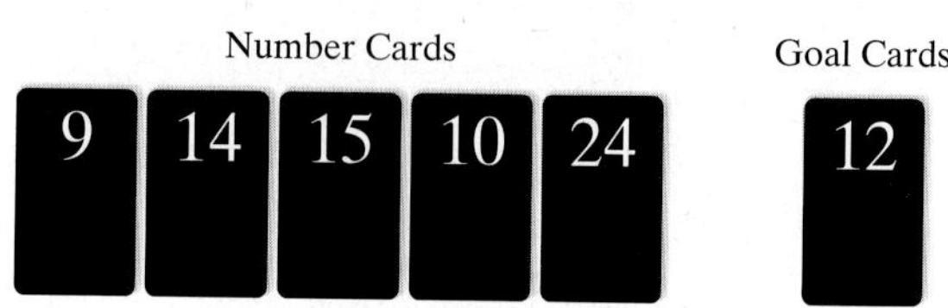

Sequence of numbers and operations to obtain goal:

$$24 \div [(15 - 14) + (10 - 9)] = 12$$

1.9 Introduction to Variables and Algebraic Expressions

OBJECTIVES

- (A) Evaluate algebraic expressions given replacement values.
- (B) Translate phrases into variable expressions.

SSM TUTOR CENTER SG CD & VIDEO MATH PRO WEB

(A) Evaluating Algebraic Expressions

Perhaps the most important quality of mathematics is that it is a science of patterns. Communicating about patterns is often made easier by using a letter to represent all the numbers fitting a pattern. We call such a letter a **variable**. For example, in Section 1.2 we presented the addition property of 0, which states that the sum of 0 and any number is that number. We might write

$$0 + 1 = 1$$
$$0 + 2 = 2$$
$$0 + 3 = 3$$
$$0 + 4 = 4$$
$$0 + 5 = 5$$
$$0 + 6 = 6$$
$$\vdots$$

continuing indefinitely. This is a pattern, and all whole numbers fit the pattern. We can communicate this pattern for all whole numbers by letting a letter, such as a, represent all whole numbers. We can then write

$$0 + a = a$$

Using variable notation is a primary goal of learning **algebra**. We now take some important first steps in beginning to use variable notation.

A combination of operations on letters (variables) and numbers is called an **algebraic expression** or simply an **expression**.

Algebraic Expressions

$3 + x \qquad 5 \cdot y \qquad 2 \cdot z - 1 + x$

If two variables or a number and a variable are next to each other, with no operation sign between them, the operation is multiplication. For example,

$2x$ means $2 \cdot x$

and

xy or $x(y)$ means $x \cdot y$

Also, the meaning of an exponent remains the same when the base is a variable. For example,

$$x^2 = \underbrace{x \cdot x}_{\text{2 factors of } x} \quad \text{and} \quad y^5 = \underbrace{y \cdot y \cdot y \cdot y \cdot y}_{\text{5 factors of } y}$$

Algebraic expressions such as $3x$ have different values depending on replacement values for x. For example, if x is 2, then $3x$ becomes

$$3x = 3 \cdot 2$$
$$= 6$$

If x is 7, then $3x$ becomes

$$3x = 3 \cdot 7$$
$$= 21$$

Replacing a variable in an expression by a number and then finding the value of the expression is called **evaluating the expression** for the variable. When finding the value of an expression, remember to follow the order of operations given in Section 1.8.

Practice Problem 1

Evaluate $x - 2$ if x is 5.

EXAMPLE 1 Evaluate $x + 7$ if x is 8.

Solution: Replace x with 8 in the expression $x + 7$.

$x + 7 = 8 + 7$ Replace x with 8.
$= 15$ Add.

When we write a statement such as "x is 5," we can use an equals symbol to represent "is" so that

x is 5 can be written as $x = 5$.

Practice Problem 2

Evaluate $y(x - 3)$ for $x = 3$ and $y = 7$.

EXAMPLE 2 Evaluate $2(x - y)$ for $x = 8$ and $y = 4$.

Solution: $2(x - y) = 2(8 - 4)$ Replace x with 8 and y with 4.
$= 2(4)$ Subtract.
$= 8$ Multiply.

Practice Problem 3

Evaluate $\frac{y + 6}{x}$ for $x = 2$ and $y = 8$.

EXAMPLE 3 Evaluate $\frac{x - 5y}{y}$ for $x = 21$ and $y = 3$.

Solution: $\frac{x - 5y}{y} = \frac{21 - 5(3)}{3}$ Replace x with 21 and y with 3.
$= \frac{21 - 15}{3}$ Multiply.
$= \frac{6}{3}$ Subtract.
$= 2$ Divide.

Practice Problem 4

Evaluate $25 - z^3 + x$ for $z = 2$ and $x = 1$.

EXAMPLE 4 Evaluate $x^2 + z - 3$ for $x = 5$ and $z = 4$.

Solution: $x^2 + z - 3 = 5^2 + 4 - 3$ Replace x with 5 and z with 4.
$= 25 + 4 - 3$ Evaluate 5^2.
$= 26$ Add and subtract from left to right.

Answers

1. 3 **2.** 0 **3.** 7 **4.** 18

Try the Concept Check in the margin.

EXAMPLE 5 The expression $\frac{5(F - 32)}{9}$ can be used to write degrees Fahrenheit F as degrees Celsius C. Find the value of this expression for $F = 86$.

Solution

$$\frac{5(F - 32)}{9} = \frac{5(86 - 32)}{9} = \frac{5(54)}{9} = \frac{270}{9} = 30$$

Thus 86°F is the same temperature as 30°C.

B Translating Phrases into Variable Expressions

To aid us in solving problems later, we practice translating verbal phrases into algebraic expressions. Certain key words and phrases suggesting addition, subtraction, multiplication, or division are reviewed next.

Addition	Subtraction	Multiplication	Division
sum	difference	product	quotient
plus	minus	times	divided by
added to	subtracted from	multiply	into
more than	less than	twice	per
increased by	decreased by	of	
total	less	double	

EXAMPLE 6 Write as an algebraic expression. Use x to represent "a number."

a. 7 increased by a number

b. 15 decreased by a number

c. The product of 2 and a number

d. The quotient of a number and 5

e. 2 subtracted from a number

Concept Check

What's wrong with the solution to the following problem?

Evaluate $3x + 2y$ for $x = 2$ and $y = 3$.

Solution:
$$3x + 2y = 3(3) + 2(2) = 9 + 4 = 13$$

Practice Problem 5

Evaluate $\frac{5(F - 32)}{9}$ for $F = 41$.

Practice Problem 6

Write as an algebraic expression. Use x to represent "a number."

a. Twice a number.

b. 8 increased by a number.

c. 10 minus a number.

d. 10 subtracted from a number.

e. The quotient of 6 and a number.

Answers

5. 5 **6. a.** $2x$ **b.** $x + 8$ **c.** $10 - x$ **d.** $x - 10$ **e.** $6 \div x$ or $\frac{6}{x}$

Concept Check:
$$3x + 2y = 3(2) + 2(3) = 6 + 6 = 12$$

Solution:

a. In words: 7 increased by a number
Translate: $7 + x$

b. In words: 15 decreased by a number
Translate: $15 - x$

c. In words: The product of 2 and a number
Translate: $2 \cdot x$ or $2x$

d. In words: The quotient of a number and 5
Translate: $x \div 5$ or $\frac{x}{5}$

e. In words: 2 subtracted from a number
Translate: $x - 2$

Helpful Hint

Remember that order is important when subtracting. Study the order of numbers and variables below.

Phrase	Translation
a number *decreased by* 5	$x - 5$
a number *subtracted from* 5	$5 - x$

Name ______________________ Section ____________ Date ____________

EXERCISE SET 1.9

 *Evaluate each following expression for $x = 2$, $y = 5$, and $z = 3$. See Examples 1 through 5.*

 **1.** $3 + 2z$

2. $7 + 3z$

3. $6xz - 5x$

4. $4yz + 2x$

5. $z - x + y$

6. $x + 5y - z$

7. $3x - z$

8. $2y + 5z$

 **9.** $y^3 - 4x$

10. $y^3 - z$

11. $2xy^2 - 6$

12. $3yz^2 + 1$

13. $8 - (y - x)$

14. $5 + (2x - 1)$

15. $y^4 + (z - x)$

16. $x^4 - (y - z)$

17. $\frac{6xy}{z}$

18. $\frac{8yz}{15}$

19. $\frac{2y - 2}{x}$

20. $\frac{6 + 3x}{z}$

21. $\frac{x + 2y}{z}$

22. $\frac{2z + 6}{3}$

23. $\frac{5x}{y} - \frac{10}{y}$

24. $\frac{70}{2y} - \frac{15}{z}$

25. $2y^2 - 4y + 3$

26. $3z^2 - z + 10$

27. $(3y - 2x)^2$

28. $(4y + 3z)^2$

29. $(xy + 1)^2$

30. $(xz - 5)^4$

31. $2y(4z - x)$

32. $3x(y + z)$

33. $xy(5 + z - x)$

34. $xz(2y + x - z)$

35. $\frac{7x + 2y}{3x}$

36. $\frac{6z + 2y}{4}$

37. The expression $16t^2$ gives the distance in feet that an object falls after t seconds. Complete the table by evaluating $16t^2$ for each given value of t.
see table

t	1	2	3	4
$16t^2$				

38. The expression $\frac{5(F-32)}{9}$ gives the equivalent degrees Celsius for F degrees Fahrenheit. Complete the table by evaluating this expression for each given value of F.

see table

F	50	59	68	77
$\frac{5(F-32)}{9}$				

B *Write each phrase as a variable expression. Use x to represent "a number." See Example 6.*

39. The sum of a number and five

40. Ten plus a number

41. The total of a number and eight

42. The difference of a number and five hundred

43. Twenty decreased by a number

44. A number less thirty

45. The product of 512 and a number

46. A number times twenty

47. A number divided by 2

48. The quotient of six and a number

49. The sum of seventeen and a number added to the product of five and the number

50. The difference of twice a number, and four

51. The product of five and a number

52. The quotient of twenty and a number, decreased by three

53. A number subtracted from 11

54. Twelve subtracted from a number

55. A number less 5

56. The product of a number and 7

57. 6 divided by a number

58. The sum of a number and 7

59. Fifty decreased by eight times a number

60. Twenty decreased by twice a number

Combining Concepts

Use a calculator to evaluate each expression for $x = 23$ and $y = 72$.

61. $x^4 - y^2$

62. $2(x + y)^2$

63. $x^2 + 5y - 112$

64. $16y - 20x + x^3$

65. If x is a whole number, which expression is the largest: $2x$, $5x$, or $\frac{x}{3}$?
Explain your answer.

66. If x is a whole number, which expression is the smallest: $2x$, $5x$, or $\frac{x}{3}$?
Explain your answer.

67. In Exercise 37, what do you notice about the value of $16t^2$ as t gets larger?

68. In Exercise 38, what do you notice about the value of $\frac{5(F - 32)}{9}$ as F gets larger?

CHAPTER 1 ACTIVITY Investigating Endangered and Threatened Species

This activity may be completed by working in groups or individually.

An **endangered** species is one that is thought to be in danger of becoming extinct throughout all or a major part of its habitat. A **threatened** species is one that may become endangered. The Division of Endangered Species at the U.S. Fish and Wildlife Service keeps close tabs on the state of threatened and endangered wildlife in the United States and around the world. The table below was compiled from data in the Division of Endangered Species' box score published on July 1, 2003. The "Total Species" column gives the total number of endangered and threatened species for each group.

1. Round each number of *endangered animal species* to the nearest ten to estimate the Animal Total.
2. Round each number of *endangered plant species* to the nearest ten to estimate the Plant Total.
3. Add the exact numbers of endangered animal species to find the exact Animal Total and record it in the table in the Endangered Species column. Add the exact numbers of endangered plant species to find the Plant Total and record it in the table in the Endangered Species column. Then find the total number of endangered species (animals and plants combined) and record this number in the table as the Grand Total in the Endangered Species column.
4. Find the Animal Total, Plant Total, and Grand Total for the Total Species column. Record these values in the table.
5. Use the data in the table to complete the Threatened Species column.
6. Write a paragraph discussing the conclusions that can be drawn from the table.
7. (Optional) The Division of Endangered Species updates its endangered/threatened species box score monthly. Visit the current box score of endangered and threatened species at http://ecos.fws.gov/tess/html/boxscore.html on the World Wide Web. How have the figures changed since July, 2003? Write a paragraph summarizing the changes in the numbers of endangered and threatened animals and plants.

Endangered and Threatened Species Worldwide

	Group	Endangered Species	Threatened Species	Total Species
Animals	Mammals	316		342
	Birds	253		273
	Reptiles	78		115
	Amphibians	20		30
	Fishes	82		126
	Clams	64		72
	Snails	22		33
	Insects	39		48
	Arachnids	12		12
	Crustaceans	18		21
	Animal Total			
Plants	Flowering Plants	578		716
	Conifers	2		5
	Ferns and Allies	24		26
	Lichens	2		2
	Plant Total			
	Grand Total			

(*Source:* U.S. Fish and Wildlife Service, Division of Endangered Species)

Chapter 1 Vocabulary Check

Fill in each blank with one of the words or phrases listed below.

place value	whole numbers	perimeter	variable
area	natural numbers	place value	exponent

1. The ______________ are 0, 1, 2, 3, ...
2. The __________ of a polygon is its distance around or the sum of the lengths of its sides.
3. The position of each digit in a number determines its ___________ .
4. An _________ is a shorthand notation for repeated multiplication of the same factor.
5. To find the ______ of a rectangle, multiply length times width.
6. The _______________ are 1, 2, 3, 4, ...
7. The position of a digit in a number determines its ___________ .
8. A letter used to represent a number is called a ________ .

CHAPTER 1 Highlights

DEFINITIONS AND CONCEPTS	**EXAMPLES**
SECTION 1.2 PLACE VALUE AND NAMES FOR NUMBERS	
The **whole numbers** are 0, 1, 2, 3, 4, 5,... The position of each digit in a number determines its **place value**. A place-value chart is shown next with the names of the periods given. 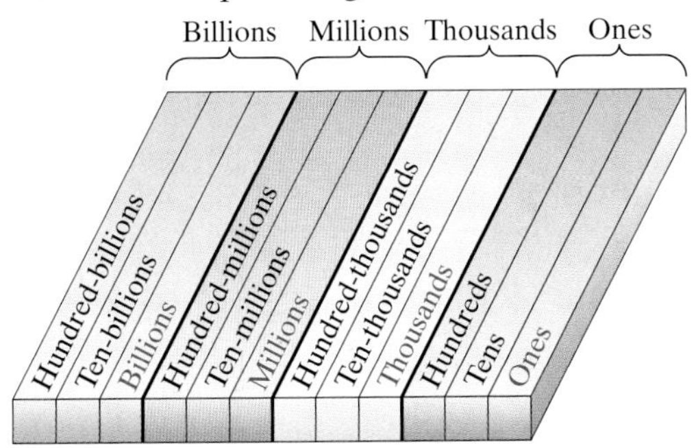	0, 14, 968, 5,268,619
To write a whole number in words, write the number in the period followed by the name of the period. The name of the ones period is not included.	9,078,651,002 is written as nine billion, seventy-eight million, six hundred fifty-one thousand, two.
SECTION 1.3 ADDING WHOLE NUMBERS	
To add whole numbers, add the digits in the ones place, then the tens place, then the hundreds place, and so on, carrying when necessary.	Find the sum: 2 1 1 2689 ← addend 1735 ← addend + 662 ← addend 5086 ← sum
The **perimeter** of a polygon is its distance around or the sum of the lengths of its sides.	Find the perimeter of the polygon shown. 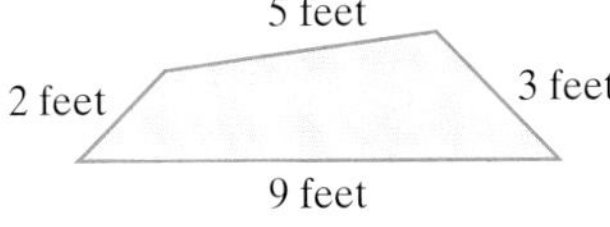 The perimeter is 5 feet + 3 feet + 9 feet + 2 feet = 19 feet.

Definitions and Concepts	**Examples**
Section 1.4 Subtracting Whole Numbers	
To subtract whole numbers, subtract the digits in the ones place, then the tens place, then the hundreds place, and so on, borrowing when necessary.	Subtract: 8 15 7984 ← minuend − 5673 ← subtrahend 2281 ← difference
Section 1.5 Rounding and Estimating	
Rounding Whole Numbers to a Given Place Value **Step 1.** Locate the digit to the right of the given place value. **Step 2.** If this digit is 5 or greater, add 1 to the digit in the given place value and replace each digit to its right with 0. **Step 3.** If this digit is less than 5, replace it and each digit to its right with 0.	Round 15,721 to the nearest thousand. 15,⑦21 Add 1 → ; Replace with zeros. Since the circled digit is 5 or greater, add 1 to the given place value and replace digits to its right with zeros. 15,721 rounded to the nearest thousand is 16,000.
Section 1.6 Multiplying Whole Numbers	
To multiply 73 and 58, for example, multiply 73 and 8, then 73 and 50. The sum of these partial products is the product of 73 and 58. Use the notation to the right.	73 ← factor × 58 ← factor 584 ← 73 × 8 3650 ← 73 × 50 4234 ← product
To find the **area** of a rectangle, multiply length times width.	Find the area of the rectangle shown. 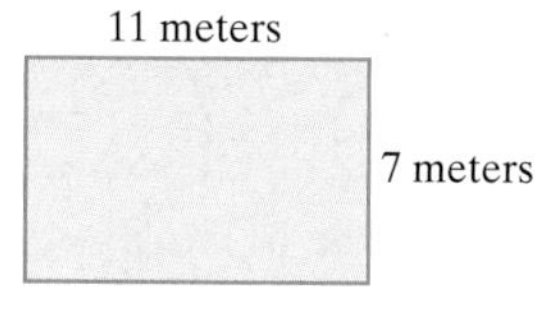 area of rectangle = length · width = (11 meters)(7 meters) = 77 square meters

DEFINITIONS AND CONCEPTS | EXAMPLES

SECTION 1.7 DIVIDING WHOLE NUMBERS

To divide larger whole numbers, use the process called **long division** as shown to the right.

$$\begin{array}{rrl} \text{divisor} \rightarrow & 507\text{ R }2 & \leftarrow \text{quotient} \\ & 14\overline{)7100} & \leftarrow \text{dividend} \\ & -70\downarrow\;\; & 5(14) = 70 \\ & 10\;\; & \text{Subtract and bring down the 0.} \\ & -\;0\downarrow & 0(14) = 0 \\ & 100 & \text{Subtract and bring down the 0.} \\ & 98 & 7(14) = 98 \\ & 2 & \text{Subtract. The remainder is 2.} \end{array}$$

To check, see that $507 \cdot 14 + 2 = 7100$.

The **average** of a list of numbers is

$$\text{average} = \frac{\text{sum of numbers}}{\text{number of numbers}}$$

Find the average of 23, 35, and 38.

$$\text{average} = \frac{23 + 35 + 38}{3} = \frac{96}{3} = 32$$

SECTION 1.8 EXPONENTS AND ORDER OF OPERATIONS

An **exponent** is a shorthand notation for repeated multiplication of the same factor.

$$3^4 = \underbrace{3 \cdot 3 \cdot 3 \cdot 3}_{\text{4 factors of 3}} = 81$$

(3 is the base; 4 is the exponent)

ORDER OF OPERATIONS

1. Do all operations within grouping symbols such as parentheses or brackets.
2. Evaluate any expressions with exponents.
3. Multiply or divide in order from left to right.
4. Add or subtract in order from left to right.

Simplify: $\dfrac{5 + 3^2}{2(7 - 6)}$

Simplify above and below the fraction bar separately.

$$\begin{aligned} \frac{5 + 3^2}{2(7 - 6)} &= \frac{5 + 9}{2(1)} && \text{Evaluate } 3^2. \text{ Subtract: } 7 - 6. \\ &= \frac{14}{2} && \text{Add. Multiply.} \\ &= 7 && \text{Divide.} \end{aligned}$$

The **area of a square** is $(\text{side})^2$.

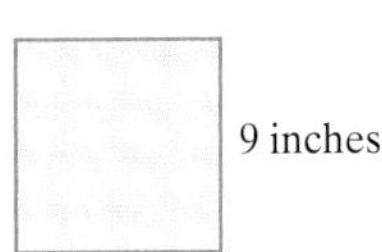

Find the area of the square shown.

$$\begin{aligned} \text{area of the square} &= (\text{side})^2 \\ &= (9 \text{ inches})^2 \\ &= 81 \text{ square inches} \end{aligned}$$

DEFINITIONS AND CONCEPTS	**EXAMPLES**
SECTION 1.9 INTRODUCTION TO VARIABLES AND ALGEBRAIC EXPRESSIONS	
A letter used to represent a number is called a **variable**.	Variables: $x, \quad y, \quad z, \quad a, \quad b$
A combination of operations on variables and numbers is called an **algebraic expression**.	Algebraic Expressions: $3 + x, \quad 7y, \quad x^3 + y - 10$
Replacing a variable in an expression by a number, and then finding the value of the expression is called **evaluating the expression** for the variable.	Evaluate $2x + y$ for $x = 22$ and $y = 4$. $2x + y = 2 \cdot 22 + 4$ Replace x with 22 and y with 4. $= 44 + 4$ Multiply. $= 48$ Add.

Name ________________ Section ________ Date ________

Chapter 1 Review

(1.2) *Determine the place value of the digit 4 in each whole number.*

1. 5480

2. 46,200,120

Write each whole number in words.

3. 5480

4. 46,200,120

Write each whole number in expanded form.

5. 6279

6. 403,225,000

Write each whole number in standard form.

7. Fifty-nine thousand, eight hundred

8. Six billion, three hundred four million

The following table shows the populations of the ten largest cities in the United States. Use this table to answer Exercises 9 through 12.

Rank	City	2000	1990	1980
1	New York, NY	8,008,278	7,322,564	7,071,639
2	Los Angeles, CA	3,694,820	3,485,398	2,968,528
3	Chicago, IL	2,896,016	2,783,726	3,005,072
4	Houston, TX	1,953,631	1,630,553	1,595,138
5	Philadelphia, PA	1,517,550	1,585,577	1,688,210
6	Phoenix, AZ	1,321,045	983,403	789,704
7	San Diego, CA	1,223,400	1,110,549	875,538
8	Dallas, TX	1,188,580	1,006,877	904,599
9	San Antonio, TX	1,144,646	935,933	785,940
10	Detroit, MI	951,270	1,027,974	1,203,368

(*Source*: U.S. Census Bureau)

9. Find the population of Houston, Texas, in 1990.

10. Find the population of Los Angeles, California, in 1980.

11. Find the increase in population for Phoenix, Arizona, from 1980 to 2000.

12. Find the decrease in population for Detroit, Michigan, from 1990 to 2000.

(1.3) *Add.*

13. $7 + 6$

14. $8 + 9$

15. $3 + 0$

16. $0 + 10$

17. $25 + 8 + 5$

18. $27 + 41$

19. $32 + 24$

20. $19 + 21$

21. $47 + 63$

22. $77 + 43$

23. 567 + 383

24. 463 + 787

25. 591 + 623 + 497

26. 5982 + 1647 + 2238

27. The distance from Chicago to New York City is 714 miles. The distance from New York City to New Delhi, India, is 7318 miles. Find the total distance from Chicago to New Delhi if traveling through New York City.

28. Susan Summerline earned salaries of $62,589, $65,340, and $69,770 during the years 1998, 1999, and 2000, respectively. Find her total earnings during those three years.

Find the perimeter of each figure.

△ **29.**

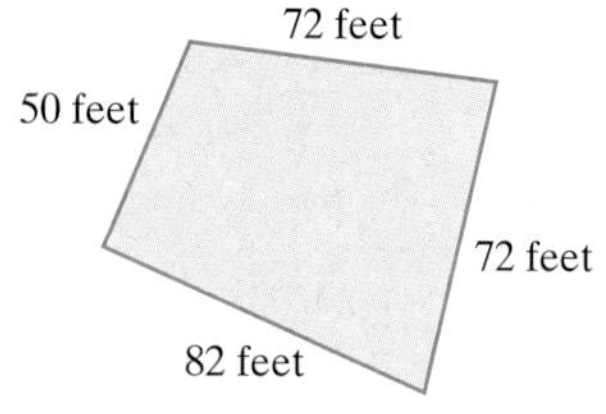

△ **30.** 11 kilometers 20 kilometers 35 kilometers

(1.4) *Subtract and then check.*

31. 42 − 9

32. 67 − 24

33. 93 − 79

34. 60 − 27

35. 599 − 237

36. 462 − 397

37. 583 − 279

38. 600 − 124

39. 4000 − 1886

40. 4268 − 3947

41. Bob Roma is proofreading the Yellow Pages for his county. If he has finished 315 pages of the total 712 pages, how many pages does he have left to proofread?

42. Shelly Winters bought a new car listed at $28,425. She received a discount of $1599 and a factory rebate of $1200. Find how much she paid for the car.

The following bar graph shows the monthly savings account balance for a freshman attending a local community college. Use this graph to answer Exercises 43 through 46.

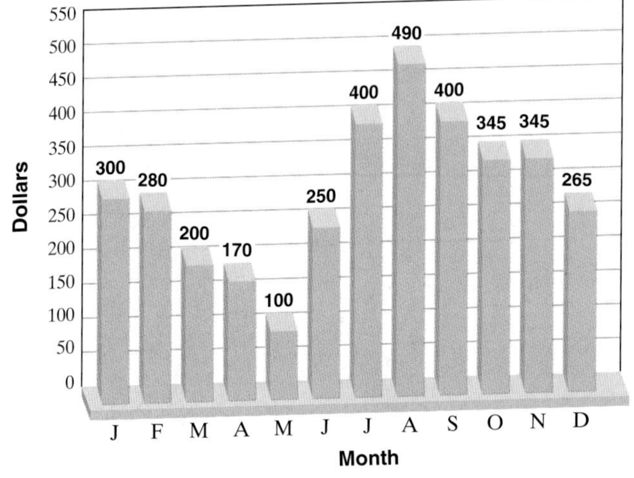

43. During what month was the balance the least?

44. During what month was the balance the greatest?

45. For what months was the balance greater than \$350?

46. For what months was the balance less than \$200?

(1.5) *Round to the given place.*

47. 93 to the nearest ten

48. 45 to the nearest ten

49. 467 to the nearest ten

50. 493 to the nearest hundred

51. 4832 to the nearest hundred

52. 57,534 to the nearest thousand

53. 49,683,712 to the nearest million

54. 768,542 to the nearest hundred-thousand

Estimate the sum or difference by rounding each number to the nearest hundred.

55. $4892 + 647 + 1876$

56. $5925 - 1787$

57. In 2001, there were 68,490,000 households in the United States subscribing to cable television services. Round this number to the nearest million.(*Source:* Nielsen Media Research-NTI)

58. In 2000, the total number of employees working for U.S. airlines was 679,967. Round this number to the nearest thousand.(*Source:* The Air Transport Association of America, Inc.)

(1.6) *Multiply.*

59. $6 \cdot 7$

60. $8 \cdot 3$

61. $5(0)$

62. $0(9)$

63. $\begin{array}{r} 47 \\ \times\ 30 \\ \hline \end{array}$

64. $\begin{array}{r} 69 \\ \times\ 42 \\ \hline \end{array}$

65. $20(8)(5)$

66. $25(9 \times 4)$

67. $\begin{array}{r} 48 \\ \times\ 77 \\ \hline \end{array}$

68. $\begin{array}{r} 77 \\ \times\ 22 \\ \hline \end{array}$

69. $49 \cdot 49 \cdot 0$

70. $62 \cdot 88 \cdot 0$

71. $\begin{array}{r} 586 \\ \times\ 29 \\ \hline \end{array}$

72. $\begin{array}{r} 242 \\ \times\ 37 \\ \hline \end{array}$

73. $\begin{array}{r} 642 \\ \times\ 177 \\ \hline \end{array}$

74. $\begin{array}{r} 347 \\ \times\ 129 \\ \hline \end{array}$

75. $\begin{array}{r} 1026 \\ \times\ 401 \\ \hline \end{array}$

76. $\begin{array}{r} 2107 \\ \times\ 302 \\ \hline \end{array}$

Estimate each product by rounding each factor to the given place.

77. $49 \cdot 32$; tens

78. $586 \cdot 357$; hundreds

79. $5231 \cdot 243$; hundreds

80. $7836 \cdot 912$; hundreds

81. One ounce of Swiss cheese contains 8 grams of fat. How many grams of fat are in 3 ounces of Swiss cheese? (*Source: Home and Garden Bulletin No. 72,* U.S. Department of Agriculture)

82. There were 5283 students enrolled at Weskan State University in the fall semester. Each paid $927 in tuition. Find the total tuition collected.

Find the area of each rectangle.

△ **83.**

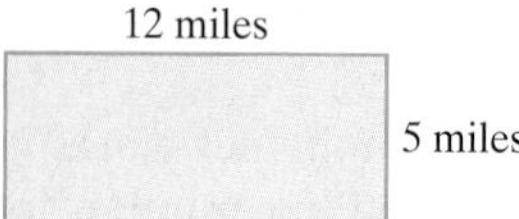

△ **84.**

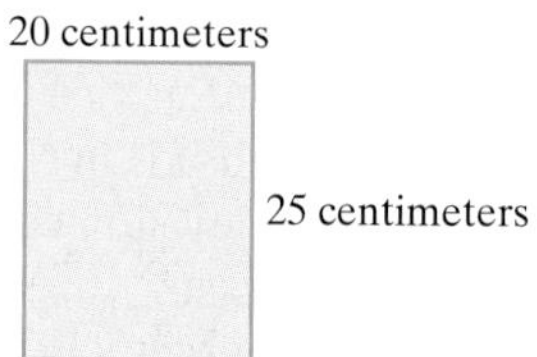

(1.7) *Divide and then check.*

85. $18 \div 6$

86. $36 \div 9$

87. $42 \div 7$

88. $25 \div 5$

89. $27 \div 5$

90. $18 \div 4$

91. $\frac{16}{0}$

92. $\frac{0}{8}$

93. $\frac{9}{9}$

94. $\frac{10}{1}$

95. $918 \div 0$

96. $0 \div 668$

97. $5\overline{)75}$

98. $8\overline{)159}$

99. $26\overline{)626}$

100. $6\overline{)336}$

101. $32\overline{)49}$

102. $19\overline{)680}$

103. $20\overline{)10{,}000}$

104. $43\overline{)909}$

105. $47\overline{)23{,}782}$

106. $30\overline{)480}$

107. $16\overline{)3192}$

108. $25\overline{)5000}$

109. One foot is 12 inches. Find how many feet there are in 5496 inches.

110. Find the average of the numbers 76, 49, 32, and 47.

(1.8) *Write using exponential notation.*

111. $7 \cdot 7 \cdot 7 \cdot 7$

112. $6 \cdot 6 \cdot 3 \cdot 3 \cdot 3$

113. $4 \cdot 2 \cdot 2 \cdot 2 \cdot 3 \cdot 3$

114. $5 \cdot 5 \cdot 7 \cdot 7 \cdot 7 \cdot 2 \cdot 2$

Simplify.

115. 7^2

116. 2^6

117. $5^3 \cdot 3^2$

118. $4^1 \cdot 10^2 \cdot 7^2$

119. $18 \div 3 + 7$

120. $12 - 8 \div 4$

121. $\dfrac{(6^2 - 3)}{3^2 + 2}$

122. $\dfrac{16 - 8}{2^3}$

123. $2 + 3[1 + (20 - 17) \cdot 3]$

124. $21 - [2^4 - (7 - 5) - 10] + 8 \cdot 2$

Find the area of each square.

125.

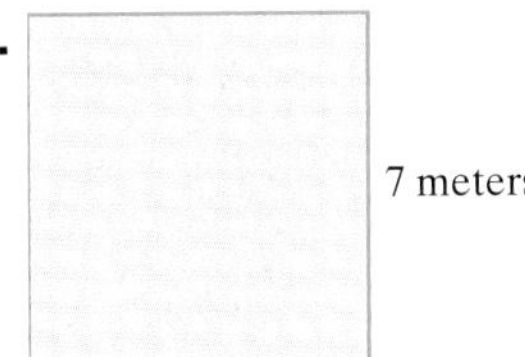

7 meters

126.

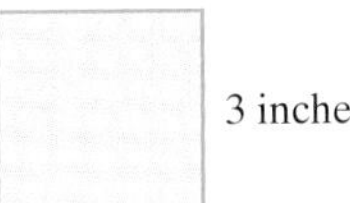

3 inches

(1.9) *Evaluate each expression for $x = 5$, $y = 0$, and $z = 2$.*

127. $\dfrac{2x}{z}$

128. $4x - 3$

129. $\dfrac{x + 7}{y}$

130. $\dfrac{y}{5x}$

131. $x^3 - 2z$

132. $\dfrac{7 + x}{3z}$

133. $(y + z)^2$

134. $\dfrac{100}{x} + \dfrac{y}{3}$

Translate each phrase into a variable expression.

135. Five subtracted from a number

136. Seven more than a number

137. Ten divided by a number

138. The product of 5 and a number

139. Complete the table by evaluating $8x^2$ for each value of x.

x	0	1	2	3
$8x^2$				

Name ________________ Section ________ Date ________

Chapter 1 Test

Remember to check your answers and use the Chapter Test Prep Video to view solutions.

Evaluate. Where it's helpful, use estimation to see if your results are reasonable.

1. $59 + 82$

2. $600 - 487$

3. $\begin{array}{r} 496 \\ \times\ 30 \\ \hline \end{array}$

4. $52{,}896 \div 69$

5. $2^3 \cdot 5^2$

6. $6^1 \cdot 2^3$

7. $98 \div 1$

8. $0 \div 49$

9. $62 \div 0$

10. $(2^4 - 5) \cdot 3$

11. $16 + 9 \div 3 \cdot 4 - 7$

12. $2[(6 - 4)^2 + (22 - 19)^2] + 10$

13. Round 52,369 to the nearest thousand.

Estimate each sum or difference by rounding each number to the nearest hundred.

14. $6289 + 5403 + 1957$

15. $4267 - 2738$

16. Twenty-nine cans of Sherwin-Williams paint cost $493. How much was each can?

17. Admission to a movie costs $7 per ticket. The Math Club has 17 members who are going to a movie together. What is the total cost of their tickets?

18. Jo McElory is looking at two new refrigerators for her apartment. One costs $599 and the other costs $725. How much more expensive is the higher-priced one?

19. Aspirin was 100 years old in 1997 and was the first U.S. drug made in tablet form. Today, people take 11 billion tablets a year for heart disease prevention and 4 billion tablets a year for headaches. How many more tablets are taken a year for heart disease prevention? (*Source:* Bayer Market Research)

Answers

1. ________
2. ________
3. ________
4. ________
5. ________
6. ________
7. ________
8. ________
9. ________
10. ________
11. ________
12. ________
13. ________
14. ________
15. ________
16. ________
17. ________
18. ________
19. ________

20. ______

21. ______

22. ______

23. a. ______

b. ______

24. ______

25. ______

26. ______

27. ______

28. ______

20. The U.S. Federal Bureau of Investigation (FBI) maintains a collection of over 250,000,000 sets of fingerprint records. Each work day, the FBI Identification Division receives over 34,000 standard fingerprint cards to add to its collection. During a standard five-day work week, how many fingerprint cards does the FBI receive? (*Source:* Federal Bureau of Investigation)

21. Evaluate $5(x^3 - 2)$ for $x = 2$.

22. Evaluate $\frac{3x - 5}{2y}$ for $x = 7$ and $y = 8$.

23. Translate the following phrases into mathematical expressions. Use x to represent "a number."

a. The product of a number and 17

b. Twice a number subtracted from 20

Find the perimeter and the area of each figure.

△ **24.** Square 5 centimeters

△ **25.** 20 yards Rectangle 10 yards

The following table shows the top grossing movies for 2001 and 2002. Use this table to answer Exercises 26 through 28.

2001	
Movie	**Gross**
Harry Potter and the Sorcerer's Stone	$317,557,891
The Lord of the Rings: The Fellowship of the Ring	$313,364,114
Shrek	$267,652,016
Monsters, Inc.	$255,870,172
Rush Hour 2	$226,138,454

(*Source:* The Internet Movie Database Ltd.)

2002	
Movie	**Gross**
Spiderman	$403,706,375
The Lord of the Rings: The Two Towers	$339,554,276
Star Wars: Episode II—Attack of the Clones	$310,675,583
Harry Potter and the Chamber of Secrets	$261,970,615
My Big Fat Greek Wedding	$182,805,123

26. Find the amount earned by the movie *Spiderman* in 2002.

27. How much more did *Lord of the Rings: The Two Towers* in 2002 earn than *Lord of the Rings: The Fellowship of the Ring* in 2001?

28. How much more did the top-grossing movie of 2002 earn than the top-grossing movie in 2001?

Integers

Whole numbers are not sufficient for representing many situations in real life. For example, to express 5° below 0° or $100 in debt, numbers less than 0 are needed. This chapter is devoted to integers, which include numbers less than 0, and to operations on integers.

CHAPTER 2

The weather on our planet is of great interest to its occupants. From snowstorms to tornadoes to hurricanes, torrential rain to severe drought, some sort of extreme weather affects nearly every portion of the Earth. But did you ever wonder about the weather on other planets? Scientists do! Information from satellites, probes, and telescopes tell of drastic atmospheric conditions of our Solar System neighbors. Pluto's distance from the Sun contributes to its frigid temperatures, some recorded as low as 338°F *below zero*. Compare that to the extreme heat of Venus, where temperatures on the surface are in excess of 800°F. In Exercises 63–66 on page 134, we will see how integers can be used to compare average daily temperatures of several planets in our Solar System.

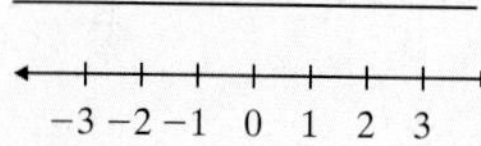

Name ______________________ Section __________ Date __________________

Chapter 2 Pretest

Answers

1. ______

(number line: −3 −2 −1 0 1 2 3)

2. ______
3. ______
4. ______
5. ______
6. ______
7. ______
8. ______
9. ______
10. ______
11. ______
12. ______
13. ______
14. ______
15. ______
16. ______
17. ______
18. ______
19. ______
20. ______
21. ______
22. ______

1. The Washington Redskins lost 22 yards on a play. Represent this quantity using an integer.

2. Graph −3, 2, and 0 on the number line.

3. Insert < or > between the given pair of numbers to make a true statement.

−31 −36

4. Simplify: $|-8|$

5. Find the opposite of −12.

6. Add: $-19 + 8$

7. Evaluate $x + y$ if $x = -6$ and $y = -11$.

8. On February 2 the temperature in Manassas, Virginia, at 7 a.m. was −9° Fahrenheit. By 8 a.m. the temperature had risen by 5°, and then between 8 a.m. and 9 a.m. it rose another 7°. What was the temperature at 9 a.m.?

9. Subtract: $-9 - (-14)$

10. Simplify: $4 - 6 - (-2) + 8$

11. Evaluate $m - n$ if $m = -5$ and $n = 10$.

12. Andrea Roberts has $88 in her checking account. She makes a deposit of $35 and writes a check for $72, and then writes another check for $55. Find the balance in her account. (Write the amount as an integer.)

13. Multiply: $(-8)(13)$

14. Divide: $\frac{-36}{-4}$

15. Evaluate $\frac{x}{y}$ if $x = -96$ and $y = -2$.

16. Evaluate xy if $x = 4$ and $y = -5$.

17. A card player had a score of −20 for each of three games. Find her total score.

Simplify.

18. -5^2

19. $\frac{9 - 17}{-6 + 2}$

20. $(-2) \cdot |-7| - (-6)$

21. Evaluate $x - y^2$ if $x = 2$ and $y = -6$.

22. Find the average of −17, 10, −9, 5, −19.

2.1 Introduction to Integers

OBJECTIVES

- A Represent real-life situations with integers.
- B Graph integers on a number line.
- C Compare two integers.
- D Find absolute value.
- E Find the opposite of a number.
- F Read bar graphs containing negative integers.

A Representing Real-Life Situations

Thus far in this text, all numbers have been 0 or greater than 0. Numbers greater than 0 are called **positive numbers**. However, sometimes situations exist that cannot be represented by a number greater than 0. For example,

To represent these situations, we need numbers less than 0.

Extending the number line to the left of 0 allows us to picture **negative numbers**, numbers that are less than 0.

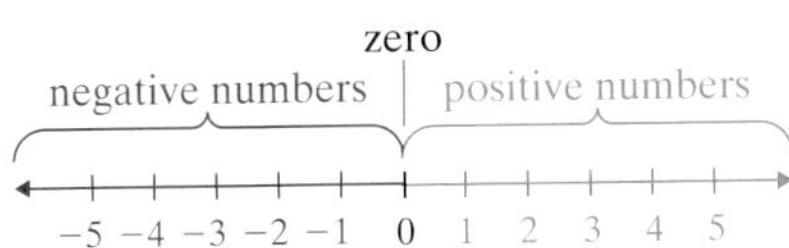

When a single + sign or no sign is in front of a number, the number is a positive number. When a single − sign is in front of a number, the number is a negative number. Together, we call positive numbers, negative numbers, and 0 the **signed numbers**.

−5 indicates "negative five."

5 and +5 both indicate "positive five."

The number 0 is neither positive nor negative.

Helpful Hint

Notice that 0 is neither positive nor negative.

Some signed numbers are integers. The **integers** consist of the numbers labeled on the number line above. The integers are

$$\ldots, -3, -2, -1, 0, 1, 2, 3, \ldots$$

Now we have numbers to represent the situations previously mentioned.

5 degrees below 0° −5°

20 feet below sea level −20 feet

EXAMPLE 1 Representing Depth with an Integer

Jack Mayfield, a miner for the Molly Kathleen Gold Mine, is presently 150 feet below the surface of the earth. Represent this position using an integer.

Solution: If 0 represents the surface of the earth, then 150 feet below the surface can be represented by −150.

Practice Problem 1

a. A deep-sea diver is 800 feet below the surface of the ocean. Represent this position using an integer.

b. A company reports a $2 million loss for the year. Represent this amount using an integer.

Answers

1. a. −800 **b.** −2 million

B Graphing Integers

EXAMPLE 2 Graph 0, −3, 2, and −2 on the number line.

Solution:

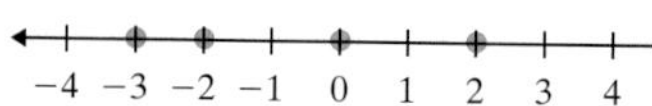

Practice Problem 2

Graph −5, −4, 3, and −3 on the number line.

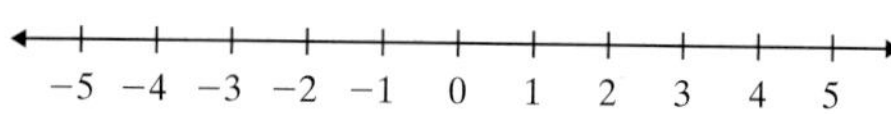

C Comparing Integers

We compare integers just as we compare whole numbers. For any two numbers graphed on a number line, the number to the **right** is the **greater number** and the number to the **left** is the **smaller number**. Recall that the inequality symbol $>$ means "is greater than" and the inequality symbol $<$ means "is less than."

Both −5 and −7 are graphed on the number line shown.

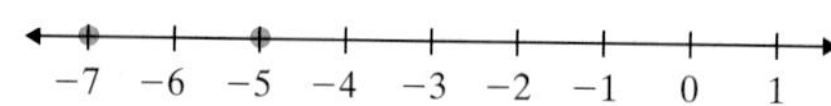

On the graph, −7 is **to the left of** −5, so −7 **is less than** −5, written as

$$-7 < -5$$

We can also write

$$-5 > -7$$

since −5 is **to the right** of −7, so −5 **is greater than** −7.

Concept Check

Is there a largest positive number? Is there a smallest negative number? Explain.

Try the Concept Check in the margin.

Practice Problem 3

Insert $<$ or $>$ between each pair of numbers to make a true statement.

a. 0 −3
b. −5 5
c. −8 −12

EXAMPLE 3 Insert $<$ or $>$ between each pair of numbers to make a true statement.

a. −7 7
b. 0 −4
c. −9 −11

Solution:

a. −7 is to the left of 7 on a number line, so $-7 < 7$.
b. 0 is to the right of −4 on a number line, so $0 > -4$.
c. −9 is to the right of −11 on a number line, so $-9 > -11$.

Helpful Hint

If you think of $<$ and $>$ as arrowheads, notice that in a true statement the arrow always points to the smaller number.

$5 > -4$ ↑ smaller number $-3 < -1$ ↑ smaller number

Answers

2.

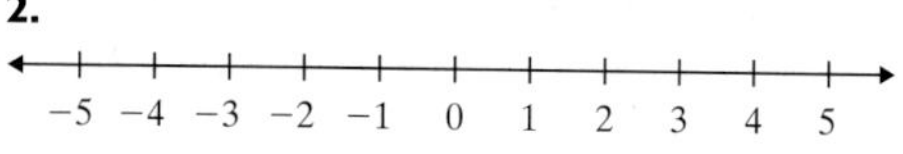

3. a. $>$ **b.** $<$ **c.** $>$

Concept Check: No.

D Finding the Absolute Value of a Number

The **absolute value** of a number is the number's distance from 0 on the number line. The symbol for absolute value is $|\ |$. For example, $|3|$ is read as "the absolute value of 3."

$|3| = 3$ because 3 is 3 units from 0.

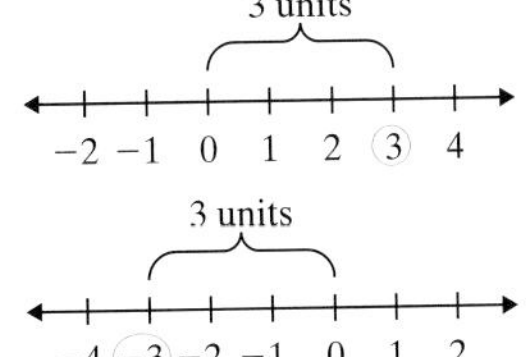

$|-3| = 3$ because -3 is 3 units from 0.

3 units

−4 −3 −2 −1 0 1 2

EXAMPLE 4 Simplify.

a. $|-2|$ **b.** $|5|$ **c.** $|0|$

Solution:

a. $|-2| = 2$ because -2 is 2 units from 0.

b. $|5| = 5$ because 5 is 5 units from 0.

c. $|0| = 0$ because 0 is 0 units from 0.

Practice Problem 4

Simplify.

a. $|-4|$ b. $|2|$

c. $|-8|$

Helpful Hint

Since the absolute value of a number is that number's *distance* from 0, the absolute value of a number is always 0 or positive. It is never negative.

$|0| = 0$ ↑ zero $\qquad |-6| = 6$ ↑ a positive number

E Finding Opposites

Two numbers that are the same distance from 0 on the number line but are on opposite sides of 0 are called **opposites**.

4 units 4 units

−5 −4 −3 −2 −1 0 1 2 3 4 5

When two numbers are opposites, we say that each is the opposite of the other. Thus **4 is the opposite of** -4 and -4 **is the opposite of 4.**

The phrase "the opposite of" is written in symbols as "$-$." For example,

The opposite of	5	is	−5
↓	↓	↓	↓
−	(5)	=	−5
The opposite of	−3	is	3
↓	↓	↓	↓
−	(−3)	=	3

Notice we just stated that

$-(-3) = 3$

If a is a number, then $-(-a) = a$.

Answers

4. a. 4 **b.** 2 **c.** 8

Practice Problem 5

Find the opposite of each number.

a. 7 b. -17

EXAMPLE 5 Find the opposite of each number.

a. 11 **b.** -2 **c.** 0

Solution:

a. The opposite of 11 is -11.

b. The opposite of -2 is $-(-2)$ or 2.

c. The opposite of 0 is 0.

Helpful Hint

Remember that 0 is neither positive nor negative.

Concept Check

True or false? The number 0 is the only number that is its own opposite.

Try the Concept Check in the margin.

Practice Problem 6

Simplify.

a. $-|-2|$ b. $-|5|$

c. $-(-11)$

EXAMPLE 6 Simplify.

a. $-(-4)$ **b.** $-|-5|$ **c.** $-|6|$

Solution:

a. $-(-4) = 4$ The opposite of negative 4 is 4.

b. $-|-5| = -5$ The opposite of the absolute value of -5 is the opposite of 5, or -5.

c. $-|6| = -6$ The opposite of the absolute value of 6 is the opposite of 6, or -6.

Practice Problem 7

Evaluate $-|x|$ if $x = -9$.

EXAMPLE 7 Evaluate $-|-x|$ if $x = -2$.

Solution: Carefully replace x with -2; then simplify.

$$-|-x| = -|-(-2)| \quad \text{Replace } x \text{ with } -2.$$

Then $-|-(-2)| = -|2| = -2$.

Answers

5. a. -7 **b.** 17

6. a. -2 **b.** -5 **c.** 11

7. -9

Concept Check: True.

F Reading Bar Graphs Containing Negative Numbers

The bar graph below shows the average temperature (in Fahrenheit) of known planets. Notice that a negative temperature is illustrated by a bar below the horizontal line representing 0°F, and a positive temperature is illustrated by a bar above the horizontal line representing 0°F.

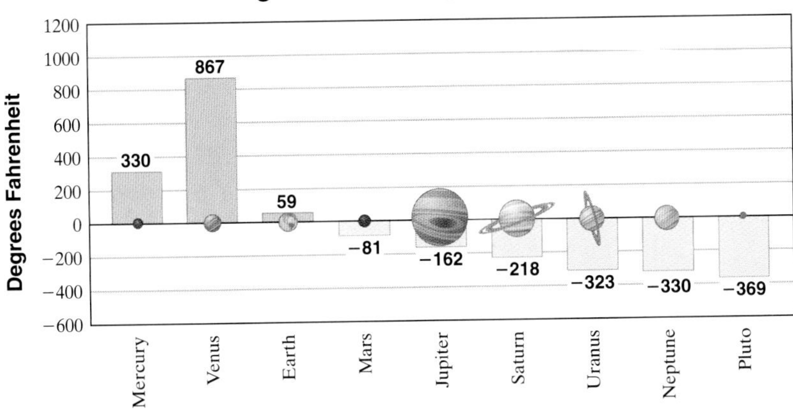

*(For some planets, the temperature given is the temperature where the atmosphere pressure equals 1 Earth atmosphere; Source: *The World Almanac*, 2003)

EXAMPLE 8. Which planet has the lowest temperature?

Solution The planet with the lowest temperature is the one that corresponds to the bar that extends the furthest in the negative direction (downward.) Pluto has the lowest temperature of -369°F. ●

Later, we will be reading graphs without labels on bars. Notice that the height of the bar can be checked or estimated by looking horizontally to the left and checking the vertical axis scale.

Practice Problem 8

Which planet has the highest temperature?

Answer

8. Venus; 867°F

STUDY SKILLS REMINDER

Are you reviewing new terms?

Remember that many of the terms used in this text may be new to you. It will be helpful to make a list of new mathematical terms and symbols as you encounter them and to review them frequently.

Focus on The Real World

Using Negative Numbers

Did you know that many countries around the world label the underground floors in a tall building with negative numbers? As we have seen in this section and will see throughout this chapter, negative numbers are very useful in many different situations where numbers less than 0 are needed. Another use of negative numbers is in pregnancy and childbirth. Just prior to labor, a baby will drop, or engage, in the mother's pelvis. Doctors and midwives use "stations," ranging from -5 to $+5$, to describe how much a baby has engaged compared to the 0 station at the entrance to the pelvic canal. A negative station means that a baby is still above the pelvic bones used for reference. A positive station describes how far the baby has moved into the pelvic canal.

Critical Thinking

Make a list of as many real-world situations as you can think of that use negative numbers in some way. Consider flipping through a newspaper or news magazine for inspiration.

Name ______________________ Section __________ Date __________

EXERCISE SET 2.1

A *Represent each quantity by an integer. See Example 1.*

1. A worker in a silver mine in Nevada works 1445 feet underground.

2. A scuba diver is swimming 35 feet below the surface of the water in the Gulf of Mexico.

3. The peak of Mount Elbert in Colorado is 14,433 feet above sea level. (*Source:* U.S. Geological Survey)

4. The lowest elevation in the United States is found at Death Valley, California, at an elevation of 282 feet below sea level. (*Source:* U.S. Geological Survey)

5. The record high temperature in Nevada is 118 degrees Fahrenheit above zero. (*Source:* National Climatic Data Center)

6. The Minnesota Viking football team lost 15 yards on a play.

7. The average depth of the Atlantic Ocean is 11,730 feet below the surface of the ocean. (*Source: 2003 World Almanac*)

8. The Dow Jones stock market average fell 317 points in one day.

9. In fiscal year 2001, US Airways posted a net loss of \$1683 million. (*Source:* US Airways)

10. For the fourth quarter in fiscal year 2002, Apple Computer, Inc., reported a net loss of \$45 million. (*Source:* Apple Computer, Inc.)

11. Two divers are exploring the bottom of a trench in the Pacific Ocean. Joe is at 135 feet below the surface of the ocean and Sara is at 157 feet below the surface. Represent each quantity by an integer and determine who is deeper in the water.

12. The temperature on one January day in Chicago was 10° below 0° Celsius. Represent this quantity by an integer and tell whether this temperature is cooler or warmer than 5° below 0° Celsius.

13. In 2002, the number of music CD singles shipped to retailers reflected an 81 percent loss from the previous year. Write an integer to represent the percent loss in CD singles shipped. (*Source:* Recording Industry Association of America)

14. In 2002, the number of music cassettes shipped to retailers reflected a 49 percent loss from the previous year. Write an integer to represent the percent loss of cassettes shipped. (*Source:* Recording Industry Association of America)

B *Graph each integer in the list on the same number line. See Example 2.*

15. 1, 2, 4, 6

−7 −6 −5 −4 −3 −2 −1 0 1 2 3 4 5 6 7

16. 3, 5, 2, 0

−7 −6 −5 −4 −3 −2 −1 0 1 2 3 4 5 6 7

17. 1, −1, −2, −4, −7

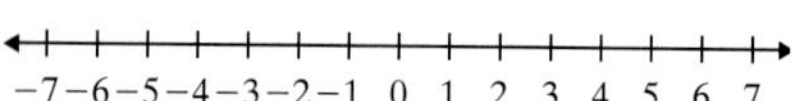

18. 2, −2, −4, 6

−7 −6 −5 −4 −3 −2 −1 0 1 2 3 4 5 6 7

19. 0, 2, 5, 7

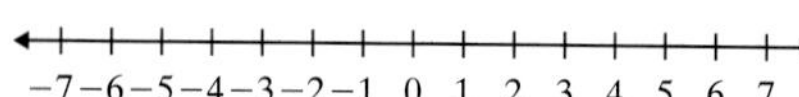

20. 0, 3, 6, 10

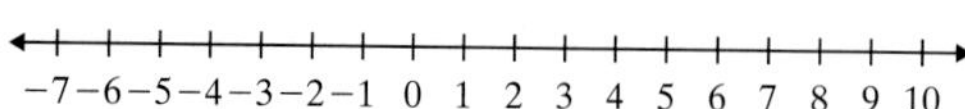

21. 0, −2, −7, −5

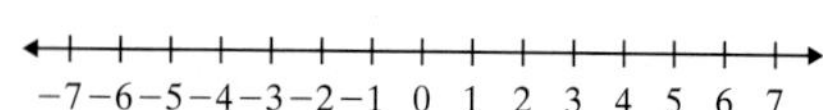

22. 0, −7, 3, −6

−7 −6 −5 −4 −3 −2 −1 0 1 2 3 4 5 6 7

C *Insert $<$ or $>$ between each pair of integers to make a true statement. See Example 3.*

23. −4 0 **24.** −8 0 **25.** −7 −5 **26.** −12 −10

27. −30 −35 **28.** −27 −29 **29.** −26 26 **30.** 13 −13

D **E** *Simplify. See Example 4.*

31. $|5|$ **32.** $|7|$ **33.** $|-8|$ **34.** $|-19|$

35. $|0|$ **36.** $|100|$ **37.** $|-5|$ **38.** $|-10|$

Find the opposite of each integer. See Example 5.

39. 5 **40.** 8 **41.** −4 **42.** −6

43. 23 **44.** 123 **45.** −10 **46.** −23

Simplify. See Example 6.

47. $|-7|$ **48.** $|-11|$ **49.** $-|20|$ **50.** $-|43|$

51. $-|-3|$ **52.** $-|-18|$ **53.** $-(-8)$ **54.** $-(-7)$

55. $|-14|$ **56.** $-(-14)$ **57.** $-(-29)$ **58.** $-|-29|$

Evaluate. See Example 7.

59. $|-x|$ if $x = -8$ **60.** $-|x|$ if $x = -8$ **61.** $-|-x|$ if $x = 3$ **62.** $-|-x|$ if $x = 7$

63. $|x|$ if $x = -23$ **64.** $|x|$ if $x = 23$ **65.** $-|x|$ if $x = 4$ **66.** $|-x|$ if $x = 1$

Insert $<$, $>$, *or* $=$ *between each pair of numbers to make a true statement.*

67. $-3 \quad -5$ **68.** $-17 \quad -6$ **69.** $|-9| \quad |-14|$

70. $|-8| \quad |-4|$ **71.** $|-33| \quad -(-33)$ **72.** $-|17| \quad -(-17)$

73. $-|-10| \quad -(-10)$ **74.** $|-24| \quad -(-24)$ **75.** $0 \quad -9$

76. $-45 \quad 0$ **77.** $|0| \quad |-9|$ **78.** $|-45| \quad |0|$

79. $-|-2| \quad -|-10|$ **80.** $-|-8| \quad -|-4|$

81. $-(-12) \quad -(-18)$ **82.** $-22 \quad -(-38)$

Write the given integers in order from least to greatest.

83. $2^2, -|3|, -(-5), -|-8|$ **84.** $|10|, 2^3, -|-5|, -(-4)$ **85.** $|-1|, -|-6|, -(-6), -|1|$

86. $1^4, -(-3), -|7|, |-20|$ **87.** $-(-2), 5^2, -10, -|-9|, |-12|$ **88.** $3^3, -|-11|, -(-10), -4, -|2|$

F *The bar graph shows elevation of selected lakes. Use this graph For Exercises 89–92.* (*Source:* U.S. Geological Survey). *See Example 8.*

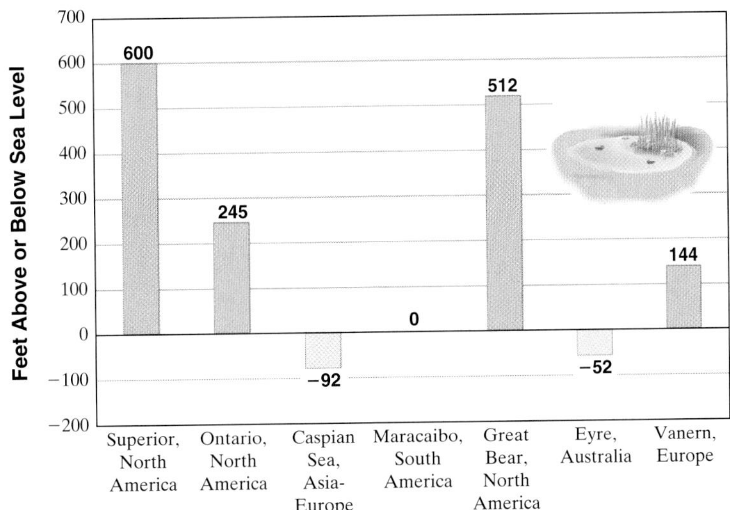

89. Which lake has an elevation at sea level?

90. Which lake shown has the lowest elevation?

91. Which lake shown has the second lowest elevation?

92. Which lake shown has the highest elevation?

Use the bar graph from Example 8 to answer exercises 93 through 96.

93. Which planet has an average temperature closest to $0°F$?

94. Which planet has a negative average temperature closest to $0°F$?

95. Which planet has an average temperature closest to $-200°F$?

96. Which planet has an average temperature closest to $-300°F$?

Review and Preview

Add. See Section 1.3.

97. $0 + 13$

98. $9 + 0$

99. $15 + 4^2 + 4$

100. $20 + 3^2 + 6$

101. $47 + 236 + 77$

102. $362 + 37 + 90$

Combining Concepts

Choose all numbers for x from each given list that make each statement true.

103. $|x| > 8$
a. 0 **b.** −5 **c.** 8 **d.** −12

104. $|x| > 4$
a. 0 **b.** 4 **c.** −1 **d.** −100

105. Evaluate: $-(-|-5|)$

106. Evaluate: $-(-|-(-7)|)$

Answer true or false for Exercises 107 through 111.

107. If $a > b$, then a must be a positive number.

108. The absolute value of a number is *always* a positive number.

109. A positive number is always greater than a negative number.

110. Zero is always less than a positive number.

111. The number $-a$ is always a negative number. (*Hint:* Read "−" as "the opposite of.")

112. Given the number line 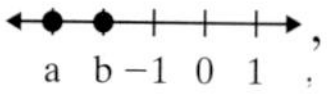

, is it true that $b < a$?

113. Write in your own words how to find the absolute value of an integer.

114. Explain how to determine which of two integers is larger.

2.2 Adding Integers

OBJECTIVES

- Ⓐ Add integers.
- Ⓑ Evaluate an algebraic expression by adding.
- Ⓒ Solve problems by adding integers.

Ⓐ Adding Integers

Adding integers can be visualized using a number line. A positive number can be represented on the number line by an arrow of appropriate length pointing to the right, and a negative number by an arrow of appropriate length pointing to the left.

Both arrows represent 2 or +2. They both point to the right and they are both 2 units long.

Both arrows represent −3. They both point to the left and they are both 3 units long.

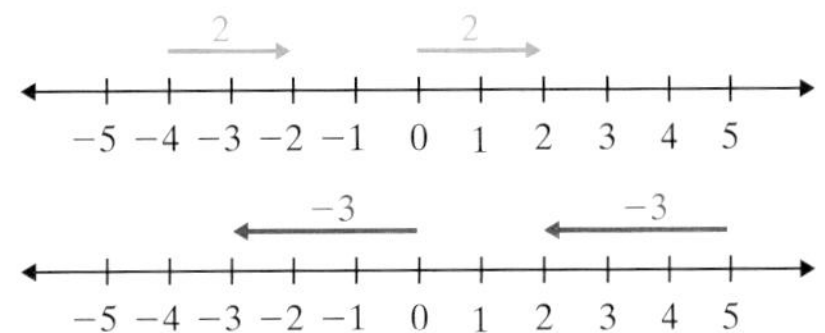

EXAMPLE 1 Add using a number line: $5 + (-2)$

Solution: To add integers on a number line, such as $5 + (-2)$, we start at 0 on the number line and draw an arrow representing 5. From the tip of this arrow, we draw another arrow representing −2. The tip of the second arrow ends at their sum, 3.

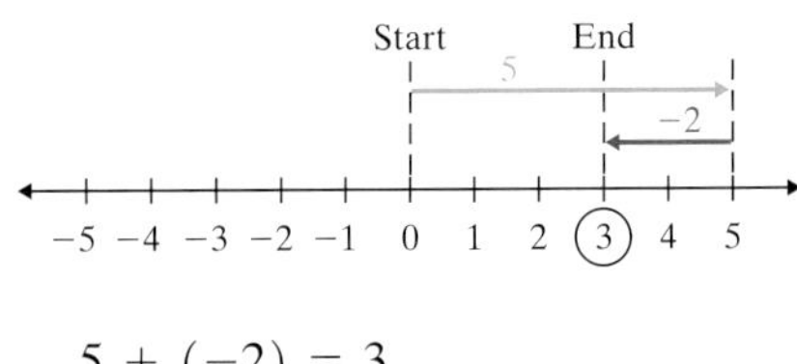

$$5 + (-2) = 3$$

Practice Problem 1

Add using a number line: $5 + (-1)$

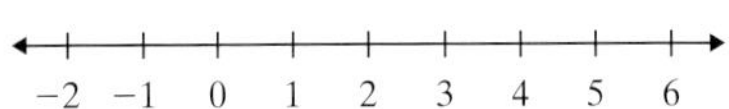

EXAMPLE 2 Add using a number line: $-1 + (-4)$

Start at 0 and draw an arrow representing −1. From the tip of this arrow, we draw another arrow representing −4. The tip of the second arrow ends at their sum, −5.

Solution:

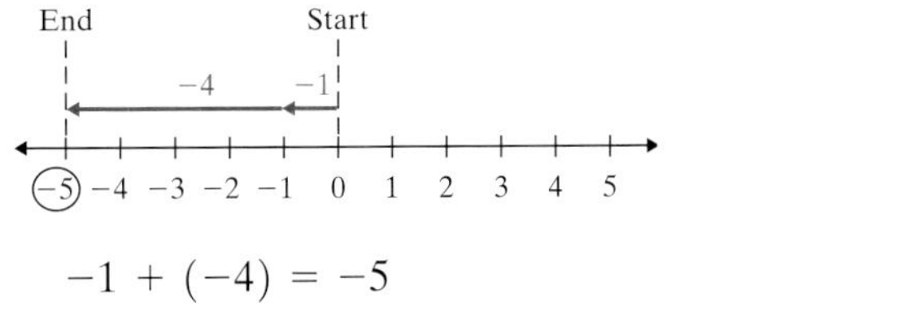

$$-1 + (-4) = -5$$

Practice Problem 2

Add using a number line: $-6 + (-2)$

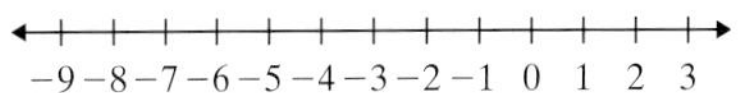

Practice Problem 3

Add using a number line: $-8 + 3$

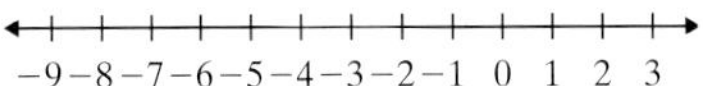

EXAMPLE 3 Add using a number line: $-7 + 3$

Solution:

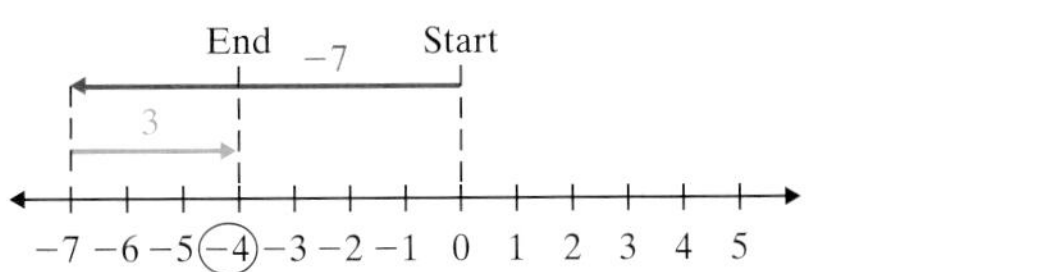

Using a number line each time we add two numbers can be time consuming. Instead, we can notice patterns in the previous examples and write rules for adding signed numbers.

Rules for adding signed numbers depend on whether we are adding numbers with the same sign or different signs. When adding two numbers with the same sign, as in Example 2, notice that the sign of the sum is the same as the sign of the addends.

Answers

1.
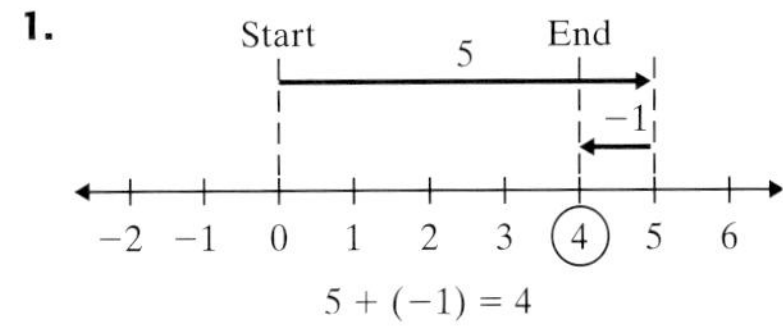

2.
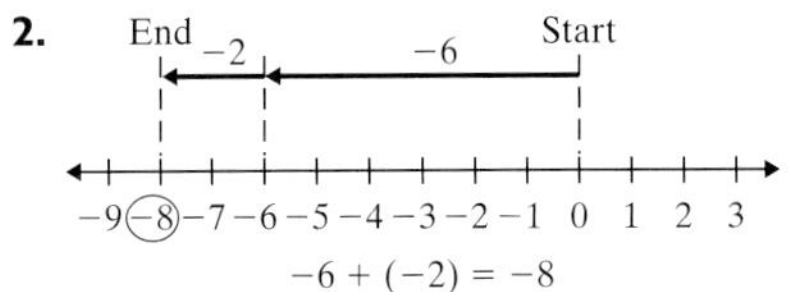

3.
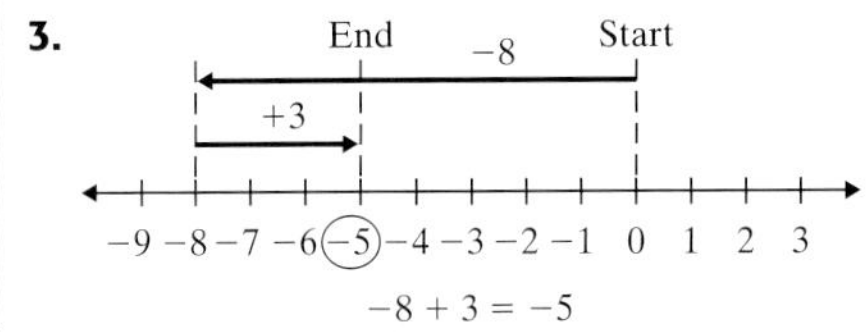

Adding Two Numbers with the Same Sign

Step 1. Add their absolute values.
Step 2. Use their common sign as the sign of the sum.

Practice Problem 4

Add: $(-3) + (-9)$

EXAMPLE 4 Add: $-2 + (-21)$

Solution:

Step 1. $|-2| = 2, |-21| = 21$, and $2 + 21 = 23$.
Step 2. Their common sign is negative, so the sum is negative:

$$-2 + (-21) = -23$$

Practice Problems 5–6

Add.

5. $-12 + (-3)$
6. $9 + 5$

EXAMPLES Add.

5. $-5 + (-1) = -6$
6. $2 + 6 = 8$

When adding two numbers with different signs, as in Examples 1 and 3, the sign of the result may be positive, negative, or the result may be 0.

Adding Two Numbers with Different Signs

Step 1. Find the larger absolute value minus the smaller absolute value.
Step 2. Use the sign of the number with the larger absolute value as the sign of the sum.

Practice Problem 7

Add: $-3 + 9$

EXAMPLE 7 Add: $-2 + 5$

Solution:

Step 1. $|-2| = 2, |5| = 5$, and $5 - 2 = 3$.
Step 2. 5 has the larger absolute value and its sign is an understood +:

$$-2 + 5 = +3 \text{ or } 3$$

Practice Problem 8

Add: $2 + (-8)$

EXAMPLE 8 Add: $3 + (-7)$

Solution:

Step 1. $|3| = 3, |-7| = 7$, and $7 - 3 = 4$.
Step 2. -7 has the larger absolute value and its sign is $-$:

$$3 + (-7) = -4$$

Practice Problems 9–11

Add.

9. $-46 + 20$
10. $8 + (-6)$
11. $-2 + 0$

EXAMPLES Add.

9. $-18 + 10 = -8$
10. $12 + (-8) = 4$
11. $0 + (-5) = -5$ The sum of 0 and any number is the number.

Recall that numbers such as 7 and -7 are called opposites. In general, the sum of a number and its opposite is always 0.

$7 + (-7) = 0$ (opposites) $\quad -26 + 26 = 0$ (opposites) $\quad 1008 + (-1008) = 0$ (opposites)

Answers

4. -12 **5.** -15 **6.** 14 **7.** 6
8. -6 **9.** -26 **10.** 2 **11.** -2

If a is a number, then

$-a$ is its opposite. Also,

$$\left.\begin{aligned} a + (-a) &= 0 \\ -a + a &= 0 \end{aligned}\right\}$$ The sum of a number and its opposite is 0.

EXAMPLES Add.

12. $-21 + 21 = 0$
13. $36 + (-36) = 0$

Try the Concept Check in the margin.

In the following examples, we add three or more integers. Remember that by the associative and commutative properties for addition, we may add numbers in any order that we wish. In Examples 14 and 15, let's add the numbers from left to right.

EXAMPLE 14 Add: $(-3) + 4 + (-11)$

Solution: $(-3) + 4 + (-11) = 1 + (-11)$
$= -10$

EXAMPLE 15 Add: $1 + (-10) + (-8) + 9$

Solution: $1 + (-10) + (-8) + 9 = -9 + (-8) + 9$
$= -17 + 9$
$= -8$

The sum will be the same if we add the numbers in any order. To see this, let's add the positive numbers together and then the negative numbers together first.

$1 + 9 = 10$ Add the positive numbers.
$(-10) + (-8) = -18$ Add the negative numbers.
$10 + (-18) = -8$ Add these results.

The sum is -8.

Helpful Hint

Don't forget that addition is commutative and associative. In other words, numbers may be added in any order.

B Evaluating Algebraic Expressions

We can continue our work with algebraic expressions by evaluating expressions given integer replacement values.

EXAMPLE 16 Evaluate $x + y$ for $x = 3$ and $y = -5$.

Solution: Replace x with 3 and y with -5 in $x + y$.

$x + y = 3 + (-5)$
$= -2$

Practice Problems 12–13

Add.
12. $15 + (-15)$
13. $-80 + 80$

Concept Check

What is wrong with the following calculation?

$6 + (-22) = 16$

Practice Problem 14

Add: $8 + (-3) + (-13)$

Practice Problem 15

Add: $5 + (-3) + 12 + (-14)$

Practice Problem 16

Evaluate $x + y$ for $x = -4$ and $y = 1$.

Answers

12. 0 **13.** 0 **14.** -8 **15.** 0 **16.** -3

Concept Check: $6 + (-22) = -16$

Practice Problem 17

Evaluate $x + y$ for $x = -11$ and $y = -6$.

EXAMPLE 17 Evaluate $x + y$ for $x = -2$ and $y = -10$.

Solution: $x + y = (-2) + (-10)$ Replace x with -2 and y with -10.

$= -12$

C Solving Problems by Adding Integers

Next, we practice solving problems that require adding integers.

Practice Problem 18

If the temperature was $-8°$ Fahrenheit at 6 a.m., and it rose 4 degrees by 7 a.m. and then rose another 7 degrees in the hour from 7 a.m. to 8 a.m., what was the temperature at 8 a.m.?

EXAMPLE 18 Calculating Temperature

On January 6, the temperature in Caribou, Maine, at 8 a.m. was $-12°$ Fahrenheit. By 9 a.m., the temperature had risen 4 degrees, and by 10 a.m. it had risen 6 degrees from the 9 a.m. temperature. What was the temperature at 10 a.m.?

Solution:

In words:	temperature at 10 a.m.	=	8 a.m. temperature	+	rise of 4°	+	rise of 6°
	↓		↓		↓		↓
Translate:	temperature at 10 a.m.	=	−12	+	(+4)	+	(+6)

$= -8 + (+6)$

$= -2$

The temperature was -2°F at 10 a.m.

CALCULATOR EXPLORATIONS

Entering Negative Numbers

To enter a negative number on a calculator, find the key marked [+/−]. (Some calculators have a key marked [CHS] and some calculators have a special key [(−)] for entering a negative sign.) To enter the number −2, for example, press the keys [2] [+/−]. The display will read [−2].

To find $-32 + (-131)$, press the keys

[32] [+/−] [+] [131] [+/−] [=] or

[(−)] [32] [+] [(−)] [131] [ENTER]

The display will read [−163]. Thus $-32 + (-131) = -163$.

Use a calculator to perform each indicated operation.

1. $-256 + 97$

2. $811 + (-1058)$

3. $6(15) + (-46)$

4. $-129 + 10(48)$

5. $-108{,}650 + (-786{,}205)$

6. $-196{,}662 + (-129{,}856)$

Answers

17. −17 **18.** 3°F

Name ______________________ Section ____________ Date ____________

Mental Math

Add.

1. $5 + 0$ **2.** $(-2) + 0$ **3.** $0 + (-35)$ **4.** $0 + 3$

5. $-12 + 12$ **6.** $48 + (-48)$ **7.** $28 + (-28)$ **8.** $-9 + 9$

EXERCISE SET 2.2

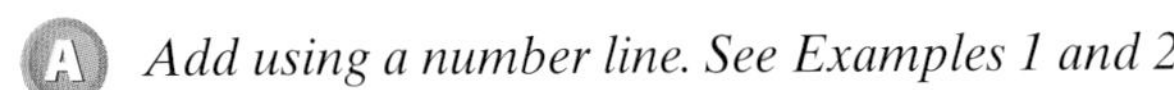

A *Add using a number line. See Examples 1 and 2.*

1. $-1 + (-6)$

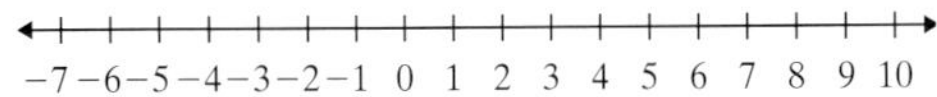

2. $9 + (-4)$

−7 −6 −5 −4 −3 −2 −1 0 1 2 3 4 5 6 7 8 9 10

3. $-4 + 7$

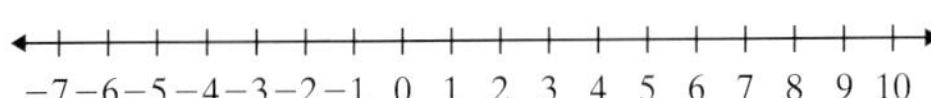

4. $10 + (-3)$

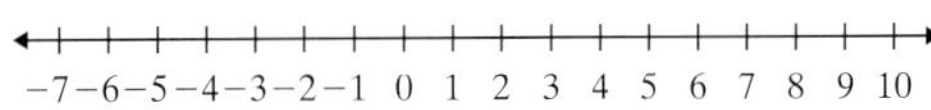

5. $-13 + 7$

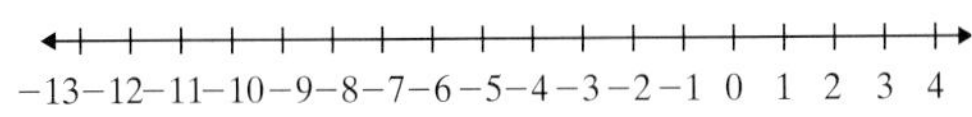

6. $-6 + (-5)$

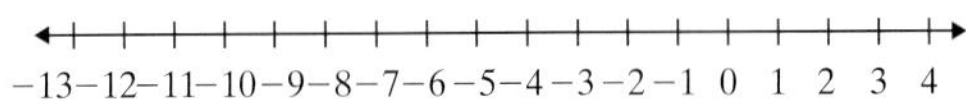

Add. See Examples 3 through 13.

7. $23 + 12$ **8.** $15 + 42$ **9.** $-6 + (-2)$ **10.** $-5 + (-4)$

11. $-43 + 43$ **12.** $-62 + 62$ **13.** $6 + (-2)$ **14.** $8 + (-3)$

15. $-6 + 8$ **16.** $-8 + 12$ **17.** $3 + (-5)$ **18.** $5 + (-9)$

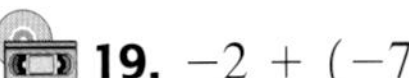

19. $-2 + (-7)$ **20.** $-6 + (-1)$ **21.** $-12 + (-12)$ **22.** $-23 + (-23)$

23. $-25 + (-32)$ **24.** $-45 + (-90)$ **25.** $-123 + (-100)$ **26.** $-500 + (-230)$

27. $-7 + 7$ **28.** $-10 + 10$ **29.** $12 + (-5)$ **30.** $24 + (-10)$

31. $-6 + 3$ **32.** $-8 + 2$ **33.** $-12 + 3$ **34.** $-15 + 5$

35. $56 + (-26)$ **36.** $89 + (-37)$ **37.** $-37 + 57$ **38.** $-25 + 65$

39. $-42 + 93$ **40.** $-64 + 164$ **41.** $34 + (-67)$ **42.** $42 + (-83)$

43. $124 + (-144)$ **44.** $325 + (-375)$ **45.** $-82 + (-43)$ **46.** $-56 + (-33)$

Add. See Examples 14 and 15.

47. $-4 + 2 + (-5)$ **48.** $-1 + 5 + (-8)$

49. $-52 + (-77) + (-117)$ **50.** $-103 + (-32) + (-27)$

51. $12 + (-4) + (-4) + 12$ **52.** $18 + (-9) + 5 + (-2)$

53. $(-10) + 14 + 25 + (-16)$ **54.** $34 + (-12) + (-11) + 213$

Add.

55. $-8 + (-14) + (-11)$ **56.** $-10 + (-6) + (-1)$

57. $-26 + 5$ **58.** $-35 + (-12)$

59. $5 + (-1) + 17$ **60.** $3 + (-23) + 6$

61. $-14 + (-31)$ **62.** $-100 + 70$

63. $13 + 14 + (-18)$ **64.** $(-45) + 22 + 20$

65. $-87 + 87$ **66.** $-87 + 0$

67. $-3 + (-8) + 12 + (-1)$ **68.** $-16 + 6 + (-14) + (-20)$

69. $0 + (-103)$ **70.** $94 + (-94)$

B *Evaluate $x + y$ for the given replacement values. See Examples 16 and 17.*

71. $x = -2$ and $y = 3$ **72.** $x = -7$ and $y = 11$ **73.** $x = -20$ and $y = -50$

74. $x = -1$ and $y = -29$ **75.** $x = 3$ and $y = -30$ **76.** $x = 13$ and $y = -17$

C *Solve. See Example 18.*

The bar graph below shows the yearly net income for Gateway, Inc. Net income is one indication of a company's health. It measures revenue (money taken in) minus cost (money spent). Use this graph for Exercises 77 through 80. (*Source*: Gateway, Inc.)

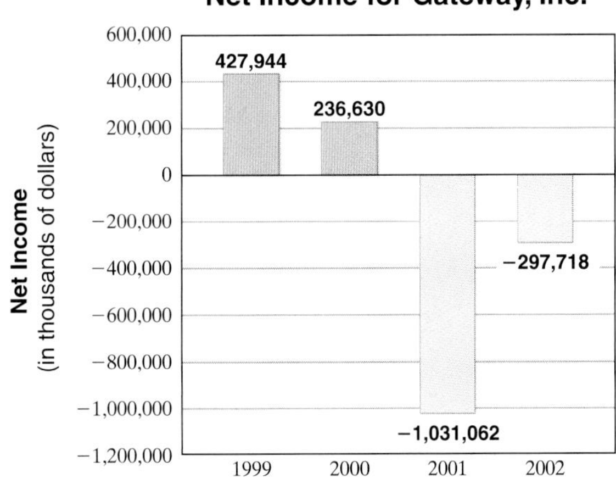

77. What was the net income (in dollars) for Gateway, Inc. in 2000?

78. What was the net income (in dollars) for Gateway, Inc. in 2001?

79. Find the total net income for years 2001 and 2002.

80. Find the total net income for all the years shown.

81. The temperature at 4 p.m. on February 2 was $-10°$ Celsius. By 11 p.m. the temperature had risen 12 degrees. Find the temperature at 11 p.m.

82. Scores in golf can be positive or negative integers. For example, a score of 3 *over* par can be represented by $+3$ and a score of 5 *under* par can be represented by -5. If Fred Couples had scores of 3 over par, 6 under par, and 7 under par for three games of golf, what was his total score?

83. A small business company reports the following net incomes. Find their sum.

Year	Net income (in dollars)
2000	$75,083
2001	−$10,412
2002	−$1,786
2003	$96,398

84. Suppose a deep-sea diver dives from the surface to 248 meters below the surface and then swims up 6 meters, down 17 meters, down another 24 meters, and then up 23 meters. Use positive and negative numbers to represent this situation. Then find the diver's depth after these movements.

In some card games, it is possible to have positive and negative scores. The table shows the scores for two teams playing a series of four card games. Use this table to answer Exercises 85 and 86.

	Game 1	Game 2	Game 3	Game 4
Team 1	−2	−13	20	2
Team 2	5	11	−7	−3

85. Find each team's total score after four games. If the winner is the team with the greater score, find the winning team.

86. Find each team's total score after three games. If the winner is the team with the greater score, which team was winning after three games?

87. The all-time record low temperature for Wyoming is −66°F, which was recorded on February 13, 1905. Kansas's all-time record low temperature is 26°F higher than Wyoming's record low. What is Kansas's record low temperature? (*Source:* National Climatic Data Center)

88. The all-time record low temperature for New York is −52°F, which occurred on February 13, 1905. In Mississippi's, the lowest temperature ever recorded is 33°F higher than New York's all-time low temperature. What is the all-time record low temperature for Mississippi? (*Source:* National Climatic Data Center)

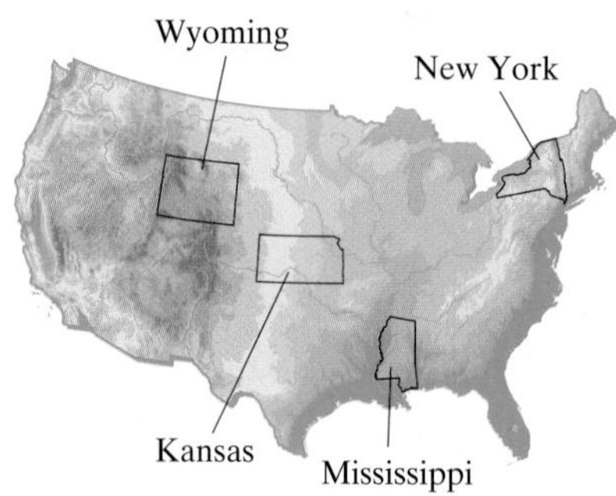

89. The difference between a country's exports and imports is called the country's *trade balance*. The United States had a trade balance of −\$230 billion in 1998, −\$370 billion in 2000, and −\$346 billion in 2001. What was the total U.S. trade balance for these years? (*Source:* U.S. Department of Commerce)

90. The U.S. trade balance for natural gas was −3538 billion cubic feet in 2000 and −3604 billion cubic feet in 2001. What was the total U.S. trade balance for natural gas for these years? (*Source:* U.S. Energy Information Administration)

Review and Preview

Subtract. See Section 1.4.

91. $44 - 0$ **92.** $91 - 0$ **93.** $52 - 52$ **94.** $103 - 103$ **95.** $87 - 59$

96. $32 - 18$ **97.** $-8602 + (-1056)$ **98.** $-7959 + 2335$ **99.** $-|-56| + |-56|$ **100.** $-|86| + (-|17|)$

Combining Concepts

For Exercises 101–104, determine whether each statement is true or false.

101. The sum of two negative numbers is always a negative number.

102. The sum of two positive numbers is always a positive number.

103. The sum of a positive number and a negative number is always a negative number.

104. The sum of zero and a negative number is always a negative number.

105. In your own words, explain how to add two negative numbers.

106. In your own words, explain how to add a positive number and a negative number.

2.3 Subtracting Integers

OBJECTIVES

- A Subtract integers.
- B Add and subtract integers.
- C Evaluate an algebraic expression by subtracting.
- D Solve problems by subtracting integers.

SSM TUTOR CENTER | SG | CD & VIDEO | MATH PRO | WEB

In Section 2.1, we discussed the opposite of an integer.

The opposite of 3 is -3.
The opposite of -6 is 6.

In this section, we use opposites to subtract integers.

A Subtracting Integers

To subtract integers, we will write the subtraction problem as an addition problem. To see how to do this, study the examples below.

$$10 - 4 = 6$$
$$10 + (-4) = 6$$

Since both expressions simplify to 6, this means that

$$10 - 4 = 10 + (-4) = 6$$

Also,

$$3 - 2 = 3 + (-2) = 1$$
$$15 - 1 = 15 + (-1) = 14$$

Thus, to subtract two numbers, we add the first number to the opposite of the second number. (The opposite of a number is also known as its **additive inverse**.)

Subtracting Two Numbers

If a and b are numbers, then $a - b = a + (-b)$.

EXAMPLES Subtract.

	subtraction ↓	=	first number ↓	+ ↓	opposite of the second number ↓	
1.	$8 - 5$	=	8	+	(-5)	$= 3$
2.	$-4 - 10$	=	-4	+	(-10)	$= -14$
3.	$6 - (-5)$	=	6	+	5	$= 11$
4.	$-11 - (-7)$	=	-11	+	7	$= -4$

Practice Problems 1–4

Subtract.

1. $12 - 7$
2. $-6 - 4$
3. $11 - (-14)$
4. $-9 - (-1)$

EXAMPLES Subtract.

5. $-10 - 5 = -10 + (-5) = -15$

6. $8 - 15 = 8 + (-15) = -7$

7. $-4 - (-5) = -4 + 5 = 1$

Practice Problems 5–7

Subtract.

5. $5 - 9$
6. $-12 - 4$
7. $-2 - (-7)$

Answers

1. 5 **2.** -10 **3.** 25 **4.** -8 **5.** -4 **6.** -16 **7.** 5

Helpful Hint:

To visualize subtraction, try the following:

The difference between 5°F and −2°F can be found by subtracting. That is,

$$5-(-2) = 5 + 2 = 7$$

Can you visually see from the thermometer on the right that there is actually 7 degrees between 5°F and −2°F?

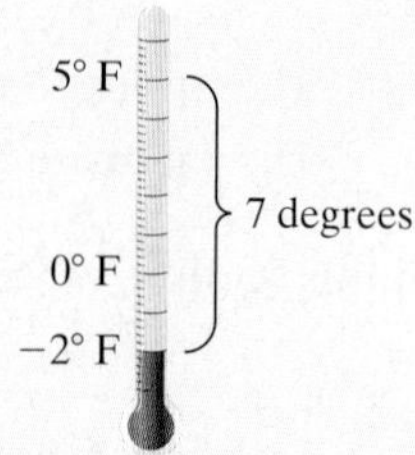

Concept Check

What is wrong with the following calculation?

$-9 - (-6) = -15$

Try the Concept Check in the margin.

Practice Problem 8

Subtract 5 from −10.

EXAMPLE 8 Subtract 7 from −3.

Solution: To subtract 7 *from* −3, we find

$$-3 - 7 = -3 + (-7) = -10$$

B Adding and Subtracting Integers

If a problem involves adding or subtracting more than two integers, we rewrite differences as sums and add. Recall that by associative and commutative properties, we may add numbers in any order. In Examples 9 and 10, we will add from left to right.

Practice Problem 9

Simplify: $-4 - 3 - 7 - (-5)$

EXAMPLE 9 Simplify: $7 - 8 - (-5) - 1$

Solution:
$$\begin{aligned} 7 - 8 - (-5) - 1 &= \underbrace{7 + (-8)} + 5 + (-1) \\ &= \underbrace{-1 + 5} + (-1) \\ &= \underbrace{4 + (-1)} \\ &= 3 \end{aligned}$$

Practice Problem 10

Simplify: $3 + (-5) - 6 - (-4)$

EXAMPLE 10 Simplify: $7 + (-12) - 3 - (-8)$

Solution:
$$\begin{aligned} 7 + (-12) - 3 - (-8) &= \underbrace{7 + (-12)} + (-3) + 8 \\ &= -5 + (-3) + 8 \\ &= -8 + 8 \\ &= 0 \end{aligned}$$

C Evaluating Expressions

Now let's practice evaluating expressions when the replacement values are integers.

Practice Problem 11

Evaluate $x - y$ for $x = -2$ and $y = 14$.

EXAMPLE 11 Evaluate $x - y$ for $x = -3$ and $y = 9$.

Solution: Replace x with −3 and y with 9 in $x - y$.

$$\begin{aligned} &x \;-\; y \\ &\downarrow \;\; \downarrow \;\; \downarrow \\ &= (-3) - 9 \\ &= (-3) + (-9) \\ &= -12 \end{aligned}$$

Answers

8. −15 **9.** −9 **10.** −4 **11.** −16

Concept Check: $-9 - (-6) = -3$

EXAMPLE 12 Evaluate $a - b$ for $a = 8$ and $b = -6$.

Solution: Watch your signs carefully!

$$\begin{aligned} & a - b \\ &= 8 - (-6) \quad \text{Replace } a \text{ with 8 and } b \text{ with } -6. \\ &= 8 + 6 \\ &= 14 \end{aligned}$$

Practice Problem 12

Evaluate $y - z$ for $y = -3$ and $z = -4$.

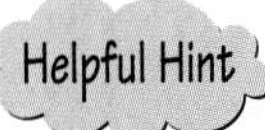

Watch carefully when replacing variables in the expression $x - y$. Make sure that all symbols are inserted and accounted for.

D Solving Problems by Subtracting Integers

Solving problems often requires subtraction of integers.

EXAMPLE 13 Finding a Change in Elevation

The highest point in the United States is the top of Mount McKinley, in Denali County, Alaska, at a height of 20,320 feet above sea level. The lowest point is Death Valley, California, which is 282 feet below sea level. How much higher is Mount McKinley than Death Valley? (*Source:* U.S. Geological Survey)

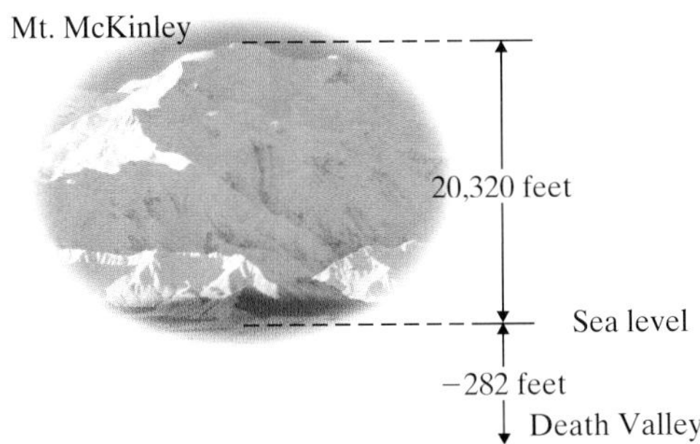

Solution:

	how much higher Mt. McKinley is	=	height of Mt. McKinley	minus	height of Death Valley
In words:	↓		↓	↓	↓
Translate:	how much higher Mt. McKinley is	=	20,320	−	(−282)

$$= 20{,}320 + 282 = 20{,}602$$

Mt. McKinley is 20,602 feet higher than Death Valley.

Practice Problem 13

The highest point in Asia is the top of Mount Everest, at a height of 29,028 feet above sea level. The lowest point is the Dead Sea, which is 1312 feet below sea level. How much higher is Mount Everest than the Dead Sea? (*Source:* National Geographic Society)

Answers

12. 1 **13.** 30,340 feet

Focus on History

MAGIC SQUARES

A magic square is a set of numbers arranged in a square table so that the sum of the numbers in each column, row, and diagonal is the same. For instance, in the magic square below, the sum of each column, row, and diagonal is 15. Notice that no number is used more than once in the magic square.

2	9	4
7	5	3
6	1	8

The properties of magic squares have been known for a very long time and once were thought to be good luck charms. The ancient Egyptians and Greeks understood their patterns. A magic square even made it into a famous work of art. The engraving titled *Melencolia I*, created by German artist Albrecht Dürer in 1514, features the following four-by-four magic square on the building behind the central figure.

16	3	2	13
5	10	11	8
9	6	7	12
4	15	14	1

CRITICAL THINKING

1. Verify that what is shown in the Dürer engraving is, in fact, a magic square. What is the common sum of the columns, rows, and diagonals?

2. Negative numbers can also be used in magic squares. Complete the following magic square:

		-2
	-1	
0		-4

3. Use the numbers $-16, -12, -8, -4, 0, 4, 8, 12,$ and 16 to form a magic square:

Name ______________________ Section __________ Date __________

EXERCISE SET 2.3

A *Perform each indicated subtraction. See Examples 1 through 8.*

1. $5 - 5$

2. $-6 - (-6)$

3. $8 - 3$

4. $5 - 2$

5. $3 - 8$

6. $2 - 5$

7. $7 - (-7)$

8. $12 - (-12)$

9. $-5 - (-8)$

10. $-25 - (-25)$

11. $-14 - 4$

12. $-2 - 42$

13. $2 - 16$

14. $8 - 9$

15. $-10 - (-10)$

16. $-5 - (-5)$

17. $-15 - (-15)$

18. $-24 - (-24)$

19. $3 - 7$

20. $4 - 12$

21. $30 - 45$

22. $29 - 56$

23. $-4 - 10$

24. $-5 - 8$

25. $-230 - 0$

26. $-15 - 0$

27. $4 - (-6)$

28. $6 - (-9)$

29. $-7 - (-3)$

30. $-12 - (-5)$

31. $-16 - (-23)$

32. $-45 - (-16)$

33. Subtract 18 from -20.

34. Subtract 10 from -22.

35. Find the difference of -20 and -3.

36. Find the difference of -8 and -13.

37. Subtract -11 from 2.

38. Subtract -50 from -50.

B *Simplify. See Examples 9 and 10.*

39. $7 - 3 - 2$

40. $8 - 4 - 1$

41. $12 - 5 - 7$

42. $30 - 7 - 12$

43. $-5 - 8 - (-12)$

44. $-10 - 6 - (-9)$

45. $-10 + (-5) - 12$

46. $-15 + (-8) - 4$

47. $12 - (-34) + (-6)$

48. $23 - (-17) + (-9)$

49. $-(-6) - 12 + (-16)$

50. $-(-9) - 7 + (-23)$

51. $-9 - (-12) + (-7) - 4$

52. $-6 - (-8) + (-12) - 7$

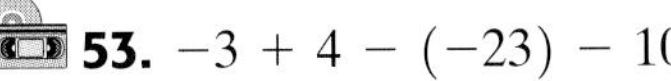

53. $-3 + 4 - (-23) - 10$

54. $5 + (-18) - (-21) - 2$

C *Evaluate $x - y$ for the given replacement values. See Examples 11 and 12.*

55. $x = -3$ and $y = 5$

56. $x = -7$ and $y = 1$

57. $x = 6$ and $y = -30$

58. $x = 9$ and $y = -2$

59. $x = -4$ and $y = -4$

60. $x = -8$ and $y = -10$

61. $x = 1$ and $y = -18$

62. $x = 14$ and $y = -12$

Recall that the bar graph from Section 2.1 below shows the average temperature in Fahrenheit of known planets. Notice that a negative temperature is illustrated by a bar below the horizontal line representing 0°F, and a positive temperature is illustrated by a bar above the horizontal line representing 0°F.

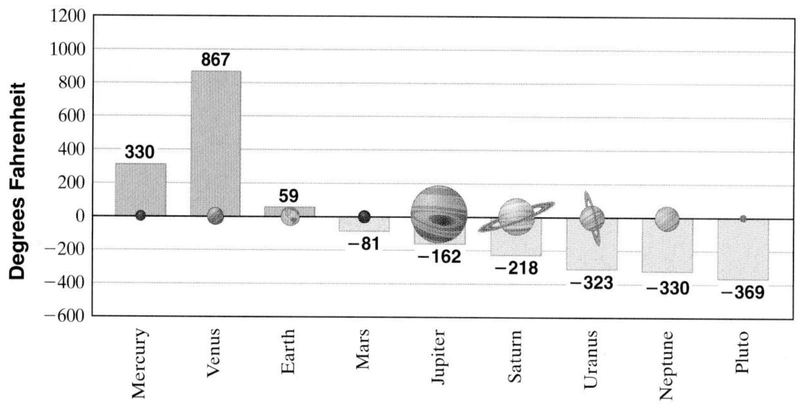

*(For some planets, the temperature given is the temperature where the atmosphere pressure equals 1 Earth atmosphere; Source: *The World Almanac*, 2003)

63. Find the difference in temperature between Earth and Pluto.

64. Find the difference in temperature between Venus and Mars.

65. Find the difference in temperature between the two plants with the lowest temperature.

66. Find the difference in temperature between Jupiter and Saturn.

D *Solve. See Example 13.*

67. The coldest temperature ever recorded on Earth was −129°F in Antarctica. The warmest temperature ever recorded was 136°F in the Sahara Desert. How many degrees warmer is 136°F than −129°F? (*Source: Questions Kids Ask*, Grolier Limited, 1991, and *The World Almanac, 2003*)

68. The coldest temperature ever recorded in the United States was −80°F in Alaska. The warmest temperature ever recorded was 134°F in California. How many degrees warmer is 134°F than −80°F? (*Source*: *The World Almanac*, 2003)

69. Aaron Aiken has $125 in his checking account. He writes a check for $117, makes a deposit of $45, and then writes another check for $69. Find the balance in his account. (Write the amount as an integer.)

70. In the card game canasta, it is possible to have a negative score. If Juan Santanilla's score is 15, what is his new score if he loses 20 points?

71. The temperature on a February morning is −6° Celsius at 6 a.m. If the temperature drops 3 degrees by 7 a.m., rises 4 degrees between 7 a.m. and 8 a.m., and then drops 7 degrees between 8 a.m. and 9 a.m., find the temperature at 9 a.m.

72. Mauna Kea in Hawaii has an elevation of 13,796 feet above sea level. The Mid-America Trench in the Pacific Ocean has an elevation of 21,857 feet below sea level. Find the difference in elevation between those two points. (*Source:* National Geographic Society and Defense Mapping Agency)

Some places on Earth lie below sea level, which is the average level of the surface of the oceans. Use this diagram to answer Exercises 73 through 76. (Source: Fantastic Book of Comparisons, Russell Ash, 1999)

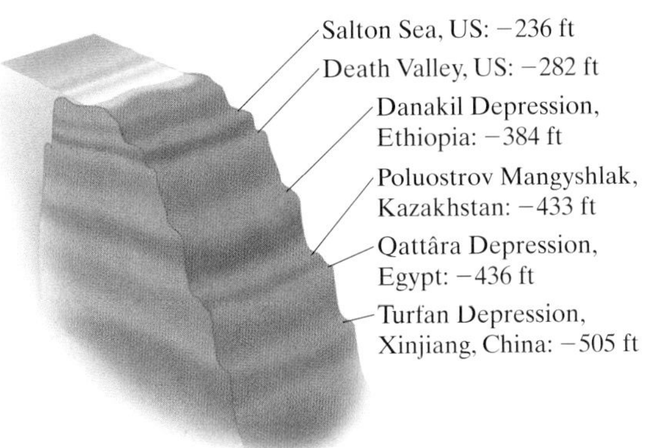

73. Find the difference in elevation between Death Valley and Quattâra Depression.

74. Find the difference in elevation between Danakil and Turfan Depressions.

75. Find the difference in elevation between the two lowest elevations shown.

76. Find the difference in elevation between the highest elevation shown and the lowest elevation shown.

The bar graph shows heights of selected lakes. For Exercises 77–80, find the difference in elevation for the lakes listed. (*Source*: U.S. Geological Survey)

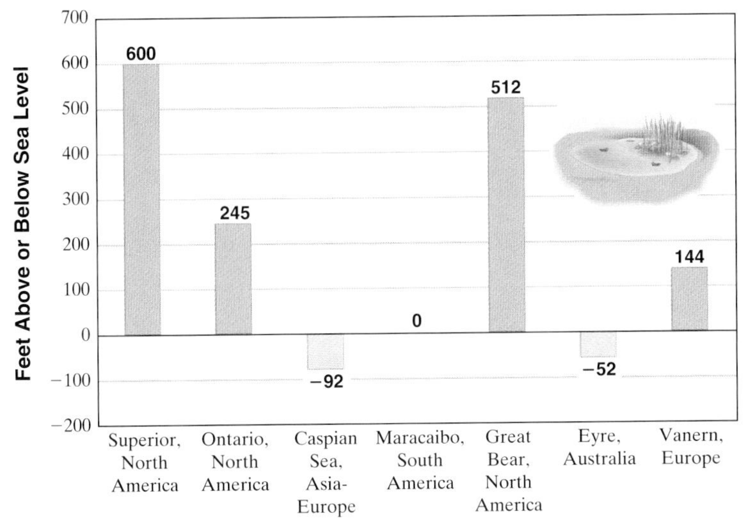

77. Lake Superior and Lake Eyre

78. Great Bear Lake and Caspian Sea

79. Lake Maracaibo and Lake Vanern

80. Lake Eyre and Caspian Sea

81. Recall that the difference between a country's exports and imports is called the country's *trade balance*. In 2001, the United States had \$731 billion in exports and \$1141 billion in imports. What was the U.S. trade balance in 2001? (*Source:* U.S. Department of Commerce)

82. In 2000, the United States exported 380 million barrels of petroleum products and imported 788 million barrels of petroleum products. What was the U.S. trade balance for petroleum products in 1997? (*Source:* U.S. Energy Information Administration)

Review and Preview

Multiply. See Section 1.6.

83. $436 \cdot 0$ **84.** $0 \cdot 86$ **85.** $436 \cdot 1$ **86.** $1 \cdot 704$ **87.** $\begin{array}{r} 23 \\ \times 46 \\ \hline \end{array}$ **88.** $\begin{array}{r} 51 \\ \times 89 \\ \hline \end{array}$

Simplify. See Section 1.8.

89. $5^2 - 6 \cdot 2 + 8$ **90.** $80 - 7 \cdot 10 + 8^2$

Combining Concepts

Evaluate each expression for the given replacement values.

91. $x - y - z$ for $x = -4$, $y = 3$, and $z = 15$

92. $x - y - z$ for $x = -14$, $y = 8$, and $z = -6$

93. $a + b - c$ for $a = -16$, $b = 14$, and $c = -22$

94. $a + b - c$ for $a = -1$, $b = -1$, and $c = 100$

Simplify. (Evaluate absolute values first.)

95. $|-3| - |-7|$

96. $|-12| - |-5|$

97. $|-6| - |6|$

98. $|-23| - |-42|$

99. $-8067 - 1129$

100. $76{,}804 - 96{,}971$

For Exercises 101 and 102, determine whether each statement is true or false.

101. $|-8 - 3| = 8 - 3$

102. $|-2 - (-6)| = |-2| - |-6|$

103. In your own words, explain how to subtract one integer from another.

104. A student explains to you that the first step to simplify $8 + 12 \cdot 5 - 100$ is to add 8 and 12. Is the student correct? Explain why or why not.

Internet Excursions

Go To: 

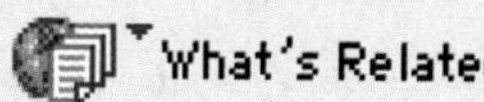

The World Wide Web address listed here will direct you to the Web site of the CNN Financial Network, or a related site. You will be able to collect data on prices of stocks traded on the New York Stock Exchange and answer the questions below.

105. Pick stocks for any four companies. You can use the Web site to look up the stock market ticker symbol for the companies. Then get a stock quote for each company. Complete the table below. For each company in the table, show that the sum of the previous closing price and the change in price give the last price.

106. For the same four companies you used in Exercise 105, show how to use the last price and the change in price to calculate the previous closing price.

Date of stock quotes:_____ Time of stock quotes: _____

Company Name	Ticker Symbol	Previous Close	Change	Last

Name ______________________ Section __________ Date __________

Integrated Review—Integers

Represent each quantity by an integer.

1. The peak of Mount Everest in Asia is 29,028 feet above sea level. (*Source:* U.S. Geological Survey)

2. The Marianas Trench in the Pacific Ocean is 35,840 feet below sea level. (*Source:* The World Almanac)

3. The deepest hole ever drilled in the Earth's crust is in Russia and its depth is over 7 miles below sea level. (*Source: Fantastic Book of Comparisons*, 1999)

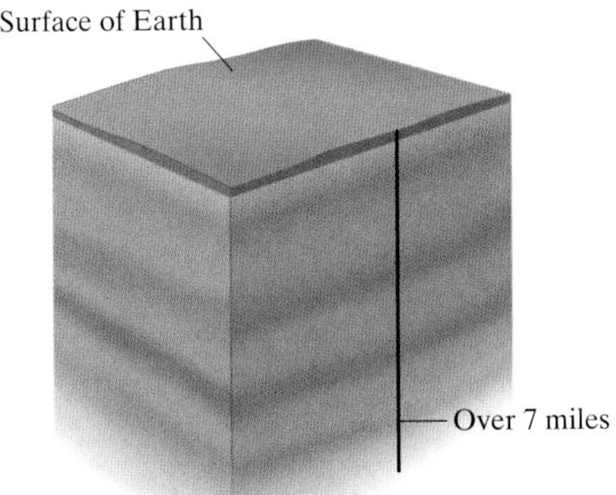

4. Graph the signed numbers on the given number line. $-4, 0, -1, 3$

−5 −4 −3 −2 −1 0 1 2 3 4 5

Insert $<$ or $>$ *between each pair of numbers to make a true statement.*

5. $0 \quad -3$

6. $-15 \quad -5$

7. $-1 \quad 1$

8. $-2 \quad -7$

Simplify.

9. $|-1|$

10. $-|-4|$

11. $|-8|$

12. $-(-5)$

Find the opposite of each number.

13. 6

14. -3

15. 89

16. 0

Add or subtract as indicated.

17. $-7 + 12$

18. $-9 + (-11)$

19. $25 + (-35)$

Answers

1. ______
2. ______
3. ______
4. see number line
5. ______
6. ______
7. ______
8. ______
9. ______
10. ______
11. ______
12. ______
13. ______
14. ______
15. ______
16. ______
17. ______
18. ______
19. ______

20. ______

21. ______

22. ______

23. ______

24. ______

25. ______

26. ______

27. ______

28. ______

29. ______

30. ______

31. ______

32. ______

33. ______

34. ______

20. $1 - 3$ **21.** $26 - (-26)$ **22.** $-2 - 1$

23. $-18 - (-102)$ **24.** $-8 + (-6) + 20$ **25.** $-11 - 7 - (-19)$

26. $-4 + (-8) - 16 - (-9)$ **27.** Subtract 14 from 26. **28.** Subtract -8 from -12.

Choose all numbers for x from each given list that make each statement true.

29. $|x| > 0$
a. 0 **b.** 18 **c.** -3 **d.** -21

30. $|x| > -5$
a. 0 **b.** 3 **c.** -1 **d.** -1000

Evaluate the expressions below for $x = -1$ and $y = 11$

31. $x + y$ **32.** $x - y$

33. $y - x$ **34.** $y + x$

2.4 Multiplying and Dividing Integers

Multiplying and dividing integers is similar to multiplying and dividing whole numbers. One difference is that we need to determine whether the result is a positive number or a negative number.

OBJECTIVES

- (A) Multiply integers.
- (B) Divide integers.
- (C) Evaluate an algebraic expression by multiplying or dividing.
- (D) Solve problems by multiplying or dividing integers.

SSM TUTOR CENTER SG CD & VIDEO MATH PRO WEB

(A) Multiplying Integers

Consider the following pattern of products.

First factor decreases by 1 each time.

$$3 \cdot 2 = 6$$
$$2 \cdot 2 = 4$$
$$1 \cdot 2 = 2$$
$$0 \cdot 2 = 0$$

Product decreases by 2 each time.

This pattern can be continued, as follows.

$$-1 \cdot 2 = -2$$
$$-2 \cdot 2 = -4$$
$$-3 \cdot 2 = -6$$

This suggests that the product of a negative number and a positive number is a negative number.

What is the sign of the product of two negative numbers? To find out, we form another pattern of products. Again, we decrease the first factor by 1 each time, but this time the second factor is negative.

$$2 \cdot (-3) = -6$$
$$1 \cdot (-3) = -3$$
$$0 \cdot (-3) = 0$$

Product increases by 3 each time.

This pattern continues as:

$$-1 \cdot (-3) = 3$$
$$-2 \cdot (-3) = 6$$
$$-3 \cdot (-3) = 9$$

This suggests that the product of two negative numbers is a positive number. Thus we can determine the sign of a product when we know the signs of the factors.

Multiplying Numbers

The product of two numbers having the same sign is a positive number.

Product of Like Signs

$$(+)(+) = +$$
$$(-)(-) = +$$

The product of two numbers having different signs is a negative number.

Product of Different Signs

$$(-)(+) = -$$
$$(+)(-) = -$$

EXAMPLES Multiply.

1. $-7 \cdot 3 = -21$
2. $-2(-5) = 10$
3. $0 \cdot (-4) = 0$
4. $10(-8) = -80$

Practice Problems 1–4

Multiply.

1. $-2 \cdot 6$
2. $-4(-3)$
3. $0 \cdot (-10)$
4. $5(-15)$

Answers

1. -12 **2.** 12 **3.** 0 **4.** -75

Recall that by the associative and commutative properties for multiplication, we may multiply numbers in any order that we wish. In Example 5, we multiply from left to right.

EXAMPLES Multiply.

5. $7(-6)(-2) = -42(-2)$
$= 84$

6. $(-2)(-3)(-4) = 6(-4)$
$= -24$

7. $(-1)(-2)(-3)(-4) = -1(-24)$ We have -24 from Example 6.
$= 24$

Practice Problems

Multiply.

5. $7(-2)(-4)$
6. $(-5)(-6)(-1)$
7. $(-2)(-5)(-6)(-1)$

Try the Concept Check in the margin.

Concept Check

What is the sign of the product of five negative numbers? Explain.

Recall from our study of exponents that $2^3 = 2 \cdot 2 \cdot 2 = 8$. We can now work with bases that are negative numbers. For example,

$$(-2)^3 = (-2)(-2)(-2) = -8$$

EXAMPLE 8 Evaluate: $(-5)^2$

Solution: Remember that $(-5)^2$ means 2 factors of -5.

$$(-5)^2 = (-5)(-5) = 25$$

Practice Problem 8

Evaluate $(-3)^4$.

Helpful Hint:

Have you noticed a pattern when multiplying signed numbers?

If we let $(-)$ represent a negative number and $(+)$ represent a positive number, then

$$(-)(-) = (+)$$
$$(-)(-)(-) = (-)$$
$$(-)(-)(-)(-) = (+)$$
$$(-)(-)(-)(-)(-) = (-)$$

The product of an even number of negative numbers is a positive result.

The product of an odd number of negative numbers is a negative result.

Notice in Example 8 the parentheses around -5 in $(-5)^2$. With these parentheses, -5 is the base that is squared. Without parentheses, such as -5^2, only the 5 is squared.

EXAMPLE 9 Evaluate: -7^2

Solution: Remember that without parentheses, only the 7 is squared.

$$-7^2 = -(7 \cdot 7) = -49$$

Practice Problem 9.

Evaluate: -9^2

Answers

5. 56 **6.** -30 **7.** 60 **8.** 81 **9.** -81

Concept Check: Negative

Make sure you understand the difference between Examples 8 and 9.

parentheses, so −5 is squared

$$(-5)^2 = (-5)(-5) = 25$$

no parentheses, so only the 7 is squared

$$-7^2 = -(7 \cdot 7) = -49$$

B Dividing Integers

Division of integers is related to multiplication of integers. The sign rules for division can be discovered by writing a related multiplication problem. For example,

$$\frac{6}{2} = 3 \quad \text{because} \quad 3 \cdot 2 = 6$$

$$\frac{-6}{2} = -3 \quad \text{because} \quad -3 \cdot 2 = -6$$

$$\frac{6}{-2} = -3 \quad \text{because} \quad -3 \cdot (-2) = 6$$

$$\frac{-6}{-2} = 3 \quad \text{because} \quad 3 \cdot (-2) = -6$$

Helpful Hint

Just as for whole numbers, division can be checked by multiplication.

Dividing Numbers

The quotient of two numbers having the same sign is a positive number.

Quotient of Like Signs

$$\frac{(+)}{(+)} = + \qquad \frac{(-)}{(-)} = +$$

The quotient of two numbers having different signs in a negative number.

Quotient of Different Signs

$$\frac{(+)}{(-)} = - \qquad \frac{(-)}{(+)} = -$$

EXAMPLES Divide.

10. $\frac{-12}{6} = -2$

11. $-20 \div (-4) = 5$

12. $\frac{48}{-3} = -16$

Try the Concept Check in the margin.

EXAMPLES Divide, if possible.

13. $\frac{0}{-5} = 0$ because $0 \cdot -5 = 0$

14. $\frac{-7}{0}$ is undefined because there is no number that gives a product of −7 when multiplied by 0.

Practice Problems 10-12

Divide.

10. $\frac{28}{-7}$ 11. $-18 \div (-2)$ 12. $\frac{-60}{10}$

Concept Check

What is wrong with the following calculation?

$$\frac{-27}{-9} = -3$$

Practice Problems 13–14

Divide, if possible.

13. $\frac{-1}{0}$ 14. $\frac{0}{-2}$

Answers

10. −4 **11.** 9 **12.** −6 **13.** undefined **14.** 0

Concept Check: $\frac{-27}{-9} = 3$

C Evaluating Expressions

Next, we practice evaluating expressions given integer replacement values.

Practice Problem 15

Evaluate xy for $x = 5$ and $y = -9$.

EXAMPLE 15 Evaluate xy for $x = -2$ and $y = 7$.

Solution: Recall that xy means $x \cdot y$.

Replace x with -2 and y with 7 in xy.

$$\begin{aligned} xy &= -2 \cdot 7 \\ &= -14 \end{aligned}$$

Practice Problem 16

Evaluate $\frac{x}{y}$ for $x = -9$ and $y = -3$.

EXAMPLE 16 Evaluate $\frac{x}{y}$ for $x = -24$ and $y = 6$.

Solution: $\frac{x}{y} = \frac{-24}{6}$ Replace x with -24 and y with 6.

$= -4$

D Solving Problems by Multiplying or Dividing Integers

Many real-life problems involve multiplication and division of integers.

Practice Problem 17

A card player had a score of -12 for each of four games. Find her total score.

EXAMPLE 17 Calculating Total Golf Score

A professional golfer finished seven strokes under par (-7) for each of three days of a tournament. What was his total score for the tournament?

Solution:

In words:	golfer's total score	=	number of days	·	score each day
	↓		↓		↓
Translate:	golfer's total score	=	3	·	(-7)
		=	-21		

The golfer's total score is -21, or 21 strokes under par.

Answers

15. -45 **16.** 3 **17.** -48

Name ______________________ Section __________ Date __________

EXERCISE SET 2.4

A *Multiply. See Examples 1 through 4.*

1. $-2(-3)$ **2.** $5(-3)$ **3.** $-4(9)$ **4.** $-7(-2)$

5. $8(-8)$ **6.** $-9(9)$ **7.** $0(-14)$ **8.** $-6(0)$

Multiply. See Example 5 through 7.

9. $6(-4)(2)$ **10.** $-2(3)(-7)$ **11.** $-1(-2)(-4)$ **12.** $8(-3)(3)$

13. $-4(4)(-5)$ **14.** $-2(-5)(-4)$ **15.** $10(-5)(0)$ **16.** $2(-1)(3)(-2)$

17. $-5(3)(-1)(-1)$ **18.** $3(0)(-4)(-8)$

Evaluate. See Examples 8 and 9.

19. -2^2 **20.** -2^4 **21.** $(-3)^3$ **22.** $(-1)^4$

23. -5^2 **24.** -4^3 **25.** $(-2)^3$ **26.** $(-3)^2$

B *Find each quotient. See Examples 10 through 14.*

27. $-24 \div 6$ **28.** $90 \div (-9)$ **29.** $\frac{-30}{6}$ **30.** $\frac{56}{-8}$

31. $\frac{-88}{-11}$ **32.** $\frac{-32}{4}$ **33.** $\frac{0}{14}$ **34.** $\frac{-13}{0}$

35. $\frac{2}{0}$ **36.** $\frac{0}{-5}$ **37.** $\frac{39}{-3}$ **38.** $\frac{-24}{-12}$

A B *Multiply or divide as indicated.*

39. $-12(0)$ **40.** $0(-100)$ **41.** $-4(3)$ **42.** $-6 \cdot 2$

43. $-9 \cdot 6$ **44.** $-12(13)$ **45.** $-7(-6)$ **46.** $-9(-5)$

 **47.** $-3(-4)(-2)$ **48.** $-7(-5)(-3)$ **49.** $(-4)^2$ **50.** $(-5)^2$

51. $-\frac{10}{5}$ **52.** $-\frac{25}{5}$ **53.** $-\frac{56}{8}$ **54.** $-\frac{49}{7}$

55. $-12 \div 3$

56. $-15 \div 3$

57. $4(-4)(-3)$

58. $6(-5)(-2)$

59. $-30(6)(-2)(-3)$

60. $-20 \cdot 5 \cdot (-5) \cdot (-3)$

61. $3 \cdot (-2) \cdot 0$

62. $-5(4)(0)$

63. $\frac{100}{-20}$

64. $\frac{45}{-9}$

65. $240 \div (-40)$

66. $480 \div (-8)$

67. $\frac{-12}{-4}$

68. $\frac{-36}{-3}$

69. -1^4

70. -2^3

71. $(-3)^5$

72. $(-9)^2$

73. $-2(3)(5)(-6)$

74. $-1(2)(7)(-3)$

75. $(-1)^{32}$

76. $(-1)^{33}$

77. $-2(-2)(-5)$

78. $-2(-2)(-3)(-2)$

79. $-42 \cdot 23$

80. $-56 \cdot 43$

81. $25 \cdot (-82)$

82. $70 \cdot (-23)$

C *Evaluate ab for the given replacement values. See Example 15.*

83. $a = -4$ and $b = 7$

84. $a = 5$ and $b = -1$

85. $a = 3$ and $b = -2$

86. $a = -9$ and $b = -6$

87. $a = -5$ and $b = -5$

88. $a = -8$ and $b = 8$

Evaluate $\frac{x}{y}$ for the given replacement values. See Example 16.

89. $x = 5$ and $y = -5$

90. $x = 9$ and $y = -3$

91. $x = -12$ and $y = 0$

92. $x = -10$ and $y = -10$

93. $x = -36$ and $y = -6$

94. $x = 0$ and $y = -5$

Evaluate xy and also $\frac{x}{y}$ for the given replacement values.

95. $x = -4$ and $y = -2$

96. $x = 20$ and $y = -5$

97. $x = 0$ and $y = -6$

98. $x = -3$ and $y = 0$

D *The graph shows melting points in degrees Celsius of selected elements. Use this graph to answer Exercises 99 through 102. See Example 17.*

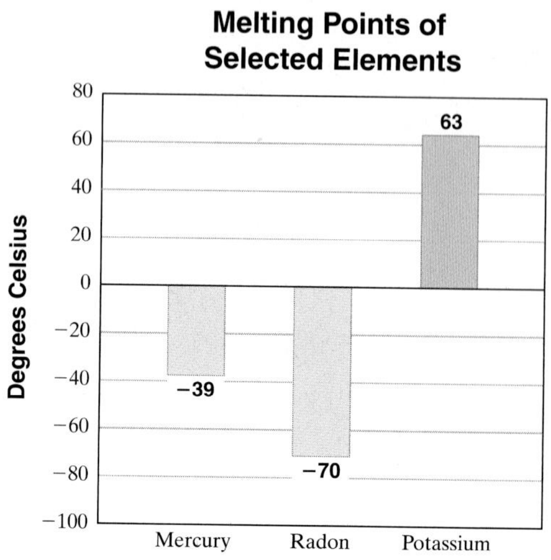

Melting Points of Selected Elements

	Degrees Celsius
Mercury	−39
Radon	−70
Potassium	63

99. The melting point of nitrogen is 3 times the melting point of radon. Find the melting point of nitrogen.

100. The melting point of rubidium is -1 times the melting point of mercury. Find the melting point of rubidium.

101. The melting point of argon is −3 times the melting point of potassium. Find the melting point of argon.

102. The melting point of strontium is −11 times the melting point of radon. Find the melting point of strontium.

Solve. See Example 17

103. A football team lost 4 yards on each of three consecutive plays. Represent the total loss as a product of integers, and find the total loss.

104. Joe Norstrom lost $400 on each of seven consecutive days in the stock market. Represent his total loss as a product of integers, and find his total loss.

105. A deep-sea diver must move up or down in the water in short steps to keep from getting a physical condition called the bends. Suppose the diver moves down from the surface in five steps of 20 feet each. Represent his movement as a product of integers, and find his final depth.

106. A weather forecaster predicts that the temperature will drop 5 degrees each hour for the next six hours. Represent this drop as a product of integers, and find the total drop in temperature.

107. For fiscal year 2001, Kmart reported a net loss of $2420 million. If this continues, what would Kmart's income be after three years (*Source*: Kmart Corporation)

108. In 2001, the earnings for a share of Kmart stock were approximately $5 per share. If an investor owned 15 shares of Kmart stock in 2001, what were his total earnings for these shares? (*Source*: Kmart corporation)

109. In 1987, there were only 27 California Condors in the entire world. In 2002, there were 192 California Condors. (*Source:* California Department of Fish and Game)

a. Find the change in the number of California Condors from 1987 to 2002.

b. Find the average change per year in the California Condor population over this period.

110. In 1997, music cassettes with a total value of $2227 million were produced by American music manufacturers. In 2001, this value had dropped to $467 million. (*Source:* Recording Industry Association of America)

a. Find the change in the value of music cassettes produced from 1997 to 2001.

b. Find the average change per year in the value of music cassettes produced over this period.

Recall that the average of a list of numbers is the sum of the numbers divided by how many numbers are in the list. Use this for Exercises 111 and 112, (We will review this again in Section 2.5.)

111. During the 2002 PGA Championship in Chaska, MN, Fred Funk had scores of −4, −2, 1 and 1 in four rounds of golf. Find his average score per tournament round. (*Source*: Professional Golf Association)

112. During the 2002 Senior PGA Championship, Fuzzy Zoeller had scores of −1, +1, and 0 for three rounds of golf. Find this average score per tournament round. (*Source*: Professional Golf Association)

Review and Preview

Perform each indicated operation. See Section 1.8.

113. $(3 \cdot 5)^2$

114. $(12 - 3)^2(18 - 10)$

115. $90 + 12^2 - 5^3$

116. $3 \cdot (7 - 4) + 2 \cdot 5^2$

117. $12 \div 4 - 2 + 7$

118. $12 \div (4 - 2) + 7$

Combining Concepts

Let a and b be positive numbers. Answer true or false for each statement.

119. $a(-b)$ is a negative number.

120. $(-a)(-b)$ is a negative number.

121. $(-a)(-a)$ is a positive number.

122. $(-a)(-a)(-a)$ is a positive number.

Let a and b be positive numbers. Determine the sign of each expression.

123. $(-a)^{30}$

124. $(-b)^{29}$

125. $(-a)^5(-b)^4$

126. $(-a)^7(-b)^5$

Without actually finding the product, write the list of numbers in Exercises 127 and 128 in order from least to greatest. For help, see a helpful hint box in this section.

127. $(-2)^{12}, (-2)^{17}, (-5)^{12}, (-5)^{17}$

128. $(-1)^{50}, (-1)^{55}, 0^{15}, (-7)^{20}, (-7)^{23}$

129. In 2001, there were 2190 commercial country music radio stations in the United States. By 2002, that number had declined to 2131. (*Source:* M Street Corporation)

a. Find the change in the number of country music radio stations from 2001 to 2002.

b. If this change continues, what will be the total change in the number of country music stations four years after 2002?

c. Based on your answer to part (b), how many country music radio stations will there be in 2006?

130. In 2000, Saturn sold a total of 260,899 cars. By 2001, the number of Saturns sold had increased to 275,425 (*Source:* Ward's Auto Info Bank)

a. Find the change in the number of Saturns sold from 2000 to 2001.

b. If this change continues, what will be the total change in the number of Saturns sold five years after 2001?

c. Based on your answer to part (b), how many Saturns would have been sold in 2006?

131. In your own words, explain how to multiply two integers.

132. In your own words, explain how to divide two integers.

133. $87^2 - (-12)^5$

134. $\dfrac{(-60 - 21)^3 - 9^4}{(-3)^6}$

Focus on Mathematical Connections

Modeling Adding and Subtracting Integers

Just as we used physical models to represent fractions and operations on whole numbers, we can use a physical model to help us add and subtract integers. In this model, we use objects called counters to represent numbers. Black counters • represent positive numbers, and red counters • represent negative numbers. The key to this model is remembering that taking a • and • together creates a neutral or zero pair. Once a neutral pair has been formed, it can be removed from or added to the model without changing the overall value.

Adding Integers

$5 + (-8)$

Begin with 5 black counters and add to that 8 red counters.

Next, form and remove neutral pairs.

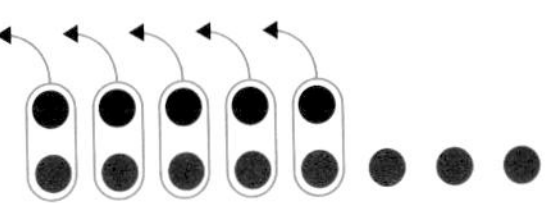

Now there are only 3 red counters left, so $5 + (-8) = -3$

Subtracting Integers

$6 - (-2)$

Begin with 6 black counters.

Subtracting a -2 indicates taking away 2 red counters. However, there are no red counters in the model at this point. Add enough neutral pairs to the model to obtain 2 red counters.

Now take away 2 red counters.

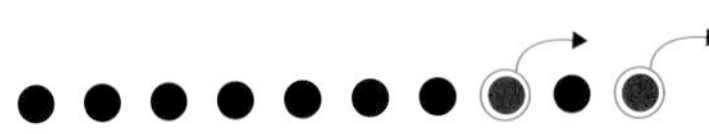

Because there are 8 black counters remaining, $6 - (-2) = 8$.

$-4 + (-3)$

Begin with 4 red counters and add to that 3 red counters.

Because this group does not contain a mixture of red and black counters, there is no need to form neutral pairs. Simply find the total number of red counters and remember that red counters represent negative numbers. So,

$$-4 + (-3) = -7$$

$-3 - 9$

Begin with 3 red counters.

Subtracting 9 indicates taking away 9 black counters, However, there are no black counters in the model at this point. Add enough neutral pairs to the model to obtain 9 black counters.

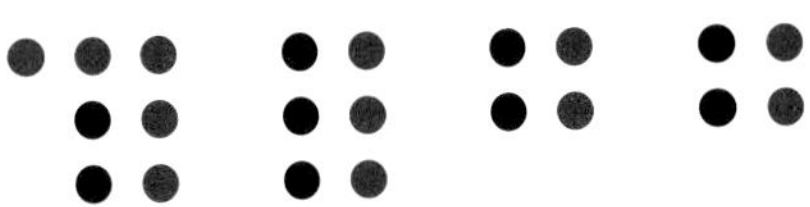

Now take away 9 black counters.

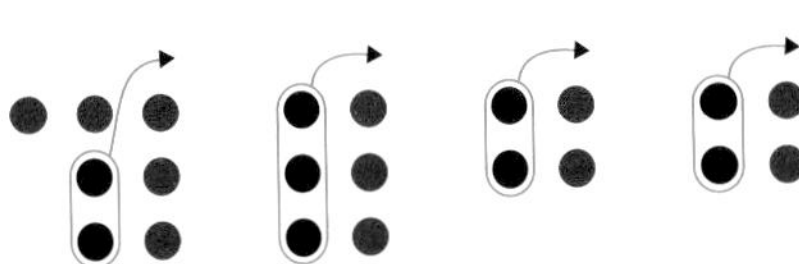

Because there are 12 red counters remaining, $-3 - 9 = -12$.

Critical Thinking

Use the counter model to perform each indicated operation.

1. $3 + 7$ **2.** $(-5) + 6$ **3.** $(-8) + (-4)$ **4.** $9 + (-11)$ **5.** $(-2) + 2$

6. $8 - 4$ **7.** $(-3) - (-1)$ **8.** $2 - (-5)$ **9.** $(-6) - 4$ **10.** $5 - 7$

STUDY SKILLS REMINDER

How are your homework assignments going?

It is so important in mathematics to keep up with homework. Why? Many concepts build on each other. Often, your understanding of a day's lecture in mathematics depends on an understanding of the previous day's material.

Remember that completing your homework assignment involves a lot more than attempting a few of the problems assigned.

To complete a homework assignment, remember these four things:

1. Attempt all of it.
2. Check it.
3. Correct it.
4. If needed, ask questions about it.

Focus on Business and Career

NET INCOME AND NET LOSS

For most businesses, a financial goal is to "make money." But what does this mean from a mathematical point of view? To find out, we must first discuss some common business terms.

- **Revenue** is the amount of money a business takes in. A company's annual revenue is the amount of money it collects during its fiscal, or business, year. For most companies, the largest source of revenue is from the sales of their products or services. For instance, a grocery store's annual revenue is the amount of money it collects during the year from selling groceries to customers. Large companies may also have revenues from interest or rentals.
- **Expenses** are the costs of doing business. For instance, a large part of a grocery store's expenses includes the cost of the food items it buys from wholesalers to resell to customers. Other expenses include salaries, mortgage payments, equipment, taxes, advertising, and so on.
- **Net income/loss** is the difference between a company's annual revenues and expenses. If the company's revenues are larger than its expenses, the difference is a positive number, and the company posts a net income for the year. Posting a net income can be interpreted as "making money." If the company's revenues are smaller than its expenses, the difference is a negative number, and the company posts a net loss for the year. Posting a net loss can be interpreted as "losing money."

Net income can also be thought of as profit. A negative profit is a net loss.

GROUP ACTIVITY

Search for corporate annual reports or articles in financial newspapers and magazines that report a company's net income or net loss. Describe what the income or loss means for the company. What are some of the factors that contributed to the net income or loss?

2.5 Order of Operations

OBJECTIVES

A. Simplify expressions by using the order of operations.

B. Evaluate an algebraic expression.

C. Find the average of a list of numbers.

A Simplifying Expressions

We first discussed the order of operations in Chapter 1. In this section, you are given an opportunity to practice using the order of operations when expressions contain integers. The rules for order of operations from Section 1.8 are repeated here.

If there are no other grouping symbols such as fraction bars or absolute values, perform operations in the following order.

Order of Operations

1. Do all operations within grouping symbols such as parentheses or brackets. (Start with the innermost set.)
2. Evaluate any expressions with exponents.
3. Multiply or divide in order from left to right.
4. Add or subtract in order from left to right.

EXAMPLES Find the value of each expression.

1. $(-3)^2 = (-3)(-3) = 9$ The base of the exponent is -3.
2. $-3^2 = -(3)(3) = -9$ The base of the exponent is 3.
3. $2 \cdot 5^2 = 2 \cdot (5 \cdot 5) = 2 \cdot 25 = 50$ The base of the exponent is 5.

Practice Problems 1–3

Find the value of each expression.

1. $(-2)^4$
2. -2^4
3. $3 \cdot 6^2$

Helpful Hint

When simplifying expressions with exponents, remember that parentheses make an important difference.

$(-3)^2$ and -3^2 **do not** mean the same thing.

$(-3)^2$ means $(-3)(-3) = 9$.

-3^2 means the opposite of $3 \cdot 3$, or -9.

Only with parentheses around it is the -3 squared.

Spend time studying Examples 1 through 3 and the Helpful Hint above. It is important to be able to determine the base of an exponent. After reading these examples, if you have trouble completing the Practice Problems, review this material again.

EXAMPLE 4 Simplify: $\dfrac{-6(2)}{-3}$

Solution: The fraction bar serves as a grouping symbol. First we multiply -6 and 2. Then we divide.

$$\frac{-6(2)}{-3} = \frac{-12}{-3}$$
$$= 4$$

Practice Problem 4

Simplify: $\dfrac{25}{5(-1)}$

EXAMPLE 5 Simplify: $\dfrac{12 - 16}{-1 + 3}$

Solution: We simplify above and below the fraction bar separately. Then we divide.

$$\frac{12 - 16}{-1 + 3} = \frac{-4}{2}$$
$$= -2$$

Practice Problem 5

Simplify: $\dfrac{-18 + 6}{-3 - 1}$

Answers

1. 16 **2.** −16 **3.** 108 **4.** −5 **5.** 3

Practice Problem 6

Simplify: $-20 + 2 \cdot 7 + 4$

Practice Problem 7

Simplify: $-2^3 + (-4)^2 + 1^5$

Practice Problem 8

Simplify: $2(2 - 8) + (-12) - 3$

Practice Problem 9

Simplify: $(-5) \cdot |-4| + (-3) + 2^3$

Practice Problem 10

Simplify: $4(-6) \div [3(5 - 7)^2]$

Concept Check

True or false? Explain your answer. The result of

$-4 \cdot (3 - 7) - 8 \cdot (9 - 6)$

is positive because there are four negative signs.

Practice Problem 11

Evaluate x^2 and $-x^2$ for $x = -12$.

Answers

6. -2 **7.** 9 **8.** -27 **9.** -15
10. -2 **11.** 144; -144

Concept Check: False; $-4 \cdot (3 - 7) - 8 \cdot (9 - 6) = -8$

EXAMPLE 6 Simplify: $3 + 4 \cdot 5 - 27$

Solution: Follow the order of operations.

$$\begin{aligned} 3 + 4 \cdot 5 - 27 &= 3 + 20 - 27 && \text{Multiply.} \\ &= 23 - 27 && \text{Add or subtract from left to right.} \\ &= -4 && \text{Subtract.} \end{aligned}$$

EXAMPLE 7 Simplify: $-4^2 + (-3)^2 - 1^3$

Solution: Follow the order of operations.

$$\begin{aligned} -4^2 + (-3)^2 - 1^3 &= -16 + 9 - 1 && \text{Simplify expressions with exponents.} \\ &= -7 - 1 && \text{Add or subtract from left to right.} \\ &= -8 \end{aligned}$$

EXAMPLE 8 Simplify: $3(4 - 7) + (-2) - 5$

Solution: Follow the order of operations.

$$\begin{aligned} 3(4 - 7) + (-2) - 5 &= 3(-3) + (-2) - 5 && \text{Simplify inside parentheses.} \\ &= -9 + (-2) - 5 && \text{Multiply.} \\ &= -11 - 5 && \text{Add or subtract from left to right.} \\ &= -16 \end{aligned}$$

EXAMPLE 9 Simplify: $(-3) \cdot |-5| - (-2) + 4^2$

Solution: Follow the order of operations.

$$\begin{aligned} (-3) \cdot |-5| - (-2) + 4^2 &= (-3) \cdot 5 - (-2) + 4^2 && \text{Write } |-5| \text{ as 5.} \\ &= (-3) \cdot 5 - (-2) + 16 && \text{Write } 4^2 \text{ as 16.} \\ &= -15 - (-2) + 16 && \text{Multiply.} \\ &= -13 + 16 && \text{Add or subtract from left to right.} \\ &= 3 \end{aligned}$$

EXAMPLE 10 Simplify: $-2[-3 + 2(-1 + 6)] - 5$

Solution: By the order of operations, we begin with the innermost set of parentheses.

$$\begin{aligned} -2[-3 + 2(-1 + 6)] - 5 &= -2[-3 + 2(5)] - 5 && \text{Write } -1 + 6 \text{ as 5.} \\ &= -2[-3 + 10] - 5 && \text{Multiply.} \\ &= -2(7) - 5 && \text{Add.} \\ &= -14 - 5 && \text{Multiply.} \\ &= -19 && \text{Subtract.} \end{aligned}$$

Try the Concept Check in the margin.

B Evaluating Expressions

Now we practice evaluating expressions.

EXAMPLE 11 Evaluate x^2 and $-x^2$ for $x = -11$.

Solution: $x^2 = (-11)^2 = (-11)(-11) = 121$

$-x^2 = -(-11)^2 = -(-11)(-11) = -121$

EXAMPLE 12 Evaluate $6z^2$ for $z = 2$ and $z = -2$.

Solution: $6z^2 = 6(2)^2 = 6(4) = 24$

$6z^2 = 6(-2)^2 = 6(4) = 24$

Practice Problem 12

Evaluate $5y^2$ for $y = 3$ and $y = -3$.

EXAMPLE 13 Evaluate $x + 2y - z$ for $x = 3$ and $y = -5$ and $z = -4$.

Solution: Replace x with 3, y with -5, z with -4 and simplify.

$x + 2y - z = 3 + 2(-5) - (-4)$ Let $x = 3$, $y = -5$, and $z = -4$.

$= 3 + (-10) + 4$ Replace $2(-5)$ with its product, -10.

$= -3$ Add.

Practice Problem 13

Evaluate $x^2 + y$ for $x = -5$ and $y = -2$.

Helpful Hint:

Remember to rewrite the subtraction sign.

EXAMPLE 14 Evaluate $7 - x^2$ for $x = -4$.

Solution: Replace x with -4 and simplify carefully!

$7 - x^2 = 7 - (-4)^2$

$= 7 - 16$ $(-4)^2 = (-4)(-4) = 16$

$= -9$ Subtract.

Practice Problem 14

Evaluate $4 - x^2$ for $x = -9$.

C Finding Averages

Recall from Chapter 1 that the average of a list of numbers is

$$\text{average} = \frac{\text{sum of numbers}}{\textit{number}\text{ of numbers}}$$

EXAMPLE 15 The graph shows the monthly normal temperatures for Barrow, Alaska. Use this graph to find the average of the temperatures for months January through May.

Practice Problem 15

Find the average the temperatures for the months October through April.

Monthly Normal Temperatures for Barrow, Alaska

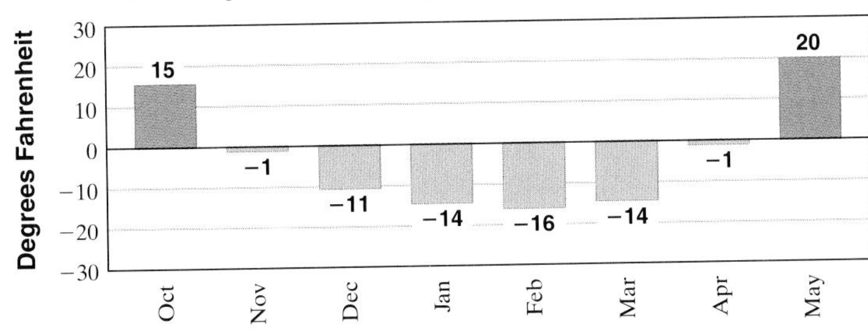

Solution: By reading the graph, we have

$$\text{average} = \frac{-14 + (-16) + (-14) + (-1) + 20}{5}$$ There are 5 months from January through May.

$$= \frac{-25}{5}$$

$$= -5$$

The average for the temperatures is $-5°$F.

Answers

12. 45; 45 **13.** 23 **14.** −77 **15.** −6.

CALCULATOR EXPLORATIONS

Simplifying an Expression Containing a Fraction Bar

Recall that even though most calculators follow the order of operations, parentheses must sometimes be inserted. For example, to simplify $\frac{-8 + 6}{-2}$ on a calculator, enter parentheses about the expression above the fraction bar so that it is simplified separately.

To simplify $\frac{-8 + 6}{-2}$, press the keys

[( ] [8] [+/−] [+] [6] [ )] [÷] [2] [+/−] [=] or

[( ] [(−)] [8] [+] [6] [ )] [÷] [(−)] [2] [ENTER]

The display will read [1].

Thus $\frac{-8 + 6}{-2} = 1$.

Use a calculator to simplify.

1. $\frac{-120 - 360}{-10}$

2. $\frac{4750}{-2 + (-17)}$

3. $\frac{-316 + (-458)}{28 + (-25)}$

4. $\frac{-234 + 86}{-18 + 16}$

STUDY SKILLS REMINDER

Are you prepared for a test on Chapter 2?

Below I have listed some *common trouble areas* for topics covered in Chapter 2. After studying for your test—but before taking your test—read these.

■ Don't forget the difference between $-(-5)$ and $-|-5|$.

$-(-5) = 5$ The opposite of −5 is 5.

$-|-5| = -5$ The opposite of the absolute value of −5 is the opposite of 5, which is −5

■ Remember how to simplify $(-7)^2$ and -7^2.

$(-7)^2 = (-7)(-7) = 49$

$-7^2 = -(7)(7) = -49$

■ Don't forget the order of operations.

$$\begin{aligned} 1 + 3(4 - 6) &= 1 + 3(-2) && \text{Simplify inside parentheses.} \\ &= 1 + (-6) && \text{Multiply.} \\ &= -5 && \text{Add.} \end{aligned}$$

Remember: This is simply a checklist of common trouble spots. For a review of Chapter 2, see the Highlights and Chapter Review at the end of this chapter.

Name ______________________ Section __________ Date __________

Mental Math

Identify the base and exponent of each expression. Do not simplify.

1. -3^2

2. $(-3)^2$

3. $4 \cdot 2^3$

4. $9 \cdot 5^6$

5. $(-7)^5$

6. -9^4

7. $5^7 \cdot 10$

8. $2^8 \cdot 11$

EXERCISE SET 2.5

 *Simplify. See Examples 1 through 10.*

1. $(-4)^3$

2. -2^4

3. -4^3

4. $(-2)^4$

5. $6 \cdot 2^2$

6. $5 \cdot 2^3$

7. $-1(-2) + 1$

8. $3 + (-8) \div 2$

9. $9 - 12 - 4$

10. $10 - 23 - 12$

11. $4 + 3(-6)$

12. $-8 + 4(3)$

13. $5(-9) + 2$

14. $7(-6) + 3$

15. $(-10) + 4 \div 2$

16. $(-12) + 6 \div 3$

17. $6 + 7 \cdot 3 - 40$

18. $5 + 9 \cdot 4 - 52$

19. $\dfrac{16 - 13}{-3}$

20. $\dfrac{20 - 15}{-1}$

21. $\dfrac{24}{10 + (-4)}$

22. $\dfrac{88}{-8 - 3}$

23. $5(-3) - (-12)$

24. $7(-4) - (-6)$

25. $(-19) - 12(3)$

26. $(-24) - 14(2)$

27. $-8 + 4^2$

28. $-12 + 3^3$

29. $[8 + (-4)]^2$

30. $[9 + (-2)]^3$

31. $8 \cdot 6 - 3 \cdot 5 + (-20)$

32. $7 \cdot 6 - 6 \cdot 5 + (-10)$

33. $16 - (-3)^4$

34. $20 - (-5)^2$

35. $|5 + 3| \cdot 2^3$

36. $|-3 + 7| \cdot 7^2$

37. $7 \cdot 8^2 + 4$

38. $10 \cdot 5^3 + 7$

39. $5^3 - (4 - 2^3)$

40. $8^2 - (5 - 2)^4$

41. $|3 - 12| \div 3$

42. $|12 - 19| \div 7$

43. $-(-2)^2$

44. $-(-2)^3$

45. $(5 - 9)^2 \div (4 - 2)^2$

46. $(2 - 7)^2 \div (4 - 3)^4$

47. $|8 - 24| \cdot (-2) \div (-2)$

48. $|3 - 15| \cdot (-4) \div (-16)$

49. $(-12 - 20) \div 16 - 25$

50. $(-20 - 5) \div 5 - 15$

51. $5(5 - 2) + (-5)^2 - 6$

52. $3 \cdot (8 - 3) + (-4) - 10$

53. $(2 - 7) \cdot (6 - 19)$

54. $(4 - 12) \cdot (8 - 17)$

55. $2 - 7 \cdot 6 - 19$

56. $4 - 12 \cdot 8 - 17$

57. $(-36 \div 6) - (4 \div 4)$

58. $(-4 \div 4) - (8 \div 8)$

59. $-5^2 - 6^2$

60. $-4^4 - 5^4$

61. $(-5)^2 - 6^2$

62. $(-4)^4 - (5)^4$

63. $(10 - 4^2)^2$

64. $(11 - 3^2)^3$

65. $2(8 - 10)^2 - 5(1 - 6)^2$

66. $-3(4 - 8)^2 + 5(14 - 16)^3$

67. $3(-10) \div [5(-3) - 7(-2)]$

68. $12 - [7 - (3 - 6)] + (2 - 3)^3$

69. $\dfrac{(-7)(-3) - (4)(3)}{3[7 \div (3 - 10)]}$

70. $\dfrac{10(-1) - (-2)(-3)}{2[-8 \div (-2 - 2)]}$

71. $-5[4 + 5(-3 + 5)] + 11$

72. $-2[1 + 3(7 - 12)] - 35$

B *Evaluate each expression for $x = -2$, $y = 4$, and $z = -1$. See Examples 11 through 14.*

73. $x + y + z$

74. $x - y - z$

75. $2x - 3y - 4z$

76. $5x - y + 4z$

77. $x^2 - y$

78. $x^2 + z$

79. $\dfrac{5y}{z}$

80. $\dfrac{4x}{y}$

Evaluate each expression for $x = -3$ and $z = -4$. See Examples 11 through 14.

81. x^2

82. z^2

83. $-z^2$

84. $-x^2$

85. $10 - x^2$

86. $3 - z^2$

87. $2x^3 - z$

88. $3z^2 - x$

C *Find the average of each list of numbers. See Example 15.*

89. $-10, 8, -4, 2, 7, -5, -12$

90. $-18, -8, -1, -1, 0, 4$

91. $-17, -26, -20, -13$

92. $-40, -20, -10, -15, -5$

Scores in golf can be 0 (also called par), a positive integer (also called above par), or a negative integer (also called below par). On the next page are scores of selected golfers from the 2003 Memorial Tournament in Dublin, Ohio. Use this graph for Exercises 93 through 98.

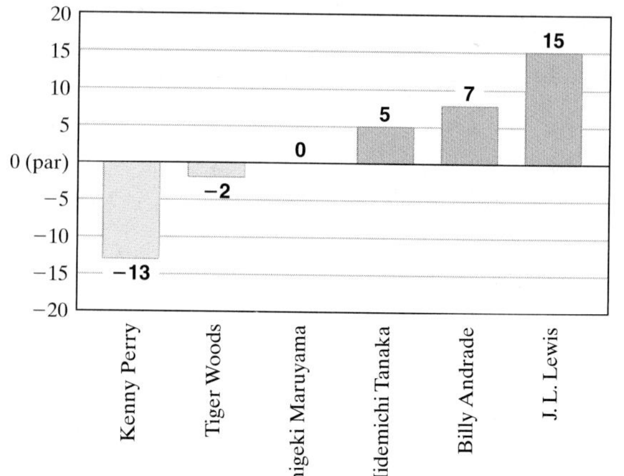

93. Find the difference between the lowest score shown and the highest score shown.

94. Find the difference between the two lowest scores.

95. Find the average of the scores shown. (*Hint*: Here, the average is the sum of the scores divided by the number of players.)

96. Find the average of the scores for Perry, Woods, Maruyama, and Lewis.

97. Can the average for these scores be greater than the highest score, 15? Explain why or why not.

98. Can the average of all the scores shown be less than the lowest score, −13. Explain why or why not.

Review and Preview

Perform each indicated operation. See Sections 1.3, 1.4, 1.6, and 1.7.

99. $45 \cdot 90$

100. $90 \div 45$

101. $90 - 45$

102. $45 + 90$

Find the perimeter of each figure. See Section 1.2.

103. Square

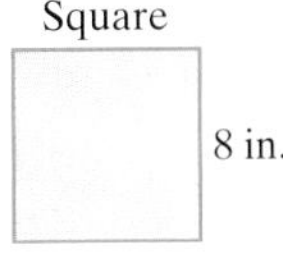

104. Parallelogram

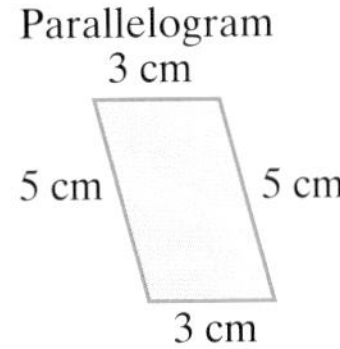

105. Rectangle

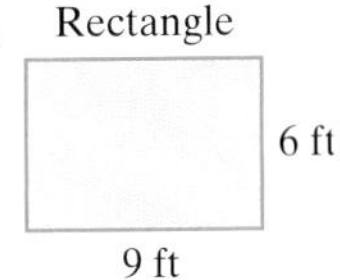

106.

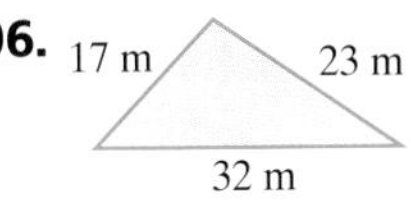

Combining Concepts

Insert parentheses where needed so that each expression evaluates to the given number.

107. $2 \cdot 7 - 5 \cdot 3$; evaluates to 12

108. $7 \cdot 3 - 4 \cdot 2$; evaluates to 34

109. $-6 \cdot 10 - 4$; evaluates to -36

110. $2 \cdot 8 \div 4 - 20$; evaluates to -36

Evaluate.

111. $(-12)^4$

112. $(-17)^5$

113. $x^3 - y^2$ for $x = 21$ and $y = -19$

114. $3x^2 + 2x - y$ for $x = -18$ and $y = 2868$

115. $(xy + z)^x$ for $x = 2$, $y = -5$, and $z = 7$

116. $5(ab + 3)^b$ for $a = -2$, $b = 3$

117. Discuss the effect parentheses have in an exponential expressions. For example, what is the difference between $(-6)^2$ and -6^2.

118. Discuss the effect parentheses have in an exponential expressions. For example, what is the difference between $(2 \cdot 4)^2$ and $2 \cdot 4^2$?

CHAPTER 2 ACTIVITY Investigating Positive and Negative Numbers

MATERIALS:

- colored thumbtacks
- coin
- cardboard
- six-sided die
- tape

Work with a partner or a small group to try the following activity. The object is to have the largest absolute value at the end of the game.

1. Attach this page to a piece of cardboard with tape. Each person should choose a different colored thumbtack as his or her playing piece. Insert each thumbtack at the starting place 0 on the number line.

2. Each player takes a turn as follows: Roll the die and flip the coin. "Heads" on the coin makes the number that lands faceup on the die positive. "Tails" on the coin makes the number that lands faceup on the die negative. Record your number along with its sign (positive or negative) in the table below. Move your thumbtack on the number line according to your number.

3. Continue taking turns, until each person has taken five turns. Verify your final position on the number line by finding the total of the integers in the table. Do your total and final position agree?

	Positive or Negative Number
Turn 1	
Turn 2	
Turn 3	
Turn 4	
Turn 5	
Total	

4. The winner is the player having the final position with the largest absolute value. Find the absolute value of your final position. Create a table listing the absolute values of each person's final position. Who won? How could you tell who won just by looking at the number line?

5. Many board games include instructions for moving playing pieces forward or backward. Make a list of games that include such instructions. Then explain how these instructions for moving forward or backward are related to positive and negative numbers.

Chapter 2 Vocabulary Check

Fill in each blank with one of the words or phrases listed below.

opposites	integers	positive integers
signed	absolute value	negative integers

1. The ______________ are integers less than 0.
2. Two numbers that are the same distance from 0 on the number line but are on opposite sides of 0 are called _________.
3. Together, positive numbers, negative numbers, and 0 are called _______ numbers.
4. The ______________ are integers greater than 0.
5. The ______________ of a number is that number's distance from 0 on the number line.
6. The ________ are $\ldots, -3, -2, -1, 0, 1, 2, 3, \ldots$.

Highlights

DEFINITIONS AND CONCEPTS | **EXAMPLES**

Section 2.1 Introduction to Integers

The **integers** are $\ldots, -3, -2, -1, 0, 1, 2, 3, \ldots$.

Integers:

$-432, \quad -10, \quad 0, \quad 15$

The **absolute value** of a number is that number's distance from 0 on the number line. The symbol for absolute value is $|\ |$.

$|-2| = 2$ (2 units)

$|2| = 2$ (2 units)

Two numbers that are the same distance from 0 on the number line but are on opposite sides of 0 are called **opposites**.

5 and -5 are opposites.

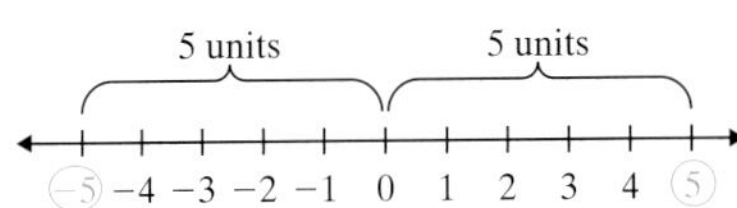

If a is a number, then $-(-a) = a$.

$-(-11) = 11, \quad -|-3| = -3$

Section 2.2 Adding Integers

Adding Two Numbers with the Same Sign

Step 1. Add their absolute values.

Step 2. Use their common sign as the sign of the sum.

Add:

$-3 + (-2) = -5$

$-7 + (-15) = -22$

Adding Two Numbers with Different Signs

Step 1. Find the larger absolute value minus the smaller absolute value.

Step 2. Use the sign of the number with the larger absolute value as the sign of the sum.

$-6 + 4 = -2$

$17 + (-12) = 5$

$-32 + (-2) + 14 = -34 + 14$

$= -20$

DEFINITIONS AND CONCEPTS	EXAMPLES
SECTION 2.3 SUBTRACTING INTEGERS	
SUBTRACTING TWO NUMBERS If a and b are numbers, then $a - b = a + (-b)$.	Subtract: $-35 - 4 = -35 + (-4) = -39$ $3 - 8 = 3 + (-8) = -5$ $-10 - (-12) = -10 + 12 = 2$ $7 - 20 - 18 - (-3) = 7 + (-20) + (-18) + (+3)$ $= -13 + (-18) + 3$ $= -31 + 3$ $= -28$
SECTION 2.4 MULTIPLYING AND DIVIDING INTEGERS	
MULTIPLYING NUMBERS The product of two numbers having the same sign is a positive number. The product of two numbers having unlike signs is a negative number.	Multiply: $(-7)(-6) = 42$ $9(-4) = -36$ Evaluate: $(-3)^2 = (-3)(-3) = 9$
DIVIDING NUMBERS The quotient of two numbers having the same sign is a positive number. The quotient of two numbers having unlike signs is a negative number.	Divide: $-100 \div (-10) = 10$ $\frac{14}{-2} = -7,\quad \frac{0}{-3} = 0,\quad \frac{22}{0}$ is undefined.
SECTION 2.5 ORDER OF OPERATIONS	
ORDER OF OPERATIONS 1. Do all operations within grouping symbols such as parentheses or brackets. 2. Evaluate any expressions with exponents. 3. Multiply or divide in order from left to right. 4. Add or subtract in order from left to right.	Simplify: $3 + 2 \cdot (-5) = 3 + (-10)$ Multiply. $= -7$ Add. $\frac{-2(5-7)}{-7+\lvert -3\rvert} = \frac{-2(-2)}{-7+3}$ $= \frac{4}{-4}$ $= -1$ $-3 + 5[2^3 - (12 - 9)] = -3 + 5[2^3 - 3]$ Subtract. $= -3 + 5[8 - 3]$ Evaluate exponent. $= -3 + 5(5)$ Subtract. $= -3 + 25$ Multiply. $= 22$ Add.

Name ________________ Section __________ Date __________

Chapter 2 Review

The map below shows selected cities and their normal high and low temperatures. Use this map as indicated throughout the review to fill-in each missing temperature in the picture.

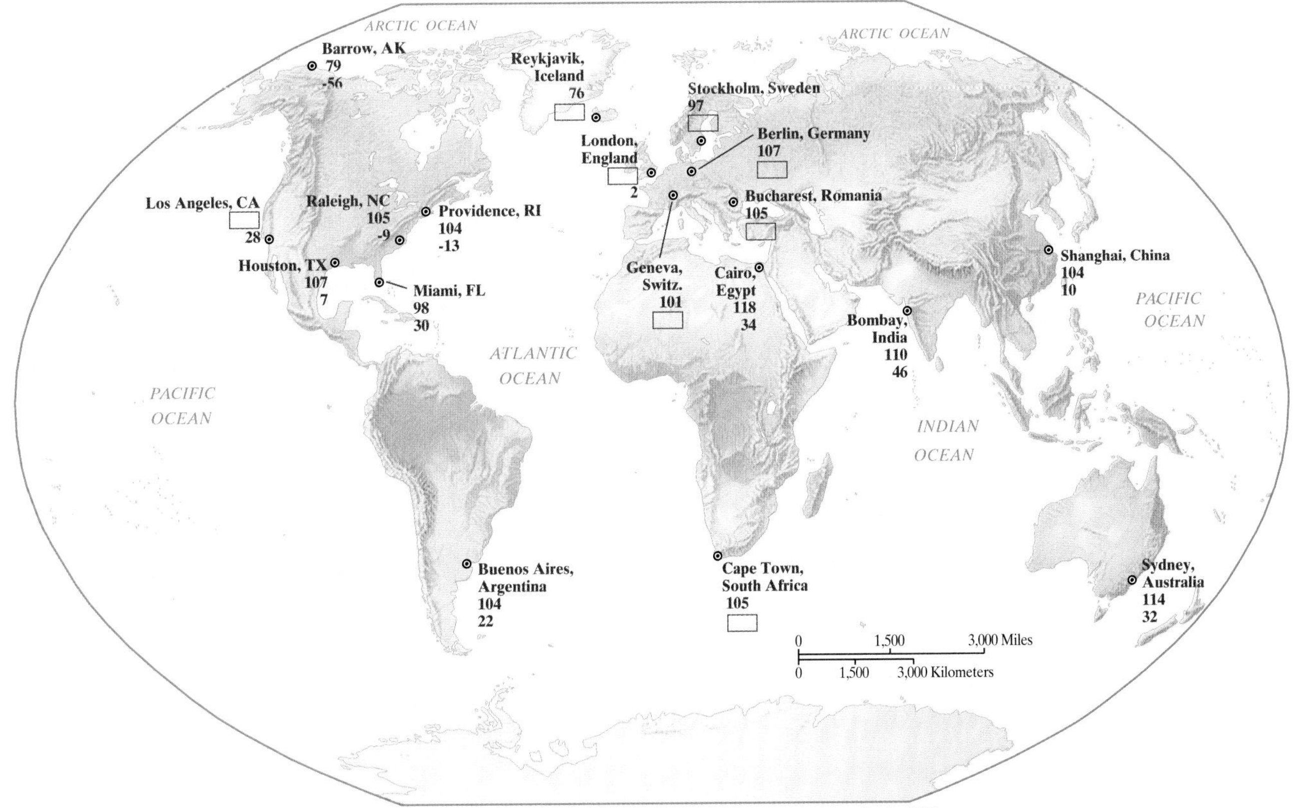

Source: World Almanac 2003; temperatures are for a 30-year period ending in 1990.

Extreme High and Low Temperatures for Selected Locations (in degrees Fahrenheit)					
	Max	**Min**		**Max**	**Min**
Berlin, Germany	107		Barrow, Alaska	79	−56
Raleigh, NC	105	−9	London, England		2
Houston, TX	107	7	Cairo, Egypt	118	34
Miami, FL	98	30	Sydney, Australia	114	32
Los Angeles, CA		28	Shanghai, China	104	10
Bucharest, Romania	105		Reykjavik, Iceland	76	
Geneva, Switzerland	101		Capetown, South Africa	105	
Providence, RI	104	−13	Buenos Aires, Argentina	104	22
Stockholm, Sweden	97		Bombay, India	110	46

(2.1) *Represent each quantity by an integer.*

1. A gold miner is working 1435 feet down in a mine.

2. A mountain peak is 7562 meters above sea level.

Graph each integer in the list on the same number line.

3. $-2, -5, 0, 5$

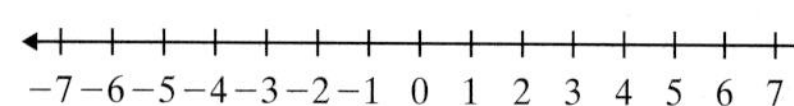

4. $-7, -1, 0, 7$

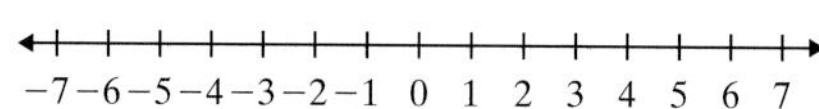

Simplify.

5. $|-12|$

6. $|0|$

7. $-|6|$

8. $-(-9)$

9. $-|-9|$

10. $-(-2)$

Insert $<$ or $>$ between each pair of integers to make a true statement.

11. $-18 \quad -20$

12. $-5 \quad 5$

13. $|-123| \quad -|-198|$

14. $8 - |-12| \quad -|-16|$

Find the opposite of each integer.

15. -12

16. $-(-3)$

Answer true or false for each statement.

17. If $a < b$, then a must be a negative number.

18. The absolute value of an integer is always 0 or a positive number.

19. A negative number is always less than a positive number.

20. If a is a negative number, then $-a$ is a positive number.

(2.2) *Add.*

21. $5 + (-3)$

22. $18 + (-4)$

23. $-12 + 16$

24. $-23 + 40$

25. $-8 + (-15)$

26. $-5 + (-17)$

27. $-24 + 3$

28. $-89 + 19$

29. $15 + (-15)$

30. $-24 + 24$

31. $-43 + (-108)$

32. $-100 + (-506)$

33. During the 2002 Kellogg-Keebler Classic, the winner, Annika Sorenstam had scores of -9, -5, and -7 for 3 rounds of golf. Find her total score (*Source*: Ladies Professional Golf Association)

34. During the 2002 LPGA Corning Classic, Joanne Morley had scores of -1, 3, -5, and 3. Find her total score. (*Source*: Ladies Professional Golf Association)

35. The temperature at 5 a.m. on a day in January was $-15°C$. By 6 a.m. the temperature had fallen 5 degrees. Find the temperature at 6 a.m.

36. A diver starts out at 127 feet below the surface and then swims downward another 23 feet. Find the diver's current depth.

For Exercises 37 and 38, use the map at the beginning of this review.

37. The high temperature for London, England is 155 degrees greater than the low temperature for Barrow, Alaska. Find the high temperature for London.

38. The high temperature for Los Angeles, California is 125 degrees greater than the low temperature for Providence, Rhode Island. Find the high temperature for Los Angeles.

(2.3) *Subtract.*

39. $12 - 4$

40. $-12 - 4$

41. $8 - 19$

42. $-8 - 19$

43. $7 - (-13)$

44. $-6 - (-14)$

45. $16 - 16$

46. $-16 - 16$

47. $-12 - (-12)$

48. $|-5| - |-12|$

49. $-(-5) - 12 + (-3)$

50. $-8 + |-12| - 10 - |-3|$

Solve.

51. Josh Weidner has \$142 in his checking account. He writes a check for \$125, makes a deposit for \$43, and then writes another check for \$85. Represent the balance in his account by an integer.

52. If the elevation of Lake Superior is 600 feet above sea level and the elevation of the Caspian Sea is 92 feet below sea level, find the difference of the elevations.

For exercises 53 and 54, use the map at the beginning of this review.

53. The low temperature for Reykjavik is 35 degrees less than the low temperature for Sydney, Australia Find the low temperature for Reykjavik.

54. The low temperature for Berlin, Germany is 14 degrees less than the low temperature for Shanghai, China. Find the low temperature for Berlin.

Answer true or false for each statement.

55. $|-5| - |-6| = 5 - 6$

56. $|-5 - (-6)| = 5 + 6$

57. If $b > a$, then $b - a$ is a positive number.

58. If $b < a$, then $b - a$ is a negative number.

(2.4) *Multiply.*

59. $-3(-7)$

60. $-6(3)$

61. $-4(16)$

62. $-5(-12)$

63. $(-5)^2$

64. $(-1)^5$

65. $12(-3)(0)$

66. $-1(6)(2)(-2)$

Divide.

67. $-15 \div 3$

68. $\frac{-24}{-8}$

69. $\frac{0}{-3}$

70. $\frac{-46}{0}$

71. $\frac{100}{-5}$

72. $\frac{-72}{8}$

73. $\frac{-38}{-1}$

74. $\frac{45}{-9}$

75. A football team lost 5 yards on each of two consecutive plays. Represent the total loss by a product of integers, and find the product.

76. A race horse bettor loss \$50 on each of four consecutive races. Represent the total loss by a product of integers, and find the product.

For exercises 77 through 80, use the map at the beginning of this review.

77. The low temperature for Bucharest, Romania is 2 times the low temperature for Raleigh, North Carolina. Find the low temperature for Bucharest.

78. The low temperature for Geneva, Switzerland is the same as the low temperature for Miami, Florida divided by -10. Find the low temperature for Geneva.

79. The low temperature for Capetown, South Africa is the same as the low temperature for Barrow, Alaska divided by -2. Find the low temperature for Capetown.

80. The low temperature for Stockholm, Sweden, is 2 times the low temperature for Providence, Rhode Island. Find the low temperature for Stockholm.

(2.5) *Simplify.*

81. $(-7)^2$

82. -7^2

83. -2^5

84. $(-2)^5$

85. $5 - 8 + 3$

86. $-3 + 12 + (-7) - 10$

87. $-10 + 3 \cdot (-2)$

88. $5 - 10 \cdot (-3)$

89. $16 \cdot (-2) + 4$

90. $3 \cdot (-12) - 8$

91. $5 + 6 \div (-3)$

92. $-6 + (-10) \div (-2)$

93. $16 + (-3) \cdot 12 \div 4$

94. $(-12) + 25 \cdot 1 \div (-5)$

95. $4^3 - (8 - 3)^2$

96. $4^3 - 90$

97. $-(-4) \cdot |-3| - 5$

98. $|5 - 1|^2 \cdot (-5)$

99. $\dfrac{(-4)(-3) - (-2)(-1)}{-10 + 5}$

100. $\dfrac{4(12 - 18)}{-10 \div (-2 - 3)}$

Find the average of each list of numbers.

101. $-18, 25, -30, 7, 0, -2$

102. $-45, -40, -30, -25$

Evaluate each expression for $x = -2$ and $y = 1$.

103. $2x - y$

104. $y^2 + x^2$

105. $\dfrac{3x}{6}$

106. $\dfrac{5y - x}{-y}$

107. x^2

108. $-x^2$

109. $7 - x^2$

110. $100 - x^3$

Name ______________________ Section __________ Date __________

Chapter 2 Test

Remember to check your answers and use the Chapter Test Prep Video to view solutions.

Simplify each expression.

1. $-5 + 8$

2. $18 - 24$

3. $5 \cdot (-20)$

4. $(-16) \div (-4)$

5. $(-18) + (-12)$

6. $-7 - (-19)$

7. $(-5) \cdot (-13)$

8. $\frac{-25}{-5}$

9. $|-25| + (-13)$

10. $14 - |-20|$

11. $|5| \cdot |-10|$

12. $\frac{|-10|}{-|-5|}$

13. $(-8) + 9 \div (-3)$

14. $-7 + (-32) - 12 + 5$

15. $(-5)^3 - 24 \div (-3)$

16. $(5 - 9)^2 \cdot (8 - 2)^3$

17. $-(-7)^2 \div 7 \cdot (-4)$

18. $3 - (8 - 2)^3$

Answers

1. ________

2. ________

3. ________

4. ________

5. ________

6. ________

7. ________

8. ________

9. ________

10. ________

11. ________

12. ________

13. ________

14. ________

15. ________

16. ________

17. ________

18. ________

19. ______________

20. ______________

21. ______________

22. ______________

23. ______________

24. ______________

25. ______________

26. ______________

27. ______________

28. ______________

29. ______________

30. ______________

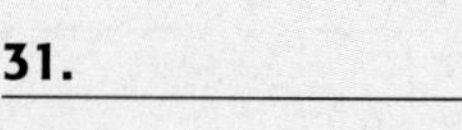

31. ______________

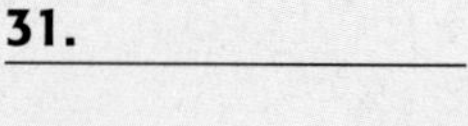

32. ______________

33. ______________

34. ______________

35. ______________

19. $-6 + (-15) \div (-3)$

20. $\frac{4}{2} - \frac{8^2}{16}$

21. $\frac{-3(-2) + 12}{-1(-4 - 5)}$

22. $\frac{|25 - 30|^2}{2(-6) + 7}$

23. $5(-8) - [6 - (2 - 4)] + (12 - 16)^2$

24. $-2^3 - 2^2$

Evaluate each expression for $x = 0$, $y = -3$, *and* $z = 2$.

25. $3x + y$

26. $|y| + |x| + |z|$

27. $\frac{3z}{2y}$

28. $2y^3$

29. $10 - y^2$

30. $7x + 3y - 4z$

31. Mary Dunstan, a diver, starts at sea level and then makes 4 successive descents of 22 feet. After the descents, what is her elevation?

32. Aaron Hawn has $129 in his checking account. He writes a check for $79, withdraws $40 from an ATM, and then deposits $35. Represent the new balance in his account by an integer.

33. Mt. Washington in New Hampshire has an elevation of 6288 feet above sea level. The Romanche Gap in the Atlantic Ocean has an elevation of 25,354 feet below sea level. Represent the difference in elevation between these two points by an integer. (*Source:* National Geographic Society and Defense Mapping Agency)

34. Lake Baykal in Siberian Russia is the deepest lake in the world with a maximum depth of 5315 feet. The elevation of the lake's surface is 1495 feet above sea level. What is the elevation (with respect to sea level) of the deepest point in the lake? (*Source:* U.S. Geological Survey)

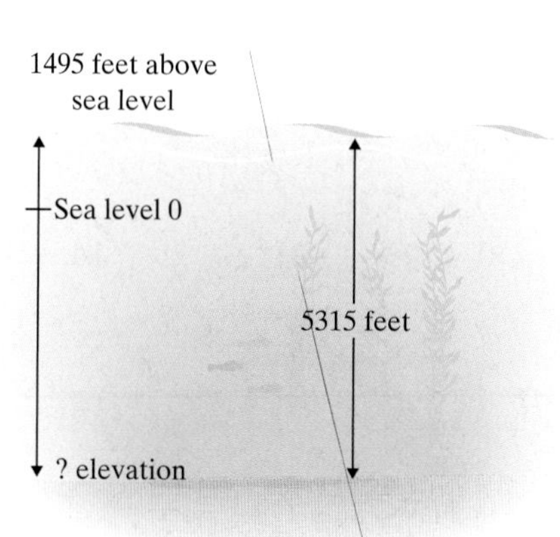

35. Find the average of $-12, -13, 0, 9$.

Name ______________________ Section ____________ Date ____________

Chapter 2 Cumulative Review

Find the place value of the digit 4 in each whole number.

1. 48,761

2. 3408

3. 249

4. 694,298

5. 524,007,656

6. 267,401,818

7. Insert $<$ or $>$ to make a true statement.
a. 5 50
b. 101 0
c. 29 27

8. Insert $<$ or $>$ to make a true statement
a. 12 4
b. 13 31
c. 82 79

9. Add: $13 + 2 + 7 + 8 + 9$

10. Add: $11 + 3 + 9 + 16$

11. Subtract: $7826 - 505$
Check by adding.

12. Subtract: $3285 - 272$
Check by adding.

13. The radius of Venus is 6052 kilometers. The radius of Mercury is 3612 kilometers less than the radius of Venus. Find the radius of Mercury. (*Source:* National Space Science Data Center)

14. C.J. Dufour wants to buy a digital camera. She has $762 in her savings account. If the camera costs $237, how much money will she have in her account after buying the camera?

15. Round 568 to the nearest ten.

16. Round 568 to the nearest hundred.

17. Round each number to the nearest hundred to find an estimated difference.

$$\begin{array}{r} 4725 \\ -2879 \\ \hline \end{array}$$

18. Round each number to the nearest thousand to find an estimated difference.

$$\begin{array}{r} 8394 \\ -2913 \\ \hline \end{array}$$

19. Rewrite each using the distributive property.
a. $3(4 + 5)$
b. $10(6 + 8)$
c. $2(7 + 3)$

20. Rewrite each using the distributive property.
a. $5(2 + 12)$
b. $9(3 + 6)$
c. $4(8 + 1)$

21. Multiply: 631×125

22. Multiply: 299×104

23. Find each quotient. Check by multiplying.
a. $42 \div 7$
b. $\frac{81}{9}$
c. $4\overline{)24}$

24. Find each quotient. Check by multiplying.
a. $\frac{35}{5}$
b. $64 \div 8$
c. $4\overline{)48}$

Answers

1. ______
2. ______
3. ______
4. ______
5. ______
6. ______
7. a. ______
b. ______
c. ______
8. a. ______
b. ______
c. ______
9. ______
10. ______
11. ______
12. ______
13. ______
14. ______
15. ______
16. ______
17. ______
18. ______
19. a. ______
b. ______
c. ______
20. a. ______
b. ______
c. ______
21. ______
22. ______
23. a. ______
b. ______
c. ______
24. a. ______
b. ______
c. ______

25. ____

26. ____

27. ____

28. ____

29. ____

30. ____

31. ____

32. ____

33. ____

34. ____

35. ____

36. ____

37. ____

38. ____

39. a. ____

b. ____

c. ____

40. a. ____

b. ____

25. Divide: $3705 \div 5$
Check by multiplying.

26. Divide: $3648 \div 8$
Check by multiplying.

27. How many boxes are needed to ship 56 pairs of Nikes to a shoe store in Texarkana if 9 pairs of shoe will fit in each shipping box?

28. Mrs. Mallory's first grade class is going to the zoo. She pays a total of $324 for 36 admission tickets. How much did each ticket cost?

Evaluate.

29. 8^2

30. 5^3

31. 7^1

32. 4^1

33. $5 \cdot 6^2$

34. $2^3 \cdot 7$

35. Simplify: $\dfrac{7 - 2 \cdot 3 + 3^2}{2^2 + 1}$

36. Simplify: $\dfrac{6^2 + 4 \cdot 4 + 2^3}{37 - 5^2}$

37. Evaluate $x + 7$ if x is 8.

38. Evaluate $5 + x$ if x is 9.

39. Simplify:

a. $|-2|$

b. $|5|$

c. $|0|$

40. Simplify:

a. $|4|$

b. $|-7|$

41. Add: $-2 + 5$

42. Add: $8 + (-3)$

43. Evaluate $a - b$ for $a = 8$ and $b = -6$.

44. Evaluate $x - y$ for $x = -2$ and $y = -7$.

45. Multiply: $-7 \cdot 3$

46. Multiply: $5(-2)$

47. Multiply: $0 \cdot (-4)$

48. Multiply: $-6 \cdot 9$

49. Simplify: $3(4 - 7) + (-2) - 5$

50. Simplify: $4 - 8(7 - 3) - (-1)$

41. ____________

42. ____________

43. ____________

44. ____________

45. ____________

46. ____________

47. ____________

48. ____________

49. ____________

50. ____________

STUDY SKILLS REMINDER

Are you getting all the mathematics help that you need?

Remember that, in addition to your instructor, there are many places to get help with your mathematics course. For example, see which of the list below are available.

- This text has an accompanying video for every section in this text and also for the chapter tests.
- The back of this book contains answers to odd-numbered exercises and selected solutions.
- MathPro is available with this text. It is a tutorial software program with lessons corresponding to each section in the text.
- A student solutions manual is available that contains worked-out solutions to odd-numbered exercises as well as solutions to every exercise in the Chapter Pretests, Integrated Reviews, Chapter Reviews, Chapter Tests, and Cumulative Reviews.
- Don't forget to check with your instructor for other local resources available to you, such as a tutoring center.

3.1 Simplifying Algebraic Expressions

Just as we can add, subtract, multiply, and divide numbers, we can add, subtract, multiply, and divide algebraic expressions. In previous sections we evaluated algebraic expressions like $x + 3, 4x$, and $x + 2y$ for particular values of the variables. In this section, we explore working with variable expressions without evaluating them. We begin with a definition of a term.

A Combining Like Terms

The addends of an algebraic expression are called the **terms** of the expression.

$x + 3$ — 2 terms

$3y^2 + (-6y) + 4$ — 3 terms

A term that is only a number has a special name. It is called a **constant term**, or simply a **constant**. A term that contains a variable is called a **variable term**.

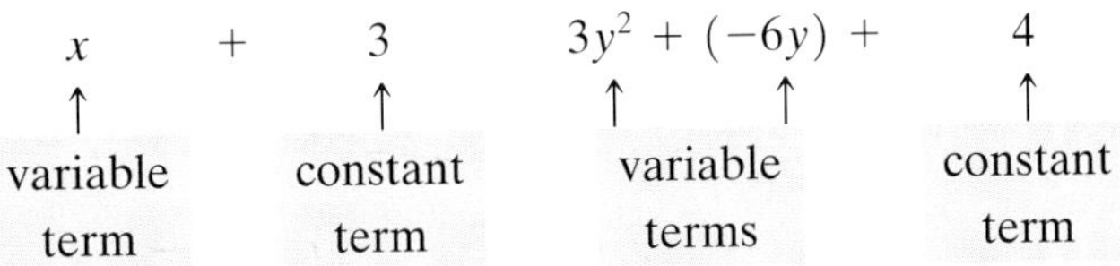

The number factor of a variable term is called the **numerical coefficient**. A numerical coefficient of 1 is usually not written.

$5x$	x or $1x$	$3y^2$	$-6y$
↑ Numerical coefficient is 5.	↑ Understood numerical coefficient is 1.	↑ Numerical coefficient is 3.	↑ Numerical coefficient is −6.

Terms that contain the same variables raised to the same exponents are called like terms.

Like Terms	Unlike Terms
$3x, \frac{1}{2}x$	$5x, x^2$
$-6y, 2y, y$	$7x, 7y$
$-5, 4$	$4y, 4$

Try the Concept Check in the margin.

A sum or difference of like terms can be simplified using the **distributive property**. Recall from Chapter 1 that the distributive property says that multiplication distributes over addition (and subtraction). For example,

$$(2 + 7)5 = 2 \cdot 5 + 7 \cdot 5$$

Using variables, we have

$$(a + b)c = a \cdot c + b \cdot c$$

If we write this from right to left, we can state the distributive property as follows.

OBJECTIVES

- **A** Use properties of numbers to combine like terms.
- **B** Use properties of numbers to multiply expressions.
- **C** Simplify expressions by multiplying and then combining like terms.
- **D** Find the perimeter and area of figures.

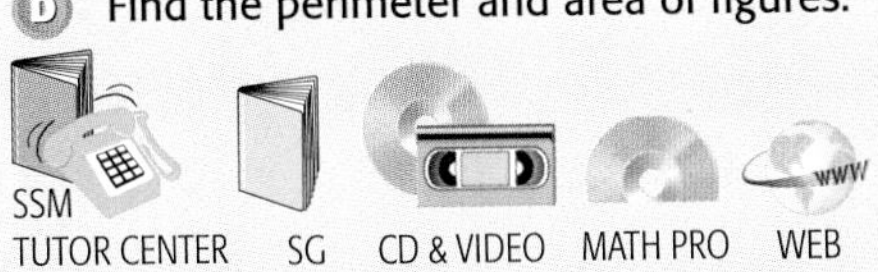

Concept Check

True or false? The terms $-7xz^2$ and $3z^2x$ are like terms. Explain.

Answer

Concept Check: True.

Distributive Property

If a, b, and c are numbers, then

$$ac + bc = (a + b)c$$

Also,

$$ac - bc = (a - b)c$$

By the distributive property then,

$$\begin{aligned} 7x + 2x &= (7 + 2)x \\ &= 9x \end{aligned}$$

We have combined like terms and the expression $7x + 2x$ simplifies to $9x$.

Practice Problem 1

Combine like terms.

a. $8m - 11m$

b. $5a + a$

c. $-y^2 + 3y^2 + 7$

EXAMPLE 1 Simplify by combining like terms.

a. $3x + 2x$

b. $y - 7y$

c. $3x^2 + 5x^2 - 2$

Solution: Add or subtract like terms.

a. $$\begin{aligned} 3x + 2x &= (3 + 2)x \\ &= 5x \end{aligned}$$

b. $$\begin{aligned} y - 7y &= 1y - 7y \\ &= (1 - 7)y \\ &= -6y \end{aligned}$$

c. $$\begin{aligned} 3x^2 + 5x^2 - 2 &= (3 + 5)x^2 - 2 \\ &= 8x^2 - 2 \end{aligned}$$

The commutative and associative properties of addition and multiplication can also help us simplify expressions. We presented these properties in Sections 1.2 and 1.5 and state them again using variables.

Properties of Addition and Multiplication

If a, b, and c are numbers, then

$a + b = b + a$ Commutative property of addition

$a \cdot b = b \cdot a$ Commutative property of multiplication

That is, the **order** of adding or multiplying two numbers can be changed without changing their sum or product.

$(a + b) + c = a + (b + c)$ Associative property of addition

$(a \cdot b) \cdot c = a \cdot (b \cdot c)$ Associative property of multiplication

That is, the **grouping** of numbers in addition or multiplication can be changed without changing their sum or product.

Answers

1. a. $-3m$ **b.** $6a$ **c.** $2y^2 + 7$

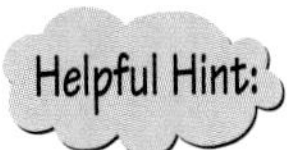

Examples of these properties are

$2 + 3 = 3 + 2$ Commutative property of addition

$7 \cdot 9 = 9 \cdot 7$ Commutative property of multiplication

$(1 + 8) + 10 = 1 + (8 + 10)$ Associative property of addition

$(4 \cdot 2) \cdot 3 = 4 \cdot (2 \cdot 3)$ Associative property of addition

EXAMPLE 2 Simplify: $2y - 6 + 4y + 8$

Solution: We begin by writing subtraction as the opposite of addition.

$$\begin{aligned} 2y - 6 + 4y + 8 &= 2y + (-6) + 4y + 8 && \text{Apply the commutative property of addition.} \\ &= 2y + 4y + (-6) + 8 \\ &= (2 + 4)y + (-6) + 8 && \text{Apply the distributive property.} \\ &= 6y + 2 && \text{Simplify.} \end{aligned}$$

Practice Problem 2

Simplify: $8m + 5 + m - 4$

EXAMPLES Simplify each expression by combining like terms.

3. $6x + 2x - 5 = 8x - 5$

4. $4x + 2 - 5x + 3 = 4x + 2 + (-5x) + 3$
$= 4x + (-5x) + 2 + 3$
$= -1x + 5$ or $-x + 5$

5. $2x - 5 + 3y + 4x - 10y + 11$
$= 2x + (-5) + 3y + 4x + (-10y) + 11$
$= 2x + 4x + 3y + (-10y) + (-5) + 11$
$= 6x - 7y + 6$

Practice Problems 3–5

Simplify each expression by combining like terms.

3. $7y + 11y - 8$
4. $2y - 6 + y + 7$
5. $-9y + 2 - 4y - 8x + 12 - x$

As we practice combining like terms, keep in mind that some of the steps may be performed mentally.

B Multiplying Expressions

We can also use properties of numbers to multiply expressions such as $3(2x)$. By the associative property of multiplication, we can write the product $3(2x)$ as $(3 \cdot 2)x$, which simplifies to $6x$.

EXAMPLES Multiply.

6. $5(3y) = (5 \cdot 3)y$ Apply the associative property of multiplication.
$= 15y$ Multiply.

7. $-2(4x) = (-2 \cdot 4)x$ Apply the associative property of multiplication.
$= -8x$ Multiply.

Practice Problems 6–7

Multiply.

6. $7(8a)$
7. $-5(9x)$

We can use the distributive property to combine like terms, which we have done, and also to multiply expressions such as $2(3 + x)$. By the distributive property, we have

$$\begin{aligned} 2(3 + x) &= 2 \cdot 3 + 2 \cdot x && \text{Apply the distributive property.} \\ &= 6 + 2x && \text{Multiply.} \end{aligned}$$

Answers

2. $9m + 1$ **3.** $18y - 8$ **4.** $3y + 1$ **5.** $-13y - 9x + 14$ **6.** $56a$ **7.** $-45x$

Practice Problem 8

Use the distributive property to multiply: $7(y + 2)$

Concept Check

What's wrong with the following? $8(a - b) = 8a - b$

Practice Problem 9

Multiply: $4(7a - 5)$

Practice Problem 10

Multiply: $6(5 - y)$

Practice Problem 11

Simplify: $5(y - 3) - 8 + y$

Practice Problem 12

Simplify: $5(2x - 3) + 7(x - 1)$

Answers

8. $7y + 14$ **9.** $28a - 20$ **10.** $30 - 6y$ **11.** $6y - 23$ **12.** $17x - 22$

Concept Check: Did not distribute the 8; $8(a - b) = 8a - 8b$

EXAMPLE 8 Use the distributive property to multiply: $6(x + 4)$

Solution: By the distributive property,

$$6(x + 4) = 6 \cdot x + 6 \cdot 4 \quad \text{Apply the distributive property.}$$
$$= 6x + 24 \quad \text{Multiply.}$$

Try the Concept Check in the margin.

EXAMPLE 9 Multiply: $-3(5a + 2)$

Solution: By the distributive property,

$$-3(5a + 2) = -3(5a) + (-3)(2) \quad \text{Apply the distributive property.}$$
$$= (-3 \cdot 5)a + (-6) \quad \text{Use the associative property and multiply.}$$
$$= -15a - 6 \quad \text{Multiply.}$$

To simplify expressions containing parentheses, we first use the distributive property and multiply.

EXAMPLE 10 Multiply: $8(x - 4)$

Solution:

$$8(x - 4) = 8 \cdot x - 8 \cdot 4$$
$$= 8x - 32$$

C Simplifying Expressions

Next we will **simplify** expressions by first using the distributive property to multiply and then **combining** any like terms.

EXAMPLE 11 Simplify: $2(3 + x) - 15$

Solution: First we use the distributive property to remove parentheses.

$$2(3 + x) - 15 = 2 \cdot 3 + 2 \cdot x + (-15) \quad \text{Apply the distributive property.}$$
$$= 6 + 2x + (-15) \quad \text{Multiply.}$$
$$= 2x + (-9) \text{ or } 2x - 9 \quad \text{Combine like terms.}$$

Helpful Hint

2 is *not* distributed to the -15 since -15 is not within the parentheses.

EXAMPLE 12 Simplify: $-2(x - 5) + 4(2x + 2)$

Solution: First we use the distributive property to remove parentheses.

$$-2(x - 5) + 4(2x + 2) = -2 \cdot x + (-2)(-5) + 4 \cdot 2x + 4 \cdot 2$$
$$= -2x + 10 + 8x + 8$$
$$= 6x + 18$$

EXAMPLE 13 Simplify: $-(x + 4) + 5x + 16$

Solution: The expression $-(x + 4)$ means $-1(x + 4)$.

$$\begin{aligned} -(x+4) + 5x + 16 &= -1(x+4) + 5x + 16 && \text{Apply the distributive property.} \\ &= -1 \cdot x + (-1)(4) + 5x + 16 \\ &= -x + (-4) + 5x + 16 && \text{Multiply.} \\ &= 4x + 12 && \text{Combine like terms.} \end{aligned}$$

Practice Problem 13

Simplify: $-(y + 1) + 3y - 12$

D Finding Perimeter and Area

EXAMPLE 14 Find the perimeter of the triangle.

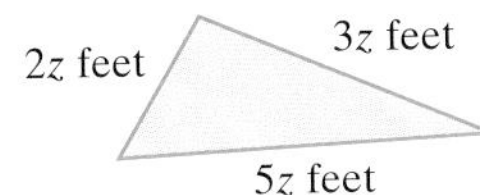

Solution: Recall that the perimeter of a figure is the distance around the figure. To find the perimeter, then, we find the sum of the lengths of the sides.

$$\begin{aligned} \text{perimeter} &= 2z + 3z + 5z \\ &= 10z \end{aligned}$$

Helpful Hint

Don't forget to insert proper units.

The perimeter is $10z$ feet.

Practice Problem 14

Find the perimeter of the square.

2x centimeters

EXAMPLE 15 Find the area of the rectangular deck.

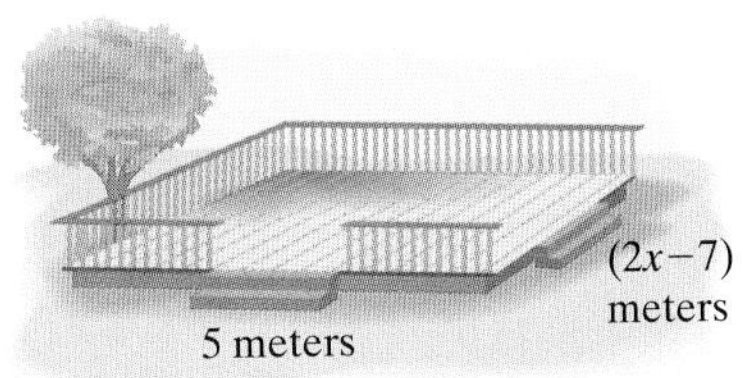

Solution: Recall how to find the area of a rectangle.

$$\begin{aligned} A &= \text{length} \cdot \text{width} \\ &= 5(2x - 7) && \text{Let length} = 5 \text{ and width} = (2x - 7). \\ &= 10x - 35 && \text{Multiply.} \end{aligned}$$

The area is $(10x - 35)$ *square* meters.

Don't forget ...

Area:	Perimeter:
• surface enclosed	• distance around
• measured in square units	• measured in units

Practice Problem 15

Find the area of the rectangular garden.

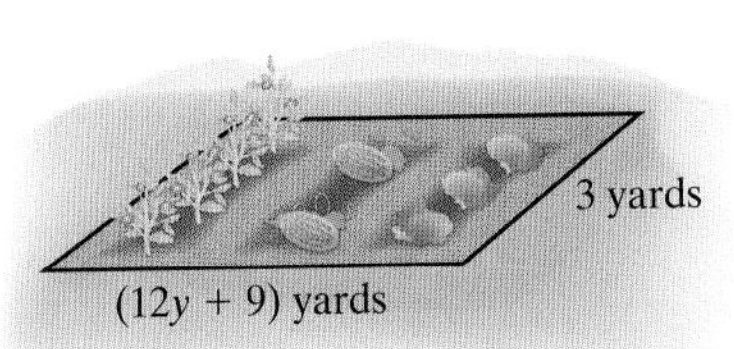

Answers

13. $2y - 13$ **14.** $8x$ centimeters
15. $(36y + 27)$ square yards

Focus on History

THE EQUAL SIGN

The most important symbol in an equation is the equal sign, =. It indicates that a statement is, in fact, an equation and not just an algebraic expression. However, the equal sign has not always been around. In fact, the = symbol has been used only since 1557 when it first appeared in a book called *The Whetstone of Witte* by English physician and mathematician Robert Recorde. Recorde's original symbol used much longer lines, ═══, and can be seen in this page from *The Whetstone of Witte*.

In his book, Recorde gave an interesting reason for his choice of symbol. He explained, "I will sette as I doe often in woorke use, a paire of parralles, or Gemowe lines of one lengthe, thus: ═══, bicause noe 2, thynges, can be moare equalle."

Prior to the appearance of the equal sign, authors of mathematics texts often used words, such as "aequales" or "gleich" to indicate equality. These words continued to be used regularly until the use of the = symbol became widely accepted in the 1700s.

The Arte

as their woorkes doe extende) to distincte it onely into twoo partes. Whereof the firste is, *when one nomber is equalle unto one other*. And the seconde is, *when one nomber is compared as equalle unto .2. other nombers.*

Alwaies willyng you to remember, that you reduce your nombers, to their leaste denominations, and smalleste formes, before you procede any farther.

And again, if your *equation* be soche, that the greateste denomination *Cosike*, be ioined to any parte of a compounde nomber, you shall tourne it so, that the nomber of the greateste signe alone, maie stande as equalle to the reste.

And this is all that neadeth to be taughte, concernyng this woorke.

Howbeit, for easie alteratiō of *equations*. I will propounde a fewe exāples, bicause the extraction of their rootes, maie the more aptly bee wroughte. And to avoide the tediouse repetition of these woordes: is equalle to: I will sette as I doe often in woorke use, a paire of paralleles, or Gemowe lines of one lengthe, thus: ═══, bicause noe .2. thynges, can be moare equalle. And now marke these nombers.

1. 14.ze. —+— .15.ꝭ ═══ 71.ꝭ.
2. 20.ze. —— .18.ꝭ ═══ .102.ꝭ.
3. 26.ʒ —+— 10ze ═══ 9.ʒ —— 10ze —+— 213.ꝭ.
4. 19.ze —+— 192.ꝭ ═══ 10ʒ —+— 108ꝭ —— 19ze.
5. 18.ze —+— 24.ꝭ. ═══ 8.ʒ. —+— 2.ze.
6. 34ʒ —— 12ze ═══ 40ze —+— 480ꝭ —— 9.ʒ.

1. In the firste there appeareth. 2. nombers, that is 14.ze.

Name ______________________ Section __________ Date __________

Mental Math

Find the numerical coefficient of each variable term.

1. $5y$
2. $-2z$
3. z
4. $3xy^2$
5. $11a$
6. $-x$

EXERCISE SET 3.1

A *Simplify each expression by combining like terms. See Examples 1 through 5.*

1. $3x + 5x$
2. $8y + 3y$
3. $5n - 9n$
4. $7z - 10z$

5. $4c + c - 7c$
6. $5b - 8b - b$
7. $5x - 7x + x - 3x$
8. $8y + y - 2y - y$
9. $4a + 3a + 6a - 8$
10. $5b - 4b + b - 15$

B *Multiply. See Examples 6 and 7.*

11. $6(5x)$
12. $4(4x)$
13. $-2(11y)$
14. $-3(21z)$

15. $12(6a)$
16. $9(7b)$

Multiply. See Examples 8 through 10.

17. $2(y + 2)$
18. $3(x + 1)$
19. $5(a - 8)$
20. $4(y - 6)$
21. $-4(3x + 7)$
22. $-8(8y + 10)$

C *Simplify each expression. First use the distributive property to multiply and remove parentheses. See Examples 11 through 13.*

23. $2(x + 4) + 7$
24. $5(6 - y) - 2$
25. $-4(6n - 5) + 3n$
26. $-3(5 - 2b) - 4b$
27. $8 + 5(3c - 1)$
28. $10 + 4(6d - 2)$
29. $3 + 6(w + 2) + w$
30. $8z + 5(6 + z) + 20$
31. $2(3x + 1) + 5(x - 2)$
32. $3(5x - 2) + 2(3x + 1)$
33. $-(5x - 1) - 10$
34. $-(2y - 6) + 10$

A B C *Simplify each expression. See Examples 1 through 13.*

35. $18y - 20y$

36. $x + 12x$

37. $z - 8z$

38. $12x - 8x$

39. $9d - 3c - d$

40. $8r + s - 7s$

41. $2y - 6 + 4y - 8$

42. $a + 4 - 7a - 5$

43. $5q + p - 6q - p$

44. $m - 8n + m + 8n$

45. $2(x + 1) + 20$

46. $5(x - 1) + 18$

47. $5(x - 7) - 8x$

48. $3(x + 2) - 11x$

49. $-5(z + 3) + 2z$

50. $-8(1 + v) + 6v$

51. $8 - x + 4x - 2 - 9x$

52. $5y - 4 + 9y - y + 15$

53. $-7(x + 5) + 5(2x + 1)$

54. $-2(x + 4) + 8(x - 1)$

55. $3r - 5r + 8 + r$

56. $6x - 4 + 2x - x + 3$

57. $-3(n - 1) - 4n$

58. $5(c + 2) + 7c$

59. $4(z - 3) + 5z - 2$

60. $8(m + 3) - 20 + m$

61. $6(2x - 1) - 12x$

62. $5(2a + 3) - 10a$

63. $-(4x - 5) + 5$

64. $-(7y - 2) + 6$

65. $-(4xy - 10) + 2(3xy + 5)$

66. $-(12ab - 10) + 5(3ab - 2)$

67. $3a + 4(a + 3)$

68. $b + 2(b - 5)$

69. $5y - 2(y - 1) + 3$

70. $3x - 4(x + 2) + 1$

D *Find the perimeter of each figure. See Example 14.*

71.

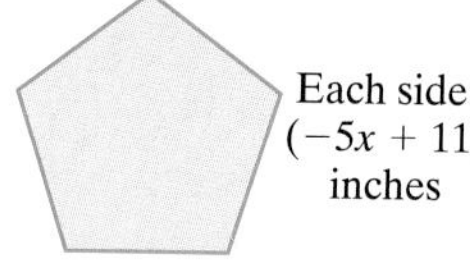

72.

Each side:
$(9y + 1)$
kilometers

△ **73.**

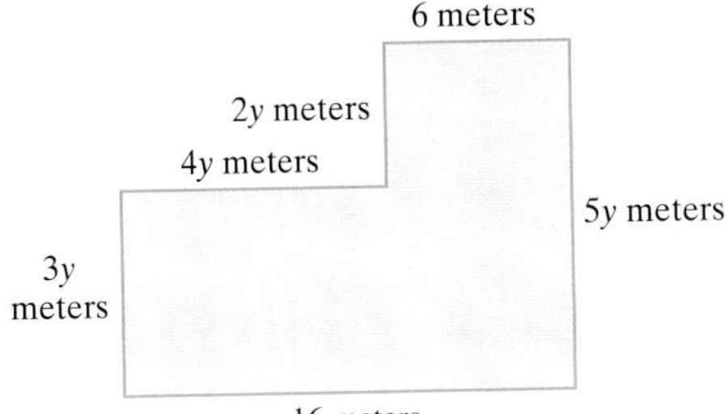

△ **74.**

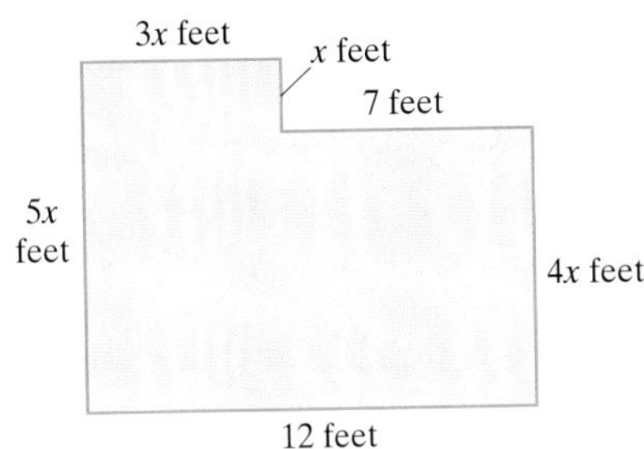

△ **75.**

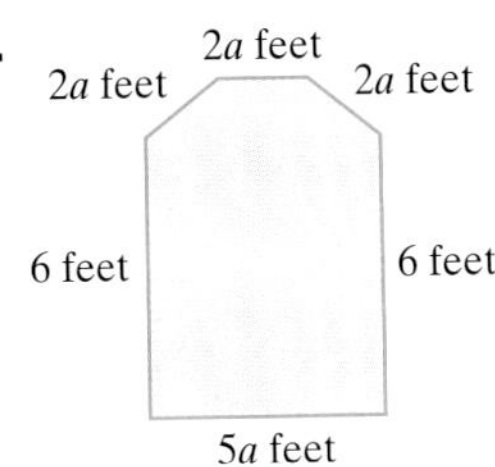

△ **76.**

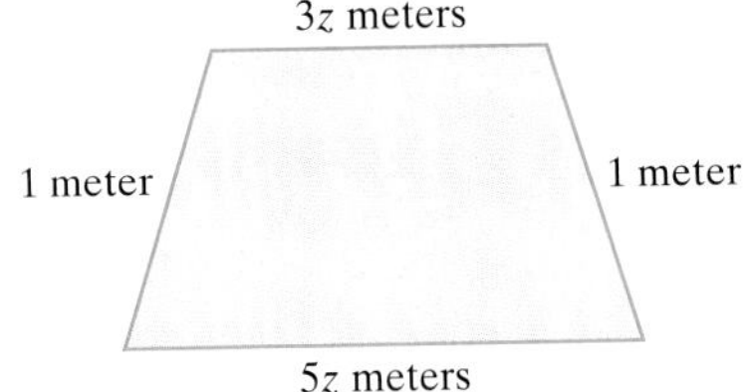

Find the area of each Square or rectangle. See Example 15.

△ **77.**

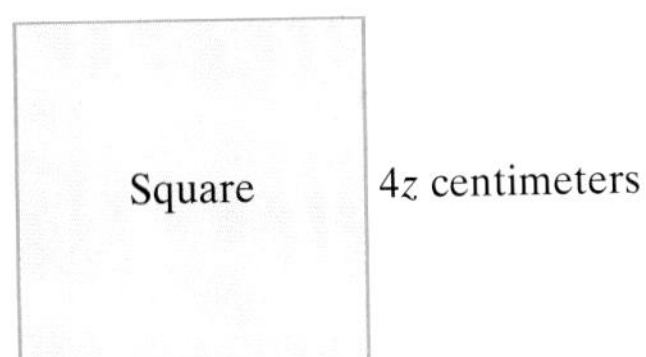

78.

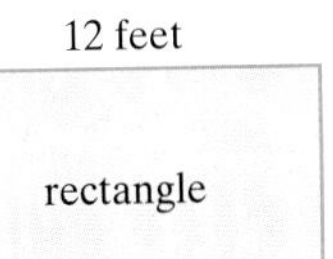

△ **79.**

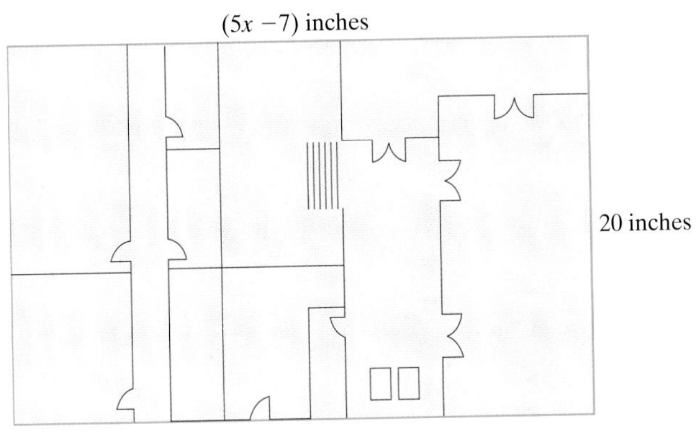

△ **80.**

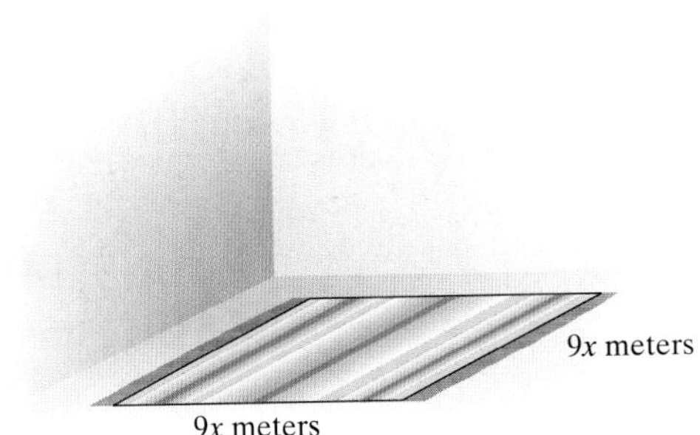

Review and Preview

Perform each indicated operation. See Sections 2.2 and 2.3.

81. $-13 + 10$

82. $-7 - (-4)$

83. $-4 - (-12)$

84. $-15 + 23$

85. $-4 + 4$

86. $8 + (-8)$

Combining Concepts

Simplify.

87. $9684q - 686 - 4860q + 12{,}960$

88. $76(268x + 592) - 2960$

89. If x is a whole number, which expression is the greatest: $2x$ or $5x$? Explain your answer.

90. If x is a whole number, which expression is the greatest: $-2x$ or $-5x$? Explain your answer.

Find the area of each figure.

91.

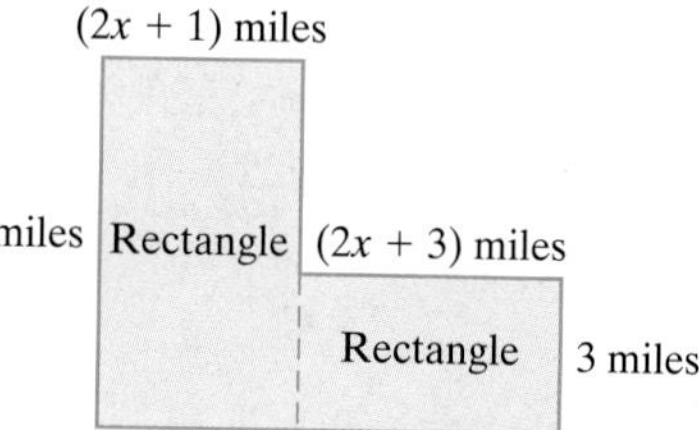

92.

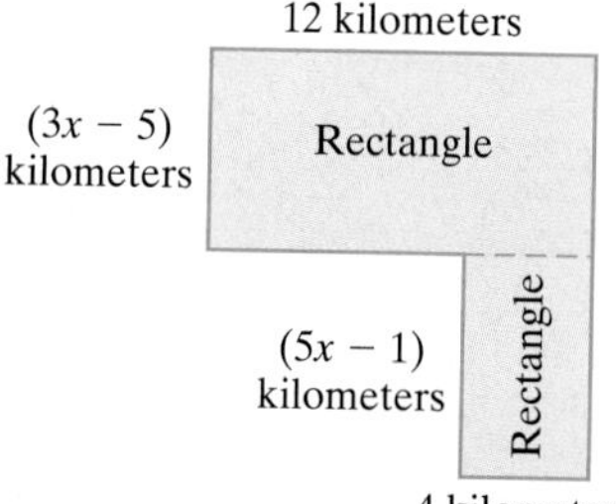

93. Explain what makes two terms "like terms."

94. Explain how to combine like terms.

3.2 Solving Equations: The Addition Property

OBJECTIVES

(A) Determine whether a given number is a solution of an equation.

(B) Use the addition property of equality to solve equations.

Frequently in this book, we have written statements like $7 + 4 = 11$ or area = length · width. Each of these statements is called an **equation**. An equation is of the form

expression = **expression**

An equation can be labeled as

$$\underbrace{x + 7}_{\text{left side}} \overset{\downarrow}{=} \underbrace{10}_{\text{right side}}$$

(A) Determining Whether a Number is a Solution

When an equation contains a variable, deciding which values of the variable make an equation a true statement is called **solving** an equation for the variable. A **solution** of an equation is a value for the variable that makes an equation a true statement. For example, 2 is a solution of the equation $x + 5 = 7$, since replacing x with 2 results in the *true* statement $2 + 5 = 7$. Similarly, 3 is not a solution of $x + 5 = 7$, since replacing x with 3 results in the *false* statement $3 + 5 = 7$.

EXAMPLE 1 Determine whether 6 is a solution of the equation $4(x - 3) = 12$.

Solution: We replace x with 6 in the equation.

$$4(x - 3) = 12$$

$$4(6 - 3) \stackrel{?}{=} 12 \quad \text{Replace } x \text{ with 6.}$$

$$4(3) \stackrel{?}{=} 12$$

$$12 = 12 \quad \text{True.}$$

Since $12 = 12$ is a true statement, 6 *is* a solution of the equation.

Practice Problem 1

Determine whether 4 is a solution of the equation $3(y - 6) = 6$.

EXAMPLE 2 Determine whether -1 is a solution of the equation $3y + 1 = 3$.

Solution:

$$3y + 1 = 3$$

$$3(-1) + 1 \stackrel{?}{=} 3$$

$$-3 + 1 \stackrel{?}{=} 3$$

$$-2 = 3 \quad \text{False.}$$

Since $-2 = 3$ is a false statement, -1 *is not* a solution of the equation.

Practice Problem 2

Determine whether -2 is a solution of the equation $-4x - 3 = 5$.

(B) Using the Addition Property to Solve Equations

To solve an equation, we will use properties of equality to write simpler equations, all equivalent to the original equation, until the final equation has the form

x = **number** or **number** = x

Answers

1. no **2.** yes

Equivalent equations have the same solution. For example, $x + 3 = 5$ and $x = 2$ are equivalent equations because they both have the same solution, 2. Since we will be writing equivalent equations, the word "number" in the equations on the previous page represents the solution of the original equation. The first property of equality to help us write simpler equations is the **addition property of equality**.

Addition Property of Equality

Let a, b, and c represent numbers.
If $a = b$, then

$$a + c = b + c \quad \text{and} \quad a - c = b - c$$

In other words, the same number may be added to or subtracted from both sides of an equation without changing the solution of the equation.

(Recall in Section 2.3 that we defined subtraction as addition of the first number and the opposite of the second number. Because of this, the Addition Property of Equality also allows us to subtract the same number from both sides.)

A good way to visualize a true equation is to picture a balanced scale. Since the scale is balanced, each side weighs the same amount. Similarly, in a true equation the expressions on each side have the same value. Picturing our balanced scale, if we add the same weight to each side, the scale remains balanced.

Practice Problem 3

Solve for y: $y - 5 = -3$

EXAMPLE 3 Solve for x: $x - 2 = 1$

Solution: To solve the equation for x, we need to rewrite the equation in the form $x =$ number. In other words, our goal is to get x alone on one side of the equation. To do so, we add 2 to both sides of the equation.

$$x - 2 = 1$$
$$x - 2 + 2 = 1 + 2 \quad \text{Add 2 to both sides of the equation.}$$
$$x = 3 \quad \text{Simplify.}$$

To check, we replace x with 3 in the *original* equation.

$$x - 2 = 1 \quad \text{Original equation}$$
$$3 - 2 \stackrel{?}{=} 1 \quad \text{Replace } x \text{ with 3.}$$
$$1 = 1 \quad \text{True.}$$

Since $1 = 1$ is a true statement, 3 is the solution of the equation.

Helpful Hint

Remember to check the solution in the *original* equation to see that it makes the equation a true statement.

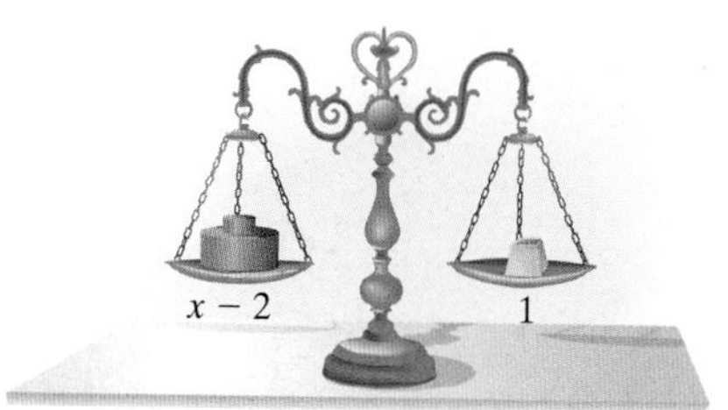

Let's visualize how we used the addition property of equality to solve the equation in Example 3. Picture the original equation $x - 2 = 1$ as a balanced scale. The left side of the equation has the same value as the right side.

Answer

3. 2

If the same weight is added to each side of a scale, the scale remains balanced. Likewise, if the same number is added to each side of an equation, the left side continues to have the same value as the right side.

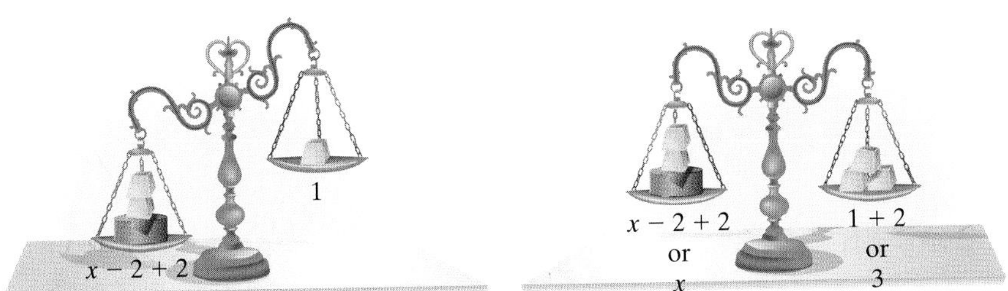

EXAMPLE 4 Solve: $-8 = x + 1$

Solution: To get x alone on one side of the equation, we subtract 1 from both sides of the equation.

$$-8 = x + 1$$
$$-8 - 1 = x + 1 - 1 \quad \text{Subtract 1 from both sides.}$$
$$-8 + (-1) = x + 1 + (-1)$$
$$-9 = x \quad \text{Simplify.}$$

Check: $-8 = x + 1$
$$-8 \stackrel{?}{=} -9 + 1 \quad \text{Replace x with } -9$$
$$-8 = -8 \quad \text{True.}$$

The solution is -9.

Practice Problem 4

Solve: $z + 9 = 1$

Helpful Hint

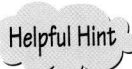

Remember that we can get the variable alone on either side of the equation. For example, the equations $-9 = x$ and $x = -9$ both have the solution of -9.

EXAMPLE 5 Solve: $7x = 6x + 4$

Solution: Subtract $6x$ from both sides so that variable terms will be on the same side of the equation.

$$7x = 6x + 4$$
$$7x - 6x = 6x + 4 - 6x \quad \text{Subtract } 6x \text{ from both sides.}$$
$$1x = 4 \quad \text{or} \quad x = 4 \quad \text{Simplify.}$$

Check to see that 4 is the solution.

Practice Problem 5

Solve: $10x = -2 + 9x$

If one or both sides of an equation can be simplified, do that first.

EXAMPLE 6 Solve: $y - 5 = -2 - 6$

Solution: First we simplify the right side of the equation.

$$y - 5 = -2 - 6$$
$$y - 5 = -8 \quad \text{Combine like terms.}$$

Practice Problem 6

Solve: $x + 6 = 1 - 3$

Answers

4. -8 **5.** -2 **6.** -8

Next we get y alone by adding 5 to both sides of the equation.

$$y - 5 + 5 = -8 + 5 \quad \text{Add 5 to both sides.}$$
$$y = -3 \quad \text{Simplify.}$$

Check to see that -3 is the solution.

Try the Concept Check in the margin.

Concept Check

What number should be added to or subtracted from both sides of the equation in order to solve the equation $-3 = y + 2$?

Practice Problem 7

Solve: $-6y - 1 + 7y = 17$

EXAMPLE 7 Solve: $5x + 2 - 4x = 7 - 9$

Solution: First we simplify each side of the equation separately.

$$5x + 2 - 4x = 7 - 9$$
$$\underbrace{5x - 4x} + 2 = \underbrace{7 - 9}$$
$$1x + 2 = -2 \quad \text{Combine like terms.}$$

To get x alone on the left side, we subtract 2 from both sides.

$$1x + 2 - 2 = -2 - 2 \quad \text{Subtract 2 from both sides.}$$
$$1x = -4 \quad \text{or} \quad x = -4 \quad \text{Simplify.}$$

Check to see that -4 is the solution.

Practice Problem 8

Solve: $13x = 4(3x - 1)$

EXAMPLE 8 Solve: $3(3x - 5) = 10x$

Solution: First we multiply on the left side to remove the parentheses.

$$3(3x - 5) = 10x$$
$$3 \cdot 3x - 3 \cdot 5 = 10x \quad \text{Use the distributive property.}$$
$$9x - 15 = 10x$$

Now we subtract $9x$ from both sides.

$$9x - 15 - 9x = 10x - 9x \quad \text{Subtract } 9x \text{ from both sides.}$$
$$-15 = 1x \quad \text{or} \quad x = -15 \quad \text{Simplify.}$$

Answers

7. 18 **8.** -4

Concept Check: Subtract 2 from both sides.

Name ______________________ Section __________ Date __________

EXERCISE SET 3.2

 *Decide whether the given number is a solution of the given equation. See Examples 1 and 2.*

1. Is 10 a solution of $x - 8 = 2$?

2. Is 9 a solution of $y - 2 = 7$?

3. Is 6 a solution of $z + 8 = 14$?

4. Is 5 a solution of $n + 11 = 16$?

5. Is -5 a solution of $x + 12 = 7$?

6. Is -7 a solution of $a + 23 = 16$?

7. Is 8 a solution of $7f = 64 - f$?

8. Is 3 a solution of $12 - k = 9$?

9. Is 0 a solution of $h - 8 = -8$?

10. Is -2 a solution of $3 + d = 0$?

11. Is 3 a solution of $4c + 2 - 3c = -1 + 6$?

12. Is 1 a solution of $2(b - 3) = 10$?

B *Solve and check the solution. See Examples 3 through 5.*

13. $a + 5 = 23$

14. $s - 7 = 15$

15. $d - 9 = 17$

16. $f + 4 = -6$

17. $7 = y - 2$

18. $-10 = z - 15$

19. $-12 = x + 4$

20. $1 = y + 7$

21. $3x = 2x + 11$

22. $5y = 4y + 12$

23. $-4 + y = 2y$

24. $8x - 4 = 9x$

Solve and check the solution. See Examples 6 through 8.

25. $x - 3 = -1 + 4$

26. $x + 7 = 2 + 3$

27. $y + 1 = -3 + 4$

28. $y - 8 = -5 - 1$

29. $-7 + 10 = m - 5$

30. $1 - 8 = n + 2$

31. $-2 - 3 = -4 + x$

32. $7 - (-10) = x - 5$

33. $2(5x - 3) = 11x$

34. $6(3x + 1) = 19x$

35. $3y = 2(y + 12)$

36. $17x = 4(4x - 6)$

37. $-8x + 4 + 9x = -1 + 7$

38. $3x - 2x + 5 = 5$

39. $2 - 2 = 5x - 4x$

40. $11 - 15 = 6x - 4 - 5x$

41. $7x + 14 - 6x = -4 - 10$

42. $-10x + 11x + 5 = 9 - 5$

Solve each equation. See Examples 3 through 8.

43. $57 = y - 16$

44. $56 = -45 + x$

45. $67 = z + 67$

46. $66 = 8 + z$

47. $x + 5 = 4 - 3$

48. $x - 7 = -14 + 10$

49. $z - 23 = -88$

50. $x + 3 = -8$

51. $7a + 7 - 6a = 20$

52. $-8a - 11 + 9a = -5$

53. $-12 + x = -15$

54. $10 + x = 12$

55. $-8 - 9 = 3x + 5 - 2x$

56. $-7 + 10 = 4x - 6 - 3x$

57. $8(3x - 2) = 25x$

58. $9(4x + 5) = 37x$

59. $7x + 7 - 6x = 10$

60. $-3 + 5x - 4x = 13$

61. $50y = 7(7y + 4)$

62. $65y = 8(8y - 9)$

Review and Preview

The trumpeter swan is the largest waterfowl in the United States. Although it was thought to be nearly extinct at the beginning of the twentieth century, recent conservation efforts have been succeeding. Use the bar graph to answer Exercises 63–66. See Section 1.3.

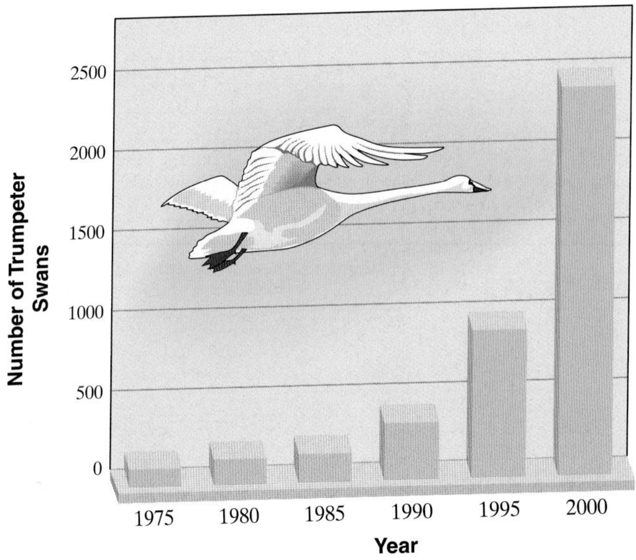

Source: *U.S. Fish and Wildlife Service*

63. Estimate the number of trumpeter swans in the Great Lakes region in 2000.

64. Estimate the number of trumpeter swans in the Great Lakes region in 1995.

65. How many more trumpeter swans in the Great Lakes region were there in 2000 than in 1985?

66. Describe any trends shown in this graph.

Simplify. See Section 1.7.

67. $\frac{8}{8}$

68. $\frac{11}{11}$

69. $\frac{-3}{-3}$

70. $\frac{-7}{-7}$

Combining Concepts

71. In your own words, explain the addition property of equality.

72. Write an equation that can be solved using the addition property of equality.

73. Are the equations below equivalent? Why or why not?

$$x + 7 = 4 + (-9)$$
$$x + 7 = 5$$

74. Write 2 equivalent equations.

Solve.

75. $x - 76{,}862 = 86{,}102$

76. $-968 + 432 = 86y - 508 - 85y$

77. $5^3 = x + 4^4$

78. $7y - 10^4 = 8y$

79. $|-13| + 3^2 = 100y - |-20| - 99y$

80. $4(x - 11) + |90| - |-86| + 2^5 = 5x$

A football team's total offense T is found by adding the total passing yardage P to the total rushing yardage R: $T = P + R$.

81. During the 2002 football season, the Green Bay Packers' total offense was 5560 yards. The Packers' rushing yardage for the season was 1933 yards. How many yards did the Packers gain by passing during the season? (*Source:* National Football League)

82. During the 2002 football season, the Denver Broncos' total offense was 6090 yards. The Broncos' passing yardage for the season was 3824 yards. How many yards did the Broncos gain by rushing during the season? (*Source:* National Football League)

In accounting, a company's annual net income, I, can be computed using the relation $I = R - E$, where R is the company's total revenues for a year and E is the company's total expenses for the year.

83. At the end of fiscal year 2002, Wal-Mart had a net income of $6671 million. During the year, Wal-Mart incurred a total of $213,141 million in expenses. What was Wal-Mart's total revenues for the year? (*Source:* Wal-Mart Stores, Inc.)

84. At the end of fiscal year 2001, Home Depot had a net income of $3044 million. During the year, Home Depot had a total of $50,559 million in expenses. What was Home Depot's total revenues for the year? (*Source:* Home Depot, Inc.)

Focus on Mathematical Connections

Modeling Equation Solving with Addition and Subtraction

We can use positive counters ● and negative counters ● to help us model the equation-solving process. We also need to use an object that represents a variable. We use small slips of paper with the variable name written on them.

Recall that taking a ● and ● together creates a neutral or zero pair. After a neutral pair has been formed, it can be removed from or added to an equation model without changing the overall value. We also need to remember that we can add or remove the same number of positive or negative counters from both sides of an equation without changing the overall value.

We can represent the equation $x + 5 = 2$ as follows:

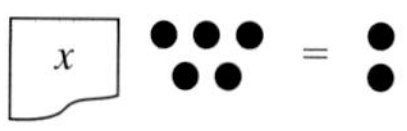

To get the variable by itself, we must remove 5 black counters from both sides of the model. Because there are only 2 counters on the right side, we must add 5 negative counters to both sides of the model. Then we can remove neutral pairs: 5 from the left side and 2 from the right side.

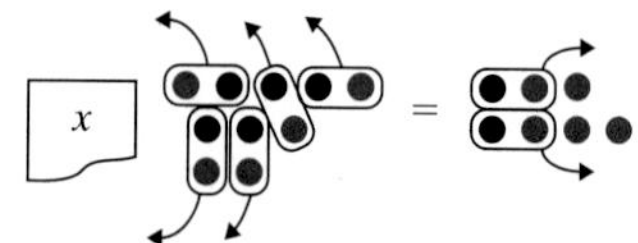

We are left with the following model, which represents the solution, $x = -3$.

Similarly, we can represent the equation $x - 4 = -6$ as follows:

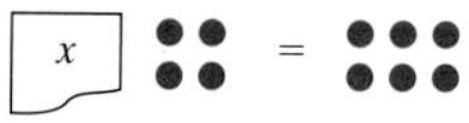

To get the variable by itself, we must remove 4 red counters from both sides of the model

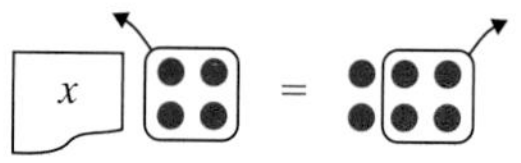

We are left with the following model, which represents the solution, $x = -2$.

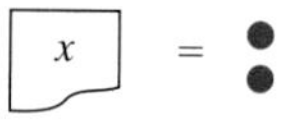

Critical Thinking

Use the counter model to solve each equation.

1. $x - 3 = -7$

2. $x - 1 = -9$

3. $x + 2 = 8$

4. $x + 4 = 5$

5. $x + 8 = 3$

6. $x - 5 = -1$

7. $x - 2 = 1$

8. $x - 5 = 10$

9. $x + 3 = -7$

10. $x + 8 = -2$

3.3 Solving Equations: The Multiplication Property

OBJECTIVES

A Use the multiplication property to solve equations.

B Translate word phrases into algebraic expressions.

A Using the Multiplication Property to Solve Equations

Although the addition property of equality is a powerful tool for helping us solve equations, it cannot help us solve all types of equations. For example, it cannot help us solve an equation such as $2x = 6$. To solve this equation, we use a second property of equality called the **multiplication property of equality**.

Multiplication Property of Equality

Let a, b, and c represent numbers and let $c \neq 0$. If $a = b$, then

$$a \cdot c = b \cdot c \quad \text{and} \quad \frac{a}{c} = \frac{b}{c}$$

In other words, both sides of an equation may be multiplied or divided by the same nonzero number without changing the solution of the equation. (We will see in Chapter 4 how the Multiplication Property also allows us to divide both sides of an equation by the same nonzero number.)

Picturing again our balanced scale, if we multiply or divide the weight on each side by the same nonzero number, the scale (or equation) remains balanced.

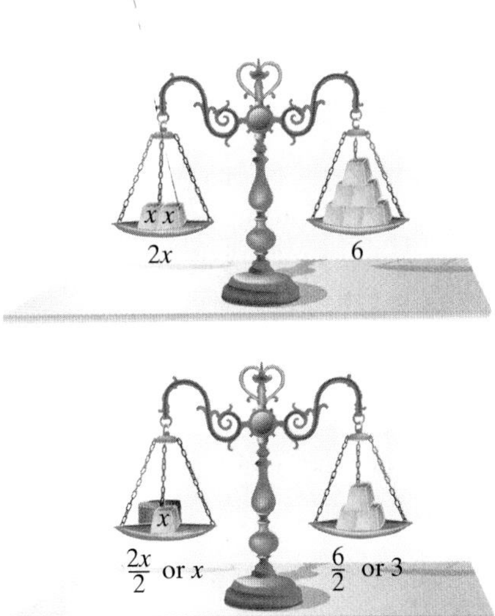

The multiplication property allows us to solve equations like

$$\frac{x}{5} = 2$$

Here, x is *divided* by 5. To get x alone, we use the multiplication property to *multiply* both sides by 5.

$$5 \cdot \frac{x}{5} = 5 \cdot 2$$

It can be shown that an expression such as $5 \cdot \frac{x}{5}$ is equivalent to $\frac{5}{5} \cdot x$, so we have that $5 \cdot \frac{x}{5} = 5 \cdot 2$ can be written as

$$\frac{5}{5} \cdot x = 5 \cdot 2$$

$$1 \cdot x = 10 \quad \text{or} \quad x = 10.$$

Practice Problem 1

Solve: $\frac{x}{-4} = 7$

EXAMPLE 1 Solve: $\frac{x}{3} = -2$

Solution: To get x alone, multiply both sides by 3.

$$\frac{x}{3} = -2$$

$$3 \cdot \frac{x}{3} = 3 \cdot (-2) \qquad \text{Multiply both sides by 3.}$$

$$\frac{3}{3} \cdot x = 3 \cdot (-2)$$

$$1x = -6 \quad \text{or} \quad x = -6 \qquad \text{Simplify.}$$

Check: Replace x with -6 in the original equation.

$$\frac{x}{3} = -2 \qquad \text{Original equation}$$

$$\frac{-6}{3} \stackrel{?}{=} -2 \qquad \text{Let } x = -6.$$

$$-2 = -2 \qquad \text{True.}$$

The solution is -6.

To solve an equation like $2x = 6$ for x, notice that 2 is *multiplied* by x. To get x alone, we use the multiplication property of equality to *divide* both sides of the equation by 2, and simplify as follows:

$$2x = 6$$

$$\frac{2 \cdot x}{2} = \frac{6}{2} \qquad \text{Divide both sides by 2.}$$

Then it can be shown that an expression such as $\frac{2 \cdot x}{2}$ is equivalent to $\frac{2}{2} \cdot x$, so

$$\frac{2 \cdot x}{2} = \frac{6}{2} \quad \text{can be written as} \quad \frac{2}{2} \cdot x = \frac{6}{2}$$

$$1 \cdot x = 3 \quad \text{or} \quad x = 3$$

Practice Problem 2

Solve: $3y = -18$

EXAMPLE 2 Solve: $-5x = 15$

Solution: To get x alone, divide both sides by -5.

$$-5x = 15 \qquad \text{Original equation}$$

$$\frac{-5x}{-5} = \frac{15}{-5} \qquad \text{Divide both sides by } -5.$$

$$\frac{-5}{-5} \cdot x = \frac{15}{-5}$$

$$1x = -3 \quad \text{or} \quad x = -3 \qquad \text{Simplify.}$$

Answers

1. -28 **2.** -6

To check, replace x with -3 in the original equation.

$$\begin{aligned} -5x &= 15 && \text{Original equation} \\ -5(-3) &\stackrel{?}{=} 15 && \text{Let } x = -3. \\ 15 &= 15 && \text{True.} \end{aligned}$$

The solution is -3.

EXAMPLE 3 Solve: $-8 = 2y$

Solution: To get y alone, divide both sides of the equation by 2.

$$\begin{aligned} -8 &= 2y \\ \frac{-8}{2} &= \frac{2y}{2} && \text{Divide both sides by 2.} \\ \frac{-8}{2} &= \frac{2}{2} \cdot y \\ -4 &= 1y \quad \text{or} \quad y = -4 \end{aligned}$$

Check to see that -4 is the solution.

Practice Problem 3

Solve: $-16 = 8x$

EXAMPLE 4 Solve: $-12x = -36$

Solution: Divide both sides of the equation by the coefficient of x, which is -12.

$$\begin{aligned} -12x &= -36 \\ \frac{-12x}{-12} &= \frac{-36}{-12} \\ \frac{-12}{-12} \cdot x &= \frac{-36}{-12} \\ x &= 3 \end{aligned}$$

To check, replace x with 3 in the original equation.

$$\begin{aligned} -12x &= -36 \\ -12(3) &\stackrel{?}{=} -36 && \text{Let } x = 3. \\ -36 &= -36 && \text{True.} \end{aligned}$$

Since $-36 = -36$ is a true statement, the solution is 3.

Practice Problem 4

Solve: $-3y = -27$

Try the Concept Check in the margin.

Concept Check

Which operation is appropriate for solving each of the following equations, addition or division?

a. $12 = x - 3$

b. $12 = 3x$

We often need to simplify one or both sides of an equation before applying the properties of equality to get the variable alone.

EXAMPLE 5 Solve: $3y - 7y = 12$

Solution: First combine like terms.

$$\begin{aligned} 3y - 7y &= 12 \\ -4y &= 12 && \text{Combine like terms.} \\ \frac{-4y}{-4} &= \frac{12}{-4} && \text{Divide both sides by } -4. \\ y &= -3 && \text{Simplify.} \end{aligned}$$

Practice Problem 5

Solve: $10 = 2m - 4m$

Answers

3. -2 **4.** 9 **5.** -5

Concept Check:
a. Addition.
b. Division.

Check: Replace y with -3.

$$3y - 7y = 12$$

$$3(-3) - 7(-3) \stackrel{?}{=} 12$$

$$-9 + 21 \stackrel{?}{=} 12$$

$$12 = 12 \quad \text{True.}$$

The solution is -3.

Practice Problem 6

Solve: $-8 + 6 = \frac{a}{3}$

EXAMPLE 6 Solve: $\frac{z}{-4} = 11 - 5$

Solution: Simplify the right side of the equation first.

$$\frac{z}{-4} = 11 - 5$$

$$\frac{z}{-4} = 6$$

Next, to get z alone, multiply both sides by -4.

$$-4 \cdot \frac{z}{-4} = -4 \cdot 6 \quad \text{Multiply both sides by } -4$$

$$\frac{-4}{-4} \cdot z = -4 \cdot 6$$

$$1z = -24 \quad \text{or} \quad z = -24$$

Check to see that -24 is the solution.

Practice Problem 7

Solve: $-4 - 10 = 4y - 5y$

EXAMPLE 7 Solve: $8x - 9x = 12 - 17$

Solution: First combine like terms on each side of the equation.

$$8x - 9x = 12 - 17$$

$$-x = -5$$

Recall that $-x$ means $-1x$ and divide both sides by -1.

$$\frac{-1x}{-1} = \frac{-5}{-1} \quad \text{Divide both sides by } -1.$$

$$x = 5 \quad \text{Simplify.}$$

Check to see that the solution is 5.

B Translating Word Phrases into Expressions

A section later in this chapter contains a formal introduction to problem solving. To prepare for this section, let's review writing phrases as algebraic expressions using the following key words and phrases as a guide:

Addition	Subtraction	Multiplication	Division	Equal Sign
sum	difference	product	quotient	equals
plus	minus	times	divided by	gives
added to	subtracted from	multiply	into	is/was
more than	less than	twice	per	yields
increased by	decreased by	of		amounts to
total	less	double		is equal to

Answers

6. -6 **7.** 14

EXAMPLE 8 Write each phrase as an algebraic expression. Use x to represent "a number."

a. a number increased by -5
b. the product of -7 and a number
c. a number less 20
d. the quotient of -18 and a number
e. a number subtracted from -2

Solution:

a. In words: a number → x; increased by → $+$; -5 → (-5)

Translate: $x + (-5)$ or $x - 5$

b. In words: the product of: -7 → -7; and → $\cdot$; a number → x

Translate: $-7 \cdot x$ or $-7x$

c. In words: a number → x; less → $-$; 20 → 20

Translate: $x - 20$

d. In words: the quotient of: -18 → -18; and → $\div$; a number → x

Translate: $-18 \div x$ or $\frac{-18}{x}$ or $-\frac{18}{x}$

e. In words: a number; subtracted from; -2

Translate: $-2 - x$

Helpful Hint

As we reviewed in Chapter 1, don't forget that order is important when subtracting. Notice the translation order of numbers and variables below.

Phrase	Translation
a number less 9	$x - 9$
a number subtracted from 9	$9 - x$

Practice Problem 8

Write each phrase as an algebraic expression. Use x to represent "a number."

a. the sum of -3 and a number
b. -5 decreased by a number
c. three times a number
d. a number subtracted from 83
e. the quotient of a number and -4

Answers

8. a. $-3 + x$ **b.** $-5 - x$ **c.** $3x$ **d.** $83 - x$ **e.** $\frac{x}{-4}$ or $-\frac{x}{4}$

Practice Problem 9

Translate each phrase into an algebraic expression. Let x be the unknown number.

a. The product of 5 and a number, decreased by 25
b. Twice the sum of a number and 3
c. The quotient of 39 and twice a number

EXAMPLE 9 Write each phrase as an algebraic expression. Let x be the unknown number.

a. Twice a number, increased by -9
b. Three times the difference of a number and 11
c. The quotient of 5 times a number and 17

Solution:

a. In words:

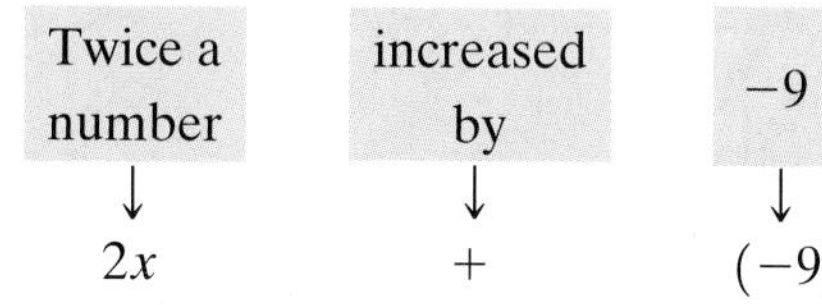

Translate: $2x + (-9)$

b. In words:

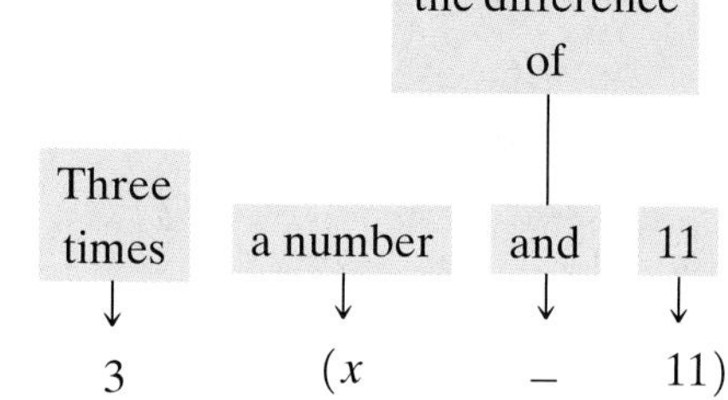

Translate: $3(x - 11)$

c. In words:

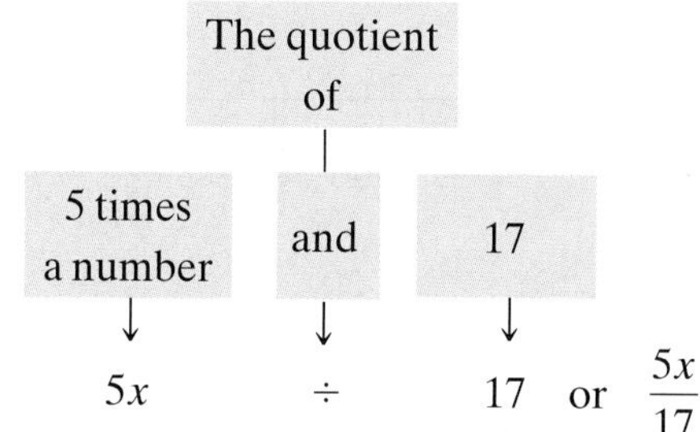

$5x \div 17$ or $\frac{5x}{17}$

STUDY SKILLS REMINDER

Are you making this class a priority?

Make sure your study habits are ones that will help you successfully complete this course. One useful habit is to make this class a priority by

- paying attention in class
- attending class
- arriving on time
- Completing assigned work on time

Answers

9. a. $5x - 25$ **b.** $2(x + 3)$ **c.** $39 \div 2x$ or $\frac{39}{2x}$

Name ______________________________ Section __________ Date __________

EXERCISE SET 3.3

A *Solve each equation. See Examples 1 through 4.*

1. $5x = 20$

2. $6y = 48$

3. $-3z = 12$

4. $-2x = 26$

5. $\frac{x}{7} = 1$

6. $\frac{y}{4} = 2$

7. $\frac{z}{-9} = 9$

8. $\frac{x}{-4} = 10$

9. $4y = 0$

10. $8x = -8$

11. $2z = -34$

12. $7y = 21$

13. $\frac{x}{-8} = -4$

14. $\frac{x}{-7} = -5$

15. $\frac{y}{-20} = 3$

16. $\frac{y}{-11} = 11$

17. $-3x = -15$

18. $-4z = -12$

19. $\frac{x}{-17} = 0$

20. $\frac{y}{-15} = 0$

Solve each equation. First combine any like terms on each side of the equation. See Examples 5 through 7.

21. $2w - 12w = 40$

22. $8y + y = 45$

23. $16 = 10t - 8t$

24. $100 = 15y + 5y$

25. $2z = 12 - 14$

26. $-3x = 11 - 2$

27. $4 - 10 = \frac{z}{-3}$

28. $20 - 22 = \frac{z}{-4}$

29. $-3x - 3x = 50 - 2$

30. $5y - 9y = -14 + (-14)$

31. $\frac{x}{5} = -26 + 16$

32. $\frac{y}{3} = 32 - 52$

Solve each equation. See Examples 1 through 7.

33. $-10x = 10$

34. $2z = -14$

35. $5x = -35$

36. $-3y = -27$

37. $0 = \frac{x}{3}$

38. $0 = \frac{x}{7}$

39. $24 = t + 3t$

40. $35 = 3r + 2r$

41. $10z - 3z = -63$

42. $x + 3x = -48$

43. $3z - 10z = -63$

44. $6y - y = 20$

45. $12 = 13y - 10y$

46. $20 = y - 6y$

47. $-4x = 20 - (-4)$

48. $6x = 5 - 35$

49. $18 - 11 = \frac{x}{-5}$

50. $9 - 14 = \frac{x}{-12}$

51. $-20 - (-50) = \frac{x}{9}$

52. $-2 - 10 = \frac{z}{10}$

53. $10p - 11p = 25$

54. $5q - 6q = 21$

55. $6x - x = 4 - 14$

56. $3x + x = 12 - 4$

57. $10 = 7t - 12t$

58. $-30 = t + 9t$

59. $5 - 5 = 3x + 2x$

60. $-42 + 20 = -2x + 13x$

61. $4r - 9r = -20$

62. $11v + 3v = 28$

63. $\frac{x}{-4} = -1 - (-8)$

64. $\frac{y}{-6} = 6 - (-1)$

65. $3w - 12w = -27$

66. $5t + 3t = -72$

67. $-36 = 9u - 10u$

68. $-50 = 4y - 5y$

69. $23x - 25x = 7 - 9$

70. $8x - 6x = 12 - 22$

B *Write each phrase as a variable expression. Use x to represent "a number." See Examples 8 and 9.*

71. The sum of -7 and a number

72. Negative eight plus a number

73. The product of -11 and a number

74. The quotient of -20 and a number, decreased by three

75. A number divided by -12

76. The quotient of six and a number, added to -80

77. Eleven subtracted from a number

78. A number subtracted from twelve

79. Negative ten decreased by 7 times a number

80. Twice a number decreased by thirty

81. The product of -13 and a number

82. Twice a number

83. The quotient of seventeen and a number, increased by -15

84. The difference of -9 times a number, and 1

85. Seven added to the product of 4 and a number

86. The product of 7 and a number, added to 100

87. Twice a number, decreased by 17

88. A number decreased by 14 and increased by 5 times the number

89. The product of -6 and the sum of a number and 15

90. Twice the sum of a number and -5

91. The quotient of 45 and the product of a number and -5

92. The quotient of ten times a number, and -4

Review and Preview

Evaluate each expression for $x = -5$. See Section 1.9.

93. $3x + 10$

94. $40x$

95. $\dfrac{x - 5}{2}$

96. $7x - 20$

97. $\dfrac{3x + 4}{x + 4}$

98. $\dfrac{2x - 4}{x - 2}$

Combining Concepts

99. Why does the multiplication property of equality not allow us to divide both sides of an equation by zero?

100. Is the equation $-x = 6$ solved for the variable? Explain why or why not.

Solve.

101. $-25x = 900$

102. $3y = -1259 - 3277$

103. $\dfrac{y}{72} = -86 - (-1029)$

104. $\dfrac{x}{-13} = 4^6 - 5^7$

105. $\dfrac{x}{-2} = 5^2 - |-10| - (-9)$

106. $\dfrac{y}{10} = (-8)^2 - |20| + (-2)^2$

For Exercises 107 through 110, the equation $d = r \cdot t$ describes the relationship between distance d in miles, rate r in miles per hour, and time t in hours.

107. The distance between New Orleans, Louisiana, and Kansas City, Missouri, by road, is approximately 780 miles. How long will it take to drive from New Orleans to Kansas City if the driver maintains a speed of 65 miles per hour? (*Source: 2003 World Almanac*)

108. The distance between Boston, Massachusetts, and Memphis, Tennessee, by road, is approximately 1320 miles. How long will it take to drive from Boston to Memphis if the driver maintains a speed of 60 miles per hour? (*Source: 2003 World Almanac*)

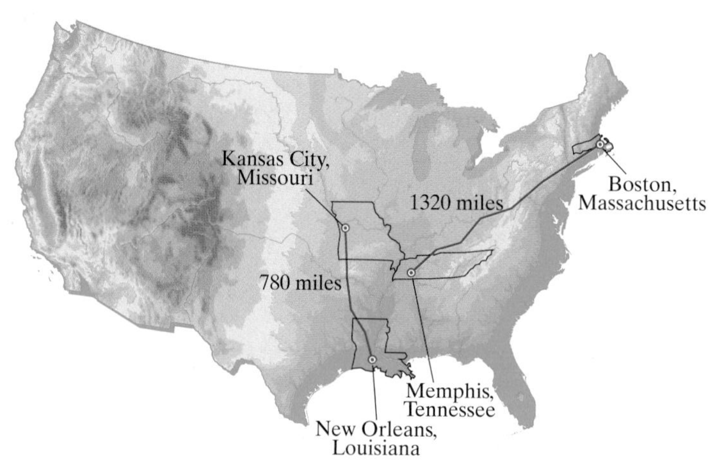

109. The distance between Toledo, Ohio, and Chicago, Illinois, by road, is approximately 232 miles. At what speed should a driver drive if he or she would like to make the trip in 4 hours? (*Source: 2003 World Almanac*)

110. The distance between St. Louis, Missouri, and Minneapolis, Minnesota, by road, is approximately 549 miles. If it took 9 hours to drive from St. Louis to Minneapolis, what was the driver's speed? (*Source: 2003 World Almanac*)

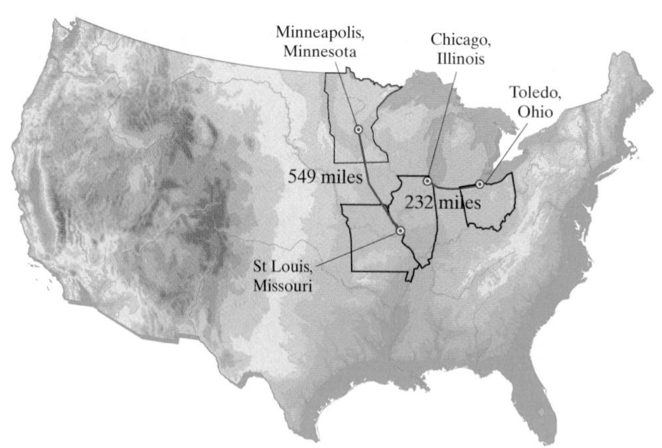

Name ______________________ Section __________ Date __________

Integrated Review—Expressions and Equations

Simplify each expression by combining like terms.

1. $7x + x$

2. $6y - 10y$

3. $2a + 5a - 9a - 2$

4. $6a - 12 - a - 14$

Multiply and simplify if possible.

5. $-2(4x + 7)$

6. $-3(2x - 10)$

7. $5(y + 2) - 20$

8. $12x + 3(x - 6) - 13$

9. Find the area of the rectangle.

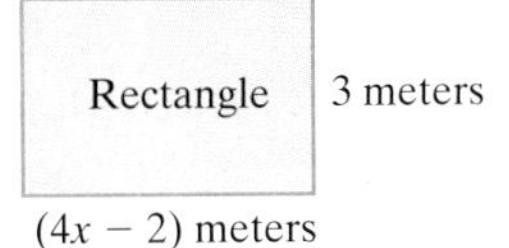

10. Find the perimeter of the triangle.

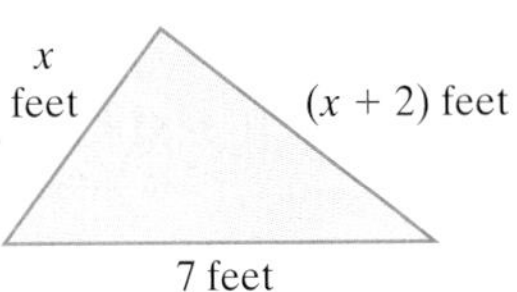

Solve and check.

11. $x + 7 = 20$

12. $-11 = x - 2$

13. $11x = 55$

14. $-7y = 0$

15. $\frac{x}{-3} = -13$

16. $\frac{z}{-5} = -11$

17. $12 = 11x - 14x$

18. $8y + 7y = -45$

19. $-3x = -15$

20. $-2m = -16$

21. $x - 12 = -45 + 23$

22. $6 - (-5) = x + 5$

Answers

1. ________
2. ________
3. ________
4. ________
5. ________
6. ________
7. ________
8. ________
9. ________
10. ________
11. ________
12. ________
13. ________
14. ________
15. ________
16. ________
17. ________
18. ________
19. ________
20. ________
21. ________
22. ________

23. ______

24. ______

25. ______

26. ______

27. ______

28. ______

29. ______

30. ______

31. ______

32. ______

33. ______

34. ______

35. ______

36. ______

23. $6(3x - 4) = 19x$

24. $25x = 6(4x - 9)$

25. $-36x - 10 + 37x = -12 - (-14)$

26. $-8 + (-14) = -80y + 20 + 81y$

27. $\frac{x}{13} = -1 - (-3)$

28. $\frac{y}{4} = 7 - 10$

29. $-8z - 2z = 26 - (-4)$

30. $-12 + (-13) = 5x - 10x$

Write each phrase as an algebraic expression. Use x to represent "a number."

31. The difference of a number and 10

32. The sum of -20 and a number

33. The product of 10 and a number

34. The quotient of 10 and a number

35. Five added to the product of -2 and a number

36. The product of -4 and the difference of a number and 1

3.4 Solving Linear Equations in One Variable

OBJECTIVES

A. Solve linear equations using the addition and multiplication properties.

B. Solve linear equations containing parentheses.

C. Write sentences as equations.

In this chapter, the equations we are solving are called **linear equations in one variable** or **first degree equations in one variable**. For example, an equation such as $5x - 2 = 6x$ is a linear equation in one variable. It is called linear or first degree because the exponent on each x is 1 and there is no variable below a fraction bar. It is an equation in one variable because it contains one variable, x.

A Solving Equations Using the Addition and Multiplication Properties

We will now solve linear equations in one variable using more than one property of equality. To solve an equation such as $2x - 6 = 18$, we first get the variable term $2x$ alone on one side of the equation.

EXAMPLE 1 Solve: $2x - 6 = 18$

Solution: We start by adding 6 to both sides to get the variable term $2x$ alone.

$2x - 6 = 18$

$2x - 6 + 6 = 18 + 6$ Add 6 to both sides.

$2x = 24$ Simplify.

To finish solving, we divide both sides by 2.

$\frac{2x}{2} = \frac{24}{2}$ Divide both sides by 2.

$x = 12$ Simplify.

Helpful Hint

Don't forget to check the proposed solution in the *original* equation.

Check: $2x - 6 = 18$

$2(12) - 6 \stackrel{?}{=} 18$ Replace x with 12 and simplify.

$24 - 6 \stackrel{?}{=} 18$

$18 = 18$ True.

The solution is 12.

Practice Problem 1

Solve: $5y + 2 = 17$

EXAMPLE 2 Solve: $20 - x = 21$

Solution: First we get the variable term alone on one side of the equation.

$20 - x = 21$

$20 - x - 20 = 21 - 20$ Subtract 20 from both sides.

$-1x = 1$ Simplify. Recall that $-x$ means $-1x$.

$\frac{-1x}{-1} = \frac{1}{-1}$ Divide both sides by -1.

$x = -1$ Simplify.

Check: $20 - x = 21$

$20 - (-1) \stackrel{?}{=} 21$

$21 = 21$ True.

The solution is -1.

Practice Problem 2

Solve: $45 = -10 - y$

Answers

1. 3 **2.** -55

Helpful Hint

Make sure you understand which property to use to solve an equation.

Addition	Understood multiplication
$x + 2 = 10$	$2x = 10$
To undo addition of 2, we subtract 2 from both sides.	To undo multiplication of 2, we divide both sides by 2.
$x + 2 - 2 = 10 - 2$ Use Addition Property of Equality.	$\frac{2x}{2} = \frac{10}{2}$ Use Multiplication Property of Equality.
$x = 8$	$x = 5$
Check: $x + 2 = 10$	Check: $2x = 10$
$8 + 2 = 10$	$2 \cdot 5 = 10$
$10 = 10$	$10 = 10$

If an equation contains variable terms on both sides, we use the addition property of equality to get all the variable terms on one side and all the constants or numbers on the other side.

Practice Problem 3

Solve: $7x + 12 = 3x - 4$

EXAMPLE 3 Solve: $3a - 6 = a + 4$

Solution:

$3a - 6 = a + 4$

$3a - 6 + 6 = a + 4 + 6$ Add 6 to both sides.

$3a = a + 10$ Simplify.

$3a - a = a + 10 - a$ Subtract a from both sides.

$2a = 10$ Simplify.

$\frac{2a}{2} = \frac{10}{2}$ Divide both sides by 2.

$a = 5$ Simplify.

Check to see that the solution is 5.

Concept Check

In Example 3, the solution is 5. If you wanted to check this solution, what equation should you use?

Try the Concept Check in the margin.

B Solving Equations Containing Parentheses

If an equation contains parentheses, we must first use the distributive property to remove them.

Practice Problem 4

Solve: $6(a - 5) = 4(a + 1)$

EXAMPLE 4 Solve: $7(x - 2) = 9x - 6$

Solution: First we apply the distributive property.

$7(x - 2) = 9x - 6$

$7x - 14 = 9x - 6$ Apply the distributive property.

Next we move variable terms to one side of the equation and constants to the other side.

$7x - 14 - 9x = 9x - 6 - 9x$ Subtract $9x$ from both sides.

$-2x - 14 = -6$ Simplify.

$-2x - 14 + 14 = -6 + 14$ Add 14 to both sides.

$-2x = 8$ Simplify.

$\frac{-2x}{-2} = \frac{8}{-2}$ Divide both sides by -2.

$x = -4$ Simplify.

Check to see that -4 is the solution.

Answers

3. -4 **4.** 17

Concept Check: $3a - 6 = a + 4$

You may want to use the following steps to solve equations.

Steps for Solving an Equation

Step 1. If parentheses are present, use the distributive property.

Step 2. Combine any like terms on each side of the equation.

Step 3. Use the addition property of equality to rewrite the equation so that variable terms are on one side of the equation and constant terms are on the other side.

Step 4. Use the multiplication property of equality to divide both sides by the numerical coefficient of the variable to solve.

Step 5. Check the solution in the *original equation*.

EXAMPLE 5 Solve: $3(2x - 6) + 6 = 0$

Solution: $3(2x - 6) + 6 = 0$

Step 1. $6x - 18 + 6 = 0$ Apply the distributive property.

Step 2. $6x - 12 = 0$ Combine like terms on the left side of the equation.

Step 3. $6x - 12 + 12 = 0 + 12$ Add 12 to both sides.

$6x = 12$ Simplify.

Step 4. $\frac{6x}{6} = \frac{12}{6}$ Divide both sides by 6.

$x = 2$ Simplify.

Check:

Step 5. $3(2x - 6) + 6 = 0$

$3(2 \cdot 2 - 6) + 6 \stackrel{?}{=} 0$

$3(4 - 6) + 6 \stackrel{?}{=} 0$

$3(-2) + 6 \stackrel{?}{=} 0$

$-6 + 6 \stackrel{?}{=} 0$

$0 = 0$ True.

The solution is 2.

Practice Problem 5

Solve: $4(x + 3) = 12$

C Writing Sentences as Equations

Next we practice translating sentences into equations. Below are key words and phrases that translate to an equal sign.

Key Words or Phrases	Examples	Symbols
equals	3 equals 2 plus 1	$3 = 2 + 1$
gives	the quotient of 10 and -5 gives -2	$\frac{10}{-5} = -2$
is/was/will be	x is 5	$x = 5$
yields	y plus 2 yields 13	$y + 2 = 13$
amounts to	twice x amounts to -30	$2x = -30$
is equal to	-24 is equal to 2 times -12	$-24 = 2(-12)$

Answer

5. 0

Practice Problem 6

Translate each sentence into an equation.

a. The difference of 110 and 80 is 30.

b. The product of 3 and the sum of -9 and 11 amounts to 6.

c. The quotient of twice 12 and -6 yields -4.

Answers

6. a. $110 - 80 = 30$ **b.** $3(-9 + 11) = 6$

c. $\frac{2(12)}{-6} = -4$

EXAMPLE 6 Translate each sentence into an equation.

a. The product of 7 and 6 is 42.

b. Twice the sum of 3 and 5 is equal to 16.

c. The quotient of -45 and 5 yields -9.

Solution:

a.

In words:	The product of 7 and 6	is	42
	↓	↓	↓
Translate:	$7 \cdot 6$	$=$	42

b.

In words:	Twice	the sum of 3 and 5	is equal to	16
	↓	↓	↓	↓
Translate:	2	$(3 + 5)$	$=$	16

c.

In words:	The quotient of -45 and 5	yields	-9
	↓	↓	↓
Translate:	$\frac{-45}{5}$	$=$	-9

CALCULATOR EXPLORATIONS

Checking Equations

A calculator can be used to check possible solutions of equations. To do this, replace the variable by the possible solution and evaluate each side of the equation separately. For example, to see whether 7 is a solution of the equation $52x = 15x + 259$, replace x with 7 and use your calculator to evaluate each side separately.

Equation: $52x = 15x + 259$

$52 \cdot 7 \stackrel{?}{=} 15 \cdot 7 + 259$ Replace x with 7.

Evaluate left side: [52] [×] [7] [=] or [ENTER]. Display: [364].

Evaluate right side: [15] [×] [7] [+] [259] [=] or [ENTER]. Display: [364].

Since the left side equals the right side, 7 is a solution of the equation $52x = 15x + 259$.

Use a calculator to determine whether the numbers given are solutions of each equation.

1. $76(x - 25) = -988$; 12

2. $-47x + 862 = -783$; 35

3. $x + 562 = 3x + 900$; -170

4. $55(x + 10) = 75x + 910$; -18

5. $29x - 1034 = 61x - 362$; -21

6. $-38x + 205 = 25x + 120$; 25

Name ______________________ Section ____________ Date ____________

EXERCISE SET 3.4

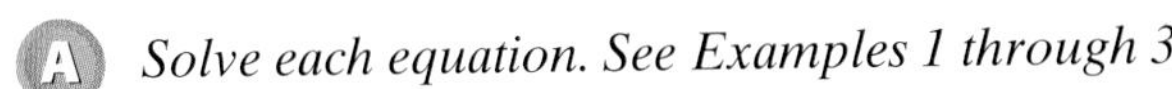

A *Solve each equation. See Examples 1 through 3.*

1. $2x - 6 = 0$

2. $3y - 12 = 0$

3. $\frac{n}{5} + 10 = 0$

4. $\frac{z}{4} + 8 = 0$

5. $6 - n = 10$

6. $7 - y = 9$

7. $10x + 15 = 6x + 3$

8. $5x - 3 = 2x - 18$

9. $3x - 7 = 4x + 5$

10. $3x + 1 = 8x - 4$

B *Solve each equation. See Examples 4 and 5.*

11. $3(x - 1) = 12$

12. $2(x + 5) = -8$

13. $-2(y + 4) = 2$

14. $-1(y + 3) = 10$

15. $35 = 17 + 3(x - 2)$

16. $22 - 42 = 4(x - 1)$

A B *Solve each equation. See Examples 1 through 5.*

17. $8 - t = 3$

18. $6 - x = 4$

19. $0 = 4x - 4$

20. $0 = 5y + 5$

21. $\frac{x}{-2} - 8 = 0$

22. $\frac{w}{-8} - 40 = 0$

23. $7 = 4c - 1$

24. $9 = 2b - 5$

25. $3r + 4 = 19$

26. $5m + 1 = 46$

27. $2x - 1 = -7$

28. $3t - 2 = -11$

29. $2 = 3z - 4$

30. $4 = 4p - 12$

31. $5x - 2 = -12$

32. $7y - 3 = -24$

33. $-7c + 1 = -20$

34. $-2b + 5 = -7$

35. $9 = \frac{x}{2} - 15$

36. $16 = \frac{y}{6} - 14$

37. $8m + 79 = -1$

38. $9a + 29 = -7$

39. $10 + 4v = -6$

40. $13 + 6b = 1$

41. $-5 = -13 - 8k$

42. $-7 = -17 - 10d$

43. $4x + 3 = 2x + 11$

44. $6y - 8 = 3y + 7$

45. $-2y - 10 = 5y + 18$

46. $7n + 5 = 12n - 10$

47. $-8n + 1 = -6n - 5$

48. $10w + 8 = w - 10$

49. $9 - 3x = 14 + 2x$

50. $4 - 7m = -3m$

51. $2(y - 3) = y - 6$

52. $3(z + 2) = 5z + 6$

53. $2t - 1 = 3(t + 7)$

54. $4 + 3c = 2(c + 2)$

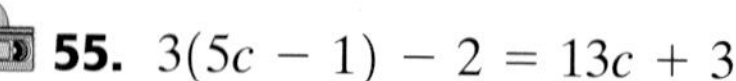

55. $3(5c - 1) - 2 = 13c + 3$

56. $4(3t + 4) - 20 = 3 + 5t$

57. $10 + 5(z - 2) = 4z + 1$

58. $14 + 4(w - 5) = 6 - 2w$

59. $7(6 + w) = 6(2 + w)$

60. $6(5 + c) = 5(c - 4)$

C *Write each sentence as an equation. See Example 6.*

61. The sum of -42 and 16 is -26.

62. The difference of -30 and 10 equals -40.

63. The product of -5 and -29 gives 145.

64. The quotient of -16 and 2 yields -8.

65. Three times the difference of -14 and 2 amounts to -48.

66. Negative 2 times the sum of 3 and 12 is -30.

67. The quotient of 100 and twice 50 is equal to 1.

68. Seventeen subtracted from -12 equals -29.

Review and Preview

The following bar graph shows the estimated number of U.S. federal individual income tax returns that will be filed electronically during the years shown. Electronically filed returns include Telefile and online returns. Use this graph to answer Exercises 69 through 72. See Section 1.3

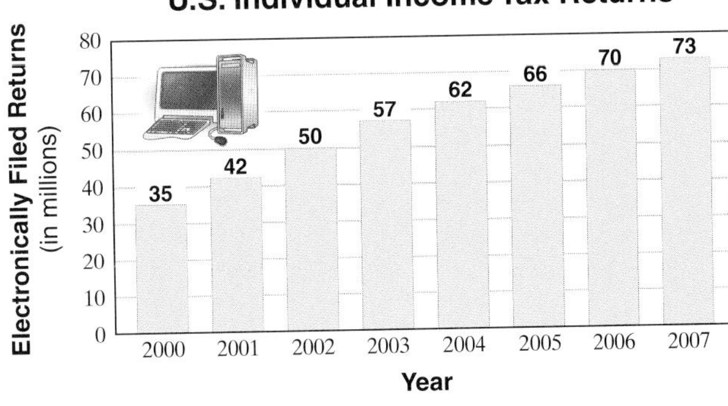

Source: IRS Compliance Research Division

69. What year shows the greatest increase in the number of electronically filed returns.

70. What year shows the smallest increase in the number of electronically filed returns.

71. By how much is the number of electronically filed returns expected to increase from 2000 to 2007?

72. Describe any trends shown in this graph.

Evaluate each expression for $x = 3$, $y = -1$, and $z = 0$. See Section 2.5.

73. $x^3 - 2xy$

74. $y^3 + 3xyz$

75. $y^5 - 4x^2$

76. $(-y)^3 + 3xyz$

77. $(2x - y)^2$

78. $(6z - 5y)^3$

79. $4xy + 6y^2$

80. $2x^2 - 7z$

Combining Concepts

Solve.

81. $(-8)^2 + 3x = 5x + 4^3$

82. $3^2 \cdot x = (-9)^3$

83. $2^3(x + 4) = 3^2(x + 4)$

84. $x + 45^2 = 54^2$

85. A classmate tries to solve $3x = 39$ by subtracting 3 from both sides of the equation. Will this step solve the equation for x? Why or why not?

86. A classmate tries to solve $2 + x = 20$ by dividing both sides by 2. Will this step solve the equation for x? Why or why not?

Focus on the Real World

SCALES

In this chapter, we have been using a scale to help illustrate true equations and the properties of equality. Scales of different types are used everyday to measure the mass or weight of objects. The scale shown is called an equal-arm balance. It was developed by the ancient Egyptians around 2500 B.C. to assist in everyday business. To use it, an object with an unknown mass is placed in one pan and objects with known masses are placed in the other pan one at a time, until the arm to which the pans are attached is perfectly level. When the pans are balanced in this way, the user knows that the object with the unknown mass has the same mass as the sum of the known masses in the other pan.

The ancient Romans invented a different kind of scale, called a steelyard, about 2000 years ago. The Romans' steelyard used a pan to hold the unknown mass on one end of the scale's arm and a known mass that slid along the other end of the arm. The position of the sliding mass over markings along the arm indicated the mass of the object in the pan. A variation of the steelyard can still be seen today in many doctors' offices.

Today, most of the scales we use on a daily basis measure weights and masses using springs, pendulums, or changes in electrical resistance. However, the most accurate scales in the world are based on the same balancing principles used in the Egyptians' equal-arm balance and the Romans' steelyard. These highly precise scales can measure accurately to a millionth of a gram.

GROUP ACTIVITY

Borrow an equal-arm balance or construct a simple one of your own. Use the scale to investigate and demonstrate the properties of equality.

3.5 Linear Equations in One Variable and Problem Solving

OBJECTIVES

A. Translate sentences into mathematical equations.

B. Use problem-solving steps to solve problems.

SSM TUTOR CENTER SG CD & VIDEO MATH PRO WEB

A Writing Sentences as Equations

Now that we have practiced solving equations for a variable, we can extend our problem-solving skills considerably. We begin by reviewing how to translate sentences into algebraic equations using the same key words and phrases table from section 3.3.

Addition	Subtraction	Multiplication	Division	Equal Sign
sum	difference	product	quotient	equals
plus	minus	times	divided by	gives
added to	subtracted from	multiply	into	is/was/will be
more than	less than	twice	per	yields
increased by	decreased by	of		amounts to
total	less	double		is equal to

EXAMPLE 1 Write each sentence as an equation. Use x to represent "a number."

a. Nine more than a number is 5.
b. Twice a number equals -10.
c. A number minus 6 amounts to 168.
d. Three times the sum of a number and 5 is -30.
e. The quotient of 8 and twice a number is equal to 2.

Solution:

a. In words: Nine | more than | a number | is | 5

Translate: $9 \quad + \quad x \quad = \quad 5$

b. In words: Twice a number | equals | -10

Translate: $2x \quad = \quad -10$

c. In words: A number | minus | 6 | amounts to | 168

Translate: $x \quad - \quad 6 \quad = \quad 168$

d. In words: Three times | the sum of a number and 5 | is | -30

Translate: $3 \quad (x + 5) \quad = \quad -30$

e. In words: The quotient of

Translate: 8 | and | twice a number | is equal to | 2

$8 \quad \div \quad 2x \quad = \quad 2$

or $\dfrac{8}{2x} = 2$

Practice Problem 1

Write each sentence as an equation. Use x to represent "a number."

a. Five times a number is 20.
b. The sum of a number and -5 yields 14.
c. Ten subtracted from a number amounts to -23.
d. Five times a number added to 7 is equal to -8.
e. The quotient of 6 and the sum of a number and 4 gives 1.

Answers

1. a. $5x = 20$ **b.** $x + (-5) = 14$ **c.** $x - 10 = -23$ **d.** $7 + 5x = -8$ **e.** $\dfrac{6}{x + 4} = 1$

B Use Problem-Solving Steps to Solve Problems

Our main purpose for studying arithmetic and algebra is to solve problems. In previous sections, we have prepared for problem solving by writing phrases as algebraic expressions and sentences as equations. We now draw upon this experience as we solve problems. The following problem-solving steps will be used throughout this text.

Problem-Solving Steps

1. UNDERSTAND the problem. During this step, become comfortable with the problem. Some ways of doing this are as follows:
   - Read and reread the problem.
   - Choose a variable to represent the unknown.
   - Construct a drawing, if possible.
   - Propose a solution and check it. Pay careful attention to how you check your proposed solution. This will help when writing an equation to model the problem.
2. TRANSLATE the problem into an equation.
3. SOLVE the equation.
4. INTERPRET the results. *Check* the proposed solution in the stated problem and *state* your conclusion.

Practice Problem 2

Translate "the sum of a number and 2 equals 6 added to three times the number" into an equation and solve.

EXAMPLE 2 Finding an Unknown Number

Twice a number, added to 3, is the same as the number minus 6. Find the number.

Solution:

1. UNDERSTAND the problem. To do so, we read and reread the problem. To help us understand the problem better, let's propose a solution and check it. Suppose that the unknown number is 10. From the stated problem, "twice 10, added to 3" is 23, but "10 minus 6" is 4. Since 23 is not equal to 4, the solution is not 10. The purpose of proposing a solution is not to guess correctly but to better understand the problem.

   Now, let's assign a variable to the unknown. We will let

   $x =$ the unknown number.

2. TRANSLATE the problem into an equation.

In words:	Twice a number	added to 3	is the same as	the number minus 6
	↓	↓	↓	↓
Translate:	$2x$	$+\ 3$	$=$	$x - 6$

3. SOLVE the equation. To solve the equation, we first subtract x from both sides.

$$2x + 3 = x - 6$$
$$2x + 3 - x = x - 6 - x \quad \text{Subtract } x \text{ from both sides.}$$
$$x + 3 = -6 \quad \text{Simplify.}$$
$$x + 3 - 3 = -6 - 3 \quad \text{Subtract 3 from both sides.}$$
$$x = -9 \quad \text{Simplify.}$$

Answer

2. -2

4. INTERPRET the results. First, *check* the proposed solution in the stated problem. Twice "−9" is −18 and −18 + 3 is −15. This is equal to the number minus 6 or "−9" −6 or −15. Then *state* your conclusion: The unknown number is −9.

Try the Concept Check in the margin.

EXAMPLE 3 Determining Voter Counts

In the 2000 Senate election in Wyoming, incumbent Craig Thomas received 100,280 *more* votes than his challenger. If a total of 204,358 votes were cast, find how many votes Craig Thomas received. (*Source: World Almanac*)

Solution:

1. UNDERSTAND the problem. We read and reread the problem. Then we assign a variable to an unknown. We use this variable to represent any other unknown quantities. We let

x = the number of challenger votes

Then

$x + 110{,}280 =$ the number of incumbent votes since Thomas received 110,280 more votes

2. TRANSLATE the problem into an equation.

In words:	challenger votes	plus	incumbent votes	is	total votes
	↓	↓	↓	↓	↓
Translate:	x	$+$	$x + 110{,}280$	$=$	$204{,}358$

3. SOLVE the equation:

$$x + x + 110{,}280 = 204{,}358$$

$$2x + 110{,}280 = 204{,}358 \quad \text{Combine like terms.}$$

$$2x + 110{,}280 - 110{,}280 = 204{,}358 - 110{,}280 \quad \text{Subtract 110,280 from both sides.}$$

$$2x = 94{,}078 \quad \text{Simplify.}$$

$$\frac{2x}{2} = \frac{94{,}078}{2} \quad \text{Divide both sides by 2.}$$

$$x = 47{,}039 \quad \text{Simplify.}$$

4. INTERPRET the results. First *Check* the proposed solution in the stated problem. Since x represents the number of votes the challenger received, the challenger received 47,039 votes. The incumbent received $x + 110{,}280 = 47{,}039 + 110{,}280 = 157{,}319$ votes. To check, notice that the total number of challenger votes and incumbent votes is $47{,}039 + 157{,}319 = 204{,}358$ votes, the given total of votes cast. Also, 157,319 is 110,280 more votes than 47,039, so the solution checks. Then, *state* your conclusion: The incumbent, Craig Thomas, received 157,319 votes.

Try the Concept Check in the margin.

Concept Check

Suppose you have solved an equation involving perimeter to find the length of a rectangular table. Explain why you would want to recheck your math if you obtain the result of −5.

Practice Problem 3

At a recent United States/Japan summit meeting, 121 delegates attended. If the United States sent 19 more delegates than Japan, find how many the United States sent.

Concept Check

When solving the problem given in Example 3, why should you be skeptical if you obtain an answer of 15,714 votes for the incumbent?

Answers

3. 70 delegates

Concept Check: Length cannot be negative.

Concept Check: Answers may vary.

Practice Problem 4

A woman's \$21,000 estate is to be divided so that her husband receives twice as much as her son. How much will each receive?

Answers

4. husband: \$14,000; son: \$7000

EXAMPLE 4 Calculating Separate Costs

Leo Leal sold a used computer system and software for \$2100, receiving four times as much money for the computer system as for the software. Find the price of each.

Solution:

1. UNDERSTAND the problem. We read and reread the problem. Then we assign a variable to an unknown. We use this variable to represent any other unknown quantities. We let

x = the software price

$4x$ = the computer system price

2. TRANSLATE the problem into an equation.

In words:	Software price	plus	computer price	is	2100
	↓	↓	↓	↓	↓
Translate:	x	$+$	$4x$	$=$	2100

3. SOLVE the equation.

$$x + 4x = 2100$$

$$5x = 2100 \quad \text{Combine like terms.}$$

$$\frac{5x}{5} = \frac{2100}{5} \quad \text{Divide both sides by 5.}$$

$$x = 420 \quad \text{Simplify.}$$

4. INTERPRET the results. *Check* the proposed solution in the stated problem. The software sold for \$420. The computer system sold for $4x = 4(\$420) = \$1680$. Since $\$420 + \$1680 = \$2100$, the total price, and \$1680 is four times \$420, the solution checks. State your conclusion: The software sold for \$420 and the computer system sold for \$1680.

Name ____________________ Section __________ Date __________

EXERCISE SET 3.5

A *Write each sentence as an equation. Use x to represent "a number." See Example 1.*

1. A number added to -5 is -7.

2. Five subtracted from a number equals 10.

3. Three times a number yields 27.

4. The quotient of 8 and a number is -2.

5. A number subtracted from -20 amounts to 104.

6. Two added to twice a number gives -14.

7. Twice the sum of a number and -1 is equal to 50.

8. Three times the difference of 7 and a number amounts to -40.

B *Solve. See Example 2.*

9. Three times a number, added to 9, is 33. Find the number.

10. Twice a number, subtracted from 60, is 20. Find the number.

11. The product of 9 and a number gives 54. Find the number.

12. The product of 7 and a number is 14. Find the number.

13. The sum of 3, 4, and a number amounts to 16. Find the number.

14. A number less 5 is 11. Find the number.

15. Seventy-two is 8 times a number added to 24. Find the number.

16. The product of 11 and a number is 121. Find the number.

17. The difference of a number and 3 is the same as 45 less the number. Find the number.

18. Sixty-four less half of 64 is equal to the sum of some number and 20. Find the number.

19. Three times the difference of some number and 5 amounts to 9. Find the number.

20. Five times the sum of some number plus 2 is equal to 8 times the number, minus 11. Find the number.

21. Eight decreased by some number equals the quotient of 15 and 5. Find the number.

22. Ten less some number is twice the sum of that number and 5. Find the number.

23. Thirteen added to the product of 3 and some number amounts to 5 times the number added to 3.

24. The product of 4 and a number is the same as 30 less twice that same number.

25. Five times a number, less 40, is 8 more than the number.

26. Thirty less a number is equal to 3 times the sum of the number and 6.

27. The difference of a number and 3 is equal to the quotient of 10 and 5.

28. The product of a number and 3 is twice the sum of that number and 5.

Solve. See Examples 3 and 4.

29. In the 2000 presidential election, George W. Bush received 5 more electoral votes than Al Gore. If a total of 527 electoral votes were cast for the two candidates, find how many votes each candidate received. (*Source:* Voter News Service)

30. Based on the 2000 Census, California has 21 more electoral votes for president than Texas. If the total number of electoral votes for these two states is 89, find the number for each state. (*Source: The World Almanac* 2003)

31. Bamboo and Pacific Kelp, a kind of sea weed, are two fast-growing plants. Bamboo grows twice as fast as kelp. If in one day both can grow a total of 54 inches, find how many inches each plant can grow in one day.

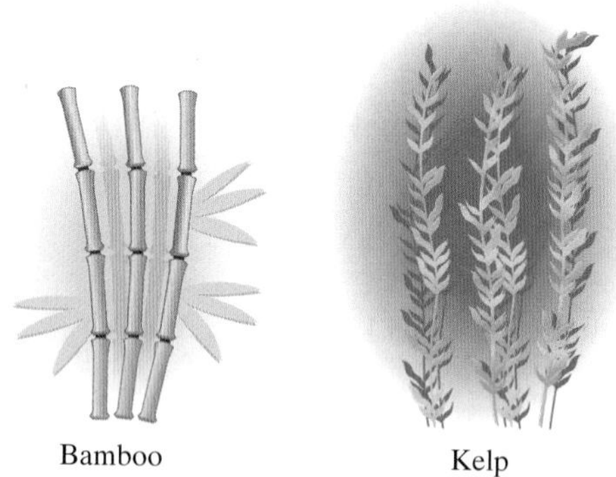

32. Norway has had three times as many rulers as Liechtenstein. If the total rulers for both countries is 56, find the number of rulers for Norway and the number for Liechtenstein.

33. The country with the most universities* is India, followed by the United States. If India has 2649 more universities than the U.S. and their combined total is 14,165, find the number of universities in India and the number in the U.S. [* Includes all further education establishments. (*Source*: The Top 10 of Everything, 2003)]

34. The average life expectancy for a man is 34 years longer than the life expectancy for a polar bear. If the total of these life expectancies is 110 years, find the life expectancy of each.

35. A Nintendo Gamecube and several games are sold for \$600. The cost of the games is 3 times as much as the cost of the Gamecube. Find the cost of the Gamecube and the cost of the games.

36. Disneyland®, in Anaheim, California, has 12 more rides than its neighboring sister park, Disney's California Adventure, which opened in 2001. Together, the two parks have 36 rides. How many rides does each park have? (*Source:* The Walt Disney Company)

37. The two NCAA stadiums with the largest capacities are Michigan Stadium (Univ. of Michigan) and Neyland Stadium (Univ. of Tennessee). Michigan Stadium has a capacity of 4647 more than Neyland. If the combined capacity for the two stadiums is 210,355, find the capacity for each stadium. (*Source:* National Collegiate Athletic Association)

38. An NHRA top fuel dragster has a top speed of 95 mph faster than an Indy Racing League car. If the combined top speeds for these two cars is 565 mph, find the top speed of each car. (*Source:* USA Today)

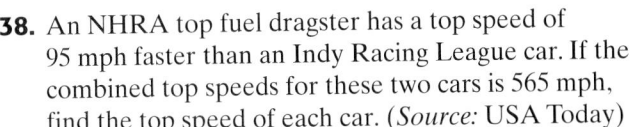

39. California contains the largest state population of native Americans. This population is three times the native American population of Washington state. If the total of these two populations is 412 thousand find the native American population in each of these two states. (*Source*: U.S. Census Bureau)

40. In 2020, China is projected to be the country with the greatest number of visiting tourists. This number is twice the number of tourists projected for Spain. If the total number of tourists for these two countries is projected to be 210 million, find the number projected for each. (*Source: The State of the World Atlas* by Dan Smith)

41. During the 2002 Women's NCAA Division I basketball championship game, the Connecticut Huskies scored 12 more points than the Oklahoma Sooners. Together, both teams scored a total of 152 points. How many points did the 2002 Champion Connecticut Huskies score during the game? (*Source:* National Collegiate Athletic Association)

42. During the 2002 Men's NCAA Division I Basketball Championship game, the Indiana Hoosiers scored 12 fewer points than the Maryland Terrapins. Together, both teams scored a total of 116 points. How many points did the 2002 Champion Maryland Terrapins score during the game? (*Source:* National Collegiate Athletic Association)

43. A crow will eat five more ounces of food a day than a finch. If together they eat 13 ounces of food, find how many ounces of food the crow consumes and how many ounces of food the finch consumes.

44. A Toyota Camry is traveling twice as fast as a Dodge truck. If their combined speed is 105 miles per hour, find the speed of the car and find the speed of the truck.

Review and Preview

Round each whole number to the given place. See Section 1.5.

45. 586 to the nearest ten

46. 82 to the nearest ten

47. 1026 to the nearest hundred

48. 52,333 to the nearest thousand

49. 2986 to the nearest thousand

50. 101,552 to the nearest hundred

Combining Concepts

51. Solve Example 3 again, but this time let x be the number of incumbent votes. Did you get the same results? Explain why or why not.

In real estate, a house's selling price P is found by adding the real estate agent's commission C to the amount A that the seller of the house receives: $P = A + C$.

52. Brianna Morley's house sold for \$230,000. Her real estate agent received a commission of \$13,800. How much did Brianna receive? (*Hint:* Substitute the known values into the equation, then solve the equation for the remaining unknown.)

53. Duncan Bostic plans to use a real estate agent to sell his house. He hopes to sell the house for \$165,000 and keep \$156,750 of that. If everything goes as he has planned, how much will his real estate agent receive as a commission?

In retailing, the retail price P of an item can be computed using the equation $P = C + M$, where C is the wholesale cost of the item and M is the amount of markup.

54. The retail price of a computer system is \$999 after a markup of \$450. What is the wholesale cost of the computer system? (*Hint:* Substitute the known values into the equation, then solve the equation for the remaining unknown.)

55. Slidell Feed and Seed sells a bag of cat food for \$12. If the store paid \$7 for the cat food, what is the markup on the cat food?

Internet Excursions

Go To:  http://www.prenhall.com/martin-gay_prealgebra  What's Related

Project Vote Smart is a nonpartisan, nonprofit organization that strives to provide the American public with unbiased information about more than 13,000 of their elected public officials as well as candidates running for public office. The Project Vote Smart goal is to help American voters make informed choices.

This World Wide Web address will provide you with access to Project Vote Smart's listing of links to official state election information, or a related site. Use this listing to connect to election result information for your state.

56. Use election results from your state's most recent governor's race to write a problem similar to Exercises 29 and 30. Exchange problems with a classmate to solve.

57. Using election results from another recent election in your state, write another problem like Exercises 29 and 30. Exchange with another student in your class to solve.

CHAPTER 3 ACTIVITY Using Equations

This activity may be completed by working in groups or individually.

A hospital nurse working second shift has been keeping track of the fluid intake and output on the chart of one of her patients, Mr. Ramirez. At the end of her shift, she totals the intakes and outputs and returns Mr. Ramirez's chart to the floor nurses' station.

Suppose you are one of the night nurses for this floor and have been assigned to Mr. Ramirez. Shortly after you come on duty, someone at the nurses' station spills coffee over several patients' charts, including the one for Mr. Ramirez. After blotting up the coffee on the chart, you notice that several entries on his chart have been smeared by the coffee spill.

MEDICAL CHART

Patient: Juan Ramirez Room: 314

Shift: Second Nurse's Initials: SRT Date: 3/27

Fluid Intake (cubic centimeters):

					Totals
Blood	500				500
Intravenous	250		100		500
Oral Fluid	300	150		100	900
Oral Meds	4 doses of		cc each		200
TOTAL					2100

Fluid Output (cubic centimeters):

				Totals
Urine	240	310		550
Emesis	120			120
Irrigation	50	25		75
TOTAL				815

1. On the Fluid Intake chart, is it possible to tell if Mr. Ramirez was given blood more than once? If so, how many times was he given blood and how can you tell?

2. On the Fluid Intake chart, is it possible to tell if Mr. Ramirez was given intravenous fluid more than once? Is it possible to tell exactly how many times Mr. Ramirez was given intravenous fluid? Explain.

3. For the Oral Fluid row of the Fluid Intake chart, notice that the third entry in that row is obliterated by the coffee stain. Let the variable x represent this third entry. Write an equation that describes the total amount of oral fluids Mr. Ramirez received during the second shift. Then solve the equation for x to find the obliterated entry.

4. For the Oral Meds row of the Fluid Intake chart, notice that the second-shift nurse indicated that Mr. Ramirez was given four equal doses of oral medication. We cannot see the size of each dose, but the chart shows that he received a total of 200 cc. Write and solve an equation to find the size of each dose.

5. Using the Fluid Output chart, write and solve an equation to find the total urine output.

6. Using the Fluid Output chart and the result from Question 5, write and solve an equation to find the obliterated entry in the Urine output row.

Chapter 3 Vocabulary Check

Fill in each blank with one of the words or phrases listed below.

Variable	Simplified	numerical coefficient	equation
terms	combined	algebraic expression	solution
like	constant	evaluating the expression	

1. An algebraic expression is __________ when all like terms have been __________ .
2. Terms that are exactly the same, except that they may have different numerical coefficients, are called _____ terms.
3. A letter used to represent a number is called a_______.
4. A combination of operations on variables and numbers is called an __________________.
5. The addends of an algebraic expression are called the ______ of the expression.
6. The number factor of a variable term is called the __________________ .
7. Replacing a variable in an expression by a number and then finding the value of the expression is called __________ ____________ for the variable.
8. A term that is a number only is called a _________.
9. An _________ is of the form expression = expression.
10. A ________ of an equation is a value for the variable that makes an equation a true statement.

CHAPTER 3 Highlights

DEFINITIONS AND CONCEPTS	EXAMPLES
SECTION 3.1 SIMPLIFYING ALGEBRAIC EXPRESSIONS	
The addends of an algebraic expression are called the **terms** of the expression.	$5x^2 + (-4x) + (-2)$ — 3 terms
The number factor of a variable term is called the **numerical coefficient**.	**Term** / **Numerical Coefficient**: 7x — 7; $-6y$ — -6; x or $1x$ — 1
Terms that are exactly the same, except that they may have different numerical coefficients, are called **like terms**.	$5x + 11x = (5 + 11)x = 16x$ like terms $y - 6y = (1 - 6)y = -5y$
An algebraic expression is **simplified** when all like terms have been **combined**.	
Use the **distributive property** to multiply an algebraic expression by a term.	Simplify: $-4(x + 2) + 3(5x - 7)$ $= -4(x) + (-4)(2) + 3(5x) + 3(-7)$ $= -4x + (-8) + 15x + (-21)$ $= 11x + (-29)$ or $11x - 29$

Definitions and Concepts / Examples

Section 3.2 Solving Equations: The Addition Property

Addition Property of Equality

Let a, b, and c represent numbers.

If $a = b$, then

$a + c = b + c$ and $a - c = b - c$

In other words, the same number may be added to or subtracted from both sides of an equation without changing the solution of the equation.

Solve: $x + 8 = 2 + (-1)$

$x + 8 = 1$ — Combine like terms.

$x + 8 - 8 = 1 - 8$ — Subract 8 from both sides.

$x = -7$ — Simplify.

The solution is -7.

Section 3.3 Solving Equations: The Multiplication Property

Multiplication Property of Equality

Let a, b, and c represent numbers and let $c \neq 0$.

If $a = b$, then

$a \cdot c = b \cdot c$ and $\frac{a}{c} = \frac{b}{c}$

In other words, both sides of an equation may be multiplied or divided by the same nonzero number without changing the solution of the equation.

Solve: $y - 7y = 30$

$-6y = 30$ — Combine like terms.

$\frac{-6y}{-6} = \frac{30}{-6}$ — Divide both sides by -6.

$y = -5$ — Simplify.

The solution is -5.

Section 3.4 Solving Linear Equations in One Variable

Steps for Solving an Equation

Step 1. If parentheses are present, use the distributive property to remove them.

Step 2. Combine any like terms on each side of the equation.

Step 3. Use the addition property of equality to rewrite the equation so that variable terms are on one side of the equation and constant terms are on the other side.

Step 4. Use the multiplication property of equality to divide both sides by the numerical coefficient of the variable to solve.

Step 5. Check the solution in the *original equation*.

Solve: $5(3x - 1) + 15 = -5$

Step 1. $15x - 5 + 15 = -5$ — Apply the distributive property.

Step 2. $15x + 10 = -5$ — Combine like terms.

Step 3. $15x + 10 - 10 = -5 - 10$ — Subtract 10 from both sides.

$15x = -15$

Step 4. $\frac{15x}{15} = \frac{-15}{15}$ — Divide both sides by 15.

$x = -1$

Step 5. Check to see that -1 is the solution.

Section 3.5 Linear Equations in One Variable and Problem Solving

Problem-Solving Steps

1. UNDERSTAND the problem. Some ways of doing this are as follows:
   - Read and reread the problem.
   - Construct a drawing, if possible.
   - Assign a variable to an unknown in the problem.
   - Propose a solution and check.

The incubation period for a golden eagle is three times the incubation period for a hummingbird. If the total of their incubation periods is 60 days, find the incubation period for each bird. (*Source: Wildlife Fact File*, International Masters Publishers)

1. UNDERSTAND the problem. Then assign a variable. Let

x = incubation period of a hummingbird

$3x$ = incubation period of a golden eagle

DEFINITIONS AND CONCEPTS

EXAMPLES

SECTION 3.5 LINEAR EQUATIONS IN ONE VARIABLE AND PROBLEM SOLVING (CONTINUED)

2. TRANSLATE the problem into an equation.

2. TRANSLATE.

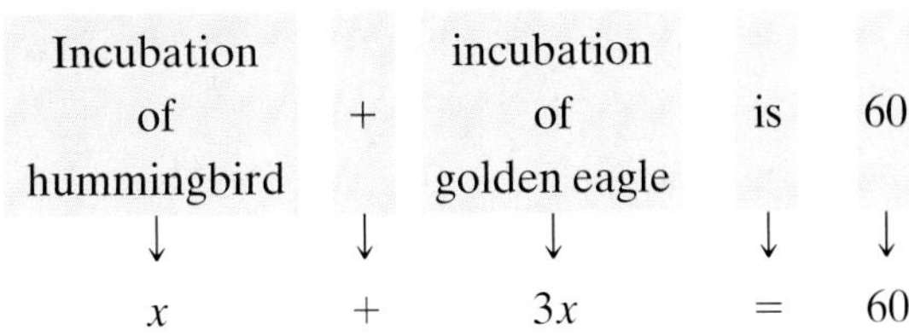

$$x + 3x = 60$$

3. SOLVE the equation.

3. SOLVE.

$$x + 3x = 60$$
$$4x = 60$$
$$\frac{4x}{4} = \frac{60}{4}$$
$$x = 15$$

4. INTERPRET the results. *Check* the proposed solution in the stated problem and *state* your conclusion.

4. INTERPRET the solution in the stated problem. The incubation period for the hummingbird is 15 days. The incubation period for the golden eagle is $3x = 3 \cdot 15 = 45$ days. Since 15 days + 45 *days* = 15 days + 45 days = 60 days and 45 is 3(15), the solution checks. State your conclusion: The incubation period for the hummingbird is 15 days. The incubation period for the golden eagle is 45 days.

As a convenient summary and for quick reference, these problem-solving steps are provided.

PROBLEM-SOLVING STEPS

1. UNDERSTAND the problem. During this step, become comfortable with the problem. Some ways of doing this are as follows:
- Read and reread the problem.
- Choose a variable to represent the unknown.
- Construct a drawing.
- Propose a solution and check it. Pay careful attention to how you check your proposed solution. This will help when writing an equation to model the problem.

2. TRANSLATE the problem into an equation.

3. SOLVE the equation.

4. INTERPRET the results. *Check* the proposed solution in the stated problem and *state* your conclusion.

STUDY SKILLS REMINDER

Are you organized?

Have you ever had trouble finding a completed assignment? When it's time to study for a test, are your notes neat and organized? Have you ever had trouble reading your own mathematics handwriting? (Be honest—I have.)

When any of these things happen, it's time to get organized. Here are a few suggestions:

Write your notes and complete your homework assignment in a notebook with pockets (spiral or ring binder.) Take class notes in this notebook, and then follow the notes with your completed homework assignment. When you receive graded papers or handouts, place them in the notebook pocket so that you will not lose them.

Remember to mark (possibly with an exclamation point) any note(s) that seem extra important to you. Also remember to mark (possibly with a question mark) any notes or homework that you are having trouble with. Don't forget to see your instructor or a math tutor to help you with the concepts or exercises that you are having trouble understanding.

Also, if you are having trouble reading your own handwriting, *slow down* and write your mathematics work clearly!

Name ____________ Section ________ Date ________

Chapter 3 Review

(3.1) *Simplify each expression by combining like terms.*

1. $3y + 7y - 15$

2. $2y - 10 - 8y$

3. $8a + a - 7 - 15a$

4. $y + 3 - 9y - 1$

Multiply.

5. $2(x + 5)$

6. $-3(y + 8)$

Simplify.

7. $7x + 3(x - 4) + x$

8. $-(3m + 2) - m - 10$

9. $3(5a - 2) - 20a + 10$

10. $6y + 3 + 2(3y - 6)$

Find the perimeter of each figure.

11.

$2x$ yards

3 yards

Rectangle

12.

Square

$5y$ meters

Find the area of each figure.

13.

$(2x - 1)$ yards

3 yards

Rectangle

14.

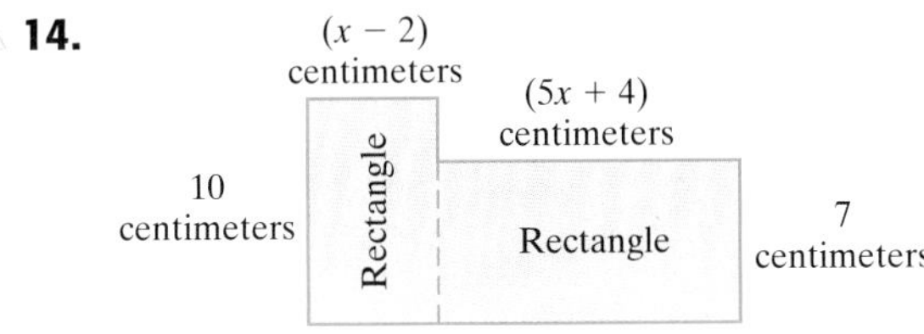

Simplify.

15. $85(7068x - 108) + 42x$

16. $-4268y + 120(63y - 32)$

(3.2)

17. Is 4 a solution of $5(2 - x) = -10$?

18. Is 0 a solution of $6y + 2 = 23 + 4y$?

Solve each equation.

19. $z - 5 = -7$

20. $3x + 10 = 4x$

21. $n + 18 = 10 - (-2)$

22. $c - 5 = -13 + 7$

23. $7x + 5 - 6x = -20$

24. $17x = 2(8x - 4)$

(3.3) *Solve each equation.*

25. $3y = -21$

26. $-8x = 72$

27. $-5n = -5$

28. $-3a = -15$

29. $\frac{x}{-6} = 2$

30. $\frac{y}{-15} = -3$

31. $-5t + 32 + 4t = 32$

32. $3z + 72 - 2z = -56$

33. $\frac{z}{4} = -8 - (-6)$

34. $-1 + (-8) = \frac{x}{5}$

35. $6y - 7y = 100 - 105$

36. $19x - 16x = 45 - 60$

Translate each phrase into an algebraic expression. Let x represent "a number."

37. The product of -5 and a number

38. Three subtracted from a number

39. The sum of -5 and a number

40. The quotient of -2 and a number

41. Eleven added to twice a number

42. The product of -5 and a number, decreased by 50

43. The quotient of 70 and the sum of a number and 6

44. Twice the difference of a number and 13

(3.4) *Solve each equation.*

45. $3x - 4 = 11$

46. $6y + 1 = 73$

47. $14 - y = -3$

48. $7 - z = 0$

49. $4z - z = -6$

50. $t - 9t = -64$

51. $2x + 5 = 7x - 100$

52. $-6x - 4 = x + 66$

53. $2x + 7 = 6x - 1$

54. $5x - 18 = -4x$

55. $\frac{x}{9} + 3 = 0$

56. $\frac{z}{-2} - 11 = 0$

57. $5(n - 3) = 7 + 3n$

58. $7(2 + x) = 4x - 1$

Write each sentence as an equation.

59. The difference of 20 and -8 is 28.

60. Five times the sum of 2 and -6 yields -20.

61. The quotient of -75 and the sum of 5 and 20 is equal to -3.

62. Nineteen subtracted from -2 amounts to -21.

(3.5) *Write each sentence as an equation using x as the variable.*

63. Twice a number minus 8 is 40.

64. Twelve subtracted from the quotient of a number and 2 is 10.

65. The difference of a number and 3 is the quotient of the number and 4.

66. The product of some number and 6 is equal to the sum of the number and 2.

Solve.

67. Five times a number subtracted from 40 is the same as three times the number. Find the number.

68. The product of a number and 3 is twice the difference of that number and 8. Find the number.

69. In an election the incumbent received 14,000 votes of the 18,500 votes cast. Of the remaining votes, the Democratic candidate received 272 more than the Independent candidate. Find how many votes the Democratic candidate received.

70. Rajiv Puri has twice as many movies on DVDs as he has video tapes. Find the number of DVDs if he has a total of 126 movie recordings.

STUDY SKILLS REMINDER

Are you prepared for a test on Chapter 3?

Below I have listed some *common trouble areas* for topics covered in Chapter 3. After studying for your test, but before taking your test, read these.

- Be careful when evaluating expressions. For example, evaluate $3x - y$ when $x = -2$ and $y = -3$.

$$\begin{aligned} 3x - y &= 3(-2) - (-3) && \text{Let } x = -2 \text{ and } y = -3. \\ &= -6 - (-3) && \text{Multiply.} \\ &= -6 + 3 && \\ &= -3 && \text{Add.} \end{aligned}$$

- Remember the distributive property.

$$\begin{aligned} 5(4x - 3) + 2 &= 5 \cdot 4x - 5 \cdot 3 + 2 && \text{Use the distributive property.} \\ &= 20x - 15 + 2 && \text{Multiply.} \\ &= 20x - 13 && \text{Combine like terms.} \end{aligned}$$

- Don't forget the steps for solving a linear equation.

$$\begin{aligned} 2(3x - 2) + 16 &= 6 && \\ 6x - 4 + 16 &= 6 && \text{Apply the distributive property.} \\ 6x + 12 &= 6 && \text{Combine like terms.} \\ 6x + 12 - 12 &= 6 - 12 && \text{Subtract 12 from both sides.} \\ 6x &= -6 && \text{Simplify.} \\ \frac{6x}{6} &= \frac{-6}{6} && \text{Divide both sides by 6.} \\ x &= -1 && \text{Simplify.} \end{aligned}$$

Remember: This is simply a checklist of common trouble areas. For a review of Chapter 3 see the Highlights and Chapter Review at the end of this chapter.

Name ____________________ Section __________ Date __________

Chapter 3 Test

Remember to check your answers and use the Chapter Test Prep Video to view solutions.

1. Simplify $7x - 5 - 12x + 10$ by combining like terms.

2. Multiply: $-2(3y + 7)$

3. Simplify: $-(3z + 2) \quad 5z - 18$

4. Write an expression that represents the perimeter of the equilateral triangle. Simplify the expression. (A triangle with three sides of equal length.)

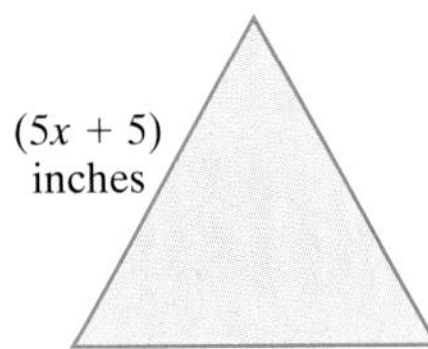

5. Write an expression that represents the area of the rectangle. Simplify the expression.

6. Solve: $9x = -90$

7. Solve: $\dfrac{y}{-4} = 4$

Solve each equation.

8. $12 = y - 3y$

9. $\dfrac{x}{2} = -5 - (-2)$

10. $5 + 4z = 37$

11. $5x + 12 - 4x - 14 = 22$

12. $-4x + 7 = 15$

13. $2(x - 6) = 0$

14. $5x - 2 = x - 10$

15. $4(5x + 3) = 2(7x + 6)$

16. Translate the following phrases into mathematical expressions. If needed, use x to represent "a number."

a. The sum of -23 and a number

b. Three times a number, subtracted from -2

17. Translate each sentence into an equation. If needed, use x to represent "a number."

a. The sum of twice 5 and -15 is -5.

b. Six added to three times a number equals -30.

Answers

1. ______
2. ______
3. ______
4. ______
5. ______
6. ______
7. ______
8. ______
9. ______
10. ______
11. ______
12. ______
13. ______
14. ______
15. ______
16. a. ______
   b. ______
17. a. ______
   b. ______

18. ______________

Solve.

18. The difference of three times a number and five times the same number is 4. Find the number.

19. In a championship basketball game, Paula Zimmerman made twice as many free throws as Maria Kaminsky. If the total number of free throws made by both women was 12, find how many free throws Paula made.

20. In a 10-kilometer race, there are 112 more men entered than women. Find the number of female runners if the total number of runners in the race is 600.

19. ______________

20. ______________

Name ______________________ Section ____________ Date ____________

Answers

Chapter 3 Cumulative Review

1. Write 106,052,447 in words.

2. Write 276,004 in words.

3. Find the perimeter of the polygon shown.

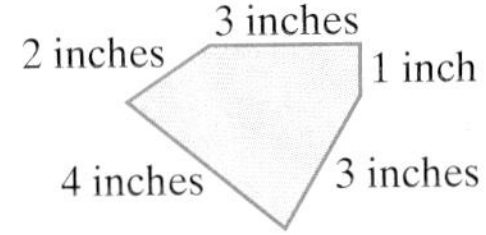

4. Find the perimeter of the rectangle shown.

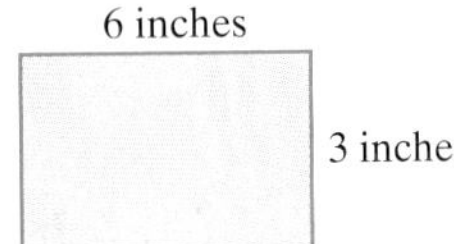

5. Subtract: $900 - 174$. Check by adding.

6. Subtract: $17{,}801 - 8216$ Check by adding.

7. Round 248,982 to the nearest hundred.

8. Round 844,497 to the nearest thousand.

9. Multiply: 25×8

10. Multiply: 395×74

11. Divide and check: $1872 \div 9$

12. Divide and check: $3956 \div 46$

13. Simplify: $2 \cdot 4 - 3 \div 3$

14. Simplify: $8 \cdot 4 + 9 \div 3$

15. Evaluate $x^2 + z - 3$ for $x = 5$ and $z = 4$.

16. Evaluate $2a^2 + 5 - c$ for $a = 2$ and $c = 3$.

17. Insert $<$ or $>$ between each pair of numbers to make a true statement.

a. -7 $\quad 7$

b. 0 $\quad -4$

c. -9 $\quad -11$

18. Insert $<$ or $>$ to make a true statement.

a. -14 $\quad 0$

b. $-(-7)$ $\quad -8$

19. Add using a number line: $5 + (-2)$

20. Add using a number line: $-3 + (-4)$

Add.

21. $-5 + (-1)$

22. $3 + (-7)$

1. ______
2. ______
3. ______
4. ______
5. ______
6. ______
7. ______
8. ______
9. ______
10. ______
11. ______
12. ______
13. ______
14. ______
15. ______
16. ______
17. a. ______
b. ______
c. ______
18. a. ______
b. ______
19. ______
20. ______
21. ______
22. ______

23.

24.

25.

26.

27.

28.

29.

30.

31.

32.

33.

34.

35.

36.

37.

38.

39.

40.

41.

42.

43.

44.

45.

46.

47.

48.

49.

50.

23. $2 + 6$

24. $21 + 15 + (-19)$

Subtract.

25. $-4 - 10$

26. $-2 - 3$

27. $6 - (-5)$

28. $19 - (-10)$

29. $-11 - (-7)$

30. $-16 - (-13)$

Divide.

31. $\frac{-12}{6}$

32. $\frac{-30}{-5}$

33. $-20 \div (-4)$

34. $26 \div (-2)$

35. $\frac{48}{-3}$

36. $\frac{-120}{12}$

Find the value of each expression.

37. $(-3)^2$

38. -2^5

39. -3^2

40. $(-5)^2$

41. Simplify:
$2y - 6 + 4y + 8$

42. Simplify:
$6x + 2 - 3x + 7$

43. Determine whether -1 is a solution of the equation $3y + 1 = 3$.

44. Determine whether 2 is a solution of $5x - 3 = 7$.

45. Solve: $-12x = -36$

46. Solve: $7y - 4y = 15$

47. Solve: $2x - 6 = 18$

48. Solve: $3a + 5 = -1$

49. Leo Leal sold a used computer system and software for $2100, receiving four times as much money for the computer system as for the software. Find the price of each.

50. Rose Daunis is thinking of a number. Two times the number, plus four is the same amount as three times the number minus seven. Find Rose's number.

Fractions

CHAPTER 4

Fractions are numbers and, like whole numbers and integers, they can be added, subtracted, multiplied, and divided. Fractions are very useful and appear frequently in everyday language, in common phrases such as "half an hour," "quarter of a pound," and "third of a cup." This chapter reviews the concept of fractions and demonstrates how to add, subtract, multiply, and divide fractions. In this chapter we also solve linear equations containing fractions.

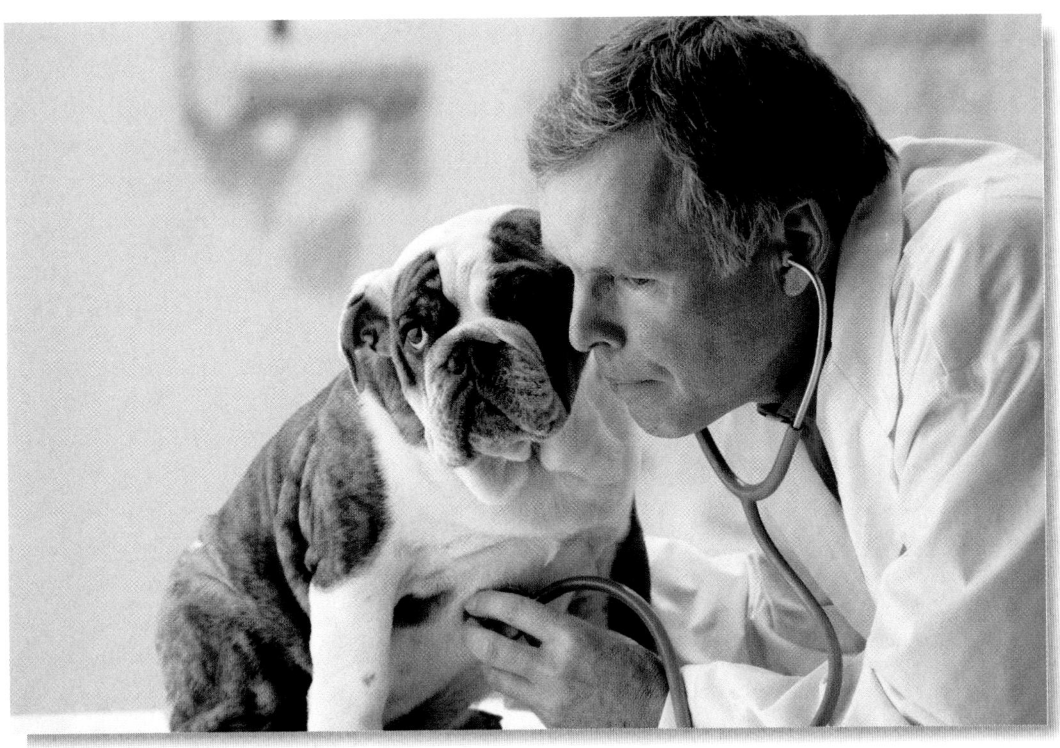

Veterinarians are doctors who diagnose, treat, and work to prevent animal diseases. According to the American Veterinary Medicine Association (AVMA), there are over 67,000 professionally active veterinarians in the United States today. Many of these work in private practice, caring for household pets or livestock. Some veterinarians work in the area of public health, mainly with the control of rabies. Research veterinarians work to establish and improve animal vaccines for diseases. There are currently 28 accredited colleges of veterinary medicine in the United States. Following graduation, the veterinarian must pass a licensing examination for the state in which he or she plans to practice. In Exercise 75 on page 270, we will see how veterinarians use fractions to compute the amount of flea medication to include in a dipping vat.

Answers

1. ______

2. ______

3. ______

4. ______

5. ______

6. ______

7. ______

8. ______

9. ______

10. ______

11. ______

12. ______

13. ______

14. ______

15. ______

16. ______

17. ______

18. ______

19. ______

20. ______

Name ______ Section ______ Date ______

Chapter 4 Pretest

1. Write a fraction to represent the shaded area of the figure.

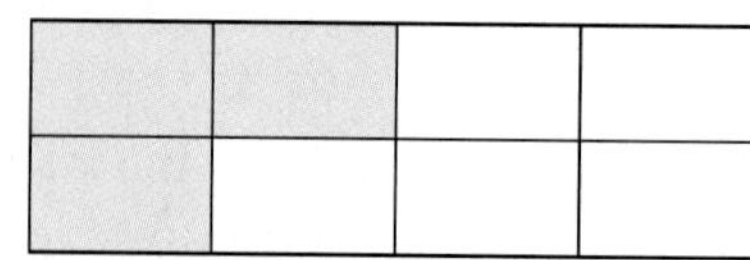

2. Write $\frac{5}{6}$ as an equivalent fraction whose denominator is 18.

3. Simplify: $\frac{0}{-4}$

4. Write the prime factorization of 140.

5. Simplify: $\frac{30}{54}$

6. There are 1000 milligrams in 1 gram. What fraction of a gram is 450 milligrams?

Perform each indicated operation and simplify.

7. $\frac{3}{4} \cdot \frac{24}{15}$

8. $\frac{5x}{7} \div 10x$

9. $\frac{8}{11} - \frac{3}{11}$

10. $-\frac{3}{10} + \frac{2}{10}$

11. Evaluate: $\left(-\frac{2}{3}\right)^3$

12. Evaluate $x \div y$ if $x = \frac{2}{9}$ and $y = -\frac{2}{3}$.

13. Add: $\frac{5}{9} + \frac{1}{12}$

14. Evaluate $x - y$ if $x = -\frac{3}{14}$ and $y = -\frac{2}{7}$.

Simplify.

15. $\frac{\frac{x}{3}}{\frac{7}{9}}$

16. $\left(\frac{2}{5}\right)^2 - 2$

17. Solve: $\frac{x}{4} + 3 = \frac{1}{8}$

18. Write $2\frac{3}{5}$ as an improper fraction.

Perform each indicated operation and simplify.

19. $3\frac{1}{5} \cdot 2\frac{3}{4}$

20. $5\frac{2}{3} + 4\frac{1}{6}$

4.1 Introduction to Fractions and Equivalent Fractions

OBJECTIVES

A. Identify the numerator and the denominator of a fraction.

B. Write a fraction to represent the shaded area of a figure.

C. Graph fractions on a number line.

D. Simplify fractions of the form $\frac{a}{a}$, $\frac{a}{1}$, and $\frac{0}{a}$.

E. Write equivalent fractions.

A Identifying Numerators and Denominators

Whole numbers are used to count whole things or units, such as cars, ball games, horses, dollars, and people. To refer to a part of a whole, fractions are used. For example, a whole baseball game is divided into nine parts called innings. If a player pitched five complete innings, the fraction $\frac{5}{9}$ can be used to show the part of a whole game he or she pitched. The 9 in the fraction $\frac{5}{9}$ is called the denominator, and it refers to the total number of equal parts (innings) in the whole game. The 5 in the fraction $\frac{5}{9}$ is called the numerator, and it tells how many of those equal parts (innings) the pitcher pitched.

$5 \leftarrow$ how many of the parts being considered
$9 \leftarrow$ number of equal parts in the whole

EXAMPLES Identify the numerator and the denominator of each fraction.

1. $\frac{3}{7}$ $\begin{matrix}\leftarrow \text{numerator} \\ \leftarrow \text{denominator}\end{matrix}$

2. $\frac{3}{7}$ $\begin{matrix}\leftarrow \text{numerator} \\ \leftarrow \text{denominator}\end{matrix}$

Practice Problems 1–2

Identify the numerator and the denominator of each fraction.

1. $\frac{9}{2}$ 2. $\frac{10y}{17}$

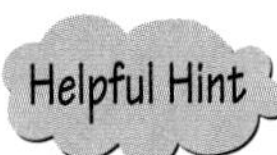

$\frac{3}{7} \leftarrow$ Remember that the bar in a fraction means division. Since division by 0 is undefined, a fraction with a denominator of 0 is undefined.

B Writing Fractions to Represent Shaded Areas of Figures

One way to become familiar with the concept of fractions is to visualize fractions with shaded figures. We can then write a fraction to represent the shaded area of the figure.

EXAMPLES Write a fraction to represent the shaded area of each figure.

3.

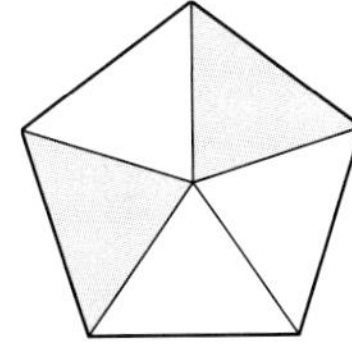

$\frac{2}{5}$ $\begin{matrix}\leftarrow \text{parts shaded} \\ \leftarrow \text{equal parts}\end{matrix}$

The figure is divided into 5 equal parts and 2 of them are shaded.

Thus, $\frac{2}{5}$ of the figure is shaded.

4.

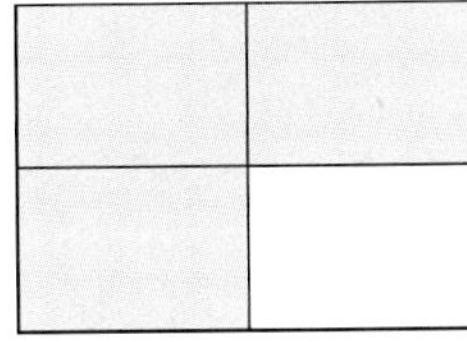

$\frac{3}{4}$ $\begin{matrix}\leftarrow \text{parts shaded} \\ \leftarrow \text{equal parts}\end{matrix}$

The figure is divided into 4 equal parts and 3 of them are shaded.

Thus, $\frac{3}{4}$ of the figure is shaded.

Practice Problems 3–4

Write a fraction to represent the shaded area of each figure.

3.

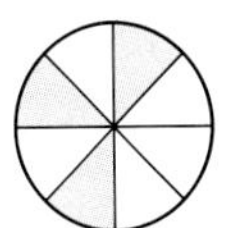

4.

Answers

1. $\frac{9}{2}$ $\begin{matrix}\leftarrow \text{numerator} \\ \leftarrow \text{denominator}\end{matrix}$ **2.** $\frac{10y}{17}$ $\begin{matrix}\leftarrow \text{numerator} \\ \leftarrow \text{denominator}\end{matrix}$ **3.** $\frac{3}{8}$ **4.** $\frac{1}{6}$

Practice Problems 5–6

Draw and shade a part of a diagram to represent each fraction.

5. $\frac{7}{11}$ of a diagram

6. $\frac{2}{3}$ of a diagram

Concept Check

Identify each as a proper fraction or improper fraction.

a. $\frac{6}{7}$ b. $\frac{13}{12}$ c. $\frac{2}{2}$ d. $\frac{99}{101}$

Answers

5. answers may vary; for example,

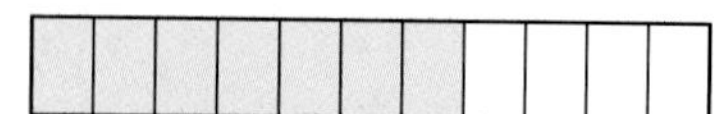

6. answers may vary; for example,

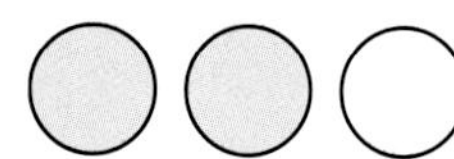

Concept Check:
a. proper **b.** improper **c.** improper **d.** proper

EXAMPLES Draw and shade a part of a diagram to represent each fraction.

5. $\frac{5}{6}$ of a diagram

We can use a geometric figure such as a rectangle and divide it into 6 equal parts. Then we will shade 5 of the equal parts.

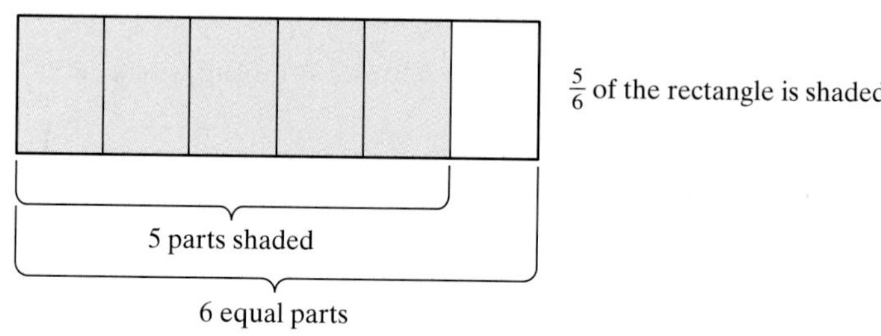

6. $\frac{3}{8}$ of a diagram

If you'd like, our diagram can consist of 8 triangles of the same size. We will shade 3 of the triangles.

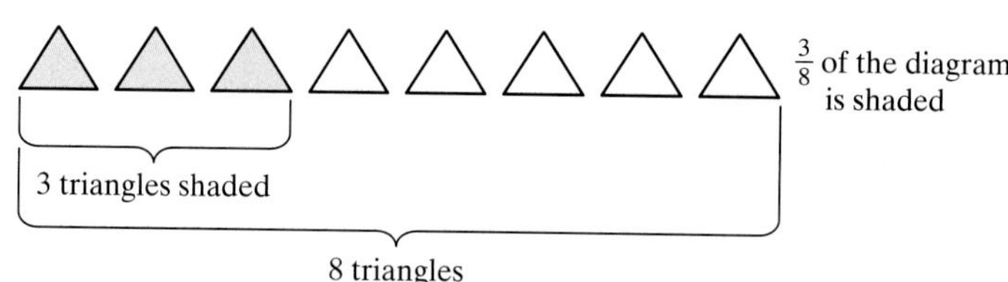

A **proper fraction** is a fraction whose numerator is less than its denominator. Proper fractions have values that are less than 1. The shaded portion of the triangle's area is represented by $\frac{2}{3}$.

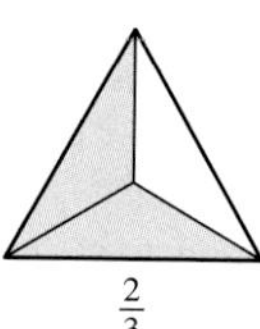

$\frac{2}{3}$

An **improper fraction** is a fraction whose numerator is greater than or equal to its denominator. Improper fractions have values that are greater than or equal to 1. The area of the shaded part of the group of circles is $\frac{9}{4}$. The shaded part of the rectangle's area is $\frac{6}{6}$.

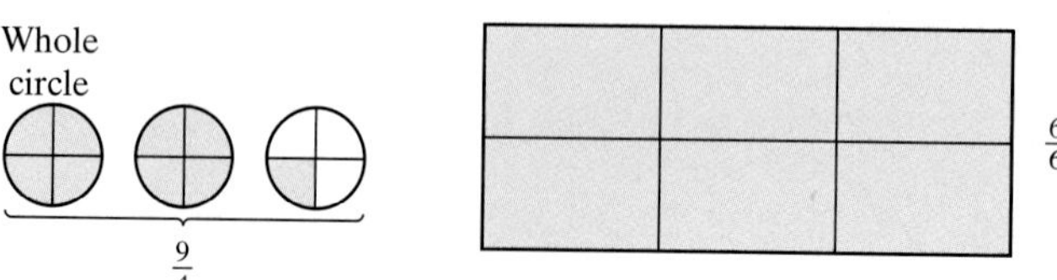

Try the Concept Check in the margin.

EXAMPLES Represent the shaded part of each figure group with an improper fraction.

7. Whole object

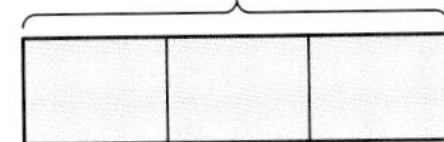
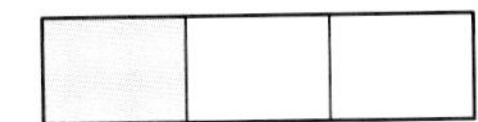

improper fraction: $\frac{4}{3}$

8.

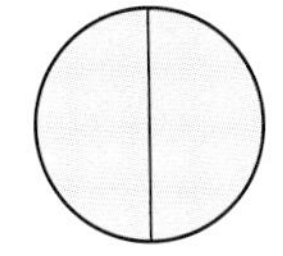
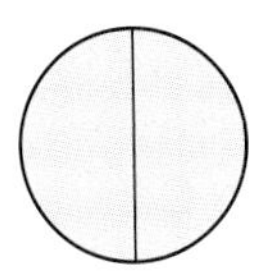
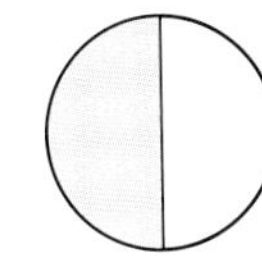

improper fraction: $\frac{5}{2}$

Practice Problems 7–8

Represent the shaded part of each figure group with an improper fraction.

7.

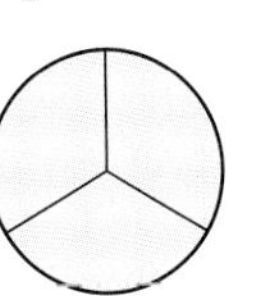
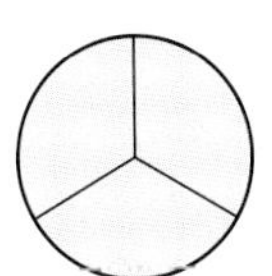
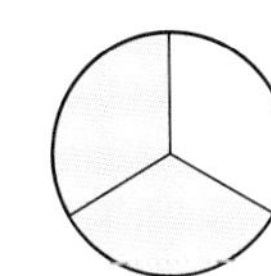

8.

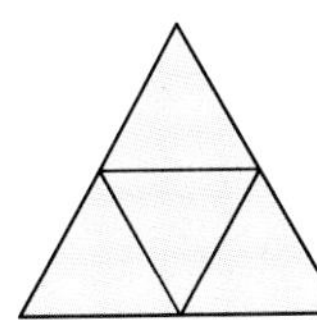
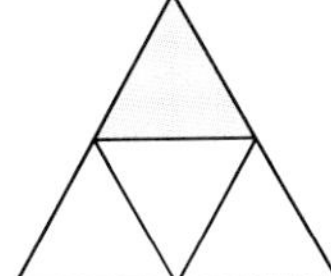

EXAMPLE 9 Writing Fractions from Real-Life Data

Of the nine planets in our solar system, two are closer to the sun than Earth is. What fraction of the planets are closer to the sun than Earth is?

Solution: The fraction closer to the sun is

$$\frac{2}{9} \quad \begin{array}{l}\leftarrow \text{number of planets that are closer} \\ \leftarrow \text{number of planets in our solar system}\end{array}$$

Thus, $\frac{2}{9}$ of the planets in our solar system are closer to the sun than Earth is.

Practice Problem 9

Of the nine planets in our solar system, seven are farther from the sun than Venus is. What fraction of the planets are farther from the sun than Venus is?

C Graphing Fractions on a Number Line

Another way to visualize fractions is to graph them on a number line. To do this, think of 1 unit on the number line as a whole. To graph $\frac{2}{5}$, for example, divide the distance from 0 to 1 into 5 equal parts. Then start at 0 and count 2 parts to the right.

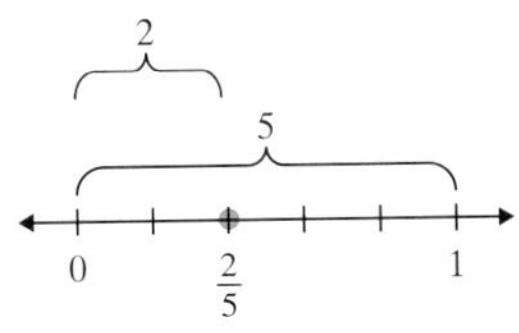

Answers

7. $\frac{8}{3}$ **8.** $\frac{5}{4}$ **9.** $\frac{7}{9}$

Practice Problem 10

Graph each proper fraction on a number line.

a. $\frac{5}{7}$ b. $\frac{2}{3}$ c. $\frac{4}{6}$

EXAMPLE 10 Graph each proper fraction on a number line.

a. $\frac{3}{4}$ **b.** $\frac{1}{2}$ **c.** $\frac{3}{6}$

Solution: **a.** Divide the distance from 0 to 1 into 4 parts. Then start at 0 and count over 3 parts.

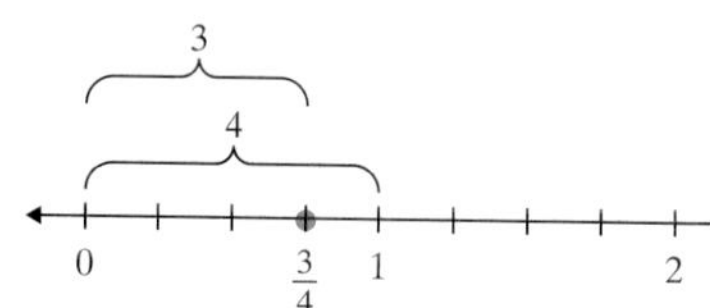

b.

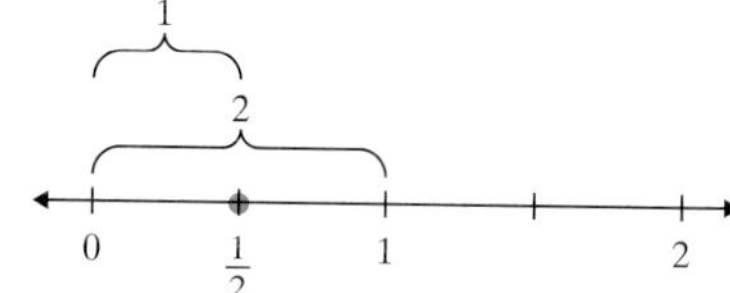

c.

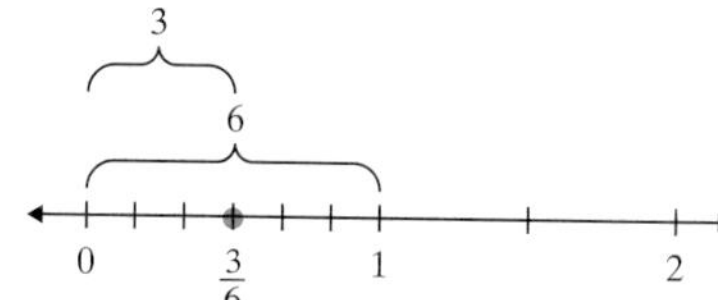

The fractions in Example 10 are all proper fractions. Notice that the value of each is less than 1. This is always true for proper fractions since the numerator of a proper fraction is less than the denominator.

Practice Problem 11

Graph each improper fraction on a number line.

a. $\frac{8}{3}$ b. $\frac{5}{4}$ c. $\frac{7}{7}$

EXAMPLE 11 Graph each improper fraction on a number line.

a. $\frac{7}{6}$ **b.** $\frac{9}{5}$ **c.** $\frac{6}{6}$ **d.** $\frac{3}{1}$

Solution:

a.

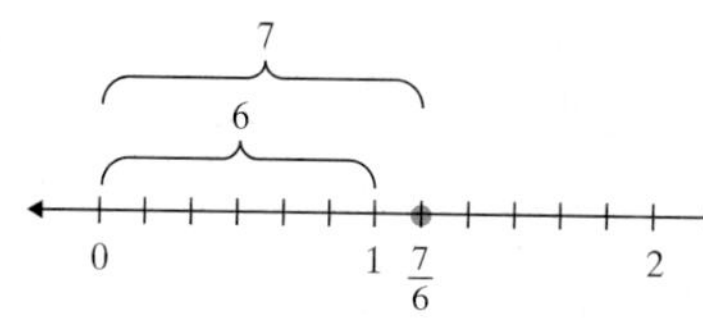

b.

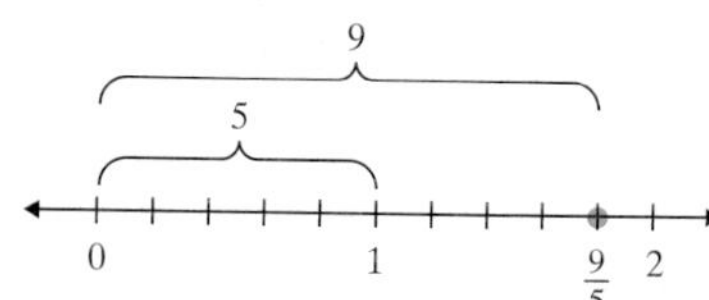

c.

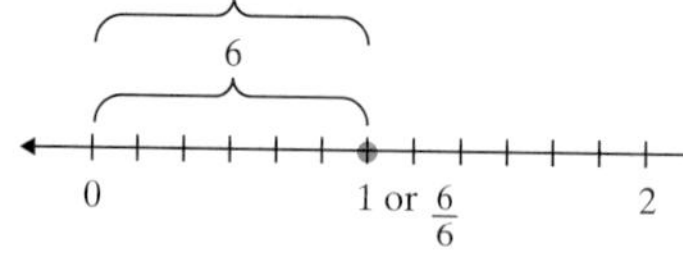

d. Each 1-unit distance has 1 equal part. Count over 3 parts.

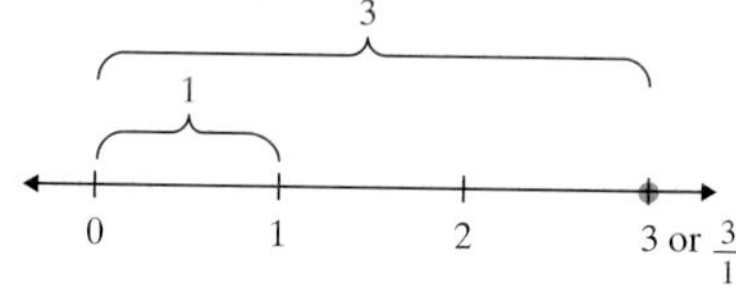

Answers

10. a. 0, $\frac{5}{7}$, 1

b. 0, $\frac{2}{3}$, 1

c. 0, $\frac{4}{6}$, 1

11. a. 0, 1, 2, $\frac{8}{3}$, 3

b. 0, 1, $\frac{5}{4}$, 2

c. 0, 1 or $\frac{7}{7}$

The fractions in Example 11 are all improper fractions. Notice that the value of each is greater than or equal to 1. This is always true since the numerator of an improper fraction is greater than or equal to the denominator.

D Simplifying $\frac{a}{a}$, $\frac{a}{1}$, and $\frac{0}{a}$.

The graphs of $\frac{6}{6}$ and $\frac{3}{1}$ from Example 11 are shown next.

Notice that the graph of $\frac{6}{6}$ is the same as the graph of 1, and the graph of $\frac{3}{1}$ is the same as 3. This makes sense if we recall that the fraction bar indicates division. Let's review some division properties for 1 and 0.

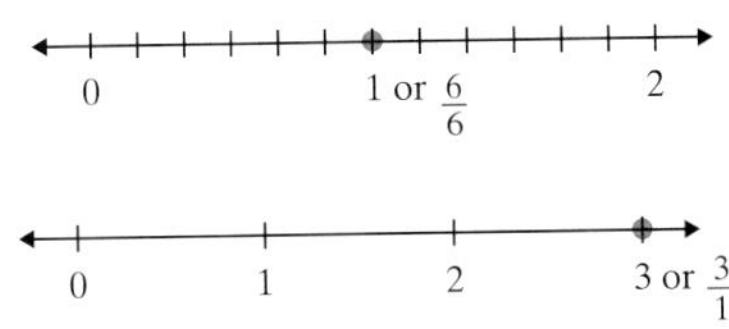

$\frac{6}{6} = 1$ because $1 \cdot 6 = 6$ $\frac{3}{1} = 3$ because $3 \cdot 1 = 3$

$\frac{0}{6} = 0$ because $0 \cdot 6 = 0$

$\frac{6}{0}$ *is undefined* because there is no number that when multiplied by 0 gives 6.

In general, we can say the following.

Let *n* be any integer.

$\frac{n}{n} = 1$ as long as n is not 0. $\frac{0}{n} = 0$ as long as n is not 0.

$\frac{n}{1} = n$ $\frac{n}{0}$ is undefined.

EXAMPLES Simplify.

12. $\frac{5}{5} = 1$ **13.** $\frac{-2}{-2} = 1$ **14.** $\frac{0}{-5} = 0$

15. $\frac{-5}{1} = -5$ **16.** $\frac{41}{1} = 41$ **17.** $\frac{19}{0}$ is undefined

Practice Problems 12–17

Simplify.

12. $\frac{9}{9}$ 13. $\frac{-6}{-6}$ 14. $\frac{0}{-1}$

15. $\frac{4}{1}$ 16. $\frac{-13}{0}$ 17. $\frac{-13}{1}$

Answers

12. 1 **13.** 1 **14.** 0 **15.** 4 **16.** undefined **17.** -13

E Writing Equivalent Fractions

When we graph $\frac{1}{3}$ and $\frac{2}{6}$ on a number line,

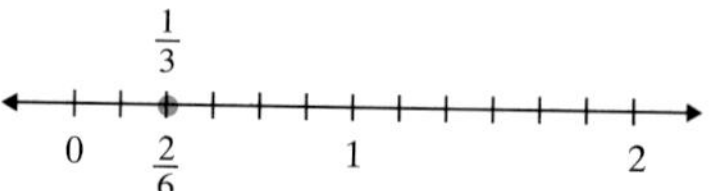

notice that both $\frac{1}{3}$ and $\frac{2}{6}$ correspond to the same point. These fractions are called **equivalent fractions**, and we write $\frac{1}{3} = \frac{2}{6}$.

Another way to visualize equivalent fractions is to use shaded figures.

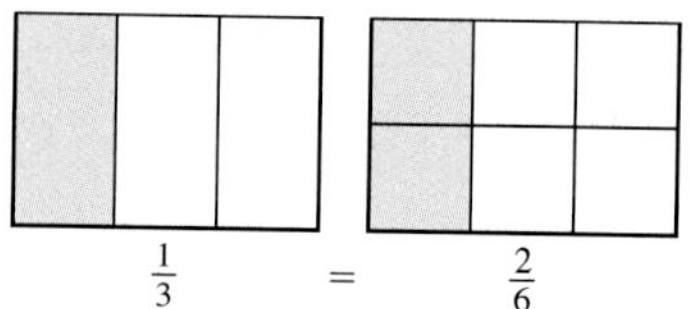

From the shaded figures, we see that equivalent fractions also represent the same portion of a whole.

Equivalent Fractions

Fractions that represent the same portion of a whole or the same point on a number line are called **equivalent fractions**.

To write equivalent fractions, we use the **fundamental property of fractions**. This property guarantees that, if we multiply or divide both the numerator and the denominator by the same nonzero number, the result is an equivalent fraction. For example, if we multiply both the numerator and denominator of $\frac{1}{3}$ by 2, the result is the equivalent fraction $\frac{2}{6}$.

$$\frac{1 \cdot 2}{3 \cdot 2} = \frac{2}{6}$$

Fundamental Property of Fractions

If *a*, *b*, and *c* are numbers, then

$$\frac{a}{b} = \frac{a \cdot c}{b \cdot c} \quad \text{and also} \quad \frac{a}{b} = \frac{a \div c}{b \div c}$$

as long as *b* and *c* are not 0. In other words, if the numerator and denominator of a fraction are multiplied or divided by the **same** nonzero number, the result is an equivalent fraction.

EXAMPLE 18 Write $\frac{2}{5}$ as an equivalent fraction whose denominator is 15.

Solution: Since $5 \cdot 3 = 15$, we use the fundamental property of fractions and multiply the numerator and denominator of $\frac{2}{5}$ by 3.

$$\frac{2}{5} = \frac{2 \cdot 3}{5 \cdot 3} = \frac{6}{15}$$

Then $\frac{2}{5}$ is equivalent to $\frac{6}{15}$. They both represent the same part of a whole. ●

EXAMPLE 19 Write $\frac{9x}{11}$ as an equivalent fraction whose denominator is 44.

Solution: Since $11 \cdot 4 = 44$, we multiply the numerator and denominator by 4.

$$\frac{9x}{11} = \frac{9x \cdot 4}{11 \cdot 4} = \frac{36x}{44}$$

Then $\frac{9x}{11}$ is equivalent to $\frac{36x}{44}$. ●

EXAMPLE 20 Write 3 as an equivalent fraction whose denominator is 7.

Solution: Recall that $3 = \frac{3}{1}$. Since $1 \cdot 7 = 7$, multiply the numerator and denominator by 7.

$$\frac{3}{1} = \frac{3 \cdot 7}{1 \cdot 7} = \frac{21}{7}$$ ●

In this chapter, we will perform operations on fractions containing variables. When the denominator of a fraction contains a variable, such as $\frac{8}{3x}$, we will assume that the variable does not represent 0. Recall that the denominator of a fraction cannot be 0.

EXAMPLE 21 Write $\frac{8}{3x}$ as an equivalent fraction whose denominator is $12x$.

Solution: Since $3x \cdot 4 = 12x$, multiply the numerator and denominator by 4.

$$\frac{8}{3x} = \frac{8 \cdot 4}{3x \cdot 4} = \frac{32}{12x}$$ ●

Try the Concept Check in the margin.

Practice Problem 18

Write $\frac{1}{4}$ as an equivalent fraction whose denominator is 20.

Practice Problem 19

Write $\frac{3x}{7}$ as an equivalent fraction with a denominator of 42.

Practice Problem 20

Write 4 as an equivalent fraction whose denominator is 6.

Practice Problem 21

Write $\frac{9}{4x}$ as an equivalent fraction with a denominator of $20x$.

Concept Check

What is the first step in writing $\frac{3}{10}$ as an equivalent fraction whose denominator is 100?

Answers

18. $\frac{5}{20}$ **19.** $\frac{18x}{42}$ **20.** $\frac{24}{6}$ **21.** $\frac{45}{20x}$

Concept Check: Answers may vary.

FOCUS ON Mathematical Connections

MODELING FRACTIONS

There are several different physical models for representing fractions.

SET MODEL

In this model, a fraction represents the portion of a set of objects that has a certain characteristic. For example, in the set of 10 shapes, 3 are hearts. That is, $\frac{3}{10}$ of the shapes are hearts.

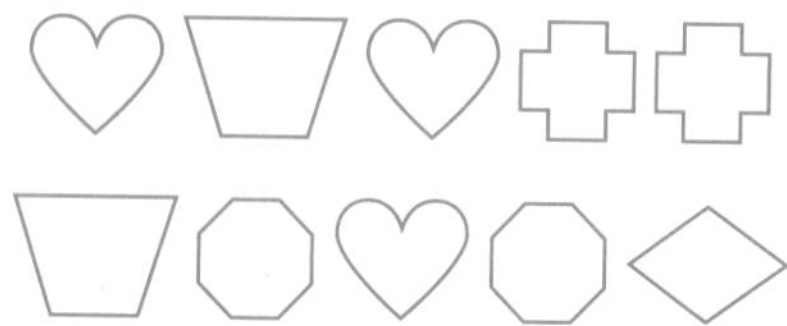

AREA MODEL

In this model, a shape is divided into a number of equal-sized regions. A fraction can be represented by shading some of the regions. For example, both of the following models represent the fraction $\frac{7}{8}$.

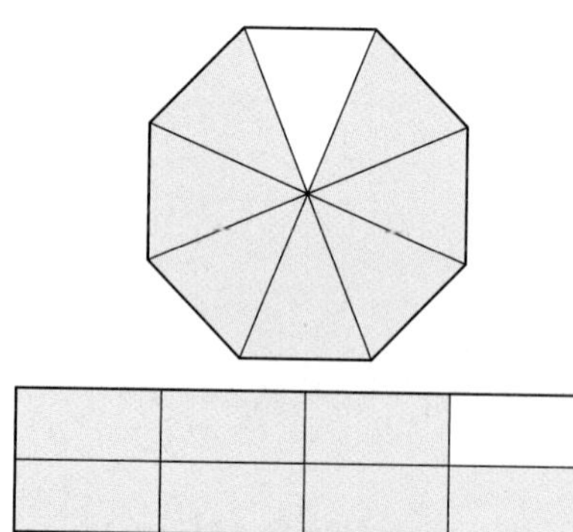

NUMBER LINE MODEL

For this model, draw a line and label a point 0 and a point to its right, 1. Now subdivide this distance from 0 to 1 depending on the denominator of the fraction that is to be represented. For example, the fraction $\frac{3}{4}$ is graphed by subdividing the portion of the number line between 0 and 1 into four equal lengths and placing a dot at the mark that represents $\frac{3}{4}$ of the distance between 0 and 1.

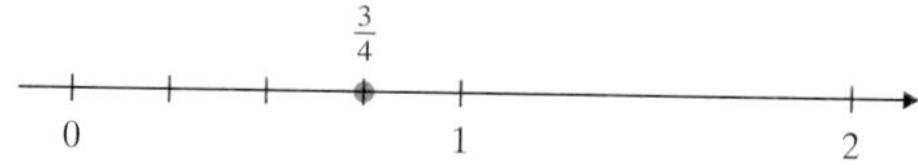

CRITICAL THINKING

1. a. Represent the fraction $\frac{5}{6}$ in two different ways using the set model.

   b. Represent the fraction $\frac{5}{6}$ in two different ways using the area model.

   c. Represent the fraction $\frac{5}{6}$ using the number line model.

2. Which model do you prefer? Why?

3. Which model do you think would be most useful for representing multiplication of fractions? Explain how this could be done.

Name ______________________ Section __________ Date ______________

Mental Math

A *Identify the numerator and the denominator of each fraction. See Examples 1 and 2.*

1. $\frac{1}{2}$

2. $\frac{1}{4}$

3. $\frac{10}{3}$

4. $\frac{53}{21}$

5. $\frac{3z}{7}$

6. $\frac{11x}{15}$

EXERCISE SET 4.1

B *Write a fraction to represent the shaded area of each figure. See Examples 3 and 4.*

1.

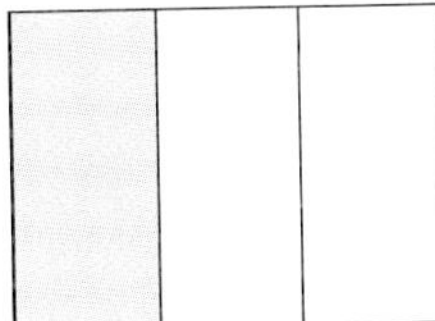

2.

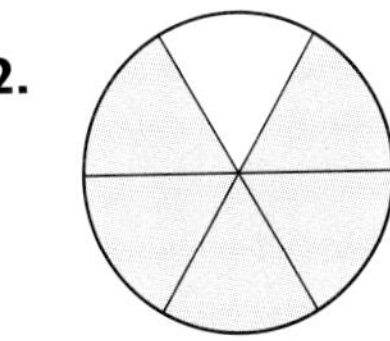

3.

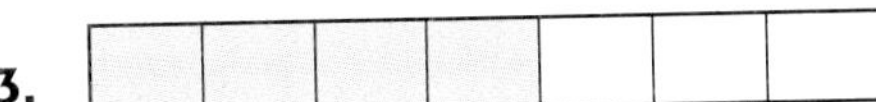

4.

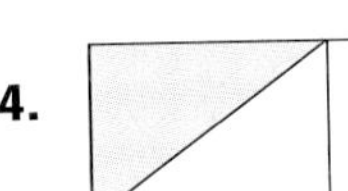

5.

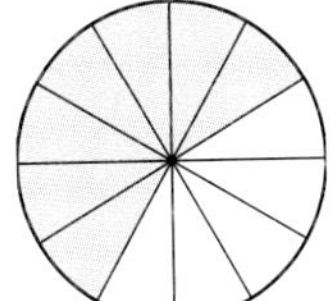

6.

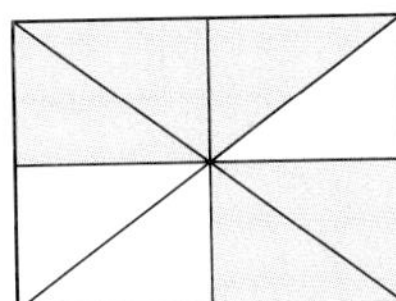

7.

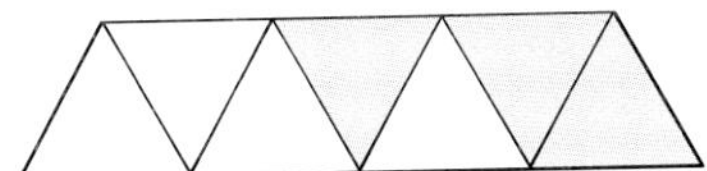

8.

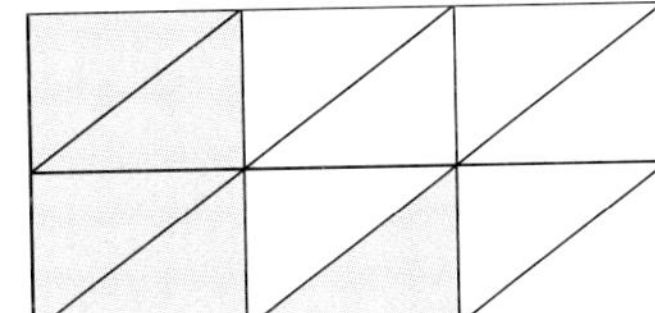

9.

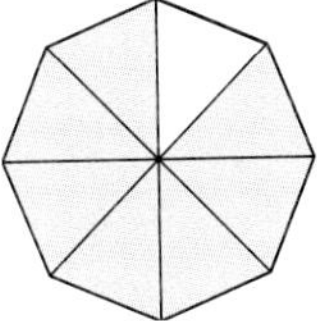

10.

Use the pizza illustration for Exercises 11 and 12.

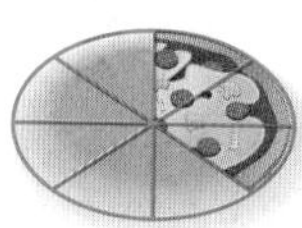

11. What fraction of the pizza is gone?

12. What fraction of the pizza remains?

Draw and shade a part of a diagram to represent each fraction. See Examples 5 and 6.

13. $\frac{5}{8}$ of a diagram

14. $\frac{1}{4}$ of a diagram

15. $\frac{1}{5}$ of a diagram

16. $\frac{3}{5}$ of a diagram

17. $\frac{6}{7}$ of a diagram

18. $\frac{7}{9}$ of a diagram

19. $\frac{4}{4}$ of a diagram

20. $\frac{6}{6}$ of a diagram

Represent the shaded area in each figure group with an improper fraction. See Examples 7 and 8.

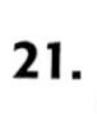

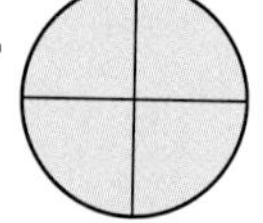 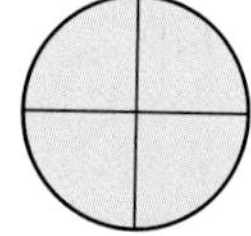 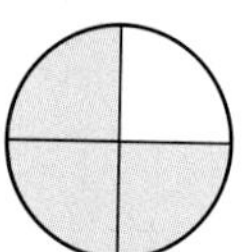

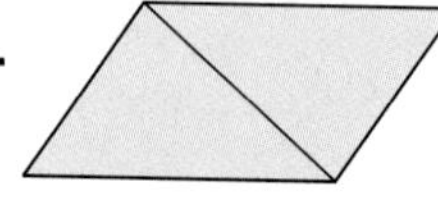 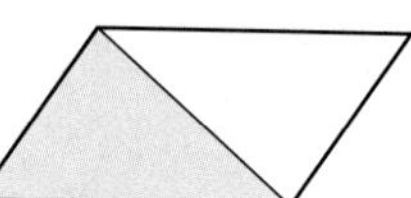

23.

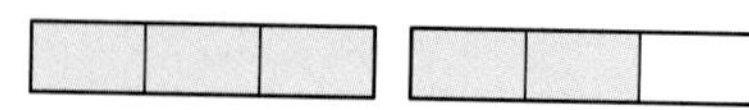

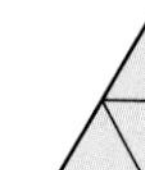 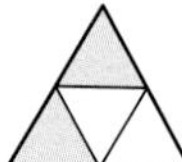

25. 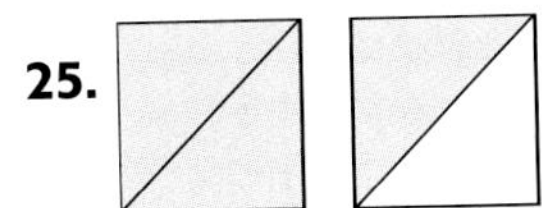

26. 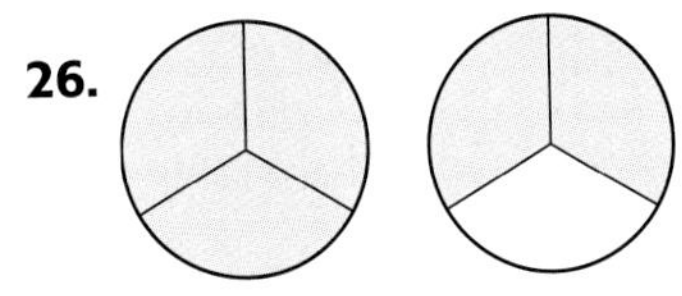

27. 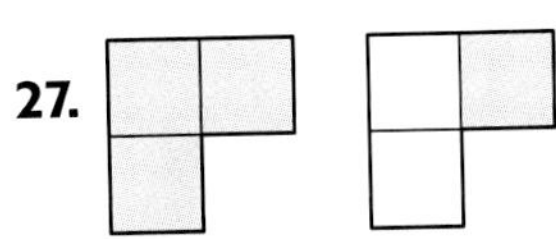

28. 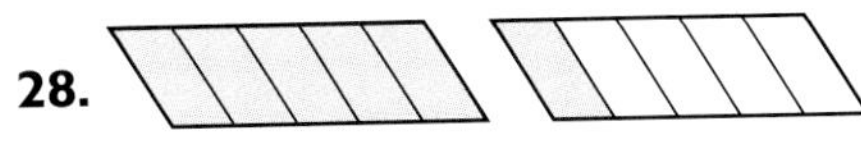

29. 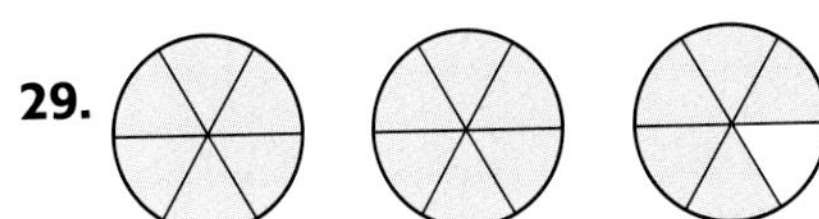

30. 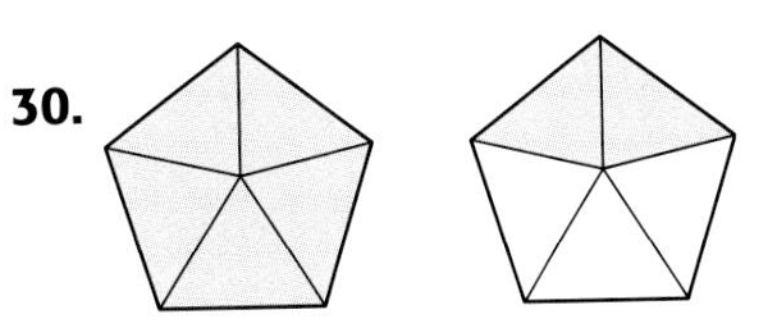

31. 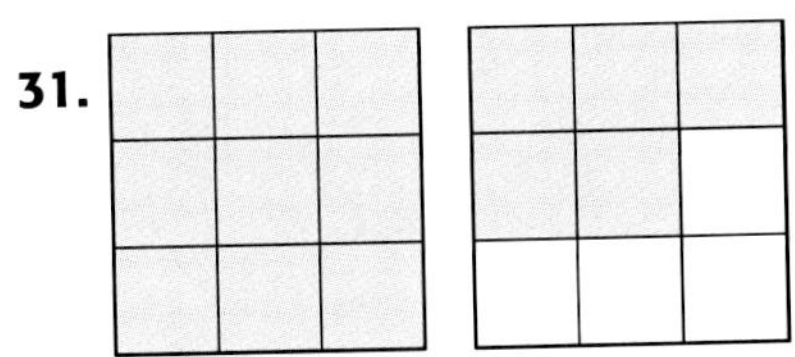

32. 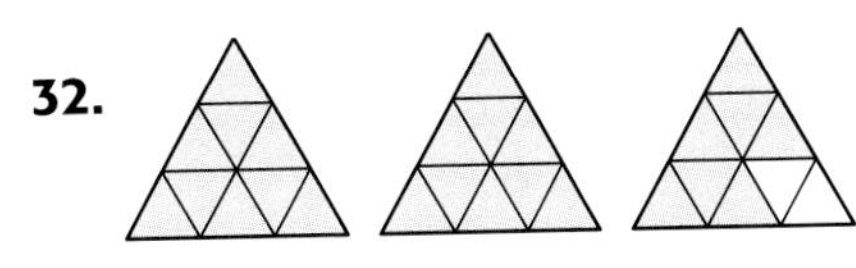

Write each fraction. See Example 9.

33. Of the 131 students at a small private school, 42 are freshmen. What fraction of the students are freshmen?

34. Of the 78 executives at a private accounting firm, 61 are men. What fraction of the executives are men?

35. From Exercise 33, how many students are *not* freshmen? What fraction of the students are *not* freshmen?

36. From Exercise 34, how many of the executives are women? What fraction of the executives are women?

37. According to a recent study, 4 out of 10 visits to U.S. hospital emergency rooms were for an injury. What fraction of emergency room visits are injury-related? (*Source*: National Center for Health Statistics)

38. The average American driver spends about 1 hour each day in a car or truck. What fraction of a day does an average American driver spend in a car or truck? (*Hint:* How many hours are in a day?) (*Source:* U.S. Department of Transportation—Federal Highway Administration)

39. As of 2003, the United States had 43 different presidents. A total of eight U.S. presidents were born in the state of Virginia, more than any other state. What fraction of U.S. presidents were born in Virginia? (*Source:* 2003 World Almanac)

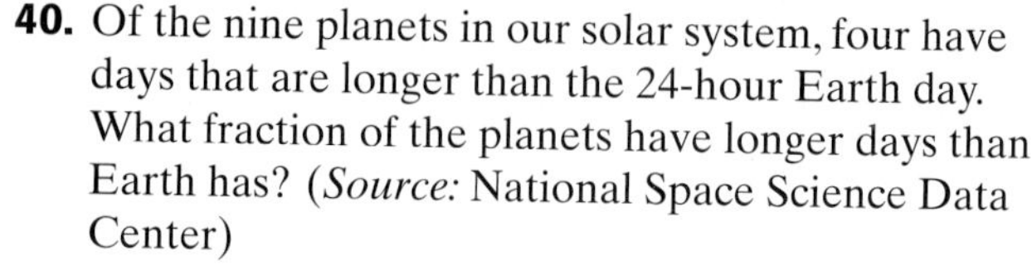

40. Of the nine planets in our solar system, four have days that are longer than the 24-hour Earth day. What fraction of the planets have longer days than Earth has? (*Source:* National Space Science Data Center)

41. The hard drive in Aaron Hawn's Computer can hold 32 gigabytes of information. He has currently used 17 gigabytes. What fraction of his hard drive has he used?

42. There are 12 inches in a foot. What fractional part of a foot does 5 inches represent?

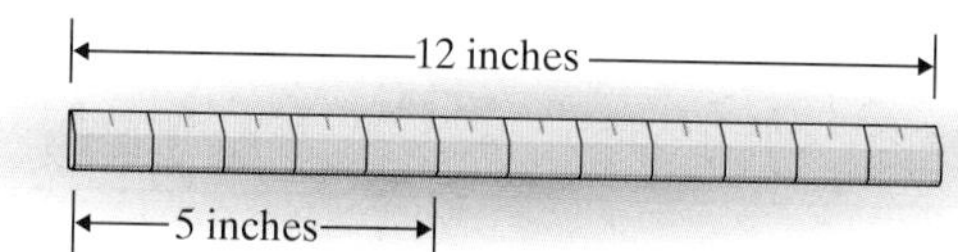

43. There are 31 days in the month of March. What fraction of the month does 11 days represent?

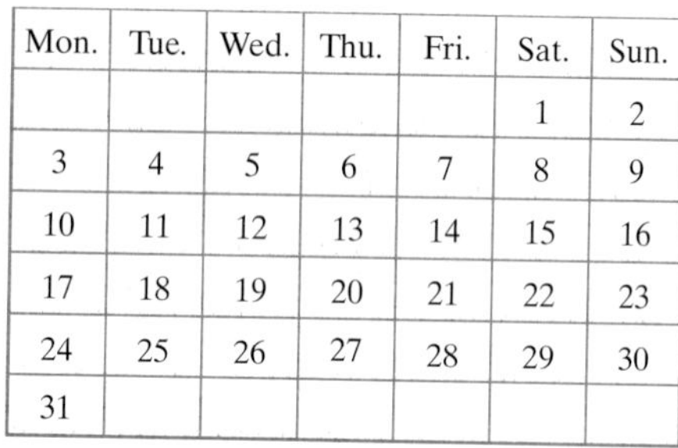

Mon.	Tue.	Wed.	Thu.	Fri.	Sat.	Sun.
					1	2
3	4	5	6	7	8	9
10	11	12	13	14	15	16
17	18	19	20	21	22	23
24	25	26	27	28	29	30
31						

44. There are 60 minutes in an hour. What fraction of an hour does 37 minutes represent?

45. In a prealgebra class containing 31 students, there are 18 freshmen, 10 sophomores, and 3 juniors. What fraction of the class is sophomores?

46. In a family with 11 children, there are 4 boys and 7 girls. What fraction of the children is girls?

47. Consumer fireworks are legal in 40 states in the United States.

a. In what fraction of the states are consumer fireworks legal?

b. In how many states are consumer fireworks illegal?

c. In what fraction of the states are consumer fireworks illegal? (*Source:* United States Fireworks Safety Council)

48. Thirty-three states in the United States contain federal Indian reservations.

a. What fraction of the states contain Indian reservations?

b. How many states do not contain Indian reservations?

c. What fraction of the states do not contain Indian reservations? (*Source:* Tiller Research, Inc., Albuquerque, NM)

C *Graph each fraction on a number line. See Examples 10 and 11.*

49. $\frac{1}{4}$

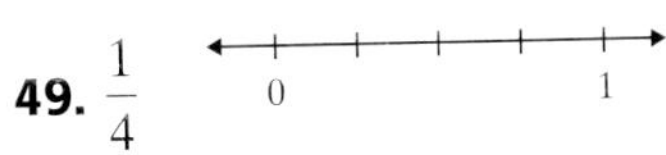

50. $\frac{1}{3}$ 0 1

51. $\frac{4}{7}$ 0 1

52. $\frac{5}{6}$ 0 1

53. $\frac{8}{5}$ 0 1 2

54. $\frac{9}{8}$ 0 1 2

55. $\frac{7}{3}$ 0 1 2 3

56. $\frac{15}{7}$ 0 1 2

57. $\frac{3}{8}$ 0 1

58. $\frac{8}{3}$ 0 1 2

D *Simplify by dividing. See Examples 12 through 17.*

59. $\frac{12}{12}$

60. $\frac{-3}{-3}$

61. $\frac{-5}{1}$

62. $\frac{-10}{1}$

63. $\frac{0}{-2}$

64. $\frac{0}{-8}$

65. $\frac{-8}{-8}$

66. $\frac{-14}{-14}$

67. $\frac{-9}{0}$

68. $\frac{-7}{0}$

69. $\frac{3}{1}$

70. $\frac{5}{5}$

 *Write each fraction as an equivalent fraction with the given denominator. See Examples 18 through 21.*

71. $\frac{4}{7}$; denominator of 35

72. $\frac{3}{5}$; denominator of 20

73. $\frac{2}{3}$; denominator of 21

74. $\frac{1}{6}$; denominator of 24

75. $\frac{2y}{5}$; denominator of 25

76. $\frac{9a}{10}$; denominator of 70

77. $\frac{1}{2}$; denominator of 30

78. $\frac{1}{3}$; denominator of 30

79. $\frac{10}{7x}$; denominator of $21x$

80. $\frac{5}{3b}$; denominator of $21b$

81. 2; denominator of 5

82. 5; denominator of 8

Write each fraction as an equivalent fraction whose denominator is 12. See Examples 18 through 21.

83. $\frac{3}{4}$

84. $\frac{4}{6}$

85. $\frac{2y}{3}$

86. $\frac{2}{3}$

87. $\frac{1}{2}$

88. $\frac{3x}{2}$

Write each fraction as an equivalent fraction whose denominator is 36x. See Example 18 through 21.

89. $\frac{4}{3}$

90. $\frac{3}{4}$

91. $\frac{5}{9}$

92. $\frac{7}{6}$

93. 1

94. 2

The table shows the fraction of the population in each country that used cell phones in a recent year. Use this table to answer Exercises 95–98.

95. Complete the table by writing each fraction as an equivalent fraction with a denominator of 100.

Country	Fraction of Population Using Cell Phones	Equivalent Fraction with a Denominator of 100
Denmark	$\frac{13}{25}$	
Finland	$\frac{39}{50}$	
Israel	$\frac{87}{100}$	$\frac{87}{100}$
Italy	$\frac{21}{25}$	
Japan	$\frac{59}{100}$	$\frac{59}{100}$
Norway	$\frac{83}{100}$	$\frac{83}{100}$
Singapore	$\frac{67}{100}$	$\frac{67}{100}$
South Korea	$\frac{6}{10}$	
Sweden	$\frac{79}{100}$	$\frac{79}{100}$
United States	$\frac{9}{20}$	

(*Source: International Telecommunication and World Almanac,* 2003)

96. Which of these countries has the largest fraction of cell phone users?

97. Which of these countries has the smallest fraction of cell phone users?

98. In which of these countries do over $\frac{3}{4}$ of the population use cell phones?

Helpful Hint

Write $\frac{3}{4}$ as an equivalent fraction with a denominator of 100.

Review and Preview

Simplify. See Section 1.8.

99. 3^2 **100.** 4^3 **101.** 5^3 **102.** 3^4

103. 7^2 **104.** 5^4 **105.** $2^3 \cdot 3$ **106.** $4^2 \cdot 5$

Combining Concepts

107. Write $\frac{2}{9}$ as an equivalent fraction whose denominator is 2088.

108. Write $\frac{3}{11}$ as an equivalent fraction whose denominator is 6479.

109. In your own words, explain how to write equivalent fractions.

110. In your own words, explain why $\frac{0}{10} = 0$ and $\frac{10}{0}$ is undefined.

Write each fraction.

111. Habitat for Humanity is a nonprofit organization that helps provide affordable housing to families in need. Habitat for Humanity does its work of building and renovating houses through 1651 local affiliates in the United States and 634 international affiliates. What fraction of the total Habitat for Humanity affiliates are located in the United States? (*Source:* Habitat for Humanity International)

112. The United States Marine Corps (USMC) has five principal training centers in California, three in North Carolina, two in South Carolina, one in Arizona, one in Hawaii, and one in Virginia. What fraction of the total USMC principal training centers are located in California? (*Source:* U.S. Department of Defense)

113. The Wendy's Corporation owns restaurants with five different names, as shown on the bar graph. What fraction of restaurants owned by Wendy's corporation are named "Wendy's" restaurants? (*Source:* The Wendy's Corporation)

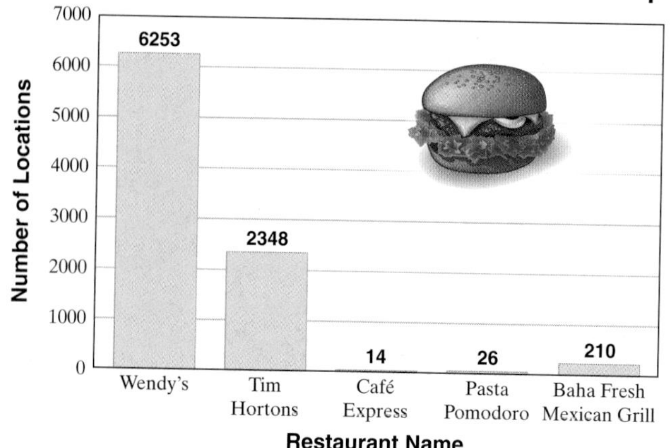

114. The Public Broadcasting Service (PBS) provides programming to the noncommercial public TV stations of the United States. The table shows a breakdown of the public television licensees by type. Each licensee operates one or more PBS member TV stations. What fraction of the public television licensees are universities or colleges? (*Source:* The Public Broadcast Service)

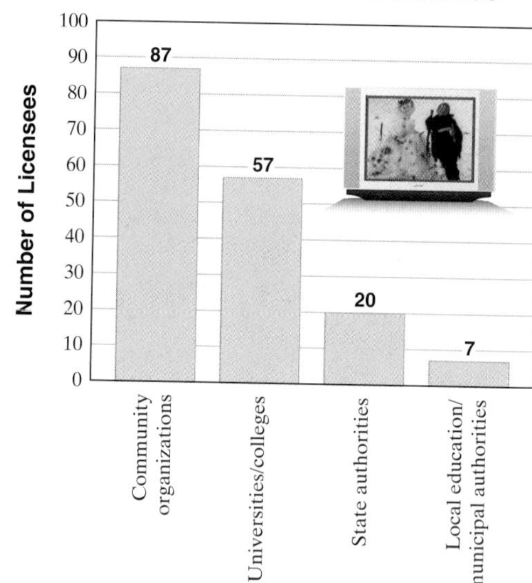

4.2 Factors and Simplest Form

A Writing a Number as a Product of Prime Numbers

Of all the equivalent ways to write a particular fraction, one special way is called **simplest form** or **lowest terms**. To help us write a fraction in simplest form, we first practice writing a number as a product of prime numbers.

A prime number is a natural number greater than 1 whose only factors are 1 and itself. The first few prime numbers are 2, 3, 5, 7, 11, 13, 17, 19, 23, 29, ...

A **composite number** is a natural number greater than 1 that is not prime.

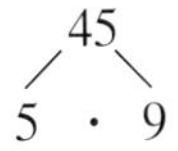

The natural number 1 is neither prime nor composite.

When a composite number is written as a product of prime numbers, this product is called the **prime factorization** of the number. For example, the prime factorization of 12 is $2 \cdot 2 \cdot 3$ because

$$12 = \underbrace{2 \cdot 2 \cdot 3}$$

This product is 12 and each number is a prime number.

Because multiplication is commutative, the order of the factors is not important. We can write the factorization $2 \cdot 2 \cdot 3$ as $2 \cdot 3 \cdot 2$ or $3 \cdot 2 \cdot 2$. Any of these is called the prime factorization of 12.

Every whole number greater than 1 has exactly one prime factorization.

Recall from Section 1.5 that since $12 = 2 \cdot 2 \cdot 3$, the numbers 2 and 3 are called *factors* of 12. A **factor** is any number that divides a number evenly (with a remainder of 0).

One method for finding the prime factorization of a number is by using a factor tree, as shown in the next example.

OBJECTIVES

A Write a number as a product of prime numbers.

B Write a fraction in simplest form.

C Solve problems by writing fractions in simplest form.

EXAMPLE 1 Write the prime factorization of 45.

Solution: We can begin by writing 45 as the product of two numbers, say 5 and 9.

```
   45
  /  \
 5  ·  9
```

The number 5 is prime but 9 is not, so we write 9 as $3 \cdot 3$.

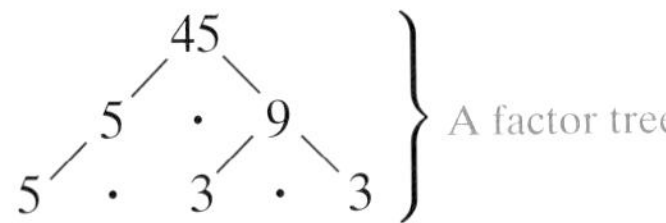

Each factor is now a prime number, so the prime factorization of 45 is $3 \cdot 3 \cdot 5$ or $3^2 \cdot 5$.

Practice Problem 1

Write the prime factorization of 28.

Answer

1. $28 = 2 \cdot 2 \cdot 7$ or $2^2 \cdot 7$

Practice Problem 2

Write the prime factorization of 60.

Helpful Hint

It makes no difference which factors you start with. The prime factorization of a number will be the same.

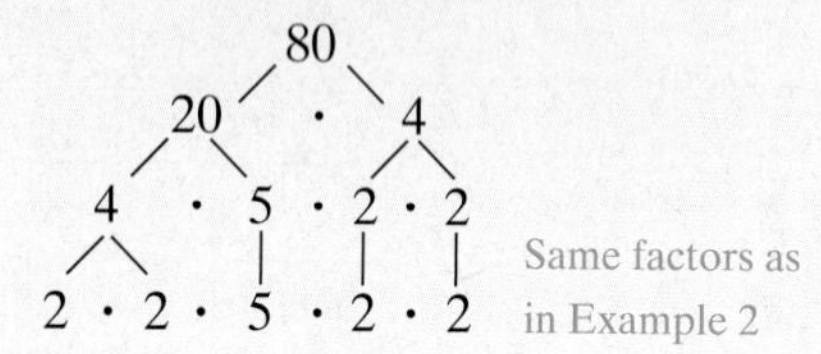

Same factors as in Example 2

EXAMPLE 2 Write the prime factorization of 80.

Solution: Write 80 as a product of two numbers. Continue this process until all factors are prime.

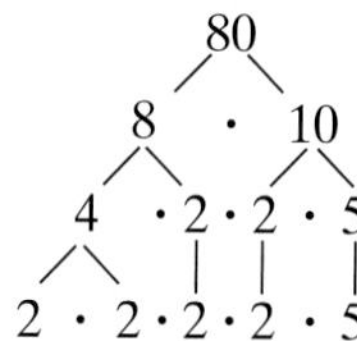

All factors are now prime, so the prime factorization of 80 is

$$2 \cdot 2 \cdot 2 \cdot 2 \cdot 5 \quad \text{or} \quad 2^4 \cdot 5.$$

There are a few quick **divisibility tests** to determine whether a number is divisible by the primes 2, 3, 5, or 10.

Divisibility Tests

A whole number is divisible by

- **2** if its last digit is 0, 2, 4, 6, or 8.

  132 is divisible by 2 since the last digit is a 2.
- **3** if the sum of its digits is divisible by 3.

  144 is divisible by 3 since $1 + 4 + 4 = 9$ is divisible by 3.
- **5** if its last digit is 0 or 5.

  1115 is divisible by 5 since the last digit is a 5.
- **10** if its last digit is 0.

  230 is divisible by 10 since the last digit is 0.

Helpful Hint

Here are a few other divisibility tests you may find interesting. A whole number is divisible by

- **4** if its last two digits are divisible by 4.

  1712 is divisible by 4.
- **6** if it is divisible by 2 and 3.

  9858 is divisible by 6.
- **9** if the sum of its digits is divisible by 9.

  5238 is divisible by 9 since $5 + 2 + 3 + 8 = 18$ is divisible by 9.

When finding the prime factorization of larger numbers, you may want to use the procedure shown in Example 3.

Practice Problem 3

Write the prime factorization of 297.

EXAMPLE 3 Write the prime factorization of 252.

Solution: For this method, we divide prime numbers into the given number. Since the ones digit of 252 is 2, we know that 252 is divisible by 2.

$$2\overline{)252}\;\; 126$$

Answers

2. $60 = 2 \cdot 2 \cdot 3 \cdot 5$ or $2^2 \cdot 3 \cdot 5$
3. $297 = 3 \cdot 3 \cdot 3 \cdot 11$ or $3^3 \cdot 11$

126 is divisible by 2 also.

$$2\overline{)252} \to 2\overline{)126} \to 63$$

63 is not divisible by 2 but is divisible by 3. Divide 63 by 3 and continue in this same manner until the quotient is a prime number.

$$2\overline{)252} \to 2\overline{)126} \to 3\overline{)63} \to 3\overline{)21} \to 7$$

Helpful Hint:

The order of choosing prime numbers does not matter. For consistency, we use the order 2, 3, 5, 7, ...

The prime factorization of 252 is $2 \cdot 2 \cdot 3 \cdot 3 \cdot 7$ or $2^2 \cdot 3^2 \cdot 7$.

Try the Concept Check in the margin.

Concept Check

True or false? The prime factorization of 72 is $2 \cdot 4 \cdot 9$. Explain your reasoning.

B Writing a Fraction in Simplest Form

We can use the prime factorization of a number to help us write a fraction in **simplest form** or **lowest terms**.

Simplest Form

A fraction is in **simplest form**, or **lowest terms**, when the numerator and the denominator have no common factors other than 1.

For example, the fraction $\frac{6}{10}$ *is not* in simplest form because 6 and 10 both have a factor of 2. That is, 2 is a common factor of 6 and 10.

We'll use the fundamental principle of fractions to divide the numerator and denominator by the common factor of 2.

$$\frac{6}{10} = \frac{6 \div 2}{10 \div 2} = \frac{3}{5}$$

The fraction $\frac{3}{5}$ is in lowest terms, since the numerator and the denominator have no common factors (other than 1).

In the future, we will write the prime factorization of the numerator and denominator to help us find common factors. Then we will divide out common factors.

Let's use the following notation to show dividing the numerator and the denominator by common factors.

$$\frac{6}{10} = \frac{2 \cdot 3}{2 \cdot 5} = \frac{2 \cdot 3}{2 \cdot 5} = \frac{3}{5}$$

Writing a Fraction in Simplest Form

To write a fraction in simplest form, write the prime factorization of the numerator and the denominator and then divide both by all common factors.

The process of writing a fraction in simplest form is called **simplifying** the fraction.

Concept Check: No; $4 = 2 \cdot 2$ and $9 = 3 \cdot 3$.

Practice Problem 4

Simplify: $\frac{30}{45}$

Practice Problem 5

Simplify: $\frac{39}{51}$

Practice Problem 6

Simplify: $\frac{45}{105y}$

Practice Problem 7

Simplify: $\frac{9a}{50a}$

Practice Problem 8

Simplify: $\frac{38}{4}$

Answers

4. $\frac{2}{3}$ **5.** $\frac{13}{17}$ **6.** $\frac{3}{7y}$ **7.** $\frac{9}{50}$ **8.** $\frac{19}{2}$

EXAMPLE 4 Simplify: $\frac{12}{20}$

Solution: First write the prime factorization of the numerator and the denominator.

$$\frac{12}{20} = \frac{2 \cdot 2 \cdot 3}{2 \cdot 2 \cdot 5}$$

Next divide the numerator and the denominator by all common factors.

$$\frac{12}{20} = \frac{2 \cdot 2 \cdot 3}{2 \cdot 2 \cdot 5} = \frac{3}{5}$$

EXAMPLE 5 Simplify: $\frac{42}{66}$

Solution: $\frac{42}{66} = \frac{2 \cdot 3 \cdot 7}{2 \cdot 3 \cdot 11} = \frac{7}{11}$

EXAMPLE 6 Simplify: $\frac{84x}{90}$

Solution: $\frac{84x}{90} = \frac{2 \cdot 2 \cdot 3 \cdot 7 \cdot x}{2 \cdot 3 \cdot 3 \cdot 5} = \frac{14x}{15}$

EXAMPLE 7 Simplify: $\frac{10y}{27y}$

Solution: $\frac{10y}{27y} = \frac{2 \cdot 5 \cdot y}{3 \cdot 3 \cdot 3 \cdot y} = \frac{10}{27}$

EXAMPLE 8 Simplify: $\frac{36}{28}$

Solution: $\frac{36}{28} = \frac{2 \cdot 2 \cdot 3 \cdot 3}{2 \cdot 2 \cdot 7} = \frac{9}{7}$

For the fraction $\frac{36}{28}$ in Example 8, you may immediately notice that a common factor of 36 and 28 is 4. If so, you may simply divide out that common factor instead of writing the prime factorizations.

$$\frac{36}{28} = \frac{4 \cdot 9}{4 \cdot 7} = \frac{9}{7}$$

The result is in simplest form. If it were not, we would repeat the same procedure until it was.

EXAMPLE 9 Simplify: $\dfrac{6x^2}{60x^3}$

Solution: Notice that 6 and 60 have a common factor of 6. Let's factor x^2 and x^3 to see what variable common factors we have,

$$\frac{6x^2}{60x^3} = \frac{6 \cdot x \cdot x}{6 \cdot 10 \cdot x \cdot x \cdot x} = \frac{1}{10x}$$

Helpful Hint

When all the factors of the numerator or denominator are divided out, don't forget that 1 still remains in that numerator or denominator. If it helps, use the following notation.

$$\frac{6x^2}{60x^3} = \frac{\overset{1}{\cancel{6}} \cdot \overset{1}{\cancel{x}} \cdot \overset{1}{\cancel{x}}}{\underset{1}{\cancel{6}} \cdot 10 \cdot \underset{1}{\cancel{x}} \cdot \underset{1}{\cancel{x}} \cdot x} = \frac{1}{10x} \quad \begin{array}{l}\leftarrow 1 \cdot 1 \cdot 1 \text{ is } 1. \\ \leftarrow 1 \cdot 10 \cdot 1 \cdot 1 \cdot x \text{ is } 10x.\end{array}$$

Try the Concept Check in the margin.

C Solving Problems by Writing Fractions in Simplest Form

Many real-life problems can be solved by writing fractions. To make the answers more clear, these fractions should be written in simplest form.

EXAMPLE 10 Calculating Fraction of Parks in Washington State

In a recent year, there were 54 national parks in the United States. Three of these parks are located in the state of Washington. What fraction of the United States' national parks can be found in Washington state? Write the fraction in simplest form. (*Source:* National Park Service)

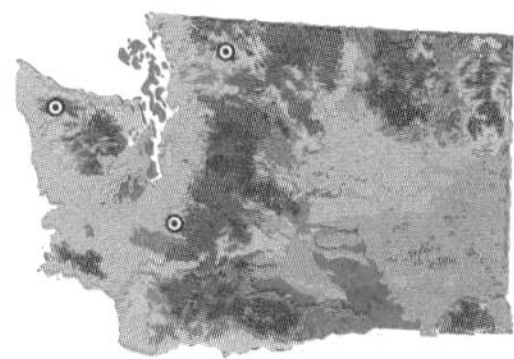

Solution: First we determine the fraction of parks found in Washington state.

$$\frac{3}{54} \quad \begin{array}{l}\leftarrow \text{national parks in Washington} \\ \leftarrow \text{total natioal parks}\end{array}$$

Next we simplify the fraction.

$$\frac{3}{54} = \frac{3}{3 \cdot 18} = \frac{1}{18}$$

Thus, $\frac{1}{18}$ of the United States' national parks are in Washington state.

Practice Problem 9

Simplify: $\dfrac{7a^3}{56a^2}$

Concept Check

Why is the following way to simplify the fraction $\frac{17}{75}$ incorrect?

$$\frac{17}{75} = \frac{1}{5}$$

Practice Problem 10

Eighty pigs were used in a recent study of olestra, a calorie-free fat substitute. A group of 12 of these pigs were fed a diet high in fat. What fraction of the pigs were fed the high-fat diet in this study? Write your answer in simplest form. (*Source:* From a study conducted by the Procter & Gamble Company)

Answers

9. $\frac{a}{8}$ **10.** $\frac{3}{20}$

Concept Check: Answers may vary.

CALCULATOR EXPLORATIONS

Simplifying Fractions

SCIENTIFIC CALCULATOR

Many calculators have a fraction key, such as [a b/c], that allows you to simplify a fraction on the calculator. For example, to simplify $\frac{324}{612}$, enter

[3] [2] [4] [a b/c] [6] [1] [2] [=]

The display will read

[9⌋17]

which represents $\frac{9}{17}$, the original fraction simplified.

GRAPHING CALCULATOR

Graphing calculators also allow you to simplify fractions. The fraction option on a graphing calculator may be found under the [MATH] menu.

To simplify $\frac{324}{612}$, enter

[3] [2] [4] [÷] [6] [1] [2] [MATH] [ENTER] [ENTER]

The display will read

[324/612 ▶ Frac 9/17]

Helpful Hint

The Calculator Explorations boxes in this chapter provide only an introduction to fraction keys on calculators. Any time you use a calculator, there are both advantages and limitations to its use. Never rely solely on your calculator. It is very important that you understand how to perform all operations on fractions by hand in order to progress through later topics. For further information, talk to your instructor.

Use your calculator to simplify each fraction.

1. $\frac{128}{224}$ **2.** $\frac{231}{396}$ **3.** $\frac{340}{459}$

4. $\frac{999}{1350}$ **5.** $\frac{810}{432}$ **6.** $\frac{315}{225}$

7. $\frac{243}{54}$ **8.** $\frac{689}{455}$

Name ______________________ Section ____________ Date ____________

Mental Math

1. Is 2430 divisible by 2? By 3? By 5?

Write the prime factorization of each number.

2. 15
3. 10
4. 6
5. 21
6. 4
7. 9
8. 14

EXERCISE SET 4.2

A *Write the prime factorization of each number. See Examples 1 through 3.*

1. 20
2. 12
3. 48
4. 75
5. 45
6. 64
7. 240
8. 128

B *Simplify each fraction. See Examples 4 through 9.*

9. $\frac{3}{12}$
10. $\frac{5}{20}$
11. $\frac{7x}{35}$
12. $\frac{9}{48z}$
13. $\frac{14}{16}$
14. $\frac{18}{4}$
15. $\frac{24a}{30a}$
16. $\frac{70y}{80xy}$
17. $\frac{35}{42}$
18. $\frac{25}{55}$
19. $\frac{30x^2}{36x}$
20. $\frac{45b}{80b^2}$
21. $\frac{16}{24}$
22. $\frac{18}{45}$
23. $\frac{45xz}{60z}$
24. $\frac{22a}{99ab}$

25. $\frac{39ab}{26a^2}$

26. $\frac{42x^2}{24xy}$

27. $\frac{63}{72}$

28. $\frac{56}{64}$

29. $\frac{21}{49}$

30. $\frac{14}{35}$

31. $\frac{24y}{40}$

32. $\frac{36}{54x}$

33. $\frac{36z}{63z}$

34. $\frac{39b}{52b}$

35. $\frac{72x^3y^2}{90xy}$

36. $\frac{24a^2b}{54ab^3}$

37. $\frac{12}{15}$

38. $\frac{18}{24}$

39. $\frac{25x^2}{40x}$

40. $\frac{36y^2}{42y}$

41. $\frac{27xy}{90y}$

42. $\frac{60y}{150yz}$

43. $\frac{36a^3bc^2}{24ab^4c^2}$

44. $\frac{60x^2yz}{36x^3y^3z^3}$

45. $\frac{40xy}{64xyz}$

46. $\frac{28abc}{60ac}$

C *Solve. Write each fraction in simplest form. See Example 10.*

47. A work shift for an employee at McDonald's consists of 8 hours. What fraction of the employee's work shift is represented by 6 hours?

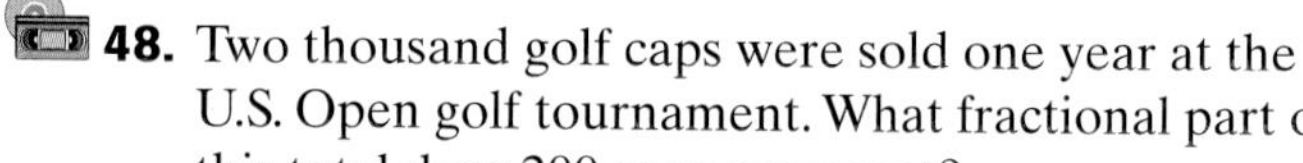

48. Two thousand golf caps were sold one year at the U.S. Open golf tournament. What fractional part of this total does 200 caps represent?

49. There are 5280 feet in a mile. What fraction of a mile is represented by 2640 feet?

50. There are 100 centimeters in 1 meter. What fraction of a meter is 20 centimeters?

51. As of the year 2000, a total of 414 individuals from around the world had flown in space. Of these, 261 were citizens of the United States. What fraction of individuals who have flown in space were Americans? (*Source:* Congressional Research Service)

52. Hallmark Cards employs 25,000 full-time employees worldwide. About 5600 employees work at the Hallmark headquarters in Kansas City, Missouri. What fraction of Hallmark employees work in Kansas City? (*Source:* Hallmark Cards,, Inc.)

53. There are 16,000 students at a local university. If 8800 are females, what fraction of the students are *male*?

54. Four out of 10 marbles are red. What fraction of marbles are *not red*?

55. Sixteen states in the United States have Ritz-Carlton hotels. (*Source:* Marriott International)

a. What fraction of states can claim at least one Ritz-Carlton hotel?

b. How many states do not have a Ritz-Carlton hotel?

c. Write the fraction of states without a Ritz-Carlton hotel.

56. As of 2003, there were 56 national parks in the United States. Eight of these parks are located in Alaska. (*Source:* National Park Service)

a. What fraction of the United States' national parks can be found in Alaska?

b. How many of the United States' national parks are found outside Alaska?

c. Write the fraction of national parks found in states other than Alaska.

57. The outer wall of the Pentagon is 24 inches wide. Ten inches is concrete, 8 inches is brick, and 6 inches is limestone. What fraction of the width is concrete? (*Source:* USA Today 1/28/2000)

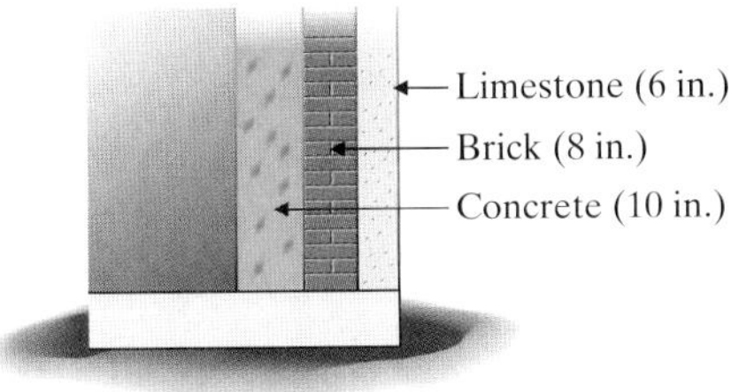

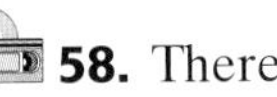 58. There are 35 students in a biology class. If 10 students made an A on the first test, what fraction of the students made an A?

The following graph is called a circle graph or pie chart. Each sector (shaped like a piece of pie) shows the fraction of entering college freshmen who expect to major in each discipline shown. The whole circle represents the entire class of college freshmen. Use this graph to answer Exercises 59–62.

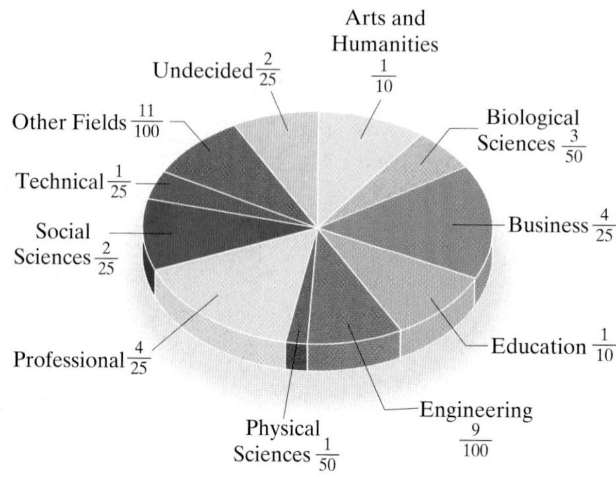

Source: Higher Education Research Institute

59. What fraction of entering college freshmen plan to major in education?

60. What fraction of entering college freshmen plan to major in social sciences?

61. Why is the Professional sector the same size as the Business sector?

62. Why is the Physical Sciences sector smaller than the Biological Sciences sector?

Review and Preview

Evaluate each expression using the given replacement numbers. See Section 2.5.

63. $\frac{x^3}{9}$ when $x = -3$

64. $\frac{y^3}{5}$ when $y = -5$

65. $2y$ when $y = -7$

66. $-5a$ when $a = -4$

67. $3z - y$ when $z = 2$ and $y = 6$

68. $7a - b$ when $a = 1$ and $b = -5$

69. $a^2 + 2b + 3$ when $a = 4$ and $b = 5$

70. $yx - z^2$ when $y = 6$, $x = 6$, and $z = 6$

Combining Concepts

71. Which fraction is closest to 0? Explain your answer.

a. $\frac{1}{2}$ b. $\frac{1}{3}$

c. $\frac{1}{4}$ d. $\frac{1}{5}$

72. Which fraction is closest to 1? Explain your answer.

a. $\frac{1}{2}$ b. $\frac{2}{3}$

c. $\frac{3}{4}$ d. $\frac{5}{6}$

Answer true or false for each statement.

73. $\frac{14}{42} = \frac{2 \cdot 7}{2 \cdot 3 \cdot 7} = \frac{0}{3}$

74. The fractions $\frac{8}{20}$, $\frac{4}{5}$, and $\frac{16}{60}$ are all equivalent.

75. A proper fraction cannot be equivalent to an improper fraction.

76. A fraction whose numerator and denominator are two different prime numbers cannot be simplified.

Simplify each fraction.

77. $\frac{372}{620}$

78. $\frac{9506}{12{,}222}$

There are generally considered to be eight basic blood types. The table shows the number of people with the various blood types in a typical group of 100 blood donors. Use this table to answer Exercises 79–83. Write each answer in simplest form.

79. What fraction of blood donors have A Rh-positive blood type?

80. What fraction of blood donors have O blood type?

81. What fraction of blood donors have AB blood type?

82. What fraction of blood donors have B blood type?

83. What fraction of blood donors have the negative Rh factor?

Distribution of Blood Types in Blood Donors

Blood Type	Number of People
O Rh-positive	37
O Rh-negative	7
A Rh-positive	36
A Rh-negative	6
B Rh-positive	9
B Rh-negative	1
AB Rh-positive	3
AB Rh-negative	1

(*Source*: American Red Cross Biomedical Services)

Use the following numbers for Exercises 84–87.

8691   786   1235   2235   85   105   22   222   900   1470

84. List the numbers divisible by both 2 and 3.

85. List the numbers that are divisible by both 3 and 5.

86. The answers to Exercise 84, are also divisible by what number? Tell why.

87. The answers to Exercise 85 are also divisible by what number? Tell why.

FOCUS ON The Real World

BLOOD AND BLOOD DONATION

Blood is the workhorse of the body. It carries to the body's tissues everything they need, from nutrients to antibodies to heat. Blood also carries away waste products like carbon dioxide. Blood contains three types of cells—red blood cells, white blood cells, and platelets—suspended in clear, watery fluid called plasma. Blood is $\frac{11}{20}$ plasma, and plasma itself is $\frac{9}{10}$ water. In the average healthy adult human, blood accounts for $\frac{1}{11}$ of a person's body weight.

Roughly every 2 seconds someone in the United States needs blood. Although only $\frac{1}{20}$ of eligible donors donate blood, the American Red Cross is still able to collect nearly 6 million volunteer donations of blood each year. This volume makes Red Cross Biomedical Services the largest supplier for blood transfusions in the United States.

The modern Red Cross blood donation program has its roots in World War II. In 1940, Dr. Charles Drew headed the Red Cross-sponsored American blood collection efforts for bombing victims in Great Britain. During that time, Dr. Drew developed techniques for separating plasma from blood cells that allowed mass production of plasma. Dr. Drew had discovered that plasma could be preserved longer than whole blood. He also found that dried plasma could be stored longer than its liquid form. By 1941, Dr. Drew had become the first medical director of the first American Red Cross Blood Bank in the United States. Plasma collected through this program saved the lives of many wounded civilians and Allied soldiers by reducing the high rate of death from shock.

GROUP ACTIVITY

Contact your local Red Cross Blood Service office. Find out how many people donated blood in your area in the past two months. Ask whether it is possible to get a breakdown of the blood donations by blood type. (For more on blood type, see Exercises 79 through 83 in Section 4.2.)

1. Research the population of the area served by your local Red Cross Blood Service office. Write the fraction of the local population who gave blood in the past two months.
2. Use the breakdown by blood type to write the fraction of donors giving each type of blood.

4.3 Multiplying and Dividing Fractions

A Multiplying Fractions

Let's use a diagram to discover how fractions are multiplied. For example, to multiply $\frac{1}{2}$ and $\frac{3}{4}$, we find $\frac{1}{2}$ of $\frac{3}{4}$. To do this, we begin with a diagram showing $\frac{3}{4}$ of rectangle's area shaded.

$\frac{3}{4}$ of the rectangle's area is shaded.

To find $\frac{1}{2}$ of $\frac{3}{4}$, we heavily shade $\frac{1}{2}$ of the part that is already shaded.

By counting smaller rectangles, we see that $\frac{3}{8}$ of the larger rectangle is now heavily shaded, so that $\frac{1}{2}$ of $\frac{3}{4}$ is $\frac{3}{8}$. This means that

$$\frac{1}{2}\cdot\frac{3}{4}=\frac{3}{8}$$

Notice that $\frac{1}{2}\cdot\frac{3}{4}=\frac{1\cdot 3}{2\cdot 4}=\frac{3}{8}$.

Notice that the numerator of the product is equal to the product of the numerators and that the denominator of the product is equal to the product of the denominators. This is how we multiply fractions.

Multiplying Two Fractions

If a, b, c, and d are numbers and b and d are not 0, then

$$\frac{a}{b}\cdot\frac{c}{d}=\frac{a\cdot c}{b\cdot d}$$

In other words, to multiply two fractions, multiply the numerators and multiply the denominators.

EXAMPLES Multiply:

1. $\frac{2}{3}\cdot\frac{5}{11}=\frac{2\cdot 5}{3\cdot 11}=\frac{10}{33}$ ← Product of numerators ← Product of denominators

2. $\frac{1}{4}\cdot\frac{1}{2}=\frac{1\cdot 1}{4\cdot 2}=\frac{1}{8}$

EXAMPLE 3 Multiply and simplify: $\frac{6}{7}\cdot\frac{14}{27}$

Solution: $\frac{6}{7}\cdot\frac{14}{27}=\frac{6\cdot 14}{7\cdot 27}$

Next, simplify by factoring into primes and dividing out common factors.

$$\frac{6\cdot 14}{7\cdot 27}=\frac{2\cdot 3\cdot 2\cdot 7}{7\cdot 3\cdot 3\cdot 3}$$

$$=\frac{4}{9}$$

OBJECTIVES

- A Multiply fractions.
- B Evaluate exponential expressions with fractional bases.
- C Divide fractions.
- D Multiply and divide given fractional replacement values.
- E Solve applications that require multiplication or division of fractions.

SSM TUTOR CENTER SG CD & VIDEO MATH PRO WEB

Practice Problems 1–2

Multiply:

1. $\frac{3}{8}\cdot\frac{5}{7}$ 2. $\frac{1}{3}\cdot\frac{1}{6}$

Practice Problem 3

Multiply and simplify: $\frac{6}{11}\cdot\frac{5}{8}$

Answers

1. $\frac{15}{56}$ **2.** $\frac{1}{18}$ **3.** $\frac{15}{44}$

Helpful Hint

In simplifying a product, it may be possible to identify common factors without actually writing the prime factorizations. For example,

$$\frac{10}{11}\cdot\frac{1}{20}=\frac{10\cdot 1}{11\cdot 20}=\frac{10\cdot 1}{11\cdot 10\cdot 2}=\frac{1}{22}$$

Practice Problem 4

Multiply and simplify: $\frac{4}{15}\cdot\frac{3}{8}$

EXAMPLE 4 Multiply and simplify: $\frac{23}{32}\cdot\frac{4}{7}$

Solution: Notice that 4 and 32 have a common factor of 4.

$$\frac{23}{32}\cdot\frac{4}{7}=\frac{23\cdot 4}{32\cdot 7}=\frac{23\cdot 4}{4\cdot 8\cdot 7}=\frac{23}{56}$$

After multiplying two fractions, *always* check to see whether the product can be simplified.

Practice Problem 5

Multiply: $\frac{1}{2}\cdot\left(-\frac{11}{28}\right)$

EXAMPLE 5 Multiply: $-\frac{1}{4}\cdot\frac{1}{2}$

Solution: Recall that the product of a negative number and a positive number is a negative number.

$$-\frac{1}{4}\cdot\frac{1}{2}=-\frac{1\cdot 1}{4\cdot 2}=-\frac{1}{8}$$

Practice Problem 6

Multiply: $\frac{9}{5}\cdot\frac{20}{12}$

EXAMPLE 6 Multiply: $\frac{13}{6}\cdot\frac{30}{26}$

Solution: $\frac{13}{6}\cdot\frac{30}{26}=\frac{13\cdot 6\cdot 5}{6\cdot 2\cdot 13}=\frac{5}{2}$

We multiply fractions in the same way if variables are involved.

Practice Problem 7

Multiply: $\frac{2}{3}\cdot\frac{3y}{2}$

EXAMPLE 7 Multiply: $\frac{3x}{4}\cdot\frac{8}{5x}$

Solution: $\frac{3x}{4}\cdot\frac{8}{5x}=\frac{3\cdot x\cdot 8}{4\cdot 5\cdot x}=\frac{3\cdot 4\cdot 2}{4\cdot 5}=\frac{6}{5}$

Helpful Hint

Recall that when the denominator of a fraction contains a variable, such as $\frac{8}{5x}$, we assume that the variable does not represent 0.

Practice Problem 8

Multiply: $\frac{a^3}{b^2}\cdot\frac{b}{a^2}$

EXAMPLE 8 Multiply: $\frac{x^2}{y}\cdot\frac{y^3}{x}$

Solution: $\frac{x^2}{y}\cdot\frac{y^3}{x}=\frac{x^2\cdot y^3}{y\cdot x}=\frac{x\cdot x\cdot y\cdot y\cdot y}{y\cdot x}=\frac{x\cdot y\cdot y}{1}=xy^2$

Answers

4. $\frac{1}{10}$ **5.** $-\frac{11}{56}$ **6.** 3 **7.** y **8.** $\frac{a}{b}$

B Evaluating Expressions with Fractional Bases

The base of an exponential expression can also be a fraction.

$$\left(\frac{1}{3}\right)^4 = \underbrace{\frac{1}{3}\cdot\frac{1}{3}\cdot\frac{1}{3}\cdot\frac{1}{3}}_{} = \frac{1\cdot1\cdot1\cdot1}{3\cdot3\cdot3\cdot3} = \frac{1}{81}$$

$\frac{1}{3}$ is a factor 4 times.

EXAMPLE 9 Evaluate.

a. $\left(\frac{2}{5}\right)^4$

b. $\left(-\frac{1}{4}\right)^2$

Solution:

a. $\left(\frac{2}{5}\right)^4 = \frac{2}{5}\cdot\frac{2}{5}\cdot\frac{2}{5}\cdot\frac{2}{5} = \frac{2\cdot2\cdot2\cdot2}{5\cdot5\cdot5\cdot5} = \frac{16}{625}$

b. $\left(-\frac{1}{4}\right)^2 = \left(-\frac{1}{4}\right)\cdot\left(-\frac{1}{4}\right) = \frac{1\cdot1}{4\cdot4} = \frac{1}{16}$

Practice Problem 9

Evaluate.

a. $\left(\frac{3}{4}\right)^3$ b. $\left(-\frac{4}{5}\right)^2$

C Dividing Fractions

Before we can divide fractions, we need to know how to find the **reciprocal** of a fraction.

Reciprocal of a Fraction

Two numbers are **reciprocals** of each other if their product is 1. The reciprocal of the fraction $\frac{a}{b}$ is $\frac{b}{a}$ because $\frac{a}{b}\cdot\frac{b}{a} = \frac{a\cdot b}{b\cdot a} = 1$.

For example,

The reciprocal of $\frac{2}{5}$ is $\frac{5}{2}$ because $\frac{2}{5}\cdot\frac{5}{2} = \frac{10}{10} = 1$.

The reciprocal of 5 is $\frac{1}{5}$ because $5\cdot\frac{1}{5} = \frac{5}{1}\cdot\frac{1}{5} = \frac{5}{5} = 1$.

The reciprocal of $-\frac{7}{11}$ is $-\frac{11}{7}$ because $-\frac{7}{11}\cdot-\frac{11}{7} = \frac{77}{77} = 1$.

Helpful Hint

Every number has a reciprocal except 0. The number 0 has no reciprocal because there is no number such that $0\cdot a = 1$.

Division of fractions has the same meaning as division of whole numbers. For example,

$10 \div 5$ means: How many 5s are there in 10?

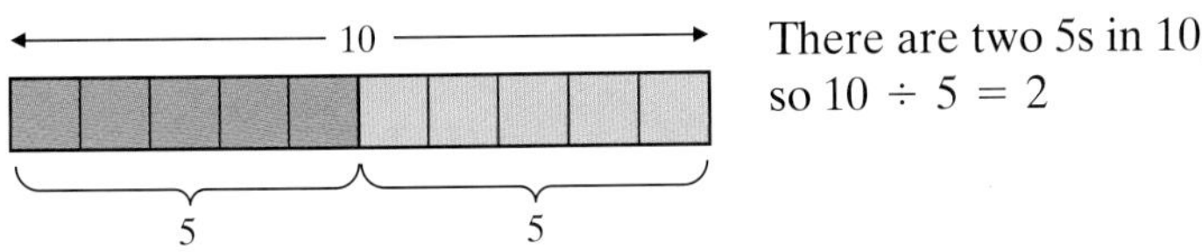

There are two 5s in 10, so $10 \div 5 = 2$

$\frac{3}{4} \div \frac{1}{8}$ means: How many $\frac{1}{8}$s are there in $\frac{3}{4}$?

Answers

9. a. $\frac{27}{64}$ **b.** $\frac{16}{25}$

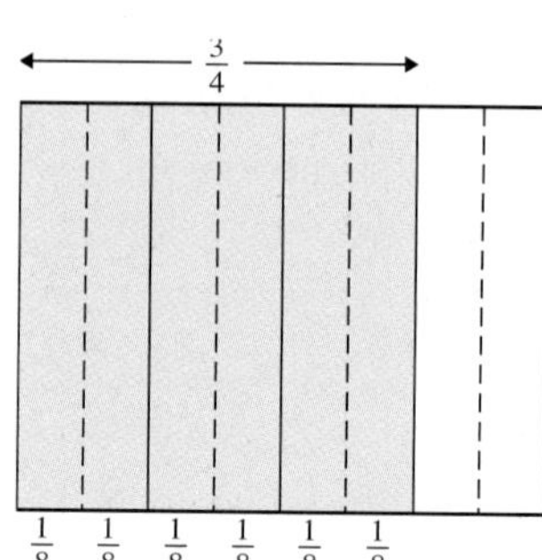

There are six $\frac{1}{8}$s in $\frac{3}{4}$, so

$$\frac{3}{4} \div \frac{1}{8} = 6$$

We use reciprocals to divide fractions.

Dividing Fractions

If b, c, and d are not 0, then $\frac{a}{b} \div \frac{c}{d} = \frac{a}{b} \cdot \frac{d}{c} = \frac{a \cdot d}{b \cdot c}$.

In other words, to divide fractions, multiply the first fraction by the reciprocal of the second fraction.

For example,

multiply by reciprocal

$$\frac{3}{4} \div \frac{1}{8} = \frac{3}{4} \cdot \frac{8}{1} = \frac{3 \cdot 8}{4 \cdot 1} = \frac{3 \cdot 2 \cdot 4}{4 \cdot 1} = \frac{6}{1} \text{ or } 6$$

Practice Problems 10–11

Divide and simplify.

10. $\frac{3}{2} \div \frac{14}{5}$

11. $\frac{4}{9} \div \frac{1}{2}$

EXAMPLES Divide and simplify.

10. $\frac{7}{8} \div \frac{2}{9} = \frac{7}{8} \cdot \frac{9}{2} = \frac{7 \cdot 9}{8 \cdot 2} = \frac{63}{16}$

11. $\frac{2}{5} \div \frac{1}{2} = \frac{2}{5} \cdot \frac{2}{1} = \frac{2 \cdot 2}{5 \cdot 1} = \frac{4}{5}$

After dividing two fractions, *always* check to see whether the result can be simplified.

Helpful Hint

When dividing by a fraction, do not look for common factors to divide out until you rewrite the division as multiplication.

Do **not** try to divide out these two 2s.

$$\frac{1}{\mathbf{2}} \div \frac{\mathbf{2}}{3} = \frac{1}{2} \cdot \frac{3}{2} = \frac{3}{4}$$

Practice Problem 12

Divide: $\frac{10}{4} \div \frac{2}{9}$

EXAMPLE 12 Divide: $-\frac{5}{16} \div -\frac{3}{4}$

Solution: Recall that the quotient (or product) of two negative numbers is a positive number.

$$-\frac{5}{16} \div -\frac{3}{4} = -\frac{5}{16} \cdot -\frac{4}{3} = \frac{5 \cdot 4}{4 \cdot 4 \cdot 3} = \frac{5}{12}$$

Answers

10. $\frac{15}{28}$ **11.** $\frac{8}{9}$ **12.** $\frac{45}{4}$

EXAMPLE 13 Divide: $\frac{2x}{3} \div 3x^2$

Solution: $\frac{2x}{3} \div 3x^2 = \frac{2x}{3} \div \frac{3x^2}{1} = \frac{2x}{3} \cdot \frac{1}{3x^2} = \frac{2 \cdot x \cdot 1}{3 \cdot 3 \cdot x \cdot x} = \frac{2}{9x}$

Try the Concept Check in the margin.

EXAMPLE 14 Simplify: $\left(\frac{4}{7} \cdot \frac{3}{8}\right) \div -\frac{3}{4}$

Solution: Remember to perform the operations inside the ( ) first.

$$\left(\frac{4}{7} \cdot \frac{3}{8}\right) \div -\frac{3}{4} = \left(\frac{4 \cdot 3}{7 \cdot 2 \cdot 4}\right) \div -\frac{3}{4} = \frac{3}{14} \div -\frac{3}{4}$$

Now divide.

$$\frac{3}{14} \div -\frac{3}{4} = \frac{3}{14} \cdot -\frac{4}{3} = -\frac{3 \cdot 2 \cdot 2}{2 \cdot 7 \cdot 3} = -\frac{2}{7}$$

D Multiplying and Dividing with Fractional Replacement Values

EXAMPLE 15 If $x = \frac{7}{8}$ and $y = -\frac{1}{3}$, evaluate **(a)** xy and **(b)** $x \div y$.

Solution: Replace x with $\frac{7}{8}$ and y with $-\frac{1}{3}$.

a. $xy = \frac{7}{8} \cdot -\frac{1}{3}$

$= -\frac{7 \cdot 1}{8 \cdot 3}$

$= -\frac{7}{24}$

b. $x \div y = \frac{7}{8} \div -\frac{1}{3}$

$= \frac{7}{8} \cdot -\frac{3}{1}$

$= -\frac{7 \cdot 3}{8 \cdot 1}$

$= -\frac{21}{8}$

EXAMPLE 16 Is $-\frac{2}{3}$ a solution of the equation $-\frac{1}{2}x = \frac{1}{3}$?

Solution: To check whether a number is a solution of an equation, recall that we replace the variable with the given number and see if a true statement results.

$-\frac{1}{2} \cdot x = \frac{1}{3}$ Recall that $-\frac{1}{2}x$ means $-\frac{1}{2} \cdot x$.

$-\frac{1}{2} \cdot -\frac{2}{3} = \frac{1}{3}$ Replace x with $-\frac{2}{3}$.

$\frac{1 \cdot 2}{2 \cdot 3} = \frac{1}{3}$ The product of two negative numbers is a positive number.

$\frac{1}{3} = \frac{1}{3}$ True.

Since we have a true statement, $-\frac{2}{3}$ is a solution.

Practice Problem 13

Divide: $\frac{3y}{4} \div 5y^3$

Concept Check

Which is the correct way to divide $\frac{3}{5}$ by $\frac{5}{12}$? Explain.

a. $\frac{3}{5} \div \frac{5}{12} = \frac{3}{5} \cdot \frac{12}{5}$

b. $\frac{3}{5} \div \frac{5}{12} = \frac{5}{3} \cdot \frac{5}{12}$

Practice Problem 14

Simplify: $\left(-\frac{2}{3} \cdot \frac{9}{14}\right) \div \frac{7}{15}$

Practice Problem 15

If $x = -\frac{3}{4}$ and $y = \frac{9}{2}$, evaluate (a) xy, and (b) $x \div y$.

Practice Problem 16

Is $-\frac{9}{8}$ a solution of the equation $2x = -\frac{9}{4}$?

Answers

13. $\frac{3}{20y^2}$ **14.** $-\frac{45}{49}$ **15. a.** $-\frac{27}{8}$ **b.** $-\frac{1}{6}$
16. yes

Concept Check: a

E Solving Applications by Multiplying and Dividing Fractions

To solve real-life problems that involve multiplying and dividing fractions, we will use our four problem-solving steps from Chapter 3. In Example 17, a new key word that implies multiplication is used. That key word is "of."

Practice Problem 17

About $\frac{1}{3}$ of all plant and animal species in the United States are at risk of becoming extinct. There are 20,439 known species of plants and animals in the United States. How many species are at risk of extinction? (*Source:* The Nature Conservancy)

EXAMPLE 17 Finding Number of Roller Coasters in an Amusement Park

Cedar Point is an amusement park located in Sandusky, Ohio. Its collection of 68 rides is the largest in the world. Of the rides, $\frac{7}{34}$ are roller coasters. How many roller coasters are in Cedar Point's collection of rides? (*Source:* Cedar Fair, L.P.)

Solution:

1. UNDERSTAND the problem. To do so, read and reread the problem. We are told that $\frac{7}{34}$ of Cedar Point's rides are roller coasters. The word "of" here means multiplication, since we are looking for the number of roller coasters.

x = number of roller coasters

2. TRANSLATE.

In words:	Number of roller coasters	is	$\frac{7}{34}$	of	total rides at Cedar Point
	↓	↓	↓	↓	↓
Translate:	x	$=$	$\frac{7}{34}$	$\cdot$	68

3. SOLVE.

$$x = \frac{7}{34} \cdot 68$$

$$= \frac{7}{34} \cdot \frac{68}{1} = \frac{7 \cdot 68}{34 \cdot 1} = \frac{7 \cdot 34 \cdot 2}{34 \cdot 1} = \frac{14}{1} \text{ or } 14$$

4. INTERPRET. *Check* your work. *State* your conclusion: The number of roller coasters at Cedar Point is 14.

Helpful Hint

To help visualize a fractional part of a whole number, look at the diagram below.

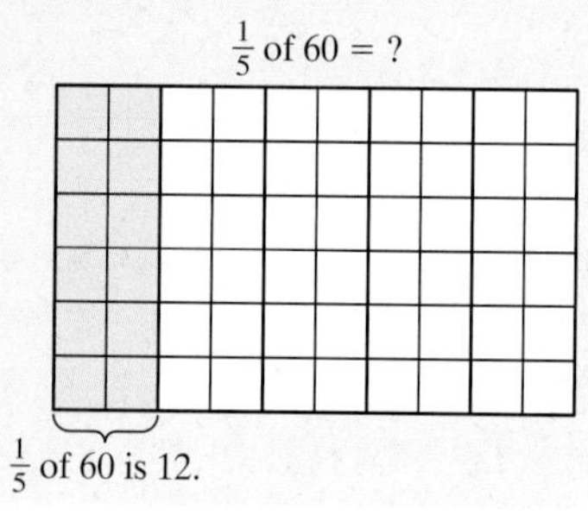

Answer

17. 6813 species

Name ______________________ Section ____________ Date ____________

Mental Math

Find each product.

1. $\frac{1}{3}\cdot\frac{2}{5}$

2. $\frac{2}{3}\cdot\frac{4}{7}$

3. $\frac{6}{5}\cdot\frac{1}{7}$

4. $\frac{7}{3}\cdot\frac{2}{3}$

5. $\frac{3}{1}\cdot\frac{3}{8}$

6. $\frac{2}{1}\cdot\frac{7}{11}$

EXERCISE SET 4.3

A *Multiply. Write the product in simplest form. See Examples 1 through 8.*

1. $\frac{7}{8}\cdot\frac{2}{3}$

2. $\frac{5}{9}\cdot\frac{7}{4}$

3. $-\frac{2}{7}\cdot\frac{5}{8}$

4. $\frac{5}{8}\cdot-\frac{1}{3}$

5. $-\frac{1}{2}\cdot-\frac{2}{15}$

6. $-\frac{3}{8}\cdot-\frac{5}{12}$

7. $\frac{18x}{20}\cdot\frac{36}{99}$

8. $\frac{5}{32}\cdot\frac{64y}{100}$

9. $3a^2\cdot\frac{1}{4}$

10. $-\frac{2}{3}\cdot 6y^3$

11. $\frac{x^3}{y^3}\cdot\frac{y^2}{x}$

12. $\frac{a}{b^3}\cdot\frac{b}{a^3}$

B *Evaluate. See Example 9.*

13. $\left(\frac{1}{5}\right)^3$

14. $\left(-\frac{1}{2}\right)^4$

15. $\left(-\frac{2}{3}\right)^2$

16. $\left(\frac{8}{9}\right)^2$

17. $\left(-\frac{2}{3}\right)^3\cdot\frac{1}{2}$

18. $\left(-\frac{3}{4}\right)^3\cdot\frac{1}{3}$

C *Divide. Write all quotients in simplest form. See Examples 10 through 13.*

19. $\frac{2}{3} \div \frac{5}{6}$

20. $\frac{5}{8} \div \frac{2}{3}$

21. $-\frac{6}{15} \div \frac{12}{5}$

22. $-\frac{4}{15} \div -\frac{8}{3}$

23. $\frac{8}{9} \div \frac{x}{2}$

24. $\frac{10}{11} \div -\frac{4}{5}$

25. $\frac{11y}{20} \div \frac{3}{11}$

26. $\frac{9z}{20} \div \frac{2}{9}$

27. $-\frac{2}{3} \div 4$

28. $-\frac{5}{6} \div 10$

29. $\frac{1}{5x} \div \frac{5}{x^2}$

30. $\frac{3}{y^2} \div \frac{9}{y}$

A **B** **C** *Perform each indicated operation. See Examples 1 through 14.*

31. $\frac{2}{3} \cdot \frac{5}{9}$

32. $\frac{8}{15} \cdot \frac{5}{32}$

33. $\frac{3x}{7} \div \frac{5}{6x}$

34. $\frac{16}{27y} \div \frac{8}{15y}$

35. $-\frac{5}{28} \cdot \frac{35}{25}$

36. $\frac{24}{45} \cdot -\frac{5}{8}$

37. $-\frac{3}{5} \div -\frac{4}{5}$

38. $-\frac{11}{16} \div -\frac{13}{16}$

39. $\left(-\frac{3}{4}\right)^2$

40. $\left(-\frac{1}{2}\right)^5$

41. $\frac{x^2}{y} \cdot \frac{y^3}{x}$

42. $\frac{b}{a^2} \cdot \frac{a^3}{b^3}$

43. $7 \div \frac{2}{11}$

44. $-100 \div \frac{1}{2}$

45. $-3x \div \frac{x^2}{12}$

46. $7x \div \frac{14x}{3}$

47. $\left(\frac{2}{7} \div \frac{7}{2}\right) \cdot \frac{3}{4}$

48. $\frac{1}{2} \cdot \left(\frac{5}{6} \div \frac{1}{12}\right)$

49. $-\frac{19}{63y} \cdot 9y^2$

50. $16a^2 \cdot -\frac{31}{24a}$

51. $-\frac{2}{3} \cdot -\frac{6}{11}$

52. $-\frac{1}{5} \cdot -\frac{6}{7}$

53. $\frac{4}{8} \div \frac{3}{16}$

54. $\frac{9}{2} \div \frac{16}{15}$

55. $\frac{21x^2}{10y} \div \frac{14x}{25y}$

56. $\frac{17y^2}{24x} \div \frac{13y}{18x}$

57. $\left(1 \div \frac{3}{4}\right) \cdot \frac{2}{3}$

58. $\left(33 \div \frac{2}{11}\right) \cdot \frac{5}{9}$

59. $\frac{a^3}{2} \div 30a^3$

60. $15c^3 \div \frac{3c^2}{5}$

61. $\frac{ab^2}{c} \cdot \frac{c}{ab}$

62. $\frac{ac}{b} \cdot \frac{b^3}{a^2c}$

63. $\left(\frac{1}{2} \cdot \frac{2}{3}\right) \div \frac{5}{6}$

64. $\left(\frac{3}{4} \cdot \frac{8}{9}\right) \div \frac{2}{5}$

65. $-\frac{4}{7} \div \left(\frac{4}{5} \cdot \frac{3}{7}\right)$

66. $\frac{5}{8} \div \left(\frac{4}{7} \cdot -\frac{5}{16}\right)$

D *Given the following replacement values, evaluate (**a**) xy and (**b**) x ÷ y. See Example 15.*

67. $x = \frac{2}{5}$ and $y = \frac{5}{6}$

a.

b.

68. $x = \frac{8}{9}$ and $y = \frac{1}{4}$

a.

b.

69. $x = -\frac{4}{5}$ and $y = \frac{9}{11}$

a.

b.

70. $x = \frac{7}{6}$ and $y = -\frac{1}{2}$

a.

b.

Determine whether the given replacement values are solutions of the given equations. See Example 16.

71. Is $-\frac{5}{18}$ a solution to $3x = -\frac{5}{6}$?

72. Is $\frac{9}{11}$ a solution to $\frac{2}{3}y = \frac{6}{11}$?

73. Is $\frac{2}{5}$ a solution to $-\frac{1}{2}z = \frac{1}{10}$?

74. Is $\frac{3}{5}$ a solution to $5x = \frac{1}{3}$?

E *Solve. See Example 17.*

75. A veterinarian's dipping vat holds 36 gallons of liquid. She usually fills it $\frac{5}{6}$ full of a medicated flea dip solution. Find how many gallons of solution are usually in the vat.

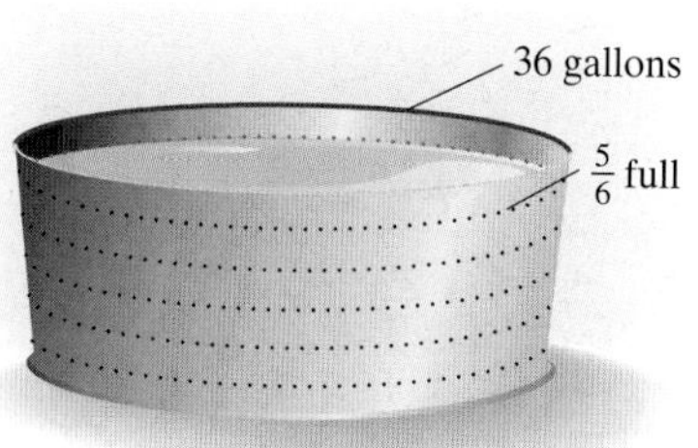

76. Each turn of a screw sinks it $\frac{3}{16}$ of an inch deeper into a piece of wood. Find how deep the screw is after 8 turns.

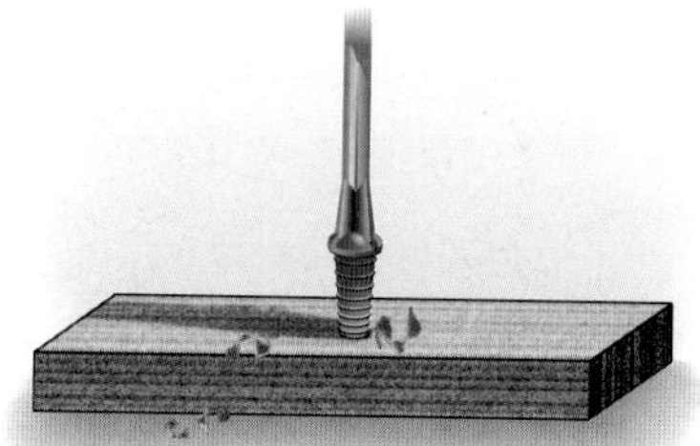

77. A special on a cruise to the Bahamas is advertised to be $\frac{2}{3}$ of the regular price. If the regular price is \$2757, what is the sale price?

78. The Gonzales recently sold their house for \$102,000, but $\frac{3}{50}$ of this amount goes to the real estate companies that helped them sell their house. How much money do the Gonzales pay to the real estate companies?

△ **79.** The radius of a circle is one-half of its diameter as shown. If the diameter of a circle is $\frac{3}{8}$ of an inch, what is its radius?

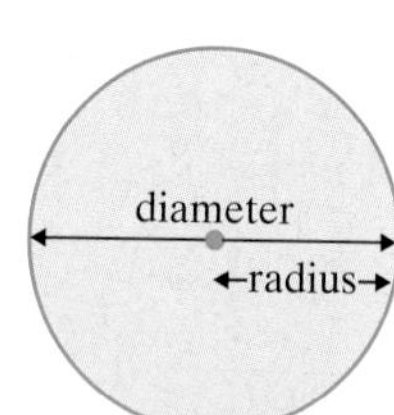

80. A recipe calls for $\frac{1}{3}$ of a cup of flour. How much flour should be used if only $\frac{1}{2}$ of the recipe is being made?

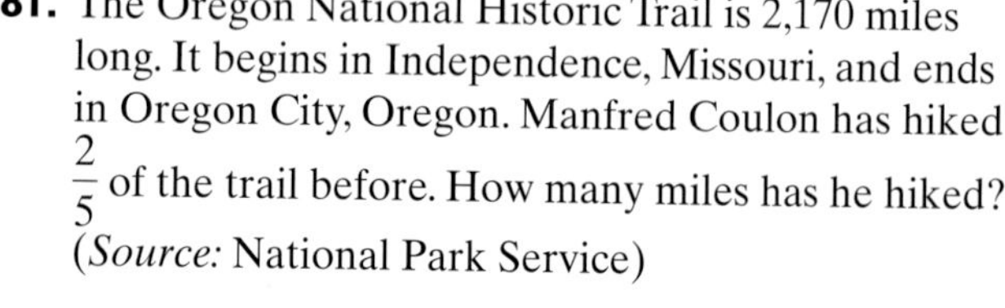

81. The Oregon National Historic Trail is 2,170 miles long. It begins in Independence, Missouri, and ends in Oregon City, Oregon. Manfred Coulon has hiked $\frac{2}{5}$ of the trail before. How many miles has he hiked? (*Source:* National Park Service)

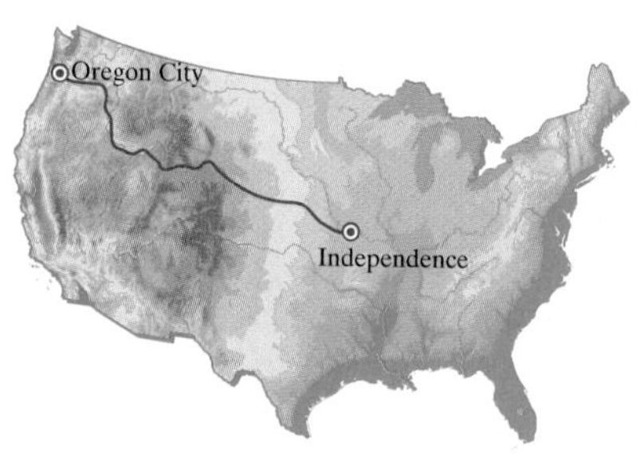

82. Movie theater owners received a total of \$7660 million in movie admission tickets, about $\frac{7}{10}$ of this amount was for R-rated movies. Find the amount of money received from R-rated movies. (*Source:* Motion Picture Association of America)

83. As part of his research, famous tornado expert Dr. T. Fujita studied approximately 31,050 tornadoes that occurred in the United States between 1916 and 1985. He found that roughly $\frac{7}{10}$ of these tornadoes occurred during April, May, June, and July. How many of these tornadoes occurred during these four months? (*Source: U.S. Tornadoes Part 1*, T. Fujita, University of Chicago)

84. Campbell Soup Company ships its soup in boxes containing 24 cans. If each can weighs about $\frac{3}{4}$ pound, find how much the contents of a box weighs.

85. An estimate for the measure of an adult's wrist is $\frac{1}{4}$ of the waist size. If Jorge has a 34-inch waist, estimate the size of his wrist.

86. An estimate for an adult's waist measurement is found by dividing the neck size (in inches) by $\frac{1}{2}$. Jock's neck measures 18 inches. Estimate his waist measurement.

When setting a post for building a deck or fence, it is recommended that $\frac{1}{3}$ of the total length of the post be buried in the ground. Find the amount of post to be buried in the ground for the given post lengths.

87. a 9-foot post

88. a 12-foot post

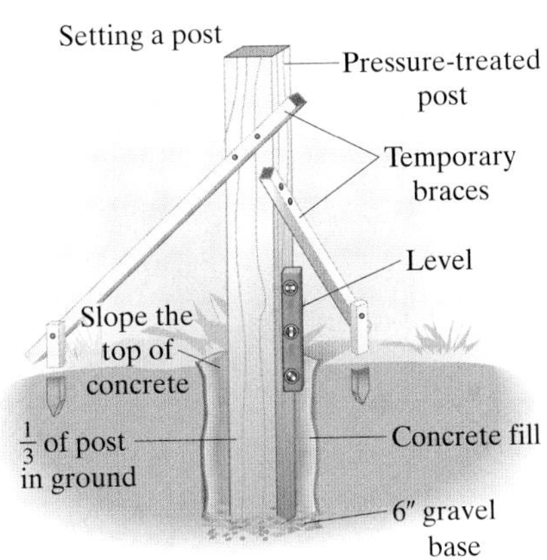

Source: *Southern Living Magazine*

89. A child needs a $\frac{1}{12}$-fluid ounce dose of a guaifenesin solution. How many doses are available in 2 fluid ounces?

90. How many $\frac{1}{3}$-ounce doses are available in a 5-ounce container of medicine.

Find the area of each rectangle. Recall that area = length · width, *or lw.*

△ **91.**

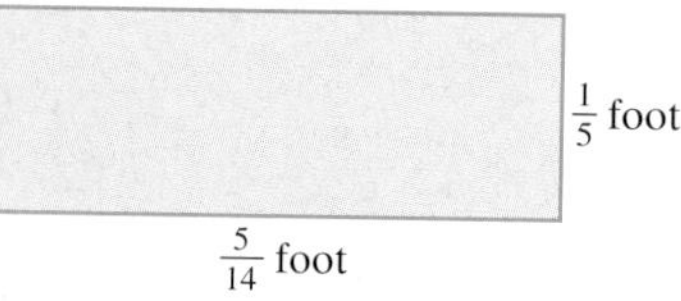

△ **92.**

$\frac{1}{2}$ mile

$\frac{3}{8}$ mile

Review and Preview

Write the prime factorization of each number. See Section 4.2.

93. 90

94. 42

95. 65

96. 72

97. 126

98. 112

Combining Concepts

99. In a recent year, approximately $\frac{7}{10}$ of U.S. households that have on-line service made on-line purchases. If 10,300,000 U.S. households have on-line services, how many of these made on-line purchases? (*Source:* Inteco Corp.)

100. Approximately $\frac{3}{25}$ of the U.S. population lives in the state of California. If the U.S. population is approximately 281,422,000 find the population of California. (*Source:* U.S. Bureau of the Census, 2000)

101. In your own words, describe how to multiply fractions.

102. In your own words, describe how to divide fractions.

103. One-third of all native flowering plant species in the United States are at risk of becoming extinct. That translates into 5144 at-risk flowering plant species. Based on this data, how many flowering plant species are native to the United States overall? (*Source*: The Nature Conservancy) (*Hint*: How many $\frac{1}{3}$s are in 5144?)

104. The FedEx fleet of aircraft includes 264 Cessnas. These Cessnas make up $\frac{66}{149}$ of the FedEx fleet. What is the size of the entire FedEx fleet of aircraft? (*Source:* Federal Express Corp.)

105. $\frac{42}{25} \cdot \frac{125}{36} \div \frac{7}{6}$

106. $\left(\frac{8}{13} \cdot \frac{39}{16} \cdot \frac{8}{9}\right)^2$

4.4 Adding and Subtracting Like Fractions and Least Common Denominator

OBJECTIVES

- (A) Add or subtract like fractions.
- (B) Add and subtract given fractional replacement values.
- (C) Solve equations containing fractions.
- (D) Solve problems by adding or subtracting like fractions.
- (E) Find the least common denominator of a list of fractions.

Fractions that have the same or a common denominator are called **like fractions**. Fractions that have different denominators are called **unlike fractions**.

Like Fractions	Unlike Fractions
$\frac{2}{5}$ and $\frac{3}{5}$	$\frac{2}{5}$ and $\frac{3}{4}$
$\frac{5}{21}, \frac{16}{21},$ and $\frac{7}{21}$	$-\frac{5}{7}$ and $\frac{5}{9}$
$-\frac{9}{15}$ and $\frac{13}{15}$	$\frac{3}{4}, \frac{9}{12},$ and $\frac{18}{24}$

(A) Adding or Subtracting Like Fractions

We can add like fractions on a number line just as we added whole numbers and integers on a number line.

To add $\frac{1}{5} + \frac{3}{5}$, start at 0 and draw an arrow $\frac{1}{5}$ of a unit long pointing to the right. From the tip of this arrow, draw an arrow $\frac{3}{5}$ of a unit long also pointing to the right. The tip of the second arrow ends at their sum, $\frac{4}{5}$.

Notice that the numerator of the sum is the sum of the numerators. Also, the denominator of the sum is the common denominator. This is how we add fractions. A similar method is used to subtract fractions.

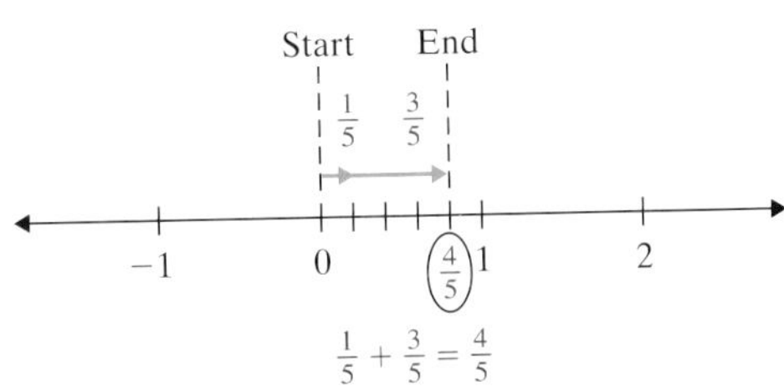

Adding or Subtracting Like Fractions (Fractions with the Same Denominator)

If a, b, and c, are numbers and b is not 0, then

$$\frac{a}{b} + \frac{c}{b} = \frac{a+c}{b} \quad \text{and also} \quad \frac{a}{b} - \frac{c}{b} = \frac{a-c}{b}$$

In other words, to add or subtract fractions with the same denominator, add or subtract their numerators and write the sum or difference over the **common** denominator.

For example,

$$\frac{1}{4} + \frac{2}{4} = \frac{1+2}{4} = \frac{3}{4}$$ Add the numerators. Keep the denominator.

$$\frac{4}{5} - \frac{2}{5} = \frac{4-2}{5} = \frac{2}{5}$$ Subtract the numerators. Keep the denominator.

Helpful Hint

As usual, don't forget to write all answers in simplest form.

EXAMPLES Add and simplify.

1. $\frac{2}{7} + \frac{3}{7} = \frac{2+3}{7} = \frac{5}{7}$ Add the numerators. Keep the common denominator.

2. $\frac{3}{16} + \frac{7}{16} = \frac{3+7}{16} = \frac{10}{16} = \frac{2 \cdot 5}{2 \cdot 8} = \frac{5}{8}$

3. $\frac{7}{8} + \frac{6}{8} + \frac{3}{8} = \frac{7+6+3}{8} = \frac{16}{8} = 2$

Try the Concept Check in the margin.

EXAMPLES Subtract and simplify.

4. $\frac{8}{9} - \frac{1}{9} = \frac{8-1}{9} = \frac{7}{9}$ ← Subtract the numerators ← Keep the common denominator.

5. $\frac{7}{8} - \frac{5}{8} = \frac{7-5}{8} = \frac{2}{8} = \frac{2}{2 \cdot 4} = \frac{1}{4}$

From our earlier work, we know that $\frac{-12}{6} = -2$ and that $\frac{12}{-6} = -2$. Also, $-\frac{12}{6} = -2$. Since all these fractions simplify to -2, we have that

$$\frac{-12}{6} = \frac{12}{-6} = -\frac{12}{6}$$

In general, the following is true:

$$\frac{-a}{b} = \frac{a}{-b} = -\frac{a}{b} \text{ as long as } b \text{ is not } 0.$$

For example, $\frac{-3}{4} = \frac{3}{-4} = -\frac{3}{4}$.

EXAMPLE 6 Add: $-\frac{11}{8} + \frac{6}{8}$

Solution: $-\frac{11}{8} + \frac{6}{8} = \frac{-11+6}{8}$

$= \frac{-5}{8}$ or $-\frac{5}{8}$

EXAMPLE 7 Subtract: $\frac{3x}{4} - \frac{7}{4}$

Solution: $\frac{3x}{4} - \frac{7}{4} = \frac{3x-7}{4}$

Recall from Section 3.1 that the terms in the numerator are unlike terms and cannot be combined.

Practice Problems 1–3

Add and simplify.

1. $\frac{5}{9} + \frac{2}{9}$
2. $\frac{5}{8} + \frac{1}{8}$
3. $\frac{10}{11} + \frac{1}{11} + \frac{7}{11}$

Concept Check

Find and correct the error in the following.

$$\frac{3}{7} + \frac{3}{7} = \frac{6}{14}$$

Practice Problems 4–5

Subtract and simplify.

4. $\frac{7}{12} - \frac{2}{12}$
5. $\frac{9}{10} - \frac{1}{10}$

Practice Problem 6

Add: $-\frac{8}{5} + \frac{4}{5}$

Practice Problem 7

Subtract: $\frac{2}{5} - \frac{3y}{5}$

Answers

1. $\frac{7}{9}$ **2.** $\frac{3}{4}$ **3.** $\frac{18}{11}$ **4.** $\frac{5}{12}$ **5.** $\frac{4}{5}$ **6.** $-\frac{4}{5}$
7. $\frac{2-3y}{5}$

Concept Check: $\frac{3}{7} + \frac{3}{7} = \frac{6}{7}$; keep the common denominator.

EXAMPLE 8 Subtract: $\frac{3}{7} - \frac{6}{7} - \frac{3}{7}$

Solution: $\frac{3}{7} - \frac{6}{7} - \frac{3}{7} = \frac{3 - 6 - 3}{7} = \frac{-6}{7}$ or $-\frac{6}{7}$

Practice Problem 8

Subtract: $\frac{4}{11} - \frac{6}{11} - \frac{3}{11}$

Helpful Hint

Recall that

$\frac{-6}{7} = -\frac{6}{7}$ $\left(\text{Also, } \frac{6}{-7} = -\frac{6}{7}, \text{ if needed.}\right)$

B Adding and Subtracting Given Fractional Replacement Values

EXAMPLE 9 Evaluate $y - x$ if $x = -\frac{3}{10}$ and $y = -\frac{8}{10}$.

Solution: Be very careful when replacing x and y with replacement values.

$$y - x = -\frac{8}{10} - \left(-\frac{3}{10}\right) \quad \text{Replace } x \text{ with } -\frac{3}{10} \text{ and } y \text{ with } -\frac{8}{10}.$$

$$= \frac{-8 - (-3)}{10}$$

$$= \frac{-5}{10} = \frac{-1 \cdot 5}{2 \cdot 5} = \frac{-1}{2} \text{ or } -\frac{1}{2}$$

Practice Problem 9

Evaluate $x + y$ if $x = -\frac{10}{12}$ and $y = \frac{5}{12}$.

C Solving Equations Containing Fractions

EXAMPLE 10 Solve: $x - \frac{1}{5} = \frac{3}{5}$

Solution: To solve, add $\frac{1}{5}$ to both sides of the equation.

$$x - \frac{1}{5} = \frac{3}{5}$$

$$x - \frac{1}{5} + \frac{1}{5} = \frac{3}{5} + \frac{1}{5}$$

$$x = \frac{4}{5}$$

To check, replace x with $\frac{4}{5}$ in the original equation.

$$x - \frac{1}{5} = \frac{3}{5}$$

$$\frac{4}{5} - \frac{1}{5} = \frac{3}{5} \quad \text{Replace } x \text{ with } \frac{4}{5}.$$

$$\frac{4 - 1}{5} = \frac{3}{5}$$

$$\frac{3}{5} = \frac{3}{5} \quad \text{True.}$$

The solution of $x - \frac{1}{5} = \frac{3}{5}$ is $\frac{4}{5}$.

Practice Problem 10

Solve: $\frac{7}{10} = x - \frac{1}{10}$

Answers

8. $-\frac{5}{11}$ **9.** $-\frac{5}{12}$ **10.** $\frac{4}{5}$

D Solving Problems by Adding and Subtracting Like Fractions

We can combine our skills in adding and subtracting like fractions with our four problem-solving steps from Chapter 3 to solve many kinds of real-life problems.

Practice Problem 11

If a piano student practices the piano $\frac{3}{4}$ of an hour in the morning and $\frac{1}{4}$ of an hour in the evening, how long did she practice that day?

EXAMPLE 11 Total Amount of an Ingredient in a Recipe

A recipe calls for $\frac{1}{3}$ of a cup of flour at the beginning and $\frac{2}{3}$ of a cup of flour later. How much total flour is needed to make that recipe?

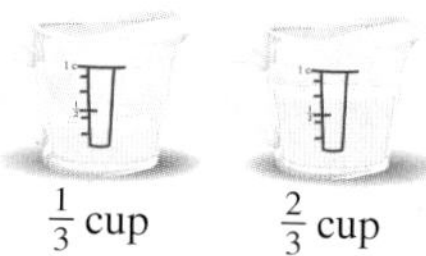

Solution:

1. UNDERSTAND the problem. To do so, read and reread the problem. Since we are finding total flour, we will add. Let x = total flour.
2. TRANSLATE.

In words:	Total flour	is	flour at the beginning	added to	flour later
	↓	↓	↓	↓	↓
Translate:	x	$=$	$\frac{1}{3}$	$+$	$\frac{2}{3}$

3. SOLVE. $x = \frac{1}{3} + \frac{2}{3}$

$$= \frac{1+2}{3} = \frac{3}{3} = 1$$

4. INTERPRET. *Check* your work. *State* your conclusion: The total flour needed for the recipe is 1 cup. ●

Practice Problem 12

A walker attends a gym that has a $\frac{1}{8}$-mile track. If he walks 9 laps on the track on Monday and 3 laps on Wednesday, how much farther did he walk on Monday than on Wednesday?

EXAMPLE 12 Calculating Distance

The distance from home to the World Gym is $\frac{7}{8}$ of a mile and from home to the Post Office is $\frac{5}{8}$ of a mile. How much farther is it from home to the World Gym than from home to the Post Office?

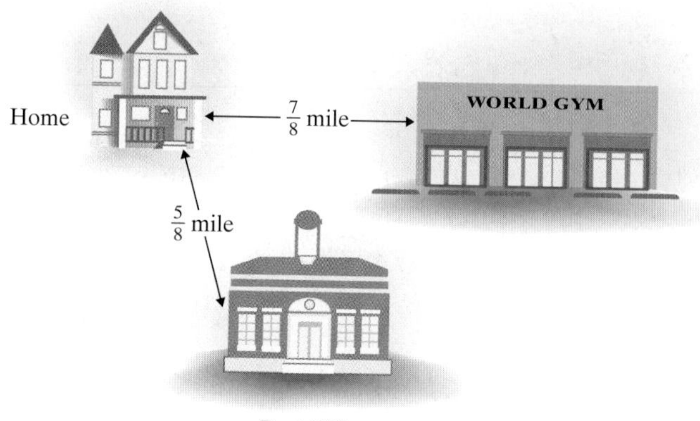

Answers

11. 1 hour **12.** $\frac{3}{4}$ mile

Solution:

1. UNDERSTAND. Read and reread the problem. The phrase "How much farther" tells us to subtract distances.

   Let x = distance farther.

2. TRANSLATE.

In words:	Distance farther	is	home to World Gym distance	minus	home to Post Office distance
	↓	↓	↓	↓	↓
Translate:	x	$=$	$\frac{7}{8}$	$-$	$\frac{5}{8}$

3. SOLVE: $x = \frac{7}{8} - \frac{5}{8}$

$$= \frac{7-5}{8} = \frac{2}{8} = \frac{2}{2 \cdot 4} = \frac{1}{4}$$

4. INTERPRET. *Check* your work. *State* your conclusion: The distance from home to the World Gym is $\frac{1}{4}$ mile farther than from home to the Post Office.

E Finding the Least Common Denominator

In the next section, we will add and subtract fractions that have different denominators. To add or subtract fractions that have unlike, or different, denominators, we first write them as equivalent fractions with a common denominator.

Although any common denominator can be used to add or subtract unlike fractions, we will use the **least common denominator (LCD)** or the **least common multiple (LCM)** of the denominators. Why? Since the LCD is the *smallest* of all common denominators, operations are usually less tedious with this number.

The **least common denominator (LCD)** of a list of fractions is the smallest positive number divisible by all the denominators in the list. (The least common denominator is also the **least common multiple (LCM) of the denominators.)**

For example, the LCD of $\frac{1}{4}$ and $\frac{3}{10}$ is 20 because 20 is the smallest positive number divisible by both 4 and 10.

Finding the LCD: Method 1

One way to find the LCD is to see whether the larger denominator is divisible by the smaller denominator. If so, the larger number is the LCD. If not, then check consecutive multiples of the larger denominator until the LCD is found.

To find the LCD for $\frac{1}{4}$ and $\frac{3}{10}$, we check to see whether 10 is a multiple of 4. No, it is not, so we check consecutive multiples of 10.

$2 \cdot 10 = 20$ 20 is divisible by 4, so LCD $= 20$.

Practice Problem 13

Find the LCD of $\frac{7}{12}$ and $\frac{2}{15}$.

EXAMPLE 13 Find the LCD of $\frac{3}{8}$ and $\frac{1}{6}$.

Solution: Is 8 divisible by 6? No, so we check multiples of 8.

$2 \cdot 8 = 16$ 16 is not divisible by 6.

$3 \cdot 8 = 24$ 24 is divisible by 6, so LCD $= 24$.

Finding the LCD: Method 2

Another way to find the LCD is to first write each denominator as a product of primes. To find the LCD of $\frac{1}{4}$ and $\frac{3}{10}$, we write

$4 = 2 \cdot 2$

$10 = 2 \cdot 5$

If the LCD is divisible by 4, it must contain the factors $2 \cdot 2$. If the LCD is divisible by 10, it must contain the factors $2 \cdot 5$. Since 4 and 10 will divide into the LCD separately, the LCD needs to contain a factor the greatest number of times that the factor appears in any **one** prime factorization.

Helpful Hint

The number 2 is a factor twice since that is the greatest number of times that 2 is a factor in the prime factorization of any one denominator.

factors of 4

LCD $= 2 \cdot 2 \cdot 5 = 20$

factors of 10

Practice Problem 14

Find the LCD of $\frac{9}{14}$ and $\frac{11}{35}$.

EXAMPLE 14 Find the LCD of $\frac{5}{24}$ and $\frac{1}{18}$.

Solution: First write each denominator as a product of primes.

$24 = 2 \cdot 2 \cdot 2 \cdot 3$

$18 = 2 \cdot 3 \cdot 3$

Write each factor the greatest number of times that it appears in any **one** prime factorization.

The greatest number of times that 2 appears is **3** times:

$24 = 2 \cdot 2 \cdot 2 \cdot 3$

The greatest number of times that 3 appears is **2** times:

$18 = 2 \cdot 3 \cdot 3$

LCD $= 2 \cdot 2 \cdot 2 \cdot 3 \cdot 3 = 72$

Notice that 72 is the smallest positive number that is divisible by both 18 and 24.

Answers

13. 70 **14.** 60

EXAMPLE 15 Find the LCD of $-\frac{2}{5}, \frac{1}{6}$, and $\frac{5}{12}$.

Solution: To help find the LCD, we write the prime factorization of each denominator and circle each different factor the greatest number of times that it appears in any one factor.

$$5 = \textcircled{5}$$
$$6 = 2 \cdot \textcircled{3}$$
$$12 = \textcircled{2 \cdot 2} \cdot 3$$
$$\text{LCD} = 2 \cdot 2 \cdot 3 \cdot 5 = 60$$

EXAMPLE 16 Find the LCD of $\frac{3}{5}, \frac{2}{x}$, and $\frac{7}{x^3}$.

Solution:
$$5 = \textcircled{5}$$
$$x = x$$
$$x^3 = \textcircled{x \cdot x \cdot x}$$
$$\text{LCD} = 5 \cdot x \cdot x \cdot x = 5x^3$$

Try the Concept Check in the margin.

STUDY SKILLS REMINDER

Have you decided to successfully complete this course?

Ask yourself if one of your current goals is to successfully complete this course.

If it is not a goal of yours, ask yourself why? One common reason is fear of failure. Amazingly enough, fear of failure alone can be strong enough to keep many of us from doing our best in any endeavor. Another common reason is that you simply haven't taken the time to make successfully completing this course one of your goals.

If you are taking this mathematics course, then successfully completing this course probably should be one of your goals. To make it a goal, start by writing this goal in your mathematics notebook. Then read or reread Section 1.1 and make a commitment to try the suggestions in this section.

If successfully completing this course is already a goal of yours, also read or reread Section 1.1 and try some suggestions in that section so that you are actively working toward your goal.

Good luck and don't forget that a positive attitude will make a big difference.

Practice Problem 15

Find the LCD of $\frac{7}{4}, \frac{7}{15}$, and $\frac{3}{10}$.

If you prefer working with exponents,

Example 15: $5 = \textcircled{5}$
$$6 = 2 \cdot \textcircled{3}$$
$$12 = \textcircled{2^2} \cdot 3$$
$$\text{LCD} = 2^2 \cdot 3 \cdot 5 = 60$$

Practice Problem 16

Find the LCD of $\frac{7}{y}$ and $\frac{6}{11}$.

Concept Check

True or false? The LCD of the fractions $\frac{1}{6}$ and $\frac{1}{8}$ is 48.

Answers

15. 60 **16.** $11y$

Concept Check: False; it is 24.

FOCUS ON Business and Career

INTENDED MAJORS

Table 1 shows the breakdown of entering college freshmen who plan to major in a professional field by specific professional major. Table 2 shows a similar breakdown for technical majors.

TABLE 1

Professional Majors	Fraction
Architecture or urban planning	$\frac{11}{155}$
Home economics	$\frac{1}{155}$
Health technology	$\frac{13}{155}$
Nursing	$\frac{7}{31}$
Pharmacy	$\frac{9}{155}$
Pre-dental, pre-medical, pre-veterinary	$\frac{39}{155}$
Therapy (occupational, physical, speech)	$\frac{36}{155}$
Other	$\frac{11}{155}$

(*Source:* Higher Education Research Institute)

TABLE 2

Technical Majors	Fraction
Building trades	$\frac{1}{8}$
Data processing, computer programming	$\frac{3}{8}$
Drafting or design	$\frac{1}{8}$
Electronics	$\frac{3}{40}$
Mechanics	$\frac{7}{40}$
Other	$\frac{1}{8}$

(*Source:* Higher Education Research Institute)

CRITICAL THINKING

1. Interpret the meaning of the first line of Table 1.
2. Interpret the meaning of the third line of Table 2.
3. Which of the professional majors is most popular? Explain your reasoning.
4. Which of the technical majors is most popular? Explain your reasoning.

GROUP ACTIVITY

Research the number of majors offered through your discipline area (such as Arts and Humanities, Biological Sciences, Business, Education, Engineering, etc.) at your school. How many entering students plan to take each major? Make a table similar to Tables 1 and 2 to display your results. Show the fraction of entering students taking each major. Do additional research to find the number of students that graduate with each major. Make another table to display your results. How does the breakdown of majors (those intended by entering students and those actually completed by graduating) change?

Name ______________________ Section __________ Date __________

Mental Math

State whether the fractions in each list are like or unlike fractions.

1. $\frac{7}{8}, \frac{7}{10}$ **2.** $\frac{2}{3}, \frac{2}{9}$ **3.** $\frac{9}{10}, \frac{1}{10}$ **4.** $\frac{8}{11}, \frac{2}{11}$

5. $\frac{2}{31}, \frac{30}{31}, \frac{19}{31}$ **6.** $\frac{3}{10}, \frac{3}{11}, \frac{3}{13}$ **7.** $\frac{5}{12}, \frac{7}{12}, \frac{12}{11}$ **8.** $\frac{1}{5}, \frac{2}{5}, \frac{4}{5}$

 *Add or subtract as indicated. See Examples 1 through 8.*

9. $\frac{3}{7} + \frac{2}{7}$ **10.** $\frac{5}{9} + \frac{2}{9}$ **11.** $\frac{10}{11} - \frac{4}{11}$ **12.** $\frac{9}{13} - \frac{5}{13}$

13. $\frac{5}{11} + \frac{2}{11}$ **14.** $\frac{4}{7} + \frac{2}{7}$ **15.** $\frac{9}{15} - \frac{1}{15}$ **16.** $\frac{3}{15} - \frac{1}{15}$

EXERCISE SET 4.4

Add or subtract as indicated. See Examples 1 through 8.

1. $-\frac{1}{2} + \frac{1}{2}$ **2.** $-\frac{3}{x} + \frac{1}{x}$ **3.** $\frac{2}{9x} + \frac{4}{9x}$ **4.** $\frac{3}{10y} + \frac{2}{10y}$

5. $-\frac{4}{13} + \frac{2}{13} + \frac{1}{13}$ **6.** $-\frac{5}{11} + \frac{1}{11} + \frac{2}{11}$ **7.** $\frac{7}{18} + \frac{3}{18} + \frac{2}{18}$ **8.** $\frac{2}{15} + \frac{4}{15} + \frac{9}{15}$

9. $\frac{1}{y} - \frac{4}{y}$ **10.** $\frac{4}{z} - \frac{7}{z}$ **11.** $\frac{7a}{4} - \frac{3}{4}$ **12.** $\frac{18b}{5} - \frac{3}{5}$

13. $\frac{1}{8} - \frac{7}{8}$

14. $\frac{1}{6} - \frac{5}{6}$

15. $\frac{20}{21} - \frac{10}{21} - \frac{17}{21}$

16. $\frac{27}{28} - \frac{5}{28} - \frac{28}{28}$

17. $\frac{9x}{15} + \frac{1x}{15}$

18. $\frac{2x}{15} - \frac{7}{15}$

19. $\frac{7x}{16} - \frac{15x}{16}$

20. $\frac{15b}{16} + \frac{7b}{16}$

21. $\frac{15}{16z} - \frac{3}{16z}$

22. $\frac{7}{16a} + \frac{15}{16a}$

23. $\frac{3}{10} - \frac{6}{10}$

24. $-\frac{6}{10} + \frac{3}{10}$

25. $\frac{15}{17} + \frac{5}{17} + \frac{14}{17}$

26. $\frac{1}{8} - \frac{15}{8} + \frac{2}{8}$

27. $\frac{9}{12} - \frac{7}{12} - \frac{10}{12}$

28. $\frac{9}{13} + \frac{10}{13} + \frac{7}{13}$

29. $\frac{x}{4} + \frac{3x}{4} - \frac{2x}{4} + \frac{x}{4}$

30. $\frac{9y}{8} + \frac{2y}{8} + \frac{5y}{8} - \frac{4y}{8}$

B *Evaluate each expression for the given replacement values. See Example 9.*

31. $x + y; x = \frac{3}{4}, y = \frac{2}{4}$

32. $x - y; x = \frac{7}{8}, y = \frac{9}{8}$

33. $x - y; x = -\frac{1}{5}, y = -\frac{3}{5}$

34. $x + y; x = -\frac{1}{6}, y = \frac{5}{6}$

35. $x - y + z; x = \frac{3}{12}, y = \frac{5}{12}, z = -\frac{7}{12}$

36. $x + y - z; x = \frac{2}{14}, y = \frac{3}{14}, z = \frac{8}{14}$

C *Solve and check. See Example 10.*

37. $x + \frac{1}{3} = -\frac{1}{3}$

38. $x + \frac{1}{9} = -\frac{7}{9}$

39. $y - \frac{3}{13} = -\frac{2}{13}$

40. $z - \frac{5}{14} = \frac{4}{14}$

41. $3x - \frac{1}{5} - 2x = \frac{1}{5} + \frac{2}{5}$

42. $5x + \frac{1}{11} - 4x = \frac{2}{11} - \frac{5}{11}$

D *Find the perimeter of each figure. Recall that the perimeter of a figure is the distance around a figure.*

△ **43.** $\frac{4}{20}$ inch, $\frac{7}{20}$ inch, $\frac{9}{20}$ inch

△ **44.** Square $\frac{1}{6}$ centimeter

△ **45.** $\frac{5}{12}$ meter Rectangle $\frac{7}{12}$ meter

△ **46.**

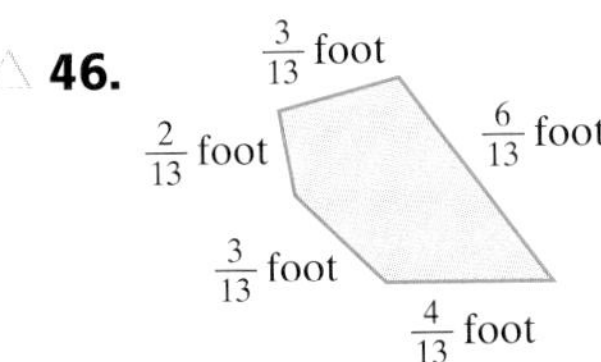

Solve. Write each answer in simplest form. See Examples 11 and 12.

47. Nathan Payne worked in his yard for $\frac{5}{8}$ of an hour on Saturday and $\frac{7}{8}$ of an hour on Sunday. How long did Nathan work in his yard over the weekend?

48. A recipe for Heavenly Hash Cake calls for $\frac{3}{4}$ cup of flour and later $\frac{1}{4}$ cup of flour. How much flour is needed to make the recipe?

The chart shows the breakdown of all U.S. employees covered by health benefits in 2002 by type of health plan. Use this chart to answer Exercises 49–52.

49. Put the health plans in order from the smallest fraction of employees covered to the largest fraction of employees covered.

50. Find the fraction of employees that are *not* covered by a Health Maintenance Organization.

Type of Health Plan	Fraction of Employees with Health Benefits
Health Maintenance Organization	$\frac{29}{100}$
Point-of-Service	$\frac{14}{100}$
Preferred Provider Organization	$\frac{50}{100}$
Traditional fee-for-service	$\frac{7}{100}$

(*Source:* William M. Mercer, Inc.)

51. Find the fraction of the employees that are *not* covered by a Preferred Provider Organization.

52. Which type of health plan is the most popular?

53. As of 2002, the fraction of states in the United States with maximum interstate highway speed limits up to and including 70 mph was $\frac{39}{50}$. The fraction of states with 70 mph speed limits was $\frac{18}{50}$. What fraction of states had speed limits that were less than 70 mph? (*Source:* National Motorists Association)

54. When people take aspirin, $\frac{31}{50}$ of the time it is used to treat some type of pain. Approximately $\frac{7}{50}$ of all aspirin use is for treating headaches. What fraction of aspirin use is for treating pain other than headaches? (*Source:* Bayer Market Research)

E *Find the LCD of each list of fractions. See Examples 13 through 16.*

55. $\frac{1}{3}, \frac{3}{4}$

56. $\frac{1}{4}, \frac{5}{6}$

57. $-\frac{2}{9}, \frac{6}{15}$

58. $-\frac{7}{12}, \frac{3}{20}$

59. $\frac{5}{12}, \frac{5}{18}$

60. $\frac{7}{12}, \frac{7}{15}$

61. $-\frac{7}{24}, -\frac{5}{x}$

62. $-\frac{11}{y}, -\frac{13}{70}$

63. $\frac{2}{25}, \frac{3}{15}, \frac{5}{6}$

64. $\frac{3}{4}, \frac{1}{6}, \frac{13}{18}$

65. $\frac{23}{18}, \frac{1}{21}$

66. $\frac{45}{24}, \frac{2}{45}$

67. $-\frac{16}{15}, -\frac{11}{25}$

68. $-\frac{22}{21}, \frac{3}{14}$

69. $\frac{1}{8}, \frac{1}{24}$

70. $\frac{1}{15}, \frac{1}{90}$

71. $\frac{8}{25}, \frac{7}{10}$

72. $\frac{7}{8}, \frac{13}{12}$

73. $-\frac{1}{a}, -\frac{7}{12}$

74. $-\frac{1}{9}, -\frac{80}{b}$

75. $\frac{4}{3}, \frac{8}{21}, \frac{3}{56}$

76. $\frac{6}{70}, \frac{11}{80}, \frac{15}{90}$

77. $\frac{12}{11}, \frac{20}{33}, \frac{12}{121}$

78. $\frac{7}{10}, \frac{8}{15}, \frac{9}{100}$

Review and Preview

Perform each indicated operation. See Sections 4.3 and 2.5.

79. $\frac{4}{5} \cdot \frac{3}{7}$

80. $\frac{5}{3} \cdot \frac{4}{9}$

81. $-2 + 10$

82. $-2(10)$

83. $\frac{2}{5} \div \frac{1}{2}$

84. $\frac{4}{7} \div \frac{1}{3}$

85. $-12 - 16$

86. $-18 - (-2)$

Combining Concepts

Perform each indicated operation.

87. $\frac{4}{11} + \frac{5}{11} - \frac{3}{11} + \frac{2}{11}$

88. $\frac{9}{12} + \frac{1}{12} - \frac{3}{12} - \frac{5}{12}$

Solve. Write each answer in simplest form.

89. A person's marital status is either single (never married), married, widowed, or divorced. Of American men over the age of 65, $\frac{38}{50}$ are married and $\frac{7}{50}$ are widowed. What fraction of American men over age 65 are either single or divorced? (*Source:* Based on data from the U.S. Bureau of the Census)

90. Trey Nguyen jogged $\frac{3}{8}$ of a mile from home and then rested. Then he continued jogging for another $\frac{3}{8}$ of a mile until he discovered his watch had fallen off. He walked back along the same path for $\frac{4}{8}$ of a mile until he found his watch. Find how far he was from his *starting point.*

91. In your own words, explain how to add like fractions.

92. In your own words, explain how to subtract like fractions.

FOCUS ON Mathematical Connections

INDUCTIVE REASONING

Inductive reasoning is the process of drawing a general conclusion from just a few observations. Many times in mathematics, we observe similarities among numbers or calculations and notice a pattern emerging. Identifying a pattern in this way uses inductive reasoning.

For example, look at the following list of numbers. The dots at the end of the list indicate that the list continues indefinitely. What do you notice?

2, 4, 6, 8, 10, 12, . . .

You probably noticed the pattern that each number in the list is 2 greater than the previous number and that all of the numbers are even numbers. If we were asked to guess the next number in the list, we could be confident that "14" would be a good response.

Let's try finding another pattern in a list of numbers.

10, 13, 18, 25, 34, 45, . . .

What do you notice? It might be useful to find the difference between successive numbers in the list. Using the differences found below, we can see that each successive difference is 2 greater than the previous difference. We can guess that the next difference will be 13, so the next number in the list is probably $45 + 13 = 58$.

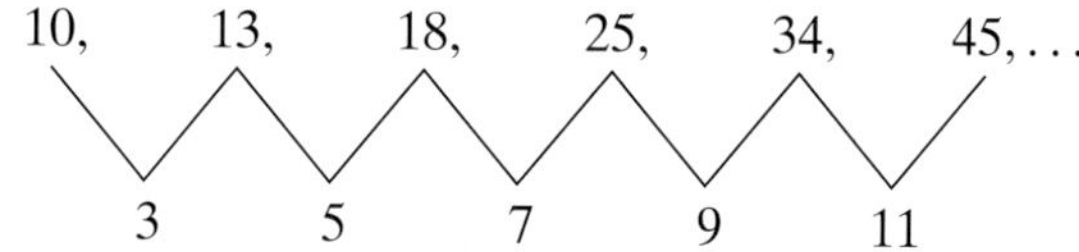

CRITICAL THINKING

Give the next two numbers in each list.

1. 5, 8, 11, 14, 17, 20, . . .

2. 100, 95, 90, 85, 80, . . .

3. $\frac{1}{2}, \frac{1}{3}, \frac{1}{4}, \frac{1}{5}, \ldots$

4. 5, 1, 6, 1, 1, 7, 1, 1, 1, 8, 1, 1, 1, . . .

5. 3, 4, 6, 9, 13, 18, . . .

6. 2, 4, 8, 16, 32, . . .

STUDY SKILLS REMINDER

What should you do the day of an exam?

On the day of an exam, try the following:

- Allow yourself plenty of time to arrive.
- Read the directions on the test carefully.
- Read each problem carefully as you take your test. Make sure that you answer the question asked.
- Watch your time and pace yourself so that you may attempt each problem on your test.
- If you have time, check your work and answers.
- Do not turn your test in early. If you have extra time, spend it double-checking your work.

Good luck!

4.5 Adding and Subtracting Unlike Fractions

OBJECTIVES

- A Add or subtract unlike fractions.
- B Write fractions in order.
- C Evaluate expressions given fractional replacement values.
- D Solve equations containing fractions.
- E Solve problems by adding or subtracting unlike fractions.

A Adding and Subtracting Unlike Fractions

In this section we add and subtract fractions with different denominators. To add or subtract these unlike fractions, first write the fractions as equivalent fractions with a common denominator and then add or subtract the like fractions. The common denominator we will use is the least common denominator (LCD).

For example, add the following unlike fractions: $\frac{3}{4} + \frac{1}{6}$. The LCD of the denominators 4 and 6 is 12. Write each fraction as an equivalent fraction with a denominator of 12.

$$\frac{3}{4} = \frac{3 \cdot 3}{4 \cdot 3} = \frac{9}{12} \quad \text{and} \quad \frac{1}{6} = \frac{1 \cdot 2}{6 \cdot 2} = \frac{2}{12}$$

Then

$$\frac{3}{4} + \frac{1}{6} = \frac{9}{12} + \frac{2}{12} = \frac{11}{12}$$

Adding or Subtracting Unlike Fractions

Step 1. Find the LCD of the denominators of the fractions.

Step 2. Write each fraction as an equivalent fraction whose denominator is the LCD.

Step 3. Add or subtract the like fractions.

Step 4. Write the sum or difference in simplest form.

EXAMPLE 1 Add: $\frac{2}{5} + \frac{4}{15}$

Solution: **Step 1.** The LCD of the denominators 5 and 15 is 15.

Step 2. $\frac{2}{5} = \frac{2 \cdot 3}{5 \cdot 3} = \frac{6}{15}$, $\quad \frac{4}{15} = \frac{4}{15}$ ← This fraction already has a denominator of 15.

Step 3. Add. $\frac{2}{5} + \frac{4}{15} = \frac{6}{15} + \frac{4}{15} = \frac{10}{15}$

Step 4. Write in simplest form.

$$\frac{10}{15} = \frac{2 \cdot 5}{3 \cdot 5} = \frac{2}{3}$$

Practice Problem 1

Add: $\frac{4}{7} + \frac{3}{14}$

EXAMPLE 2 Subtract: $\frac{2}{3} - \frac{10}{11}$

Solution: **Step 1.** The LCD of the denominators 3 and 11 is 33.

Step 2. $\frac{2}{3} = \frac{2 \cdot 11}{3 \cdot 11} = \frac{22}{33}$ and $\frac{10}{11} = \frac{10 \cdot 3}{11 \cdot 3} = \frac{30}{33}$

Step 3. Subtract.

$$\frac{2}{3} - \frac{10}{11} = \frac{22}{33} - \frac{30}{33} = \frac{22 - 30}{33} = \frac{-8}{33} \text{ or } -\frac{8}{33}$$

Step 4. $-\frac{8}{33}$ is in simplest form.

Practice Problem 2

Subtract: $\frac{3}{7} - \frac{9}{10}$

Answers

1. $\frac{11}{14}$ **2.** $-\frac{33}{70}$

Helpful Hint

Remember that $-\frac{a}{b} = \frac{a}{-b} = \frac{-a}{b}$. For example, $-\frac{8}{33} = \frac{8}{-33} = \frac{-8}{33}$.

Practice Problem 3

Add: $-\frac{1}{5} + \frac{3}{20}$

EXAMPLE 3 Add: $-\frac{1}{6} + \frac{1}{2}$

Solution: The LCD of the denominators 6 and 2 is 6.

$$-\frac{1}{6} + \frac{1}{2} = \frac{-1}{6} + \frac{1 \cdot 3}{2 \cdot 3} = \frac{-1}{6} + \frac{3}{6} = \frac{2}{6}$$

Next, simplify $\frac{2}{6}$.

$$\frac{2}{6} = \frac{2}{2 \cdot 3} = \frac{1}{3}$$

When the fractions contain variables, we add and subtract the same way.

Practice Problem 4

Add: $5 + \frac{3y}{4}$

Helpful Hint

The expression $\frac{6 - x}{3}$ from Example 4 *does not simplify* to $2 - x$. The number 3 must be a factor of both terms in the numerator (not just 6) in order to factor it out.

EXAMPLE 4 Subtract: $2 - \frac{x}{3}$

Solution: Recall that $2 = \frac{2}{1}$. The LCD of the denominators 1 and 3 is 3.

$$\frac{2}{1} - \frac{x}{3} = \frac{2 \cdot 3}{1 \cdot 3} - \frac{x}{3} = \frac{6}{3} - \frac{x}{3} = \frac{6 - x}{3}$$

The numerator $6 - x$ cannot be simplified further since 6 and $-x$ are unlike terms.

Practice Problem 5

Find: $\frac{5}{8} - \frac{1}{3} - \frac{1}{12}$

EXAMPLE 5 Find: $-\frac{3}{4} - \frac{1}{14} + \frac{6}{7}$

Solution: The LCD of 4, 14, and 7 is 28.

$$-\frac{3}{4} - \frac{1}{14} + \frac{6}{7} = -\frac{3 \cdot 7}{4 \cdot 7} - \frac{1 \cdot 2}{14 \cdot 2} + \frac{6 \cdot 4}{7 \cdot 4} = -\frac{21}{28} - \frac{2}{28} + \frac{24}{28} = \frac{1}{28}$$

Answers

3. $-\frac{1}{20}$ **4.** $\frac{20 + 3y}{4}$ **5.** $\frac{5}{24}$

Try the Concept Check in the margin.

B Writing Fractions in Order

One important application of the least common denominator is to use the LCD to help order or compare fractions.

EXAMPLE 6. Insert < or > to form a true sentence.

$$\frac{3}{4} \quad \frac{9}{11}$$

Solution: The LCD for these fractions is 44. Let's write each fraction as an equivalent fraction with a denominator of 44.

$$\frac{3}{4} = \frac{3 \cdot 11}{4 \cdot 11} = \frac{33}{44} \qquad \frac{9}{11} = \frac{9 \cdot 4}{11 \cdot 4} = \frac{36}{44}$$

Since $33 < 36$, then $\frac{33}{44} < \frac{36}{44}$ or

$$\frac{3}{4} < \frac{9}{11}$$

EXAMPLE 7. Insert < or > to form a true sentence.

$$-\frac{2}{7} \quad -\frac{1}{3}$$

Solution: The LCD is 21.

$$-\frac{2}{7} = -\frac{2 \cdot 3}{7 \cdot 3} = -\frac{6}{21} \qquad -\frac{1}{3} = -\frac{1 \cdot 7}{3 \cdot 7} = -\frac{7}{21}$$

Since $-6 > -7$, then $-\frac{6}{21} > -\frac{7}{21}$ or

$$-\frac{2}{7} > -\frac{1}{3}$$

C Evaluating Expressions Given Fractional Replacement Values

EXAMPLE 8 Evaluate $x - y$ if $x = \frac{7}{18}$ and $y = \frac{2}{9}$.

Solution: Replace x with $\frac{7}{18}$ and y with $\frac{2}{9}$ in the expression $x - y$.

$$x - y = \frac{7}{18} - \frac{2}{9}$$

The LCD of the denominators 18 and 9 is 18. Then

$$\frac{7}{18} - \frac{2}{9} = \frac{7}{18} - \frac{2 \cdot 2}{9 \cdot 2}$$

$$= \frac{7}{18} - \frac{4}{18}$$

$$= \frac{3}{18} = \frac{1}{6} \quad \text{Simplified.}$$

Concept Check

Find and correct the error in the following: $\frac{7}{12} - \frac{3}{4} = \frac{4}{8} = \frac{1}{2}$.

Practice Problem 6

Insert < or > to form a true sentence.

$$\frac{3}{8} \quad \frac{7}{20}$$

Practice Problem 7

Insert < or > to form a true sentence.

$$-\frac{17}{20} \quad -\frac{4}{5}$$

Practice Problem 8

Evaluate $x + y$ if $x = \frac{5}{11}$ and $y = \frac{4}{9}$.

Answers

6. > **7.** < **8.** $\frac{89}{99}$

Concept Check: $\frac{7}{12} - \frac{3}{4} = \frac{7}{12} - \frac{9}{12} = -\frac{2}{12} = -\frac{1}{6}$

D Solving Equations Containing Fractions

EXAMPLE 9 Solve: $x - \frac{3}{4} = \frac{1}{20}$

Solution: To get x by itself, add $\frac{3}{4}$ to both sides.

$$x - \frac{3}{4} = \frac{1}{20}$$

$$x - \frac{3}{4} + \frac{3}{4} = \frac{1}{20} + \frac{3}{4} \quad \text{Add } \frac{3}{4} \text{ to both sides.}$$

$$x = \frac{1}{20} + \frac{3 \cdot 5}{4 \cdot 5} \quad \text{The LCD of 20 and 4 is 20.}$$

$$x = \frac{1}{20} + \frac{15}{20}$$

$$x = \frac{16}{20}$$

$$x = \frac{4 \cdot 4}{4 \cdot 5} = \frac{4}{5} \quad \text{Write } \frac{16}{20} \text{ in simplest form.}$$

Check: To check, replace x with $\frac{4}{5}$ in the original equation.

$$x - \frac{3}{4} = \frac{1}{20}$$

$$\frac{4}{5} - \frac{3}{4} = \frac{1}{20} \quad \text{Replace } x \text{ with } \frac{4}{5}.$$

$$\frac{4 \cdot 4}{5 \cdot 4} - \frac{3 \cdot 5}{4 \cdot 5} = \frac{1}{20} \quad \text{The LCD of 5 and 4 is 20.}$$

$$\frac{16}{20} - \frac{15}{20} = \frac{1}{20}$$

$$\frac{1}{20} = \frac{1}{20} \quad \text{True.}$$

Thus $\frac{4}{5}$ is the solution of $x - \frac{3}{4} = \frac{1}{20}$.

Practice Problem 9

Solve: $y - \frac{2}{3} = \frac{5}{12}$

E Solving Problems by Adding or Subtracting Unlike Fractions

Very often, real-world problems involve adding or subtracting unlike fractions.

EXAMPLE 10 **Finding Total Weight**

A freight truck has $\frac{1}{4}$ ton of computers, $\frac{1}{3}$ ton of televisions, and $\frac{3}{8}$ ton of small appliances. Find the total weight of its load.

Solution: **1.** UNDERSTAND. Read and reread the problem. The phrase "total weight" tells us to add. Since the unknown is total weight, let x = total weight.

Practice Problem 10

To repair her sidewalk, a homeowner must pour small amounts of cement in three different locations. She needs $\frac{3}{5}$ of a cubic yard, $\frac{2}{10}$ of a cubic yard, and $\frac{2}{15}$ of a cubic yard for these locations. Find the total amount of concrete the homeowner needs. If she bought enough cement to mix 1 cubic yard, did she buy enough?

Answers

9. $\frac{13}{12}$ **10.** $\frac{14}{15}$ cubic yard; yes, she bought enough.

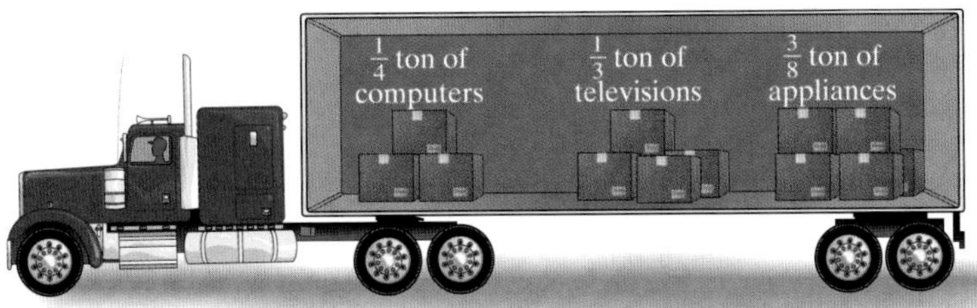

2. TRANSLATE.

In words:	Total weight	is	weight of computers	plus	weight of televisions	plus	weight of appliances
	↓	↓	↓	↓	↓	↓	↓
Translate:	x	$=$	$\frac{1}{4}$	$+$	$\frac{1}{3}$	$+$	$\frac{3}{8}$

3. SOLVE. The LCD is 24.

$$x = \frac{1}{4} + \frac{1}{3} + \frac{3}{8}$$
$$= \frac{1 \cdot 6}{4 \cdot 6} + \frac{1 \cdot 8}{3 \cdot 8} + \frac{3 \cdot 3}{8 \cdot 3}$$
$$= \frac{6}{24} + \frac{8}{24} + \frac{9}{24}$$
$$= \frac{23}{24}$$

4. INTERPRET. Check the solution. State your conclusion: The total weight of the truck's load is $\frac{23}{24}$ ton. ●

EXAMPLE 11 Calculating Flight Time

A flight from Tucson, Arizona, to Phoenix, Arizona, requires $\frac{5}{12}$ of an hour. If the plane has been flying $\frac{1}{4}$ of an hour, find how much time remains before landing.

Solution:

1. UNDERSTAND. Read and reread the problem. The phrase "how much time remains" tells us to subtract. Let $x =$ time remaining.

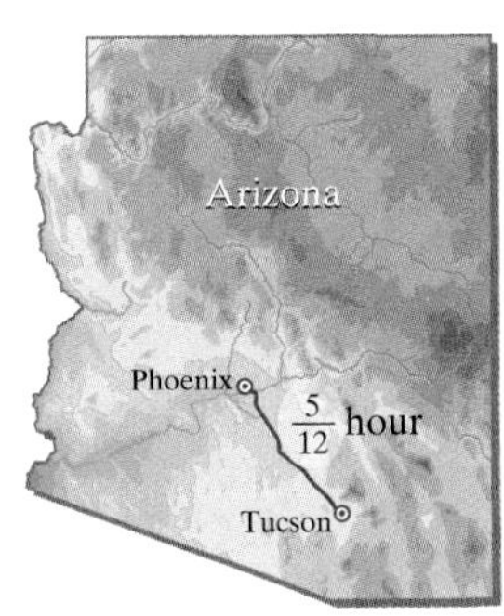

Practice Problem 11

Find the difference in length of two boards if one board is $\frac{4}{5}$ of a foot long and the other is $\frac{2}{3}$ of a foot long.

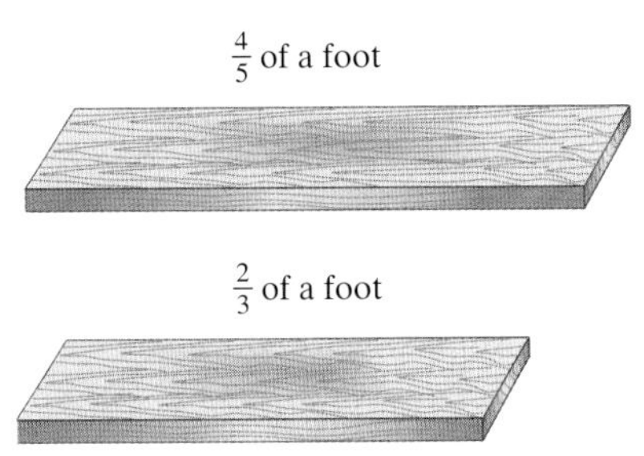

Answer

11. $\frac{2}{15}$ of a foot

2. TRANSLATE.

In words:	Time remaining	is	flight time from Tucson to Phoenix	minus	flight time already passed
	↓	↓	↓	↓	↓
Translate:	x	$=$	$\frac{5}{12}$	$-$	$\frac{1}{4}$

3. SOLVE. The LCD is 12.

$$x = \frac{5}{12} - \frac{1}{4}$$
$$= \frac{5}{12} - \frac{1 \cdot 3}{4 \cdot 3}$$
$$= \frac{5}{12} - \frac{3}{12}$$
$$= \frac{2}{12} = \frac{2}{2 \cdot 6} = \frac{1}{6}$$

4. INTERPRET. *Check* the solution. *State* your conclusion: The flight time remaining is $\frac{1}{6}$ of an hour.

CALCULATOR EXPLORATIONS

Performing Operations on Fractions

SCIENTIFIC CALCULATOR

Many calculators have a fraction key, such as [a b/c], that allows you to enter fractions, perform operations on fractions, and will give the result as a fraction. If your calculator has a fraction key, use it to calculate

$$\frac{3}{5} + \frac{4}{7}$$

Enter the keystrokes

[3] [a b/c] [5] [+] [4] [a b/c] [7] [=]

The display should read [1_6⌋35]

which represents the mixed number $1\frac{6}{35}$. Until we discuss mixed numbers in Section 4.8, let's write the result as a fraction. To convert from mixed number notation to fractional notation, press

[2nd] [d/c]

The display now reads [41⌋35]

which represents $\frac{41}{35}$, the sum in fractional notation.

GRAPHING CALCULATOR

Graphing calculators also allow you to perform operations on fractions and will give exact fractional results. The fraction option on a graphing calculator may be found under the [MATH] menu. To perform the addition above, try the keystrokes.

[3] [÷] [5] [+] [4] [÷] [7] [MATH] [ENTER] [ENTER]

The display should read

[3/5 + 4/7 ▶Frac 41/35]

Use a calculator to add the following fractions. Give each sum as a fraction.

1. $\frac{1}{16} + \frac{2}{5}$ **2.** $\frac{3}{20} + \frac{2}{25}$ **3.** $\frac{4}{9} + \frac{7}{8}$

4. $\frac{9}{11} + \frac{5}{12}$ **5.** $\frac{10}{17} + \frac{12}{19}$ **6.** $\frac{14}{31} + \frac{15}{21}$

Name ______________________ Section ____________ Date ____________

Mental Math

Find the LCD of each pair of fractions.

1. $\frac{1}{2}, \frac{2}{3}$

2. $\frac{1}{2}, \frac{3}{4}$

3. $\frac{1}{6}, \frac{5}{12}$

4. $\frac{2}{5}, \frac{7}{10}$

5. $\frac{4}{7}, \frac{1}{8}$

6. $\frac{23}{24}, \frac{1}{3}$

7. $\frac{11}{12}, \frac{3}{4}$

8. $\frac{2}{3}, \frac{3}{11}$

EXERCISE SET 4.5

A *Add or subtract as indicated. See Examples 1 through 4.*

1. $\frac{2}{3} + \frac{1}{6}$

2. $\frac{5}{6} + \frac{1}{12}$

3. $\frac{1}{2} - \frac{1}{3}$

4. $\frac{2}{3} - \frac{1}{4}$

5. $-\frac{2}{11} + \frac{2}{33}$

6. $-\frac{5}{9} + \frac{1}{3}$

7. $\frac{3x}{14} - \frac{3}{7}$

8. $\frac{2y}{5} - \frac{2}{15}$

9. $\frac{11}{35} + \frac{2}{7}$

10. $\frac{2}{5} + \frac{3}{25}$

11. $2y - \frac{5}{12}$

12. $5y - \frac{3}{20}$

13. $\frac{5}{12} - \frac{1}{9}$

14. $\frac{7}{12} - \frac{5}{18}$

15. $\frac{5}{7} + 1$

16. $-10 + \frac{7}{10}$

17. $\frac{5a}{11} + \frac{4a}{9}$

18. $\frac{7x}{18} + \frac{2x}{9}$

19. $\frac{2y}{3} - \frac{1}{6}$

20. $\frac{5}{6} - \frac{1}{12}$

21. $\frac{1}{2} + \frac{3}{x}$

22. $\frac{2}{5} + \frac{3}{x}$

23. $-\frac{2}{11} - \frac{2}{33}$

24. $-\frac{5}{9} - \frac{1}{3}$

25. $\frac{9}{14} - \frac{3}{7}$

26. $\frac{4}{5} - \frac{2}{15}$

27. $\frac{11y}{35} - \frac{2}{7}$

28. $\frac{2b}{5} - \frac{3}{25}$

29. $\frac{1}{9} - \frac{5}{12}$

30. $\frac{5}{18} - \frac{7}{12}$

31. $\frac{7}{15} - \frac{5}{12}$ **32.** $\frac{5}{8} - \frac{3}{20}$ **33.** $\frac{5}{7} - \frac{1}{8}$ **34.** $\frac{10}{13} - \frac{7}{10}$ **35.** $\frac{7}{8} + \frac{3}{16}$

36. $-\frac{7}{18} - \frac{2}{9}$ **37.** $\frac{5}{9} + \frac{3}{9}$ **38.** $\frac{4}{13} - \frac{1}{13}$ **39.** $\frac{5}{11} + \frac{y}{3}$ **40.** $\frac{5z}{13} + \frac{3}{26}$

41. $-\frac{5}{6} - \frac{3}{7}$ **42.** $\frac{1}{2} - \frac{3}{29}$ **43.** $\frac{7}{9} - \frac{1}{6}$ **44.** $\frac{9}{16} - \frac{3}{8}$ **45.** $\frac{2a}{3} + \frac{6a}{13}$

46. $\frac{3y}{4} + \frac{y}{7}$ **47.** $\frac{7}{30} - \frac{5}{12}$ **48.** $\frac{7}{30} - \frac{3}{20}$ **49.** $\frac{5}{9} + \frac{1}{y}$ **50.** $\frac{1}{12} - \frac{5}{x}$

51. $\frac{4}{5} + \frac{4}{9}$ **52.** $\frac{11}{12} - \frac{7}{24}$ **53.** $\frac{5}{9x} + \frac{1}{8}$ **54.** $\frac{3}{8} + \frac{5}{12x}$

Perform each indicated operation. See Example 5.

55. $-\frac{2}{5} + \frac{1}{3} - \frac{3}{10}$ **56.** $-\frac{1}{3} - \frac{1}{4} + \frac{2}{5}$ **57.** $\frac{x}{2} + \frac{x}{4} + \frac{2x}{16}$ **58.** $\frac{z}{4} + \frac{z}{8} + \frac{2z}{16}$

59. $\frac{6}{5} - \frac{3}{4} + \frac{1}{2}$ **60.** $\frac{6}{5} + \frac{3}{4} - \frac{1}{2}$ **61.** $-\frac{9}{12} + \frac{17}{24} - \frac{1}{6}$ **62.** $-\frac{5}{14} + \frac{3}{7} - \frac{1}{2}$

63. $\frac{3x}{8} + \frac{2x}{7} - \frac{5}{14}$ **64.** $\frac{9x}{10} - \frac{1}{2} + \frac{x}{5}$

B *Insert < or > to form a true sentence. See examples 6 and 7.*

65. $\frac{2}{7} \quad \frac{3}{10}$ **66.** $\frac{5}{9} \quad \frac{6}{11}$ **67.** $\frac{5}{6} \quad -\frac{13}{15}$

68. $-\frac{7}{8} \quad -\frac{5}{6}$ **69.** $-\frac{3}{4} \quad -\frac{11}{14}$ **70.** $-\frac{2}{9} \quad -\frac{3}{13}$

C *Evaluate each expression if $x = \frac{1}{3}$ and $y = \frac{3}{4}$. See Example 8.*

71. $x + y$

72. $x - y$

73. xy

74. $x \div y$

75. $2y + x$

76. $2x + y$

D *Solve and check. See Example 9.*

77. $x - \frac{1}{12} = \frac{5}{6}$

78. $y - \frac{8}{9} = \frac{1}{3}$

79. $\frac{2}{5} + y = -\frac{3}{10}$

80. $\frac{1}{2} + a = -\frac{3}{8}$

81. $7z + \frac{1}{16} - 6z = \frac{3}{4}$

82. $9x - \frac{2}{7} - 8x = \frac{11}{14}$

83. $-\frac{2}{9} = x - \frac{5}{6}$

84. $-\frac{1}{4} = y - \frac{7}{10}$

E *Find the perimeter of each geometric figure. (Hint: Recall that perimeter means distance around.)*

85.

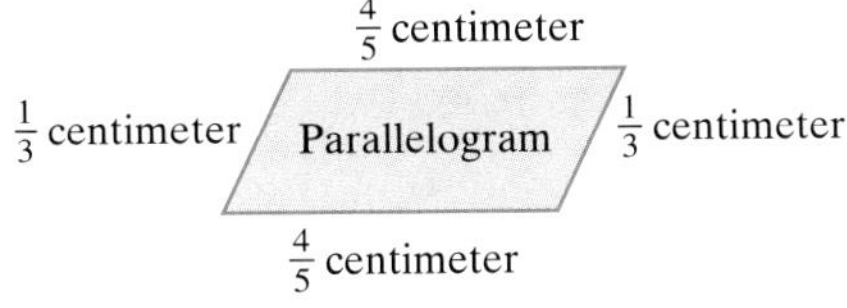

86.

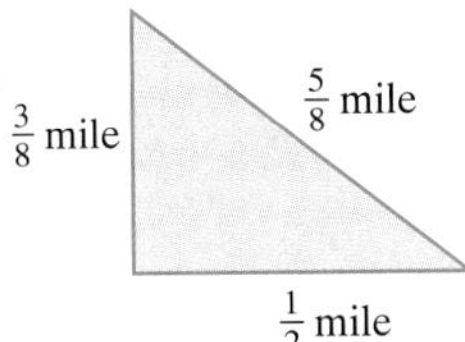

87.

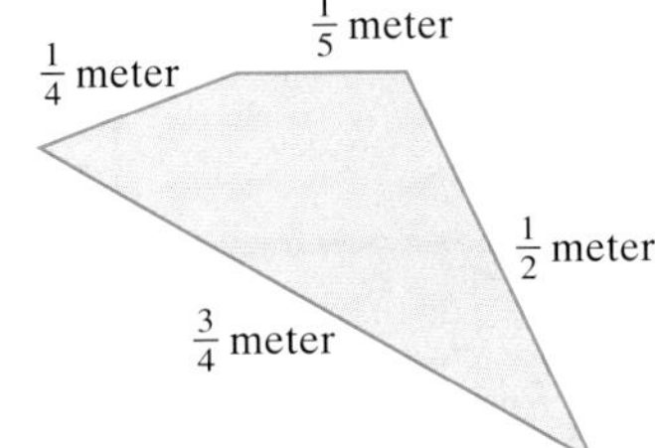

88.

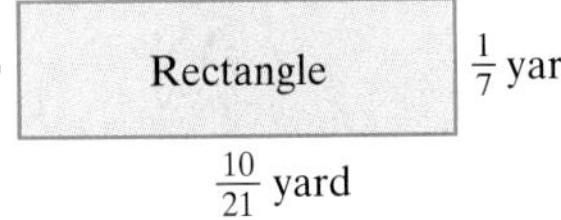

Solve. See Examples 10 and 11.

89. Killer bees have been known to chase people for up to $\frac{1}{4}$ of a mile, while domestic European honeybees will normally chase a person for no more than 100 feet, or $\frac{5}{264}$ of a mile. How much farther will a killer bee chase a person than a domestic honeybee? (*Source:* Coachella Valley Mosquito & Vector Control District)

90. The slowest mammal is the three-toed sloth from South America. The sloth has an average ground speed of $\frac{1}{10}$ mph. In the trees, it can accelerate to $\frac{17}{100}$ mph. How much faster can a sloth travel in the trees? (*Source: The Guiness Book of World Records*)

91. Given the following diagram, find L, its total length.

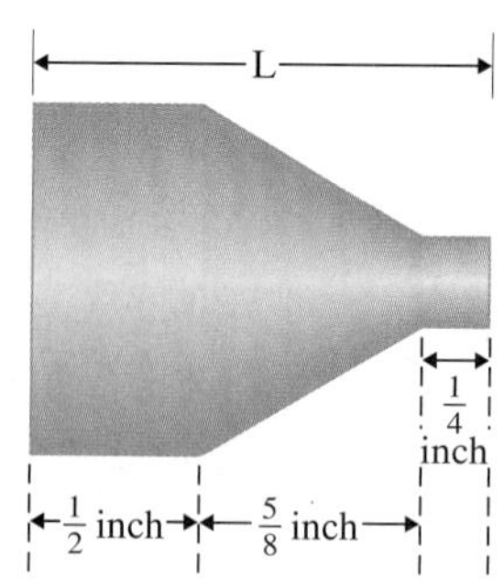

92. Given the following diagram, find W, its total width.

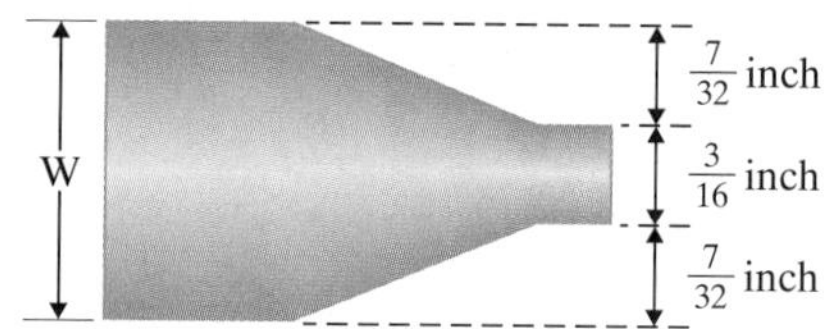

93. About $\frac{13}{20}$ of American students ages 10 to 17 name math, science, or art as their favorite subject in school. Art is the favorite subject for about $\frac{4}{25}$ of these students. For what fraction of students this age is math or science their favorite subject? (*Source:* Peter D. Hart Research Associates for the National Science Foundation)

94. Together, the United States' and Japan's postal services handle $\frac{49}{100}$ of the world's mail volume. Japan's postal service alone handles $\frac{3}{50}$ of the world's mail. What fraction of the world's mail is handled by the postal service of the United States? (*Source:* United States Postal Service)

The table gives the fraction of Americans who eat pasta at various intervals. Use this table to answer Exercises 95 and 96.

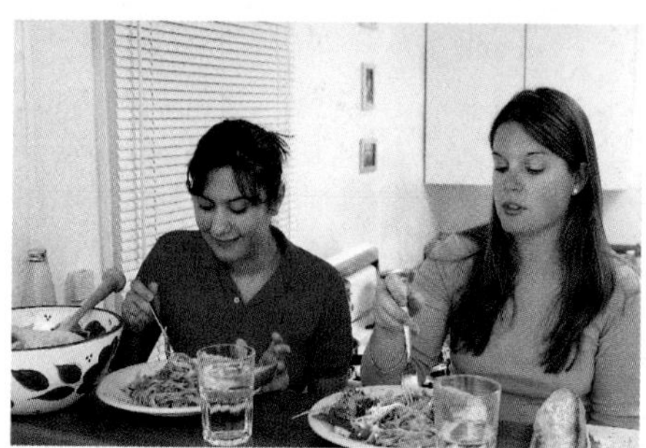

How Often Americans Eat Pasta

Frequency	Fraction
3 times per week	$\frac{31}{100}$
1 or 2 times per week	$\frac{23}{50}$
1 or 2 times per month	$\frac{17}{100}$
Less often	$\frac{3}{50}$

(*Source:* Princeton Survey Research)

95. What fraction of Americans eat pasta 1, 2, or 3 times a week?

96. What fraction of Americans eat pasta 1 or 2 times a month or less often?

The circle graph shows the fraction of American adults who drive various distances in miles in an average week. Use this graph to answer Exercises 97 and 98.

97. What fraction of American adults drive less than 50 miles in an average week?

98. What fraction of American adults drive between 50 and 149 miles in an average week?

Miles Driven by American Adults in an Average Week

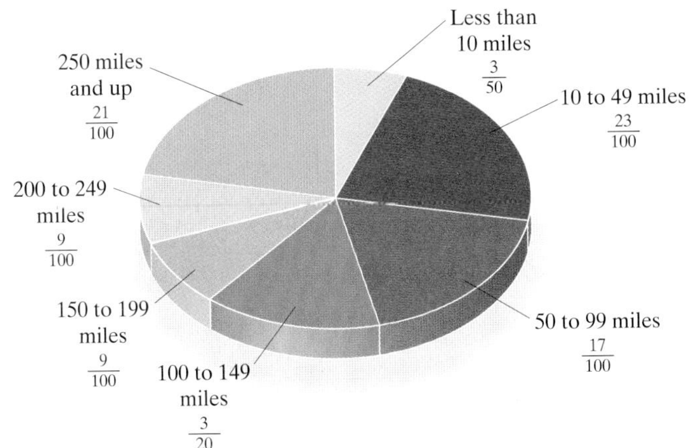

Source: Simmons

Review and Preview

Evaluate. See Section 4.3.

99. $\left(\frac{5}{6}\right)^2$

100. $\left(\frac{1}{2}\right)^2$

101. $\left(-\frac{5}{6}\right)^2$

102. $\left(-\frac{1}{2}\right)^2$

Round each number to the given place value. See Section 1.5.

103. 57,236 to the nearest hundred

104. 576 to the nearest hundred

105. 327 to the nearest ten

106. 2333 to the nearest ten

Combining Concepts

Perform each indicated operation.

107. $\frac{30}{55} + \frac{1000}{1760}$; the LCD of 55 and 1760 is 1760.

108. $\frac{19}{26} - \frac{968}{1352}$; the LCD of 26 and 1352 is 1352.

109. In your own words, describe how to add two fractions with different denominators.

110. Find the sum of the fractions in the circle graph on page 297. Did the sum surprise you? Why or why not?

Solve.

111. $x - \frac{5}{117} = \frac{71}{27}$

112. $-\frac{8}{81} = y - \frac{3}{45}$

The table shows the fraction of the world's land area occupied by each continent. Use this table to answer Exercises 113–117.

113. What fraction of the world's land area is accounted for by North and South America?

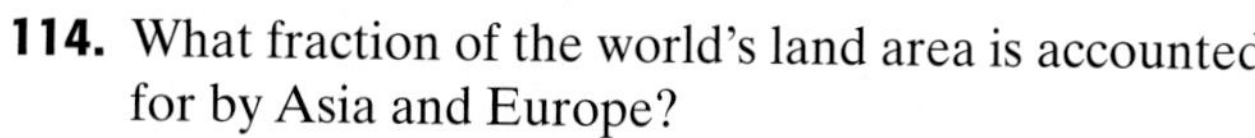

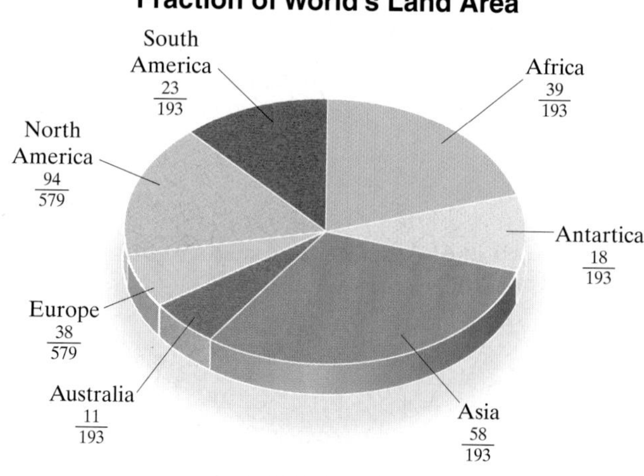

Source: *2000 World Almanac*

114. What fraction of the world's land area is accounted for by Asia and Europe?

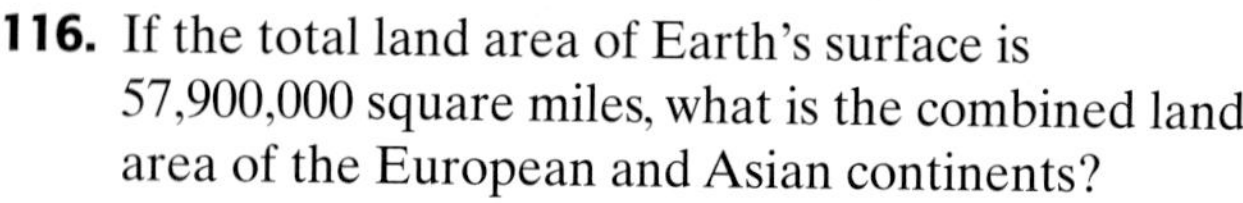

115. If the total land area of Earth's surface is 57,900,000 square miles, what is the combined land area of the North and South American continents?

116. If the total land area of Earth's surface is 57,900,000 square miles, what is the combined land area of the European and Asian continents?

117. Antarctica is generally considered to be uninhabited. What fraction of the world's land area is accounted for by inhabited continents?

Solve.

118. In 2000, about $\frac{11}{67}$ of the total weight of mail delivered by the United States Postal Service was first-class mail. That same year, about $\frac{75}{134}$ of the total weight of mail delivered by the United States Postal Service was standard mail. Which of these two categories account for a greater portion of the mail handled by weight? (*Source:* U.S. Postal Service)

119. The National Park System (NPS) in the United States includes a wide variety of park types. National military parks account for $\frac{3}{128}$ of all NPS parks, and $\frac{1}{24}$ of NPS parks are classified as national preserves. Which category, national military park or national preserve, is bigger? (*Source*: National Park Service)

120. Approximately $\frac{7}{10}$ of U.S. adults have a savings account. About $\frac{11}{25}$ of U.S. adults have a non-interest bearing checking account. Which type of banking service, savings account or non-interest checking account, do adults in the United States use more? (*Source*: Scarborough Research/USData.com, Inc.)

121. About $\frac{127}{500}$ of U.S. adults rent one or two videos per month. Approximately $\frac{31}{200}$ of U.S. adults rent three or four videos per month. Which video rental category, 1–2 videos or 3–4 videos per month, is bigger? (*Source:* Telenation/Market Facts, Inc.)

Name ____________ Section ________ Date ________

Integrated Review–Summary of Fractions and Factors

Use a fraction to represent the shaded area of each figure or figure group.

1. 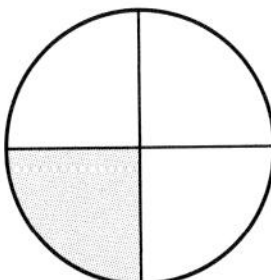

2. 

3. 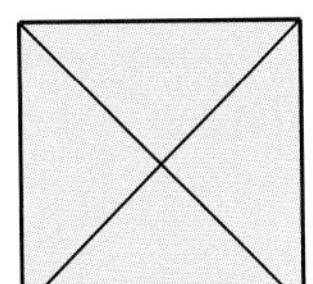

4. In a survey, 73 people out of 85 get less than 8 hours of sleep each night. What fraction of people in the survey get less than 8 hours of sleep?

Write the prime factorization of each number. Write any repeated factors using exponents.

5. 6 **6.** 70 **7.** 252

Write each fraction in simplest form.

8. $\frac{2}{14}$ **9.** $\frac{20}{24}$ **10.** $\frac{18}{38}$

11. $\frac{42}{110}$ **12.** $\frac{32}{64}$ **13.** $\frac{72}{80}$

14. Of the 50 United States, 2 states are not adjacent to any other states. What fraction of the states are not adjacent to other states? Write the fraction in simplest form.

The following summary will help you with this review of operations on fractions.

Operations on Fractions

Let a, b, c, and d be integers.

Addition: $\frac{a}{b} + \frac{c}{b} = \frac{a+c}{b}$ $(b \neq 0)$ ↑ ↑ **common denominator**

Subtraction: $\frac{a}{b} - \frac{c}{b} = \frac{a-c}{b}$ $(b \neq 0)$ ↑ ↑ **common denominator**

Multiplication: $\frac{a}{b} \cdot \frac{c}{d} = \frac{a \cdot c}{b \cdot d}$ $(b \neq 0, d \neq 0)$

Division: $\frac{a}{b} \div \frac{c}{d} = \frac{a}{b} \cdot \frac{d}{c} = \frac{a \cdot d}{b \cdot c}$ $(b \neq 0, d \neq 0, c \neq 0)$

Answers

1. ________
2. ________
3. ________
4. ________
5. ________
6. ________
7. ________
8. ________
9. ________
10. ________
11. ________
12. ________
13. ________
14. ________

15. ______

16. ______

17. ______

18. ______

19. ______

20. ______

21. ______

22. ______

23. ______

24. ______

25. ______

26. ______

27. ______

28. ______

29. ______

30. ______

31. ______

32. ______

33. ______

34. ______

Perform each indicated operation.

15. $\frac{1}{5} + \frac{3}{5}$

16. $\frac{1}{5} - \frac{3}{5}$

17. $\frac{1}{5} \cdot \frac{3}{5}$

18. $\frac{1}{5} \div \frac{3}{5}$

19. $\frac{2}{3} \div \frac{5}{6}$

20. $\frac{2}{3} \cdot \frac{5}{6}$

21. $\frac{2}{3} - \frac{5}{6}$

22. $\frac{2}{3} + \frac{5}{6}$

23. $-\frac{1}{7} \cdot -\frac{7}{18}$

24. $-\frac{4}{9} \cdot -\frac{3}{7}$

25. $-\frac{7}{8} \div 6$

26. $-\frac{9}{10} \div 5$

27. $\frac{7}{8} + \frac{1}{20}$

28. $\frac{5}{12} - \frac{1}{9}$

29. $\frac{9}{11} - \frac{2}{3}$

30. $\frac{2}{9} + \frac{1}{18}$

31. $\frac{2}{9} + \frac{1}{18} + \frac{1}{3}$

32. $\frac{3}{10} + \frac{1}{5} + \frac{6}{25}$

33. A contractor is using 18 acres of his land to sell $\frac{3}{4}$-acre lots. How many lots can he sell?

34. Suppose that the cross-section of a piece of pipe looks like the diagram shown. What is the inner diameter?

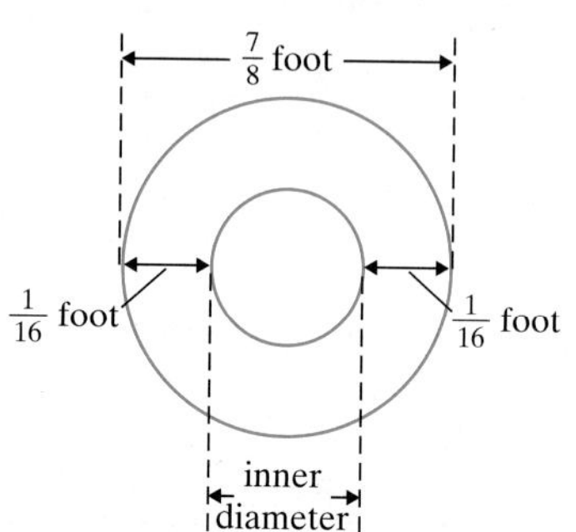

4.6 Complex Fractions and Review of Order of Operations

OBJECTIVES

A. Simplify complex fractions.

B. Review the order of operations.

A Simplifying Complex Fractions

Thus far, we have studied operations on fractions. We now practice simplifying fractions whose numerators or denominators themselves contain fractions. These fractions are called **complex fractions**.

Complex Fraction

A fraction whose numerator or denominator or both numerator and denominator contain fractions is called a **complex fraction**.

Examples of complex fractions are

$$\frac{\frac{x}{4}}{\frac{3}{2}} \qquad \frac{\frac{1}{2}+\frac{3}{8}}{\frac{3}{4}-\frac{1}{6}} \qquad \frac{\frac{y}{5}-2}{\frac{3}{10}}$$

Method 1 for Simplifying Complex Fractions

Two methods are presented to simplify complex fractions. The first method makes use of the fact that a fraction bar means division.

EXAMPLE 1 Simplify: $\dfrac{\frac{x}{4}}{\frac{3}{2}}$

Solution: Since a fraction bar means division, the complex fraction $\dfrac{\frac{x}{4}}{\frac{3}{2}}$ can be written as $\frac{x}{4} \div \frac{3}{2}$. Then divide as usual to simplify.

$$\frac{x}{4} \div \frac{3}{2} = \frac{x}{4}\cdot\frac{2}{3} \quad \text{Multiply by the reciprocal.}$$

$$= \frac{x \cdot 2}{2 \cdot 2 \cdot 3}$$

$$= \frac{x}{6}$$

Practice Problem 1

Simplify: $\dfrac{\frac{7y}{10}}{\frac{1}{5}}$

EXAMPLE 2 Simplify: $\dfrac{\frac{1}{2}+\frac{3}{8}}{\frac{3}{4}-\frac{1}{6}}$

Solution: Recall the order of operations. Since the fraction bar is considered a grouping symbol, we simplify the numerator and the denominator of the complex fraction separately. Then we divide.

$$\frac{\frac{1}{2}+\frac{3}{8}}{\frac{3}{4}-\frac{1}{6}} = \frac{\frac{1\cdot 4}{2\cdot 4}+\frac{3}{8}}{\frac{3\cdot 3}{4\cdot 3}-\frac{1\cdot 2}{6\cdot 2}} = \frac{\frac{4}{8}+\frac{3}{8}}{\frac{9}{12}-\frac{2}{12}} = \frac{\frac{7}{8}}{\frac{7}{12}}$$

Practice Problem 2

Simplify: $\dfrac{\frac{1}{2}+\frac{1}{6}}{\frac{3}{4}-\frac{2}{3}}$

Answers

1. $\frac{7y}{2}$ 2. $\frac{8}{1}$ or 8

Thus,

$$\frac{\frac{1}{2}+\frac{3}{8}}{\frac{3}{4}-\frac{1}{6}} = \frac{\frac{7}{8}}{\frac{7}{12}}$$

$$= \frac{7}{8} \div \frac{7}{12} \quad \text{Rewrite the quotient using the } \div \text{ sign.}$$

$$= \frac{7}{8} \cdot \frac{12}{7} \quad \text{Multiply by the reciprocal.}$$

$$= \frac{7 \cdot 3 \cdot 4}{2 \cdot 4 \cdot 7} \quad \text{Multiply.}$$

$$= \frac{3}{2} \quad \text{Simplify.}$$

Method 2 for Simplifying Complex Fractions

The second method for simplifying complex fractions is to multiply the numerator and the denominator of the complex fraction by the LCD of all the fractions in its numerator and its denominator. Since this LCD is divisible by all denominators, this has the effect of leaving sums and differences of integers in the numerator and the denominator. Let's use this second method to simplify the complex fraction in Example 2 again.

Practice Problem 3

Use Method 2 to simplify: $\dfrac{\frac{1}{2}+\frac{1}{6}}{\frac{3}{4}-\frac{2}{3}}$

EXAMPLE 3 Simplify: $\dfrac{\frac{1}{2}+\frac{3}{8}}{\frac{3}{4}-\frac{1}{6}}$

Solution: The complex fraction contains fractions with denominators 2, 8, 4, and 6. The LCD is 24. By the fundamental property of fractions, we can multiply the numerator and the denominator of the complex fraction by 24. Notice below that by the distributive property, this means that we multiply each term in the numerator and denominator by 24.

$$\frac{\frac{1}{2}+\frac{3}{8}}{\frac{3}{4}-\frac{1}{6}} = \frac{24\left(\frac{1}{2}+\frac{3}{8}\right)}{24\left(\frac{3}{4}-\frac{1}{6}\right)}$$

$$= \frac{\left(24 \cdot \frac{1}{2}\right) + \left(24 \cdot \frac{3}{8}\right)}{\left(24 \cdot \frac{3}{4}\right) - \left(24 \cdot \frac{1}{6}\right)} \quad \text{Apply the distributive property.}$$

$$= \frac{12 + 9}{18 - 4} \quad \text{Multiply.}$$

$$= \frac{21}{14}$$

$$= \frac{7 \cdot 3}{7 \cdot 2} = \frac{3}{2} \quad \text{Simplify.}$$

The simplified result is the same, of course, no matter which method is used.

Answer

3. $\frac{8}{1}$ or 8

EXAMPLE 4 Simplify: $\dfrac{\dfrac{y}{5} - 2}{\dfrac{3}{10}}$

Solution: Use the second method and multiply the numerator and the denominator of the complex fraction by the LCD of all fractions. Recall that $2 = \frac{2}{1}$. The LCD of the denominators 5, 1, and 10 is 10.

$$\frac{\frac{y}{5} - \frac{2}{1}}{\frac{3}{10}} = \frac{10\left(\frac{y}{5} - \frac{2}{1}\right)}{10\left(\frac{3}{10}\right)}$$ Multiply the numerator and denominator by 10.

$$= \frac{\left(10 \cdot \frac{y}{5}\right) - \left(10 \cdot \frac{2}{1}\right)}{10 \cdot \frac{3}{10}}$$ Apply the distributive property.

$$= \frac{2y - 20}{3}$$ Multiply.

Practice Problem 4

Simplify: $\dfrac{\dfrac{3}{4}}{\dfrac{x}{5} - 1}$

Helpful Hint

Don't forget to multiply the numerator and the denominator of the complex fraction by the same number—the LCD.

B Reviewing the Order of Operations

At this time, it is probably a good idea to review the order of operations on expressions containing fractions.

Order of Operations

1. Do all operations within grouping symbols such as parentheses or brackets.
2. Evaluate any expressions with exponents.
3. Multiply or divide in order from left to right.
4. Add or subtract in order from left to right.

EXAMPLE 5 Simplify: $\left(\frac{4}{5}\right)^2 - 1$

Solution: According to the order of operations, first evaluate $\left(\frac{4}{5}\right)^2$.

$$\left(\frac{4}{5}\right)^2 - 1 = \frac{16}{25} - 1$$ Write $\left(\frac{4}{5}\right)^2$ as $\frac{16}{25}$.

Next, combine the fractions. The LCD of 25 and 1 is 25.

$$\frac{16}{25} - 1 = \frac{16}{25} - \frac{25}{25}$$ Write 1 as $\frac{25}{25}$.

$$= \frac{-9}{25} \text{ or } -\frac{9}{25}$$ Subtract.

Practice Problem 5

Simplify: $\left(2 - \frac{2}{3}\right)^3$

Answers

4. $\dfrac{15}{4x - 20}$ **5.** $\dfrac{64}{27}$

EXAMPLE 6 Simplify: $\left(\frac{1}{4}+\frac{2}{3}\right)\left(\frac{11}{12}+\frac{1}{4}\right)$

Solution: First perform operations inside parentheses. Then multiply.

$$\left(\frac{1}{4}+\frac{2}{3}\right)\left(\frac{11}{12}+\frac{1}{4}\right)=\left(\frac{1\cdot 3}{4\cdot 3}+\frac{2\cdot 4}{3\cdot 4}\right)\left(\frac{11}{12}+\frac{1\cdot 3}{4\cdot 3}\right) \quad \text{Each LCD is 12.}$$

$$=\left(\frac{3}{12}+\frac{8}{12}\right)\left(\frac{11}{12}+\frac{3}{12}\right)$$

$$=\left(\frac{11}{12}\right)\left(\frac{14}{12}\right) \quad \text{Add.}$$

$$=\frac{11\cdot 2\cdot 7}{2\cdot 6\cdot 12} \quad \text{Multiply.}$$

$$=\frac{77}{72} \quad \text{Simplify.}$$

Practice Problem 6

Simplify: $\left(-\frac{1}{2}+\frac{1}{5}\right)\left(\frac{7}{8}+\frac{1}{8}\right)$

Try the Concept Check in the margin.

Concept Check

What should be done first to simplify the expression $\frac{1}{5}\cdot\frac{5}{2}-\left(\frac{2}{3}+\frac{4}{5}\right)^2$?

EXAMPLE 7 Evaluate $2x + y^2$ if $x = -\frac{1}{2}$ and $y = \frac{1}{3}$.

Solution: Replace x and y with the given values and simplify.

$$2x + y^2 = 2\left(-\frac{1}{2}\right)+\left(\frac{1}{3}\right)^2 \quad \text{Replace } x \text{ with } -\frac{1}{2} \text{ and } y \text{ with } \frac{1}{3}.$$

$$=2\left(-\frac{1}{2}\right)+\frac{1}{9} \quad \text{Write } \left(\frac{1}{3}\right)^2 \text{ as } \frac{1}{9}.$$

$$=-1+\frac{1}{9} \quad \text{Multiply.}$$

$$=-\frac{9}{9}+\frac{1}{9} \quad \text{The LCD is 9.}$$

$$=-\frac{8}{9} \quad \text{Add.}$$

Practice Problem 7

Evaluate $-\frac{3}{5} - xy$ if $x = \frac{3}{10}$ and $y = \frac{2}{3}$.

Answers

6. $-\frac{3}{10}$ **7.** $-\frac{4}{5}$

Concept Check: Add inside parentheses.

Name ______________________ Section ____________ Date ____________

EXERCISE SET 4.6

A *Simplify each complex fraction. See Examples 1 through 4.*

1. $\dfrac{\frac{1}{8}}{\frac{3}{4}}$

2. $\dfrac{\frac{2}{3}}{\frac{2}{7}}$

3. $\dfrac{\frac{9}{10}}{\frac{21}{10}}$

4. $\dfrac{\frac{14}{5}}{\frac{7}{5}}$

5. $\dfrac{\frac{2x}{27}}{\frac{4}{9}}$

6. $\dfrac{\frac{3y}{11}}{\frac{1}{2}}$

7. $\dfrac{\frac{3}{4}+\frac{2}{5}}{\frac{1}{2}+\frac{3}{5}}$

8. $\dfrac{\frac{7}{6}+\frac{2}{3}}{\frac{3}{2}-\frac{8}{9}}$

9. $\dfrac{\frac{3x}{4}}{5-\frac{1}{8}}$

10. $\dfrac{\frac{3}{10}+2}{\frac{2}{5y}}$

B *Use the order of operations to simplify each expression. See Examples 5 and 6.*

11. $\frac{1}{5}+\frac{1}{3}\cdot\frac{1}{4}$

12. $\frac{1}{2}+\frac{1}{6}\cdot\frac{1}{3}$

13. $\frac{5}{6}\div\frac{1}{3}\cdot\frac{1}{4}$

14. $\frac{7}{8}\div\frac{1}{4}\cdot\frac{1}{7}$

15. $2^2-\left(\frac{1}{3}\right)^2$

16. $3^2-\left(\frac{1}{2}\right)^2$

17. $\left(\frac{2}{9}+\frac{4}{9}\right)\left(\frac{1}{3}-\frac{9}{10}\right)$

18. $\left(\frac{1}{5}-\frac{1}{10}\right)\left(\frac{1}{5}+\frac{1}{10}\right)$

19. $\left(\frac{7}{8}-\frac{1}{2}\right)\div\frac{3}{11}$

20. $\left(-\frac{2}{3}-\frac{7}{3}\right)\div\frac{4}{9}$

21. $2\cdot\left(\frac{1}{4}+\frac{1}{5}\right)+2$

22. $\frac{2}{5}\cdot\left(5-\frac{1}{2}\right)-1$

23. $\left(\frac{3}{4}\right)^2 \div \left(\frac{3}{4} - \frac{1}{12}\right)$

24. $\left(\frac{8}{9}\right)^2 \div \left(2 - \frac{2}{3}\right)$

25. $\left(\frac{2}{3} - \frac{5}{9}\right)^2$

26. $\left(1 - \frac{2}{5}\right)^3$

27. $\left(\frac{3}{4} + \frac{1}{8}\right)^2 - \left(\frac{1}{2} + \frac{1}{8}\right)$

28. $\left(\frac{1}{6} + \frac{1}{3}\right)^3 + \left(\frac{2}{5} \cdot \frac{3}{4}\right)^2$

Evaluate each expression if $x = -\frac{1}{3}$, $y = \frac{2}{5}$, *and* $z = \frac{5}{6}$. *See Example 7.*

29. $5y - z$

30. $2z - x$

31. $\frac{x}{z}$

32. $\frac{y + x}{z}$

33. $x^2 - yz$

34. $(1 + x)(1 + z)$

A **B** *Simplify the following. See Examples 1 through 6.*

35. $\dfrac{\frac{5}{24}}{\frac{1}{12}}$

36. $\dfrac{\frac{7}{10}}{\frac{14}{25}}$

37. $\left(\frac{3}{2}\right)^3 + \left(\frac{1}{2}\right)^3$

38. $\left(\frac{5}{21} \div \frac{1}{2}\right) + \left(\frac{1}{7} \cdot \frac{1}{3}\right)$

39. $\left(-\frac{1}{3}\right)^2 + \frac{1}{3}$

40. $\left(-\frac{3}{4}\right)^2 + \frac{3}{8}$

41. $\dfrac{2 + \frac{1}{6}}{1 - \frac{4}{3}}$

42. $\dfrac{3 - \frac{1}{2}}{4 + \frac{1}{5}}$

43. $\left(1 - \frac{2}{5}\right)^2$

44. $\left(-\frac{1}{2}\right)^2 - \left(\frac{3}{4}\right)^2$

45. $\left(\frac{3}{4} - 1\right)\left(\frac{1}{8} + \frac{1}{2}\right)$

46. $\left(\frac{1}{10} + \frac{3}{20}\right)\left(\frac{1}{5} - 1\right)$

47. $\left(-\frac{2}{9}-\frac{7}{9}\right)^4$

48. $\left(\frac{5}{9}-\frac{2}{3}\right)^2$

49. $\dfrac{\frac{1}{2}-\frac{3}{8}}{\frac{3}{4}+\frac{1}{2}}$

50. $\dfrac{\frac{7}{10}+\frac{1}{2}}{\frac{4}{5}+\frac{3}{4}}$

51. $\left(\frac{3}{4}\div\frac{6}{5}\right)-\left(\frac{3}{4}\cdot\frac{6}{5}\right)$

52. $\left(\frac{1}{2}\cdot\frac{2}{7}\right)-\left(\frac{1}{2}\div\frac{2}{7}\right)$

53. $\dfrac{\frac{x}{3}+2}{5+\frac{1}{3}}$

54. $\dfrac{1-\frac{x}{4}}{2+\frac{3}{8}}$

Review and Preview

Evaluate. See Section 1.8.

55. 2^3

56. 3^2

57. 5^2

58. 2^5

Multiply. See Section 2.4.

59. $\frac{1}{3}(3x)$

60. $\frac{1}{5}(5y)$

61. $\frac{2}{3}\left(\frac{3}{2}a\right)$

62. $-\frac{9}{10}\left(-\frac{10}{9}m\right)$

Combining Concepts

63. Calculate $\frac{2^3}{3}$ and $\left(\frac{2}{3}\right)^3$. Do both of these expressions simplify to the same number? Explain why or why not.

64. Calculate $\left(\frac{1}{2}\right)^2\cdot\left(\frac{3}{4}\right)^2$ and $\left(\frac{1}{2}\cdot\frac{3}{4}\right)^2$. Do both of these expressions simplify to the same number? Explain why or why not.

Evaluate each expression if $x=\frac{3}{4}$ and $y=-\frac{4}{7}$.

65. $\dfrac{2+x}{y}$

66. $4x+y$

67. x^2+7y

68. $\dfrac{\frac{9}{14}}{x+y}$

Recall that to find the average of two numbers, find their sum and divide by 2. For example, the average of $\frac{1}{2}$ *and* $\frac{3}{4}$ *is* $\frac{\frac{1}{2}+\frac{3}{4}}{2}$. *Find the average of each pair of numbers.*

69. $\frac{1}{2}, \frac{3}{4}$

70. $\frac{3}{5}, \frac{9}{10}$

71. $\frac{1}{4}, \frac{2}{14}$

72. $\frac{5}{6}, \frac{7}{9}$

73. Two positive numbers, a and b, are graphed below. Where should the graph of their average lie?

Answer true or false for each statement.

74. It is possible for the average of two numbers to be greater than both numbers.

75. It is possible for the average of two numbers to be less than both numbers.

76. The sum of two negative fractions is always a negative number.

77. The sum of a negative fraction and a positive fraction is always a positive number.

78. It is possible for the sum of two fractions to be a whole number.

79. It is possible for the difference of two fractions to be a whole number.

80. What operation should be performed first to simplify

$$\frac{1}{5}\cdot\frac{5}{2} - \left(\frac{2}{3} + \frac{4}{5}\right)^2$$

Explain your answer.

81. A student is to evaluate $x - y$ when $x = \frac{1}{5}$ and $y = -\frac{1}{7}$. This student is asking you if he should evaluate

$\frac{1}{5} - \frac{1}{7}$. What do you tell this student and why?

4.7 Solving Equations Containing Fractions

OBJECTIVES

- (A) Solve equations containing fractions.
- (B) Review adding and subtracting fractions.

(A) Solving Equations Containing Fractions

In Chapter 3, we solved linear equations in one variable. In this section, we practice this skill by solving linear equations containing fractions. To help us solve these equations, let's review the multiplication property of equality.

Multiplication Property of Equality

Let a, b, and c be numbers, and $c \neq 0$.

If $a = b$, then $a \cdot c = b \cdot c$ and also $\frac{a}{c} = \frac{b}{c}$.

In other words, both sides of an equation may be multiplied or divided by the **same** nonzero number without changing the solution of the equation.

(We now know that we define division of a number as multiplication of its reciprocal. Because of this, the Multiplication Property of Equality also allows us to divide the same nonzero number from both sides.)

EXAMPLE 1 Solve: $\frac{1}{3}x = 7$

Solution: Recall that isolating x means that we want the coefficient of x to be 1. To do so, we use the multiplication property of equality and multiply both sides of the equation by the reciprocal of $\frac{1}{3}$, or 3. Since $\frac{1}{3} \cdot 3 = 1$, we will have isolated x.

$$\frac{1}{3}x = 7$$

$$3 \cdot \frac{1}{3}x = 3 \cdot 7 \quad \text{Multiply both sides by 3.}$$

$$1 \cdot x = 21 \text{ or } x = 21 \quad \text{Simplify.}$$

To check, replace x with 21 in the original equation.

$$\frac{1}{3}x = 7 \quad \text{Original equation}$$

$$\frac{1}{3} \cdot 21 = 7 \quad \text{Replace } x \text{ with 21.}$$

$$7 = 7 \quad \text{True.}$$

Since $7 = 7$ is a true statement, 21 is the solution of $\frac{1}{3}x = 7$.

Practice Problem 1

Solve: $\frac{1}{5}y = 2$

Answer

1. 10

Practice Problem 2

Solve: $\frac{5}{7}b = 25$

EXAMPLE 2 Solve: $\frac{3}{5}a = 9$

Solution: Multiply both sides by $\frac{5}{3}$, the reciprocal of $\frac{3}{5}$, so that the coefficient of a is 1.

$$\frac{3}{5}a = 9$$

$$\frac{5}{3}\cdot\frac{3}{5}a = \frac{5}{3}\cdot 9 \quad \text{Multiply both sides by } \frac{5}{3}.$$

$$1a = \frac{5\cdot 9}{3} \quad \text{Multiply.}$$

$$a = 15 \quad \text{Simplify.}$$

To check, replace a with 15 in the original equation.

$$\frac{3}{5}a = 9$$

$$\frac{3}{5}\cdot 15 = 9 \quad \text{Replace } a \text{ with 15.}$$

$$\frac{3\cdot 15}{5} = 9 \quad \text{Multiply.}$$

$$9 = 9 \quad \text{True.}$$

Since $9 = 9$ is true, 15 is the solution of $\frac{3}{5}a = 9$.

Practice Problem 3

Solve: $-\frac{7}{10}x = \frac{2}{5}$

EXAMPLE 3 Solve: $\frac{3}{4}x = -\frac{1}{8}$

Solution: Multiply both sides of the equation by $\frac{4}{3}$, the reciprocal of $\frac{3}{4}$.

$$\frac{3}{4}x = -\frac{1}{8}$$

$$\frac{4}{3}\cdot\frac{3}{4}x = \frac{4}{3}\cdot -\frac{1}{8} \quad \text{Multiply both sides by } \frac{4}{3}.$$

$$1x = -\frac{4\cdot 1}{3\cdot 8} \quad \text{Multiply.}$$

$$x = -\frac{1}{6} \quad \text{Simplify.}$$

To check, replace x with $-\frac{1}{6}$ in the original equation.

$$\frac{3}{4}x = -\frac{1}{8} \quad \text{Original equation.}$$

$$\frac{3}{4}\cdot -\frac{1}{6} = -\frac{1}{8} \quad \text{Replace } x \text{ with } -\frac{1}{6}.$$

$$-\frac{1}{8} = -\frac{1}{8} \quad \text{True.}$$

Since we arrived at a true statement, $-\frac{1}{6}$ is the solution of $\frac{3}{4}x = -\frac{1}{8}$.

Answers

2. 35 **3.** $-\frac{4}{7}$

EXAMPLE 4 Solve: $3y = -\frac{2}{11}$

Solution: We can either divide both sides by 3 or multiply both sides by the reciprocal of 3.

$$3y = -\frac{2}{11}$$

$$\frac{1}{3} \cdot 3y = \frac{1}{3} \cdot -\frac{2}{11} \quad \text{Multiply both sides by } \frac{1}{3}.$$

$$1y = -\frac{1 \cdot 2}{3 \cdot 11} \quad \text{Multiply.}$$

$$y = -\frac{2}{33} \quad \text{Simplify.}$$

Check to see that the solution is $-\frac{2}{33}$.

Practice Problem 4

Solve: $5x = -\frac{3}{4}$

Solving equations with fractions can be tedious. If an equation contains fractions, it is often helpful to first multiply both sides of the equation by the LCD of the fractions. This has the effect of eliminating the fractions in the equation, as shown in the next example.

EXAMPLE 5 Solve: $\frac{x}{6} + 1 = \frac{4}{3}$

Solution: First multiply both sides of the equation by the LCD of the fractions. The LCD of the denominators 6 and 3 is 6.

$$\frac{x}{6} + 1 = \frac{4}{3}$$

$$6\left(\frac{x}{6} + 1\right) = 6\left(\frac{4}{3}\right) \quad \text{Multiply both sides by 6.}$$

$$\overset{1}{\cancel{6}}\left(\frac{x}{\underset{1}{\cancel{6}}}\right) + 6(1) = \overset{2}{\cancel{6}}\left(\frac{4}{\underset{1}{\cancel{3}}}\right) \quad \text{Apply the distributive property.}$$

$$x + 6 = 8 \quad \text{Simplify.}$$

$$x + 6 + (-6) = 8 + (-6) \quad \text{Add } -6 \text{ to both sides.}$$

$$x = 2 \quad \text{Simplify.}$$

To check, replace x with 2 in the original equation.

$$\frac{x}{6} + 1 = \frac{4}{3} \quad \text{Original equation}$$

$$\frac{2}{6} + 1 = \frac{4}{3} \quad \text{Replace } x \text{ with 2.}$$

$$\frac{1}{3} + \frac{3}{3} = \frac{4}{3} \quad \text{Simplify } \frac{2}{6}. \text{ The LCD of 3 and 1 is 3.}$$

$$\frac{4}{3} = \frac{4}{3} \quad \text{True.}$$

Since we arrived at a true statement, 2 is the solution of $\frac{x}{6} + 1 = \frac{4}{3}$.

Practice Problem 5

Solve: $\frac{y}{8} + \frac{3}{4} = 2$

Answers

4. $-\frac{3}{20}$ **5.** 10

Let's review the steps for solving equations in x. An extra step is now included to handle equations containing fractions.

Solving an Equation in x

Step 1. If fractions are present, multiply both sides of the equation by the LCD of the fractions.

Step 2. If parentheses are present, use the distributive property.

Step 3. Combine any like terms on each side of the equation.

Step 4. Use the addition property of equality to rewrite the equation so that variable terms are on one side of the equation and constant terms are on the other side.

Step 5. Divide both sides of the equation by the numerical coefficient of x to solve.

Step 6. Check the answer in the **original equation.**

Practice Problem 6

Solve: $\frac{x}{5} - x = \frac{1}{5}$

EXAMPLE 6 Solve: $\frac{z}{5} - \frac{z}{3} = 6$

Solution:

$$\frac{z}{5} - \frac{z}{3} = 6$$

$$15\left(\frac{z}{5} - \frac{z}{3}\right) = 15(6) \quad \text{Multiply both sides by the LCD, 15.}$$

$$\overset{3}{\cancel{15}}\left(\frac{z}{\underset{1}{\cancel{5}}}\right) - \overset{5}{\cancel{15}}\left(\frac{z}{\underset{1}{\cancel{3}}}\right) = 15(6) \quad \text{Apply the distributive property.}$$

$$3z - 5z = 90 \quad \text{Simplify.}$$

$$-2z = 90 \quad \text{Combine like terms.}$$

$$\frac{-2z}{-2} = \frac{90}{-2} \quad \text{Divide both sides by } -2\text{, the coefficient of } z.$$

$$z = -45 \quad \text{Simplify.}$$

Helpful Hint

Don't forget to multiply *both* sides of the equation by the LCD.

To check, replace z with -45 in the **original equation** to see that a true statement results.

Practice Problem 7

Solve: $\frac{y}{2} = \frac{y}{5} + \frac{3}{2}$

EXAMPLE 7 Solve: $\frac{x}{2} = \frac{x}{3} + \frac{1}{2}$

Solution: First multiply both sides by the LCD, 6.

$$\frac{x}{2} = \frac{x}{3} + \frac{1}{2}$$

$$6\left(\frac{x}{2}\right) = 6\left(\frac{x}{3} + \frac{1}{2}\right) \quad \text{Multiply both sides by the LCD, 6.}$$

$$\overset{3}{\cancel{6}}\left(\frac{x}{\underset{1}{\cancel{2}}}\right) = \overset{2}{\cancel{6}}\left(\frac{x}{\underset{1}{\cancel{3}}}\right) + \overset{3}{\cancel{6}}\left(\frac{1}{\underset{1}{\cancel{2}}}\right) \quad \text{Apply the distributive property.}$$

$$3x = 2x + 3 \quad \text{Simplify.}$$

$$3x + (-2x) = 2x + 3 + (-2x) \quad \text{Add } (-2x) \text{ to both sides.}$$

$$x = 3 \quad \text{Simplify.}$$

To check, replace x with 3 in the original equation to see that a true statement results.

Answers

6. $-\frac{1}{4}$ **7.** 5

B Review of Adding and Subtracting Fractions

Make sure you understand the difference between **solving an equation** containing fractions and **adding or subtracting two fractions**. To solve an equation containing fractions, we use the multiplication property of equality and multiply both sides by the LCD of the fractions, thus eliminating the fractions. This method does not apply to adding or subtracting fractions. The multiplication property of equality applies only to equations. To add or subtract unlike fractions, we write each fraction as an equivalent fraction using the LCD of the fractions as the denominator. See the next example for a review.

EXAMPLE 8 Add: $\frac{x}{3} + \frac{2}{5}$

Solution: This expression is not an equation. Here, we are adding two unlike fractions. To add unlike fractions, we need to find the LCD. The LCD of the denominators 3 and 5 is 15. Write each fraction as an equivalent fraction with a denominator of 15.

$$\frac{x}{3} + \frac{2}{5} = \frac{x \cdot 5}{3 \cdot 5} + \frac{2 \cdot 3}{5 \cdot 3}$$

$$= \frac{5x}{15} + \frac{6}{15}$$

$$= \frac{5x + 6}{15}$$

Try the Concept Check in the margin.

Practice Problem 8

Subtract: $\frac{9}{10} - \frac{y}{3}$

Concept Check

Which of the following are equations and which are expressions?

a. $\frac{1}{2} + 3x = 5$

b. $\frac{2}{3}x - \frac{x}{5}$

c. $\frac{x}{12} + \frac{5x}{24}$

d. $\frac{x}{5} = \frac{1}{10}$

Answer

8. $\frac{27 - 10y}{30}$

Concept Check: equations: a, d; expressions: b, c

FOCUS ON History

THE DEVELOPMENT OF FRACTIONS

Many ancient cultures were familiar with the use of fractions and used them in everyday life. Two examples are discussed here.

- Although the ancient Egyptians were comfortable with the idea of fractions, their system of hieroglyphic numbers allowed them to only write fractions with 1 as the numerator, like $\frac{1}{3}$ or $\frac{1}{5}$. To write fractions with numerators other than 1, the Egyptians had to write a sum of fractions with 1 as the numerator. They preferred to never use the same denominator more than once in one of these sums of fractions. For example, instead of representing the fraction $\frac{2}{5}$ as the sum $\frac{1}{5} + \frac{1}{5}$ (which uses the denominator 5 more than once), they would prefer instead the sum $\frac{1}{3} + \frac{1}{15}$, which also is equal to $\frac{2}{5}$ but uses two different denominators. Similarly, $\frac{2}{13}$ would not be represented by $\frac{1}{13} + \frac{1}{13}$, as we might think would be easiest, but instead by $\frac{1}{8} + \frac{1}{52} + \frac{1}{104}$.

- The ancient Babylonians also dealt with fractions but in a more direct way than did the Egyptians. For instance, they thought of the fraction $\frac{3}{8}$ as the product of 3 and the reciprocal of 8. They compiled detailed lists of reciprocals and could calculate with fractions very easily and quickly.

Although many ancient cultures used fractions, they didn't represent them the way we do today. Fractions did not evolve into anything similar to our present-day format until the seventh century A.D. Around that time, the Hindus became the first to write fractions with the numerator above the denominator. However, Hindu mathematicians, such as Brahmagupta in A.D. 628, wrote these fractions without the fraction bar we use today. Thus, the fraction $\frac{5}{8}$ would have been written by the Hindus as $\begin{matrix}5\\8\end{matrix}$.

It wasn't until between A.D. 1100 and 1200 that the next giant step in fraction notation was made. Up to this point, Arab mathematicians had copied Hindu notation for fractions. However, they then improved this notation by inserting a horizontal bar, called a *vinculum*, between the numerator and denominator to separate the two parts of the fraction.

Today we see fractions in the form ⁴⁄₉, or 4/9 just as often as in the $\frac{4}{9}$ form, developed by the Arabs. The use of the diagonal fraction bar, called a *solidus*, was introduced with the appearance of the printing press. Typesetting fractions with a horizontal fraction bar was difficult and space-consuming because three lines of type were required for the fraction. Typesetting fractions with a diagonal fraction bar was far easier for printers because the entire fraction fit on a single line of type.

Name ______________________________ Section ____________ Date ____________

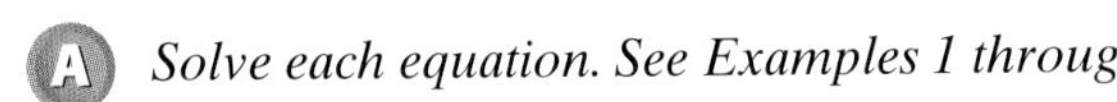

EXERCISE SET 4.7

A *Solve each equation. See Examples 1 through 4.*

1. $7x = 2$ **2.** $-5x = 4$ **3.** $\frac{1}{4}x = 3$ **4.** $\frac{1}{3}x = 6$ **5.** $\frac{2}{9}y = -6$ **6.** $\frac{4}{7}x = -8$

7. $-\frac{4}{9}z = -\frac{3}{2}$ **8.** $-\frac{11}{10}x = -\frac{2}{7}$ **9.** $7a = \frac{1}{3}$ **10.** $2z = -\frac{5}{12}$ **11.** $-3x = -\frac{6}{11}$ **12.** $-4z = -\frac{12}{25}$

Solve each equation. See Examples 5 through 7.

13. $\frac{x}{3} + 2 = \frac{7}{3}$ **14.** $\frac{x}{5} - 1 = \frac{7}{5}$ **15.** $\frac{x}{5} - \frac{5}{10} = 1$ **16.** $\frac{x}{3} - x = -6$

17. $\frac{1}{2} - \frac{3}{5} = \frac{x}{10}$ **18.** $\frac{2}{3} - \frac{1}{4} = \frac{x}{12}$ **19.** $\frac{x}{3} = \frac{x}{5} - 2$ **20.** $\frac{a}{2} = \frac{a}{7} + \frac{5}{2}$

B *Add or subtract as indicated. See Example 8.*

21. $\frac{x}{7} - \frac{4}{3}$ **22.** $-\frac{5}{9} + \frac{y}{8}$ **23.** $\frac{y}{2} + 5$ **24.** $2 + \frac{7x}{3}$ **25.** $\frac{3x}{10} + \frac{x}{6}$ **26.** $\frac{9x}{8} - \frac{5x}{6}$

A B *Solve. If no equation is given, perform the indicated operation.*

27. $\frac{3}{8}x = \frac{1}{2}$ **28.** $\frac{2}{5}y = \frac{3}{10}$ **29.** $\frac{2}{3} - \frac{x}{5} = \frac{4}{15}$ **30.** $\frac{4}{5} + \frac{x}{4} = \frac{21}{20}$

31. $\frac{9}{14}z = \frac{27}{20}$ **32.** $\frac{5}{16}a = \frac{5}{6}$ **33.** $-3m - 5m = \frac{4}{7}$ **34.** $30n - 34n = \frac{3}{20}$

35. $\frac{x}{4} + 1 = \frac{1}{4}$ **39.** $\frac{5}{9} - \frac{2}{3}$ **40.** $\frac{8}{11} - \frac{1}{2}$ **41.** $-\frac{3}{4}x = \frac{9}{2}$

36. $\frac{y}{7} - 2 = \frac{1}{7}$

37. $\frac{1}{5}y = 10$

38. $\frac{1}{4}x = -2$

42. $\frac{5}{7}y = -\frac{15}{49}$

43. $\frac{x}{2} - x = -2$

44. $\frac{y}{3} = -4 + y$

45. $-\frac{5}{8}y = \frac{3}{16} - \frac{9}{16}$

46. $-\frac{7}{9}x = -\frac{5}{18} - \frac{4}{18}$

47. $17x - 25x = \frac{1}{3}$

48. $27x - 30x = \frac{4}{9}$

49. $\frac{7}{6}x = \frac{1}{4} - \frac{2}{3}$

50. $\frac{5}{4}y = \frac{1}{2} - \frac{7}{10}$

51. $\frac{b}{4} = \frac{b}{12} + \frac{2}{3}$

52. $\frac{a}{6} = \frac{a}{3} + \frac{1}{2}$

53. $\frac{x}{3} + 2 = \frac{x}{2} + 8$

54. $\frac{y}{5} - 2 = \frac{y}{3} - 4$

Review and Preview

Perform each indicated operation. See Section 4.5.

55. $3 + \frac{1}{2}$

56. $2 + \frac{2}{3}$

57. $5 + \frac{9}{10}$

58. $1 + \frac{7}{8}$

59. $9 - \frac{5}{6}$

60. $4 - \frac{1}{5}$

Combining Concepts

61. Explain why the method for eliminating fractions in an **equation** does not apply to simplifying **expressions** containing fractions.

Solve.

62. $\frac{14}{11} + \frac{3x}{8} = \frac{x}{2}$

63. $\frac{19}{53} = \frac{353x}{1431} + \frac{23}{27}$

64. The area of the rectangle is $\frac{5}{12}$ square inches. Find its length, x.

x

$\frac{3}{4}$

4.8 Operations on Mixed Numbers

OBJECTIVES

- **A** Illustrate mixed numbers.
- **B** Write mixed numbers as improper fractions.
- **C** Write improper fractions as mixed or whole numbers.
- **D** Multiply or divide mixed or whole numbers.
- **E** Add or subtract mixed numbers.
- **F** Solve problems containing mixed numbers.
- **G** Perform operations on negative mixed numbers.

SSM TUTOR CENTER | SG | CD & VIDEO | MATH PRO | WEB

A Illustrating Mixed Numbers

Recall the graph of the improper fraction $\frac{9}{5}$.

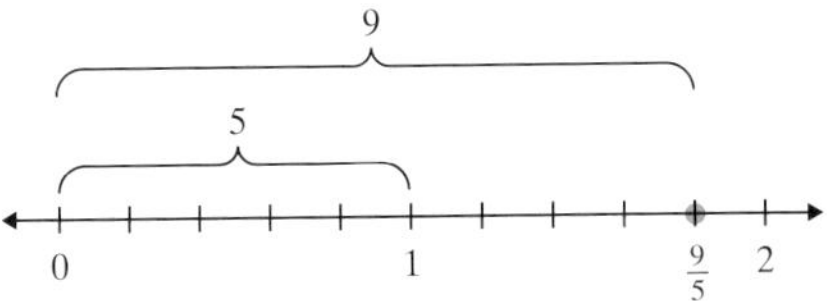

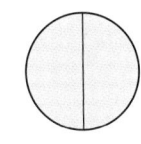

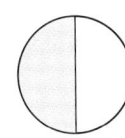

Notice that the graph of $\frac{9}{5}$ is $\frac{4}{5}$ past the number 1. Another way to name the improper fraction $\frac{9}{5}$ is by the mixed number $1\frac{4}{5}$ (read as "one and four fifths"). This **mixed number** contains a whole number and a proper fraction. The mixed number $1\frac{4}{5}$ means $1 + \frac{4}{5}$. Examples of mixed numbers are

$$2\frac{1}{4} = 2 + \frac{1}{4} \quad \text{and} \quad 5\frac{3}{8} = 5 + \frac{3}{8}$$

We can write a mixed number as an improper fraction by adding.

$$1\frac{4}{5} = 1 + \frac{4}{5} = \frac{1 \cdot 5}{1 \cdot 5} + \frac{4}{5} = \frac{5}{5} + \frac{4}{5} = \frac{9}{5}$$

We can also see that $1\frac{4}{5} = \frac{9}{5}$ by studying the following illustration.

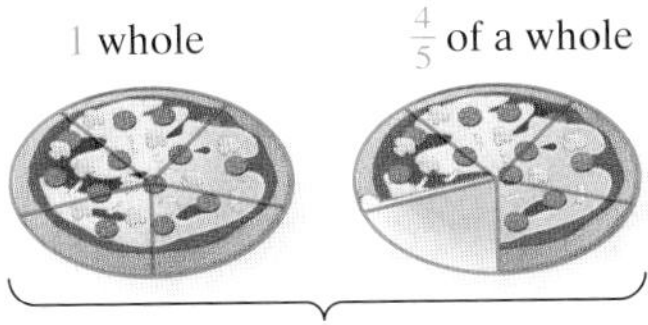

EXAMPLES Represent the shaded area of each figure group with both an improper fraction and a mixed number.

1.

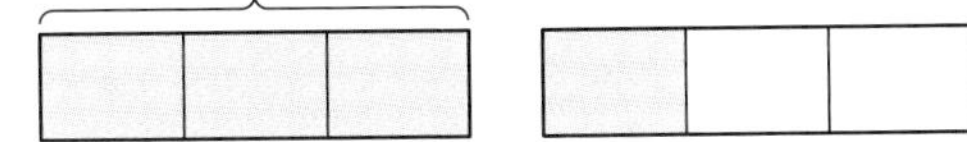

improper fraction: $\frac{4}{3}$

mixed number: $1\frac{1}{3}$

2.

improper fraction: $\frac{5}{2}$

mixed number: $2\frac{1}{2}$

Practice Problems 1–2

Represent the shaded area of each figure group with both an improper fraction and a mixed number.

1.

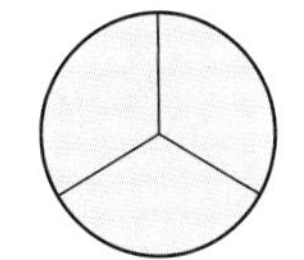
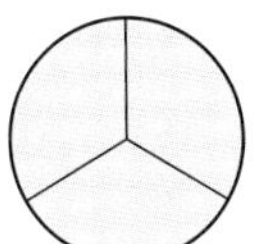
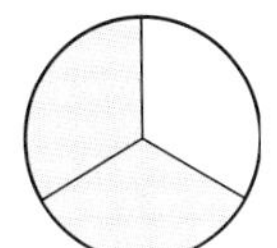

2.

Answers

1. $\frac{8}{3}$ or $2\frac{2}{3}$ **2.** $\frac{5}{4}$ or $1\frac{1}{4}$

Try the Concept Check in the margin.

B Writing Mixed Numbers as Improper Fractions

Notice from Examples 1 and 2 that mixed numbers and improper fractions were both used to represent the shaded area of the figure groups. For example,

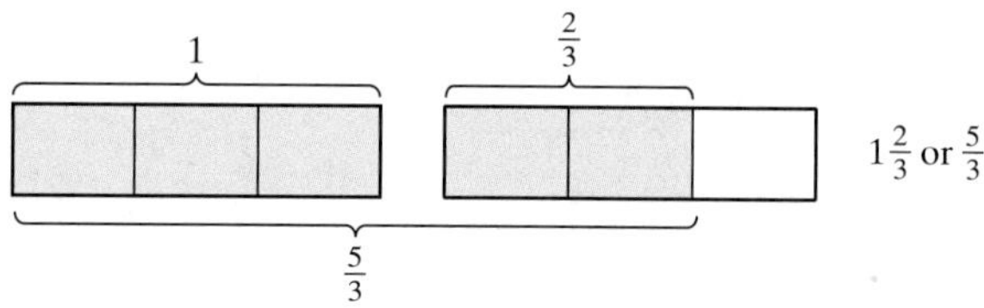

By studying these examples, we can see that a pattern occurs when writing mixed numbers as improper fractions. This pattern leads to the following method for writing a mixed number as an improper fraction.

Writing a Mixed Number as an Improper Fraction

To write a mixed number as an improper fraction,

Step 1. Multiply the whole number by the denominator of the fraction.

Step 2. Add the numerator of the fraction to the product from Step 1.

Step 3. Write the sum from Step 2 as the numerator of the improper fraction over the original denominator.

For example,

$$1\frac{2}{3} = \frac{3 \cdot 1 + 2}{3} = \frac{3 + 2}{3} = \frac{5}{3}$$

EXAMPLE 3 Write each mixed number as an improper fraction.

a. $4\frac{2}{9}$

b. $1\frac{8}{11}$

Solution:

a. $4\frac{2}{9} = \frac{9 \cdot 4 + 2}{9} = \frac{36 + 2}{9} = \frac{38}{9}$

b. $1\frac{8}{11} = \frac{11 \cdot 1 + 8}{11} = \frac{11 + 8}{11} = \frac{19}{11}$

C Writing Improper Fractions as Mixed Numbers or Whole Numbers

Just as there are times when an improper fraction is preferred, sometimes a mixed or a whole number better suits a situation. To write improper fractions as mixed or whole numbers, just remember that the fraction bar means division.

Concept Check

Is $2\frac{1}{8}$ closer to the number 2 or the number 3? If you were to estimate $2\frac{1}{8}$ by a whole number, which number would you choose? Why?

Practice Problem 3

Write each mixed number as an improper fraction.

a. $2\frac{5}{7}$ b. $5\frac{1}{3}$

c. $9\frac{3}{10}$ d. $1\frac{1}{5}$

Answers

3. a. $\frac{19}{7}$ **b.** $\frac{16}{3}$ **c.** $\frac{93}{10}$ **d.** $\frac{6}{5}$

Concept Check: 2; Answers may vary.

Writing an Improper Fraction as a Mixed Number or a Whole Number

To write an improper fraction as a mixed number or a whole number,

Step 1. Divide the denominator into the numerator.

Step 2. The whole-number part of the mixed number is the quotient. The fraction part of the mixed number is the remainder over the original denominator.

$$\text{quotient}\,\frac{\text{remainder}}{\text{original denominator}}$$

For example,

$$\frac{5}{3} = \begin{array}{r} 1 \\ 3\overline{)5} \\ \underline{3} \\ 2 \end{array} = 1\frac{2}{3}$$

← remainder
← original denominator
↑ quotient

EXAMPLE 4 Write each improper fraction as a mixed number or a whole number.

a. $\frac{30}{7}$ **b.** $\frac{16}{15}$ **c.** $\frac{84}{6}$

Solution:

a. $\begin{array}{r} 4 \\ 7\overline{)30} \\ \underline{-28} \\ 2 \end{array}$ $\frac{30}{7} = 4\frac{2}{7}$

b. $\begin{array}{r} 1 \\ 15\overline{)16} \\ \underline{-15} \\ 1 \end{array}$ $\frac{16}{15} = 1\frac{1}{15}$

c. $\begin{array}{r} 14 \\ 6\overline{)84} \\ \underline{-6} \\ 24 \\ \underline{-24} \\ 0 \end{array}$ $\frac{84}{6} = 14$ Since the remainder is 0, the result is the whole number 14.

Helpful Hint

When the remainder is 0, the improper fraction is a whole number. For example, $\frac{92}{4} = 23$.

$$\begin{array}{r} 23 \\ 4\overline{)92} \\ \underline{-8} \\ 12 \\ \underline{-12} \\ 0 \end{array}$$

Practice Problem 4

Write each fraction as a mixed number or a whole number.

a. $\frac{8}{5}$ b. $\frac{17}{6}$

c. $\frac{48}{4}$ d. $\frac{35}{4}$

e. $\frac{51}{7}$ f. $\frac{21}{20}$

Answers

4. a. $1\frac{3}{5}$ **b.** $2\frac{5}{6}$ **c.** 12 **d.** $8\frac{3}{4}$ **e.** $7\frac{2}{7}$ **f.** $1\frac{1}{20}$

Concept Check

Which of the following are equivalent to 9?

a. $7\frac{6}{3}$ b. $8\frac{4}{4}$ c. $8\frac{9}{9}$

d. $\frac{18}{2}$ e. all of these

Practice Problem 5

Multiply: $2\frac{1}{2}\cdot\frac{8}{15}$

Practice Problems 6–7

Multiply.

6. $3\frac{1}{5}\cdot 2\frac{3}{4}$ 7. $\frac{2}{3}\cdot 18$

Concept Check

a. Find the error.

$$2\frac{1}{4}\cdot\frac{1}{2} = 2\frac{1\cdot 1}{4\cdot 2} = 2\frac{1}{8}$$

b. How could you estimate the product $3\frac{1}{7}\cdot 4\frac{5}{6}$?

Practice Problems 8–10

Divide.

8. $\frac{4}{9}\div 5$ 9. $\frac{8}{15}\div 3\frac{4}{5}$ 10. $3\frac{2}{5}\div 2\frac{2}{15}$

Answers

5. $\frac{4}{3}$ or $1\frac{1}{3}$ **6.** $\frac{44}{5}$ or $8\frac{4}{5}$ **7.** $\frac{12}{1}$ or 12 **8.** $\frac{4}{45}$

9. $\frac{8}{57}$ **10.** $\frac{51}{32}$ or $1\frac{19}{32}$

Concept Check: e

Concept Check:

a. $2\frac{1}{4}\cdot\frac{1}{2} = \frac{9}{2}\cdot\frac{1}{2} = \frac{9}{4} = 2\frac{1}{4}$

b. answers may vary

Try the Concept Check in the margin.

D Multiplying or Dividing with Mixed Numbers or Whole Numbers

To multiply or divide with mixed numbers or whole numbers, first write each number as an improper fraction. (Note: If an exercise contains a mixed number, we will write the solution as a mixed number, if possible.)

EXAMPLE 5 Multiply: $3\frac{1}{3}\cdot\frac{7}{8}$

Solution: The mixed number $3\frac{1}{3}$ can be written as the fraction $\frac{10}{3}$. Then,

$$3\frac{1}{3}\cdot\frac{7}{8} = \frac{10}{3}\cdot\frac{7}{8} = \frac{2\cdot 5\cdot 7}{3\cdot 2\cdot 4} = \frac{35}{12} \text{ or } 2\frac{11}{12}$$

Don't forget that a whole number can be written as a fraction by writing the whole number over 1. For example,

$$20 = \frac{20}{1} \quad\text{and}\quad 7 = \frac{7}{1}$$

EXAMPLES Multiply.

6. $1\frac{2}{3}\cdot 2\frac{1}{4} = \frac{5}{3}\cdot\frac{9}{4} = \frac{5\cdot 9}{3\cdot 4} = \frac{5\cdot 3\cdot 3}{3\cdot 4} = \frac{15}{4}$ or $3\frac{3}{4}$

7. $\frac{3}{4}\cdot 20 = \frac{3}{4}\cdot\frac{20}{1} = \frac{3\cdot 20}{4\cdot 1} = \frac{3\cdot 4\cdot 5}{4\cdot 1} = \frac{15}{1}$ or 15

Try the Concept Check in the margin.

EXAMPLES Divide.

8. $\frac{3}{4}\div 5 = \frac{3}{4}\div\frac{5}{1} = \frac{3}{4}\cdot\frac{1}{5} = \frac{3\cdot 1}{4\cdot 5} = \frac{3}{20}$

9. $\frac{11}{18}\div 2\frac{5}{6} = \frac{11}{18}\div\frac{17}{6} = \frac{11}{18}\cdot\frac{6}{17} = \frac{11\cdot 6}{18\cdot 17}$

$= \frac{11\cdot 6}{6\cdot 3\cdot 17} = \frac{11}{51}$

10. $5\frac{2}{3}\div 2\frac{5}{9} = \frac{17}{3}\div\frac{23}{9} = \frac{17}{3}\cdot\frac{9}{23} = \frac{17\cdot 9}{3\cdot 23}$

$= \frac{17\cdot 3\cdot 3}{3\cdot 23} = \frac{51}{23}$ or $2\frac{5}{23}$

E Adding or Subtracting Mixed Numbers

We can add or subtract mixed numbers, too, by first writing each mixed number as an improper fraction. But it is often easier to add or subtract the whole-number parts and add or subtract the proper-fraction parts vertically, as shown in the next examples.

EXAMPLE 11 Add: $2\frac{1}{3} + 5\frac{3}{8}$

Solution: The LCD of 3 and 8 is 24.

$$\begin{aligned} 2\frac{1\cdot 8}{3\cdot 8} &= 2\frac{8}{24} \\ +5\frac{3\cdot 3}{8\cdot 3} &= +5\frac{9}{24} \\ & 7\frac{17}{24} \end{aligned}$$

← Add the fractions.

Add the whole numbers.

EXAMPLE 12 Add: $3\frac{4}{5} + 1\frac{4}{15}$

Solution: The LCD of 5 and 15 is 15.

$$\begin{aligned} 3\frac{4}{5} &= 3\frac{12}{15} \\ +1\frac{4}{15} &= +1\frac{4}{15} \\ & 4\frac{16}{15} \end{aligned}$$

Add the fractions, then add the whole numbers.

Notice that the fractional part is improper.

Since $\frac{16}{15}$ is —, we can write the sum as

$$4\frac{16}{15} = 4 + 1\frac{1}{15} = 5\frac{1}{15}$$

EXAMPLE 13 Add: $1\frac{4}{5} + 4 + 2\frac{1}{2}$

Solution: The LCD of 5 and 2 is 10.

$$\begin{aligned} 1\frac{4}{5} &= 1\frac{8}{10} \\ 4 &= 4 \\ +2\frac{1}{2} &= +2\frac{5}{10} \\ & 7\frac{13}{10} = 7 + 1\frac{3}{10} = 8\frac{3}{10} \end{aligned}$$

Try the Concept Check in the margin.

EXAMPLE 14 Subtract: $9\frac{3}{7} - 5\frac{4}{21}$

Solution: The LCD of 7 and 21 is 21.

$$\begin{aligned} 9\frac{3}{7} &= 9\frac{9}{21} \\ -5\frac{4}{21} &= -5\frac{4}{21} \\ & 4\frac{5}{21} \end{aligned}$$

← Subtract the fractions.

Subtract the whole numbers.

Practice Problem 11

Add: $4\frac{2}{5} + 5\frac{3}{10}$

Practice Problem 12

Add: $2\frac{5}{14} + 5\frac{6}{7}$

Practice Problem 13

Add: $10 + 2\frac{1}{7} + 3\frac{1}{5}$

Concept Check

Explain how you could estimate the following sum: $5\frac{1}{9} + 14\frac{10}{11}$.

Practice Problem 14

Subtract: $29\frac{7}{8} - 13\frac{3}{16}$

Answers

11. $9\frac{7}{10}$ **12.** $8\frac{3}{14}$ **13.** $15\frac{12}{35}$ **14.** $16\frac{11}{16}$

Concept Check: Round each mixed number to the nearest whole number and add. $5\frac{1}{9}$ rounds to 5 and $14\frac{10}{11}$ rounds to 15, and the estimated sum is $5 + 15 = 20$.

When subtracting mixed numbers, borrowing may be needed, as shown in the next example.

Practice Problem 15

Subtract: $9\frac{7}{15} - 5\frac{4}{5}$

EXAMPLE 15 Subtract: $7\frac{3}{14} - 3\frac{6}{7}$

Solution: The LCD of 7 and 14 is 14.

$$\begin{array}{rcr} 7\frac{3}{14} & = & 7\frac{3}{14} \\ -3\frac{6}{7} & = & -3\frac{12}{14} \\ \hline \end{array}$$

Notice that we cannot subtract $\frac{12}{14}$ from $\frac{3}{14}$, so we borrow from the whole number 7.

Borrow 1 from 7.

$$7\frac{3}{14} = 6 + 1\frac{3}{14} = 6 + \frac{17}{14} \text{ or } 6\frac{17}{14}$$

Now subtract.

$$\begin{array}{rcrcr} 7\frac{3}{14} & = & 7\frac{3}{14} & = & 6\frac{17}{14} \\ -3\frac{6}{7} & = & -3\frac{12}{14} & = & -3\frac{12}{14} \\ \hline & & & & 3\frac{5}{14} \end{array}$$

← Subtract the fractions.

↑ Subtract the whole numbers.

Concept Check

In the subtraction problem $5\frac{1}{4} - 3\frac{3}{4}$, $5\frac{1}{4}$ must be rewritten because $\frac{3}{4}$ cannot be subtracted from $\frac{1}{4}$. Why is it incorrect to rewrite $5\frac{1}{4}$ as $5\frac{5}{4}$?

Try the Concept Check in the margin.

Practice Problem 16

Subtract: $25 - 10\frac{2}{9}$

EXAMPLE 16 Subtract: $12 - 8\frac{3}{7}$

Solution:

$$\begin{array}{rcr} 12 & = & 11\frac{7}{7} \\ -8\frac{3}{7} & = & -8\frac{3}{7} \\ \hline & & 3\frac{4}{7} \end{array}$$

Borrow 1 from 12 and write it as $\frac{7}{7}$.

← Subtract the fractions.

↑ Subtract the whole numbers.

F Solving Problems Containing Mixed Numbers

Now that we know how to perform operations on mixed numbers, we can solve real-life problems.

Answers

15. $3\frac{2}{3}$ **16.** $14\frac{7}{9}$

Concept Check: Rewrite $5\frac{1}{4}$ as $4\frac{5}{4}$ by borrowing from the 5.

EXAMPLE 17 Finding Legal Lobster Size

Lobster fishermen must measure the upper body shells of the lobsters they catch. Lobsters that are too small are thrown back into the ocean. Each state has its own size standard for lobsters to help control the breeding stock. In 1988, Massachusetts increased its legal lobster size from $3\frac{3}{16}$ inches to $3\frac{7}{32}$ inches. How much of an increase was this? (*Source:* Peabody Essex Museum, Salem, MA)

Solution:

1. UNDERSTAND. Read and reread the problem carefully. The word "increase" found in the problem might make you think that we add to solve the problem. But the phrase "how much of an increase" tells us to subtract to find the increase. Let

   x = the increase

2. TRANSLATE.

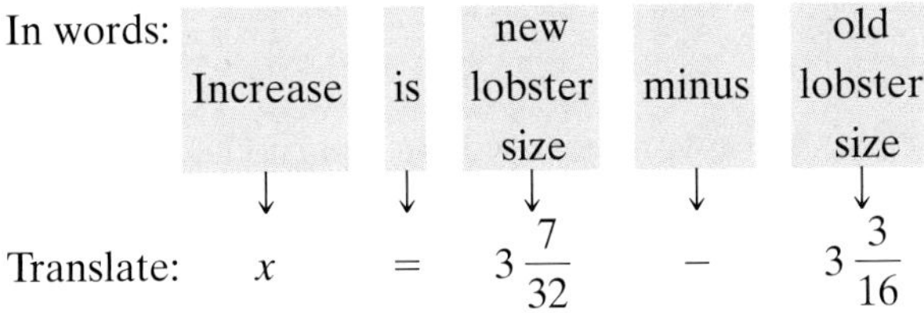

In words:	Increase	is	new lobster size	minus	old lobster size
	↓	↓	↓	↓	↓
Translate:	x	$=$	$3\frac{7}{32}$	$-$	$3\frac{3}{16}$

3. Solve. Since x is by itself on one side of the equation, the equation is solved for x. We just need to perform the subtraction.

$$\begin{array}{rcr} 3\frac{7}{32} & = & 3\frac{7}{32} \\ -3\frac{3}{16} & = & -3\frac{6}{32} \\ \hline & & \frac{1}{32} \end{array}$$

   Thus, $x = \frac{1}{32}$.

4. INTERPRET. *Check* your work. *State* your conclusion: The increase in lobster size is $\frac{1}{32}$ of an inch.

Practice Problem 17

The measurement around the trunk of a tree just below shoulder height is called its girth. The largest known American Beech tree in the United States has a girth of $23\frac{1}{4}$ feet. The largest known Sugar Maple tree in the United States has a girth of $19\frac{5}{12}$ feet. How much larger is the girth of the largest known American Beech tree than the girth of the largest known Sugar Maple tree? (*Source*: American Forests)

Answer

17. $3\frac{5}{6}$ feet

Practice Problem 18

A designer of women's clothing designs a woman's dress that requires $2\frac{1}{7}$ yards of material. How many dresses can be made from a 30-yard bolt of material?

EXAMPLE 18 Calculating Manufacturing Materials Needed

In a manufacturing process, a metal-cutting machine cuts strips $1\frac{3}{5}$ inches long from a piece of metal stock. How many such strips can be cut from a 48-inch piece of stock?

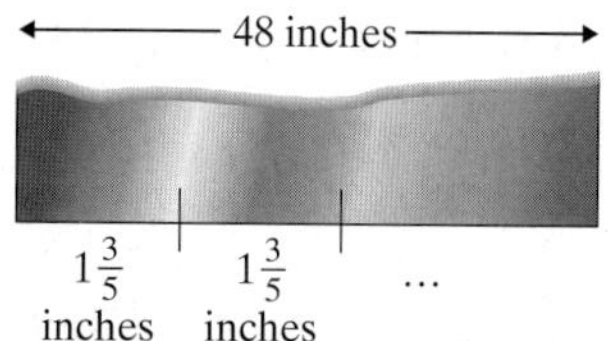

Solution:

1. UNDERSTAND the problem. To do so, read and reread the problem. Then draw a diagram.

We want to know how many $1\frac{3}{5}$s there are in 48. Let x = number of strips

2. TRANSLATE.

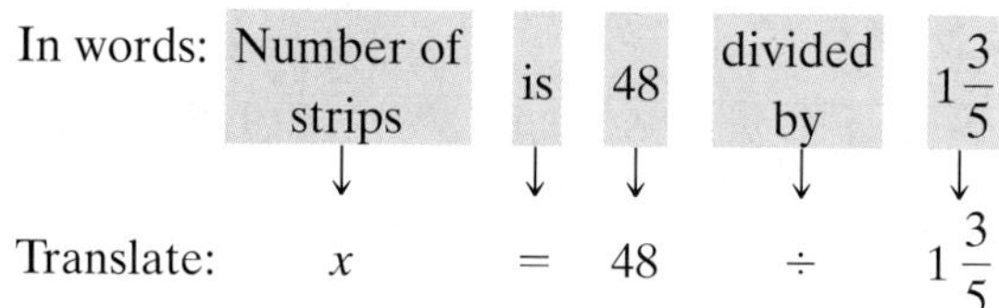

In words:	Number of strips	is	48	divided by	$1\frac{3}{5}$
	↓	↓	↓	↓	↓
Translate:	x	$=$	48	$\div$	$1\frac{3}{5}$

3. SOLVE.

$$x = 48 \div 1\frac{3}{5}$$

$$= 48 \div \frac{8}{5} = \frac{48}{1}\cdot\frac{5}{8} = \frac{48\cdot 5}{1\cdot 8} = \frac{8\cdot 6\cdot 5}{1\cdot 8} = \frac{30}{1} \text{ or } 30$$

4. INTERPRET. *Check* your work. *State* your conclusion: Thirty strips can be cut from the 48-inch piece of stock.

G Operating on Negative Mixed Numbers

To perform operations on negative mixed numbers, let's first practice writing these numbers as negative fractions and negative fractions as negative mixed numbers.

To understand negative mixed numbers, we simply need to know that, for example,

$$-3\frac{2}{5} \text{ means } -\left(3\frac{2}{5}\right)$$

Thus, to write a negative mixed number as a fraction, we do the following.

$$-3\frac{2}{5} = -\left(3\frac{2}{5}\right) = -\left(\frac{17}{5}\right) \text{ or } -\frac{17}{5}$$

$$3\frac{2}{5} = \frac{5\cdot 3 + 2}{5} = \frac{17}{5}$$

Answer

18. 14 dresses

EXAMPLES Write each as a fraction.

19. $-1\frac{7}{8} = -\frac{8 \cdot 1 + 7}{8} = -\frac{15}{8}$ Write $1\frac{7}{8}$ as an improper fraction and keep the negative sign.

20. $-23\frac{1}{2} = -\frac{2 \cdot 23 + 1}{2} = -\frac{47}{2}$ Write $23\frac{1}{2}$ as an improper fraction and keep the negative sign.

To write a negative fraction as a negative mixed number, we use a similar procedure. We simply disregard the negative sign, convert the improper fraction to a mixed number, then reinsert the negative sign.

EXAMPLES Write each as a mixed number.

21. $-\frac{22}{5} = -4\frac{2}{5}$

$$\begin{array}{r} 4 \\ 5\overline{)22} \\ -20 \\ \hline 2 \end{array} \qquad \frac{22}{5} = 4\frac{2}{5}$$

22. $-\frac{9}{4} = -2\frac{1}{4}$

$$\begin{array}{r} 2 \\ 4\overline{)9} \\ -8 \\ \hline 1 \end{array} \qquad \frac{9}{4} = 2\frac{1}{4}$$

We multiply or divide with negative mixed numbers the same way that we multiply or divide with positive mixed numbers. First, write each mixed number as a fraction.

EXAMPLES Perform the indicated operations.

23. $-4\frac{2}{5} \cdot 1\frac{3}{11} = -\frac{22}{5} \cdot \frac{14}{11} = -\frac{22 \cdot 14}{5 \cdot 11} = -\frac{2 \cdot 11 \cdot 14}{5 \cdot 11} = -\frac{28}{5}$ or $-5\frac{3}{5}$

24. $-2\frac{1}{3} \div \left(-2\frac{1}{2}\right) = -\frac{7}{3} \div \left(-\frac{5}{2}\right) = -\frac{7}{3} \cdot \left(-\frac{2}{5}\right) = \frac{7 \cdot 2}{3 \cdot 5} = \frac{14}{15}$

Helpful Hint
Recall that
$(-) \cdot (-) = +$

To add or subtract with negative mixed numbers, we must be very careful! Problems arise because recall that

$$-3\frac{2}{5} \text{ means } -\left(3\frac{2}{5}\right)$$

This means that

$$-3\frac{2}{5} = -\left(3\frac{2}{5}\right) = -\left(3 + \frac{2}{5}\right) = -3 - \frac{2}{5}$$ This can sometimes be easily overlooked.

To avoid problems, we will add or subtract negative mixed numbers by rewriting as addition and recalling how to add signed numbers.

Practice Problems

Write each as a fraction

19. $-7\frac{3}{7}$

20. $-4\frac{10}{11}$

Practice Problems

Write each as a mixed number.

21. $-\frac{29}{8}$

22. $-\frac{31}{5}$

Practice Problems

23. $3\frac{3}{4} \cdot \left(-2\frac{3}{5}\right)$

24. $-1\frac{2}{7} \div 4\frac{1}{4}$

Answers

19. $-\frac{52}{7}$ or $-7\frac{3}{7}$ **20.** $-\frac{54}{11}$ or $-4\frac{10}{11}$ **21.** $-3\frac{5}{8}$ **22.** $-6\frac{1}{5}$ **23.** $-9\frac{3}{4}$ **24.** $-\frac{36}{119}$

Practice Problem 25

Add: $7\frac{2}{3} + \left(-11\frac{3}{4}\right)$

EXAMPLE 25 Add: $6\frac{3}{5} + \left(-9\frac{7}{10}\right)$

Solution: Here we are adding two numbers with different signs. Recall that we then subtract the absolute values and keep the sign of the larger absolute value.

Since $-9\frac{7}{10}$ has the larger absolute value, the answer is negative.

First, subtract absolute values:

$$\begin{aligned} 9\frac{7}{10} &= 9\frac{7}{10} \\ -6\frac{3 \cdot 2}{5 \cdot 2} &= -6\frac{6}{10} \\ & 3\frac{1}{10} \end{aligned}$$

Thus,

$$6\frac{3}{5} + \left(-9\frac{7}{10}\right) = -3\frac{1}{10}$$ The result is negative since $-9\frac{7}{10}$ has the larger absolute value.

Practice Problem 26

Subtract: $-9\frac{2}{7} - 15\frac{11}{14}$

EXAMPLE 26 Subtract: $-11\frac{5}{6} - 20\frac{4}{9}$

Solution: Let's write as an equivalent addition: $-11\frac{5}{6} + \left(-20\frac{4}{9}\right)$. Here, we are adding two numbers with like signs. Recall that we add their absolute values and keep the common negative sign.

First, add absolute values:

$$\begin{aligned} 11\frac{5 \cdot 3}{6 \cdot 3} &= 11\frac{15}{18} \\ +20\frac{4 \cdot 2}{9 \cdot 2} &= +20\frac{8}{18} \\ & 31\frac{23}{18} \text{ or } 32\frac{5}{18} \end{aligned}$$

Since $\frac{23}{18} = 1\frac{5}{18}$

Thus,

$$-11\frac{5}{6} - 20\frac{4}{9} = -32\frac{5}{18}$$

Keep the common sign.

Answers

25. $-4\frac{1}{12}$ **26.** $-25\frac{1}{14}$

CHAPTER 4 ACTIVITY Comparing Fractions

Make seven identical copies of the strip shown in Figure 1. Cut out each strip.

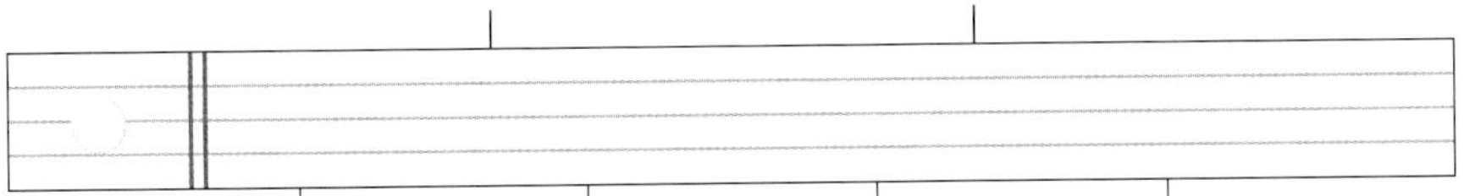

Figure 1

Figure 2

1. Fold and crease one of the strips (as shown in Figure 2) to represent two equal halves. Now unfold the strip so that it lies flat again. You have just created a fraction strip for the denominator 2.
2. Make similar fraction strips for the denominators 3, 4, 5, 6, 8, and 12 by folding. (Note: The markings along the top of Figure 1 will help you gauge equal thirds. The markings along the bottom of Figure 1 will help you gauge equal fifths.)
3. Shade one of the fraction strips to represent $\frac{1}{2}$ and another to represent $\frac{2}{5}$. Compare the fraction strips to decide which fraction is larger or whether the fractions are equal.
4. Shade one of the fraction strips to represent $\frac{3}{8}$ and another to represent $\frac{2}{3}$. Compare the fraction strips to decide which fraction is larger or whether the fractions are equal.
5. Shade one of the fraction strips to represent $\frac{3}{4}$ and another to represent $\frac{5}{6}$. Compare the fraction strips to decide which fraction is larger or whether the fractions are equal.
6. Shade one of the fraction strips to represent $\frac{2}{8}$ and another to represent $\frac{3}{12}$. (To reuse a fraction strip, just flip it over and use the blank side.) Compare the fraction strips to decide which fraction is larger or whether the fractions are equal.
7. Shade one of the fraction strips to represent $\frac{4}{5}$ and another to represent $\frac{7}{12}$. (To reuse a fraction strip just flip it over and use the blank side.) Compare the fraction strips to decide which fraction is larger or whether the fractions are equal.

CALCULATOR EXPLORATIONS

Converting Between Mixed-Number and Fraction Notation

If your calculator has a fraction key, such as [a b/c], you can use it to convert between mixed-number notation and fraction notation.

To write $13\frac{7}{16}$ as an improper fraction, press

[1] [3] [a b/c] [7] [a b/c] [1] [6] [2nd] [d/c]

The display will read

[215|16]

which represents $\frac{215}{16}$. Thus $13\frac{7}{16} = \frac{215}{16}$.

To convert $\frac{190}{13}$ to a mixed number, press

[1] [9] [0] [a b/c] [1] [3] [=]

The display will read

[14_8/13]

which represents $14\frac{8}{13}$. Thus $\frac{190}{13} = 14\frac{8}{13}$.

Write each mixed number as a fraction and each fraction as a mixed number.

1. $25\frac{5}{11}$ **2.** $67\frac{14}{15}$ **3.** $107\frac{31}{35}$

4. $186\frac{17}{21}$ **5.** $\frac{365}{14}$ **6.** $\frac{290}{13}$

7. $\frac{2769}{30}$ **8.** $\frac{3941}{17}$

Name ______________________ Section ____________ Date ____________

EXERCISE SET 4.8

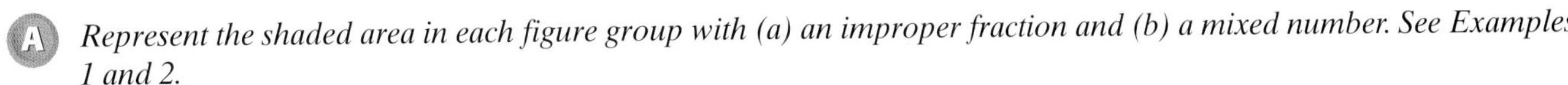

A *Represent the shaded area in each figure group with (a) an improper fraction and (b) a mixed number. See Examples 1 and 2.*

1.  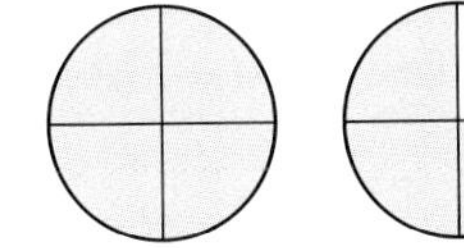

2.  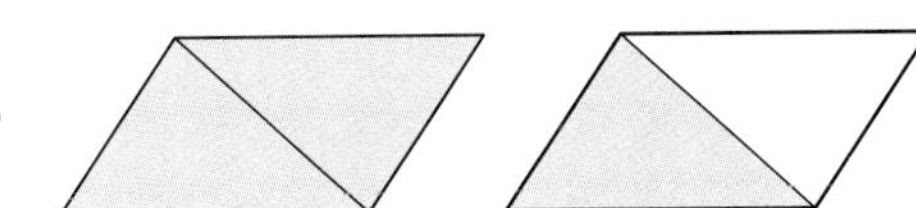

3. 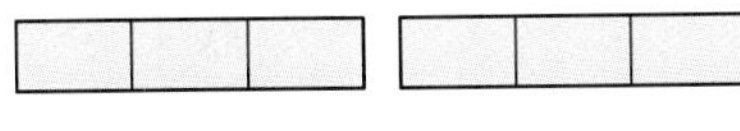 

4.   

5. 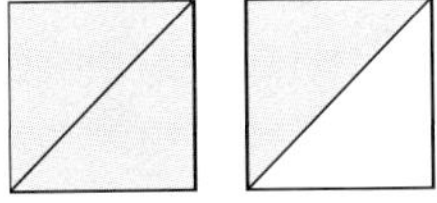

6. 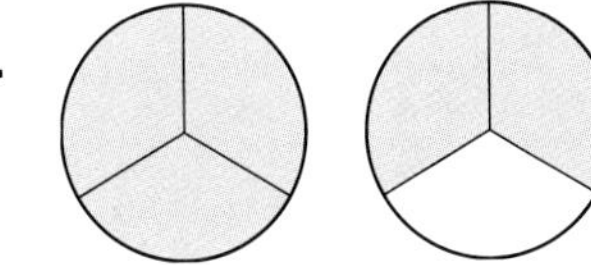

7. 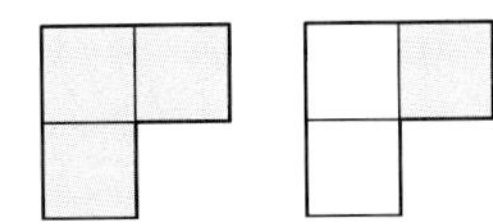

8. 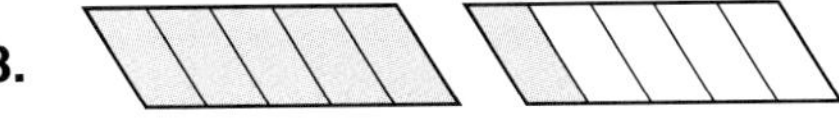

B *Write each mixed number as an improper fraction. See Example 3.*

9. $2\frac{1}{3}$

10. $6\frac{3}{4}$

 **11.** $3\frac{3}{8}$

12. $2\frac{5}{8}$

 **13.** $11\frac{6}{7}$

14. $8\frac{9}{10}$

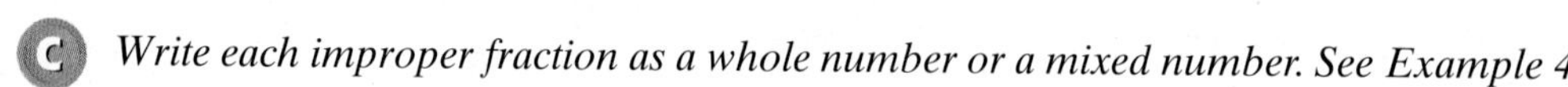

C *Write each improper fraction as a whole number or a mixed number. See Example 4.*

15. $\frac{13}{7}$

16. $\frac{42}{13}$

17. $\frac{47}{15}$

18. $\frac{65}{12}$

19. $\frac{37}{8}$

20. $\frac{42}{6}$

D *Multiply or divide. See Examples 5 through 10.*

21. $2\frac{2}{3} \cdot \frac{1}{7}$

22. $\frac{5}{9} \cdot 4\frac{1}{5}$

23. $8 \div 1\frac{5}{7}$

24. $5 \div 3\frac{3}{4}$

25. $3\frac{2}{3} \cdot 1\frac{1}{2}$

26. $2\frac{4}{5} \cdot 2\frac{5}{8}$

27. $2\frac{2}{3} \div \frac{1}{7}$

28. $\frac{5}{9} \div 4\frac{1}{5}$

E *Add. See Examples 11 through 13.*

29. $4\frac{7}{10} + 2\frac{1}{10}$

30. $7\frac{4}{9} + 3\frac{2}{9}$

31. $\begin{array}{r} 15\frac{4}{7} \\ +9\frac{11}{14} \\ \hline \end{array}$

32. $\begin{array}{r} 23\frac{3}{5} \\ +8\frac{8}{15} \\ \hline \end{array}$

33. $\begin{array}{r} 3\frac{5}{8} \\ 2\frac{1}{6} \\ +7\frac{3}{4} \\ \hline \end{array}$

34. $\begin{array}{r} 4\frac{1}{3} \\ 9\frac{2}{5} \\ +3\frac{1}{6} \\ \hline \end{array}$

Subtract. See Examples 14 through 16.

35. $4\frac{7}{10} - 2\frac{1}{10}$

36. $7\frac{4}{9} - 3\frac{2}{9}$

37. $10\frac{13}{14} - 3\frac{4}{7}$

38. $12\frac{5}{12} - 4\frac{1}{6}$

39. $9\frac{1}{5} - 8\frac{6}{25}$

40. $6\frac{2}{13} - 4\frac{7}{26}$

D E *Perform each indicated operation. See Examples 5 through 16.*

41. $2\frac{3}{4} + 1\frac{1}{4}$

42. $5\frac{5}{8} + 2\frac{3}{8}$

43. $15\frac{4}{7} - 9\frac{11}{14}$

44. $23\frac{3}{5} - 8\frac{8}{15}$

45. $3\frac{1}{9} \cdot 2$

46. $4\frac{1}{2} \cdot 3$

47. $1\frac{2}{3} \div 2\frac{1}{5}$

48. $5\frac{1}{5} \div 3\frac{1}{4}$

49. $22\frac{4}{9} + 13\frac{5}{18}$

50. $15\frac{3}{25} - 5\frac{2}{5}$

51. $5\frac{2}{3} - 3\frac{1}{6}$

52. $5\frac{3}{8} - 2\frac{13}{16}$

53. $15\frac{1}{5} + 20\frac{3}{10} + 37\frac{2}{15}$

54. $7\frac{3}{7} + 15 + 20\frac{1}{2}$

55. $6\frac{4}{7} - 5\frac{11}{14}$

56. $47\frac{5}{12} - 23\frac{19}{24}$

57. $4\frac{2}{7} \cdot 1\frac{3}{10}$

58. $6\frac{2}{3} \cdot 2\frac{3}{4}$

59.
$$\begin{array}{r} 6\frac{2}{11} \\ 3 \\ +4\frac{10}{33} \\ \hline \end{array}$$

60.
$$\begin{array}{r} 3\frac{7}{16} \\ 6\frac{1}{2} \\ +9\frac{3}{8} \\ \hline \end{array}$$

 *Solve. See Examples 17 and 18.*

The first three total eclipses of the sun visible from the Atlantic Ocean in the 21st century are listed below. The duration of each eclipse is included in the table. Use this table to answer Exercises 61–64.

Total Solar Eclipses Visible from the Atlantic Ocean

Date of Eclipse	Duration (in minutes)
June 21, 2001	$4\frac{14}{15}$
March 29, 2006	$4\frac{7}{60}$
November 3, 2013	$1\frac{2}{3}$

(*Source*: 2003 *World Almanac*)

61. What is the total duration for the three eclipses?

62. How much longer is the June 21, 2001, eclipse than the November 3, 2013, eclipse?

63. How much longer is the June 21, 2001, eclipse than the March 29, 2006, eclipse?

64. What is the total duration for the two eclipses occurring in odd-numbered years?

65. A sidewalk is built six bricks wide by laying each brick side by side. How many inches wide is the sidewalk if each brick measures $3\frac{1}{4}$ inches wide?

66. Vonshay Bartlet wants to build a bookcase with six equally spaced shelves. Disregarding the width of the shelves, he finds he has $5\frac{1}{2}$ feet of vertical space for shelves. How far apart should he space the shelves?

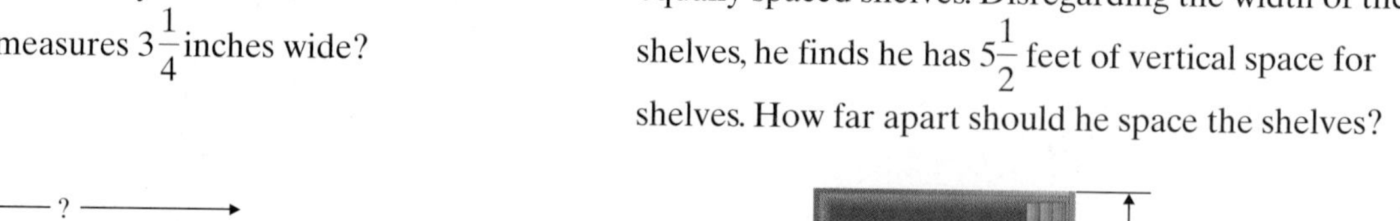

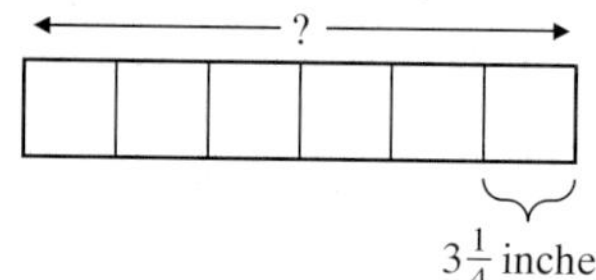

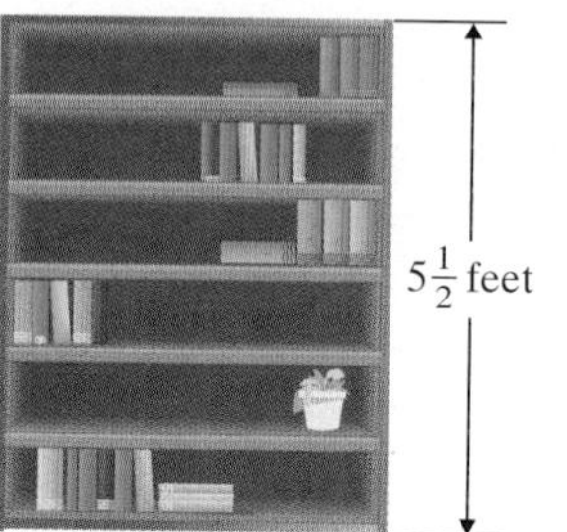

67. On March 12, 2003, the closing price for Kellogg Company stock was \$28 1/5 per share. Brian Robinson spent \$423 buying shares of Kellogg stock at the close of the stock market on that day. How many shares did he purchase? (Based on data from Commodity Systems, Inc.)

68. The nutrition label on a can of chili shows there are 26 grams of carbohydrates for each cup of chili. How many grams of carbohydrates are in a 2 1/2 cup can?

△ **69.** To prevent intruding birds, birdhouses built for Eastern Bluebirds should have an entrance hole measuring $1\frac{1}{2}$ inches in diameter. Entrance holes in bird houses for Mountain Bluebirds should measure $1\frac{9}{16}$ inches in diamater. How much wider should entrance holes for Mountain Bluebirds be than for Eastern Bluebirds? (*Source:* North American Bluebird Society)

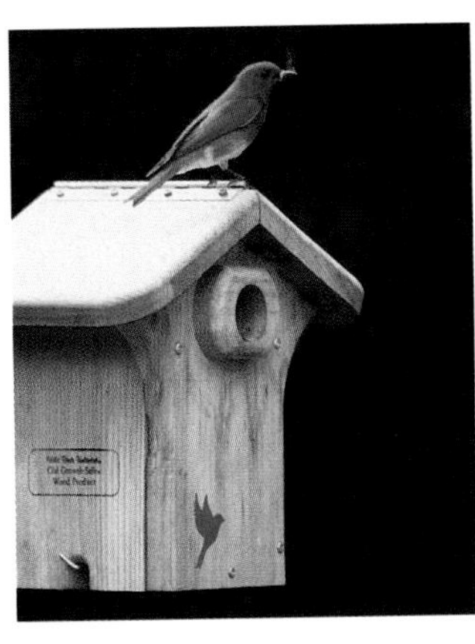

70. If the total weight allowable without overweight charges is 50 pounds and the traveler's luggage weighs $60\frac{5}{8}$ pounds, on how many pounds will the traveler's overweight charges be based?

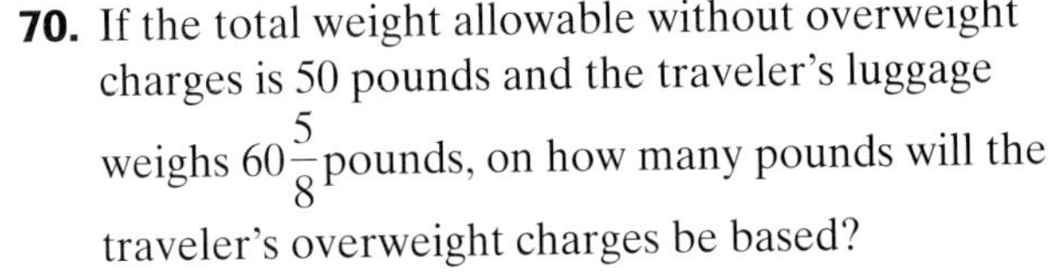

71. Charlotte Dowlin has $15\frac{2}{3}$ feet of plastic pipe. She cuts off a $2\frac{1}{2}$-foot length and then a $3\frac{1}{4}$-foot length. If she now needs a 10-foot piece of pipe, will the remaining piece do?

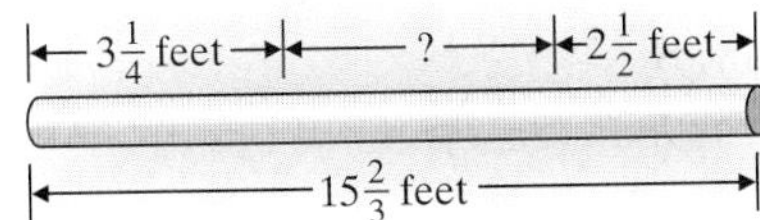

72. Shamalika Corning, a trim carpenter, cuts a board $3\frac{3}{8}$ feet long from one 6 feet long. How long is the remaining piece?

$3\frac{3}{8}$ feet
?
6 feet

73. A small airplane used $58\frac{3}{4}$ gallons of fuel to fly a $7\frac{1}{2}$ hour trip. How many gallons of fuel were used for each hour?

74. Kesha Jonston is planning a Memorial Day barbeque. She has $27\frac{3}{4}$ pounds of hamburger. How many quarter-pound hamburgers can she make?

75. The Gauge Act of 1846 set the standard gauge for U.S. railroads at $56\frac{1}{2}$ inches. (See figure.) If the standard gauge in Spain is $65\frac{9}{10}$ inches, how much wider is Spain's standard gauge than the U.S. standard gauge? (*Source:* San Diego Railroad Museum)

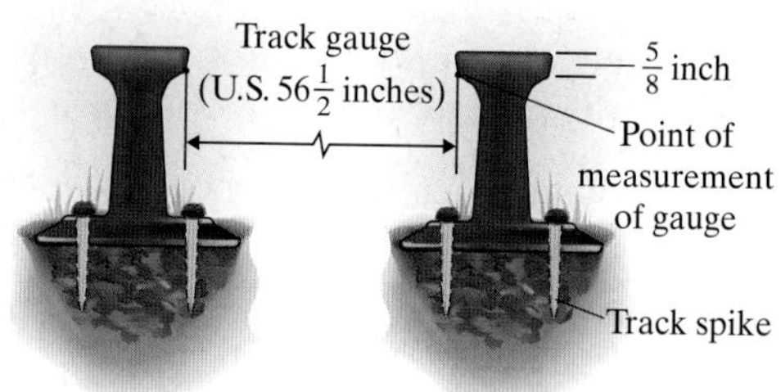

76. The standard railroad track gauge (see figure) in Spain is $65\frac{9}{10}$ inches, while in neighboring Portugal it is $65\frac{11}{20}$ inches. Which gauge is wider and by how much? (*Source:* San Diego Railroad Museum)

77. If Tucson's average rainfall is $11\frac{1}{4}$ inches and Yuma's is $3\frac{3}{5}$ inches, how much more rain, on the average, does Tucson get than Yuma?

78. A pair of crutches needs adjustment. One crutch is 43 inches and the other is $41\frac{5}{8}$ inches. Find how much the short crutch should be lengthened to make both crutches the same length.

For Exercises 79 and 80, find the area of each figure.

△ **79.**

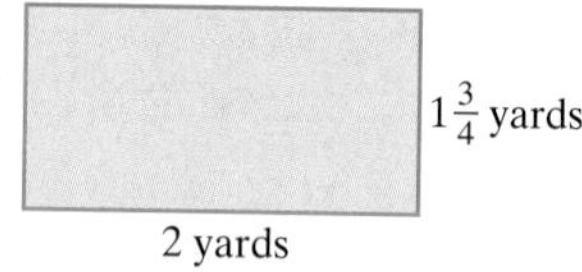

 **80.**

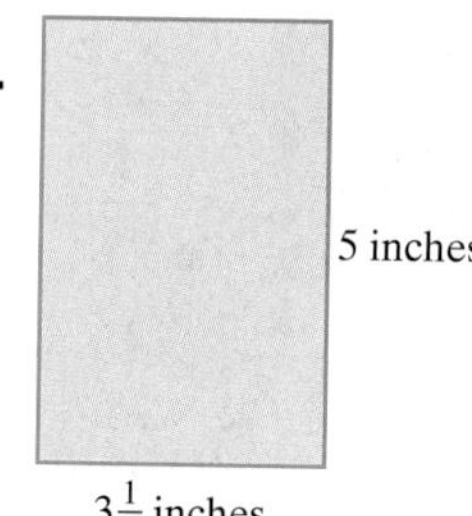

△ **81.** A model for a proposed computer chip measures $\frac{3}{4}$ inch by $1\frac{1}{4}$ inches. Find its area.

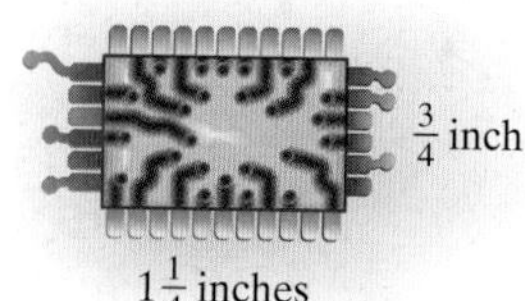

△ **82.** The Saltalamachios are planning to build a deck that measures $4\frac{1}{2}$ yards by $6\frac{1}{3}$ yards. Find the area of their proposed deck.

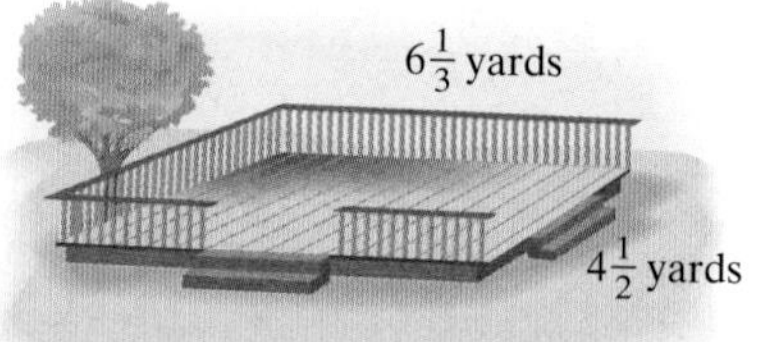

For Exercises 83 through 86, find the perimeter of each figure.

△ **83.**

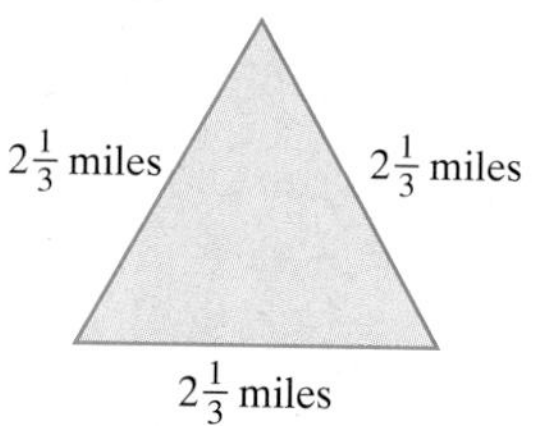

△ **84.**

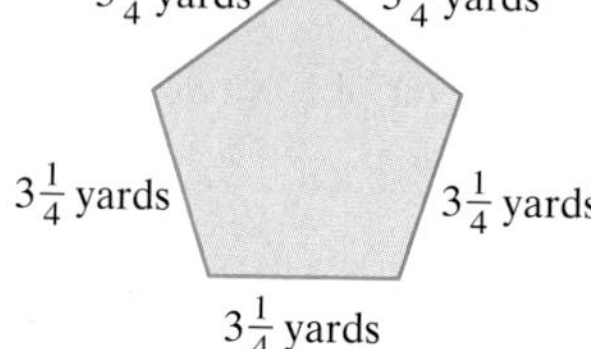

85.

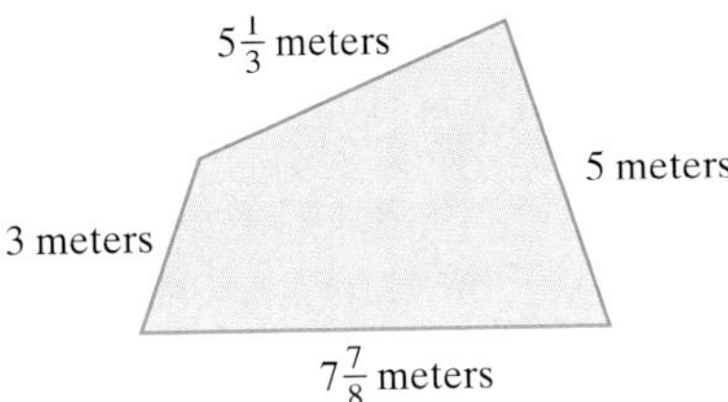

86.

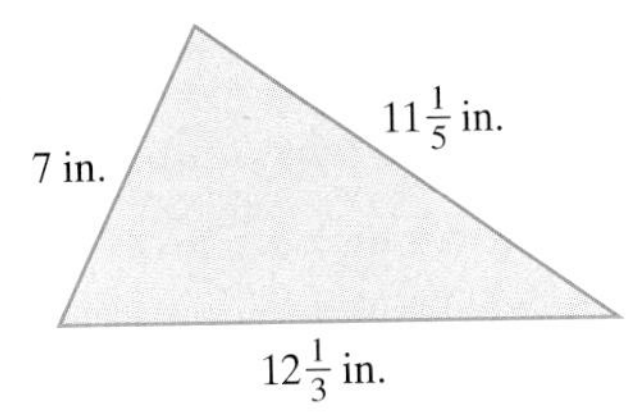

87. The life expectancy of a circulating coin is 30 years. The life expectancy of a circulating dollar bill is only $\frac{1}{20}$ as long. Find the life expectancy of circulating paper money. (*Source:* The U.S. Mint)

88. A decorative wall in Ben and Joy Lander's garden is to be built using brick that is $2\frac{3}{4}$ inches wide and a mortar joint that is $\frac{1}{2}$ inch wide. Use the diagram to find the height of the wall.

89. Located on an island in New York City's harbor, the Statue of Liberty is one of the largest statues in the world. The copper figure is $152\frac{1}{6}$ feet tall from feet to tip of torch. The figure stands on a pedestal that is $154\frac{1}{2}$ feet tall. What is the overall height of the Statue of Liberty from the base of the pedestal to the tip of the torch? (*Source: Microsoft Encarta Encyclopedia*)

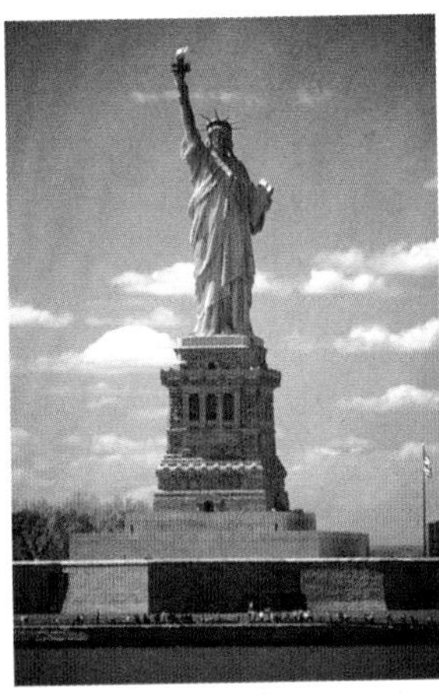

90. The record for largest rainbow trout ever caught is $42\frac{1}{8}$ pounds and was set in Alaska in 1970. The record for largest tiger trout ever caught is $20\frac{13}{16}$ pounds and was set in Michigan in 1978. How much more did the record-setting rainbow trout weigh than the record-setting tiger trout? (*Source:* International Game Fish Association)

91. The record for rainfall during a 24-hour period in Alaska is $15\frac{1}{5}$ inches. This record was set in Angoon, Alaska, in October 1982. How much rain fell per hour on average? (*Source:* National Climatic Data Center)

92. The longest floating pontoon bridge in the United States is the Evergreen Point Bridge in Seattle, Washington. It is 2526 yards long. The second-longest pontoon bridge in the United States is the Hood Canal Bridge in Point Gamble, Washington which is $2173\frac{2}{3}$ yards long. How much longer is the Evergreen Point Bridge than the Hood Canal Bridge? (*Source:* Federal Highway Administration)

G *Perform the indicated operations. See Examples 19 through 26.*

93. $-4\frac{2}{5} \cdot 2\frac{3}{10}$

94. $-3\frac{5}{6} \div \left(-3\frac{2}{3}\right)$

95. $-5\frac{1}{8} - 19\frac{3}{4}$

96. $17\frac{5}{9} + \left(-14\frac{2}{3}\right)$

97. $-31\frac{2}{15} + 17\frac{3}{20}$

98. $-1\frac{5}{7} \cdot \left(-2\frac{1}{2}\right)$

99. $1\frac{3}{4} \div \left(-3\frac{1}{2}\right)$

100. $-31\frac{7}{8} - \left(-26\frac{5}{12}\right)$

101. $11\frac{7}{8} - 13\frac{5}{6}$

102. $-20\frac{2}{5} + \left(-30\frac{3}{10}\right)$

103. $-7\frac{3}{10} \div (-100)$

104. $-4\frac{1}{4} \div 2\frac{3}{8}$

105. $-3\frac{1}{6} \cdot \left(-2\frac{3}{4}\right)$

106. $42\frac{2}{9} - 50\frac{1}{3}$

107. $-21\frac{5}{12} + \left(-10\frac{3}{4}\right)$

108. $4\frac{1}{3} \cdot \left(-5\frac{1}{4}\right)$

Review and Preview

Simplify by combining like terms. See Section 1.9.

109. $2x - 5 + 7x - 8$

110. $11 + 8y - 5 - 11y$

111. $3(y - 2) - 6y$

112. $8(3 + 2y) - 20$

Simplify. See Section 1.8.

113. $2^3 + 3^2$

114. $8 \cdot 5 + 5 \div 5$

115. $\dfrac{7 - 3}{2^2}$

116. $4^2 - 2^4$

Combining Concepts

Solve.

117. Carmen's Candy Clutch is famous for its "Nutstuff," a special blend of nuts and candy. A Supreme box of Nutstuff has $2\frac{1}{4}$ pounds of nuts and $3\frac{1}{2}$ pounds of candy. A Deluxe box has $1\frac{3}{8}$ pounds of nuts and $4\frac{1}{4}$ pounds of candy. Which box is heavier and by how much?

118. Willie Cassidie purchased three Supreme boxes and two Deluxe boxes of Nutstuff from Carmen's Candy Clutch. What is the total weight of his purchase? (See Exercise 117.)

119. Explain in your own words why $9\frac{13}{9}$ is equal to $10\frac{4}{9}$.

120. In your own words, explain how to borrow when subtracting mixed numbers.

121. Find two mixed numbers whose sum is a whole number.

122. Find two mixed numbers whose difference is a whole number.

Internet Excursions

Go To: 

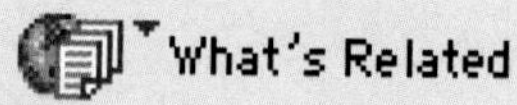

This World Wide Web site will provide access to a site called the Low Fat Vegetarian Archive, or a related site. It contains more than 2500 recipes for healthy low-fat and fat-free vegetarian dishes. Users can search the recipes by category, such as Mexican foods or desserts, or by ingredient.

123. Visit this site and find a recipe that is interesting to you. Be sure that the ingredient list contains at least two measures that are fractions (for example, $\frac{1}{3}$ cup). Print or copy the recipe. Then rewrite the list of ingredients to show the amounts that would be needed to make $\frac{1}{2}$ of the recipe.

124. Find a different recipe at this site that is interesting to you. Make sure that the recipe lists at least two fractional measures. Print or copy the recipe. Then rewrite the list of ingredients to show the amounts that would be needed to triple the recipe.

Chapter 4 Vocabulary Check

Fill in each blank with one of the words or phrases listed below.

prime factorization	simplest form	fraction	prime	equivalent	least common multiple	like
mixed number	improper	composite	reciprocals	numerator	proper	denominator

1. Fractions that have the same denominator are called _____ fractions.
2. The ____________________ is the smallest number that is a multiple of all numbers in a list of numbers.
3. _________ fractions represent the same portion of a whole.
4. A _____________ has a whole number part and a fraction part.
5. A __________ number is a whole number greater than 1 that is not prime.
6. A ________ is a number of the form $\frac{a}{b}$ where a and b are integers and b is not 0. The number a is called the _________ and the number b is called the ___________ .
7. An _________ fraction's numerator is greater than or equal to its denominator.
8. A _______ number is a whole number greater than 1 whose only divisions are 1 and itself.
9. A _______ fraction's numerator is less than its denominator.
10. Two numbers are ___________ of each other if their product is 1.
11. A fraction is in ____________ when the numerator and denominator have no common factors other than 1.
12. The ________________ of a number is that number written as a product of prime numbers.

CHAPTER 4 Highlights

DEFINITIONS AND CONCEPTS	**EXAMPLES**
SECTION 4.1 INTRODUCTION TO FRACTIONS AND EQUIVALENT FRACTIONS	
A **fraction** is a number of the form $\frac{a}{b}$, where a and b are integers and b is not 0. The number a is the **numerator** of the fraction, and the number b is the **denominator** of the fraction.	
A fraction is called a **proper fraction** if its numerator is less than its denominator.	Proper Fractions: $\frac{1}{3}, \frac{2}{5}, \frac{7}{8}, \frac{100}{101}$
A fraction is called an **improper fraction** if its numerator is greater than or equal to its denominator.	Improper Fractions: $\frac{5}{4}, \frac{2}{2}, \frac{9}{7}, \frac{101}{100}$
Fractions that represent the same portion of a whole are called **equivalent fractions**.	
FUNDAMENTAL PROPERTY OF FRACTIONS If a, b, and c are numbers, then $\frac{a}{b} = \frac{a \cdot c}{b \cdot c}$ and also $\frac{a}{b} = \frac{a \div c}{b \div c}$ as long as b and c are not 0.	Write $\frac{3}{7}$ as an equivalent fraction with a denominator of 42. $\frac{3}{7} = \frac{3 \cdot 6}{7 \cdot 6} = \frac{18}{42}$

Definitions and Concepts	**Examples**
Section 4.2 Factors and Simplest Form	
A **prime number** is a whole number greater than 1 whose only divisors are 1 and itself. A **composite number** is a whole number greater than 1 that is not prime. Every whole number greater than 1 has exactly one prime factorization.	Prime Numbers: 2, 3, 5, 7, 11, 13, 17, . . . Composite Numbers: 4, 6, 8, 9, 10, 12, 14, 15, 16, . . . Write the prime factorization of 60. $60 = 6 \cdot 10$ $= 2 \cdot 3 \cdot 2 \cdot 5$ or $2^2 \cdot 3 \cdot 5$
A fraction is in **simplest form** or **lowest terms** when the numerator and the denominator have no common factors other than 1.	Write $\frac{30x}{36x}$ in simplest form. $\frac{30x}{36x} = \frac{2 \cdot 3 \cdot 5 \cdot x}{2 \cdot 2 \cdot 3 \cdot 3 \cdot x} = \frac{5}{6}$
Section 4.3 Multiplying and Dividing Fractions	
To **multiply** two fractions, $\frac{a}{b} \cdot \frac{c}{d} = \frac{a \cdot c}{b \cdot d}, b \neq 0, d \neq 0$	Multiply: $\frac{2x}{3} \cdot \frac{5}{7} = \frac{2x \cdot 5}{3 \cdot 7} = \frac{10x}{21}$
Two numbers are **reciprocals** of each other if their product is 1.	The reciprocal of $\frac{3}{5}$ is $\frac{5}{3}$.
To **divide** two fractions, $\frac{a}{b} \div \frac{c}{d} = \frac{a}{b} \cdot \frac{d}{c} = \frac{a \cdot d}{b \cdot c}, b \neq 0, c \neq 0, d \neq 0$	Divide: $-\frac{3}{4} \div \frac{3}{8} = -\frac{3}{4} \cdot \frac{8}{3}$ $= -\frac{3 \cdot 2 \cdot 4}{4 \cdot 3} = -2$
Section 4.4 Adding and Subtracting Like Fractions and Least Common Denominator	
Fractions that have a common denominator are called **like fractions**.	Like Fractions: $-\frac{1}{3}$ and $\frac{2}{3}$; $\frac{5x}{7}$ and $\frac{6}{7}$
To add or subtract like fractions, $\frac{a}{b} + \frac{c}{b} = \frac{a + c}{b}, \frac{a}{b} - \frac{c}{b} = \frac{a - c}{b}, b \neq 0$	$\frac{10}{11} + \frac{3}{11} - \frac{8}{11} = \frac{10 + 3 - 8}{11} = \frac{5}{11}$.
The **least common denominator (LCD)** of a list of fractions is the smallest positive number divisible by all the denominators in the list.	Find the LCD of $\frac{7}{30}$ and $\frac{1}{12}$. $30 = 2 \cdot 3 \cdot 5$ $12 = 2 \cdot 2 \cdot 3$ Notice that the greatest number of times that 2 appears is 2 times. LCD $= 2 \cdot 2 \cdot 3 \cdot 5 = 60$
Section 4.5 Adding and Subtracting Unlike Fractions	
To Add or Subtract Unlike Fractions	Add: $\frac{3}{20} + \frac{2}{5}$
Step 1. Find the LCD.	**Step 1.** The LCD is 20.

Definitions and Concepts | Examples

Section 4.5 Adding and Subtracting Unlike Fractions (continued)

Step 2. Write equivalent fractions with the LCD as the denominator.

Step 3. Add or subtract the like fractions.

Step 4. Write the result in simplest form.

Step 2. $\frac{2}{5} = \frac{2 \cdot 4}{5 \cdot 4} = \frac{8}{20}$.

Step 3. $\frac{3}{20} + \frac{2}{5} = \frac{3}{20} + \frac{8}{20} = \frac{11}{20}$.

Step 4. $\frac{11}{20}$ is in simplest form.

Section 4.6 Complex Fractions and Review of Order of Operations

A fraction whose numerator or denominator or both contain fractions is called a **complex fraction**.

One method for simplifying complex fractions is to multiply the numerator and the denominator of the complex fraction by the LCD of all fractions in its numerator and its denominator.

Complex Fractions:

$$\frac{\frac{x}{4}}{\frac{7}{10}}, \quad \frac{\frac{y}{6} - 11}{\frac{4}{3}}$$

$$\frac{\frac{y}{6} - 11}{\frac{4}{3}} = \frac{6\left(\frac{y}{6} - 11\right)}{6\left(\frac{4}{3}\right)} = \frac{6\left(\frac{y}{6}\right) - 6(11)}{6\left(\frac{4}{3}\right)}$$

$$= \frac{y - 66}{8}$$

Section 4.7 Solving Equations Containing Fractions

Multiplication and Division Properties of Equality

Let a, b, and c be numbers, and $c \neq 0$.

If $a = b$, then $a \cdot c = b \cdot c$ and also $\frac{a}{c} = \frac{b}{c}$.

To Solve an Equation in x

Step 1. If fractions are present, multiply both sides of the equation by the LCD of the fractions.

Step 2. If parentheses are present, use the distributive property.

Step 3. Combine any like terms on each side of the equation.

Step 4. Use the addition property of equality to rewrite the equation so that variable terms are on one side of the equation and constant terms are on the other side.

Step 5. Divide both sides by the numerical coefficient of x to solve.

Step 6. Check the answer in the *original equation*.

Solve: $\frac{x}{15} + 2 = \frac{7}{3}$

$$15\left(\frac{x}{15} + 2\right) = 15\left(\frac{7}{3}\right) \quad \text{Multiply by the LCD 15.}$$

$$15\left(\frac{x}{15}\right) + 15 \cdot 2 = 15\left(\frac{7}{3}\right)$$

$$x + 30 = 35$$

$$x + 30 + (-30) = 35 + (-30)$$

$$x = 5$$

Check to see that 5 is the solution.

Definitions and Concepts | Examples

Section 4.8 Operations on Mixed Numbers

A **mixed number** is the sum of a whole number and a proper fraction.

Mixed Numbers: $1\frac{2}{5}$, $4\frac{7}{8}$

Writing a Mixed Number as an Improper Fraction

Step 1. Multiply the denominator of the fraction by the whole-number part. Add the numerator of the fraction to this product.

Step 2. Write this sum as the numerator of the improper fraction over the original denominator.

$$4\frac{7}{8} = \frac{8 \cdot 4 + 7}{8} = \frac{39}{8}$$

Writing an Improper Fraction as a Mixed Number or Whole Number

Step 1. Divide the denominator into the numerator.

Step 2. The whole-number part of the quotient is the whole-number part of the mixed number and the $\frac{\text{remainder}}{\text{divisor}}$ is the fractional part.

$$\frac{29}{7} = 4\frac{1}{7}$$

$$\begin{array}{r} 4 \\ 7\overline{)29} \\ -28 \\ \hline 1 \end{array}$$

To perform operations on mixed numbers, first write each mixed number as an improper fraction.

Multiply: $1\frac{3}{4} \cdot 2\frac{1}{5}$

$$\frac{7}{4} \cdot \frac{11}{5} = \frac{77}{20} = 3\frac{17}{20}$$

Addition and subtraction of mixed numbers can be performed using a vertical format.

Add: $2\frac{1}{2} + 5\frac{7}{8}$

$$\begin{array}{rcl} 2\frac{1}{2} & = & 2\frac{4}{8} \\ +5\frac{7}{8} & = & 5\frac{7}{8} \\ \hline & & 7\frac{11}{8} = 7 + 1\frac{3}{8} = 8\frac{3}{8} \end{array}$$

Name______________________ Section__________ Date__________

Chapter 4 Review

(4.1) *Represent the shaded area of each figure with a fraction.*

1. 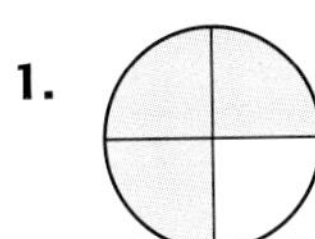

2.

Graph each fraction on a number line.

3. $\frac{7}{9}$

4. $\frac{4}{7}$

5. $\frac{5}{4}$

6. $\frac{7}{5}$

7. The United States Army offers 242 job specialties to its soldiers. Certain specialties, such as infantry and field artillery, are closed to women. There are 207 army job specialties open to women and 35 army job specialties closed to women. What fraction of army job specialties are closed to women? (*Source:* U.S. Department of the Army)

8. In the United States Armed Forces today, roughly 43 out of every 50 personnel are men. What fraction of U.S. Armed Forces are men? (*Source:* U.S. Department of Defense)

Write each fraction as an equivalent fraction with the given denominator.

9. $\frac{2}{3} = \frac{?}{30}$

10. $\frac{5}{8} = \frac{?}{56}$

11. $\frac{7a}{6} = \frac{?}{42}$

12. $\frac{9b}{4} = \frac{?}{20}$

13. $\frac{4}{5x} = \frac{?}{50x}$

14. $\frac{5}{9y} = \frac{?}{18y}$

(4.2) *Write each fraction in simplest form.*

15. $\frac{12}{28}$

16. $\frac{15}{27}$

17. $\frac{25x}{75x^2}$

18. $\frac{36y^3}{72y}$

19. $\frac{29ab}{32abc}$

20. $\frac{18xyz}{23xy}$

21. $\frac{45x^2y}{27xy^3}$

22. $\frac{42ab^2c}{30abc^3}$

23. There are 12 inches in a foot. What fractional part of a foot does 8 inches represent?

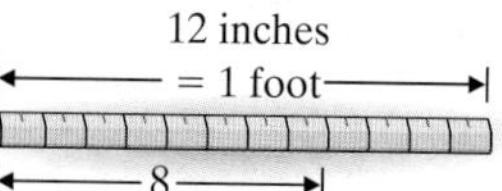

24. Six out of 15 cars are white. What fraction of cars are *not* white?

(4.3) *Multiply.*

25. $\frac{3}{5} \cdot \frac{1}{2}$

26. $-\frac{6}{7} \cdot \frac{5}{12}$

27. $\frac{7}{8x} \cdot -\frac{2}{3}$

28. $\frac{6}{15} \cdot \frac{5y}{8}$

29. $-\frac{24x}{5} \cdot -\frac{15}{8x^3}$

30. $\frac{27y^3}{21} \cdot \frac{7}{18y^2}$

31. $\left(-\frac{1}{3}\right)^3$

32. $\left(-\frac{5}{12}\right)^2$

33. $\frac{x^3}{y} \cdot \frac{y^3}{x}$

34. $\frac{ac}{b} \cdot \frac{b^2}{a^3c}$

35. Evaluate xy if $x = \frac{2}{3}$ and $y = \frac{1}{5}$.

36. Evaluate ab if $a = -7$ and $b = \frac{9}{10}$.

Divide.

37. $-\frac{3}{4} \div \frac{3}{8}$

38. $\frac{21a}{4} \div \frac{7a}{5}$

39. $\frac{18x}{5} \div \frac{2}{5x}$

40. $-\frac{9}{2} \div -\frac{1}{3}$

41. $-\frac{5}{3} \div 2y$

42. $\frac{5x^2}{y} \div \frac{10x^3}{y^3}$

43. Evaluate $x \div y$ if $x = \frac{9}{7}$ and $y = \frac{3}{4}$.

44. Evaluate $a \div b$ if $a = -5$ and $b = \frac{2}{3}$.

Find the area of each figure.

△ **45.**

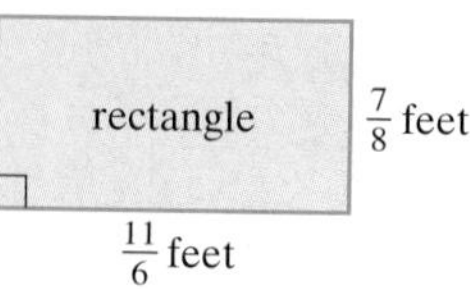

△ **46.**

square $\frac{2}{3}$ meter

(4.4) *Add or subtract as indicated.*

47. $\frac{7}{11} + \frac{3}{11}$

48. $\frac{4}{9} + \frac{2}{9}$

49. $\frac{1}{12} - \frac{5}{12}$

50. $\frac{3}{y} - \frac{1}{y}$

51. $\frac{11x}{15} + \frac{x}{15}$

52. $\frac{4y}{21} - \frac{3}{21}$

53. $\frac{4}{15} + \frac{3}{15} - \frac{2}{15}$

54. $\frac{4}{15} - \frac{3}{15} - \frac{2}{15}$

Find the LCD of each list of fractions.

55. $\frac{2}{3}, \frac{5}{x}$

56. $\frac{3}{4}, \frac{3}{8}, \frac{7}{12}$

57. Determine whether $\frac{4}{5}$ is a solution of $z + \frac{1}{5} = 1$.

58. Determine whether $\frac{3}{4}$ is a solution of $x - \frac{2}{4} = \frac{1}{4}$.

Solve.

59. If a student studies math for $\frac{3}{8}$ of an hour and geography for $\frac{1}{8}$ of an hour, find how long she studies.

60. Beryl Goldstein mixed $\frac{5}{8}$ of a gallon of water with $\frac{1}{8}$ of a gallon of punch concentrate. Then she and her friends drank $\frac{3}{8}$ of a gallon of the punch. Find how much was left.

61. One evening Mark Alorenzo did $\frac{3}{8}$ of his homework before supper, another $\frac{2}{8}$ of it while his children did their homework, and $\frac{1}{8}$ of it after his children went to bed. Find what part of his homework he did that day.

△ **62.** The Simpsons will be fencing in their land. To do this, they need to find its perimeter. Find the perimeter of their land given that it is the shape of a rectangle.

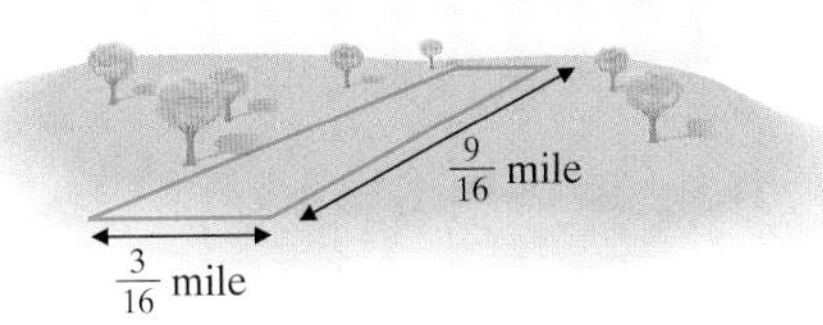

(4.5) *Add or subtract as indicated.*

63. $\frac{7}{18} + \frac{2}{9}$

64. $\frac{4}{13} - \frac{1}{26}$

65. $-\frac{1}{3} + \frac{1}{4}$

66. $-\frac{2}{3} + \frac{1}{4}$

67. $\frac{5x}{11} + \frac{2}{55}$

68. $\frac{4}{15} + \frac{b}{5}$

69. $\frac{5y}{12} - \frac{2y}{9}$

70. $\frac{7x}{18} + \frac{2x}{9}$

71. $\frac{4}{9} + \frac{5}{y}$

72. $-\frac{9}{14} - \frac{3}{7}$

73. $\frac{4}{25} + \frac{23}{75} + \frac{7}{50}$

74. $\frac{2}{3} - \frac{2}{9} - \frac{1}{6}$

Solve each equation.

75. $a - \frac{2}{3} = \frac{1}{6}$

76. $9x + \frac{1}{5} - 8x = -\frac{7}{10}$

Find the perimeter of each figure.

△ 77.

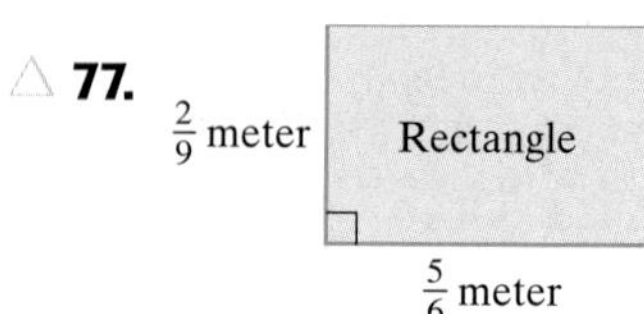

△ 78.

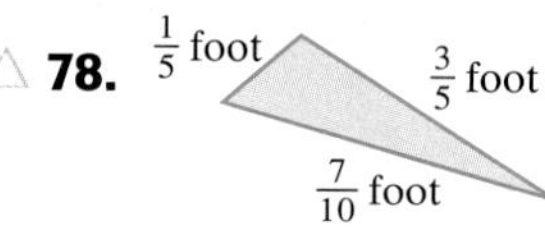

79. Determine whether $\frac{8}{11}$ is a solution of $x + \frac{1}{3} = \frac{35}{11}$.

80. Determine whether $\frac{9}{2}$ is a solution of $\frac{1}{9}y - \frac{1}{4} = \frac{1}{4}$.

81. In a group of 100 blood donors, typically $\frac{9}{25}$ have type A Rh-positive blood and $\frac{3}{50}$ have type A Rh-negative blood. What fraction have type A blood?

82. Find the difference in length of two scarves if one scarf is $\frac{5}{12}$ of a yard long and the other is $\frac{2}{3}$ of a yard long.

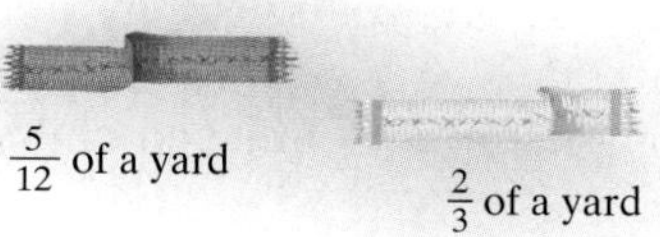

(4.6) *Simplify each complex fraction.*

83. $\dfrac{\frac{2x}{5}}{\frac{7}{10}}$

84. $\dfrac{\frac{3y}{7}}{\frac{11}{7}}$

85. $\dfrac{2 + \frac{3}{4}}{1 - \frac{1}{8}}$

86. $\dfrac{\frac{5}{6}+2}{\frac{11}{3}-1}$

87. $\dfrac{\frac{2}{5}-\frac{1}{2}}{\frac{3}{4}-\frac{7}{10}}$

88. $\dfrac{\frac{5}{6}-\frac{1}{4}}{\frac{-1}{12y}}$

Evaluate each expression if $x = \frac{1}{2}$, $y = -\frac{2}{3}$, *and* $z = \frac{4}{5}$.

89. $2x + y$

90. $\dfrac{x}{y + z}$

91. $\dfrac{x + y}{z}$

92. $x + y + z$

93. y^2

94. $x - z$

(4.7) *Solve.*

95. $-\frac{3}{5}x = 6$

96. $\frac{2}{9}y = -\frac{4}{3}$

97. $\frac{x}{7} - 3 = -\frac{6}{7}$

98. $\frac{y}{5} + 2 = \frac{11}{5}$

99. $\frac{1}{6} + \frac{x}{4} = \frac{17}{12}$

100. $\frac{x}{5} - \frac{5}{4} = \frac{x}{2} - \frac{1}{20}$

(4.8) *Write each improper fraction as a mixed number or a whole number.*

101. $\frac{15}{4}$

102. $\frac{39}{13}$

103. $\frac{7}{7}$

104. $\frac{125}{4}$

Write each mixed or whole number as an improper fraction.

105. $2\frac{1}{5}$

106. 5

107. $3\frac{8}{9}$

108. 3

Perform each indicated operation.

109. $\begin{array}{r} 31\frac{2}{7} \\ +14\frac{10}{21} \\ \hline \end{array}$

110. $\begin{array}{r} 24\frac{4}{5} \\ +35\frac{1}{5} \\ \hline \end{array}$

111. $\begin{array}{r} 69\frac{5}{22} \\ -36\frac{7}{11} \\ \hline \end{array}$

112. $\begin{array}{r} 36\frac{3}{20} \\ -32\frac{5}{6} \\ \hline \end{array}$

113. $\begin{array}{r} 29\frac{2}{9} \\ 27\frac{7}{18} \\ +54\frac{2}{3} \\ \hline \end{array}$

114. $\begin{array}{r} 7\frac{3}{8} \\ 9\frac{5}{6} \\ +3\frac{1}{12} \\ \hline \end{array}$

115. $1\frac{5}{8} \cdot \frac{2}{3}$

116. $3\frac{6}{11} \cdot \frac{5}{13}$

117. $4\frac{1}{6} \cdot 2\frac{2}{5}$

118. $5\frac{2}{3} \cdot 2\frac{1}{4}$

119. $6\frac{3}{4} \div 1\frac{2}{7}$

120. $5\frac{1}{2} \div 2\frac{1}{11}$

121. $\frac{7}{2} \div 1\frac{1}{2}$

122. $1\frac{3}{5} \div \frac{1}{4}$

123. Two packages of soup bones weigh $3\frac{3}{4}$ pounds and $2\frac{3}{5}$ pounds. Find their comined weight.

124. A ribbon $5\frac{1}{2}$ yards long is cut from a reel of ribbon with 50 yards on it. Find the length of the piece remaining on the reel.

125. The average annual snowfall at a certain ski resort is $62\frac{3}{10}$ inches. Last year it had $54\frac{1}{2}$ inches. Find how many inches below average last year's annual snowfall was.

△ **126.** Find the area of a rectangular sheet of gift wrap that is $2\frac{1}{4}$ feet by $3\frac{1}{3}$ feet.

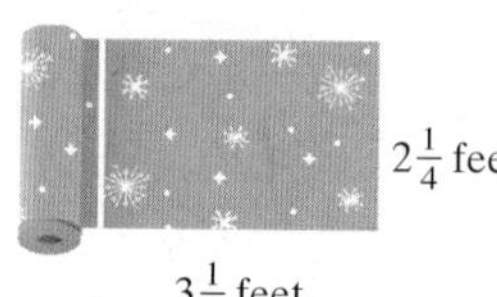

△ **127.** Find the perimeter of a sheet of shelf paper needed to fit exactly a square drawer $1\frac{1}{4}$ feet long on each side.

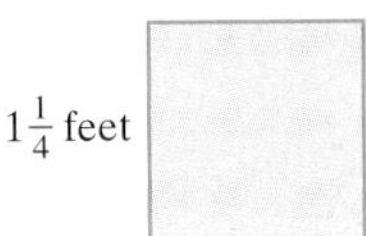

△ **128.** Find the area of the rectangle.

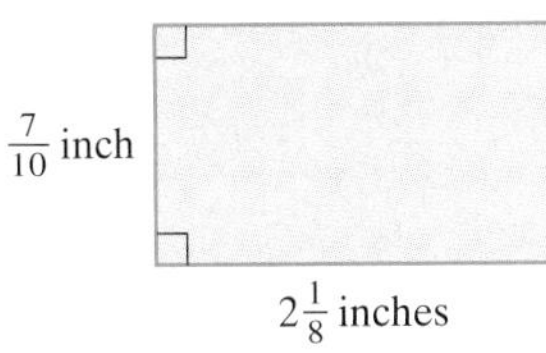

△ **129.** A flower gardener has a square flower bed $\frac{2}{3}$ meter on a side and a rectangular one that is $\frac{7}{8}$ meter by $\frac{1}{3}$ meter. Find the total perimeter of his flower beds.

130. There are 58 calories in 1 ounce of turkey. Find how many calories there are in a $3\frac{1}{2}$-ounce serving of turkey.

131. There are $3\frac{1}{3}$ grams of fat in each ounce of lean hamburger. Find how many grams of fat are in a 4-ounce hamburger.

132. Herman Heltznutt walks 5 days a week for a total distance of $5\frac{1}{4}$ miles per week. If he walks the same distance each day, find the distance he walks each day.

Perform the indicated operations.

133. $-12\frac{1}{7} + \left(-15\frac{3}{14}\right)$

134. $-3\frac{1}{5} \div \left(-2\frac{7}{10}\right)$

135. $-2\frac{1}{4} \cdot 1\frac{3}{4}$

136. $23\frac{7}{8} - 24\frac{7}{10}$

STUDY SKILLS REMINDER

Are you prepared for a test on Chapter 4?

Below I have listed some *common trouble areas* for topics covered in Chapter 4. After studying for your test—but before taking your test—read these.

Make sure you remember how to perform different operations on fractions!!! Try to add, subtract, multiply, then divide $\frac{3}{5}$ and $\frac{7}{15}$. Check your results below.

$$\frac{3}{5}+\frac{7}{15}=\frac{3\cdot 3}{5\cdot 3}+\frac{7}{15}=\frac{9}{15}+\frac{7}{15}=\frac{16}{15}=1\frac{1}{15}$$

To add or subtract, you need common denominators.

$$\frac{3}{5}-\frac{7}{15}=\frac{3\cdot 3}{5\cdot 3}-\frac{7}{15}=\frac{9}{15}-\frac{7}{15}=\frac{2}{15}$$

$$\frac{3}{5}\cdot\frac{7}{15}=\frac{3\cdot 7}{5\cdot 15}=\frac{\overset{1}{\cancel{3}}\cdot 7}{5\cdot \underset{1}{\cancel{3}}\cdot 5}=\frac{7}{25}$$

$$\frac{3}{5}\div\frac{7}{15}=\frac{3}{5}\cdot\frac{15}{7}=\frac{3\cdot 3\cdot \overset{1}{\cancel{5}}}{\underset{1}{\cancel{5}}\cdot 7}=\frac{9}{7}=1\frac{2}{7}$$

To divide, multiply by the reciprocal.

Name ______________________ Section ____________ Date ____________

Answers

Chapter 4 Test

Remember to check your answers and use the Chapter Test Prep Video to view solutions.

Write each mixed number as an improper fraction.

1. $7\frac{2}{3}$

2. $3\frac{6}{11}$

Write each improper fraction as a mixed number or a whole number.

3. $\frac{23}{5}$

4. $\frac{75}{4}$

Write each fraction in simplest form.

5. $\frac{54}{210}$

6. $-\frac{42}{70}$

Perform each indicated operation and write the answers in simplest form.

7. $\frac{4}{4} \div \frac{3}{4}$

8. $-\frac{4}{3} \cdot \frac{4}{4}$

9. $\frac{7x}{9} + \frac{x}{9}$

10. $\frac{1}{7} - \frac{3}{x}$

11. $\frac{xy^3}{z} \cdot \frac{z}{xy}$

12. $-\frac{2}{3} \cdot -\frac{8}{15}$

13. $\frac{9a}{10} + \frac{2}{5}$

14. $-\frac{8}{15y} - \frac{2}{15y}$

15. $8y^3 \div \frac{y}{3}$

1. ____________
2. ____________
3. ____________
4. ____________
5. ____________
6. ____________
7. ____________
8. ____________
9. ____________
10. ____________
11. ____________
12. ____________
13. ____________
14. ____________
15. ____________

16. ______

17. ______

18. ______

19. ______

20. ______

21. ______

22. ______

23. ______

24. ______

25. ______

26. ______

27. ______

28. ______

29. ______

30. ______

16. $5\frac{1}{4} \div \frac{7}{12}$

17. $\begin{array}{r} 3\frac{7}{8} \\ 7\frac{2}{5} \\ +2\frac{3}{4} \\ \hline \end{array}$

18. $\frac{3a}{8} \cdot \frac{16}{6a^3}$

19. $-\frac{16}{3} \div -\frac{3}{12}$

20. $3\frac{1}{3} \cdot 6\frac{3}{4}$

21. $12 \div 3\frac{1}{3}$

22. $\left(\frac{14}{5} \cdot \frac{25}{21}\right) \div 10$

23. $\frac{11}{12} - \frac{3}{8} + \frac{5}{24}$

Simplify each complex fraction.

24. $\dfrac{\frac{5x}{7}}{\frac{20x^2}{21}}$

25. $\dfrac{5 + \frac{3}{7}}{2 - \frac{1}{2}}$

Solve.

26. $-\frac{3}{8}x = \frac{3}{4}$

27. $\frac{x}{5} + x = -\frac{24}{5}$

28. $\frac{2}{3} + \frac{x}{4} = \frac{5}{12} + \frac{x}{2}$

Evaluate each expression for the given replacement values.

29. $-5x$; $x = -\frac{1}{2}$

30. $x \div y$; $x = \frac{1}{2}$, $y = 3\frac{7}{8}$

Solve.

31. A McDonald's Big Mac® sandwich has 560 calories. There are 280 calories from fat in a Big Mac. What fraction of a Big Mac's calories are from fat? (*Source:* McDonald's Corporation)

32. A carpenter cuts a piece $2\frac{3}{4}$ feet long from a cedar plank that is $6\frac{1}{2}$ feet long. How long is the remaining piece?

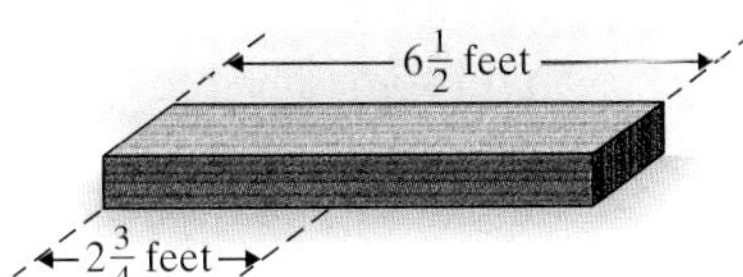

As shown in the circle graph, the market for backpacks is divided among five companies. For instance, Wilderness, Inc.'s backpack accounts for $\frac{1}{4}$ of all backpack sales. Use this graph to answer Questions 33 and 34.

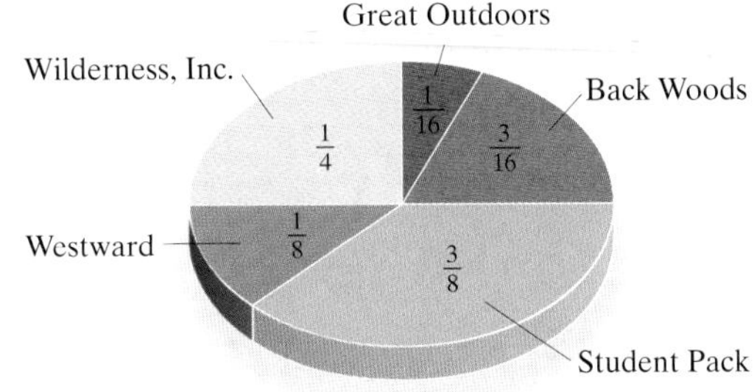

33. What fraction of backpack sales goes to Back Woods and Westward combined?

34. If a total of 500,000 backpacks are sold each year, how many backpacks does Wilderness, Inc. sell?

△ **35.** How many square yards of artificial turf are necessary to cover a football field, including the end zones and 10 yards beyond the side lines? (*Hint*: A football field measures $100 \times 53\frac{1}{3}$ yards and the end zones are 10 yards deep.)

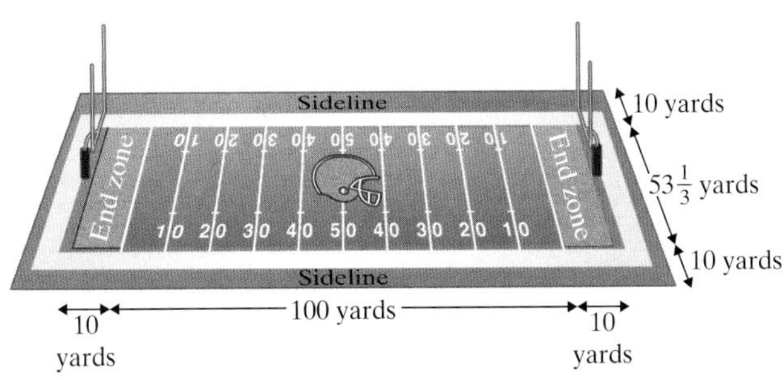

31. ____________

32. ____________

33. ____________

34. ____________

35. ____________

36. ______

37. ______

38. ______

△ **36.** Find the area of the figure.

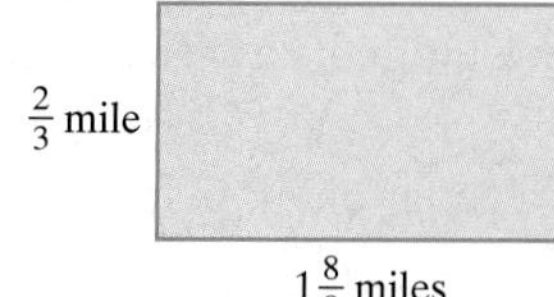

37. During a 258-mile trip, a car used $10\frac{3}{4}$ gallons of gas. How many miles could we expect the car to travel on one gallon of gas?

38. Prior to an oil spill, the stock in an oil company sold for $120 per share. As a result of the liability that the company incurred from the spill, the price per share fell to $\frac{3}{4}$ of the price before the spill. What did the stock sell for after the spill?

Name ____________ Section ________ Date ____________

Chapter 4 Cumulative Review

Write each number in words.

1. 126

2. 115

3. 3005

4. 6573

5. Add: $23 + 136$

6. Add: $587 + 44$

7. Subtract: $43 - 29$. Check by adding.

8. Subtract: $995 - 62$. Check by adding.

9. Round 278,362 to the nearest thousand.

10. Round 1436 to the nearest ten.

11. A certain computer disk can hold about 1510 thousand bytes of information. How many total bytes can 42 such disks hold?

12. On a trip across country, Daniel Daunis travels 435 miles per day. How many total miles does he travel in 3 days?

13. Divide and check: $56{,}717 \div 8$

14. Divide and check: $4558 \div 12$

Answers

1. ____________
2. ____________
3. ____________
4. ____________
5. ____________
6. ____________
7. ____________
8. ____________
9. ____________
10. ____________
11. ____________
12. ____________
13. ____________
14. ____________

15. ______

16. ______

17. ______

18. ______

19. ______

20. ______

21. ______

22. ______

23. ______

24. ______

25. ______

26. ______

27. ______

28. ______

29. ______

30. ______

Write using exponential notation.

15. $4 \cdot 4 \cdot 4$

16. $7 \cdot 7$

17. $6 \cdot 6 \cdot 6 \cdot 8 \cdot 8 \cdot 8 \cdot 8 \cdot 8$

18. $9 \cdot 9 \cdot 9 \cdot 9 \cdot 5 \cdot 5$

19. Evaluate $2(x - y)$ if $x = 8$ and $y = 4$.

20. Evaluate $8a + 3(b - 5)$ if $a = 5$ and $b = 9$.

21. Jack Mayfield, a miner for the Molly Kathleen Gold Mine, is presently 150 feet below the surface of the Earth. Represent this position using an integer.

22. The temperature on a cold day in Minneapolis, MN is 21° F below zero. Represent this temperature using an integer.

23. Add using a number line: $-7 + 3$

24. Add using a number line: $-3 + 8$

25. Simplify: $7 - 8 - (-5) - 1$

26. Simplify: $6 + (-8) - (-9) + 3$

27. Evaluate: $(-5)^2$

28. Evaluate: -2^4

29. Simplify: $\dfrac{12 - 16}{-1 + 3}$

30. Simplify: $(20 - 5^2)^2$

Multiply.

31. $5(3y)$

32. $12(3c)$

33. $-2(4x)$

34. $-7(14a)$

35. Solve: $-8 = x + 1$

36. Solve: $x - 3 = 5$

37. Solve: $8x - 9x = 12 - 17$

38. Solve: $2x + 5x = 0 - 7$

39. Translate each sentence into an equation.

a. The product of 7 and 6 is 42.
b. Twice the sum of 3 and 5 is equal to 16.
c. The quotient of -45 and 5 yields -9.

40. Translate each sentence into an equation:

a. The sum of 4 and 3 is 7.
b. Four times the difference of 5 and 2 equals 12.
c. The product of -4 and -6 yields 24.

41. Twice a number, added to 3, is the same as the number minus 6. Find the number.

42. A number is tripled, then added to 4. The result is equal to the original number minus 6. Find the number.

43. Write $\frac{9x}{11}$ as an equivalent fraction whose denominator is 44.

44. Write $\frac{2a}{3}$ as an equivalent fraction whose denominator is 15.

45. Write the prime factorization of 45.

46. Write the prime factorization of 92.

31. ______
32. ______
33. ______
34. ______
35. ______
36. ______
37. ______
38. ______
39. a. ______
b. ______
c. ______
40. a. ______
b. ______
c. ______
41. ______
42. ______
43. ______
44. ______
45. ______
46. ______

47.

48.

49.

50.

Multiply.

47. $\frac{2}{3} \cdot \frac{5}{11}$

48. $\frac{1}{7} \cdot \frac{2}{5}$

49. $\frac{1}{4} \cdot \frac{1}{2}$

50. $\frac{3}{5} \cdot \frac{1}{5}$

Decimals

C H A P T E R 5

Decimals are an important part of everyday life. For example, we use decimal numbers in our money system. One penny is 0.01 dollar, and one dime—or ten pennies, is 0.10 dollar. Among many other uses, decimals also express batting averages. A baseball player with a 0.333 batting average is a pretty good batter. Decimal numbers represent parts of a whole, just like fractions. In this chapter, we analyze the relationship between fractions and decimals.

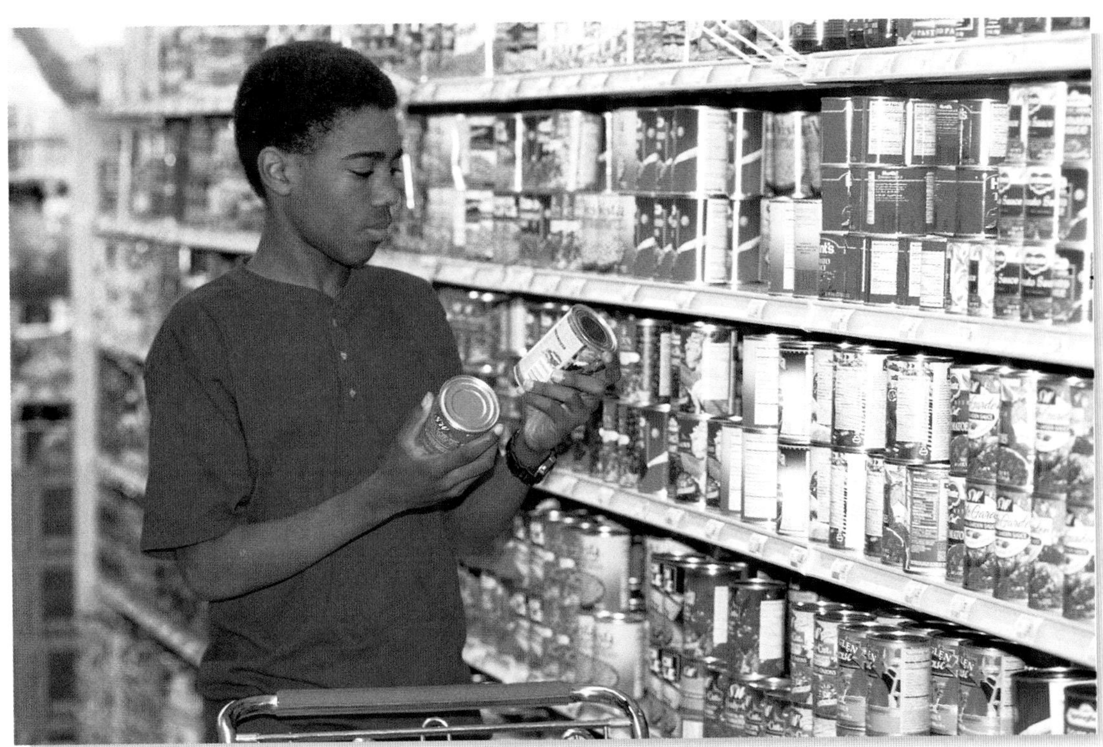

The Nutrition Labeling and Education Act (NLEA) was signed into law on November 8, 1990. It requires food manufacturers to include nutrition information on their product labels. The NLEA provides specific guidelines concerning the use of terms such as "low fat," or "high fiber." Labels contain information about portion sizes, vitamins and minerals, and sodium, fat, and cholesterol content of foods. The result of this important legislation is to help consumers make more informed and healthier food choices. In Exercise 37 on page 412, we will see how comparing decimals can help consumers to determine if a product can truly be called "fat free."

Answers

1. ______
2. ______
3. ______
4. ______
5. ______
6. ______
7. ______
8. ______
9. ______
10. ______
11. ______
12. ______
13. ______
14. ______
15. ______
16. ______
17. ______
18. ______
19. ______
20. ______

Name ______________ Section ______ Date ______________

Chapter 5 Pretest

1. Write 0.27 as a fraction.`

2. Insert <, >, or = to form a true statement. 0.205 0.213

3. Round 54.651 to the nearest tenth.

Perform each indicated operation.

4. $38.41 + 14.032 + 7.6$

5. $(-3.4)(-2.1)$

6. $(2.016)(100)$

7. $16.24 \div 0.4$

8. $\frac{891}{10{,}000}$

9. Evaluate $x - y$ for $x = 12.3$ and $y = 0.61$.

10. Simplify: $-9.8 - 6.2x - 7.9 + 1.4x$

11. Evaluate xy for $x = 4.2$ and $y = 0.03$.

△ **12.** Find the circumference of a circle whose radius is 6 inches. Then use the approximation 3.14 for π to approximate the circumference.

13. Perry Sitongia borrowed $576 from his father. He plans to pay it back over the next 20 months with equal payments. How much will each monthly payment be?

14. Simplify: $0.2(6.9 - 3.01)$

15. Write $\frac{3}{8}$ as a decimal.

16. Insert <, >, or = to form a true statement. $\frac{9}{11}$ 0.8182

17. Solve: $4(x + 0.22) = -3.4$

18. Find the square root: $\sqrt{\frac{36}{49}}$

19. Approximate $\sqrt{46}$ to the nearest hundredth.

△ **20.** Find the length of the hypotenuse of the given right triangle.

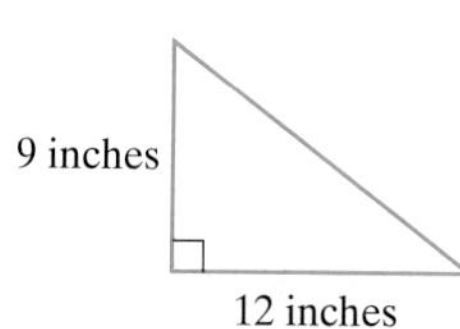

5.1 Introduction to Decimals

A Decimal Notation and Writing Decimals in Words

Like fractional notation, decimal notation is used to denote a part of a whole. Numbers written in decimal notation are called **decimal numbers**, or simply **decimals**. The decimal 17.758 has three parts.

OBJECTIVES

- A Know the meaning of place value for a decimal number and write decimals in words.
- B Write decimals in standard form.
- C Write decimals as fractions.
- D Compare decimals.
- E Round decimals to a given place value.

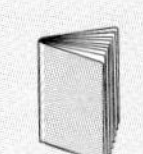

SSM TUTOR CENTER SG CD & VIDEO MATH PRO WEB

In Section 1.2, we introduced place value for whole numbers. Place names and place values for the whole-number part of a decimal number are exactly the same, as shown next. Place names and place values for the decimal part are also shown.

Place-Value Chart

	1	7	.	7	5	8		
hundreds	tens	ones	↑	tenths	hundredths	thousandths	ten-thousandths	hundred-thousandths
100	10	1	decimal point	$\frac{1}{10}$	$\frac{1}{100}$	$\frac{1}{1000}$	$\frac{1}{10,000}$	$\frac{1}{100,000}$

Notice that the value of each place is $\frac{1}{10}$ of the value of the place to its left.

For example,

$$1 \cdot \frac{1}{10} = \frac{1}{10} \qquad \text{and} \qquad \frac{1}{10} \cdot \frac{1}{10} = \frac{1}{100}$$

ones → tenths; tenths → hundredths

For the example above, 17.758, the digit 5 is in the hundredths place, so its value is 5 hundredths or $\frac{5}{100}$.

Writing or reading a decimal in words is similar to writing or reading a whole number. Use the following steps.

Writing (or Reading) a Decimal in Words

Step 1. Write the whole-number part in words.

Step 2. Write "and" for the decimal point.

Step 3. Write the decimal part in words as though it were a whole number, followed by the place value of the last digit.

EXAMPLE 1 Write each decimal in words.

a. 0.3 **b.** −5.82 **c.** 21.093

Solution:

a. Three tenths

b. Negative five and eighty-two hundredths

c. Twenty-one and ninety-three thousandths

Practice Problem 1

Write each decimal in words.

a. 0.08 b. −500.025 c. 0.0329

Answers

1. a. eight hundredths **b.** negative five hundred and twenty-five thousandths **c.** three hundred twenty-nine ten-thousandths

Practice Problem 2

Write the decimal 97.28 in words.

EXAMPLE 2 Write the decimal in the sentence in words: The Golden Jubilee Diamond is a 545.67 carat cut diamond. (*Source*: *The Guinness Book of Records*)

Solution: Five hundred forty-five and sixty-seven hundredths ●

Practice Problem 3

Write the decimal 72.1085 in words.

EXAMPLE 3 Write the decimal in the sentence in words: The oldest known fragments of Earth's crust are Zircon crystals. They were discovered in Australia and are believed to be 4.276 billion years old. (*Source*: *The Guinness Book of Records*)

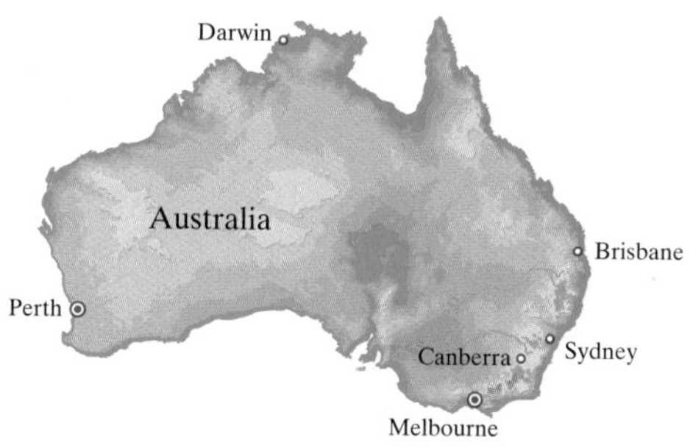

Solution: Four and two hundred seventy-six thousandths ●

Suppose that you are paying for a purchase of $368.42 at Circuit City by writing a check. Checks are usually written using the following format.

Elayn Martin-Gay 00-1110/1000 2236

Date *(Insert today's date)*

PAY TO THE ORDER OF *Circuit City* $ *368.42*

Three hundred sixty-eight and $\frac{42}{100}$ Dollars Contains Security Features. Details on Back.

For *Elayn Martin-Gay* MP

⑆0000000000⑆ 0000000000⑈ 0000

Answers

2. ninety-seven and twenty-eight hundredths
3. seventy-two and one thousand eighty-five ten-thousandths

EXAMPLE 4 Fill in the check to Camelot Music to pay for your purchase of $92.98.

Solution:

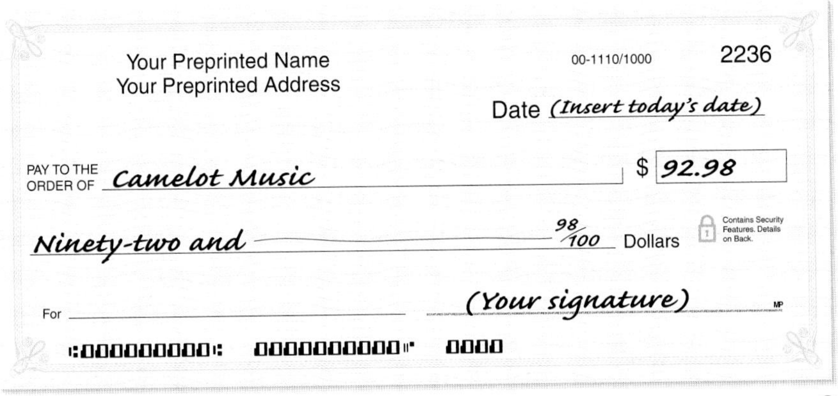

Your Preprinted Name
Your Preprinted Address
00-1110/1000 2236
Date *(Insert today's date)*
PAY TO THE ORDER OF *Camelot Music* $ *92.98*
Ninety-two and $\frac{98}{100}$ Dollars
Contains Security Features. Details on Back.
For *(Your signature)*
⑆0000000000⑆ 0000000000⑈ 0000

Practice Problem 4

Fill in the check to CLECO (Central Louisiana Electric Company) to pay for your monthy electric bill of $207.40.

Your Preprinted Name
Your Preprinted Address
00-1110/1000 2236
Date
PAY TO THE ORDER OF $
Dollars
For
⑆0000000000⑆ 0000000000⑈ 0000

B Writing Decimals in Standard Form

A decimal written in words can be written in standard form by reversing the above procedure.

EXAMPLES Write each decimal in standard form.

5. Forty-eight and twenty-six hundredths

48.26

hundredths place

6. Six and ninety-five thousandths

6.095

thousandths place

Practice Problems 5–6

Write each decimal in standard form.

5. Three hundred and ninety-six hundredths
6. Thirty-nine and forty-two thousandths

Helpful Hint

When writing a decimal from words to decimal notation, make sure the last digit is in the correct place by inserting 0s after the decimal point if necessary. For example,

Two and thirty-eight thousandths is 2.038

thousandths place

Answers

4.

Your Preprinted Name
Your Preprinted Address
00-1110/1000 2237
Date *(Current date)*
PAY TO THE ORDER OF *CLECO* $ *207.40*
Two hundred seven and $\frac{40}{100}$ Dollars
For *(Your signature)*
⑆0000000000⑆ 0000000000⑈ 0000

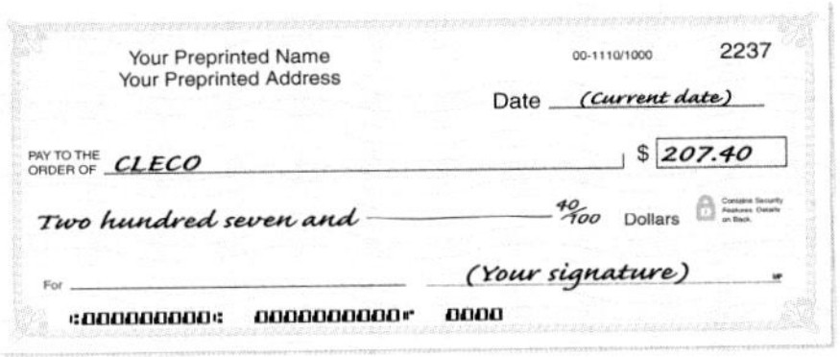

5. 300.96 **6.** 39.042

C Writing Decimals as Fractions

Once you master reading and writing decimals, writing a decimal as a fraction follows naturally.

Decimal	In Words	Fraction
0.7	seven tenths	$\frac{7}{10}$
0.51	fifty-one hundredths	$\frac{51}{100}$
0.009	nine thousandths	$\frac{9}{1000}$

Notice that the number of decimal places in a decimal number is the same as the number of zeros in the denominator of the equivalent fraction. We can use this fact to write decimals as fractions.

$$\underbrace{0.51}_{\text{2 decimal places}} = \frac{51}{\underbrace{100}_{\text{2 zeros}}} \qquad \underbrace{0.009}_{\text{3 decimal places}} = \frac{9}{\underbrace{1000}_{\text{3 zeros}}}$$

Practice Problem 7

Write 0.037 as a fraction.

EXAMPLE 7 Write 0.43 as a fraction.

Solution: $0.43 = \frac{43}{100}$

↑ 2 decimal places ↑ 2 zeros

Practice Problem 8

Write 14.97 as a mixed number.

EXAMPLE 8 Write 5.6 as a mixed number.

Solution: $5.6 = 5\frac{6}{10} = 5\frac{3}{5}$ in simplest form

↑ 1 decimal place ↑ 1 zero

Practice Problems 9–11

Write each decimal as a fraction or mixed number. Write your answer in simplest form.

9. 0.12
10. 57.8
11. −209.986

EXAMPLES Write each decimal as a fraction or mixed number. Write your answer in simplest form.

9. $0.125 = \frac{125}{1000} = \frac{1}{8}$

10. $23.5 = 23\frac{5}{10} = 23\frac{1}{2}$

11. $-105.083 = -105\frac{83}{1000}$

Answers

7. $\frac{37}{1000}$ **8.** $14\frac{97}{100}$ **9.** $\frac{3}{25}$
10. $57\frac{4}{5}$ **11.** $-209\frac{493}{500}$

D Comparing Decimals

One way to compare positive decimals is by comparing digits in corresponding places. To see why this works, let's compare 0.5 or $\frac{5}{10}$ and 0.8 or $\frac{8}{10}$. We know

$$\frac{5}{10} < \frac{8}{10} \text{ since } 5 < 8, \text{ so}$$

$$\downarrow \qquad \downarrow$$

$$0.5 < 0.8 \text{ since } 5 < 8$$

This leads to the following.

Comparing Two Positive Decimals

Compare digits in the same places from left to right. When two digits are not equal, the number with the larger digit is the larger decimal. If necessary, insert 0s after the last digit to the right of the decimal point to continue comparing.

Compare hundredths place digits

28.253 28.263

↑ ↑

5 < 6

so 28.253 < 28.263

Helpful Hint

For any decimal, writing 0s after the last digit to the right of the decimal point does not change the value of the number.

$7.6 = 7.60 = 7.600$, and so on.

When a whole number is written as a decimal, the decimal point is placed to the right of the ones digit.

$25 = 25.0 = 25.00$, and so on

EXAMPLE 12 Insert <, >, or = to form a true statement.

0.378 0.368

Solution: 0.3 78 0.3 68 The tenths places are the same.

0.3 7 8 0.3 6 8 The hundredths places are different.

Since $7 > 6$, then $0.378 > 0.368$.

EXAMPLE 13 Insert <, >, or = to form a true statement.

0.052 0.236

Solution: $0.052 < 0.236$ 0 is smaller than 2 in the tenths place.

↑ ↑

Practice Problem 12

Insert <, >, or = to form a true statement.

13.208 13.28

Practice Problem 13

Insert <, >, or = to form a true statement.

0.12 0.086

Answers

12. < **13.** >

We can also use a number line to compare decimals. This is especially helpful when comparing negative decimals. Remember, the number whose graph is to the left is smaller and the number whose graph is to the right is larger.

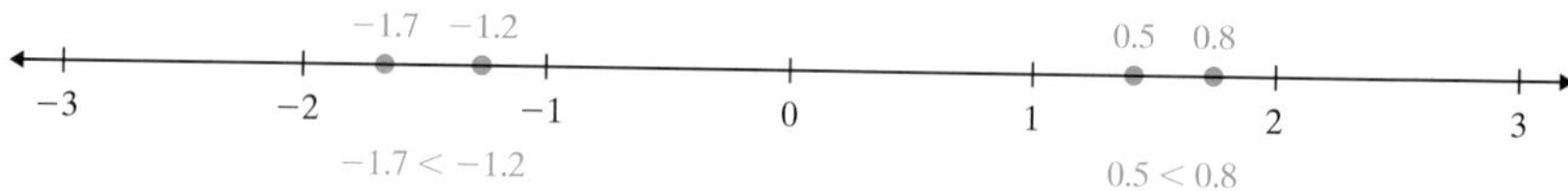

Helpful Hint:

If you have trouble comparing two negative decimals, try the following: Compare their absolute values. Then to correctly compare the negative decimals, reverse the direction of the inequality symbol.

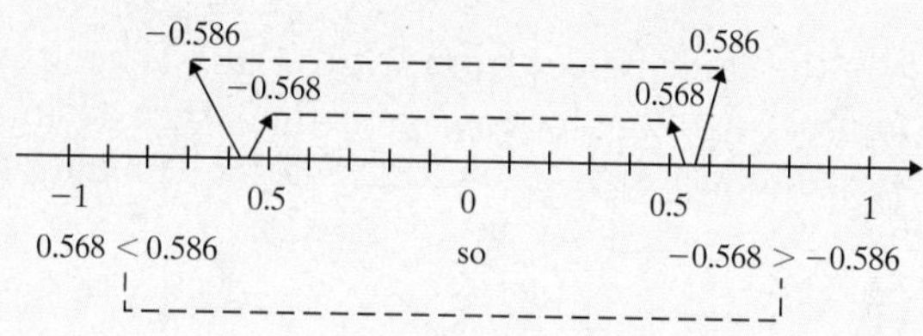

Practice Problem 14

Insert <, >, or = to form a true statement.

−0.029 −0.0209

EXAMPLE 14 Insert <, >, or = to form a true statement.

−0.0101 −0.00109

Solution: Since 0.0101 > 0.00109, then −0.0101 < −0.00109

E Rounding Decimals

We **round the decimal part** of a decimal number in nearly the same way as we round whole numbers. The only difference is that we drop digits to the right of the rounding place, instead of replacing these digits with 0s. For example,

24.954 rounded to the nearest hundredth is 24.95.

Rounding Decimals to a Place Value to the Right of the Decimal Point

Step 1. Locate the digit to the right of the given place value.

Step 2. If this digit is 5 or greater, add 1 to the digit in the given place value and drop all digits to its right. If this digit is less than 5, drop all digits to the right of the given place.

Practice Problem 15

Round 123.7817 to the nearest thousandth.

EXAMPLE 15 Round 736.2359 to the nearest tenth.

Solution:

Step 1. We locate the digit to the right of the tenths place.

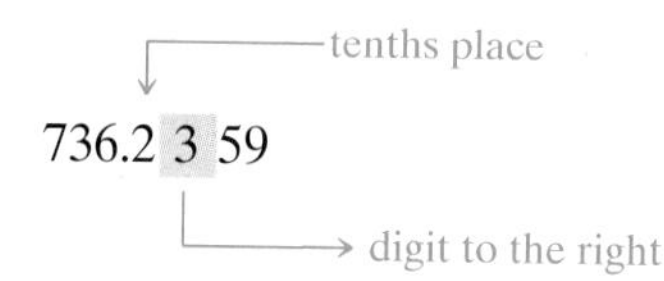

Answers

14. −0.029 < −0.0209 **15.** 123.782

Step 2. Since this digit to the right is less than 5, we drop it and all digits to its right.

Thus, 736.2359 rounded to the nearest tenth is 736.2.

The same steps for rounding can be used when the decimal is negative.

EXAMPLE 16 Round −0.027 to the nearest hundredth.

Solution:

Step 1. Locate the digit to the right of the hundredths place.

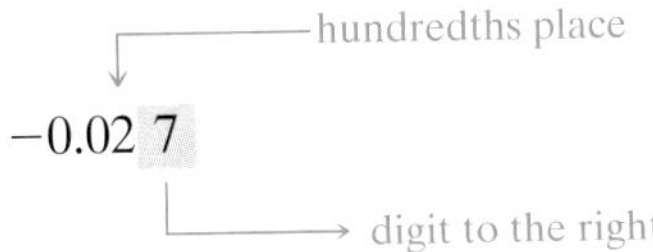

Step 2. Since this digit to the right is 5 or greater, we add 1 to the hundredths digit and drop all digits to its right.

Thus, −0.027 is −0.03 rounded to the nearest hundredth.

The following number line illustrates this rounding.

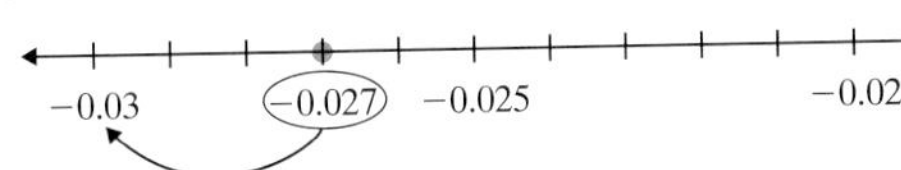

Rounding often occurs with money amounts. Since there are 100 cents in a dollar, each cent is $\frac{1}{100}$ of a dollar. This means that if we want to round to the nearest cent, we round to the nearest hundredth of a dollar.

Practice Problem 16

Round −0.072 to the nearest hundredth.

EXAMPLE 17 Finding Gasoline Prices

The price of a gallon of gasoline in Aimsville is currently $1.3279 per gallon. Round this to the nearest cent.

Solution:

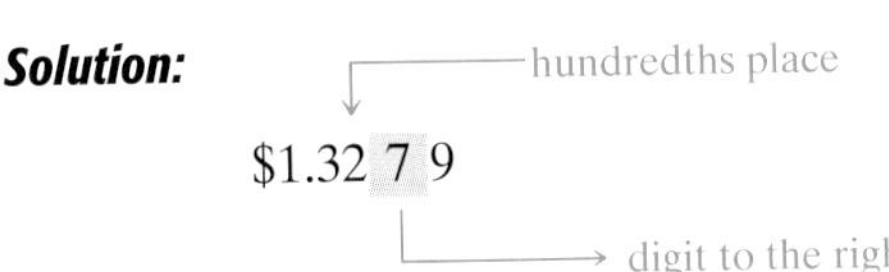

Since this digit to the right is 5 or greater, we add 1 to the hundredths digit and drop all digits to the right of the hundredths digit. Thus, the rounded gasoline price per gallon is $1.33.

Practice Problem 17

In Cititown, the price of a gallon of gasoline is $1.2789 per gallon. Round this to the nearest cent.

EXAMPLE 18 Determining State Taxable Income

A high school teacher's taxable income is $31,567.72. The tax tables in the teacher's state use amounts to the nearest dollar. Round the teacher's income to the nearest whole dollar.

Solution: Rounding to the nearest whole dollar means rounding to the ones place.

ones place

$31,567. 7 2

digit to the right

Since the digit to the right is 5 or greater, we add 1 to the digit in the ones place and drop all digits to the right of the ones place. The teacher's income rounded to the nearest dollar is $31,568.

Practice Problem 18

Water bills in Gotham City are always rounded to the nearest dime. Lois's water bill was $24.43. Round her bill to the nearest dime (tenth).

Answers

16. −0.07 **17.** $1.28 **18.** $24.40

STUDY SKILLS REMINDER

Are you satisfied with your performance on a particular quiz or exam?

If not, analyze your quiz or exam like you would a good mystery novel. Look for common themes in your errors.

Were most of your errors a result of

- *Carelessness*? If your errors were careless, did you turn in your work before the allotted time expired? If so, resolve next time to use the entire time allotted. Any extra time can be spent checking your work.
- *Running out of time*? If so, make a point to better manage your time on your next exam. A few suggestions are to work any questions that you are unsure of last and to check your work after all questions have been answered.
- *Not understanding a concept*? If so, review that concept and correct your work. Remember next time to make sure that all concepts expected to be on a quiz or exam are understood before the exam.

Name ______________________ Section ____________ Date ____________

Mental Math

Determine the place value for the 7 in each decimal or integer.

1. 70 **2.** 700 **3.** 0.7 **4.** 0.07

EXERCISE SET 5.1

A *Write each decimal in words. See Examples 1 through 3.*

1. 6.52 **2.** 7.59 **3.** 16.23 **4.** −47.65

5. −3.205 **6.** 7.495 **7.** 167.009 **8.** 233.056

Fill in each check for the described purchase. See Example 4.

9. Your monthly car loan of $321.42 to R.W. Financial.

Your Preprinted Name
Your Preprinted Address 00-1110/1000 2236
Date ______
PAY TO THE ORDER OF ______ $
______ Dollars
For ______

10. Your part of the monthly apartment rent, which is $213.70. You pay this to Amanda Dupre.

Your Preprinted Name
Your Preprinted Address 00-1110/1000 2236
Date ______
PAY TO THE ORDER OF ______ $
______ Dollars
For ______

11. Your cell phone bill of $59.68 to Bell South.

Your Preprinted Name
Your Preprinted Address 00-1110/1000 2236
Date ______
PAY TO THE ORDER OF ______ $
______ Dollars
For ______

12. Your grocery bill of $87.49 at Albertsons.

Your Preprinted Name
Your Preprinted Address 00-1110/1000 2236
Date ______
PAY TO THE ORDER OF ______ $
______ Dollars
For ______

B *Write each decimal number in standard form. See Examples 5 and 6.*

13. Six and five tenths **14.** Three and nine tenths **15.** Nine and eight hundredths

16. Twelve and six hundredths **17.** Negative five and six hundred twenty-five thousandths **18.** Negative four and three hundred ninety-nine thousandths

19. Sixty-four ten-thousandths

20. Thirty-eight ten-thousandths

21. The average annual rainfall for the state of Louisiana is sixty four and sixteen hundredths inches. (*Source:* National Climatic Data Center)

22. The United States Postal Service vehicle fleet averages nine and sixty-two hundredths miles per gallon of fuel. (*Source:* United States Postal Service)

23. In 2001, there was an average of five and four tenths on-the-job injuries for every 100 workers in the mining industry. (*Source:* Bureau of Labor Statistics)

24. The Olympic record for the 200 meter dash is nineteen and thirty two hundredths seconds, and was obtained by Michael Johnson at the 1996 Olympics in Atlanta, GA. (*Source: 2003 World Almanac*)

C *Write each decimal as a fraction or a mixed number. Write your answer in simplest form. See Examples 7 through 11.*

25. 0.3 **26.** 0.7 **27.** 0.27 **28.** 0.39 **29.** −5.47 **30.** −6.3

31. 0.048 **32.** 0.082 **33.** 7.07 **34.** 9.09 **35.** 15.802 **36.** 11.406

37. 0.3005 **38.** 0.2006 **39.** 487.32 **40.** 298.62

D *Insert* <, >, *or* = *between each pair of numbers to form a true statement. See Examples 12 through 14.*

41. 0.15 0.16 **42.** 0.12 0.15 **43.** −0.57 −0.54 **44.** −0.59 −0.52

45. 0.098 0.1 **46.** 0.0756 0.2 **47.** 0.54900 0.549 **48.** 0.98400 0.984

49. 167.908 167.980 **50.** 519.3405 519.3054 **51.** 420,000 0.000042 **52.** 0.000987 987,000

53. −1.0621 −1.07 **54.** −18.1 −18.01 **55.** −7.052 7.0052 **56.** 0.01 −0.1

57. −0.023 −0.024 **58.** −0.562 −0.652

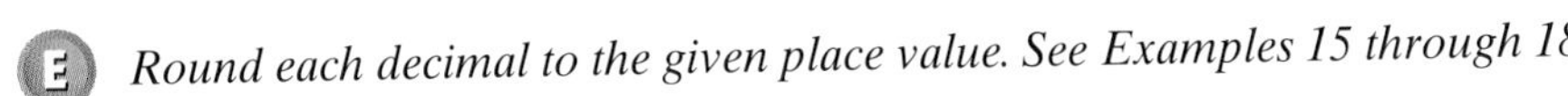

Round each decimal to the given place value. See Examples 15 through 18.

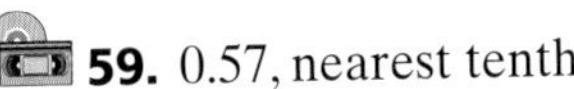

59. 0.57, nearest tenth

60. 0.54, nearest tenth

61. 0.234, nearest hundredth

62. 0.452, nearest hundredth

63. 0.5942, nearest thousandth

64. 63.4523, nearest thousandth

65. 98,207.23, nearest ten

66. 68,934.543, nearest ten

67. 12.999, nearest tenth

68. 42.5799, nearest thousandth

69. −17.667, nearest hundredth

70. −0.766, nearest hundredth

71. −0.501, nearest tenth

72. −0.602, nearest tenth

Round each money amount to the nearest cent or dollar as indicated. See Examples 17 and 18.

73. \$0.067, nearest cent

74. \$0.025, nearest cent

75. \$26.95, nearest dollar

76. \$14,769.52, nearest dollar

77. \$0.1992, nearest cent

78. \$0.7633, nearest cent

79. Which number(s) rounds to 0.26?
0.26559 0.26499 0.25786 0.25186

80. Which number(s) rounds to 0.06?
0.066 0.0586 0.0506 0.0612

Round each number to the given place value. See Examples 17 through 18.

81. Brendan Schmelter bought the DVD "Who Framed Roger Rabbit" at Target for \$15.99. Round this price to the nearest dollar.

82. The attendance at a Mets baseball game was reported to be 39,867 people. Round this number to the nearest thousand.

83. During the 2002 Boston Marathon, Margaret Okaya of Kenya was the women's winner, finishing with a time of 2.34528 hours. Round this time to the nearest hundredth (*Source: 2003 World Almanac*)

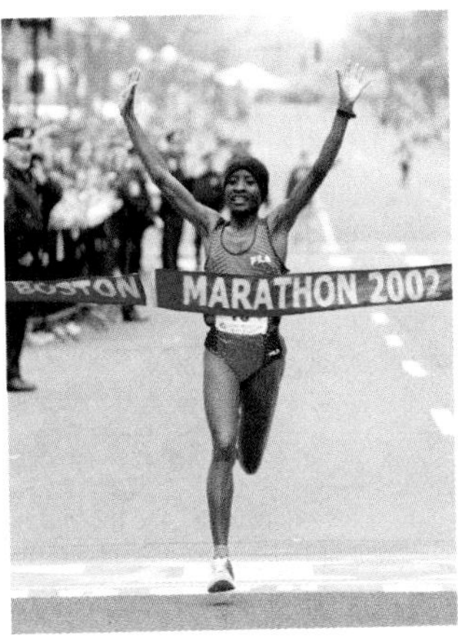

84. The population density of the state of Ohio is roughly 271.0961 people per square mile. Round this population density to the nearest tenth. (*Source:* U.S. Bureau of the Census)

85. The length of a day on Mars is 24.6229 hours. Round this figure to the nearest thousandth. (*Source:* National Space Science Data Center)

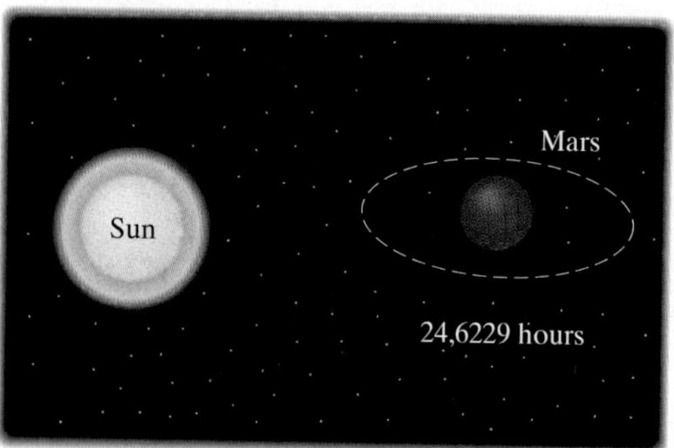

86. Venus makes a complete orbit around the sun every 224.695 days. Round this figure to the nearest whole day. (*Source:* National Space Science Data Center)

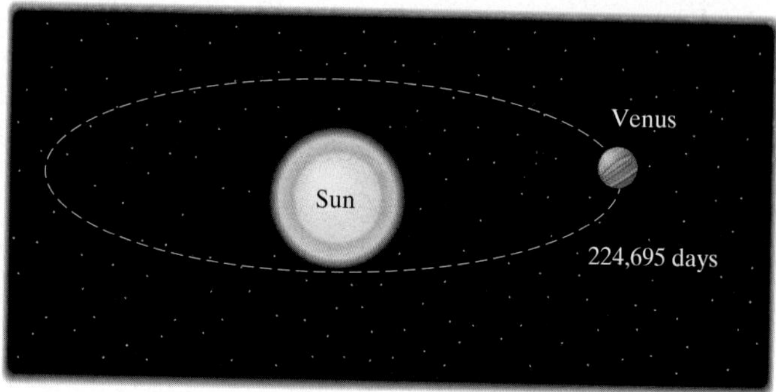

87. The official barefoot water-skiing record is 135.74 mph by Scott Pellaton. Round this figure to the nearest mile per hour. (*Source: The Guinness Book of Records*)

88. Raptor is a roller coaster at Cedar Point, an amusement park in Sandusky, Ohio. It is one of the world's tallest, fastest, and steepest inverted roller coasters. A ride on Raptor lasts about 2.267 minutes. Round this figure to the nearest tenth. (*Source:* Cedar Fair, L.P.)

89. The leading NBA scorer for the 2002 regular season was Allen Iverson of the Philadelphia 76ers. The average number of points he scored per game was 31.3833. Round this figure to the nearest whole point. (*Source*: National Basketball Association)

90. The leading WNBA scorer for the 2002 regular season was Chamique Holdsclaw of the Washington Mystics. The average number of points she scored per game was 19.85. Round this figure to the nearest whole point. (*Source:* Women's National Basketball Association)

Write these numbers from smallest to largest.

91. 0.9
0.1038
0.10299
0.1037

92. 0.01
0.0839
0.09
0.1

Review and Preview

Perform each indicated operation. See Sections 1.3 and 1.4.

93. 3452 + 2314 **94.** 8945 + 4536 **95.** 94 − 23 **96.** 82 − 47 **97.** 482 − 239 **98.** 4002 − 3897

Combining Concepts

The table gives the leading bowling averages for the Professional Bowlers Association Tour for each of the years listed. Use this table to answer Exercises 99–101.

Leading PBA Averages By Year

Year	Bowler	Average Score
1995	Mike Aulby	225.490
1996	Walter Ray Williams, Jr.	225.370
1997	Walter Ray Williams, Jr.	222.008
1998	Walter Ray Williams, Jr.	226.130
1999	Parker Bohn	228.040
2000	Chris Barnes	220.930
2001	Parker Bohn	221.546
2002	Walter Ray Williams, Jr.	224.940

(*Source:* Professional Bowlers Association)

99. What is the highest average score on the list? Which bowler achieved that average?

100. What is the lowest average score on the list? Which bowler achieved that average?

101. Make a list of the leading averages in order from greatest to least for the years shown in the table.

102. Write a 4-digit number that rounds to 26.3.

103. Write a 5-digit number that rounds to 1.7.

104. Explain how to identify the value of the 9 in the decimal 486.3297.

105. Write 0.0000203 in words.

STUDY SKILLS REMINDER

How well do you know this textbook?

See if you can answer the questions below.

1. What does the icon mean?
2. What does the icon mean?
3. What does the icon △ mean?
4. Where can you find a review for each chapter? Which answers to this review can be found in the back of your text?
5. Each chapter contains an overview of the chapter along with examples. What is this feature called?
6. Does this text contain any solutions to exercises? If so, where?
7. What help is available to you for each chapter test?

5.2 Adding and Subtracting Decimals

OBJECTIVES

- **A** Add or subtract decimals.
- **B** Evaluate expressions and check solutions with decimal replacement values.
- **C** Simplify expressions containing decimals.
- **D** Solve problems by adding or subtracting decimals.

SSM TUTOR CENTER SG CD & VIDEO MATH PRO WEB

Adding or subtracting decimals is similar to adding or subtracting whole numbers. We add or subtract digits in corresponding place values from right to left, carrying or borrowing if necessary. To make sure that digits in corresponding place values are added or subtracted, we line up the decimal points vertically.

Adding or Subtracting Decimals

Step 1. Write the decimals so that the decimal points line up vertically.

Step 2. Add or subtract the same as for whole numbers.

Step 3. Place the decimal point in the sum or difference so that it lines up vertically with the decimal points in the problem.

EXAMPLE 1 Add: 23.85 + 1.604

Solution: Line up the decimal points vertically and add as the same for whole numbers.

$$\begin{array}{r} \overset{1}{2}3.850 \\ +\ 1.604 \\ \hline 25.454 \end{array}$$

One 0 is inserted.

Place the decimal point in the sum so that all decimal points line up.

Practice Problem 1

Add.

a. 15.52 + 2.371

b. 20.06 + 17.612

c. 0.125 + 122.8

Helpful Hint

Recall that 0s may be inserted to the right of the decimal point after the last digit without changing the value of the decimal. This may be used to help line up place values when adding decimals.

$$\begin{array}{r} 3.2 \\ 15.567 \\ +0.11 \\ \hline \end{array} \quad \text{becomes} \quad \begin{array}{r} 3.\mathbf{200} \\ 15.567 \\ +\ 0.11\mathbf{0} \\ \hline 18.877 \end{array}$$

← Two 0s are inserted.

← One 0 is inserted.

EXAMPLE 2 Add: 763.7651 + 22.001 + 43.89

Solution:

$$\begin{array}{r} 763.7651 \\ 22.0010 \\ +\ 43.8900 \\ \hline 829.6561 \end{array}$$

← One 0 is inserted.

← Two 0s are inserted.

Add.

Practice Problem 2

Add.

a. 34.567 + 129.43 + 2.8903

b. 11.21 + 46.013 + 362.526

Answers

1. a. 17.891 **b.** 37.672 **c.** 122.925

2. a. 166.8873 **b.** 419.749

Helpful Hint

Don't forget that the decimal point in a whole number is after the last digit.

Practice Problem 3

Add: 27 + 0.00043

EXAMPLE 3 Add: 39 + 0.0021

Solution:

$$\begin{array}{r} 39.0000 \\ +\ 0.0021 \\ \hline 39.0021 \end{array}$$

For a whole number, place the decimal point to the right of the ones digit.

Concept Check

Find and correct the error made in the following addition of 3.05, 2.6, and 1.941.

$$\begin{array}{r} 3.05 \\ 2.6 \\ +1.941 \\ \hline 2.272 \end{array}$$

Try the Concept Check in the margin.

Practice Problem 4

Add: 8.1 + (−99.2)

EXAMPLE 4 Add: 3.62 + (−4.78)

Solution: Recall from Chapter 2 that to add two numbers with different signs we find the difference of the larger absolute value and the smaller absolute value. The sign of the answer is the same as the sign of the number with the larger absolute value.

$$\begin{array}{r} 4.78 \\ -3.62 \\ \hline 1.16 \end{array}$$

Subtract the absolute values.

Thus, $3.62 + (-4.78) = -1.16$

The sign of the number with the larger absolute value −4.78 has the larger absolute value.

Just as for whole numbers, borrowing may sometimes be needed when subtracting decimals.

Practice Problem 5

Subtract: 5.8 − 3.92

EXAMPLE 5 Subtract: 3.5 − 0.068

Solution:

$$\begin{array}{r} 3.500 \\ -0.068 \\ \hline 3.432 \end{array}$$

← Insert two 0s.

Recall that we can check a subtraction problem by adding.

$$\begin{array}{rl} 3.432 & \text{Difference} \\ +0.068 & \text{Subtrahend} \\ \hline 3.500 & \text{Minuend} \end{array}$$

Practice Problem 6

Subtract: 53 − 29.31

EXAMPLE 6 Subtract: 85 − 17.31

Solution:

$$\begin{array}{r} 85.00 \\ -17.31 \\ \hline 67.69 \end{array}$$

Check:

$$\begin{array}{r} 67.69 \\ +17.31 \\ \hline 85.00 \end{array}$$

Answers

3. 27.00043 **4.** −91.1 **5.** 1.88 **6.** 23.69

Concept Check: Decimal points were not lined up before adding.

$$\begin{array}{r} 3.050 \\ 2.600 \\ +1.941 \\ \hline 7.591 \end{array}$$

EXAMPLE 7 Subtract 3 from 6.98.

Solution:

$$\begin{array}{r} 6.98 \\ -3.00 \\ \hline 3.98 \end{array}$$ Insert two 0s.

Practice Problem 7

Subtract: 18 from 26.99.

EXAMPLE 8 Subtract: $-5.8 - 1.7$

Solution: Recall from Chapter 2 that to subtract 1.7 we add the opposite of 1.7, or -1.7. Thus

$-5.8 - 1.7 = -5.8 + (-1.7)$ To subtract, add the opposite of 1.7 which is -1.7.

Add the absolute values.

$= -7.5.$

Use the common negative sign.

Practice Problem 8

Subtract: $-3.4 - 9.6$.

EXAMPLE 9 Subtract: $-2.56 - (-4.01)$

Solution: $-2.56 - (-4.01) = -2.56 + 4.01$ To subtract, add the opposite of -4.01, which is 4.01.

Subtract the absolute values.

$= 1.45$

The answer is (understood) positive since 4.01 has the larger absolute value.

Practice Problem 9

Subtract: $-1.05 - (-7.23)$.

B Using Decimals as Replacement Values

Let's review evaluating expressions with given replacement values. This time the replacement values are decimals.

EXAMPLE 10 Evaluate $x - y$ for $x = 2.8$ and $y = 0.92$.

Solution: Replace x with 2.8 and y with 0.92 and simplify.

$x - y = 2.8 - 0.92$
$= 1.88$

$$\begin{array}{r} 2.80 \\ -0.92 \\ \hline 1.88 \end{array}$$

Practice Problem 10

Evaluate $y - z$ for $y = 11.6$ and $z = 10.8$.

EXAMPLE 11 Is 2.3 a solution of the equation $6.3 = x + 4$?

Solution: Replace x with 2.3 in the equation $6.3 = x + 4$ to see if the result is a true statement.

$6.3 = x + 4$

$6.3 = 2.3 + 4$ Replace x with 2.3.

$6.3 = 6.3$ True.

Since $6.3 = 6.3$ is a true statement, 2.3 is a solution of $6.3 = x + 4$.

Practice Problem 11

Is 12.1 a solution of the equation $y - 4.3 = 7.8$?

Answers

7. 8.99 **8.** -13 **9.** 6.18 **10.** 0.8 **11.** yes

C Simplify Expressions Containing Decimals

Practice Problem 12

Simplify by combining like terms:
$-4.3y + 7.8 - 20.1y + 14.6$

EXAMPLE 12 Simplify by combining like terms:

$$11.1x - 6.3 + 8.9x - 4.6$$

Solution:

$$\begin{aligned} 11.1x - 6.3 + 8.9x - 4.6 &= 11.1x + 8.9x + (-6.3) + (-4.6) \\ &= 20x + (-10.9) \\ &= 20x - 10.9 \end{aligned}$$

D Solving Problems by Adding or Subtracting Decimals

Decimals are very common in real-life problems.

Practice Problem 13

Find the total monthly cost of owning and operating a certain automobile given the expenses shown.

Monthly car payment:	\$536.50
Monthly insurance cost:	\$52.70
Average gasoline bill per month:	\$87.50

EXAMPLE 13 Calculating the Cost of Owning an Automobile

Find the total monthly cost of owning and operating a certain automobile given the expenses shown.

Monthly car payment:	\$256.63
Monthly insurance cost:	\$ 47.52
Average gasoline bill per month:	\$ 95.33

Solution:

1. UNDERSTAND. Read and reread the problem. The phrase "total monthly cost" tells us to add.

   Note: If you would like to assign a variable to the unknown, let x = total monthly cost. Since the translated equation will already be solved for the unknown, variables will not be used here.

2. TRANSLATE.

In words:	Total monthly cost	is	car payment	plus	insurance	plus	gasoline bill
	↓	↓	↓	↓	↓	↓	↓
Translate:	Total monthly cost	=	\$256.63	+	\$47.52	+	\$95.33

3. SOLVE.

$$\begin{array}{r} \overset{1\,1\,1}{256.63} \\ 47.52 \\ +\ 95.33 \\ \hline \$399.48 \end{array}$$

4. INTERPRET. *Check* your work. *State* your conclusion: The total monthly cost is \$399.48.

Answers

12. $-24.4y + 22.4$ **13.** \$676.70

EXAMPLE 14 Comparing Average Heights

The bar graph shows the current average heights for adults in various countries. How much taller is the average height in Denmark than the average height in the United States?

Average Adult Height

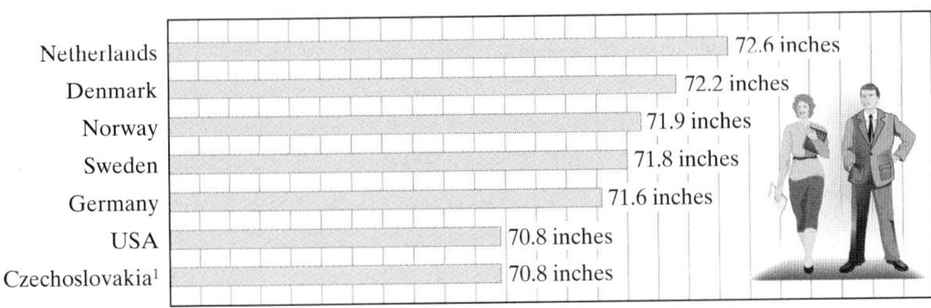

[1]Average for Czech Republic, Slovakia
Source: *USA Today*, 8/28/97

Solution:

1. UNDERSTAND. Read and reread the problem. Since we want to know "how much taller," we subtract.
2. TRANSLATE.

In words:

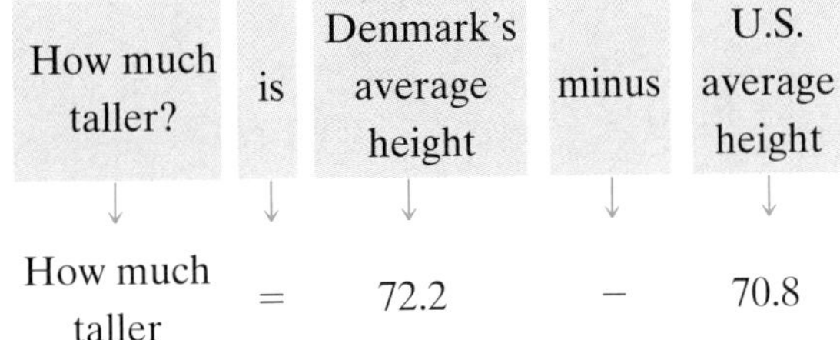

Translate: How much taller $= 72.2 - 70.8$

3. SOLVE.

$$\begin{array}{r} 72.2 \\ -70.8 \\ \hline 1.4 \end{array}$$

4. INTERPRET. *Check* your work. *State* your conclusion: The average height in Denmark is 1.4 inches more than the average height in the U.S. ●

Practice Problem 14

Use the bar graph for Example 14. How much taller is the average height in the Netherlands than the average height in Czechoslovakia?

Answer

14. 1.8 inches

CALCULATOR EXPLORATIONS

ENTERING DECIMAL NUMBERS

To enter a decimal number, find the key marked [.]. To enter the number 2.56, for example, press the keys

[2] [.] [5] [6]

The display will read [2.56].

OPERATIONS ON DECIMAL NUMBERS

Operations on decimal numbers are performed in the same way as operations on whole or signed numbers. For example, to find 8.625 − 4.29, press the keys

[8.625] [−] [4.29] [=] or [ENTER]

The display will read [4.335].

(Although entering 8.625, for example, requires pressing more than one key, we group numbers together for easier reading.)

Use a calculator to perform each indicated operation.

1. 315.782 + 12.96

2. 29.68 + 85.902

3. 6.249 − 1.0076

4. 5.238 − 0.682

5.
$$\begin{array}{r} 12.555 \\ 224.987 \\ 5.2 \\ +622.65 \\ \hline \end{array}$$

6.
$$\begin{array}{r} 47.006 \\ 0.17 \\ 313.259 \\ +139.088 \\ \hline \end{array}$$

Name ______________________ Section ____________ Date ____________

Mental Math

State each sum or difference.

1. $\begin{array}{r} 0.3 \\ +0.2 \\ \hline \end{array}$ **2.** $\begin{array}{r} 0.4 \\ +0.5 \\ \hline \end{array}$ **3.** $\begin{array}{r} 1.00 \\ +0.26 \\ \hline \end{array}$ **4.** $\begin{array}{r} 3.00 \\ +0.19 \\ \hline \end{array}$

5. $\begin{array}{r} 7.6 \\ +1.3 \\ \hline \end{array}$ **6.** $\begin{array}{r} 4.5 \\ +3.2 \\ \hline \end{array}$ **7.** $\begin{array}{r} 0.9 \\ -\ 0.3 \\ \hline \end{array}$ **8.** $\begin{array}{r} 0.6 \\ -\ 0.2 \\ \hline \end{array}$

EXERCISE SET 5.2

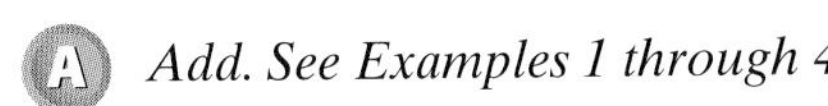

A *Add. See Examples 1 through 4.*

1. $1.3 + 2.2$

2. $2.5 + 4.1$

3. $5.7 + 1.13$

4. $2.31 + 6.4$

5. $24.6 + 2.39 + 0.0678$

6. $32.4 + 1.58 + 0.0934$

7. $\begin{array}{r} 45.023 \\ 3.006 \\ +\ 8.403 \\ \hline \end{array}$

8. $\begin{array}{r} 65.0028 \\ 5.0903 \\ +\ 6.9003 \\ \hline \end{array}$

9. $-2.6 + (-5.97)$

10. $-18.2 + (-10.8)$

11. $15.78 + (-4.62)$

12. $6.91 + (-7.03)$

Subtract. See Examples 5 through 9.

13. $8.8 - 2.3$

14. $7.6 - 2.1$

15. $18 - 2.7$

16. $28 - 3.3$

17. $\begin{array}{r} 654.9 \\ -56.67 \\ \hline \end{array}$

18. $\begin{array}{r} 863.23 \\ -39.453 \\ \hline \end{array}$

19. Subtract 6.7 from 23

20. Subtract 9.2 from 45

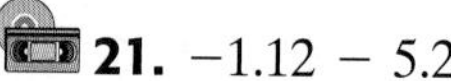

21. $-1.12 - 5.2$

22. $-8.63 - 5.6$

23. $7.7 - 14.1$

24. $10.25 - 21.76$

25. $-2.6 - (-5.7)$

26. $-9.4 - (-10.4)$

Perform each indicated operation. See Examples 1 through 9.

27. $0.9 + 2.2$

28. $0.7 + 3.4$

29. $-5.9 - 4$

30. $-6.4 - 3.4$

31. $45.67 - 20$

32. $56.89 - 30$

33. $-6.06 + 0.44$

34. $-5.05 + 0.88$

35. $900.34 - 123.45$

36. $800.74 - 463.98$

37. $3490.23 + 8493.09$

38. $600.004 + 7983.0062$

39. $\begin{array}{r} 234.89 \\ +230.67 \\ \hline \end{array}$

40. $\begin{array}{r} 734.89 \\ +640.56 \\ \hline \end{array}$

41. $50.2 - 600$

42. $40.3 - 700$

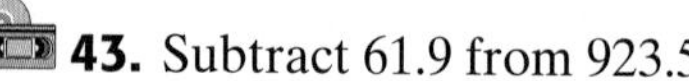

43. Subtract 61.9 from 923.5

44. Subtract 45.8 from 845.93

45. $\begin{array}{r} 100.009 \\ 6.08 \\ +\ 9.034 \\ \hline \end{array}$

46. $\begin{array}{r} 200.89 \\ 7.49 \\ +\ 62.83 \\ \hline \end{array}$

47. $\begin{array}{r} 1000 \\ -\ 123.4 \\ \hline \end{array}$

48. $\begin{array}{r} 2000 \\ -\ 327.47 \\ \hline \end{array}$

49. $-0.003 + 0.091$

50. $-0.004 + 0.085$

51. $500 - 34.098$

52. $300 - 98.345$

53. $-102.4 - 78.04$

54. $-36.2 - 10.02$

55. $-2.9 - (-1.8)$

56. $-6.5 - (-3.3)$

B *Evaluate each expression for* $x = 3.6$, $y = 5$, *and* $z = 0.21$. *See Example 10.*

57. $x + z$

58. $y + x$

59. $x - z$

60. $y - z$

61. $y - x + z$

62. $x + y + z$

Determine whether the given values are solutions to the given equations. See Example 11.

63. Is 7 a solution to $x + 2.7 = 9.3$?

64. Is 3.7 a solution to $x + 5.9 = 8.6$?

65. Is -11.4 a solution to $27.4 + y = 16$?

66. Is -22.9 a solution to $45.9 + z = 23$?

67. Is 1 a solution to $2.3 + x = 5.3 - x$?

68. Is 0.9 a solution to $1.9 - x = x + 0.1$?

C *Simplify by combining like terms. See Example 12.*

69. $30.7x + 17.6 - 23.8x - 10.7$

70. $14.2z + 11.9 - 9.6z - 15.2$

71. $-8.61 + 4.23y - 2.36 - 0.76y$

72. $-8.96x - 2.31 - 4.08x + 9.68$

D *Solve. See Examples 13 and 14.*

73. Find the total monthly cost of owning and maintaining a car given the information shown.

Monthly car payment:	\$275.36
Monthly insurance cost:	\$83.00
Average cost of gasoline per month:	\$81.60
Average maintenance cost per month:	\$14.75

74. Find the total monthly cost of owning and maintaining a car given the information shown.

Monthly car payment:	\$306.42
Monthly insurance cost:	\$53.50
Average cost of gasoline per month:	\$123.00
Average maintenance cost per month:	\$23.50

75. Gasoline was \$1.499 per gallon on one day and \$1.559 per gallon the next day. By how much did the price change?

76. A pair of eyeglasses costs a total of \$347.89. The frames of the glasses are \$97.23. How much do the lenses of the eyeglasses cost?

77. Ann-Margaret Tober bought a book for \$32.48. If she paid with two \$20 bills, what was her change?

78. Tom Mackey bought a car part for \$18.26. If he paid with two \$10 bills, what was his change?

79. Americans' consumption of sugar is on the rise. During 1990, Americans consumed an average of 136.8 pounds of sugar in its various forms such as refined white sugar, honey, and corn sweeteners. By 2000, the average American was consuming 150.1 pounds of sugar products per year. How much more sugar was the average American consuming annually in 2000 than in 1990? (*Source:* Economic Research Service, U.S. Department of Agriculture)

80. In 2002, the average wage for U.S. production workers was \$14.77 per hour. In 1997, this average wage was \$12.28 per hour. How much of an increase is this? (*Source:* Bureau of Labor Statistics)

81. The average wind speed at the weather station on Mt. Washington in New Hampshire is 35.2 miles per hour. The highest speed ever recorded at the station is 321.0 miles per hour. How much faster is the highest speed than the average wind speed? (*Source:* National Climatic Data Center)

82. The average annual rainfall in Omaha, Nebraska, is 30.22 inches. The average annual rainfall in New Orleans, Louisiana, is 61.88 inches. On average, how much more rain does New Orleans receive annually than Omaha? (*Source:* National Climatic Data Center)

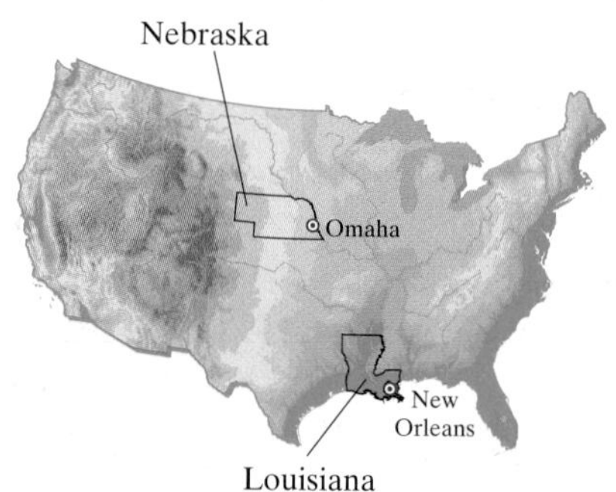

83. In October 1997, Andy Green set a new one-mile land speed record. This record was 129.567 miles per hour faster than a previous record of 633.468 set in 1983. What was Green's record-setting speed? (*Source:* United States Auto Club)

84. It costs \$3.13 to send a 2-pound package locally via parcel post at a U.S. Post Office. To send the same package as Priority Mail, it costs \$3.95. How much more does it cost to send a package as Priority Mail? (*Source*: USPS)

85. The three North America concert tours that have earned the most money are the Rolling Stones (1994) \$121.2 million, Pink Floyd (1994). \$103.5 million, and U2 (2001) \$109.7 million What was the total amount of money earned from these three concerts? (*Source:* Pollstar, Fresno, CA)

86. In 1995, the average credit-card late fee was \$12.53. In 2002, the average late fee had increased to \$27.82. By how much did the average credit-card late fee increase from 1995 to 2002? (*Source:* Consumer Action)

87. The snowiest city in the United States is Blue Canyon, California, which receives an average of 111.6 more inches of snow than the second snowiest city. The second snowiest city in the United States is Marquette, Michigan. Marquette receives an average of 129.2 inches of snow annually. How much snow does Blue Canyon receive on average each year? (*Source:* National Climatic Data Center)

88. The driest city in the world is Aswan, Egypt, which receives an average of only 0.02 inches of rain per year. Yuma, Arizona, is the driest city in the United States. Yuma receives an average of 2.63 more inches of rain each year than Aswan. What is the average annual rainfall in Yuma? (*Source:* National Climatic Data Center)

89. A landscape architect is planning a border for a flower garden shaped like a triangle. The sides of the garden measure 12.4 feet, 29.34 feet, and 25.7 feet. Find the amount of border material needed.

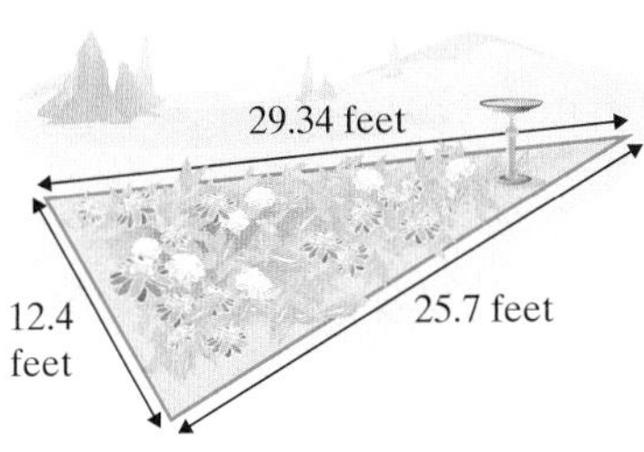

90. A contractor purchased enough buy railing to completely enclose the newly built deck shown below. Find the amount of railing purchased.

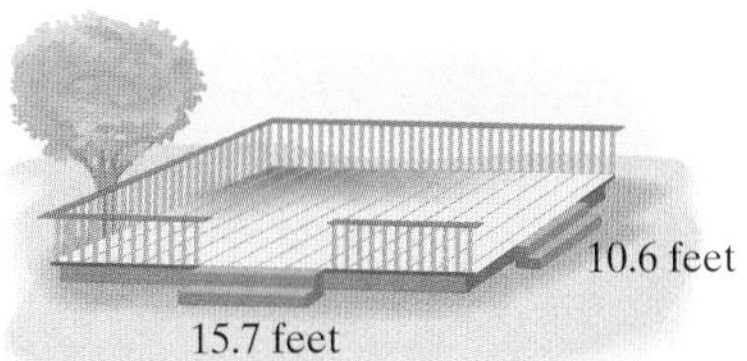

The table shows the average speeds for the Daytona 500 winners for the years shown. Use this table to answer Exercises 91–92.

Daytona 500 Winners		
Year	**Winner**	**Average Speed**
1959	Lee Petty	135.521
1969	Lee Roy Yarborough	160.875
1979	Richard Petty	143.977
1989	Darrell Waltrip	148.466
1999	Jeff Gordon	161.551

91. How much slower was the average Daytona 500 winning speed in 1989 than in 1969?

92. How much faster is the average Daytona 500 winning speed in 1999 than in 1969?

The Mercury space program had 6 manned flights. They are shown in the following table. Use this table to answer Exercises 93–94.

Year	Mission	Duration (in minutes)
1961	Redstone 3	15.467
1961	Redstone 4	15.167
1962	Atlas 6	295.383
1962	Atlas 7	296.083
1962	Atlas 8	553.183
1963	Atlas 9	2059.817

93. How many more minutes longer was the last Mercury mission than the first?

94. What is the difference in duration of the first two Mercury missions flown in 1961?

The bar graph shows the five top chocolate-consuming nations in the world. Use this table to answer Exercises 95–99.

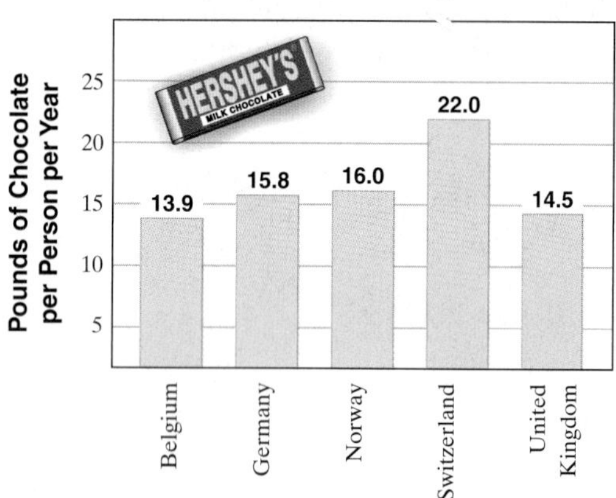

Source: Hershey Foods Corporation

95. Which country in the table has the greatest chocolate consumption per person?

96. Which country in the table has the least chocolate consumption per person?

97. How much more is the greatest chocolate consumption than the least chocolate consumption shown in the table?

98. How much more chocolate does the average German consume than the average citizen of the United Kingdom?

99. Make a new chart listing the countries and their corresponding chocolate consumption in order from greatest to least.

Review and Preview

Multiply. See Section 4.3.

100. $\left(\frac{2}{3}\right)^2$

101. $\left(\frac{1}{5}\right)^3$

102. $-\frac{12}{7}\cdot\frac{14}{3}$

103. $\frac{25}{36}\cdot\frac{24}{40}$

Combining Concepts

Let's review the values of these popular U.S. coins in order to answer the following exercises.

Penny	Nickel	Dime	Quarter
$0.01	$0.05	$0.10	$0.25

Write the value of each group of coins. To do so, it is usually easiest to start with the coin(s) of greatest value and end with the coin(s) of least value.

104.

105.

106. Name the different ways that coins can have a value of $0.17 given that you may use no more than 10 coins.

107. Name the different ways that coin(s) can have a value of $0.25 given that there are no pennies.

108. Laser beams can be used to measure the distance to the moon. One measurement showed the distance to the moon to be 256,435.235 miles. A later measurement showed that the distance is 256,436.012 miles. Find how much farther away the moon is in the second measurement compared to the first.

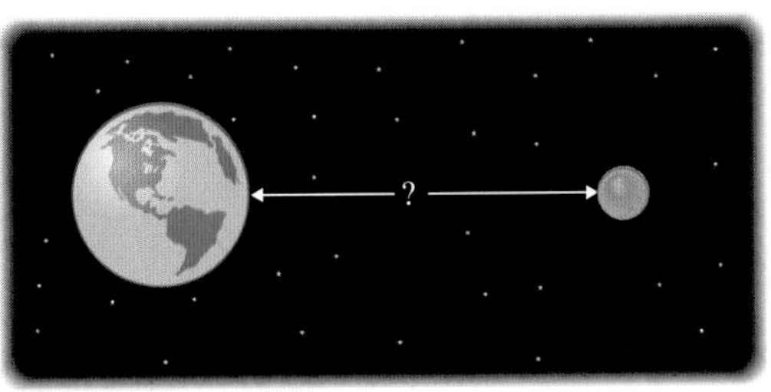

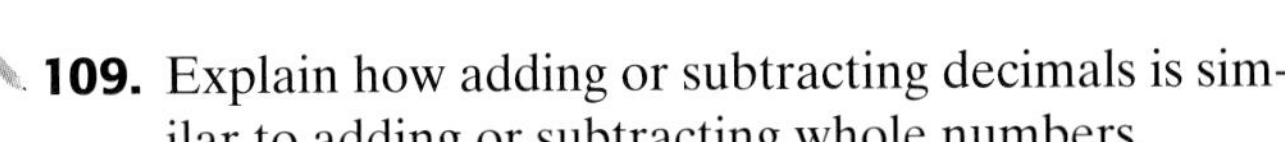

109. Explain how adding or subtracting decimals is similar to adding or subtracting whole numbers.

Combine like terms and simplify.

110. $-8.689 + 4.286x - 14.295 - 12.966x + 30.861x$

111. $14.271 - 8.968x + 1.333 - 201.815x + 101.239x$

112. Can the sum of two negative decimals ever be a positive decimal? Why or why not?

Find the unknown length in each figure.

113.

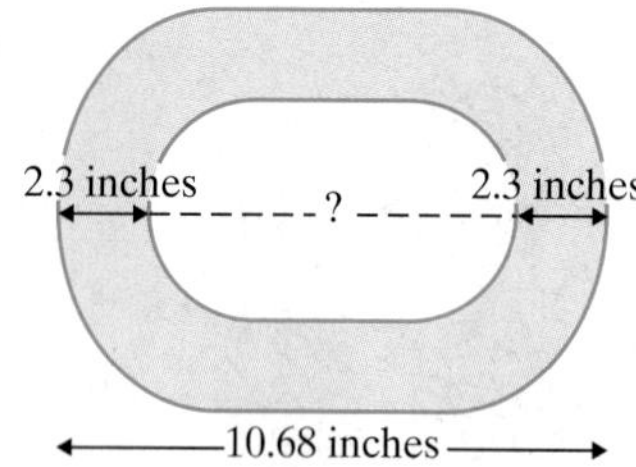

114.

5.26 meters | 7.82 meters | ? meters

17.67 meters

5.3 Multiplying Decimals and Circumference of a Circle

OBJECTIVES

- A Multiply decimals.
- B Multiply decimals by powers of 10.
- C Evaluate expressions and check solutions with decimal replacement values.
- D Find the circumference of a circle.
- E Solve problems by multiplying decimals.

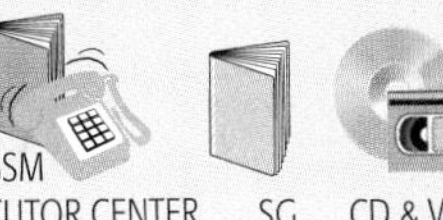

SSM TUTOR CENTER SG CD & VIDEO MATH PRO WEB

A Multiplying Decimals

Multiplying decimals is similar to multiplying whole numbers. The only difference is that we place a decimal point in the product. To discover where a decimal point is placed in the product, let's multiply 0.6×0.03. We first write each decimal as an equivalent fraction and then multiply.

$$\underset{\text{1 decimal place}}{0.6} \times \underset{\text{2 decimal places}}{0.03} = \frac{6}{10} \times \frac{3}{100} = \frac{18}{1000} = \underset{\text{3 decimal places}}{0.018}$$

Now let's multiply 0.03×0.002.

$$\underset{\text{2 decimal places}}{0.03} \times \underset{\text{3 decimal places}}{0.002} = \frac{3}{100} \times \frac{2}{1000} = \frac{6}{100{,}000} = \underset{\text{5 decimal places}}{0.00006}$$

Instead of writing decimals as fractions each time we want to multiply, we notice a pattern from these examples and state a rule that we can use.

Multiplying Decimals

Step 1. Multiply the decimals as though they were whole numbers.

Step 2. The decimal point in the product is placed so the number of decimal places in the product is equal to the *sum* of the number of decimal places in the factors.

EXAMPLE 1 Multiply: 23.6×0.78

Solution:

```
   23.6    1 decimal place
 × 0.78    2 decimal places
 ------
   1888
  16520
 ------
 18.408    3 decimal places
```

Practice Problem 1

Multiply: 45.9×0.42

EXAMPLE 2 Multiply: 0.283×0.3

Solution:

```
   0.283   3 decimal places
 ×   0.3   1 decimal place
 -------
  0.0849   4 decimal places
   ↑
   └── Insert one 0 since the product must have 4 decimal places.
```

Practice Problem 2

Multiply: 0.112×0.6

EXAMPLE 3 Multiply: 0.0531×16

Solution:

```
   0.0531   4 decimal places
 ×     16   0 decimal places
 --------
     3186
    05310
 --------
   0.8496   4 decimal places
```

Practice Problem 3

Multiply: 0.0721×48

Answers

1. 19.278 **2.** 0.0672 **3.** 3.4608

Concept Check

True or false? The number of decimal places in the product of 0.261 and 0.78 is 6. Explain.

Practice Problem 4

Multiply: $(5.4)(-1.3)$

Try the Concept Check in the margin.

EXAMPLE 4 Multiply: $(-2.6)(0.8)$

Solution: Recall that the product of a negative number and a positive number is a negative number.

$$(-2.6)(0.8) = -2.08$$

B Multiplying Decimals by Powers of 10

There are some patterns that occur when we multiply a number by a power of 10, such as 10, 100, 1000, 10,000, and so on.

$23.6951 \times 10 = 236.951$ — Move the decimal point 1 *place* to the *right*. (10: 1 zero)

$23.6951 \times 100 = 2369.51$ — Move the decimal point 2 *places* to the *right*. (100: 2 zeros)

$23.6951 \times 100{,}000 = 2{,}369{,}510.$ — Move the decimal point 5 *places* to the *right* (insert a 0). (100,000: 5 zeros)

Notice that we move the decimal point the same number of places as there are zeros in the power of 10.

Multiplying Decimals by Powers of 10 such as 10, 100, 1000, 10,000, . . .

Move the decimal point to the *right* the same number of places as there are *zeros* in the power of 10.

Practice Problems 5–7

Multiply.

5. 23.7×10
6. 203.004×100
7. 1.15×1000

EXAMPLES Multiply.

5. $7.68 \times 10 = 76.8$ — 7.68

6. $23.702 \times 100 = 2370.2$ — 23.702

7. $(-76.3)(1000) = -76{,}300$ — −76.300 Recall that the product of a negative number and a positive number is a negative number.

There are also powers of 10 that are less than 1. The decimals 0.1, 0.01, 0.001, 0.0001, and so on, are examples of powers of 10 less than 1. Notice the pattern when we multiply by these powers of 10.

$569.2 \times 0.1 = 56.92$ — Move the decimal point 1 *place* to the *left*. (0.1: 1 decimal place)

$569.2 \times 0.01 = 5.692$ — Move the decimal point 2 *places* to the *left*. (0.01: 2 decimal places)

$569.2 \times 0.0001 = 0.05692$ — Move the decimal point 4 *places* to the *left* (insert one 0). (0.0001: 4 decimal places)

Answers

4. −7.02 **5.** 237 **6.** 20,300.4 **7.** 1150

Concept Check: False; 3 decimal places + 2 decimal places is 5 decimal places in the product.

Multiplying Decimals by Powers of 10 such as 0.1, 0.01, 0.001, 0.0001, . . .

Move the decimal point to the *left* the same number of places as there are *decimal places* in the power of 10.

EXAMPLES Multiply.

8. $42.1 \times 0.1 = 4.21$

9. $76{,}805 \times 0.01 = 768.05$

10. $(-9.2)(-0.001) = 0.0092$

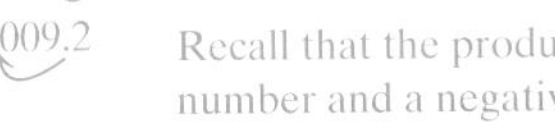

Recall that the product of a negative number and a negative number is a positive result.

Try the Concept Check in the margin.

Many times we see large numbers written, for example, in the form 281.4 million rather than in the longer standard notation. The next example shows how to interpret these numbers.

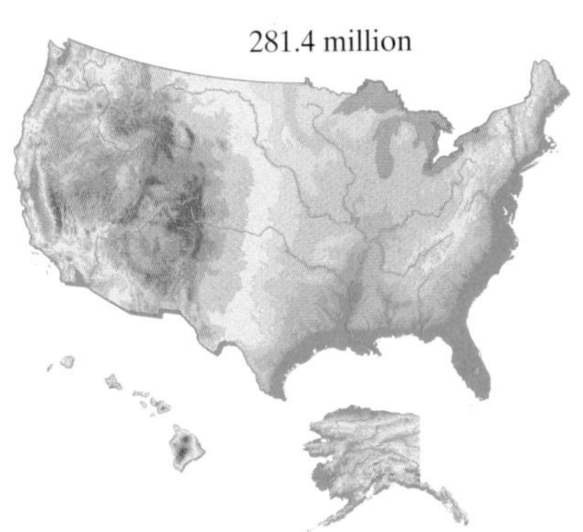

EXAMPLE 11 Population of the United States

The population of the United States is approximately 281.4 million. Write this number in standard notation. (*Source: The World Almanac*, 2003)

Solution: 281.4 million $= 281.4 \times 1$ million

$= 281.4 \times 1{,}000{,}000 = 281{,}400{,}000$

C Using Decimals as Replacement Values

Now let's practice working with variables.

EXAMPLE 12 Evaluate xy for $x = 2.3$ and $y = 0.44$.

Solution: Recall that xy means $x \cdot y$.

$xy = (2.3)(0.44)$

$$\begin{array}{r} 2.3 \\ \times\ 0.44 \\ \hline 92 \\ 920 \\ \hline 1.012 \end{array}$$

$= 1.012 \longleftarrow$

Practice Problems 8–10

Multiply.

8. 7.62×0.1
9. 1.9×0.01
10. 7682×0.001

Concept Check

True or false? 372.511 multiplied by 100 is 3.72511.

Practice Problem 11

There are 2158 thousand farms in the United States. Write this number in standard notation. (*Source: The World Almanac*, 2003)

Practice Problem 12

Evaluate $7y$ for $y = -0.028$.

Answers

8. 0.762 **9.** 0.019 **10.** 7.682 **11.** 2,158,000 **12.** −0.196

Concept Check: False.

Practice Problem 13

Is -5.5 a solution of the equation $-4x = 22$?

EXAMPLE 13 Is -9 a solution of the equation $3.7y = -3.33$?

Solution: Replace y with -9 in the equation $3.7y = -3.33$ to see if a true equation results.

$$3.7y = -3.33$$
$$3.7(-9) = -3.33 \quad \text{Replace } y \text{ with } -9.$$
$$-33.3 = -3.33 \quad \text{False.}$$

Since $-33.3 = -3.33$ is a false statement, -9 is **not** a solution of $3.7y = -3.33$.

D Finding the Circumference of a Circle

Recall that the distance around a polygon is called its perimeter. The distance around a circle is given a special name called the **circumference**, and this distance depends on the radius or the diameter of the circle.

Circumference of a Circle

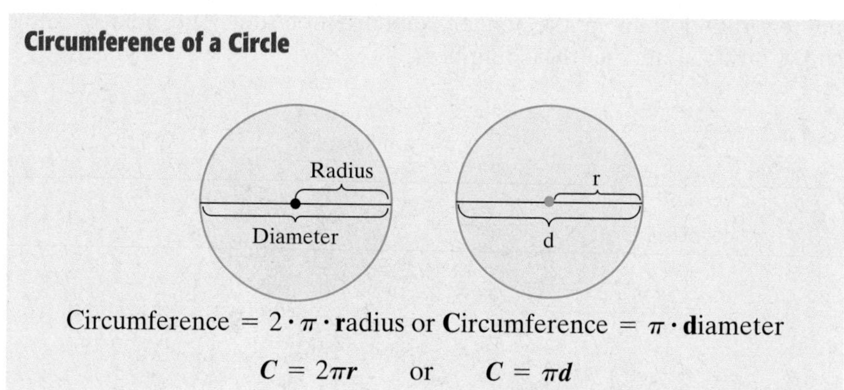

Circumference = $2 \cdot \pi \cdot$ **radius** or Circumference = $\pi \cdot$ **diameter**

$$C = 2\pi r \quad \text{or} \quad C = \pi d$$

The symbol π is the Greek letter pi, pronounced "pie." It is a constant between 3 and 4. A decimal approximation for π is 3.14. Also, a fraction approximation for π is $\frac{22}{7}$.

Practice Problem 14

Find the circumference of a circle whose radius is 11 meters. Then use the approximation 3.14 for π to approximate this circumference.

EXAMPLE 14 Circumference of a Circle

Find the circumference of a circle whose radius is 5 inches. Then use the approximation 3.14 for π to approximate the circumference.

Solution: Let $r = 5$ in the formula $C = 2\pi r$.

$$C = 2\pi r$$
$$= 2\pi \cdot 5$$
$$= 10\pi$$

5 inches

Next, replace π with the approximation 3.14.

$$C = 10\pi$$
$$\text{(is approximately)} \longrightarrow \approx 10(3.14)$$
$$= 31.4$$

The **exact** circumference or distance around the circle is 10π inches, which is **approximately** 31.4 inches.

Answers

13. yes **14.** 22π meters ≈ 69.08 meters

E Solving Problems by Multiplying Decimals

The solutions to many real-life problems are found by multiplying decimals. We continue using our four problem-solving steps to solve such problems.

EXAMPLE 15 Finding Total Cost of Materials for a Job

A college student is hired to paint a billboard with paint costing $2.49 per quart. If the job requires 3 quarts of paint, what is the total cost of the paint?

Solution:

1. UNDERSTAND. Read and reread the problem. The phrase "total cost" might make us think addition, but since this is repeated addition, let's multiply.
2. TRANSLATE.

In words:	Total cost	is	cost per quart of paint	times	number of quarts
	↓	↓	↓	↓	↓
Translate:	Total cost	=	2.49	×	3

3. SOLVE.

$$\begin{array}{r} 2.49 \\ \times \quad 3 \\ \hline 7.47 \end{array}$$

4. INTERPRET. *Check* your work. *State* your conclusion: The total cost of the paint is $7.47.

Practice Problem 15

Elaine Rehmann is fertilizing her garden. She uses 5.6 ounces of fertilizer per square yard. The garden measures 60.5 square yards. How much fertilizer does she need?

Answer

15. 338.8 ounces

FOCUS ON Mathematical Connections

MODELING MULTIPLICATION WITH DECIMAL NUMBERS

We can use an area model to represent multiplying decimal numbers. For instance, we can think of the 10 × 10 grid shown below on the left as representing 1, or a whole. We can show the product 0.5 × 0.2 by shading five tenths of the grid along one side of the square and shading two tenths of the grid along the other side.

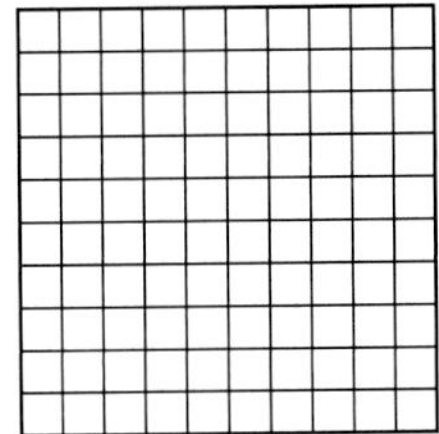

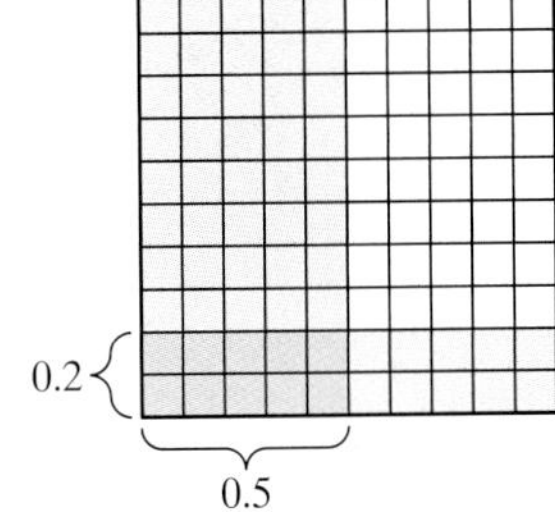

CRITICAL THINKING

1. What does the region where the shading overlaps represent? See whether you can figure it out by comparing it to the product of 0.5 × 0.2 found numerically. (Remember that one square within the grid represents one hundredth.)
2. How could you use this type of area model to represent the product of a decimal number (such as 0.6) and a whole number (such as 2)?
3. Use the model to show each of the following products:
   - **a.** 0.4 × 0.4
   - **b.** 0.9 × 0.1
   - **c.** 0.8 × 2
   - **d.** 0.7 × 1.5

FOCUS ON The Real World

McDONALD'S MENU

Today's worldwide chain of McDonald's fast-food restaurants started as a single drive-in restaurant run by the McDonald brothers, Dick and Mac, in San Bernardino, California. In the late 1940s, they decided to revamp their business, eliminating car hops in favor of counter-only service and reducing their 25-item menu to just 9 items. They focused on quick service, large volume, and low prices. The newly revamped business was hugely popular and attracted the attention of a milkshake machine salesman named Ray Kroc. Seeing the franchising potential of the McDonald brothers' fast-food formula, Ray Kroc became the exclusive McDonald's franchising agent for the entire United States in 1955. By 1956, there were 14 McDonald's restaurants, and that number had grown to 228 by 1960. Today, there are over 23,000 McDonald's restaurants worldwide—with a presence on every continent except Antarctica.

Ray Kroc's franchise McDonald's menu from 1955 is shown at the right. Local franchise owners have frequently been responsible for additions to the basic menu. The Filet-O-Fish sandwich was the first item to be added to the original menu in 1963 after it was developed by a franchise owner in Cincinnati, Ohio. The Big Mac sandwich, the invention of a Pittsburgh franchise owner, followed in 1968. And the Egg McMuffin was introduced nationally in 1973 after a Santa Barbara, California, franchisee searched for a breakfast product to offer at his restaurant. Other items that are standards on today's menu followed: the Quarter Pounder in 1972, Chicken McNuggets in 1983, and fresh tossed salads in 1987.

International McDonald's restaurants include local specialities on the menu. McDonald's restaurants in Norway include a grilled salmon sandwich called McLaks. McDonald's customers will find McSpaghetti on the menu in the Philippines, a Kiwi burger in New Zealand, and the Samurai Pork burger in Thailand.

GROUP ACTIVITY

Visit a McDonald's restaurant near you and record the present-day prices for each of the seven items on the original 1955 menu. (Unless otherwise noted, assume that sizes on the original menu are "small" for the sake of comparison with today's menu.) For each menu item, find the number of times greater that today's price is than the original price.

Original 1955 McDonald's Menu	
Hamburger	\$0.15
Cheeseburger	\$0.19
Fries	\$0.10
Small soft drink	\$0.10
Large soft drink	\$0.15
Coffee	\$0.10
Shake	\$0.20

Name ______________________ Section ____________ Date ____________

EXERCISE SET 5.3

A *Multiply. See Examples 1 through 4.*

1. $\begin{array}{r} 0.2 \\ \times\ 0.6 \\ \hline \end{array}$

2. $\begin{array}{r} 0.7 \\ \times\ 0.9 \\ \hline \end{array}$

3. $\begin{array}{r} 1.2 \\ \times\ 0.5 \\ \hline \end{array}$

4. $\begin{array}{r} 6.8 \\ \times\ 0.3 \\ \hline \end{array}$

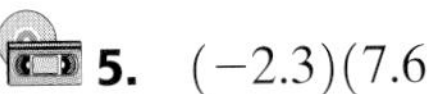

5. $(-2.3)(7.65)$

6. $(4.7)(-9.02)$

7. $(-6.89)(-5.7)$

8. $(-6.45)(-2.8)$

B *Multiply. See Examples 5 through 10.*

9. 6.5×10

10. 7.2×10

11. 6.5×0.1

12. 7.2×0.1

13. $(-7.093)(1000)$

14. $(-1.123)(1000)$

15. $(-9.83)(-0.01)$

16. $(-4.72)(-0.01)$

A **B** *Multiply. See Examples 1 through 10.*

17. $\begin{array}{r} 5.62 \\ \times\ 7.7 \\ \hline \end{array}$

18. $\begin{array}{r} 8.03 \\ \times\ 5.5 \\ \hline \end{array}$

19. $\begin{array}{r} 1.0047 \\ \times\ 8.2 \\ \hline \end{array}$

20. $\begin{array}{r} 2.0005 \\ \times\ 5.5 \\ \hline \end{array}$

21. $(147.9)(100)$

22. $(345.2)(10)$

23. $(937.62)(-0.01)$

24. $(-0.001)(562.01)$

25. 49.02×0.023

26. 30.09×0.0032

27. $(-0.023)(6.28)$

28. $(0.071)(-5.19)$

Write each number in standard notation. See Example 11.

29. The storage silos at the main Hershey chocolate factory in Hershey, Pennsylvania, can hold enough cocoa beans to make 5.5 billion Hershey's milk chocolate bars. (*Source:* Hershey Foods Corporation)

30. Of the 105.5 million American households that have televisions, 78.07 million have more than one television set. Write both numbers in standard notation. (*Source*: Nielsen Media Research)

31. About 36.4 million American households own at least one dog. (*Source:* American Pet Products Manufacturers Association)

32. The world's most popular amusement park in 2001 was Tokyo Disneyland, which had an attendance of 17.708 million visitors. (*Source: Amusement Business*, August 2002)

33. Americans lose more than 1.6 million hours each day stuck in traffic. (*Source:* United States Department of Transportation—Federal Highway Administration)

34. In 2001, the top advertiser in the United States was General Motors, who spent $3.37 billion on advertising. (*Source:* Crain Communications, Inc.)

C *Evaluate each expression for $x = 3$, $y = -0.2$, and $z = 5.7$. See Example 12.*

35. xy

36. yz

37. $xz - y$

38. $-5y + z$

39. Evaluate $2LW + 2WH + 2HL$ for $L = 10$, $W = 2.1$ and $H = 0.6$. (Surface area of a rectangular solid)

40. Evaluate $6s^2$ for $s = 2.7$. (Surface area of a cube)

Determine whether the given value is a solution of each given equation. See Example 13.

41. Is 14.2 a solution of $0.6x = 4.92$?

42. Is 1414 a solution of $100z = 14.14$?

43. Is 0.08 a solution of $-3x = -2.4$?

44. Is 17.8 a solution of $2x = 3.56$?

45. Is -4 a solution of $3.5y = -14$?

46. Is -3.6 a solution of $0.7x = -2.52$?

D *Find the circumference of each circle. Then use the approximation 3.14 for π and approximate each circumference. See Example 14.*

△ **47.**

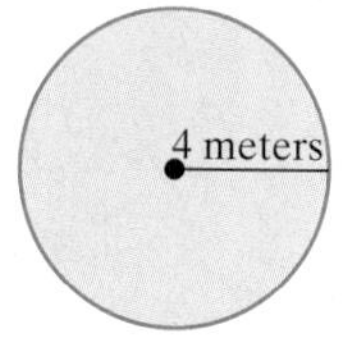

△ **48.**

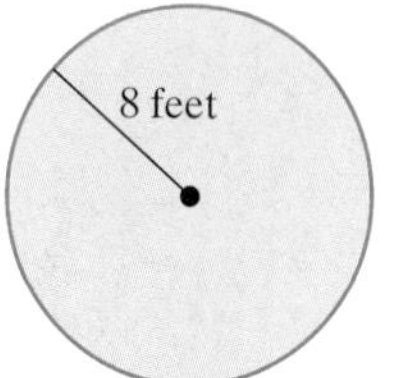

△ **49.**

10 centimeters

△ **50.** 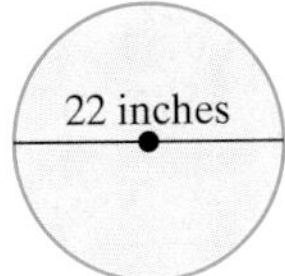

△ **51.** 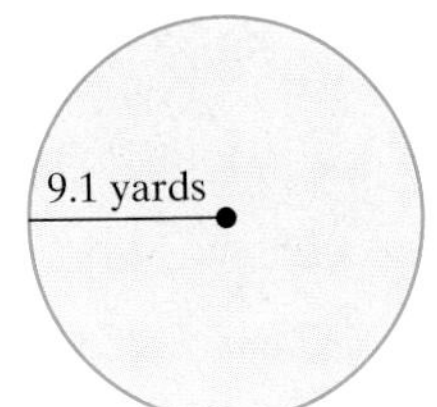

△ **52.** 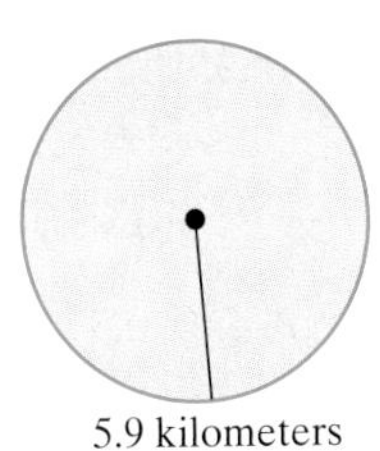

△ **53.** In 1893, the first ride called a Ferris wheel was constructed by Washington Gale Ferris. Its diameter was 250 feet. Find its circumference. Give an exact answer and an approximation using 3.14 for π. (*Source: The Handy Science Answer Book*, Visible Ink Press, 1994)

Source: *The Handy Science Answer Book*, Visible Ink Press, 1994

△ **54.** The radius of Earth is approximately 3950 miles. Find the distance around Earth at the equator. Give an exact answer and an approximation using 3.14 for π.(*Hint*: Find the circumference of a circle with radius 3950 miles.)

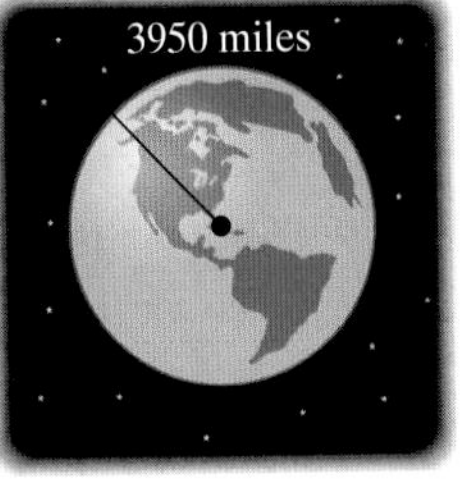

△ **55.** The London Eye, built for the Millennium celebration in London, resembles a gigantic ferris wheel with a diameter of 135 meters. If Adam Hawn rides the Eye for one revolution, find how far he travels. (*Hint*: Give an exact answer and an approximation using 3.14 for π. (*Source*: Londoneye.com)

△ **56.** The world's longest suspension bridge is the Akashi Kaikyo Bridge in Japan. This bridge has two circular caissons, which are under water foundations. If the diameter of a caisson is 80 meters, find its circumference. Give an exact answer and an approximation using 3.14 for π. (*Source*: *Scientific American*; How Things Work Today)

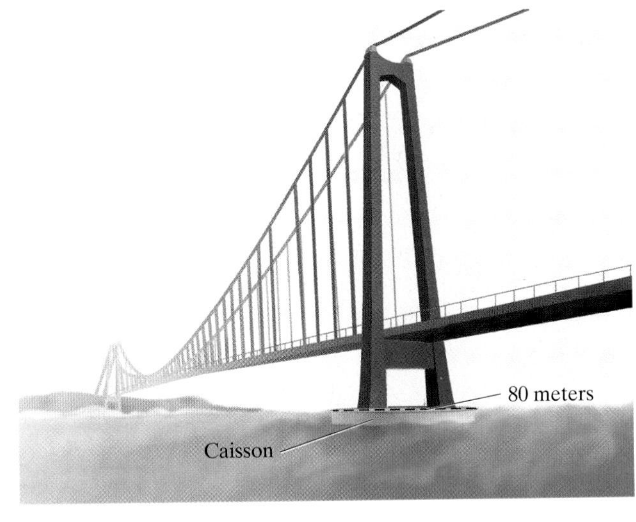

△ **57. a.** Approximate the circumference of each circle.

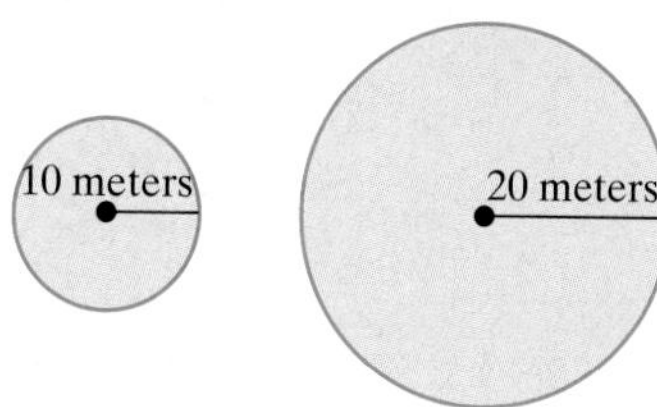

b. If the radius of a circle is doubled, is its corresponding circumference doubled?

△ **58. a.** Approximate the circumference of each circle.

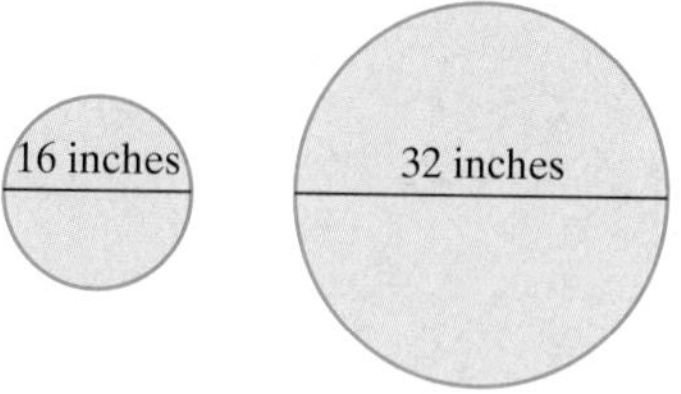

b. If the diameter of a circle is doubled, is its corresponding circumference doubled?

E *Solve. See Example 15.*

59. A 1-ounce serving of cream cheese contains 6.2 grams of saturated fat. How much saturated fat is in 4 ounces of cream cheese? (*Source: Home and Garden Bulletin No. 72*; U.S. Department of Agriculture)

60. A 3.5-ounce serving of lobster meat contains 0.1 gram of saturated fat. How much saturated fat is in 3 servings of lobster meat? (*Source:* The National Institute of Health)

61. In 2001, American farmers received an average of $2.80 per bushel of wheat. How much did a farmer receive for selling 1000 bushels of wheat? (*Source:* National Agricultural Statistics Service)

62. In 2001, American farmers received an average of $4.30 per bushel of soybeans. How much did a farmer receive for selling 10,000 bushels of soybeans? (*Source:* National Agricultural Statistics Service)

63. A meter is a unit of length in the metric system that is approximately equal to 39.37 inches. Sophia Wagner is 1.65 meters tall. Find her approximate height in inches.

64. The doorway to a room is 2.15 meters tall. Approximate this height in inches. (*Hint*: See Exercise 63.)

65. Jose Severos, an electrician for Central Power and Light, worked 40 hours last week. Calculate his pay before taxes for last week if his hourly wage is $13.88.

66. Maribel Chin, an assembly line worker, worked 20 hours last week. Her hourly rate is $8.52 per hour. Calculate her pay before taxes.

The table shows currency exchange rates for various countries on March 31, 2003. To find the amount of foreign currency equivalent to an amount of U.S. money, multiply the U.S. dollar amount by the exchange rate listed in the table. Use this table to answer Exercises 67–70.

Foreign Currency Exchange Rates	
Country	**Exchange Rate**
European Union Euro	0.9174
Australian Dollar	1.6543
British Pound	0.6333
Canadian Dollar	1.4695
Sri Lanka Rupee	96.97
Japanese Yen	118.0700

(*Source:* New York Federal Reserve Bank)

67. How many Japanese yen are equivalent to 675 U.S. dollars?

68. Suppose you wish to exchange 500 U.S. dollars into Sri Lanka Rupees. How much money, in Sri Lanka Rupees, would you receive?

69. The Goldthwaite family is taking a vacation to Canada. How much Canadian currency can they "buy" with 350 U.S. dollars?

70. A British tourist bought a $30 sweatshirt in New York City. What did he pay for the sweatshirt in British pounds?

Review and Preview

Divide. See Sections 1.7 and 4.3.

71. $130 \div 5$

72. $-495 \div -27$

73. $2016 \div 56$

74. $1863 \div 69$

75. $2920 \div 365$

76. $2916 \div 6$

77. $-\frac{24}{7} \div \frac{8}{21}$

78. $\frac{162}{25} \div \frac{9}{75}$

Combining Concepts

79. Find how far radio waves travel in 20.6 seconds. (Radio waves travel at a speed of $1.86 \times 100{,}000$ miles per second.)

80. If it takes radio waves approximately 8.3 minutes to travel from the sun to Earth, find approximately how far it is from the sun to Earth. (*Hint:* See Exercise 79.)

81. In your own words, explain how to find the number of decimal places in a product of decimal numbers.

82. In your own words, explain how to multiply by a power of 10.

83. Evaluate $4\pi r^2$ for $r = 7.68$. Round to the nearest hundredth. (Surface area of a sphere)

Internet Excursions

Go To: 

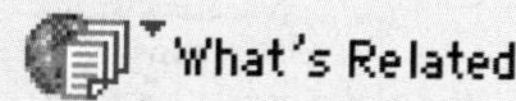

This World Wide Web address will provide you with access to a listing of current foreign currency exchange rates for the U.S. dollar, or a related site. The exchange rates in the "To United States Dollar" column are used to convert U.S. dollars to other currencies. The exchange rates in the "In United States Dollar" column are used to convert other currencies into U.S. dollars. Visit this site and use today's exchange rates to answer the following questions.

84. Convert \$650 U.S. dollars to each given currancy.
- **a.** Chinese renminbi
- **b.** Italian lira
- **c.** Swiss francs

85. Determine the cost in U.S. dollars for each item.
- **a.** A silk scarf costing 1400 Japanese yen
- **b.** A paperback book costing 9.95 Canadian dollars
- **c.** A sweater costing 50 British pounds

5.4 Dividing Decimals

OBJECTIVES

- **A** Divide decimals.
- **B** Divide decimals by powers of 10.
- **C** Evaluate expressions and check solutions with decimal replacement values.
- **D** Solve problems by dividing decimals.

A Dividing Decimals

Division of decimal numbers is similar to division of whole numbers. The only difference is the placement of a decimal point in the quotient. If the divisor is a whole number, see the example below, we divide as for whole numbers; then place the decimal point in the quotient directly above the decimal point in the dividend. Recall that division can be checked by multiplication.

EXAMPLE 1. Divide: $32\overline{)8.32}$

Solution: We divide as usual. The decimal point in the quotient is directly above the decimal point in the dividend.

$$\begin{array}{r} 0.26 \\ 32\overline{)8.32} \\ -64 \\ \hline 192 \\ -192 \\ \hline 0 \end{array}$$

0.26 ← quotient; divisor → 32, 8.32 ← dividend

Check:

$$\begin{array}{r} 0.26 \\ \times 32 \\ \hline 52 \\ 7\,80 \\ \hline 8.32 \end{array}$$

0.26 quotient; × 32 divisor; 8.32 dividend

Practice Problem 1

Divide: $48\overline{)34.08}$

If the divisor is not a whole number, we need to move the decimal point to the right until the divisor is a whole number before we divide.

$1.5\overline{)64.85}$

divisor: 1.5; dividend: 64.85

To understand how this works, let's rewrite

$1.5\overline{)64.85}$ as $\frac{64.85}{1.5}$

and then multiply the numerator and the denominator by 10.

$$\frac{64.85}{1.5} = \frac{64.85 \times 10}{1.5 \times 10} = \frac{648.5}{15}$$

which can be written as $15\overline{)648.5}$. Notice that

$1.5\overline{)64.85}$ is equivalent to $15.\overline{)648.5}$

The decimal points in the dividend and the divisor were both moved one place to the right, and the divisor is now a whole number. This procedure is summarized next.

Dividing by a Decimal

Step 1. Move the decimal point in the divisor to the right until the divisor is a whole number.

Step 2. Move the decimal point in the dividend to the right the *same number of places* as the decimal point was moved in Step 1.

Step 3. Divide. Place the decimal point in the quotient directly over the moved decimal point in the dividend.

Answer

1. 0.71

Practice Problem 2

Divide: $5.6\overline{)166.88}$

Practice Problem 3

Divide: $-2.808 \div (-104)$

Practice Problem 4

Divide: $23.4 \div 0.57$. Round the quotient to the nearest hundredth.

Answers

2. 29.8 **3.** 0.027 **4.** 41.05

EXAMPLE 2 Divide: $2.3\overline{)10.764}$

Solution: Move the decimal points in the divisor and the dividend one place to the right so that the divisor is a whole number.

$$2.3\overline{)10.764} \quad \text{becomes} \quad \begin{array}{r} 4.68 \\ 23.\overline{)107.64} \\ \underline{-92} \\ 15\,6 \\ \underline{-13\,8} \\ 1\,84 \\ \underline{-1\,84} \\ 0 \end{array}$$

To check, see that $4.68 \times 2.3 = 10.764$.

EXAMPLE 3 Divide: $-5.98 \div 115$

Solution: Recall that a negative number divided by a positive number gives a negative quotient. The divisor is a whole number, so the decimal point in the divisor is not moved.

$$\begin{array}{r} 0.052 \\ 115\overline{)5.980} \\ \underline{-5\,75} \\ 230 \\ \underline{-230} \\ 0 \end{array} \quad \leftarrow \text{Insert one 0.}$$

Thus $-5.98 \div 115 = -0.052$.

EXAMPLE 4 Divide: $17.5 \div 0.48$. Round the quotient to the nearest hundredth.

Solution: Move the decimal points in the divisor and the dividend 2 places.

$$\begin{array}{r} 36.458 \approx 36.46 \\ 48.\overline{)1750.000} \\ \underline{-144} \\ 310 \\ \underline{-288} \\ 220 \\ \underline{-192} \\ 280 \\ \underline{-240} \\ 400 \\ \underline{-384} \\ 16 \end{array}$$

hundredths place (arrow to the 5 in 36.458); approximately (arrow to ≈)

If rounding to the nearest hundredth, carry the division process out to one more decimal place, the thousandths place.

Try the Concept Check in the margin.

B Dividing Decimals by Powers of 10

There are patterns that occur when we divide decimals by powers of 10 such as 10, 100, 1000, and so on.

$$\frac{569.2}{10} = 56.92$$ Move the decimal point 1 *place* to the *left*.

1 zero

$$\frac{569.2}{10{,}000} = 0.05692$$ Move the decimal point 4 *places* to the *left*.

4 zeros

This pattern suggests the following rule.

Dividing Decimals by Powers of 10 such as 10, 100, 1000, ...

Move the decimal point of the dividend to the *left* the same number of places as there are *zeros* in the power of 10.

Notice that this is the same pattern as *multiplying* by powers of 10 such as 0.1, 0,01, or 0.001. This is because dividing by a power of 10 such as 100 is the same as multiplying by its reciprocal $\frac{1}{100}$, or 0.01. For example,

$$\frac{569.2}{10} = 569.2 \times \frac{1}{10} = 569.2 \times 0.1 = 56.92$$

EXAMPLES Divide.

5. $\frac{786}{10{,}000} = 0.0786$ Move the decimal point 4 *places* to the *left*.

4 zeros

6. $\frac{0.12}{10} = 0.012$ Move the decimal point 1 *place* to the *left*.

1 zero

Try the Concept Check in the margin.

C Using Decimals as Replacement Values

EXAMPLE 7 Evaluate $x \div y$ for $x = 2.5$ and $y = 0.05$.

Solution: Replace x with 2.5 and y with 0.05.

$$x \div y = 2.5 \div 0.05 = 50$$

$0.05\overline{)2.5}$ becomes $5\overline{)250}$ with quotient 50

Concept Check

If a quotient is to be rounded to the nearest thousandth, to what place should the division be carried out?

Practice Problems 5–6

Divide.

5. $\frac{28}{1000}$

6. $\frac{8.56}{100}$

Concept Check

Describe how to check the answer in Example 5.

Practice Problem 7

Evaluate $x \div y$ for $x = 0.035$ and $y = 0.02$.

Answers

5. 0.028 **6.** 0.0856 **7.** 1.75

Concept Check: The ten-thousandths place.

Concept Check: Answers may vary.

Practice Problem 8

Is 39 a solution of the equation $\frac{x}{100} = 3.9$?

EXAMPLE 8 Is 720 a solution of the equation $\frac{y}{100} = 7.2$?

Solution: Replace y with 720 to see if a true statement results.

$\frac{y}{100} = 7.2$ Original equation

$\frac{720}{100} = 7.2$ Replace y with 720.

$7.2 = 7.2$ True.

Since $7.2 = 7.2$ is a true statement, 720 is a solution of the equation.

D Solving Problems by Dividing Decimals

Many real-life problems involve dividing decimals.

Practice Problem 9

A bag of fertilizer covers 1250 square feet of lawn. Tim Parker's lawn measures 14,800 square feet. How many bags of fertilizer does he need? If he can buy whole bags of fertilizer only, how many whole bags does he need?

EXAMPLE 9 Calculating Materials Needed for a Job

A gallon of paint covers a 250-square-foot area. If Betty Adkins wishes to paint a wall that measures 1450 square feet, how many gallons of paint does she need? If she can buy gallon containers of paint only, how many gallon containers does she need?

Solution:

1. UNDERSTAND. Read and reread the problem. To find the number of gallons, divide 1450 by 250.
2. TRANSLATE.

In words:	Number of gallons	is	square feet	divided by	square feet per gallon
	↓	↓	↓	↓	↓
Translate:	Number of gallons	=	1450	÷	250

3. SOLVE.

$$\begin{array}{r} 5.8 \\ 250\overline{)1450.0} \\ -1250 \\ \hline 200\ 0 \\ -200\ 0 \\ \hline 0 \end{array}$$

4. INTERPRET. *Check* your work. *State* your conclusion: Betty needs 5.8 gallons of paint. If she can buy gallon containers of paint only, she needs 6 gallon containers of paint to complete the job.

Answers

8. no
9. 11.84 bags; 12 bags

Name ______________________ Section __________ Date __________

EXERCISE SET 5.4

 *Divide. See Examples 1 through 3.*

1. $0.47 \div 5$

2. $11.8 \div 2$

3. $-18 \div 0.06$

4. $20 \div (-0.04)$

 **5.** $4.756 \div 0.82$

6. $3.312 \div 0.92$

7. $-36.3 \div -5.5$

8. $-21.78 \div -2.2$

Divide, and round each quotient as indicated. See Example 4.

9. Divide 429.34 by 2.4 and round the quotient to the nearest hundred.

10. Divide 54.8 by 2.6 and round the quotient to the nearest ten.

 **11.** Divide 0.549 by 0.023 and round the quotient to the nearest hundredth.

12. Divide 0.0453 by 0.98 and round the quotient to the nearest thousandth.

13. Divide -45.23 by 0.4 and round the quotient to the nearest ten.

14. Divide -983.32 by 0.061 and round the quotient to the nearest thousand.

B *Divide. See Examples 5 and 6.*

15. $54.982 \div 100$

16. $342.54 \div 100$

17. $12.9 \div (-1000)$

18. $13.49 \div (-10{,}000)$

19. $87 \div 10$

20. $0.27 \div 10$

  *Divide. See Examples 1 through 6.*

21. $1.239 \div 3$

22. $0.54 \div 12$

23. Divide -4.2 by 0.6

24. Divide 3.6 by 0.9

25. $1.296 \div 0.27$

26. $-2.176 \div 0.34$

27. Divide 42 by 0.02

28. Divide 24 by 0.03

29. Divide -18 by -0.6

30. Divide 20 by 0.4

31. Divide 35 by 0.005

32. Divide -35 by -0.0007

33. $-1.104 \div 1.6$

34. $-2.156 \div 0.98$

35. $-2.4 \div (-100)$

36. $-86.79 \div (-1000)$

37. $\dfrac{4.615}{0.071}$

38. $\dfrac{23.8}{-0.035}$

39. Divide 0.00263 by 8.9 and round the quotient to the nearest ten-thousandth.

40. Divide 200 by 0.054 and round the quotient to the nearest thousand.

41. Divide 500 by 0.0043 and round the quotient to the nearest ten-thousand.

42. Divide 4.56 by 9.34 and round the quotient to the nearest hundredth.

C *Evaluate each expression for $x = 5.65$, $y = -0.8$, and $z = 4.52$. See Example 7.*

43. $z \div y$

44. $z \div x$

45. $x \div y$

46. $z \div 2$

47. $x \div 5$

48. $y \div 2$

Determine whether the given values are solutions of the given equations. See Example 8.

49. $\frac{x}{4} = 3.04$; $x = 12.16$

50. $\frac{y}{8} = 0.89$; $y = 7.12$

51. $\frac{x}{4.3} = 2$; $x = 0.86$

52. $\frac{z}{3.7} = 5$; $z = 185$

53. $\frac{z}{10} = 0.8$; $z = 8$

54. $\frac{x}{100} = 0.23$; $x = 23$

D *Solve. See Example 9.*

55. Dorren Schmidt pays $73.86 per month to pay back a loan of $1772.64. In how many months will the loan be paid off?

56. Josef Jones is painting the walls of a room. The walls have a total area of 546 square feet. If a quart of paint covers 52 square feet, how many quarts of paint does he need to buy?

57. The leading money winner in men's professional golf in 2001 was Tiger woods. He earned $5,687,777. Suppose he had earned this working 40 hours per week for a year. Determine his hourly wage to the nearest cent. (*Source: The World Almanac*, 2003)

58. Juanita Gomez bought unleaded gasoline for her car at $1.169 per gallon. She wanted to keep a record of how many gallons her car is using but forgot to write down how many gallons she purchased. She wrote a check for $27.71 to pay for it. How many gallons, to the nearest tenth of a gallon, did she buy?

59. A pound of fertilizer covers 39 square feet of lawn. Vivian Bulgakov's lawn measures 7883.5 square feet. How much fertilizer, to the nearest tenth of a pound, does she need to buy?

60. A page of a book contains about 1.5 kilobytes of information. If a computer disk can hold 740 kilobytes of information, how many pages of a book can be stored on one computer disk? Round to the nearest tenth of a page.

61. There are approximately 39.37 inches in 1 meter. How many meters, to the nearest tenth of a meter, are there in 200 inches?

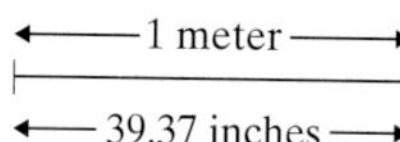

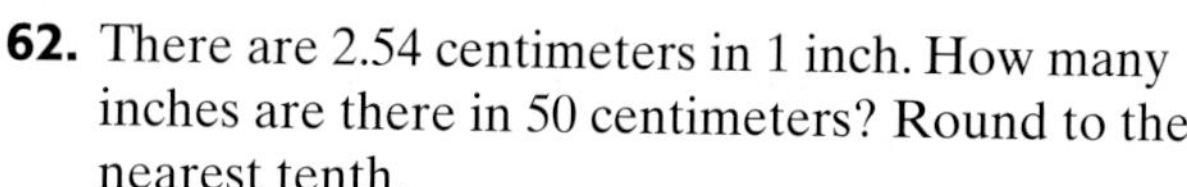

62. There are 2.54 centimeters in 1 inch. How many inches are there in 50 centimeters? Round to the nearest tenth.

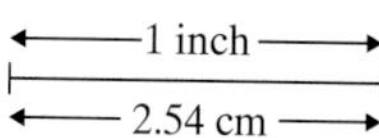

63. In the United States, an average child will wear down 730 crayons by his or her tenth birthday. Find the number of boxes of 64 crayons this is equivalent to. Round to the nearest tenth. (*Source:* Binney & Smith Inc.)

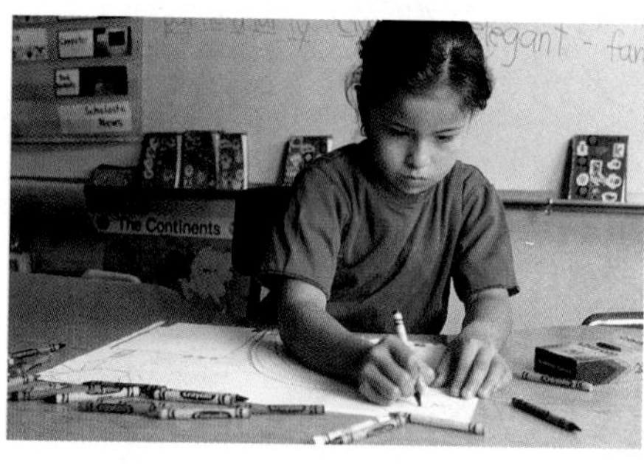

64. In 2001, American farmers received an average of \$66.90 per 100 pounds of lamb. What was the average price per pound for lambs? (*Source:* National Agricultural Statistics Service)

65. During the 24 Hours of Le Mans endurance auto race in 2002, the winning team of Frank Biela, Tom Kristensen, and Emanuele Pirro drove 3180.00 miles in 24 hours. What was their average speed in miles per hour? (*Source*: 2003 *World Almanac*)

66. In 2002, runner Khalid Khannouchi set a new world record for a marathon. His time was 2.094 hours. What was his average speed in miles per hour? Round to the nearest tenth. (*Hint*:A marathon is approximately 26.2 miles). (*Source*: International Amateur Athletic Federation)

A child is to receive a dose of 0.5 teaspoon of cough medicine every 4 hours. If the bottle contains 4 fluid ounces, answer Exercises 67 through 69.

67. A fluid ounce equals 6 teaspoons. How many teaspoons are in 4 fluid ounces?

68. The bottle of medicine contains how many doses for the child?

69. If the child takes a dose every four hours, how many days will the medicine last?

70. During the 2002 National Hockey League season, Jarome Iginia of the Calgary Flames scored 96 points. The Calgary Flames played a total of 82 games during the season. Find the average number of points scored by Jarome per game. Round to the nearest tenth.

Review and Preview

Round each decimal to the given place value. See Section 5.1.

71. 345.219, nearest hundredth

72. 902.155, nearest hundredth

73. -1000.994, nearest tenth

74. 234.1029, nearest thousandth

Use the order of operations to simplify each expression. See Section 1.8.

75. $2 + 3 \cdot 6$

76. $9 \cdot 2 + 8 \cdot 3$

77. $20 - 10 \div (-5)$

78. $(20 - 10) \div 5$

Combining Concepts

Recall from Section 1.7 that the average of a list of numbers is their total divided by how many numbers there are in the list. Use this to find the average of the numbers listed. If necessary, round to the nearest tenth.

79. 86, 78, 91, 85

80. 56, 75, 80

81. 9.6, 8.5, 4.1, 4.1, 9.8

82. 1.7, 3.8, 4.6, 8.7

83. The area of this rectangle is 38.7 square feet. If its width is 4.5 feet, find its length.

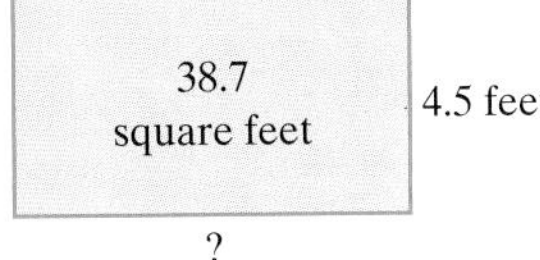

84. The perimeter of the square is 180.8 centimeters. Find the length of a side.

Perimeter is
180.8
centimeters

?

85. When dividing decimals, describe the process you use to place the decimal point in the quotient.

86. In your own words, describe how to quickly divide a number by a power of 10 such as 10, 100, 1000, etc.

To convert wind speeds in miles per hour to knots, divide by 1.15. Use this information and the Saffir-Simpson Hurricane Intensity chart below to answer Exercises 87–88. Round to the nearest tenth.

87. The chart gives wind speeds in miles per hour. What is the range of wind speeds for a Category 1 hurricane in knots?

88. What is the range of wind speeds for a Category 4 hurricane in knots?

Saffir-Simpson Hurricane Intensity Scale

Category	Wind Speed	Barometric Pressure [inches of mercury (Hg)]	Storm Surge	Damage Potential
1 (Weak)	75–95 mph	≥28.94 in.	4–5 ft	Minimal damage to vegetation
2 (Moderate)	96–110 mph	28.50–28.93 in.	6–8 ft	Moderate damage to houses
3 (Strong)	111–130 mph	27.91–28.49 in.	9–12 ft	Extensive damage to small buildings
4 (Very Strong)	131–155 mph	27.17–27.90 in.	13–18 ft	Extreme structural damage
5 (Devastating)	>155 mph	<27.17 in.	>18 ft	Catastrophic building failures possible

Name ______________________ Section ____________ Date ____________

Integrated Review—Operations on Decimals

Perform each indicated operation.

1. $1.6 + 0.97$ **2.** $3.2 + 0.85$ **3.** $9.8 - 0.9$ **4.** $10.2 - 6.7$

5. $\begin{array}{r} 0.8 \\ \times\ 0.2 \\ \hline \end{array}$ **6.** $\begin{array}{r} 0.6 \\ \times\ 0.4 \\ \hline \end{array}$ **7.** $8\overline{)2.16}$ **8.** $6\overline{)3.12}$

9. $(9.6)(-0.5)$ **10.** $(-8.7)(-0.7)$ **11.** $\begin{array}{r} 123.6 \\ -\ \ 48.04 \\ \hline \end{array}$ **12.** $\begin{array}{r} 325.2 \\ -\ 36.08 \\ \hline \end{array}$

13. $-25 + 0.026$ **14.** $0.125 + (-44)$ **15.** $29.24 \div (-3.4)$ **16.** $-10.26 \div (-1.9)$

17. -2.8×100 **18.** 1.6×1000 **19.** $\begin{array}{r} 96.21 \\ 7.028 \\ +121.7 \\ \hline \end{array}$ **20.** $\begin{array}{r} 0.268 \\ 1.93 \\ +142.881 \\ \hline \end{array}$

21. $-25.76 \div -46$ **22.** $-27.09 \div 43$ **23.** $\begin{array}{r} 12.004 \\ \times\ \ \ 2.3 \\ \hline \end{array}$ **24.** $\begin{array}{r} 28.006 \\ \times\ \ \ 5.2 \\ \hline \end{array}$

25. Subtract 4.6 from 10 **26.** Subtract 18 from 0.26 **27.** $-268.19 - 146.25$

Answers

1. ____________
2. ____________
3. ____________
4. ____________
5. ____________
6. ____________
7. ____________
8. ____________
9. ____________
10. ____________
11. ____________
12. ____________
13. ____________
14. ____________
15. ____________
16. ____________
17. ____________
18. ____________
19. ____________
20. ____________
21. ____________
22. ____________
23. ____________
24. ____________
25. ____________
26. ____________
27. ____________

28. ______

29. ______

30. ______

31. ______

32. ______

33. ______

34. ______

35. ______

36. ______

37. ______

38. ______

28. $-860.18 - 434.85$

29. $\dfrac{2.958}{-0.087}$

30. $\dfrac{-1.708}{0.061}$

31. $160 - 43.19$

32. $120 - 101.21$

33. 15.62×10

34. $15.62 \div 10$

35. $15.62 + 10$

36. $15.62 - 10$

37. According to the federal Nutrition Labeling and Education Act of 1990, a food manufacturer may label a food product "Fat Free" if it contains less than 0.5 grams of fat per serving. A new type of cookie contains 0.38 grams of fat per cookie. Can the packaging for this cookie be labeled "Fat Free"? Explain.

38. The highest NBA scorer for the 2002–2003 regular season was Tracy McGrady of the Orlando Magic. He scored a total of 2407 points in 75 games. Find his average number of points per game to the nearest tenth of a point. (*Source:* National Basketball Association)

5.5 Estimating and Order of Operations

OBJECTIVES

- (A) Estimate operations on decimals.
- (B) Simplify expressions containing decimals using the order of operations.
- (C) Evaluate expressions and check solutions given decimal replacement values.

(A) Estimating Operations on Decimals

Estimating sums, differences, products, and quotients of decimal numbers is an important skill whether you use a calculator or perform decimal operations by hand. When you can estimate results as well as calculate them, you can judge whether the calculations are reasonable. If they aren't, you know you've made an error somewhere along the way. To estimate, we round numbers.

EXAMPLE 1 Subtract: \$78.62 − \$16.85. Then estimate the difference to see whether the proposed result is reasonable by rounding each decimal to the nearest dollar and then subtracting.

Solution:

Given		Estimate
\$78.62	rounds to	\$79
−\$16.85	rounds to	−\$17
\$61.77		\$62

The estimated difference is \$62, so \$61.77 is reasonable. ●

Practice Problem 1

Subtract: \$65.34 − \$14.68. Then estimate the difference to see whether the proposed result is reasonable.

Try the Concept Check in the margin.

Concept Check

Should you estimate the sum 21.98 + 42.36 as 30 + 50 = 80? Why or why not?

EXAMPLE 2 Multiply: 28.06 × 1.95. Then estimate the product to see whether the proposed result is reasonable.

Solution:

Given	Estimate 1	Estimate 2
28.06	28	30
× 1.95	× 2	× 2
14030	56	60
252540		
280600		
54.7170		

The answer 54.7170 is reasonable. ●

Practice Problem 2

Multiply: 30.26 × 2.98. Then estimate the product to see whether the proposed solution is reasonable.

As shown in Example 2, estimated results will vary depending on what estimates are used. Notice that estimating results is a good way to see whether the decimal point has been correctly placed.

Helpful Hint

When rounding to check a calculation, you may want to round the numbers to a place value of your choosing so that your estimates are easy to compute mentally.

Answers

1. \$50.66 **2.** 90.1748

Concept Check: No; each number is rounded incorrectly. The estimate is too high.

Practice Problem 3

Divide: 713.7 ÷ 91.5. Then estimate the quotient to see whether the proposed answer is reasonable.

EXAMPLE 3 Divide: 272.356 ÷ 28.4. Then estimate the quotient to see whether the proposed result is reasonable.

Solution: **Given** **Estimate**

$$\begin{array}{r} 9.59 \\ 284.\overline{)2723.56} \\ -2556 \\ \hline 1675 \\ -1420 \\ \hline 2556 \\ -2556 \\ \hline 0 \end{array} \qquad \begin{array}{r} 9 \\ 30\overline{)270} \end{array} \quad \text{or} \quad \begin{array}{r} 9 \\ 300\overline{)2700} \end{array}$$

The estimate is 9, so 9.59 is reasonable.

Practice Problem 4

Using the figure below, estimate how much farther it is between Dodge City and Pratt than it is between Garden City and Dodge City by rounding each given distance to the nearest mile.

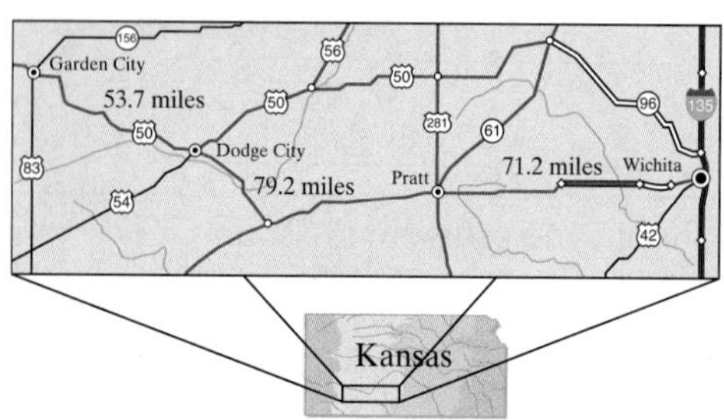

EXAMPLE 4 Estimating Distance

Use the figure in the margin to estimate the distance in miles between Garden City, Kansas, and Wichita, Kansas, by rounding each given distance to the nearest ten.

Solution:

Calculated Distance		Estimate
53.7	rounds to	50
79.2	rounds to	80
+71.2	rounds to	+70
		200

The distance between Garden City and Wichita is approximately 200 miles. (The calculated distance is 204.1 miles.)

Practice Problem 5

Use the estimation to determine whether the calculaor result is reasonable or not.

Simplify: 51.793 ÷ 13

Calculator display: 19.612

EXAMPLE 5 Use estimation to determine whether the calculator result is reasonable or not. (For example, a result that is not reasonable can occur if proper keys are not pressed.)

Simplify: 82.064 ÷ 23   Calculator display: 35.68

Solution: Round each number to the nearest 10. Since 80 ÷ 20 = 4, the calculator display 35.68 is not reasonable.

B Simplifying Expressions Containing Decimals

In the remaining examples, we will review the order of operations by simplifying expressions that contain decimals.

Order of Operations

1. Do all operations within grouping symbols such as parentheses or brackets.
2. Evaluate any expressions with exponents.
3. Multiply or divide in order from left to right.
4. Add or subtract in order from left to right.

Answers

3. 7.8 **4.** 25 miles **5.** not reasonable

EXAMPLE 6 Simplify: $-0.5(8.6 - 1.2)$

Solution: According to the order of operations, simplify inside the parentheses first.

$$\begin{aligned} -0.5(8.6 - 1.2) &= -0.5(7.4) && \text{Subtract.} \\ &= -3.7 && \text{Multiply. The product of a negative number and a positive number is a negative result.} \end{aligned}$$

Practice Problem 6

Simplify: $-8.6(3.2 - 1.8)$

EXAMPLE 7 Simplify: $(-1.3)^2 + 2.4$

Solution: Recall the meaning of an exponent.

$$\begin{aligned} (-1.3)^2 &= (-1.3)(-1.3) + 2.4 && \text{Use the definition of an exponent.} \\ &= 1.69 + 2.4 && \text{Multiply. The product of two negative numbers is a positive number} \\ &= 4.09 && \text{Add.} \end{aligned}$$

Practice Problem 7

Simplify: $(0.7)^2 + 2.1$

EXAMPLE 8 Simplify: $\dfrac{0.7 + 1.84}{0.4}$

Solution: First simplify the numerator of the fraction. Then divide.

$$\begin{aligned} \frac{0.7 + 1.84}{0.4} &= \frac{2.54}{0.4} && \text{Simplify the numerator.} \\ &= 6.35 && \text{Divide.} \end{aligned}$$

Practice Problem 8

Simplify: $\dfrac{8.78 - 2.8}{20}$

C Using Decimals as Replacement Values

EXAMPLE 9 Evaluate $-2x + 5$ for $x = 3.8$.

Solution: Replace x with 3.8 in the expression $-2x + 5$ and simplify.

$$\begin{aligned} -2x + 5 &= -2(3.8) + 5 && \text{Replace } x \text{ with 3.8.} \\ &= -7.6 + 5 && \text{Multiply.} \\ &= -2.6 && \text{Add.} \end{aligned}$$

Practice Problem 9

Evaluate $1.7y - 2$ for $y = 2.3$.

EXAMPLE 10 Determine whether -3.3 is a solution of $2x - 1.2 = -7.8$.

Solution: Replace x with -3.3 in the equation $2x - 1.2 = -7.8$ to see if a true statement results.

$$\begin{aligned} 2x - 1.2 &= -7.8 \\ 2(-3.3) - 1.2 &= -7.8 && \text{Replace } x \text{ with} \\ -6.6 - 1.2 &= -7.8 && \text{Multiply.} \\ -7.8 &= -7.8 && \text{True.} \end{aligned}$$

Since $-7.8 = -7.8$ is a true statement, -3.3 is a solution of $2x - 1.2 = -7.8$.

Practice Problem 10

Is -2.1 a solution of $3x + 7.5 = 1.2$?

Answers

6. -12.04 **7.** 2.59 **8.** 0.299 **9.** 1.91 **10.** yes

FOCUS ON Business and Career

TIME SHEETS

Many jobs require employees to fill out a weekly time sheet. A time sheet allows employees to show how much time they worked on various projects and to find how much time they worked overall. For ease of calculation, it is standard to round times on a time sheet to the nearest quarter hour. The following is an example of a time sheet.

Time Sheet for the week of: 5/9								
Employee: Darla Williams						**Employee Number: 630087**		
	Project Number	**Monday**	**Tuesday**	**Wednesday**	**Thursday**	**Friday**	**Saturday/Sunday**	**TOTALS**
1	315	5.5	1		3.25	1.5		
2	629	2.5		4	1.25	1.5	3	
3	471		3.25	2		1.75		
4	511		3.75	1.5	2.75			
5	512				0.75	2.5	3.25	
6								
7								
8								
9								
10								
	TOTALS							

CRITICAL THINKING

Find the column totals for columns Monday through Saturday/Sunday on the time sheet. What do these column totals represent?

1. Find the row totals for rows 1 through 5 on the time sheet. What do these row totals represent?
2. Find the sum of the row totals and the sum of the column totals. Are these sums the same? If so, explain why. Otherwise, explain why having different row sums and column sums should prompt you to recheck your math.
3. Darla Williams is a full-time employee who is expected to work 40 hours per week. Any time over 40 hours is considered overtime. Did Darla work any overtime this week? If so, how much?
4. Darla is paid an hourly rate of \$16.60 per hour. For overtime, she receives "time and a quarter," or 1.25 times her normal hourly pay rate. How much will Darla be paid for this week of work?
5. Describe how Darla's time was distributed among the five projects on which she worked during this week.

Name ______________________ Section __________ Date __________

EXERCISE SET 5.5

A *Perform each indicated operation. Then estimate to see whether each proposed result is reasonable. See Examples 1 through 3.*

1. 4.9 − 2.1

2. 7.3 − 3.8

3. 6 × 483.11

4. 98.2 × 405.8

5. 62.16 ÷ 14.8

6. 186.75 ÷ 24.9

7. 69.2 + 32.1 + 48.57

8. 79.52 + 21.47 + 97.2

9. 34.92 − 12.03

10. 349.87 − 251.32

Solve. See Example 4.

11. Estimate the perimeter of the rectangle by rounding each measurement to the nearest whole inch.

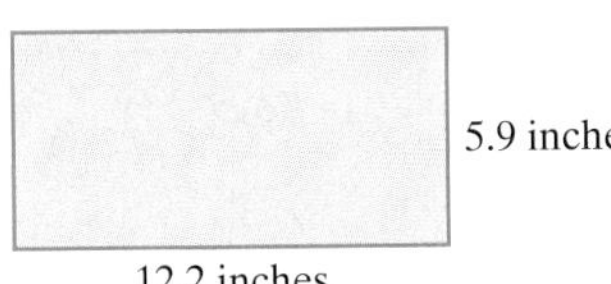

12. Joe Armani wants to put plastic edging around his rectangular flower garden to help control the weeds. The length and width are 16.2 meters and 12.9 meters, respectively. Estimate how much edging he needs by rounding each measurement to the nearest whole meter.

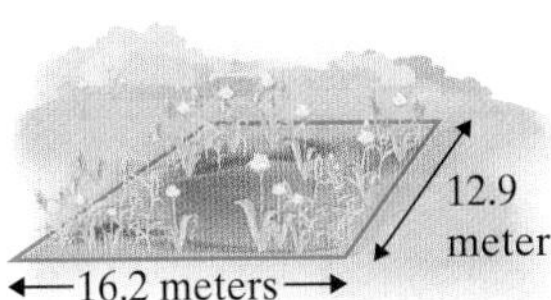

13. Estimate the perimeter of the triangle by rounding each measurement to the nearest whole foot.

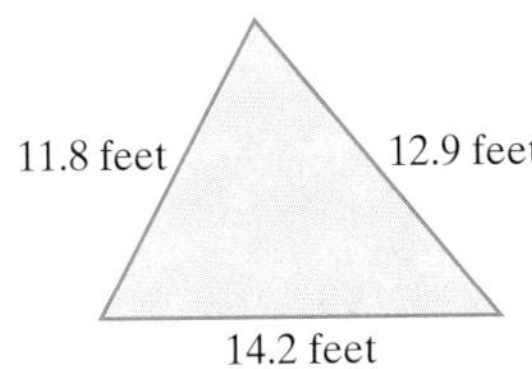

14. The area of a triangle is area $= \frac{1}{2} \cdot \text{base} \cdot \text{height}$.

Approximate the area of a triangle whose base is 21.9 centimeters and whose height is 9.9 centimeters. Round each measurement to the nearest whole centimeter.

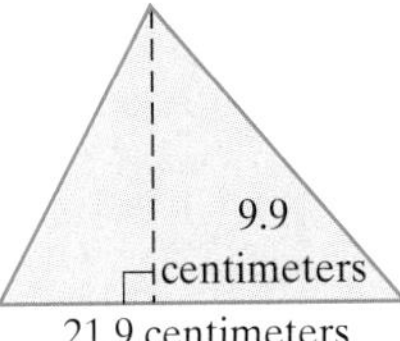

△ **15.** Use 3.14 for π to approximate the circumference of a circle whose radius is 7 meters.

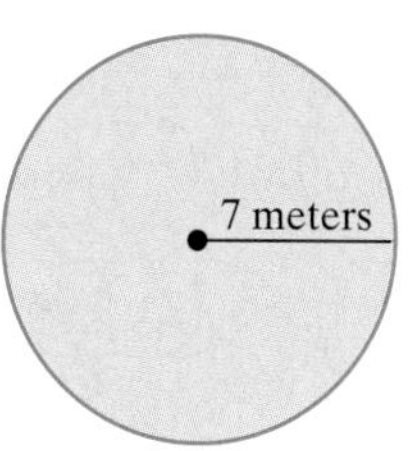

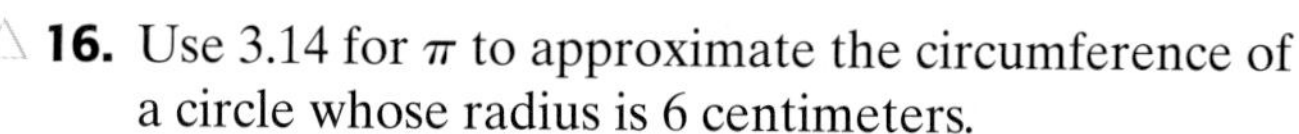

△ **16.** Use 3.14 for π to approximate the circumference of a circle whose radius is 6 centimeters.

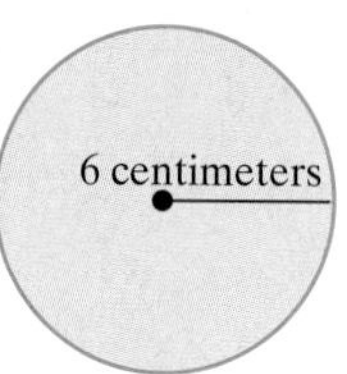

17. Mike and Sandra Hallahan are taking a trip and plan to travel about 1550 miles. Their car gets 32.8 miles to the gallon. Approximate how many gallons of gasoline their car will use by rounding 32.8 to the nearest ten. Round the result to the nearest gallon.

18. Refer to Exercise 17. Suppose gasoline costs $1.659 per gallon. Approximate how much money Mike and Sandra need for gasoline for their 1550-mile trip by rounding $1.659 to the nearest cent.

19. Robert and Kathi Hawn purchased a new car. They pay $398.79 per month on their car loan for 5 years. Approximate how much they will pay on the loan altogether by rounding the monthly payment to the nearest hundred.

20. Yoshikazu Sumo is purchasing the following groceries. He has only a ten dollar bill in his pocket. Estimate the cost of each item to see whether he has enough money. Bread, $1.49; milk, $2.09; carrots, $0.97; corn, $0.89; salt, $0.53; and butter, $2.89.

21. Estimate the total distance to the nearest mile between Grove City and Jerome by rounding each distance to the nearest mile.

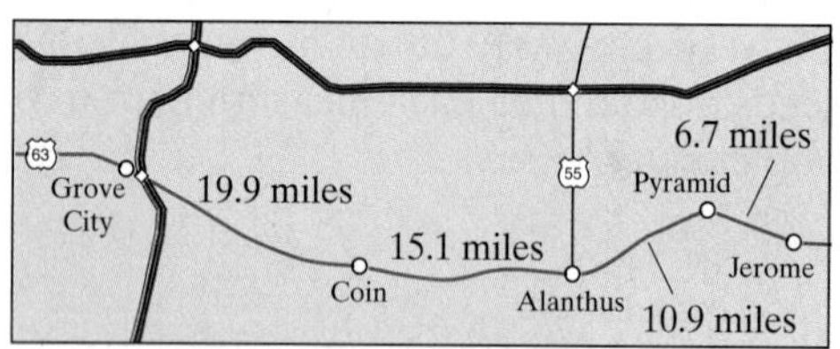

△ **22.** Estimate the perimeter of the figure shown by rounding each measurement to the nearest whole foot.

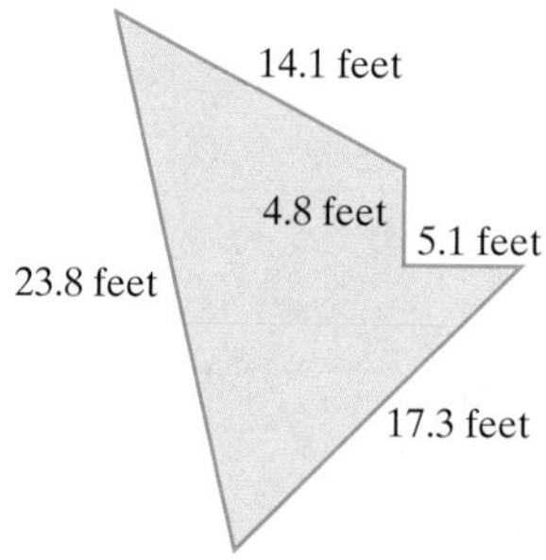

23. The all-time top six movies (that is, those that have earned the most money in the United States) along with the approximate amount of money they have earned are listed in the table. Estimate the total amount of money that these movies have earned by rounding each earning to the nearest hundred million. (*Source: Variety* magazine)

All-Time Top Six American Movies

Movie	Gross Domestic Earnings
Titanic (1997)	$600.8 million
Star Wars (1977)	$461.0 million
E.T. (1982)	$435.0 million
The Phantom Menace (1999)	$431.1 million
Spider Man (2002)	$403.7 million
Jurassic Park (1993)	$357.1 million

24. San Marino is one of the smallest countries in the world. The current San Marino population density is 1197 people per square mile. The area of San Marino is about 20 square miles. Estimate the population of San Marino by rounding 1197 to the nearest ten. (*Source: 2003 World Almanac*)

25. Micronesia is a string of islands in the western Pacific Ocean. The current Micronesia has a population density is 501 people per square mile. The area of Micronesia is about 271 square miles. Estimate the population of Micronesia by rounding both 501 and 271 to the nearest ten. (*Source: 2003 World Almanac*)

26. In 2001, American manufacturers shipped approximately 17.3 million music CD singles to retailers. The value of these shipments was about \$79.4 million. Estimate the value of an individual CD single by rounding \$79.4 and 17.3 to the nearest ten. (*Source:* Recording Industry Association of America)

Choose the best estimate. See Example 5.

27. 8.62×41.7
- **a.** 36
- **b.** 32
- **c.** 360
- **d.** 3.6

28. $1.437 + 20.69$
- **a.** 34
- **b.** 22
- **c.** 3.4
- **d.** 2.2

29. $78.6 \div 97$
- **a.** 7.86
- **b.** 0.786
- **c.** 786
- **d.** 7860

30. $302.729 - 28.697$
- **a.** 270
- **b.** 20
- **c.** 27
- **d.** 300

Use estimation to determine whether each result is reasonable or not. See Example 5.

31. 102.62×41.8 Result: 428.9516

32. $174.835 \div 47.9$ Result: 3.65

33. $1025.68 - 125.42$ Result: 900.26

34. $562.781 + 2.96$ Result: 858.781

B *Simplify each expression. See Examples 6 through 8.*

35. $(0.4)^2 - 0.1$

36. $(-100)(2.3) - 30$

37. $\frac{1 + 0.8}{-0.6}$

38. $(-0.09)^2 + 1.16$

39. $1.4(2 - 1.8)$

40. $\frac{0.29 + 1.69}{3}$

41. $4.83 \div 2.1$

42. $62.1 \div 2.7$

43. $(-2.3)^2(0.3 + 0.7)$

44. $(8.2)(100) - (8.2)(10)$

45. $(3.1 + 0.7)(2.9 - 0.9)$

46. $\frac{0.707 - 3.19}{13}$

47. $\frac{(4.5)^2}{100}$

48. $0.9(5.6 - 6.5)$

49. $\frac{7 + 0.74}{-6}$

50. $(1.5)^2 + 0.5$

Evaluate each expression for $x = 6$, $y = 0.3$, and $z = -2.4$. See Example 9.

51. z^2

52. y^2

53. $x - y$

54. $x - z$

55. $4y - z$

56. $\frac{x}{y}$

Determine whether the given value is a solution of the given equation. See Example 10.

57. Is -1.3 a solution to $7x + 2.1 = -7$?

58. Is 6.2 a solution to $5y - 8.6 = 23.9$?

59. Is -4.7 a solution to $x - 6.5 = 2x + 1.8$?

60. Is -0.9 a solution to $2x - 3.7 = x - 4.6$?

Review and Preview

Perform each indicated operation. See Sections 4.3, 4.5, and 4.8.

61. $\frac{3}{4} \cdot \frac{5}{12}$

62. $\frac{56}{23} \cdot \frac{46}{8}$

63. $-\frac{36}{56} \div \frac{30}{35}$

64. $\frac{5}{7} \div \left(-\frac{14}{10}\right)$

65. $\frac{5}{12} - \frac{1}{3}$

66. $\frac{12}{15} + \frac{5}{21}$

Combining Concepts

Simplify each expression. Then estimate to see whether your proposed result is reasonable.

67. $1.96(7.852 - 3.147)^2$

68. $(6.02)^2 + (2.06)^2$

69. Suppose that your approximation of 12,743 is 13,000 and your friend's approximation is 12,700. Who is correct? Explain your answer.

70. When simplifying $5 + 0.1(1.26 - 0.23)$, do you add, multiply, or subtract first? Explain your answer.

71. When simplifying $3(1.5)^2$, do you square first or multiply by 3 first? Explain your answer.

5.6 Fractions and Decimals

A Writing Fractions as Decimals

To write a fraction as a decimal, we interpret the fraction bar as division and find the quotient.

Writing Fractions as Decimals

To write a fraction as a decimal, divide the numerator by the denominator.

EXAMPLE 1 Write $\frac{1}{4}$ as a decimal.

Solution: $\frac{1}{4} = 1 \div 4$

$$\begin{array}{r} 0.25 \\ 4\overline{)1.00} \\ \underline{-8} \\ 20 \\ \underline{-20} \\ 0 \end{array}$$

Thus, $\frac{1}{4}$ written as a decimal is 0.25.

EXAMPLE 2 Write $-\frac{5}{8}$ as a decimal.

Solution: $-\frac{5}{8} = -(5 \div 8) = -0.625$

$$\begin{array}{r} 0.625 \\ 8\overline{)5.000} \\ \underline{-48} \\ 20 \\ \underline{-16} \\ 40 \\ \underline{-40} \\ 0 \end{array}$$

EXAMPLE 3 Write $\frac{2}{3}$ as a decimal.

Solution:

$$\begin{array}{r} 0.666\ldots \\ 3\overline{)2.000} \\ \underline{-18} \\ 20 \\ \underline{-18} \\ 20 \\ \underline{-18} \\ 2 \end{array}$$

This pattern will continue so that $\frac{2}{3} = 0.6666\ldots$.

We can place a bar over the digit 6 to indicate that it repeats.

$$\frac{2}{3} = 0.666\ldots = 0.\overline{6}$$

We can also write a decimal approximation for $\frac{2}{3}$. For example, $\frac{2}{3}$ rounded to the nearest hundredth is 0.67. This can be written as $\frac{2}{3} \approx 0.67$.

OBJECTIVES

A Write fractions as decimals.

B Compare fractions and decimals.

C Solve area problems containing fractions and decimals.

SSM TUTOR CENTER SG CD & VIDEO MATH PRO WEB

Practice Problem 1

a. Write $\frac{2}{5}$ as a decimal.

b. Write $\frac{9}{40}$ as a decimal.

Practice Problem 2

Write $-\frac{3}{8}$ as a decimal.

Practice Problem 3

a. Write $\frac{5}{6}$ as a decimal.

b. Write $\frac{2}{9}$ as a decimal.

Answers

1. a. 0.4 **b.** 0.225
2. −0.375
3. a. $0.8\overline{3} \approx 0.83$ **b.** $0.\overline{2} \approx 0.22$

Practice Problem 4

Write $\frac{4}{11}$ as a decimal. Give an exact answer and a 3 decimal place approximation.

EXAMPLE 4 Write as a decimal. Give an exact answer and a 3 decimal place approximation.

Solution:

```
     0.5454...
 11)6.0000
   -55
     50
    -44
     60
    -55
      50
     -44
       6
```

This pattern will continue, so $\frac{6}{11} = 0.\overline{54}$.

Rounded to 3 decimal places, $\frac{6}{11} \approx 0.545$.

Practice Problem 5

Write $\frac{1}{9}$ as a decimal. Round to the nearest thousandth.

EXAMPLE 5 Write $\frac{22}{7}$ as a decimal. Round to the nearest hundredth. (The fraction $\frac{22}{7}$ is an approximation for π.)

Solution:

```
     3.142 ≈ 3.14
  7)22.000
   -21
     10
    - 7
     30
    -28
      20
     -14
       6
```

Carry the division out to the thousandths place.

The fraction $\frac{22}{7}$ in decimal form is approximately 3.14.

Concept Check

Suppose you are rewriting the fraction $\frac{9}{16}$ as a decimal. How do you know you have made a mistake if you come up with 1.735?

Try the Concept Check in the margin.

B Comparing Fractions and Decimals

Now we can compare decimals and fractions by writing fractions as equivalent decimals.

Practice Problem 6

Insert <, >, or = to form a true statement.

$\frac{1}{5}$ 0.25

EXAMPLE 6 Insert <, >, or = to form a true statement.

$\frac{1}{8}$ 0.12

Answers

4. $0.\overline{36} \approx 0.364$ **5.** ≈ 0.111 **6.** <

Concept Check: Answers may vary.

Solution: First we write $\frac{1}{8}$ as an equivalent decimal. Then we compare decimal places.

$$\begin{array}{r} 0.125 \\ 8\overline{)1.000} \\ \underline{-8} \\ 20 \\ \underline{-16} \\ 40 \\ \underline{-40} \\ 0 \end{array}$$

Original numbers	$\frac{1}{8}$	0.12
Decimals	0.125	0.120
Compare	0.125 > 0.12	
Thus,	$\frac{1}{8} > 0.12$	

EXAMPLE 7 Insert <, >, or = to form a true statement.

$0.\overline{7}$ $\quad \frac{7}{9}$

Solution: We write $\frac{7}{9}$ as a decimal and then compare.

$$\begin{array}{r} 0.77\ldots = 0.\overline{7} \\ 9\overline{)7.00} \\ \underline{-63} \\ 70 \\ \underline{-63} \\ 7 \end{array}$$

Original numbers	$0.\overline{7}$	$\frac{7}{9}$
Decimals	$0.\overline{7}$	$0.\overline{7}$
Compare	$0.\overline{7} = 0.\overline{7}$	
Thus,	$0.\overline{7} = \frac{7}{9}$	

EXAMPLE 8 Write the numbers in order from smallest to largest.

$\frac{9}{20}$, $\frac{4}{9}$, 0.456

Solution:

Original numbers	$\frac{9}{20}$	$\frac{4}{9}$	0.456
Decimals	0.450	0.444. . .	0.456
Compare in order	2nd	1st	3rd

The numbers written in order are

$\frac{4}{9}$, $\frac{9}{20}$, 0.456

Practice Problem 7

Insert <, >, or = to form a true statement.

a. $\frac{1}{2}$ $\quad$ 0.54

b. $0.\overline{2}$ $\quad \frac{2}{9}$

c. $\frac{5}{7}$ $\quad$ 0.72

Practice Problem 8

Write the numbers in order from smallest to largest.

a. $\frac{1}{3}$, 0.302, $\frac{3}{8}$

b. 1.26, $1\frac{1}{4}$, $1\frac{2}{5}$

c. 0.4, 0.41, $\frac{5}{7}$

Answers

7. a. < **b.** = **c.** <

8. a. 0.302, $\frac{1}{3}$, $\frac{3}{8}$ **b.** $1\frac{1}{4}$, 1.26, $1\frac{2}{5}$

c. 0.4, 0.41, $\frac{5}{7}$

C Solving Area Problems Containing Fractions and Decimals

Sometimes real-life problems contain both fractions and decimals. In this section, we will solve such problems about area.

Practice Problem 9

Find the area of the triangle.

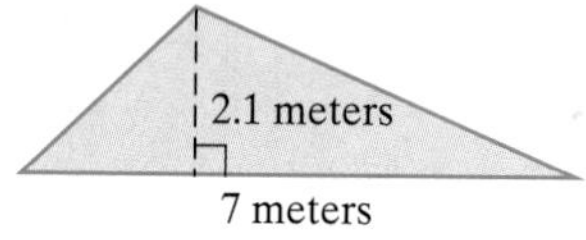

Answer

9. 7.35 square meters

EXAMPLE 9 Finding the Area of a Triangle

The area of a triangle is Area $= \frac{1}{2} \cdot \text{base} \cdot \text{height}$. Find the area of the triangle shown.

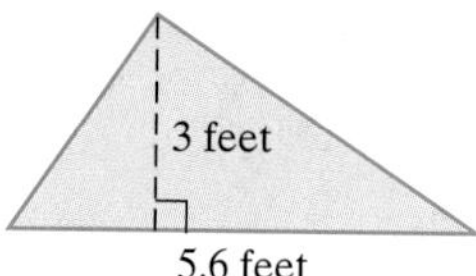

Solution:

$$\begin{aligned} \text{area} &= \frac{1}{2} \cdot \text{base} \cdot \text{height} \\ &= \frac{1}{2} \cdot 5.6 \cdot 3 && \text{Let base} = 5.6 \text{ and height} = 3. \\ &= 0.5 \cdot 5.6 \cdot 3 && \text{Write } \frac{1}{2} \text{ as the decimal } 0.5. \\ &= 8.4 \end{aligned}$$

The area of the triangle is 8.4 square feet.

STUDY SKILLS REMINDER

Are you prepared for a test on Chapter 5?

Below I have listed some *common trouble areas* for topics covered in Chapter 5. After studying for your test—but before taking your test—read these.

- Don't forget the order of operations. To simplify $0.7 + 1.3(5 - 0.1)$, should you add, subtract, or multiply first? First, perform the subtraction within parentheses, then multiply, and finally add.

$$\begin{aligned} 0.7 + 1.3(5 - 0.1) &= 0.7 + 1.3(4.9) && \text{Subtract.} \\ &= 0.7 + 6.37 && \text{Multiply.} \\ &= 7.07 && \text{Add.} \end{aligned}$$

- If you are having trouble with ordering or operations on decimals, don't forget that you can insert 0s after the last digit to the right of the decimal point as needed.

Addition	Addition with zeros inserted	Subtraction	Subtraction with zeros inserted
8.1	8.100	7	6 910 ~~7.00~~
0.6	0.600	−0.28	−0.28
+23.003	+23.003		6.72
	31.703		

Place in order from smallest to largest: 0.108, 0.18, 0.0092.
If we insert zeros, we have 0.1080, 0.1800, 0.0092.
The decimals in order are 0.0092, 0.1080, 0.1800 or 0.0092, 0.108, 0.18.

Name ______________________ Section ____________ Date ____________

EXERCISE SET 5.6

A *Write each fraction as a decimal. See Examples 1 through 5.*

1. $\frac{1}{5}$

2. $\frac{2}{5}$

3. $\frac{4}{8}$

4. $\frac{6}{8}$

5. $\frac{3}{4}$

6. $\frac{6}{5}$

7. $-\frac{2}{25}$

8. $-\frac{3}{25}$

9. $\frac{3}{8}$

10. $\frac{1}{4}$

11. $\frac{11}{12}$

12. $\frac{19}{25}$

13. $\frac{17}{40}$

14. $\frac{5}{12}$

15. $\frac{9}{20}$

16. $\frac{31}{40}$

17. $-\frac{1}{3}$

18. $-\frac{7}{9}$

19. $\frac{7}{16}$

20. $\frac{27}{16}$

21. $\frac{2}{9}$

22. $\frac{9}{11}$

23. $\frac{5}{3}$

24. $\frac{4}{11}$

Round each number as indicated.

25. Round your decimal answer to Exercise 17 to the nearest hundredth.

26. Round your decimal answer to Exercise 18 to the nearest hundredth.

27. Round your decimal answer to Exercise 19 to the nearest hundredth.

28. Round your decimal answer to Exercise 20 to the nearest hundredth.

29. Round your decimal answer to Exercise 21 to the nearest tenth.

30. Round your decimal answer to Exercise 22 to the nearest tenth.

31. Round your decimal answer to Exercise 23 to the nearest tenth.

32. Round your decimal answer to Exercise 24 to the nearest tenth.

Write each fraction as a decimal. When necessary, round all decimal answers to the nearest hundredth. See Examples 1 through 5.

33. In 2000, about $\frac{701}{1048}$ Americans used the Internet in some form. (*Source:* UCLA Center for Communications Policy)

34. About $\frac{21}{50}$ of all blood donors have type A blood. (*Source:* American Red Cross Biomedical Services)

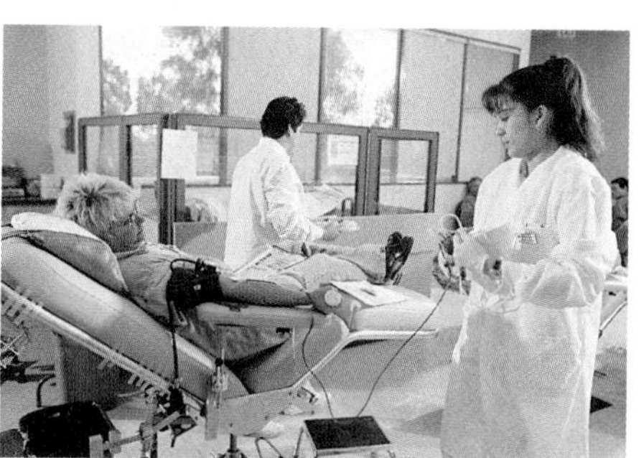

35. The National Athletic Trainers' Association is an organization dedicated to advancing the athletic training profession. Women make up approximately $\frac{11}{25}$ of its membership. (*Source:* National Athletic Trainers' Association)

36. In a recent year, approximately $\frac{29}{46}$ of all individuals who had flown in space were citizens of the United States. (*Source:* Congressional Research Service)

37. Of the U.S. mountains that are over 14,000 feet in elevation, $\frac{56}{91}$ are located in Colorado. (*Source:* U.S. Geological Survey)

38. In the United States, $\frac{1}{40}$ of all births are twins. (*Source:* Columbia Encyclopedia)

B *Insert <, >, or = between each pair of numbers to form a true statement. See Examples 6 and 7.*

39. 0.562   0.569

40. 0.983   0.988

41. 0.823   0.813

42. 0.824   0.821

43. −0.0923   −0.0932

44. −0.00536   −0.00563

45. $\frac{2}{3}$ $\frac{5}{6}$

46. $\frac{1}{9}$ $\frac{2}{17}$

47. $\frac{5}{9}$ $\frac{51}{91}$

48. $\frac{7}{12}$ $\frac{6}{11}$

49. $\frac{4}{7}$ 0.14

50. $\frac{5}{9}$ 0.557

51. 1.38 $\frac{18}{13}$

52. 0.372 $\frac{22}{59}$

53. 7.123 $\frac{456}{64}$

54. 12.713 $\frac{89}{7}$

Write each list of numbers in order from smallest to largest. See Example 8.

55. 0.34, 0.35, 0.32

56. 0.47, 0.42, 0.40

57. 0.49, 0.491, 0.498

58. 0.72, 0.727, 0.728

59. $\frac{3}{4}$, 0.78, 0.73

60. $\frac{2}{5}$, 0.49, 0.42

61. $\frac{4}{7}$, 0.453, 0.412

62. $\frac{6}{9}$, 0.663, 0.668

63. 5.23, $\frac{42}{8}$, 5.34

64. 7.56, $\frac{67}{9}$, 7.562

65. $\frac{12}{5}$, 2.37, $\frac{17}{8}$

66. $\frac{29}{16}$, 1.75, $\frac{58}{32}$

 *Find the area of each triangle or rectangle. See Example 9.*

△ **67.**

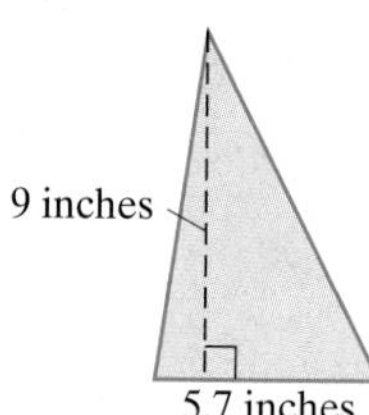

△ **68.**

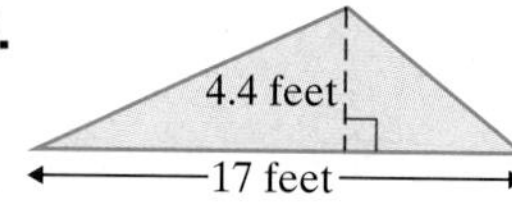

△ **69.**

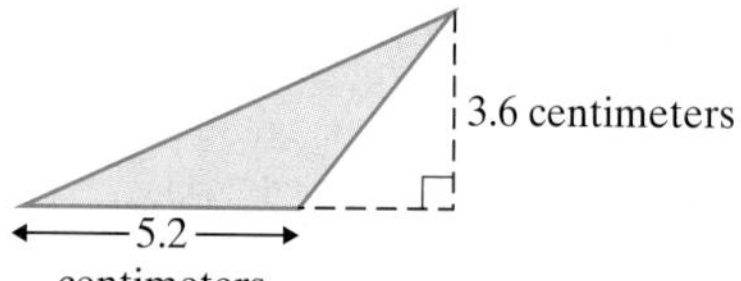

△ **70.**

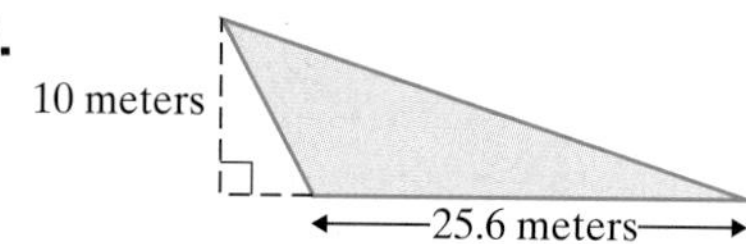

△ **71.**

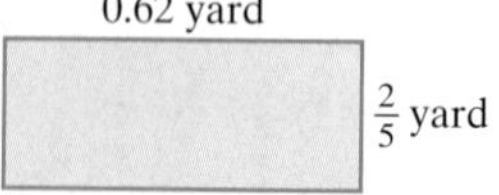

△ **72.**

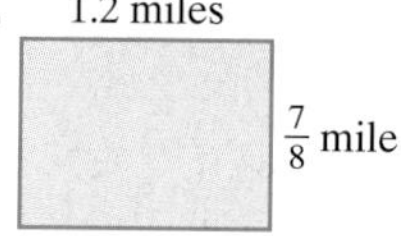

Review and Preview

Simplify. See Sections 1.8 and 4.3.

73. 2^3

74. 5^4

75. $6^2 \cdot 2$

76. $4 \cdot 3^4$

77. $\left(\frac{1}{3}\right)^4$

78. $\left(\frac{4}{5}\right)^3$

79. $\left(\frac{3}{5}\right)^2$

80. $\left(\frac{7}{2}\right)^2$

81. $\left(\frac{2}{5}\right)\left(\frac{5}{2}\right)^2$

82. $\left(\frac{2}{3}\right)\left(\frac{3}{2}\right)^2$

Combining Concepts

In 2002, there were 10,569 commercial radio stations in the United States. The most popular formats are listed in the table along with their counts. Use this graph to answer Exercises 83–86.

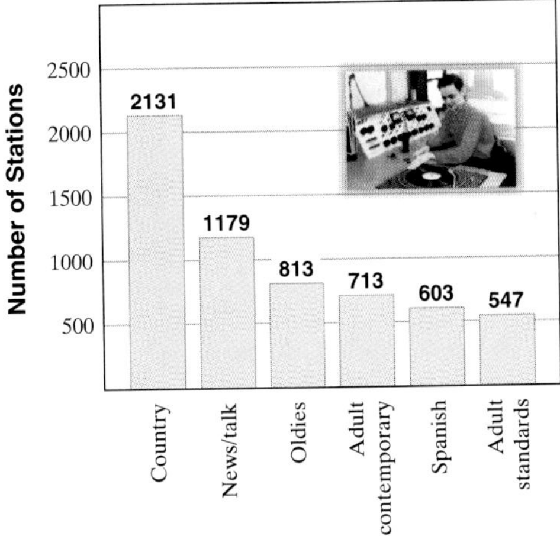

Format (Total stations: 10,569)

83. Write the fraction of radio stations with a country music format as a decimal. Round to the nearest thousandth.

84. Write the fraction of radio stations with a news/talk format as a decimal. Round to the nearest hundredth.

85. Estimate, by rounding each number in the table to the nearest hundred, the total number of stations with the top six formats in 2002.

86. Use your estimate from Exercise 85 to write the fraction of radio stations accounted for by the top six formats as a decimal. Round to the nearest hundredth.

87. Suppose that you are writing a paper on a word processor. You have been instructed to use $\frac{5}{8}$-inch margins. To change margins in the word processing software, you must enter margin widths in inches as a decimal. What number should you enter?

88. Suppose that you are using a word processor to format a table for a presentation. The first column of the table must be $3\frac{7}{16}$ inches wide. To adjust the width of the column in the word processing software, you must enter the column width in inches as a decimal. What number should you enter?

89. Describe how to determine the larger of two fractions.

Find the value of each expression. Give the result as a decimal.

90. $\frac{1}{5} - 2(7.8)$

91. $\frac{3}{4} - (9.6)(5)$

92. $8.25 - \left(\frac{1}{2}\right)^2$

93. $\left(\frac{1}{10}\right)^2 + (1.6)(2.1)$

94. $\frac{1}{4}(-9.6 - 5.2)$

95. $\frac{3}{8}(4.7 - 5.9)$

FOCUS ON History

THE GOLDEN RECTANGLE IN ART

The golden rectangle is a rectangle whose length is approximately 1.6 times its width. The early Greeks thought that a rectangle with these dimensions was the most pleasing to the eye. Examples of the golden rectangle are found in many ancient, as well as modern, works of art. For example, the Parthenon in Athens, Greece, shows the golden rectangle in many aspects of its design. Modern-era artists, including Piet Mondrian (1872–1944) and Georges Seurat (1859–1891), also frequently used the proportions of a golden rectangle in their paintings.

To test whether a rectangle is a golden rectangle, divide the rectangle's length by its width. If the result is approximately 1.6, we can consider the rectangle to be a golden rectangle. For instance, consider Mondrian's *Composition with Gray and Light Brown*, which was painted on an 80.2cm × 49.9cm canvas. Because $\frac{80.2}{49.9} \approx 1.6$, the dimensions of the canvas form a golden rectangle. In what other ways are golden rectangles connected with this painting?

Examples of golden rectangles can be found in the designs of many everyday objects. Visual artists, from architects to product and package designers, use the golden rectangle shape in such things as the face of a building, the floor of a room, the front of a food package, the front cover of a book, and even the shape of a credit card.

GROUP ACTIVITY

Find an example of a golden rectangle in a building or an everyday object. Use a ruler to measure its dimensions and verify that the length is approximately 1.6 times the width.

5.7 Equations Containing Decimals

OBJECTIVE

A Solve equations containing decimals.

A Solving Equations Containing Decimals

In this section, we continue our work with decimals and algebra by solving equations containing decimals. First, we review the steps given earlier for solving an equation.

Steps for Solving an Equation in *x*

Step 1. If fractions are present, multiply both sides of the equation by the LCD of the fractions.

Step 2. If parentheses are present, use the distributive property.

Step 3. Combine any like terms on each side of the equation.

Step 4. Use the addition property of equality to rewrite the equation so that variable terms are on one side of the equation and constant terms are on the other side.

Step 5. Divide both sides by the numerical coefficient of x to solve.

Step 6. Check the answer in the **original equation**.

EXAMPLE 1 Solve: $x - 1.5 = 8$

Solution: Steps 1 through 3 are not needed for this equation, so we begin with Step 4. To get x alone on one side of the equation, add 1.5 to both sides.

$x - 1.5 = 8$ Original equation

$x - 1.5 + 1.5 = 8 + 1.5$ Add 1.5 to both sides.

$x = 9.5$ Simplify.

Check: To check, replace x with 9.5 in the *original equation*.

$x - 1.5 = 8$ Original equation

$9.5 - 1.5 = 8$ Replace x with 9.5.

$8 = 8$ True.

Since $8 = 8$ is a true statement, 9.5 is a solution of the equation.

Practice Problem 1

Solve: $z + 0.9 = 1.3$

EXAMPLE 2 Solve: $-2y = 6.7$

Solution: Steps 1 through 4 are not needed for this equation, so we begin with Step 5. To solve for y, divide both sides by the coefficient of y, which is -2.

$-2y = 6.7$ Original equation

$\frac{-2y}{-2} = \frac{6.7}{-2}$ Divide both sides by -2.

$y = -3.35$ Simplify.

Check: To check, replace y with -3.35 in the original equation.

$-2y = 6.7$ Original equation

$-2(-3.35) = 6.7$ Replace y with -3.35.

$6.7 = 6.7$ True.

Thus -3.35 is a solution of the equation $-2y = 6.7$.

Practice Problem 2

Solve: $0.17x = -0.34$

Answers

1. 0.4 **2.** -2

Practice Problem 3

Solve: $2.9 = 1.7 + 0.3x$

EXAMPLE 3 Solve: $1.2x + 5.8 = 8.2$

Solution: We begin with Step 4 and get the variable term alone by subtracting 5.8 from both sides.

$$\begin{aligned} 1.2x + 5.8 &= 8.2 \\ 1.2x + 5.8 - 5.8 &= 8.2 - 5.8 && \text{Subtract 5.8 from both sides.} \\ 1.2x &= 2.4 && \text{Simplify.} \\ \frac{1.2x}{1.2} &= \frac{2.4}{1.2} && \text{Divide both sides by 1.2.} \\ x &= 2 && \text{Simplify.} \end{aligned}$$

To check, replace x with 2 in the original equation. The solution is 2. ●

Practice Problem 4

Solve: $8x + 4.2 = 10x + 11.6$

EXAMPLE 4 Solve: $7x + 3.2 = 4x - 1.6$

Solution:

$$\begin{aligned} 7x + 3.2 &= 4x - 1.6 \\ 7x + 3.2 - 3.2 &= 4x - 1.6 - 3.2 && \text{Subtract 3.2 from both sides.} \\ 7x &= 4x - 4.8 && \text{Simplify.} \\ 7x - 4x &= 4x - 4.8 - 4x && \text{Subtract } 4x \text{ from both sides.} \\ 3x &= -4.8 && \text{Simplify.} \\ \frac{\overset{1}{\cancel{3}}x}{\underset{1}{\cancel{3}}} &= -\frac{4.8}{3} && \text{Divide both sides by 3.} \\ x &= -1.6 && \text{Simplify.} \end{aligned}$$

Check to see that -1.6 is the solution. ●

Practice Problem 5

Solve: $6.3 - 5x = 3(x + 2.9)$

EXAMPLE 5 Solve: $5(x - 0.36) = -x + 2.4$

Solution: First use the distributive property to distribute the factor 5.

$$\begin{aligned} 5(x - 0.36) &= -x + 2.4 && \text{Original equation} \\ 5x - 1.8 &= -x + 2.4 && \text{Apply the distributive property.} \end{aligned}$$

Next, get x alone on the left side of the equation by adding 1.8 to both sides of the equation and then adding x to both sides of the equation.

$$\begin{aligned} 5x - 1.8 + 1.8 &= -x + 2.4 + 1.8 && \text{Add 1.8 to both sides.} \\ 5x &= -x + 4.2 && \text{Simplify.} \\ 5x + x &= -x + 4.2 + x && \text{Add } x \text{ to both sides.} \\ 6x &= 4.2 && \text{Simplify.} \\ \frac{6x}{6} &= \frac{4.2}{6} && \text{Divide both sides by 6.} \\ x &= 0.7 && \text{Simplify.} \end{aligned}$$

To verify that 0.7 is the solution, replace x with 0.7 in the original equation. ●

Instead of solving equations with decimals, sometimes it may be easier to first rewrite the equation so it contains integers only. Recall that multiplying a decimal by a power of 10, such as 10, 100, or 1000, has the effect of moving the decimal point to the right. We can use the multiplication property of equality to multiply both sides of the equation through by an appropriate power of 10. The resulting equivalent equation will contain integers only.

Answers

3. 4 **4.** −3.7 **5.** −0.3

EXAMPLE 6 Solve: $0.5y + 2.3 = 1.65$

Solution: Multiply the equation through by 100. This will move the decimal point in each term two places to the right.

$$0.5y + 2.3 = 1.65 \quad \text{Original equation}$$

$$100(0.5y + 2.3) = 100(1.65) \quad \text{Multiply both sides by 100.}$$

$$100(0.5y) + 100(2.3) = 100(1.65) \quad \text{Apply the distributive property.}$$

$$50y + 230 = 165 \quad \text{Simplify.}$$

Now the equation contains integers only. Finish solving by subtracting 230 from both sides.

$$50y + 230 = 165$$

$$50y + 230 - 230 = 165 - 230 \quad \text{Subtract 230 from both sides.}$$

$$50y = -65 \quad \text{Simplify.}$$

$$\frac{50y}{50} = \frac{-65}{50} \quad \text{Divide both sides by 50.}$$

$$y = -1.3 \quad \text{Simplify.}$$

Check to see that -1.3 is the solution by replacing y with -1.3 in the original equation.

Try the Concept Check in the margin.

Practice Problem 6

Solve: $0.2y + 2.6 = 4$

Concept Check

By what number would you multiply both sides of the following equation to make calculations easier? Explain your choice.

$$1.7x + 3.655 = -14.2$$

Answer

6. 7

Concept Check: Multiply by 1000.

FOCUS ON Business and Career

IN-DEMAND OCCUPATIONS

According to U.S. Bureau of Labor Statistics projections, the careers listed below will have the largest job growth in the next decade:

	Employment (numbers in thousands)		
Occupation	1998	2008	Change
1. Systems analysts	617	1194	+577
2. Retail salespersons	4056	4620	+564
3. Cashiers	3198	3754	+556
4. General managers and top executives	3362	3913	+551
5. Truck drivers, light and heavy	2970	3463	+493
6. Office clerks, general	3021	3484	+463
7. Registered nurses	2079	2530	+451
8. Computer support specialists	429	869	+440
9. Personal care and home-health aides	746	1179	+433
10. Teacher assistants	1192	1567	+375

(*Source:* Bureau of Labor Statistics, U.S. Department of Labor)

What do all of these in-demand occupations have in common? They all require a knowledge of math! For some careers like systems analysts, salespersons, cashiers, and nurses, the ways math is used on the job may be obvious. For other occupations, the use of math may not be quite as obvious. However, tasks common to many jobs, such as filling in a time sheet or a mileage log, writing up an expense report, planning a budget, figuring a bill, ordering supplies, completing a packing list, and even making a work schedule, all require math.

CRITICAL THINKING

Suppose that your college placement office is planning to publish an occupational handbook on math in popular occupations. Choose one of the occupations from the list above that interests you. Research the occupation. Then write a brief entry for the occupational handbook that describes how a person in that career would use math in his or her job. Include an example if possible.

Name ______________________ Section ____________ Date ____________

EXERCISE SET 5.7

A *Solve each equation. See Examples 1 and 2.*

1. $x + 1.2 = 7.1$

2. $y - 0.5 = 9$

3. $-5y = 2.15$

4. $-0.4x = 50$

5. $6.2 = y - 4$

6. $9.7 = x + 11.6$

7. $3.1x = -13.95$

8. $3y = -25.8$

Solve each equation. See Examples 3 through 5.

9. $-3.5x + 2.8 = -11.2$

10. $7.1 - 0.2x = 6.1$

11. $6x + 8.65 = 3x + 10$

12. $7x - 9.64 = 5x + 2.32$

13. $2(x - 1.3) = 5.8$

14. $5(x + 2.3) = 19.5$

Solve each equation by first multiplying both sides through by an appropriate power of 10 so that the equation contains integers only. See Example 6.

15. $0.4x + 0.7 = -0.9$

16. $0.7x + 0.1 = 1.5$

17. $7x - 10.8 = x$

18. $3y = 7y + 24.4$

19. $2.1x + 5 - 1.6x = 10$

20. $1.5x + 2 - 1.2x = 12.2$

Solve.

21. $y - 3.6 = 4$

22. $x + 5.7 = 8.4$

23. $-0.02x = -1.2$

24. $-9y = -0.162$

25. $6.5 = 10x + 7.2$

26. $2x - 4.2 = 8.6$

27. $2.7x - 25 = 1.2x + 5$

28. $9y - 6.9 = 6y - 11.1$

29. $200x - 0.67 = 100x + 0.81$

30. $2.3 + 500x = 600x - 0.2$

31. $3(x + 2.71) = 2x$

32. $7(x + 8.6) = 6x$

33. $8x - 5 = 10x - 8$

34. $24y - 10 = 20y - 17$

35. $1.2 + 0.3x = 0.9$

36. $1.5 = 0.4x + 0.5$

37. $-0.9x + 2.65 = -0.5x + 5.45$

38. $-50x + 0.81 = -40x - 0.48$

39. $4x + 7.6 = 2(3x - 3.2)$

40. $4(2x - 1.6) = 5x - 6.4$

41. $0.7x + 13.8 = x - 2.16$

42. $y - 5 = 0.3y + 4.1$

Review and Preview

Simplify each expression by combining like terms. If parentheses are present, first use the distributive property. See Section 3.1.

43. $2x - 6 + 4x - 10$

44. $x - 4 - x - 4$

45. $3(x - 5) + 10$

46. $9x + 4(x + 20)$

47. $5y - 1.2 - 7y + 8$

48. $7.76 + 8z - 12z + 8.91$

Combining Concepts

49. Explain in your own words the property of equality that allows us to multiply an equation through by a power of 10.

50. Construct an equation whose solution is 1.4.

Solve.

51. $-5.25x = -40.33575$

52. $7.68y = -114.98496$

53. $1.95y + 6.834 = 7.65y - 19.8591$

54. $6.11x + 4.683 = 7.51x + 18.235$

5.8 Square Roots and the Pythagorean Theorem

OBJECTIVES

(A) Find the square root of a number.

(B) Approximate square roots.

(C) Use the Pythagorean theorem.

SSM TUTOR CENTER SG CD & VIDEO MATH PRO WEB

(A) Finding Square Roots

The square of a number is the number times itself. For example,

The square of 5 is 25 because 5^2 or $5 \cdot 5 = 25$.
The square of -5 is also 25 because $(-5)^2$ or $(-5)(-5) = 25$.

The reverse process of squaring is finding a **square root**. For example,

A square root of 25 is 5 because $^2 = 25$.
A square root of 25 is also because $(-5)^2 = 25$.

Every positive number has two square roots. We see above that the square roots of 25 are 5 and -5.

We use the symbol $\sqrt{}$, called a **radical sign**, to indicate the positive square root of a nonnegative number. For example,

$\sqrt{25} = 5$ because $5^2 = 25$ and 5 is positive.
$\sqrt{9} = 3$ because $3^2 = 9$ and 3 is positive.

Square Root of a Number

The square root, $\sqrt{}$, of a positive number a is the positive number b whose square is a. In symbols,

$$\sqrt{a} = b, \qquad \text{if } b^2 = a$$

Also, $\sqrt{0} = 0$.

Helpful Hint

Remember that the radical sign $\sqrt{}$ is used to indicate the **positive square root** of a nonnegative number.

EXAMPLES Find each square root.

1. $\sqrt{49} = 7$ because $7^2 = 49$.
2. $\sqrt{36} = 6$ because $6^2 = 36$.
3. $\sqrt{1} = 1$ because $1^2 = 1$.
4. $\sqrt{81} = 9$ because $9^2 = 81$.
5. Find $\sqrt{\frac{1}{36}} = \frac{1}{6}$ because $\left(\frac{1}{6}\right)^2$ or $\frac{1}{6} \cdot \frac{1}{6} = \frac{1}{36}$.
6. Find $\sqrt{\frac{4}{25}} = \frac{2}{5}$ because $\left(\frac{2}{5}\right)^2$ or $\frac{2}{5} \cdot \frac{2}{5} = \frac{4}{25}$.

Practice Problems

Find each square root.

1. $\sqrt{100}$
2. $\sqrt{64}$
3. $\sqrt{121}$
4. $\sqrt{0}$
5. $\sqrt{\frac{1}{4}}$
6. $\sqrt{\frac{9}{16}}$

(B) Approximating Square Roots

Thus far, we have found square roots of perfect squares. Numbers like $\frac{1}{4}$, 36, $\frac{4}{25}$, and 1 are called **perfect squares** because their square root is a whole number or a fraction. A square root such as $\sqrt{5}$ cannot be written as a whole number or a fraction since 5 is not a perfect square.

Answers

1. 10 **2.** 8 **4.** 11 **3.** 0 **5.** $\frac{1}{2}$ **6.** $\frac{3}{4}$

Although $\sqrt{5}$ cannot be written as a whole number or a fraction, it can be approximated by estimating, by using a table (as in Appendix B), or by using a calculator.

Practice Problem 7

Use Appendix B or a calculator to approximate the square root of 11 to the nearest thousandth.

EXAMPLE 7 Use Appendix B or a calculator to approximate the square root of 43 to the nearest thousandth.

Solution: $\sqrt{43} \approx 6.557$

Helpful Hint

$\sqrt{43}$ is *approximately* 6.557. This means that if we multiply 6.557 by 6.557, the product is *close* to 43.

$$6.557 \times 6.557 = 42.994249$$

Practice Problem 8

Approximate $\sqrt{29}$ to the nearest thousandth.

EXAMPLE 8 Approximate $\sqrt{32}$ to the nearest thousandth.

Solution: $\sqrt{32} \approx 5.657$

C Using the Pythagorean Theorem

One important application of square roots has to do with right triangles. Recall that a **right triangle** is a triangle in which one of the angles is a right angle, or measures 90° (degrees). The **hypotenuse** of a right triangle is the side opposite the right angle. The **legs** of a right triangle are the other two sides. These are shown in the following figure. The right angle in the triangle is indicated by the small square drawn in that angle.

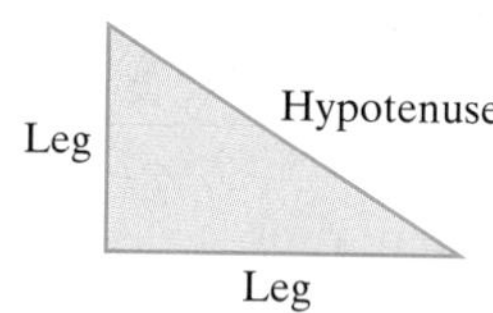

The following theorem is true for all right triangles.

Pythagorean Theorem

If a and b are the lengths of the legs of a right triangle and c is the length of the hypotenuse, then

$$a^2 + b^2 = c^2$$

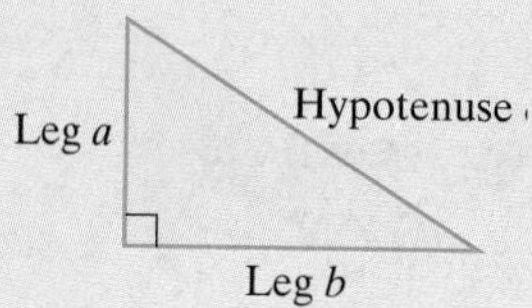

In other words, $(\text{leg})^2 + (\text{other leg})^2 = (\text{hypotenuse})^2$.

Answers

7. ≈3.317 **8.** ≈5.385

EXAMPLE 9 Find the length of the hypotenuse of the given right triangle.

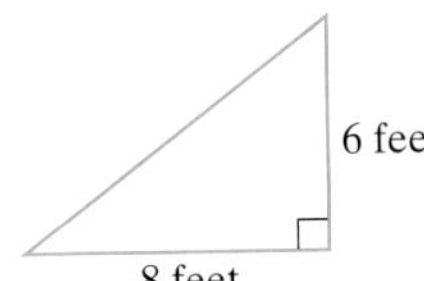

Solution: Let $a = 6$ and $b = 8$. According to the Pythagorean theorem,

$$a^2 + b^2 = c^2$$
$$6^2 + 8^2 = c^2 \quad \text{Let } a = 6 \text{ and } b = 8.$$
$$36 + 64 = c^2 \quad \text{Evaluate } 6^2 \text{ and } 8^2.$$
$$100 = c^2 \quad \text{Add.}$$

In the equation $c^2 = 100$, the solutions of c are the square roots of 100. This means that the solutions are 10 and -10. Since c represents a length, we are only interested in the positive square root of c^2.

$$c = \sqrt{100}$$
$$= 10$$

The hypotenuse is 10 feet long.

EXAMPLE 10 Approximate the length of the hypotenuse of the given right triangle. Round the length to the nearest whole unit.

Solution: Let $a = 17$ and $b = 10$.

10 meters

17 meters

$$a^2 + b^2 = c^2$$
$$17^2 + 10^2 = c^2$$
$$289 + 100 = c^2$$
$$389 = c^2$$
$$\sqrt{389} = c \text{ or } c \approx 20 \quad \text{From Appendix B or a calculator.}$$

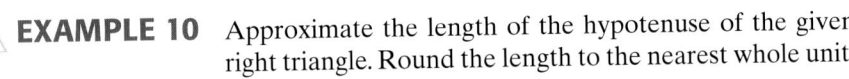

The hypotenuse is exactly $\sqrt{389}$ meters, which is approximately 20 meters.

EXAMPLE 11 Find the length of the leg in the given right triangle. Give the exact length and a two-decimal-place approximation.

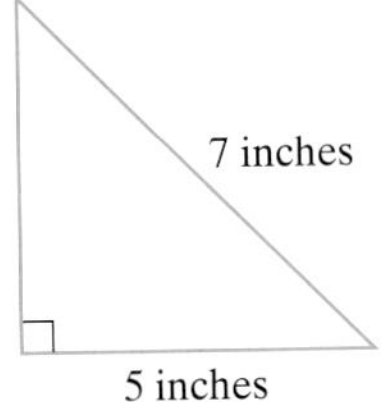

Solution: Notice that the hypotenuse measures 7 inches and that the length of one leg measures 5 inches. Thus, let $c = 7$ and a or b be 5. We will let $a = 5$.

$$a^2 + b^2 = c^2$$
$$5^2 + b^2 = 7^2 \quad \text{Let } a = 5 \text{ and } c = 7.$$
$$25 + b^2 = 49 \quad \text{Evaluate } 5^2 \text{ and } 7^2.$$
$$b^2 = 24 \quad \text{Subtract 25 from both sides.}$$
$$b = \sqrt{24} \approx 4.90$$

The length of the leg is exactly $\sqrt{24}$ inches and approximately 4.90 inches.

Try the Concept Check in the margin.

Practice Problem 9

Find the length of the hypotenuse of the given right triangle.

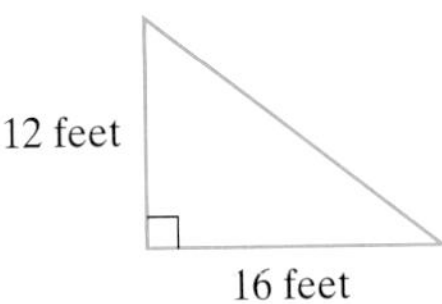

Practice Problem 10

Approximate the length of the hypotenuse of the given right triangle. Round to the nearest whole unit.

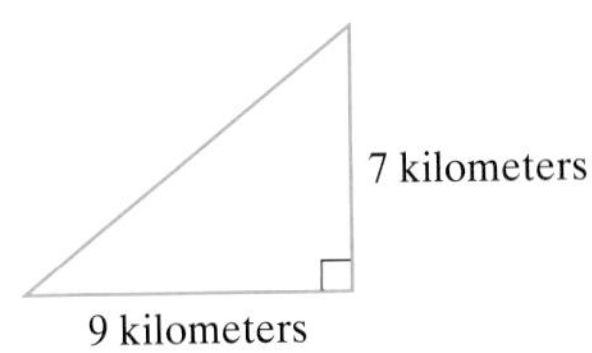

Practice Problem 11

Find the length of the leg in the given right triangle. Give the exact length and a two-decimal-place approximation.

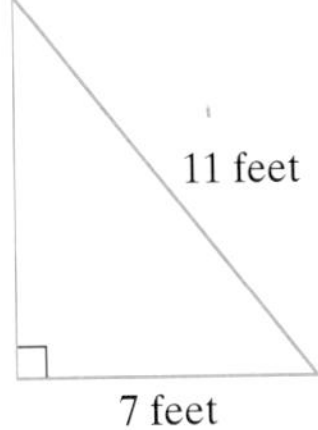

Concept Check

The following lists are the lengths of the sides of two triangles. Which set forms a right triangle?

a. 8, 15, 17

b. 24, 30, 40

Answers

9. 20 feet **10.** 11 kilometers

11. $\sqrt{72}$ feet $\approx$ 8.49 feet

Concept Check: a

Practice Problem 12

A football field is a rectangle measuring 100 yards by 53 yards. Draw a diagram and find the length of the diagonal of a football field to the nearest yard.

Answer

12. ≈ 113 yards

EXAMPLE 12 Finding the Dimensions of a Park

An inner-city park is in the shape of a square that measures 300 feet on a side. A sidewalk is to be constructed along the diagonal of the park. Find the length of the sidewalk rounded to the nearest whole foot.

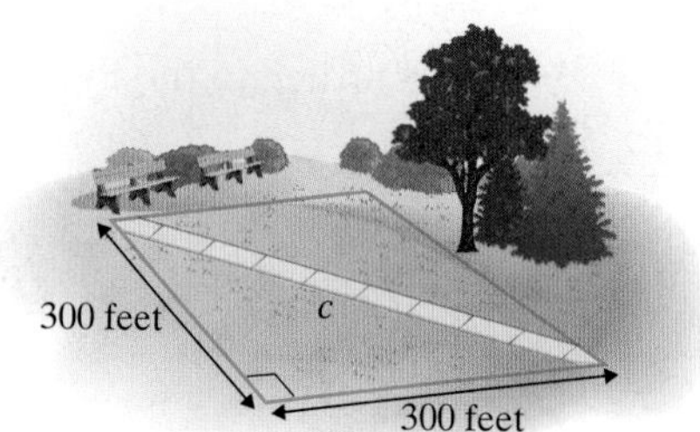

Solution: The diagonal is the hypotenuse of a right triangle, which we label c.

$$a^2 + b^2 = c^2$$

$$300^2 + 300^2 = c^2 \qquad \text{Let } a = 300 \text{ and } b = 300.$$

$$90{,}000 + 90{,}000 = c^2 \qquad \text{Evaluate } (300)^2.$$

$$180{,}000 = c^2 \qquad \text{Add.}$$

$$\sqrt{180{,}000} = c \text{ or } c \approx 424$$

The length of the sidewalk is approximately 424 feet.

CALCULATOR EXPLORATIONS

Finding Square Roots

To simplify or approximate square roots using a calculator, locate the key marked $\boxed{\sqrt{\ }}$.

To simplify $\sqrt{64}$, for example, press the keys

$\boxed{64}$ $\boxed{\sqrt{\ }}$ or $\boxed{\sqrt{\ }}$ $\boxed{64}$

The display should read $\boxed{8}$. Then

$\sqrt{64} = 8$

To *approximate* $\sqrt{10}$, press the keys

$\boxed{10}$ $\boxed{\sqrt{\ }}$ or $\boxed{\sqrt{\ }}$ $\boxed{10}$

The display should read $\boxed{3.16227766}$. This is an *approximation* for $\sqrt{10}$. A three-decimal-place approximation is

$\sqrt{10} \approx 3.162$

Is this answer reasonable? Since 10 is between the perfect squares 9 and 16, $\sqrt{10}$ is between $\sqrt{9} = 3$ and $\sqrt{16} = 4$. Our answer is reasonable since 3.162 is between 3 and 4.

Simplify.

1. $\sqrt{1024}$

2. $\sqrt{676}$

Approximate each square root. Round each answer to the nearest thousandth.

3. $\sqrt{15}$

4. $\sqrt{19}$

5. $\sqrt{97}$

6. $\sqrt{56}$

Name ______________________ Section ____________ Date ____________

EXERCISE SET 5.8

A *Find each square root. See Examples 1 through 6.*

1. $\sqrt{4}$ **2.** $\sqrt{9}$ **3.** $\sqrt{625}$ **4.** $\sqrt{16}$

5. $\sqrt{\frac{1}{81}}$ **6.** $\sqrt{\frac{1}{64}}$ **7.** $\sqrt{\frac{144}{64}}$ **8.** $\sqrt{\frac{36}{81}}$

9. $\sqrt{256}$ **10.** $\sqrt{144}$ **11.** $\sqrt{\frac{9}{4}}$ **12.** $\sqrt{\frac{121}{169}}$

B *Use Appendix B or a calculator to approximate each square root. Round the square root to the nearest thousandth. See Examples 7 and 8.*

13. $\sqrt{3}$ **14.** $\sqrt{5}$ **15.** $\sqrt{15}$ **16.** $\sqrt{17}$

17. $\sqrt{14}$ **18.** $\sqrt{18}$ **19.** $\sqrt{47}$ **20.** $\sqrt{85}$

21. $\sqrt{8}$ **22.** $\sqrt{10}$ **23.** $\sqrt{26}$ **24.** $\sqrt{35}$

25. $\sqrt{71}$ **26.** $\sqrt{62}$ **27.** $\sqrt{7}$ **28.** $\sqrt{2}$

C *Find the unknown length in each right triangle. If necessary, approximate the length to the nearest thousandth. See Examples 9 through 12.*

29.

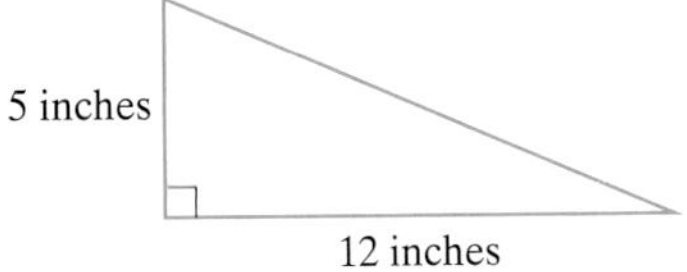

30.

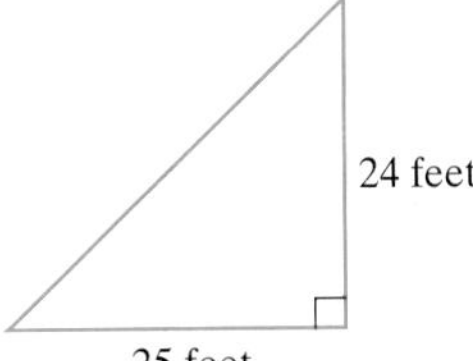

31.

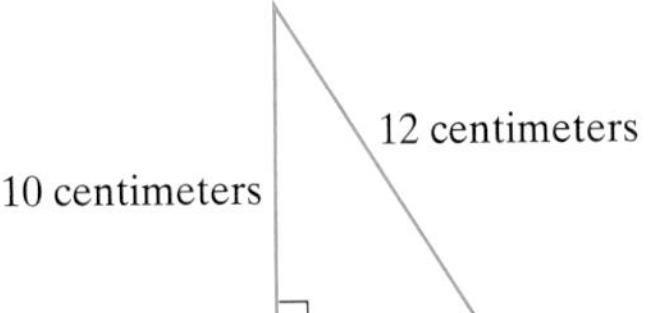

32.

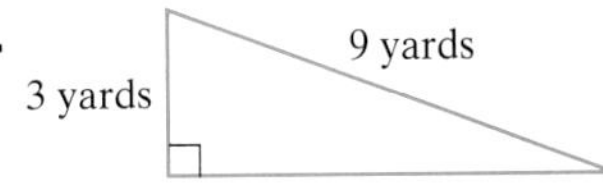

Sketch each right triangle and find the length of the side not given. If necessary, approximate the length to the nearest thousandth. See Examples 9 through 12.

△ **33.** leg = 3, leg = 4

△ **34.** leg = 9, leg = 12

△ **35.** leg = 6, hypotenuse = 10

△ **36.** leg = 48, hypotenuse = 53

△ **37.** leg = 10, leg = 14

△ **38.** leg = 32, leg = 19

△ **39.** leg = 2, leg = 16

△ **40.** leg = 27, leg = 36

△ **41.** leg = 5, hypotenuse = 13

△ **42.** leg = 45, hypotenuse = 117

△ **43.** leg = 35, leg = 28

△ **44.** leg = 30, leg = 15

△ **45.** leg = 30, leg = 30

△ **46.** leg = 110, leg = 132

△ **47.** hypotenuse = 2, leg = 1

△ **48.** hypotenuse = 7, leg = 6

Solve. See Example 12.

△ **49.** A standard city block is a square with each side measuring 100 yards. Find the length of the diagonal of a city block to the nearest hundredth yard.

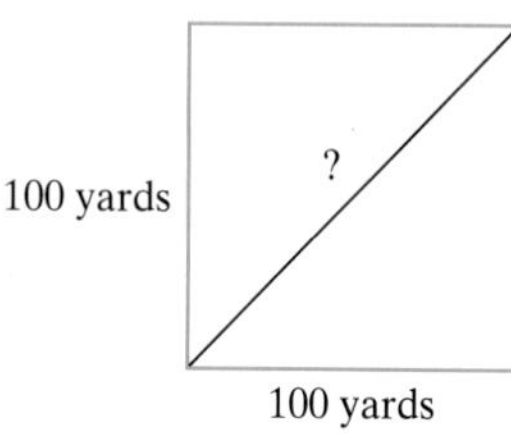

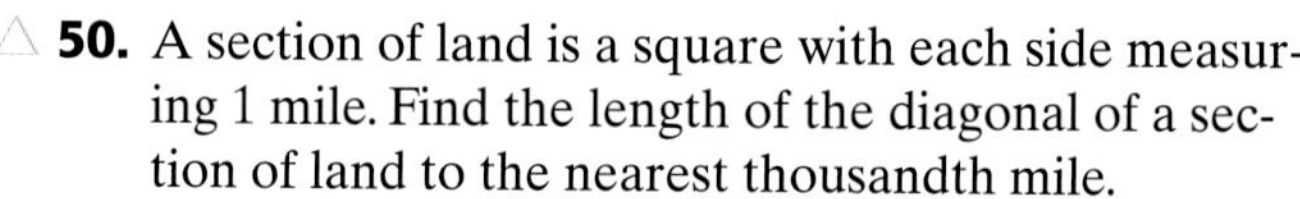
△ **50.** A section of land is a square with each side measuring 1 mile. Find the length of the diagonal of a section of land to the nearest thousandth mile.

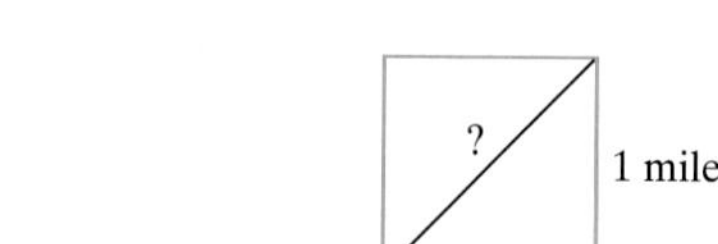

△ **51.** Find the height of the tree. Round the height to one decimal place.

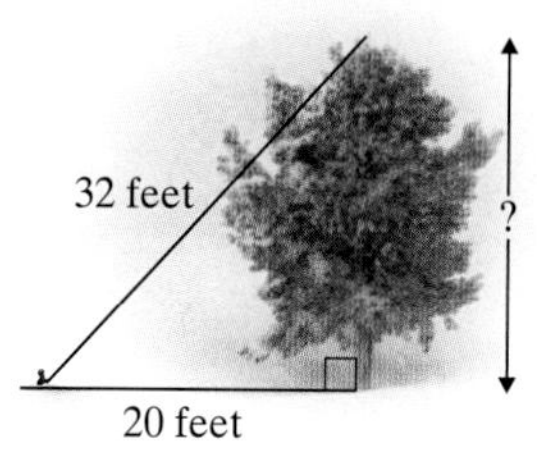

△ **52.** Find the height of the antenna. Round the height to one decimal place.

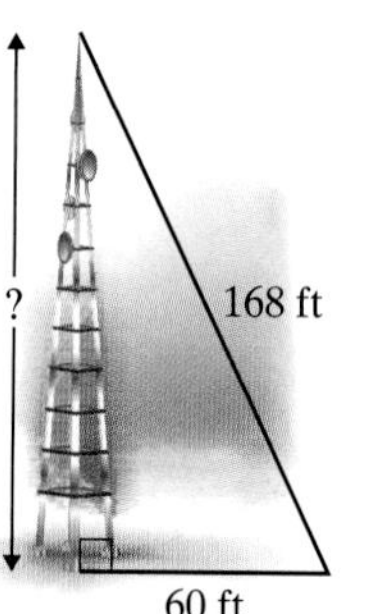

53. A football field is a rectangle that is 300 feet long by 160 feet wide. Find, to the nearest foot, the length of a straight-line run that started at one corner and went diagonally to end at the opposite corner.

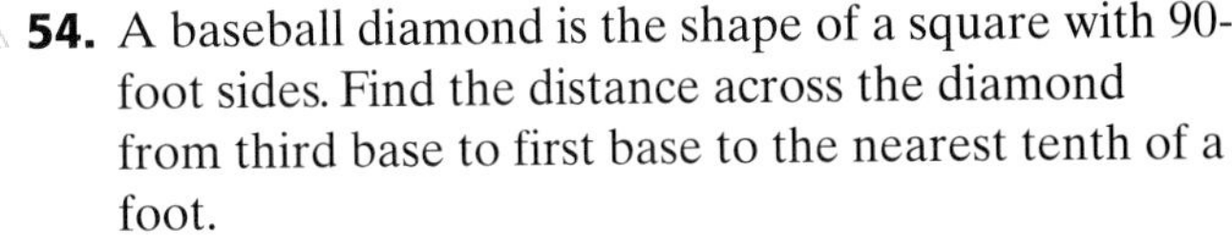

54. A baseball diamond is the shape of a square with 90-foot sides. Find the distance across the diamond from third base to first base to the nearest tenth of a foot.

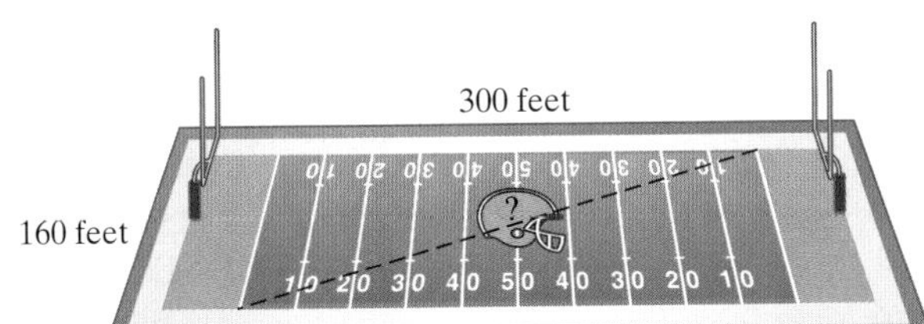

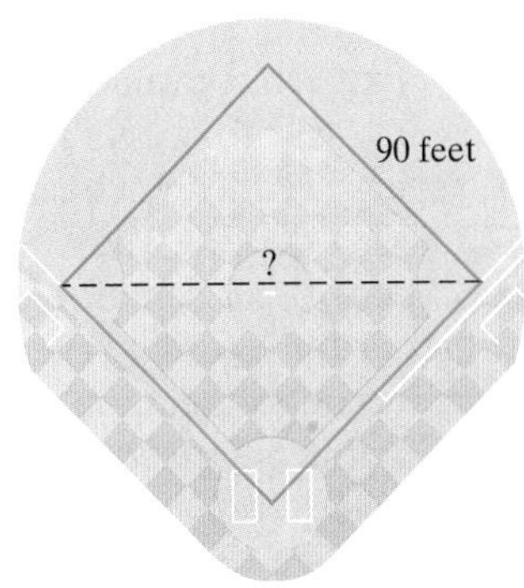

Review and Preview

Write each fraction in simplest form. See Section 4.2.

55. $\frac{10}{12}$

56. $\frac{10}{15}$

57. $\frac{24}{60}$

58. $\frac{35}{75}$

59. $\frac{30}{72}$

60. $\frac{18}{30}$

Combining Concepts

Determine what two whole numbers each square root is between without using a calculator or table. Then use a calculator or table to check.

61. $\sqrt{38}$

62. $\sqrt{27}$

63. $\sqrt{101}$

64. $\sqrt{85}$

65. Without using a calculator, explain why you know that $\sqrt{11}$ is between 3 and 4.

66. Find the exact length of x. Then give a 2-decimal place approximation.

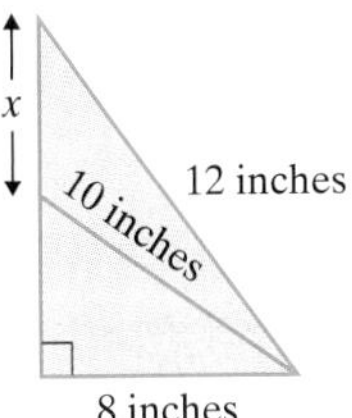

CHAPTER 5 ACTIVITY Maintaining a Checking Account

This activity may be completed by working in groups or individually.

A checking account is a convenient way of handling money and paying bills. To open a checking account, the bank or savings and loan association requires a customer to make a deposit. Then the customer receives a checkbook that contains checks, deposit slips, and a register for recording checks written and deposits made. It is important to record all payments and deposits that affect the account. It is also important to keep the checkbook balance current by subtracting checks written (and service charges) and adding deposits made (or interest earned).

About once a month checking customers receive a statement from the bank listing all activity that the account has had in the last month. The statement lists a beginning balance, all checks and deposits, any service charges made against the account, and an ending balance. Because it may take several days for checks that a customer has written to clear the banking system, the check register may list checks that do not appear on the monthly bank statement. These checks are called **outstanding checks**. Deposits that are recorded in the check register but do not appear on the statement are called **deposits in transit**. Because of these differences, it is important to balance, or reconcile, the checkbook against the monthly statement. The steps for doing so are listed at the right.

Balancing or Reconciling a Checkbook

Step 1. Place a check mark in the checkbook register next to each check and deposit listed on the monthly bank statement. Any entries in the register without a check mark are outstanding checks or deposits in transit.

Step 2. Find the ending checkbook register balance and add to it any outstanding checks and any interest paid on the account.

Step 3. From the total in Step 2, subtract any deposits in transit and any service charges.

Step 4. Compare the amount found in Step 3 with the ending balance listed on the bank statement. If they are the same, the checkbook balances with the bank statement. Be sure to update the check register with service charges and interest.

Step 5. If the checkbook does not balance, recheck the balancing process. Next, make sure that the running checkbook register balance was calculated correctly. Finally, compare the checkbook register with the statement to make sure that each check was recorded for the correct amount.

For the checkbook register and monthly bank statement given:

a. update the checkbook register

b. list the outstanding checks and deposits in transit

c. balance the checkbook—be sure to update the register with any interest or service fees

Checkbook Register

#	Date	Description	Payment	√	Deposit	Balance 425.86
114	4/1	Market Basket	30.27			
115	4/3	May's Texaco	8.50			
	4/4	Cash at ATM	50.00			
116	4/6	UNO Bookstore	121.38			
	4/7	Deposit			100.00	
117	4/9	MasterCard	84.16			
118	4/10	Blockbuster	6.12			
119	4/12	Kroger	18.72			
120	4/14	Parking sticker	18.50			
	4/15	Direct deposit			294.36	
121	4/20	Rent	395.00			
122	4/25	Student fees	20.00			
	4/28	Deposit			75.00	

FIRST NATIONAL BANK
Monthly Statement 4/30

BEGINNING BALANCE:		425.86
Date	Number	Amount
CHECKS AND ATM WITHDRAWALS		
4/3	114	30.27
4/4	ATM	50.00
4/11	117	84.16
4/13	115	8.50
4/15	119	18.72
4/22	121	395.00
DEPOSITS		
4/7		100.00
4/15	Direct deposit	294.36
SERVICE CHARGES		
Low balance fee		7.50
INTEREST		
Credited 4/30 1.15		
ENDING BALANCE:		227.22

Chapter 5 Vocabulary Check

Fill in each blank with one of the words listed below.

vertically	decimal	and	sum
denominator	numerator		

1. Like fractional notation, _________ notation is used to denote a part of a whole.
2. To write fractions as decimals, divide the ___________ by the ____________ .
3. To add or subtract decimals, write the decimals so that the decimal points line up __________ .
4. When writing decimals in words, write "_____" for the decimal point.
5. When multiplying decimals, the decimal point in the product is placed so that the number of decimal places in the product is equal to the ______ of the number of decimal places in the factors.

Chapter 5 Highlights

Definitions and Concepts | **Examples**

Section 5.1 Introduction to Decimals

Place-Value Chart

	1	7	.	7	5	8		
hundreds	tens	ones		tenths	hundredths	thousandths	ten-thousandths	hundred-thousandths
100	10	1	decimal point	$\frac{1}{10}$	$\frac{1}{100}$	$\frac{1}{1000}$	$\frac{1}{10{,}000}$	$\frac{1}{100{,}000}$

Write 3.08 in words.

Three and eight hundredths

Rounding Decimals to a Place Value to the Right of the Decimal Point

Step 1. Locate the digit to the right of the given place value.

Step 2. If this digit is 5 or greater, add 1 to the digit in the given place value and drop all digits to its right. If this digit is less than 5, drop all digits to the right of the given place value.

Round 86.1256 to the nearest hundredth.

86.1256 (2 is in the hundredths place; 5 is 5 or greater)

rounds to 86.13

Section 5.2 Adding and Subtracting Decimals

Adding or Subtracting Decimals

Step 1. Write the decimals so that the decimal points line up vertically.

Step 2. Add or subtract as for whole numbers.

Step 3. Place the decimal point in the sum or difference so that it lines up vertically with the decimal points in the problem.

Subtract: 2.8 − 1.04

$$\begin{array}{r} \overset{7\,10}{2.8\not{0}} \\ -\ 1.04 \\ \hline 1.76 \end{array}$$

DEFINITIONS AND CONCEPTS | EXAMPLES

SECTION 5.3 MULTIPLYING DECIMALS AND CIRCUMFERENCE OF A CIRCLE

Multiplying Decimals

Step 1. Multiply the decimals as though they were whole numbers.

Step 2. The decimal point in the product is placed so that the number of decimal places in the product is equal to the *sum* of the number of decimal places in the factors.

Multiply: 1.48×5.9

$$\begin{array}{r l} 1.48 & \leftarrow 2 \text{ decimal places.} \\ \times\ 5.9 & \leftarrow 1 \text{ decimal place.} \\ \hline 1332 & \\ 7400 & \\ \hline 8.732 & \leftarrow 3 \text{ decimal places.} \end{array}$$

The circumference of a circle is the distance around the circle.

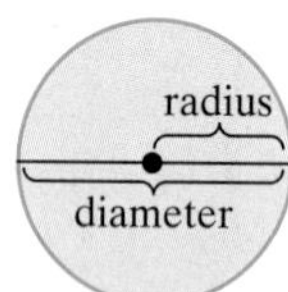

$C = \pi d$ or $C = 2\pi r$

where $\pi \approx 3.14$ or $\pi \approx \frac{22}{7}$.

or

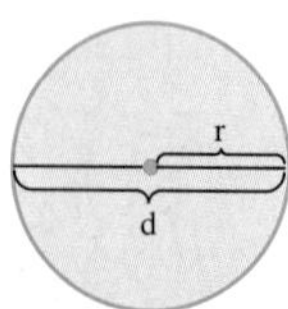

Find the exact circumference and a two-decimal-place approximation.

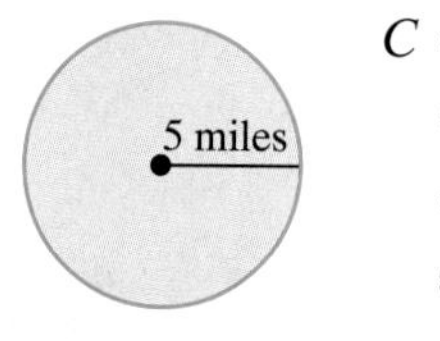

$$\begin{aligned} C &= 2\pi r \\ &= 2\pi(5) \\ &= 10\pi \\ &\approx 10(3.14) \\ &= 31.4 \end{aligned}$$

The circumference is exactly 10π miles and approximately 31.4 miles.

SECTION 5.4 DIVIDING DECIMALS

Dividing Decimals

Step 1. Move the decimal point in the divisor to the right until the divisor is a whole number.

Step 2. Move the decimal point in the dividend to the right the *same number of places* as the decimal point was moved in Step 1.

Step 3. Divide. The decimal point in the quotient is directly over the moved decimal point in the dividend.

Divide: $1.118 \div 2.6$

$$\begin{array}{r} 0.43 \\ 2.6\overline{)1.118} \\ \underline{-104} \\ 78 \\ \underline{-78} \\ 0 \end{array}$$

SECTION 5.5 ESTIMATING AND ORDER OF OPERATIONS

Order of Operations

1. Do all operations within grouping symbols such as parentheses or brackets.
2. Evaluate any expressions with exponents.
3. Multiply or divide in order from left to right.
4. Add or subtract in order from left to right.

Simplify:

$$\begin{aligned} -1.9(12.8 - 4.1) &= -1.9(8.7) && \text{Subtract.} \\ &= -16.53 && \text{Multiply.} \end{aligned}$$

DEFINITIONS AND CONCEPTS — EXAMPLES

SECTION 5.6 FRACTIONS AND DECIMALS

To write fractions as decimals, divide the numerator by the denominator.

Write $\frac{3}{8}$ as a decimal.

$$\begin{array}{r} 0.375 \\ 8\overline{)3.000} \\ \underline{-24} \\ 60 \\ \underline{-56} \\ 40 \\ \underline{-\ 40} \\ 0 \end{array}$$

SECTION 5.7 EQUATIONS CONTAINING DECIMALS

Steps for Solving an Equation in x

Step 1. If fractions are present, multiply both sides of the equation by the LCD of the fractions.

Step 2. If parentheses are present, use the distributive property.

Step 3. Combine any like terms on each side of the equation.

Step 4. Use the addition property of equality to rewrite the equation so that variable terms are on one side of the equation and constant terms are on the other side.

Step 5. Divide both sides by the numerical coefficient of x to solve.

Step 6. Check the answer in the *original equation.*

Solve:

$$3(x + 2.6) = 10.92$$

$$3x + 7.8 = 10.92 \quad \text{Apply the distributive property.}$$

$$3x + 7.8 - 7.8 = 10.92 - 7.8 \quad \text{Subtract 7.8 from both sides.}$$

$$3x = 3.12 \quad \text{Simplify.}$$

$$\frac{3x}{3} = \frac{3.12}{3} \quad \text{Divide both sides by 3.}$$

$$x = 1.04 \quad \text{Simplify.}$$

Check 1.04 in the original equation.

SECTION 5.8 SQUARE ROOTS AND THE PYTHAGOREAN THEOREM

SQUARE ROOT OF A NUMBER

The square root of a positive number a is the positive number b whose square is a. In symbols,

$$\sqrt{a} = b, \quad \text{if} \quad b^2 = a$$

Also, $\sqrt{0} = 0$.

$\sqrt{9} = 3$, $\sqrt{100} = 10$, $\sqrt{1} = 1$

PYTHAGOREAN THEOREM

If a and b are the lengths of the legs of a right triangle and c is the length of the hypotenuse, then

$$a^2 + b^2 = c^2$$

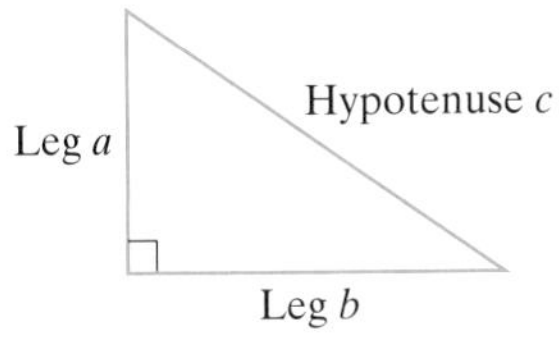

Find c.

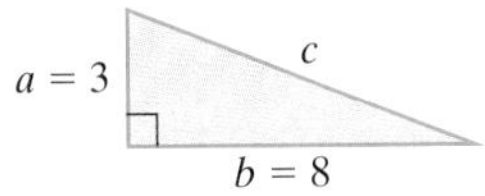

$$a^2 + b^2 = c^2$$

$$3^2 + 8^2 = c^2 \quad \text{Let } a = 3 \text{ and } b = 8.$$

$$9 + 64 = c^2 \quad \text{Multiply.}$$

$$73 = c^2 \quad \text{Simplify.}$$

$$\sqrt{73} = c \text{ or } c \approx 8.5$$

Name ______________________ Section ____________ Date ____________

Chapter 5 Review

(5.1) *Determine the place value of the number 4 in each decimal.*

1. 23.45

2. 0.000345

Write each decimal in words.

3. −23.45

4. 0.00345

5. 109.23

6. 200.000032

Write each decimal in standard form.

7. Two and fifteen hundredths

8. Negative five hundred three and one hundred two thousandths

9. Sixteen thousand twenty-five and fourteen ten-thousandths

Write each decimal as a fraction or a mixed number.

10. 0.16

11. −12.023

12. 1.0045

13. 0.00231

14. 25.25

Insert $<$, $>$, or $=$ between each pair of numbers to make a true statement.

15. 0.49 0.43

16. 0.973 0.9730

17. −402.00032 −402.000032

18. −0.230505 −0.23505

Round each decimal to the given place value.

19. 0.623, nearest tenth

20. 0.9384, nearest hundredth

21. −42.895, nearest hundredth

22. 16.34925, nearest thousandth

23. Every day in America an average of 13,490.5 people get married. Round this number to the nearest hundred.

24. A certain kind of chocolate candy bar contains 10.75 teaspoons of sugar. Convert this number to a mixed number.

(5.2) *Add.*

25. $2.4 + 7.1$

26. $3.9 + 1.2$

27. $-6.4 + (-0.88)$

28. $-19.02 + 6.98$

29. $200.49 + 16.82 + 103.002$

30. $0.00236 + 100.45 + 48.29$

Subtract.

31. $4.9 - 3.2$

32. $5.23 - 2.74$

33. $-892.1 - 432.4$

34. $0.064 - 10.2$

35. $100 - 34.98$

36. $200 - 0.00198$

37. The greatest one-day point loss on the Dow Jones Industrial Average (DJIA) occurred on September 17, 2001. On that day, the DJIA changed by -684.81 points and its value at the close of the stock market was 8920.70. What was the value of the DJIA when the stock market opened on September 17, 2001. (*Source:* Dow Jones & Co.)

38. Evaluate $x - y$ for $x = 1.2$ and $y = 6.9$.

Simplify by combining like terms.

39. $2.3x + 6.5 + 1.9x + 6.3$

40. $8.6y - 7.61 + 1.29y + 3.44$

(5.3) *Multiply.*

41. 7.2×10

42. 9.345×1000

43. -34.02×2.3

44. $-839.02 \times (-87.3)$

45. Find the exact circumference of the circle. Then use the approximation 3.14 for π and approximate the circumference.

7 meters

46. A kilometer is approximately 0.625 mile. It is 102 kilometers from Hays to Colby. Write 102 kilometers in miles to the nearest tenth mile.

(5.4) *Divide. Round the quotient to the nearest thousandth if necessary.*

47. $21 \div 0.3$

48. $-0.0063 \div 0.03$

49. $24.5 \div (-0.005)$

50. $54.98 \div 2.3$

51. $274 \div 34$

52. $-3165 \div (-20)$

53. There are approximately 3.28 feet in 1 meter. Find how many meters there are in 24 feet to the nearest tenth of a meter.

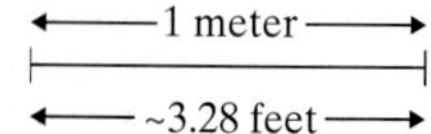

54. George Strait pays \$69.71 per month to pay back a loan of \$3136.95. In how many months will the loan be paid off?

(5.5) *Perform each indicated operation. Then estimate to see whether your proposed result is reasonable.*

55. $2.4 + 6.7 + 9.1$

56. $15.9 + 34.1$

57. $340.03 - 240.98$

58. $100 - 45.9$

59. 6.02×5.91

60. 0.205×1.72

61. $62.13 \div 1.9$

62. $601.92 \div 19.8$

63. The table shows the world's top five consumers of primary energy (in quadrillion BTU's) for the year 2000. Estimate the total energy consumption of these countries by rounding each amount to the nearest whole.

United States	98.77
China	36.67
Russia	28.07
Japan	21.77
Germany	13.98

(*Source:* Energy Information Administration)

64. Julio Tomaso is going to fertilize his lawn, a rectangle measuring 77.3 feet by 115.9 feet. Approximate the area of the lawn by rounding each measurement to the nearest foot.

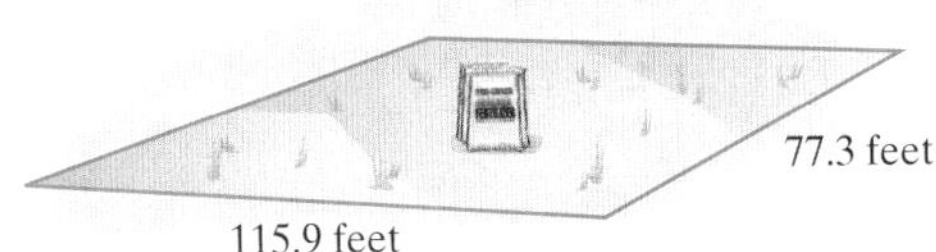

65. Estimate the cost of the items to see whether the groceries can be purchased with a \$5 bill.

Simplify each expression.

66. $(-7.6)(1.9) + 2.5$

67. $2.3^2 - 1.4$

68. $\frac{(-3.2)^2}{100}$

69. $(2.6 + 1.4)(4.5 - 3.6)$

(5.6) *Write each fraction as a decimal. Round to the nearest thousandth if necessary.*

70. $\frac{4}{5}$

71. $-\frac{12}{13}$

72. $-\frac{3}{7}$

73. $\frac{13}{60}$

74. $\frac{9}{80}$

75. $\frac{8935}{175}$

Insert $<$, $>$, or $=$ *between each pair of numbers to form a true statement.*

76. 0.3920   0.392

77. $\frac{4}{7}$ $\frac{5}{8}$

78. 0.293 $\frac{5}{17}$

79. $\frac{6}{11}$ 0.55

Write each list of numbers in order from smallest to largest.

80. 0.837, 0.839, 0.832

81. $\frac{3}{7}$, 0.42, 0.43

82. $\frac{18}{11}$, 1.63, $\frac{19}{12}$

83. $\frac{6}{7}, \frac{8}{9}, \frac{3}{4}$

Find each area.

84.

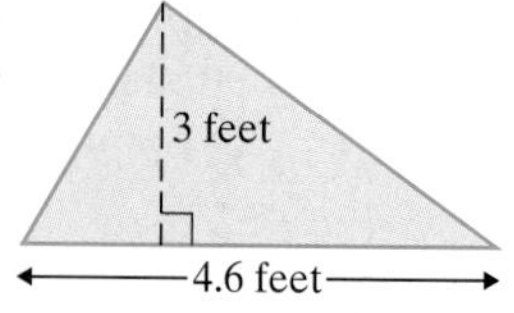

85.

2.1 inches

5.2 inches

(5.7) *Solve.*

86. $x + 3.9 = 4.2$

87. $70 = y + 22.81$

88. $2x = 17.2$

89. $-1.1y = 88$

90. $\frac{x}{4} = -0.12$

91. $6.8 = \frac{y}{5}$

92. $x + 0.78 = 1.2$

93. $0.56 = 2x$

94. $-1.3x - 9.4 = -0.4x + 8.6$

95. $3(x - 1.1) = 5x - 5.3$

(5.8) *Simplify.*

96. $\sqrt{64}$

97. $\sqrt{144}$

98. $\sqrt{36}$

99. $\sqrt{1}$

100. $\sqrt{\frac{4}{25}}$

101. $\sqrt{\frac{1}{100}}$

Find the unknown length in each given right triangle. If necessary, round to the nearest tenth.

△ **102.** leg = 12, leg = 5

△ **103.** leg = 20, leg = 21

△ **104.** leg = 9, hypotenuse = 14

△ **105.** leg = 124, hypotenuse = 155

△ **106.** leg = 66, leg = 56

△ **107.** Find the length to the nearest hundredth of the diagonal of a square that has a side of length 20 centimeters.

△ **108.** Find the height of the building rounded to the nearest tenth.

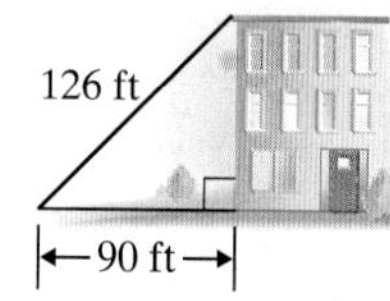

Name ______________________ Section __________ Date __________

Chapter 5 Test

Remember to check your answers and use the Chapter Test Prep Video to view solutions.

Write each decimal as indicated.

1. 45.092, in words

2. Three thousand and fifty-nine thousandths, in standard form

Perform each indicated operation. Round the result to the nearest thousandth if necessary.

3. $2.893 + 4.21 + 10.492$

4. $-47.92 - 3.28$

5. $9.83 - 30.25$

6. 10.2×4.01

7. $(-0.00843) \div (-0.23)$

Round each decimal to the indicated place value.

8. 34.8923, nearest tenth

9. 0.8623, nearest thousandth

Insert <, >, or = between each pair of numbers to form a true statement.

10. 25.0909 ⠀⠀ 25.9090

11. $\frac{4}{9}$ ⠀⠀ 0.445

Write each decimal as a fraction or a mixed number.

12. 0.345

13. -24.73

Write each fraction as a decimal. If necessary, round to the nearest thousandth.

14. $-\frac{13}{26}$

15. $\frac{16}{17}$

Simplify.

16. $(-0.6)^2 + 1.57$

17. $\frac{0.23 + 1.63}{-0.3}$

18. $2.4x - 3.6 - 1.9x - 9.8$

Find each square root and simplify. Round to the nearest thousandth if necessary.

19. $\sqrt{49}$

20. $\sqrt{157}$

21. $\sqrt{\frac{64}{100}}$

Solve.

22. $0.2x + 1.3 = 0.7$

23. $2(x + 5.7) = 6x - 3.4$

Answers

1. ______
2. ______
3. ______
4. ______
5. ______
6. ______
7. ______
8. ______
9. ______
10. ______
11. ______
12. ______
13. ______
14. ______
15. ______
16. ______
17. ______
18. ______
19. ______
20. ______
21. ______
22. ______
23. ______

24. ______

25. ______

26. ______

27. ______

28. ______

29. ______

△ **24.** Approximate to the nearest hundredth of a centimeter the length of the missing side of a right triangle with legs of 4 centimeters each.

△ **25.** Find the area.

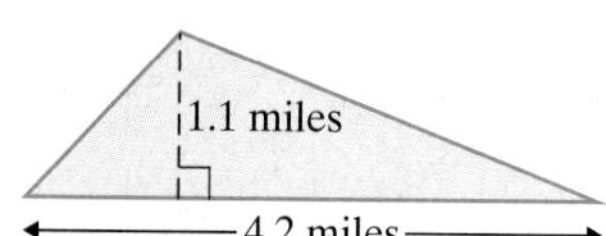

△ **26.** Vivian Thomas is going to put insecticide on her lawn to control grubworms. The lawn is a rectangle measuring 123.8 feet by 80 feet. The amount of insecticide required is 0.02 ounces per square foot. Find how much insecticide Vivian needs to purchase.

△ **27.** Find the exact circumference of the circle. Then use the approximation 3.14 for π and approximate the circumference.

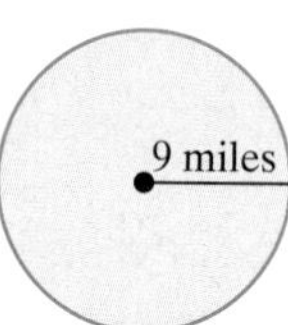

28. A CD (compact disk) holds 700 megabytes of data. A DVD (digital video disc) holds 8740 megabytes of data. How many CDs does it take to hold as much data as a DVD? Round to the nearest whole CD.

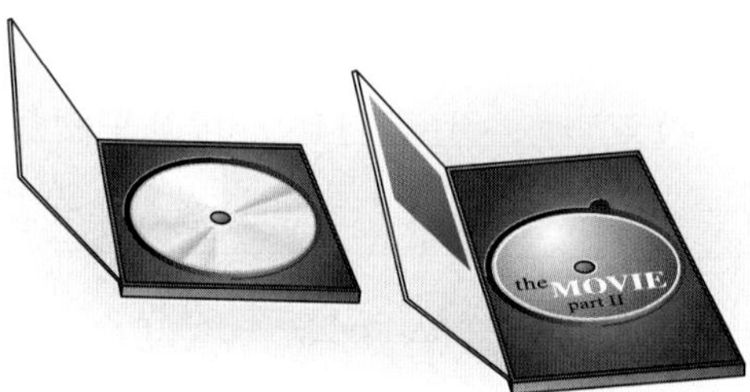

29. Estimate the total distance from Bayette to Center City by rounding each distance to the nearest mile.

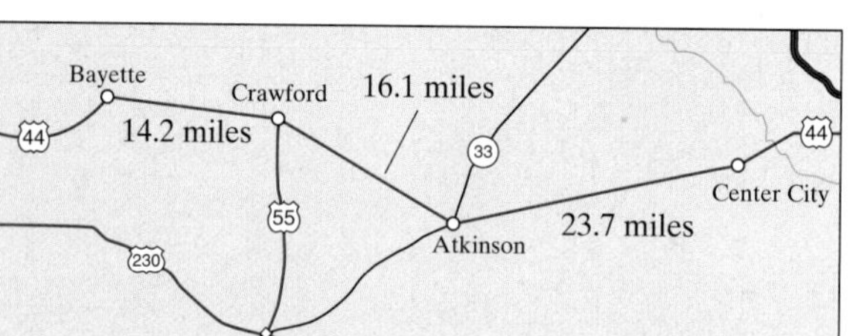

Name ______________________ Section __________ Date __________

Cumulative Review

1. Add: 34,285 + 149,761

2. Add: 5,785 + 210,199

3. Round each number to the nearest hundred to find an estimated sum.

$$\begin{array}{r} 294 \\ 625 \\ 1071 \\ +\ 349 \\ \hline \end{array}$$

4. Round each number to the nearest hundred to find the estimated sum.

$$\begin{array}{r} 186 \\ 404 \\ 853 \\ +1445 \\ \hline \end{array}$$

5. Divide: $6819 \div 17$

6. Divide: $2047 \div 14$

7. Evaluate $\frac{x - 5y}{y}$ for $x = 21$ and $y = 3$.

8. Evaluate $\frac{2a + 4}{c}$ for $a = 7$ and $c = 3$.

9. Find the opposite of each number.

a. 11 **b.** -2 **c.** 0

10. Find the opposite of each number.

a. -7 **b.** 4 **c.** -1

11. Add: $-2 + (-21)$

12. Add: $-7 + (-15)$

Find the value of each expression.

13. $2 \cdot 5^2$

14. $4 \cdot 2^3$

15. -3^2

16. $(-2)^5$

17. $(-3)^2$

18. -7^2

19. Simplify by combining like terms.

a. $3x + 2x$

b. $y - 7y$

c. $3x^2 + 5x^2 - 2$

20. Simplify by combining like terms

a. $2a - 5a$

b. $-3z^2 - z^2$

c. $4k + 2 + 7k$

21. Solve: $5x + 2 - 4x = 7 - 9$

22. Solve: $7x - 3 = 6x + 2$

23. Solve: $-5x = 15$

24. Solve: $-21 = -7x$

25. Solve: $7(x - 2) = 9x - 6$

26. Solve: $2(x + 4) = -x - 1$

Answers

1. ______
2. ______
3. ______
4. ______
5. ______
6. ______
7. ______
8. ______
9. a. ______
 b. ______
 c. ______
10. a. ______
 b. ______
 c. ______
11. ______
12. ______
13. ______
14. ______
15. ______
16. ______
17. ______
18. ______
19. a. ______
 b. ______
 c. ______
20. a. ______
 b. ______
 c. ______
21. ______
22. ______
23. ______
24. ______
25. ______
26. ______

27. ______
28. ______
29. ______
30. ______
31. ______
32. ______
33. ______
34. ______
35. ______
36. ______
37. ______
38. ______
39. ______
40. ______
41. ______
42. ______
43. ______
44. ______
45. ______
46. ______
47. ______
48. ______
49. ______
50. ______

27. In the 2000 Senate election in Wyoming, incumbent Craig Thomas received 110,280 *more* votes than his challenger. If a total of 204,358 votes were cast, find how many votes Craig Thomas received. (*Source: World Almanac*)

28. Three times a number added to 84 is 153. Find the number.

Identify the numerator and the denominator of each fraction.

29. $\frac{3}{7}$

30. $\frac{2}{5a}$

31. $\frac{13x}{5}$

32. $\frac{7y}{3x}$

33. Simplify: $\frac{12}{20}$

34. Simplify: $\frac{64}{112}$

35. Multiply: $-\frac{1}{4} \cdot \frac{1}{2}$

36. Multiply: $\left(-\frac{1}{3}\right)\left(-\frac{1}{4}\right)$

Add and simplify.

37. $\frac{2}{7} + \frac{3}{7}$

38. $\frac{2}{5} + \frac{3}{5}$

39. $\frac{7}{8} + \frac{6}{8} + \frac{3}{8}$

40. $\frac{2}{3} + \frac{1}{3} + \frac{2}{3}$

41. $\frac{2}{5} + \frac{4}{15}$

42. $\frac{3}{4} + \frac{5}{12}$

43. Simplify: $\frac{\frac{x}{4}}{\frac{3}{2}}$

44. Simplify: $\frac{\frac{2}{3}}{\frac{4x}{9}}$

45. Solve: $\frac{3}{5}a = 9$

46. Solve: $\frac{2}{3}y = 12$

47. Add: $763.7651 + 22.001 + 43.89$

48. Add: $89.27 + 14.361 + 127.2318$

49. Multiply: 23.6×0.78

50. Multiply: 43.8×0.645

Percent

CHAPTER 6

This chapter is devoted to percent, a concept used virtually every day in ordinary and business life. Understanding percent and using it efficiently depends on understanding ratios, because a percent is a ratio whose denominator is 100. We present techniques to write percents as fractions and as decimals. We then solve problems relating to interest rates, sales tax, discounts, and other real-life situations by writing percent equations.

- **6.1** Ratio and Proportion
- **6.2** Percents, Decimals, and Fractions
- **6.3** Solving Percent Problems with Equations
- **6.4** Solving Percent Problems with Proportions
- Integrated Review—Ratio, Proportion, and Percent
- **6.5** Applications of Percent
- **6.6** Percent and Problem Solving: Sales Tax, Commission, and Discount
- **6.7** Percent and Problem Solving: Interest

The Model T, developed by Henry Ford in 1908, was the world's first mass-produced automobile. It sold for \$850. In 1913, Ford Motor Company introduced interchangeable parts and the moving assembly line, and the automobile industry as we know it today was born. For almost half a century, American automobile manufacturers produced the majority of the world's motor vehicles. However, in the 1970s and 1980s, the sales of foreign auto imports in the United States began to rise. Imports reached a high in 1987, when 31.3% of all auto sales in the United States were imports. However, American auto producers started to make a comeback, reducing the sales of imported vehicles to a low of 14.9% of all such sales in 1996. Sales of imports have rebounded since then, reaching 28.2% of all auto sales in the United States in the first half of 2001. In Exercises 103–106, Section 6.2, we see some of the ways percents are used by the automobile manufacturing industry.

Answers

1. ____
2. ____
3. ____
4. ____
5. ____
6. ____
7. ____
8. ____
9. ____
10. ____
11. ____
12. ____
13. ____
14. ____
15. ____
16. ____
17. ____
18. ____
19. ____
20. ____
21. ____
22. ____
23. ____
24. ____
25. ____

Name ____ Section ____ Date ____

Chapter 6 Pretest

Write each ratio as a fraction in simplest form.

1. 5.1 to 7.9

2. \$30 to \$20

Solve.

3. $\frac{22}{x} = \frac{66}{12}$

4. $\frac{x}{8} = \frac{0.6}{2.4}$

5. On Jill Tilbert's map, 1 inch represents 20 miles. Find the distance represented by $5\frac{3}{8}$ inches on her map.

6. In a group of 100 people, 12 are female. What percent of the group is female?

7. Write 57% as a decimal.

8. Write 2.75 as a percent.

9. Write 7.5% as a fraction in simplest form.

10. Write $\frac{3}{20}$ as a percent.

Translate each question to an equation.

11. 18% of 50 is what number?

12. 4% of what number is 89?

Translate each question to a proportion.

13. 90% of what number is 82?

14. 48 is what percent of 112?

Solve.

15. What percent of 80 is 16?

16. What number is 1.5% of 220?

17. 32 is 16% of what number?

18. In a box of 250 lightbulbs, 1.2% were found to be defective. How many lightbulbs were defective?

19. The enrollment at a local high school decreased 2% over last year's enrollment of 4200. Find the decrease in enrollment and the current enrollment.

20. The sales tax on a \$499 printer is \$34.93. Find the sales tax rate.

21. Jerry Williams receives 2% commission on his sales of computer equipment. Last week his commission was \$448. Find the amount of his sales last week.

22. A television that normally sells for \$650 is on sale at 12% off. What is the discount and what is the sales price?

23. Find the simple interest after 3 years on \$600 at an interest rate of 8%.

24. \$5000 is invested at 6% compounded quarterly for 6 years. Find the total amount at the end of 6 years. Use Appendix C.

25. Find the monthly payment on a \$700 loan for 2 years if the interest on the 2-year loan is \$196.

6.1 Ratio and Proportion

OBJECTIVES

- A Write ratios as fractions.
- B Solve proportions.
- C Solve problems modeled by proportions.

A Writing Ratios as Fractions

A **ratio** is the quotient of two numbers or two quantities. For example, a percent can be thought of as a ratio, since it is the quotient of a number and 100.

$$53\% = \frac{53}{100} \quad \text{or} \quad \text{the ratio of 53 to 100}$$

Ratio

The ratio of a number a to a number b is their quotient. Ways of writing ratios are

$$a \text{ to } b, \qquad a:b, \quad \text{and} \quad \frac{a}{b}$$

Whenever possible, we will convert quantities in a ratio to the same unit of measurement.

EXAMPLE 1 Write a ratio for each phrase. Use fractional notation.

a. The ratio of 2 parts salt to 5 parts water

b. The ratio of 18 inches to 2 feet

Solution:

a. The ratio of 2 parts salt to 5 parts water is $\frac{2}{5}$.

b. First we convert to the same unit of measurement. For example,

$$2 \text{ feet} = 2 \cdot 12 \text{ inches} = 24 \text{ inches}$$

The ratio of 18 inches to 2 feet is then $\frac{18}{24}$, or $\frac{3}{4}$ in lowest terms.

Practice Problem 1

Write a ratio for each phrase. Use fractional notation.

a. The ratio of 3 parts oil to 7 parts gasoline

b. The ratio of 40 minutes to 3 hours

If a ratio compares two decimal numbers, we will write the simplified ratio as a ratio of whole numbers.

EXAMPLE 2 Write the ratio of 2.5 to 3.15 as a fraction in simplest form.

Solution: The ratio is

$$\frac{2.5}{3.15}$$

Now let's clear the ratio of decimals.

$$\frac{2.5}{3.15} = \frac{2.5 \cdot 100}{3.15 \cdot 100} = \frac{250}{315} = \frac{50}{63} \quad \text{Simplest form}$$

Practice Problem 2

Write the ratio of 1.68 to 4.8 as a fraction in simplest form.

Answers

1. a. $\frac{3}{7}$ **b.** $\frac{2}{9}$ **2.** $\frac{7}{20}$

Practice Problem 3

Given the triangle shown:

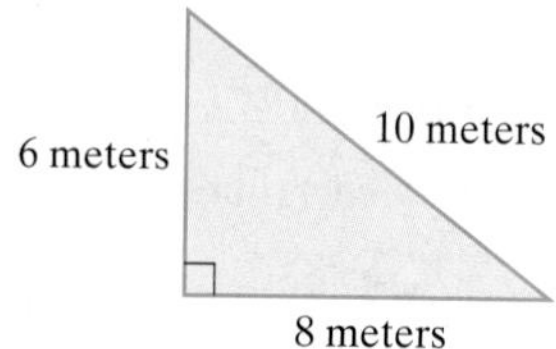

a. Find the ratio of the length of the shortest side to the length of the longest side in simplest form.

b. Find the ratio of the length of the longest side to the perimeter of the triangle in simplest form.

Concept Check

Explain why the answer $\frac{7}{5}$ would be incorrect for part (a) of Example 3.

Answers

3. a. $\frac{3}{5}$ **b.** $\frac{5}{12}$

Concept Check: $\frac{7}{5}$ is the ratio of the rectangle's length to its width.

EXAMPLE 3 Given the rectangle shown:

a. Find the ratio of its width (shorter side) to its length (longer side).

b. Find the ratio of its length to its perimeter.

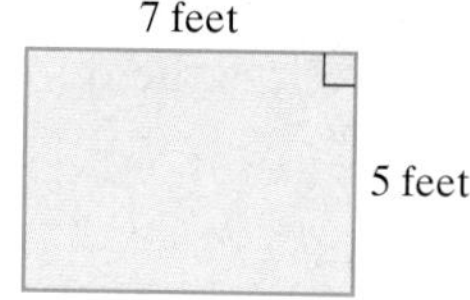

Solution:

a. The ratio of its width to its length is

$$\frac{\text{width}}{\text{length}} = \frac{5 \cancel{\text{feet}}}{7 \cancel{\text{feet}}} = \frac{5}{7}$$

b. Recall that the perimeter of the rectangle is the distance around the rectangle: $7 + 5 + 7 + 5 = 24$ feet. The ratio of its length to its perimeter is

$$\frac{\text{length}}{\text{perimeter}} = \frac{7 \cancel{\text{feet}}}{24 \cancel{\text{feet}}} = \frac{7}{24}$$

Try the Concept Check in the margin.

B Solving Proportions

Ratios can be used to form proportions. A **proportion** is a mathematical statement that two ratios are equal.

For example, the equation

$$\frac{1}{2} = \frac{4}{8}$$

is a proportion that says that the ratios $\frac{1}{2}$ and $\frac{4}{8}$ are equal.

Notice that a proportion contains four numbers. If any three numbers are known, we can solve and find the fourth number. One way to do so is to use cross products. To understand cross products, let's start with the proportion

$$\frac{a}{b} = \frac{c}{d}$$

and multiply both sides by the LCD, bd.

$$\frac{a}{b} = \frac{c}{d}$$

$$bd\left(\frac{a}{b}\right) = bd\left(\frac{c}{d}\right) \quad \text{Multiply both sides by the LCD, } bd.$$

$$ad = bc \quad \text{Simplify.}$$

cross product (ad) cross product (bc)

Notice why ad and bc are called cross products.

$$\frac{a}{b} = \frac{c}{d} \qquad \begin{matrix} bc \\ ad \end{matrix}$$

Cross Products

If $\frac{a}{b} = \frac{c}{d}$, then $ad = bc$.

EXAMPLE 4 Solve for x: $\frac{45}{x} = \frac{5}{7}$

Solution: To solve, we set cross products equal.

$$\frac{45}{x} = \frac{5}{7}$$

$$\begin{aligned} 45 \cdot 7 &= x \cdot 5 && \text{Set cross products equal.} \\ 315 &= 5x && \text{Multiply.} \\ \frac{315}{5} &= \frac{5x}{5} && \text{Divide both sides by 5.} \\ 63 &= x && \text{Simplify.} \end{aligned}$$

Check: To check, substitute 63 for x in the original proportion. The solution is 63.

EXAMPLE 5 Solve for x: $\frac{x-5}{3} = \frac{x+2}{5}$

Solution:

$$\frac{x-5}{3} = \frac{x+2}{5}$$

$$\begin{aligned} 5(x-5) &= 3(x+2) && \text{Set cross products equal.} \\ 5x - 25 &= 3x + 6 && \text{Multiply.} \\ 5x &= 3x + 31 && \text{Add 25 to both sides.} \\ 2x &= 31 && \text{Subtract } 3x \text{ from both sides.} \\ \frac{2x}{2} &= \frac{31}{2} && \text{Divide both sides by 2.} \\ x &= \frac{31}{2} \end{aligned}$$

Check: Verify that $\frac{31}{2}$ is the solution.

Try the Concept Check in the margin.

Practice Problem 4

Solve for x: $\frac{3}{8} = \frac{63}{x}$

Practice Problem 5

Solve for x: $\frac{2x+1}{7} = \frac{x-3}{5}$

Concept Check

For which of the following equations can we immediately use cross products to solve for x?

a. $\frac{2-x}{5} = \frac{1+x}{3}$

b. $\frac{2}{5} - x = \frac{1+x}{3}$

Answers

4. $x = 168$ **5.** $x = -\frac{26}{3}$

Concept Check: a

C Solving Problems Modeled by Proportions

Writing proportions is a powerful tool for solving problems in almost every field, including business, chemistry, biology, health sciences, and engineering, as well as in daily life. Given a specified ratio (or rate) of two quantities, a proportion can be used to determine an unknown quantity.

Practice Problem 6

On an architect's blueprint, 1 inch corresponds to 12 feet. How long is a wall represented by a $3\frac{1}{2}$-inch line on the blueprint?

EXAMPLE 6 Determining Distances from a Map

On a Chamber of Commerce map of Abita Springs, 5 miles corresponds to 2 inches. How many miles correspond to 7 inches?

Solution:

1. UNDERSTAND. Read and reread the problem. You may want to draw a diagram.

5 miles	5 miles	5 miles	? between 0 and 5 miles	= between 15 and 20 miles
2 inches	2 inches	2 inches	1 inch	= 7 inches

From the diagram we can see that our solution should be between 15 and 20 miles.

2. TRANSLATE. We will let n represent our unknown number. Since we are given that 5 miles corresponds to 2 inches, let's use this rate as the first fraction in our proportion. In this section, we will solve by writing proportions so that numerators have the same unit and denominators have the same unit. Since 5 miles corresponds to 2 inches as n miles corresponds to 7 inches, we have the proportion:

$$\begin{matrix}\text{miles} \rightarrow \\ \text{inches} \rightarrow\end{matrix} \frac{5}{2} = \frac{n}{7} \begin{matrix}\leftarrow \text{miles} \\ \leftarrow \text{inches}\end{matrix}$$

3. SOLVE.

$$\frac{5}{2} = \frac{n}{7}$$

$$2n = 35 \quad \text{Set cross products equal.}$$

$$\frac{2n}{2} = \frac{35}{2} \quad \text{Divide both sides by 2.}$$

$$n = 17.5 \quad \text{Simplify.}$$

4. INTERPRET. *Check* your work. This result is reasonable since it is between 15 and 20 miles. *State* your conclusion: 7 inches corresponds to 17.5 miles.

Answer

6. 42 feet

We can also solve Example 6 by writing the proportion

$$\frac{2 \text{ inches}}{5 \text{ miles}} = \frac{7 \text{ inches}}{n \text{ miles}}$$

Although other proportions may be used to solve Example 6, we will solve by writing proportions so that the numerators have the same unit measures and the denominators have the same unit measures.

EXAMPLE 7 Finding Medicine Dosage

The standard dose of an antibiotic is 4 cc (cubic centimeters) for every 25 pounds (lb) of body weight. At this rate, find the standard dose for a 140-lb woman.

Solution:

1. UNDERSTAND. Read and reread the problem. You may want to draw a diagram to estimate a reasonable solution.

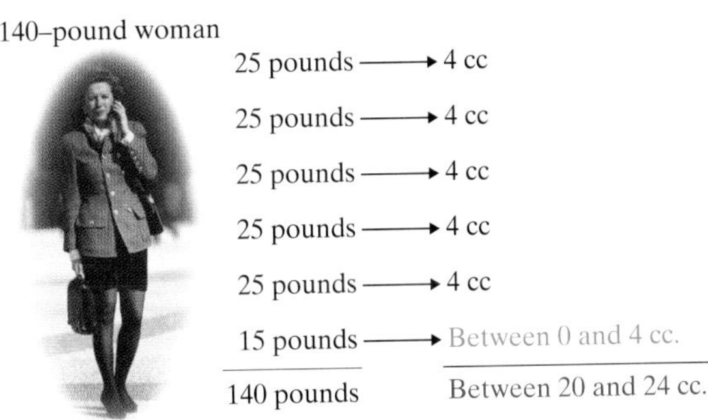

From the diagram, we can see that a reasonable solution is between 20 cc and 24 cc.

2. TRANSLATE. We will let n represent the unknown number. From the problem, we know that 4 cc is to 25 lb as n cc is to 140 lb, or

$$\begin{matrix}\text{cc} \rightarrow \\ \text{lb} \rightarrow\end{matrix} \frac{4}{25} = \frac{n}{140} \begin{matrix}\leftarrow \text{cc} \\ \leftarrow \text{lb}\end{matrix}$$

3. SOLVE.

$$\frac{4}{25} = \frac{n}{140}$$

$$\begin{aligned} 25n &= 560 && \text{Set cross products equal.} \\ \frac{25n}{25} &= \frac{560}{25} && \text{Divide both sides by 25.} \\ n &= 22.4 && \text{Simplify.} \end{aligned}$$

4. INTERPRET. *Check* your work. This result is reasonable since it is between 20 and 24 cc. *State* your conclusion: The standard dose for a 140-lb woman is 22.4 cc.

Practice Problem 7

An auto mechanic recommends that 3 ounces of isopropyl alcohol be mixed with a tankful of gas (14 gallons) to increase the octane of the gasoline for better engine performance. At this rate, how many gallons of gas can be treated with a 16-ounce bottle of alcohol?

Answer

7. $74\frac{2}{3}$ gallons

Name ______________________ Section __________ Date __________

EXERCISE SET 6.1

A *Write each ratio in fractional notation in lowest terms. See Examples 1 through 3.*

1. 2 megabytes to 15 megabytes

2. 18 disks to 41 disks

3. 10 inches to 12 inches

4. 15 miles to 40 miles

5. 5 quarts to 3 gallons

6. 8 inches to 3 feet

7. 4 nickels to 2 dollars

8. 12 quarters to 2 dollars

9. 175 centimeters to 5 meters

10. 90 centimeters to 4 meters

11. 190 minutes to 3 hours

12. 60 hours to 2 days

△ **13.** Find the ratio of the length to the width of a regulation size basketball court.

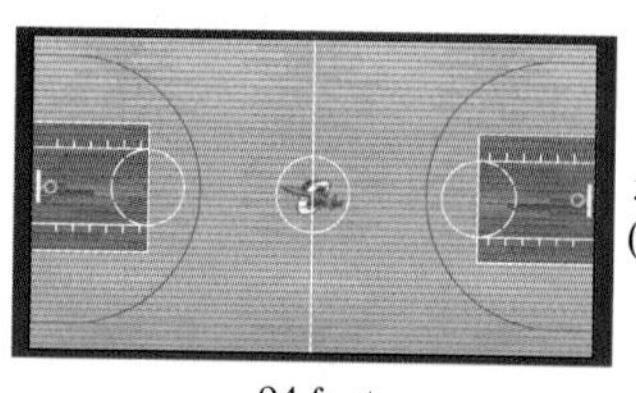

△ **14.** Find the ratio of the base to the height of the triangular mainsail.

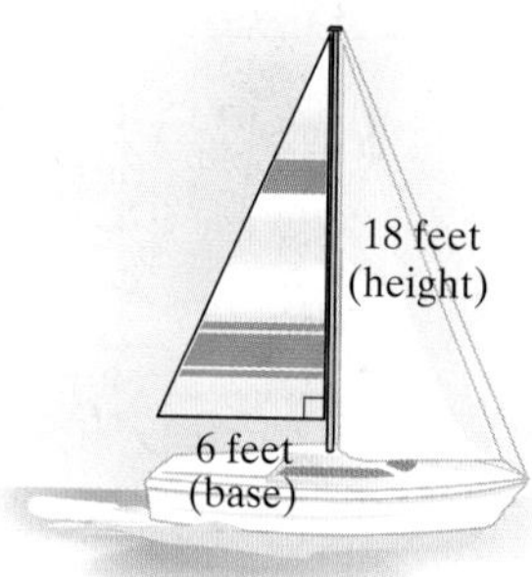

△ **15.** Find the ratio of the longest side to the perimeter of the right-triangular-shaped billboard.

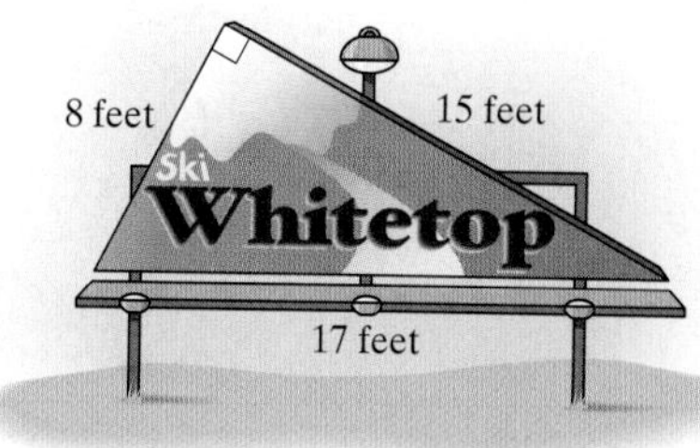

△ **16.** Find the ratio of the width to the perimeter of the rectangular vegetable garden.

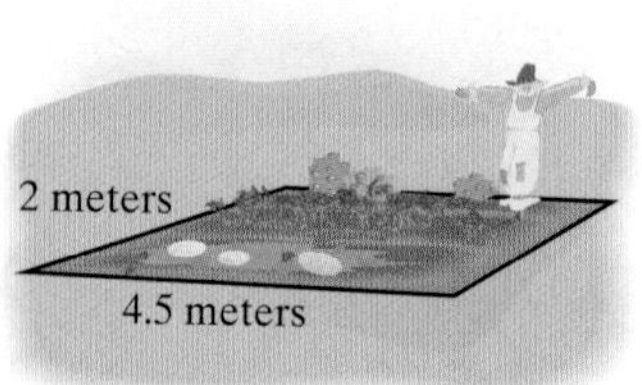

17. A large order of McDonald's french fries has 450 calories. Of this total, 200 calories are from fat. Find the ratio of calories from fat to total calories in a large order of McDonald's french fries. (*Source:* McDonald's Corporation)

18. A McDonald's Quarter Pounder® with Cheese contains 30 grams of fat. A McDonald's Grilled Chicken™ sandwich contains 20 grams of fat. Find the ratio of the amount of fat in a Quarter Pounder with Cheese to the amount of fat in a Grilled Chicken sandwich. (*Source:* McDonald's Corporation)

Blood contains three types of cells: red blood cells, white blood cells, and platelets. For approximately every 600 red blood cells in healthy humans, there are 40 platelets and 1 white blood cell. (*Source:* American Red Cross Biomedical Services) *Use this for Exercises 19 and 20.*

19. Write the ratio of red blood cells to platelet cells.

20. Write the ratio of white blood cells to red blood cells.

21. Suppose someone tells you that the ratio of 11 inches to 2 feet is $\frac{11}{2}$. How would you correct that person and explain the error?

22. Write a ratio that can be written in fractional notation as $\frac{3}{2}$.

B *Solve each proportion. See Examples 4 and 5.*

23. $\frac{2}{3} = \frac{x}{6}$

24. $\frac{x}{2} = \frac{16}{6}$

25. $\frac{x}{10} = \frac{5}{9}$

26. $\frac{9}{4x} = \frac{6}{2}$

27. $\frac{4x}{6} = \frac{7}{2}$

28. $\frac{a}{5} = \frac{3}{2}$

29. $\frac{x-3}{x} = \frac{4}{7}$

30. $\frac{y}{y-16} = \frac{5}{3}$

31. $\frac{x+1}{2x+3} = \frac{2}{3}$

32. $\frac{x+1}{x+2} = \frac{5}{3}$

33. $\frac{9}{5} = \frac{12}{3x+2}$

34. $\frac{6}{11} = \frac{27}{3x-2}$

35. $\frac{3}{x+1} = \frac{5}{2x}$

36. $\frac{7}{x-3} = \frac{8}{2x}$

37. $\frac{15}{3x-4} = \frac{5}{x}$

38. $\frac{x}{3} = \frac{2x+5}{6}$

C *Solve. See Examples 6 and 7.*

39. The ratio of the weight of an object on Earth to the weight of the same object on Pluto is 100 to 3. If an elephant weighs 4100 pounds on Earth, find the elephant's weight on Pluto.

40. If a 170-pound person weighs approximately 65 pounds on Mars, how much does a 9000-pound satellite weigh on Mars? Round to the nearest pound.

41. There are 110 calories per 28.4 grams of Crispy Rice cereal. Find how many calories are in 42.6 grams of this cereal.

42. On an architect's blueprint, 1 inch corresponds to 4 feet. Find the length of a wall represented by a line that is $3\frac{7}{8}$ inches long on the blueprint.

43. The daily supply of oxygen for one person is provided by 625 square feet of lawn. A total of 3750 square feet of lawn would provide the daily supplies of oxygen for how many people? (*Source:* Professional Lawn Care Association of America)

44. In the United States, approximately 71 million of the 200 million cars and light trucks in service have driver side air bags. In a parking lot containing 800 cars and light trucks, how many would be expected to have driver side air bags? (*Source:* Insurance Institute for Highway Safety)

45. Mary Boyd would like to estimate the height of the Statue of Liberty in New York City's harbor. According to the *2003 World Almanac*, the length of the Statue of Liberty's right arm is 42 feet. Mary's right arm is 2 feet long and her height is $5\frac{1}{3}$ feet. Use this information to estimate the height of the Statue of Liberty. (The actual height of the Statue of Liberty, from heel to top of the head, is 111 feet, 1 inch. How close is the estimated height to the actual height of the statue?)

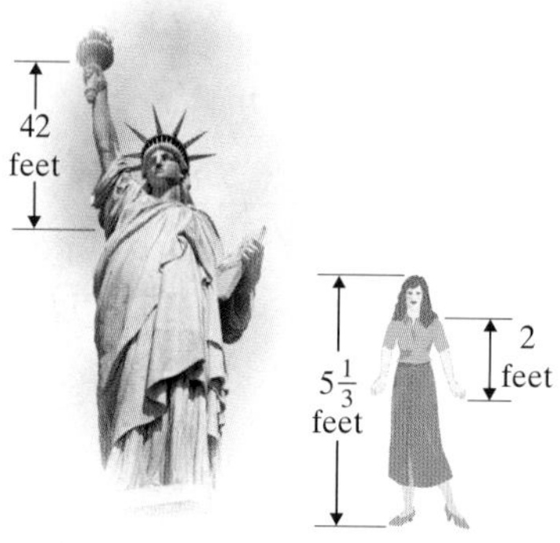

46 There are 72 milligrams of cholesterol in a 3.5 ounce serving of lobster. How much cholesterol is in 5 ounces of lobster? Round to the nearest tenth of a milligram. (*Source:* The National Institute of Health)

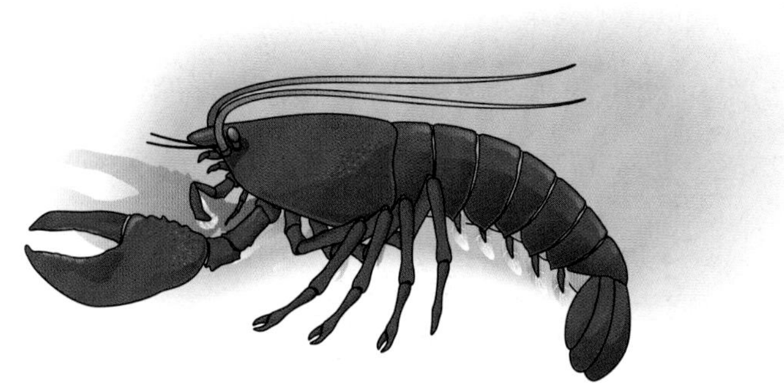

47. One pound of firmly packed brown sugar yields $2\frac{1}{4}$ cups. How many pounds of brown sugar will be required by a recipe that calls for 6 cups of firmly packed brown sugar? (*Source:* Based on data from *Family Circle* magazine)

48. Eleven out of every 25 greeting cards sold in the United States are Hallmark brand cards. If a consumer purchased 75 greeting cards in the past year, how many do you expect would have been Hallmark brand cards? (*Source:* Hallmark Cards, Inc.)

49. Most men are concerned about their health, but they continue to have poor eating habits. Two out of every 5 men blame their poor eating habits on too much fast food. In a room of 40 men, how many would you expect to blame their not eating well on fast food? (*Source*: Healthy Choice Mixed Grills survey)

50. Trump World Tower in New York City is 881 feet tall and contains 72 stories. The Empire State Building contains 102 stories. If the Empire State Building has the same number of feet per floor as the Trump World Tower, approximate its height rounded to the nearest foot. (*Source*: skyscrapers.com)

51. Medication is prescribed in 7 out of every 10 hospital emergency room visits that involve an injury. If a large urban hospital had 620 emergency room visits involving an injury in the past month, how many of these visits would you expect included a prescription for medication? (*Source:* National Center for Health Statistics)

52. Currently in the American population of people aged 65 years old and older, there are 145 women for every 100 men. In a nursing home with 280 male residents over the age of 65, how many female residents over the age of 65 would be expected? (*Source:* U.S. Bureau of the Census)

When making homemade ice cream in a hand-cranked freezer, the tub containing the ice cream mix is surrounded by a brine solution. To freeze the ice cream mix rapidly so that smooth and creamy ice cream results, the brine solution should combine crushed ice and rock salt in a ratio of 5 to 1. (*Source:* White Mountain Freezers, The Rival Company)

53. A small ice cream freezer requires 12 cups of crushed ice. How much rock salt should be mixed with the ice to create the necessary brine solution?

54. A large ice cream freezer requires $18\frac{3}{4}$ cups of crushed ice. How much rock salt will be needed?

55. The gas/oil ratio for a certain chainsaw is 50 to 1.
   a. How much oil (in gallons) should be mixed with 5 gallons of gasoline?

   b. If 1 gallon equals 128 fluid ounces, write the answer to part **a** in fluid ounces. Round to the nearest whole ounce.

56. The gas/oil ratio for a certain tractor mower is 20 to 1.
   a. How much oil (in gallons) should be mixed with 10 gallons of gas?

   b. If 1 gallon equals 4 quarts, write the answer to part **a** in quarts.

57. The adult daily dosage for a certain medicine is 150 mg (milligrams) of medicine for every 20 pounds of body weight.
   a. At this rate, find the daily dose for a man who weighs 275 pounds.

   b. If the man is to receive 500 mg of this medicine every 8 hours, is he receiving the proper dosage?

58. The adult daily dosage for a certain medicine is 80 mg (milligrams) for every 25 pounds of body weight.
   a. At this rate, find the daily dose for a woman who weighs 190 pounds.

   b. If she is to receive this medicine every 6 hours, find the amount to be given every 6 hours.

Review and Preview

Find the prime factorization of each number. See Section 4.2.

59. 15 **60.** 21 **61.** 20 **62.** 24

63. 200 **64.** 300 **65.** 32 **66.** 81

Combining Concepts

As we have seen earlier, proportions are often used in medicine dosage calculations. The exercises below have to do with liquid drug preparations, where the weight of the drug is contained in a volume of solution. The description of mg and ml below will help. We will study metric units further in Chapter 8.

mg means milligrams (A paper clip weighs about a gram. A milligram is about the weight of $\frac{1}{1000}$ *of a paper clip.)*

ml means milliliter (A liter is about a quart. A milliliter is about the amount of liquid in $\frac{1}{1000}$ *of a quart.)*

One way to solve the applications below is to set up the proportion $\frac{\text{mg}}{\text{ml}} = \frac{\text{mg}}{\text{ml}}$.

A solution strength of 15 mg of medicine in 1 ml of solution is available.

67. If a patient needs 12 mg of medicine, how many ml do you administer?

68. If a patient needs 33 mg of medicine, how many ml do you administer?

A solution strength of 8 mg of medicine in 1 ml of solution is available.

69. If a patient needs 10 mg of medicine, how many ml do you administer?

70. If a patient needs 6 mg of medicine, how many ml do you administer?

6.2 Percents, Decimals, and Fractions

OBJECTIVES

- A Know the meaning of percent.
- B Write percents as decimals.
- C Write decimals as percents.
- D Write percents as fractions.
- E Write fractions as percents.
- F Convert percents, decimals, and fractions.

A Understanding Percent

The word **percent** comes from the Latin phrase *per centum*, which means **"per 100."** For example, 53 percent (%) means 53 per 100. In the square below, 53 of the 100 squares are shaded. Thus 53% of the figure is shaded.

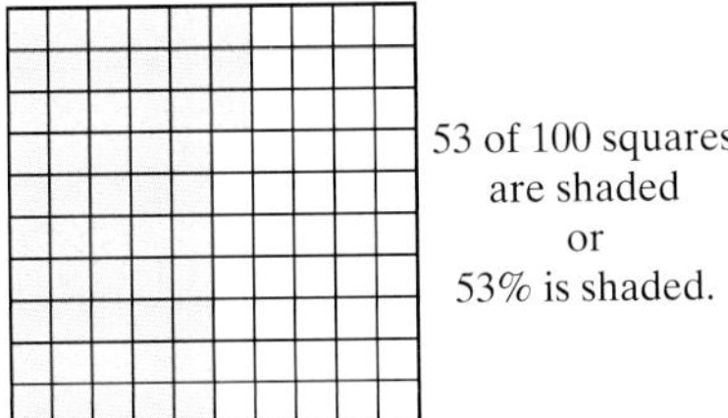

53 of 100 squares are shaded or 53% is shaded.

Since 53% means 53 per 100, 53% is the ratio of 53 to 100, or $\frac{53}{100}$.

$$53\% = \frac{53}{100}$$

Percent

Percent means **per one hundred.** The "%" symbol is used to denote percent.

Also,

$$7\% = \frac{7}{100} \quad \text{7 parts per 100 parts}$$

$$73\% = \frac{73}{100} \quad \text{73 parts per 100 parts}$$

$$109\% = \frac{109}{100} \quad \text{109 parts per 100 parts}$$

Percent is used in a variety of everyday situations. For example:

The interest rate is 5.7%.
47.4% of U.S. homes have Internet access.
The store is having a 25% off sale.
78% of us trust our local fire department.
The enrollment in community colleges has increased 141% in the last 30 years.

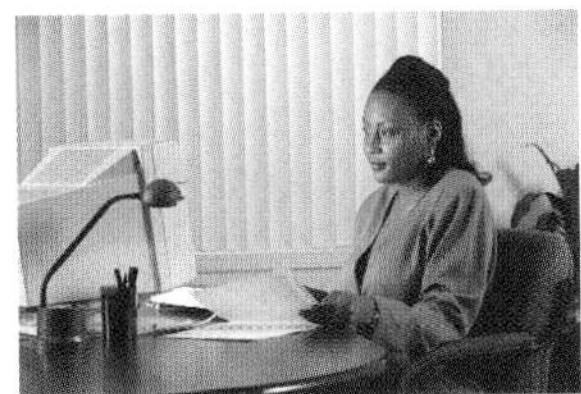

EXAMPLE 1 In a survey of 100 people, 17 people drive blue cars. What percent of people drive blue cars?

Solution: Since 17 people out of 100 drive blue cars, the fraction is $\frac{17}{100}$.

Then

$$\frac{17}{100} = 17\%$$

Practice Problem 1

Of 100 students in a club, 23 are freshmen. What percent of the students are freshmen?

Answer

1. 23%

Practice Problem 2

29 out of 100 executives are in their forties. What percent of executives are in their forties?

EXAMPLE 2 46 out of every 100 college students live at home. What percent of students live at home? (*Source:* Independent Insurance Agents of America)

Solution: $\frac{46}{100} = 46\%$

B Writing Percents as Decimals

Since percent means "per hundred," we have that

$$1\% = \frac{1}{100} = 0.01$$

To write a percent as a decimal, we can first write the percent as a fraction.

$$53\% = \frac{53}{100}$$

Now we can write the fraction as a decimal as we did in Section 5.6.

$$\frac{53}{100} = 53(0.01) = 0.53 \text{ (53-hundredths)}$$

Notice that the result is

$53\% = 53(0.01) = 0.53$ Replace the percent symbol with 0.01. Then multiply.

Writing a Percent as a Decimal

Replace the percent symbol with its decimal equivalent, 0.01; then multiply.

$$43\% = 43(0.01) = 0.43$$

Practice Problem 3

Write 89% as a decimal.

Practice Problems 4–7

Write each percent as a decimal.

4. 2.7%
5. 150%
6. 0.69%
7. 500%

EXAMPLE 3 Write 23% as a decimal.

Solution: $23\% = 23(0.01)$ Replace the percent symbol with 0.01.
$= 0.23$ Multiply.

EXAMPLES Write each percent as a decimal.

4. $4.6\% = 4.6(0.01) = 0.046$ Replace the percent symbol with 0.01. Then multiply.

5. $190\% = 190(0.01) = 1.90$ or 1.9

6. $0.74\% = 0.74(0.01) = 0.0074$

7. $100\% = 100(0.01) = 1.00$ or 1

Concept Check

Why is it incorrect to write the percent 0.033% as 3.3 in decimal form?

Try the Concept Check in the margin.

Answers

2. 29% **3.** 0.89 **4.** 0.027 **5.** 1.5 **6.** 0.0069 **7.** 5

Concept Check: To write a percent as a decimal, the decimal point should be moved two places to the left, not to the right. So the correct answer is 0.00033.

C Writing Decimals as Percents

To write a decimal as a percent, we use the result of Example 7 above. In this example, we found that 1 = 100%.

$$0.38 = 0.38(1) = 0.38(100\%) = 38\%$$

Notice that the result is

$$0.38 = 0.38(100\%) = 38.\% \text{ or } 38\% \quad \text{Multiply by 1 in the form of 100\%.}$$

Writing a Decimal as a Percent

Multiply by 1 in the form of 100%.

$$0.27 = 0.27(100\%) = 27.\% \text{ or } 27\%$$

EXAMPLE 8 Write 0.65 as a percent.

Solution: $0.65 = 0.65(100\%) = 65.\%$ Multiply by 100%.

$= 65\%$

Practice Problem 8

Write 0.19 as a percent.

EXAMPLES Write each decimal as a percent.

9. $1.25 = 1.25(100\%) = 125.\%$ or 125%

10. $0.012 = 0.012(100\%) = 001.2\%$ or 1.2%

11. $0.6 = 0.6(100\%) = 060.\%$ or 60%

Practice Problems 9–11

Write each decimal as a percent.

9. 1.75 10. 0.044 11. 0.7

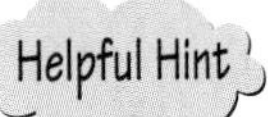

A zero was inserted as a placeholder.

Try the Concept Check in the margin.

Concept Check

Why is it incorrect to write the decimal 0.0345 as 34.5% in percent form?

Answers

8. 19% **9.** 175% **10.** 4.4% **11.** 70%

Concept Check: To change a decimal to a percent, the decimal point should be moved *only* two places to the right. So the correct answer is 3.45%.

D Writing Percents as Fractions

When we write a percent as a fraction, we usually then write the fraction in simplest form. For example, recall from the previous section that

$$50\% = \frac{50}{100}$$

Then we write the fraction in simplest form:

$$\frac{50}{100} = \frac{\overset{1}{\cancel{50}}}{2 \cdot \underset{1}{\cancel{50}}} = \frac{1}{2}$$

Writing a Percent as a Fraction

Replace the percent symbol with its fraction equivalent, $\frac{1}{100}$; then multiply. Don't forget to simplify the fraction, if possible.

$$7\% = 7 \cdot \frac{1}{100} = \frac{7}{100}$$

Practice Problems 12–16

Write each percent as a fraction in simplest form.

12. 25%
13. 2.3%
14. 150%
15. $66\frac{2}{3}\%$
16. 8%

EXAMPLES Write each percent as a fraction or mixed number in simplest form.

12. $40\% = 40 \cdot \frac{1}{100} = \frac{40}{100} = \frac{2 \cdot \overset{1}{\cancel{20}}}{5 \cdot \underset{1}{\cancel{20}}} = \frac{2}{5}$

13. $1.9\% = 1.9 \cdot \frac{1}{100} = \frac{1.9}{100}$. Next we multiply the numerator and denominator by 10.

$$\frac{1.9 \cdot 10}{100 \cdot 10} = \frac{19}{1000}$$

14. $125\% = 125 \cdot \frac{1}{100} = \frac{125}{100} = \frac{5 \cdot \overset{1}{\cancel{25}}}{4 \cdot \underset{1}{\cancel{25}}} = \frac{5}{4}$ or $1\frac{1}{4}$

15. $33\frac{1}{3}\% = 33\frac{1}{3} \cdot \frac{1}{100} = \frac{100}{3} \cdot \frac{1}{100} = \frac{\overset{1}{\cancel{100}} \cdot 1}{3 \cdot \underset{1}{\cancel{100}}} = \frac{1}{3}$

Write as an improper fraction.

16. $100\% = 100 \cdot \frac{1}{100} = \frac{100}{100} = 1$

E Writing Fractions as Percents

Recall that to write a percent as a fraction, we drop the percent symbol and divide by 100. We reverse these steps to write a fraction as a percent.

Answers

12. $\frac{1}{4}$ **13.** $\frac{23}{1000}$ **14.** $\frac{3}{2}$ **15.** $\frac{2}{3}$ **16.** $\frac{2}{25}$

Writing a Fraction as a Percent

Multiply by 1 in the form of 100%.

$$\frac{1}{8} = \frac{1}{8} \cdot 100\% = \frac{1}{8} \cdot \frac{100}{1}\% = \frac{100}{8}\% = 12\frac{1}{2}\% \quad \text{or} \quad 12.5\%$$

Helpful Hint

From Example 7, we know that

$$100\% = 1$$

Recall that when we multiply a number by 1, we are not changing the value of that number. Therefore, when we multiply a number by 100%, we are not changing its value but rather writing the number as an equivalent percent.

EXAMPLES Write each fraction or mixed number as a percent.

17. $\frac{9}{20} = \frac{9}{20} \cdot 100\% = \frac{9}{20} \cdot \frac{100}{1}\% = \frac{900}{20}\% = 45\%$

18. $\frac{2}{3} = \frac{2}{3} \cdot 100\% = \frac{2}{3} \cdot \frac{100}{1}\% = \frac{200}{3}\% = 66\frac{2}{3}\%$

19. $1\frac{1}{2} = \frac{3}{2} \cdot 100\% = \frac{3}{2} \cdot \frac{100}{1}\% = \frac{300}{2}\% = 150\%$

Try the Concept Check in the margin.

EXAMPLE 20 Write $\frac{1}{12}$ as a percent. Round to the nearest hundredth percent.

Solution:

$$\frac{1}{12} = \frac{1}{12} \cdot 100\% = \frac{1}{12} \cdot \frac{100\%}{1} = \frac{100}{12}\% \approx 8.33\%$$

("≈" means "approximately")

$$\begin{array}{r} 8.333 \approx 8.33 \\ 12\overline{)100.000} \\ -96 \\ \hline 4\,0 \\ -3\,6 \\ \hline 40 \\ -36 \\ \hline 40 \\ -36 \\ \hline 4 \end{array}$$

Thus, $\frac{1}{12}$ is approximately 8.33%.

Practice Problems 17–19

Write each fraction or mixed number as a percent.

17. $\frac{1}{2}$ 18. $\frac{7}{40}$ 19. $2\frac{1}{4}$

Concept Check

Which digit in the percent 76.4582% represents

a. A tenth percent?
b. A thousandth percent?
c. A hundredth percent?
d. A whole percent?

Practice Problem 20

Write $\frac{3}{17}$ as a percent. Round to the nearest hundredth percent.

Answers

17. 50% **18.** $17\frac{1}{2}\%$ **19.** 225% **20.** 17.65%

Concept Check: **a.** 4 **b.** 8 **c.** 5 **d.** 6

F Converting Percents, Decimals, and Fractions

Let's summarize what we have learned so far about percents, decimals, and fractions:

Summary of Converting Percents, Decimals, and Fractions

- *To write a percent as a decimal*, replace the % symbol with its decimal equivalent, 0.01; then multiply.
- *To write a percent as a fraction*, replace the % symbol with its fraction equivalent, $\frac{1}{100}$; then multiply.
- *To write a decimal or fraction as a percent*, multiply by 100%.

Practice Problem 21

A family decides to spend no more than 25% of its monthly income on rent. Write 25% as a decimal.

EXAMPLE 21 17.8% of automobile thefts in the continental United States occur in the Midwest. Write this percent as a decimal. (*Source:* The American Automobile Manufacturers Association)

Solution:

$$17.8\% = 17.8(0.01) = 0.178.$$

Thus, 17.8% written as a decimal is 0.178.

Practice Problem 22

Provincetown's budget for waste disposal increased by $1\frac{1}{4}$ times over the budget from last year. What percent increase is this?

EXAMPLE 22 An advertisement for a stereo system reads "$\frac{1}{4}$ off". What percent off is this?

Solution: Write $\frac{1}{4}$ as a percent.

$$\frac{1}{4} = \frac{1}{4} \cdot 100\% = \frac{1}{4} \cdot \frac{100}{1}\% = \frac{100}{4}\% = 25\%$$

Thus, "$\frac{1}{4}$ off" is the same as "25% off."

It is helpful to know a few basic percent conversions. Appendix A contains a handy reference of percent, decimal, and fraction equivalencies.

STUDY SKILLS REMINDER

Is your notebook still organized?

Is your notebook still organized? If it's not, it's not too late to start organizing it. Start writing your notes and completing your homework assignment in a notebook with pockets (spiral or ring binder). Take class notes in this notebook, and then follow the notes with your completed homework assignment. When you receive graded papers or handouts, place them in the notebook pocket so that you will not lose them.

Remember to mark (possibly with an exclamation point) any note(s) that seem extra important to you. Also remember to mark (possibly with a question mark) any notes or homework that you are having trouble with. Don't forget to see your instructor or a math tutor to help you with the concepts or exercises that you are having trouble understanding.

Also—don't forget to write neatly and keep a positive attitude.

Answers

21. 0.25 **22.** 125%

Name ______________________ Section ____________ Date ____________

Mental Math

Write each fraction as a percent.

1. $\frac{13}{100}$

2. $\frac{92}{100}$

3. $\frac{87}{100}$

4. $\frac{71}{100}$

5. $\frac{1}{100}$

6. $\frac{2}{100}$

EXERCISE SET 6.2

A *Solve. See Examples 1 and 2.*

1. A basketball player made 81 out of 100 attempted free throws. What percent of free throws was made?

2. In a survey of 100 people, 54 preferred chocolate syrup on their ice cream. What percent preferred chocolate syrup?

Adults were asked what type of cookie was their favorite. The circle graph shows the results for every 100 people. Use this graph to answer Exercises 3–6. See Examples 1 and 2.

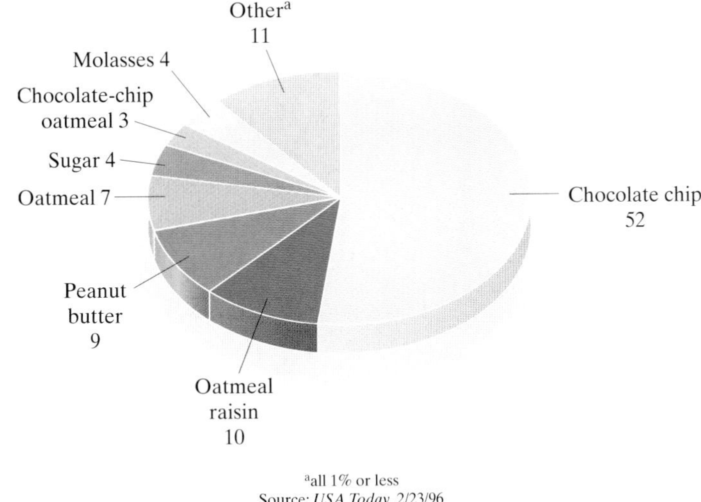

3. What percent preferred peanut butter cookies?

4. What percent preferred oatmeal raisin cookies?

5. What type of cookie was preferred by most adults? What percent preferred this type of cookie?

6. What two types of cookies were preferred by the same number of adults? What percent preferred each type?

7. 12 out of 100 adults have watched an entire infomercial. What percent is this? (*Source:* Aragon Consulting Group)

8. In 2002, 93 out of 100 elementary schools had computers. What percent is this? (*Source:* Quality Education Data, Inc.)

B *Write each percent as a decimal. See Examples 3 through 7.*

9. 48%

10. 64%

11. 6%

12. 9%

13. 100%

14. 136%

15. 61.3%

16. 52.7%

17. 2.8%

18. 1.7%

19. $64\frac{1}{4}\%$ (Hint: First write $\frac{1}{4}$ as a decimal)

20. $18\frac{3}{5}\%$

21. 300%

22. 500%

23. 32.58%

24. 72.18%

Write each percent as a decimal. See Examples 3 through 7.

25. 73.7% of the workforce in the United States works 35 hours or more per week. (*Source:* U.S. Bureau of Labor)

26. In 2001, approximately 18.4% of new luxury cars sold were silver. (*Source:* Du Pont Automotive Products)

27. The United States is the largest consumer of energy in the world, using 25% of all energy produced worldwide each year. (*Source:* Energy Information Administration)

28. Some health insurance companies pay 80% of a person's medical costs.

29. In January, 2003, 11.1% of all Southwest Airlines' flights arrived late. (*Source:* Bureau of Transportation Statistics)

30. In 2002, 88% of all public high schools in the United States owned at least one computer. (*Source:* Quality Education Data, Inc.)

31. 46.2% of registered dental hygienists in the United States have earned an associate's degree or higher. (*Source:* The American Dental Hygienists' Association)

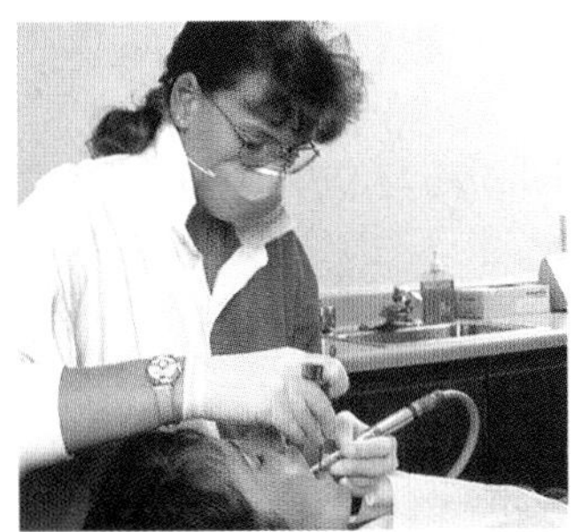

32. Video games made up 21.2% of the total toy market in the United States in 2000. (*Source:* The NPD Group Worldwide)

C *Write each decimal as a percent. See Examples 8 through 11.*

33. 3.1

34. 4.8

35. 29

36. 56

37. 0.003

38. 0.006

39. 0.22

40. 0.45

41. 0.056

42. 0.027

43. 0.3328

44. 0.1115

45. 3.00

46. 5.00

47. 0.7

48. 0.8

Write each decimal as a percent. See Examples 8 through 11.

49. The Munoz family saves 0.10 of their take-home pay.

50. The cost of an item for sale is 0.7 of the sales price.

51. In the state of California, the Hispanic population is 0.324 of the total population. (*Source:* 2003 *World Almanac*)

52. In a recent year, 0.522 of all mail delivered by the United States Postal Service was first-class mail. (*Source:* United States Postal Service)

53. People take aspirin for a variety of reasons. The most common use of aspirin is to prevent heart disease, accounting for 0.38 of all aspirin use. (*Source:* Bayer Market Research)

54. The highest state income tax rate in the state of Iowa is 0.0898 on taxable income over $51,660. (*Source: CCH State Tax Guide*)

D *Write each percent as a fraction or mixed number in simplest form. See Examples 12 through 16.*

55. 4%

56. 2%

57. 4.5%

58. 7.5%

59. 175%

60. 250%

61. 73%

62. 86%

63. 12.5%

64. 62.5%

65. 6.25%

66. 37.5%

67. $10\frac{1}{3}\%$

68. $7\frac{3}{4}\%$

69. $22\frac{3}{8}\%$

70. $15\frac{5}{8}\%$

E *Write each fraction or mixed number as a percent. See Examples 17 through 20.*

71. $\frac{3}{4}$

72. $\frac{1}{2}$

73. $\frac{7}{10}$

74. $\frac{3}{10}$

75. $\frac{2}{5}$

76. $\frac{4}{5}$

77. $\frac{59}{100}$

78. $\frac{73}{100}$

79. $\frac{17}{50}$

80. $\frac{47}{50}$

81. $\frac{3}{8}$

82. $\frac{5}{8}$

83. $\frac{5}{16}$ **84.** $\frac{7}{16}$ **85.** $\frac{2}{3}$ **86.** $\frac{1}{3}$

87. $2\frac{1}{2}$ **88.** $2\frac{1}{5}$ **89.** $1\frac{9}{10}$ **90.** $2\frac{7}{10}$

Write each fraction as a percent. Round to the nearest hundredth percent. See Example 20.

91. $\frac{7}{11}$ **92.** $\frac{5}{12}$  **93.** $\frac{4}{15}$ **94.** $\frac{10}{11}$

95. $\frac{1}{7}$ **96.** $\frac{1}{9}$ **97.** $\frac{11}{12}$ **98.** $\frac{5}{6}$

 *Complete each table. See Examples 21 and 22.*

99.

Percent	Decimal	Fraction
35%		
		$\frac{1}{5}$
	0.5	
70%		
		$\frac{3}{8}$

100.

Percent	Decimal	Fraction
	0.525	
		$\frac{3}{4}$
$66\frac{2}{3}\%$		
		$\frac{5}{6}$
100%		

101.

Percent	Decimal	Fraction
40%		
	0.235	
		$\frac{4}{5}$
$33\frac{1}{3}\%$		
		$\frac{7}{8}$
7.5%		

102.

Percent	Decimal	Fraction
50%		
		$\frac{2}{5}$
	0.25	
12.5%		
		$\frac{5}{8}$
		$\frac{7}{50}$

Solve. See Examples 21 and 22.

103. Approximately 24.9% of new full-size cars are silver, making silver the most popular new vehicle color for that class. Write this percent as a fraction. (*Source:* American Automobile Manufacturers Association)

104. In 2001, 10.6% of all new cars sold in the United States were imported from Japan. Write this percent as a fraction. (*Source:* American Automobile Manufacturers Association)

105. In 1998, $\frac{17}{200}$ of all new cars sold in the United States were imported from Japan. Write this fraction as a percent. (*Source:* Ward's Communications)

106. In 1980, $\frac{53}{250}$ of all new cars sold in the United States were imported from Japan. Write this fraction as a percent. (*Source:* American Automobile Manufacturers Association)

107. In 2001, New Mexico had the highest percent of residents not covered by any type of health insurance at 20.7%. Write this percent as a fraction. (*Source:* U.S. Bureau of the Census)

108. In 2001, 70.2% of all households with televisions subscribed to a cable television service. Write this percent as a decimal. (*Source:* Nielsen Media Research)

109. For the 2001–2002 television season, the top-rated prime-time television program was *Friends*, which had an average audience share of approximately $\frac{1}{4}$ of all those watching television during that time slot. Write this fraction as a percent. (*Source:* Nielsen Media Research)

110. Approximately $\frac{37}{50}$ of all structure fires take place in personal residences. Write this fraction as a percent. (*Source:* National Fire Protection Association)

Write each percent in this circle graph as a fraction.

World Population by Continent

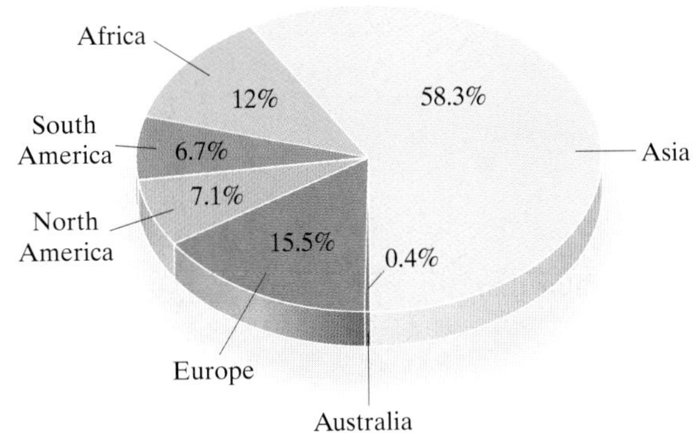

111. 0.4%

112. 58.3%

113. 12%

114. 6.7%

115. 7.1%

116. 15.5%

Review and Preview

Find the value of n. See Section 3.3.

117. $3n = 45$

118. $7n = 48$

119. $-8n = 80$

120. $-2n = 16$

121. $-6n = -72$

122. $5n = -35$

Combining Concepts

What percent of the figure is shaded?

123.

124.

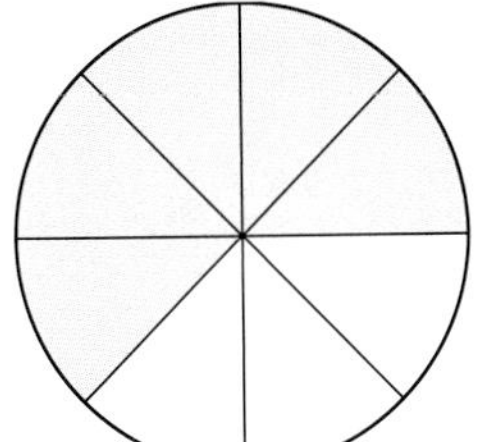

125.

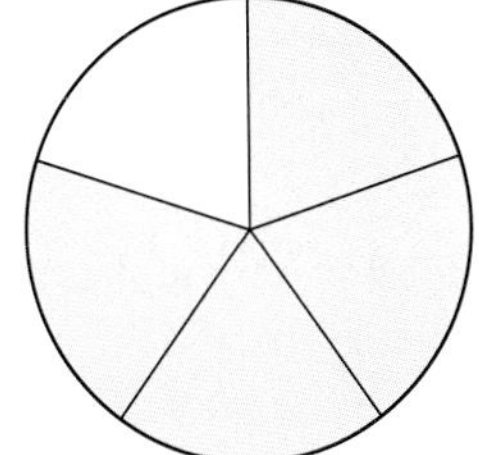

126.

Write each fraction as a decimal and then write each decimal as a percent. Round the decimal to three decimal places and the percent to the nearest tenth of a percent.

127. $\frac{850}{736}$

128. $\frac{506}{248}$

Fill in the blanks.

129. A fraction written as a percent is greater than 100% when the numerator is __________ than the denominator. (greater/less)

130. A decimal written as a percent is less than 100% when the decimal is __________ than 1. (greater/less)

131. In your own words, explain how to write a percent as a fraction.

132. In your own words, explain how to write a fraction as a decimal.

The bar graph shows the predicted fastest-growing occupations. Use this graph to answer Exercises 133–136.

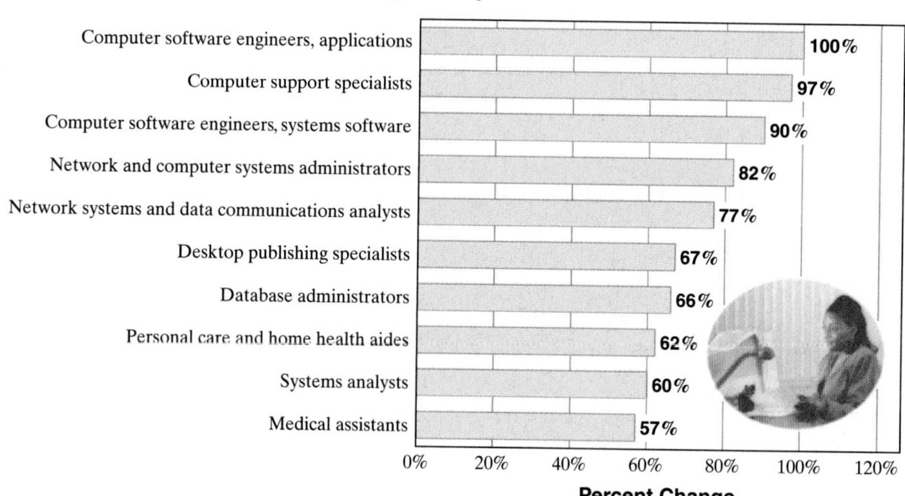

133. What occupation is predicted to be the fastest growing?

134. What occupation is predicted to be the second fastest growing?

135. Write the percent change for personal care and home health aides as a decimal.

136. Write the percent change for systems analysts as a decimal.

6.3 Solving Percent Problems with Equations

OBJECTIVES

A Write percent problems as equations.

B Solve percent problems.

Sections 6.3 and 6.4 introduce two methods for solving percent problems. It is not necessary that you study both sections. You may want to check with your instructor for further advice.

To solve percent problems in this section, we will translate the problems into mathematical statements, or equations.

A Writing Percent Problems as Equations

Recognizing key words in a percent problem is helpful in writing the problem as an equation. Three key words in the statement of a percent problem and their meanings are as follows:

of means **multiplication** (·)

is means **equals** (=)

what (or some equivalent) means **the unknown number**

In our examples, we will let the letter x stand for the unknown number.

EXAMPLE 1 Translate to an equation:

5 is what percent of 20?

Solution: 5 is what percent of 20?

$$5 = x \cdot 20$$

Practice Problem 1

Translate: 6 is what percent of 24?

Helpful Hint

Remember that an equation is simply a mathematical statement that contains an equal sign (=).

$$5 = 20x$$

↑ equal sign

EXAMPLE 2 Translate to an equation:

1.2 is 30% of what number?

Solution: 1.2 is 30% of what number?

$$1.2 = 30\% \cdot x$$

Practice Problem 2

Translate: 1.8 is 20% of what number?

EXAMPLE 3 Translate to an equation:

What number is 25% of 0.008?

Solution: What number is 25% of 0.008?

$$x = 25\% \cdot 0.008$$

Practice Problem 3

Translate: What number is 40% of 3.6?

Answers

1. $6 = x \cdot 24$ **2.** $1.8 = 20\% \cdot x$ **3.** $x = 40\% \cdot 3.6$

Practice Problems 4–6

Translate each question to an equation.

4. 42% of 50 is what number?
5. 15% of what number is 9?
6. What percent of 150 is 90?

EXAMPLES Translate each question to an equation.

4. 38% of 200 is what number?

$$38\% \cdot 200 = x$$

5. 40% of what number is 80?

$$40\% \cdot x = 80$$

6. What percent of 85 is 34?

$$x \cdot 85 = 34$$

Concept Check

In the equation $2x = 10$, what step is taken to solve the equation?

Try the Concept Check in the margin.

B Solving Percent Problems

You may have noticed by now that each percent problem has contained three numbers—in our examples, two are known and one is unknown. Each of these numbers is given a special name.

15% of 60 is 9

15% percent	·	60 base	=	9 amount

We call this equation the **percent equation.**

Percent Equation

$$\text{percent} \cdot \text{base} = \text{amount}$$

Once a percent problem has been written as a percent equation, we can use the equation to find the unknown number.

Solving Percent Equations for the Amount

Practice Problem 7

What number is 20% of 85?

EXAMPLE 7 What number is 35% of 60?

Solution:

$x = 35\% \cdot 60$ Translate to an equation.

$x = 0.35 \cdot 60$ Write 35% as 0.35.

$x = 21$ Multiply:

$$\begin{array}{r} 60 \\ \underline{0.35} \\ 300 \\ \underline{1800} \\ 21.00 \end{array}$$

Then 21 is 35% of 40. Is this reasonable? To see, round 35% to 40%. Then 40% or 0.40(60) is 24. Our result is reasonable since 21 is close to 24.

Answers

4. $42\% \cdot 50 = x$ **5.** $15\% \cdot x = 9$ **6.** $x \cdot 150 = 90$ **7.** 17

Concept Check: Divide both sides of the equation by 2.

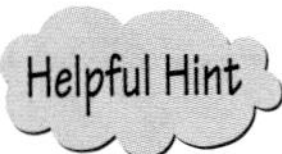

When solving a percent equation, write the percent as a decimal or fraction.

EXAMPLE 8 85% of 300 is what number?

Solution:

$$85\% \cdot 300 = x \quad \text{Translate to an equation.}$$
$$0.85 \cdot 300 = x \quad \text{Write 85\% as 0.85.}$$
$$255 = x \quad \text{Multiply: } 0.85 \cdot 300 = 255.$$

Then 85% of 300 is 255. Is this result reasonable? To see, round 85% to 90%. Then 90% of 300 or $0.90(300) = 270$, which is close to 255.

Practice Problem 8

90% of 150 is what number?

Solving Percent Equations for the Base

EXAMPLE 9 12% of what number is 0.6?

Solution:

$$12\% \cdot x = 0.6 \quad \text{Translate to an equation.}$$
$$0.12 \cdot x = 0.6 \quad \text{Write 12\% as 0.12.}$$
$$\frac{0.12 \cdot x}{0.12} = \frac{0.6}{0.12} \quad \text{Divide both sides by 0.12.}$$
$$x = 5$$

$$0.12\overline{)0.60} = 5.;\quad 60,\ 0$$

Then 12% of 5 is 0.6. Is this reasonable? To see, round 12% to 10%. Then 10% of 5 or 0.10(5) is 0.5, which is close to 0.6.

Practice Problem 9

15% of what number is 1.2?

EXAMPLE 10 13 is $6\frac{1}{2}\%$ of what number?

Solution:

$$13 = 6\frac{1}{2}\% \cdot x \quad \text{Translate to an equation.}$$
$$13 = 0.065 \cdot x \quad 6\frac{1}{2}\% = 6.5\% = 0.065.$$
$$\frac{13}{0.065} = \frac{0.065 \cdot x}{0.065} \quad \text{Divide both sides by 0.065.}$$
$$200 = x$$

$$0.065\overline{)13.000} = 200.;\quad 130,\ 0$$

Then 13 is $6\frac{1}{2}\%$ of 200. Check to see if this result is reasonable.

Practice Problem 10

27 is $4\frac{1}{2}\%$ of what number?

Answers

8. 135 **9.** 8 **10.** 600

Solving Percent Equations for the Percent

Practice Problem 11

What percent of 80 is 8?

EXAMPLE 11 What percent of 12 is 9?

Solution:

$$x \cdot 12 = 9 \quad \text{Translate to an equation.}$$

$$\frac{x \cdot 12}{12} = \frac{9}{12} \quad \text{Divide both sides by 12.}$$

$$x = 0.75$$

Next, since we are looking for percent, we write 0.75 as a percent.

$x = 75\%$

Then 75% of 12 is 9. To check, see that $75\% \cdot 12 = 9$.

Helpful Hint

If your unknown in the percent equation is percent, don't forget to convert your answer to a percent.

Practice Problem 12

35 is what percent of 25?

EXAMPLE 12 78 is what percent of 65?

Solution:

$$78 = x \cdot 65 \quad \text{Translate to an equation.}$$

$$\frac{78}{65} = \frac{x \cdot 65}{65} \quad \text{Divide both sides by 65.}$$

$$1.2 = x$$

$$120\% = x \quad \text{Write 1.2 as a percent.}$$

Then 78 is 120% of 65. Check this result.

Concept Check

Consider the problem

10 is 40% of what number?

Will the solution be greater than or less than 10? How do you know without solving the problem?

Try the Concept Check in the margin.

Helpful Hint

Use the following to see if your answers are reasonable.

100% of a number = the number

(a percent greater than 100%) of a number = a number larger than the original number

(a percent less than 100%) of a number = a number less than the original number

Answers

11. 10% **12.** 140%

Concept Check: Greater. Answers may vary.

Name ______________________ Section __________ Date __________

Mental Math

Identify the percent, the base, and the amount in each equation. Recall that percent · base = amount.

1. $42\% \cdot 50 = 21$

2. $30\% \cdot 65 = 19.5$

3. $107.5 = 125\% \cdot 86$

4. $99 = 110\% \cdot 90$

EXERCISE SET 6.3

A *Translate each question to an equation. Do not solve. See Examples 1 through 6.*

1. 15% of 72 is what number?

2. What number is 25% of 55?

3. 30% of what number is 80?

4. 0.5 is 20% of what number?

5. What percent of 90 is 20?

6. 8 is 50% of what number?

7. 1.9 is 40% of what number?

8. 72% of 63 is what number?

9. What number is 9% of 43?

10. 4.5 is what percent of 45?

B *Translate to an equation and solve. See Examples 7 and 8.*

11. 10% of 35 is what number?

12. 25% of 60 is what number?

13. What number is 14% of 52?

14. What number is 30% of 17?

Solve. See Examples 9 and 10.

15. 30 is 5% of what number?

16. 25 is 25% of what number?

17. 1.2 is 12% of what number?

18. 0.22 is 44% of what number?

Solve. See Examples 11 and 12.

19. 66 is what percent of 60?

20. 30 is what percent of 20?

21. 16 is what percent of 50?

22. 27 is what percent of 50?

Solve. See Examples 7 through 12.

23. 0.1 is 10% of what number?

24. 0.5 is 5% of what number?

25. 125% of 36 is what number?

26. 200% of 13.5 is what number?

27. 82.5 is $16\frac{1}{2}\%$ of what number?

28. 7.2 is $6\frac{1}{4}\%$ of what number?

29. 2.58 is what percent of 50?

30. 264 is what percent of 33?

31. What number is 42% of 60?

32. What number is 36% of 80?

33. What percent of 150 is 67.5?

34. What percent of 105 is 88.2?

35. 120% of what number is 42?

36. 160% of what number is 40?

Review and Preview

Find the value of n in each proportion. See Section 6.1.

37. $\frac{27}{n} = \frac{9}{10}$

38. $\frac{35}{n} = \frac{7}{5}$

39. $\frac{n}{5} = \frac{8}{11}$

40. $\frac{n}{3} = \frac{6}{13}$

Write each sentence as a proportion.

41. 17 is to 12 as n is to 20.

42. 20 is to 25 as n is to 10.

43. 8 is to 9 as 14 is to n.

44. 5 is to 6 as 15 is to n.

Combining Concepts

Solve.

45. 1.5% of 45,775 is what number?

46. What percent of 75,528 is 27,945.36?

47. 22,113 is 180% of what number?

48. In your own words, explain how to solve a percent equation.

FOCUS ON The Real World

M&M's®

In the 1930s, many American stores did not stock chocolates during the summer because, without widespread air conditioning, the chocolate tended to melt and sales declined. In 1940, Forrest E. Mars, Sr., set out to develop a chocolate candy that could be sold year-round without the melting problem. So he formed a company in Newark, New Jersey, to make bite-sized melt-proof chocolate candies encased in a thin sugar shell. Thus M&M's® Plain Chocolate Candies were born!

The first Plain M&M's colors were brown, green, orange, red, violet, and yellow. Violet was replaced by tan in 1950. When the safety of a particular type of red food coloring was publicly questioned in 1976, red was completely eliminated from the M&M's color mix to avoid alarming consumers, even though the coloring in question was never used in M&Ms. After an 11-year hiatus, red returned to the color mix in 1987. In 1995, over 10 million Americans responded to a marketing campaign asking for help in choosing a new M&M's color. Given the choices of blue, pink, purple, or no change, 54% of the respondents chose blue, and it replaced tan. The new color mix of Plain M&M's became blue, brown, green, orange, red, and yellow in the percentages shown in the circle graph. (Note: M&M continues to introduce different colors for a limited time. This may affect the outcome of the exercises below.)

Color Mix for M&M's® Plain Chocolate Candies

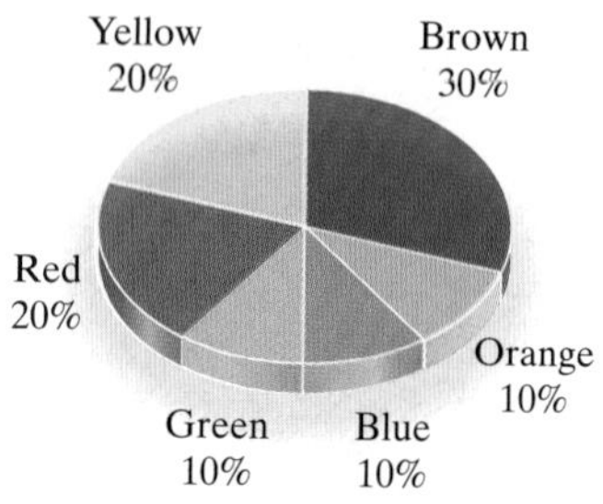

Source: Mars, Incorporated

M&M's Peanut Chocolate Candies debuted in 1954. At first, Peanut M&M's were all brown. However, in 1960, red, green, and yellow were added. Orange joined the mix in 1976, and, finally, blue was introduced in 1995. The current color mix for Peanut M&M's is shown in the circle graph.

Color Mix for M&M's® Peanut Chocolate Candies

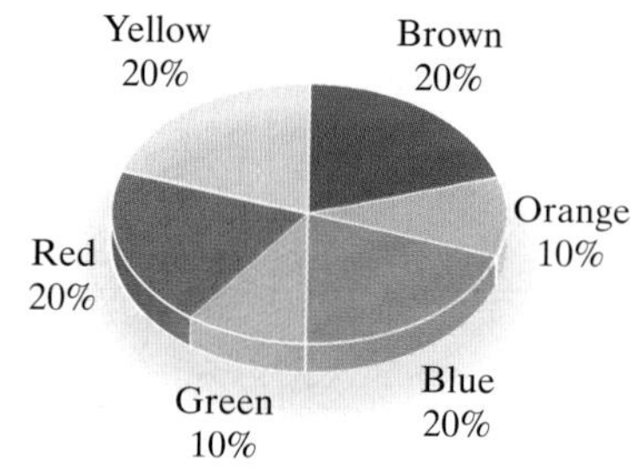

Source: Mars, Incorporated

GROUP ACTIVITY

1. Use a single-serving pouch of M&M's Plain Chocolate Candies and find the percentage of each color in the bag. How does it compare to the overall percentages officially reported?
2. Repeat Question 2 with a single-serving pouch of M&M's Peanut Chocolate Candies.
3. For each type of M&M's Candies, combine your group's color counts with those of the other groups in your class. Find the percentage of each color for the combined figures. How do these percentages compare to the official color mix?

6.4 Solving Percent Problems with Proportions

OBJECTIVES

A. Write percent problems as proportions.

B. Solve percent problems.

There is more than one method that can be used to solve percent problems. (See the note at the beginning of Section 6.3.) In the last section, we used the percent equation. In this section, we will use proportions.

A Writing Percent Problems as Proportions

To understand the proportion method, recall that 70% means the ratio of 70 to 100, or $\frac{70}{100}$.

$\frac{7}{10}$ shaded

$$70\% = \frac{70}{100} = \frac{7}{10}$$

70% or $\frac{70}{100}$ shaded

Since the ratio $\frac{70}{100}$ is equal to the ratio $\frac{7}{10}$, we have the proportion

$$\frac{7}{10} = \frac{70}{100}$$

We call this proportion the **percent proportion.** In general, we can name the parts of this proportion as follows.

Percent Proportion

$$\frac{\text{amount}}{\text{base}} = \frac{\text{percent}}{100} \quad \leftarrow \text{always } 100$$

or

$$\text{amount} \rightarrow \frac{a}{b} = \frac{p}{100} \leftarrow \text{percent}$$
$$\text{base} \rightarrow$$

When we translate percent problems to proportions, the **percent** can be identified by looking for the symbol % or the word *percent.* The **base** usually follows the word *of.* The **amount** is the part compared to the whole.

Helpful Hint

This table may be useful when identifying the parts of a proportion.

Part of Proportion	How It's Identified
Percent	% or percent
Base	Appears after *of*
Amount	Part compared to whole

Practice Problem 1

Translate to a proportion: 15% of what number is 55?

EXAMPLE 1 Translate to a proportion:

12% of what number is 47?

Solution:

percent	base	amount
12%	what number — It appears after the word *of*.	47 — It is the part compared to the whole.

$$\frac{47}{b} = \frac{12}{100}$$

(amount → 47, base → b, percent → 12)

Practice Problem 2

Translate to a proportion: 35 is what percent of 70?

EXAMPLE 2 Translate to a proportion:

101 is what percent of 200?

Solution:

amount	percent	base
101 — It is the part compared to the whole.	what percent	200 — It appears after the word *of*.

$$\frac{101}{200} = \frac{p}{100}$$

(amount → 101, base → 200, percent → p)

Practice Problem 3

Translate to a proportion: What number is 25% of 68?

EXAMPLE 3 Translate to a proportion:

What number is 90% of 45?

Solution:

amount	percent	base
What number — It is the part compared to the whole.	90%	45 — It appears after the word *of*.

$$\frac{a}{45} = \frac{90}{100}$$

(amount → a, base → 45, percent → 90)

Practice Problem 4

Translate to a proportion: 520 is 65% of what number?

EXAMPLE 4 Translate to a proportion:

238 is 40% of what number?

Solution:

amount	percent	base
238	40%	what number

$$\frac{238}{b} = \frac{40}{100}$$

Answers

1. $\frac{55}{b} = \frac{15}{100}$ 2. $\frac{35}{70} = \frac{p}{100}$ 3. $\frac{a}{68} = \frac{25}{100}$
4. $\frac{520}{b} = \frac{65}{100}$

EXAMPLE 5 Translate to a proportion:

75 is what percent of 30?

Solution: amount — percent — base

$$\frac{75}{30} = \frac{p}{100}$$

EXAMPLE 6 Translate to a proportion:

45% of 105 is what number?

Solution: percent — base — amount

$$\frac{a}{105} = \frac{45}{100}$$

Try the Concept Check in the margin.

B Solving Percent Problems

The proportions that we have written in this section contain three values that can change: the percent, the base, and the amount. If any two of these values are known, we can find the third (unknown value). To do this, we write a percent proportion and find the unknown value as we did in Section 6.3.

Try the Concept Check in the margin.

Solving Percent Proportions for the Amount

EXAMPLE 7 What number is 30% of 9?

Solution: amount — percent — base

$$\frac{a}{9} = \frac{30}{100}$$

To solve, we set cross products equal to each other.

$$\frac{a}{9} = \frac{30}{100}$$

$a \cdot 100 = 9 \cdot 30$ Set cross products equal.

$a \cdot 100 = 270$ Multiply.

$\frac{a \cdot 100}{100} = \frac{270}{100}$ Divide both sides by 100.

$a = 2.7$ Simplify.

Then 2.7 is 30% of 9.

Practice Problem 5

Translate to a proportion: 65 is what percent of 50?

Practice Problem 6

Translate to a proportion: 36% of 80 is what number?

Concept Check

When solving a percent problem with a proportion, describe how you can check the result.

Practice Problem 7

What number is 8% of 120?

Concept Check

Consider the problem

78 is what percent of 350?

Which part of the percent proportion is unknown?

a. the amount
b. the base
c. the percent

Answers

5. $\frac{65}{50} = \frac{p}{100}$ **6.** $\frac{a}{80} = \frac{36}{100}$ **7.** 9.6

Concept Check: Put the result into the proportion and check that the proportion is true.

Concept Check: c

Helpful Hint

The proportion in Example 7 contained the ratio $\frac{30}{100}$. A ratio in a proportion may be simplified before solving the proportion. The unknown number in both

$$\frac{a}{9} = \frac{30}{100} \quad \text{and} \quad \frac{a}{9} = \frac{3}{10}$$

is 2.7

Solving Percent Proportions for the Base

Practice Problem 8

75% of what number is 60?

EXAMPLE 8 150% of what number is 30?

150% → percent; what number is → base; 30 → amount

Solution:

$\frac{30}{b} = \frac{150}{100}$ Write the proportion.

$\frac{30}{b} = \frac{3}{2}$ Simplify $\frac{150}{100}$.

$30 \cdot 2 = b \cdot 3$ Set cross products equal.

$60 = b \cdot 3$ Multiply.

$\frac{60}{3} = \frac{b \cdot 3}{3}$ Divide both sides by 3.

$20 = b$ Simplify.

Then 150% of 20 is 30.

Practice Problem 9

15 is 5% of what number?

EXAMPLE 9 20.8 is 40% of what number?

20.8 → amount; 40% → percent; what number → base

Solution:

$\frac{20.8}{b} = \frac{40}{100}$ or $\frac{20.8}{b} = \frac{2}{5}$ Write the proportion and simplify $\frac{40}{100}$.

$20.8 \cdot 5 = b \cdot 2$ Set cross products equal.

$104 = b \cdot 2$ Multiply.

$\frac{104}{2} = \frac{b \cdot 2}{2}$ Divide both sides by 2.

$52 = b$ Simplify.

Then 20.8 is 40% of 52.

Answers

8. 80 **9.** 300

Solving Percent Proportions for the Percent

EXAMPLE 10 What percent of 50 is 8?

What percent → percent; 50 → base; 8 → amount

Solution:

$$\frac{8}{50} = \frac{p}{100} \text{ or } \frac{4}{25} = \frac{p}{100}$$ Write the proportion and simplify $\frac{8}{50}$.

$4 \cdot 100 = 25 \cdot p$ Set cross products equal.

$400 = 25 \cdot p$ Multiply.

$\frac{400}{25} = \frac{25 \cdot p}{25}$ Divide both sides by 25.

$16 = p$ Simplify.

Then 16% of 50 is 8.

Practice Problem 10

What percent of 40 is 5?

Helpful Hint

Recall from our percent proportion that this number already is a percent. Just keep the number the same and attach a % symbol.

EXAMPLE 11 504 is what percent of 360?

504 → amount; what percent → percent; 360 → base

Solution:

$$\frac{504}{360} = \frac{p}{100}$$

Let's choose not to simplify the ratio $\frac{504}{360}$.

$504 \cdot 100 = 360 \cdot p$ Set cross products equal.

$50{,}400 = 360 \cdot p$ Multiply.

$\frac{50{,}400}{360} = \frac{360 \cdot p}{360}$ Divide both sides by 360.

$140 = p$ Simplify.

Notice by choosing not to simplify $\frac{504}{360}$, we had larger numbers in our equation. Either way, we find that 504 is 140% of 360.

Practice Problem 11

What percent of 160 is 336?

You may have noticed the following while working examples.

Helpful Hint

Use the following to see whether your answers are reasonable.

100% of a number = the number

(a percent greater than 100%) of a number = a number larger than the original number

(a percent less than 100%) of a number = a number less than the original number

Answers

10. 12.5% **11.** 210%

FOCUS ON The Real World

HOW MUCH CAN YOU AFFORD FOR A HOUSE?

When a home buyer takes out a mortgage to buy a house, the loan is generally repaid on a monthly basis with a monthly mortgage payment. (Some banks also offer biweekly payment programs.) An important consideration in choosing a house is the amount of the monthly payment. Usually, the amount that a home buyer can afford to make as a monthly payment will dictate the house purchase price that can be afforded.

The first step in deciding how much can be afforded for a house is finding out how much income the household has each month before taxes. The Mortgage Bankers Association of American (MBAA) suggests that the monthly mortgage payment be between 25% and 28% of the total monthly income. If other long-term debts exist (such as car or education loans and long-term credit card debt repayment), the MBAA further recommends that the total of housing costs and other monthly debt payments not exceed 36% of the total monthly income.

Once the size of the monthly payment that can be afforded has been found, a mortgage payment calculator can be used to work backward to estimate the mortgage amount that will give that desired monthly payment. For example, the Interest.com Web site includes a mortgage payment calculator at http://www.interest.com/calculators/monthly-payment.shtml. (Alternatively, visit www.interest.com and navigate to "Use our mortgage calculators." Look for the calculator to calculate the monthly payment for a particular mortgage loan.) With this mortgage payment calculator, the user can input the interest rate (as a percent), the term of the loan (in years), and total home loan amount (in dollars). This information is then used to calculate the associated monthly payment. To work backward with this mortgage payment calculator to find the total loan amount that can be afforded:

- Enter the interest rate that is likely for your loan and the term of the loan in which you are interested.
- Then make a guess (perhaps $100,000?) for the total home loan amount that can be afforded.
- Have the mortgage calculator calculate the monthly payment.
- If the monthly payment that is calculated is higher than the range that can be afforded, repeat the calculation using the same interest rate and loan term but a lower value for the total home loan amount.
- If the monthly payment that is calculated is lower than the range that can be afforded, repeat the calculation using the same interest rate and loan term but a higher value for the total home loan amount.
- Repeat these calculations methodically until a monthly payment is obtained that is in the range that can be afforded. The initial principal value that gave this monthly payment amount is an estimate of the mortgage amount that can be afforded to buy a home.

GROUP ACTIVITY

1. Research current interest rates on 30-year mortgages.
2. Use the method described above to find the size of mortgages that can be afforded by households with the following total monthly incomes before taxes. (Assume in each case that the household has no other debts.) Use a loan term of 30 years and a current interest rate on a 30-year mortgage.
   a. $1500 **b.** $2000 **c.** $2500
   d. $3000 **e.** $3500 **f.** $4000
3. Create a table of your results.

Name ______________________ Section ____________ Date ____________

Mental Math

Identify the amount, the base, and the percent in each equation. Recall that $\frac{\text{amount}}{\text{base}} = \frac{\text{percent}}{100}$.

1. $\frac{12.6}{42} = \frac{30}{100}$

2. $\frac{201}{300} = \frac{67}{100}$

3. $\frac{20}{100} = \frac{102}{510}$

4. $\frac{40}{100} = \frac{248}{620}$

EXERCISE SET 6.4

A *Translate each question to a proportion. Do not solve. See Examples 1 through 6.*

1. 32% of 65 is what number?

2. What number is 5% of 125?

3. 40% of what number is 75?

4. 1.2 is 47% of what number?

5. What percent of 200 is 70?

6. 520 is 85% of what number?

7. 2.3 is 58% of what number?

8. 92% of 30 is what number?

9. What number is 19% of 130?

10. 8.2 is what percent of 82?

B *Solve. See Example 7.*

11. 10% of 55 is what number?

12. 25% of 84 is what number?

13. What number is 18% of 105?

14. What number is 40% of 29?

Solve. See Examples 8 and 9.

15. 60 is 15% of what number?

16. 75 is 75% of what number?

17. 7.8 is 78% of what number?

18. 1.1 is 44% of what number?

Solve. See Examples 10 and 11.

19. 105 is what percent of 84?

20. 77 is what percent of 44?

21. 14 is what percent of 50?

22. 37 is what percent of 50?

Solve. See Examples 7 through 11.

23. 2.9 is 10% of what number?

24. 6.2 is 5% of what number?

25. 2.4% of 80 is what number?

26. 6.5% of 120 is what number?

27. 160 is 16% of what number?

28. 30 is 6% of what number?

29. 348.6 is what percent of 166?

30. 262.4 is what percent of 82?

31. What number is 89% of 62?

32. What number is 53% of 130?

33. What percent of 8 is 3.6?

34. What percent of 5 is 1.6?

35. 140% of what number is 119?

36. 170% of what number is 221?

Review and Preview

Add or subtract as indicated. See Sections 4.4, 4.5, and 4.8.

37. $\frac{11}{16} + \frac{3}{16}$

38. $\frac{5}{8} - \frac{7}{12}$

39. $3\frac{1}{2} - \frac{11}{30}$

40. $2\frac{2}{3} + 4\frac{1}{2}$

Add or subtract as indicated. See Section 5.2.

41. $\begin{array}{r} 0.41 \\ +0.29 \\ \hline \end{array}$

42. $\begin{array}{r} 10.78 \\ 4.3 \\ +0.21 \\ \hline \end{array}$

43. $\begin{array}{r} 2.38 \\ -0.19 \\ \hline \end{array}$

44. $\begin{array}{r} 16.37 \\ -\ 2.61 \\ \hline \end{array}$

Combining Concepts

Solve. Round to the nearest tenth, if necessary.

45. What number is 22.3% of 53,862?

46. What percent of 110,736 is 88,542?

47. 8652 is 119% of what number?

48. In your own words, describe how to identify the percent, the base, and the amount in a percent problem.

Name ______________________ Section __________ Date __________

Integrated Review—Ratio, Proportion, and Percent

Write each ratio as a ratio of whole numbers using fractional notation. Write the fraction in simplest form.

1. 18 to 20 **2.** 36 to 100 **3.** 8.6 to 10 **4.** 1.6 to 4.6

Find the ratio described in each problem.

5. Find the ratio of the width to the length of the sign below.

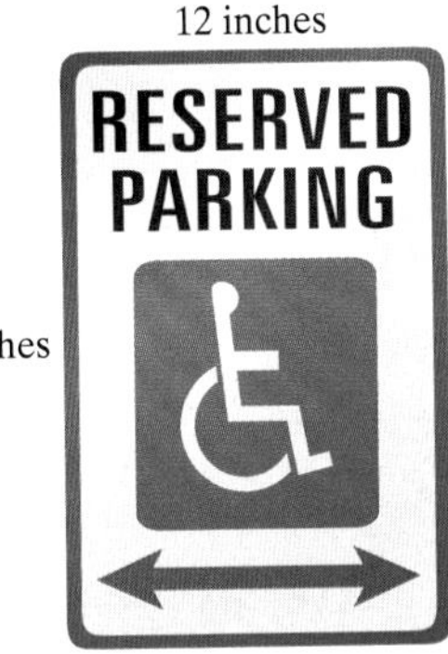

6. At the end of 2002 Lockheed Martin Corporation had \$26 hundred million in assets and \$8 hundred million in debts. Find the ratio of assets to debt. (*Source:* Lockheed Martin Corporation)

Solve.

7. $\frac{3.5}{12.5} = \frac{7}{z}$

8. $\frac{x + 7}{3} = \frac{2x}{5}$

An office uses 5 boxes of envelopes every 3 weeks.

9. Find how long a gross of envelope boxes is likely to last. (A gross of boxes is 144 boxes.) Round to the nearest week.

10. Find how many boxes should be purchased to last a month. Round to the nearest box.

Write each number as a percent.

11. 0.12 **12.** 0.68 **13.** $\frac{1}{4}$ **14.** $\frac{1}{2}$

15. 5.2 **16.** 7.8 **17.** $\frac{3}{50}$ **18.** $\frac{11}{25}$

19. $2\frac{1}{2}$ **20.** $3\frac{1}{4}$ **21.** 0.03 **22.** 0.05

Write each percent as a decimal.

23. 65% **24.** 31% **25.** 8% **26.** 7%

Answers

1. ______
2. ______
3. ______
4. ______
5. ______
6. ______
7. ______
8. ______
9. ______
10. ______
11. ______
12. ______
13. ______
14. ______
15. ______
16. ______
17. ______
18. ______
19. ______
20. ______
21. ______
22. ______
23. ______
24. ______
25. ______
26. ______

27. ______
28. ______
29. ______
30. ______
31. ______
32. ______
33. ______
34. ______
35. ______
36. ______
37. ______
38. ______
39. ______
40. ______
41. ______
42. ______
43. ______
44. ______
45. ______
46. ______
47. ______
48. ______
49. ______
50. ______

27. 142% **28.** 538% **29.** 2.9% **30.** 6.6%

Write each percent as a fraction or mixed number in simplest form.

31. 3% **32.** 8% **33.** 5.25% **34.** 12.75%

35. 38% **36.** 45% **37.** $12\frac{1}{3}\%$ **38.** $16\frac{2}{3}\%$

Solve each percent problem.

39. 12% of 70 is what number?

40. 36 is 36% of what number?

41. 212.5 is 85% of what number?

42. 66 is what percent of 55?

43. 23.8 is what percent of 85?

44. 38% of 200 is what number?

45. What number is 25% of 44?

46. What percent of 99 is 128.7?

47. What percent of 250 is 215?

48. What number is 45% of 84?

49. 63 is 42% of what number?

50. 58.9 is 95% of what number?

6.5 Applications of Percent

Ⓐ Solving Applications Involving Percent

The next few examples show just a few ways that percent occurs in real-life settings. (Each of these examples shows two ways of solving these problems. If you studied Section 6.3 only, see *Method 1.* If you studied Section 6.4 only, see *Method 2.*)

OBJECTIVES

Ⓐ Solve applications involving percent.

Ⓑ Find percent increase and percent decrease.

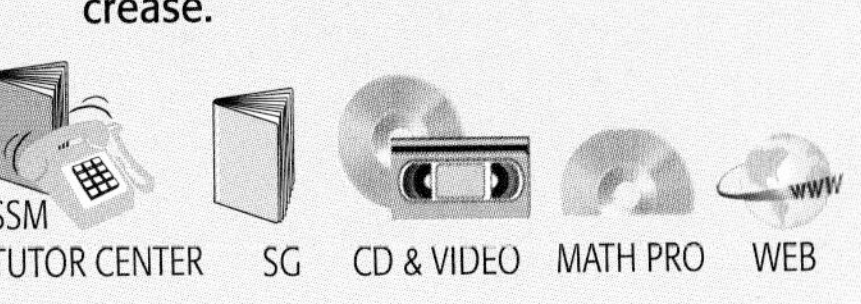

EXAMPLE 1 Finding Increases in Nursing Schools

There is a world-wide shortage of nurses that is projected to be 20% below requirements by 2020. Until 2003, there has also been a continual decline in enrollment in nursing schools.

In 2003, 2178 of the total 2593 nursing schools in the U.S. had an increase in applications or enrollment. What percent of nursing schools had an increase? Round to the nearest whole percent. (*Source*: CNN and *Nurse Week*)

Solution: *Method 1.* First, we state the problem in words.

In words: 2178 is what percent of 2593?

Translate: $2178 = x \cdot 2593$

Next, solve for x.

$$\frac{2178}{2593} = \frac{x \cdot 2593}{2593} \quad \text{Divide both sides by 2593.}$$

$$0.84 \approx x \quad \text{Round to the nearest hundredth.}$$

$$84\% \approx x \quad \text{Write as a percent.}$$

In 2003, about 84% of nursing schools had an increase in applications or enrollment.

Method 2.

In words: 2178 is what percent of 2593?

2178 → amount; what percent → percent; 2593 → base

Translate: $\frac{\text{amount} \rightarrow 2178}{\text{base} \rightarrow 2593} = \frac{p}{100} \leftarrow \text{percent}$

Next, solve for p.

$$2178 \cdot 100 = 2593 \cdot p \quad \text{Set cross products equal.}$$

$$217{,}800 = 2593 \cdot p \quad \text{Multiply.}$$

$$\frac{217{,}800}{2593} = \frac{2593 \cdot p}{2593} \quad \text{Divide both sides by 2593.}$$

$$84 \approx p$$

In 2003, about 84% of nursing schools had an increase in applications or enrollment.

Practice Problem 1

There are 106 nursing schools in Ohio. Of these schools, 61 offer RN (registered nurse) degrees. What percent of nursing schools in Ohio offer RN degrees? Round to the nearest whole percent.

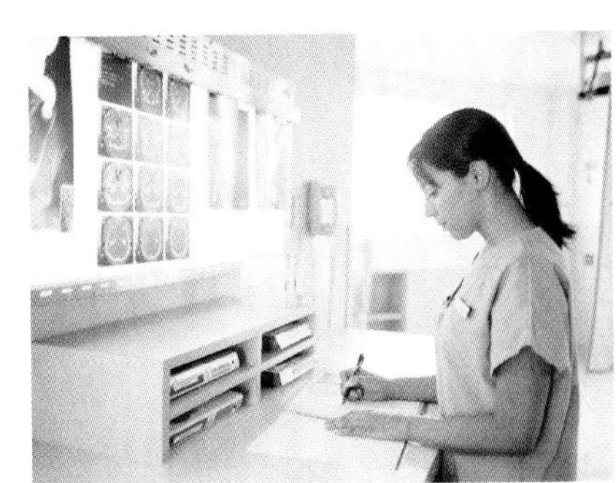

Answer

1. 58%

Practice Problem 2

The freshmen class of 775 students is 31% of all students at Euclid University. How many students go to Euclid University?

Answer

2. 2500 students

EXAMPLE 2 Finding Totals Using Percents

Mr. Buccaran, the principal at Slidell High School, counted 31 freshmen absent during a particular day. If this is 4% of the total number of freshmen, how many freshmen are there at Slidell High School?

Solution: *Method 1.* First we state the problem in words, then we translate.

In words: 31 is 4% of what number?

Translate: $31 = 4\% \cdot x$

Next, we solve for x.

$31 = 0.04 \cdot x$ Write 4% as a decimal.

$\frac{31}{0.04} = \frac{0.04 \cdot x}{0.04}$ Divide both sides by 0.04.

$775 = x$ Simplify.

There are 775 freshmen at Slidell High School.

Method 2. First we state the problem in words, then we translate.

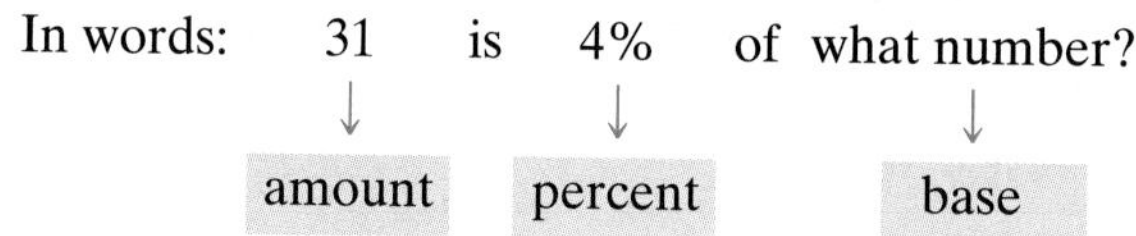
In words: 31 is 4% of what number?
amount
percent
base

Translate: amount→ $\frac{31}{b} = \frac{4}{100}$ ←percent; base→

Next we solve for b.

$31 \cdot 100 = b \cdot 4$ Set cross products equal.

$3100 = b \cdot 4$ Multiply.

$\frac{3100}{4} = \frac{b \cdot 4}{4}$ Divide both sides by 4.

$775 = b$ Simplify.

There are 775 freshmen at Slidell High School.

EXAMPLE 3 Finding Percents

Standardized nutrition labeling like the one shown has been on foods since 1994. It is recommended that no more than 30% of your calorie intake be from fat. Find what percent of the total calories shown are fat.

Fruit snacks nutrition label

Solution: *Method 1.*

In words: 10 is what percent of 80?

Translate: $10 = x \cdot 80$

Next we solve for x.

$$\frac{10}{80} = \frac{x \cdot 80}{80} \quad \text{Divide both sides by 80.}$$

$$0.125 = x \quad \text{Simplify.}$$

$$12.5\% = x \quad \text{Write 0.125 as a percent.}$$

This food contains 12.5% of its total calories as fat.

Method 2.

In words: 10 is what percent of 80?

10 → amount, what percent → percent, 80 → base

Translate: $\text{amount} \rightarrow \frac{10}{80} = \frac{p}{100} \leftarrow \text{percent}$ (base → 80)

Next we solve for p.

$$10 \cdot 100 = 80 \cdot p \quad \text{Set cross products equal.}$$

$$1000 = 80 \cdot p \quad \text{Multiply.}$$

$$\frac{1000}{80} = \frac{80 \cdot p}{80} \quad \text{Divide both sides by 80.}$$

$$12.5 = p \quad \text{Simplify.}$$

This food contains 12.5% of its total calories as fat.

Practice Problem 3

The nutrition label below is from a can of cashews. Find what percent of total calories are from fat. Round to the nearest tenth of a percent.

Answer

3. 68.4%

Practice Problem 4

From 1980 to 2000, the number of U.S. registered vehicles on the road has increased by 10%. If the number of vehicles on the road in 1980 was 122 million, find the number of vehicles on the road in 2000. (*Source:* Federal Highway Administration)

Answer

4. 134.2 million

B Finding Percent Increase and Percent Decrease

We often use percents to show how much an amount has increased or decreased.

EXAMPLE 4 Finding an Increase

From 1980 to 2000, the number of U.S. licensed drivers on the road increased by 35%. If the number of drivers on the road in 1980 was 145 million, find the number of drivers on the road in 2000. (*Source:* Federal Highway Administration)

Solution: *Method 1.* First we find the increase in drivers.

In words:	What number	is	35%	of	145?
	↓	↓	↓	↓	↓
Translate:	x	$=$	35%	$\cdot$	145

$x = 0.35 \cdot 145$ Write 35% as a decimal.

$x = 50.75$ Multiply.

The increase in drivers is 50.75 million. This means that the number of drivers in 2000 was

145 million + 50.75 million = 195.75 million drivers

Method 2. First we find the increase in drivers.

In words:	What number	is	35%	of	145?
	↓		↓		↓
	amount		percent		base

Translate: $\begin{matrix}\text{amount} \rightarrow \\ \text{base} \rightarrow\end{matrix} \frac{a}{145} = \frac{35}{100} \leftarrow \text{percent}$

$a \cdot 100 = 145 \cdot 35$ Set cross products equal.

$a \cdot 100 = 5075$ Multiply.

$\frac{a \cdot 100}{100} = \frac{5075}{100}$ Divide both sides by 100.

$a = 50.75$ Simplify.

The increase in drivers is 50.75 million. This means that the number of drivers in 2000 was

145 million + 50.75 million = 195.75 million drivers ●

Suppose that the population of a town is 10,000 people and then it increases by 2000 people. The **percent increase** is

$$\begin{matrix}\text{amount of increase} \rightarrow \\ \text{original amount} \rightarrow\end{matrix} \frac{2000}{10{,}000} = 0.2 = 20\%$$

Percent Increase

$$\text{percent increase} = \frac{\text{amount of increase}}{\text{original amount}}$$

Then write the quotient as a percent.

EXAMPLE 5 Finding Percent Increase

The number of applications for a mathematics scholarship at Yale increased from 34 to 45 in one year. What is the percent increase? Round to the nearest whole percent.

Solution: First we find the amount of increase by subtracting the original number of applicants from the new number of applicants.

$$\text{amount of increase} = 45 - 34 = 11$$

The amount of increase is 11 applicants. To find the percent increase,

$$\text{percent increase} = \frac{\text{amount of increase}}{\text{original amount}} = \frac{11}{34} \approx 0.32 = 32\%$$

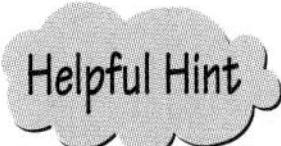

Make sure that this number in the denominator is the original number and not the new number.

The number of applications increased by about 32%.

Try the Concept Check in the margin.

Suppose that your income was \$300 a week and then it decreased by \$30. The percent decrease is

$$\begin{matrix}\text{amount of decrease} \rightarrow \\ \text{original amount} \rightarrow\end{matrix} \frac{\$30}{\$300} = 0.1 = 10\%$$

Percent Decrease

$$\text{percent decrease} = \frac{\text{amount of decrease}}{\text{original amount}}$$

Then write the quotient as a percent.

Try the Concept Check in the margin.

EXAMPLE 6 Finding Percent Decrease

In response to a decrease in sales, a company with 1500 employees reduces the number of employees to 1230. What is the percent decrease?

Solution: First we find the amount of decrease by subtracting 1230 from 1500.

$$\text{amount of decrease} = 1500 - 1230 = 270$$

The amount of decrease is 270. To find the percent decrease,

$$\text{percent decrease} = \frac{\text{amount of decrease}}{\text{original amount}} = \frac{270}{1500} = 0.18 = 18\%$$

The number of employees decreased by 18%.

Practice Problem 5

Saturday's attendance at the play *Peter Pan* increased to 333 people over Friday's attendance of 285 people. What was the percent increase in attendance? Round to the nearest tenth of a percent.

Concept Check

A student is calculating the percent increase in enrollment from 180 students one year to 200 students the next year. Explain what is wrong with the following calculations.

$$\text{amount of increase} = 200 - 180 = 20$$

$$\text{percent increase} = \frac{20}{200} = 0.1 = 10\%$$

Concept Check

An ice cream stand sold 6000 ice cream cones last summer. This year the same stand sold 5400 cones. Was there a 10% increase, a 10% decrease, or neither? Explain.

Practice Problem 6

A town with a population of 20,145 decreased to 18,430 over a 10-year period. What was the percent decrease? Round to the nearest tenth of a percent.

Answers

5. 16.8% **6.** 8.5%

Concept Check: To find the percent increase, you have to divide the amount of increase by the original amount $\left(\frac{20}{180}\right)$.

Concept Check: 10% decrease

FOCUS ON The Real World

MORTGAGES

Buying a house may be one of the most expensive purchases we make. The amount borrowed from a lending institution for real estate is called a **mortgage.** The lending institutions that normally make mortgage loans include banks, savings and loan associations, credit unions, and mortgage companies.

There are basically three items that define a mortgage loan: the principal, the loan term, and the interest rate. The principal is the dollar amount being borrowed, or financed, by the home buyers. The loan term is the length of the loan, or how long it will take to pay off the loan. The interest rate, normally expressed as a percent, governs how much must be paid for the privilege of borrowing the money.

Mortgages come in all shapes and sizes. Loan terms can range anywhere from 10 years to 15, 20, 25, 30, or even 40 years. The interest rates on shorter loans are generally lower than the interest rates on longer loans. For instance, the interest rate on a 15-year loan might be 7.25% while the interest rate on a 30-year loan is 7.5%. Interest rates also tend to be lower on loans with a smaller principal as compared with a larger principal. For example, many banks offer a jumbo mortgage loan that applies only to principals over a certain limit, generally over $227,150. The interest rates on jumbo mortgages are higher (often by about 0.25%) than on other mortgage programs.

Most lending institutions require a home buyer to make a **down payment** in cash on a home. The size of the down payment usually depends on the buyer's circumstances and the specific mortgage program chosen, but down payments generally range from 3% to 20% of the home's value. A typical down payment on a house is 10% of the purchase price. After a down payment has been chosen, the mortgage amount can be calculated by subtracting the amount of the down payment from the purchase price:

$$\text{mortgage} = \text{purchase price} - \text{down payment}$$

Besides the down payment, there are a number of initial costs related to the mortgage that must be paid. These costs are called **closing costs.** Two expensive items on the list of closing costs are the **loan origination fee** and **loan discount points.** Both of these items are generally given in "points," where each point is equal to 1% of the mortgage amount. For example, "3 points" means 3% of the mortgage amount. The loan origination fee is the fee charged by the lender to cover the costs of preparing all the loan documents. Loan discount points are prepaid interest paid at closing. Home buyers generally can choose whether or not they will pay loan discount points. Doing so lowers the interest rate on the mortgage.

$$\text{loan origination fee} = \text{mortgage} \cdot \text{points}$$
$$\text{loan discount points} = \text{mortgage} \cdot \text{points}$$

CRITICAL THINKING

Suppose you are considering buying a house with a purchase price of $140,000.

1. Find the amount of the down payment if you plan to make a 10% down payment.
2. Find the mortgage amount.
3. Calculate the loan origination fee of 1 point.
4. To get a lower interest rate on your loan, suppose you choose to pay 2.5 loan discount points at closing. How much will you be paying in loan discount points?

Name ______________________ Section __________ Date __________

EXERCISE SET 6.5

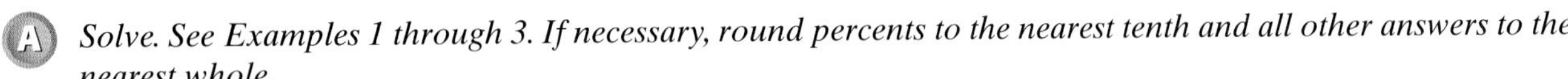

A *Solve. See Examples 1 through 3. If necessary, round percents to the nearest tenth and all other answers to the nearest whole.*

1. An inspector found 24 defective bolts during an inspection. If this is 1.5% of the total number of bolts inspected, how many bolts were inspected?

2. A daycare worker found 28 children absent one day during an epidemic of chicken pox. If this was 35% of the total number of children attending the daycare, how many children attend this daycare?

3. An owner of a repair service company estimates that for every 40 hours a repairperson is on the job, he can only bill for 75% of the hours. The remaining hours, the repairperson is idle or driving to or from a job. Determine the number of hours per 40-hour week the owner can bill for a repairperson.

4. The Hodder family paid 20% of the purchase price of a \$75,000 home as a down payment. Determine the amount of the down payment.

5. Vera Faciane earns \$2000 per month and budgets \$300 per month for food. What percent of her monthly income is spent on food?

6. Last year, Mai Toberlan bought a share of stock for \$83. She was paid a dividend of \$4.15. Determine what percent of the stock price is the dividend.

7. A manufacturer of electronic components expects 1.04% of its product to be defective. Determine the number of defective components expected in a batch of 28,350 components. Round to the nearest whole component.

8. 18% of Marvin Frank's wages are withheld for income tax. Find the amount withheld from his wages of \$3680 per month. Round to the nearest cent.

9. Of the 535 members of the 108th U.S. Congress, 73 have attended a community college. What percent of the members of the 108th Congress is this? (*Source:* American Association of Community Colleges)

10. 31.6% of all households in the United States own at least one pet dog. There are 11,250 households in Anytown. How many of these households would you expect own a dog? (*Source:* American Veterinary Medical Association)

11. There are about 98,400 female dental hygienists registered in the United States. If this represents about 98.3% of the nation's dental hygienists, find the approximate number of dental hygienists in the United States. (*Source:* The American Dental Hygienists' Association)

12. The Los Angeles County courts excused 775,130 prospective jurors from jury duty in a recent year. This represented 28% of all juror qualification affidavits sent out that year. How many juror qualification affidavits were sent out that year? (*Source:* Los Angeles Superior Court)

For each food described, find what percent of total calories is from fat. If necessary, round to the nearest tenth of a percent. See Example 3.

13.

Nutrition Facts

Serving Size 18 crackers (29g)
Servings Per Container About 9

Amount Per Serving	
Calories 120 Calories from Fat 35	
	% Daily Value*
Total Fat 4g	**6%**
Saturated Fat 0.5g	**3%**
Polyunsaturated Fat 0g	
Monounsaturated Fat 1.5g	
Cholesterol 0mg	**0%**
Sodium 220mg	**9%**
Total Carbohydrate 21g	**7%**
Dietary Fiber 2g	**7%**
Sugars 3g	
Protein 2g	

Vitamin A 0% • Vitamin C 0%
Calcium 2% • Iron 4%
Phosphorus 10%

14.

Nutrition Facts

Serving Size 28 crackers (31g)
Servings Per Container About 6

Amount Per Serving	
Calories 130 Calories from Fat 35	
	% Daily Value*
Total Fat 4g	**6%**
Saturated Fat 2g	**10%**
Polyunsaturated Fat 1g	
Monounsaturated Fat 1g	
Cholesterol 0mg	**0%**
Sodium 470mg	**20%**
Total Carbohydrate 23g	**8%**
Dietary Fiber 1g	**4%**
Sugars 4g	
Protein 2g	

Vitamin A 0% • Vitamin C 0%
Calcium 0% • Iron 2%

B *Solve. Round dollar amounts to the nearest cent and all other amounts to the nearest tenth. See Examples 4 through 6.*

15. Ace Furniture Company currently produces 6200 chairs per month. If production increases 8%, find the increase and the new number of chairs produced each month.

16. The enrollment at a local college increased 5% over last year's enrollment of 7640. Find the increase in enrollment and the current enrollment.

17. By carefully planning their meals, a family was able to decrease their weekly grocery bill by 20%. Their weekly grocery bill used to be $170. What is their new weekly grocery bill?

18. The profit of Ramone Company last year was $175,000. This year's profit decreased by 11%. Find this year's profit.

19. A car manufacturer announced that next year the price of a certain model car would increase 4.5%. This year the price is $19,286. Find the increase and the new price.

20. A union contract calls for a 6.5% salary increase for all employees. Determine the increase and the new salary that a worker currently making $28,500 under this contract can expect.

21. The population of Americans aged 65 and older was 35 million in 2000. That population is projected to increase by 80% by 2025. Find the increase and the projected 2025 population. (*Source*: Bureau of the Census)

22. The from 2000 to 2010, the number of masters degrees awarded to women is projected to increase by 8.3%. The number of women who received masters degrees in 2000 was 265,000. Find the predicted number of women to be awarded masters degrees in 2010. (*Source*: U.S. National Center for Education Statistics)

Find the amount of increase and the percent increase. See Examples 4 and 5.

	Original Amount	New Amount	Amount of Increase	Percent Increase
23.	40	50	____	____
24.	10	15	____	____
25.	85	187	____	____
26.	78	351	____	____

Find the amount of decrease and the percent decrease. See Example 6.

	Original Amount	New Amount	Amount of Decrease	Percent Decrease
27.	8	6	____	____
28.	25	20	____	____
29.	160	40	____	____
30.	200	162	____	____

Solve. Round percents to the nearest tenth, if necessary. See Examples 4 through 6.

31. There are 150 calories in a cup of whole milk and only 84 in a cup of skimmed milk. In switching to skimmed milk, find the percent decrease in number of calories.

32. In reaction to a slow economy, the number of employees at a soup company decreased from 530 to 477. What was the percent decrease in employees?

33. By changing his driving routines, Alan Miller increased his car's rate of miles per gallon from 19.5 to 23.7. Find the percent increase.

34. John Smith decided to decrease the number of calories in his diet from 3250 to 2100. Find the percent decrease.

35. The number of cable TV systems recently decreased from 10,845 to 10,700. Find the percent decrease.

36. Before taking a typing course, Geoffry Landers could type 32 words per minute. By the end of the course, he was able to type 76 words per minute. Find the percent increase.

37. In 1940, there were approximately 52.1 million sheep on farms in the United States. In 2002, this number had decreased to 6.7 million sheep. What was the percent decrease? (*Source:* National Agricultural Statistics Service)

38. In 1970, there were approximately 12.1 million milk cows on farms in the United States. In 2002, this number had decreased to 9.1 million milk cows. What was the percent decrease? (*Source:* National Agricultural Statistics Service)

39. In 1999, discarded electronics, including obsolete computer equipment, accounted for 75,000 tons of solid waste per year in Massachusetts. By 2006, discarded electronic waste is expected to increase to 300,000 tons of waste per year in the state. Find the percent increase. (*Source:* Massachusetts Department of Environmental Protection)

40. The average soft-drink size has increased from 13.1 oz to 19.9 oz over the past two decades. Find the percent increase. (*Source*: University of North Carolina at Chapel Hill, *Journal for American Medicine*)

41. In 1994, approximately 16,000 occupational therapy assistants and aides were employed in the United States. According to one survey, by 2005, this number is expected to increase to 29,000 assistants and aides. What is the percent increase? (*Source:* Bureau of Labor Statistics)

42. In 1994, approximately 206,000 medical assistants were employed in the United States. By 2005, this number is expected to increase to 327,000 medical assistants. What is the percent increase? (*Source:* Bureau of Labor Statistics)

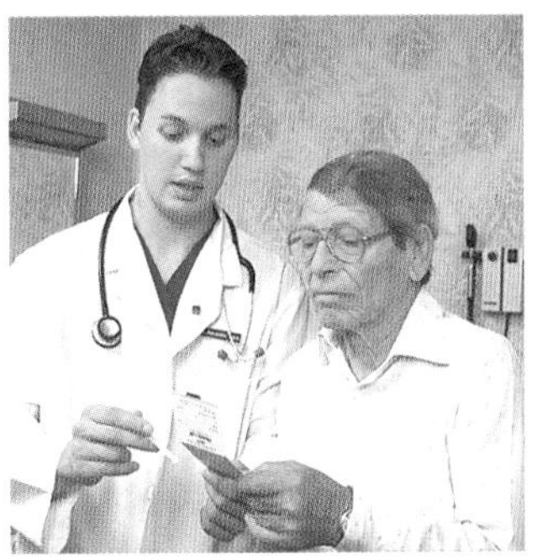

43. In 1995, 272.6 million recorded music cassettes were shipped to retailers in the United States. By 2000, this number had decreased to 76.0 million cassettes. What was the percent decrease? (*Source:* Recording Industry Association of America)

44. In 1940, the average size of a U.S. private owned farm was 174 acres. By 2000, the average size of a U.S. private owned farm had increased to 434 acres. What was the percent increase? (*Source:* National Agricultural Statistics Service)

45. In 1994, approximately 16,000,000 Americans subscribed to cellular phone service. By 2000, this number had increased to about 110,000,000 American subscribers. What was the percent increase? (*Source:* Network World, Inc.)

46. In 1998, approximately 299,000 computer engineers were employed in the United States. By 2008, this number is expected to increase to 622,000 computer engineers. What is the percent increase? (*Source:* Bureau of Labor Statistics)

47. The population of Tokyo is expected to increase from 26,518 thousand in 1994 to 28,700 thousand in 2015. Find the percent increase. (*Source:* United Nations, Department for Economic and Social Information and Policy Analysis)

48. In 1970, there were 1754 deaths from boating accidents in the United States. By 2000, the number of deaths from boating accidents had decreased to 698. What was the percent decrease? (*Source:* U.S. Coast Guard)

Review and Preview

Perform each indicated operation. See Sections 5.2 and 5.3.

49. $\begin{array}{r} 0.12 \\ \times\ 38 \\ \hline \end{array}$

50. $\begin{array}{r} 42 \\ \times\ 0.7 \\ \hline \end{array}$

51. $9.20 + 1.98$

52. $46 + 7.89$

53. $78 - 19.46$

54. $64.80 - 10.72$

Combining Concepts

55. If a number is increased by 100%, how does the increased number compare with the original number? Explain your answer.

56. In your own words, explain what is wrong with the following statement. "Last year we had 80 students attend. This year we have a 50% increase or a total of 160 students attend."

Internet Excursions

Go To:  http://www.prenhall.com/martin-gay_prealgebra  What's Related

This World Wide Web address will provide you with access to an American Savings Education Council worksheet (or a related site) for calculating a ballpark estimate of the savings needed for retirement. Print out two copies of the worksheet to use with Exercises 57–58.

57. Fill out a copy of the ballpark estimate worksheet for a person who currently is 25 years old, earns $32,000 a year, plans to retire at age 60, expects no traditional employer pension, expects $5000 per year in part-time income in retirement, and has $2000 in retirement savings. How much does this person need to save each year toward retirement?

58. Fill out a copy of the ballpark estimate worksheet for your own situation. How much do you need to save each year toward retirement?

6.6 Percent and Problem Solving: Sales Tax, Commission, and Discount

OBJECTIVES

- (A) Calculate sales tax and total price.
- (B) Calculate commissions.
- (C) Calculate discount and sale price.

(A) Calculating Sales Tax and Total Price

Percents are frequently used in the retail trade. For example, most states charge a tax on certain items when purchased. This tax is called a **sales tax,** and retail stores collect it for the state. Sales tax is almost always stated as a percent of the purchase price.

A 6% sales tax rate on a purchase of a $10.00 item gives a sales tax of

$$\text{sales tax} = 6\% \text{ of } \$10 = 0.06 \cdot \$10.00 = \$0.60$$

The total price to the customer would be

purchase price	plus	sales tax	
↓	↓	↓	
$10.00	+	$0.60 =	$10.60

This example suggests the following equations.

Sales Tax and Total Price

$$\text{sales tax} = \text{tax rate} \cdot \text{purchase price}$$
$$\text{total price} = \text{purchase price} + \text{sales tax}$$

In this section, we will round dollar amounts to the nearest cent.

EXAMPLE 1 Finding Sales Tax and Purchase Price

Find the sales tax and the total price on a purchase of an $85.50 trench coat in a city where the sales tax rate is 7.5%.

Practice Problem 1

If the sales tax rate is 6%, what is the sales tax and the total amount due on a $29.90 Goodgrip tire?

Answers

1. tax: $1.75; total: $31.65

Solution: The purchase price is \$85.50 and the tax rate is 7.5%.

sales tax	=	tax rate	·	purchase price	
↓		↓		↓	
sales tax	=	7.5%	·	\$85.50	
	=	0.075	·	\$85.5	Write 7.5% as a decimal.
	≈	\$6.41			Round to the nearest cent.

Thus

total price	=	purchase price	+	sales tax
↓		↓		↓
total price	=	\$85.50	+	\$6.41
	=	\$91.91		

The sales tax on \$85.50 is \$6.41 and the total price is \$91.91.

Try the Concept Check in the margin.

Concept Check

The purchase price of a textbook is \$50 and sales tax is 10%. If you are told by the cashier that the total price is \$75, how can you tell that a mistake has been made?

Practice Problem 2

The sales tax on a \$13,500 automobile is \$1080.00. Find the sales tax rate.

EXAMPLE 2 Finding a Sales Tax Rate

The sales tax on a \$300 printer is \$22.50. Find the sales tax rate.

Solution: Let r be the unknown sales tax rate. Then

sales tax	=	tax rate	·	purchase price
↓		↓		↓
22.50	=	r	·	300

$$\frac{22.50}{300} = \frac{r \cdot 300}{300} \quad \text{Divide both sides by 300.}$$

$$0.075 = r \quad \text{Simplify.}$$

$$7.5\% = r \quad \text{Write 0.075 as a percent.}$$

The sales tax rate is 7.5%.

B Calculating Commissions

A **wage** is payment for performing work. Hourly wage, commissions, and salary are some of the ways wages can be paid. Many people who work in sales are paid a commission. An employee who is paid a **commission** is paid a percent of his or her total sales.

Answer

2. 8%

Concept Check: Since 10% = 0.10, the sales tax is \$50 · 0.10 = \$5. The total price should have been \$55.

Commission

commission = commission rate · sales

EXAMPLE 3 Finding a Commission

Sherry Souter, a real estate broker for Wealth Investments, sold a house for \$114,000 last week. If her commission is 1.5% of the selling price of the home, find the amount of her commission.

Solution:

commission	=	commission rate	·	sales	
↓		↓		↓	
commission	=	1.5%	·	114,000	Write 1.5% as 0.015.
	=	0.015	·	114,000	Multiply.
	=	1710			

Her commission on the house is \$1710.

Practice Problem 3

Mr. Olsen is a sales representative for Miko Copiers. Last month he sold \$37,632 worth of copy equipment and supplies. What is his commission for the month if he is paid a commission of 6.6% of his total sales for the month?

EXAMPLE 4 Finding a Commission Rate

A salesperson earned \$1560 for selling \$13,000 worth of television and stereo systems. Find the commission rate.

Solution: Let r stand for the unknown commission rate. Then

commission	=	commission rate	·	sales	
↓		↓		↓	
1560	=	r	·	13,000	
$\frac{1560}{13,000}$	=	$\frac{r \cdot 13,000}{13,000}$			Divide both sides by 13,000.
0.12	=	r			Simplify.
12%	=	r			Write 0.12 as a percent.

The commission rate is 12%.

Practice Problem 4

A salesperson earns \$1290 for selling \$8600 worth of appliances. Find the commission rate.

Answers

3. \$2483.71 **4.** 15%

(C) Calculating Discount and Sale Price

Suppose that an item that normally sells for $40 is on sale for 25% off. This means that the **original price** of $40 is reduced, or **discounted** by 25% of $40, or $10. The **discount rate** is 25%, the **amount of discount** is $10, and the **sale price** is $40 − $10 or $30.

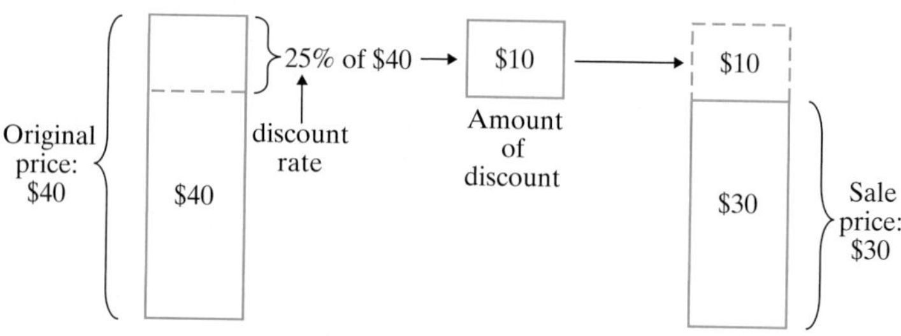

To calculate discounts and sale prices, we can use the following equations.

Discount and Sale Price

amount of discount = discount rate · original price
sale price = original price − amount of discount

Practice Problem 5

A Panasonic TV is advertised on sale for 15% off the regular price of $700. Find the discount and the sale price.

EXAMPLE 5 Finding a Discount and a Sale Price

A speaker that normally sells for $65.00 is on sale at 25% off. What is the discount and what is the sale price?

Solution: First we find the discount.

amount of discount	=	discount rate	·	original price
↓		↓		↓
amount of discount	=	25%	·	65
	=	0.25	·	65 Write 25% as 0.25.
	=	16.25 Multiply.		

The discount is $16.25. Next we find the sale price.

sale price	=	original price	−	discount
↓		↓		↓
sale price	=	$65	−	$16.25
	=	$48.75 Subtract.		

The sale price is $48.75.

Answers

5. $105; $595

Name ______________________ Section ____________ Date ____________

EXERCISE SET 6.6

 *Solve. See Examples 1 and 2.*

1. What is the sales tax on a suit priced at $150.00 if the sales tax rate is 5%?

2. If the sales tax rate is 6%, find the sales tax on a microwave oven priced at $188.

3. The purchase price of a camcorder is $799. What is the total price if the sales tax rate is 7.5%?

4. A stereo system has a purchase price of $426. What is the total price if the sales tax rate is 8%?

5. A chair and ottoman have a purchase price of $600. If the sales tax on this purchase is $54, find the sales tax rate.

6. The sales tax on the purchase of a $2500 computer is $162.50. Find the sales tax rate.

7. A computer desk sells for $220. With a sales tax rate of 8.5%, find the total price.

8. A one-half carat diamond ring is priced at $800. The sales tax rate is 6.5%. Find the total price.

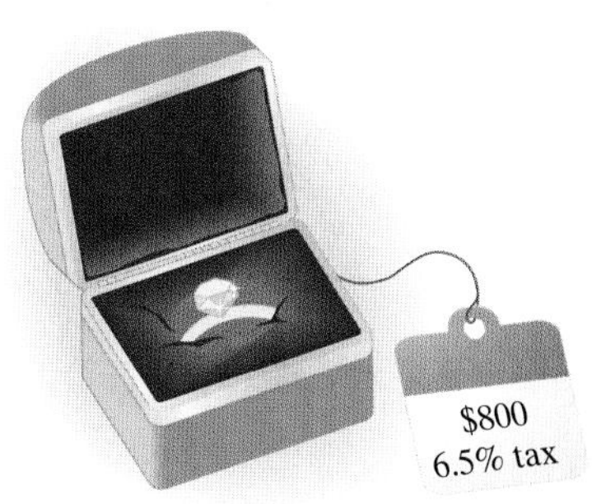

9. A gold and diamond bracelet sells for $1800. Find the total price if the sales tax rate is 6.5%.

10. The purchase price of a personal computer is $1890. If the sales tax rate is 8%, what is the total price?

11. The sales tax on the purchase of a used truck is $920. If the tax rate is 8%, find the purchase price of the truck.

12. The sales tax on the purchase of a desk is $27.50. If the tax rate is 5%, find the purchase price of the desk.

13. A cordless phone costs $90 and a battery recharger costs $15. What is the total price for purchasing these items if the sales tax rate is 7%?

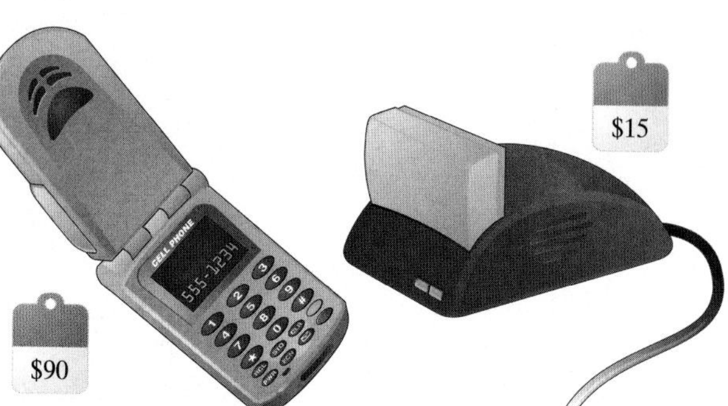

14. Ms. Warner bought a blouse for $35, a skirt for $55, and a blazer for $95. Find the total price she paid, given a sales tax rate of 6.5%.

15. The sales tax is $98.70 on a stereo sound system purchase of $1645. Find the sales tax rate.

16. The sales tax is $103.50 on a necklace purchase of $1150. Find the sales tax rate.

B *Solve. See Examples 3 and 4.*

17. Jane Moreschi, a sales representative for a large furniture warehouse, is paid a commission rate of 4%. Find her commission if she sold $1,236,856 worth of furniture last month.

18. Rosie Davis-Smith is a beauty consultant for a cosmetic business. She is paid a commission rate of 4.8%. Find her commission if she sold $1638 in cosmetics last month.

19. A salesperson earned a commission of $1380.40 for selling $9860.00 worth of paper products. Find the commission rate.

20. A salesperson earned a commission of $3575 for selling $32,500 worth of books to various book stores. Find the commission rate.

21. How much commission will Jack Pruet make on the sale of a $125,900 house if he receives 1.5% of the selling price?

22. Frankie Lopez sold $9638.00 of jewelry this week. Find her commission for the week if she is paid a commission rate of 5.6%.

23. A real estate agent earned a commission of \$2565 for selling a house. If his rate is 3%, find the selling price of the house.

24. A salesperson earned \$1750 for selling fertilizer. If her commission rate is 7%, find the selling price of the fertilizer.

C *Find the amount of discount and the sale price. See Example 5.*

	Original Price	Discount Rate	Amount of Discount	Sale Price
25.	\$68.00	10%	______	______
26.	\$47.00	20%	______	______
27.	\$96.50	50%	______	______
28.	\$110.60	40%	______	______
29.	\$215.00	35%	______	______
30.	\$370.00	25%	______	______
31.	\$21,700.00	15%	______	______
32.	\$17,800.00	12%	______	______

33. A \$300 fax machine is on sale at 15% off. Find the discount and the sale price.

34. A \$2000 designer dress is on sale at 30% off. Find the discount and the sale price.

Find the missing amounts. See Example 5.

	Original Price	Discount Rate	Amount of Discount	Sale Price
35.	$75	20%		
36.	$40	15%		
37.	$120		$39.60	
38.	$87		$8.70	
39.		40%	$370	
40.		65%	$299	

Review and Preview

Multiply. See Sections 5.3 and 5.6.

41. $2000 \cdot 0.3 \cdot 2$

42. $500 \cdot 0.08 \cdot 3$

43. $400 \cdot 0.03 \cdot 11$

44. $1000 \cdot 0.05 \cdot 5$

45. $600 \cdot 0.04 \cdot \frac{2}{3}$

46. $6000 \cdot 0.06 \cdot \frac{3}{4}$

Combining Concepts

47. A diamond necklace normally sells for $24,966. If it is first discounted by 15% and then taxed at 7.5%, find the total price.

48. A house recently sold for $562,560. The commission rate on the sale is 5.5%. If a real estate agent is to receive 60% of the commission, find the amount received by the agent.

49. Suppose that the original price of a shirt is $50. Which is better, a 60% discount or a discount of 30% followed by a discount of 35% of the reduced price. Explain your answer.

One very useful application of percent is mentally calculating a tip. Recall that to find 10% of a number, simply move the decimal point one place to the left. To find 20% of a number, just double 10% of the number. To find 15% of a number, find 10% and the add to that number half of the 10% amount. Mentally fill in the chart below. To do so, start by rounding the bill amount to the nearest dollar.

Tipping Chart

	Bill Amount	10%	15%	20%
50.	$40.21			
51.	$15.89			
52.	$72.17			
53.	$9.33			

FOCUS ON Business and Career

FASTEST-GROWING OCCUPATIONS

According to U.S. Bureau of Labor Statistics projections, the careers listed below are the top ten fastest-growing jobs, ranked by expected percent increase through the year 2010.

Occupation	Employment in 2000	Percent Increase from 2000 to 2010
Computer software engineers, applications	380,000	100%
Computer support specialists	506,000	97%
Computer software engineers, systems software	317,000	90%
Network and computer systems administrators	229,000	82%
Network systems and data communications analysts	119,000	77%
Desktop publishers	38,000	67%
Database Administrators	106,000	66%
Personal and home care aides	414,000	62%
Computer systems analysts	431,000	60%
Medical assistants	329,000	57%

(*Source*: November 2001, *Monthly Labor Review*, Bureau of Labor Statistics)

What do all of these fast-growing occupations have in common? They all require a knowledge of math! For some careers, such as desktop publishing specialists, medical assistants, and computer engineers, the ways math is used on the job may be obvious. For other occupations, the use of math may not be quite as apparent. However, tasks common to many jobs—filling in a time sheet, writing up an expense or mileage report, planning a budget, figuring a bill, ordering supplies, and even making a work schedule—all require math.

GROUP ACTIVITY

1. List the top five occupations by order of employment figures for 2000.
2. Using the 2000 employment figures and the percent increase from 2000 to 2010, find the expected 2010 employment figures for each occupation listed in the table.
3. List the top five occupations by order of employment figures for 2010. Did the order change at all from 2000? Explain.
4. How many of the occupations on the list are expected to increase by at least 80% the number of positions in 2010 than in 2000? Explain how you identified these occupations.

6.7 Percent and Problem Solving: Interest

OBJECTIVES

A. Calculate simple interest.
B. Use a compound interest table to calculate compound interest.
C. Calculate monthly payments on loans.

SSM TUTOR CENTER SG CD & VIDEO MATH PRO WEB

A Calculating Simple Interest

Interest is money charged for using other people's money. When you borrow money, you pay interest. When you loan or invest money, you earn interest. The money borrowed, loaned, or invested is called the **principal amount,** or simply **principal.** Interest is normally stated in terms of a percent of the principal for a given period of time. The **interest rate** is the percent used in computing the interest. Unless stated otherwise, *the rate is understood to be per year.* When the interest is computed on the original principal, it is called **simple interest.** Simple interest is calculated using the following equation.

Simple Interest

$$\text{simple interest} = \text{principal} \cdot \text{rate} \cdot \text{time}$$

or

$$I = P \cdot R \cdot T$$

where the rate is understood to be the rate per year and time is in years.

EXAMPLE 1 Finding Simple Interest

Find the simple interest after 2 years on \$500 at an interest rate of 12%.

Solution: In this example, $P = \$500$, $R = 12\%$, and $T = 2$ years. Replace the variables with values in the formula $I = PRT$.

$$I = P \cdot R \cdot T$$
$$I = \$500 \cdot 12\% \cdot 2 \quad \text{Let } P = \$500,\ R = 12\%, \text{ and } T = 2.$$
$$= \$500 \cdot (0.12) \cdot 2 \quad \text{Write 12\% as a decimal.}$$
$$= \$120 \quad \text{Multiply.}$$

The simple interest is \$120.

Practice Problem 1

Find the simple interest after 3 years on \$750 at an interest rate of 8%.

Try the Concept Check in the margin.

Concept Check

If \$2000 is borrowed at 12% simple interest for 14 months, is 2000 the variable T, P, or I in the simple interest formula

$I = P \cdot R \cdot T$

If time is not given in years, we need to convert the given time to years.

EXAMPLE 2 Finding Simple Interest

Ivan Borski borrowed \$2400 at 10% simple interest for 8 months to buy a used Chevy S-10. Find the simple interest he paid.

Solution: Since there are 12 months in a year, we first find what part of a year 8 months is.

$$8 \text{ months} = \frac{8}{12}\text{ year} = \frac{2}{3}\text{ year}$$

Now we find the simple interest.

$$I = P \cdot R \cdot T$$
$$= \$2400 \cdot (0.10) \cdot \frac{2}{3} \quad \text{Let } P = \$2400,\ R = 10\% \text{ or } 0.10, \text{ and } T = \tfrac{2}{3}$$
$$= \$160$$

The interest on Ivan's loan is \$160.

Practice Problem 2

Juanita Lopez borrowed \$800 for 9 months at a simple interest rate of 20%. How much interest did she pay?

Answers

1. \$180 **2.** \$120

Concept Check: 2000 is P, principal.

Try the Concept Check in the margin.

When money is borrowed, the borrower pays the original amount borrowed, the principal, as well as the interest. When money is invested, the investor receives the original amount invested, or the principal, as well as the interest. In either case, the **total amount** is the sum of the principal and the interest.

Finding the Total Amount of a Loan or Investment

total amount (paid or received) = principal + interest

Concept Check

Suppose in Example 2 you had obtained an answer of \$16,000. How would you know that you had made a mistake in this problem?

Practice Problem 3

If \$500 is borrowed at a simple interest rate of 12% for 6 months, find the total amount paid.

EXAMPLE 3 Finding the Total Amount of an Investment

An accountant invested \$2000 at a simple interest rate of 10% for 2 years. What total amount of money will she have from her investment in 2 years?

Solution: First we find her interest.

$$I = P \cdot R \cdot T$$
$$= \$2000 \cdot (0.10) \cdot 2 \quad \text{Let } P = \$2000,\ R = 10\% \text{ or } 0.10, \text{ and } T = 2.$$
$$= \$400$$

The interest is \$400.

Next we add the interest to the principal.

$$\text{total amount} = \text{principal} + \text{interest}$$
$$\text{total amount} = \$2000 + \$400$$
$$= \$2400$$

After 2 years, she will have a total amount of \$2400.

Try the Concept Check in the margin.

Concept Check

Which investment would earn more interest: an amount of money invested at 8% interest for 2 years or the same amount of money invested at 8% for 3 years? Explain.

B Calculating Compound Interest

Recall that simple interest depends on the original principal only. Another type of interest is compound interest. **Compound interest** is computed on not only the principal, but also on the interest already earned in previous compounding periods. Compound interest is used more often than simple interest in saving accounts.

Let's see how compound interest differs from simple interest. Suppose that \$2000 is invested at 7% interest **compounded annually** for 3 years. This means that interest is added to the principal at the end of each year and next year's interest is computed on this new amount. In this section, we will round dollar amounts to the nearest cent.

Amount at Beginning of Year	Principal · Rate · Time = Interest	Amount at End of Year
1st year \$2000	$\$2000 \cdot 0.07 \cdot 1 = \$140$	$\$2000 + \$140 = \$2140$
2nd year \$2140	$\$2140 \cdot 0.07 \cdot 1 = \$149.80$	$\$2140 + \$149.80 = \$2289.80$
3rd year \$2289.80	$\$2289.80 \cdot 0.07 \cdot 1 = \$160.29$	$\$2289.80 + \$160.29 = \$2450.09$

Answer

3. \$530

Concept Check: \$16,000 is too much interest. Answers may vary.

Concept Check: 8% for 3 years. Since the interest rate is the same, the longer you keep the money invested, the more interest you earn.

The compound interest earned can be found by

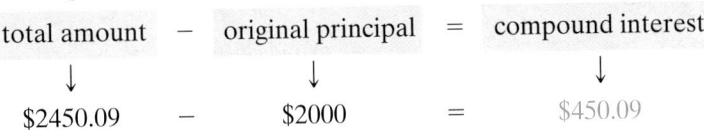

total amount	−	original principal	=	compound interest
↓		↓		↓
\$2450.09	−	\$2000	=	\$450.09

The simple interest earned would have been

principal	·	rate	·	time	=	interest
↓		↓		↓		↓
\$2000	·	0.07	·	3	=	\$420

Since compound interest earns "interest on interest," compound interest produces more interest than simple interest.

Computing compound interest using the method above can be tedious. We can use a **compound interest table** to compute interest more quickly. The compound interest table is found in Appendix C. This table gives the total compound interest and principal paid on \$1 for given rates and number of years. Then we can use the following equation to find the total amount of interest and principal.

Finding Total Amounts with Compound Interest

total amount = original principal · compound interest factor (from table)

EXAMPLE 4 Finding Total Amount Received on an Investment

\$4000 is invested at 8% compounded semiannually for 10 years. Find the total amount at the end of 10 years.

Solution: Look in Appendix C. The compound interest factor for 10 years at 8% in the compounded semiannually section is 2.19112.

total amount	=	original principal	·	compound interest factor
↓		↓		↓
total amount	=	\$4000	·	2.19112
	=	\$8764.48		

Therefore, the total amount at the end of 10 years is \$8764.48.

Practice Problem 4

\$5500 is invested at 7% compounded daily for 5 years. Find the total amount at the end of 5 years.

EXAMPLE 5 Finding Compound Interest Earned

In Example 4, we found that the total amount for \$4000 invested at 8% compounded semiannually for 10 years is \$8764.48. Find the compound interest earned.

Solution:

interest earned	=	total amount	−	original principal
↓		↓		↓
interest earned	=	\$8764.48	−	\$4000
	=	\$4764.48		

The compound interest earned is \$4764.48.

Practice Problem 5

If the total amount is \$9933.14 when \$5500 is invested, find the compound interest earned.

C Calculating a Monthly Payment

We conclude this section with a method to find the monthly payment on a loan.

Answers

4. \$7804.61 **5.** \$4433.14

Finding the Monthly Payment of a Loan

$$\text{monthly payment} = \frac{\text{principal + interest}}{\text{total number of payments}}$$

Practice Problem 6

Find the monthly payment on a $3000 3-year loan if the interest on the loan is $1123.58.

EXAMPLE 6 Finding a Monthly Payment

Find the monthly payment on a $2000 loan for 2 years. The interest on the 2-year loan is $435.88.

Solution: First we determine the total number of monthly payments. The loan is for 2 years. Since there are 12 months per year, the number of payments is 2 · 12, or 24. Now we can calculate the monthly payment.

$$\text{monthly payment} = \frac{\text{principal + interest}}{\text{total number of payments}}$$

$$\text{monthly payment} = \frac{\$2000 + \$435.88}{24}$$

$$\approx \$101.50$$

The monthly payment is $101.50.

CALCULATOR EXPLORATIONS

Compound Interest Factor

A compound interest factor may be found by using your calculator and evaluating the formula

$$\textbf{compound interest factor} = \left(1 + \frac{r}{n}\right)^{nt}$$

where r is the interest rate, t is the time in years, and n is the number of times compounded per year. For example, we stated earlier that the compound interest factor for 10 years at 8% compounded semiannually is 2.19112. Let's find this factor by evaluating the compound interest factor formula when $r = 8\%$ or 0.08, $t = 10$, and $n = 2$ (compounded semiannually means 2 times per year at six month intervals). Thus,

$$\text{compound interest factor} = \left(1 + \frac{0.08}{2}\right)^{2\cdot 10} \text{ or } \left(1 + \frac{0.08}{2}\right)^{20}$$

To evaluate, press the keys

[( ] [1] [+] [0.08] [÷] [2] [ )] [y^x] [20] [=] or [ENTER]

The display will read [2.1911231]. Rounded to five decimal places, this is 2.19112.

Find the compound interest factor. Use the table in Appendix C to check your answer.

1. 5 years, 9%, compounded quarterly
2. 15 years, 14%, compounded daily
3. 20 years, 11%, compounded annually
4. 1 year, 7%, compounded semiannually
5. Find the total amount after 4 years if $500 is invested at 6% compounded quarterly.
6. Find the total amount for 19 years if $2500 is invested at 5% compounded daily.

Answer

6. $114.54

Name ______________________ Section __________ Date __________

EXERCISE SET 6.7

A *Find the simple interest. See Examples 1 and 2.*

	Principal	Rate	Time
1.	\$200	8%	2 years
3.	\$160	11.5%	4 years
5.	\$5000	10%	$1\frac{1}{2}$ years
7.	\$375	18%	6 months
9.	\$2500	16%	21 months

	Principal	Rate	Time
2.	\$800	9%	3 years
4.	\$950	12.5%	5 years
6.	\$1500	14%	$2\frac{1}{4}$ years
8.	\$1000	10%	18 months
10.	\$775	15%	8 months

Solve. See Examples 1 through 3.

11. A company borrows \$62,500 for 2 years at a simple interest rate of 12.5% to buy an airplane. Find the total amount paid on the loan.

12. \$65,000 is borrowed to buy a house. If the simple interest rate on the 30-year loan is 10.25%, find the total amount paid on the loan.

13. A money market fund advertises a simple interest rate of 9%. Find the total amount received on an investment of \$5000 for 15 months.

14. The Real Service Company takes out a 270-day (9-month) short-term, simple interest loan of \$4500 to finance the purchase of some new equipment. If the interest rate is 14%, find the total amount that the company pays back.

15. Marsha Waide borrows $8500 and agrees to pay it back in 4 years. If the simple interest rate is 12%, find the total amount she pays back.

16. Ms. Lapchinski gives her 18-year-old daughter a graduation gift of $2000. If this money is invested at 8% simple interest for 5 years, find the total amount after 5 years.

B *Find the total amount in each compound interest account. See Example 4.*

17. $6150 is compounded semiannually at a rate of 14% for 15 years.

18. $2060 is compounded annually at a rate of 15% for 10 years.

19. $1560 is compounded daily at a rate of 8% for 5 years.

20. $1450 is compounded quarterly at a rate of 10% for 15 years.

21. $10,000 is compounded semiannually at a rate of 9% for 20 years.

22. $3500 is compounded daily at a rate of 8% for 10 years.

Find the amount of compound interest earned. See Example 5.

23. $2675 is compounded annually at a rate of 9% for 1 year.

24. $6375 is compounded semiannually at a rate of 10% for 1 year.

25. $2000 is compounded annually at a rate of 8% for 5 years.

26. $2000 is compounded semiannually at a rate of 8% for 5 years.

27. $2000 is compounded quarterly at a rate of 8% for 5 years.

28. $2000 is compounded daily at a rate of 8% for 5 years.

Solve. See Example 6.

29. A college student borrows \$1500 for 6 months to pay for a semester of school. If the interest is \$61.88, find the monthly payment.

30. Jim Tillman borrows \$1800 for 9 months. If the interest is \$148.90, find his monthly payment.

31. \$20,000 is borrowed for 4 years. If the interest on the loan is \$10,588.70, find the monthly payment.

32. \$105,000 is borrowed for 15 years. If the interest on the loan is \$181,125.00, find the monthly payment.

Review and Preview

Perform each indicated operation. See Sections 2.2 to 2.4.

33. $-5 + (-24)$

34. $7 - (-31)$

35. $(-5)(-30)$

36. $-30 \div (-5)$

37. $\dfrac{7 - 10}{3}$

38. $\dfrac{22 + (-4)}{-4 - 5}$

Combining Concepts

39. Explain how to look up the compound interest factor in the compound interest table.

40. Explain how to find the amount of interest on a compounded account.

41. Compare the following accounts: Account 1: \$1000 is invested for 10 years at a simple interest rate of 6%. Account 2: \$1000 is compounded semiannually at a rate of 6% for 10 years. Discuss how the interest is computed for each account. Determine which account earns more interest. Why?

CHAPTER 6 ACTIVITY Investigating Probability

This activity may be completed by working in groups or individually.

The measure of the chance of an event occurring is its probability. In this activity, you will investigate one way that probabilities can be estimated. You will need a cup and 30 thumbtacks.

1. Place the thumbtacks in the cup. Shake the cup and toss out the thumbtacks onto a flat surface. Count the number of tacks that land point up, and record this number in the table. Repeat for 60, 90, 120, and 150 tacks. Record your results in the second column of the table. (**Hint:** For 60 thumbtacks, count the number of tacks landing point up in two tosses of the 30 thumbtacks, etc.)

2. For each row of the table, find the fraction of tacks that landed point up. Express this fraction as a percent. Complete the third and fourth columns of the table.

3. Each of the percents you computed in Question 2 is an *estimate* of the probability that a single thumbtack will land point up when tossed. With an estimate like this, the larger the number of tack tosses used, the better the estimate of the actual probability. What do you suppose is the value of the actual probability? Explain your reasoning.

4. Combine your results for all of your tack tosses recorded in the table with the results of other students or groups in your class. Of this total number of tack tosses, compute the percent of tacks that landed point up. This is your best estimate of the probability that a tack will land point up when tossed.

5. If you tossed 200 thumbtacks, what percent would you expect to land point up? How many tacks would you expect to land point up? Use the percent (probability) you computed in Question 4 to make this calculation. What if you tossed 300 thumbtacks?

Number of Tacks	Number of Tacks Landing Point Up	Fraction of Point-Up Tacks	Percent of Point-Up Tacks
30			
60			
90			
120			
150			

STUDY SKILLS REMINDER

Are you prepared for a test on Chapter 6?

Below I have listed some *common trouble areas* for topics covered in Chapter 6. After studying for your test—but before taking your test—read these.

- Can you convert from percents to fractions or decimals and from fractions or decimals to percents?

  Percent to decimal: $7.5\% = 7.5(0.01) = 0.075$

  Percent to fraction: $11\% = 11 \cdot \frac{1}{100} = \frac{11}{100}$

  Decimal to percent: $0.36 = 0.36(100\%) = 36\%$

  Fraction to percent: $\frac{6}{7} = \frac{6}{7} \cdot 100\% = \frac{6}{7} \cdot \frac{100}{1}\% = \frac{600}{7}\%$

  $= 85\frac{5}{7}\%$

- Do you remember how to find percent increase or percent decrease? The number of CDs increased from 40 to 48. Find the percent increase.

  $$\text{percent increase} = \frac{\text{increase}}{\text{original number}} = \frac{8}{40} = 0.20 = 20\%$$

  Remember: This is simply a checklist of common trouble areas. For a review of Chapter 6, see the Highlights and Chapter Review at the end of the chapter.

Chapter 6 Vocabulary Check

Fill in each blank with one of the words or phrases listed below.

percent	of	amount	100%	compound interest	ratio
base	is	0.01	$\frac{1}{100}$	proportion	

1. In a mathematical statement, ____ usually means "multiplication."
2. In a mathematical statement, ____ means "equals."
3. _______ means "per hundred."
4. ________________ is computed not only on the principal, but also on interest already earned in previous compounding periods.
5. In the percent proportion $\frac{\_\_\_\_}{\_\_\_\_} = \frac{\text{percent}}{100}$
6. To write a decimal or fraction as a percent, multiply by _______ .
7. The decimal equivalent of the % symbol is ______ .
8. The fraction equivalent of the % symbol is ______ .
9. A __________ is a mathematical statement that two ratios are equal.
10. A _____ is the quotient of two numbers or two quantities..

Chapter 6 Highlights

CHAPTER 6

DEFINITIONS AND CONCEPTS	EXAMPLES
SECTION 6.1 RATIO AND PROPORTION	
A **ratio** is the quotient of two numbers or two quantities.	Write the ratio of 5 hours to 1 day using fractional notation. $\frac{5\text{ hours}}{1\text{ day}} = \frac{5\text{ hours}}{24\text{ hours}} = \frac{5}{24}$
A **proportion** is a mathematical statement that two ratios are equal.	$\frac{2}{3} = \frac{8}{12}$ $\quad \frac{x}{7} = \frac{15}{35}$
In the proportion $\frac{a}{b} = \frac{c}{d}$, the products ad and bc are called **cross products.**	$\frac{2}{3} = \frac{8}{12}$ → $3 \cdot 8$ or 24; → $2 \cdot 12$ or 24
If $\frac{a}{b} = \frac{c}{d}$ then $ad = bc$.	Solve: $\frac{3}{4} = \frac{x}{x-1}$ $\frac{3}{4} = \frac{x}{x-1}$ $3(x-1) = 4x$ Set cross products equal. $3x - 3 = 4x$ $-3 = x$
SECTION 6.2 PERCENTS, DECIMALS, AND FRACTIONS	
Percent means per hundred. The % symbol denotes percent.	$51\% = \frac{51}{100}$ 51 per 100 $7\% = \frac{7}{100}$ 7 per 100
To write a percent as a decimal, replace the percent symbol with its decimal equivalent, 0.01; then multiply.	$32\% = 32(0.01) = 0.32$
To write a decimal as a percent, multiply by 1 in the form of 100%.	$0.78 = 0.78(100\%)$ $= 78\%$
To write a percent as a fraction, replace the percent symbol with its fraction equivalent, $\frac{1}{100}$; then multiply. Don't forget to simplify the fraction, if possible.	$25\% = 25 \cdot \frac{1}{100} = \frac{25}{100} = \frac{\overset{1}{\cancel{25}}}{4 \cdot \underset{1}{\cancel{25}}} = \frac{1}{4}$
To write a fraction as a percent, multiply the fraction by 100%.	$\frac{1}{6} = \frac{1}{6} \cdot 100\% = \frac{1}{6} \cdot \frac{100}{1}\% = \frac{100}{6}\% = 16\frac{2}{3}\%$
SECTION 6.3 SOLVING PERCENT PROBLEMS WITH EQUATIONS	
Three key words in the statement of a percent problem are **of,** which means multiplication (·) **is,** which means equals (=) **what** (or some equivalent word or phrase), which stands for the unknown (x)	Solve. 6 is 12% of what number? $6 = 12\% \cdot x$ $6 = 0.12 \cdot x$ Write 12% as a decimal. $\frac{6}{0.12} = \frac{0.12 \cdot x}{0.12}$ Divide both sides by 0.12. $50 = x$ Thus, 6 is 12% of 50.

Definitions and Concepts | Examples

Section 6.4 Solving Percent Problems with Proportions

Percent Proportion

$$\frac{\text{amount}}{\text{base}} = \frac{\text{percent}}{100} \leftarrow \text{always 100}$$

or

$$\frac{\text{amount} \rightarrow a}{\text{base} \rightarrow b} = \frac{p}{100} \leftarrow \text{percent}$$

Solve.

20.4 is what percent of 85?

↘ ↓ ↓

amount percent base

$$\frac{\text{amount} \rightarrow 20.4}{\text{base} \rightarrow 85} = \frac{p}{100} \leftarrow \text{percent}$$

$$\begin{aligned} 20.4 \cdot 100 &= 85 \cdot p && \text{Set cross products equal.} \\ 2040 &= 85 \cdot p && \text{Multiply.} \\ \frac{2040}{85} &= \frac{85 \cdot p}{85} && \text{Divide both sides by 85.} \\ 24 &= p && \text{Simplify.} \end{aligned}$$

Thus, 20.4 is 24% of 85.

Section 6.5 Applications of Percent

Percent Increase

$$\text{percent increase} = \frac{\text{amount of increase}}{\text{original amount}}$$

Percent Decrease

$$\text{percent decrease} = \frac{\text{amount of decrease}}{\text{original amount}}$$

A town with a population of 16,480 decreased to 13,870 over a 12-year period. Find the percent decrease. Round to the nearest whole percent.

$$\begin{aligned} \text{amount of decrease} &= 16{,}480 - 13{,}870 \\ &= 2610 \end{aligned}$$

$$\begin{aligned} \text{percent decrease} &= \frac{\text{amount of decrease}}{\text{original amount}} \\ &= \frac{2610}{16{,}480} \approx 0.16 \\ &= 16\% \end{aligned}$$

The town's population decreased by about 16%.

Section 6.6 Percent and Problem Solving: Sales Tax, Commission, and Discount

Sales Tax

$$\text{sales tax} = \text{sales tax rate} \cdot \text{purchase price}$$

$$\text{total price} = \text{purchase price} + \text{sales tax}$$

Find the sales tax and the total price of a purchase of \$42.00 if the sales tax rate is 9%.

sales tax = sales tax rate · purchase price

↓ ↓ ↓

$$\begin{aligned} \text{sales tax} &= 9\% \cdot \$42 \\ &= 0.09 \cdot \$42 \\ &= \$3.78 \end{aligned}$$

The total price is

total price = purchase price + sales tax

↓ ↓ ↓

$$\begin{aligned} \text{total price} &= \$42.00 + \$3.78 \\ &= \$45.78 \end{aligned}$$

The total price is \$45.78.

Definitions and Concepts | Examples

Section 6.6 (continued)

Commission

commission = commission rate · sales

A salesperson earns a commission of 3%. Find the commission from sales of $12,500 worth of appliances.

commission	=	commission rate	·	sales
↓		↓		↓
commission	=	3%	·	$12,500
	=	0.03 · 12,500		
	=	$375		

The commission is $375.

Discount and Sale Price

amount of discount = discount rate · original price

sale price = original price − amount of discount

A suit is priced at $320 and is on sale today for 25% off. What is the sale price?

amount of discount	=	discount rate	·	original price
↓		↓		↓
amount of discount	=	25%	·	$320
	=	0.25 · 320		
	=	$80		

sale price	=	original price	−	amount of discount
↓		↓		↓
sale price	=	$320	−	$80
	=	$240		

The sale price is $240.

Section 6.7 Percent and Problem Solving: Interest

Simple Interest

interest = principal · rate · time

where the rate is understood to be per year, unless told otherwise.

Find the simple interest after 3 years on $800 at an interest rate of 5%.

interest	=	principal	·	rate	·	time
↓		↓		↓		↓
interest	=	$800	·	5%	·	3
	=	$800 · 0.05 · 3		Write 5% as 0.05.		
	=	$120		Multiply.		

The interest is $120.

Compound Interest

total amount = original principal · compound interest factor

Compound interest is computed not only on the principal, but also on interest already earned in previous compounding periods. (See Appendix C.)

$800 is invested at 5% compounded quarterly for 10 years. Find the total amount at the end of 10 years.

total amount	=	original principal	·	compound interest factor
↓		↓		↓
total amount	=	$800	·	1.64362
	≈	$1314.90		

Name ______________________ Section ____________ Date ____________

Chapter 6 Review

(6.1) *Write each phrase as a ratio in fractional notation.*

1. 20 cents to 1 dollar

2. four parts red to six parts white

Solve each proportion.

3. $\frac{x}{2} = \frac{12}{4}$

4. $\frac{20}{1} = \frac{x}{25}$

5. $\frac{32}{100} = \frac{100}{x}$

6. $\frac{20}{2} = \frac{c}{5}$

7. $\frac{2}{x - 1} = \frac{3}{x + 3}$

8. $\frac{4}{y - 3} = \frac{2}{y - 3}$

9. $\frac{y + 2}{y} = \frac{5}{3}$

10. $\frac{x - 3}{3x + 2} = \frac{2}{6}$

11. A machine can process 300 parts in 20 minutes. Find how many parts can be processed in 45 minutes.

12. As his consulting fee, Mr. Visconti charges $90.00 per day. Find how much he charges for 3 hours of consulting. Assume an 8-hour work day.

(6.2) *Solve.*

13. In a survey of 100 adults, 37 preferred pepperoni on their pizza. What percent preferred pepperoni?

14. A basketball player made 77 out of 100 attempted free throws. What percent of free throws was made?

Write each percent as a decimal.

15. 83%

16. 75%

17. 73.5%

18. 1.5%

19. 125%

20. 145%

21. 0.5%

22. 0.7%

23. 200%

24. 400%

25. 26.25%

26. 85.34%

Write each decimal as a percent.

27. 2.6

28. 0.055

29. 0.35

30. 1.02

31. 0.725

32. 0.25

33. 0.076

34. 0.085

35. 0.75

36. 0.65

37. 4.00

38. 9.00

Write each percent as a fraction or a mixed number in simplest form.

39. 1%

40. 10%

41. 25%

42. 8.5%

43. 10.2%

44. $16\frac{2}{3}\%$

45. $33\frac{1}{3}\%$

46. 110%

Write each fraction or mixed number as a percent.

47. $\frac{1}{5}$

48. $\frac{7}{10}$

49. $\frac{5}{6}$

50. $\frac{5}{8}$

51. $1\frac{2}{3}$

52. $1\frac{1}{4}$

53. $\frac{3}{5}$

54. $\frac{1}{16}$

55. More and more consumers are following the "90-10 rule," that is, eating healthy foods 90% of the time, but indulging in junk food the other 10% of the time. Write 90% and 10% as fractions. (*Source:* Grocery Manufacturers of America)

56. In medium to large firms in the United States, 96% of full-time employees receive paid vacation benefits. Write 96% as a fraction. (*Source:* U.S. Bureau of Labor Statistics)

57. During a survey recently taken, $\frac{7}{10}$ of Americans said they would give up a gym membership if it meant they could buy a new home. Write $\frac{7}{10}$ as a percent. (*Source:* KRC Research for Century 21)

58. The number of U.S. coffee drinkers drinking cappuccino and other espresso drinks has grown 150% since the early 1990s. Write 150% as a decimal. (*Source:* National Coffee Association)

(6.3) *Translate each question into an equation and solve.*

59. 1250 is 1.25% of what number?

60. What number is $33\frac{1}{3}\%$ of 24,000?

61. 124.2 is what percent of 540?

62. 22.9 is 20% of what number?

63. What number is 40% of 7500?

64. 693 is what percent of 462?

(6.4) *Translate each question into a proportion and solve.*

65. 104.5 is 25% of what number?

66. 16.5 is 5.5% of what number?

67. What number is 36% of 180?

68. 63 is what percent of 35?

69. 93.5 is what percent of 85?

70. What number is 33% of 500?

(6.5) *Solve.*

71. In a survey of 2000 people, it was found that 1320 have a microwave oven. Find the percent of people who own microwaves.

72. Of the 12,360 freshmen entering County College, 2000 are enrolled in Prealgebra. Find the percent of entering freshmen who are enrolled in Prealgebra. Round to the nearest whole percent.

73. The current charge for dumping waste in a local landfill is $16 per cubic foot. To cover new environmental costs, the charge will increase to $33 per cubic foot. Find the percent increase.

74. The number of violent crimes in a city decreased from 675 to 534. Find the percent decrease. Round to the nearest tenth of a percent.

75. This year the fund drive for a charity collected $215,000. Next year, a 4% decrease is expected. Find how much is expected to be collected in next year's drive.

76. A local union negotiated a new contract that increases the hourly pay 15% over last year's pay. The old hourly rate was $11.50. Find the new hourly rate rounded to the nearest cent.

(6.6) *Solve.*

77. If the sales tax rate is 5.5%, what is the total amount charged for a $250 coat?

78. Find the sales tax paid on a $25.50 purchase if the sales tax rate is 4.5%.

79. Russ James is a sales representative for a chemical company and is paid a commission rate of 5% on all sales. Find his commission if he sold $100,000 worth of chemicals last month.

80. Carol Sell is a sales clerk in a clothing store. She receives a commission of 7.5% on all sales. Find her commission for the week if her sales for the week were $4005. Round to the nearest cent.

81. A \$3000 mink coat is on sale for 30% off. Find the discount and the sale price.

82. A \$90 calculator is on sale for 10% off. Find the discount and the sale price.

(6.7) *Solve.*

83. Find the simple interest due on \$4000 loaned for 3 months at 12% interest.

84. Find the total amount due on an 8-month loan of \$1200 at a simple interest rate of 15%.

85. Find the total amount in an account if \$5500 is compounded annually at 12% for 15 years.

86. Find the total amount in an account if \$6000 is compounded semiannually at 11% for 10 years.

87. Find the compound interest earned if \$100 is compounded quarterly at 12% for 5 years.

88. Find the compound interest earned if \$1000 is compounded quarterly at 18% for 20 years.

Name ______________________ Section ____________ Date ____________

Chapter 6 Test

Remember to check your answers and use the Chapter Test Prep Video to view solutions.

Write each percent as a decimal.

1. 85%

2. 500%

3. 0.6%

Write each decimal as a percent.

4. 0.056

5. 6.1

6. 0.35

Write each percent as a fraction or a mixed number in simplest form.

7. 120%

8. 38.5%

9. 0.2%

Write each fraction or mixed number as a percent.

10. $\frac{11}{20}$

11. $\frac{3}{8}$

12. $1\frac{3}{4}$

13. Sales of bottled water have recently surged. Bottled water accounts for $\frac{1}{5}$ of the total noncarbonated beverage category in convenience stores. Write $\frac{1}{5}$ as a percent. (*Source:* Grocery Manufacturers of America)

14. In small firms in the United States, 64% of full-time employees receive medical insurance benefits. Write 64% as a fraction. (*Source:* U.S. Bureau of Labor Statistics)

Solve.

15. What number is 42% of 80?

16. 0.6% of what number is 7.5?

17. 567 is what percent of 756?

Answers

1. ____________

2. ____________

3. ____________

4. ____________

5. ____________

6. ____________

7. ____________

8. ____________

9. ____________

10. ____________

11. ____________

12. ____________

13. ____________

14. ____________

15. ____________

16. ____________

17. ____________

18. ______

19. ______

20. ______

21. ______

22. ______

23. ______

24. ______

25. ______

26. ______

27. ______

28. ______

29. ______

30. ______

31. ______

Solve. If necessary, round percents to the nearest tenth, dollar amounts to the nearest cent, and all other numbers to the nearest whole.

18. An alloy is 12% copper. How much copper is contained in 320 pounds of this alloy?

19. A farmer in Nebraska estimates that 20% of his potential crop, or \$11,350, has been lost to a hard freeze. Find the total value of his potential crop.

20. If the local sales tax rate is 1.25%, find the total amount charged for a stereo system priced at \$354.

21. A town's population increased from 25,200 to 26,460. Find the percent increase.

22. A \$120 framed picture is on sale for 15% off. Find the discount and the sale price.

23. Randy Nguyen is paid a commission rate of 4% on all sales. Find Randy's commission if his sales were \$9875.

24. A sales tax of \$1.53 is added to an item's price of \$152.99. Find the sales tax rate. Round to the nearest whole percent.

25. Find the simple interest earned on \$2000 saved for $3\frac{1}{2}$ years at an interest rate of 9.25%.

26. \$1365 is compounded annually at 8%. Find the total amount in the account after 5 years.

27. A couple borrowed \$400 from a bank at 13.5% for 6 months for car repairs. Find the total amount due the bank at the end of the 6-month period.

28. The number of crimes reported in New York City was 125,587 in the first half of 2001 and 118,346 in the first half of 2002. Find the percent decrease in the number of crimes in New York City from 2001 to 2002. (*Source*: Federal Bureau of Investigation. Uniform Crime Reports, January–June 2002)

29. Write the ratio \$75 to \$10 as a fraction in simplest form.

30. Solve: $\frac{5}{y+1} = \frac{4}{y+2}$

31. In a sample of 85 fluorescent bulbs, 3 were found to be defective. At this rate, how many defective bulbs should be found in 510 bulbs?

Name ______________________ Section __________ Date __________

Chapter 6 Cumulative Review

1. Multiply: 236×86

2. Multiply: 409×76

3. Subtract: 7 from -3

4. Subtract: -2 from 8.

5. Solve: $x - 2 = 1$

6. Solve: $x + 4 = 3$

7. Solve: $3(2x - 6) + 6 = 0$

8. Solve: $5(x - 2) = 3x$

9. Write 3 as an equivalent fraction whose denominator is 7.

10. Write 8 as an equivalent fraction whose denominator is 5.

11. Simplify: $\frac{84x}{90}$

12. Simplify: $\frac{10y}{32}$

13. Divide: $-\frac{5}{16} \div -\frac{3}{4}$

14. Divide: $\frac{-2}{5} \div \frac{7}{10}$

15. Evaluate $y - x$ if $x = -\frac{3}{10}$ and $y = -\frac{8}{10}$.

16. Evaluate $2x + 3y$ if $x = \frac{2}{5}$ and $y = \frac{-1}{5}$.

17. Find: $-\frac{3}{4} - \frac{1}{14} + \frac{6}{7}$

18. Find: $\frac{2}{9} + \frac{7}{15} - \frac{1}{3}$

19. Simplify: $\dfrac{\frac{1}{2} + \frac{3}{8}}{\frac{3}{4} - \frac{1}{6}}$

20. Simplify: $\dfrac{\frac{2}{3} + \frac{1}{6}}{\frac{3}{4} - \frac{3}{5}}$

21. Solve: $\frac{x}{2} = \frac{x}{3} + \frac{1}{2}$

22. Solve: $\frac{x}{2} + \frac{1}{5} = 3 - \frac{x}{5}$

23. Write each mixed number as an improper fraction.

a. $4\frac{2}{9}$ **b.** $1\frac{8}{11}$

24. Write each mixed number as an improper fraction.

a. $3\frac{2}{5}$ **b.** $6\frac{2}{7}$

Answers

1. ______
2. ______
3. ______
4. ______
5. ______
6. ______
7. ______
8. ______
9. ______
10. ______
11. ______
12. ______
13. ______
14. ______
15. ______
16. ______
17. ______
18. ______
19. ______
20. ______
21. ______
22. ______
23. a. ______
b. ______
24. a. ______
b. ______

25. ______

26. ______

27. ______

28. ______

29. ______

30. ______

31. ______

32. ______

33. ______

34. ______

35. ______

36. ______

37. ______

38. ______

39. ______

40. ______

41. ______

42. ______

43. ______

44. ______

45. ______

46. ______

Write each decimal as a fraction or mixed number. Write your answer in simplest form.

25. 0.125

26. 0.85

27. -105.083

28. 17.015

29. Subtract: $85 - 17.31$

30. Subtract: $38 - 10.06$

Multiply.

31. 7.68×10

32. 12.483×100

33. $(-76.3)(1000)$

34. -853.75×10

35. Is 720 a solution of the equation $\frac{y}{100} = 7.2$?

36. Is 470 a solution of the equation $\frac{x}{100} = 4.75$?

37. Find: $\sqrt{\frac{4}{25}}$

38. Find: $\sqrt{\frac{9}{16}}$

39. Write the ratio of 2.5 to 3.15 as a fraction in simplest form.

40. Write the ratio of 5.8 to 7.6 as a fraction in simplest form.

41. On a Chamber of Commerce map of Abita Springs, 5 miles corresponds to 2 inches. How many miles correspond to 7 inches?

42. A student doing math homework can complete 7 problems in about 6 minutes. At this rate, how many problems can be completed in 30 minutes?

Write each percent as a fraction or mixed number in simplest form.

43. 1.9%

44. 2.3%

45. $33\frac{1}{3}\%$

46. 108%

Graphing and Introduction to Statistics

CHAPTER 7

Numerical data is all around us—in newspaper and magazine articles and in television news reports. Frequently, numerical data is summarized in graphs and basic statistics. To understand this data, we must first know how to read graphs and interpret basic statistics. We learn to do just that in this chapter.

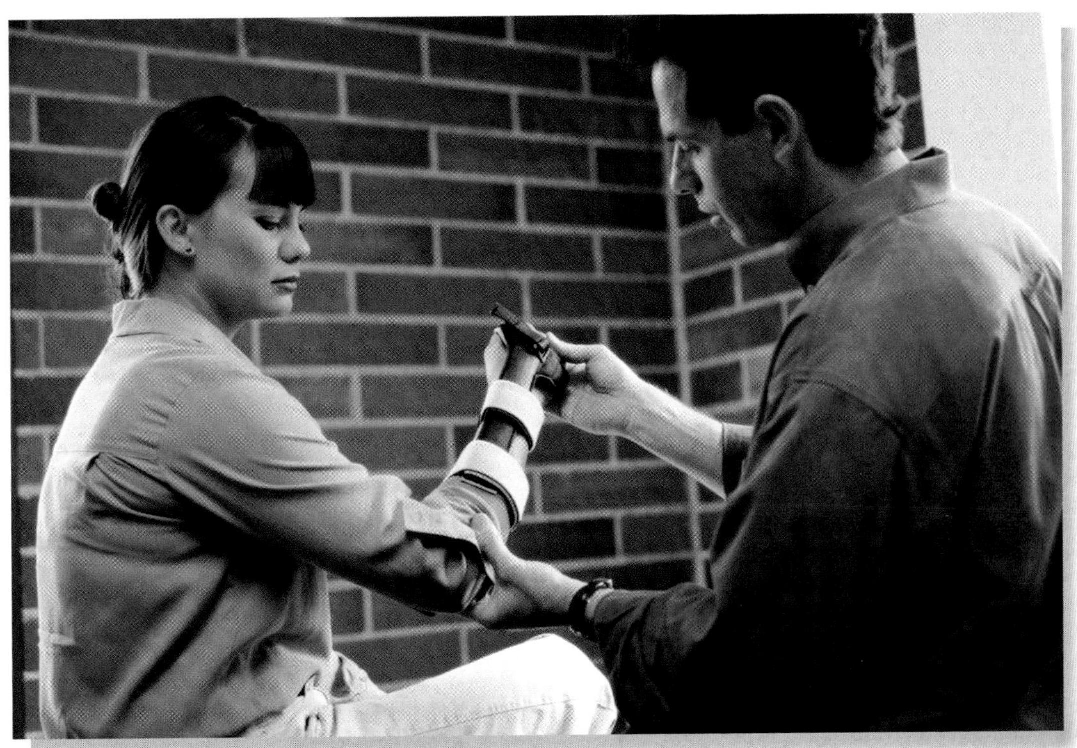

Carpal Tunnel Syndrome (CTS) is a disorder of the wrist and hand associated with pain and numbness in the fingers. It is caused by the compression of the median nerve inside the carpal tunnel, a small passage inside the wrist. Doctors can treat CTS by applying a splint to the wrist and prescribing an anti-inflammatory drug, such as ibuprofen. In some cases, however, surgery is required to enlarge the carpal tunnel and relieve the compression on the median nerve. Most people associate CTS with repetitive wrist motion, such as clicking a computer mouse. CTS can also occur as a result of fractures, arthritis, pregnancy, or diabetes. It is one of the most common work-related injuries. In Exercise 27, Section 7.2, we see how a circle graph can be used to describe the number of days of absence from work due to carpal tunnel syndrome.

Answers
1. ______
2. ______
3. ______
4. ______
5. ______
6. ______
7. ______
8. ______
9. ______
10. ______
11. ______
12. ______
13. ______

Name ______________ Section ________ Date ________

Chapter 7 Pretest

The line graph below shows the number of burglaries in a town during the months of March through September.

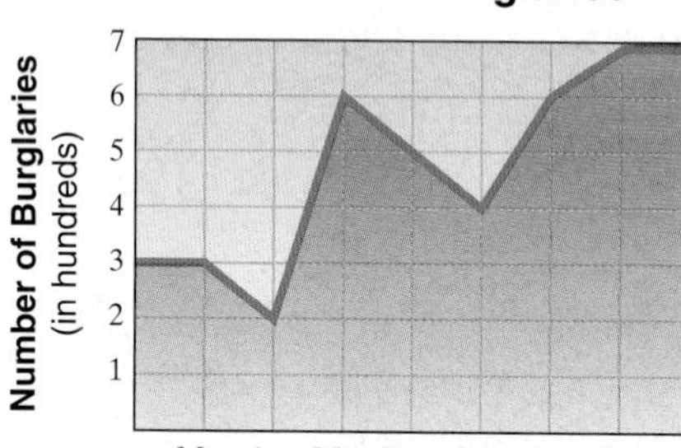

1. During which month, between March and September, did the fewest number of burglaries occur?

2. During which month, between March and September, were there 400 burglaries?

3. How many burglaries were there in September?

4. The following table shows a breakdown of an average day for Dawn Miller.

Attending college classes	4 hours
Studying	3 hours
Working	5 hours
Sleeping	8 hours
Driving	1 hour
Other	3 hours

Draw a circle graph showing this data.

Below is a list of scores from the final exam given in Mrs. Maxwell's basic college mathematics class. Use this list to complete the table below:

76 71 94 73 81 78 96 65
95 80 90 86 98 88 62 91

	Class Intervals (Scores)	Tally	Class Frequency (Number of Exams)
5.	60–69		
6.	70–79		
7.	80–89		
8.	90–99		

9. Use the table from Exercises 5 through 8 to draw a histogram.

10. Find the mean, the median, and the mode for the given set of numbers. If necessary round the mean to one decimal place.
28, 36, 81, 73, 28, 74, 31, 74, 64, 25, 74

A single die is tossed. Find the probability that the die lands showing each of the following.

11. a 4

12. a number greater than 3

13. a 3 or a 5

7.1 Reading Pictographs, Bar Graphs, and Line Graphs

OBJECTIVES

- (A) Read pictographs.
- (B) Read and construct bar graphs.
- (C) Read and construct histograms.
- (D) Read line graphs.

Often data is presented visually in a graph. In this section, we practice reading several kinds of graphs including pictographs, bar graphs, and line graphs.

(A) Reading Pictographs

A **pictograph** such as the one below is a graph in which pictures or symbols are used to represent some fixed amount. This type of graph contains a key that explains the meaning of the symbol used. An advantage of using a pictograph to display information is that comparisons can easily be made. A disadvantage of using a pictograph is that it is often hard to tell what fractional part of a symbol is shown. For example, in the pictograph below, Germany shows three complete symbols and a part of a fourth symbol. Notice that it's hard to read with any accuracy what fractional part of the 4th symbol is shown.

EXAMPLE 1 The following pictograph shows the approximate amount of nuclear energy generated by selected countries in the year 2001. Use this pictograph to answer the questions.

Nuclear Energy Generated by Selected Countries (2001)

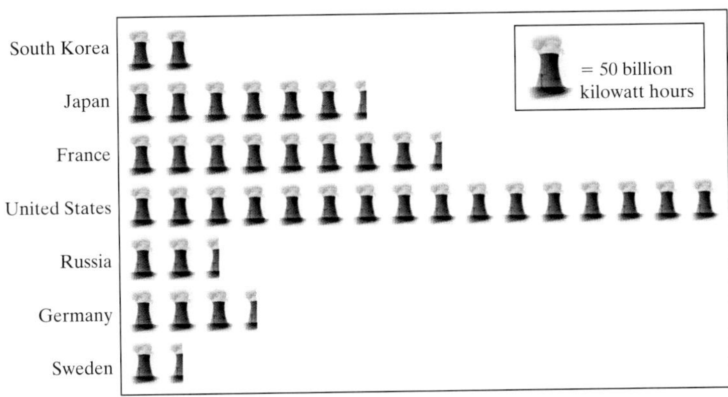

Source: Energy Information Administration

a. Approximate the amount of nuclear energy that is generated in South Korea.

b. Approximate how much more nuclear energy is generated in France than in South Korea.

Solution:

a. South Korea corresponds to 2 symbols, and each symbol represents 50 billion kilowatt hours of energy. This means that South Korea generates approximately $2 \cdot (50 \text{ billion})$ or 100 billion kilowatt hours of nuclear energy.

b. France shows about $6\frac{1}{2}$ more symbols than South Korea. This means that France generates $6\frac{1}{2} \cdot (50 \text{ billion})$ or 325 billion more kilowatt hours of nuclear energy than South Korea.

Practice Problem 1

Use the pictograph shown in Example 1 to answer the following questions:

a. Approximate the amount of nuclear energy that is generated in Sweden.

b. Approximate the total nuclear energy generated in Sweden and Russia.

(B) Reading and Constructing Bar Graphs

Another way to present data graphically is with a **bar graph.** Bar graphs can appear with vertical bars or horizontal bars. Although we have studied bar

Answers

1. a. 75 billion kilowatt hours **b.** 200 billion kilowatt hours

graphs in previous sections, we now practice reading the height of the bars contained in a bar graph. An advantage to using bar graphs is that a scale is usually included for greater accuracy. Care must be taken when reading bar graphs, as well as other types of graphs—they may be misleading, as shown later in this section.

Practice Problem 2

Use the bar graph in Example 2 to answer the following questions:

a. Approximate the number of endangered species that are insects.

b. Which category (or categories) shows the fewest endangered species?

EXAMPLE 2 The following bar graph shows the number of endangered species in 2003. Use this graph to answer the questions.

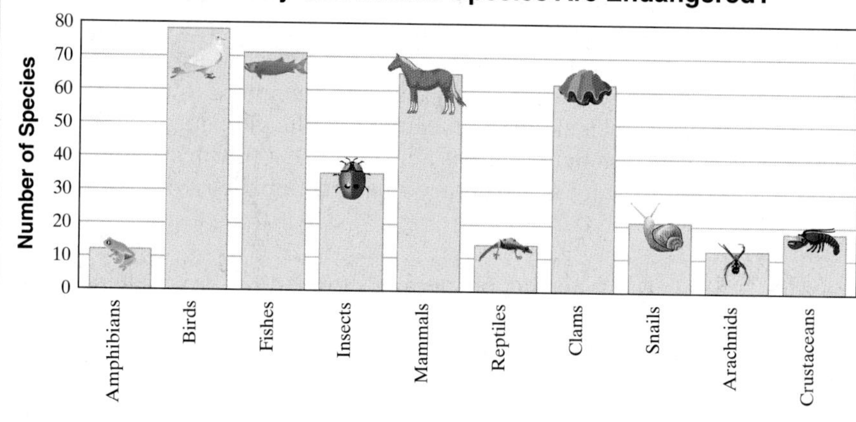

Source: U.S. Fish and Wildlife Service

a. Approximate the number of endangered species that are reptiles.

b. Which category has the most endangered species?

Solution:

a. To approximate the number of endangered species that are reptiles, we go to the top of the bar that represents reptiles. From the top of this bar, we move horizontally to the left until the scale is reached. We read the height of the bar on the scale as approximately 14. There are approximately 14 reptile species that are endangered, as shown in the next figure.

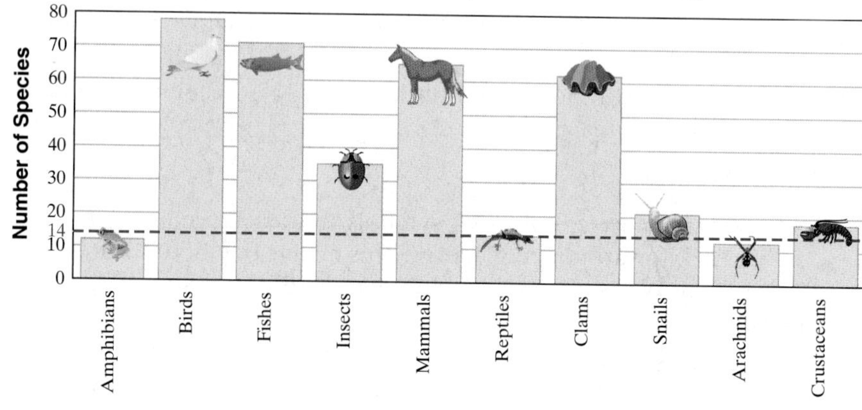

Source: U.S. Fish and Wildlife Service

b. The most endangered species is represented by the tallest bar. The tallest bar corresponds to birds. ●

Answers

2. a. 35 **b.** amphibians and arachnids

As mentioned previously, graphs can be misleading. Both graphs below show the same information, but with different scales. Special care should be taken when forming conclusions from the appearance of a graph. (Recall that the ⩘ symbol on the vertical line with the scale means that some values have been omitted.)

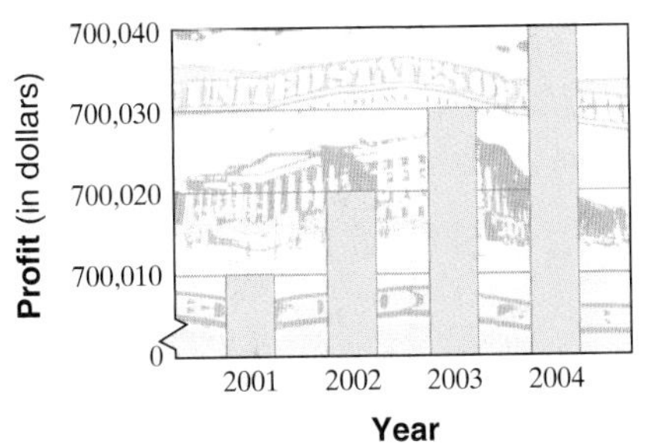

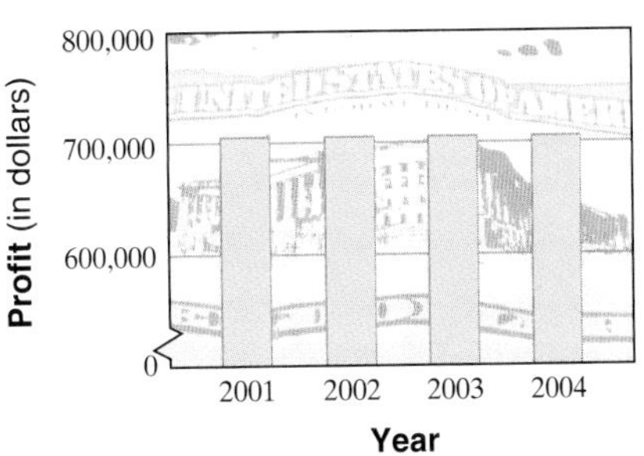

Are profits shown in the graphs above greatly increasing, or are they remaining about the same?

Next, we practice constructing a bar graph.

EXAMPLE 3 Draw a vertical bar graph using the information in the table below that gives the caffeine content of selected foods.

Average Caffeine Content of Selected Foods			
Food	**Milligrams**	**Food**	**Milligrams**
Brewed coffee (percolator, 8 ounces)	124	Instant coffee (8 ounces)	104
Brewed decaffeinated coffee (8 ounces)	3	Brewed tea (U.S. brands, 8 ounces)	64
Coca-Cola classic (8 ounces)	31	Mr. Pibb (8 ounces)	27
Dark chocolate (semi sweet, $1\frac{1}{2}$ ounces)	30	Milk chocolate (8 ounces)	9

(*Sources:* International Food Information Council and the Coca-Cola Company)

Solution: We draw and label a vertical line and a horizontal line as shown at the top of the next page. These lines are also called axes. We place the different food categories along the horizontal axis. Along the vertical axis, we place a scale.

There are many choices of scales that would be appropriate. Notice that the milligrams range from a low of 3 to a high of 124. From this information, we use a scale that starts at 0 and then shows multiples of 20 so that the scale is not too cluttered. The scale stops at 140, the smallest multiple of 20 that will allow all milligrams to be graphed. It may also be helpful to draw horizontal lines along the scale markings to help draw the vertical bars at the correct height. The finished bar graph is shown on the top of the next page.

Practice Problem 3

Draw a vertical bar graph using the information in the table about electoral votes for selected states.

Electoral Votes for President by Selected States	
State	**Electoral Votes**
Texas	34
California	55
Florida	27
Nebraska	5
Indiana	11
Georgia	15

(*Source: World Almanac 2003*)

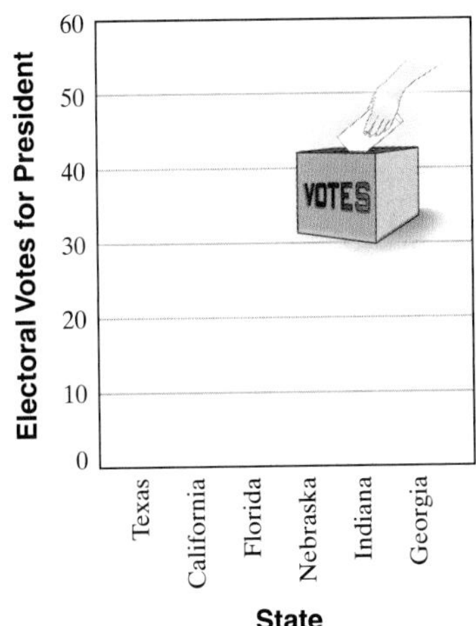

Answer

3.

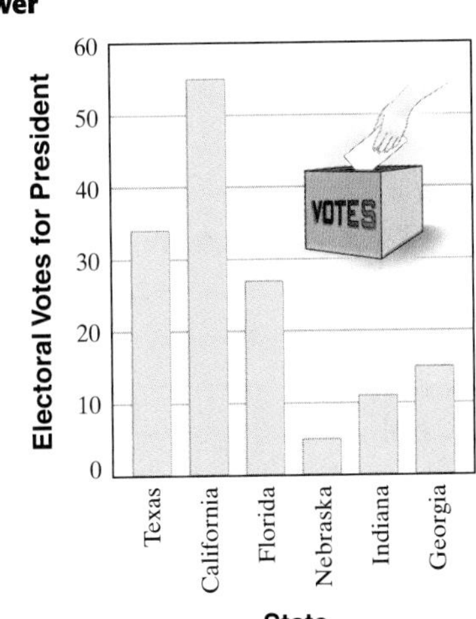

Helpful Hint

Can you see from the bar graph to the right that a scale too small may be too time-consuming to draw and too hard to read? Think of a scale that starts at 0 and shows multiples of 2. Think of the number of tick marks that would be needed.

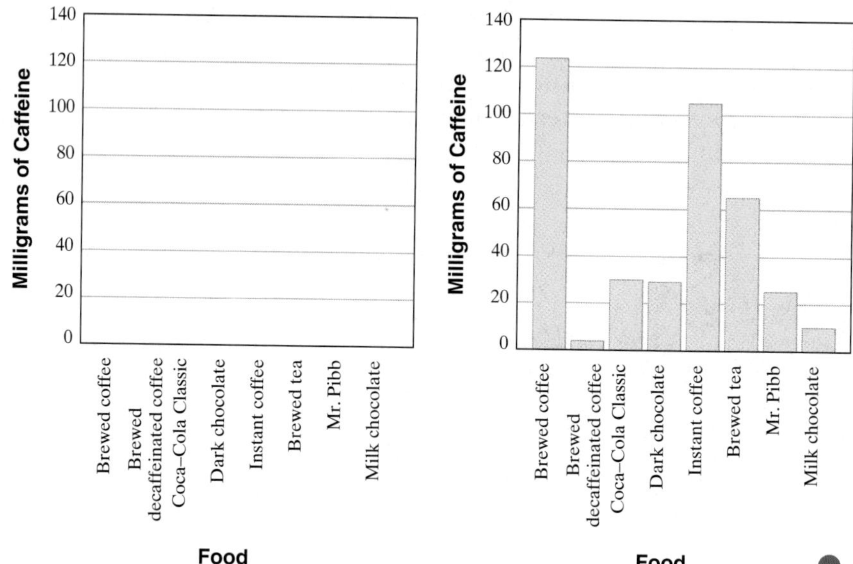

C Reading and Constructing Histograms

Suppose that the test scores of 36 students are summarized in the table below:

Student Scores	Frequency (number of students)
40–49	1
50–59	3
60–69	2
70–79	10
80–89	12
90–99	8

The results in the table can be displayed in a histogram. A **histogram** is a special bar graph. The width of each bar represents a range of numbers called a **class interval.** The height of each bar corresponds to how many times a number in the class interval occurred and is called the **class frequency.** The bars in a histogram lie side by side with no space between them.

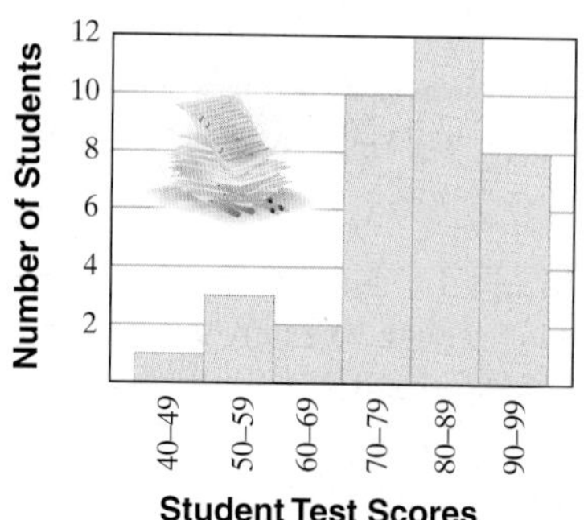

EXAMPLE 4 Use the preceding histogram to determine how many students scored 50–59 on the test.

Solution: We find the bar representing 50–59. The height of this bar is 3, which means 3 students scored 50–59 on the test.

Practice Problem 4

Use the histogram on the right to determine how many students scored 70–79 on the test.

Answer

4. 10

EXAMPLE 5 Use the preceding histogram to determine how many students scored 80 or above on the test.

Solution: We see that two different bars fit this description. There are 12 students who scored 80–89 and 8 students who scored 90–99. The sum of these two categories is 12 + 8 or 20 students. Thus, 20 students scored 80 or above on the test.

Now we will look at a way to construct histograms.

The daily high temperatures for 1 month in New Orleans, Louisiana, are recorded in the following list:

85°	90°	95°	89°	88°	94°
87°	90°	95°	92°	95°	94°
82°	92°	96°	91°	94°	92°
89°	89°	90°	93°	95°	91°
88°	90°	88°	86°	93°	89°

The data in this list have not been organized and can be hard to interpret. One way to organize the data is to place it in a **frequency distribution table.** We will do this in Example 6.

EXAMPLE 6 Complete the frequency distribution table for the preceding temperature data.

Solution: Go through the data and place a tally mark next to the class interval (in the second table column). Then count the tally marks and write each total in the third table column.

Class Intervals (Temperatures)	Tally	Class Frequency (Number of Days)
82°–84°	|	1
85°–87°	|||	3
88°–90°	~~||||~~ ~~||||~~ |	11
91°–93°	~~||||~~ ||	7
94°–96°	~~||||~~ |||	8

EXAMPLE 7 Construct a histogram from the frequency distribution table in Example 6.

Solution:

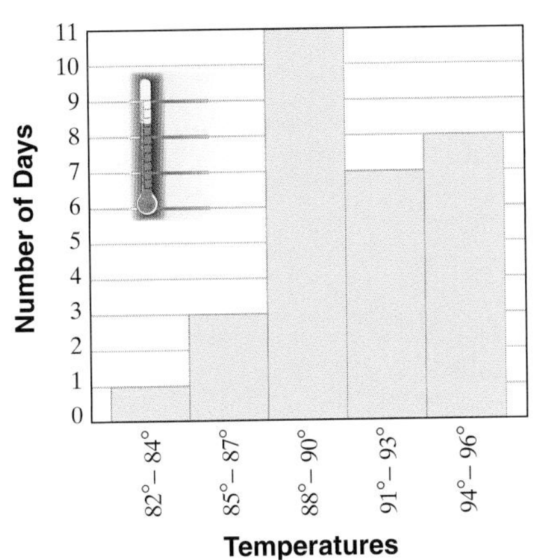

Practice Problem 5

Use the histogram for Example 4 to determine how many students scored less than 60 on the test.

Practice Problem 6

Complete the frequency distribution table for the data below. Each number represents a credit card owner's unpaid balance each month.

0	53	89	125
265	161	37	76
62	201	136	42

Class Intervals (Credit Card Balances)	Tally	Class Frequency (Number of Months)
\$0–\$49	____	____
\$50–\$99	____	____
\$100–\$149	____	____
\$150–\$199	____	____
\$200–\$249	____	____
\$250–\$299	____	____

Practice Problem 7

Construct a histogram from the frequency distribution table above.

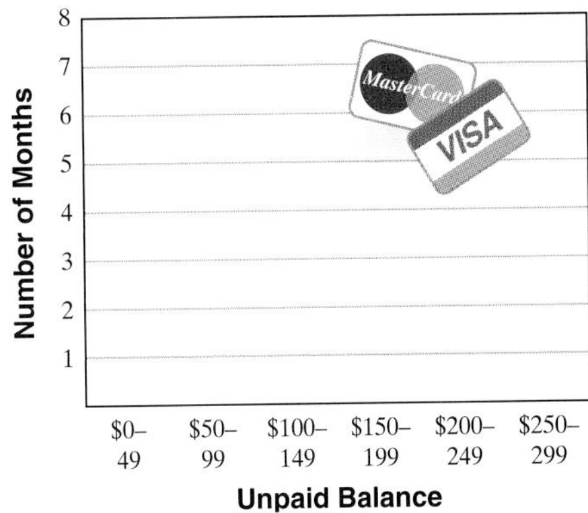

Answers

5. 4

6.

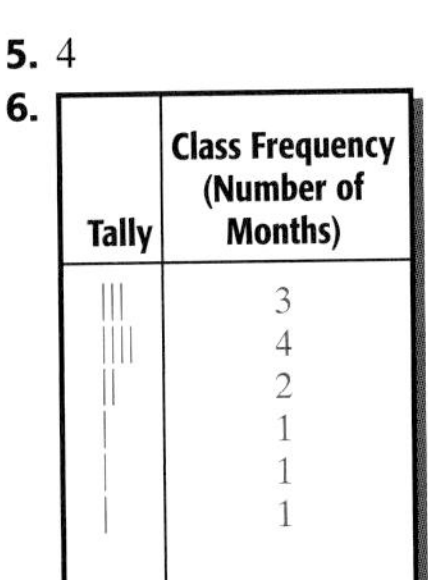

Tally	Class Frequency (Number of Months)
|||	3
||||	4
||	2
|	1
|	1
|	1

7.

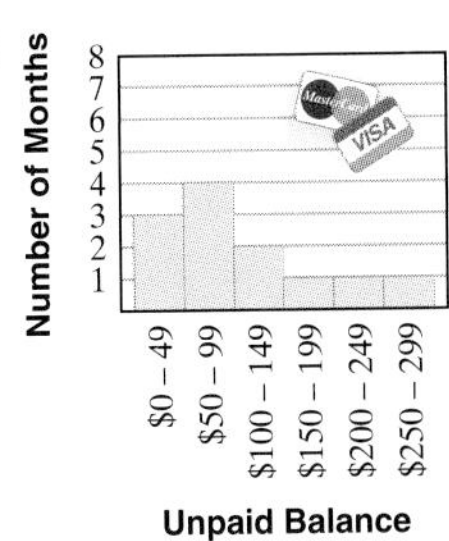

Concept Check

Which of the following sets of data is better suited to representation by a histogram? Explain.

Grade on Final	# of Students
51–60	12
61–70	18
71–80	29
81–90	23
91–100	25

Section Number	Avg. Grade on Final
150	78
151	83
152	87
153	73

Practice Problem 8

Use the temperature graph in Example 8 to answer the following questions:

a. During what month is the average daily high temperature the lowest?

b. During what month is the average daily high temperature 25°F?

c. During what months is the average daily high temperature greater than 70°F?

Answers

8. a. January **b.** December
c. June, July, and August

Concept Check: first set of data

Try the Concept Check in the margin.

D Reading Line Graphs

Another common way to display information graphically is by using a **line graph.** An advantage of a line graph is that it can be used to visualize relationships between two quantities. A line graph can also be very useful in showing change over time.

EXAMPLE 8 The following line graph shows the average daily high temperature for each month for Omaha, Nebraska. Use this graph to answer the questions.

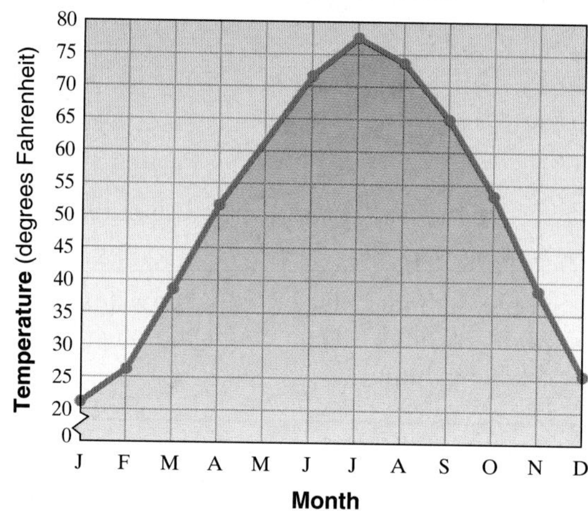

Source: National Climatic Data Center

a. During what month is the average daily high temperature the highest?

b. During what month is the average daily high temperature 65°F?

c. During what months is the average daily high temperature less than 30°F?

Solution:

a. The month with the highest temperature corresponds to the highest point on the graph. We follow the highest point downward to the horizontal month scale and see that this point corresponds to July.

b. We find the 65°F mark on the vertical axis and move to the right until a darkened point on the graph is reached. From that point, we move downward to the Month axis and read the corresponding month. For the month of September, the average daily high temperature was 65°F.

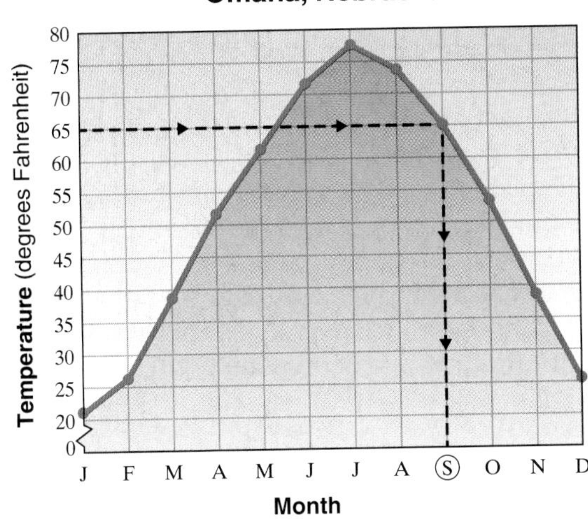

Source: National Climatic Data Center

c. To see what months the temperature is less than 30°F, we find what months correspond to darkened points that fall below the 30°F mark on the vertical axis. These months are January, February, and December.

Average Daily High Temperature in Omaha, Nebraska

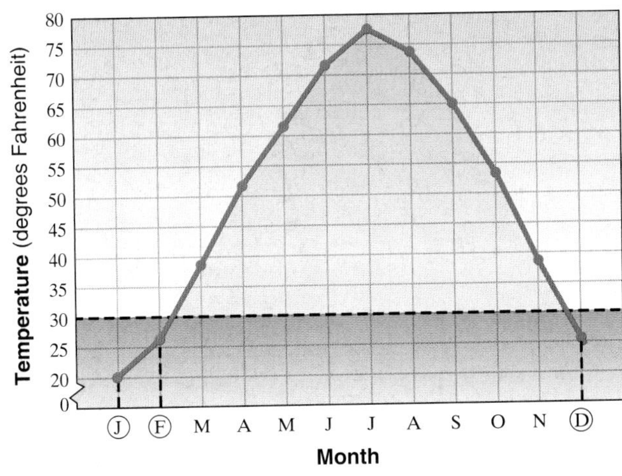

Source: National Climatic Data Center

FOCUS ON Mathematical Connections

SCATTER DIAGRAMS

In Section 7.1, we learned about presenting data visually in a graph, such as a pictograph, bar graph, or line graph. Now we learn about another type of graph, based on ordered pairs, that can be used to present data.

Data that can be represented as an ordered pair is called paired data. Many types of data collected from the real world are paired data. For instance, the total amount of rainfall a location receives each year can be written as an ordered pair of the form (year, total rainfall in inches) and is paired data. The graph of paired data as points in the rectangular coordinate system is called a **scatter diagram,** or scatter plot. Scatter diagrams can be used to look for patterns and trends in paired data.

For example, the data shown in the table is paired data that can be written as a set of ordered pairs of the form (year, number of restaurants in thousands), such as (1998, 25) and (1999, 26). A scatter diagram of the paired data is shown below. Notice that the horizontal axis is labeled "Year" to describe the x-coordinates in the ordered pairs. The vertical axis is labeled "Number of Restaurants (in thousands)" to describe the y-coordinates in the ordered pairs.

Number of McDonald's Restaurants

Year	Number of Restaurants (in thousands)
1998	25
1999	26
2000	29
2001	30
2002	31

(*Source:* McDonald's Corporation)

Number of McDonald's Restaurants Worldwide

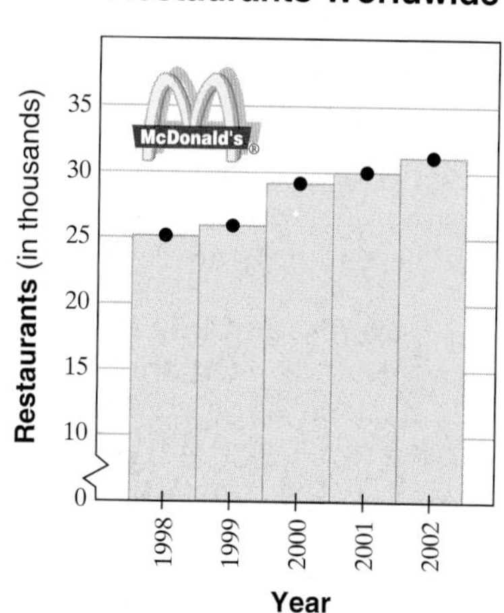

CRITICAL THINKING

The table gives the annual revenue of McDonald's restaurants in operation worldwide each year. Use the table to answer the following questions.

1. Write this paired data as a set of ordered pairs of the form (year, revenue of restaurants in millions).
2. Create a scatter diagram of the paired data.
3. What trend in the paired data does the scatter diagram show?

Annual Revenue of McDonald's Restaurants Worldwide (in millions)

Year	Annual Revenue of Restaurants (in millions)
1998	\$36
1999	\$38
2000	\$40
2001	\$41
2002	\$42

(*Source:* McDonald's Corporation)

GROUP ACTIVITY

4. Find or collect your own paired data and present it in a scatter diagram. What does the graph show?

Name ________________ Section ________ Date ________

EXERCISE SET 7.1

A *The following pictograph shows the annual automobile production by one plant for the years 1997–2003. Use this graph to answer Exercises 1 through 8. See Example 1.*

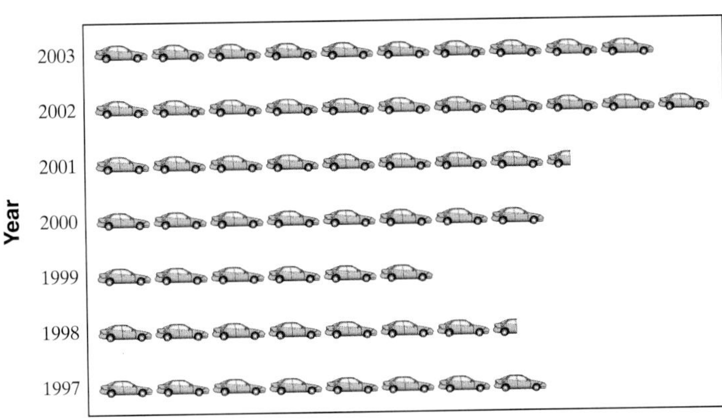

1. In what year was the greatest number of cars manufactured?

2. In what year was the least number of cars manufactured?

3. Approximate the number of cars manufactured in the year 2000.

4. Approximate the number of cars manufactured in the year 2001.

5. In what year(s) did the production of cars decrease from the previous year?

6. In what year(s) did the production of cars increase from the previous year?

7. In what year(s) were 4000 cars manufactured?

8. In what year(s) were 5500 cars manufactured?

The following pictograph shows the average number of ounces of chicken consumed per person per week in the United States. Use this graph to answer Exercises 9 through 16. See Example 1.

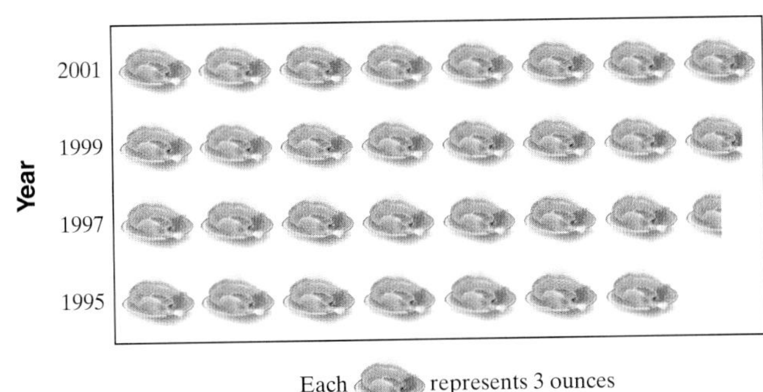

Source: National Agricultural Statistics Service

9. Approximate the number of ounces of chicken consumed per week in 1997.

10. Approximate the number of ounces of chicken consumed per week in 2001.

11. In what year(s) was the number of ounces of chicken consumed per week greater than 21 ounces?

12. In what year(s) was the number of ounces of chicken consumed per week 21 ounces or less?

13. What was the increase in average chicken consumption from 1995 to 2001?

14. What was the increase in average chicken consumption from 1997 to 2001?

15. Describe a trend in eating habits shown by this graph.

16. In 2001, did you consume less than, greater than, or about the same as the U.S. average number of ounces consumed per week?

B *The following bar graph shows the average number of people killed by tornadoes during the months of the year. Use this graph to answer Exercises 17 through 22. See Example 2.*

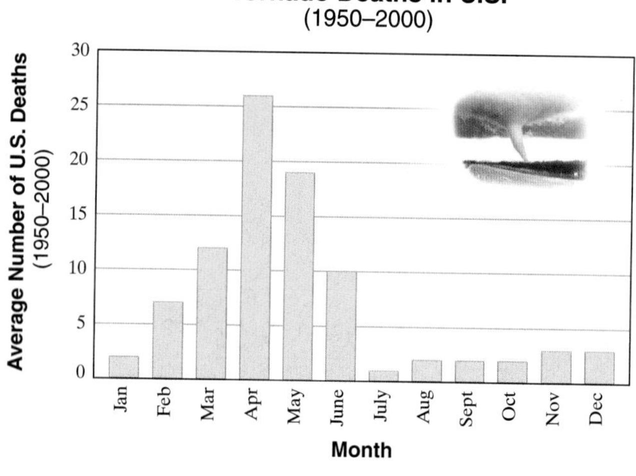

Source: Storm Prediction Center

17. In which month(s) did the most tornado-related deaths occur?

18. In which month(s) did the fewest tornado-related deaths occur?

19. Approximate the number of tornado-related deaths that occurred in May.

20. Approximate the number of tornado-related deaths that occurred in April.

21. In which month(s) did more than 5 deaths occur?

22. In which month(s) did more than 15 deaths occur?

The following horizontal bar graph shows the 2001 population of the world's largest cities (including their suburbs). Use this graph to answer Exercises 23 through 28. See Example 2.

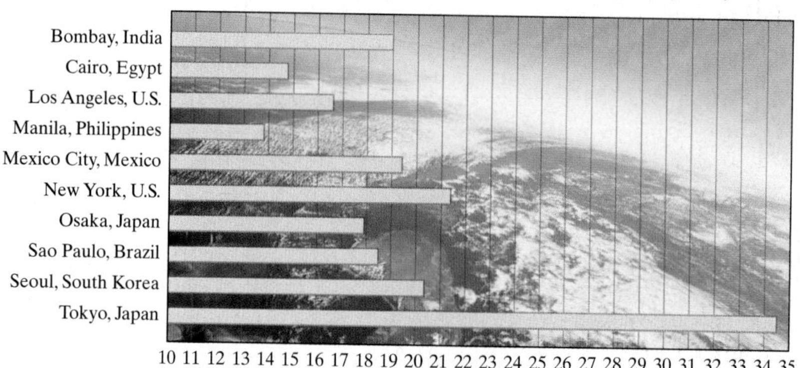

Source: Thomas Brinkhoff: *The Principal Agglomerations of the World*, http://www.citypopulation.de, 13.05.2001

23. Name the city with the largest population and estimate its population.

24. Name the city whose population is between 16 and 17 million and estimate its population.

25. Name the city in the United States with the largest population and estimate its population.

26. Name the city with the smallest population and estimate its population.

27. How much larger is Tokyo than Sao Paulo?

28. How much larger is Bombay than Manila?

Use the information given to draw a vertical bar graph. Clearly label the bars. See Example 3.

29.

Fiber Content of Selected Foods	
Food	**Grams of Total Fiber**
Kidney beans ($\frac{1}{2}$c)	4.5
Oatmeal, cooked ($\frac{3}{4}$c)	3.0
Peanut butter, chunky (2 tbsp)	1.5
Popcorn (1 c)	1.0
Potato, baked with skin (1 med)	4.0
Whole wheat bread (1 slice)	2.5

(*Sources:* American Dietetic Association and National Center for Nutrition and Dietetics)

Fiber Content of Selected Foods

Grams of Total Fiber

2
1
0

Food

30.

U.S. Restaurant Industry Annual Food and Beverage Sales	
Year	**Sales in Billions of Dollars**
1970	43
1980	120
1990	239
2001	399

(*Source:* National Restaurant Association)

U.S. Restaurant Industry Annual Food and Beverage Sales

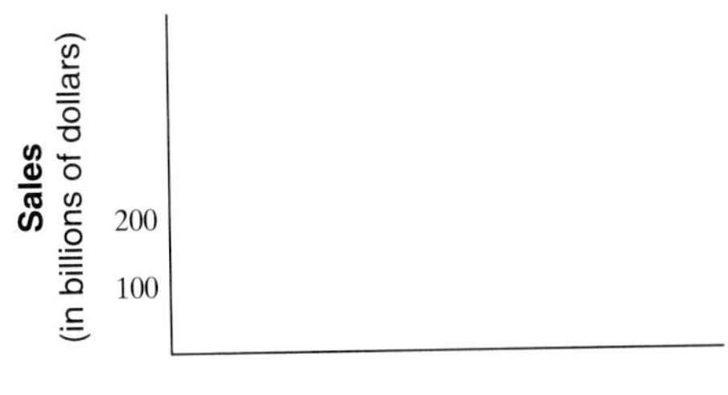

31.

Best-selling Albums of All Time (U.S. Sales)	
Album	**Estimated Sales (in millions)**
Shania Twain: *Come on Over* (1997)	19
Pink Floyd: *The Wall* (1979)	23
Michael Jackson: *Thriller* (1982)	26
AC/DC: *Back in Black* (1980)	19
Billy Joel: *Greatest Hits Volumes I & II* (1985)	21
Eagles: *Their Greatest Hits* (1976)	27
Led Zeppelin: *Led Zeppelin IV* (1971)	22

(*Source:* Recording Industry Association of America)

Best-selling Albums of All-Time
(U.S. sales)

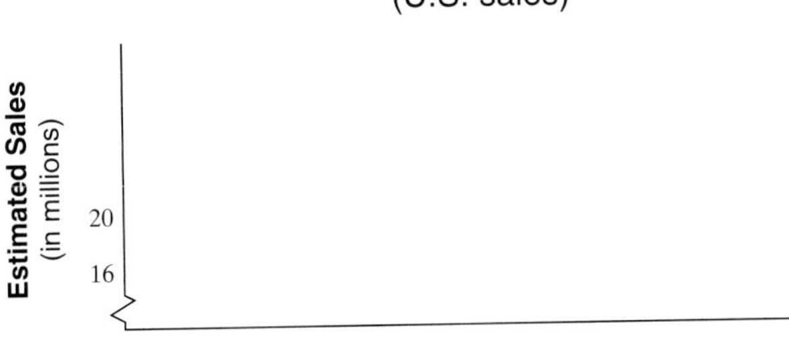

32.

Fuel Economy of the Top-Selling Vehicles in the United States for 2003*	
Vehicle (sales rank)	**Highway Fuel Economy* (in miles per gallon)**
Ford F-Series (1)	20
Chevrolet Silverado (2)	21
Dodge Ram (3)	21
Toyota Camry (4)	33
Honda Accord (5)	34
Ford Explorer (6)	22

*Maximum fuel economy available among all model trims, through June 2003
(*Sources:* U.S. Dept. of Energy, Reuters)

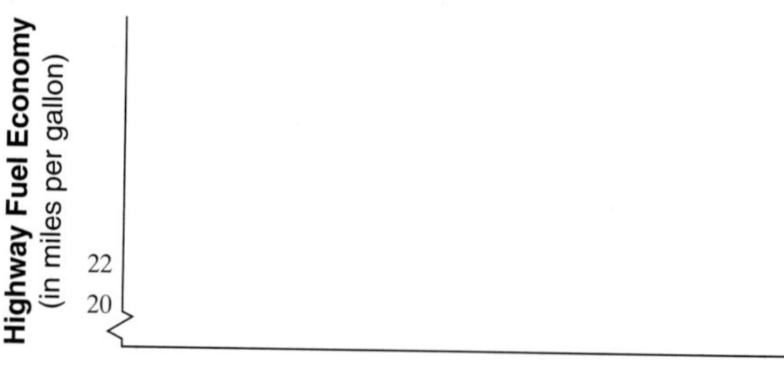

C *The following histogram shows the number of miles that each adult, from a survey of 100 adults, drives per week. Use this histogram to answer Exercises 33 through 42. See Examples 4 and 5.*

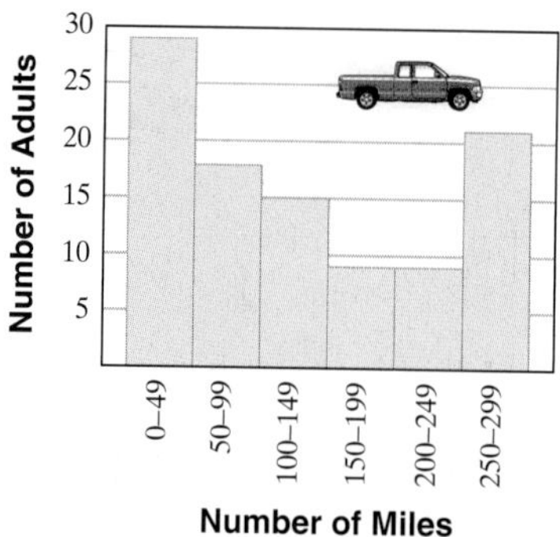

33. How many adults drive 100–149 miles per week?

34. How many adults drive 200–249 miles per week?

35. How many adults drive fewer than 150 miles per week?

36. How many adults drive 200 miles or more per week?

37. How many adults drive 100–199 miles per week?

38. How many adults drive 0–149 miles per week?

39. How many more adults drive 250–299 miles per week than 200–249 miles per week?

40. How many more adults drive 0–49 miles per week than 50–99 miles per week?

41. What is the ratio of adults who drive 150–199 miles per week to the total number of adults surveyed?

42. What is the ratio of adults who drive 50–99 miles per week to the total number of adults surveyed?

The following histogram shows the projected ages of householders for the year 2005. Use this histogram to answer Exercises 43 through 50. See Examples 4 and 5.

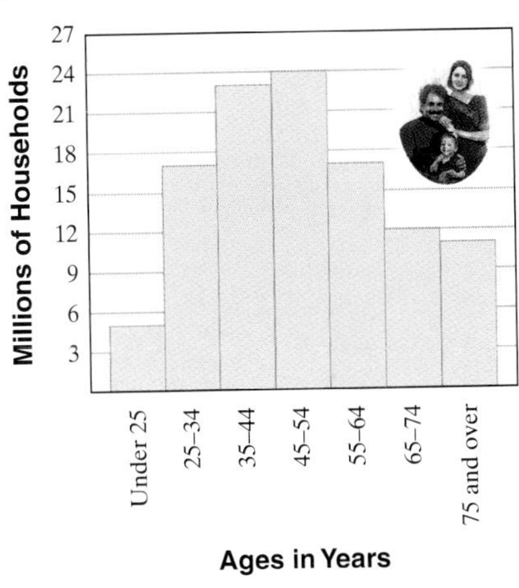

Source: U.S. Bureau of the Census, *Current Population Reports*

43. The most householders will be in what age range?

44. The least number of householders will be in what age range?

45. How many householders will be 55–64 years old?

46. How many householders will be 35–44 years old?

47. How many householders will be 44 years old or younger?

48. How many householders will be 55 years old or older?

49. Which bar represents the household you expect to be in during the year 2005?

50. How many more householders will be 45–54 years old than 55–64 years old?

The following list shows the golf scores for an amateur golfer. Use this list to complete the frequency distribution table to the right. See Example 6.

78	84	91	93	97
97	95	85	95	96
101	89	92	89	100

	Class Intervals (Scores)	Tally	Class Frequency (Number of Games)
51.	70–79	____	____
52.	80–89	____	____
53.	90–99	____	____
54.	100–109	____	____

Twenty-five people in a survey were asked to give their current checking account balances. Use the balances shown in the following list to complete the frequency distribution table to the right. See Example 6.

$53	$105	$162	$443	$109
$468	$47	$259	$316	$228
$207	$357	$15	$301	$75
$86	$77	$512	$219	$100
$192	$288	$352	$166	$292

	Class Intervals (Account Balances)	Tally	Class Frequency (Number of People)
55.	$0–$99	____	____
56.	$100–$199	____	____
57.	$200–$299	____	____
58.	$300–$399	____	____
59.	$400–$499	____	____
60.	$500–$599	____	____

61. Use the table from Exercises 51 through 54 to construct a histogram. See Example 7.

62. Use the table from Exercises 55 through 60 to construct a histogram. See Example 7.

Number of People

2

1

Account Balances

D *The following line graph shows the World Cup goals per game average during the years shown. Use this graph to answer Exercises 63 through 68. See Example 8.*

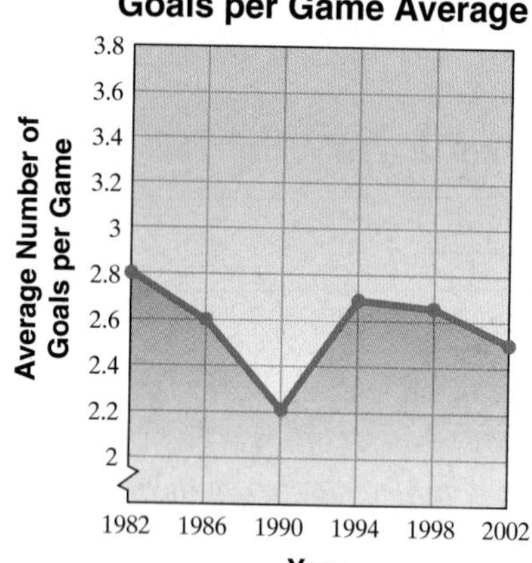

Source: *Soccer America Magazine*

63. Find the average number of goals per game in 1994.

64. Find the average number of goals per game in 2002.

65. During what year shown was the average number of goals per game the highest?

66. During what year shown was the average number of goals per game the lowest?

67. Between 1998 and 2000, did the average number of goals per game increase or decrease?

68. Between 1990 and 1994, did the average number of goals per game increase or decrease?

Review and Preview

Find each percent. See Sections 6.3 and 6.4.

69. 30% of 12

70. 45% of 120

71. 10% of 62

72. 95% of 50

Write each fraction as a percent. See Section 6.2.

73. $\frac{1}{4}$

74. $\frac{2}{5}$

75. $\frac{17}{50}$

76. $\frac{9}{10}$

Combining Concepts

The following double-line graph shows temperature highs and lows for a week. Use this graph to answer Exercises 77 through 82.

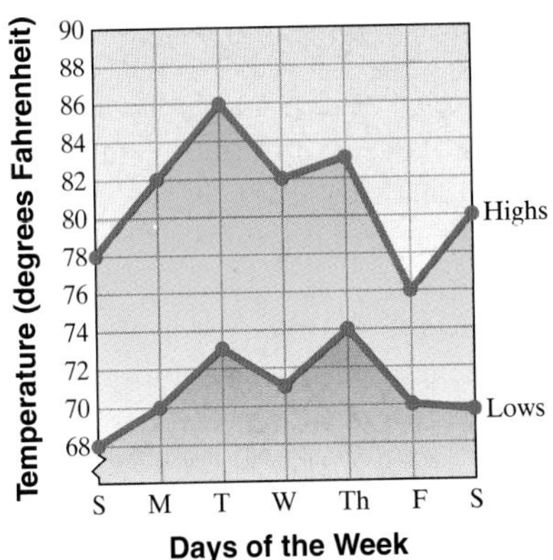

77. What was the high temperature reading on Thursday?

78. What was the low temperature reading on Thursday?

79. What day was the temperature the lowest? What was this low temperature?

80. What day of the week was the temperature the highest? What was this high temperature?

81. On what day of the week was the difference between the high temperature and the low temperature the greatest? What was this difference in temperature?

82. On what day of the week was the difference between the high temperature and the low temperature the least? What was this difference in temperature?

83. True or false? With a bar graph, the width of the bar is just as important as the height of the bar. Explain your answer.

Internet Excursions

Go To:  http://www.prenhall.com/martin-gay_prealgebra  What's Related

The Bureau of Labor Statistics, within the U.S. Department of Labor, is the principal fact-finding agency for the Federal Government in the broad field of labor economics and statistics. The World Wide Web address listed here will provide you with access to the "U.S. Economy at a Glance" Web site of the Bureau of Labor Statistics, or a related site. You will find links to graphs of various data series.

84. Visit this Web site and view the graph of "Unemployment Rate." What type of graph is this? Use the graph to estimate when the highest unemployment rate occurred during the period of time covered by the graph, and estimate that unemployment rate.

85. Visit this Web site and view the graph of "Average Hourly Earnings." What type of graph is this? Describe any trends that you see in the graph.

FOCUS ON the Real World

MISLEADING GRAPHS

Graphs are very common in magazines and in newspapers such as *USA Today*. Graphs can be a convenient way to illustrate an idea because, as the old saying goes, "A picture is worth a thousand words." However, some graphs can be deceptive, which may or may not be intentional. It is important to know some of the ways that graphs can be misleading.

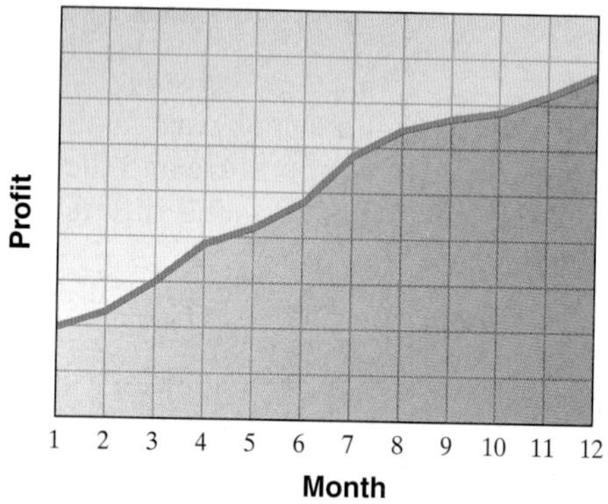

Beware of graphs like the one above. Notice that the graph shows a company's profit for various months. It appears that profit is growing quite rapidly. However, this impressive picture tells us little without knowing what units of profit are being graphed. Does the graph show profit in dollars or millions of dollars? An unethical company with profit increases of only a few pennies could use a graph like this one to make the profit increase seem much more substantial than it really is. A truthful graph describes the size of the units used along the vertical axis.

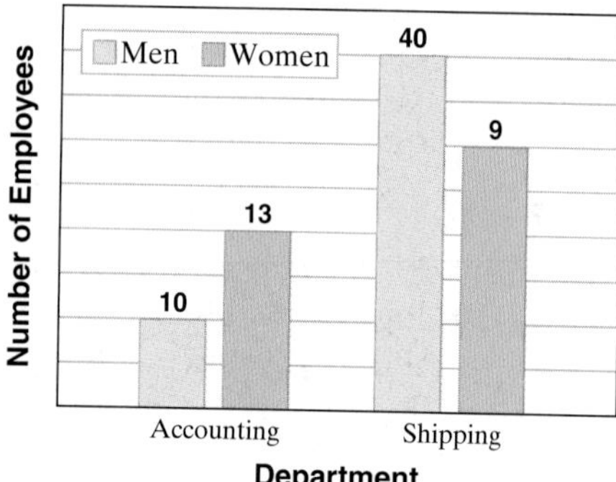

Another type of graph to watch for is one that misrepresents relationships. This can occur in both bar graphs and circle graphs. For example, the bar graph in the bottom of the left column shows the number of men and women employees in the accounting and shipping departments of a certain company. In the accounting department, the bar representing the number of women is twice as tall as the bar representing the number of men. However, the number of women (13) is not twice the number of men (10). This set of bars misrepresents the relationship between the number of men and women. Do you see how the relationship between the number of men and women in the shipping department is distorted by the heights of the bars used? A truthful graph will use bar heights or circle sectors that are proportional in size to the numbers they represent.

We have already seen that the impression a graph can give also depends on its vertical scale. Here is another example: The two graphs below represent exactly the same data. The only difference between the two graphs is the vertical scale—one shows enrollments from 246 to 260 students, and the other shows enrollments between 0 and 300 students. If you were trying to convince readers that algebra enrollment at UPH had changed drastically over the period 2000–2004, which graph would you use? Which graph do you think gives the more honest representation?

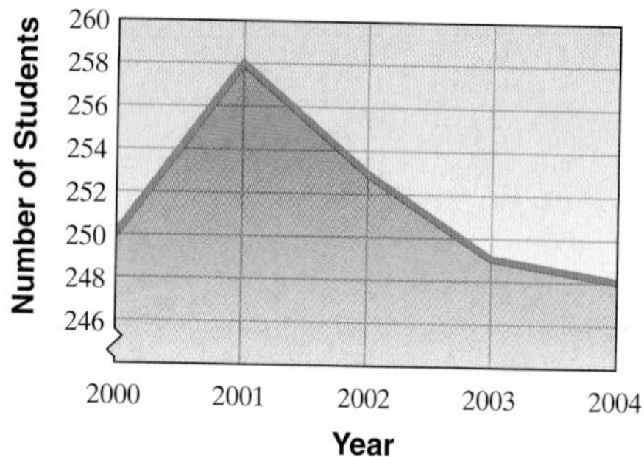

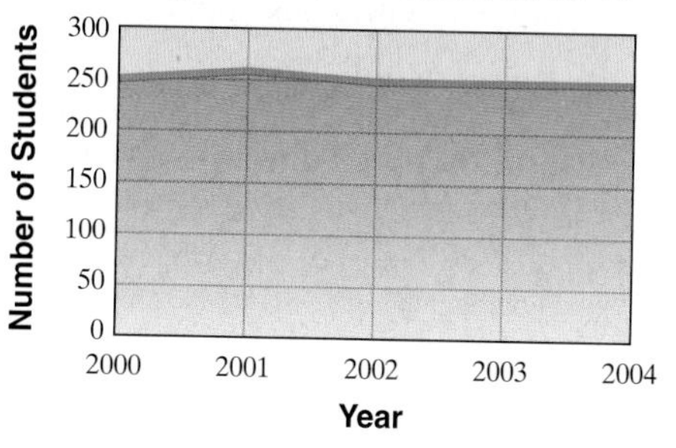

7.2 Reading Circle Graphs

OBJECTIVES

- A Read circle graphs.
- B Draw circle graphs.

SSM TUTOR CENTER SG CD & VIDEO MATH PRO WEB

A Reading Circle Graphs

In Section 6.2, the following graph was shown. This particular graph shows the favorite cookie for every 100 people.

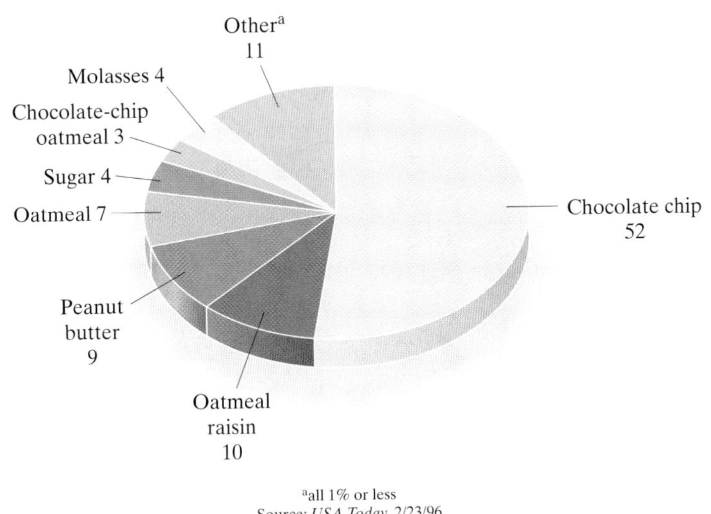

[a]all 1% or less
Source: *USA Today*, 2/23/96

Each sector of the graph (shaped like a piece of pie) shows a category and the relative size of the category. In other words, the most popular cookie is the chocolate chip cookie because it is represented by the largest sector.

EXAMPLE 1 Find the ratio of people preferring chocolate chip cookies to the total number of people. Write the ratio as a fraction in simplest form.

Solution: The ratio is

$$\frac{52 \text{ people preferring chocolate chip}}{100 \text{ people}} = \frac{52}{100} = \frac{13}{25}$$

Practice Problem 1

Find the ratio of people preferring oatmeal raisin cookies to total people. Write the ratio as a fraction in simplest form.

A circle graph is often used to show percents in different categories, with the whole circle representing 100%. For example, in 2002 the population of the United States was about 288,400,000. The following circle graph shows the percent of Americans with various numbers of working computers at home. Notice that the percents in each category sum to 100%.

Number of Working Computers at Home

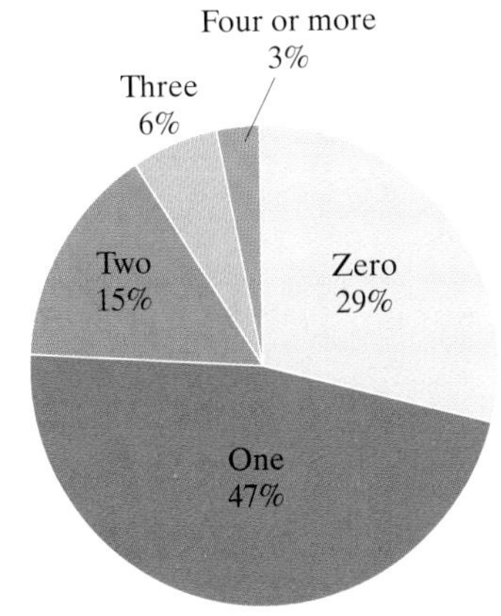

Source: UCLA Center for Communication Policy, 2003

Answer

1. $\frac{1}{10}$

Practice Problem 2

Using the circle graph for Example 2, determine the percent of Americans that have two or more working computers at home.

Practice Problem 3

Using the circle graph for Example 2, find the number of Americans that have four or more working computers at home.

Concept Check

Can the following data be represented by a circle graph? Why or why not?

Responses to the question, "In which activities are you involved?"	
Intramural sports	60%
On-campus job	42%
Fraternity/sorority	27%
Academic clubs	21%
Music programs	14%

Practice Problem 4

Use the data shown to draw a circle graph.

Freshmen	30%
Sophomores	27%
Juniors	25%
Seniors	18%

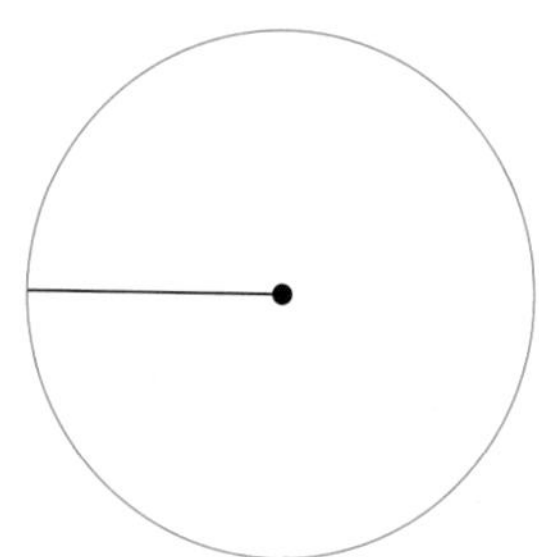

Answers

2. 24% **3.** 8,652,000 Americans **4.** see next page

Concept Check: no; the percents add up to more than 100%

EXAMPLE 2 Using the circle graph from the previous page, determine the percent of Americans that have one or more working computers at home.

Solution: To find this percent, we add the percents corresponding to one, two, three, and four or more working computers at home. The percent of Americans that have one or more working computers at home is

$$47\% + 15\% + 6\% + 3\% = 71\%$$

EXAMPLE 3 Using the circle graph for Example 2, find the *number* of Americans that have no working computers at home.

Solution: Since the *percent* of Americans with no computer is 29%, we find the *number* of Americans by finding 29% of the population. To do this, we can use the percent equation.

$$\begin{aligned} \text{amount} &= \text{percent} \cdot \text{base} \\ \text{amount} &= 0.29 \cdot 288{,}400{,}000 \\ &= 0.29(288{,}400{,}000) \\ &= 83{,}636{,}000 \end{aligned}$$

Thus, 83,636,000 Americans have no working computer at home.

Try the Concept Check in the margin.

B Drawing Circle Graphs

To draw a circle graph, we use the fact that a whole circle contains 360° (degrees).

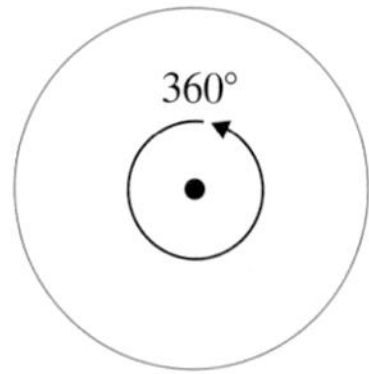

EXAMPLE 4 The following table shows the percent of U.S. armed forces personnel that are in each branch of service. (*Source:* U.S. Department of Defense)

Branch of Service	Percent
Army	33%
Navy	27%
Marine Corps	12%
Air Force	25%
Coast Guard	3%

Draw a circle graph showing this data.

Solution: First we find the number of degrees in each sector representing each branch of service. Remember that the whole circle contains 360°. (We will round degrees to the nearest whole.)

Sector	Degrees in Each Sector
Army	$33\% \times 360° = 118.8° \approx 119°$
Navy	$27\% \times 360° = 97.2° \approx 97°$
Marine Corps	$12\% \times 360° = 43.2° \approx 43°$
Air Force	$25\% \times 360° = 90° = 90°$
Coast Guard	$3\% \times 360° = 10.8° \approx 11°$

Helpful Hint

Check your calculations by finding the sum of the degrees.

$119° + 97° + 43° + 90° + 11° = 360°$.

The sum should be 360°. (It may vary only slightly because of rounding.)

Next we draw a circle and mark its center. Then we draw a line from the center of the circle to the circle itself (that is, a radius).

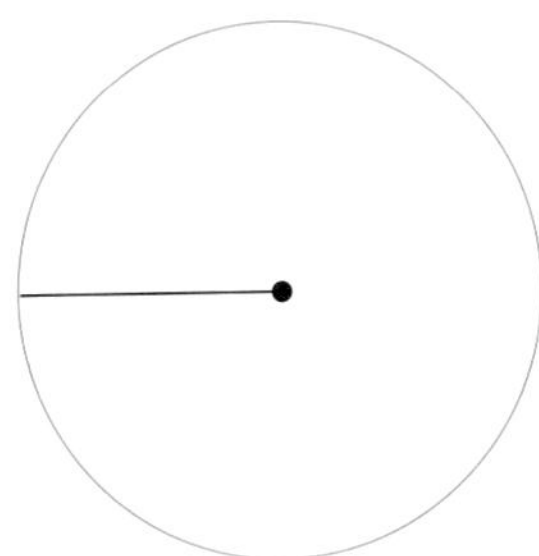

To construct the sectors, we will use a **protractor.** Recall that a protractor measures the number of degrees in an angle. We place the hole in the protractor over the center of the circle. Then we adjust the protractor so that 0° on the protractor is aligned with the line that we drew.

It makes no difference which sector we draw first. To construct the "Army" sector, we find 119° on the protractor and mark our circle. Then we remove the protractor and use this mark to draw a second line from the center to the circle itself.

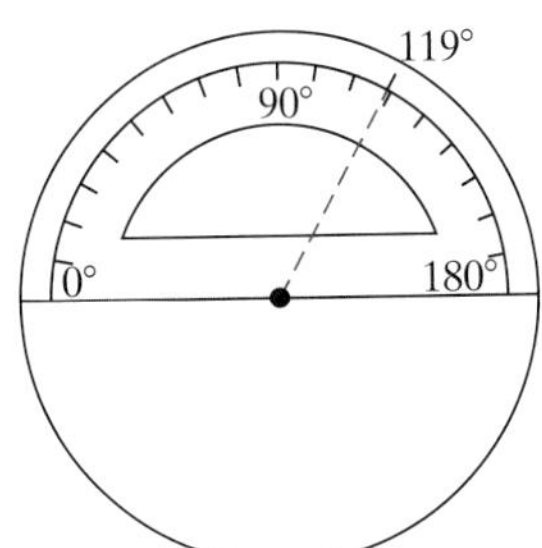

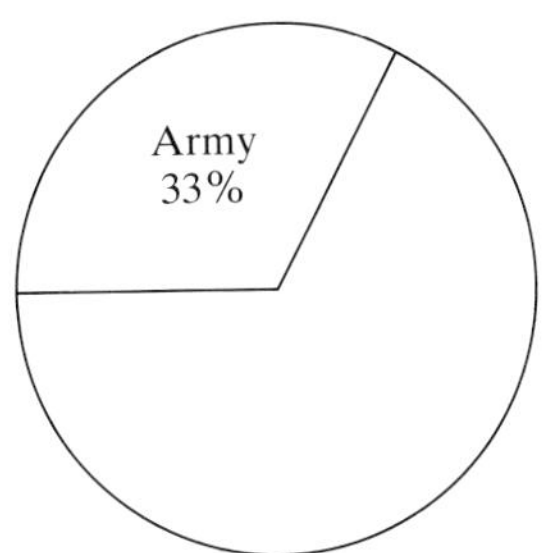

To construct the "Navy" sector, we follow the same procedure as above, except that we line up 0° with the second line we drew and mark the protractor at 97°.

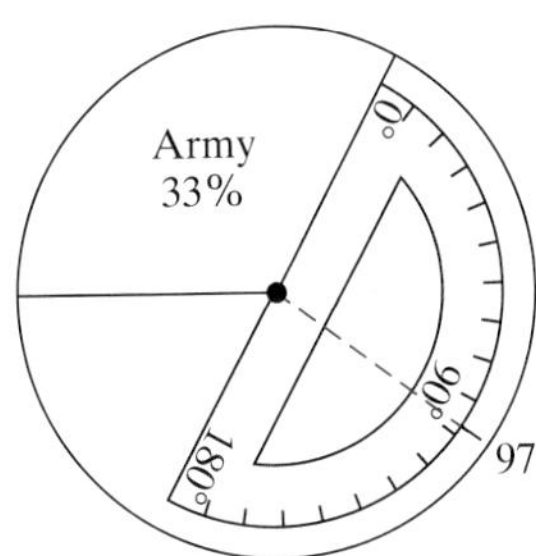

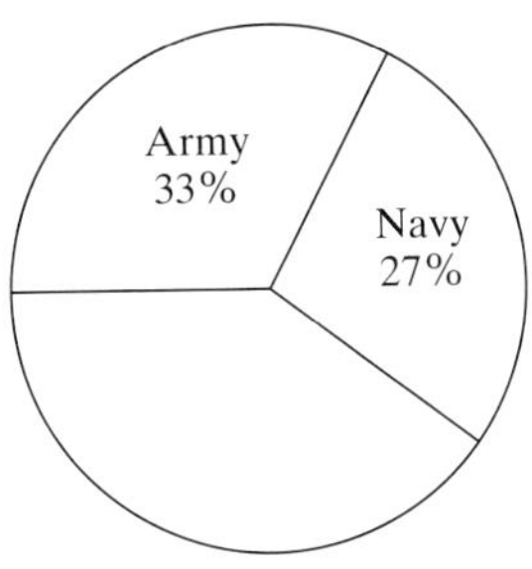

Answer

4.

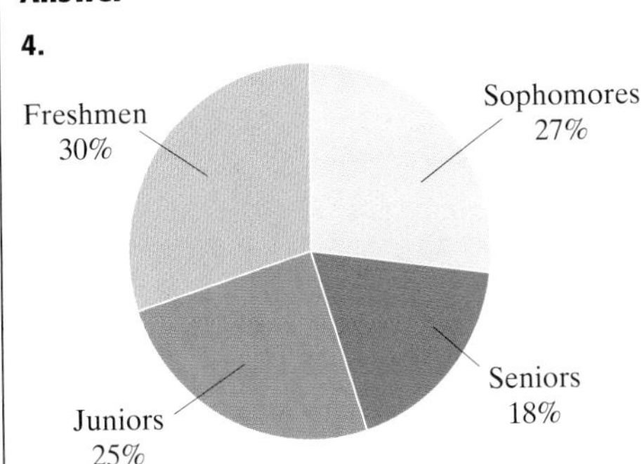

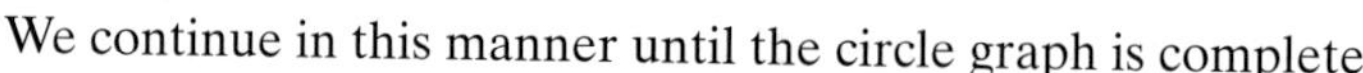
We continue in this manner until the circle graph is complete.

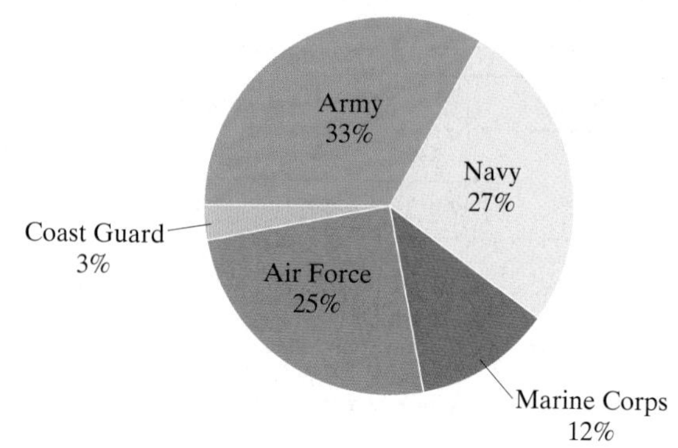

Concept Check

True or false? The larger a sector in a circle graph, the larger the percent of the total it represents. Explain your answer.

Try the Concept Check in the margin.

STUDY SKILLS REMINDER

Are you satisfied with your performance on a particular quiz or exam?

If not, don't forget to analyze your quiz or exam and look for common errors.

Were most of your errors a result of

- *Carelessness*? If your errors were careless, did you turn in your work before the allotted time expired? If so, resolve to use the entire time allotted next time. Any extra time can be spent checking your work.
- *Running out of time*? If so, make a point to better manage your time on your next exam. A few suggestions are to work any questions that you are unsure of last and to check your work after all questions have been answered.
- *Not understanding a concept*? If so, review that concept and correct your work. Remember next time to make sure that all concepts on a quiz or exam are understood before the exam.
- *Test conditions*? When studying for your test, are you placing yourself in conditions similar to test conditions? In other words, once you feel that you know the material, use a few sheets of blank paper and take a sample test. (A sample test can be one provided by your instructor or you may use the Chapter Test found at the end of each chapter.)

Concept Check: true

Name ______________________ Section ____________ Date ____________

EXERCISE SET 7.2

A *The following circle graph is a result of surveying 700 college students. They were asked where they live while attending college. Use this graph to answer Exercises 1 through 6. Write all ratios as fractions in simplest form. See Example 1.*

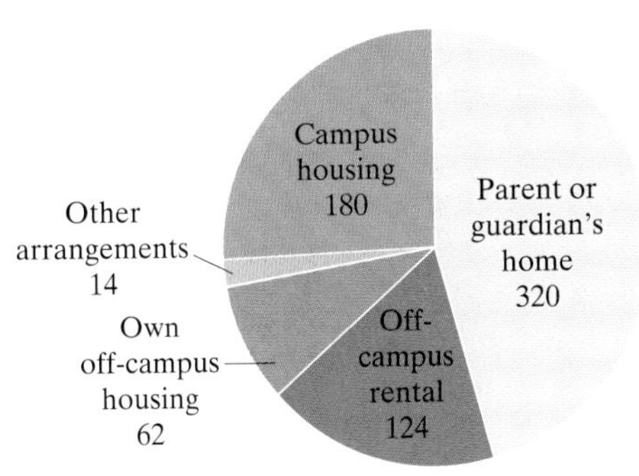

1. Where do most of these college students live?

2. Besides the category "Other Arrangements," where do the fewest of these college students live?

3. Find the ratio of students living in campus housing to total students.

4. Find the ratio of students living in off-campus rentals to total students.

5. Find the ratio of students living in campus housing to students living in a parent or guardian's home.

6. Find the ratio of students living in off-campus rentals to students living in a parent or guardian's home.

The following circle graph shows the percent of Earth's land in each of the continents. Use this graph for Exercises 7 through 14. See Example 2.

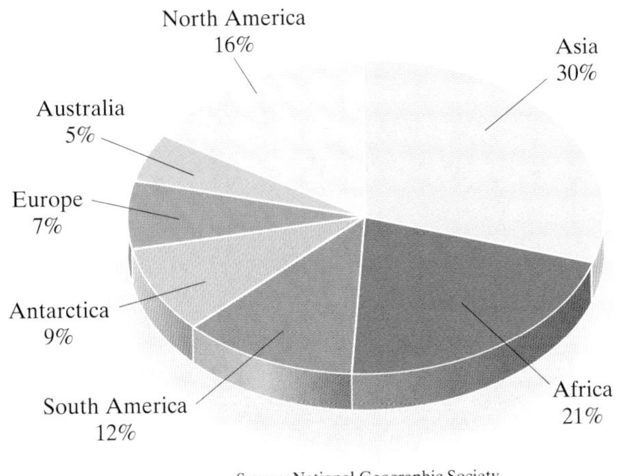

Source: National Geographic Society

7. Which continent is the largest?

8. Which continent is the smallest?

9. What percent of the land on Earth is accounted for by Asia and Europe together?

10. What percent of the land on Earth is accounted for by North and South America?

The total amount of land on Earth is approximately 57,000,000 square miles. Use the graph to find the area of the continents given in Exercises 11 through 14. See Example 3.

11. Asia

12. South America

13. Australia

14. Europe

The following circle graph shows the percent of the types of books available at Midway Memorial Library. Use this graph for Exercises 15 through 24. See Example 2.

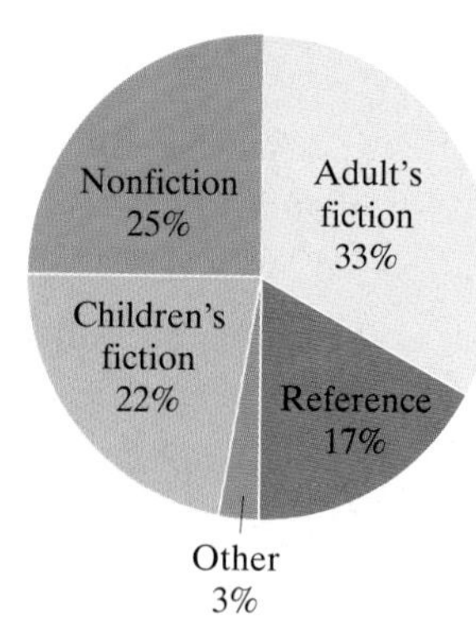

15. What percent of books are classified as some type of fiction?

16. What percent of books are nonfiction or reference?

17. What is the second-largest category of books?

18. What is the third-largest category of books?

If this library has 125,600 books, find how many books are in each category given in Exercises 19 through 24. See Example 3.

19. Nonfiction

20. Reference

21. Children's fiction

22. Adult's fiction

23. Reference or other

24. Nonfiction or other

B *Fill in the table. Round to the nearest degree. Then draw a circle graph to represent the information given in each table. (Remember: The total of the Degrees in Sector column should equal 360° or very close to 360° because of rounding.) See Example 4.*

25.

2001 U.S. Car Sales by Vehicle Origin

Country of Origin	Percent	Degrees in Sector
United States	75%	
Japan	11%	
Germany	6%	
Other Countries	8%	

(*Source:* Ward's AutoInfoBank)

26.

Size of Kellogg's Business Segments after Acquiring Keebler

Business Segment	Percent of Annual Sales	Degrees in Sector
U.S. cereal	27%	
U.S. convenience foods	43%	
International	30%	

(*Source:* Kellogg Company's Annual Report)

Name ______________________ Section __________ Date __________

Integrated Review—Reading Graphs

The following pictograph shows the average number of pounds of beef and veal consumed per person per year in the United States. Use this graph to answer Exercises 1 through 4.

Beef and Veal Consumption

Each [cow symbol] represents 10 pounds

Source: U.S. Department of Agriculture

1. Approximate the number of pounds of beef and veal consumed per person in 1995.

2. Approximate the number of pounds of beef and veal consumed per person in 1980.

3. In what year(s) was the number of pounds consumed the greatest?

4. In what year(s) was the number of pounds consumed the least?

The following bar graph shows the highest U.S. dams. Use this graph to answer Exercises 5 through 8.

Highest U.S. Dams

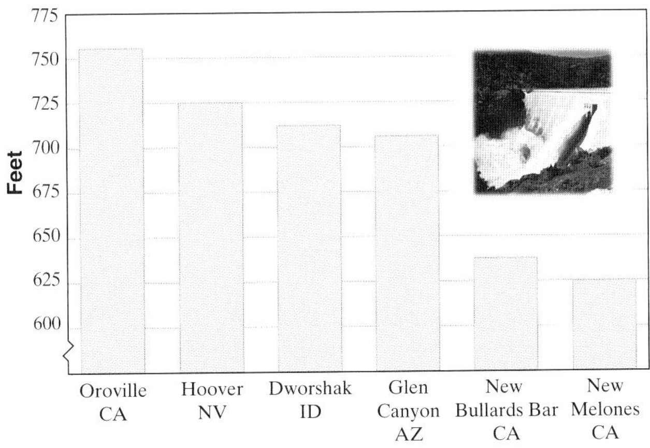

Source: Committee on Register of Dams

5. Name the U. S. dam with the greatest height and estimate its height.

6. Name the U.S. dam whose height is between 625 and 650 feet and estimate its height.

7. Estimate how much higher the Hoover Dam is than the Glen Canyon Dam.

8. How many U.S. dams have heights over 700 feet?

Answers

1. ____________

2. ____________

3. ____________

4. ____________

5. ____________

6. ____________

7. ____________

8. ____________

9. ____

10. ____

11. ____

12. ____

13. ____

14. ____

15. ____

16. ____

17. see table

18. see table

19. see table

20. see table

21. see table

22. see graph

The following line graph shows the daily high temperatures for one week in Annapolis, Maryland. Use this graph to answer Exercises 9 through 12.

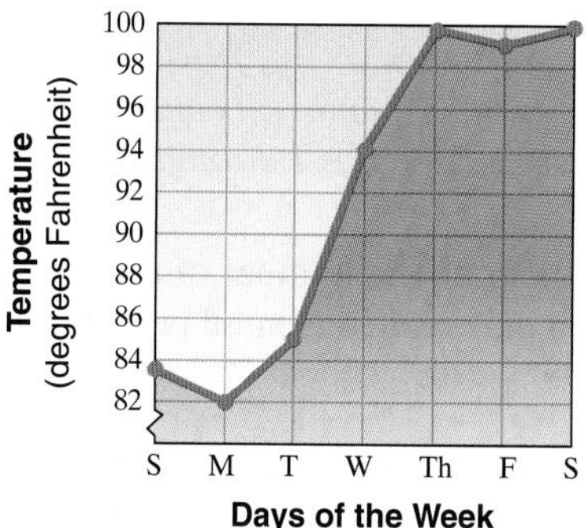

9. Name the day(s) of the week with the highest high temperature and give that temperature.

10. Name the day(s) of the week with the lowest high temperature and give that temperature.

11. On what days of the week was the high temperature less than 90° Fahrenheit?

12. On what days of the week was the temperature greater than 90° Fahrenheit?

The following circle graph shows the type of beverage milk consumed in the United States. Use this graph for Exercises 13 through 16. If a store in Kerrville, Texas, sells 200 quart containers of milk per week, estimate how many quart containers are sold in each category below.

Types of Milk Consumed

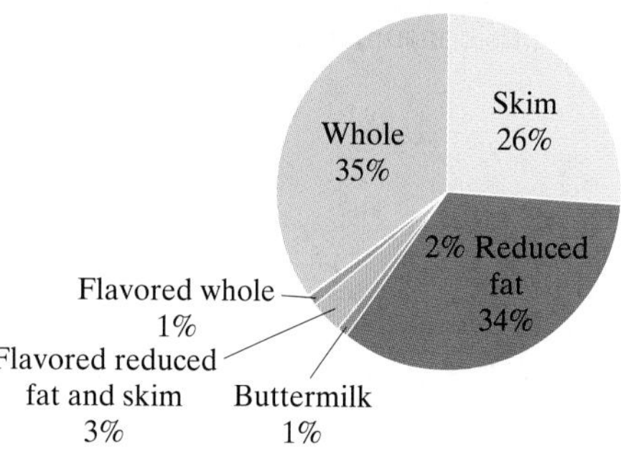

Source: U.S. Department of Agriculture

13. Whole milk

14. Skim milk

15. Buttermilk

16. Flavored reduced fat and skim milk

The following list shows weekly quiz scores for a student in basic college mathematics. Use this list to complete the frequency distribution table.

50	80	71	83	86
67	89	93	88	97
75	80	78	93	99
	53	90		

	Class Intervals (Scores)	Tally	Class Frequency (Number of Quizzes)
17.	50–59	____	____
18.	60–69	____	____
19.	70–79	____	____
20.	80–89	____	____
21.	90–99	____	____

22. Use the table from Exercises 17 through 21 to construct a histogram.

Number of Quizzes
2
1
Quiz Scores

7.3 Mean, Median, and Mode

OBJECTIVES

- A Find the mean of a list of numbers.
- B Find the median of a list of numbers.
- C Find the mode of a list of numbers.

A Finding the Mean

Sometimes we want to summarize data by displaying them in a graph, but sometimes it is also desirable to be able to describe a set of data, or a set of numbers, by a single "middle" number. Three such **measures of central tendency** are the **mean,** the **median,** and the **mode.**

The most common measure of central tendency is the mean (sometimes called the "arithmetic mean" or the "average"). Recall that we first introduced finding the average of a list of numbers in Section 1.7.

The **mean (average)** of a set of number items is the sum of the items divided by the number of items.

EXAMPLE 1 Seven students in a psychology class conducted an experiment on mazes. Each student was given a pencil and asked to successfully complete the same maze. The timed results are below.

Student	Ann	Thanh	Carlos	Jesse	Melinda	Ramzi	Dayni
Time (seconds)	13.2	11.8	10.7	16.2	15.9	13.8	18.5

a. Who completed the maze in the shortest time? Who completed the maze in the longest time?

b. Find the mean time.

c. How many students took longer than the mean time? How many students took shorter than the mean time?

Practice Problem 1

Find the mean of the following test scores: 77, 85, 86, 91, and 88.

Solution:

a. Carlos completed the maze in 10.7 seconds, the shortest time. Dayni completed the maze in 18.5 seconds, the longest time.

b. To find the mean (or average), we find the sum of the number items and divide by 7, the number of items.

$$\text{mean} = \frac{13.2 + 11.8 + 10.7 + 16.2 + 15.9 + 13.8 + 18.5}{7}$$

$$= \frac{100.1}{7} = 14.3$$

c. Three students, Jesse, Melinda, and Dayni, had times longer than the mean time. Four students, Ann, Thanh, Carlos, and Ramzi, had times shorter than the mean time.

Concept Check

Find the mean of the following set of data:

5, 10, 10, 10, 10, 15

Try the Concept Check in the margin.

Often in college, the calculation of a **grade point average** (GPA) is a **weighted mean** and is calculated as shown in Example 2.

Answer

1. 85.4

Concept Check: 10

Practice Problem 2

Find the grade point average if the following grades were earned in one semester.

Grade	Credit Hours
A	2
C	4
B	5
D	2
A	2

Answer

2. 2.73

EXAMPLE 2 The following grades were earned by a student during one semester. Find the student's grade point average.

Course	Grade	Credit Hours
College mathematics	A	3
Biology	B	3
English	A	3
PE	C	1
Social studies	D	2

Solution: To calculate the grade point average, we need to know the point values for the different possible grades. The point values of grades commonly used in colleges and universities are given below:

A: 4, B: 3, C: 2, D: 1, F: 0

Now, to find the grade point average, we multiply the number of credit hours for each course by the point value of each grade. The grade point average is the sum of these products divided by the sum of the credit hours.

Course	Grade	Point Value of Grade	Credit Hours	Point Value · Credit Hours
College mathematics	A	4	3	12
Biology	B	3	3	9
English	A	4	3	12
PE	C	2	1	2
Social studies	D	1	2	2
		Totals:	12	37

$$\text{grade point average} = \frac{37}{12} \approx 3.08 \text{ rounded to two decimal places}$$

The student earned a grade point average of 3.08. ●

B Finding the Median

You may have noticed that a very low number or a very high number can affect the mean of a list of numbers. Because of this, you may sometimes want to use another measure of central tendency. A second measure of central tendency is called the **median.** The median of a list of numbers is not affected by a low or high number in the list.

The **median** of an *ordered set* of numbers is the middle number. If the number of items is even, the median is the mean (average) of the two middle numbers.

In order to compute the median, the numbers must first be placed in order.

EXAMPLE 3 Find the median of the following list of numbers:

25, 54, 56, 57, 60, 71, 98

Solution: Because this list is in numerical order, the median is the middle number, 57.

25, 54, 56, 57, 60, 71, 98
↑
middle number

Practice Problem 3

Find the median of the list of numbers:
7, 9, 13, 23, 24, 35, 38, 41, 43

EXAMPLE 4 Find the median of the following list of scores:

67, 91, 75, 86, 55, 91

Solution: First we list the scores in numerical order and then find the middle number.

55, 67, 75, 86, 91, 91
↑
middle numbers

Since there is an even number of scores, there are two middle numbers. The median is the mean of the two middle numbers.

$$\text{median} = \frac{75 + 86}{2} = 80.5$$

The median is 80.5.

Practice Problem 4

Find the median of the list of scores:
43, 89, 78, 65, 95, 95, 88, 71

C Finding the Mode

The last common measure of central tendency is called the **mode.**

The **mode** of a set of numbers is the number that occurs most often. (It is possible for a set of numbers to have more than one mode or to have no mode.)

EXAMPLE 5 Find the mode of the list of numbers:

11, 14, 14, 16, 31, 56, 65, 77, 77, 78, 79

Solution: There are two numbers that occur the most often. They are 14 and 77. This list of numbers has two modes, 14 and 77.

Practice Problem 5

Find the mode of the list of numbers:
9, 10, 10, 13, 15, 15, 15, 17, 18, 18, 20

EXAMPLE 6 Find the median and the mode of the following list of numbers. These numbers were high temperatures for 14 consecutive days in a city in Montana.

76, 80, 85, 86, 89, 87, 82, 77, 76, 79, 82, 89, 89, 92

Practice Problem 6

Find the median and the mode of the list of numbers:
26, 31, 15, 15, 26, 30, 16, 18, 15, 35

Answers

3. 24 **4.** 83 **5.** 15 **6.** median: 22; mode: 15

Solution: First we write the numbers in numerical order.

76, 76, 77, 79, 80, 82, 82, 85, 86, 87, 89, 89, 89, 92

Since there is an even number of items, the median is the mean of the two middle numbers.

$$\text{median} = \frac{82 + 85}{2} = 83.5$$

The mode is 89, since 89 occurs most often.

Concept Check

True or false? Every set of numbers *must* have a mean, median, and mode. Explain your answer.

Try the Concept Check in the margin.

Helpful Hint

Don't forget that it is possible for a list of numbers to have no mode. For example, the list

2, 4, 5, 6, 8, 9

has no mode. There is no number or numbers that occur more often than the others.

Concept Check: false; a set of numbers may have no mode

Name ______________________ Section __________ Date __________

Mental Math

State the mean for each list of numbers.

1. 3, 5 **2.** 10, 20 **3.** 1, 3, 5 **4.** 7, 7, 7

EXERCISE SET 7.3

  C *For each set of numbers, find the mean, the median, and the mode. If necessary, round the mean to one decimal place. See Examples 1 and 3 through 6.*

1. 21, 28, 16, 42, 38

2. 42, 35, 36, 40, 50

3. 7.6, 8.2, 8.2, 9.6, 5.7, 9.1

4. 4.9, 7.1, 6.8, 6.8, 5.3, 4.9

5. 0.2, 0.3, 0.5, 0.6, 0.6, 0.9, 0.2, 0.7, 1.1

6. 0.6, 0.6, 0.8, 0.4, 0.5, 0.3, 0.7, 0.8, 0.1

7. 231, 543, 601, 293, 588, 109, 334, 268

8. 451, 356, 478, 776, 892, 500, 467, 780

The eight tallest buildings in the world are listed in the following table. Use this table to answer Exercises 9 through 14. If necessary, round results to one decimal place. See Examples 1 and 3 through 6.

Building	Height (in feet)
Petronas Tower 1, Kuala Lumpur	1483
Petronas Tower 2, Kuala Lumpur	1483
Sears Tower, Chicago	1450
Jin Mao Building, Shanghai	1381
Citic Plaza, Guangzhou	1283
Shun Hing Square, Shenzhen	1260
Empire State Building, New York	1250
Central Plaza, Hong Kong	1227

(*Source:* Council on Tall Buildings and Urban Habitat)

9. Find the mean height of the five tallest buildings.

10. Find the median height of the five tallest buildings.

11. Find the median height of the eight tallest buildings.

12. Find the mean height of the eight tallest buildings.

13. Given the building heights, explain how you know, without calculating, that the answer to Exercise 10 is more than the answer to Exercise 11.

14. Given the building heights, explain how you know, without calculating, that the answer to Exercise 12 is less than the answer to Exercise 9.

For Exercises 15 through 18, the grades are given for a student for a particular semester. Find the grade point average. If necessary, round the grade point average to the nearest hundredth. See Example 2.

15.

Grade	Credit Hours
B	3
C	3
A	4
C	4

16.

Grade	Credit Hours
D	1
F	1
C	4
B	5

17.

Grade	Credit Hours
A	3
A	3
B	4
B	1
B	2

18.

Grade	Credit Hours
B	2
B	2
A	3
C	3
B	3

During an experiment, the following times (in seconds) were recorded:

7.8, 6.9, 7.5, 4.7, 6.9, 7.0

19. Find the mean. Round to the nearest tenth.

20. Find the median.

21. Find the mode.

In a mathematics class, the following test scores were recorded for a student:

86, 95, 91, 74, 77, 85

22. Find the mean. Round to the nearest hundredth.

23. Find the median.

24. Find the mode.

The following pulse rates were recorded for a group of 15 students:

78, 80, 66, 68, 71, 64, 82, 71, 70, 65, 70, 75, 77, 86, 72

25. Find the mean.

26. Find the median.

27. Find the mode.

28. How many rates were higher than the mean?

29. How many rates were lower than the mean?

Review and Preview

Write each fraction in simplest form. See Section 4.2.

30. $\frac{12}{20}$

31. $\frac{6}{18}$

32. $\frac{4}{36}$

33. $\frac{18}{30}$

34. $\frac{35}{100}$

35. $\frac{55}{75}$

Combining Concepts

Find the missing numbers in each set of numbers.

36. 16, 18, _, _, _. The mode is 21. The median is 20.

37. _, _, _, 40, _. The mode is 35. The median is 37. The mean is 38.

38. Write a list of numbers for which you feel the median would be a better measure of central tendency than the mean.

39. Without making any computations, decide whether the median of the following list of numbers will be a whole number. Explain your reasoning.

36, 77, 29, 58, 43

7.4 Counting and Introduction to Probability

A Using a Tree Diagram

In our daily conversations, we often talk about the likelihood or the probability of a given result occurring. For example,

> The *chance* of thundershowers is 70 percent.
> What are the *odds* that the Saints will go to the Super Bowl?
> What is the *probability* that you will finish cleaning your room today?

Each of these chance happenings—thundershowers, the Saints playing in the Super Bowl, and cleaning your room today—is called an **experiment.** The possible results of an experiment are called **outcomes.** For example, flipping a coin is an experiment and the possible outcomes are heads (H) or tails (T) and are equally likely to happen.

One way to picture the outcomes of an experiment is to draw a tree diagram. Each outcome is shown on a separate branch. For example, the outcomes of tossing a coin are

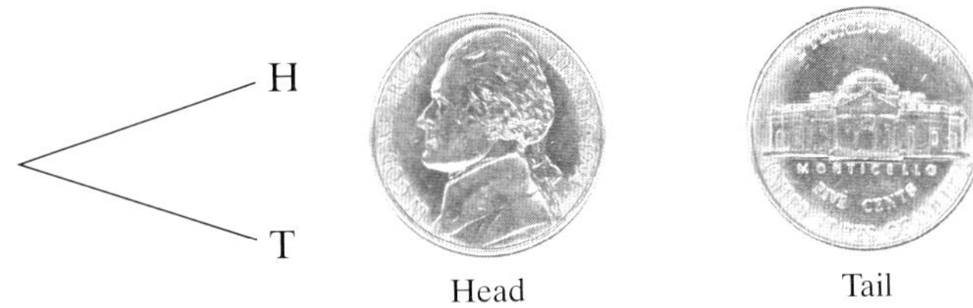

EXAMPLE 1 Draw a tree diagram for tossing a coin twice. Then use the diagram to find the number of possible outcomes.

Solution: There are 4 possible outcomes when tossing a coin twice.

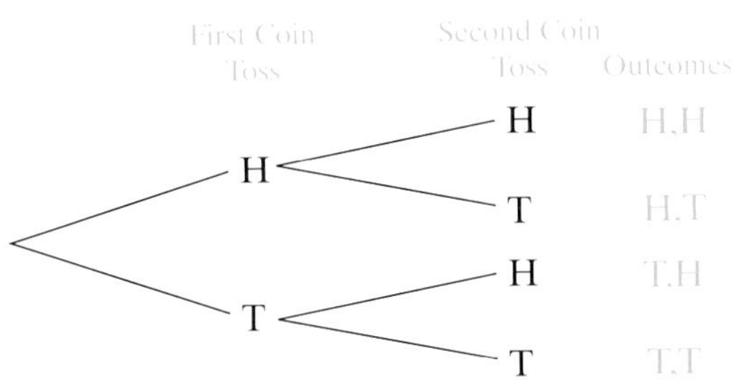

EXAMPLE 2 Draw a tree diagram for an experiment consisting of rolling a die and then tossing a coin. Then use the diagram to find the number of possible outcomes.

Die

OBJECTIVES

- A Use a tree diagram to count outcomes.
- B Find the probability of an event.

Practice Problem 1

Draw a tree diagram for tossing a coin three times. Then use the diagram to find the number of possible outcomes.

Practice Problem 2

Draw a tree diagram for an experiment consisting of tossing a coin and then rolling a die. Then use the diagram to find the number of possible outcomes.

Answers

1.

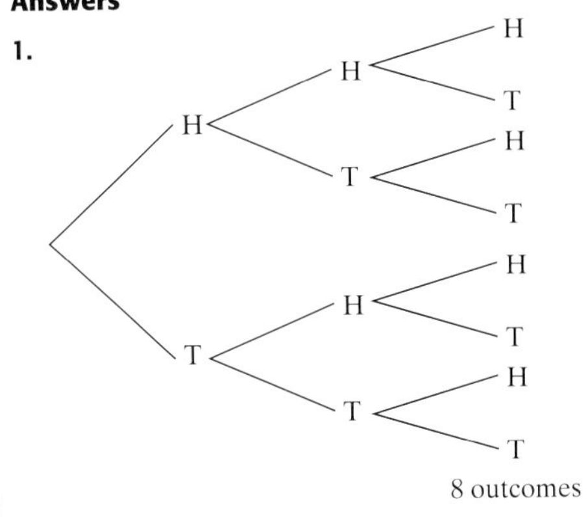

2.

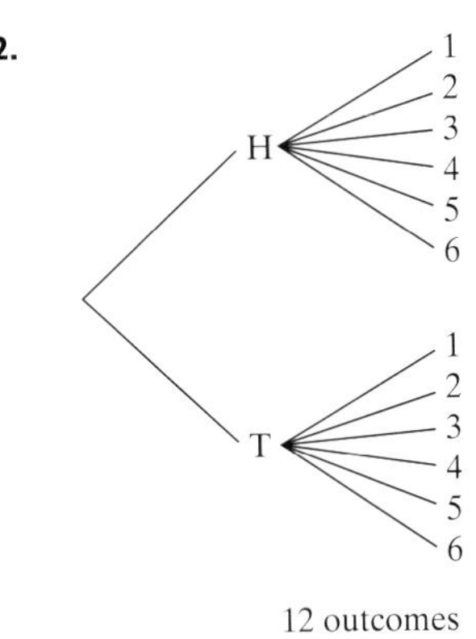

Solution: Recall that a die has six sides and that each side represents a number, 1 through 6. Each side is as likely to be rolled as another side.

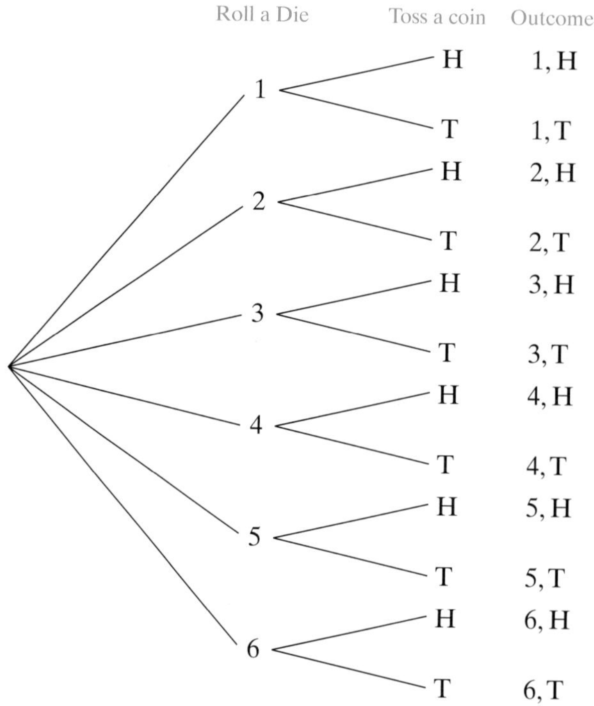

There are 12 possible outcomes for rolling a die and then tossing a coin. ●

Any number of outcomes considered together is called an **event.** For example, when tossing a coin twice, (H, H) is an event. The event is tossing heads first and tossing heads second. Another event would be tossing tails first and then heads (T, H), and so on.

B Finding the Probability of an Event

As we mentioned earlier, the probability of an event is a measure of the chance or likelihood of it occurring. For example, if a coin is tossed, what is the probability that heads occurs? Since one of two equally likely possible outcomes is heads, the probability is $\frac{1}{2}$.

The Probability of an Event

$$\text{probability of an event} = \frac{\text{number of ways that the event can occur}}{\text{number of possible outcomes}}$$

Helpful Hint

Note from the definition of probability that the probability of an event is always between 0 and 1, inclusive (i.e., including 0 and 1). A probability of 0 means that an event won't occur, and a probability of 1 means that an event is certain to occur.

EXAMPLE 3 If a coin is tossed twice, find the probability of tossing heads and then heads (H, H).

Solution: 1 way the event can occur

(H, H), (H, T), (T, H), (T, T)

4 possible outcomes

$$\text{probability} = \frac{1}{4} \quad \begin{array}{l}\text{Number of ways the event can occur} \\ \text{Number of possible outcomes}\end{array}$$

The probability of tossing heads and then heads is $\frac{1}{4}$.

Practice Problem 3

If a coin is tossed three times, find the probability of tossing heads, then tails, then tails (H, T, T).

EXAMPLE 4 If a die is rolled one time, find the probability of rolling a 3 or a 4.

Solution: Recall that there are 6 possible outcomes when rolling a die.

2 ways that the event can occur

possible outcomes: 1, 2, 3, 4, 5, 6

6 possible outcomes

$$\text{probability of a 3 or a 4} = \frac{2}{6} \quad \begin{array}{l}\text{Number of ways the event can occur} \\ \text{Number of possible outcomes}\end{array}$$

$$= \frac{1}{3} \quad \text{Simplest form}$$

Practice Problem 4

If a die is rolled one time, find the probability of rolling a 1 or a 2.

Concept Check

Suppose you have calculated a probability of $\frac{11}{9}$. How do you know that you have made an error in your calculation?

Try The Concept Check in the margin.

EXAMPLE 5 Find the probability of choosing a red marble from a box containing 1 red, 1 yellow, and 2 blue marbles.

Solution:

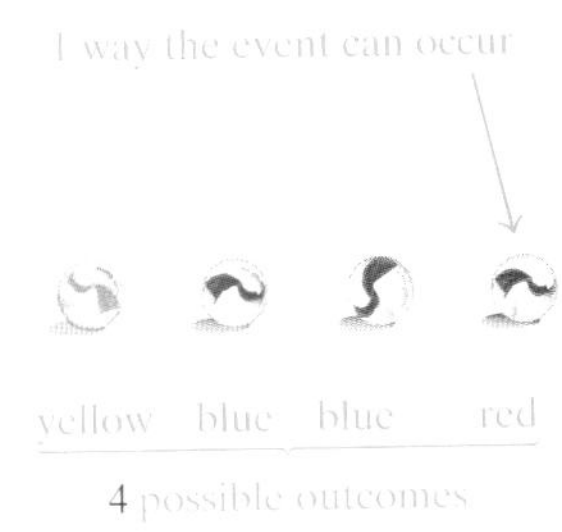

$$\text{probability} = \frac{1}{4}$$

Practice Problem 5

Use the diagram from Example 5 and find the probability of choosing a blue marble from the box.

Answers

3. $\frac{1}{8}$ **4.** $\frac{1}{3}$ **5.** $\frac{1}{2}$

Concept Check: The number of ways an event can occur can't be larger than the number of possible outcomes.

STUDY SKILLS REMINDER

Are you prepared for a test on Chapter 7?

Below I have listed some *common trouble areas* for topics covered in Chapter 7. After studying for your test—but before taking your test—read these.

- Do you remember that a set of numbers can have no mode, 1 mode, or even more than 1 mode?

  2, 5, 8, 9 — no mode

  2, 2, 8, 9 — mode: 2

  2, 2, 3, 3, 5, 7, 7 — mode: 2, 3, 7

- Do you remember how to find the median of an even-numbered set of numbers?

  2, 5, 8, 9 $\quad \dfrac{5+8}{2} = 6.5$ — The median is the average of the two "middle" numbers.

- Don't forget that the probability of an event is always between 0 and 1 inclusive (including 0 and 1).
- What is the probability of an event that won't occur? 0

  What is the probability of an event that is certain to occur? 1

Remember: This is simply a sampling of selected topics given to check your understanding. For a review of Chapter 7 in your text, see the material at the end of this chapter.

FOCUS ON Mathematical Connections

RANGE

In addition to measures of central tendency, *measures of dispersion* can also help to describe or summarize a set of data. These measures indicate how "spread out" the numbers are in a set of data. One measure of dispersion is the **range** of a set of data. The range is computed as the difference between the largest and the smallest number in the data set.

Take, for example, the data set 39, 27, 30, 66, 63, and 57. The smallest number in the set is 27 and the largest number in the set is 66. The difference between these two numbers is $66 - 27 = 39$. Thus, the range of the data is 39.

What can the range tell us about a set of data? Suppose we have two different sets of data. The median (middle number) of the first set is 18 and the median of the second set is also 18. So far, the data sets seem to be similar. But if we are then told that the range of the first set is 2 and the range of the second set is 28, we know that the two sets of data are likely to be quite different. In the first set, the numbers are clustered very closely to the median, 18. In the second set, the numbers are probably spread out much farther, covering a wider slice of values. In fact, the first set are ages of the students in a high school math class and the second set are ages of the students in a college math class (see below). We would expect the age makeup of these two classes to be quite different, and comparing the ranges of the ages confirms this.

Ages in High School Math Class

17 17 17 17 17 17 17 18 18 18 18 18 18 18 18 18 18 18 18 18 19

Median: 18 Range: 2

Ages in College Math Class:

18 18 18 18 18 18 18 18 18 18 18 18 19 19 20 21 23 27 33 41 46

Median: 18 Range: 28

GROUP ACTIVITY

Collect class grade information for two different tests or quizzes in this class or another class; both sets of grades should be for the same class. Find the mean, median, mode, and range for each set of grades. Compare statistics for each set of grades. What can you conclude?

Name ______________________ Section ____________ Date ____________

Mental Math

If a coin is tossed once, find the probability of each event.

1. The coin lands heads up.

2. The coin lands tails up.

If the spinner shown is spun once, find the probability of each event.

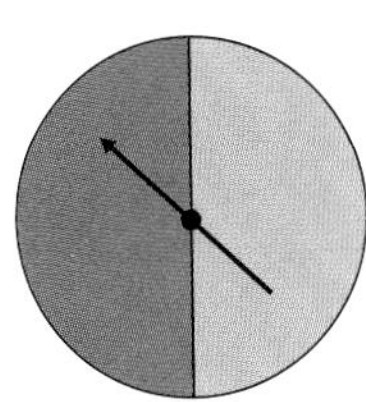

3. The spinner stops on red.

4. The spinner stops on blue.

EXERCISE SET 7.4

A *Draw a tree diagram for each experiment. Then use the diagram to find the number of possible outcomes. See Examples 1 and 2.*

1. Choosing a vowel, (a, e, i, o, u) and then a number (1, 2, or 3).

2. Choosing a number (1 or 2) and then a vowel (a, e, i, o, u).

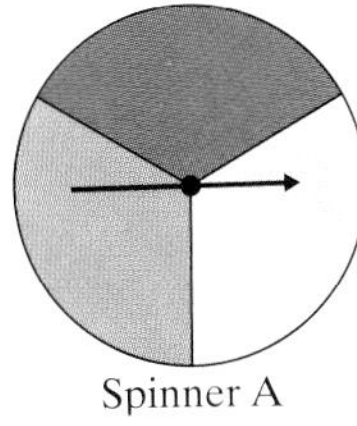

Spinner A

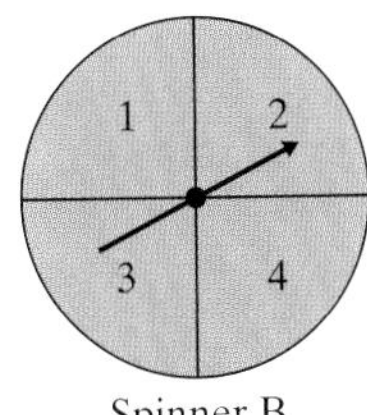

Spinner B

3. Spinning Spinner A once.

4. Spinning Spinner B once.

5. Spinning Spinner B twice.

6. Spinning Spinner A twice.

7. Spinning Spinner A and then Spinner B.

8. Spinning Spinner B and then Spinner A.

9. Tossing a coin and then spinning Spinner B.

10. Tossing a coin and then spinning Spinner A.

B *If a single die is tossed once, find the probability of each event. See Examples 3 through 5.*

11. A 5

12. A 7

13. A 1 or a 4

14. A 2 or a 3

15. An even number

16. An odd number

Suppose the spinner shown is spun once. Find the probability of each event. See Examples 3 through 5.

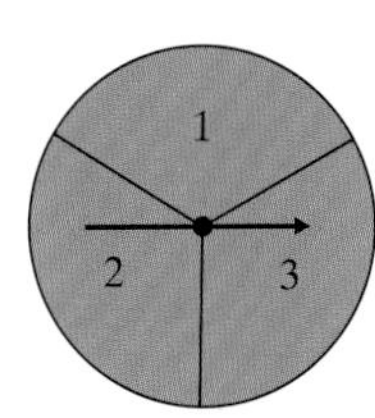

17. The result of the spin is 2.

18. The result of the spin is 3.

19. The result of the spin is an odd number.

20. The result of the spin is an even number.

If a single choice is made from the bag of marbles shown, find the probability of each event. See Examples 3 through 5.

21. A red marble is chosen.

22. A blue marble is chosen.

23. A yellow marble is chosen.

24. A green marble is chosen.

A new drug is being tested that is supposed to lower blood pressure. This drug was given to 200 people and the results are below.

Lower Blood Pressure	Higher Blood Pressure	Blood Pressure Not Changed
152	38	10

25. If a person is testing this drug, what is the probability that their blood pressure will be higher?

26. If a person is testing this drug, what is the probability that their blood pressure will be lower?

27. If a person is testing this drug, what is the probability that their blood pressure will not change?

28. What is the sum of the answers to exercises 25, 26, and 27? In your own words, explain why.

Review and Preview

Perform each indicated operation. See Sections 4.3, 4.4, and 4.5.

29. $\frac{1}{2} + \frac{1}{3}$ **30.** $\frac{7}{10} - \frac{2}{5}$ **31.** $\frac{1}{2} \cdot \frac{1}{3}$ **32.** $\frac{7}{10} \div \frac{2}{5}$ **33.** $5 \div \frac{3}{4}$ **34.** $\frac{3}{5} \cdot 10$

Combining Concepts

Recall that a deck of cards contains 52 cards. These cards consist of four suits (hearts, spades, clubs, and diamonds) of each of the following: 2, 3, 4, 5, 6, 7, 8, 9, 10, jack, queen, king, and ace. If a card is chosen from a deck of cards, find the probability of each event.

35. The king of hearts

36. The 10 of spades

37. A king

38. A 10

39. A heart

40. A club

Two dice are tossed. Find the probability of each sum of the dice. (Hint: Draw a tree diagram of the possibilities of two tosses of a die, and then find the sum of the numbers on each branch.)

41. A sum of 4

42. A sum of 11

43. A sum of 13

44. A sum of 2

45. In your own words, explain why the probability of an event cannot be greater than 1.

46. In your own words, explain when the probability of an event is 0.

CHAPTER 7 ACTIVITY Conducting a Survey

This activity may be completed by working in groups or individually.

How often have you read an article in a newspaper or in a magazine that included results from a survey or poll? Surveys seem to have become very popular ways of getting feedback on anything from a political candidate, to a new product, to services offered by a health club. In this activity, you will conduct a survey and analyze the results.

1. Conduct a survey of 30 students in one of your classes. Ask each student to report his or her age.
2. Classify each age according to the following categories: under 20, 20 to 24, 25 to 29, 30 to 39, 40 to 49, and 50 or over. Tally the number of your survey respondents that fall into each category. Make a bar graph of your results. What does this graph tell you about the ages of your survey respondents?
3. Find the average age of your survey respondents.
4. Find the median age of your survey respondents.
5. Find the mode of the ages of your survey respondents.
6. Compare the mean, median, and mode of your age data. Are these measures similar? Which is largest? Which is smallest? If there is a noticeable difference between any of these measures, can you explain why?

FOCUS ON Mathematical Connections

STEM-AND-LEAF DISPLAYS

Stem-and-leaf displays are another way to organize data. After data are logically organized, it can be much easier to draw conclusions from them.

Suppose we have collected the following set of data. It could represent the test scores for an algebra class or the pulse rates of a group of small children.

90	73	93	99	79	95	69	78	93	80
89	85	97	78	75	79	72	76	97	88
83	98	72	94	92	79	70	98	85	99

In a stem-and-leaf display, the last digit of each number forms the *leaf*, and the remaining digits to the left form the *stem*. For the first number in the list, 90, 9 is the stem and 0 is the leaf. To make the stem-and-leaf display, we write all of the stems in numerical order in a column. Then we write each leaf on the horizontal line next to its stem, aligning leaves in vertical columns. In this case, because the data range from 69 to 99, we use the stems 6, 7, 8, and 9. Each line of the display represents an interval of data; for instance, the line corresponding to the stem **7** represents all data that fall in the interval **70** to **79,** inclusive. After the data have been divided into stems and leaves on the display as shown in the table below, we simply rearrange the leaves on each line to appear in numerical order, as shown in the table in the right column.

Stem	Leaf		Stem	Leaf
6	9		6	9
7	39885926290	⟶	7	02235688999
8	095835		8	035589
9	039537784289		9	023345778899

Now that the data have been organized into a stem-and-leaf display, it is easy to answer questions about the data such as: What are the least and greatest values in the set of data? Which data interval contains the most items from the data set? How many values fall between 74 and 84? Which data value occurs most frequently in the data set? What patterns or trends do you see in the data?

CRITICAL THINKING

Make a stem-and-leaf display of the weekend emergency room admission data shown on the right. Then answer the following questions:

1. What is the difference between the least number and the greatest number of weekend ER admissions?
2. How many weekends had between 125 and 165 ER admissions?
3. What number of weekend ER admissions occurred most frequently?
4. Which interval contains the most weekend ER admissions?

Number of Emergency Room Admissions on Weekends				
198	168	117	185	159
160	177	169	112	175
170	188	137	117	145
198	169	154	163	192
167	179	155	133	121
162	188	124	145	146
128	181	198	149	140
122	162	161	180	177

Chapter 7 Vocabulary Check

Fill in each blank with one of the words or phrases listed below.

outcomes	bar	experiment	mean	tree diagram
pictograph	line	circle	median	probability
histogram		class frequency	mode	class interval

1. A ______ graph presents data using vertical or horizontal bars.
2. The ________ of a set of number items is $\frac{\text{sum of items}}{\text{number of items}}$.
3. The possible results of an experiment are the __________ .
4. A __________ is a graph in which pictures or symbols are used to visually present data.
5. The ________ of a set of numbers is the number that occurs most often.
6. A ______ graph displays information with a line that connects data points.
7. The __________ of an ordered set of numbers is the middle number.
8. A ____________ is one way to picture and count outcomes.
9. An __________ is an activity being considered, such as tossing a coin or rolling a die.
10. In a ________ graph, each section (shaped like a piece of pie) shows a category and the relative size of the category.
11. The __________ of an event is $\frac{\text{number of ways that the event can occur}}{\text{number of possible outcomes}}$.
12. A __________ is a special bar graph in which the width of each bar represents a __________ and the height of each bar represents the __________ .

CHAPTER 7 Highlights

DEFINITIONS AND CONCEPTS | **EXAMPLES**

SECTION 7.1 READING PICTOGRAPHS, BAR GRAPHS, AND LINE GRAPHS

A **pictograph** is a graph in which pictures or symbols are used to visually present data.

A **line graph** displays information with a line that connects data points.

A **bar graph** presents data using vertical or horizontal bars.

The bar graph on the right shows the number of acres of wheat harvested in 1996 for leading states.

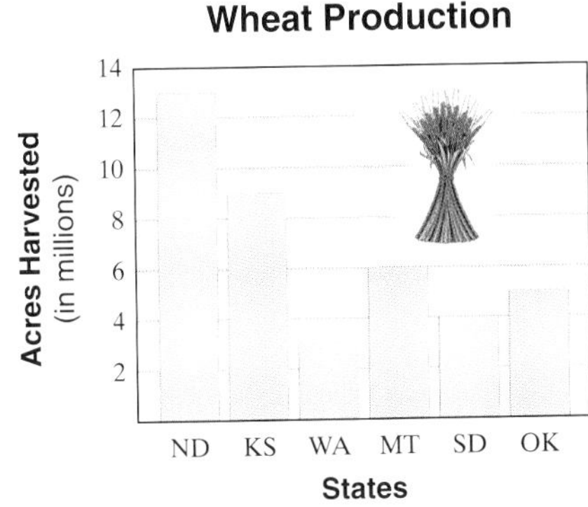

Source: U.S. Department of Agriculture

1. Approximately how many acres of wheat were harvested in Kansas?

   9,000,000 acres

2. About how many more acres of wheat were harvested in North Dakota than South Dakota?

   13 million
   − 4 million
   9 million or 9,000,000 acres

Definitions and Concepts

Examples

Section 7.1 Reading Pictographs, Bar Graphs, and Line Graphs (*continued*)

A **histogram** is a special bar graph in which the width of each bar represents a **class interval** and the height of each bar represents the **class frequency.** The histogram on the right shows student quiz scores.

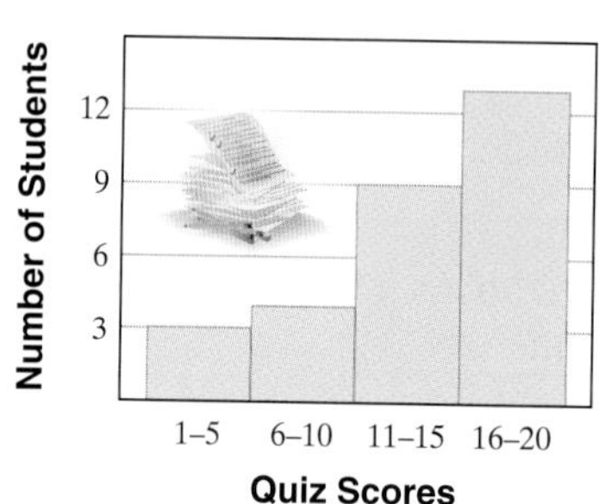

1. How many students received a score of 6–10?

4 students

2. How many students received a score of 11–20?

$9 + 13 = 22$ students

Section 7.2 Reading Circle Graphs

In a **circle graph,** each section (shaped like a piece of pie) shows a category and the relative size of the category.

The circle graph on the right classifies tornadoes by wind speed.

Tornado Wind Speeds

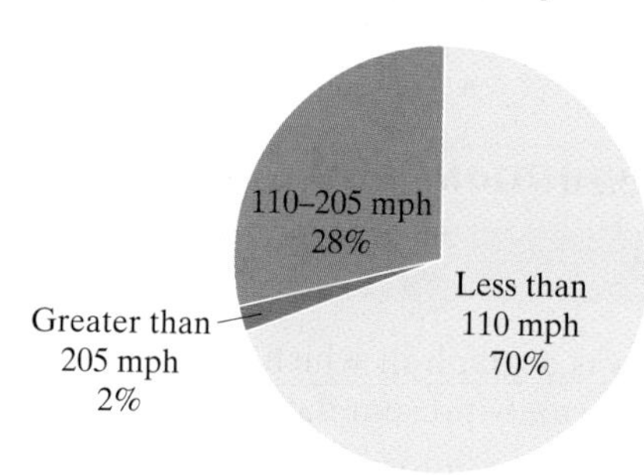

Source: National Oceanic and Atmospheric Administration

1. What percent of tornadoes have wind speeds of 110 mph or greater?

$28\% + 2\% = 30\%$

2. If there were 1235 tornadoes in the United States in 1995, how many of these might we expect to have had wind speeds less than 110 mph? Find 70% of 1235.

$70\%(1235) = 0.70(1235) = 864.5 \approx 865$

Around 865 tornadoes would be expected to have had wind speeds of less than 110 mph.

Definitions and Concepts

Examples

Section 7.3 Mean, Median, and Mode

The **mean** (or **average**) of a set of number items is

$$\text{mean} = \frac{\text{sum of items}}{\text{number of items}}$$

The **median** of an ordered set of numbers is the middle number. If the number of items is even, the median is the mean of the two middle numbers.

The **mode** of a set of numbers is the number that occurs most often. (A set of numbers may have no mode or more than one mode.)

Find the mean, median, and mode of the following set of numbers: 33, 35, 35, 43, 68, 68

$$\text{mean} = \frac{33 + 35 + 35 + 43 + 68 + 68}{6} = 47$$

The median is the mean of the two middle numbers:

$$\text{median} = \frac{35 + 43}{2} = 39$$

There are two modes because there are two numbers that occur twice:

35 and 68

Section 7.4 Counting and Introduction to Probability

An **experiment** is an activity being considered, such as tossing a coin or rolling a die. The possible results of an experiment are the **outcomes.** A **tree diagram** is one way to picture and count outcomes.

Any number of outcomes considered together is called an **event.** The **probability** of an event is a measure of the chance or likelihood of it occurring.

$$\text{probability of an event} = \frac{\text{number of ways that the event can occur}}{\text{number of possible outcomes}}$$

Draw a tree diagram for tossing a coin and then choosing a number from 1 to 4.

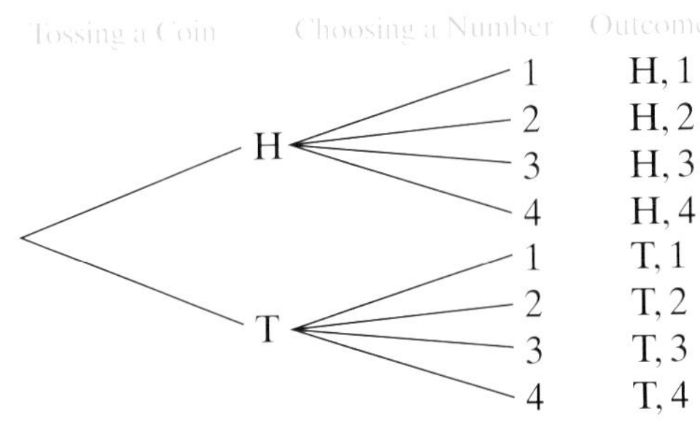

Find the probability of tossing a coin twice and tails occurring each time.

1 way the event can occur

(H,H), (H,T), (T,H), (T,T)

4 possible outcomes

$$\text{probability} = \frac{1}{4}$$

Name ______________________ Section ____________ Date ____________

Chapter 7 Review

(7.1) *The following pictograph shows the number of new homes constructed in 2000, by state. Use this graph to answer Exercises 1 through 6.*

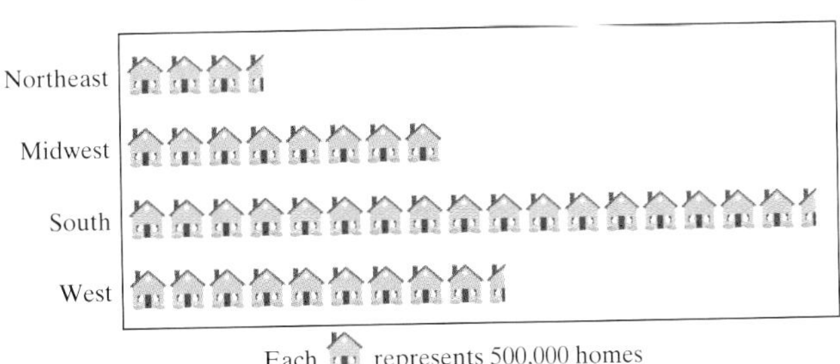

Source: U.S. Census Bureau

1. How many housing starts were there in the Midwest in 2000?

2. How many housing starts were there in the Northeast in 2000?

3. Which region had the most housing starts?

4. Which region had the fewest housing starts?

5. Which region(s) had 4,000,000 or more housing starts?

6. Which region(s) had fewer than 4,000,000 housing starts?

The following bar graph shows the percent of persons age 25 or over who completed four or more years of college. Use this graph to answer Exercises 7 through 10.

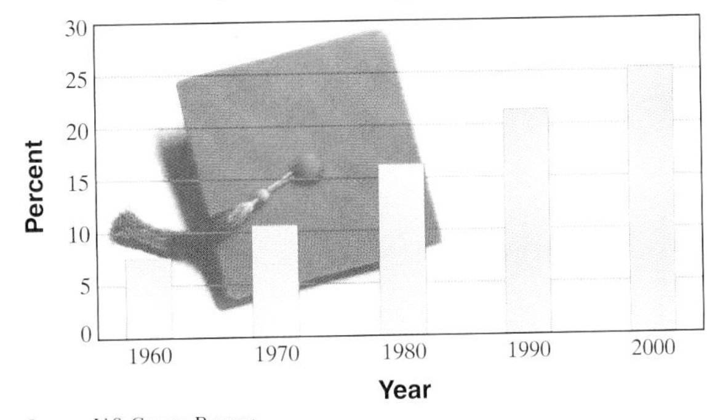

Source: U.S. Census Bureau

7. Approximate the percent of persons who completed four or more years of college in 1960.

8. What year shown had the greatest percent of persons completing four or more years of college?

9. What years shown had 15% or more of persons completing four or more years of college?

10. Describe any patterns you notice in this graph.

The following line graph shows the average price of a 30-second television advertisement during the Super Bowl for the years shown. Use this graph to answer Exercises 11 through 15.

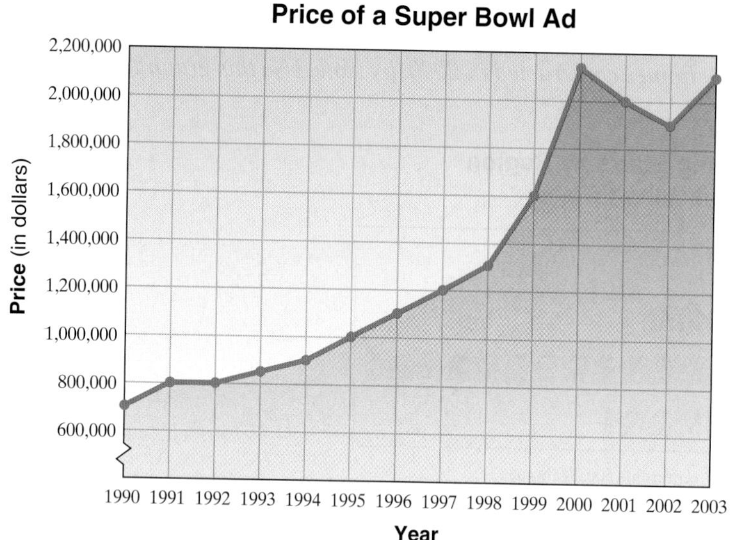

Sources: Nielsen Media Research and *Advertising Age* research

11. Approximate the price of a Super Bowl ad in 2003.

12. Approximate the price of a Super Bowl ad in 1997.

13. Between which two years did the price of a Super Bowl ad *not* increase?

14. Between which two years did the price of a Super Bowl ad increase the most?

15. During which years was the price of a Super Bowl ad *less than* $1,000,000?

The following histogram shows the hours worked per week by the employees of Southern Star Furniture. Use this histogram to answer Exercises 16 through 19.

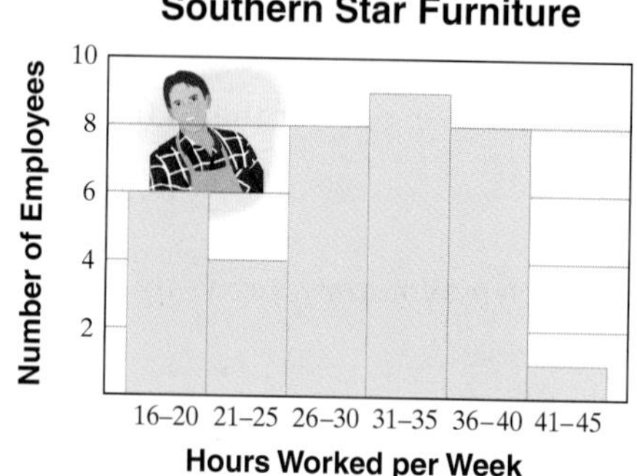

16. How many employees work 21–25 hours per week?

17. How many employees work 41–45 hours per week?

18. How many employees work 36 hours or more per week?

19. How many employees work 30 hours or less per week?

Following is a list of monthly record high temperatures for New Orleans, Louisiana. Use this list to complete the frequency distribution table below.

83	96	101	92
85	100	92	102
89	101	87	84

	Class Intervals (Temperatures)	Tally	Class Frequency (Number of Months)
20.	80°–89°	______	______
21.	90°–99°	______	______
22.	100°–109°	______	______

23. Use the table from Exercises 20, 21, and 22 to draw a histogram.

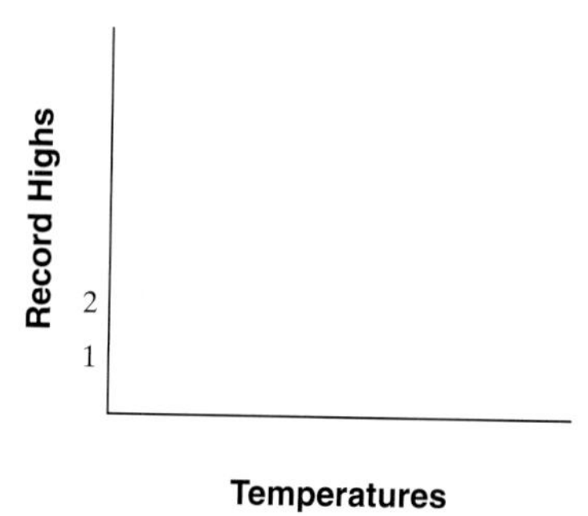

(7.2) *The following circle graph shows a family's \$4000 monthly budget. Use this graph to answer Exercises 24 through 30. Write all ratios as fractions in simplest form.*

24. What is the largest budget item?

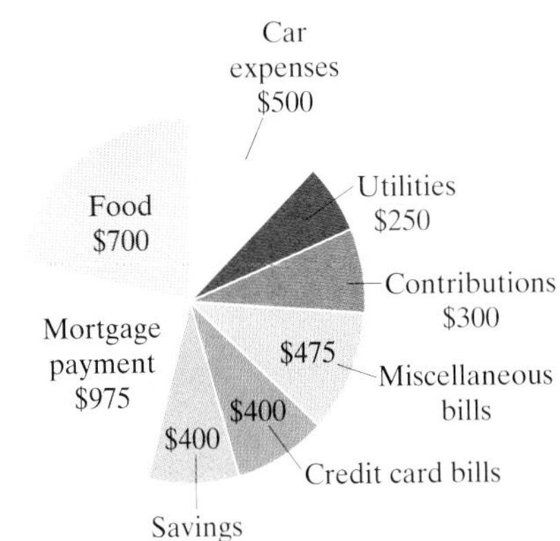

25. What is the smallest budget item?

26. How much money is budgeted for the mortgage payment and utilities?

27. How much money is budgeted for savings and contributions?

28. Find the ratio of the mortage payment to the total monthly budget.

29. Find the ratio of food to the total monthly budget.

30. Find the ratio of car expenses to food.

The following circle graph shows the percent of states with various rural interstate highway speed limits in 2000. Use this graph to determine the number of states with each speed limit in Exercises 31 through 34.

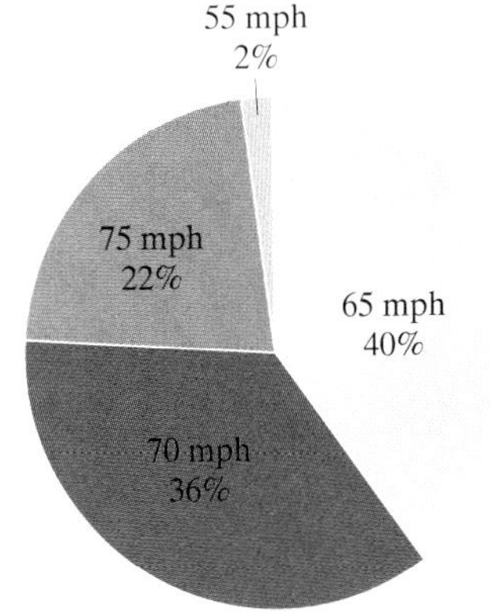

31. How many states have a rural interstate highway speed limit of 65 mph?

32. How many states have a rural interstate highway speed limit of 75 mph?

33. How many states have a rural interstate highway speed limit of 55 mph?

34. How many states have a rural interstate highway speed limit of 70 mph or 75 mph?

(7.3) *Find the mean, median, and any mode(s) for each list of numbers.*

35. 13, 23, 33, 14, 6

36. 45, 21, 60, 86, 64

37. $14,000, $20,000, $12,000, $20,000, $36,000, $45,000

38. 560, 620, 123, 400, 410, 300, 400, 780, 430, 450

For Exercises 39 and 40, the grades are given for a student for a particular semester. Find each grade point average. If necessary, round the grade point average to the nearest hundredth.

39.

Grade	Credit Hours
A	3
A	3
C	2
B	3
C	1

40.

Grade	Credit Hours
B	3
B	4
C	2
D	2
B	3

(7.4) *Draw a tree diagram for each experiment. Then use the diagram to determine the number of outcomes.*

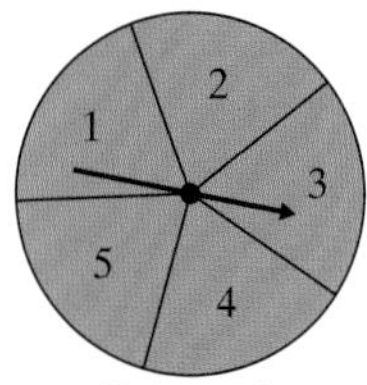

Spinner 1

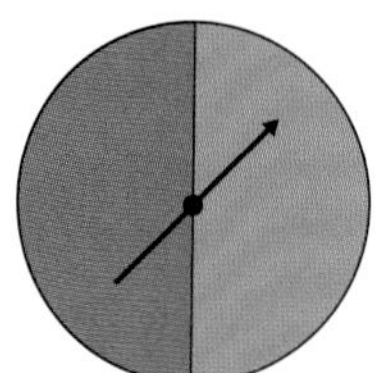
Spinner 2

41. Tossing a coin and then spinning Spinner 1

42. Spinning Spinner 2 and then tossing a coin

43. Spinning Spinner 1 twice

44. Spinning Spinner 2 twice

45. Spinning Spinner 1 and then Spinner 2

Find the probability of each event.

Die

46. Rolling a 4 on a die

47. Rolling a 3 on a die

48. Spinning a 4 on Spinner 1

49. Spinning a 3 on Spinner 1

50. Spinning either a 1, 3, or 5 on Spinner 1

51. Spinning either a 2 or a 4 on Spinner 1

STUDY SKILLS REMINDER

Tips for studying for an exam

To prepare for an exam, try the following study techniques.

- Start the study process days before your exam.
- Make sure that you are current and up-to-date on your assignments.
- If there is a topic that you are unsure of, use one of the many resources that are available to you. For example,

  See your instructor.

  Visit a learning resource center on campus where math tutors are available.

  Read the textbook material and examples on the topic.

  View a videotape on the topic.

- Reread your notes and carefully review the Chapter Highlights at the end of the chapter.
- Work the review exercises at the end of the chapter and check your answers. Make sure that you correct any missed exercises. If you have trouble on a topic, use a resource listed above.
- Find a quiet place to take the Chapter Test found at the end of the chapter. Do not use any resources when taking this sample test. This way you will have a clear indication of how prepared you are for your exam. Check your answers and make sure that you correct any missed exercises.
- Get lots of rest the night before the exam. It's hard to show how well you know the material if your brain is foggy from lack of sleep.

Good luck, and keep a positive attitude.

Name ______________________ Section __________ Date __________

Answers

1. ______

2. ______

3. ______

4. ______

5. ______

6. ______

Chapter 7 Test

Remember to check your answers and use the Chapter Test Prep Video to view solutions.

The following pictograph shows the money collected each week from a wrapping paper fundraiser. Use this graph to answer Exercises 1 through 3.

1. How much money was collected during the second week?

2. During which week was the most money collected? How much money was collected during that week?

3. What was the total money collected for the fundraiser?

The bar graph shows the normal monthly precipitation in centimeters for Chicago, Illinois. Use this graph to answer Exercises 4–6.

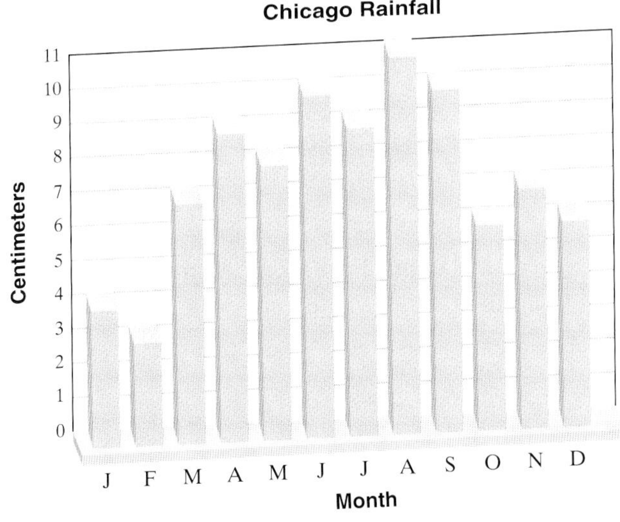

Source: U.S. National Oceanic and Atmospheric Administration, *Climatography of the United States*, No. 81

4. During which month(s) does Chicago normally have greater than 9 centimeters of rainfall?

5. During which month does Chicago normally have the least amount of rainfall? How much rain falls during that month?

6. During which month(s) does 7 centimeters of rain normally fall?

7. see graph

7. Use the information in the table to draw a bar graph. Clearly label each bar.

Countries with the Highest Newspaper Cirulations	
Country	**Average Daily Circulation (in millions)**
Japan	72
US	56
China	50
India	31
Germany	24
Russia	24
UK	19

(*Source:* World Association of Newspapers

8. ______

The result of a survey of 200 people is shown in the following circle graph. Each person was asked to tell his or her favorite type of music. Use this graph to answer Exercises 8 and 9.

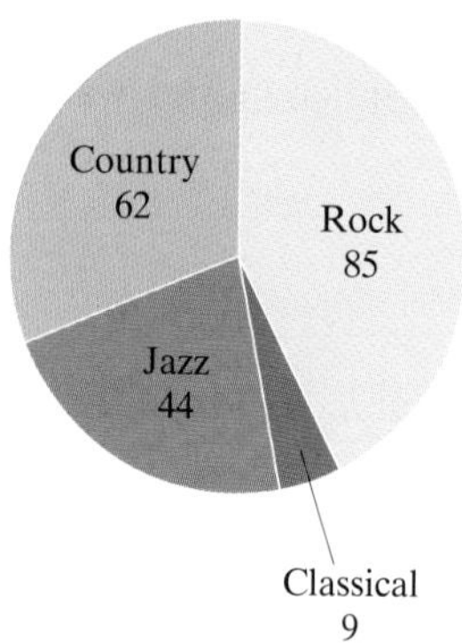

8. Find the ratio of those who prefer rock music to the total number surveyed.

9. Find the ratio of those who prefer country music to those who prefer jazz.

9. ______

The following circle graph shows the U.S. labor force employment by industry for a recent year. This graph is based on 132,000,000 people employed by these industries in the United States. Use the graph to find how many people were employed by the industries given in Exercises 10 and 11.

10. ______

U.S. Labor Force Employment by Industry

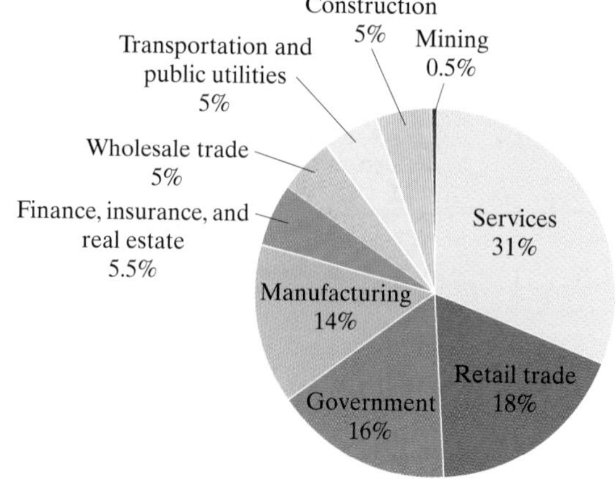

Source: Bureau of Labor Statistics

11. ______

10. Services

11. Government

A professor measures the heights of the students in her class. The results are shown in the following histogram. Use this histogram to answer Exercises 12 and 13.

12. ______

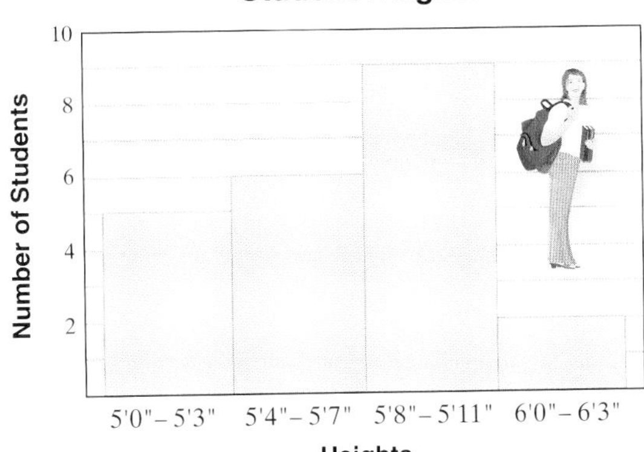

13. ______

12. How many students are 5′8″–5′11″ tall?

13. How many students are 5′7″ or shorter?

14. The history test scores of 25 students are shown below. Use these scores to complete the frequency distribution table.

70	86	81	65	92
43	72	85	69	97
82	51	75	50	68
88	83	85	77	99
77	63	59	84	90

Class Intervals (Scores)	Tally	Class Frequency (Number of Students)
40–49	____	____
50–59	____	____
60–69	____	____
70–79	____	____
80–89	____	____
90–99	____	____

15. Use the results of Exercise 14 to draw a histogram.

Number of Students

2

Scores

14. see table

15. see graph

16. ______

17. ______

18. ______

19. ______

20. ______

21. ______

22. ______

Find the mean, median, and mode of each list of numbers.

16. 26, 32, 42, 43, 49

17. 8, 10, 16, 16, 14, 12, 12, 13

Find the grade point average. If necessary, round to the nearest hundredth.

18.

Grade	Credit Hours
A	3
B	3
C	3
B	4
A	1

19. Draw a tree diagram for the experiment of spinning the spinner twice.

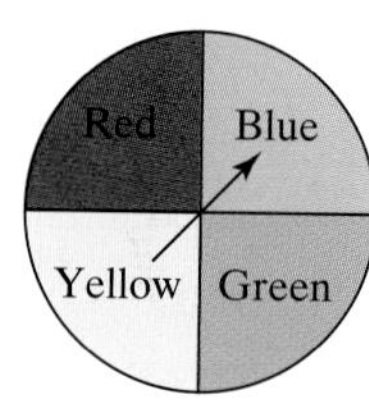

20. Draw a tree diagram for the experiment of tossing a coin twice.

Suppose that the numbers 1 through 10 are each written on the same size sheet of paper and placed in a bag. You then select one sheet of paper from the bag.

21. What is the probability of choosing a 6 from the bag?

22. What is the probability of choosing a 3 or a 4 from the bag?

Name ____________________ Section ________ Date ________

Answers

Chapter 7 Cumulative Review

1. Simplify: $4^3 + [3^2 - (10 \div 2)] - 7 \cdot 3$

2. $7^2 - [5^3 + (6 \div 3)] + 4 \cdot 2$

3. Evaluate $x - y$ for $x = -3$ and $y = 9$.

4. Evaluate $x - y$ for $x = 7$ and $y = -2$.

5. Solve: $3y - 7y = 12$

6. Solve: $2x - 6x = 24$

7. Solve: $\frac{x}{6} + 1 = \frac{4}{3}$

8. Solve: $\frac{7}{2} + \frac{a}{4} = 1$

9. Add: $2\frac{1}{3} + 5\frac{3}{8}$

10. Add: $3\frac{2}{5} + 4\frac{3}{4}$

11. Write 5.6 as a mixed number.

12. Write 2.8 as a mixed number.

13. Subtract: $3.5 - 0.068$

14. Subtract: $7.4 - 0.073$.

15. Multiply: 0.283×0.3

16. Multiply: 0.147×0.2.

17. Divide: $-5.98 \div 115$

18. Divide: $27.88 \div 205$.

19. Simplify: $(-1.3)^2 + 2.4$

20. Simplify: $(-2.7)^2$

21. Write $\frac{1}{4}$ as a decimal.

22. Write $\frac{3}{8}$ as a decimal.

23. Solve: $5(x - 0.36) = -x + 2.4$

24. Solve: $4(0.35 - x) = x - 7$

25. Approximate $\sqrt{32}$ to the nearest thousandth.

26. Approximate $\sqrt{60}$ to the nearest thousandth.

1. ______
2. ______
3. ______
4. ______
5. ______
6. ______
7. ______
8. ______
9. ______
10. ______
11. ______
12. ______
13. ______
14. ______
15. ______
16. ______
17. ______
18. ______
19. ______
20. ______
21. ______
22. ______
23. ______
24. ______
25. ______
26. ______

27. __________

28. __________

29. __________

30. __________

31. __________

32. __________

33. __________

34. __________

35. __________

36. __________

37. __________

38. __________

39. __________

40. __________

41. __________

42. __________

Write each percent as a decimal.

27. 4.6%

28. 32%

29. 0.74%

30. 2.7%

31. What number is 35% of 60?

32. What number is 40% of 36?

33. 20.8 is 40% of what number?

34. 9.5 is 25% of what number?

35. The sales tax on a $300 printer is $22.50. Find the sales tax rate.

36. The sales tax on a $2.00 yo-yo is $0.13. Find the sales tax rate.

37. Find the monthly payment on a $2000 loan for 2 years. The interest on the 2-year loan is $435.88.

38. Linda Bonnett borrows $1600 for 1 year. If the interest is $128.60, find the monthly payment.

39. Find the median of the list of scores: 67, 91, 75, 86, 55, 91

40. Find the median of the list of numbers: 43, 46, 47, 50, 52, 83

41. If a die is rolled one time, find the probability of rolling a 3 or a 4.

42. If a die is rolled once, find the probability of rolling an even number.

Geometry and Measurement

The word "geometry" is formed from the Greek words *geo*, meaning Earth, and *metron*, meaning measure. Geometry literally means to measure the Earth. In this chapter, we learn about various geometric figures and their properties such as perimeter, area, and volume. Knowledge of geometry can help us solve practical problems in real-life situations. For instance, knowing certain measures of a circular swimming pool allows us to calculate how much water it can hold.

CHAPTER

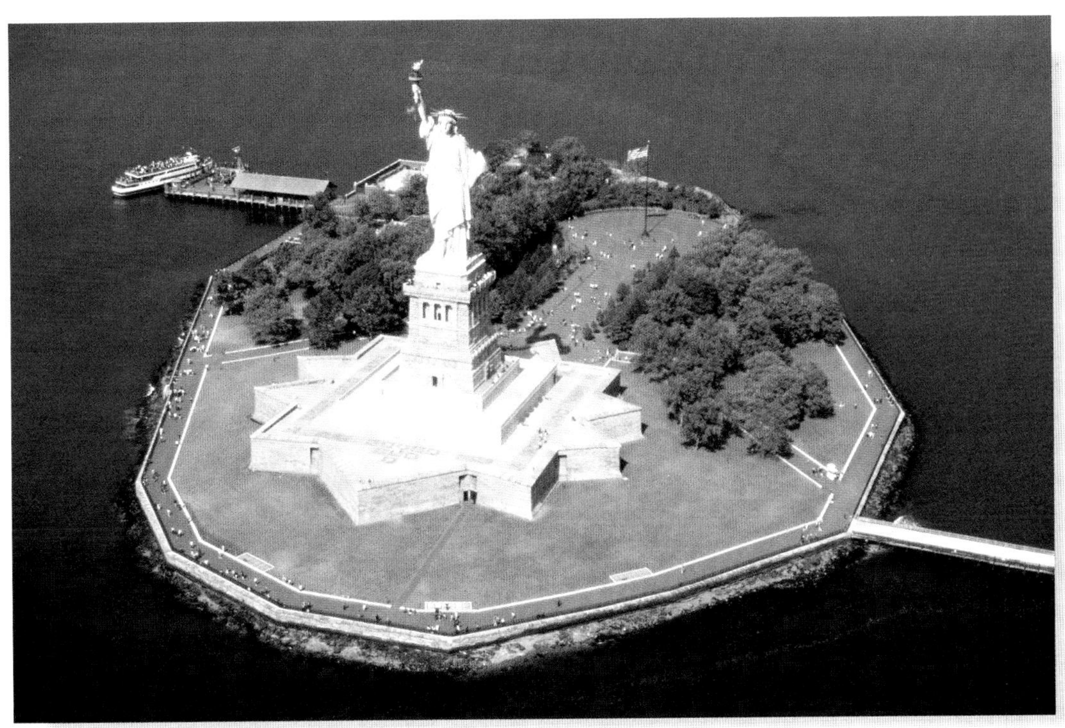

The Statue of Liberty, built in the late 1800s, was the inspiration of Frenchman Edouard-Rene Lefebvre de Laboulaye. His idea was to present a monument to the American people, from the people of France, that celebrated their common ideals of freedom and liberty. French designer and sculptor Frederic Auguste Bartholdi oversaw the building of the statue and chose its location in New York Harbor. The statue was dedicated on October 28, 1886, under its official title *Liberty Enlightening the World*. The poem "The New Colossus," by Emma Lazarus, was placed on the statue in 1903. Its verse, "Give me your tired, your poor, your huddled masses yearning to breathe free," identifies the statue as a powerful symbol of welcome to immigrants from all over the world. In Exercises 39 and 40, Section 8.2, we will explore some measurements of parts of the Statue of Liberty.

Answers

1. ____________
2. ____________
3. ____________
4. ____________
5. ____________
6. ____________
7. ____________
8. ____________
9. ____________
10. ____________
11. ____________
12. ____________
13. ____________
14. ____________
15. ____________
16. ____________
17. ____________

Name ____________ Section ____________ Date ____________

Chapter 8 Pretest

Classify each angle as acute, right, obtuse, or straight.

△ **1.**

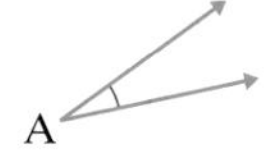

△ **2.**

B

△ **3.** Find the supplement of a 92° angle.

△ **4.** Find the measures of x, y, and z.

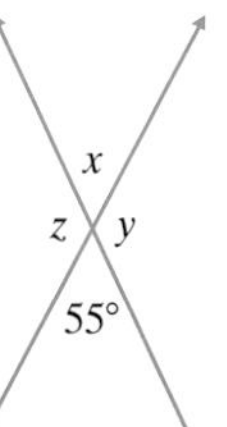

△ **5.** Find the measure of $\angle x$.

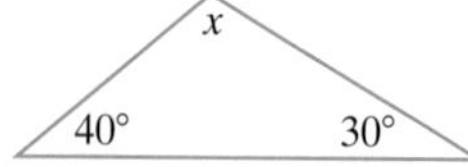

6. Convert 11 feet to yards.

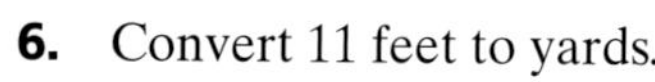

7. Convert 6,250,000 cm to kilometers.

△ **8.** Find the perimeter of a rectangle with a length of 24 inches and a width of 6 inches.

△ **9.** Find the circumference of the given circle. Use $\pi \approx 3.14$.

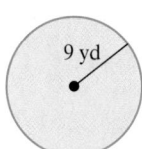

△ **10.** Find the area of the given triangle.

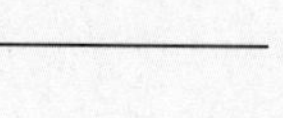

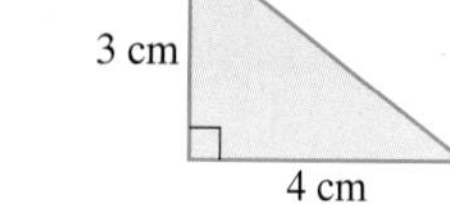

11. Divide 15 lb 8 oz by 2.

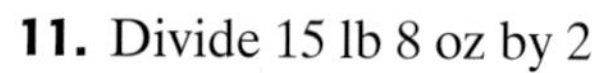

12. Subtract 2 qt from 9 gal 1 qt.

13. Convert 25 L to millileters.

14. Convert 118°F to Celsius. If necessary, round to the nearest tenth of a degree.

△ **15.** Find the ratio of the corresponding sides of the given similar triangles.

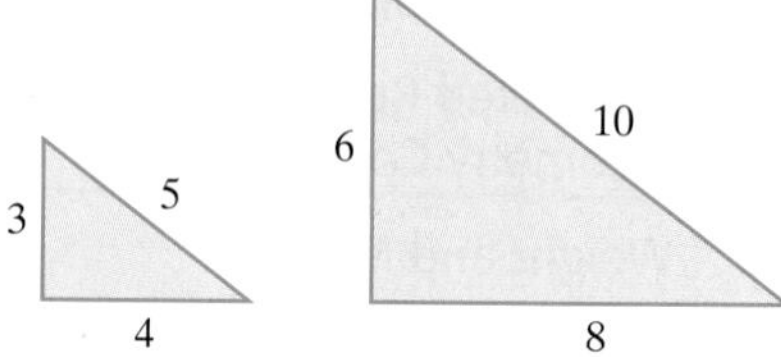

Given that the pairs of triangles are similar, find the length of the side labeled n.

△ **16.**

17. If a 40-foot tree casts a 22-foot shadow, find the length of the shadow cast by a 36-foot tree.

8.1 Lines and Angles

OBJECTIVES

- **A** Identify lines, line segments, rays, and angles.
- **B** Classify angles as acute, right, obtuse, or straight.
- **C** Identify complementary and supplementary angles.
- **D** Find measures of angles.

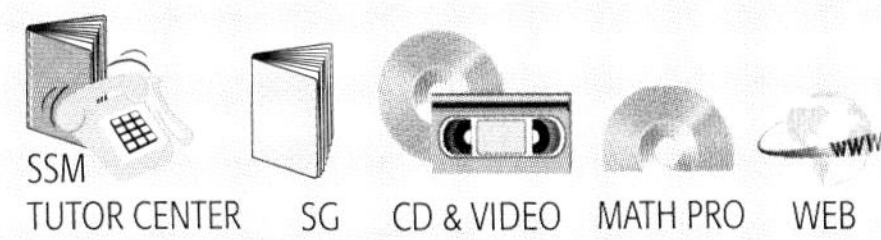

A Identify Lines, Line Segments, Rays, and Angles

Let's begin with a review of two important concepts—plane and space.

A **plane** is a flat surface that extends indefinitely in all directions. Surfaces like portions of a plane are a classroom floor or a blackboard or whiteboard.

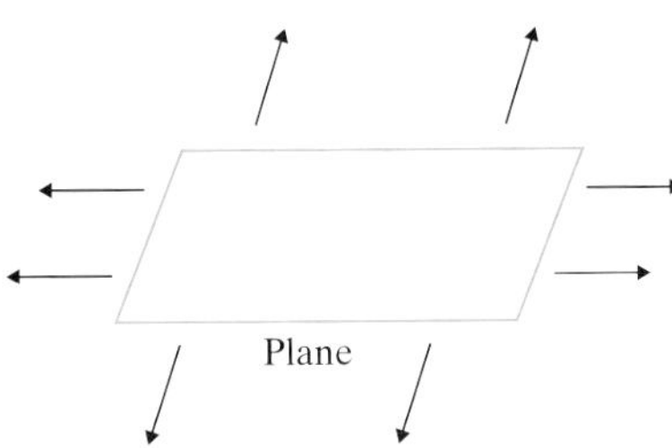

Space extends in all directions indefinitely. Examples of objects in space are houses, grains of salt, bushes, your textbook, and you.

The most basic concept of geometry is the idea of a point in space. A **point** has no length, no width, and no height, but it does have location. We represent a point by a dot, and we label points with letters.

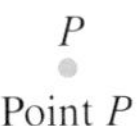

Point P

A **line** is a set of points extending indefinitely in two directions. A line has no width or height, but it does have length. We name a line by any two of its points. The line below to the left is named $\overleftrightarrow{AB}$ or $\overleftrightarrow{BA}$. A **line segment** is a piece of a line with two endpoints. The line segment below with endpoints A and B is named $\overline{AB}$ or $\overline{BA}$.

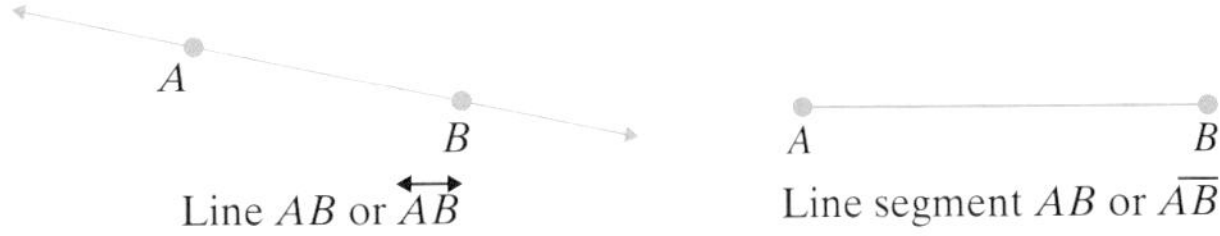

Line AB or $\overleftrightarrow{AB}$ Line segment AB or $\overline{AB}$

A **ray** is a part of a line with one endpoint. A ray extends indefinitely in one direction. To name a ray, name its endpoint first, then any other point on the ray. Below is ray AB, or $\overrightarrow{AB}$. An **angle** is made up of two rays that share the same endpoint. The common endpoint is called the **vertex.**

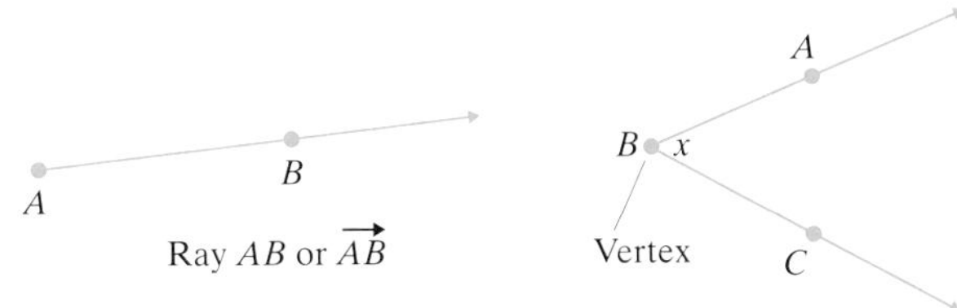

Ray AB or $\overrightarrow{AB}$

The angle in the figure above can be named

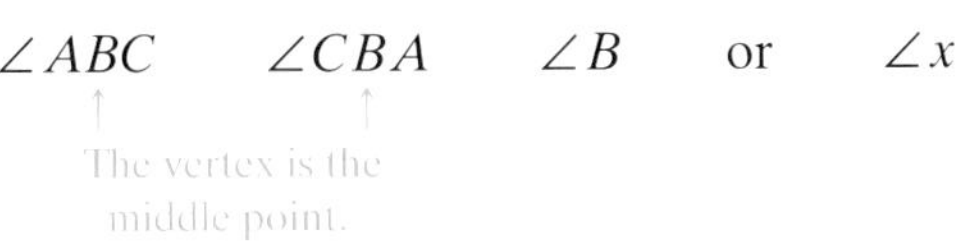

$\angle ABC$ $\angle CBA$ $\angle B$ or $\angle x$
The vertex is the middle point.

Helpful Hint

Use the vertex alone to name an angle only when there is no confusion as to what angle is being named.

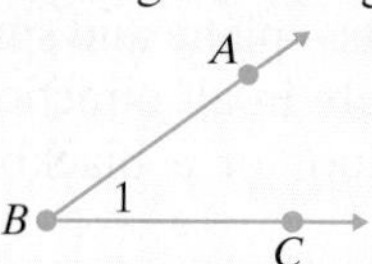

You may use ∠B to name the angle above. There is no confusion. ∠B means ∠1.

Do not use ∠B to name the angle above There is confusion. Does ∠B mean ∠1, ∠2, ∠3, or ∠4?

Rays BA and BC are **sides** of the angle.

Practice Problem 1

Identify each figure as a line, a ray, a line segment, or an angle. Then name the figure using the given points.

a. b. c. d.

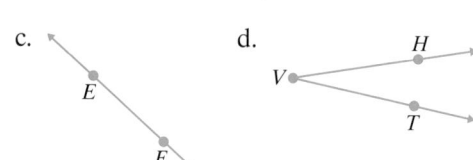

EXAMPLE 1 Identify each figure as a line, a ray, a line segment, or an angle. Then name the figure using the given points.

a.

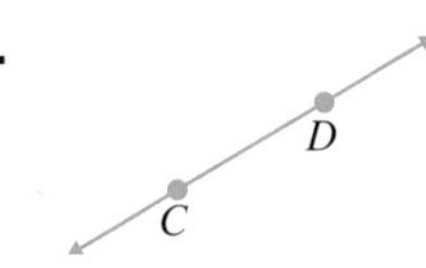

b.

c.

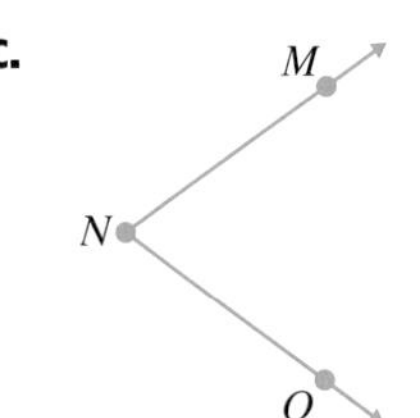

d.

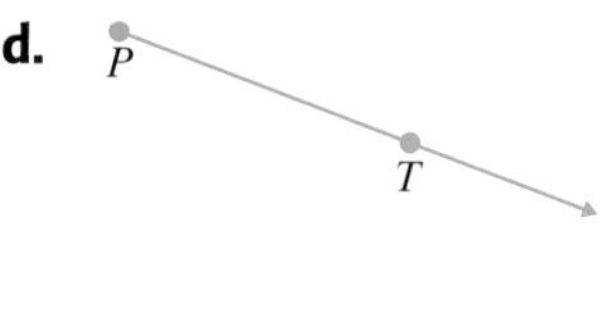

Solution: Figure (a) extends indefinitely in two directions. It is line CD or $\overleftrightarrow{CD}$. It can also be named line DC or $\overleftrightarrow{DC}$.

Figure (b) has two endpoints. It is line segment EF or $\overline{EF}$. It can also be named line segment FE or $\overline{FE}$.

Figure (c) has two rays with a common endpoint. It is an angle that can be named ∠MNO, ∠ONM, or ∠N.

Figure (d) is part of a line with one endpoint. It is ray PT or $\overrightarrow{PT}$

Practice Problem 2

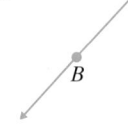

Use the figure in Example 2 to list other ways to name ∠z.

EXAMPLE 2 List other ways to name ∠y.

Solution: Two other ways to name ∠y are ∠QTR and ∠RTQ. We may *not* use the vertex alone to name this angle because three different angles have T as their vertex.

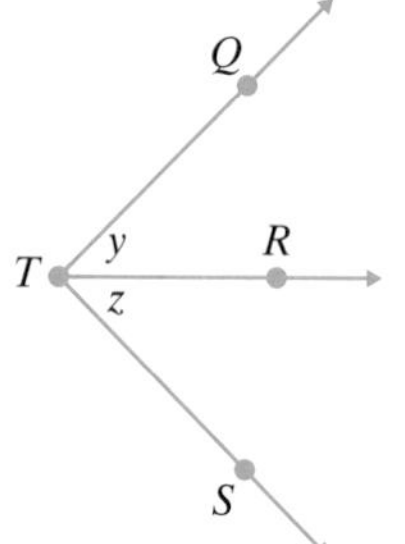

B Classifying Angles as Acute, Right, Obtuse, or Straight

An angle can be measured in **degrees.** The symbol for degrees is a small, raised circle, °. There are 360° in a full revolution, or full circle.

Answers

1. **a.** line segment; line segment RS or $\overline{RS}$ as well as SR or $\overline{SR}$.
   b. ray; ray AB or $\overrightarrow{AB}$
   c. line; line EF or $\overleftrightarrow{EF}$ as well as FE or $\overleftrightarrow{FE}$
   d. angle; ∠TVH or ∠HVT or ∠V
2. ∠RTS, ∠STR

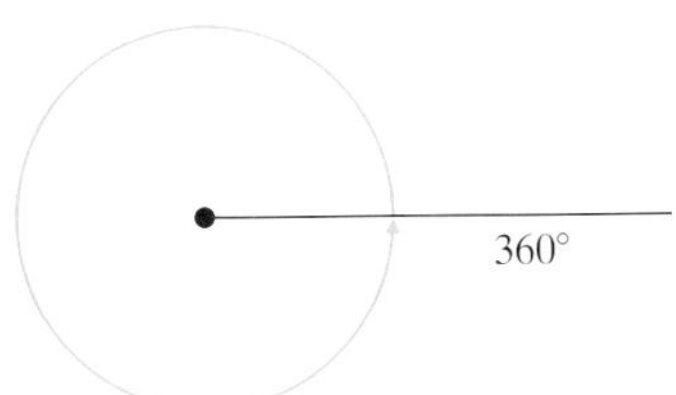

$\frac{1}{2}$ of a revolution measures $\frac{1}{2}(360°) = 180°$. An angle that measures 180° is called a **straight angle.**

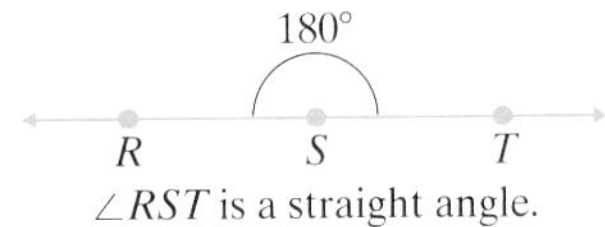

$\angle RST$ is a straight angle.

$\frac{1}{4}$ of a revolution measures $\frac{1}{4}(360°) = 90°$. An angle that measures 90° is called a **right angle.** The symbol ⊾ is used to denote a right angle.

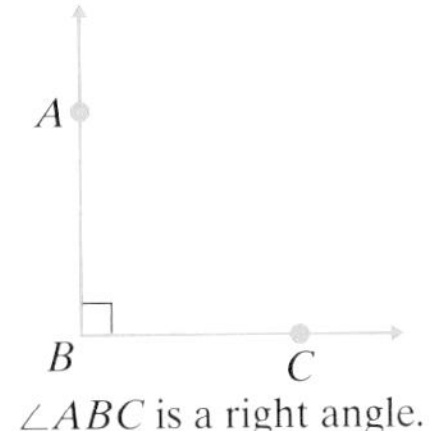

$\angle ABC$ is a right angle.

An angle whose measure is between 0° and 90° is called an **acute angle.**

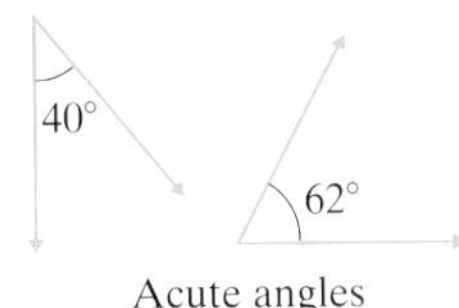

Acute angles

An angle whose measure is between 90° and 180° is called an **obtuse angle.**

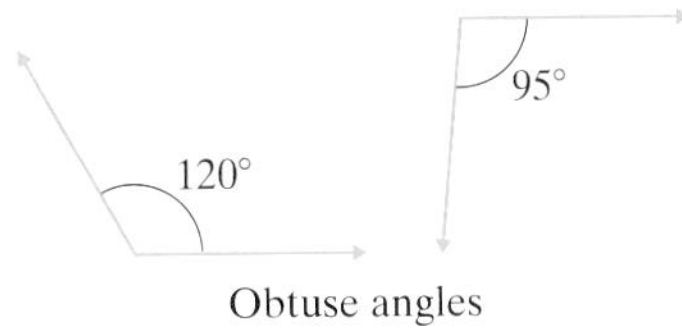

Obtuse angles

EXAMPLE 3 Classify each angle as acute, right, obtuse, or straight.

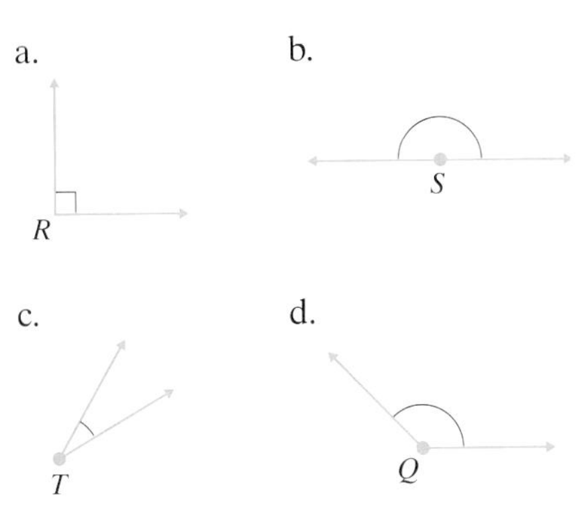

Practice Problem 3

Classify each angle as acute, right, obtuse, or straight.

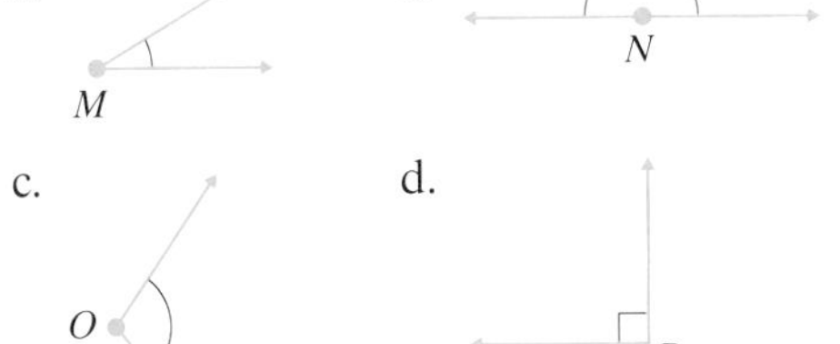

Answer

3. a. acute **b.** straight **c.** obtuse **d.** right

Solution:

a. $\angle R$ is a right angle, denoted by ⊾.

b. $\angle S$ is a straight angle.

c. $\angle T$ is an acute angle. It measures between 0° and 90°.

d. $\angle Q$ is an obtuse angle. It measures between 90° and 180°.

C Identifying Complementary and Supplementary Angles

Two angles that have a sum of 90° are called **complementary angles.** We say that each angle is the **complement** of the other.

$\angle R$ and $\angle S$ are complementary angles because

$$60° + 30° = 90°$$

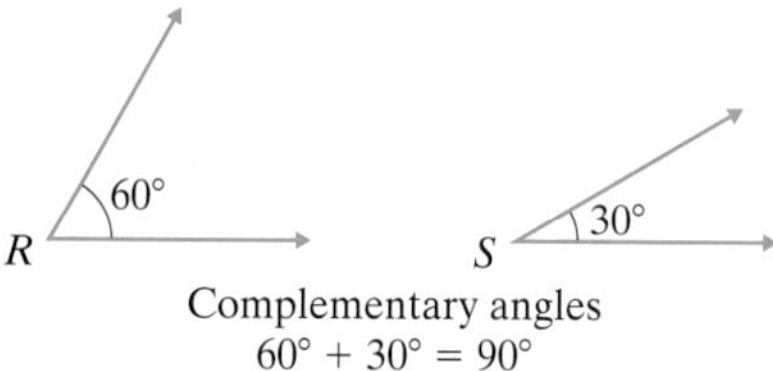

Complementary angles
60° + 30° = 90°

Two angles that have a sum of 180° are called **supplementary angles.** We say that each angle is the **supplement** of the other.

$\angle M$ and $\angle N$ are supplementary angles because

$$125° + 55° = 180°$$

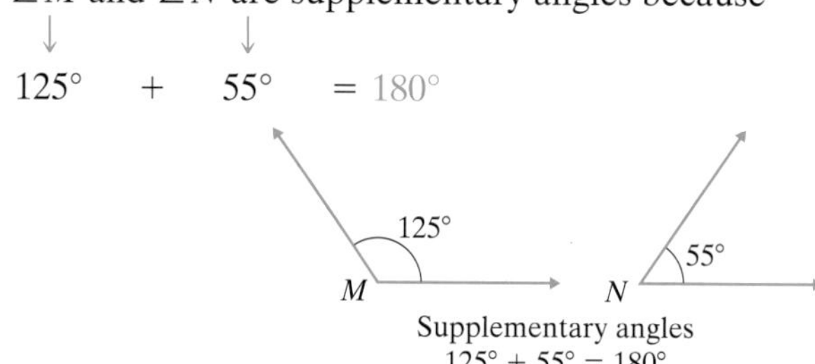

Supplementary angles
125° + 55° = 180°

Practice Problem 4

Find the complement of a 36° angle.

EXAMPLE 4 Find the complement of a 48° angle.

Solution: The complement of an angle that measures 48° is an angle that measures 90° − 48° = 42°.

Check:

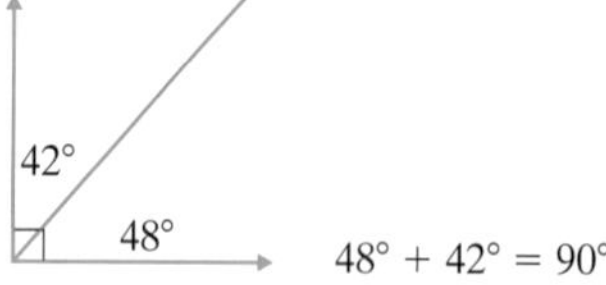

48° + 42° = 90°

Practice Problem 5

Find the supplement of an 88° angle.

EXAMPLE 5 Find the supplement of a 107° angle.

Solution: The supplement of an angle that measures 107° is an angle that measures 180° − 107° = 73°.

Check:

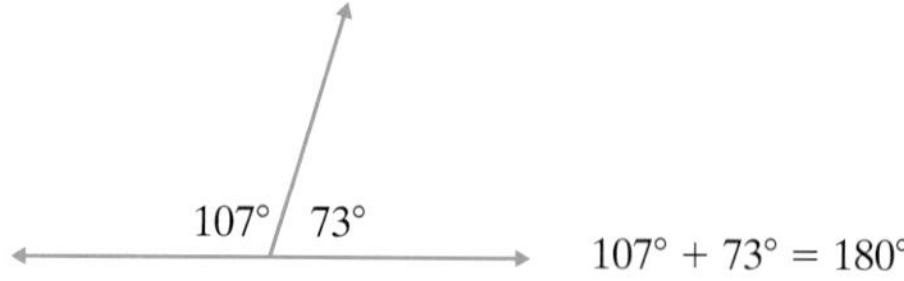

107° + 73° = 180°

Concept Check

True or false? The supplement of a 48° angle is 42°. Explain.

Try the Concept Check in the margin.

Answers

4. 54° **5.** 92°

Concept Check: False; the complement of a 48° angle is 42°; the supplement of a 48° angle is 132°.

D Finding Measures of Angles

Measures of angles can be added or subtracted to find measures of related angles.

EXAMPLE 6 Find the measure of $\angle x$.

Solution: The measure of $\angle x = 87° - 52° = 35°$

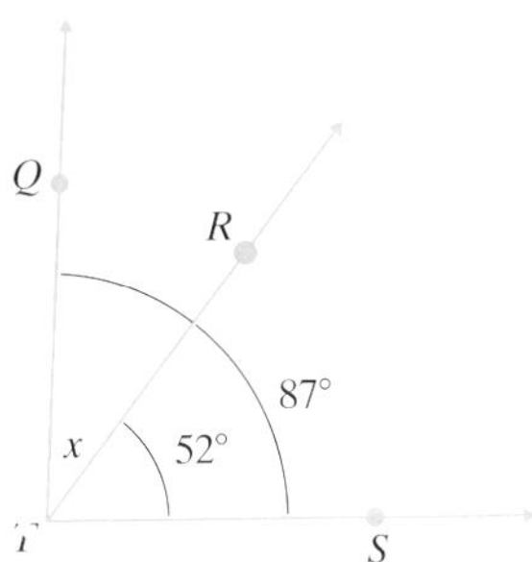

Two lines in a plane can be either parallel or intersecting. **Parallel lines** never meet. **Intersecting lines** meet at a point. The symbol is used to indicate "is parallel to." For example, in the figure $p \| q$.

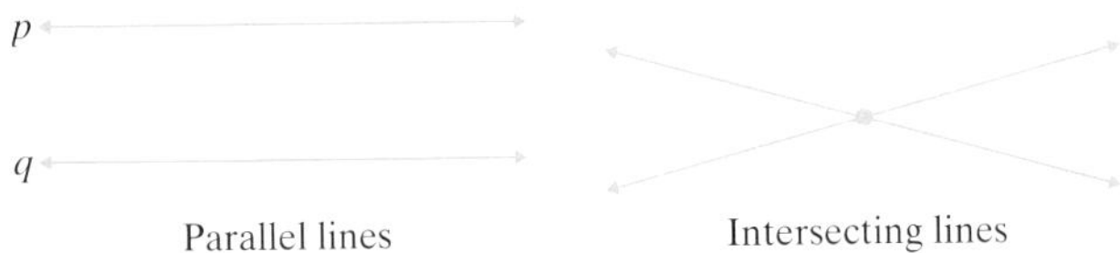

Parallel lines Intersecting lines

Some intersecting lines are perpendicular. Two lines are **perpendicular** if they form right angles when they intersect. The symbol is used to denote "is perpendicular to." For example, in the figure below, $n \perp m$.

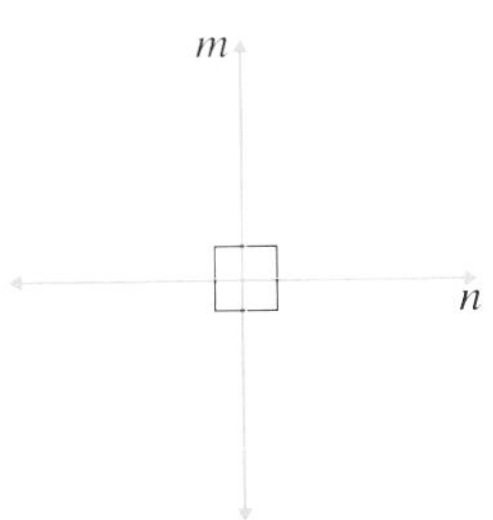

Perpendicular lines

When two lines intersect, four angles are formed. Two of these angles that are opposite each other are called **vertical angles.** Vertical angles have the same measure.

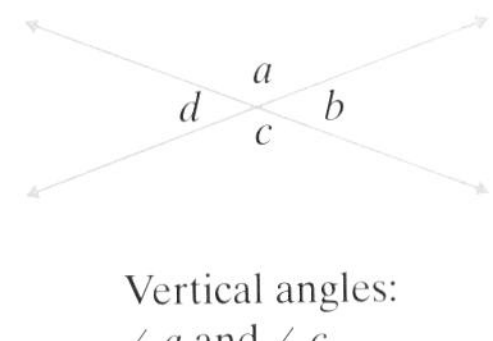

Vertical angles:
$\angle a$ and $\angle c$
$\angle d$ and $\angle b$

Practice Problem 6

Find the measure of $\angle y$.

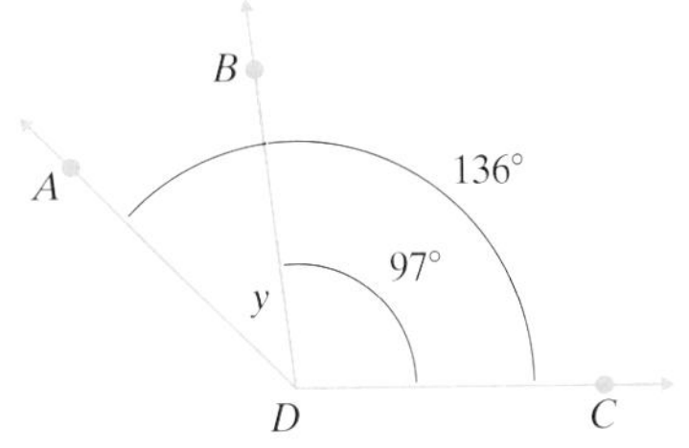

Answer

6. 39°

Practice Problem 7

Find the measure of $\angle ABY$ and $\angle ZBA$.

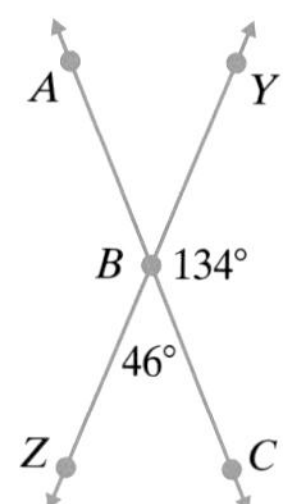

EXAMPLE 7 Find the measures of $\angle FDG$ and $\angle GDC$.

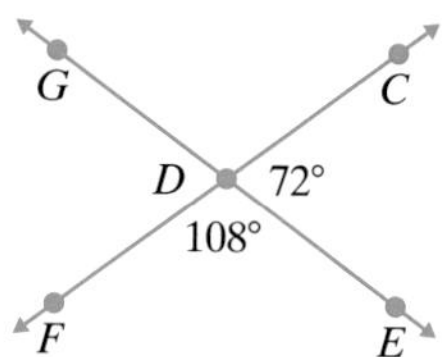

Solution: $\angle FDG$ and $\angle CDE$ are vertical angles, so they have the same measure.

measure of $\angle FDG = 72°$

$\angle GDC$ and $\angle FDE$ are vertical angles, so they have the same measure.

measure of $\angle GDC = 108°$

Notice in Example 7 that $\angle CDE$ and $\angle EDF$ share a common side, ray DE. When this happens, the angles are called *adjacent angles.*

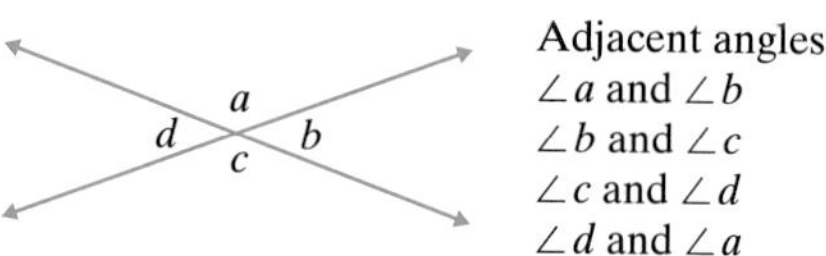

Also notice in Example 7 that adjacent angles $\angle CDE$ and $\angle EDF$ are also supplementary angles since the sum of their measures is 180°.

$72° + 108° = 180°$

This is true in general. Adjacent angles formed by intersecting lines are supplementary.

Practice Problem 8

Find the measure of $\angle a$, $\angle b$, and $\angle c$.

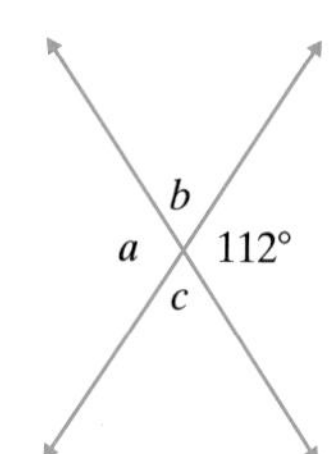

EXAMPLE 8 Find the measure of $\angle x$, $\angle y$, and $\angle z$ if the measure of $\angle t$ is 42°.

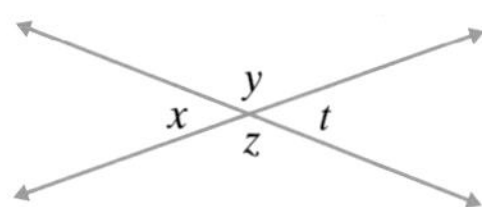

Solution: Since $\angle t$ and $\angle x$ are vertical angles, they have the same measure, so $\angle x$ measures 42°.

Since $\angle t$ and $\angle y$ are adjacent angles, their measures have a sum of 180°, so $\angle y$ measures $180° - 42° = 138°$.

Since $\angle y$ and $\angle z$ are vertical angles, they have the same measure. So $\angle z$ measures 138°.

A line that intersects two or more lines at different points is called a **transversal.** Line l is a transversal that intersects lines m and n.

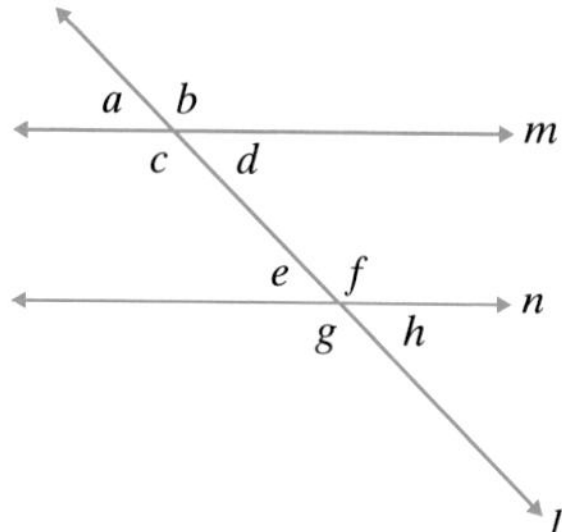

Answers

7. $\angle ABY$ measures 46°. $\angle ZBA$ measures 134°.
8. $\angle a = 112°$; $\angle b = 68°$; $\angle c = 68°$

The eight angles formed have special names. To understand these names, let's first learn what angles are called interior and what angles are called exterior by the illustration below.

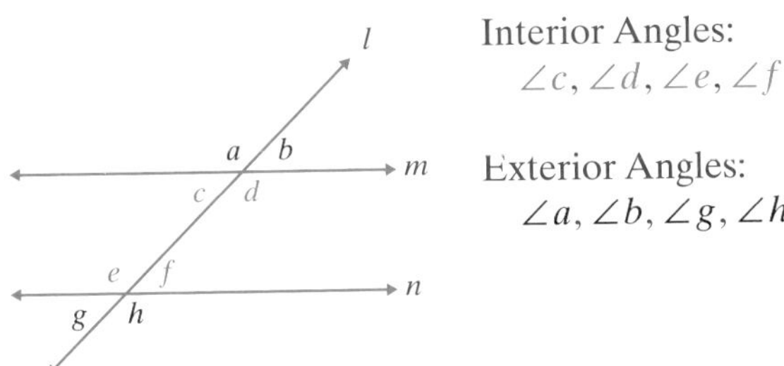

Interior Angles:
$\angle c, \angle d, \angle e, \angle f$

Exterior Angles:
$\angle a, \angle b, \angle g, \angle h$

Now that you know which angles are called interior angles, below we illustrate which pairs of these angles we call alternate interior angles.

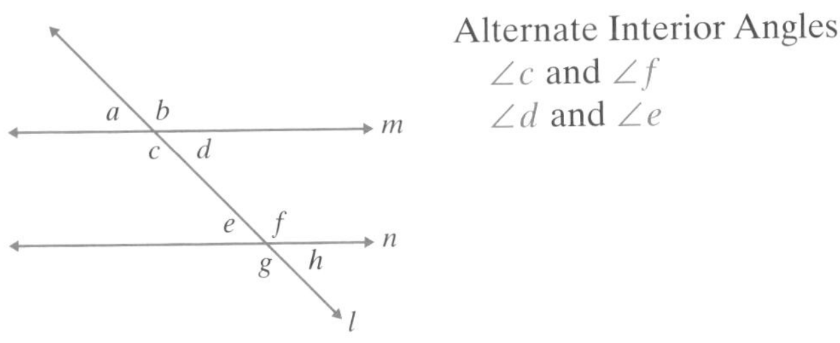

Alternate Interior Angles:
$\angle c$ and $\angle f$
$\angle d$ and $\angle e$

Another special name we give certain pairs of angles is corresponding angles. If lines m and n are cut by transveral l, then corresponding angles have the same exact position: one with part of line m as a side and one with part of line n as a side.

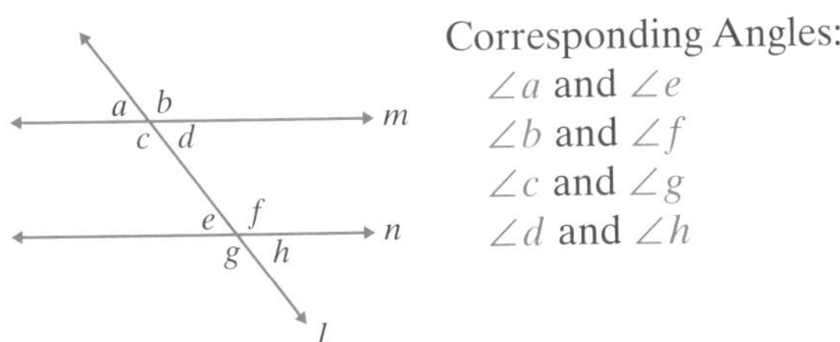

Corresponding Angles:
$\angle a$ and $\angle e$
$\angle b$ and $\angle f$
$\angle c$ and $\angle g$
$\angle d$ and $\angle h$

Note: There are other special names for angles formed by two lines cut by a transversal, such as alternate exterior angles, but we will not review them here.

When two lines cut by a transversal are *parallel*, the following are true:

Parallel Lines Cut by a Transversal

If two parallel lines are cut by a transversal, then the measures of **corresponding angles are equal** and **alternate interior angles are equal.**

Practice Problem 9

Given that $m \| n$ and that the measure of $\angle w = 40°$, find the measures of $\angle x$, $\angle y$, and $\angle z$.

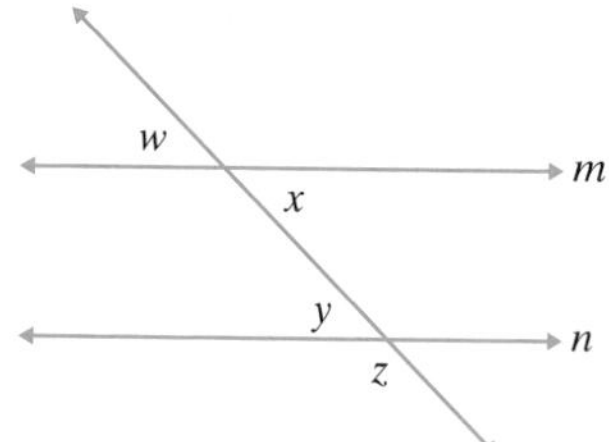

Answer

9. $\angle x = 40°$; $\angle y = 40°$; $\angle z = 140°$

EXAMPLE 9 Given that $m \| n$ and that the measure of $\angle w$ is 100°, find the measures of $\angle x$, $\angle y$, and $\angle z$.

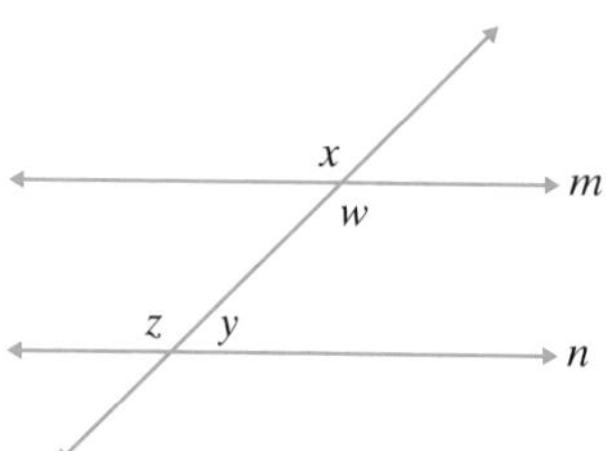

Solution:

The measure of $\angle x = 100°$. — $\angle x$ and $\angle w$ are vertical angles.

The measure of $\angle z = 100°$. — $\angle x$ and $\angle z$ are corresponding angles.

The measure of $\angle y = 180° - 100° = 80°$. — $\angle z$ and $\angle y$ are supplementary angles.

Name ______________________ Section __________ Date __________

EXERCISE SET 8.1

A *Identify each figure as a line, a ray, a line segment, or an angle. Then name the figure using the given points. See Examples 1 and 2.*

1.

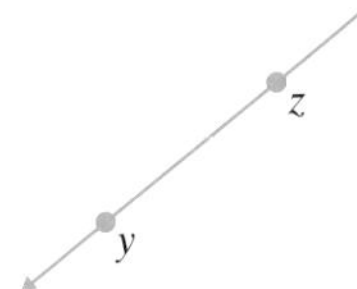

2.

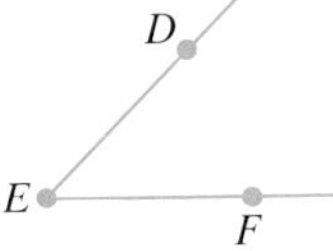

3.

4.

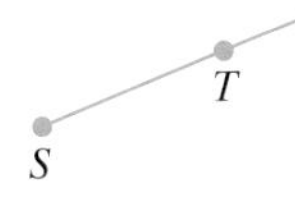

5.

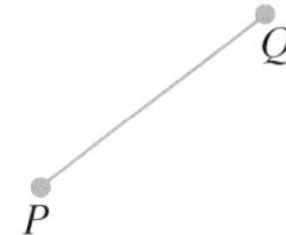

6.

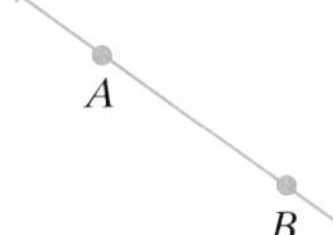

7.

8.

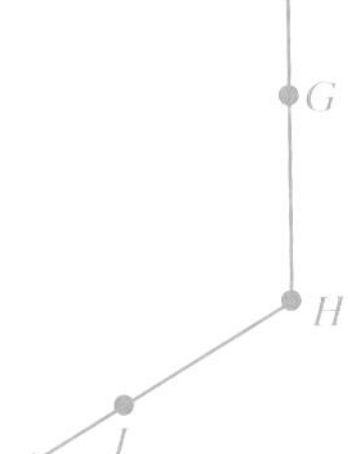

B *Find the measure of each angle in the following figure:*

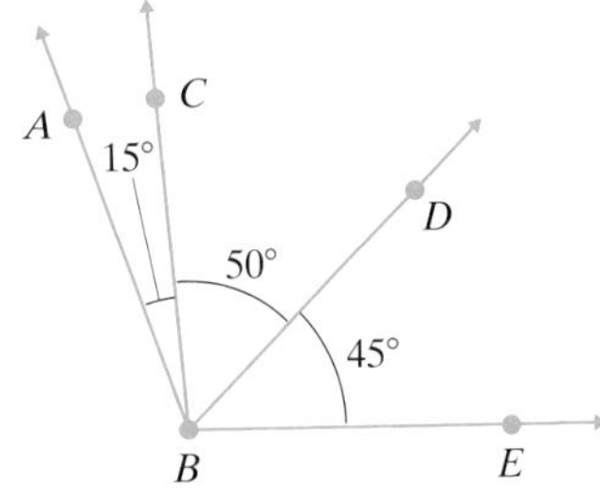

9. $\angle ABC$

10. $\angle EBD$

11. $\angle CBD$

12. $\angle CBA$

13. $\angle DBA$

14. $\angle EBC$

15. $\angle CBE$

16. $\angle ABE$

Fill in each blank. See Example 3.

17. A right angle has a measure of _____.

18. A straight angle has a measure of _____.

19. An acute angle measures between _____ and _____.

20. An obtuse angle measures between _____ and _____.

Classify each angle as acute, right, obtuse, or straight. See Example 3.

△ **21.**

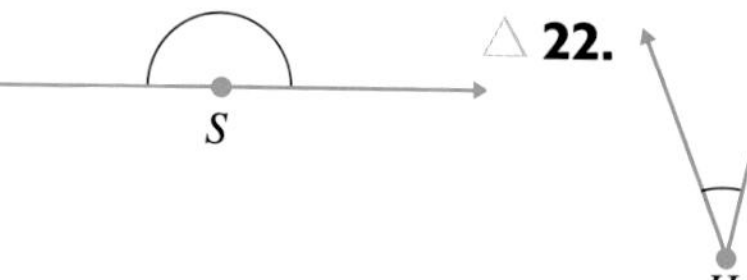

△ **22.**

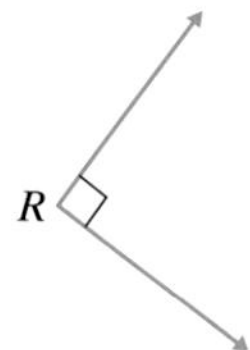

△ **23.**

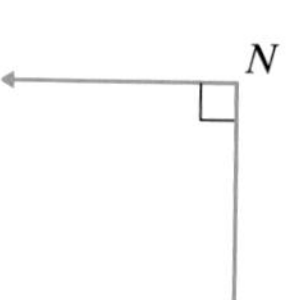

△ **24.**

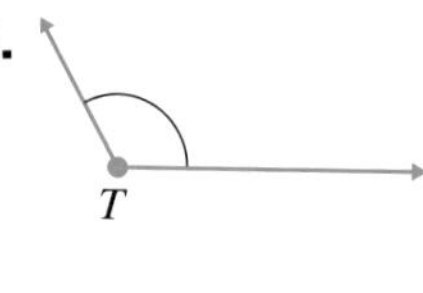

△ **25.**

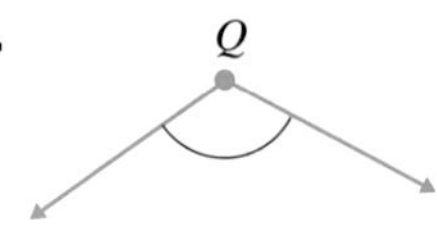

△ **26.**

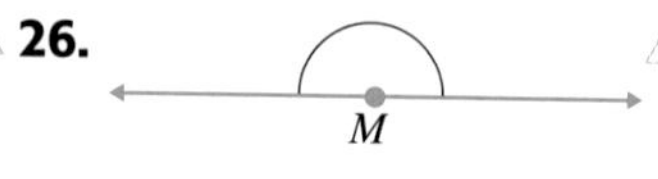

△ **27.**

△ **28.**

C *Find each complementary or supplementary angle as indicated. See Examples 4 and 5.*

△ **29.** Find the complement of a 17° angle.

△ **30.** Find the complement of an 87° angle.

△ **31.** Find the supplement of a 17° angle.

△ **32.** Find the supplement of an 87° angle.

△ **33.** Find the complement of a 48° angle.

△ **34.** Find the complement of a 22° angle.

△ **35.** Find the supplement of a 125° angle.

△ **36.** Find the supplement of a 155° angle.

△ **37.** Identify the pairs of complementary angles.

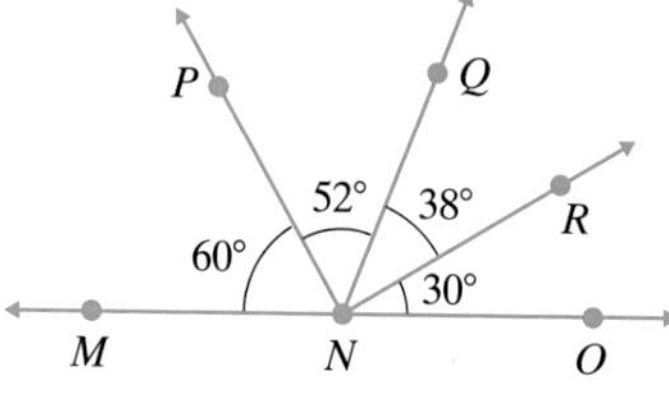

△ **38.** Identify the pairs of complementary angles.

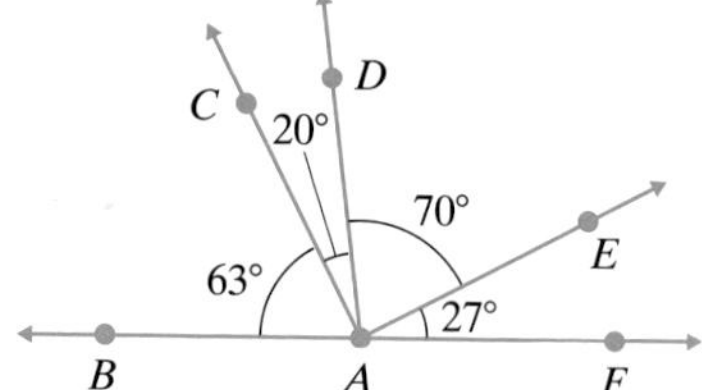

△ **39.** Identify the pairs of supplementary angles.

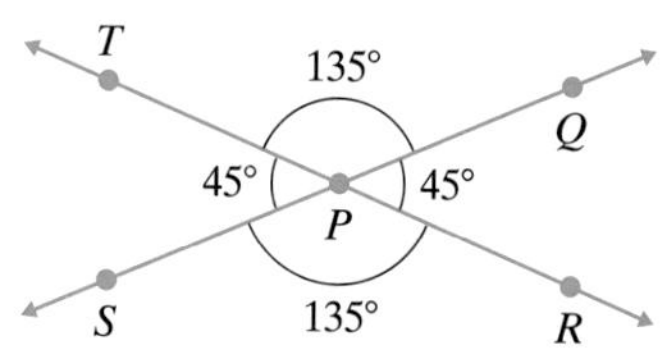

△ **40.** Identify the pairs of supplementary angles.

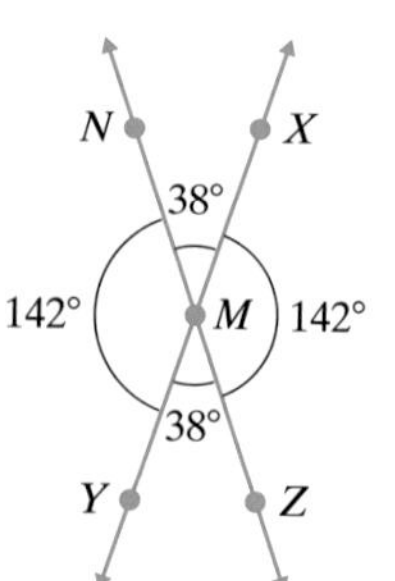

△ **41.** The angle between the two walls of the Vietnam Veterans Memorial in Washington, D.C., is 125.2°. Find the supplement of this angle. (*Source:* National Park Service)

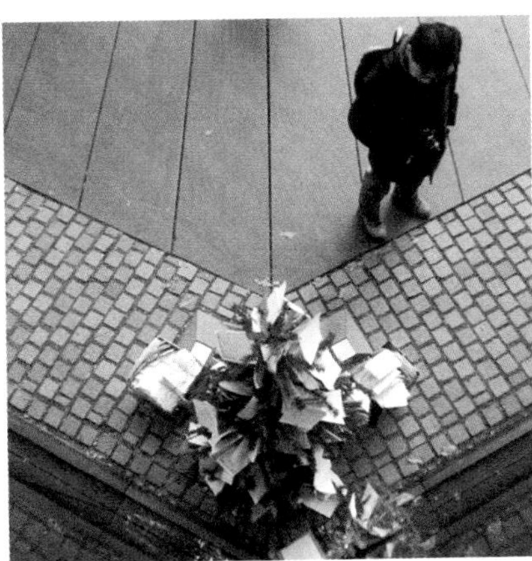

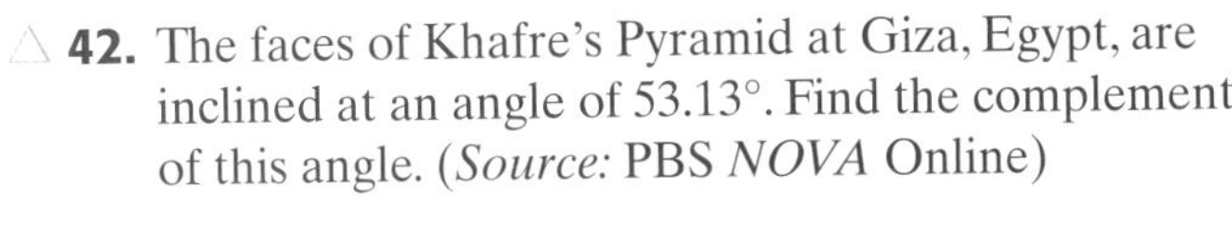

△ **42.** The faces of Khafre's Pyramid at Giza, Egypt, are inclined at an angle of 53.13°. Find the complement of this angle. (*Source:* PBS *NOVA* Online)

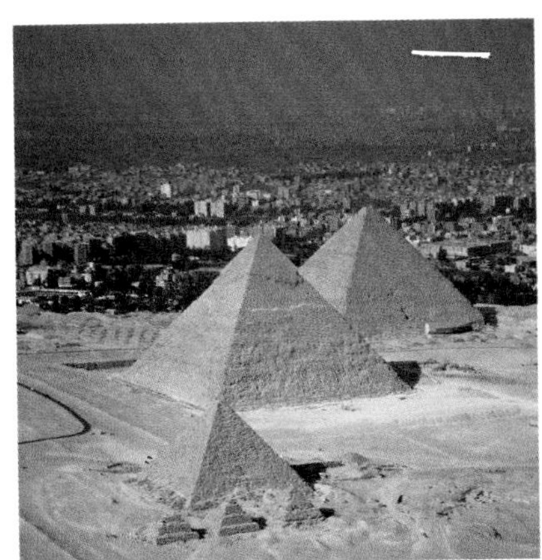

D *Find the measure of ∠x in each figure. See Example 6.*

△ **43.**

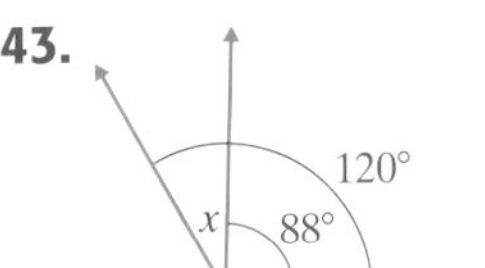

△ **44.**

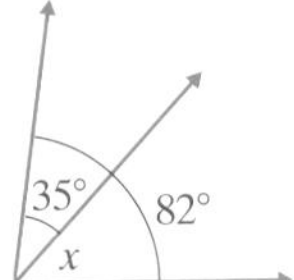

△ **45.**

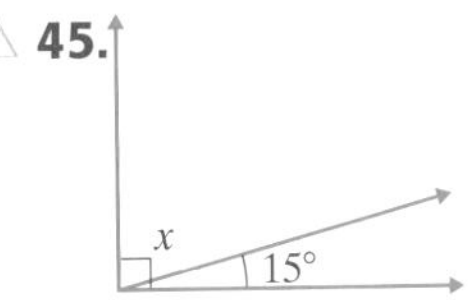

△ **46.**

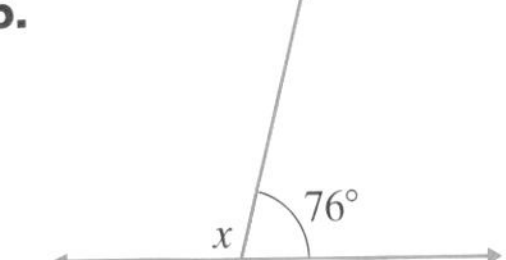

Find the measures of angles x, y, and z in each figure. See Examples 7, 8, and 9.

△ **47.**

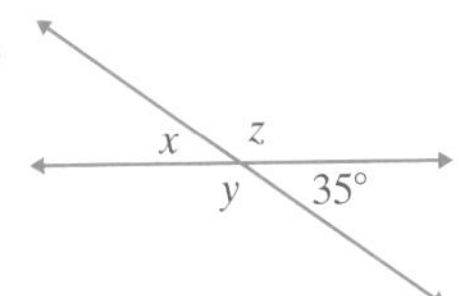

△ **48.**

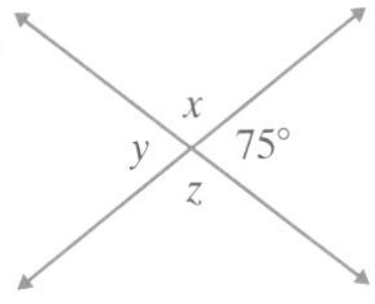

△ **49.**

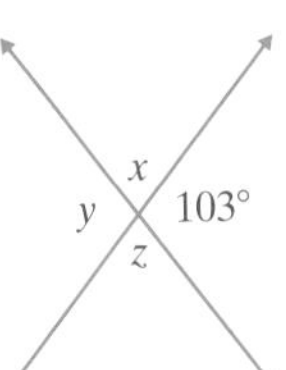

△ **50.**

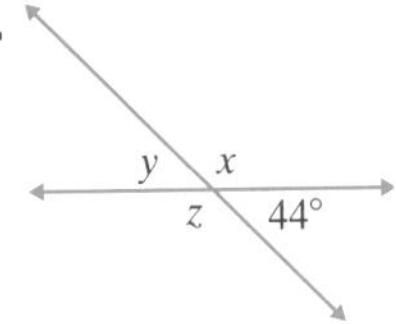

△ **51.** $m \| n$

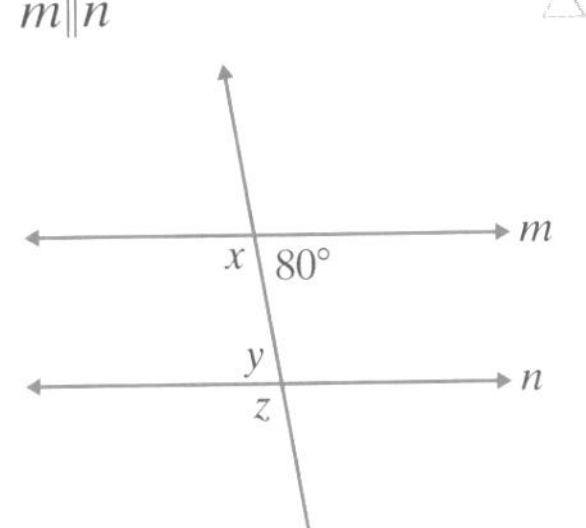

△ **52.** $m \| n$

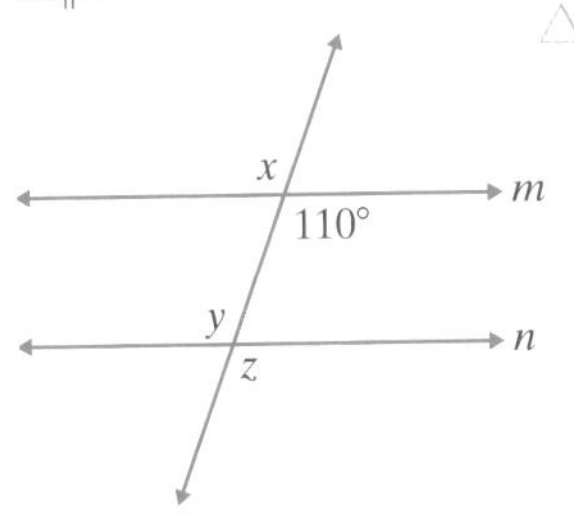

△ **53.** $m \| n$

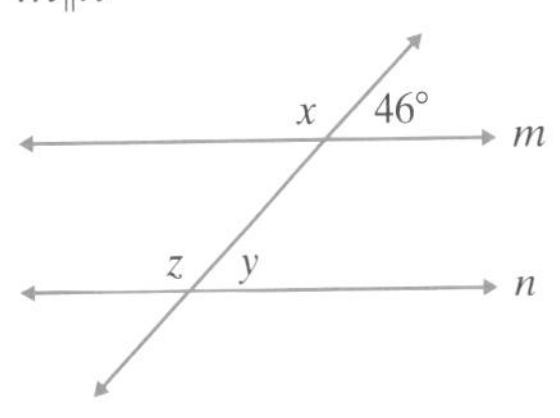

△ **54.** $m \| n$

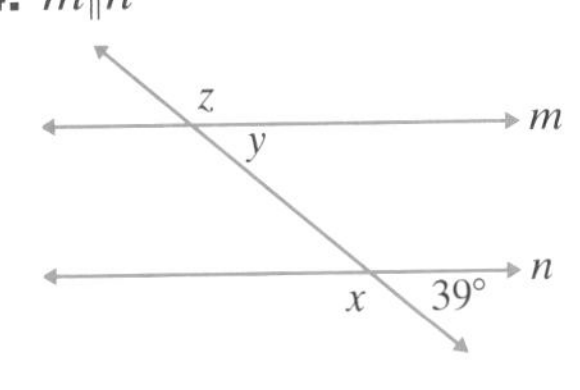

Review and Preview

Perform each indicated operation. See Sections 4.3, 4.5, and 4.8.

55. $\frac{7}{8} + \frac{1}{4}$

56. $\frac{7}{8} - \frac{1}{4}$

57. $\frac{7}{8} \cdot \frac{1}{4}$

58. $\frac{7}{8} \div \frac{1}{4}$

59. $3\frac{1}{3} - 2\frac{1}{2}$

60. $3\frac{1}{3} + 2\frac{1}{2}$

61. $3\frac{1}{3} \div 2\frac{1}{2}$

62. $3\frac{1}{3} \cdot 2\frac{1}{2}$

Combining Concepts

△ **63.** If lines m and n are parallel, find the measures of angles a through e.

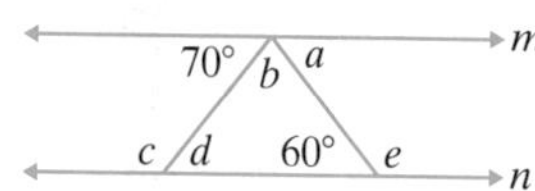

△ **64.** In your own words, describe how to find the complement and the supplement of a given angle.

△ **65.** Find two complementary angles with the same measure.

△ **66.** Can two supplementary angles both be acute?

△ **67.** Is the figure below possible? Why or why not?

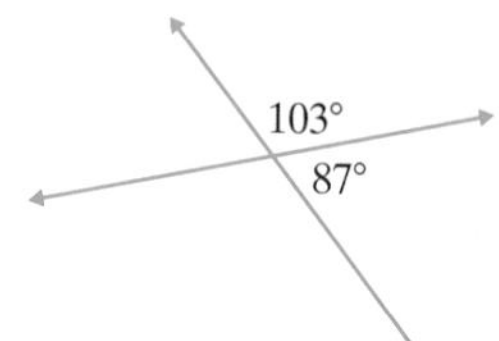

△ **68.** Below is a rectangle. List which segments, if extended, would be parallel lines.

8.2 Linear Measurement

A Defining and Converting U.S. System Units of Length

OBJECTIVES

A Define U.S. units of length and convert from one unit to another.

B Use mixed units of length.

C Perform arithmetic operations on U.S. units of length.

D Define metric units of length and convert from one unit to another.

E Perform arithmetic operations on metric units of length.

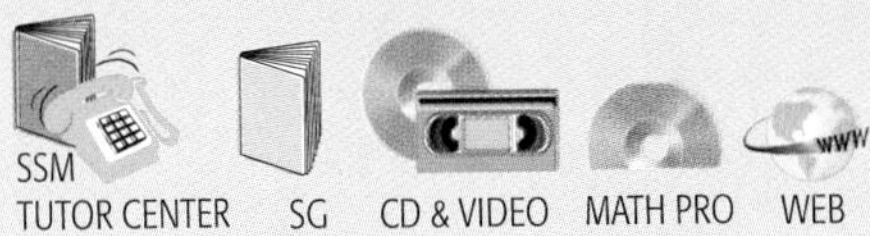

In the United States, two systems of measurement are commonly used. They are the **United States (U.S.), or English, measurement system** and the **metric system.** The U.S. measurement system is familiar to most Americans. Units such as feet, miles, ounces, and gallons are used. However, the metric system is also commonly used in fields such as medicine, sports, international marketing, and certain physical sciences. We are accustomed to buying 2-liter bottles of soft drinks, watching televised coverage of the 100-meter dash at the Olympic Games, or taking a 200-milligram dose of pain reliever.

The U.S. system of measurement uses the **inch, foot, yard,** and **mile** to measure **length.** The following is a summary of equivalencies between units of length:

U.S. Units of Length	Unit Fractions
12 inches (in.) = 1 foot (ft)	$\frac{12\text{ in.}}{1\text{ ft}} = \frac{1\text{ ft}}{12\text{ in.}} = 1$
3 feet = 1 yard (yd)	$\frac{3\text{ ft}}{1\text{ yd}} = \frac{1\text{ yd}}{3\text{ ft}} = 1$
5280 feet = 1 mile (mi)	$\frac{5280\text{ ft}}{1\text{ mi}} = \frac{1\text{ mi}}{5280\text{ ft}} = 1$

To convert from one unit of length to another, **unit fractions** may be used. A unit fraction is a fraction that equals 1. For example, since 12 in. = 1 ft, we have the unit fractions

$$\frac{12\text{ in.}}{1\text{ ft}} = \frac{1\text{ ft}}{12\text{ in.}} = 1$$

For example, to convert 48 inches to feet, we *multiply by a unit fraction* that relates feet to inches. The unit fraction should be written so that *the units we are converting to,* feet, *are in the numerator and the original units,* inches, *are in the denominator.* We do this so that like units will divide out, as shown next:

$$48\text{ in.} = \frac{48\text{ }\cancel{\text{in.}}}{1} \cdot \underbrace{\frac{1\text{ ft}}{12\text{ }\cancel{\text{in.}}}}_{\text{Unit fraction}} \quad \begin{matrix}\leftarrow \text{Units to convert to} \\ \leftarrow \text{Original units}\end{matrix}$$

$$= \frac{48 \cdot 1\text{ ft}}{1 \cdot 12}$$

$$= \frac{48\text{ ft}}{12}$$

$$= 4\text{ ft}$$

Therefore, 48 inches equals 4 feet, as seen in the diagram:

12 in.	12 in.	12 in.	12 in.	48 in. = 4 ft
1 ft	1 ft	1 ft	1 ft	

Helpful Hint:

When converting from one unit to another, select a unit fraction with the properties below:

$$\frac{\text{units you are converting to}}{\text{original units}}$$

By using this unit fraction, the original units will divide out, as wanted.

Practice Problem 1

Convert 5 feet to inches.

EXAMPLE 1 Convert 8 feet to inches.

Solution: We multiply 8 feet by a unit fraction that compares 12 inches to 1 foot. The unit fraction should be $\frac{\text{units to convert to}}{\text{original units}}$ or $\frac{12 \text{ inches}}{1 \text{ foot}}$.

$$8 \text{ ft} = \frac{8 \cancel{\text{ft}}}{1} \cdot \underbrace{\frac{12 \text{ in.}}{1 \cancel{\text{ft}}}}_{\text{Unit fraction}}$$

$$= 8 \cdot 12 \text{ in.}$$

$$= 96 \text{ in.} \quad \text{Multiply.}$$

Thus, 8 ft = 96 in., as shown in the diagram:

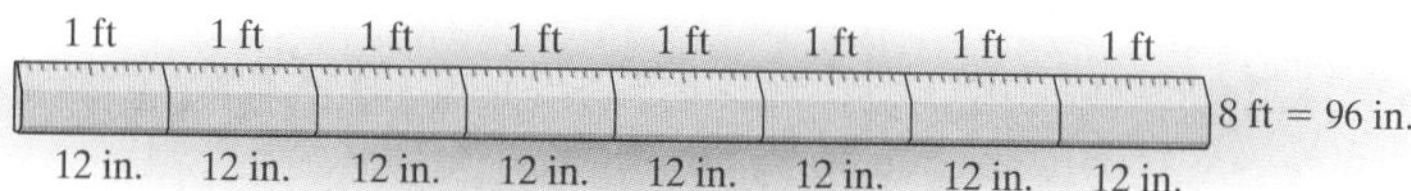

Practice Problem 2

Convert 7 yards to feet.

EXAMPLE 2 Convert 7 feet to yards.

Solution: We multiply by a unit fraction that compares 1 yard to 3 feet.

$$7 \text{ ft} = \frac{7 \cancel{\text{ft}}}{1} \cdot \frac{1 \text{ yd}}{3 \cancel{\text{ft}}} \quad \begin{array}{l} \leftarrow \text{Units to convert to} \\ \leftarrow \text{Original units} \end{array}$$

$$= \frac{7 \text{ yd}}{3}$$

$$= 2\frac{1}{3} \text{ yd} \quad \text{Divide.}$$

Thus, 7 ft = $2\frac{1}{3}$ yd.

B Using Mixed U.S. System Units of Length

Sometimes it is more meaningful to express a measurement of length with mixed units, such as 1 ft and 5 in. We usually condense this and write 1 ft 5 in.

In Example 2, we found that 7 feet was the same as $2\frac{1}{3}$ yards. The measurement can also be written as a mixture of yards and feet. That is,

7 ft = ____ yd ____ ft

Because 3 ft = 1 yd, we divide 3 into 7 to see how many whole yards are in 7 feet. The quotient is the number of yards, and the remainder is the number of feet.

$$\begin{array}{r} 2 \text{ yd } 1 \text{ ft} \\ 3\overline{)7} \quad \\ \underline{-6} \quad \\ 1 \quad \end{array}$$

Thus, 7 ft = 2 yd 1 ft, as seen in the diagram:

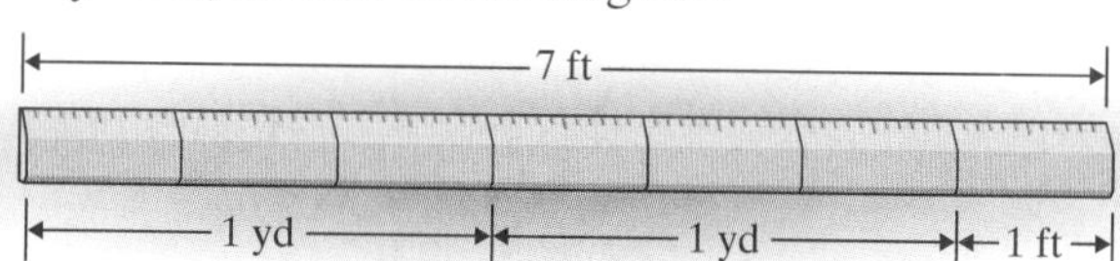

Answers

1. 60 in. **2.** 21 ft

EXAMPLE 3 Convert: 134 in. = ____ ft____ in.

Solution: Because 12 in. = 1 ft, we divide 12 into 134. The quotient is the number of feet. The remainder is the number of inches. To see why we divide 12 into 134, notice that

$$134 \text{ in.} = \frac{134 \text{ in.}}{1} \cdot \frac{1 \text{ ft}}{12 \text{ in.}} = \frac{134}{12} \text{ ft}$$

```
     11 ft 2 in.
12)134      ↑
  -12       |
   14       |
  -12       |
    2 ------
```

Thus, 134 in. = 11 ft 2 in.

Practice Problem 3

Convert: 68 in. = ____ ft____ in.

EXAMPLE 4 Convert 3 feet 7 inches to inches.

Solution: First, we will convert 3 feet to inches. Then we add 7 inches.

$$3 \text{ ft} = \frac{3 \text{ ft}}{1} \cdot \frac{12 \text{ in.}}{1 \text{ ft}} = 36 \text{ in.}$$

Then

$$3 \text{ ft } 7 \text{ in.} = 36 \text{ in.} + 7 \text{ in.} = 43 \text{ in.}$$

Practice Problem 4

Convert 5 yards 2 feet to feet.

C Performing Operations on U.S. System Units of Length

Finding sums or differences of measurements often involves converting units, as shown in the next example. Just remember that, as usual, only like units can be added or subtracted.

EXAMPLE 5 Add 3 ft 2 in. and 5 ft 11 in.

Solution: To add, we line up the similar units.

```
  3 ft  2 in.
+5 ft 11 in.
  8 ft 13 in.
```

Since 13 inches is the same as 1 ft 1 in., we have

$$8 \text{ ft } 13 \text{ in.} = 8 \text{ ft} + 1 \text{ ft } 1 \text{ in.}$$
$$= 9 \text{ ft } 1 \text{ in.}$$

Practice Problem 5

Add 4 ft 8 in. to 8 ft 11 in.

Concept Check

How could you estimate the following sum?

```
  7 yd  4 in.
+3 yd 27 in.
```

Try the Concept Check in the margin.

EXAMPLE 6 Multiply 8 ft 9 in. by 3.

Solution: By the distributive property, we multiply 8 ft by 3 and 9 in. by 3.

```
  8 ft  9 in.
×          3
 24 ft 27 in.
```

Practice Problem 6

Multiply 4 ft 7 in. by 4.

Answers

3. 5 ft 8 in. **4.** 17 ft **5.** 13 ft 7 in. **6.** 18 ft 4 in.

Concept Check: Round each to the nearest yard: 7 yd + 4 yd = 11 yd

Since 27 in. is the same as 2 ft 3 in., we simplify the product as

$$\begin{aligned} 24\text{ ft }27\text{ in.} &= 24\text{ ft} + 2\text{ ft }3\text{ in.} \\ &= 26\text{ ft }3\text{ in.} \end{aligned}$$

Practice Problem 7

Divide 18 ft 6 in. by 2.

EXAMPLE 7 Divide 24 yd 6 in. by 3.

Solution: We divide each of the units by 3.

$$\begin{array}{r} 8\text{ yd }2\text{ in.} \\ 3\overline{)24\text{ yd }6\text{ in.}} \\ \underline{-24\text{ yd}\phantom{\text{ 6 in.}}} \\ 6\text{ in.} \\ \underline{-6\text{ in.}} \\ 0 \end{array}$$

The quotient is 8 yd 2 in.

To check, see that 8 yd 2 in. multiplied by 3 is 24 yd 6 in.

Practice Problem 8

A carpenter cuts 1 ft 9 in. from a board of length 5 ft 8 in. Find the remaining length of the board.

EXAMPLE 8 Finding the Length of a Piece of Rope

A rope of length 6 yd 1 ft has 2 yd 2 ft cut from one end. Find the length of the remaining rope.

Solution: Subtract 2 yd 2 ft from 6 yd 1 ft.

beginning length	→	6 yd 1 ft
− amount cut	→	−2 yd 2 ft
remaining length		

We cannot subtract 2 ft from 1 ft, so we borrow 1 yd from the 6 yd. One yard is converted to 3 ft and combined with the 1 ft already there.

Borrow 1 yd = 3 ft — The problem now reads:

5 yd + 1 yd 3 ft

$$\begin{array}{r} \not{6}\text{ yd }1\text{ ft} \\ \underline{-\ 2\text{ yd }2\text{ ft}} \end{array} \qquad \begin{array}{r} 5\text{ yd }4\text{ ft} \\ \underline{-\ 2\text{ yd }2\text{ ft}} \\ 3\text{ yd }2\text{ ft} \end{array}$$

The remaining rope is 3 yd 2 ft long.

D Defining and Converting Metric System Units of Length

The basic unit of length in the metric system is the **meter.** A meter is slightly longer than a yard. It is approximately 39.37 inches long. Recall that a yard is 36 inches long.

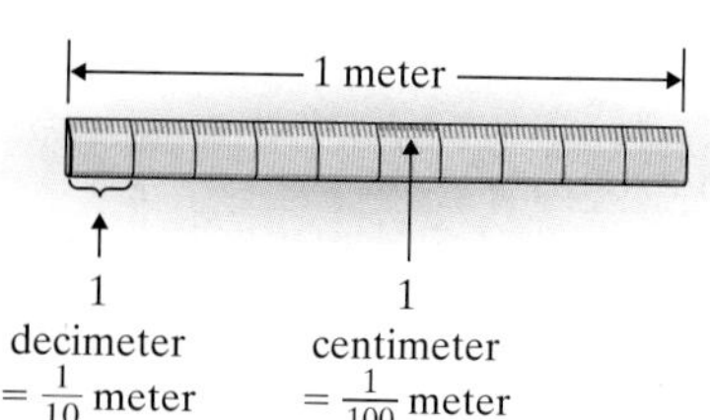

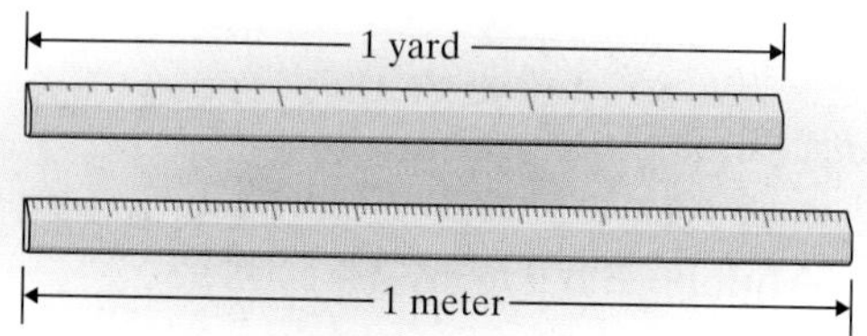

All units of length in the metric system are based on the meter. The following is a summary of the prefixes used in the metric system. Also shown are equivalencies between units of length. Like the decimal system, the metric system uses powers of 10 to define units.

Answers

7. 9 ft 3 in. **8.** 3 ft 11 in.

Prefix	Meaning	Metric Unit of Length	
kilo	1000	1 **kilo**meter (km) = 1000 meters (m)	
hecto	100	1 **hecto**meter (hm) = 100 m	
deka	10	1 **deka**meter (dam) = 10 m	
		1 meter (m) = 1 m	
deci	1/10	1 **deci**meter (dm) = 1/10 m	or 0.1 m
centi	1/100	1 **centi**meter (cm) = 1/100 m	or 0.01 m
milli	1/1000	1 **milli**meter (mm) = 1/1000 m	or 0.001 m

These same prefixes are used in the metric system for mass and capacity. The most commonly used measurements of length in the metric system are the **meter, millimeter, centimeter,** and **kilometer.**

Being comfortable with the metric units of length means gaining a "feeling" for metric lengths, just as you have a "feeling" for the length of an inch, a foot, and a mile. To help you accomplish this, study the following examples:

A millimeter is about the thickness of a large paper clip.

A centimeter is about the width of a large paper clip.

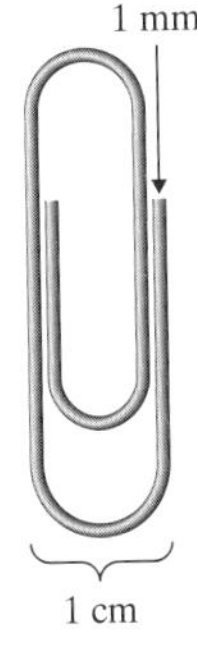

A meter is slightly longer than a yard.

A kilometer is about two-thirds of a mile.

$2\frac{1}{2}$ centimeters is about 1 inch.

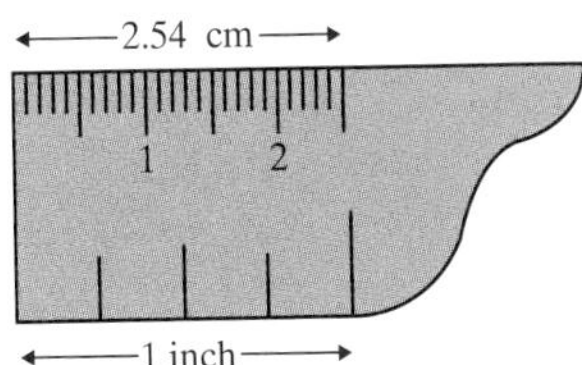

The length of this book is approximately 27.5 centimeters.

The width of this book is approximately 21.5 centimeters.

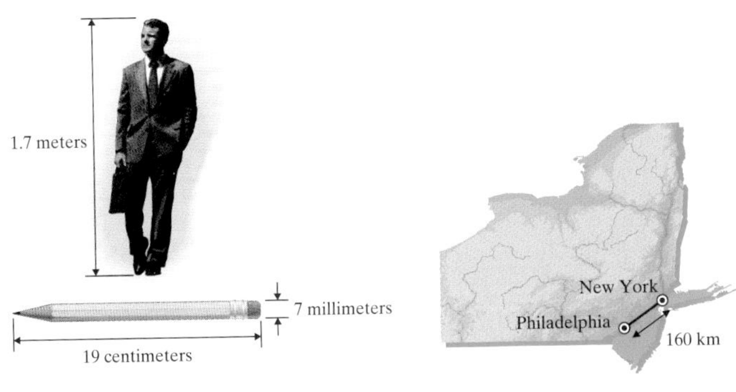

As with the U.S. system of measurement, unit fractions may be used to convert from one unit of length to another. The metric system does, however, have a distinct advantage over the U.S. system of measurement: the ease of converting from one unit of length to another. Since all units of length are powers of 10 of the meter, converting from one unit of length to another is as simple as moving the decimal point. Listing units of length in order from largest to smallest helps to keep track of how many places to move the decimal point when converting.

For example, let's convert 1200 meters to kilometers. To convert from meters to kilometers, we move along the chart below 3 units to the left, from meters to kilometers. This means that we move the decimal point 3 places to the left.

End / Start

km hm dam **m** dm cm mm

3 units to the left

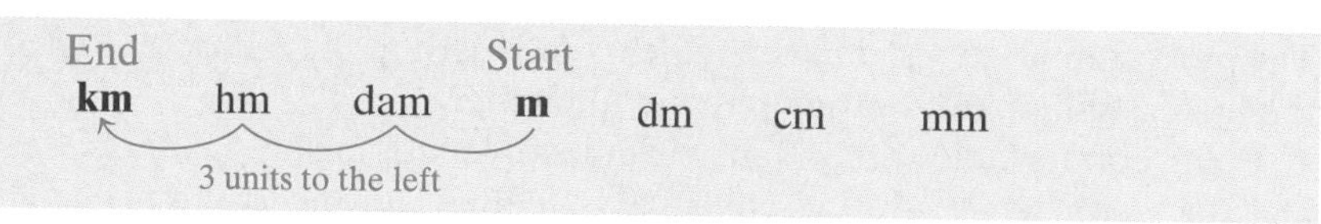

The same conversion can be made using unit fractions.

$$1200 \text{ m} = \frac{1200 \cancel{\text{m}}}{1} \cdot \underbrace{\frac{1 \text{ km}}{1000 \cancel{\text{m}}}}_{\text{Unit fraction}} = \frac{1200 \text{ km}}{1000} = 1.2 \text{ km}$$

Practice Problem 9

Convert 3.5 m to kilometers.

EXAMPLE 9 Convert 2.3 m to centimeters.

Solution: First we will convert by using a unit fraction.

$$2.3 \text{ m} = \frac{2.3 \cancel{\text{m}}}{1} \cdot \underbrace{\frac{100 \text{ cm}}{1 \cancel{\text{m}}}}_{\text{Unit fraction}} = 230 \text{ cm}$$

Now we will convert by listing the units of length in a chart and moving from meters to centimeters.

Start / End

km hm dam m dm cm mm

2 units to the right

2.30 m = 230. cm or 230 cm

2 places to the right

With either method, we get 230 cm.

Practice Problem 10

Convert 2.5 m to millimeters.

EXAMPLE 10 Convert 450,000 mm to meters.

Solution: We list the units of length in a chart and move from millimeters to meters.

End / Start

km hm dam m dm cm mm

3 units to the left

450,000 mm = 450.000 m or 450 m

3 places to the left

Try the Concept Check in the margin.

Concept Check

What is wrong with the following conversion of 150 cm to meters?

150.00 cm = 15,000 m

Answers

9. 0.0035 km **10.** 2500 mm

Concept Check: Decimal should be moved to the left: 1.5 m

E Performing Operations on Metric System Units of Length

To add, subtract, multiply, or divide with metric measurements of length, we write all numbers using the same unit of length and then add, subtract, multiply, or divide as with decimals.

EXAMPLE 11 Subtract 430 m from 1.3 km.

Solution: First we convert both measurements to kilometers or both to meters.

430 m = 0.43 km or 1.3 km = 1300 m

$$\begin{array}{r} 1.30\text{ km} \\ -\ 0.43\text{ km} \\ \hline 0.87\text{ km} \end{array} \qquad \begin{array}{r} 1300\text{ m} \\ -\ 430\text{ m} \\ \hline 870\text{ m} \end{array}$$

The difference is 0.87 km or 870 m.

Practice Problem 11

Subtract 640 m from 2.1 km.

EXAMPLE 12 Multiply 5.7 mm by 4.

Solution: Here we simply multiply the two numbers. Note that the unit of measurement remains the same.

$$\begin{array}{r} 5.7\text{ mm} \\ \times\ 4 \\ \hline 22.8\text{ mm} \end{array}$$

Practice Problem 12

Multiply 18.3 hm by 5.

EXAMPLE 13 Finding a Person's Height

Fritz Martinson was 1.2 meters tall on his last birthday. Since then, he has grown 14 centimeters. Find his current height in meters.

Solution:

original height	→	1.20 m
+ height grown	→	+ 0.14 m (Since 14 cm = 0.14 m)
current height		1.34 m

Fritz is now 1.34 meters tall.

Practice Problem 13

Doris Blackwell is knitting a scarf that is currently 0.8 meter long. If she knits an additional 45 centimeters, how long will the scarf be?

Answers

11. 1.46 km or 1460 m **12.** 91.5 hm
13. 125 cm or 1.25 m

FOCUS ON Mathematical Connections

POLYGONS

A **plane figure** is a figure that lies on a plane. Plane figures, like planes, have length and width but no thickness or depth.

A **polygon** is a closed plane figure that basically consists of three or more line segments that meet at their endpoints. A polygon is named according to the number of its sides. The table summarizes information about some basic polygons.

Polygons			
Number of Sides	**Name**	**Sum of Interior Angles**	**Example**
3	Triangle	180°	
4	Quadrilateral	360°	
5	Pentagon	540°	
6	Hexagon	720°	
7	Heptagon		
8	Octagon		
9	Nonagon		
10	Decagon		

CRITICAL THINKING

Study the sums of interior angles shown in the table for the triangle, quadrilateral, pentagon, and hexagon. What pattern do you notice? Use this pattern to complete the Sum of Interior Angles column in the table.

Name ______________________ Section ____________ Date ____________

Mental Math

Convert as indicated.

1. 12 inches to feet
2. 6 feet to yards
3. 24 inches to feet
4. 36 inches to feet
5. 36 inches to yards
6. 2 yards to inches

Determine whether the measurement in each statement is reasonable.

7. The screen of a home television set has a 30-meter diagonal.
8. A window measures 1 meter by 0.5 meter.
9. A drinking glass is made of glass 2 millimeters thick.
10. A paper clip is 4 kilometers long.
11. The distance across the Colorado River is 50 kilometers.
12. A model's hair is 30 centimeters long.

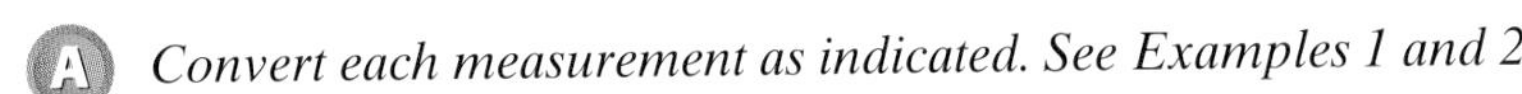

EXERCISE SET 8.2

A *Convert each measurement as indicated. See Examples 1 and 2.*

 1. 60 in. to feet

2. 84 in. to feet

 3. 12 yd to feet

4. 18 yd to feet

5. 42,240 ft to miles

6. 36,960 ft to miles

7. 102 in. to feet

8. 150 in. to feet

9. 10 ft to yards

10. 25 ft to yards

11. 6.4 mi to feet

12. 3.8 mi to feet

B *Convert each measurement as indicated. See Examples 3 and 4.*

13. 40 ft = ____ yd ____ ft

14. 100 ft = ____ yd ____ ft

15. 41 in. = ____ ft ____ in.

16. 75 in. = ____ ft ____ in.

17. 10,000 ft = ____ mi ____ ft

18. 25,000 ft = ____ mi ____ ft

19. 5 ft 2 in. = ____ in.

20. 4 ft 11 in. = ____ in.

21. 5 yd 2 ft = ____ ft

22. 7 yd 1 ft = ____ ft

23. 2 yd 1 ft = ____ in.

24. 1 yd 2 ft = ____ in.

C *Perform each indicated operation. Simplify the result if possible. See Examples 5 through 8.*

25. 5 ft 8 in. + 6 ft 7 in.

26. 9 ft 10 in. + 8 ft 4 in.

27. 12 yd 2 ft + 9 yd 2 ft

28. 16 yd 2 ft + 8 yd 1 ft

29. 24 ft 8 in. − 16 ft 3 in.

30. 15 ft 5 in. − 8 ft 2 in.

31. 16 ft 3 in. − 10 ft 9 in.

32. 14 ft 8 in. − 3 ft 11 in.

33. 6 ft 8 in. ÷ 2

34. 26 ft 10 in. ÷ 2

35. 12 yd 2 ft × 4

36. 15 yd 1 ft × 8

Solve. Remember to insert units when writing your answers. See Example 8.

37. The National Zoo maintains a small patch of bamboo, which it grows as a food supply for its pandas. Two weeks ago, the bamboo was 6 ft 10 in. tall. Since then, the bamboo has grown 3 ft 8 in. taller. How tall is the bamboo now?

38. While exploring in the Marianas Trench, a submarine probe was lowered to a point 1 mile 1400 feet below the ocean's surface. Later it was lowered an additional 1 mile 4000 feet below this point. How far is the probe below the surface of the Pacific?

39. The length of one of the Statue of Liberty's hands is 16 ft 5 in. One of the Statue's eyes is 2 ft 6 in. across. How much longer is a hand than the width of an eye? (*Source:* National Park Service)

40. The width of the Statue of Liberty's head from ear to ear is 10 ft. The height of the Statue's head from chin to cranium is 17 ft 3 in. How much taller is the Statue's head than its width? (*Source:* National Park Service)

41. The Amana Corporation stacks up its microwave ovens in a distribution warehouse. Each stack is 1 ft 9 in. wide. How far from the wall would 9 of these stacks extend?

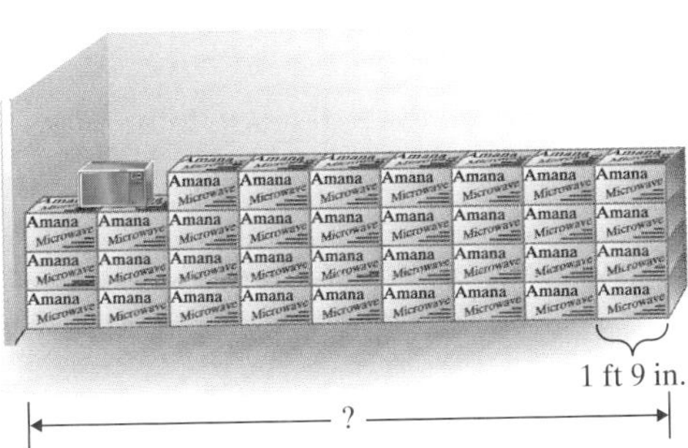

42. The highway commission is installing concrete sound barriers along a highway. Each barrier is 1 yd 2 ft long. How far will 25 barriers in a row reach?

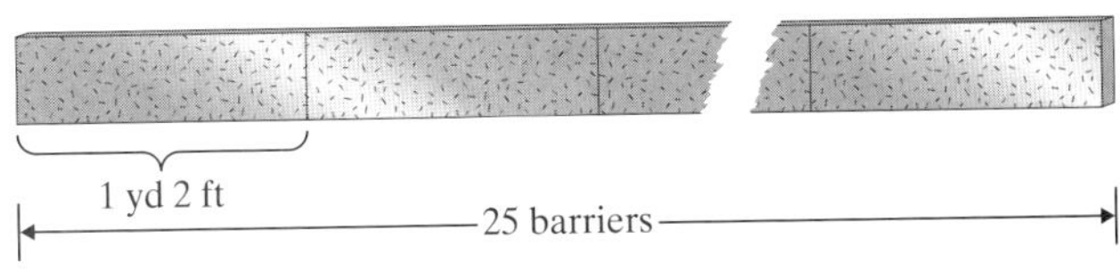

43. A carpenter needs to cut a board into thirds. If the board is 9 ft 3 in. long originally, how long will each cut piece be?

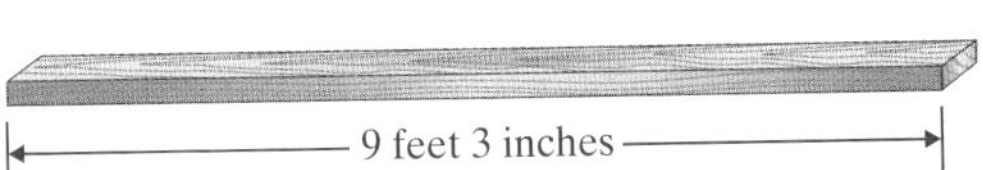

44. A wall is erected exactly halfway between two buildings that are 192 ft 8 in. apart. If the wall is 8 in. wide, how far is it from the wall to either of the buildings?

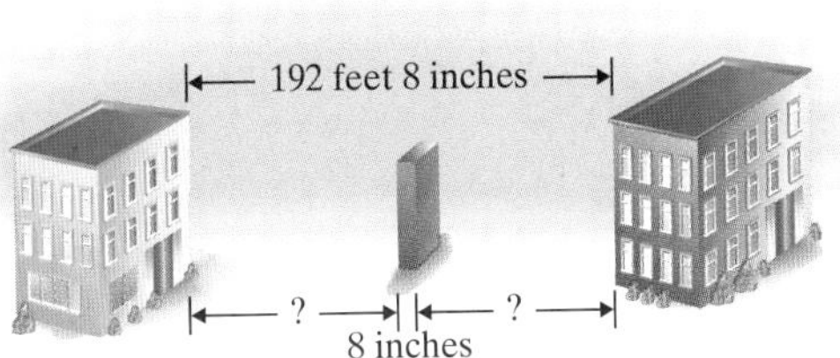

△ 45. Evelyn Pittman plans to erect a fence around her garden to keep the rabbits out. If the garden is a rectangle 24 ft 9 in. long by 18 ft 6 in. wide, what is the length of the fencing material she must purchase?

△ 46. Ronnie Hall needs new gutters for the front and *both sides* of his home. The front of the house is 50 ft. 8 in., and each side is 22 ft 9 in. wide. What length of gutter must he buy?

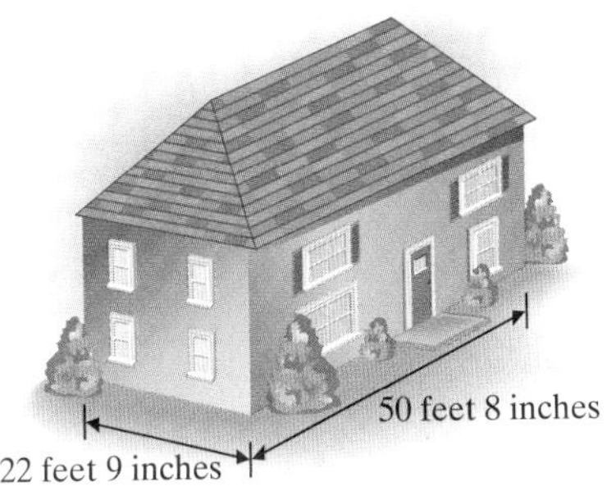

47. The world's longest Coca-Cola truck is in Sweden and is 79 feet long. How many *yards* long are 4 of these trucks? (*Source: Coca-Cola Today*)

△ 48. The world's largest Coca-Cola sign is in Arica, Chile. It is in the shape of a rectangle whose length is 400 feet and whose width is 131 feet. Find the area of the sign. (*Source: Coca-Cola Today*) (*Hint*: Recall that area of a rectangle is the product of length times width.)

D *Convert as indicated. See Examples 9 and 10.*

49. 40 m to centimeters

50. 18 m to centimeters

51. 40 mm to centimeters

52. 18 mm to centimeters

53. 300 m to kilometers

54. 400 m to kilometers

55. 1400 mm to meters

56. 6400 mm to meters

57. 1500 cm to meters

58. 6400 cm to meters

59. 8.3 cm to millimeters

60. 4.6 cm to millimeters

61. 20.1 mm to decimeters

62. 140.2 mm to decimeters

63. 0.04 m to millimeters

64. 0.2 m to millimeters

E *Perform each indicated operation. See Examples 11 and 12.*

65. 8.6 m + 0.34 m

66. 14.1 cm + 3.96 cm

67. 2.9 m + 40 mm

68. 30 cm + 8.9 m

69. 24.8 mm − 1.19 cm

70. 45.3 m − 2.16 dam

71. 15 km − 2360 m

72. 14 cm − 15 mm

73. 18.3 m × 3

74. 14.1 m × 4

75. 6.2 km ÷ 4

76. 9.6 m ÷ 5

Solve. Remember to insert units when writing your answers. See Example 13.

77. A 3.4-m rope is attached to a 5.8-m rope. However, when the ropes are tied, 8 cm of length are lost to form the knot. What is the length of the tied ropes?

78. A 2.15-m-long sash cord has become frayed at both ends so that 1 cm is trimmed from each end. How long is the remaining cord?

79. The ice on Doc Miller's pond is 5.33 cm thick. For safe skating, Doc insists that it must be 80 mm thick. How much thicker must the ice be before Doc goes skating?

80. The sediment on the bottom of the Towamencin Creek is normally 14 cm thick, but the recent flood washed away 22 mm of sediment. How thick is it now?

81. An art class is learning how to make kites. The two sticks used for each kite have lengths of 1 m and 65 cm. What total length of wood must be ordered for the sticks if 25 kites are to be built?

82. The total pages of a hardbound economics text are 3.1 cm thick. The front and back covers are each 2 mm thick. How high would a stack of 10 of these texts be?

83. A logging firm needs to cut a 67 m long redwood log into 20 equal pieces before loading it onto a truck for shipment. How long will each piece be?

84. An 18.3 m tall flagpole is mounted on a 65 cm high pedestal. How far is the top of the flagpole from the ground?

85. At one time it was believed that the fort of Basasi, on the Indian-Tibetan border, was the highest located structure at an elevation of 5.988 km above sea level. However, a settlement has been located that is 21 m higher than the fort. What is the elevation of this settlement?

86. The average American male at age 35 is 1.75 m tall. The average 65-year-old male is 48 mm shorter. How tall is the average 65-year-old male?

87. A floor tile is 22.86 cm wide. How many tiles in a row are needed to cross a room 3.429 m wide?

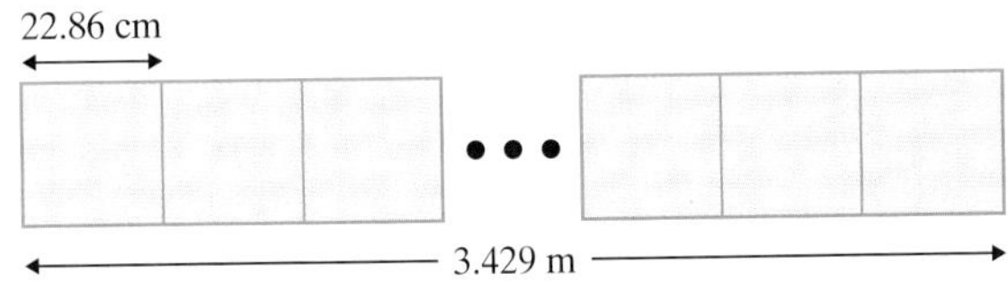

88. A standard postcard is 1.6 times longer than it is wide. If it is 9.9 cm wide, what is its length?

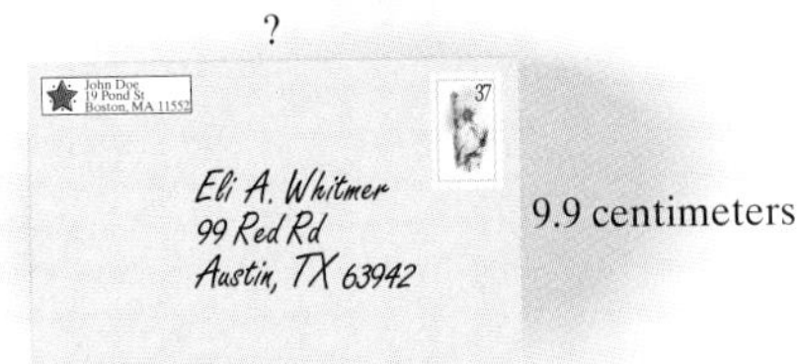

Review and Preview

Write each decimal or fraction as a percent. See Section 6.2.

89. 0.21 **90.** 0.86 **91.** $\frac{13}{100}$ **92.** $\frac{47}{100}$ **93.** $\frac{1}{4}$ **94.** $\frac{3}{20}$

Combining Concepts

95. To convert from meters to centimeters, the decimal point is moved two places to the right. Explain how this relates to the fact that the prefix *centi* means $\frac{1}{100}$.

96. Explain why conversions in the metric system are easier to make than conversions in the U.S. system of measurement.

97. Anoa Longway plans to use 26.3 meters of leftover fencing material to enclose a square garden plot for her daughter. How long will each side of the garden be?

98. A marathon is a running race over a distance of 26 mi 385 yd. If a runner runs five marathons in a year, what is the total distance he or she runs? (*Source: Microsoft Encarta Encyclopedia*)

8.3 Perimeter

A Using Formulas to Find Perimeters

Recall from Section 1.3 that the perimeter of a polygon is the distance around the polygon. This means that the perimeter of a polygon is the sum of the lengths of its sides.

EXAMPLE 1 Find the perimeter of the rectangle below.

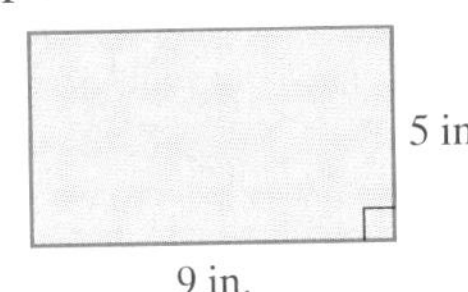

Solution: perimeter = 9 inches + 9 inches + 5 inches + 5 inches

= 28 inches

Notice that the perimeter of the rectangle in Example 1 can be written as 2 · (9 inches) + 2 · (5 inches).

↑ length ↑ width

In general, we can say that the perimeter of a rectangle is always

2 · length + 2 · width

As we have seen above and in previous sections, the perimeter of some special figures such as rectangles form patterns. These patterns are given as **formulas.** The formula for the perimeter of a rectangle is shown next:

Perimeter of a Rectangle

perimeter = 2 · length + 2 · width

In symbols, this can be written as

$P = 2l + 2w$

length

width width

length

EXAMPLE 2 Find the perimeter of a rectangle with a length of 11 inches and a width of 3 inches.

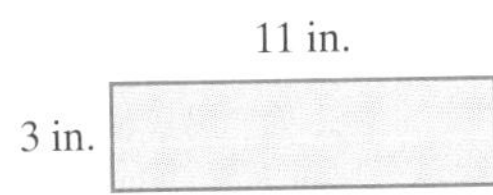

Solution: We use the formula for perimeter and replace the letters by their known measures.

$P = 2l + 2w$

$= 2 \cdot 11 \text{ in.} + 2 \cdot 3 \text{ in.}$ Replace l with 11 in. and w with 3 in.

$= 22 \text{ in.} + 6 \text{ in.}$

$= 28 \text{ in.}$

The perimeter is 28 inches.

OBJECTIVES

A Use formulas to find perimeter.

B Use formulas to find circumferences.

Practice Problem 1

Find the perimeter of the rectangular lot shown below:

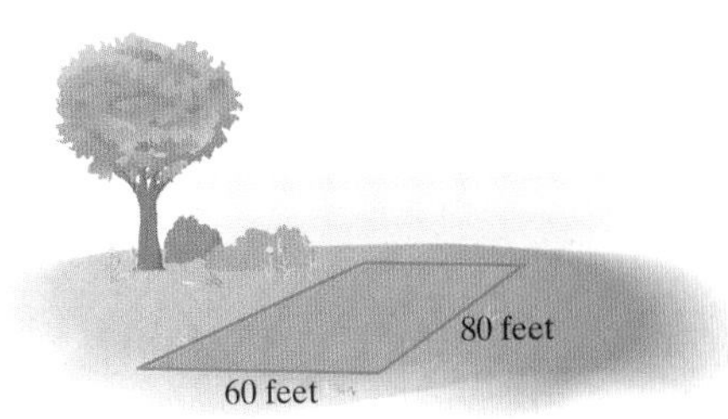

Practice Problem 2

Find the perimeter of a rectangle with a length of 22 centimeters and a width of 10 centimeters.

Answers

1. 280 ft **2.** 64 cm

Recall that a square is a special rectangle with all four sides the same length. The formula for the perimeter of a square is shown next:

Perimeter of a Square

$$\text{Perimeter} = \text{side} + \text{side} + \text{side} + \text{side}$$
$$= 4 \cdot \text{side}$$

In symbols,

$$P = 4s$$

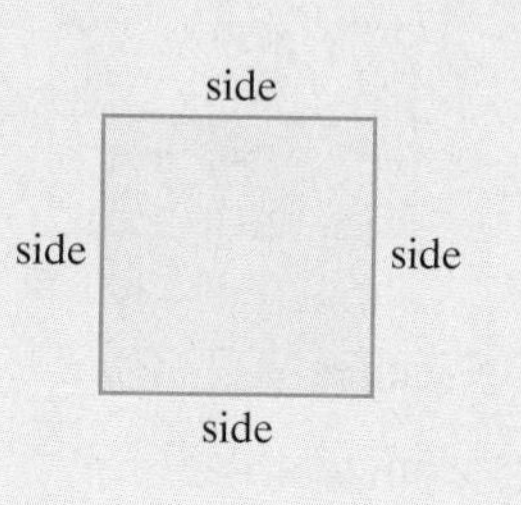

Practice Problem 3

How much fencing is needed to enclose a square field 50 yards on a side?

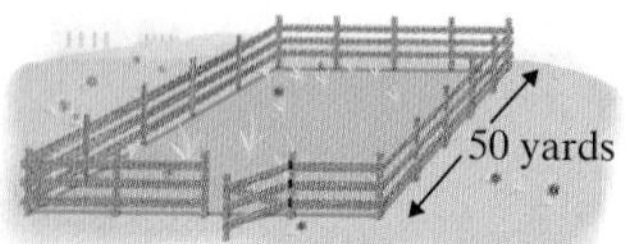

EXAMPLE 3 Finding the Perimeter of a Tabletop

Find the perimeter of a square tabletop if each side is 5 feet long.

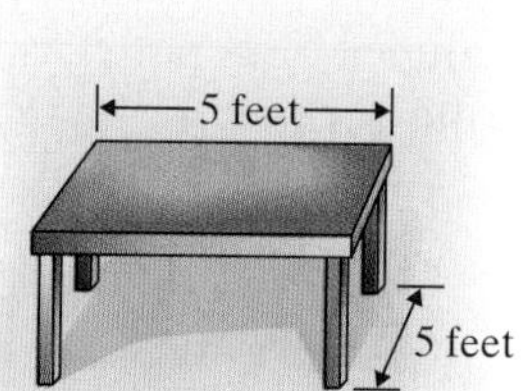

Solution: The formula for the perimeter of a square is $P = 4s$. We use this formula and replace s by 5 feet.

$$P = 4s$$
$$= 4 \cdot 5 \text{ feet}$$
$$= 20 \text{ feet}$$

The perimeter of the square tabletop is 20 feet.

The formula for the perimeter of a triangle with sides of lengths a, b, and c is given next:

Perimeter of a Triangle

$$\text{Perimeter} = \text{side } a + \text{side } b + \text{side } c$$

In symbols,

$$P = a + b + c$$

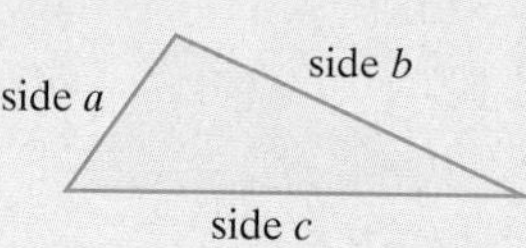

Answer

3. 200 yd

EXAMPLE 4 Find the perimeter of a triangle when the sides are 3 inches, 7 inches, and 6 inches.

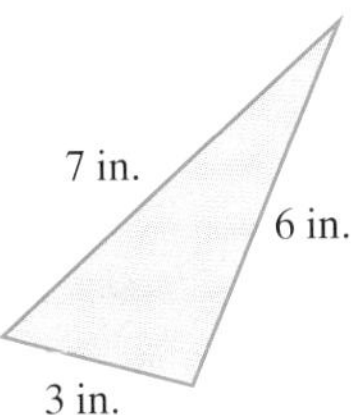

Solution: The formula is $P = a + b + c$, where a, b, and c are the lengths of the sides. Thus,

$$\begin{aligned} P &= a + b + c \\ &= 3 \text{ in.} + 7 \text{ in.} + 6 \text{ in.} \\ &= 16 \text{ in.} \end{aligned}$$

The perimeter of the triangle is 16 inches.

Recall that to find the perimeter of other polygons, we find the sum of the lengths of their sides.

EXAMPLE 5 Find the perimeter of the quadrilateral shown below:

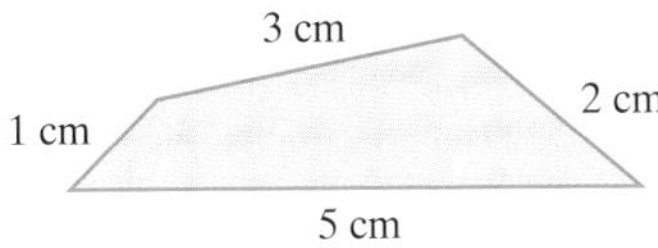

Solution: To find the perimeter, we find the sum of the lengths of its sides.

$$\text{perimeter} = 3 \text{ cm} + 2 \text{ cm} + 5 \text{ cm} + 1 \text{ cm} = 11 \text{ cm}$$

The perimeter is 11 centimeters.

EXAMPLE 6 Finding the Perimeter of a Room

Find the perimeter of the room shown below:

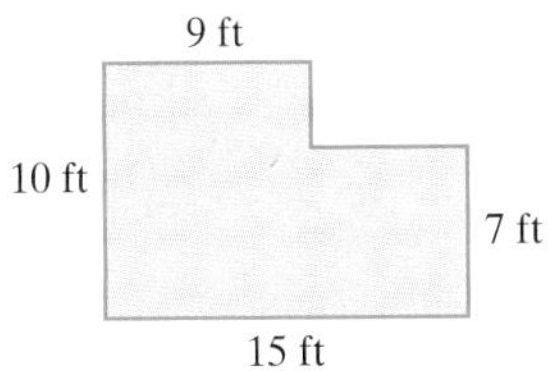

Solution: To find the perimeter of the room, we first need to find the lengths of all sides of the room.

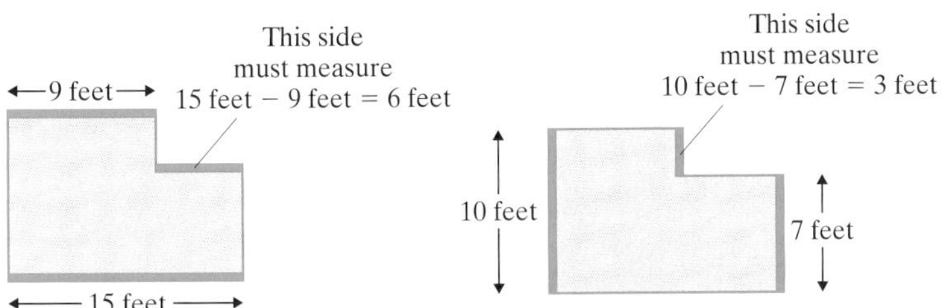

Practice Problem 4

Find the perimeter of a triangle when the sides are 5 centimeters, 9 centimeters, and 7 centimeters in length.

Practice Problem 5

Find the perimeter of the trapezoid shown.

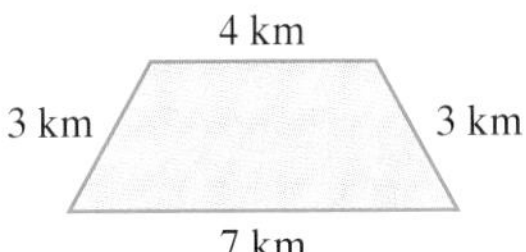

Practice Problem 6

Find the perimeter of the room shown.

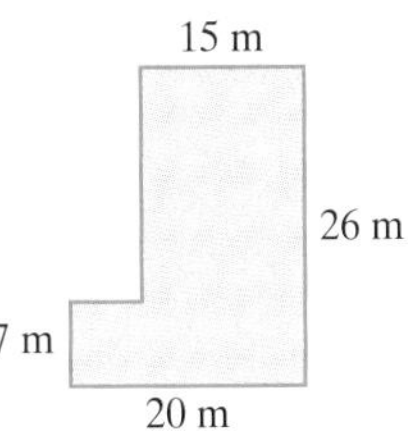

Answers

4. 21 cm **5.** 17 km **6.** 92 m

Now that we know the measures of all sides of the room, we can add the measures to find the perimeter.

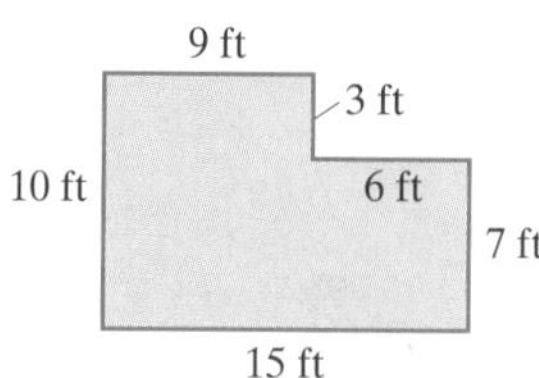

$$\text{perimeter} = 10\text{ ft} + 9\text{ ft} + 3\text{ ft} + 6\text{ ft} + 7\text{ ft} + 15\text{ ft}$$
$$= 50\text{ ft}$$

The perimeter of the room is 50 feet.

Practice Problem 7

A rectangular lot measures 60 feet by 120 feet. Find the cost to install fencing around the lot if the cost of fencing is \$1.90 per foot.

EXAMPLE 7 Calculating the Cost of Baseboard

A rectangular room measures 10 feet by 12 feet. Find the cost to hang a wallpaper border on the walls close to the ceiling if the cost of the wallpaper border is \$1.09 per foot.

Solution: First we find the perimeter of the room.

$$P = 2l + 2w$$
$$= 2 \cdot 12\text{ feet} + 2 \cdot 10\text{ feet} \quad \text{Replace } l \text{ with 12 feet and } w \text{ with 10 feet.}$$
$$= 24\text{ feet} + 20\text{ feet}$$
$$= 44\text{ feet}$$

The cost of the wallpaper is

$$\text{cost} = 1.09 \cdot 44 = 47.96$$

The cost of the wallpaper is \$47.96

B Using Formulas to Find Circumferences

Recall from Section 5.3 that the distance around a circle is called the **circumference.** This distance depends on the radius or the diameter of the circle.

The formulas for circumference are shown next:

Circumference of a Circle

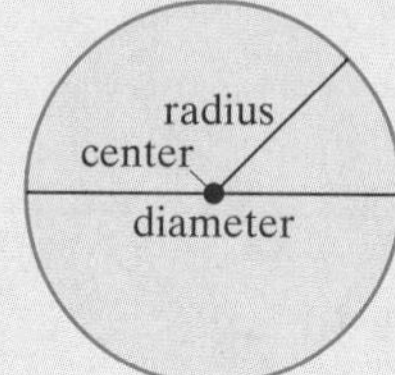

Circumference = $2 \cdot \pi \cdot$ radius or Circumference = $\pi \cdot$ diameter

In symbols,

$$C = 2\pi r \quad \text{or} \quad C = \pi d,$$

where $\pi \approx 3.14$ or $\pi \approx \frac{22}{7}$.

Answer

7. \$684

To better understand circumference and π(pi), try the following experiment. Take any can and measure its circumference and its diameter.

The can in the figure above has a circumference of 23.5 centimeters and a diameter of 7.5 centimeters. Now divide the circumference by the diameter.

$$\frac{\text{circumference}}{\text{diameter}} = \frac{23.5 \text{ cm}}{7.5 \text{ cm}} \approx 3.1$$

Try this with other sizes of cylinders and circles—you should always get a number close to 3.1. The exact ratio of circumference to diameter is π. (Recall that $\pi \approx 3.14$ or $\approx \frac{22}{7}$.)

EXAMPLE 8 Installing a Border in a Circular Spa

Mary Catherine Dooley plans to install a border of new tiling around the circumference of her circular spa. If her spa has a diameter of 14 feet, find its circumference.

Solution: Because we are given the diameter, we use the formula $C = \pi d$.

$$C = \pi d$$
$$= \pi \cdot 14 \text{ ft} \quad \text{Replace } d \text{ with 14 feet.}$$
$$= 14\pi \text{ ft}$$

The circumference of the spa is *exactly* 14π feet. By replacing π with the *approximation* 3.14, we find that the circumference is *approximately* $14 \text{ feet} \cdot 3.14 = 43.96$ feet.

Try the Concept Check in the margin.

Practice Problem 8

An irrigation device waters a circular region with a diameter of 20 yards. What is the circumference of the watered region? Use $\pi \approx 3.14$.

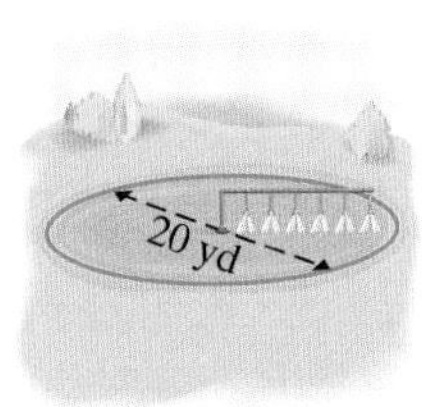

Concept Check

The distance around which figure is greater: a square with side length 5 inches or a circle with radius 3 inches?

Answer

8. 20π yd ≈ 62.8 yd

Concept Check: a square with length 5 in.

FOCUS ON The Real World

SPEEDS

A speed measures how far something travels in a given unit of time. The speed 55 miles per hour is a unit rate that can be written as $\frac{55 \text{ miles}}{1 \text{ hour}}$. Just as there are different units of measurement for length or distance, there are different units of measurement for speed as well. It is also possible to perform unit conversions on speeds. Before we learn about converting speeds, we will review units of time. The following is a summary of equivalencies between various units of time.

Units of Time	Unit Fractions
60 seconds (s) = 1 minute (min)	$\frac{60 \text{ s}}{1 \text{ min}} = \frac{1 \text{ min}}{60 \text{ s}} = 1$
60 minutes = 1 hour (h)	$\frac{60 \text{ min}}{1 \text{ h}} = \frac{1 \text{ h}}{60 \text{ min}} = 1$
3600 seconds = 1 hour	$\frac{3600 \text{ s}}{1 \text{ h}} = \frac{1 \text{ h}}{3600 \text{ s}}$

Here are some common speeds.

Speeds

Miles per hour (mph)
Miles per minute (mi/min)
Miles per second (mi/s)
Feet per second (ft/s)
Feet per minute (ft/min)
Kilometers per hour (kmph or km/h)
Kilometers per second (kmps or km/s)
Meters per second (m/s)
Knots

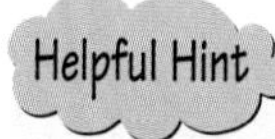

A **knot** is 1 nautical mile per hour and is a measure of speed used for ships.

1 nautical mile (nmi) ≈ 1.15 miles (mi)
1 nautical mile (nmi) ≈ 6076.12 feet (ft)

To convert from one speed to another, unit fractions may be used. To convert from mph to ft/s, first write the original speed as a unit rate. Then multiply by a unit fraction that relates miles to feet and by a unit fraction that relates hours to seconds. The unit fractions should be written so that the given units in the original rate will divide out. For example, to convert 55 mph to ft/s:

$$\begin{aligned} 55 \text{ mph} &= \frac{55 \text{ miles}}{1 \text{ hour}} = \frac{55 \text{ \sout{miles}}}{1 \text{ \sout{hour}}} \cdot \frac{5280 \text{ ft}}{1 \text{ \sout{mile}}} \cdot \frac{1 \text{ \sout{hour}}}{3600 \text{ s}} \\ &= \frac{55 \cdot 5280 \text{ ft}}{3600 \text{ s}} \\ &= \frac{290{,}400 \text{ ft}}{3600 \text{ s}} \\ &= 80\frac{2}{3} \text{ ft/s} \end{aligned}$$

GROUP ACTIVITY

1. Research the current world land speed record. Convert the speed from mph to feet per second.
2. Research the current world water speed record. Convert from mph to knots.
3. Research and then describe the Beaufort Wind Scale, its origins, and how it is used. Give the scale keyed to both miles per hour and knots. Why would both measures be useful?

Name ____________________ Section __________ Date __________

EXERCISE SET 8.3

A *Find the perimeter of each figure. See Examples 1 through 6.*

1.

15 ft Rectangle

17 ft

2.

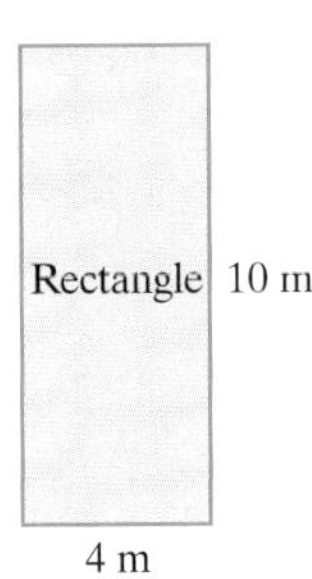

3.

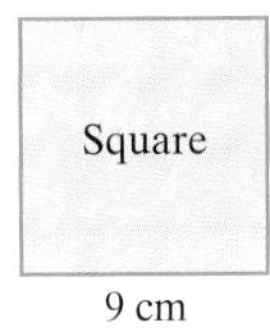

4.

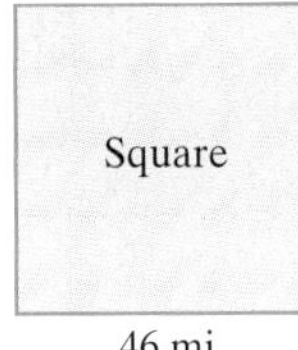

5.

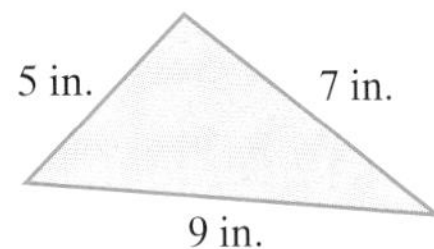

6.

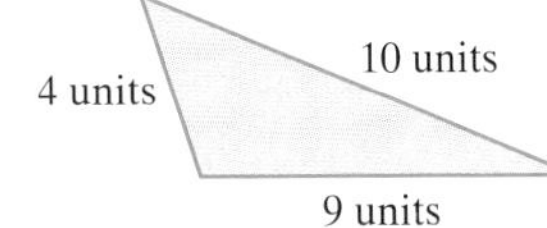

7.

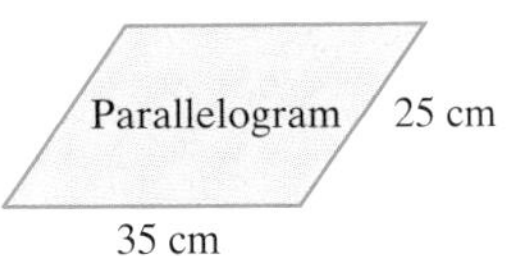

8.

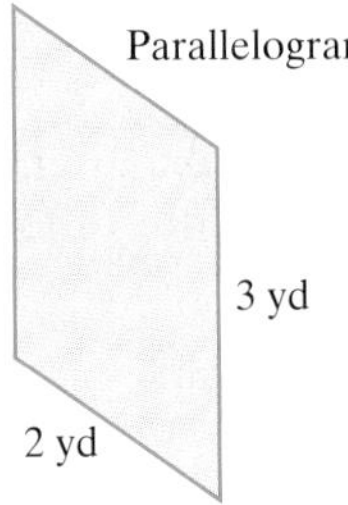

9.

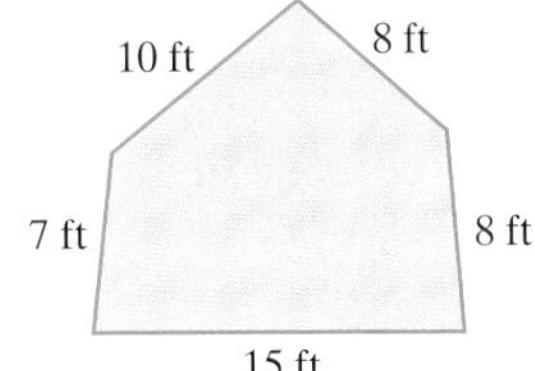

10.

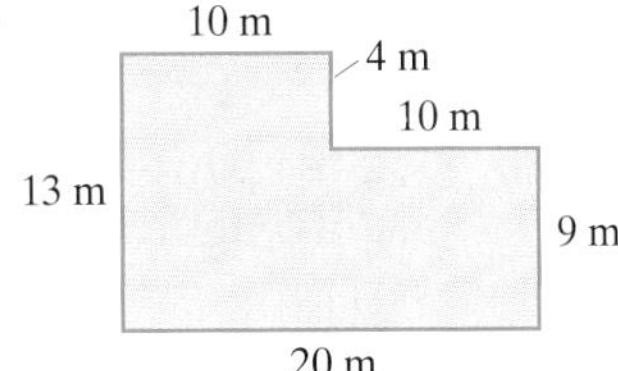

11.

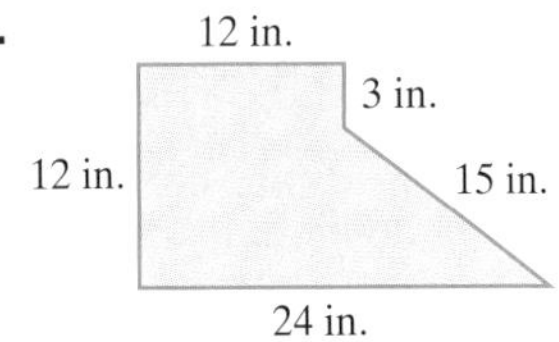

12.

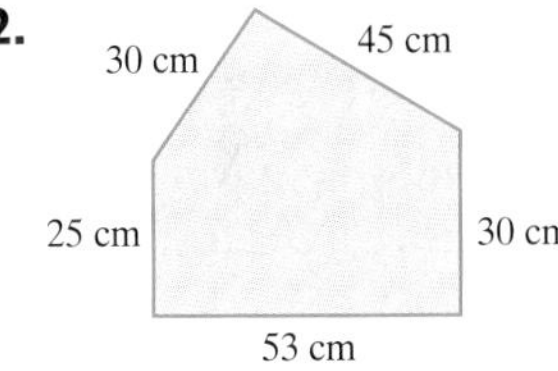

Solve. See Examples 1 through 7.

13. A polygon has sides of length 5 feet, 3 feet, 2 feet, 7 feet, and 4 feet. Find its perimeter.

14. A triangle has sides of length 8 inches, 12 inches, and 10 inches. Find its perimeter.

15. Baseboard is to be installed in a square room that measures 15 feet on one side. Find how much baseboard is needed. (*Note:* Ignore the baseboard not needed because of doors.)

16. Find how much fencing is needed to enclose a rectangular rose garden 85 feet by 15 feet.

17. If a football field is 53 yards wide and 120 yards long, what is the perimeter?

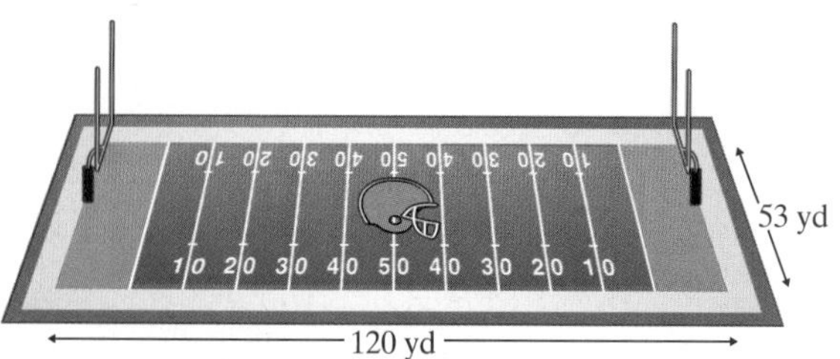

18. A stop sign has eight equal sides of length 12 inches. Find its perimeter.

19. A metal strip is being installed around a workbench that is 8 feet long and 3 feet wide. Find how much stripping is needed.

20. Find how much fencing is needed to enclose a rectangular garden 70 feet by 21 feet.

21. If the stripping in Exercise 19 costs \$3 per foot, find the total cost of the stripping.

22. If the fencing in Exercise 20 costs \$2 per foot, find the total cost of the fencing.

23. A regular hexagon has a side length of 6 inches. Find its perimeter.

24. A regular pentagon has a side length of 14 meters. Find its perimeter.

25. Find the perimeter of the top of a square compact disc case if the length of one side is 7 inches.

26. Find the perimeter of a square ceramic tile with a side of length 5 inches.

27. A rectangular room measures 6 feet by 8 feet. Find the cost of installing a strip of wallpaper around the top of the room if the wallpaper costs \$0.86 per foot.

28. A rectangular house measures 75 feet by 60 feet. Find the cost of installing gutters around the house if the cost is \$2.36 per foot.

Find the perimeter of each figure. See Example 6.

29.

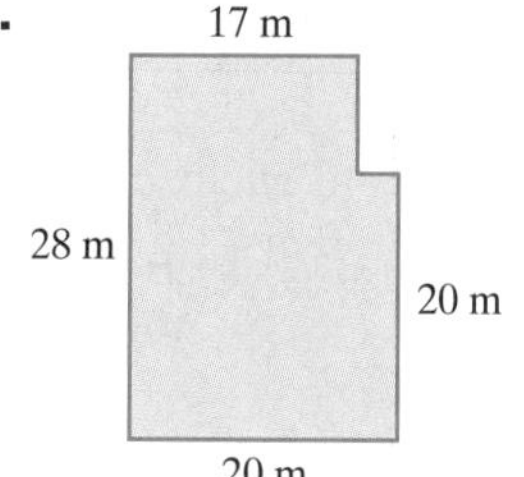

30.

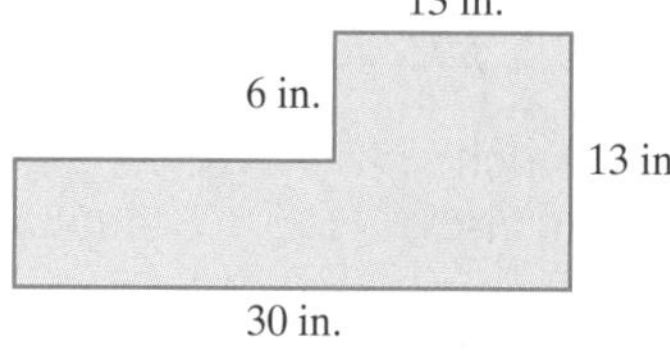

31.

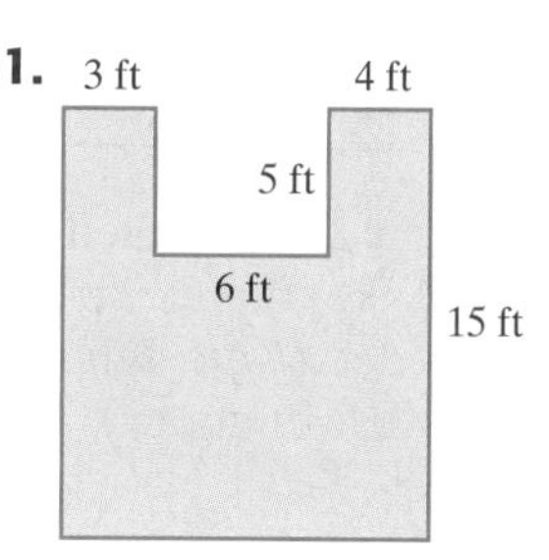

32.

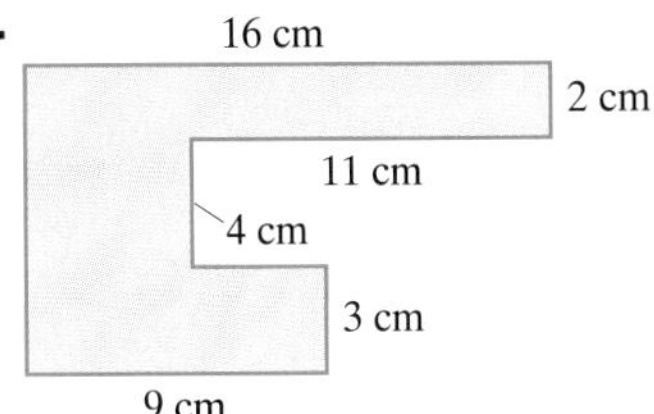

33.

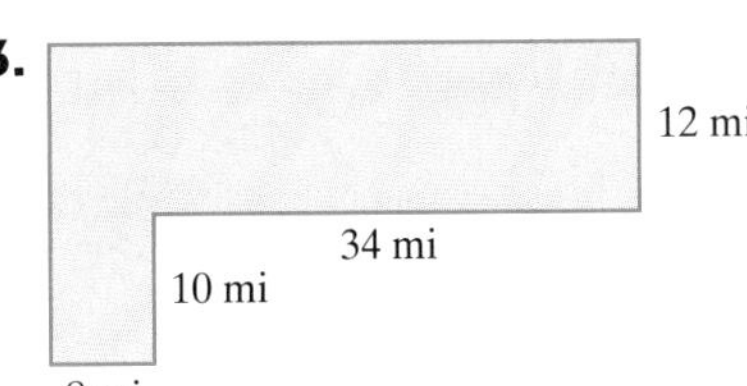

34.

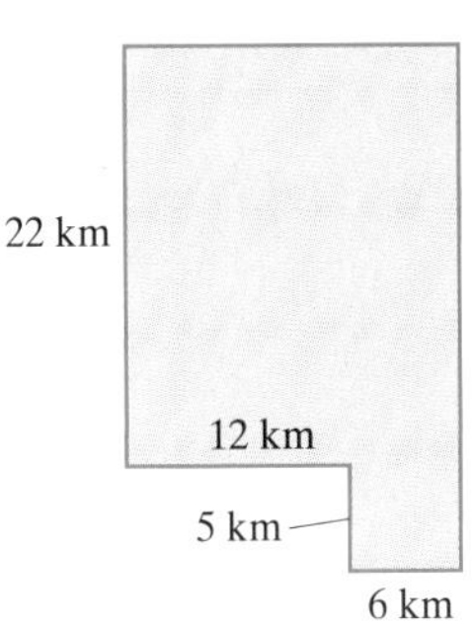

B *Find the circumference of each circle. Give the exact circumference and then an approximation. Use* $\pi \approx 3.14$. *See Example 8.*

35.

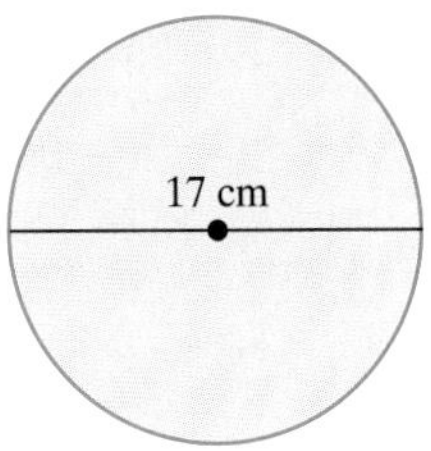

36.

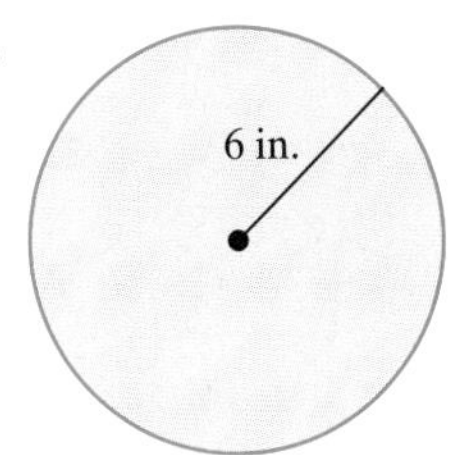

37.

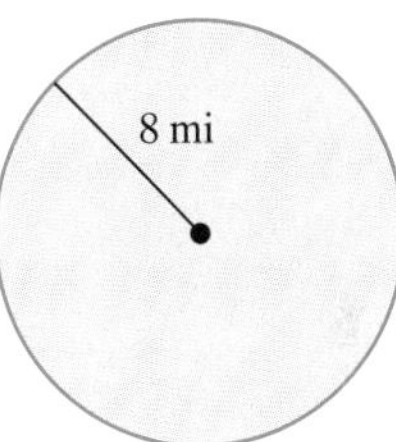

38.

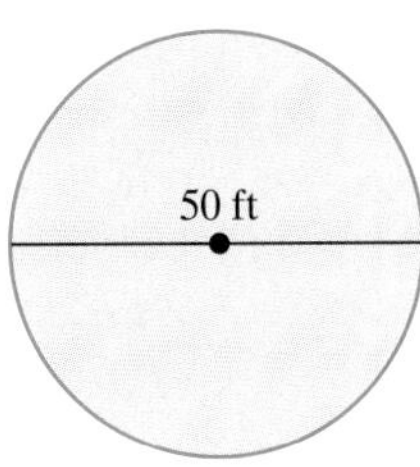

39.

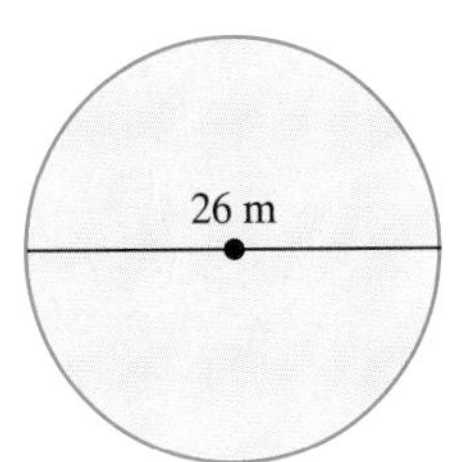

40.

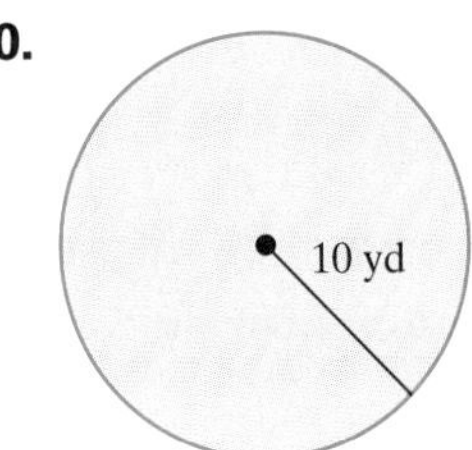

41. A circular fountain has a radius of 5 feet. Approximate the distance around the fountain. Use $\frac{22}{7}$ for π.

42. A circular walkway has a radius of 40 meters. Approximate the distance around the walkway. Use 3.14 for π.

43. Meteor Crater, near Winslow, Arizona, is about 4000 feet in diameter. Approximate the distance around the crater. Use 3.14 for π. (*Source: The Handy Science Answer Book*)

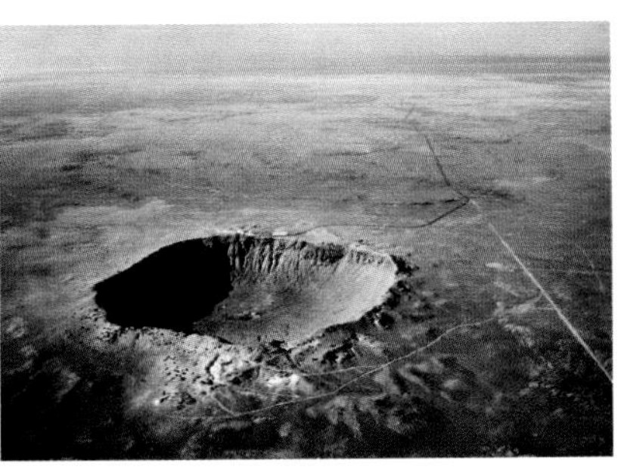

44. The largest pearl, the *Pearl of Lao-tze*, has a diameter of $5\frac{1}{2}$ inches. Approximate the distance around the pearl. Use $\frac{22}{7}$ for π. (*Source: The Guinness Book of Records*)

Review and Preview

Simplify. See Section 1.8.

45. $5 + 6 \cdot 3$

46. $25 - 3 \cdot 7$

47. $(20 - 16) \div 4$

48. $6 \cdot (8 + 2)$

49. $(18 + 8) - (12 + 4)$

50. $72 \div (2 \cdot 6)$

51. $(72 \div 2) \cdot 6$

52. $4^1 \cdot (2^3 - 8)$

Combining Concepts

53. a. Find the circumference of each circle. Approximate the circumference by using 3.14 for π.

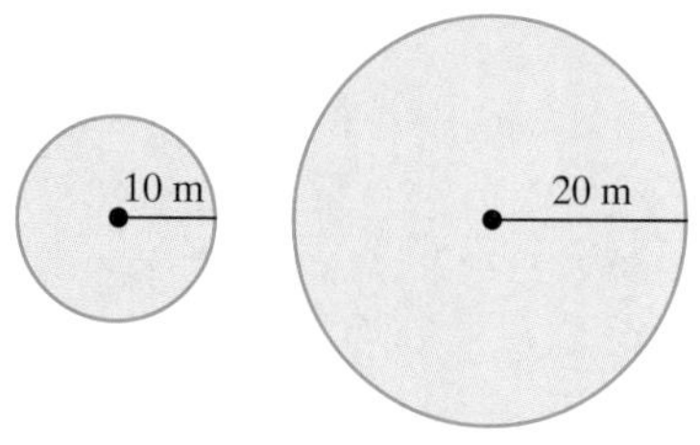

b. If the radius of a circle is doubled, is its corresponding circumference doubled?

54. a. Find the circumference of each circle. Approximate the circumference by using 3.14 for π.

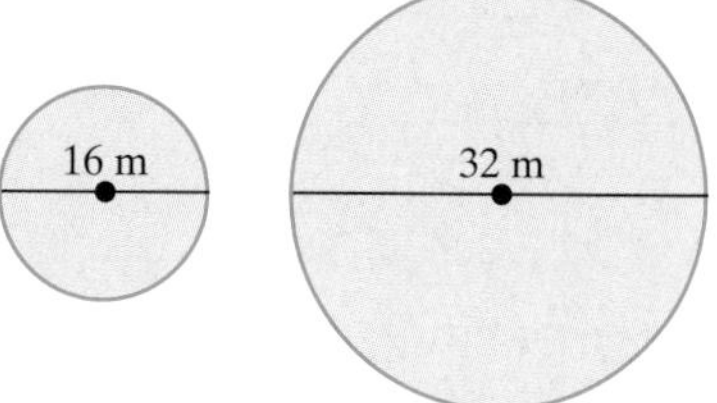

b. If the diameter of a circle is doubled, is its corresponding circumference doubled?

55. Find the perimeter of the skating rink. Give the exact answer and an approximate answer using 3.14 for π.

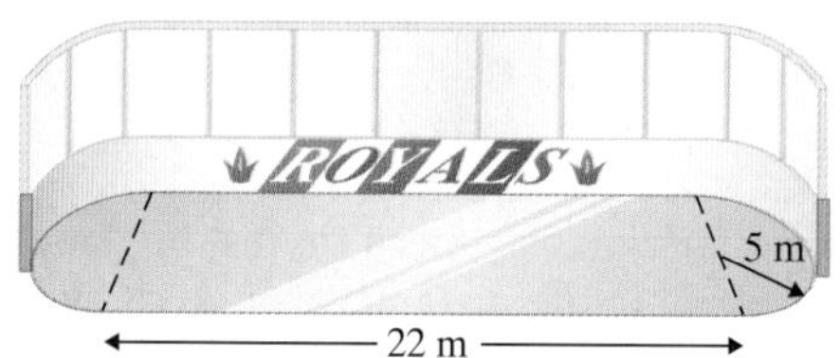

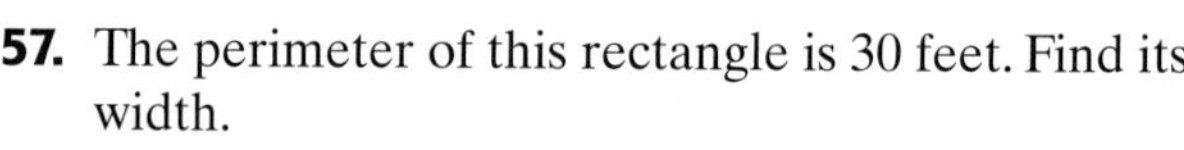

56. In your own words, explain how to find the perimeter of any polygon.

57. The perimeter of this rectangle is 30 feet. Find its width.

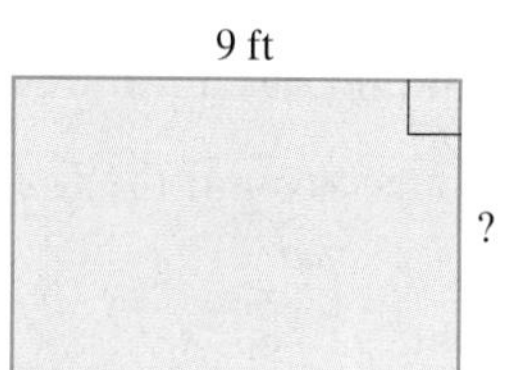

Find the perimeter. Round your results to the nearest tenth.

58.

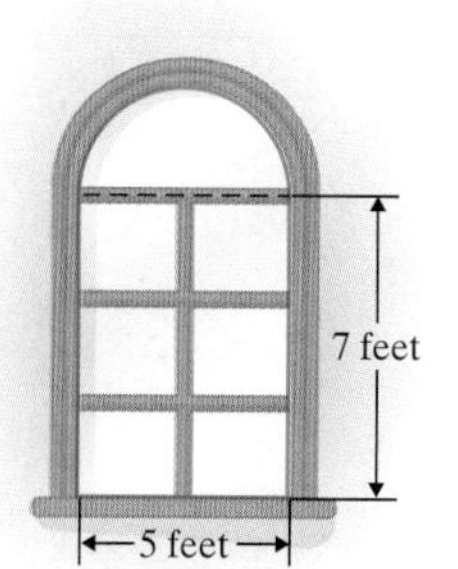

59.

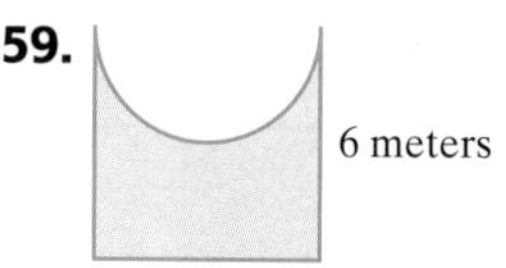

8.4 Area and Volume

OBJECTIVES

- **A** Find the area of geometric figures.
- **B** Find the volume of geometric figures.

A Finding Area

Recall that area measures the number of square units that cover the surface of a region. Thus far, we know how to find the area of a rectangle and a square. These formulas, as well as formulas for finding the areas of other common geometric figures, are given next.

Area Formulas of Common Geometric Figures

Geometric Figure | **Area Formula**

RECTANGLE

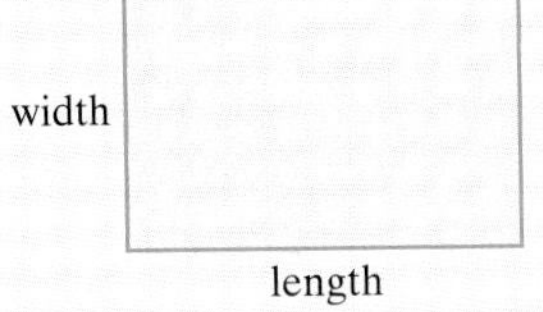

Area of a rectangle:

area = **l**ength · **w**idth

$A = lw$

SQUARE

Area of a square:

area = **s**ide · **s**ide

$A = s \cdot s = s^2$

TRIANGLE

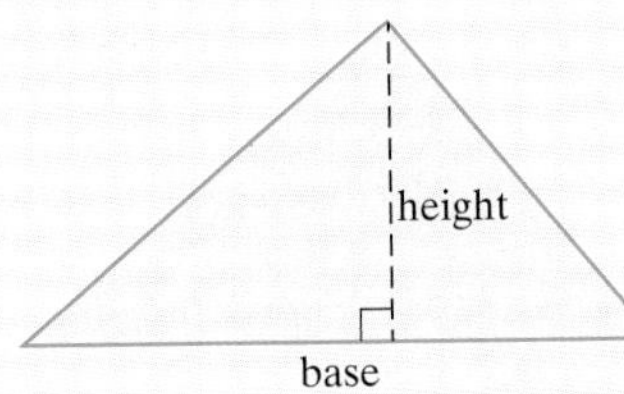

Area of a triangle:

area $= \frac{1}{2} \cdot$ **b**ase · **h**eight

$A = \frac{1}{2}bh$

PARALLELOGRAM

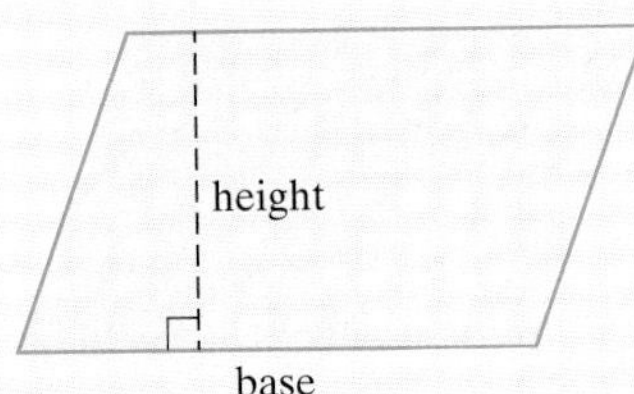

Area of a parallelogram:

area = **b**ase · **h**eight

$A = bh$

TRAPEZOID

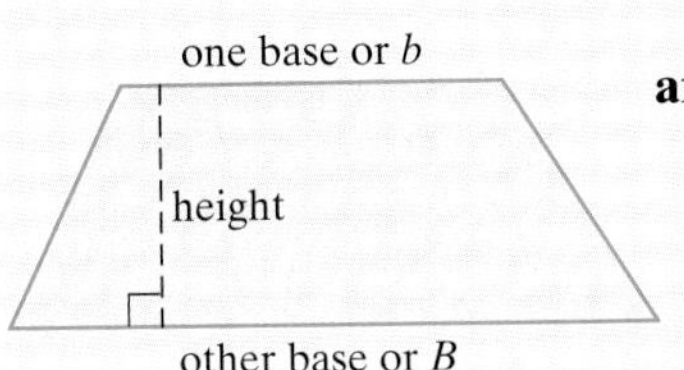

Area of a trapezoid:

area $= \frac{1}{2} \cdot$ (one **b**ase + other **b**ase) · **h**eight

$A = \frac{1}{2}(b + B)h$

Practice Problem 1

Find the area of the triangle.

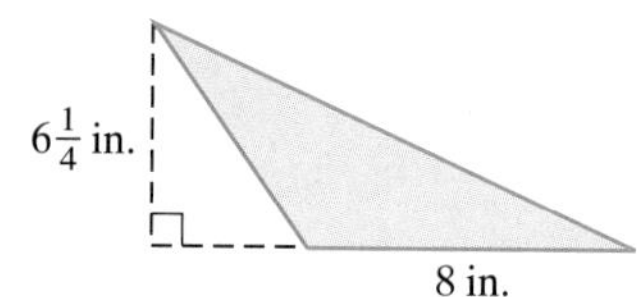

Practice Problem 2

Find the area of the trapezoid.

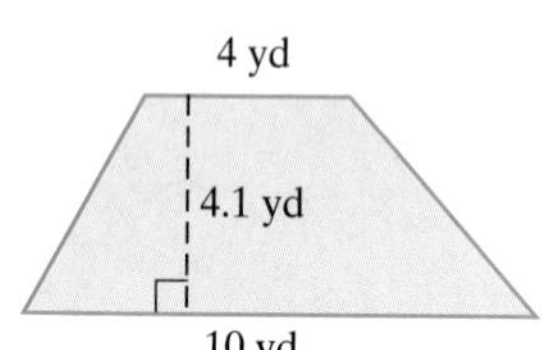

Practice Problem 3

Find the area of the figure.

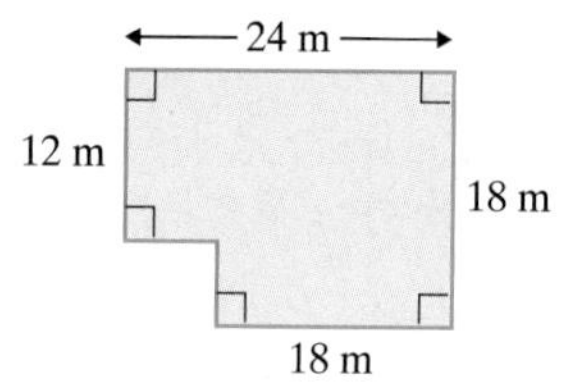

Answers

1. 25 sq in. **2.** 28.7 sq yd **3.** 396 sq m

Use these formulas for the following examples.

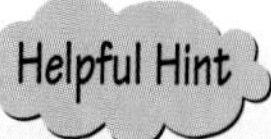

Area is always measured in square units.

EXAMPLE 1 Find the area of the triangle.

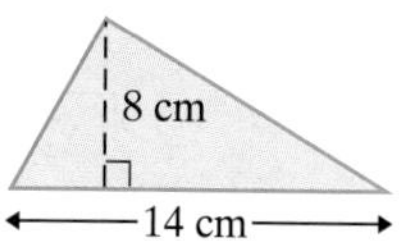

Solution:

$$A = \frac{1}{2}bh$$

$$= \frac{1}{2} \cdot 14 \text{ cm} \cdot 8 \text{ cm}$$ Replace b, base, with 14 cm and h, height, with 8 cm.

$$= \frac{\overset{1}{\cancel{2}} \cdot 7 \cdot 8}{\underset{1}{\cancel{2}}} \text{ square centimeters}$$ Write 14 as $2 \cdot 7$.

$$= 56 \text{ square centimeters}$$

The area is 56 square centimeters.

EXAMPLE 2 Find the area of the parallelogram.

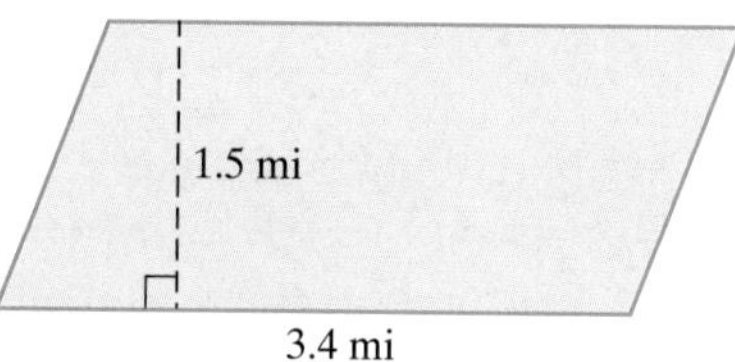

Solution:

$$A = bh$$

$$= 3.4 \text{ miles} \cdot 1.5 \text{ miles}$$ Replace b, base, with 3.4 miles and h, height, with 1.5 miles.

$$= 5.1 \text{ square miles}$$

The area is 5.1 square miles.

Helpful Hint

When finding the area of figures, check to make sure that all measurements are in the same units before calculations are made.

EXAMPLE 3 Find the area of the figure.

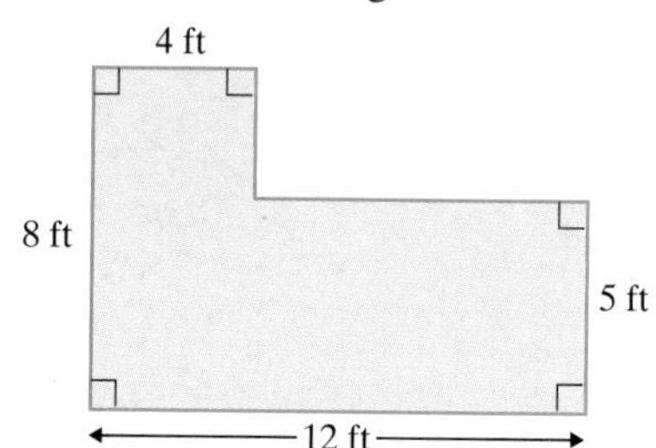

Solution: Split the figure into two rectangles. To find the area of the figure, we find the sum of the areas of the two rectangles.

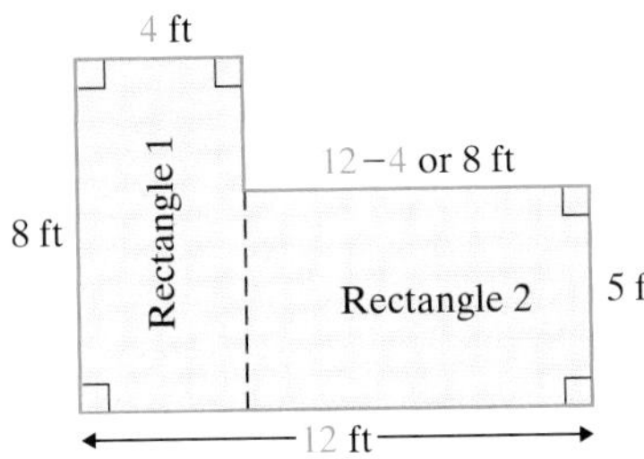

$$\begin{aligned}\text{area of Rectangle 1} &= lw\\ &= 8\text{ feet}\cdot 4\text{ feet}\\ &= 32\text{ square feet}\end{aligned}$$

Notice that the length of Rectangle 2 is 12 feet −4 feet or 8 feet.

$$\begin{aligned}\text{area of Rectangle 2} &= lw\\ &= 8\text{ feet}\cdot 5\text{ feet}\\ &= 40\text{ square feet}\end{aligned}$$

$$\begin{aligned}\text{area of the figure} &= \text{area of Rectangle 1} + \text{area of Rectangle 2}\\ &= 32\text{ square feet} + 40\text{ square feet}\\ &= 72\text{ square feet}\end{aligned}$$

Helpful Hint

The figure in Example 3 could also be split into two rectangles as shown.

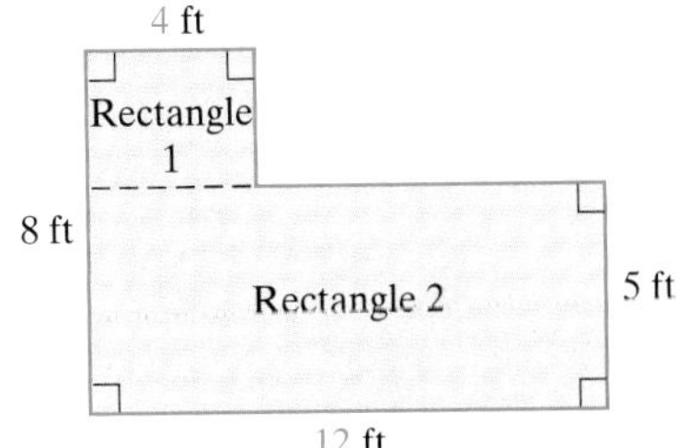

To better understand the formula for area of a circle, try the following. Cut a circle into many pieces, as shown.

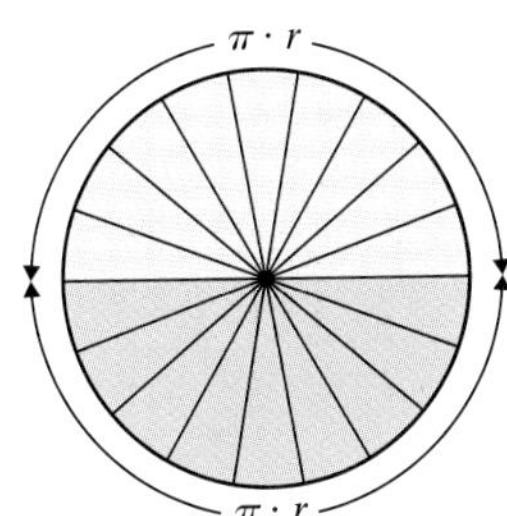

The circumference of a circle is $2\pi r$. This means that the circumference of half a circle is half of $2\pi r$, or πr.

Then unfold the two halves of the circle and place them together, as shown.

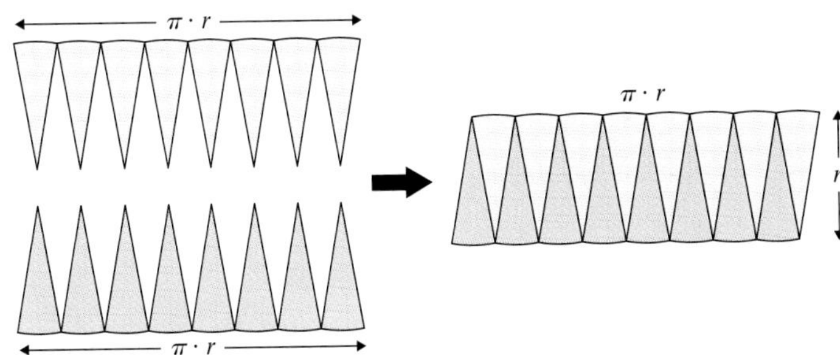

The figure on the right is almost a parallelogram with a base of πr and a height of r. The area is

$$\begin{aligned}A &= \text{base} \cdot \text{height}\\ &= (\pi r) \cdot r\\ &= \pi r^2\end{aligned}$$

This is the formula for area of a circle.

Area Formula of a Circle

Circle

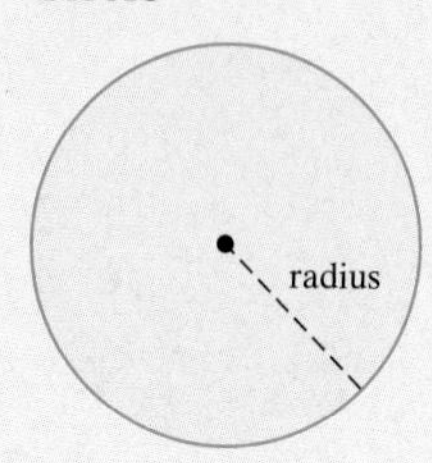

Area of a circle

area $= \pi \cdot (\text{radius})^2$

$A = \pi r^2$

(A fraction approximation for π is $\frac{22}{7}$.)

(A decimal approximation for π is 3.14.)

Practice Problem 4

Find the area of the given circle. Find the exact area and an approximation. Use 3.14 as an approximation for π.

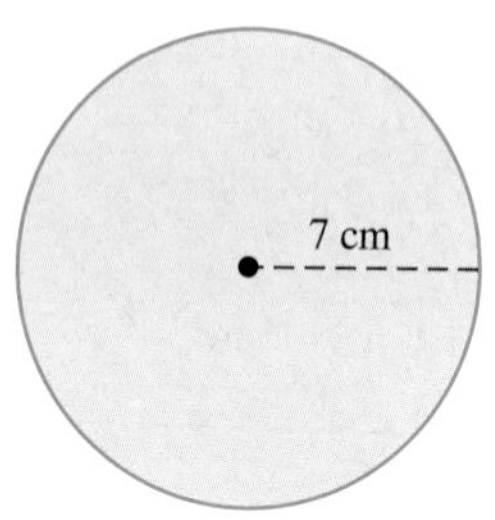

EXAMPLE 4 Find the area of a circle with a radius of 3 feet. Find the exact area and an approximation. Use 3.14 as an approximation for π.

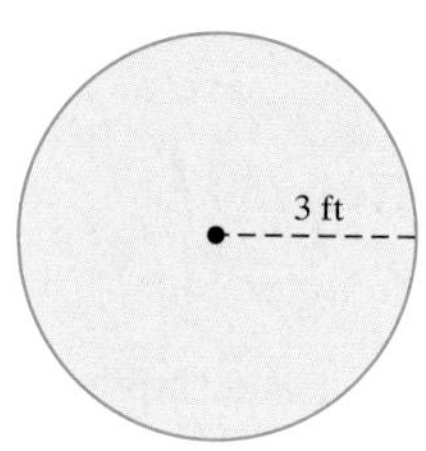

Solution: We let $r = 3$ feet and use the formula

$$A = \pi r^2$$
$$= \pi \cdot (3\text{ feet})^2 \quad \text{Replace } r \text{ with 3 feet.}$$
$$= 9 \cdot \pi \text{ square feet} \quad \text{Replace } (3\text{ feet})^2 \text{ with 9 sq ft.}$$

To approximate this area, we substitute 3.14 for π.

$$9 \cdot \pi \text{ square feet} \approx 9 \cdot 3.14 \text{ square feet}$$
$$= 28.26 \text{ square feet}$$

The *exact* area of the circle is 9π square feet, which is *approximately* 28.26 square feet.

Try the Concept Check in the margin.

Concept Check

Use estimation to decide which figure would have a larger area: a circle of diameter 10 in. or a square 10 in. long on each side.

B Finding Volume

Volume measures the number of cubic units that fill the space of a solid. The volume of a box or can, for example, is the amount of space inside. Volume can be used to describe the amount of juice in a pitcher or the amount of concrete needed to pour a foundation for a house.

The volume of a solid is the number of **cubic units** in the solid. A cubic centimeter and a cubic inch are illustrated.

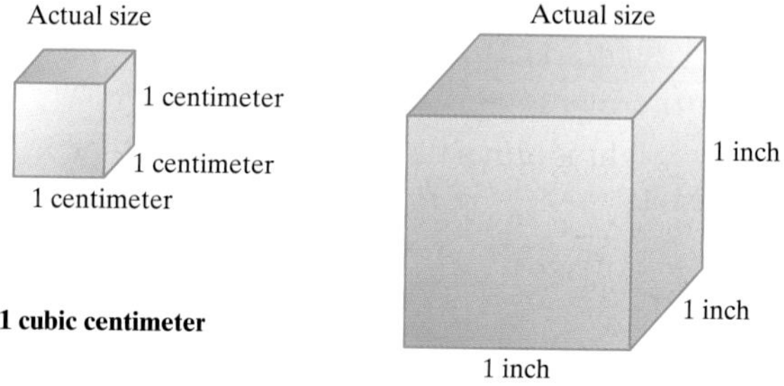

Formulas for finding the volumes of some common solids are given next.

Answers

4. 49π sq cm $\approx$ 153.86 sq cm

Concept Check: A square 10 in. long on each side would have a larger area.

Volume Formulas of Common Solids

Solid	Volume Formulas
RECTANGULAR SOLID 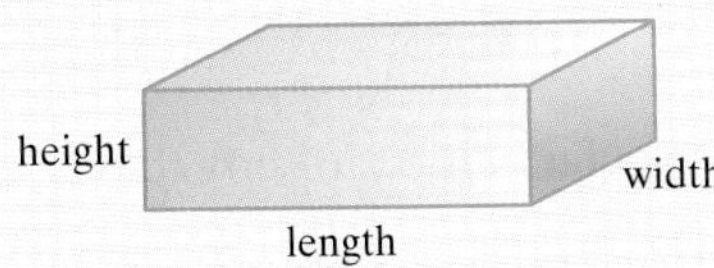	Volume of a rectangular solid: **v**olume = **l**ength · **w**idth · **h**eight $V = lwh$
CUBE 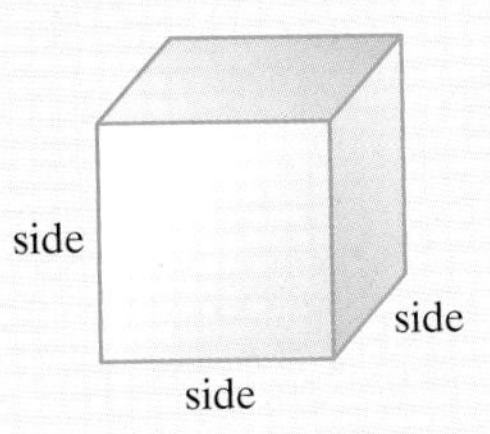	Volume of a cube: **v**olume = **s**ide · **s**ide · **s**ide $V = s \cdot s \cdot s = s^3$
SPHERE 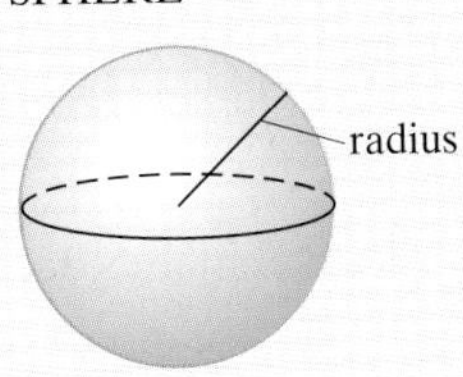	Volume of a sphere: **v**olume $= \frac{4}{3} \cdot \pi \cdot$ (**r**adius)^3 $V = \frac{4}{3}\pi r^3$
CIRCULAR CYLINDER 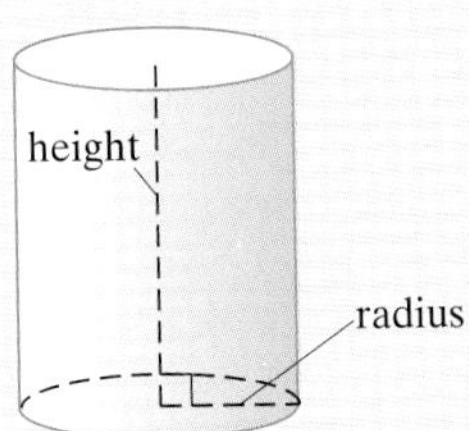	Volume of a circular cylinder: **v**olume $= \pi \cdot$ (**r**adius)^2 · **h**eight $V = \pi r^2 h$
CONE 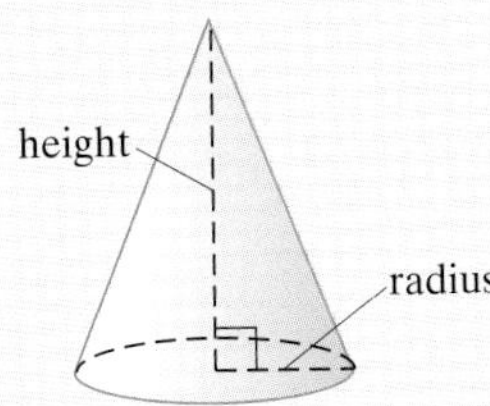	Volume of a cone: **v**olume $= \frac{1}{3} \cdot \pi \cdot$ (**r**adius)^2 · **h**eight $V = \frac{1}{3}\pi r^2 h$
SQUARE-BASED PYRAMID 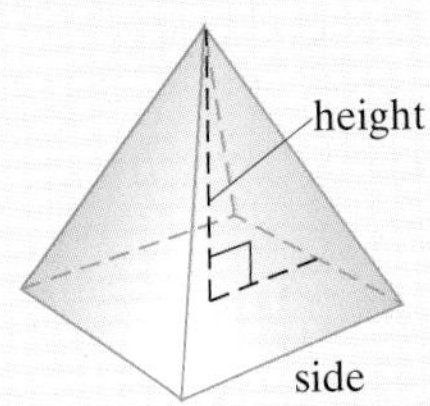	Volume of a square-based pyramid: **v**olume $= \frac{1}{3} \cdot$ (**s**ide)^2 · **h**eight $V = \frac{1}{3}s^2 h$

Helpful Hint

Volume is always measured in cubic units.

Practice Problem 5

Draw a diagram and find the volume of a rectangular box that is 5 ft long, 2 ft wide, and 4 ft deep.

EXAMPLE 5 Find the volume of a rectangular cardboard box that is 12 inches long, 6 inches wide, and 3 inches high.

Solution: $V = lwh$

$$V = 12 \text{ inches} \cdot 6 \text{ inches} \cdot 3 \text{ inches} = 216 \text{ cubic inches}$$

The volume of the rectangular box is 216 cubic inches.

Try the Concept Check in the margin.

Concept Check

Juan Lopez is calculating the volume of the rectangular solid shown. Find the error in his calculation.

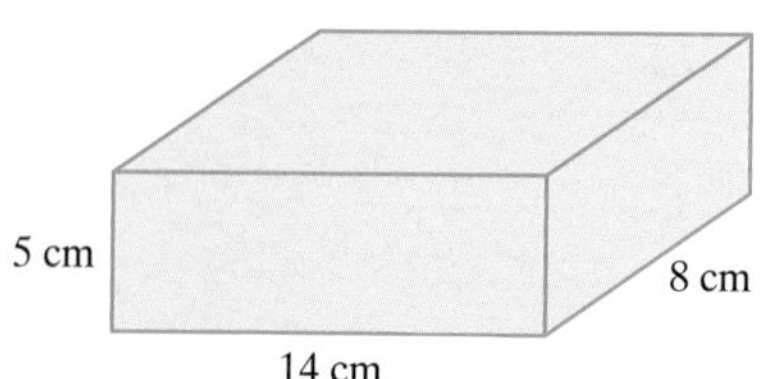

$$\begin{aligned} \text{volume} &= l + w + h \\ &= 14 + 8 + 5 \\ &= 27 \text{ cubic cm} \end{aligned}$$

EXAMPLE 6 Approximate the volume of a ball of radius 3 inches. Use the approximation $\frac{22}{7}$ for π.

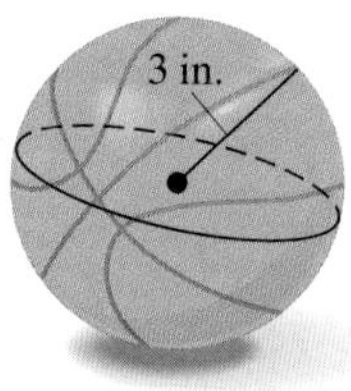

Solution: $V = \frac{4}{3}\pi r^3$

$$\approx \frac{4}{3} \cdot \frac{22}{7} \cdot (3 \text{ inches})^3$$

$$= \frac{4}{3} \cdot \frac{22}{7} \cdot 27 \text{ cubic inches}$$

$$= \frac{4 \cdot 22 \cdot \overset{1}{\cancel{3}} \cdot 9}{\underset{1}{\cancel{3}} \cdot 7} \text{ cubic inches} \quad \text{Multiply and write 27 as } 3 \cdot 9.$$

$$= \frac{792}{7} \text{ cubic inches or } 113\frac{1}{7} \text{ cubic inches}$$

The volume is *approximately* $113\frac{1}{7}$ cubic inches.

Practice Problem 6

Draw a diagram and approximate the volume of a ball of radius $\frac{1}{2}$ cm. Use $\frac{22}{7}$ for π.

Answers

5. 40 cu ft **6.** $\frac{11}{21}$ cu cm

Concept Check: volume $= l \cdot w \cdot h$
$= 14 \cdot 8 \cdot 5$
$= 560$ cu cm

EXAMPLE 7 Approximate the volume of a can that has a $3\frac{1}{2}$-inch radius and a height of 6 inches. Use $\frac{22}{7}$ for π.

Solution: Using the formula for volume of a circular cylinder, we have

$$V = \pi r^2 h$$

$$3\frac{1}{2} = \frac{7}{2}$$

$$= \pi \cdot \left(\frac{7}{2}\text{inches}\right)^2 \cdot 6 \text{ inches}$$

or approximately

$$\approx \frac{22}{7} \cdot \frac{49}{4} \cdot 6 \text{ cubic inches}$$

$$= 231 \text{ cubic inches}$$

The volume is approximately 231 cubic inches.

Practice Problem 7

Approximate the volume of a cylinder of radius 5 in. and height 7 in. Use 3.14 for π.

EXAMPLE 8 Approximate the volume of a cone that has a height of 14 centimeters and radius of 3 centimeters. Use $\frac{22}{7}$ for π.

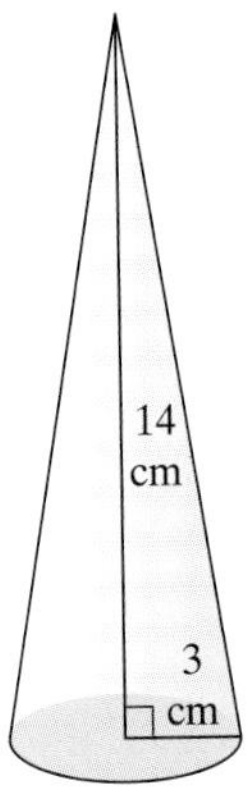

Solution: Using the formula for volume of a cone, we have

$$V = \frac{1}{3}\pi r^2 h$$

$$= \frac{1}{3} \cdot \pi \cdot (3 \text{ centimeters})^2 \cdot 14 \text{ centimeters}$$ Replace r with 3 cm and h with 14 cm.

$$= 42\pi \text{ cubic centimeters}$$

or approximately

$$\approx 42 \cdot \frac{22}{7} \text{cubic centimeters}$$

$$= 132 \text{ cubic centimeters}$$

The volume is approximately 132 cubic centimeters.

Practice Problem 8

Find the volume of a square-based pyramid that has a 3-meter side and height of 5.1 meters.

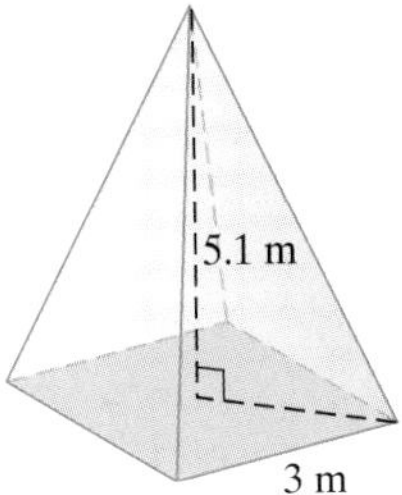

Answers

7. 549.5 cu in. **8.** 15.3 cu m

FOCUS ON The Real World

COMPUTER STORAGE CAPACITY

Have you ever heard someone describe a computer as having a 4 gig hard drive and wondered what exactly that means? In this chapter, we focused on measuring basic physical characteristics such as length, weight/mass, and capacity. However, another type of measurement that we frequently encounter in the real world deals with computer storage capacity; that is, how much information can be stored on a computer device such as a floppy diskette, data cartridge, CD-ROM, hard drive, or computer memory (such as RAM).

To understand computer storage capacities, we must first understand the building blocks of information in the world of computers. The smallest piece of information handled by a computer is called a **bit** (abbreviated "b"). Bit is short for "binary digit" and consists of either a 0 or a 1. A collection of 8 bits is called a **byte** (abbreviated "B"). A single byte represents a standard character such as a letter of the alphabet, a number, or a punctuation mark. The following table shows the relationships among various terms used to describe computer storage capacities:

1 nibble = 4 bits

1 byte (B) = 8 bits

1 kilobyte (KB) = 1024 bytes = 2^{10} bytes

1 megabyte (MB) = 1024 kilobytes = 1024^2 bytes = 2^{20} bytes

1 gigabyte (GB) = 1024 megabytes = 1024^3 bytes = 2^{30} bytes

1 terabyte (TB) = 1024 gigabytes = 1024^4 bytes = 2^{40} bytes

Note: Sometimes a megabyte is referred to as a "meg" and a gigabyte as a "gig."

GROUP ACTIVITY

Using advertisements for computer systems, find five different examples of uses of any of the computer storage capacity terms defined above. Then convert each example to both bytes and bits. Make a table to organize your results. Be sure to include an explanation of the original use of each example.

Name ______________________ Section ____________ Date ____________

EXERCISE SET 8.4

A *Find the area of each geometric figure. If the figure is a circle, give an exact area and then use the given* **approximation** *for π to approximate the area. See Examples 1 through 4.*

1.

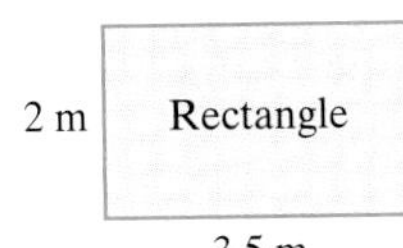

2.

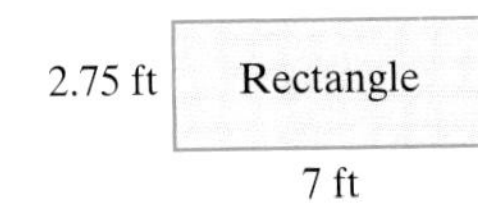

3.

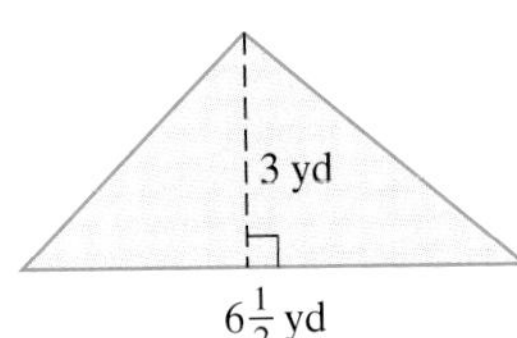

4.

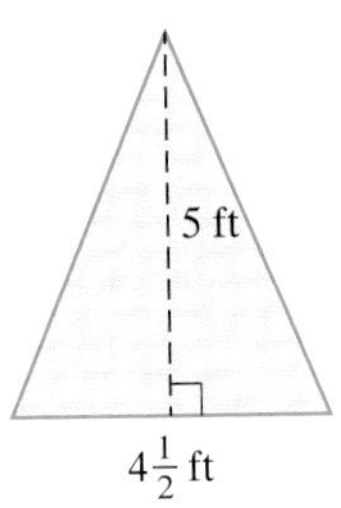

5.

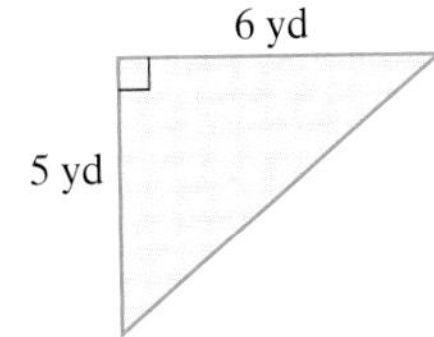

6.

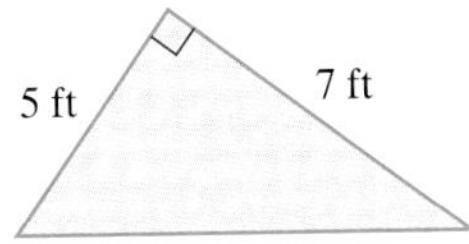

7. Use 3.14 for π.

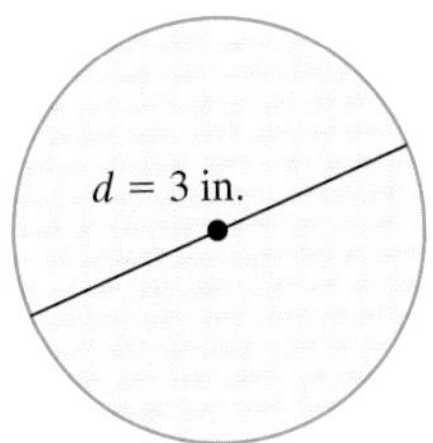

8. Use $\frac{22}{7}$ for π.

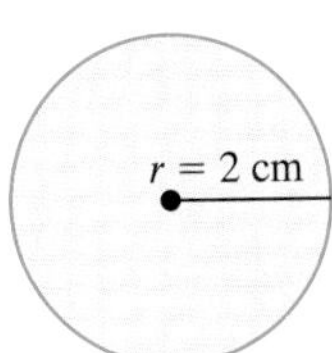

9.

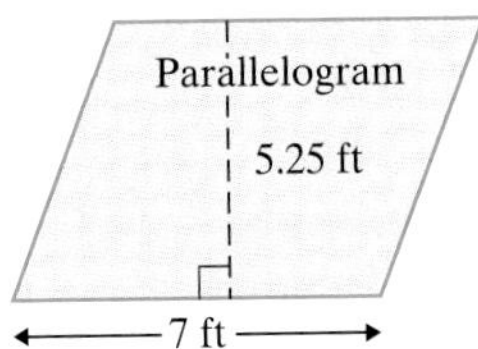

10.

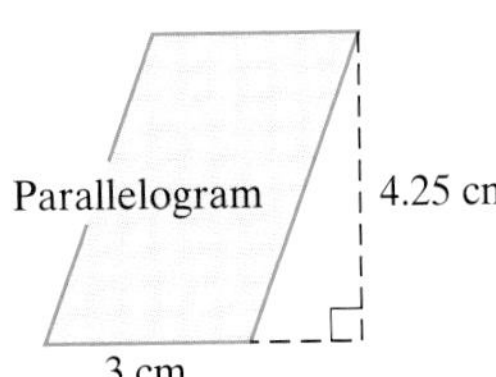

11.

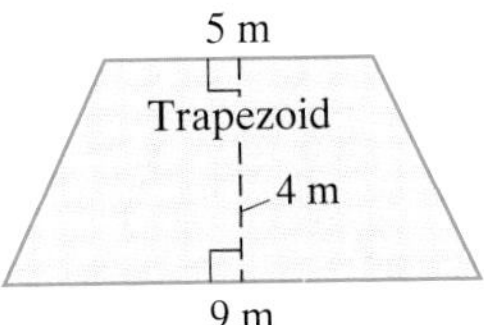

12.

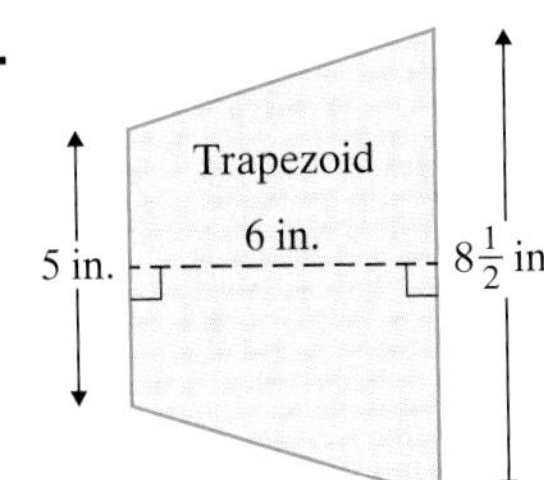

13.

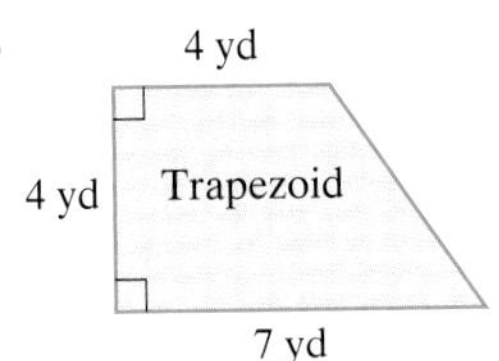

14.

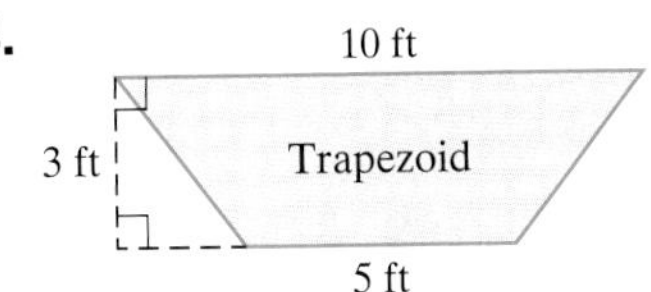

15.

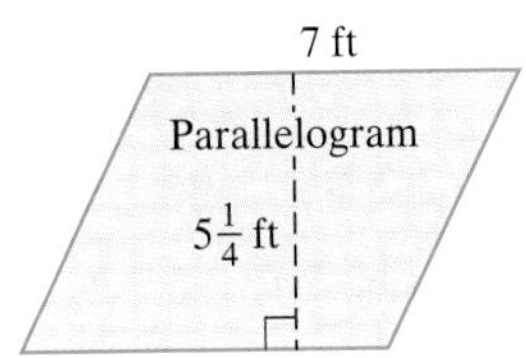

16.

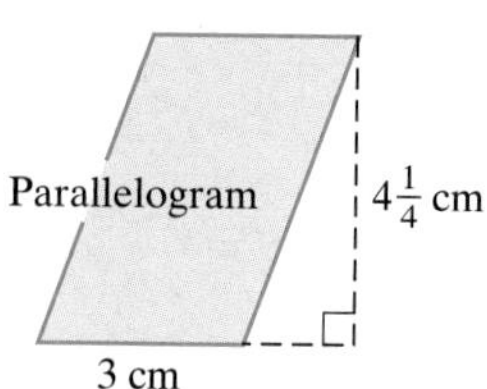

17.

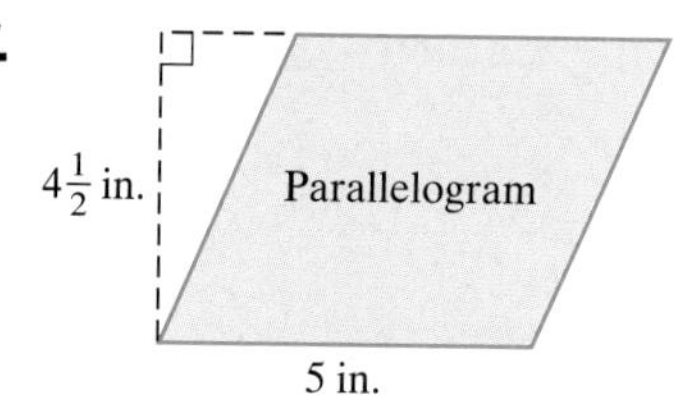

18.

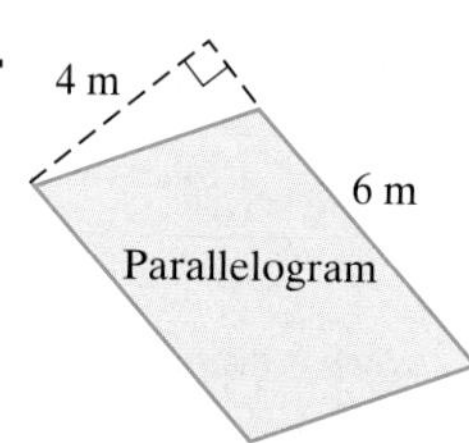

19.

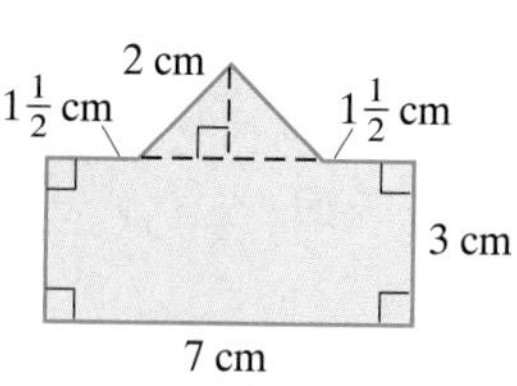

20.

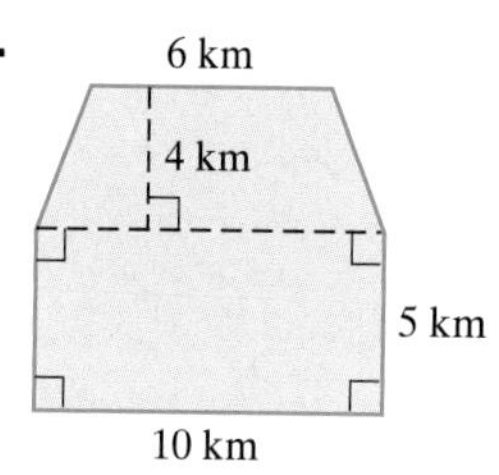

21.

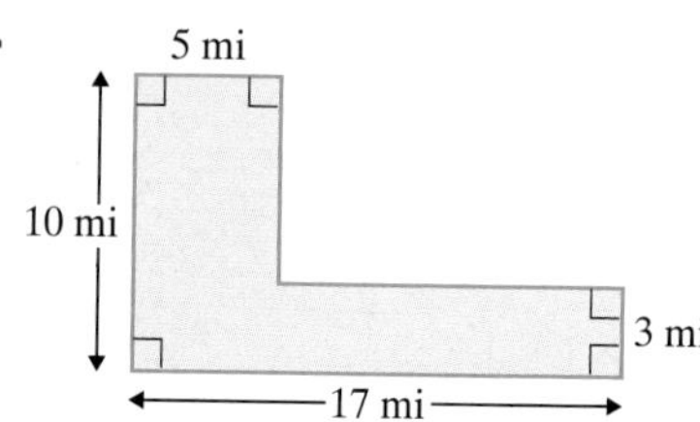

22.

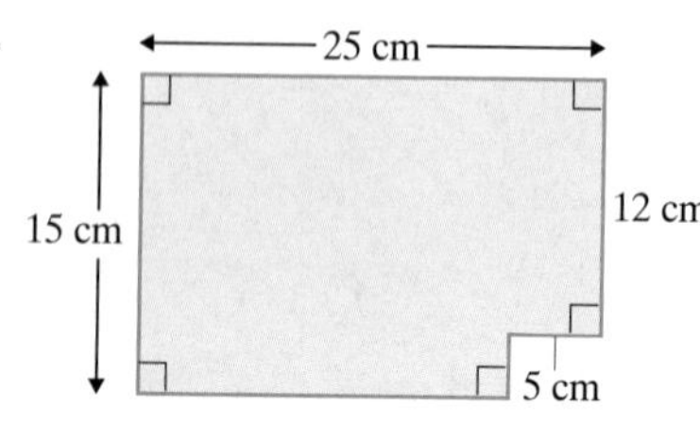

23.

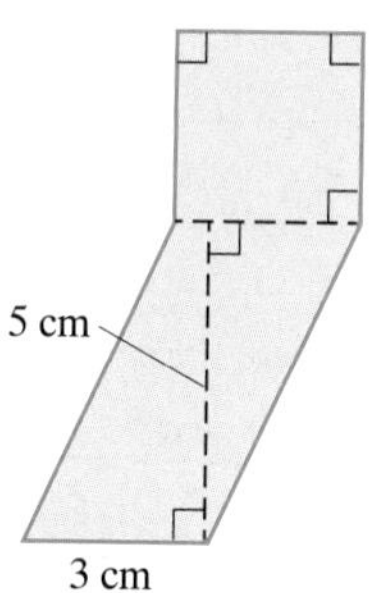

24.

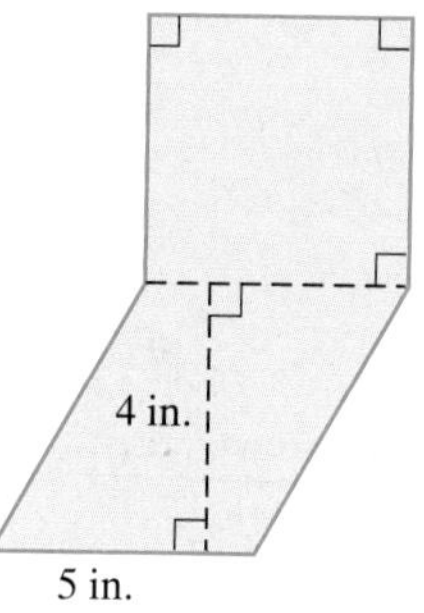

25. Use $\frac{22}{7}$ for π.

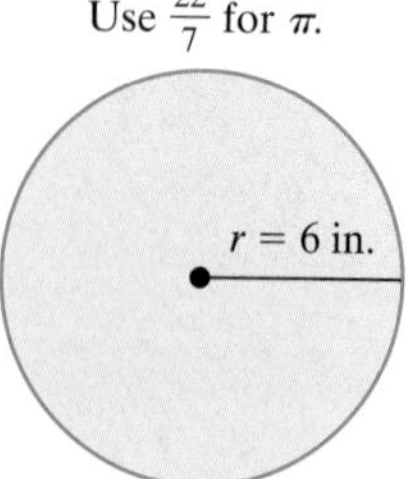

26. Use 3.14 for π.

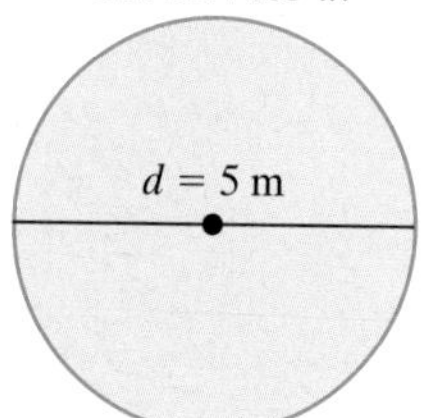

B *Find the volume of each solid. Use $\frac{22}{7}$ as an approximation for π. See Examples 5 through 8.*

27.

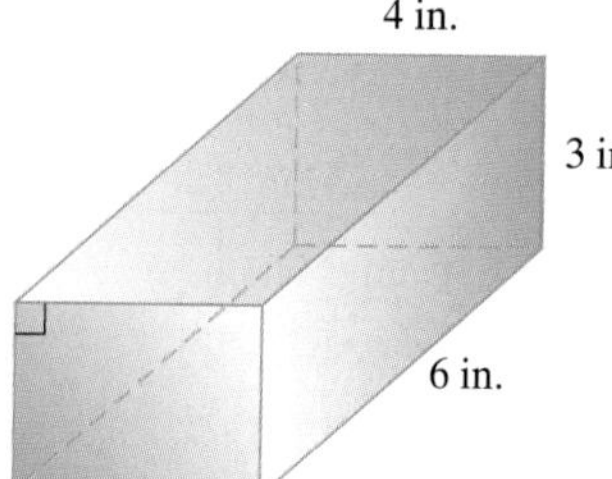

28.

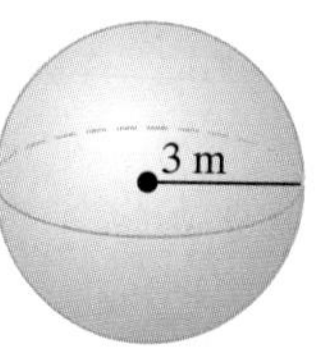

29.

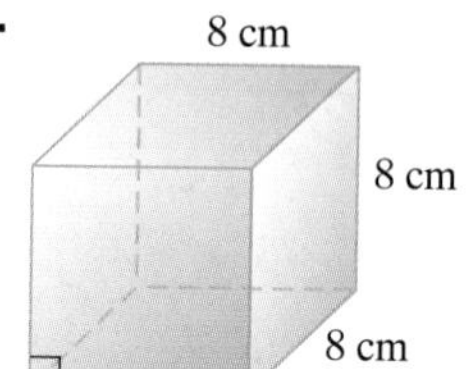

30.

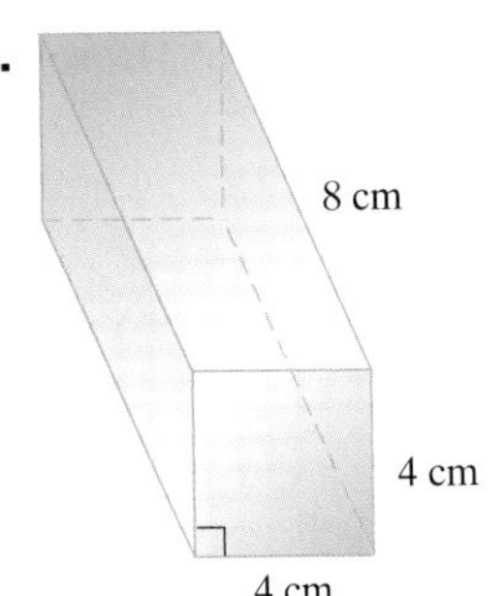

31.

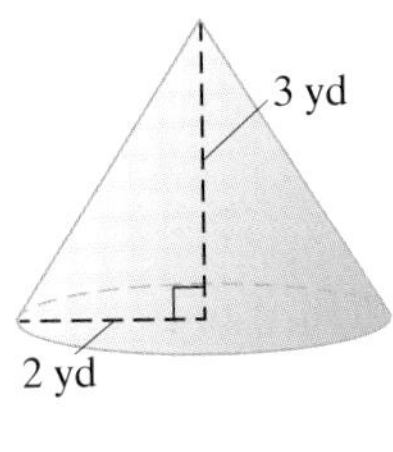

32.

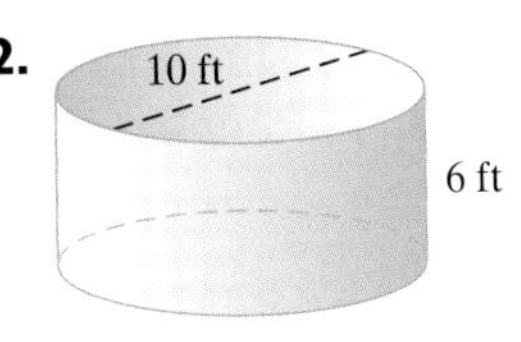

33.

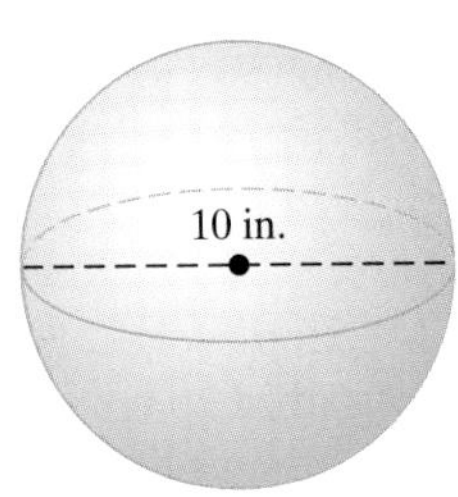

34.

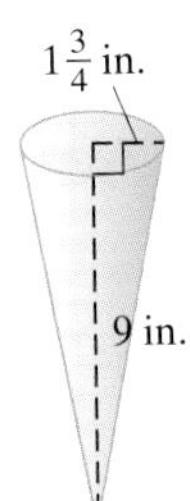

35.

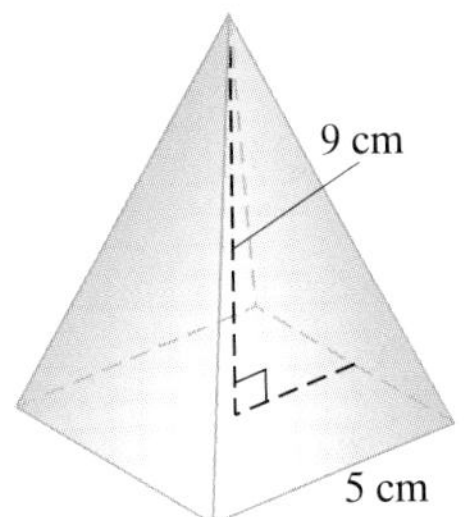

36.

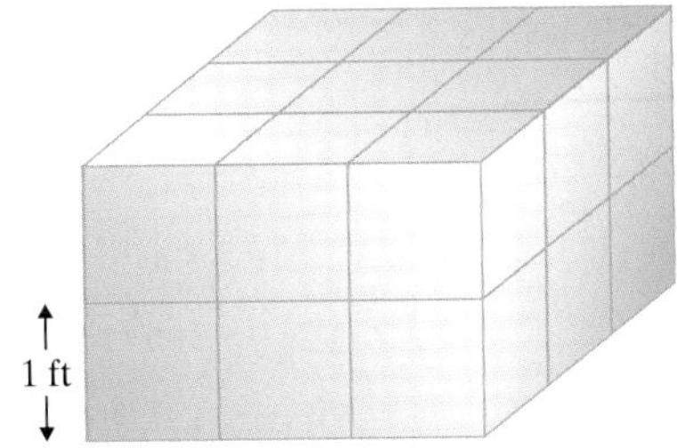

 *Solve. See Examples 1 through 8.*

 **37.** Find the volume of a cube with edges of $1\frac{1}{3}$ inches.

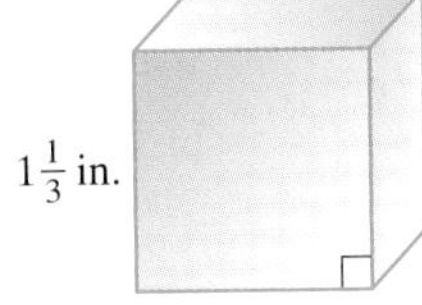

38. A water storage tank is in the shape of a cone with the pointed end down. If the radius is 14 ft and the depth of the tank is 15 ft, approximate the volume of the tank in cubic feet. Use $\frac{22}{7}$ for π.

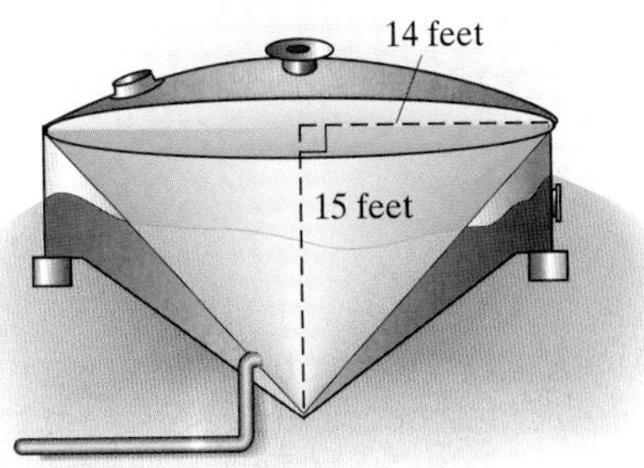

39. Find the volume of a rectangular box 2 ft by 1.4 ft by 3 ft.

40. Find the volume of a box in the shape of a cube that is 5 ft on each side.

41. The world's largest flag measures 505 feet by 225 feet. It's the U.S. "Super flag" owned by "Ski" Demski of Long Beach, California. Find its area. (*Source: Guinness World Records*)

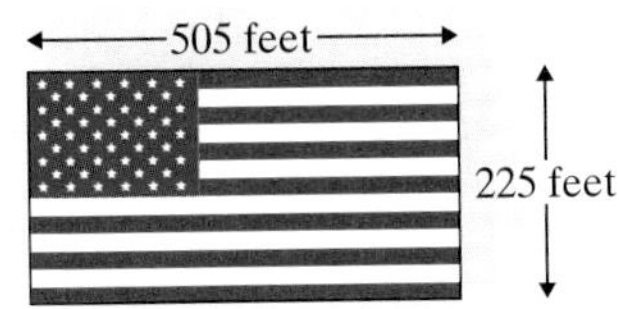

42. The longest illuminated sign is in Ramat Gan, Israel, and measures 197 feet by 66 feet. Find its area. (*Source: The Guinness Book of Records*)

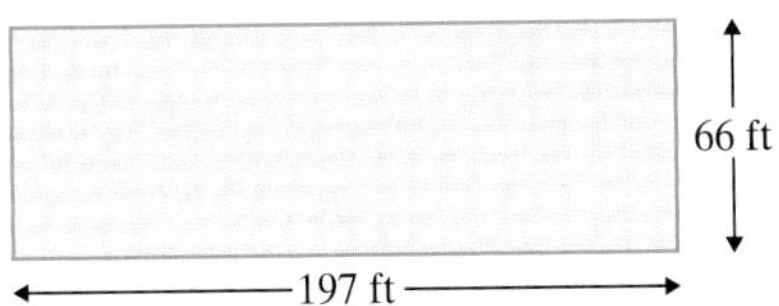

43. A drapery panel measures 6 ft by 7 ft. Find how many square feet of material are needed for *four* panels.

44. A page in this book measures 27.5 cm by 21.8 cm. Find its area.

45. A paperweight is in the shape of a square-based pyramid 20 cm tall. If an edge of the base is 12 cm, find the volume of the paperweight.

46. A bird bath is made in the shape of a hemisphere (half-sphere). If its radius is 10 in., approximate the volume. Use $\frac{22}{7}$ for π.

47. Find how many square feet of land are in the following plot:

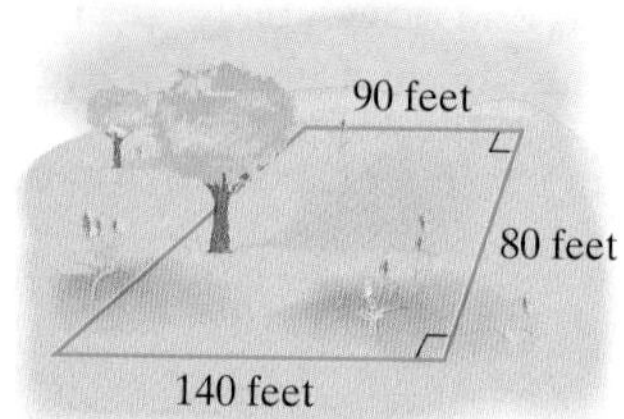

48. For Gerald Gomez to determine how much grass seed he needs to buy, he must know the size of his yard. Use the drawing to determine how many square feet are in his yard.

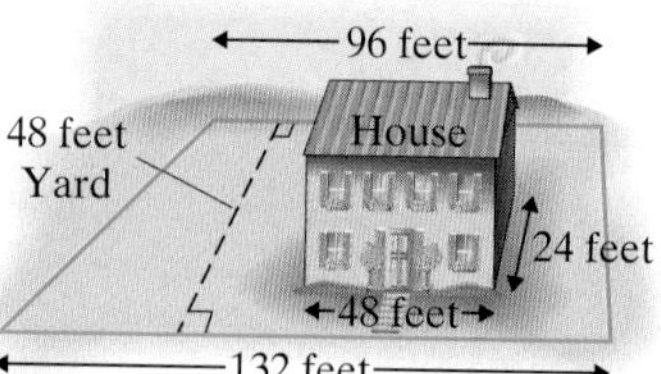

49. The shaded part of the roof shown is in the shape of a trapezoid and needs to be shingled. The number of packages of shingles to buy depends on the area. Use the dimensions given to find the area of the shaded part of the roof to the nearest whole square foot.

50. The end of the building shaded in the drawing is to be bricked. The number of bricks to buy depends on the area.

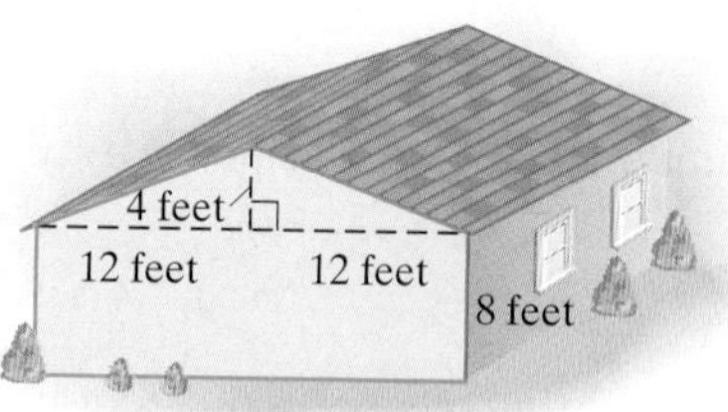

a. Find the area.

b. If the side area of each brick (including mortar room) is $\frac{1}{6}$ square foot, find the number of bricks needed to buy.

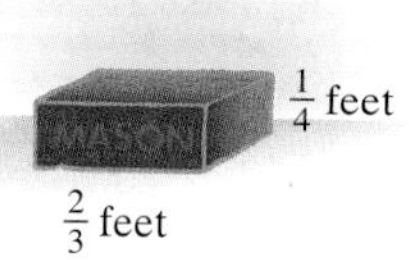

51. Find the exact volume of a sphere with a radius of 7 inches.

52. A tank is in the shape of a cylinder 8 feet tall and 3 feet in radius. Find the exact volume of the tank.

53. Find the volume of a rectangular block of ice 2 feet by $2\frac{1}{2}$ feet by $1\frac{1}{2}$ feet.

54. Find the capacity (volume in cubic feet) of a rectangular ice chest with inside measurements of 3 feet by $1\frac{1}{2}$ feet by $1\frac{3}{4}$ feet.

55. An ice cream cone with a 4-centimeter diameter and 3-centimeter depth is filled exactly level with the top of the cone. Approximate how much ice cream (in cubic centimeters) is in the cone. Use $\frac{22}{7}$ for π.

56. A child's toy is in the shape of a square-based pyramid 10 inches tall. If an edge of the base is 7 inches, find the volume of the toy.

57. Ball lightning is a rare form of lightning in which a moving white or colored luminous sphere is seen. It can last from a few seconds to a few minutes and travels at about walking pace. An average sphere size is 6 inches in diameter. Find the exact volume of a sphere with this diameter and then approximate the volume using 3.14 for π.

58. A monkey ball tree produces large green fruit in the shape of spheres. These fruits are approximately 4 inches (or 10 centimeters) in diameter and have a coarse surface. Find the exact volume of a sphere with diameter 4 inches and then approximate the volume using 3.14 for π. (Round to the nearest tenth.)

59. Find the volume of a pyramid with a square base 5 inches on a side and a height of 1.3 inches.

60. Approximate to the nearest hundredth the volume of a sphere with a radius of 2 centimeters. Use 3.14 for π.

Review and Preview

Evaluate. See Section 1.8.

61. 5^2

62. 7^2

63. 3^2

64. 20^2

65. $1^2 + 2^2$

66. $5^2 + 3^2$

67. $4^2 + 2^2$

68. $1^2 + 6^2$

Combining Concepts

Given the following situations, tell whether you are more likely to be concerned with area or perimeter.

69. ordering fencing to fence a yard

70. ordering grass seed to plant in a yard

71. buying carpet to install in a room

72. buying gutters to install on a house

73. ordering paint to paint a wall

74. ordering baseboards to install in a room

75. buying a wallpaper border to go on the walls around a room

76. buying fertilizer for your yard

77. A pizza restaurant recently advertised two specials. The first special was a 12-inch pizza for $10. The second special was two 8-inch pizzas for $9. Determine the better buy. (*Hint*: First compare the areas of the two specials and then find a price per square inch for both specials.)

78. Find the approximate area of the state of Utah.

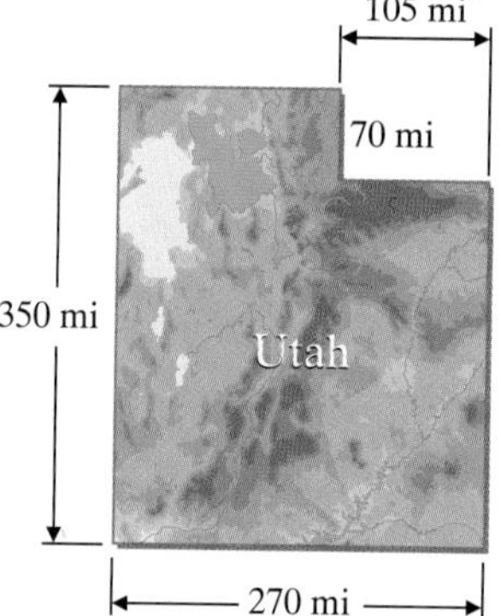

79. Find the area of a rectangle that measures 2 *feet* by 8 *inches*. Give the area in square feet and in square inches.

80. In your own words, explain why perimeter is measured in units and area is measured in square units. (*Hint*: See Section 1.6 for an introduction on the meaning of area.)

Find the area of each figure. Round results to the nearest tenth.

81.

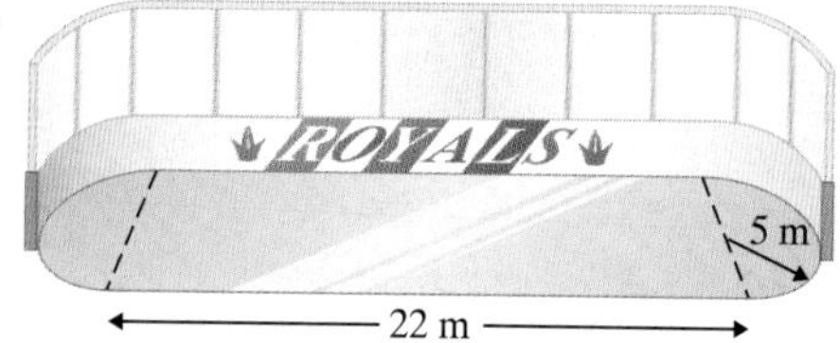

82.

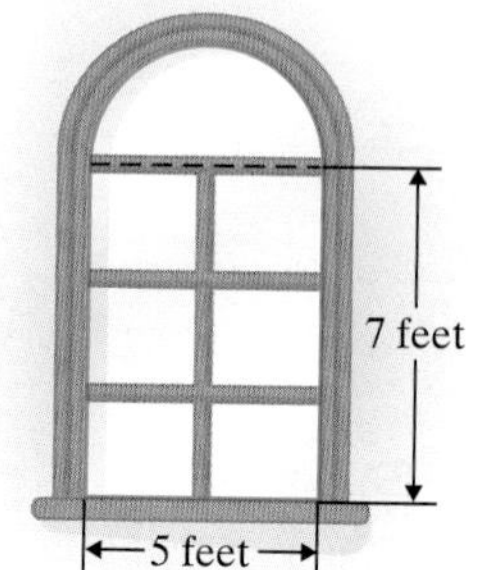

83. The Great Pyramid of Khufu at Giza is the largest of the ancient Egyptian pyramids. Its original height was 146.5 meters. The length of each side of its square base was originally 230 meters. Find the volume of the Great Pyramid of Khufu as it was originally built. Round to the nearest whole cubic meter. (*Source*: PBS *NOVA* Online)

84. The second-largest pyramid at Giza is Khafre's Pyramid. Its original height was 471 feet. The length of each side of its square base was originally 704 feet. Find the volume of Khafre's Pyramid as it was originally built. (*Source*: PBS *NOVA* Online)

85. Menkaure's Pyramid, the smallest of the three Great Pyramids at Giza, was originally 65.5 meters tall. Each of the sides of its square base was originally 344 meters long. What was the volume of Menkaure's Pyramid as it was originally built? Round to the nearest whole cubic meter. (*Source*: PBS *NOVA* Online)

86. Due to factors such as weathering and loss of outer stones, the Great Pyramid of Khufu now stands only 137 meters tall. Its square base is now only 227 meters on a side. Find the current volume of the Great Pyramid of Khufu to the nearest whole cubic meter. How much has its volume decreased since it was built? See Exercise 83 for comparison. (*Source*: PBS *NOVA* Online)

87. The centerpiece of the New England Aquarium in Boston is its Giant Ocean Tank. This exhibit is a four-story cylindrical saltwater tank containing sharks, sea turtles, stingrays, and tropical fish. The radius of the tank is 16.3 feet and its height is 32 feet (assuming that a story is 8 feet). What is the volume of the Giant Ocean Tank? Use $\pi \approx 3.14$ and round to the nearest tenth of a cubic foot. (*Source*: New England Aquarium)

88. Except for service dogs for guests with disabilities, Walt Disney World does not allow pets in its parks or hotels. However, the resort does make pet-boarding services available to guests. The pet-care kennels at Walt Disney World offer three different sizes of indoor kennels. Of these, the smaller two kennels measure

a. $2'1'' \times 1'8'' \times 1'7''$ and

b. $1'1'' \times 2' \times 8''$

What is the volume of each kennel rounded to the nearest cubic foot? Which is larger? (*Source*: Walt Disney World Resort)

89. Can you compute the volume of a rectangle? Why or why not?

90. Do two rectangles with the same perimeter have the same area? To see, find the perimeter and the area of each rectangle.

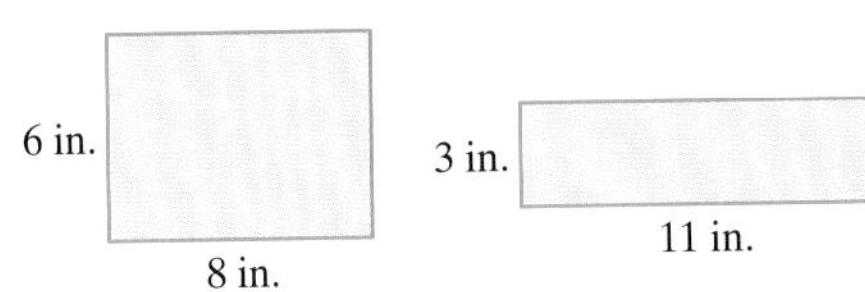

Internet Excursions

Go To: 

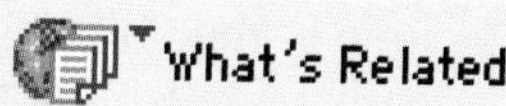

The World Wide Web address listed above will direct you to the official Web site of the U.S. Department of Defense and the Pentagon in Washington, D.C., or a related site. You will find information that helps you complete the questions below.

91. Visit this Web site and locate information on the length of each outer wall of the Pentagon. Then calculate the outer perimeter of the Pentagon.

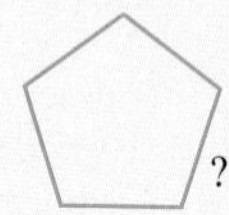

92. Notice that each of the five outer walls of the Pentagon is in the shape of a rectangle. Visit this Web site and locate information on the height of the building. Then use this height information along with the length information found in Exercise 91 to calculate the area of each outer wall of the Pentagon.

Name ______________________ Section __________ Date __________

Answers

1. ________

2. ________

3. ________

4. ________

5. ________

6. ________

7. ________

8. ________

9. ________

10. ________

11. ________

12. ________

13. ________

14. ________

15. ________

16. ________

Integrated Review—Geometry Concepts

1. Find the supplement and the complement of a 27° angle.

Find the measures of angles x, y, and z in each figure.

2.

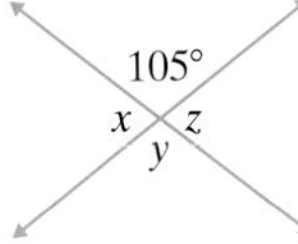

3. $m \| n$

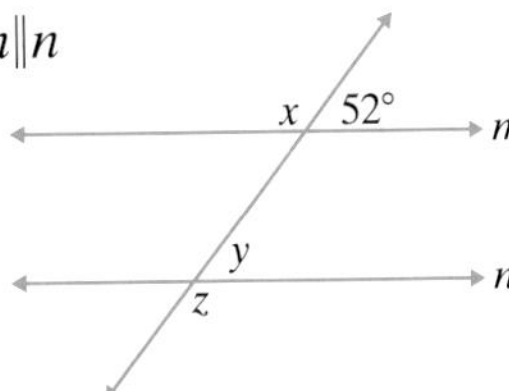

4. Recall that the sum of the measures of the angles of a triangle is 180°. Find the measure of $\angle x$.

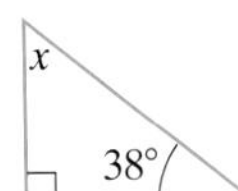

Convert each measurement of length as indicated.

5. 36 in. = ____ ft

6. 10,560 ft = ____ mi

7. 20 ft = ____ yd

8. 6 yd = ____ ft

9. 2.1 mi = ____ ft

10. 3.2 ft = ____ in.

11. 30 m = ____ cm

12. 24 mm = ____ cm

13. 2000 mm = ____ m

14. 18 m = ____ cm

15. 7.2 cm = ____ mm

16. 600 m = ____ km

17. ______

18. ______

19. ______

20. ______

21. ______

22. ______

23. ______

24. ______

25. ______

Find the perimeter (or circumference) and area of each figure. For the circle, give an exact circumference and area, then use $\pi \approx 3.14$ to approximate each. Don't forget to attach correct units.

△ **17.** Square 5 m

△ **18.**

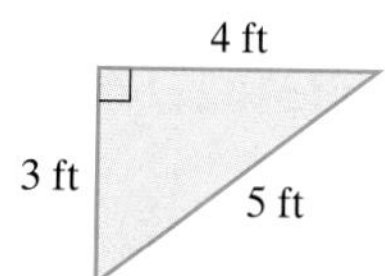

△ **19.**

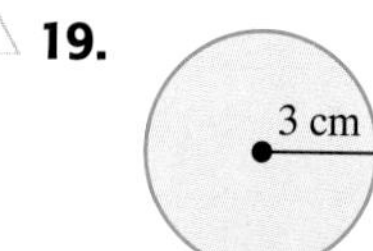

△ **20.**

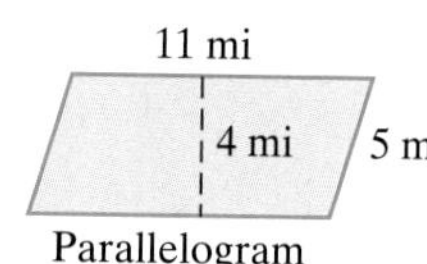

△ **21.** The smallest cathedral is in Highlandville, Missouri, and its floor is a rectangle that measures 14 feet by 17 feet. Find its perimeter and its area. (*Source: The Guinness Book of Records*)

Find the volume of each solid. Don't forget to attach correct units.

△ **22.** A cube with edges of 4 inches each

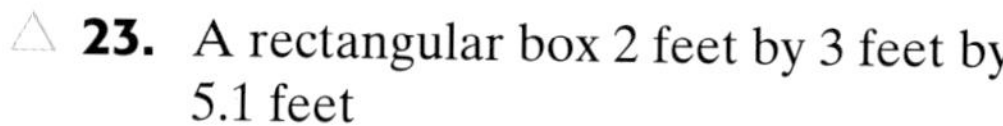

△ **23.** A rectangular box 2 feet by 3 feet by 5.1 feet

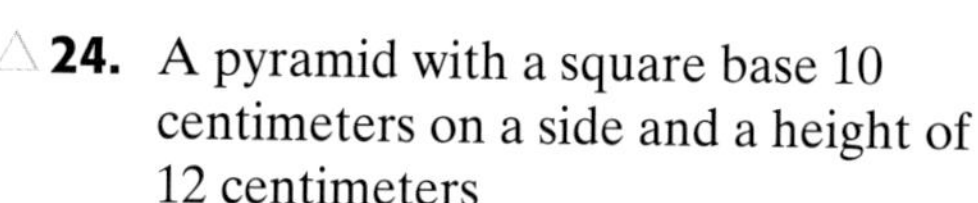

△ **24.** A pyramid with a square base 10 centimeters on a side and a height of 12 centimeters

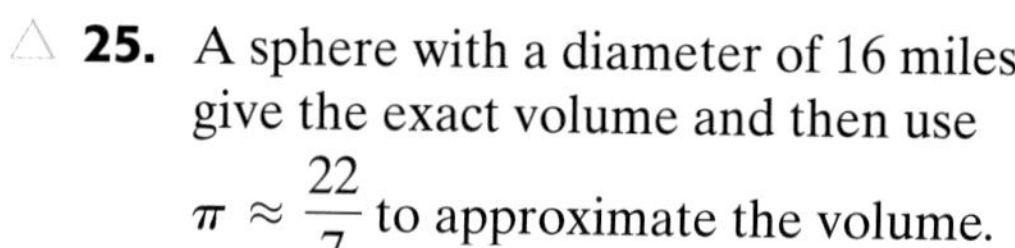

△ **25.** A sphere with a diameter of 16 miles; give the exact volume and then use $\pi \approx \frac{22}{7}$ to approximate the volume.

8.5 Weight and Mass

A Defining and Converting U.S. System Units of Weight

Whenever we talk about how heavy an object is, we are concerned with the object's **weight.** We discuss weight when we refer to a 12-ounce box of Rice Krunchies, an overweight 19-pound tabby cat, or a barge hauling 24 tons of garbage.

The most common units of weight in the U.S. measurement system are the **ounce,** the **pound,** and the **ton.** The following is a summary of equivalencies between units of weight:

U.S. Units of Weight	Unit Fractions
16 ounces (oz) = 1 pound (lb)	$\frac{16\text{ oz}}{1\text{ lb}} = \frac{1\text{ lb}}{16\text{ oz}} = 1$
2000 pounds = 1 ton	$\frac{2000\text{ lb}}{1\text{ ton}} = \frac{1\text{ ton}}{2000\text{ lb}} = 1$

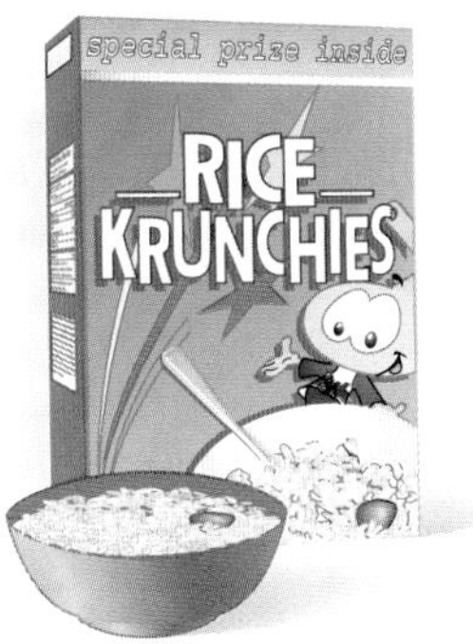

Try the Concept Check in the margin.

Unit fractions that equal 1 are used to convert between units of weight in the U.S. system. When converting using unit fractions, recall that the numerator of a unit fraction should contain the units we are converting to and the denominator should contain the original units.

To convert 40 ounces to pounds, multiply by $\frac{1\text{ lb}}{16\text{ oz}}$ ← Units to convert to / ← Original units

$$40\text{ oz} = \frac{40\text{ oz}}{1} \cdot \underbrace{\frac{1\text{ lb}}{16\text{ oz}}}_{\text{Unit fraction}}$$

$$= \frac{40\text{ lb}}{16} \quad \text{Multiply.}$$

$$= \frac{5}{2}\text{lb} \quad \text{or} \quad 2\frac{1}{2}\text{lb, as a mixed number}$$

EXAMPLE 1 Convert 9000 pounds to tons.

Solution: We multiply 9000 lb by the unit fraction

$\frac{1\text{ ton}}{2000\text{ lb}}$ ← Units to convert to / ← Original units

OBJECTIVES

A Define U.S. units of weight and convert from one unit to another.

B Perform arithmetic operations on units of weight.

C Define metric units of mass and convert from one unit to another.

D Perform arithmetic operations on units of mass.

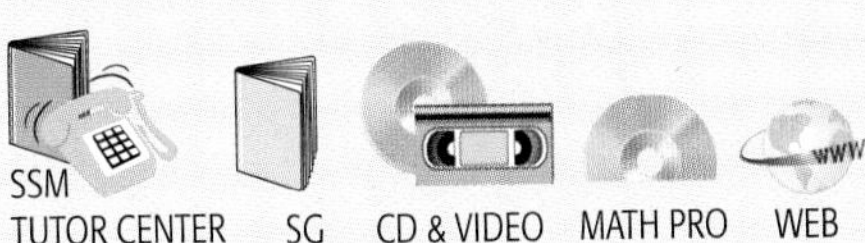

Concept Check

If you were describing the weight of a semitrailer, which type of unit would you use: ounce, pound, or ton? Why?

Practice Problem 1

Convert 4500 pounds to tons.

Answer

1. $2\frac{1}{4}$ tons

Concept Check: ton

$$9000 \text{ lb} = \frac{9000 \cancel{\text{lb}}}{1} \cdot \frac{1 \text{ ton}}{2000 \cancel{\text{lb}}} = \frac{9000 \text{ tons}}{2000} = \frac{9}{2} \text{tons or } 4\frac{1}{2} \text{tons}$$

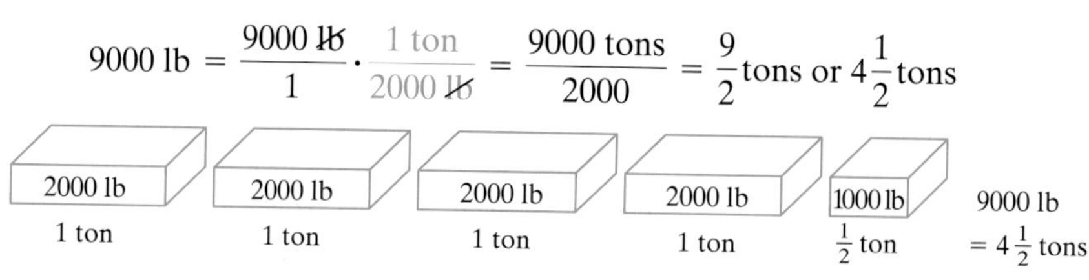

Practice Problem 2

Convert 56 ounces to pounds.

EXAMPLE 2 Convert 3 pounds to ounces.

Solution: We multiply by the unit fraction $\frac{16 \text{ oz}}{1 \text{ lb}}$ to convert from pounds to ounces.

$$3 \text{ lb} = \frac{3 \cancel{\text{lb}}}{1} \cdot \frac{16 \text{ oz}}{1 \cancel{\text{lb}}} = 3 \cdot 16 \text{ oz} = 48 \text{ oz}$$

1 lb 16 oz | 1 lb 16 oz | 1 lb 16 oz | 3 lb = 48 oz

As with length, it is sometimes useful to simplify a measurement of weight by writing it in terms of mixed units.

33 ounces = ____ lb ____ oz

Because 16 oz = 1 lb, divide 16 into 33 to see how many pounds are in 33 ounces. The quotient is the number of pounds and the remainder is the number of ounces. To see why we divide 16 into 33, notice that

$$33 \text{ oz} = 33 \cancel{\text{oz}} \cdot \frac{1 \text{ lb}}{16 \cancel{\text{oz}}} = \frac{33 \text{ lb}}{16}$$

$$\begin{array}{r} 2 \text{ lb } 1 \text{ oz} \\ 16\overline{)33} \\ -32 \\ \hline 1 \end{array}$$

Thus 33 ounces is the same as 2 lb 1 oz.

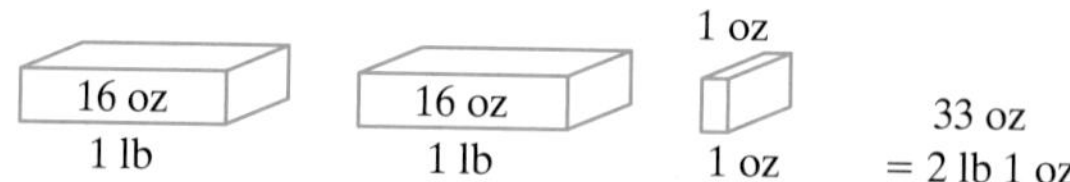

B Performing Operations on U.S. System Units of Weight

Performing arithmetic operations on units of weight works the same way as performing arithmetic operations on units of length.

Practice Problem 3

Subtract 5 tons 1200 lb from 8 tons 100 lb.

EXAMPLE 3 Subtract 3 tons 1350 lb from 8 tons 1000 lb.

Solution: To subtract, we line up similar units.

$$\begin{array}{r} 8 \text{ tons } 1000 \text{ lb} \\ -3 \text{ tons } 1350 \text{ lb} \\ \hline \end{array}$$

Since we cannot subtract 1350 lb from 1000 lb, we borrow 1 ton from the 8 tons. To do so, we write 1 ton as 2000 lb and combine it with the 1000 lb.

Answers

2. $3\frac{1}{2}$ lb **3.** 2 tons 900 lb

7 tons + 1 ton 2000 lb

$$\begin{array}{r} 8 \text{ tons } 1000 \text{ lb} \\ -3 \text{ tons } 1350 \text{ lb} \\ \hline \end{array} \quad \text{becomes} \quad \begin{array}{r} 7 \text{ tons } 3000 \text{ lb} \\ -3 \text{ tons } 1350 \text{ lb} \\ \hline 4 \text{ tons } 1650 \text{ lb} \end{array}$$

To check, see that the sum of 4 tons 1650 lb and 3 tons 1350 lb is 8 tons 1000 lb.

EXAMPLE 4 Multiply 5 lb 9 oz by 6.

Solution: We multiply 5 lb by 6 and 9 oz by 6.

$$\begin{array}{r} 5 \text{ lb } 9 \text{ oz} \\ \times \qquad 6 \\ \hline 30 \text{ lb } 54 \text{ oz} \end{array}$$

To write 54 oz as mixed units, we divide by 16 (1 lb = 16 oz).

$$\begin{array}{r} 3 \text{ lb } 6 \text{ oz} \\ 16\overline{)54} \\ -48 \\ \hline 6 \end{array}$$

Thus,

$$30 \text{ lb } 54 \text{ oz} = 30 \text{ lb} + 3 \text{ lb } 6 \text{ oz} = 33 \text{ lb } 6 \text{ oz}$$

EXAMPLE 5 Divide 9 lb 6 oz by 2.

Solution: We divide each of the units by 2.

$$\begin{array}{rl} 4 \text{ lb} & 11 \text{ oz} \\ 2\overline{)9 \text{ lb}} & 6 \text{ oz} \\ -8 & \\ \hline 1 \text{ lb} = & 16 \text{ oz} \\ & \overline{22 \text{ oz}} \end{array}$$

Divide 2 into 22 oz to get 11 oz.

To check, multiply 4 pounds 11 ounces by 2. The result is 9 pounds 6 ounces.

EXAMPLE 6 Finding the Weight of a Child

Bryan weighed 8 lb 8 oz at birth. By the time he was 1 year old, he had gained 11 lb 14 oz. Find his weight at age 1 year.

Solution:

$$\begin{array}{rcr} \text{birth weight} & \rightarrow & 8 \text{ lb } 8 \text{ oz} \\ + \text{ weight gained} & \rightarrow & + \ 11 \text{ lb } 14 \text{ oz} \\ \hline \text{total weight} & \rightarrow & 19 \text{ lb } 22 \text{ oz} \end{array}$$

Since 22 oz equals 1 lb 6 oz,

$$\begin{aligned} 19 \text{ lb } 22 \text{ oz} &= 19 \text{ lb} + 1 \text{ lb } 6 \text{ oz} \\ &= 20 \text{ lb } 6 \text{ oz} \end{aligned}$$

Bryan weighed 20 lb 6 oz on his first birthday.

Practice Problem 4

Multiply 4 lb 11 oz by 8.

Practice Problem 5

Divide 5 lb 8 oz by 4.

Practice Problem 6

A 5-lb 14-oz batch of cookies is packed into a 6-oz container before it is mailed. Find the total weight.

Answers

4. 37 lb 8 oz **5.** 1 lb 6 oz **6.** 6 lb 4 oz

C Defining and Converting Metric System Units of Mass

In scientific and technical areas, a careful distinction is made between **weight** and **mass. Weight** is really a measure of the pull of gravity. The farther from Earth an object gets, the less it weighs. However, **mass** is a measure of the amount of substance in the object and does not change. Astronauts orbiting Earth weigh much less than they weigh on Earth, but they have the same mass in orbit as they do on Earth. Here on Earth weight and mass are the same, so either term may be used.

The basic unit of mass in the metric system is the **gram.** It is defined as the mass of water contained in a cube 1 centimeter (cm) on each side.

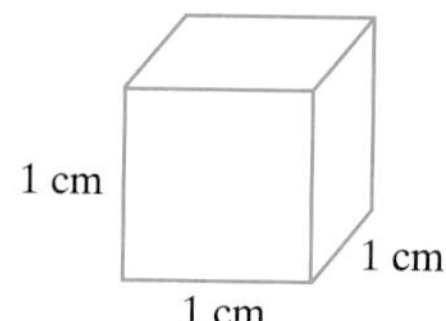

The following examples may help you get a feeling for metric masses.

A tablet contains 200 milligrams of ibuprofen.

A large paper clip weighs approximately 1 gram.

A box of crackers weighs 453 grams.

A kilogram is slightly over 2 pounds. An adult woman may weigh 60 kilograms.

The prefixes for units of mass in the metric system are the same as for units of length, as shown in the following table:

Prefix	Meaning	Metric Unit of Mass
kilo	1000	1 kilogram (kg) = 1000 grams (g)
hecto	100	1 hectogram (hg) = 100 g
deka	10	1 dekagram (dag) = 10 g
		1 gram (g) = 1 g
deci	1/10	1 decigram (dg) = 1/10 g or 0.1 g
centi	1/100	1 centigram (cg) = 1/100 g or 0.01 g
milli	1/1000	1 milligram (mg) = 1/1000 g or 0.001 g

Try the Concept Check in the margin.

The **milligram,** the **gram,** and the **kilogram** are the three most commonly used units of mass in the metric system.

As with lengths, all units of mass are powers of 10 of the gram, so converting from one unit of mass to another involves only moving the decimal point. To convert from one unit of mass to another in the metric system, list the units of mass in order from largest to smallest.

Let's convert 4300 milligrams to grams. To convert from milligrams to grams, we move along the table 3 units to the left.

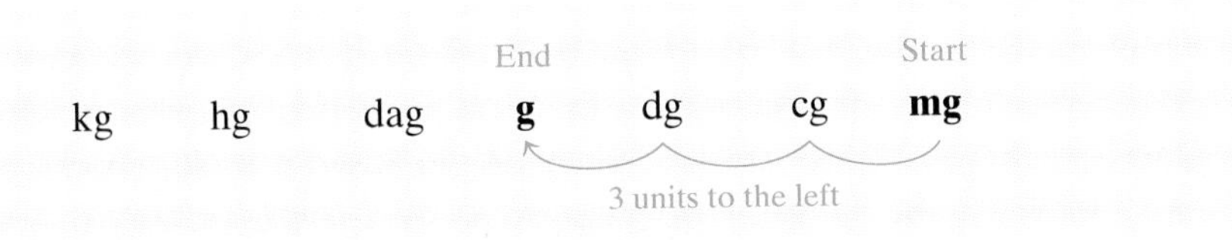

This means that we move the decimal point 3 places to the left to convert from milligrams to grams.

$$4300 \text{ mg} = 4.3 \text{ g}$$

The same conversion can be done with unit fractions.

$$4300 \text{ mg} = \frac{4300 \cancel{\text{mg}}}{1} \cdot \frac{0.001 \text{ g}}{1 \cancel{\text{mg}}}$$

$$= 4300 \cdot 0.001 \text{ g}$$

$$= 4.3 \text{ g} \quad \text{To multiply by 0.001, move the decimal point 3 places to the left.}$$

To see that this is reasonable, study the diagram:

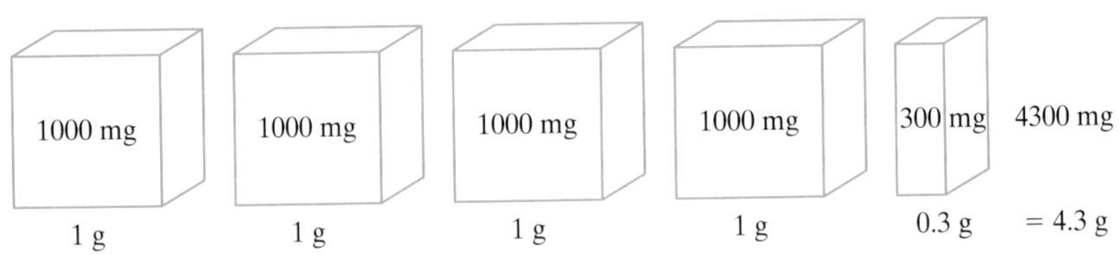

Thus $4300 \text{ mg} = 4.3 \text{ g}$

Concept Check

True or false? A decigram is larger than a dekagram. Explain.

EXAMPLE 7 Convert 3.2 kg to grams.

Solution: First we convert by using a unit fraction.

$$3.2 \text{ kg} = 3.2 \cancel{\text{kg}} \cdot \underbrace{\frac{1000 \text{ g}}{1 \cancel{\text{kg}}}}_{\text{Unit fraction}} = 3200 \text{ g}$$

Now let's list the units of mass in a chart and move from kilograms to grams.

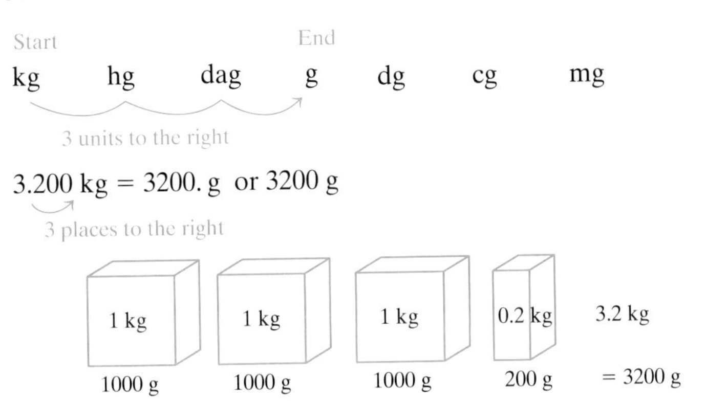

Practice Problem 7

Convert 3.41 g to milligrams.

Answer

7. 3410 mg

Concept Check: false

EXAMPLE 8 Convert 2.35 cg to grams.

Solution: We list the units of mass in a chart and move from centigrams to grams.

kg	hg	dag	g (End)	dg	cg (Start)	mg

2 units to the left

02.35 cg = 0.0235 g

2 places to the left

Practice Problem 8

Convert 56.2 cg to grams.

D Performing Operations on Metric System Units of Mass

Arithmetic operations can be performed with metric units of mass just as we performed operations with metric units of length. We convert each number to the same unit of mass and add, subtract, multiply, or divide as with decimals.

EXAMPLE 9 Subtract 5.4 dg from 1.6 g.

Solution: We convert both numbers to decigrams or to grams before subtracting.

5.4 dg = 0.54 g

$$\begin{array}{r} 1.60\text{ g} \\ -0.54\text{ g} \\ \hline 1.06\text{ g} \end{array}$$

or

1.6 g = 16 dg

$$\begin{array}{r} 16.0\text{ dg} \\ -5.4\text{ dg} \\ \hline 10.6\text{ dg} \end{array}$$

The difference is 1.06 g or 10.6 dg.

Practice Problem 9

Subtract 3.1 dg from 2.5 g.

EXAMPLE 10 Multiply 15.4 kg by 5.

Solution: We multiply the two numbers together.

$$\begin{array}{r} 15.4\text{ kg} \\ \times\quad 5 \\ \hline 77.0\text{ kg} \end{array}$$

The result is 77.0 kg.

Practice Problem 10

Multiply 12.6 kg by 4.

EXAMPLE 11 Calculating Allowable Weight in an Elevator

An elevator has a weight limit of 1400 kg. A sign posted in the elevator indicates that the maximum capacity of the elevator is 17 persons. What is the average allowable weight for each passenger, rounded to the nearest kilogram?

Practice Problem 11

Twenty-four bags of cement weigh a total of 550 kg. Find the average weight of 1 bag, rounded to the nearest kilogram.

Answers

8. 0.562 g **9.** 2.19 g or 21.9 dg **10.** 50.4 kg **11.** 23 kg

Solution: To solve, notice that the total weight of 1400 kilograms ÷ 17 = average weight

$$\begin{array}{r} 82.3\text{ kg} \approx 82\text{ kg} \\ 17\overline{)1400.0\text{ kg}} \\ -136\phantom{.0\text{ kg}} \\ 40\phantom{.0\text{ kg}} \\ -34\phantom{\text{ kg}} \\ 60\phantom{\text{ kg}} \\ -51\phantom{\text{ kg}} \\ 9\phantom{\text{ kg}} \end{array}$$

Each passenger can weigh an average of 82 kg. (Recall that a kilogram is slightly over 2 pounds, so 82 kilograms is over 164 pounds.)

FOCUS ON The Real World

THE COST OF ROAD SIGNS

There are nearly 4 million miles of streets and roads in the United States. With streets, roads, and highways comes the need for traffic control, guidance, warning, and regulation. Road signs perform many of these tasks. Just in our routine travels, we see a wide variety of road signs every day. Think how many road signs must exist on the 4 million miles of roads in the United States. Have you ever wondered how much signs like these cost?

The cost of a road sign generally depends on the type of sign. Costs for several types of signs and signposts are listed in the table. Examples of various types of signs are shown below.

Road Sign Costs	
Type of Sign	**Cost**
Regulatory, warning, marker	\$15–\$18 per square foot
Large guide	\$20–\$25 per square foot
Type of Post	**Cost**
U-channel	\$125–\$200 each
Square tube	\$10–\$15 per foot
Steel breakaway posts	\$15–\$25 per foot

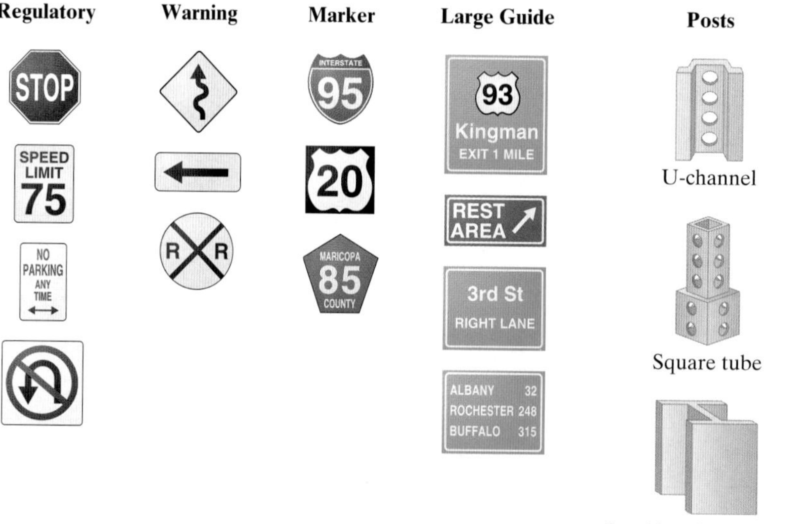

The cost of a sign is based on its area. For diamond, square, or rectangular signs, the area is found by multiplying the length (in feet) times the width (in feet). Then the area is multiplied by the cost per square foot. For signs with irregular shapes, costs are generally figured *as if* the sign were a rectangle, multiplying the height and width at the tallest and widest parts of the sign.

GROUP ACTIVITY

Locate four different kinds of road signs on or near your campus. Measure the dimensions of each sign and the height of the post on which it is mounted. Using the cost data given in the table, find the minimum and maximum costs of each sign, including its post. Summarize your results in a table, and include a sketch of each sign.

Name ______________________ Section ____________ Date ____________

Mental Math

Convert.

1. 16 ounces to pounds

2. 32 ounces to pounds

3. 1 ton to pounds

4. 4 tons to pounds

5. 1 pound to ounces

6. 6 pounds to ounces

7. 2000 pounds to tons

8. 4000 pounds to tons

Determine whether the measurement in each statement is reasonable.

9. The doctor prescribed a pill containing 2 kg of medication.

10. A full-grown cat weighs approximately 15 g.

11. A bag of flour weighs 4.5 kg.

12. A staple weighs 15 mg.

13. A professor weighs less than 150 g.

14. A car weighs 2000 mg.

EXERCISE SET 8.5

A *Convert as indicated. See Examples 1 and 2.*

1. 2 pounds to ounces

2. 3 pounds to ounces

3. 5 tons to pounds

4. 3 tons to pounds

5. 12,000 pounds to tons

6. 32,000 pounds to tons

7. 60 ounces to pounds

8. 90 ounces to pounds

9. 3500 pounds to tons

10. 9000 pounds to tons

11. 16.25 pounds to ounces

12. 14.5 pounds to ounces

13. 4.9 tons to pounds

14. 8.3 tons to pounds

15. $4\frac{3}{4}$ pounds to ounces

16. $9\frac{1}{8}$ pounds to ounces

17. 2950 pounds to the nearest tenth of a ton

18. 51 ounces to the nearest tenth of a pound

B *Perform each indicated operation. See Examples 3 through 5.*

19. 34 lb 12 oz + 18 lb 14 oz

20. 6 lb 10 oz + 10 lb 8 oz

21. 6 tons 1540 lb + 2 tons 850 lb

22. 2 tons 1575 lb + 1 ton 480 lb

23. 5 tons 1050 lb − 2 tons 875 lb

24. 4 tons 850 lb − 1 ton 260 lb

25. 12 lb 4 oz − 3 lb 9 oz

26. 45 lb 6 oz − 26 lb 10 oz

27. 5 lb 3 oz × 6

28. 2 lb 5 oz × 5

29. 6 tons 1500 lb ÷ 5

30. 5 tons 400 lb ÷ 4

Solve. Remember to insert units when writing your answers. See Example 6.

31. Doris Johnson has two open containers of Uncle Ben's rice. If she combines 1 lb 10 oz from one container with 3 lb 14 oz from the other container, how much total rice does she have?

32. Dru Mizel maintains the records of the amount of coal delivered to his department in the steel mill. In January, 3 tons 1500 lb were delivered. In February, 2 tons 1200 lb were delivered. Find the total amount delivered in these two months.

33. Carla Hamtini was amazed when she grew a 28 lb 10 oz zucchini in her garden, but later she learned that the heaviest zucchini ever grown weighed 64 lb 8 oz in Llanharry, Wales, by B. Lavery in 1990. How far below the record weight was Carla's zucchini? (*Source: The Guinness Book of Records*)

34. The heaviest baby born in good health weighed an incredible 22 lb 8 oz. He was born in Italy in September, 1955. How much heavier is this than a 7 lb 12 oz baby? (*Source: The Guinness Book of Records*)

35. The Shop 'n Bag supermarket chain ships hamburger meat by placing 10 packages of hamburger in a box, with each package weighing 3 lb 4 oz. How much will 4 boxes of hamburger weigh?

36. The Quaker Oats Company ships its 1-lb 2-oz boxes of oatmeal in cartons containing 12 boxes of oatmeal. How much will 3 such cartons weigh?

37. A carton of Del Monte Pineapple weighs 55 lb 4 oz, but 2 lb 8 oz of this weight is due to packaging. Subtract the weight of the packaging to find the actual weight of the pineapple in 4 cartons.

38. The Hormel Corporation ships cartons of canned ham weighing 43 lb 2 oz each. Of this weight, 3 lb 4 oz is due to packaging. Find the actual weight of the ham found in 3 cartons.

39. One bag of Pepperidge Farm Bordeaux cookies weighs $6\frac{3}{4}$ ounces. How many pounds will a dozen bags weigh?

40. One can of Payless Red Beets weighs $8\frac{1}{2}$ ounces. How much will eight cans weigh?

Convert as indicated. See Examples 7 and 8.

41. 500 g to kilograms

42. 650 g to kilograms

43. 4 g to milligrams

44. 9 g to milligrams

45. 25 kg to grams

46. 18 kg to grams

47. 48 mg to grams

48. 112 mg to grams

49. 6.3 g to kilograms

50. 4.9 g to kilograms

51. 15.14 g to milligrams

52. 16.23 g to milligrams

53. 4.01 kg to grams

54. 3.16 kg to grams

(D) *Perform each indicated operation. See Examples 9 and 10.*

55. 3.8 mg + 9.7 mg

56. 41.6 g + 9.8 g

57. 205 mg + 5.61 g

58. 2.1 g + 153 mg

59. 9 g − 7150 mg

60. 4 kg − 2410 g

61. 1.61 kg − 250 g

62. 6.13 g − 418 mg

63. 5.2 kg × 2.6

64. 4.8 kg × 9.3

65. 17 kg ÷ 8

66. 8.25 g ÷ 6

Solve. Remember to insert units when writing your answers. See Example 11.

67. A can of 7-Up weighs 336 grams. Find the weight in kilograms of 24 cans.

68. Guy Green normally weighs 73 kg, but he lost 2800 grams after being sick with the flu. Find Guy's new weight.

69. Sudafed is a decongestant that comes in two strengths. Regular strength contains 60 mg of medication. Extra strength contains 0.09 g of medication. How much extra medication is in the extra-strength tablet?

70. A small can of Planters sunflower seeds weighs 177 g. If each can contains 6 servings, find the weight of one serving.

71. Amanda Caucutt's doctor recommends that Amanda limit her daily intake of sodium to 0.6 gram. A one-ounce serving of Cheerios with $\frac{1}{2}$ cup of fortified skim milk contains 350 mg of sodium. How much more sodium can Amanda have after she eats a bowl of Cheerios for breakfast, assuming she intends to follow the doctor's orders?

72. A large bottle of Hire's Root Beer weighs 1900 grams. If a carton contains 6 large bottles of root beer, find the weight in kilograms of 5 cartons.

73. Three milligrams of preservatives are added to a 0.5-kg box of dried fruit. How many milligrams of preservatives are in 3 cartons of dried fruit if each carton contains 16 boxes?

74. One box of Swiss Miss Cocoa Mix weighs 0.385 kg, but 39 grams of this weight is the packaging. Find the actual weight of the cocoa in 8 boxes.

75. A carton of 12 boxes of Quaker Oats Oatmeal weighs 6.432 kg. Each box includes 26 grams of packaging material. What is the actual weight of the oatmeal in the carton?

76. The supermarket prepares hamburger in 85-gram market packages. When Leo Gonzalas gets home, he divides the package in half before refrigerating the meat. How much will each package weigh?

77. A package of Trailway's Gorp, a high-energy hiking trail mix, contains 0.3 kg of nuts, 0.15 kg of chocolate bits, and 400 grams of raisins. Find the total weight of the package.

78. The manufacturer of Anacin wants to reduce the caffeine content of its aspirin by $\frac{1}{4}$. Currently, each regular tablet contains 32 mg of caffeine. How much caffeine should be removed from each tablet?

79. A regular-size bag of Lay's potato chips weighs 198 grams. Find the weight of a dozen bags, rounded to the nearest hundredth of a kilogram.

80. A cat weighs a hefty 9 kg. The vet has recommended that this cat lose 1500 grams. How much should the cat weigh?

Review and Preview

Write each fraction as a decimal. See Section 4.7.

81. $\frac{1}{4}$

82. $\frac{1}{20}$

83. $\frac{4}{25}$

84. $\frac{3}{5}$

85. $\frac{7}{8}$

86. $\frac{3}{16}$

Combining Concepts

87. Why is the decimal point moved to the right when grams are converted to milligrams?

88. To change 8 pounds to ounces, multiply by 16. Why is this the correct procedure?

8.6 Capacity

A Defining and Converting U.S. System Units of Capacity

Units of **capacity** are generally used to measure liquids. The number of gallons of gasoline needed to fill a gas tank in a car, the number of cups of water needed in a bread recipe, and the number of quarts of milk sold each day at a supermarket are all examples of using units of capacity. The following summary shows equivalencies between units of capacity:

OBJECTIVES

- A Define U.S. units of capacity and convert from one unit to another.
- B Perform arithmetic operations on U.S. units of capacity.
- C Define metric units of capacity and convert from one unit to another.
- D Perform arithmetic operations on metric units of capacity.

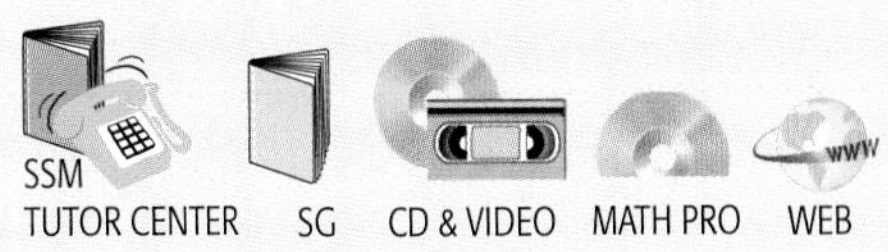

U.S. Units of Capacity

8 fluid ounces (fl oz) = 1 cup (c)
2 cups = 1 pint (pt)
2 pints = 1 quart (qt)
4 quarts = 1 gallon (gal)

Just as with units of length and weight, we can form unit fractions to convert between different units of capacity. For instance,

$$\frac{2\text{ c}}{1\text{ pt}} = \frac{1\text{ pt}}{2\text{ c}} = 1 \quad \text{and} \quad \frac{2\text{ pt}}{1\text{ qt}} = \frac{1\text{ qt}}{2\text{ pt}} = 1$$

EXAMPLE 1 Convert 9 quarts to gallons.

Solution: We multiply by the unit fraction $\frac{1\text{ gal}}{4\text{ qt}}$.

$$\begin{aligned} 9\text{ qt} &= \frac{9\text{ qt}}{1} \cdot \frac{1\text{ gal}}{4\text{ qt}} \\ &= \frac{9\text{ gal}}{4} \\ &= 2\frac{1}{4}\text{ gal} \end{aligned}$$

Thus, 9 quarts is the same as $2\frac{1}{4}$ gallons, as shown in the diagram:

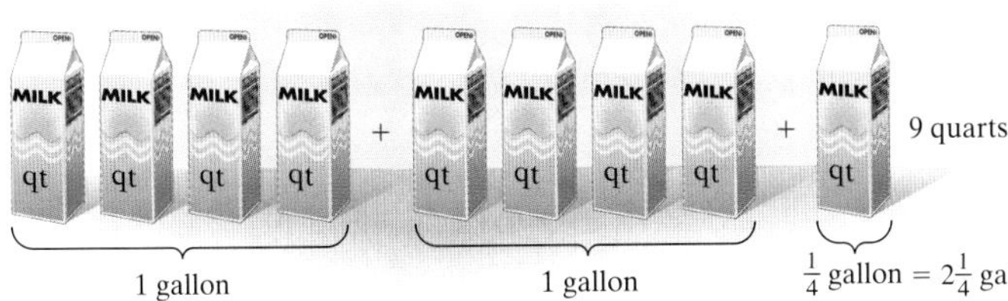

Practice Problem 1

Convert 43 pints to quarts.

EXAMPLE 2 Convert 14 cups to quarts.

Solution: Our equivalency table contains no direct conversion from cups to quarts. However, from this table we know that

$$1\text{ qt} = 2\text{ pt} = 4\text{ c}$$

so 1 qt = 4 c. Now we have the unit fraction $\frac{1\text{ qt}}{4\text{ c}}$. Thus,

$$14\text{ c} = \frac{14\text{ c}}{1} \cdot \frac{1\text{ qt}}{4\text{ c}} = \frac{7}{2}\text{qt} \quad \text{or} \quad 3\frac{1}{2}\text{qt}$$

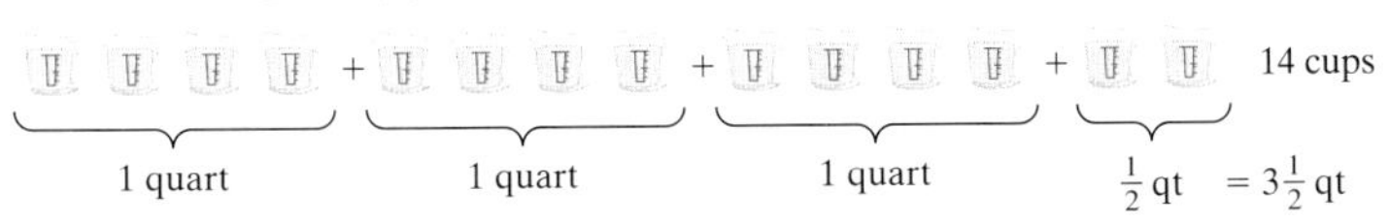

Practice Problem 2

Convert 26 quarts to cups.

Answers

1. $21\frac{1}{2}$qt **2.** 104 c

Note: Another way to solve this example is to use two unit fractions.

$$14 \text{ c} = \frac{\overset{7}{\cancel{14}} \cancel{\text{c}}}{1} \cdot \frac{1 \cancel{\text{pt}}}{\underset{1}{\cancel{2}} \cancel{\text{c}}} \cdot \frac{1 \text{ qt}}{2 \cancel{\text{pt}}} = \frac{7}{2} \text{qt} \quad \text{or} \quad 3\frac{1}{2} \text{qt}$$

Try the Concept Check in the margin.

B Performing Operations on U.S. System Units of Capacity

As is true of units of length and weight, units of capacity can be added, subtracted, multiplied, and divided.

EXAMPLE 3 Subtract 3 qt from 4 gal 2 qt.

Solution: To subtract, we line up similar units.

$$\begin{array}{r} 4 \text{ gal } 2 \text{ qt} \\ - \quad 3 \text{ qt} \\ \hline \end{array}$$

We cannot subtract 3 qt from 2 qt. We need to borrow 1 gallon from the 4 gallons, convert it to 4 quarts, and then combine it with the 2 quarts.

Borrow 1 gal = 4 qt

3 gal + (1 gal) 4 qt

$$\begin{array}{r} 4 \text{ gal } 2 \text{ qt} \\ - \quad 3 \text{ qt} \\ \hline \end{array} \quad \text{or} \quad \begin{array}{r} 3 \text{ gal } 6 \text{ qt} \\ - \quad 3 \text{ qt} \\ \hline 3 \text{ gal } 3 \text{ qt} \end{array}$$

To check, see that the sum of 3 gal 3 qt and 3 qt is 4 gal 2 qt.

EXAMPLE 4 Multiply 3 qt 1 pt by 3.

Solution: We multiply each of the units of capacity by 3.

$$\begin{array}{r} 3 \text{ qt } 1 \text{ pt} \\ \times \quad 3 \\ \hline 9 \text{ qt } 3 \text{ pt} \end{array}$$

Since 3 pints is the same as 1 quart and 1 pint, we have

$$9 \text{ qt } 3 \text{ pt} = 9 \text{ qt} + 1 \text{ qt } 1 \text{ pt} = 10 \text{ qt } 1 \text{ pt}$$

The 10 quarts can be changed to gallons by dividing by 4, since there are 4 quarts in a gallon. To see why we divide, notice that

$$10 \text{ qt} = \frac{10 \cancel{\text{qt}}}{1} \cdot \frac{1 \text{ gal}}{4 \cancel{\text{qt}}} = \frac{10}{4} \text{gal}$$

$$\begin{array}{r} 2 \text{ gal } 2 \text{ qt} \\ 4 \overline{) 10 \text{ qt}} \\ -8 \\ \hline 2 \end{array}$$

Then the product is 10 qt 1 pt or 2 gal 2 qt 1 pt.

EXAMPLE 5 Divide 3 gal 2 qt by 2.

Solution: We divide each unit of capacity by 2.

$$\begin{array}{r} 1 \text{ gal} \quad 3 \text{ qt} \\ 2 \overline{) 3 \text{ gal} \quad 2 \text{ qt}} \\ -2 \\ \hline 1 \text{ gal} = 4 \text{ qt} \\ 6 \text{ qt} \end{array} \quad 6 \text{ qt} \div 2 = 3 \text{ qt}$$

Concept Check

If 50 cups are converted to quarts, will the equivalent number of quarts be less than or greater than 50? Explain.

Practice Problem 3

Subtract 2 qt from 1 gal 1 qt.

Practice Problem 4

Multiply 2 gal 3 qt by 2.

Practice Problem 5

Divide 6 gal 1 qt by 2.

Answers

3. 3 qt **4.** 5 gal 2 qt **5.** 3 gal 1 pt

Concept Check: less than 50

EXAMPLE 6 Finding the Amount of Water in an Aquarium

An aquarium contains 6 gal 3 qt of water. If 2 gal 2 qt of water is added, what is the total amount of water in the aquarium?

Solution:

beginning water	→	6 gal 3 qt
+ water added	→	+ 2 gal 2 qt
total water	→	8 gal 5 qt

Since 5 qt = 1 gal 1 qt, we have

$$= \overbrace{8 \text{ gal}}^{8 \text{ gal}} + \overbrace{1 \text{ gal } 1 \text{ qt}}^{5 \text{ qt}}$$

$$= 9 \text{ gal } 1 \text{ qt}$$

The total amount of water is 9 gal 1 qt.

C Defining and Converting Metric System Units of Capacity

Thus far, we know that the basic unit of length in the metric system is the meter and that the basic unit of mass in the metric system is the gram. What is the basic unit of capacity? The **liter.** By definition, a **liter** is the capacity or volume of a cube measuring 10 centimeters on each side.

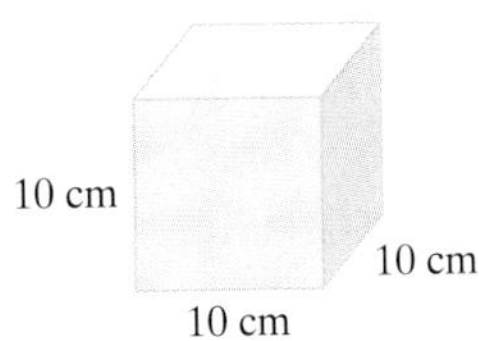

The following examples may help you get a feeling for metric capacities:

One liter of liquid is slightly more than one quart.

1 quart 1 liter

Many soft drinks are packaged in 2-liter bottles.

The metric system was designed to be a consistent system. Once again, the prefixes for metric units of capacity are the same as for metric units of length and mass, as summarized in the following table:

Practice Problem 6

A large oil drum contains 15 gal 3 qt of oil. How much will be in the drum if an additional 4 gal 3 qt of oil is poured into it?

Answer

6. 20 gal 2 qt

Prefix	Meaning	Metric Unit of Capacity	
kilo	1000	1 kiloliter	(kl) = 1000 liters (L)
hecto	100	1 hectoliter	(hl) = 100 L
deka	10	1 dekaliter	(dal) = 10 L
			1 liter (L) = 1 L
deci	1/10	1 deciliter	(dl) = 1/10 L or 0.1 L
centi	1/100	1 centiliter	(cl) = 1/100 L or 0.01 L
milli	1/1000	1 milliliter	(ml) = 1/1000 L or 0.001 L

The **milliliter** and the **liter** are the two most commonly used metric units of capacity.

Converting from one unit of capacity to another involves multiplying by powers of 10 or moving the decimal point to the left or to the right. Listing units of capacity in order from largest to smallest helps to keep track of how many places to move the decimal point when converting.

Let's convert 2.6 liters to milliliters. To convert from liters to milliliters, we move along the chart 3 units to the right.

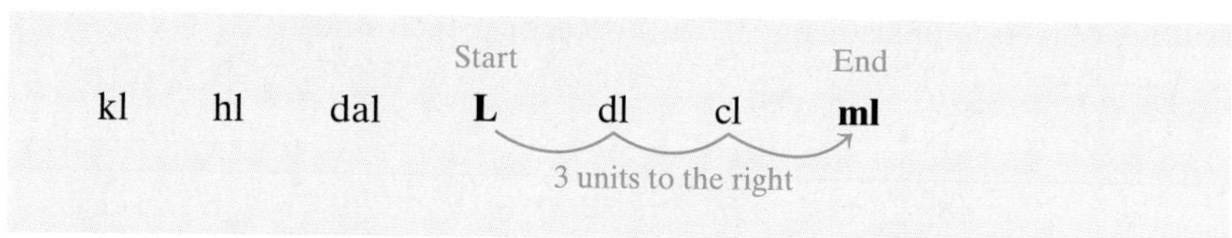

This means that we move the decimal point 3 places to the right to convert from liters to milliliters.

2.600 L = 2600. ml or 2600 ml

This same conversion can be done with unit fractions.

$$2.6 \text{ L} = \frac{2.6 \not{\text{L}}}{1} \cdot \frac{1000 \text{ ml}}{1 \not{\text{L}}}$$
$$= 2.6 \cdot 1000 \text{ ml}$$
$$= 2600 \text{ ml}$$

To multiply by 1000, move the decimal point 3 places to the right.

To visualize the result, study the diagram below:

Thus 2.6 L = 2600 ml.

Practice Problem 7

Convert 2100 ml to liters.

EXAMPLE 7 Convert 3210 ml to liters.

Solution: Let's use the unit fraction method first.

$$3210 \text{ ml} = 3210 \not{\text{ml}} \cdot \underbrace{\frac{1 \text{ L}}{1000 \not{\text{ml}}}}_{\text{Unit fraction}} = 3.21 \text{ L}$$

Now let's list the unit measures in a chart and move from milliliters to liters.

Answer

7. 2.1 L

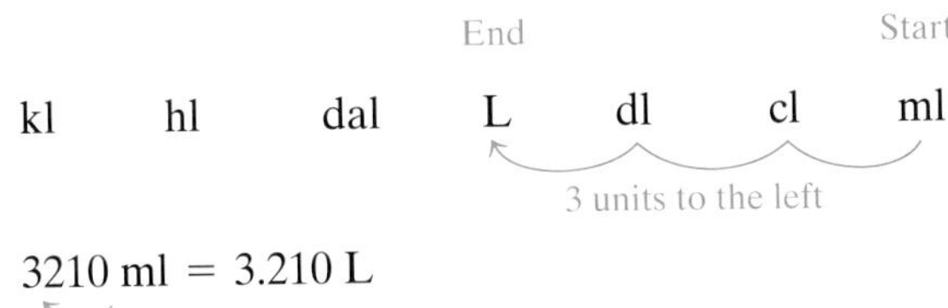

3210 ml = 3.210 L

3 places to the left

EXAMPLE 8 Convert 0.185 dl to milliliters.

Solution: We list the unit measures in a chart and move from deciliters to milliliters.

				Start		End
kl	hl	dal	L	dl	cl	ml

2 units to the right

0.185 dl = 18.5 ml

2 places to the right

(D) Performing Operations on Metric System Units of Capacity

As was true for length and weight, arithmetic operations involving metric units of capacity can also be performed. Make sure the metric units of capacity are the same before adding, subtracting, multiplying, or dividing.

EXAMPLE 9 Add 2400 ml to 8.9 L.

Solution: We must convert both to liters or both to milliliters before adding the capacities together.

2400 ml = 2.4 L or 8.9 L = 8900 ml

$$\begin{array}{r} 2.4\text{ L} \\ +8.9\text{ L} \\ \hline 11.3\text{ L} \end{array} \qquad \begin{array}{r} 2400\text{ ml} \\ +8900\text{ ml} \\ \hline 11{,}300\text{ ml} \end{array}$$

The total is 11.3 L or 11,300 ml. They both represent the same capacity.

Try the Concept Check in the margin.

EXAMPLE 10 Divide 18.08 ml by 16.

Solution:

$$\begin{array}{r} 1.13\text{ ml} \\ 16\overline{)18.08\text{ ml}} \\ -16\phantom{.08\text{ ml}} \\ \hline 2\,0\phantom{8\text{ ml}} \\ -1\,6\phantom{8\text{ ml}} \\ \hline 48\phantom{\text{ ml}} \\ -48\phantom{\text{ ml}} \\ \hline 0\phantom{\text{ ml}} \end{array}$$

The solution is 1.13 ml.

Practice Problem 8

Convert 2.13 dal to liters.

Practice Problem 9

Add 1250 ml to 2.9 L.

Concept Check

How could you estimate the following operation? Subtract 950 ml from 7.5 L.

Practice Problem 10

Divide 146.9 L by 13.

Answers

8. 21.3 L **9.** 4150 ml or 4.15 L **10.** 11.3 L

Concept Check: 950 ml = 0.95 L; round 0.95 to 1; 7.5 − 1 = 6.5 L

Practice Problem 11

If 28.6 L of water can be pumped every minute, how much water can be pumped in 85 minutes?

Answer

11. 2431 L

EXAMPLE 11 Finding the Amount of Medication a Person Has Received

A patient hooked up to an IV unit in the hospital is to receive 12.5 ml of medication every hour. How much medication does the patient receive in 3.5 hours?

Solution: We multiply 12.5 ml by 3.5.

medication per hour	→	12.5 ml
× hours	→	× 3.5
total medication		625
		375
		43.75 ml

The patient receives 43.75 ml of medication.

Name ______________________ Section ____________ Date ____________

Mental Math

Convert as indicated.

1. 2 c to pints

2. 4 c to pints

3. 4 qt to gallons

4. 8 qt to gallons

5. 2 pt to quarts

6. 6 pt to quarts

7. 8 fl oz to cups

8. 24 fl oz to cups

9. 1 pt to cups

10. 3 pt to cups

11. 1 gal to quarts

12. 2 gal to quarts

Determine whether the measurement in each statement is reasonable.

13. Clair took a dose of 2 L of cough medicine to cure her cough.

14. John drank 250 ml of milk for lunch.

15. Jeannie likes to relax in a tub filled with 3000 ml of hot water.

16. Sarah pumped 20 L of gasoline into her car yesterday.

EXERCISE SET 8.6

A *Convert each measurement as indicated. See Examples 1 and 2.*

1. 32 fluid ounces to cups

2. 16 quarts to gallons

3. 8 quarts to pints

4. 9 pints to quarts

5. 10 quarts to gallons

6. 15 cups to pints

7. 80 fluid ounces to pints

8. 18 pints to gallons

9. 2 quarts to cups

10. 3 pints to fluid ounces

11. 120 fluid ounces to quarts

12. 20 cups to gallons

13. 6 gallons to fluid ounces

14. 5 quarts to cups

15. $4\frac{1}{2}$ pints to cups

16. $6\frac{1}{2}$ gallons to quarts

17. $2\frac{3}{4}$ gallons to pints

18. $3\frac{1}{4}$ quarts to cups

B *Perform each indicated operation. See Examples 3 through 5.*

19. 4 gal 3 qt + 5 gal 2 qt

20. 2 gal 3 qt + 8 gal 3 qt

21. 1 c 5 fl oz + 2 c 7 fl oz

22. 2 c 3 fl oz + 2 c 6 fl oz

23. 3 gal − 1 gal 3 qt

24. 2 pt − 1 pt 1 c

25. 3 gal 1 qt − 1 qt 1 pt

26. 3 qt 1 c − 1 c 4 fl oz

27. 1 pt 1 c × 3

28. 1 qt 1 pt × 2

29. 8 gal 2 qt × 2

30. 6 gal 1 pt × 2

31. 9 gal 2 qt ÷ 2

32. 5 gal 6 fl oz ÷ 2

Solve. Remember to insert units when writing your answers. See Example 6.

33. A can of Hawaiian punch holds $1\frac{1}{2}$ quarts of liquid. How many fluid ounces is this?

34. Weight Watchers Double Fudge bars contain 21 fluid ounces of ice cream. How many cups of ice cream is this?

35. Many diet experts advise individuals to drink 64 ounces of water each day. How many quarts of water is this?

36. A recipe for walnut fudge cake calls for $1\frac{1}{4}$ cups of water. How many fluid ounces is this?

37. Can 5 pt 1 c of fruit punch and 2 pt 1 c of ginger ale be poured into a 1-gal container without it overflowing?

38. Three cups of prepared Jell-O are poured into 6 dessert dishes. How many fluid ounces of Jell-O are in each dish?

39. A garden tool engine requires a 30 to 1 gas to oil mixture. This means that $\frac{1}{30}$ of a gallon of oil should be mixed with 1 gallon of gas. Convert $\frac{1}{30}$ gallon to fluid ounces. Round to the nearest tenth.

40. Henning's Supermarket sells homemade soup in 1 qt 1 pt containers. How much soup is contained in three such containers?

41. A case of Pepsi Cola holds 24 cans, each of which contains 12 fluid ounces of Pepsi Cola. How many *quarts* are there in a case of Pepsi?

42. Manuela's Service Station has a drum that holds 40 gallons of oil. If 6 gallons and 3 quarts have been used, how much oil remains?

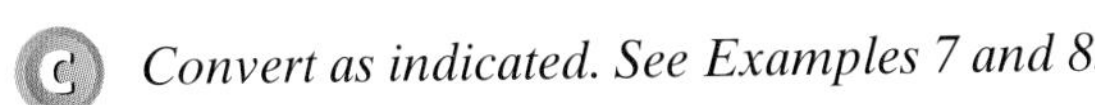

Convert as indicated. See Examples 7 and 8.

43. 5 L to milliliters

44. 8 L to milliliters

45. 4500 ml to liters

46. 3100 ml to liters

47. 410 L to kiloliters

48. 250 L to kiloliters

49. 64 ml to liters

50. 39 ml to liters

51. 0.16 kl to liters

52. 0.48 kl to liters

53. 3.6 L to milliliters

54. 1.9 L to milliliters

55. 0.16 L to kiloliters

56. 0.127 L to kiloliters

D *Perform each indicated operation. See Examples 9 and 10.*

57. 2.9 L + 19.6 L

58. 18.5 L + 4.6 L

59. 2700 ml + 1.8 L

60. 4.6 L + 1600 ml

61. 8.6 L − 190 ml

62. 4.8 L − 283 ml

63. 11,400 ml − 0.8 L

64. 6850 ml − 0.3 L

65. 480 ml × 8

66. 290 ml × 6

67. 81.2 L ÷ 0.5

68. 5.4 L ÷ 3.6

Solve. Remember to insert units when writing your answers. See Example 11.

69. Mike Schaferkotter drank 410 ml of Mountain Dew from a 2-liter bottle. How much Mountain Dew remains in the bottle?

70. A Volvo has a 54.5-L gas tank. Only 3.8 liters of gasoline still remain in the tank. How much is needed to fill it?

71. A woman added 354 ml of Prestone dry gasoline to the 18.6 L of gasoline in her car's tank. Find the total amount of gasoline in the tank.

72. Chris Peckaitis wishes to share a 2-L bottle of Coca Cola equally with 7 of his friends. How much will each person get?

73. A salesman paid $14 to fill his car with 44.3 liters of gasoline. Find the price per liter of gasoline to the nearest tenth of a cent.

74. A student carelessly misread the scale on a cylinder in the chemistry lab and added 40 cl of water to a mixture instead of 40 ml. Find the excess amount of water.

75. A large bottle of Langers Apple Juice contains 1.89 L of beverage. A smaller bottle of apple juice contains only 946 ml. How much more is in the larger bottle?

76. In a lab experiment, a student added 400 ml of salt water to 1.65 L of water. Later 320 ml of the solution was drained off. How much of the solution still remained?

Review and Preview

Write each decimal as a fraction. See Section 5.1.

77. 0.7 **78.** 0.9 **79.** 0.03 **80.** 0.007 **81.** 0.006 **82.** 0.08

Combining Concepts

83. Explain how to borrow in order to subtract 1 gal 2 qt from 3 gal 1 qt.

A cubic centimeter (cc) is the amount of space that a volume of 1 ml occupies. Because of this, we will say that 1 cc = 1 ml.

A common syringe is one with a capacity of 3 cc. Use the diagram below and give the measurement indicated by each arrow. The first measurement is done for you.

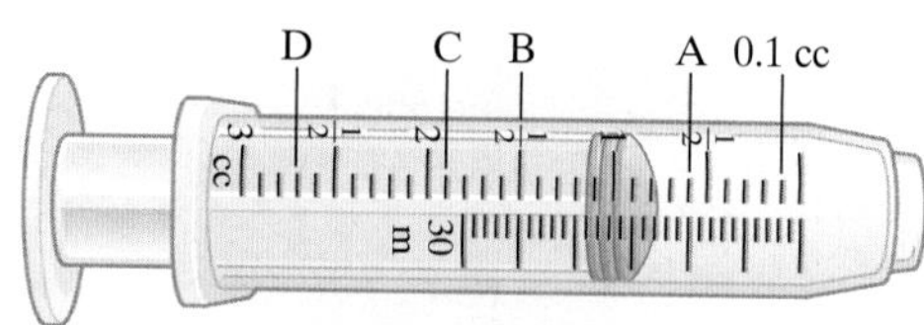

84. A **85.** B **86.** C **87.** D

In order to measure small dosages, such as for insulin, u-100 syringes are used. For these syringes, 1 cc has been divided into 100 equal units (u). Use the diagram below and give the measurement indicated by each arrow in units (u) and then cubic centimeters. Use 100 u = 1 cc *and round to the nearest hundredth.*

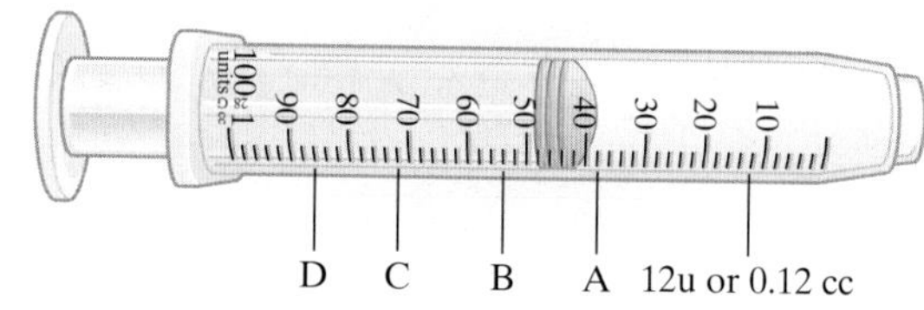

88. A **89.** B

90. C **91.** D

Internet Excursions

Go To:  http://www.prenhall.com/martin-gay_prealgebra 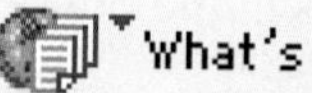 What's Related

By going to the World Wide Web address listed above, you will be directed to a site called A Dictionary of Units, or a related site, that will help you answer the questions below.

92. There are more units of capacity in use than the ones mentioned in this section. For instance, there are special measures for the capacity of dry items such as grains or fruits. Visit this Web site and locate the "U.S. System of Measurements" area. List the measure equivalencies for dry capacity. Then convert 38 pecks of apples to bushels.

93. There are also special apothecaries' measures for capacity of liquid drugs and medicines. Although metric measures are commonly used in the pharmacy industry today, awareness of these measures is still useful. Within the "U.S. System of Measurements" area of this Web site, find and list the apothecaries' measures equivalencies. Then convert 2 pints to fl drams.

8.7 Conversions between the U.S. and Metric Systems and Temperature Conversions

OBJECTIVES

Ⓐ Convert between the U.S. and Metric systems.

Ⓑ Convert between degrees Celsius and degrees Fahrenheit.

SSM TUTOR CENTER SG CD & VIDEO MATH PRO WEB

Ⓐ Converting between the U.S. and Metric Systems

The metric system probably had its beginnings in France in the 1600s, but it was the Metric Act of 1866 that made the use of this system legal (but not mandatory) in the United States. Other laws have followed that allow for a slow, but deliberate, transfer to the modernized metric system. In April, 2001, for example, the U.S. Stock Exchanges completed their change to decimal trading instead of fractions. By the end of 2009, all products sold in Europe (with some exceptions) will be required to have only metric units on their labels. (*Source*: U.S. Metric Association and National Institute of Standards and Technology)

You may be surprised at the number of everyday items we use that are already manufactured in metric units. We easily recognize 1 L and 2 L soda bottles, but what about the following?

Pencil leads (0.5 mm or 0.7 mm)
Camera film (35 mm)
Sporting events (5 km or 10 km races)
Medicines (500 mg capsules)
Labels on retail goods (dual-labeled since 1994)

Since the United States has not completely converted to the metric system, we need to practice converting from one system to the other. Below is a table of mostly approximate conversions.

Length:

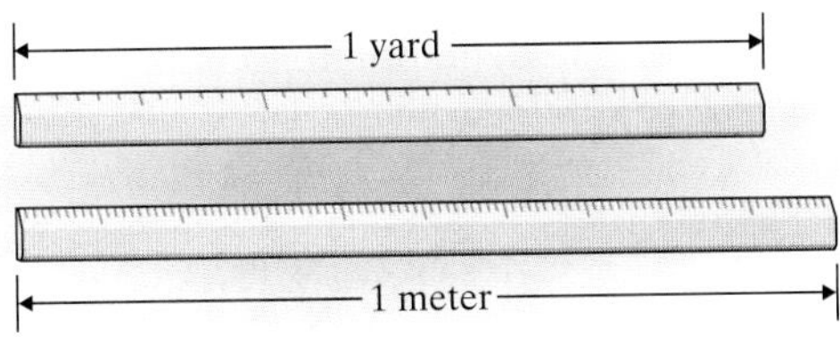

metric		U.S. System
1 m	≈	1.09 yd
1 m	≈	3.28 ft
1 km	≈	0.62 mi
2.54 cm	=	1 in.
0.30 m	≈	1 ft
1.61 km	≈	1 mi

Capacity:

metric		U.S. System
1 L	≈	1.06 qt
1 L	≈	0.26 gal
3.79 L	≈	1 gal
0.95 L	≈	1 qt
29.57 ml	≈	1 fl oz

Weight (mass):

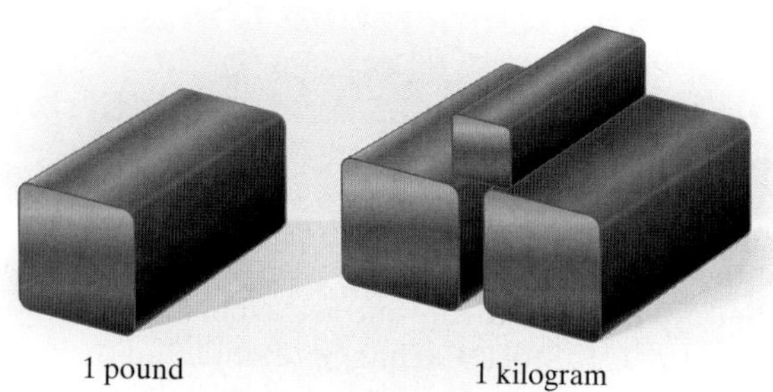

metric	U.S. System
1 kg	$\approx$ 2.20 lb
1 g	$\approx$ 0.04 oz
0.45 kg	$\approx$ 1 lb
28.35 g	$\approx$ 1 oz

There are many ways to perform these metric to U.S. Conversions. We will do so by using unit fractions.

Practice Problem 1

The center hole of a compact disk is 15 millimeters or 1.5 centimeters in diameter. Convert this length to inches. Round the result to 2 decimal places.

EXAMPLE 1 Compact Disks

Compact disks are 12 centimeters in diameter. Convert this length to inches. Round the result to two decimal places.(*Source*: usByte.com)

Solution: From our length conversion table, we know that 2.54 cm = 1 in. This fact gives us two unit fractions: $\frac{2.54\text{ cm}}{1\text{ in.}}$ and $\frac{1\text{ in.}}{2.54\text{ cm}}$. We use the unit fraction with cm in the denominator so that these units divide out.

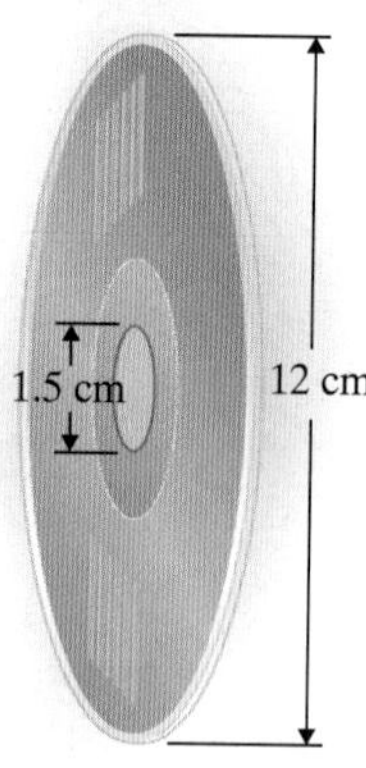

$$12\text{ cm} = \frac{12\text{ cm}}{1} \cdot \underbrace{\frac{1\text{ in.}}{2.54\text{ cm}}}_{\text{Unit fraction}} \quad \begin{array}{l}\leftarrow \text{Units to convert to} \\ \leftarrow \text{Original units}\end{array}$$

$$= \frac{12\text{ in.}}{2.54}$$

$$\approx 4.72\text{ in.} \quad \text{Divide.}$$

Thus, the diameter of a compact disk is exactly 12 cm or approximately 4.72 inches. For a dimension this size, you can use a ruler to check. Another method is to approximate. Our result, 4.72 in., is close to 5 inches. Since 1 in. is about 2.5 cm, then 5 in. is about 5(2.5 cm) = 12.5 cm, which is close to 12 cm.

Answer

1. 0.59 in.

EXAMPLE 2 Liver

The liver is your largest internal organ. It weighs about 3.5 pounds in a grown man. Convert this weight to kilograms. Round to the nearest tenth. (*Source*: *Some Body!* by Dr. Pete Rowan)

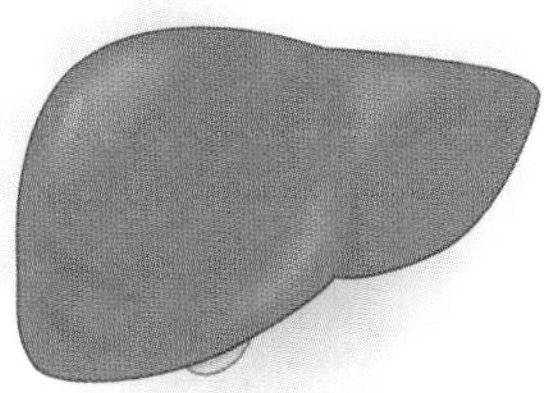

Solution:

$$3.5 \text{ lb} \approx \frac{3.5 \cancel{\text{lb}}}{1} \cdot \underbrace{\frac{0.45 \text{ kg}}{1 \cancel{\text{lb}}}}_{\text{Unit fraction}}$$

$$= 3.5(0.45 \text{ kg})$$

$$\approx 1.6 \text{ kg}$$

Thus 3.5 pounds are approximately 1.6 kilograms. From the table of conversions, we know that 1 kg $\approx$ 2.2 lb so that 0.5 kg $\approx$ 1.1 lb and adding, we have 1.5 kg $\approx$ 3.3 lb. Our result is reasonable.

Practice Problem 2

A full-grown human heart weighs about $\frac{1}{2}$ pound or 8 ounces. Convert this weight to grams. If necessary, round your result to the nearest tenth of a gram.

EXAMPLE 3 Postage Stamp

Australia converted to the metric system in 1973. In that year, four postage stamps were issued to publicize this conversion. One such stamp is shown below. Let's check the mathematics on the stamp by converting 7 fluid ounces to milliliters. Round to the nearest hundred.

Solution:

$$7 \text{ fl oz} \approx \frac{7 \cancel{\text{fl oz}}}{1} \cdot \underbrace{\frac{29.57 \text{ ml}}{1 \cancel{\text{fl oz}}}}_{\text{Unit fraction}}$$

$$= 206.99 \text{ ml}$$

Rounded to the nearest hundred, 7 fl oz $\approx$ 200 ml.

Practice Problem 3

Convert 237 ml to fluid ounces. Round to the nearest whole fluid ounce.

Answers

2. 226.8 g **3.** 8 fl oz

B Converting between Degrees Celsius and Degrees Fahrenheit

When Gabriel Fahrenheit and Anders Celsius independently established units for temperature scales, each based his unit on the heat of water the moment it boils compared to the moment it freezes. One degree Celsius is $\frac{1}{100}$ of the difference in heat. One degree Fahrenheit is $\frac{1}{180}$ of the difference in heat. Celsius arbitrarily labeled the temperature at the freezing point at 0°C, making the boiling point 100°C; Fahrenheit labeled the freezing point 32°F, making the boiling point 212°F. Water boils at 212°F or 100°C.

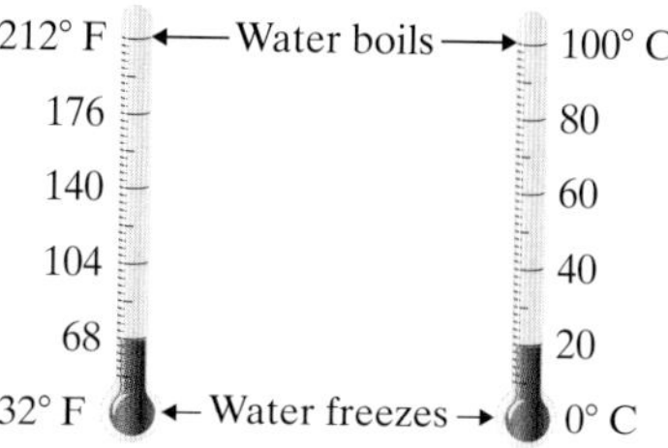

By comparing the two scales in the figure, we see that a 20°C day is as warm as a 68°F day. Similarly, a sweltering 104°F day in the Mojave desert corresponds to a 40°C day.

Concept Check

Which of the following statements is correct? Explain.

a. 6°C is below the freezing point of water.

b. 6°F is below the freezing point of water.

Try the Concept Check in the margin.

To convert from Celsius temperatures to Fahrenheit temperatures, see the box below. In this box, we use the symbol F to represent degrees Fahrenheit and the symbol C to represent degrees Celsius.

Converting Celsius to Fahrenheit

$$F = \frac{9}{5}C + 32 \quad \text{or} \quad F = 1.8C + 32$$

(To convert to Fahrenheit temperature, multiply the Celsius temperature by $\frac{9}{5}$ or 1.8, and then add 32.)

Concept Check:
a. false **b.** true

EXAMPLE 4 Convert 15°C to degrees Fahrenheit.

Solution:

$$\begin{aligned} F &= \frac{9}{5}C + 32 \\ &= \frac{9}{5} \cdot 15 + 32 \quad \text{Replace C with 15.} \\ &= 27 + 32 \quad \text{Simplify.} \\ &= 59 \quad \text{Add.} \end{aligned}$$

Thus, 15°C is equivalent to 59°F.

Practice Problem 4

Convert 50°C to degrees Fahrenheit.

EXAMPLE 5 Convert 29°C to degrees Fahrenheit.

Solution:

$$\begin{aligned} F &= 1.8C + 32 \\ &= 1.8 \cdot 29 + 32 \quad \text{Replace C with 29.} \\ &= 52.2 + 32 \quad \text{Multiply 1.8 by 29.} \\ &= 84.2 \quad \text{Add.} \end{aligned}$$

Therefore, 29°C is the same as 84.2°F.

Practice Problem 5

Convert 18°C to degrees Fahrenheit.

To convert from Fahrenheit temperatures to Celsius temperatures, see the box below. The symbol C represents degrees Celsius and the symbol F represents degrees Fahrenheit.

Converting Fahrenheit to Celsius

$$C = \frac{5}{9}(F - 32)$$

(To convert to Celsius temperature, subtract 32 from the Fahrenheit temperature, and then multiply by $\frac{5}{9}$.)

Answers

4. 122°F **5.** 64.4°F

Practice Problem 6

Convert 113°F to degrees Celsius. If necessary, round to the nearest tenth of a degree.

Practice Problem 7

During a bout with the flu, Albert's temperature reaches 102.8°F. What is his temperature measured in degrees Celsius? Round to the nearest tenth of a degree.

Concept Check

Clarissa must convert 40°F to degrees Celsius. What is wrong with her work shown below?

$$F = 1.8 \cdot C + 32$$
$$F = 1.8 \cdot 40 + 32$$
$$F = 72 + 32$$
$$F = 104$$

Answers

6. 45°C **7.** 39.3°C

Concept Check: She used the conversion for Celsius to Fahrenheit instead of Fahrenheit to Celsius.

EXAMPLE 6 Convert 114°F to degrees Celsius. If necessary, round to the nearest tenth of a degree.

Solution:

$$C = \frac{5}{9}(F - 32)$$

$$= \frac{5}{9}(114 - 32) \quad \text{Replace F with 114.}$$

$$= \frac{5}{9} \cdot (82) \quad \text{Subtract inside parentheses.}$$

$$\approx 45.6 \quad \text{Multiply.}$$

Therefore, 114°F is approximately 45.6°C.

EXAMPLE 7 Body Temperature

Normal body temperature is 98.6°F. What is this temperature in degrees Celsius?

Solution: We evaluate the formula $C = \frac{5}{9}(F - 32)$ when F is 98.6.

$$C = \frac{5}{9}(F - 32)$$

$$= \frac{5}{9}(98.6 - 32) \quad \text{Replace F with 98.6.}$$

$$= \frac{5}{9} \cdot (66.6) \quad \text{Subtract inside parentheses.}$$

$$= 37 \quad \text{Multiply.}$$

Therefore, normal body temperature is 37°C.

Try the Concept Check in the margin.

FOCUS ON History

THE DEVELOPMENT OF UNITS OF MEASURE

The earliest units of measure were based on the human body. The ancient Egyptians, Babylonians, Hebrews, and Mesopotamians used a unit of length called the **cubit,** which represents the distance between a human elbow and fingertips. For instance, in the book of Genesis in the Bible, Noah's ark is described as having a length of "three hundred cubits, its width fifty cubits, and its height thirty cubits." Other commonly used measures found in documents of these ancient cultures include the digit, hand, span, and foot. These measures are shown at the right.

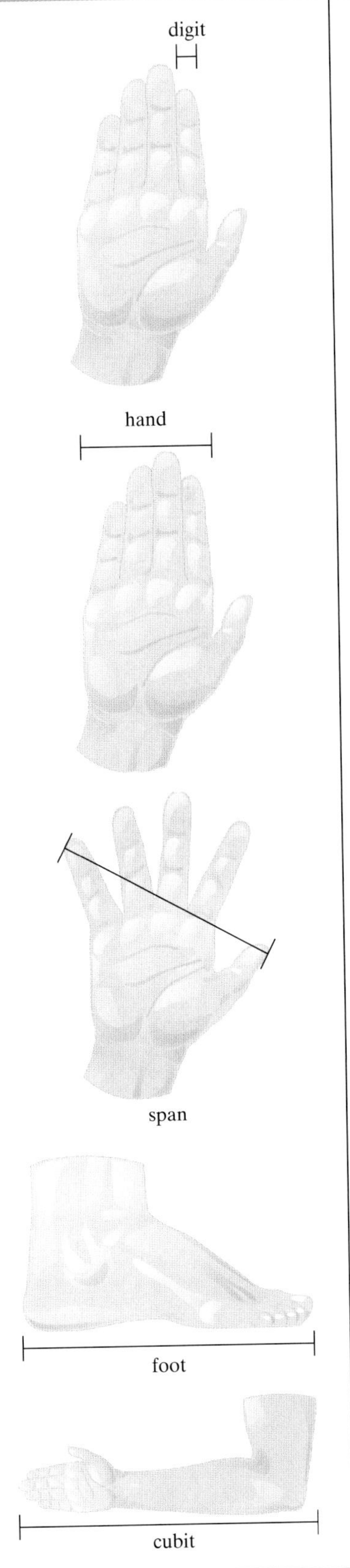

Several thousand years later, the English system of measurement also consisted of a mixed bag of body- and nature-related units of measure. A rod was the combined length of the left feet of 16 men. An inch was the distance spanned by three grains of barley. A foot was the length of the foot of the king currently in power. Around 1100 A.D. King Henry I of England decreed that a yard was the distance between the king's nose and the thumb of his outstretched arm.

Although the English system became somewhat standardized by the 13th century, the problem with it and the ancient systems of measurement was that the relationship between the various units was not necessarily easy to remember. This made it difficult to convert from one type of unit to another.

The metric system grew out of the French Revolution during the 1790s. As a reaction against the lack of consistency and utility of existing systems of measurement, the French Academy of Sciences set out to develop an easy-to-use, internationally standardized system of measurements. Originally the basic unit of the metric system, the meter, was to be one ten-millionth of the distance between the North Pole and the Equator on a line through Paris. However, it was soon discovered that this distance was nearly impossible to measure accurately. The meter has been redefined several times since the end of the French Revolution, most recently in 1983 as the distance that light travels in a vacuum during $\frac{1}{299{,}792{,}458}$ of a second.

Although the definition of the meter has evolved over time, the original idea of developing an easy-to-use system of measurements has endured. The metric system uses a standard set of prefixes for all basic unit types (length, mass, capacity, etc.) and all conversions within a unit type are based on powers of 10.

CRITICAL THINKING

Develop your own unit of measure for (a) length and (b) area. Explain how you chose your units and for what types of relative sizes of measurements they would be most useful. Discuss the advantages and disadvantages of your units of measurement. Then use your units to measure the width of your classroom desk and the area of the front cover of this textbook.

STUDY SKILLS REMINDER

Are you prepared for a test on Chapter 8?

Below I have listed some common trouble areas for topics covered in Chapter 8. After studying for your test—but before taking your test—read these.

- Don't forget the difference between complementary and supplementary angles.

  Complementary angles have a sum of 90°.

  Supplementary angles have a sum of 180°.

  The complement of a 15° angle measures 90° − 15° = 75°.

  The supplement of a 15° angle measures 180° − 15° = 165°.

- Remember:

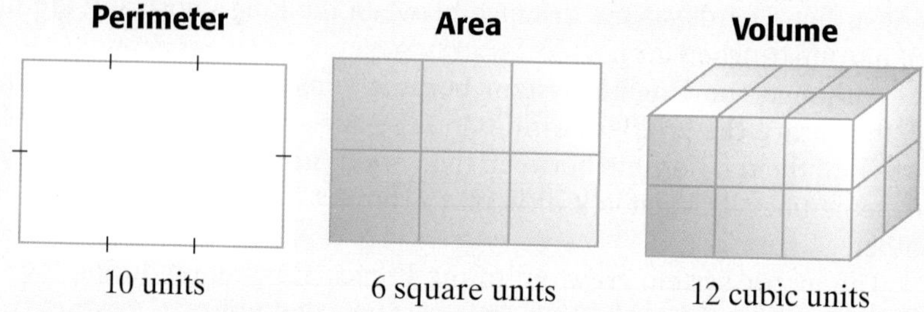

Remember: This is simply a checklist of common trouble areas. For a review of Chapter 8, see the Highlights and Chapter Review at the end of this chapter.

Name ______________________ Section __________ Date __________

Mental Math

Determine whether the measurement in each statement is reasonable.

1. A 72°F room feels comfortable.
2. Water heated to 110°F will boil.
3. Josiah has a fever if a thermometer shows his temperature to be 40°F.
4. An air temperature of 20°F on a Vermont ski slope can be expected in the winter.
5. When the temperature is 30°C outside, an overcoat is needed.
6. An air-conditioned room at 60°C feels quite chilly.
7. Barbara has a fever when a thermometer records her temperature at 40°C.
8. Water cooled to 32°C will freeze.

EXERCISE SET 8.7

A *Convert as indicated. If necessary, round answers to two decimal places. See Examples 1 through 3. Because approximations are used, your answers may vary slightly from the answers given in the back of the book.*

1. 578 milliliters to fluid ounces
2. 5 liters to quarts
3. 86 inches to centimeters
4. 86 miles to kilometers
5. 1000 grams to ounces
6. 100 kilograms to pounds
7. 93 kilometers to miles
8. 9.8 meters to feet
9. 14.5 liters to gallons
10. 150 milliliters to fluid ounces
11. 30 pounds to kilograms
12. 15 ounces to grams

Solve. If necessary, round answers to two decimal places. See Examples 1 through 3.

13. The balance beam for female gymnasts is 10 centimeters wide. Convert this width to inches.
14. In men's gymnastics, the rings are 250 centimeters from the floor. Convert this height to inches, then to feet.

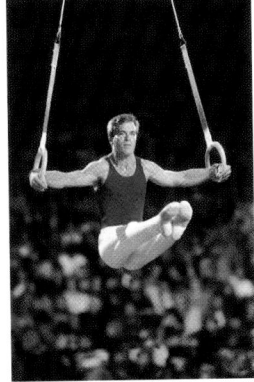

15. The speed limit is 70 miles per hour. Convert this to kilometers per hour.
16. The speed limit is 40 kilometers per hour. Convert this to miles per hour.
17. Ibuprofen comes in 200 milligram tablets. Convert this to ounces.
18. Vitamin C tablets come in 500 milligram caplets. Convert this to ounces.

19. A stone is a unit in the British customary system. Use the conversion: 14 pounds = 1 stone to check the equivalencies in this 1973 Australian stamp. Is 100 kilograms approximately 15 stone 10 pounds?

20. Convert 5 feet 11 inches to centimeters and check the conversion on this 1973 Australian stamp. Is it correct?

21. You find two soda sizes at the store. One is 12 fluid ounces and the other is 380 milliliters. Which is larger?

22. A punch recipe calls for 2 gallons of pineapple juice. You have 8 liters of pineapple juice. Do you have enough for the recipe?

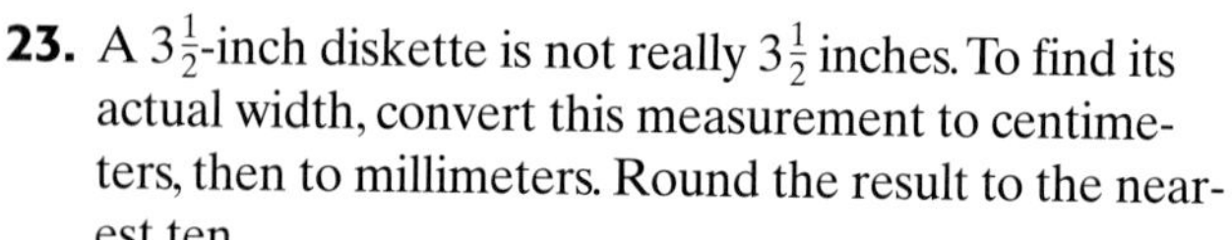

23. A $3\frac{1}{2}$-inch diskette is not really $3\frac{1}{2}$ inches. To find its actual width, convert this measurement to centimeters, then to millimeters. Round the result to the nearest ten.

24. The average two-year-old is 84 centimeters tall. Convert this to feet and inches.

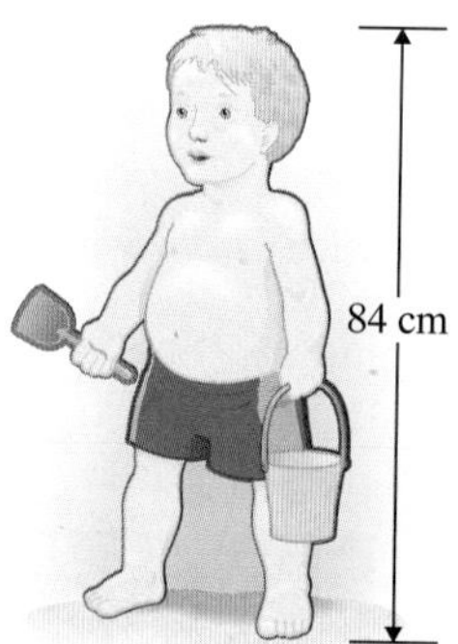

25. For an average adult, the weight of a right lung is greater than the weight of a left lung. If the right lung weighs 1.5 pounds and the left lung weighs 1.25 pounds, find the difference in grams. (*Source*: *Some Body!*)

26. The skin of an average adult weighs 9 pounds and is the heaviest organ. Find the weight in grams. (*Source*: *Some Body!*)

27. A fast sneeze has been clocked at about 167 kilometers per hour. Convert this to miles per hour. Round to the nearest whole hour.

28. The Boeing 747 has a cruising speed of about 980 kilometers per hour. Convert this to miles per hour. Round to the nearest whole hour.

29. The General Sherman giant sequoia tree has a diameter of about 8 meters at its base. Convert this to feet. (*Source*: *Fantastic Book of Comparisions*)

30. The largest crater on the near side of the moon is Billy Crater. It has a diameter of 303 kilometers, Convert this to miles. (*Source*: *Fantastic Book of Comparisions*)

31. The total length of the track on a CD is about 4.5 kilometers. Convert this to miles. Round to the nearest whole mile.

32. The distance between Mackinaw City, Michigan and Cheyenne, Wyoming is 2079 kilometers. Convert this to miles. Round to the nearest whole mile.

33. A doctor orders a dosage of 5 ml of medicine every 4 hours for 1 week. How many fluid ounces of medicine should be purchased? Round up to the next whole fluid ounce.

34. A doctor orders a dosage of 12 ml of medicine every 6 hours for 10 days. How many fluid ounces of medicine should be purchased? Round up to the next whole fluid ounce.

Without actually converting, choose the most reasonable answer.

35. A twin mattress has a width of about _____.

A. 1 m B. 100 m C. 10 m D. 1000 m

36. A pie plate has a diameter of about _____.

A. 22 m B. 22 km C. 22 cm D. 22 g

37. A liter has _____ capacity than a quart.

A. less B. greater C. the same

38. A foot is _____ a meter.

A. shorter than B. longer than C. the same length as

39. A kilogram weighs _____ a pound.

A. the same as B. less than C. greater than

40. A football field is 100 yards, which is about _____.

A. 9 m B. 90 m C. 900 m D. 9000 m

41. An $8\frac{1}{2}$-ounce glass of water has a capacity of about _____.

A. 250 L B. 25 L C. 2.5 L D. 250 ml

42. A 5-gallon gasoline can has a capacity of about _____.

A. 19 L B. 1.9 L C. 19 ml D. 1.9 ml

43. The weight of an average man is about _____.

A. 700 kg B. 7 kg C. 0.7 kg D. 70 kg

44. The weight of a pill is about _____.

A. 200 kg B. 20 kg C. 2 kg D. 200 mg

B *Convert as indicated. When necessary, round to the nearest tenth of a degree. See Examples 4 through 7.*

45. 41°F to degrees Celsius

46. 68°F to degrees Celsius

47. 104°F to degrees Celsius

48. 86°F to degrees Celsius

49. 60°C to degrees Fahrenheit

50. 80°C to degrees Fahrenheit

51. 115°C to degrees Fahrenheit

52. 35°C to degrees Fahrenheit

53. 62°F to degrees Celsius

54. 182°F to degrees Celsius

55. 142.1°F to degrees Celsius

56. 43.4°F to degrees Celsius

57. 92°C to degrees Fahrenheit

58. 75°C to degrees Fahrenheit

59. 16.3°C to degrees Fahrenheit

60. 48.6°C to degrees Fahrenheit

61. The hottest temperature ever recorded in New Mexico was 122°F. Convert this temperature to degrees Celsius. (*Source*: National Climatic Data Center)

62. The hottest temperature ever recorded in Rhode Island was 104°F. Convert this temperature to degrees Celsius. (*Source*: National Climatic Data Center)

63. A weather forecaster in Caracas predicts a high temperature of 27°C. Find this measurement in degrees Fahrenheit.

64. While driving to work, Alan Olda notices a temperature of 18°C flash on the local bank's temperature display. Find the corresponding temperature in degrees Fahrenheit.

65. At Mack Trucks' headquarters, the room temperature is to be set at 70°F, but the thermostat is calibrated in degrees Celsius. Find the temperature to be set.

66. The computer room at Merck, Sharp, and Dohm is normally cooled to 66°F. Find the corresponding temperature in degrees Celsius.

67. Najib Tan is running a fever of 100.2°F. Find his temperature as it would be shown on a Celsius thermometer.

68. William Saylor generally has a temperature of 98.2°F. Find what this temperature would be on a Celsius thermometer.

69. In a European cookbook, a recipe requires the ingredients for caramels to be heated to 118°C, but the cook has access only to a Fahrenheit thermometer. Find the temperature in degrees Fahrenheit that should be used to make the caramels.

70. The ingredients for divinity should be heated to 127°C, but the candy thermometer that Myung Kim has is calibrated to degrees Fahrenheit. Find how hot he should heat the ingredients.

71. Mark Tabbey's recipe for Yorkshire pudding calls for a 500°F oven. Find the temperature setting he should use with an oven having Celsius controls.

72. The temperature of Earth's core is estimated to be 4000°C. Find the corresponding temperature in degrees Fahrenheit.

73. The surface temperature of Venus can reach 864°F. Find this temperature in degrees Celsius.

74. In your own words, describe how to convert from degrees Celsius to degrees Fahrenheit.

Review and Preview

Perform the indicated operations. See Sections 2.2 through 2.5.

75. $-6 \cdot 4 + 5 \div (-1)$

76. $-10 \div (-2) + 9(-8)$

77. $\dfrac{-10 + 8}{-10 - 8}$

78. $\dfrac{-14 + (-1)}{-5(-3)}$

79. $3 + 5(17 - 19) - 8$

80. $1 + 4(9 - 19) + 5$

81. $3[(-1 + 5) \cdot (6 - 8)]$

82. $-5[9 - (18 - 8)]$

Combining Concepts

Body surface area (BSA) is often used to calculate dosages for some drugs. BSA is calculated in square meters using a person's weight and height.

$$\text{BSA} = \sqrt{\frac{(\text{weight in kg}) \times (\text{height in cm})}{3600}}$$

Calculate the BSA for each person. Round to the nearest hundredth.

83. An adult whose height is 182 cm and weight is 90 kg.

84. An adult whose height is 157 cm and weight is 63 kg.

85. A child whose height is 40 in. and weight is 50 kg. (Hint: Don't forget to first convert inches to centimeters)

86. A child whose height is 26 in. and weight is 13 kg.

87. An adult whose height is 60 in. and weight is 150 lb.

88. An adult whose height is 69 in. and weight is 172 lb.

89. Suppose the adult from Exercise 83 is to receive a drug that has a recommended dosage range of 10-12 mg per sq meter. Find the dosage range for the adult.

90. Suppose the child from Exercise 86 is to receive a drug that has a recommended dosage of 30 mg per sq meter. Find the dosage for the Child.

91. A handball court is a rectangle that measures 20 meters by 40 meters. Find its area in square meters and square feet.

92. A backpack measures 16 inches by 13 inches by 5 inches. Find the volume of a box with these dimensions. Find the volume in cubic inches and cubic centimeters. Round the cubic centimeters to the nearest whole cubic centimeter.

93. On February 17, 1995, in the Tokamak Fusion Test Reactor at Princeton University, the highest temperature produced in a laboratory was achieved. This temperature was 918,000,000°F. Convert this temperature to degrees Celsius. Round your answer to the nearest ten million of a degree. (*Source*: *The Guinness Book of Records*)

94. The hottest-burning substance known is carbon subnitride. Its flame at one atmospheric pressure reaches 9010°F. Convert this temperature to degrees Celsius. (*Source*: *The Guinness Book of Records*)

8.8 Congruent and Similar Triangles

OBJECTIVES

- **A** Decide whether two triangles are congruent.
- **B** Find the ratio of corresponding sides in similar triangles.
- **C** Find unknown lengths of sides in similar triangles.
- **D** Solve problems containing similar triangles.

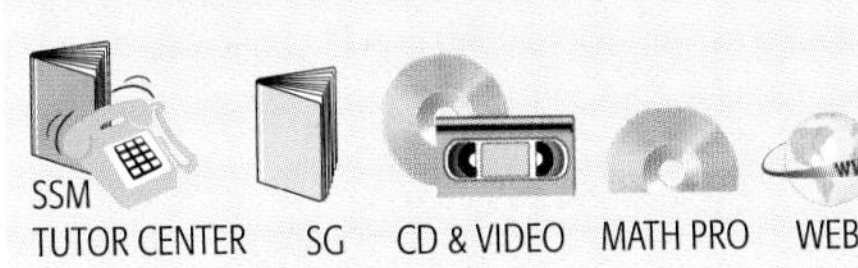

A Deciding Whether Two Triangles Are Congruent

Two triangles are **congruent** when they have the same shape and the same size. The triangles below are congruent. Angles A and D have the same measure, so they are called corresponding angles and they are marked the same way. Both have an arc with 1 tic mark. Angles C and F are corresponding angles. They have the same measure and are both marked by an arc with 2 tic marks. Finally, angles B and E are corresponding angles. They have the same measure and are both marked by an arc. In congruent triangles, the measures of corresponding angles are **equal** and the lengths of corresponding sides are **equal.**

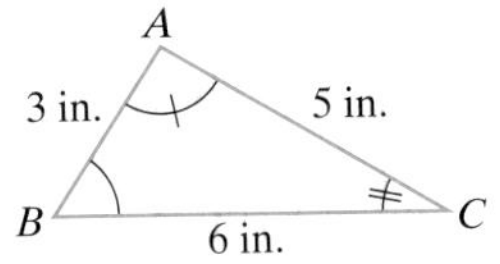

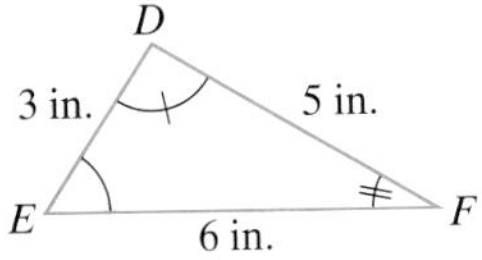

Now let's talk about corresponding sides. Sides AB and DE (denoted $\overline{AB}$ and $\overline{DE}$) are corresponding because they lie between corresponding angles. For the same reason, $\overline{AC}$ and $\overline{DF}$ are corresponding sides, and finally $\overline{BC}$ and $\overline{EF}$ are also corresponding sides.

Angles with equal measure: $\angle A$ and $\angle D$, $\angle B$ and $\angle E$, $\angle C$ and $\angle F$
Sides with equal length: $\overline{AB}$ and $\overline{DE}$, $\overline{BC}$ and $\overline{EF}$, $\overline{AC}$ and $\overline{DF}$

Any one of the following may be used to determine whether two triangles are congruent:

Congruent Triangles

Angle-Side-Angle (ASA)
If the measures of two angles of a triangle equal the measures of two angles of another triangle, and the lengths of the sides between each pair of angles are equal, the triangles are congruent.

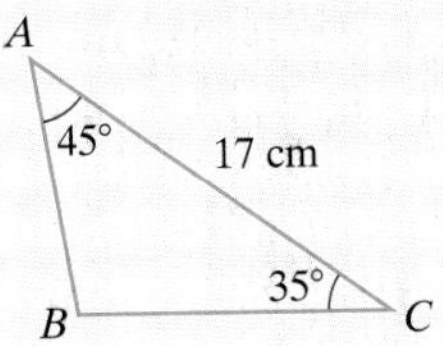

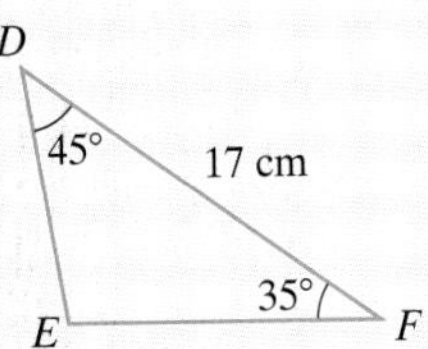

For example, these two triangles are congruent by Angle-Side-Angle.

Side-Side-Side (SSS)
If the lengths of the three sides of a triangle equal the lengths of the corresponding sides of another triangle, the triangles are congruent.

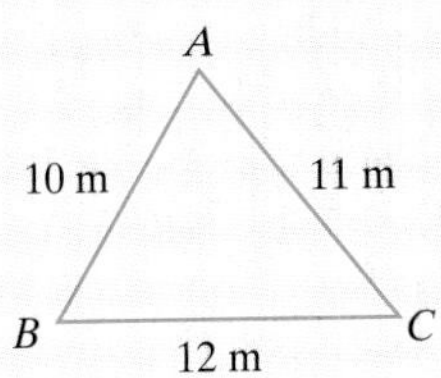

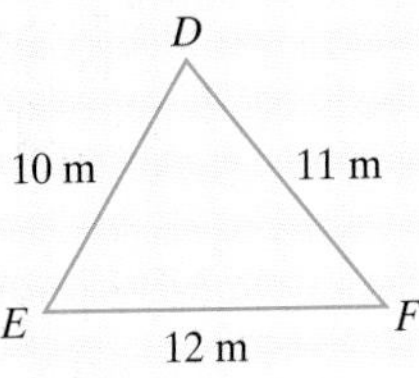

For example, these two triangles are congruent by Side-Side-Side.

Congruent Triangles, *continued*

Side-Angle-Side (SAS)

If the lengths of two sides of a triangle equal the lengths of corresponding sides of another triangle, and the measures of the angles between each pair of sides are equal, the triangles are congruent.

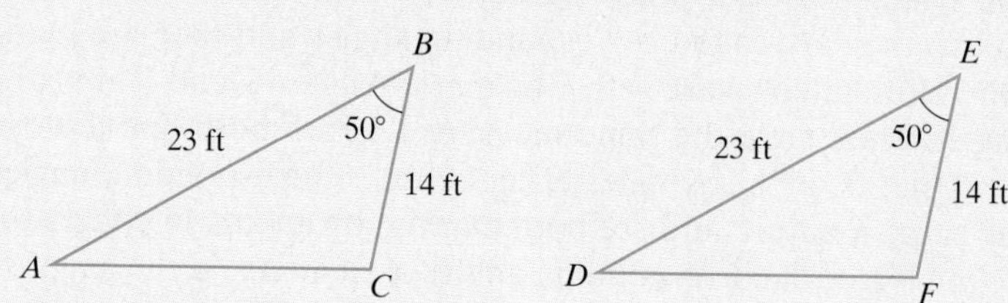

For example, these two triangles are congruent by Side-Angle-Side.

Practice Problem 1

Determine whether triangle MNO is congruent to triangle RQS.

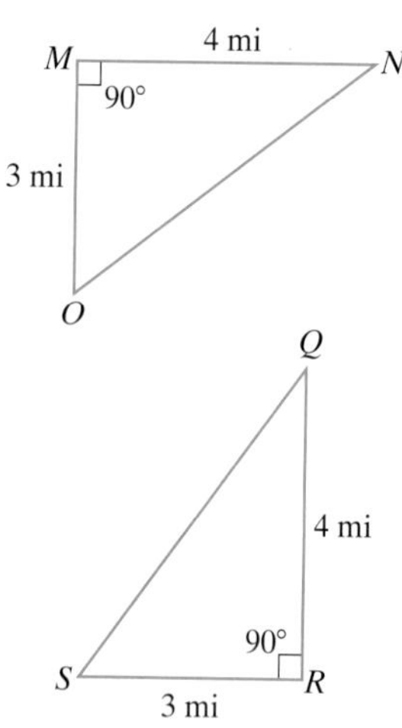

Answer

1. congruent

EXAMPLE 1 Determine whether triangle ABC is congruent to triangle DEF.

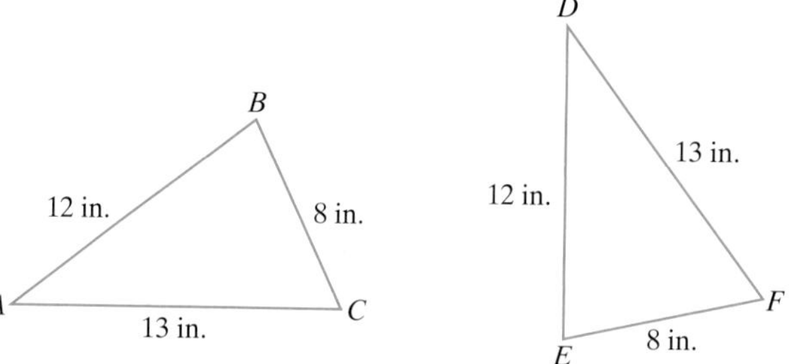

Solution: Since the lengths of all three sides of triangle ABC equal the lengths of all three sides of triangle DEF, the triangles are congruent.

In Example 1, notice that as soon as we know that the two triangles are congruent, we know that all three corresponding angles also have equal measure.

Angles with equal measure: B and E, C and F, A and D

B Finding the Ratios of Corresponding Sides in Similar Triangles

Two triangles are **similar** when they have the same shape but not necessarily the same size. In similar triangles, the measures of corresponding angles are **equal** and corresponding sides are **in proportion.** The following triangles are similar:

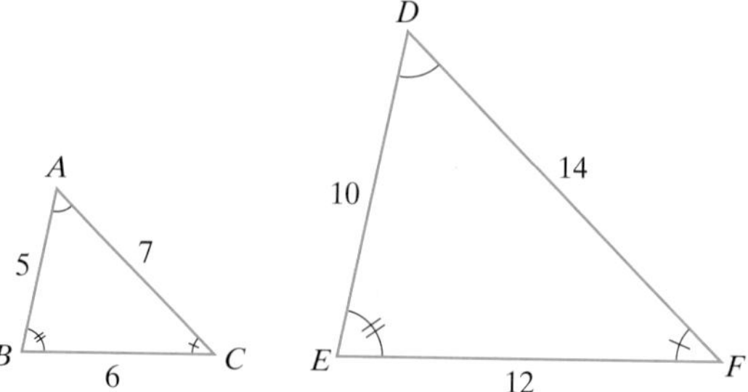

Since these triangles are similar, the measures of corresponding angles are equal.

Angles with equal measure: $\angle A$ and $\angle D$, $\angle B$ and $\angle E$, $\angle C$ and $\angle F$

Also, the lengths of corresponding sides are in proportion. This means that the ratios of the lengths of corresponding sides are equal.

Sides with lengths in proportion:

$$\frac{AB}{DE} = \frac{5}{10} = \frac{1}{2}, \frac{BC}{EF} = \frac{6}{12} = \frac{1}{2}, \frac{CA}{FD} = \frac{7}{14} = \frac{1}{2}$$

The ratio of corresponding sides is $\frac{1}{2}$.

EXAMPLE 2 Find the ratio of corresponding sides $\overline{AB}$ and $\overline{DE}$ for the similar triangles ABC and DEF.

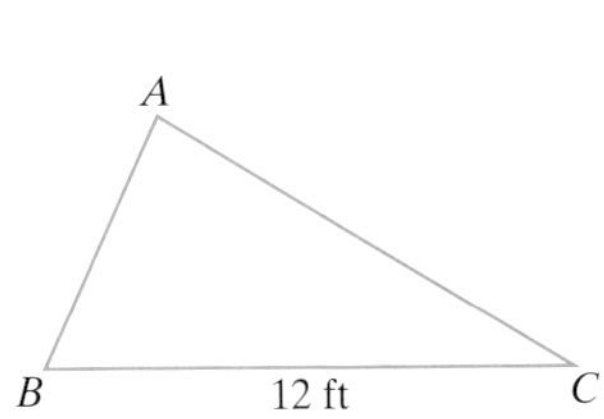

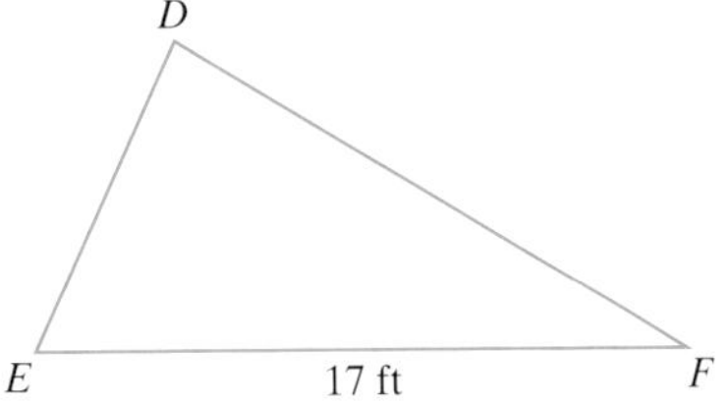

Solution: We are given the lengths of two corresponding sides. Their ratio is

$$\frac{12 \text{ feet}}{17 \text{ feet}} = \frac{12}{17}$$

C Finding Unknown Lengths of Sides in Similar Triangles

Because the ratios of lengths of corresponding sides are equal, we can use proportions to find unknown lengths in similar triangles.

Try the Concept Check in the margin.

EXAMPLE 3 Given that the triangles are similar, find the unknown length n.

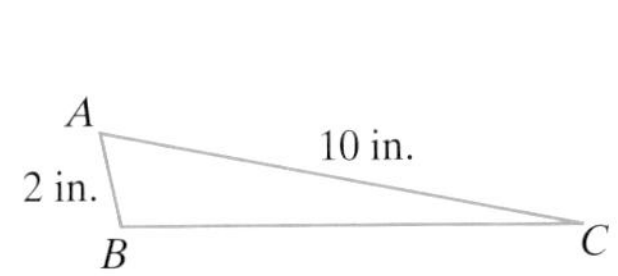

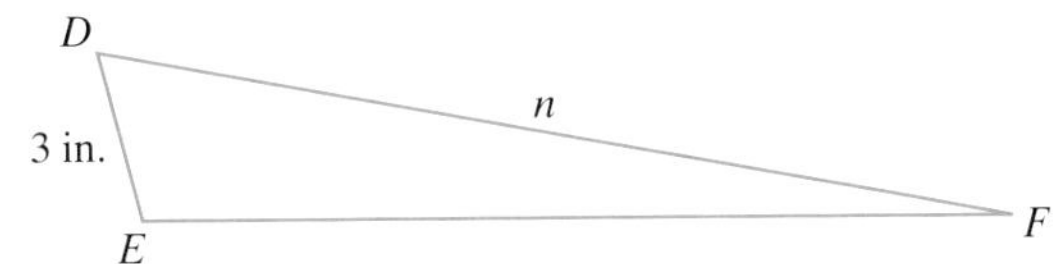

Solution: Since the triangles are similar, corresponding sides are in proportion. Thus, the ratio of 2 to 3 is the same as the ratio of 10 to n. If we write this in symbols, we have

$$\frac{2}{3} = \frac{10}{n}$$

To find the unknown length n, we set cross products equal.

$$\frac{2}{3} = \frac{10}{n}$$

$$30 = 2n \quad \text{Set cross products equal.}$$

$$\frac{30}{2} = \frac{2n}{2} \quad \text{Divide both sides by 2.}$$

$$15 = n$$

The unknown length n is 15 inches.

Practice Problem 2

Find the ratio of corresponding sides $\overline{QR}$ and $\overline{XY}$ for the similar triangles QRS and XYZ.

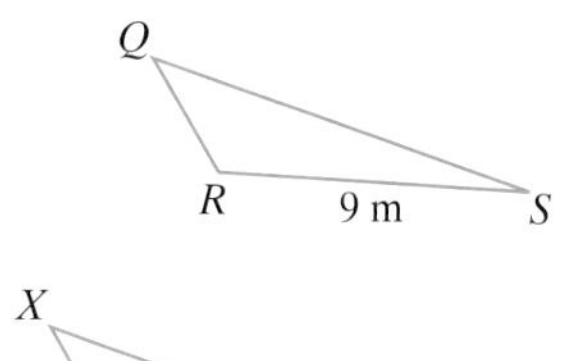

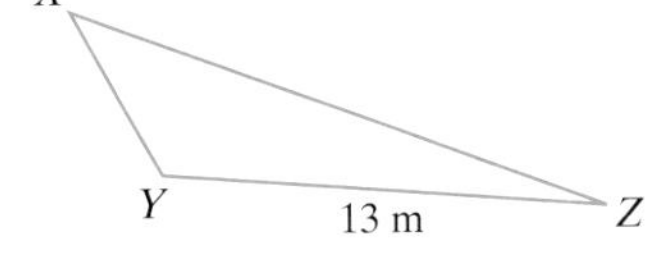

Practice Problem 3

Given that the triangles are similar, find the unknown length n.

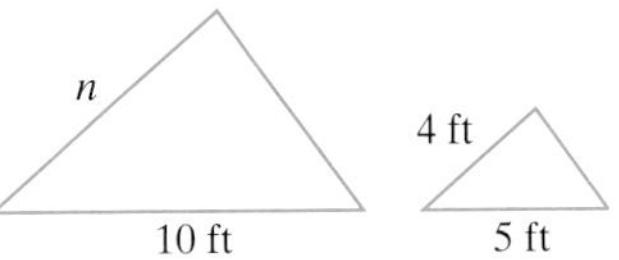

Concept Check

The following two triangles are similar. Which vertices of the first triangle appear to correspond to which vertices of the second triangle?

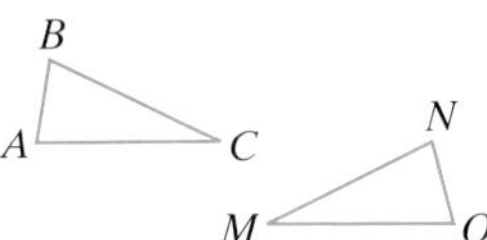

Answers

2. $\frac{9}{13}$ **3.** $n = 8$ ft

Concept Check: A corresponds to O; B corresponds to N; C corresponds to M (or A corresponds to N; B corresponds to O).

D Solving Problems Containing Similar Triangles

Many applications involve a diagram containing similar triangles. Surveyors, astronomers, and many other professionals use ratios of similar triangles frequently in their work.

EXAMPLE 4 Finding the Height of a Tree

Mel Rose is a 6-foot-tall park ranger who needs to know the height of a particular tree. He notices that when the shadow of the tree is 69 feet long, his own shadow is 9 feet long. Find the height of the tree.

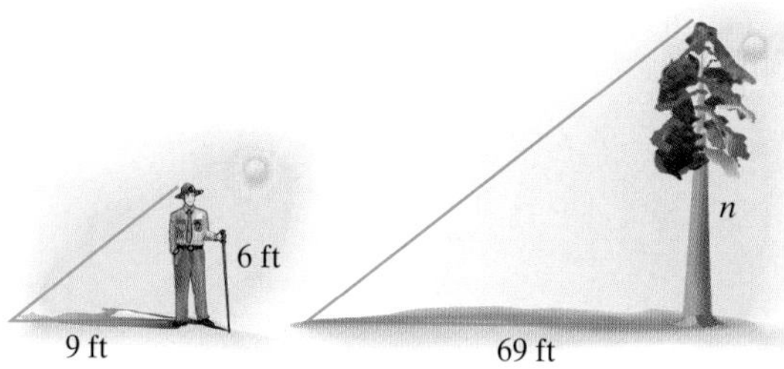

Solution:

1. UNDERSTAND. Read and reread the problem. Notice that the triangle formed by the sun's rays, Mel, and his shadow is similar to the triangle formed by the sun's rays, the tree, and its shadow.
2. TRANSLATE. Write a proportion from the similar triangles formed.

$$\begin{array}{r} \text{Mel's height} \rightarrow \\ \text{height of tree} \rightarrow \end{array} \frac{6}{n} = \frac{9}{69} \begin{array}{l} \leftarrow \text{length of Mel's shadow} \\ \leftarrow \text{length of tree's shadow} \end{array}$$

$$\text{or} \quad \frac{6}{n} = \frac{3}{23} \quad \text{Write in simplest form.}$$

3. SOLVE for n.

$$\frac{6}{n} = \frac{3}{23}$$

$$3n = 138 \quad \text{Set cross products equal.}$$

$$\frac{3n}{3} = \frac{138}{3} \quad \text{Divide both sides by 3.}$$

$$n = 46$$

4. INTERPRET. *Check* to see that replacing n with 46 in the proportion makes the proportion true. *State* your conclusion: The height of the tree is 46 feet.

Practice Problem 4

Tammy Shultz, a firefighter, needs to estimate the height of a burning building. She estimates the length of her shadow to be 8 feet long and the length of the building's shadow to be 200 feet long. Find the height of the building if she is 5 feet tall.

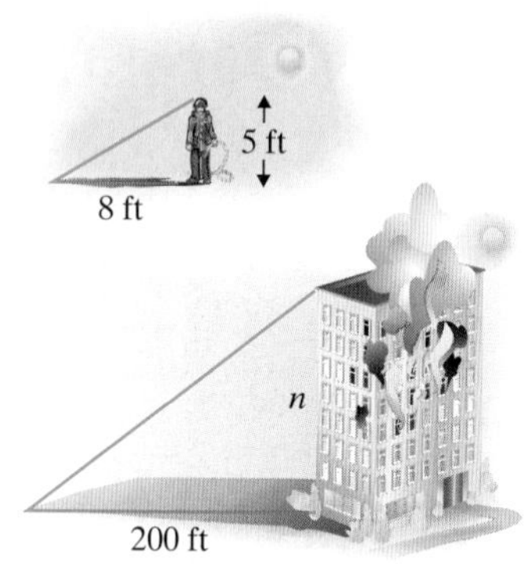

Answer

4. 125 feet

Name ______________________ Section __________ Date __________

EXERCISE SET 8.8

A *Determine whether each pair of triangles is congruent. If congruent, state the reason why, such as SSS, SAS, or ASA. See Example 1.*

1.

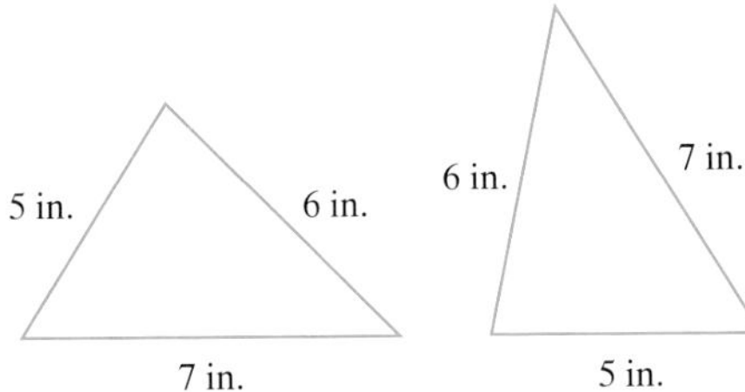

2.

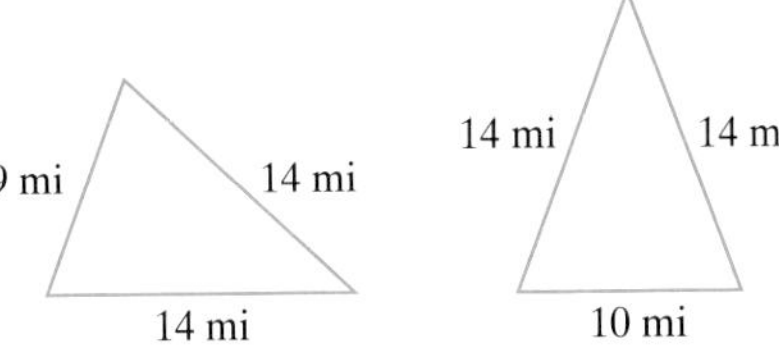

3.

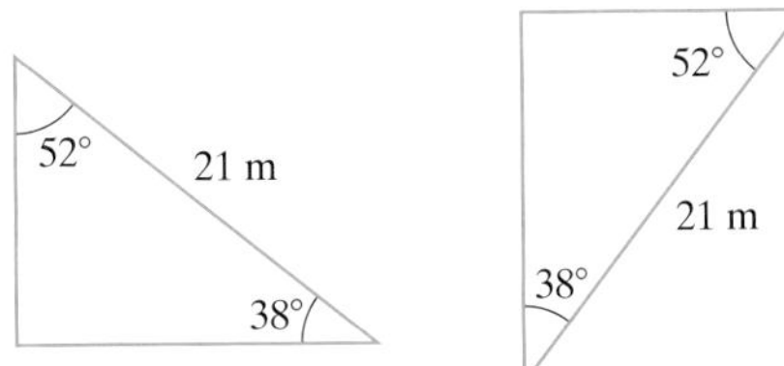

4.

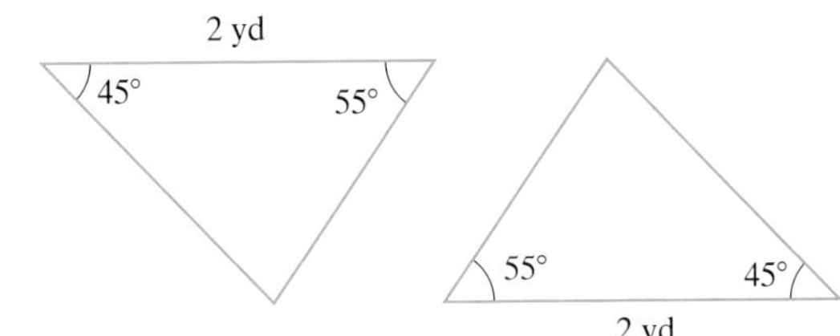

B *Find the ratio of the corresponding sides of the given similar triangles. See Example 2.*

5.

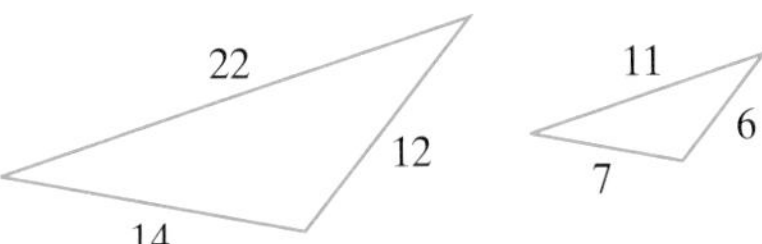

6.

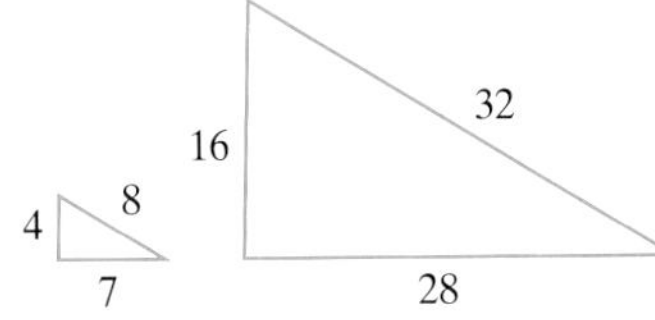

7.

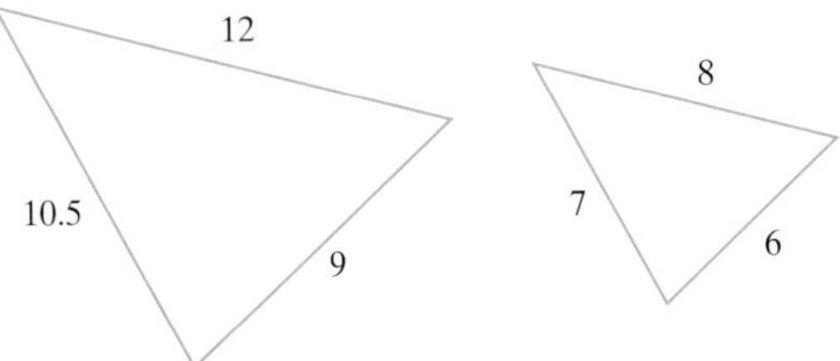

8.

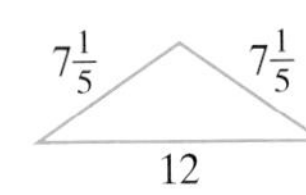

C *Given that the pairs of triangles are similar, find the length of the side labeled n. See Example 3.*

9.

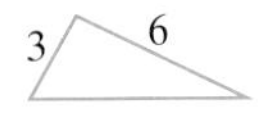

10.

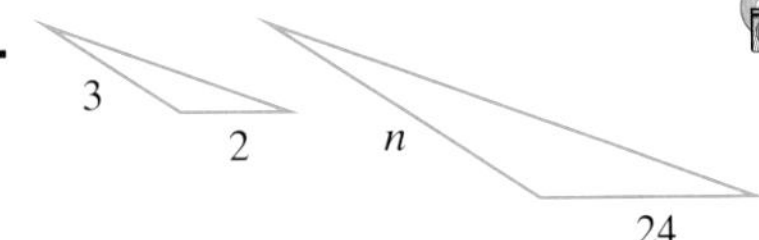

11.

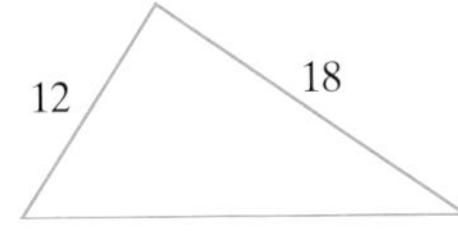

12. 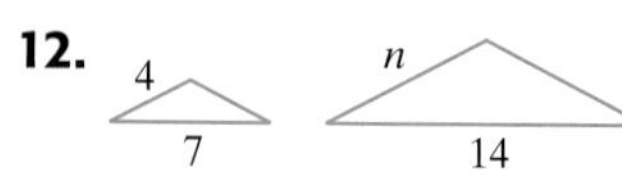

13. 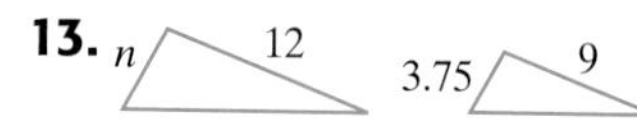

14. 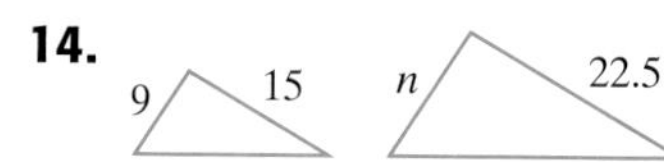

15. 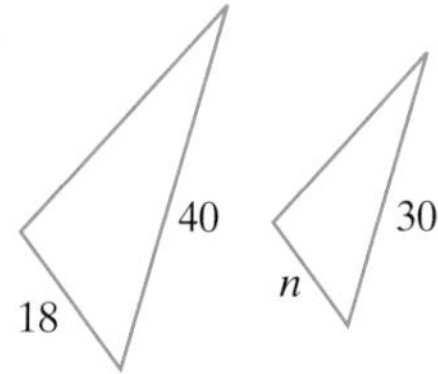

16. 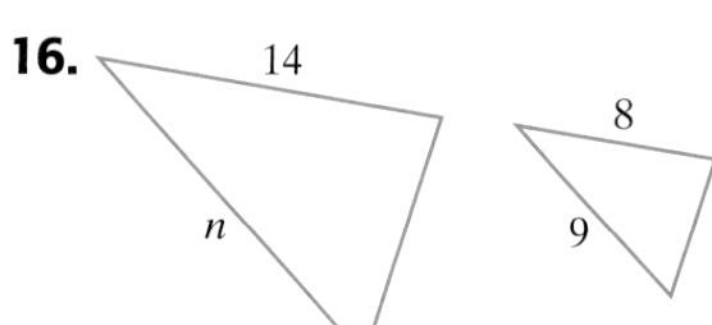

17. 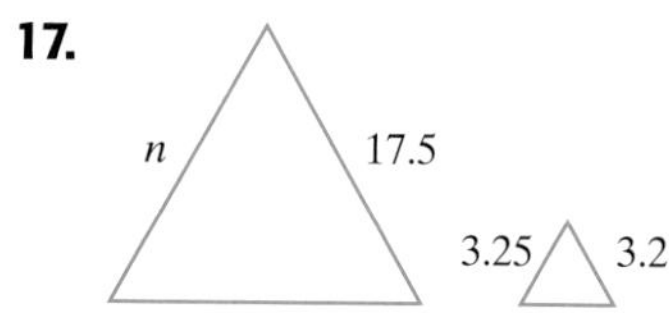

18. 21.6 n 7.2 9.6

19. 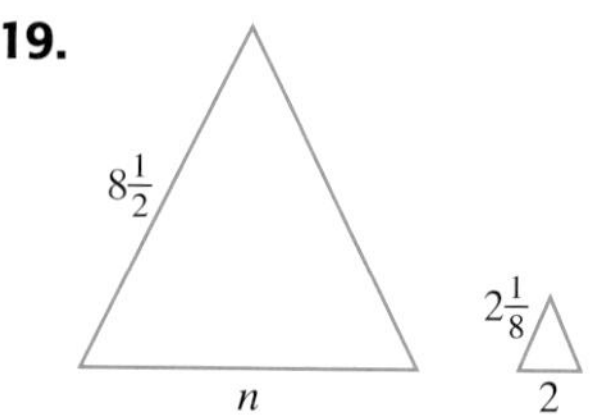

20. 9 n 6 9

21. 16 34 10 n

22. 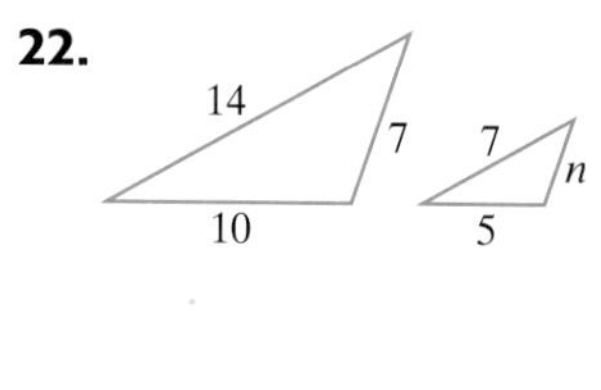

D *Solve. See Example 4.*

23. Given the following diagram, approximate the height of the First National Center in Oklahoma City, OK. (*Source: The World Almanac,* 2003)

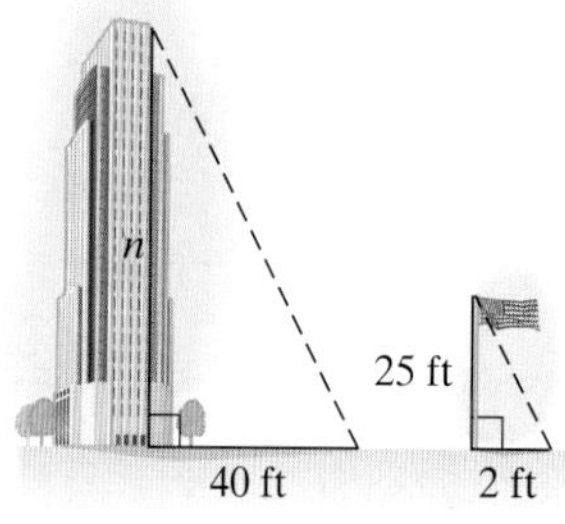
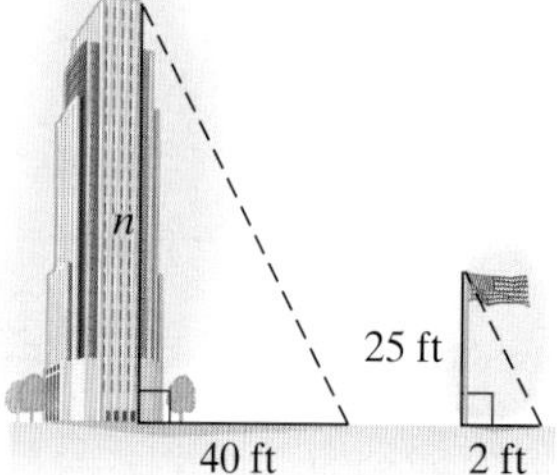

24. The tallest tree standing today is a redwood located near Ukiah, California. Given the following diagram, approximate its height. (*Source: Guinness World Record,* 2003)

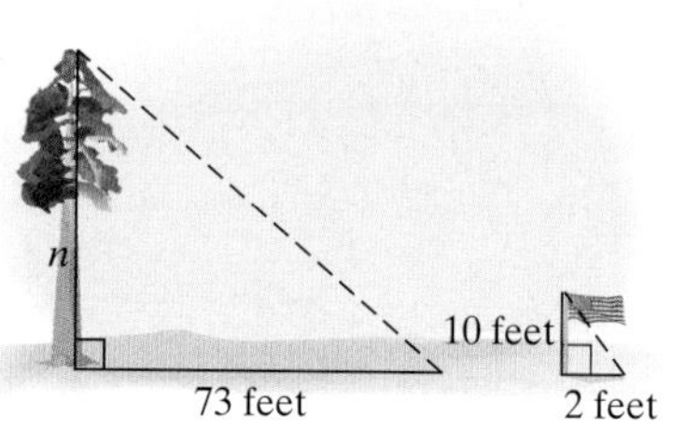

25. Samantha Black, a 5-foot-tall park ranger, needs to know the height of a tree. She notices that when the shadow of the tree is 48 feet long, her shadow is 4 feet long. Find the height of the tree.

26. Lloyd White, a firefighter, needs to estimate the height of a burning building. He estimates the length of his shadow to be 9 feet long and the length of the building's shadow to be 75 feet long. Find the approximate height of the building if he is 6 feet tall.

27. If a 30-foot tree casts an 18-foot shadow, find the length of the shadow cast by a 24-foot tree.

28. If a 24-foot flagpole casts a 32-foot shadow, find the length of the shadow cast by a 44-foot antenna. Round to the nearest tenth.

Review and Preview

Find the average of each list of numbers. See Section 1.7.

29. 14, 17, 21, 18

30. 87, 84, 93

31. 76, 79, 88

32. 7, 8, 4, 6, 3, 8

Combining Concepts

Given that the pairs of triangles are similar, find the length of the side labeled n. Round your results to 1 decimal place.

33.

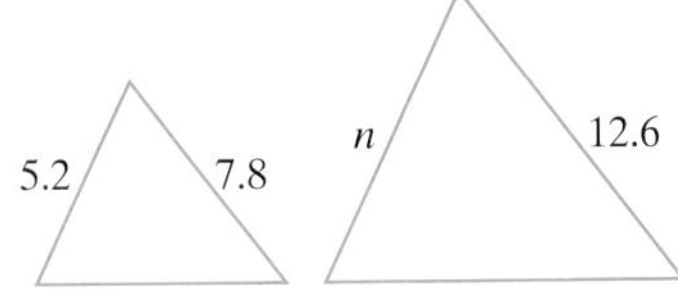

34.

35. In your own words, describe any differences in similar triangles and congruent triangles.

36. The print on a particular page measures 7 inches by 9 inches where 7 inches is the width. A printing shop is to copy the page and reduce the print so that its length is 5 inches. What will its width be? Will the print now fit on a 3 inch by 5 inch index card?

37. Ben and Joyce Lander draw a rectangular deck on their house plans. Joyce measures the deck drawing on the plans to be 3 inches by $4\frac{1}{2}$ inches. If the scale on the drawing is $\frac{1}{4}$ in. = 1 foot, find the dimensions of the deck they want built.

CHAPTER 8 ACTIVITY Map Reading

MATERIALS:

- ruler
- string
- calculator

This activity may be completed by working in groups or individually.

Investigate the route you would take from Santa Rosa, New Mexico, to San Antonio, New Mexico. Use the map in the figure to answer the following questions. You may find that using string to match the roads on the map is useful when measuring distances.

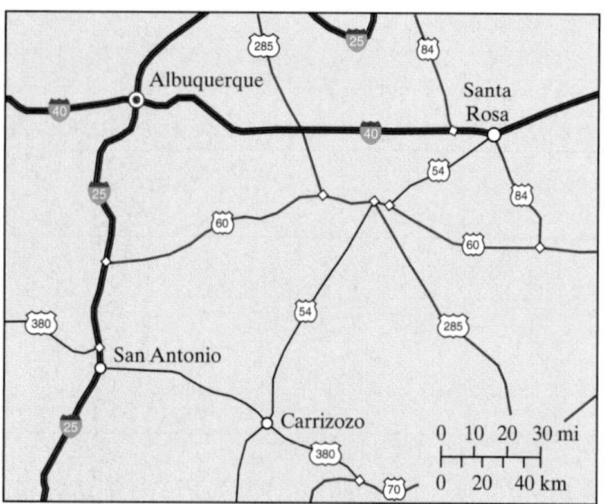

1. How many miles is it from Santa Rosa to San Antonio via Interstate 40 and Interstate 25? Convert this distance to kilometers.

2. How many miles is it from Santa Rosa to San Antonio via U.S. 54 and U.S. 380? Convert this distance to kilometers.

3. Assume that the speed limit on Interstates 40 and 25 is 65 miles per hour. How long would the trip take if you took this route and traveled 65 miles per hour the entire trip?

4. At what average speed would you have to travel on the U.S. routes to make the trip from Santa Rosa to San Antonio in the same amount of time that it would take on the interstate routes? Do you think this speed is reasonable on this route? Explain your reasoning.

5. Discuss in general the factors that might affect your decision among the different routes.

6. Explain which route you would choose in this case and why.

STUDY SKILLS REMINDER

How are your homework assignments going?

By now, you should have good homework habits. If not, it's never too late to begin. Why is it so important in mathematics to keep up with homework? You probably now know the answer to that question. You have probably realized by now that many concepts in mathematics build on each other. Your understanding of one chapter in mathematics usually depends on your understanding of the previous chapter's material.

Don't forget that completing your homework assignment involves a lot more than attempting a few of the problems assigned.

To complete a homework assignment, remember these four things:

1. Attempt all of it.
2. Check it.
3. Correct it.
4. If needed, ask questions about it.

Chapter 8 Vocabulary Check

Fill in each blank with one of the words or phrases listed below.

transversal	line segment	obtuse	straight	adjacent	right	volume	area
acute	perimeter	vertical	supplementary	ray	angle	line	complementary
vertex	mass	unit fractions	gram	weight	meter	liter	congruent
similar							

1. ______ is a measure of the pull of gravity.

2. ______ is a measure of the amount of substance in an object. This measure does not change.

3. The basic unit of length in the metric system is the ______.

4. To convert from one unit of length to another, ______ may be used.

5. A ______ is the basic unit of mass in the metric system.

6. The ______ is the basic unit of capacity in the metric system.

7. A ______ is a piece of a line with two endpoints.

8. Two angles that have a sum of 90° are called ______ angles.

9. A ______ is a set of points extending indefinitely in two directions.

10. The ______ of a polygon is the distance around the polygon.

11. An ______ is made up of two rays that share the same endpoint. The common endpoint is called the ______.

12. ______ measures the amount of surface of a region.

13. A ______ is a part of a line with one endpoint. A ray extends indefinitely in one direction.

14. A line that intersects two or more lines at different points is called a ______.

15. An angle that measures 180° is called a ______ angle.

16. The measure of the space of a solid is called its ______.

17. When two lines intersect, four angles are formed. Two of these angles that are opposite each other are called ______ angles.

18. Two of the angles from #17 that share a common side are called ______ angles.

19. An angle whose measure is between 90° and 180° is called an ______ angle.

20. An angle that measures 90° is called a ______ angle.

21. An angle whose measure is between 0° and 90° is called an ______ angle.

22. Two angles that have a sum of 180° are called ______ angles.

23. ______ triangles have the same shape and the same size.

24. ______ triangles have exactly the same shape but not necessarily the same size.

CHAPTER 8 Highlights

DEFINITIONS AND CONCEPTS | **EXAMPLES**

SECTION 8.1 LINES AND ANGLES

A **line** is a set of points extending indefinitely in two directions. A line has no width or height, but it does have length. We name a line by any two of its points.

Line AB or $\overleftrightarrow{AB}$

A **line segment** is a piece of a line with two endpoints.

Line segment AB or $\overline{AB}$

Ray AB or $\overrightarrow{AB}$

A **ray** is a part of a line with one endpoint. A ray extends indefinitely in one direction.

An **angle** is made up of two rays that share the same endpoint. The common endpoint is called the **vertex.**

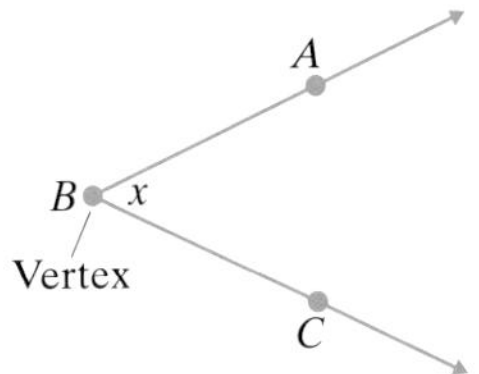

SECTION 8.2 LINEAR MEASUREMENT

To convert from one unit of length to another, unit **fractions** may be used.

The unit fraction should be in the form

$$\frac{\text{units converting to}}{\text{original units}}$$

$$\frac{12 \text{ inches}}{1 \text{ foot}}, \quad \frac{1 \text{ foot}}{12 \text{ inches}}, \quad \frac{3 \text{ feet}}{1 \text{ yard}}$$

Convert 6 feet to inches.

$$6 \text{ feet} = \frac{6 \cancel{\text{feet}}}{1} \cdot \frac{12 \text{ inches}}{1 \cancel{\text{foot}}} \quad \begin{array}{l} \leftarrow \text{units converting to} \\ \leftarrow \text{original units} \end{array}$$

$$= 6 \cdot 12 \text{ inches}$$

$$= 72 \text{ inches}$$

LENGTH: U.S. SYSTEM OF MEASUREMENT

12 inches (in.) = 1 foot (ft)
3 feet = 1 yard (yd)
5280 feet = 1 mile (mi)

The basic unit of length in the metric system is the **meter.** A meter is slightly longer than a yard.

Convert 3650 centimeters to meters.

$$3650 \text{ cm} = \frac{3650 \cancel{\text{cm}}}{1} \cdot \frac{0.01 \text{ m}}{1 \cancel{\text{cm}}} = 36.5 \text{ m}$$

or

km hm dam m (End) dm cm (Start) mm

2 units to the left

36.50 cm = 36.5 m

2 places to the left

LENGTH: METRIC SYSTEM OF MEASUREMENT

Prefix	Meaning	Metric Unit of Length
kilo	1000	1 **kilo**meter (km) = 1000 meters (m)
hecto	100	1 **hecto**meter (hm) = 100 m
deka	10	1 **deka**meter (dam) = 10 m
		1 meter (m) = 1 m
deci	1/10	1 **deci**meter (dm) = 1/10 m or 0.1 m
centi	1/100	1 **centi**meter (cm) = 1/100 m or 0.01 m
milli	1/1000	1 **milli**meter (mm) = 1/1000 m or 0.001 m

Definitions and Concepts — Examples

Section 8.3 Perimeter

Perimeter Formulas

Rectangle: $P = 2l + 2w$

Square: $P = 4s$

Triangle: $P = a + b + c$

Circumference of a Circle: $C = 2\pi r$ or $C = \pi d$

where $\pi \approx 3.14$ or $\pi \approx \frac{22}{7}$

Find the perimeter of the rectangle.

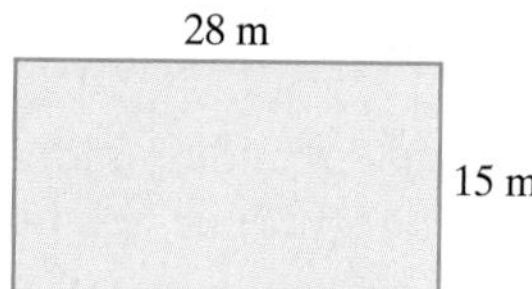

$$\begin{aligned} P &= 2l + 2w \\ &= 2 \cdot 28 \text{ meters} + 2 \cdot 15 \text{ meters} \\ &= 56 \text{ meters} + 30 \text{ meters} \\ &= 86 \text{ meters} \end{aligned}$$

The perimeter is 86 meters.

Section 8.4 Area and Volume

Area Formulas

Rectangle: $A = lw$

Square: $A = s^2$

Triangle: $A = \frac{1}{2}bh$

Parallelogram: $A = bh$

Trapezoid: $A = \frac{1}{2}(b + B)h$

Circle: $A = \pi r^2$

Volume Formulas

Rectangular Solid: $V = lwh$

Cube: $V = s^3$

Sphere: $V = \frac{4}{3}\pi r^3$

Right Circular Cylinder: $V = \pi r^2 h$

Cone: $V = \frac{1}{3}\pi r^2 h$

Square-Based Pyramid: $V = \frac{1}{3}s^2 h$

Find the area of the square.

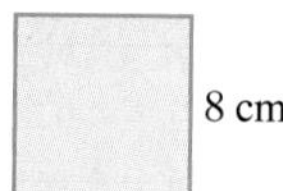

$$\begin{aligned} A &= s^2 \\ &= (8 \text{ centimeters})^2 \\ &= 64 \text{ square centimeters} \end{aligned}$$

The area of the square is 64 square centimeters.

Find the volume of the sphere. Use $\frac{22}{7}$ for π.

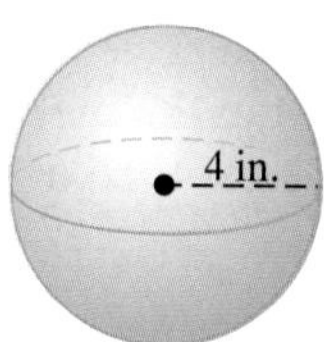

$$\begin{aligned} V &= \frac{4}{3}\pi r^3 \\ &\approx \frac{4}{3} \cdot \frac{22}{7} \cdot (4 \text{ inches})^3 \\ &= \frac{4 \cdot 22 \cdot 64}{3 \cdot 7} \text{ cubic inches} \\ &= \frac{5632}{21} \text{ or } 268\frac{4}{21} \text{ cubic inches} \end{aligned}$$

Definitions and Concepts | Examples

Section 8.5 Weight and Mass

Weight is really a measure of the pull of gravity.

Mass is a measure of the amount of substance in the object and does not change.

Weight: U.S. System of Measurement

16 ounces (oz) = 1 pound (lb)

2000 pounds = 1 ton

Convert 5 pounds to ounces.

$$5\text{ lb} = \frac{5\text{ }\cancel{\text{lb}}}{1} \cdot \frac{16\text{ oz}}{1\text{ }\cancel{\text{lb}}} = 80\text{ oz}$$

A **gram** is the basic unit of mass in the metric system. It is the mass of water contained in a cube 1 centimeter on each side. A large paper clip weighs about 1 gram.

Convert 260 grams to kilograms.

$$260\text{ g} = \frac{260\text{ }\cancel{\text{g}}}{1} \cdot \frac{1\text{ kg}}{1000\text{ }\cancel{\text{g}}} = 0.26\text{ kg}$$

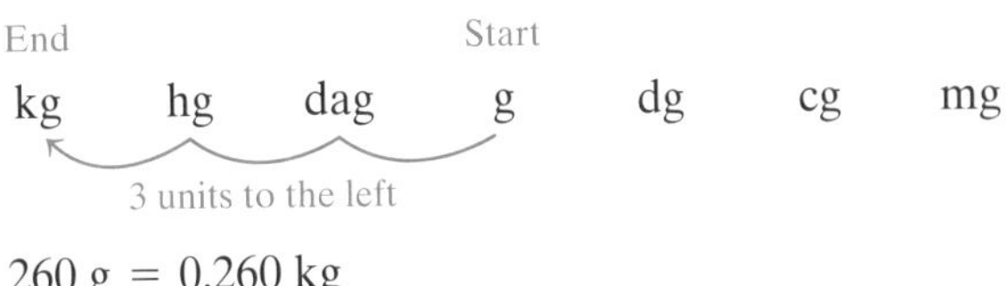

260 g = 0.260 kg

3 places to the left

Mass: Metric System of Measurement

Prefix	Meaning	Metric Unit of Mass
kilo	1000	1 kilogram (kg) = 1000 grams (g)
hecto	100	1 hectogram (hg) = 100 g
deka	10	1 dekagram (dag) = 10 g
		1 gram (g) = 1 g
deci	1/10	1 decigram (dg) = 1/10 g or 0.1 g
centi	1/100	1 centigram (cg) = 1/100 g or 0.01 g
milli	1/1000	1 milligram (mg) = 1/1000 g or 0.001 g

Section 8.6 Capacity

Capacity: U.S. System of Measurement

8 fluid ounces (fl oz) = 1 cup (c)

2 cups = 1 pint (pt)

2 pints = 1 quart (qt)

4 quarts = 1 gallon (gal)

Convert 5 pints to gallons.

1 gal = 4 qt = 8 pt

$$5\text{ pt} = \frac{5\text{ }\cancel{\text{pt}}}{1} \cdot \frac{1\text{ gal}}{8\text{ }\cancel{\text{pt}}} = \frac{5}{8}\text{ gal}$$

The **liter** is the basic unit of capacity in the metric system. It is the capacity or volume of a cube measuring 10 centimeters on each side. A liter of liquid is slightly more than 1 quart.

Convert 1.5 liters to milliliters.

$$1.5\text{ L} = \frac{1.5\text{ }\cancel{\text{L}}}{1} \cdot \frac{1000\text{ ml}}{1\text{ }\cancel{\text{L}}} = 1500\text{ ml}$$

or

Start　　End

kl　hl　dal　L　dl　cl　ml

3 units to the right

1.500 L = 1500 mL

3 places to the right

(continued)

Definitions and Concepts | Examples

Section 8.6 Capacity (continued)

Capacity: Metric System of Measurement

Prefix	Meaning	Metric Unit of Capacity
kilo	1000	1 kiloliter (kl) = 1000 liters (L)
hecto	100	1 hectoliter (hl) = 100 L
deka	10	1 dekaliter (dal) = 10 L
		1 liter (L) = 1 L
deci	1/10	1 deciliter (dl) = 1/10 L or 0.1 L
centi	1/100	1 centiliter (cl) = 1/100 L or 0.01 L
milli	1/1000	1 milliliter (ml) = 1/1000 L or 0.001 L

Section 8.7 Conversions between the U.S. and Metric Systems and Temperature Conversions

To convert between systems, use approximate unit fractions.

Convert 7 feet to meters.

$$7 \text{ ft} \approx \frac{7 \cancel{\text{ft}}}{1} \cdot \frac{0.30 \text{ m}}{1 \cancel{\text{ft}}} = 2.1 \text{ m}$$

Convert 8 liters to quarts.

$$8 \text{ L} \approx \frac{8 \cancel{\text{L}}}{1} \cdot \frac{1.06 \text{ qt}}{1 \cancel{\text{L}}} = 8.48 \text{ qt}$$

Convert 363 grams to ounces.

$$363 \text{ g} \approx \frac{363 \cancel{\text{g}}}{1} \cdot \frac{0.04 \text{ oz}}{1 \cancel{\text{g}}} = 14.52 \text{ oz}$$

Celsius to Fahrenheit

$$F = \frac{9}{5}C + 32 \quad \text{or} \quad F = 1.8C + 32$$

Convert 35°C to degrees Fahrenheit.

$$F = \frac{9}{5} \cdot 35 + 32 = 63 + 32 = 95$$

35°C = 95°F

Fahrenheit to Celsius

$$C = \frac{5}{9}(F - 32)$$

Convert 50°F to degrees Celsius.

$$C = \frac{5}{9} \cdot (50 - 32) = \frac{5}{9} \cdot (18) = 10$$

50°F = 10°C

DEFINITIONS AND CONCEPTS — EXAMPLES

SECTION 8.8 CONGRUENT AND SIMILAR TRIANGLES

Congruent triangles have the same shape and the same size. Corresponding angles are equal, and corresponding sides are equal.

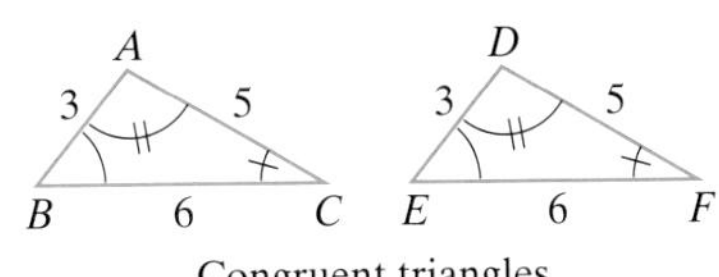

Congruent triangles

Similar triangles have exactly the same shape but not necessarily the same size. Corresponding angles are equal, and the ratios of the lengths of corresponding sides are equal.

Similar triangles

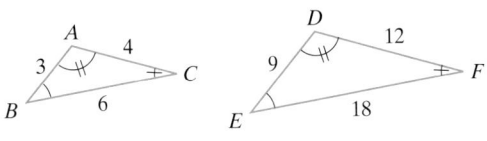

$$\frac{AB}{DE} = \frac{3}{9} = \frac{1}{3}, \quad \frac{BC}{EF} = \frac{6}{18} = \frac{1}{3},$$

$$\frac{AC}{DF} = \frac{4}{12} = \frac{1}{3}$$

STUDY SKILLS REMINDER

Do you remember what to do the day of an exam?

On the day of an exam, don't forget to try the following:

- Allow yourself plenty of time to arrive.
- Read the directions on the test carefully.
- Read each problem carefully as you take your test. Make sure that you answer the question asked.
- Watch your time and pace yourself so that you may attempt each problem on your test.
- If you have time, check your work and answers.
- Do not turn your test in early. If you have extra time, spend it double-checking your work.

Good luck!

Name ______________________ Section __________ Date __________

Chapter 8 Review

(8.1) *Classify each angle as acute, right, obtuse, or straight.*

1.

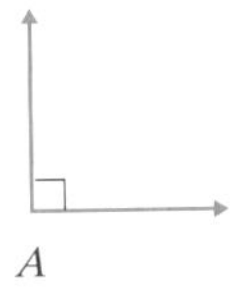

2.

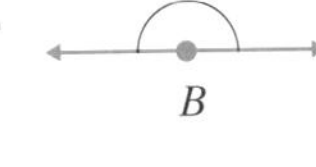

3.

4.

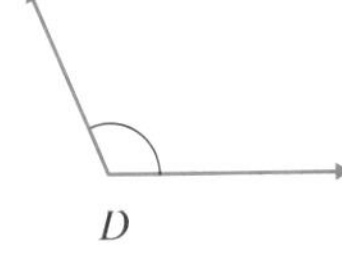

5. Find the complement of a 25° angle.

6. Find the supplement of a 105° angle.

7. Find the supplement of a 72° angle.

8. Find the complement of a 1° angle.

Find the measure of x in each figure.

9.

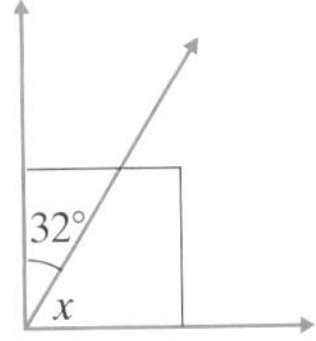

10.

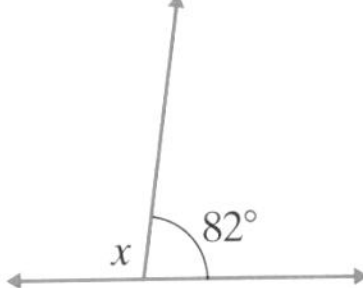

11.

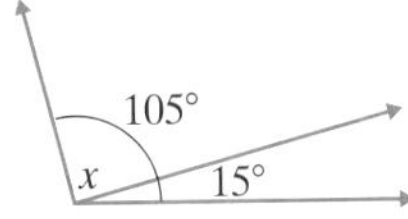

12.

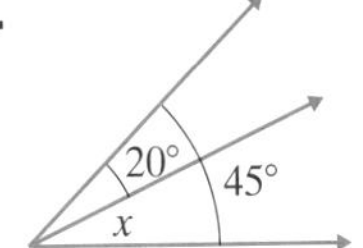

13. Identify the pairs of supplementary angles.

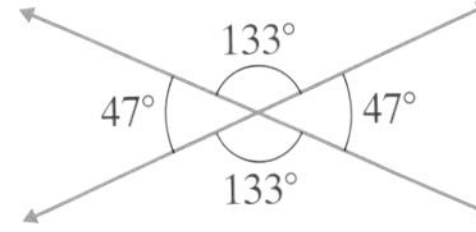

14. Identify the pairs of complementary angles.

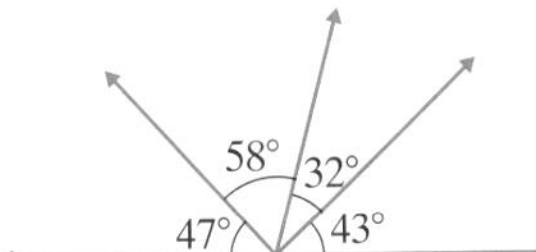

Find the measures of angles x, y, and z in each figure.

15.

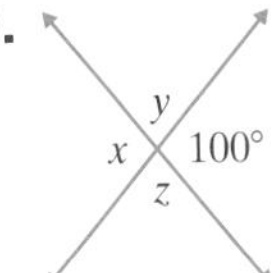

16.

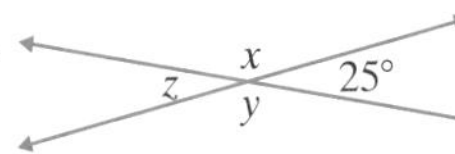

17. $m \parallel n$

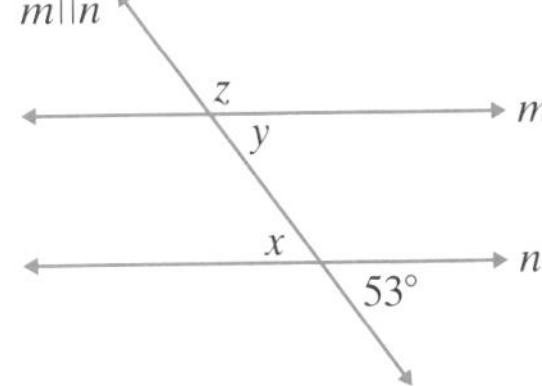

18. $m \parallel n$

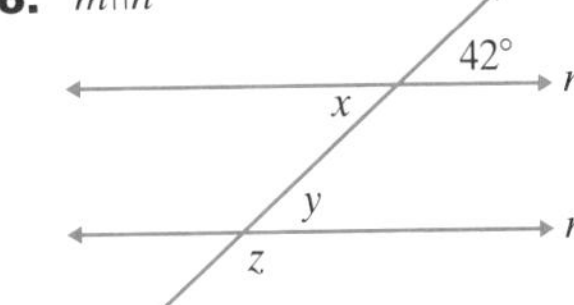

8.2 *Convert.*

19. 108 in. to feet

20. 72 ft to yards

21. 2.5 mi to feet

22. 6.25 ft to inches

23. 52 ft = ____ yd ____ ft

24. 46 in. = ____ ft ____ in.

25. 42 m to centimeters

26. 82 cm to millimeters

27. 12.18 mm to meters

28. 2.31 m to kilometers

Perform each indicated operation.

29. 4 yd 2 ft + 16 yd 2 ft

30. 12 ft 1 in. − 4 ft 8 in.

31. 8 ft 3 in. × 5

32. 7 ft 4 in. ÷ 2

33. 8 cm + 15 mm

34. 4 m + 126 cm

35. 9.3 km − 183 m

36. 4100 mm − 3 m

Solve.

37. A bolt of cloth contains 333 yd 1 ft of cotton ticking. Find the amount of material that remains after 163 yd 2 ft is removed from the bolt.

38. The local ambulance corps plans to award 20 framed certificates of valor to some of its outstanding members. If each frame requires 6 ft 4 in. of framing material, how much material is needed for all the frames?

39. The trip from Philadelphia to Washington, DC, is 217 km. Four friends agree to share the driving equally. How far must each drive on this round-trip vacation?

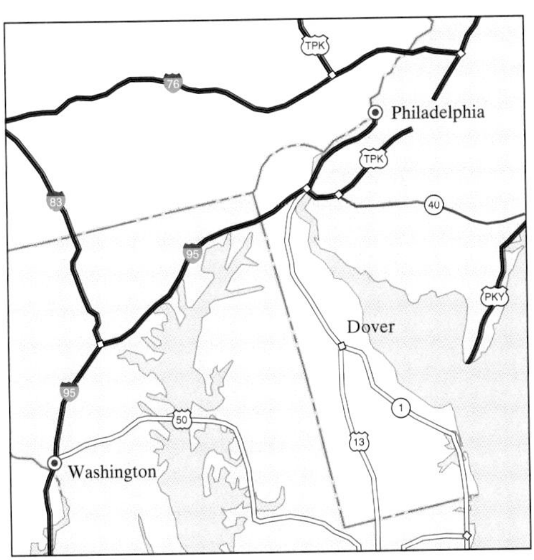

△ **40.** The college has ordered that NO SMOKING signs be placed above the doorway of each classroom. Each sign is 0.8 m long and 30 cm wide. Find the area of each sign. (*Hint*: Recall that the area of a rectangle = length · width.)

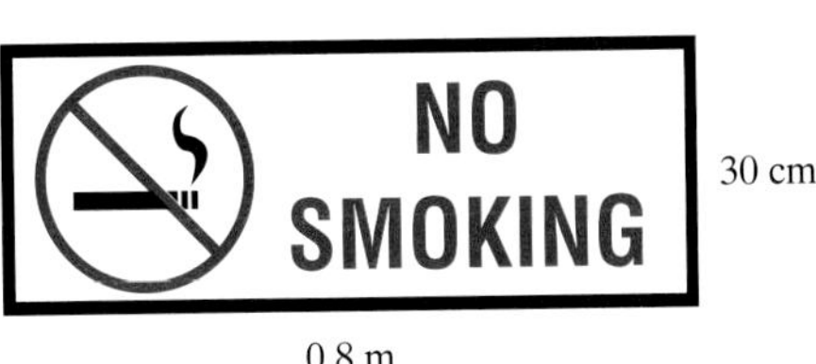

△ **(8.3)** *Find the perimeter of each figure.*

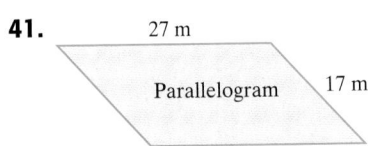

41. 27 m; 17 m; Parallelogram

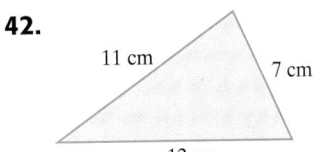

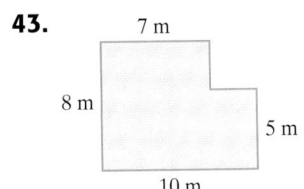

42. 11 cm; 7 cm; 12 cm

43. 7 m; 8 m; 5 m; 10 m

44. 5 ft; 4 ft; 3 ft; 11 ft; 22 ft

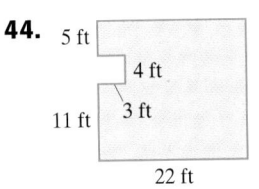

Solve.

45. Find the perimeter of a rectangular sign that measures 6 feet by 10 feet.

46. Find the perimeter of a town square that measures 110 feet on a side.

Find the circumference of each circle. Use $\pi \approx 3.14$.

47.

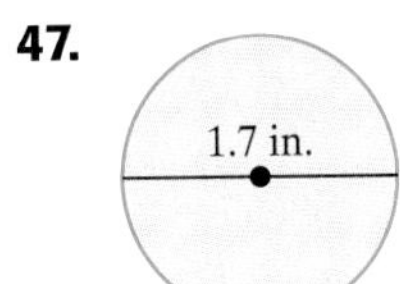

48.

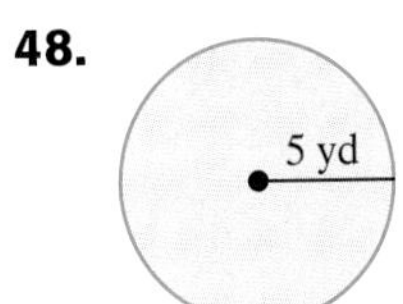

(8.4) *Find the area of each figure. For the circles, find the exact area and then use* $\pi \approx 3.14$ *to approximate the area.*

49.

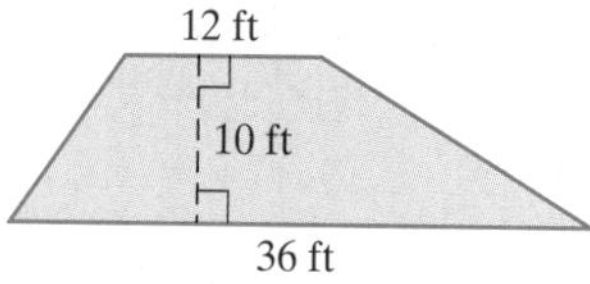

50.

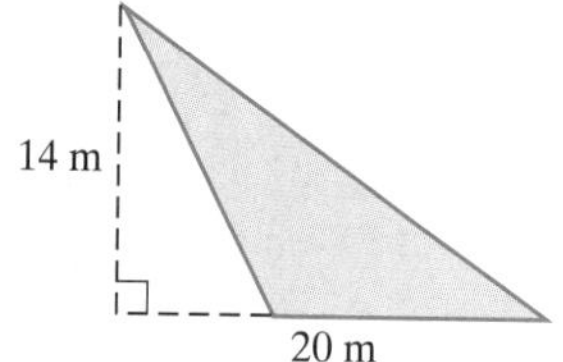

51.

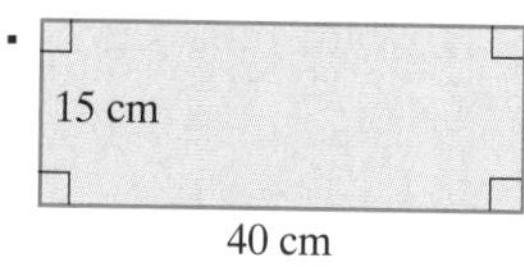

52.

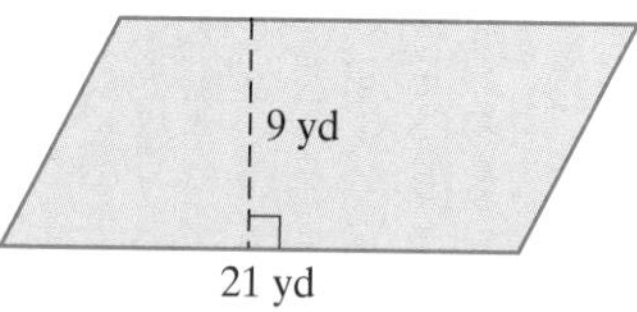

53.

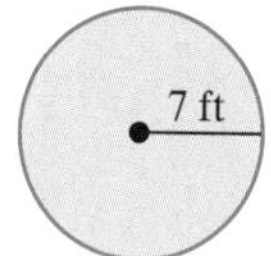

54.

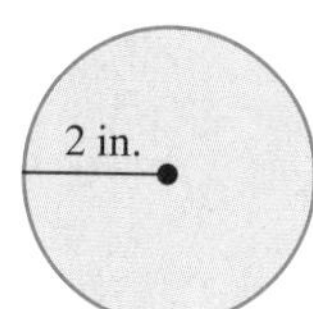

55.

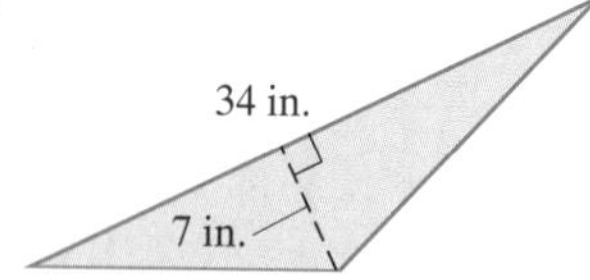

56.

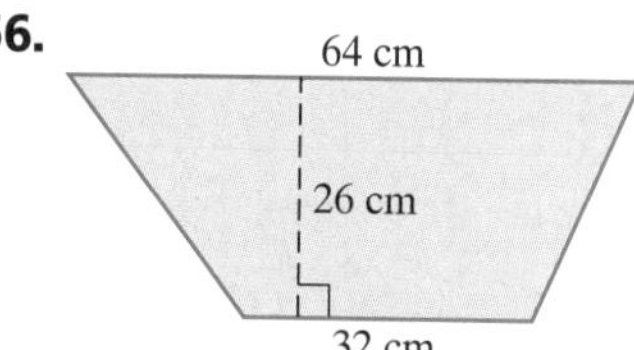

57.

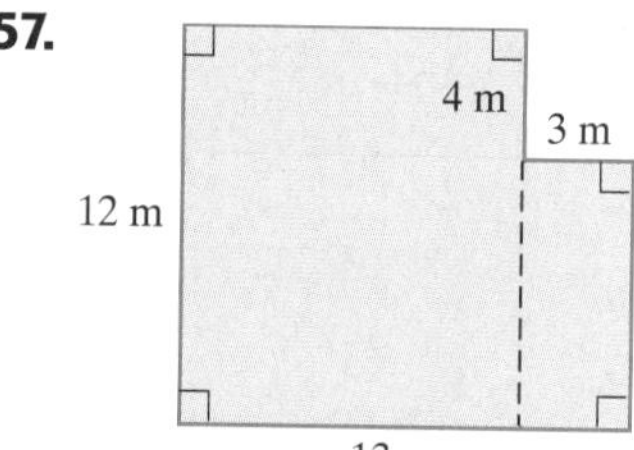

58. The amount of sealer necessary to seal a driveway depends on the area. Find the area of a rectangular driveway 36 feet by 12 feet.

59. Find how much carpet is needed to cover the floor of the room shown.

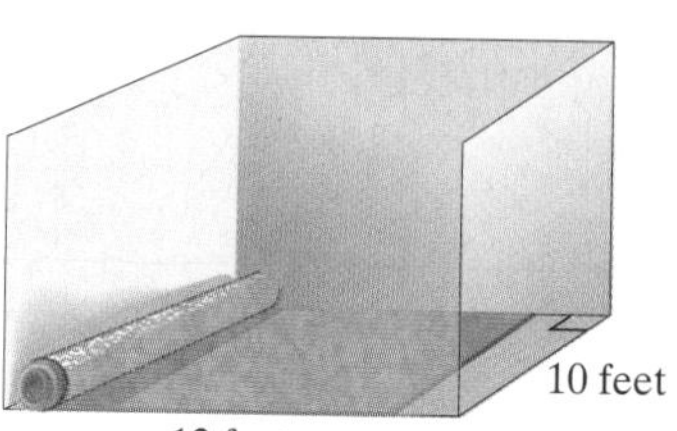

Find the volume of each solid. For Exercises 62 and 63, use $\pi \approx \frac{22}{7}$.

60.

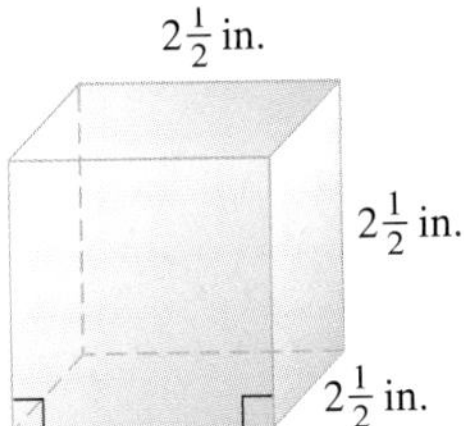

61.

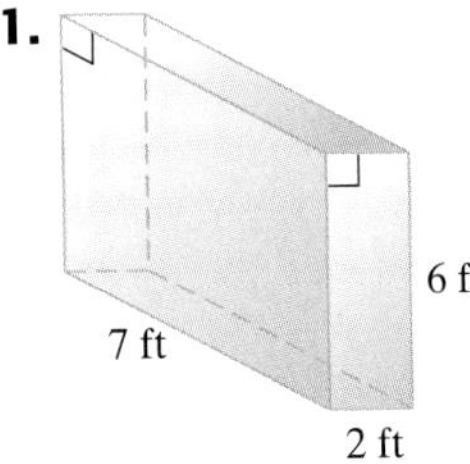

62.

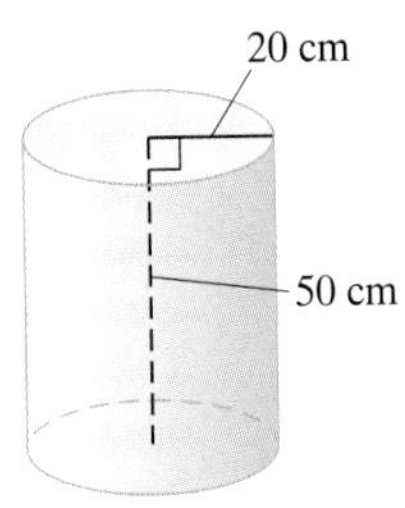

63.

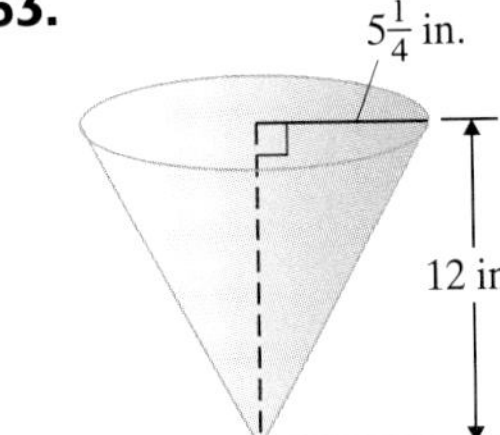

64. Find the volume of a pyramid with a square base 2 feet on a side and a height of 2 feet.

65. Approximate the volume of a tin can 8 inches high and 3.5 inches for the radius. Use $\pi \approx 3.14$

66. A chest has 3 drawers. If each drawer has inside measurements of $2\frac{1}{2}$ feet by $1\frac{1}{2}$ feet by $\frac{2}{3}$ feet, find the total volume of the 3 drawers.

67. A cylindrical canister for a shop vacuum is 2 feet tall and 1 foot for the *diameter*. Find its exact volume.

68. Find the volume of air in a rectangular room 15 feet by 12 feet with a 7-foot ceiling.

69. A mover has two boxes left for packing. Both are cubical, one 3 feet on a side and the other 1.2 feet on a side. Find their combined volume.

(8.5) *Convert.*

70. 66 oz to pounds

71. 2.3 tons to pounds

72. 52 oz = ____ lb ____ oz

73. 8200 lb = ____ tons ____ lb

74. 1400 mg to grams

75. 40 kg to grams

76. 2.1 hg to dekagrams

77. 0.03 mg to decigrams

Perform each indicated operation.

78. 6 lb 5 oz − 2 lb 12 oz

79. 5 tons 1600 lb + 4 tons 1200 lb

80. 6 tons 2250 lb ÷ 3

81. 8 lb 6 oz × 4

82. 1300 mg + 3.6 g

83. 4.8 kg + 4200 g

84. 9.3 g − 1200 mg

85. 6.3 kg × 8

Solve the following.

86. Donshay Berry ordered 1 lb 12 oz of soft-center candies and 2 lb 8 oz of chewy-center candies for his party. Find the total weight of the candy ordered.

87. Four local townships jointly purchase 38 tons 300 lb of cinders to spread on their roads during an ice storm. Determine the weight of the cinders each township receives if they share the purchase equally.

88. Linda Holden ordered 8.3 kg of whole wheat flour from the health food store, but she received 450 g less. How much flour did she actually receive?

89. Eight friends spent a weekend in the Poconos tapping maple trees and preparing 9.3 kg of maple syrup. Find the weight each friend receives if they share the syrup equally.

(8.6) *Convert.*

90. 16 pt to quarts

91. 40 fl oz to cups

92. 6.75 gal to quarts

93. 8.5 pt to cups

94. 9 pt = ____ qt ____ pt

95. 15 qt = ____ gal ____ qt

96. 3.8 L to milliliters

97. 4.2 ml to deciliters

98. 14 hl to kiloliters

99. 30.6 L to centiliters

Perform each indicated operation.

100. 1 qt 1 pt + 3 qt 1 pt

101. 3 gal 2 qt 1 pt × 2

102. 0.946 L − 210 ml

103. 6.1 L + 9400 ml

Solve.

104. Carlos Perez prepares 4 gal 2 qt of iced tea for a block party. During the first 30 minutes of the party, 1 gal 3 qt of the tea is consumed. How much iced tea remains?

105. A recipe for soup stock calls for 1 c 4 fl oz of beef broth. How much should be used if the recipe is cut in half?

106. Each bottle of Kiwi liquid shoe polish holds 85 ml of the polish. Find the number of liters of shoe polish contained in 8 boxes if each box contains 16 bottles.

107. Ivan Miller wants to pour three separate containers of saline solution into a single vat with a capacity of 10 liters. Will 6 liters of solution in the first container combined with 1300 milliliters in the second container and 2.6 liters in the third container fit into the larger vat?

(8.7) *Convert as indicated. If necessary, round to two decimal places. Because approximations are used, your answers may vary slightly from the answers given in the back of the book.*

108. 7 meters to feet

109. 11.5 yards to meters

110. 17.5 liters to gallons

111. 7.8 liters to quarts

112. 15 ounces to grams

113. 23 pounds to kilograms

114. A 100-meter dash is being held today. How many yards is this?

115. If a person weighs 82 kilograms, how many pounds is this?

116. How many quarts are contained in a 3-liter bottle of cola?

117. A compact disk is 1.2 mm thick. Find the height (in inches) of 50 disks.

Convert. Round to the nearest tenth of a degree, if necessary.

118. 245°C to degrees Fahrenheit

119. 160°C to degrees Fahrenheit

120. 42°C to degrees Fahrenheit

121. 86°C to degrees Fahrenheit

122. 93.2°F to degrees Celsius

123. 51.8°F to degrees Celsius

124. 41.3°F to degrees Celsius

125. 80°F to degrees Celsius

Solve. Round to the nearest tenth of a degree, if necessary.

126. A sharp dip in the jet stream caused the temperature in New Orleans to drop to 35°F. Find the corresponding temperature in degrees Celsius.

127. A recipe for meat loaf calls for a 165°C oven. Find the setting used if the oven has a Fahrenheit thermometer.

(8.8) *Given that the pairs of triangles are similar, find the unknown length x.*

△ **128.**

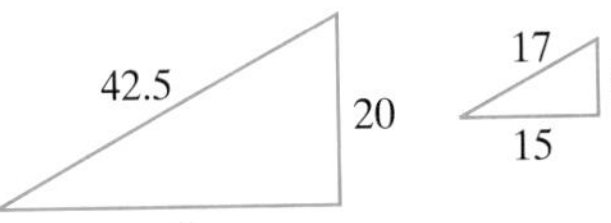

△ **129.**

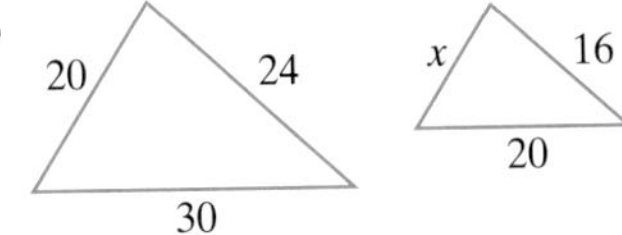

△ **130.**

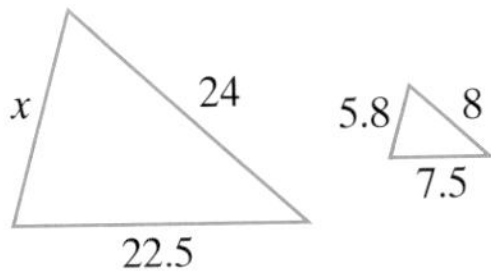

△ **131.**

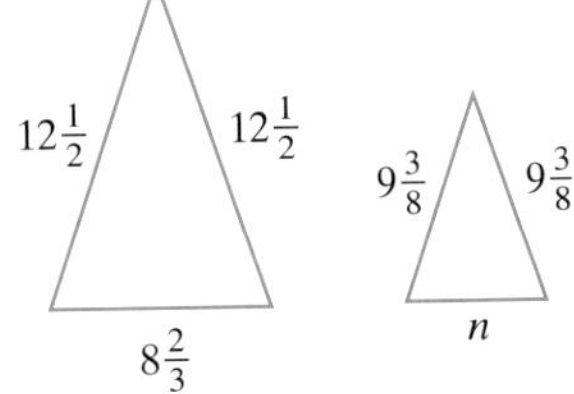

Solve.

132. A housepainter needs to estimate the height of a condominium. He estimates the length of his shadow to be 7 feet long and the length of the building's shadow to be 42 feet long. Find the height of the building if the housepainter is $5\frac{1}{2}$ feet tall.

△ **133.** Santa's elves are making a triangular sail for a toy sailboat. The toy sail is to be the same shape as a real sailboat's sail. Use the following diagram to find the unknown lengths x and y.

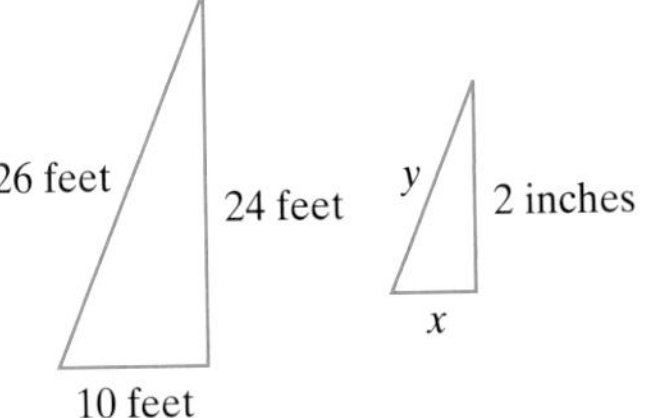

Determine whether each pair of triangles is congruent. If congruent, state the reason why, such as SSS, SAS, or ASA.

134.

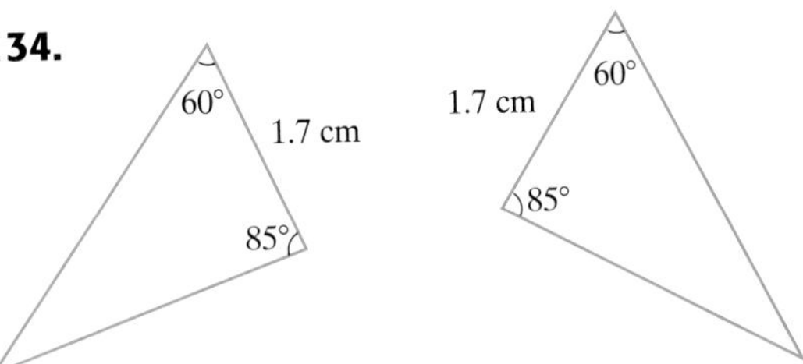

135.

STUDY SKILLS REMINDER

How are you doing?

If you haven't done so yet, take a few moments and think about how you are doing in this course. Are you working toward your goal of successfully completing this course? Is your performance on homework, quizzes, and tests satisfactory? If not, you might want to see your instructor to see if he/she has any suggestions on how you can improve your performance. Let me once again remind you that, in addition to your instructor, there are many places to get help with your mathematics course. A few suggestions are below.

- This text has an accompanying video lesson for every text section.
- The back of this book contains answers to odd-numbered exercises and selected solutions.
- MathPro is available with this text. It is a tutorial software program with lessons corresponding to each text section.
- There is a student solutions manual available that contains worked-out solutions to odd-numbered exercises as well as solutions to every exercise in the Chapter Pretests, Integrated Reviews, Chapter Reviews, Chapter Tests, and Cumulative Reviews.
- Don't forget to check with your instructor for other local resources available to you, such as a tutoring center.

Name ____________________ Section ____________ Date ____________

Chapter 8 Test

Remember to check your answers and use the Chapter Test Prep Video to view solutions.

1. Find the complement of a 78° angle.

2. Find the supplement of a 124° angle.

3. Find the measure of $\angle x$.

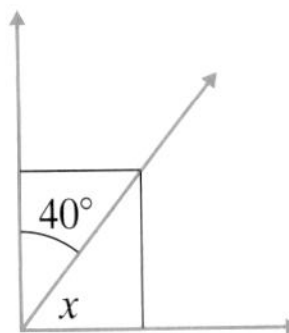

Find the measure of x, y, and z in each figure.

4.

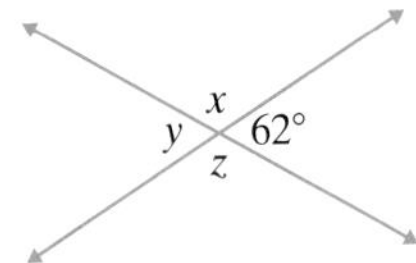

5.

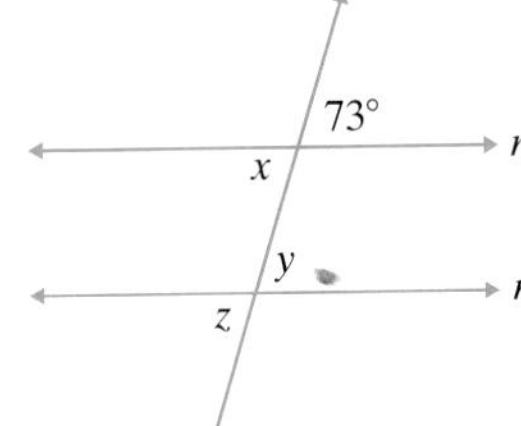

Find the unknown diameter or radius as indicated.

6.

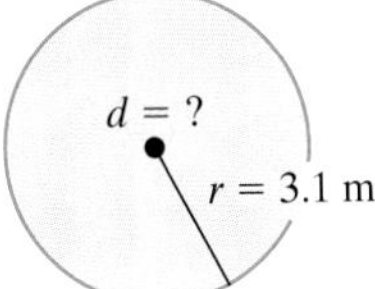

7.

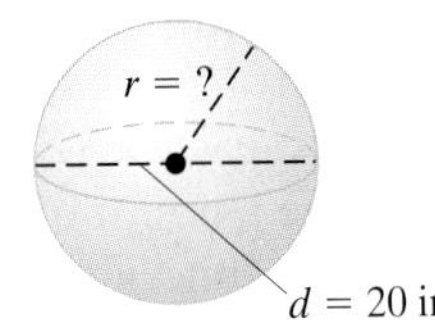

Find the perimeter (or circumference) and area of each figure. For the circle, give the exact value and then use $\pi \approx 3.14$.

8.

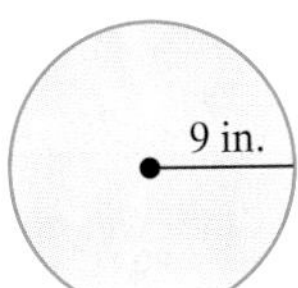

9.

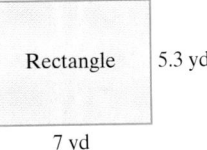

Rectangle

5.3 yd

7 yd

10.

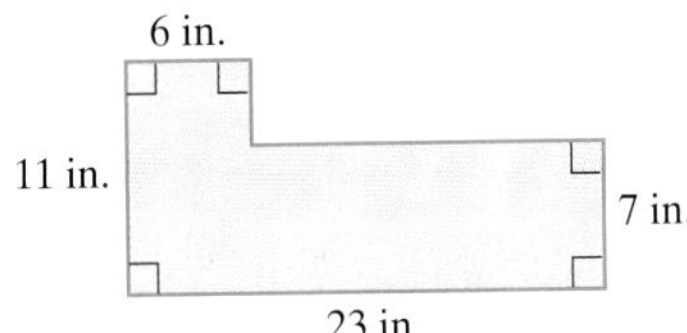

Answers

1. ____________

2. ____________

3. ____________

4. ____________

5. ____________

6. ____________

7. ____________

8. ____________

9. ____________

10. ____________

11. ______

12. ______

13. ______

14. ______

15. ______

16. ______

17. ______

18. ______

19. ______

Find the volume of each solid. For the cylinder, use $\pi \approx \frac{22}{7}$.

△ **11.**

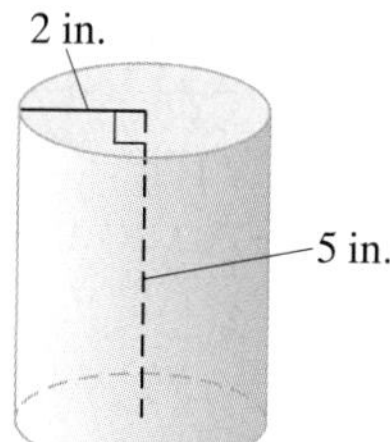

△ **12.**

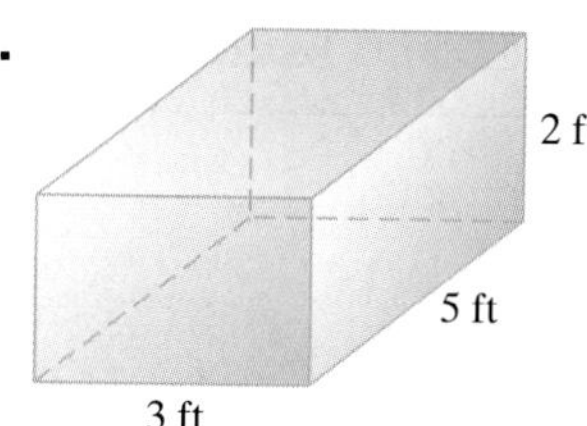

Solve.

△ **13.** Find the perimeter of a square frame with a side length of 4 inches.

△ **14.** How much soil is needed to fill a rectangular hole 3 feet by 3 feet by 2 feet?

△ **15.** Vivian Thomas is going to put insecticide on her lawn to control grubworms. The lawn is a rectangle measuring 123.8 feet by 80 feet. The amount of insecticide required is 0.02 ounces per square foot. Find how much insecticide Vivian needs to purchase.

Convert. If necessary, round to 3 decimal places.

16. 280 in. to feet and inches

17. $2\frac{1}{2}$ gal to quarts

18. 30 oz to pounds

19. 2.8 tons to pounds

20. 2.4 km to meters

21. 3.6 cm to millimeters

22. 4.3 dg to grams

23. 0.83 L to milliliters

24. 7 kg to pounds

25. 8.5 in. to centimeters

Convert. Round to the nearest tenth of a degree, if necessary.

26. 84°F to degrees Celsius

27. 12.6°C to degrees Fahrenheit

28. The sugar maples in front of Bette MacMillan's house are 8.4 meters tall. Because they interfere with the phone lines, the telephone company plans to remove the top third of the trees. How tall will the maples be after they are cut back?

29. A total of 15 gal 1 qt of oil has been removed from a 20-gallon drum. How much oil still remains in the container?

30. The engineer in charge of the bridge construction said that the span of the bridge would be 88 m. But the actual construction required it to be 340 cm longer. Find the span of the bridge.

31. If 2 ft 9 in. of material is used to manufacture one scarf, how much material is needed for 6 scarves?

20. ____________

21. ____________

22. ____________

23. ____________

24. ____________

25. ____________

26. ____________

27. ____________

28. ____________

29. ____________

30. ____________

31. ____________

32. ______

33. ______

34. ______

35. ______

36. ______

37. ______

32. The largest ice cream sundae ever made in the U.S. was assembled in Anaheim, California, in 1985. This giant sundae used 4667 gallons of ice cream. How many pints of ice cream were used?

33. A piece of candy weighs 5 grams. How many ounces is this?

34. A 5-gallon container holds how many liters?

35. A 5-kilometer race is being held today. How many miles is this?

△ **36.** Given that the following triangles are similar, find the unknown length n.

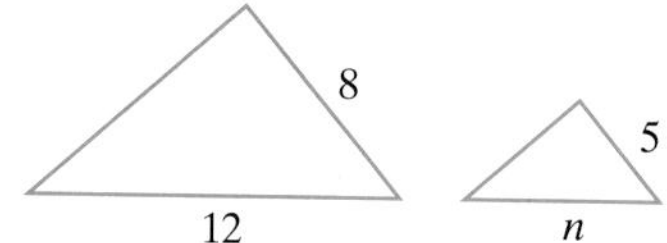

△ **37.** Tamara Watford, a surveyor, needs to estimate the height of a tower. She estimates the length of her shadow to be 4 feet long and the length of the tower's shadow to be 48 feet long. Find the height of the tower if she is $5\frac{3}{4}$ feet tall.

$5\frac{3}{4}$ ft

4 ft

?

48 ft

Name ____________ Section ________ Date ________

Cumulative Review

1. Solve: $3a - 6 = a + 4$

2. Solve: $2x + 1 = 3x - 5$

3. Evaluate:

   a. $\left(\frac{2}{5}\right)^4$ b. $\left(-\frac{1}{4}\right)^2$

4. Evaluate:

   a. $\left(-\frac{1}{3}\right)^3$ b. $\left(\frac{3}{7}\right)^2$

5. Add: $1\frac{4}{5} + 4 + 2\frac{1}{2}$

6. Add: $2\frac{1}{3} + 4\frac{2}{5} + 3$

7. Simplify by combining like terms:

   $11.1x - 6.3 + 8.9x - 4.6$

8. Simplify by combining like terms:

   $2.5y + 3.7 - 1.3y - 1.9$

9. Simplify: $\dfrac{0.7 + 1.84}{0.4}$

10. Simplify: $\dfrac{0.12 + 0.96}{0.5}$

11. Insert $<$, $>$, or $=$ to form a true statement.

    $0.\overline{7} \quad \frac{7}{9}$

12. Insert $<$, $>$, or $=$ to form a true statement.

    $0.43 \quad \frac{2}{5}$

Answers

1. ____________

2. ____________

3. a. ____________

   b. ____________

4. a. ____________

   b. ____________

5. ____________

6. ____________

7. ____________

8. ____________

9. ____________

10. ____________

11. ____________

12. ____________

13. ____________

14. ____________

15. ____________

16. ____________

17. a. ____________

b. ____________

18. a. ____________

b. ____________

19. ____________

20. ____________

21. ____________

22. ____________

13. Solve: $0.5y + 2.3 = 1.65$

14. Solve: $0.4x - 9.3 = 2.7$

△ **15.** An inner city park is in the shape of a square that measures 300 feet on a side. Find the length of the diagonal of the park rounded to the nearest whole foot.

△ **16.** A rectangular field is 200 feet by 125 feet. Find the length of the diagonal of the field, rounded to the nearest whole foot.

△ **17.** Given the rectangle shown:

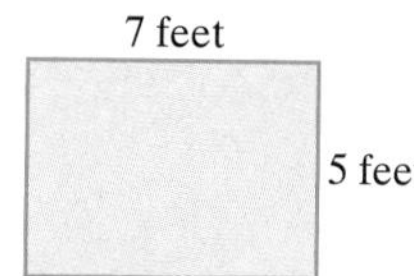

a. Find the ratio of its width to its length.

b. Find the ratio of its length to its perimeter.

△ **18.** A square is 9 inches by 9 inches.

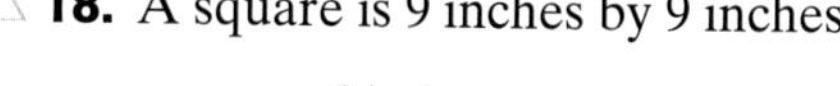

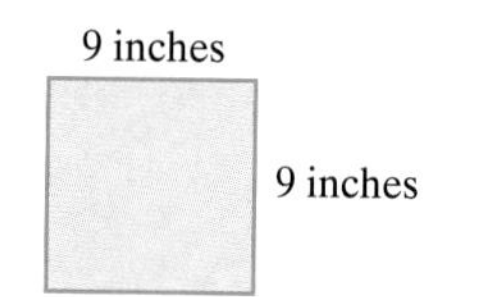

a. Find the ratio of a side to its perimeter.

b. Find the ratio of its perimeter to its area.

19. The standard dose of an antibiotic is 4 cc (cubic centimeters) for every 25 pounds (lb) of body weight. At this rate, find the standard dose for a 140-lb woman.

20. A recipe that makes 2 pie crusts calls for 3 cups of flour. How much flour is needed to make 5 pie crusts?

21. In a survey of 100 people, 17 people drive blue cars. What percent of people drive blue cars?

22. Of 100 shoppers surveyed at a mall, 17 paid for their purchases using only cash. What percent of shoppers used only cash to pay for their purchases?

23. 13 is $6\frac{1}{2}\%$ of what number?

24. 54 is $4\frac{1}{2}\%$ of what number?

25. What number is 30% of 9?

26. What number is 42% of 30?

27. The number of applications for a mathematics scholarship at Yale increased from 34 to 45 in one year. What is the percent increase? Round to the nearest whole percent.

28. The price of a gallon of paint rose from \$15 to \$19. Find the percent increase, rounded to the nearest whole percent.

29. Find the sales tax and the total price on a purchase of an \$85.50 trench coat in a city where the sales tax rate is 7.5%.

30. A sofa has a purchase price of \$375. If the sales tax rate is 8%, find the amount of sales tax and the total cost of the sofa.

31. Find the median of the list of numbers: 25, 54, 56, 57, 60, 71, 98

32. Find the median in the list of scores. 60, 95, 89, 72, 83

33. Find the probability of choosing a red marble from a box containing 1 red, 1 yellow, and 2 blue marbles.

34. Find the probability of choosing a nickel at random in a coin purse that contains 2 pennies, 2 nickels, and 3 quarters.

23. ____________

24. ____________

25. ____________

26. ____________

27. ____________

28. ____________

29. ____________

30. ____________

31. ____________

32. ____________

33. ____________

34. ____________

35.

36.

37.

38.

39.

40.

41.

42.

35. Find the complement of a 48° angle.

36. Find the supplement of a 137° angle.

37. Convert 8 feet to inches.

38. Convert 7 yards to feet.

39. Subtract 3 tons 1350 lb from 8 tons 1000 lb.

40. Add 8 lb 15 oz to 9 lb 3 oz.

41. Convert 114°F to degrees Celsius. If necessary, round to the nearest tenth of a degree.

42. Convert 86°F to degrees Celsius. If necessary, round to the nearest tenth of a degree.

Equations, Inequalities, and Problem Solving

CHAPTER 9

In this chapter, we solve equations and inequalities. Once we know how to solve equations and inequalities, we may solve word problems. Of course, problem solving is an integral topic in algebra and its discussion is continued throughout this text.

Crayola crayons have a long history starting in 1903. In that year, cousins Edwin Binney and C. Harold Smith produced their first Crayola product: a box of eight petroleum-based paraffin crayons, in the colors red, orange, yellow, green, blue, violet, brown, and black, that sold for 5¢. The "Crayola" brand name was coined by Edwin's wife, Alice, by combining the French word *craie* (chalk) with *ola* (short for oleaginous). Today, the Crayola brand name is recognized by 99% of Americans, and Crayola crayons are available in 120 colors. In Section 9.4, Exercise 17, you will find the number of votes cast for America's two favorite Crayola crayon colors.

Answers

1. ______
2. ______
3. ______
4. ______
5. ______
6. ______
7. ______
8. ______
9. ______
10. ______
11. ______
12. ______
13. ______
14. ______
15. ______
16. ______
17. ______
18. ______
19. ______
20. ______

Name ______________________ Section __________ Date __________

Chapter 9 Pretest

Simplify.

1. $5(2x + 9) - 3$

2. $-5(2y - 3) + 1$

Solve.

3. $3 - x = -12$

4. $12 - (5 - 4b) = 9 + 3b$

5. $\frac{2}{3}m = -8$

6. $-7 - 3y = 17 + 5y$

7. $3(1 - 4x) + 2(5x) = 9$

8. $0.20x + 0.15(60) = 0.75(18)$

9. $2(x - 1) = 2x + 5$

10. Three times the sum of a number and -2 is the same as 2 more than the number. Find the number.

11. Find two consecutive even integers such that three times the smaller is 16 more than twice the larger.

12. Substitute the given values into the given formula and solve for the unknown variable.

$V = \frac{1}{3}Ah; V = 60, h = 4$

△ **13.** If the area of a right-triangularly shaped sign is 18 square feet and its height is 4 feet, find the base of the sign.

14. Solve the given formula for the specified variable.

$2x + y = 8$ for y

Name the property illustrated by each statement.

15. $7 \cdot 6 = 6 \cdot 7$

16. $(8 + x) + 4 = 8 + (x + 4)$

17. $0 + 78 = 78$

Solve each inequality. Graph the solutions.

18. $-4 + x \leq 2$

−10 −8 −6 −4 −2 0 2 4 6 8 10

19. $-\frac{3}{2}y > 6$

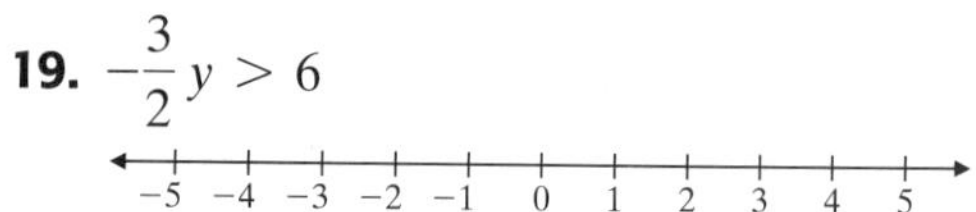

20. $-5x + 3 \leq 4(x - 6)$

−5 −4 −3 −2 −1 0 1 2 3 4 5

9.1 Symbols and Sets of Numbers

OBJECTIVES

- **A** Define the meaning of the symbols $=$, $\neq$, $<$, $>$, $\leq$, and $\geq$.
- **B** Translate sentences into mathematical statements.
- **C** Identify integers, rational numbers, irrational numbers, and real numbers.

Throughout the previous chapters, we have studied different sets of numbers. In this section, we review these sets of numbers. We also introduce a few new sets of numbers in order to show the relationships among these commom sets of real numbers. We begin with a review of the set of natural numbers and the set of whole numbers and how we use symbols to compare these numbers. A **set** is a collection of objects, each of which is called a **member** or **element** of the set. A pair of brace symbols { } encloses the list of elements and is translated as "the set of" or "the set containing."

Natural Numbers

$\{1, 2, 3, 4, 5, 6, \dots\}$

Whole Numbers

$\{0, 1, 2, 3, 4, 5, 6, \dots\}$

Helpful Hint

The three dots (an ellipsis) at the end of the list of elements of a set means that the list continues in the same manner indefinitely.

A Equality and Inequality Symbols

Picturing natural numbers and whole numbers on a number line helps us to see the order of the numbers. Symbols can be used to describe in writing the order of two quantities. We will use equality symbols and inequality symbols to compare quantities.

Below is a review of these symbols. The letters a and b are used to represent quantities. Letters such as a and b that are used to represent numbers or quantities are called **variables.**

		Meaning
Equality symbol:	$a = b$	a is equal to b.
Inequality symbols:	$a \neq b$	a is not equal to b.
	$a < b$	a is less than b.
	$a > b$	a is greater than b.
	$a \leq b$	a is less than or equal to b.
	$a \geq b$	a is greater than or equal to b.

These symbols may be used to form **mathematical statements** such as

$2 = 2$ and $2 \neq 6$

Recall that on the number line, we see that a number **to the right of** another number is **larger.** Similarly, a number **to the left of** another number is **smaller.** For example, 3 is to the left of 5 on the number line, which means that 3 is less than 5, or $3 < 5$. Similarly, 2 is to the right of 0 on the number line, which means 2 is greater than 0, or $2 > 0$. Since 0 is to the left of 2, we can also say that 0 is less than 2, or $0 < 2$.

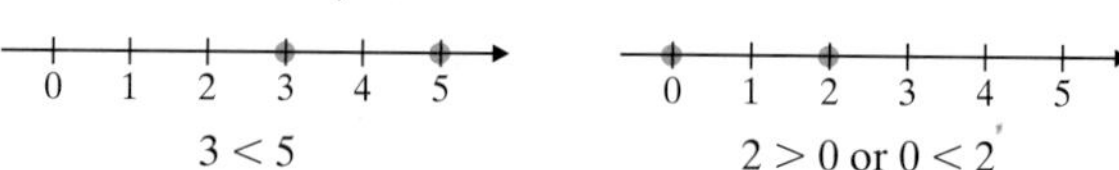

Helpful Hint

Notice that $2 > 0$ has exactly the same meaning as $0 < 2$. Switching the order of the numbers and reversing the "direction of the inequality symbol" does not change the meaning of the statement.

$5 > 3$ has the same meaning as $3 < 5$.

Also notice that when the statement is true, the inequality arrow points to the smaller number.

Practice Problems 1–6

Determine whether each statement is true or false.

1. $8 < 6$
2. $100 > 10$
3. $21 \leq 21$
4. $21 \geq 21$
5. $0 \leq 5$
6. $25 \geq 22$

EXAMPLES Determine whether each statement is true or false.

1. $2 < 3$ True. Since 2 is to the left of 3 on the number line

2. $72 > 27$ True. Since 72 is to the right of 27 on the number line

3. $8 \geq 8$ True. Since $8 = 8$ is true

4. $8 \leq 8$ True. Since $8 = 8$ is true

5. $23 \leq 0$ False. Since neither $23 < 0$ nor $23 = 0$ is true

6. $0 \leq 23$ True. Since $0 < 23$ is true

Helpful Hint

If either $3 < 3$ or $3 = 3$ is true, then $3 \leq 3$ is true.

B Translating Sentences into Mathematical Statements

Now, let's use the symbols discussed above to translate sentences into mathematical statements.

Practice Problem 7

Translate each sentence into a mathematical statement.

a. Fourteen is greater than or equal to fourteen.
b. Zero is less than five.
c. Nine is not equal to ten.

EXAMPLE 7 Translate each sentence into a mathematical statement.

a. Nine is less than or equal to eleven.
b. Eight is greater than one.
c. Three is not equal to four.

Solution:

a. nine / is less than or equal to / eleven
↓ ↓ ↓
$9 \leq 11$

b. eight / is greater than / one
↓ ↓ ↓
$8 > 1$

Answers

1. false **2.** true **3.** true **4.** true **5.** true **6.** true **7. a.** $14 \geq 14$ **b.** $0 < 5$ **c.** $9 \neq 10$

c. three is not equal to four

$$\begin{array}{ccc} \text{three} & \text{is not equal to} & \text{four} \\ \downarrow & \downarrow & \downarrow \\ 3 & \neq & 4 \end{array}$$

C Identifying Common Sets of Numbers

Whole numbers are not sufficient to describe many situations in the real world. For example, quantities smaller than zero must sometimes be represented, such as temperatures less than 0 degrees.

We can place numbers less than zero on the number line as follows: Numbers less than 0 are to the left of 0 and are labeled −1, −2, −3, and so on. The numbers we have labeled on the number line below are called the set of **integers.**

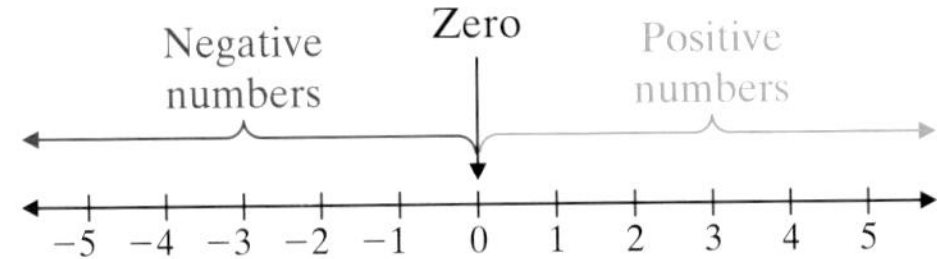

Integers to the left of 0 are called **negative integers;** integers to the right of 0 are called **positive integers.** The integer 0 is neither positive nor negative.

Integers

$\{\ldots, -3, -2, -1, 0, 1, 2, 3, \ldots\}$

Helpful Hint

A − sign, such as the one in −1, tells us that the number is to the left of 0 on the number line.

−1 is read "negative one."

A + sign or no sign tells us that a number lies to the right of 0 on the number line. For example, 3 and +3 both mean positive three.

EXAMPLE 8

Use an integer to express the number in the following. "The lowest temperature ever recorded at South Pole Station, Antarctica, occurred during the month of June. The record-low temperature was 117 degrees below zero." (*Source*: The National Oceanic and Atmospheric Administration)

Solution: The integer −117 represents 117 degrees below zero.

Practice Problem 8

Use an integer to express the number in the following. "The lowest altitude in North America is found in Death Valley, California. Its altitude is 282 feet below sea level." (*Source: The World Almanac*)

Answer

8. −282

A problem with integers in real-life settings arises when quantities are smaller than some integer but greater than the next smallest integer. On the number line, these quantities may be visualized by points between integers. Some of these quantities between integers can be represented as a quotient of integers. For example,

The point on the number line halfway between 0 and 1 can be represented by $\frac{1}{2}$, a quotient of integers.

The point on the number line halfway between 0 and -1 can be represented by $-\frac{1}{2}$. Other quotients of integers and their graphs are shown in the margin.

$-\frac{7}{4}$, $-\frac{1}{2}$, $\frac{1}{2}$, $\frac{4}{3}$, $\frac{5}{2}$

-5 -4 -3 -2 -1 0 1 2 3 4 5

These numbers, each of which can be represented as a quotient of integers, are examples of rational numbers. It's not possible to list the set of rational numbers using the notation that we have been using. For this reason, we will use a different notation.

Rational Numbers

$$\left\{\frac{a}{b}\middle|\, a \text{ and } b \text{ are integers and } b \neq 0\right\}$$

We read this set as "the set of numbers $\frac{a}{b}$ such that a and b are integers and **b is not 0**."

We commonly refer to rational numbers as fractions.

Notice that every integer is also a rational number since each integer can be written as a quotient of integers. For example, the integer 5 is also a rational number since $5 = \frac{5}{1}$. For the rational number $\frac{5}{1}$, recall that the top number, 5, is called the numerator and the bottom number, 1, is called the denominator.

Let's practice **graphing** numbers on a number line.

EXAMPLE 9 Graph the numbers on the number line.

$$-\frac{4}{3},\quad \frac{1}{4},\quad \frac{3}{2},\quad 2\frac{1}{8},\quad 3.5$$

Solution: To help graph the improper fractions in the list, we first write them as mixed numbers.

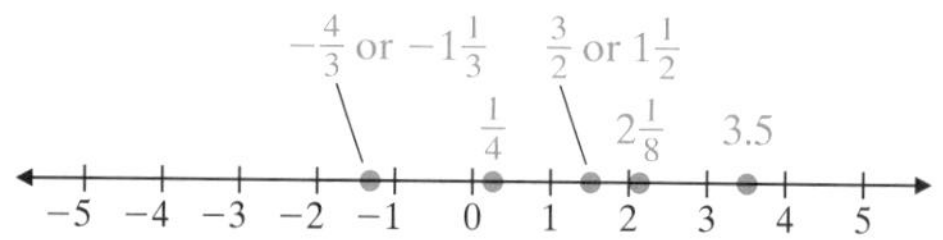

Practice Problem 9

Graph the numbers on the number line.

$$-2.5,\quad -\frac{2}{3},\quad \frac{1}{5},\quad \frac{5}{4},\quad 2.25$$

-5 -4 -3 -2 -1 0 1 2 3 4 5

Answer

9. -2.5, $-\frac{2}{3}$, $\frac{1}{5}$, $\frac{5}{4}$, 2.25

-5 -4 -3 -2 -1 0 1 2 3 4 5

Every rational number has a point on the number line that corresponds to it. But not every point on the number line corresponds to a rational number. Those points that do not correspond to rational numbers correspond instead to **irrational numbers.**

Irrational Numbers

{nonrational numbers that correspond to points on the number line}

An irrational number that you have probably seen is π. Also, $\sqrt{2}$, the length of the diagonal of the square shown in the margin, is an irrational number.

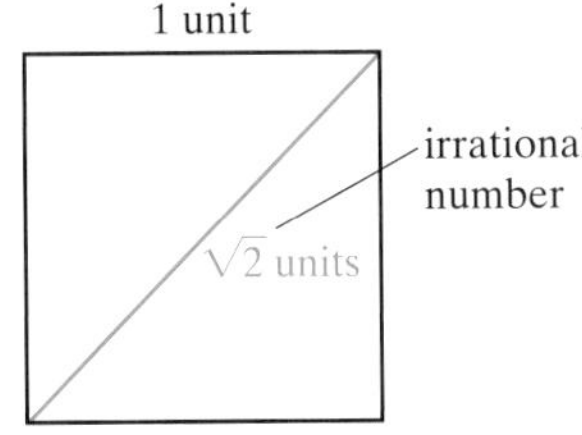

Both rational and irrational numbers can be written as decimal numbers. The decimal equivalent of a rational number will either terminate or repeat in a pattern. For example, upon dividing we find that

$$\frac{3}{4} = 0.75 \quad \text{(Decimal number terminates or ends.)}$$

$$\frac{2}{3} = 0.66666\ldots \quad \text{(Decimal number repeats in a pattern.)}$$

The decimal representation of an irrational number will neither terminate nor repeat.

The set of numbers, each of which corresponds to a point on the number line, is called the set of **real numbers.** One and only one point on the number line corresponds to each real number.

Real Numbers

{Numbers that correspond to points on the number line}

Several different sets of numbers have been discussed in this section. The following diagram shows the relationships among these sets of real numbers. Notice that, together, the rational numbers and the irrational numbers make up the real numbers.

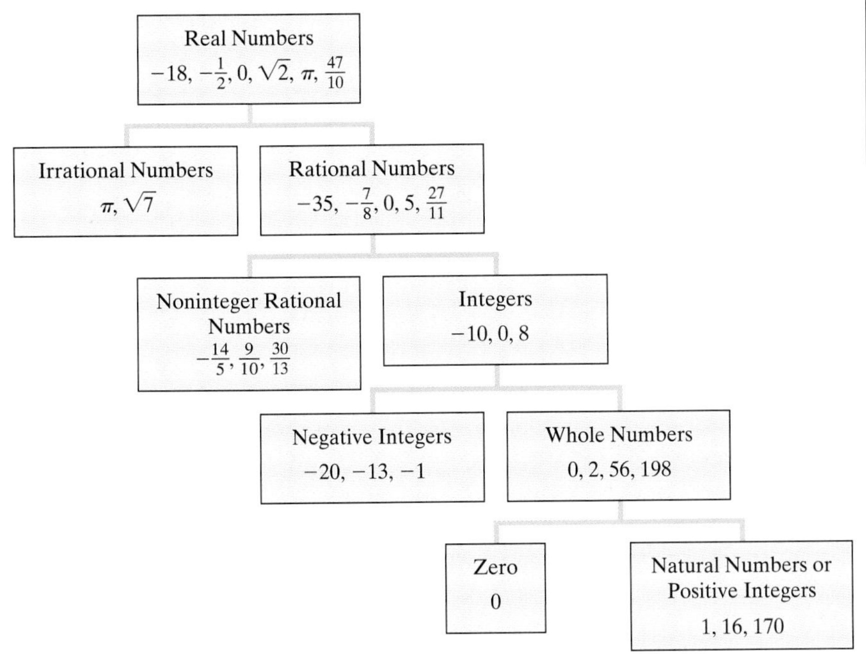

Practice Problem 10

Given the set $\{-100, -\frac{2}{5}, 0, \pi, 6, 913\}$, list the numbers in this set that belong to the set of:

a. Natural numbers
b. Whole numbers
c. Integers
d. Rational numbers
e. Irrational numbers
f. Real numbers

Answers

10. a. 6, 913 **b.** 0, 6, 913 **c.** $-100, 0, 6, 913$ **d.** $-100, -\frac{2}{5}, 0, 6, 913$ **e.** π **f.** all numbers in the given set

EXAMPLE 10

Given the set $\{-2, 0, \frac{1}{4}, 112, -3, 11, \sqrt{2}\}$, list the numbers in this set that belong to the set of:

a. Natural numbers
b. Whole numbers
c. Integers
d. Rational numbers
e. Irrational numbers
f. Real numbers

Solution:

a. The natural numbers are 11 and 112.
b. The whole numbers are 0, 11, and 112.
c. The integers are $-3, -2, 0, 11$, and 112.
d. Recall that integers are rational numbers also. The rational numbers are $-3, -2, 0, \frac{1}{4}, 11$, and 112.
e. The irrational number is $\sqrt{2}$.
f. The real numbers are all numbers in the given set. ●

STUDY SKILLS REMINDER

Are you getting all the mathematics help that you need?

Remember that in addition to your instructor, there are many places to get help with your mathematics course. For example, see the list below.

- There is an accompanying video lesson for every section in this text.
- The back of this book contains answers to odd-numbered exercises as well as answers to every exercise in the Chapter Pretests, Integrated Reviews, Chapter Reviews, Chapter Tests, and Cumulative Reviews. The back of the book also contains selected solutions.
- MathPro is available with this text. It is a tutorial software program with lessons corresponding to each section in the text.
- There is a student solutions manual available that contains worked-out solutions to odd-numbered exercises as well as solutions to every exercise in the Chapter Pretests, Integrated Reviews, Chapter Reviews, Chapter Tests, and Cumulative Reviews.
- Check with your instructor for other local resources available to you, such as a tutor center.

Name ______________________ Section __________ Date ____________

EXERCISE SET 9.1

A *Insert $<$, $>$, or $=$ in the space between the paired numbers to make each statement true. See Examples 1 through 6.*

1. 4   10

2. 8   5

3. 7   3

4. 9   15

5. 6.26   6.26

6. 2.13   1.13

7. 0   7

8. 20   0

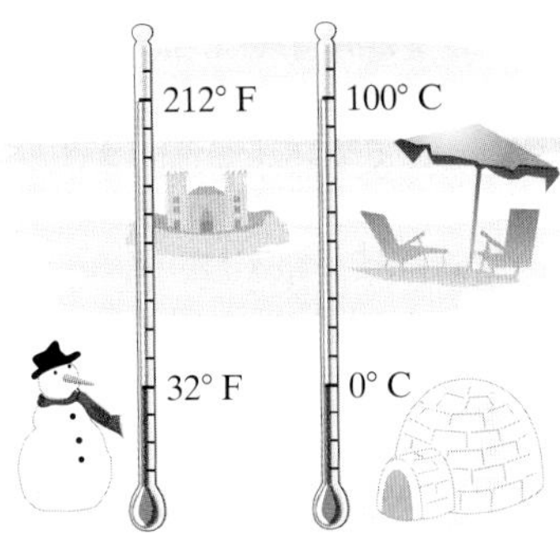

9. The freezing point of water is 32° Fahrenheit. The boiling point of water is 212° Fahrenheit. Write an inequality statement using $<$ or $>$ comparing the numbers 32 and 212.

10. The freezing point of water is 0° Celsius. The boiling point of water is 100° Celsius. Write an inequality statement using $<$ or $>$ comparing the numbers 0 and 100.

Determine whether each statement is true or false. See Examples 1 through 6.

11. $11 \leq 11$

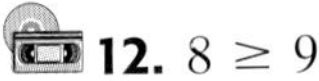

12. $8 \geq 9$

13. $10 > 11$

14. $17 > 16$

15. $3 + 8 \geq 3(8)$

16. $8 \cdot 8 \leq 8 \cdot 7$

17. $7 > 0$

18. $4 < 7$

△ **19.** An angle measuring 30° and an angle measuring 45° are shown. Use the inequality symbols $\leq$ or $\geq$ to write a statement comparing the numbers 30 and 45.

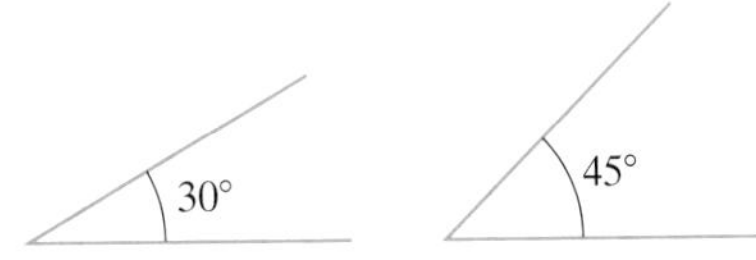

△ **20.** The sum of the measures of the angles of a triangle is 180°. The sum of the measures of the angles of a parallelogram is 360°. Use the inequality symbols $\leq$ or $\geq$ to write a statement comparing the numbers 360 and 180.

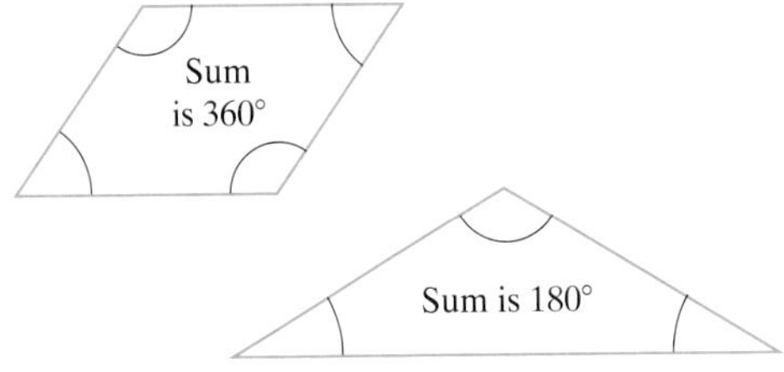

Rewrite each inequality so that the inequality symbol points in the opposite direction and the resulting statement has the same meaning as the given one.

21. $25 \geq 20$ **22.** $-13 \leq 13$ **23.** $0 < 6$ **24.** $5 > 3$ **25.** $-10 > -12$ **26.** $-4 < -2$

B *Write each sentence as a mathematical statement. See Example 7.*

27. Seven is less than eleven.

28. Twenty is greater than two.

29. Five is greater than or equal to four.

30. Negative ten is less than or equal to thirty-seven.

31. Fifteen is not equal to negative two.

32. Negative seven is not equal to seven.

C *Use integers to represent the values in each statement. See Example 8.*

33. The highest elevation in California is Mt. Whitney with an altitude of 14,494 feet. The lowest elevation in California is Death Valley with an altitude of 282 feet below sea level. (*Source:* U.S. Geological Survey)

34. Driskill Mountain, in Louisiana, has an altitude of 535 feet. New Orleans, Louisiana, lies 8 feet below sea level. (*Source:* U.S. Geological Survey)

35. From 1990 to 2000, the population of Washington, D.C., decreased by 34,841. (*Source:* U.S. Census Bureau)

36. From 1999 to 2000, the enrollment in public and private U.S. high schools increased by 143,000 students. (*Source:* National Center for Education Statistics)

37. Gretchen Bertani deposited \$475 in her savings account. She later withdrew \$195.

38. David Lopez was deep-sea diving. During his dive, he ascended 17 feet and later descended 15 feet.

Graph each set of numbers on the number line. See Example 9.

39. $-4, 0, 2, 5$

−5 −4 −3 −2 −1 0 1 2 3 4 5

40. $-3, 0, 1, 5$

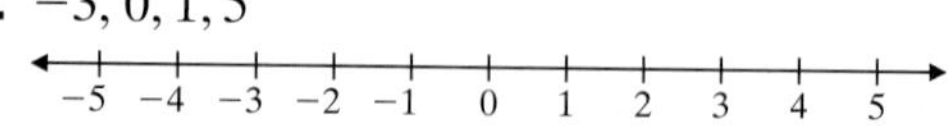

41. $-2, 4, \frac{1}{2}, -\frac{1}{4}$

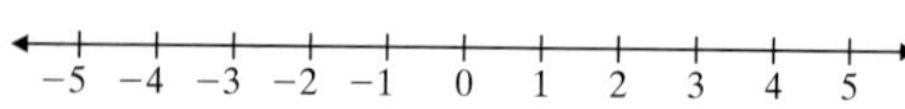

42. $-5, 3, -\frac{1}{2}, \frac{1}{4}$

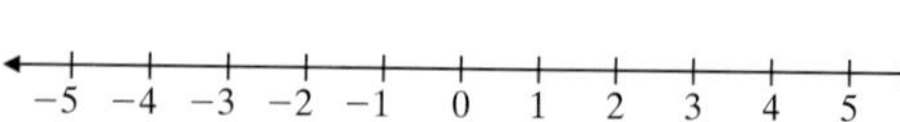

43. $-2.5, \frac{7}{4}, 3.25, -\frac{3}{2}$

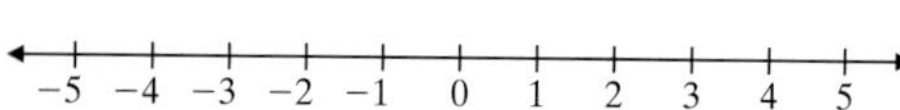

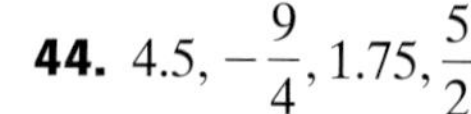

44. $4.5, -\frac{9}{4}, 1.75, \frac{5}{2}$

−5 −4 −3 −2 −1 0 1 2 3 4 5

Tell which set or sets each number belongs to: natural numbers, whole numbers, integers, rational numbers, irrational numbers, and real numbers. See Example 10.

45. 0

46. $\frac{1}{4}$

47. -7

48. $-\frac{1}{7}$

49. 265

50. 7941

51. $\frac{2}{3}$

52. $\sqrt{3}$

Determine whether each statement is true or false.

53. Every rational number is also an integer.

54. Every natural number is positive.

55. 0 is a real number.

56. Every whole number is an integer.

57. Every negative number is also a rational number.

58. Every rational number is also a real number.

59. Every real number is also a rational number.

60. $\frac{1}{2}$ is an integer.

Review and Preview

Insert $<$, $>$, or $=$ in the appropriate space to make each statement true. See Section 2.1.

61. $|-5| \quad -4$

62. $0 \quad |0|$

63. $|-1| \quad |1|$

64. $\left|\frac{2}{5}\right| \quad \left|-\frac{2}{5}\right|$

65. $|-2| \quad |-3|$

66. $-5.00 \quad |-5.0|$

67. $|0| \quad |-8|$

68. $|-12| \quad \frac{24}{2}$

Combining Concepts

Tell whether each statement is true or false.

69. $\frac{1}{2} < \frac{1}{3}$

70. $\frac{3}{6} \geq \frac{1}{2}$

71. $|-5.3| \geq |5.3|$

72. $-1\frac{1}{2} > -\frac{1}{2}$

73. $-9.6 > -9.1$

74. $-7.3 < -7.1$

75. $-\frac{2}{3} \leq -\frac{1}{5}$

76. $-\frac{5}{6} > -\frac{1}{6}$

This graph shows the number of visitors, in millions, at the three most popular U.S. national parks. Each bar represents a different park in the year 2000 and the height of each bar represents the number of visitors (in millions) for that particular park in the year 2000. Use this graph to answer Exercises 77 through 80.

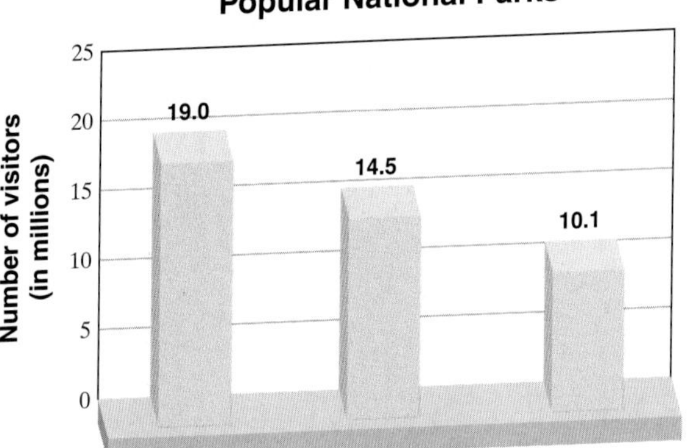

77. What is the park with the most visitors?

78. What is the number of visitors for the Great Smoky Mountains National Park?

79. Write an inequality statement using $\leq$ or $\geq$ comparing the number of visitors for Blue Ridge Parkway and Golden Gate National Recreation Area.

80. Write an inequality statement using $<$ or $>$ comparing the number of visitors for Great Smoky Mountains National Park and Blue Ridge Parkway.

The apparent magnitude of a star is the measure of its brightness as seen by someone on Earth. The smaller the apparent magnitude, the brighter the star. Below, the apparent magnitudes of some stars are listed. Use this table to answer Exercises 81 through 86.

Star	Apparent Magnitude	Star	Apparent Magnitude
Arcturus	−0.04	Spica	0.98
Sirius	−1.46	Rigel	0.12
Vega	0.03	Regulus	1.35
Antares	0.96	Canopus	−0.72
Sun	−26.7	Hadar	0.61

(*Source: Norton's 2000.0: Star Atlas and Reference Handbook*, 18th ed., Longman Group, UK, 1989)

81. The apparent magnitude of the Sun is −26.7. The apparent magnitude of the star Arcturus is −0.04. Write an inequality statement comparing the numbers −0.04 and −26.7.

82. The apparent magnitude of Antares is 0.96. The apparent magnitude of Spica is 0.98. Write an inequality statement comparing the numbers 0.96 and 0.98.

83. Which is brighter, the Sun or Arcturus?

84. Which is dimmer, Antares or Spica?

85. Which star listed is the brightest?

86. Which star listed is the dimmest?

87. In your own words, explain how to find the absolute value of a number.

88. Give an example of a real-life situation that can be described with integers but not with whole numbers.

9.2 Properties of Real Numbers

A Using the Commutative and Associative Properties

In this section we review names of properties of real numbers with which we are already familiar. Throughout this section, the variables a, b, and c represent real numbers.

We know that order does not matter when adding numbers. For example, we know that $7 + 5$ is the same as $5 + 7$. This property is given a special name—the **commutative property of addition.** We also know that order does not matter when multiplying numbers. For example, we know that $-5(6) = 6(-5)$. This property means that multiplication is commutative also and is called the **commutative property of multiplication.**

Commutative Properties

Addition: $a + b = b + a$

Multiplication: $a \cdot b = b \cdot a$

These properties state that the *order* in which any two real numbers are added or multiplied does not change their sum or product. For example, if we let $a = 3$ and $b = 5$, then the commutative properties guarantee that

$$3 + 5 = 5 + 3 \quad \text{and} \quad 3 \cdot 5 = 5 \cdot 3$$

Helpful Hint

Is subtraction also commutative? Try an example. Is $3 - 2 = 2 - 3$? **No!** The left side of this statement equals 1; the right side equals -1. There is no commutative property of subtraction. Similarly, there is no commutative property for division. For example, $10 \div 2$ does not equal $2 \div 10$.

EXAMPLE 1 Use a commutative property to complete each statement.

a. $x + 5 =$ ______ **b.** $3 \cdot x =$ ______

Solution:

a. $x + 5 = 5 + x$ By the commutative property of addition

b. $3 \cdot x = x \cdot 3$ By the commutative property of multiplication

Try the Concept Check in the margin.

Let's now discuss grouping numbers. We know that when we add three numbers, the way in which they are grouped or associated does not change their sum. For example, we know that $2 + (3 + 4) = 2 + 7 = 9$. This result is the same if we group the numbers differently. In other words, $(2 + 3) + 4 = 5 + 4 = 9$, also. Thus, $2 + (3 + 4) = (2 + 3) + 4$. This property is called the **associative property of addition.**

We also know that changing the grouping of numbers when multiplying does not change their product. For example, $2 \cdot (3 \cdot 4) = (2 \cdot 3) \cdot 4$ (check it). This is the **associative property of multiplication.**

OBJECTIVES

A Use the commutative and associative properties.

B Use the distributive property.

C Use the identity and inverse properties.

SSM TUTOR CENTER SG CD & VIDEO MATH PRO WEB

Practice Problem 1

Use a commutative property to complete each statement.

a. $7 \cdot y =$ ________

b. $4 + x =$ ________

Concept Check

Which of the following pairs of actions are commutative?

a. "raking the leaves" and "bagging the leaves"

b. "putting on your left glove" and "putting on your right glove"

c. "putting on your coat" and "putting on your shirt"

d. "reading a novel" and "reading a newspaper"

Answers

1. a. $y \cdot 7$ **b.** $x + 4$

Concept Check: b, d

Associative Properties

Addition: $(a + b) + c = a + (b + c)$

Multiplication: $(a \cdot b) \cdot c = a \cdot (b \cdot c)$

These properties state that the way in which three numbers are *grouped* does not change their sum or their product.

EXAMPLE 2 Use an associative property to complete each statement.

a. $5 + (4 + 6) =$ _____

b. $(-1 \cdot 2) \cdot 5 =$ _____

Solution:

a. $5 + (4 + 6) = (5 + 4) + 6$ By the associative property of addition

b. $(-1 \cdot 2) \cdot 5 = -1 \cdot (2 \cdot 5)$ By the associative property of multiplication

Practice Problem 2

Use an associative property to complete each statement.

a. $5 \cdot (-3 \cdot 6) =$ _______

b. $(-2 + 7) + 3 =$ _______

Helpful Hint

Remember the difference between the commutative properties and the associative properties. The commutative properties have to do with the *order* of numbers and the associative properties have to do with the *grouping* of numbers.

EXAMPLES

Determine whether each statement is true by an associative property or a commutative property.

3. $(7 + 10) + 4 = (10 + 7) + 4$ Since the order of two numbers was changed and their grouping was not, this is true by the commutative property of addition.

4. $2 \cdot (3 \cdot 1) = (2 \cdot 3) \cdot 1$ Since the grouping of the numbers was changed and their order was not, this is true by the associative property of multiplication.

Practice Problems 3–4

Determine whether each statement is true by an associative property or a commutative property.

3. $5 \cdot (4 \cdot 7) = 5 \cdot (7 \cdot 4)$

4. $-2 + (4 + 9) = (-2 + 4) + 9$

Let's now illustrate how these properties can help us simplify expressions.

EXAMPLES Simplify each expression.

5. $10 + (x + 12) = 10 + (12 + x)$ By the commutative property of addition

$= (10 + 12) + x$ By the associative property of addition

$= 22 + x$ Add.

6. $-3(7x) = (-3 \cdot 7)x$ By the associative property of multiplication

$= -21x$ Multiply.

Practice Problems 5–6

Simplify each expression.

5. $(-3 + x) + 17$

6. $4(5x)$

B Using the Distributive Property

The **distributive property of multiplication over addition** is used repeatedly throughout algebra. It is useful because it allows us to write a product as a sum or a sum as a product.

We know that $7(2 + 4) = 7(6) = 42$. Compare that with $7(2) + 7(4) = 14 + 28 = 42$. Since both original expressions equal 42, they must equal each other, or

$$7(2 + 4) = 7(2) + 7(4)$$

This is an example of the distributive property. The product on the left side of the equal sign is equal to the sum on the right side. We can think of the 7 as being distributed to each number inside the parentheses.

Answers

2. a. $(5 \cdot -3) \cdot 6$ **b.** $-2 + (7 + 3)$ **3.** commutative **4.** associative **5.** $14 + x$ **6.** $20x$

Distributive Property of Multiplication Over Addition

$$a(b + c) = ab + ac$$

Since multiplication is commutative, this property can also be written as

$$(b + c)a = ba + ca$$

The distributive property can also be extended to more than two numbers inside the parentheses. For example,

$$\begin{aligned} 3(x + y + z) &= 3(x) + 3(y) + 3(z) \\ &= 3x + 3y + 3z \end{aligned}$$

Since we define subtraction in terms of addition, the distributive property is also true for subtraction. For example,

$$\begin{aligned} 2(x - y) &= 2(x) - 2(y) \\ &= 2x - 2y \end{aligned}$$

EXAMPLES

Use the distributive property to write each expression without parentheses. Then simplify the result.

7. $2(x + y) = 2(x) + 2(y)$
$= 2x + 2y$

8. $-5(-3 + 2z) = -5(-3) + (-5)(2z)$
$= 15 - 10z$

9. $5(x + 3y - z) = 5(x) + 5(3y) - 5(z)$
$= 5x + 15y - 5z$

10. $-1(2 - y) = (-1)(2) - (-1)(y)$
$= -2 + y$

11. $-(3 + x - w) = -1(3 + x - w)$
$= (-1)(3) + (-1)(x) - (-1)(w)$
$= -3 - x + w$

12. $4(3x + 7) + 10 = 4(3x) + 4(7) + 10$ Apply the distributive property.
$= 12x + 28 + 10$ Multiply.
$= 12x + 38$ Add.

Helpful Hint

Notice in Example 11 that $-(3 + x - w)$ is first rewritten as $-1(3 + x - w)$.

The distributive property can also be used to write a sum as a product.

EXAMPLES

Use the distributive property to write each sum as a product.

13. $8 \cdot 2 + 8 \cdot x = 8(2 + x)$

14. $7s + 7t = 7(s + t)$

C Using the Identity and Inverse Properties

Next, we present the **identity properties.**

The number 0 is called the identity for addition because when 0 is added to any real number, the result is the same real number. In other words, the *identity* of the real number is not changed.

Practice Problems 7–12

Use the distributive property to write each expression without parentheses. Then simplify the result.

7. $5(x + y)$
8. $-3(2 + 7x)$
9. $4(x + 6y - 2z)$
10. $-1(3 - a)$
11. $-(8 + a - b)$
12. $9(2x + 4) + 9$

Practice Problems 13–14

Use the distributive property to write each sum as a product.

13. $9 \cdot 3 + 9 \cdot y$
14. $4x + 4y$

Answers

7. $5x + 5y$ **8.** $-6 - 21x$ **9.** $4x + 24y - 8z$ **10.** $-3 + a$ **11.** $-8 - a + b$ **12.** $18x + 45$ **13.** $9(3 + y)$ **14.** $4(x + y)$

The number 1 is called the identity for multiplication because when a real number is multiplied by 1, the result is the same real number. In other words, the *identity* of the real number is not changed.

Identities for Addition and Multiplication

0 is the identity element for addition.

$$a + 0 = a \quad \text{and} \quad 0 + a = a$$

1 is the identity element for multiplication.

$$a \cdot 1 = a \quad \text{and} \quad 1 \cdot a = a$$

Notice that 0 is the *only* number that can be added to any real number with the result that the sum is the same real number. Also, 1 is the *only* number that can be multiplied by any real number with the result that the product is the same real number.

Opposites (also called **additive inverses**) are introduced in Section 2.1. Two numbers are called additive inverses or opposites if their sum is 0. The additive inverse or opposite of 6 is -6 because $6 + (-6) = 0$. The additive inverse or opposite of -5 is 5 because $-5 + 5 = 0$.

Reciprocals (also called **multiplicative inverses**) are introduced in Section 4.3. Two nonzero numbers are called reciprocals or multiplicative inverses if their product is 1. The reciprocal or multiplicative inverse of $\frac{2}{3}$ is $\frac{3}{2}$ because $\frac{2}{3} \cdot \frac{3}{2} = 1$. Likewise, the reciprocal of -5 is $-\frac{1}{5}$ because $-5\left(-\frac{1}{5}\right) = 1$.

Additive or Multiplicative Inverses

The numbers a and $-a$ are additive inverses or opposites of each other because their sum is 0; that is,

$$a + (-a) = 0$$

The numbers b and $\frac{1}{b}$ (for $b \neq 0$) are reciprocals or multiplicative inverses of each other because their product is 1; that is,

$$b \cdot \frac{1}{b} = 1$$

Try the Concept Check in the margin.

EXAMPLES Name the property illustrated by each true statement.

15. $3 \cdot y = y \cdot 3$ — Commutative property of multiplication (order changed)

16. $(x + 7) + 9 = x + (7 + 9)$ — Associative property of addition (grouping changed)

17. $(b + 0) + 3 = b + 3$ — Identity element for addition

18. $2 \cdot (z \cdot 5) = 2 \cdot (5 \cdot z)$ — Commutative property of multiplication (order changed)

19. $-2 \cdot \left(-\frac{1}{2}\right) = 1$ — Multiplicative inverse property

20. $-2 + 2 = 0$ — Additive inverse property

21. $-6 \cdot (y \cdot 2) = (-6 \cdot 2) \cdot y$ — Commutative and associative properties of multiplication (order and grouping changed)

Concept Check

Which of the following is the

a. opposite of $-\frac{3}{10}$, and which is the

b. reciprocal of $-\frac{3}{10}$?

$$1, -\frac{10}{3}, \frac{3}{10}, 0, \frac{10}{3}, -\frac{3}{10}$$

Practice Problems 15–21

Name the property illustrated by each true statement.

15. $5 + (-5) = 0$
16. $12 + y = y + 12$
17. $-4 \cdot (6 \cdot x) = (-4 \cdot 6) \cdot x$
18. $6 + (z + 2) = 6 + (2 + z)$
19. $3\left(\frac{1}{3}\right) = 1$
20. $(x + 0) + 23 = x + 23$
21. $(7 \cdot y) \cdot 10 = y \cdot (7 \cdot 10)$

Answers

15. additive inverse property **16.** commutative property of addition **17.** associative property of multiplication **18.** commutative property of addition **19.** multiplicative inverse property **20.** identity element for addition **21.** commutative and associative properties of multiplication

Concept Check:

a. $\frac{3}{10}$ **b.** $-\frac{10}{3}$

Name ______________________ Section __________ Date __________

EXERCISE SET 9.2

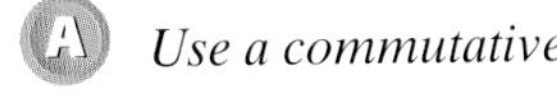 *Use a commutative property to complete each statement. See Examples 1 and 3.*

1. $x + 16 =$ ________ **2.** $4 + y =$ ________ **3.** $-4 \cdot y =$ ________ **4.** $-2 \cdot x =$ ________

5. $xy =$ ________ **6.** $ab =$ ________ **7.** $2x + 13 =$ ________ **8.** $19 + 3y =$ ________

Use an associative property to complete each statement. See Examples 2 and 4.

9. $(xy) \cdot z =$ ________ **10.** $3 \cdot (x \cdot y) =$ ________ **11.** $2 + (a + b) =$ ________

12. $(y + 4) + z =$ ________ **13.** $4 \cdot (ab) =$ ________ **14.** $(-3y) \cdot z =$ ________

15. $(a + b) + c =$ ________ **16.** $6 + (r + s) =$ ________

Use the commutative and associative properties to simplify each expression. See Examples 5 and 6.

17. $8 + (9 + b)$ **18.** $(r + 3) + 11$ **19.** $4(6y)$ **20.** $2(42x)$

21. $\frac{1}{5}(5y)$ **22.** $\frac{1}{8}(8z)$ **23.** $(13 + a) + 13$ **24.** $7 + (x + 4)$

25. $-9(8x)$ **26.** $-3(12y)$ **27.** $\frac{3}{4}\left(\frac{4}{3}s\right)$ **28.** $\frac{2}{7}\left(\frac{7}{2}r\right)$

29. Write an example that shows that division is not commutative.

30. Write an example that shows that subtraction is not commutative.

B *Use the distributive property to write each expression without parentheses. Then simplify the result. See Examples 7 through 12.*

31. $4(x + y)$

32. $7(a + b)$

33. $9(x - 6)$

34. $11(y - 4)$

35. $2(3x + 5)$

36. $5(7 + 8y)$

37. $7(4x - 3)$

38. $3(8x - 1)$

39. $3(6 + x)$

40. $2(x + 5)$

41. $-2(y - z)$

42. $-3(z - y)$

43. $-7(3y + 5)$

44. $-5(2r + 11)$

45. $5(x + 4m + 2)$

46. $8(3y + z - 6)$

47. $-4(1 - 2m + n)$

48. $-4(4 + 2p + 5)$

49. $-(5x + 2)$

50. $-(9r + 5)$

51. $-(r - 3 - 7p)$

52. $-(q - 2 + 6r)$

53. $\frac{1}{2}(6x + 8)$

54. $\frac{1}{4}(4x - 2)$

55. $-\frac{1}{3}(3x - 9y)$

56. $-\frac{1}{5}(10a - 25b)$

57. $3(2r + 5) - 7$

58. $10(4s + 6) - 40$

59. $-9(4x + 8) + 2$

60. $-11(5x + 3) + 10$

61. $-4(4x + 5) - 5$

62. $-6(2x + 1) - 1$

Use the distributive property to write each sum as a product. See Examples 13 and 14.

63. $4 \cdot 1 + 4 \cdot y$

64. $14 \cdot z + 14 \cdot 5$

65. $11x + 11y$

66. $9a + 9b$

67. $(-1) \cdot 5 + (-1) \cdot x$

68. $(-3)a + (-3)y$

69. $30a + 30b$

70. $25x + 25y$

 *Find the additive inverse or opposite of each of the following numbers. See Example 20.*

71. 16

72. 14

73. -8

74. -3

75. $-(-1.2)$

76. $-(7.9)$

77. $-|-2|$

78. $-|-9|$

Find the multiplicative inverse or reciprocal of each number. See Example 19.

79. $\frac{2}{3}$

80. $\frac{3}{4}$

81. $-\frac{5}{6}$

82. $-\frac{7}{8}$

83. $3\frac{5}{6}$

84. $2\frac{3}{5}$

85. -2

86. -5

Name the properties illustrated by each true statement. See Examples 15 through 21.

87. $3 \cdot 5 = 5 \cdot 3$

88. $4(3 + 8) = 4 \cdot 3 + 4 \cdot 8$

89. $2 + (x + 5) = (2 + x) + 5$

90. $(x + 9) + 3 = (9 + x) + 3$

91. $9(3 + 7) = 9 \cdot 3 + 9 \cdot 7$

92. $1 \cdot 9 = 9$

93. $(4 \cdot y) \cdot 9 = 4 \cdot (y \cdot 9)$

94. $6 \cdot \frac{1}{6} = 1$

95. $0 + 6 = 6$

96. $(a + 9) + 6 = a + (9 + 6)$

97. $-4(y + 7) = -4 \cdot y + (-4) \cdot 7$

98. $(11 + r) + 8 = (r + 11) + 8$

99. $-4 \cdot (8 \cdot 3) = (8 \cdot -4) \cdot 3$

100. $r + 0 = r$

Review and Preview

Evaluate each expression for the given values. See Section 2.5.

101. If $x = -1$ and $y = 3$, find $y - x^2$

102. If $g = 0$ and $h = -4$, find $gh - h^2$

103. If $a = 2$ and $b = -5$, find $a - b^2$

104. If $x = -3$, find $x^3 - x^2 + 4$

105. If $y = -5$ and $z = 0$, find $yz - y^2$

106. If $x = -2$, find $x^3 - x^2 - x$

Combining Concepts

Fill in the table with the opposite (additive inverse), the reciprocal (multiplicative inverse), or the expression. Assume that the value of each expression is not 0.

	107.	108.	109.	110.	111.	112.
Expression	8	$-\frac{2}{3}$	x	$4y$		
Opposite						$7x$
Reciprocal					$\frac{1}{2x}$	

Name the property illustrated by each step.

113. a. $\triangle + (\square + \bigcirc) = (\square + \bigcirc) + \triangle$

b. $= (\bigcirc + \square) + \triangle$

c. $= \bigcirc + (\square + \triangle)$

114. a. $(x + y) + z = x + (y + z)$

b. $= (y + z) + x$

c. $= (z + y) + x$

115. Explain why 0 is called the identity element for addition.

116. Explain why 1 is called the identity element for multiplication.

Determine which pairs of actions are commutative.

117. "taking a test" and "studying for the test"

118. "putting on your shoes" and "putting on your socks"

119. "putting on your left shoe" and "putting on your right shoe"

120. "reading the sports section" and "reading the comics section"

9.3 Further Solving Linear Equations

OBJECTIVES

A. Apply the general strategy for solving a linear equation.
B. Solve equations containing fractions or decimals.
C. Recognize identities and equations with no solution.

A Solving Linear Equations

Let's begin with a formal definition of a linear equation in one variable.

A **linear equation in one variable** can be written in the form

$$Ax + B = C$$

where $A, B,$ and C, are real numbers and $A \neq 0$.

We now combine our knowledge from the previous chapters and review solving linear equations.

To Solve Linear Equations in One Variable

Step 1. If an equation contains fractions, multiply both sides by the LCD to clear the equation of fractions.
Step 2. Use the distributive property to remove parentheses if they occur.
Step 3. Simplify each side of the equation by combining like terms.
Step 4. Get all variable terms on one side and all numbers on the other side by using the addition property of equality.
Step 5. Get the variable alone by using the multiplication property of equality.
Step 6. Check the solution by substituting it into the original equation.

EXAMPLE 1 Solve: $4(2x - 3) + 7 = 3x + 5$

Solution: There are no fractions, so we begin with Step 2.

$$4(2x - 3) + 7 = 3x + 5$$

Step 2. $8x - 12 + 7 = 3x + 5$ Apply the distributive property.
Step 3. $8x - 5 = 3x + 5$ Combine like terms.
Step 4. Get all variable terms on the same side of the equation by subtracting $3x$ from both sides and then adding 5 to both sides.

$8x - 5 - 3x = 3x + 5 - 3x$ Subtract $3x$ from both sides.
$5x - 5 = 5$ Simplify.
$5x - 5 + 5 = 5 + 5$ Add 5 to both sides.
$5x = 10$ Simplify.

Step 5. Use the multiplication property of equality to get x alone.

$\frac{5x}{5} = \frac{10}{5}$ Divide both sides by 5.
$x = 2$ Simplify.

Step 6. Check.

$4(2x - 3) + 7 = 3x + 5$ Original equation
$4[2(2) - 3] + 7 \stackrel{?}{=} 3(2) + 5$ Replace x with 2.
$4(4 - 3) + 7 \stackrel{?}{=} 6 + 5$
$4(1) + 7 \stackrel{?}{=} 11$
$4 + 7 \stackrel{?}{=} 11$
$11 = 11$ True

The solution is 2.

Practice Problem 1

Solve: $5(3x - 1) + 2 = 12x + 6$

For Your Review:

Addition Property of Equality

If $a, b,$ and c are real numbers, then

$$a = b \quad \text{and} \quad a + c = b + c$$

are equivalent equations.
Since subtraction is defined in terms of addition, we may also subtract the same number from both sides of an equation.

Multiplication Property of Equality

If a, b, and c are real numbers and $c \neq 0$, then

$$a = b \quad \text{and} \quad ac = bc$$

are equivalent equations.
Since division is defined in terms of multiplication, we may also divide both sides of an equation by the same nonzero number.

Answer
1. $x = 3$

Helpful Hint

When checking solutions, use the original written equation.

Practice Problem 2

Solve: $9(5 - x) = -3x$

EXAMPLE 2 Solve: $8(2 - t) = -5t$

Solution: First, we apply the distributive property.

$$8(2 - t) = -5t$$

Step 2. $16 - 8t = -5t$ Use the distributive property.

Step 4. $16 - 8t + 8t = -5t + 8t$ Add $8t$ to both sides.

$16 = 3t$ Combine like terms.

Step 5. $\frac{16}{3} = \frac{3t}{3}$ Divide both sides by 3.

$\frac{16}{3} = t$ Simplify.

Step 6. Check.

$8(2 - t) = -5t$ Original equation

$8\left(2 - \frac{16}{3}\right) \stackrel{?}{=} -5\left(\frac{16}{3}\right)$ Replace t with $\frac{16}{3}$.

$8\left(\frac{6}{3} - \frac{16}{3}\right) \stackrel{?}{=} -\frac{80}{3}$ The LCD is 3.

$8\left(-\frac{10}{3}\right) \stackrel{?}{=} -\frac{80}{3}$ Subtract fractions.

$-\frac{80}{3} = -\frac{80}{3}$ True

The solution is $\frac{16}{3}$.

B Solving Equations Containing Fractions or Decimals

If an equation contains fractions, we can clear the equation of fractions by multiplying both sides by the LCD of all denominators. By doing this, we avoid working with time-consuming fractions.

Practice Problem 3

Solve: $\frac{5}{2}x - 1 = \frac{3}{2}x - 4$

EXAMPLE 3 Solve: $\frac{x}{2} - 1 = \frac{2}{3}x - 3$

Solution: We begin by clearing fractions. To do this, we multiply both sides of the equation by the LCD of 2 and 3, which is 6.

$$\frac{x}{2} - 1 = \frac{2}{3}x - 3$$

Step 1. $6\left(\frac{x}{2} - 1\right) = 6\left(\frac{2}{3}x - 3\right)$ Multiply both sides by the LCD, 6.

Step 2. $6\left(\frac{x}{2}\right) - 6(1) = 6\left(\frac{2}{3}x\right) - 6(3)$ Apply the distributive property.

$3x - 6 = 4x - 18$ Simplify.

Helpful Hint

Don't forget to multiply *each* term by the LCD.

Answers

2. $x = \frac{15}{2}$ **3.** $x = -3$

There are no longer grouping symbols and no like terms on either side of the equation, so we continue with Step 4.

$$3x - 6 = 4x - 18$$

Step 4. $3x - 6 - 3x = 4x - 18 - 3x$ Subtract $3x$ from both sides.

$-6 = x - 18$ Simplify.

$-6 + 18 = x - 18 + 18$ Add 18 to both sides.

$12 = x$ Simplify.

Step 5. The variable is now alone, so there is no need to apply the multiplication property of equality.

Step 6. Check.

$\frac{x}{2} - 1 = \frac{2}{3}x - 3$ Original equation

$\frac{12}{2} - 1 \stackrel{?}{=} \frac{2}{3} \cdot 12 - 3$ Replace x with 12.

$6 - 1 \stackrel{?}{=} 8 - 3$ Simplify.

$5 = 5$ True

The solution is 12.

EXAMPLE 4 Solve: $\frac{2(a + 3)}{3} = 6a + 2$

Solution: We clear the equation of fractions first.

$$\frac{2(a + 3)}{3} = 6a + 2$$

Step 1. $3 \cdot \frac{2(a + 3)}{3} = 3(6a + 2)$ Clear the fraction by multiplying both sides by the LCD, 3.

Step 2. Next, we use the distributive property and remove parentheses.

$2a + 6 = 18a + 6$ Apply the distributive property.

Step 4. $2a + 6 - 6 = 18a + 6 - 6$ Subtract 6 from both sides.

$2a = 18a$

$2a - 18a = 18a - 18a$ Subtract $18a$ from both sides.

$-16a = 0$

Step 5. $\frac{-16a}{-16} = \frac{0}{-16}$ Divide both sides by -16.

$a = 0$ Write the fraction in simplest form.

Step 6. To check, replace a with 0 in the original equation. The solution is 0.

When solving a problem about money, you may need to solve an equation containing decimals. If you choose, you may multiply to clear the equation of decimals.

Practice Problem 4

Solve: $\frac{3(x - 2)}{5} = 3x + 6$

Answer

4. $x = -3$

Practice Problem 5

Solve:
$0.06x - 0.10(x - 2) = -0.02(8)$

EXAMPLE 5 Solve: $0.25x + 0.10(x - 3) = 0.05(22)$

Solution: First we clear this equation of decimals by multiplying both sides of the equation by 100. Recall that multiplying a decimal number by 100 has the effect of moving the decimal point 2 places to the right.

$$0.25x + 0.10(x - 3) = 0.05(22)$$

Step 1. $0.25x + 0.10(x - 3) = 0.05(22)$ Multiply both sides by 100.

$$25x + 10(x - 3) = 5(22)$$

Step 2. $25x + 10x - 30 = 110$ Apply the distributive property.

Step 3. $35x - 30 = 110$ Combine like terms.

Step 4. $35x - 30 + 30 = 110 + 30$ Add 30 to both sides.

$35x = 140$ Combine like terms.

Step 5. $\frac{35x}{35} = \frac{140}{35}$ Divide both sides by 35.

$$x = 4$$

Step 6. To check, replace x with 4 in the original equation. The solution is 4.

C Recognizing Identities and Equations with No Solution

So far, each equation that we have solved has had a single solution. However, not every equation in one variable has a single solution. Some equations have no solution, while others have an infinite number of solutions. For example,

$$x + 5 = x + 7$$

has no solution since no matter which **real number** we replace x with, the equation is false.

real number + 5 = same real number + 7 FALSE

On the other hand,

$$x + 6 = x + 6$$

has infinitely many solutions since x can be replaced by any real number and the equation is always true.

real number + 6 = same real number + 6 TRUE

The equation $x + 6 = x + 6$ is called an **identity**. The next few examples illustrate special equations like these.

Answer

5. $x = 9$

EXAMPLE 6 Solve: $-2(x - 5) + 10 = -3(x + 2) + x$

Solution: $-2(x - 5) + 10 = -3(x + 2) + x$

$$-2x + 10 + 10 = -3x - 6 + x \quad \text{Apply the distributive property on both sides.}$$

$$-2x + 20 = -2x - 6 \quad \text{Combine like terms.}$$

$$-2x + 20 + 2x = -2x - 6 + 2x \quad \text{Add } 2x \text{ to both sides.}$$

$$20 = -6 \quad \text{Combine like terms.}$$

The final equation contains no variable terms, and there is no value for x that makes $20 = -6$ a true equation. We conclude that there is **no solution** to this equation.

EXAMPLE 7 Solve: $3(x - 4) = 3x - 12$

Solution: $3(x - 4) = 3x - 12$

$$3x - 12 = 3x - 12 \quad \text{Apply the distributive property.}$$

The left side of the equation is now identical to the right side. Every real number may be substituted for x and a true statement will result. We arrive at the same conclusion if we continue.

$$3x - 12 = 3x - 12$$

$$3x - 12 + 12 = 3x - 12 + 12 \quad \text{Add 12 to both sides.}$$

$$3x = 3x \quad \text{Combine like terms.}$$

$$3x - 3x = 3x - 3x \quad \text{Subtract } 3x \text{ from both sides.}$$

$$0 = 0$$

Again, one side of the equation is identical to the other side. Thus, $3(x - 4) = 3x - 12$ is an **identity** and **every real number** is a solution.

Try the Concept Check in the margin.

Practice Problem 6

Solve: $5(2 - x) + 8x = 3(x - 6)$

Practice Problem 7

Solve: $-6(2x + 1) - 14 = -10(x + 2) - 2x$

Concept Check

Suppose you have simplified several equations and obtain the following results. What can you conclude about the solutions to the original equation?

a. $7 = 7$ b. $x = 0$ c. $7 = -4$

Answers

6. no solution **7.** Every real number is a solution.

Concept Check: **a.** Every real number is a solution. **b.** The solution is 0. **c.** There is no solution.

CALCULATOR EXPLORATIONS

Checking Equations

We can use a calculator to check possible solutions of equations. To do this, replace the variable by the possible solution and evaluate both sides of the equation separately.

Equation: $3x - 4 = 2(x + 6)$ Solution: $x = 16$

$3x - 4 = 2(x + 6)$ Original equation

$3(16) - 4 \stackrel{?}{=} 2(16 + 6)$ Replace x with 16.

Now evaluate each side with your calculator.

Evaluate left side: [3] [×] [16] [−] [4] [=] or [ENTER] Display: [44]

Evaluate right side: [2] [(] [16] [+] [6] [)] [=] or [ENTER] Display: [44]

Since the left side equals the right side, the equation checks.

Use a calculator to check the possible solutions to each equation.

1. $2x = 48 + 6x;\quad x = -12$

2. $-3x - 7 = 3x - 1;\quad x = -1$

3. $5x - 2.6 = 2(x + 0.8);\quad x = 4.4$

4. $-1.6x - 3.9 = -6.9x - 25.6;\quad x = 5$

5. $\frac{564x}{4} = 200x - 11(649);\quad x = 121$

6. $20(x - 39) = 5x - 432;\quad x = 23.2$

Name ______________________ Section ________ Date ________

EXERCISE SET 9.3

 *Solve each equation. See Examples 1 and 2.*

1. $-4y + 10 = -2(3y + 1)$
2. $-3x + 1 = -2(x - 2)$
3. $9x - 8 = 10 + 15x$
4. $15x - 5 = 7 + 12x$
5. $-2(3x - 4) = 2x$
6. $-(5x - 10) = 5x$
7. $4(2n - 1) = (6n + 4) + 1$
8. $4(4y + 2) = 2(1 + 6y) + 8$
9. $5(2x - 1) - 2(3x) = 1$
10. $3(2 - 5x) + 4(6x) = 12$
11. $6(x - 3) + 10 = -8$
12. $-4(2 + n) + 9 = 1$
13. $8 - 2(a - 1) = 7 + a$
14. $5 - 6(2 + b) = b - 14$
15. $4x + 3 = 2x + 11$
16. $6y - 8 = 3y + 7$
17. $-2y - 10 = 5y + 18$
18. $7n + 5 = 10n - 10$
19. $-3(t - 5) + 2t = 5t - 4$
20. $-(4a - 7) - 5a = 10 + a$
21. $5y + 2(y - 6) = 4(y + 1) - 2$
22. $9x + 3(x - 4) = 10(x - 5) + 7$

B *Solve each equation. See Examples 3 through 5.*

23. $\frac{3}{4}x - \frac{1}{2} = 1$

24. $\frac{2}{3}x + \frac{5}{3} = \frac{5}{3}$

25. $x + \frac{5}{4} = \frac{3}{4}x$

26. $\frac{7}{8}x + \frac{1}{4} = \frac{3}{4}x$

27. $\frac{x}{2} - 1 = \frac{x}{5} + 2$

28. $\frac{x}{5} - 2 = \frac{x}{3}$

29. $\frac{6(3 - z)}{5} = -z$

30. $\frac{4(5 - w)}{3} = -w$

31. $0.06 - 0.01(x + 1) = -0.02(2 - x)$

32. $-0.01(5x + 4) = 0.04 - 0.01(x + 4)$

33. $\frac{3(x - 5)}{2} = \frac{2(x + 5)}{3}$

34. $\frac{5(x - 1)}{4} = \frac{3(x + 1)}{2}$

35. $0.50x + 0.15(70) = 0.25(142)$

36. $0.40x + 0.06(30) = 0.20(49)$

37. $0.12(y - 6) + 0.06y = 0.08y - 0.07(10)$

38. $0.60(z - 300) + 0.05z = 0.70z - 0.41(500)$

39. $\frac{2(x + 1)}{4} = 3x - 2$

40. $\frac{3(y + 3)}{5} = 2y + 6$

41. $x + \frac{7}{6} = 2x - \frac{7}{6}$

42. $\frac{5}{2}x - 1 = x + \frac{1}{4}$

43. $\frac{9}{2} + \frac{5}{2}y = 2y - 4$

44. $3 - \frac{1}{2}x = 5x - 8$

45. Explain the difference between simplifying an expression and solving an equation.

46. When solving an equation, if an equivalent equation is $0 = 5$, what can we conclude? If an equivalent equation is $-2 = -2$, what can we conclude?

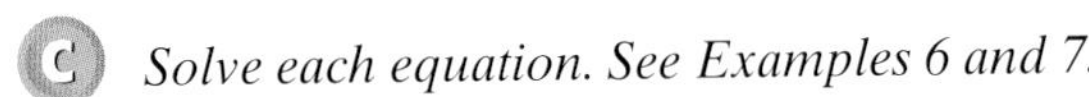

Solve each equation. See Examples 6 and 7.

47. $5x - 5 = 2(x + 1) + 3x - 7$

48. $3(2x - 1) + 5 = 6x + 2$

49. $\frac{x}{4} + 1 = \frac{x}{4}$

50. $\frac{x}{3} - 2 = \frac{x}{3}$

51. $3x - 7 = 3(x + 1)$

52. $2(x - 5) = 2x + 10$

53. $2(x + 3) - 5 = 5x - 3(1 + x)$

54. $4(2 + x) + 1 = 7x - 3(x - 2)$

55. On your own, construct an equation for which every real number is a solution.

56. On your own, construct an equation that has no solution.

Review and Preview

Write each algebraic expression described. See Section 3.5.

57. A plot of land is in the shape of a triangle. If one side is x meters, a second side is $(2x - 3)$ meters, and a third side is $(3x - 5)$ meters, express the perimeter of the lot as a simplified expression in x.

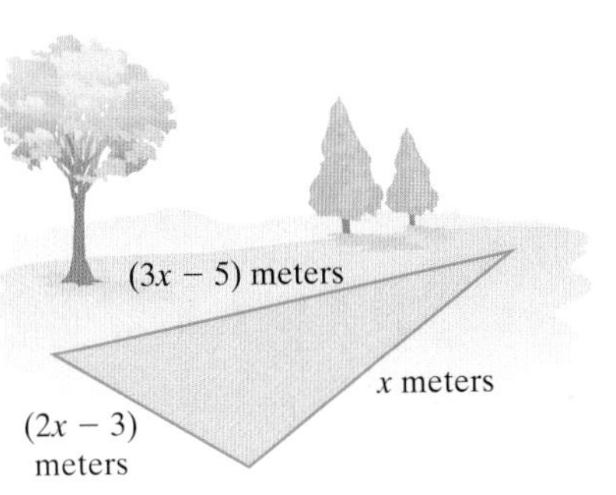

58. A portion of a board has length x feet. The other part has length $(7x - 9)$ feet. Express the total length of the board as a simplified expression in x.

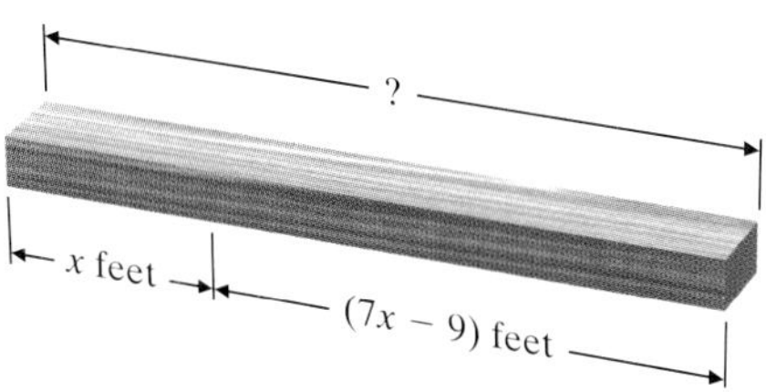

Write each phrase as an algebraic expression. Use x for the unknown number.

59. A number subtracted from -8

60. Three times a number

61. The sum of -3 and twice a number

62. The difference of 8 and twice a number

63. The product of 9 and the sum of a number and 20

64. The quotient of -12 and the difference of a number and 3

Combining Concepts

Solve.

65. $1000(7x - 10) = 50(412 + 100x)$

66. $1000(x + 40) = 100(16 + 7x)$

67. $0.035x + 5.112 = 0.010x + 5.107$

68. $0.127x - 2.685 = 0.027x - 2.38$

69. The perimeter of a geometric figure is the sum of the lengths of its sides. If the perimeter of the following pentagon (five-sided figure) is 28 centimeters, find the length of each side.

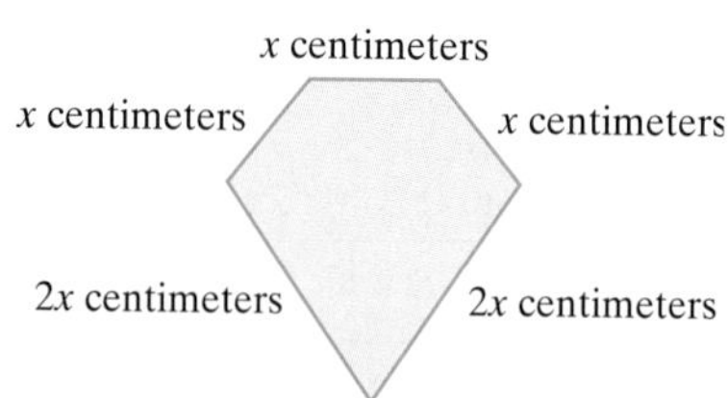

70. The perimeter of the following triangle is 35 meters. Find the length of each side.

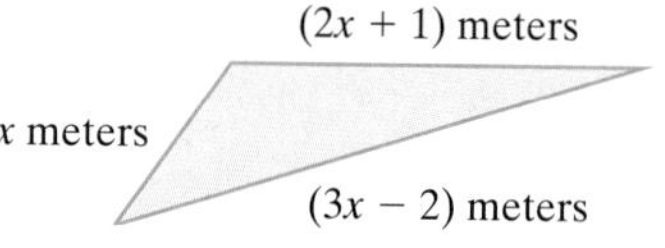

Name ______ Section ______ Date ______

Answers

Integrated Review—Real Numbers and Solving Linear Equations

Tell which set or sets each number belongs to: natural numbers, whole numbers, integers, rational numbers, irrational numbers, and real numbers.

1. 0

2. 143

3. $\frac{3}{8}$

4. 1

5. -13

6. $\frac{9}{10}$

7. $-\frac{1}{9}$

8. $\sqrt{5}$

Remove parentheses and simplify each expression.

9. $7(d - 3) + 10$

10. $9(z + 7) - 15$

11. $-4(3y - 4) + 12y$

12. $-3(2x + 5) - 6x$

Solve. Feel free to use the steps given in Section 9.3.

13. $2x - 7 = 6x - 27$

14. $3 + 8y = 3y - 2$

15. $-3a + 6 + 5a = 7a - 8a$

16. $4b - 8 - b = 10b - 3b$

17. $-\frac{2}{3}x = \frac{5}{9}$

18. $-\frac{3}{8}y = -\frac{1}{16}$

19. $10 = -6n + 16$

20. $-5 = -2m + 7$

1. ______
2. ______
3. ______
4. ______
5. ______
6. ______
7. ______
8. ______
9. ______
10. ______
11. ______
12. ______
13. ______
14. ______
15. ______
16. ______
17. ______
18. ______
19. ______
20. ______

21. ____________

22. ____________

23. ____________

24. ____________

25. ____________

26. ____________

27. ____________

28. ____________

29. ____________

30. ____________

31. ____________

32. ____________

21. $3(5c - 1) - 2 = 13c + 3$

22. $4(3t + 4) - 20 = 3 + 5t$

23. $\frac{2(z + 3)}{3} = 5 - z$

24. $\frac{3(w + 2)}{4} = 2w + 3$

25. $-2(2x - 5) = -3x + 7 - x + 3$

26. $-4(5x - 2) = -12x + 4 - 8x + 4$

27. $0.02(6t - 3) = 0.04(t - 2) + 0.02$

28. $0.03(m + 7) = 0.02(5 - m) + 0.03$

29. $-3y = \frac{4(y - 1)}{5}$

30. $-4x = \frac{5(1 - x)}{6}$

31. $\frac{5}{3}x - \frac{7}{3} = x$

32. $\frac{7}{5}n + \frac{3}{5} = -n$

9.4 Further Problem Solving

In this section, we review our problem-solving steps first introduced in Section 3.5 and continue to solve problems that are modeled by linear equations.

OBJECTIVE

A Translate a problem to an equation, then use the equation to solve the problem.

SSM TUTOR CENTER SG CD & VIDEO MATH PRO WEB

General Strategy for Problem Solving

1. UNDERSTAND the problem. During this step, become comfortable with the problem. Some ways of doing this are:
   Read and reread the problem.
   Choose a variable to represent the unknown.
   Construct a drawing.
   Propose a solution and check. Pay careful attention to how you check your proposed solution. This will help when writing an equation to model the problem.
2. TRANSLATE the problem into an equation.
3. SOLVE the equation.
4. INTERPRET the results: *Check* the proposed solution in the stated problem and *state* your conclusion.

A Translating and Solving Problems

Much of problem solving involves a direct translation from a sentence to an equation.

EXAMPLE 1 Finding an Unknown Number

Twice the sum of a number and 4 is the same as four times the number, decreased by 12. Find the number.

Solution:

1. UNDERSTAND. Read and reread the problem. If we let
   x = the unknown number, then
   "the sum of a number and 4" translates to "$x + 4$" and
   "four times the number" translates to "$4x$"
2. TRANSLATE.

twice	sum of a number and 4	is the same as	four times the number	decreased by	12
↓	↓	↓	↓	↓	↓
2	$(x + 4)$	=	$4x$	−	12

3. SOLVE

$$2(x + 4) = 4x - 12$$
$$2x + 8 = 4x - 12 \quad \text{Apply the distributive property.}$$
$$2x + 8 - 4x = 4x - 12 - 4x \quad \text{Subtract } 4x \text{ from both sides.}$$
$$-2x + 8 = -12$$
$$-2x + 8 - 8 = -12 - 8 \quad \text{Subtract 8 from both sides.}$$
$$-2x = -20$$
$$\frac{-2x}{-2} = \frac{-20}{-2} \quad \text{Divide both sides by } -2.$$
$$x = 10$$

Practice Problem 1

Three times the difference of a number and 5 is the same as twice the number decreased by 3. Find the number.

Answer

1. The number is 12.

4. INTERPRET.

Check: Check this solution in the problem as it was originally stated. To do so, replace "number" with 10. Twice the sum of "10" and 4 is 28, which is the same as 4 times "10" decreased by 12.

State: The number is 10.

Practice Problem 2

An 18-foot wire is to be cut so that the longer piece is 5 times longer than the shorter piece. Find the length of each piece.

EXAMPLE 2 Finding the Length of a Board

A 10-foot board is to be cut into two pieces so that the longer piece is 4 times the shorter. Find the length of each piece.

Solution:

1. UNDERSTAND the problem. To do so, read and reread the problem. You may also want to propose a solution. For example, if 3 feet represents the length of the shorter piece, then $4(3) = 12$ feet is the length of the longer piece, since it is 4 times the length of the shorter piece. This guess gives a total board length of 3 feet + 12 feet = 15 feet, which is too long. However, the purpose of proposing a solution is not to guess correctly, but to help better understand the problem and how to model it.

In general, if we let

x = length of shorter piece, then
$4x$ = length of longer piece

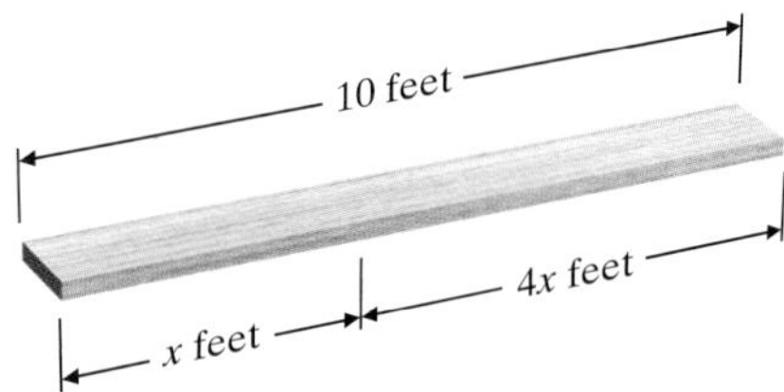

2. TRANSLATE the problem. First, we write the equation in words.

length of shorter piece	added to	length of longer piece	equals	total length of board
↓	↓	↓	↓	↓
x	$+$	$4x$	$=$	10

3. SOLVE.

$$x + 4x = 10$$
$$5x = 10 \quad \text{Combine like terms.}$$
$$\frac{5x}{5} = \frac{10}{5} \quad \text{Divide both sides by 5.}$$
$$x = 2$$

4. INTERPRET.

Check: Check the solution in the stated problem. If the shorter piece of board is 2 feet, the longer piece is $4 \cdot (2 \text{ feet}) = 8$ feet and the sum of the two pieces is 2 feet + 8 feet = 10 feet.

Answer

2. shorter piece = 3 ft; longer piece = 15 ft

State: The shorter piece of board is 2 feet and the longer piece of board is 8 feet.

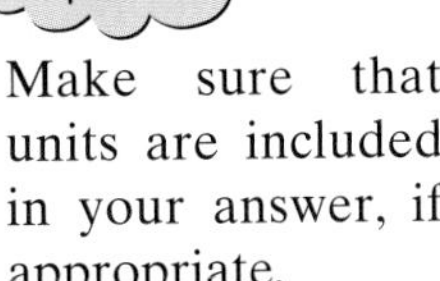

Make sure that units are included in your answer, if appropriate.

EXAMPLE 3 Finding the Number of Republican and Democratic Senators

In the 107th Congress, the U.S. House of Representatives had a total of 430 Democrats and Republicans. There were 10 more Republican representatives than Democratic. Find the number of representatives from each party. (*Source:* Office of the Clerk of the U.S. House of Representatives)

Solution:

1. UNDERSTAND the problem. Read and reread the problem. Let's suppose that there are 200 Democratic representatives. Since there are 10 more Republicans than Democrats, there must be $200 + 10 = 210$ Republicans. The total number of Democrats and Republicans is then $200 + 210 = 410$. This is incorrect since the total should be 430, but we now have a better understanding of the problem.

In general, if we let

$$x = \text{number of Democrats, then}$$
$$x + 10 = \text{number of Republicans}$$

2. TRANSLATE the problem. First, we write the equation in words.

number of Democrats	added to	number of Republicans	equals	430
↓	↓	↓	↓	↓
x	$+$	$(x + 10)$	$=$	430

3. SOLVE.

$$x + (x + 10) = 430$$
$$2x + 10 = 430 \quad \text{Combine like terms.}$$
$$2x + 10 - 10 = 430 - 10 \quad \text{Subtract 10 from both sides.}$$
$$2x = 420$$
$$\frac{2x}{2} = \frac{420}{2} \quad \text{Divide both sides by 2.}$$
$$x = 210$$

4. INTERPRET.

Check: If there are 210 Democratic representatives, then there are $210 + 10 = 220$ Republican representatives. The total number of representatives is then $210 + 220 = 430$. The results check.

State: There are 210 Democratic and 220 Republican representatives in the 107th Congress.

Practice Problem 3

Through the year 2000, the state of California had 22 more electoral votes for president than the state of Texas. If the total electoral votes for these two states was 86, find the number of electoral votes for each state.

Answer

3. Texas = 32 electoral votes; California = 54 electoral votes

Practice Problem 4

Enterprise Car Rental charges a daily rate of \$34 plus \$0.20 per mile. Suppose that you rent a car for a day and your bill (before taxes) is \$104. How many miles did you drive?

EXAMPLE 4 Calculating Cellular Phone Usage

A local cellular phone company charges Elaine Chapoton \$50 per month and \$0.36 per minute of phone use in her usage category. If Elaine was charged \$99.68 for a month's cellular phone use, determine the number of whole minutes of phone use.

Solution:

1. UNDERSTAND. Read and reread the problem. Let's propose that Elaine uses the phone for 70 minutes. Pay careful attention as to how we calculate her bill. For 70 minutes of use, Elaine's phone bill will be \$50 plus \$0.36 per minute of use. This is $\$50 + 0.36(70) = \$75.20$, less than \$99.68. We now understand the problem and know that the number of minutes is greater than 70.

If we let

x = number of minutes, then

$0.36x$ = charge per minute of phone use

2. TRANSLATE.

\$50	added to	minute charge	is equal to	\$99.68
↓	↓	↓	↓	↓
50	+	$0.36x$	=	99.68

3. SOLVE.

$$50 + 0.36x = 99.68$$

$$50 + 0.36x - 50 = 99.68 - 50 \quad \text{Subtract 50 from both sides.}$$

$$0.36x = 49.68 \quad \text{Simplify.}$$

$$\frac{0.36x}{0.36} = \frac{49.68}{0.36} \quad \text{Divide both sides by 0.36.}$$

$$x = 138 \quad \text{Simplify.}$$

4. INTERPRET.

Check: If Elaine spends 138 minutes on her cellular phone, her bill is $\$50 + \$0.36(138) = \$99.68$.

State: Elaine spent 138 minutes on her cellular phone this month.

Practice Problem 5

The measure of the second angle of a triangle is twice the measure of the smallest angle. The measure of the third angle of the triangle is three times the measure of the smallest angle. Find the measures of the angles.

Answers

4. 350 miles **5.** smallest = 30°; second = 60°; third = 90°

EXAMPLE 5 Finding Angle Measures

If the two walls of the Vietnam Veterans Memorial in Washington, D.C., were connected, an isosceles triangle would be formed. The measure of the third angle is 97.5° more than the measure of either of the other two equal angles. Find the measure of the third angle. (*Source:* National Park Service)

Solution:

1. UNDERSTAND. Read and reread the problem. We then draw a diagram (recall that an isosceles triangle has two angles with the same measure) and let

x = degree measure of one angle
x = degree measure of the second equal angle
$x + 97.5$ = degree measure of the third angle

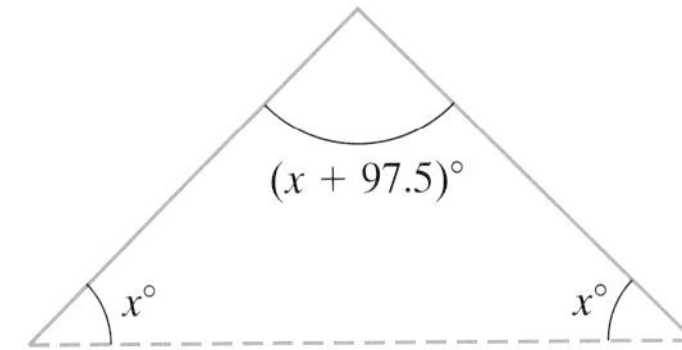

2. TRANSLATE. Recall that the sum of the measures of the angles of a triangle equals 180.

measure of first angle		measure of second angle	measure of third angle	equals	180
↓		↓	↓	↓	↓
x	$+$	x	$+ (x + 97.5)$	$=$	180

3. SOLVE.

$$x + x + (x + 97.5) = 180$$
$$3x + 97.5 = 180 \quad \text{Combine like terms.}$$
$$3x + 97.5 - 97.5 = 180 - 97.5 \quad \text{Subtract 97.5 from both sides.}$$
$$3x = 82.5$$
$$\frac{3x}{3} = \frac{82.5}{3} \quad \text{Divide both sides by 3.}$$
$$x = 27.5$$

4. INTERPRET.

Check: If $x = 27.5$, then the measure of the third angle is $x + 97.5 = 125$. The sum of the angles is then $27.5 + 27.5 + 125 = 180$, the correct sum.

State: The third angle measures 125°.*

*(The two walls actually meet at an angle of 125 degrees 12 minutes. The measurement of 97.5° given in the problem is an approximation.)

Some of our applications in the exercise set have to do with consecutive integers. To prepare for these exercises, let's practice writing expressions for consecutive integers.

Practice Problem 6

If x is the first of two consecutive integers, express the sum of the first and the second integer in terms of x. Simplify if possible.

EXAMPLE 6 Writing an Expression for Consecutive Integers

If x is the first of three consecutive integers, express the sum of the three integers in terms of x. Simplify if possible.

Solution: An example of three consecutive integers is 7, 8, and 9.

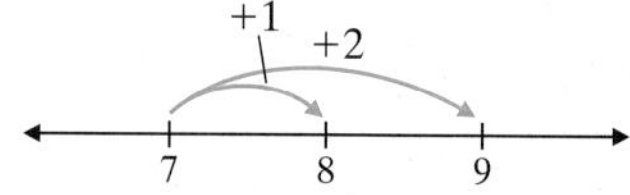

The second consecutive integer is always 1 more than the first, and the third consecutive integer is 2 more than the first. If x is the first of three consecutive integers, the three consecutive integers are x, $x + 1$, and $x + 2$.

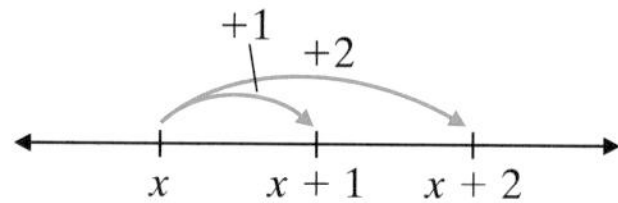

Their sum is shown below.

In words:	first integer	+	second integer	+	third integer
Translate:	x	+	$(x + 1)$	+	$(x + 2)$

This simplifies to $3x + 3$.

Study these examples of consecutive even and consecutive odd integers.

Consecutive even integers:

Example:

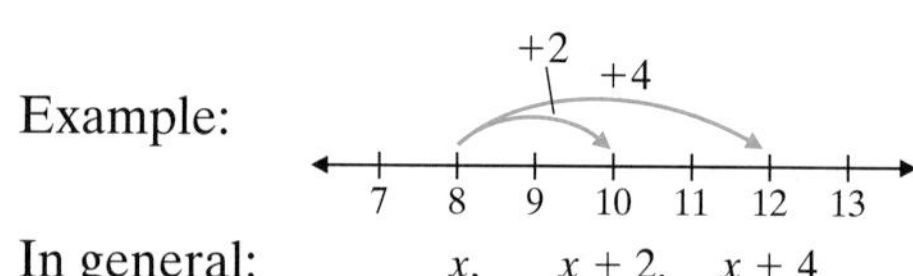

In general: x, $x + 2$, $x + 4$

Consecutive odd integers:

Example:

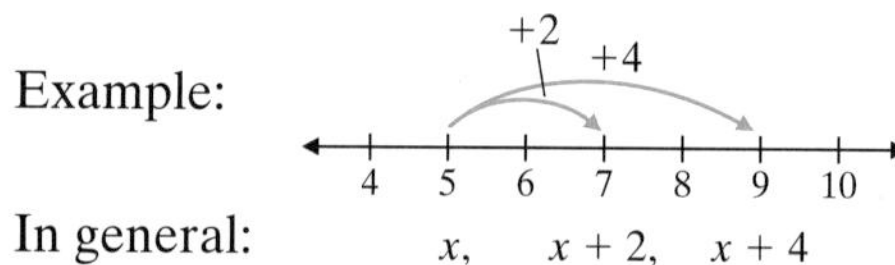

In general: x, $x + 2$, $x + 4$

Helpful Hint

If x is an odd integer, then $x + 2$ is the next odd integer. This 2 simply means that odd integers are always 2 units from each other.

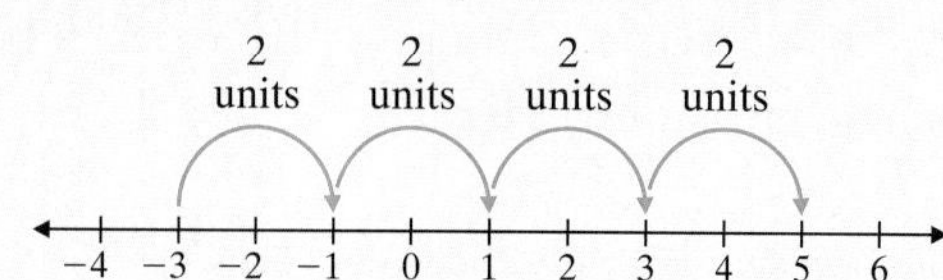

Answer

6. $2x + 1$

Name ______________________ Section __________ Date __________

EXERCISE SET 9.4

 *Solve. See Example 1.*

1. The sum of twice a number and $\frac{1}{5}$ is equal to the difference between three times the number and $\frac{4}{5}$. Find the number.

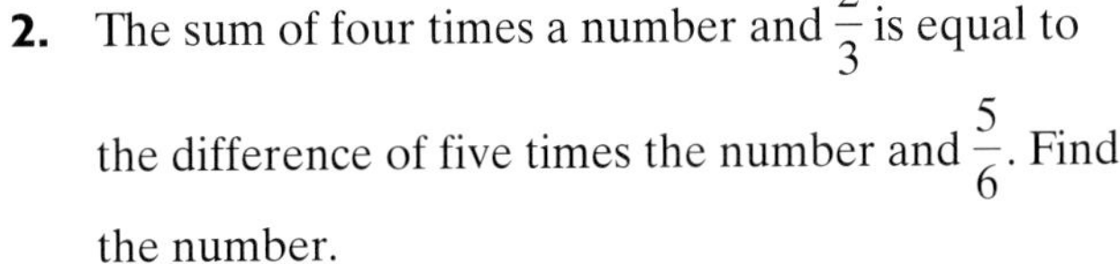

2. The sum of four times a number and $\frac{2}{3}$ is equal to the difference of five times the number and $\frac{5}{6}$. Find the number.

3. Twice the difference of a number and 8 is equal to three times the sum of the number and 3. Find the number.

4. Five times the sum of a number and -1 is the same as 6 times the number. Find the number.

5. The product of twice a number and three is the same as the difference of five times the number and $\frac{3}{4}$. Find the number.

6. If the difference of a number and four is doubled, the result is $\frac{1}{4}$ less than the number. Find the number.

7. If the sum of a number and five is tripled, the result is one less than twice the number. Find the number.

8. Twice the sum of a number and six equals three times the sum of the number and four. Find the number.

Solve. See Examples 2 through 5.

9. The governor of Michigan makes $83,400 more per year than the governor of Oregon. If the total of their salaries is $270,600, find the salary of each. (*Source: The World Almanac, 2003*)

10. In the 2000 Summer Olympics, the United States Team won 12 more gold medals than the China Team. If the total number of gold medals for both is 68, find the number of gold medals that each team won. (*Source: The World Almanac, 2003*)

11. A 40-inch board is to be cut into three pieces so that the second piece is twice as long as the first piece and the third piece is 5 times as long as the first piece. If x represents the length of the first piece, find the lengths of all three pieces.

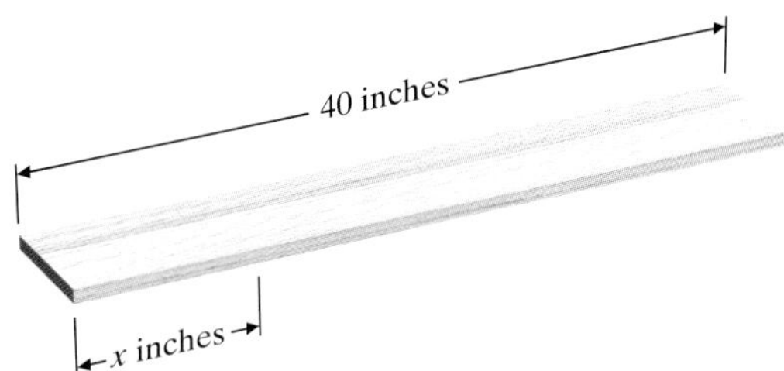

12. A 21-foot beam is to be divided so that the longer piece is 1 foot more than 3 times the shorter piece. If x represents the length of the shorter piece, find the lengths of both pieces.

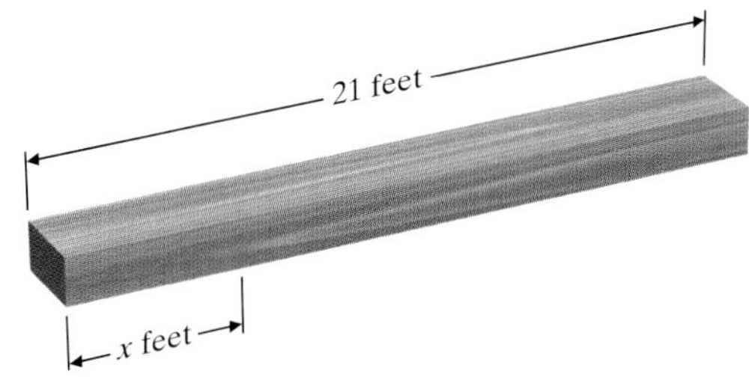

13. A car rental agency advertised renting a Buick Century for \$24.95 per day and \$0.29 per mile. If you rent this car for 2 days, how many whole miles can you drive on a \$100 budget?

14. A plumber gave an estimate for the renovation of a kitchen. Her hourly pay is \$27 per hour and the plumber's parts will cost \$80. If her total estimate is \$404, how many hours does she expect this job to take?

△ 15. The flag of Equatorial Guinea contains an isosceles triangle. (Recall that an isosceles triangle contains two angles with the same measure.) If the measure of the third angle of the triangle is 30° more than twice the measure of either of the other two angles, find the measure of each angle of the triangle. (*Hint:* Recall that the sum of the measures of the angles of a triangle is 180°.)

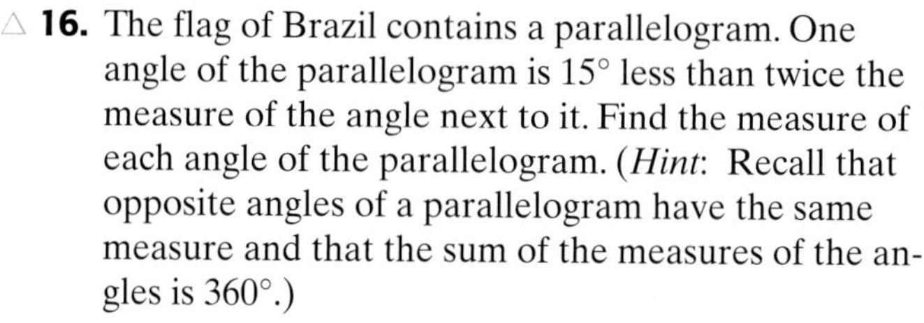

△ 16. The flag of Brazil contains a parallelogram. One angle of the parallelogram is 15° less than twice the measure of the angle next to it. Find the measure of each angle of the parallelogram. (*Hint*: Recall that opposite angles of a parallelogram have the same measure and that the sum of the measures of the angles is 360°.)

17. In the Crayola Color Census 2000, Crayola crayon users were asked to vote for their favorite Crayola color. The color blue was ranked first, followed by cerulean in second place. Blue received 3366 more votes than cerulean. Together, both colors received a total of 19,278 votes. Find the number of votes each color received. (*Source:* Binney & Smith, Inc.)

18. In 2001 the U.S. poverty threshold for a family of four with two children was \$501 more than the poverty threshold for the same family in 2000. The sum of the poverty thresholds for 2000 and 2001 was \$35,707. What was the poverty threshold in each year? (*Source:* U.S. Census Bureau)

△ 19. Two angles are supplementary if their sum is 180°. One angle measures three times the measure of a smaller angle. If x represents the measure of the smaller angle and these two angles are supplementary, find the measure of each angle.

$x°$

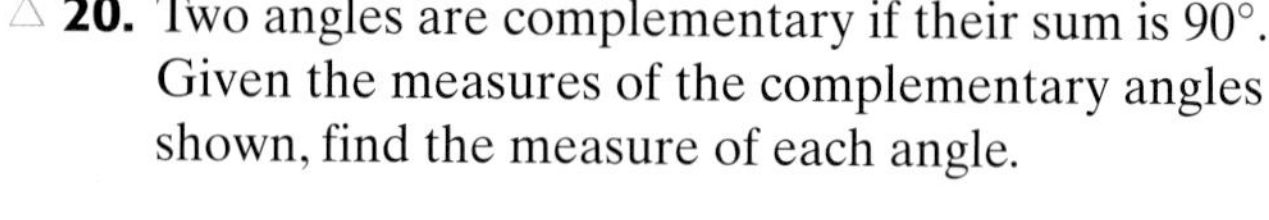

△ 20. Two angles are complementary if their sum is 90°. Given the measures of the complementary angles shown, find the measure of each angle.

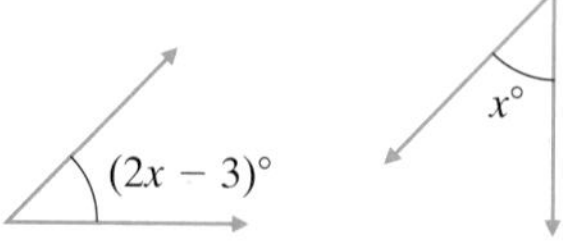

21. A 17-foot piece of string is cut into two pieces so that one piece is 2 feet longer than twice the shorter piece. If the shorter piece is x feet long, find the lengths of both pieces.

22. An 18-foot wire is to be cut so that the longer piece is 5 times longer than the shorter piece. Find the length of each piece.

23. On April 7, 2001, the Mars Odyssey spacecraft was launched, beginning a multi-year mission to observe and map the planet Mars. Mars Odyssey was launched on Boeing's Delta II 7925 launch vehicle using nine strap-on solid rocket motors. Each solid rocket motor has a height that is 8 meters more than 5 times its diameter. If the sum of the height and the diameter for a single solid rocket motor is 14 meters, find each dimension. (*Source:* NASA)

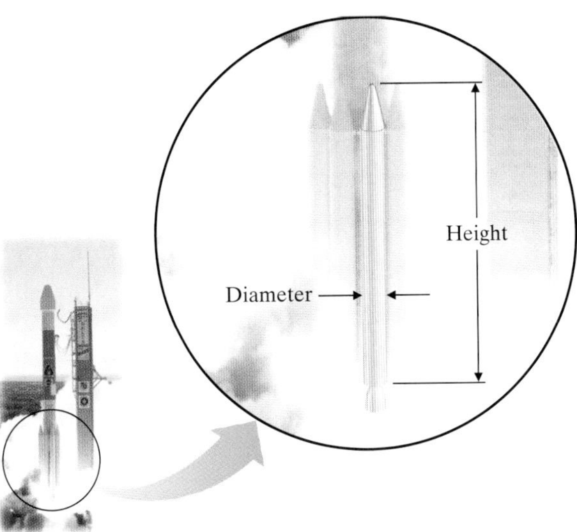

24. Over the past few years the satellite Voyager II has passed by the planets Saturn, Uranus, and Neptune, continually updating information about these planets, including the number of moons for each. Uranus is now believed to have 13 more moons than Neptune. Also, Saturn is now believed to have 2 more than twice the number of moons of Neptune. If the total number of moons for these planets is 47, find the number of moons for each planet. (*Source:* National Space Science Data Center)

25. The area of the Sahara Desert is 7 times the area of the Gobi Desert. If the sum of their areas is 4,000,000 square miles, find the area of each desert.

26. The largest meteorite in the world is the Hoba West located in Namibia. Its weight is 3 times the weight of the Armanty meteorite located in Outer Mongolia. If the sum of their weights is 88 tons, find the weight of each.

The graph below shows the states with the highest tourism budgets for a recent year. Use the graph for Exercises 27 through 32.

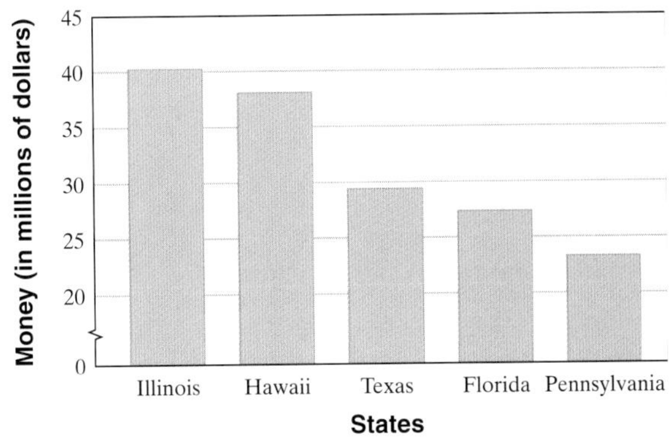

Source: Travel Industry Association of America

27. In your own words, describe what the word *tourism* means.

28. Which state spends the most money on tourism?

29. Which states spend between \$25 and \$30 million on tourism?

30. The states of Texas and Florida spend a total of \$56.6 million for tourism. The state of Texas spends \$2.2 million more than the state of Florida. Find the amount that each state spends on tourism.

31. The states of Hawaii and Pennsylvania spend a total of \$60.9 million for tourism. The state of Hawaii spends \$8.1 million less than twice the amount of money that the state of Pennsylvania spends. Find the amount that each state spends on tourism.

32. Compare the heights of the bars in the graph with your results of Exercises 30 and 31. Are your answers reasonable?

33. The Pentagon in Washington, D.C., is the headquarters for the U.S. Department of Defense. The Pentagon is also the world's largest office building in terms of ground space with a floor area of over 6.5 million square feet. This is three times the floor area of the Empire State Building. About how much floor space does the Empire State Building have? Round to the nearest tenth of a million.

34. Hertz Car Rental charges a daily rate of \$39 plus \$0.20 per mile for a certain car. Suppose that you rent that car for a day and your bill (before taxes) is \$95. How many miles did you drive?

Solve. These applications have to do with consecutive integers. For a review of consecutive integers, see Example 6.

35. On April 1, 2001, Notre Dame defeated Purdue in the 2001 NCAA Division I Women's Basketball Championship. The two teams' final scores for the game were two consecutive even integers whose sum was 134. Find each final score. (*Source:* National Collegiate Athletic Association)

36. The number of counties in California and the number of counties in Montana are consecutive even integers whose sum is 114. If California has more counties than Montana, how many counties does each state have? (*Source: The World Almanac and Book of Facts 2003*)

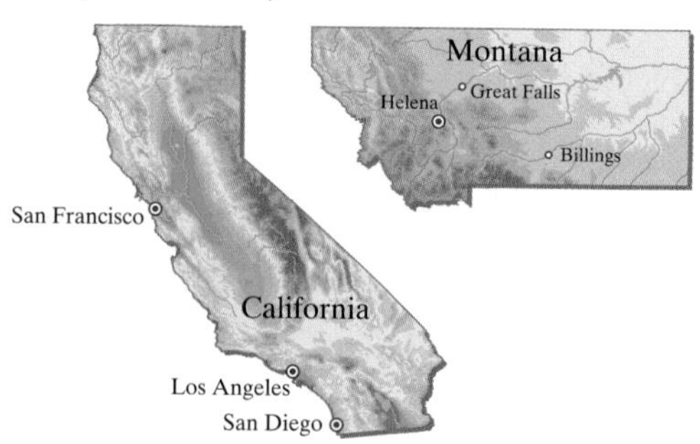

37. In the 2000 Summer Olympics, China won more medals than Australia, who won more medals than Germany. If the number of medals won by each country is three consecutive integers whose sum is 174, how many medals did each country win? (*Source: The World Almanac and Book of Facts 2003*)

38. To make an international telephone call, you need the code for the country you are calling. The codes for Mali Republic, Côte d'Ivoire, and Niger are three consecutive odd integers whose sum is 675. Find the code for each country.

△ 39. The measures of the angles of a triangle are 3 consecutive even integers. Find the measure of each angle.

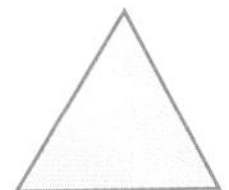

△ 40. A quadrilateral is a polygon with 4 sides. The sum of the measures of the 4 angles in a quadrilateral is 360°. If the measures of the angles of a quadrilateral are consecutive odd integers, find the measures.

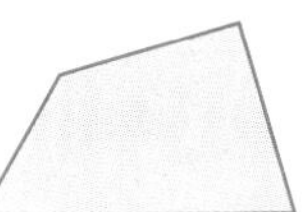

Review and Preview

Translate each sentence into an equation. See Sections 9.1 and 9.4.

41. Half of the difference of a number and one is thirty-seven.

42. Five times the opposite of a number is the number plus sixty.

43. If three times the sum of a number and 2 is divided by 5, the quotient is 0.

44. If the sum of a number and 9 is subtracted from 50, the result is 0.

Evaluate each expression for the given values. See Section 1.9.

45. $2W + 2L$; $W = 7$ and $L = 10$

46. $\frac{1}{2}Bh$; $B = 14$ and $h = 22$

47. πr^2; $r = 15$

48. $r \cdot t$; $r = 15$ and $t = 2$

Combining Concepts

49. Give an example of how you recently solved a problem using mathematics.

50. In your own words, explain why a solution of a word problem should be checked using the original wording of the problem and not the equation written from the wording.

△ **51.** The golden rectangle is a rectangle whose length is approximately 1.6 times its width. The early Greeks thought that a rectangle with these dimensions was the most pleasing to the eye and examples of the golden rectangle are found in many early works of art. For example, the Parthenon in Athens contains many examples of golden rectangles. Mike Hallahan would like to plant a rectangular garden in the shape of a golden rectangle. If he has 78 feet of fencing available, find the dimensions of the garden.

The length-width rectangle approximates the golden rectangle as well as the width-height rectangle.

△ **53.** Examples of golden rectangles can be found today in architecture and manufacturing packaging. Find an example of a golden rectangle in your home. A few suggestions: the front face of a book, the floor of a room, the front of a box of food.

△ **52.** It is thought that for about 75% of adults, a rectangle in the shape of the golden rectangle is the most pleasing to the eye. Draw 3 rectangles, one in the shape of the golden rectangle, and poll your class. Do the results agree with the percentage given above?

9.5 Formulas and Problem Solving

OBJECTIVES

A. Use formulas to solve problems.

B. Solve a formula or equation for one of its variables.

SSM TUTOR CENTER SG CD & VIDEO MATH PRO WEB

A Using Formulas to Solve Problems

A **formula** describes a known relationship among quantities. Many formulas are given as equations. For example, the formula

$$d = r \cdot t$$

stands for the relationship

$$\text{distance} = \text{rate} \cdot \text{time}$$

Let's look at one way that we can use this formula.

If we know we traveled a distance of 100 miles at a rate of 40 miles per hour, we can replace the variables d and r in the formula $d = rt$ and find our travel time, t.

$d = rt$ Formula

$100 = 40t$ Replace d with 100 and r with 40.

To solve for t, we divide both sides of the equation by 40.

$\frac{100}{40} = \frac{40t}{40}$ Divide both sides by 40.

$\frac{5}{2} = t$ Simplify.

The travel time was $\frac{5}{2}$ hours, or $2\frac{1}{2}$ hours.

In this section, we solve problems that can be modeled by known formulas. We use the same problem-solving strategy that was introduced in the previous section.

EXAMPLE 1 Finding Time Given Rate and Distance

A glacier is a giant mass of rocks and ice that flows downhill like a river. Portage Glacier in Alaska is about 6 miles, or 31,680 *feet*, long and moves 400 *feet* per year. Icebergs are created when the front end of the glacier flows into Portage Lake. How long does it take for ice at the head (beginning) of the glacier to reach the lake?

Practice Problem 1

A family is planning their vacation to visit relatives. They will drive from Cincinnati, Ohio to Rapid City, South Dakota, a distance of 1180 miles. They plan to average a rate of 50 miles per hour. How much time will they spend driving?

Answer

1. 23.6 hours

Solution:

1. UNDERSTAND. Read and reread the problem. The appropriate formula needed to solve this problem is the distance formula, $d = rt$. To become familiar with this formula, let's find the distance that ice traveling at a rate of 400 feet per year travels in 100 years. To do so, we let time t be 100 years and rate r be the given 400 feet per year, and substitute these values into the formula $d = rt$. We then have that distance $d = 400(100) = 40{,}000$ feet. Since we are interested in finding how long it takes ice to travel 31,680 feet, we now know that it is less than 100 years.

   Since we are using the formula $d = rt$, we let

   $t =$ the time in years for ice to reach the lake

   $r =$ rate or speed of ice

   $d =$ distance from beginning of glacier to lake

2. TRANSLATE. To translate to an equation, we use the formula $d = rt$ and let distance $d = 31{,}680$ feet and rate $r = 400$ feet per year.

$$d = r \cdot t$$

$$31{,}680 = 400 \cdot t \qquad \text{Let } d = 31{,}680 \text{ and } r = 400.$$

3. SOLVE. Solve the equation for t. To solve for t, divide both sides by 400.

$$\frac{31{,}680}{400} = \frac{400 \cdot t}{400} \qquad \text{Divide both sides by 400.}$$

$$79.2 = t \qquad \text{Simplify.}$$

4. INTERPRET.

Check: To check, substitute 79.2 for t and 400 for r in the distance formula and check to see that the distance is 31,680 feet.

State: It takes 79.2 years for the ice at the head of Portage Glacier to reach the lake.

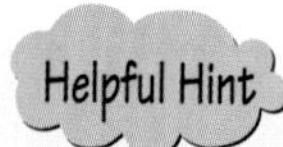

Don't forget to include units, if appropriate.

Practice Problem 2

A wood deck is being built behind a house. The width of the deck must be 18 feet because of the shape of the house. If there is 450 square feet of decking material, find the length of the deck.

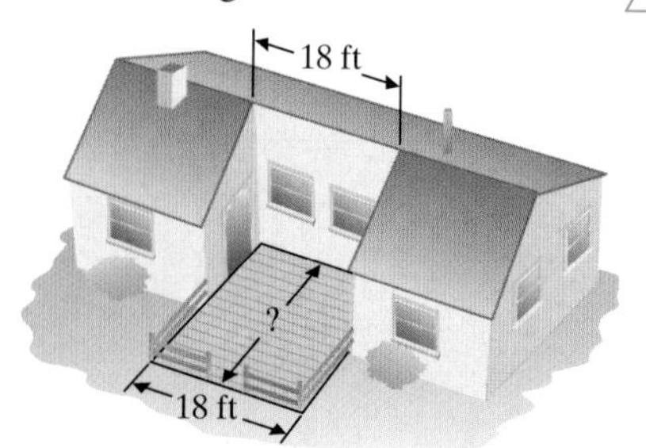

Answer

2. 25 feet

EXAMPLE 2 Calculating the Length of a Garden

Charles Pecot can afford enough fencing to enclose a rectangular garden with a perimeter of 140 feet. If the width of his garden is to be 30 feet, find the length.

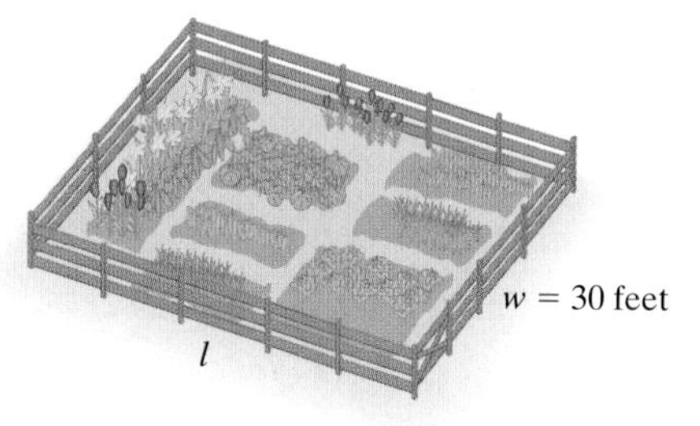

Solution:

1. UNDERSTAND. Read and reread the problem. The formula needed to solve this problem is the formula for the perimeter of a rectangle, $P = 2l + 2w$. Before continuing, let's become familar with this formula.

   l = the length of the rectangular garden
   w = the width of the rectangular garden
   P = perimeter of the garden

2. TRANSLATE. To translate to an equation, we use the formula $P = 2l + 2w$ and let perimeter $P = 140$ feet and width $w = 30$ feet.

   $P = 2l + 2w$ Let $P = 140$ and $w = 30$.

   $140 = 2l + 2(30)$

3. SOLVE.

   $140 = 2l + 2(30)$
   $140 = 2l + 60$ Multiply 2(30).
   $140 - 60 = 2l + 60 - 60$ Subtract 60 from both sides.
   $80 = 2l$ Combine like terms.
   $40 = l$ Divide both sides by 2.

4. INTERPRET.

Check: Substitute 40 for l and 30 for w in the perimeter formula and check to see that the perimeter is 140 feet.

State: The length of the rectangular garden is 40 feet.

Solving a Formula for a Variable

We say that the formula

$$d = rt$$

is solved for d because d is alone on one side of the equation and the other side contains no d's. Suppose that we have a large number of problems to solve where we are given distance d and rate r and asked to find time t. In this case, it may be easier to first solve the formula $d = rt$ for t. To solve for t, we divide both sides of the equation by r.

$$d = rt$$

$$\frac{d}{r} = \frac{rt}{r} \quad \text{Divide both sides by } r.$$

$$\frac{d}{r} = t \quad \text{Simplify.}$$

To solve a formula or an equation for a specified variable, we use the same steps as for solving a linear equation. These steps are listed next.

To Solve Equations for a Specified Variable

Step 1. If an equation contains fractions, multiply both sides by the LCD to clear the equation of fractions.

Step 2. Use the distributive property to remove parentheses if they occur.

Step 3. Simplify each side of the equation by combining like terms.

Step 4. Get all terms containing the specified variable on one side and all other terms on the other side by using the addition property of equality.

Step 5. Get the specified variable alone by using the multiplication property of equality.

Practice Problem 3

Solve $C = 2\pi r$ for r. (This formula is used to find the circumference C of a circle given its radius r.)

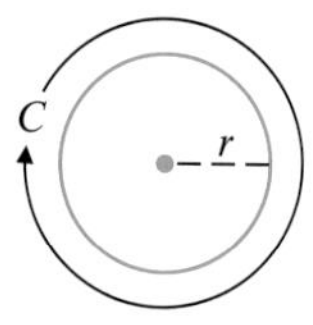

EXAMPLE 3 Solve $V = lwh$ for l.

Solution: This formula is used to find the volume of a box. To solve for l, we divide both sides by wh.

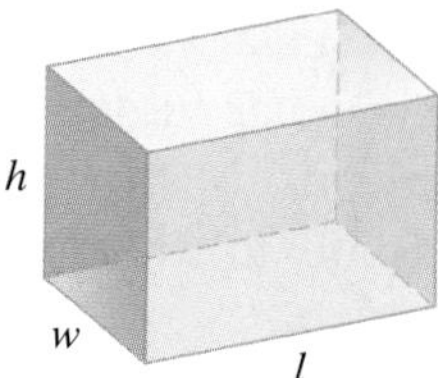

$$V = lwh$$

$$\frac{V}{wh} = \frac{lwh}{wh} \quad \text{Divide both sides by } wh.$$

$$\frac{V}{wh} = l \quad \text{Simplify.}$$

Since we have l alone on one side of the equation, we have solved for l in terms of V, w, and h. Remember that it does not matter on which side of the equation we get the variable alone.

Practice Problem 4 

Solve $P = 2l + 2w$ for w.

EXAMPLE 4 Solve $y = mx + b$ for x.

Solution: First we get mx alone by subtracting b from both sides.

$$y = mx + b$$

$$y - b = mx + b - b \quad \text{Subtract } b \text{ from both sides.}$$

$$y - b = mx \quad \text{Combine like terms.}$$

Next we solve for x by dividing both sides by m.

$$\frac{y-b}{m} = \frac{mx}{m}$$

$$\frac{y-b}{m} = x \quad \text{Simplify.}$$

Practice Problem 5

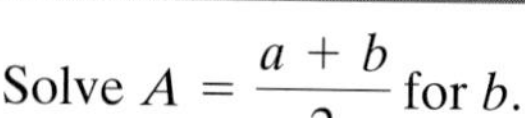

Solve $A = \frac{a+b}{2}$ for b.

EXAMPLE 5 Solve $A = \frac{bh}{2}$ for h.

Solution: First let's clear the equation of fractions by multiplying both sides by 2.

$$A = \frac{bh}{2}$$

$$2 \cdot A = 2\left(\frac{bh}{2}\right) \quad \text{Multiply both sides by 2 to clear fractions.}$$

$$2A = bh$$

$$\frac{2A}{b} = \frac{bh}{b} \quad \text{Divide both sides by } b \text{ to get } h \text{ alone.}$$

$$\frac{2A}{b} = h \quad \text{Simplify.}$$

Answers

3. $r = \frac{C}{2\pi}$ **4.** $w = \frac{P - 2l}{2}$

5. $b = 2A - a$

Name ____________________ Section __________ Date __________

EXERCISE SET 9.5

A *Substitute the given values into each given formula and solve for the unknown variable. See Examples 1 and 2.*

△ **1.** $A = bh$; $A = 45$, $b = 15$
(Area of a parallelogram)

2. $d = rt$; $d = 195$, $t = 3$
(Distance formula)

△ **3.** $S = 4lw + 2wh$; $S = 102$, $l = 7$, $w = 3$
(Surface area of a special rectangular box)

△ **4.** $V = lwh$; $l = 14$, $w = 8$, $h = 3$
(Volume of a rectangular box)

△ **5.** $A = \frac{1}{2}(B + b)h$; $A = 180$, $B = 11$, $b = 7$
(Area of a trapezoid)

△ **6.** $A = \frac{1}{2}(B + b)h$; $A = 60$, $B = 7$, $b = 3$
(Area of a trapezoid)

△ **7.** $P = a + b + c$; $P = 30$, $a = 8$, $b = 10$
(Perimeter of a triangle)

△ **8.** $V = \frac{1}{3}Ah$; $V = 45$, $h = 5$
(Volume of a pyramid)

△ **9.** $C = 2\pi r$; $C = 15.7$ (use the approximation 3.14 for π)
(Circumference of a circle)

△ **10.** $A = \pi r^2$; $r = 4.5$ (use the approximation 3.14 for π)
(Area of a circle)

11. $I = PRT$; $I = 3750$, $P = 25{,}000$, $R = 0.05$
(Simple interest formula)

12. $I = PRT$; $I = 1{,}056{,}000$, $R = 0.055$, $T = 6$
(Simple interest formula)

△ **13.** $V = \frac{1}{3}\pi r^2 h$; $V = 565.2$, $r = 6$ (use the approximation 3.14 for π)
(Volume of a cone)

△ **14.** $V = \frac{4}{3}\pi r^3$; $r = 3$ (use the approximation 3.14 for π)
(Volume of a sphere)

Solve. See Examples 1 and 2.

△ **15.** The world's largest sign for Coca-Cola is located in Arica, Chile. The rectangular sign has a length of 400 feet and has an area of 52,400 square feet. Find the width of the sign. (*Source:* Fabulous Facts about Coca-Cola, Atlanta, GA)

△ **16.** The length of a rectangular garden is 6 meters. If 21 meters of fencing are required to fence the garden, find its width.

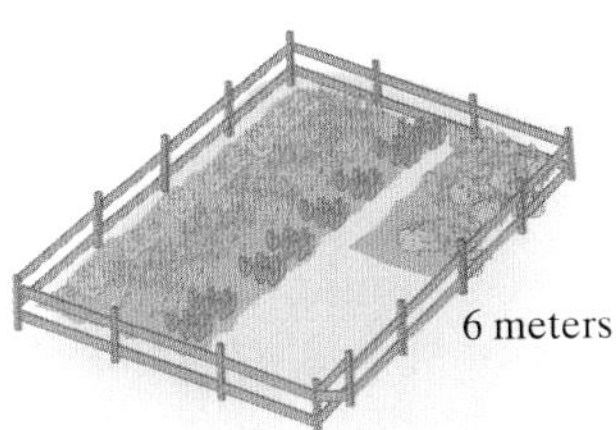

17. The Cat is a high-speed catamaran auto ferry that operates between Bar Harbor, Maine, and Yarmouth, Nova Scotia. The Cat can make the trip in about $2\frac{1}{2}$ hours at a speed of 55 mph. About how far apart are Bar Harbor and Yarmouth? (*Source:* Bay Ferries)

18. A family is planning their vacation to Disney World. They will drive from a small town outside New Orleans, Louisiana, to Orlando, Florida, a distance of 700 miles. They plan to average a rate of 55 miles per hour. How long will this trip take?

19. The highest temperature ever recorded in Europe was 122°F in Seville, Spain, in August 1881. Convert this record high temperature to Celsius. (*Source:* National Climatic Data Center)

20. The lowest temperature ever recorded in Oceania was −10°C at the Haleakala Summit in Maui, Hawaii, in January 1961. Convert this record low temperature to Fahrenheit. (*Source:* National Climatic Data Center)

△ 21. Piranha fish require 1.5 cubic feet of water per fish to maintain a healthy environment. Find the maximum number of piranhas you could put in a tank measuring 8 feet by 3 feet by 6 feet.

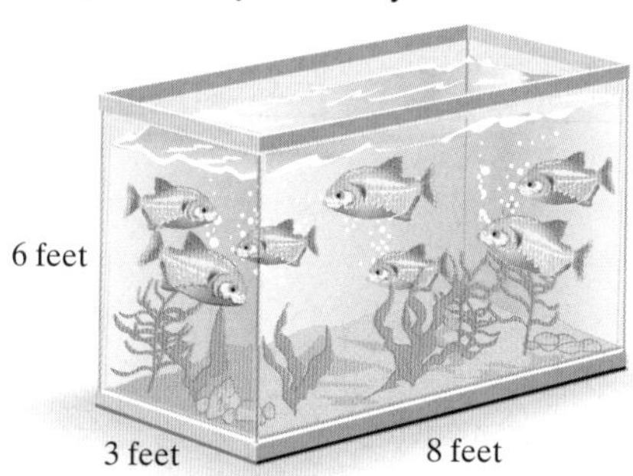

△ 22. Find how many goldfish you can put in a cylindrical tank whose diameter is 8 meters and whose height is 3 meters if each goldfish needs 2 cubic meters of water.

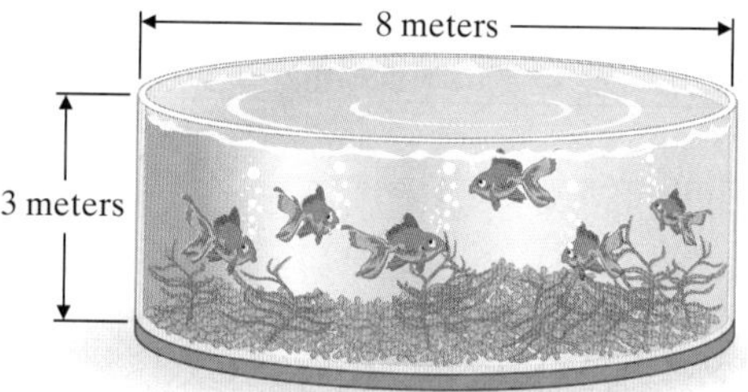

△ 23. The longest runway at Los Angeles International Airport has the shape of a rectangle and an area of 1,813,500 square feet. This runway is 150 feet wide. How long is the runway? (*Source:* Los Angeles World Airports)

24. Beaumont, Texas, is about 150 miles from Toledo Bend. If Leo Miller leaves Beaumont at 4 a.m. and averages 45 mph, when should he arrive at Toledo Bend?

25. The X-30 is a new "space plane" being developed that will skim the edge of space at 4000 miles per hour. Neglecting altitude, if the circumference of the Earth is approximately 25,000 miles, how long will it take for the X-30 to travel around the Earth?

26. In the United States, the longest hang glider flight was a 303-mile, $8\frac{1}{2}$-hour flight from New Mexico to Kansas. What was the average rate during this flight?

△ **27.** A lawn is in the shape of a trapezoid with a height of 60 feet and bases of 70 feet and 130 feet. How many bags of fertilizer must be purchased to cover the lawn if each bag covers 4000 square feet?

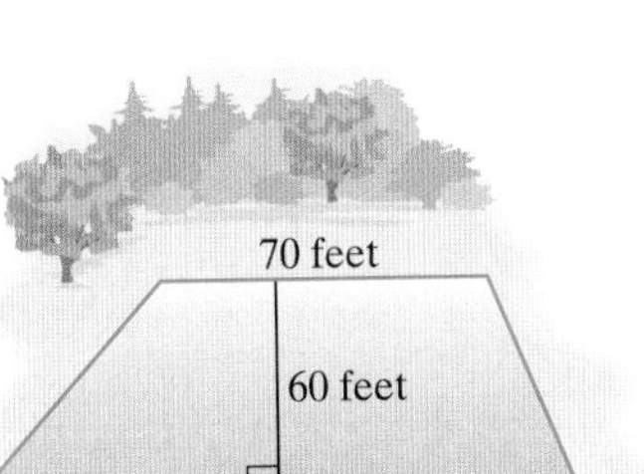

△ **28.** If the area of a right-triangularly shaped sail is 20 square feet and its base is 5 feet, find the height of the sail.

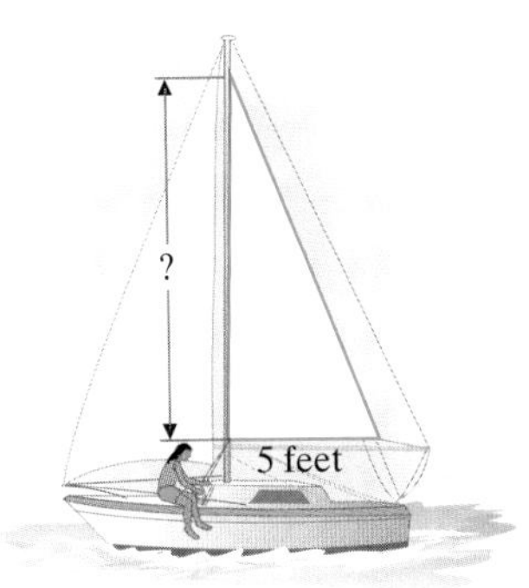

△ **29.** The CART Fed Ex Championship Series is an open-wheeled race car competition based in the United States. A CART car has a maximum length of 199 inches, a maximum width of 78.5 inches, and a maximum height of 33 inches. When the CART series travels to another country for a grand prix, teams must ship their cars. Find the volume of the smallest shipping crate needed to ship a CART car of maximum dimensions. (*Source:* Championship Auto Racing Teams, Inc.)

CART Racing Car

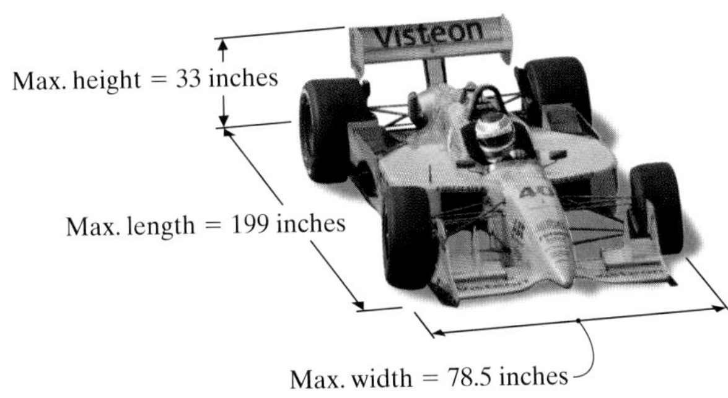

30. On a road course, a CART car's speed can average up to around 105 mph. Based on this speed, how long would it take a CART driver to travel from Los Angeles to New York City, a distance of about 2810 miles by road, without stopping? Round to the nearest tenth of an hour.

△ **31.** Maria's Pizza sells one 16-inch cheese pizza or two 10-inch cheese pizzas for \$9.99. Determine which size gives more pizza.

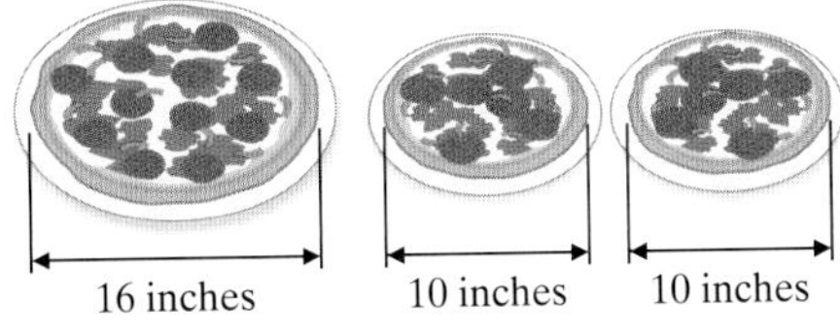

△ **32.** Find how much rope is needed to wrap the Earth at the equator, if the radius of the Earth is 4000 miles. (*Hint*: Use 3.14 for π and the formula for circumference.)

33. Dry ice is a name given to solidified carbon dioxide. At $-78.5°C$ it changes directly from a solid to a gas. Convert this temperature to degrees Fahrenheit.

34. Lightning bolts can reach a temperature of 50,000°F. Convert this temperature to degrees Celsius.

35. The distance from the Sun to the Earth is approximately 93,000,000 miles. If light travels at a rate of 186,000 miles per second, how long does it take light from the Sun to reach us?

36. Light travels at a rate of 186,000 miles per second. If our Moon is 238,860 miles from the Earth, how long does it take light reflected off the Moon to reach us? (Round to the nearest tenth of a second.)

37. The Hoberman Sphere is a toy ball that expands and contracts. When it is completely closed, it has a diameter of 9.5 inches. Find the volume of the Hoberman Sphere when it is completely closed. Use 3.14 for π. Round to the nearest whole cubic inch. (*Source:* Hoberman Designs, Inc.)

38. When the Hoberman Sphere (see Exercise 37) is completely expanded, its diameter is 30 inches. Find the volume of the Hoberman Sphere when it is completely expanded. Use 3.14 for π. Round to the nearest whole cubic inch. (*Source:* Hoberman Designs, Inc.)

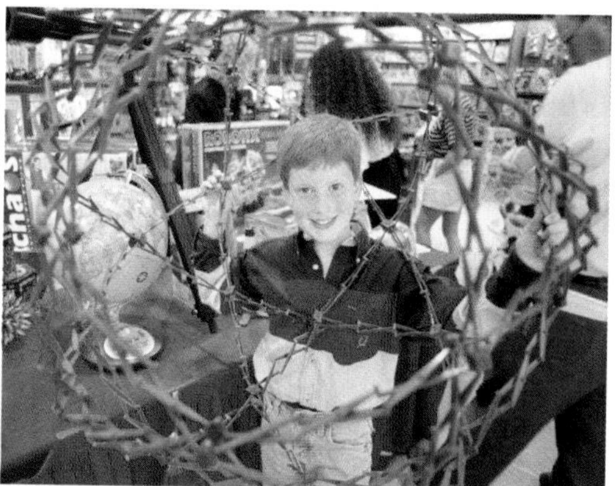

39. The average temperature on the planet Mercury is 167°C. Convert this temperature to degrees Fahrenheit. (*Source:* National Space Science Data Center)

40. The average temperature on the planet Jupiter is −227°F. Convert this temperature to degrees Celsius. Round to the nearest degree. (*Source:* National Space Science Data Center)

41. Bolts of lightning can travel at 270,000 miles per second. How many times can a lightning bolt travel around the world in one second? (See Exercise 32. Round to the nearest tenth.)

42. A glacier is a giant mass of rocks and ice that flows downhill like a river. Exit Glacier, near Seward, Alaska, moves at a rate of 20 inches a day. Find the distance in feet the glacier moves in a year. (Assume 365 days a year. Round to 2 decimal places.)

43. Flying fish do not *actually* fly, but glide. They have been known to travel a distance of 1300 feet at a rate of 20 miles per hour. How many seconds did it take to travel this distance? (*Hint:* First convert miles per hour to feet per second. Recall that 1 mile = 5280 feet. Round to the nearest tenth of a second.)

△ 44. Stalactites join stalagmites to form columns. A column found at Natural Bridge Caverns near San Antonio, Texas, rises 15 feet and has a *diameter* of only 2 inches. Find the volume of this column in cubic inches. (*Hint*: Use the formula for volume of a cylinder and use 3.14 for π.)

45. A Japanese "bullet" train set a new world record for train speed at 552 kilometers per hour during a manned test run on the Yamanashi Maglev Test Line in April 1999. The Yamanashi Maglev Test Line is 42.8 kilometers long. How many *minutes* would a test run on the Yamanashi Line last at this record-setting speed? Round to the nearest hundredth of a minute. (*Source:* Japan Railways Central Co.)

46. In 1983, the Hawaiian volcano Kilauea began erupting in a series of episodes still occurring at the time of this writing. At times, the lava flows advanced at speeds of up to 0.5 kilometer per hour. In 1983 and 1984 lava flows destroyed 16 homes in the Royal Gardens subdivision, about 6 km away from the eruption site. Roughly how long did it take the lava to reach Royal Gardens? Round to the nearest hour. (*Source:* U.S. Geological Survey Hawaiian Volcano Observatory)

B *Solve each formula for the specified variable. See Examples 3 through 5.*

47. $f = 5gh$ for h

△ **48.** $C = 2\pi r$ for r

△ **49.** $V = LWH$ for W

50. $T = mnr$ for n

51. $3x + y = 7$ for y

52. $-x + y = 13$ for y

53. $A = P + PRT$ for R

54. $A = P + PRT$ for T

△ **55.** $V = \frac{1}{3}Ah$ for A

56. $D = \frac{1}{4}fk$ for k

△57. $P = a + b + c$ for a

58. $PR = s_1 + s_2 + s_3 + s_4$ for s_3

△59. $S = 2\pi rh + 2\pi r^2$ for h

△60. $S = 4lw + 2wh$ for h

Review and Preview

Write each percent as a decimal. See Section 6.2.

61. 32%

62. 8%

63. 200%

64. 0.5%

Write each decimal as a percent. See Section 6.2.

65. 0.17

66. 0.03

67. 7.2

68. 5

Combining Concepts

Solve.

69. $N = R + \frac{V}{G}$ for V
(Urban forestry: tree plantings per year)

70. $B = \frac{F}{P - V}$ for V
(Business: break-even point)

△71. The formula $V = LWH$ is used to find the volume of a box. If the length of a box is doubled, the width is doubled, and the height is doubled, how does this affect the volume? Explain your answer.

△72. The formula $A = bh$ is used to find the area of a parallelogram. If the base of a parallelogram is doubled and its height is doubled, how does this affect the area? Explain your answer.

73. Find the temperature at which the Celsius measurement and Fahrenheit measurement are the same number.

9.6 Linear Inequalities and Problem Solving

In Section 9.1, we reviewed these inequality symbols and their meanings:

$<$ means "is less than" $\leq$ means "is less than or equal to"
$>$ means "is greater than" $\geq$ means "is greater than or equal to"

An **inequality** is a statement that contains one of the symbols above.

Equations	Inequalities
$x = 3$	$x \leq 3$
$5n - 6 = 14$	$5n - 6 \geq 14$
$12 = 7 - 3y$	$12 \leq 7 - 3y$
$\frac{x}{4} - 6 = 1$	$\frac{x}{4} - 6 > 1$

OBJECTIVES

- **A** Graph inequalities on a number line.
- **B** Use the addition property of inequality to solve inequalities.
- **C** Use the multiplication property of inequality to solve inequalities.
- **D** Use both properties to solve inequalities.
- **E** Solve problems modeled by inequalities.

A Graphing Inequalities on a Number Line

Recall that the single solution to the equation $x = 3$ is 3. The solutions of the inequality $x \leq 3$ include 3 and *all real numbers less than 3* (for example, -10, $\frac{1}{2}$, 2, and 2.9). Because we can't list all numbers less than 3, we show instead a picture of the solutions by graphing them.

To graph $x \leq 3$, we shade the numbers to the left of 3 since they are less than 3. Then we place a closed circle on the point representing 3. The closed circle indicates that 3 *is* a solution: 3 *is* less than or equal to 3.

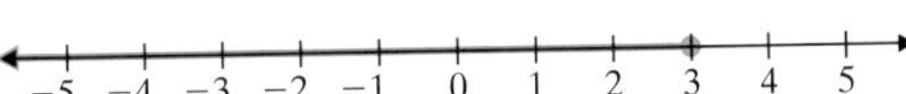

To graph $x < 3$, we shade the numbers to the left of 3. Then we place an open circle on the point representing 3. The open circle indicates that 3 *is not* a solution: 3 *is not* less than 3.

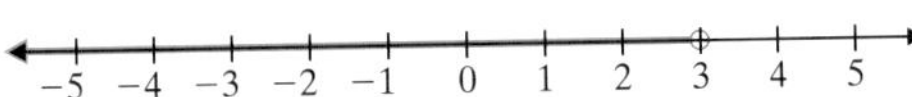

EXAMPLE 1 Graph: $x \geq -1$

Solution: We place a closed circle at -1 since the inequality symbol is $\geq$ and -1 is greater than or equal to -1. Then we shade to the right of -1.

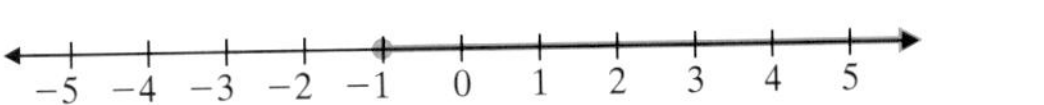

Practice Problem 1

Graph: $x \geq -2$

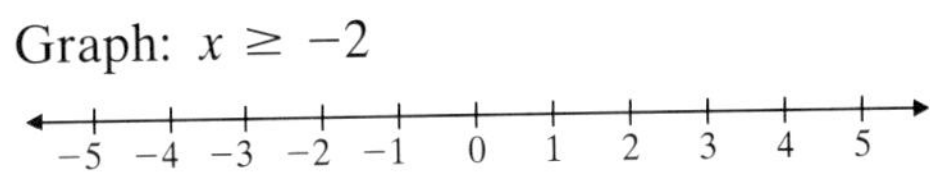

EXAMPLE 2 Graph: $-1 > x$

Solution: Recall from Chapter 1 that $-1 > x$ means the same as $x < -1$, shown below.

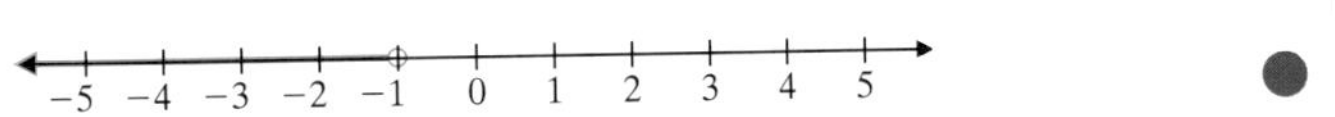

Practice Problem 2

Graph: $5 > x$

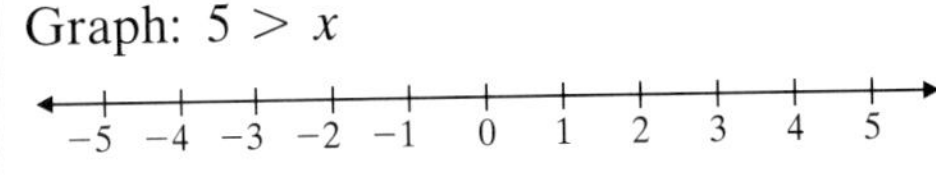

B Using the Addition Property

When solutions of a linear inequality are not immediately obvious, they are found through a process similar to the one used to solve a linear equation. Our goal is to get the variable alone on one side of the inequality. We use properties of inequality similar to properties of equality.

Answers

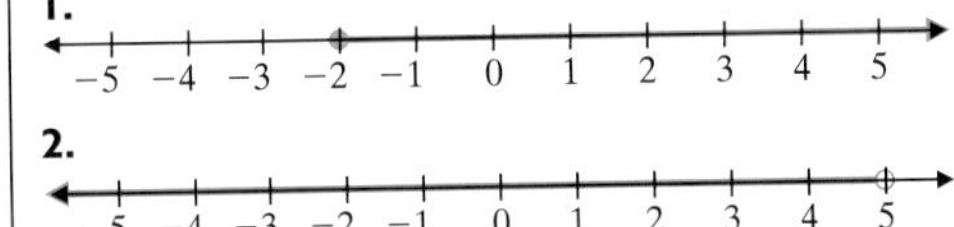

Addition Property of Inequality

If a, b, and c are real numbers, then

$$a < b \quad \text{and} \quad a + c < b + c$$

are equivalent inequalities.

This property also holds true for subtracting values, since subtraction is defined in terms of addition. In other words, adding or subtracting the same quantity from both sides of an inequality does not change the solutions of the inequality.

Practice Problem 3

Solve $x - 6 \geq -11$. Graph the solutions.

−5 −4 −3 −2 −1 0 1 2 3 4 5

EXAMPLE 3 Solve $x + 4 \leq -6$. Graph the solutions.

Solution: To solve for x, subtract 4 from both sides of the inequality.

$x + 4 \leq -6$	Original inequality
$x + 4 - 4 \leq -6 - 4$	Subtract 4 from both sides.
$x \leq -10$	Simplify.

The graph of the solutions is shown below.

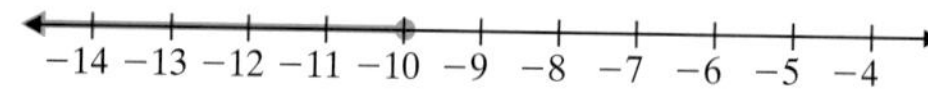

Helpful Hint

Notice that any number less than or equal to -10 is a solution to $x \leq -10$. For example, solutions include

$$-10, \quad -200, \quad -11\frac{1}{2}, \quad -7\pi, \quad -\sqrt{130}, \quad \text{and} \quad -50.3$$

C Using the Multiplication Property

An important difference between solving linear equations and solving linear inequalities is shown when we multiply or divide both sides of an inequality by a nonzero real number. For example, start with the true statement $6 < 8$ and multiply both sides by 2. As we see below, the resulting inequality is also true.

$6 < 8$	True
$2(6) < 2(8)$	Multiply both sides by 2.
$12 < 16$	True

But if we start with the same true statement $6 < 8$ and multiply both sides by -2, the resulting inequality is not a true statement.

$6 < 8$	True
$-2(6) < -2(8)$	Multiply both sides by -2.
$-12 < -16$	False

Notice, however, that if we reverse the direction of the inequality symbol, the resulting inequality is true.

$-12 < -16$	False
$-12 > -16$	True

This demonstrates the multiplication property of inequality.

Answer

3. $x \geq -5$

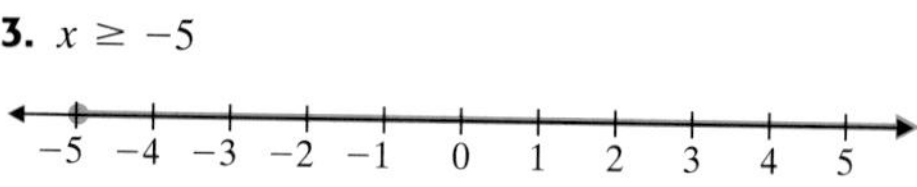

Multiplication Property of Inequality

1. If a, b, and c are real numbers, and c is **positive**, then
   $a < b$ and $ac < bc$
   are equivalent inequalities.
2. If a, b, and c are real numbers, and c is **negative**, then
   $a < b$ and $ac > bc$
   are equivalent inequalities.

Because division is defined in terms of multiplication, this property also holds true when dividing both sides of an inequality by a nonzero number: If we multiply or divide both sides of an inequality by a negative number, **the direction of the inequality sign must be reversed for the inequalities to remain equivalent**.

Try the Concept Check in the margin.

EXAMPLE 4 Solve $-2x \le -4$. Graph the solutions.

Solution: Remember to reverse the direction of the inequality symbol when dividing by a negative number.

$-2x \le -4$

$\frac{-2x}{-2} \ge \frac{-4}{-2}$ Divide both sides by -2 and reverse the inequality sign.

$x \ge 2$ Simplify.

The graph of the solutions is shown.

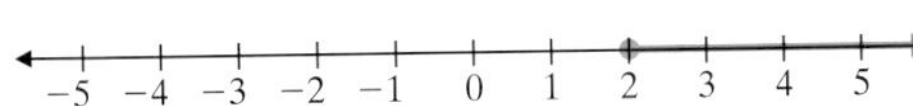

EXAMPLE 5 Solve $2x < -4$. Graph the solutions.

Solution: $2x < -4$

$\frac{2x}{2} < \frac{-4}{2}$ Divide both sides by 2. Do not reverse the inequality sign.

$x < -2$ Simplify.

The graph of the solutions is shown.

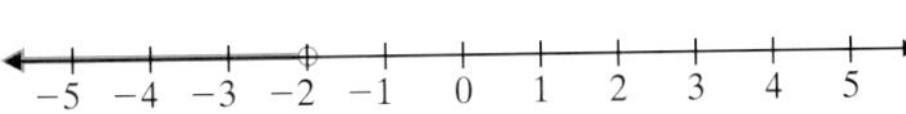

Since we cannot list all solutions to an inequality such as $x < -2$, we will use the set notation $\{x|x < -2\}$. Recall from Section 9.1 that this is read "the set of all x such that x is less than -2." We will use this notation when solving inequalities.

Concept Check

Fill in the blank with $<$, $>$, $\le$, or $\ge$.

a. Since $-8 < -4$, then $3(-8)$ ___ $3(-4)$.

b. Since $5 \ge -2$, then $\frac{5}{-7}$ ___ $\frac{-2}{-7}$.

c. If $a < b$, then $2a$ ___ $2b$.

d. If $a \ge b$, then $\frac{a}{-3}$ ___ $\frac{b}{-3}$.

Practice Problem 4

Solve $-3x \le 12$. Graph the solutions.

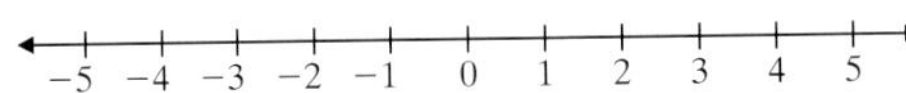

Practice Problem 5

Solve $5x > -20$. Graph the solutions.

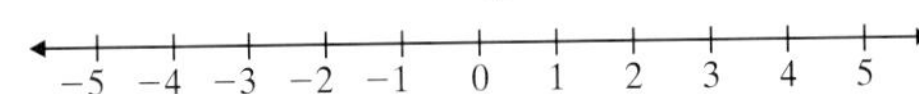

Answers

4. $x \ge -4$,

5. $x > -4$,

Concept Check: **a.** $<$, **b.** $\le$, **c.** $<$, **d.** $\le$

D Using Both Properties of Inequality

The following steps may be helpful when solving inequalities. Notice that these steps are similar to the ones given in Section 9.3 for solving equations.

To Solve Inequalities in One Variable

Step 1. If an inequality contains fractions, multiply both sides by the LCD to clear the inequality of fractions.
Step 2. Use the distributive property to remove parentheses if they occur.
Step 3. Simplify each side of the inequality by combining like terms.
Step 4. Get all variable terms on one side and all numbers on the other side by using the addition property of inequality.
Step 5. Get the variable alone by using the multiplication property of inequality.

Helpful Hint

Don't forget that if both sides of an inequality are multiplied or divided by a negative number, the direction of the inequality sign must be reversed.

EXAMPLE 6 Solve $-4x + 7 \geq -9$. Graph the solution set.

Solution:

$$-4x + 7 \geq -9$$

$$-4x + 7 - 7 \geq -9 - 7 \quad \text{Subtract 7 from both sides.}$$

$$-4x \geq -16 \quad \text{Simplify.}$$

$$\frac{-4x}{-4} \leq \frac{-16}{-4} \quad \text{Divide both sides by } -4 \text{ and reverse the direction of the inequality sign.}$$

$$x \leq 4 \quad \text{Simplify.}$$

The graph of the solution set $\{x|x \leq 4\}$ is shown.

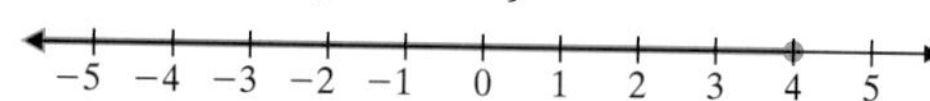

EXAMPLE 7 Solve $-5x + 7 < 2(x - 3)$. Graph the solution set.

Solution:

$$-5x + 7 < 2(x - 3)$$

$$-5x + 7 < 2x - 6 \quad \text{Apply the distributive property.}$$

$$-5x + 7 - 2x < 2x - 6 - 2x \quad \text{Subtract } 2x \text{ from both sides.}$$

$$-7x + 7 < -6 \quad \text{Combine like terms.}$$

$$-7x + 7 - 7 < -6 - 7 \quad \text{Subtract 7 from both sides.}$$

$$-7x < -13 \quad \text{Combine like terms.}$$

$$\frac{-7x}{-7} > \frac{-13}{-7} \quad \text{Divide both sides by } -7 \text{ and reverse the direction of the inequality sign.}$$

$$x > \frac{13}{7} \quad \text{Simplify.}$$

The graph of the solution set $\left\{x|x > \frac{13}{7}\right\}$ is shown.

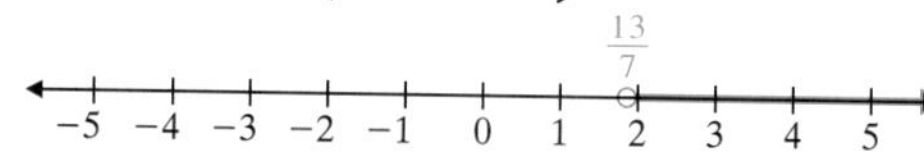

Practice Problem 6

Solve $-3x + 11 \leq -13$. Graph the solution set.

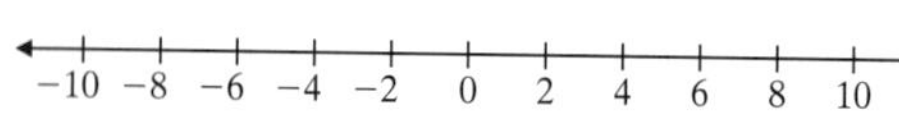

Practice Problem 7

Solve $-6x - 3 > -4(x + 1)$. Graph the solution set.

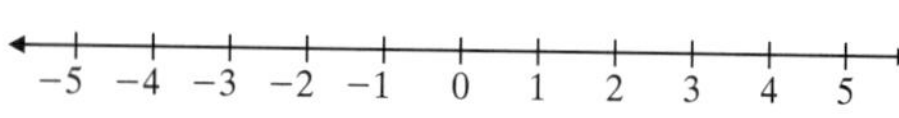

Answers

6. $\{x|x \geq 8\}$,

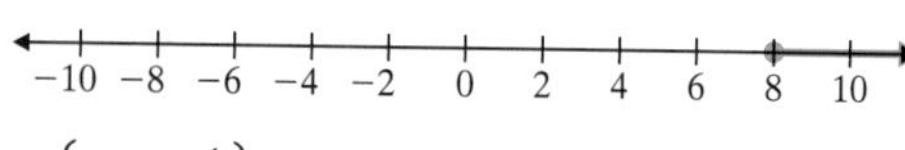

7. $\left\{x|x < \frac{1}{2}\right\}$,

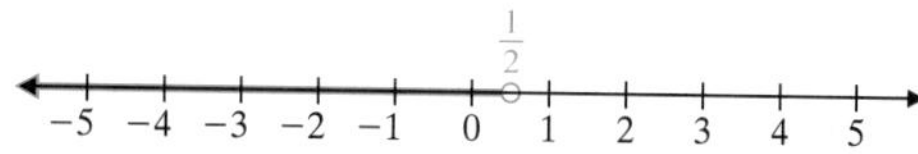

EXAMPLE 8 Solve: $2(x - 3) - 5 \le 3(x + 2) - 18$

Solution:

$$\begin{aligned}
2(x - 3) - 5 &\le 3(x + 2) - 18 \\
2x - 6 - 5 &\le 3x + 6 - 18 && \text{Apply the distributive property.} \\
2x - 11 &\le 3x - 12 && \text{Combine like terms.} \\
-x - 11 &\le -12 && \text{Subtract } 3x \text{ from both sides.} \\
-x &\le -1 && \text{Add 11 to both sides.} \\
\frac{-x}{-1} &\ge \frac{-1}{-1} && \text{Divide both sides by } -1 \text{ and reverse the direction of the inequality sign.} \\
x &\ge 1 && \text{Simplify.}
\end{aligned}$$

The solution set is $\{x|x \ge 1\}$.

Practice Problem 8

Solve: $3(x + 5) - 1 \ge 5(x - 1) + 7$

 Solving Problems Modeled by Inequalities

Problems containing words such as "at least," "at most," "between," "no more than," and "no less than" usually indicate that an inequality should be solved instead of an equation. In solving applications involving linear inequalities, we use the same procedure we used to solve applications involving linear equations.

EXAMPLE 9 Budgeting for a Wedding

Marie Chase and Jonathan Edwards are having their wedding reception at the Gallery reception hall. They may spend at most \$1000 for the reception. If the reception hall charges a \$100 cleanup fee plus \$14 per person, find the greatest number of people that they can invite and still stay within their budget. (Note: The bride and groom are admitted free of charge.)

Solution:

1. UNDERSTAND. Read and reread the problem. Suppose that 50 people attend the reception. The cost is then $\$100 + \$14(50) = \$100 + \$700 = \$800$.

   Let $x =$ the number of people who attend the reception.

2. TRANSLATE.

cleanup fee	+	cost per person	times	number of people	must be less than or equal to	\$1000
↓		↓	↓	↓	↓	↓
100	+	14	·	x	$\le$	1000

3. SOLVE.

$$\begin{aligned}
100 + 14x &\le 1000 \\
14x &\le 900 && \text{Subtract 100 from both sides.} \\
x &\le 64\frac{2}{7} && \text{Divide both sides by 14.}
\end{aligned}$$

4. INTERPRET.

Check: Since x represents the number of people, we round down to the nearest whole, or 64. Notice that if 64 people attend, the cost is $\$100 + \$14(64) = \$996$. If 65 people attend, the cost is $\$100 + \$14(65) = \$1010$, which is more than the given \$1000.

State: Marie Chase and Jonathan Edwards can invite at most 64 people to the reception.

Practice Problem 9

Alex earns \$600 per month plus 4% of all his sales. Find the minimum sales that will allow Alex to earn at least \$3000 per month.

Answers

8. $\{x|x \le 6\}$ **9.** \$60,000

FOCUS ON Mathematical Connections

SEQUENCES

The ratio of the lengths of the sides of the golden rectangle is approximately 1.6. This value is also known as the golden ratio. The actual value of the golden ratio is $\frac{1 + \sqrt{5}}{2} \approx 1.618033989\ldots$, which is an irrational number. Interestingly enough, there is a very simple sequence, or ordered list of numbers, that can be used to approximate the golden ratio. The sequence is called the Fibonacci sequence, and it is very easy to construct. Start with the first two terms of the sequence, the numbers 1 and 1. The next term in the sequence is the sum of the two previous terms, and so on. For example,

$$1, \quad 1, \quad \overset{1+1=}{2}, \quad \overset{1+2=}{3}, \quad \overset{2+3=}{5}, \quad \overset{3+5=}{8}, \quad \overset{5+8=}{13}, \quad \overset{8+13=}{21}, \quad \overset{13+21=}{34}, \quad \overset{21+34=}{55}, \ldots$$

Now, let's look at the ratios of a Fibonacci sequence term to the previous one:

$$\frac{1}{1} = 1, \quad \frac{2}{1} = 2, \quad \frac{3}{2} = 1.5, \quad \frac{5}{3} \approx 1.66667, \quad \frac{8}{5} = 1.6, \quad \frac{13}{8} = 1.625, \quad \frac{21}{13} \approx 1.615, \quad \frac{34}{21} \approx 1.619$$

Notice that each successive ratio is a bit closer than the one before it to the value of the golden ratio, $1.618033989\ldots$.

CRITICAL THINKING

Generate a few additional terms of the Fibonacci sequence and extend the list of ratios of terms. How much closer to the value of the golden ratio can you get?

Name ____________ Section ________ Date ________

Mental Math

Solve each inequality.

1. $5x > 10$ **2.** $4x < 20$ **3.** $2x \geq 16$ **4.** $9x \leq 63$

Decide which number listed is not a solution to each given inequality.

5. $x \geq -3;\quad -3, 0, -5, \pi$

6. $x < 6;\quad -6, |-6|, 0, -3.2$

7. $x < 4.01;\quad 4, -4.01, 4.1, -4.1$

8. $x \geq -3;\quad -4, -3, -2, -(-2)$

EXERCISE SET 9.6

A *Graph each on a number line. See Examples 1 and 2.*

1. $x \leq -1$

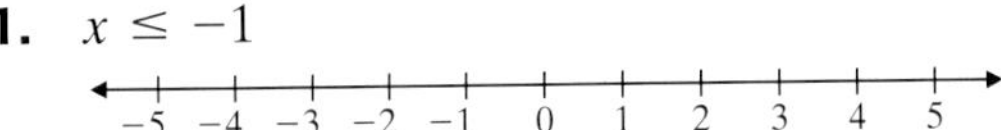

2. $y < 0$

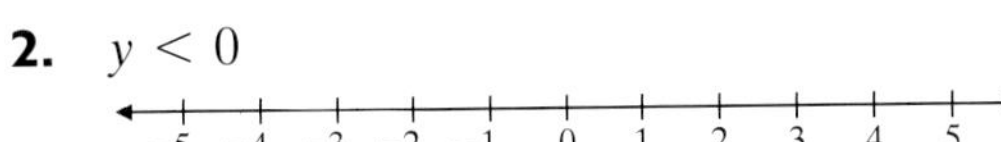

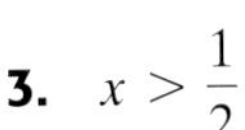
3. $x > \frac{1}{2}$

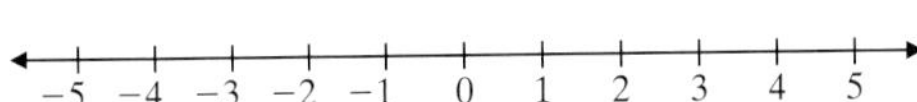

4. $z \geq -\frac{2}{3}$

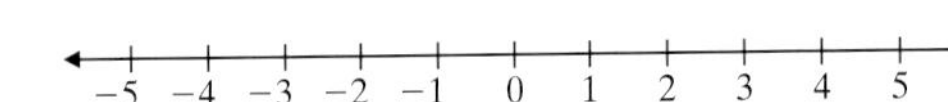

5. $y < 4$

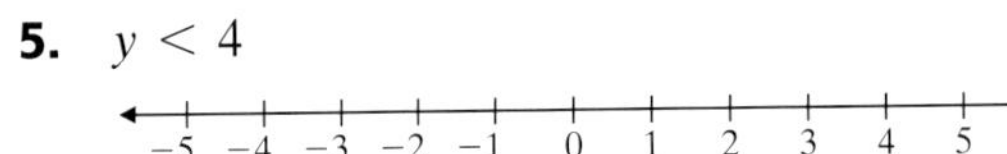

6. $x > 3$

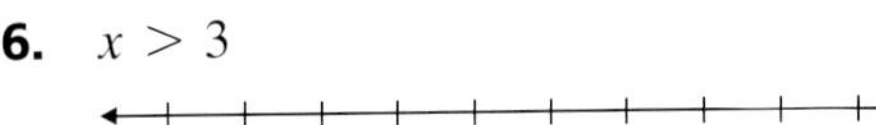

7. $-2 \leq m$

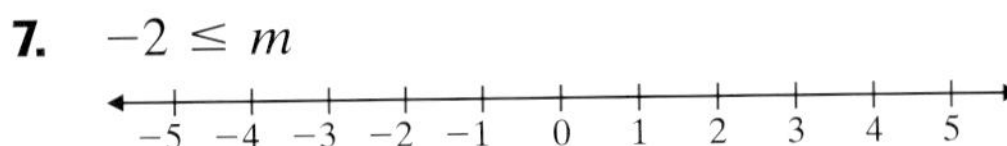

8. $-5 \geq x$

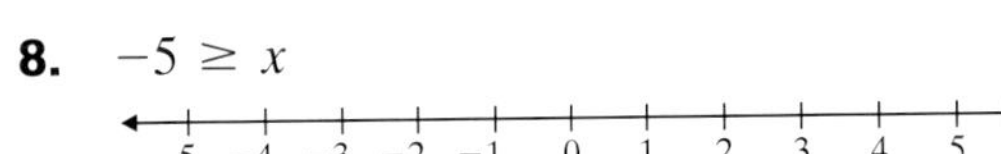

B *Solve each inequality. Graph the solution set. See Example 3.*

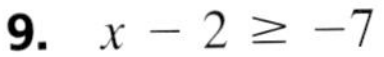
9. $x - 2 \geq -7$

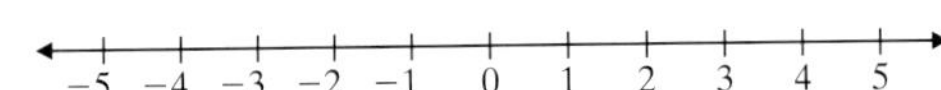

10. $x + 4 \leq 1$

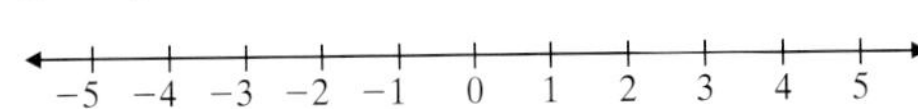

11. $-9 + y < 0$

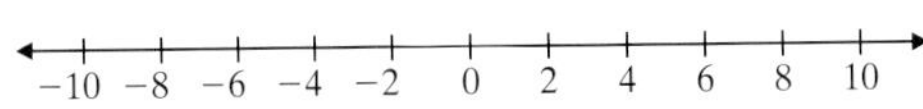

12. $-3 + m > 5$

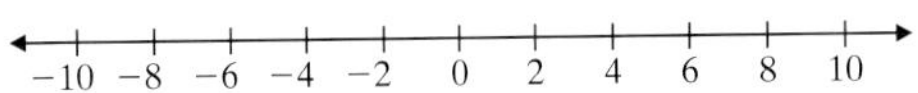

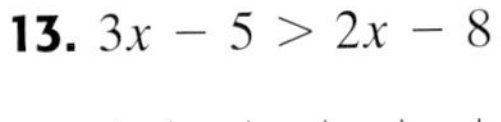
13. $3x - 5 > 2x - 8$

14. $3 - 7x \geq 10 - 8x$

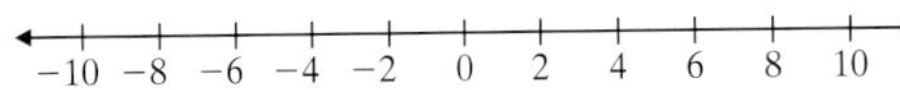

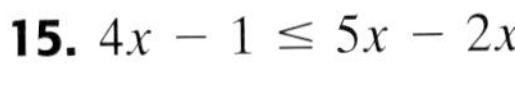
15. $4x - 1 \leq 5x - 2x$

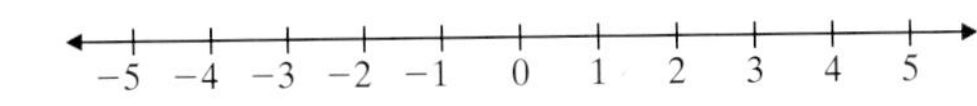

16. $7x + 3 < 9x - 3x$

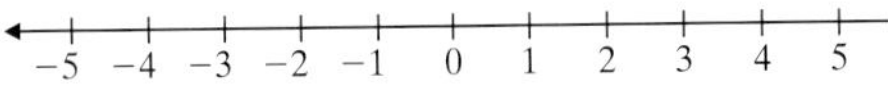

C *Solve each inequality. Graph the solution set. See Examples 4 and 5.*

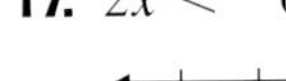
17. $2x < -6$

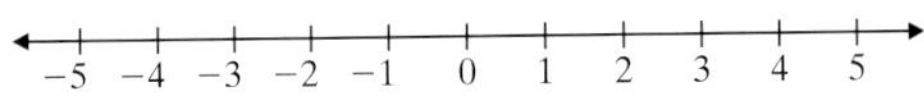

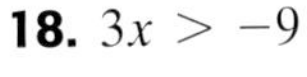
18. $3x > -9$

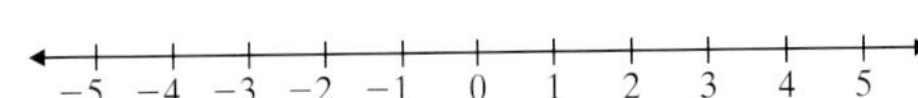

19. $-8x \leq 16$

20. $-5x < 20$

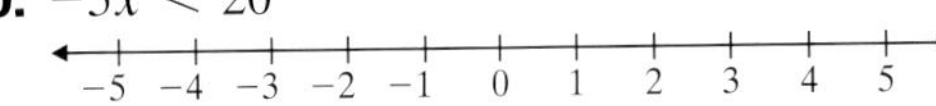

21. $-x > 0$

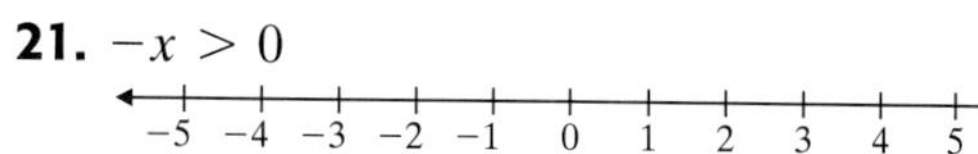

22. $-y \geq 0$

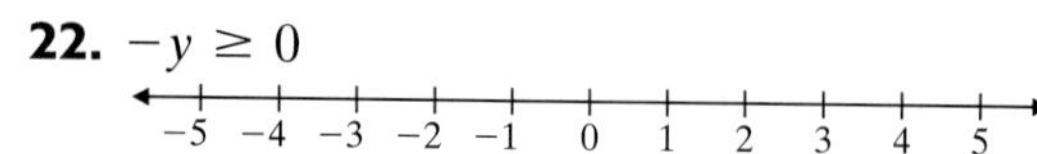

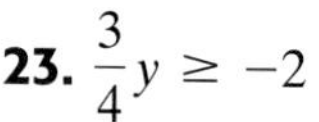

23. $\frac{3}{4}y \geq -2$

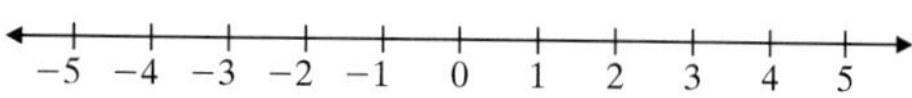

24. $\frac{5}{6}x \leq -8$

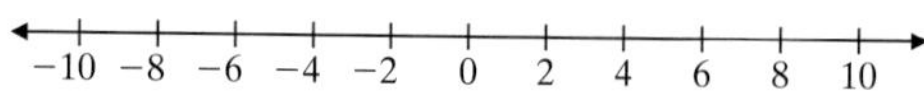

25. $-0.6y < -1.8$

26. $-0.3x > -2.4$

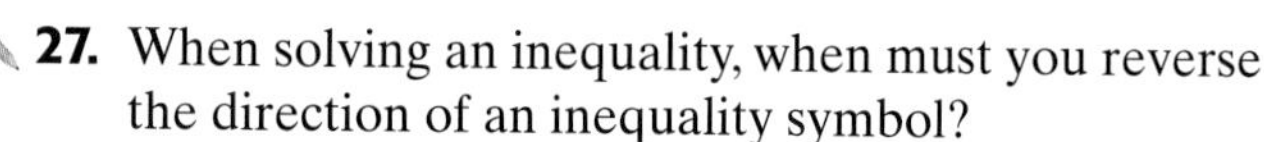

27. When solving an inequality, when must you reverse the direction of an inequality symbol?

28. If both sides of the inequality $-3x < -30$ are divided by 3, do you reverse the direction of the inequality symbol? Why or why not?

D *Solve each inequality. See Examples 6 through 8.*

29. $3x - 7 < 6x + 2$

30. $2x - 1 \geq 4x - 5$

31. $5x - 7x \leq x + 2$

32. $4 - x < 8x + 2x$

33. $-6x + 2 \geq 2(5 - x)$

34. $-7x + 4 > 3(4 - x)$

35. $4(3x - 1) \leq 5(2x - 4)$

36. $3(5x - 4) \leq 4(3x - 2)$

37. $3(x + 2) - 6 > -2(x - 3) + 14$

38. $7(x - 2) + x \leq -4(5 - x) - 12$

39. $-2(x - 4) - 3x < -(4x + 1) + 2x$

40. $-5(1 - x) + x \leq -(6 - 2x) + 6$

41. $\frac{1}{2}(x - 5) < \frac{1}{3}(2x - 1)$

42. $\frac{1}{4}(x + 4) < \frac{1}{5}(2x + 3)$

43. $-5x + 4 \leq -4(x - 1)$

44. $-6x + 2 < -3(x + 4)$

E *Solve the following. See Example 9.*

45. Six more than twice a number is greater than negative fourteen. Find all numbers that make this statement true.

46. Five times a number increased by one is less than or equal to ten. Find all such numbers.

△ **47.** The perimeter of a rectangle is to be no greater than 100 centimeters and the width must be 15 centimeters. Find the maximum length of the rectangle.

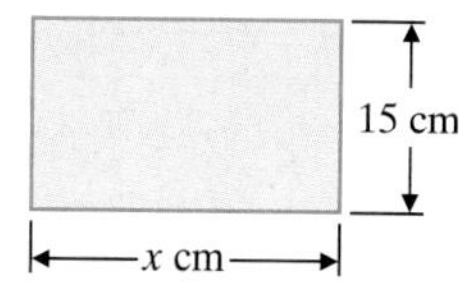

△ **48.** One side of a triangle is four times as long as another side, and the third side is 12 inches long. If the perimeter can be no longer than 87 inches, find the maximum lengths of the other two sides.

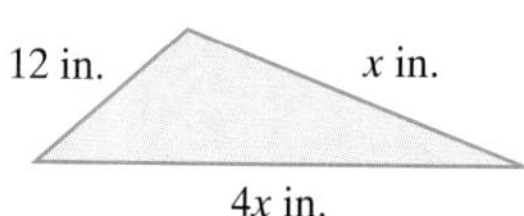

49. Ben Holladay bowled 146 and 201 in his first two games. What must he bowl in his third game to have an average of at least 180? (*Hint:* The average of a list of numbers is their sum divided by the number of numbers in the list.)

50. On an NBA team the two forwards measure 6′8″ and 6′6″ tall and the two guards measure 6′0″ and 5′9″ tall. How tall a center should they hire if they wish to have a starting team average height of at least 6′5″?

Review and Preview

Evaluate each expression. See Section 1.8.

51. 3^4

52. 4^3

53. 1^8

54. 0^7

55. $\left(\frac{7}{8}\right)^2$

56. $\left(\frac{2}{3}\right)^3$

This graph shows the growth in the number of Starbucks coffee bar locations from 1996 through 2002. The height of the graph for each year shown corresponds to the number of Starbucks locations. Use this graph to answer Exercises 57 through 62.

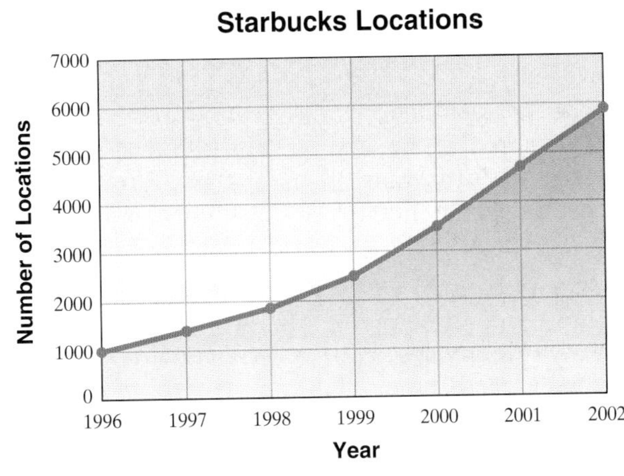

57. How many Starbucks locations were there in 1996?

58. How many Starbucks locations were there in 2002?

59. Between which two years did the greatest increase in the number of Starbucks locations occur?

60. In what year were there approximately 3500 Starbucks locations?

61. During what year did the number of Starbucks locations rise above 4000?

62. During what year did the number of Starbucks locations rise above 2000?

Combining Concepts

Solve.

63. Eric Daly has scores of 75, 83, and 85 on his history tests. Use an inequality to find the scores he can make on his final exam to receive a B in the class. The final exam counts as two tests, and a B is received if the final course average is greater than or equal to 80.

64. Maria Lipco has scores of 85, 95, and 92 on her algebra tests. Use an inequality to find the scores she can make on her final exam to receive an A in the course. The final exam counts as three tests, and an A is received if the final course average is greater than or equal to 90. Round to one decimal place.

Internet Excursions

Go To: 
http://www.prenhall.com/martin-gay_prealgebra
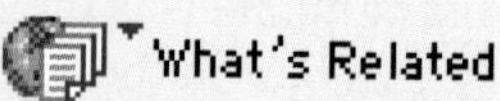
What's Related

The National Climatic Data Center (NCDC) is the world's largest active archive of weather data. The given World Wide Web address will provide you with access to NCDC's Climate Visualization site, or a related site, where you can find data on precipitation, temperature, and drought conditions for any state. Most states are subdivided into geographical divisions. Data from 1895 through the present are available for many states.

Choose the state in which your college or university is located. Select temperature as the parameter, and then choose a year. The graph that is then generated will show the average monthly temperatures for that year for all divisions of the state you specified. At the bottom of the Web page, note the link to the data file listing all the values shown in the graphs. Use the information in these Web pages to complete the following exercises.

65. Fill in the following blanks. In the year ________, the lowest average monthly temperature of ________ °F occurred in Division ________ of the state of ________. Let F represent the average monthly temperature in Fahrenheit degrees. Write an inequality that describes the minimum average monthly temperature in Fahrenheit degrees in your state for this year. Then use the relationship $F = \frac{9}{5}C + 32$ to convert this inequality to one that describes the minimum average monthly temperature in degrees Celsius.

66. Fill in the following blanks. In the year ________, the highest average monthly temperature of ________ °F occurred in Division ________ of the state of ________. Let F represent the average monthly temperature in Fahrenheit degrees. Write an inequality that describes the maximum average monthly temperature in Fahrenheit degrees in your state for this year. Then use the relationship $F = \frac{9}{5}C + 32$ to convert this inequality to one that describes the maximum average monthly temperature in degrees Celsius.

CHAPTER 9 ACTIVITY Investigating Averages

MATERIALS:

- small rubber ball or crumpled paper ball
- bucket or waste can

This activity may be completed by working in groups or individually.

1. Try shooting the ball into the bucket or waste can 5 times. Record your results below.

Shots Made	Shots Missed

2. Find your shooting percent for the 5 shots (that is, the percent of the shots you actually made out of the number you tried).

3. Suppose you are going to try an additional 5 shots. How many of the next 5 shots will you have to make to have a 50% shooting percent for all 10 shots? An 80% shooting percent?

4. Did you solve an equation in Question 3? If so, explain what you did. If not, explain how you could use an equation to find the answers.

5. Now suppose you are going to try an additional 22 shots. How many of the next 22 shots will you have to make to have at least a 50% shooting percent for all 27 shots? At least a 70% shooting percent?

6. Choose one of the sports played at your college that is currently in season. How many regular-season games are scheduled? What is the team's current percentage of games won?

7. Suppose the team has a goal of finishing the season with a winning percentage better than 110% of their current wins. At least how many of the remaining games must they win to achieve their goal?

Chapter 9 VOCABULARY CHECK

Fill in each blank with one of the words or phrases listed below.

like terms	numerical coefficient	linear equation in one variable
equivalent equations	formula	proportion
linear inequality in one variable	ratio	inequality symbols

1. Terms with the same variables raised to exactly the same powers are called ____________
2. A ______________________________ can be written in the form $ax + b = c$.
3. Equations that have the same solution are called ______________________
4. An equation that describes a known relationship among quantities is called a __________
5. A ________________________________ can be written in the form $ax + b < c$, (or $>, \leq, \geq$).
6. The ______________________ of a term is its numerical factor.
7. A ________ is the quotient of two numbers or two quantities.
8. A _____________ is a mathematical statement that two ratios are equal.
9. The symbols $\neq$, <, and > are called ____________________.

CHAPTER 9 Highlights

DEFINITIONS AND CONCEPTS	**EXAMPLES**
Section 9.1 Symbols and Sets of Numbers	
A **set** is a collection of objects, called **elements**, enclosed in braces.	$\{a, c, e\}$
Natural numbers: $\{1, 2, 3, 4, \ldots\}$ **Whole numbers**: $\{0, 1, 2, 3, 4, \ldots\}$ **Integers**: $\{\ldots, -3, -2, -1, 0, 1, 2, 3, \ldots\}$ **Rational numbers**: {real numbers that can be expressed as a quotient of integers} **Irrational numbers**: {real numbers that cannot be expressed as a quotient of integers} **Real numbers**: {all numbers that correspond to a point on the number line}	Given the set $\left\{-3.4, \sqrt{3}, 0, \frac{2}{3}, 5, -4\right\}$ list the numbers that belong to the set of Natural numbers $\;5$ Whole numbers $\;0,\ 5$ Integers $\;-4, 0, 5$ Rational numbers $\;-3.4, 0, \frac{2}{3}, 5, -4$ Irrational numbers $\;\sqrt{3}$ Real numbers $\;-3.4, \sqrt{3}, 0, \frac{2}{3}, 5, -4$
A line used to picture numbers is called a **number line**.	−5 −4 −3 −2 −1 0 1 2 3 4 5
The **absolute value** of a real number a denoted by $\|a\|$ is the distance between a and 0 on the number line.	$\|5\| = 5 \quad \|0\| = 0 \quad \|-2\| = 2$
SYMBOLS: $=$ is equal to $\neq$ is not equal to $>$ is greater than $<$ is less than $\leq$ is less than or equal to $\geq$ is greater than or equal to	$-7 = -7$ $3 \neq -3$ $4 > 1$ $1 < 4$ $6 \leq 6$ $18 \geq -\frac{1}{3}$

Definitions and Concepts	**Examples**
Section 9.2 Properties of Real Numbers	
Commutative Properties Addition: $a + b = b + a$ Multiplication: $a \cdot b = b \cdot a$	$3 + (-7) = -7 + 3$ $-8 \cdot 5 = 5 \cdot (-8)$
Associative Properties Addition: $(a + b) + c = a + (b + c)$ Multiplication: $(a \cdot b) \cdot c = a \cdot (b \cdot c)$	$(5 + 10) + 20 = 5 + (10 + 20)$ $(-3 \cdot 2) \cdot 11 = -3 \cdot (2 \cdot 11)$
Two numbers whose product is 1 are called **multiplicative inverses** or **reciprocals**. The reciprocal of a nonzero number a is $\frac{1}{a}$ because $a \cdot \frac{1}{a} = 1$.	The reciprocal of 3 is $\frac{1}{3}$. The reciprocal of $-\frac{2}{5}$ is $-\frac{5}{2}$.
Distributive Property $a(b + c) = a \cdot b + a \cdot c$	$5(6 + 10) = 5 \cdot 6 + 5 \cdot 10$ $-2(3 + x) = -2 \cdot 3 + (-2)(x)$
Identities $a + 0 = a$ $\quad 0 + a = a$ $a \cdot 1 = a$ $\quad 1 \cdot a = a$	$5 + 0 = 5$ $\quad 0 + (-2) = -2$ $-14 \cdot 1 = -14$ $\quad 1 \cdot 27 = 27$
Inverses Additive or opposite: $a + (-a) = 0$ Multiplicative or reciprocal: $b \cdot \frac{1}{b} = 1, \quad b \neq 0$	$7 + (-7) = 0$ $3 \cdot \frac{1}{3} = 1$

Section 9.3 Further Solving Linear Equations		
To Solve Linear Equations	Solve: $\frac{5(-2x + 9)}{6} + 3 = \frac{1}{2}$	
1. Clear the equation of fractions.	**1.** $6 \cdot \frac{5(-2x + 9)}{6} + 6 \cdot 3 = 6 \cdot \frac{1}{2}$	
2. Remove any grouping symbols such as parentheses.	**2.** $5(-2x + 9) + 18 = 3$ $-10x + 45 + 18 = 3$	Apply the distributive property.
3. Simplify each side by combining like terms.	**3.** $-10x + 63 = 3$	Combine like terms.
4. Get all variable terms on one side and all numbers on the other side by using the addition property of equality.	**4.** $-10x + 63 - 63 = 3 - 63$ $-10x = -60$	Subtract 63.
5. Get the variable alone by using the multiplication property of equality.	**5.** $\frac{-10x}{-10} = \frac{-60}{-10}$ $x = 6$	Divide by -10.
6. Check the solution by substituting it into the original equation.		

DEFINITIONS AND CONCEPTS

EXAMPLES

Section 9.4 Further Problem Solving

PROBLEM-SOLVING STEPS

The height of the Hudson volcano in Chile is twice the height of the Kiska volcano in the Aleutian Islands. If the sum of their heights is 12,870 feet, find the height of each.

1. UNDERSTAND the problem.

1. Read and reread the problem. Guess a solution and check your guess.
   Let x be the height of the Kiska volcano. Then $2x$ is the height of the Hudson volcano.

2. TRANSLATE the problem.

2.

height of Kiska	added to	height of Hudson	is	12,870
↓	↓	↓	↓	↓
x	$+$	$2x$	$=$	12,870

3. SOLVE the equation.

3. $x + 2x = 12{,}870$
   $3x = 12{,}870$
   $x = 4290$

4. INTERPRET the results.

4. *Check:* If x is 4290, then $2x$ is $2(4290)$ or 8580. Their sum is $4290 + 8580$ or 12,870, the required amount.
   State: The Kiska volcano is 4290 feet high, and the Hudson volcano is 8580 feet high.

Section 9.5 Formulas and Problem Solving

An equation that describes a known relationship among quantities is called a **formula**.

To solve a formula for a specified variable, use the same steps as for solving a linear equation. Treat the specified variable as the only variable of the equation.

$A = lw$ (area of a rectangle)
$I = PRT$ (simple interest)

Solve $P = 2l + 2w$ for l.

$$P = 2l + 2w$$

$$P - 2w = 2l + 2w - 2w \quad \text{Subtract } 2w.$$

$$P - 2w = 2l$$

$$\frac{P - 2w}{2} = \frac{2l}{2} \quad \text{Divide by 2.}$$

$$\frac{P - 2w}{2} = l$$

DEFINITIONS AND CONCEPTS — EXAMPLES

Section 9.6 Linear Inequalities and Problem Solving

Properties of inequalities are similar to properties of equations. Don't forget that if you multiply or divide both sides of an inequality by the same *negative* number, you must reverse the direction of the inequality symbol.

$-2x \leq 4$

$\frac{-2x}{-2} \geq \frac{4}{-2}$ Divide by -2; reverse the inequality symbol.

$x \geq -2$

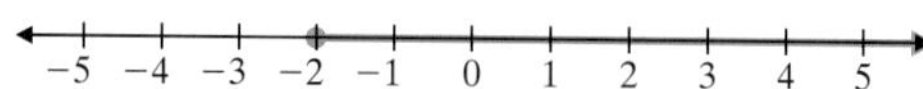

TO SOLVE LINEAR INEQUALITIES

1. Clear the inequality of fractions.
2. Remove grouping symbols.
3. Simplify each side by combining like terms.
4. Write all variable terms on one side and all numbers on the other side using the addition property of inequality.
5. Get the variable alone by using the multiplication property of inequality.

Solve: $3(x + 2) \leq -2 + 8$

1. $3(x + 2) \leq -2 + 8$ No fractions to clear.

2. $3x + 6 \leq -2 + 8$ Apply the distributive property.

3. $3x + 6 \leq 6$ Combine like terms.

4. $3x + 6 - 6 \leq 6 - 6$ Subtract 6.

$3x \leq 0$

5. $\frac{3x}{3} \leq \frac{0}{3}$ Divide by 3.

$x \leq 0$

The solution set is $\{x|x \leq 0\}$.

−5 −4 −3 −2 −1 0 1 2 3 4 5

Name ______________________ Section __________ Date __________

Chapter 9 Review

(9.1) *Insert <, >, or = in the appropriate space to make each statement true.*

1. $8 \quad 10$

2. $7 \quad 2$

3. $-4 \quad -5$

4. $\frac{12}{2} \quad -8$

5. $|-7| \quad |-8|$

6. $|-9| \quad -9$

7. $-|-1| \quad -1$

8. $|-14| \quad -(-14)$

9. $1.2 \quad 1.02$

10. $-\frac{3}{2} \quad -\frac{3}{4}$

Translate each statement into symbols.

11. Four is greater than or equal to negative three.

12. Six is not equal to five.

13. 0.03 is less than 0.3.

14. New York City has 155 museums and 400 art galleries. Write an inequality statement comparing the numbers 155 and 400. (*Source*: Absolute Trivia.com)

Given the sets of numbers below, list the numbers in each set that also belong to the set of:

a. Natural numbers
b. Whole numbers
c. Integers
d. Rational numbers
e. Irrational numbers
f. Real numbers

15. $\left\{-6, 0, 1, 1\frac{1}{2}, 3, \pi, 9.62\right\}$

16. $\left\{-3, -1.6, 2, 5, \frac{11}{2}, 15.1, \sqrt{5}, 2\pi\right\}$

The following chart shows the gains and losses in dollars of Density Oil and Gas stock for a particular week. Use this chart to answer Exercises 17 and 18.

Day	Gain or Loss (in dollars)
Monday	+1
Tuesday	−2
Wednesday	+5
Thursday	+1
Friday	−4

17. Which day showed the greatest loss?

18. Which day showed the greatest gain?

(9.2) *Name the property illustrated in each equation.*

19. $-6 + 5 = 5 + (-6)$

20. $6 \cdot 1 = 6$

21. $3(8 - 5) = 3 \cdot 8 + 3 \cdot (-5)$

22. $4 + (-4) = 0$

23. $2 + (3 + 9) = (2 + 3) + 9$

24. $2 \cdot 8 = 8 \cdot 2$

25. $6(8 + 5) = 6 \cdot 8 + 6 \cdot 5$

26. $(3 \cdot 8) \cdot 4 = 3 \cdot (8 \cdot 4)$

27. $4 \cdot \frac{1}{4} = 1$

28. $8 + 0 = 8$

(9.3) *Solve each equation.*

29. $5x + 25 = 20$

30. $5x - 6 + x = 4x$

31. $-y + 4y = -y$

32. $-5x + \frac{3}{7} = \frac{10}{7}$

33. $\frac{5}{3}x + 4 = \frac{2}{3}x$

34. $-(5x + 1) = -7x + 3$

35. $-4(2x + 1) = -5x + 5$

36. $-6(2x - 5) = -3(9 + 4x)$

37. $3(8y - 1) = 6(5 + 4y)$

38. $\frac{3(2 - z)}{5} = z$

39. $\frac{4(n + 2)}{5} = -n$

40. $0.5(2n - 3) - 0.1 = 0.4(6 + 2n)$

41. $-9 - 5a = 3(6a - 1)$

42. $\frac{5(c + 1)}{6} = 2c - 3$

43. $\frac{2(8 - a)}{3} = 4 - 4a$

44. $200(70x - 3560) = -179(150x - 19{,}300)$

45. $1.72y - 0.04y = 0.42$

46. Write the sum of three consecutive integers as an expression in x. Let x be the first even integer.

(9.4) *Solve each of the following.*

47. The height of the Washington Monument is 50.5 inches more than 10 times the length of a side of its square base. If the sum of these two dimensions is 7327 inches, find the height of the Washington Monument. (*Source:* National Park Service)

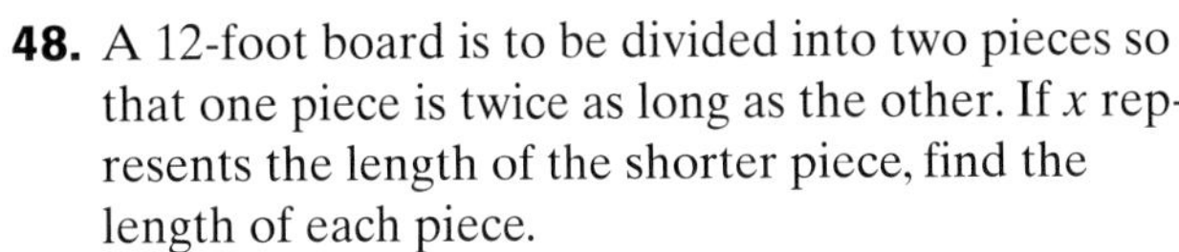

48. A 12-foot board is to be divided into two pieces so that one piece is twice as long as the other. If x represents the length of the shorter piece, find the length of each piece.

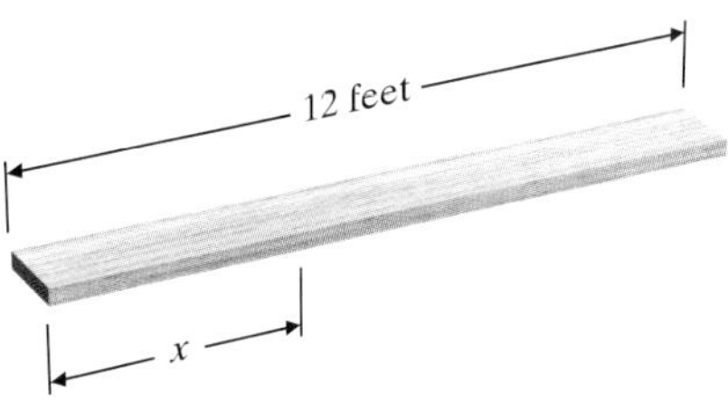

49. In March 2001, Kellogg Company acquired Keebler Foods Company. After the merger, the total number of Kellogg and Keebler manufacturing plants was 53. The number of Kellogg plants was one less than twice the number of Keebler plants. How many of each type of plant were there? (*Source: Kellogg Company 2000 Annual Report*)

50. Find three consecutive integers whose sum is negative 114.

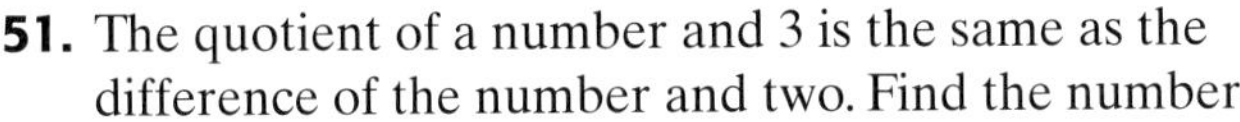

51. The quotient of a number and 3 is the same as the difference of the number and two. Find the number.

52. Double the sum of a number and 6 is the opposite of the number. Find the number.

(9.5) *Substitute the given values into the given formulas and solve for the unknown variable.*

53. $P = 2l + 2w$; $P = 46, l = 14$

54. $V = lwh$; $V = 192, l = 8, w = 6$

Solve each equation for the indicated variable.

55. $y = mx + b$ for m

56. $r = vst - 5$ for s

57. $2y - 5x = 7$ for x

58. $3x - 6y = -2$ for y

△ **59.** $C = \pi D$ for π

△ **60.** $C = 2\pi r$ for π

△ **61.** A swimming pool holds 900 cubic meters of water. If its length is 20 meters and its height is 3 meters, find its width.

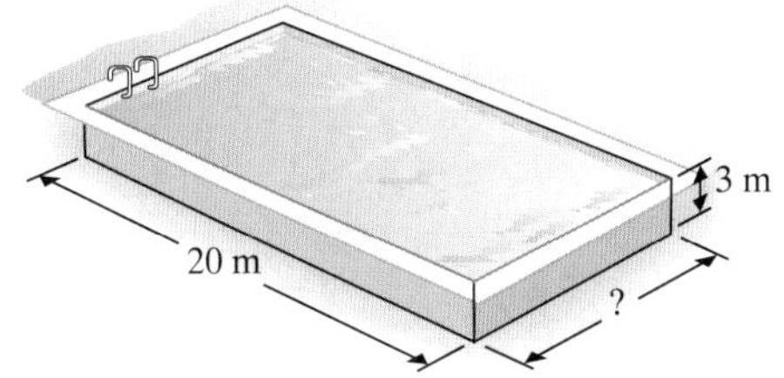

62. On April 28, 2001, the highest temperature recorded in the United States was 104°F, which occurred in Death Valley, California. Convert this temperature to degrees Celsius. (*Source:* National Weather Service)

(9.6) *Graph on a number line.*

63. $x \leq -2$

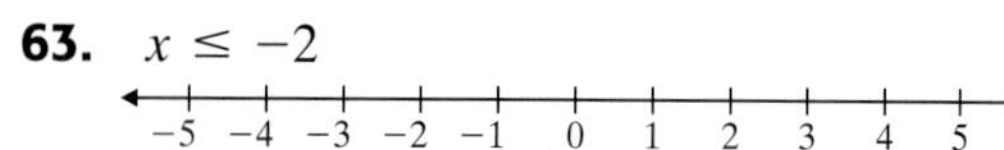

64. $x > 0$

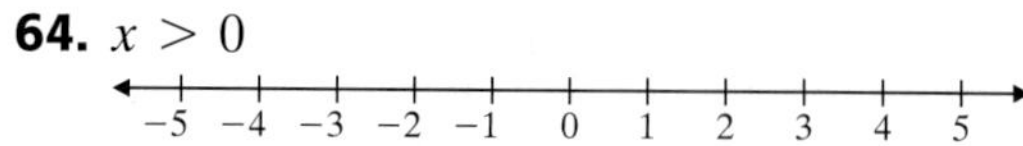

Solve each inequality.

65. $x - 5 \leq -4$

66. $x + 7 > 2$

67. $-2x \geq -20$

68. $-3x > 12$

69. $5x - 7 > 8x + 5$

70. $x + 4 \geq 6x - 16$

71. $\frac{2}{3}y > 6$

72. $-0.5y \leq 7.5$

73. $-2(x - 5) > 2(3x - 2)$

74. $4(2x - 5) \leq 5x - 1$

75. Carol Abolafia earns $175 per week plus a 5% commission on all her sales. Find the minimum amount of sales she must make to ensure that she earns at least $300 per week.

76. Joseph Barrow shot rounds of 76, 82, and 79 golfing. What must he shoot on his next round so that his average will be below 80?

Name ______________________ Section __________ Date __________

Answers

1. ______
2. ______
3. a. ______
b. ______
c. ______
d. ______
e. ______
f. ______
4. ______
5. ______
6. ______
7. ______
8. ______
9. ______
10. ______
11. ______
12. ______
13. ______
14. ______
15. ______
16. ______
17. ______
18. ______

Chapter 9 Test

Remember to check your answers and use the Chapter Test Prep Video to view solutions.

Translate each statement into symbols.

1. The absolute value of negative seven is greater than five.

2. The sum of nine and five is greater than or equal to four.

3. Given

$\{-5, -1, \frac{1}{4}, 0, 1, 7, 11.6, \sqrt{7}, 3\pi\}$, list the numbers in this set that also belong to the set of:

a. Natural numbers

b. Whole numbers

c. Integers

d. Rational numbers

e. Irrational numbers

f. Real numbers

Identify the property illustrated by each expression.

4. $8 + (9 + 3) = (8 + 9) + 3$

5. $6 \cdot 8 = 8 \cdot 6$

6. $-6(2 + 4) = -6 \cdot 2 + (-6) \cdot 4$

7. $\frac{1}{6}(6) = 1$

8. Find the opposite of -9.

9. Find the reciprocal of $-\frac{1}{3}$.

Use the distributive propety to write each expression without parentheses. Then simplify if possible.

10. $7 + 2(5y - 3)$

11. $4(x - 2) - 3(2x - 6)$

Solve each equation.

12. $4(n - 5) = -(4 - 2n)$

13. $-2(x - 3) = x + 5 - 3x$

14. $4z + 1 - z = 1 + z$

15. $\frac{2(x + 6)}{3} = x - 5$

16. $\frac{1}{2} - x + \frac{3}{2} = x - 4$

17. $-0.3(x - 4) + x = 0.5(3 - x)$

18. $-4(a + 1) - 3a = -7(2a - 3)$

19. ______

20. ______

21. ______

22. ______

23. ______

24. ______

25. ______

26. ______

27. ______

28. ______

Solve each application.

19. A number increased by two-thirds of the number is 35. Find the number.

△ **20.** A gallon of water seal covers 200 square feet. How many gallons are needed to paint two coats of water seal on a deck that measures 20 feet by 35 feet?

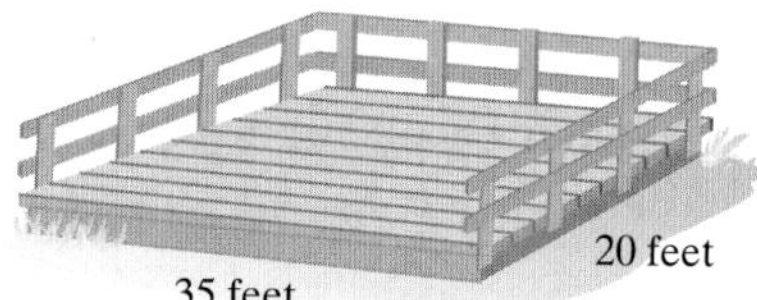

21. Find the value of x if $y = -14$, $m = -2$, and $b = -2$ in the formula $y = mx + b$.

Solve the equation for the indicated variable.

22. $V = \pi r^2 h$ for h

23. $3x - 4y = 10$ for y

Solve the inequality. Graph the solution set.

24. $3x - 5 > 7x + 3$

−5 −4 −3 −2 −1 0 1 2 3 4 5

Solve each inequality.

25. $-0.3x \geq 2.4$

26. $-5(x - 1) + 6 \leq -3(x + 4) + 1$

27. $\dfrac{2(5x + 1)}{3} > 2$

28. New York State has more public libraries than any other state. It has 650 more public libraries than Indiana does. If the total number of public libraries for these states is 1504, find the number of public libraries in New York and the number in Indiana. (*Source: The World Almanac and Book of Facts*)

Name ______________ Section ________ Date ________

Answers

Chapter 9 Cumulative Review

Simplify.

1. a. $-(-4)$ **b.** $-|-5|$ **c.** $-|6|$

2. Insert $<$, $>$, or $=$ in the appropriate space to make each statement true.

a. $|0| \quad 2$

b. $|-5| \quad 5$

c. $|-3| \quad |-2|$

d. $|5| \quad |6|$

e. $|-7| \quad |6|$

3. Simplify $-2[-3 + 2(-1 + 6)] - 5$

4. Simplify the expression $\dfrac{3 + |4 - 3| + 2^2}{6 - 3}$.

Add.

5. $-18 + 10$ **6.** $(-8) + (-11)$ **7.** $12 + (-8)$

8. $(-2) + 10$ **9.** $(-3) + 4 + (-11)$ **10.** $0.2 + (-0.5)$

Perform the indicated operation.

11. $\dfrac{2}{3} - \dfrac{10}{11}$ **12.** $\dfrac{2}{5} - \dfrac{39}{40}$ **13.** $2 - \dfrac{x}{3}$ **14.** $5 + \dfrac{x}{2}$

Evaluate. Let $x = -\dfrac{1}{2}$ and $y = \dfrac{1}{3}$.

15. $2x + y^2$ **16.** $2y + x^2$

Perform the indicated operation.

17. $-4\frac{2}{5} \cdot 1\frac{3}{11}$ **18.** $-2\frac{1}{2} \cdot (-2\frac{1}{2})$ **19.** $-2\frac{1}{3} \div (-2\frac{1}{2})$

20. $3\frac{3}{5} \div (-3\frac{1}{3})$ **21.** $(-1.3)^2 + 2.4$ **22.** $1.2(7.3 - 9.3)$

1. a. ________

b. ________

c. ________

2. a. ________

b. ________

c. ________

d. ________

e. ________

3. ________

4. ________

5. ________

6. ________

7. ________

8. ________

9. ________

10. ________

11. ________

12. ________

13. ________

14. ________

15. ________

16. ________

17. ________

18. ________

19. ________

20. ________

21. ________

22. ________

23.
24.
25.
26.
27.
28.
29.
30.
31.
32.
33.
34.
35.
36.
37.
38.
39.
40.
41.
42.
43.
44.
45.
46.
47.
48.
49.
50.

23. $\sqrt{49}$ **24.** $\sqrt{64}$ **25.** $\sqrt{\frac{4}{25}}$ **26.** $\sqrt{\frac{9}{100}}$

Solve for x.

27. $1.2x + 5.8 = 8.2$ **28.** $1.3x - 2.6 = -9.1$

Determine whether each statement is true or false.

29. $8 \geq 8$ **30.** $-4 \leq -4$ **31.** $8 \leq 8$ **32.** $-4 \geq -4$

33. $23 \leq 0$ **34.** $-8 \leq 0$ **35.** $0 \leq 23$ **36.** $0 \leq -8$

Use the distributive property to write each expression without parentheses. Then simplify the result.

37. $-5(-3 + 2z)$ **38.** $-4(2x-1)$ **39.** $4(3x + 7) + 10$ **40.** $9 + 2\,(5x + 6)$

41. Solve: $\dfrac{2(a + 3)}{3} = 6a + 2$ **42.** Solve: $\dfrac{x}{2} + \dfrac{x}{5} = 3$

43. In the 107th Congress, the U.S. House of Representatives had a total of 430 Democrats and Republicans. There were 10 more Republican representatives than Democratic. Find the number of representatives from each party. (*Source:* Office of the Clerk of the U.S. House of Representatives)

44. A glacier is a giant mass of rocks and ice that flows downhill like a river. Portage Glacier in Alaska is about 6 miles, or 31,680 feet, long and moves 400 feet per year. Icebergs are created when the front end of the glacier flows into Portage Lake. How long does it take for ice at the head (beginning) of the glacier to reach the lake?

45. Three times the sum of a number and 2 is the same as 9 times the number

46. Solve: $V = lwh$ for w.

47. Graph $-1 > x$.

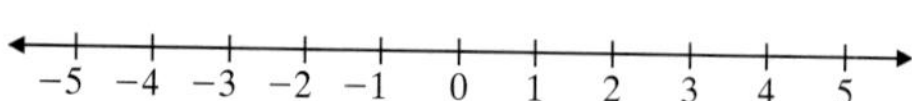

48. Solve: $-3x < -30$

49. Solve $2(x - 3) - 5 \leq 3(x + 2) - 18$

50. Solve: $10 + x < 6x - 10$

Exponents and Polynomials

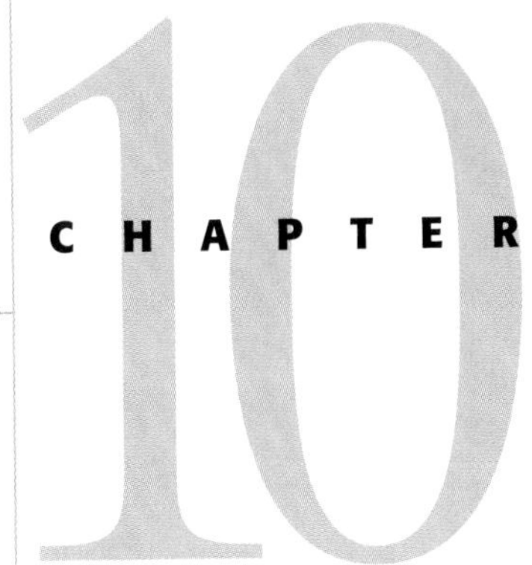

Recall from Chapter 1 that an exponent is a shorthand notation for repeated factors. This chapter explores additional concepts about exponents and exponential expressions. An especially useful type of exponential expression is a polynomial. Polynomials model many real-world phenomena. Our goal in this chapter is to become proficient with operations on polynomials.

Niagara Falls is visited by millions of tourists each year. Located between Niagara Falls, New York, and Niagara Falls, Ontario, Canada, on the Niagara River linking Lake Erie and Lake Ontario, the Falls consist of the American Falls, Bridal Veil Falls, and the Canadian, or Horseshoe, Falls. Together, they are known simply as Niagara Falls. The Falls were formed about 12,000 years ago and have since receded upstream 7 miles to their present location. Together, the Falls are roughly 3660 feet wide along their brinks. Water flowing over Niagara Falls drops about 167 feet to the river below and has the potential of generating 4.4 million kilowatts in hydroelectric power. In In Section 10.2, Exercise 99 we will use exponents, through scientific notation, to compute the phenomenal volume of water flowing over Niagara Falls in an hour.

Answers

1. ______
2. ______
3. ______
4. ______
5. ______
6. ______
7. ______
8. ______
9. ______
10. ______
11. ______
12. ______
13. ______
14. ______
15. ______
16. ______
17. ______
18. ______
19. ______
20. ______

Name ______ Section ______ Date ______

Chapter 10 Pretest

1. Evaluate: $\left(-\frac{3}{4}\right)^2$

Simplify.

2. $(4y^6)(2y^7)$

3. $\frac{a^9b^{16}}{a^{12}b^5}$

4. $4^0 + 2x^0$

5. $\left(-\frac{1}{6}\right)^{-3}$

6. $\left(\frac{m^{-2}n}{m^6n^{-8}}\right)^{-2}$

7. $12x^2 + 3x - 5 - 8x^2 + 7x$

8. Express in scientific notation: 0.000000814

9. Find the degree of the following polynomial. $8x - 4x^5 + 6x^3 + 10$

10. Find the value of the given polynomial when $x = -1$. $-3x^3 + 2x^2 - 4$

Perform the indicated operations.

11. $(4x^2 - 3x + 9) + (6x^2 + 3x - 8)$

12. $(6y^2 - 4) - (-3y^2 + 5y - 1)$

13. $(2a^2 + 3ab - 7b^2) - (3a^2 + 3ab + 9b^2)$

14. $\left(-\frac{2}{7}n^6\right)\left(\frac{21}{16}n^3\right)$

15. $-2t^2(3t^5 + 4t^3 - 8)$

16. $(2y - 1)(5y + 6)$

17. $(7a - 5)^2$

18. $(4b + 9)(4b - 9)$

19. $\frac{16p^4 - 8p^3 + 20p^2}{4p}$

20. $\frac{5x^2 - 28x - 12}{x - 6}$

10.1 Exponents

OBJECTIVES

- (A) Evaluate exponential expressions.
- (B) Use the product rule for exponents.
- (C) Use the power rule for exponents.
- (D) Use the power rules for products and quotients.
- (E) Use the quotient rule for exponents, and define a number raised to the 0 power.
- (F) Decide which rule(s) to use to simplify an expression.

(A) Evaluating Exponential Expressions

In this section, we continue our work with integer exponents. As we reviewed in Section 1.8, for example,

$$2 \cdot 2 \cdot 2 \cdot 2 \cdot 2 = 2^5$$

The exponent 5 tells us how many times that 2 is a factor. The expression 2^5 is called an **exponential expression.** It is also called the fifth **power** of 2, or we can say that 2 is **raised** to the fifth power.

$$5^6 = \underbrace{5 \cdot 5 \cdot 5 \cdot 5 \cdot 5 \cdot 5}_{6 \text{ factors; each factor is } 5} \quad \text{and} \quad (-3)^4 = \underbrace{(-3) \cdot (-3) \cdot (-3) \cdot (-3)}_{4 \text{ factors; each factor is } -3}$$

The base of an exponential expression is the repeated factor. The exponent is the number of times that the base is used as a factor.

$$\overset{\text{exponent or power}}{\underset{\text{base}}{a^n}} = \underbrace{a \cdot a \cdot a \ldots a}_{n \text{ factors of } a}$$

Practice Problems 1–6

Evaluate (find the value of) each expression.

1. 3^4
2. 7^1
3. $(-2)^3$
4. -2^3
5. $\left(\frac{2}{3}\right)^2$
6. $5 \cdot 6^2$

EXAMPLES Evaluate (find the value of) each expression.

1. $2^3 = 2 \cdot 2 \cdot 2 = 8$
2. $3^1 = 3$. To raise 3 to the first power means to use 3 as a factor only once. When no exponent is shown, the exponent is assumed to be 1.
3. $(-4)^2 = (-4)(-4) = 16$
4. $-4^2 = -(4 \cdot 4) = -16$
5. $\left(\frac{1}{2}\right)^4 = \frac{1}{2} \cdot \frac{1}{2} \cdot \frac{1}{2} \cdot \frac{1}{2} = \frac{1}{16}$
6. $4 \cdot 3^2 = 4 \cdot 9 = 36$

Notice how similar -4^2 is to $(-4)^2$ in the examples above. The difference between the two is the parentheses. In $(-4)^2$, the parentheses tell us that the base, or the repeated factor, is -4. In -4^2, only 4 is the base.

Helpful Hint

Be careful when identifying the base of an exponential expression. Pay close attention to the use of parentheses.

$(-3)^2$	-3^2	$2 \cdot 3^2$
The base is -3.	The base is 3.	The base is 3.
$(-3)^2 = (-3)(-3) = 9$	$-3^2 = -(3 \cdot 3) = -9$	$2 \cdot 3^2 = 2 \cdot 3 \cdot 3 = 18$

An exponent has the same meaning whether the base is a number or a variable. If x is a real number and n is a positive integer, then x^n is the product of n factors, each of which is x.

$$x^n = \underbrace{x \cdot x \cdot x \cdot x \cdot x \ldots x}_{n \text{ factors; each factor is } x}$$

Answers

1. 81 **2.** 7 **3.** -8 **4.** -8 **5.** $\frac{4}{9}$ **6.** 180

Practice Problem 7

Evaluate each expression for the given value of x.

a. $3x^2$ when x is 4

b. $\frac{x^4}{-8}$ when x is -2

EXAMPLE 7 Evaluate each expression for the given value of x.

a. $2x^3$ when x is 5 **b.** $\frac{9}{x^2}$ when x is -3

Solution:

a. When x is 5, $2x^3 = 2 \cdot 5^3$

$= 2 \cdot (5 \cdot 5 \cdot 5)$

$= 2 \cdot 125$

$= 250$

b. When x is -3, $\frac{9}{x^2} = \frac{9}{(-3)^2}$

$= \frac{9}{(-3)(-3)}$

$= \frac{9}{9} = 1$

B Using the Product Rule

Exponential expressions can be multiplied, divided, added, subtracted, and themselves raised to powers. By our definition of an exponent,

$$5^4 \cdot 5^3 = \underbrace{(5 \cdot 5 \cdot 5 \cdot 5)}_{4 \text{ factors of } 5} \cdot \underbrace{(5 \cdot 5 \cdot 5)}_{3 \text{ factors of } 5}$$

$$= \underbrace{5 \cdot 5 \cdot 5 \cdot 5 \cdot 5 \cdot 5 \cdot 5}_{7 \text{ factors of } 5}$$

$$= 5^7$$

Also,

$$x^2 \cdot x^3 = (x \cdot x) \cdot (x \cdot x \cdot x)$$

$$= x \cdot x \cdot x \cdot x \cdot x$$

$$= x^5$$

In both cases, notice that the result is exactly the same if the exponents are added.

$$5^4 \cdot 5^3 = 5^{4+3} = 5^7 \quad \text{and} \quad x^2 \cdot x^3 = x^{2+3} = x^5$$

This suggests the following rule.

Product Rule for Exponents

If m and n are positive integers and a is a real number, then

$$a^m \cdot a^n = a^{m+n}$$

For example,

$$3^5 \cdot 3^7 = 3^{5+7} = 3^{12}$$

In other words, to multiply two exponential expressions with the **same base,** we keep the base and add the exponents. We call this **simplifying** the exponential expression.

Practice Problems 8–12

Use the product rule to simplify each expression.

8. $7^3 \cdot 7^2$
9. $x^4 \cdot x^9$
10. $r^5 \cdot r$
11. $s^6 \cdot s^2 \cdot s^3$
12. $(-3)^9 \cdot (-3)$

Answers

7. a. 48 **b.** -2 **8.** 7^5 **9.** x^{13} **10.** r^6 **11.** s^{11} **12.** $(-3)^{10}$

EXAMPLES Use the product rule to simplify each expression.

8. $4^2 \cdot 4^5 = 4^{2+5} = 4^7$

9. $x^2 \cdot x^5 = x^{2+5} = x^7$

10. $y^3 \cdot y = y^3 \cdot y^1$

$= y^{3+1}$

$= y^4$

Helpful Hint

Don't forget that if no exponent is written, it is assumed to be 1.

11. $y^3 \cdot y^2 \cdot y^7 = y^{3+2+7} = y^{12}$

12. $(-5)^7 \cdot (-5)^8 = (-5)^{7+8} = (-5)^{15}$

EXAMPLE 13 Use the product rule to simplify $(2x^2)(-3x^5)$.

Solution: Recall that $2x^2$ means $2 \cdot x^2$ and $-3x^5$ means $-3 \cdot x^5$.

$$\begin{aligned}(2x^2)(-3x^5) &= 2 \cdot x^2 \cdot -3 \cdot x^5 && \text{Remove parentheses.}\\ &= 2 \cdot -3 \cdot x^2 \cdot x^5 && \text{Group factors with common bases.}\\ &= -6x^7 && \text{Simplify.}\end{aligned}$$

Practice Problem 13

Use the product rule to simplify $(6x^3)(-2x^9)$.

Helpful Hint

These examples will remind you of the difference between adding and multiplying terms.

Addition

$5x^3 + 3x^3 = (5 + 3)x^3 = 8x^3$ By the distributive property

$7x + 4x^2 = 7x + 4x^2$ Cannot be combined.

Multiplication

$(5x^3)(3x^3) = 5 \cdot 3 \cdot x^3 \cdot x^3 = 15x^{3+3} = 15x^6$ By the product rule

$(7x)(4x^2) = 7 \cdot 4 \cdot x \cdot x^2 = 28x^{1+2} = 28x^3$ By the product rule

C Using the Power Rule

Exponential expressions can themselves be raised to powers. Let's try to discover a rule that simplifies an expression like $(x^2)^3$. By the definition of a^n,

$$(x^2)^3 = (x^2)(x^2)(x^2) \quad (x^2)^3 \text{ means 3 factors of } (x^2).$$

which can be simplified by the product rule for exponents.

$$(x^2)^3 = (x^2)(x^2)(x^2) = x^{2+2+2} = x^6$$

Notice that the result is exactly the same if we multiply the exponents.

$$(x^2)^3 = x^{2 \cdot 3} = x^6$$

The following rule states this result.

Power Rule for Exponents

If m and n are positive integers and a is a real number, then

$$(a^m)^n = a^{mn}$$

For example,

$$(7^2)^5 = 7^{2 \cdot 5} = 7^{10}$$

In other words, to raise an exponential expression to a power, we keep the base and multiply the exponents.

EXAMPLES Use the power rule to simplify each expression.

14. $(5^3)^6 = 5^{3 \cdot 6} = 5^{18}$

15. $(y^8)^2 = y^{8 \cdot 2} = y^{16}$

Practice Problems 14–15

Use the power rule to simplify each expression.

14. $(9^4)^{10}$ 15. $(z^6)^3$

Answers

13. $-12x^{12}$ **14.** 9^{40} **15.** z^{18}

Helpful Hint

Take a moment to make sure that you understand when to apply the product rule and when to apply the power rule.

Product Rule → Add Exponents
$x^5 \cdot x^7 = x^{5+7} = x^{12}$
$y^6 \cdot y^2 = y^{6+2} = y^8$

Power Rule → Multiply Exponents
$(x^5)^7 = x^{5\cdot 7} = x^{35}$
$(y^6)^2 = y^{6\cdot 2} = y^{12}$

D Using the Power Rules for Products and Quotients

When the base of an exponential expression is a product, the definition of a^n still applies. For example, simplify $(xy)^3$ as follows.

$(xy)^3 = (xy)(xy)(xy)$ — $(xy)^3$ means 3 factors of (xy).

$= x \cdot x \cdot x \cdot y \cdot y \cdot y$ — Group factors with common bases.

$= x^3y^3$ — Simplify.

Notice that to simplify the expression $(xy)^3$, we raise each factor within the parentheses to a power of 3.

$$(xy)^3 = x^3y^3$$

In general, we have the following rule.

Power of a Product Rule

If n is a positive integer and a and b are real numbers, then

$$(ab)^n = a^nb^n$$

For example:

$$(3x)^5 = 3^5x^5$$

In other words, to raise a product to a power, we raise each factor to the power.

EXAMPLES Simplify each expression.

16. $(st)^4 = s^4 \cdot t^4 = s^4t^4$ — Use the power of a product rule.

17. $(2a)^3 = 2^3 \cdot a^3 = 8a^3$ — Use the power of a product rule.

18. $(-5x^2y^3z)^2 = (-5)^2 \cdot (x^2)^2 \cdot (y^3)^2 \cdot (z^1)^2$ — Use the power of a product rule.

$= 25x^4y^6z^2$ — Use the power rule for exponents.

Practice Problems 16–18

Simplify each expression.

16. $(xy)^7$

17. $(3y)^4$

18. $(-2p^4q^2r)^3$

Answers

16. x^7y^7 **17.** $81y^4$ **18.** $-8p^{12}q^6r^3$

Let's see what happens when we raise a quotient to a power. For example, we simplify $\left(\frac{x}{y}\right)^3$ as follows.

$$\left(\frac{x}{y}\right)^3 = \left(\frac{x}{y}\right)\left(\frac{x}{y}\right)\left(\frac{x}{y}\right) \qquad \left(\frac{x}{y}\right)^3 \text{ means 3 factors of } \left(\frac{x}{y}\right).$$

$$= \frac{x \cdot x \cdot x}{y \cdot y \cdot y} \qquad \text{Multiply fractions.}$$

$$= \frac{x^3}{y^3} \qquad \text{Simplify.}$$

Notice that to simplify the expression, $\left(\frac{x}{y}\right)^3$, we raise both the numerator and the denominator to a power of 3.

$$\left(\frac{x}{y}\right)^3 = \frac{x^3}{y^3}$$

In general, we have the following rule.

Power of a Quotient Rule

If n is a positive integer and a and c are real numbers, then

$$\left(\frac{a}{c}\right)^n = \frac{a^n}{c^n}, \quad c \neq 0$$

For example:

$$\left(\frac{y}{7}\right)^3 = \frac{y^3}{7^3}$$

In other words, to raise a quotient to a power, we raise both the numerator and the denominator to the power.

EXAMPLES Simplify each expression.

19. $\left(\frac{m}{n}\right)^7 = \frac{m^7}{n^7}, \quad n \neq 0$ Use the power of a quotient rule.

20. $\left(\frac{2x^4}{3y^5}\right)^4 = \frac{2^4 \cdot (x^4)^4}{3^4 \cdot (y^5)^4}$ Use the power of a quotient rule.

$= \frac{16x^{16}}{81y^{20}}, \quad y \neq 0$ Use the power rule for exponents.

E Using the Quotient Rule and Defining the Zero Exponent

Another pattern for simplifying exponential expressions involves quotients.

$$\frac{x^5}{x^3} = \frac{x \cdot x \cdot x \cdot x \cdot x}{x \cdot x \cdot x}$$

$$= \frac{x \cdot x \cdot x \cdot x \cdot x}{x \cdot x \cdot x}$$

$$= 1 \cdot 1 \cdot 1 \cdot x \cdot x$$

$$= x \cdot x$$

$$= x^2$$

Practice Problems 19–20

Simplify each expression.

19. $\left(\frac{r}{s}\right)^6$ 20. $\left(\frac{5x^6}{9y^3}\right)^2$

Answers

19. $\frac{r^6}{s^6}, \quad s \neq 0$ **20.** $\frac{25x^{12}}{81y^6}, \quad y \neq 0$

Notice that the result is exactly the same if we subtract exponents of the common bases.

$$\frac{x^5}{x^3} = x^{5-3} = x^2$$

The following rule states this result in a general way.

Quotient Rule for Exponents

If m and n are positive integers and a is a real number, then

$$\frac{a^m}{a^n} = a^{m-n}, \quad a \neq 0$$

For example,

$$\frac{x^6}{x^2} = x^{6-2} = x^4, \quad x \neq 0$$

In other words, to divide one exponential expression by another with a common base, we keep the base and subtract the exponents.

Practice Problems 21–24

Simplify each quotient.

21. $\frac{y^7}{y^3}$ 22. $\frac{5^9}{5^6}$

23. $\frac{(-2)^{14}}{(-2)^{10}}$ 24. $\frac{7a^4b^{11}}{ab}$

EXAMPLES Simplify each quotient.

21. $\frac{x^5}{x^2} = x^{5-2} = x^3$ Use the quotient rule.

22. $\frac{4^7}{4^3} = 4^{7-3} = 4^4 = 256$ Use the quotient rule.

23. $\frac{(-3)^5}{(-3)^2} = (-3)^3 = -27$ Use the quotient rule.

24. $\frac{2x^5y^2}{xy} = 2 \cdot \frac{x^5}{x^1} \cdot \frac{y^2}{y^1}$

$= 2 \cdot (x^{5-1}) \cdot (y^{2-1})$ Use the quotient rule.

$= 2x^4y^1$ or $2x^4y$

Let's now give meaning to an expression such as x^0. To do so, we will simplify $\frac{x^3}{x^3}$ in two ways and compare the results.

$\frac{x^3}{x^3} = x^{3-3} = x^0$ Apply the quotient rule.

$\frac{x^3}{x^3} = \frac{x \cdot x \cdot x}{x \cdot x \cdot x} = 1$ Apply the fundamental principle for fractions.

Since $\frac{x^3}{x^3} = x^0$ and $\frac{x^3}{x^3} = 1$, we define that $x^0 = 1$ as long as x is not 0.

Answers

21. y^4 **22.** 125 **23.** 16 **24.** $7a^3b^{10}$

Zero Exponent

$a^0 = 1$, as long as a is not 0.

For example, $5^0 = 1$.

In other words, a base raised to the 0 power is 1, as long as the base is not 0.

EXAMPLES Simplify each expression.

25. $3^0 = 1$

26. $(5x^3y^2)^0 = 1$

27. $(-4)^0 = 1$

28. $-4^0 = -1 \cdot 4^0 = -1 \cdot 1 = -1$

Try the Concept Check in the margin.

F Deciding Which Rule to Use

Let's practice deciding which rule to use to simplify.

EXAMPLE 29 Simplify each expression.

a. $x^7 \cdot x^4$

b. $\left(\frac{1}{2}\right)^4$

c. $(9y^5)^2$

Solution:

a. Here, we have a product, so we use the product rule to simplify.

$$x^7 \cdot x^4 = x^{7+4} = x^{11}$$

b. This is a quotient raised to a power, so we use the power of a quotient rule.

$$\left(\frac{1}{2}\right)^4 = \frac{1^4}{2^4} = \frac{1}{16}$$

c. This is a product raised to a power, so we use the power of a product rule.

$$(9y^5)^2 = 9^2(y^5)^2 = 81y^{10}$$

Practice Problems 25–28

Simplify each expression.

25. 8^0 26. $(2r^2s)^0$

27. $(-5)^0$ 28. -5^0

Concept Check

Suppose you are simplifying each expression. Tell whether you would *add* the exponents, *subtract* the exponents, *multiply* the exponents, *divide* the exponents, or *none of these*.

a. $(x^{63})^{21}$ b. $\frac{y^{15}}{y^3}$

c. $z^{16} + z^8$ d. $w^{45} \cdot w^9$

Practice Problem 29

Simplify each expression.

a. $\frac{x^7}{x^4}$ b. $(3y^4)^4$ c. $\left(\frac{x}{4}\right)^3$

Answers

25. 1 **26.** 1 **27.** 1 **28.** -1

29. a. x^3 **b.** $81y^{16}$ **c.** $\frac{x^3}{64}$

Concept Check: **a.** multiply **b.** subtract **c.** none of these **d.** add

STUDY SKILLS REMINDER

Are you satisfied with your performance on a particular quiz or exam?

If not, analyze your quiz or exam like you would a good mystery novel. Look for common themes in your errors.

Were most of your errors a result of

- *Carelessness?* If your errors were careless, did you turn in your work before the allotted time expired? If so, resolve next time to use the entire time allotted. Any extra time can be spent checking your work.
- *Running out of time?* If so, make a point to better manage your time on your next exam. A few suggestions are to work any questions that you are unsure of last and to check your work after all questions have been answered.
- *Not understanding a concept?* If so, review that concept and correct your work. Remember next time to make sure that all concepts on a quiz or exam are understood before the exam.

Name ______________ Section ________ Date ________

Mental Math

State the bases and the exponents for each expression.

1. 3^2 **2.** 5^4 **3.** $(-3)^6$ **4.** -3^7 **5.** -4^2

6. $(-4)^3$ **7.** $5 \cdot 3^4$ **8.** $9 \cdot 7^6$ **9.** $5x^2$ **10.** $(5x)^2$

EXERCISE SET 10.1

A *Evaluate each expression. See Examples 1 through 6.*

1. 7^2 **2.** -3^2 **3.** $(-5)^1$ **4.** $(-3)^2$ **5.** -2^4 **6.** -4^3

7. $(-2)^4$ **8.** $(-4)^3$ **9.** $\left(\frac{1}{3}\right)^3$ **10.** $\left(-\frac{1}{9}\right)^2$ **11.** $7 \cdot 2^4$ **12.** $9 \cdot 1^2$

13. Explain why $(-5)^4 = 625$, while $-5^4 = -625$.

14. Explain why $5 \cdot 4^2 = 80$, while $(5 \cdot 4)^2 = 400$.

Evaluate each expression with the given replacement values. See Example 7.

15. x^2 when $x = -2$

16. x^3 when $x = -2$

17. $5x^3$ when $x = 3$

18. $4x^2$ when $x = -1$

19. $2xy^2$ when $x = 3$ and $y = 5$

20. $-4x^2y^3$ when $x = 2$ and $y = -1$

21. $\frac{2z^4}{5}$ when $z = -2$

22. $\frac{10}{3y^3}$ when $y = 5$

B *Use the product rule to simplify each expression. Write the results using exponents. See Examples 8 through 13.*

23. $x^2 \cdot x^5$ **24.** $y^2 \cdot y$ **25.** $(-3)^3 \cdot (-3)^9$ **26.** $(-5)^7 \cdot (-5)^6$

27. $(5y^4)(3y)$ **28.** $(-2z^3)(-2z^2)$ **29.** $(4z^{10})(-6z^7)(z^3)$ **30.** $(12x^5)(-x^6)(x^4)$

△ **31.** The rectangle below has width $4x^2$ feet and length $5x^3$ feet. Find its area.

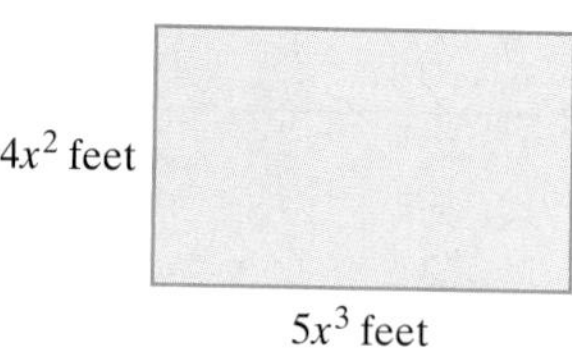

△ **32.** The parallelogram below has base length $9y^7$ meters and height $2y^{10}$ meters. Find its area.

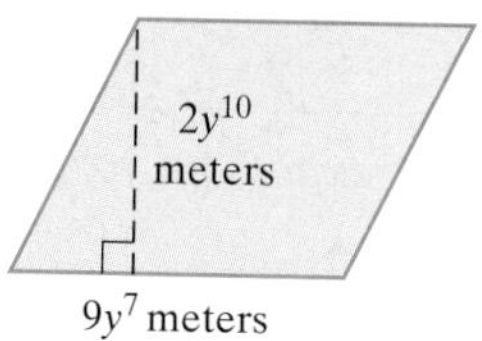

C D *Use the power rule and the power of a product or quotient rule to simplify each expression. See Examples 14 through 20.*

33. $(x^9)^4$

34. $(y^7)^5$

35. $(pq)^7$

36. $(ab)^6$

37. $(2a^5)^3$

38. $(4x^6)^2$

39. $\left(\frac{m}{n}\right)^9$

40. $\left(\frac{xy}{7}\right)^2$

41. $(x^2y^3)^5$

42. $(a^4b)^7$

43. $\left(\frac{-2xz}{y^5}\right)^2$

44. $\left(\frac{y^4}{-3z^3}\right)^3$

△ **45.** The square shown has sides of length $8z^5$ decimeters. Find its area.

△ **46.** Given the circle below with radius $5y$ centimeters, find its area. Do not approximate π.

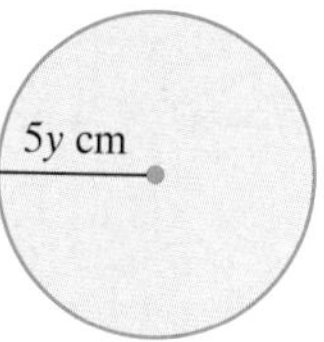

△ **47.** The vault below is in the shape of a cube. If each side is $3y^4$ feet, find its volume.

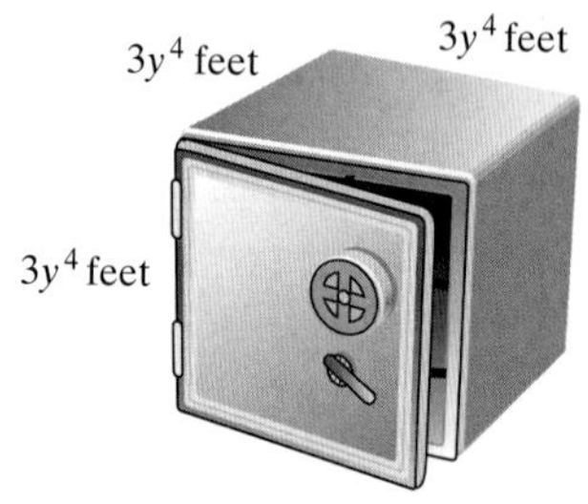

△ **48.** The silo shown is in the shape of a cylinder. If its radius is $4x$ meters and its height is $5x^3$ meters, find its volume. Do not approximate π.

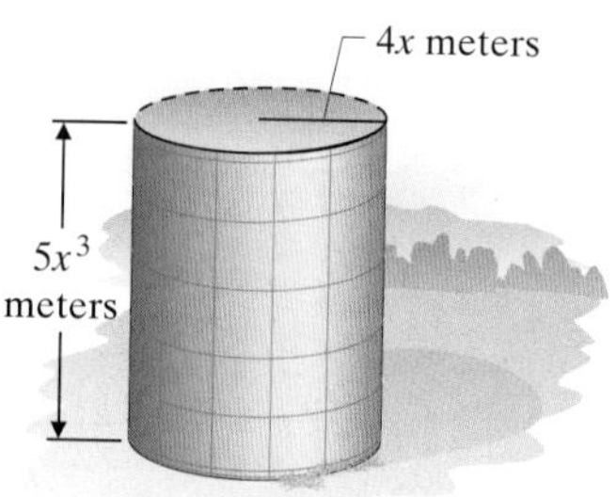

E *Use the quotient rule and simplify each expression. See Examples 21 through 24.*

49. $\frac{x^3}{x}$

50. $\frac{y^{10}}{y^9}$

51. $\frac{(-2)^5}{(-2)^3}$

52. $\frac{(-5)^{14}}{(-5)^{11}}$

53. $\frac{p^7q^{20}}{pq^{15}}$

54. $\frac{x^8y^6}{xy^5}$

55. $\frac{7x^2y^6}{14x^2y^3}$

56. $\frac{9a^4b^7}{3ab^2}$

Simplify each expression. See Examples 25 through 28.

57. $(2x)^0$

58. $-4x^0$

59. $-2x^0$

60. $(4y)^0$

61. $5^0 + y^0$

62. $-3^0 + 4^0$

63. In your own words, explain why $5^0 = 1$.

64. In your own words, explain when $(-3)^n$ is positive and when it is negative.

F *Simplify each expression. See Example 29.*

65. -5^2

66. $(-5)^2$

67. $\left(\frac{1}{4}\right)^3$

68. $\left(\frac{2}{3}\right)^3$

69. $\frac{z^{12}}{z^4}$

70. $\frac{b^4}{b}$

71. $(9xy)^2$

72. $(2ab)^5$

73. $(6b)^0$

74. $(5ab)^0$

75. $2^3 + 2^5$

76. $7^2 - 7^0$

77. b^4b^2

78. y^4y^1

79. $a^2a^3a^4$

80. $x^2x^{15}x^9$

81. $(2x^3)(-8x^4)$

82. $(3y^4)(-5y)$

83. $(4a)^3$

84. $(2ab)^4$

85. $(-6xyz^3)^2$

86. $(-3xy^2a^3b)^3$

87. $\left(\frac{3y^5}{6x^4}\right)^3$

88. $\left(\frac{2ab}{6yz}\right)^4$

89. $\frac{3x^5}{x^4}$

90. $\frac{5x^9}{x^3}$

91. $\frac{2x^3y^2z}{xyz}$

92. $\frac{x^{12}y^{13}}{x^5y^7}$

Review and Preview

Subtract. See Section 9.3.

93. $5 - 7$

94. $9 - 12$

95. $3 - (-2)$

96. $5 - (-10)$

97. $-11 - (-4)$

98. $-15 - (-21)$

Combining Concepts

△ **99.** The formula $V = x^3$ can be used to find the volume V of a cube with side length x. Find the volume of a cube with side length 7 meters. (Volume is measured in cubic units.)

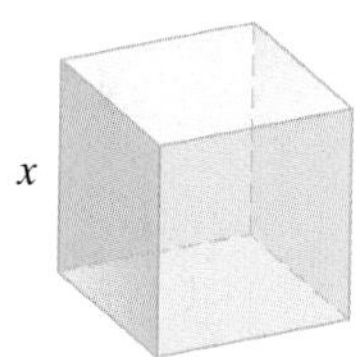

△ **100.** The formula $S = 6x^2$ can be used to find the surface area S of a cube with side length x. Find the surface area of a cube with side length 5 meters. (Surface area is measured in square units.)

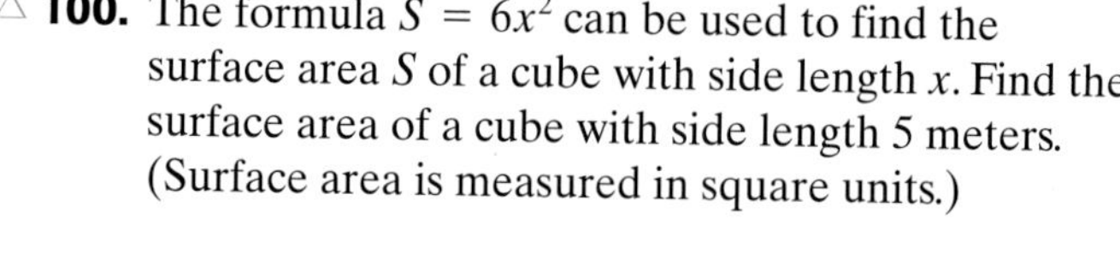

△ **101.** To find the amount of water that a swimming pool in the shape of a cube can hold, do we use the formula for volume of the cube or surface area of the cube? (See Exercises 99 and 100.)

△ **102.** To find the amount of material needed to cover an ottoman in the shape of a cube, do we use the formula for volume of the cube or surface area of the cube? (See Exercises 99 and 100.)

Simplify each expression. Assume that variables represent positive integers.

103. $x^{5a}x^{4a}$

104. $b^{9a}b^{4a}$

105. $(a^b)^5$

106. $(2a^{4b})^4$

107. $\dfrac{x^{9a}}{x^{4a}}$

108. $\dfrac{y^{15b}}{y^{6b}}$

109. Suppose you borrow money for 6 months. If the interest rate is compounded monthly, the formula $A = P\left(1 + \dfrac{r}{12}\right)^6$ gives the total amount A to be repaid at the end of 6 months. For a loan of $P = \$1000$ and interest rate of 9% ($r = 0.09$), how much money will you need to pay off the loan?

110. On August 24, 2003, the Federal Reserve discount rate was set at 2%. (*Source:* Federal Reserve Board) The discount rate is the interest rate at which banks can borrow money from the Federal Reserve System. Suppose a bank needs to borrow money from the Federal Reserve System for 3 months. If the interest is compounded monthly, the formula $A = P\left(1 + \dfrac{r}{12}\right)^3$ gives the total amount A to be repaid at the end of 3 months. For a loan of $P = \$500{,}000$ and interest rate of $r = 0.02$, how much money will the bank repay to the Federal Reserve at the end of 3 months? Round to the nearest dollar.

10.2 Negative Exponents and Scientific Notation

OBJECTIVES

- **A** Simplify expressions containing negative exponents.
- **B** Use the rules and definitions for exponents to simplify exponential expressions.
- **C** Write numbers in scientific notation.
- **D** Convert numbers in scientific notation to standard form.

SSM TUTOR CENTER SG CD & VIDEO MATH PRO WEB

A Simplifying Expressions Containing Negative Exponents

Our work with exponential expressions so far has been limited to exponents that are positive integers or 0. Here we will also give meaning to an expression like x^{-3}.

Suppose that we wish to simplify the expression $\frac{x^2}{x^5}$. If we use the quotient rule for exponents, we subtract exponents:

$$\frac{x^2}{x^5} = x^{2-5} = x^{-3}, \quad x \neq 0$$

But what does x^{-3} mean? Let's simplify $\frac{x^2}{x^5}$ using the definition of a^n.

$$\frac{x^2}{x^5} = \frac{x \cdot x}{x \cdot x \cdot x \cdot x \cdot x}$$

$$= \frac{x \cdot x}{x \cdot x \cdot x \cdot x \cdot x}$$ Divide numerator and denominator by common factors by applying the fundamental principle for fractions.

$$= \frac{1}{x^3}$$

If the quotient rule is to hold true for negative exponents, then x^{-3} must equal $\frac{1}{x^3}$.

From this example, we state the definition for negative exponents.

Negative Exponents

If a is a real number other than 0 and n is an integer, then

$$a^{-n} = \frac{1}{a^n}$$

For example,

$$x^{-3} = \frac{1}{x^3}$$

In other words, another way to write a^{-n} is to take its reciprocal and change the sign of its exponent.

EXAMPLES Simplify by writing each expression with positive exponents only.

1. 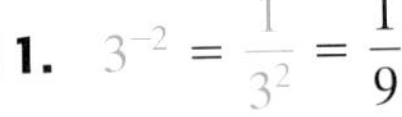 $3^{-2} = \frac{1}{3^2} = \frac{1}{9}$ Use the definition of negative exponent.

2. 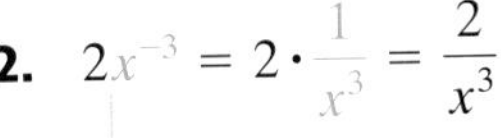 $2x^{-3} = 2 \cdot \frac{1}{x^3} = \frac{2}{x^3}$ Use the definition of negative exponent.

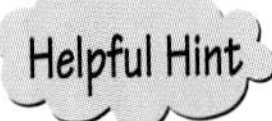

Don't forget that since there are no parentheses, only x is the base for the exponent -3.

3. $2^{-1} + 4^{-1} = \frac{1}{2} + \frac{1}{4} = \frac{2}{4} + \frac{1}{4} = \frac{3}{4}$

4. $(-2)^{-4} = \frac{1}{(-2)^4} = \frac{1}{(-2)(-2)(-2)(-2)} = \frac{1}{16}$

Practice Problems 1–4

Simplify by writing each expression with positive exponents only.

1. 5^{-3}
2. $7x^{-4}$
3. $5^{-1} + 3^{-1}$
4. $(-3)^{-4}$

Answers

1. $\frac{1}{125}$ **2.** $\frac{7}{x^4}$ **3.** $\frac{8}{15}$ **4.** $\frac{1}{81}$

Helpful Hint

A negative exponent *does not affect* the sign of its base.

Remember: Another way to write a^{-n} is to take its reciprocal and change the sign of its exponent: $a^{-n} = \frac{1}{a^n}$. For example,

$$x^{-2} = \frac{1}{x^2}, \qquad 2^{-3} = \frac{1}{2^3} \text{ or } \frac{1}{8}$$

$$\frac{1}{y^{-4}} = \frac{1}{\frac{1}{y^4}} = y^4, \qquad \frac{1}{5^{-2}} = 5^2 \text{ or } 25$$

Practice Problems 5–8

Simplify each expression. Write each result using positive exponents only.

5. $\left(\frac{6}{7}\right)^{-2}$ 6. $\frac{x}{x^{-4}}$

7. $\frac{y^{-9}}{z^{-5}}$ 8. $\frac{y^{-4}}{y^6}$

Answers

5. $\frac{49}{36}$ **6.** x^5 **7.** $\frac{z^5}{y^9}$ **8.** $\frac{1}{y^{10}}$

EXAMPLES Simplify each expression. Write each result using positive exponents only.

5. $\left(\frac{2}{3}\right)^{-3} = \frac{2^{-3}}{3^{-3}} = \frac{3^3}{2^3} = \frac{27}{8}$ Use the negative exponent rule.

6. $\frac{y}{y^{-2}} = \frac{y^1}{y^{-2}} = y^{1-(-2)} = y^3$ Use the quotient rule.

7. $\frac{p^{-4}}{q^{-9}} = \frac{q^9}{p^4}$ Use the negative exponent rule.

8. $\frac{x^{-5}}{x^7} = x^{-5-7} = x^{-12} = \frac{1}{x^{12}}$

B Simplifying Exponential Expressions

All the previously stated rules for exponents apply for negative exponents also. Here is a summary of the rules and definitions for exponents.

Summary of Exponent Rules

If m and n are integers and a, b, and c are real numbers, then:

Product rule for exponents:	$a^m \cdot a^n = a^{m+n}$
Power rule for exponents:	$(a^m)^n = a^{m \cdot n}$
Power of a product:	$(ab)^n = a^n b^n$
Power of a quotient:	$\left(\frac{a}{c}\right)^n = \frac{a^n}{c^n}, \quad c \neq 0$
Quotient rule for exponents:	$\frac{a^m}{a^n} = a^{m-n}, \quad a \neq 0$
Zero exponent:	$a^0 = 1, \quad a \neq 0$
Negative exponent:	$a^{-n} = \frac{1}{a^n}, \quad a \neq 0$

EXAMPLES

Simplify each expression. Write each result using positive exponents only.

9. $\frac{(x^3)^4 x}{x^7} = \frac{x^{12} \cdot x}{x^7} = \frac{x^{12+1}}{x^7} = \frac{x^{13}}{x^7} = x^{13-7} = x^6$ Use the power rule.

10. $\left(\frac{3a^2}{b}\right)^{-3} = \frac{3^{-3}(a^2)^{-3}}{b^{-3}}$ Raise each factor in the numerator and the denominator to the -3 power.

$= \frac{3^{-3}a^{-6}}{b^{-3}}$ Use the power rule.

$= \frac{b^3}{3^3a^6}$ Use the negative exponent rule.

$= \frac{b^3}{27a^6}$ Write 3^3 as 27.

11. $(y^{-3}z^6)^{-6} = (y^{-3})^{-6}(z^6)^{-6}$ Raise each factor to the -6 power.

$= y^{18}z^{-36} = \frac{y^{18}}{z^{36}}$

12. $\frac{(2x)^5}{x^3} = \frac{2^5 \cdot x^5}{x^3} = 2^5 \cdot x^{5-3} = 32x^2$ Raise each factor in the numerator to the fifth power.

13. $\frac{x^{-7}}{(x^4)^3} = \frac{x^{-7}}{x^{12}} = x^{-7-12} = x^{-19} = \frac{1}{x^{19}}$

14. $(5y^3)^{-2} = 5^{-2}(y^3)^{-2}$ Raise each factor to the -2 power.

$= 5^{-2}y^{-6} = \frac{1}{5^2y^6} = \frac{1}{25y^6}$

15. $\frac{(2xy)^{-3}}{(x^2y^3)^2} = \frac{2^{-3}x^{-3}y^{-3}}{(x^2)^2 \cdot (y^3)^2} = \frac{2^{-3}x^{-3}y^{-3}}{x^4y^6}$

$= 2^{-3}x^{-3-4}y^{-3-6} = 2^{-3}x^{-7}y^{-9}$

$= \frac{1}{2^3x^7y^9}$ or $\frac{1}{8x^7y^9}$

Practice Problems 9–15

Simplify each expression. Write each result using positive exponents only.

9. $\frac{(x^5)^3x}{x^4}$

10. $\left(\frac{9x^3}{y}\right)^{-2}$

11. $(a^{-4}b^7)^{-5}$

12. $\frac{(2x)^4}{x^8}$

13. $\frac{y^{-10}}{(y^5)^4}$

14. $(4a^2)^{-3}$

15. $\frac{(3x^{-2}y)^{-2}}{4x^7y}$

C Writing Numbers in Scientific Notation

Both very large and very small numbers frequently occur in many fields of science. For example, the distance between the Sun and the planet Pluto is approximately 5,906,000,000 kilometers, and the mass of a proton is approximately 0.00000000000000000000000165 gram. It can be tedious to write these numbers in this standard decimal notation, so **scientific notation** is used as a convenient shorthand for expressing very large and very small numbers.

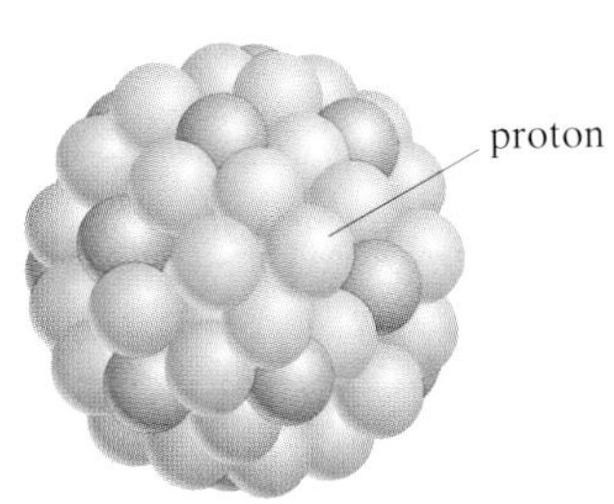

Mass of proton is approximately 0.00000000000000000000000165 gram

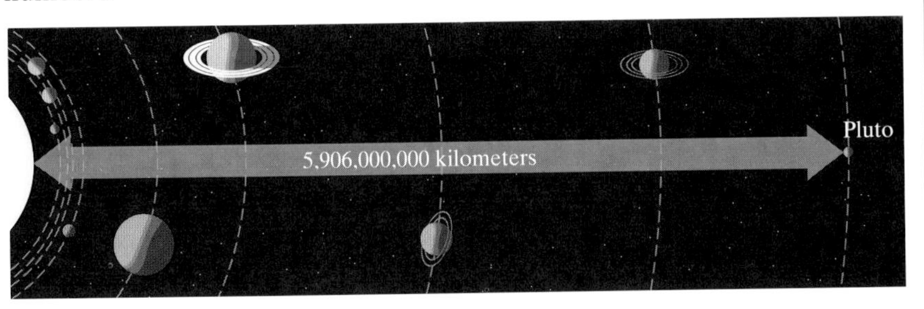

Answers

9. x^{12} **10.** $\frac{y^2}{81x^6}$ **11.** $\frac{a^{20}}{b^{35}}$ **12.** $\frac{16}{x^4}$ **13.** $\frac{1}{y^{30}}$ **14.** $\frac{1}{64a^6}$ **15.** $\frac{1}{36x^3y^3}$

Scientific Notation

A positive number is written in scientific notation if it is written as the product of a number a, where $1 \leq a < 10$, and an integer power r of 10: $a \times 10^r$.

The following numbers are written in scientific notation. The $\times$ sign for multiplication is used as part of the notation.

2.03×10^2 7.362×10^7 5.906×10^9 (Distance between the Sun and Pluto)

1×10^{-3} 8.1×10^{-5} 1.65×10^{-24} (Mass of a proton)

The following steps are useful when writing numbers in scientific notation.

To Write a Number in Scientific Notation

Step 1. Move the decimal point in the original number so that the new number has a value between 1 and 10.

Step 2. Count the number of decimal places the decimal point is moved in Step 1. If the original number is 10 or greater, the count is positive. If the original number is less than 1, the count is negative.

Step 3. Multiply the new number in Step 1 by 10 raised to an exponent equal to the count found in Step 2.

Practice Problem 16

Write each number in scientific notation.

a. 420,000 b. 0.00017
c. 9,060,000,000 d. 0.000007

EXAMPLE 16 Write each number in scientific notation.

a. 367,000,000 **b.** 0.000003
c. 20,520,000,000 **d.** 0.00085

Solution:

a. Step 1. Move the decimal point until the number is between 1 and 10.

367,000,000 (8 places)

Step 2. The decimal point is moved 8 places and the original number is 10 or greater, so the count is positive 8.

Step 3. $367{,}000{,}000 = 3.67 \times 10^8$.

b. Step 1. Move the decimal point until the number is between 1 and 10.

0.000003 (6 places)

Step 2. The decimal point is moved 6 places and the original number is less than 1, so the count is -6.

Step 3. $0.000003 = 3.0 \times 10^{-6}$

c. $20{,}520{,}000{,}000 = 2.052 \times 10^{10}$

d. $0.00085 = 8.5 \times 10^{-4}$

D Converting Numbers to Standard Form

A number written in scientific notation can be rewritten in standard form. For example, to write 8.63×10^3 in standard form, recall that $10^3 = 1000$.

$$8.63 \times 10^3 = 8.63(1000) = 8630$$

Notice that the exponent on the 10 is positive 3, and we moved the decimal point 3 places to the right.

Answers

16. a. 4.2×10^5 **b.** 1.7×10^{-4} **c.** 9.06×10^9 **d.** 7×10^{-6}

To write 7.29×10^{-3} in standard form, recall that $10^{-3} = \frac{1}{10^3} = \frac{1}{1000}$.

$$7.29 \times 10^{-3} = 7.29\left(\frac{1}{1000}\right) = \frac{7.29}{1000} = 0.00729$$

The exponent on the 10 is negative 3, and we moved the decimal to the left 3 places.

In general, **to write a scientific notation number in standard form,** move the decimal point the same number of places as the exponent on 10. If the exponent is positive, move the decimal point to the right; if the exponent is negative, move the decimal point to the left.

Try the Concept Check in the margin.

Concept Check

Which number in each pair is larger?

a. 7.8×10^3 or 2.1×10^5
b. 9.2×10^{-2} or 2.7×10^4
c. 5.6×10^{-4} or 6.3×10^{-5}

EXAMPLE 17

Write each number in standard notation, without exponents.

a. 1.02×10^5 **b.** 7.358×10^{-3}
c. 8.4×10^7 **d.** 3.007×10^{-5}

Solution:

a. Move the decimal point 5 places to the right.

$1.02 \times 10^5 = 102{,}000.$

b. Move the decimal point 3 places to the left.

$7.358 \times 10^{-3} = 0.007358$

c. $8.4 \times 10^7 = 84{,}000{,}000.$ 7 places to the right

d. $3.007 \times 10^{-5} = 0.00003007$ 5 places to the left

Practice Problem 17

Write the numbers in standard notation, without exponents.

a. 3.062×10^{-4} b. 5.21×10^4
c. 9.6×10^{-5} d. 6.002×10^6

Performing operations on numbers written in scientific notation makes use of the rules and definitions for exponents.

EXAMPLE 18

Perform each indicated operation. Write each result in standard decimal notation.

a. $(8 \times 10^{-6})(7 \times 10^3)$

b. $\dfrac{12 \times 10^2}{6 \times 10^{-3}}$

Solution:

a. $(8 \times 10^{-6})(7 \times 10^3) = 8 \cdot 7 \cdot 10^{-6} \cdot 10^3$
$= 56 \times 10^{-3}$
$= 0.056$

b. $\dfrac{12 \times 10^2}{6 \times 10^{-3}} = \dfrac{12}{6} \times 10^{2-(-3)} = 2 \times 10^5 = 200{,}000$

Practice Problem 18

Perform each indicated operation. Write each result in standard decimal notation.

a. $(9 \times 10^7)(4 \times 10^{-9})$

b. $\dfrac{8 \times 10^4}{2 \times 10^{-3}}$

Answers

17. a. 0.0003062 **b.** 52,100 **c.** 0.000096 **d.** 6,002,000 **18. a.** 0.36 **b.** 40,000,000

Concept Check: **a.** 2.1×10^5 **b.** 2.7×10^4 **c.** 5.6×10^{-4}

CALCULATOR EXPLORATIONS

Scientific Notation

To enter a number written in scientific notation on a scientific calculator, locate the scientific notation key, which may be marked [EE] or [EXP]. To enter 3.1×10^7, press [3.1] [EE] [7]. The display should read [3.1 07].

Enter each number written in scientific notation on your calculator.

1. 5.31×10^3
2. -4.8×10^{14}
3. 6.6×10^{-9}
4. -9.9811×10^{-2}

Multiply each of the following on your calculator. Notice the form of the result.

5. $3{,}000{,}000 \times 5{,}000{,}000$
6. $230{,}000 \times 1{,}000$

Multiply each of the following on your calculator. Write the product in scientific notation.

7. $(3.26 \times 10^6)(2.5 \times 10^{13})$
8. $(8.76 \times 10^{-4})(1.237 \times 10^9)$

Name ______________________ Section __________ Date __________

Mental Math

State each expression using positive exponents only.

1. $5x^{-2}$ **2.** $3x^{-3}$ **3.** $\frac{1}{y^{-6}}$ **4.** $\frac{1}{x^{-3}}$ **5.** $\frac{4}{y^{-3}}$ **6.** $\frac{16}{y^{-7}}$

EXERCISE SET 10.2

A *Simplify each expression. Write each result using positive exponents only. See Examples 1 through 8.*

1. 4^{-3} **2.** 6^{-2} **3.** $7x^{-3}$ **4.** $(7x)^{-3}$ **5.** $\left(-\frac{1}{4}\right)^{-3}$ **6.** $\left(-\frac{1}{8}\right)^{-2}$

7. $3^{-1} + 2^{-1}$ **8.** $4^{-1} + 4^{-2}$ **9.** $\frac{1}{p^{-3}}$ **10.** $\frac{1}{q^{-5}}$ **11.** $\frac{p^{-5}}{q^{-4}}$ **12.** $\frac{r^{-5}}{s^{-2}}$

13. $\frac{x^{-2}}{x}$ **14.** $\frac{y}{y^{-3}}$ **15.** $\frac{z^{-4}}{z^{-7}}$ **16.** $\frac{x^{-4}}{x^{-1}}$ **17.** $2^0 + 3^{-1}$ **18.** $4^{-2} - 4^{-3}$

19. $(-3)^{-2}$ **20.** $(-2)^{-6}$ **21.** $\frac{-1}{p^{-4}}$ **22.** $\frac{-1}{y^{-6}}$ **23.** $-2^0 - 3^0$ **24.** $5^0 + (-5)^0$

B *Simplify each expression. Write each result using positive exponents only. See Examples 9 through 15.*

25. $\frac{x^2x^5}{x^3}$ **26.** $\frac{y^4y^5}{y^6}$ **27.** $\frac{p^2p}{p^{-1}}$ **28.** $\frac{y^3y}{y^{-2}}$ **29.** $\frac{(m^5)^4m}{m^{10}}$ **30.** $\frac{(x^2)^8x}{x^9}$

31. $\frac{r}{r^{-3}r^{-2}}$ **32.** $\frac{p}{p^{-3}q^{-5}}$ **33.** $(x^5y^3)^{-3}$ **34.** $(z^5x^5)^{-3}$ **35.** $\frac{(x^2)^3}{x^{10}}$ **36.** $\frac{(y^4)^2}{y^{12}}$

37. $\frac{(a^5)^2}{(a^3)^4}$ **38.** $\frac{(x^2)^5}{(x^4)^3}$ **39.** $\frac{8k^4}{2k}$ **40.** $\frac{27r^4}{3r^6}$ **41.** $\frac{-6m^4}{-2m^3}$ **42.** $\frac{15a^4}{-15a^5}$

43. $\frac{-24a^6b}{6ab^2}$ **44.** $\frac{-5x^4y^5}{15x^4y^2}$ **45.** $\frac{6x^2y^3}{-7xy^5}$ **46.** $\frac{-8xa^2b}{-5xa^5b}$ **47.** $(a^{-5}b^2)^{-6}$ **48.** $(4^{-1}x^5)^{-2}$

49. $\left(\frac{x^{-2}y^4}{x^3y^7}\right)^2$ **50.** $\left(\frac{a^5b}{a^7b^{-2}}\right)^{-3}$ **51.** $\frac{4^2z^{-3}}{4^3z^{-5}}$ **52.** $\frac{3^{-1}x^4}{3^3x^{-7}}$ **53.** $\frac{2^{-3}x^{-4}}{2^2x}$ **54.** $\frac{5^{-1}z^7}{5^{-2}z^9}$

55. $\frac{7ab^{-4}}{7^{-1}a^{-3}b^2}$ **56.** $\frac{6^{-5}x^{-1}y^2}{6^{-2}x^{-4}y^4}$ **57.** $\left(\frac{a^{-5}b}{ab^3}\right)^{-4}$ **58.** $\left(\frac{r^{-2}s^{-3}}{r^{-4}s^{-3}}\right)^{-3}$ **59.** $\frac{(xy^3)^5}{(xy)^{-4}}$ **60.** $\frac{(rs)^{-3}}{(r^2s^3)^2}$

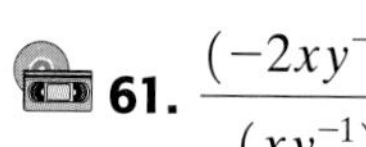
61. $\dfrac{(-2xy^{-3})^{-3}}{(xy^{-1})^{-1}}$

62. $\dfrac{(-3x^2y^2)^{-2}}{(xyz)^{-2}}$

63. $\dfrac{(a^4b^{-7})^{-5}}{(5a^2b^{-1})^{-2}}$

64. $\dfrac{(a^6b^{-2})^4}{(4a^{-3}b^{-3})^3}$

65. Find the volume of the cube.

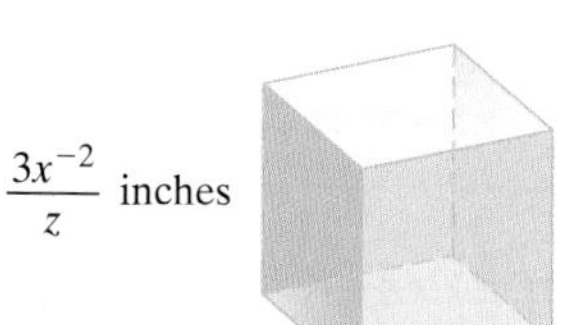

66. Find the area of the triangle.

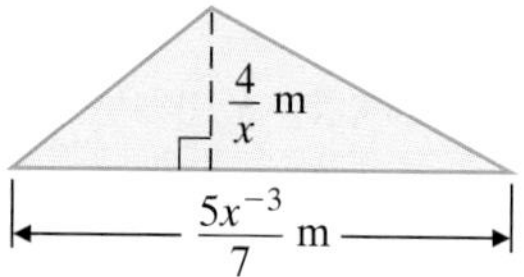

C *Write each number in scientific notation. See Example 16.*

67. 78,000

68. 9,300,000,000

69. 0.00000167

70. 0.00000017

71. 0.00635

72. 0.00194

73. 1,160,000

74. 700,000

75. The temperature of the Sun at its core is 15,600,000 degrees Kelvin. Write 15,600,000 in scientific notation. (*Source:* Students for the Exploration and Development of Space)

76. Google.com is an Internet search engine that allows users to search over 1,346,966,000 Web pages. Write 1,346,966,000 in scientific notation. (*Source:* Google, Inc.)

77. At this writing, the world's largest optical telescopes are the twin Keck Telescopes located near the summit of Mauna Kea in Hawaii. The elevation of the Keck Telescopes is about 13,600 feet above sea level. Write 13,600 in scientific notation. (*Source:* W. M. Keck Observatory)

78. More than 2,000,000,000 pencils are manufactured in the U.S. annually. Write this number in scientific notation. (*Source:* AbsoluteTrivia.com)

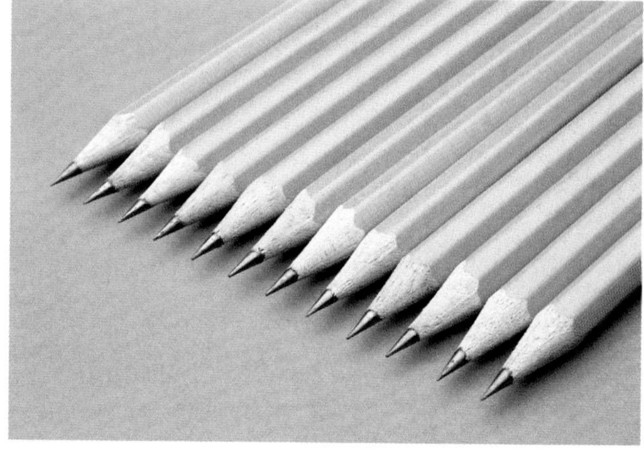

79. In May 2001, the population of the United States was roughly 284,000,000. Write 284,000,000 in scientific notation. (*Source:* U.S. Census Bureau)

80. Pioneer 10 became the first spacecraft to leave the solar system eleven years after it was launched in 1972. When it was contacted in April 2001 to verify that it could still transmit a radio signal, Pioneer 10 was 7,290,000,000 miles from Earth. Write 7,290,000,000 in scientific notation. (*Source:* NASA Ames Research Center)

D *Write each number in standard notation. See Example 17.*

81. 8.673×10^{-10}

82. 9.056×10^{-4}

83. 3.3×10^{-2}

84. 4.8×10^{-6}

85. 2.032×10^{4}

86. 9.07×10^{10}

87. The mass of an atom of the uranium isotope U-238 is 3.97×10^{-22} grams. Write this number in standard notation.

88. The mass of a hydrogen atom is 1.7×10^{-24} grams. Write this number in standard notation.

89. Each second, the Sun converts 7.0×10^{8} tons of hydrogen into helium and energy in the form of gamma rays. Write this number in standard notation. (*Source:* Students for the Exploration and Development of Space)

90. In chemistry, Avogadro's number is the number of atoms in one mole of an element. Avogadro's number is $6.02214199 \times 10^{23}$. Write this number in standard notation. (*Source:* National Institute of Standards and Technology)

Evaluate each expression using exponential rules. Write each result in standard notation. See Example 18.

91. $(1.2 \times 10^{-3})(3 \times 10^{-2})$

92. $(2.5 \times 10^{6})(2 \times 10^{-6})$

93. $(4 \times 10^{-10})(7 \times 10^{-9})$

94. $(5 \times 10^{6})(4 \times 10^{-8})$

95. $\dfrac{8 \times 10^{-1}}{16 \times 10^{5}}$

96. $\dfrac{25 \times 10^{-4}}{5 \times 10^{-9}}$

97. $\dfrac{1.4 \times 10^{-2}}{7 \times 10^{-8}}$

98. $\dfrac{0.4 \times 10^{5}}{0.2 \times 10^{11}}$

99. Although the actual amount varies by season and time of day, the average volume of water that flows over Niagara Falls (the American and Canadian falls combined) each second is 7.5×10^{5} gallons. How much water flows over Niagara Falls in an hour? Write the result in scientific notation. (*Hint:* 1 hour equals 3600 seconds) (*Source:* niagarafallslive.com)

100. A beam of light travels 9.460×10^{12} kilometers per year. How far does light travel in 10,000 years? Write the result in scientific notation.

Review and Preview

Simplify each expression by combining any like terms. See Section 9.2.

101. $3x - 5x + 7$

102. $7w + w - 2w$

103. $y - 10 + y$

104. $-6z + 20 - 3z$

105. $7x + 2 - 8x - 6$

106. $10y - 14 - y - 14$

Combining Concepts

Simplify each expression. Write each result in standard notation.

107. $(2.63 \times 10^{12})(-1.5 \times 10^{-10})$

108. $(6.785 \times 10^{-4})(4.68 \times 10^{10})$

Light travels at a rate of 1.86×10^5 *miles per second. Use this information and the distance formula* $d = r \cdot t$ *to answer Exercises 109 and 110.*

109. If the distance from the Moon to the Earth is 238,857 miles, find how long it takes the reflected light of the Moon to reach the Earth. (Round to the nearest tenth of a second.)

110. If the distance from the Sun to the Earth is 93,000,000 miles, find how long it takes the light of the Sun to reach the Earth. (Round to the nearest tenth of a second.)

Simplify each expression. Assume that variables represent positive integers.

111. $a^{-4m} \cdot a^{5m}$

112. $(x^{-3s})^3$

113. $(3y^{2z})^3$

114. $a^{4m+1} \cdot a^4$

115. It was stated earlier that for an integer n,

$$x^{-n} = \frac{1}{x^n}, \quad x \neq 0$$

Explain why x may not equal 0.

116. Determine whether each statement is true or false.

a. $5^{-1} < 5^{-2}$

b. $\left(\frac{1}{5}\right)^{-1} < \left(\frac{1}{5}\right)^{-2}$

c. $a^{-1} < a^{-2}$ for all nonzero numbers.

Internet Excursions

Go To: 

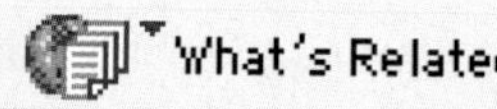

The Bureau of the Public Debt is part of the U.S. Department of the Treasury. The Bureau of the Public Debt borrows the money needed to run the federal government and keeps track of the debt. The given World Wide Web address will provide you with access to the Bureau of the Public Debt's Web Site, or a related site, where you can find the current size of the U.S. public debt to the penny. This site is updated daily. It also lists the amount of the public debt for the past month, as well as for selected dates in prior months and years.

117. Find the size of the public debt that is listed most recently on the Bureau of the Public Debt's Web site. Use the information on the Web site to record the debt amount and its date. Then write the debt amount in scientific notation, rounded to the nearest hundredth.
Debt amount: ______________
Date: ______________
Scientific notation (rounded to nearest hundredth): ______________

118. Look up the size of the public debt at the end of the month six months ago. Use the information on the Web site to record the debt amount and its date. Then write the debt amount in scientific notation, rounded to the nearest hundredth.
Debt amount: ______________
Date: ______________
Scientific notation (rounded to nearest hundredth): ______________

10.3 Introduction to Polynomials

OBJECTIVES

- A. Define term and coefficient of a term.
- B. Define polynomial, monomial, binomial, trinomial, and degree.
- C. Evaluate polynomials for given replacement values.
- D. Simplify a polynomial by combining like terms.
- E. Simplify a polynomial in several variables.

SSM TUTOR CENTER SG CD & VIDEO MATH PRO WEB

A Defining Term and Coefficient

In this section, we introduce a special algebraic expression called a polynomial. Let's first review some definitions presented in Section 3.1.

Recall that a term is a number or the product of a number and variables raised to powers. The terms of an expression are separated by plus signs. The terms of the expression $4x^2 + 3x$ are $4x^2$ and $3x$. The terms of the expression $9x^4 - 7x - 1$, or $9x^4 + (-7x) + (-1)$, are $9x^4$, $-7x$, and -1.

Expression	Terms
$4x^2 + 3x$	$4x^2, 3x$
$9x^4 - 7x - 1$	$9x^4, -7x, -1$
$7y^3$	$7y^3$

The **numerical coefficient** of a term, or simply the **coefficient,** is the numerical factor of each term. If no numerical factor appears in the term, then the coefficient is understood to be 1. If the term is a number only, it is called a **constant term** or simply a **constant.**

Term	Coefficient
x^5	1
$3x^2$	3
$-4x$	-4
$-x^2y$	-1
3 (constant)	3

EXAMPLE 1

Complete the table for the expression $7x^5 - 8x^4 + x^2 - 3x + 5$.

Term	Coefficient
x^2	
	-8
$-3x$	
	7
5	

Solution: The completed table is shown below.

Term	Coefficient
x^2	1
$-8x^4$	-8
$-3x$	-3
$7x^5$	7
5	5

Practice Problem 1

Complete the table for the expression $-6x^6 + 4x^5 + 7x^3 - 9x^2 - 1$.

Term	Coefficient
$7x^3$	
	-9
$-6x^6$	
	4
-1	

Answer

1. term: $-9x^2$; $4x^5$ coefficient: 7; -6; -1

B Defining Polynomial, Monomial, Binomial, Trinomial, and Degree

Now we are ready to define what we mean by a polynomial.

Polynomial

A **polynomial in x** is a finite sum of terms of the form ax^n, where a is a real number and n is a whole number.

For example,

$$x^5 - 3x^3 + 2x^2 - 5x + 1$$

is a polynomial in x. Notice that this polynomial is written in **descending powers** of x because the powers of x decrease from left to right. (Recall that the term 1 can be thought of as $1x^0$.)

On the other hand,

$$x^{-5} + 2x - 3$$

is **not** a polynomial because one of its terms contains a variable with an exponent, -5, that is not a whole number.

Types of Polynomials

A **monomial** is a polynomial with exactly one term.
A **binomial** is a polynomial with exactly two terms.
A **trinomial** is a polynomial with exactly three terms.

The following are examples of monomials, binomials, and trinomials. Each of these examples is also a polynomial.

Polynomials			
Monomials	**Binomials**	**Trinomials**	**More Than Three Terms**
ax^2	$x + y$	$x^2 + 4xy + y^2$	$5x^3 - 6x^2 + 3x - 6$
$-3z$	$3p + 2$	$x^5 + 7x^2 - x$	$-y^5 + y^4 - 3y^3 - y^2 + y$
4	$4x^2 - 7$	$-q^4 + q^3 - 2q$	$x^6 + x^4 - x^3 + 1$

Each term of a polynomial has a degree. The **degree of a term in one variable** is the exponent on the variable.

Practice Problem 2

Identify the degree of each term of the trinomial $-15x^3 + 2x^2 - 5$.

EXAMPLE 2

Identify the degree of each term of the trinomial $12x^4 - 7x + 3$.

Solution: The term $12x^4$ has degree 4.

The term $-7x$ has degree 1 since $-7x$ is $-7x^1$.

The term 3 has degree 0 since 3 is $3x^0$.

Each polynomial also has a degree.

Answer

2. 3; 2; 0

Degree of a Polynomial

The **degree of a polynomial** is the greatest degree of any term of the polynomial.

EXAMPLE 3

Find the degree of each polynomial and tell whether the polynomial is a monomial, binomial, trinomial, or none of these.

a. $-2t^2 + 3t + 6$ **b.** $15x - 10$ **c.** $7x + 3x^3 + 2x^2 - 1$

Solution:

a. The degree of the trinomial $-2t^2 + 3t + 6$ is 2, the greatest degree of any of its terms.

b. The degree of the binomial $15x - 10$ or $15x^1 - 10$ is 1.

c. The degree of the polynomial $7x + 3x^3 + 2x^2 - 1$ is 3.

Practice Problem 3

Find the degree of each polynomial and tell whether the polynomial is a monomial, binomial, trinomial, or none of these.

a. $-6x + 14$

b. $9x - 3x^6 + 5x^4 + 2$

c. $10x^2 - 6x - 6$

C Evaluating Polynomials

Polynomials have different values depending on the replacement values for the variables. When we find the value of a polynomial for a given replacement value, we are evaluating the polynomial for that value.

EXAMPLE 4 Evaluate each polynomial when $x = -2$.

a. $-5x + 6$

b. $3x^2 - 2x + 1$

Solution:

a. $-5x + 6 = -5(-2) + 6$ Replace x with -2.
$= 10 + 6$
$= 16$

b. $3x^2 - 2x + 1 = 3(-2)^2 - 2(-2) + 1$ Replace x with -2.
$= 3(4) + 4 + 1$
$= 12 + 4 + 1$
$= 17$

Practice Problem 4

Evaluate each polynomial when $x = -1$.

a. $-2x + 10$

b. $6x^2 + 11x - 20$

Many physical phenomena can be modeled by polynomials.

EXAMPLE 5 Finding Free-Fall Time

The CN Tower in Toronto, Ontario, is 1821 feet tall and is the world's tallest self-supporting structure. An object is dropped from the top of this building. Neglecting air resistance, the height in feet of the object at time t seconds is given by the polynomial $-16t^2 + 1821$. Find the height of the object when $t = 1$ second and when $t = 10$ seconds. (*Source:* World Almanac)

Solution: To find each height, we evaluate the polynomial when $t = 1$ and when $t = 10$.

$-16t^2 + 1821 = -16(1)^2 + 1821$ Replace t with 1.
$= -16(1) + 1821$
$= -16 + 1821$
$= 1805$

Practice Problem 5

Find the height of the object in Example 5 when $t = 3$ seconds and when $t = 7$ seconds.

Answers

3. a. binomial, 1 **b.** none of these, 6 **c.** trinomial, 2 **4. a.** 12 **b.** -25 **5.** 1677 ft; 1037 ft

The height of the object at 1 second is 1805 feet.

$$-16t^2 + 1821 = -16(10)^2 + 1821 \quad \text{Replace } t \text{ with 10.}$$
$$= -16(100) + 1821$$
$$= -1600 + 1821$$
$$= 221$$

The height of the object at 10 seconds is 221 feet.

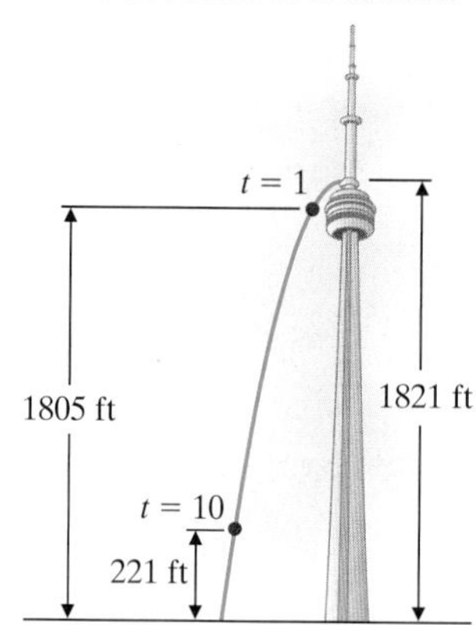

D Simplifying Polynomials by Combining Like Terms

We can simplify polynomials with like terms by combining the like terms. Recall that like terms are terms that contain exactly the same variables raised to exactly the same powers.

Like Terms	Unlike Terms
$5x^2, -7x^2$	$3x, 3y$
$y, 2y$	$-2x^2, -5x$
$\frac{1}{2}a^2b, -a^2b$	$6st^2, 4s^2t$

Only like terms can be combined. We combine like terms by applying the distributive property.

Practice Problems 6–10

Simplify each polynomial by combining any like terms.

6. $-6y + 8y$
7. $14y^2 + 3 - 10y^2 - 9$
8. $7x^3 + x^3$
9. $23x^2 - 6x - x - 15$
10. $\frac{2}{7}x^3 - \frac{1}{4}x + 2 - \frac{1}{2}x^3 + \frac{3}{8}x$

Answers

6. $2y$ **7.** $4y^2 - 6$ **8.** $8x^3$
9. $23x^2 - 7x - 15$ **10.** $-\frac{3}{14}x^3 + \frac{1}{8}x + 2$

EXAMPLES Simplify each polynomial by combining any like terms.

6. $-3x + 7x = (-3 + 7)x = 4x$

7. $11x^2 + 5 + 2x^2 - 7 = 11x^2 + 2x^2 + 5 - 7$
$$= 13x^2 - 2$$

8. $9x^3 + x^3 = 9x^3 + 1x^3$ Write x^3 as $1x^3$.
$$= 10x^3$$

9. $5x^2 + 6x - 9x - 3 = 5x^2 - 3x - 3$ Combine like terms $6x$ and $-9x$.

10. $\frac{2}{5}x^4 + \frac{2}{3}x^3 - x^2 + \frac{1}{10}x^4 - \frac{1}{6}x^3$
$$= \left(\frac{2}{5} + \frac{1}{10}\right)x^4 + \left(\frac{2}{3} - \frac{1}{6}\right)x^3 - x^2$$
$$= \left(\frac{4}{10} + \frac{1}{10}\right)x^4 + \left(\frac{4}{6} - \frac{1}{6}\right)x^3 - x^2$$
$$= \frac{5}{10}x^4 + \frac{3}{6}x^3 - x^2$$
$$= \frac{1}{2}x^4 + \frac{1}{2}x^3 - x^2$$

 **EXAMPLE 11** Write a polynomial that describes the total area of the squares and rectangles shown below. Then simplify the polynomial.

Solution:

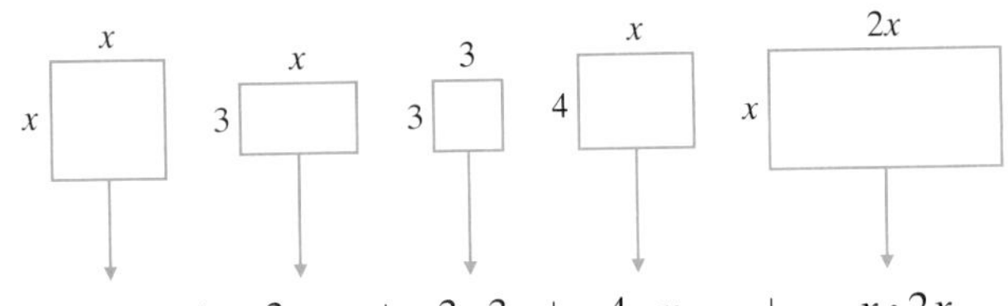

Area:

$$x \cdot x + 3 \cdot x + 3 \cdot 3 + 4 \cdot x + x \cdot 2x$$

$= x^2 + 3x + 9 + 4x + 2x^2$ Recall that the area of a rectangle is length times width.

$= 3x^2 + 7x + 9$ Combine like terms.

 Simplifying Polynomials Containing Several Variables

A polynomial may contain more than one variable. One example is

$$5x + 3xy^2 - 6x^2y^2 + x^2y - 2y + 1$$

We call this expression a polynomial in several variables.

The **degree of a term** with more than one variable is the sum of the exponents on the variables. The **degree of the polynomial** in several variables is still the greatest degree of the terms of the polynomial.

EXAMPLE 12

Identify the degrees of the terms and the degree of the polynomial $5x + 3xy^2 - 6x^2y^2 + x^2y - 2y + 1$.

Solution: To organize our work, we use a table.

Terms of Polynomial	Degree of Term	Degree of Polynomial
$5x$	1	
$3xy^2$	1 + 2 or 3	
$-6x^2y^2$	2 + 2 or 4	4 (highest degree)
x^2y	2 + 1 or 3	
$-2y$	1	
1	0	

To simplify a polynomial containing several variables, we combine any like terms.

EXAMPLES Simplify each polynomial by combining any like terms.

13. $3xy - 5y^2 + 7xy - 9x^2 = (3 + 7)xy - 5y^2 - 9x^2$
$= 10xy - 5y^2 - 9x^2$

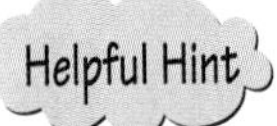

This term can be written as $10xy$ or $10yx$.

14. $9a^2b - 6a^2 + 5b^2 + a^2b - 11a^2 + 2b^2$
$= 10a^2b - 17a^2 + 7b^2$

Practice Problem 11

Write a polynomial that describes the total area of the squares and rectangles shown below. Then simplify the polynomial.

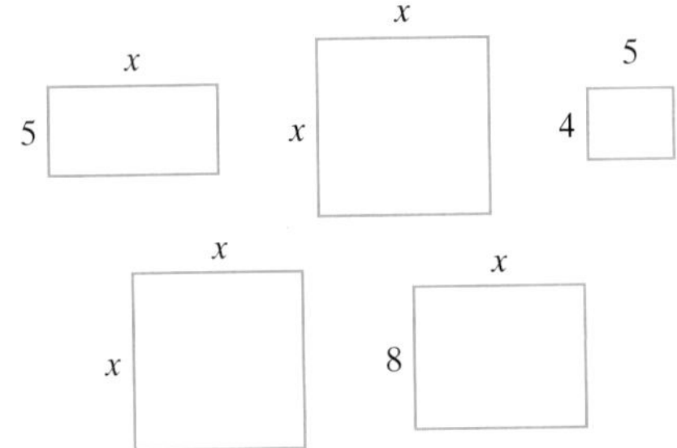

Practice Problem 12

Identify the degrees of the terms and the degree of the polynomial $-2x^3y^2 + 4 - 8xy + 3x^3y + 5xy^2$.

Practice Problems 13–14

Simplify each polynomial by combining any like terms.

13. $11ab - 6a^2 - ba + 8b^2$
14. $7x^2y^2 + 2y^2 - 4y^2x^2 + x^2 - y^2 + 5x^2$

Answers

11. $2x^2 + 13x + 20$ **12.** 5, 0, 2, 4, 3; 5
13. $10ab - 6a^2 + 8b^2$ **14.** $3x^2y^2 + y^2 + 6x^2$

FOCUS ON The Real World

SPACE EXPLORATION

From scientific observations on Earth, we know that Saturn, the second largest planet in our solar system, is a giant ball of gas surrounded by rings and orbited by 19 moons. We also know that Saturn has a diameter of 120,000 kilometers and a mass of 569,000,000,000,000,000,000,000,000 kilograms. But what is Saturn like below its outer layer of clouds? What are Saturn's rings made of? What is the surface of Saturn's largest moon, Titan, like? Could life ever be supported on Titan?

NASA is hoping to answer these questions and more about the sixth planet from the sun in our solar system with its \$3,400,000,000 Cassini mission. The Cassini spacecraft is scheduled to arrive in orbit around Saturn in June 2004. The goal of this mission is to study Saturn, its rings, and its moons. The Cassini spacecraft will also launch the Huygens probe to study Titan.

The Cassini mission began on October 15, 1997, with the launch of a mighty Titan IV booster rocket. The entire launch vehicle including rocket fuel weighed more than 2,000,000 pounds before launch. The Titan IV flung Cassini into space at a speed of 14,400 kilometers per hour. To take advantage of something called *gravity assist*, Cassini is taking a roundabout path to Saturn past Venus (twice), Earth, and Jupiter. Altogether, Cassini will travel 3,540,000,000 kilometers before reaching Saturn, a planet that is only 1,430,000,000 kilometers from the Sun.

Once Cassini reaches Saturn, it will begin collecting all kinds of data about the planet and its moons. Over the course of Cassini's mission, it will collect over 2,000,000,000,000 bits of scientific information, or about the same amount of data in 800 sets of the *Encyclopedia Britannica*. About once per day, Cassini will use its 4-meter antenna to transmit the latest data that it has collected back to Earth at a frequency of 8,400,000,000 cycles per second. For comparison, the FM band on a radio is centered around 100,000,000 cycles per second. It will take from 70 to 90 minutes for Cassini's transmissions to reach Earth and by then the signals will be very weak. The power of the signal transmitted by the spacecraft is 20 watts but, even with the huge antennas used on Earth, only 0.0000000000000001 watt can be received. (*Source:* based on data from National Aeronautics and Space Administration)

CRITICAL THINKING

1. Make a list of the numbers (other than those in dates) used in the article. Rewrite each number in scientific notation.
2. What are the advantages of scientific notation?
3. What are the disadvantages of scientific notation?
4. In your opinion, how large or small should a number be to make using scientific notation worthwhile?

Name ______________________ Section __________ Date __________

EXERCISE SET 10.3

A *Complete each table for each polynomial. See Example 1.*

1. $x^2 - 3x + 5$

Term	Coefficient
x^2	
	-3
5	

2. $2x^3 - x + 4$

Term	Coefficient
	2
$-x$	
4	

3. $-5x^4 + 3.2x^2 + x - 5$

Term	Coefficient
$-5x^4$	
$3.2x^2$	
x	
-5	

4. $9.7x^7 - 3x^5 + x^3 - \frac{1}{4}x^2$

Term	Coefficient
$9.7x^7$	
$-3x^5$	
x^3	
$-\frac{1}{4}x^2$	

B *Find the degree of each polynomial and determine whether it is a monomial, binomial, trinomial, or none of these. See Examples 2 and 3.*

5. $x + 2$

6. $-6y + y^2 + 4$

7. $9m^3 - 5m^2 + 4m - 8$

8. $5a^2 + 3a^3 - 4a^4$

9. $12x^4 - x^2 - 12x^2$

10. $7r^2 + 2r - 3r^5$

11. $3z - 5$

12. $5y + 2$

13. Describe how to find the degree of a term.

14. Describe how to find the degree of a polynomial.

15. Explain why xyz is a monomial while $x + y + z$ is a trinomial.

16. Explain why the degree of the term $5y^3$ is 3 and the degree of the polynomial $2y + y + 2y$ is 1.

 *Evaluate each polynomial when* **(a)** $x = 0$ *and* **(b)** $x = -1$. *See Examples 4 and 5.*

17. $x + 6$

18. $2x - 10$

19. $x^2 - 5x - 2$

20. $x^2 - 4$

21. $x^3 - 15$

22. $-2x^3 + 3x^2 - 6$

A rocket is fired upward from the ground with an initial velocity of 200 feet per second. Neglecting air resistance, the height of the rocket at any time t can be described in feet by the polynomial $-16t^2 + 200t$. *Find the height of the rocket at the time given in Exercises 23 through 26. See Example 5.*

23. $t = 1$ second

24. $t = 5$ seconds

25. $t = 7.6$ seconds

26. $t = 10.3$ seconds

27. The number of wireless telephone subscribers (in millions) x years after 1990 is given by the polynomial $0.97x^2 - 0.91x + 7.46$ for 1993 through 2000. Use this model to predict the number of wireless telephone subscribers in 2005 ($x = 15$). (*Source:* Based on data from Cellular Telecommunications & Internet Association)

28. The annual per capita consumption of chicken in pounds in the United States x years after 1990 is given by the polynomial $0.08x^3 - 1.19x^2 + 6.45x + 69.93$ for 1991 through 2000. Use this model to predict the per capita consumption of chicken in 2003 ($x = 13$). (*Source:* Based on data from U.S. Department of Agriculture, Economic Research Service)

D *Simplify each expression by combining like terms. See Examples 6 through 10.*

29. $14x^2 + 9x^2$

30. $18x^3 - 4x^3$

31. $15x^2 - 3x^2 - y$

32. $12k^3 - 9k^3 + 11$

33. $8s - 5s + 4s$

34. $5y + 7y - 6y$

35. $0.1y^2 - 1.2y^2 + 6.7 - 1.9$

36. $7.6y + 3.2y^2 - 8y - 2.5y^2$

Recall that the perimeter of a figure such as the ones shown in Exercises 37 and 38 is the sum of the lengths of its sides. Write each perimeter as a polynomial. Then simplify the polynomial.

37.

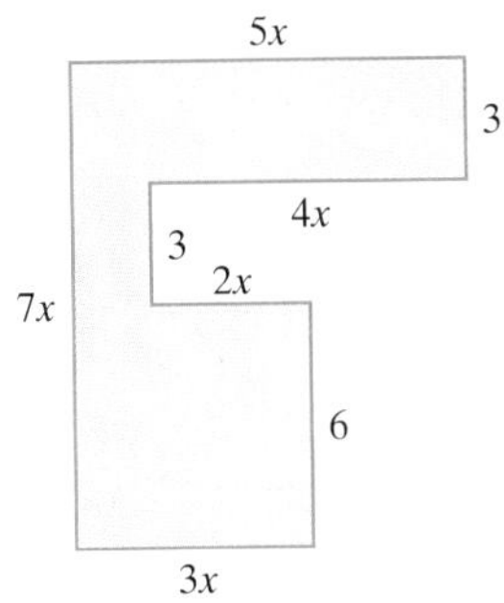

38.

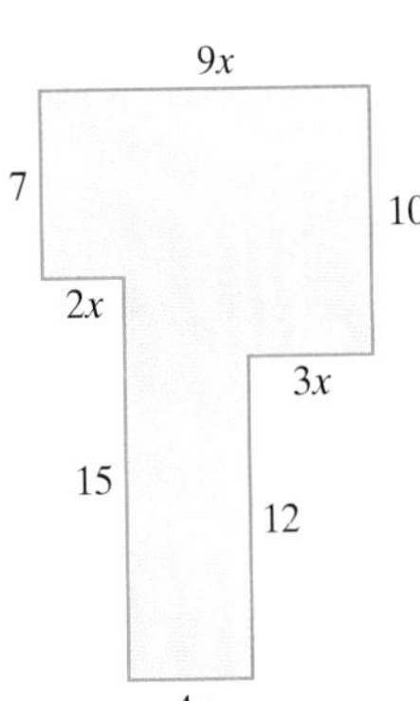

Write a polynomial that describes the total area of each set of rectangles and squares shown in Exercises 39 and 40. Then simplify the polynomial. See Example 11.

39.

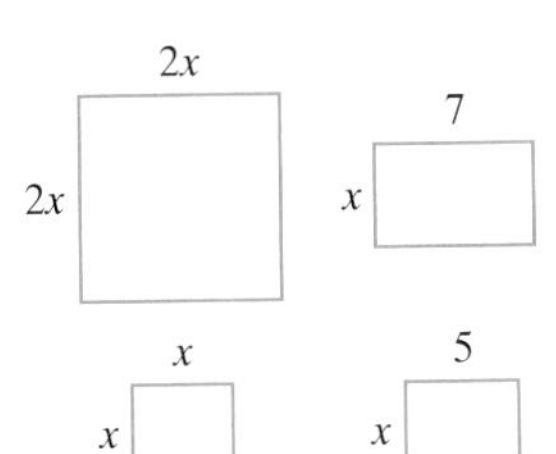

40.

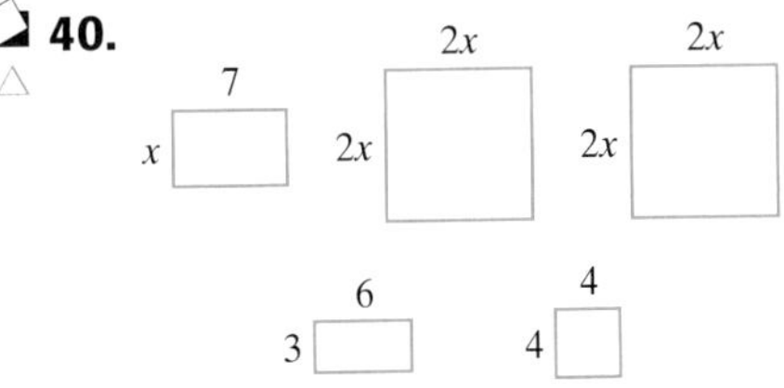

E *Identify the degrees of the terms and the degree of the polynomial. See Example 12.*

41. $9ab - 6a + 5b - 3$

42. $y^4 - 6y^3x + 2x^2y^2 - 5y^2 + 3$

43. $x^3y - 6 + 2x^2y^2 + 5y^3$

44. $2a^2b + 10a^4b - 9ab + 6$

Simplify each polynomial by combining any like terms. See Examples 13 and 14.

45. $3ab - 4a + 6ab - 7a$

46. $-9xy + 7y - xy - 6y$

47. $4x^2 - 6xy + 3y^2 - xy$

48. $3a^2 - 9ab + 4b^2 - 7ab$

49. $5x^2y + 6xy^2 - 5yx^2 + 4 - 9y^2x$

50. $17a^2b - 16ab^2 + 3a^3 + 4ba^3 - b^2a$

51. $14y^3 - 9 + 3a^2b^2 - 10 - 19b^2a^2$

52. $18x^4 + 2x^3y^3 - 1 - 2y^3x^3 - 17x^4$

Review and Preview

Simplify each expression. See Section 9.2.

53. $4 + 5(2x + 3)$

54. $9 - 6(5x + 1)$

55. $2(x - 5) + 3(5 - x)$

56. $-3(w + 7) + 5(w + 1)$

Combining Concepts

57. Explain why the height of the rocket in Exercises 23 through 26 increases and then decreases as time passes.

58. Approximate (to the nearest tenth of a second) how long before the rocket in Exercises 23 through 26 hits the ground.

Simplify each polynomial by combining like terms.

59. $1.85x^2 - 3.76x + 9.25x^2 + 10.76 - 4.21x$

60. $7.75x + 9.16x^2 - 1.27 - 14.58x^2 - 18.34$

10.4 Adding and Subtracting Polynomials

A Adding Polynomials

To add polynomials, we use commutative and associative properties and then combine like terms. To see if you are ready to add polynomials,

Try the Concept Check in the margin.

To Add Polynomials

To add polynomials, combine all like terms.

EXAMPLES Add.

1. $(4x^3 - 6x^2 + 2x + 7) + (5x^2 - 2x)$

$= 4x^3 - 6x^2 + 2x + 7 + 5x^2 - 2x$ Remove parentheses.

$= 4x^3 + (-6x^2 + 5x^2) + (2x - 2x) + 7$ Combine like terms.

$= 4x^3 - x^2 + 7$ Simplify.

2. $(-2x^2 + 5x - 1)$ and $(-2x^2 + x + 3)$ translates to

$(-2x^2 + 5x - 1) + (-2x^2 + x + 3)$

$= -2x^2 + 5x - 1 - 2x^2 + x + 3$ Remove parentheses.

$= (-2x^2 - 2x^2) + (5x + 1x) + (-1 + 3)$ Combine like terms.

$= -4x^2 + 6x + 2$ Simplify.

Polynomials can be added vertically if we line up like terms underneath one another.

EXAMPLE 3 Add $(7y^3 - 2y^2 + 7)$ and $(6y^2 + 1)$ using a vertical format.

Solution: Vertically line up like terms and add.

$$\begin{array}{r} 7y^3 - 2y^2 + 7 \\ 6y^2 + 1 \\ \hline 7y^3 + 4y^2 + 8 \end{array}$$

B Subtracting Polynomials

To subtract one polynomial from another, recall the definition of subtraction. To subtract a number, we add its opposite: $a - b = a + (-b)$. To subtract a polynomial, we also add its opposite. Just as $-b$ is the opposite of b, $-(x^2 + 5)$ is the opposite of $(x^2 + 5)$.

EXAMPLE 4 Subtract: $(5x - 3) - (2x - 11)$

Solution: From the definition of subtraction, we have

$(5x - 3) - (2x - 11) = (5x - 3) + [-(2x - 11)]$ Add the opposite.

$= (5x - 3) + (-2x + 11)$ Apply the distributive property.

$= 5x - 3 - 2x + 11$ Remove parentheses.

$= 3x + 8$ Combine like terms.

OBJECTIVES

- A Add polynomials.
- B Subtract polynomials.
- C Add or subtract polynomials in one variable.
- D Add or subtract polynomials in several variables.

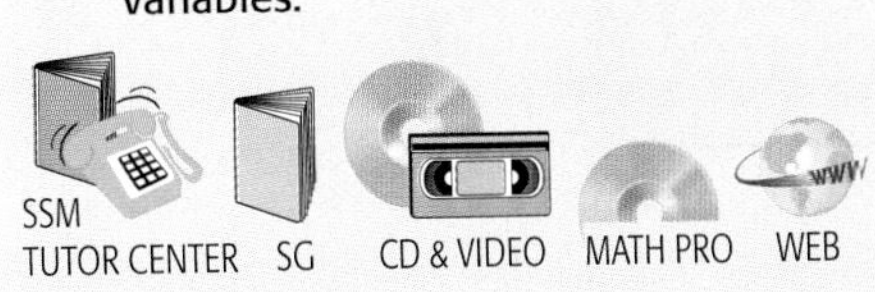

Concept Check

When combining like terms in the expression $5x - 8x^2 - 8x$, which of the following is the proper result?

a. $-11x^2$ b. $-3x - 8x^2$
c. $-11x$ d. $-11x^4$

Practice Problems 1–2

Add.

1. $(3x^5 - 7x^3 + 2x - 1) + (3x^3 - 2x)$
2. $(5x^2 - 2x + 1)$ and $(-6x^2 + x - 1)$

Practice Problem 3

Add $(9y^2 - 6y + 5)$ and $(4y + 3)$ using a vertical format.

To Subtract Polynomials

To subtract two polynomials, change the signs of the terms of the polynomial being subtracted and then add.

Practice Problem 4

Subtract: $(9x + 5) - (4x - 3)$

Answers

1. $3x^5 - 4x^3 - 1$ **2.** $-x^2 - x$
3. $9y^2 - 2y + 8$ **4.** $5x + 8$

Concept Check: **b**

Practice Problem 5

Subtract:
$(4x^3 - 10x^2 + 1) - (-4x^3 + x^2 - 11)$

EXAMPLE 5 Subtract: $(2x^3 + 8x^2 - 6x) - (2x^3 - x^2 + 1)$

Solution: First, we change the sign of each term of the second polynomial; then we add.

$$\begin{aligned}&(2x^3 + 8x^2 - 6x) - (2x^3 - x^2 + 1)\\ &= (2x^3 + 8x^2 - 6x) + (-2x^3 + x^2 - 1)\\ &= 2x^3 + 8x^2 - 6x - 2x^3 + x^2 - 1\\ &= 2x^3 - 2x^3 + 8x^2 + x^2 - 6x - 1\\ &= 9x^2 - 6x - 1 \qquad \text{Combine like terms.}\end{aligned}$$

Just as polynomials can be added vertically, so can they be subtracted vertically.

Practice Problem 6

Subtract $(6y^2 - 3y + 2)$ from $(2y^2 - 2y + 7)$ using a vertical format.

EXAMPLE 6

Subtract $(5y^2 + 2y - 6)$ from $(-3y^2 - 2y + 11)$ using a vertical format.

Solution: Arrange the polynomials in a vertical format, lining up like terms.

$$\begin{array}{r} -3y^2 - 2y + 11 \\ -(5y^2 + 2y - 6) \\ \hline \end{array} \qquad \begin{array}{r} -3y^2 - 2y + 11 \\ -5y^2 - 2y + 6 \\ \hline -8y^2 - 4y + 17 \end{array}$$

Helpful Hint

Don't forget to change the sign of each term in the polynomial being subtracted.

C Adding and Subtracting Polynomials in One Variable

Let's practice adding and subtracting polynomials in one variable.

Practice Problem 7

Subtract $(3x + 1)$ from the sum of $(4x - 3)$ and $(12x - 5)$.

EXAMPLE 7 Subtract $(5z - 7)$ from the sum of $(8z + 11)$ and $(9z - 2)$.

Solution: Notice that $(5z - 7)$ is to be subtracted **from** a sum. The translation is

$$\begin{aligned}&[(8z + 11) + (9z - 2)] - (5z - 7)\\ &= 8z + 11 + 9z - 2 - 5z + 7 \qquad \text{Remove grouping symbols.}\\ &= 8z + 9z - 5z + 11 - 2 + 7 \qquad \text{Group like terms.}\\ &= 12z + 16 \qquad \text{Combine like terms.}\end{aligned}$$

D Adding and Subtracting Polynomials in Several Variables

Now that we know how to add or subtract polynomials in one variable, we can also add and subtract polynomials in several variables.

Practice Problems 8–9

Add or subtract as indicated.

8. $(2a^2 - ab + 6b^2) - (-3a^2 + ab - 7b^2)$
9. $(5x^2y^2 + 3 - 9x^2y + y^2) - (-x^2y^2 + 7 - 8xy^2 + 2y^2)$

Answers

5. $8x^3 - 11x^2 + 12$ **6.** $-4y^2 + y + 5$
7. $13x - 9$ **8.** $5a^2 - 2ab + 13b^2$
9. $6x^2y^2 - 4 - 9x^2y + 8xy^2 - y^2$

EXAMPLES Add or subtract as indicated.

8. $(3x^2 - 6xy + 5y^2) + (-2x^2 + 8xy - y^2)$
$= 3x^2 - 6xy + 5y^2 - 2x^2 + 8xy - y^2$
$= x^2 + 2xy + 4y^2$ Combine like terms.

9. $(9a^2b^2 + 6ab - 3ab^2) - (5b^2a + 2ab - 3 - 9b^2)$ Change the sign of each term of the polynomial being subtracted.
$= 9a^2b^2 + 6ab - 3ab^2 - 5b^2a - 2ab + 3 + 9b^2$
$= 9a^2b^2 + 4ab - 8ab^2 + 9b^2 + 3$ Combine like terms.

Name ____________________ Section __________ Date __________

EXERCISE SET 10.4

 *Add. See Examples 1 through 3.*

1. $(3x + 7) + (9x + 5)$

2. $(3x^2 + 7) + (3x^2 + 9)$

3. $(-7x + 5) + (-3x^2 + 7x + 5)$

4. $(3x - 8) + (4x^2 - 3x + 3)$

5. $(-5x^2 + 3) + (2x^2 + 1)$

6. $(-y - 2) + (3y + 5)$

 **7.** $(-3y^2 - 4y) + (2y^2 + y - 1)$

8. $(7x^2 + 2x - 9) + (-3x^2 + 5)$

Add using a vertical format. See Example 3.

9. $\begin{array}{r} 3t^2 + 4 \\ \underline{5t^2 - 8} \end{array}$

10. $\begin{array}{r} 7x^3 + 3 \\ \underline{2x^3 + 1} \end{array}$

11. $\begin{array}{r} 10a^3 - 8a^2 + 9 \\ \underline{5a^3 + 9a^2 + 7} \end{array}$

12. $\begin{array}{r} 2x^3 - 3x^2 + x - 4 \\ \underline{5x^3 + 2x^2 - 3x + 2} \end{array}$

B *Subtract. See Examples 4 and 5.*

13. $(2x + 5) - (3x - 9)$

14. $(5x^2 + 4) - (-2y^2 + 4)$

15. $3x - (5x - 9)$

16. $4 - (-y - 4)$

17. $(2x^2 + 3x - 9) - (-4x + 7)$

18. $(-7x^2 + 4x + 7) - (-8x + 2)$

19. $(-7y^2 + 5) - (-8y^2 + 12)$

20. $(4 + 5a) - (-a - 5)$

21. $(5x + 8) - (-2x^2 - 6x + 8)$

22. $(-6y^2 + 3y - 4) - (9y^2 - 3y)$

Subtract using a vertical format. See Example 6.

23. $\begin{array}{r} 4z^2 - 8z + 3 \\ \underline{-\ (6z^2 + 8z - 3)} \end{array}$

24. $\begin{array}{r} 7a^2 - 9a + 6 \\ \underline{-(11a^2 - 4a + 2)} \end{array}$

25. $\begin{array}{r} 5u^5 - 4u^2 + 3u - 7 \\ \underline{-(3u^5 + 6u^2 - 8u + 2)} \end{array}$

26. $\begin{array}{r} 5x^3 - 4x^2 + 6x - 2 \\ \underline{-(3x^3 - 2x^2 - x - 4)} \end{array}$

C *Add or subtract as indicated. See Example 7.*

27. $(3x + 5) + (2x - 14)$

28. $(9x - 1) - (5x + 2)$

29. $(7y + 7) - (y - 6)$

30. $(14y + 12) + (-3y - 5)$

31. $(x^2 + 2x + 1) - (3x^2 - 6x + 2)$

32. $(5y^2 - 3y - 1) - (2y^2 + y + 1)$

33. $(3x^2 + 5x - 8) + (5x^2 + 9x + 12) - (x^2 - 14)$

34. $(-a^2 + 1) - (a^2 - 3) + (5a^2 - 6a + 7)$

Perform each indicated operation. See Examples 2, 6, and 7.

35. Subtract $4x$ from $7x - 3$.

36. Subtract y from $y^2 - 4y + 1$.

37. Add $(4x^2 - 6x + 1)$ and $(3x^2 + 2x + 1)$.

38. Add $(-3x^2 - 5x + 2)$ and $(x^2 - 6x + 9)$.

39. Subtract $(5x + 7)$ from $(7x^2 + 3x + 9)$.

40. Subtract $(5y^2 + 8y + 2)$ from $(7y^2 + 9y - 8)$.

41. Subtract $(4y^2 - 6y - 3)$ from the sum of $(8y^2 + 7)$ and $(6y + 9)$.

42. Subtract $(4x^2 - 2x + 2)$ from the sum of $(x^2 + 7x + 1)$ and $(7x + 5)$.

43. Subtract $(3x^2 - 4)$ from the sum of $(x^2 - 9x + 2)$ and $(2x^2 - 6x + 1)$.

44. Subtract $(y^2 - 9)$ from the sum of $(3y^2 + y + 4)$ and $(2y^2 - 6y - 10)$.

D *Add or subtract as indicated. See Examples 8 and 9.*

45. $(9a + 6b - 5) + (-11a - 7b + 6)$

46. $(3x - 2 + 6y) + (7x - 2 - y)$

47. $(4x^2 + y^2 + 3) - (x^2 + y^2 - 2)$

48. $(7a^2 - 3b^2 + 10) - (-2a^2 + b^2 - 12)$

49. $(x^2 + 2xy - y^2) + (5x^2 - 4xy + 20y^2)$

50. $(a^2 - ab + 4b^2) + (6a^2 + 8ab - b^2)$

51. $(11r^2s + 16rs - 3 - 2r^2s^2) - (3sr^2 + 5 - 9r^2s^2)$

52. $(3x^2y - 6xy + x^2y^2 - 5) - (11x^2y^2 - 1 + 5yx^2)$

Review and Preview

Multiply. See Section 10.1.

53. $3x(2x)$

54. $-7x(x)$

55. $(12x^3)(-x^5)$

56. $6r^3(7r^{10})$

57. $10x^2(20xy^2)$

58. $-z^2y(11zy)$

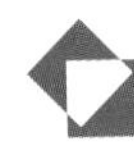

Combining Concepts

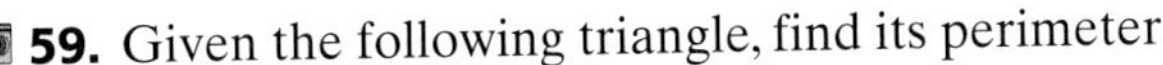

59. Given the following triangle, find its perimeter.

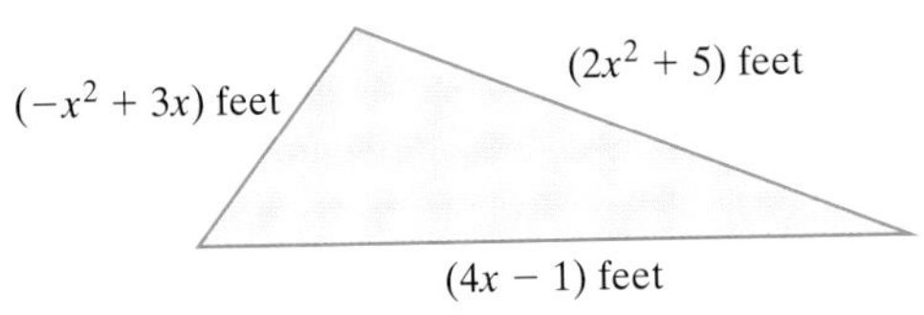

60. Given the following quadrilateral, find its perimeter.

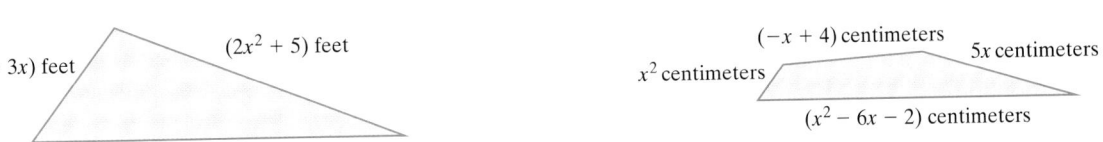

$(-x + 4)$ centimeters
$5x$ centimeters
x^2 centimeters
$(x^2 - 6x - 2)$ centimeters

61. A wooden beam is $(4y^2 + 4y + 1)$ meters long. If a piece $(y^2 - 10)$ meters is cut, express the length of the remaining piece of beam as a polynomial in y.

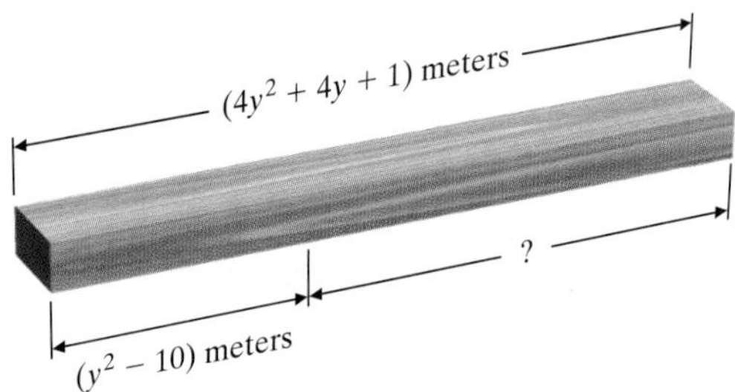

62. A piece of quarter-round molding is $(13x - 7)$ inches long. If a piece $(2x + 2)$ inches is removed, express the length of the remaining piece of molding as a polynomial in x.

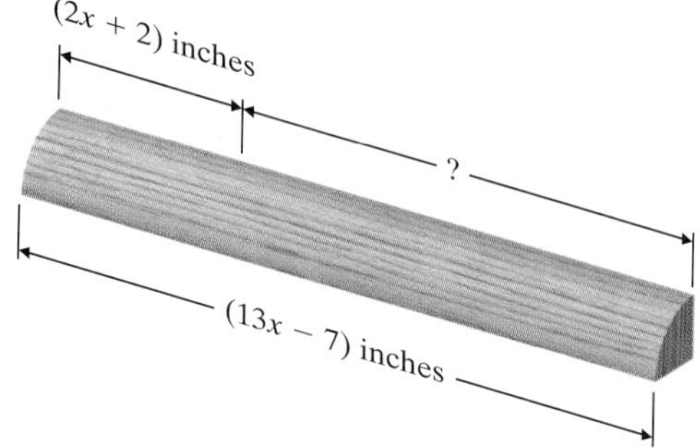

Perform each indicated operation.

63. $[(1.2x^2 - 3x + 9.1) - (7.8x^2 - 3.1 + 8)] + (1.2x - 6)$

64. $[(7.9y^4 - 6.8y^3 + 3.3y) + (6.1y^3 - 5)] - (4.2y^4 + 1.1y - 1)$

65. The polynomial $-0.26x^2 + 4.58x + 14$ represents the number of Americans (in millions) enrolled in individual-practice-association HMOs during 1993–2001. The polynomial $-0.30x^2 + 5.74x + 21.72$ represents the number of Americans (in millions) enrolled in all other types of HMOs during 1993–2001. In both polynomials, x represents the number of years after 1993. Find a polynomial for the total enrollment (in millions) in HMOs of all kinds during this period. (*Source:* Based on data from the Public Health Service)

66. The polynomial $0.015x^2 + 0.002x + 1.128$ represents the sales of electricity (in trillion kilowatt hours) in the U.S. residential sector during 1998–2000. The polynomial $-0.0075x^2 + 0.0615x + 2.113$ represents the sales of electricity (in trillion kilowatt hours) in all other U.S. sectors during 1998–2000. In both polynomials, x represents the number of years after 1998. Find a polynomial for the total sales of electricity (in trillion kilowatt hours) to all sectors in the United States during this period. (Source: Based on data from the Energy Information Administration)

67. Simplify each expression by performing the indicated operation. Explain how you arrived at each answer.

a. $x + x$

b. $x \cdot x$

c. $-x - x$

d. $(-x)(-x)$

10.5 Multiplying Polynomials

OBJECTIVES

- (A) Multiply monomials.
- (B) Multiply a monomial by a polynomial.
- (C) Multiply two polynomials.
- (D) Multiply polynomials vertically.

(A) Multiplying Monomials

Recall from Section 10.1 that to multiply two monomials such as $(-5x^3)$ and $(-2x^4)$, we use the associative and commutative properties and regroup. Remember also that to multiply exponential expressions with a common base, we add exponents.

$$(-5x^3)(-2x^4) = (-5)(-2)(x^3 \cdot x^4)$$ Use the commutative and associative properties.

$$= 10x^7$$ Multiply.

EXAMPLES Multiply.

1. $6x \cdot 4x = (6 \cdot 4)(x \cdot x)$ Use the commutative and associative properties.
$= 24x^2$ Multiply.

2. $-7x^2 \cdot 2x^5 = (-7 \cdot 2)(x^2 \cdot x^5)$
$= -14x^7$

3. $(-12x^5)(-x) = (-12x^5)(-1x)$
$= (-12)(-1)(x^5 \cdot x)$
$= 12x^6$

Practice Problems 1–3

Multiply.

1. $10x \cdot 9x$
2. $8x^3(-11x^7)$
3. $(-5x^4)(-x)$

(B) Multiplying Monomials by Polynomials

To multiply a monomial such as $7x$ by a trinomial such as $x^2 + 2x + 5$, we use the distributive property.

EXAMPLES Multiply.

4. $7x(x^2 + 2x + 5) = 7x(x^2) + 7x(2x) + 7x(5)$ Apply the distributive property.
$= 7x^3 + 14x^2 + 35x$ Multiply.

5. $5x(2x^3 + 6) = 5x(2x^3) + 5x(6)$ Apply the distributive property.
$= 10x^4 + 30x$ Multiply.

6. $-3x^2(5x^2 + 6x - 1)$
$= (-3x^2)(5x^2) + (-3x^2)(6x) + (-3x^2)(-1)$ Apply the distributive property.
$= -15x^4 - 18x^3 + 3x^2$ Multiply.

Practice Problems 4–6

Multiply.

4. $4x(x^2 + 4x + 3)$
5. $8x(7x^4 + 1)$
6. $-2x^3(3x^2 - x + 2)$

(C) Multiplying Two Polynomials

We also use the distributive property to multiply two binomials.

EXAMPLE 7 Multiply: $(3x + 2)(2x - 5)$

Solution:

$(3x + 2)(2x - 5) = 3x(2x - 5) + 2(2x - 5).$ Use the distributive property.

$= 3x(2x) + 3x(-5) + 2(2x) + 2(-5)$
$= 6x^2 - 15x + 4x - 10$ Multiply.
$= 6x^2 - 11x - 10$ Combine like terms.

This idea can be expanded so that we can multiply any two polynomials.

Practice Problem 7

Multiply: $(4x + 5)(3x - 4)$

Answers

1. $90x^2$ **2.** $-88x^{10}$ **3.** $5x^5$ **4.** $4x^3 + 16x^2 + 12x$ **5.** $56x^5 + 8x$ **6.** $-6x^5 + 2x^4 - 4x^3$ **7.** $12x^2 - x - 20$

To Multiply Two Polynomials

Multiply each term of the first polynomial by each term of the second polynomial, and then combine like terms.

Practice Problems 8–9

Multiply.

8. $(3x - 2y)^2$
9. $(x + 3)(2x^2 - 5x + 4)$

EXAMPLES Multiply.

8. $(2x - y)^2$

$= (2x - y)(2x - y)$

$= 2x(2x) + 2x(-y) + (-y)(2x) + (-y)(-y)$

$= 4x^2 - 2xy - 2xy + y^2$ Multiply.

$= 4x^2 - 4xy + y^2$ Combine like terms.

9. $(t + 2)(3t^2 - 4t + 2)$

$= t(3t^2) + t(-4t) + t(2) + 2(3t^2) + 2(-4t) + 2(2)$

$= 3t^3 - 4t^2 + 2t + 6t^2 - 8t + 4$

$= 3t^3 + 2t^2 - 6t + 4$ Combine like terms.

D Multiplying Polynomials Vertically

Another convenient method for multiplying polynomials is to multiply vertically, similar to the way we multiply real numbers. This method is shown in the next examples.

Practice Problem 10

Multiply vertically:

$(3y^2 + 1)(y^2 - 4y + 5)$

EXAMPLE 10 Multiply vertically: $(2y^2 + 5)(y^2 - 3y + 4)$

Solution:

$$\begin{array}{rrrrrl} & & y^2 & -3y & +4 & \\ & & 2y^2 & & +5 & \\ \hline & & 5y^2 & -15y & +20 & \text{Multiply } y^2 - 3y + 4 \text{ by } 5 \\ 2y^4 & -6y^3 & +8y^2 & & & \text{Multiply } y^2 - 3y + 4 \text{ by } 2y^2 \\ \hline 2y^4 & -6y^3 & +13y^2 & -15y & +20 & \text{Combine like terms.} \end{array}$$

Practice Problem 11

Find the product of $(4x^2 - x - 1)$ and $(3x^2 + 6x - 2)$ using a vertical format.

EXAMPLE 11

Find the product of $(2x^2 - 3x + 4)$ and $(x^2 + 5x - 2)$ using a vertical format.

Solution: First, we arrange the polynomials in a vertical format. Then we multiply each term of the second polynomial by each term of the first polynomial.

$$\begin{array}{rrrrrl} & & 2x^2 & -3x & +4 & \\ & & x^2 & +5x & -2 & \\ \hline & & -4x^2 & +6x & -8 & \text{Multiply } 2x^2 - 3x + 4 \text{ by } -2. \\ & 10x^3 & -15x^2 & +20x & & \text{Multiply } 2x^2 - 3x + 4 \text{ by } 5x. \\ 2x^4 & -3x^3 & +4x^2 & & & \text{Multiply } 2x^2 - 3x + 4 \text{ by } x^2. \\ \hline 2x^4 & +7x^3 & -15x^2 & +26x & -8 & \text{Combine like terms.} \end{array}$$

Answers

8. $9x^2 - 12xy + 4y^2$
9. $2x^3 + x^2 - 11x + 12$
10. $3y^4 - 12y^3 + 16y^2 - 4y + 5$
11. $12x^4 + 21x^3 - 17x^2 - 4x + 2$

Name ______________________ Section __________ Date __________

Mental Math

Find each product.

1. $x^3 \cdot x^5$
2. $x^2 \cdot x^6$
3. $y^4 \cdot y$
4. $y^9 \cdot y$
5. $x^7 \cdot x^7$
6. $x^{11} \cdot x^{11}$

EXERCISE SET 10.5

 *Multiply. See Examples 1 through 3.*

1. $8x^2 \cdot 3x$
2. $6x \cdot 3x^2$
3. $(-3.1x^3)(4x^9)$
4. $(-5.2x^4)(3x^4)$
5. $(-x^3)(-x)$
6. $(-x^6)(-x)$
7. $\left(-\frac{1}{3}y^2\right)\left(\frac{2}{5}y\right)$
8. $\left(-\frac{3}{4}y^7\right)\left(\frac{1}{7}y^4\right)$
9. $(2x)(-3x^2)(4x^5)$
10. $(x)(5x^4)(-6x^7)$

B *Multiply. See Examples 4 through 6.*

11. $3x(2x + 5)$
12. $2x(6x + 3)$
13. $7x(x^2 + 2x - 1)$
14. $5y(y^2 + y - 10)$
15. $-2a(a + 4)$
16. $-3a(2a + 7)$
17. $3x(2x^2 - 3x + 4)$
18. $4x(5x^2 - 6x - 10)$
19. $3a(a^2 + 2)$
20. $x^3(x + 12)$
21. $-2a^2(3a^2 - 2a + 3)$
22. $-4b^2(3b^3 - 12b^2 - 6)$
23. $3x^2y(2x^3 - x^2y^2 + 8y^3)$
24. $4xy^2(7x^3 + 3x^2y^2 - 9y^3)$

25. The area of the largest rectangle below is $x(x + 3)$. Find another expression for this area by finding the sum of the areas of the smaller rectangles.

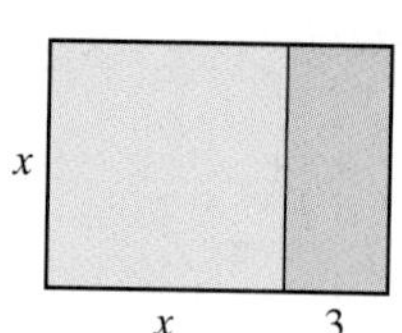

26. Write an expression for the area of the largest rectangle below in two different ways.

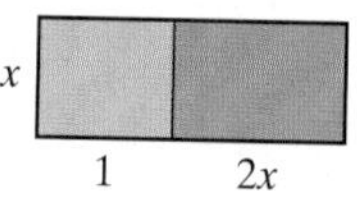

C *Multiply. See Examples 7 through 9.*

27. $(x + 4)(x + 3)$

28. $(x + 2)(x + 9)$

29. $(a + 7)(a - 2)$

30. $(y - 10)(y + 11)$

31. $\left(x + \frac{2}{3}\right)\left(x - \frac{1}{3}\right)$

32. $\left(x + \frac{3}{5}\right)\left(x - \frac{2}{5}\right)$

33. $(3x^2 + 1)(4x^2 + 7)$

34. $(5x^2 + 2)(6x^2 + 2)$

35. $(4x - 3)(3x - 5)$

36. $(8x - 3)(2x - 4)$

37. $(1 - 3a)(1 - 4a)$

38. $(3 - 2a)(2 - a)$

39. $(2y - 4)^2$

40. $(6x - 7)^2$

41. $(x - 2)(x^2 - 3x + 7)$

42. $(x + 3)(x^2 + 5x - 8)$

43. $(x + 5)(x^3 - 3x + 4)$

44. $(a + 2)(a^3 - 3a^2 + 7)$

45. $(2a - 3)(5a^2 - 6a + 4)$

46. $(3 + b)(2 - 5b - 3b^2)$

47. $(7xy - y)^2$

48. $(x^2 - 4)^2$

49. The area of the figure below is $(x + 2)(x + 3)$. Find another expression for this area by finding the sum of the areas of the smaller rectangles.

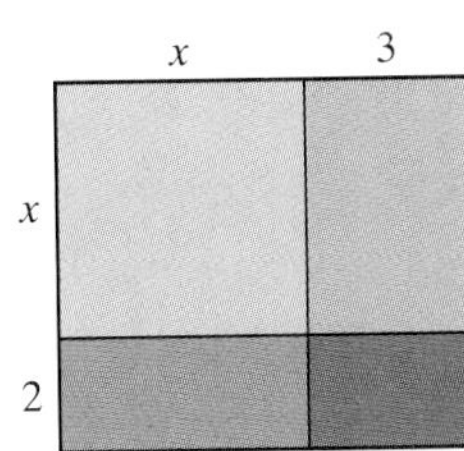

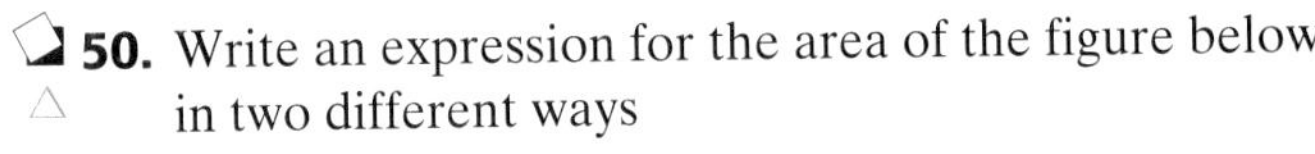

50. Write an expression for the area of the figure below in two different ways

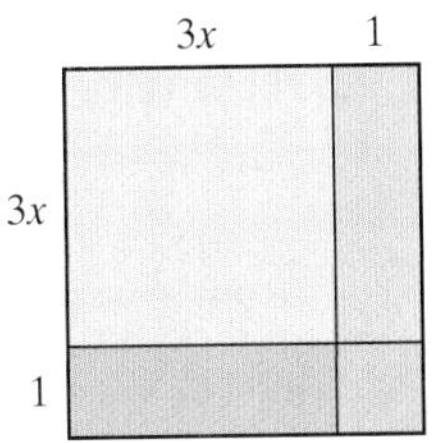

D *Multiply vertically. See Examples 10 and 11.*

51. $(2x - 11)(6x + 1)$

52. $(4x - 7)(5x + 1)$

53. $(x + 3)(2x^2 + 4x - 1)$

54. $(4x - 5)(8x^2 + 2x - 4)$

55. $(x^2 + 5x - 7)(x^2 - 7x - 9)$

56. $(3x^2 - x + 2)(x^2 + 2x + 1)$

Review and Preview

Perform each indicated operation. See Section 10.1.

57. $(5x)^2$

58. $(4p)^2$

59. $(-3y^3)^2$

60. $(-7m^2)^2$

For income tax purposes, Rob Calcutta, the owner of Copy Services, uses a method called ***straight-line depreciation*** *to show the depreciated (or decreased) value of a copy machine he recently purchased. Rob assumes that he can use the machine for 7 years. The graph below shows the depreciated (or decreased) value of the machine over the years. Use this graph to answer Exercises 61 through 66. See Section 7.1.*

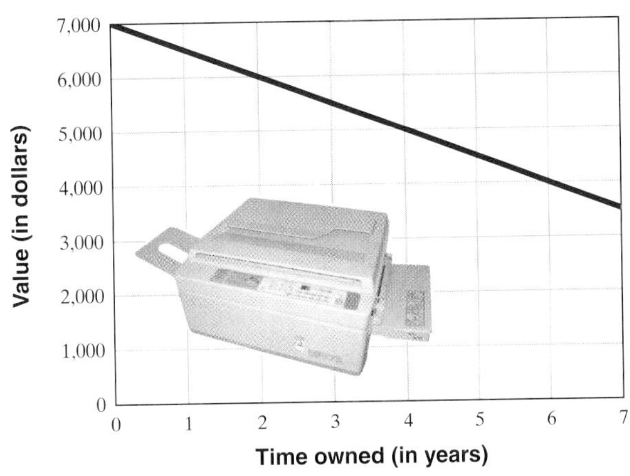

61. What was the purchase price of the copy machine? (*Hint*: This is when time owned is 0 years.)

62. What is the depreciated value of the machine in 7 years?

63. What loss in value occurred during the first year?

64. What loss in value occurred during the second year?

65. Why do you think this method of depreciating is called straight-line depreciation?

66. Why is the line tilted downward?

Combining Concepts

Express as the product of polynomials. Then multiply.

△ **67.** Find the area of the rectangle.

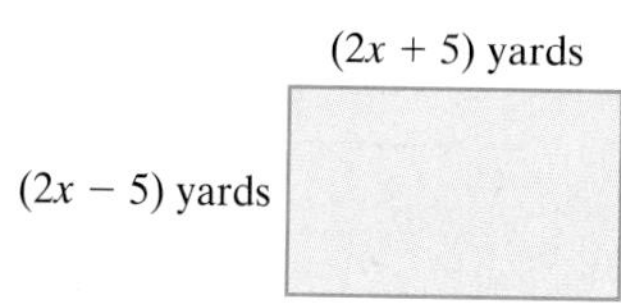

△ **68.** Find the area of the square field.

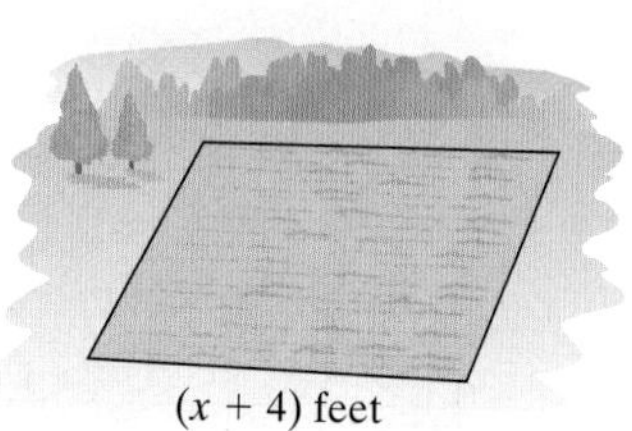

△ **69.** Find the area of the triangle.

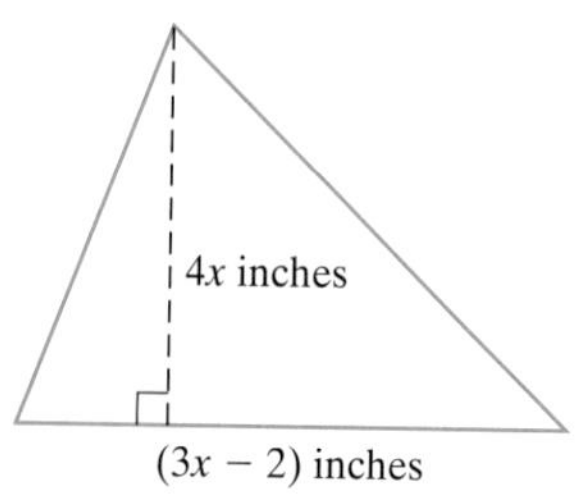

△ **70.** Find the volume of the cube-shaped glass block.

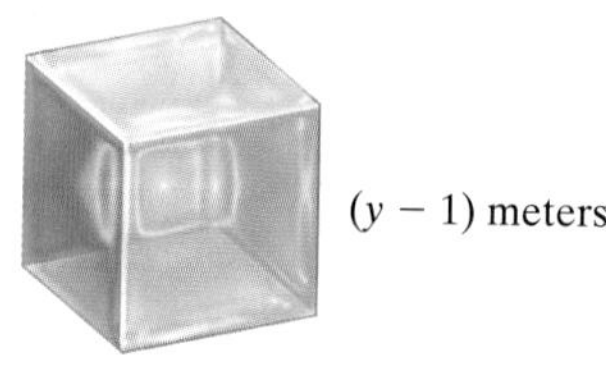

71. Perform each indicated operation. Explain the difference between the two expressions.

a. $(3x + 5) + (3x + 7)$

b. $(3x + 5)(3x + 7)$

72. Evaluate each of the following.

a. $(2 + 3)^2; 2^2 + 3^2$

b. $(8 + 10)^2; 8^2 + 10^2$

Does $(a + b)^2 = a^2 + b^2$ no matter what the values of a and b are? Why or why not?

73. Multiply each of the following polynomials.

a. $(a + b)(a - b)$

b. $(2x + 3y)(2x - 3y)$

c. $(4x + 7)(4x - 7)$

d. Can you make a general statement about all products of the form $(x + y)(x - y)$?

10.6 Special Products

OBJECTIVES

- A Multiply two binomials using the FOIL method.
- B Square a binomial.
- C Multiply the sum and difference of two terms.
- D Use special products to multiply binomials.

SSM TUTOR CENTER SG CD & VIDEO MATH PRO WEB

A Using the FOIL Method

In this section, we multiply binomials using special products. First, we introduce a special order for multiplying binomials called the FOIL order or method. We demonstrate by multiplying $(3x + 1)$ by $(2x + 5)$.

The FOIL Method

F stands for the product of the **First** terms. $(3x + 1)(2x + 5)$ $(3x)(2x) = 6x^2$ F

O stands for the product of the **Outer** terms. $(3x + 1)(2x + 5)$ $(3x)(5) = 15x$ O

I stands for the product of the **Inner** terms. $(3x + 1)(2x + 5)$ $(1)(2x) = 2x$ I

L stands for the product of the **Last** terms. $(3x + 1)(2x + 5)$ $(1)(5) = 5$ L

$$\begin{aligned}(3x + 1)(2x + 5) &= \overset{F}{6x^2} + \overset{O}{15x} + \overset{I}{2x} + \overset{L}{5} \\ &= 6x^2 + 17x + 5 \quad \text{Combine like terms.}\end{aligned}$$

Let's practice multiplying binomials using the FOIL method.

EXAMPLE 1 Multiply: $(x - 3)(x + 4)$

Solution:

$$\begin{aligned}(x - 3)(x + 4) &= \overset{F}{(x)(x)} + \overset{O}{(x)(4)} + \overset{I}{(-3)(x)} + \overset{L}{(-3)(4)} \\ &= x^2 + 4x - 3x - 12 \\ &= x^2 + x - 12 \quad \text{Combine like terms.}\end{aligned}$$

Practice Problem 1

Multiply: $(x + 7)(x - 5)$

EXAMPLE 2 Multiply: $(5x - 7)(x - 2)$

Solution:

$$\begin{aligned}(5x - 7)(x - 2) &= \overset{F}{5x(x)} + \overset{O}{5x(-2)} + \overset{I}{(-7)(x)} + \overset{L}{(-7)(-2)} \\ &= 5x^2 - 10x - 7x + 14 \\ &= 5x^2 - 17x + 14 \quad \text{Combine like terms.}\end{aligned}$$

Practice Problem 2

Multiply: $(6x - 1)(x - 4)$

EXAMPLE 3 Multiply: $(y^2 + 6)(2y - 1)$

Solution: $(y^2 + 6)(2y - 1) = \overset{F}{2y^3} - \overset{O}{1y^2} + \overset{I}{12y} - \overset{L}{6}$

Notice in this example that there are no like terms that can be combined, so the product is $2y^3 - y^2 + 12y - 6$.

Practice Problem 3

Multiply: $(2y^2 + 3)(y - 4)$

Answers

1. $x^2 + 2x - 35$ **2.** $6x^2 - 25x + 4$
3. $2y^3 - 8y^2 + 3y - 12$

B Squaring Binomials

An expression such as $(3y + 1)^2$ is called the square of a binomial. Since $(3y + 1)^2 = (3y + 1)(3y + 1)$, we can use the FOIL method to find this product.

Practice Problem 4

Multiply: $(2x + 9)^2$

EXAMPLE 4 Multiply: $(3y + 1)^2$

Solution:

$$
\begin{aligned}
(3y + 1)^2 &= (3y + 1)(3y + 1) \\
&= \overset{F}{(3y)(3y)} + \overset{O}{(3y)(1)} + \overset{I}{1(3y)} + \overset{L}{1(1)} \\
&= 9y^2 + 3y + 3y + 1 \\
&= 9y^2 + 6y + 1
\end{aligned}
$$

Notice the pattern that appears in Example 4.

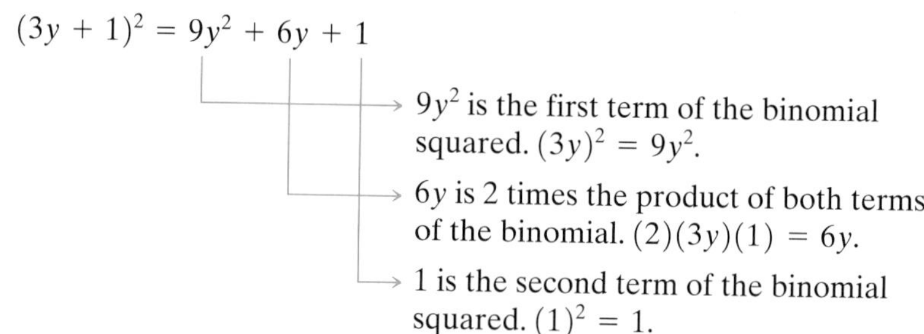

$(3y + 1)^2 = 9y^2 + 6y + 1$
$9y^2$ is the first term of the binomial squared. $(3y)^2 = 9y^2$.
$6y$ is 2 times the product of both terms of the binomial. $(2)(3y)(1) = 6y$.
1 is the second term of the binomial squared. $(1)^2 = 1$.

This pattern leads to the formulas below, which can be used when squaring a binomial. We call these **special products.**

Squaring a Binomial

A binomial squared is equal to the square of the first term plus or minus twice the product of both terms plus the square of the second term.

$$(a + b)^2 = a^2 + 2ab + b^2$$

$$(a - b)^2 = a^2 - 2ab + b^2$$

This product can be visualized geometrically.

a a^2 ab
$a + b$
b ab b^2
a b
$a + b$

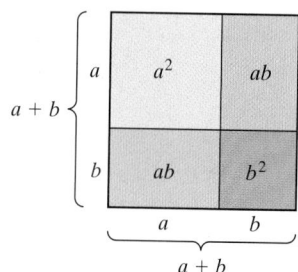

The area of the large square is side · side.
Area $= (a + b)(a + b) = (a + b)^2$
The area of the large square is also the sum of the areas of the smaller rectangles.
Area $= a^2 + ab + ab + b^2 = a^2 + 2ab + b^2$
Thus, $(a + b)^2 = a^2 + 2ab + b^2$.

Answer

4. $4x^2 + 36x + 81$

EXAMPLES Use a special product to square each binomial.

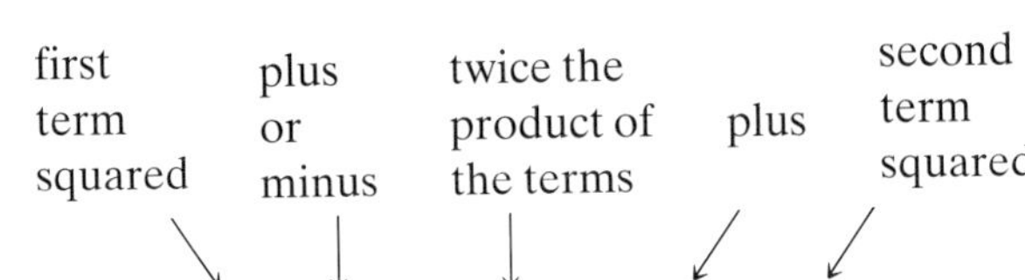

5. $(t + 2)^2 = t^2 + 2(t)(2) + 2^2 = t^2 + 4t + 4$
6. $(p - q)^2 = p^2 - 2(p)(q) + q^2 = p^2 - 2pq + q^2$
7. $(2x + 5)^2 = (2x)^2 + 2(2x)(5) + 5^2 = 4x^2 + 20x + 25$
8. $(x^2 - 7y)^2 = (x^2)^2 - 2(x^2)(7y) + (7y)^2 = x^4 - 14x^2y + 49y^2$

Helpful Hint

Notice that

$(a + b)^2 \neq a^2 + b^2$ The middle term $2ab$ is missing.

$(a + b)^2 = (a + b)(a + b) = a^2 + 2ab + b^2$

Likewise,

$(a - b)^2 \neq a^2 - b^2$

$(a - b)^2 = (a - b)(a - b) = a^2 - 2ab + b^2$

C Multiplying the Sum and Difference of Two Terms

Another special product is the product of the sum and difference of the same two terms, such as $(x + y)(x - y)$. Finding this product by the FOIL method, we see a pattern emerge.

$$(x + y)(x - y) = \overset{F}{x^2} - \overset{O}{xy} + \overset{I}{xy} - \overset{L}{y^2}$$
$$= x^2 - y^2$$

Notice that the two middle terms subtract out. This is because the **O**uter product is the opposite of the **I**nner product. Only the **difference of squares** remains.

Multiplying the Sum and Difference of Two Terms

The product of the sum and difference of two terms is the square of the first term minus the square of the second term.

$$(a + b)(a - b) = a^2 - b^2$$

EXAMPLES Use a special product to multiply.

first term squared — minus — second term squared

9. $(x + 4)(x - 4) = x^2 - 4^2 = x^2 - 16$
10. $(6t + 7)(6t - 7) = (6t)^2 - 7^2 = 36t^2 - 49$
11. $\left(x - \frac{1}{4}\right)\left(x + \frac{1}{4}\right) = x^2 - \left(\frac{1}{4}\right)^2 = x^2 - \frac{1}{16}$
12. $(2p - q)(2p + q) = (2p)^2 - q^2 = 4p^2 - q^2$
13. $(3x^2 - 5y)(3x^2 + 5y) = (3x^2)^2 - (5y)^2 = 9x^4 - 25y^2$

Practice Problems 5–8

Use a special product to square each binomial.

5. $(y + 3)^2$
6. $(r - s)^2$
7. $(6x + 5)^2$
8. $(x^2 - 3y)^2$

Practice Problems 9–13

Use a special product to multiply.

9. $(x + 7)(x - 7)$
10. $(4y + 5)(4y - 5)$
11. $\left(x - \frac{1}{3}\right)\left(x + \frac{1}{3}\right)$
12. $(3a - b)(3a + b)$
13. $(2x^2 - 6y)(2x^2 + 6y)$

Answers

5. $y^2 + 6y + 9$ **6.** $r^2 - 2rs + s^2$ **7.** $36x^2 + 60x + 25$ **8.** $x^4 - 6x^2y + 9y^2$ **9.** $x^2 - 49$ **10.** $16y^2 - 25$ **11.** $x^2 - \frac{1}{9}$ **12.** $9a^2 - b^2$ **13.** $4x^4 - 36y^2$

Concept Check

Match each expression on the left to the equivalent expression or expressions in the list on the right.

$(a + b)^2$ a. $(a + b)(a + b)$
$(a + b)(a - b)$ b. $a^2 - b^2$
c. $a^2 + b^2$
d. $a^2 - 2ab + b^2$
e. $a^2 + 2ab + b^2$

Practice Problems 14–16

Use a special product to multiply.

14. $(7x - 1)^2$
15. $(5y + 3)(2y - 5)$
16. $(2a - 1)(2a + 1)$

Answers

14. $49x^2 - 14x + 1$ **15.** $10y^2 - 19y - 15$
16. $4a^2 - 1$

Concept Check: a or e, b

Try the Concept Check in the margin.

D Using Special Products

Let's now practice using our special products on a variety of multiplication problems. This practice will help us recognize when to apply what special product formula.

EXAMPLES Use a special product to multiply.

14. $(x - 9)(x + 9)$ This is the sum and difference of the same two terms.

$= x^2 - 9^2 = x^2 - 81$

15. $(3y + 2)^2$ This is a binomial squared.

$= (3y)^2 + 2(3y)(2) + 2^2$
$= 9y^2 + 12y + 4$

16. $(6a + 1)(a - 7)$ No special product applies. Use the FOIL method.

F O I L
$= 6a \cdot a + 6a(-7) + 1 \cdot a + 1(-7)$
$= 6a^2 - 42a + a - 7$
$= 6a^2 - 41a - 7$

Name ______________________ Section __________ Date __________

EXERCISE SET 10.6

A *Multiply using the FOIL method. See Examples 1 through 3.*

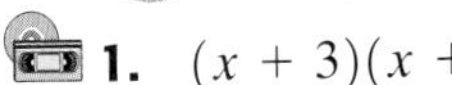

1. $(x + 3)(x + 4)$ **2.** $(x + 5)(x + 1)$ **3.** $(x - 5)(x + 10)$ **4.** $(y - 12)(y + 4)$

5. $(5x - 6)(x + 2)$ **6.** $(3y - 5)(2y - 7)$ **7.** $(y - 6)(4y - 1)$ **8.** $(2x - 9)(x - 11)$

9. $(2x + 5)(3x - 1)$ **10.** $(6x + 2)(x - 2)$ **11.** $(y^2 + 7)(6y + 4)$ **12.** $(y^2 + 3)(5y + 6)$

13. $\left(x - \frac{1}{3}\right)\left(x + \frac{2}{3}\right)$ **14.** $\left(x - \frac{2}{5}\right)\left(x + \frac{1}{5}\right)$ **15.** $(4 - 3a)(2 - 5a)$ **16.** $(3 - 2a)(6 - 5a)$

17. $(x + 5y)(2x - y)$ **18.** $(x + 4y)(3x - y)$

B *Multiply. See Examples 4 through 8.*

19. $(x + 2)^2$ **20.** $(x + 7)^2$ **21.** $(2x - 1)^2$ **22.** $(7x - 3)^2$ **23.** $(3a - 5)^2$

24. $(5a + 2)^2$ **25.** $(x^2 + 5)^2$ **26.** $(x^2 + 3)^2$ **27.** $\left(y - \frac{2}{7}\right)^2$ **28.** $\left(y - \frac{3}{4}\right)^2$

29. $(2a - 3)^2$ **30.** $(5b - 4)^2$ **31.** $(5x + 9)^2$ **32.** $(6s + 2)^2$ **33.** $(3x - 7y)^2$

34. $(4s - 2y)^2$ **35.** $(4m + 5n)^2$ **36.** $(3n + 5m)^2$

37. Using your own words, explain how to square a binomial such as $(a + b)^2$.

38. Explain how to find the product of two binomials using the FOIL method.

C *Multiply. See Examples 9 through 13.*

39. $(a - 7)(a + 7)$ **40.** $(b + 3)(b - 3)$ **41.** $(x + 6)(x - 6)$ **42.** $(x - 8)(x + 8)$

43. $(3x - 1)(3x + 1)$ **44.** $(4x - 5)(4x + 5)$ **45.** $(x^2 + 5)(x^2 - 5)$ **46.** $(a^2 + 6)(a^2 - 6)$

47. $(2y^2 - 1)(2y^2 + 1)$ **48.** $(3x^2 + 1)(3x^2 - 1)$ **49.** $(4 - 7x)(4 + 7x)$ **50.** $(8 - 7x)(8 + 7x)$

51. $\left(3x - \frac{1}{2}\right)\left(3x + \frac{1}{2}\right)$ **52.** $\left(10x + \frac{2}{7}\right)\left(10x - \frac{2}{7}\right)$ **53.** $(9x + y)(9x - y)$ **54.** $(2x - y)(2x + y)$

55. $(2m + 5n)(2m - 5n)$ **56.** $(5m + 4n)(5m - 4n)$

D *Multiply. See Examples 14 through 16.*

57. $(a + 5)(a + 4)$

58. $(a + 5)(a + 7)$

59. $(a - 7)^2$

60. $(b - 2)^2$

61. $(4a + 1)(3a - 1)$

62. $(6a + 7)(6a + 5)$

63. $(x + 2)(x - 2)$

64. $(x - 10)(x + 10)$

65. $(3a + 1)^2$

66. $(4a + 2)^2$

67. $(x + y)(4x - y)$

68. $(3x + 2)(4x - 2)$

69. $\left(a - \frac{1}{2}y\right)\left(a + \frac{1}{2}y\right)$

70. $\left(\frac{a}{2} + 4y\right)\left(\frac{a}{2} - 4y\right)$

71. $(3b + 7)(2b - 5)$

72. $(3y - 13)(y - 3)$

73. $(x^2 + 10)(x^2 - 10)$

74. $(x^2 + 8)(x^2 - 8)$

75. $(4x + 5)(4x - 5)$

76. $(3x + 5)(3x - 5)$

77. $(5x - 6y)^2$

78. $(4x - 9y)^2$

79. $(2r - 3s)(2r + 3s)$

80. $(6r - 2x)(6r + 2x)$

Review and Preview

Simplify each expression. See Section 10.1.

81. $\frac{50b^{10}}{70b^5}$

82. $\frac{60y^6}{80y^2}$

83. $\frac{8a^{17}b^5}{-4a^7b^{10}}$

84. $\frac{-6a^8y}{3a^4y}$

85. $\frac{2x^4y^{12}}{3x^4y^4}$

86. $\frac{-48ab^6}{32ab^3}$

Combining Concepts

Express each as a product of polynomials in x. Then multiply and simplify.

△ **87.** Find the area of the square rug if its side is $(2x + 1)$ feet.

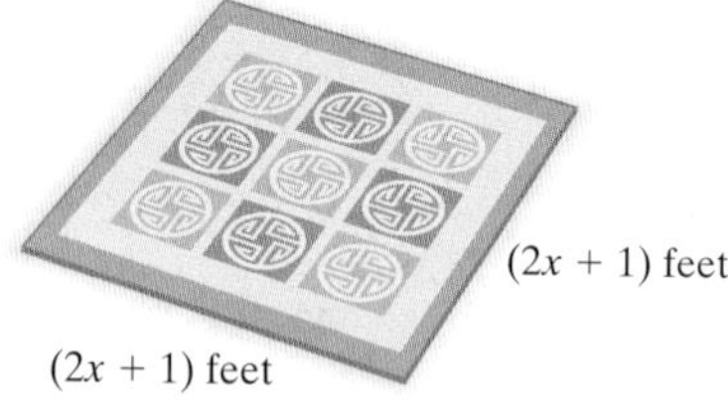

△ **88.** Find the area of the rectangular canvas if its length is $(3x - 2)$ inches and its width is $(x - 4)$ inches.

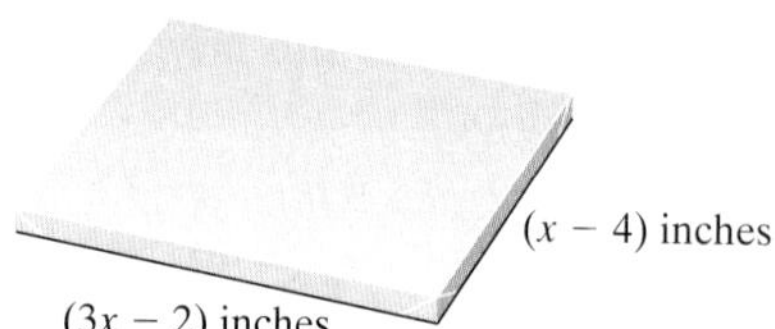

△ **89.** Find the area of the shaded region.

(5x − 3) meters

(x + 1) m

(5x − 3) meters

△ **90.** Find the area of the shaded region.

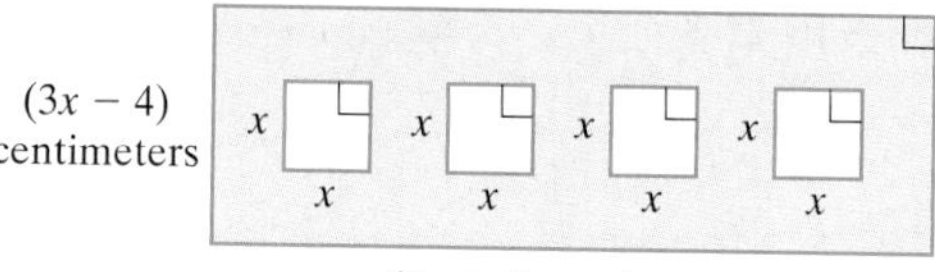

Name ______________ Section ________ Date ________

Integrated Review—Exponents and Operations on Polynomials

Perform operations and simplify.

1. $(5x^2)(7x^3)$

2. $(4y^2)(8y^7)$

3. -4^2

4. $(-4)^2$

5. $(x - 5)(2x + 1)$

6. $(3x - 2)(x + 5)$

7. $(x - 5) + (2x + 1)$

8. $(3x - 2) + (x + 5)$

9. $\frac{7x^9y^{12}}{x^3y^{10}}$

10. $\frac{20a^2b^8}{14a^2b^2}$

11. $(12m^7n^6)^2$

12. $(4y^9z^{10})^3$

13. $(4y - 3)(4y + 3)$

14. $(7x - 1)(7x + 1)$

15. $(x^{-7}y^5)^9$

16. $(3^{-1}x^9)^3$

Answers

1. ______
2. ______
3. ______
4. ______
5. ______
6. ______
7. ______
8. ______
9. ______
10. ______
11. ______
12. ______
13. ______
14. ______
15. ______
16. ______

17. ______

18. ______

19. ______

20. ______

21. ______

22. ______

23. ______

24. ______

25. ______

26. ______

27. ______

28. ______

29. ______

30. ______

31. ______

32. ______

17. $(7x^2 - 2x + 3) - (5x^2 + 9)$

18. $(10x^2 + 7x - 9) - (4x^2 - 6x + 2)$

19. $0.7y^2 - 1.2 + 1.8y^2 - 6y + 1$

20. $7.8x^2 - 6.8x - 3.3 + 0.6x^2 - 9$

21. $(x + 4)^2$

22. $(y - 9)^2$

23. $(x + 4) + (x + 4)$

24. $(y - 9) + (y - 9)$

25. $7x^2 - 6xy + 4(y^2 - xy)$

26. $5a^2 - 3ab + 6(b^2 - a^2)$

27. $(x - 3)(x^2 + 5x - 1)$

28. $(x + 1)(x^2 - 3x - 2)$

29. $(2x - 7)(3x + 10)$

30. $(5x - 1)(4x + 5)$

31. $(2x - 7)(x^2 - 6x + 1)$

32. $(5x - 1)(x^2 + 2x - 3)$

10.7 Dividing Polynomials

OBJECTIVES

Ⓐ Divide a polynomial by a monomial.

Ⓑ Use long division to divide a polynomial by a polynomial other than a monomial.

SSM TUTOR CENTER SG CD & VIDEO MATH PRO WEB

Ⓐ Dividing by a Monomial

To divide a polynomial by a monomial, recall addition of fractions. Fractions that have a common denominator are added by adding the numerators:

$$\frac{a}{c} + \frac{b}{c} = \frac{a + b}{c}$$

If we read this equation from right to left and let a, b, and c be monomials, $c \neq 0$, we have the following.

To Divide a Polynomial by a Monomial

Divide each term of the polynomial by the monomial.

$$\frac{a + b}{c} = \frac{a}{c} + \frac{b}{c}, \quad c \neq 0$$

Throughout this section, we assume that denominators are not 0.

EXAMPLE 1 Divide: $6m^2 + 2m$ by $2m$

Solution: We begin by writing the quotient in fraction form. Then we divide each term of the polynomial $6m^2 + 2m$ by the monomial $2m$.

$$\frac{6m^2 + 2m}{2m} = \frac{6m^2}{2m} + \frac{2m}{2m}$$

$$= 3m + 1 \quad \text{Simplify.}$$

Check: To check, we multiply.

$$2m(3m + 1) = 2m(3m) + 2m(1) = 6m^2 + 2m$$

The quotient $3m + 1$ checks.

Try the Concept Check in the margin.

EXAMPLE 2 Divide: $\frac{9x^5 - 12x^2 + 3x}{3x^2}$

Solution:

$$\frac{9x^5 - 12x^2 + 3x}{3x^2} = \frac{9x^5}{3x^2} - \frac{12x^2}{3x^2} + \frac{3x}{3x^2} \quad \text{Divide each term by } 3x^2.$$

$$= 3x^3 - 4 + \frac{1}{x} \quad \text{Simplify.}$$

Notice that the quotient is not a polynomial because of the term $\frac{1}{x}$. This expression is called a rational expression—we will study rational expressions in Chapter 12. Although the quotient of two polynomials is not always a polynomial, we may still check by multiplying.

Check:

$$3x^2\left(3x^3 - 4 + \frac{1}{x}\right) = 3x^2(3x^3) - 3x^2(4) + 3x^2\left(\frac{1}{x}\right)$$

$$= 9x^5 - 12x^2 + 3x$$

Practice Problem 1

Divide: $25x^3 + 5x^2$ by $5x^2$

Concept Check

In which of the following is $\frac{x + 5}{5}$ simplified correctly?

a. $\frac{x}{5} + 1$ b. x c. $x + 1$

Practice Problem 2

Divide: $\frac{30x^7 + 10x^2 - 5x}{5x^2}$

Answers

1. $5x + 1$ **2.** $6x^5 + 2 - \frac{1}{x}$

Concept Check: **a**

Practice Problem 3

Divide: $\dfrac{12x^3y^3 - 18xy + 6y}{3xy}$

EXAMPLE 3 Divide: $\dfrac{8x^2y^2 - 16xy + 2x}{4xy}$

Solution:

$$\frac{8x^2y^2 - 16xy + 2x}{4xy} = \frac{8x^2y^2}{4xy} - \frac{16xy}{4xy} + \frac{2x}{4xy} \quad \text{Divide each term by } 4xy.$$

$$= 2xy - 4 + \frac{1}{2y} \quad \text{Simplify.}$$

Check: $4xy\left(2xy - 4 + \dfrac{1}{2y}\right) = 4xy(2xy) - 4xy(4) + 4xy\left(\dfrac{1}{2y}\right)$

$$= 8x^2y^2 - 16xy + 2x$$

B Dividing by a Polynomial Other Than a Monomial

To divide a polynomial by a polynomial other than a monomial, we use a process known as long division. Polynomial long division is similar to number long division, so we review long division by dividing 13 into 3660.

Helpful Hint

Recall that 3660 is called the dividend.

```
     281
13)3660
   26↓        2·13 = 26
   106        Subtract and bring down the next digit in the dividend.
   104↓       8·13 = 104
    20        Subtract and bring down the next digit in the dividend.
    13        1·13 = 13
     7        Subtract. There are no more digits to bring down, so the remainder is 7.
```

The quotient is 281 R 7, which can be written as $281\frac{7}{13}$ ← remainder / ← divisor.

Recall that division can be checked by multiplication. To check this division problem, we see that

$13 \cdot 281 + 7 = 3660$, the dividend.

Now we demonstrate long division of polynomials.

Practice Problem 4

Divide: $x^2 + 12x + 35$ by $x + 5$

EXAMPLE 4 Divide $x^2 + 7x + 12$ by $x + 3$ using long division.

Solution:

```
              x
x + 3)x² + 7x + 12     How many times does x divide x²? x²/x = x.
    ⁻x² ⁻ 3x    ↓      Multiply: x(x + 3).
          4x + 12      Subtract and bring down the next term.
```

To subtract, change the signs of these terms and add.

Now we repeat this process.

```
              x + 4
x + 3)x² + 7x + 12     How many times does x divide 4x? 4x/x = 4.
      x² + 3x
           4x + 12
         ⁻4x ⁻ 12      Multiply: 4(x + 3).
                0      Subtract. The remainder is 0.
```

To subtract, change the signs of these terms and add.

The quotient is $x + 4$.

Answers

3. $4x^2y^2 - 6 + \dfrac{2}{x}$ **4.** $x + 7$

Check: We check by multiplying.

	divisor	·	quotient	+	remainder	=	dividend
or	↓		↓		↓		↓
	$(x + 3)$	·	$(x + 4)$	+	0	=	$x^2 + 7x + 12$

The quotient checks.

EXAMPLE 5 Divide $6x^2 + 10x - 5$ by $3x - 1$ using long division.

Solution:

$$\begin{array}{r} 2x + 4 \\ 3x - 1 \overline{) 6x^2 + 10x - 5} \\ \underline{6x^2 - 2x} \\ 12x - 5 \\ \underline{12x - 4} \\ -1 \end{array}$$

$\frac{6x^2}{3x} = 2x$, so $2x$ is a term of the quotient.

Multiply: $2x(3x - 1)$.

Subtract and bring down the next term.

$\frac{12x}{3x} = 4$. Multiply: $4(3x - 1)$.

Subtract. The remainder is -1.

Thus $(6x^2 + 10x - 5)$ divided by $(3x - 1)$ is $(2x + 4)$ with a remainder of -1. This can be written as follows.

$$\frac{6x^2 + 10x - 5}{3x - 1} = 2x + 4 + \frac{-1}{3x - 1} \begin{array}{l} \leftarrow \text{remainder} \\ \leftarrow \text{divisor} \end{array}$$

Check: To check, we multiply $(3x - 1)(2x + 4)$. Then we add the remainder, -1, to this product.

$$\begin{aligned}(3x - 1)(2x + 4) + (-1) &= (6x^2 + 12x - 2x - 4) - 1 \\ &= 6x^2 + 10x - 5\end{aligned}$$

The quotient checks.

Notice that the division process is continued until the degree of the remainder polynomial is less than the degree of the divisor polynomial.

EXAMPLE 6 Divide: $\dfrac{4x^2 + 7 + 8x^3}{2x + 3}$

Solution: Before we begin the division process, we rewrite $4x^2 + 7 + 8x^3$ as $8x^3 + 4x^2 + 0x + 7$. Notice that we have written the polynomial in descending order and have represented the missing x term by $0x$.

$$\begin{array}{r} 4x^2 - 4x + 6 \\ 2x + 3 \overline{) 8x^3 + 4x^2 + 0x + 7} \\ \underline{8x^3 + 12x^2} \\ -8x^2 + 0x \\ \underline{-8x^2 - 12x} \\ 12x + 7 \\ \underline{12x + 18} \\ -11 \end{array}$$

Remainder.

Thus, $\dfrac{4x^2 + 7 + 8x^3}{2x + 3} = 4x^2 - 4x + 6 + \dfrac{-11}{2x + 3}$.

Practice Problem 5

Divide: $6x^2 + 7x - 5$ by $2x - 1$

Practice Problem 6

Divide: $\dfrac{5 - x + 9x^3}{3x + 2}$

Answers

5. $3x + 5$ **6.** $3x^2 - 2x + 1 + \dfrac{3}{3x + 2}$

Practice Problem 7

Divide: $x^3 - 1$ by $x - 1$

EXAMPLE 7 Divide $x^3 - 8$ by $x - 2$.

Solution: Notice that the polynomial $x^3 - 8$ is missing an x^2 term and an x term. We'll represent these terms by inserting $0x^2$ and $0x$.

$$\begin{array}{r} x^2 + 2x + 4 \\ x - 2 \overline{)x^3 + 0x^2 + 0x - 8} \\ \underline{x^3 - 2x^2 } \\ 2x^2 + 0x \\ \underline{2x^2 - 4x } \\ 4x - 8 \\ \underline{4x - 8} \\ 0 \end{array}$$

Thus, $\dfrac{x^3 - 8}{x - 2} = x^2 + 2x + 4$.

Check: To check, see that $(x^2 + 2x + 4)(x - 2) = x^3 - 8$.

Answer

7. $x^2 + x + 1$

FOCUS ON History

EXPONENTIAL NOTATION

The French mathematician and philosopher René Descartes (1596–1650) is generally credited with devising the system of exponents that we use in math today. His book *La Géométrie* was the first to show successive powers of an unknown quantity x as x, xx, x^3, x^4, x^5, and so on. No one knows why Descartes preferred to write xx instead of x^2. However, the use of xx for the square of the quantity x continued to be popular. Those who used the notation defended it by saying that xx takes up no more space when written than x^2 does.

Before Descartes popularized the use of exponents to indicate powers, other less convenient methods were used. Some mathematicians preferred to write out the Latin words *quadratus* and *cubus* whenever they wanted to indicate that a quantity was to be raised to the second power or the third power. Other mathematicians used the abbreviations of *quadratus* and *cubus*, Q and C, to indicate second and third powers of a quantity.

Name ______________________ Section __________ Date __________

Mental Math

Simplify each expression.

1. $\frac{a^6}{a^4}$
2. $\frac{y^2}{y}$
3. $\frac{a^3}{a}$
4. $\frac{p^8}{p^3}$
5. $\frac{k^5}{k^2}$
6. $\frac{k^7}{k^5}$

EXERCISE SET 10.7

A *Perform each division. See Examples 1 through 3.*

1. $\frac{20x^2 + 5x + 9}{5}$
2. $\frac{8x^3 - 4x^2 + 6x + 2}{2}$
3. $\frac{12x^4 + 3x^2}{x}$
4. $\frac{15x^2 - 9x^5}{x}$
5. $\frac{15p^3 + 18p^2}{3p}$
6. $\frac{14m^2 - 27m^3}{7m}$
7. $\frac{-9x^4 + 18x^5}{6x^5}$
8. $\frac{6x^5 + 3x^4}{3x^4}$
9. $\frac{-9x^5 + 3x^4 - 12}{3x^3}$
10. $\frac{6a^2 - 4a + 12}{-2a^2}$
11. $\frac{4x^4 - 6x^3 + 7}{-4x^4}$
12. $\frac{-12a^3 + 36a - 15}{3a}$
13. $\frac{a^2b^2 - ab^3}{ab}$
14. $\frac{m^3n^2 - mn^4}{mn}$
15. $\frac{2x^2y + 8x^2y^2 - xy^2}{2xy}$
16. $\frac{11x^3y^3 - 33xy + x^2y^2}{11xy}$

B *Find each quotient using long division. See Examples 4 and 5.*

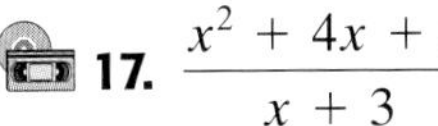

17. $\frac{x^2 + 4x + 3}{x + 3}$
18. $\frac{x^2 + 7x + 10}{x + 5}$
19. $\frac{2x^2 + 13x + 15}{x + 5}$
20. $\frac{3x^2 + 8x + 4}{x + 2}$
21. $\frac{2x^2 - 7x + 3}{x - 4}$
22. $\frac{3x^2 - x - 4}{x - 1}$
23. $\frac{8x^2 + 6x - 27}{2x - 3}$
24. $\frac{18w^2 + 18w - 8}{3w + 4}$
25. $\frac{9a^3 - 3a^2 - 3a + 4}{3a + 2}$
26. $\frac{4x^3 + 12x^2 + x - 12}{2x + 3}$
27. $\frac{2b^3 + 9b^2 + 6b - 4}{b + 4}$
28. $\frac{2x^3 + 3x^2 - 3x + 4}{x + 2}$
29. $\frac{8x^2 + 10x + 1}{2x + 1}$
30. $\frac{3x^2 + 17x + 7}{3x + 2}$
31. $\frac{2x^3 + 2x^2 - 17x + 8}{x - 2}$
32. $\frac{4x^3 + 11x^2 - 8x - 10}{x + 3}$

Find each quotient using long division. Don't forget to write the polynomials in descending order and fill in any missing terms. See Examples 6 and 7.

33. $\frac{x^3 - 27}{x - 3}$
34. $\frac{x^3 + 64}{x + 4}$
35. $\frac{1 - 3x^2}{x + 2}$
36. $\frac{7 - 5x^2}{x + 3}$

37. $\dfrac{-4b + 4b^2 - 5}{2b - 1}$

38. $\dfrac{-3y + 2y^2 - 15}{2y + 5}$

Review and Preview

Fill in each blank. See Sections 10.1 and 10.2.

39. $12 = 4 \cdot$ ______

40. $12 = 2 \cdot$ ______

41. $20 = -5 \cdot$ ______

42. $20 = -4 \cdot$ ______

43. $9x^2 = 3x \cdot$ ______

44. $9x^2 = 9x \cdot$ ______

45. $36x^2 = 4x \cdot$ ______

46. $36x^2 = 2x \cdot$ ______

Combining Concepts

Divide.

47. $\dfrac{x^5 + x^2}{x^2 + x}$

48. $\dfrac{x^6 - x^4}{x^3 + 1}$

Solve.

△ **49.** The perimeter of a square is $(12x^3 + 4x - 16)$ feet. Find the length of its side.

Perimeter is $(12x^3 + 4x - 16)$ feet

△ **50.** The volume of the swimming pool shown is $(36x^5 - 12x^3 + 6x^2)$ cubic feet. If its height is $2x$ feet and its width is $3x$ feet, find its length.

△ **51.** The area of the following parallelogram is $(10x^2 + 31x + 15)$ square meters. If its base is $(5x + 3)$ meters, find its height.

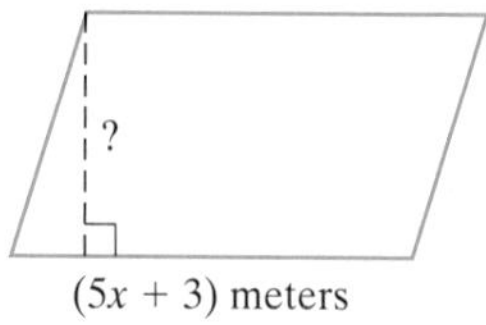

△ **52.** The area of the top of the Ping-Pong table shown is $(49x^2 + 70x - 200)$ square inches. If its length is $(7x + 20)$ inches, find its width.

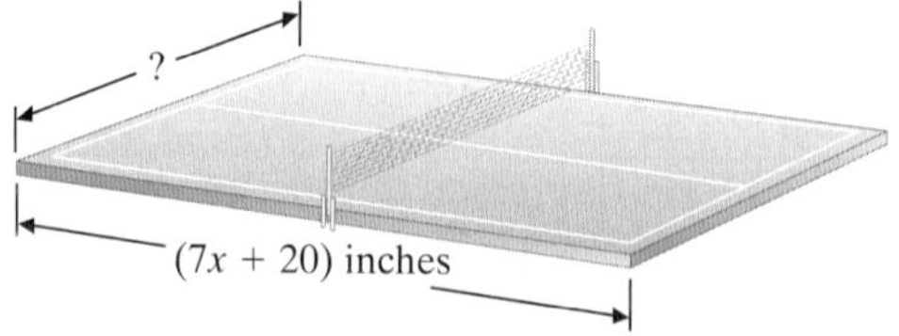

53. Explain how to check a polynomial long division result when the remainder is 0.

54. Explain how to check a polynomial long division result when the remainder is not 0.

CHAPTER 10 ACTIVITY Modeling with Polynomials

MATERIALS:

- Calculator

This activity may be completed by working in groups or individually.

The polynomial model $-40x^2 + 140x + 8393$ gives the average daily total supply of motor gasoline (in thousand barrels per day) in the United States for the period 1998–2000. The polynomial model $7x^2 + 22x + 8082$ gives the average daily supply of domestically produced motor gasoline (in thousand barrels per day) in the United States for the same period. In both models, x is the number of years after 1998. The other source of motor gasoline in the United States, contributing to the total supply, is imported motor gasoline. (*Source:* Based on data from the Energy Information Administration)

1. Use the given polynomials to complete the following table showing the average daily supply (both total and domestic) over the period 1998–2000 by evaluating each polynomial at the given values of x. Then subtract each value in the fourth column from the corresponding value in the third column. Record the result in the last column, titled "Difference." What do you think these values represent?

Year	x	Average Daily Total Supply (thousand barrels per day)	Average Daily Domestic Supply (thousand barrels per day)	Difference
1998	0			
1999	1			
2000	2			

2. Use the polynomial models to find a new polynomial model representing the average daily supply of imported motor gasoline. Then evaluate your new polynomial model to complete the accompanying table.

Year	x	Average Daily Imported Supply (thousand barrels per day)
1998	0	
1999	1	
2000	2	

3. Compare the values in the last column of the table in question 1 to the values in the last column of the table in question 2. What do you notice? What can you conclude?

4. Make a bar graph of the data in the table in question 2. Describe what you see.

Average Daily Imported Supply (thousand barrels a day)

0

STUDY SKILLS REMINDER

What should you do on the day of an exam?

On the day of an exam, try the following:

- Allow yourself plenty of time to arrive at your classroom.
- Read the directions on the test carefully.
- Read each problem carefully as you take your test. Make sure that you answer the question asked.
- Watch your time and pace yourself so that you may attempt each problem on your test.
- If you have time, check your work and answers.
- Do not turn your test in early. If you have extra time, spend it double-checking your work.

Good luck!

FOCUS ON History

NEGATIVE EXPONENTS

Negative exponents were the invention of the English mathematician John Wallis (1616–1703). His book *Arithmetica Infinitorum*, published in 1656, begins with proofs of the laws of exponents. He extended these to cover negative exponents as well and showed that $x^0 = 1$, $x^{-1} = \frac{1}{x}$, $x^{-2} = \frac{1}{x^2}$, and so on. He also showed that, in general, x^{-n} represents the reciprocal of x^n.

Not long after Wallis published his *Arithmetica Infinitorum*, Sir Issac Newton (1642–1727) adopted Wallis's definition and use of negative exponents. Newton's widely circulated mathematical and scientific writings helped the use of negative exponents become universally accepted.

Chapter 10 VOCABULARY CHECK

Fill in each blank with one of the words or phrases listed below.

term	coefficient	monomial	binomial	trinomial
polynomials	degree of a term	degree of a polynomial	FOIL	

1. A ________ is a number or the product of numbers and variables raised to powers.
2. The ________ method may be used when multiplying two binomials.
3. A polynomial with exactly 3 terms is called a ____________.
4. The ________________________ is the greatest degree of any term of the polynomial.
5. A polynomial with exactly 2 terms is called a ____________.
6. The _____________ of a term is its numerical factor.
7. The __________________ is the sum of the exponents on the variables in the term.
8. A polynomial with exactly 1 term is called a ____________.
9. Monomials, binomials, and trinomials are all examples of ______________.

10 CHAPTER Highlights

DEFINITIONS AND CONCEPTS	EXAMPLES
Section 10.1 Exponents	
a^n means the product of n factors, each of which is a.	$3^2 = 3 \cdot 3 = 9$ $(-5)^3 = (-5)(-5)(-5) = -125$ $\left(\frac{1}{2}\right)^4 = \frac{1}{2}\cdot\frac{1}{2}\cdot\frac{1}{2}\cdot\frac{1}{2} = \frac{1}{16}$
Let m and n be integers and no denominators be 0. **Product Rule:** $a^m \cdot a^n = a^{m+n}$	$x^2 \cdot x^7 = x^{2+7} = x^9$
Power Rule: $(a^m)^n = a^{mn}$	$(5^3)^8 = 5^{3\cdot 8} = 5^{24}$
Power of a Product Rule: $(ab)^n = a^n b^n$	$(7y)^4 = 7^4 y^4$
Power of a Quotient Rule: $\left(\frac{a}{b}\right)^n = \frac{a^n}{b^n}$	$\left(\frac{x}{8}\right)^3 = \frac{x^3}{8^3}$
Quotient Rule: $\frac{a^m}{a^n} = a^{m-n}$	$\frac{x^9}{x^4} = x^{9-4} = x^5$
Zero Exponent: $a^0 = 1, a \neq 0$	$5^0 = 1$; $x^0 = 1, x \neq 0$
Section 10.2 Negative Exponents and Scientific Notation	
If $a \neq 0$ and n is an integer, $a^{-n} = \frac{1}{a^n}$	$3^{-2} = \frac{1}{3^2} = \frac{1}{9}$; $5x^{-2} = \frac{5}{x^2}$ Simplify: $\left(\frac{x^{-2}y}{x^5}\right)^{-2} = \frac{x^4y^{-2}}{x^{-10}}$ $= x^{4-(-10)}y^{-2}$ $= \frac{x^{14}}{y^2}$
A positive number is written in scientific notation if it is written as the product of a number a, where $1 \leq a < 10$, and an integer power r of 10. $a \times 10^r$	$1200.0 = 1.2 \times 10^3$ $0.000000568 = 5.68 \times 10^{-7}$

Definitions and Concepts	**Examples**

Section 10.3 Introduction to Polynomials

A **term** is a number or the product of a number and variables raised to powers.

$-5x, 7a^2b, \frac{1}{4}y^4, 0.2$

The **numerical coefficient** or **coefficient** of a term is its numerical factor.

Term	Coefficient
$7x^2$	7
y	1
$-a^2b$	-1

A **polynomial** is a finite sum of terms of the form ax^n where a is a real number and n is a whole number.

$5x^3 - 6x^2 + 3x - 6$ (Polynomial)

A **monomial** is a polynomial with exactly 1 term.

$\frac{5}{6}y^3$ (Monomial)

A **binomial** is a polynomial with exactly 2 terms.

$-0.2a^2b - 5b^2$ (Binomial)

A **trinomial** is a polynomial with exactly 3 terms.

$3x^2 - 2x + 1$ (Trinomial)

The **degree of a polynomial** is the greatest degree of any term of the polynomial.

Polynomial	Degree
$5x^2 - 3x + 2$	2
$7y + 8y^2z^3 - 12$	$2 + 3 = 5$

Section 10.4 Adding and Subtracting Polynomials

To add polynomials, combine like terms.

Add.

$$(7x^2 - 3x + 2) + (-5x - 6)$$
$$= 7x^2 - 3x + 2 - 5x - 6$$
$$= 7x^2 - 8x - 4$$

To subtract two polynomials, change the signs of the terms of the second polynomial, and then add.

Subtract.

$$(17y^2 - 2y + 1) - (-3y^3 + 5y - 6)$$
$$= (17y^2 - 2y + 1) + (3y^3 - 5y + 6)$$
$$= 17y^2 - 2y + 1 + 3y^3 - 5y + 6$$
$$= 3y^3 + 17y^2 - 7y + 7$$

Section 10.5 Multiplying Polynomials

To multiply two polynomials, multiply each term of one polynomial by each term of the other polynomial, and then combine like terms.

Multiply.

$$(2x + 1)(5x^2 - 6x + 2)$$
$$= 2x(5x^2 - 6x + 2) + 1(5x^2 - 6x + 2)$$
$$= 10x^3 - 12x^2 + 4x + 5x^2 - 6x + 2$$
$$= 10x^3 - 7x^2 - 2x + 2$$

Definitions and Concepts | Examples

Section 10.6 Special Products

The **FOIL method** may be used when multiplying two binomials.

Multiply: $(5x - 3)(2x + 3)$

First, Last, Inner, Outer

$(5x - 3)(2x + 3)$

$$= \overset{F}{(5x)(2x)} + \overset{O}{(5x)(3)} + \overset{I}{(-3)(2x)} + \overset{L}{(-3)(3)}$$
$$= 10x^2 + 15x - 6x - 9$$
$$= 10x^2 + 9x - 9$$

Squaring a Binomial

$$(a + b)^2 = a^2 + 2ab + b^2$$

$$(a - b)^2 = a^2 - 2ab + b^2$$

Square each binomial.

$$(x + 5)^2 = x^2 + 2(x)(5) + 5^2$$
$$= x^2 + 10x + 25$$

$$(3x - 2y)^2 = (3x)^2 - 2(3x)(2y) + (2y)^2$$
$$= 9x^2 - 12xy + 4y^2$$

Multiplying the Sum and Difference of Two Terms

$$(a + b)(a - b) = a^2 - b^2$$

Multiply.

$$(6y + 5)(6y - 5) = (6y)^2 - 5^2$$
$$= 36y^2 - 25$$

Section 10.7 Dividing Polynomials

To divide a polynomial by a monomial,

$$\frac{a + b}{c} = \frac{a}{c} + \frac{b}{c}, c \neq 0$$

Divide.

$$\frac{15x^5 - 10x^3 + 5x^2 - 2x}{5x^2}$$
$$= \frac{15x^5}{5x^2} - \frac{10x^3}{5x^2} + \frac{5x^2}{5x^2} - \frac{2x}{5x^2}$$
$$= 3x^3 - 2x + 1 - \frac{2}{5x}$$

To divide a polynomial by a polynomial other than a monomial, use long division.

$$\begin{array}{r} 5x - 1 + \dfrac{-4}{2x + 3} \\ 2x + 3\overline{)10x^2 + 13x - 7} \\ \underline{10x^2 + 15x} \\ -2x - 7 \\ \underline{-2x - 3} \\ -4 \end{array}$$

or $5x - 1 - \dfrac{4}{2x + 3}$

STUDY SKILLS REMINDER

Are you prepared for a test on Chapter 10?

Below is a list of some *common trouble areas* for topics covered in Chapter 10. After studying for your test—but before taking your test—read these.

- Do you know that a negative exponent does not make the base a negative number? For example,

$$3^{-2} = \frac{1}{3^2} = \frac{1}{9}$$

- Make sure you remember that x has an understood coefficient of 1 and an understood exponent of 1. For example,

$$2x + x = 2x + 1x = 3x; \quad x^5 \cdot x = x^5 \cdot x^1 = x^6$$

- Do you know the difference between $5x^2$ and $(5x)^2$?

$5x^2$ is $5 \cdot x^2$; $(5x)^2 = 5^2 \cdot x^2$ or $25 \cdot x^2$

- Can you evaluate $x^2 - x$ when $x = -2$?

$$x^2 - x = (-2)^2 - (-2) = 4 - (-2) = 4 + 2 = 6$$

- Can you subtract $5x^2 + 1$ from $3x^2 - 6$?

$$(3x^2 - 6) - (5x^2 + 1) = 3x^2 - 6 - 5x^2 - 1 = -2x^2 - 7$$

- Make sure you are familiar with squaring a binomial.

$$(3x - 4)^2 = (3x)^2 - 2(3x)(4) + 4^2 = 9x^2 - 24x + 16$$

or

$$(3x - 4)^2 = (3x - 4)(3x - 4) = 9x^2 - 24x + 16$$

Remember: This is simply a checklist of common trouble areas. For a review of Chapter 10, see the Highlights and Chapter Review.

Name ______________________ Section __________ Date __________

Chapter 10 Review

(10.1) *State the base and the exponent for each expression.*

1. 3^2 **2.** $(-5)^4$ **3.** -5^4 **4.** x^6

Evaluate each expression.

5. 8^3 **6.** $(-6)^2$ **7.** -6^2 **8.** $-4^3 - 4^0$ **9.** $(3b)^0$ **10.** $\frac{8b}{8b}$

Simplify each expression.

11. $y^2 \cdot y^7$ **12.** $x^9 \cdot x^5$ **13.** $(2x^5)(-3x^6)$ **14.** $(-5y^3)(4y^4)$ **15.** $(x^4)^2$

16. $(y^3)^5$ **17.** $(3y^6)^4$ **18.** $(2x^3)^3$ **19.** $\frac{x^9}{x^4}$ **20.** $\frac{z^{12}}{z^5}$

21. $\frac{a^5b^4}{ab}$ **22.** $\frac{x^4y^6}{xy}$ **23.** $\frac{12xy^6}{3x^4y^{10}}$ **24.** $\frac{2x^7y^8}{8xy^2}$ **25.** $5a^7(2a^4)^3$

26. $(2x)^2(9x)$ **27.** $(-5a)^0 + 7^0 + 8^0$ **28.** $8x^0 + 9^0$

Simplify the given expression and choose the correct result.

29. $\left(\frac{3x^4}{4y}\right)^3$

a. $\frac{27x^{64}}{64y^3}$

b. $\frac{27x^{12}}{64y^3}$

c. $\frac{9x^{12}}{12y^3}$

d. $\frac{3x^{12}}{4y^3}$

30. $\left(\frac{5a^6}{b^3}\right)^2$

a. $\frac{10a^{12}}{b^6}$

b. $\frac{25a^{36}}{b^9}$

c. $\frac{25a^{12}}{b^6}$

d. $25a^{12}b^6$

(10.2) *Simplify each expression.*

31. 7^{-2} **32.** -7^{-2} **33.** $2x^{-4}$ **34.** $(2x)^{-4}$

35. $\left(\frac{1}{5}\right)^{-3}$ **36.** $\left(\frac{-2}{3}\right)^{-2}$ **37.** $2^0 + 2^{-4}$ **38.** $6^{-1} - 7^{-1}$

Simplify each expression. Assume that variables in an exponent represent positive integers only. Write each answer using positive exponents only.

39. $\dfrac{x^5}{x^{-3}}$

40. $\dfrac{z^4}{z^{-4}}$

41. $\dfrac{r^{-3}}{r^{-4}}$

42. $\dfrac{y^{-2}}{y^{-5}}$

43. $\left(\dfrac{bc^{-2}}{bc^{-3}}\right)^4$

44. $\left(\dfrac{x^{-3}y^{-4}}{x^{-2}y^{-5}}\right)^{-3}$

45. $\dfrac{x^{-4}y^{-6}}{x^2y^7}$

46. $\dfrac{a^5b^{-5}}{a^{-5}b^5}$

Write each number in scientific notation.

47. 0.00027

48. 0.8868

49. 80,800,000

50. −868,000

51. In January 2002, the United States imported approximately 1,302,079,000 kilograms of coffee. Write this number in scientific notation. (*Source:* International Coffee Organization)

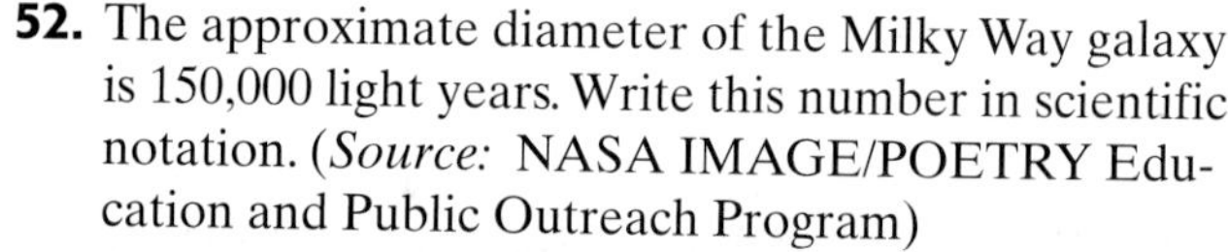

52. The approximate diameter of the Milky Way galaxy is 150,000 light years. Write this number in scientific notation. (*Source:* NASA IMAGE/POETRY Education and Public Outreach Program)

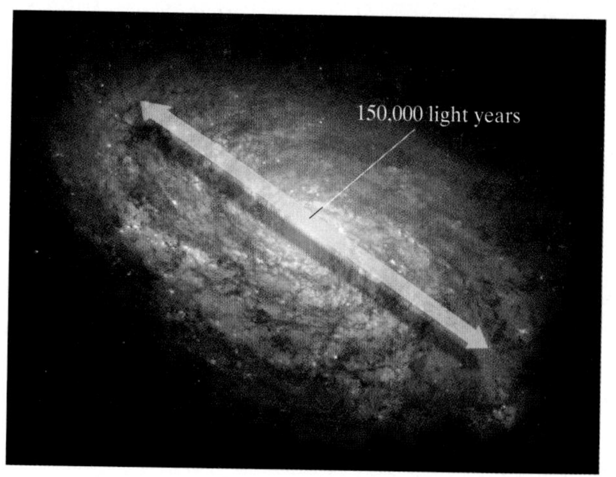

Write each number in standard form.

53. 8.67×10^5

54. 3.86×10^{-3}

55. 8.6×10^{-4}

56. 8.936×10^5

57. The volume of the planet Jupiter is 1.43128×10^{15} cubic kilometers. Write this number in standard notation. (*Source:* National Space Science Data Center)

58. An angstrom is a unit of measure, equal to 1×10^{-10} meter, used for measuring wavelengths or the diameters of atoms. Write this number in standard notation. (*Source:* National Institute of Standards and Technology)

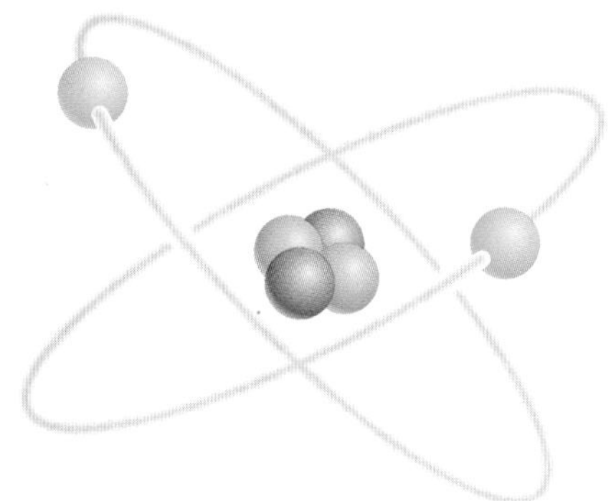

Simplify. Express each result in standard form.

59. $(8 \times 10^4)(2 \times 10^{-7})$

60. $\dfrac{8 \times 10^4}{2 \times 10^{-7}}$

(10.3) *Find the degree of each polynomial.*

61. $y^5 + 7x - 8x^4$

62. $9y^2 + 30y + 25$

63. $-14x^2y - 28x^2y^3 - 42x^2y^2$

64. $6x^2y^2z^2 + 5x^2y^3 - 12xyz$

△ **65.** The surface area of a box with a square base and a height of 5 units is given by the polynomial $2x^2 + 20x$. Fill in the table below by evaluating $2x^2 + 20x$ for the given values of x.

x	1	3	5.1	10
$2x^2 + 20x$				

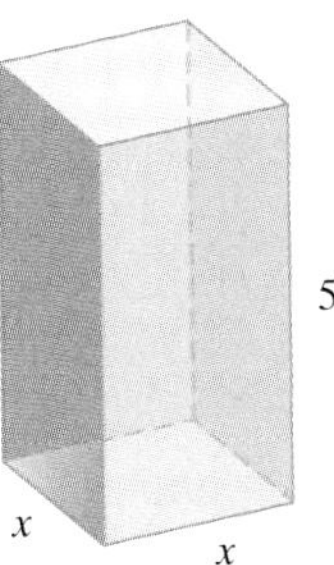

Combine like terms in each expression.

66. $7a^2 - 4a^2 - a^2$

67. $9y + y - 14y$

68. $6a^2 + 4a + 9a^2$

69. $21x^2 + 3x + x^2 + 6$

70. $4a^2b - 3b^2 - 8q^2 - 10a^2b + 7q^2$

71. $2s^{14} + 3s^{13} + 12s^{12} - s^{10}$

(10.4) *Add or subtract as indicated.*

72. $(3x^2 + 2x + 6) + (5x^2 + x)$

73. $(2x^5 + 3x^4 + 4x^3 + 5x^2) + (4x^2 + 7x + 6)$

74. $(-5y^2 + 3) - (2y^2 + 4)$

75. $(2m^7 + 3x^4 + 7m^6) - (8m^7 + 4m^2 + 6x^4)$

76. $(3x^2 - 7xy + 7y^2) - (4x^2 - xy + 9y^2)$

77. Add $(-9x^2 + 6x + 2)$ and $(4x^2 - x - 1)$.

78. Subtract $(4x^2 + 8x - 7)$ from the sum of $(x^2 + 7x + 9)$ and $(x^2 + 4)$.

(10.5) *Multiply each expression.*

79. $6(x + 5)$

80. $9(x - 7)$

81. $4(2a + 7)$

82. $9(6a - 3)$

83. $-7x(x^2 + 5)$

84. $-8y(4y^2 - 6)$

85. $-2(x^3 - 9x^2 + x)$

86. $-3a(a^2b + ab + b^2)$

87. $(3a^3 - 4a + 1)(-2a)$

88. $(6b^3 - 4b + 2)(7b)$

89. $(2x + 2)(x - 7)$

90. $(2x - 5)(3x + 2)$

91. $(4a - 1)(a + 7)$

92. $(6a - 1)(7a + 3)$

93. $(x + 7)(x^3 + 4x - 5)$

94. $(x + 2)(x^5 + x + 1)$

95. $(x^2 + 2x + 4)(x^2 + 2x - 4)$

96. $(x^3 + 4x + 4)(x^3 + 4x - 4)$

97. $(x + 7)^3$

98. $(2x - 5)^3$

(10.6) *Use special products to multiply each of the following.*

99. $(x + 7)^2$

100. $(x - 5)^2$

101. $(3x - 7)^2$

102. $(4x + 2)^2$

103. $(5x - 9)^2$

104. $(5x + 1)(5x - 1)$

105. $(7x + 4)(7x - 4)$

106. $(a + 2b)(a - 2b)$

107. $(2x - 6)(2x + 6)$

108. $(4a^2 - 2b)(4a^2 + 2b)$

Express each as a product of polynomials in x. Then multiply and simplify.

△ **109.** Find the area of the square if its side is $(3x - 1)$ meters.

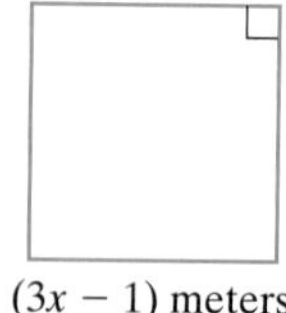

△ **110.** Find the area of the rectangle.

(x − 1) miles

(5x + 2) miles

(10.7) *Divide.*

111. $\dfrac{x^2 + 21x + 49}{7x^2}$

112. $\dfrac{5a^3b - 15ab^2 + 20ab}{-5ab}$

113. $(a^2 - a + 4) \div (a - 2)$

114. $(4x^2 + 20x + 7) \div (x + 5)$

115. $\dfrac{a^3 + a^2 + 2a + 6}{a - 2}$

116. $\dfrac{9b^3 - 18b^2 + 8b - 1}{3b - 2}$

117. $\dfrac{4x^4 - 4x^3 + x^2 + 4x - 3}{2x - 1}$

118. $\dfrac{-10x^2 - x^3 - 21x + 18}{x - 6}$

△ **119.** The area of the rectangle below is $(15x^3 - 3x^2 + 60)$ square feet. If its length is $3x^2$ feet, find its width.

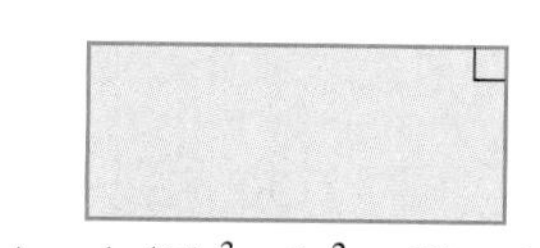

Area is $(15x^3 - 3x^2 + 60)$ sq feet

△ **120.** The perimeter of the equilateral triangle below is $(21a^3b^6 + 3a - 3)$ units. Find the length of a side.

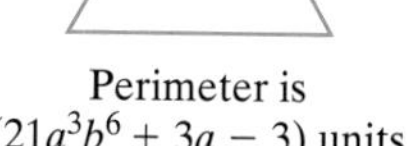

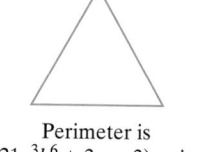

Perimeter is $(21a^3b^6 + 3a - 3)$ units

Name ______________________ Section __________ Date __________

Answers

1. ______
2. ______
3. ______
4. ______
5. ______
6. ______
7. ______
8. ______
9. ______
10. ______
11. ______
12. ______
13. ______
14. ______
15. ______
16. ______
17. ______
18. ______
19. ______
20. ______

Chapter 10 Test

Remember to check your answers and use the Chapter Test Prep Video to view solutions.

Evaluate each expression.

1. 2^5 **2.** $(-3)^4$ **3.** -3^4 **4.** 4^{-3}

Simplify each exponential expression.

5. $(3x^2)(-5x^9)$ **6.** $\dfrac{y^7}{y^2}$ **7.** $\dfrac{r^{-8}}{r^{-3}}$

Simplify each expression. Write the result using only positive exponents.

8. $\left(\dfrac{x^2y^3}{x^3y^{-4}}\right)^2$ **9.** $\dfrac{6^2x^{-4}y^{-1}}{6^3x^{-3}y^7}$

Express each number in scientific notation.

10. 563,000 **11.** 0.0000863

Write each number in standard form.

12. 1.5×10^{-3} **13.** 6.23×10^4

14. Simplify. Write the answer in standard form.
$(1.2 \times 10^5)(3 \times 10^{-7})$

15. Find the degree of the following polynomial.
$4xy^2 + 7xy + x^3y - 2$

16. Simplify by combining like terms.
$5x^2 + 4x - 7x^2 + 11 + 8x$

Perform each indicated operation.

17. $(8x^3 + 7x^2 + 4x - 7) + (8x^3 - 7x - 6)$

18.
$$\begin{array}{r} 5x^3 + x^2 + 5x - 2 \\ -\ (8x^3 - 4x^2 + x - 7) \\ \hline \end{array}$$

19. Subtract $(4x + 2)$ from the sum of $(8x^2 + 7x + 5)$ and $(x^3 - 8)$.

20. Multiply: $(3x + 7)(x^2 + 5x + 2)$

21. ____

22. ____

23. ____

24. ____

25. see table ____

26. ____

____

27. ____

28. ____

29. ____

21. Multiply: $3x^2(2x^2 - 3x + 7)$

22. Use the FOIL method to multiply $(x + 7)(3x - 5)$.

Use special products to multiply each of the following.

23. $(4x - 2)^2$

24. $(x^2 - 9b)(x^2 + 9b)$

25. When it is completed, the Suyong Bay Tower in Pusan, Korea, will be the world's tallest building at 1516 feet tall. Neglecting air resistance, the height of an object dropped from this building at time t seconds is given by the polynomial $-16t^2 + 1516$. Find the height of the object at the given times below. (*Source:* Council on Tall Buildings and Urban Habitat)

t	0 Seconds	3 Seconds	6 Seconds	9 Seconds
$-16t^2 + 1516$				

26. Find the area of the top of the table. Express the area as a product, then multiply and simplify.

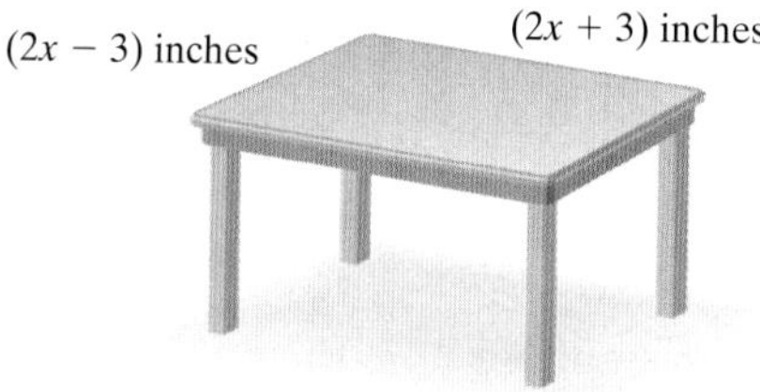

Divide.

27. $\dfrac{4x^2 + 24xy - 7x}{8xy}$

28. $(x^2 + 7x + 10) \div (x + 5)$

29. $\dfrac{27x^3 - 8}{3x + 2}$

Name ______________________ Section __________ Date __________

Answers

Chapter 10 Cumulative Review

Evaluate.

1. 8^2 **2.** 5^3 **3.** 2^5 **4.** 3^3

Write an algebraic expression. Use x to represent "a number".

5. **a.** 7 increased by a number
b. 15 decreased by a number
c. The product of 2 and a number
d. The quotient of a number and 5
e. 2 subtract from a number

6. **a.** The sum of a number and 3
b. The product of 3 and a number
c. Twice a number
d. 10 decreased by a number
e. 5 times a number increased by 7

Use the distributive property to remove parentheses. Then simplify if possible.

7. $2(3 + x) - 15$ **8.** $5(x + 2)$

9. $-2(x - 5) + 4(2x + 2)$ **10.** $-2(y + 0.3z - 1)$

Solve.

11. $y - 5 = -2 - 6$ **12.** $\frac{y}{7} = 20$

13. $7(x - 2) = 9x - 6$ **14.** $6(2a - 1) - (11a + 6) = 7$

15. $\frac{3}{5}a = 9$

1. ______
2. ______
3. ______
4. ______
5. ______
a. ______
b. ______
c. ______
d. ______
e. ______
6. ______
a. ______
b. ______
c. ______
d. ______
e. ______
7. ______
8. ______
9. ______
10. ______
11. ______
12. ______
13. ______
14. ______
15. ______

16. ______

17. ______

18. ______

19. ______

20. ______

21. ______

22. ______

23. ______

24. ______

25. ______

26. ______

27. ______

28. ______

29. a. ______

b. ______

c. ______

d. ______

e. ______

f. ______

30. ______

31. ______

16. $\frac{2}{3}y = 16$ **17.** $3y = -\frac{2}{11}$ **18.** $5y = -\frac{1}{5}$

Write each decimal as a fraction or a mixed number.

19. 0.125 **20.** 0.250

21. 23.5 **22.** 10.75

23. −105.083 **24.** −31.07

Evaluate $x - y$ for the given replacement values.

25. $x = 2.8$, $y = 0.92$ **26.** $x = -1.2$, $y = 7.6$

Evaluate xy for the given replacement values.

27. $x = 2.3$, $y = 0.44$ **28.** $x = -6.1$, $y = 0.5$

29. Given the set $\left\{-2, 0, \frac{1}{4}, 112, -3, 11, \sqrt{2}\right\}$, list the numbers in this set that belong to the set of:

a. Natural numbers **b.** Whole numbers **c.** Integers
d. Rational numbers **e.** Irrational numbers **f.** Real numbers

30. Tell which set(s) the number $-2\frac{1}{2}$ belongs to. Use the sets of numbers from Exercise 29.

31. Solve: $0.25x + 0.10(x - 3) = 0.05(22)$

32. Solve: $0.6x - 10 = 1.4x - 14$

33. Twice the sum of a number and 4 is the same as four times the number, decreased by 12. Find the number.

34. Three times the difference of a number and 2 is the same as five times the number, decreased by 10.

△ 35. Charles Pecot can afford enough fencing to enclose a rectangular garden with a perimeter of 140 feet. If the width of his garden is to be 30 feet, find the length.

36. A house is in the shape of a rectangle and has 2016 square feet of living area. If the length is 63 feet, find its width.

37. Solve: $-4x + 7 \geq -9$. Graph the solutions.

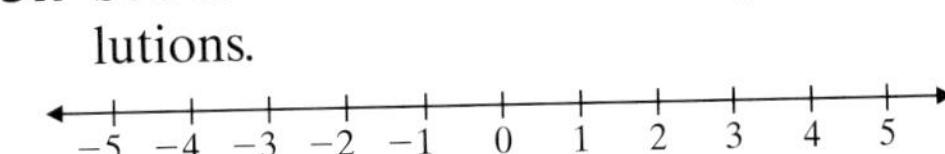

38. Solve: $3x + 4 \geq 2x - 6$

39. Simplify each expression.

a. $x^7 \cdot x^4$

b. $\left(\frac{1}{2}\right)^4$

c. $(9y^5)^2$

40. Simplify each expression.

a. $y \cdot y^5$

b. $\left(\frac{2}{3}\right)^3$

c. $(8x^3)^2$

Simplify the following expressions. Write each result using positive exponents only.

41. $\left(\frac{3a^2}{b}\right)^{-3}$

42. $\left(\frac{2x^3}{y}\right)^{-2}$

43. $(5y^3)^{-2}$

44. $(3x^7)^{-3}$

Simplify each polynomial by combining any like terms.

45. $9x^3 + x^3$

46. $9x^3 - x^3$

47. $5x^2 + 6x - 9x - 3$

48. $2x - x^2 + 5x - 4x^2$

49. Multiply: $7x(x^2 + 2x + 5)$

50. Multiply: $-2x(x^2 - x + 1)$

32. ______
33. ______
34. ______
35. ______
36. ______
37. ______
38. ______
39. a. ______
b. ______
c. ______
40. a. ______
b. ______
c. ______
41. ______
42. ______
43. ______
44. ______
45. ______
46. ______
47. ______
48. ______
49. ______
50. ______

51. ______

52. ______

51. Divide: $\dfrac{9x^5 - 12x^2 + 3x}{3x^2}$

52. Divide: $\dfrac{4x^7 - 12x^2 + 2x}{2x}$

Factoring Polynomials

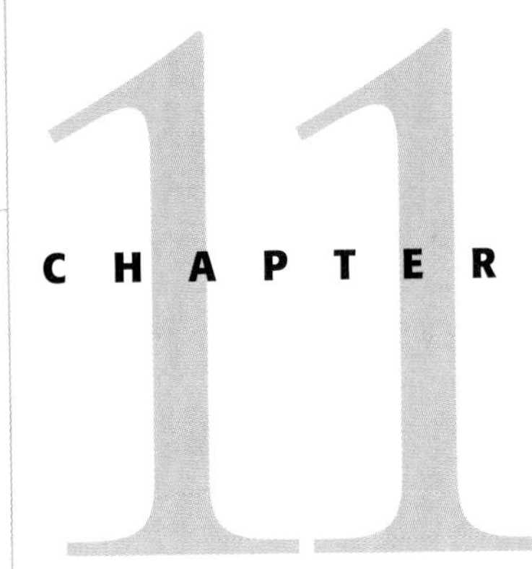

In Chapter 10, we learned how to multiply polynomials. Now we will deal with an operation that is the reverse process of multiplying—factoring. Factoring is an important algebraic skill because it allows us to write a sum as a product. As we will see in Sections 11.6 and 11.7, factoring can be used to solve equations other than linear equations. In Chapter 12, we will also use factoring to simplify and perform arithmetic operations on rational expressions.

When completed, the Petronas Twin Towers in Kuala Lumpur, Malaysia, became the world's tallest building. At a height of 1483 feet, these Islamic-influenced towers beat out the previous record holder, the Sears Tower in Chicago, by 33 feet. Each of the twin towers has 88 stories, and together they have over 32,000 windows. This colossal building complex was designed by American architect Cesar Pelli and cost over \$1.2 billion to build. In Section 11.5, Exercise 98, a polynomial expression for the height of an object dropped from the Petronas Twin Towers is factored.

Answers

1. ______
2. ______
3. ______
4. ______
5. ______
6. ______
7. ______
8. ______
9. ______
10. ______
11. ______
12. ______
13. ______
14. ______
15. ______
16. ______
17. ______
18. ______
19. ______
20. ______

Name ______ Section ______ Date ______

Chapter 11 Pretest

Factor each polynomial completely. If a polynomial cannot be factored, write "prime."

1. $2x^3y - 6x^2y^2$

2. $xy + 6x - 4y - 24$

3. $a^2 + 8a + 12$

4. $m^2 + 4m - 3$

5. $3x^3 - 18x^2 + 15x$

6. $2x^2 + 5x - 12$

7. $14x^2 + 63x + 70$

8. $24b^2 - 25b + 6$

9. $15y^2 + 38y + 7$

10. $x^2 + 24x + 144$

11. $4x^2 - 12xy + 9y^2$

12. $a^2 - 49b^2$

13. $1 - 64t^2$

14. $25b^2 + 4$

15. Fill in the blank so that $x^2 +$ ______ $x + 81$ is a perfect square trinomial.

Solve each equation.

16. $(x - 12)(x + 5) = 0$

17. $y^2 - 13y = 0$

18. $2m^3 - 2m^2 - 24m = 0$

△ **19.** The length of a rectangle is 7 inches more than its width. Its area is 120 square inches. Find the dimensions of the rectangle.

20. The sum of a number and its square is 240. Find the number.

11.1 The Greatest Common Factor

In the product $2 \cdot 3 = 6$, the numbers 2 and 3 are called **factors** of 6 and $2 \cdot 3$ is a **factored form** of 6. This is true of polynomials also. Since $(x + 2)(x + 3) = x^2 + 5x + 6$, then $(x + 2)$ and $(x + 3)$ are factors of $x^2 + 5x + 6$, and $(x + 2)(x + 3)$ is a factored form of the polynomial.

The process of writing a polynomial as a product is called **factoring** the polynomial.

Study the examples below and look for a pattern.

Try the Concept Check in the margin.

Multiplying: $5(x^2 + 3) = 5x^2 + 15$ $\quad 2x(x - 7) = 2x^2 - 14x$

Factoring: $5x^2 + 15 = 5(x^2 + 3)$ $\quad 2x^2 - 14x = 2x(x - 7)$

Do you see that factoring is the reverse process of multiplying?

$$x^2 + 5x + 6 = (x + 2)(x + 3)$$

(factoring →; ← multiplying)

A Finding the Greatest Common Factor

The first step in factoring a polynomial is to see whether the terms of the polynomial have a common factor. If there is one, we can write the polynomial as a product by **factoring out** the common factor. We will usually factor out the *greatest* common factor (GCF).

The **greatest common factor (GCF) of a list of terms** is the product of the GCF of the numerical coefficients and the GCF of the variable factors.

$$20x^2y^2 = 2 \cdot 2 \cdot 5 \cdot x \cdot x \cdot y \cdot y$$
$$6xy^3 = 2 \cdot 3 \cdot x \cdot y \cdot y \cdot y$$
$$\text{GCF} = 2 \cdot x \cdot y \cdot y = 2xy^2$$

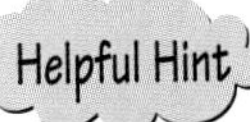

Notice below that the GCF of a list of terms contains the smallest exponent on each common variable.

The GCF of x^5y^6, x^2y^7 and x^3y^4 is x^2y^4. (Smallest exponent on x. Smallest exponent on y.)

EXAMPLE 1 Find the greatest common factor of each list of terms.

a. $6x^2$, $10x^3$, and $-8x$
b. $-18y^2$, $-63y^3$, and $27y^4$
c. a^3b^2, a^5b, and a^6b^2

Solution:

a.
$$6x^2 = 2 \cdot 3 \cdot x^2$$
$$10x^3 = 2 \cdot 5 \cdot x^3$$
$$-8x = -1 \cdot 2 \cdot 2 \cdot 2 \cdot x^1$$
$$\text{GCF} = 2 \cdot x^1 \text{ or } 2x$$

The GCF of x^2, x^3, and x is x.

OBJECTIVES

A Find the greatest common factor of a list of terms.

B Factor out the greatest common factor from the terms of a polynomial.

C Factor by grouping.

Concept Check

Multiply: $2(x - 4)$
What do you think the result of factoring $2x - 8$ would be? Why?

Practice Problem 1

Find the greatest common factor of each list of terms.

a. $6x^2$, $9x^4$, and $-12x^5$
b. $-16y$, $-20y^6$, and $40y^4$
c. a^5b^4, ab^3, and a^3b^2

Answers

1. a. $3x^2$ **b.** $4y$ **c.** ab^2

Concept Check: $2x - 8$; The result would be $2(x - 4)$ because factoring is the reverse process of multiplying.

b. $-18y^2 = -1 \cdot 2 \cdot 3 \cdot 3 \cdot y^2$
$-63y^3 = -1 \cdot 3 \cdot 3 \cdot 7 \cdot y^3$
$27y^4 = 3 \cdot 3 \cdot 3 \cdot y^4$
$\text{GCF} = 3 \cdot 3 \cdot y^2$ or $9y^2$

→ The GCF of y^2, y^3, and y^4 is y^2.

c. The GCF of a^3, a^5, and a^6 is a^3.
The GCF of b^2, b, and b^2 is b. Thus,
the GCF of a^3b^2, a^5b, and a^6b^2 is a^3b.

B Factoring Out the Greatest Common Factor

To factor a polynomial such as $8x + 14$, we first see whether the terms have a greatest common factor other than 1. In this case, they do: The GCF of $8x$ and 14 is 2.

We factor out 2 from each term by writing each term as a product of 2 and the term's remaining factors.

$$8x + 14 = 2 \cdot 4x + 2 \cdot 7$$

Using the distributive property, we can write

$$8x + 14 = 2 \cdot 4x + 2 \cdot 7$$
$$= 2(4x + 7)$$

Thus, a factored form of $8x + 14$ is $2(4x + 7)$. We can check by multiplying:

$$2(4x + 7) = 2 \cdot 4x + 2 \cdot 7 = 8x + 14.$$

Try the Concept Check in the margin.

Helpful Hint

A factored form of $8x + 14$ is *not*

$$2 \cdot 4x + 2 \cdot 7$$

Although the *terms* have been factored (written as a product), the *polynomial* $8x + 14$ has not been factored. A factored form of $8x + 14$ is the *product* $2(4x + 7)$.

Concept Check

Which of the following is/are factored form(s) of $6t + 18$?

a. 6
b. $6 \cdot t + 6 \cdot 3$
c. $6(t + 3)$
d. $3(t + 6)$

EXAMPLE 2

Factor each polynomial by factoring out the greatest common factor (GCF).

a. $6t + 18$ **b.** $y^5 - y^7$

Solution:

a. The GCF of terms $6t$ and 18 is 6. Thus,

$$6t + 18 = 6 \cdot t + 6 \cdot 3$$
$$= 6(t + 3) \quad \text{Apply the distributive property.}$$

We can check our work by multiplying 6 and $(t + 3)$.
$6(t + 3) = 6 \cdot t + 6 \cdot 3 = 6t + 18$, the original polynomial.

b. The GCF of y^5 and y^7 is y^5. Thus,

$$y^5 - y^7 = (y^5)1 - (y^5)y^2$$
$$= y^5(1 - y^2)$$

Helpful Hint

Don't forget the 1.

Practice Problem 2

Factor each polynomial by factoring out the greatest common factor (GCF).

a. $10y + 25$
b. $x^4 - x^9$

Answers

2. a. $5(2y + 5)$ **b.** $x^4(1 - x^5)$

Concept Check: c

EXAMPLE 3 Factor: $-9a^5 + 18a^2 - 3a$

Solution:

$$-9a^5 + 18a^2 - 3a = (3a)(-3a^4) + (3a)(6a) + (3a)(-1)$$
$$= 3a(-3a^4 + 6a - 1)$$

Helpful Hint

Don't forget the -1.

In Example 3, we could have chosen to factor out $-3a$ instead of $3a$. If we factor out $-3a$, we have

$$-9a^5 + 18a^2 - 3a = (-3a)(3a^4) + (-3a)(-6a) + (-3a)(1)$$
$$= -3a(3a^4 - 6a + 1)$$

Notice the changes in signs when factoring out $-3a$.

EXAMPLES Factor.

4. $6a^4 - 12a = 6a(a^3 - 2)$

5. $\frac{3}{7}x^4 + \frac{1}{7}x^3 - \frac{5}{7}x^2 = \frac{1}{7}x^2(3x^2 + x - 5)$

6. $15p^2q^4 + 20p^3q^5 + 5p^3q^3 = 5p^2q^3(3q + 4pq^2 + p)$

EXAMPLE 7 Factor: $5(x + 3) + y(x + 3)$

Solution: The binomial $(x + 3)$ is present in both terms and is the greatest common factor. We use the distributive property to factor out $(x + 3)$.

$$5(x + 3) + y(x + 3) = (x + 3)(5 + y)$$

C Factoring by Grouping

Once the GCF is factored out, we can often continue to factor the polynomial, using a variety of techniques. We discuss here a technique called **factoring by grouping.** This technique can be used to factor some polynomials with four terms.

EXAMPLE 8 Factor $xy + 2x + 3y + 6$ by grouping.

Solution: The GCF of the first two terms is x, and the GCF of the last two terms is 3.

$$xy + 2x + 3y + 6 = (xy + 2x) + (3y + 6)$$
$$= x(y + 2) + 3(y + 2)$$

Helpful Hint

Notice that this is *not* a factored form of the original polynomial. It is a sum, not a product.

Next we factor out the common binomial factor, $(y + 2)$.

$$x(y + 2) + 3(y + 2) = (y + 2)(x + 3)$$

Practice Problem 3

Factor: $-10x^3 + 8x^2 - 2x$

Practice Problems 4–6

Factor.

4. $4x^3 + 12x$

5. $\frac{2}{5}a^5 - \frac{4}{5}a^3 + \frac{1}{5}a^2$

6. $6a^3b + 3a^3b^2 + 9a^2b^4$

Practice Problem 7

Factor: $7(p + 2) + q(p + 2)$

Practice Problem 8

Factor $ab + 7a + 2b + 14$ by grouping.

Answers

3. $-2x(5x^2 - 4x + 1)$ **4.** $4x(x^2 + 3)$
5. $\frac{1}{5}a^2(2a^3 - 4a + 1)$ **6.** $3a^2b(2a + ab + 3b^3)$
7. $(p + 2)(7 + q)$ **8.** $(b + 7)(a + 2)$

Check: Multiply $(y + 2)$ by $(x + 3)$.

$$(y + 2)(x + 3) = xy + 2x + 3y + 6,$$

the original polynomial.

Thus, the factored form of $xy + 2x + 3y + 6$ is the product $(y + 2)(x + 3)$.

Practice Problems 9–10

Factor by grouping.

9. $28x^3 - 7x^2 + 12x - 3$
10. $2xy + 5y^2 - 4x - 10y$

EXAMPLES Factor by grouping.

9. $15x^3 - 10x^2 + 6x - 4$

$= (15x^3 - 10x^2) + (6x - 4)$

$= 5x^2(3x - 2) + 2(3x - 2)$ Factor each group.

$= (3x - 2)(5x^2 + 2)$ Factor out the common factor, $(3x - 2)$.

10. $3x^2 + 4xy - 3x - 4y$

$= (3x^2 + 4xy) + (-3x - 4y)$ Factor each group. A -1 is factored from the second pair of terms so that there is a common factor, $(3x + 4y)$.

$= x(3x + 4y) - 1(3x + 4y)$

$= (3x + 4y)(x - 1)$ Factor out the common factor, $(3x + 4y)$.

Practice Problems 11–13

Factor by grouping.

11. $4x^3 + x - 20x^2 - 5$
12. $2x - 2 + x^3 - 3x^2$
13. $3xy - 4 + x - 12y$

EXAMPLES Factor by grouping.

11. $3x^3 - 2x - 9x^2 + 6$ Factor each group. A -3 is factored from the second pair of terms so that there is a common factor, $(3x^2 - 2)$.

$= x(3x^2 - 2) - 3(3x^2 - 2)$

$= (3x^2 - 2)(x - 3)$ Factor out the common factor, $(3x^2 - 2)$.

12. $5x - 10 + x^3 - x^2 = 5(x - 2) + x^2(x - 1)$

There is no common binomial factor that can now be factored out. No matter how we rearrange the terms, no grouping will lead to a common factor. Thus, this polynomial is not factorable by grouping.

13. $3xy + 2 - 3x - 2y$

Notice that the first two terms have no common factor other than 1. However, if we rearrange these terms, a grouping emerges that does lead to a common factor.

$3xy + 2 - 3x - 2y$

$= (3xy - 3x) + (-2y + 2)$

$= 3x(y - 1) - 2(y - 1)$ Factor -2 from the second group.

$= (y - 1)(3x - 2)$ Factor out the common factor, $(y - 1)$.

Helpful Hint

Throughout this chapter, we will be factoring polynomials. Even when the instructions do not so state, it is always a good idea to check your answers by multiplying.

Answers

9. $(4x - 1)(7x^2 + 3)$ **10.** $(2x + 5y)(y - 2)$
11. $(4x^2 + 1)(x - 5)$ **12.** can't be factored
13. $(3y + 1)(x - 4)$

Name ______________________ Section __________ Date __________

Mental Math

Find the GCF of each pair of integers.

1. 2, 16 **2.** 3, 18 **3.** 6, 15 **4.** 20, 15 **5.** 14, 35 **6.** 27, 36

EXERCISE SET 11.1

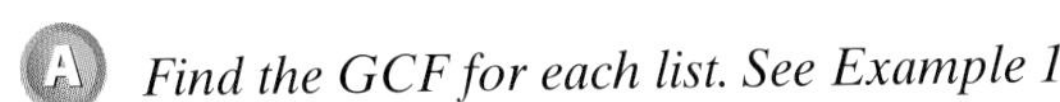

A *Find the GCF for each list. See Example 1.*

1. y^2, y^4, y^7

2. x^3, x^2, x^3

3. $x^{10}y^2, xy^2, x^3y^3$

4. p^7q, p^8q^2, p^9q^3

5. $8x, 4$

6. $9y, y$

7. $12y^4, 20y^3$

8. $32x, 18x^2$

9. $-10x^2, 15x^3$

10. $-21x^3, 14x$

11. $12x^3, -6x^4, 3x^5$

12. $15y^2, 5y^7, -20y^3$

13. $-18x^2y, 9x^3y^3, 36x^3y$

14. $7x, -21x^2y^2, 14xy$

B *Factor out the GCF from each polynomial. See Examples 2 through 7.*

15. $3a + 6$

16. $18a + 12$

17. $30x - 15$

18. $42x - 7$

19. $x^3 + 5x^2$

20. $y^5 - 6y^4$

21. $6y^4 - 2y$

22. $5x^2 + 10x^6$

23. $32xy - 18x^2$

24. $10xy - 15x^2$

25. $4x - 8y + 4$

26. $7x + 21y - 7$

27. $6x^3 - 9x^2 + 12x$

28. $12x^3 + 16x^2 - 8x$

29. $a^7b^6 - a^3b^2 + a^2b^5 - a^2b^2$

30. $x^9y^6 + x^3y^5 - x^4y^3 + x^3y^3$

31. $5x^3y - 15x^2y + 10xy$

32. $14x^3y + 7x^2y - 7xy$

33. $8x^5 + 16x^4 - 20x^3 + 12$

34. $9y^6 - 27y^4 + 18y^2 + 6$

35. $\frac{1}{3}x^4 + \frac{2}{3}x^3 - \frac{4}{3}x^5 + \frac{1}{3}x$

36. $\frac{2}{5}y^7 - \frac{4}{5}y^5 + \frac{3}{5}y^2 - \frac{2}{5}y$

37. $y(x + 2) + 3(x + 2)$

38. $z(y + 4) + 3(y + 4)$

39. $8(x + 2) - y(x + 2)$

40. $x(y^2 + 1) - 3(y^2 + 1)$

41. Construct a binomial whose greatest common factor is $5a^3$. (*Hint:* Multiply $5a^3$ by a binomial whose terms contain no common factor other than 1. $5a^3(\square + \square)$.)

42. Construct a trinomial whose greatest common factor is $2x^2$. See the hint for Exercise 41.

C *Factor each four-term polynomial by grouping. See Examples 8 through 13.*

43. $x^3 + 2x^2 + 5x + 10$

44. $x^3 + 4x^2 + 3x + 12$

45. $5x + 15 + xy + 3y$

46. $xy + y + 2x + 2$

47. $6x^3 - 4x^2 + 15x - 10$

48. $16x^3 - 28x^2 + 12x - 21$

49. $2y - 8 + xy - 4x$

50. $6x - 42 + xy - 7y$

51. $2x^3 + x^2 + 8x + 4$

52. $2x^3 - x^2 - 10x + 5$

53. $4x^2 - 8xy - 3x + 6y$

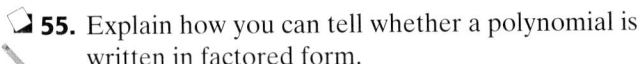

54. $5xy - 15x - 6y + 18$

55. Explain how you can tell whether a polynomial is written in factored form.

56. Construct a four-term polynomial that can be factored by grouping.

Review and Preview

Multiply. See Section 10.5.

57. $(x + 2)(x + 5)$

58. $(y + 3)(y + 6)$

59. $(b + 1)(b - 4)$

60. $(x - 5)(x + 10)$

Fill in the chart by finding two numbers that have the given product and sum. The first column is filled in for you.

		61.	62.	63.	64.	65.	66.	67.	68.
Two Numbers	4, 7								
Their Product	28	12	20	8	16	−10	−9	−24	−36
Their Sum	11	8	9	−9	−10	3	0	−5	−5

Combining Concepts

Factor out the GCF from each polynomial. Then factor by grouping.

69. $12x^2y - 42x^2 - 4y + 14$

70. $90 + 15y^2 - 18x - 3xy^2$

Write an expression for the area of each shaded region. Then write the expression as a factored polynomial.

△ **71.**

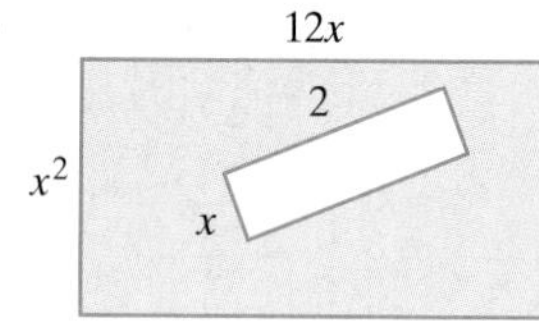

△ **72.**

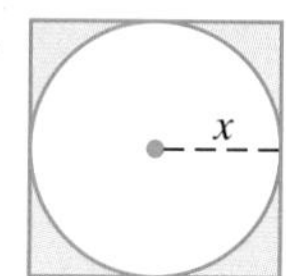

Write an expression for the length of each rectangle. (Hint: Factor the area binomial and recall that Area = width · length.)

△ **73.**

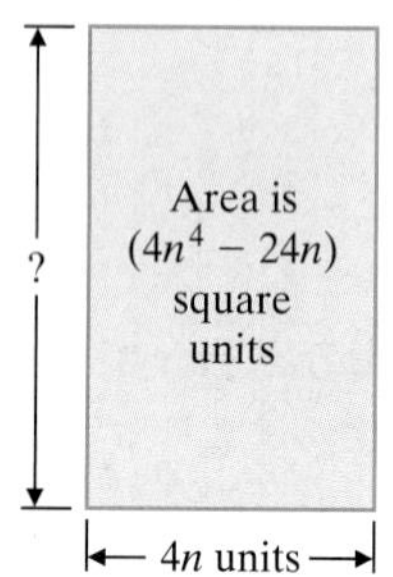

△ **74.**

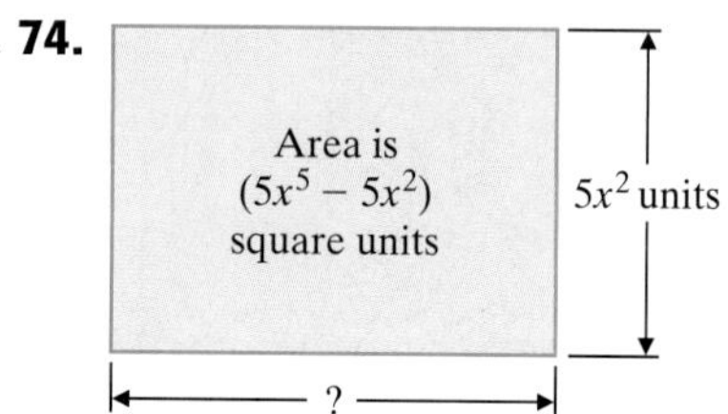

75. The nonresidential sales of electricity (in billion kilowatt hours) in the United States during 1998–2000 can be approximated by the polynomial $-8x^2 + 60x + 2000$, where x is the number of years after 1998. (*Source:* Energy Information Administration)

a. Find the amount of electricity sold in 1998. To do so, let $x = 0$ and evaluate $-8x^2 + 60x + 2000$.

b. Find the amount of electricity sold in 2000.

c. Factor the polynomial $-8x^2 + 60x + 2000$.

76. The average daily total supply of motor gasoline (in thousand barrels per day) in the United States for the period 1998–2000 can be approximated by the polynomial $-40x^2 + 140x + 8000$, where x is the number of years after 1998. (*Source:* Energy Information Administration)

a. Find the average daily total supply of motor gasoline in 1999. To do so, let $x = 1$ and evaluate $-40x^2 + 140x + 8000$.

b. Find the average daily total supply of motor gasoline in 2000.

c. Factor the polynomial $-40x^2 + 140x + 8000$.

11.2 Factoring Trinomials of the Form $x^2 + bx + c$

OBJECTIVES

A. Factor trinomials of the form $x^2 + bx + c$.

B. Factor out the greatest common factor and then factor a trinomial of the form $x^2 + bx + c$.

SSM TUTOR CENTER SG CD & VIDEO MATH PRO WEB

A Factoring Trinomials of the Form $x^2 + bx + c$

In this section, we factor trinomials of the form $x^2 + bx + c$, such as

$$x^2 + 4x + 3, \quad x^2 - 8x + 15, \quad x^2 + 4x - 12, \quad \text{and} \quad r^2 - r - 42$$

Notice that for these trinomials, the coefficient of the squared variable is 1.

Recall that factoring means to write as a product and that factoring and multiplying are reverse processes. Using the FOIL method of multiplying binomials, we have the following.

$$\begin{aligned}(x + 3)(x + 1) &= \overset{F}{x^2} + \overset{O}{1x} + \overset{I}{3x} + \overset{L}{3} \\ &= x^2 + 4x + 3\end{aligned}$$

Thus, a factored form of $x^2 + 4x + 3$ is $(x + 3)(x + 1)$.

Notice that the product of the first terms of the binomials is $x \cdot x = x^2$, the first term of the trinomial. Also, the product of the last two terms of the binomials is $3 \cdot 1 = 3$, the third term of the trinomial. The sum of these same terms is $3 + 1 = 4$, the coefficient of the middle, x, term of the trinomial.

The product of these numbers is 3.

$$x^2 + 4x + 3 = (x + 3)(x + 1)$$

The sum of these numbers is 4.

Many trinomials, such as the one above, factor into two binomials. To factor $x^2 + 7x + 10$, let's assume that it factors into two binomials and begin by writing two pairs of parentheses. The first term of the trinomial is x^2, so we use x and x as the first terms of the binomial factors.

$$x^2 + 7x + 10 = (x + \square)(x + \square)$$

To determine the last term of each binomial factor, we look for two integers whose product is 10 and whose sum is 7. The integers are 2 and 5. Thus,

$$x^2 + 7x + 10 = (x + 2)(x + 5)$$

To see if we have factored correctly, we multiply.

$$\begin{aligned}(x + 2)(x + 5) &= x^2 + 5x + 2x + 10 \\ &= x^2 + 7x + 10 \quad \text{Combine like terms.}\end{aligned}$$

Helpful Hint

Since multiplication is commutative, the factored form of $x^2 + 7x + 10$ can be written as either $(x + 2)(x + 5)$ or $(x + 5)(x + 2)$.

To Factor a Trinomial of the Form $x^2 + bx + c$

The product of these numbers is c.

$$x^2 + bx + c = (x + \square)(x + \square)$$

The sum of these numbers is b.

Practice Problem 1

Factor: $x^2 + 9x + 20$

EXAMPLE 1 Factor: $x^2 + 7x + 12$

Solution: We begin by writing the first terms of the binomial factors.

$$(x + \square)(x + \square)$$

Next we look for two numbers whose product is 12 and whose sum is 7. Since our numbers must have a positive product and a positive sum, we look at pairs of positive factors of 12 only.

Factors of 12	Sum of Factors
1, 12	13
2, 6	8
3, 4	7

Correct sum, so the numbers are 3 and 4.

$$x^2 + 7x + 12 = (x + 3)(x + 4)$$

Check: Multiply $(x + 3)$ by $(x + 4)$.

Practice Problem 2

Factor each trinomial.

a. $x^2 - 13x + 22$

b. $x^2 - 27x + 50$

EXAMPLE 2 Factor: $x^2 - 8x + 15$

Solution: Again, we begin by writing the first terms of the binomials.

$$(x + \square)(x + \square)$$

Now we look for two numbers whose product is 15 and whose sum is -8. Since our numbers must have a positive product and a negative sum, we look at pairs of negative factors of 15 only.

Factors of 15	Sum of Factors
$-1, -15$	-16
$-3, -5$	-8

Correct sum, so the numbers are -3 and -5.

$$x^2 - 8x + 15 = (x - 3)(x - 5)$$

Practice Problem 3

Factor: $x^2 + 5x - 36$

EXAMPLE 3 Factor: $x^2 + 4x - 12$

Solution: $x^2 + 4x - 12 = (x + \square)(x + \square)$

Answers

1. $(x + 4)(x + 5)$ **2. a.** $(x - 2)(x - 11)$ **b.** $(x - 2)(x - 25)$ **3.** $(x + 9)(x - 4)$

We look for two numbers whose product is -12 and whose sum is 4. Since our numbers must have a negative product, we look at pairs of factors with opposite signs.

Factors of -12	Sum of Factors
$-1, 12$	11
$1, -12$	-11
$-2, 6$	4
$2, -6$	-4
$-3, 4$	1
$3, -4$	-1

Correct sum, so the numbers are -2 and 6.

$$x^2 + 4x - 12 = (x - 2)(x + 6)$$

EXAMPLE 4 Factor: $r^2 - r - 42$

Solution: Because the variable in this trinomial is r, the first term of each binomial factor is r.

$$r^2 - r - 42 = (r + \square)(r + \square)$$

Now we look for two numbers whose product is -42 and whose sum is -1, the numerical coefficient of r. The numbers are 6 and -7. Therefore,

$$r^2 - r - 42 = (r + 6)(r - 7)$$

Practice Problem 4

Factor each trinomial.

a. $q^2 - 3q - 40$

b. $y^2 + 2y - 48$

EXAMPLE 5 Factor: $a^2 + 2a + 10$

Solution: Look for two numbers whose product is 10 and whose sum is 2. Neither 1 and 10 nor 2 and 5 give the required sum, 2. We conclude that $a^2 + 2a + 10$ is not factorable with integers. A polynomial such as $a^2 + 2a + 10$ is called a **prime polynomial.**

Practice Problem 5

Factor: $x^2 + 6x + 15$

EXAMPLE 6 Factor: $x^2 + 5xy + 6y^2$

Solution: $x^2 + 5xy + 6y^2 = (x + \square)(x + \square)$

Recall that the middle term $5xy$ is the same as $5yx$. Notice that $5y$ is the "coefficient" of x. We then look for two terms whose product is $6y^2$ and whose sum is $5y$. The terms are $2y$ and $3y$ because $2y \cdot 3y = 6y^2$ and $2y + 3y = 5y$. Therefore,

$$x^2 + 5xy + 6y^2 = (x + 2y)(x + 3y)$$

Practice Problem 6

Factor each trinomial.

a. $x^2 + 6xy + 8y^2$

b. $a^2 - 13ab + 30b^2$

EXAMPLE 7 Factor: $x^4 + 5x^2 + 6$

Solution: As usual, we begin by writing the first terms of the binomials. Since the greatest power of x in this polynomial is x^4, we write

$$(x^2 + \square)(x^2 + \square) \quad \text{since } x^2 \cdot x^2 = x^4$$

Now we look for two factors of 6 whose sum is 5. The numbers are 2 and 3. Thus,

$$x^4 + 5x^2 + 6 = (x^2 + 2)(x^2 + 3)$$

Practice Problem 7

Factor: $x^4 + 8x^2 + 12$

Answers

4. a. $(q - 8)(q + 5)$ **b.** $(y + 8)(y - 6)$
5. prime polynomial **6. a.** $(x + 2y)(x + 4y)$
b. $(a - 3b)(a - 10b)$ **7.** $(x^2 + 6)(x^2 + 2)$

The following sign patterns may be useful when factoring trinomials.

Helpful Hint

A positive constant in a trinomial tells us to look for two numbers with the same sign. The sign of the coefficient of the middle term tells us whether the signs are both positive or both negative.

both positive; same sign

$x^2 + 10x + 16 = (x + 2)(x + 8)$

both negative; same sign

$x^2 - 10x + 16 = (x - 2)(x - 8)$

A negative constant in a trinomial tells us to look for two numbers with opposite signs.

opposite signs

$x^2 + 6x - 16 = (x + 8)(x - 2)$

opposite signs

$x^2 - 6x - 16 = (x - 8)(x + 2)$

B Factoring Out the Greatest Common Factor

Remember that the first step in factoring any polynomial is to factor out the greatest common factor (if there is one other than 1 or -1).

EXAMPLE 8 Factor: $3m^2 - 24m - 60$

Solution: First we factor out the greatest common factor, 3, from each term.

$$3m^2 - 24m - 60 = 3(m^2 - 8m - 20)$$

Now we factor $m^2 - 8m - 20$ by looking for two factors of -20 whose sum is -8. The factors are -10 and 2. Therefore, the complete factored form is

$$3m^2 - 24m - 60 = 3(m + 2)(m - 10)$$

Helpful Hint

Remember to write the common factor 3 as part of the factored form.

Practice Problem 8

Factor each trinomial.

a. $x^3 + 3x^2 - 4x$

b. $4x^2 - 24x + 36$

EXAMPLE 9 Factor: $2x^4 - 26x^3 + 84x^2$

Solution:

$$2x^4 - 26x^3 + 84x^2 = 2x^2(x^2 - 13x + 42) \quad \text{Factor out common factor of } 2x^2.$$
$$= 2x^2(x - 6)(x - 7) \quad \text{Factor } x^2 - 13x + 42.$$

Practice Problem 9

Factor: $5x^5 - 25x^4 - 30x^3$

Answers

8. a. $x(x + 4)(x - 1)$ **b.** $4(x - 3)(x - 3)$
9. $5x^3(x + 1)(x - 6)$

Name ______________ Section ________ Date ________

Mental Math

Complete each factored form.

1. $x^2 + 9x + 20 = (x + 4)(x \quad)$ **2.** $x^2 + 12x + 35 = (x + 5)(x \quad)$ **3.** $x^2 - 7x + 12 = (x - 4)(x \quad)$

4. $x^2 - 13x + 22 = (x - 2)(x \quad)$ **5.** $x^2 + 4x + 4 = (x + 2)(x \quad)$ **6.** $x^2 + 10x + 24 = (x + 6)(x \quad)$

EXERCISE SET 11.2

A *Factor each trinomial completely. If a polynomial can't be factored, write "prime." See Examples 1 through 7.*

1. $x^2 + 7x + 6$ **2.** $x^2 + 6x + 8$ **3.** $x^2 - 10x + 9$ **4.** $x^2 - 6x + 9$ **5.** $x^2 - 3x - 18$

6. $x^2 - x - 30$ **7.** $x^2 + 3x - 70$ **8.** $x^2 + 4x - 32$ **9.** $x^2 + 5x + 2$ **10.** $x^2 - 7x + 5$

11. $x^2 + 8xy + 15y^2$ **12.** $x^2 + 6xy + 8y^2$ **13.** $a^4 - 2a^2 - 15$ **14.** $y^4 - 3y^2 - 70$

15. Write a polynomial that factors as $(x - 3)(x + 8)$.

16. To factor $x^2 + 13x + 42$, think of two numbers whose ________ is 42 and whose ________ is 13.

Complete each sentence in your own words.

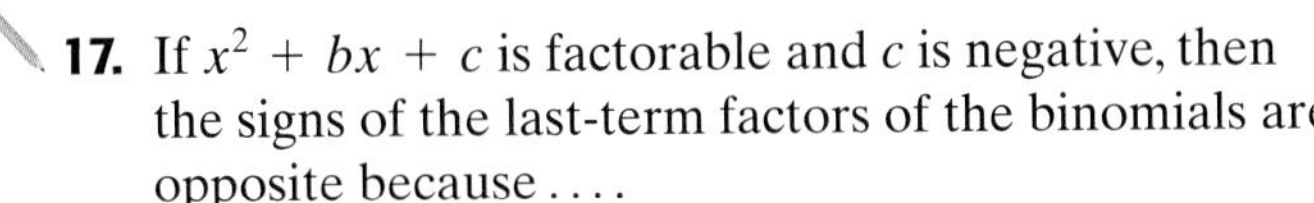

17. If $x^2 + bx + c$ is factorable and c is negative, then the signs of the last-term factors of the binomials are opposite because . . . .

18. If $x^2 + bx + c$ is factorable and c is positive, then the signs of the last-term factors of the binomials are the same because . . . .

B *Factor each trinomial completely. See Examples 1 through 9.*

19. $2z^2 + 20z + 32$ **20.** $3x^2 + 30x + 63$ **21.** $2x^3 - 18x^2 + 40x$ **22.** $x^3 - x^2 - 56x$

23. $x^2 - 3xy - 4y^2$ **24.** $x^2 - 4xy - 77y^2$ **25.** $x^2 + 15x + 36$ **26.** $x^2 + 19x + 60$

27. $x^2 - x - 2$ **28.** $x^2 - 5x - 14$ **29.** $r^2 - 16r + 48$ **30.** $r^2 - 10r + 21$

31. $x^2 + xy - 2y^2$ **32.** $x^2 - xy - 6y^2$ **33.** $3x^2 + 9x - 30$ **34.** $4x^2 - 4x - 48$

35. $3x^2 - 60x + 108$ **36.** $2x^2 - 24x + 70$ **37.** $x^2 - 18x - 144$ **38.** $x^2 + x - 42$

39. $r^2 - 3r + 6$

40. $x^2 + 4x - 10$

41. $x^2 - 8x + 15$

42. $x^2 - 9x + 14$

43. $6x^3 + 54x^2 + 120x$

44. $3x^3 + 3x^2 - 126x$

45. $4x^2y + 4xy - 12y$

46. $3x^2y - 9xy + 45y$

47. $x^2 - 4x - 21$

48. $x^2 - 4x - 32$

49. $x^2 + 7xy + 10y^2$

50. $x^2 - 3xy - 4y^2$

51. $64 + 24t + 2t^2$

52. $50 + 20t + 2t^2$

53. $x^3 - 2x^2 - 24x$

54. $x^3 - 3x^2 - 28x$

55. $2t^5 - 14t^4 + 24t^3$

56. $3x^6 + 30x^5 + 72x^4$

57. $5x^3y - 25x^2y^2 - 120xy^3$

58. $3x^2 - 6xy - 72y^2$

Review and Preview

Multiply. See Section 10.5.

59. $(2x + 1)(x + 5)$

60. $(3x + 2)(x + 4)$

61. $(5y - 4)(3y - 1)$

62. $(4z - 7)(7z - 1)$

63. $(a + 3)(9a - 4)$

64. $(y - 5)(6y + 5)$

Combining Concepts

Write the perimeter of each rectangle as a simplified polynomial. Then factor the polynomial.

△ 65.

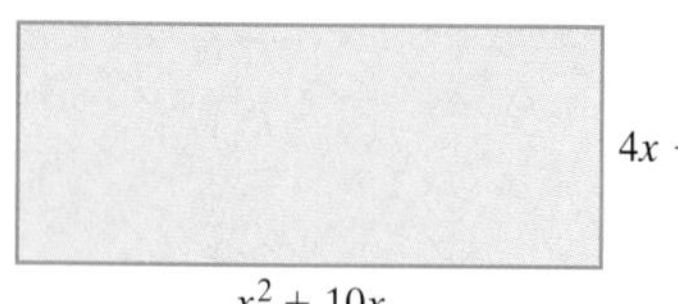

△ 66.

Factor each trinomial completely.

67. $y^2(x + 1) - 2y(x + 1) - 15(x + 1)$

68. $z^2(x + 1) - 3z(x + 1) - 70(x + 1)$

Find a positive value of c so that each trinomial is factorable.

69. $y^2 - 4y + c$

70. $n^2 - 16n + c$

Find a positive value of b so that each trinomial is factorable.

71. $x^2 + bx + 15$

72. $y^2 + by + 20$

Factor each trinomial. [Hint: Notice that $x^{2n} + 4x^n + 3$ factors as $(x^n + 1)(x^n + 3)$.]

73. $x^{2n} + 5x^n + 6$

74. $x^{2n} + 8x^n - 20$

11.3 Factoring Trinomials of the Form $ax^2 + bx + c$

OBJECTIVES

A Factor trinomials of the form $ax^2 + bx + c$, where $a \neq 1$.

B Factor out the GCF before factoring a trinomial of the form $ax^2 + bx + c$.

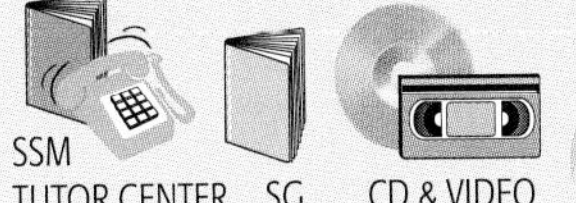

A Factoring Trinomials of the Form $ax^2 + bx + c$

In this section, we factor trinomials of the form $ax^2 + bx + c$, such as

$$3x^2 + 11x + 6, \quad 8x^2 - 22x + 5, \quad \text{and} \quad 2x^2 + 13x - 7$$

Notice that the coefficient of the squared variable in these trinomials is a number other than 1. We will factor these trinomials using a trial-and-check method based on our work in the last section.

To begin, let's review the relationship between the numerical coefficients of the trinomial and the numerical coefficients of its factored form. For example, since $(2x + 1)(x + 6) = 2x^2 + 13x + 6$, the factored form of $2x^2 + 13x + 6$ is

$$2x^2 + 13x + 6 = (2x + 1)(x + 6)$$

Notice that $2x$ and x are factors of $2x^2$, the first term of the trinomial. Also, 6 and 1 are factors of 6, the last term of the trinomial, as shown:

$$2x^2 + 13x + 6 = (2x + 1)(x + 6)$$

$2x \cdot x$

$1 \cdot 6$

Also notice that $13x$, the middle term, is the sum of the following products:

$$2x^2 + 13x + 6 = (2x + 1)(x + 6)$$

$$\begin{array}{r} 1x \\ + 12x \\ \hline 13x \end{array}$$ Middle term

Let's use this pattern to factor $5x^2 + 7x + 2$. First, we find factors of $5x^2$. Since all numerical coefficients in this trinomial are positive, we will use factors with positive numerical coefficients only. Thus, the factors of $5x^2$ are $5x$ and x. Let's try these factors as first terms of the binomials. Thus far, we have

$$5x^2 + 7x + 2 = (5x + \square)(x + \square)$$

Next, we need to find positive factors of 2. Positive factors of 2 are 1 and 2. Now we try possible combinations of these factors as second terms of the binomials until we obtain a middle term of $7x$.

$$(5x + 1)(x + 2) = 5x^2 + 11x + 2$$

$$\begin{array}{r} 1x \\ + 10x \\ \hline 11x \end{array}$$ → **Incorrect** middle term

Let's try switching factors 2 and 1.

$$(5x + 2)(x + 1) = 5x^2 + 7x + 2$$

$$\begin{array}{r} 2x \\ + 5x \\ \hline 7x \end{array}$$ → **Correct** middle term

Thus the factored form of $5x^2 + 7x + 2$ is $(5x + 2)(x + 1)$. To check, we multiply $(5x + 2)$ and $(x + 1)$. The product is $5x^2 + 7x + 2$.

Practice Problem 1

Factor each trinomial.

a. $4x^2 + 12x + 5$
b. $5x^2 + 27x + 10$

Concept Check

Do the terms of $3x^2 + 29x + 18$ have a common factor? Without multiplying, decide which of the following factored forms could not be a factored form of $3x^2 + 29x + 18$.

a. $(3x + 18)(x + 1)$
b. $(3x + 2)(x + 9)$
c. $(3x + 6)(x + 3)$
d. $(3x + 9)(x + 2)$

Practice Problem 2

Factor each trinomial.

a. $6x^2 - 5x + 1$
b. $2x^2 - 11x + 12$

Answers

1. a. $(2x + 5)(2x + 1)$ **b.** $(5x + 2)(x + 5)$
2. a. $(3x - 1)(2x - 1)$ **b.** $(2x - 3)(x - 4)$

Concept Check: no; a, c, d

EXAMPLE 1 Factor: $3x^2 + 11x + 6$

Solution: Since all numerical coefficients are positive, we use factors with positive numerical coefficients. We first find factors of $3x^2$.

$$\text{Factors of } 3x^2\text{:} \quad 3x^2 = 3x \cdot x$$

If factorable, the trinomial will be of the form

$$3x^2 + 11x + 6 = (3x + \square)(x + \square)$$

Next we factor 6.

$$\text{Factors of 6:} \quad 6 = 1 \cdot 6, \qquad 6 = 2 \cdot 3$$

Now we try combinations of factors of 6 until a middle term of $11x$ is obtained. Let's try 1 and 6 first.

$$(3x + 1)(x + 6) = 3x^2 + 19x + 6$$

$$\begin{array}{r} 1x \\ +\ 18x \\ \hline 19x \end{array} \longrightarrow \textbf{Incorrect} \text{ middle term}$$

Now let's next try 6 and 1.

$$(3x + 6)(x + 1)$$

Before multiplying, notice that the terms of the factor $3x + 6$ have a common factor of 3. The terms of the original trinomial $3x^2 + 11x + 6$ have no common factor other than 1, so the terms of its factors will also contain no common factor other than 1. This means that $(3x + 6)(x + 1)$ is not the correct factored form.

Next let's try 2 and 3 as last terms.

$$(3x + 2)(x + 3) = 3x^2 + 11x + 6$$

$$\begin{array}{r} 2x \\ +\ 9x \\ \hline 11x \end{array} \longrightarrow \textbf{Correct} \text{ middle term}$$

Thus the factored form of $3x^2 + 11x + 6$ is $(3x + 2)(x + 3)$. ●

Helpful Hint

If the terms of a trinomial have no common factor (other than 1), then the terms of each of its binomial factors will contain no common factor (other than 1).

Try the Concept Check in the margin.

EXAMPLE 2 Factor: $8x^2 - 22x + 5$

Solution: Factors of $8x^2$: $8x^2 = 8x \cdot x$, $\quad 8x^2 = 4x \cdot 2x$

We'll try $8x$ and x.

$$8x^2 - 22x + 5 = (8x + \square)(x + \square)$$

Since the middle term, $-22x$, has a negative numerical coefficient, we factor 5 into negative factors.

Factors of 5: $5 = -1 \cdot -5$

Let's try -1 and -5.

$$(8x - 1)(x - 5) = 8x^2 - 41x + 5$$

$-1x$
$+ (-40x)$
$-41x$ → **Incorrect** middle term

Now let's try -5 and -1.

$$(8x - 5)(x - 1) = 8x^2 - 13x + 5$$

$-5x$
$+ (-8x)$
$-13x$ → **Incorrect** middle term

Don't give up yet! We can still try other factors of $8x^2$. Let's try $4x$ and $2x$ with -1 and -5.

$$(4x - 1)(2x - 5) = 8x^2 - 22x + 5$$

$-2x$
$+ (-20x)$
$-22x$ → **Correct** middle term

The factored form of $8x^2 - 22x + 5$ is $(4x - 1)(2x - 5)$.

EXAMPLE 3 Factor: $2x^2 + 13x - 7$

Solution: Factors of $2x^2$: $2x^2 = 2x \cdot x$

Factors of -7: $-7 = -1 \cdot 7$, $-7 = 1 \cdot -7$

We try possible combinations of these factors:

$(2x + 1)(x - 7) = 2x^2 - 13x - 7$ **Incorrect** middle term

$(2x - 1)(x + 7) = 2x^2 + 13x - 7$ **Correct** middle term

The factored form of $2x^2 + 13x - 7$ is $(2x - 1)(x + 7)$.

EXAMPLE 4 Factor: $10x^2 - 13xy - 3y^2$

Solution: Factors of $10x^2$: $10x^2 = 10x \cdot x$, $10x^2 = 2x \cdot 5x$

Factors of $-3y^2$: $-3y^2 = -3y \cdot y$, $-3y^2 = 3y \cdot -y$

We try some combinations of these factors:

$(10x - 3y)(x + y) = 10x^2 + 7xy - 3y^2$

$(x + 3y)(10x - y) = 10x^2 + 29xy - 3y^2$

$(5x + 3y)(2x - y) = 10x^2 + xy - 3y^2$

$(2x - 3y)(5x + y) = 10x^2 - 13xy - 3y^2$ **Correct** middle term

The factored form of $10x^2 - 13xy - 3y^2$ is $(2x - 3y)(5x + y)$.

Practice Problem 3

Factor each trinomial.

a. $35x^2 + 4x - 4$

b. $4x^2 + 3x - 7$

Practice Problem 4

Factor each trinomial.

a. $14x^2 - 3xy - 2y^2$

b. $12a^2 - 16ab - 3b^2$

Answers

3. a. $(5x + 2)(7x - 2)$ **b.** $(4x + 7)(x - 1)$

4. a. $(7x + 2y)(2x - y)$ **b.** $(6a + b)(2a - 3b)$

B Factoring Out the Greatest Common Factor

Don't forget that the first step in factoring any polynomial is to look for a common factor to factor out.

Practice Problem 5

Factor each trinomial.

a. $3x^3 + 17x^2 + 10x$

b. $6xy^2 + 33xy - 18x$

EXAMPLE 5 Factor: $24x^4 + 40x^3 + 6x^2$

Solution: Notice that all three terms have a common factor of $2x^2$. Thus we factor out $2x^2$ first.

$$24x^4 + 40x^3 + 6x^2 = 2x^2(12x^2 + 20x + 3)$$

Next we factor $12x^2 + 20x + 3$.

Factors of $12x^2$: $12x^2 = 4x \cdot 3x$, $12x^2 = 12x \cdot x$, $12x^2 = 6x \cdot 2x$

Since all terms in the trinomial have positive numerical coefficients, we factor 3 using positive factors only.

Factors of 3: $3 = 1 \cdot 3$

We try some combinations of the factors.

$$2x^2(4x + 3)(3x + 1) = 2x^2(12x^2 + 13x + 3)$$
$$2x^2(12x + 1)(x + 3) = 2x^2(12x^2 + 37x + 3)$$
$$2x^2(2x + 3)(6x + 1) = 2x^2(12x^2 + 20x + 3) \quad \text{Correct middle term}$$

The factored form of $24x^4 + 40x^3 + 6x^2$ is $2x^2(2x + 3)(6x + 1)$.

Don't forget to include the common factor in the factored form.

When the term containing the squared variable has a negative coefficient, you may want to first factor out a common factor of -1.

Practice Problem 6

Factor: $-5x^2 - 19x + 4$

EXAMPLE 6 Factor: $-6x^2 - 13x + 5$

Solution: We begin by factoring out a common factor of -1.

$$-6x^2 - 13x + 5 = -1(6x^2 + 13x - 5) \quad \text{Factor out } -1.$$
$$= -1(3x - 1)(2x + 5) \quad \text{Factor } 6x^2 + 13x - 5.$$

Answers

5. a. $x(3x + 2)(x + 5)$ **b.** $3x(2y - 1)(y + 6)$

6. $-1(x + 4)(5x - 1)$

Name ______________________ Section __________ Date __________

EXERCISE SET 11.3

 *Complete each factored form.*

1. $5x^2 + 22x + 8 = (5x + 2)(\quad)$

2. $2y^2 + 15y + 25 = (2y + 5)(\quad)$

3. $50x^2 + 15x - 2 = (5x + 2)(\quad)$

4. $6y^2 + 11y - 10 = (2y + 5)(\quad)$

5. $20x^2 - 7x - 6 = (5x + 2)(\quad)$

6. $8y^2 - 2y - 55 = (2y + 5)(\quad)$

Factor each trinomial completely. See Examples 1 through 4.

7. $2x^2 + 13x + 15$

8. $3x^2 + 8x + 4$

9. $8y^2 - 17y + 9$

10. $21x^2 - 41x + 10$

11. $2x^2 - 9x - 5$

12. $36r^2 - 5r - 24$

13. $20r^2 + 27r - 8$

14. $3x^2 + 20x - 63$

15. $10x^2 + 17x + 3$

16. $2x^2 + 7x + 5$

17. $x + 3x^2 - 2$

18. $y + 8y^2 - 9$

19. $6x^2 - 13xy + 5y^2$

20. $8x^2 - 14xy + 3y^2$

21. $15x^2 - 16x - 15$

22. $25x^2 - 5x - 6$

23. $-9x + 20 + x^2$

24. $-7x + 12 + x^2$

25. $2x^2 - 7x - 99$

26. $2x^2 + 7x - 72$

27. $-27t + 7t^2 - 4$

28. $4t^2 - 7 - 3t$

29. $3a^2 + 10ab + 3b^2$

30. $2a^2 + 11ab + 5b^2$

31. $49x^2 - 7x - 2$

32. $3x^2 + 10x - 8$

33. $18x^2 - 9x - 14$

34. $42a^2 - 43a + 6$

 *Factor each trinomial completely. See Examples 1 through 6.*

35. $12x^3 + 11x^2 + 2x$

36. $8a^3 + 14a^2 + 3a$

37. $21x^2 - 48x - 45$

38. $12x^2 - 14x - 10$

39. $7x + 12x^2 - 12$

40. $16x + 15x^2 - 15$

41. $6x^2y^2 - 2xy^2 - 60y^2$

42. $8x^2y + 34xy - 84y$

43. $4x^2 - 8x - 21$

44. $6x^2 - 11x - 10$

45. $3x^2 - 42x + 63$

46. $5x^2 - 75x + 60$

47. $8x^2 + 6x - 27$

48. $-x^2 + 4x + 21$

49. $-x^2 + 2x + 24$

50. $54a^2 + 39ab - 8b^2$

51. $4x^3 - 9x^2 - 9x$

52. $6x^3 - 31x^2 + 5x$

53. $24x^2 - 58x + 9$

54. $36x^2 + 55x - 14$

55. $40a^2b + 9ab - 9b$

56. $24y^2x + 7yx - 5x$

57. $15x^4 + 19x^2 + 6$

58. $6x^3 - 28x^2 + 16x$

59. $6y^3 - 8y^2 - 30y$

60. $12x^3 - 34x^2 + 24x$

61. $10x^3 + 25x^2y - 15xy^2$

62. $42x^4 - 99x^3y - 15x^2y^2$

63. $-14x^2 + 39x - 10$

64. $-15x^2 + 26x - 8$

The following graph shows the national unemployment rate for the United States for the months February 2003 through July 2003. Use the graph to answer Exercises 65 through 68. See Section 7.1.

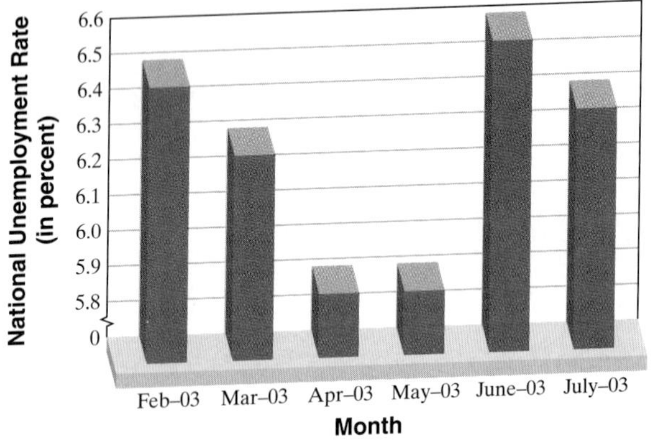

Source: U.S. Bureau of Labor Statistics

65. During which month(s) was the unemployment rate 5.8%?

66. During which month was the unemployment rate the highest?

67. By how much did the unemployment rate change from February to July?

68. Describe any trend you notice from this graph.

Combining Concepts

Factor each trinomial completely.

69. $4x^2(y - 1)^2 + 10x(y - 1)^2 + 25(y - 1)^2$

70. $3x^2(a + 3)^3 - 28x(a + 3)^3 + 25(a + 3)^3$

71. $-12x^3y^2 + 3x^2y^2 + 15xy^2$
(*Hint:* Begin by factoring out $-3xy^2$.)

Find a positive value of b so that each trinomial is factorable.

72. $3x^2 + bx - 5$

73. $2z^2 + bz - 7$

Find a positive value of c so that each trinomial is factorable.

74. $5x^2 + 7x + c$

75. $3x^2 - 8x + c$

76. In your own words, describe the steps you will use to factor a trinomial.

11.4 Factoring Trinomials of the Form $ax^2 + bx + c$, by Grouping

OBJECTIVE

A Use the grouping method to factor trinomials of the form $ax^2 + bx + c$.

SSM TUTOR CENTER SG CD & VIDEO MATH PRO WEB

A Using the Grouping Method

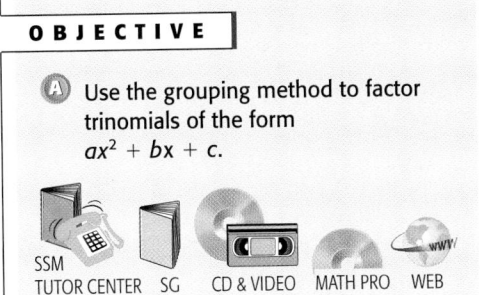

There is an alternative method that can be used to factor trinomials of the form $ax^2 + bx + c, a \neq 1$. This method is called the **grouping method** because it uses factoring by grouping as we learned in Section 11.1.

To see how this method works, let's multiply the following:

$$(2x + 1)(3x + 5) = 6x^2 + 10x + 3x + 5$$

$$10 \cdot 3 = 30$$

$$6 \cdot 5 = 30$$

$$= 6x^2 + 13x + 5$$

Notice that the product of the coefficients of the first and last terms is $6 \cdot 5 = 30$. This is the same as the product of the coefficients of the two middle terms, $10 \cdot 3 = 30$.

Let's use this pattern to write $2x^2 + 11x + 12$ as a four-term polynomial. We will then factor the polynomial by grouping.

$$2x^2 + 11x + 12 \quad \text{Find two numbers whose product is } 2 \cdot 12 = 24$$

$$= 2x^2 + \square x + \square x + 12 \quad \text{and whose sum is 11.}$$

Since we want a positive product and a positive sum, we consider pairs of positive factors of 24 only.

Factors of 24	Sum of Factors	
1, 24	25	
2, 12	14	
3, 8	11	Correct sum

The factors are 3 and 8. Now we use these factors to write the middle term $11x$ as $3x + 8x$ (or $8x + 3x$). We replace $11x$ with $3x + 8x$ in the original trinomial and then we can factor by grouping.

$$\begin{aligned} 2x^2 + 11x + 12 &= 2x^2 + 3x + 8x + 12 \\ &= (2x^2 + 3x) + (8x + 12) && \text{Group the terms.} \\ &= x(2x + 3) + 4(2x + 3) && \text{Factor each group.} \\ &= (2x + 3)(x + 4) && \text{Factor out } (2x + 3). \end{aligned}$$

In general, we have the following procedure.

To Factor Trinomials by Grouping

Step 1. Factor out a greatest common factor, if there is one other than 1.
Step 2. For the resulting trinomial $ax^2 + bx + c$, find two numbers whose product is $a \cdot c$ and whose sum is b.
Step 3. Write the middle term, bx, using the factors found in Step 2.
Step 4. Factor by grouping.

Practice Problem 1

Factor each trinomial by grouping.

a. $3x^2 + 14x + 8$

b. $12x^2 + 19x + 5$

EXAMPLE 1 Factor $8x^2 - 14x + 5$ by grouping.

Solution:

Step 1. The terms of this trinomial contain no greatest common factor other than 1.

Step 2. This trinomial is of the form $ax^2 + bx + c$ with $a = 8$, $b = -14$, and $c = 5$. Find two numbers whose product is $a \cdot c$ or $8 \cdot 5 = 40$, and whose sum is b or -14. The numbers are -4 and -10.

Step 3. Write $-14x$ as $-4x - 10x$ so that

$$8x^2 - 14x + 5 = 8x^2 - 4x - 10x + 5$$

Step 4. Factor by grouping.

$$\begin{aligned} 8x^2 - 4x - 10x + 5 &= 4x(2x - 1) - 5(2x - 1) \\ &= (2x - 1)(4x - 5) \end{aligned}$$

Practice Problem 2

Factor each trinomial by grouping.

a. $6x^2y - 7xy - 5y$

b. $30x^2 - 26x + 4$

EXAMPLE 2 Factor $6x^2 - 2x - 20$ by grouping.

Solution:

Step 1. First factor out the greatest common factor, 2.

$$6x^2 - 2x - 20 = 2(3x^2 - x - 10)$$

Step 2. Next notice that $a = 3$, $b = -1$, and $c = -10$ in the resulting trinomial. Find two numbers whose product is $a \cdot c$ or $3(-10) = -30$ and whose sum is b, -1. The numbers are -6 and 5.

Step 3. $3x^2 - x - 10 = 3x^2 - 6x + 5x - 10$

Step 4.

$$\begin{aligned} &= 3x(x - 2) + 5(x - 2) \\ &= (x - 2)(3x + 5) \end{aligned}$$

The factored form of $6x^2 - 2x - 20 = 2(x - 2)(3x + 5)$.

Don't forget to include the common factor of 2.

Answers

1. a. $(x + 4)(3x + 2)$ **b.** $(4x + 5)(3x + 1)$
2. a. $y(2x + 1)(3x - 5)$ **b.** $2(5x - 1)(3x - 2)$

Name ______________________ Section __________ Date __________

EXERCISE SET 11.4

A *Factor each polynomial by grouping. Notice that Step 3 has already been done in these exercises. See Examples 1 and 2.*

1. $x^2 + 3x + 2x + 6$

2. $x^2 + 5x + 3x + 15$

3. $x^2 - 4x + 7x - 28$

4. $x^2 - 6x + 2x - 12$

5. $y^2 + 8y - 2y - 16$

6. $z^2 + 10z - 7z - 70$

7. $3x^2 + 4x + 12x + 16$

8. $2x^2 + 5x + 14x + 35$

9. $8x^2 - 5x - 24x + 15$

10. $4x^2 - 9x - 32x + 72$

11. $5x^4 - 3x^2 + 25x^2 - 15$

12. $2y^4 - 10y^2 + 7y^2 - 35$

Factor each trinomial by grouping. Exercises 13–16 are broken into parts to help you get started. See Examples 1 and 2.

13. $6x^2 + 11x + 3$
- **a.** Find two numbers whose product is $6 \cdot 3 = 18$ and whose sum is 11.
- **b.** Write $11x$ using the factors from part (a).
- **c.** Factor by grouping.

14. $8x^2 + 14x + 3$
- **a.** Find two numbers whose product is $8 \cdot 3 = 24$ and whose sum is 14.
- **b.** Write $14x$ using the factors from part (a).
- **c.** Factor by grouping.

15. $15x^2 - 23x + 4$
- **a.** Find two numbers whose product is $15 \cdot 4 = 60$ and whose sum is -23.
- **b.** Write $-23x$ using the factors from part (a).
- **c.** Factor by grouping.

16. $6x^2 - 13x + 5$
- **a.** Find two numbers whose product is $6 \cdot 5 = 30$ and whose sum is -13.
- **b.** Write $-13x$ using the factors from part (a).
- **c.** Factor by grouping.

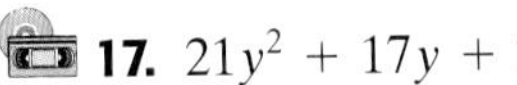

17. $21y^2 + 17y + 2$

18. $15x^2 + 11x + 2$

19. $7x^2 - 4x - 11$

20. $8x^2 - x - 9$

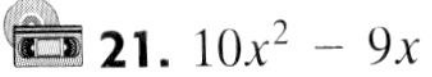

21. $10x^2 - 9x + 2$

22. $30x^2 - 23x + 3$

23. $2x^2 - 7x + 5$

24. $2x^2 - 7x + 3$

25. $12x + 4x^2 + 9$

26. $20x + 25x^2 + 4$

27. $4x^2 - 8x - 21$

28. $6x^2 - 11x - 10$

29. $10x^2 - 23x + 12$

30. $21x^2 - 13x + 2$

31. $2x^3 + 13x^2 + 15x$

32. $3x^3 + 8x^2 + 4x$

33. $16y^2 - 34y + 18$

34. $4y^2 - 2y - 12$

35. $-13x + 6 + 6x^2$

36. $-25x + 12 + 12x^2$

37. $54a^2 - 9a - 30$

38. $30a^2 + 38a - 20$

39. $20a^3 + 37a^2 + 8a$

40. $10a^3 + 17a^2 + 3a$

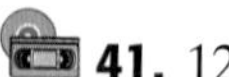

41. $12x^3 - 27x^2 - 27x$

42. $30x^3 - 155x^2 + 25x$

Review and Preview

Multiply. See Section 10.6.

43. $(x - 2)(x + 2)$

44. $(y - 5)(y + 5)$

45. $(y + 4)(y + 4)$

46. $(x + 7)(x + 7)$

47. $(9z + 5)(9z - 5)$

48. $(8y + 9)(8y - 9)$

49. $(4x - 3)^2$

50. $(2z - 1)^2$

Combining Concepts

Factor each polynomial by grouping.

51. $x^{2n} + 2x^n + 3x^n + 6$
(*Hint:* Don't forget that $x^{2n} = x^n \cdot x^n$.)

52. $x^{2n} + 6x^n + 10x^n + 60$

53. $3x^{2n} + 16x^n - 35$

54. $12x^{2n} - 40x^n + 25$

55. In your own words, explain how to factor a trinomial by grouping.

11.5 Factoring Perfect Square Trinomials and the Difference of Two Squares

OBJECTIVES

- A Recognize perfect square trinomials.
- B Factor perfect square trinomials.
- C Factor the difference of two squares.

SSM TUTOR CENTER SG CD & VIDEO MATH PRO WEB

A Recognizing Perfect Square Trinomials

A trinomial that is the square of a binomial is called a **perfect square trinomial.** For example,

$$(x+3)^2 = (x+3)(x+3)$$
$$= x^2 + 6x + 9$$

Thus $x^2 + 6x + 9$ is a perfect square trinomial.

In Chapter 10, we discovered special product formulas for squaring binomials.

$$(a+b)^2 = a^2 + 2ab + b^2 \quad \text{and} \quad (a-b)^2 = a^2 - 2ab + b^2$$

Because multiplication and factoring are reverse processes, we can now use these special products to help us factor perfect square trinomials. If we reverse these equations, we have the following.

Factoring Perfect Square Trinomials

$$a^2 + 2ab + b^2 = (a+b)^2$$
$$a^2 - 2ab + b^2 = (a-b)^2$$

To use these equations to help us factor, we must first be able to recognize a perfect square trinomial. A trinomial is a perfect square when

1. two terms, a^2 and b^2, are squares and
2. another term is $2 \cdot a \cdot b$ or $-2 \cdot a \cdot b$. That is, this term is twice the product of a and b, or its opposite.

EXAMPLE 1 Decide whether $x^2 + 8x + 16$ is a perfect square trinomial.

Solution:

1. Two terms, x^2 and 16, are squares $(16 = 4^2)$.
2. Twice the product of x and 4 is the other term of the trinomial.

$$2 \cdot x \cdot 4 = 8x$$

Thus, $x^2 + 8x + 16$ is a perfect square trinomial.

Practice Problem 1

Decide whether each trinomial is a perfect square trinomial.

a. $x^2 + 12x + 36$
b. $x^2 + 20x + 100$

EXAMPLE 2 Decide whether $4x^2 + 10x + 9$ is a perfect square trinomial.

Solution:

1. Two terms, $4x^2$ and 9, are squares.

$$4x^2 = (2x)^2 \quad \text{and} \quad 9 = 3^2$$

2. Twice the product of $2x$ and 3 is *not* the other term of the trinomial.

$$2 \cdot 2x \cdot 3 = 12x, \textit{not } 10x$$

The trinomial is *not* a perfect square trinomial.

Practice Problem 2

Decide whether each trinomial is a perfect square trinomial.

a. $9x^2 + 20x + 25$
b. $4x^2 + 8x + 11$

Answers

1. a. yes **b.** yes **2. a.** no **b.** no

Practice Problem 3

Decide whether each trinomial is a perfect square trinomial.

a. $25x^2 - 10x + 1$

b. $9x^2 - 42x + 49$

Practice Problem 4

Factor: $x^2 + 16x + 64$

Practice Problem 5

Factor: $9r^2 + 24rs + 16s^2$

Practice Problem 6

Factor: $9n^2 - 6n + 1$

Practice Problem 7

Factor: $9x^2 + 15x + 4$

Answers

3. a. yes **b.** yes **4.** $(x + 8)^2$
5. $(3r + 4s)^2$ **6.** $(3n - 1)^2$
7. $(3x + 1)(3x + 4)$

EXAMPLE 3 Decide whether $9x^2 - 12x + 4$ is a perfect square trinomial.

Solution:

1. Two terms, $9x^2$ and 4, are squares.

$$9x^2 = (3x)^2 \quad \text{and} \quad 4 = 2^2$$

2. Twice the product of $3x$ and 2 is the opposite of the other term of the trinomial.

$$2 \cdot 3x \cdot 2 = 12x, \text{ the opposite of } -12x$$

Thus, $9x^2 - 12x + 4$ is a perfect square trinomial.

B Factoring Perfect Square Trinomials

Now that we can recognize perfect square trinomials, we are ready to factor them.

EXAMPLE 4 Factor: $x^2 + 12x + 36$

Solution:

$$x^2 + 12x + 36 = x^2 + 2 \cdot x \cdot 6 + 6^2 \qquad 36 = 6^2 \text{ and } 12x = 2 \cdot x \cdot 6$$

$$a^2 + 2 \cdot a \cdot b + b^2$$

$$= (x + 6)^2$$

$$(a + b)^2$$

EXAMPLE 5 Factor: $25x^2 + 20xy + 4y^2$

Solution:

$$25x^2 + 20xy + 4y^2 = (5x)^2 + 2 \cdot 5x \cdot 2y + (2y)^2$$
$$= (5x + 2y)^2$$

EXAMPLE 6 Factor: $4m^2 - 4m + 1$

Solution:

$$4m^2 - 4m + 1 = (2m)^2 - 2 \cdot 2m \cdot 1 + 1^2$$

$$a^2 - 2 \cdot a \cdot b + b^2$$

$$= (2m - 1)^2$$

$$(a - b)^2$$

EXAMPLE 7 Factor: $25x^2 + 50x + 9$

Solution: Notice that this trinomial is not a perfect square trinomial.

$$25x^2 = (5x)^2, 9 = 3^2$$

but

$$2 \cdot 5x \cdot 3 = 30x$$

and $30x$ is not the middle term $50x$.

Although $25x^2 + 50x + 9$ is not a perfect square trinomial, it is factorable. Using techniques we learned in Section 11.3, we find that

$$25x^2 + 50x + 9 = (5x + 9)(5x + 1)$$

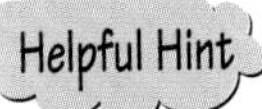

A perfect square trinomial can also be factored by the methods found in Sections 11.2 through 11.4.

EXAMPLE 8 Factor: $162x^3 - 144x^2 + 32x$

Solution: Don't forget to first look for a common factor. There is a greatest common factor of $2x$ in this trinomial.

$$\begin{aligned}162x^3 - 144x^2 + 32x &= 2x(81x^2 - 72x + 16)\\ &= 2x[(9x)^2 - 2\cdot 9x\cdot 4 + 4^2]\\ &= 2x(9x - 4)^2\end{aligned}$$

Practice Problem 8

Factor: $12x^3 - 84x^2 + 147x$

C Factoring the Difference of Two Squares

In Chapter 10, we discovered another special product, the product of the sum and difference of two terms a and b:

$$(a + b)(a - b) = a^2 - b^2$$

Reversing this equation gives us another factoring pattern, which we use to factor the difference of two squares.

Factoring the Difference of Two Squares

$$a^2 - b^2 = (a + b)(a - b)$$

Let's practice using this pattern.

EXAMPLES Factor each binomial.

9. $x^2 - 4 = x^2 - 2^2 = (x + 2)(x - 2)$

$$\uparrow \quad \uparrow \qquad \uparrow \quad \uparrow \; \uparrow \quad \uparrow$$
$$a^2 - b^2 = (a + b)(a - b)$$

10. $y^2 - 25 = y^2 - 5^2 = (y + 5)(y - 5)$

11. $y^2 - \frac{4}{9} = y^2 - \left(\frac{2}{3}\right)^2 = \left(y + \frac{2}{3}\right)\left(y - \frac{2}{3}\right)$

12. $x^2 + 4$

Note that the binomial $x^2 + 4$ is the *sum* of two squares since we can write $x^2 + 4$ as $x^2 + 2^2$. We might try to factor using $(x + 2)(x + 2)$ or $(x - 2)(x - 2)$. But when we multiply to check, we find that neither factoring is correct.

$$(x + 2)(x + 2) = x^2 + 4x + 4$$
$$(x - 2)(x - 2) = x^2 - 4x + 4$$

In both cases, the product is a trinomial, not the required binomial. In fact, $x^2 + 4$ is a prime polynomial.

Practice Problems 9–12

Factor each binomial.

9. $x^2 - 9$ 10. $a^2 - 16$

11. $c^2 - \frac{9}{25}$ 12. $s^2 + 9$

Helpful Hint

After the greatest common factor has been removed, the *sum* of two squares cannot be factored further using real numbers.

Answers

8. $3x(2x - 7)^2$ **9.** $(x - 3)(x + 3)$ **10.** $(a - 4)(a + 4)$ **11.** $\left(c - \frac{3}{5}\right)\left(c + \frac{3}{5}\right)$ **12.** prime polynomial

Practice Problems 13–15

Factor each difference of two squares.

13. $9s^2 - 1$
14. $16x^2 - 49y^2$
15. $p^4 - 81$

EXAMPLES Factor each difference of two squares.

13. $4x^2 - 1 = (2x)^2 - 1^2 = (2x + 1)(2x - 1)$

14. $25a^2 - 9b^2 = (5a)^2 - (3b)^2 = (5a + 3b)(5a - 3b)$

15. $y^4 - 16 = (y^2)^2 - 4^2$

$= (y^2 + 4)(y^2 - 4)$ Factor the difference of two squares.

$= (y^2 + 4)(y + 2)(y - 2)$ Factor the difference of two squares.

Helpful Hint

1. Don't forget to first see whether there's a greatest common factor (other than 1) that can be factored out.
2. Factor completely. In other words, check to see whether any factors can be factored further (as in Example 15).

Practice Problems 16–18

Factor completely.

16. $9x^3 - 25x$
17. $48x^4 - 3$
18. $-9x^2 + 100$

EXAMPLES Factor completely.

16. $4x^3 - 49x = x(4x^2 - 49)$ Factor out the common factor, x.

$= x[(2x)^2 - 7^2]$

$= x(2x + 7)(2x - 7)$ Factor the difference of two squares.

17. $162x^4 - 2 = 2(81x^4 - 1)$ Factor out the common factor, 2.

$= 2(9x^2 + 1)(9x^2 - 1)$ Factor the difference of two squares.

$= 2(9x^2 + 1)(3x + 1)(3x - 1)$ Factor the difference of two squares.

18. $-49x^2 + 16 = -1(49x^2 - 16)$ Factor out -1.

$= -1(7x + 4)(7x - 4)$ Factor the difference of two squares.

Answers

13. $(3s - 1)(3s + 1)$
14. $(4x - 7y)(4x + 7y)$
15. $(p^2 + 9)(p + 3)(p - 3)$
16. $x(3x - 5)(3x + 5)$
17. $3(4x^2 + 1)(2x + 1)(2x - 1)$
18. $-1(3x - 10)(3x + 10)$

GRAPHING CALCULATOR EXPLORATIONS

Graphing

A graphing calculator is a convenient tool for evaluating an expression at a given replacement value. For example, let's evaluate $x^2 - 6x$ when $x = 2$. To do so, store the value 2 in the variable x and then enter and evaluate the algebraic expression.

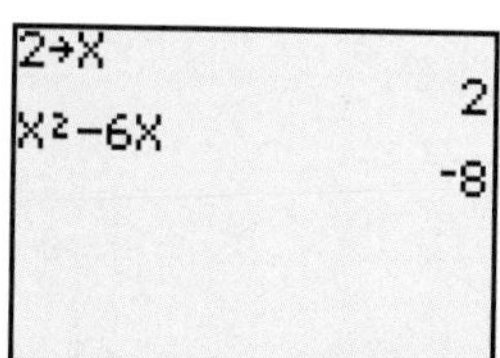

The value of $x^2 - 6x$ when $x = 2$ is -8. You may want to use this method for evaluating expressions as you explore the following.

We can use a graphing calculator to explore factoring patterns numerically. Use your calculator to evaluate $x^2 - 2x + 1$, $x^2 - 2x - 1$, and $(x - 1)^2$ for each value of x given in the table. What do you observe?

	$x^2 - 2x + 1$	$x^2 - 2x - 1$	$(x - 1)^2$
$x = 5$			
$x = -3$			
$x = 2.7$			
$x = -12.1$			
$x = 0$			

Notice in each case that $x^2 - 2x - 1 \neq (x - 1)^2$. Because for each x in the table the value of $x^2 - 2x + 1$ and the value of $(x - 1)^2$ are the same, we might guess that $x^2 - 2x + 1 = (x - 1)^2$. We can verify our guess algebraically with multiplication:

$$(x - 1)(x - 1) = x^2 - x - x + 1 = x^2 - 2x + 1$$

Name ______________________ Section __________ Date __________

Mental Math

State each number as a square.

1. 1 **2.** 25 **3.** 81 **4.** 64 **5.** 9 **6.** 100

State each term as a square.

7. $9x^2$ **8.** $16y^2$ **9.** $25a^2$ **10.** $81b^2$ **11.** $36p^4$ **12.** $4q^4$

EXERCISE SET 11.5

A *Determine whether each trinomial is a perfect square trinomial. See Examples 1 through 3.*

1. $x^2 + 16x + 64$ **2.** $x^2 + 22x + 121$ **3.** $y^2 + 5y + 25$ **4.** $y^2 + 4y + 16$

5. $m^2 - 2m + 1$ **6.** $p^2 - 4p + 4$ **7.** $a^2 - 16a + 49$ **8.** $n^2 - 20n + 144$

9. $4x^2 + 12xy + 8y^2$ **10.** $25x^2 + 20xy + 2y^2$ **11.** $25a^2 - 40ab + 16b^2$ **12.** $36a^2 - 12ab + b^2$

13. Fill in the blank so that $x^2 +$ ______ $x + 16$ is a perfect square trinomial.

14. Fill in the blank so that $9x^2 +$ ______ $x + 25$ is a perfect square trinomial.

B *Factor each trinomial completely. See Examples 4 through 8.*

15. $x^2 + 22x + 121$ **16.** $x^2 + 18x + 81$ **17.** $x^2 - 16x + 64$ **18.** $x^2 - 12x + 36$

19. $16a^2 - 24a + 9$ **20.** $25x^2 + 20x + 4$ **21.** $x^4 + 4x^2 + 4$ **22.** $m^4 + 10m^2 + 25$

23. $2n^2 - 28n + 98$ **24.** $3y^2 - 6y + 3$ **25.** $16y^2 + 40y + 25$ **26.** $9y^2 + 48y + 64$

27. $x^2y^2 - 10xy + 25$ **28.** $4x^2y^2 - 28xy + 49$ **29.** $m^3 + 18m^2 + 81m$ **30.** $y^3 + 12y^2 + 36y$

31. $1 + 6x^2 + x^4$ **32.** $1 + 16x^2 + x^4$ **33.** $9x^2 - 24xy + 16y^2$ **34.** $25x^2 - 60xy + 36y^2$

35. $x^2 + 14xy + 49y^2$ **36.** $x^2 + 10xy + 25y^2$

37. Describe a perfect square trinomial.

38. Write a perfect square trinomial that factors as $(x + 3y)^2$.

C *Factor each binomial completely. See Examples 9 through 18.*

39. $x^2 - 4$ **40.** $x^2 - 36$ **41.** $81 - p^2$ **42.** $100 - t^2$ **43.** $-4r^2 + 1$

44. $-9t^2 + 1$  **45.** $9x^2 - 16$ **46.** $36y^2 - 25$ 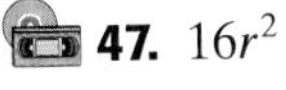 **47.** $16r^2 + 1$ **48.** $49y^2 + 1$

49. $-36 + x^2$ **50.** $-1 + y^2$ **51.** $m^4 - 1$ **52.** $n^4 - 16$

53. $x^2 - 169y^2$ **54.** $x^2 - 225y^2$ **55.** $18r^2 - 8$ **56.** $32t^2 - 50$ **57.** $9xy^2 - 4x$

58. $16xy^2 - 64x$

59. $25y^4 - 100y^2$

60. $xy^3 - 9xyz^2$

61. $x^3y - 4xy^3$

62. $36x^2 - 64$

63. $225a^2 - 81b^2$

64. $144 - 81x^2$

65. $12x^2 - 27$

66. $25y^2 - 9$

67. $49a^2 - 16$

68. $121 - 100x^2$

69. $169a^2 - 49b^2$

70. $x^2y^2 - 1$

71. $16 - a^2b^2$

72. $x^2 - \frac{1}{4}$

73. $y^2 - \frac{1}{16}$

74. $49 - \frac{9}{25}m^2$

75. $100 - \frac{4}{81}n^2$

76. What binomial multiplied by $(x - 6)$ gives the difference of two squares?

77. What binomial multiplied by $(5 + y)$ gives the difference of two squares?

Review and Preview

Solve each equation. See Section 9.3.

78. $x - 6 = 0$

79. $y + 5 = 0$

80. $2m + 4 = 0$

81. $3x - 9 = 0$

82. $5z - 1 = 0$

83. $4a + 2 = 0$

Solve each of the following. See Section 9.5.

△ **84.** A suitcase has a volume of 960 cubic inches. Find x.

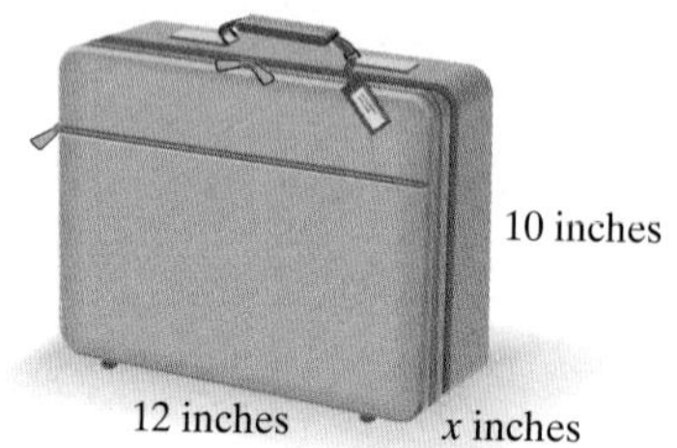

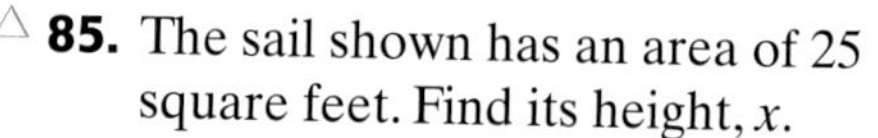

△ **85.** The sail shown has an area of 25 square feet. Find its height, x.

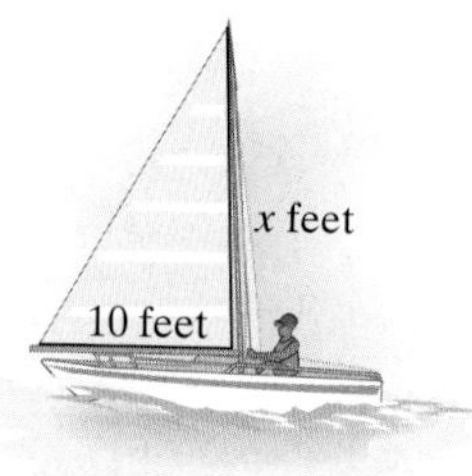

Combining Concepts

Factor each expression completely.

86. $(x + 2)^2 - y^2$

87. $(y - 6)^2 - z^2$

88. $a^2(b - 4) - 16(b - 4)$

89. $m^2(n + 8) - 9(n + 8)$

90. $(x^2 + 6x + 9) - 4y^2$ (*Hint:* Factor the trinomial in parentheses first.)

91. $(x^2 + 2x + 1) - 36y^2$

92. $x^{2n} - 100$

93. $x^{2n} - 81$

The area of the largest square in the figure is $(a + b)^2$. Use this figure to answer Exercises 94 and 95.

94. Write the area of the largest square as the sum of the areas of the smaller squares and rectangles.

95. What factoring formula from this section is visually represented by this square?

96. An object is dropped from the top of Pittsburgh's USX Tower, which is 841 feet tall. (*Source: World Almanac* research) The height of the object after t seconds is given by the expression $841 - 16t^2$.

a. Find the height of the object after 2 seconds.

b. Find the height of the object after 5 seconds.

c. To the nearest whole second, estimate when the object hits the ground.

d. Factor $841 - 16t^2$.

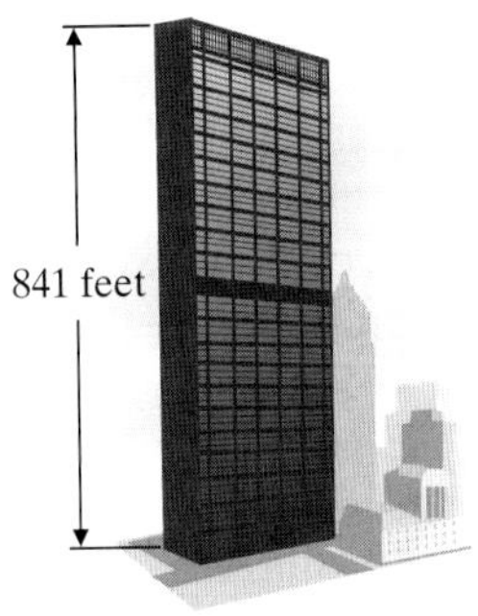

97. A worker on the top of the Aetna Life Building in San Francisco accidentally drops a bolt. The Aetna Life Building is 529 feet tall. (*Source: World Almanac* research) The height of the bolt after t seconds is given by the expression $529 - 16t^2$.

a. Find the height of the bolt after 1 second.

b. Find the height of the bolt after 4 seconds.

c. To the nearest whole second, estimate when the bolt hits the ground.

d. Factor $529 - 16t^2$.

98. At this writing, the world's tallest building is the Petronas Twin Towers in Kuala Lumpur, Malaysia, at a height of 1483 feet. (*Source:* Council on Tall Buildings and Urban Habitat) Suppose a worker is suspended 39 feet below the tip of the pinnacle atop one of the towers, at a height of 1444 feet above the ground. If the worker accidentally drops a bolt, the height of the bolt after t seconds is given by the expression $1444 - 16t^2$.

a. Find the height of the bolt after 3 seconds.

b. Find the height of the bolt after 7 seconds.

c. To the nearest whole second, estimate when the bolt hits the ground.

d. Factor $1444 - 16t^2$.

99. A performer with the Moscow Circus is planning a stunt involving a free fall from the top of the Moscow State University building, which is 784 feet tall. (*Source:* Council on Tall Buildings and Urban Habitat) Neglecting air resistance, the performer's height above gigantic cushions positioned at ground level after t seconds is given by the expression $784 - 16t^2$.

a. Find the performer's height after 2 seconds.

b. Find the performer's height after 5 seconds.

c. To the nearest whole second, estimate when the performer reaches the cushions positioned at ground level.

d. Factor $784 - 16t^2$.

FOCUS ON Mathematical Connections

NUMBER THEORY

By now, you have realized that being able to write a number as the product of prime numbers is very useful in the process of factoring polynomials. You probably also know at least a few numbers that are prime (such as 2, 3, and 5). But what about the other prime numbers? When we come across a number, how will we know if it is a prime number? Apparently, the ancient Greek mathematician Eratosthenes had a similar question because in the third century B.C. he devised a simple method for identifying primes. The method is called the Sieve of Eratosthenes because it "sifts out" the primes in a list of numbers. The Sieve of Eratosthenes is generally considered to be the most useful for identifying primes less than 1,000,000.

Here's how the sieve works: suppose you want to find the prime numbers in the first n natural numbers. Write the numbers, in order, from 2 to n. We know that 2 is prime, so circle it. Now, cross out each number greater than 2 that is a multiple of 2. Consider the next number in the list that is not crossed out. This number is 3, which we know is prime. Circle 3 and then cross out all multiples of 3 in the remainder of the list. Continue considering each uncircled number in the list. Once you have reached the largest prime number less than or equal to $\sqrt{n}$, and you have eliminated all its multiples in the remainder of the list, you can stop. Circle all the numbers left in the list that have not yet been circled. Now all of the circled numbers in the list are prime numbers. The list below demonstrates the Sieve of Eratosthenes on the numbers 2 through 30. Because $\sqrt{30} \approx 5.477$, we need only check and eliminate the multiples of primes up to and including 5, which is the largest prime less than or equal to the square root of 30.

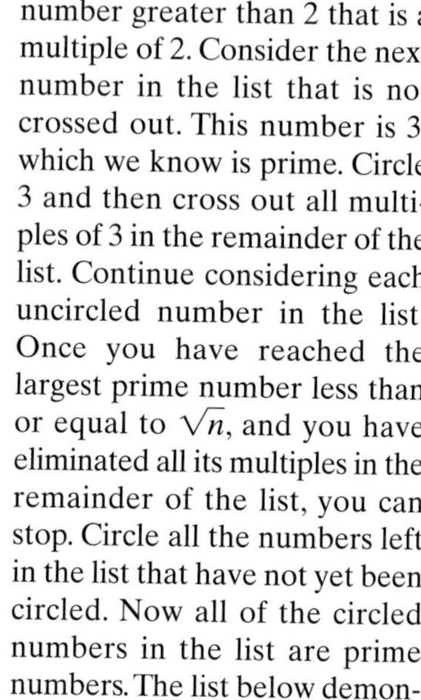

	(2)	(3)	~~4~~	(5)	~~6~~	(7)	~~8~~	~~9~~	~~10~~
(11)	~~12~~	(13)	~~14~~	~~15~~	~~16~~	(17)	~~18~~	(19)	~~20~~
~~21~~	~~22~~	(23)	~~24~~	~~25~~	~~26~~	~~27~~	~~28~~	(29)	~~30~~

We can see that the prime numbers less than 30 are 2, 3, 5, 7, 11, 13, 17, 19, 23, and 29.

GROUP ACTIVITY

Work with your group to identify the prime numbers less than 300 using the Sieve of Eratosthenes. What is the largest prime number that you will need to check in this process?

Name ______________________ Section __________ Date __________

Answers

1. ________
2. ________
3. ________
4. ________
5. ________
6. ________
7. ________
8. ________
9. ________
10. ________
11. ________
12. ________
13. ________
14. ________
15. ________
16. ________
17. ________
18. ________
19. ________
20. ________
21. ________
22. ________
23. ________
24. ________
25. ________
26. ________
27. ________
28. ________
29. ________
30. ________
31. ________
32. ________
33. ________
34. ________
35. ________
36. ________

Integrated Review—Choosing a Factoring Strategy

The following steps may be helpful when factoring polynomials.

To Factor a Polynomial

Step 1. Are there any common factors? If so, factor out the GCF.

Step 2. How many terms are in the polynomial?

- **a.** Two terms: Is it the difference of two squares? $a^2 - b^2 = (a - b)(a + b)$
- **b.** Three terms: Try one of the following.
  - **i.** Perfect square trinomial: $a^2 + 2ab + b^2 = (a + b)^2$
    $a^2 - 2ab + b^2 = (a - b)^2$
  - **ii.** If not a perfect square trinomial, factor using the methods presented in Sections 11.2 through 11.4.
- **c.** Four terms: Try factoring by grouping.

Step 3. See if any factors in the factored polynomial can be factored further.

Step 4. Check by multiplying.

Factor each polynomial completely.

1. $x^2 + x - 12$

2. $x^2 - 10x + 16$

3. $x^2 - x - 6$

4. $x^2 + 2x + 1$

5. $x^2 - 6x + 9$

6. $x^2 + x - 2$

7. $x^2 + x - 6$

8. $x^2 + 7x + 12$

9. $x^2 - 7x + 10$

10. $x^2 - x - 30$

11. $2x^2 - 98$

12. $3x^2 - 75$

13. $x^2 + 3x + 5x + 15$

14. $3y - 21 + xy - 7x$

15. $x^2 + 6x - 16$

16. $x^2 - 3x - 28$

17. $4x^3 + 20x^2 - 56x$

18. $6x^3 - 6x^2 - 120x$

19. $12x^2 + 34x + 24$

20. $8a^2 + 6ab - 5b^2$

21. $4a^2 - b^2$

22. $x^2 - 25y^2$

23. $28 - 13x - 6x^2$

24. $20 - 3x - 2x^2$

25. $x^2 - 2x + 4$

26. $a^2 + a - 3$

27. $6y^2 + y - 15$

28. $4x^2 - x - 5$

29. $18x^3 - 63x^2 + 9x$

30. $12a^3 - 24a^2 + 4a$

31. $16a^2 - 56a + 49$

32. $25p^2 - 70p + 49$

33. $14 + 5x - x^2$

34. $3 - 2x - x^2$

35. $3x^4y + 6x^3y - 72x^2y$

36. $2x^3y + 8x^2y^2 - 10xy^3$

37. ______
38. ______
39. ______
40. ______
41. ______
42. ______
43. ______
44. ______
45. ______
46. ______
47. ______
48. ______
49. ______
50. ______
51. ______
52. ______
53. ______
54. ______
55. ______
56. ______
57. ______
58. ______
59. ______
60. ______
61. ______
62. ______
63. ______
64. ______
65. ______
66. ______
67. ______
68. ______
69. ______
70. ______
71. ______
72. ______
73. ______
74. ______
75. ______
76. ______

37. $12x^3y + 243xy$

38. $6x^3y^2 + 8xy^2$

39. $2xy - 72x^3y$

40. $2x^3 - 18x$

41. $x^3 + 6x^2 - 4x - 24$

42. $x^3 - 2x^2 - 36x + 72$

43. $6a^3 + 10a^2$

44. $4n^2 - 6n$

45. $3x^3 - x^2 + 12x - 4$

46. $x^3 - 2x^2 + 3x - 6$

47. $6x^2 + 18xy + 12y^2$

48. $12x^2 + 46xy - 8y^2$

49. $5(x + y) + x(x + y)$

50. $7(x - y) + y(x - y)$

51. $14t^2 - 9t + 1$

52. $3t^2 - 5t + 1$

53. $3x^2 + 2x - 5$

54. $7x^2 + 19x - 6$

55. $1 - 8a - 20a^2$

56. $1 - 7a - 60a^2$

57. $x^4 - 10x^2 + 9$

58. $x^4 - 13x^2 + 36$

59. $x^2 - 23x + 120$

60. $y^2 + 22y + 96$

61. $x^2 - 14x - 48$

62. $16a^2 - 56ab + 49b^2$

63. $25p^2 - 70pq + 49q^2$

64. $7x^2 + 24xy + 9y^2$

65. $-x^2 - x + 30$

66. $-x^2 + 6x - 8$

67. $3rs - s + 12r - 4$

68. $x^3 - 2x^2 + 3x - 6$

69. $4x^2 - 8xy - 3x + 6y$

70. $4x^2 - 2xy - 7yz + 14xz$

71. $x^2 + 9xy - 36y^2$

72. $3x^2 + 10xy - 8y^2$

73. $x^4 - 14x^2 - 32$

74. $x^4 - 22x^2 - 75$

75. Explain why it makes good sense to factor out the GCF first, before using other methods of factoring.

76. The sum of two squares usually does not factor. Is the sum of two squares $9x^2 + 81y^2$ factorable?

11.6 Solving Quadratic Equations by Factoring

In this section, we introduce a new type of equation—the **quadratic equation.**

Quadratic Equation

A quadratic equation is one that can be written in the form

$$ax^2 + bx + c = 0$$

where a, b, and c are real numbers and $a \neq 0$.

Some examples of quadratic equations are shown below.

$$3x^2 + 5x + 6 = 0 \qquad x^2 = 9 \qquad y^2 + y = 1$$

The form $ax^2 + bx + c = 0$ is called the **standard form** of a quadratic equation. The quadratic equation $3x^2 + 5x + 6 = 0$ is the only equation above that is in standard form.

Quadratic equations model many real-life situations. For example, let's suppose an object is dropped from the top of a 256-foot cliff and we want to know how long before the object strikes the ground. The answer to this question is found by solving the quadratic equation $-16t^2 + 256 = 0$. (See Example 1 in Section 11.7.)

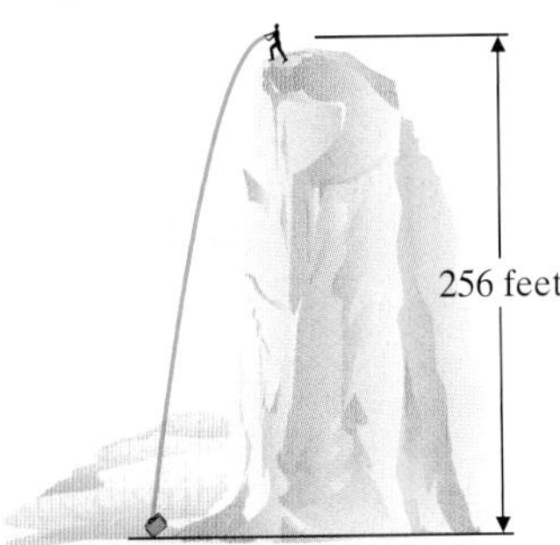

OBJECTIVES

A Solve quadratic equations by factoring.

B Solve equations with degree greater than 2 by factoring.

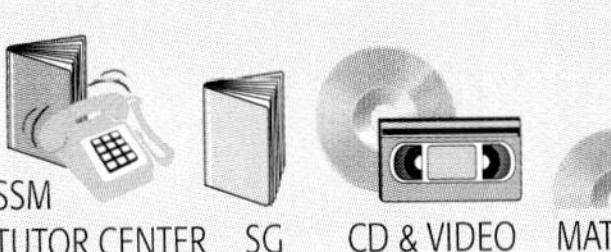

A Solving Quadratic Equations by Factoring

Some quadratic equations can be solved by making use of factoring and the **zero factor property.**

Zero Factor Property

If a and b are real numbers and if $ab = 0$, then $a = 0$ or $b = 0$.

In other words, if the product of two numbers is 0, then at least one of the numbers must be 0.

EXAMPLE 1 Solve: $(x - 3)(x + 1) = 0$

Solution: If this equation is to be a true statement, then either the factor $x - 3$ must be 0 or the factor $x + 1$ must be 0. In other words, either

$$x - 3 = 0 \qquad \text{or} \qquad x + 1 = 0$$

If we solve these two linear equations, we have

$$x = 3 \qquad \text{or} \qquad x = -1$$

Practice Problem 1

Solve: $(x - 7)(x + 2) = 0$

Answer

1. 7 and -2

Thus, 3 and -1 are both solutions of the equation $(x - 3)(x + 1) = 0$. To check, we replace x with 3 in the original equation. Then we replace x with -1 in the original equation.

Check:

$$(x - 3)(x + 1) = 0$$
$$(3 - 3)(3 + 1) \stackrel{?}{=} 0 \quad \text{Replace } x \text{ with 3.}$$
$$0(4) = 0 \quad \text{True}$$

$$(x - 3)(x + 1) = 0$$
$$(-1 - 3)(-1 + 1) \stackrel{?}{=} 0 \quad \text{Replace } x \text{ with } -1.$$
$$(-4)(0) = 0 \quad \text{True}$$

The solutions are 3 and -1.

Helpful Hint

The zero factor property says that *if a product is 0, then a factor is 0.*

If $a \cdot b = 0$, then $a = 0$ or $b = 0$.
If $x(x + 5) = 0$, then $x = 0$ or $x + 5 = 0$.
If $(x + 7)(2x - 3) = 0$, then $x + 7 = 0$ or $2x - 3 = 0$.

Use this property only when the product is 0. For example, if $a \cdot b = 8$, we do not know the value of a or b. The values may be $a = 2, b = 4$ or $a = 8, b = 1$, or any other two numbers whose product is 8.

Practice Problem 2

Solve: $(x - 10)(3x + 1) = 0$

EXAMPLE 2 Solve: $(x - 5)(2x + 7) = 0$

Solution: The product is 0. By the zero factor property, this is true only when a factor is 0. To solve, we set each factor equal to 0 and solve the resulting linear equations.

$$(x - 5)(2x + 7) = 0$$
$$x - 5 = 0 \quad \text{or} \quad 2x + 7 = 0$$
$$x = 5 \qquad 2x = -7$$
$$x = -\frac{7}{2}$$

Check: Let $x = 5$.

$$(x - 5)(2x + 7) = 0$$
$$(5 - 5)(2 \cdot 5 + 7) \stackrel{?}{=} 0 \quad \text{Replace } x \text{ with 5.}$$
$$0 \cdot 17 \stackrel{?}{=} 0$$
$$0 = 0 \quad \text{True}$$

Let $x = -\frac{7}{2}$.

$$(x - 5)(2x + 7) = 0$$
$$\left(-\frac{7}{2} - 5\right)\left(2\left(-\frac{7}{2}\right) + 7\right) \stackrel{?}{=} 0 \quad \text{Replace } x \text{ with } -\frac{7}{2}.$$
$$\left(-\frac{17}{2}\right)(-7 + 7) \stackrel{?}{=} 0$$
$$\left(-\frac{17}{2}\right) \cdot 0 \stackrel{?}{=} 0$$
$$0 = 0 \quad \text{True}$$

The solutions are 5 and $-\frac{7}{2}$.

Answer

2. 10 and $-\frac{1}{3}$

EXAMPLE 3 Solve: $x(5x - 2) = 0$

Solution: $x(5x - 2) = 0$

$$x = 0 \quad \text{or} \quad 5x - 2 = 0 \qquad \text{Use the zero factor property.}$$
$$5x = 2$$
$$x = \frac{2}{5}$$

Check these solutions in the original equation. The solutions are 0 and $\frac{2}{5}$.

Practice Problem 3

Solve each equation.

a. $y(y + 3) = 0$

b. $x(4x - 3) = 0$

EXAMPLE 4 Solve: $x^2 - 9x - 22 = 0$

Solution: One side of the equation is 0. However, to use the zero factor property, one side of the equation must be 0 *and* the other side must be written as a product (must be factored). Thus, we must first factor this polynomial.

$$x^2 - 9x - 22 = 0$$
$$(x - 11)(x + 2) = 0 \qquad \text{Factor.}$$

Now we can apply the zero factor property.

$$x - 11 = 0 \quad \text{or} \quad x + 2 = 0$$
$$x = 11 \qquad\qquad x = -2$$

Check:

Let $x = 11$.

$$x^2 - 9x - 22 = 0$$
$$11^2 - 9 \cdot 11 - 22 \stackrel{?}{=} 0$$
$$121 - 99 - 22 \stackrel{?}{=} 0$$
$$22 - 22 \stackrel{?}{=} 0$$
$$0 = 0 \quad \text{True}$$

Let $x = -2$.

$$x^2 - 9x - 22 = 0$$
$$(-2)^2 - 9(-2) - 22 \stackrel{?}{=} 0$$
$$4 + 18 - 22 \stackrel{?}{=} 0$$
$$22 - 22 \stackrel{?}{=} 0$$
$$0 = 0 \quad \text{True.}$$

The solutions are 11 and −2.

Practice Problem 4

Solve: $x^2 - 3x - 18 = 0$

EXAMPLE 5 Solve: $x^2 - 9x = -20$

Solution: First we rewrite the equation in standard form so that one side is 0. Then we factor the polynomial.

$$x^2 - 9x = -20$$
$$x^2 - 9x + 20 = 0 \qquad \text{Write in standard form by adding 20 to both sides.}$$
$$(x - 4)(x - 5) = 0 \qquad \text{Factor.}$$

Next we use the zero factor property and set each factor equal to 0.

$$x - 4 = 0 \quad \text{or} \quad x - 5 = 0 \qquad \text{Set each factor equal to 0.}$$
$$x = 4 \qquad\qquad x = 5 \qquad \text{Solve.}$$

Check: Check these solutions in the original equation. The solutions are 4 and 5.

Practice Problem 5

Solve: $x^2 - 14x = -24$

Answers

3. a. 0 and −3 **b.** 0 and $\frac{3}{4}$ **4.** 6 and −3 **5.** 12 and 2

The following steps may be used to solve a quadratic equation by factoring.

To Solve Quadratic Equations by Factoring

Step 1. Write the equation in standard form so that one side of the equation is 0.

Step 2. Factor the quadratic equation completely.

Step 3. Set each factor containing a variable equal to 0.

Step 4. Solve the resulting equations.

Step 5. Check each solution in the original equation.

Since it is not always possible to factor a quadratic polynomial, not all quadratic equations can be solved by factoring. Other methods of solving quadratic equations are presented in Chapter 16.

Practice Problem 6

Solve each equation.

a. $x(x - 4) = 5$

b. $x(3x + 7) = 6$

EXAMPLE 6 Solve: $x(2x - 7) = 4$

Solution: First we write the equation in standard form; then we factor.

$$x(2x - 7) = 4$$

$$2x^2 - 7x = 4 \quad \text{Multiply.}$$

$$2x^2 - 7x - 4 = 0 \quad \text{Write in standard form.}$$

$$(2x + 1)(x - 4) = 0 \quad \text{Factor.}$$

$$2x + 1 = 0 \quad \text{or} \quad x - 4 = 0 \quad \text{Set each factor equal to zero.}$$

$$2x = -1 \qquad x = 4 \quad \text{Solve.}$$

$$x = -\frac{1}{2}$$

Check the solutions in the original equation. The solutions are $-\frac{1}{2}$ and 4. ●

Helpful Hint

To solve the equation $x(2x - 7) = 4$, do **not** set each factor equal to 4. Remember that to apply the zero factor property, one side of the equation must be 0 and the other side of the equation must be in factored form.

B Solving Equations with Degree Greater than Two by Factoring

Some equations with degree greater than 2 can be solved by factoring and then using the zero factor property.

Practice Problem 7

Solve: $2x^3 - 18x = 0$

EXAMPLE 7 Solve: $3x^3 - 12x = 0$

Solution: To factor the left side of the equation, we begin by factoring out the greatest common factor, $3x$.

$$3x^3 - 12x = 0$$

$$3x(x^2 - 4) = 0 \quad \text{Factor out the GCF, } 3x.$$

$$3x(x + 2)(x - 2) = 0 \quad \text{Factor } x^2 - 4 \text{, a difference of two squares.}$$

$$3x = 0 \quad \text{or} \quad x + 2 = 0 \quad \text{or} \quad x - 2 = 0 \quad \text{Set each factor equal to 0.}$$

$$x = 0 \qquad x = -2 \qquad x = 2 \quad \text{Solve.}$$

Answers

6. a. 5 and −1 **b.** $\frac{2}{3}$ and −3 **7.** 0, 3, and −3

Thus, the equation $3x^3 - 12x = 0$ has three solutions: 0, −2, and 2.

Check: Replace x with each solution in the original equation.

Let $x = 0$.	Let $x = -2$.	Let $x = 2$.
$3(0)^3 - 12(0) \stackrel{?}{=} 0$	$3(-2)^3 - 12(-2) \stackrel{?}{=} 0$	$3(2)^3 - 12(2) \stackrel{?}{=} 0$
$0 = 0$	$3(-8) + 24 \stackrel{?}{=} 0$	$3(8) - 24 \stackrel{?}{=} 0$
True	$0 = 0$ True	$0 = 0$ True

The solutions are 0, −2, and 2.

EXAMPLE 8 Solve: $(5x - 1)(2x^2 + 15x + 18) = 0$

Solution:

$$(5x - 1)(2x^2 + 15x + 18) = 0$$

$$(5x - 1)(2x + 3)(x + 6) = 0 \quad \text{Factor the trinomial.}$$

$$5x - 1 = 0 \quad \text{or} \quad 2x + 3 = 0 \quad \text{or} \quad x + 6 = 0 \quad \text{Set each factor equal to 0.}$$

$$5x = 1 \qquad 2x = -3 \qquad x = -6 \quad \text{Solve.}$$

$$x = \frac{1}{5} \qquad x = -\frac{3}{2}$$

Check each solution in the original equation. The solutions are $\frac{1}{5}$, $-\frac{3}{2}$, and −6.

Practice Problem 8

Solve: $(x + 3)(3x^2 - 20x - 7) = 0$

Answer

8. $-3, -\frac{1}{3}$, and 7

STUDY SKILLS REMINDER

How well do you know your textbook?

See if you can answer the questions below.

1. What does the [video icon] icon mean?
2. What does the [pencil icon] icon mean?
3. What does the △ icon mean?
4. Where can you find a review for each chapter? What answers to this review can be found in the back of your text?
5. Each chapter contains an overview of the chapter along with examples. What is this feature called?
6. Does this text contain any solutions to exercises? If so, where?

FOCUS ON The Real World

FINDING PRIME NUMBERS

Now that you have discovered a way to identify relatively small prime numbers with the Sieve of Eratosthenes, perhaps you are wondering whether there are very large primes. The answer is yes. Another ancient Greek mathematician, Euclid, proved that there are an infinite number of primes. Thus, there do exist huge numbers that are prime. Researchers call prime numbers with more than 1000 digits *titanic primes*.

Knowing that very large prime numbers exist and finding them, and proving that they are in fact prime are two very different things. The Great Internet Mersenne Prime Search, or GIMPS, is a distributed computing project whose aim is finding large prime numbers. Founded in 1996, GIMPS now utilizes a network of over 21,500 personal computers that perform prime-locating arithmetic computations during each computer's "spare" computing time. These computers are owned by over twelve thousand different individuals, businesses, and schools around the world who donate computing time to the project. Together, this vast and varied array of individual personal computers form a virtual supercomputer capable of making 720 billion calculations per second.

The GIMPS project has been highly successful. Since its founding, it has been responsible for finding the five most recent largest-known prime numbers. At the time that this book was written, the largest-known prime number is $2^{13,466,917} - 1$. It was discovered by the GIMPS project on November 14, 2001. This record prime has 4,053,946 digits! It may be in a special class of prime numbers called the Mersenne primes, named after a 17th-century French monk and mathematician, Father Marin Mersenne. A Mersenne prime is a prime number of the form $2^p - 1$, where p is a positive integer. For instance, $2^2 - 1 = 3$, $2^3 - 1 = 7$, and $2^5 - 1 = 31$ are all Mersenne primes. This new record prime could be only the 39th known Mersenne prime.

GROUP ACTIVITIES

1. The following World Wide Web site contains information on the current status of the largest-known prime numbers: http://www.utm.edu/research/primes/largest.html. Visit this site and report on the five largest-known prime numbers. Be sure to include information about the numbers themselves (their form, whether they are Mersenne primes, how many digits, and so on), who found them, when they were found, and how they were found (if possible). Have any primes larger than $2^{13,466,917} - 1$ been found? If so, how many?
2. Research and report on the uses of prime numbers. Explain why there is an ongoing search for the largest-known prime number.

STUDY SKILLS REMINDER

Are you satisfied with your performance in this course thus far?

If not, ask yourself the following questions:

- Am I attending all class periods and arriving on time?
- Am I working and checking my homework assignments?
- Am I getting help when I need it?
- In addition to my instructor, am I using the supplements to this text that could help me? For example, the tutorial video lessons? MathPro, the tutorial software?
- Am I satisfied with my performance on quizzes and tests?

If you answered no to *any* of these questions, read or reread Section 1.1 for suggestions in these areas. Also, you may want to contact your instructor for additional feedback.

Name ______________________ Section __________ Date __________

Mental Math

Solve each equation by inspection.

1. $(a - 3)(a - 7) = 0$
2. $(a - 5)(a - 2) = 0$
3. $(x + 8)(x + 6) = 0$
4. $(x + 2)(x + 3) = 0$
5. $(x + 1)(x - 3) = 0$
6. $(x - 1)(x + 2) = 0$

EXERCISE SET 11.6

A *Solve each equation. See Examples 1 through 3.*

1. $(x - 2)(x + 1) = 0$
2. $(x + 3)(x + 2) = 0$
3. $(x - 6)(x - 7) = 0$
4. $(x + 4)(x - 10) = 0$
5. $(x + 9)(x + 17) = 0$
6. $(x - 11)(x - 1) = 0$
7. $x(x + 6) = 0$
8. $x(x - 7) = 0$
9. $3x(x - 8) = 0$
10. $2x(x + 12) = 0$
11. $(2x + 3)(4x - 5) = 0$
12. $(3x - 2)(5x + 1) = 0$
13. $(2x - 7)(7x + 2) = 0$
14. $(9x + 1)(4x - 3) = 0$
15. $\left(x - \frac{1}{2}\right)\left(x + \frac{1}{3}\right) = 0$
16. $\left(x + \frac{2}{9}\right)\left(x - \frac{1}{4}\right) = 0$
17. $(x + 0.2)(x + 1.5) = 0$
18. $(x + 1.7)(x + 2.3) = 0$
19. Write a quadratic equation that has two solutions, 6 and -1. Leave the polynomial in the equation in factored form.
20. Write a quadratic equation that has two solutions, 0 and -2. Leave the polynomial in the equation in factored form.

Solve each equation. See Examples 4 through 6.

21. $x^2 - 13x + 36 = 0$
22. $x^2 + 2x - 63 = 0$
23. $x^2 + 2x - 8 = 0$
24. $x^2 - 5x + 6 = 0$
25. $x^2 - 7x = 0$
26. $x^2 - 3x = 0$
27. $x^2 + 20x = 0$
28. $x^2 + 15x = 0$
29. $x^2 = 16$
30. $x^2 = 9$
31. $x^2 - 4x = 32$
32. $x^2 - 5x = 24$

33. $x(3x - 1) = 14$

34. $x(4x - 11) = 3$

35. $3x^2 + 19x - 72 = 0$

36. $36x^2 + x - 21 = 0$

B *Solve each equation. See Examples 7 and 8.*

37. $4x^3 - x = 0$

38. $4y^3 - 36y = 0$

39. $4(x - 7) = 6$

40. $5(3 - 4x) = 9$

41. $(4x - 3)(16x^2 - 24x + 9) = 0$

42. $(2x + 5)(4x^2 - 10x + 25) = 0$

43. $4y^2 - 1 = 0$

44. $4y^2 - 81 = 0$

45. $(2x + 3)(2x^2 - 5x - 3) = 0$

46. $(2x - 9)(x^2 + 5x - 36) = 0$

47. $x^2 - 15 = -2x$

48. $x^2 - 26 = -11x$

49. $5x^2 - 6x - 8 = 0$

50. $12x^2 + 7x - 12 = 0$

51. $30x^2 - 11x = 30$

52. $9x^2 + 6x = -2$

53. $6y^2 - 22y - 40 = 0$

54. $3x^2 - 6x - 9 = 0$

55. $(y - 2)(y + 3) = 6$

56. $(y - 5)(y - 2) = 28$

57. $x^3 - 12x^2 + 32x = 0$

58. $x^3 - 14x^2 + 49x = 0$

59. Write a quadratic equation in standard form that has two solutions, 5 and 7.

60. Write an equation that has three solutions, 0, 1, and 2.

Review and Preview

Perform each indicated operation. Write all results in lowest terms. See Section 4.3 through 4.5.

61. $\frac{3}{5} + \frac{4}{9}$

62. $\frac{2}{3} + \frac{3}{7}$

63. $\frac{7}{10} - \frac{5}{12}$

64. $\frac{5}{9} - \frac{5}{12}$

65. $\frac{4}{5} \cdot \frac{7}{8}$

66. $\frac{3}{7} \cdot \frac{12}{17}$

Combining Concepts

67. Explain the error and solve correctly:

$$x(x - 2) = 8$$
$$x = 8 \quad \text{or} \quad x - 2 = 8$$
$$x = 10$$

68. Explain the error and solve correctly:

$$(x - 4)(x + 2) = 0$$
$$x = -4 \quad \text{or} \quad x = 2$$

69. A compass is accidentally thrown upward and out of an air balloon at a height of 300 feet. The height, y, of the compass at time x is given by the equation $y = -16x^2 + 20x + 300$.

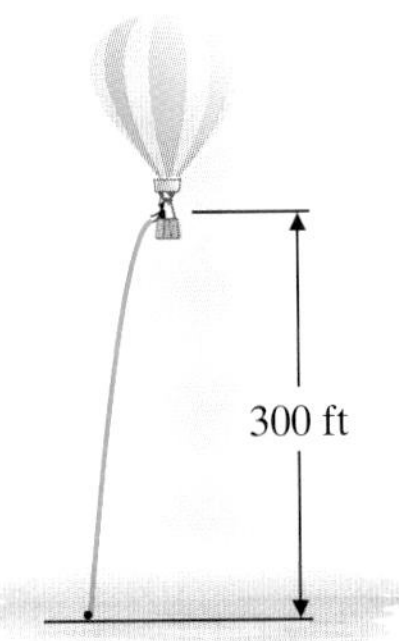

a. Find the height of the compass at the given times by filling in the table below.

Time, x (in seconds)	0	1	2	3	4	5	6
Height, y (in feet)							

b. Use the table to determine when the compass strikes the ground.

c. Use the table to approximate the maximum height of the compass.

70. A rocket is fired upward from the ground with an initial velocity of 100 feet per second. The height, y, of the rocket at any time x is given by the equation $y = -16x^2 + 100x$.

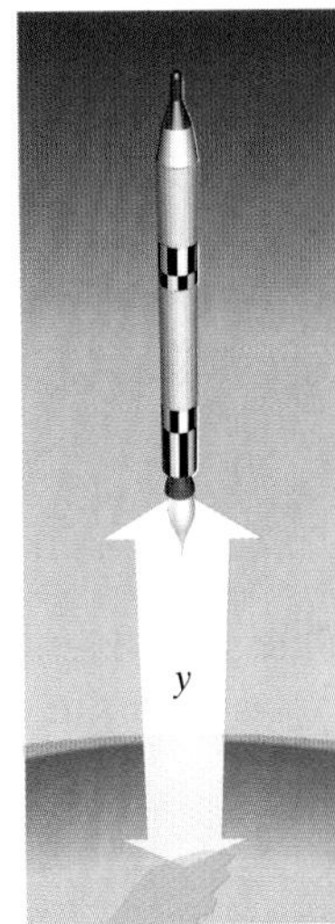

a. Find the height of the rocket at the given times by filling in the table below.

Time, x (in seconds)	0	1	2	3	4	5	6	7
Height, y (in feet)								

b. Use the table to help approximate when the rocket strikes the ground to the nearest tenth of a second.

c. Use the table to approximate the maximum height of the rocket.

Solve each equation.

71. $(x - 3)(3x + 4) = (x + 2)(x - 6)$

72. $(2x - 3)(x + 6) = (x - 9)(x + 2)$

73. $(2x - 3)(x + 8) = (x - 6)(x + 4)$

74. $(x + 6)(x - 6) = (2x - 9)(x + 4)$

FOCUS ON Mathematical Connections

GEOMETRY

Factoring polynomials can be visualized using areas of rectangles. To see this, let's first find the areas of the following squares and rectangles. (Recall that Area = Length · Width.)

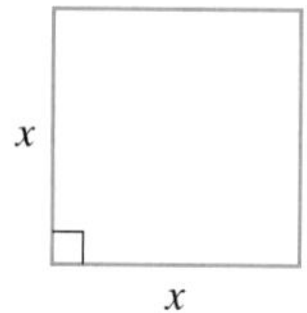

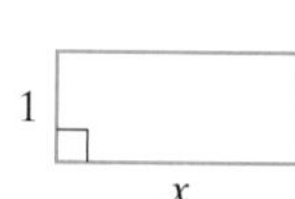

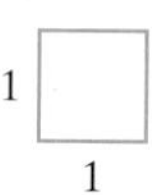

To use these areas to visualize factoring the polynomial $x^2 + 3x + 2$, for example, use the shapes below to form a rectangle. The factored form is found by reading the length and the width of the rectangle as shown below.

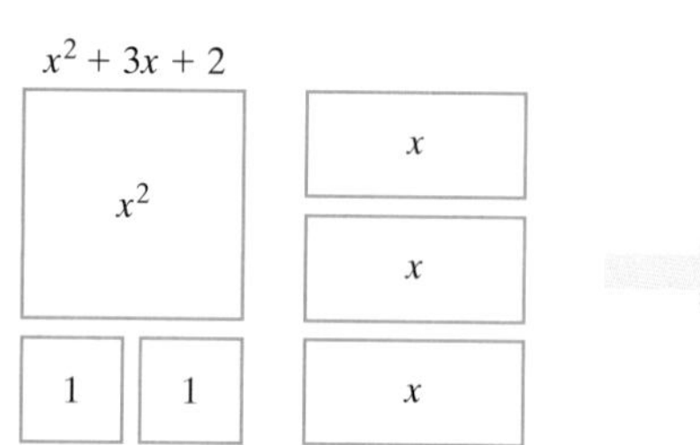

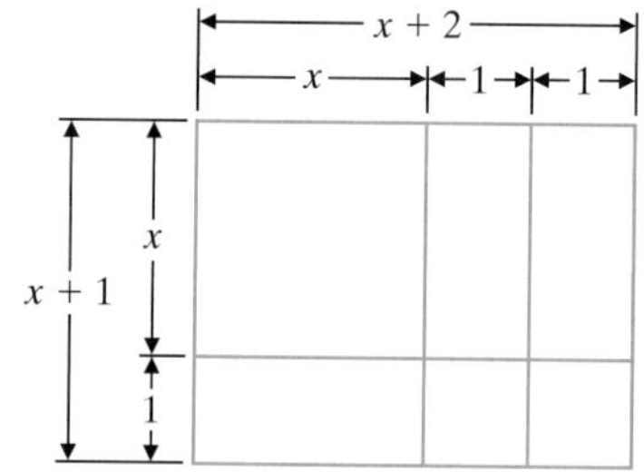

Thus, $x^2 + 3x + 2 = (x + 2)(x + 1)$.

Try using this method to visualize the factored form of each polynomial below.

GROUP ACTIVITY

Work in a group and use tiles to find the factored form of the polynomials below. (Tiles can be hand made from index cards.)

1. $x^2 + 6x + 5$

2. $x^2 + 5x + 6$

3. $x^2 + 5x + 4$

4. $x^2 + 4x + 3$

5. $x^2 + 6x + 9$

6. $x^2 + 4x + 4$

11.7 Quadratic Equations and Problem Solving

A Solving Problems Modeled by Quadratic Equations

Some problems may be modeled by quadratic equations. To solve these problems, we use the same problem-solving steps that were introduced in Section 9.4. When solving these problems, keep in mind that a solution of an equation that models a problem may not be a solution to the problem. For example, a person's age or the length of a rectangle is always a positive number. Thus we discard solutions that do not make sense as solutions of the problem.

EXAMPLE 1 Finding Free-Fall Time

For a TV commercial, a piece of luggage is dropped from a cliff 256 feet above the ground to show the durability of the luggage. Neglecting air resistance, the height h in feet of the luggage above the ground after t seconds is given by the quadratic equation

$$h = -16t^2 + 256$$

Find how long it takes for the luggage to hit the ground.

Solution:

1. UNDERSTAND. Read and reread the problem. Then draw a picture of the problem.

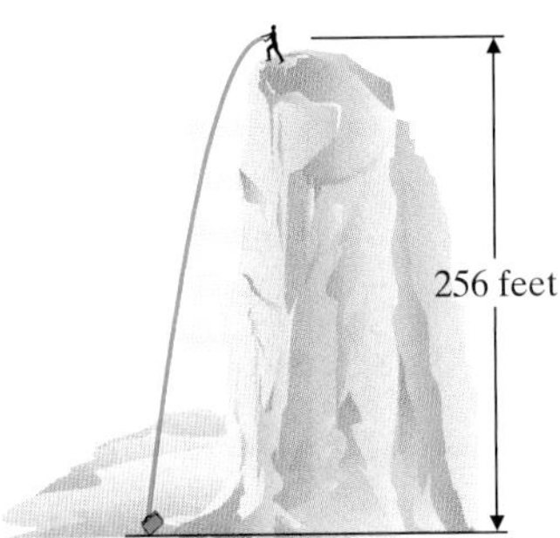

The equation $h = -16t^2 + 256$ models the height of the falling luggage at time t. Familiarize yourself with this equation by finding the height of the luggage at $t = 1$ second and $t = 2$ seconds.

When $t = 1$ second, the height of the suitcase is

$h = -16(1)^2 + 256 = 240$ feet.

When $t = 2$ seconds, the height of the suitcase is

$h = -16(2)^2 + 256 = 192$ feet.

2. TRANSLATE. To find how long it takes the luggage to hit the ground, we want to know the value of t for which the height $h = 0$.

$$0 = -16t^2 + 256$$

3. SOLVE. We solve the quadratic equation by factoring.

$$0 = -16t^2 + 256$$
$$0 = -16(t^2 - 16)$$
$$0 = -16(t - 4)(t + 4)$$
$$t - 4 = 0 \quad \text{or} \quad t + 4 = 0$$
$$t = 4 \qquad\qquad t = -4$$

4. INTERPRET. Since the time t cannot be negative, the proposed solution is 4 seconds.

Check: Verify that the height of the luggage when t is 4 seconds is 0.

When $t = 4$ seconds, $h = -16(4)^2 + 256 = -256 + 256 = 0$ feet.

OBJECTIVE

A Solve problems that can be modeled by quadratic equations.

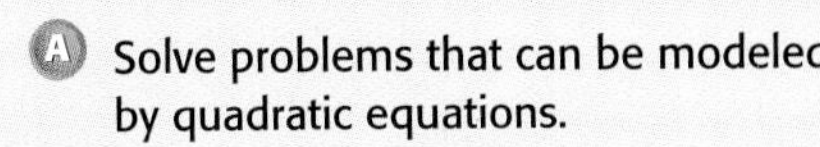

Practice Problem 1

An object is dropped from the roof of a 144-foot-tall building. Neglecting air resistance, the height h in feet of the object above ground after t seconds is given by the quadratic equation

$$h = -16t^2 + 144$$

Find how long it takes the object to hit the ground.

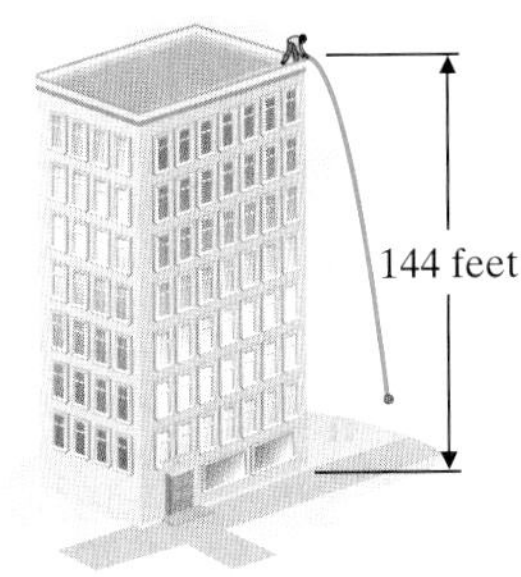

Answer

1. 3 seconds

State: The solution checks and the luggage hits the ground 4 seconds after it is dropped.

Practice Problem 2

The square of a number minus twice the number is 63. Find the number.

EXAMPLE 2 Finding a Number

The square of a number plus three times the number is 70. Find the number.

Solution:

1. UNDERSTAND. Read and reread the problem. Suppose that the number is 5. The square of 5 is 5^2 or 25. Three times 5 is 15. Then $25 + 15 = 40$, not 70, so the number must be greater than 5. Remember, the purpose of proposing a number, such as 5, is to better understand the problem. Now that we do, we will let $x =$ the number.
2. TRANSLATE.

the square of a number	plus	three times the number	is	70
↓	↓	↓	↓	↓
x^2	$+$	$3x$	$=$	70

3. SOLVE.

$$x^2 + 3x = 70$$
$$x^2 + 3x - 70 = 0 \quad \text{Subtract 70 from both sides.}$$
$$(x + 10)(x - 7) = 0 \quad \text{Factor.}$$
$$x + 10 = 0 \quad \text{or} \quad x - 7 = 0 \quad \text{Set each factor equal to 0.}$$
$$x = -10 \qquad x = 7 \quad \text{Solve.}$$

4. INTERPRET.

Check: The square of -10 is $(-10)^2$, or 100. Three times -10 is $3(-10)$ or -30. Then $100 + (-30) = 70$, the correct sum, so -10 checks.

The square of 7 is 7^2 or 49. Three times 7 is $3(7)$, or 21. Then $49 + 21 = 70$, the correct sum, so 7 checks.

State: There are two numbers. They are -10 and 7.

Practice Problem 3

The length of a rectangle is 5 feet more than its width. The area of the rectangle is 176 square feet. Find the length and the width of the rectangle.

EXAMPLE 3 Finding the Dimensions of a Sail

The height of a triangular sail is 2 meters less than twice the length of the base. If the sail has an area of 30 square meters, find the length of its base and the height.

Solution:

1. UNDERSTAND. Read and reread the problem. Since we are finding the length of the base and the height, we let

   $x =$ the length of the base

   Since the height is 2 meters less than twice the base,

   $2x - 2 =$ the height

   An illustration is shown on the next page.
2. TRANSLATE. We are given that the area of the triangle is 30 square meters, so we use the formula for area of a triangle.

area of triangle	$=$	$\frac{1}{2}$	$\cdot$	base	$\cdot$	height
↓		↓		↓		↓
30	$=$	$\frac{1}{2}$	$\cdot$	x	$\cdot$	$(2x - 2)$

Answers

2. 9 and -7 **3.** length $= 16$ ft; width $= 11$ ft

3. SOLVE. Now we solve the quadratic equation.

$$30 = \frac{1}{2}x(2x - 2)$$

$$30 = x^2 - x \quad \text{Multiply.}$$

$$x^2 - x - 30 = 0 \quad \text{Write in standard form.}$$

$$(x - 6)(x + 5) = 0 \quad \text{Factor.}$$

$$x - 6 = 0 \quad \text{or} \quad x + 5 = 0 \quad \text{Set each factor equal to 0.}$$

$$x = 6 \qquad x = -5$$

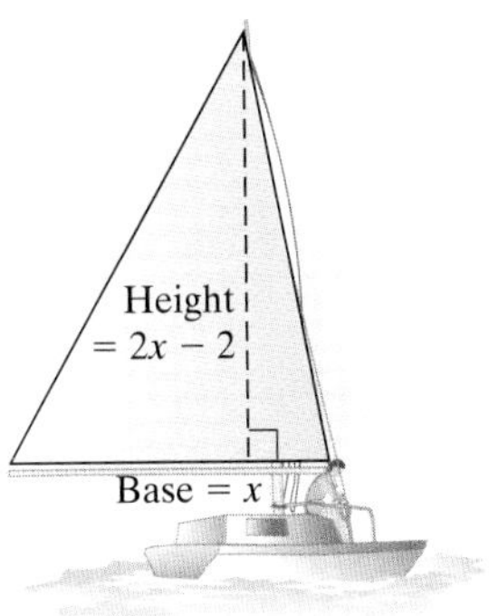

4. INTERPRET. Since x represents the length of the base, we discard the solution -5. The base of a triangle cannot be negative. The base is then 6 feet and the height is $2(6) - 2 = 10$ feet.

Check: To check this problem, we recall that $\frac{1}{2}\text{base} \cdot \text{height} = \text{area}$, or

$$\frac{1}{2}(6)(10) = 30 \quad \text{The required area}$$

State: The base of the triangular sail is 6 meters and the height is 10 meters.

The next example makes use of the **Pythagorean theorem** and consecutive integers. Before we review this theorem, recall that a **right triangle** is a triangle that contains a 90° or right angle. The **hypotenuse** of a right triangle is the side opposite the right angle and is the longest side of the triangle. The **legs** of a right triangle are the other sides of the triangle.

Pythagorean Theorem

In a right triangle, the sum of the squares of the lengths of the two legs is equal to the square of the length of the hypotenuse.

$$(\text{leg})^2 + (\text{leg})^2 = (\text{hypotenuse})^2 \quad \text{or} \quad a^2 + b^2 = c^2$$

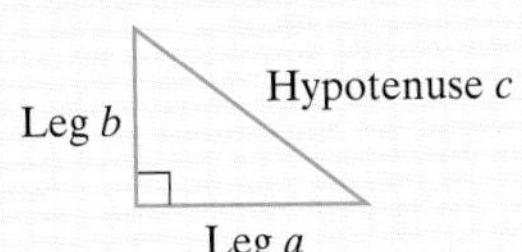

Study the following diagrams for a review of consecutive integers.

Examples

If x is the first integer, then consecutive integers are $x, x + 1, x + 2, \ldots$

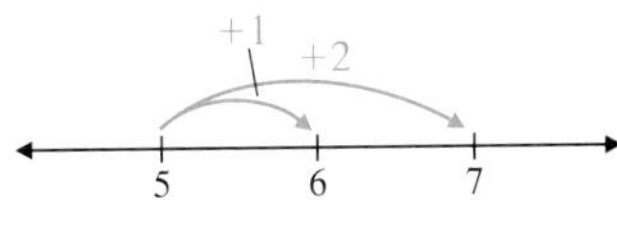

If x is the first even integer, then consecutive even integers are $x, x + 2, x + 4, \ldots$

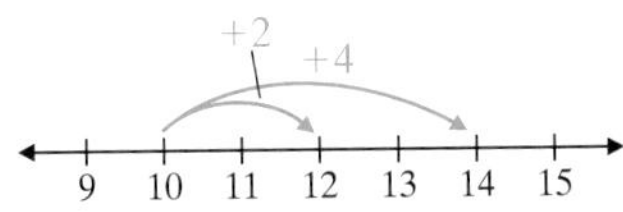

If x is the first odd integer, then consecutive odd integers are $x, x + 2, x + 4, \ldots$

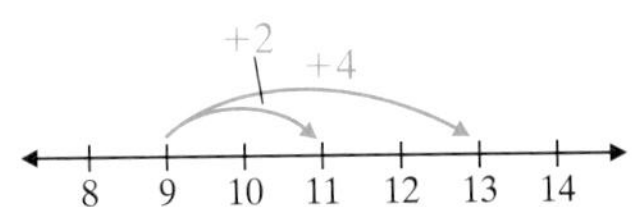

Practice Problem 4

Solve.

a. Find two consecutive odd integers whose product is 23 more than their sum.

b. The length of one leg of a right triangle is 7 meters less than the length of the other leg. The length of the hypotenuse is 13 meters. Find the lengths of the legs.

Answers

4. a. 5 and 7 or −5 and −3 **b.** 5 meters, 12 meters

EXAMPLE 4 Finding the Dimensions of a Triangle

Find the lengths of the sides of a right triangle if the lengths can be expressed as three consecutive even integers.

Solution:

1. UNDERSTAND. Read and reread the problem. Let's suppose that the length of one leg of the right triangle is 4 units. Then the other leg is the next even integer, or 6 units, and the hypotenuse of the triangle is the next even integer, or 8 units. Remember that the hypotenuse is the longest side. Let's see if a triangle with sides of these lengths forms a right triangle. To do this, we check to see whether the Pythagorean theorem holds true.

$$4^2 + 6^2 \stackrel{?}{=} 8^2$$
$$16 + 36 \stackrel{?}{=} 64$$
$$52 = 64 \quad \text{False.}$$

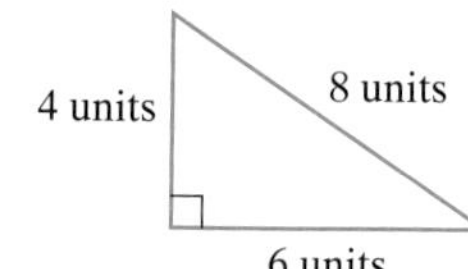

Our proposed numbers do not check, but we now have a better understanding of the problem.

We let $x, x + 2$, and $x + 4$ be three consecutive even integers. Since these integers represent lengths of the sides of a right triangle, we have the following.

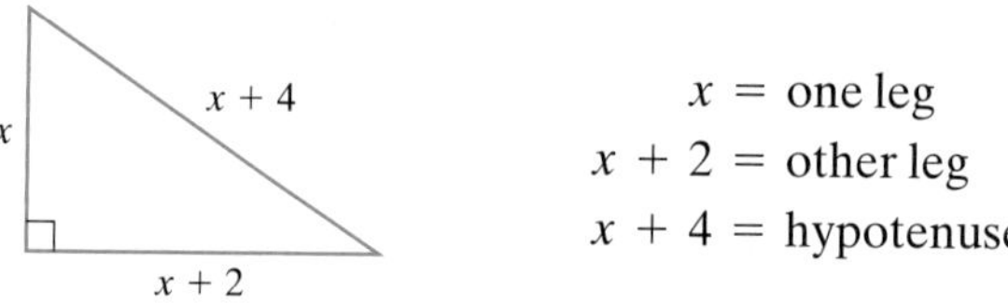

$$x = \text{one leg}$$
$$x + 2 = \text{other leg}$$
$$x + 4 = \text{hypotenuse (longest side)}$$

2. TRANSLATE. By the Pythagorean theorem, we have that

$$(\text{hypotenuse})^2 = (\text{leg})^2 + (\text{leg})^2$$
$$(x + 4)^2 = (x)^2 + (x + 2)^2$$

3. SOLVE. Now we solve the equation.

$$(x + 4)^2 = x^2 + (x + 2)^2$$
$$x^2 + 8x + 16 = x^2 + x^2 + 4x + 4 \quad \text{Multiply.}$$
$$x^2 + 8x + 16 = 2x^2 + 4x + 4 \quad \text{Combine like terms.}$$
$$x^2 - 4x - 12 = 0 \quad \text{Write in standard form.}$$
$$(x - 6)(x + 2) = 0 \quad \text{Factor.}$$
$$x - 6 = 0 \quad \text{or} \quad x + 2 = 0 \quad \text{Set each factor equal to 0.}$$
$$x = 6 \qquad x = -2$$

4. INTERPRET. We discard $x = -2$ since length cannot be negative. If $x = 6$, then $x + 2 = 8$ and $x + 4 = 10$.

Check: Verify that

$$(\text{hypotenuse})^2 = (\text{leg})^2 + (\text{leg})^2$$
$$10^2 \stackrel{?}{=} 6^2 + 8^2$$
$$100 \stackrel{?}{=} 36 + 64$$
$$100 = 100 \quad \text{True}$$

State: The sides of the right triangle have lengths 6 units, 8 units, and 10 units.

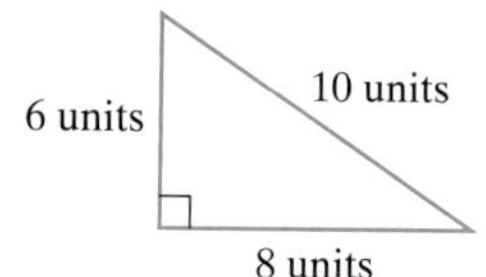

Name ______________________ Section __________ Date __________

EXERCISE SET 11.7

A *See Examples 1 through 4 for all exercises. For Exercises 1 through 6, represent each given condition using a single variable, x.*

△ **1.** The length and width of a rectangle whose length is 4 centimeters more than its width

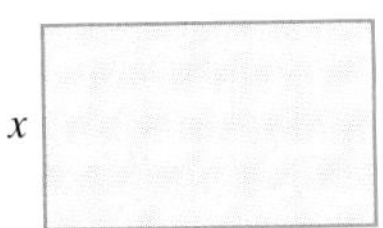

△ **2.** The length and width of a rectangle whose length is twice its width

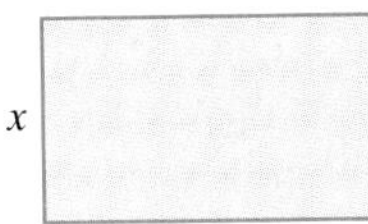

3. Two consecutive odd integers

4. Two consecutive even integers

△ **5.** The base and height of a triangle whose height is one more than four times its base

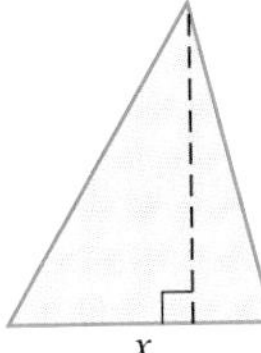

△ **6.** The base and height of a trapezoid whose base is three less than five times its height

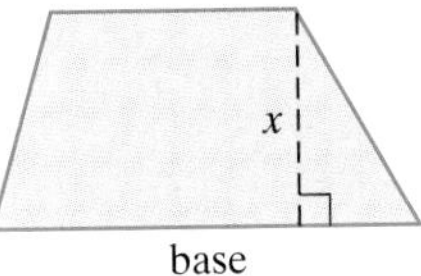

Use the information given to find the dimensions of each figure.

△ **7.**

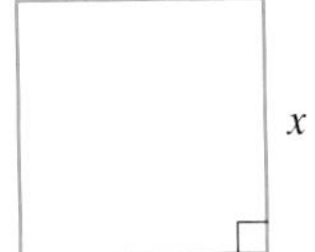

The *area* of the square is 121 square units. Find the length of its sides.

△ **8.**

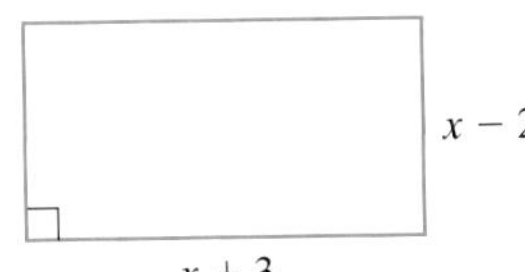

The *area* of the rectangle is 84 square inches. Find its length and width.

△ **9.**

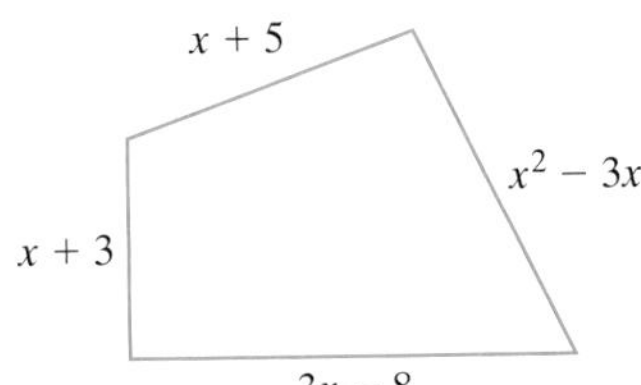

The *perimeter* of the quadrilateral is 120 centimeters. Find the lengths of the sides.

△ **10.**

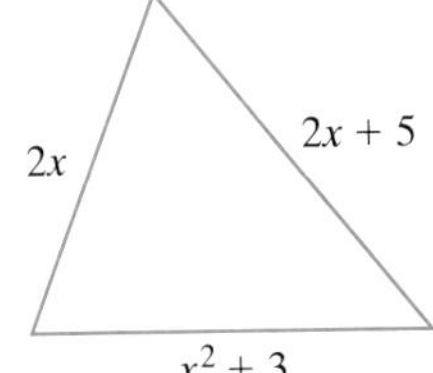

The *perimeter* of the triangle is 85 feet. Find the lengths of its sides.

△ **11.**

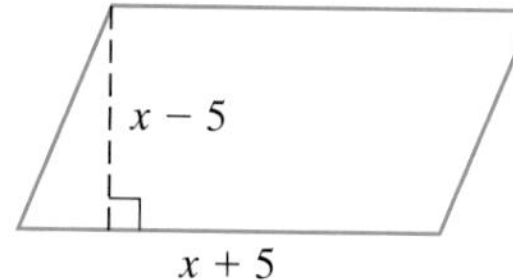

The *area* of the parallelogram is 96 square miles. Find its base and height.

△ **12.**

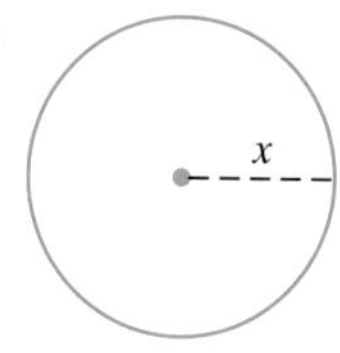

The *area* of the circle is 25π square kilometers. Find its radius.

Solve.

13. An object is thrown upward from the top of an 80-foot building with an initial velocity of 64 feet per second. The height h of the object after t seconds is given by the quadratic equation $h = -16t^2 + 64t + 80$. When will the object hit the ground?

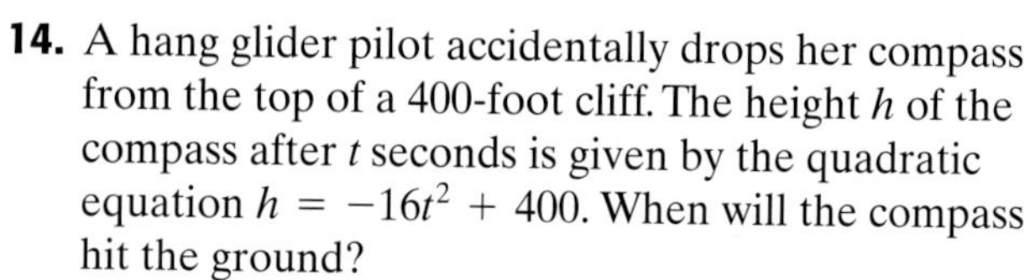

14. A hang glider pilot accidentally drops her compass from the top of a 400-foot cliff. The height h of the compass after t seconds is given by the quadratic equation $h = -16t^2 + 400$. When will the compass hit the ground?

15. The length of a rectangle is 7 centimeters less than twice its width. Its area is 30 square centimeters. Find the dimensions of the rectangle.

16. The length of a rectangle is 9 inches more than its width. Its area is 112 square inches. Find the dimensions of the rectangle.

△ *The equation $D = \frac{1}{2}n(n - 3)$ gives the number of diagonals D for a polygon with n sides. For example, a polygon with 6 sides has $D = \frac{1}{2}\cdot 6(6 - 3)$ or $D = 9$ diagonals. (See if you can count all 9 diagonals. Some are shown in the figure.) Use this equation, $D = \frac{1}{2}n(n - 3)$, for Exercises 17 through 20.*

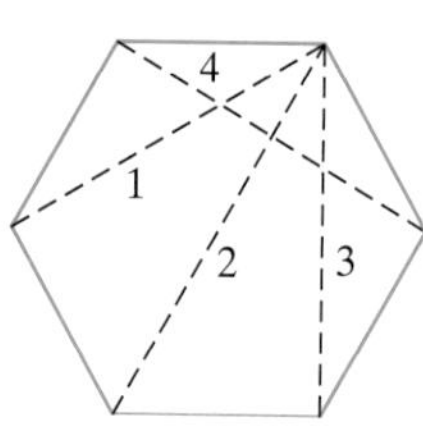

17. Find the number of diagonals for a polygon that has 12 sides.

18. Find the number of diagonals for a polygon that has 15 sides.

19. Find the number of sides n for a polygon that has 35 diagonals.

20. Find the number of sides n for a polygon that has 14 diagonals.

Solve.

21. The sum of a number and its square is 132. Find the number.

22. The sum of a number and its square is 182. Find the number.

△ **23.** Two boats travel at a right angle to each other after leaving the same dock at the same time. One hour later the boats are 17 miles apart. If one boat travels 7 miles per hour faster than the other boat, find the rate of each boat.

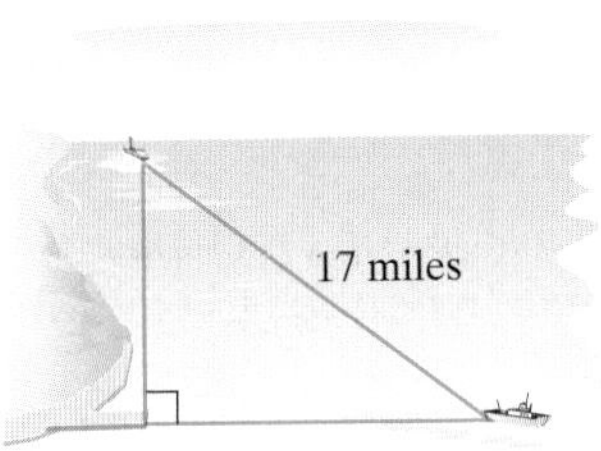

△ **24.** The side of a square equals the width of a rectangle. The length of the rectangle is 6 meters longer than its width. The sum of the areas of the square and the rectangle is 176 square meters. Find the side of the square.

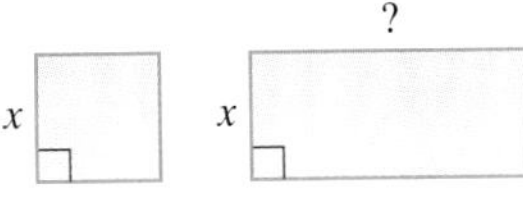

25. The sum of two numbers is 20, and the sum of their squares is 218. Find the numbers.

26. The sum of two numbers is 25, and the sum of their squares is 325. Find the numbers.

△ **27.** If the sides of a square are increased by 3 inches, the area becomes 64 square inches. Find the length of the sides of the original square.

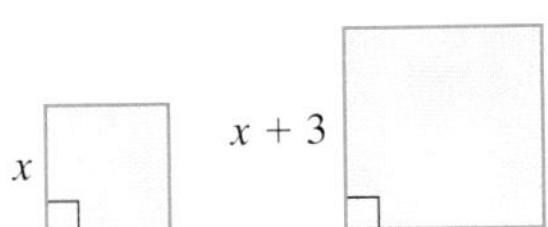

△ **28.** If the sides of a square are increased by 5 meters, the area becomes 100 square meters. Find the length of the sides of the original square.

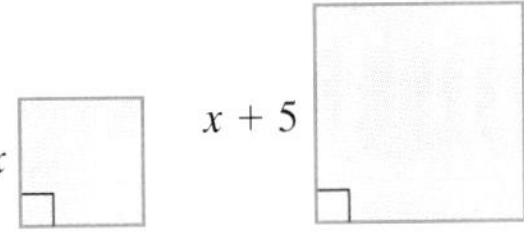

△ **29.** One leg of a right triangle is 4 millimeters longer than the smaller leg and the hypotenuse is 8 millimeters longer than the smaller leg. Find the lengths of the sides of the triangle.

△ **30.** One leg of a right triangle is 9 centimeters longer than the other leg and the hypotenuse is 45 centimeters. Find the lengths of the legs of the triangle.

△ **31.** The length of the base of a triangle is twice its height. If the area of the triangle is 100 square kilometers, find the height.

x

$2x$

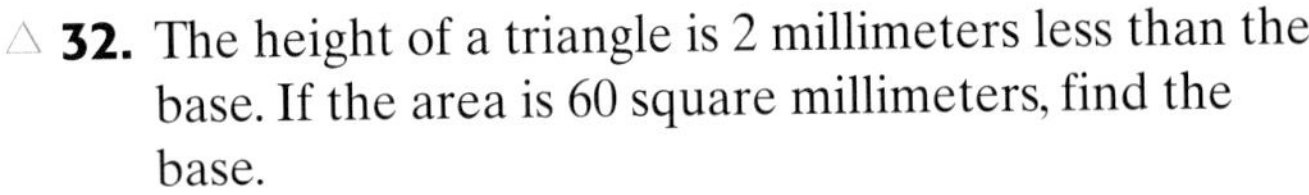

△ **32.** The height of a triangle is 2 millimeters less than the base. If the area is 60 square millimeters, find the base.

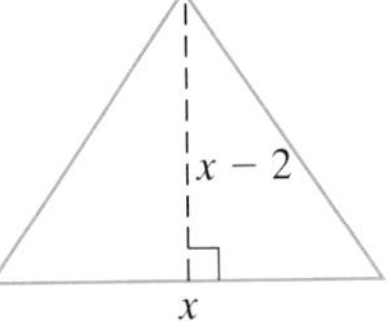

△ **33.** Find the length of the shorter leg of a right triangle if the longer leg is 12 feet more than the shorter leg and the hypotenuse is 12 feet less than twice the shorter leg.

△ **34.** Find the length of the shorter leg of a right triangle if the longer leg is 10 miles more than the shorter leg and the hypotenuse is 10 miles less than twice the shorter leg.

35. An object is dropped from 39 feet below the tip of the pinnacle atop one of the 1483-foot-tall Petronas Twin Towers in Kuala Lumpor, Malaysia. (*Source:* Council on Tall Buildings and Urban Habitat) The height h of the object after t seconds is given by the equation $h = -16t^2 + 1444$. Find how many seconds pass before the object reaches the ground.

36. An object is dropped from the top of 311 South Wacker Drive, a 961-foot-tall office building in Chicago. (*Source:* Council on Tall Buildings and Urban Habitat) The height h of the object after t seconds is given by the equation $h = -16t^2 + 961$. Find how many seconds pass before the object reaches the ground.

37. At the end of 2 years, P dollars invested at an interest rate r compounded annually increases to an amount, A dollars, given by

$$A = P(1 + r)^2$$

Find the interest rate if \$100 increased to \$144 in 2 years.

38. At the end of 2 years, P dollars invested at an interest rate r compounded annually increases to an amount, A dollars, given by

$$A = P(1 + r)^2$$

Find the interest rate if \$2000 increased to \$2420 in 2 years.

△ **39.** Find the dimensions of a rectangle whose width is 7 miles less than its length and whose area is 120 square miles.

△ **40.** Find the dimensions of a rectangle whose width is 2 inches less than half its length and whose area is 160 square inches.

41. If the cost, C, for manufacturing x units of a certain product is given by $C = x^2 - 15x + 50$, find the number of units manufactured at a cost of \$9500.

42. If a switchboard handles n telephones, the number C of telephone connections it can make simultaneously is given by the equation $C = \frac{n(n-1)}{2}$. Find how many telephones are handled by a switchboard making 120 telephone connections simultaneously.

Review and Preview

The following double line graph shows a comparison of the amount of land (in thousand acres) operated by farms in Alabama during the years shown with the amount of land operated by farms in North Carolina. Use this graph to answer Exercises 43 through 49. See Section 7.1.

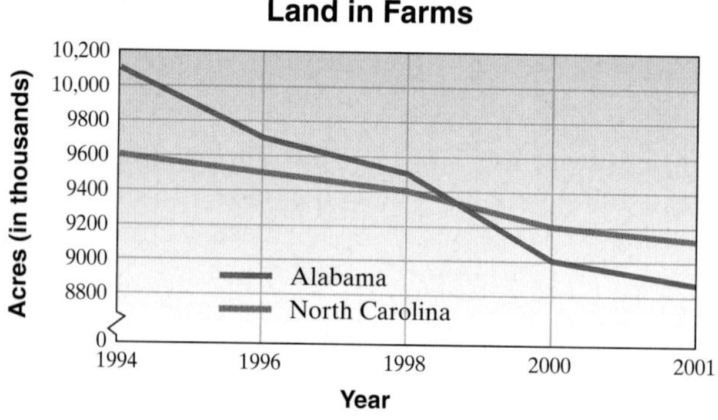

Source: U.S. Department of Agriculture, National Agricultural Statistics Service

43. Approximate the amount of land operated by farms in North Carolina in 1994.

44. Approximate the amount of land operated by farms in North Carolina in 1996.

45. Approximate the amount of land operated by farms in Alabama in 1998.

46. Approximate the amount of land operated by farms in Alabama in 2001.

47. Approximate the year that the colored lines in this graph intersect.

48. In your own words, explain the meaning of the point of intersection in the graph.

49. Describe the trends shown in this graph and speculate as to why these trends have occurred.

Combining Concepts

△ **50.** According to the International America's Cup Class (IACC) rule, a sailboat competing in the America's Cup match must have a 110-foot-tall mast and a combined mainsail and jib sail area of 3000 square feet. (*Source:* America's Cup Organizing Committee) A design for an IACC-class sailboat calls for the mainsail to be 60% of the combined sail area. If the height of the triangular mainsail is 28 feet more than twice the length of the boom, find the length of the boom and the height of the mainsail.

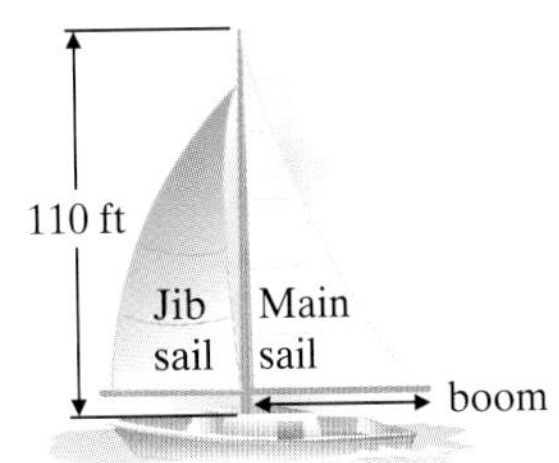

△ **51.** A rectangular pool is surrounded by a walk 4 meters wide. The pool is 6 meters longer than its width. If the total area is 576 square meters more than the area of the pool, find the dimensions of the pool.

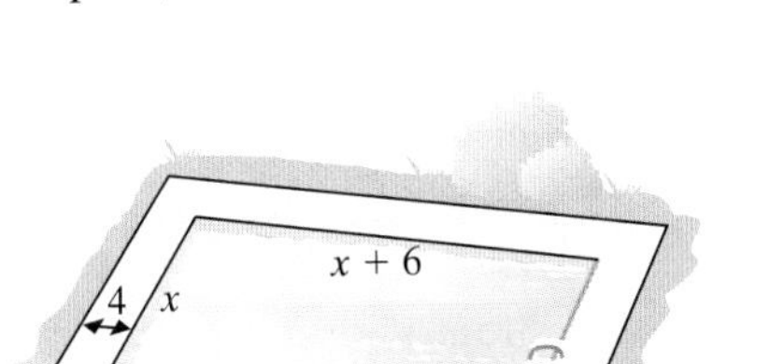

△ **52.** A rectangular garden is surrounded by a walk of uniform width. The area of the garden is 180 square yards. If the dimensions of the garden plus the walk are 16 yards by 24 yards, find the width of the walk.

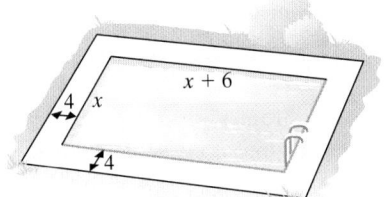

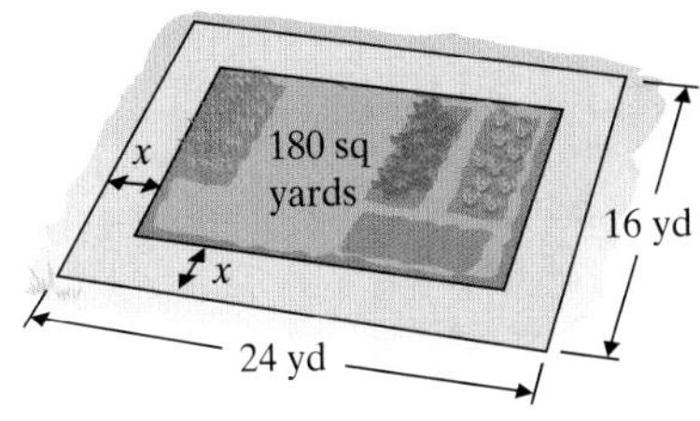

Internet Excursions

Go To:  http://www.prenhall.com/martin-gay_prealgebra 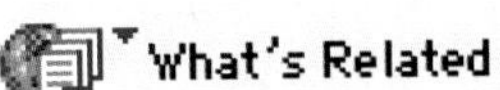

The United States Postal Service uses five-digit ZIP codes to simplify mail distribution. Each ZIP code corresponds to a unique post office location. If you know a ZIP code, you can look up the city and state associated with it by visiting the above World Wide Web address where you will gain access to the United States Postal Service's City/State/ZIP Code Associations Web site, or a related site.

53. The first three digits in the ZIP codes of two cities are 704. The remaining two-digit numbers in each ZIP code have a sum of 54, and the sum of their squares is 1556. What are the two ZIP codes? Use the given link to look up the cities and states associated with these ZIP codes.

54. The first three digits in the ZIP codes of two cities are 347. The remaining two-digit numbers in each ZIP code have a sum of 40, and the sum of their squares is 962. What are the two ZIP codes? Use the given link to look up the cities and states associated with these ZIP codes.

CHAPTER 11 ACTIVITY Factoring

Choosing Among Building Options: Shaunesa has just had a 10-foot-by-15-foot, in-ground swimming pool installed in her backyard. She has $3000 left from the building project that she would like to spend on surrounding the pool with a patio of constant width (see the figure). She has talked to several local suppliers about options for building this patio and must choose among the following.

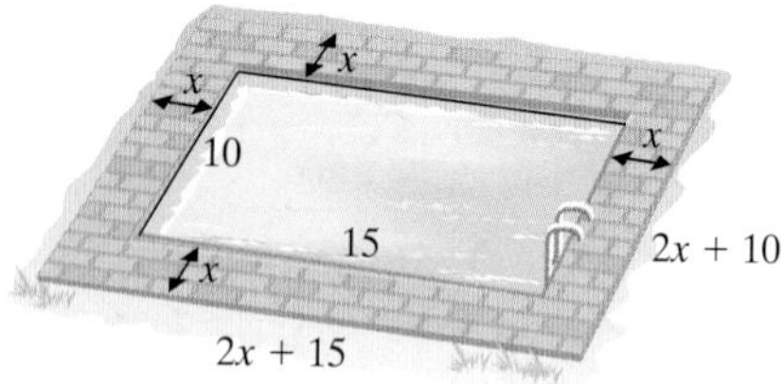

Option	Material	Price
A	Poured cement	$5 per square foot
B	Brick	$7.50 per square foot plus a $30 flat fee for delivering the bricks
C	Outdoor carpeting	$4.50 per square foot plus $10.86 per foot of the pool's perimeter to install an edging

1. Find the area of the swimming pool.
2. Write an algebraic expression for the total area of the region containing both the pool and patio.
3. Use subtraction to find an algebraic expression for the area of just the patio (not including the pool).
4. Find the perimeter of the swimming pool alone.
5. For each patio material option, write an algebraic expression for the total cost of installing the patio based on its area and the given price information.
6. If Shaunesa plans to spend the entire $3000 she has saved for the patio, how wide can the patio in option A be?
7. If Shaunesa plans to spend the entire $3000 she has saved for the patio, how wide can the patio in option B be?
8. If Shaunesa plans to spend the entire $3000 she has saved for the patio, how wide can the patio in option C be?
9. Which option should Shaunesa choose? Why? Discuss the pros and cons of each option.

Chapter 11 VOCABULARY CHECK

Fill in each blank with one of the words or phrases listed below.

factoring | quadratic equation | perfect square trinomial
greatest common factor

1. An equation that can be written in the form $ax^2 + bx + c = 0$ (with a not 0) is called a ________________.
2. ____________ is the process of writing an expression as a product.
3. The ________________________ of a list of terms is the product of all common factors.
4. A trinomial that is the square of some binomial is called a ________________________.

CHAPTER 11 Highlights

DEFINITIONS AND CONCEPTS	EXAMPLES
Section 11.1 The Greatest Common Factor	
Factoring is the process of writing an expression as a product.	Factor: $6 = 2 \cdot 3$ Factor: $x^2 + 5x + 6 = (x + 2)(x + 3)$
The GCF of a list of variable terms contains the smallest exponent on each common variable.	The GCF of z^5, z^3, and z^{10} is z^3.
The GCF of a list of terms is the product of all common factors.	Find the GCF of $8x^2y, 10x^3y^2$, and $50x^2y^3$. $8x^2y = 2 \cdot 2 \cdot 2 \cdot x^2 \cdot y$ $10x^3y^2 = 2 \cdot 5 \cdot x^3 \cdot y^2$ $50x^2y^3 = 2 \cdot 5 \cdot 5 \cdot x^2 \cdot y^3$ $\text{GCF} = 2 \cdot x^2 \cdot y$ or $2x^2y$
TO FACTOR BY GROUPING **Step 1.** Group the terms into two groups of two terms. **Step 2.** Factor out the GCF from each group. **Step 3.** If there is a common binomial factor, factor it out. **Step 4.** If not, rearrange the terms and try Steps 1–3 again.	Factor: $10ax + 15a - 6xy - 9y$ **Step 1.** $(10ax + 15a) + (-6xy - 9y)$ **Step 2.** $5a(2x + 3) - 3y(2x + 3)$ **Step 3.** $(2x + 3)(5a - 3y)$
Section 11.2 Factoring Trinomials of the Form $x^2 + bx + c$	
The sum of these numbers is b. $x^2 + bx + c = (x + \square)(x + \square)$ The product of these numbers is c.	Factor: $x^2 + 7x + 12$ $3 + 4 = 7$ $\quad 3 \cdot 4 = 12$ $x^2 + 7x + 12 = (x + 3)(x + 4)$
Section 11.3 Factoring Trinomials of the Form $ax^2 + bx + c$	
To factor $ax^2 + bx + c$, try various combinations of factors of ax^2 and c until a middle term of bx is obtained when checking.	Factor: $3x^2 + 14x - 5$ Factors of $3x^2$: $3x, x$ Factors of -5: $-1, 5$ and $1, -5$. $(3x - 1)(x + 5)$ $-1x$ $+ 15x$ $14x$ Correct middle term

DEFINITIONS AND CONCEPTS — EXAMPLES

Section 11.4 Factoring Trinomials of the Form $ax^2 + bx + c$, by Grouping

TO FACTOR $ax^2 + bx + c$ BY GROUPING

Step 1. Find two numbers whose product is $a \cdot c$ and whose sum is b.

Step 2. Rewrite bx, using the factors found in Step 1.

Step 3. Factor by grouping.

Factor: $3x^2 + 14x - 5$

Step 1. Find two numbers whose product is $3 \cdot (-5)$ or -15 and whose sum is 14. They are 15 and -1.

Step 2. $3x^2 + 14x - 5$
$= 3x^2 + 15x - 1x - 5$

Step 3. $= 3x(x + 5) - 1(x + 5)$
$= (x + 5)(3x - 1)$

Section 11.5 Factoring Perfect Square Trinomials and the Difference of Two Squares

A **perfect square trinomial** is a trinomial that is the square of some binomial.

PERFECT SQUARE TRINOMIAL = SQUARE OF BINOMIAL

$$x^2 + 4x + 4 = (x + 2)^2$$
$$25x^2 - 10x + 1 = (5x - 1)^2$$

Factoring Perfect Square Trinomials

$$a^2 + 2ab + b^2 = (a + b)^2$$
$$a^2 - 2ab + b^2 = (a - b)^2$$

Factor.

$$x^2 + 6x + 9 = x^2 + 2(x \cdot 3) + 3^2 = (x + 3)^2$$
$$4x^2 - 12x + 9 = (2x)^2 - 2(2x \cdot 3) + 3^2 = (2x - 3)^2$$

Difference of Two Squares

$$a^2 - b^2 = (a + b)(a - b)$$

Factor.

$$x^2 - 9 = x^2 - 3^2 = (x + 3)(x - 3)$$

Section 11.6 Solving Quadratic Equations by Factoring

A **quadratic equation** is an equation that can be written in the form $ax^2 + bx + c = 0$ with a not 0.

The form $ax^2 + bx + c = 0$ is called the **standard form** of a quadratic equation.

Quadratic Equation	Standard Form
$x^2 = 16$	$x^2 - 16 = 0$
$y = -2y^2 + 5$	$2y^2 + y - 5 = 0$

Zero Factor Property

If a and b are real numbers and if $ab = 0$, then $a = 0$ or $b = 0$.

If $(x + 3)(x - 1) = 0$, then $x + 3 = 0$ or $x - 1 = 0$.

TO SOLVE QUADRATIC EQUATIONS BY FACTORING

Step 1. Write the equation in standard form so that one side of the equation is 0.

Step 2. Factor completely.

Step 3. Set each factor containing a variable equal to 0.

Step 4. Solve the resulting equations.

Step 5. Check solutions in the original equation.

Solve: $3x^2 = 13x - 4$

Step 1. $3x^2 - 13x + 4 = 0$

Step 2. $(3x - 1)(x - 4) = 0$

Step 3. $3x - 1 = 0$ or $x - 4 = 0$

Step 4. $3x = 1$ $\qquad x = 4$

$x = \frac{1}{3}$

Step 5. Check both $\frac{1}{3}$ and 4 in the original equation.

Definitions and Concepts | **Examples**

Section 11.7 Quadratic Equations and Problem Solving

Problem-Solving Steps

A garden is in the shape of a rectangle whose length is two feet more than its width. If the area of the garden is 35 square feet, find its dimensions.

1. UNDERSTAND the problem.

1. Read and reread the problem. Guess a solution and check your guess. Draw a diagram.

   Let x be the width of the rectangular garden. Then $x + 2$ is the length.

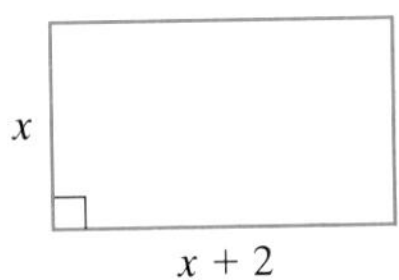

2. TRANSLATE.

2. length · width = area

$$(x + 2) \cdot x = 35$$

3. SOLVE.

3.
$$(x + 2)x = 35$$
$$x^2 + 2x - 35 = 0$$
$$(x - 5)(x + 7) = 0$$
$$x - 5 = 0 \quad \text{or} \quad x + 7 = 0$$
$$x = 5 \qquad\qquad x = -7$$

4. INTERPRET.

4. Discard the solution of -7 since x represents width.

   Check: If x is 5 feet, then $x + 2 = 5 + 2 = 7$ feet. The area of a rectangle whose width is 5 feet and whose length is 7 feet is (5 feet)(7 feet) or 35 square feet.

   State: The garden is 5 feet by 7 feet.

STUDY SKILLS REMINDER

Are you prepared for a test on Chapter 11?

Below is a list of some *common trouble areas* for topics covered in Chapter 11. After studying for your test—but before taking your test—read these.

- Don't forget that the first step to factor any polynomial is to first factor out any common factors.

  $$9x^2 - 36 = 9(x^2 - 4) = 9(x + 2)(x - 2)$$

- Can you completely factor $x^4 - 24x^2 - 25$?

  $$\begin{aligned} x^4 - 24x^2 - 25 &= (x^2 - 25)(x^2 + 1) \\ &= (x + 5)(x - 5)(x^2 + 1) \end{aligned}$$

- Remember that to use the zero factor property to solve a quadratic equation, one side of the equation must be 0 and the other side must be a factored polynomial.

  $$\begin{aligned} x(x - 2) &= 3 && \text{Cannot use zero factor property.} \\ x^2 - 2x - 3 &= 0 \\ (x - 3)(x + 1) &= 0 && \text{Now we can use zero factor property.} \end{aligned}$$

  $$x - 3 = 0 \quad \text{or} \quad x + 1 = 0$$
  $$x = 3 \quad \text{or} \quad x = -1$$

Remember: This is simply a sampling of selected topics given to check your understanding. For a review of Chapter 11 in your text, see the material at the end of this chapter.

Name ______________________ Section __________ Date __________

Chapter 11 Review

(11.1) *Complete each factoring.*

1. $6x^2 - 15x = 3x(\qquad)$

2. $4x^5 + 2x - 10x^4 = 2x(\qquad\qquad)$

Factor out the GCF from each polynomial.

3. $5m + 30$

4. $20x^3 + 12x^2 + 24x$

5. $3x(2x + 3) - 5(2x + 3)$

6. $5x(x + 1) - (x + 1)$

Factor each polynomial by grouping.

7. $3x^2 - 3x + 2x - 2$

8. $6x^2 + 10x - 3x - 5$

9. $3a^2 + 9ab + 3b^2 + ab$

(11.2) *Factor each trinomial.*

10. $x^2 + 6x + 8$

11. $x^2 - 11x + 24$

12. $x^2 + x + 2$

13. $x^2 - 5x - 6$

14. $x^2 + 2x - 8$

15. $x^2 + 4xy - 12y^2$

16. $x^2 + 8xy + 15y^2$

17. $72 - 18x - 2x^2$

18. $32 + 12x - 4x^2$

19. $5y^3 - 50y^2 + 120y$

20. To factor $x^2 + 2x - 48$, think of two numbers whose product is ________ and whose sum is ________.

21. What is the first step to factoring $3x^2 + 15x + 30$?

(11.3) or (11.4) *Factor each trinomial.*

22. $2x^2 + 13x + 6$

23. $4x^2 + 4x - 3$

24. $6x^2 + 5xy - 4y^2$

25. $x^2 - x + 2$

26. $2x^2 - 23x - 39$

27. $18x^2 - 9xy - 20y^2$

28. $10y^3 + 25y^2 - 60y$

Write the perimeter of each figure as a simplified polynomial. Then factor each polynomial.

△ **29.**

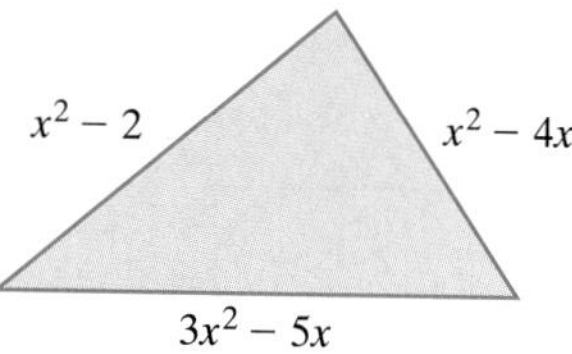

△ **30.**

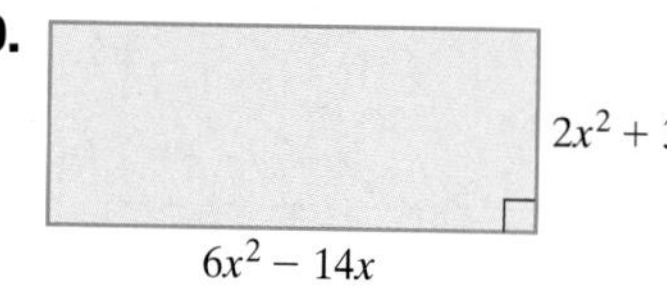

(11.5) *Determine whether each polynomial is a perfect square trinomial.*

31. $x^2 + 6x + 9$ **32.** $x^2 + 8x + 64$ **33.** $9m^2 - 12m + 16$ **34.** $4y^2 - 28y + 49$

Determine whether each binomial is a difference of two squares.

35. $x^2 - 9$ **36.** $x^2 + 16$ **37.** $4x^2 - 25y^2$ **38.** $9a^3 - 1$

Factor each polynomial completely.

39. $x^2 - 81$ **40.** $x^2 + 12x + 36$ **41.** $4x^2 - 9$ **42.** $9t^2 - 25s^2$

43. $16x^2 + y^2$ **44.** $n^2 - 18n + 81$ **45.** $3r^2 + 36r + 108$ **46.** $9y^2 - 42y + 49$

47. $5m^8 - 5m^6$ **48.** $4x^2 - 28xy + 49y^2$ **49.** $3x^2y + 6xy^2 + 3y^3$ **50.** $16x^4 - 1$

(11.6) *Solve each equation.*

51. $(x + 6)(x - 2) = 0$ **52.** $3x(x + 1)(7x - 2) = 0$ **53.** $4(5x + 1)(x + 3) = 0$

54. $x^2 + 8x + 7 = 0$ **55.** $x^2 - 2x - 24 = 0$ **56.** $x^2 + 10x = -25$

57. $x(x - 10) = -16$

58. $(3x - 1)(9x^2 + 3x + 1) = 0$

59. $56x^2 - 5x - 6 = 0$

60. $m^2 = 6m$

61. $r^2 = 25$

62. Write a quadratic equation that has the two solutions 4 and 5.

(11.7) *Use the given information to choose the correct dimensions.*

△ **63.** The perimeter of a rectangle is 24 inches. The length is twice the width. Find the dimensions of the rectangle.
- **a.** 5 inches by 7 inches
- **b.** 5 inches by 10 inches
- **c.** 4 inches by 8 inches
- **d.** 2 inches by 10 inches

△ **64.** The area of a rectangle is 80 meters. The length is one more than three times the width. Find the dimensions of the rectangle.
- **a.** 8 meters by 10 meters
- **b.** 4 meters by 13 meters
- **c.** 4 meters by 20 meters
- **d.** 5 meters by 16 meters

Use the given information to find the dimensions of each figure.

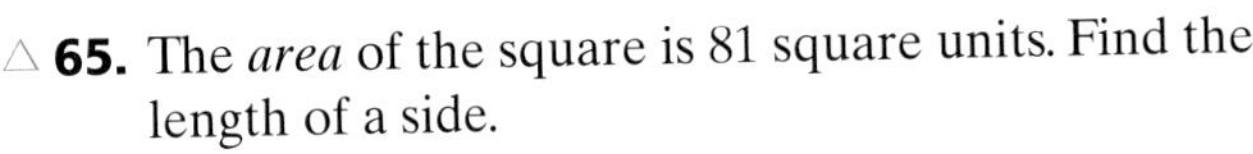

△ **65.** The *area* of the square is 81 square units. Find the length of a side.

△ **66.** The *perimeter* of the quadrilateral is 47 units. Find the lengths of the sides.

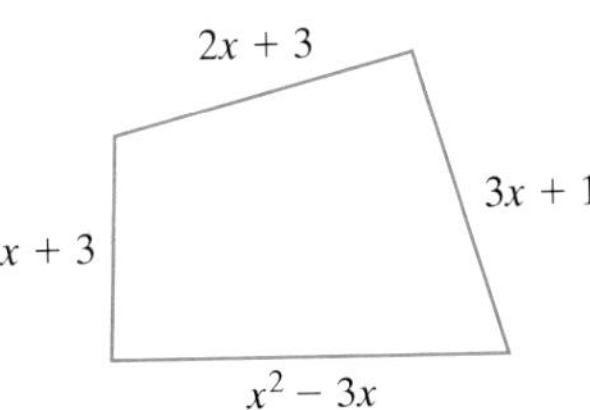

Solve.

△ **67.** A flag for a local organization is in the shape of a rectangle whose length is 15 inches less than twice its width. If the area of the flag is 500 square inches, find its dimensions.

△ **68.** The base of a triangular sail is four times its height. If the area of the triangle is 162 square yards, find the base.

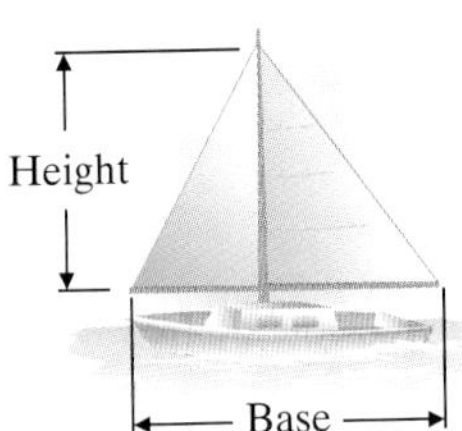

69. Find two consecutive positive integers whose product is 380.

70. A rocket is fired from the ground with an initial velocity of 440 feet per second. Its height h after t seconds is given by the equation $h = -16t^2 + 440t$.

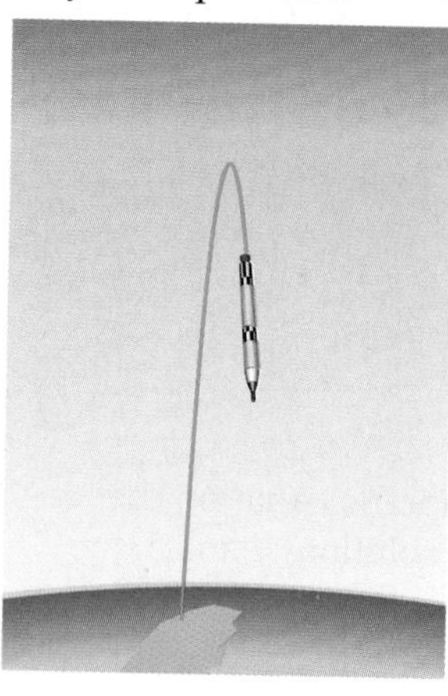

a. Find how many seconds pass before the rocket reaches a height of 2800 feet. Explain why two answers are obtained.

b. Find how many seconds pass before the rocket reaches the ground again.

△ **71.** An architect's squaring instrument is in the shape of a right triangle. Find the length of the longer leg of the right triangle if the hypotenuse is 8 centimeters longer than the longer leg and the shorter leg is 8 centimeters shorter than the longer leg.

Name ______ Section ______ Date ______

Answers

1. ______
2. ______
3. ______
4. ______
5. ______
6. ______
7. ______
8. ______
9. ______
10. ______
11. ______
12. ______
13. ______
14. ______
15. ______
16. ______
17. ______
18. ______
19. ______

Chapter 11 Test

Remember to check your answers and use the Chapter Test Prep Video to view solutions.

Factor each polynomial completely. If a polynomial cannot be factored, write "prime."

1. $9x^2 - 3x$

2. $x^2 + 11x + 28$

3. $y^2 + 22y + 121$

4. $4(a + 3) - y(a + 3)$

5. $y^2 - 8y - 48$

6. $3a^2 + 3ab - 7a - 7b$

7. $3x^2 - 5x + 2$

8. $180 - 5x^2$

9. $3x^3 - 21x^2 + 30x$

10. $6t^2 - t - 5$

11. $xy^2 - 7y^2 - 4x + 28$

12. $x - x^5$

13. $x^2 + 14xy + 24y^2$

Solve each equation.

14. $x^2 + 5x = 14$

15. $x(x + 6) = 7$

16. $3x(2x - 3)(3x + 4) = 0$

17. $5t^3 - 45t = 0$

18. $t^2 - 2t - 15 = 0$

19. $(x - 1)(3x^2 - x - 2) = 0$

20. ______

21. ______

22. ______

23. ______

Solve.

△ **20.** A deck for a home is in the shape of a triangle. The length of the base of the triangle is 9 feet longer than its height. If the area of the triangle is 68 square feet find the length of the base.

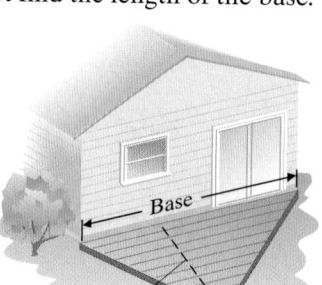

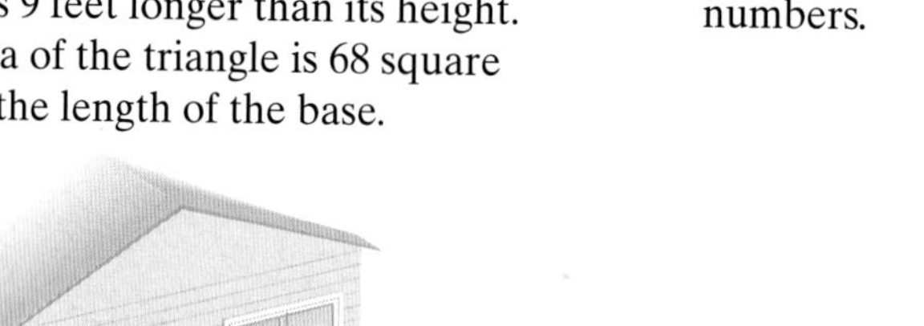

21. The sum of two numbers is 17, and the sum of their squares is 145. Find the numbers.

22. A window washer is suspended 38 feet below the roof of the 1127-foot-tall John Hancock Center in Chicago. (*Source:* Council on Tall Buildings and Urban Habitat) If the window washer drops an object from this height, the object's height h after t seconds is given by the equation $h = -16t^2 + 1089$. Find how many seconds pass before the object reaches the ground.

△ **23.** Find the lengths of the sides of a right triangle if the hypotenuse is 10 centimeters longer than the shorter leg and 5 centimeters longer than the longer leg.

Name ______________________ Section __________ Date __________

Answers

1. ______
2. ______
3. ______
4. ______
5. ______
6. ______
7. ______
8. ______
9. ______
10. ______
11. ______
12. ______
13. ______
14. ______
15. ______
16. ______

Chapter 11 Cumulative Review

1. Evaluate $x + 2y - z$ for $x = 3$, $y = -5$, and $z = -4$.

2. Evaluate $5x - y$ for $x = -2$ and $y = 4$.

3. Evaluate $7 - x^2$ for $x = -4$.

4. Evaluate $x^3 - y^3$ for $x = -2$ and $y = 4$.

Simplify.

5. $2y - 6 + 4y + 8$

6. $-5a - 3 + a + 2$

7. $4x + 2 - 5x + 3$

8. $2.3x + 5x - 6$

9. Decide whether 6 is a solution of the equation $4(x - 3) = 12$

10. Decide whether 2 is a solution of $3x + 10 = 8x$.

Solve.

11. $3(3x - 5) = 10x$

12. $2(7x - 1) = 15x$

13. $-12x = -36$

14. $-7x = 42$

15. $\frac{x - 5}{3} = \frac{x + 2}{5}$

16. $\frac{x + 1}{2} = \frac{x - 7}{3}$

17. a. ______

b. ______

c. ______

18.a. ______

b. ______

19. ______

20. ______

21. ______

22. ______

23. ______

24. ______

25. ______

26. ______

27. ______

28. ______

29. ______

30. ______

31. ______

32. ______

17. Translate each sentence into a mathematical statement.

a. Nine is less than or equal to eleven.

b. Eight is greater than one.

c. Three is not equal to four.

18. Translate each sentence into a mathematical statement.

a. Five is greater than or equal to one.

b. Two is not equal to negative four.

19. Solve: $3(x - 4) = 3x - 12$

20. Solve: $2(x + 5) = 2x + 9$

21. Solve for l: $V = lwh$

22. Solve for x: $3x - 2y = 5$

Simplify each expression.

23. $(5^3)^6$ **24.** $(7^9)^2$ **25.** $(y^8)^2$ **26.** $(x^{11})^3$

Simplify the following expressions. Write each result using positive exponents only.

27. $\dfrac{(x^3)^4 x}{x^7}$ **28.** $\dfrac{(y^3)^9 \cdot y}{y^6}$ **29.** $(y^{-3}z^6)^{-6}$

30. $(xy^{-4})^{-2}$ **31.** $\dfrac{x^{-7}}{(x^4)^3}$ **32.** $\dfrac{(y^2)^5}{y^{-3}}$

Simplify each polynomial by combining any like terms.

33. $-3x + 7x$ **34.** $y + y$ **35.** $11x^2 + 5 + 2x^2 - 7$

36. $8y - y^2 + 4y - y^2$ **37.** Multiply: $(2x - y)^2$ **38.** Multiply: $(3x + 1)^2$

Use a special product to square each binomial.

39. $(t + 2)^2$ **40.** $(x - 4)^2$ **41.** $(x^2 - 7y)^2$ **42.** $(x^2 + 7y)^2$

43. Divide: $\dfrac{8x^2y^2 - 16xy + 2x}{4xy}$ **44.** Divide: $\dfrac{20a^2b^3 - 5ab + 10b}{5ab}$

Factor.

45. $5(x + 3) + y(x + 3)$ **46.** $9(y - 2) + x(y - 2)$

47. $x^4 + 5x^2 + 6$ **48.** $x^4 - 4x^2 - 5$

49. $6x^2 - 2x - 20$ **50.** $10x^2 + 25x + 10$

51. For a TV commercial, a piece of luggage is dropped from a cliff 256 feet above the ground to show the durability of the luggage. Neglecting air resistance, the height h in feet of the luggage above the ground after t seconds is given by the quadratic equation

$h = -16t^2 + 256$

Find how long it takes for the luggage to hit the ground.

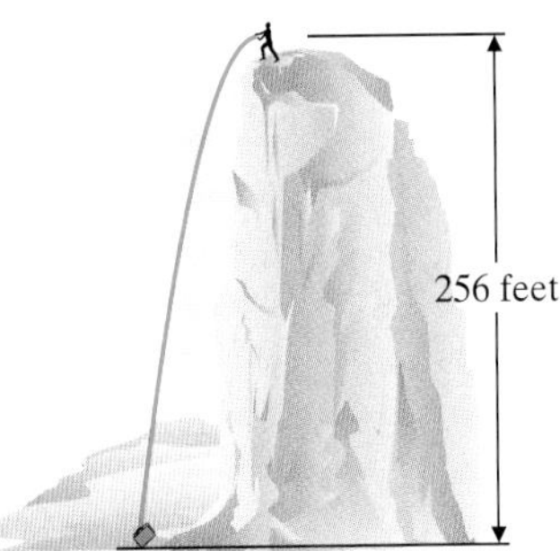

52. The square of a number plus twice the number is 120.

find the number

33. __________

34. __________

35. __________

36. __________

37. __________

38. __________

39. __________

40. __________

41. __________

42. __________

43. __________

44. __________

45. __________

46. __________

47. __________

48. __________

49. __________

50. __________

51. __________

52. __________

Rational Expressions

In this chapter, we expand our knowledge of algebraic expressions to include algebraic fractions, called **rational expressions.** We explore the operations of addition, subtraction, multiplication, and division using principles similar to the principles for numerical fractions.

Rational expressions can be found in formulas for many real-world situations. In Section 12.1, Exercise 71, you will find a rational expression used in a formula to calculate a baseball player's slugging percentage. Slugging percentage is a statistic that gauges a hitter's power. In essence, slugging percentage gives the average number of bases a hitter takes per batting attempt. A strong batter should be able to hit the ball farther, allowing him or her to get farther around the bases on a hit, which will result in a higher slugging percentage. Babe Ruth holds the record for slugging percentage in a single major league season with 0.8472 in 1920, as well as the record for career slugging percentage with 0.6897 spanning his major league career from 1914 to 1935. More recently, single-season slugging percentage leaders in the major leagues have included Todd Helton (0.6983) in 2000, Larry Walker (0.7100) in 1999, and Mark McGwire (0.7525) in 1998.

Answers

1. ______
2. ______
3. ______
4. ______
5. ______
6. ______
7. ______
8. ______
9. ______
10. ______
11. ______
12. ______
13. ______
14. ______
15. ______
16. ______
17. ______
18. ______
19. ______

Name ______ Section ______ Date ______

Chapter 12 Pretest

1. Find any real numbers for which the following expression is undefined.

$$\frac{x + 2}{x^2 - 9x - 10}$$

2. Simplify: $\dfrac{4x + 32}{x^2 + 10x + 16}$

3. Find the LCD for the following rational expressions.

$$\frac{1}{5x + 10}, \frac{3}{2x^2 + 10x + 12}$$

Perform the indicated operations and simplify if possible.

4. $\dfrac{y^2 - 8y + 7}{2y - 14} \cdot \dfrac{6y + 18}{y^2 + 2y - 3}$

5. $\dfrac{5x^3}{x^2 - 25} \div \dfrac{x^6}{(x + 5)^2}$

6. $\dfrac{b}{b^2 - 9b - 22} + \dfrac{2}{b^2 - 9b - 22}$

7. $\dfrac{3}{x - 1} - 4$

8. $\dfrac{2}{x - 5} - \dfrac{7}{5 - x}$

9. $\dfrac{x}{x^2 - 16} + \dfrac{3}{x^2 - 7x + 12}$

Solve each equation.

10. $\dfrac{5}{b} + \dfrac{3}{5} = \dfrac{4}{5b}$

11. $9 + \dfrac{7}{d - 7} = \dfrac{d}{d - 7}$

12. $\dfrac{4y + 5}{y^2 + 5y + 6} + \dfrac{3}{y + 3} = \dfrac{2}{y + 2}$

13. Solve the equation for the indicated variable. $\dfrac{2A}{b} = h$; for b.

Simplify each complex fraction.

14. $\dfrac{\dfrac{12m^3}{5n^2}}{\dfrac{4m^6}{25n^8}}$

15. $\dfrac{16 - \dfrac{1}{a^2}}{\dfrac{4}{a} + \dfrac{1}{a^2}}$

△ **16.** Given that the two triangles are similar, find x.

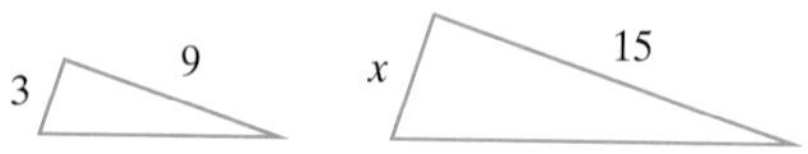

17. A number added to the product of 10 and the reciprocal of the number equals 7. Find the number.

18. Sonya can wash the windows in her house in 5 hours. Her daughter completes the same job in 8 hours. Find how long it takes if they work together.

19. Tom flies his airplane 495 miles with a tail wind of 25 miles per hour. Against the wind, he flies only 405 miles in the same amount of time. Find the rate of the plane in still air.

12.1 Simplifying Rational Expressions

OBJECTIVES

A. Find the value of a rational expression given a replacement number.

B. Identify values for which a rational expression is undefined.

C. Simplify, or write rational expressions in lowest terms.

A Evaluating Rational Expressions

A rational number is a number that can be written as a quotient of integers. A *rational expression* is also a quotient; it is a quotient of polynomials. Examples are

$$\frac{2}{3},\quad \frac{3y^3}{8},\quad \frac{-4p}{p^3+2p+1},\quad \text{and}\quad \frac{5x^2-3x+2}{3x+7}$$

Rational Expression

A **rational expression** is an expression that can be written in the form

$$\frac{P}{Q}$$

where P and Q are polynomials and $Q \neq 0$.

Rational expressions have different numerical values depending on what values replace the variables.

EXAMPLE 1

Find the numerical value of $\frac{x+4}{2x-3}$ for each replacement value.

a. $x = 5$ **b.** $x = -2$

Solution:

a. We replace each x in the expression with 5 and then simplify.

$$\frac{x+4}{2x-3} = \frac{5+4}{2(5)-3} = \frac{9}{10-3} = \frac{9}{7}$$

b. We replace each x in the expression with -2 and then simplify.

$$\frac{x+4}{2x-3} = \frac{-2+4}{2(-2)-3} = \frac{2}{-7} \quad \text{or} \quad -\frac{2}{7}$$

Practice Problem 1

Find the value of $\frac{x-3}{5x+1}$ for each replacement value.

a. $x = 4$

b. $x = -3$

In the example above, we wrote $\frac{2}{-7}$ as $-\frac{2}{7}$. For a negative fraction such as $\frac{2}{-7}$, recall from Section 1.7 that

$$\frac{2}{-7} = \frac{-2}{7} = -\frac{2}{7}$$

In general, for any fraction,

$$\frac{-a}{b} = \frac{a}{-b} = -\frac{a}{b}, \quad b \neq 0$$

This is also true for rational expressions. For example,

$$\underbrace{\frac{-(x+2)}{x}} = \frac{x+2}{-x} = -\frac{x+2}{x}$$

Notice the parentheses.

Answers

1. a. $\frac{1}{21}$ **b.** $\frac{-6}{-14} = \frac{3}{7}$

B Identifying When a Rational Expression Is Undefined

In the preceding box, notice that we wrote $b \neq 0$ for the denominator b. The denominator of a rational expression must not equal 0 since division by 0 is not defined. This means we must be careful when replacing the variable in a rational expression by a number. For example, suppose we replace x with 5 in the rational expression $\frac{2 + x}{x - 5}$. The expression becomes

$$\frac{2 + x}{x - 5} = \frac{2 + 5}{5 - 5} = \frac{7}{0}$$

But division by 0 is undefined. Therefore, in this expression we can allow x to be any real number *except* 5. **A rational expression is undefined for values that make the denominator 0.**

Practice Problem 2

Are there any values for x for which each rational expression is undefined?

a. $\frac{x}{x + 2}$ b. $\frac{x - 3}{x^2 + 5x + 4}$

c. $\frac{x^2 - 3x + 2}{5}$

EXAMPLE 2

Are there any values for x for which each expression is undefined?

a. $\frac{x}{x - 3}$ **b.** $\frac{x^2 + 2}{x^2 - 3x + 2}$ **c.** $\frac{x^3 - 6x^2 - 10x}{3}$

Solution: To find values for which a rational expression is undefined, we find values that make the denominator 0.

a. The denominator of $\frac{x}{x - 3}$ is 0 when $x - 3 = 0$ or when $x = 3$. Thus, when $x = 3$, the expression $\frac{x}{x - 3}$ is undefined.

b. We set the denominator equal to 0.

$$x^2 - 3x + 2 = 0$$

$$(x - 2)(x - 1) = 0 \quad \text{Factor.}$$

$$x - 2 = 0 \quad \text{or} \quad x - 1 = 0 \quad \text{Set each factor equal to 0.}$$

$$x = 2 \qquad\qquad x = 1 \quad \text{Solve.}$$

Thus, when $x = 2$ or $x = 1$, the denominator $x^2 - 3x + 2$ is 0. So the rational expression $\frac{x^2 + 2}{x^2 - 3x + 2}$ is undefined when $x = 2$ or when $x = 1$.

c. The denominator of $\frac{x^3 - 6x^2 - 10x}{3}$ is never 0, so there are no values of x for which this expression is undefined.

C Simplifying Rational Expressions

A fraction is said to be written in lowest terms or simplest form when the numerator and denominator have no common factors other than 1 (or −1). For example, the fraction $\frac{7}{10}$ is written in lowest terms since the numerator and denominator have no common factors other than 1 (or −1).

The process of writing a rational expression in lowest terms or simplest form is called **simplifying** a rational expression. The following **fundamental principle of rational expressions** is used to simplify a rational expression.

Fundamental Principle of Rational Expressions

If $\frac{P}{Q}$ is a rational expression and R is a nonzero polynomial, then

$$\frac{PR}{QR} = \frac{P}{Q}$$

Answers

2. a. $x = -2$ **b.** $x = -4, x = -1$ **c.** no

Simplifying a rational expression is similar to simplifying a fraction.

Simplify: $\frac{15}{20}$

$$\frac{15}{20} = \frac{3 \cdot 5}{2 \cdot 2 \cdot 5} \quad \text{Factor the numerator and the denominator.}$$

$$= \frac{3 \cdot 5}{2 \cdot 2 \cdot 5} \quad \text{Look for common factors.}$$

$$= \frac{3}{2 \cdot 2} = \frac{3}{4} \quad \text{Apply the fundamental principle.}$$

Simplify: $\frac{x^2 - 9}{x^2 + x - 6}$

$$\frac{x^2 - 9}{x^2 + x - 6} = \frac{(x - 3)(x + 3)}{(x - 2)(x + 3)} \quad \text{Factor the numerator and the denominator.}$$

$$= \frac{(x - 3)(x + 3)}{(x - 2)(x + 3)} \quad \text{Look for common factors.}$$

$$= \frac{x - 3}{x - 2} \quad \text{Apply the fundamental principle.}$$

Thus, the rational expression $\frac{x^2 - 9}{x^2 + x - 6}$ has the same value as the rational expression $\frac{x - 3}{x - 2}$ for all values of x except 2 and -3. (Remember that when x is 2, the denominator of both rational expressions is 0 and when x is -3, the original rational expression has a denominator of 0.)

As we simplify rational expressions, we will assume that the simplified rational expression is equal to the original rational expression for all real numbers except those for which either denominator is 0. The following steps may be used to simplify rational expressions.

To Simplify a Rational Expression

Step 1. Completely factor the numerator and denominator.

Step 2. Apply the fundamental principle of rational expressions to divide out common factors.

EXAMPLE 3 Simplify: $\frac{5x - 5}{x^3 - x^2}$

Solution: To begin, we factor the numerator and denominator if possible. Then we apply the fundamental principle.

$$\frac{5x - 5}{x^3 - x^2} = \frac{5(x - 1)}{x^2(x - 1)} = \frac{5}{x^2}$$

Practice Problem 3

Simplify: $\frac{x^4 + x^3}{5x + 5}$

EXAMPLE 4 Simplify: $\frac{x^2 + 8x + 7}{x^2 - 4x - 5}$

Solution: We factor the numerator and denominator and then apply the fundamental principle.

$$\frac{x^2 + 8x + 7}{x^2 - 4x - 5} = \frac{(x + 7)(x + 1)}{(x - 5)(x + 1)} = \frac{x + 7}{x - 5}$$

Practice Problem 4

Simplify: $\frac{x^2 + 11x + 18}{x^2 + x - 2}$

Answers

3. $\frac{x^3}{5}$ **4.** $\frac{x + 9}{x - 1}$

Practice Problem 5

Simplify: $\dfrac{x^2 + 10x + 25}{x^2 + 5x}$

EXAMPLE 5 Simplify: $\dfrac{x^2 + 4x + 4}{x^2 + 2x}$

Solution: We factor the numerator and denominator and then apply the fundamental principle.

$$\frac{x^2 + 4x + 4}{x^2 + 2x} = \frac{(x + 2)(x + 2)}{x(x + 2)} = \frac{x + 2}{x}$$

Helpful Hint

When simplifying a rational expression, the fundamental principle applies to **common *factors*, not common *terms*.**

$$\frac{x \cdot (x + 2)}{x \cdot x} = \frac{x + 2}{x}$$

Common factors. These can be divided out.

$$\frac{x + 2}{x}$$

Common terms. Fundamental principle does not apply. This is in simplest form.

Try the Concept Check in the margin.

Practice Problem 6

Simplify: $\dfrac{x + 5}{x^2 - 25}$

EXAMPLE 6 Simplify: $\dfrac{x + 9}{x^2 - 81}$

Solution: We factor and then apply the fundamental principle.

$$\frac{x + 9}{x^2 - 81} = \frac{x + 9}{(x + 9)(x - 9)} = \frac{1}{x - 9}$$

Practice Problem 7

Simplify each rational expression.

a. $\dfrac{x + 4}{4 + x}$ b. $\dfrac{x - 4}{4 - x}$

Concept Check

Recall that the fundamental principle applies to common factors only. Which of the following are *not* true? Explain why.

a. $\dfrac{3 - 1}{3 + 5} = -\dfrac{1}{5}$

b. $\dfrac{2x + 10}{2} = x + 5$

c. $\dfrac{37}{72} = \dfrac{3}{2}$

d. $\dfrac{2x + 3}{2} = x + 3$

EXAMPLE 7 Simplify each rational expression.

a. $\dfrac{x + y}{y + x}$ **b.** $\dfrac{x - y}{y - x}$

Solution:

a. The expression $\dfrac{x + y}{y + x}$ can be simplified by using the commutative property of addition to rewrite the denominator $y + x$ as $x + y$.

$$\frac{x + y}{y + x} = \frac{x + y}{x + y} = 1$$

b. The expression $\dfrac{x - y}{y - x}$ can be simplified by recognizing that $y - x$ and $x - y$ are opposites. In other words, $y - x = -1(x - y)$. We proceed as follows:

$$\frac{x - y}{y - x} = \frac{1 \cdot (x - y)}{(-1)(x - y)} = \frac{1}{-1} = -1$$

Answers

5. $\dfrac{x + 5}{x}$ **6.** $\dfrac{1}{x - 5}$ **7. a.** 1 **b.** −1

Concept Check: a, c, d

Name ______________________ Section __________ Date __________

Mental Math

Find any real numbers for which each rational expression is undefined. See Example 2.

1. $\frac{x + 5}{x}$ **2.** $\frac{x^2 - 5x}{x - 3}$ **3.** $\frac{x^2 + 4x - 2}{x(x - 1)}$ **4.** $\frac{x + 2}{(x - 5)(x - 6)}$

EXERCISE SET 12.1

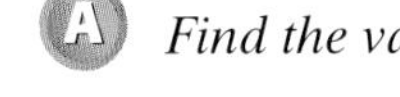

A *Find the value of the following expressions when $x = 2$, $y = -2$, and $z = -5$. See Example 1.*

1. $\frac{x + 5}{x + 2}$ **2.** $\frac{x + 8}{2x + 5}$ **3.** $\frac{y^3}{y^2 - 1}$

4. $\frac{z}{z^2 - 5}$ **5.** $\frac{x^2 + 8x + 2}{x^2 - x - 6}$ **6.** $\frac{x + 5}{x^2 + 4x - 8}$

7. The total revenue R from the sale of a popular music compact disc is approximately given by the equation

$$R = \frac{150x^2}{x^2 + 3}$$

where x is the number of years since the CD has been released and revenue R is in millions of dollars.

a. Find the total revenue generated by the end of the first year.

b. Find the total revenue generated by the end of the second year. Round to the nearest tenth.

c. Find the total revenue generated in the second year only.

8. For a certain model fax machine, the manufacturing cost C per machine is given by the equation

$$C = \frac{250x + 10{,}000}{x}$$

where x is the number of fax machines manufactured and cost C is in dollars per machine.

a. Find the cost per fax machine when manufacturing 100 fax machines.

b. Find the cost per fax machine when manufacturing 1000 fax machines.

c. Does the cost per machine decrease or increase when more machines are manufactured? Explain why this is so.

B *Find any real numbers for which each rational expression is undefined. See Example 2.*

9. $\frac{7}{2x}$ **10.** $\frac{3}{5x}$ **11.** $\frac{x + 3}{x + 2}$ **12.** $\frac{5x + 1}{x - 3}$ **13.** $\frac{4x^2 + 9}{2x - 8}$

14. $\frac{9x^3 + 4x}{15x + 45}$ **15.** $\frac{9x^3 + 4}{15x + 30}$ **16.** $\frac{19x^3 + 2}{x^3 - x}$ **17.** $\frac{x^2 - 5x - 2}{4}$ **18.** $\frac{9y^5 + y^3}{9}$

19. Explain why the denominator of a fraction or a rational expression must not equal 0.

20. Does $\frac{(x - 3)(x + 3)}{x - 3}$ have the same value as $x + 3$ for all real numbers? Explain why or why not.

Ⓒ *Simplify each expression. See Examples 3 through 7.*

21. $\frac{2}{8x + 16}$

22. $\frac{3}{9x + 6}$

23. $\frac{x - 2}{x^2 - 4}$

24. $\frac{x + 5}{x^2 - 25}$

25. $\frac{2x - 10}{3x - 30}$

26. $\frac{3x - 12}{4x - 16}$

27. $\frac{x + 7}{7 + x}$

28. $\frac{y + 9}{9 + y}$

29. $\frac{x - 7}{7 - x}$

30. $\frac{y - 9}{9 - y}$

31. $\frac{-5a - 5b}{a + b}$

32. $\frac{7x + 35}{x^2 + 5x}$

33. $\frac{x + 5}{x^2 - 4x - 45}$

34. $\frac{x - 3}{x^2 - 6x + 9}$

35. $\frac{5x^2 + 11x + 2}{x + 2}$

36. $\frac{12x^2 + 4x - 1}{2x + 1}$

37. $\frac{x + 7}{x^2 + 5x - 14}$

38. $\frac{x - 10}{x^2 - 17x + 70}$

39. $\frac{2x^2 + 3x - 2}{2x - 1}$

40. $\frac{4x^2 + 24x}{x + 6}$

41. $\frac{x^2 + 7x + 10}{x^2 - 3x - 10}$

42. $\frac{2x^2 + 7x - 4}{x^2 + 3x - 4}$

43. $\frac{3x^2 + 7x + 2}{3x^2 + 13x + 4}$

44. $\frac{4x^2 - 4x + 1}{2x^2 + 9x - 5}$

45. $\frac{2x^2 - 8}{4x - 8}$

46. $\frac{5x^2 - 500}{35x + 350}$

47. $\frac{11x^2 - 22x^3}{6x - 12x^2}$

48. $\frac{16r^2 - 4s^2}{4r - 2s}$

49. $\frac{2 - x}{x - 2}$

50. $\frac{7 - y}{y - 7}$

51. $\frac{x^2 - 1}{x^2 - 2x + 1}$

52. $\frac{x^2 - 16}{x^2 - 8x + 16}$

53. $\frac{m^2 - 6m + 9}{m^2 - 9}$

54. $\frac{m^2 - 4m + 4}{m^2 + m - 6}$

Review and Preview

Perform each indicated operation. See Section 4.3.

55. $\frac{1}{3} \cdot \frac{9}{11}$

56. $\frac{5}{27} \cdot \frac{2}{5}$

57. $\frac{5}{6} \cdot \frac{10}{11} \cdot \frac{2}{3}$

58. $\frac{4}{3} \cdot \frac{1}{7} \cdot \frac{10}{13}$

59. $\frac{1}{3} \div \frac{1}{4}$

60. $\frac{7}{8} \div \frac{1}{2}$

61. $\frac{13}{20} \div \frac{2}{9}$

62. $\frac{8}{15} \div \frac{5}{8}$

Combining Concepts

Simplify each expression. Each exercise contains a four-term polynomial that should be factored by grouping.

63. $\frac{x^2 + xy + 2x + 2y}{x + 2}$

64. $\frac{ab + ac + b^2 + bc}{b + c}$

65. $\frac{5x + 15 - xy - 3y}{2x + 6}$

66. $\frac{xy - 6x + 2y - 12}{y^2 - 6y}$

67. Explain how to write a fraction in lowest terms.

68. Explain how to write a rational expression in lowest terms.

69. A company's gross profit margin P can be computed with the formula $P = \frac{R - C}{R}$, where R = the company's revenue and C = the cost of goods sold. For fiscal year 2003, consumer electronics retailer Best Buy had revenues of \$20.9 billion and cost of goods sold of \$15.7 billion. (*Source:* Best Buy 2003 Annual Report) What was Best Buy's gross profit margin in 2003? Express the answer as a percent rounded to the nearest tenth of a percent.

70. During a storm, water treatment engineers monitor how quickly rain is falling. If too much rain comes too fast, there is a danger of sewers backing up. A formula that gives the rainfall intensity i in millimeters per hour for a certain strength storm in eastern Virginia is

$$i = \frac{5840}{t + 29}$$

where t is the duration of the storm in minutes. What rainfall intensity should engineers expect for a storm of this strength in eastern Virginia that lasts for 80 minutes? Round answer to one decimal place.

71. A baseball player's slugging percentage S can be calculated with the following formula:

$S = \frac{h + d + 2t + 3r}{b}$, where h = number of hits, d = number of doubles, t = number of triples, r = number of home runs, and b = number of at bats. During the 2003 season, Manny Ramirez of the Boston Red Sox had 472 at bats, 150 hits, 30 doubles, 1 triple, and 31 home runs. (*Source:* Major League Baseball) Calculate Ramirez's 2003 slugging percentage. Round to the nearest tenth of a percent.

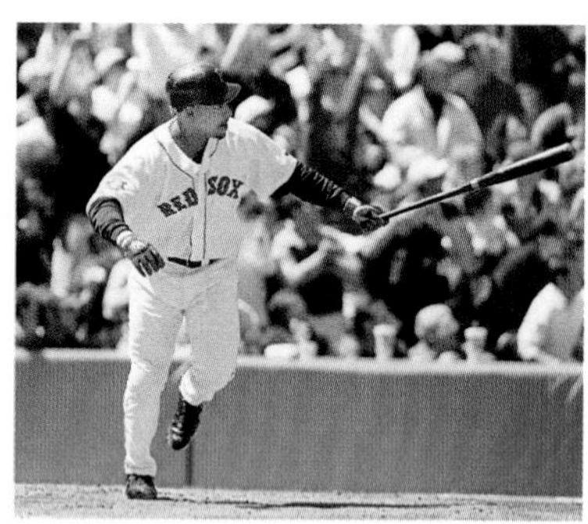

72. Anthropologists and forensic scientists use a measure called the cephalic index to help classify skulls. The cephalic index of a skull with width W and length L from front to back is given by the formula

$$C = \frac{100W}{L}$$

A long skull has an index value less than 75, a medium skull has an index value between 75 and 85, and a broad skull has an index value over 85. Find the cephalic index of a skull that is 5 inches wide and 6.4 inches long. Classify the skull.

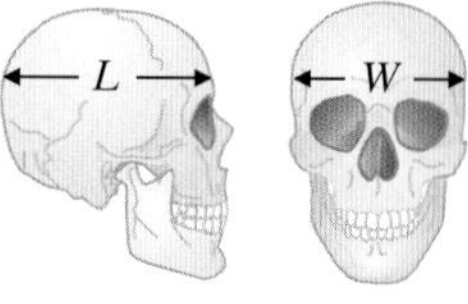

12.2 Multiplying and Dividing Rational Expressions

OBJECTIVES

- **A** Multiply rational expressions.
- **B** Divide rational expressions.
- **C** Multiply and divide rational expressions.
- **D** Convert between units of measure.

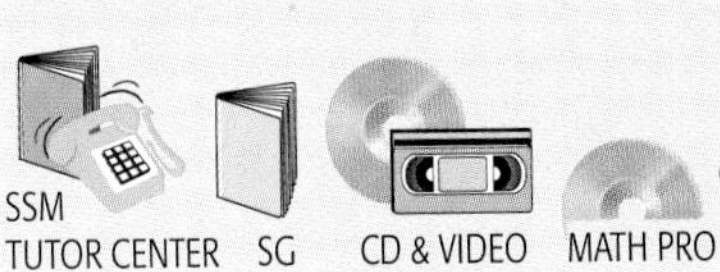

A Multiplying Rational Expressions

Just as simplifying rational expressions is similar to simplifying number fractions, multiplying and dividing rational expressions is similar to multiplying and dividing number fractions.

Multiply: $\frac{3}{5} \cdot \frac{10}{11}$	Multiply: $\frac{x-3}{x+5} \cdot \frac{2x+10}{x^2-9}$

Multiply numerators and then multiply denominators.

$\frac{3}{5} \cdot \frac{10}{11} = \frac{3 \cdot 10}{5 \cdot 11}$	$\frac{x-3}{x+5} \cdot \frac{2x+10}{x^2-9} = \frac{(x-3) \cdot (2x+10)}{(x+5) \cdot (x^2-9)}$

Simplify by factoring numerators and denominators.

$= \frac{3 \cdot 2 \cdot 5}{5 \cdot 11}$	$= \frac{(x-3) \cdot 2(x+5)}{(x+5)(x+3)(x-3)}$

Apply the fundamental principle of rational expressions.

$= \frac{3 \cdot 2}{11}$ or $\frac{6}{11}$	$= \frac{2}{x+3}$

Multiplying Rational Expressions

If $\frac{P}{Q}$ and $\frac{R}{S}$ are rational expressions, then

$$\frac{P}{Q} \cdot \frac{R}{S} = \frac{PR}{QS}$$

To multiply rational expressions, multiply the numerators and then multiply the denominators.

EXAMPLE 1 Multiply.

a. $\frac{25x}{2} \cdot \frac{1}{y^3}$ **b.** $\frac{-7x^2}{5y} \cdot \frac{3y^5}{14x^2}$

Solution: To multiply rational expressions, we first multiply the numerators and then multiply the denominators of both expressions. Then we write the product in lowest terms.

a. $\frac{25x}{2} \cdot \frac{1}{y^3} = \frac{25x \cdot 1}{2 \cdot y^3} = \frac{25x}{2y^3}$

The expression $\frac{25x}{2y^3}$ is in lowest terms.

b. $\frac{-7x^2}{5y} \cdot \frac{3y^5}{14x^2} = \frac{-7x^2 \cdot 3y^5}{5y \cdot 14x^2}$

Practice Problem 1

Multiply.

a. $\frac{16y}{3} \cdot \frac{1}{x^2}$ b. $\frac{-5a^3}{3b^3} \cdot \frac{2b^2}{15a}$

Answers

1. a. $\frac{16y}{3x^2}$ **b.** $-\frac{2a^2}{9b}$

The expression $\frac{-7x^2 \cdot 3y^5}{5y \cdot 14x^2}$ is not in lowest terms, so we factor the numerator and the denominator and apply the fundamental principle.

$$= \frac{-1 \cdot 7 \cdot 3 \cdot x^2 \cdot y \cdot y^4}{5 \cdot 2 \cdot 7 \cdot x^2 \cdot y}$$

$$= -\frac{3y^4}{10}$$

When multiplying rational expressions, it is usually best to factor each numerator and denominator first. This will help us when we apply the fundamental principle to write the product in lowest terms.

Practice Problem 2

Multiply: $\frac{6x+6}{7} \cdot \frac{14}{x^2-1}$

EXAMPLE 2 Multiply: $\frac{x^2+x}{3x} \cdot \frac{6}{5x+5}$

Solution:

$$\frac{x^2+x}{3x} \cdot \frac{6}{5x+5} = \frac{x(x+1)}{3x} \cdot \frac{2 \cdot 3}{5(x+1)}$$ Factor numerators and denominators.

$$= \frac{x(x+1) \cdot 2 \cdot 3}{3x \cdot 5(x+1)}$$ Multiply.

$$= \frac{2}{5}$$ Apply the fundamental principle.

The following steps may be used to multiply rational expressions.

To Multiply Rational Expressions

Step 1. Completely factor numerators and denominators.

Step 2. Multiply numerators and multiply denominators.

Step 3. Simplify or write the product in lowest terms by applying the fundamental principle to all common factors.

Concept Check

Which of the following is a true statement?

a. $\frac{1}{3} \cdot \frac{1}{2} = \frac{1}{5}$ b. $\frac{2}{x} \cdot \frac{5}{x} = \frac{10}{x}$

c. $\frac{3}{x} \cdot \frac{1}{2} = \frac{3}{2x}$

d. $\frac{x}{7} \cdot \frac{x+5}{4} = \frac{2x+5}{28}$

Try the Concept Check in the margin.

Practice Problem 3

Multiply: $\frac{4x+8}{7x^2-14x} \cdot \frac{3x^2-5x-2}{9x^2-1}$

EXAMPLE 3 Multiply: $\frac{3x+3}{5x^2-5x} \cdot \frac{2x^2+x-3}{4x^2-9}$

Solution:

$$\frac{3x+3}{5x^2-5x} \cdot \frac{2x^2+x-3}{4x^2-9} = \frac{3(x+1)}{5x(x-1)} \cdot \frac{(2x+3)(x-1)}{(2x-3)(2x+3)}$$ Factor.

$$= \frac{3(x+1)(2x+3)(x-1)}{5x(x-1)(2x-3)(2x+3)}$$ Multiply.

$$= \frac{3(x+1)}{5x(2x-3)}$$ Simplify.

Answers

2. $\frac{12}{x-1}$ **3.** $\frac{4(x+2)}{7x(3x-1)}$

Concept Check: c

B Dividing Rational Expressions

We can divide by a rational expression in the same way we divide by a number fraction. Recall that to divide by a fraction, we multiply by its reciprocal.

For example, to divide $\frac{3}{2}$ by $\frac{7}{8}$, we multiply $\frac{3}{2}$ by $\frac{8}{7}$.

$$\frac{3}{2} \div \frac{7}{8} = \frac{3}{2} \cdot \frac{8}{7} = \frac{3 \cdot 4 \cdot 2}{2 \cdot 7} = \frac{12}{7}$$

Helpful Hint

Don't forget how to find reciprocals. The reciprocal of $\frac{a}{b}$ is $\frac{b}{a}$, $a \neq 0$, $b \neq 0$.

Dividing Rational Expressions

If $\frac{P}{Q}$ and $\frac{R}{S}$ are rational expressions and $\frac{R}{S}$ is not 0, then

$$\frac{P}{Q} \div \frac{R}{S} = \frac{P}{Q} \cdot \frac{S}{R} = \frac{PS}{QR}$$

To divide two rational expressions, multiply the first rational expression by the reciprocal of the second rational expression.

EXAMPLE 4 Divide: $\frac{3x^3}{40} \div \frac{4x^3}{y^2}$

Solution:

$$\frac{3x^3}{40} \div \frac{4x^3}{y^2} = \frac{3x^3}{40} \cdot \frac{y^2}{4x^3}$$ Multiply by the reciprocal of $\frac{4x^3}{y^2}$.

$$= \frac{3x^3y^2}{160x^3}$$

$$= \frac{3y^2}{160}$$ Simplify.

Practice Problem 4

Divide: $\frac{7x^2}{6} \div \frac{x}{2y}$

EXAMPLE 5 Divide: $\frac{(x-1)(x+2)}{10} \div \frac{2x+4}{5}$

Solution:

$$\frac{(x-1)(x+2)}{10} \div \frac{2x+4}{5} = \frac{(x-1)(x+2)}{10} \cdot \frac{5}{2x+4}$$ Multiply by the reciprocal of $\frac{2x+4}{5}$.

$$= \frac{(x-1)(x+2) \cdot 5}{5 \cdot 2 \cdot 2 \cdot (x+2)}$$ Factor and multiply.

$$= \frac{x-1}{4}$$ Simplify.

Practice Problem 5

Divide: $\frac{(2x+3)(x-4)}{6} \div \frac{3x-12}{2}$

The following may be used to divide by a rational expression.

To Divide by a Rational Expression

Multiply by its reciprocal.

Answers

4. $\frac{7xy}{3}$ **5.** $\frac{2x+3}{9}$

Practice Problem 6

Divide: $\frac{10x + 4}{x^2 - 4} \div \frac{5x^3 + 2x^2}{x + 2}$

EXAMPLE 6 Divide: $\frac{6x + 2}{x^2 - 1} \div \frac{3x^2 + x}{x - 1}$

Solution:

$$\frac{6x + 2}{x^2 - 1} \div \frac{3x^2 + x}{x - 1} = \frac{6x + 2}{x^2 - 1} \cdot \frac{x - 1}{3x^2 + x}$$ Multiply by the reciprocal.

$$= \frac{2(3x + 1)(x - 1)}{(x + 1)(x - 1) \cdot x(3x + 1)}$$ Factor and multiply.

$$= \frac{2}{x(x + 1)}$$ Simplify.

Practice Problem 7

Divide: $\frac{3x^2 - 10x + 8}{7x - 14} \div \frac{9x - 12}{21}$

EXAMPLE 7 Divide: $\frac{2x^2 - 11x + 5}{5x - 25} \div \frac{4x - 2}{10}$

Solution:

$$\frac{2x^2 - 11x + 5}{5x - 25} \div \frac{4x - 2}{10} = \frac{2x^2 - 11x + 5}{5x - 25} \cdot \frac{10}{4x - 2}$$ Multiply by the reciprocal.

$$= \frac{(2x - 1)(x - 5) \cdot 2 \cdot 5}{5(x - 5) \cdot 2(2x - 1)}$$ Factor and multiply.

$$= \frac{1}{1} \text{ or } 1$$ Simplify.

C Multiplying and Dividing Rational Expressions

Let's make sure that we understand the difference between multiplying and dividing rational expressions.

Rational Expressions	
Multiplication	Multiply the numerators and multiply the denominators.
Division	Multiply by the reciprocal of the divisor.

Practice Problem 8

Multiply or divide as indicated.

a. $\frac{x + 3}{x} \cdot \frac{7}{x + 3}$

b. $\frac{x + 3}{x} \div \frac{7}{x + 3}$

c. $\frac{3 - x}{x^2 + 6x + 5} \cdot \frac{2x + 10}{x^2 - 7x + 12}$

EXAMPLE 8 Multiply or divide as indicated.

a. $\frac{x - 4}{5} \cdot \frac{x}{x - 4}$ **b.** $\frac{x - 4}{5} \div \frac{x}{x - 4}$

c. $\frac{x^2 - 4}{2x + 6} \cdot \frac{x^2 + 4x + 3}{2 - x}$

Solution:

a. $\frac{x - 4}{5} \cdot \frac{x}{x - 4} = \frac{(x - 4) \cdot x}{5 \cdot (x - 4)} = \frac{x}{5}$

b. $\frac{x - 4}{5} \div \frac{x}{x - 4} = \frac{x - 4}{5} \cdot \frac{x - 4}{x} = \frac{(x - 4)^2}{5x}$

c. $\frac{x^2 - 4}{2x + 6} \cdot \frac{x^2 + 4x + 3}{2 - x} = \frac{(x - 2)(x + 2) \cdot (x + 1)(x + 3)}{2(x + 3) \cdot (2 - x)}$ Factor and multiply.

Answers

6. $\frac{2}{x^2(x - 2)}$ **7.** 1 **8. a.** $\frac{7}{x}$ **b.** $\frac{(x + 3)^2}{7x}$ **c.** $-\frac{2}{(x + 1)(x - 4)}$

Recall from Section 12.1 that $x - 2$ and $2 - x$ are opposites. This means that $\frac{x-2}{2-x} = -1$. Thus,

$$\frac{(x-2)(x+2)\cdot(x+1)(x+3)}{2(x+3)\cdot(2-x)} = \frac{-1(x+2)(x+1)}{2}$$
$$= -\frac{(x+2)(x+1)}{2}$$

D Converting between Units of Measure

Now that we know how to multiply fractions and rational expressions, we can use this knowledge to help us convert between units of measure. To do so, we will use **unit fractions**. A unit fraction is a fraction that equals 1. For example, since 12 in. = 1 ft, we have the unit fractions

$$\frac{12 \text{ in.}}{1 \text{ ft}} = 1 \quad \text{and} \quad \frac{1 \text{ ft}}{12 \text{ in.}} = 1$$

EXAMPLE 9 Converting from Cubic Feet to Cubic Yards

The largest building in the world by volume is The Boeing Company's Everett, Washington, factory complex where Boeing's wide-body jetliners, the 747, 767, and 777, are built. The volume of this factory complex is 472,370,319 cubic feet. Find the volume of this Boeing facility in cubic yards. (*Source:* The Boeing Company)

Solution: There are 27 cubic feet in 1 cubic yard. (See the diagram.)

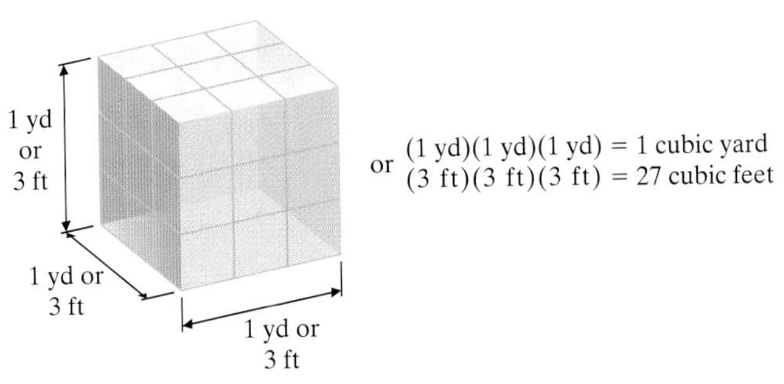

$$472{,}370{,}319 \text{ cu ft} = 472{,}370{,}319 \cancel{\text{cu ft}} \cdot \frac{1 \text{ cu yd}}{27 \cancel{\text{cu ft}}}$$
$$= \frac{472{,}370{,}319}{27} \text{ cu yd}$$
$$= 17{,}495{,}197 \text{ cu yd}$$

Practice Problem 9

The largest casino in the world is the Foxwoods Resort Casino in Ledyard, CT. The gaming area for this casino is 21,444 *square yards*. Find the size of the gaming area in *square feet*. (*Source: The Guinness Book of Records*)

Answer

9. 192,996 sq ft

Helpful Hint

When converting among units of measurement, if possible, write the unit fraction so that **the numerator contains the units you are converting to** and **the denominator contains the original units.**

$$48 \text{ in.} = \frac{48 \text{ in.}}{1} \cdot \underbrace{\frac{1 \text{ ft}}{12 \text{ in.}}}_{\text{Unit fraction}} \begin{matrix} \leftarrow \text{Units converting to} \\ \leftarrow \text{Original units} \end{matrix}$$

$$= \frac{48}{12} \text{ ft} = 4 \text{ ft}$$

Practice Problem 10

Man has been timed at a speed of approximately 40.9 feet per second. Convert this to miles per hour. Round to the nearest tenth. (*Source: World Almanac and Book of Facts*)

EXAMPLE 10 Converting from Feet per Second to Miles per Hour

At the 2000 Summer Olympics, U.S. athlete Marian Jones won the gold medal in the women's 100-meter track event. She ran the distance at an average speed of 30.5 feet per second. Convert this speed to miles per hour. (*Source: World Almanac and Book of Facts, 2003*)

Solution: Recall that 1 mile = 5280 feet and 1 hour = 3600 seconds$(60 \cdot 60)$.

Unit fractions

$$30.5 \text{ feet/second} = \frac{30.5 \text{ feet}}{1 \text{ second}} \cdot \frac{3600 \text{ seconds}}{1 \text{ hour}} \cdot \frac{1 \text{ mile}}{5280 \text{ feet}}$$

$$= \frac{30.5 \cdot 3600}{5280} \text{ miles/hour}$$

$$\approx 20.8 \text{ miles/hour (rounded to the nearest tenth)}$$

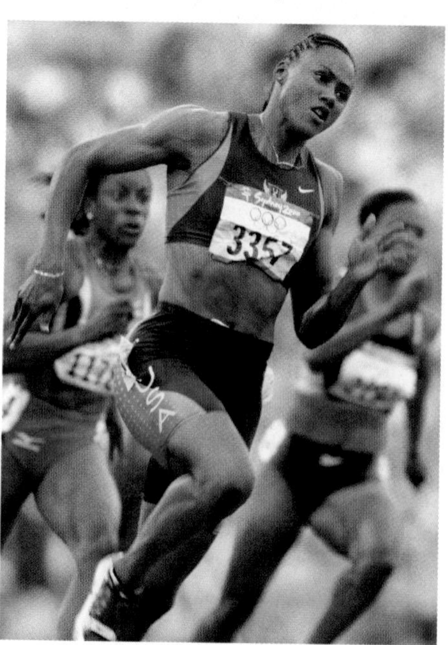

Answer

10. 27.9 mph

Name ________________ Section ________ Date ________

Mental Math

Find each product. See Example 1.

1. $\frac{2}{y} \cdot \frac{x}{3}$

2. $\frac{3x}{4} \cdot \frac{1}{y}$

3. $\frac{5}{7} \cdot \frac{y^2}{x^2}$

4. $\frac{x^5}{11} \cdot \frac{4}{z^3}$

5. $\frac{9}{x} \cdot \frac{x}{5}$

6. $\frac{y}{7} \cdot \frac{3}{y}$

EXERCISE SET 12.2

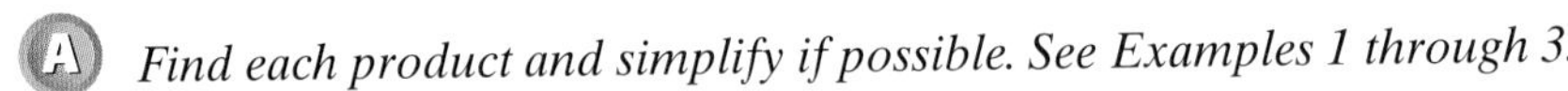

A *Find each product and simplify if possible. See Examples 1 through 3.*

1. $\frac{3x}{y^2} \cdot \frac{7y}{4x}$

2. $\frac{9x^2}{y} \cdot \frac{4y}{3x^2}$

3. $\frac{8x}{2} \cdot \frac{x^5}{4x^2}$

4. $\frac{6x^2}{10x^3} \cdot \frac{5x}{12}$

5. $-\frac{5a^2b}{30a^2b^2} \cdot b^3$

6. $-\frac{9x^3y^2}{18xy^5} \cdot y^3$

7. $\frac{x}{2x - 14} \cdot \frac{x^2 - 7x}{5}$

8. $\frac{4x - 24}{20x} \cdot \frac{5}{x - 6}$

9. $\frac{6x + 6}{5} \cdot \frac{10}{36x + 36}$

10. $\frac{x^2 + x}{8} \cdot \frac{16}{x + 1}$

11. $\frac{m^2 - n^2}{m + n} \cdot \frac{m}{m^2 - mn}$

12. $\frac{(m - n)^2}{m + n} \cdot \frac{m}{m^2 - mn}$

13. $\frac{x^2 - 25}{x^2 - 3x - 10} \cdot \frac{x + 2}{x}$

14. $\frac{a^2 + 6a + 9}{a^2 - 4} \cdot \frac{a + 3}{a - 2}$

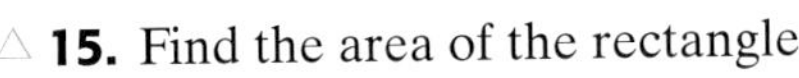

△ **15.** Find the area of the rectangle.

$\frac{2x}{x^2 - 25}$ feet

$\frac{x + 5}{9x}$ feet

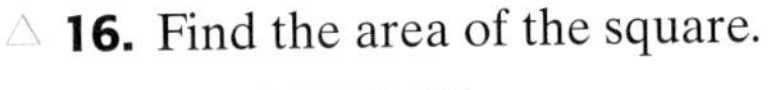

△ **16.** Find the area of the square.

$\frac{2x}{5x^2 + 3x}$ meters

B *Find each quotient and simplify. See Examples 4 through 7.*

17. $\frac{5x^7}{2x^5} \div \frac{10x}{4x^3}$

18. $\frac{9y^4}{6y} \div \frac{y^2}{3}$

19. $\frac{8x^2}{y^3} \div \frac{4x^2y^3}{6}$

20. $\frac{7a^2b}{3ab^2} \div \frac{21a^2b^2}{14ab}$

21. $\frac{(x-6)(x+4)}{4x} \div \frac{2x-12}{8x^2}$

22. $\frac{(x+3)^2}{5} \div \frac{5x+15}{25}$

23. $\frac{3x^2}{x^2-1} \div \frac{x^5}{(x+1)^2}$

24. $\frac{x+1}{(x+1)(2x+3)} \div \frac{20}{2x+3}$

25. $\frac{m^2-n^2}{m+n} \div \frac{m}{m^2+nm}$

26. $\frac{(m-n)^2}{m+n} \div \frac{m^2-mn}{m}$

27. $\frac{x+2}{7-x} \div \frac{x^2-5x+6}{x^2-9x+14}$

28. $(x-3) \div \frac{x^2+3x-18}{x}$

29. $\frac{x^2+7x+10}{x-1} \div \frac{x^2+2x-15}{x-1}$

30. $\frac{a^2-b^2}{9} \div \frac{3b-3a}{27x^2}$

C *Multiply or divide as indicated. See Example 8.*

31. $\frac{5x-10}{12} \div \frac{4x-8}{8}$

32. $\frac{6x+6}{5} \div \frac{3x+3}{10}$

33. $\frac{x^2+5x}{8} \cdot \frac{9}{3x+15}$

34. $\frac{3x^2+12x}{6} \cdot \frac{9}{2x+8}$

35. $\frac{7}{6p^2+q} \div \frac{14}{18p^2+3q}$

36. $\frac{3x+6}{20} \div \frac{4x+8}{8}$

37. $\frac{3x+4y}{x^2+4xy+4y^2} \cdot \frac{x+2y}{2}$

38. $\frac{x^2-y^2}{3x^2+3xy} \cdot \frac{3x^2+6x}{3x^2-2xy-y^2}$

39. $\frac{(x+2)^2}{x-2} \div \frac{x^2-4}{2x-4}$

40. $\frac{x^2-4}{2y} \div \frac{2-x}{6xy}$

41. $\frac{3y}{3-x} \div \frac{12xy}{x^2-9}$

42. $\frac{x+3}{x^2-9} \div \frac{5x+15}{(x-3)^2}$

43. $\dfrac{a^2 + 7a + 12}{a^2 + 5a + 6} \cdot \dfrac{a^2 + 8a + 15}{a^2 + 5a + 4}$

44. $\dfrac{b^2 + 2b - 3}{b^2 + b - 2} \cdot \dfrac{b^2 - 4}{b^2 + 6b + 8}$

D *Convert as indicated. See Examples 9 and 10.*

45. 10 square feet = ______ square inches.

46. 1008 square inches = ______ square feet.

47. 90,000 cubic inches = ______ cubic yards (round to the nearest hundredth).

48. 2 cubic yards = ______ cubic inches.

49. 50 miles per hour = ______ feet per second (round to the nearest whole).

50. 10 feet per second = ______ miles per hour (round to the nearest tenth).

51. The Pentagon, headquarters for the Department of Defense, contains 3,705,793 square feet of office and storage space. Convert this to square yards. Round to the nearest square yard. (*Source:* U.S. Department of Defense)

52. The world's tallest building, the Petronas Twin Towers in Kuala Lumpur, Malaysia, has 408,742 square yards of floor space. Convert this to square feet. (*Source:* KLCC Group)

53. In October 1997, driver Andy Green set the current world land speed record of 763 miles per hour in a specially designed car, Thrust SSC. Find this speed in feet per second. Round to the nearest whole. (*Source:* Federation International de L'Automobile)

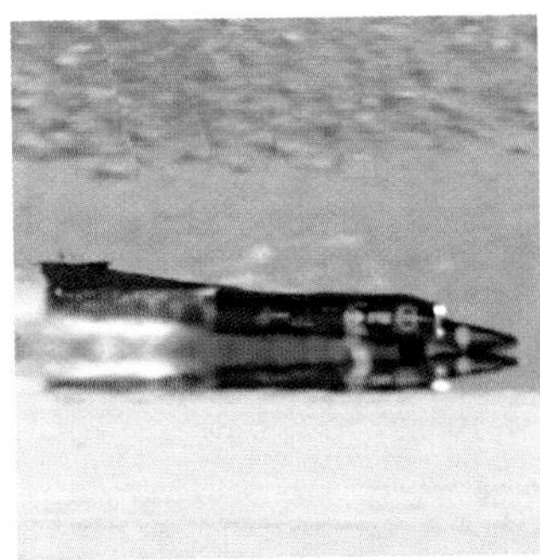

54. Maurice Greene of the United States holds the current world record for the men's 100-meter track event. In 1999, he covered the distance at an average speed of 33.5 feet per second. Convert this speed to miles per hour. Round to the nearest tenth. (*Source:* International Amateur Athletic Federation)

Review and Preview

Perform each indicated operation. See Section 4.4.

55. $\dfrac{1}{5} + \dfrac{4}{5}$

56. $\dfrac{3}{15} + \dfrac{6}{15}$

57. $\dfrac{9}{9} - \dfrac{19}{9}$

58. $\dfrac{4}{3} - \dfrac{8}{3}$

59. $\dfrac{6}{5} + \left(\dfrac{1}{5} - \dfrac{8}{5}\right)$

60. $-\dfrac{3}{2} + \left(\dfrac{1}{2} - \dfrac{3}{2}\right)$

Combining Concepts

Multiply or divide as indicated.

61. $\left(\dfrac{x^2 - y^2}{x^2 + y^2} \div \dfrac{x^2 - y^2}{3x}\right) \cdot \dfrac{x^2 + y^2}{6}$

62. $\left(\dfrac{x^2 - 9}{x^2 - 1} \cdot \dfrac{x^2 + 2x + 1}{2x^2 + 9x + 9}\right) \div \dfrac{2x + 3}{1 - x}$

63. $\left(\dfrac{2a + b}{b^2} \cdot \dfrac{3a^2 - 2ab}{ab + 2b^2}\right) \div \dfrac{a^2 - 3ab + 2b^2}{5ab - 10b^2}$

64. $\left(\dfrac{x^2y^2 - xy}{4x - 4y} \div \dfrac{3y - 3x}{8x - 8y}\right) \cdot \dfrac{y - x}{8}$

65. In your own words, explain how you multiply rational expressions.

66. Explain how dividing rational expressions is similar to dividing rational numbers.

67. On August 24, 2003, 1 euro was equivalent to 1.09 U.S. dollar. If you had wanted to exchange \$2000 U.S. to euros on that day for a European vacation, how much would you have received? Round to the nearest hundredth. (*Source:* International Monetary Fund)

68. An environmental technician finds that warm water from an industrial process is being discharged into a nearby pond at a rate of 30 gallons per minute. Plant regulations state that the flow rate should be no more than 0.1 cubic feet per second. Is the flow rate of 30 gallons per minute in violation of the plant regulations? (*Hint:* 1 cubic foot is equivalent to 7.48 gallons.)

12.3 Adding and Subtracting Rational Expressions with the Same Denominator and Least Common Denominators

OBJECTIVES

- **A** Add and subtract rational expressions with common denominators.
- **B** Find the least common denominator of a list of rational expressions.
- **C** Write a rational expression as an equivalent expression whose denominator is given.

SSM TUTOR CENTER SG CD & VIDEO MATH PRO WEB

A Adding and Subtracting Rational Expressions with the Same Denominator

Like multiplication and division, addition and subtraction of rational expressions is similar to addition and subtraction of rational numbers. In this section, we add and subtract rational expressions with a common denominator.

Add: $\frac{6}{5} + \frac{2}{5}$

Add: $\frac{9}{x+2} + \frac{3}{x+2}$

Add the numerators and place the sum over the common denominator.

$$\frac{6}{5} + \frac{2}{5} = \frac{6+2}{5}$$
$$= \frac{8}{5} \quad \text{Simplify.}$$

$$\frac{9}{x+2} + \frac{3}{x+2} = \frac{9+3}{x+2}$$
$$= \frac{12}{x+2} \quad \text{Simplify.}$$

Adding and Subtracting Rational Expressions with Common Denominators

If $\frac{P}{R}$ and $\frac{Q}{R}$ are rational expressions, then

$$\frac{P}{R} + \frac{Q}{R} = \frac{P+Q}{R} \quad \text{and} \quad \frac{P}{R} - \frac{Q}{R} = \frac{P-Q}{R}$$

To add or subtract rational expressions, add or subtract numerators and place the sum or difference over the common denominator.

EXAMPLE 1 Add: $\frac{5m}{2n} + \frac{m}{2n}$

Solution:

$$\frac{5m}{2n} + \frac{m}{2n} = \frac{5m+m}{2n} \quad \text{Add the numerators.}$$
$$= \frac{6m}{2n} \quad \text{Simplify the numerator by combining like terms.}$$
$$= \frac{3m}{n} \quad \text{Simplify by applying the fundamental principle.}$$

Practice Problem 1

Add: $\frac{8x}{3y} + \frac{x}{3y}$

EXAMPLE 2 Subtract: $\frac{2y}{2y-7} - \frac{7}{2y-7}$

Solution:

$$\frac{2y}{2y-7} - \frac{7}{2y-7} = \frac{2y-7}{2y-7} \quad \text{Subtract the numerators.}$$
$$= \frac{1}{1} \text{ or } 1 \quad \text{Simplify.}$$

Practice Problem 2

Subtract: $\frac{3x}{3x-7} - \frac{7}{3x-7}$

Answers

1. $\frac{3x}{y}$ **2.** 1

Practice Problem 3

Subtract: $\dfrac{2x^2 + 5x}{x + 2} - \dfrac{4x + 6}{x + 2}$

EXAMPLE 3 Subtract: $\dfrac{3x^2 + 2x}{x - 1} - \dfrac{10x - 5}{x - 1}$

Solution:

$$\frac{3x^2 + 2x}{x - 1} - \frac{10x - 5}{x - 1} = \frac{3x^2 + 2x - (10x - 5)}{x - 1}$$ Subtract the numerators. Notice the parentheses.

$$= \frac{3x^2 + 2x - 10x + 5}{x - 1}$$ Use the distributive property.

$$= \frac{3x^2 - 8x + 5}{x - 1}$$ Combine like terms.

$$= \frac{(x - 1)(3x - 5)}{x - 1}$$ Factor.

$$= 3x - 5$$ Simplify.

Helpful Hint

Notice how the numerator $10x - 5$ was subtracted in Example 3.

This − sign applies to the entire numerator of $10x - 5$.

So parentheses are inserted here to indicate this.

$$\frac{3x^2 + 2x}{x - 1} - \frac{10x - 5}{x - 1} = \frac{3x^2 + 2x - (10x - 5)}{x - 1}$$

B Finding the Least Common Denominator

Recall from Chapter 4 that to add and subtract fractions with different denominators, we first find a least common denominator (LCD). Then we write all fractions as equivalent fractions with the LCD.

For example, suppose we add $\frac{8}{3}$ and $\frac{2}{5}$. The LCD of denominators 3 and 5 is 15, since 15 is the smallest number that both 3 and 5 divide into evenly. So we rewrite each fraction so that its denominator is 15. (Notice how we apply the fundamental principle.)

$$\frac{8}{3} + \frac{2}{5} = \frac{8(5)}{3(5)} + \frac{2(3)}{5(3)} = \frac{40}{15} + \frac{6}{15} = \frac{40 + 6}{15} = \frac{46}{15}$$

To add or subtract rational expressions with different denominators, we also first find an LCD and then write all rational expressions as equivalent expressions with the LCD. The **least common denominator (LCD) of a list of rational expressions** is a polynomial of least degree whose factors include all the factors of the denominators in the list.

To Find the Least Common Denominator (LCD)

Step 1. Factor each denominator completely.

Step 2. The least common denominator (LCD) is the product of all unique factors found in Step 1, each raised to a power equal to the greatest number of times that the factor appears in any one factored denominator.

Answer

3. $2x - 3$

EXAMPLE 4 Find the LCD for each pair.

a. $\frac{1}{8}, \frac{3}{22}$ **b.** $\frac{7}{5x}, \frac{6}{15x^2}$

Solution:

a. We start by finding the prime factorization of each denominator.

$$8 = 2^3 \quad \text{and}$$
$$22 = 2 \cdot 11$$

Next we write the product of all the unique factors, each raised to a power equal to the greatest number of times that the factor appears.

The greatest number of times that the factor 2 appears is 3.
The greatest number of times that the factor 11 appears is 1.

$$\text{LCD} = 2^3 \cdot 11^1 = 8 \cdot 11 = 88$$

b. We factor each denominator.

$$5x = 5 \cdot x \quad \text{and}$$
$$15x^2 = 3 \cdot 5 \cdot x^2$$

The greatest number of times that the factor 5 appears is 1.
The greatest number of times that the factor 3 appears is 1.
The greatest number of times that the factor x appears is 2.

$$\text{LCD} = 3^1 \cdot 5^1 \cdot x^2 = 15x^2$$

EXAMPLE 5 Find the LCD of $\frac{7x}{x + 2}$ and $\frac{5x^2}{x - 2}$.

Solution: The denominators $x + 2$ and $x - 2$ are completely factored already. The factor $x + 2$ appears once and the factor $x - 2$ appears once.

$$\text{LCD} = (x + 2)(x - 2)$$

EXAMPLE 6 Find the LCD of $\frac{6m^2}{3m + 15}$ and $\frac{2}{(m + 5)^2}$.

Solution: We factor each denominator.

$$3m + 15 = 3(m + 5)$$
$$(m + 5)^2 = (m + 5)^2 \quad \text{This denominator is already factored.}$$

The greatest number of times that the factor 3 appears is 1.
The greatest number of times that the factor $m + 5$ appears *in any one denominator* is 2.

$$\text{LCD} = 3(m + 5)^2$$

Try the Concept Check in the margin.

EXAMPLE 7 Find the LCD of $\frac{t - 10}{t^2 - t - 6}$ and $\frac{t + 5}{t^2 + 3t + 2}$.

Solution:

$$t^2 - t - 6 = (t - 3)(t + 2)$$
$$t^2 + 3t + 2 = (t + 1)(t + 2)$$
$$\text{LCD} = (t - 3)(t + 2)(t + 1)$$

Practice Problem 4

Find the LCD for each pair.

a. $\frac{2}{9}, \frac{7}{15}$ b. $\frac{5}{6x^3}, \frac{11}{8x^5}$

Practice Problem 5

Find the LCD of $\frac{3a}{a + 5}$ and $\frac{7a}{a - 5}$.

Practice Problem 6

Find the LCD of $\frac{7x^2}{(x - 4)^2}$ and $\frac{5x}{3x - 12}$.

Concept Check

Choose the correct LCD of $\frac{x}{(x + 1)^2}$ and $\frac{5}{x + 1}$.

a. $x + 1$ b. $(x + 1)^2$
c. $(x + 1)^3$ d. $5x(x + 1)^2$

Practice Problem 7

Find the LCD of $\frac{y + 5}{y^2 + 2y - 3}$ and $\frac{y + 4}{y^2 - 3y + 2}$.

Answers

4. a. 45 **b.** $24x^5$ **5.** $(a + 5)(a - 5)$
6. $3(x - 4)^2$ **7.** $(y + 3)(y - 2)(y - 1)$
Concept Check: b

Practice Problem 8

Find the LCD of $\frac{6}{x-4}$ and $\frac{9}{4-x}$.

EXAMPLE 8 Find the LCD of $\frac{2}{x-2}$ and $\frac{10}{2-x}$.

Solution: The denominators $x - 2$ and $2 - x$ are opposites. That is, $2 - x = -1(x - 2)$. We can use either $x - 2$ or $2 - x$ as the LCD.

$$\text{LCD} = x - 2 \quad \text{or} \quad \text{LCD} = 2 - x$$

C Writing Equivalent Rational Expressions

Next we practice writing a rational expression as an equivalent rational expression with a given denominator. To do this, we apply the fundamental principle, which says that $\frac{PR}{QR} = \frac{P}{Q}$, or equivalently that $\frac{P}{Q} = \frac{PR}{QR}$. This can be seen by recalling that multiplying an expression by 1 produces an equivalent expression. In other words,

$$\frac{P}{Q} = \frac{P}{Q} \cdot 1 = \frac{P}{Q} \cdot \frac{R}{R} = \frac{PR}{QR}$$

Practice Problem 9

Write the rational expression as an equivalent rational expression with the given denominator.

$$\frac{2x}{5y} = \frac{\quad}{20x^2y^2}$$

EXAMPLE 9 Write the rational expression as an equivalent rational expression with the given denominator.

$$\frac{4b}{9a} = \frac{\quad}{27a^2b}$$

Solution: We can ask ourselves: "What do we multiply $9a$ by to get $27a^2b$?" The answer is $3ab$, since $9a(3ab) = 27a^2b$. So we multiply the numerator and denominator by $3ab$.

$$\frac{4b}{9a} = \frac{4b(3ab)}{9a(3ab)} = \frac{12ab^2}{27a^2b}$$

Practice Problem 10

Write the rational expression as an equivalent rational expression with the given denominator.

$$\frac{3}{x^2 - 25} = \frac{\quad}{(x+5)(x-5)(x-3)}$$

EXAMPLE 10 Write the rational expression as an equivalent rational expression with the given denominator.

$$\frac{5}{x^2 - 4} = \frac{\quad}{(x-2)(x+2)(x-4)}$$

Solution: First we factor the denominator $x^2 - 4$ as $(x - 2)(x + 2)$. If we multiply the original denominator $(x - 2)(x + 2)$ by $x - 4$, the result is the new denominator $(x - 2)(x + 2)(x - 4)$. Thus, we multiply the numerator and the denominator by $x - 4$.

$$\frac{5}{x^2-4} = \frac{5}{(x-2)(x+2)} = \frac{5(x-4)}{(x-2)(x+2)(x-4)}$$
$$= \frac{5x-20}{(x-2)(x+2)(x-4)}$$

Answers

8. $(x - 4)$ or $(4 - x)$ **9.** $\frac{8x^3y}{20x^2y^2}$

10. $\frac{3x-9}{(x+5)(x-5)(x-3)}$

Name ______________________ Section __________ Date __________

Mental Math

Perform each indicated operation.

1. $\frac{2}{3} + \frac{1}{3}$

2. $\frac{5}{11} + \frac{1}{11}$

3. $\frac{3x}{9} + \frac{4x}{9}$

4. $\frac{3y}{8} + \frac{2y}{8}$

5. $\frac{8}{9} - \frac{7}{9}$

6. $\frac{14}{12} - \frac{3}{12}$

7. $\frac{7y}{5} + \frac{10y}{5}$

8. $\frac{12x}{7} - \frac{4x}{7}$

EXERCISE SET 12.3

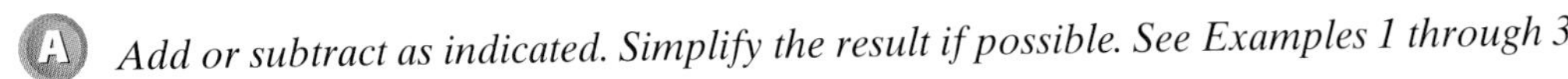

A *Add or subtract as indicated. Simplify the result if possible. See Examples 1 through 3.*

1. $\frac{a}{13} + \frac{9}{13}$

2. $\frac{x+1}{7} + \frac{6}{7}$

3. $\frac{4m}{3n} + \frac{5m}{3n}$

4. $\frac{3p}{2} + \frac{11p}{2}$

5. $\frac{4m}{m-6} - \frac{24}{m-6}$

6. $\frac{8y}{y-2} - \frac{16}{y-2}$

7. $\frac{9}{3+y} + \frac{y+1}{3+y}$

8. $\frac{9}{y+9} + \frac{y}{y+9}$

9. $\frac{5x+4}{x-1} - \frac{2x+7}{x-1}$

10. $\frac{x^2+9x}{x+7} - \frac{4x+14}{x+7}$

11. $\frac{a}{a^2+2a-15} - \frac{3}{a^2+2a-15}$

12. $\frac{3y}{y^2+3y-10} - \frac{6}{y^2+3y-10}$

13. $\frac{2x+3}{x^2-x-30} - \frac{x-2}{x^2-x-30}$

14. $\frac{3x-1}{x^2+5x-6} - \frac{2x-7}{x^2+5x-6}$

△ **15.** A square has a side of length $\frac{5}{x-2}$ meters. Express its perimeter as a rational expression.

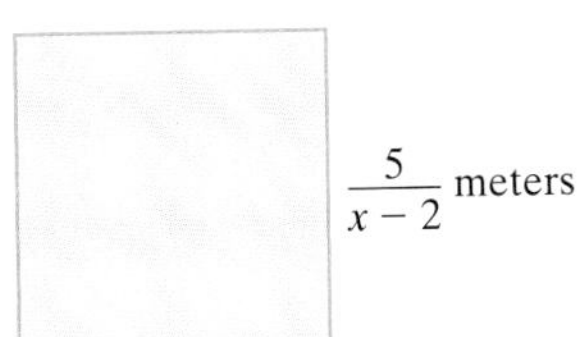

△ **16.** A trapezoid has sides of the indicated lengths. Find its perimeter.

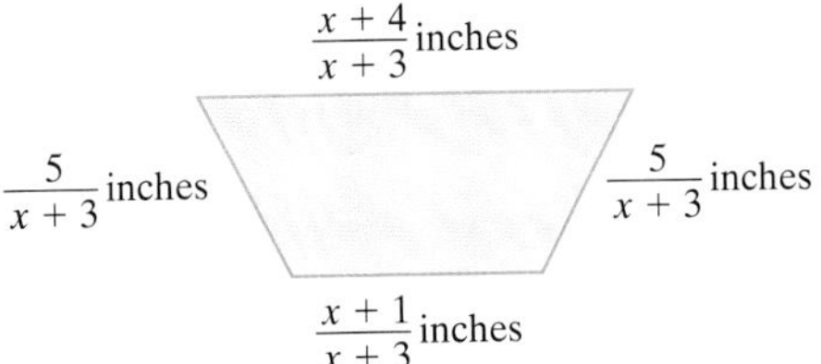

17. In your own words, describe how to add or subtract two rational expressions with the same denominators.

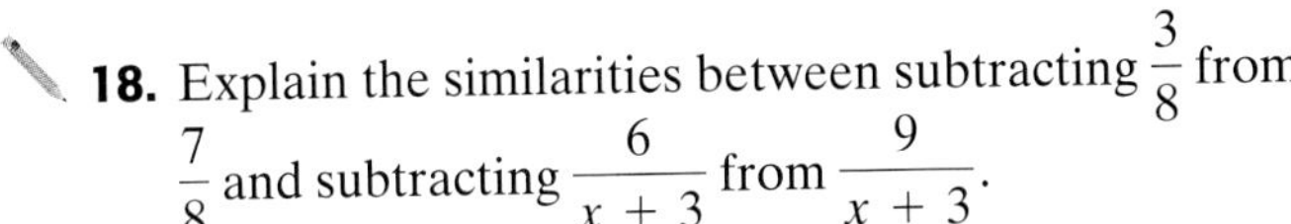

18. Explain the similarities between subtracting $\frac{3}{8}$ from $\frac{7}{8}$ and subtracting $\frac{6}{x+3}$ from $\frac{9}{x+3}$.

B *Find the LCD for each list of rational expressions. See Examples 4 through 8.*

19. $\frac{19}{2x}, \frac{5}{4x^3}$

20. $\frac{17x}{4y^5}, \frac{2}{8y}$

21. $\frac{9}{8x}, \frac{3}{2x + 4}$

22. $\frac{1}{6y}, \frac{3x}{4y + 12}$

23. $\frac{2}{x + 3}, \frac{5}{x - 2}$

24. $\frac{-6}{x - 1}, \frac{4}{x + 5}$

25. $\frac{x}{x + 6}, \frac{10}{3x + 18}$

26. $\frac{12}{x + 5}, \frac{x}{4x + 20}$

27. $\frac{1}{3x + 3}, \frac{8}{2x^2 + 4x + 2}$

28. $\frac{19x + 5}{4x - 12}, \frac{3}{2x^2 - 12x + 18}$

29. $\frac{5}{x - 8}, \frac{3}{8 - x}$

30. $\frac{2x + 5}{3x - 7}, \frac{5}{7 - 3x}$

31. $\frac{5x + 1}{x^2 + 3x - 4}, \frac{3x}{x^2 + 2x - 3}$

32. $\frac{4}{x^2 + 4x + 3}, \frac{4x - 2}{x^2 + 10x + 21}$

33. Write some instructions to help a friend who is having difficulty finding the LCD of two rational expressions.

34. Explain why the LCD of the rational expressions $\frac{7}{x + 1}$ and $\frac{9x}{(x + 1)^2}$ is $(x + 1)^2$ and not $(x + 1)^3$.

C *Rewrite each rational expression as an equivalent rational expression with the given denominator. See Examples 9 and 10.*

35. $\frac{3}{2x} = \frac{}{4x^2}$

36. $\frac{3}{9y^5} = \frac{}{72y^9}$

37. $\frac{6}{3a} = \frac{}{12ab^2}$

38. $\frac{17a}{4y^2x} = \frac{}{32y^3x^2z}$

39. $\frac{9}{x + 3} = \frac{}{2(x + 3)}$

40. $\frac{4x + 1}{3x + 6} = \frac{}{3y(x + 2)}$

41. $\frac{9a + 2}{5a + 10} = \frac{}{5b(a + 2)}$

42. $\frac{5 + y}{2x^2 + 10} = \frac{}{4(x^2 + 5)}$

43. $\frac{x}{x^3 + 6x^2 + 8x} = \frac{}{x(x + 4)(x + 2)(x + 1)}$

44. $\frac{5x}{x^2 + 2x - 3} = \frac{}{(x - 1)(x - 5)(x + 3)}$

45. $\frac{9y - 1}{15x^2 - 30} = \frac{}{30x^2 - 60}$

46. $\frac{6}{x^2 - 9} = \frac{}{(x + 3)(x - 3)(x + 2)}$

Review and Preview

Perform each indicated operation. See Section 4.5.

47. $\frac{2}{3} + \frac{5}{7}$

48. $\frac{9}{10} - \frac{3}{5}$

49. $\frac{2}{6} - \frac{3}{4}$

50. $\frac{11}{15} + \frac{5}{9}$

51. $\frac{1}{12} + \frac{3}{20}$

52. $\frac{7}{30} + \frac{3}{18}$

Combining Concepts

53. You are throwing a barbecue and you want to make sure that you purchase the same number of hot dogs as hot dog buns. Hot dogs come 8 to a package and hot dog buns come 12 to a package. What is the least number of each type of package you should buy?

54. The planet Mercury revolves around the sun in 88 Earth days. It takes Jupiter 4332 Earth days to make one revolution around the sun. (*Source:* National Space Science Data Center) If the two planets are aligned as shown in the figure, how long will it take for them to align again?

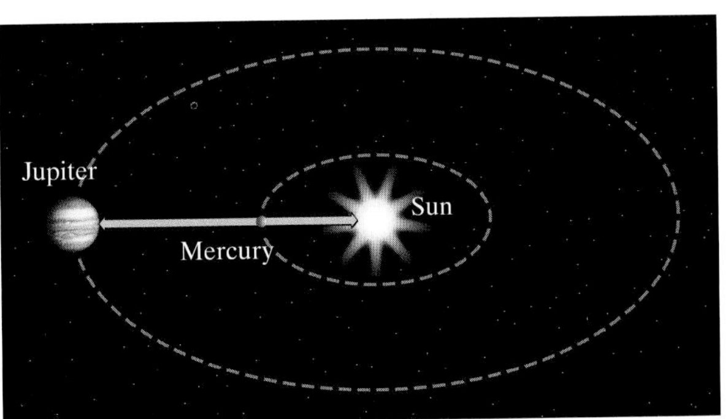

55. An algebra student approaches you with a problem. He's tried to subtract two rational expressions, but his result does not match the book's. Check to see if the student has made an error. If so, correct his work shown below.

$$\frac{2x-6}{x-5} - \frac{x+4}{x-5}$$
$$= \frac{2x-6-x+4}{x-5}$$
$$= \frac{x-2}{x-5}$$

FOCUS ON Business and Career

Fast-Growing Careers

According to U.S. Bureau of Labor Statistics projections, the careers listed below will have the largest job growth in the next decade.

	Employment (Numbers in thousands)		
Occupation	1998	2008	Change
1. Systems analysts	617	1,194	+577
2. Retail salespersons	4,056	4,620	+563
3. Cashiers	3,198	3,754	+556
4. General managers and top executives	3,362	3,913	+551
5. Truck drivers, light and heavy	2,970	3,463	+493
6. Office clerks, general	3,021	3,484	+463
7. Registered nurses	2,079	2,530	+451
8. Computer support specialists	429	869	+439
9. Personal care and home health aides	746	1,179	+433
10. Teacher assistants	1,192	1,567	+375

(*Source:* Bureau of Labor Statistics, U.S. Department of Labor)

What do all of these in-demand occupations have in common? They all require a knowledge of math! For some careers like systems analysts, salespersons, cashiers, and nurses, the ways math is used on the job may be obvious. For other occupations, the use of math may not be quite as obvious. However, tasks common to many jobs like filling in a time sheet or a mileage log, writing up an expense report, planning a budget, figuring a bill, ordering supplies, completing a packing list, and even making a work schedule all require math.

Critical Thinking

Suppose that your college placement office is planning to publish an occupational handbook on math in popular occupations. Choose one of the occupations from the list above that interests you. Research the occupation. Then write a brief entry for the occupational handbook that describes how a person in that career would use math in his or her job. Include an example if possible.

12.4 Adding and Subtracting Rational Expressions with Different Denominators

OBJECTIVE

A Add and subtract rational expressions with different denominators.

A Adding and Subtracting Rational Expressions with Different Denominators

In the previous section, we practiced all the skills we need to add and subtract rational expressions with different denominators. The steps are as follows.

To Add or Subtract Rational Expressions with Different Denominators

Step 1. Find the LCD of the rational expressions.

Step 2. Rewrite each rational expression as an equivalent expression whose denominator is the LCD found in Step 1.

Step 3. Add or subtract numerators and write the sum or difference over the common denominator.

Step 4. Simplify or write the rational expression in lowest terms.

EXAMPLE 1 Perform each indicated operation.

a. $\frac{a}{4} - \frac{2a}{8}$ **b.** $\frac{3}{10x^2} + \frac{7}{25x}$

Solution:

a. First, we must find the LCD. Since $4 = 2^2$ and $8 = 2^3$, the LCD $= 2^3 = 8$. Next we write each fraction as an equivalent fraction with the denominator 8, and then we subtract.

$$\frac{a}{4} - \frac{2a}{8} = \frac{a(2)}{4(2)} - \frac{2a}{8} = \frac{2a}{8} - \frac{2a}{8} = \frac{2a - 2a}{8} = \frac{0}{8} = 0$$

b. Since $10x^2 = 2 \cdot 5 \cdot x \cdot x$ and $25x = 5 \cdot 5 \cdot x$, the LCD $= 2 \cdot 5^2 \cdot x^2 = 50x^2$. We write each fraction as an equivalent fraction with a denominator of $50x^2$.

$$\frac{3}{10x^2} + \frac{7}{25x} = \frac{3(5)}{10x^2(5)} + \frac{7(2x)}{25x(2x)}$$

$$= \frac{15}{50x^2} + \frac{14x}{50x^2}$$

$$= \frac{15 + 14x}{50x^2}$$ Add numerators. Write the sum over the common denominator.

EXAMPLE 2 Subtract: $\frac{6x}{x^2 - 4} - \frac{3}{x + 2}$

Solution: Since $x^2 - 4 = (x + 2)(x - 2)$, the LCD $= (x - 2)(x + 2)$. We write equivalent expressions with the LCD as denominators.

Practice Problem 1

Perform each indicated operation.

a. $\frac{y}{5} - \frac{3y}{15}$ b. $\frac{5}{8x} + \frac{11}{10x^2}$

Practice Problem 2

Subtract: $\frac{10x}{x^2 - 9} - \frac{5}{x + 3}$

Answers

1. a. 0 **b.** $\frac{25x + 44}{40x^2}$ **2.** $\frac{5}{x - 3}$

$$\frac{6x}{x^2-4}-\frac{3}{x+2}=\frac{6x}{(x-2)(x+2)}-\frac{3(x-2)}{(x+2)(x-2)}$$

$$=\frac{6x-3(x-2)}{(x+2)(x-2)}$$ Subtract numerators. Write the difference over the common denominator.

$$=\frac{6x-3x+6}{(x+2)(x-2)}$$ Apply the distributive property in the numerator.

$$=\frac{3x+6}{(x+2)(x-2)}$$ Combine like terms in the numerator.

Next we factor the numerator to see if this rational expression can be simplified.

$$\frac{3x+6}{(x+2)(x-2)}=\frac{3(x+2)}{(x+2)(x-2)}$$ Factor.

$$=\frac{3}{x-2}$$ Apply the fundamental principle to simplify.

Practice Problem 3

Add: $\frac{5}{7x}+\frac{2}{x+1}$

EXAMPLE 3 Add: $\frac{2}{3t}+\frac{5}{t+1}$

Solution: The LCD is $3t(t+1)$. We write each rational expression as an equivalent rational expression with a denominator of $3t(t+1)$.

$$\frac{2}{3t}+\frac{5}{t+1}=\frac{2(t+1)}{3t(t+1)}+\frac{5(3t)}{(t+1)(3t)}$$

$$=\frac{2(t+1)+5(3t)}{3t(t+1)}$$ Add numerators. Write the sum over the common denominator.

$$=\frac{2t+2+15t}{3t(t+1)}$$ Apply the distributive property in the numerator.

$$=\frac{17t+2}{3t(t+1)}$$ Combine like terms in the numerator.

Practice Problem 4

Subtract: $\frac{10}{x-6}-\frac{15}{6-x}$

EXAMPLE 4 Subtract: $\frac{7}{x-3}-\frac{9}{3-x}$

Solution: To find a common denominator, we notice that $x-3$ and $3-x$ are opposites. That is, $3-x=-(x-3)$. We write the denominator $3-x$ as $-(x-3)$ and simplify.

$$\frac{7}{x-3}-\frac{9}{3-x}=\frac{7}{x-3}-\frac{9}{-(x-3)}$$

$$=\frac{7}{x-3}-\frac{-9}{x-3}$$ Apply $\frac{a}{-b}=\frac{-a}{b}$.

$$=\frac{7-(-9)}{x-3}$$ Subtract numerators. Write the difference over the common denominator.

$$=\frac{16}{x-3}$$

Practice Problem 5

Add: $2+\frac{x}{x+5}$

EXAMPLE 5 Add: $1+\frac{m}{m+1}$

Solution: Recall that 1 is the same as $\frac{1}{1}$. The LCD of $\frac{1}{1}$ and $\frac{m}{m+1}$ is $m+1$.

Answers

3. $\frac{19x+5}{7x(x+1)}$ **4.** $\frac{25}{x-6}$ **5.** $\frac{3x+10}{x+5}$

$$1 + \frac{m}{m+1} = \frac{1}{1} + \frac{m}{m+1}$$ Write 1 as $\frac{1}{1}$.

$$= \frac{1(m+1)}{1(m+1)} + \frac{m}{m+1}$$ Multiply both the numerator and the denominator of $\frac{1}{1}$ by $m + 1$.

$$= \frac{m + 1 + m}{m+1}$$ Add numerators. Write the sum over the common denominator.

$$= \frac{2m + 1}{m+1}$$ Combine like terms in the numerator.

EXAMPLE 6 Subtract: $\frac{3}{2x^2 + x} - \frac{2x}{6x + 3}$

Solution: First, we factor the denominators.

$$\frac{3}{2x^2 + x} - \frac{2x}{6x + 3} = \frac{3}{x(2x + 1)} - \frac{2x}{3(2x + 1)}$$

The LCD is $3x(2x + 1)$. We write equivalent expressions with denominators of $3x(2x + 1)$.

$$\frac{3}{x(2x + 1)} - \frac{2x}{3(2x + 1)} = \frac{3(3)}{x(2x + 1)(3)} - \frac{2x(x)}{3(2x + 1)(x)}$$

$$= \frac{9 - 2x^2}{3x(2x + 1)}$$ Subtract numerators. Write the difference over the common denominator.

Practice Problem 6

Subtract: $\frac{4}{3x^2 + 2x} - \frac{3x}{12x + 8}$

EXAMPLE 7 Add: $\frac{2x}{x^2 + 2x + 1} + \frac{x}{x^2 - 1}$

Solution: First we factor the denominators.

$$\frac{2x}{x^2 + 2x + 1} + \frac{x}{x^2 - 1}$$

$$= \frac{2x}{(x + 1)(x + 1)} + \frac{x}{(x + 1)(x - 1)}$$

Rewrite each expression with LCD $(x + 1)(x + 1)(x - 1)$.

$$= \frac{2x(x - 1)}{(x + 1)(x + 1)(x - 1)} + \frac{x(x + 1)}{(x + 1)(x - 1)(x + 1)}$$

$$= \frac{2x(x - 1) + x(x + 1)}{(x + 1)^2(x - 1)}$$ Add numerators. Write the sum over the common denominator.

$$= \frac{2x^2 - 2x + x^2 + x}{(x + 1)^2(x - 1)}$$ Apply the distributive property in the numerator.

$$= \frac{3x^2 - x}{(x + 1)^2(x - 1)} \text{ or } \frac{x(3x - 1)}{(x + 1)^2(x - 1)}$$

The numerator was factored as a last step to see if the rational expression could be simplified further. Since there are no factors common to the numerator and the denominator, we can't simplify further.

Practice Problem 7

Add: $\frac{6x}{x^2 + 4x + 4} + \frac{x}{x^2 - 4}$

Answers

6. $\frac{16 - 3x^2}{4x(3x + 2)}$ **7.** $\frac{x(7x - 10)}{(x + 2)^2(x - 2)}$

STUDY SKILLS REMINDER

How are you doing?

If you haven't done so yet, take a few moments and think about how you are doing in this course. Are you working toward your goal of successfully completing this course? Is your performance on homework, quizzes, and tests satisfactory? If not, you might want to see your instructor to see if he/she has any suggestions on how you can improve your performance. Let me once again remind you that, in addition to your instructor, there are many places to get help with your mathematics course. A few suggestions are below.

- This text has an accompanying video lesson for every section in this text.
- The back of this book contains answers to odd-numbered exercises and selected solutions.
- MathPro is available with this text. It is a tutorial software program with lessons corresponding to each section in the text.
- There is a student solutions manual available that contains worked-out solutions to odd-numbered exercises as well as solutions to every exercise in the Chapter Pretests, Integrated Reviews, Chapter Reviews, Chapter Tests, and Cumulative Reviews.
- Don't forget to check with your instructor for other local resources available to you, such as a tutor center.

Name ____________ Section ________ Date ________

EXERCISE SET 12.4

A *Perform each indicated operation. Simplify if possible. See Examples 1 through 7.*

1. $\frac{4}{2x} + \frac{9}{3x}$

2. $\frac{15}{7a} + \frac{8}{6a}$

3. $\frac{15a}{b} + \frac{6b}{5}$

4. $\frac{4c}{d} - \frac{8x}{5}$

5. $\frac{3}{x} + \frac{5}{2x^2}$

6. $\frac{14}{3x^2} + \frac{6}{x}$

7. $\frac{6}{x+1} + \frac{10}{2x+2}$

8. $\frac{8}{x+4} - \frac{3}{3x+12}$

9. $\frac{3}{x+2} - \frac{1}{x^2-4}$

10. $\frac{15}{2x-4} + \frac{x}{x^2-4}$

11. $\frac{3}{4x} + \frac{8}{x-2}$

12. $\frac{5}{y^2} - \frac{y}{2y+1}$

13. $\frac{6}{x-3} + \frac{8}{3-x}$

14. $\frac{9}{x-3} + \frac{9}{3-x}$

15. $\frac{-8}{x^2-1} - \frac{7}{1-x^2}$

16. $\frac{-9}{25x^2-1} + \frac{7}{1-25x^2}$

17. $\frac{5}{x} + 2$

18. $\frac{7}{x^2} - 5x$

19. $\frac{5}{x-2} + 6$

20. $\frac{6y}{y+5} + 1$

21. $\frac{y+2}{y+3} - 2$

22. $\frac{7}{2x-3} - 3$

23. $\frac{-x+2}{x} - \frac{x-6}{4x}$

24. $\frac{-y+1}{y} - \frac{2y-5}{3y}$

25. $\frac{5x}{x+2} - \frac{3x-4}{x+2}$

26. $\frac{7x}{x-3} - \frac{4x+9}{x-3}$

27. $\frac{3x^4}{x} - \frac{4x^2}{x^2}$

28. $\frac{5x}{6} + \frac{15x^2}{2}$

29. $\frac{1}{x+3} - \frac{1}{(x+3)^2}$

30. $\frac{5x}{(x-2)^2} - \frac{3}{x-2}$

31. $\frac{4}{5b} + \frac{1}{b-1}$

32. $\frac{1}{y+5} + \frac{2}{3y}$

33. $\frac{2}{m} + 1$

34. $\frac{6}{x} - 1$

35. $\frac{6}{1-2x} - \frac{4}{2x-1}$

36. $\dfrac{10}{3n - 4} - \dfrac{5}{4 - 3n}$

37. $\dfrac{7}{(x + 1)(x - 1)} + \dfrac{8}{(x + 1)^2}$

38. $\dfrac{5x + 2}{(x + 1)(x + 5)} - \dfrac{2}{x + 5}$

39. $\dfrac{x}{x^2 - 1} - \dfrac{2}{x^2 - 2x + 1}$

40. $\dfrac{x}{x^2 - 4} - \dfrac{5}{x^2 - 4x + 4}$

41. $\dfrac{3a}{2a + 6} - \dfrac{a - 1}{a + 3}$

42. $\dfrac{1}{x + y} - \dfrac{y}{x^2 - y^2}$

43. $\dfrac{y - 1}{2y + 3} + \dfrac{3}{(2y + 3)^2}$

44. $\dfrac{x - 6}{5x + 1} + \dfrac{6}{(5x + 1)^2}$

45. $\dfrac{5}{2 - x} + \dfrac{x}{2x - 4}$

46. $\dfrac{-1}{a - 2} + \dfrac{4}{4 - 2a}$

47. $\dfrac{-7}{y^2 - 3y + 2} - \dfrac{2}{y - 1}$

48. $\dfrac{2}{x^2 + 4x + 4} + \dfrac{1}{x + 2}$

49. $\dfrac{13}{x^2 - 5x + 6} - \dfrac{5}{x - 3}$

50. $\dfrac{27}{y^2 - 81} + \dfrac{3}{2(y + 9)}$

51. $\dfrac{x + 8}{x^2 - 5x - 6} + \dfrac{x + 1}{x^2 - 4x - 5}$

52. $\dfrac{x}{x^2 + 12x + 20} - \dfrac{1}{x^2 + 8x - 20}$

53. In your own words, explain how to add two rational expressions with different denominators.

54. In your own words, explain how to subtract two rational expressions with different denominators.

Review and Preview

Solve each linear or quadratic equation. See Sections 9.3 and 11.6.

55. $3x + 5 = 7$

56. $5x - 1 = 8$

57. $2x^2 - x - 1 = 0$

58. $4x^2 - 9 = 0$

59. $4(x + 6) + 3 = -3$

60. $2(3x + 1) + 15 = -7$

Combining Concepts

Perform each indicated operation.

61. $\frac{3}{x} - \frac{2x}{x^2 - 1} + \frac{5}{x + 1}$

62. $\frac{5}{x - 2} + \frac{7x}{x^2 - 4} - \frac{11}{x}$

63. $\frac{5}{x^2 - 4} + \frac{2}{x^2 - 4x + 4} - \frac{3}{x^2 - x - 6}$

64. $\frac{8}{x^2 + 6x + 5} - \frac{3x}{x^2 + 4x - 5} + \frac{2}{x^2 - 1}$

65. $\frac{9}{x^2 + 9x + 14} - \frac{3x}{x^2 + 10x + 21} + \frac{x + 4}{x^2 + 5x + 6}$

66. $\frac{x + 10}{x^2 - 3x - 4} - \frac{8}{x^2 + 6x + 5} - \frac{9}{x^2 + x - 20}$

67. A board of length $\frac{3}{x + 4}$ inches was cut into two pieces. If one piece is $\frac{1}{x - 4}$ inches, express the length of the other board as a rational expression.

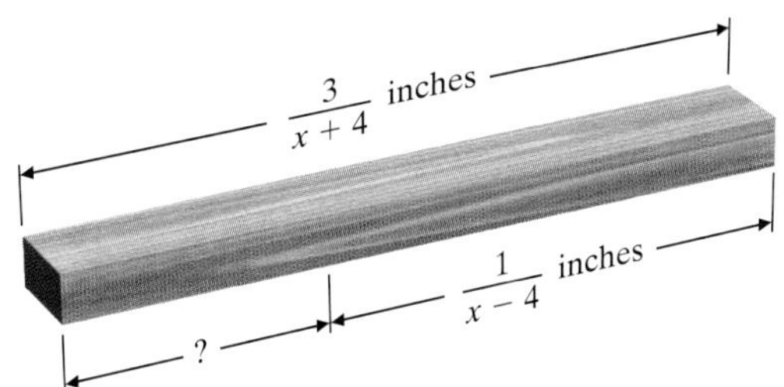

△ 68. The length of a rectangle is $\frac{3}{y - 5}$ feet, while its width is $\frac{2}{y}$ feet. Find its perimeter and then find its area.

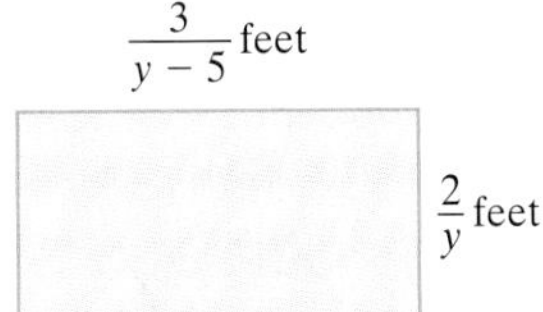

69. In ice hockey, penalty killing percentage is a statistic calculated as $1 - \frac{G}{P}$, where G = opponent's power play goals and P = opponent's power play opportunities. Simplify this expression.

70. The dose of medicine prescribed for a child depends on the child's age A in years and the adult dose D for the medication. Two expressions that give a child's dose are Young's Rule, $\frac{DA}{A + 12}$, and Cowling's Rule, $\frac{D(A + 1)}{24}$. Find an expression for the difference in the doses given by these expressions.

71. Explain when the LCD of the rational expressions in a sum is the product of the denominators.

72. Explain when the LCD is the same as one of the denominators of a rational expression to be added or subtracted.

FOCUS ON History

Epigram of Diophantus

One of the great algebraists of ancient times was a man named Diophantus. Little is known of his life other than that he lived and worked in Alexandria. Some historians believe he lived during the first century of the Christian era, about the time of Nero. The only clue to his personal life is the following epigram found in a collection called the Palatine Anthology.

> God granted him youth for a sixth of his life and added a twelfth part to this. He clothed his cheeks in down. He lit him the light of wedlock after a seventh part, and five years after his marriage, He granted him a son. Alas, lateborn wretched child. After attaining the measure of half his father's life, cruel fate overtook him, thus leaving Diophantus during the last four years of his life only such consolation as the science of numbers. How old was Diophantus at his death?*

*From *The Nature and Growth of Modern Mathematics*, Edna Kramer, 1970, Fawcett Premier Books, Vol. 1, pages 107–108.

We are looking for Diophantus' age when he died, so let x represent that age. If we sum the parts of his life, we should get the total age.

Parts of his life:

- $\frac{1}{6}\cdot x + \frac{1}{12}\cdot x$ is the time of his youth.
- $\frac{1}{7}\cdot x$ is the time between his youth and when he married.
- 5 years is the time between his marriage and the birth of his son.
- $\frac{1}{2}\cdot x$ is the time Diophantus had with his son.
- 4 years is the time between his son's death and his own.

The sum of these parts should equal Diophantus' age when he died.

$$\frac{1}{6}\cdot x + \frac{1}{12}\cdot x + \frac{1}{7}\cdot x + 5 + \frac{1}{2}\cdot x + 4 = x$$

Critical Thinking

1. Solve the epigram by solving the equation.
2. How old was Diophantus when his son was born? How old was the son when he died?
3. Solve the following epigram:

   I was four when my mother packed my lunch and sent me off to school. Half my life was spent in school and another sixth was spent on a farm. Alas, hard times befell me. My crops and cattle fared poorly and my land was sold. I returned to school for 3 years and have spent one tenth of my life teaching. How old am I?

Group Activity

4. Write an epigram describing your life. Be sure that none of the time periods in your epigram overlap. Exchange epigrams with a partner to solve and check.

12.5 Solving Equations Containing Rational Expressions

OBJECTIVES

- (A) Solve equations containing rational expressions.
- (B) Solve equations containing rational expressions for a specified variable.

(A) Solving Equations Containing Rational Expressions

In Chapter 4, we solved equations containing fractions. In this section, we continue the work we began in Chapter 4 by solving equations containing rational expressions. For example,

$$\frac{x}{5} + \frac{x+2}{9} = 8 \quad \text{and} \quad \frac{x+1}{9x-5} = \frac{2}{3x}$$

are equations containing rational expressions. To solve equations such as these, we use the multiplication property of equality to clear the equation of fractions by multiplying both sides of the equation by the LCD.

EXAMPLE 1 Solve: $\frac{x}{2} + \frac{8}{3} = \frac{1}{6}$

Solution: The LCD of denominators 2, 3, and 6 is 6, so we multiply both sides of the equation by 6.

$$6\left(\frac{x}{2} + \frac{8}{3}\right) = 6\left(\frac{1}{6}\right)$$

$$6\left(\frac{x}{2}\right) + 6\left(\frac{8}{3}\right) = 6\left(\frac{1}{6}\right) \quad \text{Use the distributive property.}$$

$$3 \cdot x + 16 = 1 \quad \text{Multiply and simplify.}$$

$$3x = -15 \quad \text{Subtract 16 from both sides.}$$

$$x = -5 \quad \text{Divide both sides by 3.}$$

Check: To check, we replace x with -5 in the original equation.

$$\frac{-5}{2} + \frac{8}{3} \stackrel{?}{=} \frac{1}{6} \quad \text{Replace } x \text{ with } -5.$$

$$\frac{1}{6} = \frac{1}{6} \quad \text{True}$$

This number checks, so the solution is -5.

Practice Problem 1

Solve: $\frac{x}{4} + \frac{4}{5} = \frac{1}{20}$

Make sure that *each* term is multiplied by the LCD.

EXAMPLE 2 Solve: $\frac{t-4}{2} - \frac{t-3}{9} = \frac{5}{18}$

Solution: The LCD of denominators 2, 9, and 18 is 18, so we multiply both sides of the equation by 18.

$$18\left(\frac{t-4}{2} - \frac{t-3}{9}\right) = 18\left(\frac{5}{18}\right)$$

$$18\left(\frac{t-4}{2}\right) - 18\left(\frac{t-3}{9}\right) = 18\left(\frac{5}{18}\right) \quad \text{Use the distributive property.}$$

$$9(t-4) - 2(t-3) = 5 \quad \text{Simplify.}$$

$$9t - 36 - 2t + 6 = 5 \quad \text{Use the distributive property.}$$

$$7t - 30 = 5 \quad \text{Combine like terms.}$$

$$7t = 35$$

$$t = 5 \quad \text{Solve for } t.$$

Practice Problem 2

Solve: $\frac{x+2}{3} - \frac{x-1}{5} = \frac{1}{15}$

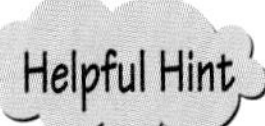

Multiply *each* term by 18.

Answers

1. $x = -3$ **2.** $x = -6$

Check:

$$\frac{t-4}{2} - \frac{t-3}{9} = \frac{5}{18}$$

$$\frac{5-4}{2} - \frac{5-3}{9} \stackrel{?}{=} \frac{5}{18} \quad \text{Replace } t \text{ with 5.}$$

$$\frac{1}{2} - \frac{2}{9} \stackrel{?}{=} \frac{5}{18} \quad \text{Simplify.}$$

$$\frac{5}{18} = \frac{5}{18} \quad \text{True}$$

The solution is 5.

Recall from Section 12.1 that a rational expression is defined for all real numbers except those that make the denominator of the expression 0. This means that if an equation contains *rational expressions with variables in the denominator*, we must be certain that the proposed solution does not make the denominator 0. If replacing the variable with the proposed solution makes the denominator 0, the rational expression is undefined and this proposed solution must be rejected.

Practice Problem 3

Solve: $2 + \frac{6}{x} = x + 7$

EXAMPLE 3 Solve: $3 - \frac{6}{x} = x + 8$

Solution: In this equation, 0 cannot be a solution because if x is 0, the rational expression $\frac{6}{x}$ is undefined. The LCD is x, so we multiply both sides of the equation by x.

$$x\left(3 - \frac{6}{x}\right) = x(x + 8)$$

$$x(3) - x\left(\frac{6}{x}\right) = x \cdot x + x \cdot 8 \quad \text{Use the distributive property.}$$

Helpful Hint

Multiply *each* term by x.

$$3x - 6 = x^2 + 8x \quad \text{Simplify.}$$

Now we write the quadratic equation in standard form and solve for x.

$$0 = x^2 + 5x + 6$$

$$0 = (x + 3)(x + 2) \quad \text{Factor.}$$

$$x + 3 = 0 \quad \text{or} \quad x + 2 = 0 \quad \text{Set each factor equal to 0 and solve.}$$

$$x = -3 \qquad x = -2$$

Notice that neither -3 nor -2 makes the denominator in the original equation equal to 0.

Check: To check these solutions, we replace x in the original equation by -3, and then by -2.

If $x = -3$:

$$3 - \frac{6}{x} = x + 8$$

$$3 - \frac{6}{-3} \stackrel{?}{=} -3 + 8$$

$$3 - (-2) \stackrel{?}{=} 5$$

$$5 = 5 \quad \text{True}$$

If $x = -2$:

$$3 - \frac{6}{x} = x + 8$$

$$3 - \frac{6}{-2} \stackrel{?}{=} -2 + 8$$

$$3 - (-3) \stackrel{?}{=} 6$$

$$6 = 6 \quad \text{True}$$

Both -3 and -2 are solutions.

Answer

3. $x = -6, x = 1$

The following steps may be used to solve an equation containing rational expressions.

To Solve an Equation Containing Rational Expressions

Step 1. Multiply both sides of the equation by the LCD of all rational expressions in the equation.

Step 2. Remove any grouping symbols and solve the resulting equation.

Step 3. Check the solution in the original equation.

EXAMPLE 4 Solve: $\dfrac{4x}{x^2 + x - 30} + \dfrac{2}{x - 5} = \dfrac{1}{x + 6}$

Solution: The denominator $x^2 + x - 30$ factors as $(x + 6)(x - 5)$. The LCD is then $(x + 6)(x - 5)$, so we multiply both sides of the equation by this LCD.

$$(x + 6)(x - 5)\left(\frac{4x}{x^2 + x - 30} + \frac{2}{x - 5}\right) = (x + 6)(x - 5)\left(\frac{1}{x + 6}\right) \quad \text{Multiply by the LCD.}$$

$$(x + 6)(x - 5) \cdot \frac{4x}{x^2 + x - 30} + (x + 6)(x - 5) \cdot \frac{2}{x - 5} \quad \text{Use the distributive property.}$$

$$= (x + 6)(x - 5) \cdot \frac{1}{x + 6}$$

$$4x + 2(x + 6) = x - 5 \quad \text{Simplify.}$$

$$4x + 2x + 12 = x - 5 \quad \text{Use the distributive property.}$$

$$6x + 12 = x - 5 \quad \text{Combine like terms.}$$

$$5x = -17$$

$$x = -\frac{17}{5} \quad \text{Divide both sides by 5.}$$

Check: Check by replacing x with $-\frac{17}{5}$ in the original equation. The solution is $-\frac{17}{5}$.

Practice Problem 4

Solve: $\dfrac{2}{x + 3} + \dfrac{3}{x - 3} = \dfrac{-2}{x^2 - 9}$

EXAMPLE 5 Solve: $\dfrac{2x}{x - 4} = \dfrac{8}{x - 4} + 1$

Solution: Multiply both sides by the LCD, $x - 4$.

$$(x - 4)\left(\frac{2x}{x - 4}\right) = (x - 4)\left(\frac{8}{x - 4} + 1\right) \quad \text{Multiply by the LCD.}$$

$$(x - 4) \cdot \frac{2x}{x - 4} = (x - 4) \cdot \frac{8}{x - 4} + (x - 4) \cdot 1 \quad \text{Use the distributive property.}$$

$$2x = 8 + (x - 4) \quad \text{Simplify.}$$

$$2x = 4 + x$$

$$x = 4$$

Notice that 4 makes the denominator 0 in the original equation. Therefore, 4 is *not* a solution and this equation has *no solution*.

Practice Problem 5

Solve: $\dfrac{5x}{x - 1} = \dfrac{5}{x - 1} + 3$

Answers

4. $x = -1$ **5.** No solution

Try the Concept Check in the margin.

Helpful Hint

As we can see from Example 5, it is important to check the proposed solution(s) in the original equation.

EXAMPLE 6 Solve: $x + \frac{14}{x-2} = \frac{7x}{x-2} + 1$

Solution: Notice the denominators in this equation. We can see that 2 can't be a solution. The LCD is $x - 2$, so we multiply both sides of the equation by $x - 2$.

$$(x-2)\left(x + \frac{14}{x-2}\right) = (x-2)\left(\frac{7x}{x-2} + 1\right)$$

$$(x-2)(x) + (x-2)\left(\frac{14}{x-2}\right) = (x-2)\left(\frac{7x}{x-2}\right) + (x-2)(1)$$

$$x^2 - 2x + 14 = 7x + x - 2 \quad \text{Simplify.}$$

$$x^2 - 2x + 14 = 8x - 2 \quad \text{Combine like terms.}$$

$$x^2 - 10x + 16 = 0 \quad \text{Write the quadratic equation in standard form.}$$

$$(x-8)(x-2) = 0 \quad \text{Factor.}$$

$$x - 8 = 0 \quad \text{or} \quad x - 2 = 0 \quad \text{Set each factor equal to 0.}$$

$$x = 8 \qquad x = 2 \quad \text{Solve.}$$

As we have already noted, 2 can't be a solution of the original equation. So we need only replace x with 8 in the original equation. We find that 8 is a solution; the only solution is 8.

B Solving Equations for a Specified Variable

The last example in this section is an equation containing several variables, and we are directed to solve for one of the variables. The steps used in the preceding examples can be applied to solve equations for a specified variable as well.

EXAMPLE 7 Solve $\frac{1}{a} + \frac{1}{b} = \frac{1}{x}$ for x.

Solution: (This type of equation often models a work problem, as we shall see in the next section.) The LCD is abx, so we multiply both sides by abx.

$$abx\left(\frac{1}{a} + \frac{1}{b}\right) = abx\left(\frac{1}{x}\right)$$

$$abx\left(\frac{1}{a}\right) + abx\left(\frac{1}{b}\right) = abx \cdot \frac{1}{x}$$

$$bx + ax = ab \quad \text{Simplify.}$$

$$x(b + a) = ab \quad \text{Factor out } x \text{ from each term on the left side.}$$

$$\frac{x(b+a)}{b+a} = \frac{ab}{b+a} \quad \text{Divide both sides by } b + a.$$

$$x = \frac{ab}{b+a} \quad \text{Simplify.}$$

This equation is now solved for x.

Concept Check

When can we clear fractions by multiplying through by the LCD?

a. When adding or subtracting rational expressions
b. When solving an equation containing rational expressions
c. Both of these
d. Neither of these

Practice Problem 6

Solve: $x - \frac{6}{x+3} = \frac{2x}{x+3} + 2$

Practice Problem 7

Solve $\frac{1}{a} + \frac{1}{b} = \frac{1}{x}$ for a.

Answers

6. $x = 4$ **7.** $a = \frac{bx}{b-x}$

Concept Check: b

Name ______________________ Section __________ Date __________

Mental Math

Solve each equation for the variable.

1. $\frac{x}{5} = 2$
2. $\frac{x}{8} = 4$
3. $\frac{z}{6} = 6$
4. $\frac{y}{7} = 8$

EXERCISE SET 12.5

 *Solve each equation and check each solution. See Examples 1 and 2.*

1. $\frac{x}{5} + 3 = 9$
2. $\frac{x}{5} - 2 = 9$
3. $\frac{x}{2} + \frac{5x}{4} = \frac{x}{12}$
4. $\frac{x}{6} + \frac{4x}{3} = \frac{x}{18}$
5. $2 - \frac{8}{x} = 6$
6. $5 + \frac{4}{x} = 1$
7. $2 + \frac{10}{x} = x + 5$
8. $6 + \frac{5}{y} = y - \frac{2}{y}$
9. $\frac{a}{5} = \frac{a-3}{2}$
10. $\frac{b}{5} = \frac{b+2}{6}$
11. $\frac{x-3}{5} + \frac{x-2}{2} = \frac{1}{2}$
12. $\frac{a+5}{4} + \frac{a+5}{2} = \frac{a}{8}$

Solve each equation and check each answer. See Examples 3 through 6.

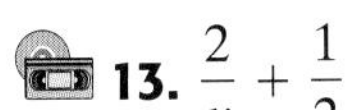

13. $\frac{2}{y} + \frac{1}{2} = \frac{5}{2y}$
14. $\frac{6}{3y} + \frac{3}{y} = 1$
15. $\frac{11}{2x} + \frac{2}{3} = \frac{7}{2x}$
16. $\frac{5}{3} - \frac{3}{2x} = \frac{3}{2}$
17. $2 + \frac{3}{a-3} = \frac{a}{a-3}$
18. $\frac{2y}{y-2} - \frac{4}{y-2} = 4$
19. $\frac{3}{2a-5} = -1$
20. $\frac{6}{4-3x} = -3$
21. $\frac{y}{y+4} + \frac{4}{y+4} = 3$
22. $\frac{5y}{y+1} - \frac{3}{y+1} = 4$
23. $\frac{a}{a-6} = \frac{-2}{a-1}$
24. $\frac{5}{x-6} = \frac{x}{x-2}$
25. $\frac{2x}{x+2} - 2 = \frac{x-8}{x-2}$
26. $\frac{4y}{y-3} - 3 = \frac{3y-1}{y+3}$
27. $\frac{4y}{y-4} + 5 = \frac{5y}{y-4}$

28. $\frac{2a}{a+2} - 5 = \frac{7a}{a+2}$

29. $\frac{2}{x-2} + 1 = \frac{x}{x+2}$

30. $1 + \frac{3}{x+1} = \frac{x}{x-1}$

31. $\frac{t}{t-4} = \frac{t+4}{6}$

32. $\frac{15}{x+4} = \frac{x-4}{x}$

33. $\frac{x+1}{3} - \frac{x-1}{6} = \frac{1}{6}$

34. $\frac{3x}{5} - \frac{x-6}{3} = -\frac{2}{5}$

35. $\frac{y}{2y+2} + \frac{2y-16}{4y+4} = \frac{2y-3}{y+1}$

36. $\frac{1}{x+2} = \frac{4}{x^2-4} - \frac{1}{x-2}$

37. $\frac{4r-4}{r^2+5r-14} + \frac{2}{r+7} = \frac{1}{r-2}$

38. $\frac{3}{x+3} = \frac{12x+19}{x^2+7x+12} - \frac{5}{x+4}$

39. $\frac{x+1}{x+3} = \frac{x^2-11x}{x^2+x-6} - \frac{x-3}{x-2}$

40. $\frac{2t+3}{t-1} - \frac{2}{t+3} = \frac{5-6t}{t^2+2t-3}$

B *Solve each equation for the indicated variable. See Example 7.*

41. $R = \frac{E}{I}$ for I (Electronics: resistance of a circuit)

△ 42. $\frac{A}{W} = L$ for W (Geometry: area of a rectangle)

43. $T = \frac{V}{Q}$ for Q (Water purification: settling time)

44. $T = \frac{2U}{B+E}$ for B (Merchandising: stock turnover rate)

45. $i = \frac{A}{t+B}$ for t (Hydrology: rainfall intensity)

46. $C = \frac{D(A+1)}{24}$ for A (Medicine: Cowling's Rule for child's dose)

47. $N = R + \frac{V}{G}$ for G (Urban forestry: tree plantings per year)

48. $B = \frac{705w}{h^2}$ for w (Health: body-mass index)

△ **49.** $\frac{C}{\pi r} = 2$ for r (Geometry: circumference of a circle)

50. $W = \frac{CE^2}{2}$ for C (Electronics: energy stored in a capacitor)

51. $\frac{1}{y} + \frac{1}{3} = \frac{1}{x}$ for x

52. $\frac{1}{5} + \frac{2}{y} = \frac{1}{x}$ for x

Review and Preview

Write each phrase as an expression.

53. The reciprocal of x

54. The reciprocal of $x + 1$

55. The reciprocal of x, added to the reciprocal of 2

56. The reciprocal of x, subtracted from the reciprocal of 5

Answer each question.

57. If a tank is filled in 3 hours, what part of the tank is filled in 1 hour?

58. If a strip of beach is cleaned in 4 hours, what part of the beach is cleaned in 1 hour?

Combining Concepts

Recall that two angles are supplementary if the sum of their measures is 180°. Find the measures of the supplementary angles.

△ **59.**

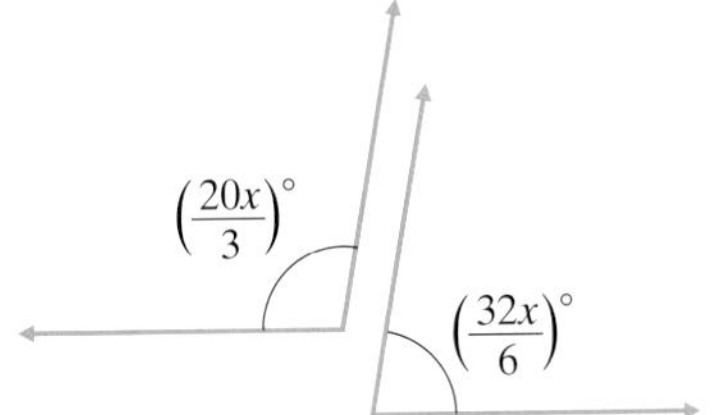

△ **60.**

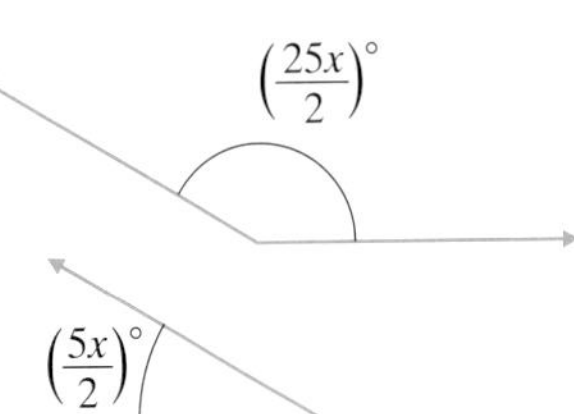

Recall that two angles are complementary if the sum of their measures is 90°. Find the measures of the complementary angles.

△ **61.**

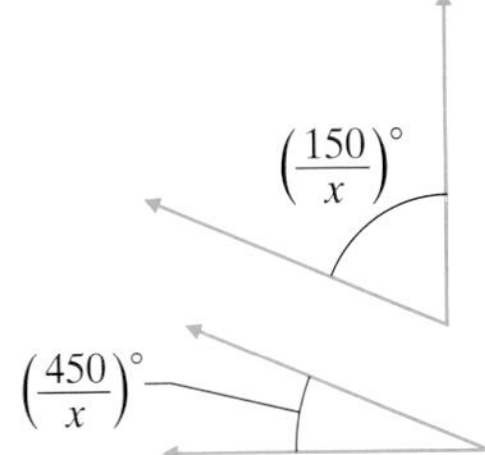

△ **62.**

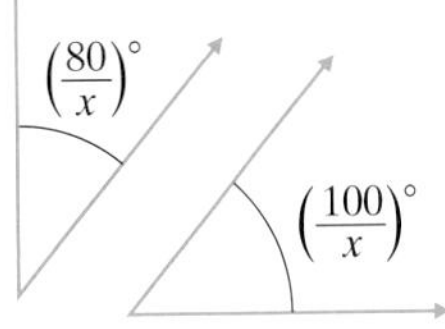

Solve each equation.

63. $\dfrac{4}{a^2 + 4a + 3} + \dfrac{2}{a^2 + a - 6} - \dfrac{3}{a^2 - a - 2} = 0$

64. $\dfrac{-4}{a^2 + 2a - 8} + \dfrac{1}{a^2 + 9a + 20} = \dfrac{-4}{a^2 + 3a - 10}$

65. When adding the expressions in $\dfrac{3x}{2} + \dfrac{x}{4}$, can you multiply each term by 4? Why or why not?

66. When solving the equation $\dfrac{3x}{2} + \dfrac{x}{4} = 1$, can you multiply both sides of the equation by 4? Why or why not?

STUDY SKILLS REMINDER

Are you prepared for a test on Chapter 12?

Below I have listed *a common trouble* area for topics covered in Chapter 12. After studying for your test, but before taking your test, read this.

Do you know the differences between how to perform operations such as $\dfrac{4}{x} + \dfrac{2}{3}$ or $\dfrac{4}{x} \div \dfrac{2}{x}$ and how to solve an equation such as $\dfrac{4}{x} + \dfrac{2}{3} = 1$?

$$\frac{\mathbf{4}}{\boldsymbol{x}} + \frac{\mathbf{2}}{\mathbf{3}} = \frac{4 \cdot 3}{x \cdot 3} + \frac{2 \cdot x}{3 \cdot x} = \frac{12}{3x} + \frac{2x}{3x} = \frac{12 + 2x}{3x} \text{ or } \frac{2(6 + x)}{3x}, \text{ the sum.}$$

Addition—write each expression as an equivalent expression with the same LCD.

$$\frac{\mathbf{4}}{\boldsymbol{x}} \div \frac{\mathbf{2}}{\boldsymbol{x}} = \frac{4}{x} \cdot \frac{x}{2} = \frac{4 \cdot x}{x \cdot 2} = \frac{4}{2} = 2, \text{ the quotient.}$$

Division—multiply the first rational expression by the reciprocal of the second.

$\dfrac{\mathbf{4}}{\boldsymbol{x}} + \dfrac{\mathbf{2}}{\mathbf{3}} = 1$	Equation to be solved.
$3x\left(\dfrac{4}{x} + \dfrac{2}{3}\right) = 3x \cdot 1$	Multiply both sides of the equation by the LCD $3x$.
$3x\left(\dfrac{4}{x}\right) + 3x\left(\dfrac{2}{3}\right) = 3x \cdot 1$	Use the distributive property.
$12 + 2x = 3x$	Multiply and simplify.
$12 = x$	Subtract $2x$ from both sides.

The solution is 12.

For more examples and exercises see the Integrated Review on the next page.

Name ____________ Section ________ Date ________

Answers

Integrated Review—Summary of Rational Expressions

It is important to know the difference between performing operations with rational expressions and solving an equation containing rational expressions. Study the examples below.

Performing Operations with Rational Expressions

Adding: $\frac{1}{x} + \frac{1}{x+5} = \frac{1\cdot(x+5)}{x(x+5)} + \frac{1\cdot x}{x(x+5)} = \frac{x+5+x}{x(x+5)} = \frac{2x+5}{x(x+5)}$

Subtracting: $\frac{3}{x} - \frac{5}{x^2y} = \frac{3\cdot xy}{x\cdot xy} - \frac{5}{x^2y} = \frac{3xy-5}{x^2y}$

Multiplying: $\frac{2}{x}\cdot\frac{5}{x-1} = \frac{2\cdot 5}{x(x-1)} = \frac{10}{x(x-1)}$

Dividing: $\frac{4}{2x+1} \div \frac{x-3}{x} = \frac{4}{2x+1}\cdot\frac{x}{x-3} = \frac{4x}{(2x+1)(x-3)}$

Solving an Equation Containing Rational Expressions

To solve an equation containing rational expressions, we clear the equation of fractions by multiplying both sides by the LCD.

$$\frac{3}{x} - \frac{5}{x-1} = \frac{1}{x(x-1)}$$ Note that x can't be 0 or 1.

$$x(x-1)\left(\frac{3}{x}\right) - x(x-1)\left(\frac{5}{x-1}\right) = x(x-1)\cdot\frac{1}{x(x-1)}$$ Multiply both sides by the LCD.

$$3(x-1) - 5x = 1$$ Simplify.

$$3x - 3 - 5x = 1$$ Use the distributive property.

$$-2x - 3 = 1$$ Combine like terms.

$$-2x = 4$$ Add 3 to both sides.

$$x = -2$$ Divide both sides by -2.

Determine whether each of the following is an equation or an expression. If it is an equation, solve it for its variable. If it is an expression, perform the indicated operation.

1. $\frac{1}{x} + \frac{2}{3}$

2. $\frac{3}{a} + \frac{5}{6}$

3. $\frac{1}{x} + \frac{2}{3} = \frac{3}{x}$

4. $\frac{3}{a} + \frac{5}{6} = 1$

5. $\frac{2}{x+1} - \frac{1}{x}$

6. $\frac{4}{x-3} - \frac{1}{x}$

7. $\frac{2}{x+1} - \frac{1}{x} = 1$

8. $\frac{4}{x-3} - \frac{1}{x} = \frac{6}{x(x-3)}$

1. ____________
2. ____________
3. ____________
4. ____________
5. ____________
6. ____________
7. ____________
8. ____________

9. ______

10. ______

11. ______

12. ______

13. ______

14. ______

15. ______

16. ______

17. ______

18. ______

19. ______

20. ______

21. ______

22. ______

23. ______

24. ______

9. $\dfrac{15x}{x+8} \cdot \dfrac{2x+16}{3x}$

10. $\dfrac{9z+5}{15} \cdot \dfrac{5z}{81z^2-25}$

11. $\dfrac{2x+1}{x-3} + \dfrac{3x+6}{x-3}$

12. $\dfrac{4p-3}{2p+7} + \dfrac{3p+8}{2p+7}$

13. $\dfrac{x+5}{7} = \dfrac{8}{2}$

14. $\dfrac{1}{2} = \dfrac{x+1}{8}$

15. $\dfrac{5a+10}{18} \div \dfrac{a^2-4}{10a}$

16. $\dfrac{9}{x^2-1} \div \dfrac{12}{3x+3}$

17. $\dfrac{x+2}{3x-1} + \dfrac{5}{(3x-1)^2}$

18. $\dfrac{4}{(2x-5)^2} + \dfrac{x+1}{2x-5}$

19. $\dfrac{x-7}{x} - \dfrac{x+2}{5x}$

20. $\dfrac{9}{x^2-4} + \dfrac{2}{x+2} = \dfrac{-1}{x-2}$

21. $\dfrac{3}{x+3} = \dfrac{5}{x^2-9} - \dfrac{2}{x-3}$

22. $\dfrac{10x-9}{x} - \dfrac{x-4}{3x}$

23. Explain the difference between solving an equation such as $\dfrac{x}{2} + \dfrac{3}{4} = \dfrac{x}{4}$ for x and performing an operation such as adding $\dfrac{x}{2} + \dfrac{3}{4}$.

24. When solving an equation such as $\dfrac{y}{4} = \dfrac{y}{2} - \dfrac{1}{4}$, we may multiply all terms by 4. When subtracting two rational expressions such as $\dfrac{y}{2} - \dfrac{1}{4}$, we may not. Explain why.

12.6 Rational Equations and Problem Solving

OBJECTIVES

- A Solve problems about numbers.
- B Solve problems about work.
- C Solve problems about distance, rate, and time.
- D Solve problems about similar triangles.

A Solving Problems about Numbers

In this section, we solve problems that can be modeled by equations containing rational expressions. To solve these problems, we use the same problem-solving steps that were first introduced in Section 9.4. In our first example, our goal is to find an unknown number.

EXAMPLE 1 Finding an Unknown Number

The quotient of a number and 6, minus $\frac{5}{3}$ is the quotient of the number and 2. Find the number.

Solution:

1. UNDERSTAND. Read and reread the problem. Suppose that the unknown number is 2, then we see if the quotient of 2 and 6, or $\frac{2}{6}$, minus $\frac{5}{3}$ is equal to the quotient of 2 and 2, or $\frac{2}{2}$.

$$\frac{2}{6} - \frac{5}{3} = \frac{1}{3} - \frac{5}{3} = -\frac{4}{3}, \text{ not } \frac{2}{2}$$

Don't forget that the purpose of a proposed solution is to better understand the problem.

Let $x =$ the unknown number.

2. TRANSLATE.

In words:	the quotient of x and 6	minus	$\frac{5}{3}$	is	the quotient of x and 2
	↓	↓	↓	↓	↓
Translate:	$\frac{x}{6}$	$-$	$\frac{5}{3}$	$=$	$\frac{x}{2}$

3. SOLVE. Here, we solve the equation $\frac{x}{6} - \frac{5}{3} = \frac{x}{2}$. We begin by multiplying both sides of the equation by the LCD 6.

$$6\left(\frac{x}{6} - \frac{5}{3}\right) = 6\left(\frac{x}{2}\right)$$

$$6\left(\frac{x}{6}\right) - 6\left(\frac{5}{3}\right) = 6\left(\frac{x}{2}\right) \quad \text{Apply the distributive property.}$$

$$x - 10 = 3x \quad \text{Simplify.}$$

$$-10 = 2x \quad \text{Subtract } x \text{ from both sides.}$$

$$-\frac{10}{2} = \frac{2x}{2} \quad \text{Divide both sides by 2.}$$

$$-5 = x \quad \text{Simplify.}$$

4. INTERPRET.

Check: To check, we verify that "the quotient of -5 and 6 minus $\frac{5}{3}$ is the quotient of -5 and 2, or $-\frac{5}{6} - \frac{5}{3} = -\frac{5}{2}$. The statement is true.

State: The unknown number is -5.

Practice Problem 1

The quotient of a number and 2, minus $\frac{1}{3}$ is the quotient of the number and 6.

Answer

1. $x = 1$

B Solving Problems about Work

The next example is often called a work problem. Work problems usually involve people or machines doing a certain task.

Practice Problem 2

Andrew and Timothy Larson volunteer at a local recycling plant. Andrew can sort a batch of recyclables in 2 hours alone while his brother Timothy needs 3 hours to complete the same job. If they work together, how long will it take them to sort one batch?

EXAMPLE 2 Finding Work Rates

Sam Waterton and Frank Schaffer work in a plant that manufactures automobiles. Sam can complete a quality control tour of the plant in 3 hours while his assistant, Frank, needs 7 hours to complete the same job. The regional manager is coming to inspect the plant facilities, so both Sam and Frank are directed to complete a quality control tour together. How long will this take?

Solution:

1. UNDERSTAND. Read and reread the problem. The key idea here is the relationship between the **time** (hours) it takes to complete the job and the **part of the job** completed in 1 unit of time (hour). For example, if the **time** it takes Sam to complete the job is 3 hours, the **part of the job** he can complete in 1 hour is $\frac{1}{3}$.

   Similarly, Frank can complete $\frac{1}{7}$ of the job in 1 hour.

   Let $x =$ the **time** in hours it takes Sam and Frank to complete the job together.

   Then $\frac{1}{x} =$ the **part of the job** they complete in 1 hour.

	Hours to Complete Total Job	Part of Job Completed in 1 Hour
Sam	3	$\frac{1}{3}$
Frank	7	$\frac{1}{7}$
Together	x	$\frac{1}{x}$

2. TRANSLATE.

   In words:

part of job Sam completed in 1 hour	added to	part of job Frank completed in 1 hour	is equal to	part of job they completed together in 1 hour
↓	↓	↓	↓	↓

   Translate: $\frac{1}{3} + \frac{1}{7} = \frac{1}{x}$

Answer

2. $1\frac{1}{5}$ hours

3. SOLVE. Here, we solve the equation $\frac{1}{3} + \frac{1}{7} = \frac{1}{x}$. We begin by multiplying both sides of the equation by the LCD, $21x$.

$$21x\left(\frac{1}{3}\right) + 21x\left(\frac{1}{7}\right) = 21x\left(\frac{1}{x}\right)$$

$$7x + 3x = 21 \quad \text{Simplify.}$$

$$10x = 21$$

$$x = \frac{21}{10} \text{ or } 2\frac{1}{10} \text{ hours}$$

4. INTERPRET.

Check: Our proposed solution is $2\frac{1}{10}$ hours. This proposed solution is reasonable since $2\frac{1}{10}$ hours is more than half of Sam's time and less than half of Frank's time. Check this solution in the originally *stated* problem.

State: Sam and Frank can complete the quality control tour in $2\frac{1}{10}$ hours.

C Solving Problems about Distance, Rate, and Time

Next we look at a problem solved by the distance/rate/time formula.

EXAMPLE 3 Finding Speeds of Vehicles

A car travels 180 miles in the same time that a truck travels 120 miles. If the car's speed is 20 miles per hour faster than the truck's, find the car's speed and the truck's speed.

Solution:

1. UNDERSTAND. Read and reread the problem. Suppose that the truck's speed is 45 miles per hour. Then the car's speed is 20 miles per hour more, or 65 miles per hour.

   We are given that the car travels 180 miles in the same time that the truck travels 120 miles. To find the time it takes the car to travel 180 miles, we use the formula $d = rt$ solved for t: $\frac{d}{r} = t$.

Car's Time

$$t = \frac{d}{r} = \frac{180}{65} = 2\frac{50}{65} = 2\frac{10}{13} \text{ hours}$$

Truck's Time

$$t = \frac{d}{r} = \frac{120}{45} = 2\frac{30}{45} = 2\frac{2}{3} \text{ hours}$$

Since the times are not the same, our proposed solution is not correct. But we have a better understanding of the problem.

Let $x =$ the speed of the truck.

Since the car's speed is 20 miles per hour faster than the truck's, then

$x + 20 =$ the speed of the car

Practice Problem 3

A car travels 280 miles in the same time that a motorcycle travels 240 miles. If the car's speed is 10 miles per hour more than the motorcycle's, find the speed of the car and the speed of the motorcycle.

Answer

3. car: 70 mph; motorcycle: 60 mph

Use the formula $d = r \cdot t$ or **d**istance = **r**ate · **t**ime. Prepare a chart to organize the information in the problem.

	Distance	= Rate	· Time
Truck	120	x	$\frac{120}{x}$ ← distance / ← rate
Car	180	$x + 20$	$\frac{180}{x+20}$ ← distance / ← rate

Helpful Hint
If $d = r \cdot t$, then $t = \frac{d}{r}$ or *time* $= \frac{distance}{rate}$.

2. TRANSLATE. Since the car and the truck traveled the same amount of time, we have the following.

In words: car's time = truck's time

Translate: $\frac{180}{x + 20} = \frac{120}{x}$

3. SOLVE. We begin by multiplying both sides of the equation by the LCD, $x(x + 20)$, or cross multiplying.

$$\frac{180}{x + 20} = \frac{120}{x}$$

$$180x = 120(x + 20)$$

$$180x = 120x + 2400 \quad \text{Use the distributive property.}$$

$$60x = 2400 \quad \text{Subtract } 120x \text{ from both sides.}$$

$$x = 40 \quad \text{Divide both sides by 60.}$$

4. INTERPRET. The speed of the truck is 40 miles per hour. The speed of the car must then be $x + 20$ or 60 miles per hour.

Check: Find the time it takes the car to travel 180 miles and the time it takes the truck to travel 120 miles.

Car's Time

$$t = \frac{d}{r} = \frac{180}{60} = 3 \text{ hours}$$

Truck's Time

$$t = \frac{d}{r} = \frac{120}{40} = 3 \text{ hours}$$

Since both travel the same amount of time, the proposed solution is correct.

State: The car's speed is 60 miles per hour and the truck's speed is 40 miles per hour.

D Solving Problems about Similar Triangles

Recall that **similar triangles** have the same shape but not necessarily the same size. In similar triangles, the measures of corresponding angles are equal, and corresponding sides are in proportion.

If triangle ABC and triangle XYZ shown on the next page are similar, then we know that the measure of angle A = the measure of angle X, the measure of angle B = the measure of angle Y, and the measure of angle C = the measure of angle Z. We also know that corresponding sides are in proportion: $\frac{a}{x} = \frac{b}{y} = \frac{c}{z}$.

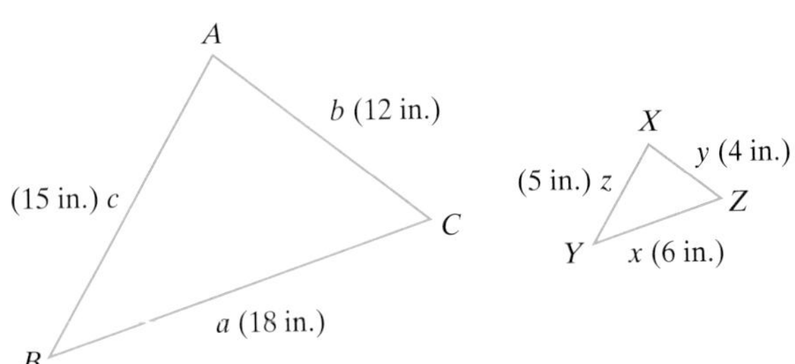

In this section, we will position similar triangles so that they have the same orientation.

To show that corresponding sides are in proportion for the triangles above, we write the ratios of the corresponding sides.

$$\frac{a}{x} = \frac{18}{6} = 3 \qquad \frac{b}{y} = \frac{12}{4} = 3 \qquad \frac{c}{z} = \frac{15}{5} = 3$$

△ EXAMPLE 4 Finding the Length of a Side of a Triangle

If the following two triangles are similar, find the missing length x.

2 yards　10 yards　3 yards　x yards

Solution: Since the triangles are similar, their corresponding sides are in proportion and we have

$$\frac{2}{3} = \frac{10}{x}$$

To solve, we multiply both sides by the LCD, $3x$, or cross multiply.

$$2x = 30$$
$$x = 15 \quad \text{Divide both sides by 2.}$$

The missing length x is 15 yards.

Practice Problem 4 △

If the following two triangles are similar, find the missing length x.

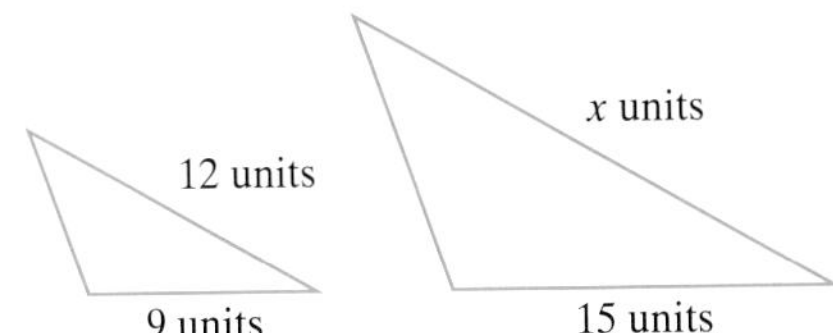

Answer

4. $x = 20$ units

FOCUS ON Business and Career

MORTGAGES

A loan to buy a house or other property is called a **mortgage**. When you are thinking of getting a mortgage to buy a house, it is helpful to know how much your monthly mortgage payment will be. One way to calculate the monthly payment P is to use the formula

$$P = \frac{\frac{Ar}{12}}{1 - \frac{1}{\left(1 + \frac{r}{12}\right)^{12t}}}$$

where A = the amount of the mortgage, r = the annual interest rate (written as a decimal), and t = the loan term in years. Try the exercises to the right and below.

CRITICAL THINKING

1. The average interest rate for a 30-year fixed mortgage in the United States in August 2003 was 6.44%. (*Source:* HSH Associates) Suppose you had borrowed $90,000 to buy a house in August 2003. If your loan term was 30 years, calculate your monthly mortgage payment.

2. The average interest rate for a 15-year fixed mortgage in the United States in August 2003 was 5.76%. (*Source:* HSH Associates) Suppose you had borrowed $80,000 to buy a house in August 2003. If your loan term was 15 years, calculate your monthly mortgage payment.

Another way to calculate a monthly mortgage payment is to use one of the many sites on the World Wide Web that offer an interactive mortgage calculator. For instance, by visiting the given World Wide Web address, you will be able to access the Interest.com Web site, or a related site, where you can calculate a monthly mortgage payment by entering the annual interest rate as a percent, the term of the loan in years, and the total home loan amount. Use the site below to solve the given exercises.

Internet Excursions

Go To: 

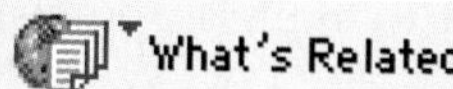

3. Suppose you would like to borrow $55,000 to buy a vacation cottage. If the annual interest rate is 6.85% and you plan to take out a 15-year loan, what will be your monthly mortgage payment?

4. Suppose you would like to borrow $120,000 to buy a house. If the annual interest rate is 7.12% and you plan to take out a 20-year loan, what will be your monthly mortgage payment?

Name ______________________ Section ________ Date __________

Mental Math

Without solving algebraically, select the best choice for each exercise.

1. One person can complete a job in 7 hours. A second person can complete the same job in 5 hours. How long will it take them to complete the job if they work together?
a. more than 7 hours
b. between 5 and 7 hours
c. less than 5 hours

2. One inlet pipe can fill a pond in 30 hours. A second inlet pipe can fill the same pond in 25 hours. How long before the pond is filled if both inlet pipes are on?
a. less than 25 hours
b. between 25 and 30 hours
c. more than 30 hours

EXERCISE SET 12.6

 *Solve. See Example 1.*

1. Three times the reciprocal of a number equals 9 times the reciprocal of 6. Find the number.

2. Twelve divided by the sum of a number and 2 equals the quotient of 4 and the difference of the number and 2. Find the number.

3. If twice a number added to 3 is divided by the number plus 1, the result is three halves. Find the number.

4. A number added to the product of 6 and the reciprocal of the number equals -5. Find the number.

5. Two divided by the difference of a number and 3, minus 4 divided by the number plus 3, equals 8 times the reciprocal of the difference of the number squared and 9. What is the number?

6. If 15 times the reciprocal of a number is added to the ratio of 9 times the number minus 7 and the number plus 2, the result is 9. What is the number?

7. One-fourth equals the quotient of a number and 8. Find the number.

8. Four times a number added to 5 is divided by 6. The result is $\frac{7}{2}$. Find the number.

B *Solve. See Example 2.*

9. Smith Engineering found that an experienced surveyor can survey a roadbed in 4 hours. An apprentice surveyor needs 5 hours to survey the same stretch of road. If the two work together, find how long it takes them to complete the job.

10. An experienced bricklayer can construct a small wall in 3 hours. The apprentice can complete the job in 6 hours. Find how long it takes if they work together.

11. In 2 minutes, a conveyor belt can move 300 pounds of recyclable aluminum from the delivery truck to a storage area. A smaller belt can move the same quantity of cans the same distance in 6 minutes. If both belts are used, find how long it takes to move the cans to the storage area.

12. Find how long it takes the conveyor belts described in Exercise 11 to move 1200 pounds of cans. (*Hint:* Think of 1200 pounds as four 300-pound jobs.)

13. Marcus and Tony work for Lombardo's Pipe and Concrete. Mr. Lombardo is preparing an estimate for a customer. He knows that Marcus can lay a slab of concrete in 6 hours. Tony can lay the same size slab in 4 hours. If both work on the job and the cost of labor is $45.00 per hour, decide what the labor estimate should be.

14. Mr. Dodson can paint his house by himself in 4 days. His son will need an additional day to complete the job if he works by himself. If they work together, find how long it takes to paint the house.

15. One custodian can clean a suite of offices in 3 hours. When a second worker is asked to join the regular custodian, the job takes only $1\frac{1}{2}$ hours. How long would it take the second worker to do the same job alone?

16. One person can proofread copy for a small newspaper in 4 hours. If a second proofreader is also employed, the job can be done in $2\frac{1}{2}$ hours. How long would it take the second proofreader to do the same job alone?

17. One pipe fills a storage pond in 20 hours. A second pipe fills the same pond in 15 hours. When a third pipe is added and all three are used to fill the pond, it takes only 6 hours. Find how long it would take the third pipe alone to do the job.

18. One pump fills a tank 3 times as fast as another pump. If the pumps work together, they fill the tank in 21 minutes. How long would it take each pump alone to fill the tank?

C *Solve. See Example 3.*

19. A runner begins her workout by jogging to the park, a distance of 3 miles. She then jogs home at the same speed but along a different route. This return trip is 9 miles and her time is one hour longer. Complete the accompanying chart and use it to find the runner's jogging speed.

	Distance =	Rate ·	Time
Trip to park	3		x
Return trip	9		$x + 1$

20. A marketing manager travels 1080 miles in a corporate jet and then an additional 240 miles by car. If the car ride takes one hour longer than the jet ride, and if the rate of the jet is 6 times the rate of the car, find the time the manager travels by jet and find the time the manager travels by car.

21. A cyclist rode the first 20-mile portion of his workout at a constant speed. For the 16-mile cooldown portion of his workout, he reduced his speed by 2 miles per hour. Each portion of the workout took the same time. Find the cyclist's speed during the first portion and find his speed during the cooldown portion.

22. A tractor-trailer travels 300 miles through the flatland in the same amount of time that it travels 180 miles through mountains. The rate of the tractor-trailer is 20 miles per hour slower in the mountains than in the flatland. Find both the flatland rate and mountain rate.

23. A boat can travel 9 miles upstream in the same amount of time it takes to travel 11 miles downstream. If the current of the river is 3 miles per hour, complete the chart below and use it to find the speed r of the boat in still water.

	Distance =	Rate ·	Time
Upstream	9	$r - 3$	
Downstream	11	$r + 3$	

24. A pilot flies 630 miles with a tail wind of 35 miles per hour. Against the wind, she flies only 455 miles in the same amount of time. Find the rate of the plane in still air.

25. A cyclist rides 16 miles per hour on level ground on a still day. He finds that he rides 48 miles with the wind behind him in the same amount of time that he rides 16 miles into the wind. Find the rate of the wind.

26. The current on a portion of the Mississippi River is 3 miles per hour. A barge can go 6 miles upstream in the same amount of time it takes to go 10 miles downstream. Find the speed of the boat in still water.

27. While road testing a new make of car, the editor of a consumer magazine finds that she can go 10 miles into a 3-mile-per-hour wind in the same amount of time she can go 11 miles with a 3-mile-per-hour wind behind her. Find the speed of the car in still air.

28. A fisherman on Pearl River rows 9 miles downstream in the same amount of time he rows 3 miles upstream. If the current is 6 miles per hour, find how long it takes him to cover the 12 miles.

D *Given that the following pairs of triangles are similar, find each missing length. See Example 4.*

△ 29.

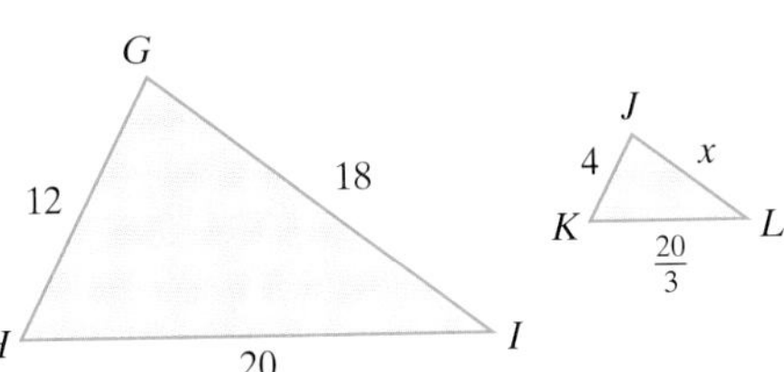

△ 30.

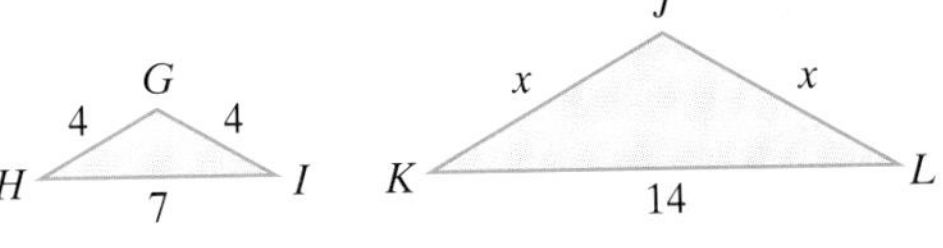

△ 31.

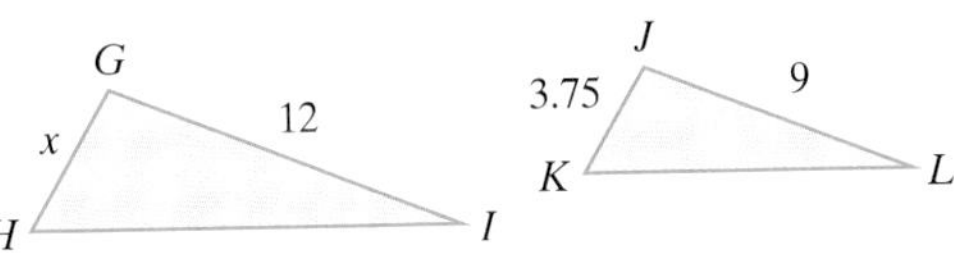

△ 32.

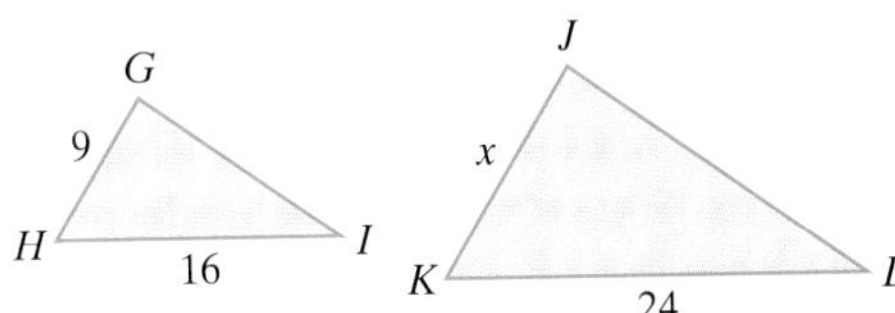

△ 33.

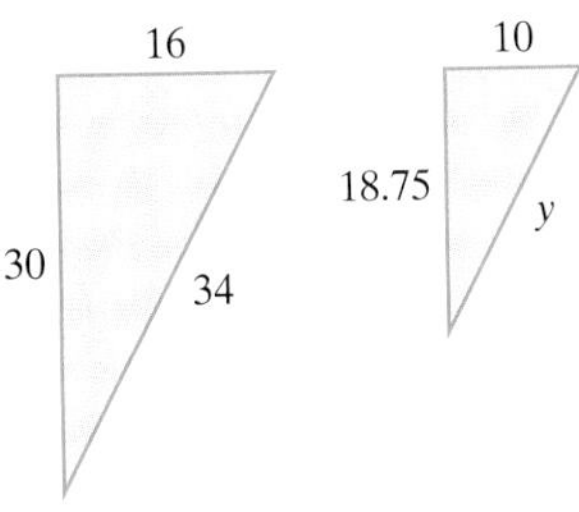

△ 34.

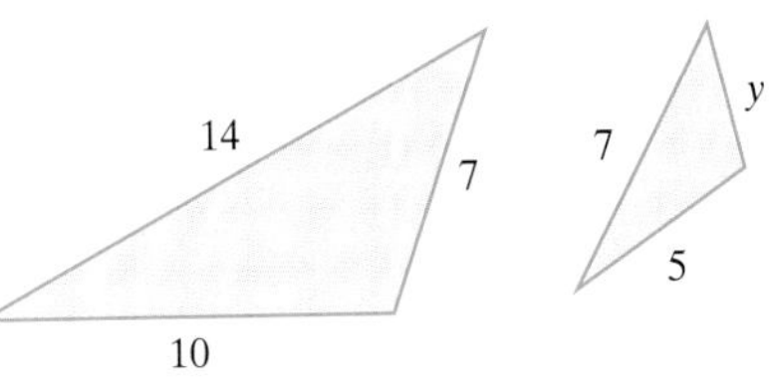

△ 35.

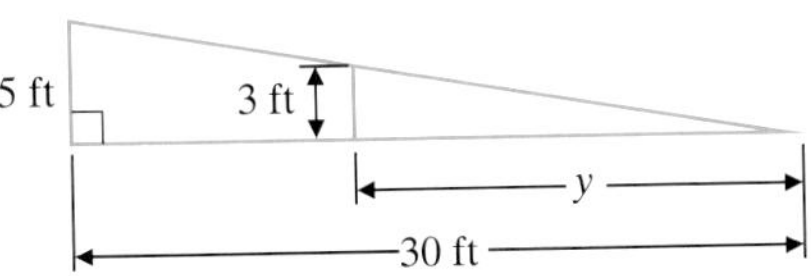

△ 36.

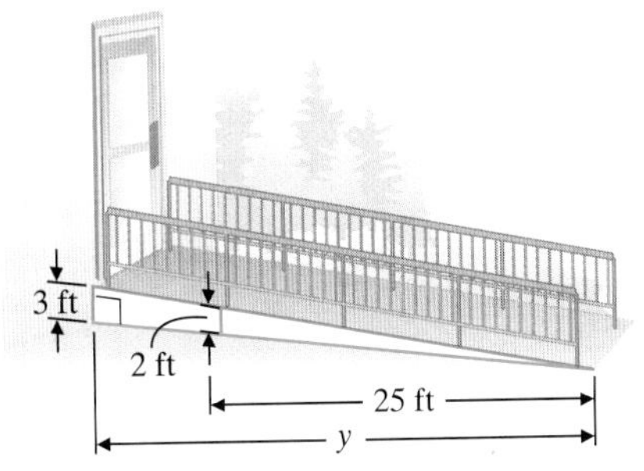

△ **37.** An architect is completing the plans for a triangular deck. Use the diagram below to find the missing dimension.

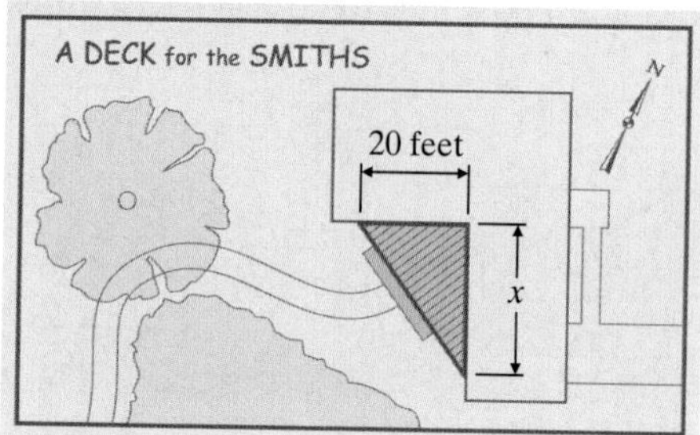

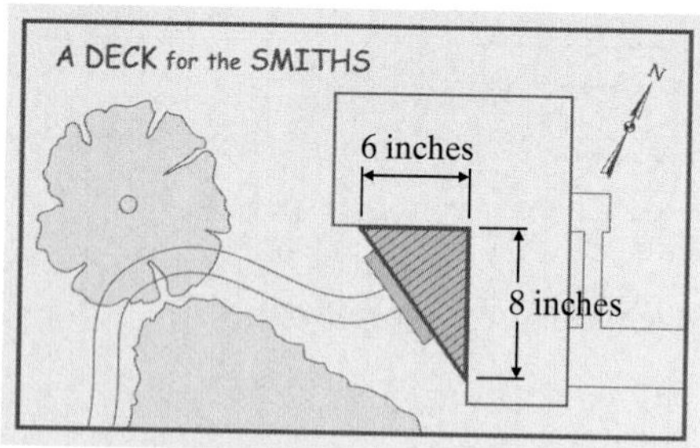

△ **38.** A student wishes to make a small model of a triangular mainsail in order to study the effects of wind on the sail. The smaller model will be the same shape as a regular size sailboat's mainsail. Use the following diagram to find the missing dimensions.

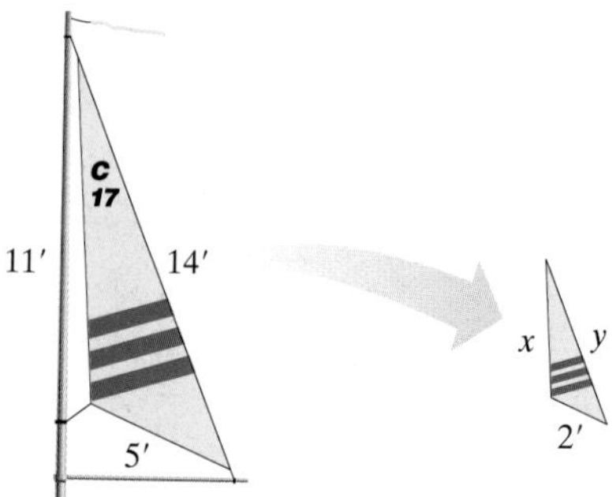

Review and Preview

Simplify. Follow the circled steps in the order shown. See Section 4.6.

39. $\dfrac{\frac{3}{4}+\frac{1}{4}}{\frac{3}{8}+\frac{13}{8}}$ ① Add. ← ③ Divide. ② Add.

40. $\dfrac{\frac{9}{5}+\frac{6}{5}}{\frac{17}{6}+\frac{7}{6}}$ ① Add. ← ③ Divide. ② Add.

41. $\dfrac{\frac{2}{5}+\frac{1}{5}}{\frac{7}{10}+\frac{7}{10}}$ ① Add. ← ③ Divide. ② Add.

42. $\dfrac{\frac{1}{4}+\frac{5}{4}}{\frac{3}{8}+\frac{7}{8}}$ ① Add. ← ③ Divide. ② Add.

Combining Concepts

43. Brazilians Helio Castroneves and Bruno Junqueira placed first and fifth, respectively, in the 2001 Indianapolis 500. The track is 2.5 miles long. When traveling at their fastest lap speeds, Junqueira drove 2.459 miles in the same time that Castroneves completed an entire 2.5-mile lap. Castroneves' fastest lap speed was 3.6 mph faster than Junqueira's fastest lap speed. Find each driver's fastest lap speed. Round each speed to the nearest tenth. (*Source:* Indy Racing League)

44. A hyena spots a giraffe 0.5 mile away and begins running toward it. The giraffe starts running away from the hyena just as the hyena begins running toward it. A hyena can run at a speed of 40 mph and a giraffe can run at 32 mph. How long will it take for the hyena to overtake the giraffe? (*Source: The World Almanac and Book of Facts, 2003*)

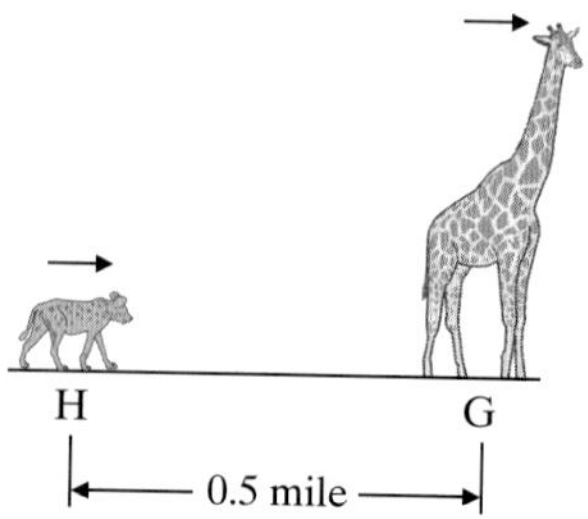

45. Person A can complete a job in 5 hours, and person B can complete the same job in 3 hours. Without solving algebraically, discuss reasonable and unreasonable answers for how long it would take them to complete the job together.

12.7 Simplifying Complex Fractions

OBJECTIVES

A. Simplify complex fractions using method 1.

B. Simplify complex fractions using method 2.

Recall from Section 4.6 that a rational expression whose numerator or denominator or both numerator and denominator contain fractions is called a **complex rational expression** or a **complex fraction.** Some examples are

$$\frac{4}{2-\frac{1}{2}} \qquad \frac{\frac{3}{2}}{\frac{4}{7}-x} \qquad \frac{\frac{1}{x+2}}{x+2-\frac{1}{x}}$$

← Numerator of complex fraction
← Main fraction bar
← Denominator of complex fraction

Our goal in this section is to review writing complex fractions in simplest form. A complex fraction is in simplest form when it is in the form $\frac{P}{Q}$, where P and Q are polynomials that have no common factors.

A Simplifying Complex Fractions–Method 1

Just as in Section 4.6, two methods of simplifying complex fractions are represented. The first method presented uses the fact that the main fraction bar indicates division.

Method 1: To Simplify a Complex Fraction

Step 1. Add or subtract fractions in the numerator or denominator so that the numerator is a single fraction and the denominator is a single fraction.

Step 2. Perform the indicated division by multiplying the numerator of the complex fraction by the reciprocal of the denominator of the complex fraction.

Step 3. Write the rational expression in lowest terms.

EXAMPLE 1 Simplify the complex fraction $\frac{\frac{5}{8}}{\frac{2}{3}}$.

Solution: Since the numerator and denominator of the complex fraction are already single fractions, we proceed to step 2: Perform the indicated division by multiplying the numerator $\frac{5}{8}$ by the reciprocal of the denominator $\frac{2}{3}$.

$$\frac{\frac{5}{8}}{\frac{2}{3}} = \frac{5}{8}\cdot\frac{3}{2} = \frac{15}{16}$$

The reciprocal of $\frac{2}{3}$ is $\frac{3}{2}$.

Practice Problem 1

Simplify the complex fraction $\frac{\frac{3}{7}}{\frac{5}{9}}$.

EXAMPLE 2 Simplify: $\dfrac{\frac{2}{3}+\frac{1}{5}}{\frac{2}{3}-\frac{2}{9}}$

Solution: We simplify the expressions above and below the main fraction bar separately. First we add $\frac{2}{3}$ and $\frac{1}{5}$ to obtain a single fraction in the numerator. Then we subtract $\frac{2}{9}$ from $\frac{2}{3}$ to obtain a single fraction in the denominator.

Practice Problem 2

Simplify: $\dfrac{\frac{3}{4}-\frac{2}{3}}{\frac{1}{2}+\frac{3}{8}}$

Answers

1. $\frac{27}{35}$ 2. $\frac{2}{21}$

$$\frac{\frac{2}{3}+\frac{1}{5}}{\frac{2}{3}-\frac{2}{9}}=\frac{\frac{2(5)}{3(5)}+\frac{1(3)}{5(3)}}{\frac{2(3)}{3(3)}-\frac{2}{9}}$$ The LCD of the numerator's fractions is 15. The LCD of the denominator's fractions is 9

$$=\frac{\frac{10}{15}+\frac{3}{15}}{\frac{6}{9}-\frac{2}{9}}$$ Simplify.

$$=\frac{\frac{13}{15}}{\frac{4}{9}}$$ Add the numerator's fractions. Subtract the denominator's fractions.

Next we perform the indicated division by multiplying the numerator of the complex fraction by the reciprocal of the denominator of the complex fraction.

$$\frac{\frac{13}{15}}{\frac{4}{9}}=\frac{13}{15}\cdot\frac{9}{4}$$ The reciprocal of $\frac{4}{9}$ is $\frac{9}{4}$.

$$=\frac{13\cdot 3\cdot 3}{3\cdot 5\cdot 4}=\frac{39}{20}$$

Practice Problem 3

Simplify: $\dfrac{\frac{2}{5}-\frac{1}{x}}{\frac{x}{10}-\frac{1}{3}}$

EXAMPLE 3 Simplify: $\dfrac{\frac{1}{z}-\frac{1}{2}}{\frac{1}{3}-\frac{z}{6}}$

Solution: Subtract to get a single fraction in the numerator and a single fraction in the denominator of the complex fraction.

$$\frac{\frac{1}{z}-\frac{1}{2}}{\frac{1}{3}-\frac{z}{6}}=\frac{\frac{2}{2z}-\frac{z}{2z}}{\frac{2}{6}-\frac{z}{6}}$$ The LCD of the numerator's fractions is $2z$. The LCD of the denominator's fractions is 6.

$$=\frac{\frac{2-z}{2z}}{\frac{2-z}{6}}$$

$$=\frac{2-z}{2z}\cdot\frac{6}{2-z}$$ Multiply by the reciprocal of $\frac{2-z}{6}$.

$$=\frac{2\cdot 3\cdot(2-z)}{2\cdot z\cdot(2-z)}$$ Factor.

$$=\frac{3}{z}$$ Write in lowest terms.

B Simplifying Complex Fractions—Method 2

Next we study a second method for simplifying complex fractions. In this method, we multiply the numerator and the denominator of the complex fraction by the LCD of all fractions in the complex fraction.

Answer

3. $\dfrac{6(2x-5)}{x(3x-10)}$

Method 2: To Simplify a Complex Fraction

Step 1. Find the LCD of all the fractions in the complex fraction.
Step 2. Multiply both the numerator and the denominator of the complex fraction by the LCD from Step 1.
Step 3. Perform the indicated operations and write the result in lowest terms.

We use method 2 to rework Example 2.

EXAMPLE 4 Simplify: $\dfrac{\frac{2}{3}+\frac{1}{5}}{\frac{2}{3}-\frac{2}{9}}$

Solution: The LCD of $\frac{2}{3}, \frac{1}{5}, \frac{2}{3}$, and $\frac{2}{9}$ is 45, so we multiply the numerator and the denominator of the complex fraction by 45. Then we perform the indicated operations, and write in lowest terms.

$$\frac{\frac{2}{3}+\frac{1}{5}}{\frac{2}{3}-\frac{2}{9}} = \frac{45\left(\frac{2}{3}+\frac{1}{5}\right)}{45\left(\frac{2}{3}-\frac{2}{9}\right)}$$

$$= \frac{45\left(\frac{2}{3}\right)+45\left(\frac{1}{5}\right)}{45\left(\frac{2}{3}\right)-45\left(\frac{2}{9}\right)} \quad \text{Apply the distributive property.}$$

$$= \frac{30+9}{30-10} = \frac{39}{20} \quad \text{Simplify.}$$

Helpful Hint

The same complex fraction was simplified using two different methods in Examples 2 and 4. Notice that the simplified results are the same.

Practice Problem 4

Use method 2 to simplify the complex fraction in Practice Problem 2:

$$\frac{\frac{3}{4}-\frac{2}{3}}{\frac{1}{2}+\frac{3}{8}}$$

EXAMPLE 5 Simplify: $\dfrac{\frac{x+1}{y}}{\frac{x}{y}+2}$

Solution: The LCD of $\frac{x+1}{y}$ and $\frac{x}{y}$ is y, so we multiply the numerator and the denominator of the complex fraction by y.

$$\frac{\frac{x+1}{y}}{\frac{x}{y}+2} = \frac{y\left(\frac{x+1}{y}\right)}{y\left(\frac{x}{y}+2\right)}$$

$$= \frac{y\left(\frac{x+1}{y}\right)}{y\left(\frac{x}{y}\right)+y\cdot 2} \quad \text{Apply the distributive property in the denominator.}$$

$$= \frac{x+1}{x+2y} \quad \text{Simplify.}$$

Practice Problem 5

Simplify: $\dfrac{1+\frac{x}{y}}{\frac{2x+1}{y}}$

Answers

4. $\frac{2}{21}$ **5.** $\frac{y+x}{2x+1}$

Practice Problem 6

Simplify: $\dfrac{\dfrac{5}{6y} + \dfrac{y}{x}}{\dfrac{y}{3} - x}$

Answer

6. $\dfrac{5x + 6y^2}{2yx(y - 3x)}$

EXAMPLE 6 Simplify: $\dfrac{\dfrac{x}{y} + \dfrac{3}{2x}}{\dfrac{x}{2} + y}$

Solution: The LCD of $\dfrac{x}{y}, \dfrac{3}{2x}, \dfrac{x}{2}$, and $\dfrac{y}{1}$ is $2xy$, so we multiply both the numerator and the denominator of the complex fraction by $2xy$.

$$\frac{\dfrac{x}{y} + \dfrac{3}{2x}}{\dfrac{x}{2} + y} = \frac{2xy\left(\dfrac{x}{y} + \dfrac{3}{2x}\right)}{2xy\left(\dfrac{x}{2} + y\right)}$$

$$= \frac{2xy\left(\dfrac{x}{y}\right) + 2xy\left(\dfrac{3}{2x}\right)}{2xy\left(\dfrac{x}{2}\right) + 2xy(y)}$$ Apply the distributive property.

$$= \frac{2x^2 + 3y}{x^2y + 2xy^2}$$

$$\text{or } \frac{2x^2 + 3y}{xy(x + 2y)}$$

STUDY SKILLS REMINDER

Is your notebook still organized?

Is your notebook still organized? If it's not, it's not too late to start organizing it. Start writing your notes and completing your homework assignment in a notebook with pockets (spiral or ring binder). Take class notes in this notebook, and then follow the notes with your completed homework assignment. When you receive graded papers or handouts, place them in the notebook pocket so that you will not lose them.

Remember to mark (possibly with an exclamation point) any note that seems extra important to you. Also remember to mark (possibly with a question mark) any notes or homework that you are having trouble with. Don't forget to see your instructor or a math tutor to help you with the concepts or exercises that you are having trouble understanding.

Also—don't forget to write neatly.

Name ______________________ Section __________ Date __________

EXERCISE SET 12.7

 **B** *Simplify each complex fraction. See Examples 1 through 6.*

1. $\dfrac{\frac{1}{2}}{\frac{3}{4}}$

2. $\dfrac{\frac{1}{8}}{-\frac{5}{12}}$

3. $\dfrac{-\frac{4x}{9}}{-\frac{2x}{3}}$

4. $\dfrac{-\frac{6y}{11}}{\frac{4y}{9}}$

5. $\dfrac{-\frac{5}{12x^2}}{\frac{25}{16x^3}}$

6. $\dfrac{-\frac{7}{8y}}{\frac{21}{4y}}$

 **7.** $\dfrac{\frac{1}{3}}{\frac{1}{2}-\frac{1}{4}}$

8. $\dfrac{\frac{7}{10}-\frac{3}{5}}{\frac{1}{2}}$

9. $\dfrac{2+\frac{7}{10}}{1+\frac{3}{5}}$

10. $\dfrac{4-\frac{11}{12}}{5+\frac{1}{4}}$

11. $\dfrac{\frac{m}{n}-1}{\frac{m}{n}+1}$

12. $\dfrac{\frac{x}{2}+2}{\frac{x}{2}-2}$

13. $\dfrac{\frac{1}{5}-\frac{1}{x}}{\frac{7}{10}+\frac{1}{x^2}}$

14. $\dfrac{\frac{1}{y^2}+\frac{2}{3}}{\frac{1}{y}-\frac{5}{6}}$

15. $\dfrac{1+\frac{1}{y-2}}{y+\frac{1}{y-2}}$

16. $\dfrac{x-\frac{1}{2x+1}}{1-\frac{x}{2x+1}}$

17. $\dfrac{\frac{4y-8}{16}}{\frac{6y-12}{4}}$

18. $\dfrac{\frac{7y+21}{3}}{\frac{3y+9}{8}}$

19. $\dfrac{\frac{x}{y}+1}{\frac{x}{y}-1}$

20. $\dfrac{\frac{3}{5y}+8}{\frac{3}{5y}-8}$

21. $\dfrac{1}{2 + \dfrac{1}{3}}$

22. $\dfrac{3}{1 - \dfrac{4}{3}}$

23. $\dfrac{\dfrac{ax + ab}{x^2 - b^2}}{\dfrac{x + b}{x - b}}$

24. $\dfrac{\dfrac{m + 2}{m - 2}}{\dfrac{2m + 4}{m^2 - 4}}$

25. $\dfrac{\dfrac{8}{x + 4} + 2}{\dfrac{12}{x + 4} - 2}$

26. $\dfrac{\dfrac{25}{x + 5} + 5}{\dfrac{3}{x + 5} - 5}$

27. $\dfrac{\dfrac{s}{r} + \dfrac{r}{s}}{\dfrac{s}{r} - \dfrac{r}{s}}$

28. $\dfrac{\dfrac{2}{x} + \dfrac{x}{2}}{\dfrac{2}{x} - \dfrac{x}{2}}$

29. Explain how to simplify a complex fraction using method 1.

30. Explain how to simplify a complex fraction using method 2.

Review and Preview

Use the bar graph below to answer Exercises 31 through 34. See Section 7.1.

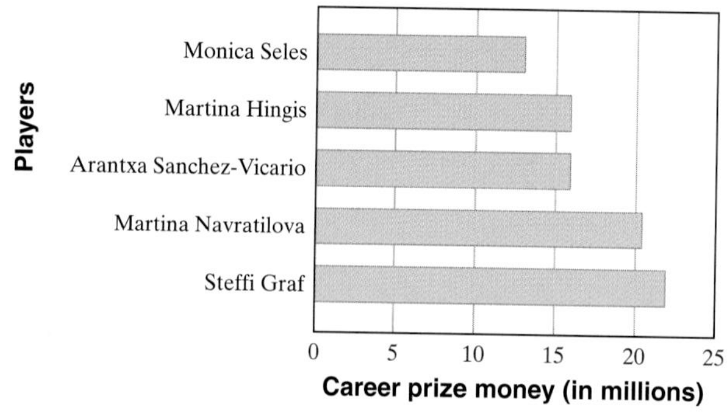

Source: Sanex WTA Tour Media Information System

31. Which women's tennis player has earned the most prize money in her career?

32. Estimate how much more prize money Arantxa Sanchez-Vicario has earned over her career than Monica Seles.

33. Which of the players shown have earned less than \$18 million in prize money over their careers?

34. During her career, Martina Navratilova won over 300 singles and doubles tournament titles. Assuming her prize money was earned only for tournament titles, how much prize money did she earn per tournament title, on average?

Combining Concepts

To find the average of two numbers, we find their sum and divide by 2. For example, the average of 65 and 81 is found by simplifying $\frac{65 + 81}{2}$. *This simplifies to* $\frac{146}{2} = 73$.

35. Find the average of $\frac{1}{3}$ and $\frac{3}{4}$.

36. Write the average of $\frac{3}{n}$ and $\frac{5}{n^2}$ as a simplified rational expression.

37. In electronics, when two resistors R_1 (read R sub 1) and R_2 (read R sub 2) are connected in parallel, the total resistance is given by the complex fraction

$$\frac{1}{\frac{1}{R_1} + \frac{1}{R_2}}.$$

Simplify this expression.

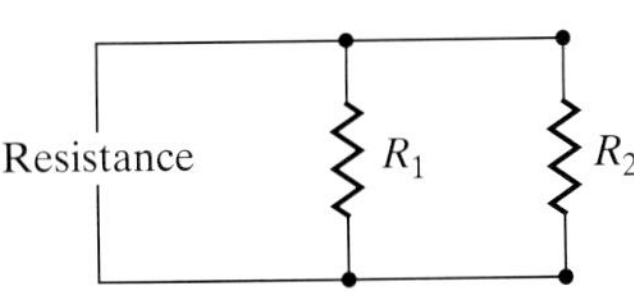

38. Astronomers occasionally need to know the day of the week a particular date fell on. The complex fraction

$$\frac{J + \frac{3}{2}}{7}$$

where J is the *Julian day number*, is used to make this calculation. Simplify this expression.

Simplify each of the following. First, write each expression without negative exponents. Then simplify the complex fraction. The first step has been completed for Exercise 39.

39. $\frac{x^{-1} + 2^{-1}}{x^{-2} - 4^{-1}} = \frac{\frac{1}{x} + \frac{1}{2}}{\frac{1}{x^2} - \frac{1}{4}}$

40. $\frac{3^{-1} - x^{-1}}{9^{-1} - x^{-2}}$

41. $\frac{y^{-2}}{1 - y^{-2}}$

42. $\frac{4 + x^{-1}}{3 + x^{-1}}$

CHAPTER 12 ACTIVITY

Analyzing a Table

MATERIALS:

- Calculator

This activity may be completed by working in groups or individually.

A person's body-mass index (BMI) can be used as an indicator of whether the person should lose weight. BMI can be calculated with the formula $B = \frac{705w}{h^2}$ where w is weight in pounds and h is height in inches. Doctors recommend that body-mass index values fall between 19 and 25. BMI values of 25 to 29.9 are considered overweight, and BMI values of 30 and over are considered obese.

1. Use the BMI formula to complete the table for the given combinations of height and weight. (*Hint*: You may want to use spreadsheet software or a calculator that has a table feature to help fill in the table.)
2. Use the table to find the BMI for (a) a person who is 66 inches tall and weighs 160 pounds, (b) a person who is 62 inches tall and weighs 120 pounds, (c) a person who is 70 inches tall and weighs 170 pounds. Do these BMI values indicate that any of these people should try to lose weight?
3. Examine the table. What pattern do you notice as you look across the rows? What pattern do you notice as you look down the columns?
4. Mark the table to show weight and height combinations with corresponding BMI values that fall within the recommended range.
5. Why would having a table like this on hand be beneficial to a doctor? Explain.

Height

Weight	60	62	64	66	68	70	72	74
100								
110								
120								
130								
140								
150								
160								
170								
180								
190								
200								

Chapter 12 VOCABULARY CHECK

Fill in each blank with one of the terms listed below.

rational expression   complex fraction

1. A ________________ is an expression that can be written in the form $\frac{P}{Q}$, where P and Q are polynomials and Q is not 0.

2. In a ________________ the numerator or denominator or both may contain fractions.

Highlights

DEFINITIONS AND CONCEPTS	EXAMPLES
Section 12.1 Simplifying Rational Expressions	
A **rational expression** is an expression that can be written in the form $\frac{P}{Q}$, where P and Q are polynomials and Q does not equal 0.	$\frac{7y^3}{4}, \frac{x^2+6x+1}{x-3}, \frac{-5}{s^3+8}$
To find values for which a rational expression is undefined, find values for which the denominator is 0.	Find any values for which the expression $\frac{5y}{y^2-4y+3}$ is undefined. $y^2 - 4y + 3 = 0$ Set the denominator equal to 0. $(y-3)(y-1) = 0$ Factor. $y - 3 = 0$ or $y - 1 = 0$ Set each factor equal to 0. $y = 3$ $y = 1$ Solve. The expression is undefined when y is 3 and when y is 1.
FUNDAMENTAL PRINCIPLE OF RATIONAL EXPRESSIONS If P, Q, and R are polynomials, and Q and R are not 0, then $\frac{PR}{QR} = \frac{P}{Q}$	By the fundamental principle, $\frac{(x-3)(x+1)}{x(x+1)} = \frac{x-3}{x}$ as long as $x \neq 0$ and $x \neq -1$.
TO SIMPLIFY A RATIONAL EXPRESSION **Step 1.** Factor the numerator and denominator. **Step 2.** Apply the fundamental principle to divide out common factors.	Simplify: $\frac{4x+20}{x^2-25}$ $\frac{4x+20}{x^2-25} = \frac{4(x+5)}{(x+5)(x-5)} = \frac{4}{x-5}$

Definitions and Concepts | Examples

Section 12.2 Multiplying and Dividing Rational Expressions

To Multiply Rational Expressions

Step 1. Factor numerators and denominators.

Step 2. Multiply numerators and multiply denominators.

Step 3. Write the product in lowest terms.

$$\frac{P}{Q}\cdot\frac{R}{S}=\frac{PR}{QS}$$

Multiply: $\frac{4x+4}{2x-3}\cdot\frac{2x^2+x-6}{x^2-1}$

$$\frac{4x+4}{2x-3}\cdot\frac{2x^2+x-6}{x^2-1}$$
$$=\frac{4(x+1)}{2x-3}\cdot\frac{(2x-3)(x+2)}{(x+1)(x-1)}$$
$$=\frac{4(x+1)(2x-3)(x+2)}{(2x-3)(x+1)(x-1)}$$
$$=\frac{4(x+2)}{x-1}$$

To divide by a rational expression, multiply by the reciprocal.

$$\frac{P}{Q}\div\frac{R}{S}=\frac{P}{Q}\cdot\frac{S}{R}=\frac{PS}{QR}$$

Divide: $\frac{15x+5}{3x^2-14x-5}\div\frac{15}{3x-12}$

$$\frac{15x+5}{3x^2-14x-5}\div\frac{15}{3x-12}$$
$$=\frac{5(3x+1)}{(3x+1)(x-5)}\cdot\frac{3(x-4)}{(3\cdot 5)}$$
$$=\frac{x-4}{x-5}$$

Section 12.3 Adding and Subtracting Rational Expressions with the Same Denominator and Least Common Denominators

To add or subtract rational expressions with the same denominator, add or subtract numerators, and place the sum or difference over the common denominator.

$$\frac{P}{R}+\frac{Q}{R}=\frac{P+Q}{R}$$
$$\frac{P}{R}-\frac{Q}{R}=\frac{P-Q}{R}$$

Perform each indicated operation.

$$\frac{5}{x+1}+\frac{x}{x+1}=\frac{5+x}{x+1}$$

$$\frac{2y+7}{y^2-9}-\frac{y+4}{y^2-9}$$
$$=\frac{(2y+7)-(y+4)}{y^2-9}$$
$$=\frac{2y+7-y-4}{y^2-9}$$
$$=\frac{y+3}{(y+3)(y-3)}$$
$$=\frac{1}{y-3}$$

To Find the Least Common Denominator (LCD)

Step 1. Factor the denominators.

Step 2. The LCD is the product of all unique factors, each raised to a power equal to the greatest number of times that it appears in any one factored denominator.

Find the LCD for

$$\frac{7x}{x^2+10x+25}\quad\text{and}\quad\frac{11}{3x^2+15x}$$

$$x^2+10x+25=(x+5)(x+5)$$
$$3x^2+15x=3x(x+5)$$
$$\text{LCD}=3x(x+5)(x+5)\text{ or }3x(x+5)^2$$

Definitions and Concepts — Examples

Section 12.4 Adding and Subtracting Rational Expressions with Different Denominators

To Add or Subtract Rational Expressions With Different Denominators

Step 1. Find the LCD.

Step 2. Rewrite each rational expression as an equivalent expression whose denominator is the LCD.

Step 3. Add or subtract numerators and place the sum or difference over the common denominator.

Step 4. Write the result in lowest terms.

Perform the indicated operation.

$$\frac{9x+3}{x^2-9}-\frac{5}{x-3}$$
$$=\frac{9x+3}{(x+3)(x-3)}-\frac{5}{x-3}$$

LCD is $(x+3)(x-3)$.

$$=\frac{9x+3}{(x+3)(x-3)}-\frac{5(x+3)}{(x-3)(x+3)}$$
$$=\frac{9x+3-5(x+3)}{(x+3)(x-3)}$$
$$=\frac{9x+3-5x-15}{(x+3)(x-3)}$$
$$=\frac{4x-12}{(x+3)(x-3)}$$
$$=\frac{4(x-3)}{(x+3)(x-3)}=\frac{4}{x+3}$$

Section 12.5 Solving Equations Containing Rational Expressions

To Solve an Equation Containing Rational Expressions

Step 1. Multiply both sides of the equation by the LCD of all rational expressions in the equation.

Step 2. Remove any grouping symbols and solve the resulting equation.

Step 3. Check the solution in the original equation.

Solve: $\frac{5x}{x+2}+3=\frac{4x-6}{x+2}$ The LCD is $x+2$.

$$(x+2)\left(\frac{5x}{x+2}+3\right)=(x+2)\left(\frac{4x-6}{x+2}\right)$$
$$(x+2)\left(\frac{5x}{x+2}\right)+(x+2)(3)=(x+2)\left(\frac{4x-6}{x+2}\right)$$
$$5x+3x+6=4x-6$$
$$4x=-12$$
$$x=-3$$

The solution checks; the solution is -3.

Section 12.6 Rational Equations and Problem Solving

Problem-Solving Steps

1. UNDERSTAND. Read and reread the problem.

A small plane and a car leave Kansas City, Missouri, and head for Minneapolis, Minnesota, a distance of 450 miles. The speed of the plane is 3 times the speed of the car, and the plane arrives 6 hours ahead of the car. Find the speed of the car.

Let x = the speed of the car.

Then $3x$ = the speed of the plane.

	Distance =	Rate ·	Time
Car	450	x	$\frac{450}{x}\left(\frac{\text{distance}}{\text{rate}}\right)$
Plane	450	$3x$	$\frac{450}{3x}\left(\frac{\text{distance}}{\text{rate}}\right)$

Definitions and Concepts	Examples

Section 12.6 Rational Equations and Problem Solving *(continued)*

2. TRANSLATE.

In words: plane's time + 6 hours = car's time

Translate: $\frac{450}{3x} + 6 = \frac{450}{x}$

3. SOLVE.

$$\frac{450}{3x} + 6 = \frac{450}{x}$$

$$3x\left(\frac{450}{3x}\right) + 3x(6) = 3x\left(\frac{450}{x}\right)$$

$$450 + 18x = 1350$$

$$18x = 900$$

$$x = 50$$

4. INTERPRET.

Check the solution by replacing x with 50 in the original equation. **State** the conclusion: The speed of the car is 50 miles per hour.

Section 12.7 Simplifying Complex Fractions

Method 1: To Simplify a Complex Fraction

Step 1. Add or subtract fractions in the numerator and the denominator of the complex fraction.

Step 2. Perform the indicated division.

Step 3. Write the result in lowest terms.

Simplify:

$$\frac{\frac{1}{x} + 2}{\frac{1}{x} - \frac{1}{y}} = \frac{\frac{1}{x} + \frac{2x}{x}}{\frac{y}{xy} - \frac{x}{xy}}$$

$$= \frac{\frac{1 + 2x}{x}}{\frac{y - x}{xy}}$$

$$= \frac{1 + 2x}{x} \cdot \frac{xy}{y - x}$$

$$= \frac{y(1 + 2x)}{y - x}$$

Method 2. To Simplify a Complex Fraction

Step 1. Find the LCD of all fractions in the complex fraction.

Step 2. Multiply the numerator and the denominator of the complex fraction by the LCD.

Step 3. Perform the indicated operations and write the result in lowest terms.

$$\frac{\frac{1}{x} + 2}{\frac{1}{x} - \frac{1}{y}} = \frac{xy\left(\frac{1}{x} + 2\right)}{xy\left(\frac{1}{x} - \frac{1}{y}\right)}$$

$$= \frac{xy\left(\frac{1}{x}\right) + xy(2)}{xy\left(\frac{1}{x}\right) - xy\left(\frac{1}{y}\right)}$$

$$= \frac{y + 2xy}{y - x} \text{ or } \frac{y(1 + 2x)}{y - x}$$

Name ______________________ Section __________ Date __________

Chapter 12 Review

(12.1) *Find any real number for which each rational expression is undefined.*

1. $\dfrac{x + 5}{x^2 - 4}$

2. $\dfrac{5x + 9}{4x^2 - 4x - 15}$

Find the value of each rational expression when $x = 5, y = 7,$ and $z = -2$.

3. $\dfrac{2 - z}{z + 5}$

4. $\dfrac{x^2 + xy - y^2}{x + y}$

Simplify each rational expression.

5. $\dfrac{2x + 6}{x^2 + 3x}$

6. $\dfrac{3x - 12}{x^2 - 4x}$

7. $\dfrac{x + 2}{x^2 - 3x - 10}$

8. $\dfrac{x + 4}{x^2 + 5x + 4}$

9. $\dfrac{x^3 - 4x}{x^2 + 3x + 2}$

10. $\dfrac{5x^2 - 125}{x^2 + 2x - 15}$

11. $\dfrac{x^2 - x - 6}{x^2 - 3x - 10}$

12. $\dfrac{x^2 - 2x}{x^2 + 2x - 8}$

Simplify each expression. First, factor the four-term polynomials by grouping.

13. $\dfrac{x^2 + xa + xb + ab}{x^2 - xc + bx - bc}$

14. $\dfrac{x^2 + 5x - 2x - 10}{x^2 - 3x - 2x + 6}$

(12.2) *Perform each indicated operation and simplify.*

15. $\dfrac{15x^3y^2}{z} \cdot \dfrac{z}{5xy^3}$

16. $\dfrac{-y^3}{8} \cdot \dfrac{9x^2}{y^3}$

17. $\dfrac{x^2 - 9}{x^2 - 4} \cdot \dfrac{x - 2}{x + 3}$

18. $\dfrac{2x + 5}{x - 6} \cdot \dfrac{2x}{-x + 6}$

19. $\dfrac{x^2 - 5x - 24}{x^2 - x - 12} \div \dfrac{x^2 - 10x + 16}{x^2 + x - 6}$

20. $\dfrac{4x + 4y}{xy^2} \div \dfrac{3x + 3y}{x^2y}$

21. $\dfrac{x^2 + x - 42}{x - 3} \cdot \dfrac{(x - 3)^2}{x + 7}$

22. $\dfrac{2a + 2b}{3} \cdot \dfrac{a - b}{a^2 - b^2}$

23. $\dfrac{x^2 - 9x + 14}{x^2 - 5x + 6} \cdot \dfrac{x + 2}{x^2 - 5x - 14}$

24. $(x - 3) \cdot \dfrac{x}{x^2 + 3x - 18}$

25. $\dfrac{2x^2 - 9x + 9}{8x - 12} \div \dfrac{x^2 - 3x}{2x}$

26. $\dfrac{x^2 - y^2}{x^2 + xy} \div \dfrac{3x^2 - 2xy - y^2}{3x^2 + 6x}$

(12.3) *Perform each indicated operation and simplify.*

27. $\dfrac{x}{x^2 + 9x + 14} + \dfrac{7}{x^2 + 9x + 14}$

28. $\dfrac{x}{x^2 + 2x - 15} + \dfrac{5}{x^2 + 2x - 15}$

29. $\dfrac{4x - 5}{3x^2} - \dfrac{2x + 5}{3x^2}$

30. $\dfrac{9x + 7}{6x^2} - \dfrac{3x + 4}{6x^2}$

Find the LCD of each pair of rational expressions.

31. $\dfrac{x + 4}{2x}, \dfrac{3}{7x}$

32. $\dfrac{x - 2}{x^2 - 5x - 24}, \dfrac{3}{x^2 + 11x + 24}$

Rewrite each rational expression as an equivalent expression whose denominator is the given polynomial.

33. $\dfrac{5}{7x} = \dfrac{\quad}{14x^3y}$

34. $\dfrac{9}{4y} = \dfrac{\quad}{16y^3x}$

35. $\dfrac{x + 2}{x^2 + 11x + 18} = \dfrac{\quad}{(x + 2)(x - 5)(x + 9)}$

36. $\dfrac{3x - 5}{x^2 + 4x + 4} = \dfrac{\quad}{(x + 2)^2(x + 3)}$

(12.4) *Perform each indicated operation and simplify.*

37. $\dfrac{4}{5x^2} - \dfrac{6}{y}$

38. $\dfrac{2}{x - 3} - \dfrac{4}{x - 1}$

39. $\dfrac{x + 7}{x + 3} - \dfrac{x - 3}{x + 7}$

40. $\dfrac{4}{x + 3} - 2$

41. $\dfrac{3}{x^2 + 2x - 8} + \dfrac{2}{x^2 - 3x + 2}$

42. $\dfrac{2x - 5}{6x + 9} - \dfrac{4}{2x^2 + 3x}$

43. $\dfrac{x - 1}{x^2 - 2x + 1} - \dfrac{x + 1}{x - 1}$

44. $\dfrac{x - 1}{x^2 + 4x + 4} + \dfrac{x - 1}{x + 2}$

Find the perimeter and the area of each figure.

 **45.**

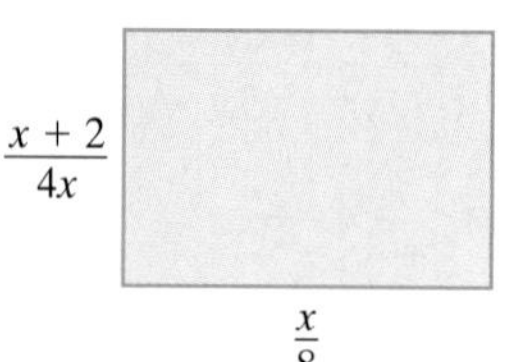

△ **46.**

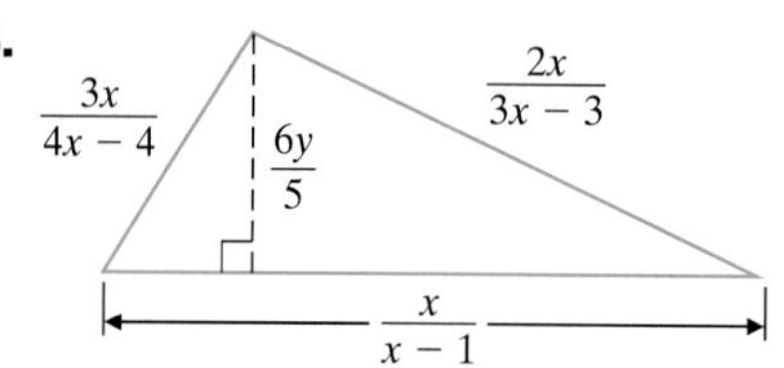

(12.5) *Solve each equation.*

47. $\frac{x+4}{9} = \frac{5}{9}$

48. $\frac{n}{10} = 9 - \frac{n}{5}$

49. $\frac{5y-3}{7} = \frac{15y-2}{28}$

50. $\frac{2}{x+1} - \frac{1}{x-2} = -\frac{1}{2}$

51. $\frac{1}{a+3} + \frac{1}{a-3} = -\frac{5}{a^2-9}$

52. $\frac{y}{2y+2} + \frac{2y-16}{4y+4} = \frac{y-3}{y+1}$

53. $\frac{4}{x+3} + \frac{8}{x^2-9} = 0$

54. $\frac{2}{x-3} - \frac{4}{x+3} = \frac{8}{x^2-9}$

55. $\frac{x-3}{x+1} - \frac{x-6}{x+5} = 0$

56. $x + 5 = \frac{6}{x}$

Solve each equation for the indicated variable.

57. $\frac{4A}{5b} = x^2$ for b

58. $\frac{x}{7} + \frac{y}{8} = 10$ for y

(12.6) *Solve each problem.*

59. Five times the reciprocal of a number equals the sum of $\frac{3}{2}$ the reciprocal of the number and $\frac{7}{6}$. What is the number?

60. The reciprocal of a number equals the reciprocal of the difference of 4 and the number. Find the number.

61. A car travels 90 miles in the same time that a car traveling 10 miles per hour slower travels 60 miles. Find the speed of each car.

62. The current in a bayou near Lafayette, Louisiana, is 4 miles per hour. A paddle boat travels 48 miles upstream in the same amount of time it takes to travel 72 miles downstream. Find the speed of the boat in still water.

63. When Mark and Maria manicure Mr. Stergeon's lawn, it takes them 5 hours. If Mark works alone, it takes 7 hours. Find how long it takes Maria alone.

64. It takes pipe A 20 days to fill a fish pond. Pipe B takes 15 days. Find how long it takes both pipes together to fill the pond.

Given that the pairs of triangles are similar, find each missing length x.

△ **65.**

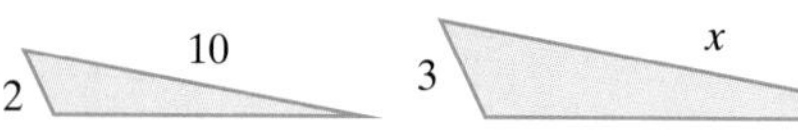

△ **66.**

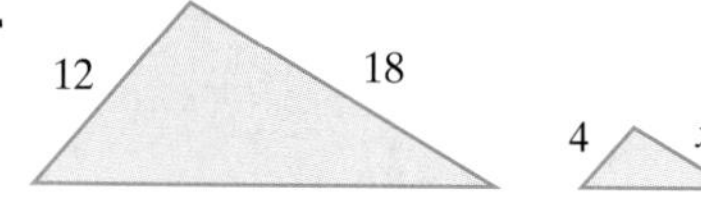

△ **67.**

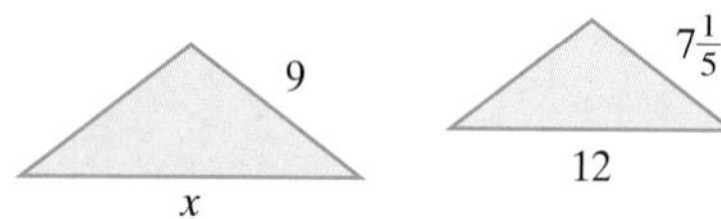

△ **68.**

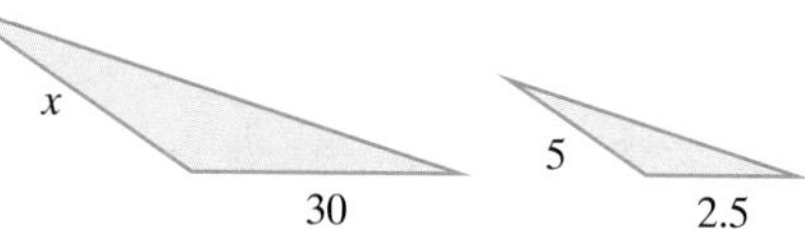

(12.7) *Simplify each complex fraction.*

69. $\dfrac{\dfrac{5x}{27}}{-\dfrac{10xy}{21}}$

70. $\dfrac{\dfrac{8x}{x^2-9}}{\dfrac{4}{x+3}}$

71. $\dfrac{\dfrac{3}{5}+\dfrac{2}{7}}{\dfrac{1}{5}+\dfrac{5}{6}}$

72. $\dfrac{2+\dfrac{1}{x^2}}{\dfrac{1}{x}+\dfrac{2}{x^2}}$

73. $\dfrac{3-\dfrac{1}{y}}{2-\dfrac{1}{y}}$

74. $\dfrac{\dfrac{6}{x+2}+4}{\dfrac{8}{x+2}-4}$

Name ______________________ Section __________ Date __________

Chapter 12 Test

Remember to check your answers and use the Chapter Test Prep Video to view solutions.

1. Find any real numbers for which the following expression is undefined.
$\dfrac{x+5}{x^2+4x+3}$

2. For a certain computer desk, the manufacturing cost C per desk (in dollars) is
$C = \dfrac{100x + 3000}{x}$
where x is the number of desks manufactured.
   a. Find the average cost per desk when manufacturing 200 computer desks.
   b. Find the average cost per desk when manufacturing 1000 computer desks.

Simplify each rational expression.

3. $\dfrac{3x-6}{5x-10}$

4. $\dfrac{x+10}{x^2-100}$

5. $\dfrac{x+6}{x^2+12x+36}$

6. $\dfrac{7-x}{x-7}$

7. $\dfrac{2m^3-2m^2-12m}{m^2-5m+6}$

8. $\dfrac{y-x}{x^2-y^2}$

Perform each indicated operation and simplify if possible.

9. $\dfrac{3}{x-1}\cdot(5x-5)$

10. $\dfrac{y^2-5y+6}{2y+4}\cdot\dfrac{y+2}{2y-6}$

11. $\dfrac{6}{2x+5}-\dfrac{5+x}{2x+5}$

12. $\dfrac{5a}{a^2-a-6}-\dfrac{2}{a-3}$

13. $\dfrac{6}{x^2-1}+\dfrac{3}{x+1}$

14. $\dfrac{x^2-9}{x^2-3x}\div\dfrac{xy+5x+3y+15}{2x+10}$

15. $\dfrac{x+2}{x^2+11x+18}+\dfrac{5}{x^2-3x-10}$

Answers

1. ______
2. a. ______
   b. ______
3. ______
4. ______
5. ______
6. ______
7. ______
8. ______
9. ______
10. ______
11. ______
12. ______
13. ______
14. ______
15. ______

16. ______

17. ______

18. ______

19. ______

20. ______

21. ______

22. ______

23. ______

24. ______

25. ______

Solve each equation.

16. $\frac{4}{y} - \frac{5}{3} = \frac{-1}{5}$

17. $\frac{5}{y+1} = \frac{4}{y+2}$

18. $\frac{a}{a-3} = \frac{3}{a-3} - \frac{3}{2}$

19. $x - \frac{14}{x-1} = 4 - \frac{2x}{x-1}$

Simplify each complex fraction.

20. $\dfrac{\dfrac{5x^2}{yz^2}}{\dfrac{10x}{z^3}}$

21. $\dfrac{5 - \dfrac{1}{y^2}}{\dfrac{1}{y} + \dfrac{2}{y^2}}$

△ **22.** Given that the two triangles are similar, find x.

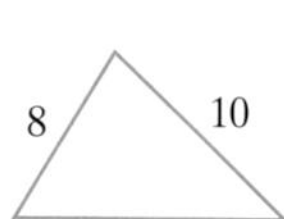

23. A number plus five times its reciprocal is equal to six. Find the number.

24. A pleasure boat traveling down the Red River takes the same time to go 14 miles upstream as it takes to go 16 miles downstream. If the current of the river is 2 miles per hour, find the speed of the boat in still water.

25. An inlet pipe can fill a tank in 12 hours. A second pipe can fill the tank in 15 hours. If both pipes are used, find how long it takes to fill the tank.

Name ______________________ Section ________ Date ________

Answers

Chapter 12 Cumulative Review

Write each fraction or mixed number as a percent.

1. $\frac{9}{20}$ **2.** $\frac{4}{5}$ **3.** $\frac{2}{3}$ **4.** $\frac{1}{9}$ **5.** $1\frac{1}{2}$ **6.** $3\frac{3}{4}$

Name the property illustrated by each true statement.

7. $3 \cdot y = y \cdot 3$

8. $3 + y = y + 3$

9. $(x + 7) + 9 = x + (7 + 9)$

10. $(x \cdot 7) \cdot 9 = x \cdot (7 \cdot 9)$

11. A 10-foot board is to be cut into two pieces so that the longer piece is 4 times longer than the shorter. Find the length of each piece.

12. A 45-meter piece of rope is to be cut into three pieces. Two pieces are to have equal length and the third piece is to be 3 feet shorter than the longest pieces. Find the length of each piece.

13. Solve $y = mx + b$ for x.

14. Solve $y = mx + b$ for b.

15. Solve $x + 4 \leq -6$. Graph the solutions.

−10 −8 −6 −4 −2 0 2 4 6 8 10

16. Solve $x - 1 > -10$

Simplify each quotient.

17. $\frac{x^5}{x^2}$ **18.** $\frac{x^9}{x}$ **19.** $\frac{4^7}{4^3}$ **20.** $\frac{7^{12}}{7^4}$

21. $\frac{(-3)^5}{(-3)^2}$ **22.** $\frac{(-4)^9}{(-4)^7}$ **23.** $\frac{2x^5y^2}{xy}$ **24.** $\frac{13a^5b}{a^4}$

1. ______
2. ______
3. ______
4. ______
5. ______
6. ______
7. ______
8. ______
9. ______
10. ______
11. ______
12. ______
13. ______
14. ______
15. ______
16. ______
17. ______
18. ______
19. ______
20. ______
21. ______
22. ______
23. ______
24. ______

25.

26.

27.

28.

29.

30.

31.

32.

33.

34.

35.

36.

37.

38.

39.

40.

41.

42.

43.

44.

Simplify by writing each expression with positive exponents only.

25. $2x^{-3}$ **26.** $9x^{-2}$ **27.** $(-2)^{-4}$ **28.** $(-3)^{-3}$

Multiply.

29. $5x(2x^3 + 6)$

30. $3y(4y^2 - 2)$

31. $-3x^2(5x^2 + 6x - 1)$

32. $-5y(7y^2 - 3y + 1)$

33. Divide: $\dfrac{4x^2 + 7 + 8x^3}{2x + 3}$

34. Divide: $\dfrac{6x^2 - 7x + 4}{2x+1}$

Factor .

35. $x^2 + 7x + 12$

36. $x^2 + 17x + 70$

37. $25x^2 + 20xy + 4y^2$

38. $36a^2 - 48ab + 16b^2$

39. Solve: $x^2 - 9x - 22 = 0$

40. Solve: $x^2 + 2x - 15 = 0$

41. Multiply: $\dfrac{x^2 + x}{3x} \cdot \dfrac{6}{5x + 5}$

42. Divide: $\dfrac{3x - 12}{2} \div \dfrac{5x - 20}{4x}$

43. Subtract: $\dfrac{3x^2 + 2x}{x - 1} - \dfrac{10x - 5}{x - 1}$

44. Add: $\dfrac{4x}{x + 2} + \dfrac{8}{x + 2}$

45. Subtract: $\dfrac{6x}{x^2 - 4} - \dfrac{3}{x + 2}$

46. Add: $\dfrac{2}{x - 3} + \dfrac{5x}{x^2 - 9}$

47. Solve: $\dfrac{t - 4}{2} - \dfrac{t - 3}{9} = \dfrac{5}{18}$

48. $\dfrac{y}{2} - \dfrac{y}{4} = \dfrac{1}{6}$

49. Sam Waterton and Frank Schaffer work in a plant that manufactures automobiles. Sam can complete a quality control tour of the plant in 3 hours while his assistant, Frank, needs 7 hours to complete the same job. The regional manager is coming to inspect the plant facilities, so both Sam and Frank are directed to complete a quality control tour together. How long will this take?

50. One machine can complete a task in 18 hours. A second machine can do the same task in 12 hours. How long would it take to complete the task if both machines are able to work on it?

51. Simplify: $\dfrac{\dfrac{1}{z} - \dfrac{1}{2}}{\dfrac{1}{3} - \dfrac{z}{6}}$

52. Simplify: $\dfrac{\dfrac{x}{9} - \dfrac{1}{x}}{1 + \dfrac{3}{x}}$

45. ____________

46. ____________

47. ____________

48. ____________

49. ____________

50. ____________

51. ____________

52. ____________

Graphing Equations and Inequalities

CHAPTER 13

In Chapter 9 we solved and graphed the solutions of linear equations and inequalities in one variable on number lines. Now we define and present techniques for solving and graphing linear equations and inequalities in two variables on grids. Two-variable equations lead directly to the concept of *function*, perhaps the most important concept in all mathematics. Functions are introduced in Section 13.6.

Ninety-three percent of American children go trick-or-treating at Halloween. Each year, Americans spend roughly \$2 billion on Halloween candy, including 20 million pounds of the ever-popular candy corn. While that means that Americans purchase approximately 8.3 billion kernels of candy corn to help celebrate Halloween, adults prefer to hand out chocolate to trick-or-treaters. According to a recent survey, nearly 80% of adults said they planned to give out chocolate treats at Halloween. The survey also revealed that SNICKERS® Bars are the top chocolate pick for distributing to little beggars on All Saints' Eve. In Exercise 20, Section 13.1, we will examine the trends in Halloween candy sales.

Answers

1. see graph

2.

3. see graph

4. see graph

5. see graph

6.

7.

8.

9.

10.

11.

12.

13.

14.

15. a.

b.

c.

16.

Name ______ Section ______ Date ______

Chapter 13 Pretest

1. Plot each ordered pair: $(-4, 3)$, $(0, -2)$, and $(5, 0)$

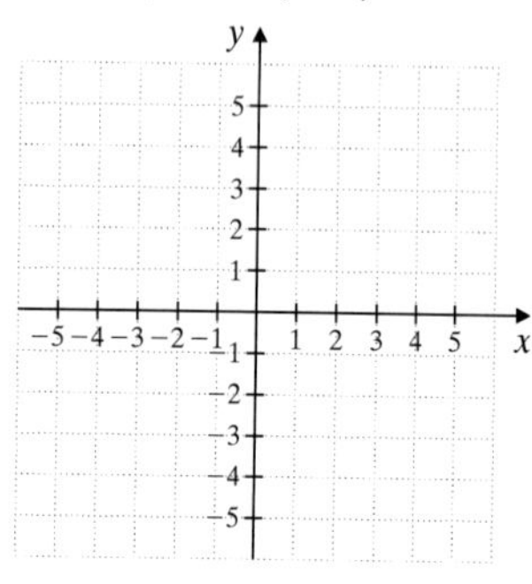

2. Complete the ordered pair solution for the given linear equation.
$8x - 3y = 2; (-2, \ )$

Graph.

3. $3x - y = 6$

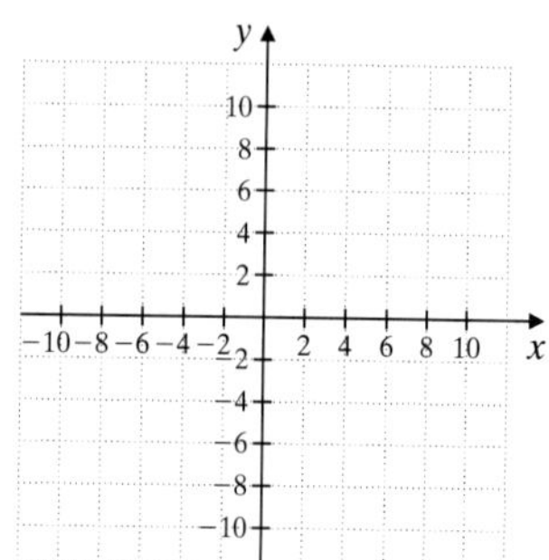

4. The line with x-intercept: $(-1, 0)$, y-intercept: $(0, 4)$

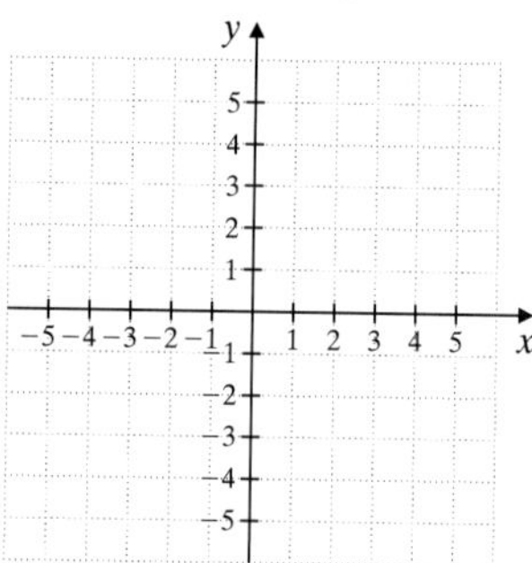

5. $3x + 2y \leq 6$

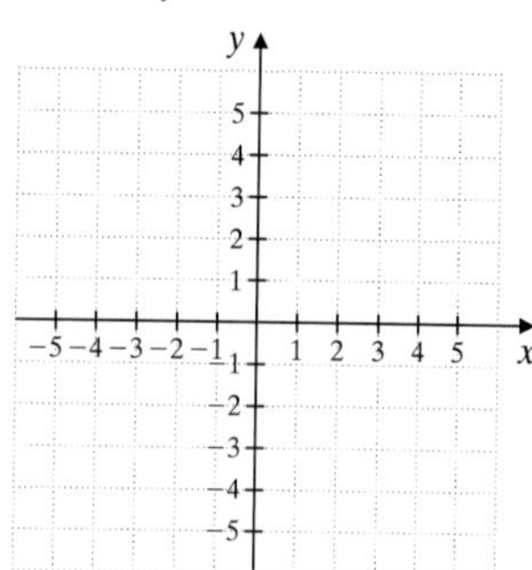

Find the slope of each line.

6. Passes through $(-7, 8)$ and $(3, 5)$

7. $4x - 5y = 20$

8. $x = 10$

Find the equation of each line. Write the equation in the form $Ax + By = C$.

9. Slope $-\frac{1}{3}$; passes through $(3, -6)$

10. Passes through the origin and $(-8, 1)$

11. Slope $\frac{2}{7}$; y-intercept $(0, 14)$

12. Find the domain and range of the given relation: $\{(-3, 8), (7, -1), (0, 6), (2, -1)\}$

Determine whether each relation is a function.

13. $\{(1, 7), (-8, 7), (6, 3), (9, 2)\}$

14. $\{(0, 4), (1, 3), (2, -5), (1, 10), (-2, -8)\}$

15. Given the function $f(x) = -3x + 8$, find each function value.
a. $f(-1)$ **b.** $f(0)$ **c.** $f(10)$

16. y varies directly as x. If $y = 16$ when $x = 24$, find y when x is 20.

13.1 The Rectangular Coordinate System

OBJECTIVES

A. Plot ordered pairs of numbers on the rectangular coordinate system.
B. Graph paired data to create a scatter diagram.
C. Find the missing coordinate of an ordered pair solution, given one coordinate of the pair.

In Sections 7.1 and 7.2, we learned how to read graphs. The broken line graph below shows the relationship between time spent smoking a cigarette and pulse rate. The horizontal line or axis shows time in minutes and the vertical line or axis shows the pulse rate in heartbeats per minute. Notice in this graph that there are two numbers associated with each point of the graph. For example, 15 minutes after "lighting up," the pulse rate is 80 beats per minute. If we agree to write the time first and the pulse rate second, we can say there is a point on the graph corresponding to the **ordered pair** of numbers (15, 80). A few more ordered pairs are shown alongside their corresponding points.

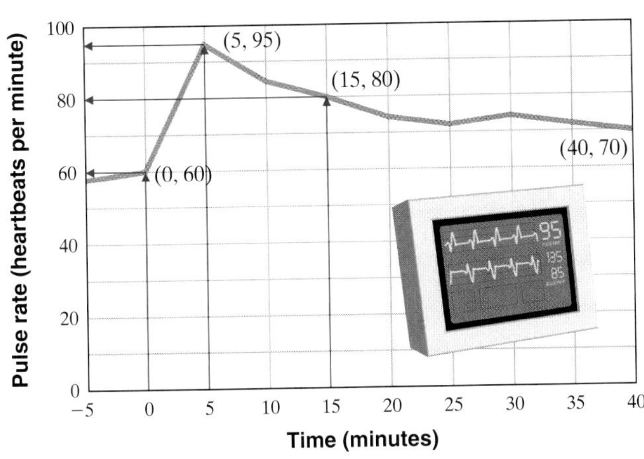

A Plotting Ordered Pairs of Numbers

In general, we use the idea of ordered pairs to describe the location of a point in a plane (such as a piece of paper). We start with a horizontal and a vertical axis. Each axis is a number line, and for the sake of consistency we construct our axes to intersect at the 0 coordinate of both. This point of intersection is called the **origin.** Notice that these two number lines or axes divide the plane into four regions called **quadrants.** The quadrants are usually numbered with Roman numerals as shown. The axes are not considered to be in any quadrant.

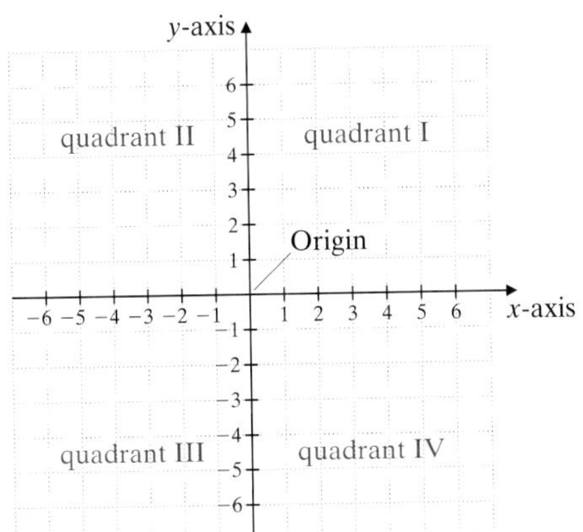

It is helpful to label axes, so we label the horizontal axis the ***x*-axis** and the vertical axis the ***y*-axis.** We call the system described above the **rectangular coordinate system,** or the **coordinate plane.** Just as with other graphs shown, we can then describe the locations of points by ordered pairs of numbers. We list the horizontal ***x*-axis** measurement first and the vertical ***y*-axis** measurement second.

To plot or graph the point corresponding to the ordered pair

(a, b),

we start at the origin. We then move a units left or right (right if a is positive, left if a is negative). From there, we move b units up or down (up if b is positive, down if b is negative). For example, to plot the point corresponding to the ordered pair $(3, 2)$, we start at the origin, move 3 units right, and from there move 2 units up. (See the figure below.) The x-value, 3, is also called the **x-coordinate** and the y-value, 2, is also called the **y-coordinate.** From now on, we will call the point with coordinates $(3, 2)$ simply the point $(3, 2)$. The point $(-2, 5)$ is graphed below also.

y-axis
(−2, 5)
5 units up
(3, 2)
2 units up
2 units left
3 units right
x-axis

Concept Check

Is the graph of the point $(-5, 1)$ in the same location as the graph of the point $(1, -5)$? Explain.

Practice Problem 1

On a single coordinate system, plot each ordered pair. State in which quadrant, if any, each point lies.

a. $(4, 2)$ b. $(-1, -3)$
c. $(2, -2)$ d. $(-5, 1)$
e. $(0, 3)$ f. $(3, 0)$
g. $(0, -4)$ h. $\left(-2\frac{1}{2}, 0\right)$

Answers

1.

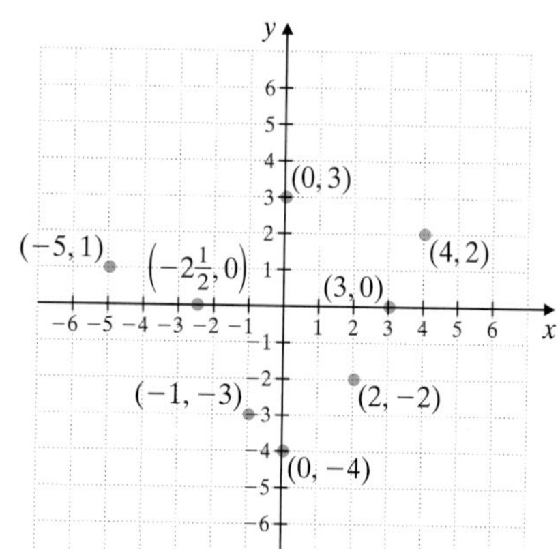

a. Point $(4, 2)$ lies in quadrant I.
b. Point $(-1, -3)$ lies in quadrant III.
c. Point $(2, -2)$ lies in quadrant IV.
d. Point $(-5, 1)$ lies in quadrant II.
e.–h. Points $(0, 3)$, $(3, 0)$, $(0, -4)$, and $\left(-2\frac{1}{2}, 0\right)$ lie on axes, so they are not in any quadrant.

Concept Check: The graph of point $(-5, 1)$ lies in quadrant II and the graph of point $(1, -5)$ lies in quadrant IV. They are *not* in the same location.

Helpful Hint

Don't forget that **each ordered pair corresponds to exactly one point in the plane and that each point in the plane corresponds to exactly one ordered pair.**

Try the Concept Check in the margin.

EXAMPLE 1 On a single coordinate system, plot each ordered pair. State in which quadrant, if any, each point lies.

a. $(5, 3)$ **b.** $(-2, -4)$ **c.** $(1, -2)$ **d.** $(-5, 3)$
e. $(0, 0)$ **f.** $(0, 2)$ **g.** $(-5, 0)$ **h.** $\left(0, -5\frac{1}{2}\right)$

Solution:

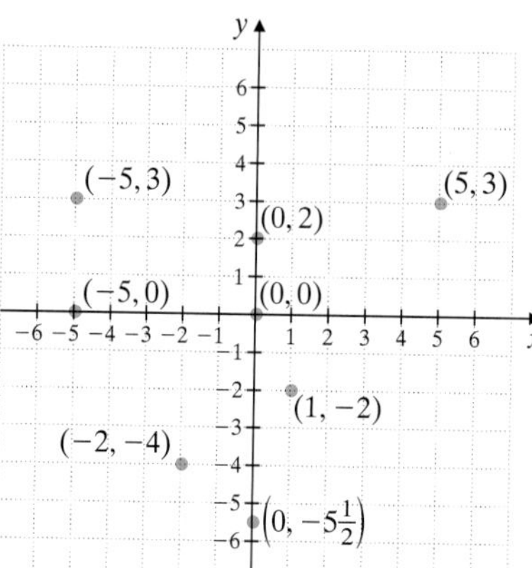

a. Point $(5, 3)$ lies in quadrant I.
b. Point $(-2, -4)$ lies in quadrant III.
c. Point $(1, -2)$ lies in quadrant IV.
d. Point $(-5, 3)$ lies in quadrant II.
e.–h. Points $(0, 0)$, $(0, 2)$, $(-5, 0)$, and $\left(0, -5\frac{1}{2}\right)$ lie on axes, so they are not in any quadrant.

Try the Concept Check in the margin.

From Example 1, notice that the y-coordinate of any point on the x-axis is 0. For example, the point $(-5, 0)$ lies on the x-axis. Also, the x-coordinate of any point on the y-axis is 0. For example, the point $(0, 2)$ lies on the y-axis.

B Creating Scatter Diagrams

Data that can be represented as ordered pairs are called **paired data.** Many types of data collected from the real world are paired data. For instance, the annual measurements of a child's height can be written as ordered pairs of the form (year, height in inches) and are paired data. The graph of paired data as points in the rectangular coordinate system is called a **scatter diagram.** Scatter diagrams can be used to look for patterns and trends in paired data.

EXAMPLE 2 The table gives the annual net sales for Wal-Mart Stores for the years shown. (*Source:* Wal-Mart Stores, Inc.)

Year	Wal-Mart Net Sales (in billions of dollars)
1999	138
2000	165
2001	191
2002	218
2003	245

a. Write this paired data as a set of ordered pairs of the form (year, revenue in billions of dollars).

b. Create a scatter diagram of the paired data.

c. What trend in the paired data does the scatter diagram show?

Solution:

a. The ordered pairs are $(1999, 138)$, $(2000, 165)$, $(2001, 191)$, $(2002, 218)$, and $(2003, 245)$,

b. We begin by plotting the ordered pairs. Because the x-coordinate in each ordered pair is a year, we label the x-axis "Year" and mark the horizontal axis with the years given. Then we label the y-axis or vertical axis "Wal-Mart Net Sales (in billions of dollars)." In this case, it is convenient to mark the vertical axis in multiples of 25, starting with 0. Since no net sale is less than 125, we use the notation ⚡ to skip to 125, then proceed by multiples of 25.

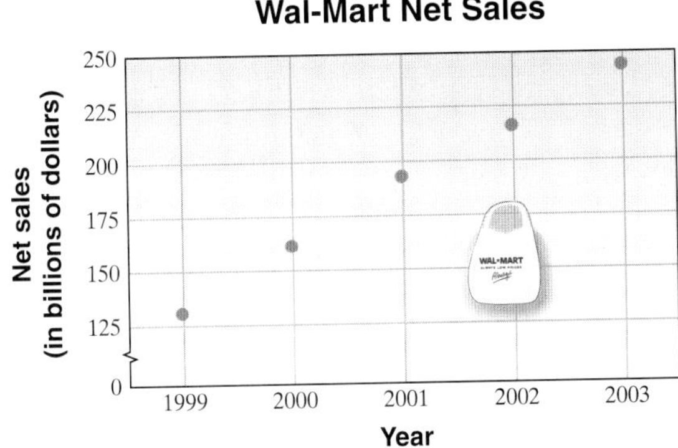

c. The scatter diagram shows that Wal-Mart net sales steadily increased over the years 1999–2003.

Concept Check

For each description of a point in the rectangular coordinate system, write an ordered pair that represents it.

a. Point A is located three units to the left of the y-axis and five units above the x-axis.

b. Point B is located six units below the origin.

Practice Problem 2

The table gives the number of tornadoes that have occurred in the United States for the years shown. (*Source:* Storm Prediction Center, National Weather Service)

Year	Tornadoes
1997	1148
1998	1424
1999	1343
2000	997
2001	1216
2002	941

a. Write this paired data as a set of ordered pairs of the form (year, tornadoes).

b. Create a scatter diagram of the paired data.

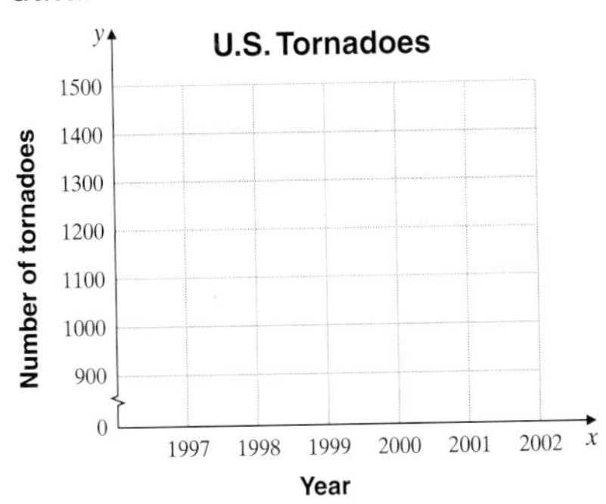

c. What trend in the paired data, if any, does the scatter diagram show?

Answers

2. a. $(1997, 1148)$, $(1998, 1424)$, $(1999, 1343)$, $(2000, 997)$, $(2001, 1216)$, $(2002, 941)$

b.

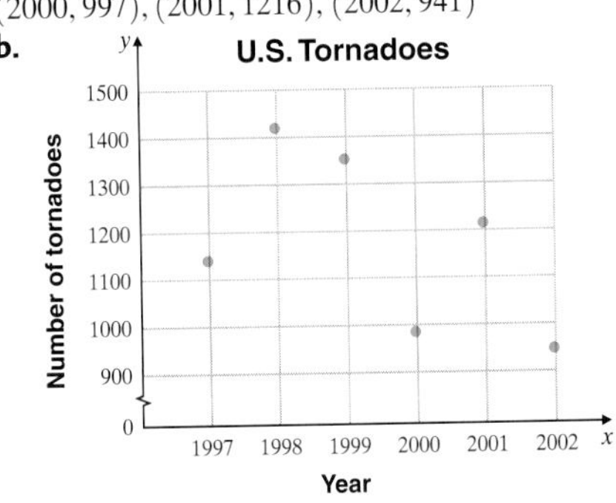

c. The number of tornadoes varies greatly from year to year.

Concept Check: **a.** $(-3, 5)$ **b.** $(0, -6)$

C Completing Ordered Pair Solutions

Let's see how we can use ordered pairs to record solutions of equations containing two variables. An equation in one variable such as $x + 1 = 5$ has one solution, which is 4: the number 4 is the value of the variable x that makes the equation true.

An equation in two variables, such as $2x + y = 8$, has solutions consisting of two values, one for x and one for y. For example, $x = 3$ and $y = 2$ is a solution of $2x + y = 8$ because, if x is replaced with 3 and y with 2, we get a true statement.

$$\begin{aligned} 2x + y &= 8 \\ 2(3) + 2 &\stackrel{?}{=} 8 \\ 8 &= 8 \quad \text{True} \end{aligned}$$

The solution $x = 3$ and $y = 2$ can be written as $(3, 2)$, an ordered pair of numbers.

In general, an ordered pair is a **solution** of an equation in two variables if replacing the variables by the values of the ordered pair results in a *true statement*. For example, another ordered pair solution of $2x + y = 8$ is $(5, -2)$. Replacing x with 5 and y with -2 results in a true statement.

$$\begin{aligned} 2x + y &= 8 \\ 2(5) + (-2) &\stackrel{?}{=} 8 \quad \text{Replace } x \text{ with 5 and } y \text{ with } -2. \\ 10 - 2 &\stackrel{?}{=} 8 \\ 8 &= 8 \quad \text{True} \end{aligned}$$

Practice Problem 3

Complete each ordered pair so that it is a solution to the equation $x + 2y = 8$.

a. $(0, \ )$
b. $(\ , 3)$
c. $(-4, \ )$

EXAMPLE 3 Complete each ordered pair so that it is a solution to the equation $3x + y = 12$.

a. $(0, \ )$ **b.** $(\ , 6)$ **c.** $(-1, \ )$

Solution:

a. In the ordered pair $(0, \ )$, the x-value is 0. We let $x = 0$ in the equation and solve for y.

$$\begin{aligned} 3x + y &= 12 \\ 3(0) + y &= 12 \quad \text{Replace } x \text{ with 0.} \\ 0 + y &= 12 \\ y &= 12 \end{aligned}$$

The completed ordered pair is $(0, 12)$.

b. In the ordered pair $(\ , 6)$, the y-value is 6. We let $y = 6$ in the equation and solve for x.

$$\begin{aligned} 3x + y &= 12 \\ 3x + 6 &= 12 \quad \text{Replace } y \text{ with 6.} \\ 3x &= 6 \quad \text{Subtract 6 from both sides.} \\ x &= 2 \quad \text{Divide both sides by 3.} \end{aligned}$$

The ordered pair is $(2, 6)$.

c. In the ordered pair $(-1, \ )$, the x-value is -1. We let $x = -1$ in the equation and solve for y.

$$\begin{aligned} 3x + y &= 12 \\ 3(-1) + y &= 12 \quad \text{Replace } x \text{ with } -1. \\ -3 + y &= 12 \\ y &= 15 \quad \text{Add 3 to both sides.} \end{aligned}$$

The ordered pair is $(-1, 15)$.

Answers

3. a. $(0, 4)$ **b.** $(2, 3)$ **c.** $(-4, 6)$

Solutions of equations in two variables can also be recorded in a **table of paired values,** as shown in the next example.

EXAMPLE 4 Complete the table for the equation $y = 3x$.

	x	y
a.	-1	
b.		0
c.		-9

Solution:

a. We replace x with -1 in the equation and solve for y.

$$y = 3x$$
$$y = 3(-1) \quad \text{Let } x = -1.$$
$$y = -3$$

The ordered pair is $(-1, -3)$.

b. We replace y with 0 in the equation and solve for x.

$$y = 3x$$
$$0 = 3x \quad \text{Let } y = 0.$$
$$0 = x \quad \text{Divide both sides by 3.}$$

The ordered pair is $(0, 0)$.

c. We replace y with -9 in the equation and solve for x.

$$y = 3x$$
$$-9 = 3x \quad \text{Let } y = -9.$$
$$-3 = x \quad \text{Divide both sides by 3.}$$

The ordered pair is $(-3, -9)$. The completed table is shown to the right.

x	y
-1	-3
0	0
-3	-9

EXAMPLE 5 Complete the table for the equation $y = 3$.

x	y
-2	
0	
-5	

Solution: The equation $y = 3$ is the same as $0x + y = 3$. No matter what value we replace x by, y always equals 3. The completed table is shown to the right.

x	y
-2	3
0	3
-5	3

Practice Problem 4

Complete the table for the equation $y = -2x$.

	x	y
a.	-3	
b.		0
c.		10

Practice Problem 5

Complete the table for the equation $x = 5$.

x	y
	-2
	0
	4

Answers

4.

	x	y
a.	-3	6
b.	0	0
c.	-5	10

5.

x	y
5	-2
5	0
5	4

By now, you have noticed that equations in two variables often have more than one solution. We discuss this more in the next section.

A table showing ordered pair solutions may be written vertically, or horizontally as shown in the next example.

Practice Problem 6

A company purchased a fax machine for \$400. The business manager of the company predicts that the fax machine will be used for 7 years and the value in dollars y of the machine in x years is $y = -50x + 400$. Complete the table.

x	1	2	3	4	5	6	7
y							

EXAMPLE 6 A small business purchased a computer for \$2000. The business predicts that the computer will be used for 5 years and the value in dollars y of the computer in x years is $y = -300x + 2000$. Complete the table.

x	0	1	2	3	4	5
y						

Solution: To find the value of y when x is 0, we replace x with 0 in the equation. We use this same procedure to find y when x is 1 and when x is 2.

When $x = 0$,

$y = -300x + 2000$
$y = -300 \cdot 0 + 2000$
$y = 0 + 2000$
$y = 2000$

When $x = 1$,

$y = -300x + 2000$
$y = -300 \cdot 1 + 2000$
$y = -300 + 2000$
$y = 1700$

When $x = 2$,

$y = -300x + 2000$
$y = -300 \cdot 2 + 2000$
$y = -600 + 2000$
$y = 1400$

We have the ordered pairs $(0, 2000)$, $(1, 1700)$, and $(2, 1400)$. This means that in 0 years the value of the computer is \$2000, in 1 year the value of the computer is \$1700, and in 2 years the value is \$1400. To complete the table of values, we continue the procedure for $x = 3$, $x = 4$, and $x = 5$.

When $x = 3$,

$y = -300x + 2000$
$y = -300 \cdot 3 + 2000$
$y = -900 + 2000$
$y = 1100$

When $x = 4$,

$y = -300x + 2000$
$y = -300 \cdot 4 + 2000$
$y = -1200 + 2000$
$y = 800$

When $x = 5$,

$y = -300x + 2000$
$y = -300 \cdot 5 + 2000$
$y = -1500 + 2000$
$y = 500$

The completed table is shown below.

x	0	1	2	3	4	5
y	2000	1700	1400	1100	800	500

The ordered pair solutions recorded in the completed table for Example 6 are another set of paired data. They are graphed next. Notice that this scatter diagram gives a visual picture of the decrease in value of the computer.

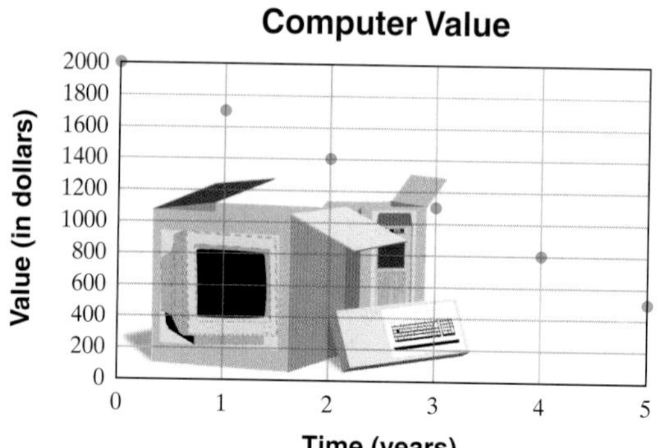

x	y
0	2000
1	1700
2	1400
3	1100
4	800
5	500

Answer

6.

x	1	2	3	4	5	6	7
y	350	300	250	200	150	100	50

Name ______________________ Section __________ Date __________

Mental Math

Give two ordered pair solutions for each linear equation.

1. $x + y = 10$

2. $x + y = 6$

EXERCISE SET 13.1

A *Plot each ordered pair. State in which quadrant, if any, each point lies. See Example 1.*

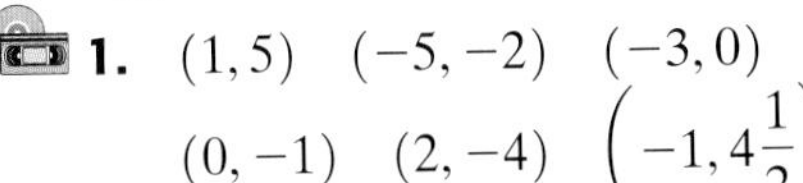

1. $(1, 5)$ $(-5, -2)$ $(-3, 0)$ $(0, -1)$ $(2, -4)$ $\left(-1, 4\frac{1}{2}\right)$

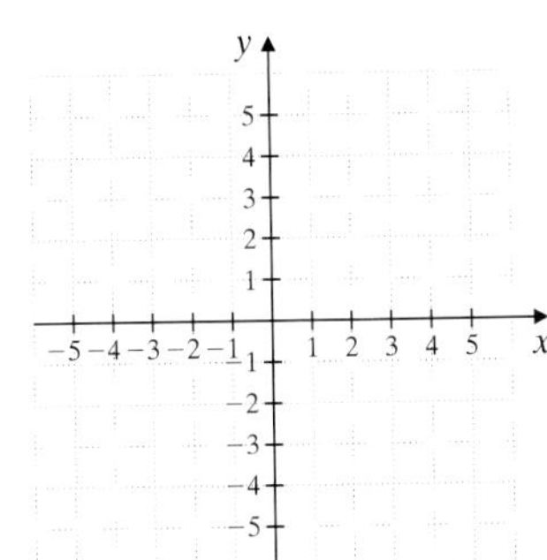

2. $(2, 4)$ $(0, 2)$ $(-2, 1)$ $(-3, -3)$ $\left(3\frac{3}{4}, 0\right)$ $(5, -4)$

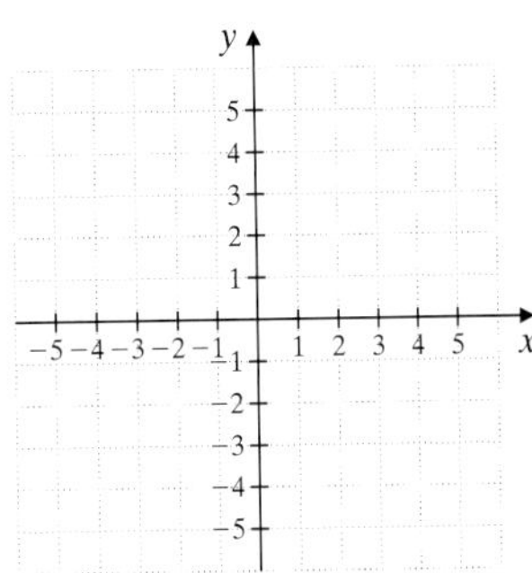

3. When is the graph of the ordered pair (a, b) the same as the graph of the ordered pair (b, a)?

4. In your own words, describe how to plot an ordered pair.

Find the x- and y-coordinates of each labeled point. See Example 1.

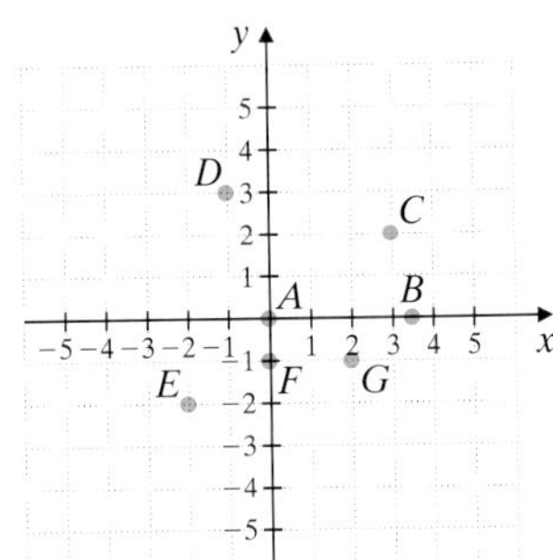

5. A **6.** B **7.** C

8. D **9.** E **10.** F

11. G

12. A **13.** B **14.** C

15. D **16.** E **17.** F

18. G

B *Solve. See Example 2.*

19. The table shows the average farm size (in acres) in the United States during the years shown. (*Source:* National Agricultural Statistics Service)

Year	Average Farm Size (in acres)
1995	438
1996	438
1997	436
1998	435
1999	432
2000	434
2001	436

a. Write this paired data as a set of ordered pairs of the form (year, average farm size).

b. Create a scatter diagram of the paired data. Be sure to label the axes appropriately.

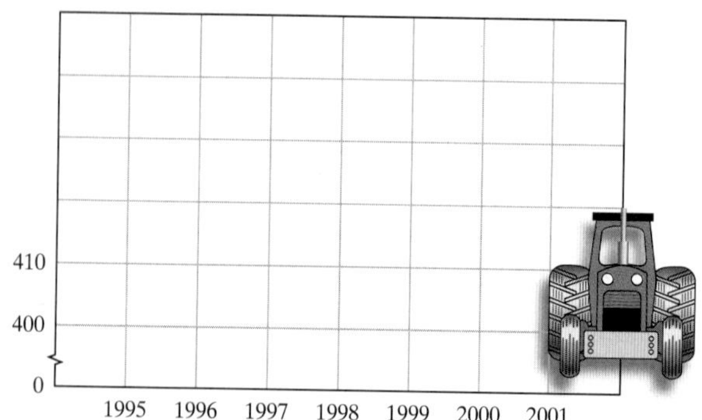

20. In the United States, Halloween is the top holiday for candy sales. The table shows the sales of chocolate and candy at Halloween (in billions of dollars) for the years shown. (*Source:* National Confectioners Association)

Year	Halloween Candy Sales (in billions of dollars)
1996	1.66
1997	1.71
1998	1.79
1999	1.90
2000	1.99
2001	1.99
2002	2.03

a. Write this paired data as a set of ordered pairs of the form (year, Halloween candy sales).

b. Create a scatter diagram of the paired data. Be sure to label the axes appropriately.

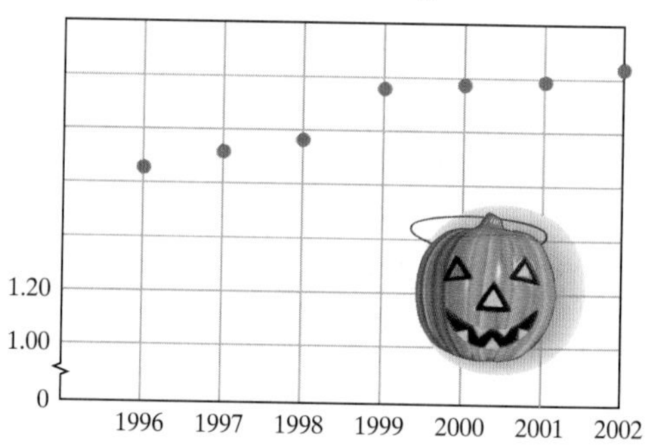

c. What trend in the paired data does the scatter diagram show?

21. The table shows the distance from the equator (in miles) and the average annual snowfall (in inches) for each of eight selected U.S. cities. (*Sources*: National Climatic Data Center, Wake Forest University Albatross Project)

City	Distance from Equator (in miles)	Average Annual Snowfall (in inches)
Atlanta, GA	2313	2
Austin, TX	2085	1
Baltimore, MD	2711	21
Chicago, IL	2869	39
Detroit, MI	2920	42
Juneau, AK	4038	99
Miami, FL	1783	0
Winston-Salem, NC	2493	9

a. Write this paired data as a set of ordered pairs of the form (distance from equator, average annual snowfall).

b. Create a scatter diagram of the paired data. Be sure to label the axes appropriately.

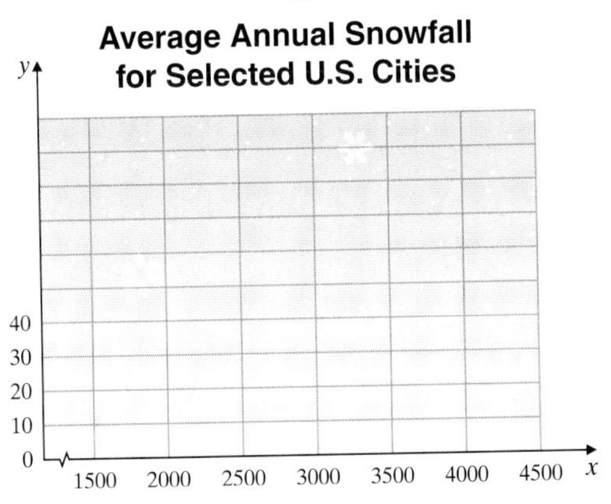

c. What trend in the paired data does the scatter diagram show?

22. The table shows the opening day payroll (in millions of dollars) and the number of games won by the ten Major League Baseball teams with the highest payrolls during the 2000 baseball season. (*Sources*: The Associated Press, Major League Baseball)

Team	2000 Opening Day Payroll (in millions of dollars)	2000 Games Won
NY Yankees	92.5	87
Los Angeles Dodgers	88.1	86
Atlanta Braves	84.5	95
Baltimore Orioles	81.4	74
Arizona Diamondbacks	81.0	85
NY Mets	79.5	94
Boston Red Sox	77.9	85
Cleveland Indians	75.9	90
Texas Rangers	70.8	71
Tampa Bay Devil Rays	62.8	69

a. Write this paired data as a set of ordered pairs of the form (opening day payroll, games won).

b. Create a scatter diagram of the paired data. Be sure to label the axes appropriately.

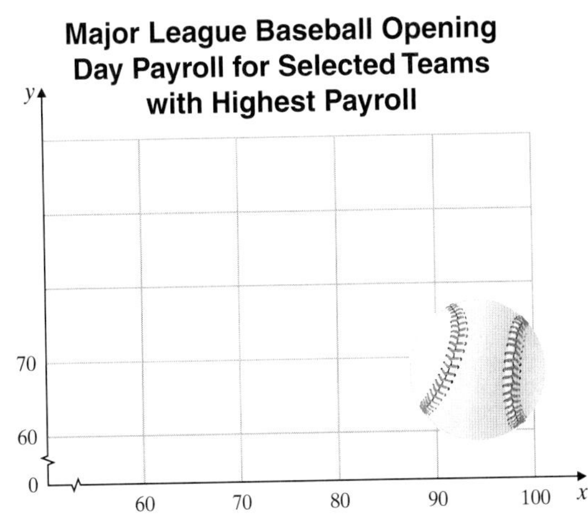

23. Minh, a psychology student, kept a record of how much time she spent studying for each of her 20-point psychology quizzes and her score on each quiz.

Hours Spent Studying	Quiz Score
0.50	10
0.75	12
1.00	15
1.25	16
1.50	18
1.50	19
1.75	19
2.00	20

a. Write each paired data as an ordered pair of the form (hours spent studying, quiz score).

b. Create a scatter diagram of the paired data. Be sure to label the axes appropriately.

Minh's Chart for Psychology

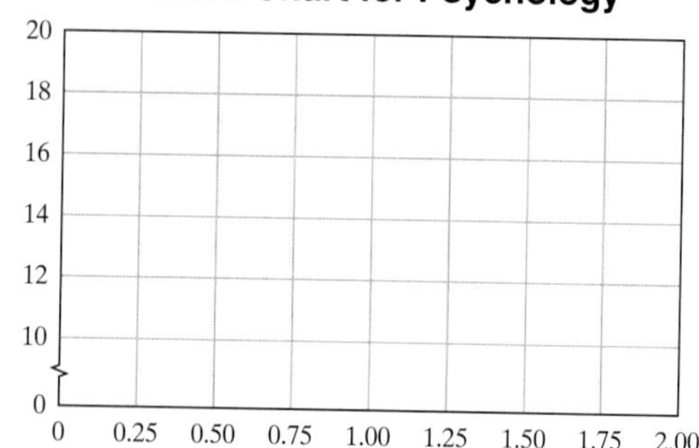

c. What might Minh conclude from the scatter diagram?

24. A local lumberyard uses quantity pricing. The table shows the price per board for different amounts of lumber purchased.

Price per Board (in dollars)	Number of Boards Purchased
8.00	1
7.50	10
6.50	25
5.00	50
2.00	100

a. Write each paired data as an ordered pair of the form (price per board, number of boards purchased).

b. Create a scatter diagram of the paired data. Be sure to label the axes appropriately.

Lumberyard Board Pricing

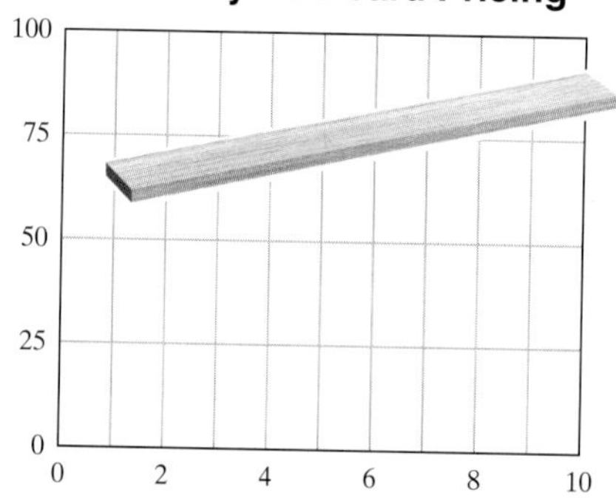

c. What trend in the paired data does the scatter diagram show?

C *Complete each ordered pair so that it is a solution of the given linear equation. See Example 3.*

25. $x - 4y = 4; (\ , -2), (4, \ )$

26. $x - 5y = -1; (\ , -2), (4, \ )$

27. $3x + y = 9; (0, \ ), (\ , 0)$

28. $x + 5y = 15; (0, \ ), (\ , 0)$

29. $y = -7; (11, \ ), (\ , -7)$

30. $x = \frac{1}{2}; (\ , 0), \left(\frac{1}{2}, \ \right)$

Complete the table of ordered pairs for each linear equation. See Examples 4 through 6.

31. $x + 3y = 6$

x	y
0	
	0
	1

32. $2x + y = 4$

x	y
0	
	0
	2

33. $2x - y = 12$

x	y
0	
	−2
3	

34. $-5x + y = 10$

x	y
	0
	5
0	

35. $2x + 7y = 5$

x	y
0	
	0
	1

36. $x - 6y = 3$

x	y
0	
1	
	−1

Complete the table of ordered pairs for each equation. Then plot the ordered pair solutions. See Examples 4 through 6.

37. $x = 3$

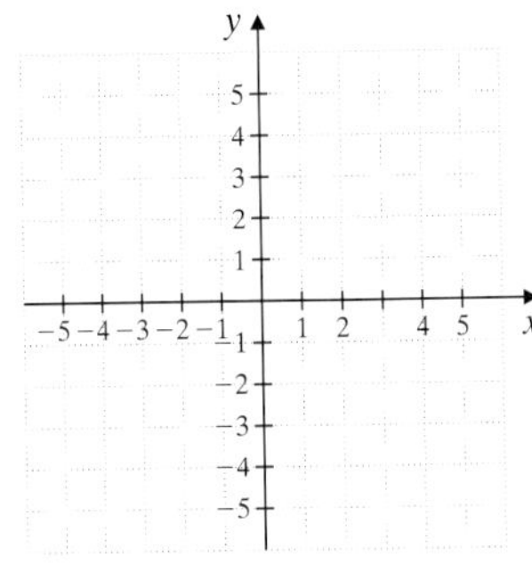

x	y
	0
	−0.5
	$\frac{1}{4}$

38. $y = -1$

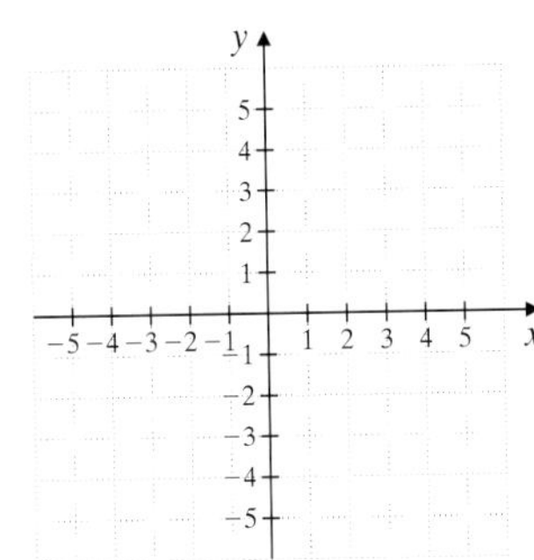

x	y
−2	
0	
−1	

39. $x = -5y$

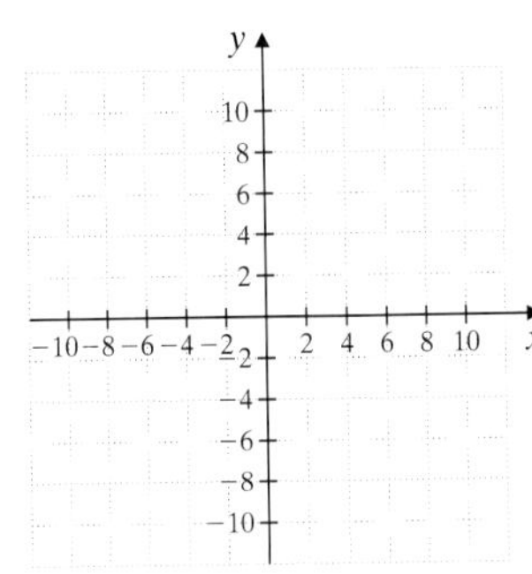

x	y
	0
	1
10	

40. $y = -3x$

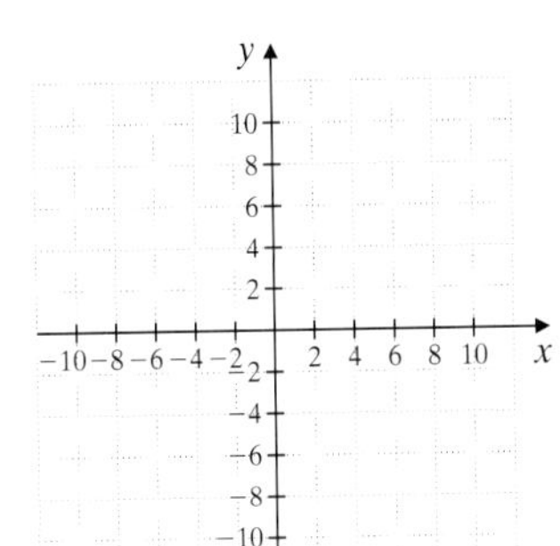

x	y
0	
−2	
	9

Solve. See Example 6.

41. The cost in dollars y of producing x computer desks is given by $y = 80x + 5000$.

a. Complete the table.

x	100	200	300
y			

b. Find the number of computer desks that can be produced for \$8600. (*Hint*: Find x when $y = 8600$.)

42. The hourly wage y of an employee at a certain production company is given by $y = 0.25x + 9$ where x is the number of units produced by the employee in an hour.

a. Complete the table.

x	0	1	5	10
y				

b. Find the number of units that an employee must produce each hour to earn an hourly wage of \$12.25. (*Hint*: Find x when $y = 12.25$.)

Review and Preview

Solve each equation for y. See Section 9.3.

43. $x + y = 5$

44. $x - y = 3$

45. $2x + 4y = 5$

46. $5x + 2y = 7$

47. $10x = -5y$

48. $4y = -8x$

Combining Concepts

△ **49.** Find the perimeter of the rectangle whose vertices are the points with coordinates $(-1, 5)$, $(3, 5)$, $(3, -4)$, and $(-1, -4)$.

△ **50.** Find the area of the rectangle whose vertices are the points with coordinates $(5, 2)$, $(5, -6)$, $(0, -6)$, and $(0, 2)$.

The scatter diagram below shows Walt Disney Company's annual revenues. The horizontal axis represents the number of years after 1996.

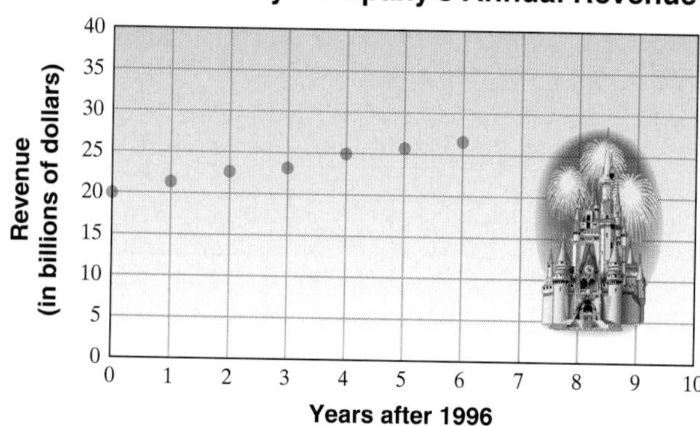

51. Estimate the revenues for years 1, 2, 3, and 4.

52. Use a straight edge or ruler and this scatter diagram to predict Disney's revenue in the year 2007.

53. Discuss any similarities in the graphs of the ordered pair solutions for Exercises 37–40.

54. The percent y of recorded music sales that were in cassette format from 1995 through 2001 is given by $y = -3.12x + 22.67$. In the equation, x represents the number of years after 1995. (*Source:* Recording Industry Association of America)

a. Complete the table.

x	1	3	5
y			

b. Find the year in which approximately 17% of recorded music sales were cassettes. (*Hint*: Find x when $y = 17$ and round to the nearest whole number.)

55. The amount y of land operated by farms in the United States (in million acres) from 1990 through 2000 is given by $y = -4.22x + 985.02$. In the equation, x represents the number of years after 1990. (*Source:* National Agricultural Statistics Service)

a. Complete the table.

x	4	7	10
y			

b. Find the year in which there were approximately 947 million acres of land operated by farms. (*Hint*: Find x when $y = 947$ and round to the nearest whole number.)

FOCUS ON The Real World

Reading a Map

How do you find a location on a map? Most maps we use today have a grid that is based on the rectangular coordinate system we use in algebra. After finding the coordinates of cities and other landmarks from the map index, the grid can help us find places on the map. To eliminate confusion, many maps use letters to label the grid along one edge and numbers along the other. However, the coordinates are still pairs of numbers and letters. For instance, the coordinates for Toledo on the map are A-2.

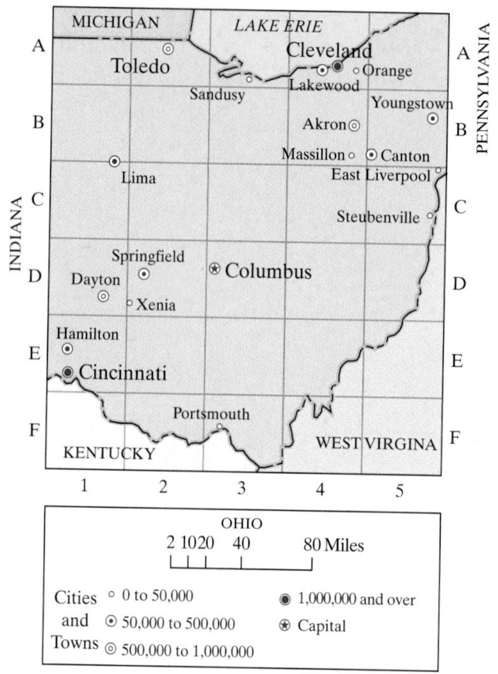

Critical Thinking

1. Find the coordinates of the following cities: Hamilton, Columbus, Youngstown, and Cincinnati.
2. What cities correspond to the following coordinates: F-3, A-3, B-4, and D-2?
3. How are the map's coordinate system and the rectangular coordinate system we use in algebra the same? How are they different? What are the advantages of each?

13.2 Graphing Linear Equations

OBJECTIVE

Ⓐ Graph a linear equation by finding and plotting ordered pair solutions.

In the previous section, we found that equations in two variables may have more than one solution. For example, both $(2, 2)$ and $(0, 4)$ are solutions of the equation $x + y = 4$. In fact, this equation has an infinite number of solutions. Other solutions include $(-2, 6)$, $(4, 0)$, and $(6, -2)$. Notice the pattern that appears in the graph of these solutions.

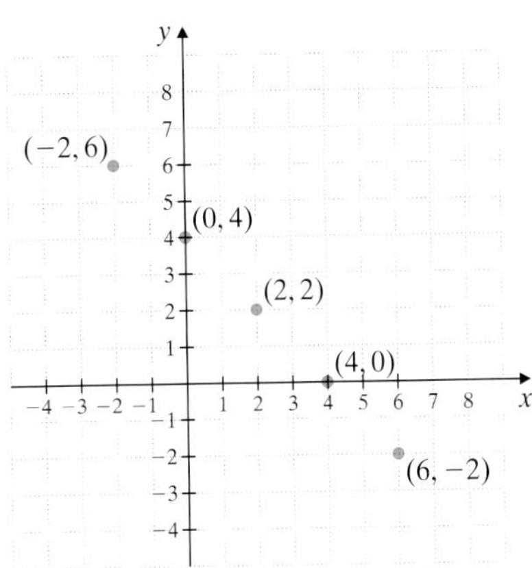

These solutions all appear to lie on the same line, as seen in the second graph. It can be shown that every ordered pair solution of the equation corresponds to a point on this line, and every point on this line corresponds to an ordered pair solution. Thus, we say that this line is the **graph of the equation** $x + y = 4$. Notice that we can only show a part of a line on a graph. The arrowheads on each end of the line below remind us that the line actually extends indefinitely in both directions.

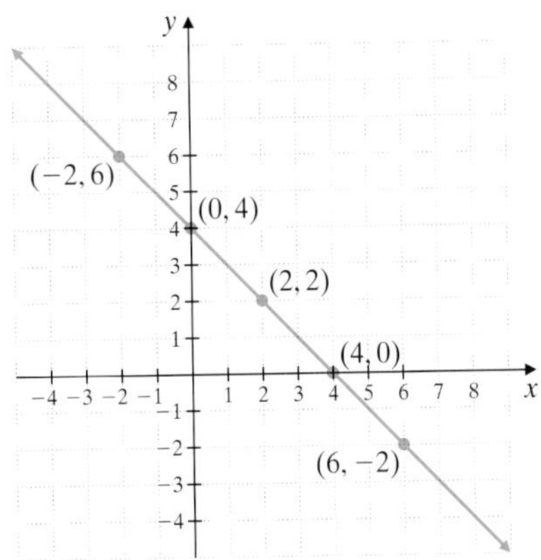

The equation $x + y = 4$ is called a *linear equation in two variables* and *the graph of every linear equation in two variables is a straight line.*

Linear Equation in Two Variables

A **linear equation in two variables** is an equation that can be written in the form

$$Ax + By = C$$

where A, B, and C are real numbers and A and B are not both 0. **The graph of a linear equation in two variables is a straight line.**

The form $Ax + By = C$ is called **standard form.** Following are examples of linear equations in two variables.

$$2x + y = 8 \qquad -2x = 7y \qquad y = \frac{1}{3}x + 2 \qquad y = 7$$

A Graphing Linear Equations

From geometry, we know that a straight line is determined by just two points. Thus, to graph a linear equation in two variables we need to find just two of its infinitely many solutions. Once we do so, we plot the solution points and draw the line connecting the points. Usually, we find a third solution as well, as a check.

Practice Problem 1

Graph the linear equation $x + 3y = 6$.

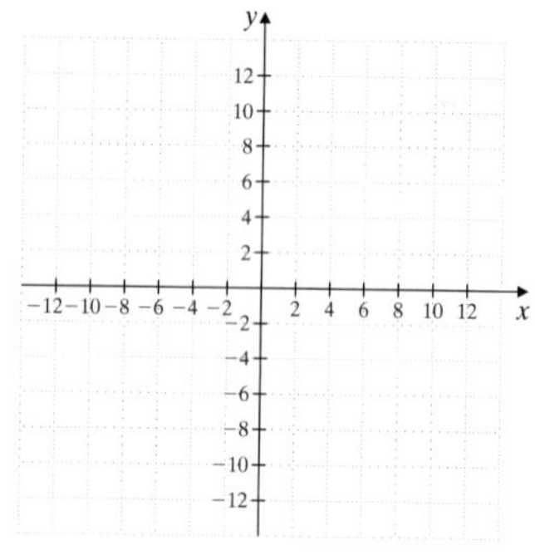

EXAMPLE 1 Graph the linear equation $2x + y = 5$.

Solution: To graph this equation, we find three ordered pair solutions of $2x + y = 5$. To do this, we choose a value for one variable, x or y, and solve for the other variable. For example, if we let $x = 1$, then $2x + y = 5$ becomes

$$\begin{aligned} 2x + y &= 5 \\ 2(1) + y &= 5 &&\text{Replace } x \text{ with 1.} \\ 2 + y &= 5 &&\text{Multiply.} \\ y &= 3 &&\text{Subtract 2 from both sides.} \end{aligned}$$

Since $y = 3$ when $x = 1$, the ordered pair $(1, 3)$ is a solution of $2x + y = 5$. Next, we let $x = 0$.

$$\begin{aligned} 2x + y &= 5 \\ 2(0) + y &= 5 &&\text{Replace } x \text{ with 0.} \\ 0 + y &= 5 \\ y &= 5 \end{aligned}$$

The ordered pair $(0, 5)$ is a second solution.

The two solutions found so far allow us to draw the straight line that is the graph of all solutions of $2x + y = 5$. However, we will find a third ordered pair as a check. Let $y = -1$.

$$\begin{aligned} 2x + y &= 5 \\ 2x + (-1) &= 5 &&\text{Replace } y \text{ with } -1. \\ 2x - 1 &= 5 \\ 2x &= 6 &&\text{Add 1 to both sides.} \\ x &= 3 &&\text{Divide both sides by 2.} \end{aligned}$$

The third solution is $(3, -1)$. These three ordered pair solutions are listed in the table and plotted on the coordinate plane. The graph of $2x + y = 5$ is the line through the three points.

x	y
1	3
0	5
3	−1

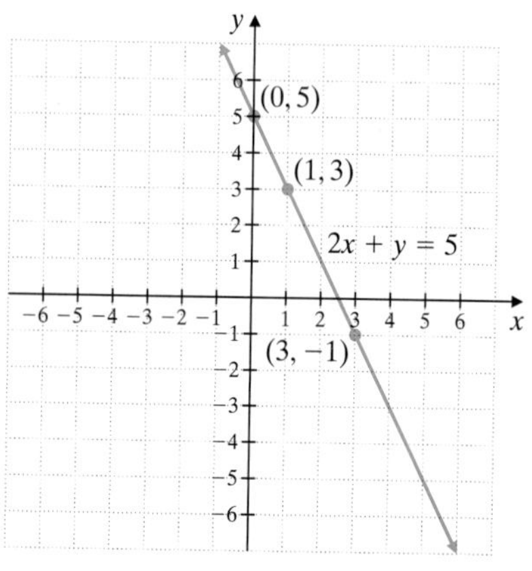

Answer

1.

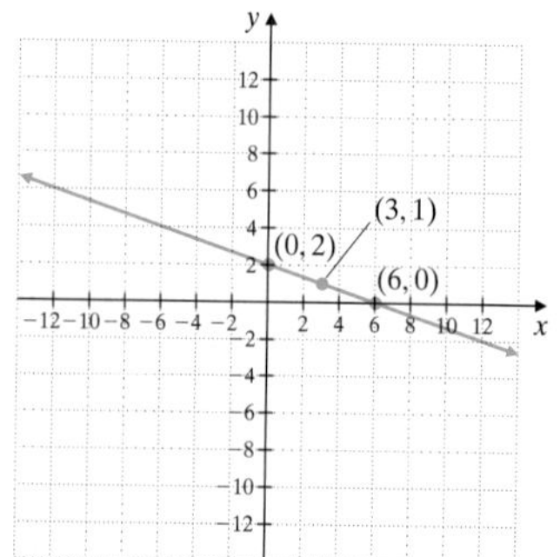

Helpful Hint

All three points should fall on the same straight line. If not, check your ordered pair solutions for a mistake.

EXAMPLE 2 Graph the linear equation $-5x + 3y = 15$.

Solution: We find three ordered pair solutions of $-5x + 3y = 15$.

Let $x = 0$.

$$\begin{aligned} -5x + 3y &= 15 \\ -5 \cdot 0 + 3y &= 15 \\ 0 + 3y &= 15 \\ 3y &= 15 \\ y &= 5 \end{aligned}$$

Let $y = 0$.

$$\begin{aligned} -5x + 3y &= 15 \\ -5x + 3 \cdot 0 &= 15 \\ -5x + 0 &= 15 \\ -5x &= 15 \\ x &= -3 \end{aligned}$$

Let $x = -2$.

$$\begin{aligned} -5x + 3y &= 15 \\ -5 \cdot -2 + 3y &= 15 \\ 10 + 3y &= 15 \\ 3y &= 5 \\ y &= \frac{5}{3} \end{aligned}$$

The ordered pairs are $(0, 5)$, $(-3, 0)$, and $\left(-2, \frac{5}{3}\right)$. The graph of $-5x + 3y = 15$ is the line through the three points.

x	y
0	5
−3	0
−2	$\frac{5}{3} = 1\frac{2}{3}$

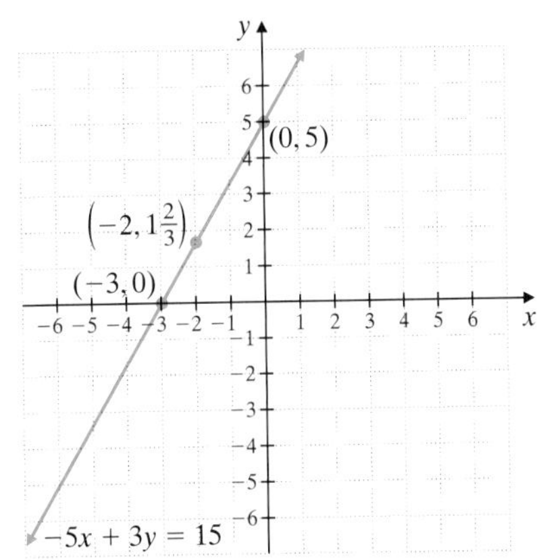

EXAMPLE 3 Graph the linear equation $y = 3x$.

Solution: We find three ordered pair solutions. Since this equation is solved for y, we'll choose three x values.

If $x = 2$, $y = 3 \cdot 2 = 6$.

If $x = 0$, $y = 3 \cdot 0 = 0$.

If $x = -1$, $y = 3 \cdot -1 = -3$.

Next, we plot the ordered pair solutions and draw a line through the plotted points. The line is the graph of $y = 3x$. Every point on the graph represents an ordered pair solution of the equation and every ordered pair solution is a point on this line.

Practice Problem 2

Graph the linear equation $-2x + 4y = 8$.

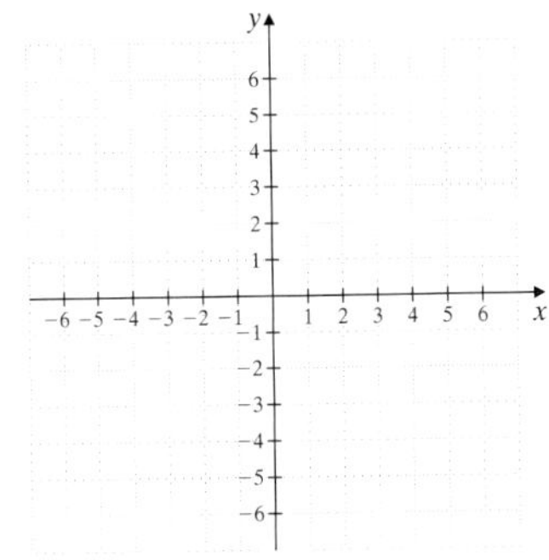

Practice Problem 3

Graph the linear equation $y = 2x$.

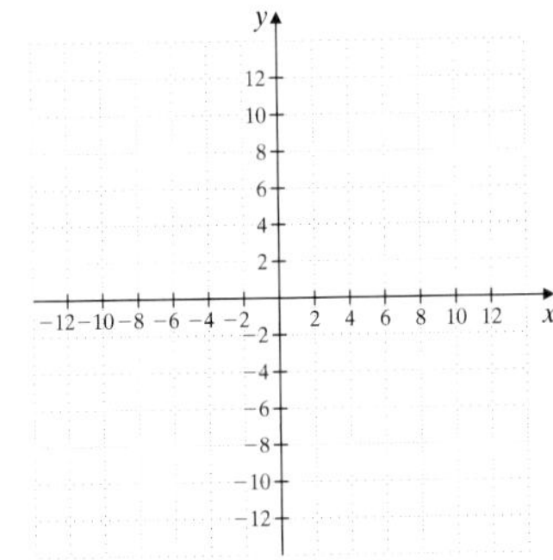

Answers

2.

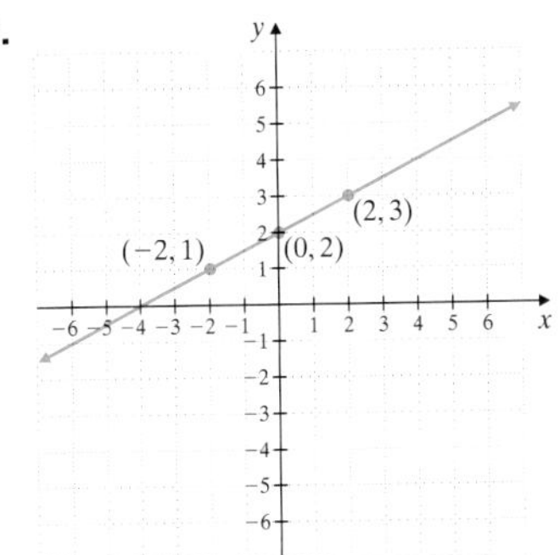

3. See page 1048.

x	y
2	6
0	0
−1	−3

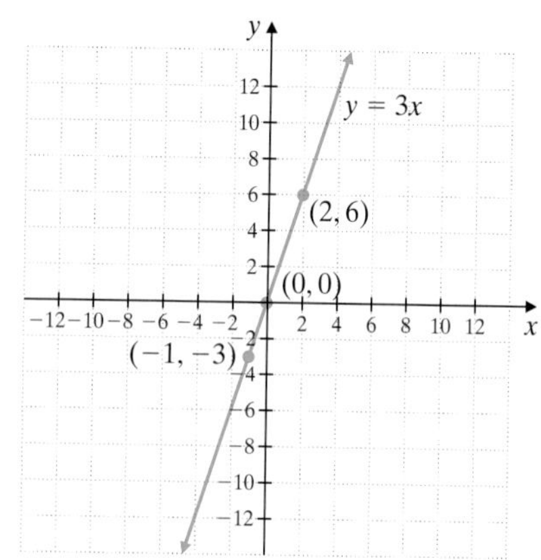

Practice Problem 4

Graph the linear equation $y = -\frac{1}{2}x$.

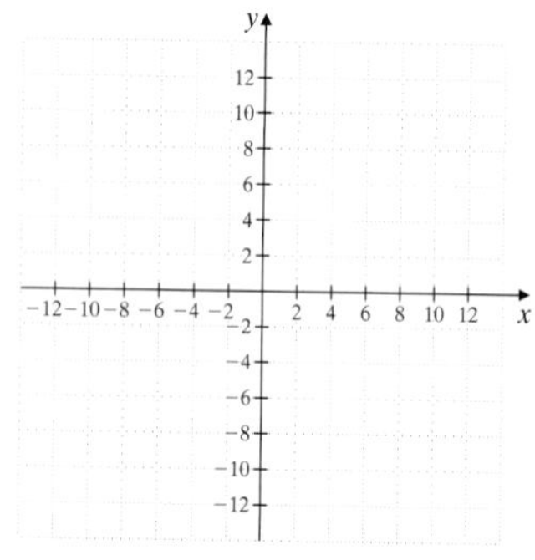

EXAMPLE 4 Graph the linear equation $y = -\frac{1}{3}x$.

Solution: We find three ordered pair solutions, plot the solutions, and draw a line through the plotted solutions. To avoid fractions, we'll choose x values that are multiples of 3 to substitute into the equation.

If $x = 6$, then $y = -\frac{1}{3} \cdot 6 = -2$.

If $x = 0$, then $y = -\frac{1}{3} \cdot 0 = 0$.

If $x = -3$, then $y = -\frac{1}{3} \cdot -3 = 1$.

x	y
6	−2
0	0
−3	1

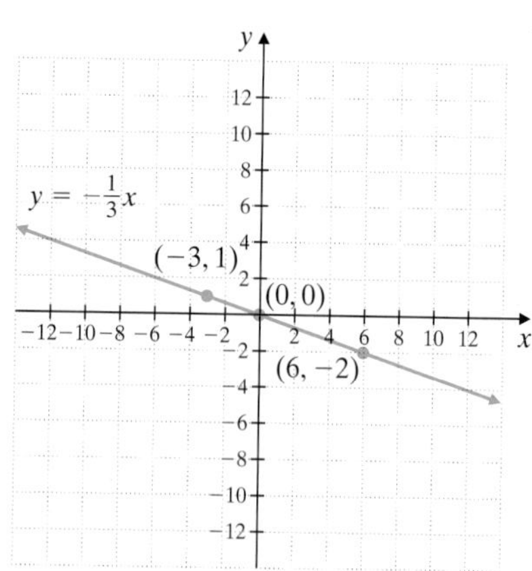

Let's compare the graphs in Examples 3 and 4. The graph of $y = 3x$ tilts upward (as we follow the line from left to right) and the graph of $y = -\frac{1}{3}x$ tilts downward (as we follow the line from left to right). Also notice that both lines go through the origin or that $(0, 0)$ is an ordered pair solution of both equations. We will learn more about the tilt, or slope, of a line in Section 13.4.

EXAMPLE 5 Graph the linear equation $y = 3x + 6$ and compare this graph with the graph of $y = 3x$ in Example 3.

Answers

3.

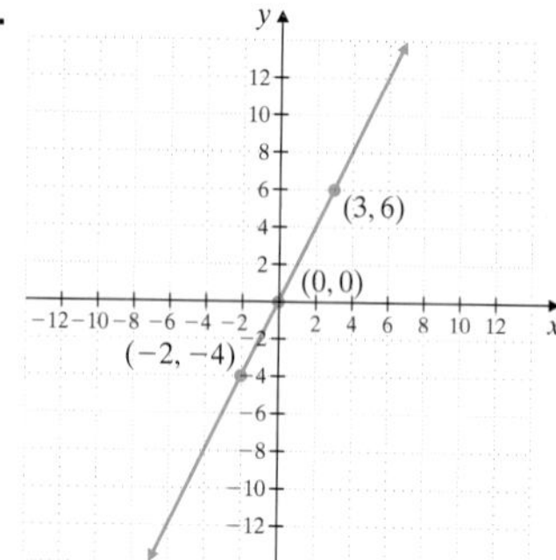

4.

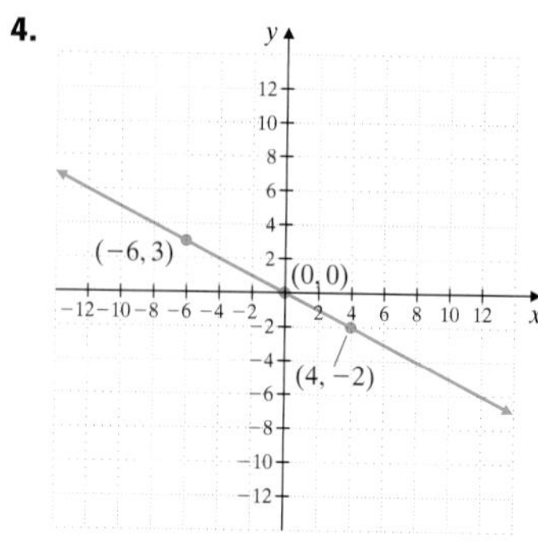

Solution: We find three ordered pair solutions, plot the solutions, and draw a line through the plotted solutions. We choose x-values and substitute them into the equation $y = 3x + 6$.

If $x = -3$, then $y = 3(-3) + 6 = -3$.
If $x = 0$, then $y = 3(0) + 6 = 6$.
If $x = 1$, then $y = 3(1) + 6 = 9$.

x	y
−3	−3
0	6
1	9

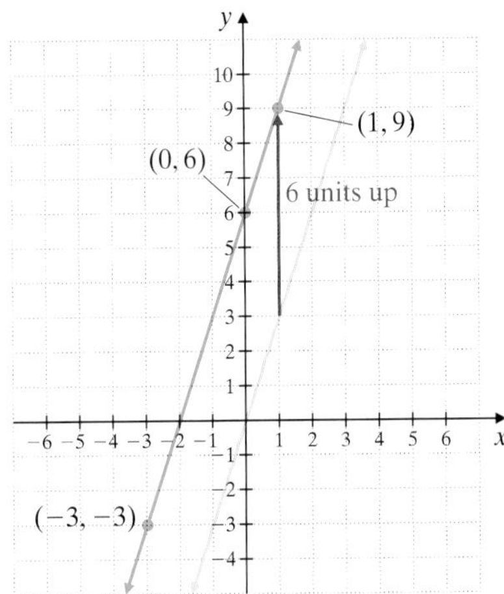

The most startling similarity is that both graphs appear to have the same upward tilt as we move from left to right. Also, the graph of $y = 3x$ crosses the y-axis at the origin, while the graph of $y = 3x + 6$ crosses the y-axis at 6. It appears that the graph of $y = 3x + 6$ is the same as the graph of $y = 3x$ except that the graph of $y = 3x + 6$ is moved 6 units upward.

EXAMPLE 6 Graph the linear equation $y = -2$.

Solution: Recall from Section 13.1 that the equation $y = -2$ is the same as $0x + y = -2$. No matter what value we replace x with, y is -2.

x	y
0	−2
3	−2
−2	−2

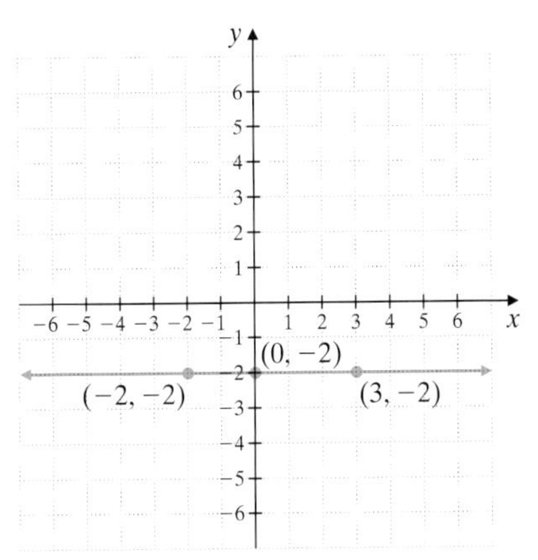

Notice that the graph of $y = -2$ is a horizontal line.

Practice Problem 5

Graph the linear equation $y = 2x + 3$ and compare this graph with the graph of $y = 2x$ in Practice Problem 3.

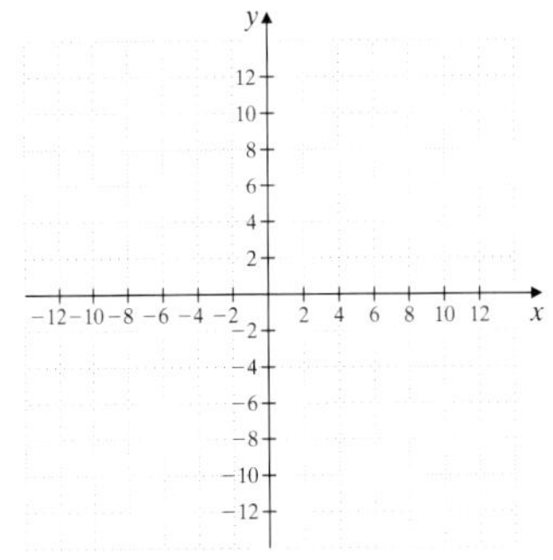

Practice Problem 6

Graph the linear equation $x = 3$.

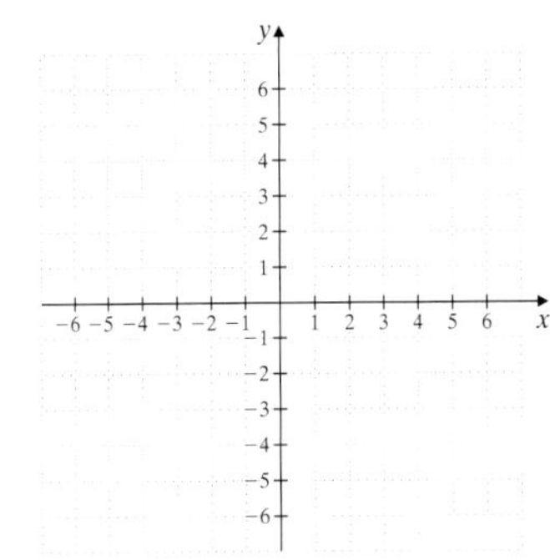

Answers

5.

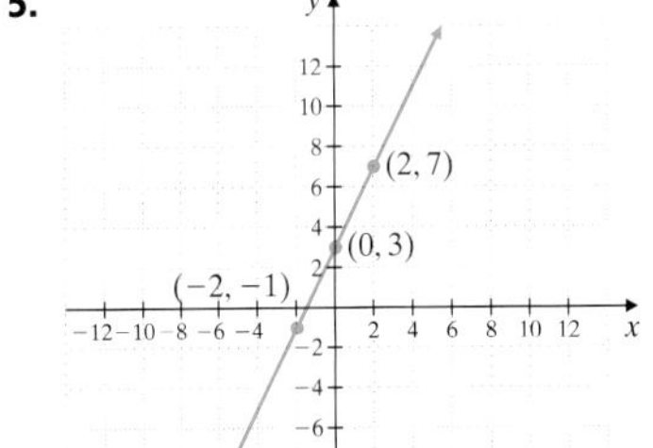

Same as the graph of $y = 2x$ except that the graph of $y = 2x + 3$ is moved 3 units upward.

6.

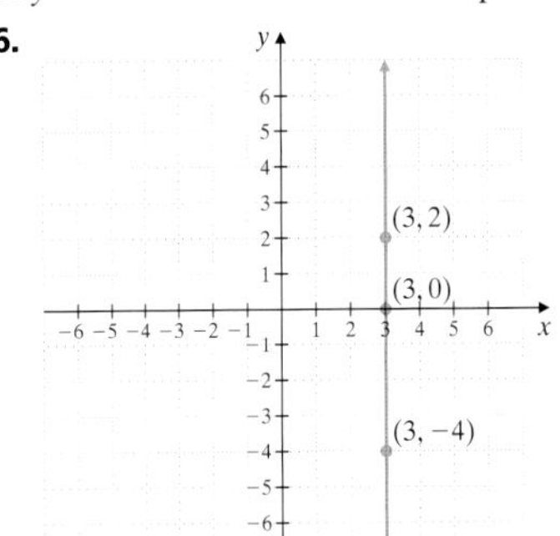

GRAPHING CALCULATOR EXPLORATIONS

In this section, we begin an optional study of graphing calculators and graphing software packages for computers. These graphers use the same point plotting technique that was introduced in this section. The advantage of this graphing technology is, of course, that graphing calculators and computers can find and plot ordered pair solutions much faster than we can. Note, however, that the features described in these boxes may not be available on all graphing calculators.

The rectangular screen where a portion of the rectangular coordinate system is displayed is called a **window.** We call it a **standard window** for graphing when both the x- and y-axes show coordinates between -10 and 10. This information is often displayed in the window menu on a graphing calculator as follows.

Xmin = −10
Xmax = 10
Xscl = 1 — The scale on the x-axis is one unit per tick mark.
Ymin = −10
Ymax = 10
Yscl = 1 — The scale on the y-axis is one unit per tick mark.

To use a graphing calculator to graph the equation $y = 2x + 3$, press the [Y=] key and enter the keystrokes [2] [x] [+] [3]. The top row should now read $Y_1 = 2x + 3$. Next press the [GRAPH] key, and the display should look like this:

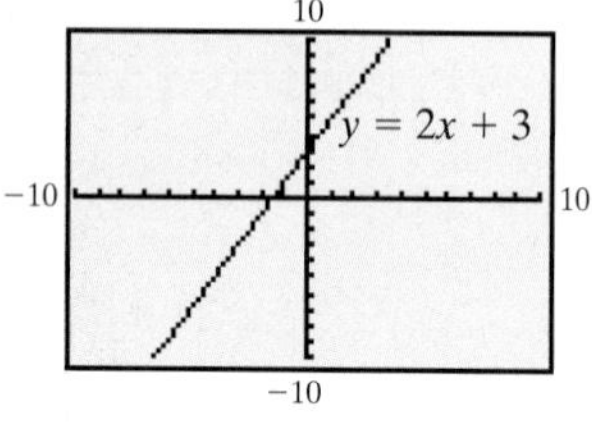

Use a standard window and graph the following linear equations. (Unless otherwise stated, use a standard window when graphing.)

1. $y = -3x + 7$

2. $y = -x + 5$

3. $y = 2.5x - 7.9$

4. $y = -1.3x + 5.2$

5. $y = -\dfrac{3}{10}x + \dfrac{32}{5}$

6. $y = \dfrac{2}{9}x - \dfrac{22}{3}$

Name ____________________ Section __________ Date __________

EXERCISE SET 13.2

 *For each equation, find three ordered pair solutions by completing the table. Then use the ordered pairs to graph the equation. See Examples 1 through 6.*

1. $x - y = 6$

x	y
	0
4	
	−1

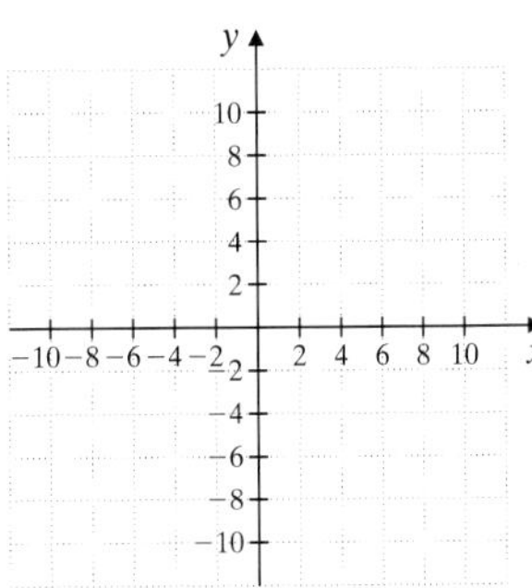

2. $x - y = 4$

x	y
0	
	2
−1	

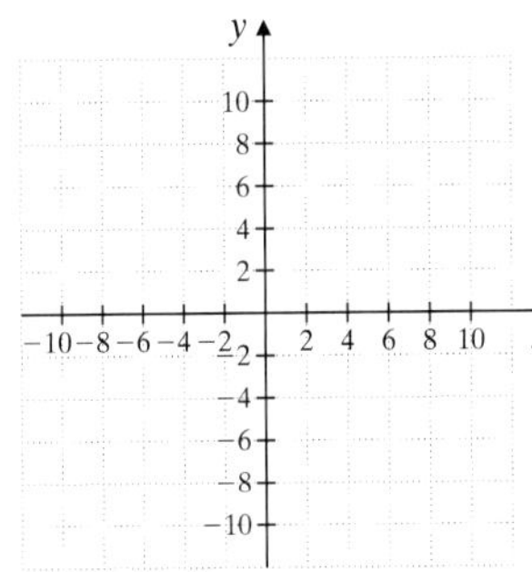

3. $y = -4x$

x	y
1	
0	
−1	

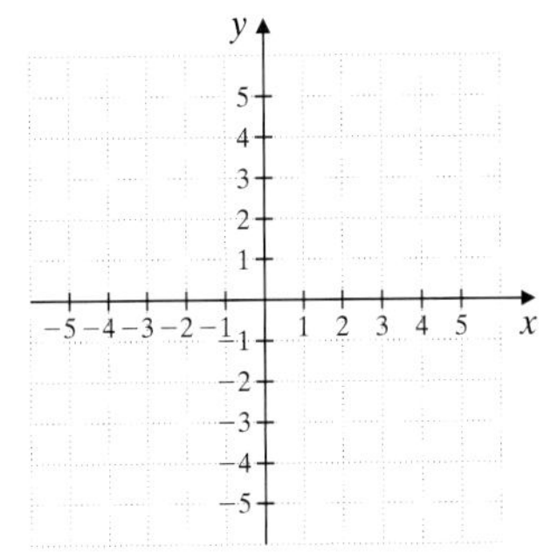

4. $y = -5x$

x	y
1	
0	
−1	

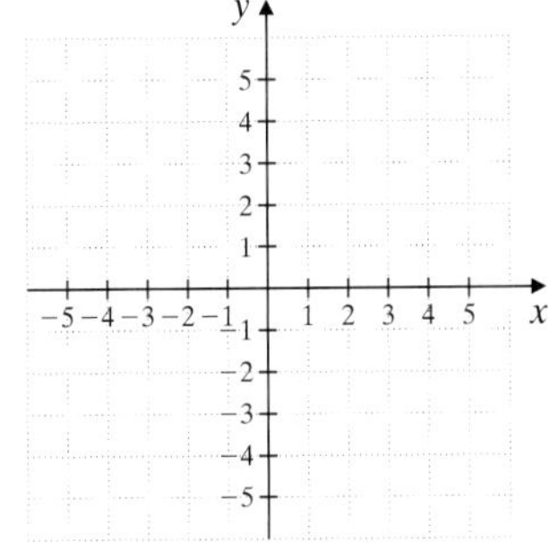

5. $y = \frac{1}{3}x$

x	y
0	
6	
−3	

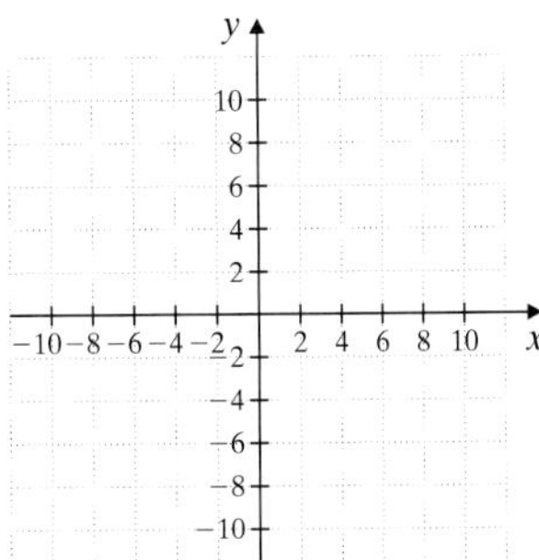

6. $y = \frac{1}{2}x$

x	y
0	
−4	
2	

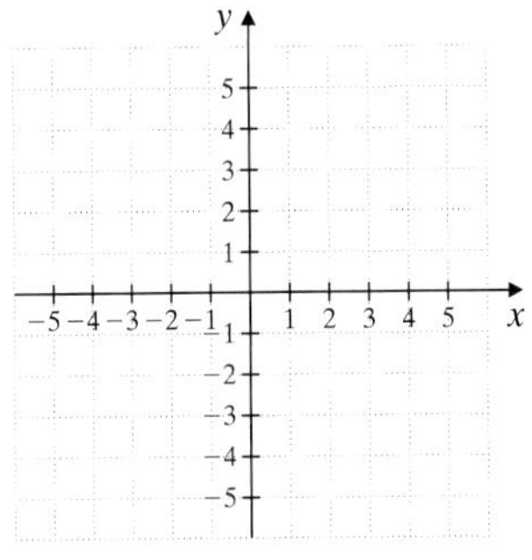

7. $y = -4x + 3$

x	y
0	
1	
2	

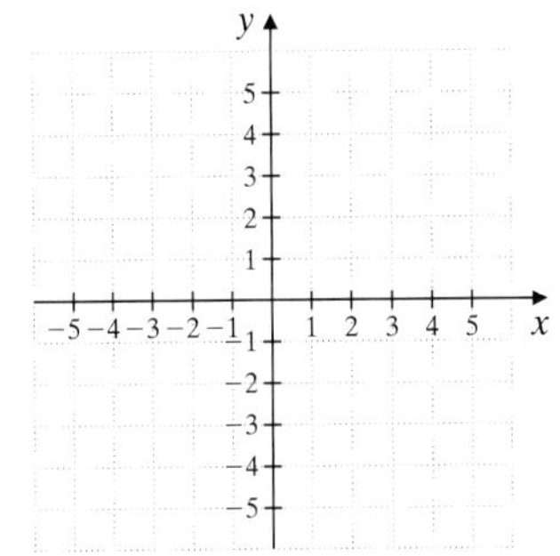

8. $y = -5x + 2$

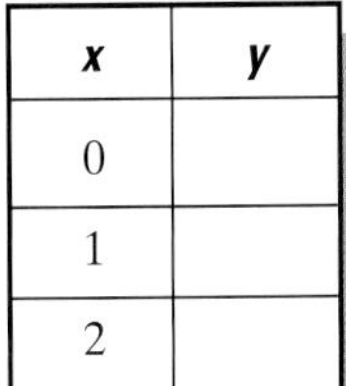

x	y
0	
1	
2	

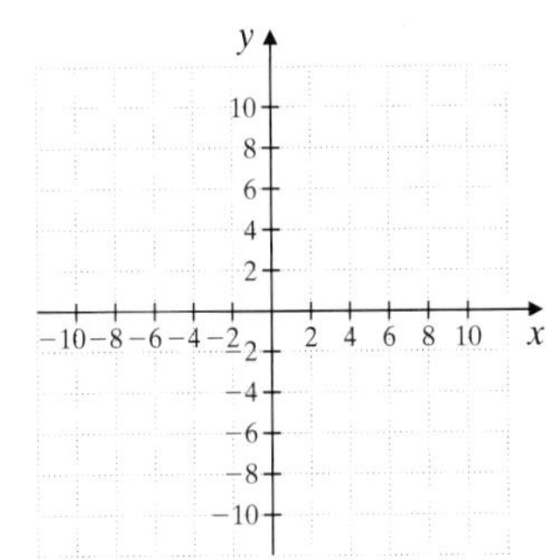

Graph each linear equation. See Examples 1 through 6.

9. $x + y = 1$

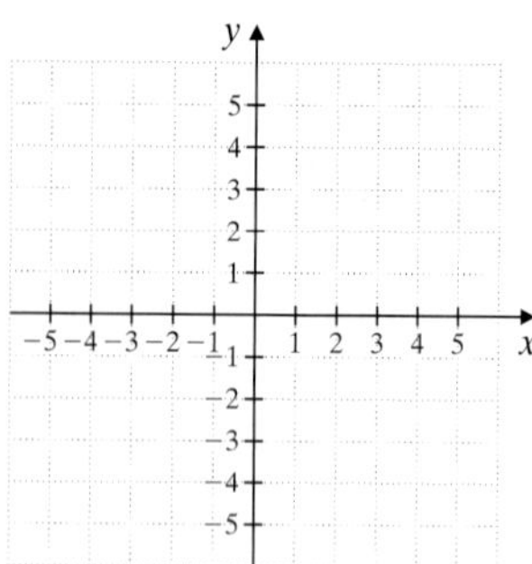

10. $x + y = 7$

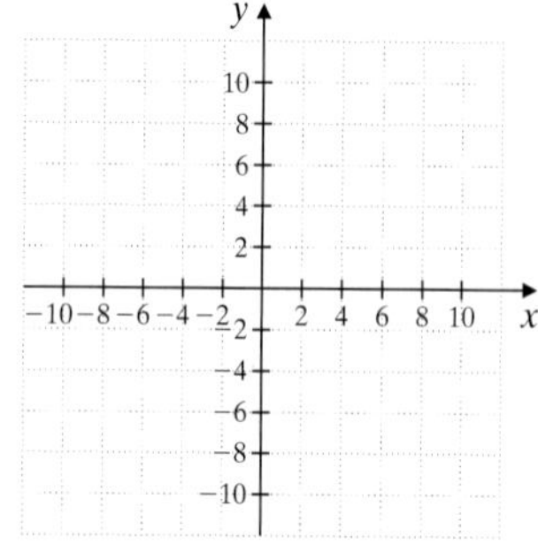

11. $-x + y = 6$

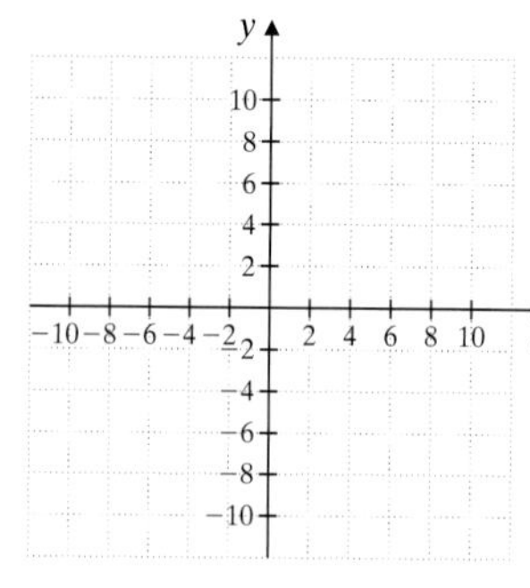

12. $x - y = -2$

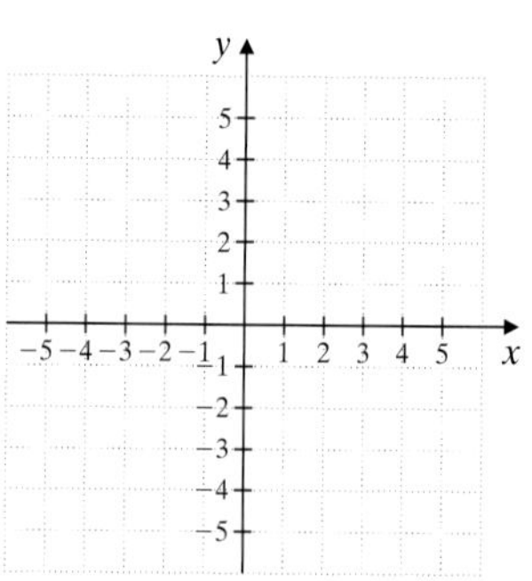

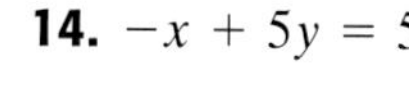

13. $x - 2y = 6$

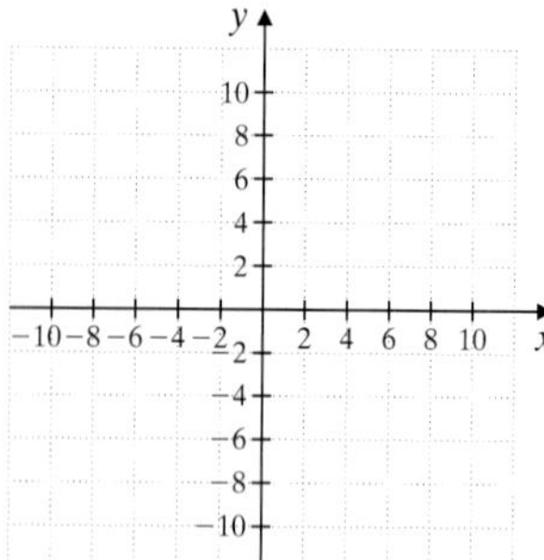

14. $-x + 5y = 5$

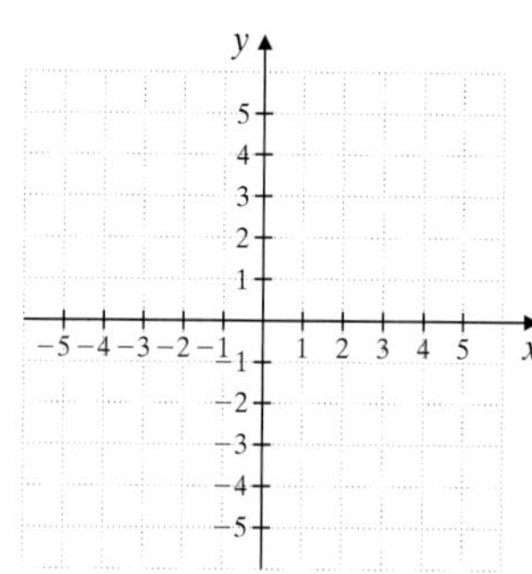

15. $y = 6x + 3$

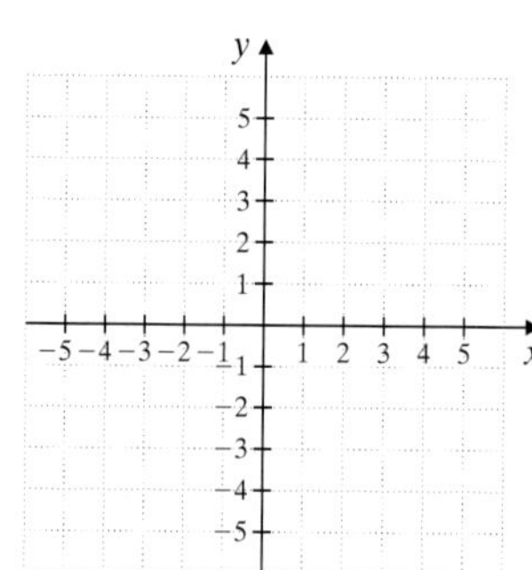

16. $y = -2x + 7$

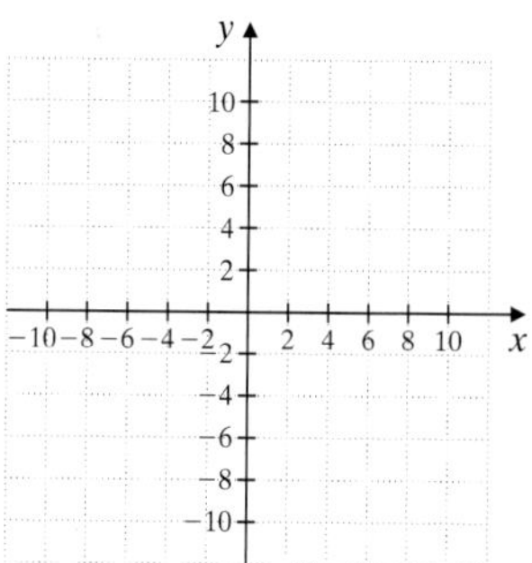

17. $x = -4$

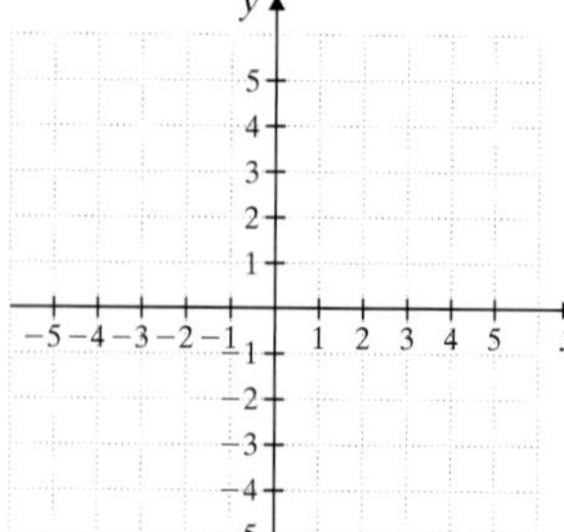

18. $y = 5$

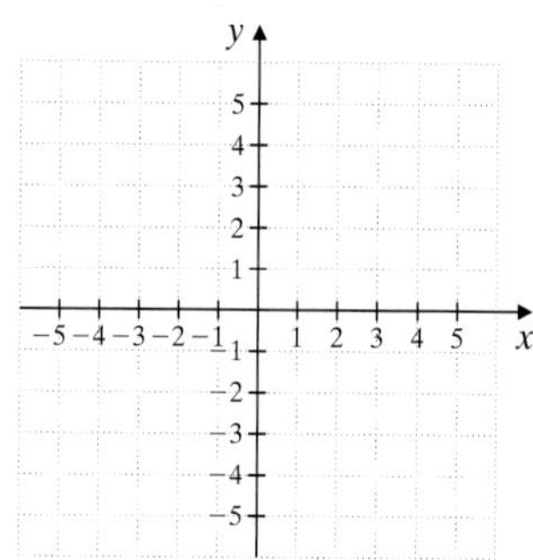

19. $y = 3$

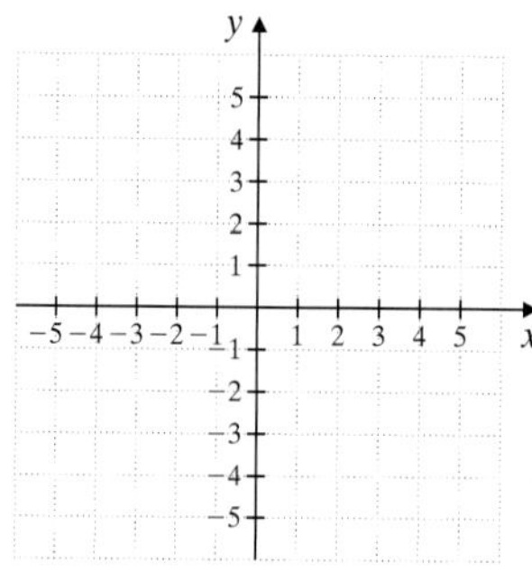

20. $x = -1$

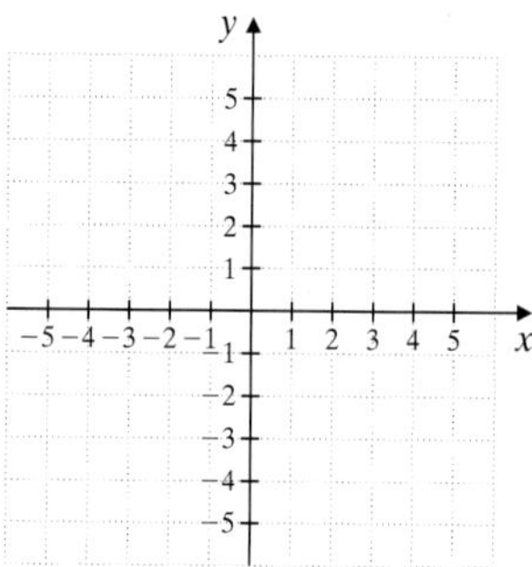

21. $y = x$

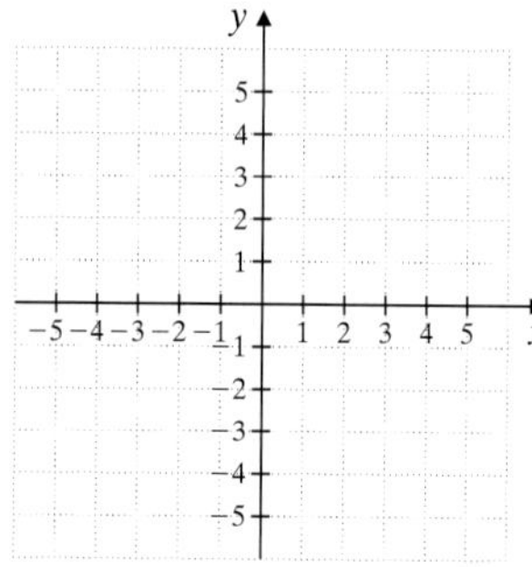

22. $y = -x$

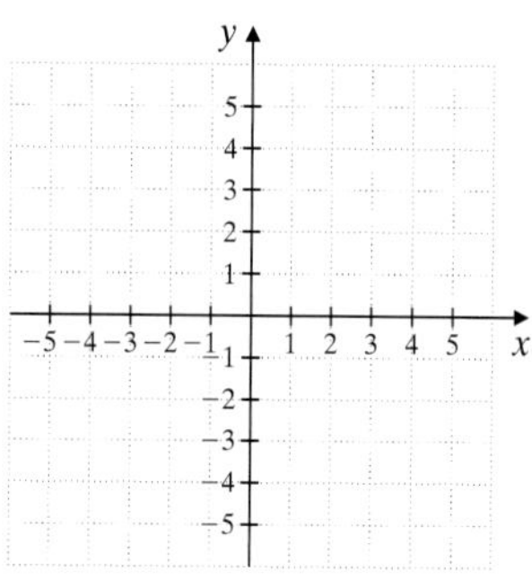

23. $y = 5x$

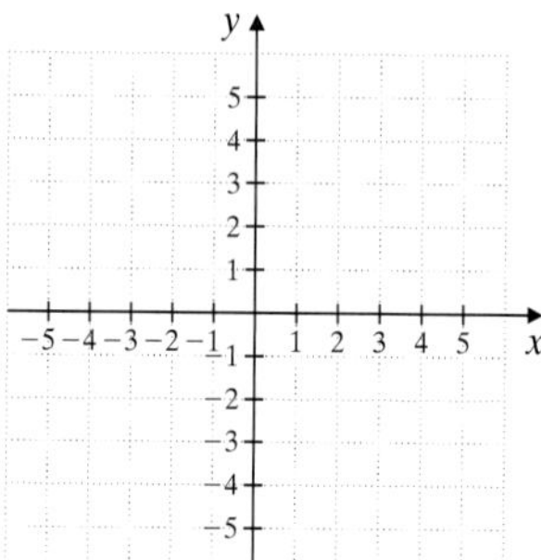

24. $y = 4x$

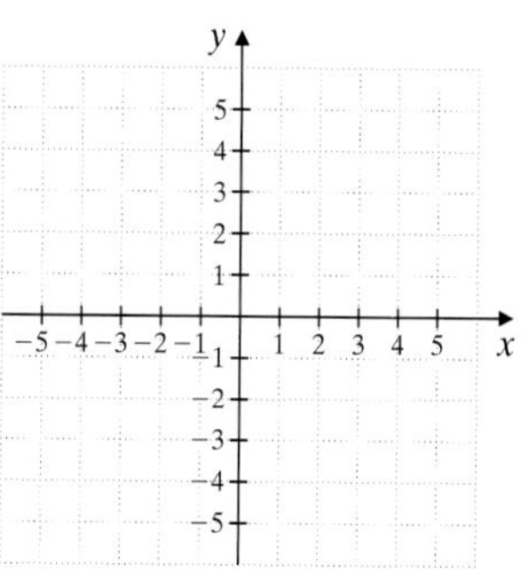

25. $x + 3y = 9$

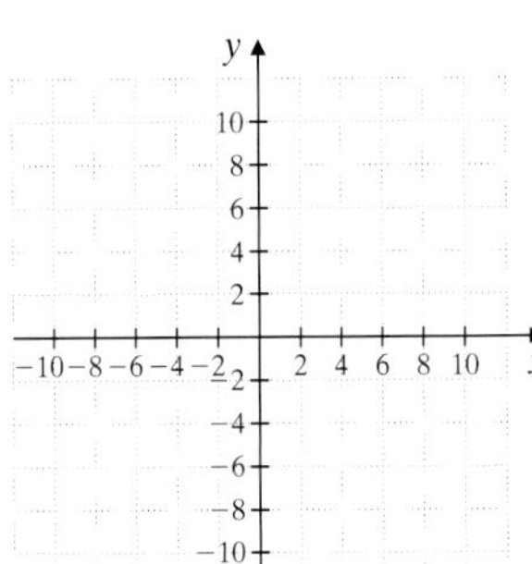

26. $2x + y = 2$

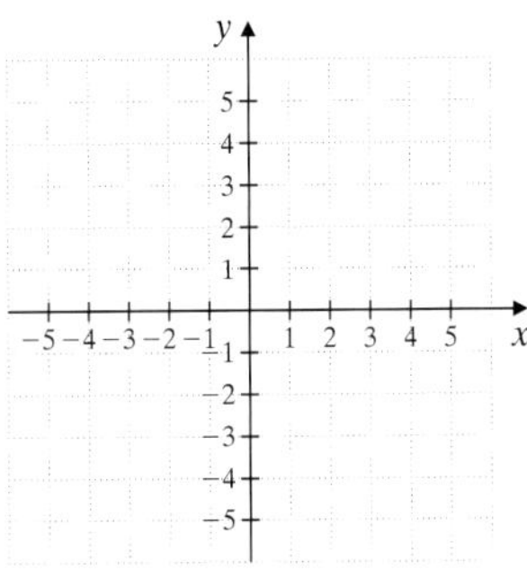

27. $y = \frac{1}{2}x - 1$

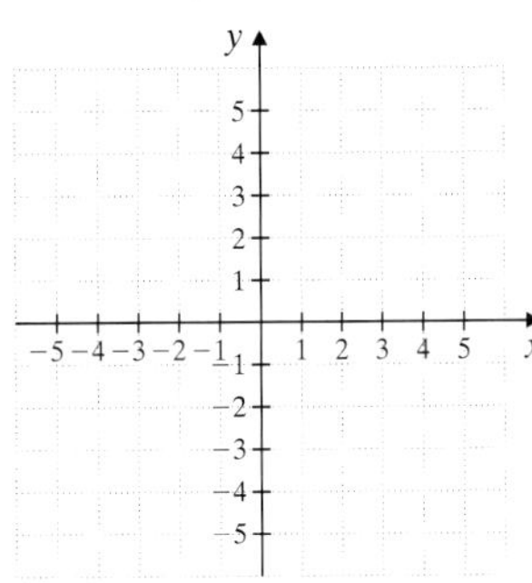

28. $y = \frac{1}{4}x + 3$

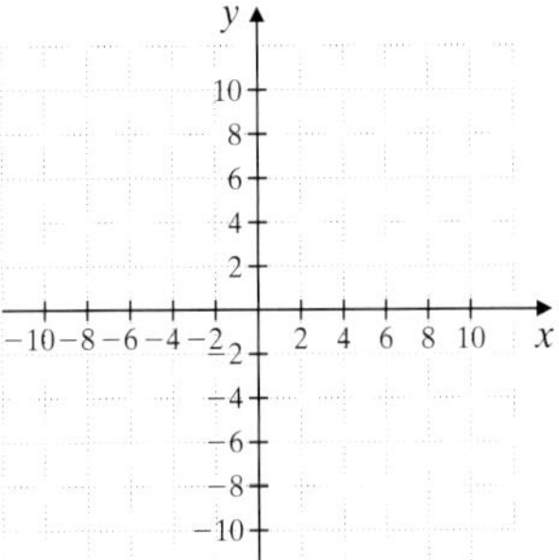

29. $3x - 2y = 12$

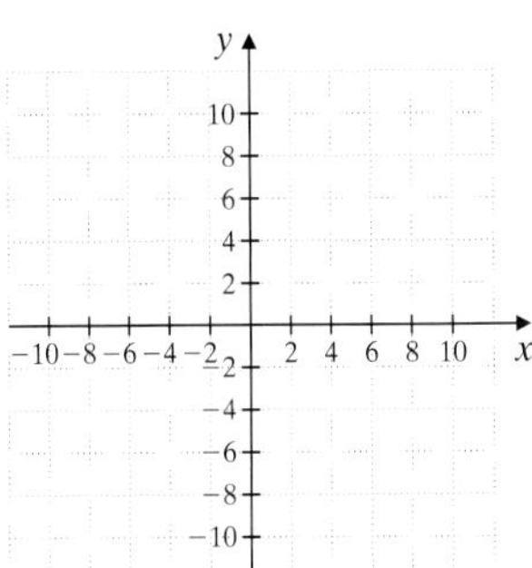

30. $2x - 7y = 14$

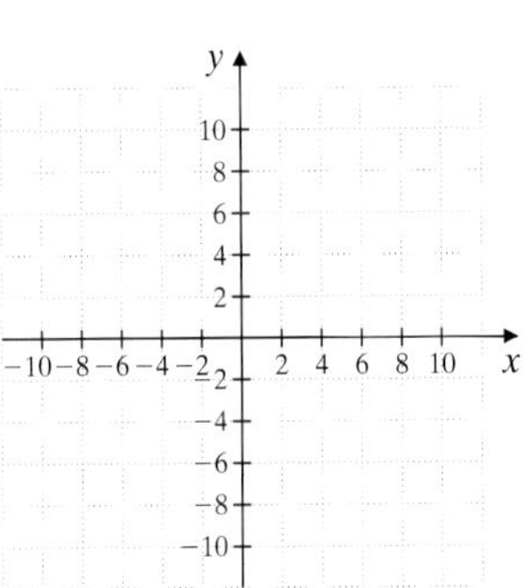

Graph each pair of linear equations on the same set of axes. Discuss how the graphs are similar and how they are different. See Example 5.

31. $y = 5x$
$y = 5x + 4$

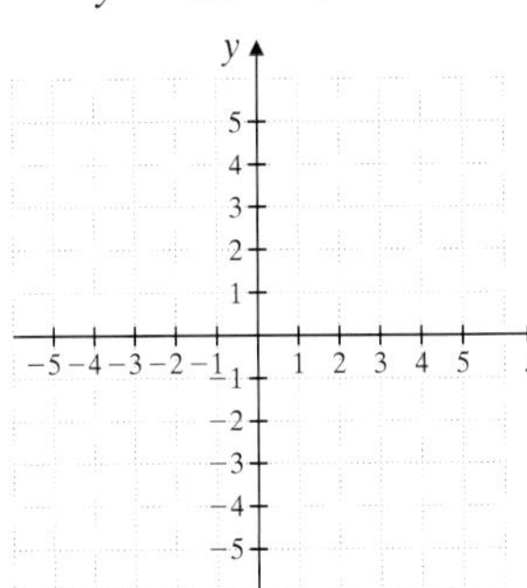

32. $y = 2x$
$y = 2x + 5$

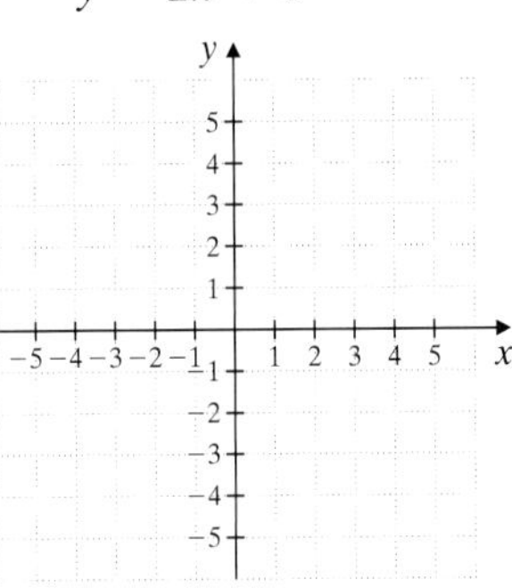

33. $y = -2x$
$y = -2x - 3$

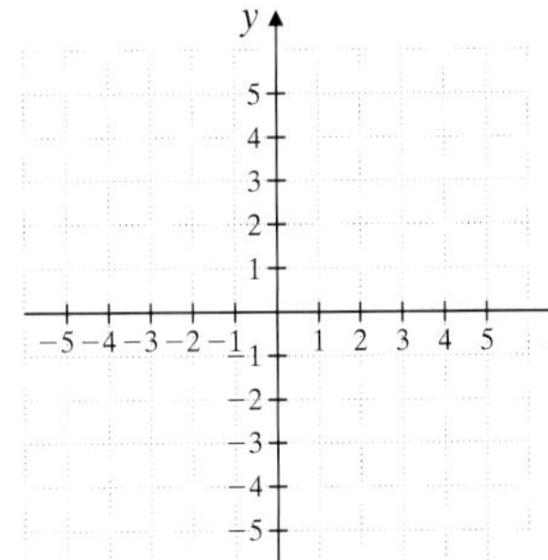

34. $y = x$
$y = x - 7$

Review and Preview

△ **35.** The coordinates of three vertices of a rectangle are $(-2, 5)$, $(4, 5)$, and $(-2, -1)$. Find the coordinates of the fourth vertex. See Section 13.1.

△ **36.** The coordinates of two vertices of a square are $(-3, -1)$ and $(2, -1)$. Find the coordinates of two pairs of points possible for the third and fourth vertices. See Section 13.1.

Complete each table. See Section 13.1.

37. $x - y = -3$

x	y
0	
	0

38. $y - x = 5$

x	y
0	
	0

39. $y = 2x$

x	y
0	
	0

40. $x = -3y$

x	y
0	
	0

Combining Concepts

41. Graph the nonlinear equation $y = x^2$ by completing the table shown. Plot the ordered pairs and connect them with a smooth curve.

x	y
0	
1	
−1	
2	
−2	

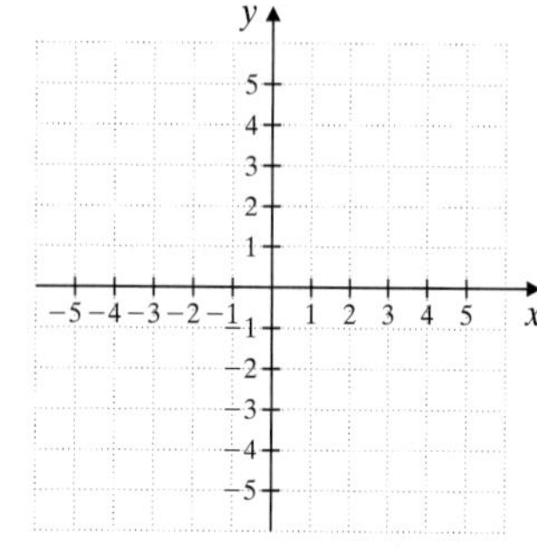

42. Graph the nonlinear equation $y = |x|$ by completing the table shown. Plot the ordered pairs and connect them. This curve is "V" shaped.

x	y
0	
1	
−1	
2	
−2	

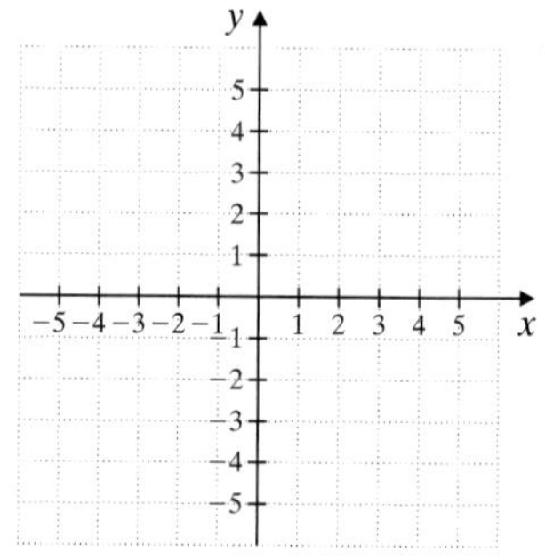

△ **43.** The perimeter of the trapezoid is 22 centimeters. Write a linear equation in two variables for the perimeter. Find y if x is 3 centimeters.

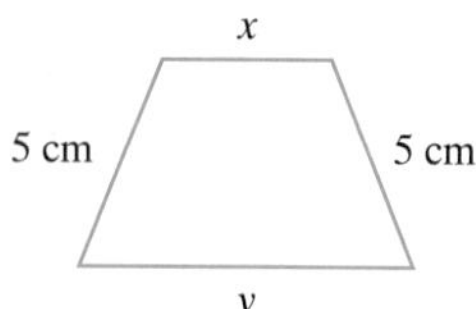

44. If (a, b) is an ordered pair solution of $x + y = 5$, is (b, a) also a solution? Explain why or why not.

45. One of the top five occupations in terms of growth in the next few years is expected to be registered nursing. The number of people y in thousands employed as registered nurses in the United States can be estimated by the linear equation $y = 45x + 2214$ where x is the number of years after 2001. (*Source:* Based on data from the Bureau of Labor Statistics)

a. Graph the linear equation. The break in the vertical axis means that the numbers between 0 and 2200 have been skipped.

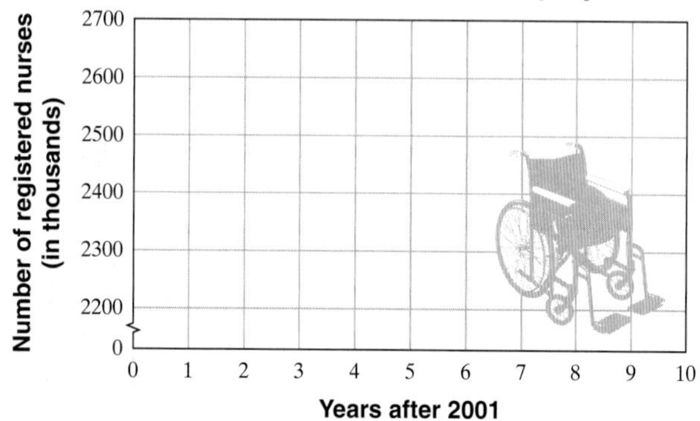

b. Does the point $(6, 2484)$ lie on the line? If so, what does this ordered pair mean?

46. Head Start is a comprehensive child development program serving young children in low-income families. The number of children y (in thousands) enrolled in Head Start from 1992–2002 can be approximated by the linear equation $y = 25x + 664$, where x is the number of years after 1992. (*Source:* Head Start Bureau, the Administration on Children, Youth and Families)

a. Graph the linear equation.

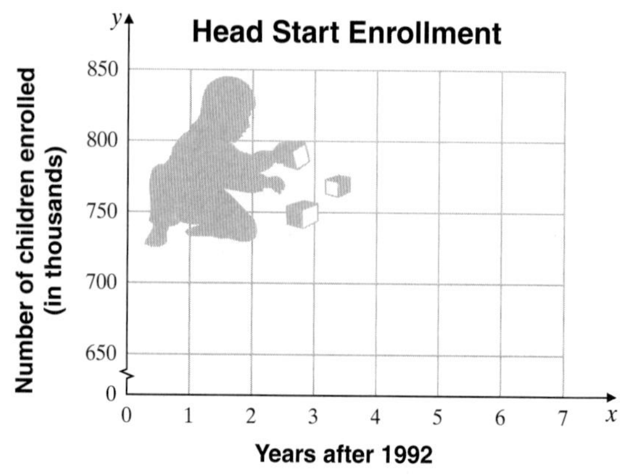

b. Does the point $(4, 764)$ lie on the line? If so, what does this ordered pair mean?

47. The number of U.S. households y in millions that have at least one television set can be estimated by the linear equation $y = 1.43x + 95$ where x is the number of years after 1995. (*Source:* Nielsen Media Research)

a. Graph the linear equation.

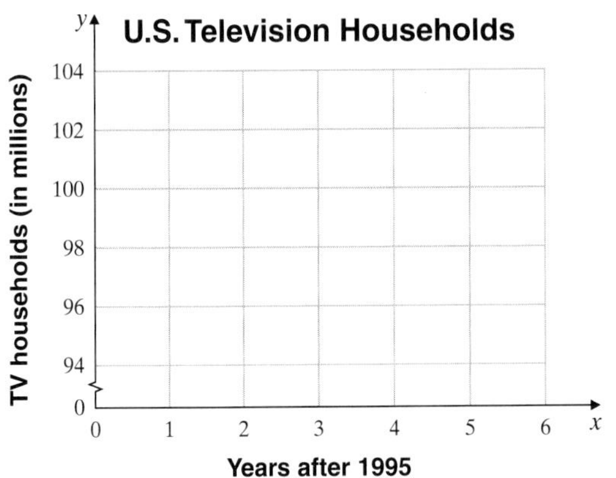

b. Complete the ordered pair (5,).

c. Write a sentence explaining the meaning of the ordered pair found in part b.

48. The restaurant industry is busier than ever. The yearly revenue for restaurants in the U.S. can be estimated by $y = 11.58x + 348$ where x is the number of years after 1998 and y is the revenue in billions of dollars. (*Source:* National Restaurant Assn.)

a. Graph the linear equation.

b. Complete the ordered pair (3,).

c. Write a sentence explaining the meaning of the ordered pair found in part b.

STUDY SKILLS REMINDER

Tips for studying for an exam

To prepare for an exam, try the following study techniques.

- Start the study process days before your exam.
- Make sure that you are current and up-to-date on your assignments.
- If there is a topic that you are unsure of, use one of the many resources that are available to you. For example,

  See your instructor.

  Visit a learning resource center on campus where math tutors are available.

  Read the textbook material and examples on the topic.

  View a videotape on the topic.
- Reread your notes and carefully review the Chapter Highlights at the end of the chapter.
- Work the review exercises at the end of the chapter and check your answers. Make sure that you correct any missed exercises. If you have trouble on a topic, use a resource listed above.
- Find a quiet place to take the Chapter Test found at the end of the chapter. Do not use any resources when taking this sample test. This way you will have a clear indication of how prepared you are for your exam. Check your answers and make sure that you correct any missed exercises.
- Get lots of rest the night before the exam. It's hard to show how well you know the material if your brain is foggy from lack of sleep.

Good luck and keep a positive attitude.

13.3 Intercepts

OBJECTIVES

(A) Identify intercepts of a graph.

(B) Graph a linear equation by finding and plotting intercept points.

(C) Identify and graph vertical and horizontal lines.

(A) Identifying Intercepts

The graph of $y = 4x - 8$ is shown below. Notice that this graph crosses the y-axis at the point $(0, -8)$. This point is called the **y-intercept.** Likewise the graph crosses the x-axis at $(2, 0)$. This point is called the **x-intercept.**

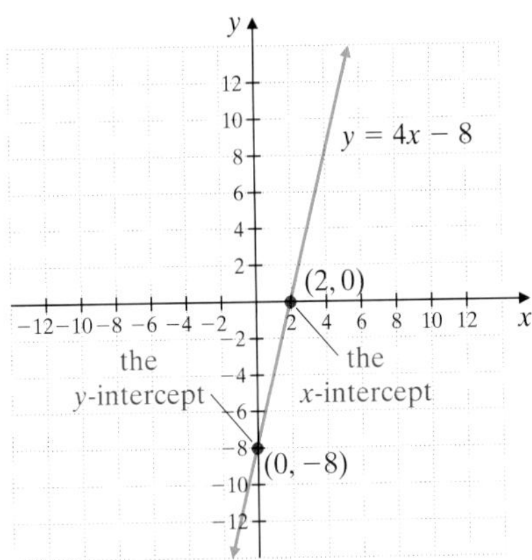

The intercepts are $(2, 0)$ and $(0, -8)$.

Helpful Hint

If a graph crosses the x-axis at $(-3, 0)$ and the y-axis at $(0, 7)$, then

$$\underbrace{(-3, 0)}_{\uparrow \; x\text{-intercept}} \qquad \underbrace{(0, 7)}_{\uparrow \; y\text{-intercept}}$$

Notice that for the y-intercept, the x-value is 0 and for the x-intercept, the y-value is 0.

EXAMPLES Identify the x- and y-intercepts.

1.

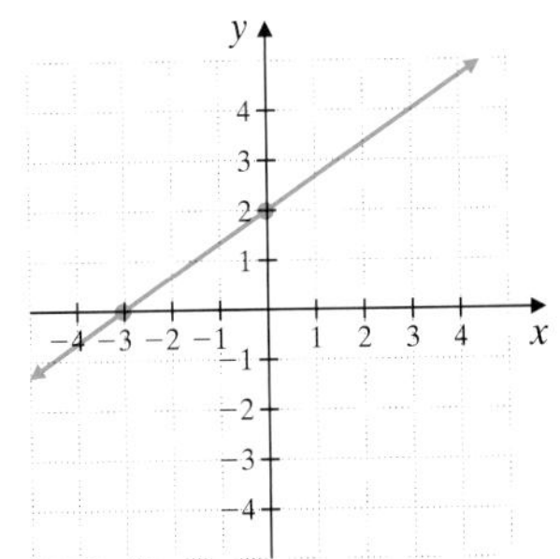

Solution:

x-intercept: $(-3, 0)$

y-intercept: $(0, 2)$

Practice Problem 1

Identify the x- and y-intercepts.

1.

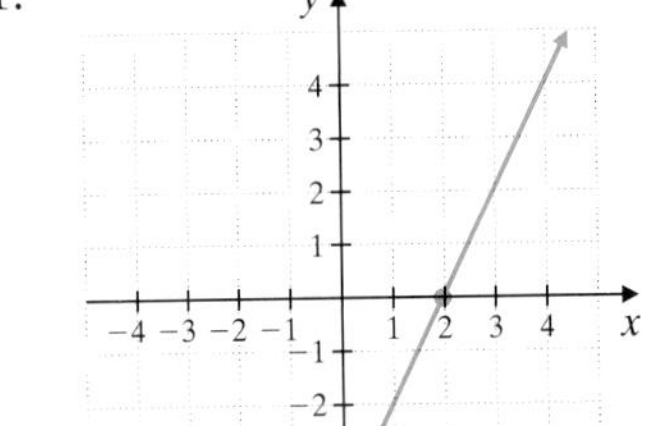

Answer

1. x-intercept: $(2, 0)$; y-intercept: $(0, -4)$

Practice Problems 2-3

Identify the x- and y-intercepts.

2.

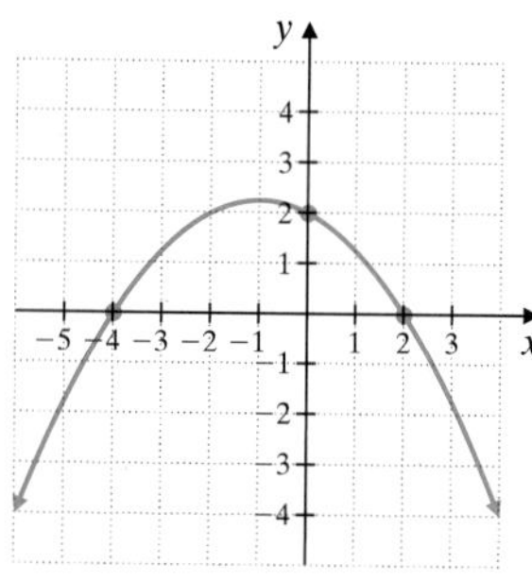

3.

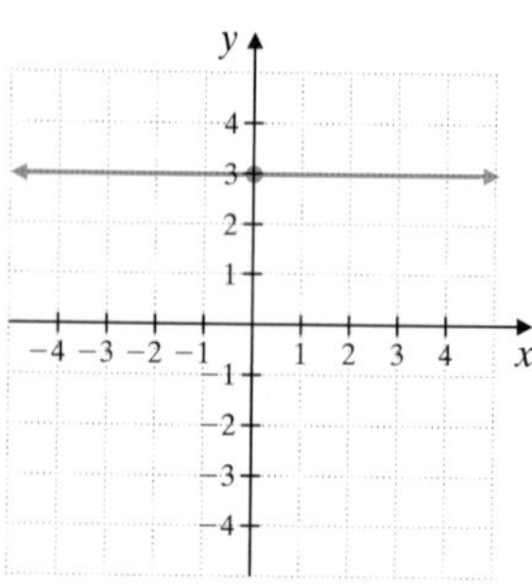

Practice Problem 4

Graph $2x - y = 4$ by finding and plotting its intercepts.

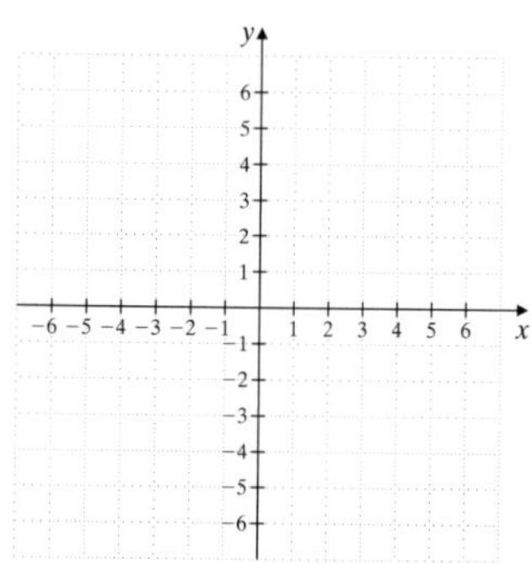

Answers

2. x-intercepts: $(-4, 0)$, $(2, 0)$; y-intercept: $(0, 2)$
3. x-intercept: none; y-intercept: $(0, 3)$
4. See page 1059.

2.

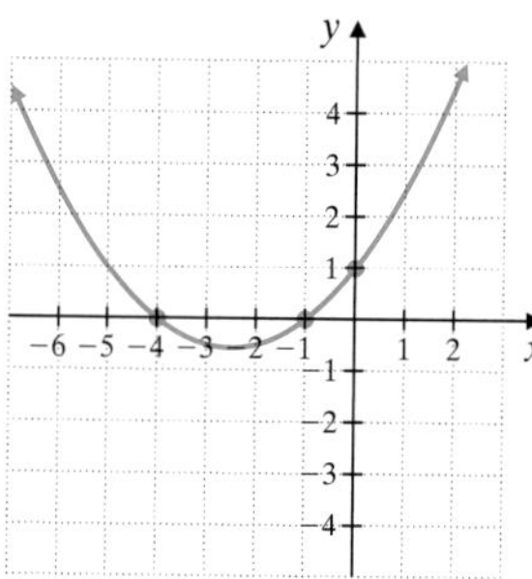

Solution:

x-intercepts: $(-4, 0)(-1, 0)$
y-intercept: $(0, 1)$

3.

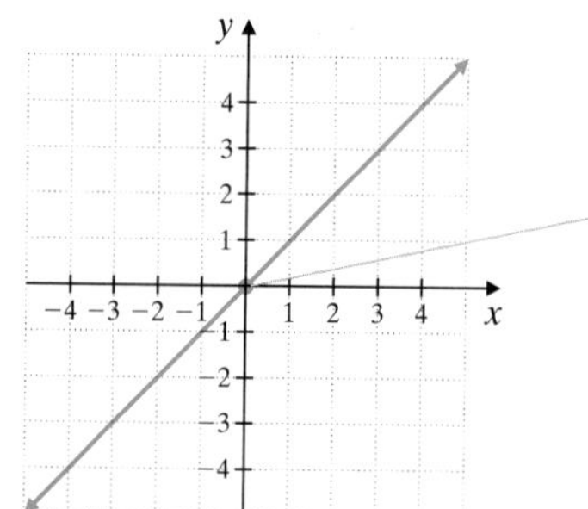

Helpful Hint

Notice that any time $(0, 0)$ is a point of a graph, then it is an x-intercept and a y-intercept.

Solution:

x-intercept: $(0, 0)$
y-intercept: $(0, 0)$

Here, the x- and y-intercept happen to be the same point.

B Finding and Plotting Intercepts

Given an equation of a line, we can usually find intercepts easily since one coordinate is 0.

One way to find the y-intercept of a line from its equation is to let $x = 0$, since a point on the y-axis has an x-coordinate of 0. To find the x-intercept of a line, let $y = 0$, since a point on the x-axis has a y-coordinate of 0.

Finding *x*- and *y*-Intercepts

To find the x-intercept, let $y = 0$ and solve for x.
To find the y-intercept, let $x = 0$ and solve for y.

EXAMPLE 4 Graph $x - 3y = 6$ by finding and plotting its intercepts.

Solution: We let $y = 0$ to find the x-intercept and $x = 0$ to find the y-intercept.

Let $y = 0$.	Let $x = 0$.
$x - 3y = 6$	$x - 3y = 6$
$x - 3(0) = 6$	$0 - 3y = 6$
$x - 0 = 6$	$-3y = 6$
$x = 6$	$y = -2$

The x-intercept is $(6, 0)$ and the y-intercept is $(0, -2)$. We find a third ordered pair solution to check our work. If we let $y = -1$, then $x = 3$. We plot the points $(6, 0)$, $(0, -2)$, and $(3, -1)$. The graph of $x - 3y = 6$ is the line drawn through these points as shown.

x	y
6	0
0	−2
3	−1

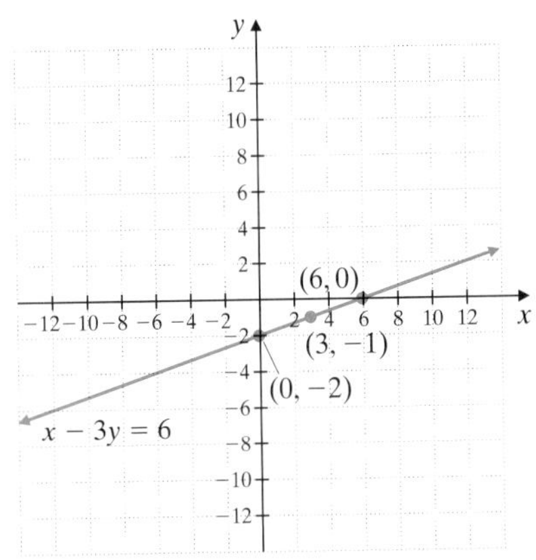

EXAMPLE 5 Graph $x = -2y$ by finding and plotting its intercepts.

Solution: We let $y = 0$ to find the x-intercept and $x = 0$ to find the y-intercept.

Let $y = 0$.	Let $x = 0$.
$x = -2y$	$x = -2y$
$x = -2(0)$	$0 = -2y$
$x = 0$	$0 = y$

Both the x-intercept and y-intercept are $(0, 0)$. In other words, when $x = 0$, then $y = 0$, which gives the ordered pair $(0, 0)$. Also, when $y = 0$, then $x = 0$, which gives the same ordered pair $(0, 0)$. This happens when the graph passes through the origin. Since two points are needed to determine a line, we must find at least one more ordered pair that satisfies $x = -2y$. We let $y = -1$ to find a second ordered pair solution and let $y = 1$ as a checkpoint.

Let $y = -1$.	Let $y = 1$.
$x = -2(-1)$	$x = -2(1)$
$x = 2$	$x = -2$

The ordered pairs are $(0, 0)$, $(2, -1)$, and $(-2, 1)$. We plot these points to graph $x = -2y$.

x	y
0	0
2	−1
−2	1

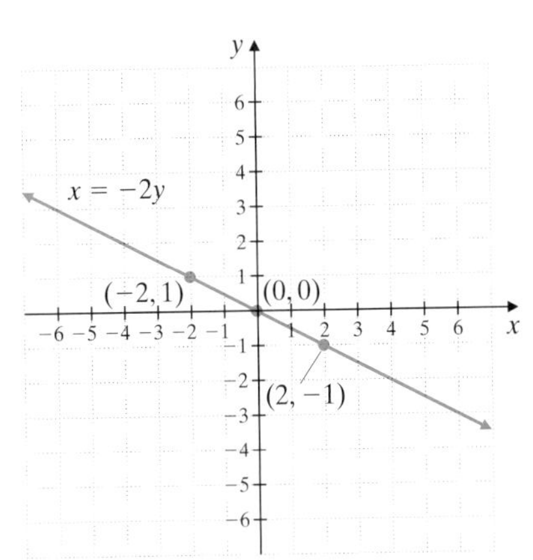

Practice Problem 5

Graph $y = 3x$ by finding and plotting its intercepts.

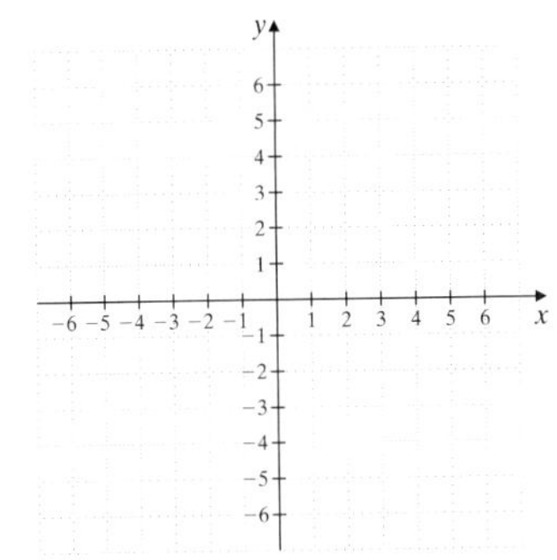

Answers

4.

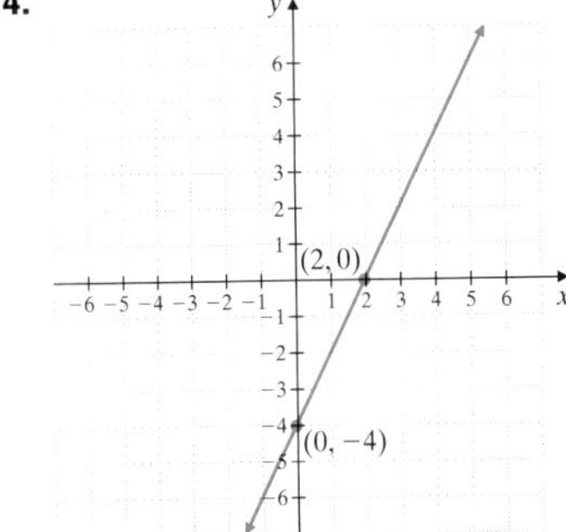

5.

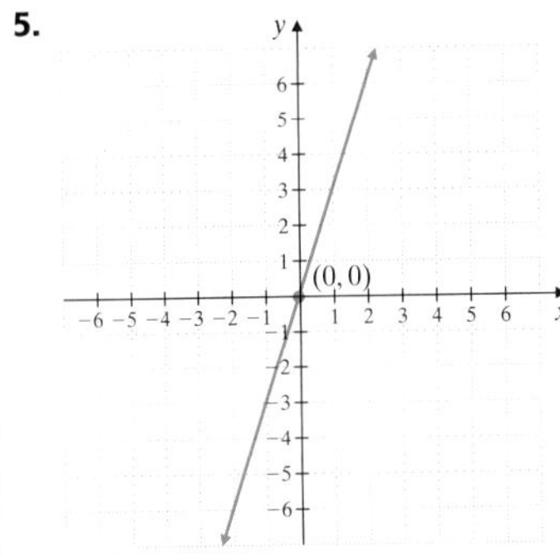

Ⓒ Graphing Vertical and Horizontal Lines

The equation $x = 2$, for example, is a linear equation in two variables because it can be written in the form $x + 0y = 2$. The graph of this equation is a vertical line, as shown in the next example.

Practice Problem 6

Graph: $x = -3$

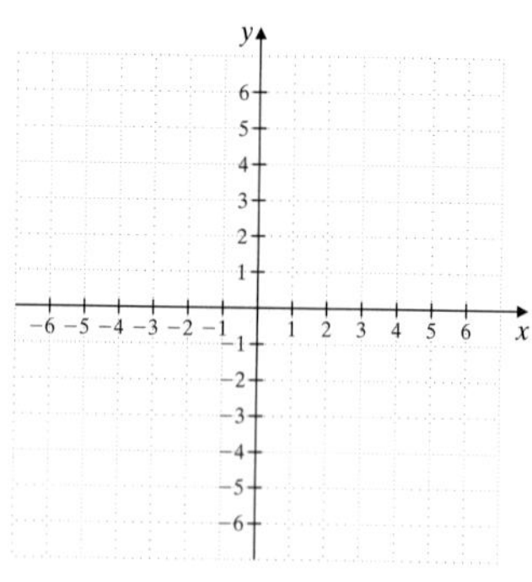

EXAMPLE 6 Graph: $x = 2$

Solution: The equation $x = 2$ can be written as $x + 0y = 2$. For any y-value chosen, notice that x is 2. No other value for x satisfies $x + 0y = 2$. Any ordered pair whose x-coordinate is 2 is a solution of $x + 0y = 2$. We will use the ordered pair solutions $(2, 3)$, $(2, 0)$, and $(2, -3)$ to graph $x = 2$.

x	y
2	3
2	0
2	−3

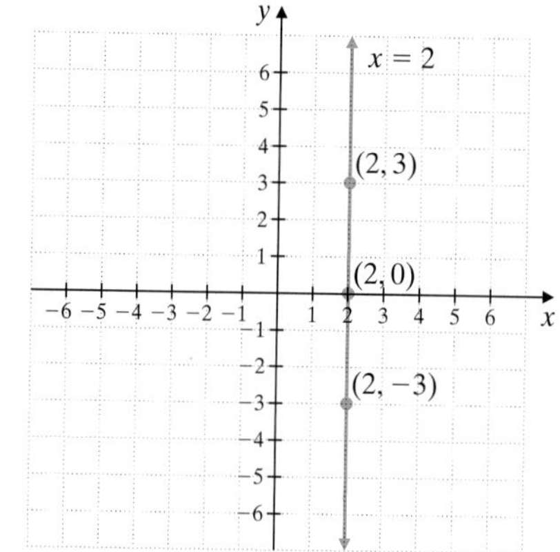

The graph is a vertical line with x-intercept 2. Note that this graph has no y-intercept because x is never 0.

In general, we have the following.

Vertical Lines

The graph of $x = c$, where c is a real number, is a vertical line with x-intercept $(c, 0)$.

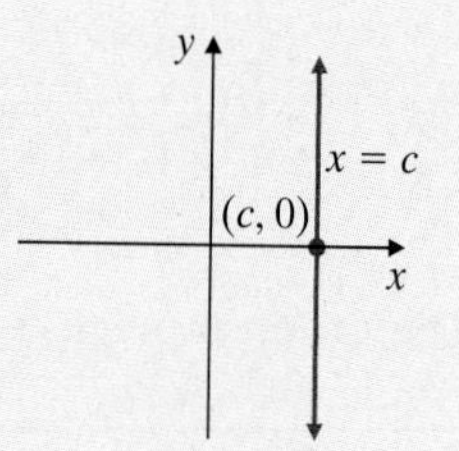

Answer

6.

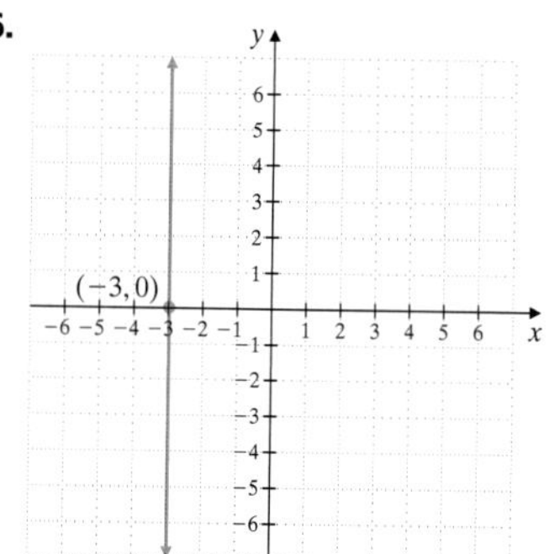

EXAMPLE 7 Graph: $y = -3$

Solution: The equation $y = -3$ can be written as $0x + y = -3$. For any x-value chosen, y is -3. If we choose 4, 1, and -2 as x-values, the ordered pair solutions are $(4, -3)$, $(1, -3)$, and $(-2, -3)$. We use these ordered pairs to graph $y = -3$. The graph is a horizontal line with y-intercept -3 and no x-intercept.

x	y
4	-3
1	-3
-2	-3

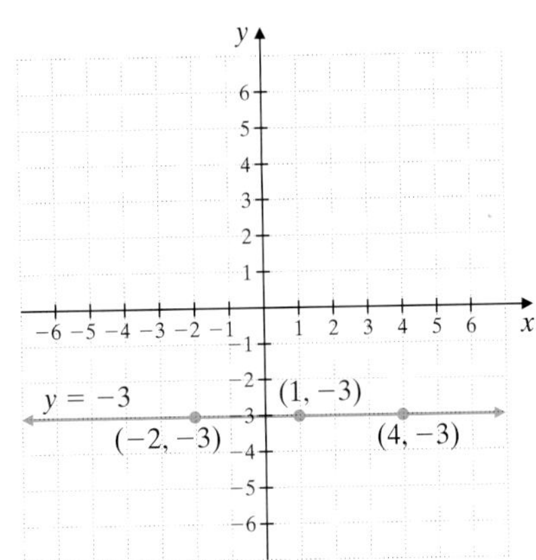

In general, we have the following.

Horizontal Lines

The graph of $y = c$, where c is a real number, is a horizontal line with y-intercept $(0, c)$.

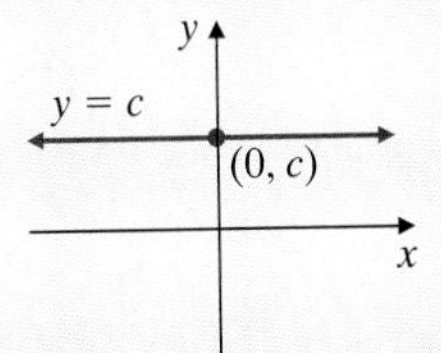

Practice Problem 7

Graph: $y = 4$

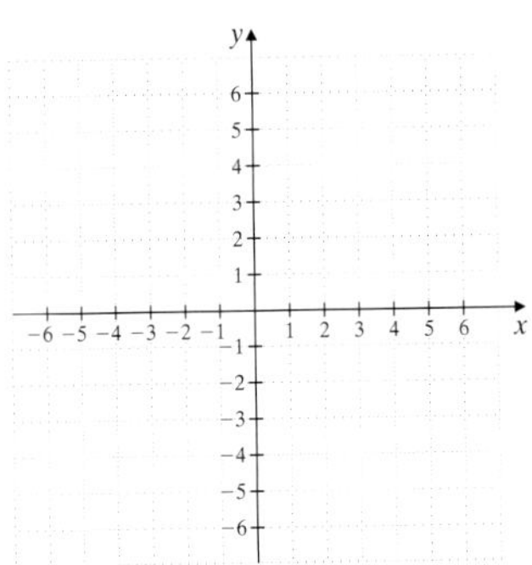

Answer

7.

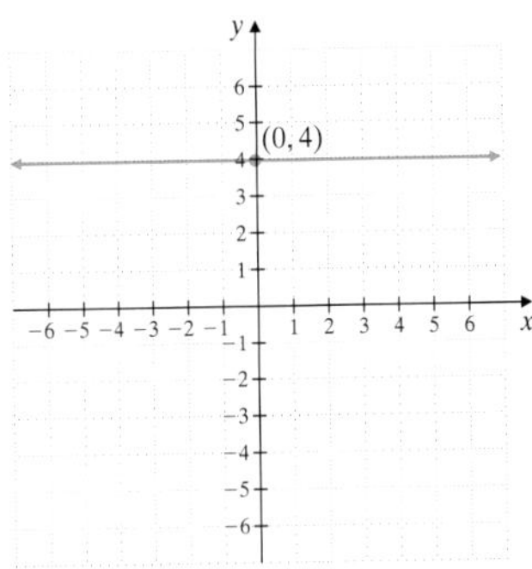

GRAPHING CALCULATOR EXPLORATIONS

You may have noticed that to use the Y= key on a graphing calculator to graph an equation, the equation must be solved for y. For example, to graph $2x + 3y = 7$, we solve this equation for y.

$$2x + 3y = 7$$

$$3y = -2x + 7 \quad \text{Subtract } 2x \text{ from both sides.}$$

$$\frac{3y}{3} = -\frac{2x}{3} + \frac{7}{3} \quad \text{Divide both sides by 3.}$$

$$y = -\frac{2}{3}x + \frac{7}{3} \quad \text{Simplify.}$$

To graph $2x + 3y = 7$ or $y = -\frac{2}{3}x + \frac{7}{3}$, press the Y= key and enter

$$Y_1 = -\frac{2}{3}x + \frac{7}{3}$$

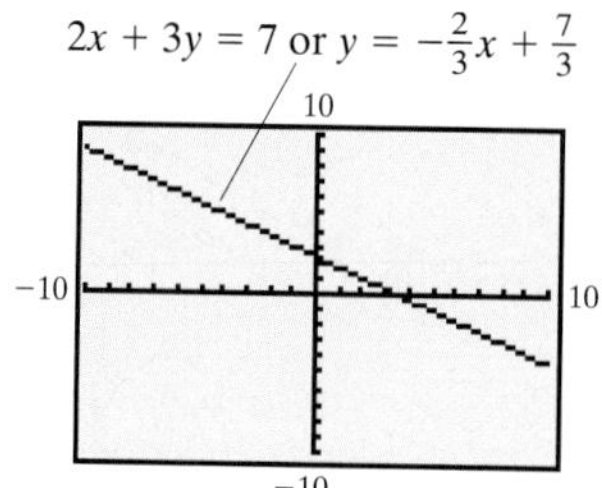

Graph each linear equation.

1. $x = 3.78y$

2. $-2.61y = x$

3. $-2.2x + 6.8y = 15.5$

4. $5.9x - 0.8y = -10.4$

Name ______________________ Section __________ Date __________

Mental Math

Answer the following true or false.

1. The graph of $x = 2$ is a horizontal line.

2. All lines have an x-intercept *and* a y-intercept.

3. The graph of $y = 4x$ contains the point $(0, 0)$.

4. The graph of $x + y = 5$ has an x-intercept of $(5, 0)$ and a y-intercept of $(0, 5)$.

EXERCISE SET 13.3

A *Identify the intercepts. See Examples 1 through 3.*

1.

2.

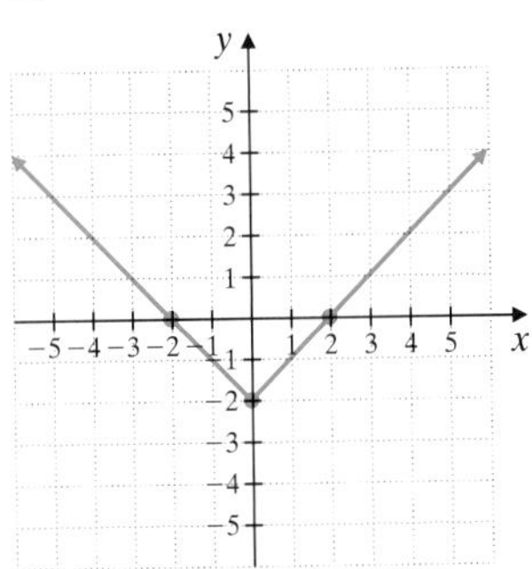

3.

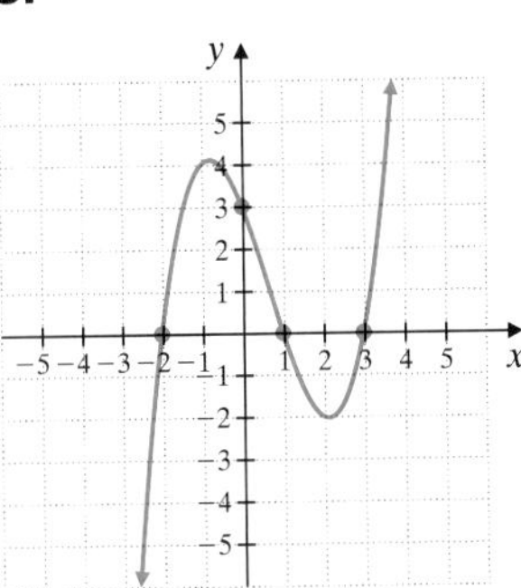

4.

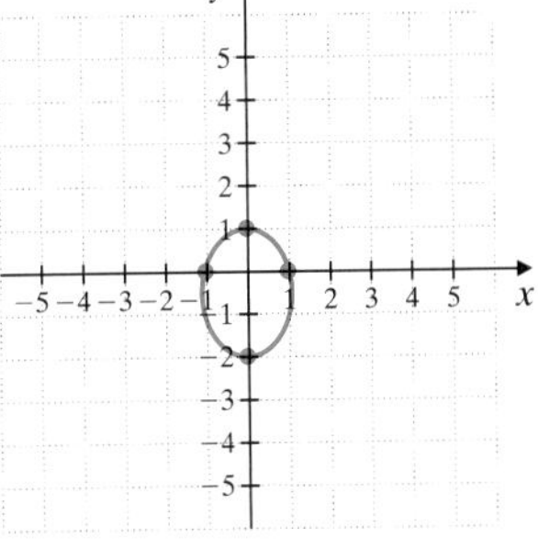

5. What is the greatest number of x- and y-intercepts that a line can have?

6. What is the smallest number of x- and y-intercepts that a line can have?

7. What is the smallest number of x- and y-intercepts that a circle can have?

8. What is the greatest number of x- and y-intercepts that a circle can have?

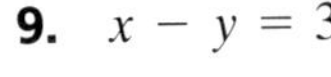

B *Graph each linear equation by finding and plotting its intercepts. See Examples 4 and 5.*

9. $x - y = 3$

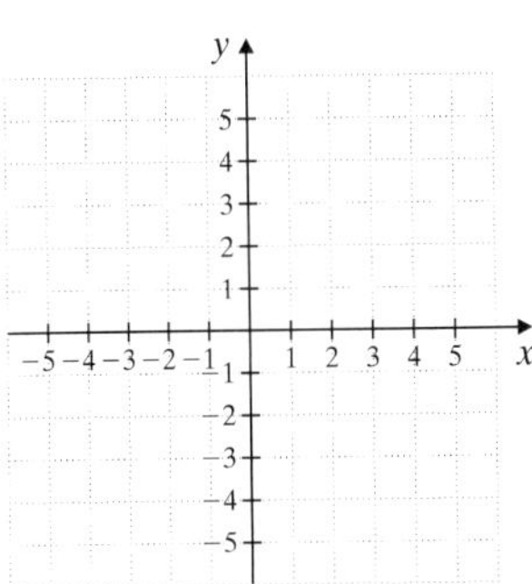

10. $x - y = -4$

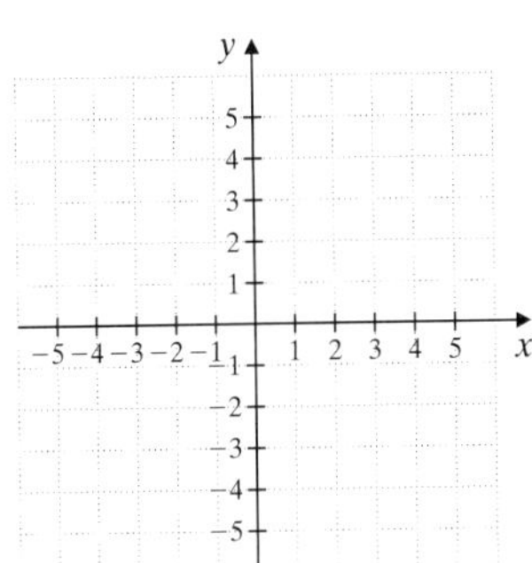

11. $x = 5y$

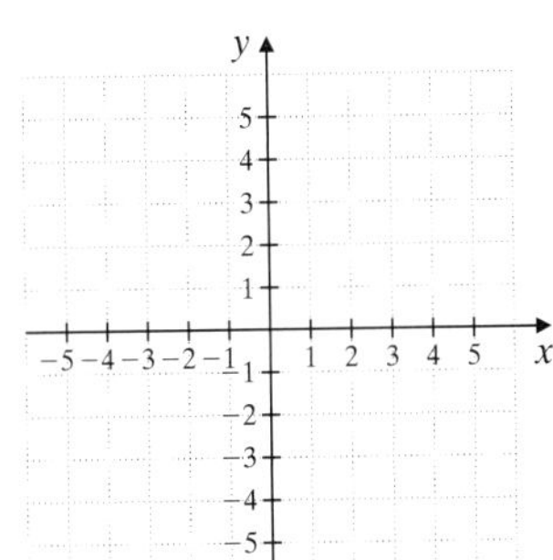

12. $2x = y$

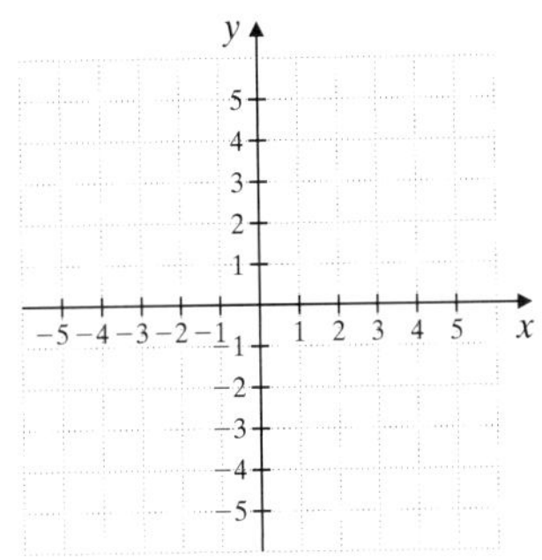

13. $-x + 2y = 6$

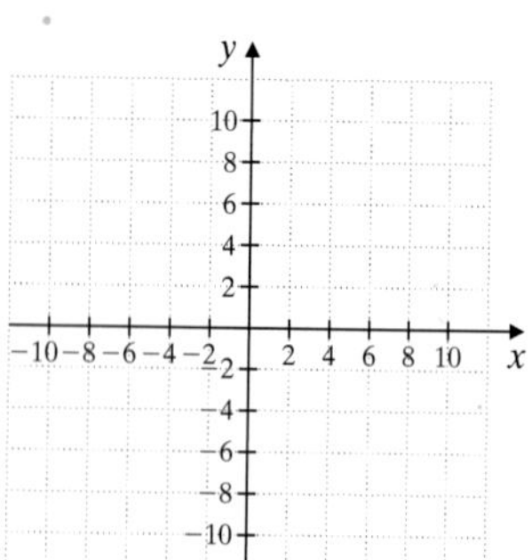

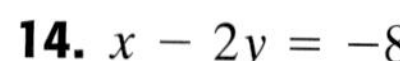

14. $x - 2y = -8$

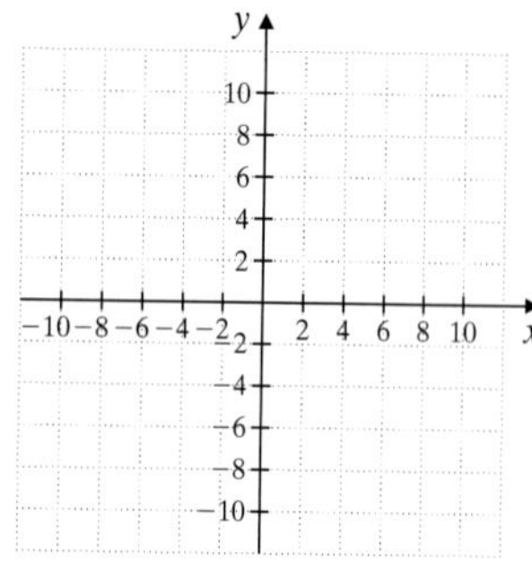

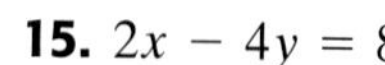

15. $2x - 4y = 8$

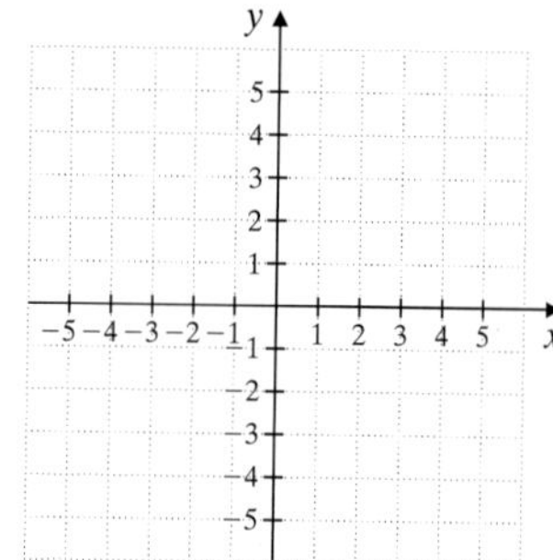

16. $2x + 3y = 6$

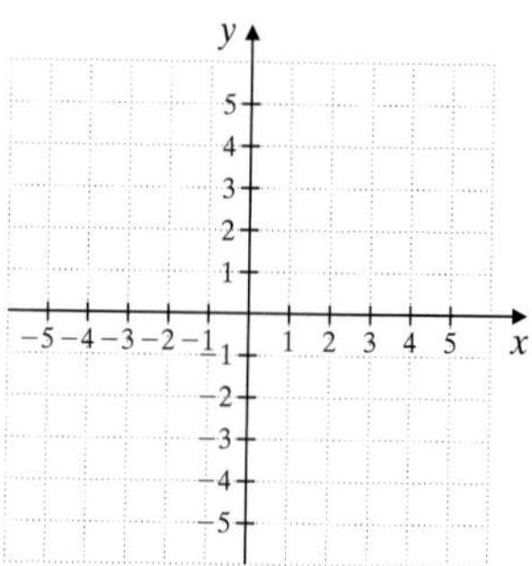

17. $x = 2y$

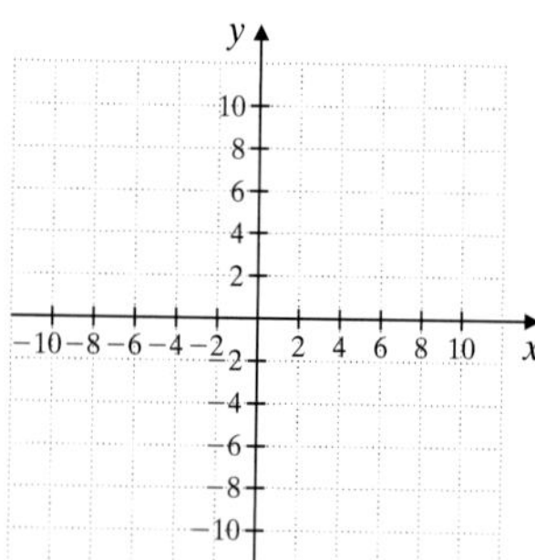

18. $y = -2x$

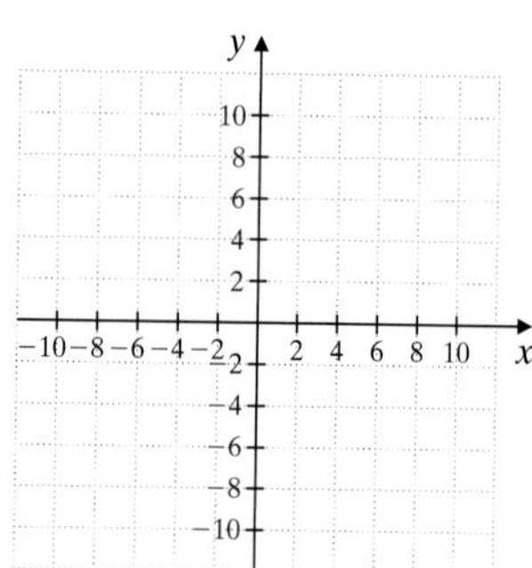

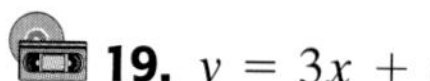

19. $y = 3x + 6$

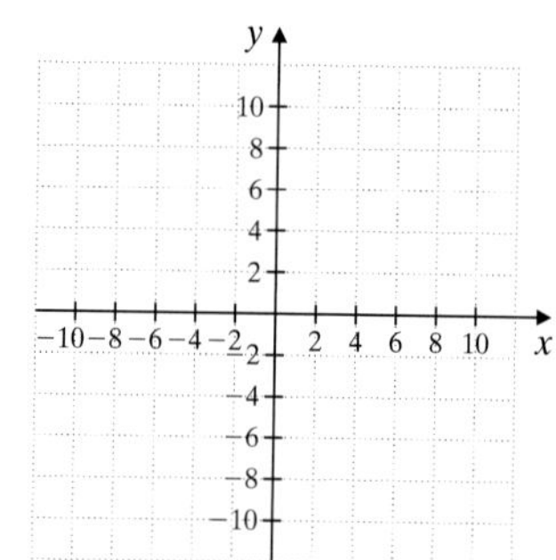

20. $y = 2x + 10$

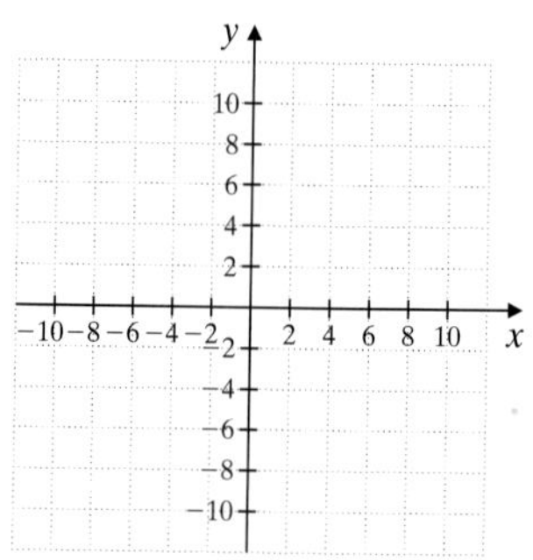

21. $x = y$

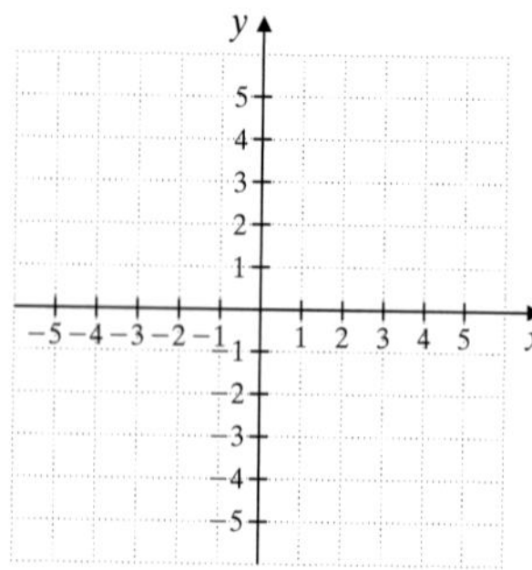

22. $x = -y$

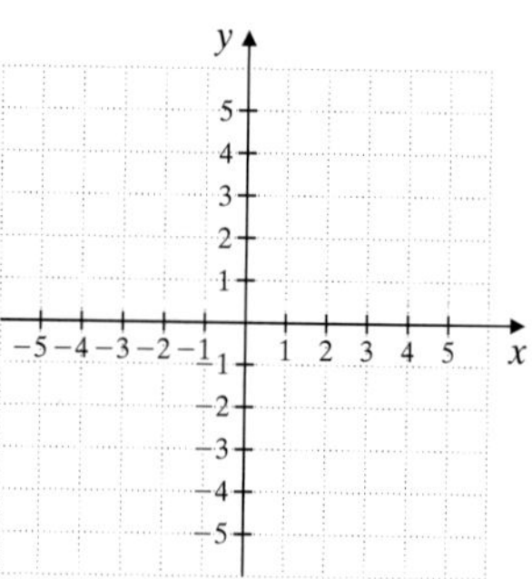

23. $x + 8y = 8$

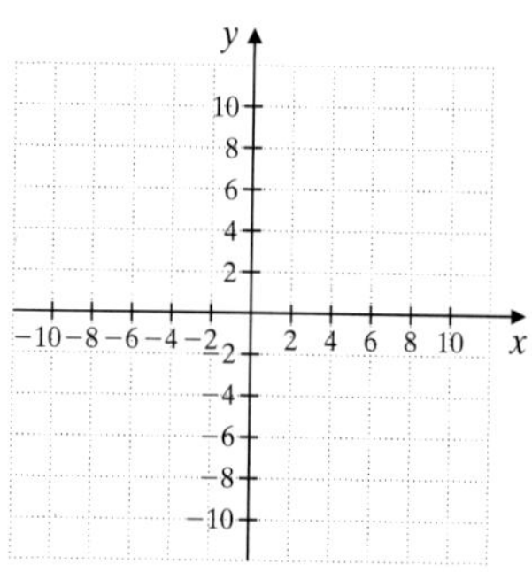

24. $x + 3y = 9$

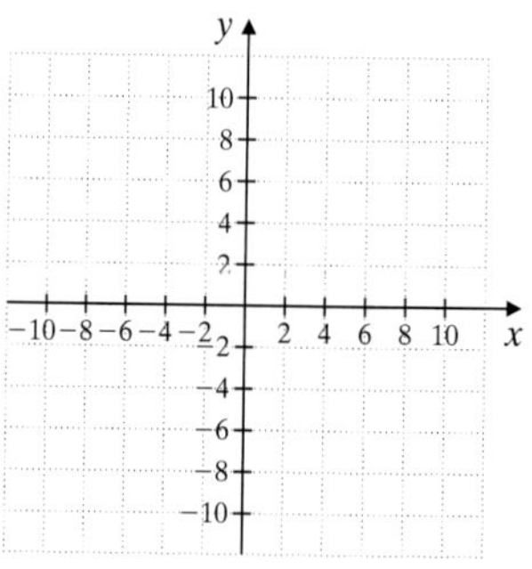

25. $5 = 6x - y$

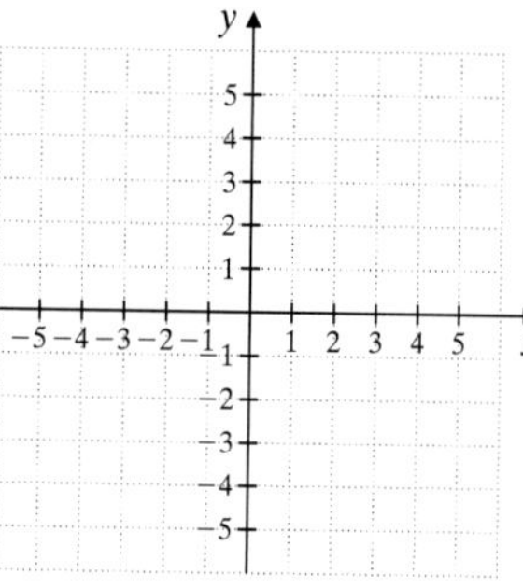

26. $4 = x - 3y$

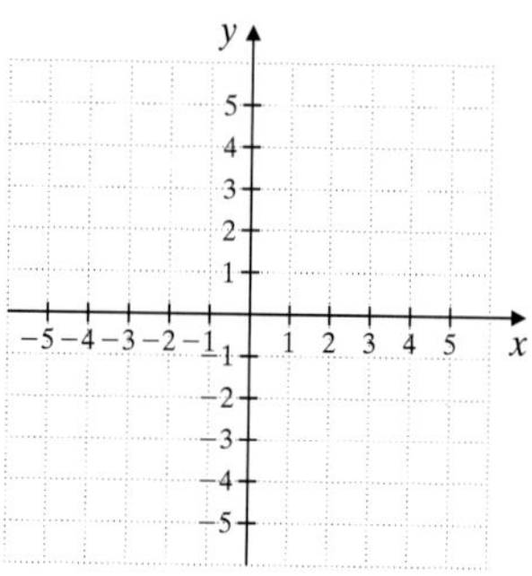

27. $-x + 10y = 11$

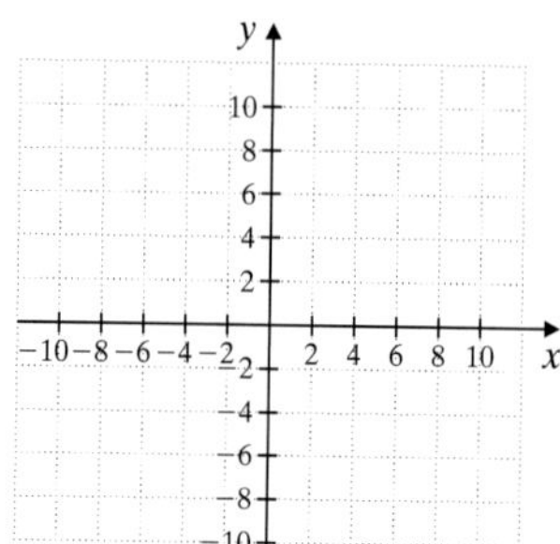

28. $-x + 9 = -y$

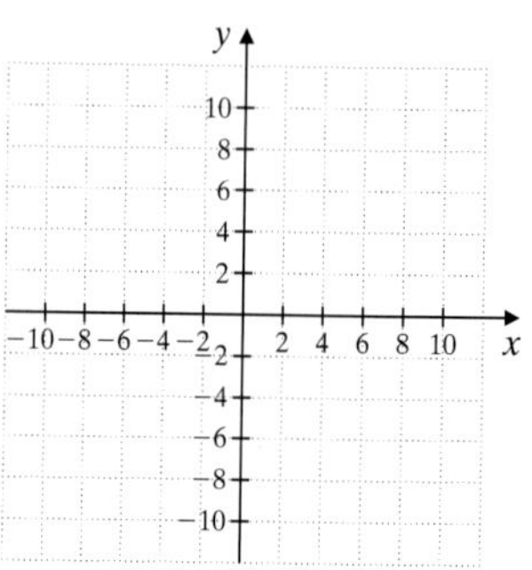

 *Graph each linear equation. See Examples 6 and 7.*

29. $x = -1$

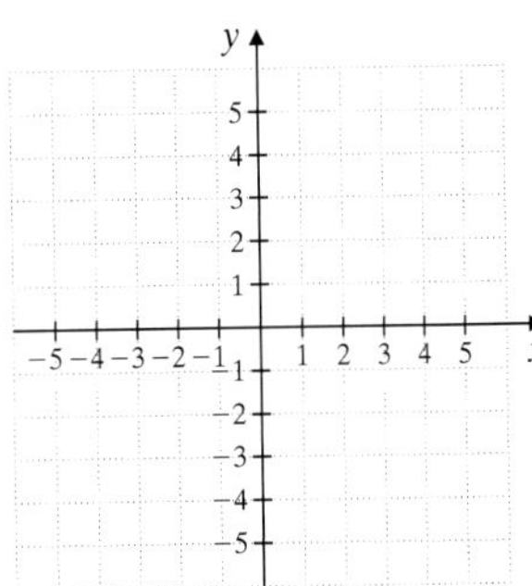

30. $y = 5$

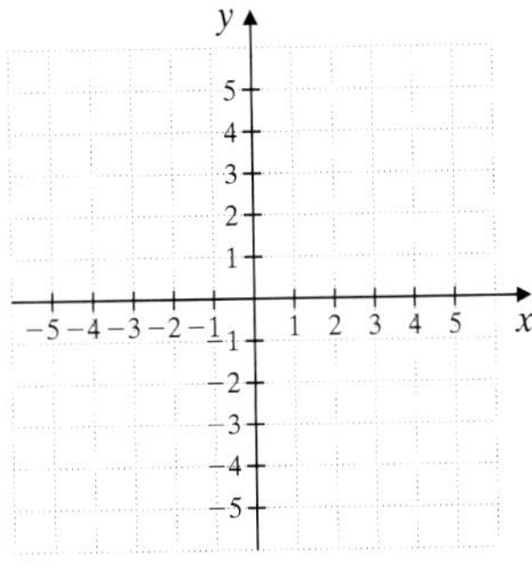

31. $y = 0$

32. $x = 0$

33. $y + 7 = 0$

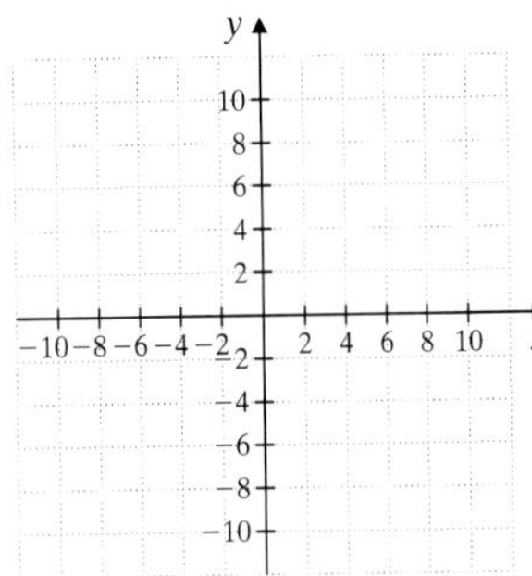

34. $x - 2 = 0$

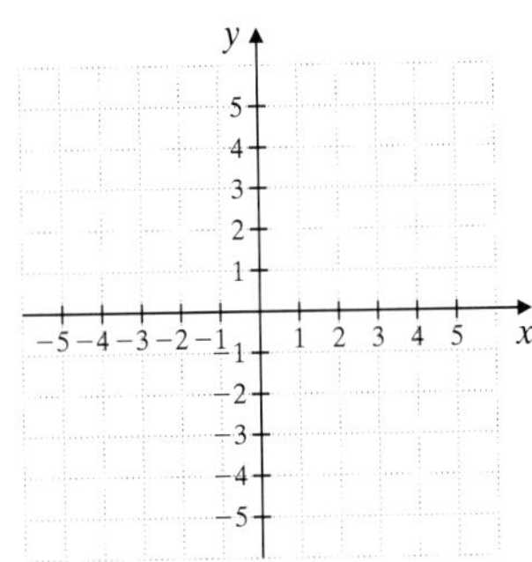

35. $x + 3 = 0$

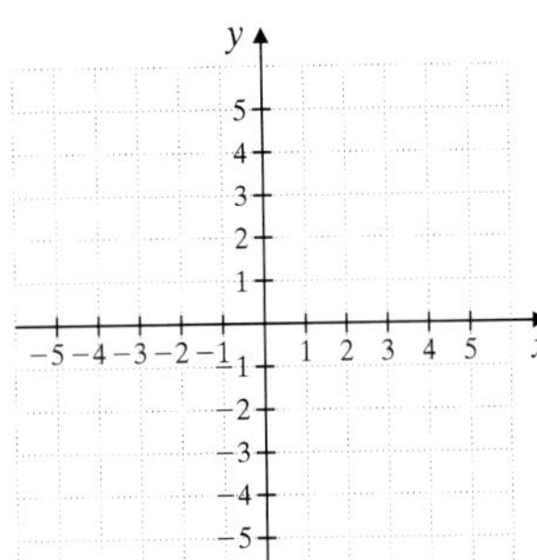

36. $y - 6 = 0$

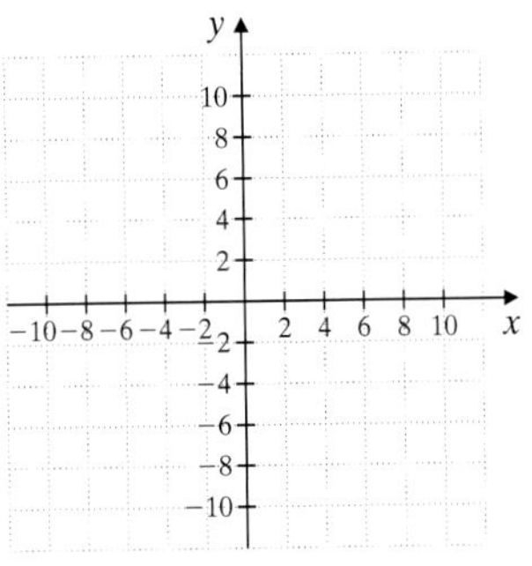

Review and Preview

Simplify. See Sections 2.2 through 2.4 and 4.2.

37. $\dfrac{-6 - 3}{2 - 8}$

38. $\dfrac{4 - 5}{-1 - 0}$

39. $\dfrac{-8 - (-2)}{-3 - (-2)}$

40. $\dfrac{12 - 3}{10 - 9}$

41. $\dfrac{0 - 6}{5 - 0}$

42. $\dfrac{2 - 2}{3 - 5}$

Combining Concepts

Match each equation with its graph.

43. $y = 3$

44. $y = 2x + 2$

45. $x = 3$

46. $y = 2x + 3$

A.

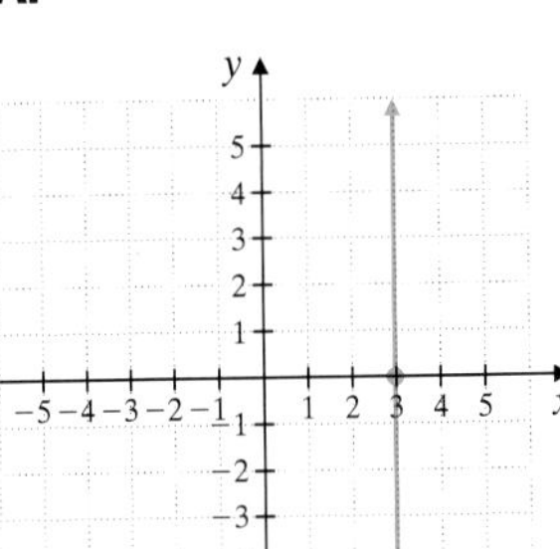

B.

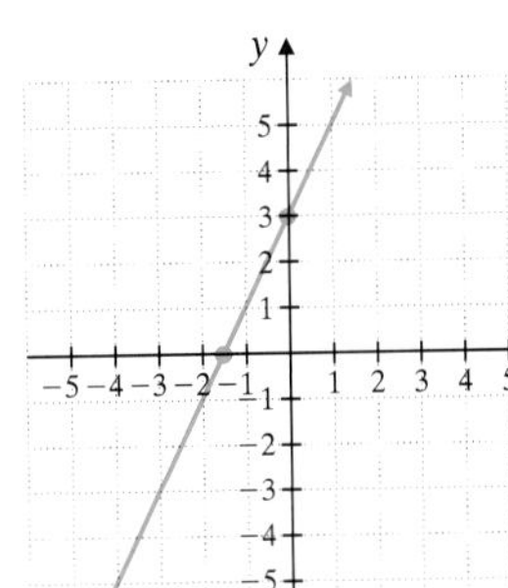

C.

D.

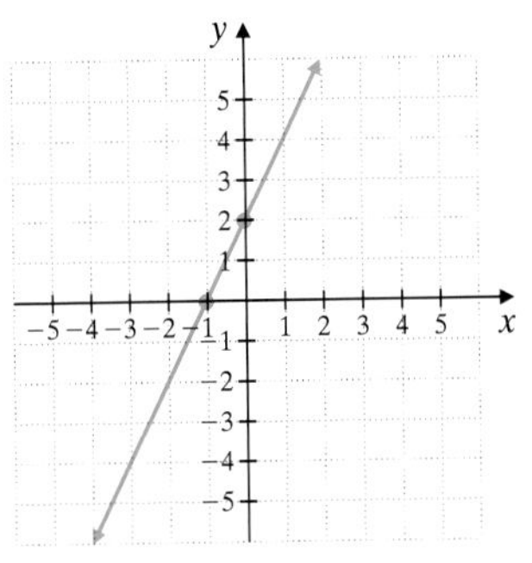

47. Discuss whether a vertical line ever has a y-intercept.

48. Discuss whether a horizontal line ever has an x-intercept.

49. The production supervisor at Alexandra's Office Products finds that it takes 3 hours to manufacture a particular office chair and 6 hours to manufacture an office desk. A total of 1200 hours is available to produce office chairs and desks of this style. The linear equation that models this situation is $3x + 6y = 1200$, where x represents the number of chairs produced and y the number of desks manufactured.

   a. Complete the ordered pair solution $(0,\ \ )$ of this equation. Describe the manufacturing situation that corresponds to this solution.

   b. Complete the ordered pair solution $(\ \ , 0)$ of this equation. Describe the manufacturing situation that corresponds to this solution.

   c. If 50 desks are manufactured, find the greatest number of chairs that can be made.

Two lines in the same plane that do not intersect are called ***parallel lines.***

50. Use your own graph paper to draw a line parallel to the line $x = 5$ that intersects the x-axis at 1. What is the equation of this line?

51. Use your own graph paper to draw a line parallel to the line $y = -1$ that intersects the y-axis at -4. What is the equation of this line?

52. The number of music cassettes y (in millions) shipped to retailers in the United States from 1995 through 2001 can be modeled by the equation $y = -36.8x + 263.77$, where x represents the number of years after 1995. (*Source:* Recording Industry Association of America)

   a. Find the x-intercept of this equation. (Round to the nearest tenth.)

   b. What does this x-intercept mean?

53. The number of Disney Stores y for the years 1996–2000 can by modeled by the equation $y = 51.6x + 560.2$, where x represents the number of years after 1996. (*Source: The Walt Disney Company Fact Book 2000*)

   a. Find the y-intercept of this equation.

   b. What does this y-intercept mean?

13.4 Slope

OBJECTIVES

A. Find the slope of a line given two points of the line.
B. Find the slope of a line given its equation.
C. Find the slopes of horizontal and vertical lines.
D. Compare the slopes of parallel and perpendicular lines.
E. Solve problems of slope.

SSM TUTOR CENTER SG CD & VIDEO MATH PRO WEB

A Finding the Slope of a Line Given Two Points

Thus far, much of this chapter has been devoted to graphing lines. You have probably noticed by now that a key feature of a line is its slant or steepness. In mathematics, the slant or steepness of a line is formally known as its **slope.** We measure the slope of a line by the ratio of vertical change (rise) to the corresponding horizontal change (run) as we move along the line.

On the line below, for example, suppose that we begin at the point $(1, 2)$ and move to the point $(4, 6)$. The vertical change is the change in y-coordinates: $6 - 2$ or 4 units. The corresponding horizontal change is the change in x-coordinates: $4 - 1 = 3$ units. The ratio of these changes is

$$\text{slope} = \frac{\text{change in } y \text{ (vertical change or rise)}}{\text{change in } x \text{ (horizontal change or run)}} = \frac{4}{3}$$

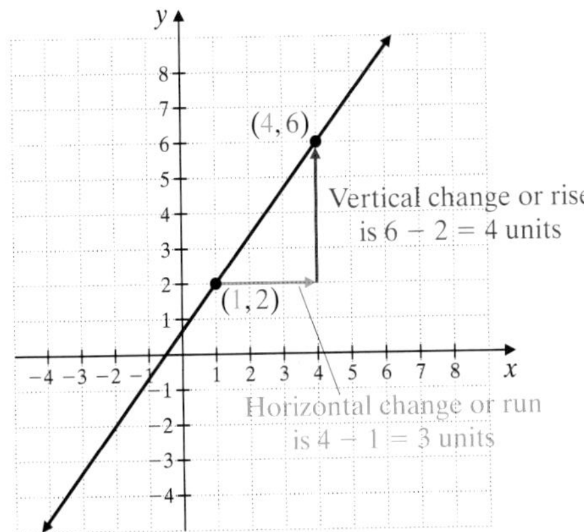

The slope of this line, then, is $\frac{4}{3}$. This means that for every 4 units of change in y-coordinates, there is a corresponding change of 3 units in x-coordinates.

Helpful Hint

It makes no difference what two points of a line are chosen to find its slope. The slope of a line is the same everywhere on the line.

Slope of a Line

The slope m of the line containing the points (x_1, y_1) and (x_2, y_2) is given by

$$m = \frac{\text{rise}}{\text{run}} = \frac{\text{change in } y}{\text{change in } x} = \frac{y_2 - y_1}{x_2 - x_1}, \quad \text{as long as } x_2 \neq x_1$$

Practice Problem 1

Find the slope of the line through $(-2, 3)$ and $(4, -1)$. Graph the line.

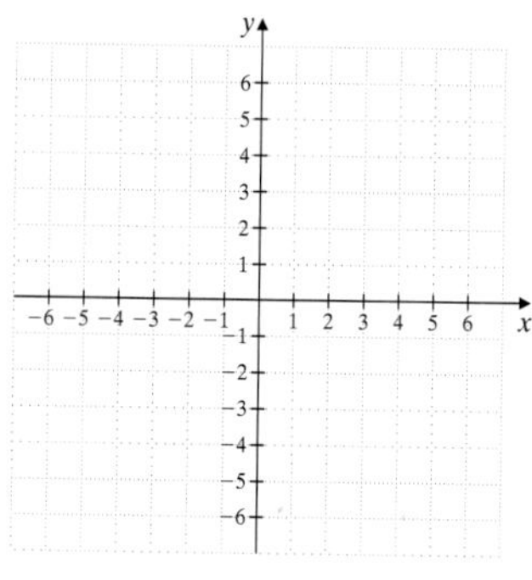

Concept Check

The points $(-2, -5)$, $(0, -2)$, $(4, 4)$, and $(10, 13)$ all lie on the same line. Work with a partner and verify that the slope is the same no matter which points are used to find slope.

Answers

1. $-\frac{2}{3}$

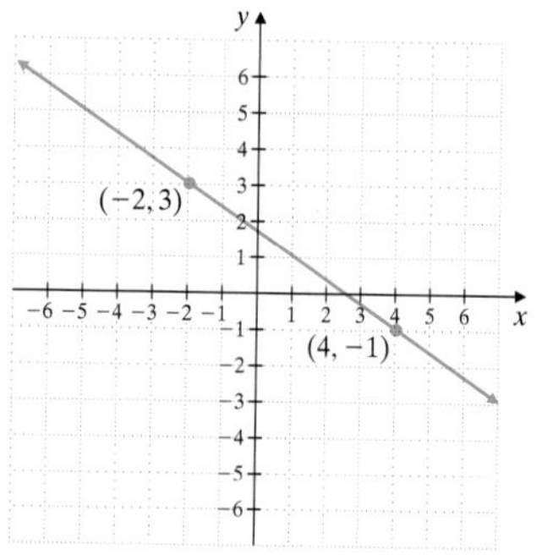

Concept Check: $m = \frac{3}{2}$

EXAMPLE 1

Find the slope of the line through $(-1, 5)$ and $(2, -3)$. Graph the line.

Solution: Let (x_1, y_1) be $(-1, 5)$ and (x_2, y_2) be $(2, -3)$. Then, by the definition of slope, we have the following.

$$m = \frac{y_2 - y_1}{x_2 - x_1}$$

$$= \frac{-3 - 5}{2 - (-1)}$$

$$= \frac{-8}{3} = -\frac{8}{3}$$

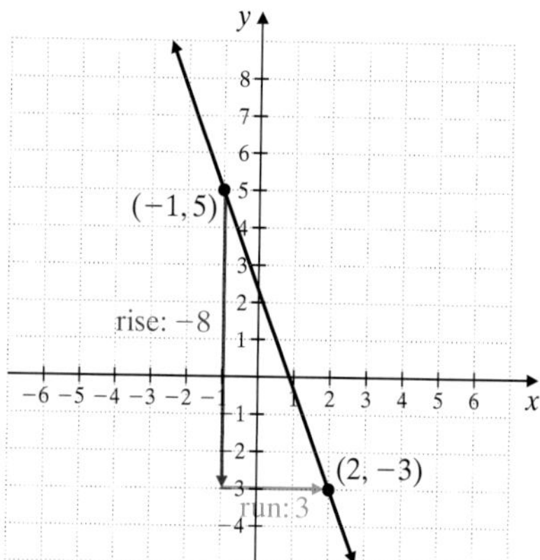

The slope of the line is $-\frac{8}{3}$.

In Example 1, we could just as well have identified (x_1, y_1) with $(2, -3)$ and (x_2, y_2) with $(-1, 5)$. It makes no difference which point is called (x_1, y_1) or (x_2, y_2).

Try the Concept Check in the margin.

Helpful Hint

When finding the slope of a line through two given points, it makes no difference which given point is called (x_1, y_1) and which is called (x_2, y_2). However, once an x-coordinate is called x_1, make sure its corresponding y-coordinate is called y_1.

EXAMPLE 2

Find the slope of the line through $(-1, -2)$ and $(2, 4)$. Graph the line.

Solution: Let (x_1, y_1) be $(2, 4)$ and (x_2, y_2) be $(-1, -2)$.

$$m = \frac{y_2 - y_1}{x_2 - x_1} = \frac{-2 - 4}{-1 - 2} = \frac{-6}{-3} = 2$$

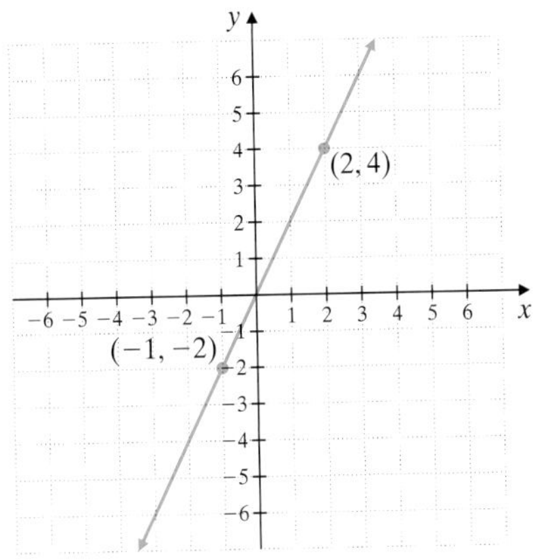

The slope is 2.

Practice Problem 2

Find the slope of the line through $(-2, 1)$ and $(3, 5)$. Graph the line.

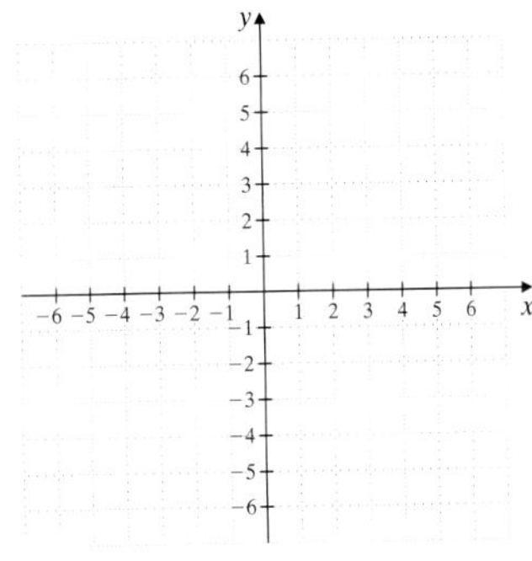

Try the Concept Check in the margin.

Notice that the slope of the line in Example 1 is negative, and the slope of the line in Example 2 is positive. Let your eye follow the line with negative slope from left to right and notice that the line "goes down." If you follow the line with positive slope from left to right, you will notice that the line "goes up." This is true in general.

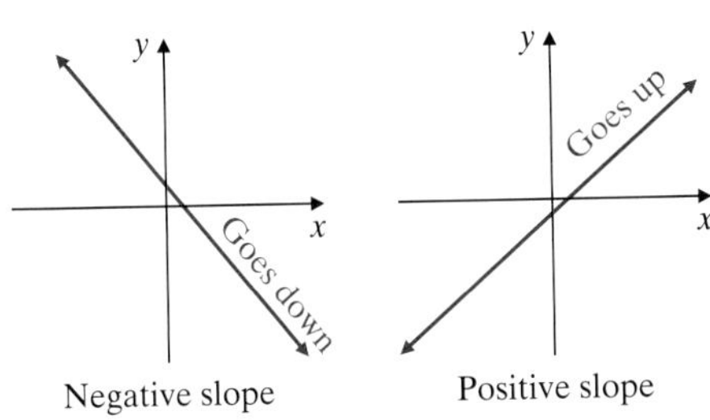

Concept Check

What is wrong with the following slope calculation for the points $(3, 5)$ and $(-2, 6)$?

$$m = \frac{5 - 6}{-2 - 3} = \frac{-1}{-5} = \frac{1}{5}$$

B Finding the Slope of a Line Given Its Equation

As we have seen, the slope of a line is defined by two points on the line. Thus, if we know the equation of a line, we can find its slope by finding two of its points. For example, let's find the slope of the line

$$y = 3x - 2$$

To find two points, we can choose two values for x and substitute to find corresponding y-values. If $x = 0$, for example, $y = 3 \cdot 0 - 2$ or $y = -2$. If $x = 1$, $y = 3 \cdot 1 - 2$ or $y = 1$. This gives the ordered pairs $(0, -2)$ and $(1, 1)$. Using the definition for slope, we have

$$m = \frac{1 - (-2)}{1 - 0} = \frac{3}{1} = 3 \quad \text{The slope is 3.}$$

Notice that the slope, 3, is the same as the coefficient of x in the equation $y = 3x - 2$. This is true in general.

Answers

2. $\frac{4}{5}$

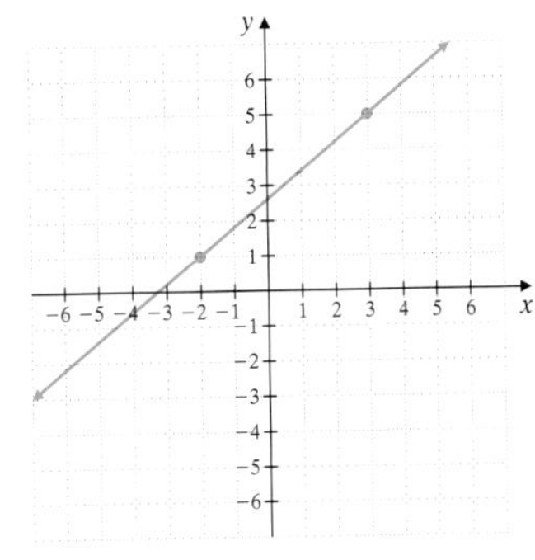

Concept Check: $m = \frac{5 - 6}{3 - (-2)} = \frac{-1}{5} = -\frac{1}{5}$

If a linear equation is solved for y, the coefficient of x is the line's slope. In other words, the slope of the line given by $y = mx + b$ is m, the coefficient of x.

Practice Problem 3

Find the slope of the line $5x + 4y = 10$.

EXAMPLE 3 Find the slope of the line $-2x + 3y = 11$.

Solution: When we solve for y, the coefficient of x is the slope.

$$-2x + 3y = 11$$

$$3y = 2x + 11 \quad \text{Add } 2x \text{ to both sides.}$$

$$y = \frac{2}{3}x + \frac{11}{3} \quad \text{Divide both sides by 3.}$$

The slope is $\frac{2}{3}$.

C Finding Slopes of Horizontal and Vertical Lines

Practice Problem 4

Find the slope of $y = 3$.

EXAMPLE 4 Find the slope of the line $y = -1$.

Solution: Recall that $y = -1$ is a horizontal line with y-intercept -1. To find the slope, we find two ordered pair solutions of $y = -1$, knowing that solutions of $y = -1$ must have a y-value of -1. We will use $(2, -1)$ and $(-3, -1)$. We let (x_1, y_1) be $(2, -1)$ and (x_2, y_2) be $(-3, -1)$.

$$m = \frac{y_2 - y_1}{x_2 - x_1} = \frac{-1 - (-1)}{-3 - 2} = \frac{0}{-5} = 0$$

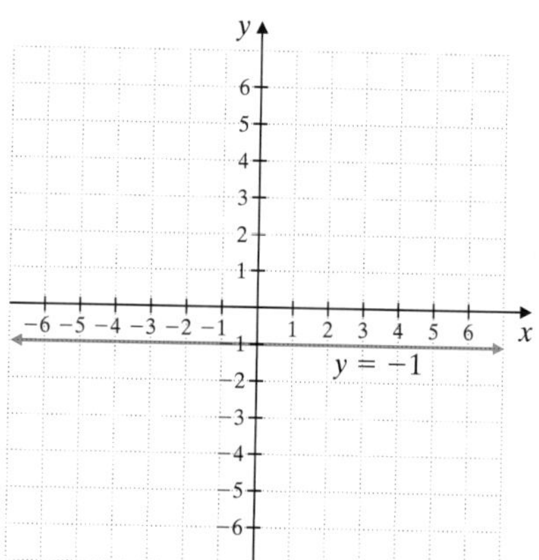

The slope of the line $y = -1$ is 0. Since the y-values will have a difference of 0 for every horizontal line, we can say that all **horizontal lines have a slope of 0.**

Practice Problem 5

Find the slope of the line $x = -2$.

EXAMPLE 5 Find the slope of the line $x = 5$.

Solution: Recall that the graph of $x = 5$ is a vertical line with x-intercept 5. To find the slope, we find two ordered pair solutions of $x = 5$. Ordered pair solutions of $x = 5$ must have an x-value of 5. We will use $(5, 0)$ and $(5, 4)$. We let $(x_1, y_1) = (5, 0)$ and $(x_2, y_2) = (5, 4)$.

$$m = \frac{y_2 - y_1}{x_2 - x_1} = \frac{4 - 0}{5 - 5} = \frac{4}{0}$$

Answers

3. $-\frac{5}{4}$ **4.** 0 **5.** undefined slope

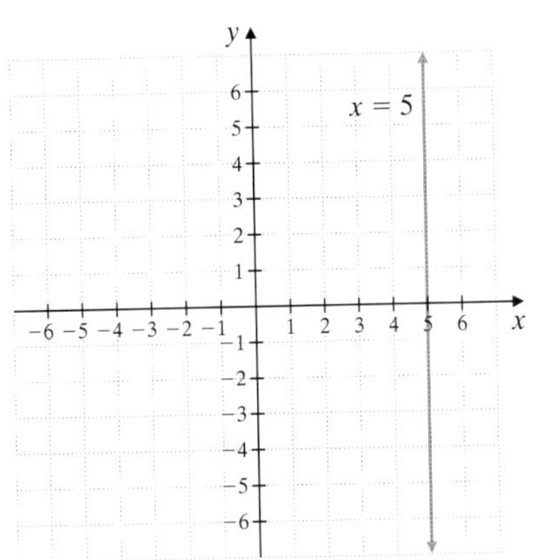

Since $\frac{4}{0}$ is undefined, we say the slope of the vertical line $x = 5$ is undefined.

Since the x-values will have a difference of 0 for every vertical line, we can say that all **vertical lines have undefined slope.**

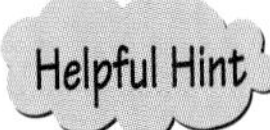

Slope of 0 and undefined slope are not the same. Vertical lines have undefined slope, while horizontal lines have a slope of 0.

Here is a general review of slope.
Slope m of the line through (x_1, y_1) and (x_2, y_2) is given by the equation

$$m = \frac{y_2 - y_1}{x_2 - x_1}.$$

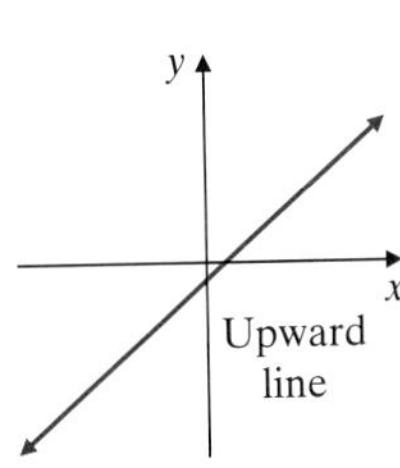

Positive slope: $m > 0$

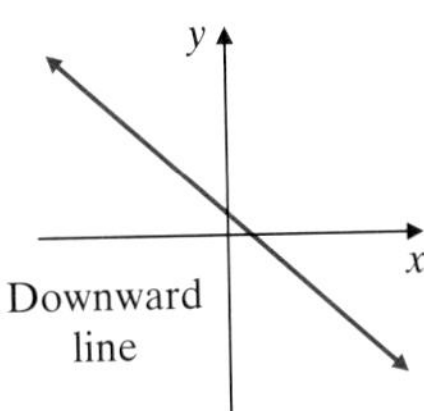

Negative slope: $m < 0$

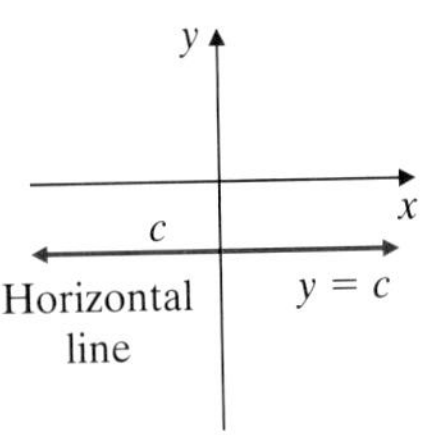

Zero slope: $m = 0$

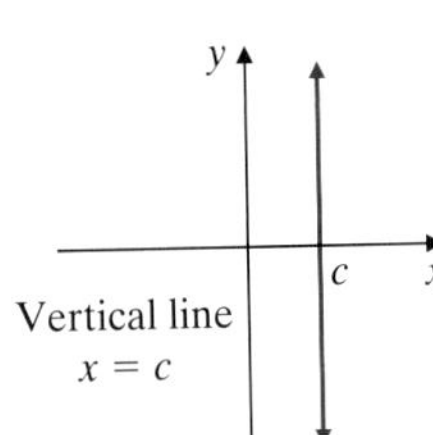

No slope or undefined slope

D Slopes of Parallel and Perpendicular Lines

Two lines in the same plane are **parallel** if they do not intersect. Slopes of lines can help us determine whether lines are parallel. Since parallel lines have the same steepness, it follows that they have the same slope.

For example, the graphs of

$$y = -2x + 4$$

and

$$y = -2x - 3$$

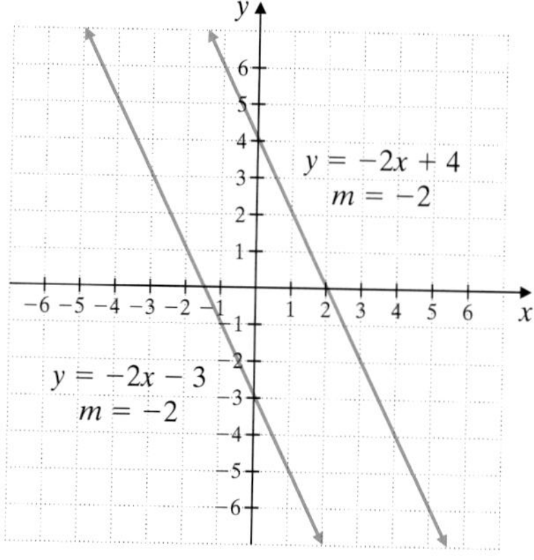

are shown. These lines have the same slope, −2. They also have different y-intercepts, so the lines are parallel. (If the y-intercepts were the same also, the lines would be the same.)

Parallel Lines

Nonvertical parallel lines have the same slope and different y-intercepts.

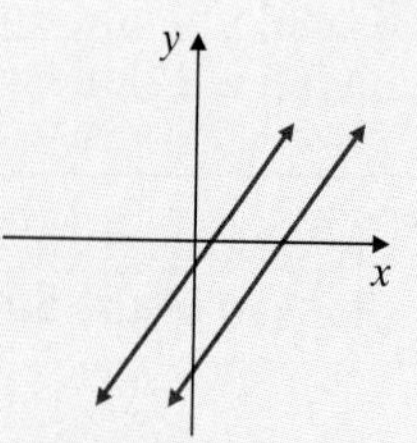

Two lines are **perpendicular** if they lie in the same plane and meet at a 90° (right) angle. How do the slopes of perpendicular lines compare? The product of the slopes of two perpendicular lines is −1.

For example, the graphs of

$$y = 4x + 1$$

and

$$y = -\frac{1}{4}x - 3$$

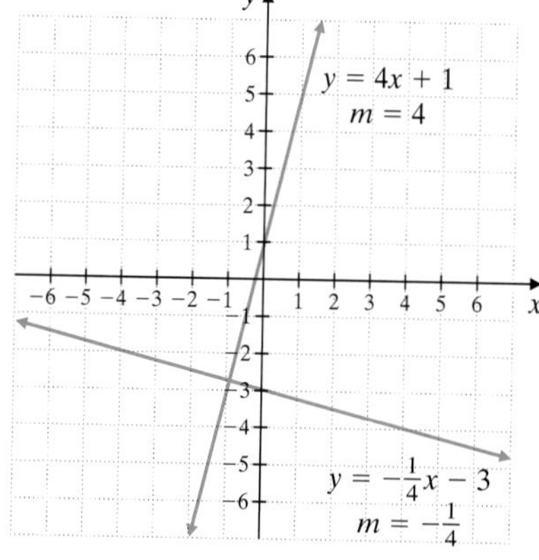

are shown. The slopes of the lines are 4 and $-\frac{1}{4}$. Their product is $4\left(-\frac{1}{4}\right) = -1$, so the lines are perpendicular.

Perpendicular Lines

If the product of the slopes of two lines is -1, then the lines are perpendicular. (Two nonvertical lines are perpendicular if the slope of one is the negative reciprocal of the slope of the other.)

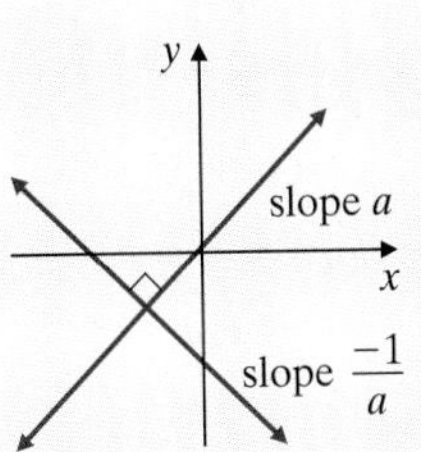

Helpful Hint

Here are examples of numbers that are negative (opposite) reciprocals.

Number	Negative Reciprocal	Their product is -1.
$\frac{2}{3}$	$-\frac{3}{2}$	$\frac{2}{3} \cdot -\frac{3}{2} = -\frac{6}{6} = -1$
-5 or $-\frac{5}{1}$	$\frac{1}{5}$	$-5 \cdot \frac{1}{5} = -\frac{5}{5} = -1$

Helpful Hint

Here are a few important points about vertical and horizontal lines.

- Two distinct vertical lines are parallel.
- Two distinct horizontal lines are parallel.
- A horizontal line and a vertical line are always perpendicular.

△ EXAMPLE 6

Determine whether each pair of lines is parallel, perpendicular, or neither.

a. $y = -\frac{1}{5}x + 1$
$2x + 10y = 3$

b. $x + y = 3$
$-x + y = 4$

c. $3x + y = 5$
$2x + 3y = 6$

Solution:

a. The slope of the line $y = -\frac{1}{5}x + 1$ is $-\frac{1}{5}$. We find the slope of the second line by solving its equation for y.

$$
\begin{aligned}
2x + 10y &= 3 \\
10y &= -2x + 3 && \text{Subtract } 2x \text{ from both sides.} \\
y &= \frac{-2}{10}x + \frac{3}{10} && \text{Divide both sides by 10.} \\
y &= -\frac{1}{5}x + \frac{3}{10} && \text{Simplify.}
\end{aligned}
$$

The slope of this line is $-\frac{1}{5}$ also. Since the lines have the same slope and different y-intercepts, they are parallel, as shown on the next page.

Practice Problem 6 △

Determine whether each pair of lines is parallel, perpendicular, or neither.

a. $x + y = 5$
$2x + y = 5$

b. $5y = 2x - 3$
$5x + 2y = 1$

c. $y = 2x + 1$
$4x - 2y = 8$

Answers

6. a. neither **b.** perpendicular **c.** parallel

b. To find each slope, we solve each equation for y.

$$x + y = 3 \qquad\qquad -x + y = 4$$
$$y = -x + 3 \qquad\qquad y = x + 4$$

The slope is -1. The slope is 1.

The slopes are not the same, so the lines are not parallel. Next we check the product of the slopes: $(-1)(1) = -1$. Since the product is -1, the lines are perpendicular, as shown in the figure.

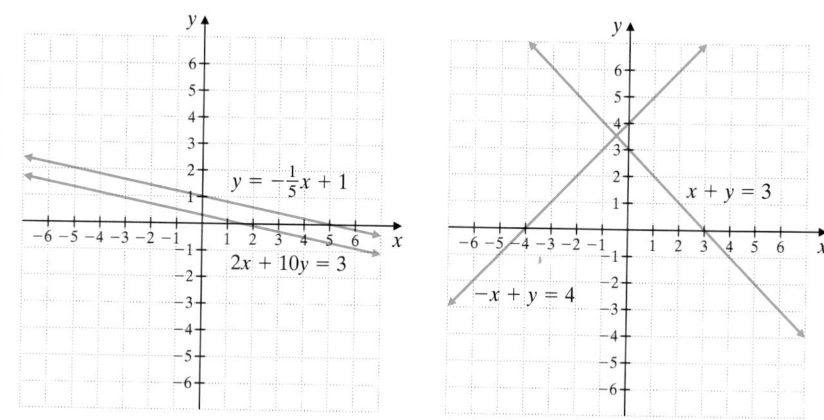

c. We solve each equation for y to find each slope. The slopes are -3 and $-\frac{2}{3}$. The slopes are not the same and their product is not -1. Thus, the lines are neither parallel nor perpendicular. ●

Concept Check

Write the equations of three parallel lines.

Try the Concept Check in the margin.

E Solving Problems of Slope

There are many real-world applications of slope. For example, the pitch of a roof used by builders and architects is its slope. The pitch of the roof is $\frac{7}{10}\left(\frac{\text{rise}}{\text{run}}\right)$. This means that the roof rises vertically 7 feet for every horizontal 10 feet.

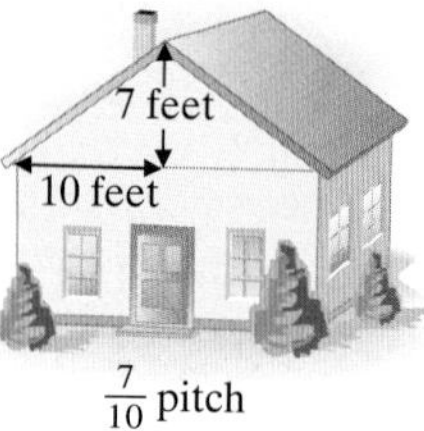

$\frac{7}{10}$ pitch

The grade of a road is its slope written as a percent. A 7% grade, as shown below, means that the road rises (or falls) 7 feet for every horizontal 100 feet. $\left(\text{Recall that } 7\% = \frac{7}{100}.\right)$

$\frac{7}{100}$ = 7% grade

7 feet

100 feet

Concept Check: for example, $y = 2x - 3, y = 2x - 1, y = 2x$

EXAMPLE 7 Finding the Grade of a Road

At one part of the road to the summit of Pike's Peak, the road rises 15 feet for a horizontal distance of 250 feet. Find the grade of the road.

Solution: Recall that the grade of a road is its slope written as a percent.

$$\text{grade} = \frac{\text{rise}}{\text{run}} = \frac{15}{250} = 0.06 = 6\%$$

15 feet

250 feet

The grade is 6%.

Slope can also be interpreted as a rate of change. In other words, slope tells us how fast y is changing with respect to x.

EXAMPLE 8 Finding the Slope of a Line

The following graph shows the cost y (in cents) of an in-state long-distance telephone call in Massachusetts where x is the length of the call in minutes. Find the slope of the line and attach the proper units for the rate of change. Then write a sentence explaining the meaning of slope in this application.

Solution: Use $(2, 48)$ and $(5, 81)$ to calculate slope.

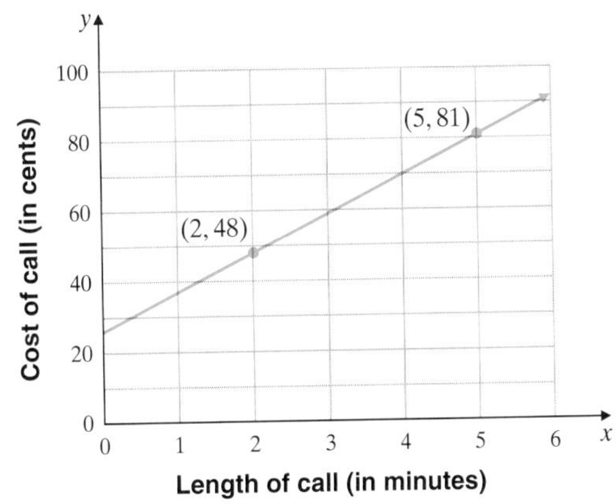

$$m = \frac{81 - 48}{5 - 2} = \frac{33}{3} = \frac{11}{1}\ \frac{\text{cents}}{\text{minute}}$$

This means that the rate of change of a phone call is 11 cents per 1 minute or the cost of the phone call increases 11 cents per minute.

Practice Problem 7

Find the grade of the road shown.

Practice Problem 8

Find the slope of the line and write the slope as a rate of change. This graph represents annual food and drink sales y (in billions of dollars) for year x.

U.S. Restaurant Food & Drink Sales

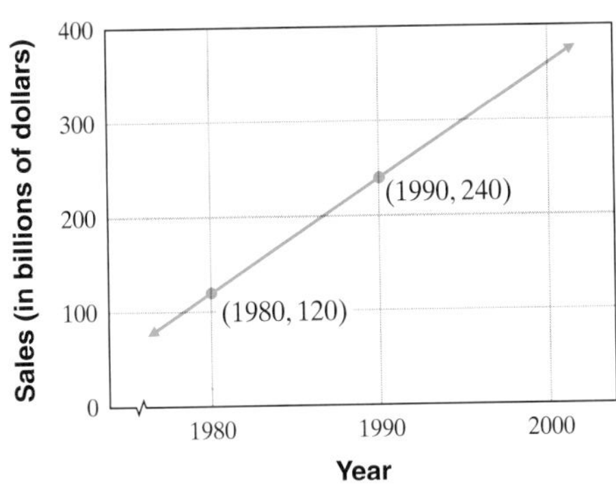

Source: National Restaurant Assn.

Answers

7. 15% **8.** $m = 12$; Each year the sales of food and drink from restaurants increases by $12 billion.

GRAPHING CALCULATOR EXPLORATIONS

It is possible to use a graphing calculator and sketch the graph of more than one equation on the same set of axes. This feature can be used to see parallel lines with the same slope. For example, graph the equations $y = \frac{2}{5}x$, $y = \frac{2}{5}x + 7$, and $y = \frac{2}{5}x - 4$ on the same set of axes. To do so, press the Y= key and enter the equations on the first three lines.

$$Y_1 = \left(\frac{2}{5}\right)x$$

$$Y_2 = \left(\frac{2}{5}\right)x + 7$$

$$Y_3 = \left(\frac{2}{5}\right)x - 4$$

The displayed equations should look like:

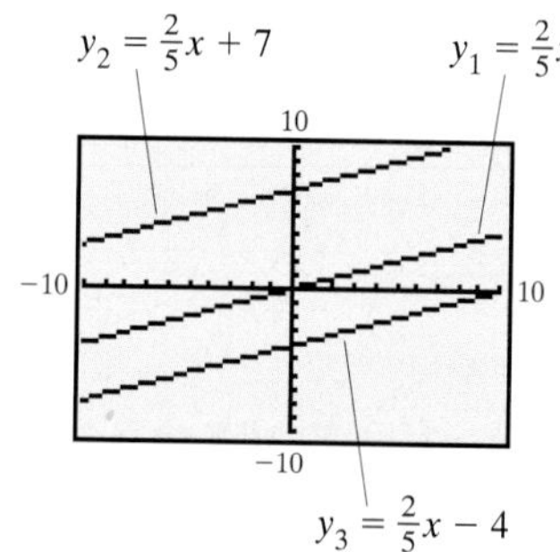

These lines are parallel as expected since they all have a slope of $\frac{2}{5}$. The graph of $y = \frac{2}{5}x + 7$ is the graph of $y = \frac{2}{5}x$ moved 7 units upward with a y-intercept of 7. Also, the graph of $y = \frac{2}{5}x - 4$ is the graph of $y = \frac{2}{5}x$ moved 4 units downward with a y-intercept of −4.

Graph the parallel lines on the same set of axes. Describe the similarities and differences in their graphs.

1. $y = 3.8x$, $y = 3.8x - 3$, $y = 3.8x + 9$

2. $y = -4.9x$, $y = -4.9x + 1$, $y = -4.9x + 8$

3. $y = \frac{1}{4}x$, $y = \frac{1}{4}x + 5$, $y = \frac{1}{4}x - 8$

4. $y = -\frac{3}{4}x$, $y = -\frac{3}{4}x - 5$, $y = -\frac{3}{4}x + 6$

Name ______________________ Section ________ Date ____________

Mental Math

Decide whether a line with the given slope is upward-sloping, downward-sloping, horizontal, or vertical.

1. $m = \frac{7}{6}$ **2.** $m = -3$ **3.** $m = 0$ **4.** m is undefined

EXERCISE SET 13.4

Use the points shown on each graph to find the slope of each line. See Examples 1 and 2.

1.

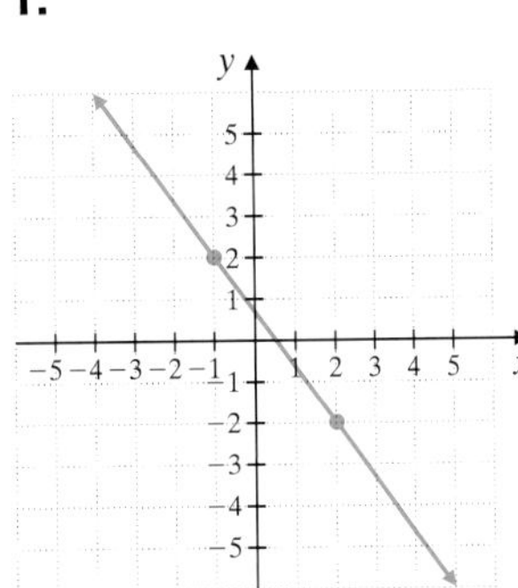

2.

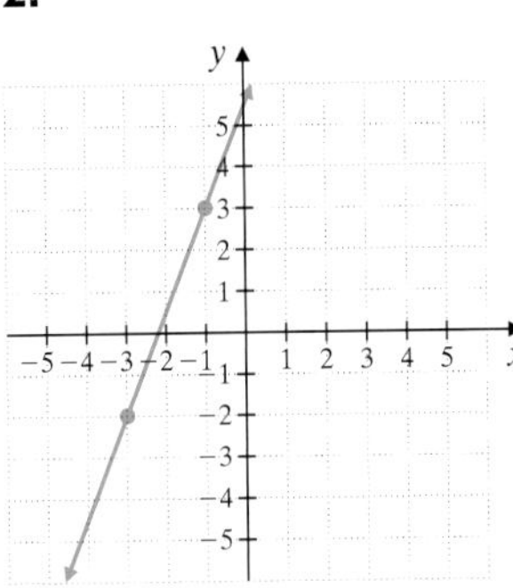

3.

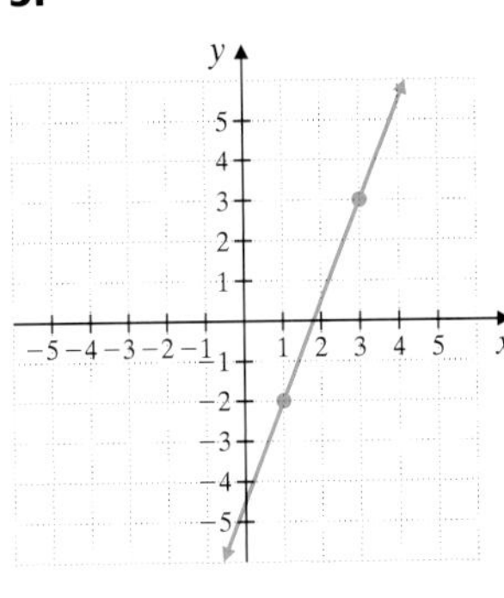

4.

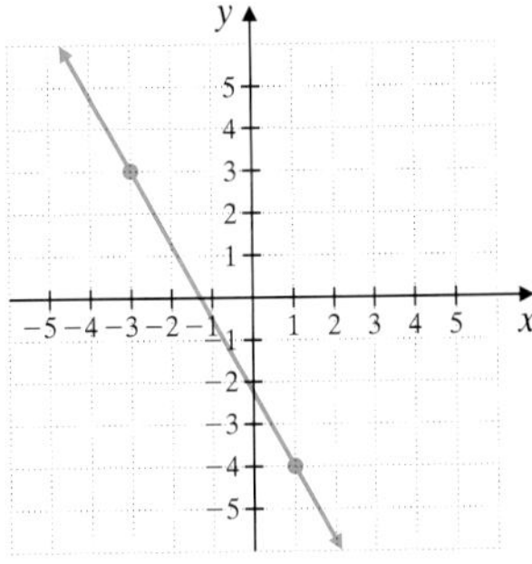

Find the slope of the line that passes through the given points. See Examples 1 and 2.

5. $(0, 0)$ and $(7, 8)$

6. $(-1, 5)$ and $(0, 0)$

7. $(-1, 5)$ and $(6, -2)$

8. $(-1, 9)$ and $(-3, 4)$

9. $(1, 4)$ and $(5, 3)$

10. $(3, 1)$ and $(2, 6)$

11. $(-2, 8)$ and $(1, 6)$

12. $(4, -3)$ and $(2, 2)$

13. $(5, 1)$ and $(-2, 1)$

14. $(5, 4)$ and $(5, 0)$

For each graph, determine which line has the larger slope.

15.

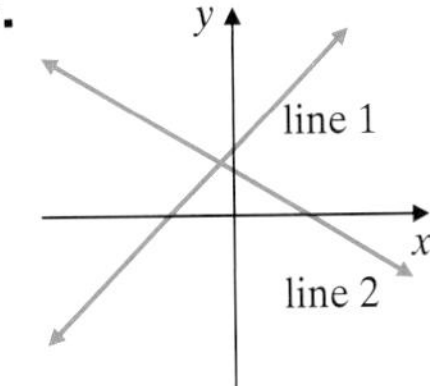

16.

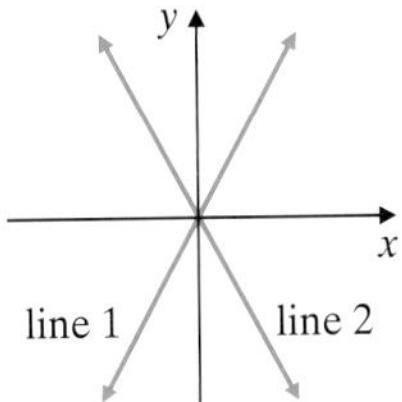

17.

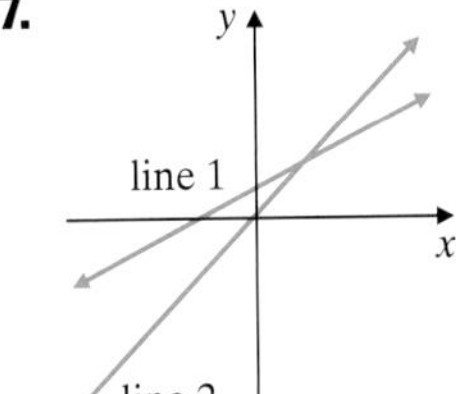

18.

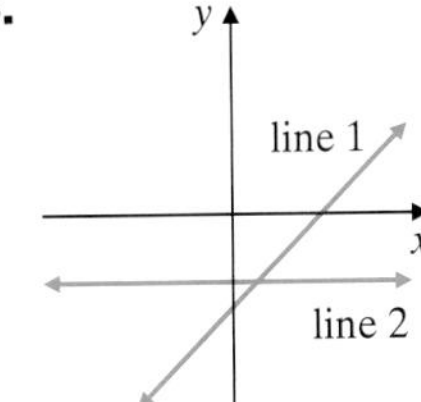

B *Find the slope of each line. See Example 3.*

19. $y = 5x - 2$

20. $y = -2x + 6$

21. $2x + y = 7$

22. $-5x + y = 10$

23. $2x - 3y = 10$

24. $-3x - 4y = 6$

25. $x = 2y$

26. $x = -4y$

C *Find the slope of each line. See Examples 4 and 5.*

27.

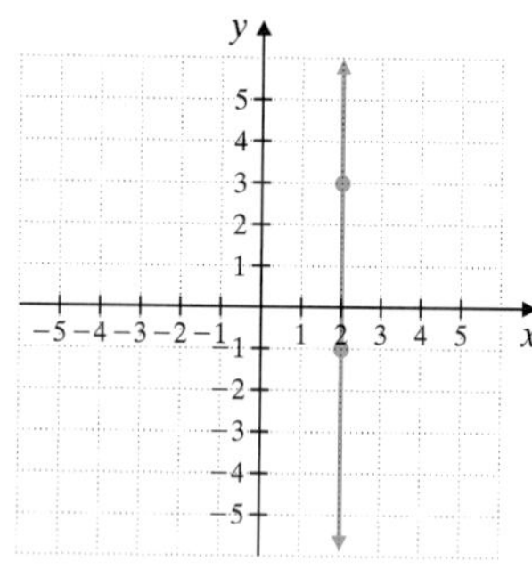

28.

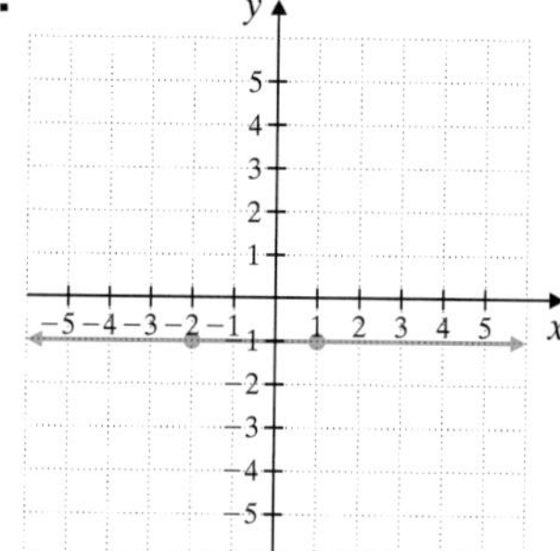

29. $x = 1$

30. $y = -2$

31. $y = -3$

32. $x = 5$

△ **D** *Determine whether each pair of lines is parallel, perpendicular, or neither. See Example 6.*

33. $x - 3y = -6$
$3x - y = 0$

34. $-5x + y = -6$
$x + 5y = 5$

35. $10 + 3x = 5y$
$5x + 3y = 1$

36. $y = 4x - 2$
$4x + y = 5$

37. $6x = 5y + 1$
$-12x + 10y = 1$

38. $-x + 2y = -2$
$2x = 4y + 3$

△ *Find the slope of the line that is (**a**) parallel and (**b**) perpendicular to the line through each pair of points. See Example 6.*

39. $(-3, -3)$ and $(0, 0)$

40. $(6, -2)$ and $(1, 4)$

41. $(-8, -4)$ and $(3, 5)$

42. $(6, -1)$ and $(-4, -10)$

E *The pitch of a roof is its slope. Find the pitch of each roof shown. See Example 7.*

43.

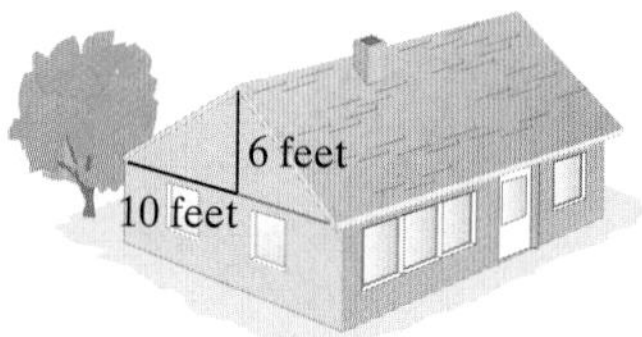

44.

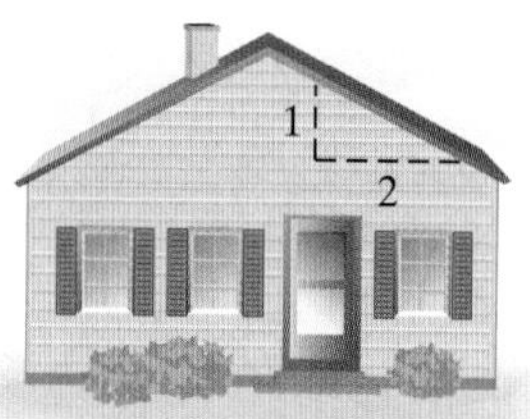

The grade of a road is its slope written as a percent. Find the grade of each road shown. See Example 7.

45.

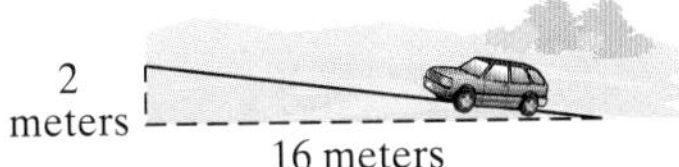

46.

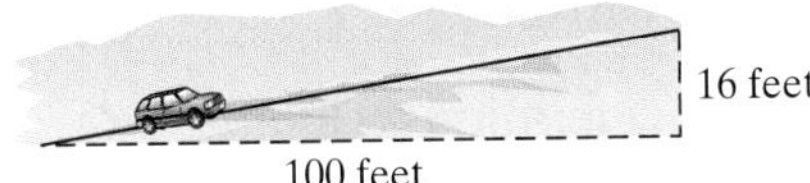

47. One of Japan's superconducting "bullet" trains is researched and tested at the Yamanashi Maglev Test Line near Otsuki City. The steepest section of the track has a rise of 2580 meters for a horizontal distance of 6450 meters. What is the grade for this section of track? (*Source:* Japan Railways Central Co.)

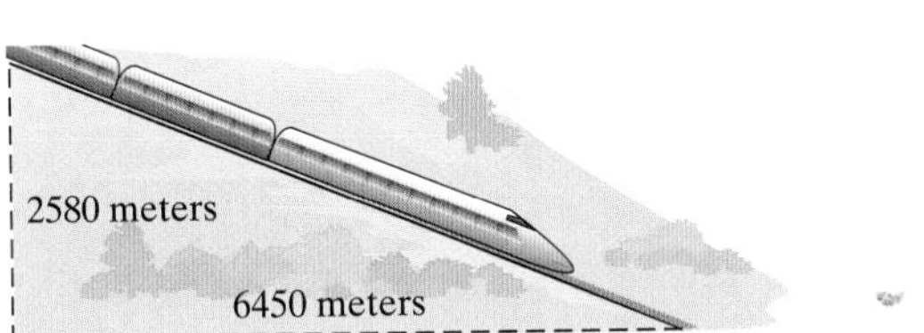

48. The steepest street is Baldwin Street in Dunedin, New Zealand. It has a maximum rise of 10 meters for a horizontal distance of 12.66 meters. Find the grade for this section of road. Round to the nearest whole percent. (*Source: The Guinness Book of Records*)

49. Professional plumbers suggest that a sewer pipe should rise 0.25 inch for every horizontal foot. Find the recommended slope for a sewer pipe. Round to the nearest hundredth.

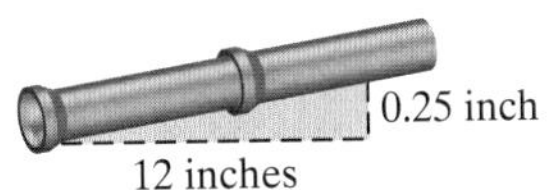

50. According to federal regulations, a wheelchair ramp should rise no more than 1 foot for a horizontal distance of 12 feet. Write the slope as a grade. Round to the nearest tenth of a percent.

Find the slope of each line and write the slope as a rate of change. Don't forget to attach the proper units. See Example 8.

51. This graph approximates the number of U.S. Internet users y (in millions) for year x.

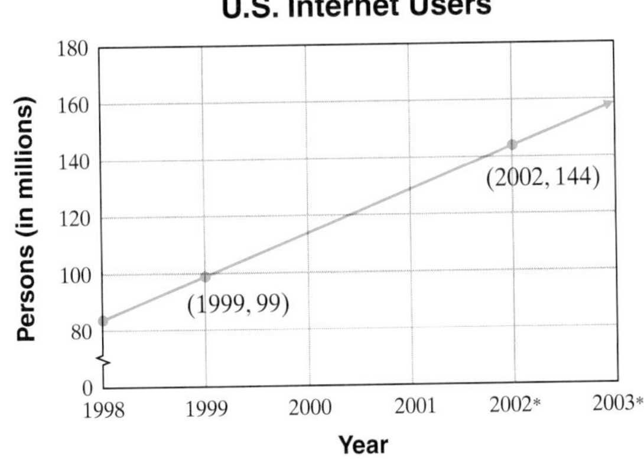

52. This graph approximates the total number of cosmetic surgeons for year x.

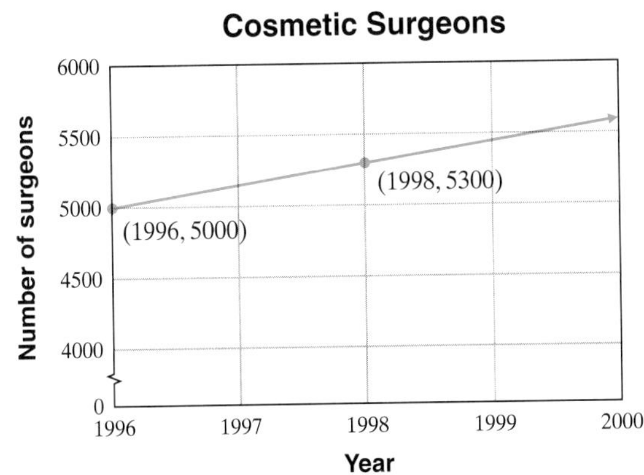

53. The graph below shows the total cost y (in dollars) of owning and operating a compact car where x is the number of miles driven.

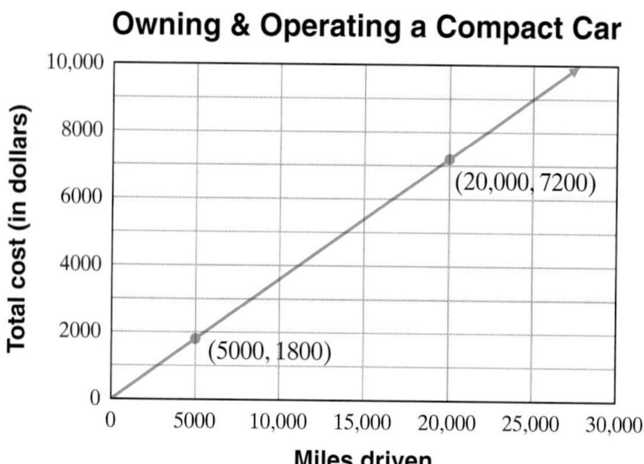

Source: Federal Highway Administration

54. The graph below shows the total cost y (in dollars) of owning and operating a full-size pickup truck, where x is the number of miles driven.

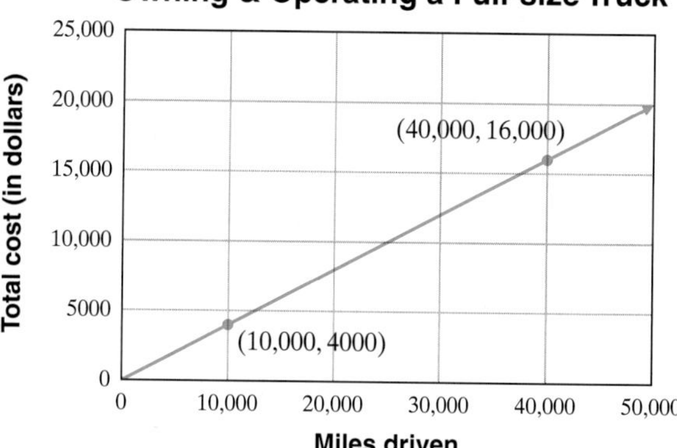

Source: Federal Highway Administration

Review and Preview

Solve each equation for y. See Section 9.5.

55. $y - (-6) = 2(x - 4)$ **56.** $y - 7 = -9(x - 6)$ **57.** $y - 1 = -6(x - (-2))$ **58.** $y - (-3) = 4(x - (-5))$

Combining Concepts

Match each line with its slope.

A. $m = 0$ **B.** undefined slope **C.** $m = 3$

D. $m = 1$ **E.** $m = -\frac{1}{2}$ **F.** $m = -\frac{3}{4}$

59.

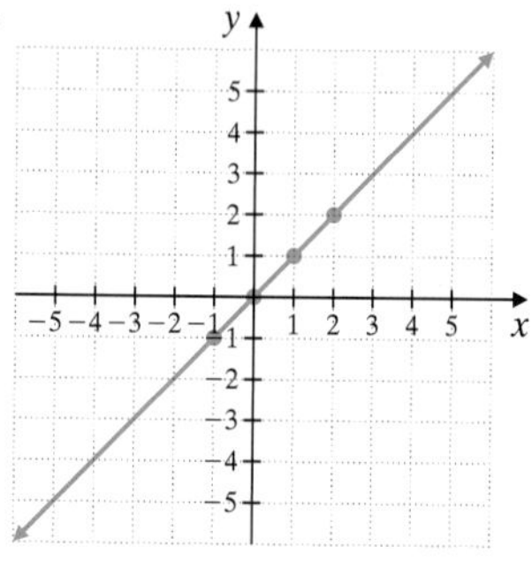

60.

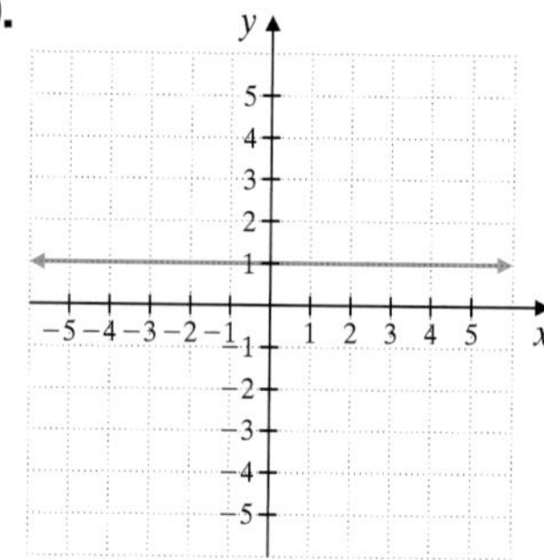

61.

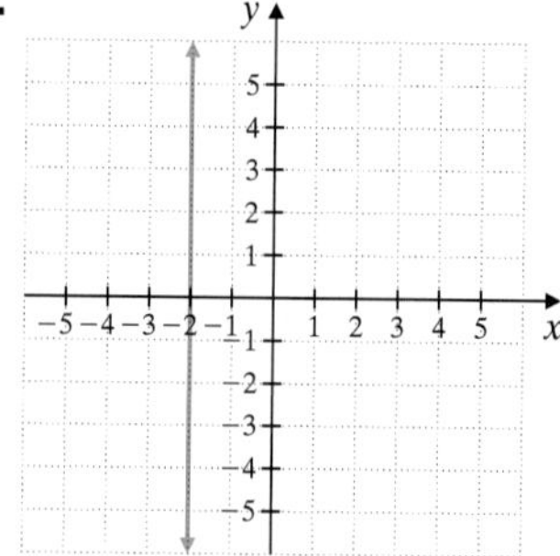

62.

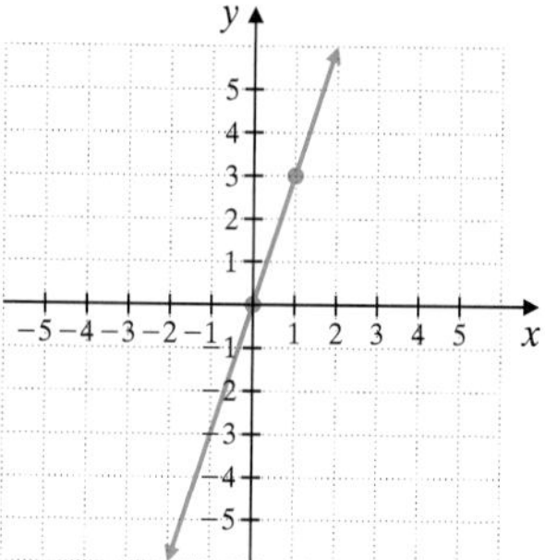

63.

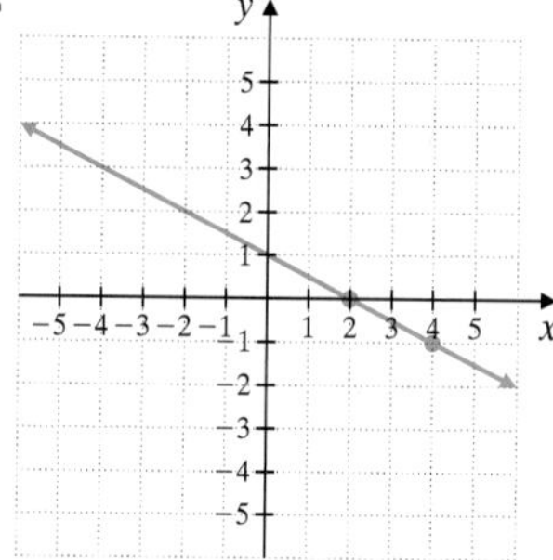

64.

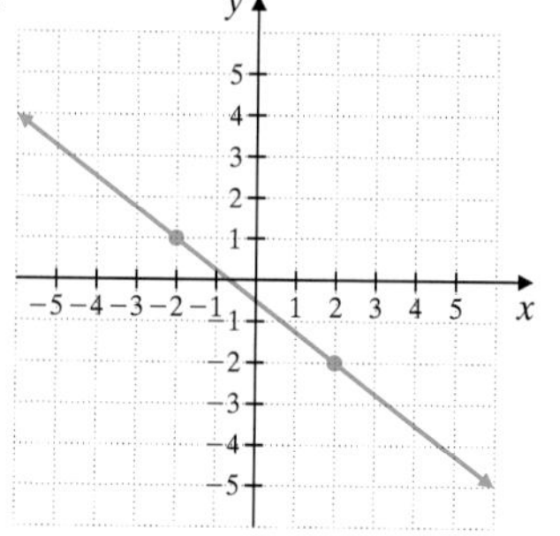

The following line graph shows the average fuel economy (in miles per gallon) by passenger automobiles produced during each of the model years shown. Use this graph to answer Exercises 65 through 70.

65. What was the average fuel economy (in miles per gallon) for automobiles produced during 1995?

66. Find the decrease in average fuel economy for automobiles for the years 1998 to 2000.

67. During which of the model years shown was average fuel economy the lowest?
What was the average fuel economy for that year?

68. During which of the model years shown was average fuel economy the highest?
What was the average fuel economy for that year?

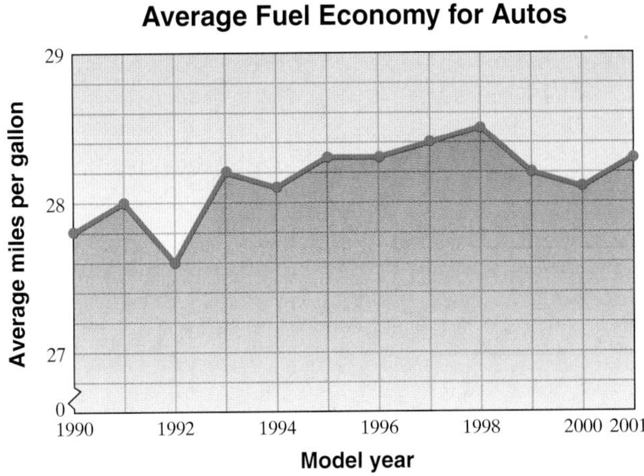

Source: U.S. Environmental Protection Agency, Office of Transportation and Air Quality

69. What line segment has the greatest slope?

70. What line segment has the smallest slope?

71. Find x so that the pitch of the roof is $\frac{1}{3}$.

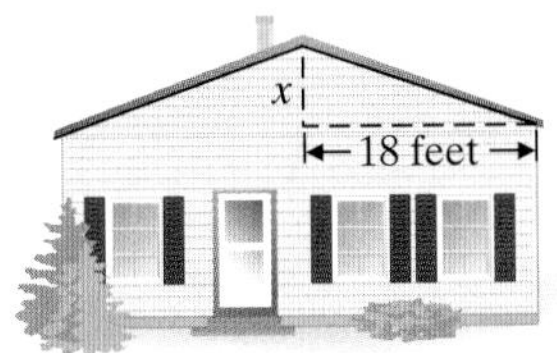

72. Find x so that the pitch of the roof is $\frac{2}{5}$.

73. The average price of an acre of U.S. farmland was \$782 in 1994. In 2001, the price of an acre rose to approximately \$1,132. (*Source:* National Agricultural Statistics Service)

a. Write two ordered pairs of the form (year, price of acre).

b. Find the slope of the line through the two points.

c. Write a sentence explaining the meaning of the slope as a rate of change.

74. There were approximately 13,290 kidney transplants performed in the United States in 2000. In 2002, the number of kidney transplants performed in the United States rose to 14,774. (*Source:* Organ Procurement and Transplantation Network)

a. Write two ordered pairs of the form (year, number of kidney transplants).

b. Find the slope of the line between the two points.

c. Write a sentence explaining the meaning of the slope as a rate of change.

75. In the years 1998 through 2000, the number of admissions to the movie theater in the U.S. and Canada can be modeled by the linear equation $y = -30x + 1485$ where x is years after 1998 and y is admissions in millions. (*Source:* Motion Picture Assn. of America)

a. Find the y-intercept of this line.

b. Write a sentence explaining the meaning of this intercept.

c. Find the slope of this line.

d. Write a sentence explaining the meaning of the slope as a rate of change.

76. The table below shows weight and fuel economy information for selected Dodge and Ford 2001 model year passenger vehicles. The linear equation $y = -0.003x + 34.7$ models the relationship between weight and fuel economy for these vehicles, where x = weight in pounds and y = combined city/highway fuel economy in miles per gallon. (*Sources*: Automotive Information Center, U.S. Environmental Protection Agency)

2001 Model	Weight (in pounds)	Combined Fuel Economy (in miles per gallon)
DODGE		
Caravan SE	3908	22
Grand Caravan ES	4252	20
Intrepid Sedan	3480	23
Dodge Neon	2635	27
Stratus Sedan SE	3227	23
FORD		
Focus SE Sedan	2564	31
Mustang Coupe	3114	23
Taurus SE Sedan	3354	22
Crown Victoria Sedan	3946	20
Escape XLS	2991	25
Expedition XLT	4891	18
Excursion XLT	6650	15
Windstar	4058	20

a. Find the y-intercept of this line.

b. Find the slope of this line.

c. Write a sentence explaining the meaning of the slope as a rate of change.

△ **77.** Show that a triangle with vertices at the points $(1, 1)$, $(-4, 4)$, and $(-3, 0)$ is a right triangle.

△ **78.** Show that the quadrilateral with vertices $(1, 3)$, $(2, 1)$, $(-4, 0)$, and $(-3, -2)$ is a parallelogram.

Find the slope of the line through the given points.

79. $(2.1, 6.7)$ and $(-8.3, 9.3)$

80. $(-3.8, 1.2)$ and $(-2.2, 4.5)$

81. $(2.3, 0.2)$ and $(7.9, 5.1)$

82. $(14.3, -10.1)$ and $(9.8, -2.9)$

83. The graph of $y = -\frac{1}{3}x + 2$ has a slope of $-\frac{1}{3}$. The graph of $y = -2x + 2$ has a slope of -2. The graph of $y = -4x + 2$ has a slope of -4. Graph all three equations on a single coordinate system. As the absolute value of the slope becomes larger, how does the steepness of the line change?

84. The graph of $y = \frac{1}{2}x$ has a slope of $\frac{1}{2}$. The graph of $y = 3x$ has a slope of 3. The graph of $y = 5x$ has a slope of 5. Graph all three equations on a single coordinate system. As slope becomes larger, how does the steepness of the line change?

13.5 Equations of Lines

OBJECTIVES

A. Use the slope-intercept form to write an equation of a line.
B. Use the slope-intercept form to graph a linear equation.
C. Use the point-slope form to find an equation of a line given its slope and a point of the line.
D. Use the point-slope form to find an equation of a line given two points of the line.
E. Use the point-slope form to solve problems.

We know that when a linear equation is solved for y, the coefficient of x is the slope of the line. For example, the slope of the line whose equation is $y = 3x + 1$ is 3. In this equation, $y = 3x + 1$, what does 1 represent? To find out, let $x = 0$ and watch what happens.

$$y = 3x + 1$$
$$y = 3 \cdot 0 + 1 \qquad \text{Let } x = 0.$$
$$y = 1$$

We now have the ordered pair $(0, 1)$, which means that 1 is the y-intercept.

This is true in general. To see this, let $x = 0$ and solve for y in $y = mx + b$.

$$y = m \cdot 0 + b \qquad \text{Let } x = 0.$$
$$y = b$$

We obtain the ordered pair $(0, b)$, which means that point is the y-intercept.

The form $y = mx + b$ is appropriately called the **slope-intercept form** of a linear equation. (m: slope; b: y-intercept)

Slope-Intercept Form

When a linear equation in two variables is written in slope-intercept form,

$$y = mx + b$$

then m is the slope of the line and $(0, b)$ is the y-intercept of the line.

A Using the Slope-Intercept Form to Write an Equation

The slope-intercept form can be used to write the equation of a line when we know its slope and y-intercept.

EXAMPLE 1

Find an equation of the line with y-intercept $(0, -3)$ and slope of $\frac{1}{4}$.

Solution: We are given the slope and the y-intercept. We let $m = \frac{1}{4}$ and $b = -3$ and write the equation in slope-intercept form, $y = mx + b$.

$$y = mx + b$$
$$y = \frac{1}{4}x + (-3) \qquad \text{Let } m = \frac{1}{4} \text{ and } b = -3.$$
$$y = \frac{1}{4}x - 3 \qquad \text{Simplify.}$$

Practice Problem 1

Find an equation of the line with y-intercept -2 and slope of $\frac{3}{5}$.

B Using the Slope-Intercept Form to Graph an Equation

We also can use the slope-intercept form of the equation of a line to graph a linear equation.

Answer

1. $y = \frac{3}{5}x - 2$

Practice Problem 2

Graph the equation $y = \frac{2}{3}x - 4$.

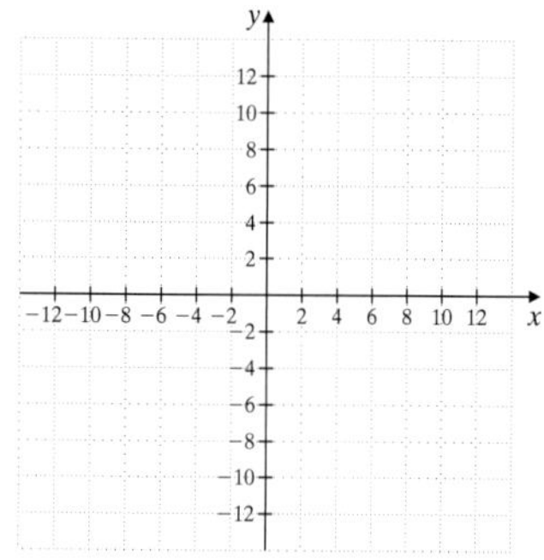

Practice Problem 3

Use the slope-intercept form to graph $3x + y = 2$.

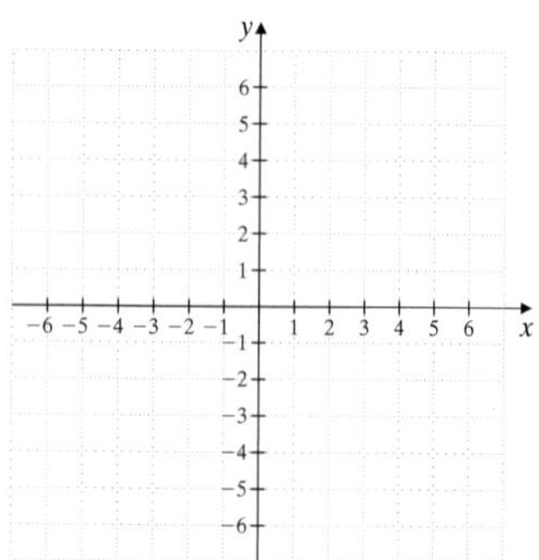

Answers

2.

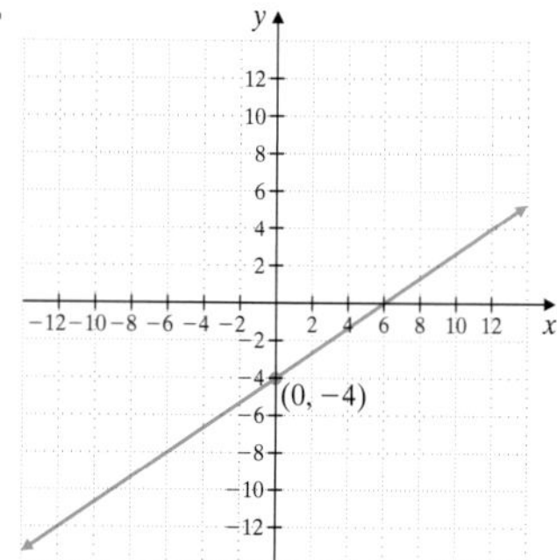

3.

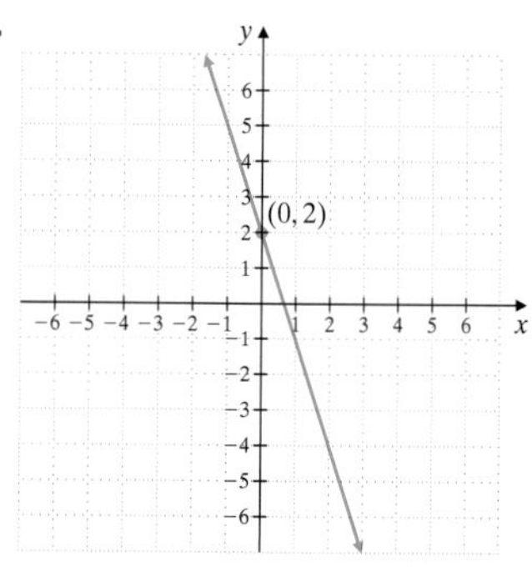

EXAMPLE 2 Use the slope-intercept form to graph the equation

$$y = \frac{3}{5}x - 2$$

Solution: Since the equation $y = \frac{3}{5}x - 2$ is written in slope-intercept form $y = mx + b$, the slope of its graph is $\frac{3}{5}$ and the y-intercept is $(0, -2)$. To graph this equation, we begin by plotting the point $(0, -2)$. From this point, we can find another point of the graph by using the slope $\frac{3}{5}$ and recalling that slope is $\frac{\text{rise}}{\text{run}}$. We start at the y-intercept and move 3 units up since the numerator of the slope is 3; then we move 5 units to the right since the denominator of the slope is 5. We stop at the point $(5, 1)$. The line through $(0, -2)$ and $(5, 1)$ is the graph of $y = \frac{3}{5}x - 2$.

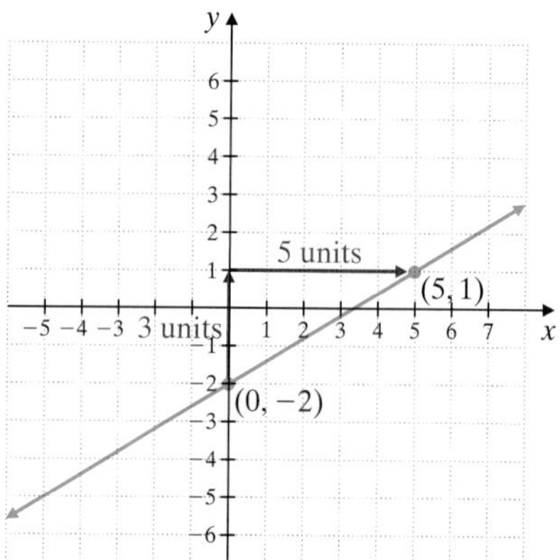

EXAMPLE 3

Use the slope-intercept form to graph the equation $4x + y = 1$.

Solution: First we write the given equation in slope-intercept form.

$$4x + y = 1$$
$$y = -4x + 1$$

The graph of this equation will have slope -4 and y-intercept $(0, 1)$. To graph this line, we first plot the point $(0, 1)$. To find another point of the graph, we use the slope -4, which can be written as $\frac{-4}{1}$ $\left(\frac{4}{-1}\text{ could also be used}\right)$. We start at the point $(0, 1)$ and move 4 units down (since the numerator of the slope is -4), and then 1 unit to the right (since the denominator of the slope is 1).

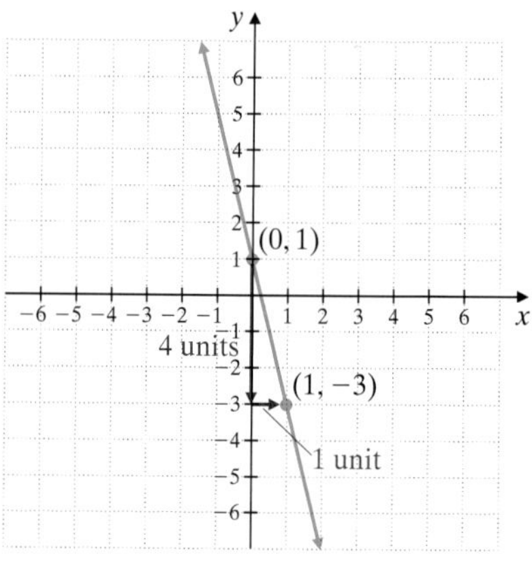

We arrive at the point $(1, -3)$. The line through $(0, 1)$ and $(1, -3)$ is the graph of $4x + y = 1$.

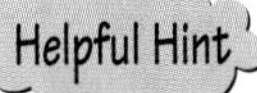

In Example 3, if we interpret the slope of -4 as $\frac{4}{-1}$, we arrive at $(-1, 5)$ for a second point. Notice that this point is also on the line.

C Writing an Equation Given Its Slope and a Point

Thus far, we have seen that we can write an equation of a line if we know its slope and y-intercept. We can also write an equation of a line if we know its slope and any point on the line. To see how we do this, let m represent slope and (x_1, y_1) represent the point on the line. Then if (x, y) is any other point of the line, we have that

$$\frac{y - y_1}{x - x_1} = m$$

$$y - y_1 = m(x - x_1) \quad \text{Multiply both sides by } (x - x_1).$$

This is the **point-slope form** of the equation of a line.

Point-Slope Form of the Equation of a Line

The point-slope form of the equation of a line is $y - y_1 = m(x - x_1)$, where m is the slope of the line and (x_1, y_1) is a point on the line.

EXAMPLE 4

Find an equation of the line passing through $(-1, 5)$ with slope -2. Write the equation in the form $Ax + By = C$.

Solution: Since the slope and a point on the line are given, we use point-slope form $y - y_1 = m(x - x_1)$ to write the equation. Let $m = -2$ and $(-1, 5) = (x_1, y_1)$.

$$y - y_1 = m(x - x_1)$$
$$y - 5 = -2[x - (-1)] \quad \text{Let } m = -2 \text{ and } (x_1, y_1) = (-1, 5).$$
$$y - 5 = -2(x + 1) \quad \text{Simplify.}$$
$$y - 5 = -2x - 2 \quad \text{Use the distributive property.}$$
$$y = -2x + 3 \quad \text{Add 5 to both sides.}$$
$$2x + y = 3 \quad \text{Add } 2x \text{ to both sides.}$$

Practice Problem 4

Find an equation of the line that passes through $(2, -4)$ with slope -3. Write the equation in the form $Ax + By = C$.

D Writing an Equation Given Two Points

We can also find the equation of a line when we are given any two points of the line.

EXAMPLE 5 Find an equation of the line through $(2, 5)$ and $(-3, 4)$. Write the equation in the form $Ax + By = C$.

Solution: First we find the slope of the line. Let (x_1, y_1) be $(2, 5)$ and (x_2, y_2) be $(-3, 4)$.

$$m = \frac{y_2 - y_1}{x_2 - x_1} = \frac{4 - 5}{-3 - 2} = \frac{-1}{-5} = \frac{1}{5}$$

Practice Problem 5

Find an equation of the line through $(1, 3)$ and $(5, -2)$. Write the equation in the form $Ax + By = C$.

Answers

4. $3x + y = 2$ **5.** $5x + 4y = 17$

Next we use the slope $\frac{1}{5}$ and either one of the given points to write the equation in point-slope form. We use $(2, 5)$. Let $x_1 = 2$, $y_1 = 5$, and $m = \frac{1}{5}$.

$$y - y_1 = m(x - x_1) \quad \text{Use point-slope form.}$$

$$y - 5 = \frac{1}{5}(x - 2) \quad \text{Let } x_1 = 2, y_1 = 5, \text{ and } m = \frac{1}{5}.$$

$$5(y - 5) = 5 \cdot \frac{1}{5}(x - 2) \quad \text{Multiply both sides by 5 to clear fractions.}$$

$$5y - 25 = x - 2 \quad \text{Use the distributive property and simplify.}$$

$$-x + 5y - 25 = -2 \quad \text{Subtract } x \text{ from both sides.}$$

$$-x + 5y = 23 \quad \text{Add 25 to both sides.}$$

Helpful Hint

When you multiply both sides of the equation from Example 5, $-x + 5y = 23$ by -1, it becomes $x - 5y = -23$. Both $-x + 5y = 23$ and $x - 5y = -23$ are in the form $Ax + By = C$ and both are equations of the same line.

E Using the Point-Slope Form to Solve Problems

Problems occurring in many fields can be modeled by linear equations in two variables. The next example is from the field of marketing and shows how consumer demand of a product depends on the price of the product.

EXAMPLE 6

The Whammo Company has learned that by pricing a newly released Frisbee at \$6, sales will reach 2000 Frisbees per day. Raising the price to \$8 will cause the sales to fall to 1500 Frisbees per day.

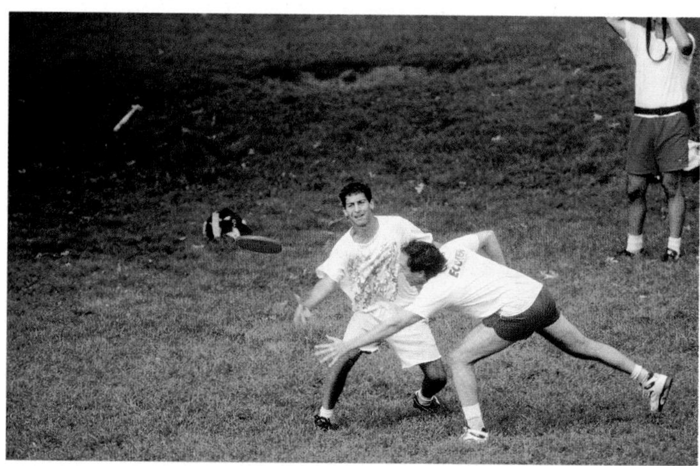

a. Assume that the relationship between sales price and number of Frisbees sold is linear and write an equation describing this relationship. Write the equation in slope-intercept form. Use ordered pairs of the form (sales price, number sold).

b. Predict the daily sales of Frisbees if the price is \$7.50.

Practice Problem 6

The Pool Entertainment Company learned that by pricing a new pool toy at \$10, local sales will reach 200 a week. Lowering the price to \$9 will cause sales to rise to 250 a week.

a. Assume that the relationship between sales price and number of toys sold is linear, and write an equation describing this relationship. Write the equation in slope-intercept form. Use ordered pairs of the form (sales price, number sold).

b. Predict the weekly sales of the toy if the price is \$7.50.

Answers

6. a. $y = -50x + 700$ **b.** 325

Solution:

a. We use the given information and write two ordered pairs. Our ordered pairs are $(6, 2000)$ and $(8, 1500)$. To use the point-slope form to write an equation, we find the slope of the line that contains these points.

$$m = \frac{2000 - 1500}{6 - 8} = \frac{500}{-2} = -250$$

Next we use the slope and either one of the points to write the equation in point-slope form. We use $(6, 2000)$.

$$y - y_1 = m(x - x_1) \quad \text{Use point-slope form.}$$
$$y - 2000 = -250(x - 6) \quad \text{Let } x_1 = 6, y_1 = 2000, \text{ and } m = -250.$$
$$y - 2000 = -250x + 1500 \quad \text{Use the distributive property.}$$
$$y = -250x + 3500 \quad \text{Write in slope-intercept form.}$$

b. To predict the sales if the price is \$7.50, we find y when $x = 7.50$.

$$y = -250x + 3500$$
$$y = -250(7.50) + 3500 \quad \text{Let } x = 7.50.$$
$$y = -1875 + 3500$$
$$y = 1625$$

If the price is \$7.50, sales will reach 1625 Frisbees per day.

We could have solved Example 6 by using ordered pairs of the form (number sold, sales price).

Here is a summary of our discussion on linear equations thus far.

Forms of Linear Equations

$Ax + By = C$	**Standard form** of a linear equation. A and B are not both 0.
$y = mx + b$	**Slope-intercept form** of a linear equation. The slope is m and the y-intercept is $(0, b)$.
$y - y_1 = m(x - x_1)$	**Point-slope form** of a linear equation. The slope is m and (x_1, y_1) is a point on the line.
$y = c$	**Horizontal line** The slope is 0 and the y-intercept is $(0, c)$.
$x = c$	**Vertical line** The slope is undefined and the x-intercept is $(c, 0)$.

Parallel and Perpendicular Lines

Nonvertical parallel lines have the same slope.
The product of the slopes of two nonvertical perpendicular lines is -1.

GRAPHING CALCULATOR EXPLORATIONS

A graphing calculator is a very useful tool for discovering patterns. To discover the change in the graph of a linear equation caused by a change in slope, try the following. Use a standard window and graph a linear equation in the form $y = mx + b$. Recall that the graph of such an equation will have slope m and y-intercept $(0, b)$.

First graph $y = x + 3$. To do so, press the Y= key and enter $Y_1 = x + 3$. Notice that this graph has slope 1 and that the y-intercept is 3. Next, on the same set of axes, graph $y = 2x + 3$ and $y = 3x + 3$ by pressing Y= and entering $Y_2 = 2x + 3$ and $Y_3 = 3x + 3$.

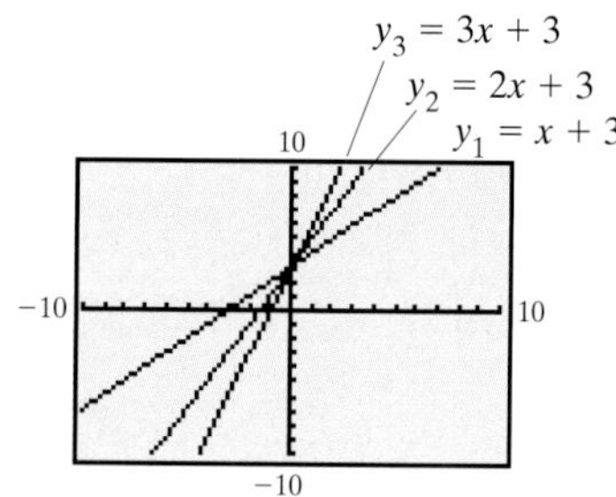

Notice the difference in the graph of each equation as the slope changes from 1 to 2 to 3. How would the graph of $y = 5x + 3$ appear? To see the change in the graph caused by a change in negative slope, try graphing $y = -x + 3$, $y = -2x + 3$, and $y = -3x + 3$ on the same set of axes.

Use a graphing calculator to graph the following equations. For each exercise, graph the first equation and use its graph to predict the appearance of the other equations. Then graph the other equations on the same set of axes and check your prediction.

1. $y = x$; $y = 6x$, $y = -6x$

2. $y = -x$; $y = -5x$, $y = -10x$

3. $y = \frac{1}{2}x + 2$; $y = \frac{3}{4}x + 2$, $y = x + 2$

4. $y = x + 1$; $y = \frac{5}{4}x + 1$, $y = \frac{5}{2}x + 1$

Name ______________________ Section __________ Date __________

Mental Math

Use the equation to identify the slope and the y-intercept of the graph of each equation.

1. $y = 2x - 1$

2. $y = -7x + 3$

3. $y = x + \frac{1}{3}$

4. $y = -x - \frac{2}{9}$

5. $y = \frac{5}{7}x - 4$

6. $y = -\frac{1}{4}x + \frac{3}{5}$

Use the equation to identify the slope and a point of the line.

7. $y - 8 = 3(x - 4)$

8. $y - 1 = 5(x - 2)$

9. $y + 3 = -2(x - 10)$

10. $y + 6 = -7(x - 2)$

11. $y = \frac{2}{5}(x + 1)$

12. $y = \frac{3}{7}(x + 4)$

EXERCISE SET 13.5

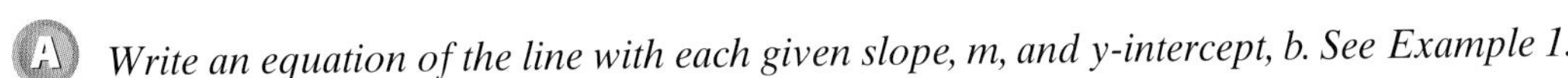

A *Write an equation of the line with each given slope, m, and y-intercept, b. See Example 1.*

1. $m = 5, b = 3$

2. $m = 2, b = \frac{3}{4}$

3. $m = \frac{2}{3}, b = 0$

4. $m = 0, b = -2$

5. $m = -\frac{1}{5}, b = \frac{1}{9}$

6. $m = -3, b = -3$

B *Use the slope-intercept form to graph each equation. See Examples 2 and 3.*

7. $y = 2x + 1$

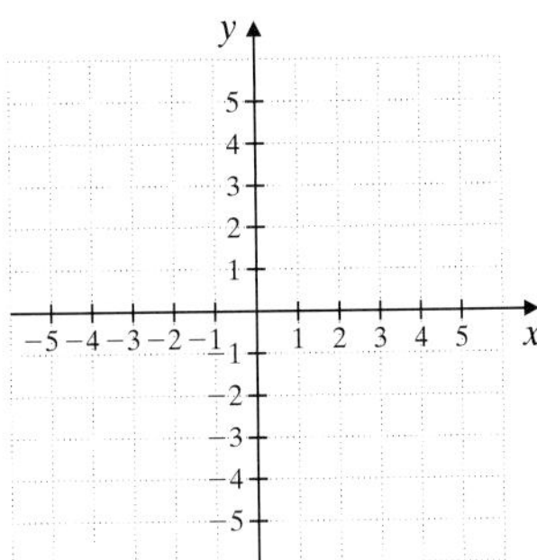

8. $y = -4x - 1$

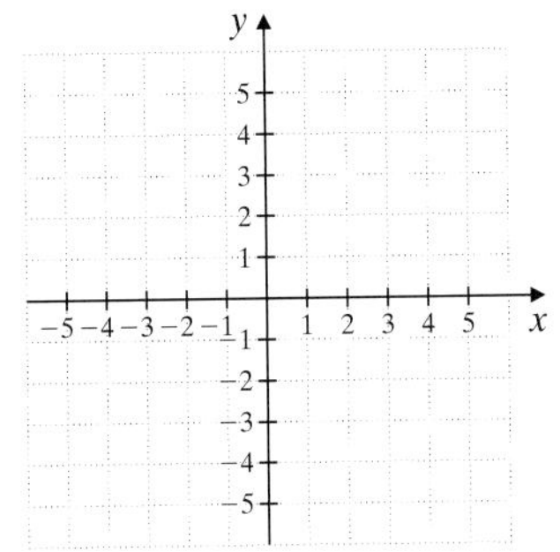

9. $y = \frac{2}{3}x + 5$

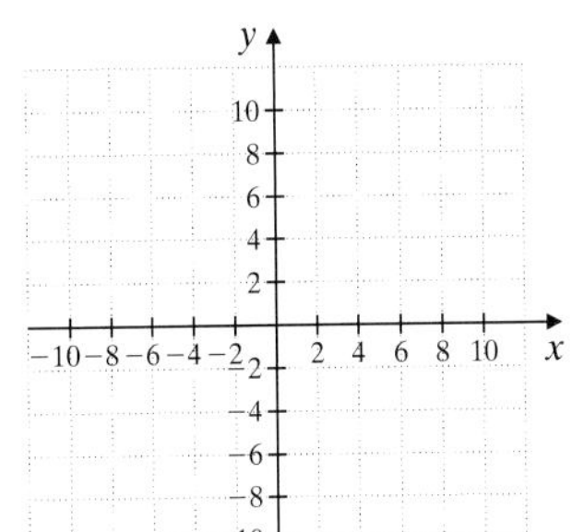

10. $y = \frac{1}{4}x - 3$

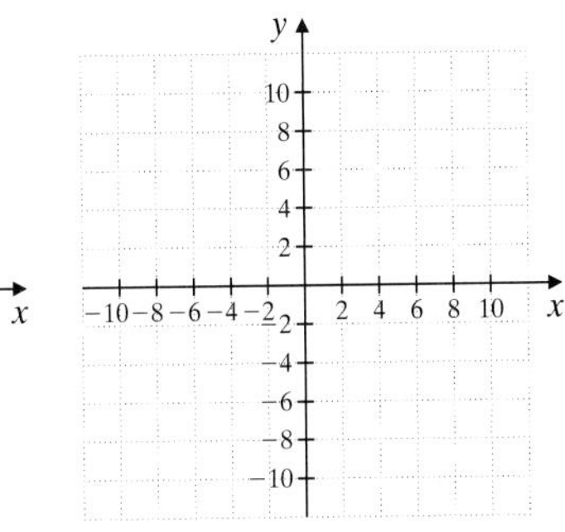

11. $y = -5x$

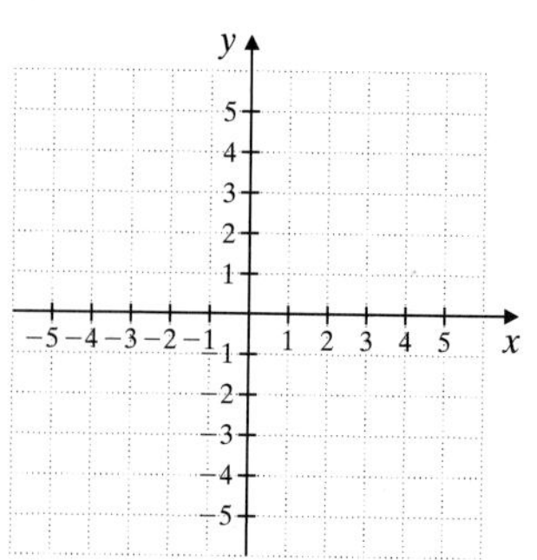

12. $y = 6x$

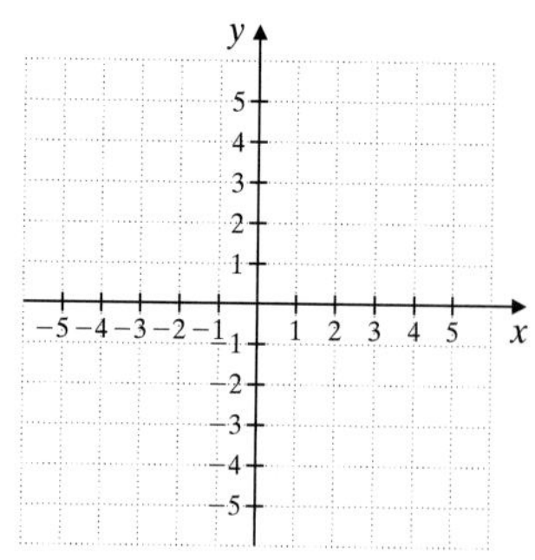

13. $4x + y = 6$

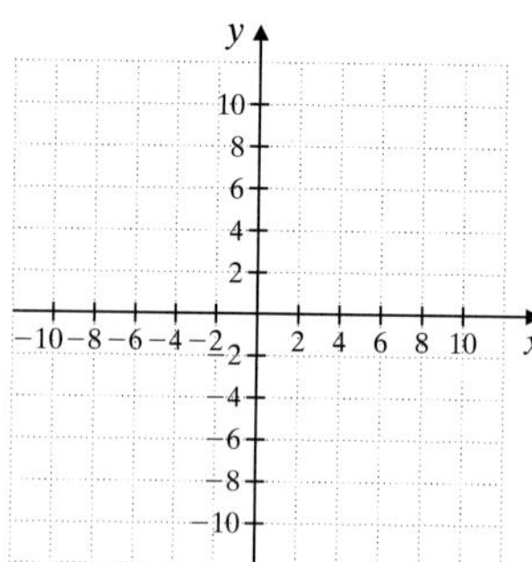

14. $-3x + y = 2$

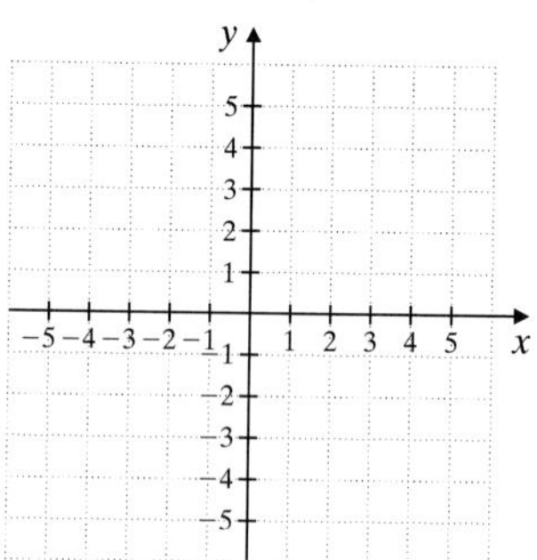

15. $4x - 7y = -14$

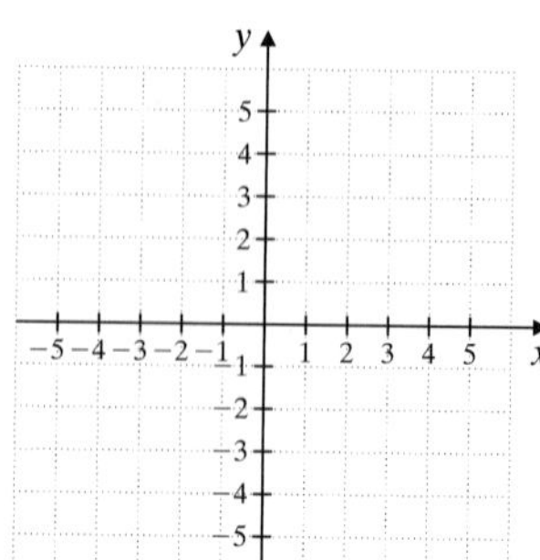

16. $3x - 4y = 4$

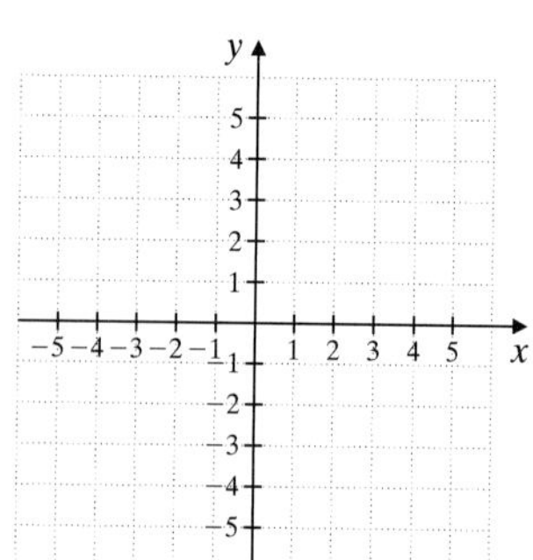

C *Find an equation of each line with the given slope that passes through the given point. Write the equation in the form $Ax + By = C$. See Example 4.*

17. $m = 6;\quad (2, 2)$

18. $m = 4;\quad (1, 3)$

19. $m = -8;\quad (-1, -5)$

20. $m = -2;\quad (-11, -12)$

21. $m = \frac{1}{2};\quad (5, -6)$

22. $m = \frac{2}{3};\quad (-8, 9)$

23. $m = -\frac{1}{2};\quad (-3, 0)$

24. $m = -\frac{1}{5};\quad (4, 0)$

D *Find an equation of the line passing through each pair of points. Write the equation in the form $Ax + By = C$. See Example 5.*

25. $(3, 2)$ and $(5, 6)$

26. $(6, 2)$ and $(8, 8)$

27. $(-1, 3)$ and $(-2, -5)$

28. $(-4, 0)$ and $(6, -1)$

29. $(2, 3)$ and $(-1, -1)$

30. $(0, 0)$ and $\left(\frac{1}{2}, \frac{1}{3}\right)$

31. $(10, 7)$ and $(7, 10)$

32. $(5, -6)$ and $(-6, 5)$

33. $(10, 7)$ and $(7, 10)$

34. $(5, -6)$ and $(-6, 5)$

35. $(-8, 1)$ and $(0, 0)$

36. $(2, 3)$ and $(0, 0)$

E *Solve. Assume each exercise describes a linear relationship. When writing a linear equation in Exercises 37 through 46, write the equation in slope-intercept form. See Example 6.*

37. A rock is dropped from the top of a 400-foot cliff. After 1 second, the rock is traveling 32 feet per second. After 3 seconds, the rock is traveling 96 feet per second.

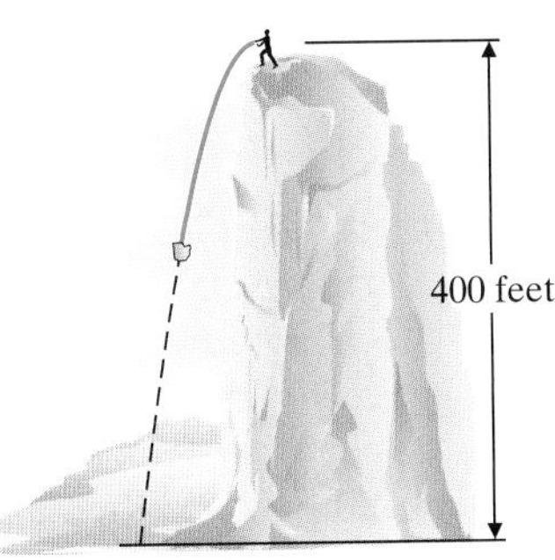

a. Assume that the relationship between time and speed is linear and write an equation describing this relationship. Use ordered pairs of the form (time, speed).

b. Use this equation to determine the speed of the rock 4 seconds after it was dropped.

38. A Hawaiian fruit company is studying the sales of a pineapple sauce to see if this product is to be continued. At the end of its first year, profits on this product amounted to \$30,000. At the end of the fourth year, profits were \$66,000.

a. Assume that the relationship between years on the market and profit is linear and write an equation describing this relationship. Use ordered pairs of the form (years on the market, profit).

b. Use this equation to predict the profit at the end of 7 years.

39. In 2000 there were 7590 electric-powered vehicles in use in the United States. In 1998 only 5242 electric vehicles were being used. (*Source:* U.S. Energy Information Administration)

a. Write an equation describing the relationship between time and number of electric-powered vehicles. Use ordered pairs of the form (years past 1998, number of vehicles).

b. Use this equation to predict the number of electric-powered vehicles in use in 2009.

40. In 2000 there were 483 thousand eating establishments in the United States. In 1996, there were 457 thousand eating establishments.

a. Write an equation describing the relationship between time and number of eating establishments. Use ordered pairs of the form (years past 1996, number of eating establishments in thousands).

b. Use this equation to predict the number of eating establishments in 2008.

41. In 2000, the U.S. population per square mile of land area was 79.6. In 1990, this person per square mile population was 70.3.

a. Write an equation describing the relationship between year and person per square mile. Use ordered pairs of the form (years past 1990, person per square mile).

b. Use this equation to predict the person per square mile population in 2007.

42. In 1996 there were 135 thousand apparel and accessory stores in the United States. In 2001 there were a total of 152 thousand apparel and accessory stores. (*Source:* U.S. Bureau of the Census, *County Business Patterns*, annual)

a. Write an equation describing this relationship. Use ordered pairs of the form (years past 1996, number of stores in thousands).

b. Use this equation to predict the number of apparel and accessory stores in 2007.

43. In 1995, the sales of battery-operated ride-on toys in the United States were $191 million. In 2000, the sales of these ride-on toys had risen to $260 million. (*Source:* Toy Manufacturers of America, Inc.)

a. Write two ordered pairs of the form (years after 1995, sales of battery-operated ride-on toys) for this situation.

b. The relationship between years after 1995 and sales of battery-operated ride-on toys is linear over this period. Use the ordered pairs from part (a) to write an equation of the line relating year to sales of battery-operated ride-on toys.

c. Use the linear equation from part (b) to estimate the sales of ride-on toys in 1999.

44. In 2000, crude oil production by OPEC countries was 29.2 million barrels per day. In 2002, OPEC crude oil production had decreased to about 28.7 million barrels per day. (*Source:* Energy Information Administration)

a. Write two ordered pairs of the form (years after 2000, crude oil production) for this situation.

b. Assume that the relationship between years after 2000 and crude oil production is linear over this period. Use the ordered pairs from part (a) to write an equation of the line relating year to crude oil production.

c. Use the linear equation from part (b) to estimate the crude oil production by OPEC countries in 2001.

45. The Pool Fun Company has learned that by pricing a newly released Fun Noodle at \$3, sales will reach 10,000 Fun Noodles per day during the summer. Raising the price to \$5 will cause the sales to fall to 8000 Fun Noodles per day.

a. Assume that the relationship between sales price and number of Fun Noodles sold is linear and write an equation describing this relationship. Use ordered pairs of the form (sales price, number sold).

b. Predict the daily sales of Fun Noodles if the price is \$3.50.

46. The value of a building bought in 1990 may be depreciated (or decreased) as time passes for income tax purposes. Seven years after the building was bought, this value was \$165,000 and 12 years after it was bought, this value was \$140,000.

a. If the relationship between number of years past 1990 and the depreciated value of the building is linear, write an equation describing this relationship. Use ordered pairs of the form (years past 1990, value of building).

b. Use this equation to estimate the depreciated value of the building in 2010.

Review and Preview

Find the value of $x^2 - 3x + 1$ for each given value of x. See Section 9.3.

47. 2 **48.** 5 **49.** -1 **50.** -3

For each graph, determine whether any x-values correspond to two or more y-values. See Section 13.1.

51.

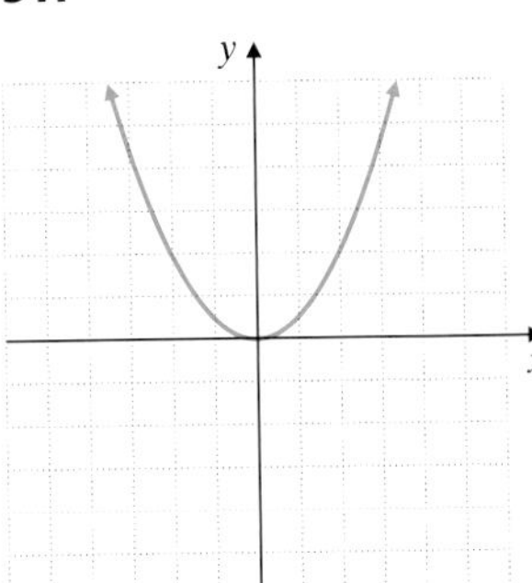

52.

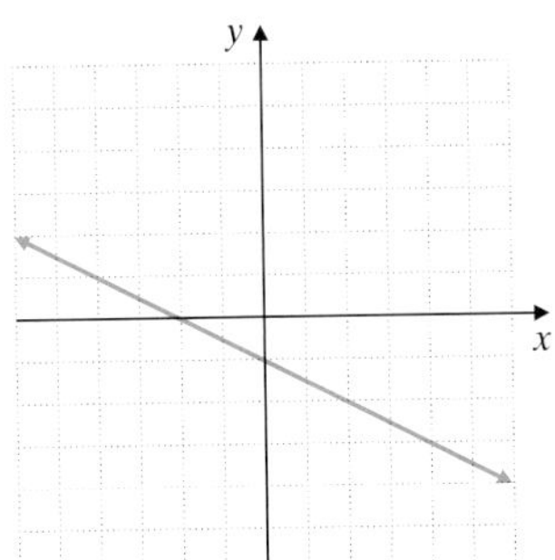

53.

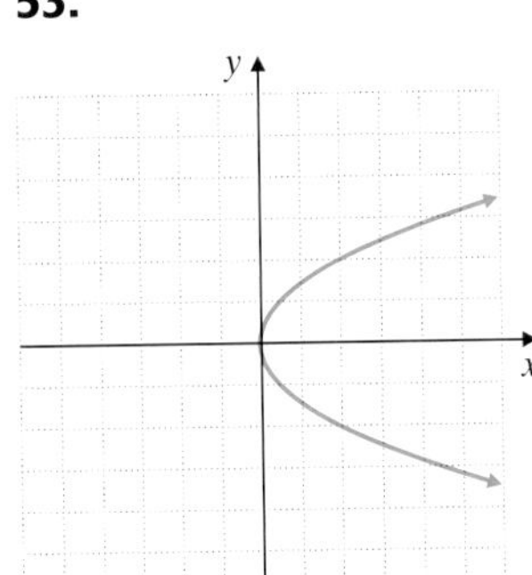

54.

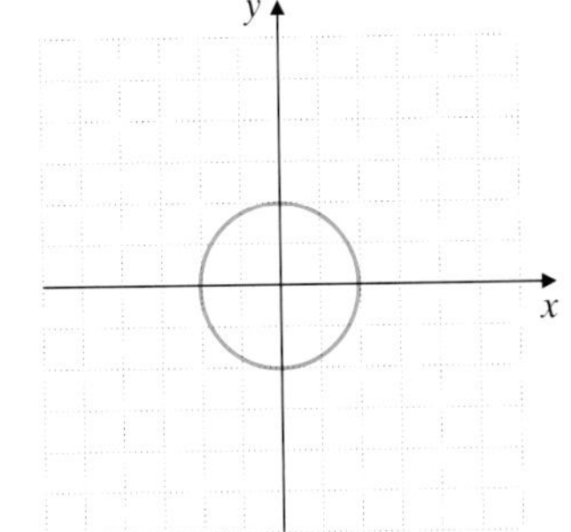

Combining Concepts

Match each linear equation with its graph.

55. $y = 2x + 1$ **56.** $y = -x + 1$ **57.** $y = -3x - 2$ **58.** $y = \frac{5}{3}x - 2$

A.

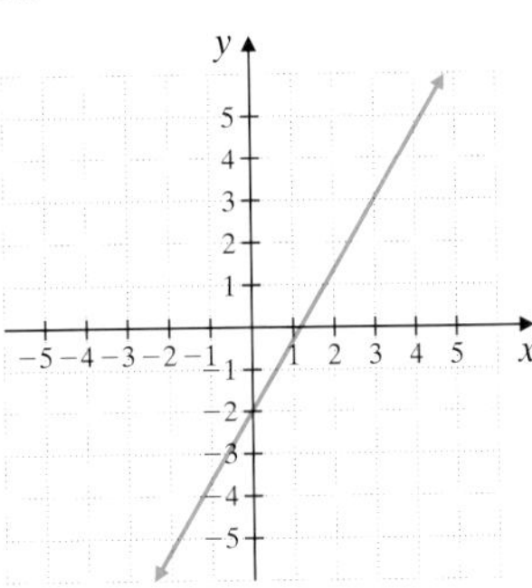

B.

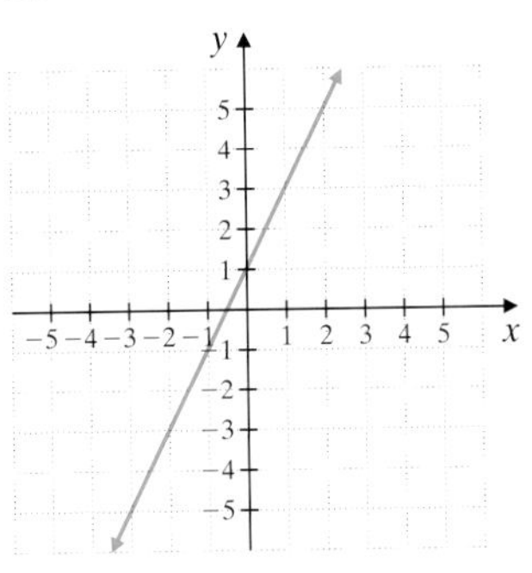

C.

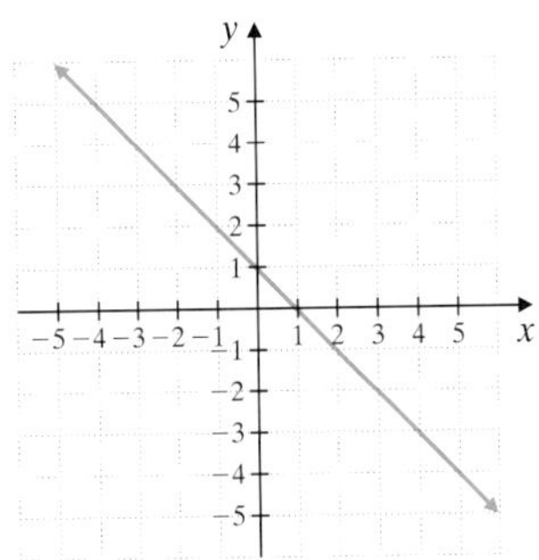

D.

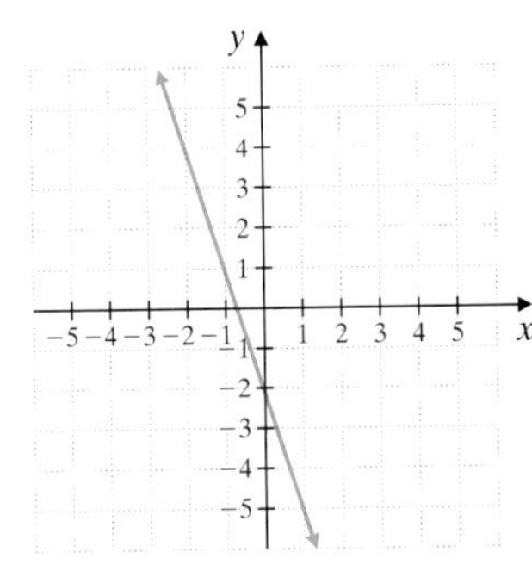

59. Write an equation of the line that contains the point $(-1, 2)$ and has the same slope as the line $y = 3x - 1$.

60. Write an equation of the line that contains the point $(4, 0)$ and has the same slope as the line $y = -2x + 3$.

61. Write an equation in standard form of the line that contains the point $(-1, 2)$ and is
a. parallel to the line $y = 3x - 1$.
b. perpendicular to the line $y = 3x - 1$.

62. Write an equation in standard form of the line that contains the point $(4, 0)$ and is
a. parallel to the line $y = -2x + 3$.
b. perpendicular to the line $y = -2x + 3$.

Internet Excursions

Go To: 

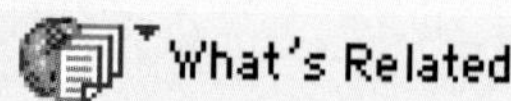

The National Center for Education Statistics (NCES) is the federal agency responsible for collecting data on the state of education in the United States. The given World Wide Web address will provide you with access to the NCES's "Encyclopedia of Education Stats" Web site, or a related site, where you will have access to a wide variety of education statistics. Browse the links to find two different sets of paired data on education that are interesting to you. Then answer the questions below.

63. For one set of data that you have found, write down two ordered pairs. Describe what the ordered pairs represent. Then use the ordered pairs to find an equation for the line passing through these two points.

64. For your other set of data, write the data as a set of ordered pairs. Describe what the ordered pairs represent. Make a scatter diagram of the data. Do you think that a linear equation would represent the data well? Explain.

Name ______________________ Section ________ Date ________

Answers

Integrated Review—Summary of Linear Equations

Find the slope of each line.

1.

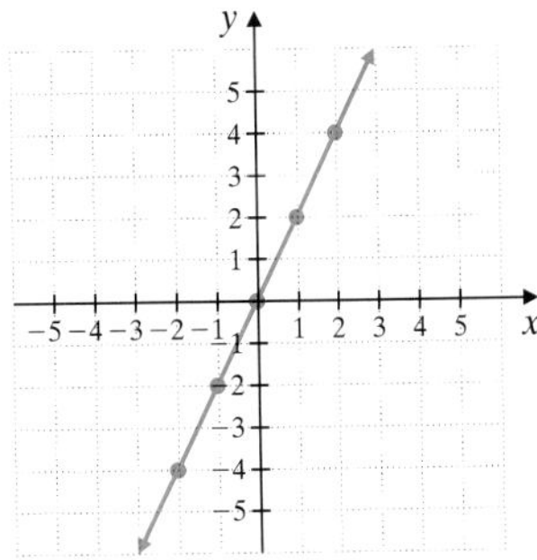

2.

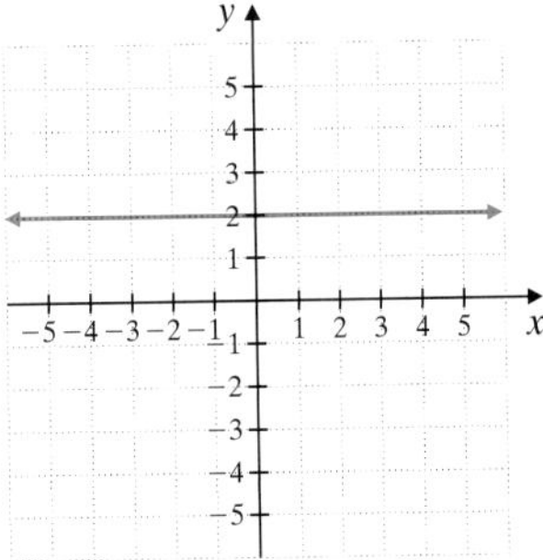

3.

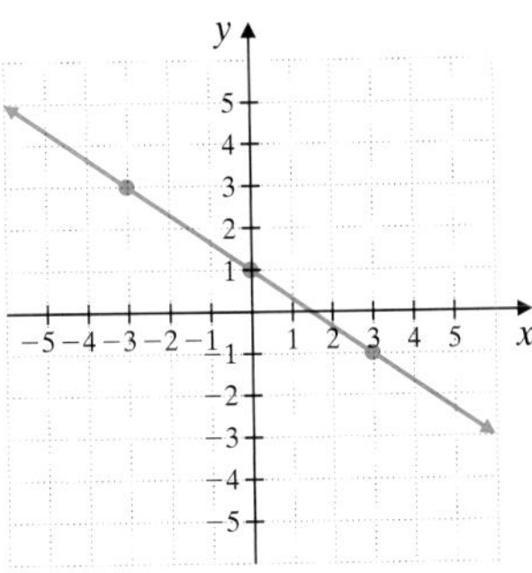

4.

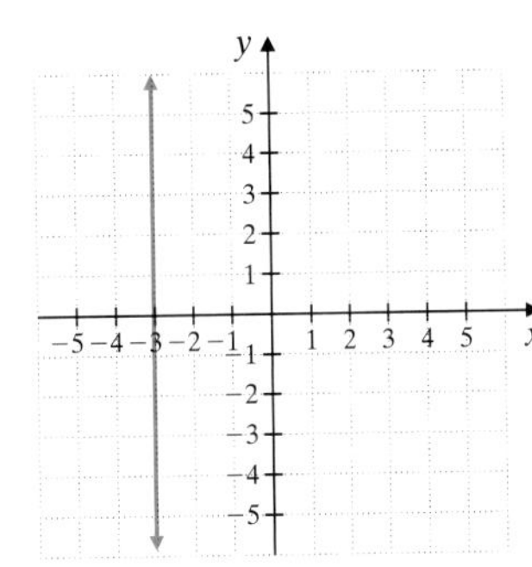

Graph each linear equation.

5. $y = -2x$

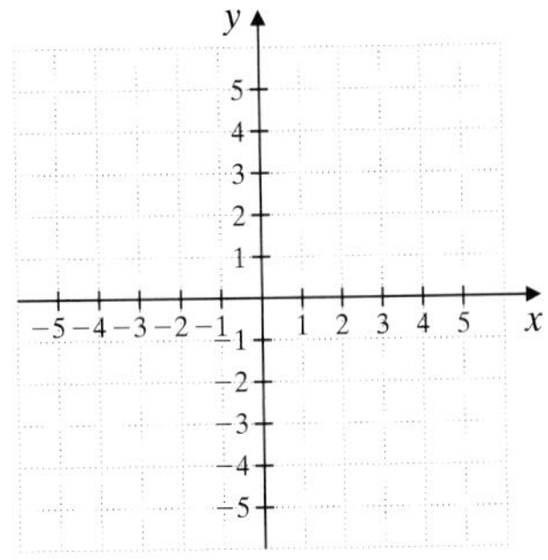

6. $x + y = 3$

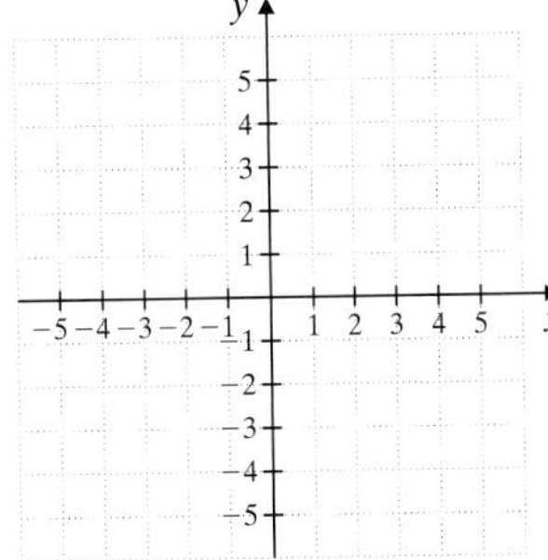

7. $x = -1$

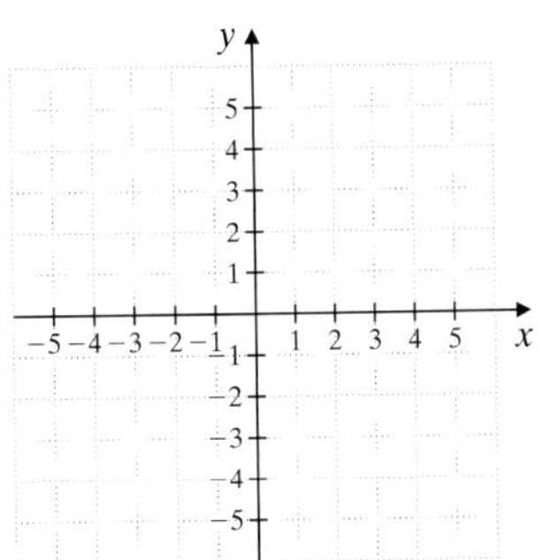

8. $y = 4$

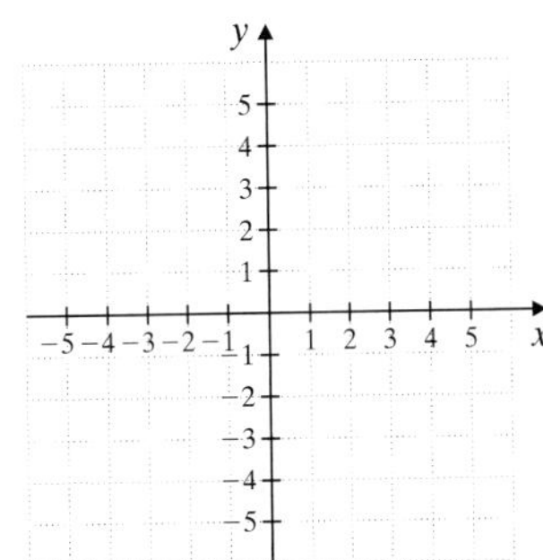

1. ____________

2. ____________

3. ____________

4. ____________

5. see graph

6. see graph

7. see graph

8. see graph

9. see graph

10. see graph

11.

12.

13.

14.

15.

16.

17.

18.

19.

20.

21.

22. a.

b.

c.

9. $x - 2y = 6$

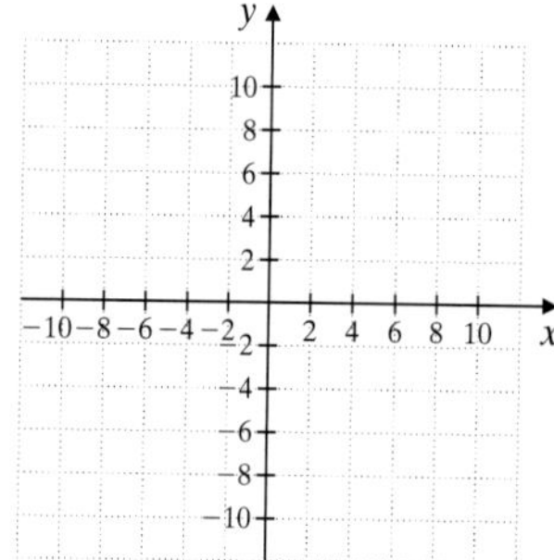

10. $y = 3x + 2$

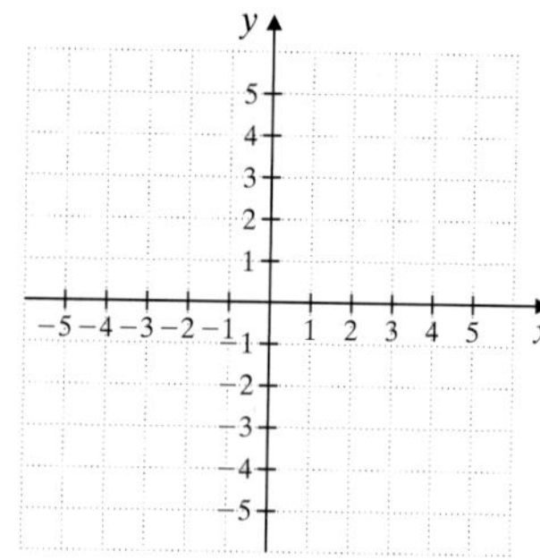

Find the slope of each line by writing the equation in slope-intercept form.

11. $y = 3x - 1$ **12.** $y = -6x + 2$ **13.** $7x + 2y = 11$ **14.** $2x - y = 0$

Find the slope of each line.

15. $x = 2$

16. $y = -4$

17. Write an equation of the line with slope $m = 2$ and y-intercept $b = -\frac{1}{3}$.

18. Find an equation of the line with slope $m = -4$ that passes through the point $(-1, 3)$. Write the equation in the form $Ax + By = C$.

19. Find an equation of the line that passes through the points $(2, 0)$ and $(-1, -3)$. Write the equation in the form $Ax + By = C$.

Determine whether each pair of lines is parallel, perpendicular, or neither.

20. $6x - y = 7$
$2x + 3y = 4$

21. $3x - 6y = 4$
$y = -2x$

22. Seventy-five percent of U.S. households own an outdoor barbecue grill. In 1997, the number of grill units shipped was 11.6 million. In 2001, the number of grill units shipped was 15.4 million. (*Source:* Barbecue Industry Assn.)

a. Write two ordered pairs of the form (year, millions of grill units shipped).
b. Find the slope of the line between the two points.
c. Write a sentence explaining the meaning of the slope as a rate of change.

13.6 Introduction to Functions

OBJECTIVES

- A Identify relations, domains, and ranges.
- B Identify functions.
- C Use the vertical line test.
- D Use function notation.

A Identifying Relations, Domains, and Ranges

In this chapter, we have studied paired data in the form of ordered pairs. For example, when we list an ordered pair such as $(3, 1)$, we are saying that when x is 3, then y is 1. In other words $x = 3$ and $y = 1$ are related to each other.

For this reason, we call a set of ordered pairs a **relation.** The set of all x-coordinates is called the **domain** of a relation, and the set of all y-coordinates is called the **range** of a relation.

EXAMPLE 1

Find the domain and the range of the relation $\{(0, 2), (3, 3), (-1, 0), (3, -2)\}$.

Solution: The domain is the set of all x-values, or $\{-1, 0, 3\}$, and the range is the set of all y-values, or $\{-2, 0, 2, 3\}$.

Practice Problem 1

Find the domain and range of the relation $\{(-3, 5), (-3, 1), (4, 6), (7, 0)\}$.

B Identifying Functions

Paired data occur often in real-life applications. Some special sets of paired data, or ordered pairs, are called *functions*.

Function

A **function** is a set of ordered pairs in which each x-coordinate has exactly one y-coordinate.

In other words, a function cannot have two ordered pairs with the same x-coordinate but different y-coordinates.

EXAMPLE 2 Which of the following relations are also functions?

a. $\{(-1, 1), (2, 3), (7, 3), (8, 6)\}$
b. $\{(0, -2), (1, 5), (0, 3), (7, 7)\}$

Solution:

a. Although the ordered pairs $(2, 3)$ and $(7, 3)$ have the same y-value, each x-value is assigned to only one y-value, so this set of ordered pairs is a function.

b. The x-value 0 is paired with two y-values, -2 and 3, so this set of ordered pairs is not a function.

Practice Problem 2

Which of the following relations are also functions?

a. $\{(2, 5), (-3, 7), (4, 5), (0, -1)\}$
b. $\{(1, 4), (6, 6), (1, -3), (7, 5)\}$

Relations and functions can be described by graphs of their ordered pairs.

Answers

1. domain: $\{-3, 4, 7\}$; range: $\{0, 1, 5, 6\}$
2. a. a function **b.** not a function

Practice Problem 3

Which graph is the graph of a function?

a.

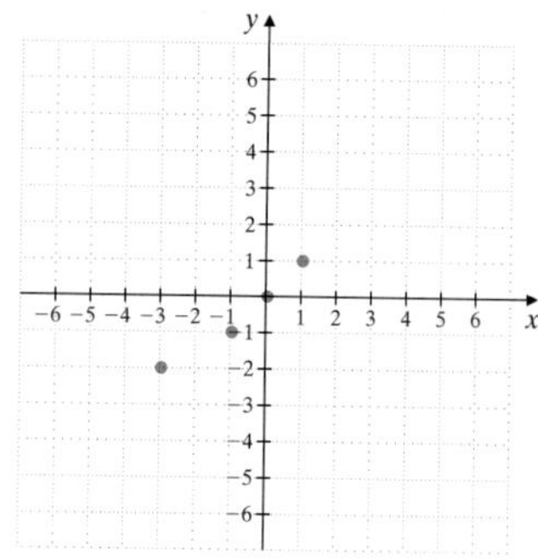

b.

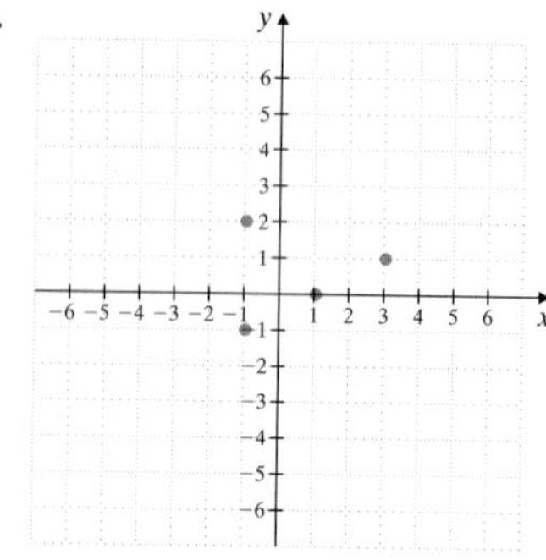

EXAMPLE 3 Which graph is the graph of a function?

a.

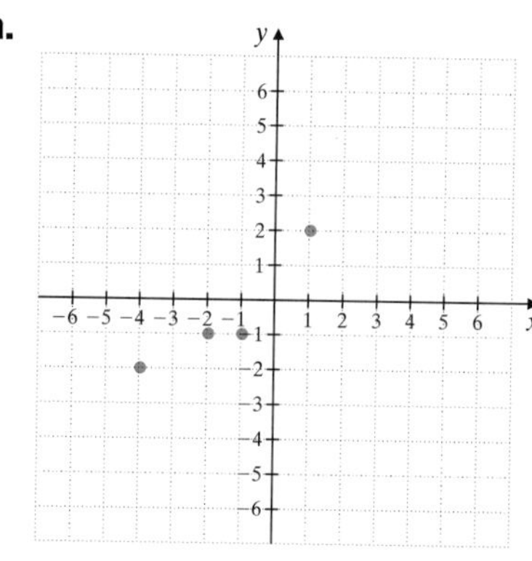

b.

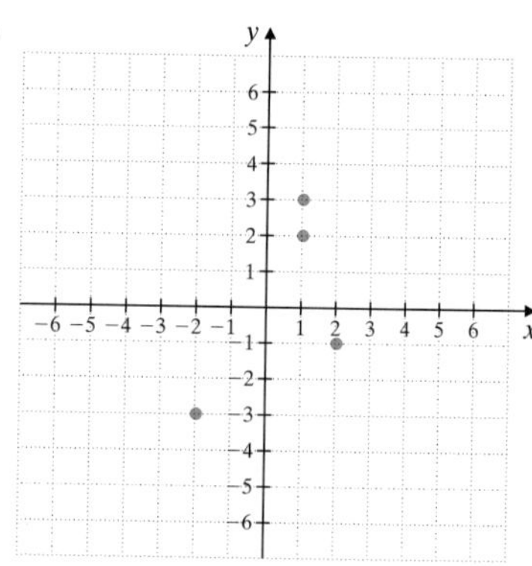

Solution:

a. This is the graph of the relation $\{(-4,-2), (-2,-1), (-1,-1), (1,2)\}$. Each x-coordinate has exactly one y-coordinate, so this is the graph of a function.

b. This is the graph of the relation $\{(-2,-3), (1,2), (1,3), (2,-1)\}$. The x-coordinate 1 is paired with two y-coordinates, 2 and 3, so this is not the graph of a function.

C Using the Vertical Line Test

The graph in Example 3(b) was not the graph of a function because the x-coordinate 1 was paired with two y-coordinates, 2 and 3. Notice that when an x-coordinate is paired with more than one y-coordinate, a vertical line can be drawn that will intersect the graph at more than one point. We can use this fact to determine whether a relation is also a function. We call this the vertical line test.

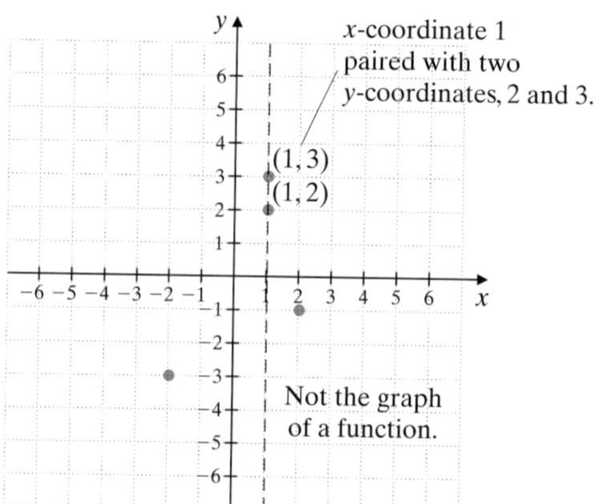

Vertical Line Test

If a vertical line can be drawn so that it intersects a graph more than once, the graph is not the graph of a function. (If no such vertical line can be drawn, the graph is that of a function.)

This vertical line test works for all types of graphs on the rectangular coordinate system.

Answers

3. a. a function **b.** not a function

EXAMPLE 4

Use the vertical line test to determine whether each graph is the graph of a function.

a.

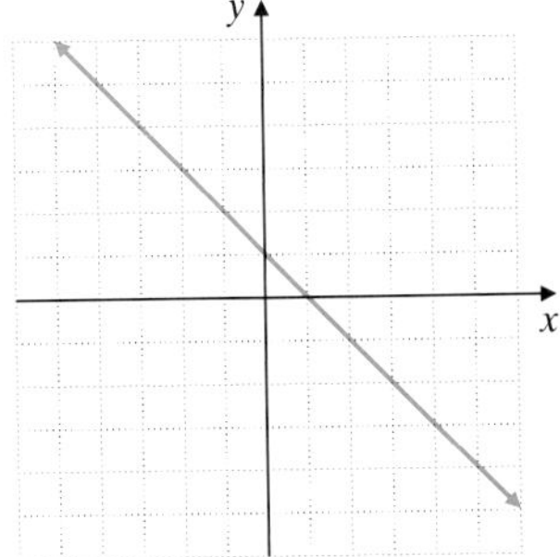

b.

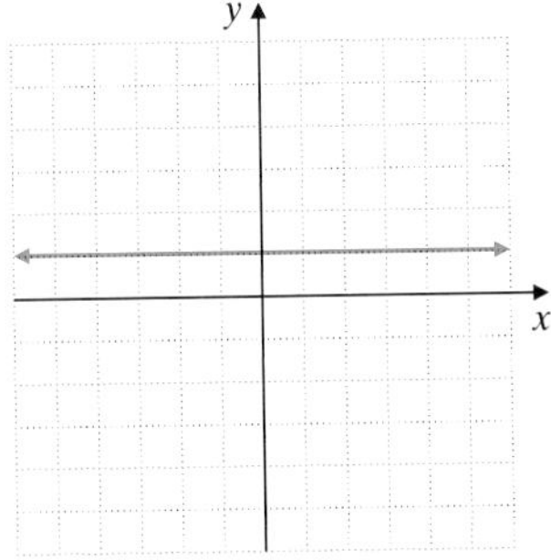

c.

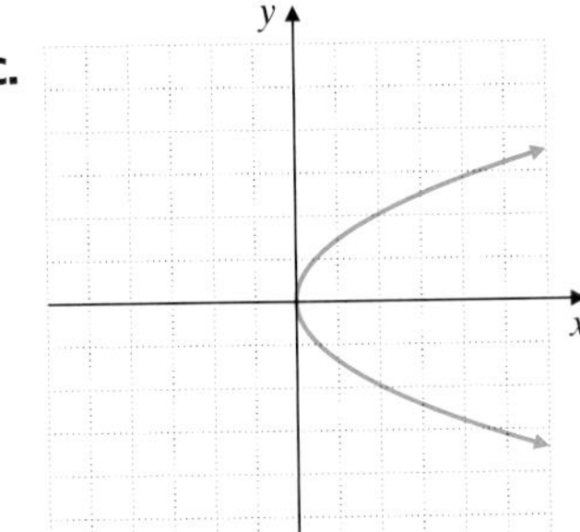

d.

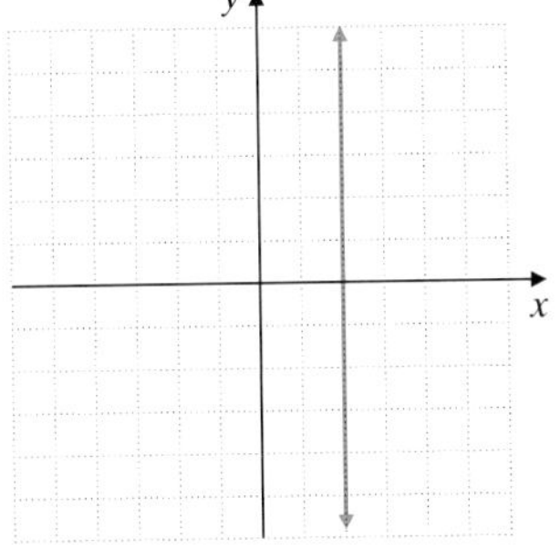

Solution:

a. This graph is the graph of a function since no vertical line will intersect this graph more than once.

b. This graph is also the graph of a function; no vertical line will intersect it more than once.

c. This graph is not the graph of a function. Vertical lines can be drawn that intersect the graph in two points. An example of one is shown.

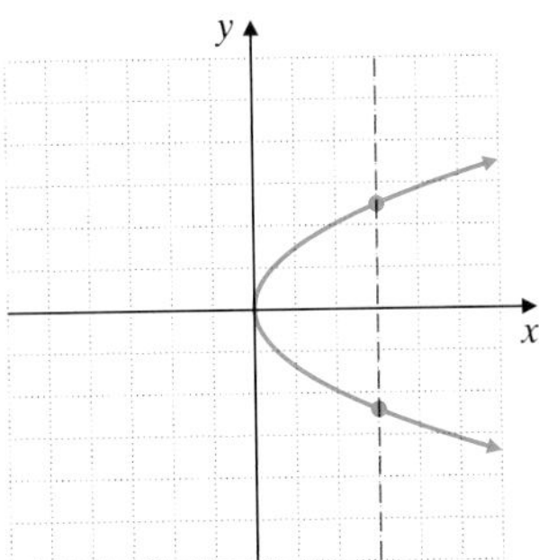

d. This graph is not the graph of a function. A vertical line can be drawn that intersects this line at every point.

Examples of functions can often be found in magazines, newspapers, books, and other printed material in the form of tables or graphs such as that in Example 5.

Practice Problem 4

Determine whether each graph is the graph of a function.

a.

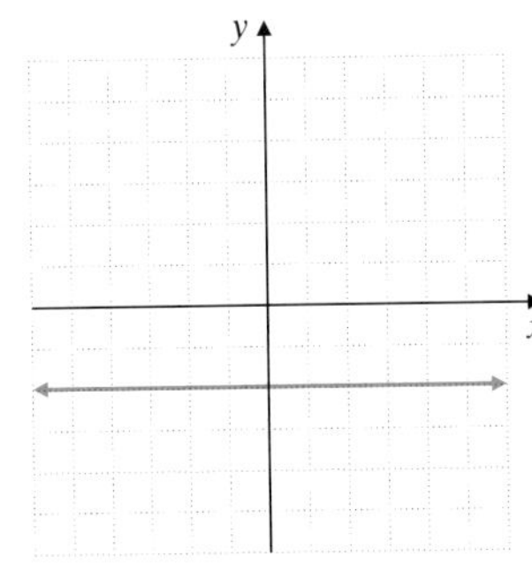

b.

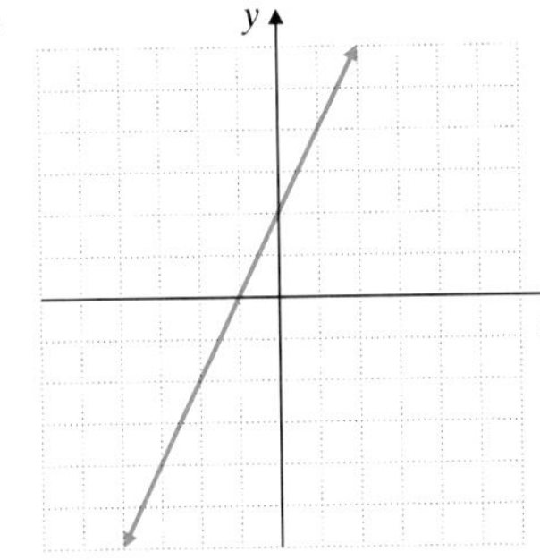

c.

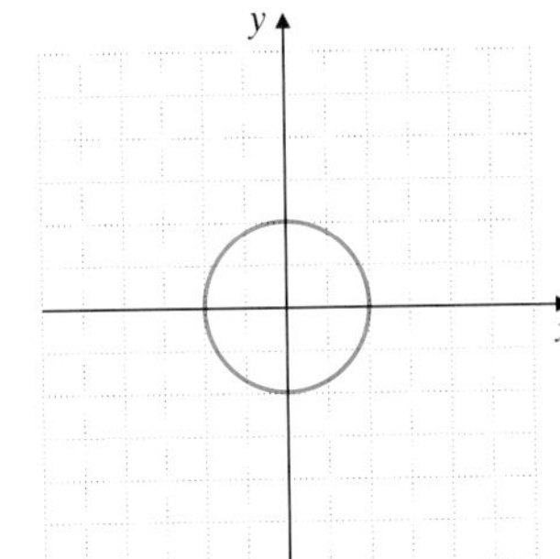

d.

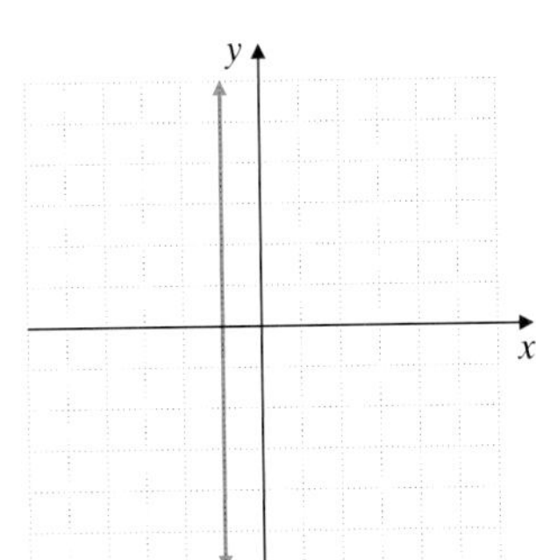

Answers

4. a. a function **b.** a function **c.** not a function **d.** not a function

Practice Problem 5

Use the graph in Example 5 to answer the questions.

a. Approximate the time of sunrise on March 1.

b. Approximate the date(s) when the sun rises at 6 A.M.

EXAMPLE 5

The graph shows the sunrise time for Indianapolis, Indiana, for the year. Use this graph to answer the questions.

a. Approximate the time of sunrise on February 1.

b. Approximate the date(s) when the sun rises at 5 A.M.

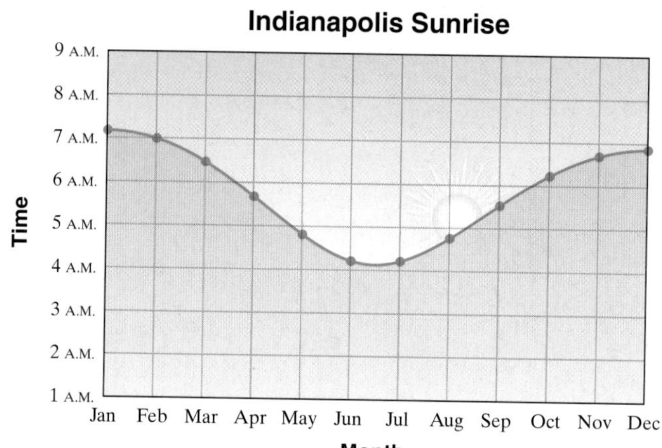

Source: Wolff World Atlas

c. Is this the graph of a function?

Solution:

a. To approximate the time of sunrise on February 1, we find the mark on the horizontal axis that corresponds to February 1. From this mark, we move vertically upward until the graph is reached. From that point on the graph, we move horizontally to the left until the vertical axis is reached. The vertical axis there reads 7 A.M. as shown below.

b. To approximate the date(s) when the sun rises at 5 A.M., we find 5 A.M. on the time axis and move horizontally to the right. Notice that we will hit the graph at two points, corresponding to two dates for which the sun rises at 5 A.M. We follow both points on the graph vertically downward until the horizontal axis is reached. The sun rises at 5 A.M. at approximately the end of the month of April and the middle of the month of August.

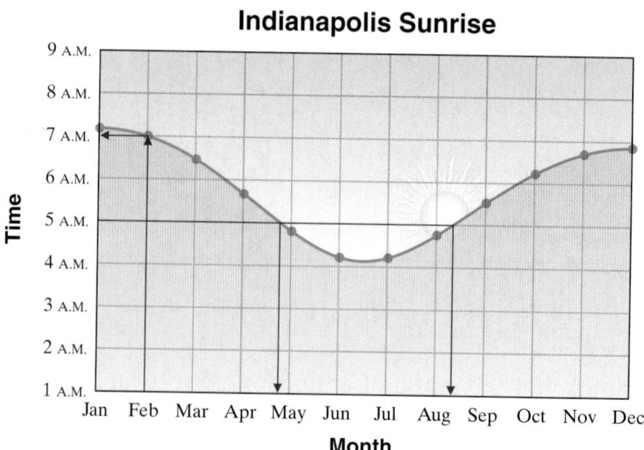

Source: Wolff World Atlas

c. The graph is the graph of a function since it passes the vertical line test. In other words, for every day of the year in Indianapolis, there is exactly one sunrise time.

Answers

5. a. 6:30 A.M. **b.** middle of March and middle of September

D Using Function Notation

The graph of the linear equation $y = 2x + 1$ passes the vertical line test, so we say that $y = 2x + 1$ is a function. In other words, $y = 2x + 1$ gives us a rule for writing ordered pairs where every x-coordinate is paired with at most one y-coordinate.

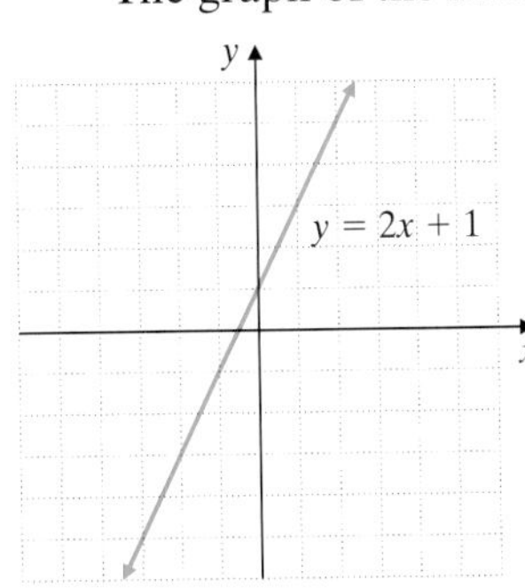

We often use letters such as f, g, and h to name functions. For example, the symbol $f(x)$ means *function of* x and is read "f of x." This notation is called **function notation.** The equation $y = 2x + 1$ can be written as $f(x) = 2x + 1$ using function notation, and these equations mean the same thing. In other words $y = f(x)$.

The notation $f(1)$ means to replace x with 1 and find the resulting y or function value. Since

$$f(x) = 2x + 1$$

then

$$f(1) = 2(1) + 1 = 3$$

This means that, when $x = 1$, y or $f(x) = 3$, and we have the ordered pair $(1, 3)$. Now let's find $f(2)$, $f(0)$, and $f(-1)$.

$f(x) = 2x + 1$	$f(x) = 2x + 1$	$f(x) = 2x + 1$
$f(2) = 2(2) + 1$	$f(0) = 2(0) + 1$	$f(-1) = 2(-1) + 1$
$= 4 + 1$	$= 0 + 1$	$= -2 + 1$
$= 5$	$= 1$	$= -1$
Ordered Pair: $(2, 5)$	$(0, 1)$	$(-1, -1)$

Helpful Hint

Note that $f(x)$ is a special symbol in mathematics used to denote a function. The symbol $f(x)$ is read "f of x." It does **not** mean $f \cdot x$ (f times x).

EXAMPLE 6 Given $g(x) = x^2 - 3$, find the following and list the corresponding ordered pair.

a. $g(2)$ **b.** $g(-2)$ **c.** $g(0)$

Solution:

a. $g(x) = x^2 - 3$	**b.** $g(x) = x^2 - 3$	**c.** $g(x) = x^2 - 3$
$g(2) = 2^2 - 3$	$g(-2) = (-2)^2 - 3$	$g(0) = 0^2 - 3$
$= 4 - 3$	$= 4 - 3$	$= 0 - 3$
$= 1$	$= 1$	$= -3$
Ordered Pair: $(2, 1)$	$(-2, 1)$	$(0, -3)$

Try the Concept Check in the margin.

Practice Problem 6

Given $f(x) = x^2 + 1$, find the following and list the corresponding ordered pair.

a. $f(1)$ b. $f(-3)$ c. $f(0)$

Concept Check

Suppose that the value of a function f is -7 when the function is evaluated at 2. Write this situation in function notation.

Answers

6. a. 2; $(1, 2)$ **b.** 10; $(-3, 10)$ **c.** 1; $(0, 1)$

Concept Check: $f(2) = -7$

FOCUS ON The Real World

ROAD GRADES

Have you ever driven on a hilly highway and seen a sign like the one below? The 7% on the sign refers to the grade of the road. The grade of a road is the same as its slope given as a percent. A 7 percent grade means that for every rise of 7 units there is a run of 100 units. The type of units doesn't matter as long as they are the same. For instance, we could say that a 7 percent grade represents a rise of 7 feet for every run of 100 feet or a rise of 7 meters for every run of 100 meters, and so on.

$$7\% = 0.07 = \frac{7}{100} \qquad 7\% \text{ grade} = \frac{\text{rise of } 7}{\text{run of } 100}$$

Most highways are designed to have grades of 6 percent or less. If a portion of a highway has a grade that is steeper than 6 percent, a sign is usually posted giving the grade and the number of miles for the grade. Truck drivers need to know when the road is particularly steep. They may need to take precautions such as using a different gear, reducing their speed, or testing their brakes.

Here is a sampling of road grades:

- A portion of the John Scott Highway in Steubenville, Ohio, has a 10% grade. (*Source:* Ohio Department of Transportation)
- Joaquin Road in Portola Valley, California, has an average grade of 15%. (*Source:* Western Wheelers Bicycle Club)
- The steepest grade in Seattle, Washington, is 26% on East Roy Street between 25th Avenue and 26th Avenue. (*Source: Seattle Post-Intelligencer*, Nov. 21, 1994)
- The steepest street in Pittsburgh, Pennsylvania, is Canton Avenue with a 37% grade. (*Source:* Pittsburgh Department of Public Works)
- The steepest street in the world is Baldwin Street in Dunedin, New Zealand. Its maximum grade is 79%. (*Source: Guinness Book of Records*, 1996)

COOPERATIVE LEARNING ACTIVITY

Try to find a road sign with a percent grade warning or the name and grade of a steep road in your area. Describe its slope and make a scale drawing to represent its grade.

Name ____________________ Section __________ Date __________

EXERCISE SET 13.6

A *Find the domain and the range of each relation. See Example 1.*

1. $\{(2,4),(0,0),(-7,10),(10,-7)\}$

2. $\{(3,-6),(1,4),(-2,-2)\}$

3. $\{(0,-2),(1,-2),(5,-2)\}$

4. $\{(5,0),(5,-3),(5,4),(5,3)\}$

B *Determine whether each relation is also a function. See Example 2.*

5. $\{(1,1),(2,2),(-3,-3),(0,0)\}$

6. $\{(1,2),(3,2),(4,2)\}$

7. $\{(-1,0),(-1,6),(-1,8)\}$

8. $\{(11,6),(-1,-2),(0,0),(3,-2)\}$

C *Use the vertical line test to determine whether each graph is the graph of a function. See Examples 3 and 4.*

9.

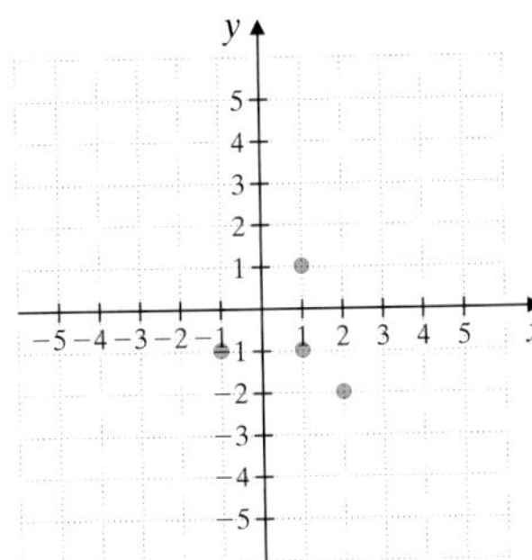

10.

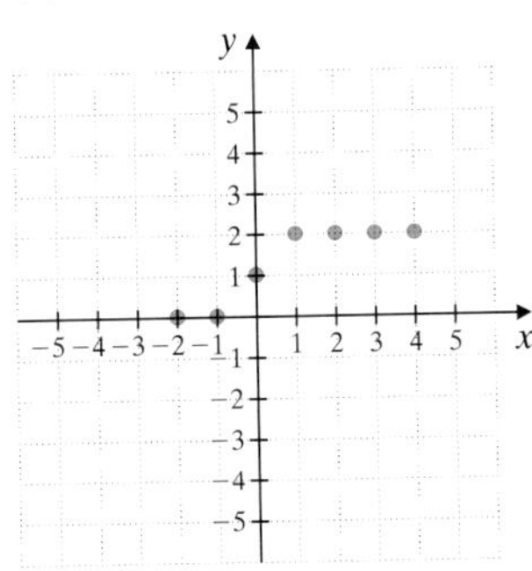

11.

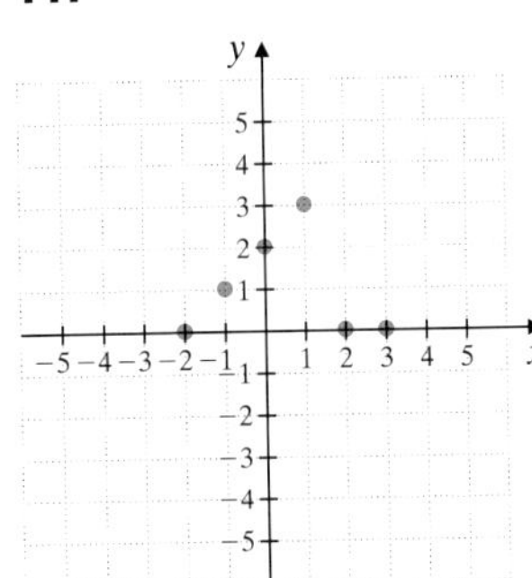

12.

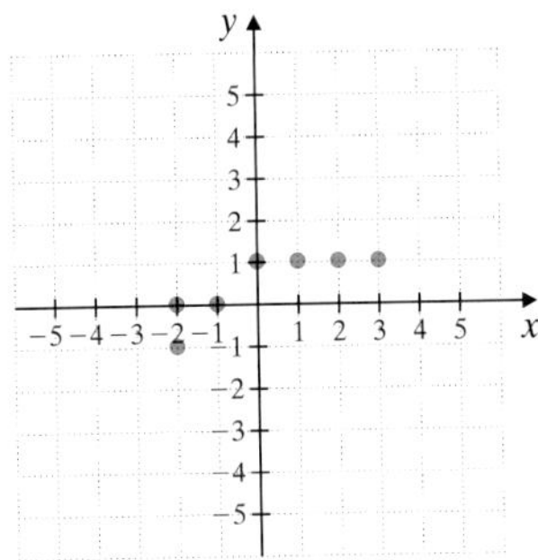

13.

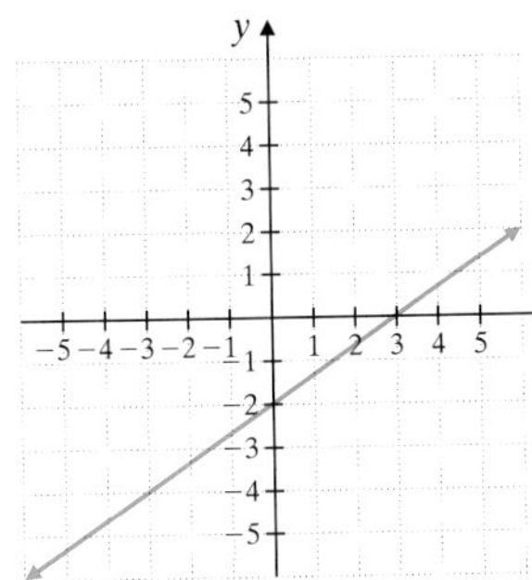

14.

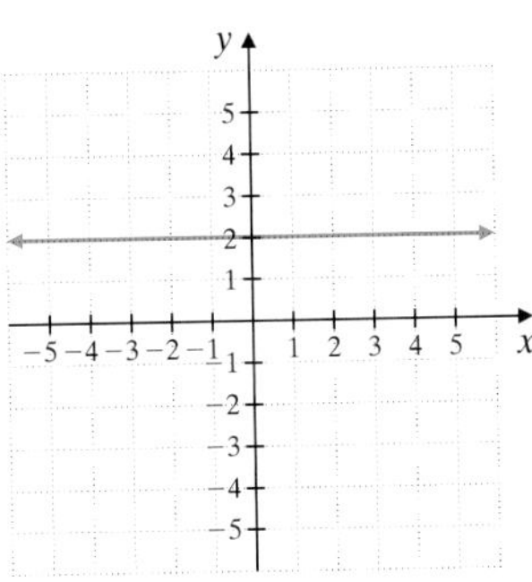

15.

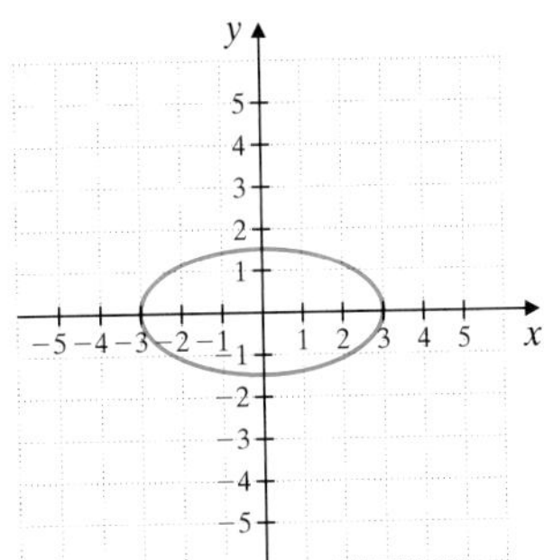

16.

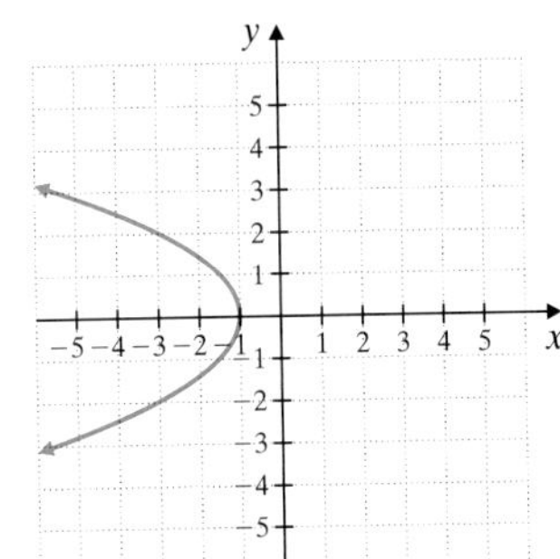

Use the graph in Example 5 to answer Exercises 17 through 20.

17. Approximate the time of sunrise on September 1 in Indianapolis.

18. Approximate the date(s) when the sun rises in Indianapolis at 7 A.M.

19. Describe the change in sunrise over the year for Indianapolis.

20. When, in Indianapolis, is the earliest sunrise? What point on the graph does this correspond to?

The graph shows the sunset times for Seward, Alaska. Use this graph to answer Exercises 21 through 26.

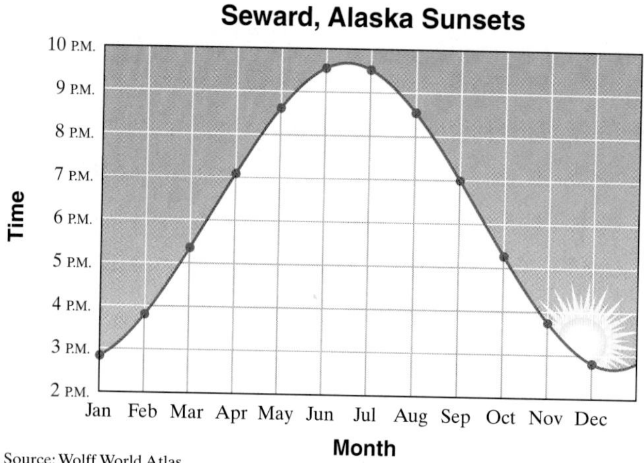

21. Approximate the time of sunset on June 1.

22. Approximate the time of sunset on November 1.

23. Approximate the date(s) when the sunset is 3 P.M.

24. Approximate the date(s) when the sunset is 9 P.M.

25. Is this graph the graph of a function? Why or why not?

26. Do you think a graph of sunset times for any location will always be a function? Why or why not?

This graph shows the U.S. hourly minimum wage for each year shown. Use this graph to answer Exercises 27 through 32.

27. Approximate the minimum wage at the beginning of 1997.

28. Approximate the minimum wage at the beginning of 1999.

29. Approximate the year when the minimum wage increased to over $4.00 per hour.

30. Approximate the year when the minimum wage increased to over $5.00 per hour.

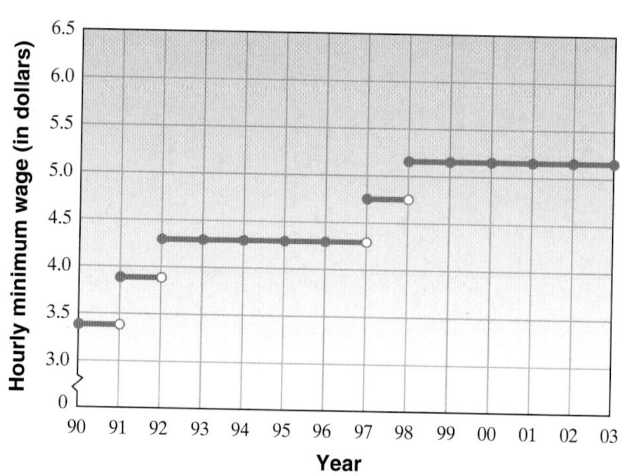

31. Is this graph the graph of a function? Why or why not?

32. Do you think that a similar graph of your hourly wage every year (whether you are working or not) will be the graph of a function? Why or why not?

D *Find $f(-2)$, $f(0)$, and $f(3)$ for each function. See Example 6.*

33. $f(x) = 2x - 5$

34. $f(x) = 3 - 7x$

35. $f(x) = x^2 + 2$

36. $f(x) = x^2 - 4$

37. $f(x) = 3x$

38. $f(x) = -3x$

39. $f(x) = |x|$

40. $f(x) = |2 - x|$

Find $h(-1)$, $h(0)$, and $h(4)$ for each function. See Example 6.

41. $h(x) = -5x$

42. $h(x) = -3x$

43. $h(x) = 2x^2 + 3$

44. $h(x) = 3x^2$

Review and Preview

Solve each inequality. See Section 9.6.

45. $2x + 5 < 7$

46. $3x - 1 \geq 11$

47. $-x + 6 \leq 9$

48. $-2x + 3 > 3$

Find the perimeter of each figure. See Section 12.4.

△ **49.**

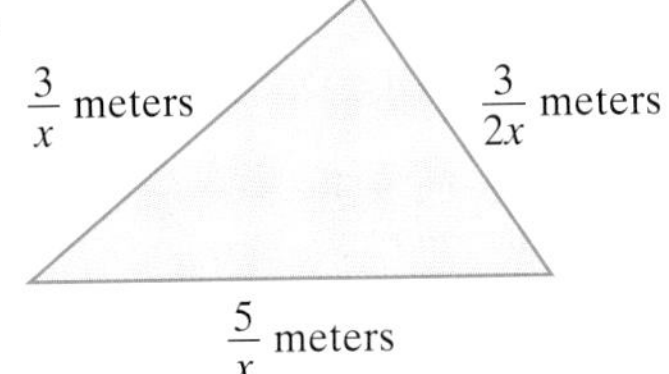

△ **50.**

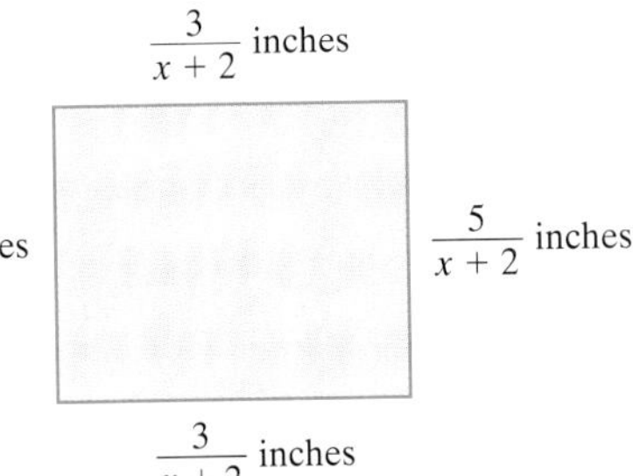

Combining Concepts

51. Forensic scientists use the function

$$f(x) = 2.59x + 47.24$$

to estimate the height of a woman given the length x of her femur bone.

a. Estimate the height of a woman whose femur measures 46 centimeters.

b. Estimate the height of a woman whose femur measures 39 centimeters.

52. The dosage in milligrams of Ivermectin, a heartworm preventive for a dog who weighs x pounds, is given by the function

$$f(x) = \frac{136}{25}x$$

a. Find the proper dosage for a dog that weighs 35 pounds.

b. Find the proper dosage for a dog that weighs 70 pounds.

53. In your own words define (a) function; (b) domain; (c) range.

54. Explain the vertical line test and how it is used.

55. Since $y = x + 7$ is a function, rewrite the equation using function notation.

STUDY SKILLS REMINDER

Do you remember what to do on the day of an exam?

On the day of an exam, don't forget to try the following:

- Allow yourself plenty of time to arrive.
- Read the directions on the test carefully.
- Read each problem carefully as you take your test. Make sure that you answer the question asked.
- Watch your time and pace yourself so that you may attempt each problem on your test.
- If you have time, check your work and answers.
- Do not turn your test in early. If you have extra time, double-check your work.

Good luck!

13.7 Graphing Linear Inequalities in Two Variables

OBJECTIVES

A Determine whether an ordered pair is a solution of a linear inequality in two variables.

B Graph a linear inequality in two variables.

Recall that a linear equation in two variables is an equation that can be written in the form $Ax + By = C$ where $A, B,$ and C are real numbers and A and B are not both 0. A **linear inequality in two variables** is an inequality that can be written in one of the forms

$$Ax + By < C \qquad Ax + By \le C$$
$$Ax + By > C \qquad Ax + By \ge C$$

where $A, B,$ and C are real numbers and A and B are not both 0.

A Determining Solutions of Linear Inequalities in Two Variables

Just as for linear equations in x and y, an ordered pair is a **solution** of an inequality in x and y if replacing the variables with the coordinates of the ordered pair results in a true statement.

EXAMPLE 1

Determine whether each ordered pair is a solution of the equation $2x - y < 6$.

a. $(5, -1)$ **b.** $(2, 7)$

Solution:

a. We replace x with 5 and y with -1 and see if a true statement results.

$$2x - y < 6$$
$$2(5) - (-1) < 6 \quad \text{Replace } x \text{ with 5 and } y \text{ with } -1.$$
$$10 + 1 < 6$$
$$11 < 6 \quad \text{False}$$

The ordered pair $(5, -1)$ is not a solution since $11 < 6$ is a false statement.

b. We replace x with 2 and y with 7 and see if a true statement results.

$$2x - y < 6$$
$$2(2) - (7) < 6 \quad \text{Replace } x \text{ with 2 and } y \text{ with 7.}$$
$$4 - 7 < 6$$
$$-3 < 6 \quad \text{True}$$

The ordered pair $(2, 7)$ is a solution since $-3 < 6$ is a true statement.

Practice Problem 1

Determine whether each ordered pair is a solution of $x - 4y > 8$.

a. $(-3, 2)$ b. $(9, 0)$

B Graphing Linear Inequalities in Two Variables

The linear equation $x - y = 1$ is graphed next. Recall that all points on the line correspond to ordered pairs that satisfy the equation $x - y = 1$.

Notice the line defined by $x - y = 1$ divides the rectangular coordinate system plane into 2 sides. All points on one side of the line satisfy the inequality $x - y < 1$ and all points on the other side satisfy the inequality $x - y > 1$. The graph on the next page shows a few examples of this.

Answers

1. a. no **b.** yes

$x - y$	< 1	
$1 - 3$	< 1	True
$-2 - 1$	< 1	True
$-4 - (-4)$	< 1	True

$x - y$	> 1	
$4 - 1$	> 1	True
$2 - (-2)$	> 1	True
$0 - (-4)$	> 1	True

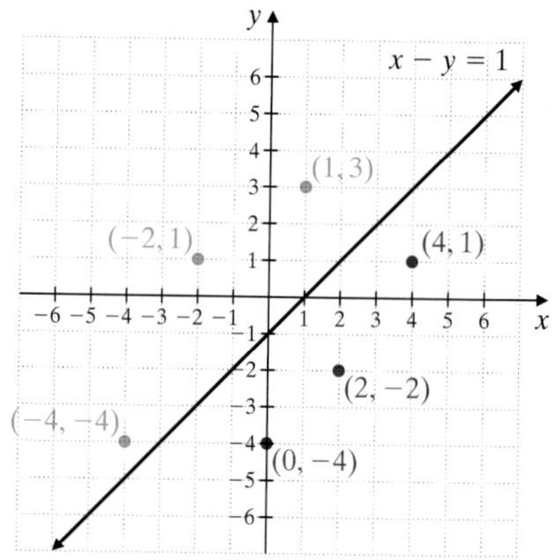

The graph of $x - y < 1$ is the region shaded blue and the graph of $x - y > 1$ is the region shaded red below.

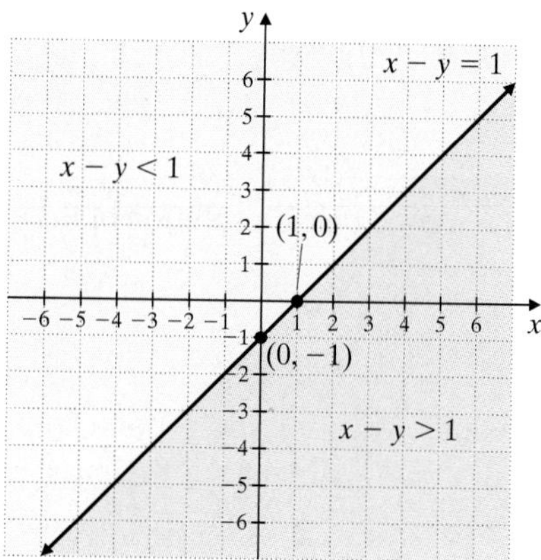

The region to the left of the line and the region to the right of the line are called **half-planes.** Every line divides the plane (similar to a sheet of paper extending indefinitely in all directions) into two half-planes; the line is called the **boundary.**

Recall that the inequality $x - y \le 1$ means

$$x - y = 1 \quad \text{or} \quad x - y < 1$$

Thus, the graph of $x - y \le 1$ is the half-plane $x - y < 1$ along with the boundary line $x - y = 1$.

To Graph a Linear Inequality in Two Variables

Step 1. Graph the boundary line found by replacing the inequality sign with an equal sign. If the inequality sign is $>$ or $<$, graph a dashed boundary line (indicating that the points on the line are not solutions of the inequality). If the inequality sign is $\ge$ or $\le$, graph a solid boundary line (indicating that the points on the line are solutions of the inequality).

Step 2. Choose a point, *not* on the boundary line, as a test point. Substitute the coordinates of this test point into the *original* inequality.

Step 3. If a true statement is obtained in Step 2, shade the half-plane that contains the test point. If a false statement is obtained, shade the half-plane that does not contain the test point.

EXAMPLE 2 Graph: $x + y < 7$

Solution:

Step 1. First we graph the boundary line by graphing the equation $x + y = 7$. We graph this boundary as a *dashed line* because the inequality sign is $<$, and thus the points on the line are not solutions of the inequality $x + y < 7$.

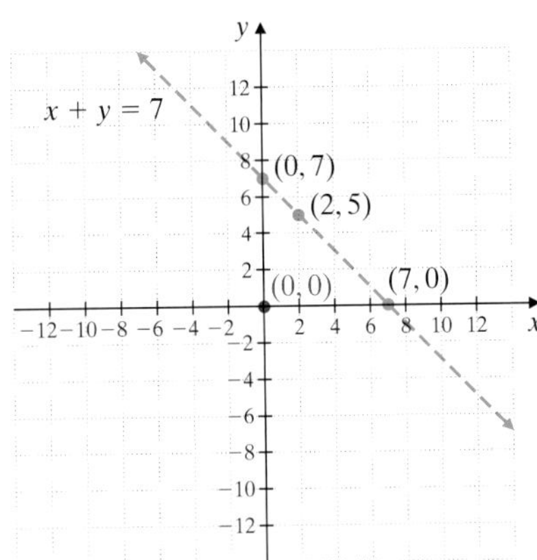

Step 2. Next we choose a test point, being careful *not* to choose a point on the boundary line. We choose $(0, 0)$, and substitute the coordinates of $(0, 0)$ into $x + y < 7$.

$x + y < 7$ Original inequality

$0 + 0 < 7$ Replace x with 0 and y with 0.

$0 < 7$ True

Step 3. Since the result is a true statement, $(0, 0)$ is a solution of $x + y < 7$, and every point in the same half-plane as $(0, 0)$ is also a solution. To indicate this, we shade the entire half-plane containing $(0, 0)$, as shown.

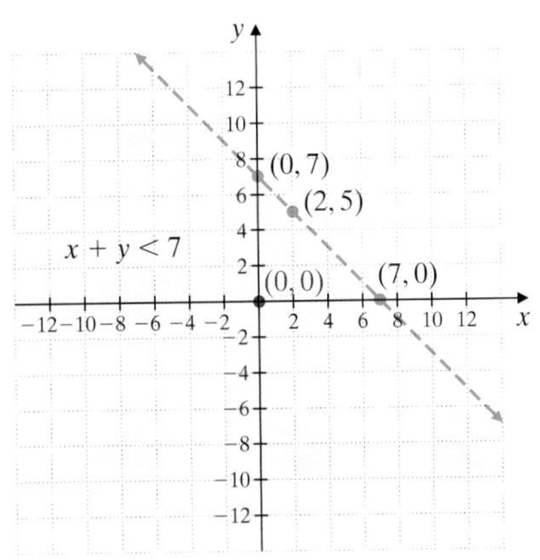

EXAMPLE 3 Graph: $2x - y \geq 3$

Solution:

Step 1. We graph the boundary line by graphing $2x - y = 3$. We draw this line as a solid line because the inequality sign is $\geq$, and thus the points on the line are solutions of $2x - y \geq 3$.

Step 2. Once again, $(0, 0)$ is a convenient test point since it is not on the boundary line.

Practice Problem 2

Graph: $x - y > 3$

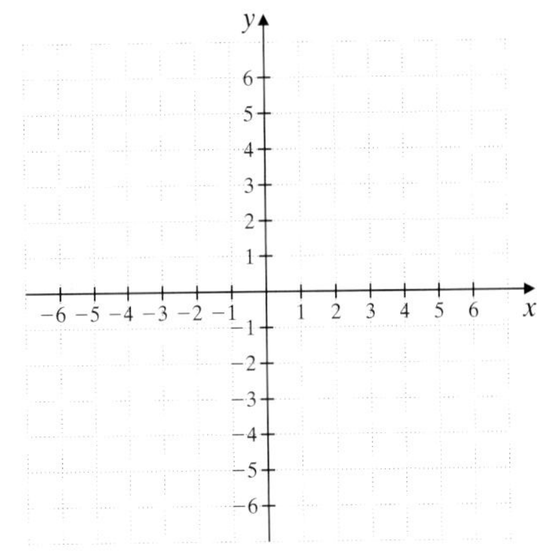

Practice Problem 3

Graph: $x - 4y \leq 4$

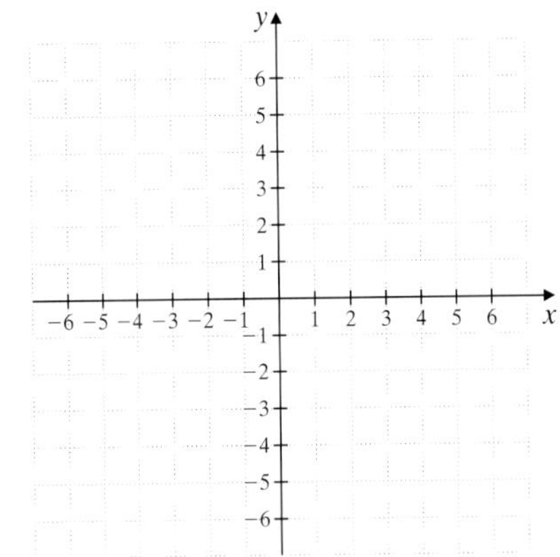

Answers

2.

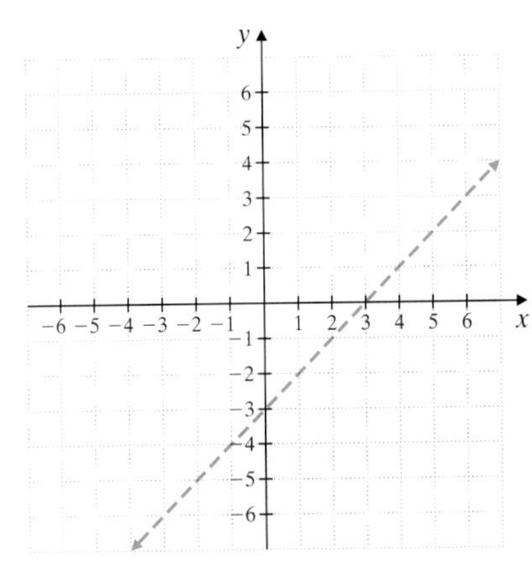

3. See page 1110.

Practice Problem 4

Graph: $y < 3x$

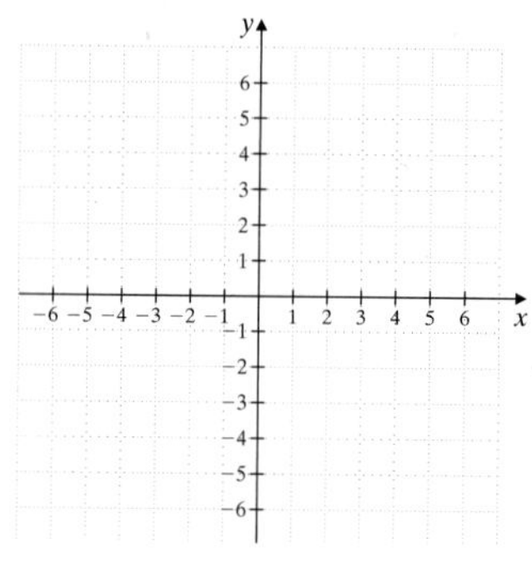

Answers

3.

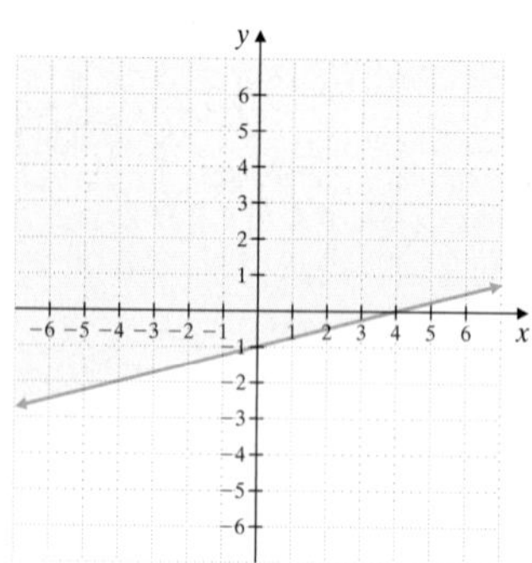

4.

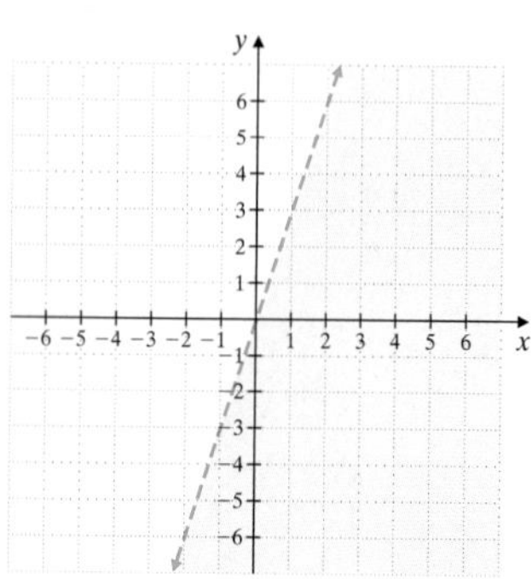

We substitute 0 for x and 0 for y into the original inequality.

$2x - y \geq 3$

$2(0) - 0 \geq 3$ Let $x = 0$ and $y = 0$.

$0 \geq 3$ False

Step 3. Since the statement is false, no point in the half-plane containing $(0, 0)$ is a solution. Therefore, we shade the half-plane that does not contain $(0, 0)$. Every point in the shaded half-plane and every point on the boundary line is a solution of $2x - y \geq 3$.

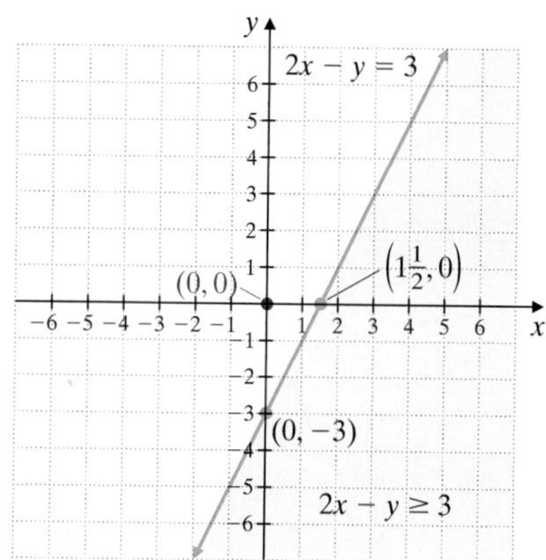

Helpful Hint

When graphing an inequality, make sure the test point is substituted into the **original inequality.** For Example 3, we substituted the test point $(0, 0)$ into the **original inequality** $2x - y \geq 3$, *not* $2x - y = 3$.

EXAMPLE 4 Graph: $x > 2y$

Solution:

Step 1. We find the boundary line by graphing $x = 2y$. The boundary line is a dashed line since the inequality symbol is $>$.

Step 2. We cannot use $(0, 0)$ as a test point because it is a point on the boundary line. We choose instead $(0, 2)$.

$x > 2y$

$0 > 2(2)$ Let $x = 0$ and $y = 2$.

$0 > 4$ False

Step 3. Since the statement is false, we shade the half-plane that does not contain the test point $(0, 2)$, as shown.

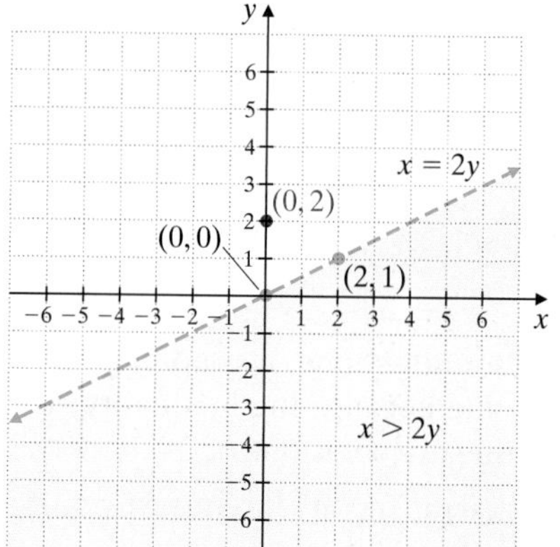

EXAMPLE 5 Graph: $5x + 4y \le 20$

Solution: We graph the solid boundary line $5x + 4y = 20$ and choose $(0, 0)$ as the test point.

$$5x + 4y \le 20$$
$$5(0) + 4(0) \le 20 \quad \text{Let } x = 0 \text{ and } y = 0.$$
$$0 \le 20 \quad \text{True}$$

We shade the half-plane that contains $(0, 0)$, as shown.

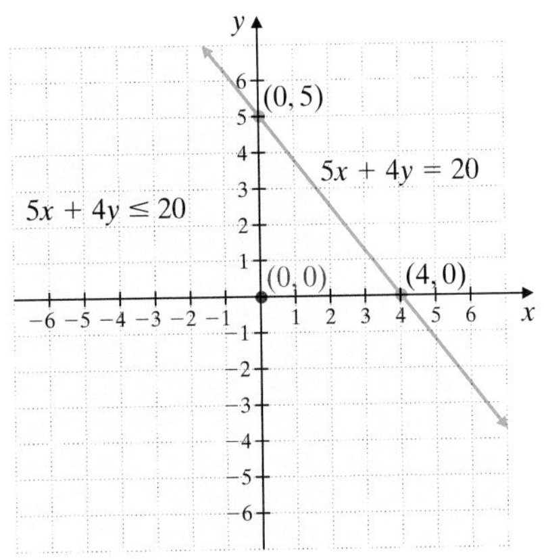

EXAMPLE 6 Graph: $y > 3$

Solution: We graph the dashed boundary line $y = 3$ and choose $(0, 0)$ as the test point. (Recall that the graph of $y = 3$ is a horizontal line with y-intercept 3.)

$$y > 3$$
$$0 > 3 \quad \text{Let } y = 0.$$
$$0 > 3 \quad \text{False}$$

We shade the half-plane that does not contain $(0, 0)$, as shown.

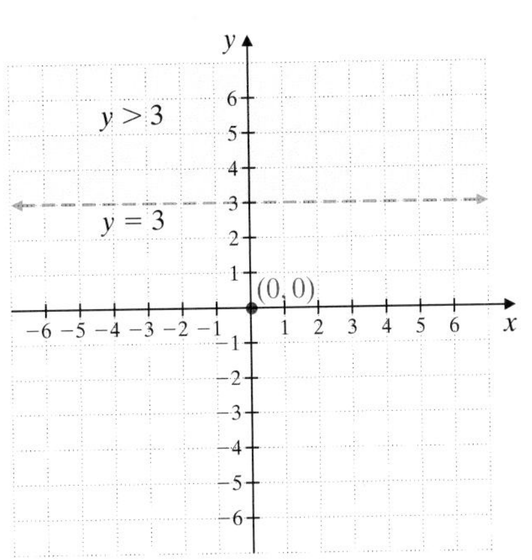

Practice Problem 5

Graph: $3x + 2y \ge 12$

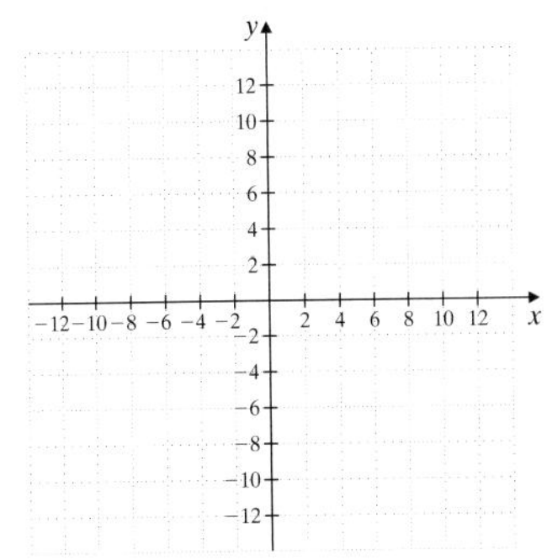

Practice Problem 6

Graph: $x < 2$

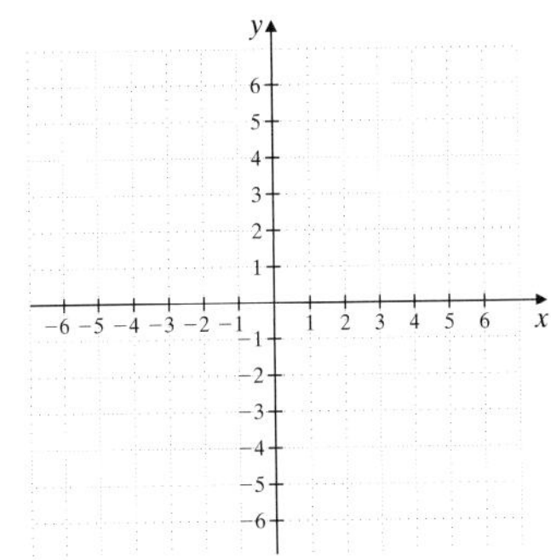

Answers

5.

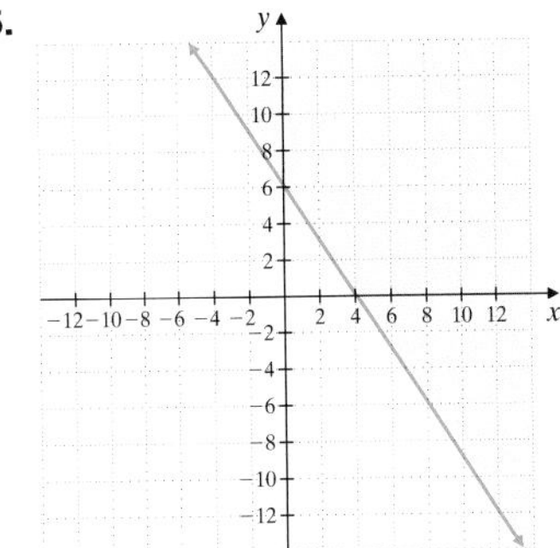

6.

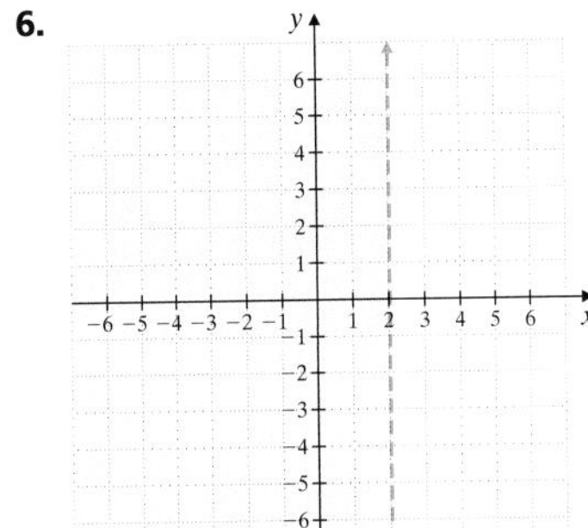

Practice Problem 7

Graph: $y \geq \frac{1}{4}x + 3$

EXAMPLE 7 Graph: $y \leq \frac{2}{3}x - 4$

Solution: Graph the solid boundary line $y = \frac{2}{3}x - 4$. This equation is in slope-intercept form with slope $\frac{2}{3}$ and y-intercept -4.

We use this information to graph the line. Then we choose $(0, 0)$ as our test point.

$$y < \frac{2}{3}x - 4$$

$$0 < \frac{2}{3} \cdot 0 - 4$$

$$0 < -4 \qquad \text{False}$$

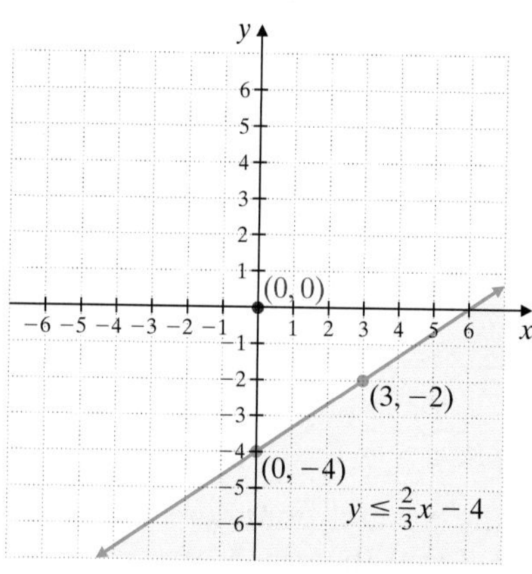

We shade the half-plane that does not contain $(0, 0)$, as shown.

Answer

7.

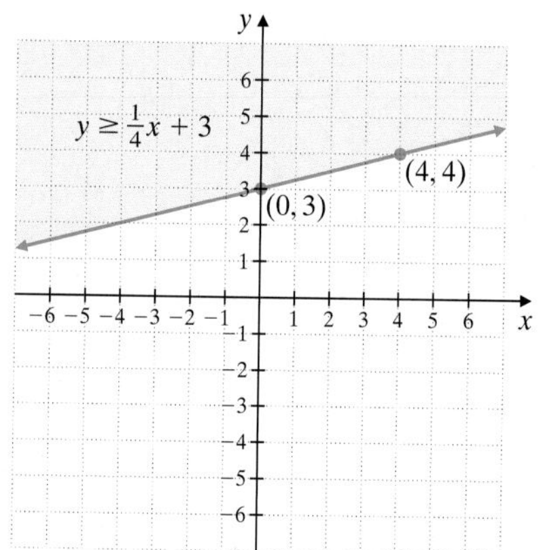

Name ______________________ Section __________ Date __________

Mental Math

State whether the graph of each inequality includes its corresponding boundary line.

1. $y \geq x + 4$ **2.** $x - y > -7$ **3.** $y \geq x$ **4.** $x > 0$

Decide whether $(0, 0)$ is a solution of each given inequality.

5. $x + y > -5$ **6.** $2x + 3y < 10$ **7.** $x - y \leq -1$ **8.** $\frac{2}{3}x + \frac{5}{6}y > 4$

EXERCISE SET 13.7

A *Determine whether the ordered pairs given are solutions of the linear inequality in two variables. See Example 1.*

1. $x - y > 3$; $(0, 3)$, $(2, -1)$ **2.** $y - x < -2$; $(2, 1)$, $(5, -1)$ **3.** $3x - 5y \leq -4$; $(2, 3)$, $(-1, -1)$

4. $2x + y \geq 10$; $(0, 11)$, $(5, 0)$ **5.** $x < -y$; $(0, 2)$, $(-5, 1)$ **6.** $y > 3x$; $(0, 0)$, $(1, 4)$

B *Graph each inequality. See Examples 2 through 7.*

7. $x + y \leq 1$

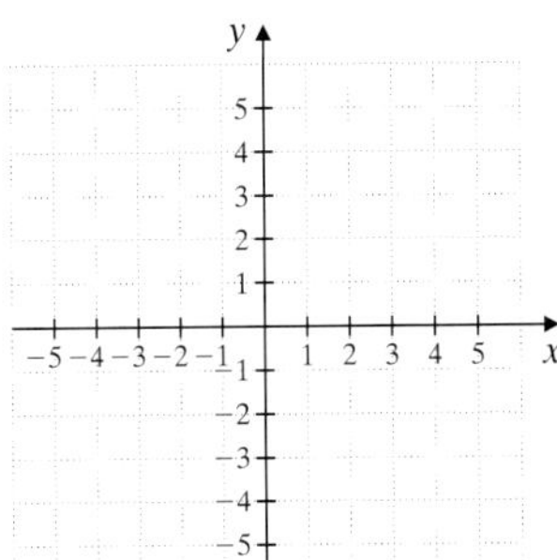

8. $x + y \geq -2$

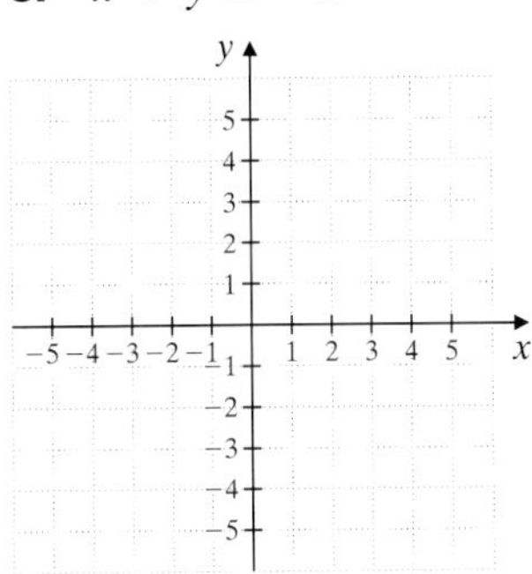

9. $2x - y > -4$

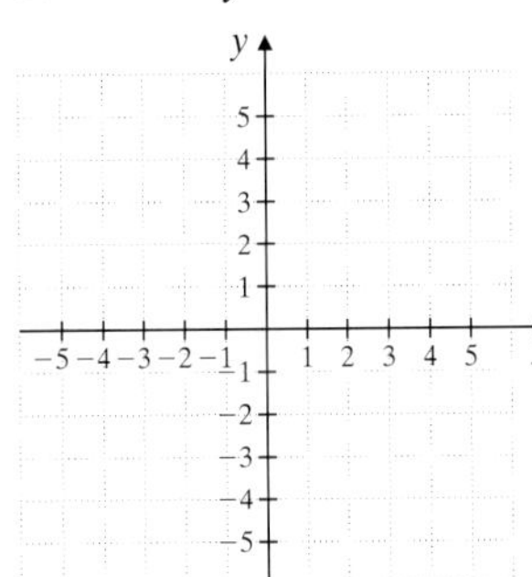

10. $x - 3y < 3$

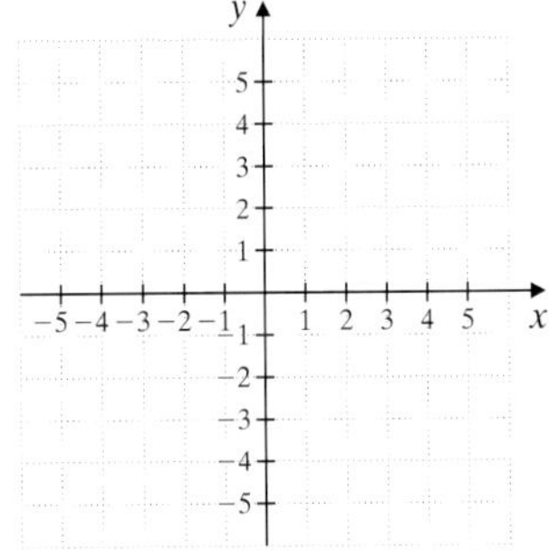

11. $y > 2x$

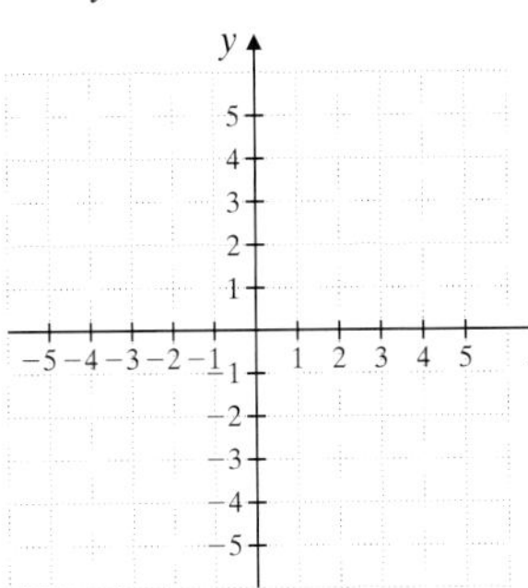

12. $y < 3x$

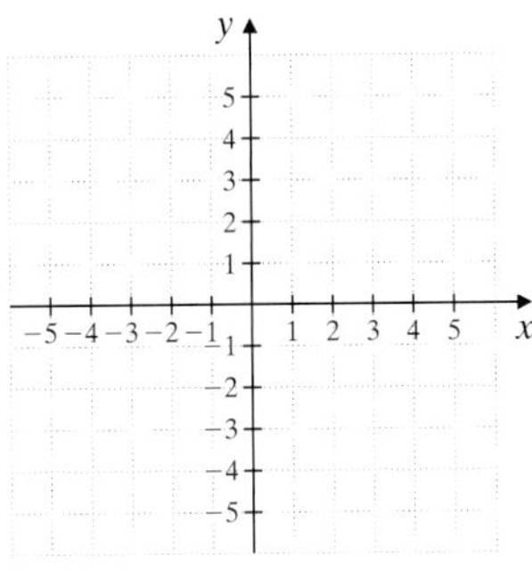

13. $x \leq -3y$

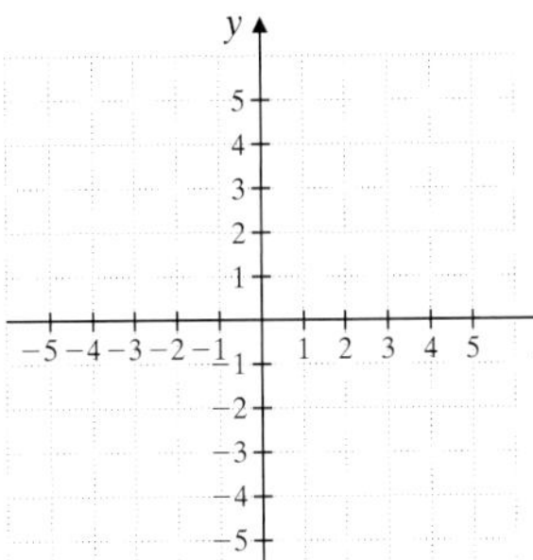

14. $x \geq -2y$

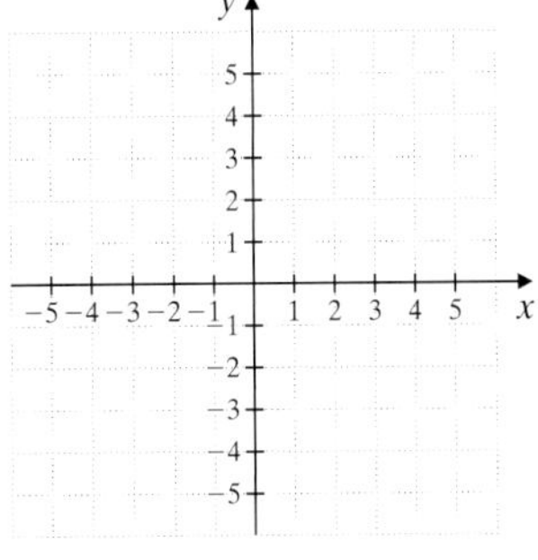

15. $y \geq x + 5$

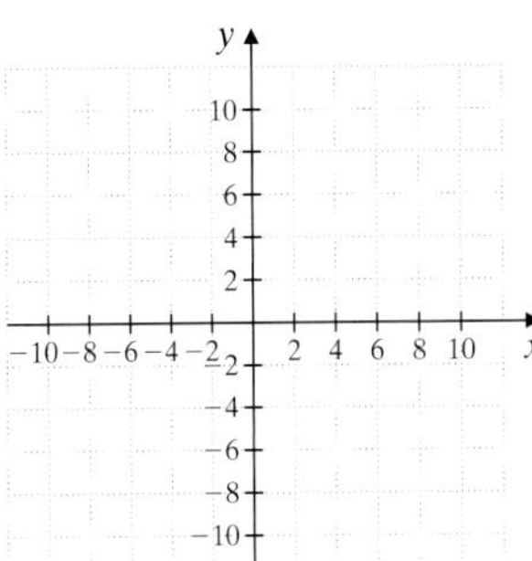

16. $y \leq x + 1$

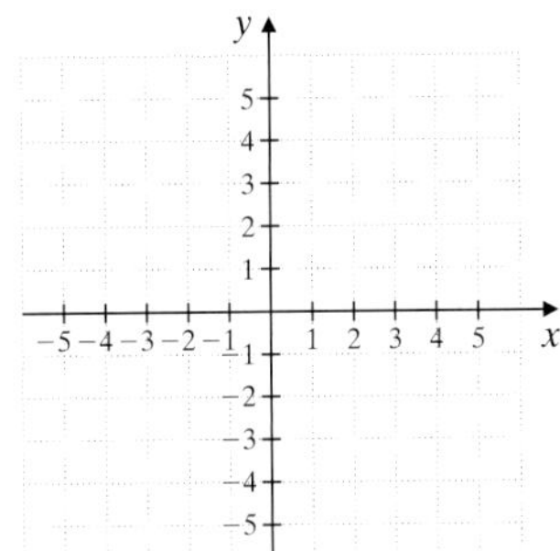

17. $y < 4$

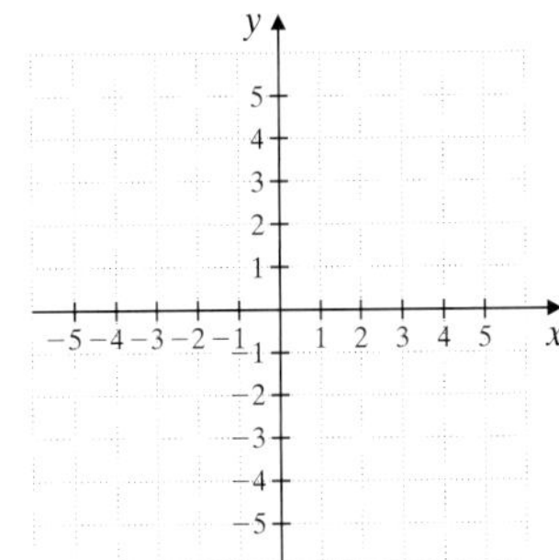

18. $y > 2$

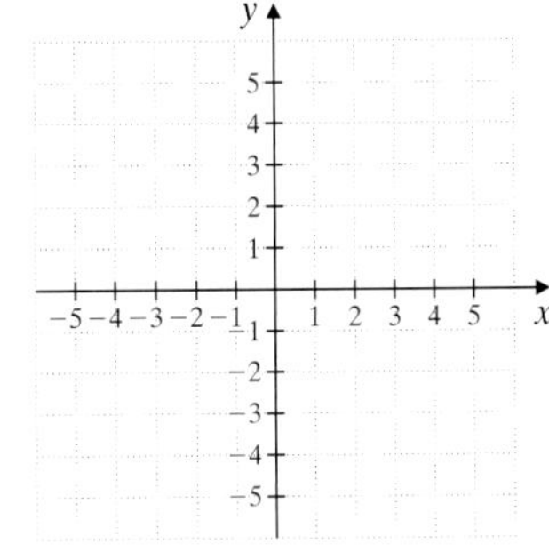

19. $x \geq -3$

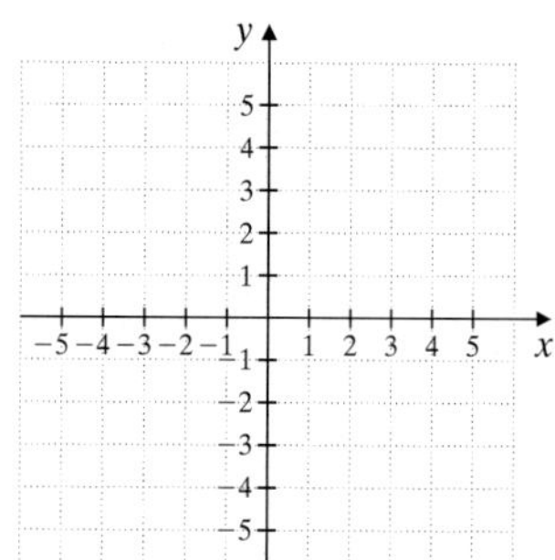

20. $x \leq -1$

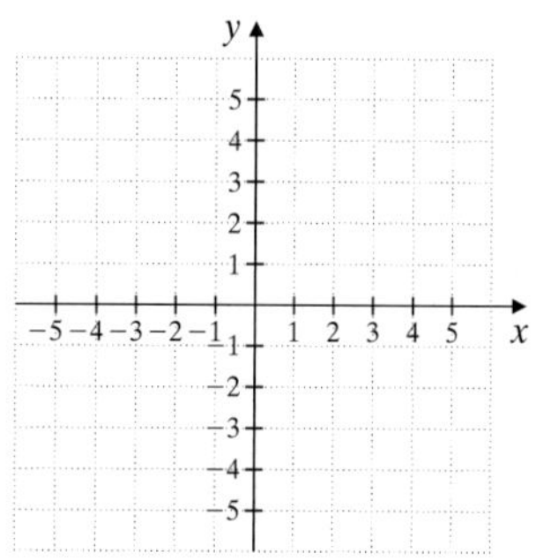

21. $5x + 2y \leq 10$

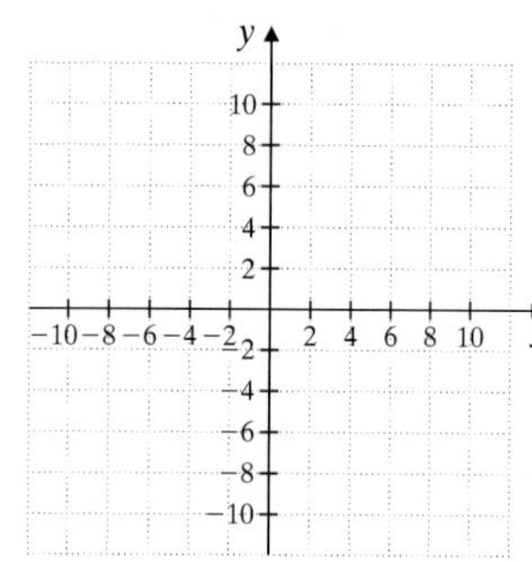

22. $4x + 3y \geq 12$

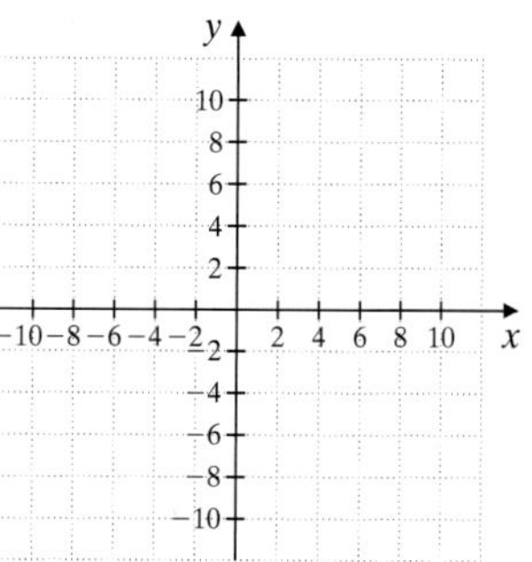

23. $x > y$

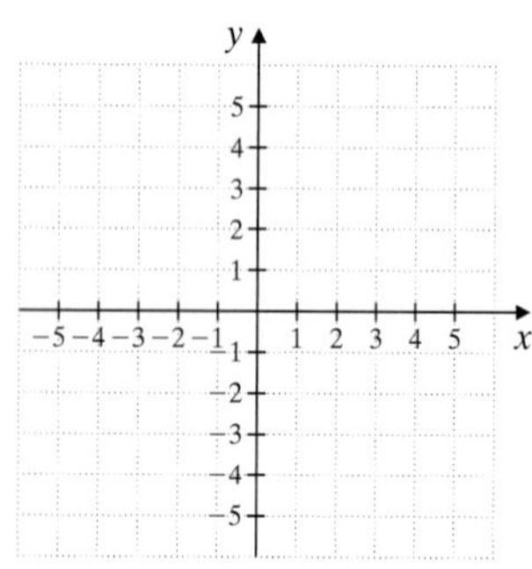

24. $x \leq -y$

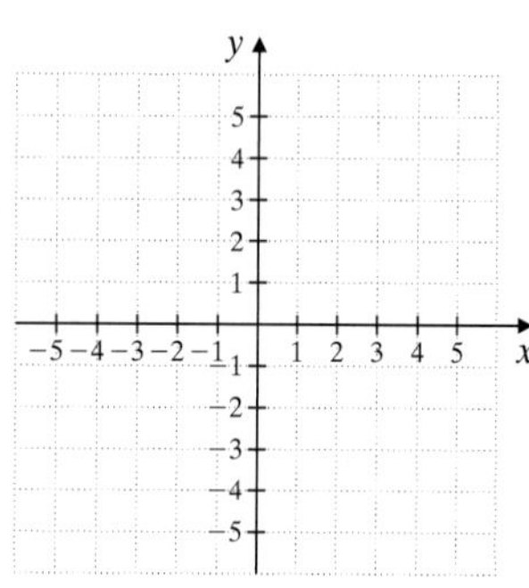

25. $x - y \leq 6$

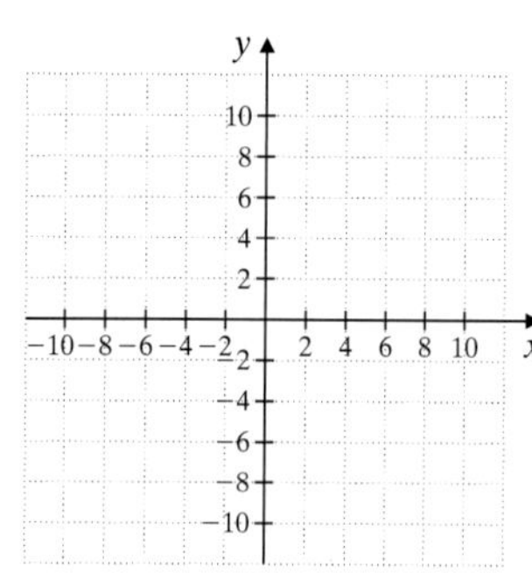

26. $x - y > 10$

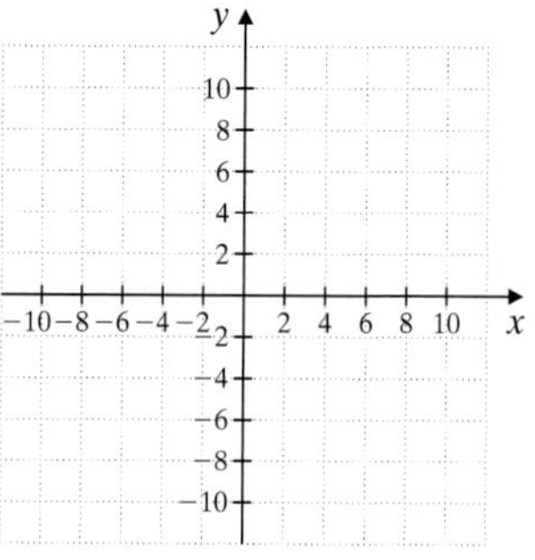

27. $x \geq 0$

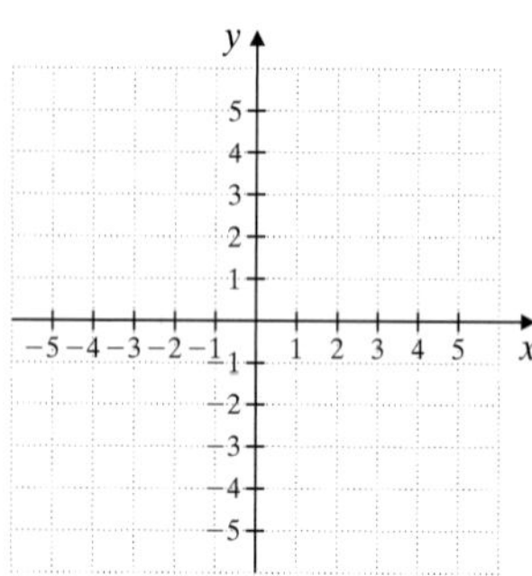

28. $y \leq 0$

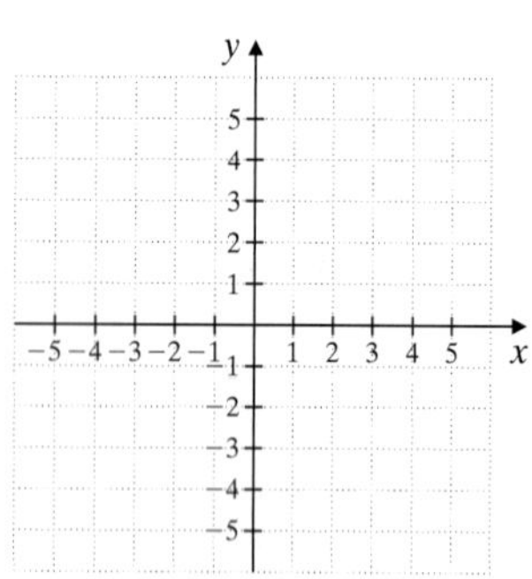

29. $2x + 7y > 5$

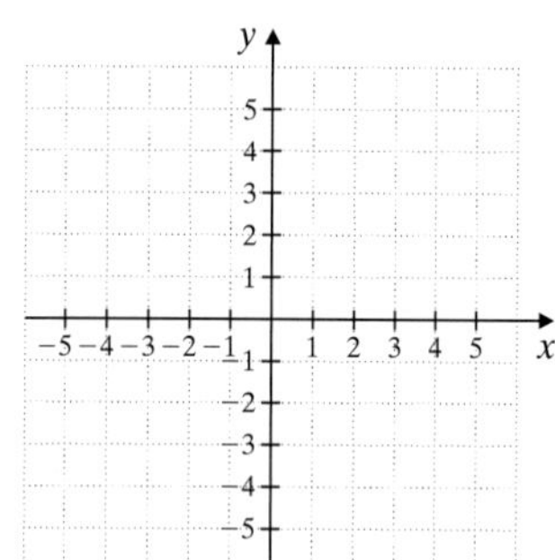

30. $3x + 5y \leq -2$

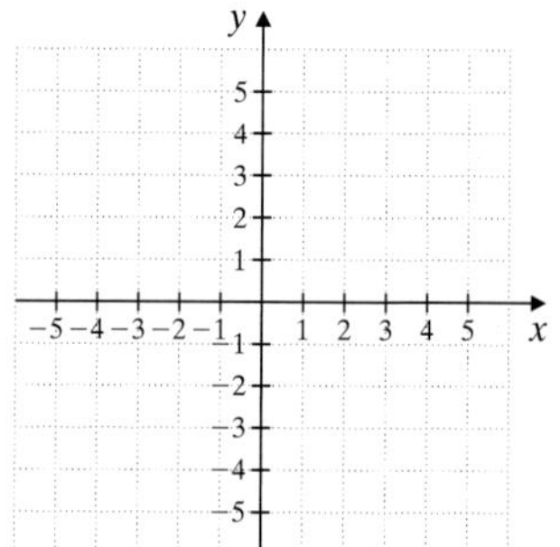

31. $y \geq \frac{1}{2}x - 4$

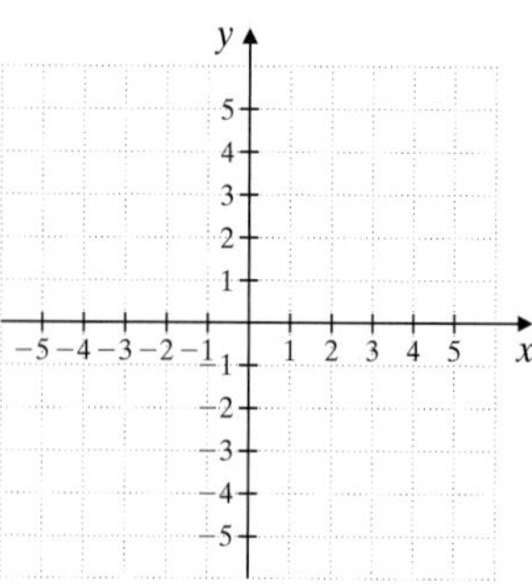

32. $y < \frac{2}{5}x - 3$

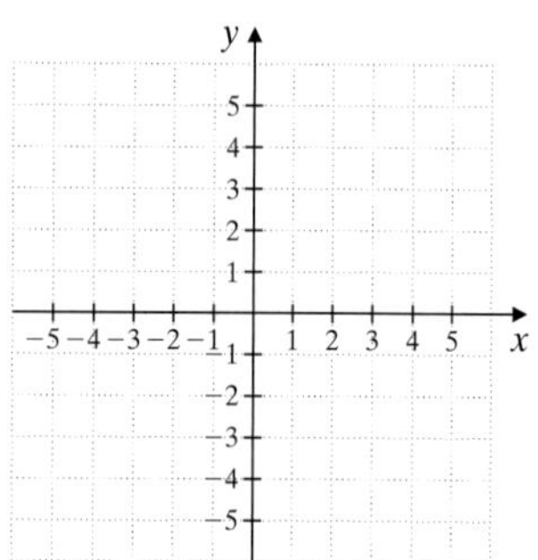

Review and Preview

Approximate the coordinates of each point of intersection. See Section 13.1.

33.

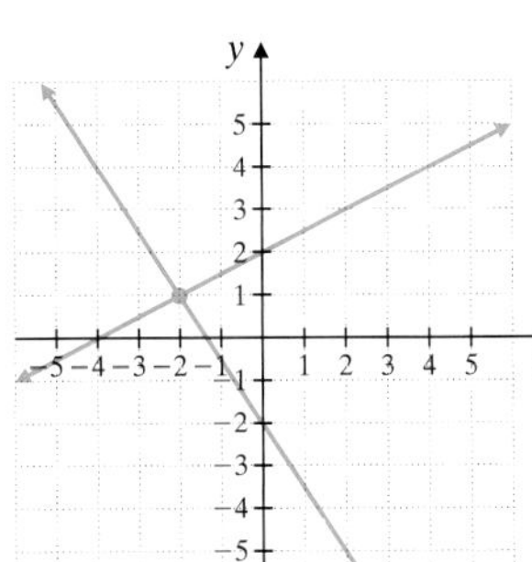

34.

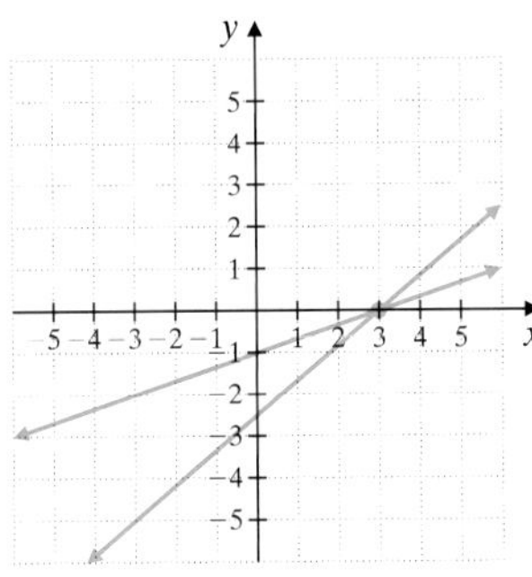

35.

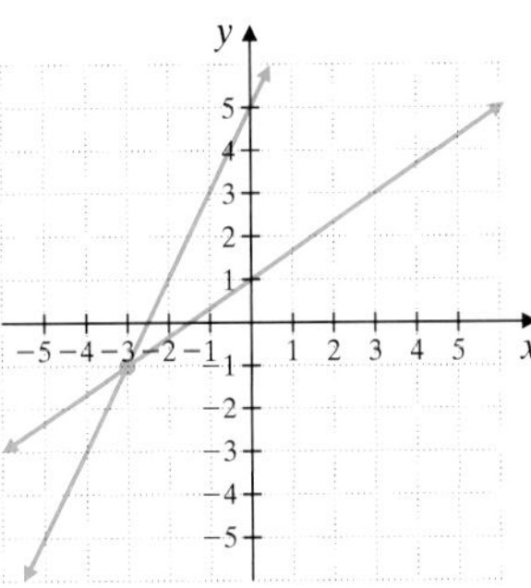

36.

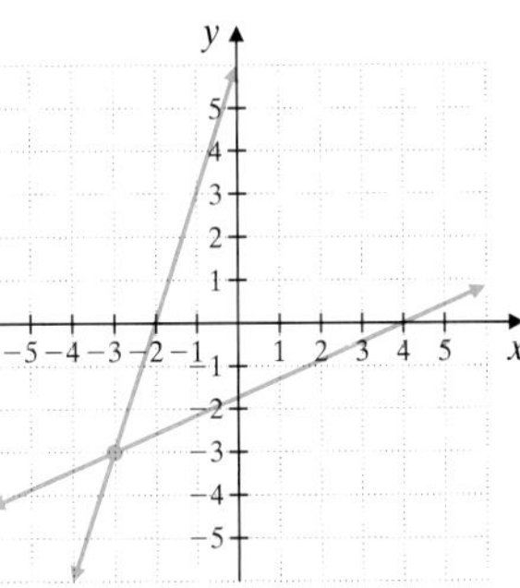

Combining Concepts

Match each inequality with its graph.

A. $x > 2$

B. $y < 2$

C. $y \le 2x$

D. $y \le -3x$

37.

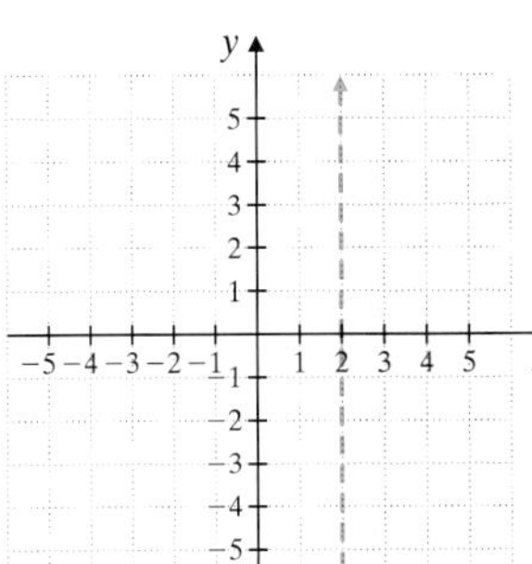

38.

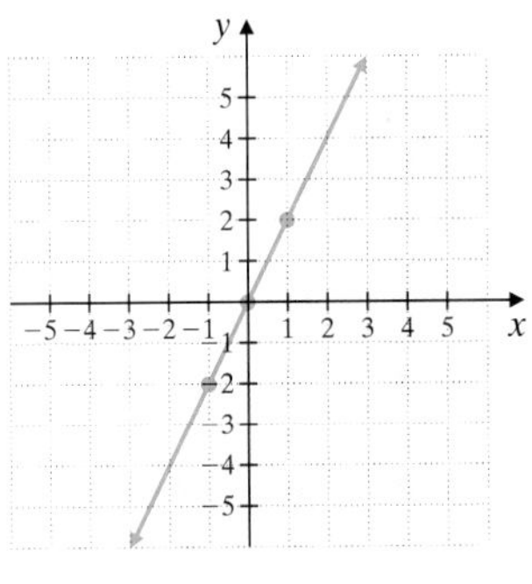

39.

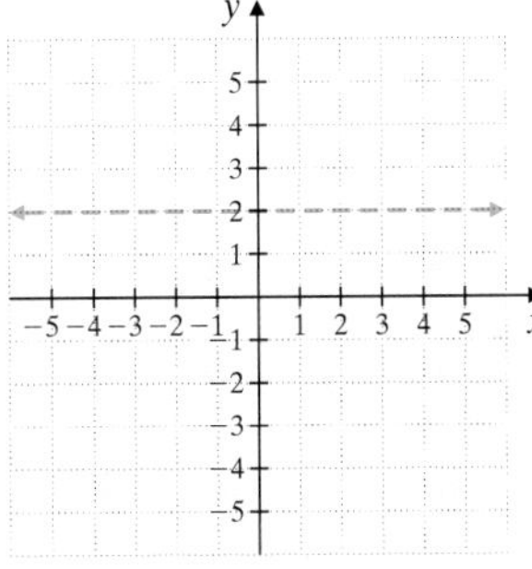

40.

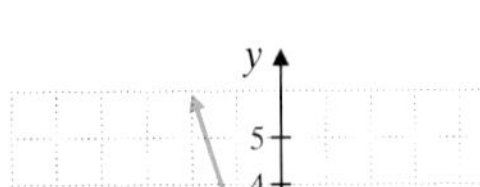

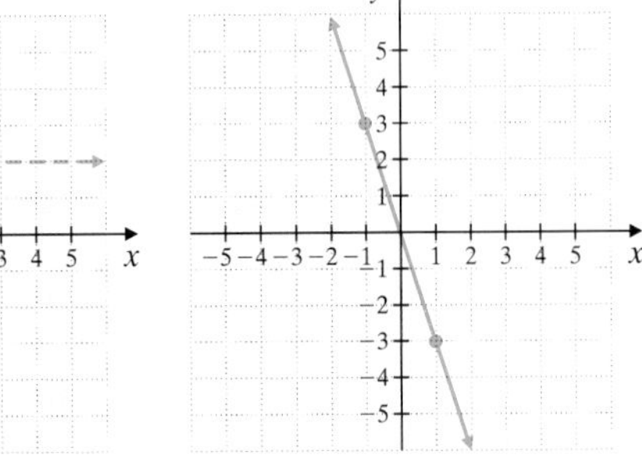

41. Explain why a point on the boundary line should not be chosen as the test point.

42. Write an inequality whose solutions are all points of numbers whose sum is at least 13.

43. It's the end of the budgeting period for Dennis Fernandes and he has \$500 left in his budget for car rental expenses. He plans to spend this budget on a sales trip throughout southern Texas. He will rent a car that costs \$30 per day and \$0.15 per mile and he can spend no more than \$500.

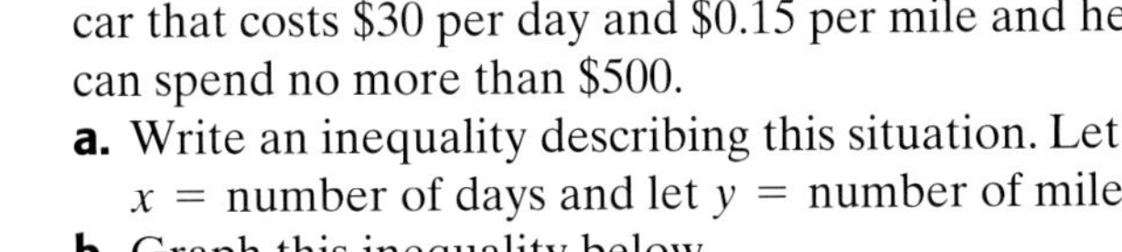

a. Write an inequality describing this situation. Let x = number of days and let y = number of miles.

b. Graph this inequality below.

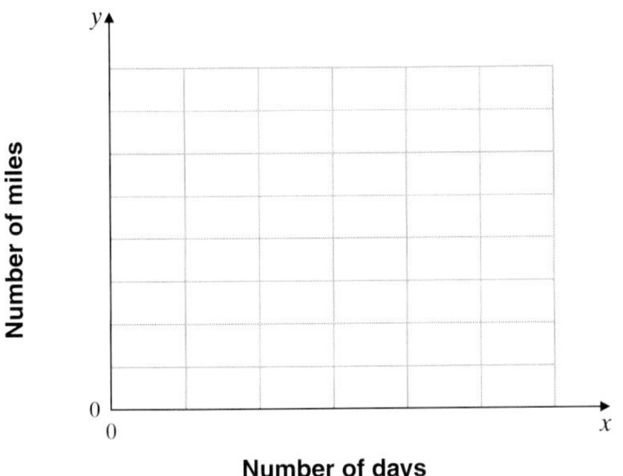

c. Why is the grid showing quadrant I only?

44. Scott Sambracci and Sara Thygeson are planning their wedding. They have calculated that they want the cost of their wedding ceremony x plus the cost of their reception y to be no more than \$5000.

a. Write an inequality describing this relationship.

b. Graph this inequality below.

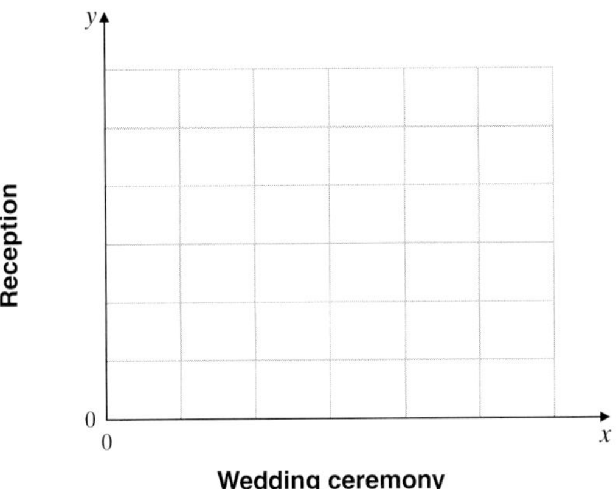

c. Why is the grid showing quadrant I only?

STUDY SKILLS REMINDER

Are you prepared for a test on Chapter 13?

Below I have listed some common trouble areas for topics covered in Chapter 13. After studying for your test—but before taking your test—read these.

- If you are having trouble with graphing, you might want to ask your instructor if you can use graph paper on your test. This will save you time and keep your graphs neat.
- Don't forget that the graph of an ordered pair is a *single* point in the rectangular coordinate system.
- Make sure you remember that to find the slope of a linear equation using its equation, *first* solve the equation for y. *Then* the coefficient of x is its slope.

$$2x + 3y = 7$$

$$3y = -2x + 7 \quad \text{Subtract } 2x \text{ from both sides.}$$

$$\frac{3y}{3} = -\frac{2}{3}x + \frac{7}{3} \quad \text{Divide both sides by 3.}$$

$$y = -\frac{2}{3}x + \frac{7}{3}$$

$\frac{7}{3}$: y-intercept; $-\frac{2}{3}$: slope

- Remember that a point that is an x-intercept will have a y-value of 0 and a point that is a y-intercept will have an x-value of 0. Also—the point $(0, 0)$ will be both an x- and y-intercept.
- If you studied functions, remember that $f(x)$ *does not* mean $f \cdot x$. It is a special function notation. If $f(x) = x^2 - 6$, then $f(-3) = (-3)^2 - 6 = 9 - 6 = 3$.

13.8 Direct and Inverse Variation

OBJECTIVES

- **A** Solve problems involving direct variation.
- **B** Solve problems involving inverse variation.
- **C** Solve problems involving other types of direct and inverse variation.
- **D** Solve applications of variation.

In Chapter 9, we studied linear equations in two variables. Recall that such an equation can be written in the form $Ax + By = C$, where A and B are not both 0. Also recall that the graph of a linear equation in two variables is a line. In this section, we begin by looking at a particular family of linear equations—those that can be written in the form

$$y = kx$$

where k is a constant. This family of equations is called *direct variation.*

A Solving Direct Variation Problems

Let's suppose that you are earning \$7.25 per hour at a part-time job. The amount of money you earn depends on the number of hours you work. This is illustrated by the following table:

Hours worked	0	1	2	3	4
Money Earned (before deductions)	0	7.25	14.50	21.75	29.00

and so on

In general, to calculate your earnings (before deductions), multiply the constant \$7.25 by the number of hours you work. If we let y represent the amount of money earned and x represent the number of hours worked, we get the direct variation equation

$$y = 7.25 \cdot x$$

earnings = \$7.25 · hours worked

Notice that in this direct variation equation, as the number of hours increases, the pay increases as well.

Direct Variation

y varies directly as x, or **y is directly proportional to x,** if there is a nonzero constant k such that

$$y = kx$$

The number k is called the **constant of variation** or the **constant of proportionality.**

In our direct variation example, $y = 7.25x$, the constant of variation is 7.25.

Let's use the previous table to graph $y = 7.25x$. We begin our graph at the ordered-pair solution $(0, 0)$. Why? We assume that the least amount of hours worked is 0. If 0 hours are worked, then the pay is \$0.

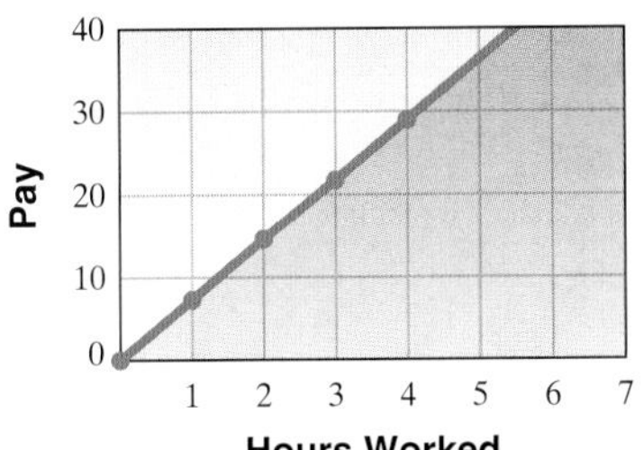

As illustrated in this graph, a direct variation equation $y = kx$ is linear. Also notice that $y = 7.25x$ is a function since its graph passes the vertical line test.

Practice Problem 1

Write a direct variation equation that satisfies:

x	4	$\frac{1}{2}$	1.5	6
y	8	1	3	12

EXAMPLE 1 Write a direct variation equation of the form $y = kx$ that satisfies the ordered pairs in the table below.

x	2	9	1.5	-1
y	6	27	4.5	-3

Solution: We are given that there is a direct variation relationship between x and y. This means that

$$y = kx$$

By studying the given values, you may be able to mentally calculate k. If not, to find k, we simply substitute one given ordered pair into this equation and solve for k. We'll use the given pair $(2, 6)$.

$$y = kx$$
$$6 = k \cdot 2$$
$$\frac{6}{2} = \frac{k \cdot 2}{2}$$
$$3 = k \quad \text{Solve for } k.$$

Since $k = 3$, we have the equation $y = 3x$.

To check, see that each given y is 3 times the given x.

Let's try another type of direct variation example.

Practice Problem 2

Suppose that y varies directly as x. If y is 15 when x is 45, find the constant of variation and the direct variation equation. Then find y when x is 3.

EXAMPLE 2 Suppose that y varies directly as x. If k is 17 when x is 34, find the constant of variation and the direct variation equation. Then find y when x is 12.

Solution: Let's use the same method as in Example 1 to find k. Since we are told that y varies directly as x, we know the relationship is of the form

$$y = kx$$

Let $y = 17$ and $x = 34$ and solve for k.

$$17 = k \cdot 34$$
$$\frac{17}{34} = \frac{k \cdot 34}{34}$$
$$\frac{1}{2} = k \quad \text{Solve for } k.$$

Thus, the constant of variation is $\frac{1}{2}$ and the equation is $y = \frac{1}{2}x$.

To find y when $x = 12$, use $y = \frac{1}{2}x$ and replace x with 12.

$$y = \frac{1}{2}x$$
$$y = \frac{1}{2} \cdot 12 \quad \text{Replace } x \text{ with 12.}$$
$$y = 6$$

Thus, when x is 12, y is 6.

Let's review a few facts about linear equations of the form $y = kx$.

Answers

1. $y = 2x$ **2.** $y = \frac{1}{3}x; y = 1$

Direct Variation: $y = kx$

- There is a direct variation relationship between x and y.
- The graph is a line.
- The line will always go through the origin $(0, 0)$. Why?

$$\text{Let } x = 0. \text{ Then } y = k \cdot 0 \text{ or } y = 0.$$

- The slope of the graph of $y = kx$ is k, the constant of variation. Why? Remember that the slope of an equation of the form $y = mx + b$ is m, the coefficient of x.
- The equation $y = kx$ describes a function. Each x has a unique y and its graph passes the vertical line test.

EXAMPLE 3 The line is the graph of a direct variation equation. Find the constant of variation and the direct variation equation.

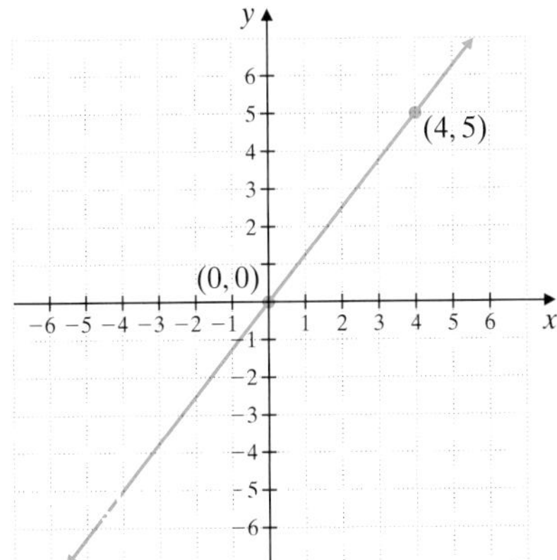

Solution: Recall that k, the constant of variation, is the same as the slope of the line. Thus, to find k, we use the slope formula and find slope.

Using the given points $(0, 0)$, and $(4, 5)$, we have

$$\text{slope} = \frac{5 - 0}{4 - 0} = \frac{5}{4}$$

Thus, $k = \dfrac{5}{4}$ and the variation equation is $y = \dfrac{5}{4}x$.

Practical Problem 3

Find the constant of variation and the direct variation equation for the line below.

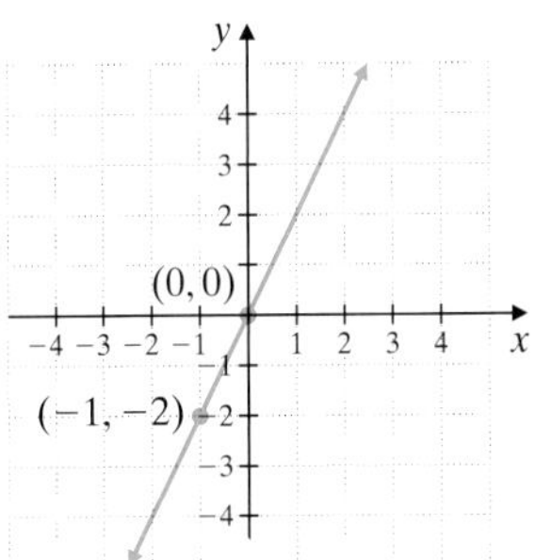

B Solving Inverse Variation Problems

In this section, we introduce another type of variation called inverse variation.

Let's suppose you need to drive a distance of 40 miles. You know that the faster you drive the distance, the sooner you arrive at your destination. Recall that there is a mathematical relationship between distance, rate, and time. It is $d = r \cdot t$. In our example, distance is a constant 40 miles, so we have $40 = r \cdot t$ or $t = \dfrac{40}{r}$.

For example, if you drive 10 mph, the time to drive the 40 miles is

$$t = \frac{40}{r} = \frac{40}{10} = 4 \text{ hours}$$

If you drive 20 mph, the time is

$$t = \frac{40}{r} = \frac{40}{20} = 2 \text{ hours}$$

Answer

3. $y = 2x$

Again, notice that as speed increases, time decreases. Below are some ordered-pair solutions of $t = \frac{40}{r}$ and its graph.

rate (mph)	r	5	10	20	40	60	80
time (hr)	t	8	4	2	1	$\frac{2}{3}$	$\frac{1}{2}$

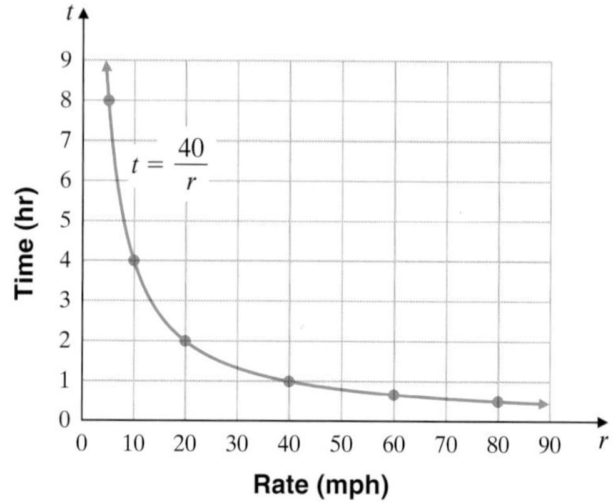

Notice that the graph of this variation is not a line, but it passes the vertical line test so $t = \frac{40}{r}$ does describe a function. This is an example of inverse variation.

Inverse Variation

***y* varies inversely as *x*,** or ***y* is inversely proportional to *x*,** if there is a nonzero constant k such that

$$y = \frac{k}{x}$$

The number k is called the **constant of variation** or the **constant of proportionality.**

In our inverse variation example, $t = \frac{40}{r}$ or $y = \frac{40}{x}$, the constant of variation is 40.

We can immediately see differences and similarities in direct variation and inverse variation.

Direct variation	$y = kx$	linear equation	both
Inverse variation	$y = \frac{k}{x}$	rational equation	Functions

Remember from Chapter 12 that $y = \frac{k}{x}$ is a rational equation and not a linear equation. Also notice that because x is in the denominator, x can be any value except 0.

We can still derive an inverse variation equation from a table of values.

Practical Problem 4

Write an inverse variation equation of the form $y = \frac{k}{x}$ that satisfies:

x	4	10	40	−2
y	5	2	$\frac{1}{2}$	−10

EXAMPLE 4 Write an inverse variation equation of the form $y = \frac{k}{x}$ that satisfies the ordered pairs in the table below.

x	2	4	$\frac{1}{2}$
y	6	3	24

Solution: Since there is an inverse variation relationship between x and y, we know that $y = \frac{k}{x}$.

Answer

4. $y = \frac{20}{x}$

To find k, choose one given ordered pair and substitute the values into the equation. We'll use (2, 6).

$$y = \frac{k}{x}$$

$$6 = \frac{k}{2}$$

$$2 \cdot 6 = 2 \cdot \frac{k}{2} \quad \text{Multiply both sides by 2.}$$

$$12 = k \quad \text{Solve.}$$

Since $k = 12$, we have the equation $y = \dfrac{12}{x}$.

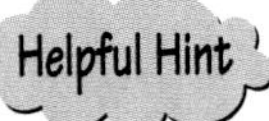

Multiply both sides of the inverse variation relationship equation $y = \dfrac{k}{x}$ by x (as long as x is not 0), and we have $xy = k$. This means that if y varies inversely as x, their product is always the constant of variation k. For an example of this, check the table above.

x	2	4	$\frac{1}{2}$
y	6	3	24

$2 \cdot 6 = 12$ $\quad 4 \cdot 3 = 12$ $\quad \frac{1}{2} \cdot 24 = 12$

EXAMPLE 5 Suppose that y varies inversely as x. If $y = 0.02$ when $x = 75$, find the constant of variation and the inverse variation equation. Then find y when x is 30.

Solution: Since y varies inversely as x, the constant of variation may be found by simply finding the product of the given x and y.

$$k = xy = 75(0.02) = 1.5$$

To check, we will use the inverse variation equation

$$y = \frac{k}{x}$$

Let $y = 0.02$ and $x = 75$ and solve for k.

$$0.02 = \frac{k}{75}$$

$$75(0.02) = 75 \cdot \frac{k}{75} \quad \text{Multiply both sides by 75.}$$

$$1.5 = k \quad \text{Solve for } k.$$

Thus, the constant of variation is 1.5 and the equation is $y = \dfrac{1.5}{x}$. To find y when $x = 30$, use $y = \dfrac{1.5}{x}$ and replace x with 30.

$$y = \frac{1.5}{x}$$

$$y = \frac{1.5}{30} \quad \text{Replace } x \text{ with 30.}$$

$$y = 0.05$$

Thus, when x is 30, y is 0.05.

Practice Problem 5

Suppose that y varies inversely as x. If y is 4 when x is 0.8, find the constant of variation and the direct variation equation. Then find y when x is 20.

Answer

5. $y = \dfrac{3.2}{x}$; $y = 0.16$

C Solving Other Types of Direct and Inverse Variation Problems

It is possible for y to vary directly or inversely as powers of x.

Direct and Inverse Variation as nth Powers of x

y varies directly as a power of x if there is a nonzero constant k and a natural number n such that

$$y = kx^n$$

y varies inversely as a power of x if there is a nonzero constant k and a natural number n such that

$$y = \frac{k}{x^n}$$

Practice Problem 6

The area of a circle varies directly as the square of its radius. A circle with radius 7 inches has an area of 49π square inches. Find the area of a circle whose radius is 4 feet.

EXAMPLE 6 The surface area of a cube A varies directly as the square of a length of its sides. If A is 54 when s is 3, find A when $s = 4.2$.

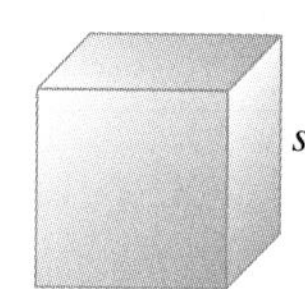

Solution: Since the surface area A varies directly as the square of side s, we have

$$A = ks^2$$

To find k, let $A = 54$ and $s = 3$.

$A = k \cdot s^2$

$54 = k \cdot 3^2$ Let $A = 54$ and $s = 3$.

$54 = 9k$ $3^2 = 9$.

$6 = k$ Divide by 9.

The formula for surface area of a cube is then

$A = 6s^2$ where s is the length of a side.

To find the surface area when $s = 4.2$, substitute.

$A = 6s^2$

$A = 6 \cdot (4.2)^2$

$A = 105.84$

The surface area of a cube whose side measures 4.2 units is 105.84 sq units.

Answer

6. 16π sq ft

D Solving Applications of Variation

There are many real-life applications of direct and inverse variation.

EXAMPLE 7 The weight of a body w varies inversely with the square of its distance from the center of Earth d. If a person weighs 160 pounds on the surface of Earth, what is the person's weight 200 miles above the surface? (Assume that the radius of Earth is 4000 miles.)

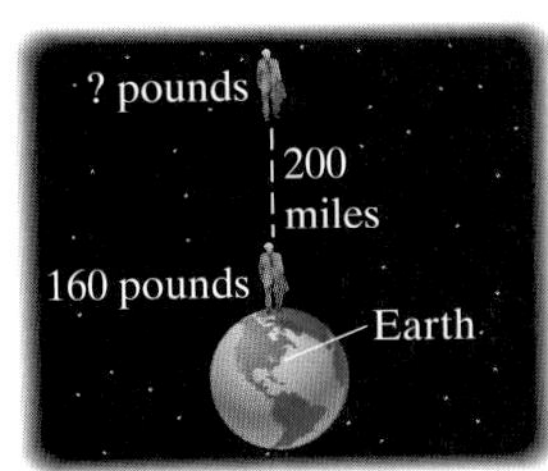

Solution:

1. UNDERSTAND. Make sure you read and reread the problem.
2. TRANSLATE. Since we are told that weight w varies inversely with the square of its distance from the center of Earth, d, we have

$$w = \frac{k}{d^2}$$

3. SOLVE. To solve the problem, we first find k. To do so, use the fact that the person weighs 160 pounds on Earth's surface, which is a distance of 4000 miles from the Earth's center.

$$w = \frac{k}{d^2}$$

$$160 = \frac{k}{(4000)^2}$$

$$2{,}560{,}000{,}000 = k$$

Thus, we have $w = \dfrac{2{,}560{,}000{,}000}{d^2}$.

Since we want to know the person's weight 200 miles above the Earth's surface, we let $d = 4200$ and find w.

$$w = \frac{2{,}560{,}000{,}000}{d^2}$$

$$w = \frac{2{,}560{,}000{,}000}{(4200)^2}$$ A person 2 miles above the Earth's surface is 4200 miles from the Earth's center.

$$w \approx 145$$ Simplify.

4. INTERPRET. *Check*: Your answer is reasonable since the further a person is from Earth, the less the person weighs. *State*: Thus, 200 miles above the surface of the Earth, a 160-pound person weighs approximately 145 pounds.

Practice Problem 7

The distance d that an object falls is directly proportional to the square of the time of the fall, t. If an object falls 144 feet in 3 seconds, find how far the object falls in 5 seconds.

Answer

7. 400 feet

STUDY SKILLS REMINDER

How are your homework assignments going?

By now, you should have good homework habits. If not, it's never too late to begin. Why is it so important in mathematics to keep up with homework? You probably now know the answer to that question. You may have realized by now that many concepts in mathematics build on each other. Your understanding of one chapter in mathematics usually depends on your understanding of the previous chapter's material.

Don't forget that completing your homework assignment involves a lot more than attempting a few of the problems assigned.

To complete a homework assignment, remember these four things:

1. Attempt all of it.
2. Check it.
3. Correct it.
4. If needed, ask questions about it.

Name ____________________ Section __________ Date __________

Mental Math

State whether each equation represents direct or indirect variation.

1. $y = 5x$ **2.** $y = \dfrac{5}{x}$ **3.** $y = \dfrac{7}{x^2}$ **4.** $y = 6.5x^4$

5. $y = \dfrac{11}{x}$ **6.** $y = 18x$ **7.** $y = 12x^2$ **8.** $y = \dfrac{20}{x^3}$

EXERCISE SET 13.8

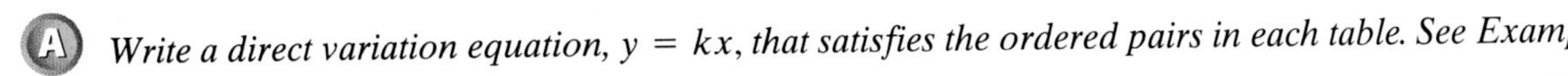

A *Write a direct variation equation, $y = kx$, that satisfies the ordered pairs in each table. See Example 1.*

1.

x	0	6	10
y	0	3	5

2.

x	0	2	−1	3
y	0	14	−7	21

3.

x	−2	2	4	5
y	−12	12	24	30

4.

x	3	9	−2	12
y	1	3	$-\frac{2}{3}$	4

Write a direct variation equation, $y = kx$, that describes each graph. See Example 3.

5.

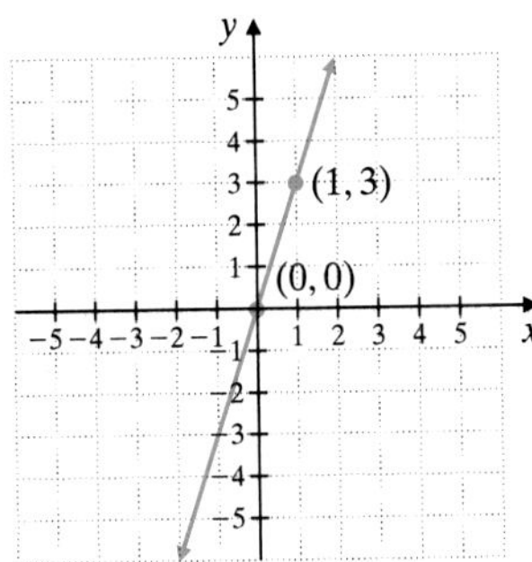

6.

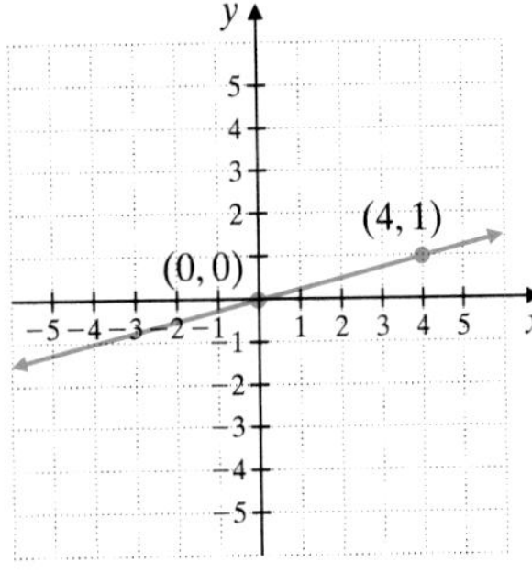

7.

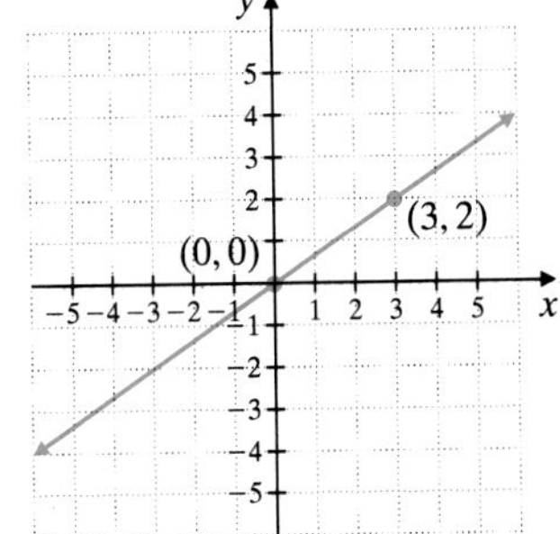

8.

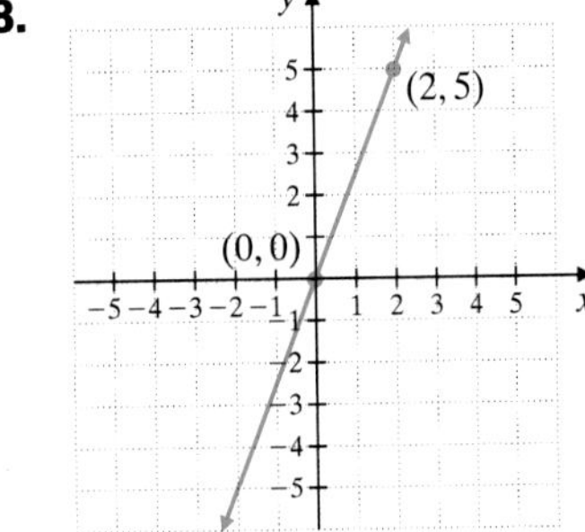

B *Write an inverse variation equation, $y = \frac{k}{x}$, that satisfies the ordered pairs in each table. See Example 4.*

9.

x	1	−7	3.5	−2
y	7	−1	2	−3.5

10.

x	2	−11	4	−4
y	11	−2	5.5	−5.5

11.

x	10	$\frac{1}{2}$	$-\frac{1}{2}$
y	0.05	1	$-\frac{1}{3}$

12.

x	4	$\frac{1}{5}$	−8
y	0.1	2	−0.05

Write an equation to describe each variation. Use k for the constant of proportionality. See Examples 1 through 5.

13. y varies directly as x

14. a varies directly as b

15. h varies inversely as t

16. s varies inversely as t

17. z varies directly as x^2

18. p varies inversely as x^2

19. y varies inversely as z^3

20. x varies directly as y^4

21. x varies inversely as $\sqrt{y}$

22. y varies directly as d^2

Solve See Examples 2, 5, and 6.

23. y varies directly as x. If $y = 20$ when $x = 5$, find y when x is 10.

24. y varies directly as x. If $y = 27$ when $x = 3$, find y when x is 2.

25. y varies inversely as x. If $y = 5$ when $x = 60$, find y when x is 100.

26. y varies inversely as x. If $y = 200$ when $x = 5$, find y when x is 4.

27. z varies directly as x^2. If $z = 96$ when $x = 4$, find z when $x = 3$.

28. s varies directly as t^3. If $s = 270$ when $t = 3$, find s when $x = 1$.

29. a varies inversely as b^3. If $a = \frac{3}{2}$ when $b = 2$, find a when b is 3.

30. p varies inversely as q^2. If $p = \frac{5}{16}$ when $q = 8$, find p when $q = \frac{1}{2}$.

C **D** *Solve. See Examples 1 through 7.*

31. Your paycheck (before deductions) varies directly as the number of hours you work. If your paycheck is \$112.50 for 18 hours, find your pay for 10 hours.

32. If your paycheck (before deductions) is \$244.50 for 30 hours, find your pay for 34 hours. See Exercise 31.

33. The cost of manufacturing a certain type of headphone varies inversely as the number of headphones increases. If 5000 headphones can be manufactured for \$9.00 each, find the cost to manufacture 7500 headphones.

34. The cost of manufacturing a certain composition notebook varies inversely as the number of notebooks increases. If 10,000 notebooks can be manufactured for \$0.50 each, find the cost to manufacture 18,000 notebooks.

35. The distance a spring stretches varies directly with the weight attached to the spring. If a 60-pound weight stretches the spring 4 inches, find the distance that an 80-pound weight stretches the spring.

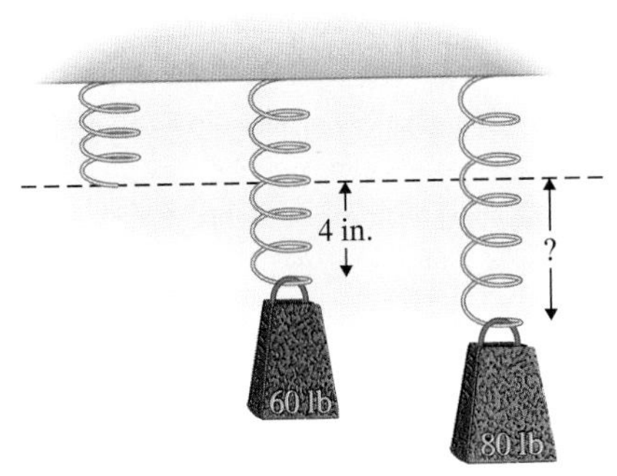

36. If a 30-pound weight stretches a spring 10 inches, find the distance a 20-pound weight stretches the spring. (See Exercise 35.)

37. The weight of an object varies inversely as the square of its distance from the *center* of the Earth. If a person weighs 180 pounds on Earth's surface, what is his weight 10 miles above the surface of the Earth? (Assume that the Earth's radius is 4000 miles.)

38. For a constant distance, the rate of travel varies inversely as the time traveled. If a family travels 55 mph and arrives at a destination in 4 hours, how long will the return trip take traveling at 60 mph?

39. The distance d that an object falls is directly proportional to the square of the time of the fall, t. A person who is parachuting for the first time is told to wait ten seconds before opening the parachute. If the person falls 64 feet in 2 seconds, find how far he falls in 10 seconds.

40. The distance needed for a car to stop, d, is directly proportional to the square of its rate of travel, r. Under certain driving conditions, a car traveling 60 mph needs 300 feet to stop. With these same driving conditions, how long does it take a car to stop if the car is traveling 30 mph when the brakes are applied?

Review and Preview

Add the equations. See Section 3.2.

41. $\begin{aligned} -3x + 4y &= 7 \\ \underline{3x - 2y} &\underline{= 9} \end{aligned}$

42. $\begin{aligned} x - y &= -9 \\ \underline{-x - y} &\underline{= -14} \end{aligned}$

43. $\begin{aligned} 5x - 0.4y &= 0.7 \\ \underline{-9x + 0.4y} &\underline{= -0.2} \end{aligned}$

44. $\begin{aligned} 1.9x - 2y &= 5.7 \\ \underline{-1.9x - 0.1y} &\underline{= 2.3} \end{aligned}$

Combining Concepts

45. Suppose that y varies directly as x. If x is tripled, what is the effect on y?

46. Suppose that y varies directly as x^2. If x is tripled, what is the effect on y?

47. The period of a pendulum p (the time of one complete back-and-forth swing) varies directly with the square root of its length, ℓ. If the length of the pendulum is quadrupled, what is the effect on the period, p?

48. For a constant distance, the rate of travel r varies inversely with the time traveled, t. If a car traveling 100 mph completes a test track in 6 minutes, find the rate needed to complete the same test track in 4 minutes. (*Hint*: Convert minutes to hours.)

CHAPTER 13 ACTIVITY Finding a Linear Model

This activity may be completed by working in groups or individually.

The following table shows the estimated number of foreign visitors (in millions) to the United States for the years 2000 through 2003.

Year	Foreign Visitors to the United States (in millions)
2000	51.5
2001	54.2
2002	56.9
2003	59.6

(*Source:* Tourism Industries/International Trade Administration, U.S. Department of Commerce)

1. Make a scatter diagram of the paired data in the table.
2. Use what you have learned in this chapter to write an equation of the line representing the paired data in the table. Explain how you found the equation, and what each variable represents.
3. What is the slope of your line? What does the slope mean in this context?
4. Use your linear equation to predict the number of foreign visitors to the United States in 2010.
5. Compare your linear equation to that found by other students or groups. Is it the same, similar, or different? How?
6. Compare your prediction from question 3 to that of other students or groups. Describe what you find.

Chapter 13 VOCABULARY CHECK

Fill in each blank with one of the words listed below.

y-axis	x-axis	solution	linear	standard	slope-intercept	relation
x-intercept	y-intercept	y	x	slope	point-slope	function
domain	range	direct	inverse			

1. An ordered pair is a __________ of an equation in two variables if replacing the variables by the coordinates of the ordered pair results in a true statement.
2. The vertical number line in the rectangular coordinate system is called the ________.
3. A ________ equation can be written in the form $Ax + By = C$.
4. A(n) __________ is a point of the graph where the graph crosses the x-axis.
5. The form $Ax + By = C$ is called __________ form.
6. A(n) __________ is a point of the graph where the graph crosses the y-axis.
7. A set of ordered pairs that assigns to each x-value exactly one y-value is called a __________.
8. The equation $y = 7x - 5$ is written in ______________ form.
9. The set of all x-coordinates of a relation is called the ________ of the relation.
10. The set of all y-coordinates of a relation is called the ________ of the relation.
11. A set of ordered pairs is called a __________.
12. The equation $y + 1 = 7(x - 2)$ is written in ____________ form.
13. To find an x-intercept of a graph, let _____ = 0.
14. The horizontal number line in the rectangular coordinate system is called the ________.
15. To find a y-intercept of a graph, let _____ = 0.
16. The ________ of a line measures the steepness or tilt of a line.
17. The equation $y = kx$ is an example of ________ variation.
18. The equation $y = \frac{k}{x}$ is an example of ________ variation.

13 CHAPTER Highlights

DEFINITIONS AND CONCEPTS | **EXAMPLES**

Section 13.1 The Rectangular Coordinate System

The **rectangular coordinate system** consists of a plane and a vertical and a horizontal number line intersecting at their 0 coordinates. The vertical number line is called the **y-axis** and the horizontal number line is called the **x-axis.** The point of intersection of the axes is called the **origin.**

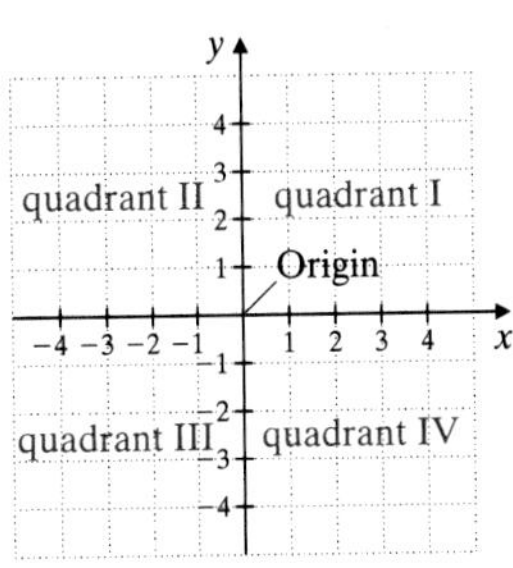

To **plot** or **graph** an ordered pair means to find its corresponding point on a rectangular coordinate system.

To plot or graph an ordered pair such as $(3, -2)$, start at the origin. Move 3 units to the right and from there, 2 units down.

To plot or graph $(-3, 4)$, start at the origin. Move 3 units to the left and from there, 4 units up.

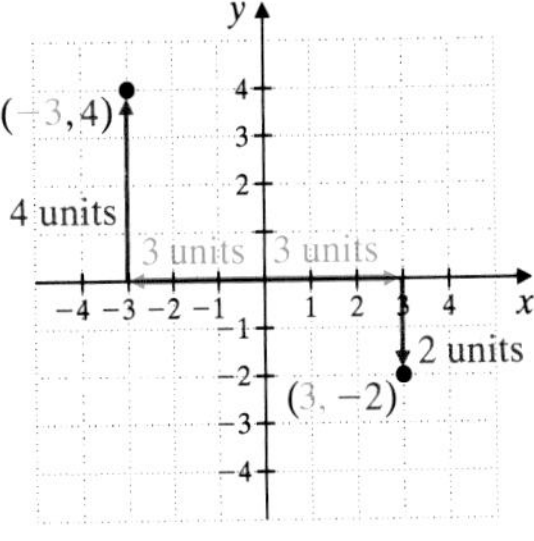

An ordered pair is a **solution** of an equation in two variables if replacing the variables with the coordinates of the ordered pair results in a true statement.

DEFINITIONS AND CONCEPTS — EXAMPLES

Section 13.1 The Rectangular Coordinate System *(continued)*

If one coordinate of an ordered pair solution is known, the other value can be determined by substitution.

Complete the ordered pair (0,) for the equation $x - 6y = 12$.

$$x - 6y = 12$$

$$0 - 6y = 12 \quad \text{Let } x = 0.$$

$$\frac{-6y}{-6} = \frac{12}{-6} \quad \text{Divide by } -6.$$

$$y = -2$$

The ordered pair solution is $(0, -2)$.

Section 13.2 Graphing Linear Equations

A **linear equation in two variables** is an equation that can be written in the form $Ax + By = C$, where A and B are not both 0. The form $Ax + By = C$ is called **standard form.**

$3x + 2y = -6 \qquad x = -5$

$y = 3 \qquad y = -x + 10$

$x + y = 10$ is in standard form.

To graph a linear equation in two variables, find three ordered pair solutions. Plot the solution points and draw the line connecting the points.

Graph: $x - 2y = 5$

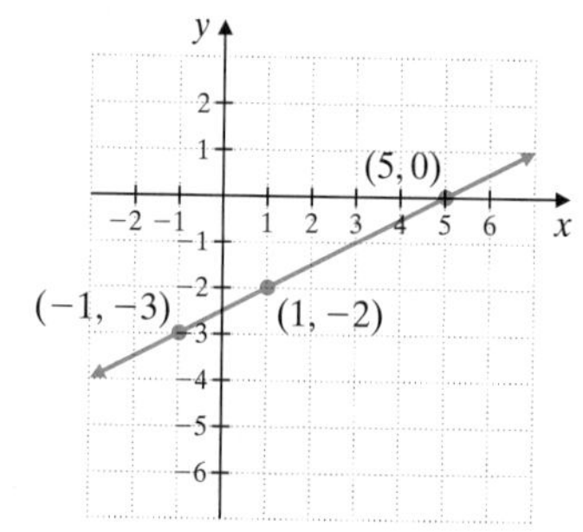

x	y
5	0
1	−2
−1	−3

Section 13.3 Intercepts

An **intercept** of a graph is a point where the graph intersects an axis. If a graph intersects the x-axis at a, then $(a, 0)$ is the **x-intercept.** If a graph intersects the y-axis at b, then $(0, b)$ is the **y-intercept.**

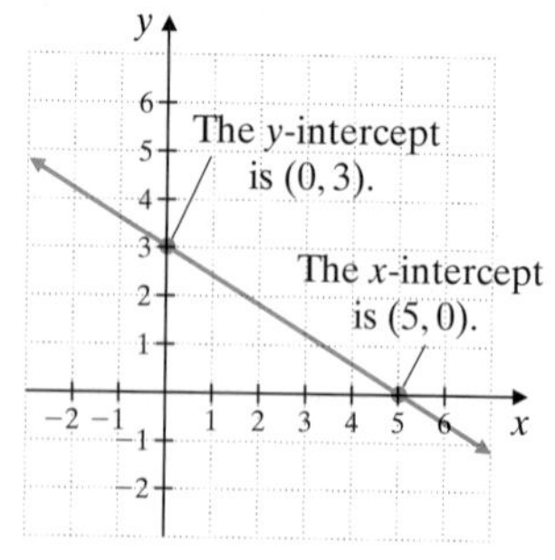

To find the x-intercept, let $y = 0$ and solve for x.
To find the y-intercept, let $x = 0$ and solve for y.

Find the intercepts for $2x - 5y = -10$.

If $y = 0$, then

$$2x - 5 \cdot 0 = -10$$

$$2x = -10$$

$$\frac{2x}{2} = \frac{-10}{2}$$

$$x = -5$$

If $x = 0$, then

$$2 \cdot 0 - 5y = -10$$

$$-5y = -10$$

$$\frac{-5y}{-5} = \frac{-10}{-5}$$

$$y = 2$$

DEFINITIONS AND CONCEPTS | **EXAMPLES**

Section 13.3 Intercepts *(continued)*

The x-intercept is $(-5, 0)$. The y-intercept is $(0, 2)$.

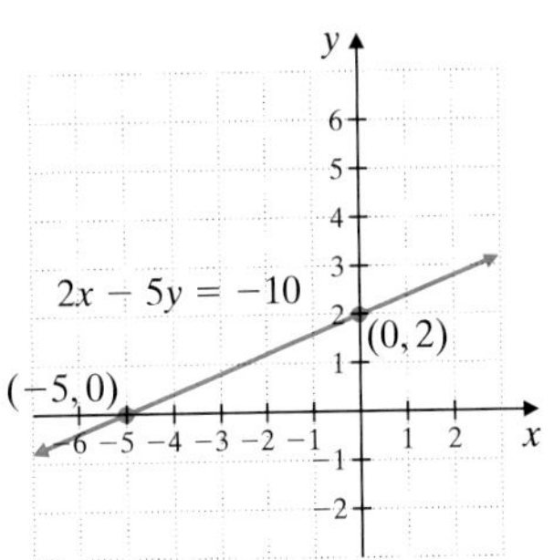

The graph of $x = c$ is a vertical line with x-intercept $(c, 0)$.
The graph of $y = c$ is a horizontal line with y-intercept $(0, c)$.

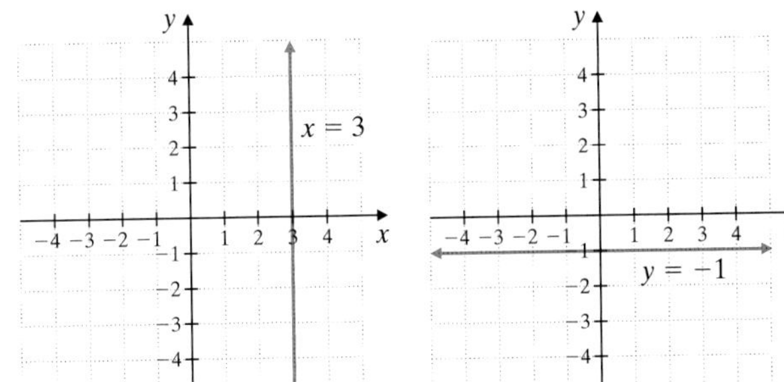

Section 13.4 Slope

The **slope** m of the line through points (x_1, y_1) and (x_2, y_2) is given by

$$m = \frac{y_2 - y_1}{x_2 - x_1} \quad \text{as long as } x_2 \neq x_1$$

A horizontal line has slope 0.
The slope of a vertical line is undefined.
Nonvertical parallel lines have the same slope.
Two nonvertical lines are perpendicular if the slope of one is the negative reciprocal of the slope of the other.

The slope of the line through points $(-1, 6)$ and $(-5, 8)$ is

$$m = \frac{y_2 - y_1}{x_2 - x_1} = \frac{8 - 6}{-5 - (-1)} = \frac{2}{-4} = -\frac{1}{2}$$

The slope of the line $y = -5$ is 0.
The line $x = 3$ has undefined slope.

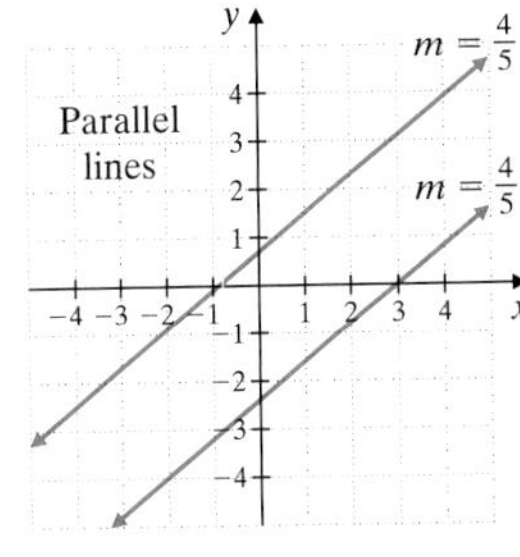

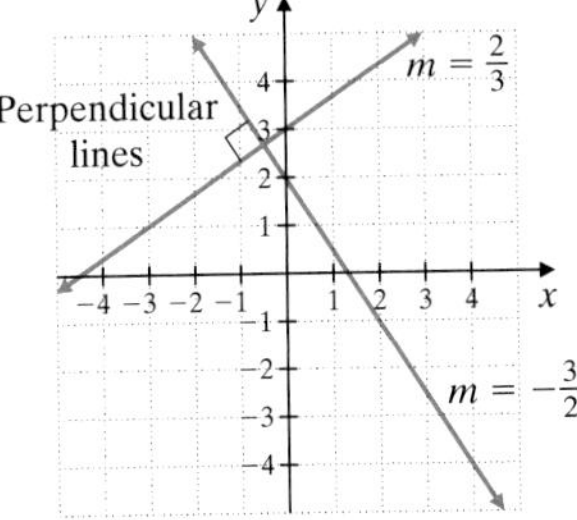

Section 13.5 Equations of Lines

SLOPE-INTERCEPT FORM

$$y = mx + b$$

m is the slope of the line.
$(0, b)$ is the y-intercept.

Find the slope and the y-intercept of the line $2x + 3y = 6$.
Solve for y: $2x + 3y = 6$

$3y = -2x + 6$ Subtract $2x$.

$y = -\frac{2}{3}x + 2$ Divide by 3.

The slope of the line is $-\frac{2}{3}$ and the y-intercept is $(0, 2)$.

DEFINITIONS AND CONCEPTS | EXAMPLES

Section 13.5 Equations of Lines *(continued)*

POINT-SLOPE FORM

$$y - y_1 = m(x - x_1)$$

m is the slope.
(x_1, y_1) is a point of the line.

Find an equation of the line with slope $\frac{3}{4}$ that contains the point $(-1, 5)$.

$$y - 5 = \frac{3}{4}(x - (-1))$$

$4(y - 5) = 3(x + 1)$ Multiply by 4.

$4y - 20 = 3x + 3$ Distribute.

$-3x + 4y = 23$ Subtract $3x$ and add 20.

Section 13.6 Introduction to Functions

A set of ordered pairs is a **relation.** The set of all x-coordinates is called the **domain** of the relation and the set of all y-coordinates is called the **range** of the relation.

The domain of the relation

$$\{(0, 5), (2, 5), (4, 5), (5, -2)\}$$

is $\{0, 2, 4, 5\}$. The range is $\{-2, 5\}$.

A **function** is a set of ordered pairs that assigns to each x-value exactly one y-value.

VERTICAL LINE TEST

If a vertical line can be drawn so that it intersects a graph more than once, the graph is not the graph of a function. (If no such line can be drawn, the graph is that of a function.)

Which are graphs of functions?

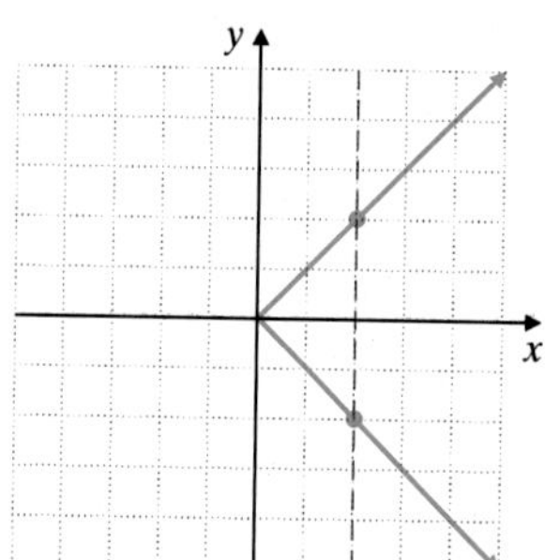

This graph is not the graph of a function.

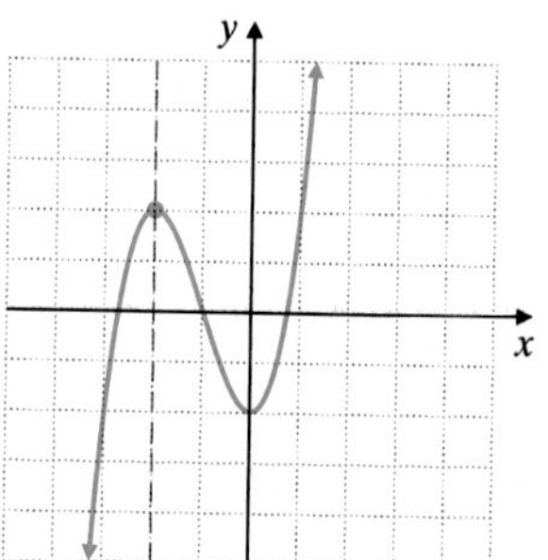

This graph is the graph of a function.

The symbol $f(x)$ means **function of *x*.** This notation is called **function notation.**

If $f(x) = 3x - 7$, then

$$f(-1) = 3(-1) - 7$$
$$= -3 - 7$$
$$= -10$$

Section 13.7 Graphing Linear Inequalities in Two Variables

A **linear inequality in two variables** is an inequality that can be written in one of these forms:

$Ax + By < C$ $\quad$ $Ax + By \le C$

$Ax + By > C$ $\quad$ $Ax + By \ge C$

where A and B are not both 0.

$2x - 5y < 6$ $\quad$ $x \ge -5$

$y > -8x$ $\quad$ $y \le 2$

TO GRAPH A LINEAR INEQUALITY

1. Graph the boundary line by graphing the related equation. Draw the line solid if the inequality symbol is $\leq$ or $\geq$. Draw the line dashed if the inequality symbol is $<$ or $>$.
2. Choose a test point not on the line. Substitute its coordinates into the original inequality.
3. If the resulting inequality is true, shade the half-plane that contains the test point. If the inequality is not true, shade the half-plane that does not contain the test point.

Graph: $2x - y \leq 4$

1. Graph $2x - y = 4$. Draw a solid line because the inequality symbol is $\leq$.
2. Check the test point $(0, 0)$ in the original inequality, $2x - y \leq 4$.

$$2 \cdot 0 - 0 \leq 4 \quad \text{Let } x = 0 \text{ and } y = 0.$$

$$0 \leq 4 \quad \text{True}$$

3. The inequality is true, so shade the half-plane containing $(0, 0)$ as shown.

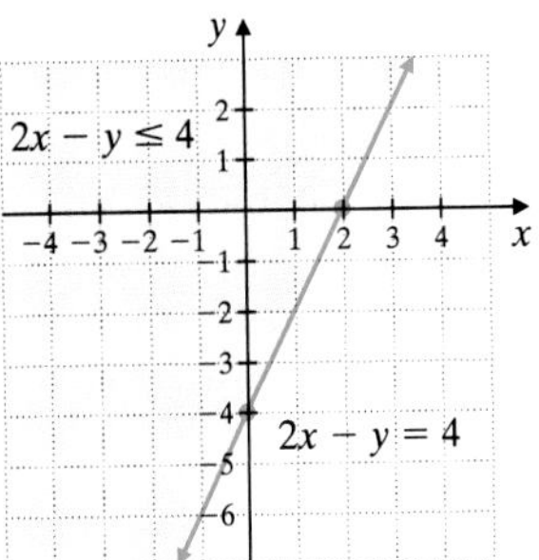

Section 13.8 Direct and Inverse Variation

y **varies directly as** *x*, or *y* is **directly proportional to** *x*, if there is a nonzero constant *k* such that

$$y = kx$$

y **varies inversely as** *x*, or *y* is **inversely proportional to** *x*, if there is a nonzero constant *k* such that

$$y = \frac{k}{x}$$

The circumference of a circle *C* varies directly as its radius r.

$$C = \underbrace{2\pi}_{k} r$$

Pressure *P* varies inversely with volume *V*.

$$P = \frac{k}{V}$$

FOCUS ON The Real World

MISLEADING GRAPHS

Graphs are very common in magazines and in newspapers such as *USA Today*. Graphs can be a convenient way to get an idea across because, as the old saying goes, "a picture is worth a thousand words." However, some graphs can be deceptive, which may or may not be intentional. It is important to know some of the ways that graphs can be misleading.

Beware of graphs like the one at the right. Notice that the graph shows a company's profit for various months. It appears that profit is growing quite rapidly. However, this impressive picture tells us little without knowing what units of profit are being graphed. Does the graph show profit in dollars or millions of dollars? An unethical company with profit increases of only a few pennies could use a graph like this one to make the profit increase seem much more substantial than it really is. A truthful graph describes the size of the units used along the vertical axis.

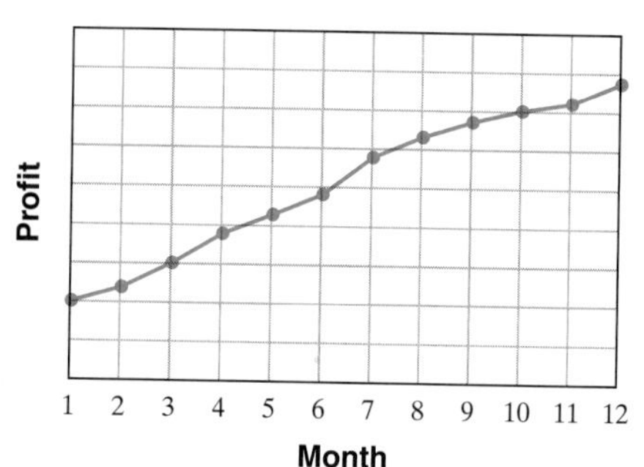

Another type of graph to watch for is one that misrepresents relationships. For example, the bar graph at the right shows the number of men and women employees in the accounting and shipping departments of a certain company. In the accounting department, the bar representing the number of women is shown twice as tall as the bar representing the number of men. However, the number of women (13) is not twice the number of men (10). This set of bars misrepresents the relationship between the number of men and women. Do you see how the relationship between the number of men and women in the shipping department is distorted by the heights of the bars used? A truthful graph will use bar heights that are proportional to the numbers they represent.

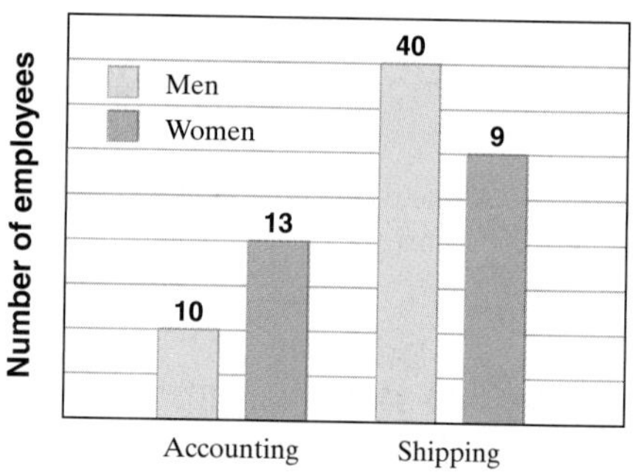

The impression a graph can give also depends on its vertical scale. The two graphs below represent exactly the same data. The only difference between the two graphs is the vertical scale—one shows enrollments from 246 to 260 students and the other shows enrollments between 0 and 300 students. If you were trying to convince readers that algebra enrollment at UPH had changed drastically over the period 1999–2003, which graph would you use? Which graph do you think gives the more honest representation?

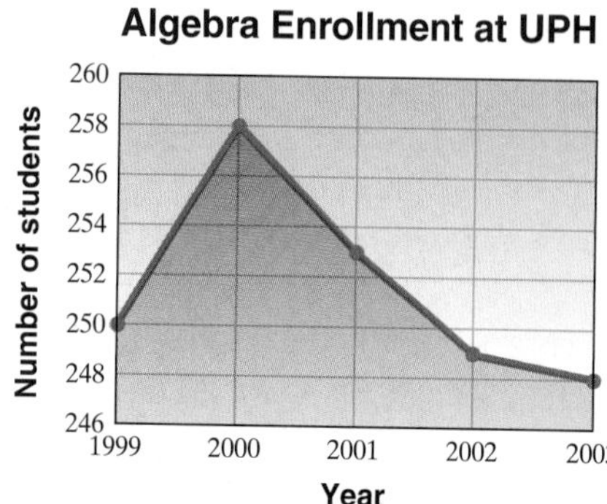

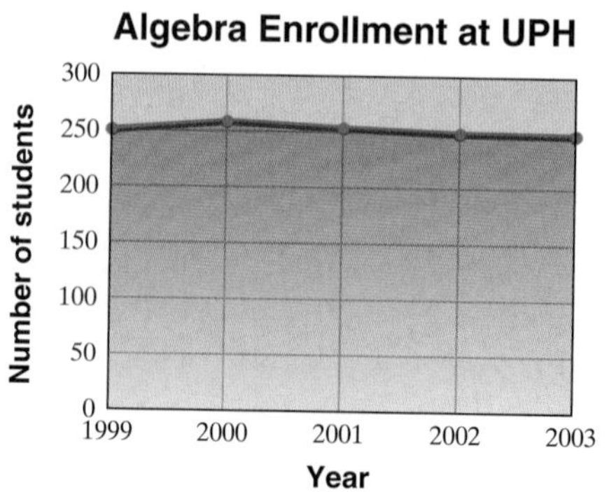

Name ______________________ Section __________ Date __________

Chapter 13 Review

(13.1) *Plot each pair on the same rectangular coordinate system.*

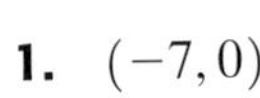
1. $(-7, 0)$

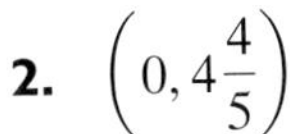
2. $\left(0, 4\frac{4}{5}\right)$

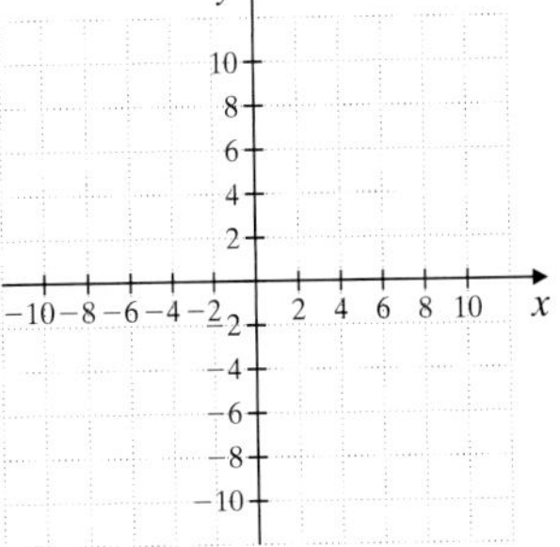

3. $(-2, -5)$

4. $(1, -3)$

5. $(0.7, 0.7)$

6. $(-6, 4)$

Complete each ordered pair so that it is a solution of the given equation.

7. $-2 + y = 6x$; $(7, \quad)$

8. $y = 3x + 5$; $(\quad, -8)$

Complete the table of values for each given equation.

9. $9 = -3x + 4y$

x	y
	0
	3
9	

10. $y = 5$

x	y
7	
−7	
0	

11. $x = 2y$

x	y
	0
	5
	−5

12. The cost in dollars of producing x compact disc holders is given by $y = 5x + 2000$.

a. Complete the table.

x	1	100	1000
y			

b. Find the number of compact disc holders that can be produced for $6430.

(13.2) *Graph each linear equation.*

13. $x - y = 1$

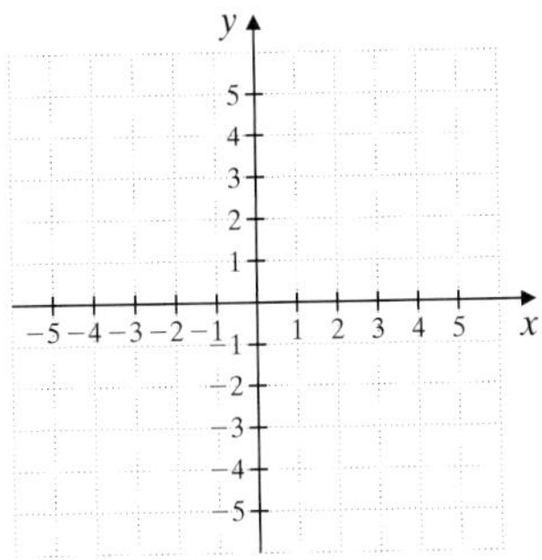

14. $x + y = 6$

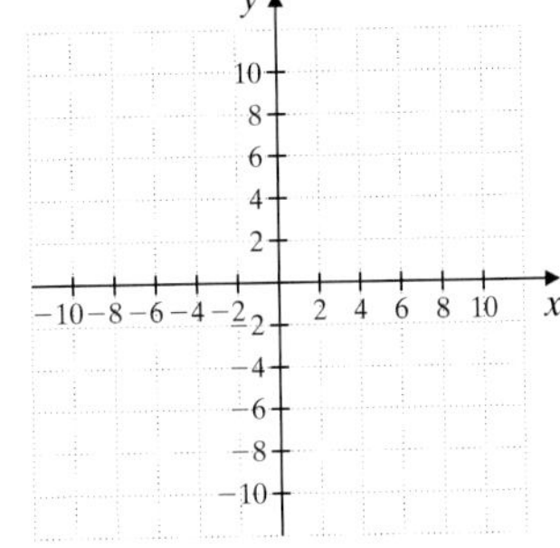

15. $x - 3y = 12$

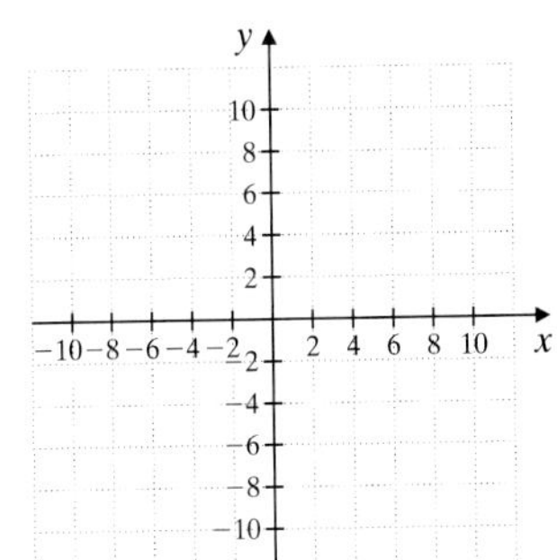

16. $5x - y = -8$

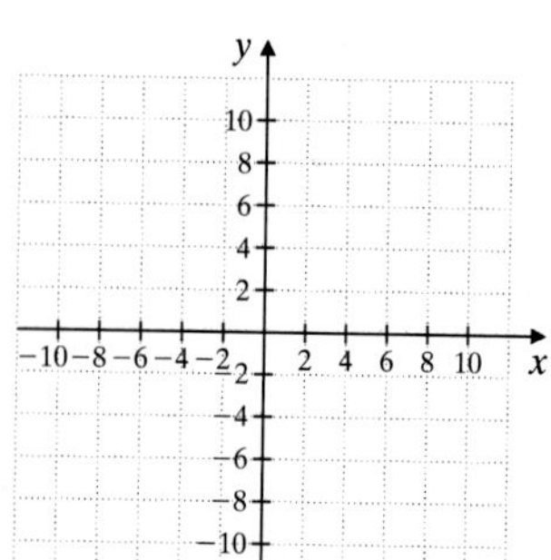

17. $x = 3y$

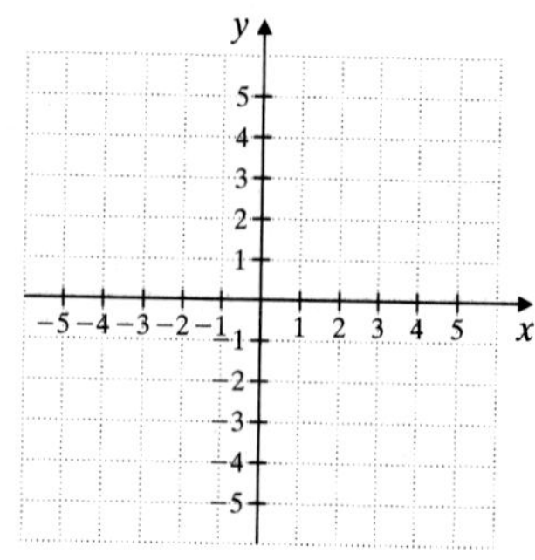

18. $y = -2x$

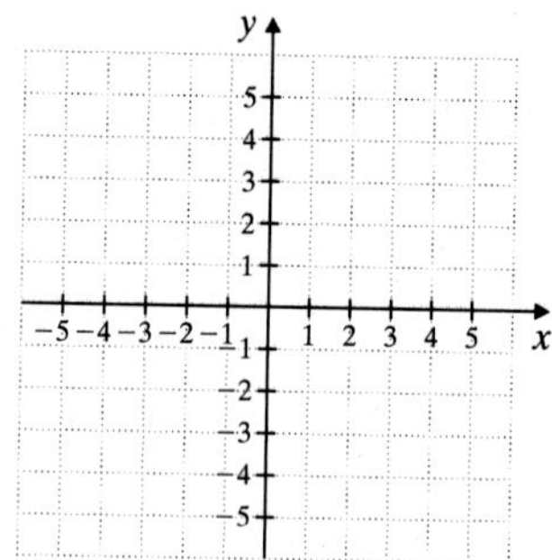

(13.3) *Identify the intercepts in each graph.*

19.

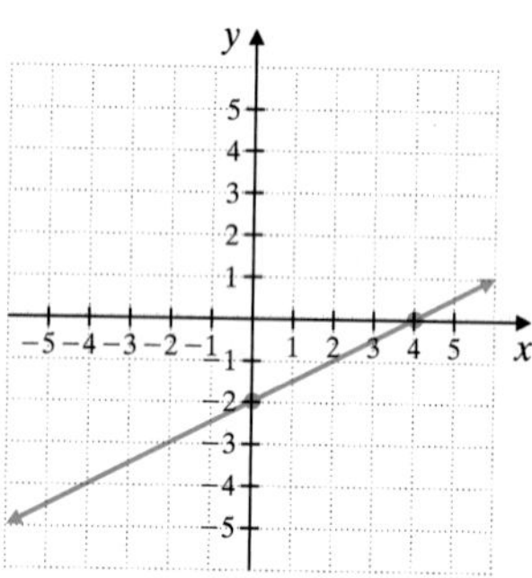

20.

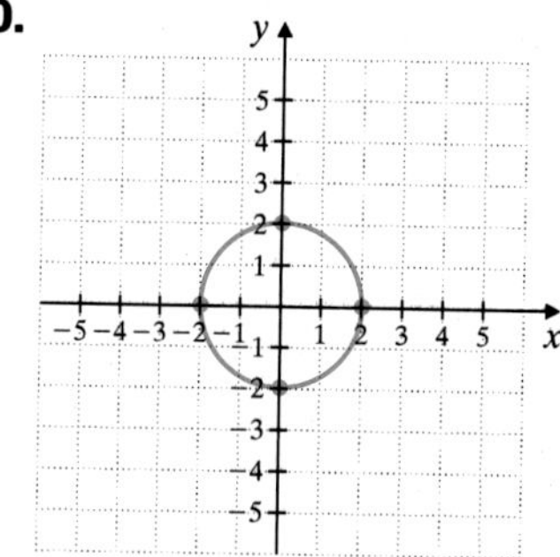

Graph each linear equation.

21. $y = -3$

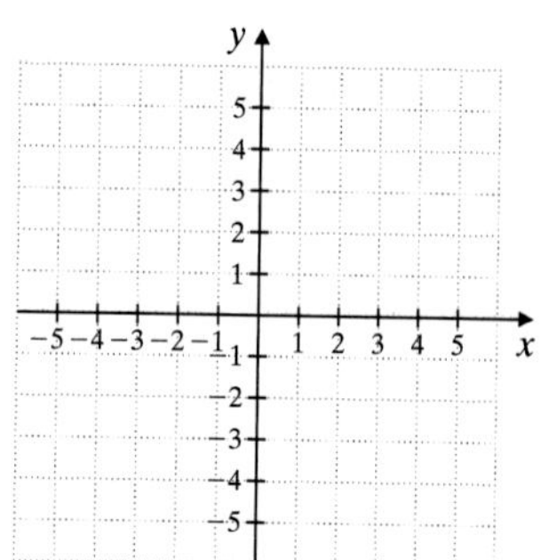

22. $x = 5$

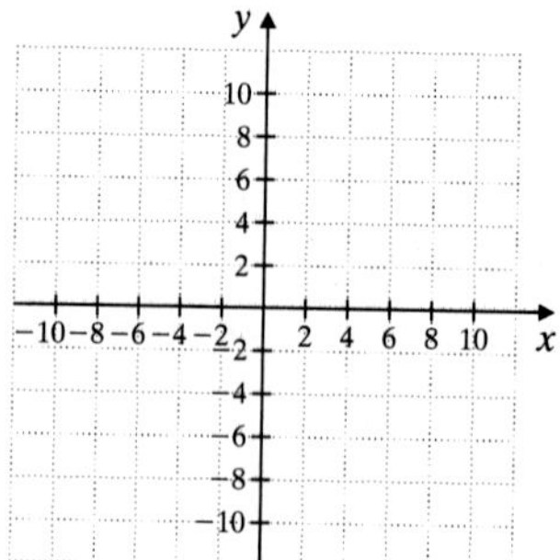

Find the intercepts for each equation.

23. $x - 3y = 12$

24. $-4x + y = 8$

(13.4) *Find the slope of each line.*

25.

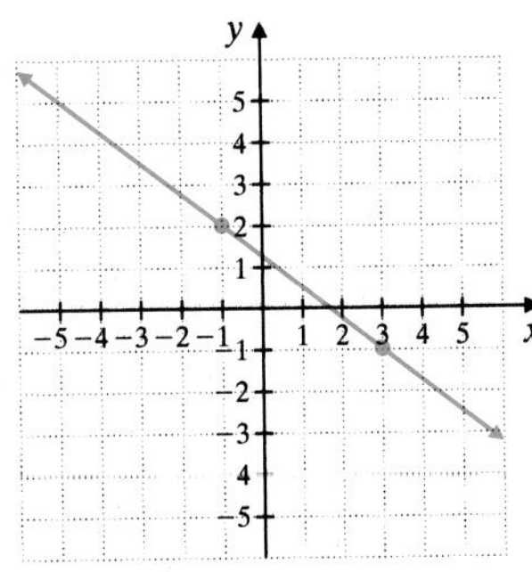

26.

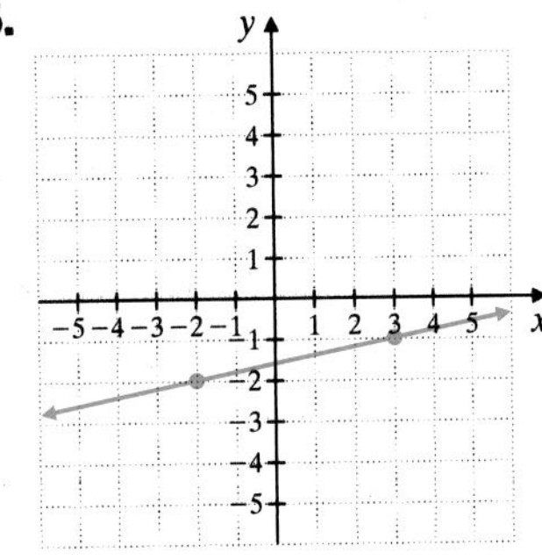

Match each line with its slope.

A.

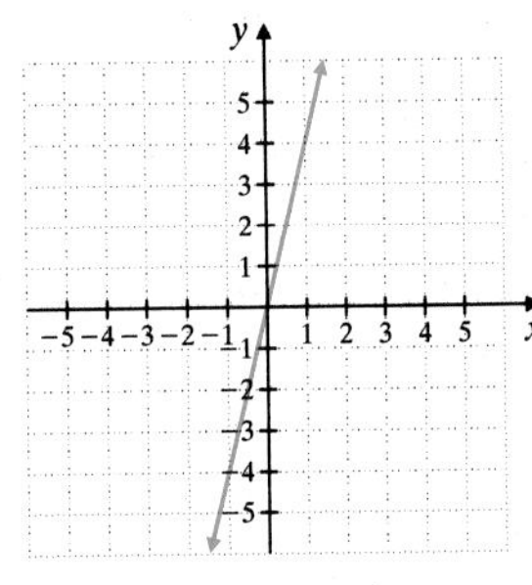

B.

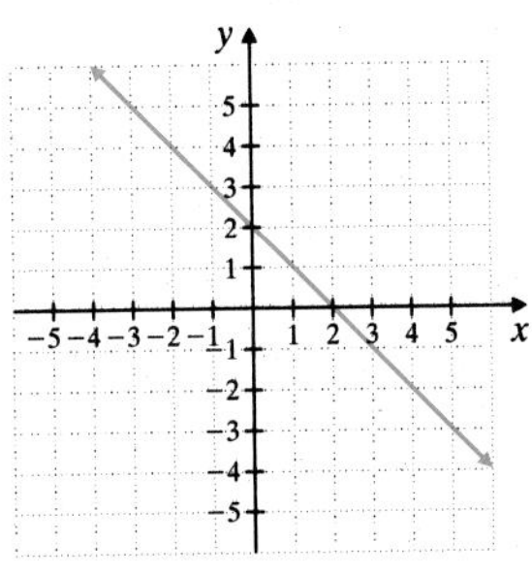

C.

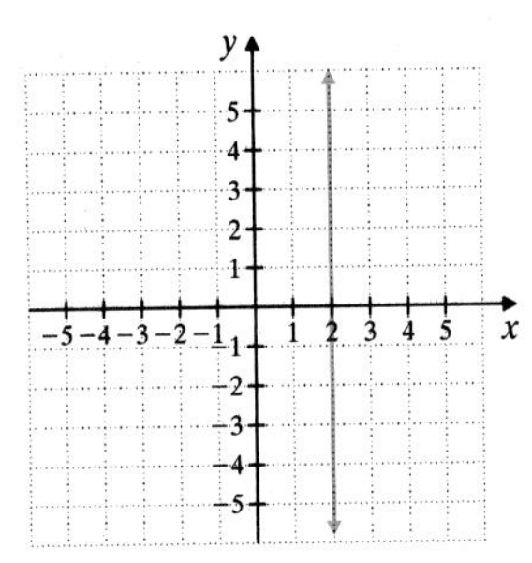

D.

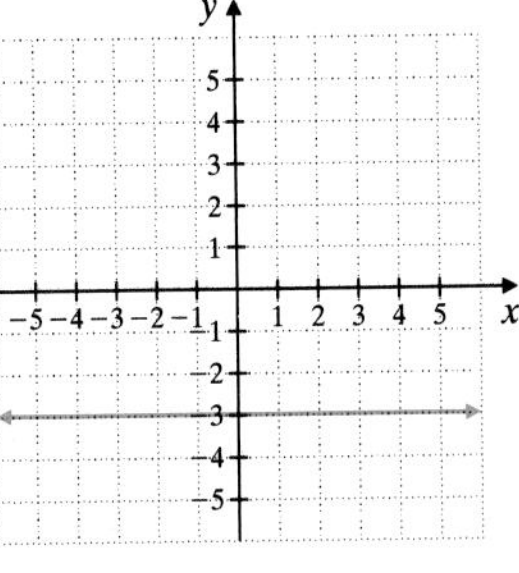

27. $m = 0$

28. $m = -1$

29. undefined slope

30. $m = 4$

Find the slope of the line that passes through each pair of points.

31. $(2, 5)$ and $(6, 8)$

32. $(4, 7)$ and $(1, 2)$

33. $(1, 3)$ and $(-2, -9)$

34. $(-4, 1)$ and $(3, -6)$

Find the slope of each line.

35. $y = 3x + 7$

36. $x - 2y = 4$

37. $y = -2$

38. $x = 0$

Determine whether each pair of lines is parallel, perpendicular, or neither.

39. $x - y = -6$

$x + y = 3$

40. $3x + y = 7$

$-3x - y = 10$

41. $y = 4x + \frac{1}{2}$

$4x + 2y = 1$

Find the slope of each line and write the slope as a rate of change. Don't forget to attach the proper units.

42. The graph below approximates the number of U.S. persons y (in millions) who have a bachelor's degree or higher per year x.

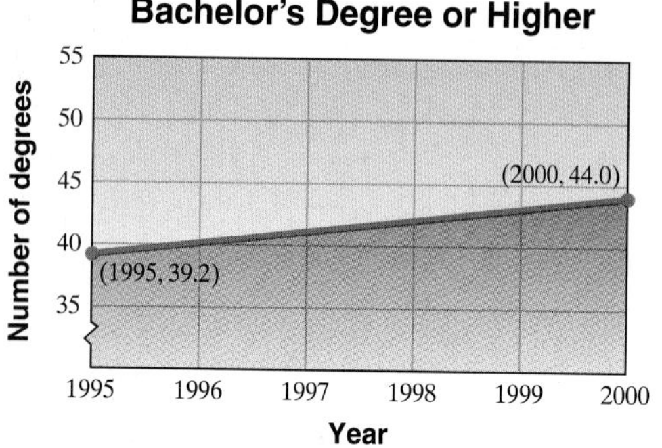

43. The graph below approximates the number of U.S. travelers y (in millions) that are vacationing per year x.

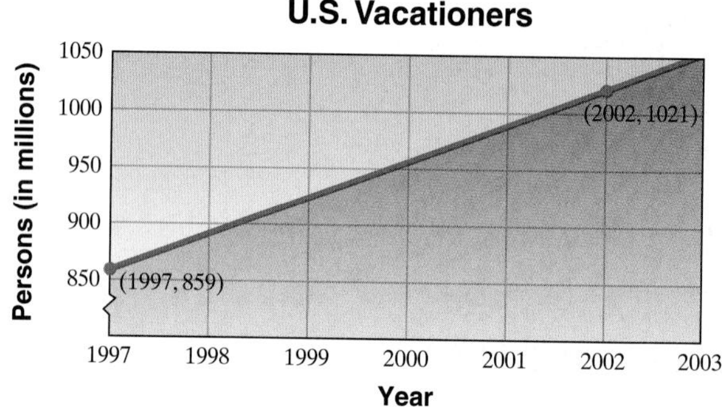

(13.5) *Determine the slope and the y-intercept of the graph of each equation.*

44. $3x + y = 7$

45. $x - 6y = -1$

Write an equation of each line.

46. slope -5; y-intercept $\left(0, \frac{1}{2}\right)$

47. slope $\frac{2}{3}$; y-intercept $(0, 6)$

Match each equation with its graph.

48. $y = 2x + 1$

49. $y = -4x$

50. $y = 2x$

51. $y = 2x - 1$

A.

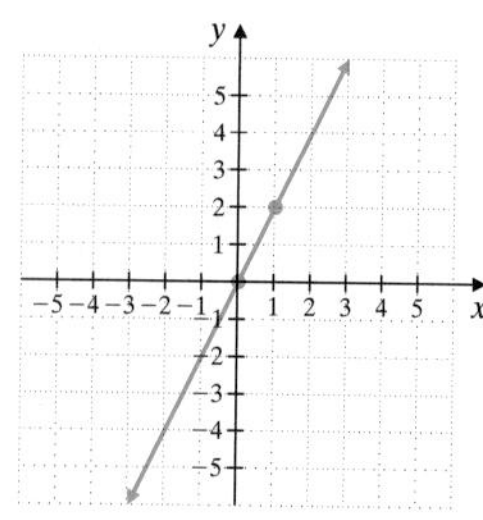

B.

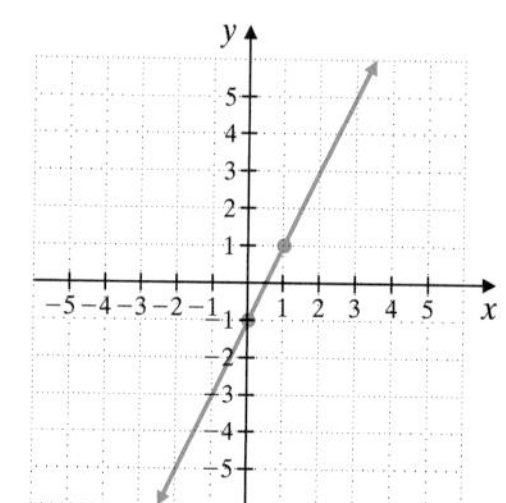

C.

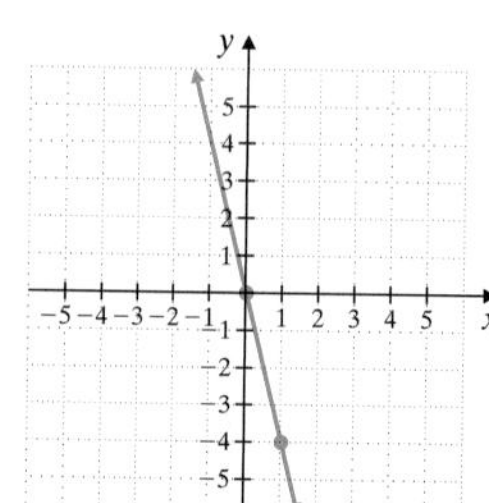

D.

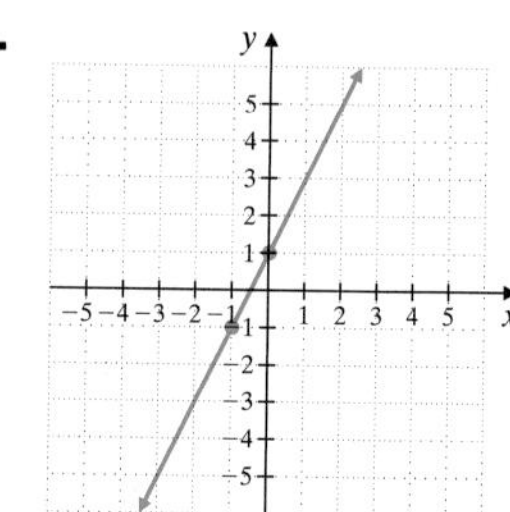

Write an equation of the line with the given slope that passes through the given point. Write the equation in the form $Ax + By = C$.

52. $m = 4; (2, 0)$

53. $m = -3; (0, -5)$

54. $m = \frac{3}{5}; (1, 4)$

55. $m = -\frac{1}{3}; (-3, 3)$

Write an equation of the line passing through each pair of points. Write the equation in the form $Ax + By = C$.

56. $(1, 7)$ and $(2, -7)$

57. $(-2, 5)$ and $(-4, 6)$

(13.6) *Determine whether each relation or graph is a function.*

58. $\{(7, 1), (7, 5), (2, 6)\}$

59. $\{(0, -1), (5, -1), (2, 2)\}$

60.

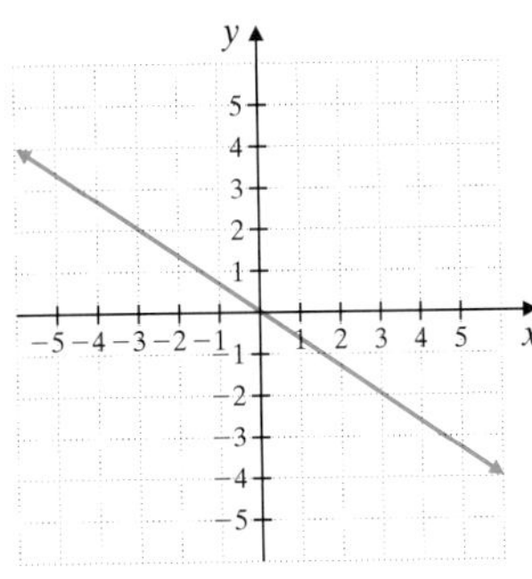

61.

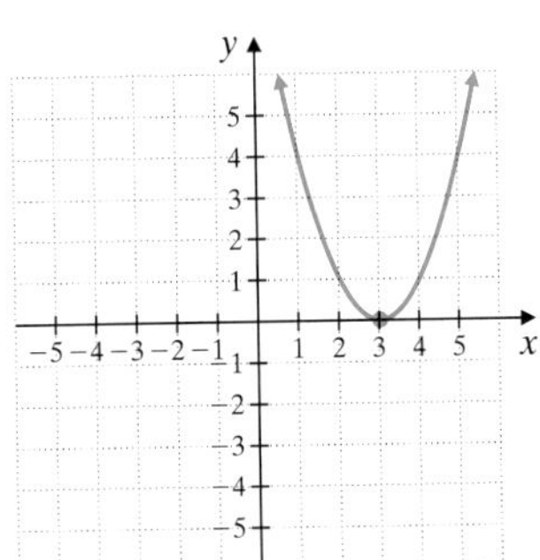

62.

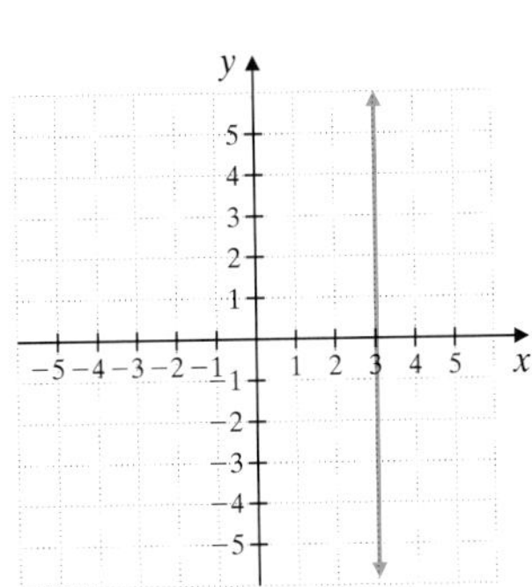

63.

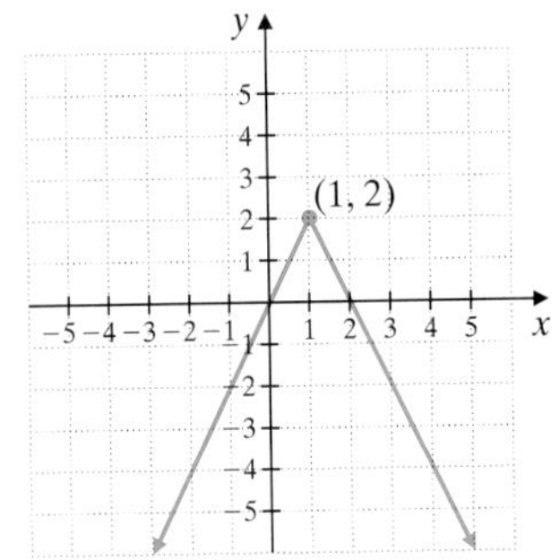

64. Find the indicated function value for the function, $f(x) = -2x + 6$.

a. $f(0)$

b. $f(-2)$

c. $f\left(\frac{1}{2}\right)$

(13.7) *Graph each inequality.*

65. $x + 6y < 6$

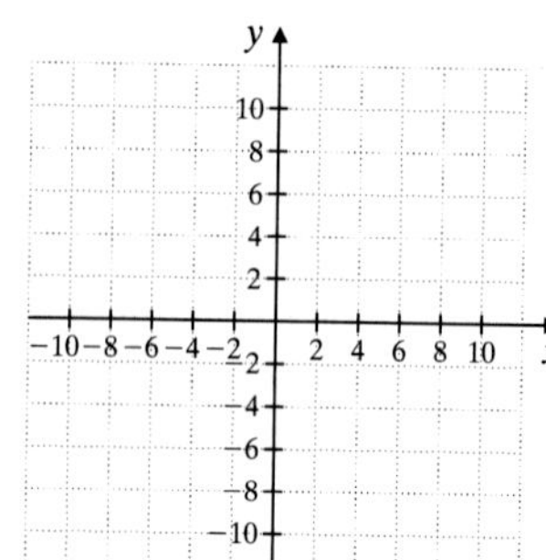

66. $x + y > -2$

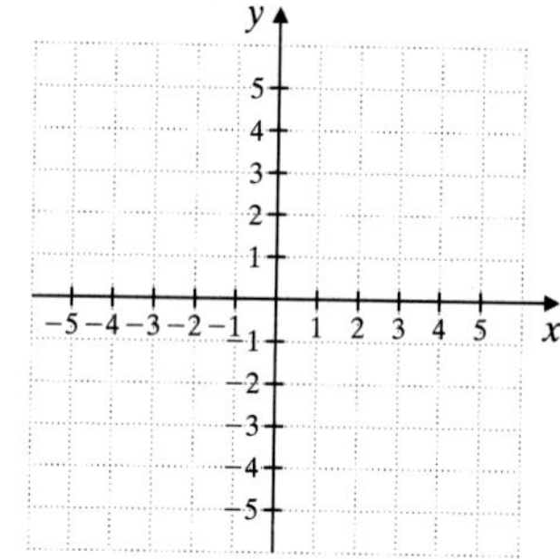

67. $y \geq -7$

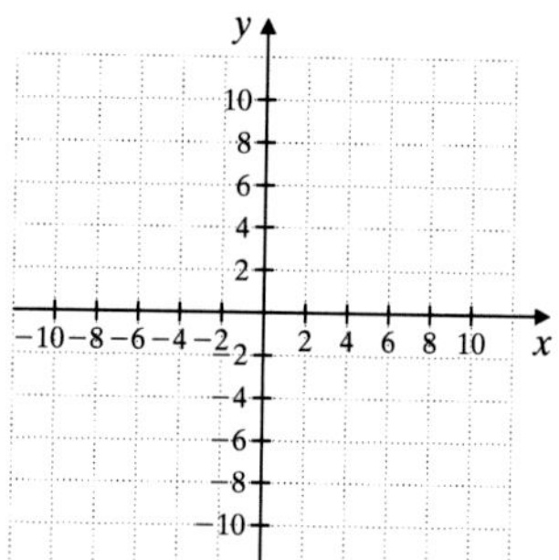

68. $y \leq -4$

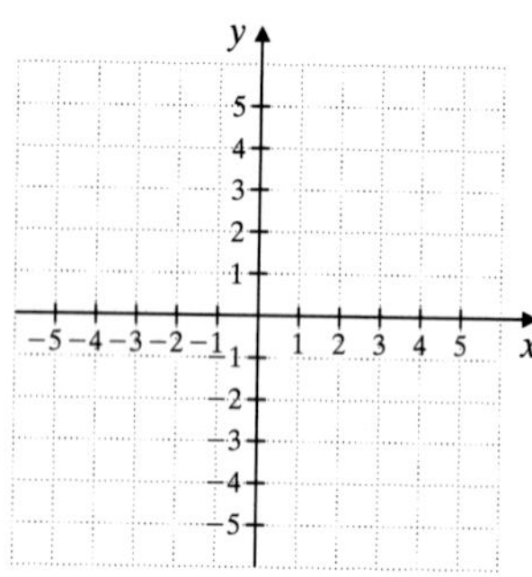

69. $-x \leq y$

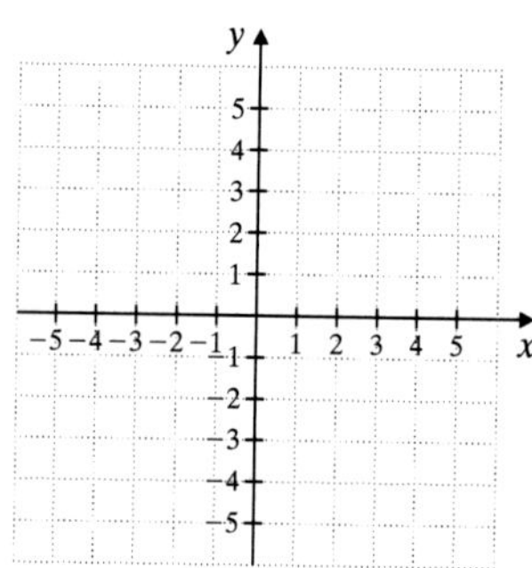

70. $x \geq -y$

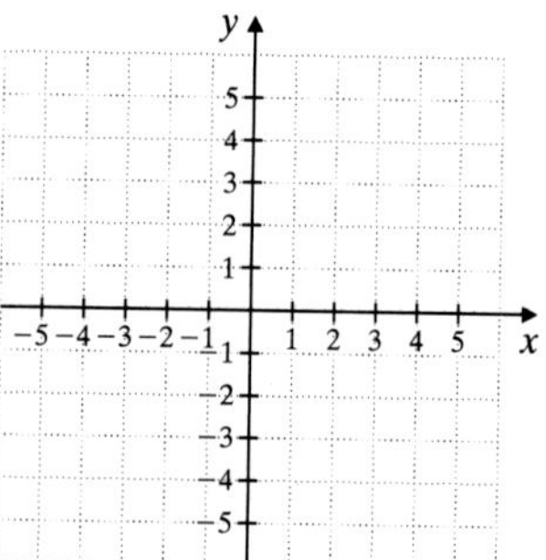

(13.8) *Solve.*

71. y varies directly as x. If $y = 40$ when $x = 4$, find y when x is 11.

72. y varies inversely as x. If $y = 4$ when $x = 6$, find y when x is 48.

73. y varies inversely as x^3. If $y = 12.5$ when $x = 2$, find y when x is 3.

74. y varies directly as x^2. If $y = 175$ when $x = 5$, find y when $x = 10$.

75. The cost of manufacturing a certain medicine varies inversely as the amount of medicine manufactured increases. If 3000 milliliters can be manufactured for $6600, find the cost to manufacture 5000 milliliters.

76. The distance a spring stretches varies directly with the weight attached to the spring. If a 150-pound weight stretches the spring 8 inches, find the distance that a 90-pound weight stretches the spring.

Name ______________________ Section ____________ Date ____________

Chapter 13 Test

Remember to check your answers and use the Chapter Test Prep Video to view solutions.

1. The table gives the percent of new mothers who returned to work after having their child. (*Source:* Census Bureau)

Year	Percent of New Mothers Returning to Work
1980	38
1984	47
1988	51
1992	54
1996	59
2000	55

a. Write this data as a set of ordered pairs of the form (year, percent of new mothers returning to work).

b. Create a scatter diagram of the data. Be sure to label the axes properly.

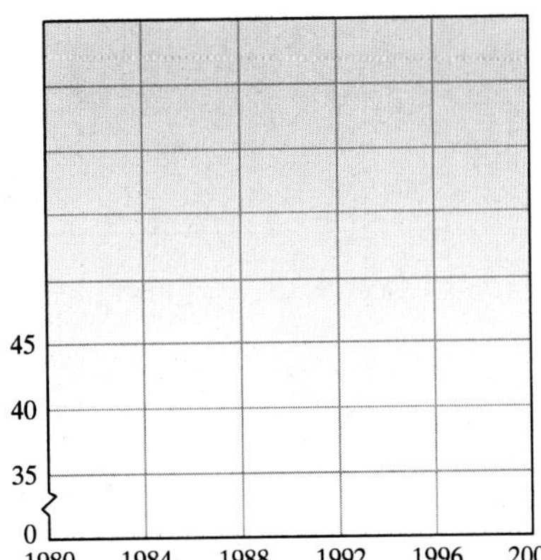

Graph.

2. $2x + y = 8$

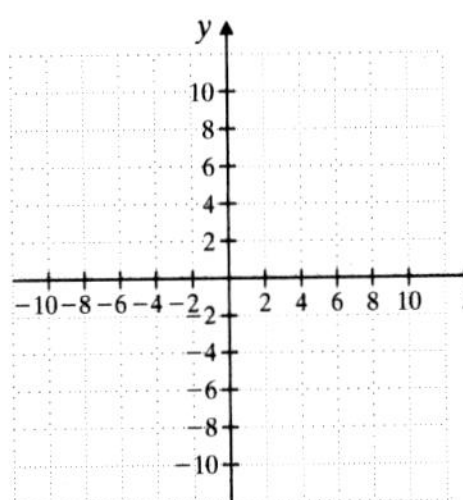

3. $5x - 7y = 10$

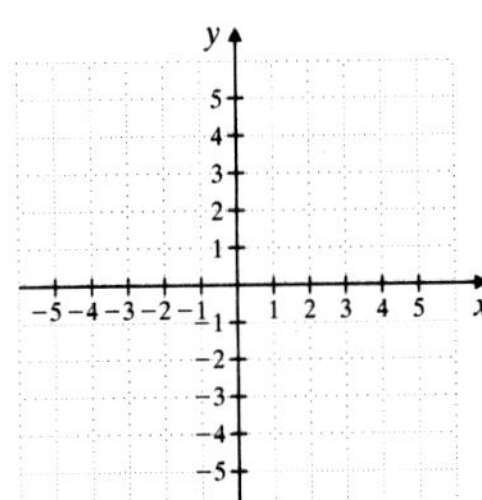

4. $y = -1$

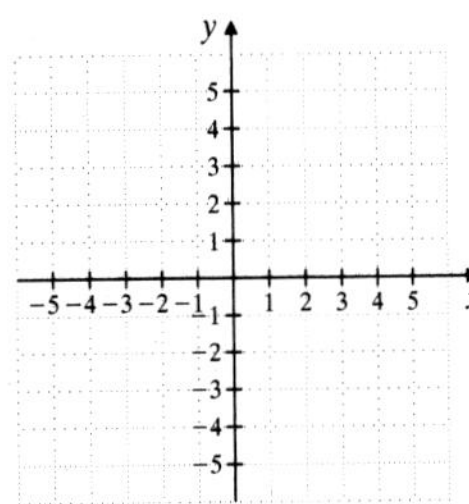

5. $x - 3 = 0$

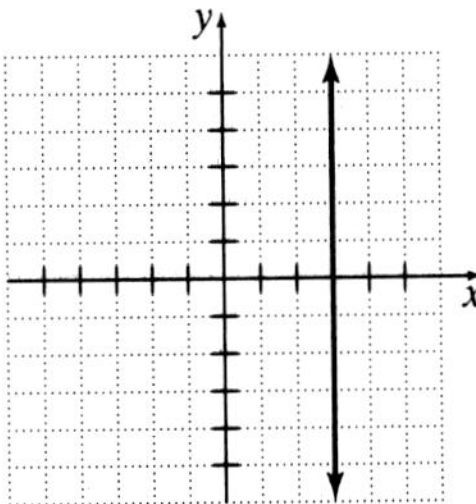

6. $y \geq -4x$

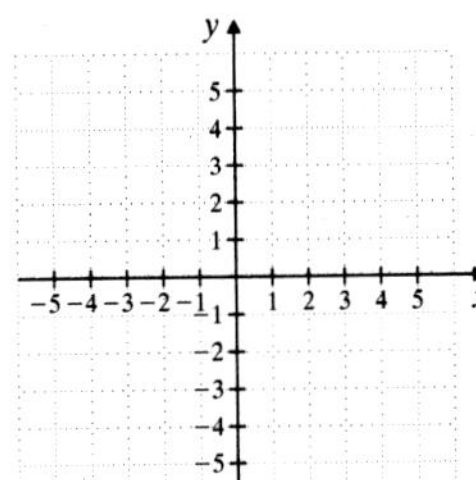

7. $2x - 3y > -6$

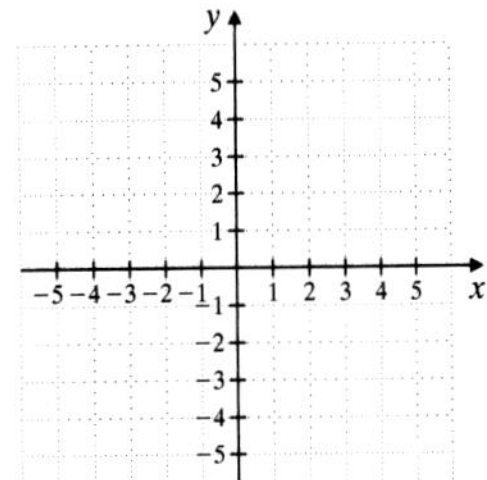

Find the slope of each line.

8.

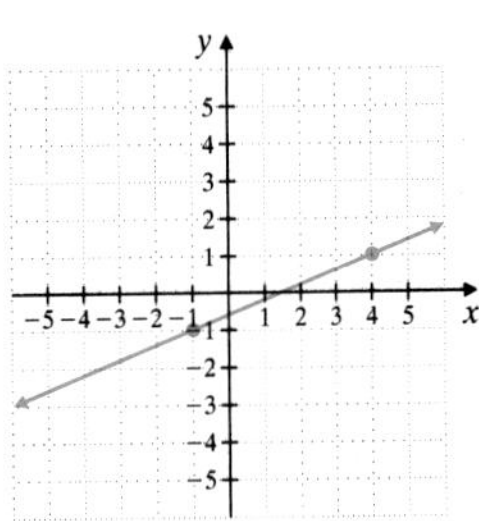

9.

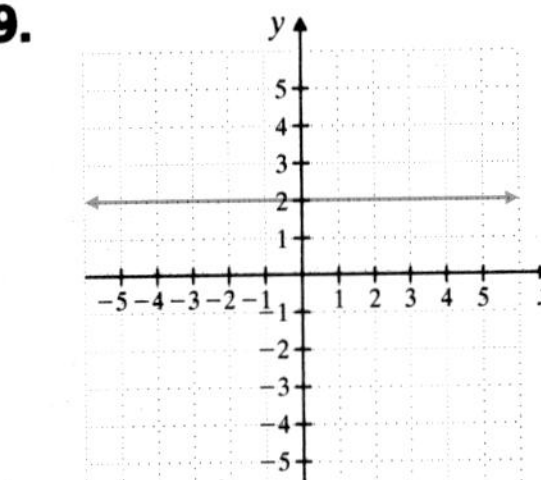

Answers

1. a. ____________

b. see graph

2. see graph

3. see graph

4. see graph

5. see graph

6. see graph

7. see graph

8. ____________

9. ____________

10. ______
11. ______
12. ______
13. ______
14. ______
15. ______
16. ______
17. ______
18. ______
19. ______
20. ______
21. ______
22. ______
23. a. ______
b. ______
c. ______
24. a. ______
b. ______
c. ______
25. ______
26. ______
27. ______
28. ______

10. Passes through $(6, -5)$ and $(-1, 2)$

11. $-3x + y = 5$

12. $x = 6$

13. Determine the slope and y-intercept of the graph of $7x - 3y = 2$

14. Determine whether the graphs of $y = 2x - 6$ and $-4x = 2y$ are parallel lines, perpendicular lines, or neither.

Find the equation of each line. Write the equation in the form $Ax + By = C$.

15. Slope $-\frac{1}{4}$, passes through $(2, 2)$

16. Passes through the origin and $(6, -7)$

17. Passes through $(2, -5)$ and $(1, 3)$

18. Slope $\frac{1}{8}$; y-intercept 12

Determine whether each relation is a function.

19. $\{(-1, 2), (-2, 4), (-3, 6), (-4, 8)\}$

20. $\{(-3, -3), (0, 5), (-3, 2), (0, 0)\}$

21. The graph shown in Exercise 8.

22. The graph shown in Exercise 9.

Find the indicated function values for each function.

23. $f(x) = 2x - 4$

a. $f(-2)$

b. $f(0.2)$

c. $f(0)$

24. $h(x) = x^3 - x$

a. $h(-1)$

b. $h(0)$

c. $h(4)$

25. The perimeter of the parallelogram below is 42 meters. Write a linear equation in two variables for the perimeter. Use this equation to find x when y is 8.

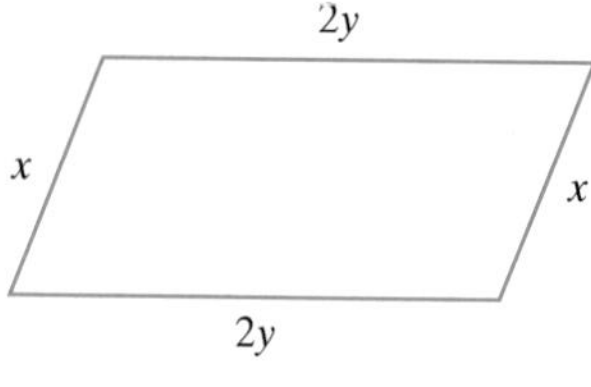

26. This graph approximates the movie ticket sales y (in millions) for the year x. Find the slope of the line and write the slope as a rate of change. Don't forget to attach the proper units.

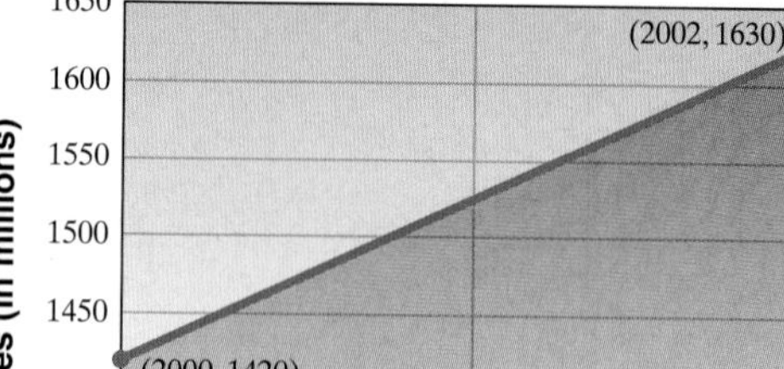

Source: National Association of Theater Owners

27. y varies directly as x. If $y = 10$ when $x = 15$, find y when x is 42.

28. y varies inversely as x^2. If $y = 8$ when $x = 5$, find y when x is 15.

Name ____________________ Section __________ Date __________

Chapter 13 Cumulative Review

1. The state of Colorado is in the shape of a rectangle whose length is about 380 miles and whose width is about 280 miles. Find its area.

2. In a pecan orchard, there are 21 trees in each row and 7 rows of trees. How many pecan trees are there?

3. Add: $1 + (-10) + (-8) + 9$

4. Add: $-2 + (-7) + 3 + (-4)$

5. Write $\frac{8}{3x}$ as an equivalent fraction whose denominator is $12x$.

6. Write $\frac{3}{2c}$ as an equivalent fraction with denominator $8c$.

7. Subtract: $12 - 8\frac{3}{7}$

8. Subtract: $15 - 4\frac{2}{5}$

9. Evaluate $-2x + 5$ for $x = 3.8$.

10. Evaluate $6x - 1$ for $x = -2.1$.

11. Write $\frac{22}{7}$ as a decimal. Round to the nearest hundredth.

12. Write $\frac{37}{19}$ as a decimal. Round to the nearest thousandth.

13. Solve $2x < -4$. Graph the solutions.

−5 −4 −3 −2 −1 0 1 2 3 4 5

14. Solve $3x \leq -9$. Graph the solution set.

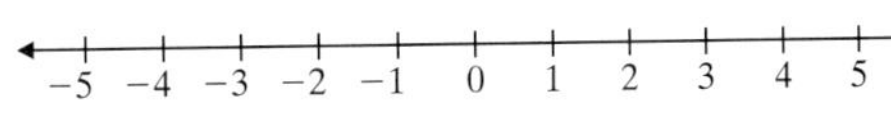

Find the degree of each polynomial and tell whether the polynomial is a monomial, binomial, trinomial, or none of these.

15. **a.** $-2t^2 + 3t + 6$
    b. $15x - 10$
    c. $7x + 3x^3 + 2x^2 - 1$

16. **a.** $-7y + 2$
    b. $8x - x^2 - 1$
    c. $9y^3 - 6y + 2 + y^2$

17. Add: $(-2x^2 + 5x - 1)$ and $(-2x^2 + x + 3)$

18. Add: $(9x - 5)$ and $(x^2 - 6x + 5)$

19. Multiply: $(3y + 1)^2$

20. Multiply: $(2x - 5)^2$

21. Factor: $-9a^5 + 18a^2 - 3a$

22. Factor: $2x^5 - x^3$

23. Factor: $x^2 + 4x - 12$

24. Factor: $x^2 + 4x - 21$

25. Factor: $8x^2 - 22x + 5$

26. Factor: $15x^2 + x - 2$

Answers

1. ______
2. ______
3. ______
4. ______
5. ______
6. ______
7. ______
8. ______
9. ______
10. ______
11. ______
12. ______
13. ______
14. ______
15. a. ______
 b. ______
 c. ______
16. a. ______
 b. ______
 c. ______
17. ______
18. ______
19. ______
20. ______
21. ______
22. ______
23. ______
24. ______
25. ______
26. ______

27.
28.
29.
30.
31.
32.
33.
34.
35.
36.
37.
38.
39. a.
b.
c.
40. a.
b.
c.
41. see graph
42. see graph
43.
44.
45.
46.
47. a.
b.
c.
48. a.
b.
c.

27. Solve: $x^2 - 9x = -20$

28. Solve: $x^2 - 9x = -14$

29. Divide: $\dfrac{2x^2 - 11x + 5}{5x - 25} \div \dfrac{4x - 2}{10}$

30. Multiply: $\dfrac{3x^2 + 17x - 6}{5x + 5} \cdot \dfrac{2x + 2}{4x + 24}$

Write the rational expression as an equivalent rational expression with the given denominator.

31. $\dfrac{4b}{9a} = \dfrac{}{27a^2b}$

32. $\dfrac{7x}{11y} = \dfrac{}{99x^2y^2}$

33. Add: $1 + \dfrac{m}{m + 1}$

34. Subtract: $1 - \dfrac{m}{m + 1}$

35. Solve: $3 - \dfrac{6}{x} = x + 8$

36. Solve: $2 + \dfrac{10}{x} = x + 5$

37. Simplify: $\dfrac{\dfrac{x + 1}{y}}{\dfrac{x}{y} + 2}$

38. Simplify: $\dfrac{\dfrac{x}{2} + 2}{\dfrac{x}{2} - 2}$

Complete each ordered pair solution so that it is a solution of each equation.

39. $3x + y = 12$

a. $(0, \ )$

b. $(\ , 6)$

c. $(-1, \ )$

40. $-x + 4y = -20$

a. $(0, \ )$

b. $(\ , 0)$

c. $(\ , -2)$

41. Graph: $2x + y = 5$

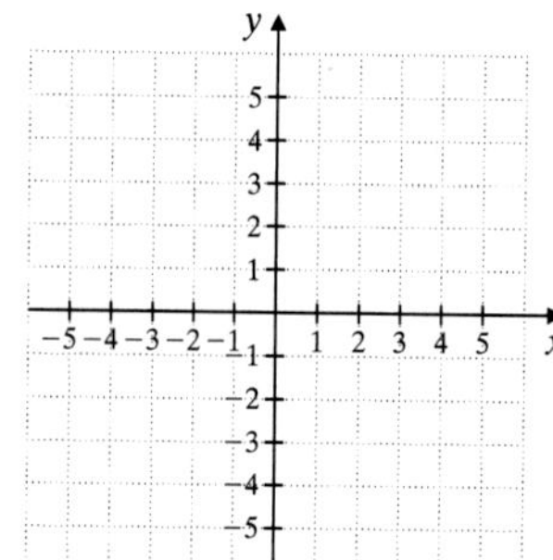

42. Graph: $y = -2x$

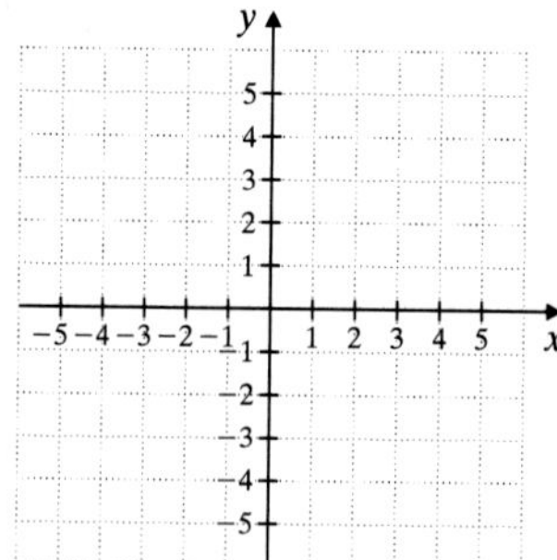

43. Find the slope of the line:
$-2x + 3y = 11$

44. Find the slope of the line:
$7x + 4y = 10$

45. Find an equation of the line passing through $(-1, 5)$ with slope -2. Write the equation in the form $Ax + By = C$.

46. Find an equation of the line passing through $(2, -7)$ with slope -5. Write the equation in the form $Ax + By = C$.

47. Given $g(x) = x^2 - 3$, find each function value and list the corresponding ordered pair.

a. $g(2)$ **b.** $g(-2)$ **c.** $g(0)$

48. Given $f(x) = 3x^2 + 2$, find each function value and list the corresponding ordered pair.

a. $f(0)$ **b.** $f(4)$ **c.** $f(-1)$

Systems of Equations

In Chapter 13, we graphed equations containing two variables. As we have seen, equations like these are often needed to represent relationships between two different quantities. There are also many opportunities to compare and contrast two such equations, called a **system of equations.** This chapter presents **linear systems** and ways we solve these systems and apply them to real-life situations.

Today, television is a fact of life. In 1946, a year after World War II ended, the television industry came to life. Broadcasting was initially dominated by two radio companies, Columbia Broadcasting System (CBS) and National Broadcasting Company (NBC). Originally, the Federal Communications Commission (FCC) restricted television broadcasts in the U.S. to the very high frequency (VHF) channels 2–13. The use of channels 14–83, the ultra-high frequency (UHF) channels, was banned. Competition for the VHF channels was fierce, and together three major networks. ABC (American Broadcasting Company), CBS, and NBC, controlled 95% of viewership well into the 1970s. The FCC gradually began making UHF channels available for broadcasting and the number of television stations (particularly UHF stations) exploded. Exercise 53 in Section 14.3 uses a system of equations to find the year in which the number of UHF stations caught up to and surpassed the number of VHF stations in the U.S.

Answers

1. see graph
2. see graph
3.
4.
5.
6.
7.
8.
9.
10.
11.
12.
13.
14.
15.
16.
17.
18.
19.
20.

Name ______ Section ______ Date ______

Chapter 14 Pretest

Solve each system by graphing.

1. $\begin{cases} x + y = 5 \\ x - y = 7 \end{cases}$

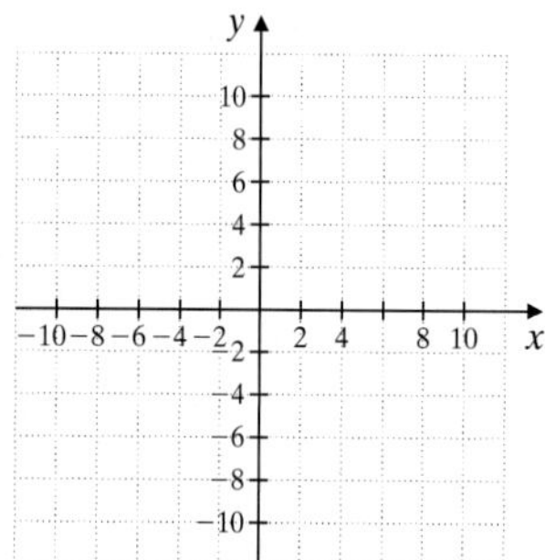

2. $\begin{cases} y = 4x \\ x = 1 \end{cases}$

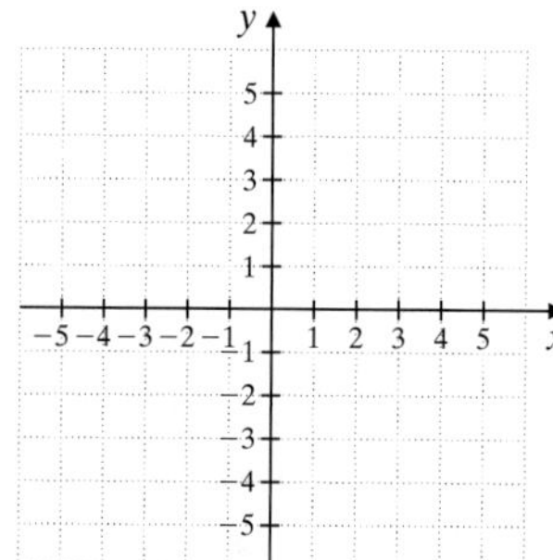

Solve each system using the substitution method.

3. $\begin{cases} x + y = 6 \\ x = 3y - 2 \end{cases}$

4. $\begin{cases} 5x + y = 13 \\ 4x - 5y = 22 \end{cases}$

5. $\begin{cases} 7y = x - 6 \\ 2x + 3y = -5 \end{cases}$

6. $\begin{cases} 4x = y + 6 \\ 8x - 2y = 12 \end{cases}$

7. $\begin{cases} x - 5 = 3y \\ 6y - 2x = 10 \end{cases}$

8. $\begin{cases} 4x + 6y = -14 \\ 6x + y = -1 \end{cases}$

9. $\begin{cases} \frac{1}{5}x - y = 3 \\ x - 5y = 15 \end{cases}$

10. $\begin{cases} y = 3x + 7 \\ y = 10x + 21 \end{cases}$

Solve each system by the addition method.

11. $\begin{cases} 2x + y = 11 \\ 3x - y = 29 \end{cases}$

12. $\begin{cases} 4x - 3y = 13 \\ 5x - 9y = 53 \end{cases}$

13. $\begin{cases} 6x + 8y = 92 \\ 5x - 3y = 9 \end{cases}$

14. $\begin{cases} 3x - 4y = 7 \\ -9x + 12y = 21 \end{cases}$

15. $\begin{cases} \frac{x}{2} + \frac{y}{3} = 2 \\ \frac{x}{6} - \frac{y}{4} = 5 \end{cases}$

16. $\begin{cases} 6x + 10y = -4 \\ -x + y = -1 \end{cases}$

17. $\begin{cases} 2x = 8 - 3y \\ 9y = 24 - 6x \end{cases}$

18. $\begin{cases} 11x = 5y + 30 \\ 3x + 4y = -24 \end{cases}$

Solve.

19. Two numbers have a sum of 97 and a difference of 65. Find the numbers.

20. Find the measures of two complementary angles if one angle is 6° less than twice the other.

14.1 Solving Systems of Linear Equations by Graphing

OBJECTIVES

- **A** Decide whether an ordered pair is a solution of a system of linear equations.
- **B** Solve a system of linear equations by graphing.
- **C** Identify special systems: those with no solution and those with an infinite number of solutions.

A **system of linear equations** consists of two or more linear equations. In this section, we focus on solving systems of linear equations containing two equations in two variables. Examples of such linear systems are

$$\begin{cases} 3x - 3y = 0 \\ x = 2y \end{cases} \quad \begin{cases} x - y = 0 \\ 2x + y = 10 \end{cases} \quad \begin{cases} y = 7x - 1 \\ y = 4 \end{cases}$$

A Deciding Whether an Ordered Pair Is a Solution

A **solution** of a system of two equations in two variables is an ordered pair of numbers that is a solution of both equations in the system.

EXAMPLE 1 Determine whether $(12, 6)$ is a solution of the system

$$\begin{cases} 2x - 3y = 6 \\ x = 2y \end{cases}$$

Solution: To determine whether $(12, 6)$ is a solution of the system, we replace x with 12 and y with 6 in both equations.

$2x - 3y = 6$	First equation	$x = 2y$	Second equation
$2(12) - 3(6) \stackrel{?}{=} 6$	Let $x = 12$ and $y = 6$.	$12 \stackrel{?}{=} 2(6)$	Let $x = 12$ and $y = 6$.
$24 - 18 \stackrel{?}{=} 6$	Simplify.	$12 = 12$	True
$6 = 6$	True		

Since $(12, 6)$ is a solution of both equations, it is a solution of the system.

Practice Problem 1

Determine whether $(3, 9)$ is a solution of the system

$$\begin{cases} 5x - 2y = -3 \\ y = 3x \end{cases}$$

EXAMPLE 2 Determine whether $(-1, 2)$ is a solution of the system

$$\begin{cases} x + 2y = 3 \\ 4x - y = 6 \end{cases}$$

Solution: We replace x with -1 and y with 2 in both equations.

$x + 2y = 3$	First equation	$4x - y = 6$	Second equation
$-1 + 2(2) \stackrel{?}{=} 3$	Let $x = -1$ and $y = 2$.	$4(-1) - 2 \stackrel{?}{=} 6$	Let $x = -1$ and $y = 2$.
$-1 + 4 \stackrel{?}{=} 3$	Simplify.	$-4 - 2 \stackrel{?}{=} 6$	Simplify.
$3 = 3$	True	$-6 = 6$	False

$(-1, 2)$ is not a solution of the second equation, $4x - y = 6$, so it is not a solution of the system.

Practice Problem 2

Determine whether $(3, -2)$ is a solution of the system

$$\begin{cases} 2x - y = 8 \\ x + 3y = 4 \end{cases}$$

B Solving Systems of Equations by Graphing

Since a solution of a system of two equations in two variables is a solution common to both equations, it is also a point common to the graphs of both equations. Let's practice finding solutions of both equations in a system—that is, solutions of the system—by graphing and identifying points of intersection.

Answers

1. $(3, 9)$ is a solution of the system **2.** $(3, -2)$ is not a solution of the system

Practice Problem 3

Solve the system of equations by graphing.

$$\begin{cases} -3x + y = -10 \\ x - y = 6 \end{cases}$$

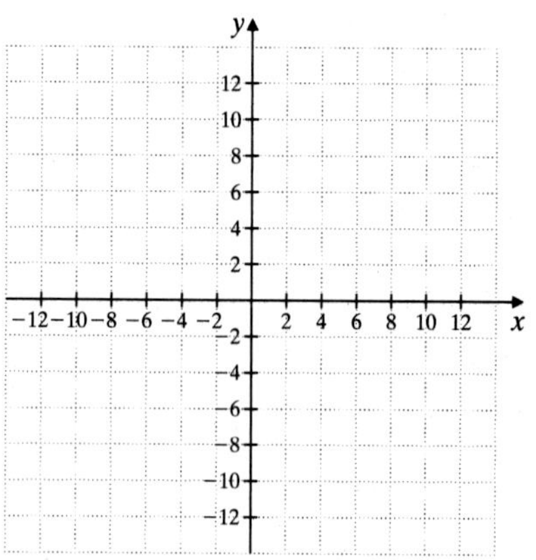

Practice Problem 4

Solve the system of equations by graphing.

$$\begin{cases} x + 3y = -1 \\ y = 1 \end{cases}$$

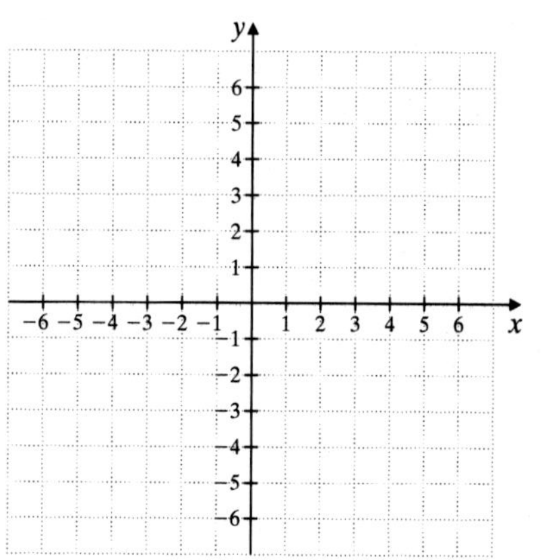

Answers

3. please see page 1150 **4.** please see page 1150

EXAMPLE 3 Solve the system of equations by graphing.

$$\begin{cases} -x + 3y = 10 \\ x + y = 2 \end{cases}$$

Solution: On a single set of axes, graph each linear equation.

$-x + 3y = 10$

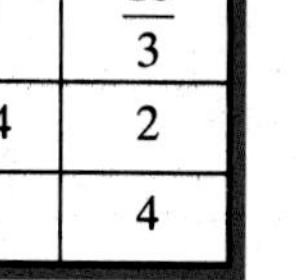

x	y
0	$\frac{10}{3}$
−4	2
2	4

$x + y = 2$

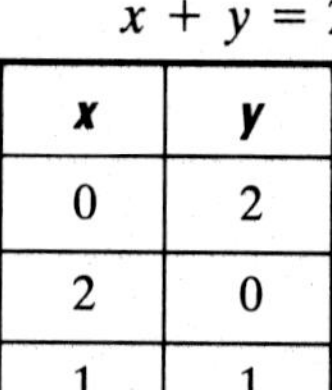

x	y
0	2
2	0
1	1

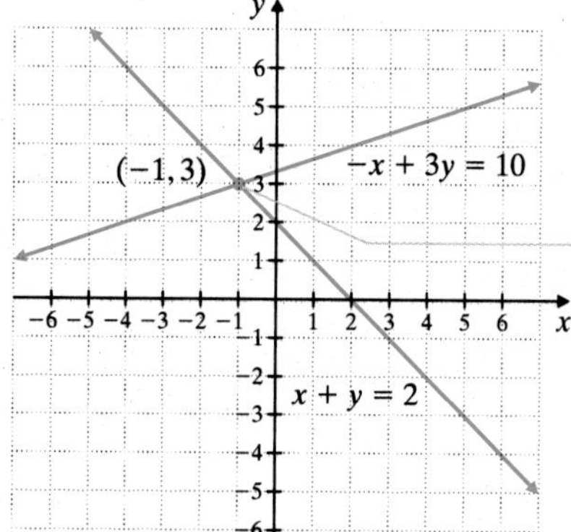

Helpful Hint

The point of intersection gives the solution of the system.

The two lines appear to intersect at the point $(-1, 3)$. To check, we replace x with -1 and y with 3 in both equations.

$-x + 3y = 10$	First equation	$x + y = 2$	Second equation
$-(-1) + 3(3) \stackrel{?}{=} 10$	Let $x = -1$ and $y = 3$.	$-1 + 3 \stackrel{?}{=} 2$	Let $x = -1$ and $y = 3$.
$1 + 9 \stackrel{?}{=} 10$	Simplify.	$2 = 2$	True
$10 = 10$	True		

$(-1, 3)$ checks, so it is the solution of the system.

Helpful Hint

Neatly drawn graphs can help when "guessing" the solution of a system of linear equations by graphing.

EXAMPLE 4 Solve the system of equations by graphing.

$$\begin{cases} 2x + 3y = -2 \\ x = 2 \end{cases}$$

Solution: We graph each linear equation on a single set of axes.

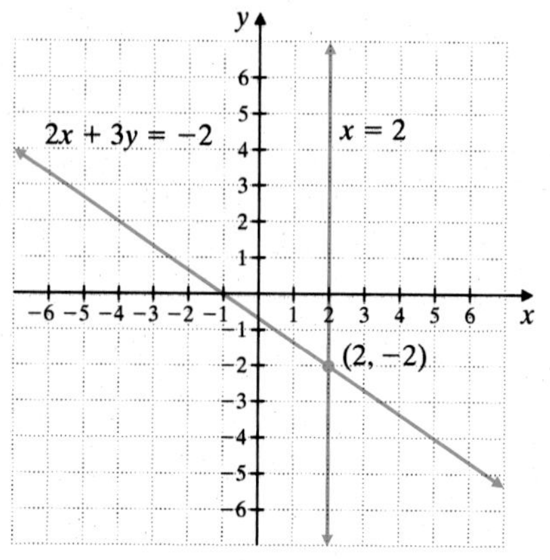

The two lines appear to intersect at the point $(2, -2)$. To determine whether $(2, -2)$ is the solution, we replace x with 2 and y with -2 in both equations.

$2x + 3y = -2$	First equation	$x = 2$	Second equation
$2(2) + 3(-2) \stackrel{?}{=} -2$	Let $x = 2$ and $y = -2$.	$2 = 2$	Let $x = 2$.
$4 + (-6) \stackrel{?}{=} -2$	Simplify.	$2 = 2$	True
$-2 = -2$	True		

Since a true statement results in both equations, $(2, -2)$ is the solution of the system.

C Identifying Special Systems of Linear Equations

Not all systems of linear equations have a single solution. Some systems have no solution and some have an infinite number of solutions.

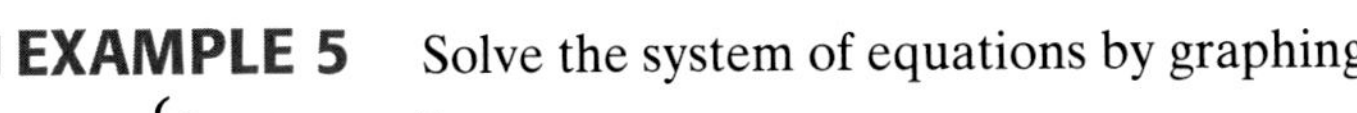

EXAMPLE 5 Solve the system of equations by graphing.

$$\begin{cases} 2x + y = 7 \\ 2y = -4x \end{cases}$$

Solution: We graph the two equations in the system. The equations in slope-intercept form are $y = -2x + 7$ and $y = -2x$. Notice from the equations that the lines have the same slope, -2, and different y-intercepts. This means that the lines are parallel.

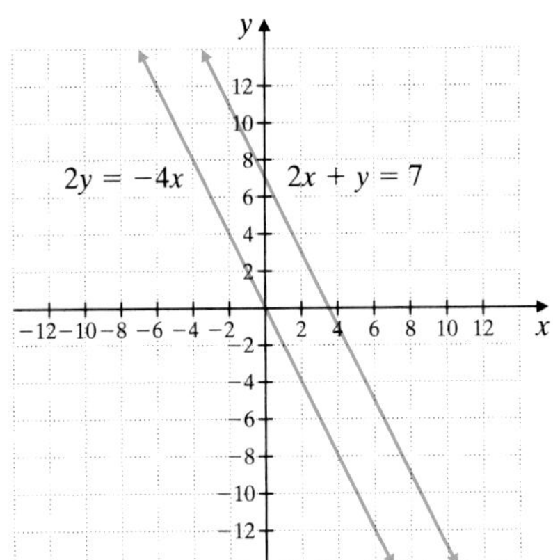

Since the lines are parallel, they do not intersect. This means that the system has *no solution*.

EXAMPLE 6 Solve the system of equations by graphing.

$$\begin{cases} x - y = 3 \\ -x + y = -3 \end{cases}$$

Solution: We graph each equation.

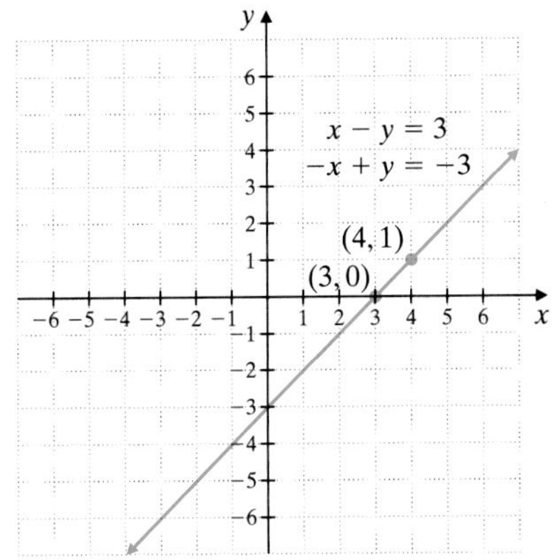

Practice Problem 5

Solve the system of equations by graphing.

$$\begin{cases} 3x - y = 6 \\ 6x = 2y \end{cases}$$

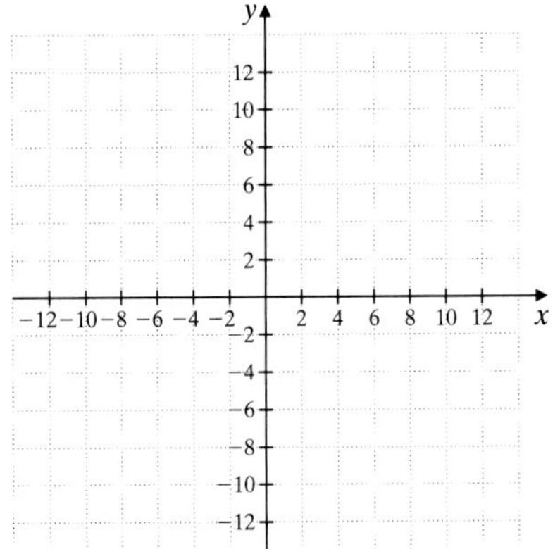

Practice Problem 6

Solve the system of equations by graphing.

$$\begin{cases} 3x + 4y = 12 \\ 9x + 12y = 36 \end{cases}$$

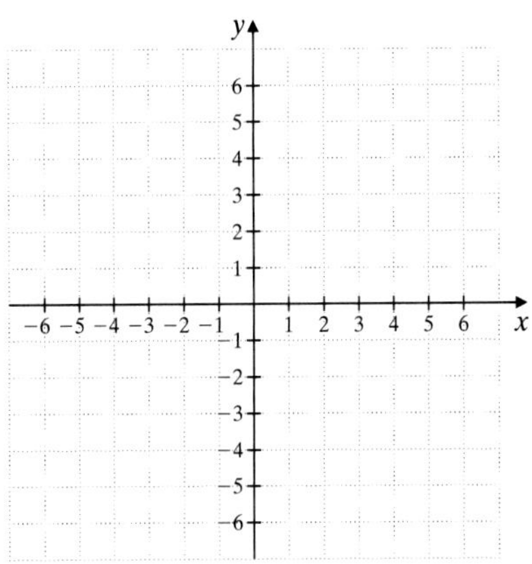

Answers

5. please see page 1150 **6.** please see page 1150

Answers

3. (2, −4)

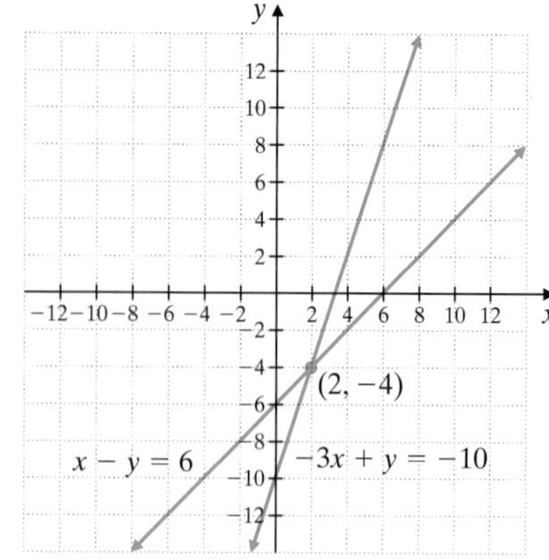

4. (−4, 1)

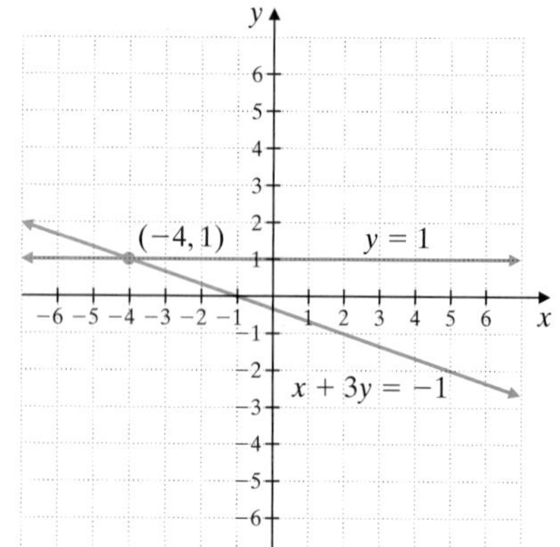

5. no solution

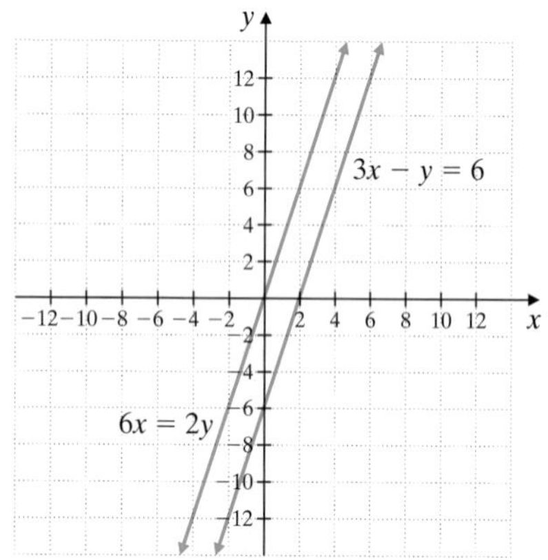

6. infinite number of solutions

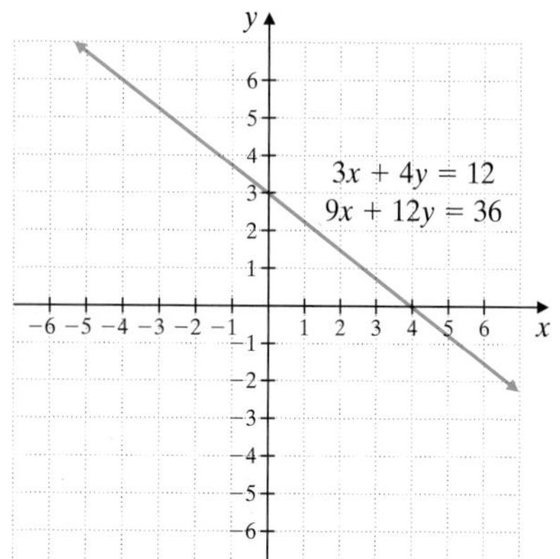

The graphs of the equations are the same line. To see this, notice that if both sides of the first equation in the system are multiplied by −1, the result is the second equation.

$$x - y = 3 \quad \text{First equation}$$
$$-1(x - y) = -1(3) \quad \text{Multiply both sides by } -1.$$
$$-x + y = -3 \quad \text{Simplify. This is the second equation.}$$

This means that the system has an infinite number of solutions. Any ordered pair that is a solution of one equation is a solution of the other and is then a solution of the system.

Examples 5 and 6 are special cases of systems of linear equations. A system that has no solution is said to be an **inconsistent system.** If the graphs of the two equations of a system are identical, we call the equations **dependent equations.**

As we have seen, three different situations can occur when graphing the two lines associated with the equations in a linear system. These situations are shown in the figures.

One point of intersection: one solution

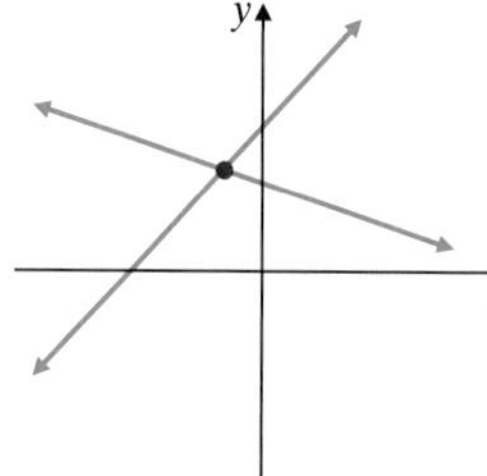

Consistent system
(at least one solution)
Independent equations
(graphs of equations differ)

Parallel lines: no solution

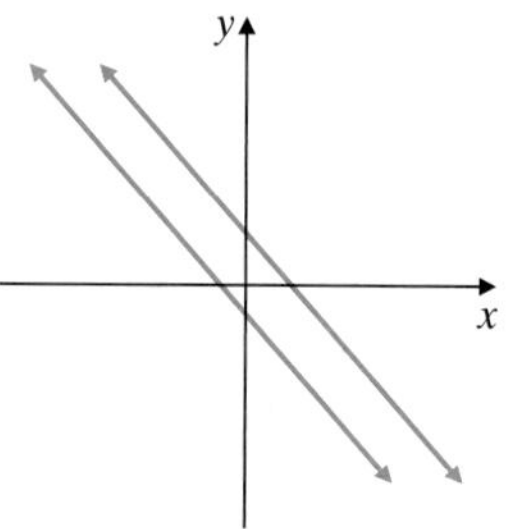

Inconsistent system
(no solution)
Independent equations
(graphs of equations differ)

Same line: infinite number of solutions

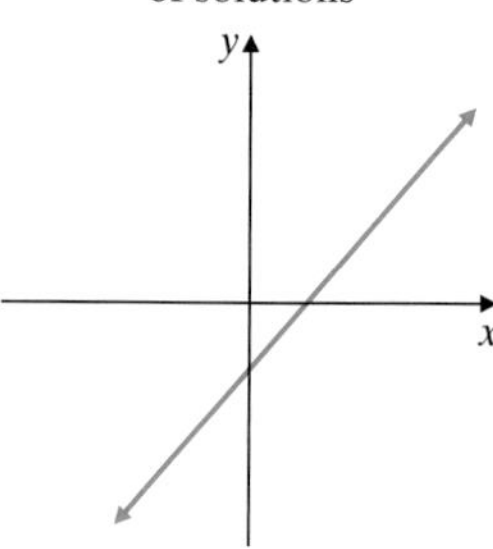

Consistent system
(at least one solution)
Dependent equations
(graphs of equations identical)

GRAPHING CALCULATOR EXPLORATIONS

A graphing calculator may be used to approximate solutions of systems of equations. For example, to approximate the solution of the system

$$\begin{cases} y = -3.14x - 1.35 \\ y = 4.88x + 5.25, \end{cases}$$

first graph each equation on the same set of axes. Then use the intersect feature of your calculator to approximate the point of intersection.

The approximate point of intersection is $(-0.82, 1.23)$.

Solve each system of equations. Approximate the solutions to two decimal places.

1. $\begin{cases} y = -2.68x + 1.21 \\ y = 5.22x - 1.68 \end{cases}$

2. $\begin{cases} y = 4.25x + 3.89 \\ y = -1.88x + 3.21 \end{cases}$

3. $\begin{cases} 4.3x - 2.9y = 5.6 \\ 8.1x + 7.6y = -14.1 \end{cases}$

4. $\begin{cases} -3.6x - 8.6y = 10 \\ -4.5x + 9.6y = -7.7 \end{cases}$

FOCUS ON Business and Career

ASSESSING JOB OFFERS

When you finish your present course of study, you will probably look for a job. When your job search has paid off and you receive a job offer, how will you decide whether to take the job? How do you decide between two or more job offers? These decisions are an important part of the job search and may not be easy to make. To evaluate the job offer, you should consider the nature of the work involved in the job, the type of company or organization that has offered the job, and the salary and benefits offered by the employer. You may also need to compare the compensation packages of two or more job offers. The following hints on assessing a job's compensation package were included in the Bureau of Labor Statistics' *Occupational Outlook Handbook*, 2000–01 edition.

> ***Salaries and benefits.*** Wait for the employer to introduce these subjects. Some companies will not talk about pay until they have decided to hire you. In order to know if their offer is reasonable, you need a rough estimate of what the job should pay. You may have to go to several sources for this information. Try to find family, friends, or acquaintances who recently were hired in similar jobs. Ask your teachers and the staff in placement offices about starting pay for graduates with your qualifications. Help-wanted ads in newspapers sometimes give salary ranges for similar positions. Check the library or your school's career center for salary surveys such as those conducted by the National Association of Colleges and Employers or various professional associations.
>
> If you are considering the salary and benefits for a job in another geographic area, make allowances for differences in the cost of living, which may be significantly higher in a large metropolitan area than in a smaller city, town, or rural area.
>
> You also should learn the organization's policy regarding overtime. Depending on the job, you may or may not be exempt from laws requiring the employer to compensate you for overtime. Find out how many hours you will be expected to work each week and whether you receive overtime pay or compensatory time off for working more than the specified number of hours in a week.
>
> Also take into account that the starting salary is just that—the start. Your salary should be reviewed on a regular basis; many organizations do it every year. How much can you expect to earn after 1, 2, or 3 or more years? An employer cannot be specific about the amount of pay if it includes commissions and bonuses.
>
> Benefits can also add a lot to your base pay, but they vary widely. Find out exactly what the benefit package includes and how much of the costs you must bear.
>
> National, state, and metropolitan area data from the National Compensation Survey, which integrates data from three existing Bureau of Labor Statistics programs—the Employment Cost Index, the Occupational Compensation Survey, and the Employee Benefits Survey—are available from:
>
> Bureau of Labor Statistics, Office of Compensation and Working Conditions, 2 Massachusetts Ave. NE, Room 4130, Washington, DC 20212-0001. Telephone: (202) 691-6199.
> Internet: http://stats.bls.gov/comhome.htm
>
> Data on earnings by detailed occupation from the Occupational Employment Statistics (OES) Survey are available from:
>
> Bureau of Labor Statistics, Office of Employment and Unemployment Statistics, Occupational Employment Statistics, 2 Massachusetts Ave. NE, Room 4840, Washington, DC 20212-0001. Telephone: (202) 691-6569.
> Internet: http://stats.bls.gov/oeshome.htm

CRITICAL THINKING

1. Suppose you have been searching for a position as an electronics sales associate. You have received two job offers. The first job pays a monthly salary of \$2000 per month plus a commission of 4% on all sales made. The second pays a monthly salary of \$2300 per month plus a commission of 2% on all sales made. At what level of monthly sales would the jobs pay the same amount? Based only on the given information about the jobs, which job would you choose? Why? What other information would you want to have about the jobs before making a decision?

2. Suppose you have been searching for an entry-level bookkeeping position. You have received two job offers. The first company offers you a starting hourly wage of \$7.50 per hour and says that each year entry-level workers receive a raise of \$0.75 per hour. The second company offers you a starting hourly wage of \$8.50 per hour and says that you can expect a \$0.50 per hour raise each year. After how many years will the two jobs pay the same hourly wage? Based only on the given information about the jobs, which job would you choose? Why? What other information would you want to have about the jobs before making a decision?

Name ______________________ Section __________ Date __________

Mental Math

Each rectangular coordinate system shows the graph of the equations in a system of equations. Use each graph to determine the number of solutions for each associated system. If the system has only one solution, give its coordinates.

1.

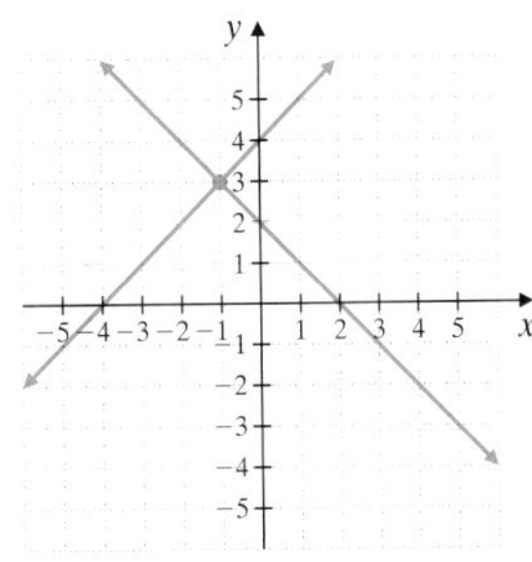

2.

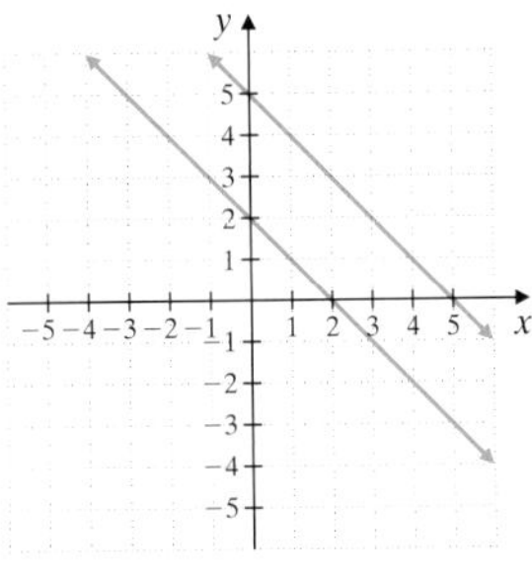

3.

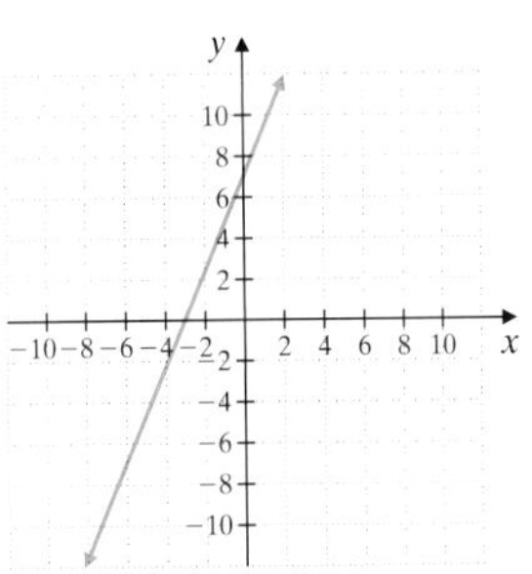

4.

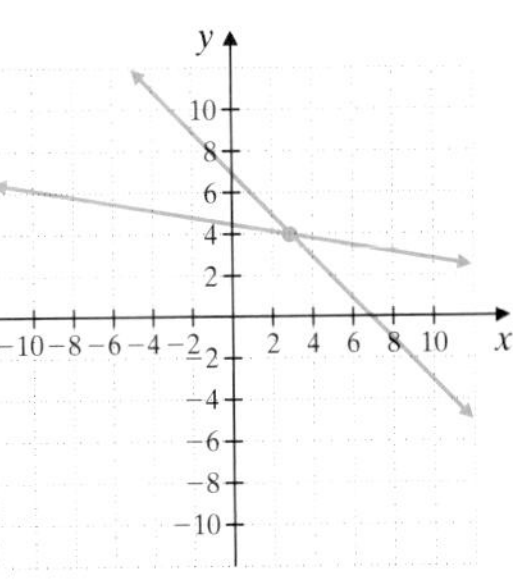

5.

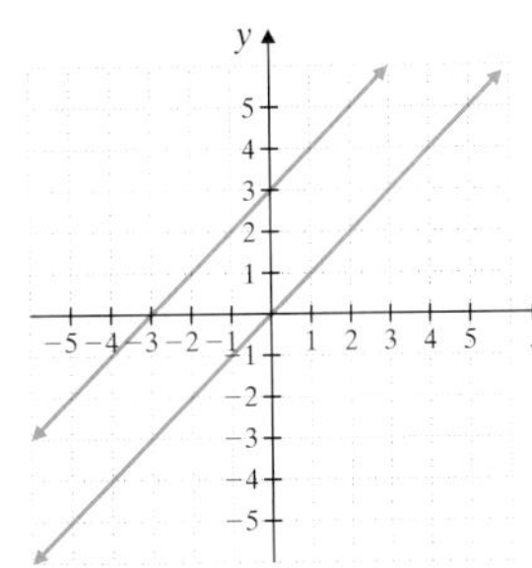

6.

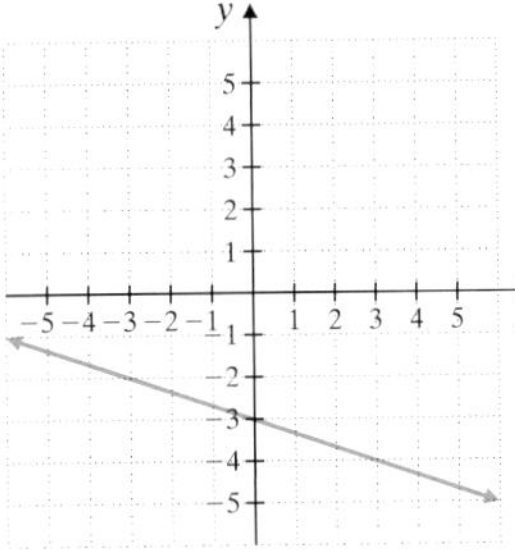

7.

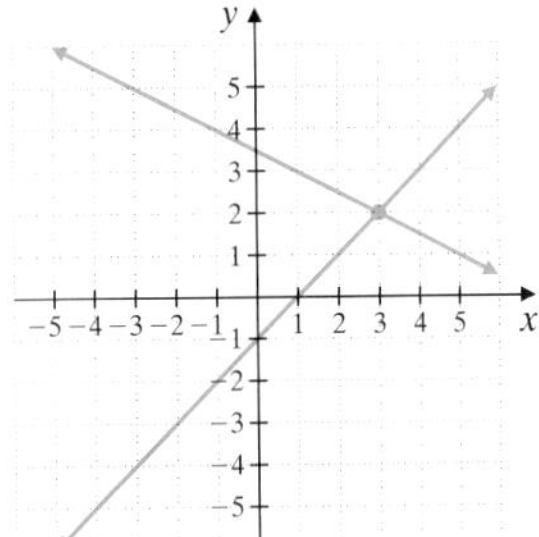

8.

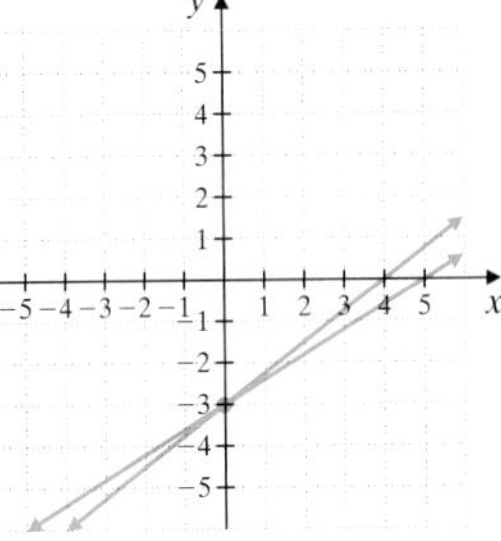

EXERCISE SET 14.1

A *Determine whether each ordered pair is a solution of the system of linear equations. See Examples 1 and 2.*

1. $\begin{cases} x + y = 8 \\ 3x + 2y = 21 \end{cases}$

a. $(2, 4)$

b. $(5, 3)$

2. $\begin{cases} 2x + y = 5 \\ x + 3y = 5 \end{cases}$

a. $(5, 0)$

b. $(2, 1)$

3. $\begin{cases} 3x - y = 5 \\ x + 2y = 11 \end{cases}$

a. $(3, 4)$

b. $(0, -5)$

4. $\begin{cases} 2x - 3y = 8 \\ x - 2y = 6 \end{cases}$

a. $(-2, -4)$

b. $(7, 2)$

5. $\begin{cases} 2y = 4x \\ 2x - y = 0 \end{cases}$

a. $(-3, -6)$

b. $(0, 0)$

6. $\begin{cases} 4x = 1 - y \\ x - 3y = -8 \end{cases}$

a. $(0, 1)$

b. $(1, -3)$

B **C** *Solve each system of linear equations by graphing. See Examples 3 through 6.*

7. $\begin{cases} x + y = 4 \\ x - y = 2 \end{cases}$

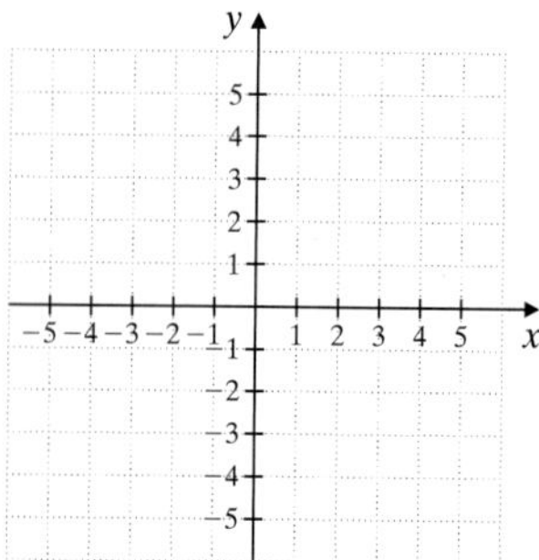

8. $\begin{cases} x + y = 3 \\ x - y = 5 \end{cases}$

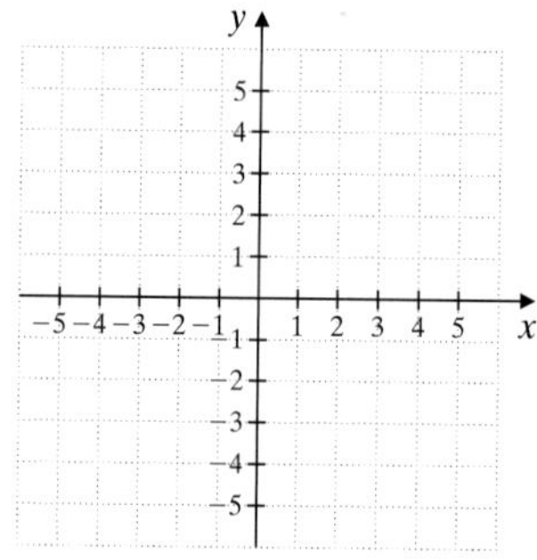

9. $\begin{cases} x + y = 6 \\ -x + y = -6 \end{cases}$

10. $\begin{cases} x + y = 1 \\ -x + y = -3 \end{cases}$

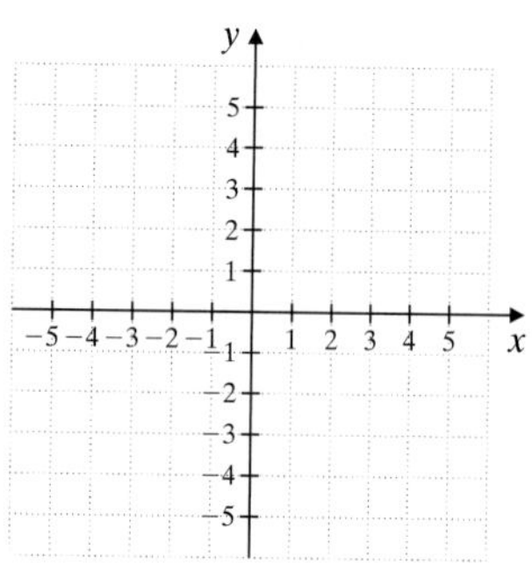

11. $\begin{cases} y = 2x \\ 3x - y = -2 \end{cases}$

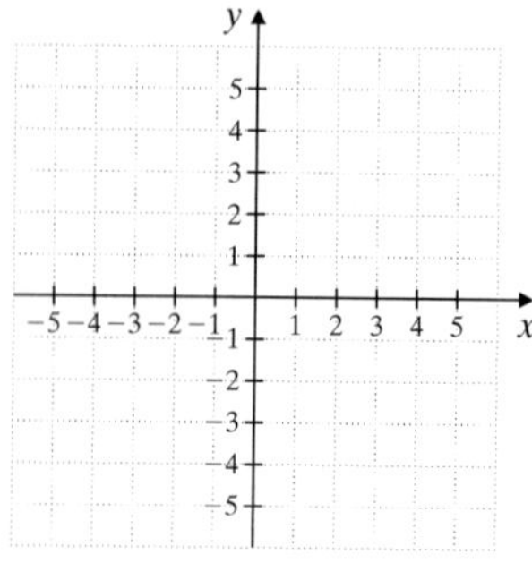

12. $\begin{cases} y = -3x \\ 2x - y = -5 \end{cases}$

13. $\begin{cases} y = x + 1 \\ y = 2x - 1 \end{cases}$

14. $\begin{cases} y = 3x - 4 \\ y = x + 2 \end{cases}$

15. $\begin{cases} 2x + y = 0 \\ 3x + y = 1 \end{cases}$

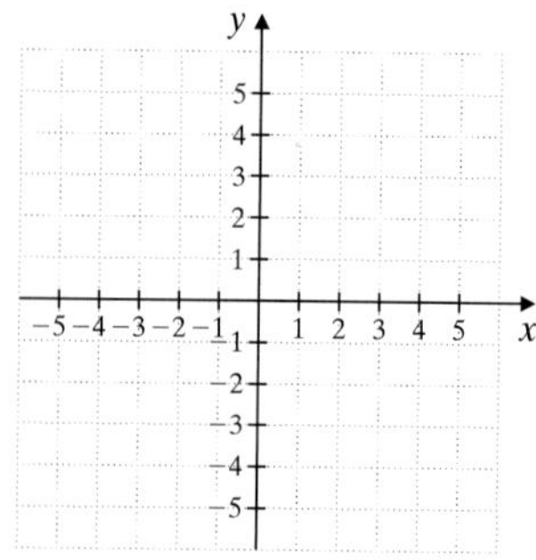

16. $\begin{cases} 2x + y = 1 \\ 3x + y = 0 \end{cases}$

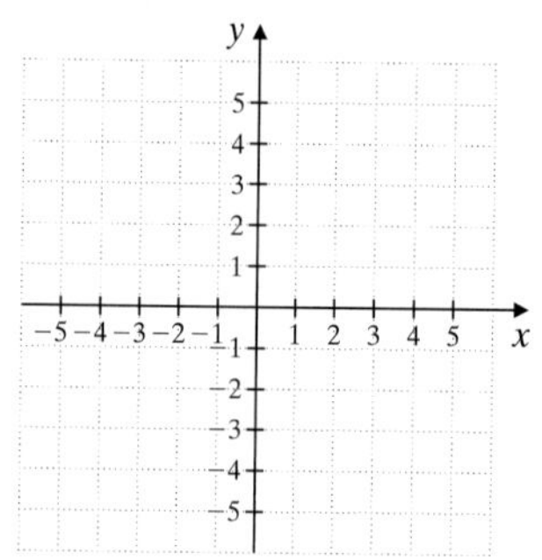

17. $\begin{cases} y = -x - 1 \\ y = 2x + 5 \end{cases}$

18. $\begin{cases} y = x - 1 \\ y = -3x - 5 \end{cases}$

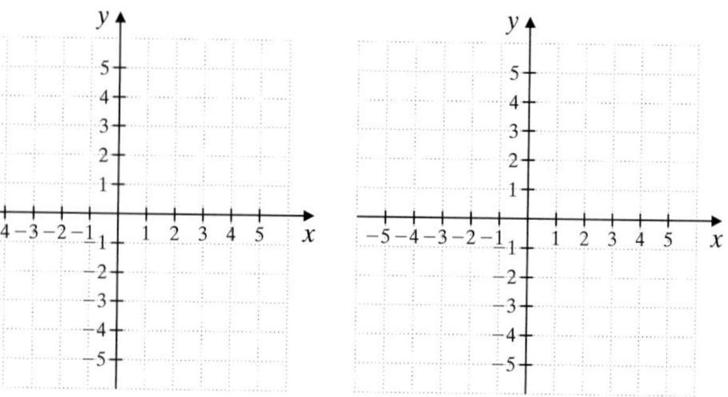

19. $\begin{cases} 2x - y = 6 \\ y = 2 \end{cases}$

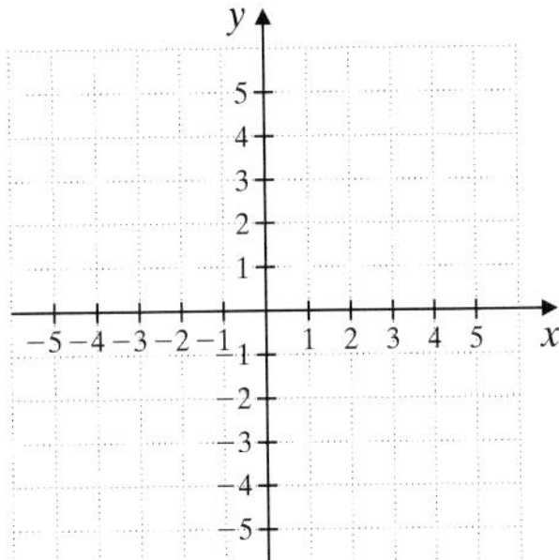

20. $\begin{cases} x + y = 5 \\ x = 4 \end{cases}$

21. 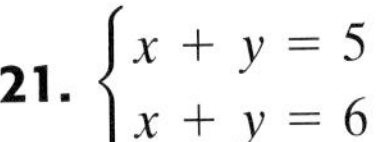 $\begin{cases} x + y = 5 \\ x + y = 6 \end{cases}$

22. $\begin{cases} x - y = 4 \\ x - y = 1 \end{cases}$

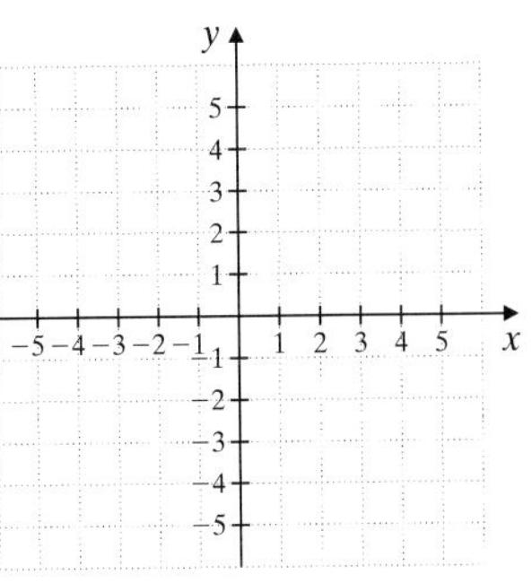

23. $\begin{cases} 2x + y = 4 \\ x + y = 2 \end{cases}$

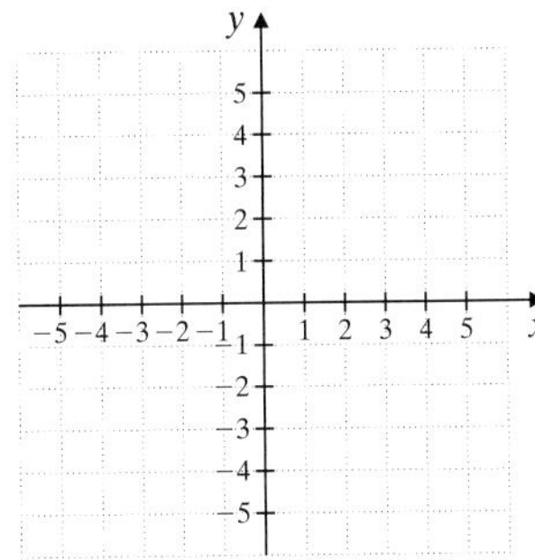

24. $\begin{cases} y + 2x = 3 \\ 4x = 2 - 2y \end{cases}$

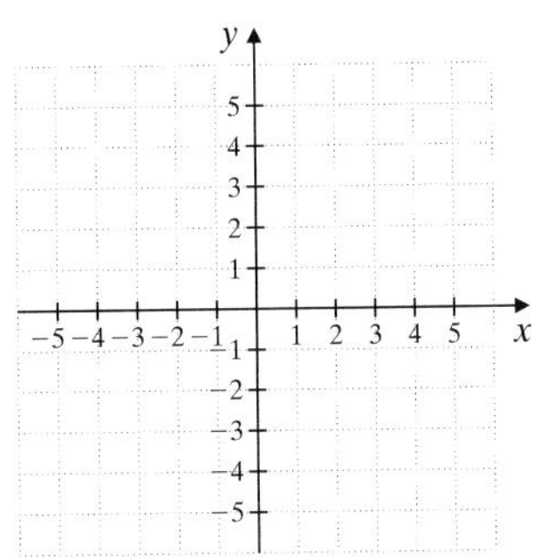

25. $\begin{cases} x - 2y = 2 \\ 3x + 2y = -2 \end{cases}$

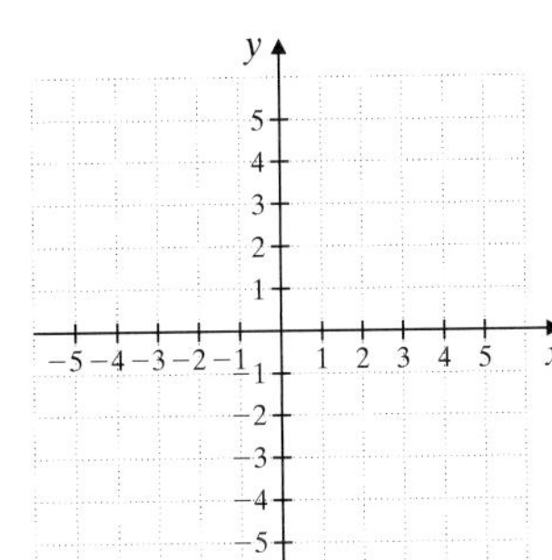

26. $\begin{cases} x + 3y = 7 \\ 2x - 3y = -4 \end{cases}$

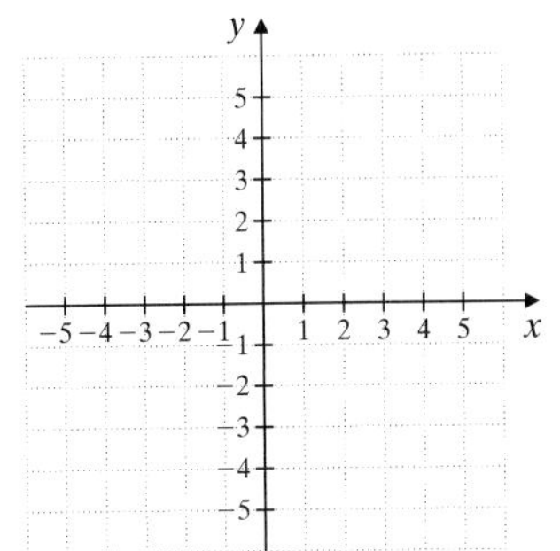

27. $\begin{cases} y - 3x = -2 \\ 6x - 2y = 4 \end{cases}$

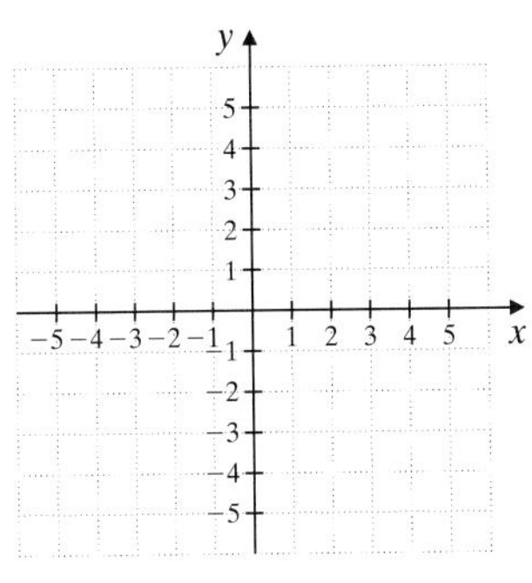

28. $\begin{cases} x - 2y = -6 \\ -2x + 4y = 12 \end{cases}$

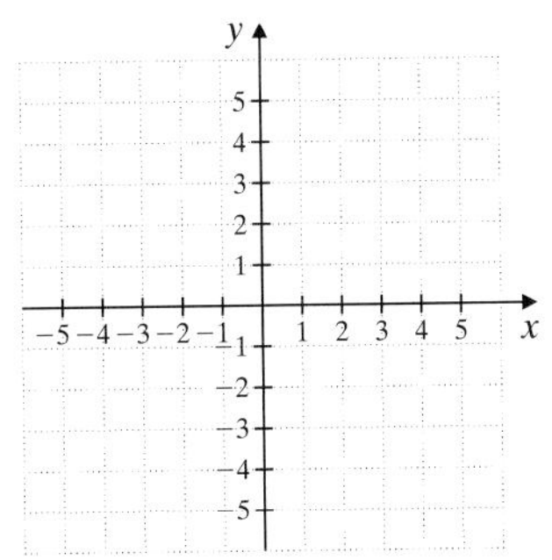

29. $\begin{cases} x = 3 \\ y = -1 \end{cases}$

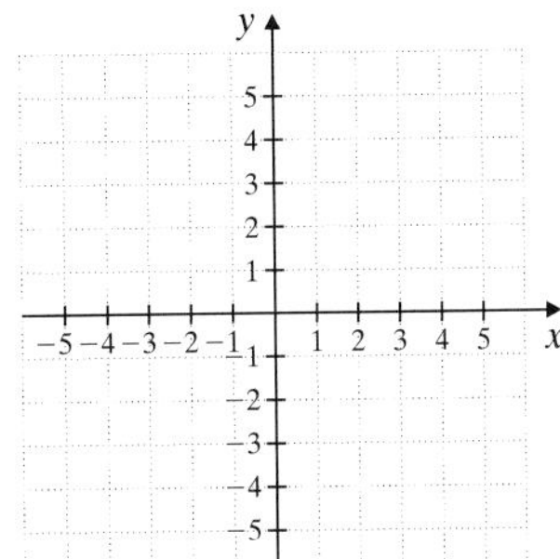

30. $\begin{cases} x = -5 \\ y = 3 \end{cases}$

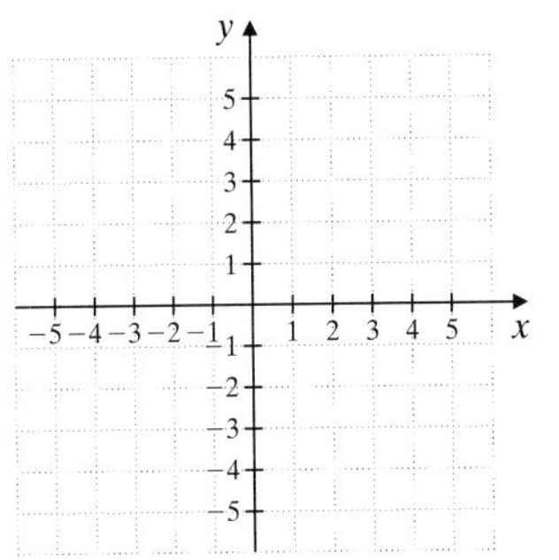

31. $\begin{cases} y = x - 2 \\ y = 2x + 3 \end{cases}$

32. $\begin{cases} y = x + 5 \\ y = -2x - 4 \end{cases}$

33. $\begin{cases} 2x - 3y = -2 \\ -3x + 5y = 5 \end{cases}$

34. $\begin{cases} 4x - y = 7 \\ 2x - 3y = -9 \end{cases}$

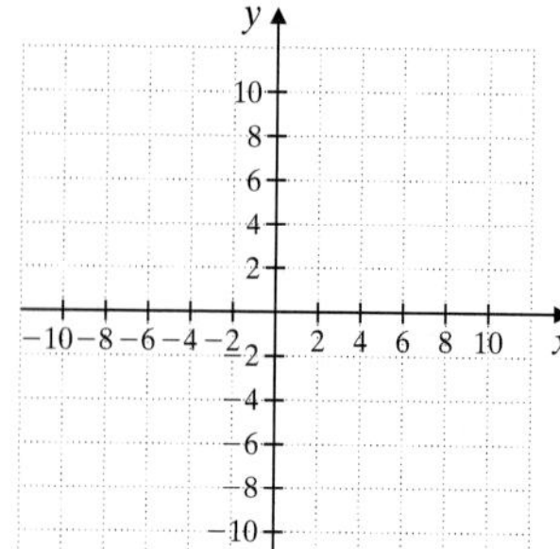

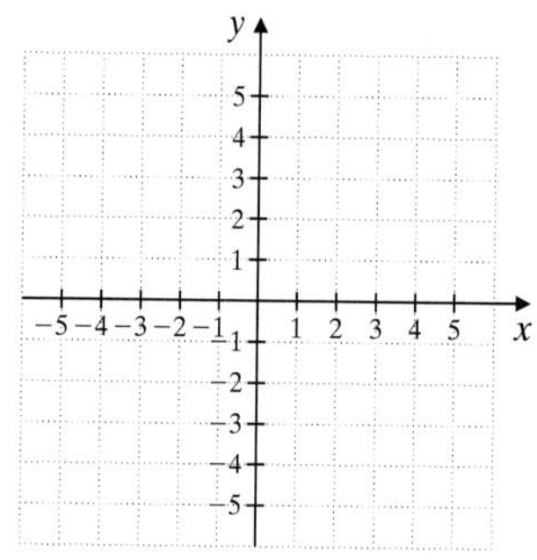

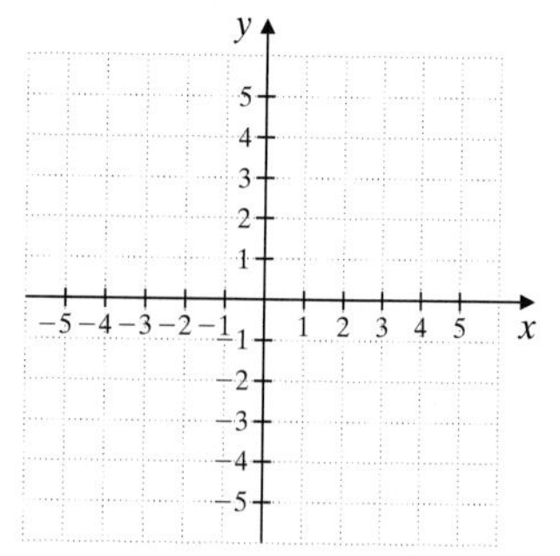

35. Draw a graph of two linear equations whose associated system has the solution $(-1, 4)$.

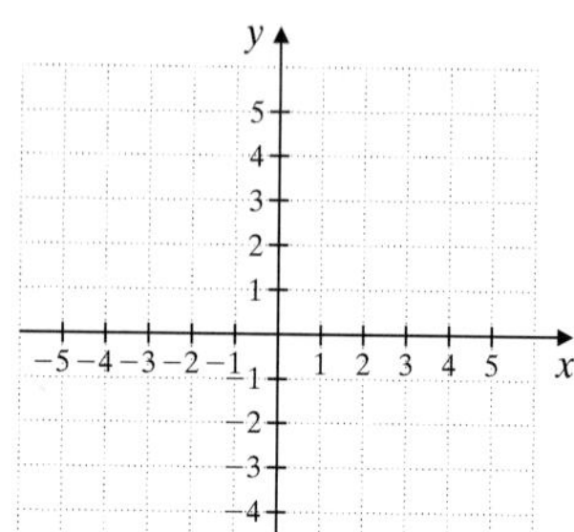

36. Draw a graph of two linear equations whose associated system has the solution $(3, -2)$.

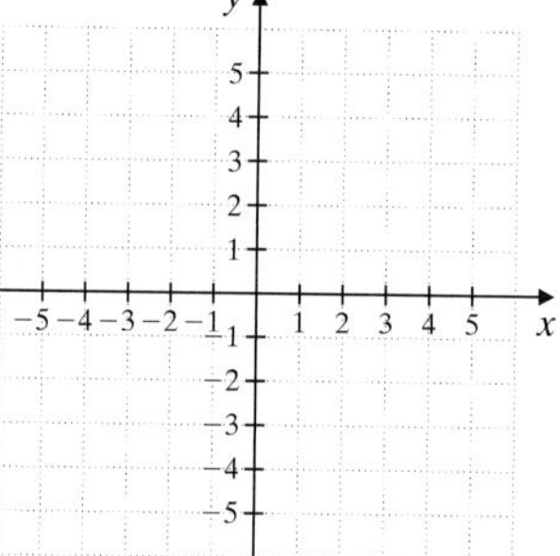

37. Draw a graph of two linear equations whose associated system has no solution.

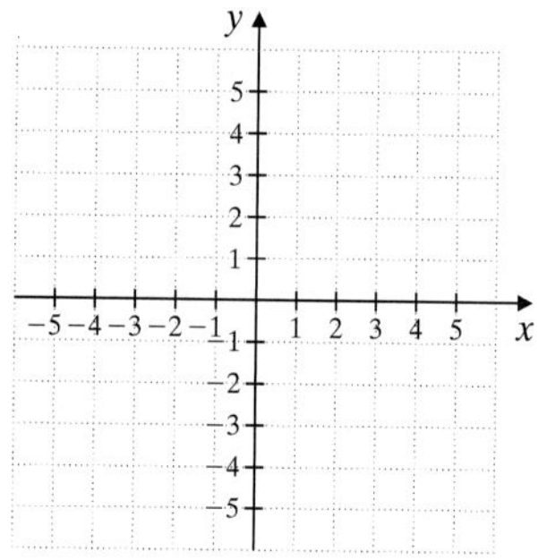

38. Draw a graph of two linear equations whose associated system has an infinite number of solutions.

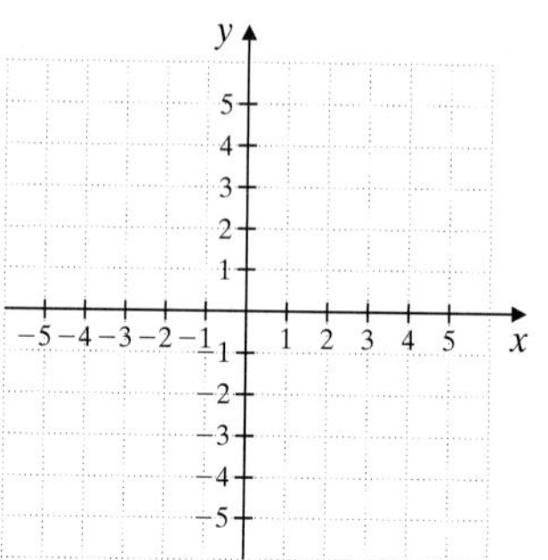

The double line graph below shows the number of pounds of fishery products from U.S. domestic catch and from imports. Use this graph to answer Exercises 39 and 40.

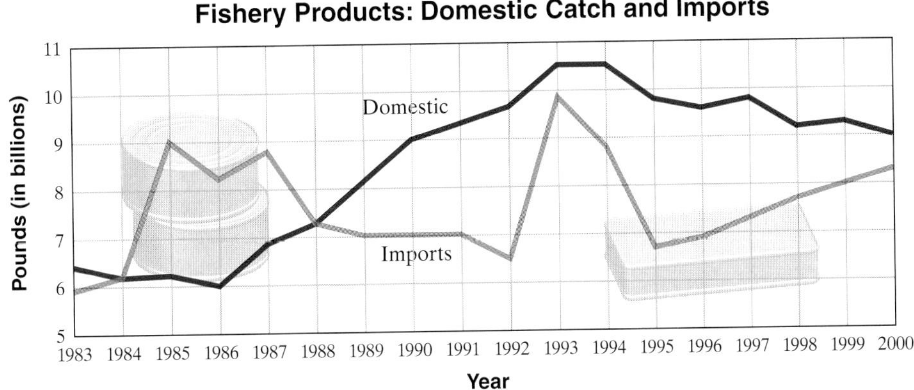

Source: U.S. Bureau of the Census, *Statistical Abstract of the United States*: 2002, 115th ed., Washington, DC, 1995.

39. In what year(s) was the number of pounds of imported fishery products equal to the number of pounds of domestic catch?

40. In what year(s) was the number of pounds of imported fishery products less than or equal to the number of pounds of domestic catch?

The double line graph below shows the number of Kmart stores vs. the number of Wal-Mart and Wal-Mart Supercenter stores. Use this graph to answer Exercises 41 and 42.

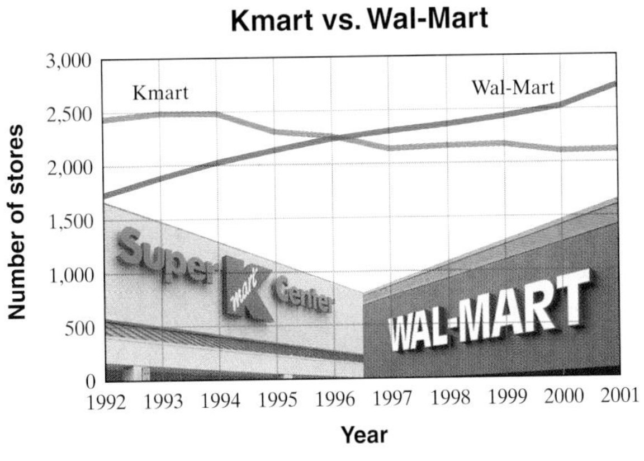

Sources: Kmart Corporation, Wal-Mart Stores, Inc.

41. In what year was the number of Kmart stores approximately equal to the number of Wal-Mart stores?

42. In what years was the number of Wal-Mart stores greater than the number of Kmart stores?

43. In the next column are tables of values for two linear equations.

a. Find a solution of the corresponding system.

b. Graph several ordered pairs from each table and sketch the two lines.

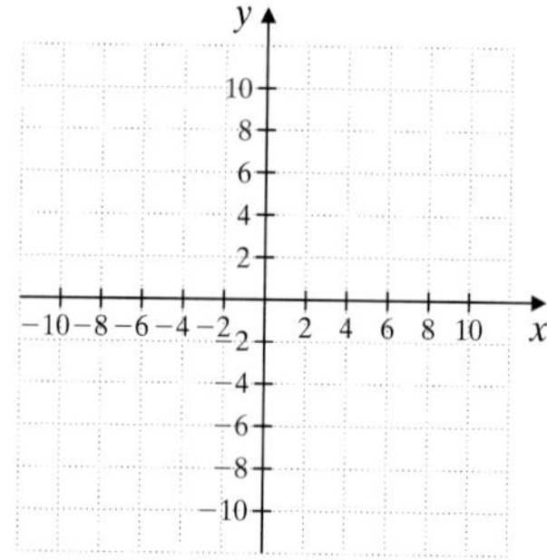

x	y
1	3
2	5
3	7
4	9
5	11

x	y
1	6
2	7
3	8
4	9
5	10

c. Does your graph confirm the solution from part (**a**)?

Review and Preview

Solve each equation. See Section 9.3.

44. $5(x - 3) + 3x = 1$

45. $-2x + 3(x + 6) = 17$

46. $4\left(\frac{y + 1}{2}\right) + 3y = 0$

47. $-y + 12\left(\frac{y - 1}{4}\right) = 3$

48. $8a - 2(3a - 1) = 6$

49. $3z - (4z - 2) = 9$

Combining Concepts

50. Construct a system of two linear equations that has $(2, 5)$ as a solution.

51. Construct a system of two linear equations that has $(0, 1)$ as a solution.

52. The ordered pair $(-2, 3)$ is a solution of the three linear equations below:

$$x + y = 1$$
$$2x - y = -7$$
$$x + 3y = 7$$

If each equation has a distinct graph, describe the graph of all three equations on the same axes.

53. Explain how to use a graph to determine the number of solutions of a system.

14.2 Solving Systems of Linear Equations by Substitution

OBJECTIVE

A Use the substitution method to solve a system of linear equations.

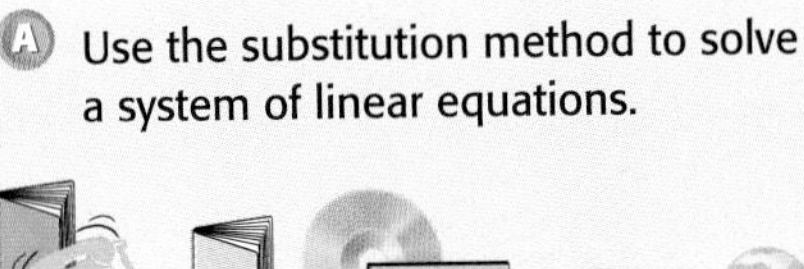

A Using the Substitution Method

You may have suspected by now that graphing alone is not an accurate way to solve a system of linear equations. For example, a solution of $\left(\frac{1}{2}, \frac{2}{9}\right)$ is unlikely to be read correctly from a graph. In this section, we discuss a second, more accurate method for solving systems of equations. This method is called the **substitution method** and is introduced in the next example.

EXAMPLE 1 Solve the system:

$$\begin{cases} 2x + y = 10 & \text{First equation} \\ x = y + 2 & \text{Second equation} \end{cases}$$

Solution: The second equation in this system is $x = y + 2$. This tells us that x and $y + 2$ have the same value. This means that we may substitute $y + 2$ for x in the first equation.

$$2x + y = 10 \quad \text{First equation}$$
$$2(y + 2) + y = 10 \quad \text{Substitute } y + 2 \text{ for } x \text{ since } x = y + 2.$$

Notice that this equation now has one variable, y. Let's now solve this equation for y.

Helpful Hint

Don't forget the distributive property.

$$\begin{aligned} 2(y + 2) + y &= 10 & \\ 2y + 4 + y &= 10 & \text{Use the distributive property.} \\ 3y + 4 &= 10 & \text{Combine like terms.} \\ 3y &= 6 & \text{Subtract 4 from both sides.} \\ y &= 2 & \text{Divide both sides by 3.} \end{aligned}$$

Now we know that the y-value of the ordered pair solution of the system is 2. To find the corresponding x-value, we replace y with 2 in the equation $x = y + 2$ and solve for x.

$$\begin{aligned} x &= y + 2 & \\ x &= 2 + 2 & \text{Let } y = 2. \\ x &= 4 & \end{aligned}$$

The solution of the system is the ordered pair $(4, 2)$. Since an ordered pair solution must satisfy both linear equations in the system, we could have chosen the equation $2x + y = 10$ to find the corresponding x-value. The resulting x-value is the same.

Check: We check to see that $(4, 2)$ satisfies both equations of the original system.

First Equation

$$\begin{aligned} 2x + y &= 10 & \\ 2(4) + 2 &\stackrel{?}{=} 10 & \\ 10 &= 10 & \text{True} \end{aligned}$$

Second Equation

$$\begin{aligned} x &= y + 2 & \\ 4 &\stackrel{?}{=} 2 + 2 & \text{Let } x = 4 \text{ and } y = 2. \\ 4 &= 4 & \text{True} \end{aligned}$$

Practice Problem 1

Use the substitution method to solve the system:

$$\begin{cases} 2x + 3y = 13 \\ x = y + 4 \end{cases}$$

Answer

1. $(5, 1)$

The solution of the system is $(4, 2)$.
A graph of the two equations shows the two lines intersecting at the point $(4, 2)$.

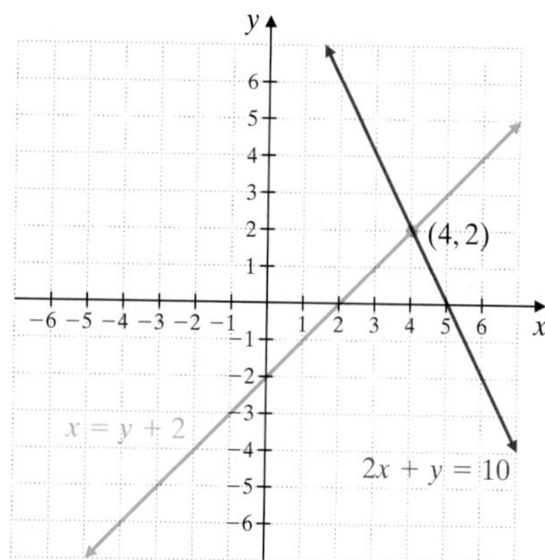

Practice Problem 2

Use the substitution method to solve the system:

$$\begin{cases} 4x - y = 2 \\ y = 5x \end{cases}$$

EXAMPLE 2 Solve the system:

$$\begin{cases} 5x - y = -2 \\ y = 3x \end{cases}$$

Solution: The second equation is solved for y in terms of x. We substitute $3x$ for y in the first equation.

$$5x - y = -2 \quad \text{First equation}$$
$$5x - (3x) = -2$$

Now we solve for x.

$$5x - 3x = -2$$
$$2x = -2 \quad \text{Combine like terms.}$$
$$x = -1 \quad \text{Divide both sides by 2.}$$

The x-value of the ordered pair solution is -1. To find the corresponding y-value, we replace x with -1 in the equation $y = 3x$.

$$y = 3x$$
$$y = 3(-1) \quad \text{Let } x = -1.$$
$$y = -3$$

Check to see that the solution of the system is $(-1, -3)$.

To solve a system of equations by substitution, we first need an equation solved for one of its variables.

Practice Problem 3

Solve the system:

$$\begin{cases} 3x + y = 5 \\ 3x - 2y = -7 \end{cases}$$

Answers

2. $(-2, -10)$ **3.** $\left(\frac{1}{3}, 4\right)$

EXAMPLE 3 Solve the system:

$$\begin{cases} x + 2y = 7 \\ 2x + 2y = 13 \end{cases}$$

Solution: We choose one of the equations and solve for x or y. We will solve the first equation for x by subtracting $2y$ from both sides.

$$x + 2y = 7 \quad \text{First equation}$$
$$x = 7 - 2y \quad \text{Subtract } 2y \text{ from both sides.}$$

Since $x = 7 - 2y$, we now substitute $7 - 2y$ for x in the second equation and solve for y.

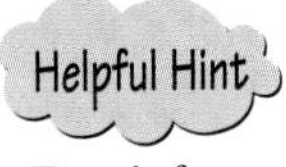

Don't forget to insert parentheses when substituting $7 - 2y$ for x.

$$\begin{aligned} 2x + 2y &= 13 && \text{Second equation} \\ 2(7 - 2y) + 2y &= 13 && \text{Let } x = 7 - 2y. \\ 14 - 4y + 2y &= 13 && \text{Apply the distributive property.} \\ 14 - 2y &= 13 && \text{Simplify.} \\ -2y &= -1 && \text{Subtract 14 from both sides.} \\ y &= \frac{1}{2} && \text{Divide both sides by } -2. \end{aligned}$$

To find x, we let $y = \frac{1}{2}$ in the equation $x = 7 - 2y$.

$$\begin{aligned} x &= 7 - 2y \\ x &= 7 - 2\left(\frac{1}{2}\right) && \text{Let } y = \frac{1}{2}. \\ x &= 7 - 1 \\ x &= 6 \end{aligned}$$

Check the solution in both equations of the original system. The solution is $\left(6, \frac{1}{2}\right)$.

The following steps summarize how to solve a system of equations by the substitution method.

To Solve a System of Two Linear Equations by the Substitution Method

Step 1. Solve one of the equations for one of its variables.
Step 2. Substitute the expression for the variable found in Step 1 into the other equation.
Step 3. Solve the equation from Step 2 to find the value of one variable.
Step 4. Substitute the value found in Step 3 in any equation containing both variables to find the value of the other variable.
Step 5. Check the proposed solution in the original system.

Try the Concept Check in the margin.

EXAMPLE 4 Solve the system:

$$\begin{cases} 7x - 3y = -14 \\ -3x + y = 6 \end{cases}$$

Solution: To avoid introducing fractions, we will solve the second equation for y.

$$\begin{aligned} -3x + y &= 6 && \text{Second equation} \\ y &= 3x + 6 \end{aligned}$$

Concept Check

As you solve the system

$$\begin{cases} 2x + y = -5 \\ x - y = 5 \end{cases}$$

you find that $y = -5$. Is this the solution of the system?

Practice Problem 4

Solve the system:

$$\begin{cases} 5x - 2y = 6 \\ -3x + y = -3 \end{cases}$$

Answers

4. $(0, -3)$

Concept Check: no, the solution will be an ordered pair

Next, we substitute $3x + 6$ for y in the first equation.

$$7x - 3y = -14 \quad \text{First equation}$$
$$7x - 3(3x + 6) = -14 \quad \text{Let } y = 3x + 6.$$
$$7x - 9x - 18 = -14 \quad \text{Use the distributive property.}$$
$$-2x - 18 = -14 \quad \text{Simplify.}$$
$$-2x = 4 \quad \text{Add 18 to both sides.}$$
$$\frac{-2x}{-2} = \frac{4}{-2} \quad \text{Divide both sides by } -2.$$
$$x = -2$$

To find the corresponding y-value, we substitute -2 for x in the equation $y = 3x + 6$. Then $y = 3(-2) + 6$ or $y = 0$. The solution of the system is $(-2, 0)$. Check this solution in both equations of the system.

Helpful Hint

When solving a system of equations by the substitution method, begin by solving an equation for one of its variables. If possible, solve for a variable that has a coefficient of 1 or -1 to avoid working with time-consuming fractions.

Practice Problem 5

Solve the system:

$$\begin{cases} -x + 3y = 6 \\ y = \frac{1}{3}x + 2 \end{cases}$$

EXAMPLE 5 Solve the system: $\begin{cases} \frac{1}{2}x - y = 3 \\ x = 6 + 2y \end{cases}$

Solution: The second equation is already solved for x in terms of y. Thus we substitute $6 + 2y$ for x in the first equation and solve for y.

$$\frac{1}{2}x - y = 3 \quad \text{First equation}$$
$$\frac{1}{2}(6 + 2y) - y = 3 \quad \text{Let } x = 6 + 2y.$$
$$3 + y - y = 3 \quad \text{Apply the distributive property.}$$
$$3 = 3 \quad \text{Simplify.}$$

Arriving at a true statement such as $3 = 3$ indicates that the two linear equations in the original system are equivalent. This means that their graphs are identical, as shown in the figure. There is an infinite number of solutions to the system, and any solution of one equation is also a solution of the other.

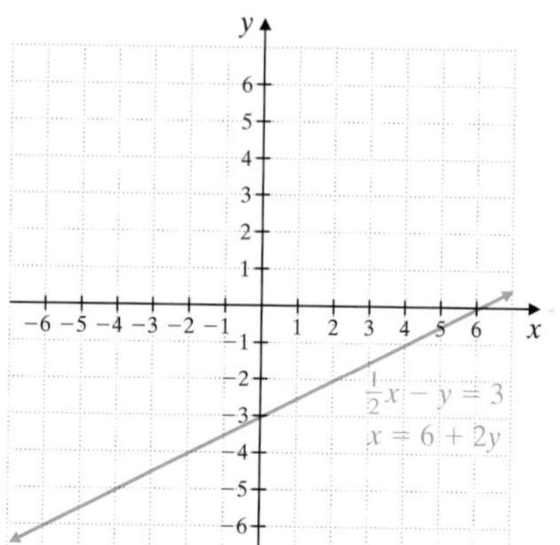

Answer

5. infinite number of solutions

EXAMPLE 6 Solve the system:

$$\begin{cases} 6x + 12y = 5 \\ -4x - 8y = 0 \end{cases}$$

Solution: We choose the second equation and solve for y.

$$-4x - 8y = 0 \quad \text{Second equation}$$
$$-8y = 4x \quad \text{Add } 4x \text{ to both sides.}$$
$$\frac{-8y}{-8} = \frac{4x}{-8} \quad \text{Divide both sides by } -8.$$
$$y = -\frac{1}{2}x \quad \text{Simplify.}$$

Now we replace y with $-\frac{1}{2}x$ in the first equation.

$$6x + 12y = 5 \quad \text{First equation}$$
$$6x + 12\left(-\frac{1}{2}x\right) = 5 \quad \text{Let } y = -\frac{1}{2}x.$$
$$6x + (-6x) = 5 \quad \text{Simplify.}$$
$$0 = 5 \quad \text{Combine like terms.}$$

The false statement $0 = 5$ indicates that this system has no solution. The graph of the linear equations in the system is a pair of parallel lines, as shown in the figure.

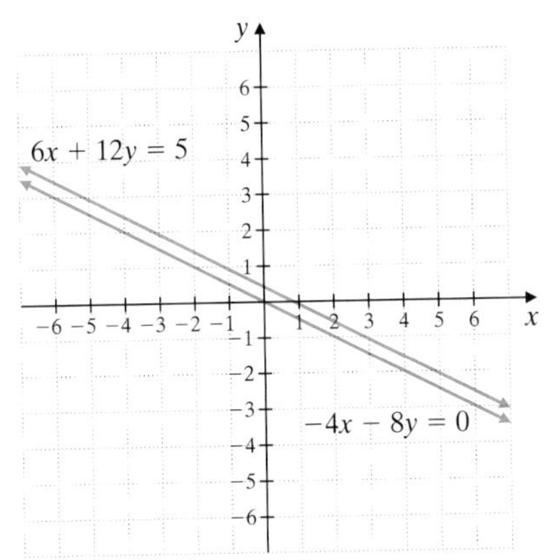

Try the Concept Check in the margin.

Practice Problem 6

Solve the system:

$$\begin{cases} 2x - 3y = 6 \\ -4x + 6y = -12 \end{cases}$$

Concept Check

Describe how the graphs of the equations in a system appear if the system has

a. no solution

b. one solution

c. an infinite number of solutions

Answers

6. infinite number of solutions

Concept Check: **a.** parallel lines **b.** intersect at one point **c.** identical graphs

FOCUS ON Business and Career

BREAK-EVEN POINT

When a business sells a new product, it generally does not start making a profit right away. There are usually many expenses associated with creating a new product. These expenses might include an advertising blitz to introduce the product to the public. These start-up expenses might also include the cost of market research and product development or any brand-new equipment needed to manufacture the product. Start-up costs like these are generally called *fixed costs* because they don't depend on the number of items manufactured. Expenses that depend on the number of items manufactured, such as the cost of materials and shipping, are called *variable costs*. The total cost of manufacturing the new product is given by the cost equation: Total cost = Fixed costs + Variable costs.

For instance, suppose a greeting card company is launching a new line of greeting cards. The company spent \$7000 doing product research and development for the new line and spent \$15,000 on advertising the new line. The company does not need to buy any new equipment to manufacture the cards, but the paper and ink needed to make each card will cost \$0.20 per card. The total cost y in dollars for manufacturing x cards is $y = 22{,}000 + 0.20x$.

Once a business sets a price for the new product, the company can find the product's expected *revenue*. Revenue is the amount of money the company takes in from the sales of its product. The revenue from selling a product is given by the revenue equation: Revenue = Price per item × Number of items sold.

For instance, suppose that the card company plans to sell its new cards for \$1.50 each. The revenue y, in dollars, that the company can expect to receive from the sales of x cards is $y = 1.50x$.

If the total cost and revenue equations are graphed on the same coordinate system, the graphs should intersect. The point of intersection is where total cost equals revenue and is called the *break-even point*. The break-even point gives the number of items x that must be manufactured and sold for the company to recover its expenses. If fewer than this number of items are produced and sold, the company loses money. If more than this number of items are produced and sold, the company makes a profit. In the case of the greeting card company, approximately 16,923 cards must be manufactured and sold for the company to break even on this new card line. The total cost and revenue of producing and selling 16,923 cards is the same. It is approximately \$25,385.

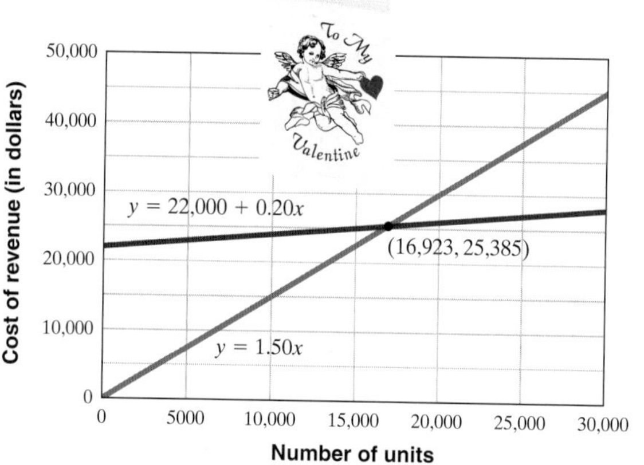

GROUP ACTIVITY

Suppose your group is starting a small business near your campus.

a. Choose a business and decide what campus-related product or service you will provide.

b. Research the fixed costs of starting up such a business.

c. Research the variable costs of producing such a product or providing such a service.

d. Decide how much you would charge per unit of your product or service.

e. Find a system of equations for the total cost and revenue of your product or service.

f. How many units of your product or service must be sold before your business will break even?

Name ______________________ Section __________ Date __________

EXERCISE SET 14.2

A *Solve each system of equations by the substitution method. See Examples 1 and 2.*

1. $\begin{cases} x + y = 3 \\ x = 2y \end{cases}$

2. $\begin{cases} x + y = 20 \\ x = 3y \end{cases}$

3. $\begin{cases} x + y = 6 \\ y = -3x \end{cases}$

4. $\begin{cases} x + y = 6 \\ y = -4x \end{cases}$

5. $\begin{cases} 3x + 2y = 16 \\ x = 3y - 2 \end{cases}$

6. $\begin{cases} 2x + 3y = 18 \\ x = 2y - 5 \end{cases}$

7. $\begin{cases} 3x - 4y = 10 \\ x = 2y \end{cases}$

8. $\begin{cases} 3x - 4y = 10 \\ y = 2x \end{cases}$

9. $\begin{cases} y = 3x + 1 \\ 4y - 8x = 12 \end{cases}$

10. $\begin{cases} y = 2x + 3 \\ 5y - 7x = 18 \end{cases}$

11. $\begin{cases} y = 2x + 9 \\ y = 7x + 10 \end{cases}$

12. $\begin{cases} y = 5x - 3 \\ y = 8x + 4 \end{cases}$

Solve each system of equations by the substitution method. See Examples 3 through 6.

13. $\begin{cases} x + 2y = 6 \\ 2x + 3y = 8 \end{cases}$

14. $\begin{cases} x + 3y = -5 \\ 2x + 2y = 6 \end{cases}$

15. $\begin{cases} 2x - 5y = 1 \\ 3x + y = -7 \end{cases}$

16. $\begin{cases} 4x + 2y = 5 \\ 2x + y = -4 \end{cases}$

17. $\begin{cases} 2y = x + 2 \\ 6x - 12y = 0 \end{cases}$

18. $\begin{cases} 3y = x + 6 \\ 4x + 12y = 0 \end{cases}$

19. $\begin{cases} 4x + y = 11 \\ 2x + 5y = 1 \end{cases}$

20. $\begin{cases} 3x + y = -14 \\ 4x + 3y = -22 \end{cases}$

21. $\begin{cases} 2x - 3y = -9 \\ 3x = y + 4 \end{cases}$

22. $\begin{cases} 8x - 3y = -4 \\ 7x = y + 3 \end{cases}$

23. $\begin{cases} 6x - 3y = 5 \\ x + 2y = 0 \end{cases}$

24. $\begin{cases} 10x - 5y = -21 \\ x + 3y = 0 \end{cases}$

25. $\begin{cases} 3x - y = 1 \\ 2x - 3y = 10 \end{cases}$

26. $\begin{cases} 2x - y = -7 \\ 4x - 3y = -11 \end{cases}$

27. $\begin{cases} -x + 2y = 10 \\ -2x + 3y = 18 \end{cases}$

28. $\begin{cases} -x + 3y = 18 \\ -3x + 2y = 19 \end{cases}$

29. $\begin{cases} 5x + 10y = 20 \\ 2x + 6y = 10 \end{cases}$

30. $\begin{cases} 2x + 4y = 6 \\ 5x + 10y = 15 \end{cases}$

31. $\begin{cases} 3x + 6y = 9 \\ 4x + 8y = 16 \end{cases}$

32. $\begin{cases} 6x + 3y = 12 \\ 9x + 6y = 15 \end{cases}$

33. $\begin{cases} \frac{1}{3}x - y = 2 \\ x - 3y = 6 \end{cases}$

34. $\begin{cases} \frac{1}{4}x - 2y = 1 \\ x - 8y = 4 \end{cases}$

35. Explain how to identify a system with no solution when using the substitution method.

36. Occasionally, when using the substitution method, we obtain the equation $0 = 0$. Explain how this result indicates that the graphs of the equations in the system are identical.

Review and Preview

Write equivalent equations by multiplying both sides of each given equation by the given nonzero number. See Section 9.3.

37. $3x + 2y = 6$ by -2

38. $-x + y = 10$ by 5

39. $-4x + y = 3$ by 3

40. $5a - 7b = -4$ by -4

Add the binomials. See Section 10.4.

41. $\begin{array}{r} 3n + 6m \\ \underline{2n - 6m} \end{array}$

42. $\begin{array}{r} -2x + 5y \\ \underline{2x + 11y} \end{array}$

43. $\begin{array}{r} -5a - 7b \\ \underline{5a - 8b} \end{array}$

44. $\begin{array}{r} 9q + p \\ \underline{-9q - p} \end{array}$

Combining Concepts

Solve each system by the substitution method. First simplify each equation by combining like terms.

45. $\begin{cases} -5y + 6y = 3x + 2(x - 5) - 3x + 5 \\ 4(x + y) - x + y = -12 \end{cases}$

46. $\begin{cases} 5x + 2y - 4x - 2y = 2(2y + 6) - 7 \\ 3(2x - y) - 4x = 1 + 9 \end{cases}$

Use a graphing calculator to solve each system.

47. $\begin{cases} y = 5.1x + 14.56 \\ y = -2x - 3.9 \end{cases}$

48. $\begin{cases} y = 3.1x - 16.35 \\ y = -9.7x + 28.45 \end{cases}$

49. $\begin{cases} 3x + 2y = 14.04 \\ 5x + y = 18.5 \end{cases}$

50. $\begin{cases} x + y = -15.2 \\ -2x + 5y = -19.3 \end{cases}$

51. For the years 1970 through 2001, the annual percentage y of U.S. households that used fuel oil to heat their homes is given by the equation $y = -0.49x + 24.75$, where x is the number of years after 1970. For the same period the annual percentage y of U.S. households that used electricity to heat their homes is given by the equation $y = 0.71x + 9.19$, where x is the number of years after 1970. (*Source:* U.S. Census Bureau, American Housing Survey Branch)

a. Use the substitution method to solve this system of equations. (Round your final results to the nearest whole numbers.)

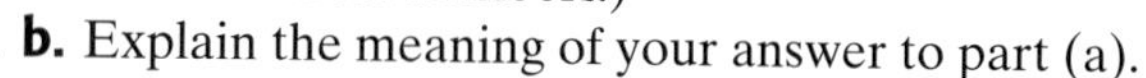

b. Explain the meaning of your answer to part (a).

c. Sketch a graph of the system of equations. Write a sentence describing the use of fuel oil and electricity for heating homes between 1970 and 2001.

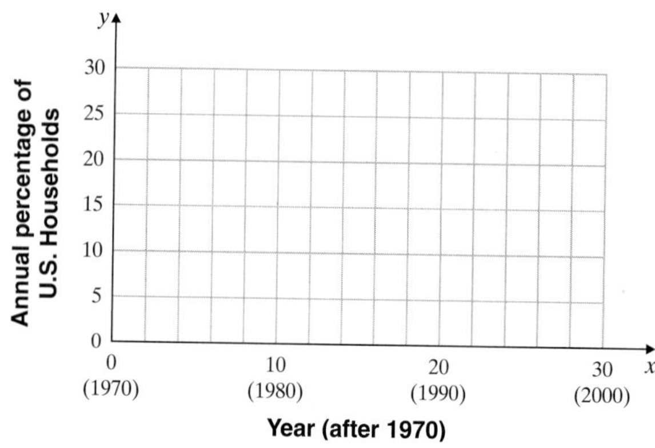

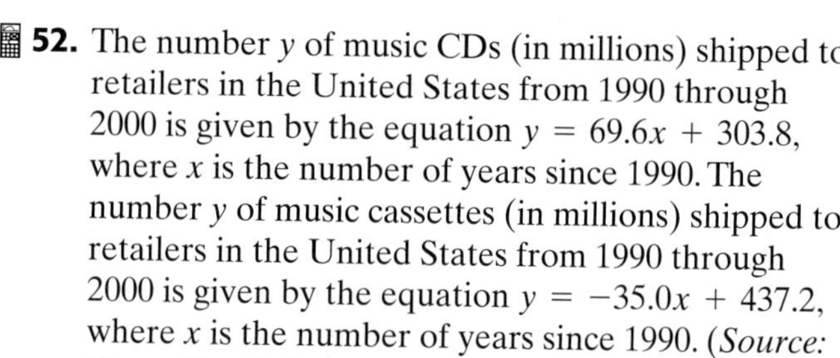

52. The number y of music CDs (in millions) shipped to retailers in the United States from 1990 through 2000 is given by the equation $y = 69.6x + 303.8$, where x is the number of years since 1990. The number y of music cassettes (in millions) shipped to retailers in the United States from 1990 through 2000 is given by the equation $y = -35.0x + 437.2$, where x is the number of years since 1990. (*Source:* Recording Industry Association of America)

a. Use the substitution method to solve this system of equations. (Round x to the nearest tenth and y to the nearest whole.)

b. Explain the meaning of your answer to part (a).

c. Sketch a graph of the system of equations. Write a sentence describing the trends in the popularity of these two types of music formats.

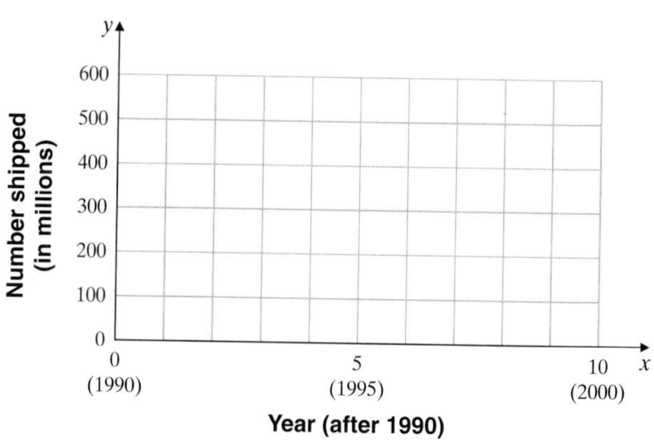

14.3 Solving Systems of Linear Equations by Addition

OBJECTIVE

(A) Use the addition method to solve a system of linear equations.

SSM TUTOR CENTER SG CD & VIDEO MATH PRO WEB

(A) Using the Addition Method

We have seen that substitution is an accurate method for solving a system of linear equations. Another accurate method is the **addition** or **elimination method.** The addition method is based on the addition property of equality: Adding equal quantities to both sides of an equation does not change the solution of the equation. In symbols,

$$\text{if } A = B \text{ and } C = D, \text{ then } A + C = B + D$$

EXAMPLE 1 Solve the system: $\begin{cases} x + y = 7 \\ x - y = 5 \end{cases}$

Solution: Since the left side of each equation is equal to its right side, we are adding equal quantities when we add the left sides of the equations together and the right sides of the equations together. This adding eliminates the variable y and gives us an equation in one variable, x. We can then solve for x.

$$\begin{aligned} x + y &= 7 && \text{First equation} \\ x - y &= 5 && \text{Second equation} \\ \hline 2x \quad &= 12 && \text{Add the equations to eliminate } y. \\ x &= 6 && \text{Divide both sides by 2.} \end{aligned}$$

The x-value of the solution is 6. To find the corresponding y-value, we let $x = 6$ in either equation of the system. We will use the first equation.

$$\begin{aligned} x + y &= 7 && \text{First equation} \\ 6 + y &= 7 && \text{Let } x = 6. \\ y &= 1 && \text{Solve for } y. \end{aligned}$$

The solution is $(6, 1)$.

Check: Check the solution in both equations.

First Equation

$$\begin{aligned} x + y &= 7 \\ 6 + 1 &\stackrel{?}{=} 7 && \text{Let } x = 6 \text{ and } y = 1. \\ 7 &= 7 && \text{True} \end{aligned}$$

Second Equation

$$\begin{aligned} x - y &= 5 \\ 6 - 1 &\stackrel{?}{=} 5 && \text{Let } x = 6 \text{ and } y = 1. \\ 5 &= 5 && \text{True} \end{aligned}$$

Thus, the solution of the system is $(6, 1)$ and the graphs of the two equations intersect at the point $(6, 1)$ as shown.

Practice Problem 1

Use the addition method to solve the system: $\begin{cases} x + y = 13 \\ x - y = 5 \end{cases}$

Answer

1. $(9, 4)$

$x + y = 7$

$(6, 1)$

$x - y = 5$

Practice Problem 2

Solve the system: $\begin{cases} 2x - y = -6 \\ -x + 4y = 17 \end{cases}$

EXAMPLE 2 Solve the system: $\begin{cases} -2x + y = 2 \\ -x + 3y = -4 \end{cases}$

Solution: If we simply add these two equations, the result is still an equation in two variables. However, our goal is to eliminate one of the variables so that we have an equation in the other variable. To do this, notice what happens if we multiply *both sides* of the first equation by -3. We are allowed to do this by the multiplication property of equality. Then the system

$$\begin{cases} -3(-2x + y) = -3(2) \\ -x + 3y = -4 \end{cases} \quad \text{simplifies to} \quad \begin{cases} 6x - 3y = -6 \\ -x + 3y = -4 \end{cases}$$

When we add the resulting equations, the y variable is eliminated.

$$\begin{aligned} 6x - 3y &= -6 \\ -x + 3y &= -4 \\ \hline 5x \qquad &= -10 \quad \text{Add.} \\ x &= -2 \quad \text{Divide both sides by 5.} \end{aligned}$$

To find the corresponding y-value, we let $x = -2$ in either of the original equations. We use the first equation of the original system.

$$\begin{aligned} -2x + y &= 2 \quad \text{First equation} \\ -2(-2) + y &= 2 \quad \text{Let } x = -2. \\ 4 + y &= 2 \\ y &= -2 \end{aligned}$$

Check the ordered pair $(-2, -2)$ in both equations of the *original* system. The solution is $(-2, -2)$.

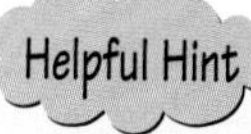

When finding the second value of an ordered pair solution, any equation equivalent to one of the original equations in the system may be used.

Answer

2. $(-1, 4)$

In Example 2, the decision to multiply the first equation by -3 was no accident. **To eliminate a variable** when adding two equations, **the coefficient**

of the variable in one equation must be the opposite of its coefficient in the other equation.

Helpful Hint

Be sure to multiply *both sides* of an equation by a chosen number when solving by the addition method. A common mistake is to multiply only the side containing the variables.

EXAMPLE 3 Solve the system: $\begin{cases} 2x - y = 7 \\ 8x - 4y = 1 \end{cases}$

Solution: When we multiply both sides of the first equation by -4, the resulting coefficient of x is -8. This is the opposite of 8, the coefficient of x in the second equation. Then the system

$$\begin{cases} -4(2x - y) = -4(7) \\ 8x - 4y = 1 \end{cases} \quad \text{simplifies to}$$

Helpful Hint

Don't forget to multiply both sides by -4.

$$\begin{cases} -8x + 4y = -28 \\ 8x - 4y = 1 \end{cases}$$
$$0 = -27 \quad \text{Add the equations.}$$

When we add the equations, both variables are eliminated and we have $0 = -27$, a false statement. This means that the system has no solution. The equations, if graphed, represent parallel lines.

EXAMPLE 4 Solve the system: $\begin{cases} 3x - 2y = 2 \\ -9x + 6y = -6 \end{cases}$

Solution: First we multiply both sides of the first equation by 3 and then we add the resulting equations.

$$\begin{cases} 3(3x - 2y) = 3 \cdot 2 \\ -9x + 6y = -6 \end{cases} \quad \text{simplifies to} \quad \begin{cases} 9x - 6y = 6 \\ -9x + 6y = -6 \end{cases} \quad \text{Add the equations.}$$
$$0 = 0$$

Both variables are eliminated and we have $0 = 0$, a true statement. This means that the system has an infinite number of solutions.

Try the Concept Check in the margin.

EXAMPLE 5 Solve the system: $\begin{cases} 3x + 4y = 13 \\ 5x - 9y = 6 \end{cases}$

Solution: We can eliminate the variable y by multiplying the first equation by 9 and the second equation by 4. Then we add the resulting equations.

$$\begin{cases} 9(3x + 4y) = 9(13) \\ 4(5x - 9y) = 4(6) \end{cases} \quad \text{simplifies to} \quad \begin{cases} 27x + 36y = 117 \\ 20x - 36y = 24 \end{cases}$$
$$47x = 141 \quad \text{Add the equations.}$$
$$x = 3 \quad \text{Solve for } x.$$

To find the corresponding y-value, we let $x = 3$ in one of the original equations of the system. Doing so in any of these equations will give $y = 1$. Check to see that $(3, 1)$ satisfies each equation in the original system. The solution is $(3, 1)$.

Practice Problem 3

Solve the system: $\begin{cases} x - 3y = -2 \\ -3x + 9y = 5 \end{cases}$

Practice Problem 4

Solve the system: $\begin{cases} 2x + 5y = 1 \\ -4x - 10y = -2 \end{cases}$

Concept Check

Suppose you are solving the system

$$\begin{cases} 3x + 8y = -5 \\ 2x - 4y = 3 \end{cases}$$

You decide to use the addition method by multiplying both sides of the second equation by 2. In which of the following was the multiplication performed correctly? Explain.

a. $4x - 8y = 3$ b. $4x - 8y = 6$

Practice Problem 5

Solve the system: $\begin{cases} 4x + 5y = 14 \\ 3x - 2y = -1 \end{cases}$

Answers

3. no solution **4.** infinite number of solutions **5.** $(1, 2)$

Concept Check: b

If we had decided to eliminate x instead of y in Example 5, the first equation could have been multiplied by 5 and the second by -3. Try solving the original system this way to check that the solution is $(3, 1)$.

The following steps summarize how to solve a system of linear equations by the addition method.

To Solve a System of Two Linear Equations by the Addition Method

Step 1. Rewrite each equation in standard form $Ax + By = C$.

Step 2. If necessary, multiply one or both equations by a nonzero number so that the coefficients of a chosen variable in the system are opposites.

Step 3. Add the equations.

Step 4. Find the value of one variable by solving the resulting equation from Step 3.

Step 5. Find the value of the second variable by substituting the value found in Step 4 into either of the original equations.

Step 6. Check the proposed solution in the original system.

Try the Concept Check in the margin.

EXAMPLE 6 Solve the system: $\begin{cases} -x - \dfrac{y}{2} = \dfrac{5}{2} \\ \dfrac{x}{6} - \dfrac{y}{2} = 0 \end{cases}$

Solution: We begin by clearing each equation of fractions. To do so, we multiply both sides of the first equation by the LCD 2 and both sides of the second equation by the LCD 6. Then the system

$$\begin{cases} 2\left(-x - \dfrac{y}{2}\right) = 2\left(\dfrac{5}{2}\right) \\ 6\left(\dfrac{x}{6} - \dfrac{y}{2}\right) = 6(0) \end{cases} \quad \text{simplifies to} \quad \begin{cases} -2x - y = 5 \\ x - 3y = 0 \end{cases}$$

We can now eliminate the variable x by multiplying the second equation by 2.

$$\begin{cases} -2x - y = 5 \\ 2(x - 3y) = 2 \cdot 0 \end{cases} \quad \text{simplifies to} \quad \begin{cases} -2x - y = 5 \\ 2x - 6y = 0 \end{cases}$$

$$-7y = 5 \quad \text{Add the equations.}$$

$$y = -\frac{5}{7} \quad \text{Solve for } y.$$

To find x, we could replace y with $-\frac{5}{7}$ in one of the equations with two variables. Instead, let's go back to the simplified system and multiply by appropriate factors to eliminate the variable y and solve for x. To do this, we multiply the first equation by -3. Then the system

$$\begin{cases} -3(-2x - y) = -3(5) \\ x - 3y = 0 \end{cases} \quad \text{simplifies to} \quad \begin{cases} 6x + 3y = -15 \\ x - 3y = 0 \end{cases}$$

$$7x = -15 \quad \text{Add the equations.}$$

$$x = -\frac{15}{7} \quad \text{Solve for } x.$$

Check the ordered pair $\left(-\frac{15}{7}, -\frac{5}{7}\right)$ in both equations of the original system. The solution is $\left(-\frac{15}{7}, -\frac{5}{7}\right)$.

Concept Check

Suppose you are solving the system

$$\begin{cases} -4x + 7y = 6 \\ x + 2y = 5 \end{cases}$$

by the addition method.

a. What step(s) should you take if you wish to eliminate x when adding the equations?

b. What step(s) should you take if you wish to eliminate y when adding the equations?

Practice Problem 6

Solve the system: $\begin{cases} -\dfrac{x}{3} + y = \dfrac{4}{3} \\ \dfrac{x}{2} - \dfrac{5}{2}y = -\dfrac{1}{2} \end{cases}$

Answers

6. $\left(-\frac{17}{2}, -\frac{3}{2}\right)$

Concept Check: **a.** multiply the second equation by 4, **b.** possible answer: multiply the first equation by -2 and the second equation by 7

Name ______________________ Section __________ Date __________

EXERCISE SET 14.3

Solve each system of equations by the addition method. See Example 1.

1. $\begin{cases} 3x + y = 5 \\ 6x - y = 4 \end{cases}$

2. $\begin{cases} 4x + y = 13 \\ 2x - y = 5 \end{cases}$

3. $\begin{cases} x - 2y = 8 \\ -x + 5y = -17 \end{cases}$

4. $\begin{cases} x - 2y = -11 \\ -x + 5y = 23 \end{cases}$

5. $\begin{cases} 3x + 2y = 11 \\ 5x - 2y = 29 \end{cases}$

6. $\begin{cases} 4x + 2y = 2 \\ 3x - 2y = 12 \end{cases}$

7. $\begin{cases} x + y = 6 \\ x - y = 6 \end{cases}$

8. $\begin{cases} x - y = 1 \\ -x + 2y = 0 \end{cases}$

Solve each system of equations by the addition method. See Examples 2 through 5.

9. $\begin{cases} 3x + y = -11 \\ 6x - 2y = -2 \end{cases}$

10. $\begin{cases} 4x + y = -13 \\ 6x - 3y = -15 \end{cases}$

11. $\begin{cases} x + 5y = 18 \\ 3x + 2y = -11 \end{cases}$

12. $\begin{cases} x + 4y = 14 \\ 5x + 3y = 2 \end{cases}$

13. $\begin{cases} 2x - 5y = 4 \\ 3x - 2y = 4 \end{cases}$

14. $\begin{cases} 6x - 5y = 7 \\ 4x - 6y = 7 \end{cases}$

15. $\begin{cases} 2x + 3y = 0 \\ 4x + 6y = 3 \end{cases}$

16. $\begin{cases} -x + 5y = -1 \\ 3x - 15y = 3 \end{cases}$

17. $\begin{cases} 3x + y = 4 \\ 9x + 3y = 6 \end{cases}$

18. $\begin{cases} 2x + y = 6 \\ 4x + 2y = 12 \end{cases}$

19. $\begin{cases} 3x - 2y = 7 \\ 5x + 4y = 8 \end{cases}$

20. $\begin{cases} 6x - 5y = 25 \\ 4x + 15y = 13 \end{cases}$

21. $\begin{cases} \frac{2}{3}x + 4y = -4 \\ 5x + 6y = 18 \end{cases}$

22. $\begin{cases} \frac{3}{2}x + 4y = 1 \\ 9x + 24y = 5 \end{cases}$

23. $\begin{cases} 4x - 6y = 8 \\ 6x - 9y = 12 \end{cases}$

24. $\begin{cases} 9x - 3y = 12 \\ 12x - 4y = 18 \end{cases}$

25. $\begin{cases} 8x = -11y - 16 \\ 2x + 3y = -4 \end{cases}$

26. $\begin{cases} 10x + 3y = -12 \\ 5x = -4y - 16 \end{cases}$

27. When solving a system of equations by the addition method, how do we know when the system has no solution?

28. Explain why the addition method might be preferred over the substitution method for solving the system $\begin{cases} 2x - 3y = 5 \\ 5x + 2y = 6. \end{cases}$

Solve each system of equations by the addition method. See Example 6.

29. $\begin{cases} \frac{x}{3} + \frac{y}{6} = 1 \\ \frac{x}{2} - \frac{y}{4} = 0 \end{cases}$

30. $\begin{cases} \frac{x}{2} + \frac{y}{8} = 3 \\ x - \frac{y}{4} = 0 \end{cases}$

31. $\begin{cases} x - \frac{y}{3} = -1 \\ -\frac{x}{2} + \frac{y}{8} = \frac{1}{4} \end{cases}$

32. $\begin{cases} 2x - \frac{3y}{4} = -3 \\ x + \frac{y}{9} = \frac{13}{3} \end{cases}$

33. $\begin{cases} \dfrac{x}{3} - y = 2 \\ -\dfrac{x}{2} + \dfrac{3y}{2} = -3 \end{cases}$

34. $\begin{cases} \dfrac{x}{2} + \dfrac{y}{4} = 1 \\ -\dfrac{x}{4} - \dfrac{y}{8} = 1 \end{cases}$

35. $\begin{cases} \dfrac{3}{5}x - y = -\dfrac{4}{5} \\ 3x + \dfrac{y}{2} = -\dfrac{9}{5} \end{cases}$

36. $\begin{cases} 3x + \dfrac{7}{2}y = \dfrac{3}{4} \\ -\dfrac{x}{2} + \dfrac{5}{3}y = -\dfrac{5}{4} \end{cases}$

37. $\begin{cases} 3.5x + 2.5y = 17 \\ -1.5x - 7.5y = -33 \end{cases}$

38. $\begin{cases} -2.5x - 6.5y = 47 \\ 0.5x - 4.5y = 37 \end{cases}$

39. $\begin{cases} 0.02x + 0.04y = 0.09 \\ -0.1x + 0.3y = 0.8 \end{cases}$

40. $\begin{cases} 0.04x - 0.05y = 0.105 \\ 0.2x - 0.6y = 1.05 \end{cases}$

Review and Preview

Rewrite each sentence using mathematical symbols. Do not solve the equations. See Sections 9.1 and 9.4.

41. Twice a number, added to 6, is 3 less than the number.

42. The sum of three consecutive integers is 66.

43. Three times a number, subtracted from 20, is 2.

44. Twice the sum of 8 and a number is the difference of the number and 20.

45. The product of 4 and the sum of a number and 6 is twice the number.

46. If the quotient of twice a number and 7 is subtracted from the reciprocal of the number, the result is 2.

Combining Concepts

47. Use the system of linear equations below to answer the questions.

$$\begin{cases} x + y = 5 \\ 3x + 3y = b \end{cases}$$

a. Find the value of b so that the system has an infinite number of solutions.

b. Find a value of b so that there are no solutions to the system.

48. Use the system of linear equations below to answer the questions.

$$\begin{cases} x + y = 4 \\ 2x + by = 8 \end{cases}$$

a. Find the value of b so that the system has an infinite number of solutions.

b. Find a value of b so that the system has a single solution.

Solve each system by the addition method.

49. $\begin{cases} 2x + 3y = 14 \\ 3x - 4y = -69.1 \end{cases}$

50. $\begin{cases} 5x - 2y = -19.8 \\ -3x + 5y = -3.7 \end{cases}$

51. Suppose you are solving the system

$$\begin{cases} 3x + 8y = -5 \\ 2x - 4y = 3. \end{cases}$$

You decide to use the addition method by multiplying both sides of the second equation by 2. In which of the following was the multiplication performed correctly? Explain.

a. $4x - 8y = 3$
b. $4x - 8y = 6$

52. Suppose you are solving the system

$$\begin{cases} -2x - y = 0 \\ -2x + 3y = 6. \end{cases}$$

You decide to use the addition method by multiplying both sides of the first equation by 3, then adding the resulting equation to the second equation. Which of the following is the correct sum? Explain.

a. $-8x = 6$
b. $-8x = 9$

53. Commercial broadcast television stations can be divided into VHF stations (channels 2 through 13) and UHF stations (channels 14 through 83). The number y of VHF stations in the United States from 1980 through 2001 is given by the equation $3x - y = -514$, where x is the number of years since 1980. The number y of UHF stations in the United States from 1980 through 2001 is given by the equation $-23x + y = 249$, where x is the number of years since 1980. (*Source:* Television Bureau of Advertising, Inc.)

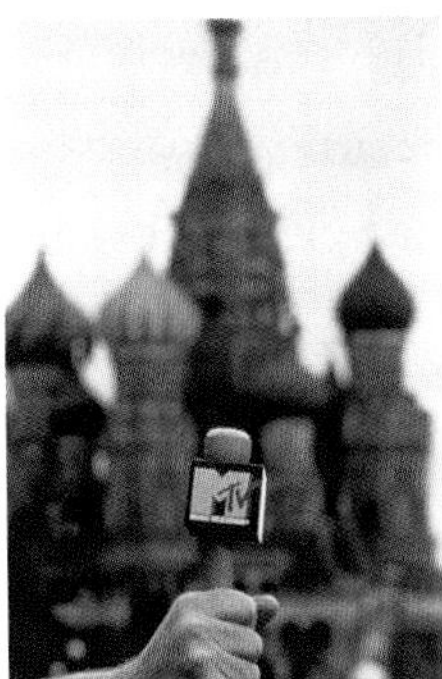

a. Use the addition method to solve this system of equations. (Round your final results to the nearest whole numbers.)
b. Interpret your solution from part (a).
c. During which years were there more UHF commercial television stations than VHF stations?

54. In recent years, the number of daily newspapers printed as morning editions has been increasing and the number of daily newspapers printed as evening editions has been decreasing. The number y of daily morning newspapers in existence from 1990 through 1999 is given by the equation $40x - 2y = -1120$, where x is the number of years since 1990. The number y of daily evening newspapers in existence from 1990 through 1999 is given by the equation $364x + 10y = 10{,}747$, where x is the number of years since 1990. (*Source: Editor & Publisher International Year Book*, annual, Editor & Publisher Co., New York, NY)

a. Use the addition method to find the year in which the number of morning newspapers equaled the number of evening newspapers. (Round to the nearest whole.)
b. How many of each type of newspaper were in existence in that year?

FOCUS ON History

Everyday Math in Ancient China

The oldest known arithmetic book is a Chinese textbook called *Nine Chapters on the Mathematical Art*. No one knows for sure who wrote this text or when it was first written. Experts believe that it was a collection of works written by many different people. It was probably written over the course of several centuries. Even though no one knows the original date of *Nine Chapters*, we do know that it existed in 213 B.C. In that year, all of the original copies of *Nine Chapters*, along with many other books, were burned when the first emperor of the Qin Dynasty (221–206 B.C.) tried to erase all traces of previous rulers and dynasties.

The Qin Emperor was not quite successful in destroying all of the *Nine Chapters*. Pieces of the text were found and many Chinese mathematicians filled in the missing material. In 263 A.D., the Chinese mathematician Liu Hui wrote a summary of *Nine Chapters*, adding his own solutions to its problems. Liu Hui's version was studied in China for over a thousand years. At one point, the Chinese government even adopted *Nine Chapters* as the official study aid for university students to use when preparing for civil service exams.

Nine Chapters is a guide to everyday math in ancient China. It contains a total of 246 problems covering widely encountered problems like field measurement, rice exchange, fair taxation, and construction. It includes the earliest known use of negative numbers and shows the first development of solving systems of linear equations. The following problem appears in Chapter 7, "Excess and Deficiency," of *Nine Chapters*. (Note: A *wen* is a unit of currency.)

> A certain number of people are purchasing some chickens together. If each person contributes 9 wen, there is an excess of 11 wen. If each person contributes just 6 wen, there is a deficiency of 16 wen. Find the number of people and the total price of the chickens. (Adapted from *The History of Mathematics: An Introduction*, second edition, David M. Burton, 1991, Wm. C. Brown Publishers, p. 164)

Critical Thinking

The information in the excess/deficiency problem from *Nine Chapters* can be translated into two equations in two variables. Let c represent the total price of the chickens, and let x represent the number of people pooling their money to buy the chickens. In this situation, an excess of 11 wen can be interpreted as 11 more than the price of the chickens. A deficiency of 16 wen can be interpreted as 16 less than the price of the chickens.

1. Use what you have learned so far in this book about translating sentences into equations to write two equations in two variables for the excess/deficiency problem.
2. Solve the problem from *Nine Chapters* by solving the system of equations you wrote in Question 1. How many people pooled their money? What was the price of the chickens?
3. Write a modern-day excess/deficiency problem of your own.

Name ______________________ Section ________ Date ________

Answers

1. ________
2. ________
3. ________
4. ________
5. ________
6. ________
7. ________
8. ________
9. ________
10. ________
11. ________
12. ________

Integrated Review—Summary of Solving Systems of Equations

Solve each system by either the addition method or the substitution method.

1. $\begin{cases} 2x - 3y = -11 \\ y = 4x - 3 \end{cases}$

2. $\begin{cases} 4x - 5y = 6 \\ y = 3x - 10 \end{cases}$

3. $\begin{cases} x + y = 3 \\ x - y = 7 \end{cases}$

4. $\begin{cases} x - y = 20 \\ x + y = -8 \end{cases}$

5. $\begin{cases} x + 2y = 1 \\ 3x + 4y = -1 \end{cases}$

6. $\begin{cases} x + 3y = 5 \\ 5x + 6y = -2 \end{cases}$

7. $\begin{cases} y = x + 3 \\ 3x - 2y = -6 \end{cases}$

8. $\begin{cases} y = -2x \\ 2x - 3y = -16 \end{cases}$

9. $\begin{cases} y = 2x - 3 \\ y = 5x - 18 \end{cases}$

10. $\begin{cases} y = 6x - 5 \\ y = 4x - 11 \end{cases}$

11. $\begin{cases} x + \frac{1}{6}y = \frac{1}{2} \\ 3x + 2y = 3 \end{cases}$

12. $\begin{cases} x + \frac{1}{3}y = \frac{5}{12} \\ 8x + 3y = 4 \end{cases}$

13. ______

14. ______

15. ______

16. ______

17. ______

18. ______

19. ______

20. ______

13. $\begin{cases} x - 5y = 1 \\ -2x + 10y = 3 \end{cases}$

14. $\begin{cases} -x + 2y = 3 \\ 3x - 6y = -9 \end{cases}$

15. $\begin{cases} 0.2x - 0.3y = -0.95 \\ 0.4x + 0.1y = 0.55 \end{cases}$

16. $\begin{cases} 0.08x - 0.04y = -0.11 \\ 0.02x - 0.06y = -0.09 \end{cases}$

17. $\begin{cases} x = 3y - 7 \\ 2x - 6y = -14 \end{cases}$

18. $\begin{cases} y = \dfrac{x}{2} - 3 \\ 2x - 4y = 0 \end{cases}$

19. Which method, substitution or addition, would you prefer to use to solve the system below? Explain your reasoning.

$$\begin{cases} 3x + 2y = -2 \\ y = -2x \end{cases}$$

20. Which method, substitution or addition, would you prefer to use to solve the system below? Explain your reasoning.

$$\begin{cases} 3x - 2y = -3 \\ 6x + 2y = 12 \end{cases}$$

14.4 Systems of Linear Equations and Problem Solving

OBJECTIVE

A Use a system of equations to solve problems.

SSM TUTOR CENTER SG CD & VIDEO MATH PRO WEB

A Using a System of Equations for Problem Solving

Many of the word problems solved earlier with one-variable equations can also be solved with two equations in two variables. We use the same problem-solving steps that have been used throughout this text. The only difference is that two variables are assigned to represent the two unknown quantities and that the problem is translated into two equations.

Problem-Solving Steps

1. UNDERSTAND the problem. During this step, become comfortable with the problem. Some ways of doing this are to

   Read and reread the problem.

   Choose two variables to represent the two unknowns.

   Construct a drawing.

   Propose a solution and check. Pay careful attention to how you check your proposed solution. This will help when writing equations to model the problem.
2. TRANSLATE the problem into two equations.
3. SOLVE the system of equations.
4. INTERPRET the results: *Check* the proposed solution in the stated problem and *state* your conclusion.

EXAMPLE 1 Finding Unknown Numbers

Find two numbers whose sum is 37 and whose difference is 21.

Solution:

1. UNDERSTAND. Read and reread the problem. Suppose that one number is 20. If their sum is 37, the other number is 17 because $20 + 17 = 37$. Is their difference 21? No; $20 - 17 = 3$. Our proposed solution is incorrect, but we now have a better understanding of the problem.

   Since we are looking for two numbers, we let

   x = first number and

   y = second number
2. TRANSLATE. Since we have assigned two variables to this problem, we translate our problem into two equations.

In words:	two numbers whose sum	is	37
	↓	↓	↓
Translate:	$x + y$	$=$	37

In words:	two numbers whose difference	is	21
	↓	↓	↓
Translate:	$x - y$	$=$	21

Practice Problem 1

Find two numbers whose sum is 50 and whose difference is 22.

Answer

1. 36 and 14

3. SOLVE. Now we solve the system.

$$\begin{cases} x + y = 37 \\ x - y = 21 \end{cases}$$

Notice that the coefficients of the variable y are opposites. Let's then solve by the addition method and begin by adding the equations.

$$\begin{aligned} x + y &= 37 \\ x - y &= 21 \\ \hline 2x \quad &= 58 \end{aligned}$$ Add the equations.

$$x = \frac{58}{2} = 29$$ Divide both sides by 2.

Now we let $x = 29$ in the first equation to find y.

$$x + y = 37$$ First equation

$$29 + y = 37$$

$$y = 37 - 29 = 8$$

4. INTERPRET. The solution of the system is $(29, 8)$.

Check: Notice that the sum of 29 and 8 is $29 + 8 = 37$, the required sum. Their difference is $29 - 8 = 21$, the required difference.
State: The numbers are 29 and 8.

Practice Problem 2

Admission prices at a local weekend fair were \$5 for children and \$7 for adults. The total money collected was \$3379, and 587 people attended the fair. How many children and how many adults attended the fair?

EXAMPLE 2 Solving a Problem about Prices

The Cirque du Soleil show Alegria is performing locally. Matinee admission for 4 adults and 2 children is \$374, while admission for 2 adults and 3 children is \$285.

a. What is the price of an adult's ticket?

b. What is the price of a child's ticket?

c. Suppose that a special rate of \$1000 is offered for groups of 20 persons. Should a group of 4 adults and 16 children use the group rate? Why or why not?

Solution:

1. UNDERSTAND. Read and reread the problem and guess a solution. Let's suppose that the price of an adult's ticket is \$50 and the price of a child's ticket is \$40. To check our proposed solution, let's see if admission for 4 adults and 2 children is \$374. Admission for 4 adults is 4(\$50) or \$200 and admission for 2 children is 2(\$40) or \$80. This gives a total admission of \$200 + \$80 = \$280, not the required \$374. Again though, we have

Answer

2. 365 children and 222 adults

accomplished the purpose of this process: We have a better understanding of the problem. To continue, we let

A = the price of an adult's ticket and
C = the price of a child's ticket

2. TRANSLATE. We translate the problem into two equations using both variables.

In words:	admission for 4 adults	and	admission for 2 children	is	$374
	↓	↓	↓	↓	↓
Translate:	$4A$	$+$	$2C$	$=$	374

In words:	admission for 2 adults	and	admission for 3 children	is	$285
	↓	↓	↓	↓	↓
Translate:	$2A$	$+$	$3C$	$=$	285

3. SOLVE. We solve the system.

$$\begin{cases} 4A + 2C = 374 \\ 2A + 3C = 285 \end{cases}$$

Since both equations are written in standard form, we solve by the addition method. First we multiply the second equation by -2 so that when we add the equations we eliminate the variable A. Then the system

$$\begin{cases} 4A + 2C = 374 \\ -2(2A + 3C) = -2(285) \end{cases} \text{ simplifies to } \begin{cases} 4A + 2C = 374 \\ -4A - 6C = -570 \end{cases}$$

Add the equations. $\quad -4C = -196$

$$-4C = -196$$

$$C = \frac{-196}{-4} = 49 \text{ or } \$49\text{, the children's ticket price.}$$

To find A, we replace C with 49 in the first equation.

$$4A + 2C = 374 \quad \text{First equation}$$
$$4A + 2(49) = 374 \quad \text{Let } C = 49.$$
$$4A + 98 = 374$$
$$4A = 276$$
$$A = \frac{276}{4} = 69 \text{ or } \$69\text{, the adult's ticket price.}$$

4. INTERPRET.

Check: Notice that 4 adults and 2 children will pay

4($69) + 2($49) = $276 + $98 = $374, the required amount. Also, the price for 2 adults and 3 children is 2($69) + 3($49) = $138 + $147 = $285, the required amount.

State: Answer the three original questions.

a. Since $A = 69$, the price of an adult's ticket is $69.

b. Since $C = 49$, the price of a child's ticket is $49.

c. The regular admission price for 4 adults and 16 children is

$$4(\$69) + 16(\$49) = \$276 + \$784$$
$$= \$1060$$

This is \$60 more than the special group rate of \$1000, so they should request the group rate. ●

Practice Problem 3

Two cars are 440 miles apart and traveling toward each other. They meet in 3 hours. If one car's speed is 10 miles per hour faster than the other car's speed, find the speed of each car.

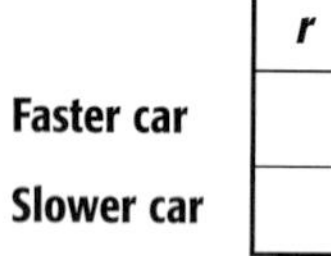

	r	$\cdot\ t$	$= d$
Faster car			
Slower car			

Answer

3. One car's speed is $68\frac{1}{3}$ mph and the other car's speed is $78\frac{1}{3}$ mph.

EXAMPLE 3 Finding Rates

Betsy Beasley and Alfredo Drizarry live 15 miles away from each other. They decide to meet one day by walking toward one another. After 2 hours they meet. If Betsy walks one mile per hour faster than Alfredo, find both walking speeds.

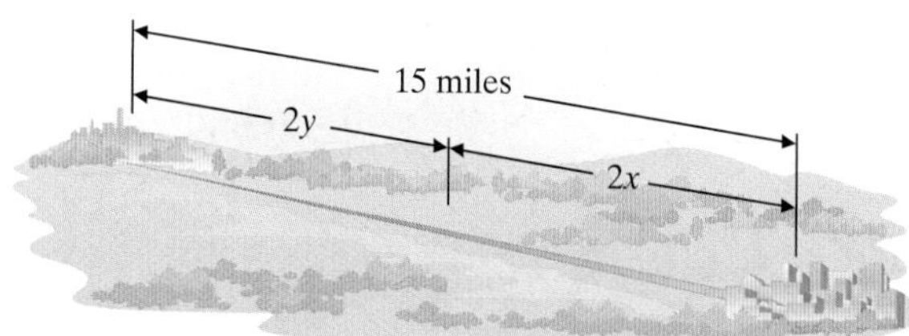

Solution:

1. UNDERSTAND. Read and reread the problem. Let's propose a solution and use the formula $d = r \cdot t$ to check. Suppose that Betsy's rate is 4 miles per hour. Since Betsy's rate is 1 mile per hour faster, Alfredo's rate is 3 miles per hour. To check, see if they can walk a total of 15 miles in 2 hours. Betsy's distance is rate $\cdot$ time $= 4(2) = 8$ miles and Alfredo's distance is rate $\cdot$ time $= 3(2) = 6$ miles. Their total distance is 8 miles + 6 miles = 14 miles, not the required 15 miles. Now that we have a better understanding of the problem, let's model it with a system of equations.

First, we let

$x =$ Alfredo's rate in miles per hour and

$y =$ Betsy's rate in miles per hour

Now we use the facts stated in the problem and the formula $d = rt$ to fill in the following chart.

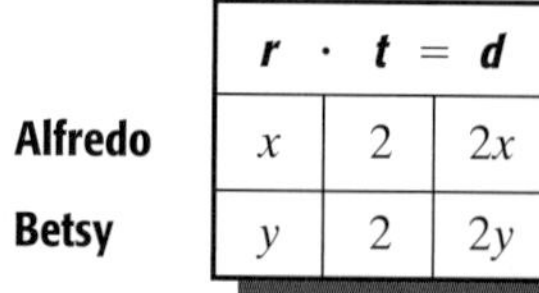

	r	$\cdot\ t$	$= d$
Alfredo	x	2	$2x$
Betsy	y	2	$2y$

2. TRANSLATE. We translate the problem into two equations using both variables.

In words:	Alfredo's distance	+	Betsy's distance	=	15
	↓		↓		↓
Translate:	$2x$	+	$2y$	=	15

In words:	Betsy's rate	is	1 mile per hour faster than Alfredo's
	↓	↓	↓
Translate:	y	=	$x + 1$

3. SOLVE. The system of equations we are solving is

$$\begin{cases} 2x + 2y = 15 \\ y = x + 1 \end{cases}$$

Let's use substitution to solve the system since the second equation is solved for y.

$$2x + 2y = 15 \quad \text{First equation}$$

$$2x + 2(x + 1) = 15 \quad \text{Replace } y \text{ with } x + 1.$$

$$2x + 2x + 2 = 15$$

$$4x = 13$$

$$x = \frac{13}{4} = 3.25$$

$$y = x + 1 = 3.25 + 1 = 4.25$$

4. INTERPRET. Alfredo's proposed rate is 3.25 miles per hour and Betsy's proposed rate is 4.25 miles per hour.

Check: Use the formula $d = rt$ and find that in 2 hours, Alfredo's distance is $(3.25)(2)$ miles or 6.5 miles. In 2 hours, Betsy's distance is $(4.25)(2)$ miles or 8.5 miles. The total distance walked is 6.5 miles + 8.5 miles or 15 miles, the given distance.

State: Alfredo walks at a rate of 3.25 miles per hour and Betsy walks at a rate of 4.25 miles per hour.

EXAMPLE 4 Finding Amounts of Solutions

Eric Daly, a chemistry teaching assistant, needs 10 liters of a 20% saline solution (salt water) for his 2 p.m. laboratory class. Unfortunately, the only mixtures on hand are a 5% saline solution and a 25% saline solution. How much of each solution should he mix to produce the 20% solution?

Solution:

1. UNDERSTAND. Read and reread the problem. Suppose that we need 4 liters of the 5% solution. Then we need $10 - 4 = 6$ liters of the 25% solution. To see if this gives us 10 liters of a 20% saline solution, let's find the amount of pure salt in each solution.

	concentration rate ↓	×	amount of solution ↓	=	amount of pure salt ↓
5% solution:	0.05	×	4 liters	=	0.2 liters
25% solution:	0.25	×	6 liters	=	1.5 liters
20% solution:	0.20	×	10 liters	=	2 liters

Since 0.2 liters + 1.5 liters = 1.7 liters, not 2 liters, our proposed solution is incorrect. But we have gained some insight into how to model and check this problem.

We let

x = number of liters of 5% solution and

y = number of liters of 25% solution

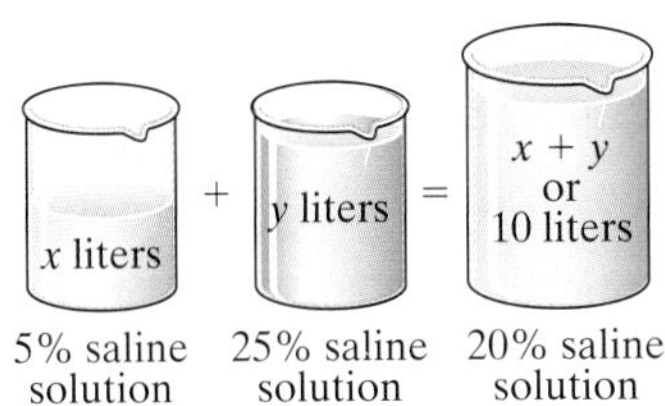

Practice Problem 4

Barb Hayes, a pharmacist, needs 50 liters of a 60% alcohol solution. She currently has available a 20% solution and a 70% solution. How many liters of each must she use to make the needed 50 liters of 60% alcohol solution?

Answer

4. 10 liters of the 20% alcohol solution and 40 liters of the 70% alcohol solution

Now we use a table to organize the given data.

	Concentration Rate	Liters of Solution	Liters of Pure Salt
First solution	5%	x	$0.05x$
Second solution	25%	y	$0.25y$
Mixture needed	20%	10	$(0.20)(10)$

2. TRANSLATE. We translate into two equations using both variables.

In words:	liters of 5% solution	+	liters of 25% solution	=	10
	↓		↓		↓
Translate:	x	+	y	=	10

In words:	salt in 5% solution	+	salt in 25% solution	=	salt in mixture
	↓		↓		↓
Translate:	$0.05x$	+	$0.25y$	=	$(0.20)(10)$

3. SOLVE. Here we solve the system

$$\begin{cases} x + y = 10 \\ 0.05x + 0.25y = 2 \end{cases}$$

To solve by the addition method, we first multiply the first equation by -25 and the second equation by 100. Then the system

$$\begin{cases} -25(x + y) = -25(10) \\ 100(0.05x + 0.25y) = 100(2) \end{cases} \quad \text{simplifies to} \quad \begin{cases} -25x - 25y = -250 \\ 5x + 25y = 200 \\ \hline -20x = -50 \quad \text{Add.} \\ x = 2.5 \end{cases}$$

To find y, we let $x = 2.5$ in the first equation of the original system.

$$x + y = 10$$
$$2.5 + y = 10 \quad \text{Let } x = 2.5.$$
$$y = 7.5$$

4. INTERPRET. Thus, we propose that Eric needs to mix 2.5 liters of 5% saline solution with 7.5 liters of 25% saline solution.

Check: Notice that $2.5 + 7.5 = 10$, the required number of liters. Also, the sum of the liters of salt in the two solutions equals the liters of salt in the required mixture:

$$0.05(2.5) + 0.25(7.5) = 0.20(10)$$
$$0.125 + 1.875 = 2$$

State: Eric needs 2.5 liters of the 5% saline solution and 7.5 liters of the 25% solution.

Try the Concept Check in the margin.

Concept Check

Suppose you mix an amount of a 30% acid solution with an amount of a 50% acid solution. Which of the following acid strengths would be possible for the resulting acid mixture?

a. 22% b. 44% c. 63%

Answer

Concept Check: b

Name ______________________ Section __________ Date __________

Mental Math

Without actually solving each problem, choose the correct solution by deciding which choice satisfies the given conditions.

1. The length of a rectangle is 3 feet longer than the width. The perimeter is 30 feet. Find the dimensions of the rectangle.
a. length = 8 feet; width = 5 feet
b. length = 8 feet; width = 7 feet
c. length = 9 feet; width = 6 feet

2. An isosceles triangle, a triangle with two sides of equal length, has a perimeter of 20 inches. Each of the equal sides is one inch longer than the third side. Find the lengths of the three sides.
a. 6 inches, 6 inches, and 7 inches
b. 7 inches, 7 inches, and 6 inches
c. 6 inches, 7 inches, and 8 inches

3. Two computer disks and three notebooks cost \$17. However, five computer disks and four notebooks cost \$32. Find the price of each.
a. notebook = \$4; computer disk = \$3
b. notebook = \$3; computer disk = \$4
c. notebook = \$5; computer disk = \$2

4. Two music CDs and four music cassette tapes cost a total of \$40. However, three music CDs and five cassette tapes cost \$55. Find the price of each.
a. CD = \$12; cassette = \$4
b. CD = \$15; cassette = \$2
c. CD = \$10; cassette = \$5

5. Kesha has a total of 100 coins, all of which are either dimes or quarters. The total value of the coins is \$13.00. Find the number of each type of coin.
a. 80 dimes; 20 quarters
b. 20 dimes; 44 quarters
c. 60 dimes; 40 quarters

6. Yolanda has 28 gallons of saline solution available in two large containers at her pharmacy. One container holds three times as much as the other container. Find the capacity of each container.
a. 15 gallons; 5 gallons
b. 20 gallons; 8 gallons
c. 21 gallons; 7 gallons

EXERCISE SET 14.4

Write a system of equations describing each situation. Do not solve the system. See Example 1.

1. Two numbers add up to 15 and have a difference of 7.

2. The total of two numbers is 16. The first number plus 2 more than 3 times the second equals 18.

3. Keiko has a total of \$6500, which she has invested in two accounts. The larger account is \$800 greater than the smaller account.

4. Dominique has four times as much money in his savings account as in his checking account. The total amount is \$2300.

Solve. See Example 1.

5. Two numbers total 83 and have a difference of 17. Find the two numbers.

6. The sum of two numbers is 76 and their difference is 52. Find the two numbers.

7. A first number plus twice a second number is 8. Twice the first number plus the second totals 25. Find the numbers.

8. One number is 4 more than twice the second number. Their total is 25. Find the numbers.

9. The highest scorer during the WNBA 2003 regular season was Lauren Jackson of the Seattle Storm. Over the season, Jackson scored 142 more points than the second-highest scorer, Chamique Holdsclaw of the Washington Mystics. Together, Jackson and Holdsclaw scored 1174 points during the 2003 regular season. How many points did each player score over the course of the season? (*Source:* Women's National Basketball Association)

10. Milan Hejduk of the Colorado Avalanche was the NHL's leading goal scorer during the 2002–2003 regular season. Glen Murray of the Boston Bruins, who was ranked fifth for goals, scored 6 fewer goals than Hejduk. Together, these two players made a total of 94 goals during the 2002–2003 regular season. How many goals each did Murray and Hejduk make? (*Source:* National Hockey League)

Solve. See Example 2.

11. Ann Marie Jones has been pricing Amtrak train fares for a group trip to New York. Three adults and four children must pay \$159. Two adults and three children must pay \$112. Find the price of an adult's ticket, and find the price of a child's ticket.

12. Last month, Jerry Papa purchased five cassettes and two compact discs at Wall-to-Wall Sound for \$65. This month he bought three cassettes and four compact discs for \$81. Find the price of each cassette, and find the price of each compact disc.

13. Johnston and Betsy Waring have a jar containing 80 coins, all of which are either quarters or nickels. The total value of the coins is \$14.60. How many of each type of coin do they have?

14. Art and Bette Meish purchased 40 stamps, a mixture of 32¢ and 19¢ stamps. Find the number of each type of stamp if they spent \$12.15.

15. David and Jacquelyn Bick own 30 shares of General Electric Co. stock and 55 shares of The Ohio Art Company (makers of Etch A Sketch and other toys). At the close of the markets on a particular day, their stock portfolio consisting of these two stocks was worth \$2348.10. The closing price of General Electric stock was \$35.77 more per share than the closing price of The Ohio Art Company stock on that day. What was the closing price of each stock on that day? (*Source:* Bridge Information Services)

16. Caitlin Jackson has an investment in Polaroid and AOL Time Warner stock. On a particular day, Polaroid stock closed at \$3 per share and AOL Time Warner stock closed at \$52.80 per share. Caitlin's portfolio made up of these two stocks was worth \$3255 at the end of the day. If Caitlin owns 31 more shares of AOL Time Warner stock than Polaroid stock, how many shares of each type of stock does she own? (*Source:* Bridge Information Services)

17. Cyril and Anoa Nantambu operate a small construction and supply company. In July they charged the Shaffers \$1702.50 for 65 hours of labor and 3 tons of material. In August the Shaffers paid \$1349 for 49 hours of labor and $2\frac{1}{2}$ tons of material. Find the cost per hour of labor and the cost per ton of material.

18. Joan Gundersen rented a car from Hertz, which rents its cars for a daily fee plus an additional charge per mile driven. Joan recalls that a car rented for 5 days and driven for 300 miles cost her \$178, while a car rented for 4 days and driven for 500 miles cost \$197. Find the daily fee, and find the mileage charge.

Solve. See Example 3.

19. Pratap Puri rowed 18 miles down the Delaware River in 2 hours, but the return trip took him $4\frac{1}{2}$ hours. Find the rate Pratap can row in still water, and find the rate of the current.
Let x = rate Pratap can row in still water and
y = rate of the current

	d =	r ·	t
Downstream	18	$x + y$	2
Upstream	18	$x - y$	$4\frac{1}{2}$

20. The Jonathan Schultz family took a canoe 10 miles down the Allegheny River in 1 hour and 15 minutes. After lunch it took them 4 hours to return. Find the rate of the current.
Let x = rate the family can row in still water and
y = rate of the current

	d =	r ·	t
Downstream	10	$x + y$	$1\frac{1}{4}$
Upstream	10	$x - y$	4

21. Dave and Sandy Hartranft are frequent flyers with Delta Airlines. They often fly from Philadelphia to Chicago, a distance of 780 miles. On one particular trip they fly into the wind, and the flight takes 2 hours. The return trip, with the wind behind them, only takes $1\frac{1}{2}$ hours. Find the speed of the wind and find the speed of the plane in still air.

22. With a strong wind behind it, a United Airlines jet flies 2400 miles from Los Angeles to Orlando in 4 hours and 45 minutes. The return trip takes 6 hours, as the plane flies into the wind. Find the speed of the plane in still air, and find the wind speed to the nearest tenth of a mile per hour.

23. Jim Williamson began a 186-mile bicycle trip to build up stamina for a triathlete competition. Unfortunately, his bicycle chain broke, so he finished the trip walking. The whole trip took 6 hours. If Jim walks at a rate of 4 miles per hour and rides at 40 mph, find the amount of time he spent on the bicycle.

24. In Canada, eastbound and westbound trains travel along the same track, with sidings to pull onto to avoid accidents. Two trains are now 150 miles apart, with the westbound train traveling twice as fast as the eastbound train. A warning must be issued to pull one train onto a siding or else the trains will crash in $1\frac{1}{4}$ hours. Find the speed of the eastbound train and the speed of the westbound train.

Solve. See Example 4.

25. Dorren Schmidt is a chemist with Gemco Pharmaceutical. She needs to prepare 12 ounces of a 9% hydrochloric acid solution. Find the amount of a 4% solution and the amount of a 12% solution she should mix to get this solution.

Concentration Rate	Liters of Solution	Liters of Pure Acid
0.04	x	$0.04x$
0.12	y	?
0.09	12	?

26. Elise Everly is preparing 15 liters of a 25% saline solution. Elise has two other saline solutions with strengths of 40% and 10%. Find the amount of 40% solution and the amount of 10% solution she should mix to get 15 liters of a 25% solution.

Concentration Rate	Liters of Solution	Liters of Pure Salt
0.40	x	$0.40x$
0.10	y	?
0.25	15	?

27. Wayne Osby blends coffee for a local coffee café. He needs to prepare 200 pounds of blended coffee beans selling for \$3.95 per pound. He intends to do this by blending together a high-quality bean costing \$4.95 per pound and a cheaper bean costing \$2.65 per pound. To the nearest pound, find how much high-quality coffee bean and how much cheaper coffee bean he should blend.

28. Macadamia nuts cost an astounding \$16.50 per pound, but research by an independent firm says that mixed nuts sell better if macadamias are included. The standard mix costs \$9.25 per pound. Find how many pounds of macadamias and how many pounds of the standard mix should be combined to produce 40 pounds that will cost \$10 per pound. Find the amounts to the nearest tenth of a pound.

Solve. See Examples 1 through 4.

△ **29.** Recall that two angles are complementary if their sum is 90°. Find the measures of two complementary angles if one angle is twice the other.

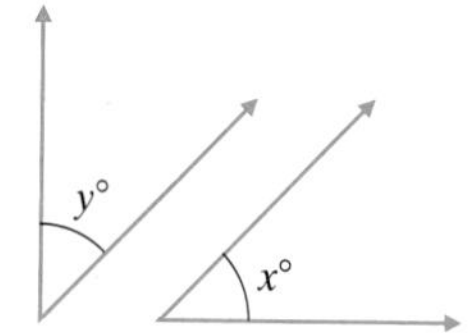

△ **30.** Recall that two angles are supplementary if their sum is 180°. Find the measures of two supplementary angles if one angle is 20° more than four times the other.

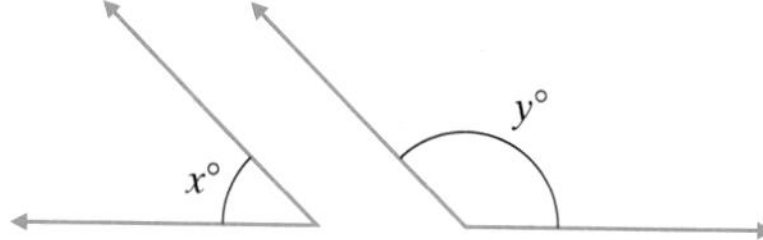

△ **31.** Find the measures of two complementary angles if one angle is 10° more than three times the other.

△ **32.** Find the measures of two supplementary angles if one angle is 18° more than twice the other.

33. Carrie and Raymond McCormick had a pottery stand at the annual Skippack Craft Fair. They sold some of their pottery at the original price of \$9.50 each, but later decreased the price of each by \$2. If they sold all 90 pieces and took in \$721, find how many they sold at the original price and how many they sold at the reduced price.

34. Trinity Church held its annual spaghetti supper and fed a total of 387 people. They charged \$6.80 for adults and half-price for children. If they took in \$2444.60, find how many adults and how many children attended the supper.

35. The Santa Fe National Historic Trail is approximately 1200 miles between Old Franklin, Missouri, and Santa Fe, New Mexico. Suppose that a group of hikers start from each town and walk the trail toward each other. They meet after a total hiking time of 240 hours. If one group travels $\frac{1}{2}$ mile per hour slower than the other group, find the rate of each group. (*Source:* National Park Service)

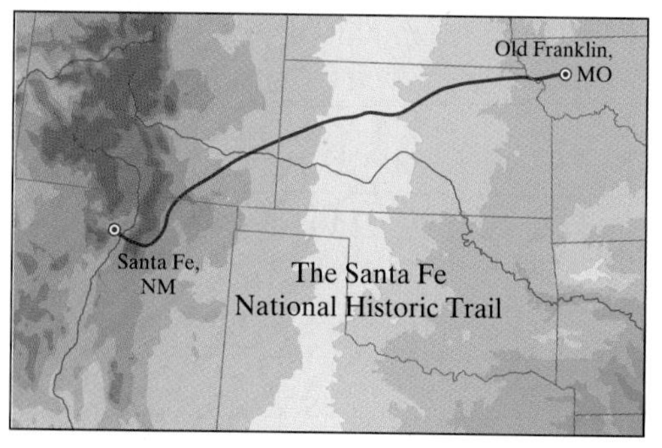

36. California 1 South is a historic highway that stretches 123 miles along the coast from Monterey to Morro Bay. Suppose that two cars start driving this highway, one from each town. They meet after 3 hours. Find the rate of each car if one car travels 1 mile per hour faster than the other car. (*Source: National Geographic*)

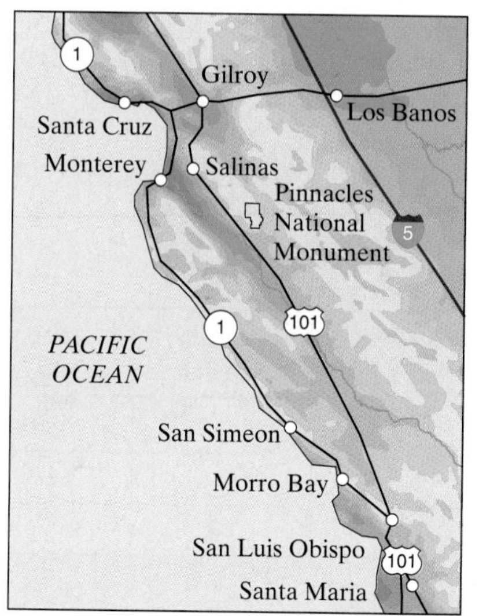

37. A 30% solution of fertilizer is to be mixed with a 60% solution of fertilizer in order to get 150 gallons of a 50% solution. How many gallons of the 30% solution and 60% solution should be mixed?

38. A 10% acid solution is to be mixed with a 50% acid solution in order to get 120 ounces of a 20% acid solution. How many ounces of the 10% solution and 50% solution should be mixed?

39. Traffic signs are regulated by the *Manual on Uniform Traffic Control Devices* (MUTCD). According to this manual, if the sign below is placed on a freeway, its perimeter must be 144 inches. Also, its length is 12 inches longer than its width. Find the dimensions of this sign.

40. According to the MUTCD (see Exercise 39), this sign must have a perimeter of 60 inches. Also, its length must be 6 inches longer than its width. Find the perimeter of this sign.

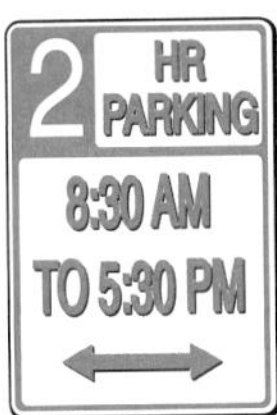

Review and Preview

Find the square of each expression. For example, the square of 7 is 7^2 or 49. The square of $5x$ is $(5x)^2$ or $25x^2$. See Section 10.1.

41. 4 **42.** 3 **43.** $6x$ **44.** $11y$ **45.** $10y^3$ **46.** $8x^5$

Combining Concepts

△**47.** Dale and Sharon Mahnke have decided to fence off a garden plot behind their house, using their house as the "fence" along one side of the garden. The length (which runs parallel to the house) is 3 feet less than twice the width. Find the dimensions if 33 feet of fencing is used along the three sides requiring it.

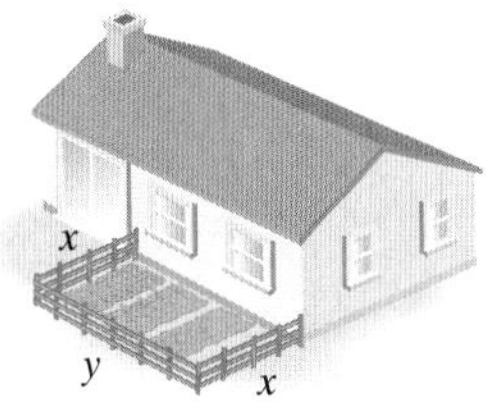

△**48.** Judy McElroy plans to erect 152 feet of fencing around her rectangular horse pasture. A river bank serves as one side length of the rectangle. If each width is 4 feet longer than half the length, find the dimensions.

Internet Excursions

Go To:  http://www.prenhall.com/martin-gay_prealgebra 

Major League Soccer (MLS) had its inaugural season in 1996, and by 2003 had expanded to 10 teams. The given World Wide Web address will provide you with access to the Major League Soccer site, or a related site, for a listing of statistics for the current regular season and links to statistics for past seasons.

49. Using actual data for the current (or most recent) season, write a problem similar to Exercises 9 and 10 involving statistics for MLS leading scorers. When you have finished writing your problem, trade with another student in your class and solve each other's problem. Then check each other's work.

50. Using actual data for the current (or most recent) season, write a problem similar to Exercises 9 and 10 involving statistics for MLS game attendance (found under Regular Season League Stats). When you have finished writing your problem, trade with another student in your class and solve each other's problem. Then check each other's work.

CHAPTER 14 ACTIVITY

Analyzing the Courses of Ships

MATERIALS:

- Ruler, graphing calculator (optional)

This activity may be completed by working in groups or individually.

From overhead photographs or satellite imagery of ships on the ocean, defense analysts can tell a lot about a ship's immediate course by looking at its wake. Assuming that two ships will maintain their present courses, it is possible to extend their paths, based on the wakes visible in the photograph, and find possible points of collision.

Investigate the courses and possibility of collision of the two ships shown in the figure. Assume that the ships will maintain their present courses.

1. Using each ship's wake as a guide, extend the paths of the ships on the figure. Estimate the coordinates of the point of intersection of the ships' courses from the grid. If the ships continue in these courses, they could possibly collide at the point of intersection of their paths.

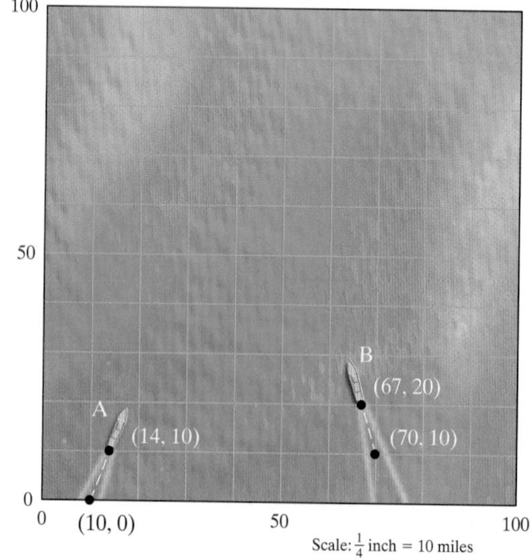

2. Using the coordinates labeled on each ship's wake, find a linear equation that describes each path.

3. (Optional) Use a graphing calculator to graph both equations in the same window. Use the Intersect or Trace feature to estimate the point of intersection of the two paths. Compare this estimate to your estimate in Question 1.

4. Solve the system of two linear equations using one of the methods in this chapter. The solution is the point of intersection of the two paths. Compare your answer to your estimates from Questions 1 and 3.

5. Plot the point of intersection you found in Question 4 on the figure. Use the figure's scale to find each ship's distance from this point of collision by measuring from the bow (tip) of each ship with a ruler. Suppose that the speed of ship A is r_1 and the speed of ship B is r_2. Given the present positions and courses of the two ships, find a relationship between their speeds that would ensure their collision.

Chapter 14 VOCABULARY CHECK

Fill in each blank with one of the words or phrases listed below.

system of linear equations	solution	consistent	independent
dependent	inconsistent	substitution	addition

1. In a system of linear equations in two variables, if the graphs of the equations are the same, the equations are ____________ equations.
2. Two or more linear equations are called a ____________________.
3. A system of equations that has at least one solution is called a(n) ____________ system.
4. A __________ of a system of two equations in two variables is an ordered pair of numbers that is a solution of both equations in the system.
5. Two algebraic methods for solving systems of equations are __________ and ____________.
6. A system of equations that has no solution is called a(n) ____________ system.
7. In a system of linear equations in two variables, if the graphs of the equations are different, the equations are ____________ equations.

CHAPTER 14 Highlights

DEFINITIONS AND CONCEPTS | **EXAMPLES**

Section 14.1 Solving Systems of Linear Equations by Graphing

A **system of linear equations** consists of two or more linear equations.

$$\begin{cases} 2x + y = 6 \\ x = -3y \end{cases} \quad \begin{cases} -3x + 5y = 10 \\ x - 4y = -2 \end{cases}$$

A **solution** of a system of two equations in two variables is an ordered pair of numbers that is a solution of both equations in the system.

Determine whether $(-1, 3)$ is a solution of the system.

$$\begin{cases} 2x - y = -5 \\ x = 3y - 10 \end{cases}$$

Replace x with -1 and y with 3 in both equations.

$$2x - y = -5$$
$$2(-1) - 3 \stackrel{?}{=} -5$$
$$-5 = -5 \quad \text{True}$$

$$x = 3y - 10$$
$$-1 \stackrel{?}{=} 3(3) - 10$$
$$-1 = -1 \quad \text{True}$$

$(-1, 3)$ is a solution of the system.

Graphically, a solution of a system is a point common to the graphs of both equations.

Solve by graphing: $\begin{cases} 3x - 2y = -3 \\ x + y = 4 \end{cases}$

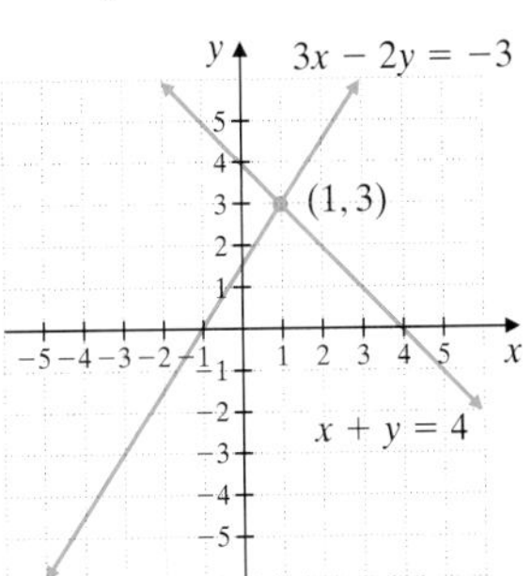

DEFINITIONS AND CONCEPTS | EXAMPLES

Section 14.1 Solving Systems of Linear Equations by Graphing *(continued)*

Three different situations can occur when graphing the two lines associated with the equations in a linear system.

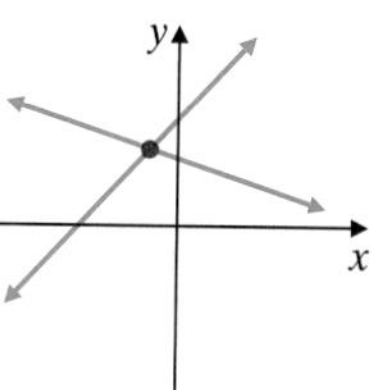

One point of intersection; one solution

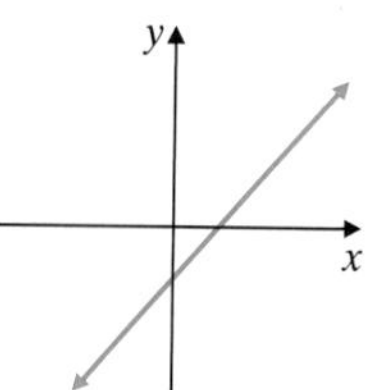

Same line; infinite number of solutions

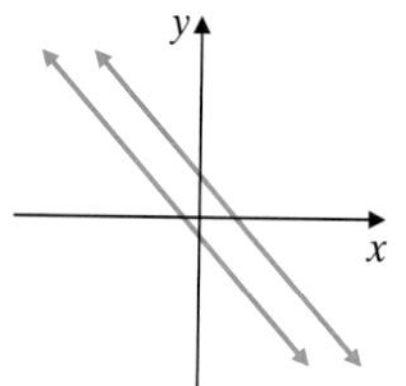

Parallel lines: no solution

Section 14.2 Solving Systems of Linear Equations by Substitution

TO SOLVE A SYSTEM OF LINEAR EQUATIONS BY THE SUBSTITUTION METHOD

Step 1. Solve one equation for a variable.

Step 2. Substitute the expression for the variable into the other equation.

Step 3. Solve the equation from Step 2 to find the value of one variable.

Step 4. Substitute the value from Step 3 in either original equation to find the value of the other variable.

Step 5. Check the solution in both original equations.

Solve by substitution.

$$\begin{cases} 3x + 2y = 1 \\ x = y - 3 \end{cases}$$

Substitute $y - 3$ for x in the first equation.

$$3x + 2y = 1$$
$$3(y - 3) + 2y = 1$$
$$3y - 9 + 2y = 1$$
$$5y = 10$$
$$y = 2 \quad \text{Divide by 5.}$$

To find x, substitute 2 for y in $x = y - 3$ so that $x = 2 - 3$ or -1. The solution $(-1, 2)$ checks.

Section 14.3 Solving Systems of Linear Equations by Addition

TO SOLVE A SYSTEM OF LINEAR EQUATIONS BY THE ADDITION METHOD

Step 1. Rewrite each equation in standard form $Ax + By = C$.

Step 2. Multiply one or both equations by a nonzero number so that the coefficients of a variable are opposites.

Step 3. Add the equations.

Step 4. Find the value of one variable by solving the resulting equation.

Step 5. Substitute the value from Step 4 into either original equation to find the value of the other variable.

Solve by addition.

$$\begin{cases} x - 2y = 8 \\ 3x + y = -4 \end{cases}$$

Multiply both sides of the first equation by -3.

$$\begin{cases} -3x + 6y = -24 \\ 3x + y = -4 \end{cases}$$
$$7y = -28 \quad \text{Add.}$$
$$y = -4 \quad \text{Divide by 7.}$$

To find x, let $y = -4$ in an original equation.

$$x - 2(-4) = 8 \quad \text{First equation}$$
$$x + 8 = 8$$
$$x = 0$$

Definitions and Concepts | Examples

Section 14.3 Solving Systems of Linear Equations by Addition *(continued)*

Step 6. Check the solution in both original equations.

The solution $(0, -4)$ checks.

If solving a system of linear equations by substitution or addition yields a true statement such as $-2 = -2$, then the graphs of the equations in the system are identical and there is an infinite number of solutions of the system.

Solve: $\begin{cases} 2x - 6y = -2 \\ x = 3y - 1 \end{cases}$

Substitute $3y - 1$ for x in the first equation.

$$\begin{aligned} 2(3y - 1) - 6y &= -2 \\ 6y - 2 - 6y &= -2 \\ -2 &= -2 \quad \text{True} \end{aligned}$$

The system has an infinite number of solutions.

If solving a system of linear equations yields a false statement such as $0 = 3$, the graphs of the equations in the system are parallel lines and the system has no solution.

Solve: $\begin{cases} 5x - 2y = 6 \\ -5x + 2y = -3 \end{cases}$

$$0 = 3 \quad \text{False}$$

The system has no solution.

Section 14.4 Systems of Linear Equations and Problem Solving

Problem-Solving Steps

1. UNDERSTAND. Read and reread the problem.

Two angles are supplementary if their sum is 180°. The larger of two supplementary angles is three times the smaller, decreased by twelve. Find the measure of each angle. Let

x = measure of smaller angle and

y = measure of larger angle

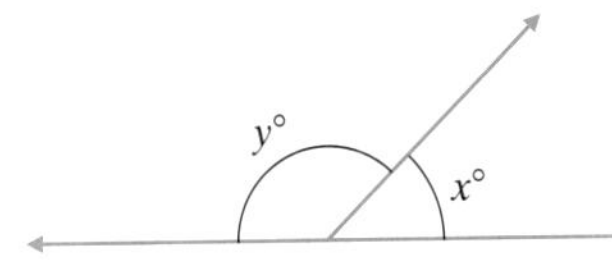

2. TRANSLATE.

In words:

	the sum of supplementary angles	is	180°
	↓	↓	↓
Translate:	$x + y$	$=$	180

In words:	larger angle	is	3 times smaller	decreased by	12
	↓	↓	↓	↓	↓
Translate:	y	$=$	$3x$	$-$	12

3. SOLVE.

Solve the system.

$$\begin{cases} x + y = 180 \\ y = 3x - 12 \end{cases}$$

Use the substitution method and replace y with $3x - 12$ in the first equation.

$$\begin{aligned} x + y &= 180 \\ x + (3x - 12) &= 180 \\ 4x &= 192 \\ x &= 48 \end{aligned}$$

4. INTERPRET.

Since $y = 3x - 12$, then $y = 3 \cdot 48 - 12$ or 132.

The solution checks. The smaller angle measures 48° and the larger angle measures 132°.

STUDY SKILLS REMINDER

Are you prepared for a test on Chapter 14?

Below I have listed some common trouble areas for topics covered in Chapter 14. After studying for your test—but before taking your test—read these.

- If you are having trouble drawing a neat graph, remember to ask your instructor if you can use graph paper on your test. This will save you time and keep your graphs neat.
- Do you remember how to check solutions of systems of equations? If $(-1, 5)$ is a solution of the system

$$\begin{cases} 3x - y = -8 \\ -x + y = 6 \end{cases},$$

then the ordered pair will make *both* equations a true statement.

$3x - y = -8$		$-x + y = 6$	
$3(-1) - 5 = -8$	Let $x = -1$ and $y = 5$.	$-(-1) + 5 = 6$	Let $x = -1$ and $y = 5$.
$-8 = -8$	True	$6 = 6$	True

Remember: This is simply a list of a few common trouble areas. For a review of Chapter 14, see the Highlights and Chapter Review at the end of this chapter.

Name ______________________ Section __________ Date __________

Chapter 14 Review

(14.1) *Determine whether each ordered pair is a solution of the system of linear equations.*

1. $\begin{cases} 2x - 3y = 12 \\ 3x + 4y = 1 \end{cases}$

a. $(12, 4)$

b. $(3, -2)$

2. $\begin{cases} 4x + y = 0 \\ -8x - 5y = 9 \end{cases}$

a. $\left(\frac{3}{4}, -3\right)$

b. $(-2, 8)$

3. $\begin{cases} 5x - 6y = 18 \\ 2y - x = -4 \end{cases}$

a. $(-6, -8)$

b. $3, \frac{5}{2}$

4. $\begin{cases} 2x + 3y = 1 \\ 3y - x = 4 \end{cases}$

a. $(2, 2)$

b. $(-1, 1)$

Solve each system of equations by graphing.

5. $\begin{cases} x + y = 5 \\ x - y = 1 \end{cases}$

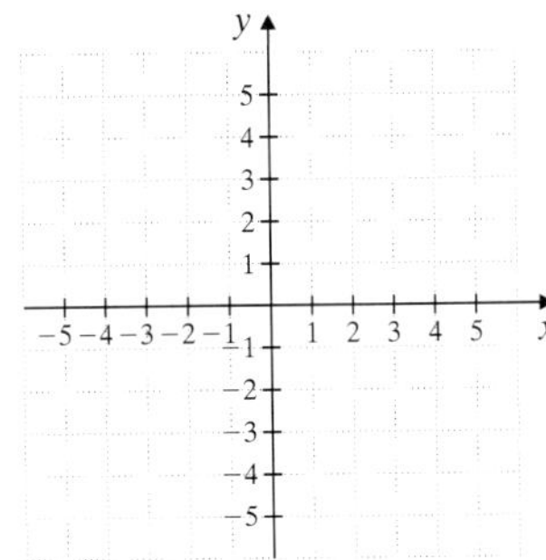

6. $\begin{cases} x + y = 3 \\ x - y = -1 \end{cases}$

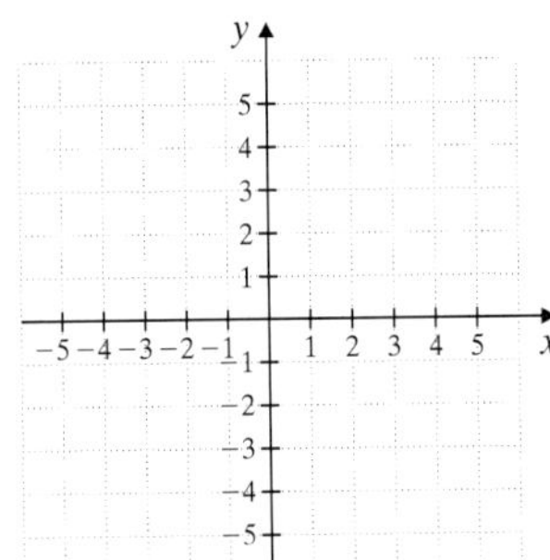

7. $\begin{cases} x = 5 \\ y = -1 \end{cases}$

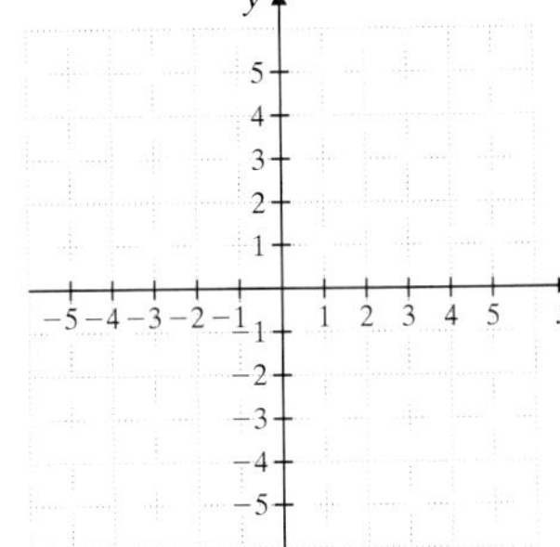

8. $\begin{cases} x = -3 \\ y = 2 \end{cases}$

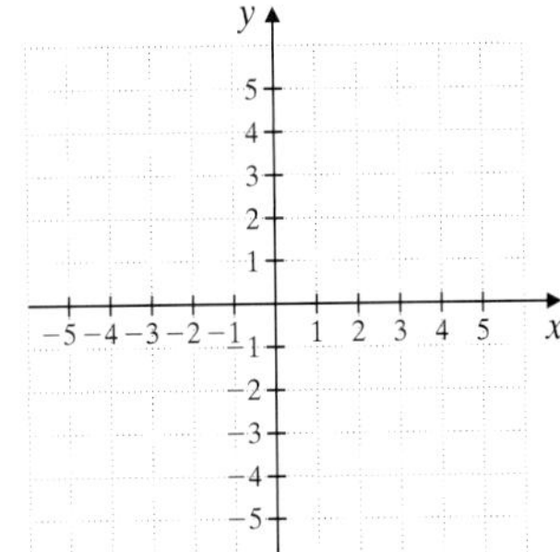

9. $\begin{cases} 2x + y = 5 \\ x = -3y \end{cases}$

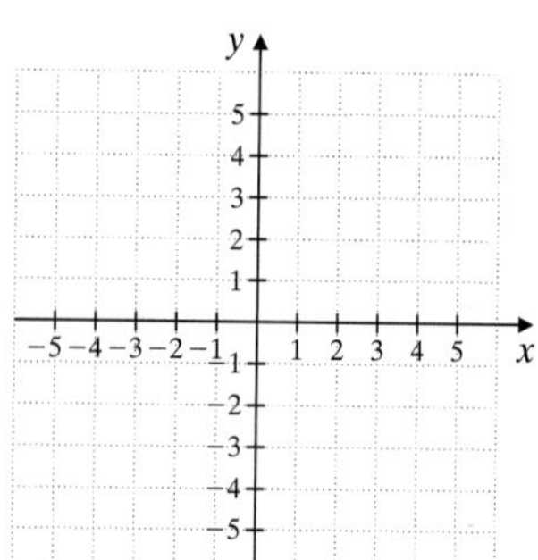

10. $\begin{cases} 3x + y = -2 \\ y = -5x \end{cases}$

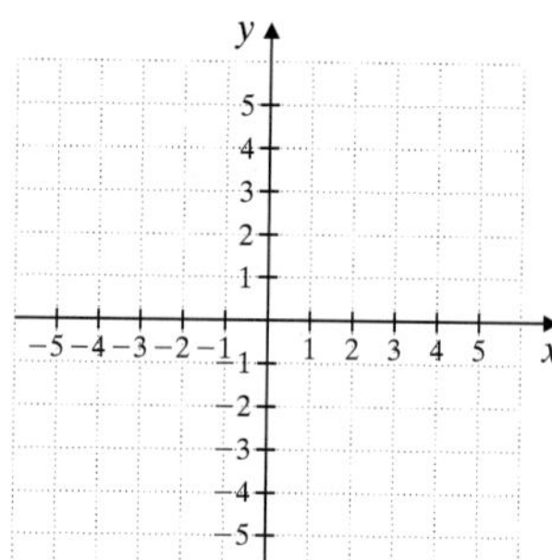

11. $\begin{cases} y = 2x + 4 \\ y = -x - 5 \end{cases}$

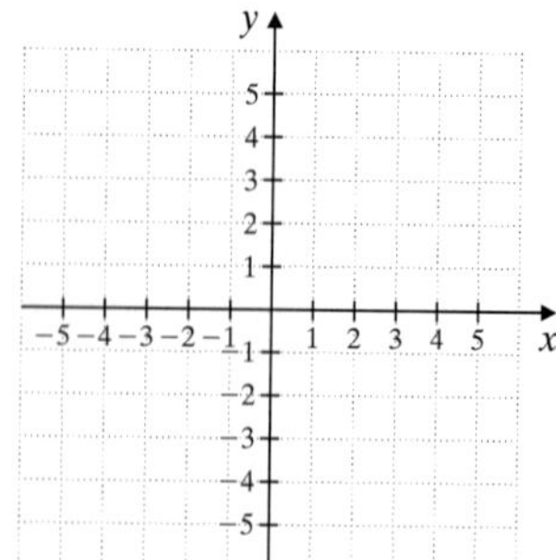

12. $\begin{cases} y = x - 5 \\ y = -2x + 2 \end{cases}$

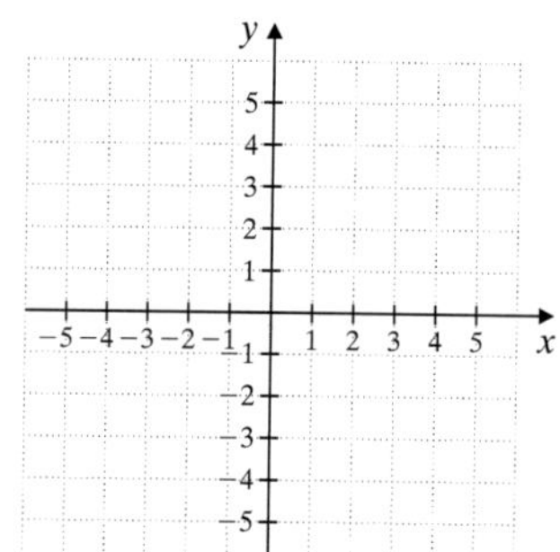

13. $\begin{cases} y = 3x \\ -6x + 2y = 6 \end{cases}$

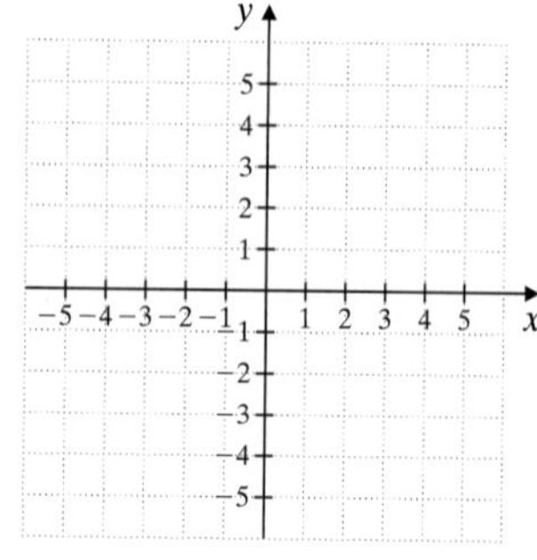

14. $\begin{cases} x - 2y = 2 \\ -2x + 4y = -4 \end{cases}$

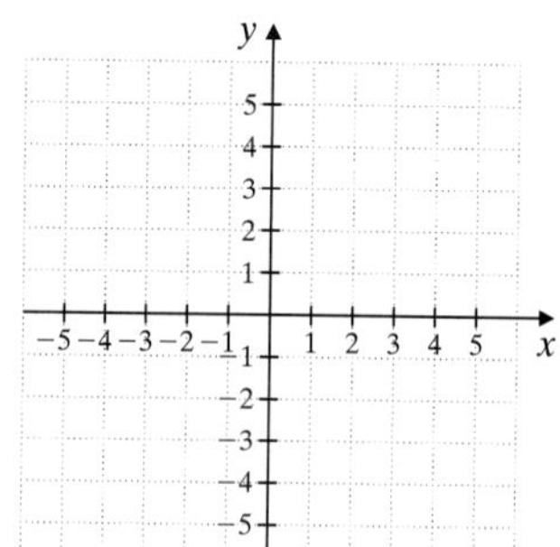

(14.2) *Solve each system of equations by the substitution method.*

15. $\begin{cases} x = 2y \\ 2x - 3y = 2 \end{cases}$

16. $\begin{cases} x = 5y \\ x - 4y = 1 \end{cases}$

17. $\begin{cases} y = 2x + 6 \\ 3x - 2y = -11 \end{cases}$

18. $\begin{cases} y = 3x - 7 \\ 2x - 3y = 7 \end{cases}$

19. $\begin{cases} x + 3y = -3 \\ 2x + y = 4 \end{cases}$

20. $\begin{cases} 3x + y = 11 \\ x + 2y = 12 \end{cases}$

21. $\begin{cases} 4y = 2x - 3 \\ x - 2y = 4 \end{cases}$

22. $\begin{cases} 2x = 3y - 18 \\ x + 4y = 2 \end{cases}$

23. $\begin{cases} x + y = 6 \\ y = -x - 4 \end{cases}$

24. $\begin{cases} -3x + y = 6 \\ y = 3x + 2 \end{cases}$

(14.3) *Solve each system of equations by the addition method.*

25. $\begin{cases} x + y = 14 \\ x - y = 18 \end{cases}$

26. $\begin{cases} x + y = 9 \\ x - y = 13 \end{cases}$

27. $\begin{cases} 2x + 3y = -6 \\ x - 3y = -12 \end{cases}$

28. $\begin{cases} 4x + y = 15 \\ -4x + 3y = -19 \end{cases}$

29. $\begin{cases} 2x - 3y = -15 \\ x + 4y = 31 \end{cases}$

30. $\begin{cases} x - 5y = -22 \\ 4x + 3y = 4 \end{cases}$

31. $\begin{cases} 2x - 6y = -1 \\ -x + 3y = \frac{1}{2} \end{cases}$

32. $\begin{cases} -4x - 6y = 8 \\ 2x + 3y = -3 \end{cases}$

33. $\begin{cases} \frac{3}{4}x + \frac{2}{3}y = 2 \\ x + \frac{y}{3} = 6 \end{cases}$

34. $\begin{cases} \frac{2}{5}x + \frac{3}{4}y = 1 \\ x + 3y = -2 \end{cases}$

35. $\begin{cases} 10x + 2y = 0 \\ 3x + 5y = 33 \end{cases}$

36. $\begin{cases} 0.6x - 0.3y = -1.5 \\ 0.04x - 0.02y = -0.1 \end{cases}$

(14.4) *Solve each problem by writing and solving a system of linear equations.*

37. The sum of two numbers is 16. Three times the larger number decreased by the smaller number is 72. Find the two numbers.

38. The Forrest Theater can seat a total of 360 people. They take in \$15,150 when every seat is sold. If orchestra section tickets cost \$45 and balcony tickets cost \$35, find the number of seats in the orchestra section and the number of seats in the balcony.

39. A riverboat can head 340 miles upriver in 19 hours, but the return trip takes only 14 hours. Find the current of the river and find the speed of the ship in still water to the nearest tenth of a mile.

	d =	***r*** ·	***t***
Upriver	340	$x - y$	19
Downriver	340	$x + y$	14

40. Sam and Cynthia Abney invested \$9000 one year ago. Part of the money was invested at 6% and the rest at 10%. If the total interest earned in one year was \$652.80, find how much was invested at each rate.

△ **41.** Ancient Greeks thought that a picture had the most pleasing dimensions if the length was approximately 1.6 times as long as the width. This ratio is known as the Golden Ratio. If Sandreka Walker has 6 feet of framing material, find the dimensions of the largest frame she can make that satisfies the Golden Ratio. Find the dimensions to the nearest hundredth of a foot.

42. Find the amount of a 6% acid solution and the amount of a 14% acid solution Pat Mayfield should combine to prepare 50 cc (cubic centimeters) of a 12% solution.

43. A deli charges $3.80 for a breakfast of three eggs and four strips of bacon. The charge is $2.75 for two eggs and three strips of bacon. Find the cost of each egg and the cost of each strip of bacon.

44. An exercise enthusiast alternates between jogging and walking. He traveled 15 miles during the past 3 hours. He jogs at a rate of 7.5 miles per hour and walks at a rate of 4 miles per hour. Find how much time, to the nearest hundredth of an hour, he actually spent jogging and how much time he spent walking.

STUDY SKILLS REMINDER

Are you satisfied with your performance on a particular quiz or exam?

If not, don't forget to analyze your quiz or exam and look for common errors.

Were most of your errors a result of

- *Carelessness*? If your errors were careless, did you turn in your work before the allotted time expired? If so, resolve next time to use the entire time allotted. Any extra time can be spent checking your work.
- *Running out of time*? If so, make a point to better manage your time on your next exam. A few suggestions are to work any questions that you are unsure of last and to check your work after all questions have been answered.
- *Not understanding a concept*? If so, review that concept and correct your work. Remember next time to make sure that all concepts on a quiz or exam are understood before the exam.

Name ______________________ Section ________ Date ________

Chapter 14 Test

Remember to check your answers and use the Chapter Test Prep Video to view solutions.

Solve the system by graphing.

1. $\begin{cases} y - x = 6 \\ y + 2x = -6 \end{cases}$

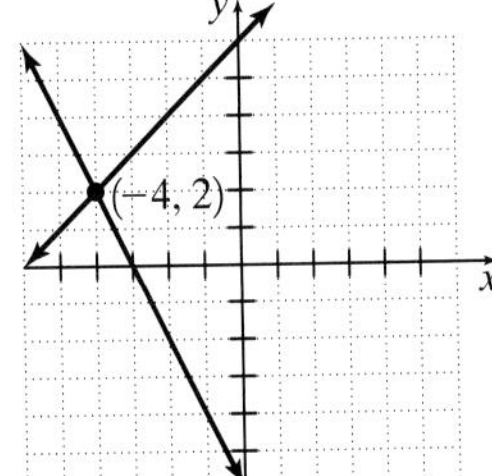

Solve each system by the substitution method.

2. $\begin{cases} 4x + 3y = 1 \\ y = -3x + 7 \end{cases}$

3. $\begin{cases} 3x - 2y = -14 \\ x + 3y = -1 \end{cases}$

Solve each system by the addition method.

4. $\begin{cases} x + y = 28 \\ x - y = 12 \end{cases}$

5. $\begin{cases} 3x + 5y = 2 \\ 2x - 3y = 14 \end{cases}$

Solve.

6. $\begin{cases} x - y = 4 \\ x - 2y = 11 \end{cases}$

7. $\begin{cases} 8x - 4y = 12 \\ y = 2x - 3 \end{cases}$

8. $\begin{cases} 3(2x + y) = 4x + 20 \\ x - 2y = 3 \end{cases}$

9. $\begin{cases} \frac{1}{2}x + 2y = -\frac{15}{4} \\ 4x = -y \end{cases}$

10. $\begin{cases} x - \frac{2}{3}y = 3 \\ -2x + 3y = 10 \end{cases}$

11. $\begin{cases} 0.01x - 0.06y = -0.23 \\ 0.2x + 0.4y = 0.2 \end{cases}$

12. $\begin{cases} 4x - 6y = 7 \\ -2x + 3y = 0 \end{cases}$

13. $\begin{cases} \frac{x - 3}{2} = \frac{2 - y}{4} \\ \frac{7 - 2x}{3} = \frac{y}{2} \end{cases}$

Answers

1. see graph
2. ________
3. ________
4. ________
5. ________
6. ________
7. ________
8. ________
9. ________
10. ________
11. ________
12. ________
13. ________

14. ______

15. ______

16. ______

17. ______

Solve each problem by writing and using a system of linear equations.

14. Two numbers have a sum of 124 and a difference of 32. Find the numbers.

15. Find the amount of a 12% saline solution a lab assistant should add to 80 cc (cubic centimeters) of a 22% saline solution in order to have a 16% solution.

16. Don has invested \$4000, part at 5% simple annual interest and the rest at 9%. Find how much he invested at each rate if the total interest after 1 year is \$311.

17. Although the number of farms in the U.S. is still decreasing, small farms are making a comeback. Texas and Missouri are the states with the most number of farms. Texas has 116 thousand more farms than Missouri and the total number of farms for these two states is 336 thousand. Find the number of farms for each state.

Name ________________ Section ________ Date ________

Answers

Chapter 14 Cumulative Review

Subtract.

1. $8 - 15$

2. $4 - 7$

3. $-4 - (-5)$

4. $3 - (-2)$

5. Solve: $7x = 6x + 4$

6. Solve: $4x = -2 + 3x$

7. Translate to an equation: 1.2 is 30% of what number?

8. Translate to an equation: 9 is 45% of what number?

9. What percent of 50 is 8?

10. What percent of 16 is 4?

11. Mr. Buccaran, the principal at Slidell High School, counted 31 freshmen absent during a particular day. If this is 4% of the total number of freshmen, how many freshmen are there at Slidell High School?

12. 2% of the apples in a shipment are rotten. If there are 29 rotten apples, how many apples are in the shipment?

13. Solve:
$-2(x - 5) + 10 = -3(x + 2) + x$

14. Solve: $4(4y + 2) = 2(1 + 6y) + 8$

15. Solve $-5x + 7 < 2(x - 3)$. Graph the solution set.

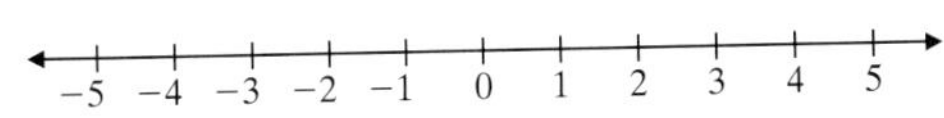

16. Solve $-7x + 4 \le 3(4 - x)$. Graph the solution set.

−4 −3 −2 −1 0 1 2 3 4

Simplify each expression.

17. $\left(\dfrac{m}{n}\right)^7$

18. $(-5x^3)(-7x^4)$

19. $\left(\dfrac{2x^4}{3y^5}\right)^4$

20. $\left(\dfrac{5x^2}{4y^3}\right)^2$

21. Subtract: $(2x^3 + 8x^2 - 6x) - (2x^3 - x^2 + 1)$

22. Subtract: $(7x + 1) - (-x - 3)$

1. ________
2. ________
3. ________
4. ________
5. ________
6. ________
7. ________
8. ________
9. ________
10. ________
11. ________
12. ________
13. ________
14. ________
15. ________
16. ________
17. ________
18. ________
19. ________
20. ________
21. ________
22. ________

23. ______

24. ______

25. ______

26. ______

27. ______

28. ______

29. ______

30. ______

31. ______

32. ______

33. ______

34. ______

35. ______

36. ______

37. ______

38. ______

39. ______

40. ______

41. ______

42. ______

23. Divide $6x^2 + 10x - 5$ by $3x - 1$.

24. Divide $3x^2 - x - 4$ by $x - 1$.

25. Solve: $x(2x - 7) = 4$

26. Solve: $x(x - 5) = 24$

△ **27.** Find the lengths of the sides of a right triangle if the lengths can be expressed by three consecutive even integers.

28. The sum of a number and its square is 132. Find the number.

29. Subtract: $\dfrac{2y}{2y - 7} - \dfrac{7}{2y - 7}$

30. Add: $\dfrac{x^2 + 3}{x + 9} + \dfrac{9x - 3}{x + 9}$

31. Find the slope of the line $y = -1$.

32. Find the slope of the line $x = 2$.

33. Find an equation of the line through $(2, 5)$ and $(-3, 4)$. Write the equation in the form $Ax + By = C$.

34. Find an equation of the line through $(5, -6)$ and $(-6, 5)$. Write the equation in the form $Ax + By = C$.

35. Find the domain and the range of the relation $\{(0, 2), (3, 3), (-1, 0), (3, -2)\}$.

36. Find the domain and the range of the relation $\{(2, 3), (2, 0), (2, -2), (2, 4)\}$.

37. Solve the system: $\begin{cases} x + 2y = 7 \\ 2x + 2y = 13 \end{cases}$

38. Solve the system: $\begin{cases} 3y = x + 6 \\ 4x + 12y = 0 \end{cases}$

39. Solve the system: $\begin{cases} -x - \dfrac{y}{2} = \dfrac{5}{2} \\ \dfrac{x}{6} - \dfrac{y}{2} = 0 \end{cases}$

40. Solve the system: $\begin{cases} x - \dfrac{3y}{8} = -\dfrac{3}{2} \\ x + \dfrac{y}{9} = \dfrac{13}{3} \end{cases}$

41. Find two numbers whose sum is 37 and whose difference is 21.

42. The sum of two numbers is 75 and their difference is 9. Find the two numbers.

Roots and Radicals

Having spent the last chapter studying equations, we return now to algebraic expressions. We expand on our skills of operating on expressions—adding, subtracting, multiplying, dividing, and raising to powers—to include finding roots. Just as subtraction is defined by addition and division by multiplication, finding roots is defined by raising to powers. As we master finding roots, we will work with equations that contain roots and solve problems that can be modeled by such equations.

Nearly 85% of the gold recovered by humankind in all of recorded history is still in use. It's likely that the gold we use today in electronics or jewelry was once mined by the ancient Egyptians or retrieved from the New World by Christopher Columbus in the name of Spain. Gold's scarcity and durability made it a natural form of currency. By the 17th century, merchants who had accumulated gold were finding it bulky to transport and difficult to store safely. They began leaving their gold with goldsmiths in return for a receipt. Trading gold receipts having the same value as the gold itself became a popular way to conduct business and the idea of gold-backed paper money was born. In Section 15.1, Exercise 83, roots are used to find the dimensions of a cube representing all of the gold found by humankind since the beginning of recorded history.

Answers

1. ______
2. ______
3. ______
4. ______
5. ______
6. ______
7. ______
8. ______
9. ______
10. ______
11. ______
12. ______
13. ______
14. ______
15. ______
16. ______
17. ______
18. ______
19. ______
20. ______

Name ______ Section ______ Date ______

Chapter 15 Pretest

Simplify the following. Indicate if the expression is not a real number. Assume that x represents a positive number.

1. $-\sqrt{49}$

2. $\sqrt{\frac{4}{25}}$

3. $\sqrt[3]{-64}$

4. $\sqrt{120}$

5. $\sqrt{\frac{24}{y^6}}$

6. $\sqrt[3]{112}$

Perform each indicated operation.

7. $\sqrt{15} + 2\sqrt{15} - 6\sqrt{15}$

8. $3\sqrt{12} - 2\sqrt{27}$

9. $\sqrt{\frac{7}{4}} + \sqrt{\frac{7}{25}}$

10. $\sqrt{6} \cdot \sqrt{18}$

11. $\sqrt{2}(\sqrt{14} - \sqrt{5})$

12. $(\sqrt{y} - 3)^2$

13. $\frac{\sqrt{56x^5}}{\sqrt{2x^3}}$

Rationalize each denominator.

14. $\sqrt{\frac{5}{11}}$

15. $\frac{16}{\sqrt{2a}}$

16. $\frac{3}{2 - \sqrt{x}}$

Solve each of the following radical equations.

17. $\sqrt{x} + 9 = 16$

18. $\sqrt{x + 4} = \sqrt{x} + 1$

△ **19.** Find the length of the unknown leg of the right triangle. Give an exact answer.

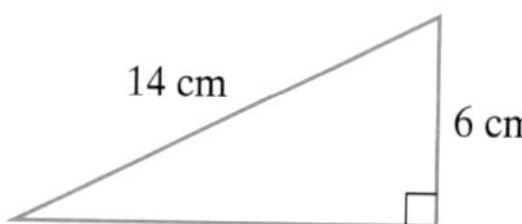

△ **20.** The formula $r = \sqrt{\frac{S}{4\pi}}$ can be used to find the radius of a sphere given its surface area S. Use this formula to approximate the radius of a sphere if its surface area is 80 square inches. Round to two decimal places.

15.1 Introduction to Radicals

OBJECTIVES

A. Find square roots.
B. Find cube roots.
C. Find nth roots.
D. Approximate square roots.
E. Simplify radicals containing variables.

SSM TUTOR CENTER SG CD & VIDEO MATH PRO WEB

A Finding Square Roots

In this section, we define finding the **root** of a number by its reverse operation, raising a number to a power. We begin with squares and square roots.

The square of 5 is $5^2 = 25$.
The square of -5 is $(-5)^2 = 25$
The square of $\frac{1}{2}$ is $\left(\frac{1}{2}\right)^2 = \frac{1}{4}$.

The reverse operation of squaring a number is finding the **square root** of a number. For example,

A square root of 25 is 5, because $5^2 = 25$.
A square root of 25 is also -5, because $(-5)^2 = 25$.
A square root of $\frac{1}{4}$ is $\frac{1}{2}$, because $\left(\frac{1}{2}\right)^2 = \frac{1}{4}$.

In general, the number b is a square root of a number a if $b^2 = a$.

The symbol $\sqrt{}$ is used to denote the **positive** or **principal square root** of a number. For example,

$\sqrt{25} = 5$ only, since $5^2 = 25$ and 5 is positive.

The symbol $-\sqrt{}$ is used to denote the **negative square root.** For example,

$-\sqrt{25} = -5$.

The symbol $\sqrt{}$ is called a **radical** or **radical sign.** The expression within or under a radical sign is called the **radicand.** An expression containing a radical is called a **radical expression.**

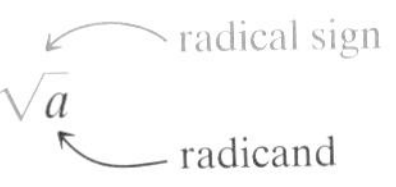

Square Root

If a is a positive number, then

$\sqrt{a}$ is the **positive square root** of a and
$-\sqrt{a}$ is the **negative square root** of a.

Also, $\sqrt{0} = 0$.

EXAMPLES Find each square root.

1. $\sqrt{36} = 6$, because $6^2 = 36$ and 6 is positive.
2. $\sqrt{64} = 8$, because $8^2 = 64$ and 8 is positive.
3. $-\sqrt{25} = -5$. The negative sign in front of the radical indicates the negative square root of 25.
4. $\sqrt{\frac{9}{100}} = \frac{3}{10}$ because $\left(\frac{3}{10}\right)^2 = \frac{9}{100}$ and $\frac{3}{10}$ is positive.
5. $\sqrt{0} = 0$ because $0^2 = 0$.

Practice Problems 1–5

Find each square root.

1. $\sqrt{100}$
2. $\sqrt{9}$
3. $-\sqrt{36}$
4. $\sqrt{\frac{25}{81}}$
5. $\sqrt{1}$

Is the square root of a negative number a real number? For example, is $\sqrt{-4}$ a real number? To answer this question, we ask ourselves, is there a real number whose square is -4? Since there is no real number whose square is -4, we say that $\sqrt{-4}$ is not a real number. In general,

Answers

1. 10 **2.** 3 **3.** -6 **4.** $\frac{5}{9}$ **5.** 1

A square root of a negative number is not a real number.

B Finding Cube Roots

We can find roots other than square roots. For example, since $2^3 = 8$, we call 2 the **cube root** of 8. In symbols, we write

$\sqrt[3]{8} = 2$ The number 3 is called the **index.**

Also,

$\sqrt[3]{27} = 3$ Since $3^3 = 27$

$\sqrt[3]{-64} = -4$ Since $(-4)^3 = -64$

Notice that unlike the square root of a negative number, the cube root of a negative number is a real number. This is so because while we cannot find a real number whose *square* is negative, we *can* find a real number whose *cube* is negative. In fact, the cube of a negative number is a negative number. Therefore, the cube root of a negative number is a negative number.

Practice Problems 6–8

Find each cube root.

6. $\sqrt[3]{27}$
7. $\sqrt[3]{-8}$
8. $\sqrt[3]{\frac{1}{64}}$

EXAMPLES Find each cube root.

6. $\sqrt[3]{1} = 1$ because $1^3 = 1$.

7. $\sqrt[3]{-27} = -3$ because $(-3)^3 = -27$.

8. $\sqrt[3]{\frac{1}{125}} = \frac{1}{5}$ because $\left(\frac{1}{5}\right)^3 = \frac{1}{125}$.

C Finding *n*th Roots

Just as we can raise a real number to powers other than 2 or 3, we can find roots other than square roots and cube roots. In fact, we can take the *n*th root of a number where *n* is any natural number. An ***n*th root** of a number *a* is a number whose *n*th power is *a*.

In symbols, the *n*th root of *a* is written as $\sqrt[n]{a}$. Recall that *n* is called the **index.** The index 2 is usually omitted for square roots.

Concept Check

Which of the following is a real number?

a. $\sqrt{-64}$
b. $\sqrt[4]{-64}$
c. $\sqrt[5]{-64}$
d. $\sqrt[6]{-64}$

Helpful Hint

If the index is even, as it is in $\sqrt{}$, $\sqrt[4]{}$, $\sqrt[6]{}$, and so on, the radicand must be nonnegative for the root to be a real number. For example,

$\sqrt[4]{81} = 3$ but $\sqrt[4]{-81}$ is not a real number.

$\sqrt[6]{64} = 2$ but $\sqrt[6]{-64}$ is not a real number.

Try the Concept Check in the margin.

Practice Problems 9–12

Find each root.

9. $\sqrt[4]{-16}$
10. $\sqrt[5]{-1}$
11. $\sqrt[4]{81}$
12. $\sqrt[6]{-64}$

EXAMPLES Find each root.

9. $\sqrt[4]{16} = 2$ because $2^4 = 16$ and 2 is positive.

10. $\sqrt[5]{-32} = -2$ because $(-2)^5 = -32$.

11. $-\sqrt[3]{8} = -2$ because $\sqrt[3]{8} = 2$.

12. $\sqrt[4]{-81}$ is not a real number since the index 4 is even and the radicand -81 is negative. In other words, there is no real number that when raised to the 4th power gives -81.

Answers

6. 3 **7.** -2 **8.** $\frac{1}{4}$ **9.** not a real number **10.** -1 **11.** 3 **12.** not a real number

Concept Check: c

D Approximating Square Roots

Recall that numbers such as 1, 4, 9, 25, and $\frac{4}{25}$ are called **perfect squares,** since $1^2 = 1$, $2^2 = 4$, $3^2 = 9$, $5^2 = 25$, and $\left(\frac{2}{5}\right)^2 = \frac{4}{25}$. Square roots of perfect square radicands simplify to rational numbers.

What happens when we try to simplify a root such as $\sqrt{3}$? Since 3 is not a perfect square, $\sqrt{3}$ is not a rational number. It cannot be written as a quotient of integers. It is called an **irrational number** and we can find a decimal **approximation** of it. To find decimal approximations, use a calculator or an appendix. (For calculator help, see the next example or the box at the end of this section.)

EXAMPLE 13

Use a calculator or an appendix to approximate $\sqrt{3}$ to three decimal places.

Solution: We may use an appendix or a calculator to approximate $\sqrt{3}$. To use a calculator, find the square root key $\boxed{\sqrt{}}$.

$$\sqrt{3} \approx 1.732050808$$

To three decimal places, $\sqrt{3} \approx 1.732$.

Practice Problem 13

Use a calculator or Appendix B to approximate $\sqrt{10}$ to three decimal places.

E Simplifying Radicals Containing Variables

Radicals can also contain variables. To simplify radicals containing variables, special care must be taken. To see how we simplify $\sqrt{x^2}$, let's look at a few examples in this form.

If $x = 3$, we have $\sqrt{3^2} = \sqrt{9} = 3$, or x.

If x is 5, we have $\sqrt{5^2} = \sqrt{25} = 5$, or x.

From these two examples, you may think that $\sqrt{x^2}$ simplifies to x. Let's now look at an example where x is a negative number. If $x = -3$, we have $\sqrt{(-3)^2} = \sqrt{9} = 3$, not -3, our original x. To make sure that $\sqrt{x^2}$ simplifies to a nonnegative number, we have the following.

For any real number a,

$$\sqrt{a^2} = |a|.$$

Thus,

$$\sqrt{x^2} = |x|,$$
$$\sqrt{(-8)^2} = |-8| = 8$$
$$\sqrt{(7y)^2} = |7y|, \qquad \text{and so on.}$$

To avoid this, for the rest of the chapter we assume that **if a variable appears in the radicand of a radical expression, it represents positive numbers only.** Then

$\sqrt{x^2} = |x| = x$ since x is a positive number.

$\sqrt{y^2} = y$ Because $(y)^2 = y^2$

$\sqrt{x^8} = x^4$ Because $(x^4)^2 = x^8$

$\sqrt{9x^2} = 3x$ Because $(3x)^2 = 9x^2$

Answer

13. 3.162

Practice Problems 14–17

Simplify each expression. Assume that all variables represent positive numbers.

14. $\sqrt{x^8}$
15. $\sqrt{x^{20}}$
16. $\sqrt{4x^6}$
17. $\sqrt[3]{8y^{12}}$

Answers

14. x^4 **15.** x^{10} **16.** $2x^3$ **17.** $2y^4$

EXAMPLES

Simplify each expression. Assume that all variables represent positive numbers.

14. $\sqrt{x^2} = x$ because $(x)^2 = x^2$.
15. $\sqrt{x^6} = x^3$ because $(x^3)^2 = x^6$.
16. $\sqrt[3]{27y^6} = 3y^2$ because $(3y^2)^3 = 27y^6$.
17. $\sqrt{16x^{16}} = 4x^8$ because $(4x^8)^2 = 16x^{16}$.

CALCULATOR EXPLORATIONS

To simplify or approximate square roots using a calculator, locate the key marked [$\sqrt{\ }$]. To simplify $\sqrt{25}$ using a scientific calculator, press [25] [$\sqrt{\ }$]. The display should read [5]. To simplify $\sqrt{25}$ using a graphing calculator, press [$\sqrt{\ }$] [25] [ENTER].

To approximate $\sqrt{30}$, press [30] [$\sqrt{\ }$] (or [$\sqrt{\ }$] [30]). The display should read [5.4772256]. This is an approximation for $\sqrt{30}$. A three-decimal-place approximation is

$$\sqrt{30} \approx 5.477$$

Is this answer reasonable? Since 30 is between perfect squares 25 and 36, $\sqrt{30}$ is between $\sqrt{25} = 5$ and $\sqrt{36} = 6$. The calculator result is then reasonable since 5.4772256 is between 5 and 6.

Use a calculator to approximate each expression to three decimal places. Decide whether each result is reasonable.

1. $\sqrt{7}$ **2.** $\sqrt{14}$ **3.** $\sqrt{11}$

4. $\sqrt{200}$ **5.** $\sqrt{82}$ **6.** $\sqrt{46}$

Many scientific calculators have a key, such as [$\sqrt[x]{y}$], that can be used to approximate roots other than square roots. To approximate these roots using a graphing calculator, look under the [MATH] menu or consult your manual. To use a [$\sqrt[x]{y}$] key to find $\sqrt[3]{8}$, press [3] [$\sqrt[x]{y}$] [8] (press [ENTER] if needed.) The display should read [2].

Use a calculator to approximate each expression to three decimal places. Decide whether each result is reasonable.

7. $\sqrt[3]{40}$ **8.** $\sqrt[3]{71}$ **9.** $\sqrt[4]{20}$

10. $\sqrt[4]{15}$ **11.** $\sqrt[5]{18}$ **12.** $\sqrt[6]{2}$

Name ______________________ Section ____________ Date ____________

EXERCISE SET 15.1

 *Find each square root. See Examples 1 through 5.*

1. $\sqrt{16}$

2. $\sqrt{9}$

3. $\sqrt{81}$

4. $\sqrt{49}$

5. $\sqrt{\frac{1}{25}}$

6. $\sqrt{\frac{1}{64}}$

7. $-\sqrt{100}$

8. $-\sqrt{36}$

9. $\sqrt{-4}$

10. $\sqrt{-25}$

11. $-\sqrt{121}$

12. $-\sqrt{49}$

13. $\sqrt{\frac{9}{25}}$

14. $\sqrt{\frac{4}{81}}$

15. $\sqrt{900}$

16. $\sqrt{400}$

17. $\sqrt{144}$

18. $\sqrt{169}$

19. $\sqrt{\frac{1}{100}}$

20. $\sqrt{\frac{1}{121}}$

B *Find each cube root. See Examples 6 through 8.*

21. $\sqrt[3]{125}$

22. $\sqrt[3]{64}$

23. $\sqrt[3]{-64}$

24. $\sqrt[3]{-27}$

25. $-\sqrt[3]{8}$

26. $-\sqrt[3]{27}$

27. $\sqrt[3]{\frac{1}{8}}$

28. $\sqrt[3]{\frac{1}{64}}$

29. $\sqrt[3]{-125}$

30. $\sqrt[3]{-27}$

31. Explain why the square root of a negative number is not a real number.

32. Explain why the cube root of a negative number is a real number.

C *Find each root. See Examples 9 through 12.*

33. $\sqrt[5]{32}$

34. $\sqrt[4]{81}$

35. $\sqrt[4]{-16}$

36. $\sqrt{-9}$

37. $-\sqrt[4]{625}$

38. $-\sqrt[5]{32}$

39. $\sqrt[6]{1}$

40. $\sqrt[5]{1}$

D *Approximate each square root to three decimal places. See Example 13.*

41. $\sqrt{7}$

42. $\sqrt{10}$

43. $\sqrt{12}$

44. $\sqrt{19}$

45. $\sqrt{37}$

46. $\sqrt{27}$

47. $\sqrt{136}$

48. $\sqrt{8}$

49. A standard baseball diamond is a square with 90-foot sides connecting the bases. The distance from home plate to second base is $90 \cdot \sqrt{2}$ feet. Approximate $\sqrt{2}$ to two decimal places and use your result to approximate the distance $90 \cdot \sqrt{2}$ feet.

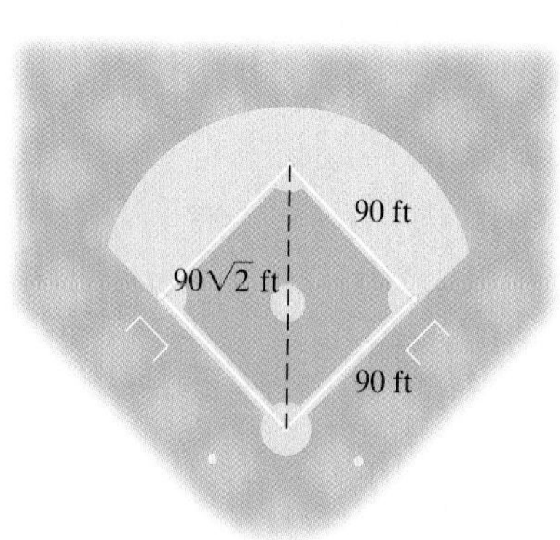

50. The roof of the warehouse shown needs to be shingled. The total area of the roof is exactly $480\sqrt{29}$ square feet. Approximate this area to the nearest whole number.

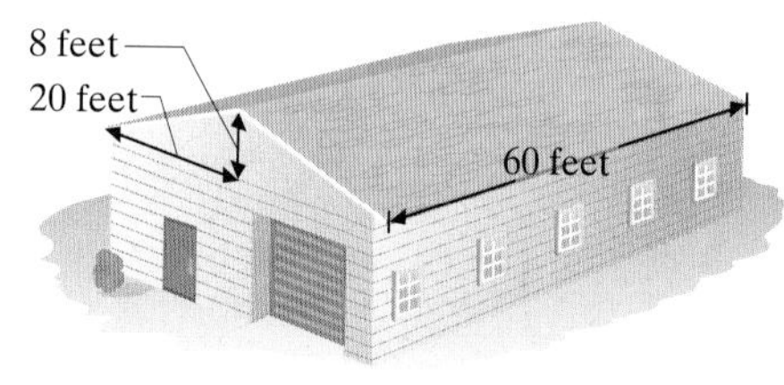

E *Find each root. Assume that all variables represent positive numbers. See Examples 14 through 17.*

51. $\sqrt{z^2}$

52. $\sqrt{y^{10}}$

53. $\sqrt{x^4}$

54. $\sqrt{x^6}$

55. $\sqrt{9x^8}$

56. $\sqrt{36x^{12}}$

57. $\sqrt{81x^2}$

58. $\sqrt{100z^4}$

59. $\sqrt{a^2b^4}$

60. $\sqrt{x^{12}y^{20}}$

61. $\sqrt{16a^6b^4}$

62. $\sqrt{4m^{14}n^2}$

Review and Preview

Write each integer as a product of two integers such that one of the factors is a perfect square. For example, we can write $18 = 9 \cdot 2$, *where 9 is a perfect square. See Section 2.4.*

63. 50

64. 8

65. 32

66. 75

67. 28

68. 44

69. 27

70. 90

Combining Concepts

The length of a side of a square is given by the expression $\sqrt{A}$ *where A is the square's area. Use this expression for Exercises 71 through 74. Be sure to attach the appropriate units.*

△ **71.** The area of a square is 49 square miles. Find the length of a side of the square.

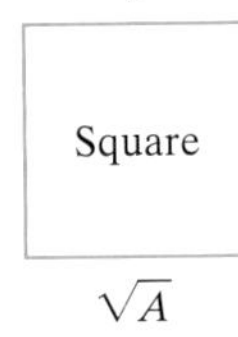

△ **72.** The area of a square is $\frac{1}{81}$ square meters. Find the length of a side of the square.

△ **73.** The base of a Sharp mini disc player is in the shape of a square with area 9.61 square inches. Find the length of a side. (*Source: Guinness World Records*)

△ **74.** A parking lot is in the shape of a square with area 2500 square yards. Find the length of a side.

75. Simplify $\sqrt{\sqrt{81}}$.

76. Simplify $\sqrt[3]{\sqrt[3]{1}}$.

77. Graph $y = \sqrt{x}$. (Complete the table below, plot the ordered pair solutions, and draw a smooth curve through the points. Remember that since the radicand cannot be negative, this particular graph begins at the point with coordinates (0, 0).)

x	y	
0	0	
1		
3		(approximate)
4		
9		

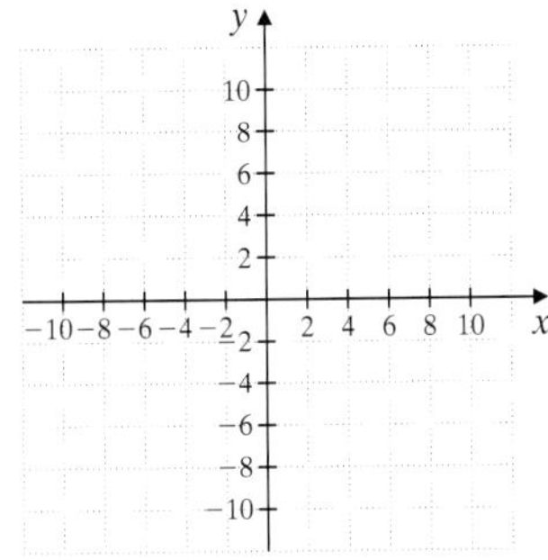

78. Graph $y = \sqrt[3]{x}$. (Complete the table below, plot the ordered pair solutions, and draw a smooth curve through the points.)

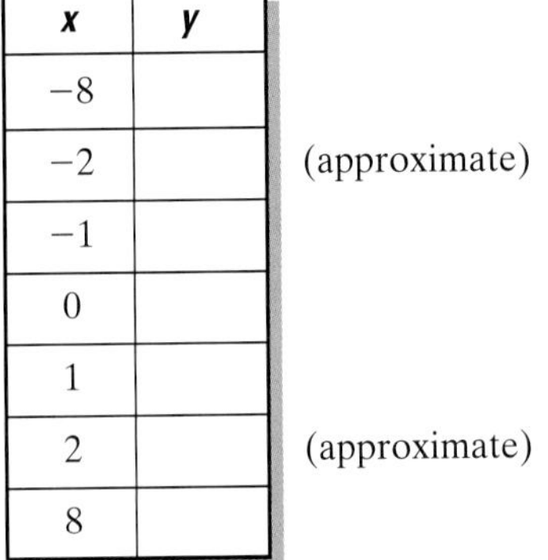

x	y	
−8		
−2		(approximate)
−1		
0		
1		
2		(approximate)
8		

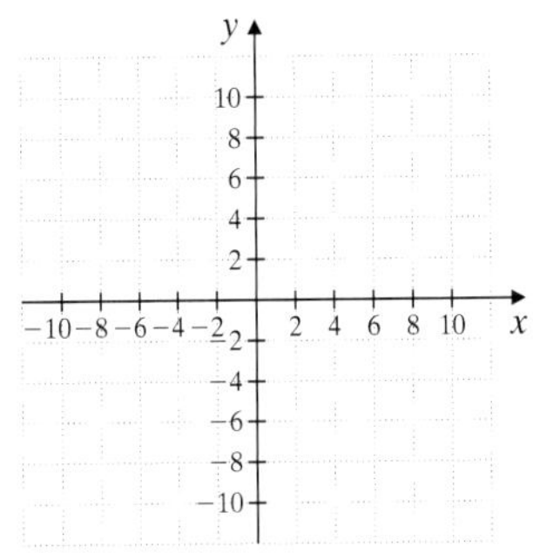

 *Use a graphing calculator and graph each function. Observe the graph from left to right and give the ordered pair that corresponds to the "beginning" of the graph. Then tell why the graph starts at that point.*

79. $y = \sqrt{x - 2}$

80. $y = \sqrt{x + 3}$

81. $y = \sqrt{x + 4}$

82. $y = \sqrt{x - 5}$

83. If the amount of gold discovered by humankind could be assembled in one place, it would make a cube with a volume of 195,112 cubic feet. Each side of the cube would be $\sqrt[3]{195,112}$ feet long. How long would one side of the cube be? (*Source: Reader's Digest*)

15.2 Simplifying Radicals

OBJECTIVES

- A Use the product rule to simplify radicals.
- B Use the quotient rule to simplify radicals.
- C Use both rules to simplify radicals containing variables.
- D Simplify roots other than square roots.

SSM TUTOR CENTER SG CD & VIDEO MATH PRO WEB

A Simplifying Radicals Using the Product Rule

A square root is simplified when the radicand contains no perfect square factors (other than 1). For example, $\sqrt{20}$ is not simplified because $\sqrt{20} = \sqrt{4 \cdot 5}$ and 4 is a perfect square.

To begin simplifying square roots, we notice the following pattern.

$$\sqrt{9 \cdot 16} = \sqrt{144} = 12$$
$$\sqrt{9} \cdot \sqrt{16} = 3 \cdot 4 = 12$$

Since both expressions simplify to 12, we can write

$$\sqrt{9 \cdot 16} = \sqrt{9} \cdot \sqrt{16}$$

This suggests the following product rule for square roots.

Product Rule for Square Roots

If $\sqrt{a}$ and $\sqrt{b}$ are real numbers, then

$$\sqrt{a \cdot b} = \sqrt{a} \cdot \sqrt{b}$$

In other words, the square root of a product is equal to the product of the square roots.

To simplify $\sqrt{20}$, for example, we factor 20 so that one of its factors is a perfect square factor.

$\sqrt{20} = \sqrt{4 \cdot 5}$ Factor 20.

$= \sqrt{4} \cdot \sqrt{5}$ Use the product rule.

$= 2\sqrt{5}$ Write $\sqrt{4}$ as 2.

The notation $2\sqrt{5}$ means $2 \cdot \sqrt{5}$. Since the radicand 5 has no perfect square factor other than 1, the expression $2\sqrt{5}$ is in simplest form.

Helpful Hint

A radical expression in simplest form *does not mean* a decimal approximation. The simplest form of a radical expression is an exact form and may still contain a radical.

$\underbrace{\sqrt{20} = 2\sqrt{5}}_{\text{exact}}$ $\qquad$ $\sqrt{20} \approx \underbrace{4.47}_{\text{decimal approximation}}$

EXAMPLES Simplify.

1. $\sqrt{54} = \sqrt{9 \cdot 6}$ Factor 54 so that one factor is a perfect square. 9 is a perfect square.

$= \sqrt{9} \cdot \sqrt{6}$ Use the product rule.

$= 3\sqrt{6}$ Write $\sqrt{9}$ as 3.

2. $\sqrt{12} = \sqrt{4 \cdot 3}$ Factor 12 so that one factor is a perfect square. 4 is a perfect square.

$= \sqrt{4} \cdot \sqrt{3}$ Use the product rule.

$= 2\sqrt{3}$ Write $\sqrt{4}$ as 2.

Practice Problems 1–4

Simplify.

1. $\sqrt{40}$
2. $\sqrt{18}$
3. $\sqrt{700}$
4. $\sqrt{15}$

Answers

1. $2\sqrt{10}$ **2.** $3\sqrt{2}$ **3.** $10\sqrt{7}$ **4.** $\sqrt{15}$

3. $\sqrt{200} = \sqrt{100 \cdot 2}$ Factor 200 so that one factor is a perfect square. 100 is a perfect square.

$= \sqrt{100} \cdot \sqrt{2}$ Use the product rule.

$= 10\sqrt{2}$ Write $\sqrt{100}$ as 10.

4. $\sqrt{35}$ The radicand 35 contains no perfect square factors other than 1. Thus $\sqrt{35}$ is in simplest form.

In Example 3, 100 is the largest perfect square factor of 200. What happens if we don't use the largest perfect square factor? Although using the largest perfect square factor saves time, the result is the same no matter what perfect square factor is used. For example, it is also true that $200 = 4 \cdot 50$. Then

$$\begin{aligned}\sqrt{200} &= \sqrt{4} \cdot \sqrt{50} \\ &= 2 \cdot \sqrt{50}\end{aligned}$$

Since $\sqrt{50}$ is not in simplest form, we continue.

$$\begin{aligned}\sqrt{200} &= 2 \cdot \sqrt{50} \\ &= 2 \cdot \sqrt{25} \cdot \sqrt{2} \\ &= 2 \cdot 5 \cdot \sqrt{2} \\ &= 10\sqrt{2}\end{aligned}$$

B Simplifying Radicals Using the Quotient Rule

Next, let's examine the square root of a quotient.

$$\sqrt{\frac{16}{4}} = \sqrt{4} = 2$$

Also,

$$\frac{\sqrt{16}}{\sqrt{4}} = \frac{4}{2} = 2$$

Since both expressions equal 2, we can write

$$\sqrt{\frac{16}{4}} = \frac{\sqrt{16}}{\sqrt{4}}$$

This suggests the following quotient rule.

Quotient Rule for Square Roots

If $\sqrt{a}$ and $\sqrt{b}$ are real numbers and $b \neq 0$, then

$$\sqrt{\frac{a}{b}} = \frac{\sqrt{a}}{\sqrt{b}}$$

In other words, the square root of a quotient is equal to the quotient of the square roots.

EXAMPLES Use the quotient rule to simplify.

5. $\sqrt{\frac{25}{36}} = \frac{\sqrt{25}}{\sqrt{36}} = \frac{5}{6}$

6. $\sqrt{\frac{3}{64}} = \frac{\sqrt{3}}{\sqrt{64}} = \frac{\sqrt{3}}{8}$

Practice Problems 5–7

Use the quotient rule to simplify.

5. $\sqrt{\frac{16}{81}}$

6. $\sqrt{\frac{2}{25}}$

7. $\sqrt{\frac{45}{49}}$

Answers

5. $\frac{4}{9}$ **6.** $\frac{\sqrt{2}}{5}$ **7.** $\frac{3\sqrt{5}}{7}$

7. $\sqrt{\dfrac{40}{81}} = \dfrac{\sqrt{40}}{\sqrt{81}}$ Use the quotient rule.

$= \dfrac{\sqrt{4}\cdot\sqrt{10}}{9}$ Use the product rule and write $\sqrt{81}$ as 9.

$= \dfrac{2\sqrt{10}}{9}$ Write $\sqrt{4}$ as 2.

C Simplifying Radicals Containing Variables

Recall that $\sqrt{x^6} = x^3$ because $(x^3)^2 = x^6$. If a variable radicand has an odd exponent, we write the exponential expression so that one factor is the greatest even power contained in the expression. Then we use the product rule to simplify.

EXAMPLES

Simplify each radical. Assume that all variables represent positive numbers.

8. $\sqrt{x^5} = \sqrt{x^4\cdot x} = \sqrt{x^4}\cdot\sqrt{x} = x^2\sqrt{x}$

9. $\sqrt{8y^2} = \sqrt{4\cdot 2\cdot y^2} = \sqrt{4y^2\cdot 2} = \sqrt{4y^2}\cdot\sqrt{2} = 2y\sqrt{2}$

10. $\sqrt{\dfrac{45}{x^6}} = \dfrac{\sqrt{45}}{\sqrt{x^6}} = \dfrac{\sqrt{9\cdot 5}}{x^3} = \dfrac{\sqrt{9}\cdot\sqrt{5}}{x^3} = \dfrac{3\sqrt{5}}{x^3}$

Practice Problems 8–10

Simplify each radical. Assume that all variables represent positive numbers.

8. $\sqrt{x^{11}}$
9. $\sqrt{18x^4}$
10. $\sqrt{\dfrac{27}{x^8}}$

D Simplifying Roots Other Than Square Roots

The product and quotient rules also apply to roots other than square roots. For example, to simplify cube roots, we look for perfect cube factors of the radicand. Recall that 8 is a perfect cube since $2^3 = 8$. Therefore, to simplify $\sqrt[3]{48}$, we factor 48 as $8\cdot 6$.

$\sqrt[3]{48} = \sqrt[3]{8\cdot 6}$ Factor 48.

$= \sqrt[3]{8}\cdot\sqrt[3]{6}$ Use the product rule.

$= 2\sqrt[3]{6}$ Write $\sqrt[3]{8}$ as 2.

$2\sqrt[3]{6}$ is in simplest form since the radicand 6 contains no perfect cube factors other than 1.

EXAMPLES

Simplify each radical.

11. $\sqrt[3]{54} = \sqrt[3]{27\cdot 2} = \sqrt[3]{27}\cdot\sqrt[3]{2} = 3\sqrt[3]{2}$

12. $\sqrt[3]{18}$ The number 18 contains no perfect cube factors, so $\sqrt[3]{18}$ cannot be simplified further.

13. $\sqrt[3]{\dfrac{7}{8}} = \dfrac{\sqrt[3]{7}}{\sqrt[3]{8}} = \dfrac{\sqrt[3]{7}}{2}$

14. $\sqrt[3]{\dfrac{40}{27}} = \dfrac{\sqrt[3]{40}}{\sqrt[3]{27}} = \dfrac{\sqrt[3]{8\cdot 5}}{3} = \dfrac{\sqrt[3]{8}\cdot\sqrt[3]{5}}{3} = \dfrac{2\sqrt[3]{5}}{3}$

Practice Problems 11–14

Simplify each radical.

11. $\sqrt[3]{40}$
12. $\sqrt[3]{50}$
13. $\sqrt[3]{\dfrac{10}{27}}$
14. $\sqrt[3]{\dfrac{81}{8}}$

Answers

8. $x^5\sqrt{x}$ **9.** $3x^2\sqrt{2}$ **10.** $\dfrac{3\sqrt{3}}{x^4}$ **11.** $2\sqrt[3]{5}$ **12.** $\sqrt[3]{50}$ **13.** $\dfrac{\sqrt[3]{10}}{3}$ **14.** $\dfrac{3\sqrt[3]{3}}{2}$

FOCUS ON The Real World

ESCAPE VELOCITY

Each planet in the solar system has a minimum speed that an object must reach to escape the planet's pull of gravity. This speed is called the **escape velocity.** For instance, Earth's escape velocity is 11.19 kilometers per second. A rocket launched from Earth's surface must be going at least 11.19 kilometers per second to leave the planet for outer space. If the rocket goes slower than 11.19 kilometers per second, it will either fall back to Earth or go into orbit around Earth.

The escape velocity v from a planet with mass m and radius r is given by the formula

$$v = \sqrt{\frac{2Gm}{r}}$$

where G is a constant called the *universal constant of gravitation*. The escape velocity of each planet in our solar system, along with its mass and radius, is given in the table.

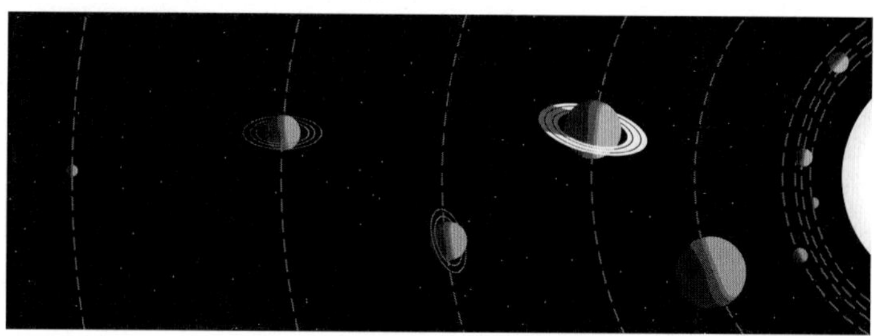

Planet	Mass (kilograms)	Radius (kilometers)	Escape Velocity (kilometers per second)
Mercury	3.302×10^{23}	2440	4.30
Venus	4.869×10^{24}	6052	10.36
Earth	5.974×10^{24}	6371	11.19
Moon	7.349×10^{22}	1738	2.38
Mars	6.419×10^{23}	3390	5.03
Jupiter	1.899×10^{27}	69,911	59.50
Saturn	5.685×10^{26}	58,232	35.50
Uranus	8.683×10^{25}	25,362	21.30
Neptune	1.024×10^{26}	24,624	23.50
Pluto	1.250×10^{22}	1137	1.10

(*Source*: National Space Science Data Center)

CRITICAL THINKING

1. List the planets in order of decreasing mass.
2. List the planets in order of decreasing escape velocity.
3. What do you notice about your lists from Questions 1 and 2? In general, what could you conclude about the relationship between mass and escape velocity?
4. Use the formula for escape velocity and the data in the table to estimate the value of the universal constant of gravitation G. Explain how you found your estimate.

Name ______________________ Section __________ Date __________

Mental Math

Simplify each radical. Assume that all variables represent positive numbers.

1. $\sqrt{4 \cdot 9}$
2. $\sqrt{9 \cdot 36}$
3. $\sqrt{x^2}$
4. $\sqrt{y^4}$
5. $\sqrt{0}$
6. $\sqrt{1}$
7. $\sqrt{25x^4}$
8. $\sqrt{49x^2}$

EXERCISE SET 15.2

A *Use the product rule to simplify each radical. See Examples 1 through 4.*

1. $\sqrt{20}$
2. $\sqrt{44}$
3. $\sqrt{18}$
4. $\sqrt{45}$
5. $\sqrt{50}$
6. $\sqrt{28}$

7. $\sqrt{33}$
8. $\sqrt{98}$
9. $\sqrt{60}$
10. $\sqrt{90}$
11. $\sqrt{180}$
12. $\sqrt{150}$
13. $\sqrt{52}$
14. $\sqrt{75}$

B *Use the quotient rule and the product rule to simplify each radical. See Examples 5 through 7.*

15. $\sqrt{\frac{8}{25}}$
16. $\sqrt{\frac{63}{16}}$
17. $\sqrt{\frac{27}{121}}$
18. $\sqrt{\frac{24}{169}}$
19. $\sqrt{\frac{9}{4}}$
20. $\sqrt{\frac{100}{49}}$
21. $\sqrt{\frac{125}{9}}$
22. $\sqrt{\frac{27}{100}}$
23. $\sqrt{\frac{11}{36}}$
24. $\sqrt{\frac{30}{49}}$
25. $-\sqrt{\frac{27}{144}}$
26. $-\sqrt{\frac{84}{121}}$

C *Simplify each radical. Assume that all variables represent positive numbers. See Examples 8 through 10.*

27. $\sqrt{x^7}$
28. $\sqrt{y^3}$
29. $\sqrt{x^{13}}$
30. $\sqrt{y^{17}}$
31. $\sqrt{75x^2}$
32. $\sqrt{72y^2}$
33. $\sqrt{96x^4}$
34. $\sqrt{40y^{10}}$
35. $\sqrt{\frac{12}{y^2}}$
36. $\sqrt{\frac{63}{x^4}}$
37. $\sqrt{\frac{9x}{y^2}}$
38. $\sqrt{\frac{6y^2}{x^4}}$
39. $\sqrt{\frac{88}{x^4}}$
40. $\sqrt{\frac{x^{11}}{81}}$

D *Simplify each radical. See Examples 11 through 14.*

41. $\sqrt[3]{24}$ **42.** $\sqrt[3]{81}$ **43.** $\sqrt[3]{250}$ **44.** $\sqrt[3]{40}$ **45.** $\sqrt[3]{\frac{5}{64}}$ **46.** $\sqrt[3]{\frac{32}{125}}$

47. $\sqrt[3]{\frac{7}{8}}$ **48.** $\sqrt[3]{\frac{10}{27}}$ **49.** $\sqrt[3]{\frac{15}{64}}$ **50.** $\sqrt[3]{\frac{4}{27}}$ **51.** $\sqrt[3]{80}$ **52.** $\sqrt[3]{108}$

Review and Preview

Perform each indicated operation. See Sections 10.3 and 10.4.

53. $6x + 8x$ **54.** $(6x)(8x)$ **55.** $(2x + 3)(x - 5)$

56. $(2x + 3) + (x - 5)$ **57.** $9y^2 - 9y^2$ **58.** $(9y^2)(-8y^2)$

Combining Concepts

Simplify each radical. Assume that all variables represent positive numbers.

59. $\sqrt{x^6y^3}$ **60.** $\sqrt{98x^5y^4}$

61. $\sqrt{x^2 + 4x + 4}$ (*Hint*: Factor the trinomial first.) **62.** $\sqrt[3]{-8x^6}$

63. If a cube is to have a volume of 80 cubic inches, then each side must be $\sqrt[3]{80}$ inches long. Simplify the radical representing the side length.

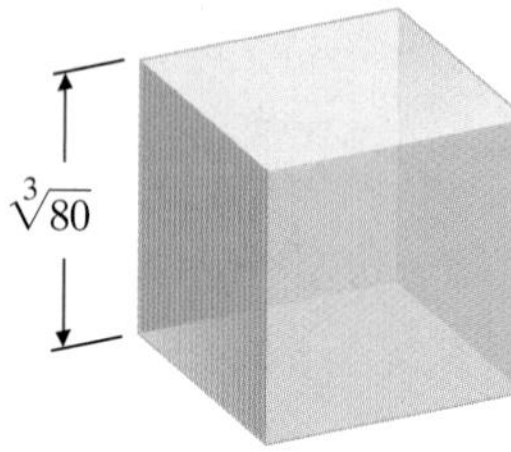

64. Jeannie Boswell is swimming across a 40-foot-wide river, trying to head straight across to the opposite shore. However, the current is strong enough to move her downstream 100 feet by the time she reaches land. (See the figure.) Because of the current, the actual distance she swam is $\sqrt{11,600}$ feet. Simplify this radical.

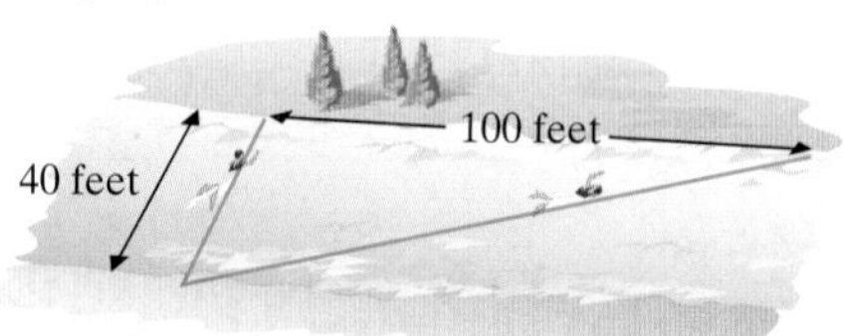

The length of a side of a cube is given by the expression $\sqrt{\frac{A}{6}}$ *where A is the cube's surface area. Use this expression for Exercises 65 through 68. Be sure to attach the appropriate units.*

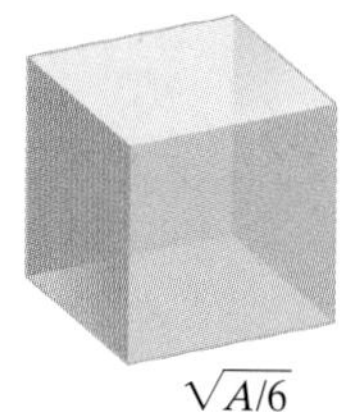

△ **65.** The surface area of a cube is 120 square inches. Find the exact length of a side of the cube.

△ **66.** The surface area of a cube is 594 square feet. Find the exact length of a side of the cube.

△ **67.** The Borg space ship on *Star Trek: The Next Generation* is in the shape of a cube. Suppose a model of this ship has a surface area of 150 square inches. Find the length of a side of the ship.

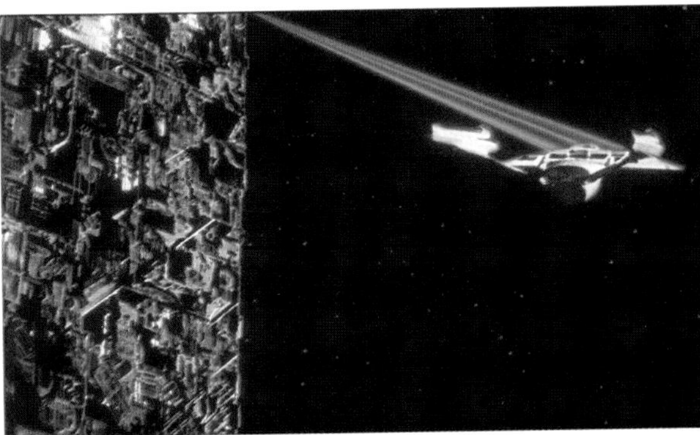

△ **68.** A shipping crate in the shape of a cube is to be constructed. If the crate is to have a surface area of 486 square feet, find the length of a side of the crate.

The cost C in dollars per day to operate a small delivery service is given by $C = 100\sqrt[3]{n} + 700$, *where n is the number of deliveries per day.*

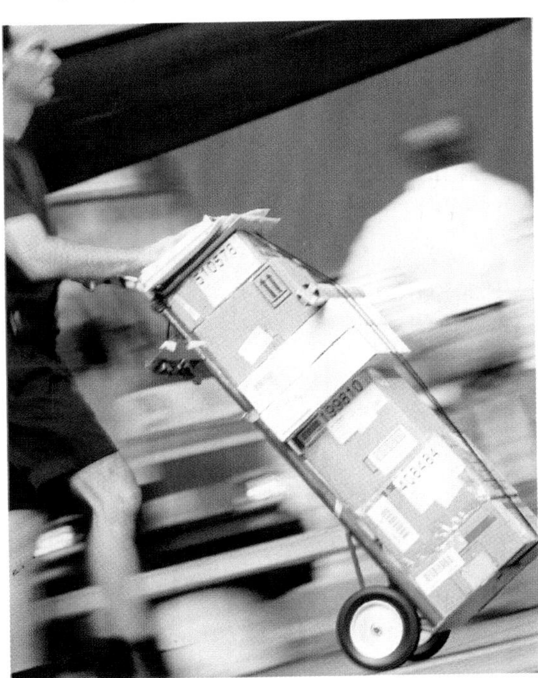

69. Find the cost if the number of deliveries is 1000.

70. Approximate the cost if the number of deliveries is 500.

71. By using replacement values for a and b, show that $\sqrt{a^2 + b^2}$ does not equal $a + b$.

72. By using replacement values for a and b, show that $\sqrt{a + b}$ does not equal $\sqrt{a} + \sqrt{b}$.

The Mosteller formula for calculating body surface area is $B = \sqrt{\frac{hw}{3600}}$, *where B is an individual's body surface area in square meters, h is the individual's height in centimeters, and w is the individual's weight in kilograms. Use this formula in Exercises 73 and 74. Round answers to the nearest tenth.*

73. Find the body surface area of a person who is 169 cm tall and weighs 64 kilograms.

74. Approximate the body surface area of a person who is 183 cm tall and weighs 85 kilograms.

FOCUS ON Mathematical Connections

Graphing and the Distance Formula

One application of radicals is finding the distance between two points in the coordinate plane. This can be very useful in graphing.

The distance d between two points with coordinates (x_1, y_1) and (x_2, y_2) is given by the **distance formula** $d = \sqrt{(x_2 - x_1)^2 + (y_2 - y_1)^2}$. Suppose we want to find the distance between the two points $(-1, 9)$ and $(3, 5)$. We can use the distance formula with $(x_1, y_1) = (-1, 9)$ and $(x_2, y_2) = (3, 5)$. Then we have

$$\begin{aligned} d &= \sqrt{(x_2 - x_1)^2 + (y_2 - y_1)^2} \\ &= \sqrt{[3 - (-1)]^2 + (5 - 9)^2} \\ &= \sqrt{(4)^2 + (-4)^2} \\ &= \sqrt{16 + 16} \\ &= \sqrt{32} = 4\sqrt{2} \end{aligned}$$

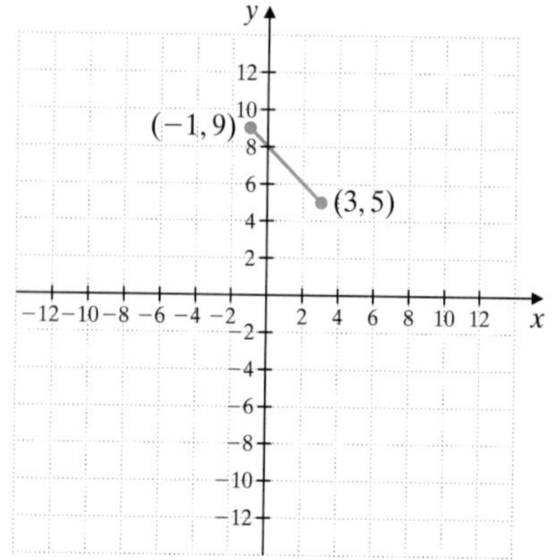

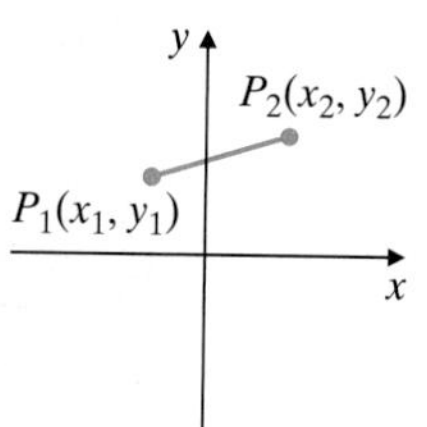

The distance between the two points is exactly $4\sqrt{2}$ units or approximately 5.66 units.

Group Activity

Brainstorm to come up with several disciplines or activities in which the distance formula might be useful. Make up an example that shows how the distance formula would be used in one of the activities on your list. Then present your example to the rest of the class.

15.3 Adding and Subtracting Radicals

OBJECTIVES

A Add or subtract like radicals.

B Simplify radical expressions, and then add or subtract any like radicals.

A Adding and Subtracting Radicals

Recall that to combine like terms, we use the distributive property.

$$5x + 3x = (5 + 3)x = 8x$$

The distributive property can also be applied to expressions containing the same radicals. For example,

$$5\sqrt{2} + 3\sqrt{2} = (5 + 3)\sqrt{2} = 8\sqrt{2}$$

Also,

$$9\sqrt{5} - 6\sqrt{5} = (9 - 6)\sqrt{5} = 3\sqrt{5}$$

Radical terms such as $5\sqrt{2}$ and $3\sqrt{2}$ are **like radicals,** as are $9\sqrt{5}$ and $6\sqrt{5}$. Like radicals have the same index and the same radicand.

EXAMPLES Add or subtract as indicated.

1. $4\sqrt{5} + 3\sqrt{5} = (4 + 3)\sqrt{5} = 7\sqrt{5}$
2. $\sqrt{10} - 6\sqrt{10} = 1\sqrt{10} - 6\sqrt{10} = (1 - 6)\sqrt{10} = -5\sqrt{10}$
3. $2\sqrt{6} + 2\sqrt{5}$ cannot be simplified further since the radicands are not the same.
4. $\sqrt{15} + \sqrt{15} = 1\sqrt{15} + 1\sqrt{15} = (1 + 1)\sqrt{15} = 2\sqrt{15}$

Practice Problems 1–4

Add or subtract as indicated.

1. $6\sqrt{11} + 9\sqrt{11}$
2. $\sqrt{7} - 3\sqrt{7}$
3. $\sqrt{2} + \sqrt{2}$
4. $3\sqrt{3} - 3\sqrt{2}$

Try the Concept Check in the margin.

Concept Check

Which is true?

a. $2 + 3\sqrt{5} = 5\sqrt{5}$
b. $2\sqrt{3} + 2\sqrt{7} = 2\sqrt{10}$
c. $\sqrt{3} + \sqrt{5} = \sqrt{8}$
d. None of the above is true. In each case, the left-hand side cannot be simplified further.

B Simplifying Radicals before Adding or Subtracting

At first glance, it appears that the expression $\sqrt{50} + \sqrt{8}$ cannot be simplified further because the radicands are different. However, the product rule can be used to simplify each radical, and then further simplification might be possible.

Answers

1. $15\sqrt{11}$ **2.** $-2\sqrt{7}$ **3.** $2\sqrt{2}$ **4.** $3\sqrt{3} - 3\sqrt{2}$

Concept Check: d

Practice Problems 5–8

Add or subtract by first simplifying each radical.

5. $\sqrt{27} + \sqrt{75}$
6. $3\sqrt{20} - 7\sqrt{45}$
7. $\sqrt{36} - \sqrt{48} - 4\sqrt{3} - \sqrt{9}$
8. $\sqrt{9x^4} - \sqrt{36x^3} + \sqrt{x^3}$

Answers

5. $8\sqrt{3}$ **6.** $-15\sqrt{5}$ **7.** $3 - 8\sqrt{3}$ **8.** $3x^2 - 5x\sqrt{x}$

EXAMPLES Add or subtract by first simplifying each radical.

5. $\sqrt{50} + \sqrt{8} = \sqrt{25\cdot 2} + \sqrt{4\cdot 2}$ Factor radicands.

$= \sqrt{25}\cdot\sqrt{2} + \sqrt{4}\cdot\sqrt{2}$ Use the product rule.

$= 5\sqrt{2} + 2\sqrt{2}$ Simplify $\sqrt{25}$ and $\sqrt{4}$.

$= 7\sqrt{2}$ Add like radicals.

6. $7\sqrt{12} - \sqrt{75} = 7\sqrt{4\cdot 3} - \sqrt{25\cdot 3}$ Factor radicands.

$= 7\sqrt{4}\cdot\sqrt{3} - \sqrt{25}\cdot\sqrt{3}$ Use the product rule.

$= 7\cdot 2\sqrt{3} - 5\sqrt{3}$ Simplify $\sqrt{4}$ and $\sqrt{25}$.

$= 14\sqrt{3} - 5\sqrt{3}$ Multiply.

$= 9\sqrt{3}$ Subtract like radicals.

7. $\sqrt{25} - \sqrt{27} - 2\sqrt{18} - \sqrt{16}$

$= 5 - \sqrt{9\cdot 3} - 2\sqrt{9\cdot 2} - 4$ Factor radicands and simplify $\sqrt{25}$ and $\sqrt{16}$.

$= 5 - \sqrt{9}\cdot\sqrt{3} - 2\sqrt{9}\cdot\sqrt{2} - 4$ Use the product rule.

$= 5 - 3\sqrt{3} - 2\cdot 3\sqrt{2} - 4$ Simplify.

$= 1 - 3\sqrt{3} - 6\sqrt{2}$ Write $5 - 4$ as 1 and $2\cdot 3$ as 6.

8. $2\sqrt{x^2} - \sqrt{25x} + \sqrt{x}$

$= 2x - \sqrt{25}\cdot\sqrt{x} + \sqrt{x}$ Write $\sqrt{x^2}$ as x and use the product rule.

$= 2x - 5\sqrt{x} + 1\sqrt{x}$ Simplify.

$= 2x - 4\sqrt{x}$ Add like radicals.

Name ______________________ Section __________ Date __________

Mental Math

Simplify each expression by combining like radicals.

1. $3\sqrt{2} + 5\sqrt{2}$
2. $3\sqrt{5} + 7\sqrt{5}$
3. $5\sqrt{x} + 2\sqrt{x}$
4. $8\sqrt{x} + 3\sqrt{x}$
5. $5\sqrt{7} - 2\sqrt{7}$
6. $8\sqrt{6} - 5\sqrt{6}$

EXERCISE SET 15.3

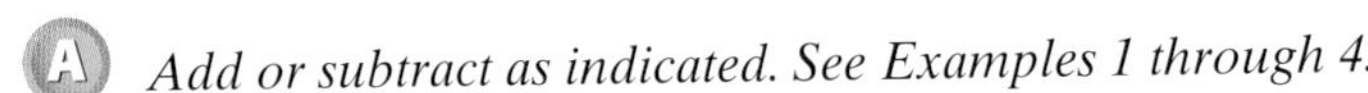

A *Add or subtract as indicated. See Examples 1 through 4.*

1. $4\sqrt{3} - 8\sqrt{3}$
2. $\sqrt{5} - 9\sqrt{5}$
3. $3\sqrt{6} + 8\sqrt{6} - 2\sqrt{6} - 5$
4. $12\sqrt{2} - 3\sqrt{2} + 8\sqrt{2} + 10$
5. $6\sqrt{5} - 5\sqrt{5} + \sqrt{2}$
6. $4\sqrt{3} + \sqrt{5} - 3\sqrt{3}$
7. $2\sqrt{3} + 5\sqrt{3} - \sqrt{3}$
8. $8\sqrt{14} + 2\sqrt{14} + 4$
9. $2\sqrt{2} - 7\sqrt{2} - 6$
10. $5\sqrt{7} + 2 - 11\sqrt{7}$
11. $12\sqrt{5} - \sqrt{5} - 4\sqrt{5}$
12. $\sqrt{5} + \sqrt{15}$
13. $\sqrt{5} + \sqrt{5}$
14. $4 + 8\sqrt{2} - 9$
15. $6 - 2\sqrt{3} - \sqrt{3}$
16. $8 - \sqrt{2} - 5\sqrt{2}$
17. In your own words, describe like radicals.
18. In the expression $\sqrt{5} + 2 - 3\sqrt{5}$, explain why 2 and -3 cannot be combined.

B *Add or subtract by first simplifying each radical and then combining any like radicals. Assume that all variables represent positive numbers. See Examples 5 through 8.*

19. $\sqrt{12} + \sqrt{27}$
20. $\sqrt{50} + \sqrt{18}$
21. $\sqrt{45} + 3\sqrt{20}$
22. $5\sqrt{32} - \sqrt{72}$
23. $2\sqrt{54} - \sqrt{20} + \sqrt{45} - \sqrt{24}$
24. $2\sqrt{8} - \sqrt{128} + \sqrt{48} + \sqrt{18}$
25. $4x - 3\sqrt{x^2} + \sqrt{x}$
26. $x - 6\sqrt{x^2} + 2\sqrt{x}$
27. $\sqrt{25x} + \sqrt{36x} - 11\sqrt{x}$
28. $\sqrt{9x} - \sqrt{16x} + 2\sqrt{x}$
29. $3\sqrt{x^3} - x\sqrt{4x}$
30. $\sqrt{16x} - \sqrt{x^3}$
31. $\sqrt{75} + \sqrt{48}$
32. $2\sqrt{80} - \sqrt{45}$
33. $\sqrt{8} + \sqrt{9} + \sqrt{18} + \sqrt{81}$
34. $\sqrt{6} + \sqrt{16} + \sqrt{24} + \sqrt{25}$
35. $\sqrt{\frac{5}{9}} + \sqrt{\frac{5}{81}}$
36. $\sqrt{\frac{3}{64}} + \sqrt{\frac{3}{16}}$
37. $\sqrt{\frac{3}{4}} - \sqrt{\frac{3}{64}}$
38. $\sqrt{\frac{2}{25}} + \sqrt{\frac{2}{9}}$
39. $2\sqrt{45} - 2\sqrt{20}$
40. $5\sqrt{18} + 2\sqrt{32}$
41. $\sqrt{35} - \sqrt{140}$
42. $\sqrt{15} - \sqrt{135}$

43. $3\sqrt{9x} + 2\sqrt{x}$

44. $5\sqrt{x} + 4\sqrt{4x}$

45. $\sqrt{9x^2} + \sqrt{81x^2} - 11\sqrt{x}$

46. $\sqrt{100x^2} + 3\sqrt{x} - \sqrt{36x^2}$

47. $\sqrt{3x^3} + 3x\sqrt{x}$

48. $x\sqrt{4x} + \sqrt{9x^3}$

49. $\sqrt{32x^2} + \sqrt{32x^2} + \sqrt{4x^2}$

50. $\sqrt{18x^2} + \sqrt{24x^3} + \sqrt{2x^2}$

51. $\sqrt{40x} + \sqrt{40x^4} - 2\sqrt{10x} - \sqrt{5x^4}$

52. $\sqrt{72x^2} + \sqrt{54x} - x\sqrt{50} - 3\sqrt{2x}$

Review and Preview

Square each binomial. See Section 10.5.

53. $(x + 6)^2$

54. $(3x + 2)^2$

55. $(2x - 1)^2$

56. $(x - 5)^2$

Solve each system of linear equations. See Section 14.2.

57. $\begin{cases} x = 2y \\ x + 5y = 14 \end{cases}$

58. $\begin{cases} y = -5x \\ x + y = 16 \end{cases}$

Combining Concepts

△ **59.** Find the perimeter of the rectangular picture frame.

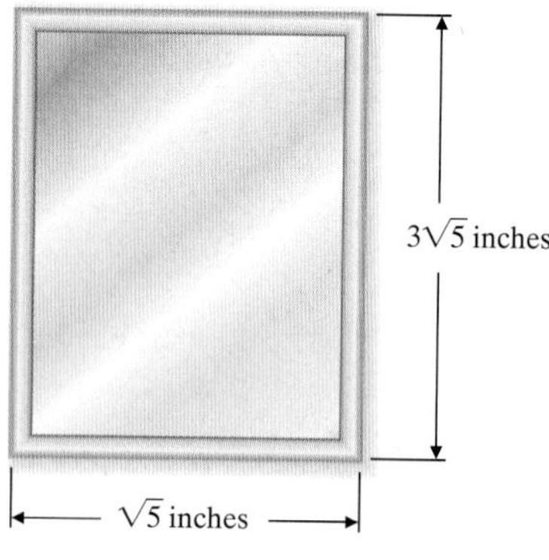

△ **60.** Find the perimeter of the plot of land.

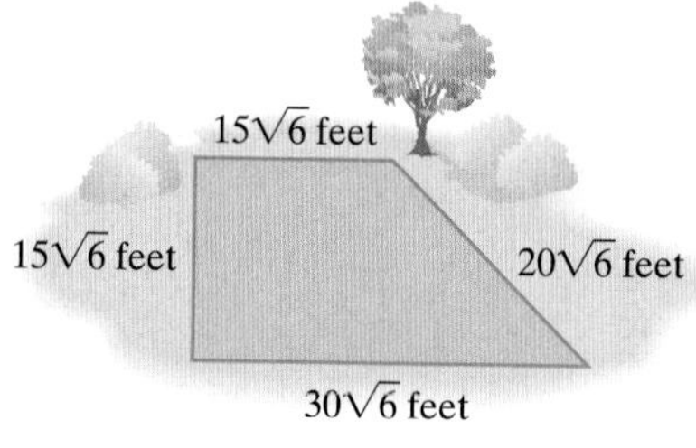

△ **61.** A water trough is to be made of wood. Each of the two triangular end pieces has an area of $\frac{3\sqrt{27}}{4}$ square feet. The two side panels are both rectangular. In simplest radical form, find the total area of the wood needed.

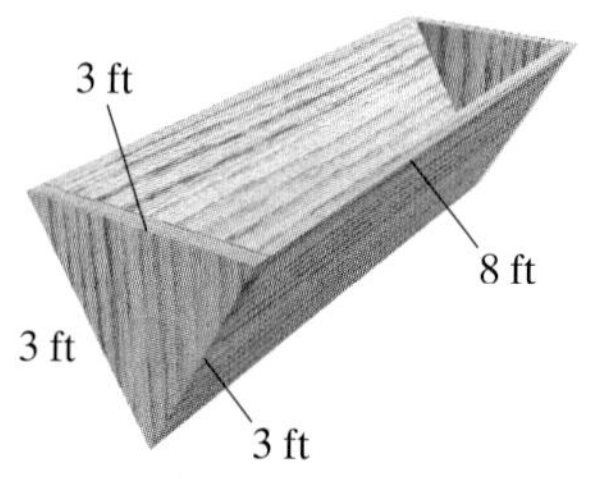

62. Eight wooden braces are to be attached along the diagonals of the vertical sides of a storage bin. Each of four of these diagonals has a length of $\sqrt{52}$ feet, while each of the other four has a length of $\sqrt{80}$ feet. In simplest radical form, find the total length of the wood needed for these braces.

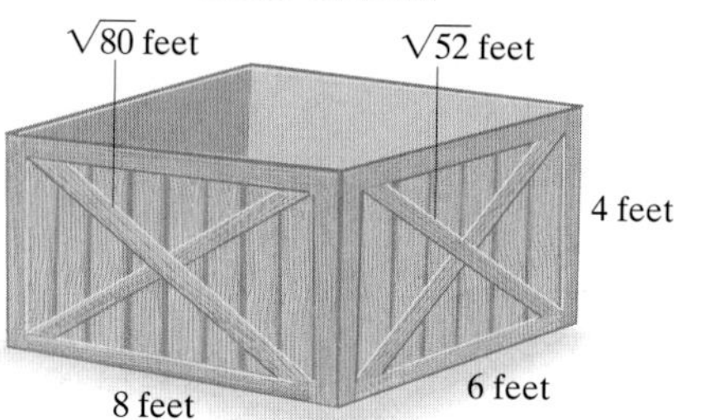

15.4 Multiplying and Dividing Radicals

OBJECTIVES

- (A) Multiply radicals.
- (B) Divide radicals.
- (C) Rationalize denominators.
- (D) Rationalize using conjugates.

SSM TUTOR CENTER SG CD & VIDEO MATH PRO WEB

(A) Multiplying Radicals

In Section 15.2, we used the product and quotient rules for radicals to help us simplify radicals. In this section, we use these rules to simplify products and quotients of radicals.

Product Rule for Radicals

If $\sqrt{a}$ and $\sqrt{b}$ are real numbers, then

$$\sqrt{a} \cdot \sqrt{b} = \sqrt{a \cdot b}$$

In other words, the product of the square roots of two numbers is the square root of the product of the two numbers. For example,

$$\sqrt{3} \cdot \sqrt{2} = \sqrt{3 \cdot 2} = \sqrt{6}$$

EXAMPLES Multiply. Then simplify each product if possible.

1. $\sqrt{7} \cdot \sqrt{3} = \sqrt{7 \cdot 3}$
$= \sqrt{21}$

2. $\sqrt{3} \cdot \sqrt{15} = \sqrt{45}$ Use the product rule.
$= \sqrt{9 \cdot 5}$ Factor the radicand.
$= \sqrt{9} \cdot \sqrt{5}$ Use the product rule.
$= 3\sqrt{5}$ Simplify $\sqrt{9}$.

3. $\sqrt{2x^3} \cdot \sqrt{6x} = \sqrt{2x^3 \cdot 6x}$ Use the product rule.
$= \sqrt{12x^4}$ Multiply.
$= \sqrt{4x^4 \cdot 3}$ Write $12x^4$ so that one factor is a perfect square.
$= \sqrt{4x^4} \cdot \sqrt{3}$ Use the product rule.
$= 2x^2\sqrt{3}$ Simplify.

Practice Problems 1–3

Multiply. Then simplify each product if possible.

1. $\sqrt{5} \cdot \sqrt{2}$
2. $\sqrt{6} \cdot \sqrt{3}$
3. $\sqrt{10x} \cdot \sqrt{2x}$

When multiplying radical expressions containing more than one term, we use the same techniques we use to multiply other algebraic expressions with more than one term.

EXAMPLE 4 Multiply.

a. $\sqrt{5}(\sqrt{5} - \sqrt{2})$
b. $(\sqrt{x} + \sqrt{2})(\sqrt{3} - \sqrt{2})$

Solution:

a. Using the distributive property, we have

$$\sqrt{5}(\sqrt{5} - \sqrt{2}) = \sqrt{5} \cdot \sqrt{5} - \sqrt{5} \cdot \sqrt{2}$$
$$= \sqrt{25} - \sqrt{10}$$
$$= 5 - \sqrt{10}$$

b. Using the FOIL method of multiplication, we have

$$(\sqrt{x} + \sqrt{2})(\sqrt{3} - \sqrt{2})$$

F O I L

$= \sqrt{x} \cdot \sqrt{3} - \sqrt{x} \cdot \sqrt{2} + \sqrt{2} \cdot \sqrt{3} - \sqrt{2} \cdot \sqrt{2}$
$= \sqrt{3x} - \sqrt{2x} + \sqrt{6} - \sqrt{4}$ Use the product rule.
$= \sqrt{3x} - \sqrt{2x} + \sqrt{6} - 2$ Simplify.

Practice Problem 4

Multiply.

a. $\sqrt{7}(\sqrt{7} - \sqrt{3})$
b. $(\sqrt{x} + \sqrt{5})(\sqrt{x} - \sqrt{3})$

Answers

1. $\sqrt{10}$ **2.** $3\sqrt{2}$ **3.** $2x\sqrt{5}$ **4. a.** $7 - \sqrt{21}$ **b.** $x - \sqrt{3x} + \sqrt{5x} - \sqrt{15}$

Concept Check

Identify the true statement(s).

a. $\sqrt{7} \cdot \sqrt{7} = 7$
b. $\sqrt{2} \cdot \sqrt{3} = 6$
c. $\sqrt{131} \cdot \sqrt{131} = 131$
d. $\sqrt{5x} \cdot \sqrt{5x} = 5x$ (Here x is a positive number.)

From Example 4, we found that

$$\sqrt{5} \cdot \sqrt{5} = 5 \quad \text{and} \quad \sqrt{2} \cdot \sqrt{2} = 2$$

This is true in general.

If a is a positive number,

$$\sqrt{a} \cdot \sqrt{a} = a$$

Try the Concept Check in the margin.

The special product formulas also can be used to multiply expressions containing radicals.

Practice Problem 5

Multiply.

a. $(\sqrt{3} + 6)(\sqrt{3} - 6)$
b. $(\sqrt{5x} + 4)^2$

EXAMPLE 5 Multiply.

a. $(\sqrt{5} - 7)(\sqrt{5} + 7)$
b. $(\sqrt{7x} + 2)^2$

Solution:

a. $(\sqrt{5} - 7)(\sqrt{5} + 7) = (\sqrt{5})^2 - 7^2 = 5 - 49 = -44$ Recall that $(a - b)(a + b) = a^2 - b^2$.

b. $(\sqrt{7x} + 2)^2 = (\sqrt{7x})^2 + 2(\sqrt{7x})(2) + (2)^2 = 7x + 4\sqrt{7x} + 4$ Recall that $(a + b)^2 = a^2 + 2ab + b^2$.

B Dividing Radicals

To simplify quotients of rational expressions, we use the quotient rule.

Quotient Rule for Radicals

If $\sqrt{a}$ and $\sqrt{b}$ are real numbers and $b \neq 0$, then

$$\frac{\sqrt{a}}{\sqrt{b}} = \sqrt{\frac{a}{b}}$$

Practice Problems 6–8

Divide. Then simplify the quotient if possible.

6. $\dfrac{\sqrt{15}}{\sqrt{3}}$

7. $\dfrac{\sqrt{90}}{\sqrt{2}}$

8. $\dfrac{\sqrt{75x^3}}{\sqrt{5x}}$

EXAMPLES Divide. Then simplify the quotient if possible.

6. $\dfrac{\sqrt{14}}{\sqrt{2}} = \sqrt{\dfrac{14}{2}} = \sqrt{7}$

7. $\dfrac{\sqrt{100}}{\sqrt{5}} = \sqrt{\dfrac{100}{5}} = \sqrt{20} = \sqrt{4 \cdot 5} = \sqrt{4} \cdot \sqrt{5} = 2\sqrt{5}$

8. $\dfrac{\sqrt{12x^3}}{\sqrt{3x}} = \sqrt{\dfrac{12x^3}{3x}} = \sqrt{4x^2} = 2x$

C Rationalizing Denominators

It is sometimes easier to work with radical expressions if the denominator does not contain a radical. To get rid of the radical in the denominator of a radical expression, we use the fact that we can multiply the numerator and the denominator of a fraction by the same nonzero number without changing the value of the expression. This is the same as multiplying the fraction by 1. For example, to get rid of the radical in the denominator of $\dfrac{\sqrt{5}}{\sqrt{2}}$, we multiply the numerator and the denominator by $\sqrt{2}$. Then

Answers

5. a. -33 **b.** $5x + 8\sqrt{5x} + 16$ **6.** $\sqrt{5}$
7. $3\sqrt{5}$ **8.** $x\sqrt{15}$

Concept Check: a, c, d

$$\frac{\sqrt{5}}{\sqrt{2}} = \frac{\sqrt{5} \cdot \sqrt{2}}{\sqrt{2} \cdot \sqrt{2}} = \frac{\sqrt{10}}{2}$$

This process is called **rationalizing** the denominator.

EXAMPLE 9 Rationalize the denominator of $\frac{2}{\sqrt{7}}$.

Solution: To get rid of the radical in the denominator of $\frac{2}{\sqrt{7}}$, we multiply the numerator and the denominator by $\sqrt{7}$.

$$\frac{2}{\sqrt{7}} = \frac{2 \cdot \sqrt{7}}{\sqrt{7} \cdot \sqrt{7}} = \frac{2\sqrt{7}}{7}$$

●

EXAMPLE 10 Rationalize the denominator of $\frac{\sqrt{5}}{\sqrt{12}}$.

Solution: We can multiply the numerator and denominator by $\sqrt{12}$, but see what happens if we simplify first.

$$\frac{\sqrt{5}}{\sqrt{12}} = \frac{\sqrt{5}}{\sqrt{4 \cdot 3}} = \frac{\sqrt{5}}{2\sqrt{3}}$$

To rationalize the denominator now, we multiply the numerator and the denominator by $\sqrt{3}$.

$$\frac{\sqrt{5}}{2\sqrt{3}} = \frac{\sqrt{5} \cdot \sqrt{3}}{2\sqrt{3} \cdot \sqrt{3}} = \frac{\sqrt{15}}{2 \cdot 3} = \frac{\sqrt{15}}{6}$$

●

EXAMPLE 11 Rationalize the denominator of $\sqrt{\frac{1}{18x}}$.

Solution: First we simplify.

$$\sqrt{\frac{1}{18x}} = \frac{\sqrt{1}}{\sqrt{18x}} = \frac{1}{\sqrt{9} \cdot \sqrt{2x}} = \frac{1}{3\sqrt{2x}}$$

Now to rationalize the denominator, we multiply the numerator and denominator by $\sqrt{2x}$.

$$\frac{1}{3\sqrt{2x}} = \frac{1 \cdot \sqrt{2x}}{3\sqrt{2x} \cdot \sqrt{2x}} = \frac{\sqrt{2x}}{3 \cdot 2x} = \frac{\sqrt{2x}}{6x}$$

●

D Rationalizing Denominators Using Conjugates

To rationalize a denominator that is a sum or a difference, such as the denominator in

$$\frac{2}{4 + \sqrt{3}}$$

we multiply the numerator and the denominator by $4 - \sqrt{3}$. The expressions $4 + \sqrt{3}$ and $4 - \sqrt{3}$ are called conjugates of each other. When a radical expression such as $4 + \sqrt{3}$ is multiplied by its conjugate $4 - \sqrt{3}$, the product simplifies to an expression that contains no radicals.

In general, the expressions $a + b$ and $a - b$ are **conjugates** of each other.

$$(a + b)(a - b) = a^2 - b^2$$

$$(4 + \sqrt{3})(4 - \sqrt{3}) = 4^2 - (\sqrt{3})^2 = 16 - 3 = 13$$

Then

$$\frac{2}{4 + \sqrt{3}} = \frac{2(4 - \sqrt{3})}{(4 + \sqrt{3})(4 - \sqrt{3})} = \frac{2(4 - \sqrt{3})}{13}$$

Practice Problem 9

Rationalize the denominator of $\frac{5}{\sqrt{3}}$.

Practice Problem 10

Rationalize the denominator of $\frac{\sqrt{7}}{\sqrt{20}}$.

Practice Problem 11

Rationalize the denominator of $\sqrt{\frac{2}{45x}}$.

Answers

9. $\frac{5\sqrt{3}}{3}$ **10.** $\frac{\sqrt{35}}{10}$ **11.** $\frac{\sqrt{10x}}{15x}$

Practice Problem 12

Rationalize the denominator of $\frac{3}{1 + \sqrt{7}}$.

EXAMPLE 12 Rationalize the denominator of $\frac{2}{1 + \sqrt{3}}$.

Solution: We multiply the numerator and the denominator of this fraction by the conjugate of $1 + \sqrt{3}$, that is, by $1 - \sqrt{3}$.

$$\frac{2}{1 + \sqrt{3}} = \frac{2(1 - \sqrt{3})}{(1 + \sqrt{3})(1 - \sqrt{3})}$$

$$= \frac{2(1 - \sqrt{3})}{1^2 - (\sqrt{3})^2}$$

$$= \frac{2(1 - \sqrt{3})}{1 - 3}$$

$$= \frac{2(1 - \sqrt{3})}{-2}$$

$$= -\frac{2(1 - \sqrt{3})}{2} \qquad \frac{a}{-b} = -\frac{a}{b}$$

$$= -1(1 - \sqrt{3}) \qquad \text{Simplify.}$$

$$= -1 + \sqrt{3} \qquad \text{Multiply.}$$

Helpful Hint

Don't forget that $(\sqrt{3})^2 = \sqrt{3} \cdot \sqrt{3} = 3$

Practice Problem 13

Rationalize the denominator of $\frac{\sqrt{2} + 5}{\sqrt{2} - 1}$.

EXAMPLE 13 Rationalize the denominator of $\frac{\sqrt{5} + 4}{\sqrt{5} - 1}$.

Solution:

$$\frac{\sqrt{5} + 4}{\sqrt{5} - 1} = \frac{(\sqrt{5} + 4)(\sqrt{5} + 1)}{(\sqrt{5} - 1)(\sqrt{5} + 1)}$$

Multiply the numerator and denominator by $\sqrt{5} + 1$, the conjugate of $\sqrt{5} - 1$.

$$= \frac{5 + \sqrt{5} + 4\sqrt{5} + 4}{5 - 1} \qquad \text{Multiply.}$$

$$= \frac{9 + 5\sqrt{5}}{4} \qquad \text{Simplify.}$$

Practice Problem 14

Rationalize the denominator of $\frac{7}{2 - \sqrt{x}}$.

EXAMPLE 14 Rationalize the denominator of $\frac{3}{1 + \sqrt{x}}$.

Solution:

$$\frac{3}{1 + \sqrt{x}} = \frac{3(1 - \sqrt{x})}{(1 + \sqrt{x})(1 - \sqrt{x})}$$

Multiply the numerator and denominator by $1 - \sqrt{x}$, the conjugate of $1 + \sqrt{x}$.

$$= \frac{3(1 - \sqrt{x})}{1 - x}$$

Answers

12. $\frac{-1 + \sqrt{7}}{2}$ **13.** $7 + 6\sqrt{2}$

14. $\frac{7(2 + \sqrt{x})}{4 - x}$

Name ______________________ Section __________ Date __________

Mental Math

Multiply. Assume that all variables represent positive numbers.

1. $\sqrt{2} \cdot \sqrt{3}$

2. $\sqrt{5} \cdot \sqrt{7}$

3. $\sqrt{1} \cdot \sqrt{6}$

4. $\sqrt{7} \cdot \sqrt{x}$

5. $\sqrt{10} \cdot \sqrt{y}$

6. $\sqrt{x} \cdot \sqrt{y}$

EXERCISE SET 15.4

A *Multiply and simplify. Assume that all variables represent positive real numbers. See Examples 1 through 5.*

1. $\sqrt{8} \cdot \sqrt{2}$

2. $\sqrt{3} \cdot \sqrt{12}$

3. $\sqrt{10} \cdot \sqrt{5}$

4. $\sqrt{2} \cdot \sqrt{14}$

5. $\sqrt{6} \cdot \sqrt{6}$

6. $\sqrt{10} \cdot \sqrt{10}$

7. $\sqrt{2x} \cdot \sqrt{2x}$

8. $\sqrt{5y} \cdot \sqrt{5y}$

9. $(2\sqrt{5})^2$

10. $(3\sqrt{10})^2$

11. $(6\sqrt{x})^2$

12. $(8\sqrt{y})^2$

13. $\sqrt{3y} \cdot \sqrt{6x}$

14. $\sqrt{21y} \cdot \sqrt{3x}$

15. $\sqrt{2xy^2} \cdot \sqrt{8xy}$

16. $\sqrt{18x^2y^2} \cdot \sqrt{2x^2y}$

17. $\sqrt{2}(\sqrt{5} + 1)$

18. $\sqrt{3}(\sqrt{2} - 1)$

19. $\sqrt{10}(\sqrt{2} + \sqrt{5})$

20. $\sqrt{6}(\sqrt{3} + \sqrt{2})$

21. $\sqrt{6}(\sqrt{5} + \sqrt{7})$

22. $\sqrt{10}(\sqrt{3} - \sqrt{7})$

23. $(\sqrt{3} + 6)(\sqrt{3} - 6)$

24. $(\sqrt{5} + 2)(\sqrt{5} - 2)$

25. $(\sqrt{3} + \sqrt{5})(\sqrt{2} - \sqrt{5})$

26. $(\sqrt{7} + \sqrt{5})(\sqrt{2} - \sqrt{5})$

27. $(2\sqrt{11} + 1)(\sqrt{11} - 6)$

28. $(5\sqrt{3} + 2)(\sqrt{3} - 1)$

29. $(\sqrt{x} + 6)(\sqrt{x} - 6)$

30. $(\sqrt{y} + 5)(\sqrt{y} - 5)$

31. $(\sqrt{x} - 7)^2$

32. $(\sqrt{x} + 4)^2$

33. $(\sqrt{6y} + 1)^2$

34. $(\sqrt{3y} - 2)^2$

B *Divide and simplify. Assume that all variables represent positive real numbers. See Examples 6 through 8.*

35. $\frac{\sqrt{32}}{\sqrt{2}}$

36. $\frac{\sqrt{40}}{\sqrt{10}}$

37. $\frac{\sqrt{21}}{\sqrt{3}}$

38. $\frac{\sqrt{55}}{\sqrt{5}}$

39. $\frac{\sqrt{90}}{\sqrt{5}}$

40. $\frac{\sqrt{96}}{\sqrt{8}}$

41. $\frac{\sqrt{75y^5}}{\sqrt{3y}}$

42. $\frac{\sqrt{24x^7}}{\sqrt{6x}}$

43. $\frac{\sqrt{150}}{\sqrt{2}}$

44. $\frac{\sqrt{120}}{\sqrt{3}}$

45. $\frac{\sqrt{72y^5}}{\sqrt{3y^3}}$

46. $\frac{\sqrt{54x^3}}{\sqrt{2x}}$

47. $\frac{\sqrt{24x^3y^4}}{\sqrt{2xy}}$

48. $\frac{\sqrt{96x^5y^3}}{\sqrt{3x^2y}}$

C *Rationalize each denominator and simplify. Assume that all variables represent positive real numbers. See Examples 9 through 11.*

49. $\frac{\sqrt{3}}{\sqrt{5}}$

50. $\frac{\sqrt{2}}{\sqrt{3}}$

51. $\frac{7}{\sqrt{2}}$

52. $\frac{8}{\sqrt{11}}$

53. $\frac{1}{\sqrt{6y}}$

54. $\frac{1}{\sqrt{10z}}$

55. $\sqrt{\frac{5}{18}}$ **56.** $\sqrt{\frac{7}{12}}$ **57.** $\sqrt{\frac{3}{x}}$ **58.** $\sqrt{\frac{5}{x}}$ **59.** $\sqrt{\frac{1}{8}}$ **60.** $\sqrt{\frac{1}{27}}$

61. $\sqrt{\frac{2}{15}}$ **62.** $\sqrt{\frac{11}{14}}$ **63.** $\sqrt{\frac{3}{20}}$ **64.** $\sqrt{\frac{3}{50}}$ **65.** $\frac{3x}{\sqrt{2x}}$ **66.** $\frac{5y}{\sqrt{3y}}$

67. $\frac{8y}{\sqrt{5}}$ **68.** $\frac{7x}{\sqrt{2}}$ **69.** $\sqrt{\frac{y}{12x}}$ **70.** $\sqrt{\frac{x}{20y}}$

D *Rationalize each denominator and simplify. Assume that all variables represent positive real numbers. See Examples 12 through 14.*

71. $\frac{3}{\sqrt{2} + 1}$ **72.** $\frac{6}{\sqrt{5} + 2}$ **73.** $\frac{4}{2 - \sqrt{5}}$ **74.** $\frac{2}{\sqrt{10} - 3}$

75. $\frac{\sqrt{5} + 1}{\sqrt{6} - \sqrt{5}}$ **76.** $\frac{\sqrt{3} + 1}{\sqrt{3} - \sqrt{2}}$ **77.** $\frac{\sqrt{3} + 1}{\sqrt{2} - 1}$ **78.** $\frac{\sqrt{2} - 2}{2 - \sqrt{3}}$

79. $\frac{5}{2 + \sqrt{x}}$ **80.** $\frac{9}{3 + \sqrt{x}}$ **81.** $\frac{3}{\sqrt{x} - 4}$ **82.** $\frac{4}{\sqrt{x} - 1}$

Review and Preview

Solve each equation. See Sections 9.3 and 11.6.

83. $x + 5 = 7^2$

84. $2y - 1 = 3^2$

85. $4z^2 + 6z - 12 = (2z)^2$

86. $16x^2 + x + 9 = (4x)^2$

87. $9x^2 + 5x + 4 = (3x + 1)^2$

88. $x^2 + 3x + 4 = (x + 2)^2$

Combining Concepts

△ **89.** Find the area of a rectangular room whose length is $13\sqrt{2}$ meters and width is $5\sqrt{6}$ meters.

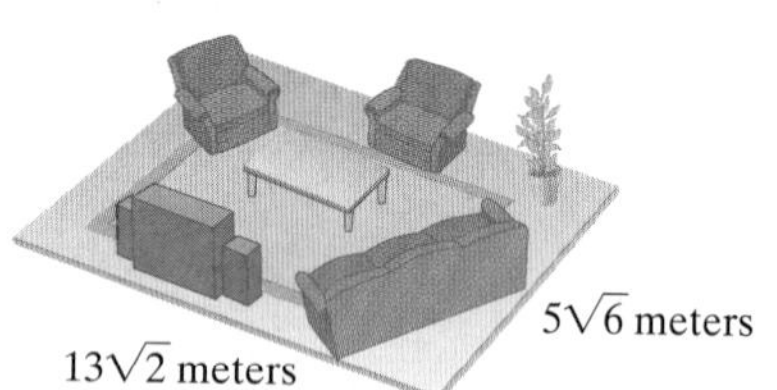

△ **90.** Find the volume of a microwave oven whose length is $\sqrt{3}$ feet, width is $\sqrt{2}$ feet, and height is $\sqrt{2}$ feet.

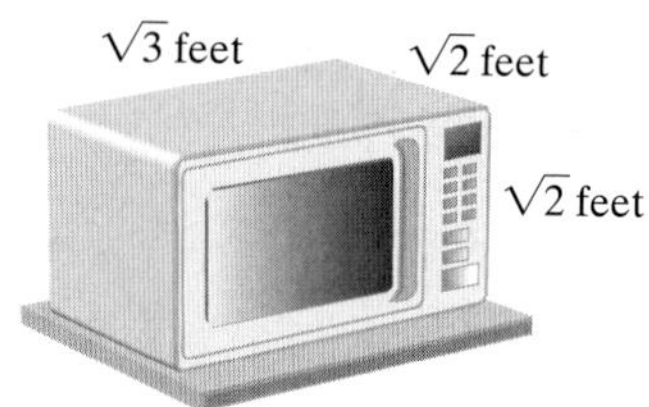

91. When rationalizing the denominator of $\frac{\sqrt{2}}{\sqrt{3}}$, explain why both the numerator and the denominator must be multiplied by $\sqrt{3}$.

△ **92.** If a circle has area A, then the formula for the radius r of the circle is

$$r = \sqrt{\frac{A}{\pi}}$$

Rationalize the denominator of this expression.

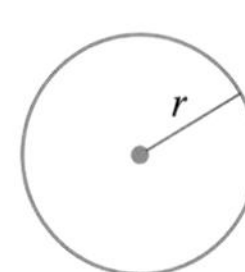

It is often more convenient to work with a radical expression whose numerator is rationalized. Rationalize the numerator of each expression by multiplying the numerator and denominator by the conjugate of the numerator.

93. $\frac{\sqrt{3} + 1}{\sqrt{2} - 1}$

94. $\frac{\sqrt{2} - 2}{2 - \sqrt{3}}$

Name ______________________ Section ________ Date ____________

Integrated Review—Simplifying Radicals

Simplify. Assume that all variables represent positive numbers.

1. $\sqrt{36}$ **2.** $\sqrt{48}$ **3.** $\sqrt{x^4}$ **4.** $\sqrt{y^7}$

5. $\sqrt{16x^2}$ **6.** $\sqrt{18x^{11}}$ **7.** $\sqrt[3]{8}$ **8.** $\sqrt[4]{81}$

9. $\sqrt[3]{-27}$ **10.** $\sqrt{-4}$ **11.** $\sqrt{\frac{11}{9}}$ **12.** $\sqrt[3]{\frac{7}{64}}$

13. $-\sqrt{16}$ **14.** $-\sqrt{25}$ **15.** $\sqrt{\frac{9}{49}}$ **16.** $\sqrt{\frac{1}{64}}$

17. $\sqrt{a^8b^2}$ **18.** $\sqrt{x^{10}y^{20}}$ **19.** $\sqrt{25m^6}$ **20.** $\sqrt{9n^{16}}$

Add or subtract as indicated.

21. $5\sqrt{7} + \sqrt{7}$ **22.** $\sqrt{50} - \sqrt{8}$

23. $5\sqrt{2} - 5\sqrt{3}$ **24.** $2\sqrt{x} + \sqrt{25x} - \sqrt{36x} + 3x$

Answers

1. ________
2. ________
3. ________
4. ________
5. ________
6. ________
7. ________
8. ________
9. ________
10. ________
11. ________
12. ________
13. ________
14. ________
15. ________
16. ________
17. ________
18. ________
19. ________
20. ________
21. ________
22. ________
23. ________
24. ________

25. ______

26. ______

27. ______

28. ______

29. ______

30. ______

31. ______

32. ______

33. ______

34. ______

35. ______

36. ______

37. ______

38. ______

39. ______

40. ______

41. ______

42. ______

Multiply and simplify if possible.

25. $\sqrt{2} \cdot \sqrt{15}$ **26.** $\sqrt{3} \cdot \sqrt{3}$ **27.** $(2\sqrt{7})^2$ **28.** $(3\sqrt{5})^2$

29. $\sqrt{3}(\sqrt{11} + 1)$ **30.** $\sqrt{6}(\sqrt{3} - 2)$ **31.** $\sqrt{8y} \cdot \sqrt{2y}$

32. $\sqrt{15x^2} \cdot \sqrt{3x^2}$ **33.** $(\sqrt{x} - 5)(\sqrt{x} + 2)$ **34.** $(3 + \sqrt{2})^2$

Divide and simplify if possible.

35. $\frac{\sqrt{8}}{\sqrt{2}}$ **36.** $\frac{\sqrt{45}}{\sqrt{15}}$ **37.** $\frac{\sqrt{24x^5}}{\sqrt{2x}}$ **38.** $\frac{\sqrt{75a^4b^5}}{\sqrt{5ab}}$

Rationalize each denominator.

39. $\sqrt{\frac{1}{6}}$ **40.** $\frac{x}{\sqrt{20}}$ **41.** $\frac{4}{\sqrt{6} + 1}$ **42.** $\frac{\sqrt{2} + 1}{\sqrt{x} - 5}$

15.5 Solving Equations Containing Radicals

OBJECTIVES

A. Solve radical equations by using the squaring property of equality once.

B. Solve radical equations by using the squaring property of equality twice.

A Using the Squaring Property of Equality Once

In this section, we solve **radical equations** such as

$$\sqrt{x+3} = 5 \quad \text{and} \quad \sqrt{2x+1} = \sqrt{3x}$$

Radical equations contain variables in the radicand. To solve these equations, we rely on the following squaring property.

The Squaring Property of Equality

If $a = b$, then $a^2 = b^2$.

Unfortunately, this squaring property does not guarantee that all solutions of the new equation are solutions of the original equation. For example, if we square both sides of the equation

$$x = 2$$

we have

$$x^2 = 4$$

This new equation has two solutions, 2 and -2, while the original equation $x = 2$ has only one solution. For this reason, we must **always check proposed solutions of radical equations in the original equation.**

EXAMPLE 1 Solve: $\sqrt{x+3} = 5$

Solution: To solve this radical equation, we use the squaring property of equality and square both sides of the equation.

$$\sqrt{x+3} = 5$$
$$(\sqrt{x+3})^2 = 5^2 \quad \text{Square both sides.}$$
$$x + 3 = 25 \quad \text{Simplify.}$$
$$x = 22 \quad \text{Subtract 3 from both sides.}$$

Check: We replace x with 22 in the original equation.

Don't forget to check the proposed solutions of radical equations in the original equation.

$$\sqrt{x+3} = 5 \quad \text{Original equation}$$
$$\sqrt{22+3} \stackrel{?}{=} 5 \quad \text{Let } x = 22.$$
$$\sqrt{25} \stackrel{?}{=} 5$$
$$5 = 5 \quad \text{True}$$

Since a true statement results, 22 is the solution.

Practice Problem 1

Solve: $\sqrt{x-2} = 7$

EXAMPLE 2 Solve: $\sqrt{x} + 6 = 4$

Solution: First we write the equation so that the radical is by itself on one side of the equation.

$$\sqrt{x} + 6 = 4$$
$$\sqrt{x} = -2 \quad \text{Subtract 6 from both sides to get the radical by itself.}$$

Practice Problem 2

Solve: $\sqrt{x} + 9 = 2$

Answers

1. $x = 51$ **2.** no solution

Normally we would now square both sides. Recall, however, that $\sqrt{x}$ is the principal or nonnegative square root of x so that $\sqrt{x}$ cannot equal -2 and thus this equation has no solution. We arrive at the same conclusion if we continue by applying the squaring property.

$$\sqrt{x} = -2$$

$$(\sqrt{x})^2 = (-2)^2 \quad \text{Square both sides.}$$

$$x = 4 \quad \text{Simplify.}$$

Check: We replace x with 4 in the original equation.

$$\sqrt{x} + 6 = 4 \quad \text{Original equation}$$

$$\sqrt{4} + 6 \stackrel{?}{=} 4 \quad \text{Let } x = 4.$$

$$2 + 6 = 4 \quad \text{False}$$

Since 4 *does not* satisfy the original equation, this equation has no solution. ●

Example 2 makes it very clear that we *must* check proposed solutions in the original equation to determine if they are truly solutions. If a proposed solution does not work, we say that the value is an **extraneous solution.**

The following steps can be used to solve radical equations containing square roots.

To Solve a Radical Equation Containing Square Roots

Step 1. Arrange terms so that one radical is by itself on one side of the equation. That is, isolate a radical.

Step 2. Square both sides of the equation.

Step 3. Simplify both sides of the equation.

Step 4. If the equation still contains a radical term, repeat Steps 1 through 3.

Step 5. Solve the equation.

Step 6. Check all solutions in the original equation for extraneous solutions.

Practice Problem 3

Solve: $\sqrt{6x - 1} = \sqrt{x}$

EXAMPLE 3 Solve: $\sqrt{x} = \sqrt{5x - 2}$

Solution: Each of the radicals is already isolated, since each is by itself on one side of the equation. So we begin solving by squaring both sides.

$$\sqrt{x} = \sqrt{5x - 2} \quad \text{Original equation}$$

$$(\sqrt{x})^2 = (\sqrt{5x - 2})^2 \quad \text{Square both sides.}$$

$$x = 5x - 2 \quad \text{Simplify.}$$

$$-4x = -2 \quad \text{Subtract } 5x \text{ from both sides.}$$

$$x = \frac{-2}{-4} = \frac{1}{2} \quad \text{Divide both sides by } -4 \text{ and simplify.}$$

Answer

3. $x = \frac{1}{5}$

Check: We replace x with $\frac{1}{2}$ in the original equation.

$$\sqrt{x} = \sqrt{5x - 2} \quad \text{Original equation}$$

$$\sqrt{\frac{1}{2}} \stackrel{?}{=} \sqrt{5 \cdot \frac{1}{2} - 2} \quad \text{Let } x = \frac{1}{2}.$$

$$\sqrt{\frac{1}{2}} \stackrel{?}{=} \sqrt{\frac{5}{2} - 2} \quad \text{Multiply.}$$

$$\sqrt{\frac{1}{2}} \stackrel{?}{=} \sqrt{\frac{5}{2} - \frac{4}{2}} \quad \text{Write 2 as } \frac{4}{2}.$$

$$\sqrt{\frac{1}{2}} = \sqrt{\frac{1}{2}} \quad \text{True}$$

This statement is true, so the solution is $\frac{1}{2}$.

EXAMPLE 4 Solve: $\sqrt{4y^2 + 5y - 15} = 2y$

Solution: The radical is already isolated, so we start by squaring both sides.

$$\sqrt{4y^2 + 5y - 15} = 2y$$

$$\left(\sqrt{4y^2 + 5y - 15}\right)^2 = (2y)^2 \quad \text{Square both sides.}$$

$$4y^2 + 5y - 15 = 4y^2 \quad \text{Simplify.}$$

$$5y - 15 = 0 \quad \text{Subtract } 4y^2 \text{ from both sides.}$$

$$5y = 15 \quad \text{Add 15 to both sides.}$$

$$y = 3 \quad \text{Divide both sides by 5.}$$

Check: We replace y with 3 in the original equation.

$$\sqrt{4y^2 + 5y - 15} = 2y \quad \text{Original equation}$$

$$\sqrt{4 \cdot 3^2 + 5 \cdot 3 - 15} \stackrel{?}{=} 2 \cdot 3 \quad \text{Let } y = 3.$$

$$\sqrt{4 \cdot 9 + 15 - 15} \stackrel{?}{=} 6 \quad \text{Simplify.}$$

$$\sqrt{36} \stackrel{?}{=} 6$$

$$6 = 6 \quad \text{True}$$

This statement is true, so the solution is 3.

EXAMPLE 5 Solve: $\sqrt{x + 3} - x = -3$

Solution: First we isolate the radical by adding x to both sides. Then we square both sides.

$$\sqrt{x + 3} - x = -3$$

$$\sqrt{x + 3} = x - 3 \quad \text{Add } x \text{ to both sides.}$$

$$(\sqrt{x + 3})^2 = (x - 3)^2 \quad \text{Square both sides.}$$

$$x + 3 = x^2 - 6x + 9$$

Helpful Hint

Don't forget that $(x - 3)^2 = (x - 3)(x - 3) = x^2 - 6x + 9$

Practice Problem 4

Solve: $\sqrt{9y^2 + 2y - 10} = 3y$

Practice Problem 5

Solve: $\sqrt{x + 1} - x = -5$

Answers

4. $y = 5$ **5.** $x = 8$

To solve the resulting quadratic equation, we write the equation in standard form by subtracting x and 3 from both sides.

$$x + 3 = x^2 - 6x + 9$$
$$3 = x^2 - 7x + 9 \quad \text{Subtract } x \text{ from both sides.}$$
$$0 = x^2 - 7x + 6 \quad \text{Subtract 3 from both sides.}$$
$$0 = (x - 6)(x - 1) \quad \text{Factor.}$$
$$0 = x - 6 \quad \text{or} \quad 0 = x - 1 \quad \text{Set each factor equal to zero.}$$
$$6 = x \qquad 1 = x \quad \text{Solve for } x.$$

Check: We replace x with 6 and then x with 1 in the original equation.

Let $x = 6$.

$$\sqrt{x + 3} - x = -3$$
$$\sqrt{6 + 3} - 6 \stackrel{?}{=} -3$$
$$\sqrt{9} - 6 \stackrel{?}{=} -3$$
$$3 - 6 \stackrel{?}{=} -3$$
$$-3 = -3 \quad \text{True}$$

Let $x = 1$.

$$\sqrt{x + 3} - x = -3$$
$$\sqrt{1 + 3} - 1 \stackrel{?}{=} -3$$
$$\sqrt{4} - 1 \stackrel{?}{=} -3$$
$$2 - 1 \stackrel{?}{=} -3$$
$$1 = -3 \quad \text{False}$$

Since replacing x with 1 resulted in a false statement, 1 is an extraneous solution. The only solution is 6.

B Using the Squaring Property of Equality Twice

If a radical equation contains two radicals, we may need to use the squaring property twice.

Practice Problem 6

Solve: $\sqrt{x} + 3 = \sqrt{x + 15}$

EXAMPLE 6 Solve: $\sqrt{x - 4} = \sqrt{x} - 2$

Solution:

$$\sqrt{x - 4} = \sqrt{x} - 2$$
$$(\sqrt{x - 4})^2 = (\sqrt{x} - 2)^2 \quad \text{Square both sides.}$$
$$x - 4 = x - 4\sqrt{x} + 4$$
$$-8 = -4\sqrt{x}$$
$$2 = \sqrt{x} \quad \text{Divide both sides by } -4.$$
$$4 = x \quad \text{Square both sides again.}$$

Helpful Hint

$$(\sqrt{x} - 2)^2 = (\sqrt{x} - 2)(\sqrt{x} - 2)$$
$$= \sqrt{x} \cdot \sqrt{x} - 2\sqrt{x} - 2\sqrt{x} + 4$$
$$= x - 4\sqrt{x} + 4$$

Check the proposed solution in the original equation. The solution is 4.

Answer

6. $x = 1$

Name ______________________ Section ____________ Date ____________

EXERCISE SET 15.5

Solve each equation. See Examples 1 through 3.

1. $\sqrt{x} = 9$

2. $\sqrt{x} = 4$

3. $\sqrt{x + 5} = 2$

4. $\sqrt{x + 12} = 3$

5. $\sqrt{2x + 6} = 4$

6. $\sqrt{3x + 7} = 5$

7. $\sqrt{x} - 2 = 5$

8. $4\sqrt{x} - 7 = 5$

9. $3\sqrt{x} + 5 = 2$

10. $3\sqrt{x} + 5 = 8$

11. $\sqrt{x + 6} + 1 = 3$

12. $\sqrt{x + 5} + 2 = 5$

13. $\sqrt{2x + 1} + 3 = 5$

14. $\sqrt{3x - 1} + 4 = 1$

15. $\sqrt{x} + 3 = 7$

16. $\sqrt{x} + 5 = 10$

17. $\sqrt{x + 6} + 5 = 3$

18. $\sqrt{2x - 1} + 7 = 1$

19. $\sqrt{4x - 3} = \sqrt{x + 3}$

20. $\sqrt{5x - 4} = \sqrt{x + 8}$

21. $\sqrt{x} = \sqrt{3x - 8}$

22. $\sqrt{x} = \sqrt{4x - 3}$

23. $\sqrt{4x} = \sqrt{2x + 6}$

24. $\sqrt{5x + 6} = \sqrt{8x}$

Solve each equation. See Examples 4 and 5.

25. $\sqrt{9x^2 + 2x - 4} = 3x$

26. $\sqrt{4x^2 + 3x - 9} = 2x$

27. $\sqrt{16x^2 - 3x + 6} = 4x$

28. $\sqrt{9x^2 - 2x + 8} = 3x$

29. $\sqrt{16x^2 + 2x + 2} = 4x$

30. $\sqrt{4x^2 + 3x - 2} = 2x$

31. $\sqrt{2x^2 + 6x + 9} = 3$

32. $\sqrt{3x^2 + 6x + 4} = 2$

33. $\sqrt{x + 7} = x + 5$

34. $\sqrt{x + 5} = x - 1$

35. $\sqrt{x} = x - 6$

36. $\sqrt{x} = x + 6$

37. $\sqrt{2x + 1} = x - 7$

38. $\sqrt{2x + 5} = x - 5$

39. $x = \sqrt{2x - 2} + 1$

40. $\sqrt{1 - 8x} + 2 = x$

41. $\sqrt{1 - 8x} - x = 4$

42. $\sqrt{3x + 7} - x = 3$

43. $\sqrt{2x + 5} - 1 = x$

44. $x = \sqrt{4x - 7} + 1$

Solve each equation. See Example 6.

45. $\sqrt{x - 7} = \sqrt{x} - 1$

46. $\sqrt{x} + 2 = \sqrt{x + 24}$

47. $\sqrt{x} + 3 = \sqrt{x + 15}$

48. $\sqrt{x - 8} = \sqrt{x} - 2$

49. $\sqrt{x + 8} = \sqrt{x} + 2$

50. $\sqrt{x} + 1 = \sqrt{x + 15}$

Review and Preview

Translate each sentence into an equation and then solve. See Section 9.4.

51. If 8 is subtracted from the product of 3 and x, the result is 19. Find x.

52. If 3 more than x is subtracted from twice x, the result is 11. Find x.

53. The length of a rectangle is twice the width. The perimeter is 24 inches. Find the length.

54. The length of a rectangle is 2 inches longer than the width. The perimeter is 24 inches. Find the length.

Combining Concepts

55. The formula $b = \sqrt{\frac{V}{2}}$ can be used to determine the length b of a side of the base of a square-based pyramid with height 6 units and volume V cubic units.

a. Find the length of the side of the base that produces a pyramid with each volume. (Round to the nearest tenth of a unit.)

V	20	200	2000
b			

b. Notice in the table that volume V has been increased by a factor of 10 each time. Does the corresponding length b of a side increase by a factor of 10 each time also?

56. The formula $r = \sqrt{\frac{V}{2\pi}}$ can be used to determine the radius r of a cylinder with height 2 units and volume V cubic units.

a. Find the radius needed to manufacture a cylinder with each volume. (Round to the nearest tenth of a unit.)

V	10	100	1000
r			

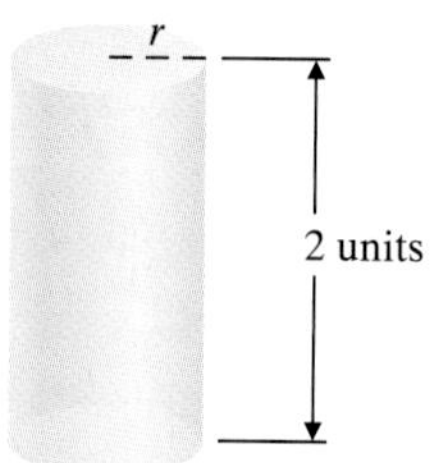

b. Notice in the table that volume V has been increased by a factor of 10 each time. Does the corresponding radius increase by a factor of 10 each time also?

57. Explain why proposed solutions of radical equations must be checked in the original equation.

Graphing calculators can be used to solve equations. To solve $\sqrt{x-2} = x - 5$, for example, graph $y_1 = \sqrt{x-2}$ and $y_2 = x - 5$ on the same set of axes. Use the Trace and Zoom features or an Intersect feature to find the point of intersection of the graphs. The x-value of the point is the solution of the equation. Use a graphing calculator to solve the equations below. Approximate solutions to the nearest hundredth.

58. $\sqrt{x-2} = x - 5$

59. $\sqrt{x+1} = 2x - 3$

60. $-\sqrt{x+4} = 5x - 6$

61. $-\sqrt{x+5} = -7x + 1$

15.6 Radical Equations and Problem Solving

OBJECTIVES

Ⓐ Use the Pythagorean theorem to solve problems.

Ⓑ Solve problems using formulas containing radicals.

Ⓐ Using the Pythagorean Theorem

Applications of radicals can be found in geometry, finance, science, and other areas of technology. Our first application involves the Pythagorean theorem, which gives a formula that relates the lengths of the three sides of a right triangle. We first studied the Pythagorean theorem in Chapter 5 and we review it here.

The Pythagorean Theorem

If a and b are lengths of the legs of a right triangle and c is the length of the hypotenuse, then $a^2 + b^2 = c^2$.

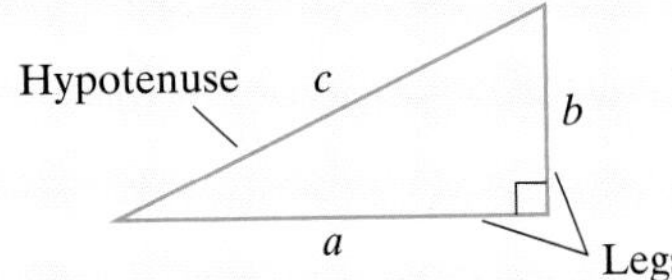

EXAMPLE 1

Find the length of the hypotenuse of a right triangle whose legs are 6 inches and 8 inches long.

Solution: Because this is a right triangle, we use the Pythagorean theorem. We let $a = 6$ inches and $b = 8$ inches. Length c must be the length of the hypotenuse.

$a^2 + b^2 = c^2$ Use the Pythagorean theorem.

$6^2 + 8^2 = c^2$ Substitute the lengths of the legs.

$36 + 64 = c^2$ Simplify.

$100 = c^2$

6 inches

Hypotenuse

8 inches

Since c represents a length, we know that c is positive and is the principal square root of 100.

$100 = c^2$

$\sqrt{100} = c$ Use the definition of principal square root.

$10 = c$ Simplify.

The hypotenuse has a length of 10 inches. ●

Practice Problem 1

Find the length of the hypotenuse of the right triangle shown.

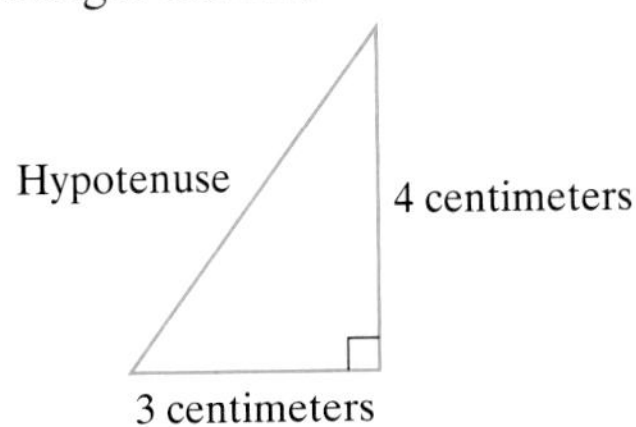

EXAMPLE 2

Find the length of the leg of the right triangle shown. Give the exact length and a two-decimal-place approximation.

Solution: We let $a = 2$ meters and b be the unknown length of the other leg. The hypotenuse is $c = 5$ meters.

$a^2 + b^2 = c^2$ Use the Pythagorean theorem.

$2^2 + b^2 = 5^2$ Let $a = 2$ and $c = 5$.

$4 + b^2 = 25$

$b^2 = 21$

$b = \sqrt{21} \approx 4.58$ meters

2 meters

5 meters

Leg

The length of the leg is exactly $\sqrt{21}$ meters or approximately 4.58 meters. ●

Practice Problem 2

Find the length of the leg of the right triangle shown. Give the exact length and a two-decimal-place approximation.

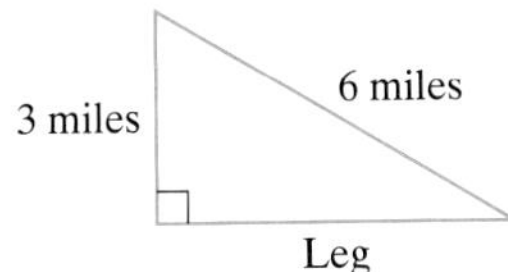

Answers

1. 5 cm **2.** $3\sqrt{3}$ mi; 5.20 mi

Practice Problem 3

Evan Saacks wants to determine the distance at certain points across a pond on his property. He is able to measure the distances shown on the following diagram. Find how wide the pond is to the nearest tenth of a foot.

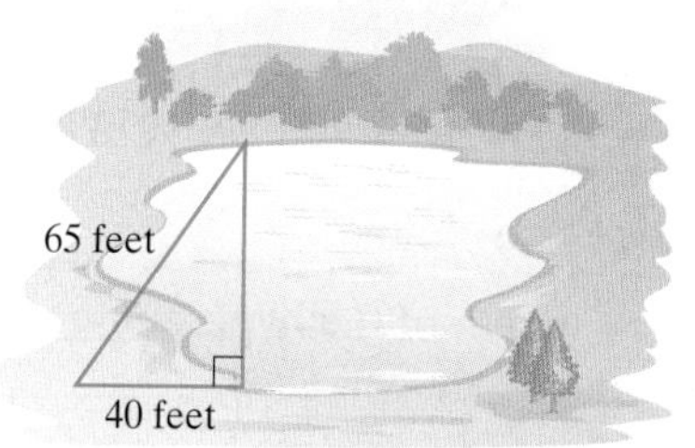

Answer

3. 51.2 ft

EXAMPLE 3 Finding a Distance

A surveyor must determine the distance across a lake at points P and Q as shown in the figure. To do this, she finds a third point R perpendicular to line PQ. If the length of $\overline{PR}$ is 320 feet and the length of $\overline{QR}$ is 240 feet, what is the distance across the lake? Approximate this distance to the nearest whole foot.

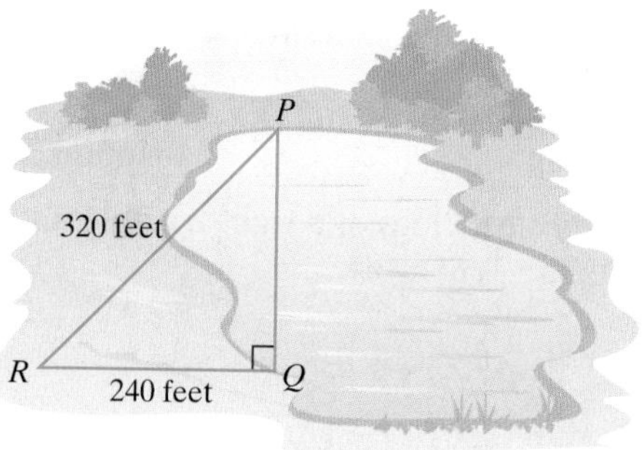

Solution:

1. UNDERSTAND. Read and reread the problem. We will set up the problem using the Pythagorean theorem. By creating a line perpendicular to line PQ, the surveyor deliberately constructed a right triangle. The hypotenuse, $\overline{PR}$, has a length of 320 feet, so we let $c = 320$ in the Pythagorean theorem. The side $\overline{QR}$ is one of the legs, so we let $a = 240$ and $b =$ the unknown length.

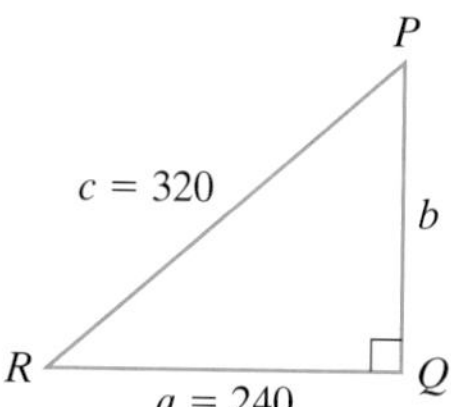

2. TRANSLATE.

$a^2 + b^2 = c^2$ Use the Pythagorean theorem.

$240^2 + b^2 = 320^2$ Let $a = 240$ and $c = 320$.

3. SOLVE.

$57{,}600 + b^2 = 102{,}400$

$b^2 = 44{,}800$ Subtract 57,600 from both sides.

$b = \sqrt{44{,}800}$ Use the definition of principal square root.

4. INTERPRET.

Check: See that $240^2 + (\sqrt{44{,}800})^2 = 320^2$.

State: The distance across the lake is *exactly* $\sqrt{44{,}800}$ feet. The surveyor can now use a calculator to find that $\sqrt{44{,}800}$ feet is *approximately* 211.6601 feet, so the distance across the lake is roughly 212 feet. ●

B Using Formulas Containing Radicals

The Pythagorean theorem is an extremely important result in mathematics and should be memorized. But there are other applications involving formulas containing radicals that are not quite as well known, such as the velocity formula used in the next example.

EXAMPLE 4 Finding the Velocity of an Object

A formula used to determine the velocity v, in feet per second, of an object after it has fallen a certain height (neglecting air resistance) is $v = \sqrt{2gh}$, where g is the acceleration due to gravity and h is the height the object has fallen. On Earth, the acceleration g due to gravity is approximately 32 feet per second per second. Find the velocity of a watermelon after it has fallen 5 feet.

Solution: We are told that $g = 32$ feet per second per second. To find the velocity v when $h = 5$ feet, we use the velocity formula.

$$v = \sqrt{2gh} \quad \text{Use the velocity formula.}$$

$$= \sqrt{2 \cdot 32 \cdot 5} \quad \text{Substitute known values.}$$

$$= \sqrt{320}$$

$$= 8\sqrt{5} \quad \text{Simplify the radicand.}$$

The velocity of the watermelon after it falls 5 feet is *exactly* $8\sqrt{5}$ feet per second, or *approximately* 17.9 feet per second.

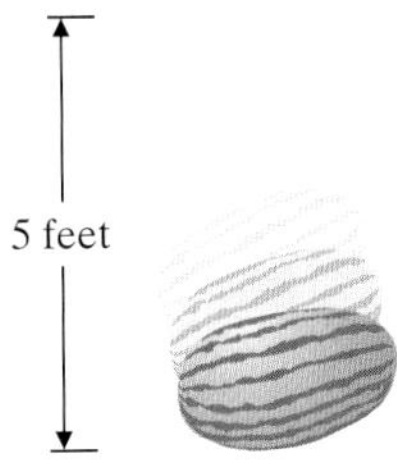

Practice Problem 4

Use the formula from Example 4 and find the velocity of an object after it has fallen 20 feet.

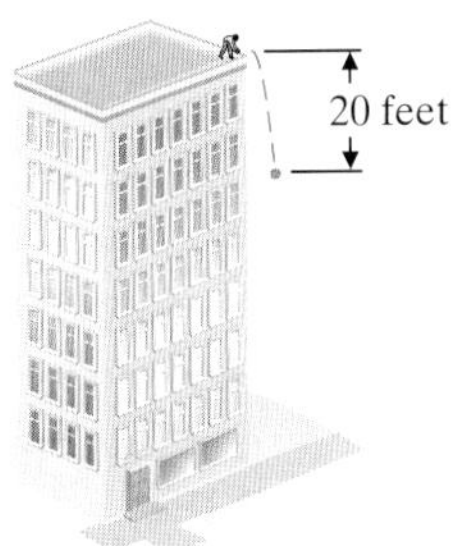

Answer

4. $16\sqrt{5}$ ft per sec ≈ 35.8 ft per sec

STUDY SKILLS REMINDER

Are you prepared for a test on Chapter 15?

Below I have listed some *common trouble areas* for topics covered in Chapter 15. After studying for your test—but before taking your test—read these.

- Do you understand the difference between $\sqrt{3} \cdot \sqrt{2}$ and $\sqrt{3} + \sqrt{2}$?

  $\sqrt{3} \cdot \sqrt{2} = \sqrt{3 \cdot 2} = \sqrt{6}$

  $\sqrt{3} + \sqrt{2}$ cannot be simplified further. The terms are unlike terms.

- Do you understand the difference between rationalizing the denominator of $\frac{\sqrt{3}}{\sqrt{7}}$ and rationalizing the denominator of $\frac{\sqrt{3}}{\sqrt{7} + 1}$?

  $$\frac{\sqrt{3}}{\sqrt{7}} = \frac{\sqrt{3} \cdot \sqrt{7}}{\sqrt{7} \cdot \sqrt{7}} = \frac{\sqrt{21}}{7}$$

  $$\frac{\sqrt{3}}{\sqrt{7} + 1} = \frac{\sqrt{3}(\sqrt{7} - 1)}{(\sqrt{7} + 1)(\sqrt{7} - 1)}$$

  $$= \frac{\sqrt{3}(\sqrt{7} - 1)}{7 - \sqrt{7} + \sqrt{7} - 1} = \frac{\sqrt{3}(\sqrt{7} - 1)}{6}$$

- To solve an equation containing a radical, don't forget to first isolate the radical

  $\sqrt{x} - 10 = -4$

  $\sqrt{x} = 6$ Isolate the radical.

  $(\sqrt{x})^2 = 6^2$ Square both sides.

  $x = 36$ Simplify.

Make sure you check the proposed solution in the original equation.

Remember: This is simply a listing of a few common trouble areas. For a review of Chapter 15, see the Highlights and Chapter Review at the end of the chapter.

Name ______________________ Section __________ Date __________

EXERCISE SET 15.6

Use the Pythagorean theorem to find the length of the unknown side of each right triangle. Give an exact answer and a two-decimal-place approximation. See Examples 1 and 2.

1.

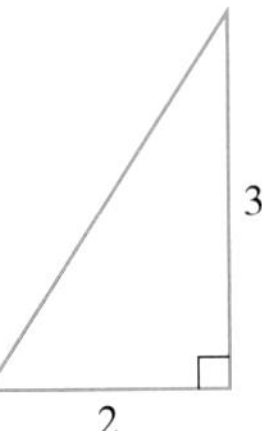

2.

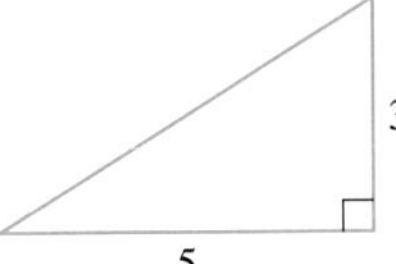

3.

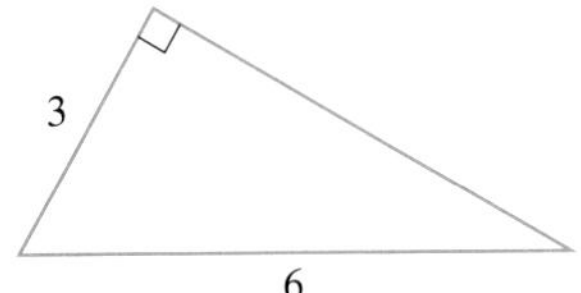

4.

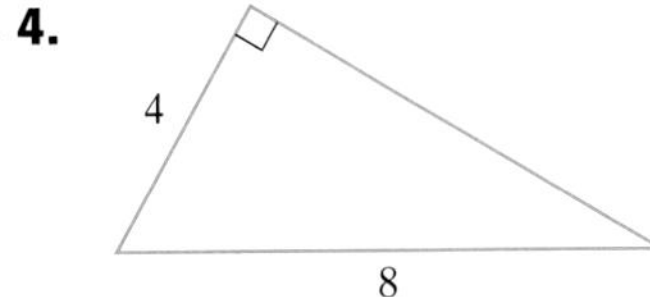

5.

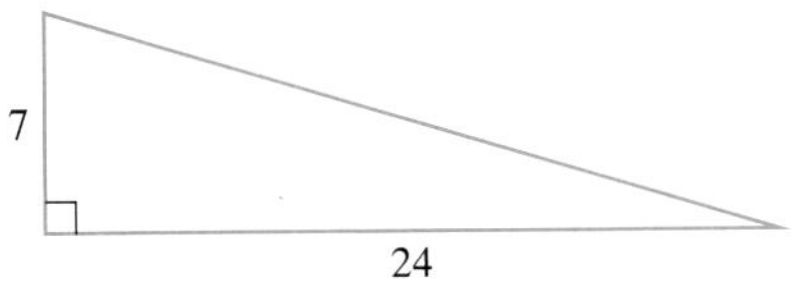

6.

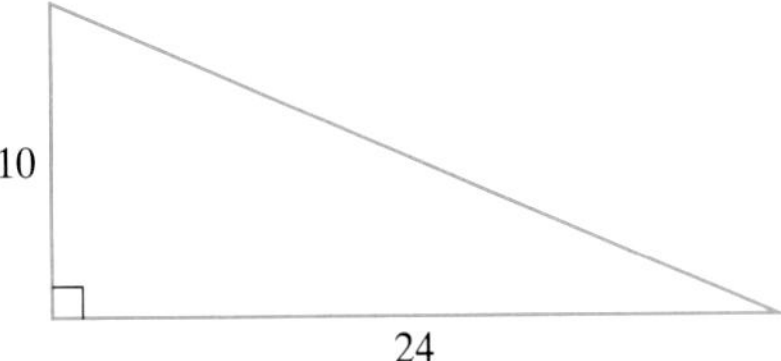

7.

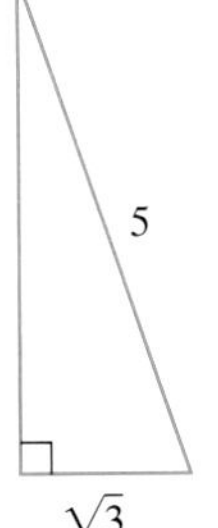

8.

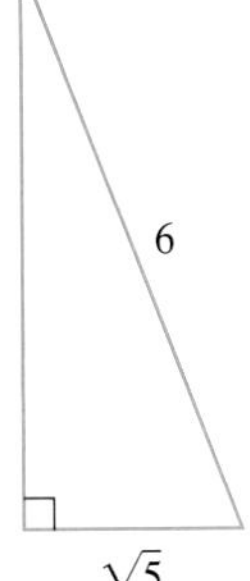

9. △

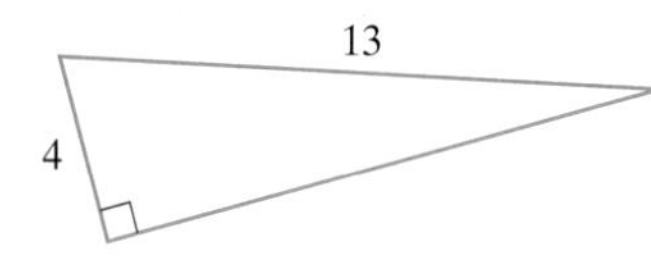

△ **10.**

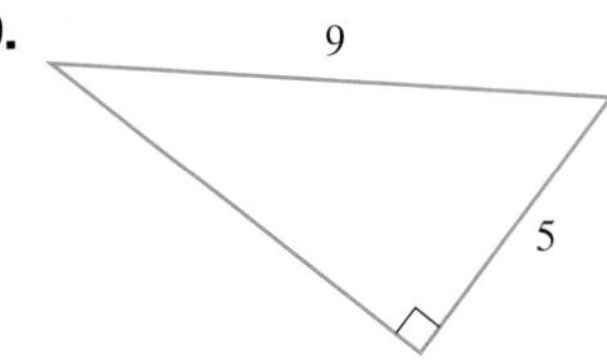

Find the length of the unknown side of each right triangle with sides a, b, and c, where c is the hypotenuse. See Examples 1 and 2. Give an exact answer and a two-decimal-place approximation.

△ **11.** $a = 4, b = 5$

△ **12.** $a = 2, b = 7$

△ **13.** $b = 2, c = 6$

△ **14.** $b = 1, c = 5$

△ **15.** $a = \sqrt{10}, c = 10$

△ **16.** $a = \sqrt{7}, c = \sqrt{35}$

Solve each problem. See Example 3.

17. △ A wire is used to anchor a 20-foot-high pole. One end of the wire is attached to the top of the pole. The other end is fastened to a stake five feet away from the bottom of the pole. Find the length of the wire, to the nearest tenth of a foot.

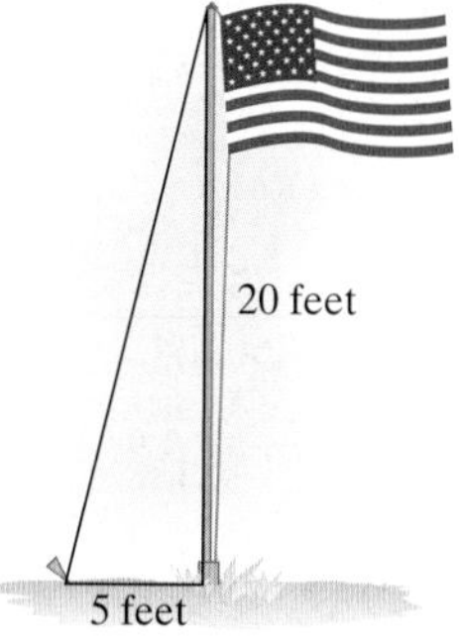

△ **18.** Jim Spivey needs to connect two underground pipelines, which are offset by 3 feet, as pictured in the diagram. Neglecting the joints needed to join the pipes, find the length of the shortest possible connecting pipe rounded to the nearest hundredth of a foot.

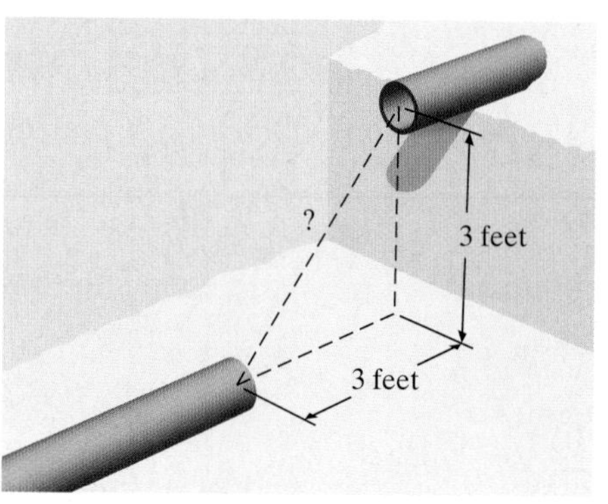

△ **19.** Robert Weisman needs to attach a diagonal brace to a rectangular frame in order to make it structurally sound. If the framework is 6 feet by 10 feet, find how long the brace needs to be to the nearest tenth of a foot.

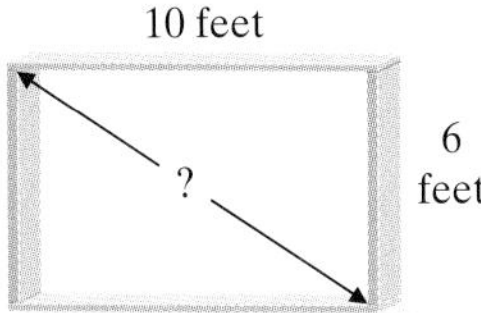

△ **20.** Elizabeth Kaster is flying a kite. She let out 80 feet of string and attached the string to a stake in the ground. The kite is now directly above her brother Mike, who is 32 feet away from Elizabeth. Find the height of the kite to the nearest foot.

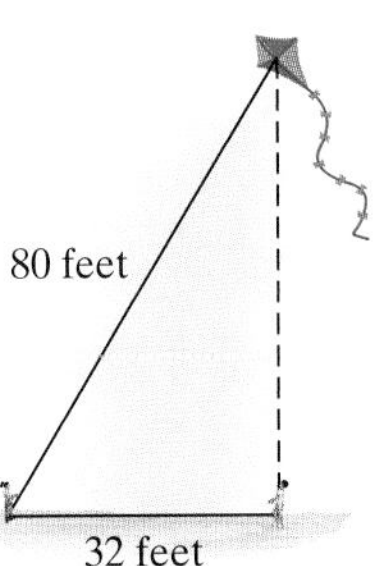

B *Solve each problem. See Example 4.*

△ **21.** For a square-based pyramid, the formula $b = \sqrt{\frac{3V}{h}}$ describes the relationship between the length b of one side of the base, the volume V, and the height h. Find the volume if each side of the base is 6 feet long, and the pyramid is 2 feet high.

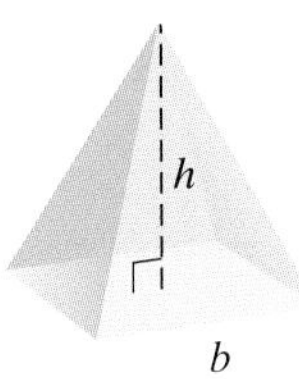

22. The formula $t = \frac{\sqrt{d}}{4}$ relates the distance d, in feet, that an object falls in t seconds, assuming that air resistance does not slow down the object. Find how long, to the nearest hundredth of a second, it takes an object to reach the ground from the top of the Sears Tower in Chicago, a distance of 1730 feet. (*Source*: Council on Tall Buildings and Urban Habitat)

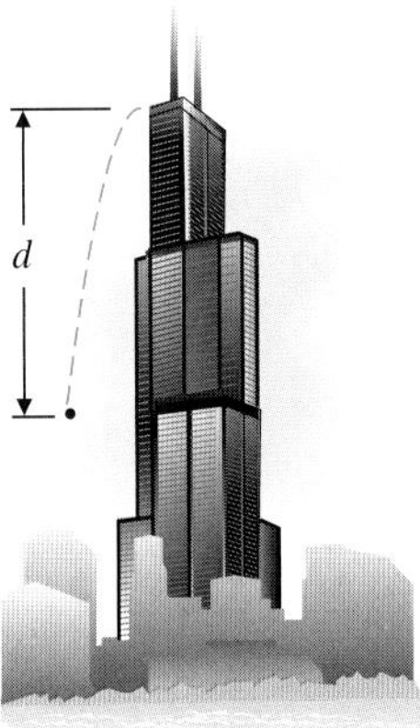

23. Police use the formula $s = \sqrt{30fd}$ to estimate the speed s of a car just before it skidded. In this formula, the speed s is measured in miles per hour, d represents the distance the car skidded in feet and f represents the coefficient of friction. The value of f depends on the type of road surface, and for wet concrete f is 0.35. Find how fast a car was moving if it skidded 280 feet on wet concrete. Round your result to the nearest mile per hour.

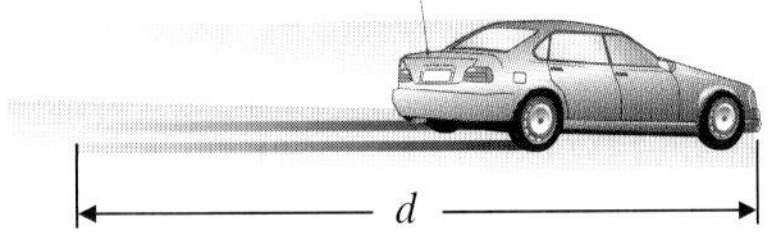

24. The coefficient of friction of a certain dry road is 0.95. Use the formula in Exercise 23 to find how far a car will skid on this dry road if it is traveling at a rate of 60 mph. Round the length to the nearest foot.

25. The formula $v = \sqrt{2.5r}$ can be used to estimate the maximum safe velocity, v, in miles per hour, at which a car can travel if it is driven along a curved road with a **radius of curvature** r in feet. Find the maximum safe speed to the nearest whole number if a cloverleaf exit on an expressway has a radius of curvature of 300 feet.

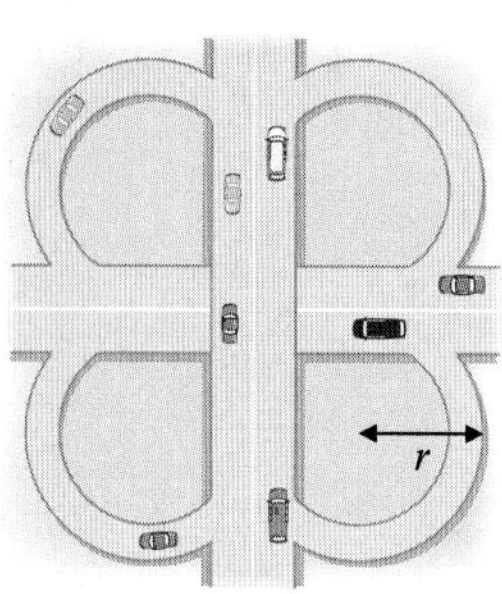

26. Use the formula from Exercise 25 to find the radius of curvature if the safe velocity is 30 mph.

△ **27.** The maximum distance d in kilometers that you can see from a height of h meters is given by $d = 3.5\sqrt{h}$. Find how far you can see from the top of the Bank One Tower in Indianapolis, a height of 285.4 meters. Round to the nearest tenth of a kilometer. (*Source: World Almanac and Book of Facts*)

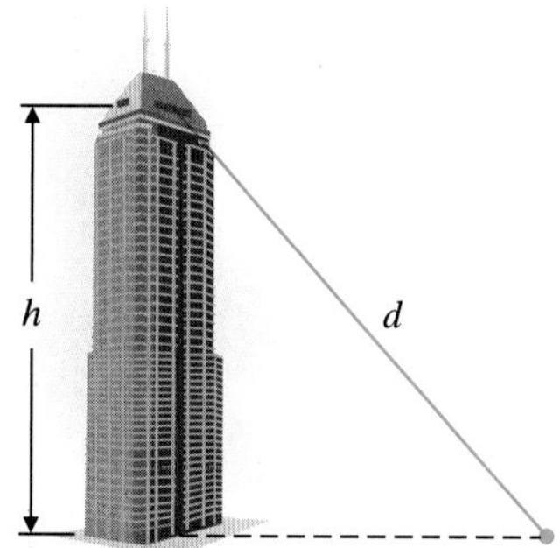

△ **28.** Use the formula from Exercise 27 to find how far you can see from the top of the Chase Tower Building in Houston, Texas, a height of 305 meters. Round to the nearest tenth of a kilometer. (*Source*: Council on Tall Buildings and Urban Habitat)

Review and Preview

Find two numbers whose square is the given number. See Section 15.1.

29. 9 **30.** 25 **31.** 100 **32.** 49 **33.** 64 **34.** 121

Combining Concepts

For each triangle, find the length of x.

△ **35.**

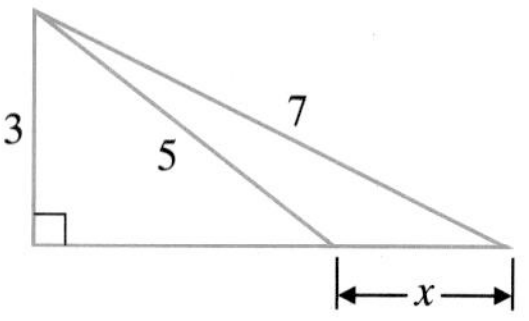

△ **36.**

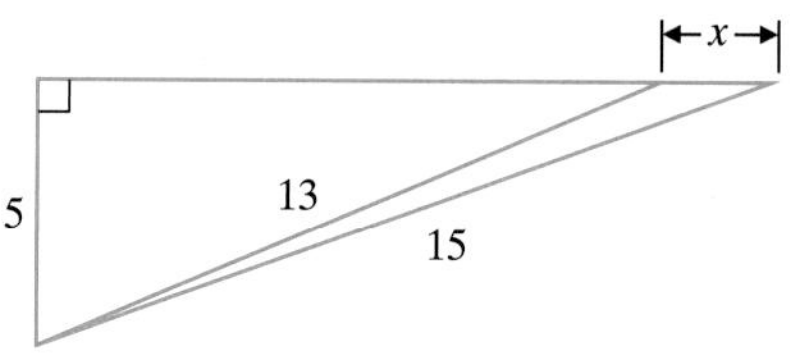

△ **37.** Mike and Sandra Hallahan leave the seashore at the same time. Mike drives northward at a rate of 30 miles per hour, while Sandra drives west at 60 mph. Find how far apart they are after 3 hours to the nearest mile.

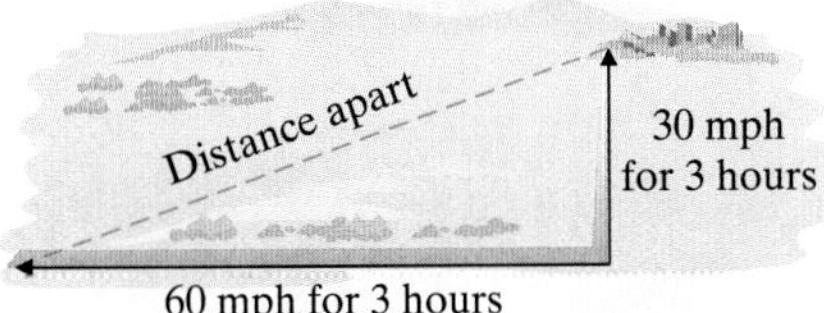

△ **38.** Railroad tracks are invariably made up of relatively short sections of rail connected by expansion joints. To see why this construction is necessary, consider a single rail 100 feet long (or 1200 inches). On an extremely hot day, suppose it expands 1 inch in the hot sun to a new length of 1201 inches. Theoretically, the track would bow upward as pictured.

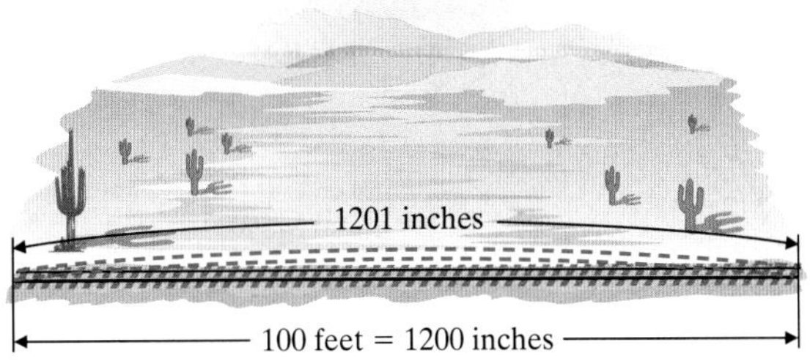

Let us approximate the bulge in the railroad this way.

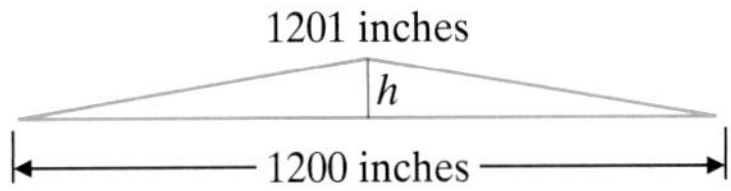

Calculate the height h of the bulge to the nearest tenth of an inch.

39. Based on the results of Exercise 38, explain why railroads use short sections of rail connected by expansion joints.

Internet Excursions

Go To: http://www.prenhall.com/martin-gay_prealgebra

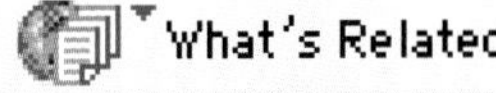

The Central Intelligence Agency of the United States publishes an annual factbook as a basic reference for 267 nations around the world. It includes a variety of information for each nation, such as descriptions of and data on the nation's geography, people, government, transportation, economy, communications, and defense. The given World Wide Web address will provide you with access to the CIA Factbook, or a related site. You can look up data for any country. The section on geography in a country's listing gives information on the highest and lowest points in the country.

40. Choose any country in North America. Look up that country on the CIA Factbook Web site. Using data in the geography section, complete the following information: Country: ______; Highest point: ______; Elevation: ______. Use the formula $d = 3.5\sqrt{h}$, where d is the maximum distance (in kilometers) that you can see from a height of h meters above the surface of the Earth, to find the distance that can be seen from the highest point in the country that you chose.

41. Choose any country in Asia. Look up that country on the CIA Factbook Web site. Using data in the geography section, complete the following information: Country: ______; Highest point: ______; Elevation: ______. Use the formula from Exercise 40 to find the distance that can be seen from the highest point in the country that you chose.

CHAPTER 15 ACTIVITY Investigating the Dimensions of Cylinders

MATERIALS:

- calculator
- several empty cans of different sizes
- 2-cup (16-fluid-ounce) transparent measuring cup with metric markings (in milliliters)
- metric ruler
- water

This activity may be completed by working in groups or individually.

The radius r of a cylinder is related to its volume V (in cubic units) and its height h by the formula $r = \sqrt{\frac{V}{\pi h}}$. You will investigate the radii of several cylindrical cans by completing the following table.

Can	Volume (ml)	Height (cm)	Calculated Radius (cm)	Measured Radius (cm)
A				
B				
C				
D				

1. For each can, measure its volume by filling it with water and pouring the water into the measuring cup. Find the volume of the water in milliliters (ml). Record the volumes of the cans in the table. (Remember that 1 ml $=$ 1 cm^3).
2. For each can, use a ruler to measure its height in centimeters (cm). Record the heights in the table.
3. Use the formula $r = \sqrt{\frac{V}{\pi h}}$ to calculate an estimate of each can's radius. Record these calculated radii in the table.
4. Try to measure the radius of each can and record these measured radii in the table. $\left(\text{Remember that radius} = \frac{1}{2}\text{diameter.}\right)$
5. How close are the values of the calculated radius and the measured radius of each can? What factors could account for the differences?

Chapter 15 VOCABULARY CHECK

Fill in each blank with one of the words or phrases listed below.

index	radicand	like radicals
rationalizing the denominator	conjugate	
principal square root	radical	

1. The expressions $5\sqrt{x}$ and $7\sqrt{x}$ are examples of ______________.
2. In the expression $\sqrt[3]{45}$ the number 3 is the ________, the number 45 is the ___________, and $\sqrt{}$ is called the __________ sign.
3. The ____________ of $(a + b)$ is $(a - b)$.
4. The ____________________ of 25 is 5.
5. The process of eliminating the radical in the denominator of a radical expression is called ____________________________.

CHAPTER 15 Highlights

DEFINITIONS AND CONCEPTS | **EXAMPLES**

Section 15.1 Introduction to Radicals

The **positive or principal square root** of a positive number a is written as $\sqrt{a}$. The **negative square root** of a is written as $-\sqrt{a}$. $\sqrt{a} = b$ only if $b^2 = a$ and $b > 0$.

$\sqrt{25} = 5$ $\qquad$ $\sqrt{100} = 10$

$-\sqrt{9} = -3$ $\qquad$ $\sqrt{\dfrac{4}{49}} = \dfrac{2}{7}$

A square root of a negative number is not a real number.

$\sqrt{-4}$ is not a real number.

The **cube root** of a real number a is written as $\sqrt[3]{a}$ and $\sqrt[3]{a} = b$ only if $b^3 = a$.

$\sqrt[3]{64} = 4$ $\qquad$ $\sqrt[3]{-8} = -2$

The ***n*th root** of a number a is written as $\sqrt[n]{a}$ and $\sqrt[n]{a} = b$ only if $b^n = a$.

$\sqrt[4]{81} = 3$

$\sqrt[5]{-32} = -2$

The natural number n is called the **index,** the symbol $\sqrt{}$ is called a **radical,** and the expression within the radical is called the **radicand.**

(Note: If the index is even, the radicand must be nonnegative for the root to be a real number.)

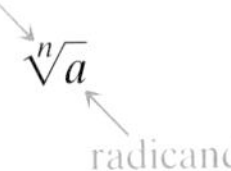

Section 15.2 Simplifying Radicals

PRODUCT RULE FOR RADICALS

If $\sqrt{a}$ and $\sqrt{b}$ are real numbers, then

$$\sqrt{a \cdot b} = \sqrt{a} \cdot \sqrt{b}$$

A square root is in **simplified form** if the radicand contains no perfect square factors other than 1. To simplify a square root, factor the radicand so that one of its factors is a perfect square factor.

$$\sqrt{45} = \sqrt{9 \cdot 5} = \sqrt{9} \cdot \sqrt{5} = 3\sqrt{5}$$

QUOTIENT RULE FOR RADICALS

If $\sqrt{a}$ and $\sqrt{b}$ are real numbers and $b \neq 0$, then

$$\sqrt{\frac{a}{b}} = \frac{\sqrt{a}}{\sqrt{b}}$$

$$\sqrt{\frac{18}{x^6}} = \frac{\sqrt{9 \cdot 2}}{\sqrt{x^6}} = \frac{\sqrt{9} \cdot \sqrt{2}}{x^3} = \frac{3\sqrt{2}}{x^3}$$

Definitions and Concepts	**Examples**
Section 15.3 Adding and Subtracting Radicals	
Like radicals are radical expressions that have the same index and the same radicand.	$5\sqrt{2}, -7\sqrt{2}, \sqrt{2}$
To **combine like radicals** use the distributive property.	$2\sqrt{7} - 13\sqrt{7} = (2 - 13)\sqrt{7} = -11\sqrt{7}$ $\sqrt{8} + \sqrt{50} = 2\sqrt{2} + 5\sqrt{2} = 7\sqrt{2}$
Section 15.4 Multiplying and Dividing Radicals	
The product and quotient rules for radicals may be used to simplify products and quotients of radicals.	Perform each indicated operation and simplify. Multiply. $\sqrt{2}\cdot\sqrt{8} = \sqrt{16} = 4$ $(\sqrt{3x} + 1)(\sqrt{5} - \sqrt{3})$ $= \sqrt{15x} - \sqrt{9x} + \sqrt{5} - \sqrt{3}$ $= \sqrt{15x} - 3\sqrt{x} + \sqrt{5} - \sqrt{3}$ Divide. $\frac{\sqrt{20}}{\sqrt{2}} = \sqrt{\frac{20}{2}} = \sqrt{10}$
The process of eliminating the radical in the denominator of a radical expression is called **rationalizing the denominator.**	Rationalize the denominator. $\frac{5}{\sqrt{11}} = \frac{5\cdot\sqrt{11}}{\sqrt{11}\cdot\sqrt{11}} = \frac{5\sqrt{11}}{11}$
The **conjugate** of $a + b$ is $a - b$.	The conjugate of $2 + \sqrt{3}$ is $2 - \sqrt{3}$.
To rationalize a denominator that is a sum or difference of radicals, multiply the numerator and the denominator by the conjugate of the denominator.	Rationalize the denominator. $\frac{5}{6 - \sqrt{5}} = \frac{5(6 + \sqrt{5})}{(6 - \sqrt{5})(6 + \sqrt{5})}$ $= \frac{5(6 + \sqrt{5})}{36 - 5}$ $= \frac{5(6 + \sqrt{5})}{31}$
Section 15.5 Solving Equations Containing Radicals	
To Solve a Radical Equation Containing Square Roots **Step 1.** Get one radical by itself on one side of the equation. **Step 2.** Square both sides of the equation. **Step 3.** Simplify both sides of the equation. **Step 4.** If the equation still contains a radical term, repeat Steps 1 through 3. **Step 5.** Solve the equation. **Step 6.** Check solutions in the original equation.	Solve: $\sqrt{2x - 1} - x = -2$ $\sqrt{2x - 1} = x - 2$ $(\sqrt{2x - 1})^2 = (x - 2)^2$ Square both sides. $2x - 1 = x^2 - 4x + 4$ $0 = x^2 - 6x + 5$ $0 = (x - 1)(x - 5)$ Factor. $x - 1 = 0$ or $x - 5 = 0$ $x = 1$ $x = 5$ Solve. Check both proposed solutions in the original equation. Here, 5 checks but 1 does not. The only solution is 5.

DEFINITIONS AND CONCEPTS | **EXAMPLES**

Section 15.6 Radical Equations and Problem Solving

PROBLEM-SOLVING STEPS

1. UNDERSTAND. Read and reread the problem.

A rain gutter is to be mounted on the eaves of a house 15 feet above the ground. A garden is adjacent to the house so that the closest a ladder can be placed to the house is 6 feet. How long a ladder is needed for installing the gutter?

Let $x =$ the length of the ladder.

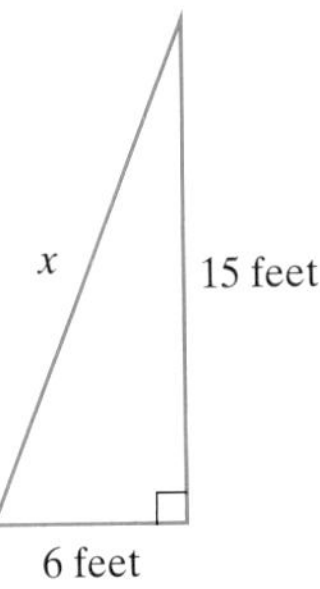

2. TRANSLATE.

Here, we use the Pythagorean theorem. The unknown length x is the hypotenuse.

In words:

$$(\text{leg})^2 \quad + \quad (\text{leg})^2 \quad = \quad (\text{hypotenuse})^2$$

3. SOLVE.

Translate:

$$6^2 + 15^2 = x^2$$
$$36 + 225 = x^2$$
$$261 = x^2$$
$$\sqrt{261} = x \quad \text{or} \quad x = 3\sqrt{29}$$

4. INTERPRET.

Check and state. The ladder needs to be $3\sqrt{29}$ feet or approximately 16.2 feet long.

STUDY SKILLS REMINDER

Are you prepared for your final exam?

To prepare for your final exam, try the following study techniques.

- Review the material that you will be responsible for on your exam. Also check your notebook for any lecture notes that you highlighted.
- Review any formulas that you may need to memorize.
- Check to see if your instructor or math department will be conducting a final exam review.
- Check with your instructor to see whether there are final exams from previous semesters/quarters that are available to students for study.
- Use your previously taken tests as a practice final exam. To do so, rewrite the test questions in mixed order on blank sheets of paper. This will help you prepare for exam conditions.
- If you are unsure of a few topics, see your instructor or visit a learning lab for further assistance. Also, viewing the video segment of a troublesome section will help.
- If you need further exercises to work, try the chapter tests and cumulative reviews at the end of appropriate chapters.

Good luck!

Name ______________________ Section __________ Date ____________

Chapter 15 Review

(15.1) *Find each root.*

1. $\sqrt{81}$

2. $-\sqrt{49}$

3. $\sqrt[3]{27}$

4. $\sqrt[4]{16}$

5. $-\sqrt{\frac{9}{64}}$

6. $\sqrt{\frac{36}{81}}$

7. $\sqrt[4]{16}$

8. $\sqrt[3]{-8}$

9. Which radical(s) is not a real number?

a. $\sqrt{4}$

b. $-\sqrt{4}$

c. $\sqrt{-4}$

d. $\sqrt[3]{-4}$

10. Which radical(s) is not a real number?

a. $\sqrt{-5}$

b. $\sqrt[3]{-5}$

c. $\sqrt[4]{-5}$

d. $\sqrt[5]{-5}$

Find each root. Assume that all variables represent positive numbers.

11. $\sqrt{x^{12}}$

12. $\sqrt{x^8}$

13. $\sqrt{9y^2}$

14. $\sqrt{25x^4}$

(15.2) *Simplify each expression using the product rule. Assume that all variables represent positive numbers.*

15. $\sqrt{40}$

16. $\sqrt{24}$

17. $\sqrt{54}$

18. $\sqrt{88}$

19. $\sqrt{x^5}$

20. $\sqrt{y^7}$

21. $\sqrt{20x^2}$

22. $\sqrt{50y^4}$

23. $\sqrt[3]{54}$

24. $\sqrt[3]{88}$

Simplify each expression using the quotient rule. Assume that all variables represent positive numbers.

25. $\sqrt{\frac{18}{25}}$

26. $\sqrt{\frac{75}{64}}$

27. $-\sqrt{\frac{50}{9}}$

28. $-\sqrt{\frac{12}{49}}$

29. $\sqrt{\frac{11}{x^2}}$

30. $\sqrt{\frac{7}{y^4}}$

31. $\sqrt{\frac{y^5}{100}}$

32. $\sqrt{\frac{x^3}{81}}$

(15.3) *Add or subtract by combining like radicals.*

33. $5\sqrt{2} - 8\sqrt{2}$

34. $\sqrt{3} - 6\sqrt{3}$

35. $6\sqrt{5} + 3\sqrt{6} - 2\sqrt{5} + \sqrt{6}$

36. $-\sqrt{7} + 8\sqrt{2} - \sqrt{7} - 6\sqrt{2}$

Add or subtract by simplifying each radical and then combining like terms. Assume that all variables represent positive numbers.

37. $\sqrt{28} + \sqrt{63} + \sqrt{56}$

38. $\sqrt{75} + \sqrt{48} - \sqrt{16}$

39. $\sqrt{\frac{5}{9}} - \sqrt{\frac{5}{36}}$

40. $\sqrt{\frac{11}{25}} + \sqrt{\frac{11}{16}}$

41. $\sqrt{45x^2} + 3\sqrt{5x^2} - 7x\sqrt{5} + 10$

42. $\sqrt{50x} - 9\sqrt{2x} + \sqrt{72x} - \sqrt{3x}$

(15.4) *Multiply and simplify if possible. Assume that all variables represent positive numbers.*

43. $\sqrt{3} \cdot \sqrt{6}$

44. $\sqrt{5} \cdot \sqrt{15}$

45. $\sqrt{2}(\sqrt{5} - \sqrt{7})$

46. $\sqrt{5}(\sqrt{11} + \sqrt{3})$

47. $(\sqrt{3} + 2)(\sqrt{6} - 5)$

48. $(\sqrt{5} + 1)(\sqrt{5} - 3)$

49. $(\sqrt{x} - 2)^2$

50. $(\sqrt{y} + 4)^2$

Divide and simplify if possible. Assume that all variables represent positive numbers.

51. $\frac{\sqrt{27}}{\sqrt{3}}$

52. $\frac{\sqrt{20}}{\sqrt{5}}$

53. $\frac{\sqrt{160}}{\sqrt{8}}$

54. $\frac{\sqrt{96}}{\sqrt{3}}$

55. $\frac{\sqrt{30x^6}}{\sqrt{2x^3}}$

56. $\frac{\sqrt{54x^5y^2}}{\sqrt{3xy^2}}$

Rationalize each denominator and simplify.

57. $\frac{\sqrt{2}}{\sqrt{11}}$

58. $\frac{\sqrt{3}}{\sqrt{13}}$

59. $\sqrt{\frac{5}{6}}$

60. $\sqrt{\frac{7}{10}}$

61. $\frac{1}{\sqrt{5x}}$

62. $\frac{5}{\sqrt{3y}}$

63. $\sqrt{\frac{3}{x}}$

64. $\sqrt{\frac{6}{y}}$

65. $\dfrac{3}{\sqrt{5} - 2}$

66. $\dfrac{8}{\sqrt{10} - 3}$

67. $\dfrac{\sqrt{2} + 1}{\sqrt{3} - 1}$

68. $\dfrac{\sqrt{3} - 2}{\sqrt{5} + 2}$

69. $\dfrac{10}{\sqrt{x} + 5}$

70. $\dfrac{8}{\sqrt{x} - 1}$

(15.5) *Solve each radical equation.*

71. $\sqrt{2x} = 6$

72. $\sqrt{x + 3} = 4$

73. $\sqrt{x} + 3 = 8$

74. $\sqrt{x} + 8 = 3$

75. $\sqrt{2x + 1} = x - 7$

76. $\sqrt{3x + 1} = x - 1$

77. $\sqrt{x} + 3 = \sqrt{x + 15}$

78. $\sqrt{x - 5} = \sqrt{x} - 1$

(15.6) *Use the Pythagorean theorem to find the length of each unknown side. Give an exact answer and a two-decimal-place approximation.*

△ **79.**

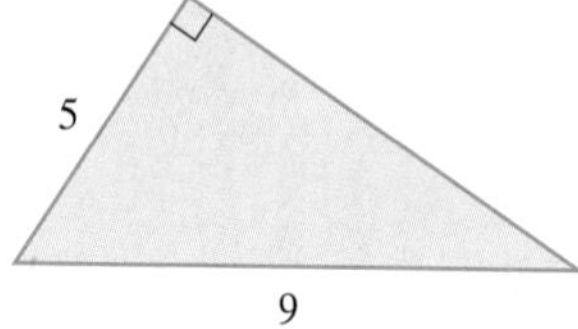

△ **80.**

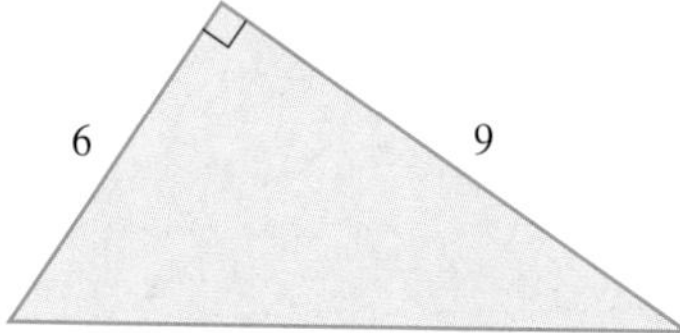

△ **81.** Romeo is standing 20 feet away from the wall below Juliet's balcony during a school play. Juliet is on the balcony, 12 feet above the ground. Find how far apart Romeo and Juliet are.

△ **82.** The diagonal of a rectangle is 10 inches long. If the width of the rectangle is 5 inches, find the length of the rectangle.

Use the formula $r = \sqrt{\dfrac{S}{4\pi}}$, *where* r = *the radius of a sphere and* S = *the surface area of the sphere, for Exercises 83 and 84.*

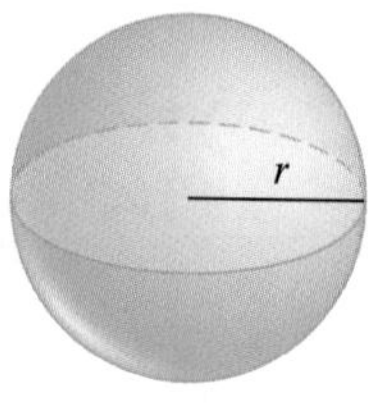

△ **83.** Find the radius of a sphere to the nearest tenth of an inch if the surface area is 72 square inches.

△ **84.** Find the exact surface area of a sphere if its radius is 6 inches. (Do not approximate π.)

Name ______________________ Section __________ Date __________

Chapter 15 Test

Remember to check your answers and use the Chapter Test Prep Video to view solutions.

Simplify each radical. Indicate if the radical is not a real number. Assume that x represents a positive number.

1. $\sqrt{16}$

2. $\sqrt[3]{-125}$

3. $\sqrt[4]{81}$

4. $\sqrt{\frac{9}{16}}$

5. $\sqrt[4]{-81}$

6. $\sqrt{x^{10}}$

Simplify each radical. Assume that all variables represent positive numbers.

7. $\sqrt{54}$

8. $\sqrt{92}$

9. $\sqrt{y^7}$

10. $\sqrt{24x^8}$

11. $\sqrt[3]{27}$

12. $\sqrt[3]{16}$

13. $\sqrt{\frac{5}{16}}$

14. $\sqrt{\frac{y^3}{25}}$

Perform each indicated operation. Assume that all variables represent positive numbers.

15. $\sqrt{13} + \sqrt{13} - 4\sqrt{13}$

16. $\sqrt{18} - \sqrt{75} + 7\sqrt{3} - \sqrt{8}$

17. $\sqrt{\frac{3}{4}} + \sqrt{\frac{3}{25}}$

18. $\sqrt{7} \cdot \sqrt{14}$

19. $\sqrt{5}(\sqrt{5} + 2\sqrt{7})$

Answers

1. ______
2. ______
3. ______
4. ______
5. ______
6. ______
7. ______
8. ______
9. ______
10. ______
11. ______
12. ______
13. ______
14. ______
15. ______
16. ______
17. ______
18. ______
19. ______

20. ______

21. ______

22. ______

23. ______

24. ______

25. ______

26. ______

27. ______

28. ______

29. ______

30. ______

31. ______

20. $(2\sqrt{x} + 3)(2\sqrt{x} - 3)$

21. $\frac{\sqrt{50}}{\sqrt{10}}$

22. $\frac{\sqrt{40x^4}}{\sqrt{2x}}$

Rationalize each denominator. Assume that all variables represent positive numbers.

23. $\sqrt{\frac{2}{3}}$

24. $\frac{8}{\sqrt{5y}}$

25. $\frac{8}{\sqrt{6} + 2}$

26. $\frac{1}{3 - \sqrt{x}}$

Solve each radical equation.

27. $\sqrt{x} + 8 = 11$

28. $\sqrt{3x - 6} = \sqrt{x + 4}$

29. $\sqrt{2x - 2} = x - 5$

△ **30.** Find the length of the unknown leg of the right triangle shown. Give an exact answer.

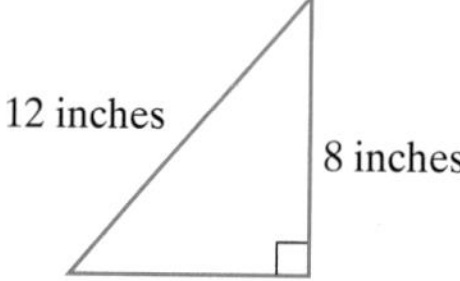

△ **31.** The formula $r = \sqrt{\frac{A}{\pi}}$ can be used to find the radius r of a circle given its area A. Use this formula to approximate the radius of the given circle. Round to two decimal places.

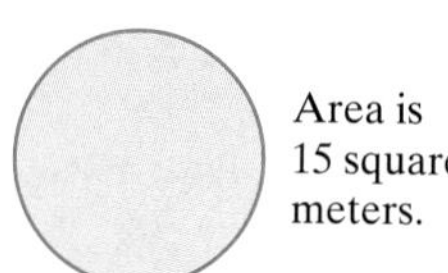

Name ______________________ Section __________ Date __________

Answers

Chapter 15 Cumulative Review

1. Round 736.2359 to the nearest tenth.

2. Round 328.174 to the nearest tenth.

3. Add: $23.85 + 1.604$

4. Add: $12.762 + 4.29$

5. Is -9 a solution of the equation $3.7y = -3.33$?

6. Is 6 a solution of the equation $2.8x = 16.8$?

7. Find: $\sqrt{\frac{1}{36}}$

8. Find: $\sqrt{\frac{4}{25}}$

9. Solve: $4(2x - 3) + 7 = 3x + 5$

10. Solve: $3(2 - 5x) + 24x = 12$

11. Write the following numbers in standard notation, without exponents.
    - **a.** 1.02×10^5
    - **b.** 7.358×10^{-3}
    - **c.** 8.4×10^7
    - **d.** 3.007×10^{-5}

12. Write the following numbers in standard notation, without exponents.
    - **a.** 8.26×10^4
    - **b.** 9.9×10^{-2}
    - **c.** 1.002×10^5
    - **d.** 8.039×10^{-3}

13. Multiply: $(3x + 2)(2x - 5)$

14. Multiply: $(5x - 1)(4x + 1)$

15. Factor $xy + 2x + 3y + 6$ by grouping.

16. Factor: $16x^3 - 28x^2 + 12x - 21$

17. Factor: $3x^2 + 11x + 6$

18. Factor: $9x^2 - 5x - 4$

Are there any values for x for which each expression is undefined?

19. **a.** $\frac{x}{x - 3}$
    b. $\frac{x^2 + 2}{x^2 - 3x + 2}$
    c. $\frac{x^3 - 6x^2 - 10x}{3}$

20. **a.** $\frac{x - 3}{x}$
    b. $\frac{x + 1}{5}$
    c. $\frac{x^2 - 3}{x^2 - 4}$

Answers:
1. ______
2. ______
3. ______
4. ______
5. ______
6. ______
7. ______
8. ______
9. ______
10. ______
11. a. ______ b. ______ c. ______ d. ______
12. a. ______ b. ______ c. ______ d. ______
13. ______
14. ______
15. ______
16. ______
17. ______
18. ______
19. a. ______ b. ______ c. ______
20. a. ______ b. ______ c. ______

21.
22.
23. a.
b.
24. a.
b.
25.
26.
27. see graph
28. see graph
29.
30.
31.
32.
33.
34.
35.
36.
37.
38.
39.
40.
41.
42.
43.
44.
45.
46.

21. Simplify: $\dfrac{x^2 + 4x + 4}{x^2 + 2x}$

22. Simplify: $\dfrac{16x^2 - 4y^2}{4x - 2y}$

Perform each indicated operation.

23. a. $\dfrac{a}{4} - \dfrac{2a}{8}$

b. $\dfrac{3}{10x^2} + \dfrac{7}{25x}$

24. a. $\dfrac{x}{5} - \dfrac{3x}{10}$

b. $\dfrac{9}{12a^2} + \dfrac{5}{16a}$

25. Solve: $\dfrac{4x}{x^2 + x - 30} + \dfrac{2}{x - 5} = \dfrac{1}{x + 6}$

26. Solve: $\dfrac{3}{x + 3} = \dfrac{12x + 19}{x^2 + 7x + 12} - \dfrac{5}{x + 4}$

27. Graph $y = -3$.

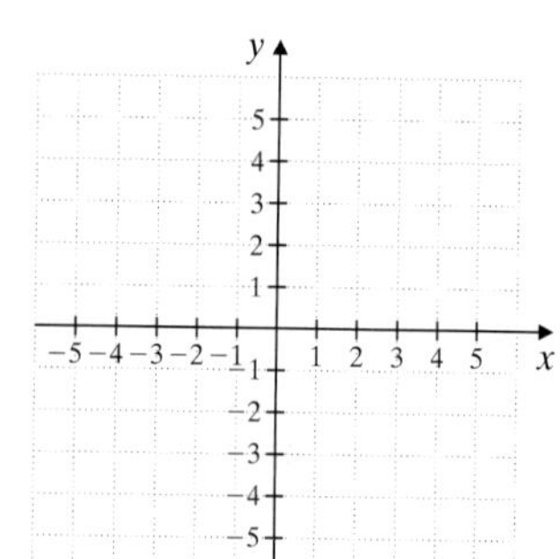

28. Graph $x = 2$.

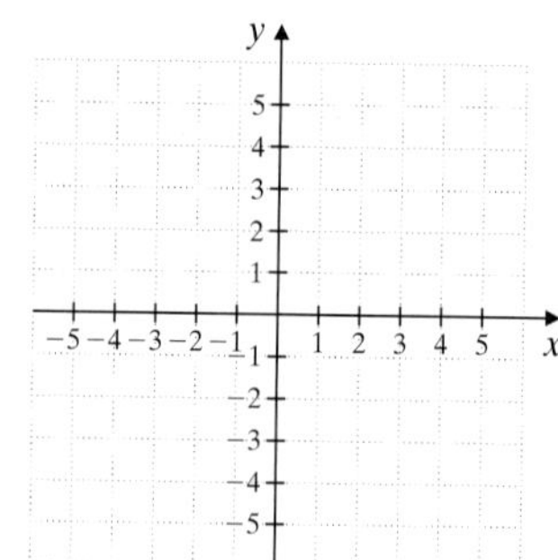

Find each cube root.

29. $\sqrt[3]{1}$

30. $\sqrt[3]{8}$

31. $\sqrt[3]{-27}$

32. $\sqrt[3]{-8}$

33. $\sqrt[3]{\dfrac{1}{125}}$

34. $\sqrt[3]{\dfrac{27}{64}}$

Simplify.

35. $\sqrt{54}$

36. $\sqrt{40}$

37. $\sqrt{200}$

38. $\sqrt{125}$

Add or subtract by first simplifying each radical.

39. $7\sqrt{12} - \sqrt{75}$

40. $\sqrt{75} + \sqrt{48}$

41. $2\sqrt{x^2} - \sqrt{25x} + \sqrt{x}$

42. $5\sqrt{x^2} + \sqrt{36x} + \sqrt{49x^2}$

Rationalize each denominator.

43. $\dfrac{2}{\sqrt{7}}$

44. $\dfrac{4}{\sqrt{5}}$

45. Solve: $\sqrt{x} = \sqrt{5x - 2}$

46. Solve: $\sqrt{x + 5} = x - 1$

Quadratic Equations

An important part of the study of algebra is learning to use methods for solving equations. Starting in Chapter 3, we presented techniques for solving linear equations in one variable. In Chapter 11, we solved quadratic equations in one variable by factoring the quadratic expressions. We now present other methods for solving quadratic equations in one variable.

It's a bird—it's a plane—no, . . . it's a Goodyear blimp! These widely recognized blimps are one of the best-known corporate symbols in the United States. Since 1925, the Goodyear Tire and Rubber Company has maintained a fleet of helium-filled lighter-than-air ships to carry out advertising and public relations functions. Today, Goodyear's fleet includes five blimps: Spirit of Goodyear based in Akron, Ohio; Stars & Stripes based in Pompano Beach, Florida; Eagle based in Carson, California; Spirit of Europe based in Wolverhampton, England; and Spirit of the Americas based in Sao Paulo, Brazil. The three U.S.-based Goodyear blimps measure 192 feet long and have top speeds of 50 mph in the air. Together, these five airships fly over 400,000 miles each year in their roles as aerial ambassadors for Goodyear. In Section 16.3, Exercise 62, we will use a method called the **quadratic formula** to predict when the net sales of the Goodyear Tire and Rubber Company will reach a certain level.

Answers

1. ________
2. ________
3. ________
4. ________
5. ________
6. ________
7. ________
8. ________
9. ________
10. ________
11. ________
12. ________
13. ________
14. ________
15. see graph
16. see graph
17. see graph
18. see graph

Name ________________ Section ________ Date ________

Chapter 16 Pretest

Solve by factoring.

1. $a^2 - 6a = 0$

2. $2x^2 - 11x = 6$

Solve using the square root property.

3. $b^2 = 144$

4. $(2y - 7)^2 = 24$

Solve by completing the square.

5. $x^2 - 14x + 48 = 0$

6. $3x^2 - 5x = 2$

Solve using the quadratic formula.

7. $x^2 - 6x - 27 = 0$

8. $m^2 - \frac{7}{4}m - \frac{3}{2} = 0$

Solve by the most appropriate method.

9. $(2x + 3)(x - 1) = 6$

10. $(5x + 3)^2 = 18$

11. $8x^2 + 18x + 9 = 0$

12. $m^2 - 6m = -3$

13. $\frac{1}{4}x^2 + x - \frac{1}{8} = 0$

14. $(y + 7)^2 - 5 = 0$

Graph the quadratic equations. Label the vertex and the intercept points with their coordinates.

15. $y = -3x^2$

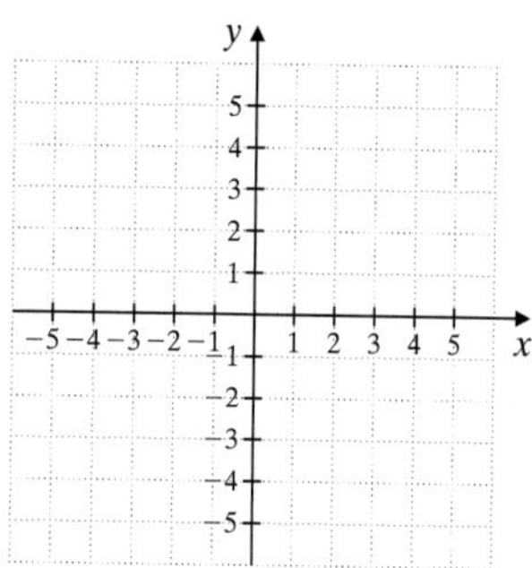

16. 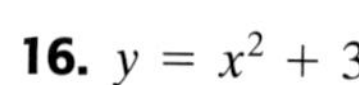

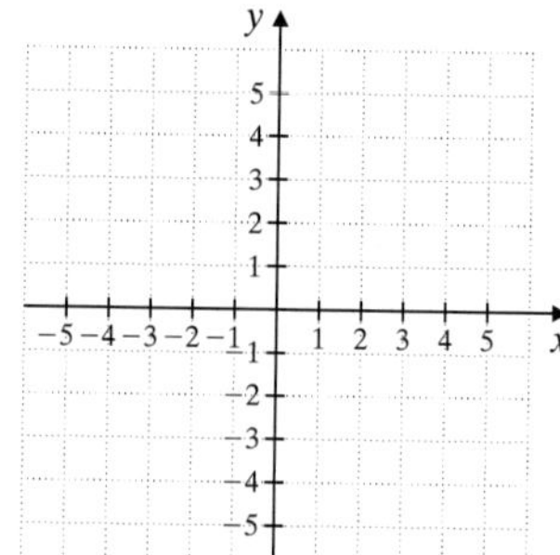

17. $y = x^2 + 4x$

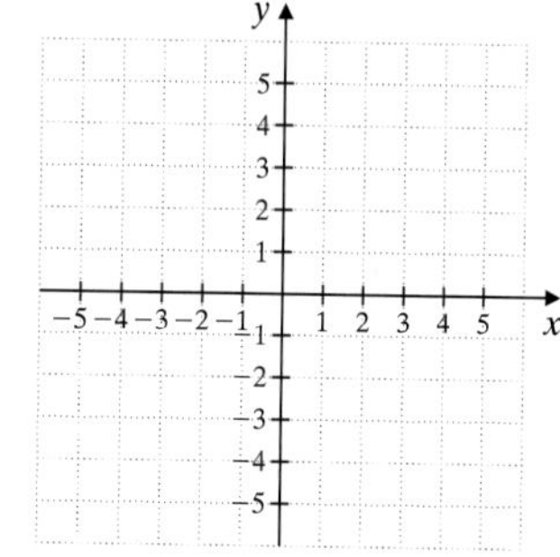

18. 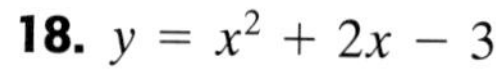

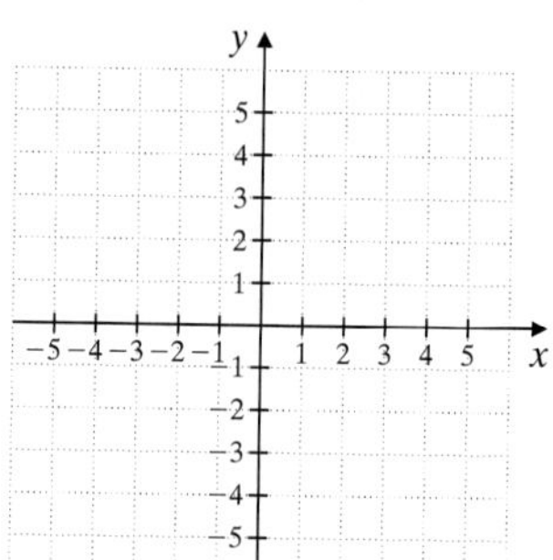

16.1 Solving Quadratic Equations by the Square Root Property

OBJECTIVES

- (A) Review factoring to solve quadratic equations.
- (B) Use the square root property to solve quadratic equations.
- (C) Use the square root property to solve applications.

Recall that a quadratic equation is an equation that can be written in the form

$$ax^2 + bx + c = 0$$

where a, b, and c are real numbers and $a \neq 0$.

(A) Solving Quadratic Equations by Factoring

To solve quadratic equations by factoring, we use the **zero factor property:** If the product of two numbers is zero, then at least one of the two numbers is zero. Examples 1 and 2 review the process of solving quadratic equations by factoring.

EXAMPLE 1 Solve: $x^2 - 4 = 0$

Solution:

$$x^2 - 4 = 0$$

$$(x + 2)(x - 2) = 0 \quad \text{Factor.}$$

$$x + 2 = 0 \quad \text{or} \quad x - 2 = 0 \quad \text{Use the zero factor property.}$$

$$x = -2 \qquad x = 2 \quad \text{Solve each equation.}$$

The solutions are -2 and 2.

Practice Problem 1

Solve: $x^2 - 25 = 0$

EXAMPLE 2 Solve: $3y^2 + 13y = 10$

Solution: Recall that to use the zero factor property, one side of the equation must be 0 and the other side must be factored.

$$3y^2 + 13y = 10$$

$$3y^2 + 13y - 10 = 0 \quad \text{Subtract 10 from both sides.}$$

$$(3y - 2)(y + 5) = 0 \quad \text{Factor.}$$

$$3y - 2 = 0 \quad \text{or} \quad y + 5 = 0 \quad \text{Use the zero factor property.}$$

$$3y = 2 \qquad y = -5 \quad \text{Solve each equation.}$$

$$y = \frac{2}{3}$$

The solutions are $\frac{2}{3}$ and -5.

Practice Problem 2

Solve: $2x^2 - 3x = 9$

(B) Using the Square Root Property

Consider solving Example 1, $x^2 - 4 = 0$, another way. First, add 4 to both sides of the equation.

$$x^2 - 4 = 0$$

$$x^2 = 4 \quad \text{Add 4 to both sides.}$$

Now we see that the value for x must be a number whose square is 4. Therefore $x = \sqrt{4} = 2$ or $x = -\sqrt{4} = -2$. This reasoning is an example of the square root property.

Answers

1. 5 and -5 **2.** $-\frac{3}{2}$ and 3

Square Root Property

If $x^2 = a$ for $a \geq 0$, then

$$x = \sqrt{a} \quad \text{or} \quad x = -\sqrt{a}$$

Practice Problem 3

Use the square root property to solve $x^2 - 16 = 0$.

EXAMPLE 3 Use the square root property to solve $x^2 - 9 = 0$.

Solution: First we solve for x^2 by adding 9 to both sides.

$$x^2 - 9 = 0$$
$$x^2 = 9 \quad \text{Add 9 to both sides.}$$

Next we use the square root property.

$$x = \sqrt{9} \quad \text{or} \quad x = -\sqrt{9}$$
$$x = 3 \qquad\qquad x = -3$$

Check:

$$x^2 - 9 = 0 \quad \text{Original equation} \qquad x^2 - 9 = 0 \quad \text{Original equation}$$
$$3^2 - 9 \stackrel{?}{=} 0 \quad \text{Let } x = 3. \qquad (-3)^2 - 9 \stackrel{?}{=} 0 \quad \text{Let } x = -3.$$
$$0 = 0 \quad \text{True} \qquad 0 = 0 \quad \text{True}$$

The solutions are 3 and −3.

Practice Problem 4

Use the square root property to solve $3x^2 = 11$.

EXAMPLE 4 Use the square root property to solve $2x^2 = 7$.

Solution: First we solve for x^2 by dividing both sides by 2. Then we use the square root property.

$$2x^2 = 7$$
$$x^2 = \frac{7}{2} \quad \text{Divide both sides by 2.}$$
$$x = \sqrt{\frac{7}{2}} \quad \text{or} \quad x = -\sqrt{\frac{7}{2}} \quad \text{Use the square root property.}$$
$$x = \frac{\sqrt{7} \cdot \sqrt{2}}{\sqrt{2} \cdot \sqrt{2}} \qquad x = -\frac{\sqrt{7} \cdot \sqrt{2}}{\sqrt{2} \cdot \sqrt{2}} \quad \text{Rationalize the denominator.}$$
$$x = \frac{\sqrt{14}}{2} \qquad x = -\frac{\sqrt{14}}{2} \quad \text{Simplify.}$$

Remember to check both solutions in the original equation. The solutions are $\frac{\sqrt{14}}{2}$ and $-\frac{\sqrt{14}}{2}$.

Practice Problem 5

Use the square root property to solve $(x - 4)^2 = 49$.

EXAMPLE 5 Use the square root property to solve $(x - 3)^2 = 16$.

Solution: Instead of x^2, here we have $(x - 3)^2$. But the square root property can still be used.

$$(x - 3)^2 = 16$$
$$x - 3 = \sqrt{16} \quad \text{or} \quad x - 3 = -\sqrt{16} \quad \text{Use the square root property.}$$
$$x - 3 = 4 \qquad x - 3 = -4 \quad \text{Write } \sqrt{16} \text{ as 4 and } -\sqrt{16} \text{ as } -4.$$
$$x = 7 \qquad x = -1 \quad \text{Solve.}$$

Answers

3. 4 and −4 **4.** $\frac{\sqrt{33}}{3}$ and $-\frac{\sqrt{33}}{3}$

5. 11 and −3

Check:

$(x-3)^2 = 16$ Original equation
$(7-3)^2 \stackrel{?}{=} 16$ Let $x = 7$.
$4^2 \stackrel{?}{=} 16$ Simplify.
$16 = 16$ True

$(x-3)^2 = 16$ Original equation
$(-1-3)^2 \stackrel{?}{=} 16$ Let $x = -1$.
$(-4)^2 \stackrel{?}{=} 16$ Simplify.
$16 = 16$ True

Both 7 and −1 are solutions.

EXAMPLE 6 Use the square root property to solve $(x+1)^2 = 8$.

Solution: $(x+1)^2 = 8$

$x + 1 = \sqrt{8}$ or $x + 1 = -\sqrt{8}$ Use the square root property.
$x + 1 = 2\sqrt{2}$ $x + 1 = -2\sqrt{2}$ Simplify the radical.
$x = -1 + 2\sqrt{2}$ $x = -1 - 2\sqrt{2}$ Solve for x.

Check both solutions in the original equation. The solutions are $-1 + 2\sqrt{2}$ and $-1 - 2\sqrt{2}$. This can be written compactly as $-1 \pm 2\sqrt{2}$. The notation $\pm$ is read as "plus or minus."

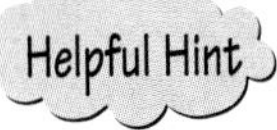

read "plus or minus"

The notation $-1 \pm \sqrt{5}$, for example, is just a shorthand notation for both $-1 + \sqrt{5}$ and $-1 - \sqrt{5}$.

EXAMPLE 7 Use the square root property to solve $(x-1)^2 = -2$.

Solution: This equation has no real solution because the square root of −2 is not a real number.

EXAMPLE 8 Use the square root property to solve $(5x-2)^2 = 10$.

Solution: $(5x-2)^2 = 10$

$5x - 2 = \sqrt{10}$ or $5x - 2 = -\sqrt{10}$ Use the square root property.
$5x = 2 + \sqrt{10}$ $5x = 2 - \sqrt{10}$ Add 2 to both sides.
$x = \frac{2 + \sqrt{10}}{5}$ $x = \frac{2 - \sqrt{10}}{5}$ Divide both sides by 5.

Check both solutions in the original equation. The solutions are $\frac{2 + \sqrt{10}}{5}$ and $\frac{2 - \sqrt{10}}{5}$, which can be written as $\frac{2 \pm \sqrt{10}}{5}$.

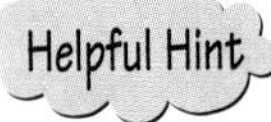

For some applications and graphing purposes, decimal approximations of exact solutions to quadratic equations may be desired.

Exact solutions from Example 8		**Decimal approximations**
$\frac{2 + \sqrt{10}}{5}$	$\approx$	1.032
$\frac{2 - \sqrt{10}}{5}$	$\approx$	−0.232

Practice Problem 6

Use the square root property to solve $(x-5)^2 = 18$.

Practice Problem 7

Use the square root property to solve $(x+3)^2 = -5$.

Practice Problem 8

Use the square root property to solve $(4x+1)^2 = 15$.

Answers

6. $5 \pm 3\sqrt{2}$ **7.** no real solution
8. $\frac{-1 \pm \sqrt{15}}{4}$

C Solving Applications Using the Square Root Property

Many real-world applications are modeled by quadratic equations.

EXAMPLE 9 Finding the Length of Time of a Dive

The record for the highest dive into a lake was made by Harry Froboess of Switzerland. In 1936 he dove 394 feet from the airship Hindenburg into Lake Constance. To the nearest tenth of a second, how long did his dive take? (*Source: The Guinness Book of Records*)

Solution:

1. UNDERSTAND. To approximate the time of the dive, we use the formula $h = 16t^2$* where t is time in seconds and h is the distance in feet traveled by a free-falling body or object. For example, to find the distance traveled in 1 second, or 3 seconds, we let $t = 1$ and then $t = 3$.

   If $t = 1, h = 16(1)^2 = 16 \cdot 1 = 16$ feet
   If $t = 3, h = 16(3)^2 = 16 \cdot 9 = 144$ feet

   Since a body travels 144 feet in 3 seconds, we now know the dive of 394 feet lasted longer than 3 seconds.
2. TRANSLATE. Use the formula $h = 16t^2$, let the distance $h = 394$, and we have the equation $394 = 16t^2$.
3. SOLVE. To solve $394 = 16t^2$ for t, we will use the square root property.

$$394 = 16t^2$$

$$\frac{394}{16} = t^2 \quad \text{Divide both sides by 16.}$$

$$24.625 = t^2 \quad \text{Simplify.}$$

$$\sqrt{24.625} = t \quad \text{or} \quad -\sqrt{24.625} = t \quad \text{Use the square root property.}$$

$$5.0 \approx t \quad \text{or} \quad -5.0 \approx t \quad \text{Approximate.}$$

4. INTERPRET.

Check: We reject the solution -5.0 since the length of the dive is not a negative number.
State: The dive lasted approximately 5 seconds.

*The formula $h = 16t^2$ does not take into account air resistance.

Name ______________________ Section __________ Date __________

EXERCISE SET 16.1

 *Solve each equation by factoring. See Examples 1 and 2.*

1. $k^2 - 9 = 0$ **2.** $k^2 - 49 = 0$ **3.** $m^2 + 2m = 15$ **4.** $m^2 + 6m = 7$ **5.** $2x^2 - 32 = 0$

6. $3p^4 - 9p^3 = 0$ **7.** $4a^2 - 36 = 0$ **8.** $7a^2 - 175 = 0$ **9.** $x^2 + 7x = -10$ **10.** $x^2 + 10x = -24$

 *Use the square root property to solve each quadratic equation. See Examples 3 and 4.*

 **11.** $x^2 = 64$ **12.** $x^2 = 121$ **13.** $x^2 = 21$ **14.** $x^2 = 22$ **15.** $x^2 = \frac{1}{25}$

16. $x^2 = \frac{1}{16}$ **17.** $x^2 = -4$ **18.** $x^2 = -25$ **19.** $3x^2 = 13$ **20.** $5x^2 = 2$

21. $7x^2 = 4$ **22.** $2x^2 = 9$ **23.** $x^2 - 2 = 0$ **24.** $x^2 - 15 = 0$

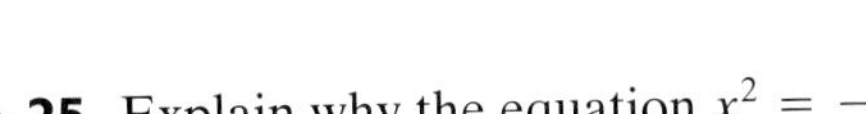

25. Explain why the equation $x^2 = -9$ has no real solution.

26. Explain why the equation $x^2 = 9$ has two solutions.

Use the square root property to solve each quadratic equation. See Examples 5 through 8.

27. $(x - 5)^2 = 49$ **28.** $(x + 2)^2 = 25$ **29.** $(x + 2)^2 = 7$ **30.** $(x - 7)^2 = 2$

31. $\left(m - \frac{1}{2}\right)^2 = \frac{1}{4}$ **32.** $\left(m + \frac{1}{3}\right)^2 = \frac{1}{9}$ **33.** $(p + 2)^2 = 10$ **34.** $(p - 7)^2 = 13$

35. $(3y + 2)^2 = 100$ **36.** $(4y - 3)^2 = 81$ **37.** $(z - 4)^2 = -9$ **38.** $(z + 7)^2 = -20$

39. $(2x - 11)^2 = 50$ **40.** $(3x - 17)^2 = 28$ **41.** $(3x - 7)^2 = 32$ **42.** $(5x - 11)^2 = 54$

C *Solve. See Example 9. For Exercises 43 through 46, use the formula from $h = 16t^2$ Example 9.*

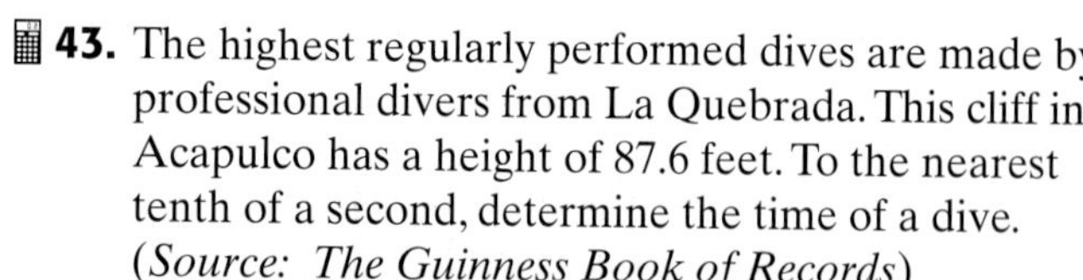

43. The highest regularly performed dives are made by professional divers from La Quebrada. This cliff in Acapulco has a height of 87.6 feet. To the nearest tenth of a second, determine the time of a dive. (*Source: The Guinness Book of Records*)

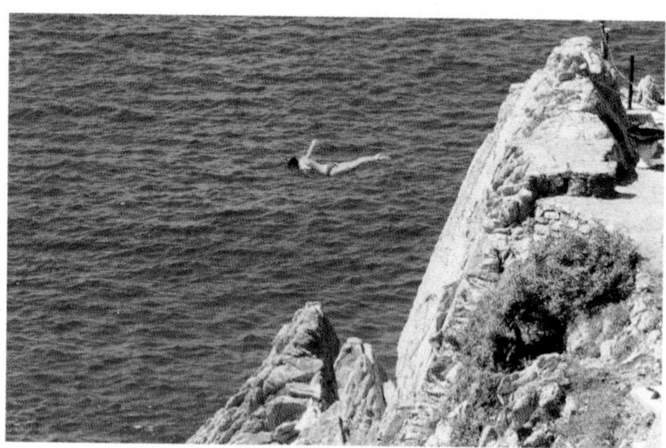

44. In 1988, Eddie Turner saved Frank Fanan, who became unconscious after an injury while jumping out of an airplane. Fanan fell 11,136 feet before Turner pulled his ripcord. To the nearest tenth of a second, determine the time of Fanan's unconscious free-fall.

45. In New Mexico, Joseph Kittinger fell 16 miles before opening his parachute on August 16, 1960. To the nearest tenth of a second, how long did Kittinger free-fall before opening his parachute? (*Hint:* First convert 16 miles to feet. Use 1 mile = 5280 feet.) (*Source: The Guinness Book of Records*)

46. In the Ukraine, Elvira Fomitcheva fell 9 miles 1056 feet before opening her parachute on October 26, 1977. To the nearest tenth of a second, how long did she free-fall before opening her parachute? (Use the hint from Exercise 45.) (*Source: The Guinness Book of Records*)

The formula for the area of a square is $A = s^2$ where s is the length of a side. Use this formula for Exercises 47 through 50. For each exercise, give an exact answer and a two decimal place approximation.

△ **47.** If the area of a square is 20 square inches, find the length of a side.

△ **48.** If the area of a square is 32 square meters, find the length of a side.

△ **49.** The Washington Monument has a square base whose area is approximately 3039 square feet. Find the length of a side. (*Source: The World Almanac*, 2003)

△ **50.** A giant strawberry shortcake was made by the Greater Plant City Chamber of Commerce in Plant City, Florida. Its base had an area of 827 square feet. If the base was in the shape of a square, find the length of a side. (*Source: The Guinness Book of Records*)

Review and Preview

Factor each perfect square trinomial. See Section 11.5.

51. $x^2 + 6x + 9$

52. $y^2 + 10y + 25$

53. $x^2 - 4x + 4$

54. $x^2 - 20x + 100$

Combining Concepts

Solve each quadratic equation by first factoring the perfect square trinomial on the left side. Then apply the square root property.

55. $x^2 + 4x + 4 = 16$

56. $y^2 - 10y + 25 = 11$

57. The area of a circle is found by the equation $A = \pi r^2$. If the area A of a certain circle is 36π square inches, find its radius r.

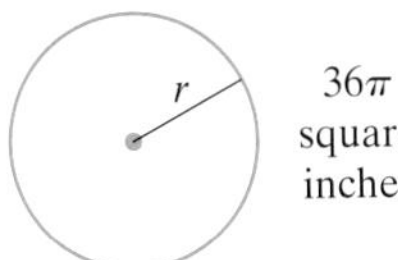

58. A 27-inch TV is advertised in the local paper. If 27 inches is the measure of the diagonal of the picture tube, use the Pythagorean theorem and the figure below to find the measures of the sides of the picture tube.

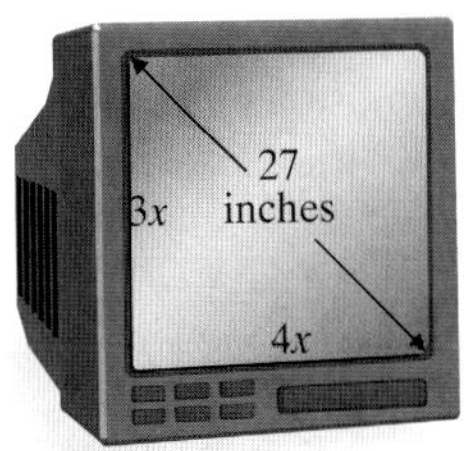

59. Neglecting air resistance, the distance d in feet that an object falls in t seconds is given by the equation $d = 16t^2$. If a sandblaster drops his goggles from a bridge 400 feet from the water below, find how long it takes for the goggles to hit the water.

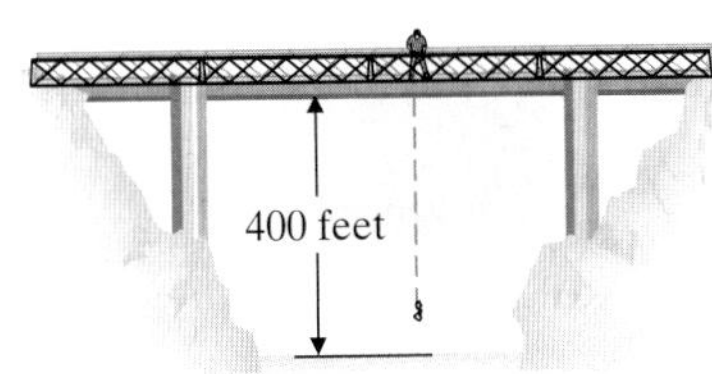

Solve each quadratic equation by using the square root property. Use a calculator and round each solution to the nearest hundredth.

60. $x^2 = 1.78$

61. $(x - 1.37)^2 = 5.71$

62. The number y of Target stores open for business from 2000 through 2002 is given by the equation $y = 9(x + 3.7)^2 + 852$, where $x = 0$ represents the year 2000. Assume that this trend continues and find the year in which there will be 2000 Target stores open for business. (*Hint:* Replace y with 2000 in the equation, solve for x and round to the nearest whole.) (*Source:* Based on data from Target Corporation)

63. World cotton production y (in thousand metric tons) from 2001 through 2003 is given by the equation $y = 1944(x - 0.914)^2 + 19{,}143$, where $x = 0$ represents the year 2001. Assume that this trend continues and find the year in which there will be 30,000 thousand metric tons. (*Hint:* Replace y with 30,000 in the equation, solve for x and round to the nearest whole.) (*Source:* Based on data from the U. S. Department of Agriculture, Foreign Agricultural Service)

FOCUS ON the Real World

USES OF PARABOLAS

We learned in Chapter 13 that the graph of an equation in two variables of the form $y = mx + b$ is a straight line. Later in this chapter, we will find that the graph of a quadratic equation in two variables of the form $y = ax^2 + bx + c$ is a shape called a **parabola.** The figure to the right shows the general shape of a parabola.

The shape of a parabola shows up in many situations, both natural and human-made, in the world around us.

NATURAL SITUATIONS

- **Hurricanes** The paths of many hurricanes are roughly shaped like a parabola. In the northern hemisphere, hurricanes generally begin moving to the northwest. Then, as they move farther from the equator, they swing around to head in a northeastern direction.

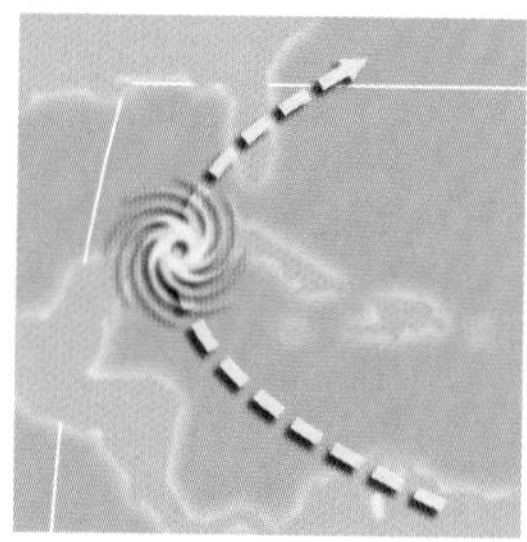

- **Projectiles** The force of the earth's gravity acts on a projectile launched into the air. The resulting path of the projectile, anything from a bullet to a football, is generally shaped like a parabola.

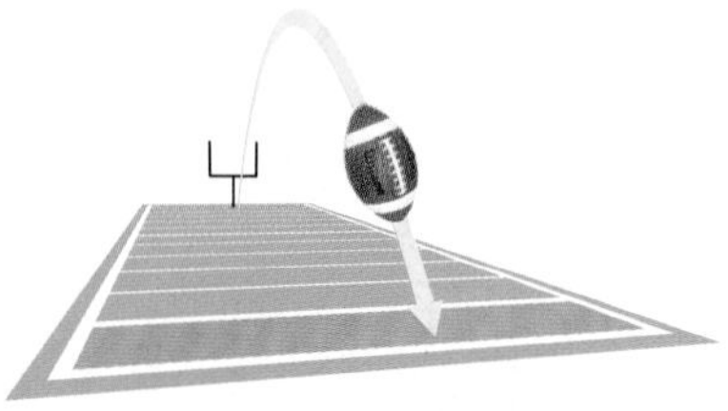

- **Orbits** There are several different possible shapes of orbits of satellites, planets, moons, and comets in outer space. One of the possible types of orbits is in the shape of a parabola. A parabolic orbit is most often seen with comets.

HUMAN-MADE SITUATIONS

- **Telescopes** Because a parabola has nice reflecting properties, its shape is used in many kinds of telescopes. The largest non-steerable radio telescope is Arecibo Observatory in Puerto Rico. This telescope consists of a huge parabolic dish built into a valley. The dish is about 1000 feet across.

- **Training Astronauts** Astronauts must be able to work in zero-gravity conditions on missions in space. However, it's nearly impossible to escape the force of gravity on earth. To help astronauts train to work in weightlessness, a specially modified jet can be flown in a parabolic path. At the top of the parabola, weightlessness can be simulated for up to 30 seconds at a time.

- **Architecture** The reinforced concrete arches used in many modern buildings are based on the shape of a parabola.

- **Music** The design of the modern flute incorporates a parabolic head joint.

16.2 Solving Quadratic Equations by Completing the Square

OBJECTIVES

A Solve quadratic equations of the form $x^2 + bx + c = 0$ by completing the square.

B Solve quadratic equations of the form $ax^2 + bx + c = 0$ by completing the square.

A Completing the Square to Solve $x^2 + bx + c = 0$

In the last section, we used the square root property to solve equations such as

$$(x + 1)^2 = 8 \quad \text{and} \quad (5x - 2)^2 = 3$$

Notice that one side of each equation is a quantity squared and that the other side is a constant. To solve

$$x^2 + 2x = 4$$

notice that if we add 1 to both sides of the equation, the left side is a perfect square trinomial that can be factored.

$$x^2 + 2x + 1 = 4 + 1 \quad \text{Add 1 to both sides.}$$
$$(x + 1)^2 = 5 \quad \text{Factor.}$$

Now we can solve this equation as we did in the previous section by using the square root property.

$$x + 1 = \sqrt{5} \quad \text{or} \quad x + 1 = -\sqrt{5} \quad \text{Use the square root property.}$$
$$x = -1 + \sqrt{5} \qquad x = -1 - \sqrt{5} \quad \text{Solve.}$$

The solutions are $-1 \pm \sqrt{5}$.

Adding a number to $x^2 + 2x$ to form a perfect square trinomial is called **completing the square** on $x^2 + 2x$.

In general, we have the following:

Completing the Square

To complete the square on $x^2 + bx$, add $\left(\frac{b}{2}\right)^2$. To find $\left(\frac{b}{2}\right)^2$, **find half the coefficient of x, and then square the result.**

EXAMPLE 1 Solve $x^2 + 6x + 3 = 0$ by completing the square.

Solution: First we get the variable terms alone by subtracting 3 from both sides of the equation.

$$x^2 + 6x + 3 = 0$$
$$x^2 + 6x = -3 \quad \text{Subtract 3 from both sides.}$$

Next we find half the coefficient of the x-term, and then square it. We add this result to *both sides* of the equation. This will make the left side a perfect square trinomial. The coefficient of x is 6, and half of 6 is 3. So we add 3^2 or 9 to both sides.

$$x^2 + 6x + 9 = -3 + 9 \quad \text{Complete the square.}$$
$$(x + 3)^2 = 6 \quad \text{Factor the trinomial } x^2 + 6x + 9.$$
$$x + 3 = \sqrt{6} \quad \text{or} \quad x + 3 = -\sqrt{6} \quad \text{Use the square root property.}$$
$$x = -3 + \sqrt{6} \qquad x = -3 - \sqrt{6} \quad \text{Subtract 3 from both sides.}$$

Check by substituting $-3 + \sqrt{6}$ and $-3 - \sqrt{6}$ in the original equation. The solutions are $-3 \pm \sqrt{6}$.

Practice Problem 1

Solve $x^2 + 8x + 1 = 0$ by completing the square.

Answer

1. $-4 \pm \sqrt{15}$

Helpful Hint

Remember, when completing the square, add the number that completes the square to **both sides of the equation.**

Practice Problem 2

Solve $x^2 - 14x = -32$ by completing the square.

EXAMPLE 2 Solve $x^2 - 10x = -14$ by completing the square.

Solution: The variable terms are already alone on one side of the equation. The coefficient of x is -10. Half of -10 is -5, and $(-5)^2 = 25$. So we add 25 to both sides.

$$x^2 - 10x = -14$$
$$x^2 - 10x + 25 = -14 + 25$$

Helpful Hint

Add 25 to *both* sides of the equation.

$(x - 5)^2 = 11$ Factor the trinomial and simplify $-14 + 25$.

$x - 5 = \sqrt{11}$ or $x - 5 = -\sqrt{11}$ Use the square root property.

$x = 5 + \sqrt{11}$ $\quad x = 5 - \sqrt{11}$ Add 5 to both sides.

The solutions are $5 \pm \sqrt{11}$.

B Completing the Square to Solve $ax^2 + bx + c = 0$

The method of completing the square can be used to solve *any* quadratic equation whether the coefficient of the squared variable is 1 or not. When the coefficient of the squared variable is not 1, we first divide both sides of the equation by the coefficient of the squared variable so that the new coefficient is 1. Then we complete the square.

Practice Problem 3

Solve $4x^2 - 16x - 9 = 0$ by completing the square.

EXAMPLE 3 Solve $4x^2 - 8x - 5 = 0$ by completing the square.

Solution: Since the coefficient of x^2 is 4, not 1, we first divide both sides of the equation by 4 so that the coefficient of x^2 is 1.

$4x^2 - 8x - 5 = 0$

$x^2 - 2x - \frac{5}{4} = 0$ Divide both sides by 4.

$x^2 - 2x = \frac{5}{4}$ Get the variable terms alone on one side of the equation.

The coefficient of x is -2. Half of -2 is -1, and $(-1)^2 = 1$. So we add 1 to both sides.

$x^2 - 2x + 1 = \frac{5}{4} + 1$

$(x - 1)^2 = \frac{9}{4}$ Factor $x^2 - 2x + 1$ and simplify $\frac{5}{4} + 1$.

$x - 1 = \sqrt{\frac{9}{4}}$ or $x - 1 = -\sqrt{\frac{9}{4}}$ Use the square root property.

$x = 1 + \frac{3}{2}$ $\quad x = 1 - \frac{3}{2}$ Add 1 to both sides and simplify the radical.

$x = \frac{5}{2}$ $\quad x = -\frac{1}{2}$ Simplify.

Both $\frac{5}{2}$ and $-\frac{1}{2}$ are solutions.

Answers

2. $7 \pm \sqrt{17}$ **3.** $\frac{9}{2}$ and $-\frac{1}{2}$

The following steps may be used to solve a quadratic equation in x by completing the square.

To Solve a Quadratic Equation in x by Completing the Square

Step 1. If the coefficient of x^2 is 1, go to Step 2. If not, divide both sides of the equation by the coefficient of x^2.
Step 2. Get all terms with variables on one side of the equation and constants on the other side.
Step 3. Find half the coefficient of x and then square the result. Add this number to both sides of the equation.
Step 4. Factor the resulting perfect square trinomial.
Step 5. Use the square root property to solve the equation.

EXAMPLE 4 Solve $2x^2 + 6x = -7$ by completing the square.

Solution: The coefficient of x^2 is not 1. We divide both sides by 2, the coefficient of x^2.

$$2x^2 + 6x = -7$$

$$x^2 + 3x = -\frac{7}{2} \quad \text{Divide both sides by 2.}$$

$$x^2 + 3x + \frac{9}{4} = -\frac{7}{2} + \frac{9}{4} \quad \text{Add } \left(\frac{3}{2}\right)^2 \text{ or } \frac{9}{4} \text{ to both sides.}$$

$$\left(x + \frac{3}{2}\right)^2 = -\frac{5}{4} \quad \text{Factor the left side and simplify the right.}$$

There is no real solution to this equation since the square root of a negative number is not a real number.

Practice Problem 4

Solve $2x^2 + 10x = -13$ by completing the square.

EXAMPLE 5 Solve $2x^2 = 10x + 1$ by completing the square.

Solution: First we divide both sides of the equation by 2, the coefficient of x^2.

$$2x^2 = 10x + 1$$

$$x^2 = 5x + \frac{1}{2} \quad \text{Divide both sides by 2.}$$

Next we get the variable terms alone by subtracting $5x$ from both sides.

$$x^2 - 5x = \frac{1}{2}$$

$$x^2 - 5x + \frac{25}{4} = \frac{1}{2} + \frac{25}{4} \quad \text{Add } \left(-\frac{5}{2}\right)^2 \text{ or } \frac{25}{4} \text{ to both sides.}$$

$$\left(x - \frac{5}{2}\right)^2 = \frac{27}{4} \quad \text{Factor the left side and simplify the right side.}$$

$$x - \frac{5}{2} = \sqrt{\frac{27}{4}} \quad \text{or} \quad x - \frac{5}{2} = -\sqrt{\frac{27}{4}} \quad \text{Use the square root property.}$$

$$x - \frac{5}{2} = \frac{3\sqrt{3}}{2} \qquad x - \frac{5}{2} = -\frac{3\sqrt{3}}{2} \quad \text{Simplify.}$$

$$x = \frac{5}{2} + \frac{3\sqrt{3}}{2} \qquad x = \frac{5}{2} - \frac{3\sqrt{3}}{2}$$

The solutions are $\frac{5 \pm 3\sqrt{3}}{2}$.

Practice Problem 5

Solve $2x^2 = -3x + 2$ by completing the square.

Answers

4. no real solution **5.** $\frac{1}{2}$ and -2

STUDY SKILLS REMINDER

Are you prepared for your final exam?

To prepare for your final exam, try the following study techniques.

- Review the material that you will be responsible for on your exam. Also check your notebook for any lecture notes that you highlighted.
- Review any formulas that you may need to memorize.
- Check to see if your instructor or math department will be conducting a final exam review.
- Check with your instructor to see whether there are final exams from previous semesters/quarters that are available to students for study.
- Use your previously taken tests as a practice final exam. To do so, rewrite the test questions in mixed order on blank sheets of paper. This will help you prepare for exam conditions.
- If you are unsure of a few topics, see your instructor or visit a learning lab for further assistance. Also, viewing the video segment of a troublesome section will help.
- If you need further exercises to work, try the chapter tests at the end of appropriate chapters.

Name ____________________ Section __________ Date __________

Mental Math

Determine the number to add to make each expression a perfect square trinomial.

1. $p^2 + 8p$

2. $p^2 + 6p$

3. $x^2 + 20x$

4. $x^2 + 18x$

5. $y^2 + 14y$

6. $y^2 + 2y$

EXERCISE SET 16.2

A *Solve each quadratic equation by completing the square. See Examples 1 and 2.*

1. $x^2 + 8x = -12$

2. $x^2 - 10x = -24$

3. $x^2 + 2x - 5 = 0$

4. $z^2 + 6z - 9 = 0$

5. $x^2 - 6x = 0$

6. $y^2 + 4y = 0$

7. $z^2 + 5z = 7$

8. $x^2 - 7x = 5$

9. $x^2 - 2x - 1 = 0$

10. $x^2 - 4x + 2 = 0$

11. $y^2 + 5y + 4 = 0$

12. $y^2 - 5y + 6 = 0$

13. $x^2 + 6x - 25 = 0$

14. $x^2 - 6x + 7 = 0$

15. $x^2 - 3x - 3 = 0$

16. $x^2 - 9x + 3 = 0$

17. $x(x + 3) = 18$

18. $x(x - 3) = 18$

B *Solve each quadratic equation by completing the square. See Examples 3 through 5.*

19. $3x^2 - 6x = 24$

20. $2x^2 + 18x = -40$

21. $5x^2 + 10x + 6 = 0$

22. $3x^2 - 12x + 14 = 0$

23. $2x^2 = 6x + 5$

24. $4x^2 = -20x + 3$

25. $2y^2 + 8y + 5 = 0$

26. $3z^2 + 6z + 4 = 0$

27. $2y^2 - 3y + 1 = 0$

28. $2y^2 - y - 1 = 0$

29. In your own words, describe a perfect square trinomial.

30. Describe how to find the number to add to $x^2 - 7x$ to make a perfect square trinomial.

Review and Preview

Simplify each expression. See Section 15.2.

31. $\frac{3}{4} - \sqrt{\frac{25}{16}}$

32. $\frac{3}{5} + \sqrt{\frac{16}{25}}$

33. $\frac{1}{2} - \sqrt{\frac{9}{4}}$

34. $\frac{9}{10} - \sqrt{\frac{49}{100}}$

Simplify each expression. See Section 15.4.

35. $\frac{6 + 4\sqrt{5}}{2}$

36. $\frac{10 - 20\sqrt{3}}{2}$

37. $\frac{3 - 9\sqrt{2}}{6}$

38. $\frac{12 - 8\sqrt{7}}{16}$

Combining Concepts

39. Find a value of k that will make $x^2 + kx + 16$ a perfect square trinomial.

40. Find a value of k that will make $x^2 + kx + 25$ a perfect square trinomial.

41. Retail sales y (in millions of dollars) for bookstores in the United States from 2000 through 2002 are given by the equation $y = -195x^2 + 602x + 15{,}375$. In this equation, x is the number of years after 2000. Assume that this trend continues and predict the year after 2000 in which the retail sales for U.S. bookstores will be \$12,500 million. (*Source:* Based on data from the U.S. Bureau of the Census, Monthly Retail Surveys Branch)

42. The average price of gold y (in dollars per ounce) from 1996 through 2000 is given by the equation $y = 10x^2 - 67x + 389$. In this equation, x is the number of years after 1996. Assume that this trend continues and find the year after 2000 in which the price of gold will be \$1025 per ounce. (*Source:* Based on data from Platinum Guild International (USA) Inc.)

Recall that a graphing calculator may be used to solve an equation. For example, to solve $x^2 + 8x = -12$ (Exercise 1), graph

$y_1 = x^2 + 8x$ *(left side of equation) and*

$y_2 = -12$ *(right side of equation)*

The x-coordinate of the point of intersection of the graphs is the solution. Use a graphing calculator and solve each equation. Round solutions to the nearest hundredth.

43. Exercise 1

44. Exercise 2

45. Exercise 23

46. Exercise 8

16.3 Solving Quadratic Equations by the Quadratic Formula

OBJECTIVE

(A) Use the quadratic formula to solve quadratic equations.

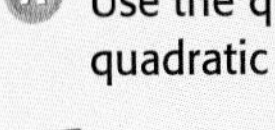

SSM TUTOR CENTER SG CD & VIDEO MATH PRO WEB

(A) Using the Quadratic Formula

We can use the technique of completing the square to develop a formula to find solutions of any quadratic equation. We develop and use the **quadratic formula** in this section.

Recall that a quadratic equation in **standard form** is

$$ax^2 + bx + c = 0, \quad a \neq 0$$

To develop the quadratic formula, let's complete the square for this quadratic equation in standard form.

First we divide both sides of the equation by the coefficient of x^2 and then get the variable terms alone on one side of the equation.

$$x^2 + \frac{b}{a}x + \frac{c}{a} = 0 \qquad \text{Divide by } a\text{; recall that } a \text{ cannot be 0.}$$

$$x^2 + \frac{b}{a}x = -\frac{c}{a} \qquad \text{Get the variable terms alone on one side of the equation.}$$

The coefficient of x is $\frac{b}{a}$. Half of $\frac{b}{a}$ is $\frac{b}{2a}$ and $\left(\frac{b}{2a}\right)^2 = \frac{b^2}{4a^2}$. So we add $\frac{b^2}{4a^2}$ to both sides of the equation.

$$x^2 + \frac{b}{a}x + \frac{b^2}{4a^2} = -\frac{c}{a} + \frac{b^2}{4a^2} \qquad \text{Add } \frac{b^2}{4a^2} \text{ to both sides.}$$

$$\left(x + \frac{b}{2a}\right)^2 = -\frac{c}{a} + \frac{b^2}{4a^2} \qquad \text{Factor the left side.}$$

$$\left(x + \frac{b}{2a}\right)^2 = -\frac{4ac}{4a^2} + \frac{b^2}{4a^2} \qquad \text{Multiply } -\frac{c}{a} \text{ by } \frac{4a}{4a} \text{ so that both terms on the right side have a common denominator.}$$

$$\left(x + \frac{b}{2a}\right)^2 = \frac{b^2 - 4ac}{4a^2} \qquad \text{Simplify the right side.}$$

Now we use the square root property.

$$x + \frac{b}{2a} = \sqrt{\frac{b^2 - 4ac}{4a^2}} \quad \text{or} \quad x + \frac{b}{2a} = -\sqrt{\frac{b^2 - 4ac}{4a^2}} \qquad \text{Use the square root property.}$$

$$x + \frac{b}{2a} = \frac{\sqrt{b^2 - 4ac}}{2a} \qquad x + \frac{b}{2a} = -\frac{\sqrt{b^2 - 4ac}}{2a} \qquad \text{Simplify the radical.}$$

$$x = -\frac{b}{2a} + \frac{\sqrt{b^2 - 4ac}}{2a} \qquad x = -\frac{b}{2a} - \frac{\sqrt{b^2 - 4ac}}{2a} \qquad \text{Subtract } \frac{b}{2a} \text{ from both sides.}$$

$$x = \frac{-b + \sqrt{b^2 - 4ac}}{2a} \qquad x = \frac{-b - \sqrt{b^2 - 4ac}}{2a} \qquad \text{Simplify.}$$

The solutions are $\frac{-b \pm \sqrt{b^2 - 4ac}}{2a}$. This final equation is called the **quadratic formula** and gives the solutions of any quadratic equation.

Quadratic Formula

If a, b, and c are real numbers and $a \neq 0$, a quadratic equation written in the form $ax^2 + bx + c = 0$ has solutions

$$x = \frac{-b \pm \sqrt{b^2 - 4ac}}{2a}$$

Helpful Hint

Don't forget that to correctly identify a, b, and c in the quadratic formula, you should write the equation in standard form.

Quadratic Equations in Standard Form

$5x^2 - 6x + 2 = 0$	$a = 5, b = -6, c = 2$
$4y^2 - 9 = 0$	$a = 4, b = 0, c = -9$
$x^2 + x = 0$	$a = 1, b = 1, c = 0$
$\sqrt{2}x^2 + \sqrt{5}x + \sqrt{3} = 0$	$a = \sqrt{2}, b = \sqrt{5}, c = \sqrt{3}$

Practice Problem 1

Solve $2x^2 - x - 5 = 0$ using the quadratic formula.

EXAMPLE 1 Solve $3x^2 + x - 3 = 0$ using the quadratic formula.

Solution: This equation is in standard form with $a = 3, b = 1$, and $c = -3$. By the quadratic formula, we have

$$x = \frac{-b \pm \sqrt{b^2 - 4ac}}{2a}$$

$$x = \frac{-1 \pm \sqrt{1^2 - 4 \cdot 3 \cdot (-3)}}{2 \cdot 3} \quad \text{Let } a = 3, b = 1, \text{ and } c = -3.$$

$$= \frac{-1 \pm \sqrt{1 + 36}}{6} \quad \text{Simplify.}$$

$$= \frac{-1 \pm \sqrt{37}}{6}$$

Check both solutions in the original equation. The solutions are $\frac{-1 + \sqrt{37}}{6}$ and $\frac{-1 - \sqrt{37}}{6}$.

Practice Problem 2

Solve $3x^2 + 8x = 3$ using the quadratic formula.

EXAMPLE 2 Solve $2x^2 - 9x = 5$ using the quadratic formula.

Solution: First we write the equation in standard form by subtracting 5 from both sides.

$$2x^2 - 9x = 5$$
$$2x^2 - 9x - 5 = 0$$

Next we note that $a = 2, b = -9$, and $c = -5$. We substitute these values into the quadratic formula.

$$x = \frac{-b \pm \sqrt{b^2 - 4ac}}{2a}$$

$$x = \frac{-(-9) \pm \sqrt{(-9)^2 - 4 \cdot 2 \cdot (-5)}}{2 \cdot 2} \quad \text{Substitute in the formula.}$$

$$= \frac{9 \pm \sqrt{81 + 40}}{4} \quad \text{Simplify.}$$

$$= \frac{9 \pm \sqrt{121}}{4} = \frac{9 \pm 11}{4}$$

Notice that the fraction bar is under the entire numerator of $-b \pm \sqrt{b^2 - 4ac}$.

Answers

1. $\frac{1 + \sqrt{41}}{4}$ and $\frac{1 - \sqrt{41}}{4}$ **2.** $\frac{1}{3}$ and -3

Then,

$$x = \frac{9 - 11}{4} = -\frac{1}{2} \quad \text{or} \quad x = \frac{9 + 11}{4} = 5$$

Check $-\frac{1}{2}$ and 5 in the original equation. Both $-\frac{1}{2}$ and 5 are solutions.

The following steps may be useful when solving a quadratic equation by the quadratic formula.

To Solve a Quadratic Equation by the Quadratic Formula

Step 1. Write the quadratic equation in standard form: $ax^2 + bx + c = 0$.

Step 2. If necessary, clear the equation of fractions to simplify calculations.

Step 3. Identify a, b, and c.

Step 4. Replace a, b, and c in the quadratic formula with the identified values, and simplify.

Try the Concept Check in the margin.

EXAMPLE 3 Solve $7x^2 = 1$ using the quadratic formula.

Solution: First we write the equation in standard form by subtracting 1 from both sides.

$$7x^2 = 1$$
$$7x^2 - 1 = 0$$

Next we replace a, b, and c with the identified values: $a = 7, b = 0, c = -1$.

$$x = \frac{0 \pm \sqrt{0^2 - 4 \cdot 7 \cdot (-1)}}{2 \cdot 7} \quad \text{Substitute in the formula.}$$

$$= \frac{\pm \sqrt{28}}{14} \quad \text{Simplify.}$$

$$= \frac{\pm 2\sqrt{7}}{14}$$

$$= \pm \frac{\sqrt{7}}{7}$$

The solutions are $\frac{\sqrt{7}}{7}$ and $-\frac{\sqrt{7}}{7}$.

Notice that we could have solved the equation $7x^2 = 1$ in Example 3 by dividing both sides by 7 and then using the square root property. We solved the equation by the quadratic formula to show that this formula can be used to solve any quadratic equation.

EXAMPLE 4 Solve $x^2 = -x - 1$ using the quadratic formula.

Solution: First we write the equation in standard form.

$$x^2 + x + 1 = 0$$

Concept Check

For the quadratic equation $2x^2 - 5 = 7x$, if $a = 2$ and $c = -5$ in the quadratic formula, the value of b is which of the following?

a. $\frac{7}{2}$

b. 7

c. -5

d. -7

Practice Problem 3

Solve $5x^2 = 2$ using the quadratic formula.

Practice Problem 4

Solve $x^2 = -2x - 3$ using the quadratic formula.

Answers

3. $\frac{\sqrt{10}}{5}$ and $-\frac{\sqrt{10}}{5}$ **4.** no real solution

Concept Check: d

Next we replace a, b, and c in the quadratic formula with $a = 1, b = 1$, and $c = 1$.

$$x = \frac{-1 \pm \sqrt{1^2 - 4 \cdot 1 \cdot 1}}{2 \cdot 1} \quad \text{Substitute in the formula.}$$

$$= \frac{-1 \pm \sqrt{-3}}{2} \quad \text{Simplify.}$$

There is no real number solution because $\sqrt{-3}$ is not a real number.

Practice Problem 5

Solve $\frac{1}{3}x^2 - x = 1$ using the quadratic formula.

EXAMPLE 5 Solve $\frac{1}{2}x^2 - x = 2$ using the quadratic formula.

Solution: We write the equation in standard form and then clear the equation of fractions by multiplying both sides by the LCD, 2.

$$\frac{1}{2}x^2 - x = 2$$

$$\frac{1}{2}x^2 - x - 2 = 0 \quad \text{Write in standard form.}$$

$$x^2 - 2x - 4 = 0 \quad \text{Multiply both sides by 2.}$$

Here, $a = 1, b = -2$, and $c = -4$, so we substitute these values into the quadratic formula.

$$x = \frac{-(-2) \pm \sqrt{(-2)^2 - 4 \cdot 1 \cdot (-4)}}{2 \cdot 1}$$

$$= \frac{2 \pm \sqrt{20}}{2} = \frac{2 \pm 2\sqrt{5}}{2} \quad \text{Simplify.}$$

$$= \frac{2(1 \pm \sqrt{5})}{2} = 1 \pm \sqrt{5} \quad \text{Factor and simplify.}$$

The solutions are $1 - \sqrt{5}$ and $1 + \sqrt{5}$.

Notice that in Example 5, although we cleared the equation of fractions, the coefficients $a = \frac{1}{2}, b = -1$, and $c = -2$ will give the same results.

Helpful Hint

When simplifying an expression such as

$$\frac{3 \pm 6\sqrt{2}}{6}$$

first factor out a common factor from the terms of the numerator and then simplify.

$$\frac{3 \pm 6\sqrt{2}}{6} = \frac{3(1 \pm 2\sqrt{2})}{2 \cdot 3} = \frac{1 \pm 2\sqrt{2}}{2}$$

Answer

5. $\frac{3 \pm \sqrt{21}}{2}$

Name ______________________ Section __________ Date __________

Mental Math

Identify the value of a, b, and c in each quadratic equation.

1. $2x^2 + 5x + 3 = 0$ **2.** $5x^2 - 7x + 1 = 0$ **3.** $10x^2 - 13x - 2 = 0$

4. $x^2 + 3x - 7 = 0$ **5.** $x^2 - 6 = 0$ **6.** $9x^2 - 4 = 0$

EXERCISE SET 16.3

Use the quadratic formula to solve each quadratic equation. See Examples 1 through 4.

1. $x^2 - 3x + 2 = 0$ **2.** $x^2 - 5x - 6 = 0$ 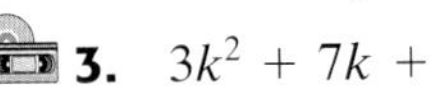 **3.** $3k^2 + 7k + 1 = 0$ **4.** $7k^2 + 3k - 1 = 0$

5. $49x^2 - 4 = 0$ **6.** $25x^2 - 15 = 0$ **7.** $5z^2 - 4z + 3 = 0$ **8.** $3x^2 + 2x + 1 = 0$

9. $y^2 = 7y + 30$ **10.** $y^2 = 5y + 36$ **11.** $2x^2 = 10$ **12.** $5x^2 = 15$

13. $m^2 - 12 = m$ **14.** $m^2 - 14 = 5m$ **15.** $3 - x^2 = 4x$ **16.** $10 - x^2 = 2x$

17. $6x^2 + 9x = 2$ **18.** $3x^2 - 9x = 8$ **19.** $7p^2 + 2 = 8p$ **20.** $11p^2 + 2 = 10p$

21. $a^2 - 6a + 2 = 0$ **22.** $a^2 - 10a + 19 = 0$ **23.** $2x^2 - 6x + 3 = 0$ **24.** $5x^2 - 8x + 2 = 0$

25. $3x^2 = 1 - 2x$ **26.** $5y^2 = 4 - y$ **27.** $4y^2 = 6y + 1$ **28.** $6z^2 + 3z + 2 = 0$

29. $20y^2 = 3 - 11y$

30. $2z^2 = z + 3$

31. $x^2 + x + 2 = 0$

32. $k^2 + 2k + 5 = 0$

Use the quadratic formula to solve each quadratic equation. See Example 5.

33. $3p^2 - \frac{2}{3}p + 1 = 0$

34. $\frac{5}{2}p^2 - p + \frac{1}{2} = 0$

35. $\frac{m^2}{2} = m + \frac{1}{2}$

36. $\frac{m^2}{2} = 3m - 1$

37. $4p^2 + \frac{3}{2} = -5p$

38. $4p^2 + \frac{3}{2} = 5p$

39. $5x^2 = \frac{7}{2}x + 1$

40. $2x^2 = \frac{5}{2}x + \frac{7}{2}$

41. $28x^2 + 5x + \frac{11}{4} = 0$

42. $\frac{2}{3}x^2 - 2x - \frac{2}{3} = 0$

43. $5z^2 - 2z = \frac{1}{5}$

44. $9z^2 + 12z = -1$

Review and Preview

Graph the following linear equations in two variables. See Section 13.2.

45. $y = -3$

46. $x = 4$

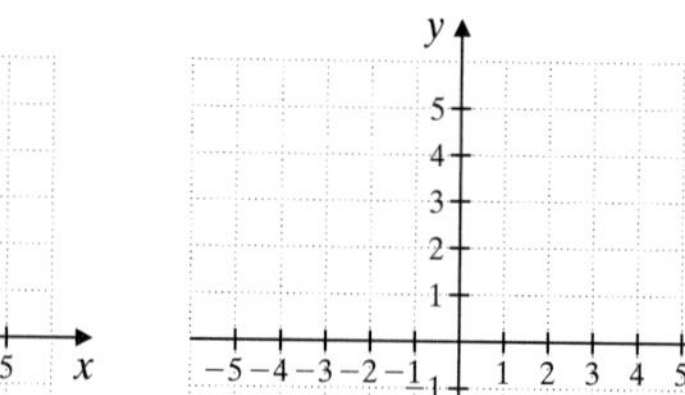

47. $y = 3x - 2$

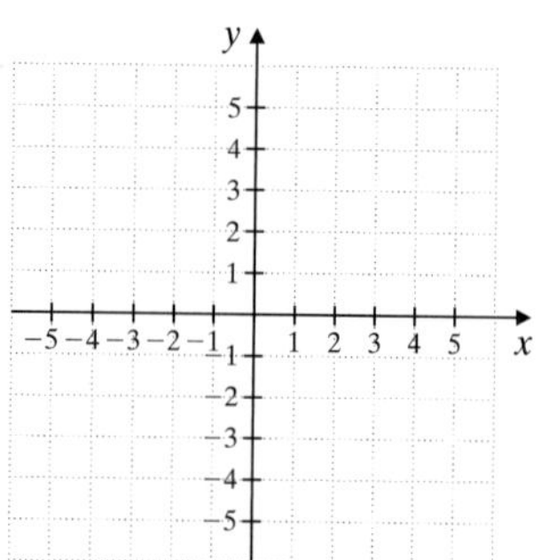

48. $y = 2x + 3$

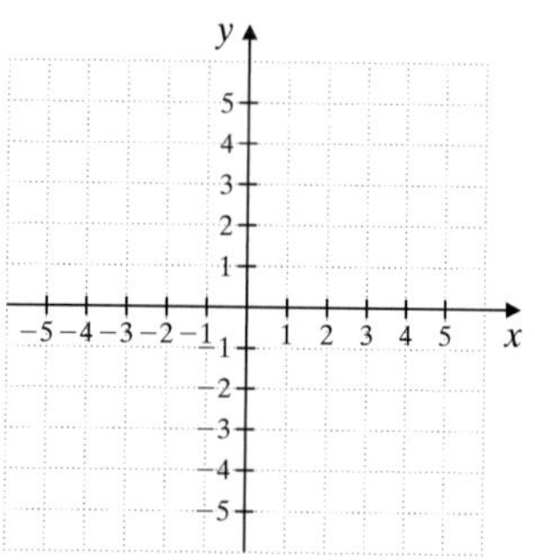

Find the length of the unknown side of each triangle.

△ **49.**

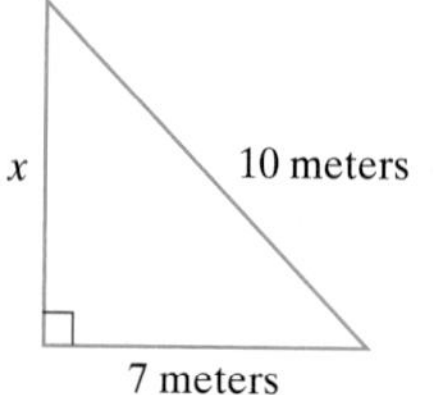

△ **50.**

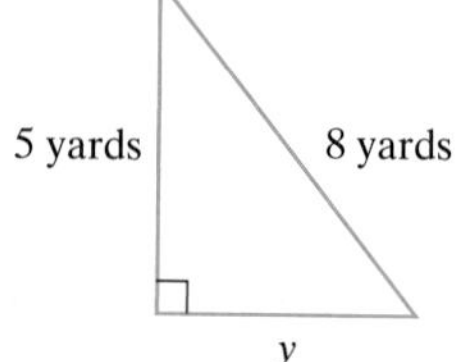

Combining Concepts

△ **51.** The largest chocolate bar was a Cadbury's Dairy Milk bar that weighed 1.1 tons. The bar had a base area of 35 square feet and its length was five feet longer than its width. Find the length and width of the bar rounded to one decimal place. (*Source: The Guinness Book of Records*)

△ **52.** The area of a rectangular conference room table is 95 square feet. If its length is six feet longer than its width, find the dimensions of the table. Round each dimension to the nearest tenth.

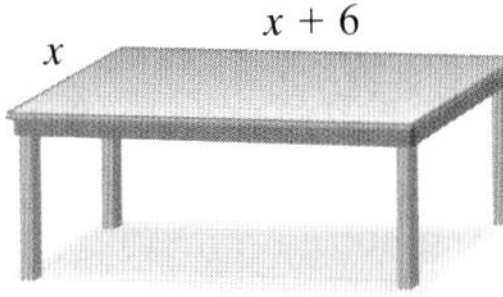

Solve each equation using the quadratic formula.

53. $x^2 + 3\sqrt{2}x - 5 = 0$

54. $y^2 - 2\sqrt{5}y - 1 = 0$

55. Explain how the quadratic formula is developed and why it is useful.

Use the quadratic formula and a calculator to solve each equation. Round solutions to the nearest tenth.

56. $x^2 + x = 15$

57. $y^2 - y = 11$

58. $1.2x^2 - 5.2x - 3.9 = 0$

59. $7.3z^2 + 5.4z - 1.1 = 0$

A rocket is launched from the top of an 80-foot cliff with an initial velocity of 120 feet per second. The height, h, of the rocket after t seconds is given by the equation $h = -16t^2 + 120t + 80$.

60. How long after the rocket is launched will it be 30 feet from the ground? Round to the nearest tenth of a second.

61. How long after the rocket is launched will it strike the ground? Round to the nearest tenth of a second. (*Hint*: The rocket will strike the ground when its height $h = 0$.)

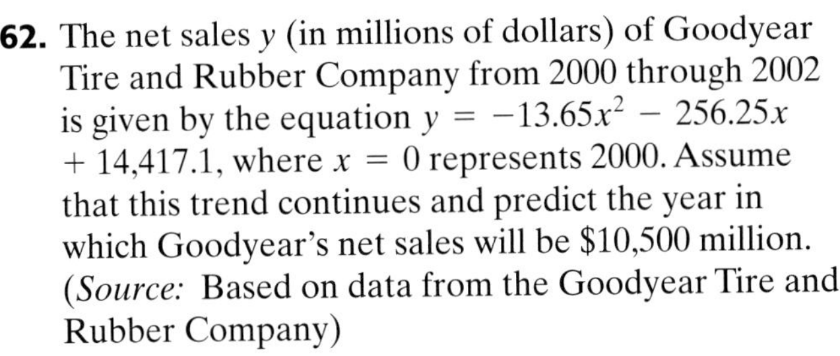

62. The net sales y (in millions of dollars) of Goodyear Tire and Rubber Company from 2000 through 2002 is given by the equation $y = -13.65x^2 - 256.25x + 14{,}417.1$, where $x = 0$ represents 2000. Assume that this trend continues and predict the year in which Goodyear's net sales will be \$10,500 million. (*Source:* Based on data from the Goodyear Tire and Rubber Company)

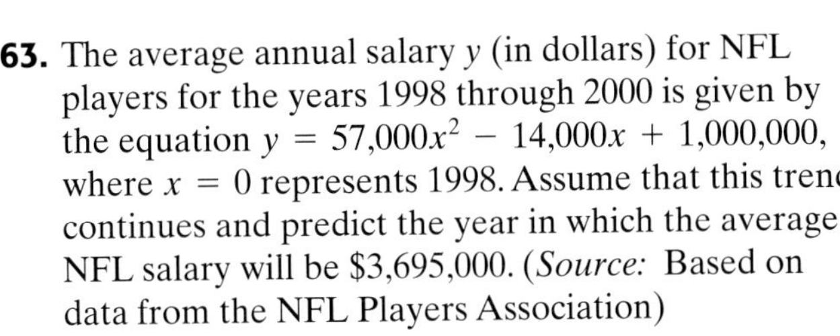

63. The average annual salary y (in dollars) for NFL players for the years 1998 through 2000 is given by the equation $y = 57{,}000x^2 - 14{,}000x + 1{,}000{,}000$, where $x = 0$ represents 1998. Assume that this trend continues and predict the year in which the average NFL salary will be \$3,695,000. (*Source:* Based on data from the NFL Players Association)

FOCUS ON Business and Career

MODELING THE SIZE OF THE FEDEX AIR FLEET

Launched in 1973, Federal Express was the first company to offer overnight package delivery service in the United States. Today, FedEx is the world's largest express transportation company. FedEx's 148,000 employees worldwide deliver over 3.3 million packages every business day to residences and businesses in 211 different countries. A ground fleet of over 45,500 delivery vehicles and the world's second largest air fleet, capable of carrying 26.5 million pounds daily, are needed to handle this high volume of packages.

The number of aircraft y in the constantly expanding FedEx air fleet from 1999 to 2003 can be modeled by the equation $y = -2.3x^2 + 10.5x + 639.9$, where $x = 0$ represents 1999. (*Source*: Based on data from FedEx Corporation)

GROUP ACTIVITY

- Use the given quadratic equation to complete the table of values.
- When was the FedEx air fleet the greatest during this period? When was the FedEx air fleet the smallest during this period?
- Use the model to predict the year in which the FedEx air fleet consists of 600 aircraft.

Year	x	y
1999	0	
2000		
2001		
2002		
2003		

Internet Excursions

Go To:  http://www.prenhall.com/martin-gay_prealgebra  What's Related

Most publicly held corporations publish an annual report summarizing annual earnings and financial positions, as well as various operating data, for their stockholders. Many corporations make their annual reports available on their Web sites. FedEx is one such corporation. The given World Wide Web address will provide you with access to the FedEx Web site, or a related site, where you will find a link to FedEx's annual reports.

1. Select FedEx's most recent annual report. Look for the section of the annual report titled "Selected Consolidated Financial Data," and then scan for aircraft fleet data under the heading "Other Operating Data." Compare the actual size of the FedEx air fleet for the years 1999 through 2003 to values listed in the table that were given by the model for the same period. How well do you think the given model fits the actual data? Explain. (*Hint*: You may need to consult earlier years' annual reports to find the actual air fleet sizes for all the required years.)

2. Use the model to predict the size of FedEx's air fleet for the most recent year listed in the current annual report. How does this prediction compare to the actual value for this year?

Name ______________________ Section __________ Date __________

Answers

1. ____
2. ____
3. ____
4. ____
5. ____
6. ____
7. ____
8. ____
9. ____
10. ____
11. ____
12. ____
13. ____
14. ____
15. ____
16. ____
17. ____
18. ____

Integrated Review—Summary of Quadratic Equations

An important skill in mathematics is learning when to use one technique in favor of another. We now practice this by deciding which method to use when solving quadratic equations. Although both the quadratic formula and completing the square can be used to solve any quadratic equation, the quadratic formula is usually less tedious and thus preferred. The following steps may be used to solve a quadratic equation.

To Solve a Quadratic Equation

Step 1. If the equation is in the form $ax^2 = c$ or $(ax + b)^2 = c$, use the square root property and solve. If not, go to Step 2.

Step 2. Write the equation in standard form: $ax^2 + bx + c = 0$.

Step 3. Try to solve the equation by the factoring method. If not possible, go to Step 4.

Step 4. Solve the equation by the quadratic formula.

Choose and use a method to solve each equation.

1. $5x^2 - 11x + 2 = 0$

2. $5x^2 + 13x - 6 = 0$

3. $x^2 - 1 = 2x$

4. $x^2 + 7 = 6x$

5. $a^2 = 20$

6. $a^2 = 72$

7. $x^2 - x + 4 = 0$

8. $x^2 - 2x + 7 = 0$

9. $3x^2 - 12x + 12 = 0$

10. $5x^2 - 30x + 45 = 0$

11. $9 - 6p + p^2 = 0$

12. $49 - 28p + 4p^2 = 0$

13. $4y^2 - 16 = 0$

14. $3y^2 - 27 = 0$

15. $x^4 - 3x^3 + 2x^2 = 0$

16. $x^3 + 7x^2 + 12x = 0$

17. $(2z + 5)^2 = 25$

18. $(3z - 4)^2 = 16$

19. ______
20. ______
21. ______
22. ______
23. ______
24. ______
25. ______
26. ______
27. ______
28. ______
29. ______
30. ______
31. ______
32. ______
33. ______
34. ______
35. ______
36. ______
37. ______
38. ______
39. ______
40. ______
41. ______

19. $30x = 25x^2 + 2$

20. $12x = 4x^2 + 4$

21. $\frac{2}{3}m^2 - \frac{1}{3}m - 1 = 0$

22. $\frac{5}{8}m^2 + m - \frac{1}{2} = 0$

23. $x^2 - \frac{1}{2}x - \frac{1}{5} = 0$

24. $x^2 + \frac{1}{2}x - \frac{1}{8} = 0$

25. $4x^2 - 27x + 35 = 0$

26. $9x^2 - 16x + 7 = 0$

27. $(7 - 5x)^2 = 18$

28. $(5 - 4x)^2 = 75$

29. $3z^2 - 7z = 12$

30. $6z^2 + 7z = 6$

31. $x = x^2 - 110$

32. $x = 56 - x^2$

33. $\frac{3}{4}x^2 - \frac{5}{2}x - 2 = 0$

34. $x^2 - \frac{6}{5}x - \frac{8}{5} = 0$

35. $x^2 - 0.6x + 0.05 = 0$

36. $x^2 - 0.1x - 0.06 = 0$

37. $10x^2 - 11x + 2 = 0$

38. $20x^2 - 11x + 1 = 0$

39. $\frac{1}{2}z^2 - 2z + \frac{3}{4} = 0$

40. $\frac{1}{5}z^2 - \frac{1}{2}z - 2 = 0$

41. Explain how you will decide what method to use when solving quadratic equations.

16.4 Graphing Quadratic Equations in Two Variables

OBJECTIVES

Ⓐ Graph quadratic equations of the form $y = ax^2$.

Ⓑ Graph quadratic equations of the form $y = ax^2 + bx + c$.

SSM TUTOR CENTER SG CD & VIDEO MATH PRO WEB

Recall from Section 13.2 that the graph of a linear equation in two variables $Ax + By = C$ is a straight line. In this section, we will find that the graph of a quadratic equation in the form $y = ax^2 + bx + c$ is a parabola.

Ⓐ Graphing $y = ax^2$

We begin our work by graphing $y = x^2$. To do so, we will find and plot ordered pair solutions of this equation. Let's select a few values for x, find the corresponding y-values, and record them in a table of values to keep track. Then we can plot the points corresponding to these solutions.

If $x = -3$, then $y = (-3)^2$, or 9.
If $x = -2$, then $y = (-2)^2$, or 4.
If $x = -1$, then $y = (-1)^2$, or 1.
If $x = 0$, then $y = 0^2$, or 0.
If $x = 1$, then $y = 1^2$, or 1.
If $x = 2$, then $y = 2^2$, or 4.
If $x = 3$, then $y = 3^2$, or 9.

x	y
−3	9
−2	4
−1	1
0	0
1	1
2	4
3	9

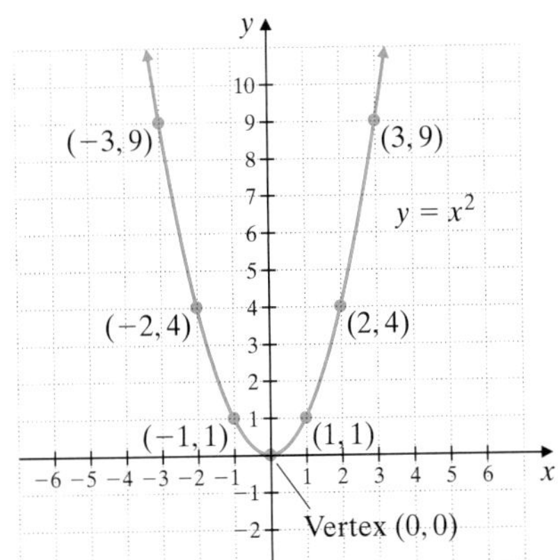

The graph of $y = x^2$ is a smooth curve through the plotted points. This curve is called a **parabola.** The lowest point on a parabola opening upward is called the **vertex.** The vertex is $(0, 0)$ for the parabola $y = x^2$. If we fold the graph paper along the y-axis, the two pieces of the parabola match perfectly. For this reason, we say the graph is **symmetric about the y-axis,** and we call the y-axis the **line of symmetry.**

Notice that the parabola that corresponds to the equation $y = x^2$ opens upward. This happens when the coefficient of x^2 is positive. In the equation $y = x^2$, the coefficient of x^2 is 1. Example 1 shows the graph of a quadratic equation where the coefficient of x^2 is negative.

EXAMPLE 1 Graph: $y = -2x^2$

Solution: We begin by selecting x-values and calculating the corresponding y-values. Then we plot the ordered pairs found and draw a smooth curve through those points. Notice that when the coefficient of x^2 is negative, the

Practice Problem 1

Graph: $y = -3x^2$

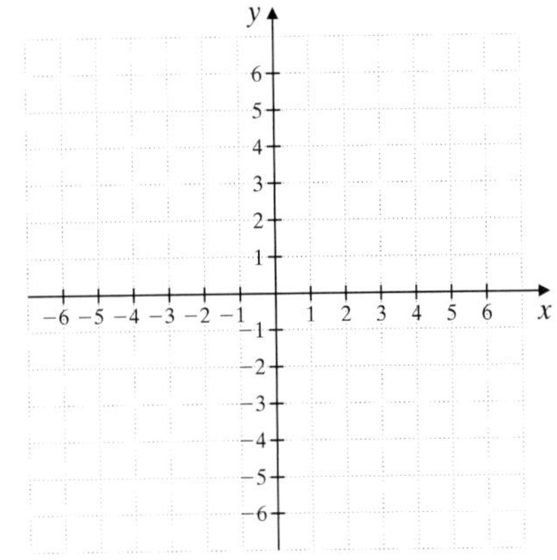

Answer

1.

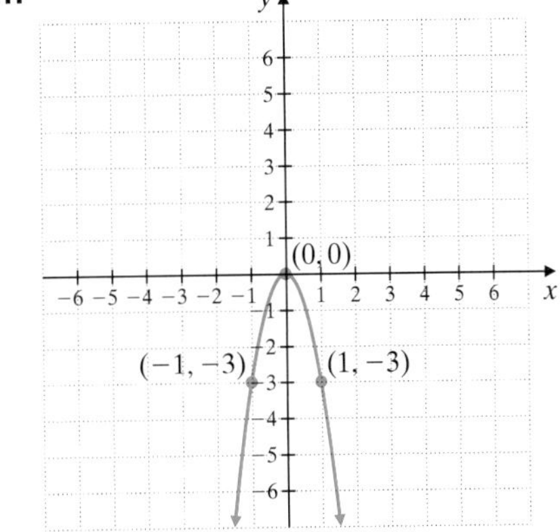

corresponding parabola opens downward. When a parabola opens downward, the vertex is the highest point of the parabola. The vertex of this parabola is $(0, 0)$.

$y = -2x^2$

x	y
0	0
1	−2
2	−8
3	−18
−1	−2
−2	−8
−3	−18

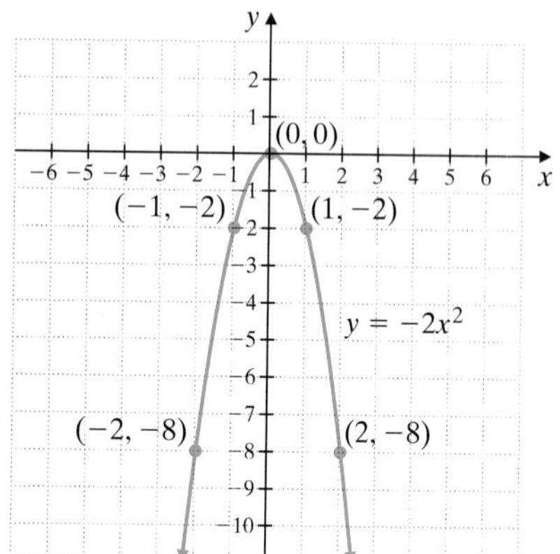

B Graphing $y = ax^2 + bx + c$

Just as for linear equations, we can use x- and y-intercepts to help graph quadratic equations. Recall from Chapter 13 that an x-intercept is the point where the graph crosses the x-axis. A y-intercept is the point where the graph crosses the y-axis.

Helpful Hint

Recall that:
To find x-intercepts, let $y = 0$ and solve for x.
To find y-intercepts, let $x = 0$ and solve for y.

Practice Problem 2

Graph: $y = x^2 - 9$

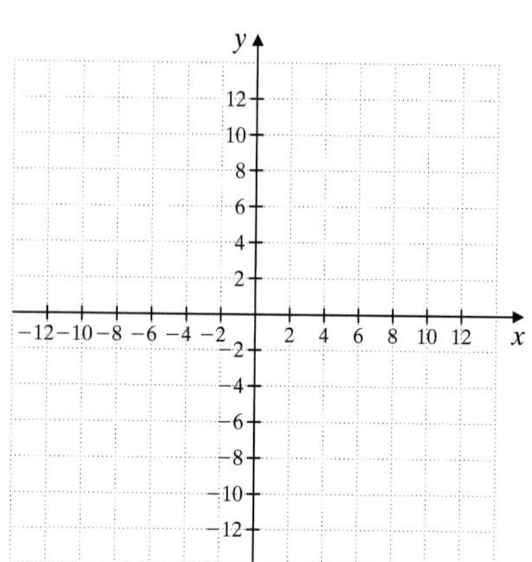

Answer

2.

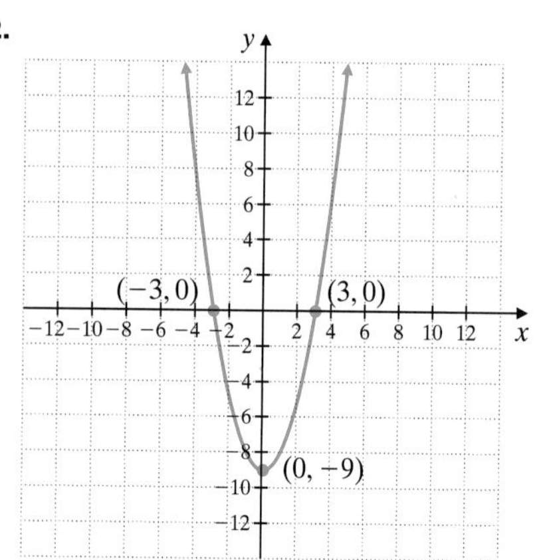

EXAMPLE 2 Graph: $y = x^2 - 4$

Solution: If we write this equation as $y = x^2 + 0x + (-4)$, we can see that it is in the form $y = ax^2 + bx + c$. To graph it, we first find the intercepts. To find the y-intercept, we let $x = 0$. Then

$$y = 0^2 - 4 = -4$$

To find x-intercepts, we let $y = 0$.

$$0 = x^2 - 4$$
$$0 = (x - 2)(x + 2)$$
$$x - 2 = 0 \quad \text{or} \quad x + 2 = 0$$
$$x = 2 \qquad\qquad x = -2$$

Thus far, we have the y-intercept $(0, -4)$ and the x-intercepts $(2, 0)$ and $(-2, 0)$. Now we can select additional x-values, find the corresponding y-values, plot the points, and draw a smooth curve through the points.

$y = x^2 - 4$

x	y
0	-4
1	-3
2	0
3	5
-1	-3
-2	0
-3	5

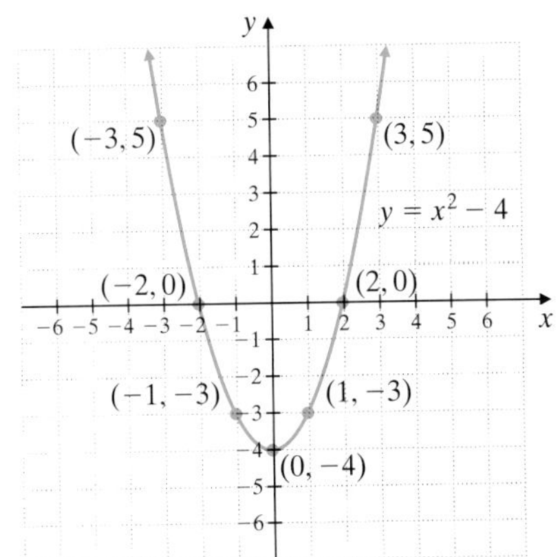

Notice that the vertex of this parabola is $(0, -4)$.

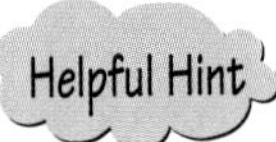

For the graph of $y = ax^2 + bx + c$,
If a is positive, the parabola opens upward.
If a is negative, the parabola opens downward.

Try the Concept Check in the margin.

Thus far, we have accidentally stumbled upon the vertex of each parabola that we have graphed. It would be helpful if we could first find the vertex of a parabola. Next we would determine whether the parabola opens upward or downward. Finally we would calculate additional points such as x- and y-intercepts as needed. In fact, there is a formula that may be used to find the vertex of a parabola.

Vertex Formula

The vertex of the parabola $y = ax^2 + bx + c$ has x-coordinate

$$\frac{-b}{2a}$$

The corresponding y-coordinate of the vertex is obtained by substituting the x-coordinate into the equation and finding y.

One way to develop this formula is to notice that the x-value of the vertex of the parabolas that we are considering lies halfway between its x-intercepts. We will not show the development of this formula here.

EXAMPLE 3 Graph: $y = x^2 - 6x + 8$

Solution: In the equation $y = x^2 - 6x + 8$, $a = 1$ and $b = -6$. The x-coordinate of the vertex is

$$\frac{-b}{2a} = \frac{-(-6)}{2 \cdot 1} = 3 \quad \text{Use the vertex formula, } \frac{-b}{2a}.$$

Concept Check

For which of the following graphs of $y = ax^2 + bx + c$ would the value of a be negative?

a. 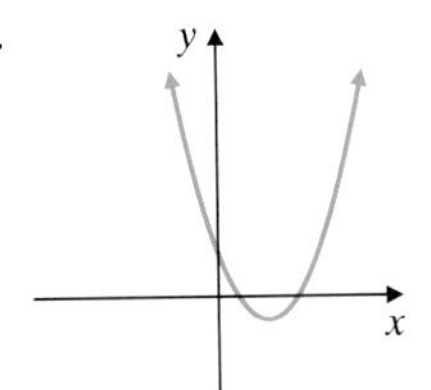

b. 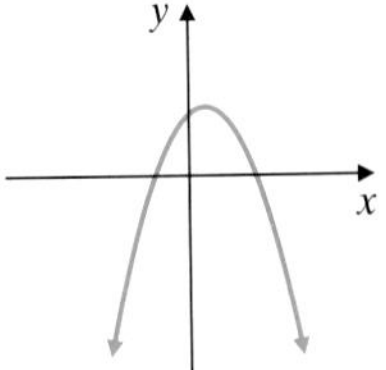

Practice Problem 3

Graph: $y = x^2 - 2x - 3$

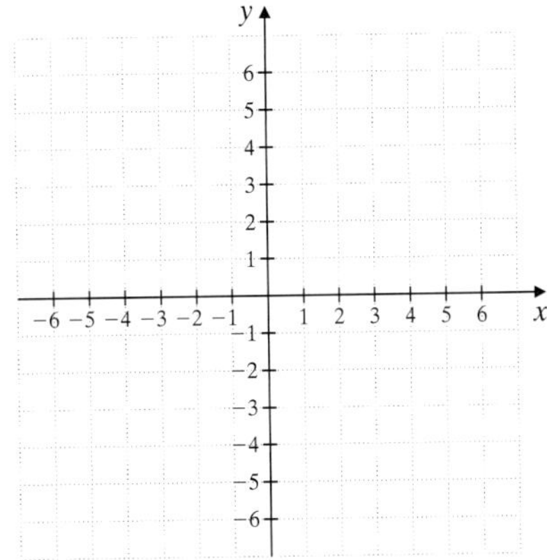

Answer

3.

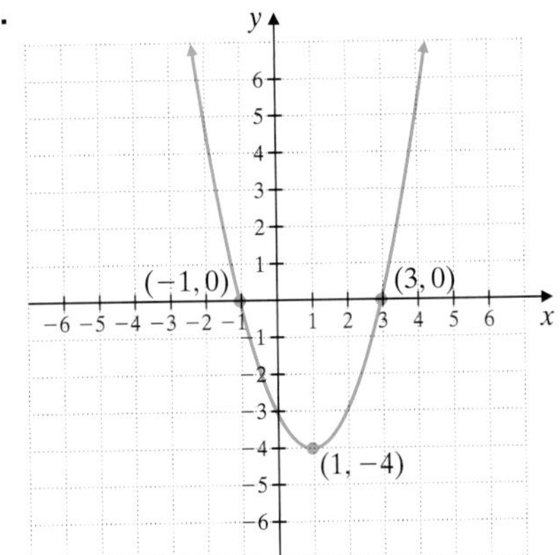

Concept Check: b

To find the corresponding y-coordinate, we let $x = 3$ in the original equation.

$$y = x^2 - 6x + 8 = 3^2 - 6 \cdot 3 + 8 = -1$$

The vertex is $(3, -1)$ and the parabola opens upward since a is positive. We now find and plot the intercepts.

To find the x-intercepts, we let $y = 0$.

$$0 = x^2 - 6x + 8$$

We factor the expression $x^2 - 6x + 8$ to find $(x - 4)(x - 2) = 0$. The x-intercepts are $(4, 0)$ and $(2, 0)$.

If we let $x = 0$ in the original equation, then $y = 8$ gives us the y-intercept $(0, 8)$. Now we plot the vertex $(3, -1)$ and the intercepts $(4, 0)$, $(2, 0)$, and $(0, 8)$. Then we can sketch the parabola. These and two additional points are shown in the table.

x	y
3	−1
4	0
2	0
0	8
1	3
5	3

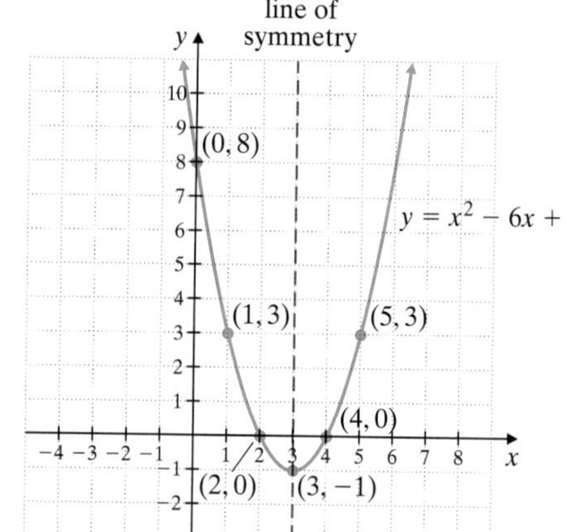

EXAMPLE 4 Graph: $y = x^2 + 2x - 5$

Solution: In the equation $y = x^2 + 2x - 5$, $a = 1$ and $b = 2$. Using the vertex formula, we find that the x-coordinate of the vertex is

$$x = \frac{-b}{2a} = \frac{-2}{2 \cdot 1} = -1$$

The y-coordinate is

$$y = (-1)^2 + 2(-1) - 5 = -6$$

Thus the vertex is $(-1, -6)$.

To find the x-intercepts, we let $y = 0$.

$$0 = x^2 + 2x - 5$$

This cannot be solved by factoring, so we use the quadratic formula.

$$x = \frac{-2 \pm \sqrt{2^2 - 4(1)(-5)}}{2 \cdot 1} \quad \text{Let } a = 1, b = 2, \text{ and } c = -5.$$

$$x = \frac{-2 \pm \sqrt{24}}{2}$$

$$x = \frac{-2 \pm 2\sqrt{6}}{2} \quad \text{Simplify the radical.}$$

$$x = \frac{2(-1 \pm \sqrt{6})}{2} = -1 \pm \sqrt{6}$$

Practice Problem 4

Graph: $y = x^2 - 3x + 1$

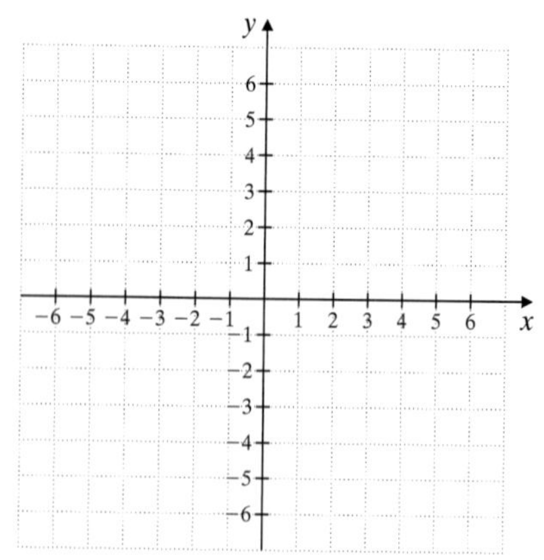

Answer

4.

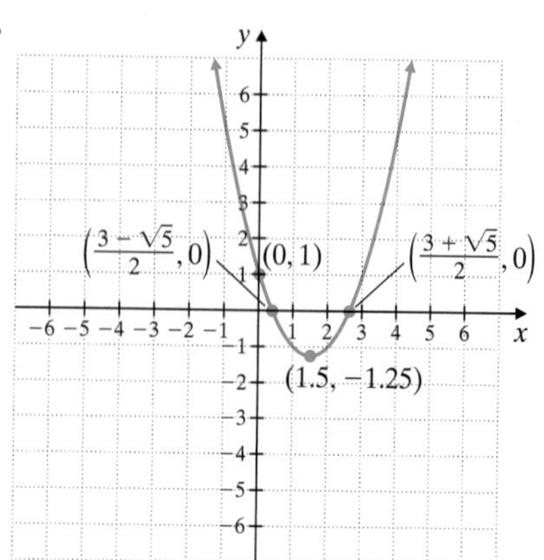

The x-intercepts are $(-1 + \sqrt{6}, 0)$ and $(-1 - \sqrt{6}, 0)$. We use a calculator to approximate these so that we can easily graph these intercepts.

$$-1 + \sqrt{6} \approx 1.4 \qquad \text{and} \qquad -1 - \sqrt{6} \approx -3.4$$

To find the y-intercept, we let $x = 0$ in the original equation and find that $y = -5$. Thus the y-intercept is $(0, -5)$.

x	y
-1	-6
$-1 + \sqrt{6}$	0
$-1 - \sqrt{6}$	0
0	-5
-2	-5

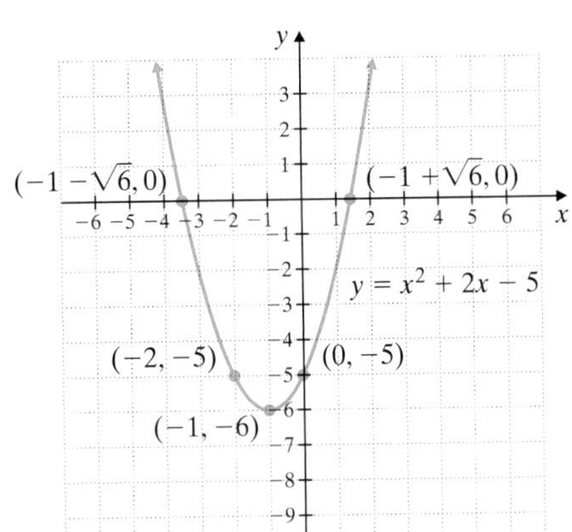

Helpful Hint

Notice that the number of x-intercepts of the graph of the parabola $y = ax^2 + bx + c$ is the same as the number of real solutions of $0 = ax^2 + bx + c$.

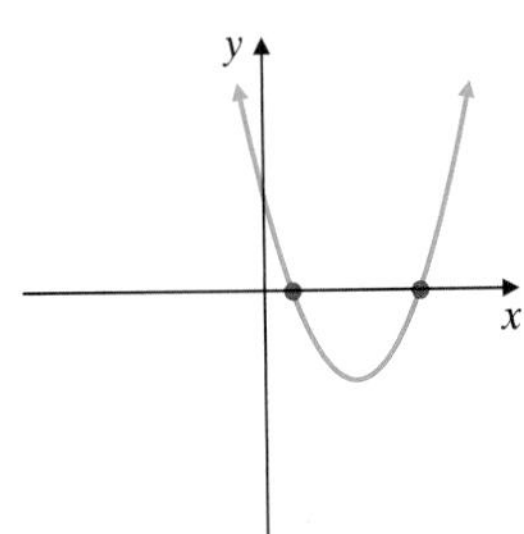

$y = ax^2 + bx + c$
$a > 0$
Two x-intercepts
Two real solutions of
$0 = ax^2 + bx + c$

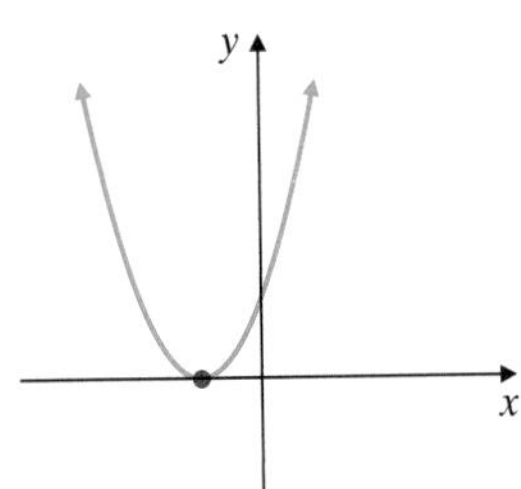

$y = ax^2 + bx + c$
$a > 0$
One x-intercept
One real solution of
$0 = ax^2 + bx + c$

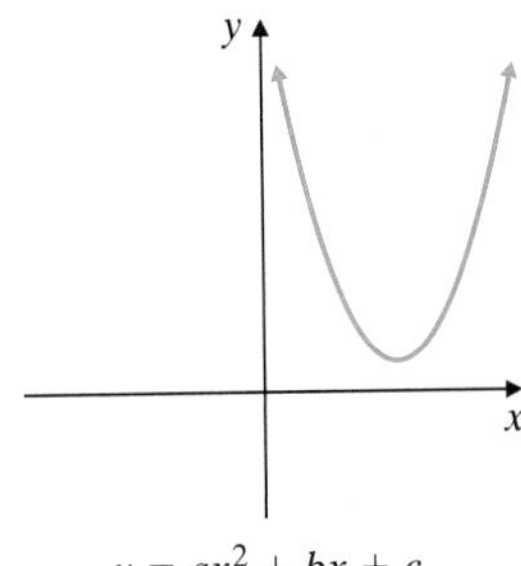

$y = ax^2 + bx + c$
$a > 0$
No x-intercepts
No real solutions of
$0 = ax^2 + bx + c$

GRAPHING CALCULATOR EXPLORATIONS

Recall that a graphing calculator may be used to solve quadratic equations. The x-intercepts of the graph of $y = ax^2 + bx + c$ are solutions of $0 = ax^2 + bx + c$. To solve $x^2 - 7x - 3 = 0$, for example, graph $y_1 = x^2 - 7x - 3$. The x-intercepts of the graph are the solutions of the equation.

Use a graphing calculator to solve each quadratic equation. Round solutions to two decimal places.

1. $x^2 - 7x - 3 = 0$

2. $2x^2 - 11x - 1 = 0$

3. $-1.7x^2 + 5.6x - 3.7 = 0$

4. $-5.8x^2 + 2.3x - 3.9 = 0$

5. $5.8x^2 - 2.6x - 1.9 = 0$

6. $7.5x^2 - 3.7x - 1.1 = 0$

Name ______________________ Section __________ Date __________

EXERCISE SET 16.4

A *Graph each quadratic equation by finding and plotting ordered pair solutions. See Example 1.*

1. $y = 2x^2$

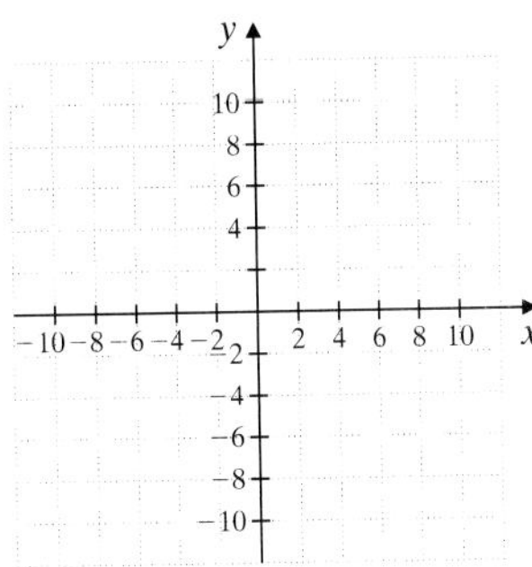

2. $y = 3x^2$

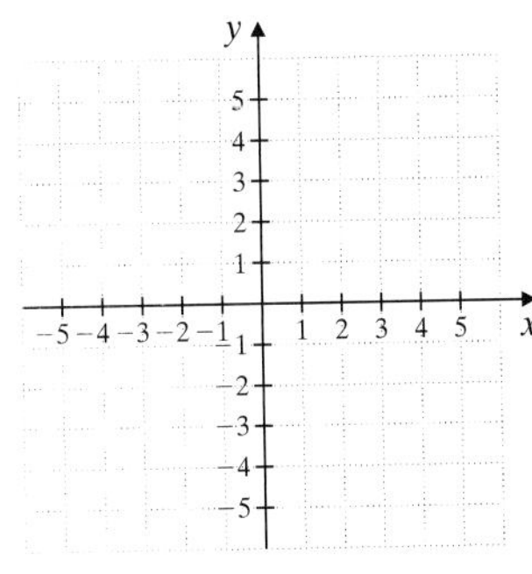

3. $y = -x^2$

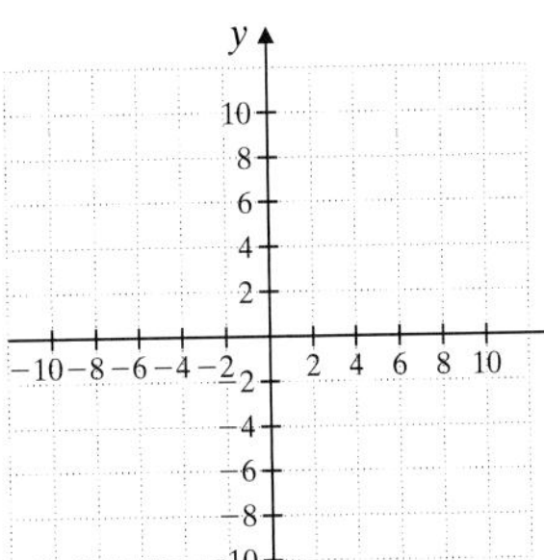

4. $y = -4x^2$

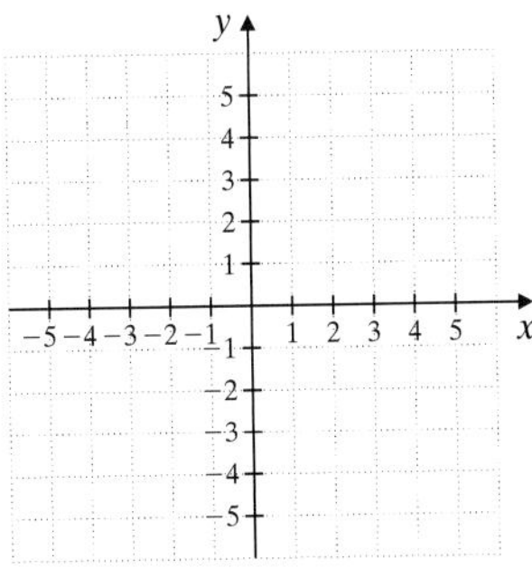

5. $y = \frac{1}{3}x^2$

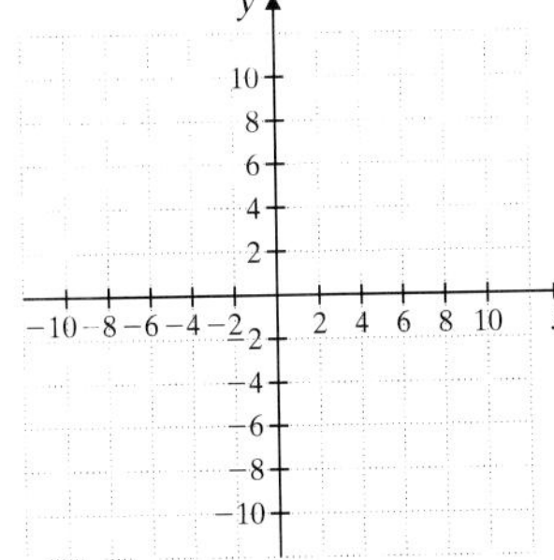

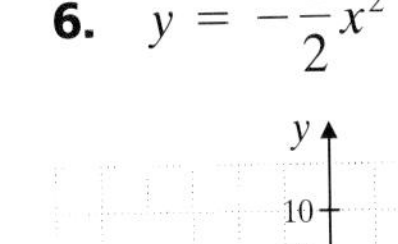

6. $y = -\frac{1}{2}x^2$

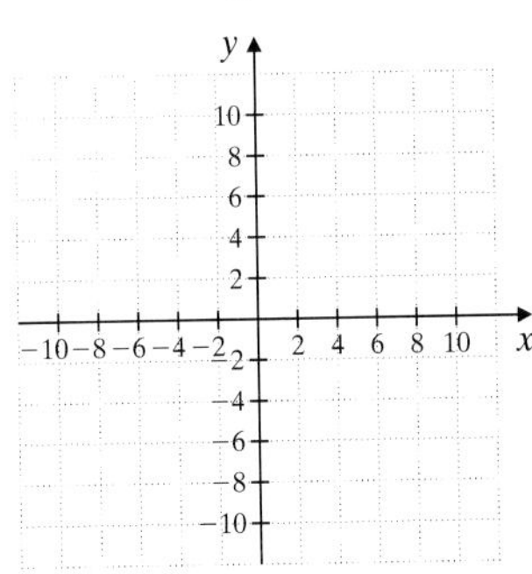

B *Sketch the graph of each equation. Identify the vertex and the intercepts. See Examples 2 through 4.*

7. $y = x^2 - 1$

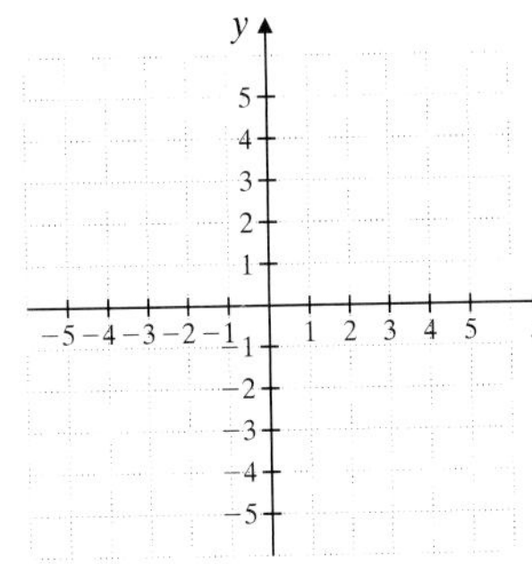

8. $y = x^2 - 16$

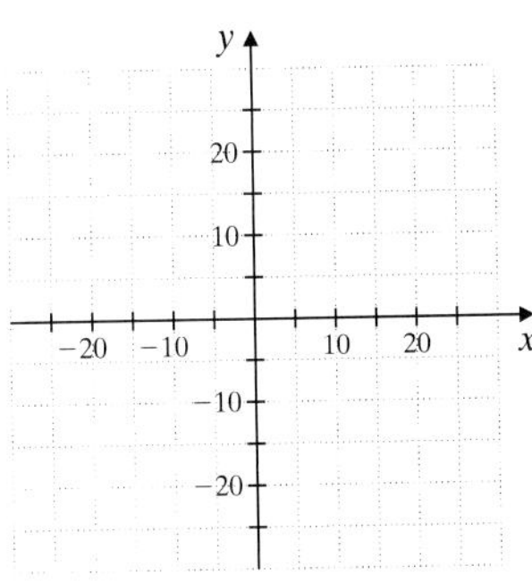

9. $y = x^2 + 4$

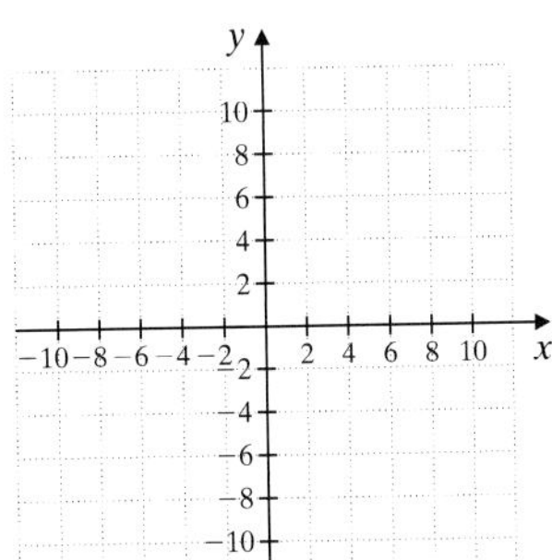

10. $y = x^2 + 9$

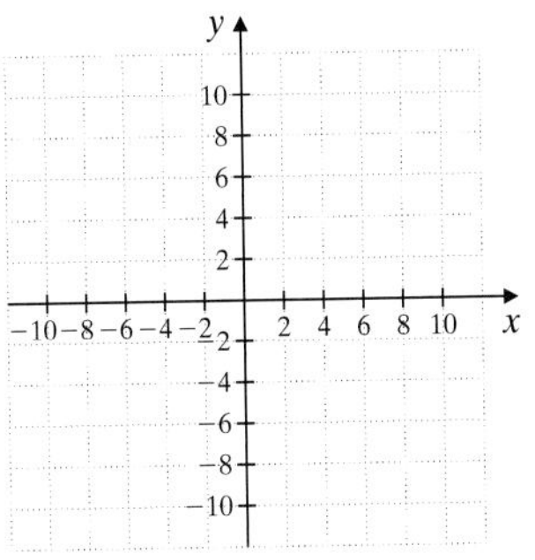

11. $y = x^2 + 6x$

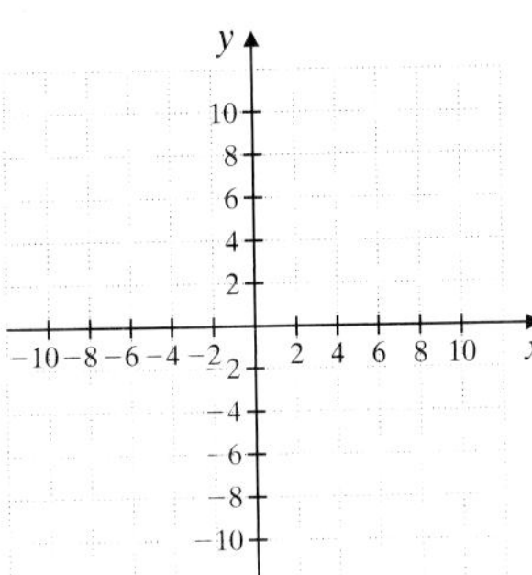

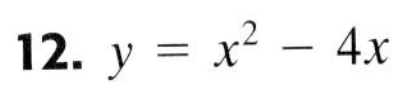

12. $y = x^2 - 4x$

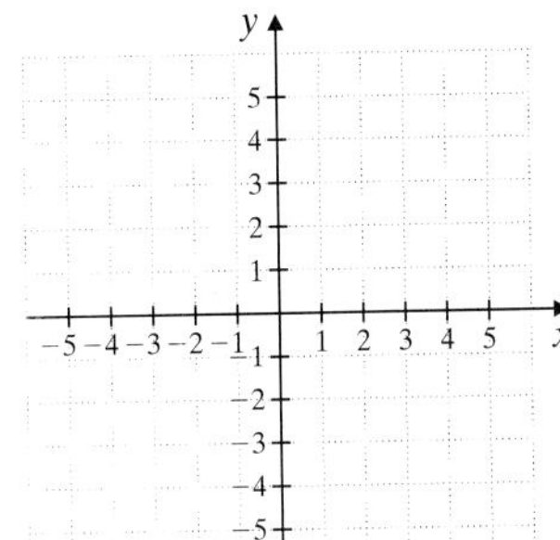

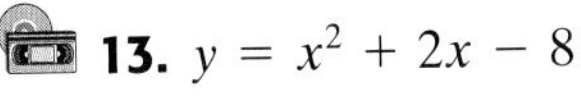

13. $y = x^2 + 2x - 8$

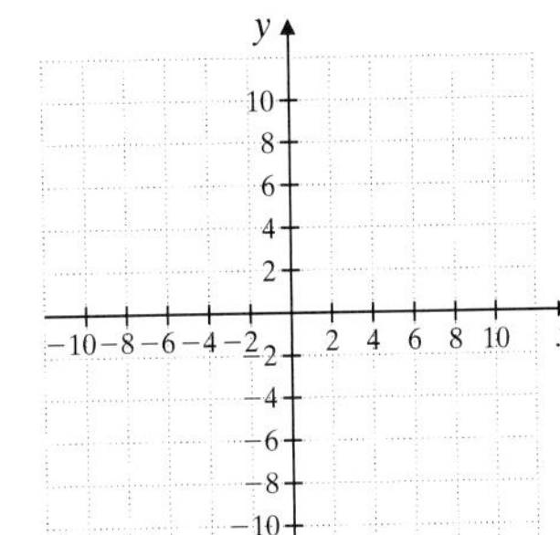

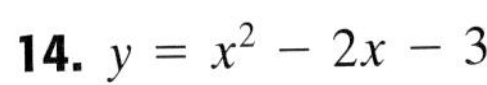

14. $y = x^2 - 2x - 3$

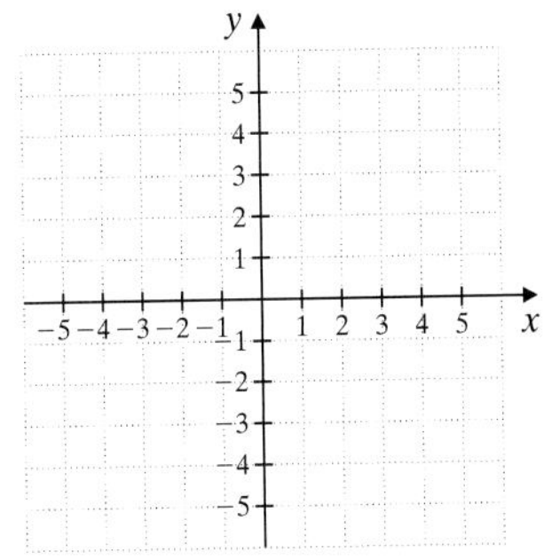

15. $y = -x^2 + x + 2$

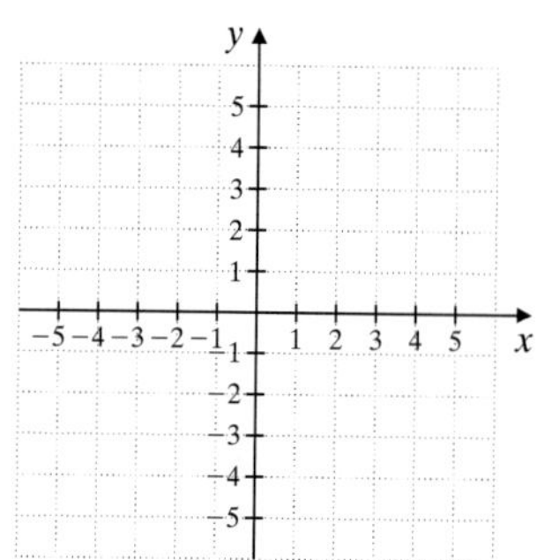

16. $y = -x^2 - 2x - 1$

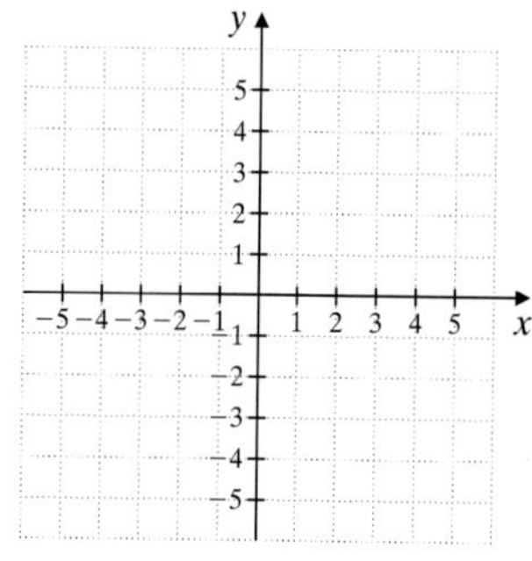

17. $y = x^2 + 5x + 4$

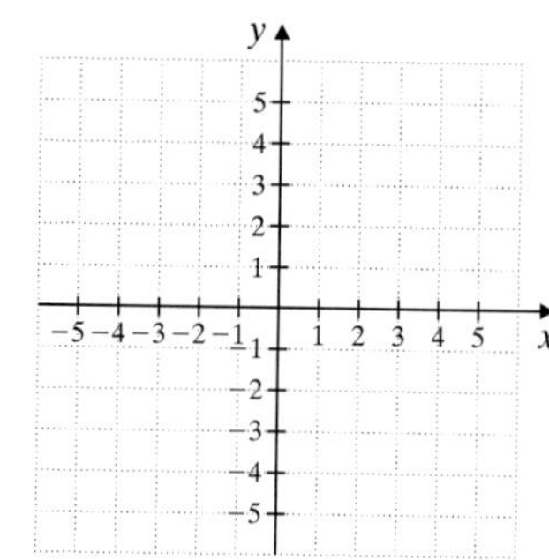

18. $y = x^2 + 7x + 10$

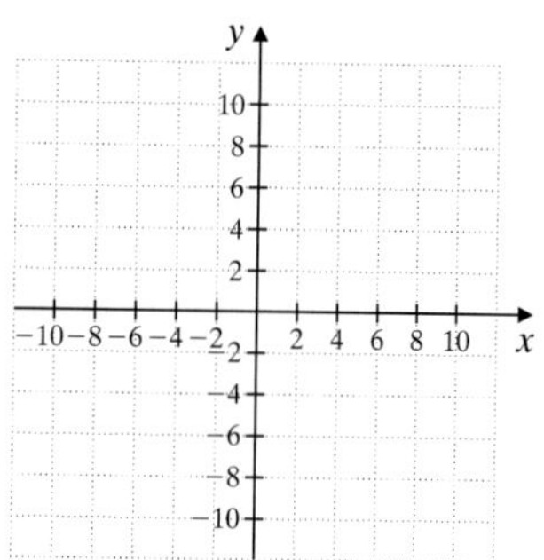

19. $y = x^2 - 4x + 5$

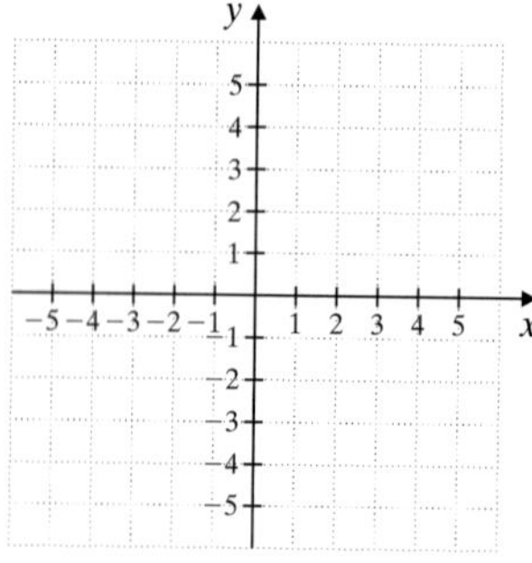

20. $y = x^2 - 6x + 10$

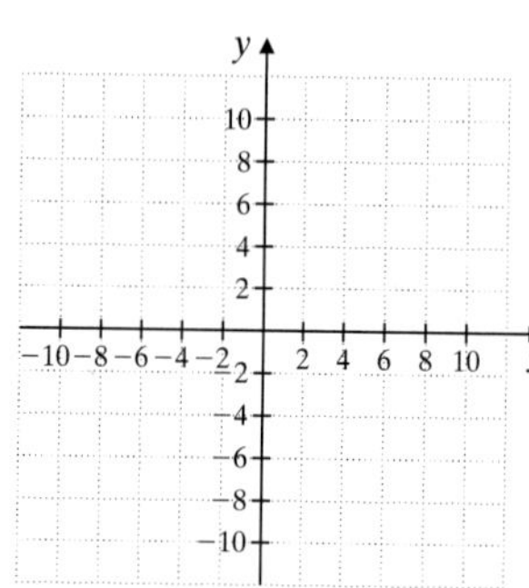

21. $y = 2 - x^2$

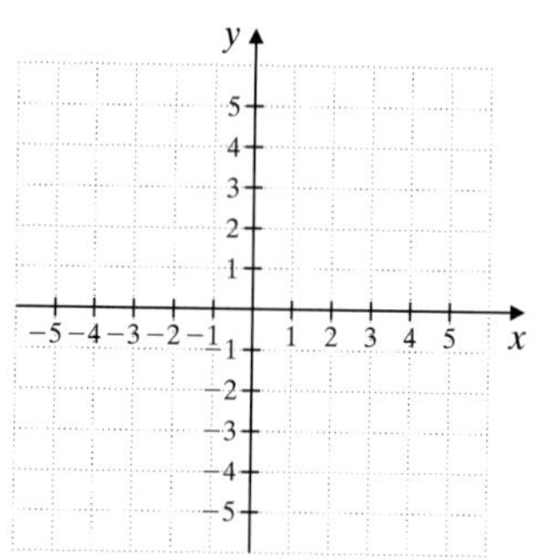

22. $y = 3 - x^2$

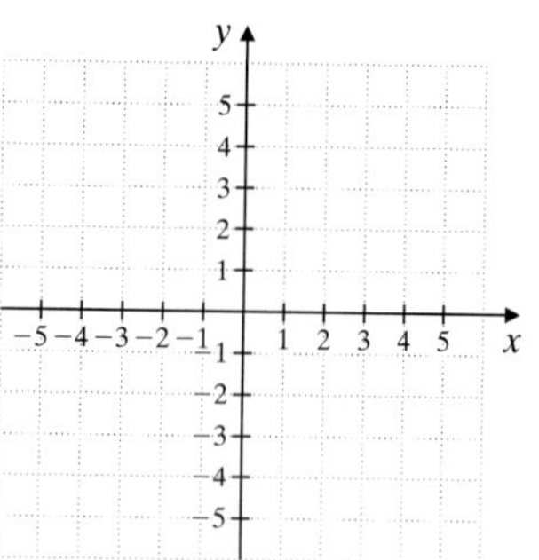

23. $y = 2x^2 - 11x + 5$

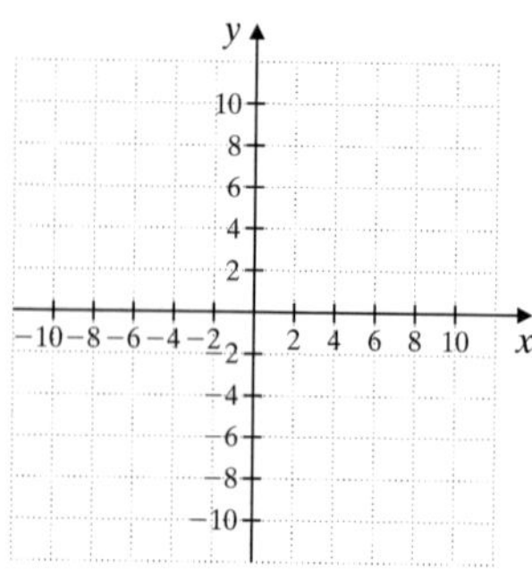

24. $y = 2x^2 + x - 3$

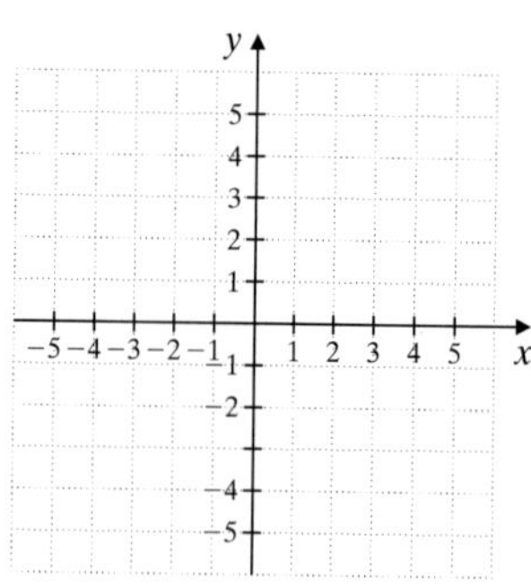

25. $y = -x^2 + 4x - 3$

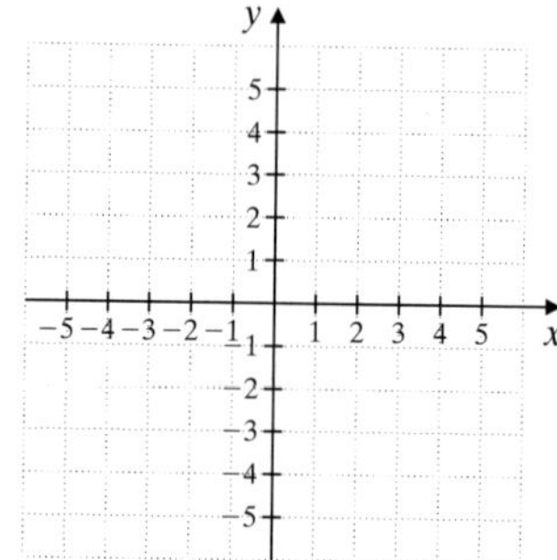

26. $y = -x^2 + 6x - 8$

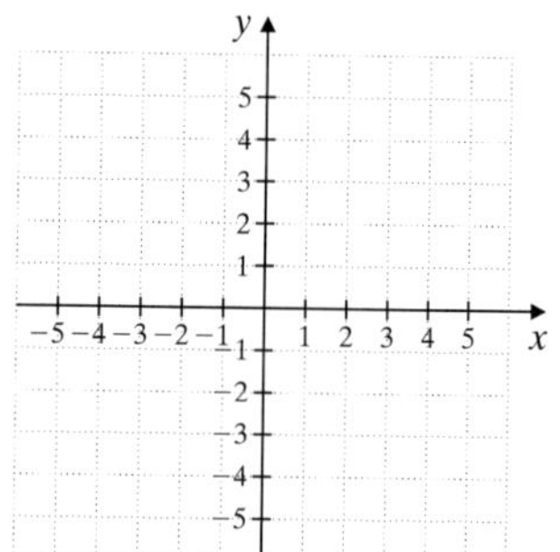

Review and Preview

Simplify each complex fraction. See Section 12.7.

27. $\dfrac{\frac{1}{7}}{\frac{2}{5}}$

28. $\dfrac{\frac{3}{8}}{\frac{1}{7}}$

29. $\dfrac{\frac{1}{x}}{\frac{2}{x^2}}$

30. $\dfrac{\frac{x}{5}}{\frac{2}{x}}$

31. $\dfrac{2x}{1 - \frac{1}{x}}$

32. $\dfrac{x}{x - \frac{1}{x}}$

33. $\dfrac{\frac{a - b}{2b}}{\frac{b - a}{8b^2}}$

34. $\dfrac{\frac{2a^2}{a - 3}}{\frac{a}{3 - a}}$

Combining Concepts

35. The height h of a fireball launched from a Roman candle with an initial velocity of 128 feet per second is given by the equation $h = -16t^2 + 128t$, where t is time in seconds after launch.

Use the graph of this equation to answer each question.

a. Estimate the maximum height of the fireball.

b. Estimate the time when the fireball is at its maximum height.

c. Estimate the time when the fireball returns to the ground.

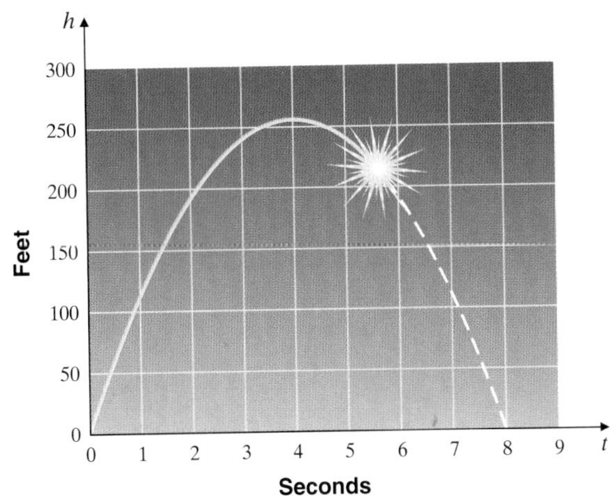

Match the graph of each equation of the form $y = ax^2 + bx + c$ with the given description.

36. $a > 0$, two x-intercepts

37. $a < 0$, one x-intercept

38. $a < 0$, no x-intercept

39. $a > 0$, no x-intercept

40. $a > 0$, one x-intercept

41. $a < 0$, two x-intercepts

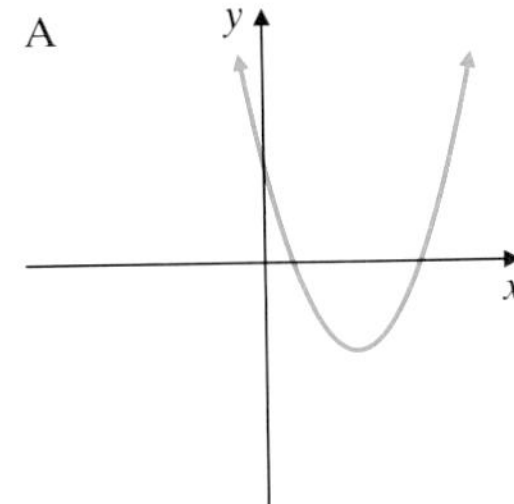

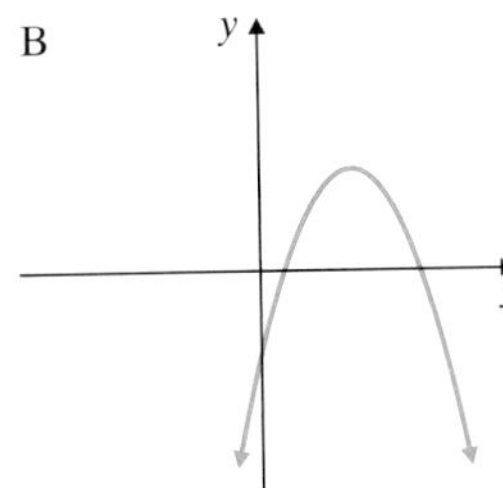

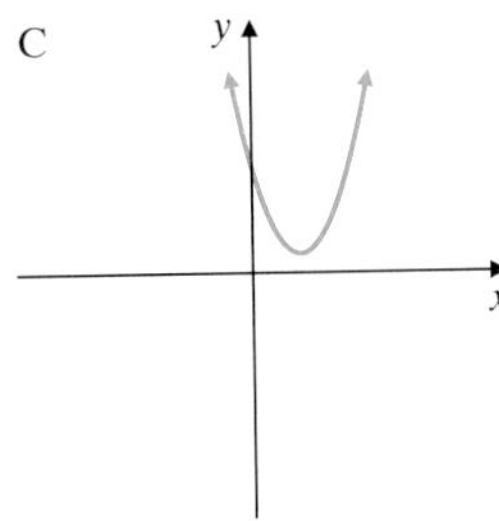

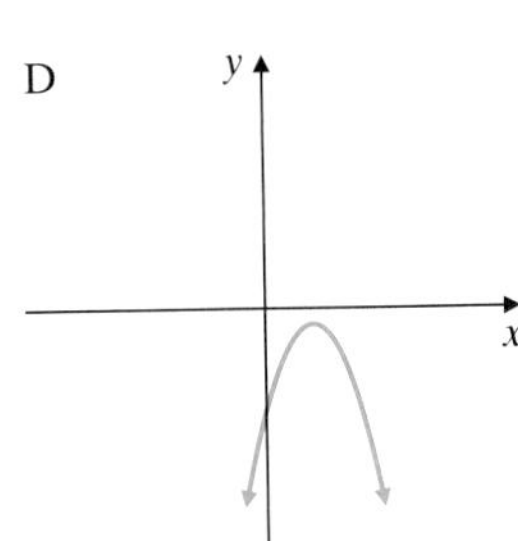

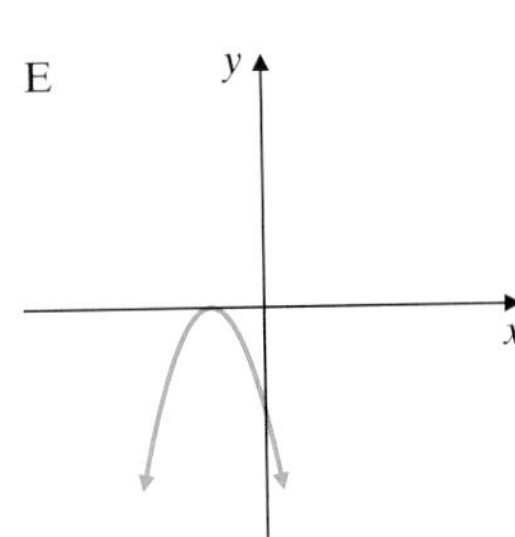

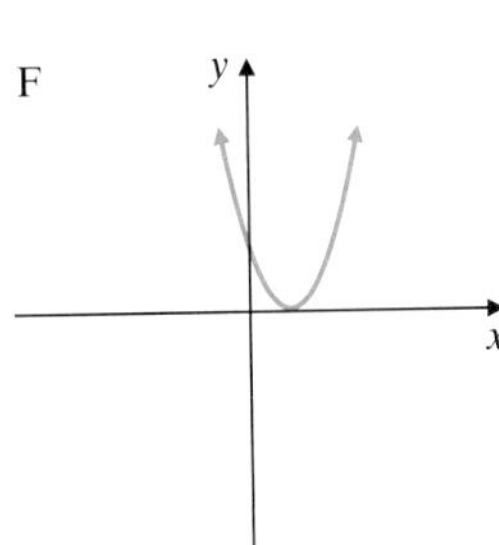

42. Determine the maximum number and the minimum number of x-intercepts for a parabola. Explain your answers.

CHAPTER 16 ACTIVITY Modeling a Physical Situation

Figure 1

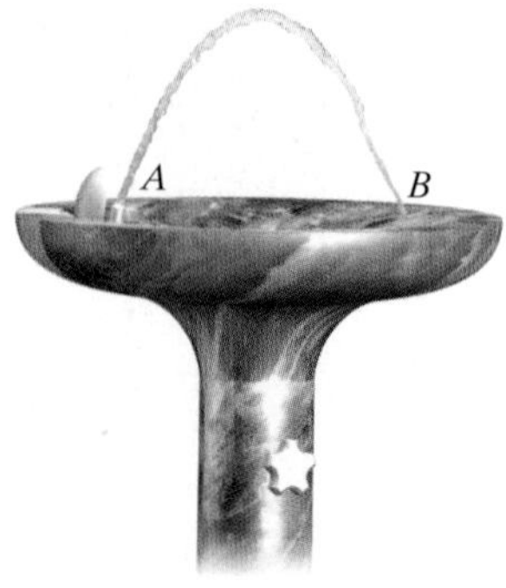

Figure 2

This activity may be completed by working in groups or individually.

Model the physical situation of the parabolic path of water from a water fountain. For simplicity, use the given Figure 2 to investigate the following questions.

Data Table		
	x	**y**
Point A	0	0
Point B		
Point V		

1. Collect data for the x-intercepts of the parabolic path. Let points A and B in Figure 2 be on the x-axis and let the coordinates of point A be $(0, 0)$. Use a ruler to measure the distance between points A and B **on Figure 2** to the nearest even one-tenth centimeter, and use this information to determine the coordinates of point B. Record this data in the data table, (*Hint*: If the distance from A to B measures 8 one-tenth centimeters, then the coordinates of point B are $(8, 0)$.)

2. Next, collect data for the vertex V of the parabolic path. What is the relationship between the x-coordinate of the vertex and the x-intercepts found in Question 1? What is the line of symmetry? To locate point V in Figure 2, find the midpoint of the line segment joining points A and B and mark point V on the path of water directly above the midpoint. To approximate the y-coordinate of the vertex, use a ruler to measure its distance from the x-axis to the nearest one-tenth centimeter. Record this data in the data table.

3. Plot the points from the data table on a rectangular coordinate system. Sketch the parabolic curve through your points A, B, and V.

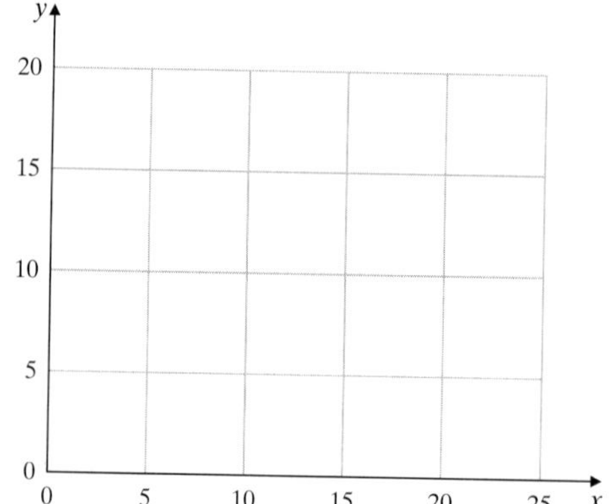

4. Which of the following models best fits the data you collected? Explain your reasoning.
   a. $y = 16x + 18$
   b. $y = -13x^2 + 20x$
   c. $y = 0.13x^2 - 2.6x$
   d. $y = -0.13x^2 + 2.6x$

5. (Optional) Enter your data into a graphing calculator and use the quadratic curve-fitting feature to find a model for your data. How does the model compare with your selection from Question 4?

Chapter 16 VOCABULARY CHECK

Fill in each blank with one of the words or phrases listed below.

square root	vertex
completing the square	quadratic

1. If $x^2 = a$, then $x = \sqrt{a}$ or $x = -\sqrt{a}$. This property is called the ______________ property.
2. The formula $\dfrac{-b}{2a}$ where $y = ax^2 + bx + c$ is called the __________ formula.
3. The process of solving a quadratic equation by writing it in the form $(x + a)^2 = c$ is called ______________________.
4. The formula $x = \dfrac{-b \pm \sqrt{b^2 - 4ac}}{2a}$ is called the ____________ formula.

CHAPTER 16 Highlights

DEFINITIONS AND CONCEPTS | **EXAMPLES**

Section 16.1 Solving Quadratic Equations by the Square Root Property

SQUARE ROOT PROPERTY

If $x^2 = a$ for $a \geq 0$, then $x = \sqrt{a}$ or $x = -\sqrt{a}$.

Solve the equation.

$$(x - 1)^2 = 15$$

$$x - 1 = \sqrt{15} \quad \text{or} \quad x - 1 = -\sqrt{15}$$

$$x = 1 + \sqrt{15} \qquad x = 1 - \sqrt{15}$$

Section 16.2 Solving Quadratic Equations by Completing the Square

TO SOLVE A QUADRATIC EQUATION BY COMPLETING THE SQUARE

Step 1. If the coefficient of x^2 is not 1, divide both sides of the equation by the coefficient.

Step 2. Get all terms with variables alone on one side.

Step 3. Complete the square by adding the square of half of the coefficient of x to both sides.

Step 4. Factor the perfect square trinomial.

Step 5. Use the square root property to solve.

Solve $2x^2 + 12x - 10 = 0$ by completing the square.

$$\frac{2x^2}{2} + \frac{12x}{2} - \frac{10}{2} = \frac{0}{2} \qquad \text{Divide by 2.}$$

$$x^2 + 6x - 5 = 0 \qquad \text{Simplify.}$$

$$x^2 + 6x = 5 \qquad \text{Add 5 to both sides.}$$

The coefficient of x is 6. Half of 6 is 3 and $3^2 = 9$. Add 9 to both sides.

$$x^2 + 6x + 9 = 5 + 9$$

$$(x + 3)^2 = 14 \qquad \text{Factor.}$$

$$x + 3 = \sqrt{14} \quad \text{or} \quad x + 3 = -\sqrt{14}$$

$$x = -3 + \sqrt{14} \quad x = -3 - \sqrt{14}$$

DEFINITIONS AND CONCEPTS | **EXAMPLES**

Section 16.3 Solving Quadratic Equations by the Quadratic Formula

QUADRATIC FORMULA

If a, b, and c are real numbers and $a \neq 0$, the quadratic equation $ax^2 + bx + c = 0$ has solutions

$$x = \frac{-b \pm \sqrt{b^2 - 4ac}}{2a}$$

Identify a, b, and c in the quadratic equation

$$4x^2 - 6x = 5$$

First, subtract 5 from both sides.

$$4x^2 - 6x - 5 = 0$$

$a = 4, b = -6$, and $c = -5$.

TO SOLVE A QUADRATIC EQUATION BY THE QUADRATIC FORMULA

Step 1. Write the equation in standard form: $ax^2 + bx + c = 0$.

Step 2. If necessary, clear the equation of fractions.

Step 3. Identify a, b, and c.

Step 4. Replace a, b, and c in the quadratic formula with the identified values, and simplify.

Solve $3x^2 - 2x - 2 = 0$.

In this equation, $a = 3, b = -2$, and $c = -2$.

$$x = \frac{-(-2) \pm \sqrt{(-2)^2 - 4(3)(-2)}}{2 \cdot 3}$$

$$= \frac{2 \pm \sqrt{4 - (-24)}}{6}$$

$$= \frac{2 \pm \sqrt{28}}{6} = \frac{2 \pm \sqrt{4 \cdot 7}}{6} = \frac{2 \pm 2\sqrt{7}}{6}$$

$$= \frac{2(1 \pm \sqrt{7})}{2 \cdot 3} = \frac{1 \pm \sqrt{7}}{3}$$

Section 16.4 Graphing Quadratic Equations in Two Variables

The graph of a quadratic equation $y = ax^2 + bx + c, a \neq 0$, is called a **parabola.** The lowest point on a parabola opening upward or the highest point on a parabola opening downward is called the **vertex.** The vertical line through the vertex is the **line of symmetry.**

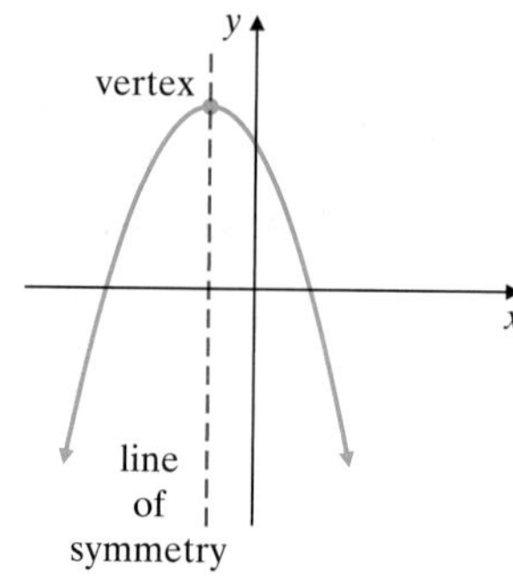

VERTEX FORMULA

The vertex of the parabola $y = ax^2 + bx + c$ has x-coordinate $\frac{-b}{2a}$. To find the corresponding y-coordinate, substitute the x-coordinate into the original equation and solve for y.

Graph: $y = 2x^2 - 6x + 4$

The x-coordinate of the vertex is

$$x = \frac{-b}{2a} = \frac{-(-6)}{2(2)} = \frac{6}{4} = \frac{3}{2}$$

The y-coordinate is

$$y = 2\left(\frac{3}{2}\right)^2 - 6\left(\frac{3}{2}\right) + 4 = 2\left(\frac{9}{4}\right) - 9 + 4$$

$$= -\frac{1}{2}$$

The vertex is $\left(\frac{3}{2}, -\frac{1}{2}\right)$.

The y-intercept is

$$y = 2 \cdot 0^2 - 6 \cdot 0 + 4 = 4$$

The x-intercepts are

$$0 = 2x^2 - 6x + 4$$

$$0 = 2(x - 2)(x - 1)$$

$$x = 2 \quad \text{or} \quad x = 1$$

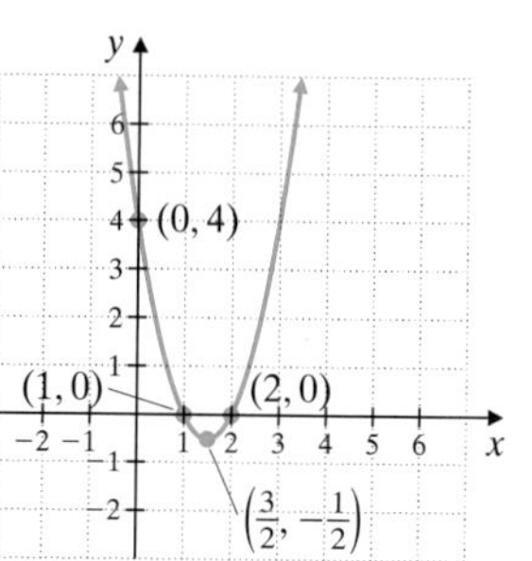

Name ____________________ Section __________ Date __________

Chapter 16 Review

(16.1) *Solve each quadradic equation by factoring.*

1. $(x - 4)(5x + 3) = 0$

2. $(x + 7)(3x + 4) = 0$

3. $3m^2 - 5m = 2$

4. $7m^2 + 2m = 5$

5. $6x^3 - 54x = 0$

6. $2x^2 - 8 = 0$

Use the square root property to solve each quadratic equation.

7. $x^2 = 36$

8. $x^2 = 81$

9. $k^2 = 50$

10. $k^2 = 45$

11. $(x - 11)^2 = 49$

12. $(x + 3)^2 = 100$

13. $(4p + 2)^2 = 100$

14. $(3p + 6)^2 = 81$

Solve. For Exercises 15 and 16, use the formula $h = 16t^2$.

15. If Kara Washington dives from a height of 100 feet, how long before she hits the water?

16. How long does a 5-mile free-fall take? Round your result to the nearest tenth of a second.

(16.2) *Solve each quadratic equation by completing the square.*

17. $x^2 + 4x = 1$

18. $x^2 - 8x = 3$

19. $x^2 - 6x + 7 = 0$

20. $x^2 + 6x + 7 = 0$

21. $2y^2 + y - 1 = 0$

22. $y^2 + 3y - 1 = 0$

(16.3) *Use the quadratic formula to solve each quadratic equation.*

23. $x^2 - 10x + 7 = 0$

24. $x^2 + 4x - 7 = 0$

25. $2x^2 + x - 1 = 0$

26. $x^2 + 3x - 1 = 0$

27. $9x^2 + 30x + 25 = 0$

28. $16x^2 - 72x + 81 = 0$

29. $15x^2 + 2 = 11x$

30. $15x^2 + 2 = 13x$

31. $2x^2 + x + 5 = 0$

32. $7x^2 - 3x + 1 = 0$

33. The average price of silver (in cents per ounce) from 1998 to 2001 is given by the equation $y = -8x^2 - 13x + 552$. In this equation, x is the number of years since 1998. Assume that this trend continues and find the year after 1998 in which the price of silver will be 180 cents per ounce. (*Source*: U.S. Bureau of Mines)

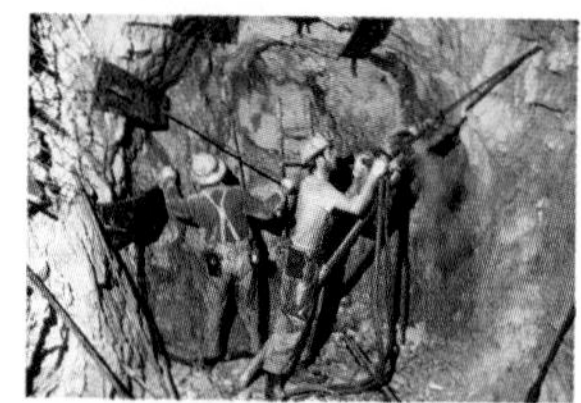

34. The average price of platinum (in dollars per ounce) from 1998 to 2001 is given by the equation $y = -5x^2 + 79.4x + 357.4$. In this equation, x is the number of years since 1998. Assume that this trend continues and find the year after 2001 in which the price of platinum will be 450 dollars per ounce. (*Source*: U.S. Bureau of Mines)

(16.4) *Graph each quadratic equation and find and plot any intercept points.*

35. $y = 3x^2$

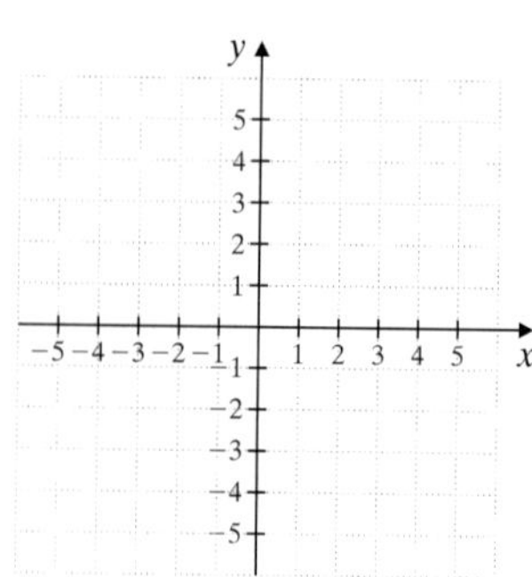

36. $y = -\frac{1}{2}x^2$

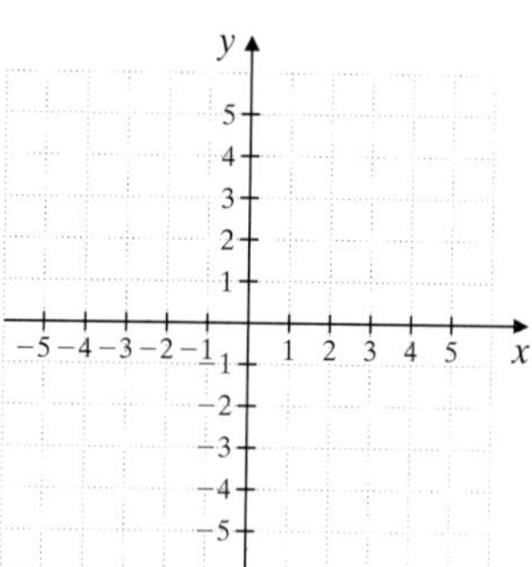

Graph each quadratic equation. Label the vertex and the intercept points with their coordinates.

37. $y = x^2 - 25$

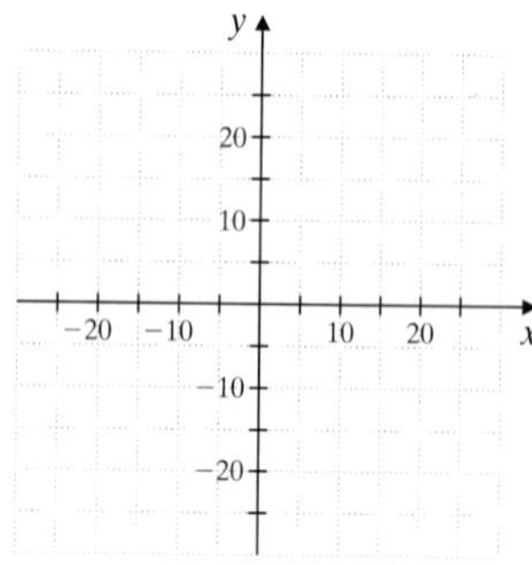

38. $y = x^2 - 36$

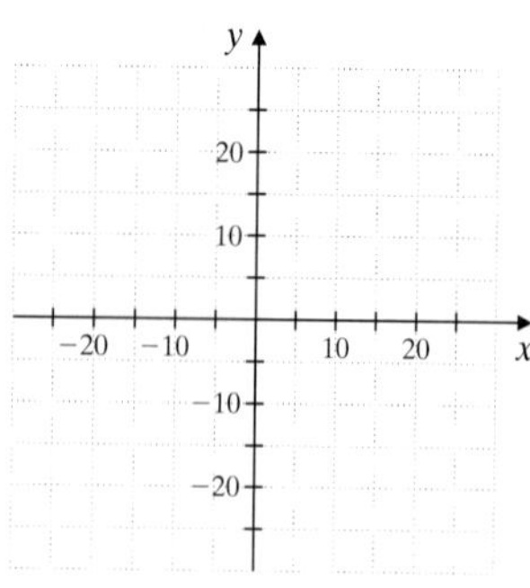

39. $y = x^2 + 3$

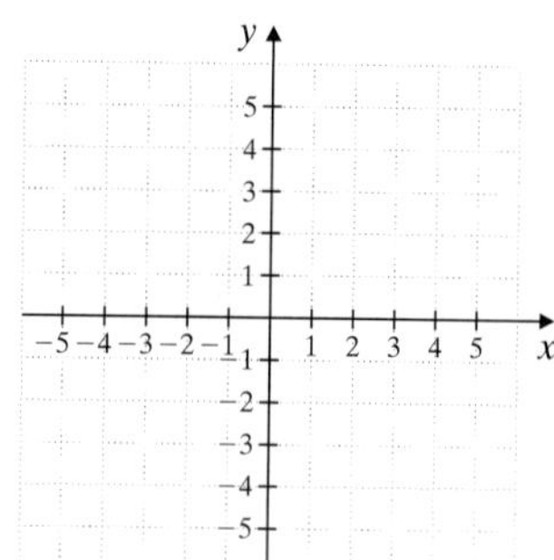

40. $y = x^2 + 8$

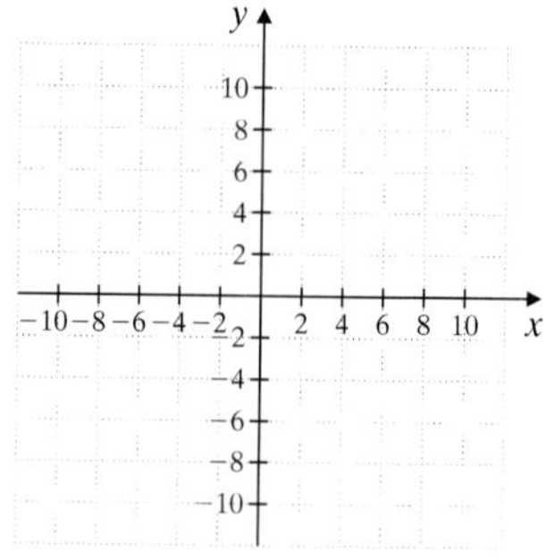

41. $y = -4x^2 + 8$

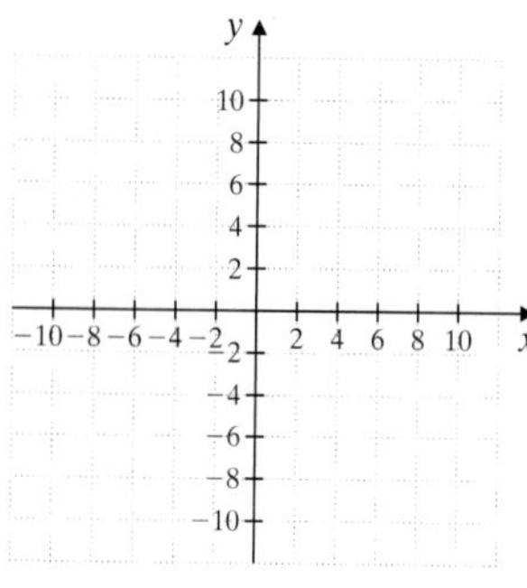

42. $y = -3x^2 + 9$

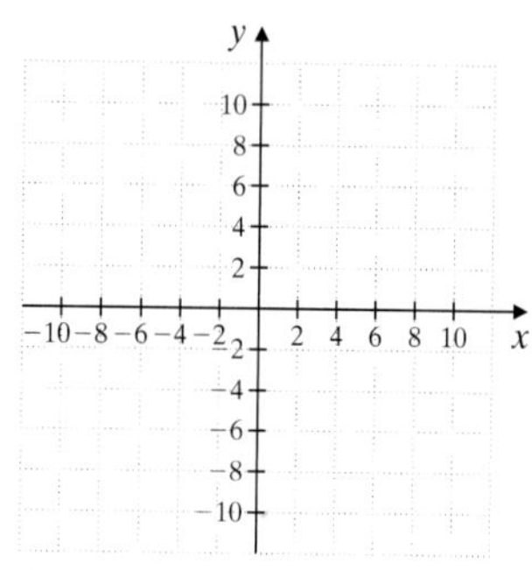

43. $y = x^2 + 3x - 10$

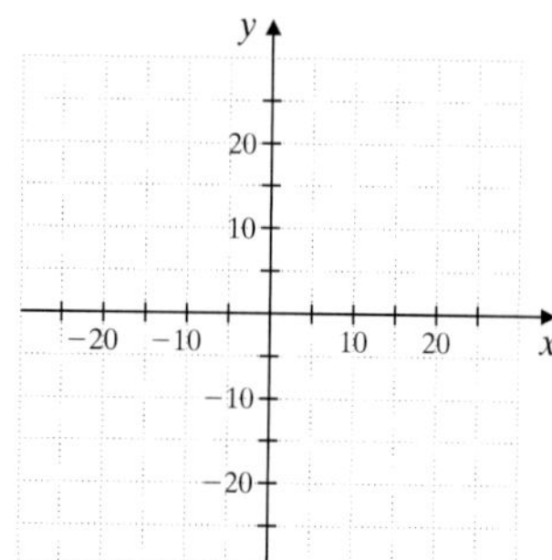

44. $y = x^2 + 3x - 4$

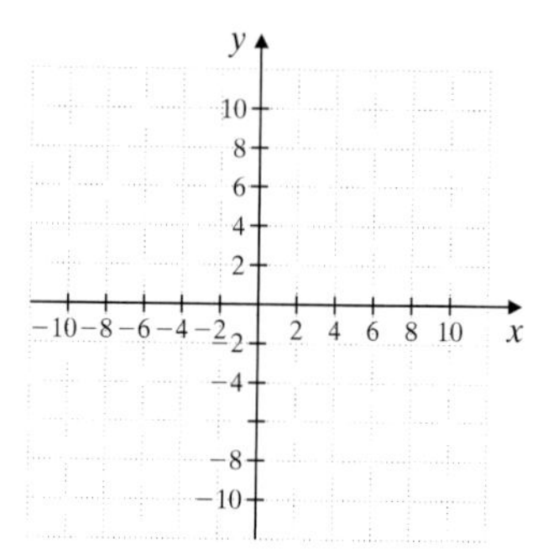

45. $y = -x^2 - 5x - 6$

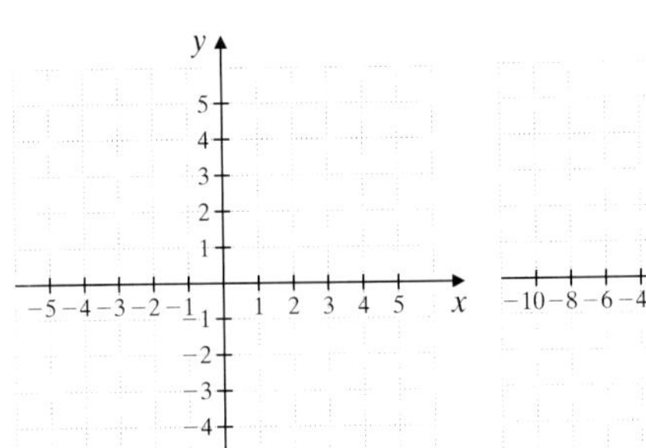

46. $y = -x^2 + 4x + 8$

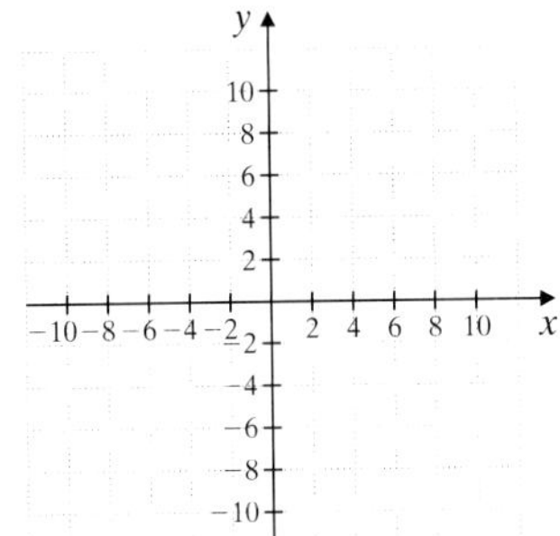

47. $y = 2x^2 - 11x - 6$

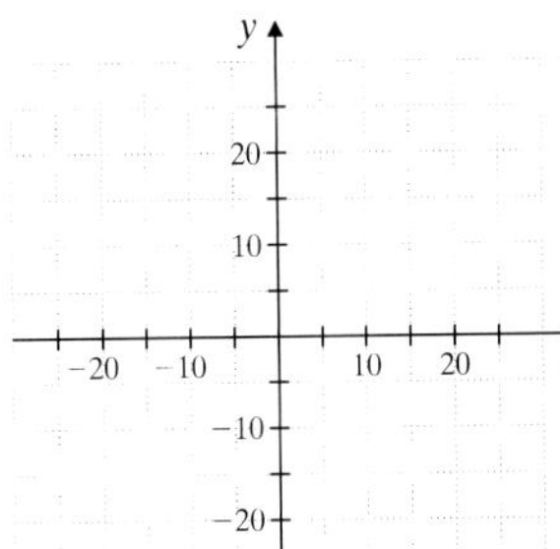

48. $y = 3x^2 - x - 2$

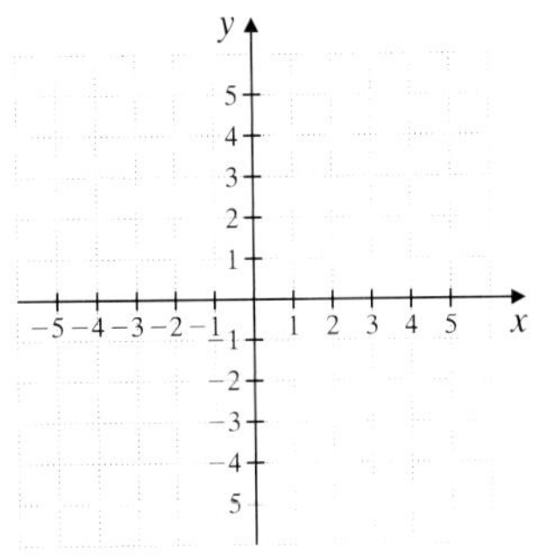

Quadratic equations in the form $y = ax^2 + bx + c$ are graphed below. Determine the number of real solutions for the related equation $0 = ax^2 + bx + c$ from each graph. List the solutions.

49.

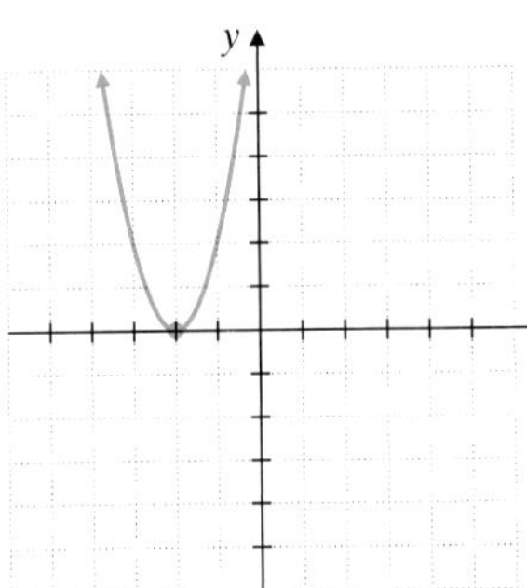

50.

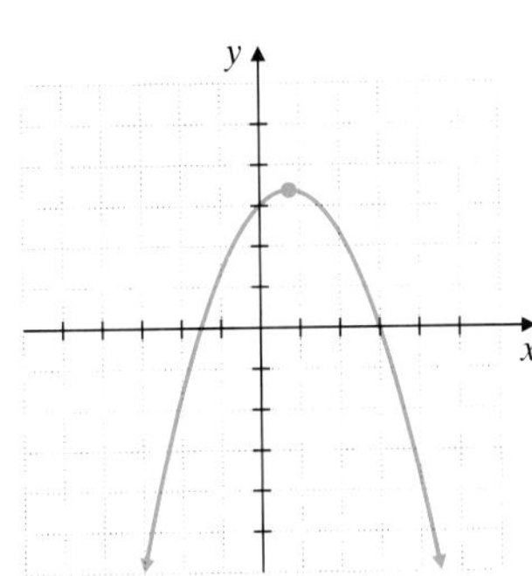

51.

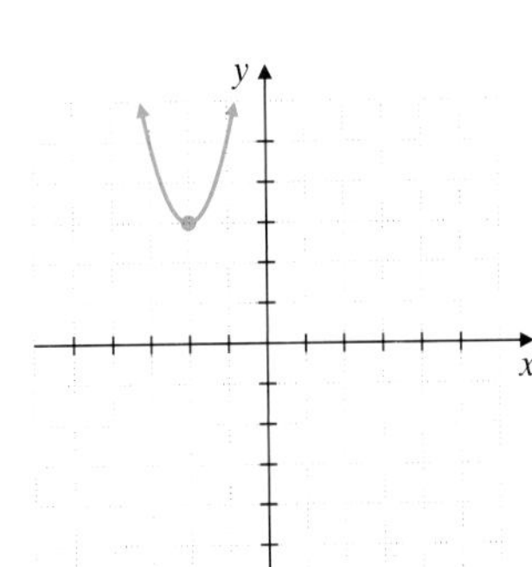

52.

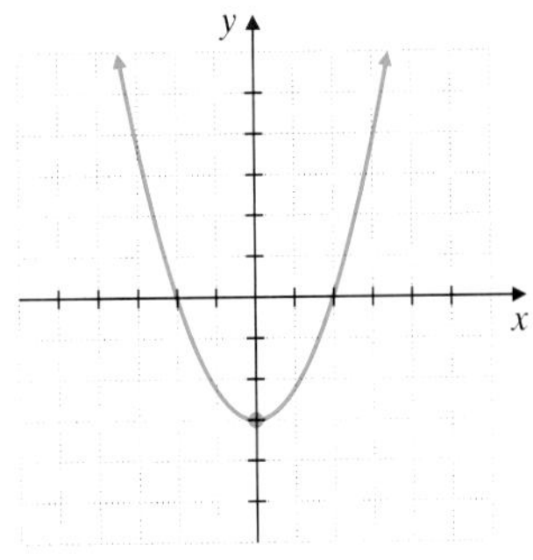

Match each quadratic equation with its graph.

53. $y = 2x^2$

54. $y = -x^2$

55. $y = x^2 + 4x + 4$

56. $y = x^2 + 5x + 4$

A

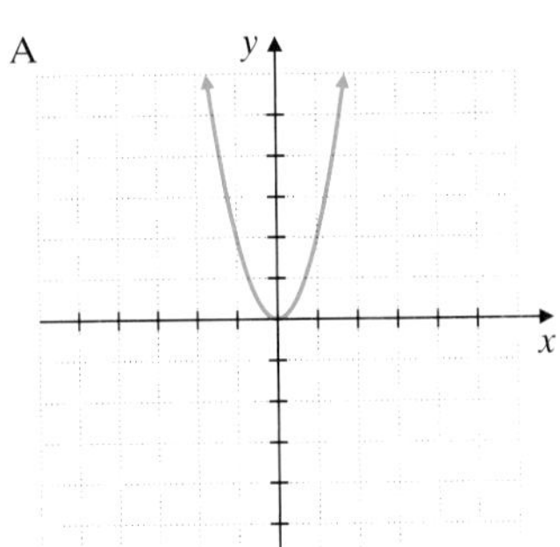

B

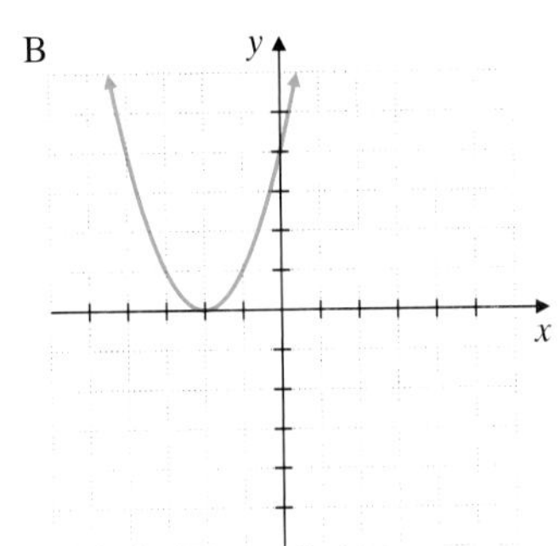

C

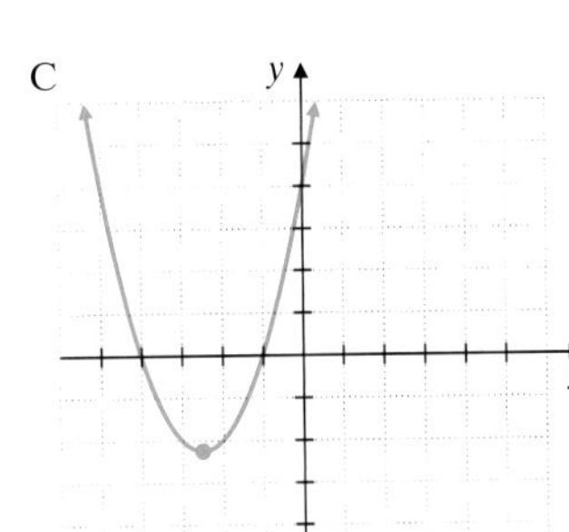

D

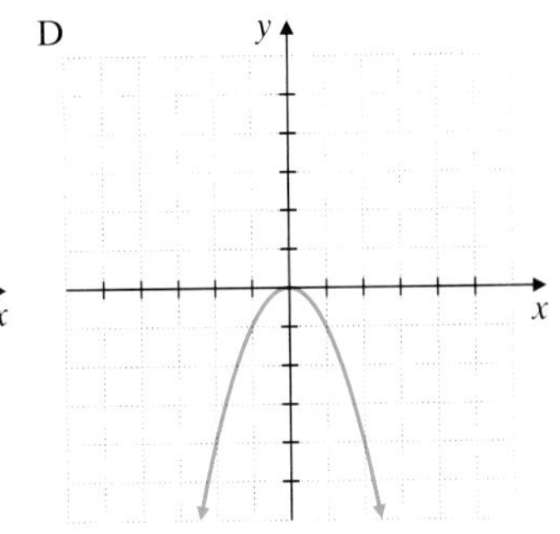

STUDY SKILLS REMINDER

Are you prepared for a test on Chapter 16?

Below I have listed some common trouble areas for topics covered in Chapter 16. After studying for your test—but before taking your test—read these.

- Don't forget that to use the square root property, one side of your equation should be a squared variable or variable expression.

Solve: $3x^2 = 15$

$x^2 = 5$ Divide both sides by 3 to isolate x^2.

$x = \sqrt{5}$ or $x = -\sqrt{5}$ Use the square root property.

- Remember that to identify a, b, and c for the quadratic formula, write the quadratic equation in standard form: $ax^2 + bx + c = 0$

Solve: $x^2 = -x + 1$

$x^2 + x - 1 = 0$ Write in standard form.

Here, $a = 1, b = 1$, and $c = -1$.

$$x = \frac{-1 \pm \sqrt{1^2 - 4(1)(-1)}}{2(1)} = \frac{-1 \pm \sqrt{5}}{2}$$

Remember: This is simply a listing of a few common trouble areas. For a review of Chapter 16, see the Highlights and Chapter Review at the end of this chapter.

Name ______________________ Section ________ Date ________

Answers

1. ______
2. ______
3. ______
4. ______
5. ______
6. ______
7. ______
8. ______
9. ______
10. ______
11. ______
12. ______
13. ______

Chapter 16 Test

Remember to check your answers and use the Chapter Test Prep Video to view solutons

Solve using the square root property.

1. $5k^2 = 80$

2. $(3m - 5)^2 = 8$

Solve by completing the square.

3. $x^2 - 26x + 160 = 0$

4. $3x^2 + 12x - 4 = 0$

Solve using the quadratic formula.

5. $x^2 - 3x - 10 = 0$

6. $p^2 - \frac{5}{3}p - \frac{1}{3} = 0$

Solve by the most appropriate method.

7. $(3x - 5)(x + 2) = -6$

8. $(3x - 1)^2 = 16$

9. $3x^2 - 7x - 2 = 0$

10. $x(x + 6) = 7$

11. $3x^2 - 7x + 2 = 0$

12. $2x^2 - 6x + 1 = 0$

13. $9x^3 = x$

14. see graph

15. see graph

16.

Graph each quadratic equation. Label the vertex and the intercept points with their coordinates.

14. $y = -3x^2$

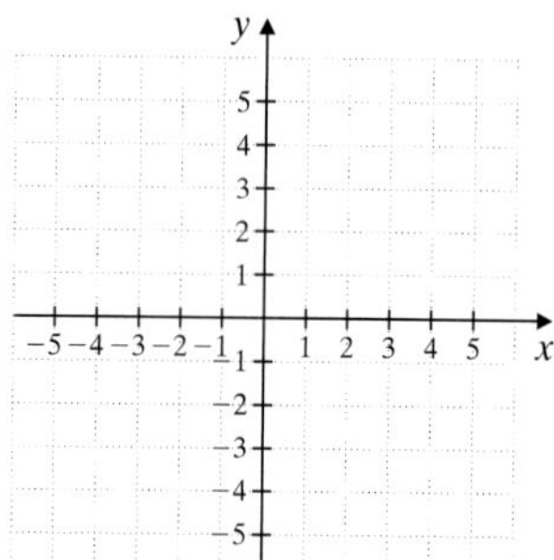

15. $y = x^2 - 7x + 10$

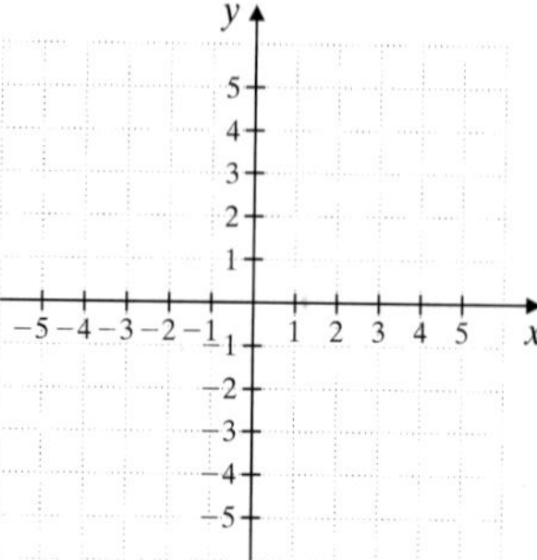

Solve.

16. The highest dive from a diving board by a woman was made by Lucy Wardle of the United States. She dove from a height of 120.75 feet at Ocean Park, Hong Kong, in 1985. To the nearest tenth of a second, how long did the dive take? Use the formula $h = 16t^2$. (*Source: The Guinness Book of Records*)

Name ______________________ Section __________ Date __________

Chapter 16 Cumulative Review

Divide.

1. $\dfrac{786}{10{,}000}$

2. $\dfrac{818}{1000}$

3. $\dfrac{0.12}{10}$

4. $\dfrac{5.03}{100}$

5. Using the circle graph shown, determine the percent of Americans that have one or more working computers at home.

6. Using the circle graph in Exercise 5, find the percent of Americans that have fewer than 3 working computers at home.

Number of Working Computers at Home

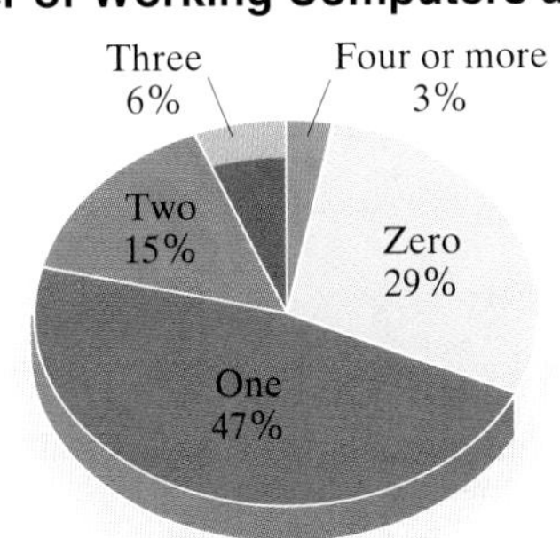

Source: UCLA Center for Communication Policy, 2003

7. Find the perimeter of a rectangle with a length of 11 inches and a width of 3 inches.

8. Find the perimeter of a triangular yard whose sides are 6 feet, 8 feet, and 11 feet.

9. Find the area of the parallelogram.

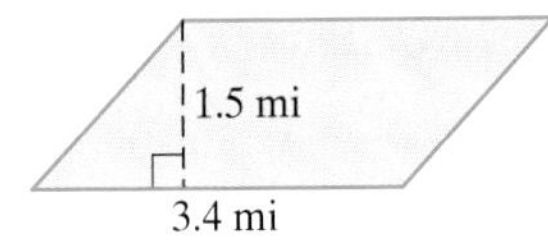

10. Find the area of the triangle.

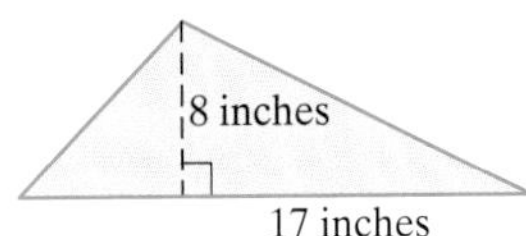

11. Convert 3210 ml to liters.

12. Convert 4321 cl to liters.

13. Solve: $8(2 - t) = -5t$

14. Solve: $\frac{5}{2}x - 1 = x + \frac{1}{4}$

Simplify the following expressions.

15. 3^0

16. -2^0

17. Factor: $r^2 - r - 42$

18. Factor: $y^2 + 3y - 70$

19. Factor: $10x^2 - 13xy - 3y^2$

20. Factor: $72x^2 - 35xy + 3y^2$

21. Factor: $8x^2 - 14x + 5$ by grouping.

22. Factor: $15x^2 - 4x - 4$ by grouping.

23. Factor completely.

a. $4x^3 - 49x$

b. $162x^4 - 2$

24. Factor completely.

a. $9x^3 - x$

b. $5x^4 - 5$

25. Solve: $(5x - 1)(2x^2 + 15x + 18) = 0$

26. Solve: $(x + 4)(40x^2 - 34x + 3) = 0$

Answers

1. ____
2. ____
3. ____
4. ____
5. ____
6. ____
7. ____
8. ____
9. ____
10. ____
11. ____
12. ____
13. ____
14. ____
15. ____
16. ____
17. ____
18. ____
19. ____
20. ____
21. ____
22. ____
23. a. ____
    b. ____
24. a. ____
    b. ____
25. ____
26. ____

27. ______

28. ______

29. ______

30. ______

31. see table

32. see table

33. a. ______ b. ______ c. ______

34. a. ______ b. ______ c. ______

35. a. ______ b. ______

36. a. ______ b. ______

37. ______

38. ______

39. ______

40. ______

41. ______

42. ______

43. ______

44. ______

45. ______

46. ______

47. ______

48. ______

27. Simplify: $\dfrac{x^2 + 8x + 7}{x^2 - 4x - 5}$

28. Simplify: $\dfrac{x^2 - 6x + 5}{x^2 + 6x - 7}$

29. The quotient of a number and 6, minus $\frac{5}{3}$ is the quotient of the number and 2. Find the number.

30. Four times a number, added to 5 is divided by 6. The result is $\frac{7}{2}$. Find the number.

31. Complete the table for the equation $y = 3x$.

	x	y
a.	-1	
b.		0
c.		-9

32. Complete the table for the equation $y = -2x + 7$.

	x	y
a.	0	
b.		0
c.	5	

Determine whether each pair of lines is parallel, perpendicular, or neither.

33. a. $y = -\frac{1}{5}x + 1$
$2x + 10y = 3$

b. $x + y = 3$
$-x + y = 4$

c. $3x + y = 5$
$2x + 3y = 6$

34. a. $y = -2x + 3$
$y = -2x + 5$

b. $-2x + y = 3$
$x + 2y = 9$

c. $3x - 2y = -8$
$3x + 2y = 1$

35. Which of the following relations are also functions?

a. $\{(-1,1), (2,3), (7,3), (8,6)\}$

b. $\{(0,-2), (1,5), (0,3), (7,7)\}$

36. Which of the following relations are also functions?

a. $\{(-2,3), (-5,7), (9,3), (0,0)\}$

b. $\left\{(9,-1), \left(0,\frac{1}{2}\right), (2,-1), (9,0)\right\}$

37. Solve the system:

$$\begin{cases} 2x + y = 10 \\ x = y + 2 \end{cases}$$

38. Solve the system:

$$\begin{cases} 8x - 3y = -4 \\ y = 7x - 3 \end{cases}$$

Find each square root.

39. $\sqrt{36}$

40. $\sqrt{81}$

41. $\sqrt{\dfrac{9}{100}}$

42. $\sqrt{\dfrac{16}{25}}$

Rationalize the denominator.

43. $\dfrac{2}{1 + \sqrt{3}}$

44. $\dfrac{7}{\sqrt{5} - 2}$

Use the square root property to solve.

45. $(x - 3)^2 = 16$

46. $(x + 4)^2 = 9$

47. Solve $\frac{1}{2}x^2 - x = 2$ by using the quadratic formula.

48. Solve $2x^2 = \frac{5}{2}x + \frac{7}{2}$ by using the quadratic formula.

APPENDIX A

Percents, Decimals, and Fractions

Percent, Decimal, and Fraction Equivalent		
Percent	**Decimal**	**Fraction**
1%	0.01	$\frac{1}{100}$
5%	0.05	$\frac{1}{20}$
10%	0.1	$\frac{1}{10}$
12.5% or $12\frac{1}{2}$%	0.125	$\frac{1}{8}$
$16.\overline{6}$% or $16\frac{2}{3}$%	$0.1\overline{6}$	$\frac{1}{6}$
20%	0.2	$\frac{1}{5}$
25%	0.25	$\frac{1}{4}$
30%	0.3	$\frac{3}{10}$
$33.\overline{3}$% or $33\frac{1}{3}$%	$0.\overline{3}$	$\frac{1}{3}$
37.5% or $37\frac{1}{2}$%	0.375	$\frac{3}{8}$
40%	0.4	$\frac{2}{5}$
50%	0.5	$\frac{1}{2}$
60%	0.6	$\frac{3}{5}$
62.5% or $62\frac{1}{2}$%	0.625	$\frac{5}{8}$
$66.\overline{6}$% or $66\frac{2}{3}$%	$0.\overline{6}$	$\frac{2}{3}$
70%	0.7	$\frac{7}{10}$
75%	0.75	$\frac{3}{4}$
80%	0.8	$\frac{4}{5}$
$83.\overline{3}$% or $83\frac{1}{3}$%	$0.8\overline{3}$	$\frac{5}{6}$
87.5% or $87\frac{1}{2}$%	0.875	$\frac{7}{8}$
90%	0.9	$\frac{9}{10}$
100%	1.0	1
110%	1.1	$1\frac{1}{10}$
125%	1.25	$1\frac{1}{4}$
$133.\overline{3}$% or $133\frac{1}{3}$%	$1.\overline{3}$	$1\frac{1}{3}$
150%	1.5	$1\frac{1}{2}$
$166.\overline{6}$% or $166\frac{2}{3}$%	$1.\overline{6}$	$1\frac{2}{3}$
175%	1.75	$1\frac{3}{4}$
200%	2.0	2

APPENDIX B

Table of Squares and Square Roots

Squares and Square Roots

n	n^2	$\sqrt{n}$	n	n^2	$\sqrt{n}$
1	1	1.000	51	2601	7.141
2	4	1.414	52	2704	7.211
3	9	1.732	53	2809	7.280
4	16	2.000	54	2916	7.348
5	25	2.236	55	3025	7.416
6	36	2.449	56	3136	7.483
7	49	2.646	57	3249	7.550
8	64	2.828	58	3364	7.616
9	81	3.000	59	3481	7.681
10	100	3.162	60	3600	7.746
11	121	3.317	61	3721	7.810
12	144	3.464	62	3844	7.874
13	169	3.606	63	3969	7.937
14	196	3.742	64	4096	8.000
15	225	3.873	65	4225	8.062
16	256	4.000	66	4356	8.124
17	289	4.123	67	4489	8.185
18	324	4.243	68	4624	8.246
19	361	4.359	69	4761	8.307
20	400	4.472	70	4900	8.367
21	441	4.583	71	5041	8.426
22	484	4.690	72	5184	8.485
23	529	4.796	73	5329	8.544
24	576	4.899	74	5476	8.602
25	625	5.000	75	5625	8.660
26	676	5.099	76	5776	8.718
27	729	5.196	77	5929	8.775
28	784	5.292	78	6084	8.832
29	841	5.385	79	6241	8.888
30	900	5.477	80	6400	8.944
31	961	5.568	81	6561	9.000
32	1024	5.657	82	6724	9.055
33	1089	5.745	83	6889	9.110
34	1156	5.831	84	7056	9.165
35	1225	5.916	85	7225	9.220
36	1296	6.000	86	7396	9.274
37	1369	6.083	87	7569	9.327
38	1444	6.164	88	7744	9.381
39	1521	6.245	89	7921	9.434
40	1600	6.325	90	8100	9.487
41	1681	6.403	91	8281	9.539
42	1764	6.481	92	8464	9.592
43	1849	6.557	93	8649	9.644
44	1936	6.633	94	8836	9.695
45	2025	6.708	95	9025	9.747
46	2116	6.782	96	9216	9.798
47	2209	6.856	97	9409	9.849
48	2304	6.928	98	9604	9.899
49	2401	7.000	99	9801	9.950
50	2500	7.071	100	10,000	10.000

APPENDIX C

Compound Interest Table

Compounded Annually

	5%	6%	7%	8%	9%	10%	11%	12%	13%	14%	15%	16%	17%	18%
1 year	1.05000	1.06000	1.07000	1.08000	1.09000	1.10000	1.11000	1.12000	1.13000	1.14000	1.15000	1.16000	1.17000	1.18000
5 years	1.27628	1.33823	1.40255	1.46933	1.53862	1.61051	1.68506	1.76234	1.84244	1.92541	2.01136	2.10034	2.19245	2.28776
10 years	1.62889	1.79085	1.96715	2.15892	2.36736	2.59374	2.83942	3.10585	3.39457	3.70722	4.04556	4.41144	4.80683	5.23384
15 years	2.07893	2.39656	2.75903	3.17217	3.64248	4.17725	4.78459	5.47357	6.25427	7.13794	8.13706	9.26552	10.53872	11.97375
20 years	2.65330	3.20714	3.86968	4.66096	5.60441	6.72750	8.06231	9.64629	11.52309	13.74349	16.36654	19.46076	23.10560	27.39303

Compounded Semiannually

	5%	6%	7%	8%	9%	10%	11%	12%	13%	14%	15%	16%	17%	18%
1 year	1.05063	1.06090	1.07123	1.08160	1.09203	1.10250	1.11303	1.12360	1.13423	1.14490	1.15563	1.16640	1.17723	1.18810
5 years	1.28008	1.34392	1.41060	1.48024	1.55297	1.62889	1.70814	1.79085	1.87714	1.96715	2.06103	2.15892	2.26098	2.36736
10 years	1.63862	1.80611	1.98979	2.19112	2.41171	2.65330	2.91776	3.20714	3.52365	3.86968	4.24785	4.66096	5.11205	5.60441
15 years	2.09757	2.42726	2.80679	3.24340	3.74532	4.32194	4.98395	5.74349	6.61437	7.61226	8.75496	10.06266	11.55825	13.26768
20 years	2.68506	3.26204	3.95926	4.80102	5.81636	7.03999	8.51331	10.28572	12.41607	14.97446	18.04424	21.72452	26.13302	31.40942

Compounded Quarterly

	5%	6%	7%	8%	9%	10%	11%	12%	13%	14%	15%	16%	17%	18%
1 year	1.05095	1.06136	1.07186	1.08243	1.09308	1.10381	1.11462	1.12551	1.13648	1.14752	1.15865	1.16986	1.18115	1.19252
5 years	1.28204	1.34686	1.41478	1.48595	1.56051	1.63862	1.72043	1.80611	1.89584	1.98979	2.08815	2.19112	2.29891	2.41171
10 years	1.64362	1.81402	2.00160	2.20804	2.43519	2.68506	2.95987	3.26204	3.59420	3.95926	4.36038	4.80102	5.28497	5.81636
15 years	2.10718	2.44322	2.83182	3.28103	3.80013	4.39979	5.09225	5.89160	6.81402	7.87809	9.10513	10.51963	12.14965	14.02741
20 years	2.70148	3.29066	4.00639	4.87544	5.93015	7.20957	8.76085	10.64089	12.91828	15.67574	19.01290	23.04980	27.93091	33.83010

Compounded Daily

	5%	6%	7%	8%	9%	10%	11%	12%	13%	14%	15%	16%	17%	18%
1 year	1.05127	1.06183	1.07250	1.08328	1.09416	1.10516	1.11626	1.12747	1.13880	1.15024	1.16180	1.17347	1.18526	1.19716
5 years	1.28400	1.34983	1.41902	1.49176	1.56823	1.64861	1.73311	1.82194	1.91532	2.01348	2.11667	2.22515	2.33918	2.45906
10 years	1.64866	1.82203	2.01362	2.22535	2.45933	2.71791	3.00367	3.31946	3.66845	4.05411	4.48031	4.95130	5.47178	6.04696
15 years	2.11689	2.45942	2.85736	3.31968	3.85678	4.48077	5.20569	6.04786	7.02625	8.16288	9.48335	11.01738	12.79950	14.86983
20 years	2.71810	3.31979	4.05466	4.95216	6.04831	7.38703	9.02202	11.01883	13.45751	16.43582	20.07316	24.51533	29.94039	36.56577

APPENDIX D

Review of Geometric Figures

Plane Figures Have Length and Width But No Thickness or Depth.		
Name	**Description**	**Figure**
Polygon	Union of three or more coplanar line segments that intersect with each other only at each end point, with each end point shared by two segments.	
Triangle	Polygon with three sides (sum of measures of three angles is 180°).	
Scalene Triangle	Triangle with no sides of equal length.	
Isosceles Triangle	Triangle with two sides of equal length.	
Equilateral Triangle	Triangle with all sides of equal length.	
Right Triangle	Triangle that contains a right angle.	leg, hypotenuse, leg
Quadrilateral	Polygon with four sides (sum of measures of four angles is 360°).	
Trapezoid	Quadrilateral with exactly one pair of opposite sides parallel.	base, leg, parallel sides, leg, base
Isosceles Trapezoid	Trapezoid with legs of equal length.	
Parallelogram	Quadrilateral with both pairs of opposite sides parallel.	

(*continued*)

Name	Description	Figure
Rhombus	Parallelogram with all sides of equal length.	
Rectangle	Parallelogram with four right angles.	
Square	Rectangle with all sides of equal length.	
Circle	All points in a plane the same distance from a fixed point called the **center**.	radius, center, diameter

Solid Figures Have Length, Width, and Height or Depth.		
Name	**Description**	**Figure**
Rectangular Solid	A solid with six sides, all of which are rectangles.	
Cube	A rectangular solid whose six sides are squares.	
Sphere	All points the same distance from a fixed point called the **center**.	radius, center
Right Circular Cylinder	A cylinder consisting of two circular bases that are perpendicular to its altitude.	
Right Circular Cone	A cone with a circular base that is perpendicular to its altitude.	

APPENDIX E

Review of Volume and Surface Area

A convex solid is a set of points, S, not all in one plane, such that for any two points A and B in S, all points between A and B are also in S. In this appendix, we will find the volume and surface area of special types of solids called polyhedrons. A solid formed by the intersection of a finite number of planes is called a **polyhedron.** The box below is an example of a polyhedron.

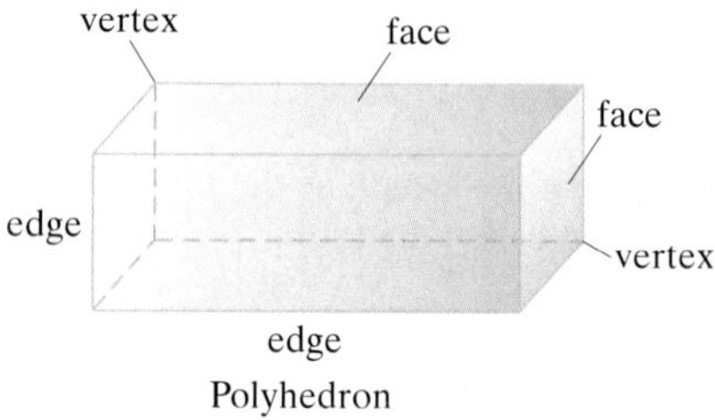

Polyhedron

Each of the plane regions of the polyhedron is called a **face** of the polyhedron. If the intersection of two faces is a line segment, this line segment is an **edge** of the polyhedron. The intersections of the edges are the **vertices** of the polyhedron.

Volume is a measure of the space of a solid. The volume of a box or can, for example, is the amount of space inside. Volume can be used to describe the amount of juice in a pitcher or the amount of concrete needed to pour a foundation for a house.

The volume of a solid is the number of **cubic units** in the solid. A cubic centimeter and a cubic inch are illustrated.

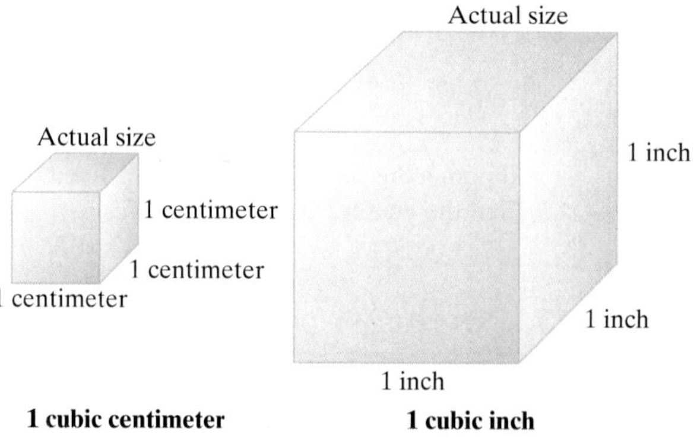

1 cubic centimeter **1 cubic inch**

The **surface area** of a polyhedron is the sum of the areas of the faces of the polyhedron. For example, each face of the cube to the left above has an area of 1 square centimeter. Since there are 6 faces of the cube, the sum of the areas of the faces is 6 square centimeters. Surface area can be used to describe the amount of material needed to cover a solid. Surface area is measured in square units.

Formulas for finding the volumes, V, and surface areas, SA, of some common solids are given next.

Volume and Surface Area Formulas of Common Solids

Solid	Formulas
RECTANGULAR SOLID 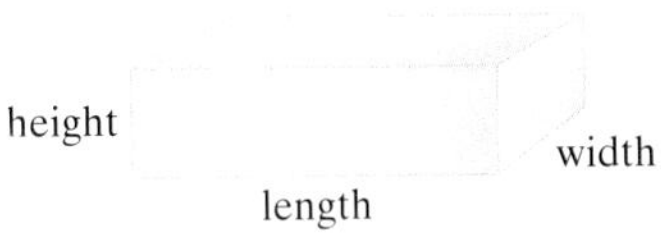	$V = lwh$ $SA = 2lh + 2wh + 2lw$ where h = height, w = width, l = length
CUBE 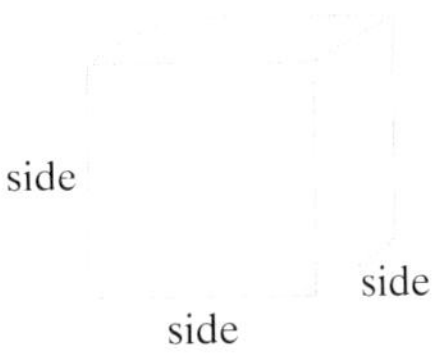	$V = s^3$ $SA = 6s^2$ where s = side
SPHERE 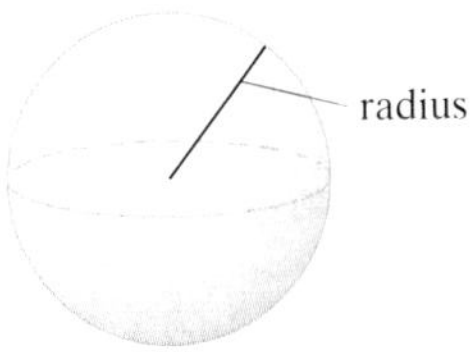	$V = \frac{4}{3}\pi r^3$ $SA = 4\pi r^2$ where r = radius
CIRCULAR CYLINDER 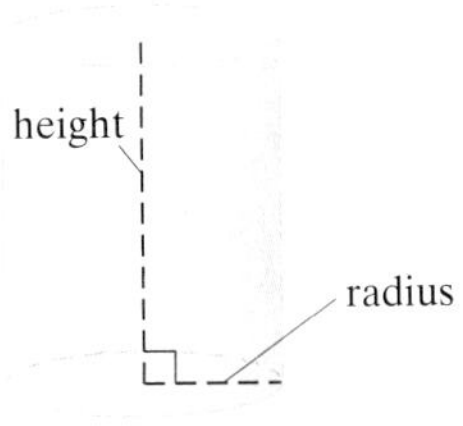	$V = \pi r^2 h$ $SA = 2\pi rh + 2\pi r^2$ where h = height, r = radius
CONE 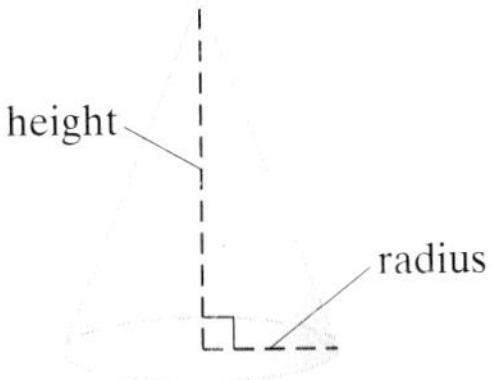	$V = \frac{1}{3}\pi r^2 h$ $SA = \pi r\sqrt{r^2 + h^2} + \pi r^2$ where h = height, r = radius
SQUARE-BASED PYRAMID 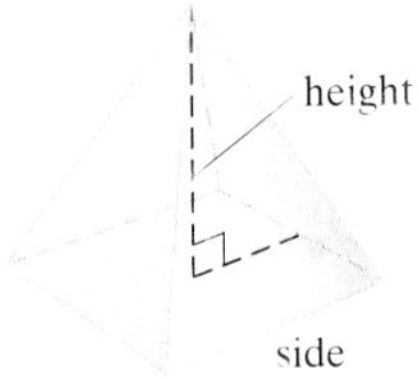	$V = \frac{1}{3}s^2 h$ $SA = B + \frac{1}{2}pl$ where B = area of base; p = perimeter of base, h = height, s = side, l = slant height

Helpful Hint

Volume is measured in cubic units. Surface area is measured in square units.

EXAMPLE 1

Find the volume and surface area of a rectangular box that is 12 inches long, 6 inches wide, and 3 inches high.

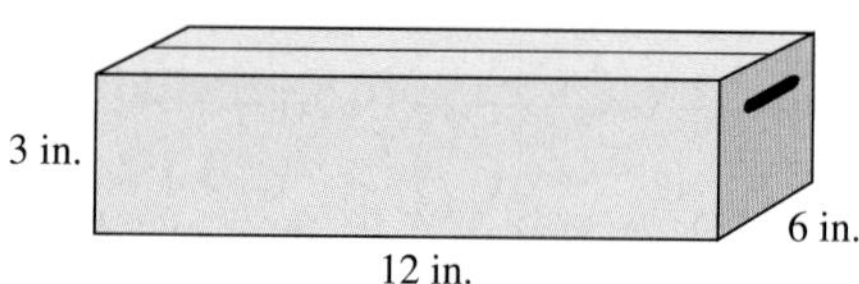

Solution: Let $h = 3$ in., $l = 12$ in., and $w = 6$ in.

$$V = lwh$$

$$V = 12 \text{ inches} \cdot 6 \text{ inches} \cdot 3 \text{ inches} = 216 \text{ cubic inches}$$

The volume of the rectangular box is 216 cubic inches.

$$\begin{aligned} SA &= 2lh + 2wh + 2lw \\ &= 2(12 \text{ in.})(3 \text{ in.}) + 2(6 \text{ in.})(3 \text{ in.}) + 2(12 \text{ in.})(6 \text{ in.}) \\ &= 72 \text{ sq in.} + 36 \text{ sq in.} + 144 \text{ sq in.} \\ &= 252 \text{ sq in.} \end{aligned}$$

The surface area of rectangular box is 252 square inches.

EXAMPLE 2

Find the volume and surface area of a ball that has a radius of 2 inches. Give the exact volume and surface area and then use the approximation $\frac{22}{7}$ for π.

Solution:

$V = \frac{4}{3}\pi r^3$ — Formula for volume of a sphere.

$V = \frac{4}{3} \cdot \pi(2 \text{ in.})^3$ — Let $r = 2$ inches.

$= \frac{32}{3}\pi \text{ cu in.}$ — Simplify.

$\approx \frac{32}{3} \cdot \frac{22}{7} \text{ cu in.}$ — Approximate π with $\frac{22}{7}$.

$= \frac{704}{21} \text{ or } 33\frac{11}{21} \text{ cu in.}$

The volume of the sphere is exactly $\frac{32}{3}\pi$ cubic inches or approximately $33\frac{11}{21}$ cubic inches.

$$
\begin{aligned}
SA &= 4\pi r^2 && \text{Formula for surface area.}\\
SA &= 4 \cdot \pi(2\,\text{in.})^2 && \text{Let } r = 2 \text{ inches.}\\
&= 16\pi \text{ sq in.} && \text{Simplify.}\\
&\approx 16 \cdot \frac{22}{7} \text{ sq in.} && \text{Approximate } \pi \text{ with } \frac{22}{7}.\\
&= \frac{352}{7} \text{ or } 50\frac{2}{7} \text{ sq in.}
\end{aligned}
$$

The surface area of the sphere is exactly 16π square inches or approximately $50\frac{2}{7}$ square inches.

Name ______________________ Section __________ Date __________

APPENDIX E EXERCISE SET

Find the volume and surface area of each solid. See Examples 1 and 2. For formulas that contain π, give an exact answer and then approximate using $\frac{22}{7}$ for π.

1.

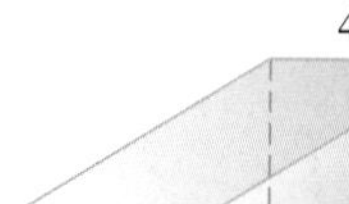

2.

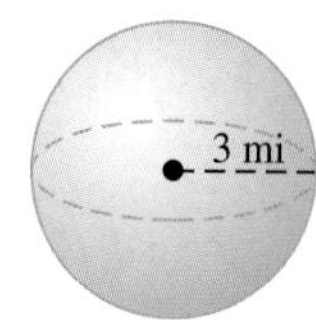

3.

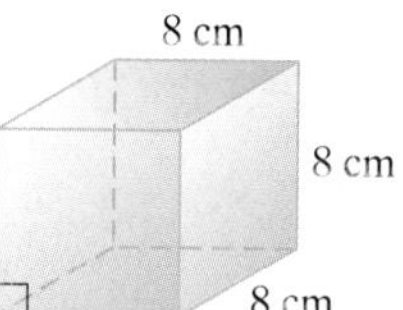

4.

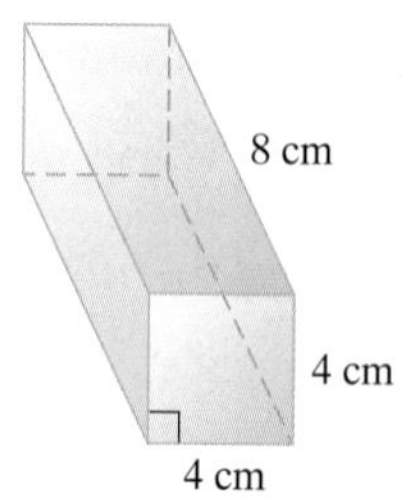

5. (For surface area, use 3.14 for π.)

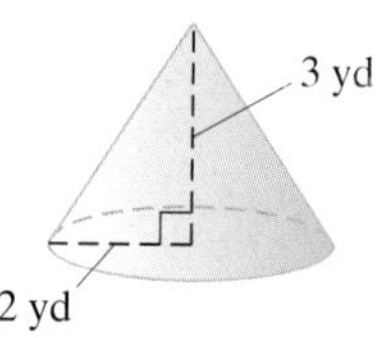

6.

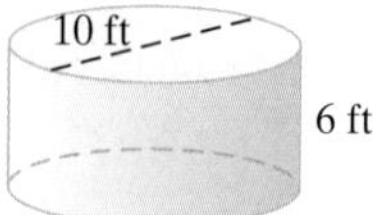

7.

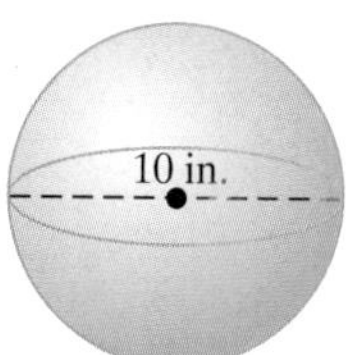

8. Find the volume only.

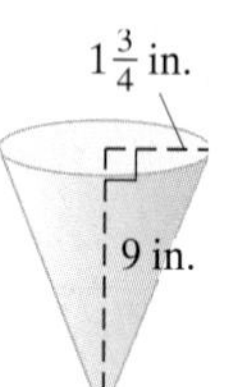

9.

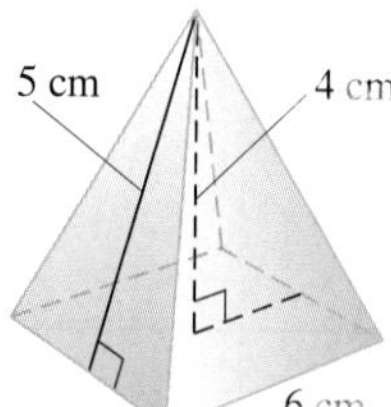

10.

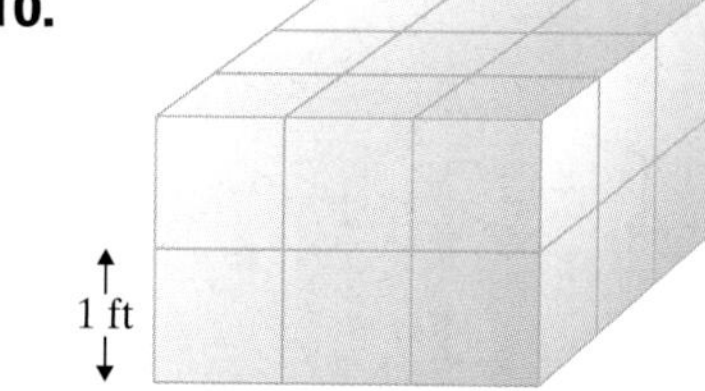

Solve.

11. Find the volume of a cube with edges of $1\frac{1}{3}$ inches.

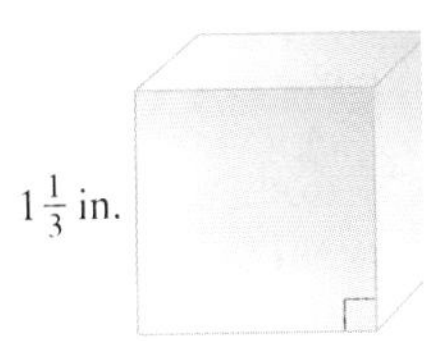

12. A water storage tank is in the shape of a cone with the pointed end down. If the radius is 14 ft and the depth of the tank is 15 ft, approximate the volume of the tank in cubic feet. Use $\frac{22}{7}$ for π.

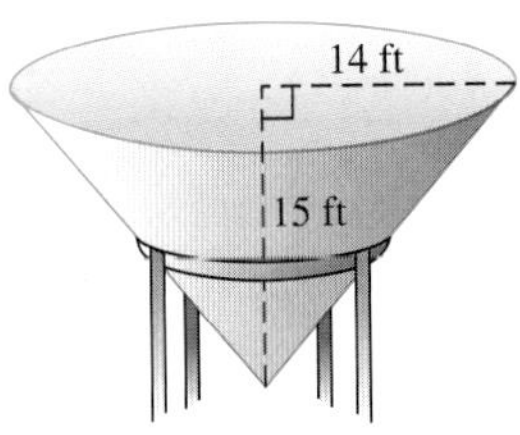

13. Find the surface area of a rectangular box 2 ft by 1.4 ft by 3 ft.

14. Find the surface area of a box in the shape of a cube that is 5 ft on each side.

15. Find the volume of a pyramid with a square base 5 in. on a side and a height of 1.3 in.

16. Approximate to the nearest hundredth the volume of a sphere with a radius of 2 cm. Use 3.14 for π.

17. A paperweight is in the shape of a square-based pyramid 20 cm tall. If an edge of the base is 12 cm, find the volume of the paperweight.

18. A bird bath is made in the shape of a hemisphere (half-sphere). If its radius is 10 in., approximate the volume. Use $\frac{22}{7}$ for π.

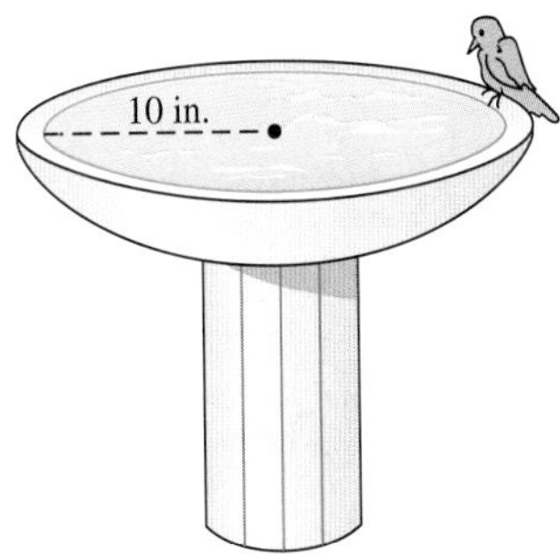

19. Find the exact surface area of a sphere with a radius of 7 in.

20. A tank is in the shape of a cylinder 8 ft tall and 3 ft in radius. Find the exact surface area of the tank.

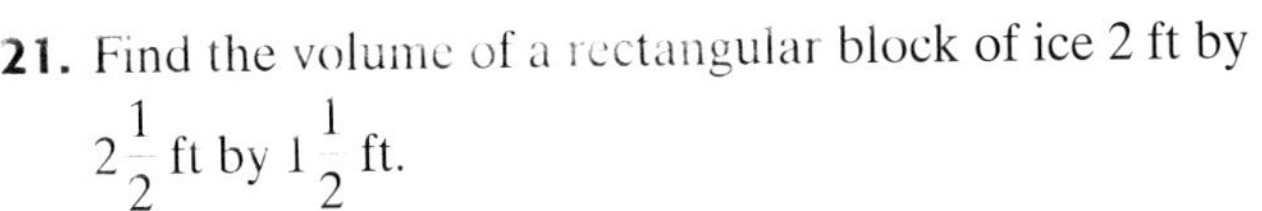

21. Find the volume of a rectangular block of ice 2 ft by $2\frac{1}{2}$ ft by $1\frac{1}{2}$ ft.

22. Find the capacity (volume in cubic feet) of a rectangular ice chest with inside measurements of 3 ft by $1\frac{1}{2}$ ft by $1\frac{3}{4}$ ft.

23. An ice cream cone with a 4-cm diameter and 3-cm depth is filled exactly level with the top of the cone. Approximate how much ice cream (in cubic centimeters) is in the cone. Use $\frac{22}{7}$ for π.

24. A child's toy is in the shape of a square-based pyramid 10 in. tall. If an edge of the base is 7 in., find the volume of the toy.

APPENDIX F

Factoring Sums and Differences of Cubes

Although the sum of two squares usually does not factor, the sum or difference of two cubes can be factored and reveal factoring patterns. The pattern for the sum of cubes can be checked by multiplying the binomial $x + y$ and the trinomial $x^2 - xy + y^2$. The pattern for the difference of two cubes can be checked by multiplying the binomial $x - y$ by the trinomial $x^2 + xy + y^2$.

SUM OR DIFFERENCE OF TWO CUBES

$$a^3 + b^3 = (a + b)(a^2 - ab + b^2)$$

$$a^3 - b^3 = (a - b)(a^2 + ab + b^2)$$

EXAMPLE 1 Factor $x^3 + 8$.

Solution:

First, write the binomial in the form $a^3 + b^3$.

$$x^3 + 8 = x^3 + 2^3 \qquad \text{Write in the form } a^3 + b^3.$$

If we replace a with x and b with 2 in the formula above, we have

$$x^3 + 2^3 = (x + 2)[x^2 - (x)(2) + 2^2]$$
$$= (x + 2)(x^2 - 2x + 4)$$

●

Helpful Hint

When factoring sums or differences of cubes, notice the sign patterns.

same sign

$$x^3 + y^3 = (x + y)(x^2 - xy + y^2)$$

opposite sign   always positive

same sign

$$x^3 - y^3 = (x - y)(x^2 + xy + y^2)$$

opposite sign   always positive

EXAMPLE 2 Factor $y^3 - 27$.

Solution:

$$\begin{aligned} y^3 - 27 &= y^3 - 3^3 && \text{Write in the form } a^3 - b^3. \\ &= (y - 3)[y^2 + (y)(3) + 3^2] \\ &= (y - 3)(y^2 + 3y + 9) \end{aligned}$$

EXAMPLE 3 Factor $64x^3 + 1$.

Solution:

$$\begin{aligned} 64x^3 + 1 &= (4x)^3 + 1^3 \\ &= (4x + 1)[(4x)^2 - (4x)(1) + 1^2] \\ &= (4x + 1)(16x^2 - 4x + 1) \end{aligned}$$

EXAMPLE 4 Factor $54a^3 - 16b^3$.

Solution: Remember to factor out common factors first before using other factoring methods.

$$\begin{aligned} 54a^3 - 16b^3 &= 2(27a^3 - 8b^3) && \text{Factor out the GCF 2.} \\ &= 2[(3a)^3 - (2b)^3] && \text{Difference of two cubes.} \\ &= 2(3a - 2b)[(3a)^2 + (3a)(2b) + (2b)^2] \\ &= 2(3a - 2b)(9a^2 + 6ab + 4b^2) \end{aligned}$$

Name ______________________ Section __________ Date __________

APPENDIX F EXERCISE SET

Factor the sum or difference of two cubes. See Examples 1 through 4.

1. $a^3 + 27$

2. $b^3 - 8$

3. $8a^3 + 1$

4. $64x^3 - 1$

5. $5k^3 + 40$

6. $6r^3 - 162$

7. $x^3y^3 - 64$

8. $8x^3 - y^3$

9. $x^3 + 125$

10. $a^3 - 216$

11. $24x^4 - 81xy^3$

12. $375y^6 - 24y^3$

Factor the binomials completely.

13. $27 - t^3$

14. $125 + r^3$

15. $8r^3 - 64$

16. $54r^3 + 2$

17. $t^3 - 343$

18. $s^3 + 216$

19. $s^3 - 64t^3$

20. $8t^3 + s^3$

APPENDIX G

Systems of Linear Inequalities

A **solution of a system of linear inequalities** is an ordered pair that satisfies each inequality in the system. The set of all such ordered pairs is the solution set of the system. Graphing this set gives us a picture of the solution set. We can graph a system of inequalities by graphing each inequality in the system and identifying the region of overlap.

EXAMPLE 1 Graph the solution of the system: $\begin{cases} 3x \geq y \\ x + 2y \leq 8 \end{cases}$

Solution: We begin by graphing each inequality on the same set of axes. The graph of the solution of the system is the region contained in the graphs of both inequalities. It is their intersection.

First, graph $3x \geq y$. The boundary line is the graph of $3x = y$. Sketch a solid boundary line since the inequality $3x \geq y$ means $3x > y$ or $3x = y$. The test point $(1, 0)$ satisfies the inequality, so shade the half-plane that includes $(1, 0)$.

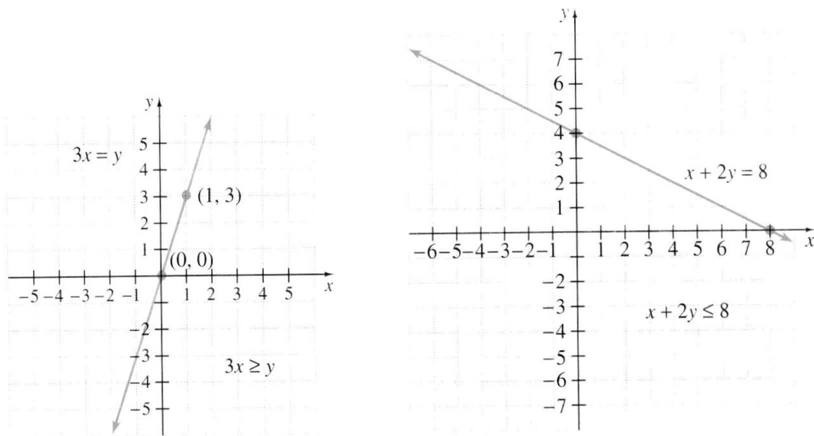

Next, sketch a solid boundary line $x + 2y = 8$ on the same set of axes. The test point $(0, 0)$ satisfies the inequality $x + 2y \leq 8$, so shade the half-plane that includes $(0, 0)$. (For clarity, the graph of $x + 2y \leq 8$ is shown on a separate set of axes.)

An ordered pair solution of the system must satisfy both inequalities. These solutions are points that lie in both shaded regions. The solution of the system is the darkest shaded region. This solution includes parts of both boundary lines.

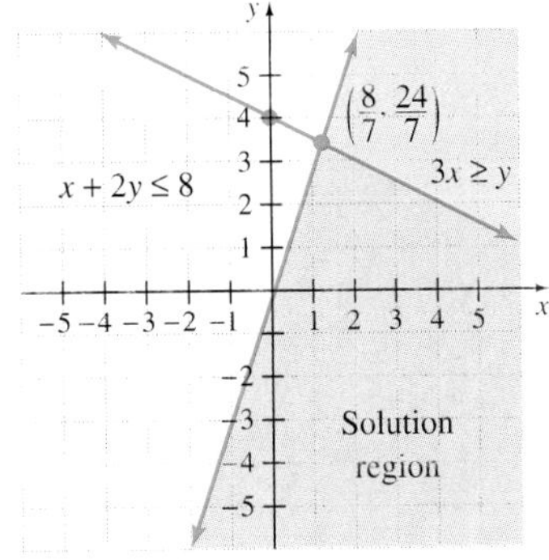

In linear programming, it is sometimes necessary to find the coordinates of the **corner point**: the point at which the two boundary lines intersect. To find the point of intersection, solve the related linear system

$$\begin{cases} 3x = y \\ x + 2y = 8 \end{cases}$$

by the substitution method or the addition method. The lines intersect at $\left(\frac{8}{7}, \frac{24}{7}\right)$, the corner point of the graph.

GRAPHING THE SOLUTION OF A SYSTEM OF LINEAR INEQUALITIES

Step 1. Graph each inequality in the system on the same set of axes.
Step 2. The solutions of the system are the points common to the graphs of all the inequalities in the system.

EXAMPLE 2 Graph the solution of the system $\begin{cases} x - y < 2 \\ x + 2y > -1 \end{cases}$

Solution: Graph both inequalities on the same set of axes. Both boundary lines are dashed lines since the inequality symbols are $<$ and $>$. The solution of the system is the region shown by the darkest shading. In this example, the boundary lines are not a part of the solution.

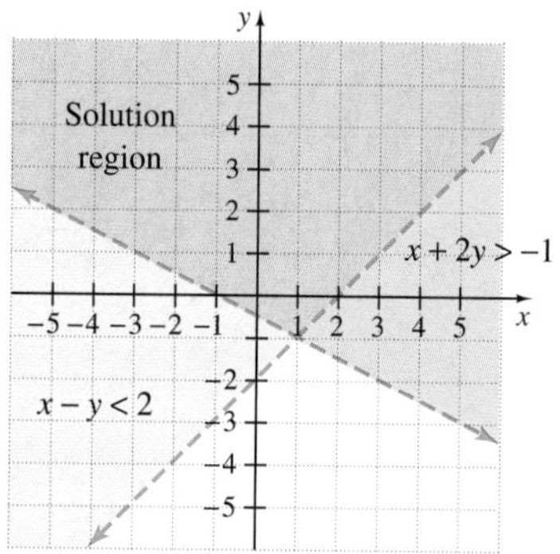

EXAMPLE 3 Graph the solution of the system $\begin{cases} -3x + 4y < 12 \\ x \geq 2 \end{cases}$

Solution: Graph both inequalities on the same set of axes.

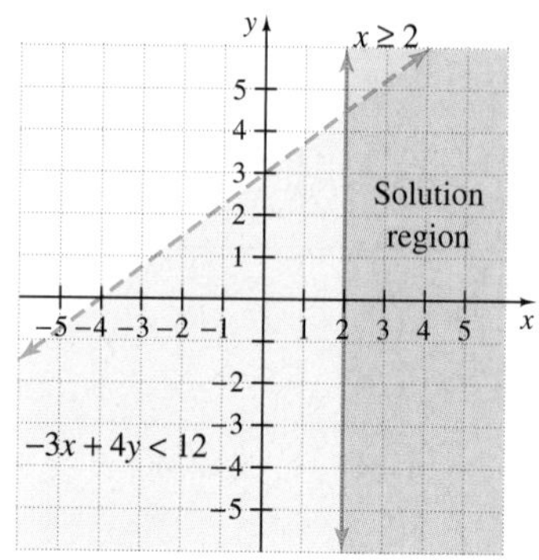

The solution of the system is the darkest shaded region, including a portion of the line $x = 2$.

APPENDIX G EXERCISE SET

Graph the solutions of each system of linear inequalities. See Examples 1 through 3.

1. $\begin{cases} y \geq x + 1 \\ y \geq 3 - x \end{cases}$

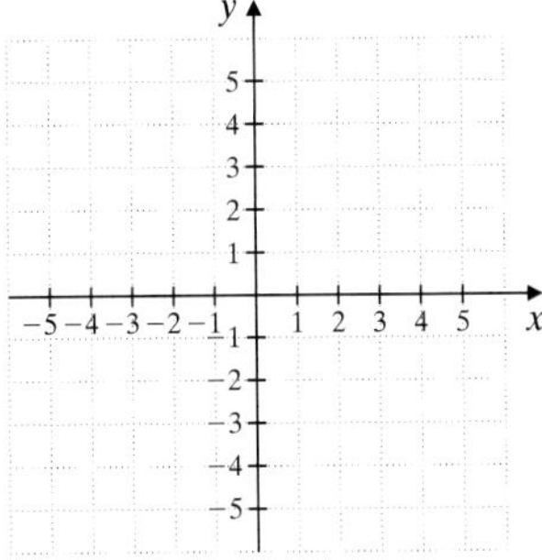

2. $\begin{cases} y \geq x - 3 \\ y \geq -1 - x \end{cases}$

3. $\begin{cases} y < 3x - 4 \\ y \leq x + 2 \end{cases}$

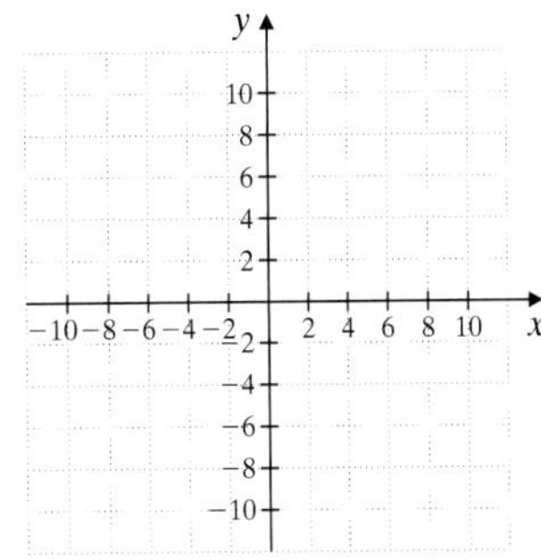

4. $\begin{cases} y \leq 2x + 1 \\ y > x + 2 \end{cases}$

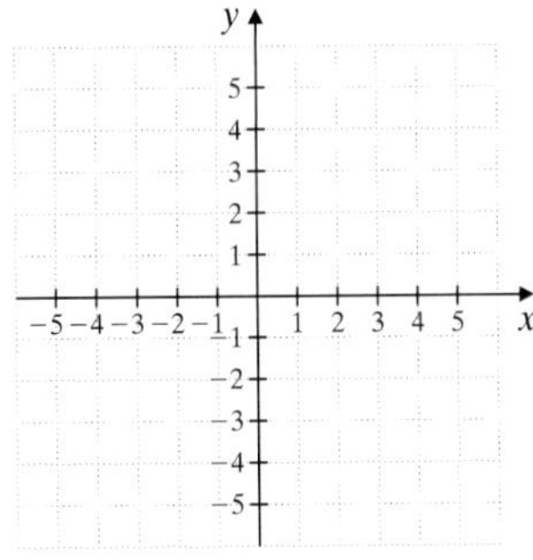

5. $\begin{cases} y \leq -2x - 2 \\ y \geq x + 4 \end{cases}$

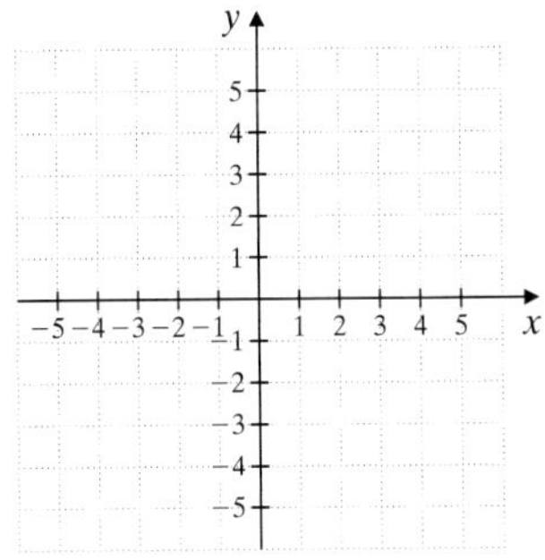

6. $\begin{cases} y \leq 2x + 4 \\ y \geq -x - 5 \end{cases}$

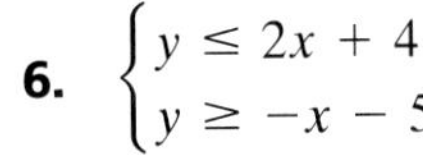

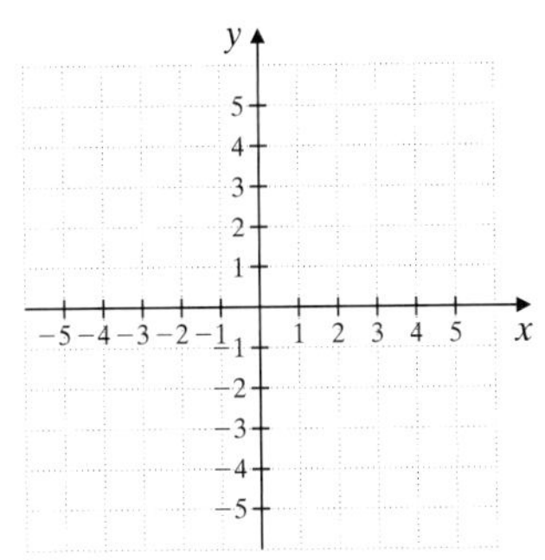

7. $\begin{cases} y \geq -x + 2 \\ y \leq 2x + 5 \end{cases}$

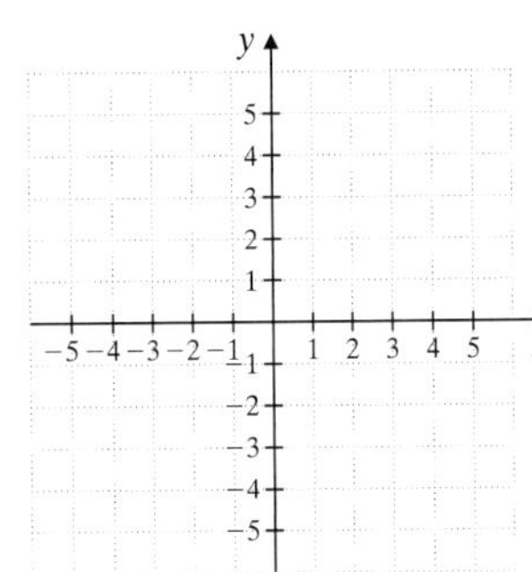

8. $\begin{cases} y \geq x - 5 \\ y \leq -3x + 3 \end{cases}$

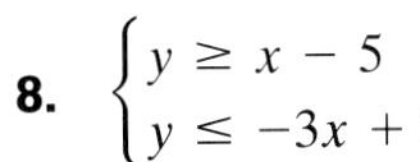

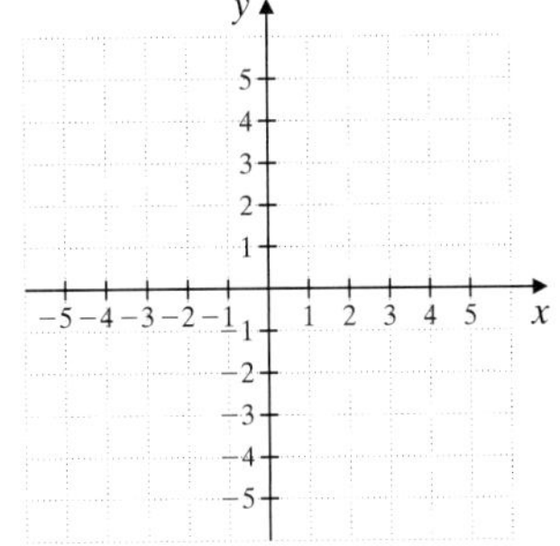

9. $\begin{cases} x \geq 3y \\ x + 3y \leq 6 \end{cases}$

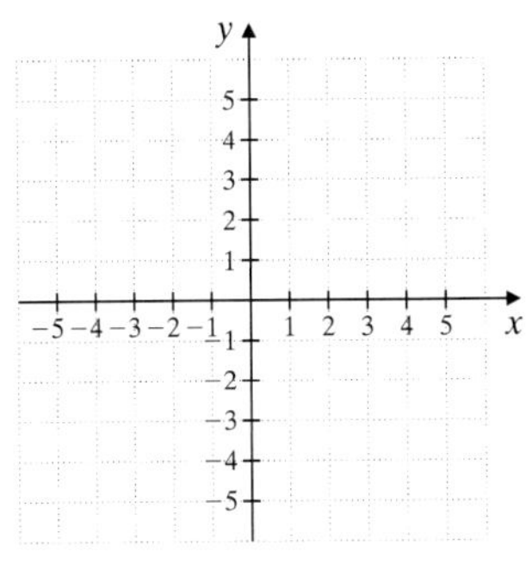

10. $\begin{cases} -2x < y \\ x + 2y < 3 \end{cases}$

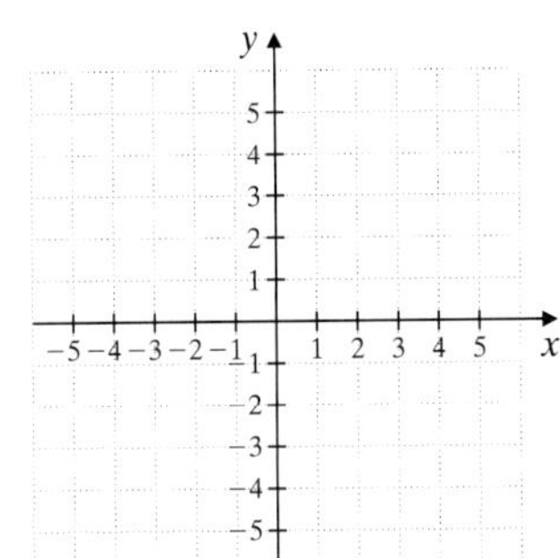

11. $\begin{cases} y + 2x \geq 0 \\ 5x - 3y \leq 12 \end{cases}$

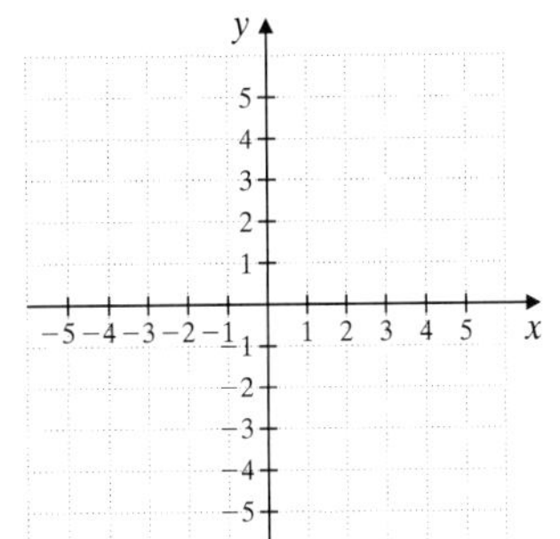

12. $\begin{cases} y + 2x \leq 0 \\ 5x + 3y \geq -2 \end{cases}$

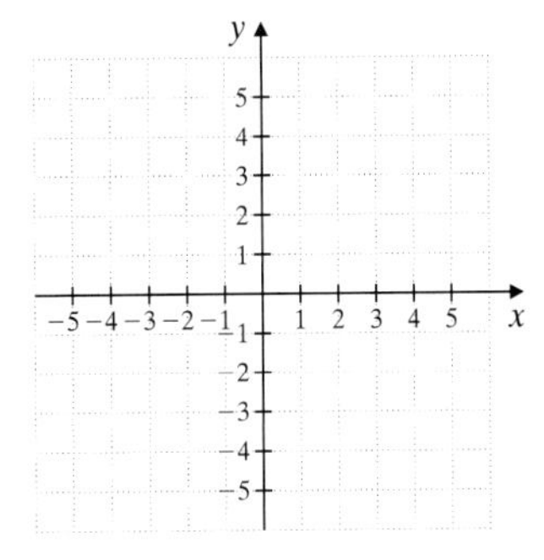

Graph the solutions of each system of linear inequalities. See Examples 1 through 3.

13. $\begin{cases} 3x - 4y \geq -6 \\ 2x + y \leq 7 \end{cases}$

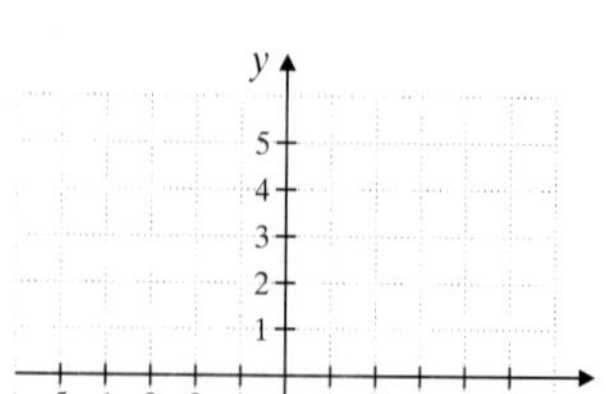

14. $\begin{cases} 4x - y \geq -2 \\ 2x + 3y \leq -8 \end{cases}$

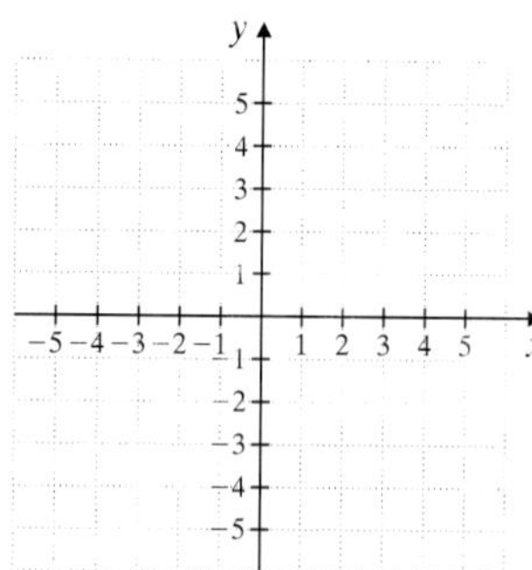

15. $\begin{cases} x \leq 2 \\ y \geq -3 \end{cases}$

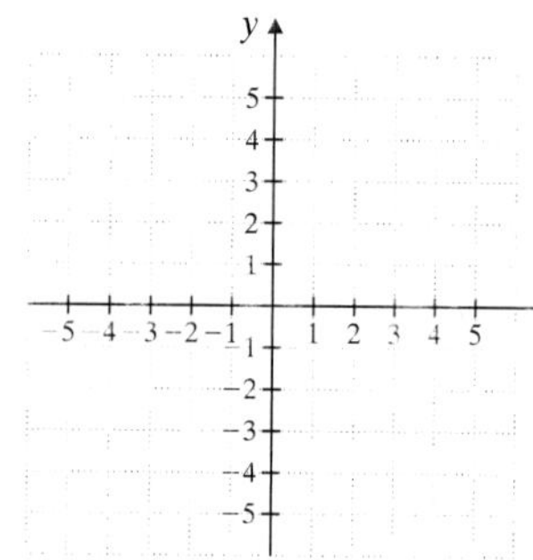

16. $\begin{cases} x \geq -3 \\ y \geq -2 \end{cases}$

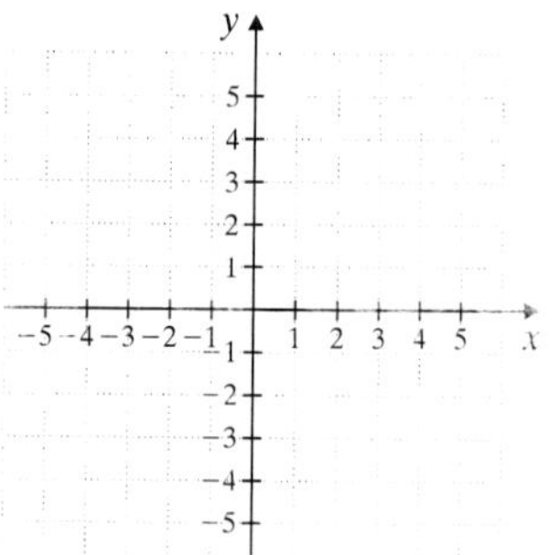

17. $\begin{cases} y \geq 1 \\ x < -3 \end{cases}$

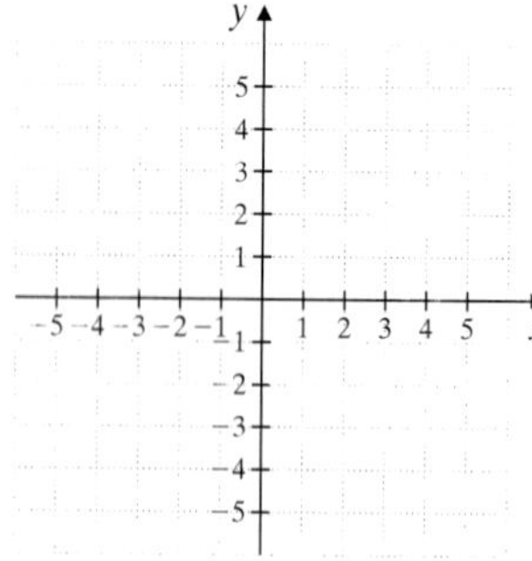

18. $\begin{cases} y > 2 \\ x \geq -1 \end{cases}$

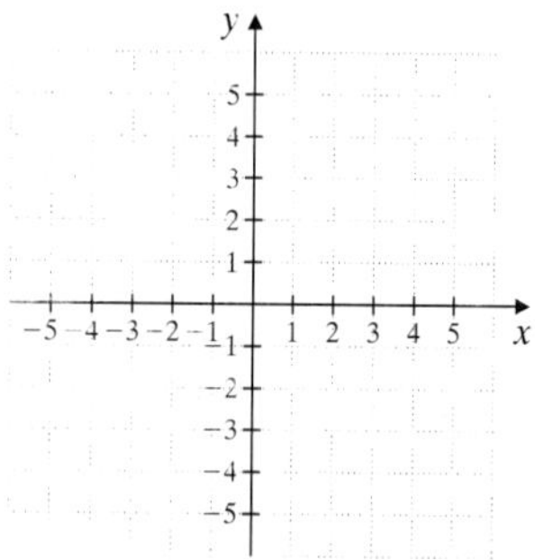

19. $\begin{cases} 2x + 3y < -8 \\ x \geq -4 \end{cases}$

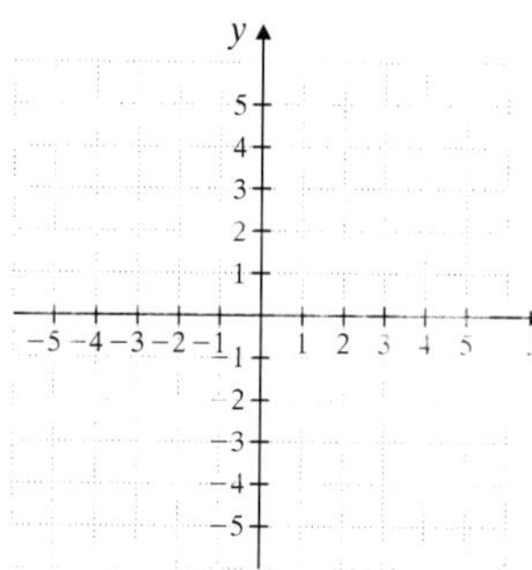

20. $\begin{cases} 3x + 2y \leq 6 \\ x < 2 \end{cases}$

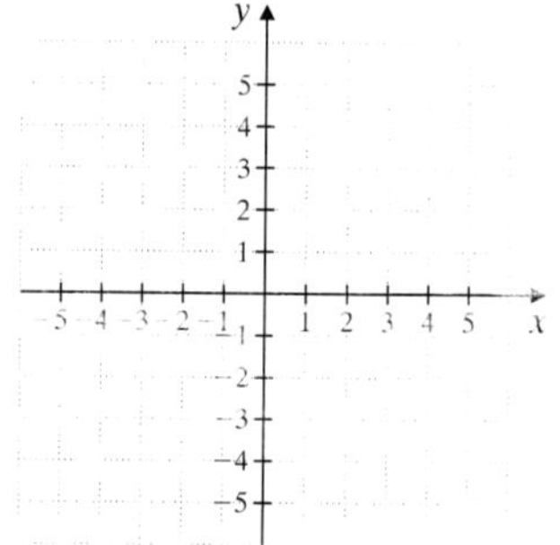

21 $\begin{cases} 2x - 5y \leq 9 \\ y \leq -3 \end{cases}$

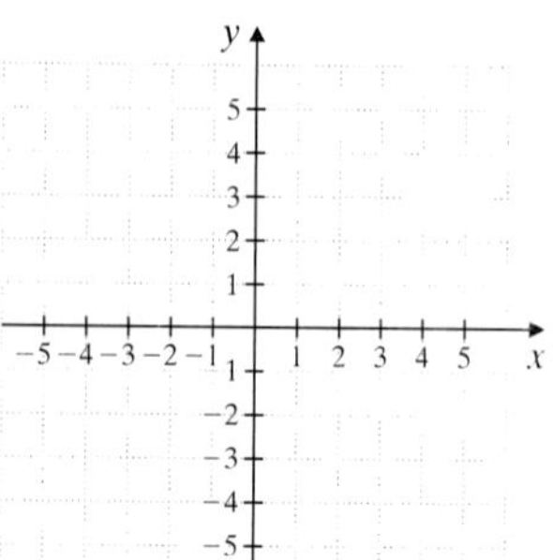

22. $\begin{cases} 2x + 5y \leq -10 \\ y \geq 1 \end{cases}$

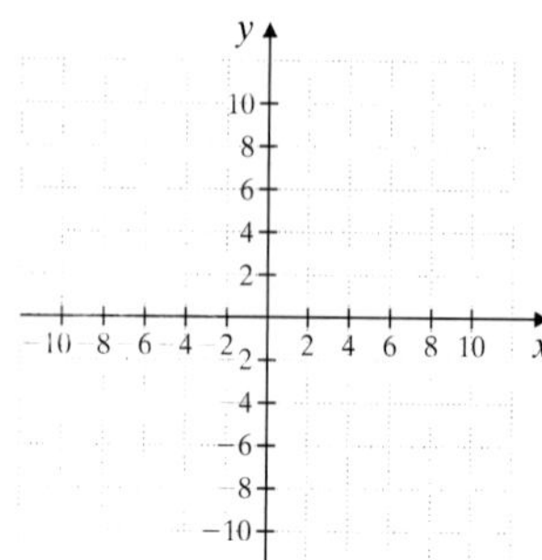

23. $\begin{cases} y \geq \frac{1}{2}x + 2 \\ y \leq \frac{1}{2}x - 3 \end{cases}$

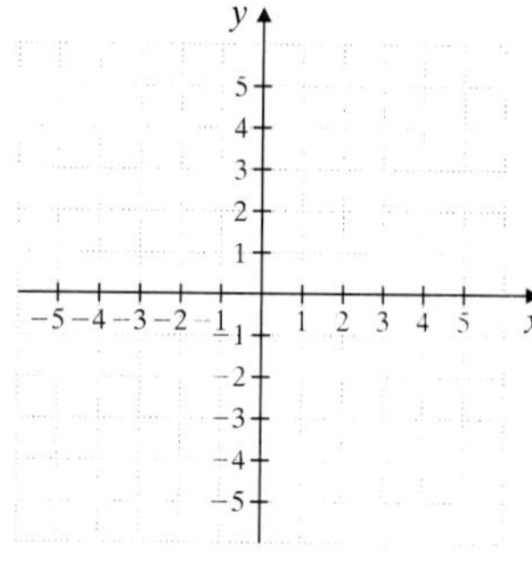

24. $\begin{cases} y \geq \frac{-3}{2}x + 3 \\ y < \frac{-3}{2}x + 6 \end{cases}$

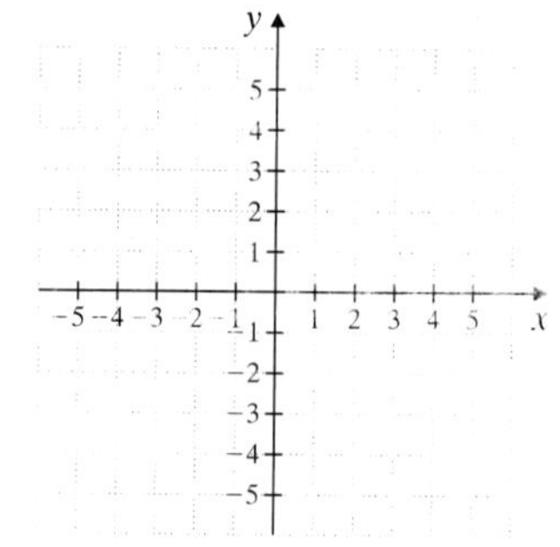

For each system of inequalities, choose the corresponding graph.

25. $\begin{cases} y < 5 \\ x > 3 \end{cases}$

26. $\begin{cases} y > 5 \\ x < 3 \end{cases}$

27. $\begin{cases} y \le 5 \\ x < 3 \end{cases}$

28. $\begin{cases} y > 5 \\ x \ge 3 \end{cases}$

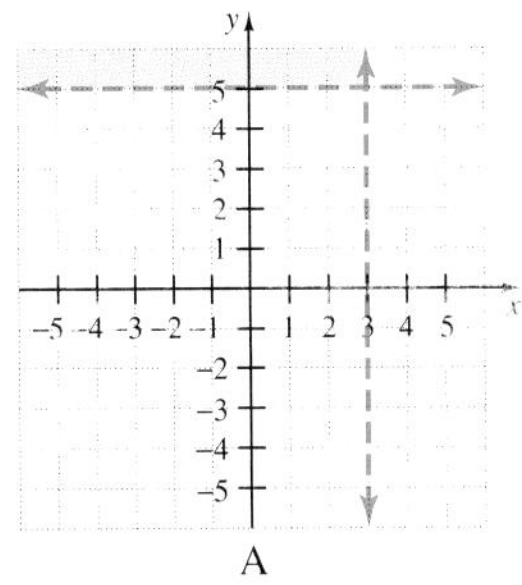
A

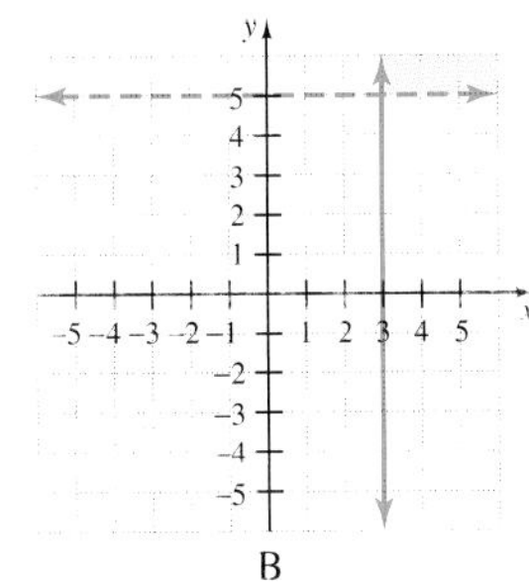
B

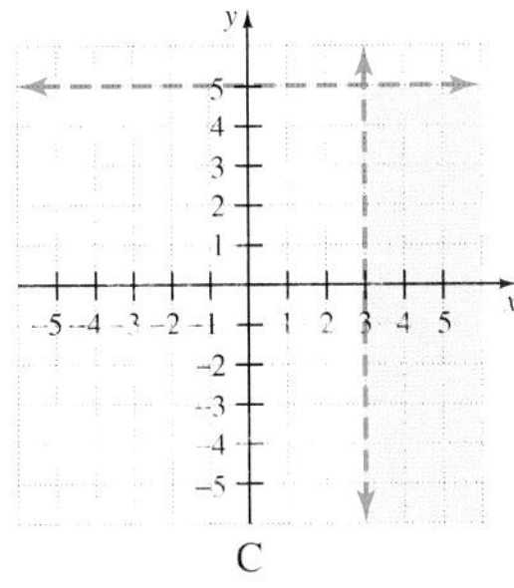
C

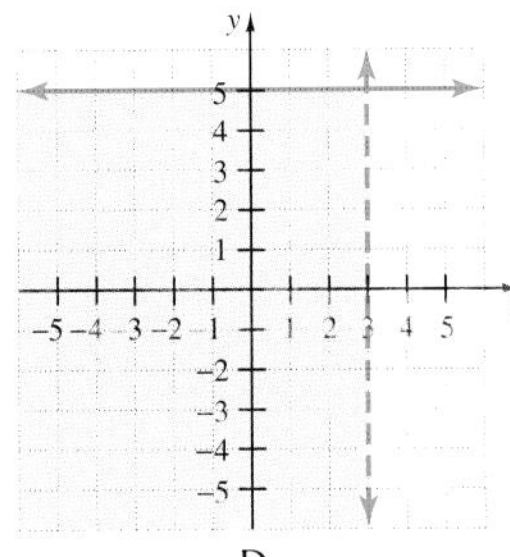
D

Answers to Selected Exercises

Chapter 1 WHOLE NUMBERS AND INTRODUCTION TO ALGEBRA

CHAPTER 1 PRETEST

1. hundreds; 1.2A **2.** twenty-three thousand, four hundred ninety; 1.2B **3.** 87; 1.3A **4.** 3717; 1.6B **5.** 626; 1.4A **6.** 32; 1.4B **7.** 136 pages; 1.4C **8.** 9050; 1.5A **9.** 3100; 1.5B **10.** $9 \cdot 3 + 9 \cdot 11$; 1.6A **11.** 25 in.; 1.3B **12.** 184 sq yd; 1.6C **13.** 576 seats; 1.6D **14.** 243; 1.7A **15.** 446 R 9; 1.7B **16.** 39; 1.7D **17.** $9880; 1.7C **18.** 9^7; 1.8A **19.** 2401; 1.8B **20.** 39; 1.8C **21.** 3; 1.9A **22.** $2x + 6$; 1.9B

EXERCISE SET 1.2

1. tens **3.** thousands **5.** hundred-thousands **7.** millions **9.** five thousand, four hundred twenty **11.** twenty-six thousand, nine hundred ninety **13.** one million, six hundred twenty thousand **15.** fifty-three million, five hundred twenty thousand, one hundred seventy **17.** sixty-three thousand, nine hundred sixty **19.** one thousand, four hundred eighty-three **21.** twelve thousand, six hundred sixty-two **23.** thirteen million, six hundred thousand **25.** three thousand, eight hundred ninety-five **27.** 6508 **29.** 29,900 **31.** 6,504,019 **33.** 3,000,014 **35.** 220 **37.** 755 **39.** 73,500,000 **41.** 1815 **43.** 1262 **45.** 45,000 **47.** $400 + 6$ **49.** $5000 + 200 + 90$ **51.** $60,000 + 2000 + 400 + 7$ **53.** $30,000 + 600 + 80$ **55.** $30,000,000 + 9,000,000 + 600,000 + 80,000$ **57.** $1000 + 6$ **59.** $<$ **61.** $>$ **63.** $>$ **65.** $>$ **67.** four thousand **69.** $4000 + 100 + 40 + 5$ **71.** Nile **73.** Labrador retriever; one hundred sixty-five thousand, nine hundred seventy **75.** Golden retriever **77.** 7632 **79.** Answers may vary. **81.** Canton

CALCULATOR EXPLORATIONS

1. 134 **3.** 340 **5.** 2834

MENTAL MATH

1. 12 **3.** 9000 **5.** 1620

EXERCISE SET 1.3

1. 36 **3.** 92 **5.** 49 **7.** 5399 **9.** 117 **11.** 71 **13.** 117 **15.** 25 **17.** 62 **19.** 212 **21.** 94 **23.** 910 **25.** 8273 **27.** 11,926 **29.** 1884 **31.** 16,717 **33.** 1110 **35.** 8999 **37.** 35,901 **39.** 612,389 **41.** 29 in. **43.** 25 ft **45.** 24 in. **47.** 8 yd **49.** 6684 ft **51.** 340 ft **53.** 263,700 motorcycles **55.** 28,000 people **57.** 8545 stores **59.** 2425 ft **61.** 13,255 mi **63.** 7485 **65.** 10,413 **67.** Texas **69.** 678 stores **71.** 1495 stores **73.** Answers may vary. **75.** 166,510,192 **77.** The computation is incorrect; answers may vary.

CALCULATOR EXPLORATIONS

1. 770 **3.** 109 **5.** 8978

MENTAL MATH

1. 7 **3.** 5 **5.** 0 **7.** 400 **9.** 500

EXERCISE SET 1.4

1. 44 **3.** 60 **5.** 265 **7.** 254 **9.** 545 **11.** 600 **13.** 25 **15.** 45 **17.** 146 **19.** 288 **21.** 168 **23.** 6 **25.** 447 **27.** 5723 **29.** 504 **31.** 89 **33.** 79 **35.** 39,914 **37.** 32,711 **39.** 5041 **41.** 31,213 **43.** 4 **45.** 20 **47.** 7 **49.** 72 **51.** 88 **53.** 264 pages **55.** 4 million sq km **57.** 6065 ft **59.** $175 **61.** 358 mi **63.** $239 **65.** 173 points **67.** 37,035 boxers **69.** Jo; 271 votes **71.** 3,044,452 people **73.** 5920 sq ft **75.** Lake Pontchartrain Bridge; 2175 ft **77.** Atlanta Hartsfield International **79.** 5842 thousand **81.** General Motors Corp. **83.** $1,389,100,000 **85.** $21,784,800,000 **87.** $5269 - 2385 = 2884$ **89.** Answers may vary.

EXERCISE SET 1.5

1. 630 **3.** 640 **5.** 790 **7.** 400 **9.** 1100 **11.** 43,000 **13.** 248,700 **15.** 36,000 **17.** 100,000 **19.** 60,000,000 **21.** 5280; 5300; 5000 **23.** 9440; 9400; 9000 **25.** 14,880; 14,900; 15,000 **27.** 11,000 **29.** 38,000 points **31.** 120,000,000 bushels **33.** 18,000 women **35.** 130 **37.** 380 **39.** 5500 **41.** 300 **43.** 8500 **45.** correct **47.** incorrect **49.** correct **51.** correct **53.** $3100 **55.** 80 mi **57.** 6000 ft **59.** 1,400,000 people **61.** 14,000,000 votes **63.** 154,000 children **65.** 1100 mi **67.** $3,400,000,000 **69.** $2,000,000,000 **71.** 8550 **73.** 1,549,999 **75.** 21,900 mi **77.** Answers may vary.

Calculator Explorations

1. 3456 **3.** 15,322 **5.** 272,291

Mental Math

1. 24 **3.** 0 **5.** 0 **7.** 87

Exercise Set 1.6

1. $4 \cdot 3 + 4 \cdot 9$ **3.** $2 \cdot 4 + 2 \cdot 6$ **5.** $10 \cdot 11 + 10 \cdot 7$ **7.** 252 **9.** 1872 **11.** 1362 **13.** 5310 **15.** 4172 **17.** 10,857 **19.** 11,326 **21.** 24,800 **23.** 0 **25.** 5900 **27.** 59,232 **29.** 142,506 **31.** 1,821,204 **33.** 456,135 **35.** 64,790 **37.** 199,548 **39.** 240,000 **41.** 300,000 **43.** 26,100 **45.** 3600 **47.** 63 sq m **49.** 390 sq ft **51.** 375 cal **53.** $1890 **55.** 192 cans **57.** 9900 sq ft **59.** 56,000 sq ft **61.** 5828 pixels **63.** 2000 characters **65.** 1280 cal **67.** 71,343 mi **69.** 506 windows **71.** 21,700,000 qt **73.** $6,335,000,000 **75.** 50 students **77.** apple and orange **79.** 2, 9 **81.** Answers may vary. **83.** 1938 points

Calculator Explorations

1. 53 **3.** 62 **5.** 261 **7.** 0

Mental Math

1. 5 **3.** 9 **5.** 0 **7.** 9 **9.** 1 **11.** 5 **13.** undefined **15.** 7 **17.** 0 **19.** 8

Exercise Set 1.7

1. 12 **3.** 37 **5.** 338 **7.** 16 R 2 **9.** 563 R 1 **11.** 37 R 1 **13.** 265 R 1 **15.** 49 **17.** 13 **19.** 97 R 40 **21.** 206 **23.** 506 **25.** 202 R 7 **27.** 45 **29.** 98 R 100 **31.** 202 R 15 **33.** 202 **35.** 58 students **37.** $252,000 **39.** 415 bushels **41.** 89 bridges **43.** Yes, she needs 176 ft; she has 9 ft left over **45.** 24 touchdowns **47.** 1760 yd **49.** 26 **51.** 498 **53.** 79 **55.** 16° **57.** $2,957,500,000 **59.** increase; answers may vary. **61.** No; answers may vary.

Integrated Review

1. 148 **2.** 6555 **3.** 1620 **4.** 562 **5.** 79 **6.** undefined **7.** 9 **8.** 1 **9.** 0 **10.** 0 **11.** 0 **12.** 3 **13.** 2433 **14.** 12,895 **15.** 213 R 3 **16.** 79,317 **17.** 27 **18.** 9 **19.** 138 **20.** 276 **21.** 1099 R 2 **22.** 111 R 1 **23.** 663 R 6 **24.** 1076 R 60 **25.** 1024 **26.** 9899 **27.** 30,603 **28.** 47,500 **29.** 0 **30.** undefined **31.** 0 **32.** 0 **33.** 86 **34.** 22 **35.** 8630; 8600; 9000 **36.** 1550; 1600; 2000 **37.** 10,900; 10,900; 11,000 **38.** 432,200; 432,200; 432,000 **39.** perimeter: 20 ft; area: 25 sq ft **40.** perimeter: 42 in.; area: 98 sq in. **41.** 26 mi **42.** 26 m **43.** 3987 mi **44.** 43 muscles **45.** 206 cases with 12 cans left over; yes **46.** $17,820

Calculator Explorations

1. 729 **3.** 1024 **5.** 2048 **7.** 2526 **9.** 4295 **11.** 8

Exercise Set 1.8

1. 3^4 **3.** 7^8 **5.** 12^3 **7.** $6^2 \cdot 5^3$ **9.** $9^3 \cdot 8$ **11.** $3 \cdot 2^5$ **13.** $3 \cdot 2^2 \cdot 5^3$ **15.** 25 **17.** 125 **19.** 64 **21.** 1024 **23.** 7 **25.** 243 **27.** 256 **29.** 64 **31.** 81 **33.** 729 **35.** 100 **37.** 10,000 **39.** 10 **41.** 1920 **43.** 729 **45.** 21 **47.** 8 **49.** 29 **51.** 4 **53.** 17 **55.** 46 **57.** 28 **59.** 10 **61.** 7 **63.** 4 **65.** 14 **67.** 72 **69.** 2 **71.** 35 **73.** 4 **75.** undefined **77.** 52 **79.** 44 **81.** 12 **83.** 13 **85.** 400 sq mi **87.** 64 sq cm **89.** 10,000 sq m **91.** $(2 + 3) \cdot 6 - 2$ **93.** $24 \div (3 \cdot 2) + 2 \cdot 5$ **95.** 1260 ft **97.** 6,384,814

Exercise Set 1.9

1. 9 **3.** 26 **5.** 6 **7.** 3 **9.** 117 **11.** 94 **13.** 5 **15.** 626 **17.** 20 **19.** 4 **21.** 4 **23.** 0 **25.** 33 **27.** 121 **29.** 121 **31.** 100 **33.** 60 **35.** 4 **37.** 16, 64, 144, 256 **39.** $x + 5$ **41.** $x + 8$ **43.** $20 - x$ **45.** $512x$ **47.** $\frac{x}{2}$ **49.** $5x + (17 + x)$ **51.** $5x$ **53.** $11 - x$ **55.** $x - 5$ **57.** $6 \div x$ or $\frac{6}{x}$ **59.** $50 - 8x$ **61.** 274,657 **63.** 777 **65.** $5x$ **67.** As t gets larger, $16t^2$ gets larger.

Chapter 1 Review

1. hundreds **2.** ten millions **3.** five thousand, four hundred eighty **4.** forty-six million, two hundred thousand, one hundred twenty **5.** $6000 + 200 + 70 + 9$ **6.** $400,000,000 + 3,000,000 + 200,000 + 20,000 + 5000$ **7.** 59,800 **8.** 6,304,000,000 **9.** 1,630,553 **10.** 2,968,528 **11.** 531,341 **12.** 76,704 **13.** 13 **14.** 17 **15.** 3 **16.** 10 **17.** 38

18. 68 **19.** 56 **20.** 40 **21.** 110 **22.** 120 **23.** 950 **24.** 1250 **25.** 1711 **26.** 9867 **27.** 8032 mi **28.** $197,699 **29.** 276 ft **30.** 66 km **31.** 33 **32.** 43 **33.** 14 **34.** 33 **35.** 362 **36.** 65 **37.** 304 **38.** 476 **39.** 2114 **40.** 321 **41.** 397 pages **42.** $25,626 **43.** May **44.** August **45.** July, August, September **46.** April, May **47.** 90 **48.** 50 **49.** 470 **50.** 500 **51.** 4800 **52.** 58,000 **53.** 50,000,000 **54.** 800,000 **55.** 7400 **56.** 4100 **57.** 68,000,000 **58.** 680,000 **59.** 42 **60.** 24 **61.** 0 **62.** 0 **63.** 1410 **64.** 2898 **65.** 800 **66.** 900 **67.** 3696 **68.** 1694 **69.** 0 **70.** 0 **71.** 16,994 **72.** 8954 **73.** 113,634 **74.** 44,763 **75.** 411,426 **76.** 636,314 **77.** 1500 **78.** 240,000 **79.** 1,040,000 **80.** 7,020,000 **81.** 24 g **82.** $4,897,341 **83.** 60 sq mi **84.** 500 sq cm **85.** 3 **86.** 4 **87.** 6 **88.** 5 **89.** 5 R 2 **90.** 4 R 2 **91.** undefined **92.** 0 **93.** 1 **94.** 10 **95.** undefined **96.** 0 **97.** 15 **98.** 19 R 7 **99.** 24 R 2 **100.** 56 **101.** 1 R 17 **102.** 35 R 15 **103.** 500 **104.** 21 R 6 **105.** 506 **106.** 16 **107.** 199 R 8 **108.** 200 **109.** 458 ft **110.** 51 **111.** 7^4 **112.** $6^2 \cdot 3^3$ **113.** $4 \cdot 2^3 \cdot 3^2$ **114.** $5^2 \cdot 7^3 \cdot 2^2$ **115.** 49 **116.** 64 **117.** 1125 **118.** 19,600 **119.** 13 **120.** 10 **121.** 3 **122.** 1 **123.** 32 **124.** 33 **125.** 49 sq m **126.** 9 sq in. **127.** 5 **128.** 17 **129.** undefined **130.** 0 **131.** 121 **132.** 2 **133.** 4 **134.** 20 **135.** $x - 5$ **136.** $x + 7$ **137.** $10 \div x$ **138.** $5x$ **139.** 0, 8, 32, 72

Chapter 1 Test

1. 141 **2.** 113 **3.** 14,880 **4.** 766 R 42 **5.** 200 **6.** 48 **7.** 98 **8.** 0 **9.** undefined **10.** 33 **11.** 21 **12.** 36 **13.** 52,000 **14.** 13,700 **15.** 1600 **16.** $17 **17.** $119 **18.** $126 **19.** 7 billion tablets **20.** 170,000 cards **21.** 30 **22.** 1 **23. a.** $17x$ **b.** $20 - 2x$ **24.** 20 cm, 25 sq cm **25.** 60 yd, 200 sq yd **26.** $403,706,375 **27.** $26,190,162 **28.** $86,148,484

Chapter 2 Integers

Chapter 2 Pretest

1. −22; 2.1 A **2.** 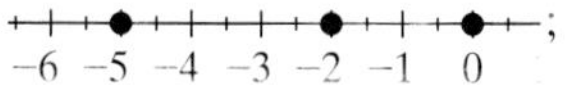

; 2.1B **3.** >; 2.1 C **4.** 8; 2.1D **5.** 12; 2.1E **6.** −11; 2.2A **7.** −17; 2.2B **8.** 3° F; 2.2C **9.** 5; 2.3A **10.** 8; 2.3B **11.** −15; 2.3C **12.** −4; 2.3D **13.** −104; 2.4A **14.** 9; 2.4B **15.** 48; 2.4C **16.** −20; 2.4C **17.** −60; 2.4D **18.** −25; 2.5A **19.** 2; 2.5A **20.** −8; 2.5A **21.** −34; 2.5B **22.** −6; 2.5C

Exercise Set 2.1

1. −1445 **3.** +14,433 **5.** +118 **7.** −11,730 **9.** −1683 million **11.** −135; −157; Sara **13.** −81

15. 

17. 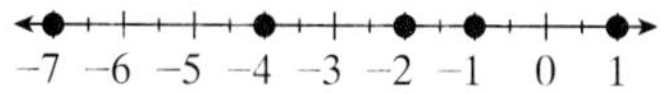

19. 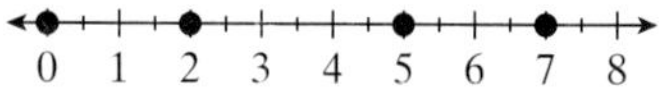

21. 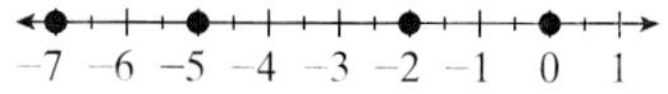

23. < **25.** < **27.** > **29.** < **31.** 5 **33.** 8 **35.** 0 **37.** 5 **39.** −5 **41.** 4 **43.** −23 **45.** 10 **47.** 7 **49.** −20 **51.** −3 **53.** 8 **55.** 14 **57.** 29 **59.** 8 **61.** −3 **63.** 23 **65.** −4 **67.** > **69.** < **71.** = **73.** < **75.** > **77.** < **79.** > **81.** < **83.** $-|-8|, -|3|, 2^2, -(-5)$ **85.** $-|-6|, -|1|, |-1|, -(-6)$ **87.** $-10, -|-9|, -(-2), |-12|, 5^2$ **89.** Maracaibo Lake **91.** Eyre Lake **93.** Earth **95.** Saturn **97.** 13 **99.** 35 **101.** 360 **103.** d **105.** 5 **107.** false **109.** true **111.** false **113.** Answers may vary.

Calculator Explorations

1. −159 **3.** 44 **5.** −894,855

Mental Math

1. 5 **3.** −35 **5.** 0 **7.** 0

Exercise Set 2.2

1. 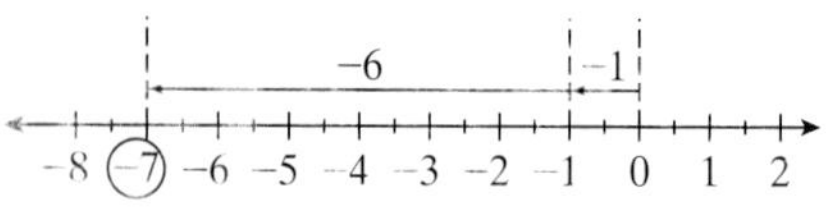

3. 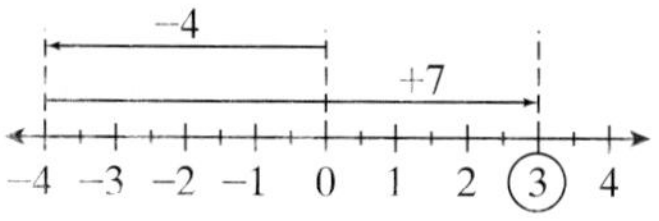

5. 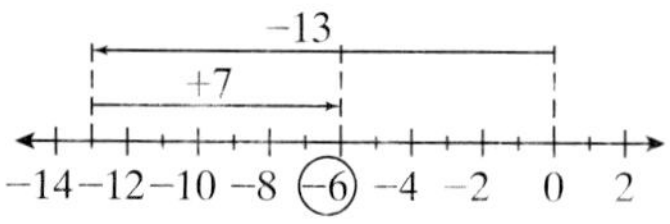

7. 35 **9.** −8 **11.** 0 **13.** 4 **15.** 2 **17.** −2 **19.** −9 **21.** −24 **23.** −57 **25.** −223 **27.** 0 **29.** 7 **31.** −3 **33.** −9 **35.** 30 **37.** 20 **39.** 51 **41.** −33 **43.** −20 **45.** −125 **47.** −7 **49.** −246 **51.** 16 **53.** 13 **55.** −33 **57.** −21 **59.** 21 **61.** −45 **63.** 9 **65.** 0 **67.** 0 **69.** −103 **71.** 1 **73.** −70 **75.** −27 **77.** $236,630,000 **79.** −$1,328,780,000 **81.** 2° C **83.** $159,283 **85.** Team 1:7; Team 2:6; winning team: Team 1 **87.** −40°F **89.** −$946 billion **91.** 44 **93.** 0 **95.** 28 **97.** −9658 **99.** 0 **101.** true **103.** false **105.** Answers may vary.

Exercise Set 2.3

1. 0 **3.** 5 **5.** −5 **7.** 14 **9.** 3 **11.** −18 **13.** −14 **15.** 0 **17.** 0 **19.** −4 **21.** −15 **23.** −14 **25.** −230 **27.** 10 **29.** −4 **31.** 7 **33.** −38 **35.** −17 **37.** 13 **39.** 2 **41.** 0 **43.** −1 **45.** −27 **47.** 40 **49.** −22 **51.** −8 **53.** 14 **55.** −8 **57.** 36 **59.** 0 **61.** 19 **63.** 428 degrees **65.** 39 degrees **67.** 265° F **69.** −$16 **71.** −12°C **73.** 154 ft **75.** 69 ft **77.** 652 ft **79.** 144 ft **81.** −$410 billion **83.** 0 **85.** 436 **87.** 1058 **89.** 21 **91.** −22 **93.** 20 **95.** −4 **97.** 0 **99.** −9196 **101.** false **103.** Answers may vary. **105.** Answers may vary.

Integrated Review

1. +29,028 **2.** −35,840 **3.** −7 **4.** 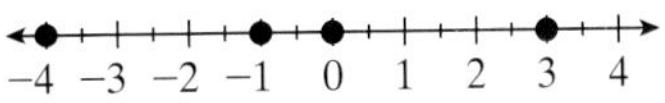

5. > **6.** < **7.** < **8.** > **9.** 1 **10.** −4 **11.** 8 **12.** 5 **13.** −6 **14.** 3 **15.** −89 **16.** 0 **17.** 5 **18.** −20 **19.** −10 **20.** −2 **21.** 52 **22.** −3 **23.** 84 **24.** 6 **25.** 1 **26.** −19 **27.** 12 **28.** −4 **29.** b, c, d **30.** a, b, c, d **31.** 10 **32.** −12 **33.** 12 **34.** 10

Exercise Set 2.4

1. 6 **3.** −36 **5.** −64 **7.** 0 **9.** −48 **11.** −8 **13.** 80 **15.** 0 **17.** −15 **19.** −4 **21.** −27 **23.** −25 **25.** −8 **27.** −4 **29.** −5 **31.** 8 **33.** 0 **35.** undefined **37.** −13 **39.** 0 **41.** −12 **43.** −54 **45.** 42 **47.** −24 **49.** 16 **51.** −2 **53.** −7 **55.** −4 **57.** 48 **59.** −1080 **61.** 0 **63.** −5 **65.** −6 **67.** 3 **69.** −1 **71.** −243 **73.** 180 **75.** 1 **77.** −20 **79.** −966 **81.** −2050 **83.** −28 **85.** −6 **87.** 25 **89.** −1 **91.** undefined **93.** 6 **95.** 8; 2 **97.** 0; 0 **99.** −210°C **101.** −189°C **103.** $(-4)(3) = -12$; a loss of 12 yd **105.** $(-20)(5) = -100$; a depth of 100 ft **107.** −$7260 million **109. a.** 165 condors **b.** 11 condors per year. **111.** −1 **113.** 225 **115.** 109 **117.** 8 **119.** true **121.** true **123.** positive **125.** negative **127.** $(-5)^{17}, (-2)^{17}, (-2)^{12}, (-5)^{12}$ **129. a.** −59 radio stations **b.** −236 radio stations **c.** 1895 radio stations **131.** Answers may vary. **133.** 256,401

Calculator Explorations

1. 48 **3.** −258

Mental Math

1. base: 3; exponent: 2 **3.** base: 2; exponent: 3 **5.** base: −7; exponent: 5 **7.** base: 5; exponent: 7

Exercise Set 2.5

1. −64 **3.** −64 **5.** 24 **7.** 3 **9.** −7 **11.** −14 **13.** −43 **15.** −8 **17.** −13 **19.** −1 **21.** 4 **23.** −3 **25.** −55 **27.** −8 **29.** 16 **31.** 13 **33.** −65 **35.** 64 **37.** 452 **39.** 129 **41.** 3 **43.** −4 **45.** 4 **47.** 16 **49.** −27 **51.** 34 **53.** 65 **55.** −59 **57.** −7 **59.** −61 **61.** −11 **63.** 36 **65.** −117 **67.** 30 **69.** −3 **71.** −59 **73.** 1 **75.** −12 **77.** 0 **79.** −20 **81.** 9 **83.** −16 **85.** 1 **87.** −50 **89.** −2 **91.** −19 **93.** 28 points **95.** 2 points **97.** No; answers may vary. **99.** 4050 **101.** 45 **103.** 32 in. **105.** 30 ft **107.** $2 \cdot (7 - 5) \cdot 3$ **109.** $-6 \cdot (10 - 4)$ **111.** 20,736 **113.** 8900 **115.** 9 **117.** Answers may vary.

Chapter 2 Review

1. −1435 **2.** +7562 **3.** 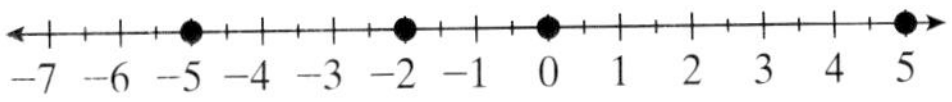

4. 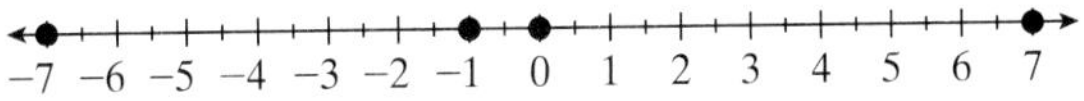

5. 12 **6.** 0 **7.** −6 **8.** 9 **9.** −9 **10.** 2 **11.** > **12.** < **13.** > **14.** > **15.** 12 **16.** −3 **17.** false **18.** true **19.** true **20.** true **21.** 2 **22.** 14 **23.** 4 **24.** 17 **25.** −23 **26.** −22 **27.** −21 **28.** −70 **29.** 0 **30.** 0 **31.** −151 **32.** −606 **33.** −21 **34.** 0 **35.** −20°C **36.** 150 ft below the surface **37.** 99° F **38.** 112° F **39.** 8 **40.** −16 **41.** −11 **42.** −27 **43.** 20 **44.** 8 **45.** 0 **46.** −32 **47.** 0 **48.** −7 **49.** −10 **50.** −9 **51.** −25 **52.** 692 ft **53.** −3°F **54.** −4°F **55.** true **56.** false **57.** true **58.** true **59.** 21 **60.** −18 **61.** −64 **62.** 60 **63.** 25 **64.** −1 **65.** 0 **66.** 24 **67.** −5 **68.** 3 **69.** 0 **70.** undefined **71.** −20 **72.** −9 **73.** 38 **74.** −5 **75.** $(-5)(2) = -10$ **76.** $(-50)(4) = -200$ **77.** −18°F **78.** −3°F **79.** 28° F **80.** −26°F **81.** 49 **82.** −49 **83.** −32 **84.** −32 **85.** 0 **86.** −8 **87.** −16 **88.** 35 **89.** −28 **90.** −44 **91.** 3 **92.** −1 **93.** 7 **94.** −17 **95.** 39 **96.** −26 **97.** 7 **98.** −80 **99.** −2 **100.** −12 **101.** −3 **102.** −35 **103.** −5 **104.** 5 **105.** −1 **106.** −7 **107.** 4 **108.** −4 **109.** 3 **110.** 108

Chapter 2 Test

1. 3 **2.** −6 **3.** −100 **4.** 4 **5.** −30 **6.** 12 **7.** 65 **8.** 5 **9.** 12 **10.** −6 **11.** 50 **12.** −2 **13.** −11 **14.** −46 **15.** −117 **16.** 3456 **17.** 28 **18.** −213 **19.** −1 **20.** −2 **21.** 2 **22.** −5 **23.** −32 **24.** −12 **25.** −3 **26.** 5 **27.** −1 **28.** −54 **29.** 1 **30.** −17 **31.** 88 ft below sea level **32.** 45 **33.** 31,642 **34.** 3820 ft below sea level **35.** −4

Cumulative Review

1. ten-thousands; Sec. 1.2, Ex. 1 **2.** hundreds **3.** tens; Sec. 1.2, Ex. 2 **4.** thousands **5.** millions; Sec. 1.2, Ex. 3 **6.** hundred-thousands **7. a.** $<$ **b.** $>$ **c.** $>$; Sec. 1.2, Ex. 12 **8. a.** $>$ **b.** $<$ **c.** $>$ **9.** 39; Sec. 1.3, Ex. 3 **10.** 39 **11.** 7321; Sec. 1.4, Ex. 2 **12.** 3013 **13.** 2440 km; Sec. 1.4, Ex. 5 **14.** \$525 **15.** 570; Sec. 1.5, Ex. 1 **16.** 600 **17.** 1800; Sec. 1.5, Ex. 5 **18.** 5000 **19. a.** $3 \cdot 4 + 3 \cdot 5$ **b.** $10 \cdot 6 + 10 \cdot 8$ **c.** $2 \cdot 7 + 2 \cdot 3$; Sec. 1.6, Ex. 2 **20. a.** $5 \cdot 2 + 5 \cdot 12$ **b.** $9 \cdot 3 + 9 \cdot 6$ **b.** $4 \cdot 8 + 4 \cdot 1$ **21.** 78,875; Sec. 1.6, Ex. 5 **22.** 31,096 **23. a.** 6 **b.** 9 **c.** 6; Sec. 1.7, Ex. 1 **24. a.** 7 **b.** 8 **c.** 12 **25.** 741; Sec. 1.7, Ex. 4 **26.** 456 **27.** 7 boxes; Sec. 1.7, Ex. 11 **28.** \$9 **29.** 64; Sec. 1.8, Ex. 5 **30.** 125 **31.** 7; Sec. 1.8, Ex. 6 **32.** 4 **33.** 180; Sec. 1.8, Ex. 8 **34.** 56 **35.** 2; Sec. 1.8, Ex. 13 **36.** 5 **37.** 15; Sec. 1.9, Ex. 1 **38.** 14 **39. a.** 2 **b.** 5 **c.** 0; Sec. 2.1, Ex. 4 **40. a.** 4 **b.** 7 **41.** 3; Sec. 2.2, Ex. 7 **42.** 5 **43.** 14; Sec. 2.3, Ex. 12 **44.** 5 **45.** -21; Sec. 2.4, Ex. 1 **46.** -10 **47.** 0; Sec. 2.4, Ex. 3 **48.** -54 **49.** -16; Sec. 2.5, Ex. 8 **50.** -27

Chapter 3 Solving Equations and Problem Solving

Chapter 3 Pretest

1. $15x + 4$; 3.1A **2.** $5x + 5$; 3.1C **3.** $56b$; 3.1B **4.** $-10y + 35$; 3.1B **5.** $(13a + 17)$ in.; 3.1D **6.** $(12x - 3)$ sq ft; 3.1D **7.** not a solution; 3.2A **8.** -12; 3.2B **9.** -9; 3.2B **10.** -7; 3.3A **11.** 11; 3.3A **12.** $2(x - 12)$; 3.3B **13.** $3x$; 3.3B **14.** 9; 3.4A **15.** 2; 3.4A **16.** 1; 3.4B **17.** -3; 3.4B **18.** $\frac{54}{-6} = -9$; 3.4C **19.** $x - 5 = 12$; 3.5A **20.** 7; 3.5B

Mental Math

1. 5 **3.** 1 **5.** 11

Exercise Set 3.1

1. $8x$ **3.** $-4n$ **5.** $-2c$ **7.** $-4x$ **9.** $13a - 8$ **11.** $30x$ **13.** $-22y$ **15.** $72a$ **17.** $2y + 4$ **19.** $5a - 40$ **21.** $-12x - 28$ **23.** $2x + 15$ **25.** $-21n + 20$ **27.** $15c + 3$ **29.** $7w + 15$ **31.** $11x - 8$ **33.** $-5x - 9$ **35.** $-2y$ **37.** $-7z$ **39.** $8d - 3c$ **41.** $6y - 14$ **43.** $-q$ **45.** $2x + 22$ **47.** $-3x - 35$ **49.** $-3z - 15$ **51.** $-6x + 6$ **53.** $3x - 30$ **55.** $-r + 8$ **57.** $-7n + 3$ **59.** $9z - 14$ **61.** -6 **63.** $-4x + 10$ **65.** $2xy + 20$ **67.** $7a + 12$ **69.** $3y + 5$ **71.** $(-25x + 55)$ in. **73.** $(14y + 22)$ m **75.** $(11a + 12)$ ft **77.** $16z^2$ sq cm **79.** $(100x - 140)$ sq in. **81.** -3 **83.** 8 **85.** 0 **87.** $4824q + 12{,}274$ **89.** Answers may vary. **91.** $(20x + 16)$ sq mi **93.** Answers may vary.

Exercise Set 3.2

1. yes **3.** yes **5.** yes **7.** yes **9.** yes **11.** yes **13.** 18 **15.** 26 **17.** 9 **19.** -16 **21.** 11 **23.** -4 **25.** 6 **27.** 0 **29.** 8 **31.** -1 **33.** -6 **35.** 24 **37.** 2 **39.** 0 **41.** -28 **43.** 73 **45.** 0 **47.** -4 **49.** -65 **51.** 13 **53.** -3 **55.** -22 **57.** -16 **59.** 3 **61.** 28 **63.** 2430 **65.** about 2200 **67.** 1 **69.** 1 **71.** Answers may vary. **73.** No; answers may vary. **75.** 162,964 **77.** -131 **79.** 42 **81.** 3627 yd **83.** \$219,812 million

Exercise Set 3.3

1. 4 **3.** -4 **5.** 7 **7.** -81 **9.** 0 **11.** -17 **13.** 32 **15.** -60 **17.** 5 **19.** 0 **21.** -4 **23.** 8 **25.** -1 **27.** 18 **29.** -8 **31.** -50 **33.** -1 **35.** -7 **37.** 0 **39.** 6 **41.** -9 **43.** 9 **45.** 4 **47.** -6 **49.** -35 **51.** 270 **53.** -25 **55.** -2 **57.** -2 **59.** 0 **61.** 4 **63.** -28 **65.** 3 **67.** 36 **69.** 1 **71.** $-7 + x$ **73.** $-11x$ **75.** $\frac{x}{-12}$ or $-\frac{x}{12}$ **77.** $x - 11$ **79.** $-10 - 7x$ **81.** $-13x$ **83.** $\frac{17}{x} + (-15)$ **85.** $4x + 7$ **87.** $2x - 17$ **89.** $-6(x + 15)$ **91.** $\frac{45}{-5x}$ **93.** -5 **95.** -5 **97.** 11 **99.** Answers may vary. **101.** -36 **103.** 67,896 **105.** -48 **107.** 12 hr **109.** 58 mph

Integrated Review

1. $8x$ **2.** $-4y$ **3.** $-2a - 2$ **4.** $5a - 26$ **5.** $-8x - 14$ **6.** $-6x + 30$ **7.** $5y - 10$ **8.** $15x - 31$ **9.** $(12x - 6)$ sq m **10.** $(2x + 9)$ ft **11.** 13 **12.** -9 **13.** 5 **14.** 0 **15.** 39 **16.** 55 **17.** -4 **18.** -3 **19.** 5 **20.** 8 **21.** -10 **22.** 6 **23.** -24 **24.** -54 **25.** 12 **26.** -42 **27.** 26 **28.** -12 **29.** -3 **30.** 5 **31.** $x - 10$ **32.** $-20 + x$ **33.** $10x$ **34.** $\frac{10}{x}$ **35.** $-2x + 5$ **36.** $-4(x - 1)$

Calculator Explorations

1. yes **3.** no **5.** yes

Exercise Set 3.4

1. 3 **3.** −50 **5.** −4 **7.** −3 **9.** −12 **11.** 5 **13.** −5 **15.** 8 **17.** 5 **19.** 1 **21.** −16 **23.** 2 **25.** 5 **27.** −3 **29.** 2 **31.** −2 **33.** 3 **35.** 48 **37.** −10 **39.** −4 **41.** −1 **43.** 4 **45.** −4 **47.** 3 **49.** −1 **51.** 0 **53.** −22 **55.** 4 **57.** 1 **59.** −30 **61.** $-42 + 16 = -26$ **63.** $-5(-29) = 145$ **65.** $3(-14 - 2) = -48$ **67.** $\frac{100}{2(50)} = 1$ **69.** 2002 **71.** 38 million returns **73.** 33 **75.** −37 **77.** 49 **79.** −6 **81.** 0 **83.** −4 **85.** No; answers may vary.

Exercise Set 3.5

1. $-5 + x = -7$ **3.** $3x = 27$ **5.** $-20 - x = 104$ **7.** $2[x + (-1)] = 50$ **9.** 8 **11.** 6 **13.** 9 **15.** 6 **17.** 24 **19.** 8 **21.** 5 **23.** 5 **25.** 12 **27.** 5 **29.** Bush: 266 votes; Gore: 261 votes **31.** bamboo: 36 in.; kelp: 18 in. **33.** India: 8407; U.S.: 5758 **35.** Gamecube: \$150; games: \$450 **37.** Michigan Stadium, 107,501; Neyland Stadium, 102,854 **39.** California, 309 thousand; Washington, 103 thousand **41.** 82 points **43.** crow: 9 oz; finch: 4 oz **45.** 590 **47.** 1000 **49.** 3000 **51.** Answers may vary. **53.** \$8250 **55.** \$5 **57.** Answers may vary.

Chapter 3 Review

1. $10y - 15$ **2.** $-6y - 10$ **3.** $-6a - 7$ **4.** $-8y + 2$ **5.** $2x + 10$ **6.** $-3y - 24$ **7.** $11x - 12$ **8.** $-4m - 12$ **9.** $-5a + 4$ **10.** $12y - 9$ **11.** $(4x + 6)$ yd **12.** $20y$ m **13.** $(6x - 3)$ sq yd **14.** $(45x + 8)$ sq cm **15.** $600{,}822x - 9180$ **16.** $3292y - 3840$ **17.** yes **18.** no **19.** −2 **20.** 10 **21.** −6 **22.** −1 **23.** −25 **24.** −8 **25.** −7 **26.** −9 **27.** 1 **28.** 5 **29.** −12 **30.** 45 **31.** 0 **32.** −128 **33.** −8 **34.** −45 **35.** 5 **36.** −5 **37.** $-5x$ **38.** $x - 3$ **39.** $-5 + x$ **40.** $\frac{-2}{x}$ **41.** $2x + 11$ **42.** $-5x - 50$ **43.** $\frac{70}{x + 6}$ **44.** $2(x - 13)$ **45.** 5 **46.** 12 **47.** 17 **48.** 7 **49.** −2 **50.** 8 **51.** 21 **52.** −10 **53.** 2 **54.** 2 **55.** −27 **56.** −22 **57.** 11 **58.** −5 **59.** $20 - (-8) = 28$ **60.** $5[2 + (-6)] = -20$ **61.** $\frac{-75}{5 + 20} = -3$ **62.** $-2 - 19 = -21$ **63.** $2x - 8 = 40$ **64.** $\frac{x}{2} - 12 = 10$ **65.** $x - 3 = \frac{x}{4}$ **66.** $6x = x + 2$ **67.** 5 **68.** −16 **69.** 2386 votes **70.** 84 DVDs

Chapter 3 Test

1. $-5x + 5$ **2.** $-6y - 14$ **3.** $-8z - 20$ **4.** $(15x + 15)$ in. **5.** $(12x - 4)$ sq m **6.** −10 **7.** −16 **8.** −6 **9.** −6 **10.** 8 **11.** 24 **12.** −2 **13.** 6 **14.** −2 **15.** 0 **16. a.** $-23 + x$ **b.** $-2 - 3x$ **17. a.** $2 \cdot 5 + (-15) = -5$ **b.** $3x + 6 = -30$ **18.** −2 **19.** 8 free throws **20.** 244 women

Cumulative Review

1. one hundred six million, fifty-two thousand, four hundred forty-seven; Sec. 1.2, Ex. 6 **2.** two hundred seventy-six thousand, four **3.** 13 in.; Sec. 1.3, Ex. 5 **4.** 18 in. **5.** 726; Sec. 1.4, Ex. 4 **6.** 9585 **7.** 249,000; Sec. 1.5, Ex. 3 **8.** 844,000 **9.** 200; Sec. 1.6, Ex. 3 **10.** 29,230 **11.** 208; Sec. 1.7, Ex. 5 **12.** 86 **13.** 7; Sec. 1.8, Ex. 9 **14.** 35 **15.** 26; Sec. 1.9, Ex. 4 **16.** 10 **17. a.** < **b.** > **c.** >; Sec. 2.1, Ex. 3 **18. a.** < **b.** > **19.** 3; Sec. 2.2, Ex. 1 **20.** −7 **21.** −6; Sec. 2.2, Ex. 5 **22.** −4 **23.** 8; Sec. 2.2, Ex. 6 **24.** 17 **25.** −14; Sec. 2.3, Ex. 2 **26.** −5 **27.** 11; Sec. 2.3, Ex. 3 **28.** 29 **29.** −4; Sec. 2.3, Ex. 4 **30.** −3 **31.** −2; Sec. 2.4, Ex. 10 **32.** 6 **33.** 5; Sec. 2.4, Ex. 11 **34.** −13 **35.** −16; Sec. 2.4, Ex. 12 **36.** −10 **37.** 9; Sec. 2.5, Ex. 1 **38.** −32 **39.** −9; Sec. 2.5, Ex. 2 **40.** 25 **41.** $6y + 2$; Sec. 3.1, Ex. 2 **42.** $3x + 9$ **43.** not a solution; Sec. 3.2, Ex. 2 **44.** solution **45.** 3; Sec. 3.3, Ex. 4 **46.** 5 **47.** 12; Sec. 3.4, Ex. 1 **48.** −2 **49.** software, \$420; computer system, \$1680; Sec. 3.5, Ex. 4 **50.** 11

Chapter 4 Fractions

Chapter 4 Pretest

1. $\frac{3}{8}$; 4.1 B **2.** $\frac{15}{18}$; 4.1 E **3.** 0; 4.1 D **4.** $2 \cdot 2 \cdot 5 \cdot 7$ or $2^2 \cdot 5 \cdot 7$; 4.2A **5.** $\frac{5}{9}$; 4.2B **6.** $\frac{9}{20}$ of a gram; 4.2C **7.** $\frac{6}{5}$; 4.3 A **8.** $\frac{1}{14}$; 4.3 C **9.** $\frac{5}{11}$; 4.4 A **10.** $-\frac{1}{10}$; 4.4A **11.** $-\frac{8}{27}$; 4.3B **12.** $-\frac{1}{3}$; 4.3D **13.** $\frac{23}{36}$; 4.5A **14.** $\frac{1}{14}$; 4.5C **15.** $\frac{3x}{7}$; 4.6A **16.** $-\frac{46}{25}$; 4.6B **17.** $-\frac{23}{2}$; 4.7A **18.** $\frac{13}{5}$; 4.8B **19.** $\frac{44}{5}$ or $8\frac{4}{5}$; 4.8D **20.** $9\frac{5}{6}$; 4.8E

MENTAL MATH

1. numerator = 1; denominator = 2 **3.** numerator = 10; denominator = 3 **5.** numerator = $3z$; denominator = 7

EXERCISE SET 4.1

1. $\frac{1}{3}$ **3.** $\frac{4}{7}$ **5.** $\frac{7}{12}$ **7.** $\frac{3}{7}$ **9.** $\frac{7}{8}$ **11.** $\frac{5}{8}$ **13.** **15.** **17.** **19.** **21.** $\frac{11}{4}$ **23.** $\frac{11}{3}$ **25.** $\frac{3}{2}$ **27.** $\frac{4}{3}$ **29.** $\frac{17}{6}$ **31.** $\frac{14}{9}$ **33.** $\frac{42}{131}$ of the students **35.** 89; $\frac{89}{131}$ of the students **37.** $\frac{4}{10}$ of the visits **39.** $\frac{8}{43}$ of the presidents **41.** $\frac{17}{32}$ of the hard drive **43.** $\frac{11}{31}$ of the month **45.** $\frac{10}{31}$ of the class **47. a.** $\frac{40}{50}$ of the states **b.** 10 states **c.** $\frac{10}{50}$ of the states **49.** **51.**

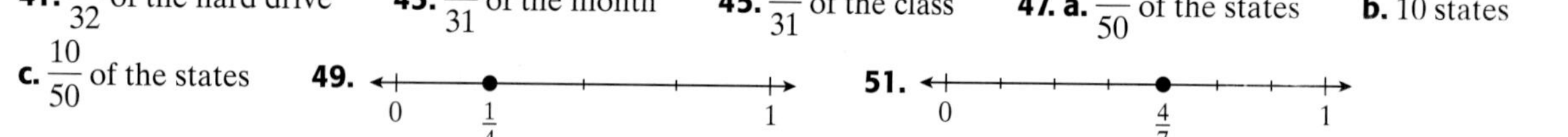

53. 0, 1, $\frac{8}{5}$, 2 **55.** 0, 1, 2, $\frac{7}{3}$, 3 **57.** 0, $\frac{3}{8}$, 1

59. 1 **61.** −5 **63.** 0 **65.** 1 **67.** undefined **69.** 3 **71.** $\frac{20}{35}$ **73.** $\frac{14}{21}$ **75.** $\frac{10y}{25}$ **77.** $\frac{15}{30}$ **79.** $\frac{30}{21x}$ **81.** $\frac{10}{5}$ **83.** $\frac{9}{12}$ **85.** $\frac{8y}{12}$ **87.** $\frac{6}{12}$ **89.** $\frac{48x}{36x}$ **91.** $\frac{20x}{36x}$ **93.** $\frac{36x}{36x}$ **95.** $\frac{52}{100}; \frac{78}{100}; \frac{87}{100}; \frac{84}{100}; \frac{59}{100}; \frac{83}{100}; \frac{67}{100}; \frac{60}{100}; \frac{79}{100}; \frac{45}{100}$ **97.** United States **99.** 9 **101.** 125 **103.** 49 **105.** 24 **107.** $\frac{464}{2088}$ **109.** Answers may vary. **111.** $\frac{1651}{2285}$ of the affiliates **113.** $\frac{6253}{8851}$ of the restaurants

CALCULATOR EXPLORATIONS

1. $\frac{4}{7}$ **3.** $\frac{20}{27}$ **5.** $\frac{15}{8}$ **7.** $\frac{9}{2}$

MENTAL MATH

1. yes, yes, yes **3.** $2 \cdot 5$ **5.** $3 \cdot 7$ **7.** 3^2

EXERCISE SET 4.2

1. $2^2 \cdot 5$ **3.** $2^4 \cdot 3$ **5.** $3^2 \cdot 5$ **7.** $2^4 \cdot 3 \cdot 5$ **9.** $\frac{1}{4}$ **11.** $\frac{x}{5}$ **13.** $\frac{7}{8}$ **15.** $\frac{4}{5}$ **17.** $\frac{5}{6}$ **19.** $\frac{5x}{6}$ **21.** $\frac{2}{3}$ **23.** $\frac{3x}{4}$ **25.** $\frac{3b}{2a}$ **27.** $\frac{7}{8}$ **29.** $\frac{3}{7}$ **31.** $\frac{3y}{5}$ **33.** $\frac{4}{7}$ **35.** $\frac{4x^2y}{5}$ **37.** $\frac{4}{5}$ **39.** $\frac{5x}{8}$ **41.** $\frac{3x}{10}$ **43.** $\frac{3a^2}{2b^3}$ **45.** $\frac{5}{8z}$ **47.** $\frac{3}{4}$ of a shift **49.** $\frac{1}{2}$ mi **51.** $\frac{29}{46}$ of individuals **53.** $\frac{9}{20}$ of the students **55. a.** $\frac{8}{25}$ of the states **b.** 34 states **c.** $\frac{17}{25}$ of the states **57.** $\frac{5}{12}$ of the width **59.** $\frac{1}{10}$ **61.** Answers may vary. **63.** −3 **65.** −14 **67.** 0 **69.** 29 **71.** d **73.** false **75.** true **77.** $\frac{3}{5}$ **79.** $\frac{9}{25}$ of the donors **81.** $\frac{1}{25}$ of the donors **83.** $\frac{3}{20}$ of the donors **85.** 2235, 105, 900, 1470 **87.** 15; Answers may vary.

MENTAL MATH

1. $\frac{2}{15}$ **3.** $\frac{6}{35}$ **5.** $\frac{9}{8}$

EXERCISE SET 4.3

1. $\frac{7}{12}$ **3.** $-\frac{5}{28}$ **5.** $\frac{1}{15}$ **7.** $\frac{18x}{55}$ **9.** $\frac{3a^2}{4}$ **11.** $\frac{x^2}{y}$ **13.** $\frac{1}{125}$ **15.** $\frac{4}{9}$ **17.** $-\frac{4}{27}$ **19.** $\frac{4}{5}$ **21.** $-\frac{1}{6}$ **23.** $\frac{16}{9x}$

25. $\frac{121y}{60}$ **27.** $-\frac{1}{6}$ **29.** $\frac{x}{25}$ **31.** $\frac{10}{27}$ **33.** $\frac{18x^2}{35}$ **35.** $-\frac{1}{4}$ **37.** $\frac{3}{4}$ **39.** $\frac{9}{16}$ **41.** xy^2 **43.** $\frac{77}{2}$ **45.** $-\frac{36}{x}$ **47.** $\frac{3}{49}$ **49.** $-\frac{19y}{7}$ **51.** $\frac{4}{11}$ **53.** $\frac{8}{3}$ **55.** $\frac{15x}{4}$ **57.** $\frac{8}{9}$ **59.** $\frac{1}{60}$ **61.** b **63.** $\frac{2}{5}$ **65.** $-\frac{5}{3}$ **67. a.** $\frac{1}{3}$ **b.** $\frac{12}{25}$ **69. a.** $-\frac{36}{55}$ **b.** $-\frac{44}{45}$ **71.** yes **73.** no **75.** 30 gal **77.** \$1838 **79.** $\frac{3}{16}$ in. **81.** 868 mi **83.** 21,735 tornados **85.** $\frac{17}{2}$ in. **87.** 3 ft **89.** 24 **91.** $\frac{1}{14}$ sq ft **93.** $2 \cdot 3 \cdot 3 \cdot 5$ or $2 \cdot 3^2 \cdot 5$ **95.** $5 \cdot 13$ **97.** $2 \cdot 3 \cdot 3 \cdot 7$ or $2 \cdot 3^2 \cdot 7$ **99.** 7,210,000 households **101.** Answers may vary. **103.** 15,432 species **105.** 5

Mental Math

1. unlike **3.** like **5.** like **7.** unlike **9.** $\frac{5}{7}$ **11.** $\frac{6}{11}$ **13.** $\frac{7}{11}$ **15.** $\frac{8}{15}$

Exercise Set 4.4

1. 0 **3.** $\frac{2}{3x}$ **5.** $-\frac{1}{13}$ **7.** $\frac{2}{3}$ **9.** $-\frac{3}{y}$ **11.** $\frac{7a-3}{4}$ **13.** $-\frac{3}{4}$ **15.** $-\frac{1}{3}$ **17.** $\frac{2x}{3}$ **19.** $-\frac{x}{2}$ **21.** $\frac{3}{4z}$ **23.** $-\frac{3}{10}$ **25.** 2 **27.** $-\frac{2}{3}$ **29.** $\frac{3x}{4}$ **31.** $\frac{5}{4}$ **33.** $\frac{2}{5}$ **35.** $-\frac{3}{4}$ **37.** $-\frac{2}{3}$ **39.** $\frac{1}{13}$ **41.** $\frac{4}{5}$ **43.** 1 in. **45.** 2 m **47.** $\frac{3}{2}$ hr **49.** Traditional fee-for-service, Point-of-Service, Health Maintenance Organization, Preferred Provider Organization **51.** $\frac{50}{100} = \frac{1}{2}$ of the employees **53.** $\frac{21}{50}$ of the states **55.** 12 **57.** 45 **59.** 36 **61.** $24x$ **63.** 150 **65.** 126 **67.** 75 **69.** 24 **71.** 50 **73.** $12a$ **75.** 168 **77.** 363 **79.** $\frac{12}{35}$ **81.** 8 **83.** $\frac{4}{5}$ **85.** -28 **87.** $\frac{8}{11}$ **89.** $\frac{1}{10}$ of men **91.** Answers may vary.

Calculator Explorations

1. $\frac{37}{80}$ **3.** $\frac{95}{72}$ **5.** $\frac{394}{323}$

Mental Math

1. 6 **3.** 12 **5.** 56 **7.** 12

Exercise Set 4.5

1. $\frac{5}{6}$ **3.** $\frac{1}{6}$ **5.** $-\frac{4}{33}$ **7.** $\frac{3x-6}{14}$ **9.** $\frac{3}{5}$ **11.** $\frac{24y-5}{12}$ **13.** $\frac{11}{36}$ **15.** $\frac{12}{7}$ **17.** $\frac{89a}{99}$ **19.** $\frac{4y-1}{6}$ **21.** $\frac{x+6}{2x}$ **23.** $-\frac{8}{33}$ **25.** $\frac{3}{14}$ **27.** $\frac{11y-10}{35}$ **29.** $-\frac{11}{36}$ **31.** $\frac{1}{20}$ **33.** $\frac{33}{56}$ **35.** $\frac{17}{16}$ **37.** $\frac{8}{9}$ **39.** $\frac{15+11y}{33}$ **41.** $-\frac{53}{42}$ **43.** $\frac{11}{18}$ **45.** $\frac{44a}{39}$ **47.** $-\frac{11}{60}$ **49.** $\frac{5y+9}{9y}$ **51.** $\frac{56}{45}$ **53.** $\frac{40+9x}{72x}$ **55.** $-\frac{11}{30}$ **57.** $\frac{7x}{8}$ **59.** $\frac{19}{20}$ **61.** $-\frac{5}{24}$ **63.** $\frac{37x-20}{56}$ **65.** $<$ **67.** $>$ **69.** $>$ **71.** $\frac{13}{12}$ **73.** $\frac{1}{4}$ **75.** $\frac{11}{6}$ **77.** $\frac{11}{12}$ **79.** $-\frac{7}{10}$ **81.** $\frac{11}{16}$ **83.** $\frac{11}{18}$ **85.** $\frac{34}{15}$ cm **87.** $\frac{17}{10}$ m **89.** $\frac{61}{264}$ mi **91.** $\frac{11}{8}$ in. **93.** $\frac{49}{100}$ of students **95.** $\frac{77}{100}$ of Americans **97.** $\frac{29}{100}$ of adults **99.** $\frac{25}{36}$ **101.** $\frac{25}{36}$ **103.** 57,200 **105.** 330 **107.** $\frac{49}{44}$ **109.** Answers may vary. **111.** $\frac{938}{351}$ **113.** $\frac{163}{579}$ of the land area **115.** 16,300,000 sq mi **117.** $\frac{175}{193}$ of land area **119.** national preserves **121.** 1-2 videos per mo

Integrated Review

1. $\frac{1}{4}$ **2.** $\frac{3}{6}$ or $\frac{1}{2}$ **3.** $\frac{7}{4}$ **4.** $\frac{73}{85}$ **5.** $2 \cdot 3$ **6.** $2 \cdot 5 \cdot 7$ **7.** $2^2 \cdot 3^2 \cdot 7$ **8.** $\frac{1}{7}$ **9.** $\frac{5}{6}$ **10.** $\frac{9}{19}$ **11.** $\frac{21}{55}$ **12.** $\frac{1}{2}$ **13.** $\frac{9}{10}$ **14.** $\frac{1}{25}$ **15.** $\frac{4}{5}$ **16.** $-\frac{2}{5}$ **17.** $\frac{3}{25}$ **18.** $\frac{1}{3}$ **19.** $\frac{4}{5}$ **20.** $\frac{5}{9}$ **21.** $-\frac{1}{6}$ **22.** $\frac{3}{2}$ **23.** $\frac{1}{18}$ **24.** $\frac{4}{21}$ **25.** $-\frac{7}{48}$ **26.** $-\frac{9}{50}$ **27.** $\frac{37}{40}$ **28.** $\frac{11}{36}$ **29.** $\frac{5}{33}$ **30.** $\frac{5}{18}$ **31.** $\frac{11}{18}$ **32.** $\frac{37}{50}$ **33.** 24 lots **34.** $\frac{3}{4}$ ft

Exercise Set 4.6

1. $\frac{1}{6}$ **3.** $\frac{3}{7}$ **5.** $\frac{x}{6}$ **7.** $\frac{23}{22}$ **9.** $\frac{2x}{13}$ **11.** $\frac{17}{60}$ **13.** $\frac{5}{8}$ **15.** $\frac{35}{9}$ **17.** $-\frac{17}{45}$ **19.** $\frac{11}{8}$ **21.** $\frac{29}{10}$ **23.** $\frac{27}{32}$ **25.** $\frac{1}{81}$ **27.** $\frac{9}{64}$ **29.** $\frac{7}{6}$ **31.** $-\frac{2}{5}$ **33.** $-\frac{2}{9}$ **35.** $\frac{5}{2}$ **37.** $\frac{7}{2}$ **39.** $\frac{4}{9}$ **41.** $-\frac{13}{2}$ **43.** $\frac{9}{25}$ **45.** $-\frac{5}{32}$ **47.** 1 **49.** $\frac{1}{10}$ **51.** $-\frac{11}{40}$ **53.** $\frac{x+6}{16}$ **55.** 8 **57.** 25 **59.** $1x$ or x **61.** $1a$ or a **63.** No; answers may vary. **65.** $-\frac{77}{16}$ **67.** $-\frac{55}{16}$ **69.** $\frac{5}{8}$ **71.** $\frac{11}{56}$ **73.** halfway between a and b **75.** false **77.** false **79.** true **81.** No; answers may vary.

Exercise Set 4.7

1. $\frac{2}{7}$ **3.** 12 **5.** -27 **7.** $\frac{27}{8}$ **9.** $\frac{1}{21}$ **11.** $\frac{2}{11}$ **13.** 1 **15.** $\frac{15}{2}$ **17.** -1 **19.** -15 **21.** $\frac{3x-28}{21}$ **23.** $\frac{y+10}{2}$ **25.** $\frac{7x}{15}$ **27.** $\frac{4}{3}$ **29.** 2 **31.** $\frac{21}{10}$ **33.** $-\frac{1}{14}$ **35.** -3 **37.** 50 **39.** $-\frac{1}{9}$ **41.** -6 **43.** 4 **45.** $\frac{3}{5}$ **47.** $-\frac{1}{24}$ **49.** $-\frac{5}{14}$ **51.** 4 **53.** -36 **55.** $\frac{7}{2}$ **57.** $\frac{59}{10}$ **59.** $\frac{49}{6}$ **61.** Answers may vary. **63.** -2

Calculator Explorations

1. $\frac{280}{11}$ **3.** $\frac{3776}{35}$ **5.** $26\frac{1}{14}$ **7.** $92\frac{3}{10}$

Exercise Set 4.8

1. a. $\frac{11}{4}$ **b.** $2\frac{3}{4}$ **3. a.** $\frac{11}{3}$ **b.** $3\frac{2}{3}$ **5. a.** $\frac{3}{2}$ **b.** $1\frac{1}{2}$ **7. a.** $\frac{4}{3}$ **b.** $1\frac{1}{3}$ **9.** $\frac{7}{3}$ **11.** $\frac{27}{8}$ **13.** $\frac{83}{7}$ **15.** $1\frac{6}{7}$ **17.** $3\frac{2}{15}$ **19.** $4\frac{5}{8}$ **21.** $\frac{8}{21}$ **23.** $4\frac{2}{3}$ **25.** $5\frac{1}{2}$ **27.** $18\frac{2}{3}$ **29.** $6\frac{4}{5}$ **31.** $25\frac{5}{14}$ **33.** $13\frac{13}{24}$ **35.** $2\frac{3}{5}$ **37.** $7\frac{5}{14}$ **39.** $\frac{24}{25}$ **41.** 4 **43.** $5\frac{11}{14}$ **45.** $6\frac{2}{9}$ **47.** $\frac{25}{33}$ **49.** $35\frac{13}{18}$ **51.** $2\frac{1}{2}$ **53.** $72\frac{19}{30}$ **55.** $\frac{11}{14}$ **57.** $5\frac{4}{7}$ **59.** $13\frac{16}{33}$ **61.** $10\frac{43}{60}$ min **63.** $\frac{49}{60}$ min **65.** $\frac{39}{2}$ or $19\frac{1}{2}$ in. **67.** 15 shares **69.** $\frac{1}{16}$ in. **71.** No, she will be $\frac{1}{12}$ of a foot short. **73.** $7\frac{5}{6}$ gal **75.** $9\frac{2}{5}$ in. **77.** $7\frac{13}{20}$ in. **79.** $\frac{7}{2}$ or $3\frac{1}{2}$ sq yd **81.** $\frac{15}{16}$ sq in. **83.** 7 mi **85.** $21\frac{5}{24}$ m **87.** $1\frac{1}{2}$ yr **89.** $306\frac{2}{3}$ ft **91.** $\frac{19}{30}$ in. **93.** $-10\frac{3}{25}$ **95.** $-24\frac{7}{8}$ **97.** $-13\frac{59}{60}$ **99.** $-\frac{1}{2}$ **101.** $-1\frac{23}{24}$ **103.** $\frac{73}{1000}$ **105.** $8\frac{17}{24}$ **107.** $-32\frac{1}{6}$ **109.** $9x-13$ **111.** $-3y-6$ **113.** 17 **115.** 1 **117.** Supreme is heavier by $\frac{1}{8}$ lb **119.** Answers may vary. **121.** Answers may vary. **123.** Answers may vary.

Chapter 4 Review

1. $\frac{3}{4}$ **2.** $\frac{2}{5}$ **3.** 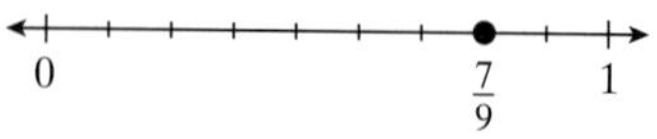 number line from 0 to 1 with point at $\frac{7}{9}$ **4.** 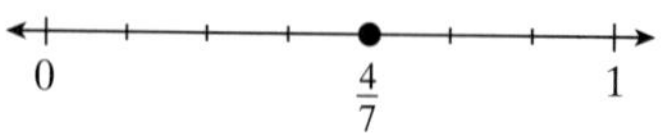 number line from 0 to 1 with point at $\frac{4}{7}$

5. 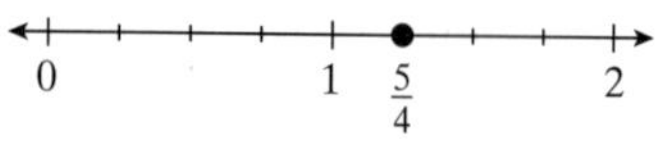 number line from 0 to 2 with point at $\frac{5}{4}$ **6.** 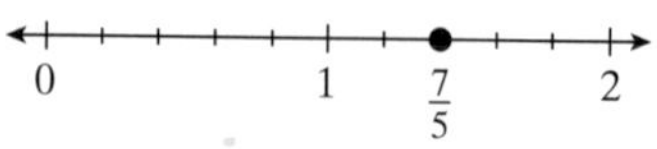 number line from 0 to 2 with point at $\frac{7}{5}$ **7.** $\frac{35}{242}$ of job specialties **8.** $\frac{43}{50}$ of personnel

9. 20 **10.** 35 **11.** $49a$ **12.** $45b$ **13.** 40 **14.** 10 **15.** $\frac{3}{7}$ **16.** $\frac{5}{9}$ **17.** $\frac{1}{3x}$ **18.** $\frac{y^2}{2}$ **19.** $\frac{29}{32c}$ **20.** $\frac{18z}{23}$ **21.** $\frac{5x}{3y^2}$ **22.** $\frac{7b}{5c^2}$ **23.** $\frac{2}{3}$ of a foot **24.** $\frac{3}{5}$ of the cars **25.** $\frac{3}{10}$ **26.** $-\frac{5}{14}$ **27.** $-\frac{7}{12x}$ **28.** $\frac{y}{4}$ **29.** $\frac{9}{x^2}$ **30.** $\frac{y}{2}$ **31.** $-\frac{1}{27}$ **32.** $\frac{25}{144}$ **33.** x^2y^2 **34.** $\frac{b}{a^2}$ **35.** $\frac{2}{15}$ **36.** $-\frac{63}{10}$ **37.** -2 **38.** $\frac{15}{4}$ **39.** $9x^2$ **40.** $\frac{27}{2}$ **41.** $-\frac{5}{6y}$ **42.** $\frac{y^2}{2x}$ **43.** $\frac{12}{7}$ **44.** $-\frac{15}{2}$ **45.** $\frac{77}{48}$ sq ft **46.** $\frac{4}{9}$ sq m **47.** $\frac{10}{11}$ **48.** $\frac{2}{3}$ **49.** $-\frac{1}{3}$ **50.** $\frac{2}{y}$ **51.** $\frac{4x}{5}$ **52.** $\frac{4y-3}{21}$ **53.** $\frac{1}{3}$ **54.** $-\frac{1}{15}$ **55.** $3x$ **56.** 24 **57.** yes **58.** yes **59.** $\frac{1}{2}$ hr **60.** $\frac{3}{8}$ gal **61.** $\frac{3}{4}$ of his homework

62. $\frac{3}{2}$ mi **63.** $\frac{11}{18}$ **64.** $\frac{7}{26}$ **65.** $-\frac{1}{12}$ **66.** $-\frac{5}{12}$ **67.** $\frac{25x + 2}{55}$ **68.** $\frac{4 + 3b}{15}$ **69.** $\frac{7y}{36}$ **70.** $\frac{11x}{18}$ **71.** $\frac{4y + 45}{9y}$ **72.** $-\frac{15}{14}$ **73.** $\frac{91}{150}$ **74.** $\frac{5}{18}$ **75.** $\frac{5}{6}$ **76.** $-\frac{9}{10}$ **77.** $\frac{19}{9}$ m **78.** $\frac{3}{2}$ ft **79.** no **80.** yes **81.** $\frac{21}{50}$ of the donors **82.** $\frac{1}{4}$ yd **83.** $\frac{4x}{7}$ **84.** $\frac{3y}{11}$ **85.** $\frac{22}{7}$ **86.** $\frac{17}{16}$ **87.** -2 **88.** $-7y$ **89.** $\frac{1}{3}$ **90.** $\frac{15}{4}$ **91.** $-\frac{5}{24}$ **92.** $\frac{19}{30}$ **93.** $\frac{4}{9}$ **94.** $-\frac{3}{10}$ **95.** -10 **96.** -6 **97.** 15 **98.** 1 **99.** 5 **100.** -4 **101.** $3\frac{3}{4}$ **102.** 3 **103.** 1 **104.** $31\frac{1}{4}$ **105.** $\frac{11}{5}$ **106.** $\frac{5}{1}$ **107.** $\frac{35}{9}$ **108.** $\frac{3}{1}$ **109.** $45\frac{16}{21}$ **110.** 60 **111.** $32\frac{13}{22}$ **112.** $3\frac{19}{60}$ **113.** $111\frac{5}{18}$ **114.** $20\frac{7}{24}$ **115.** $1\frac{1}{12}$ **116.** $1\frac{4}{11}$ **117.** 10 **118.** $12\frac{3}{4}$ **119.** $5\frac{1}{4}$ **120.** $2\frac{29}{46}$ **121.** $2\frac{1}{3}$ **122.** $6\frac{2}{5}$ **123.** $6\frac{7}{20}$ lb **124.** $44\frac{1}{2}$ yd **125.** $7\frac{4}{5}$ in. **126.** $7\frac{1}{2}$ sq ft **127.** 5 ft **128.** $\frac{119}{80}$ or $1\frac{39}{80}$ sq in. **129.** $5\frac{1}{12}$ m **130.** 203 calories **131.** $13\frac{1}{3}$ g **132.** $\frac{21}{20}$ or $1\frac{1}{20}$ mi **133.** $-27\frac{5}{14}$ **134.** $1\frac{5}{27}$ **135.** $-3\frac{15}{16}$ **136.** $-\frac{33}{40}$

Chapter 4 Test

1. $\frac{23}{3}$ **2.** $\frac{39}{11}$ **3.** $4\frac{3}{5}$ **4.** $18\frac{3}{4}$ **5.** $\frac{9}{35}$ **6.** $-\frac{3}{5}$ **7.** $\frac{4}{3}$ **8.** $-\frac{4}{3}$ **9.** $\frac{8x}{9}$ **10.** $\frac{x - 21}{7x}$ **11.** y^2 **12.** $\frac{16}{45}$ **13.** $\frac{9a + 4}{10}$ **14.** $-\frac{2}{3y}$ **15.** $24y^2$ **16.** 9 **17.** $14\frac{1}{40}$ **18.** $\frac{1}{a^2}$ **19.** $\frac{64}{3}$ **20.** $22\frac{1}{2}$ **21.** $3\frac{3}{5}$ **22.** $\frac{1}{3}$ **23.** $\frac{3}{4}$ **24.** $\frac{3}{4x}$ **25.** $\frac{76}{21}$ **26.** -2 **27.** -4 **28.** 1 **29.** $\frac{5}{2}$ **30.** $\frac{4}{31}$ **31.** $\frac{1}{2}$ of the calories **32.** $3\frac{3}{4}$ ft **33.** $\frac{5}{16}$ of the sales **34.** 125,000 backpacks **35.** 8800 sq yd **36.** $\frac{34}{27}$ or $1\frac{7}{27}$ sq mi **37.** 24 mi **38.** $90 per share

Cumulative Review

1. one hundred twenty-six; Sec. 1.2, Ex. 4 **2.** one hundred fifteen **3.** three thousand five; Sec. 1.2, Ex. 5 **4.** six thousand five hundred seventy-three **5.** 159; Sec. 1.3, Ex. 1 **6.** 631 **7.** 14; Sec. 1.4, Ex. 3 **8.** 933 **9.** 278,000; Sec. 1.5, Ex. 2 **10.** 1440 **11.** 63,420 thousand bytes; Sec. 1.6, Ex. 7 **12.** 1305 mi **13.** 7089 R 5; Sec. 1.7, Ex. 7 **14.** 379 R 10 **15.** 4^3; Sec. 1.8, Ex. 1 **16.** 7^2 **17.** $6^3 \cdot 8^5$; Sec. 1.8, Ex. 4 **18.** $9^4 \cdot 5^2$ **19.** 8; Sec. 1.9, Ex. 2 **20.** 52 **21.** -150; Sec. 2.1, Ex. 1 **22.** -21 **23.** -4; Sec. 2.2, Ex. 3 **24.** 5 **25.** 3; Sec. 2.3, Ex. 9 **26.** 10 **27.** 25; Sec. 2.4, Ex. 8 **28.** -16 **29.** -2; Sec. 2.5, Ex. 5 **30.** 25 **31.** $15y$; Sec. 3.1, Ex. 6 **32.** $36c$ **33.** $-8x$; Sec. 3.1, Ex. 7 **34.** $-98a$ **35.** -9; Sec. 3.2, Ex. 4 **36.** 8 **37.** 5; Sec. 3.3, Ex. 7 **38.** -1 **39. a.** $7 \cdot 6 = 42$ **b.** $2(3 + 5) = 16$ **c.** $\frac{-45}{5} = -9$; Sec. 3.4, Ex. 6 **40. a.** $4 + 3 = 7$ **b.** $4(5 - 2) = 12$ **c.** $(-4)(-6) = 24$ **41.** -9; Sec. 3.5, Ex. 2 **42.** -5 **43.** $\frac{36x}{44}$; Sec. 4.1, Ex. 19 **44.** $\frac{10a}{15}$ **45.** $3 \cdot 3 \cdot 5$ or $3^2 \cdot 5$; Sec. 4.2, Ex. 1 **46.** $2 \cdot 2 \cdot 23$ or $2^2 \cdot 23$ **47.** $\frac{10}{33}$; Sec. 4.3, Ex. 1 **48.** $\frac{2}{35}$ **49.** $\frac{1}{8}$; Sec. 4.3, Ex. 2 **50.** $\frac{3}{25}$

Chapter 5 Decimals

Chapter 5 Pretest

1. $\frac{27}{100}$; 5.1C **2.** $<$; 5.1D **3.** 54.7; 5.1E **4.** 60.042; 5.2A **5.** 7.14; 5.3A **6.** 201.6; 5.3B **7.** 40.6; 5.4A **8.** 0.0891; 5.4B **9.** 11.69; 5.2B **10.** $-17.7 - 4.8x$; 5.2C **11.** 0.126; 5.3C **12.** 12π in. ≈ 37.68 in.; 5.3D **13.** $28.80; 5.4D **14.** 0.778; 5.5B **15.** 0.375; 5.6A **16.** $<$; 5.6B **17.** -1.07; 5.7A **18.** $\frac{6}{7}$; 5.8A **19.** 6.78; 5.8B **20.** 15 in.; 5.8C

Mental Math

1. tens **3.** tenths

Exercise Set 5.1

1. six and fifty-two hundredths **3.** sixteen and twenty-three hundredths **5.** negative three and two hundred five thousandths

7. one hundred sixty-seven and nine thousandths

9.

Preprinted Name
Preprinted Address
Current date
DATE

PAY TO THE ORDER OF R.W. Financial $ 321.42

Three hundred twenty-one and 42/100 DOLLARS

FOR ______ Signature

11.

Preprinted Name
Preprinted Address
Current date
DATE

PAY TO THE ORDER OF Bell South $ 59.68

Fifty-nine and 68/100 DOLLARS

FOR ______ Signature

13. 6.5 **15.** 9.08 **17.** −5.625 **19.** 0.0064 **21.** 64.16 **23.** 5.4 **25.** $\frac{3}{10}$ **27.** $\frac{27}{100}$ **29.** $-5\frac{47}{100}$ **31.** $\frac{6}{125}$ **33.** $7\frac{7}{100}$ **35.** $15\frac{401}{500}$ **37.** $\frac{601}{2000}$ **39.** $487\frac{8}{25}$ **41.** < **43.** < **45.** < **47.** = **49.** < **51.** > **53.** > **55.** < **57.** > **59.** 0.6 **61.** 0.23 **63.** 0.594 **65.** 98,210 **67.** 13.0 **69.** −17.67 **71.** −0.5 **73.** \$0.07 **75.** \$27 **77.** \$0.20 **79.** 0.26499; 0.25786 **81.** \$16 **83.** 2.35 hr **85.** 24.623 hr **87.** 136 mph **89.** 31 points **91.** 0.10299, 0.1037, 0.1038, 0.9 **93.** 5766 **95.** 71 **97.** 243 **99.** 228.040; Parker Bohn **101.** 228.040; 226.130; 225.490; 225.370; 224.940; 222.008; 221.546; 220.930 **103.** Answers may vary. **105.** two hundred three ten-millionths

Calculator Explorations

1. 328.742 **3.** 5.2414 **5.** 865.392

Mental Math

1. 0.5 **3.** 1.26 **5.** 8.9 **7.** 0.6

Exercise Set 5.2

1. 3.5 **3.** 6.83 **5.** 27.0578 **7.** 56.432 **9.** −8.57 **11.** 11.16 **13.** 6.5 **15.** 15.3 **17.** 598.23 **19.** 16.3 **21.** −6.32 **23.** −6.4 **25.** 3.1 **27.** 3.1 **29.** −9.9 **31.** 25.67 **33.** −5.62 **35.** 776.89 **37.** 11,983.32 **39.** 465.56 **41.** −549.8 **43.** 861.6 **45.** 115.123 **47.** 876.6 **49.** 0.088 **51.** 465.902 **53.** −180.44 **55.** −1.1 **57.** 3.81 **59.** 3.39 **61.** 1.61 **63.** no **65.** yes **67.** no **69.** $6.9x + 6.9$ **71.** $-10.97 + 3.47y$ **73.** \$454.71 **75.** \$0.06 **77.** \$7.52 **79.** 13.3 lb **81.** 285.8 mph **83.** 763.035 mph **85.** \$334.4 million **87.** 240.8 in. **89.** 67.44 ft **91.** 12.409 mph **93.** 2044.35 min **95.** Switzerland **97.** 8.1 lb **99.**

Country	Pounds of Chocolate Per Person
Switzerland	22.0
Norway	16.0
German	15.8
United Kingdom	14.5
Belgium	13.9

101. $\frac{1}{125}$ **103.** $\frac{5}{12}$ **105.** \$1.02 **107.** 5 nickels; 2 dimes and 1 nickel; 1 dime and 3 nickels; 1 quarter **109.** Answers may vary. **111.** $-109.544x + 15.604$ **113.** 6.08 in.

Exercise Set 5.3

1. 0.12 **3.** 0.6 **5.** −17.595 **7.** 39.273 **9.** 65 **11.** 0.65 **13.** −7093 **15.** 0.0983 **17.** 43.274 **19.** 8.23854 **21.** 14,790 **23.** −9.3762 **25.** 1.12746 **27.** −0.14444 **29.** 5,500,000,000 chocolate bars **31.** 36,400,000 households **33.** 1,600,000 hr **35.** −0.6 **37.** 17.3 **39.** 56.52 **41.** no **43.** no **45.** yes **47.** 8π m ≈ 24.12 m **49.** 10π cm ≈ 31.4 cm **51.** 18.2π yd ≈ 57.148 yd **53.** 250π ft ≈ 785 ft **55.** 135π m ≈ 423.9 m **57. a.** 62.8 m; 125.6 m **b.** yes **59.** 24.8 g **61.** \$2800 **63.** 64.9605 in. **65.** \$555.20 **67.** 79,697.25 yen **69.** 514 Canadian dollars **71.** 26 **73.** 36 **75.** 8 **77.** −9 **79.** 3,831,600 mi **81.** Answers may vary. **83.** 740.82 **85.** Answers may vary.

Exercise Set 5.4

1. 0.094 **3.** −300 **5.** 5.8 **7.** 6.6 **9.** 200 **11.** 23.87 **13.** −110 **15.** 0.54982 **17.** −0.0129 **19.** 8.7 **21.** 0.413 **23.** −7 **25.** 4.8 **27.** 2100 **29.** 30 **31.** 7000 **33.** −0.69 **35.** 0.024 **37.** 65 **39.** 0.0003 **41.** 120,000 **43.** −5.65 **45.** −7.0625 **47.** 1.13 **49.** yes **51.** no **53.** yes **55.** 24 mo **57.** \$2734.51 **59.** 202.1 lb **61.** 5.1 m **63.** 11.4 boxes **65.** 132.5 mph **67.** 24 tsp **69.** 8 days **71.** 345.22 **73.** −1001.0 **75.** 20 **77.** 22 **79.** 85 **81.** 7.2 **83.** 8.6 ft **85.** Answers may vary. **87.** 65.2 – 82.6 knots

Integrated Review

1. 2.57 **2.** 4.05 **3.** 8.9 **4.** 3.5 **5.** 0.16 **6.** 0.24 **7.** 0.27 **8.** 0.52 **9.** −4.8 **10.** 6.09 **11.** 75.56 **12.** 289.12 **13.** −24.974 **14.** −43.875 **15.** −8.6 **16.** 5.4 **17.** −280 **18.** 1600 **19.** 224.938 **20.** 145.079 **21.** 0.56 **22.** −0.63 **23.** 27.6092 **24.** 145.6312 **25.** 5.4 **26.** −17.74 **27.** −414.44 **28.** −1295.03 **29.** −34 **30.** −28 **31.** 116.81 **32.** 18.79 **33.** 156.2 **34.** 1.562 **35.** 25.62 **36.** 5.62 **37.** Yes; answers may vary. **38.** 32.1 points

Exercise Set 5.5

1. 2.8 **3.** 2898.66 **5.** 4.2 **7.** 149.87 **9.** 22.89 **11.** 36 in. **13.** 39 ft **15.** 43.96 m **17.** 52 gal **19.** $24,000 **21.** 53 mi **23.** $2700 million **25.** 135,000 people **27.** c **29.** b **31.** not reasonable **33.** reasonable **35.** 0.06 **37.** −3 **39.** 0.28 **41.** 2.3 **43.** 5.29 **45.** 7.6 **47.** 0.2025 **49.** −1.29 **51.** 5.76 **53.** 5.7 **55.** 3.6 **57.** yes **59.** no **61.** $\frac{5}{16}$ **63.** $-\frac{3}{4}$ **65.** $\frac{1}{12}$ **67.** 43.388569 **69.** Answers may vary. **71.** Square; answers may vary.

Exercise Set 5.6

1. 0.2 **3.** 0.5 **5.** 0.75 **7.** −0.08 **9.** 0.375 **11.** $0.91\overline{6}$ **13.** 0.425 **15.** 0.45 **17.** $0.\overline{3}$ **19.** 0.4375 **21.** $0.\overline{2}$ **23.** $1.\overline{6}$ **25.** 0.33 **27.** 0.44 **29.** 0.2 **31.** 1.7 **33.** 0.67 **35.** 0.44 **37.** 0.62 **39.** $<$ **41.** $>$ **43.** $>$ **45.** $<$ **47.** $<$ **49.** $>$ **51.** $<$ **53.** $<$ **55.** 0.32, 0.34, 0.35 **57.** 0.49, 0.491, 0.498 **59.** 0.73, $\frac{3}{4}$, 0.78 **61.** 0.412, 0.453, $\frac{4}{7}$ **63.** 5.23, $\frac{42}{8}$, 5.34 **65.** $\frac{17}{8}$, 2.37, $\frac{12}{5}$ **67.** 25.65 sq in. **69.** 9.36 sq cm **71.** 0.248 sq yd **73.** 8 **75.** 72 **77.** $\frac{1}{81}$ **79.** $\frac{9}{25}$ **81.** $\frac{5}{2}$ **83.** 0.202 **85.** 5900 stations **87.** 0.625 **89.** Answers may vary. **91.** −47.25 **93.** 3.37 **95.** −0.45

Exercise Set 5.7

1. 5.9 **3.** −0.43 **5.** 10.2 **7.** −4.5 **9.** 4 **11.** 0.45 **13.** 4.2 **15.** −4 **17.** 1.8 **19.** 10 **21.** 7.6 **23.** 60 **25.** −0.07 **27.** 20 **29.** 0.0148 **31.** −8.13 **33.** 1.5 **35.** −1 **37.** −7 **39.** 7 **41.** 53.2 **43.** $6x - 16$ **45.** $3x - 5$ **47.** $-2y + 6.8$ **49.** Answers may vary. **51.** 7.683 **53.** 4.683

Calculator Explorations

1. 32 **3.** 3.873 **5.** 9.849

Exercise Set 5.8

1. 2 **3.** 25 **5.** $\frac{1}{9}$ **7.** $\frac{12}{8} = \frac{3}{2}$ **9.** 16 **11.** $\frac{3}{2}$ **13.** 1.732 **15.** 3.873 **17.** 3.742 **19.** 6.856 **21.** 2.828 **23.** 5.099 **25.** 8.426 **27.** 2.646 **29.** 13 in. **31.** 6.633 cm **33.** 5 **35.** 8 **37.** 17.205 **39.** 16.125 **41.** 12 **43.** 44.822 **45.** 42.426 **47.** 1.732 **49.** 141.42 yd **51.** 25.0 ft **53.** 340 ft **55.** $\frac{5}{6}$ **57.** $\frac{2}{5}$ **59.** $\frac{5}{12}$ **61.** 6, 7 **63.** 10, 11 **65.** Answers may vary.

Chapter 5 Review

1. tenths **2.** hundred-thousandths **3.** negative twenty-three and forty-five hundredths **4.** three hundred forty-five hundred-thousandths **5.** one hundred nine and twenty-three hundredths **6.** two hundred and thirty-two millionths **7.** 2.15 **8.** −503.102 **9.** 16,025.0014 **10.** $\frac{4}{25}$ **11.** $-12\frac{23}{1000}$ **12.** $1\frac{9}{2000}$ **13.** $\frac{231}{100{,}000}$ **14.** $25\frac{1}{4}$ **15.** $>$ **16.** $=$ **17.** $<$ **18.** $>$ **19.** 0.6 **20.** 0.94 **21.** −42.90 **22.** 16.349 **23.** 13,500 people **24.** $10\frac{3}{4}$ tsp **25.** 9.5 **26.** 5.1 **27.** −7.28 **28.** −12.04 **29.** 320.312 **30.** 148.74236 **31.** 1.7 **32.** 2.49 **33.** −1324.5 **34.** −10.136 **35.** 65.02 **36.** 199.99802 **37.** 9605.51 points **38.** −5.7 **39.** $4.2x + 12.8$ **40.** $9.89y - 4.17$ **41.** 72 **42.** 9345 **43.** −78.246 **44.** 73,246.446 **45.** 14π m $\approx$ 43.96 m **46.** 63.8 mi **47.** 70 **48.** −0.21 **49.** −4900 **50.** 23.904 **51.** 8.059 **52.** 158.25 **53.** 7.3 m **54.** 45 mo **55.** 18.2 **56.** 50 **57.** 99.05 **58.** 54.1 **59.** 35.5782 **60.** 0.3526 **61.** 32.7 **62.** 30.4 **63.** 200 quadrillion Btu **64.** 8932 sq ft **65.** Yes, $5 is enough. **66.** −11.94 **67.** 3.89 **68.** 0.1024 **69.** 3.6 **70.** 0.8 **71.** −0.923 **72.** −0.429 **73.** 0.217 **74.** 0.113 **75.** 51.057 **76.** $=$ **77.** $<$ **78.** $<$ **79.** $<$ **80.** 0.832, 0.837, 0.839 **81.** 0.42, $\frac{3}{7}$, 0.43 **82.** $\frac{19}{12}$, 1.63, $\frac{18}{11}$ **83.** $\frac{3}{4}, \frac{6}{7}, \frac{8}{9}$ **84.** 6.9 sq ft **85.** 5.46 sq in. **86.** 0.3 **87.** 47.19 **88.** 8.6 **89.** −80 **90.** −0.48 **91.** 34 **92.** 0.42 **93.** 0.28 **94.** −20 **95.** 1 **96.** 8

97. 12 **98.** 6 **99.** 1 **100.** $\frac{2}{5}$ **101.** $\frac{1}{10}$ **102.** 13 **103.** 29 **104.** 10.7 **105.** 93 **106.** 86.6 **107.** 28.28 cm **108.** 88.2 ft

Chapter 5 Test

1. forty-five and ninety-two thousandths **2.** 3000.059 **3.** 17.595 **4.** −51.20 **5.** −20.42 **6.** 40.902 **7.** 0.037 **8.** 34.9 **9.** 0.862 **10.** < **11.** < **12.** $\frac{69}{200}$ **13.** $-24\frac{73}{100}$ **14.** −0.5 **15.** 0.941 **16.** 1.93 **17.** −6.2 **18.** $0.5x - 13.4$ **19.** 7 **20.** 12.530 **21.** $\frac{8}{10} = \frac{4}{5}$ **22.** −3 **23.** 3.7 **24.** 5.66 cm **25.** 2.31 sq mi **26.** 198.08 oz **27.** 18π mi $\approx$ 56.52 mi **28.** 12 CDs **29.** 54 mi

Cumulative Review

1. 184,046; Sec. 1.3, Ex. 2 **2.** 215,984 **3.** 2300; Sec. 1.5, Ex. 4 **4.** 2900 **5.** 401 R 2; Sec. 1.7, Ex. 8 **6.** 146 R 3 **7.** 2; Sec. 1.9, Ex. 3 **8.** 6 **9. a.** −11 **b.** 2 **c.** 0; Sec. 2.1, Ex. 5 **10. a.** 7 **b.** −4 **c.** 1 **11.** −23; Sec. 2.2, Ex. 4 **12.** −22 **13.** 50; Sec. 2.5, Ex. 3 **14.** 32 **15.** −9; Sec. 2.5, Ex. 4 **16.** −32 **17.** 9; Sec. 2.5, Ex. 1 **18.** −9 **19. a.** $5x$ **b.** $-6y$ **c.** $8x^2 - 2$; Sec. 3.1, Ex. 1 **20. a.** $-3a$ **b.** $-4z^2$ **c.** $11k + 2$ **21.** −4; Sec. 3.2, Ex. 7 **22.** 5 **23.** −3; Sec. 3.3, Ex. 2 **24.** 3 **25.** −4; Sec. 3.4, Ex. 4 **26.** −3 **27.** 47,039 votes; Sec. 3.5, Ex. 3 **28.** 23 **29.** numerator: 3; denominator: 7; Sec. 4.1, Ex. 1 **30.** numerator; 2: denominator; $5a$ **31.** numerator: $13x$; denominator: 5; Sec. 4.1, Ex. 2 **32.** numerator; $7y$: denominator; $3x$ **33.** $\frac{3}{5}$; Sec. 4.2, Ex. 4 **34.** $\frac{4}{7}$ **35.** $-\frac{1}{8}$; Sec. 4.3, Ex. 5 **36.** $\frac{1}{12}$ **37.** $\frac{5}{7}$; Sec. 4.4, Ex. 1 **38.** 1 **39.** 2; Sec. 4.4, Ex. 3 **40.** $1\frac{2}{3}$ **41.** $\frac{2}{3}$; Sec. 4.5, Ex. 1 **42.** $1\frac{1}{6}$ **43.** $\frac{x}{6}$; Sec. 4.6, Ex. 1 **44.** $\frac{3}{2x}$ **45.** 15; Sec. 4.7, Ex. 2 **46.** 18 **47.** 829.6561; Sec. 5.2, Ex. 2 **48.** 230.8628 **49.** 18.408; Sec. 5.3, Ex. 1 **50.** 28.251

Chapter 6 Percent

Chapter 6 Pretest

1. $\frac{51}{79}$; 6.1A **2.** $\frac{3}{2}$; 6.1A **3.** 4; 6.1B **4.** 2; 6.1B **5.** $107\frac{1}{2}$ mi; 6.1C **6.** 12%; 6.2A **7.** 0.57; 6.2B **8.** 275%; 6.2C **9.** $\frac{3}{40}$; 6.2D **10.** 15%; 6.2E **11.** $18\% \cdot 50 = x$; 6.3A **12.** $4\% \cdot x = 89$; 6.3A **13.** $\frac{82}{b} = \frac{90}{100}$; 6.4A **14.** $\frac{48}{112} = \frac{p}{100}$; 6.4A **15.** 20%; 6.3B, 6.4B **16.** 3.3; 6.3B, 6.4B **17.** 200; 6.3B, 6.4B **18.** 3 lightbulbs; 6.5A **19.** decrease of 84 students; current enrollment: 4116 students; 6.5B **20.** 7%; 6.6A **21.** \$22,400; 6.6B **22.** discount: \$78, sales price: \$572; 6.6C **23.** \$144; 6.7A **24.** \$7147.52; 6.7B **25.** \$37.33; 6.7C

Section 6.1

1. $\frac{2}{15}$ **3.** $\frac{5}{6}$ **5.** $\frac{5}{12}$ **7.** $\frac{1}{10}$ **9.** $\frac{7}{20}$ **11.** $\frac{19}{18}$ **13.** $\frac{47}{25}$ **15.** $\frac{17}{40}$ **17.** $\frac{4}{9}$ **19.** $\frac{15}{1}$ **21.** Answers may vary. **23.** 4 **25.** $\frac{50}{9}$ **27.** $\frac{21}{4}$ **29.** 7 **31.** −3 **33.** $\frac{14}{9}$ **35.** 5 **37.** no solution **39.** 123 lb **41.** 165 cal **43.** 6 people **45.** 112 ft; 11-in. difference **47.** $2\frac{2}{3}$ lb **49.** 16 men **51.** 434 emergency room visits **53.** 2.4 cups **55. a.** 0.1 gal **b.** 13 fluid ounces **57. a.** 2062.5 mg **b.** no **59.** $3 \cdot 5$ **61.** $2^2 \cdot 5$ **63.** $2^3 \cdot 5^2$ **65.** 2^5 **67.** 0.8 ml **69.** 1.25 ml

Mental Math

1. 13% **3.** 87% **5.** 1%

EXERCISE SET 6.2

1. 81% **3.** 9% **5.** chocolate chip; 52% **7.** 12% **9.** 0.48 **11.** 0.06 **13.** 1 **15.** 0.613 **17.** 0.028 **19.** 0.6425 **21.** 3 **23.** 0.3258 **25.** 0.737 **27.** 0.25 **29.** 0.111 **31.** 0.462 **33.** 310% **35.** 2900% **37.** 0.3% **39.** 22% **41.** 5.6% **43.** 33.28% **45.** 300% **47.** 70% **49.** 10% **51.** 32.4% **53.** 38% **55.** $\frac{1}{25}$ **57.** $\frac{9}{200}$ **59.** $1\frac{3}{4}$ **61.** $\frac{73}{100}$ **63.** $\frac{1}{8}$ **65.** $\frac{1}{16}$ **67.** $\frac{31}{300}$ **69.** $\frac{179}{800}$ **71.** 75% **73.** 70% **75.** 40% **77.** 59% **79.** 34% **81.** $37\frac{1}{2}\%$ **83.** $31\frac{1}{4}\%$ **85.** $66\frac{2}{3}\%$ **87.** 250% **89.** 190% **91.** 63.64% **93.** 26.67% **95.** 14.29% **97.** 91.67% **99.** 0.35, $\frac{7}{20}$; 20%, 0.2; 50%, $\frac{1}{2}$; 0.7, $\frac{7}{10}$; 37.5%, 0.375 **101.** 0.4, $\frac{2}{5}$; $23\frac{1}{2}\%$, $\frac{47}{200}$; 80%, 0.8; $0.\overline{3}$, $\frac{1}{3}$; 87.5%, 0.875; 0.075, $\frac{3}{40}$ **103.** $\frac{249}{1000}$ **105.** 8.5% **107.** $\frac{207}{1000}$ **109.** 25% **111.** $\frac{1}{250}$ **113.** $\frac{3}{25}$ **115.** $\frac{71}{1000}$ **117.** 15 **119.** −10 **121.** 12 **123.** 75% **125.** 80% **127.** 1.155; 115.5% **129.** greater **131.** Answers may vary. **133.** Computer software engineers, applications **135.** 0.62

MENTAL MATH

1. percent: 42; base: 50; amount: 21 **3.** percent: 125; base: 86; amount: 107.5

EXERCISE SET 6.3

1. $15\% \cdot 72 = x$ **3.** $30\% \cdot x = 80$ **5.** $x \cdot 90 = 20$ **7.** $1.9 = 40\% \cdot x$ **9.** $x = 9\% \cdot 43$ **11.** 3.5 **13.** 7.28 **15.** 600 **17.** 10 **19.** 110% **21.** 32% **23.** 1 **25.** 45 **27.** 500 **29.** 5.16% **31.** 25.2 **33.** 45% **35.** 35 **37.** 30 **39.** $3\frac{7}{11}$ **41.** $\frac{17}{12} = \frac{n}{20}$ **43.** $\frac{8}{9} = \frac{14}{n}$ **45.** 686.625 **47.** 12,285

MENTAL MATH

1. amount: 12.6; base: 42; percent 30 **3.** amount: 102; base: 510; percent: 20

EXERCISE SET 6.4

1. $\frac{a}{65} = \frac{32}{100}$ **3.** $\frac{75}{b} = \frac{40}{100}$ **5.** $\frac{70}{200} = \frac{p}{100}$ **7.** $\frac{2.3}{b} = \frac{58}{100}$ **9.** $\frac{a}{130} = \frac{19}{100}$ **11.** 5.5 **13.** 18.9 **15.** 400 **17.** 10 **19.** 125% **21.** 28% **23.** 29 **25.** 1.92 **27.** 1000 **29.** 210% **31.** 55.18 **33.** 45% **35.** 85 **37.** $\frac{7}{8}$ **39.** $3\frac{2}{15}$ **41.** 0.7 **43.** 2.19 **45.** 12,011.2 **47.** 7270.6

INTEGRATED REVIEW

1. $\frac{9}{10}$ **2.** $\frac{9}{25}$ **3.** $\frac{43}{50}$ **4.** $\frac{8}{23}$ **5.** $\frac{2}{3}$ **6.** $\frac{13}{4}$ **7.** 25 **8.** 35 **9.** 86 weeks **10.** 7 boxes **11.** 12% **12.** 68% **13.** 25% **14.** 50% **15.** 520% **16.** 780% **17.** 6% **18.** 44% **19.** 250% **20.** 325% **21.** 3% **22.** 5% **23.** 0.65 **24.** 0.31 **25.** 0.08 **26.** 0.07 **27.** 1.42 **28.** 5.38 **29.** 0.029 **30.** 0.066 **31.** $\frac{3}{100}$ **32.** $\frac{2}{25}$ **33.** $\frac{21}{400}$ **34.** $\frac{51}{400}$ **35.** $\frac{19}{50}$ **36.** $\frac{9}{20}$ **37.** $\frac{37}{300}$ **38.** $\frac{1}{6}$ **39.** 8.4 **40.** 100 **41.** 250 **42.** 120% **43.** 28% **44.** 76 **45.** 11 **46.** 130% **47.** 86% **48.** 37.8 **49.** 150 **50.** 62

EXERCISE SET 6.5

1. 1600 bolts **3.** 30 hr **5.** 15% **7.** 295 components **9.** 13.6% **11.** 100,102 dental hygienists **13.** 29.2% **15.** 496 chairs; 6696 chairs **17.** $136 **19.** $867.87; $20,153.87 **21.** 28 million; 63 million **23.** 10; 25% **25.** 102; 120% **27.** 2; 25% **29.** 120; 75% **31.** 44% **33.** 21.5% **35.** 1.3% **37.** 87.1% **39.** 300% **41.** 81.3% **43.** 72.1% **45.** 587.5% **47.** 8.2% **49.** 4.56 **51.** 11.18 **53.** 58.54 **55.** The increased number is double the original number. **57.** Answers may vary.

Exercise Set 6.6

1. \$7.50 **3.** \$858.93 **5.** 9% **7.** \$238.70 **9.** \$1917 **11.** \$11,500 **13.** \$112.35 **15.** 6% **17.** \$49,474.24 **19.** 14% **21.** \$1888.50 **23.** \$85,500 **25.** \$6.80; \$61.20 **27.** \$48.25; \$48.25 **29.** \$75.25; \$139.75 **31.** \$3255.00; \$18,445.00 **33.** \$45; \$255 **35.** \$15; \$60 **37.** 33%; \$80.40 **39.** \$925; \$555 **41.** 1200 **43.** 132 **45.** 16 **47.** \$22,812.68 **49.** A discount of 60% is better. **51.** \$1.60; \$2.40; \$3.20 **53.** \$0.90; \$1.35; \$1.80

Calculator Explorations

1. 1.56051 **3.** 8.06231 **5.** \$634.50

Exercise Set 6.7

1. \$32 **3.** \$73.60 **5.** \$750 **7.** \$33.75 **9.** \$700 **11.** \$78,125 **13.** \$5562.50 **15.** \$12,580 **17.** \$46,815.40 **19.** \$2327.15 **21.** \$58,163.60 **23.** \$240.75 **25.** \$938.66 **27.** \$971.90 **29.** \$260.31 **31.** \$637.26 **33.** -29 **35.** 150 **37.** -1 **39.** Answers may vary. **41.** Answers may vary.

Chapter 6 Review

1. $\frac{1}{5}$ **2.** $\frac{2}{3}$ **3.** 6 **4.** 500 **5.** 312.5 **6.** 50 **7.** 9 **8.** no solution **9.** 3 **10.** no solution **11.** 675 parts **12.** \$33.75 **13.** 37% **14.** 77% **15.** 0.83 **16.** 0.75 **17.** 0.735 **18.** 0.015 **19.** 1.25 **20.** 1.45 **21.** 0.005 **22.** 0.007 **23.** 2.00 or 2 **24.** 4.00 or 4 **25.** 0.2625 **26.** 0.8534 **27.** 260% **28.** 5.5% **29.** 35% **30.** 102% **31.** 72.5% **32.** 25% **33.** 7.6% **34.** 8.5% **35.** 75% **36.** 65% **37.** 400% **38.** 900% **39.** $\frac{1}{100}$ **40.** $\frac{1}{10}$ **41.** $\frac{1}{4}$ **42.** $\frac{17}{200}$ **43.** $\frac{51}{500}$ **44.** $\frac{1}{6}$ **45.** $\frac{1}{3}$ **46.** $1\frac{1}{10}$ **47.** 20% **48.** 70% **49.** $83\frac{1}{3}\%$ **50.** 62.5% **51.** $166\frac{2}{3}\%$ **52.** 125% **53.** 60% **54.** 6.25% **55.** $\frac{9}{10}, \frac{1}{10}$ **56.** $\frac{24}{25}$ **57.** 70% **58.** 1.5 **59.** 100,000 **60.** 8000 **61.** 23% **62.** 114.5 **63.** 3000 **64.** 150% **65.** 418 **66.** 300 **67.** 64.8 **68.** 180% **69.** 110% **70.** 165 **71.** 66% **72.** 16% **73.** 106.25% **74.** 20.9% **75.** \$206,400 **76.** \$13.23 **77.** \$263.75 **78.** \$1.15 **79.** \$5000 **80.** \$300.38 **81.** discount: \$900; sale price: \$2100 **82.** discount: \$9; sale price: \$81 **83.** \$120 **84.** \$1320 **85.** \$30,104.64 **86.** \$17,506.56 **87.** \$80.61 **88.** \$32,830.10

Chapter 6 Test

1. 0.85 **2.** 5 **3.** 0.006 **4.** 5.6% **5.** 610% **6.** 35% **7.** $1\frac{1}{5}$ **8.** $\frac{77}{200}$ **9.** $\frac{1}{500}$ **10.** 55% **11.** 37.5% **12.** 175% **13.** 20% **14.** $\frac{16}{25}$ **15.** 33.6 **16.** 1250 **17.** 75% **18.** 38.4 lb **19.** \$56,750 **20.** \$358.43 **21.** 5% **22.** discount: \$18; sale price: \$102 **23.** \$395 **24.** 1% **25.** \$647.50 **26.** \$2005.64 **27.** \$427 **28.** 5.8% **29.** $\frac{15}{2}$ **30.** -6 **31.** 18 bulbs

Cumulative Review

1. 20,296; Sec. 1.6, Ex. 4 **2.** 31,084 **3.** -10; Sec. 2.3, Ex. 8 **4.** 10 **5.** 3; Sec. 3.2, Ex. 3 **6.** -1 **7.** 2; Sec. 3.4, Ex. 5 **8.** 5 **9.** $\frac{21}{7}$; Sec. 4.1, Ex. 20 **10.** $\frac{40}{5}$ **11.** $\frac{14x}{15}$; Sec. 4.2, Ex. 6 **12.** $\frac{5y}{16}$ **13.** $\frac{5}{12}$; Sec. 4.3, Ex. 12 **14.** $-\frac{4}{7}$ **15.** $-\frac{1}{2}$; Sec. 4.4, Ex. 9 **16.** $\frac{1}{5}$ **17.** $\frac{1}{28}$; Sec. 4.5, Ex. 5 **18.** $\frac{16}{45}$ **19.** $\frac{3}{2}$; Sec. 4.6, Ex. 2 **20.** $\frac{50}{9}$ **21.** 3; Sec. 4.7, Ex. 7 **22.** 4 **23. a.** $\frac{38}{9}$ **b.** $\frac{19}{11}$; Sec. 4.8, Ex. 3 **24. a.** $\frac{17}{5}$ **b.** $\frac{44}{7}$ **25.** $\frac{1}{8}$; Sec. 5.1, Ex. 9 **26.** $\frac{17}{20}$ **27.** $-105\frac{83}{1000}$; Sec. 5.1, Ex. 11 **28.** $17\frac{3}{200}$ **29.** 67.69; Sec. 5.2, Ex. 6 **30.** 27.94 **31.** 76.8; Sec. 5.3, Ex. 5 **32.** 1248.3 **33.** $-76{,}300$; Sec. 5.3, Ex. 7 **34.** -8537.5 **35.** yes; Sec. 5.4, Ex. 8 **36.** no **37.** $\frac{2}{5}$; Sec. 5.8, Ex. 6 **38.** $\frac{3}{4}$ **39.** $\frac{50}{63}$; Sec. 6.1, Ex. 2 **40.** $\frac{29}{38}$ **41.** 17.5 mi; Sec. 6.1, Ex. 6 **42.** 35 **43.** $\frac{19}{1000}$; Sec. 6.2, Ex. 13 **44.** $\frac{23}{1000}$ **45.** $\frac{1}{3}$; Sec. 6.2, Ex. 15 **46.** $1\frac{2}{25}$

Chapter 7 Graphing and Introduction to Statistics

Chapter 7 Pretest

1. April; 7.1D **2.** July; 7.1D **3.** 700; 7.1D

4. 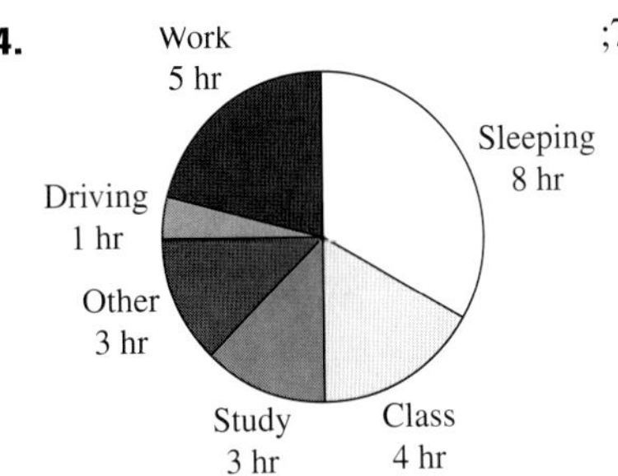

; 7.2B

5. ||; 2; 7.1C **6.** ||||; 4; 7.1C

7. ||||; 4; 7.1C **8.** 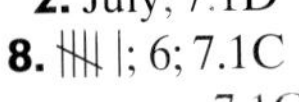 𝍸 |; 6; 7.1C

9. 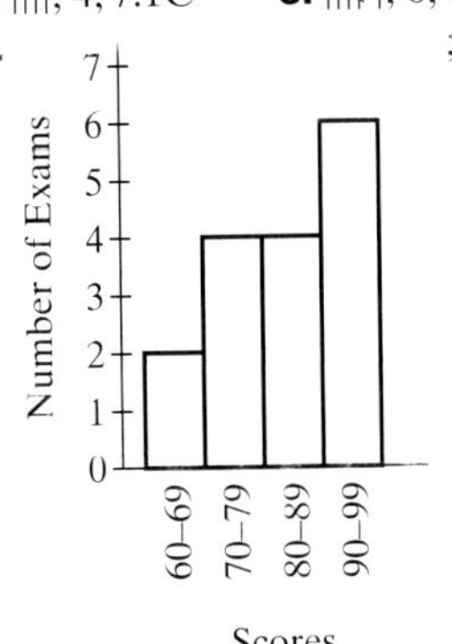

; 7.1C

10. mean: 53.5 median: 64; mode: 74; 7.3 ABC **11.** $\frac{1}{6}$; 7.4B **12.** $\frac{1}{2}$; 7.4B **13.** $\frac{1}{3}$; 7.4B

Exercise Set 7.1

1. 2002 **3.** 4000 cars **5.** 1998, 1999, 2003 **7.** 1997, 2000 **9.** 22.5 oz **11.** 1997, 1999, 2001 **13.** 3 oz/wk **15.** consumption of chicken is increasing **17.** April **19.** 19 deaths **21.** February, March, April, May, June **23.** Tokyo; 34.5 million or 34,500,000 **25.** New York; 21.4 million or 21,400,000 **27.** 16 million or 16,000,000

29. 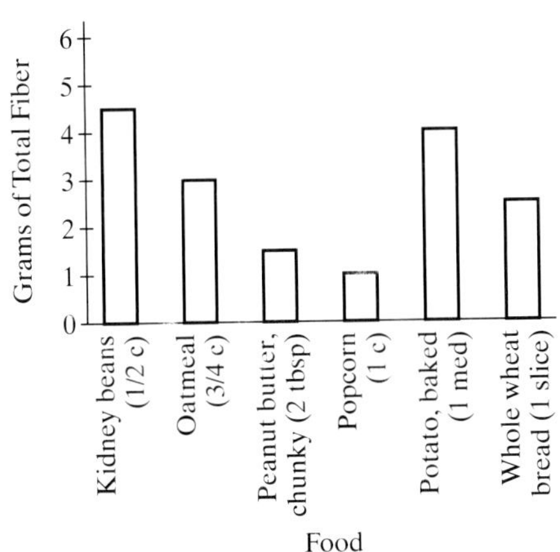

31. 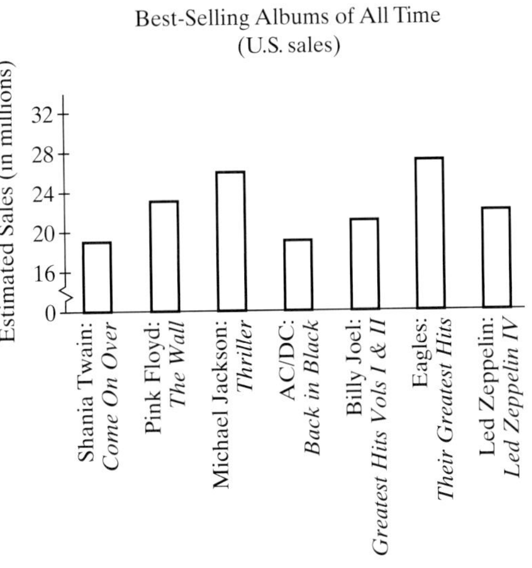

33. 15 adults **35.** 61 adults **37.** 24 adults **39.** 12 adults **41.** $\frac{9}{100}$ **43.** 45–54 **45.** 17 million householders **47.** 45 million householders **49.** Answers may vary.

51. |; 1 **53.** 𝍸 |||; 8 **55.** 𝍸 |; 6 **57.** 𝍸 |; 6 **59.** ||; 2

61. 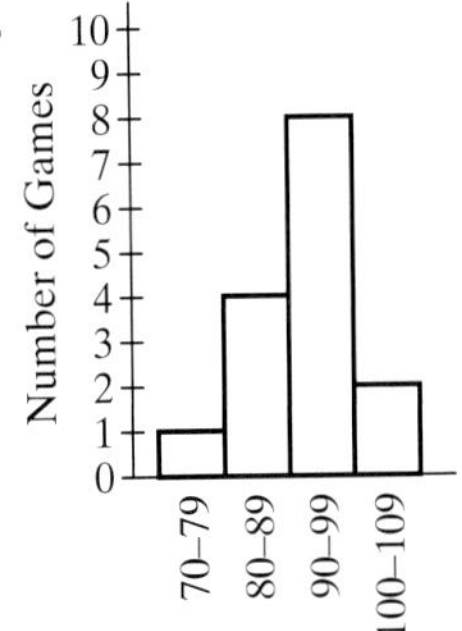

63. 2.7 goals **65.** 1982 **67.** decrease **69.** 3.6 **71.** 6.2 **73.** 25% **75.** 34% **77.** 83°F **79.** Sunday; 68°F **81.** Tuesday; 13°F **83.** Answers may vary. **85.** Answers may vary.

Exercise Set 7.4

1. a: 1, 2, 3; e: 1, 2, 3; i: 1, 2, 3; o: 1, 2, 3; u: 1, 2, 3
15 outcomes

3. Red, Blue, Yellow
3 outcomes

5. 1: 1, 2, 3, 4; 2: 1, 2, 3, 4; 3: 1, 2, 3, 4; 4: 1, 2, 3, 4
16 outcomes

7. Red: 1, 2, 3, 4; Blue: 1, 2, 3, 4; Yellow: 1, 2, 3, 4
12 outcomes

9. H: 1, 2, 3, 4; T: 1, 2, 3, 4
8 outcomes

11. $\frac{1}{6}$ **13.** $\frac{1}{3}$ **15.** $\frac{1}{2}$ **17.** $\frac{1}{3}$ **19.** $\frac{2}{3}$ **21.** $\frac{1}{7}$ **23.** $\frac{2}{7}$ **25.** $\frac{19}{100}$ **27.** $\frac{1}{20}$ **29.** $\frac{5}{6}$ **31.** $\frac{1}{6}$ **33.** $\frac{20}{3}$ or $6\frac{2}{3}$
35. $\frac{1}{52}$ **37.** $\frac{1}{13}$ **39.** $\frac{1}{4}$ **41.** $\frac{1}{12}$ **43.** 0 **45.** Answers may vary.

Chapter 7 Review

1. 4,000,000 **2.** 1,750,000 **3.** South **4.** Northeast **5.** Midwest, South, and West **6.** Northeast **7.** 7.5% **8.** 2000
9. 1980, 1990, 2000 **10.** Answers may vary. **11.** $2,100,000 **12.** $1,200,000 **13.** 1991 and 1992, 2000 and 2001, 2001 and 2002
14. 1999 and 2000 **15.** 1990, 1991, 1992, 1993, 1994 **16.** 4 employees **17.** 1 employee **18.** 9 employees
19. 18 employees **20.** 卌; 5 **21.** |||; 3 **22.** ||||; 4 **23.**

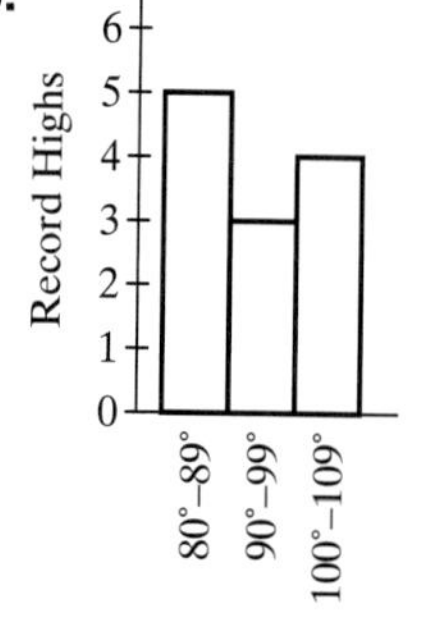

24. mortgage payment **25.** utilities
26. $1225 **27.** $700 **28.** $\frac{39}{160}$ **29.** $\frac{7}{40}$ **30.** $\frac{5}{7}$
31. 20 states **32.** 11 states **33.** 1 state **34.** 29 states
35. mean: 17.8; median: 14; no mode
36. mean: 55.2; median: 60; no mode
37. mean: $24,500; median: $20,000; mode: $20,000
38. mean: 447.3; median: 420; mode: 400
39. 3.25 **40.** 2.57

41.

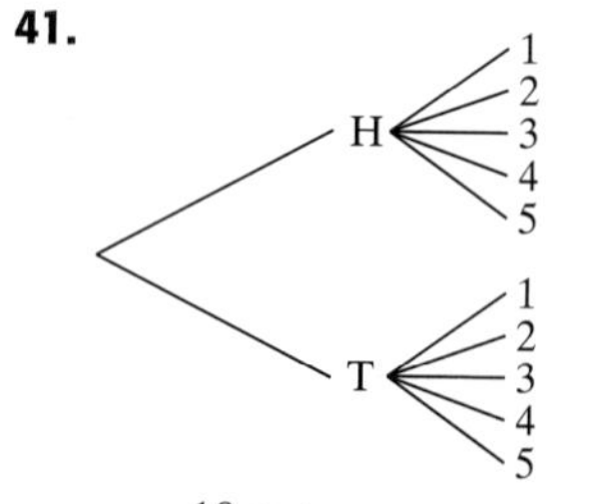

10 outcomes

42.

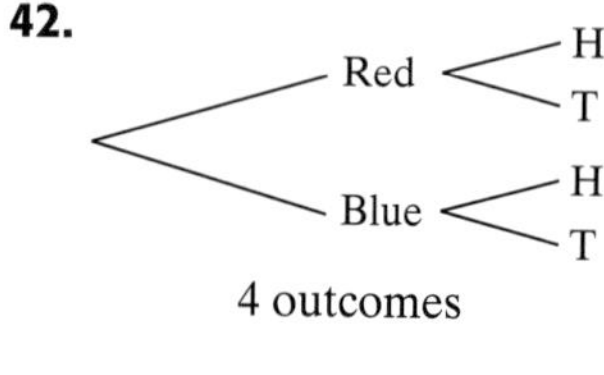

4 outcomes

43.

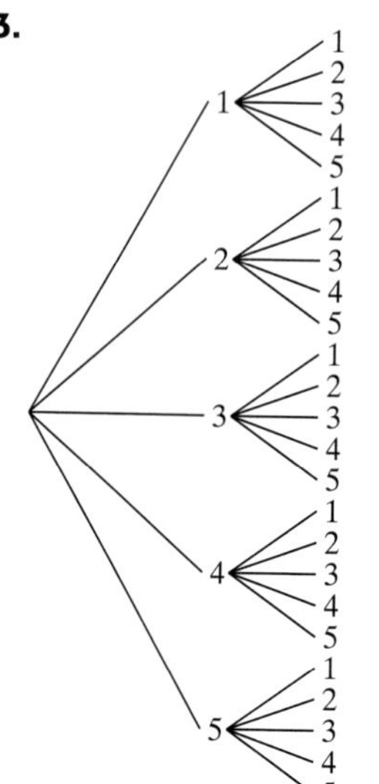

25 outcomes

44.

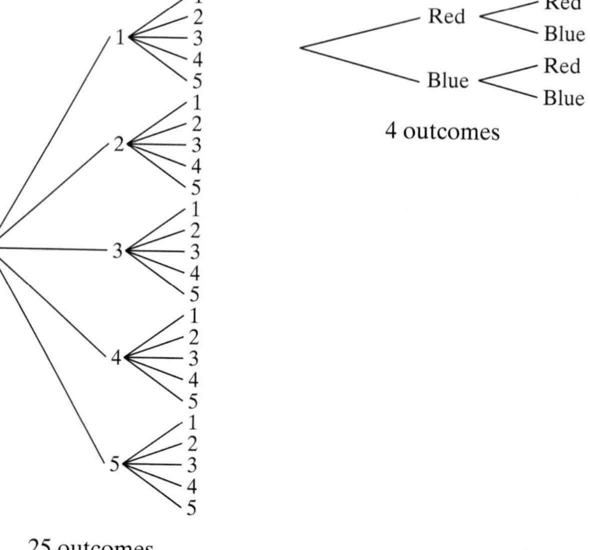

4 outcomes

45.

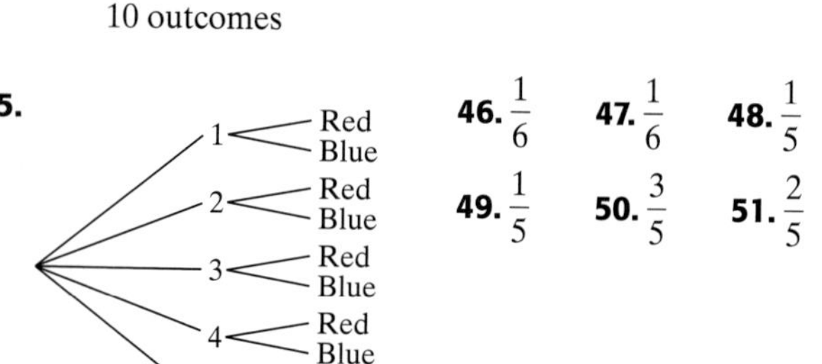

10 outcomes

46. $\frac{1}{6}$ **47.** $\frac{1}{6}$ **48.** $\frac{1}{5}$
49. $\frac{1}{5}$ **50.** $\frac{3}{5}$ **51.** $\frac{2}{5}$

Exercise Set 7.2

1. parent or guardian's home **3.** $\frac{9}{35}$ **5.** $\frac{9}{16}$ **7.** Asia **9.** 37% **11.** 17,100,000 sq mi **13.** 2,850,000 sq mi **15.** 55% **17.** nonfiction **19.** 31,400 books **21.** 27,632 books **23.** 25,120 books

25. 270°; 40°; 22°; 29°

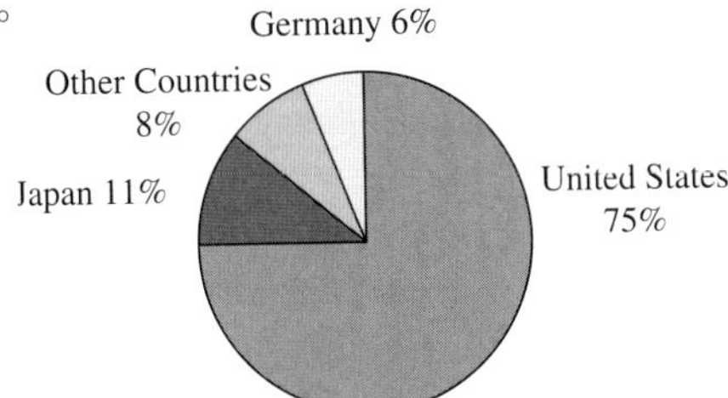

27. 25°; 65°; 50°; 220°

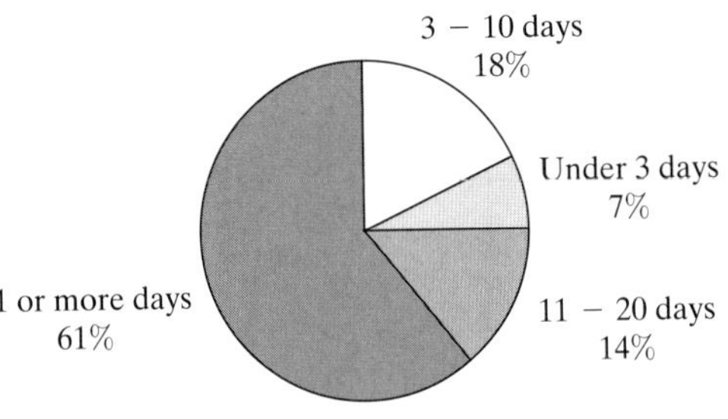

29. $2^2 \cdot 5$ **31.** $2^3 \cdot 5$ **33.** $5 \cdot 17$ **35.** Pacific; answers may vary. **37.** 129,600,002 sq km **39.** 55,542,858 sq km **41.** 602 respondents **43.** 2198 respondents **45.** $\frac{301}{837}$ **47.** No; Answers may vary.

Integrated Review

1. 69 lb **2.** 78 lb **3.** 1985 **4.** 1995 and 2000 **5.** Oroville Dam; 755 ft **6.** New Bullards Bar Dam; 635 ft **7.** 15 ft **8.** 4 dams **9.** Thursday and Saturday; 100° F **10.** Monday; 82° F **11.** Sunday, Monday, and Tuesday **12.** Wednesday, Thursday, Friday, and Saturday **13.** 70 qt containers **14.** 52 qt containers **15.** 2 qt containers **16.** 6 qt containers **17.** ||; 2 **18.** |; 1 **19.** |||; 3 **20.** ~~||||~~ |; 6 **21.** ~~||||~~; 5 **22.**

Number of Quizzes

Quiz Scores: 50–59, 60–69, 70–79, 80–89, 90–99

Mental Math

1. 4 **3.** 3

Exercise Set 7.3

1. mean: 29; median: 28; no mode **3.** mean: 8.1; median: 8.2; mode: 8.2 **5.** mean: 0.6; median: 0.6; mode: 0.2 and 0.6 **7.** mean: 370.9; median: 313.5; no mode **9.** 1416 ft **11.** 1332 ft **13.** Answers may vary. **15.** 2.79 **17.** 3.46 **19.** 6.8 **21.** 6.9 **23.** 85.5 **25.** 73 **27.** 70 and 71 **29.** 9 rates **31.** $\frac{1}{3}$ **33.** $\frac{3}{5}$ **35.** $\frac{11}{15}$ **37.** 35, 35, 37, 43 **39.** Yes; Answers may vary.

Mental Math

1. $\frac{1}{2}$ **3.** $\frac{1}{2}$

CHAPTER 7 TEST

1. \$225 **2.** 3rd week; \$350 **3.** \$1100 **4.** June, August, September **5.** February, 3 cm **6.** March and November

7.

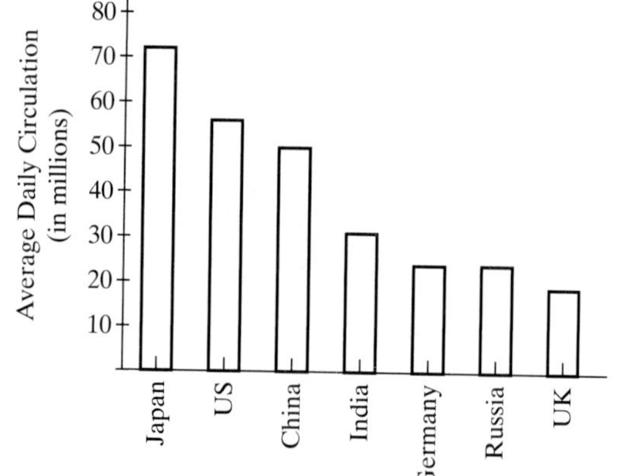

8. $\frac{17}{40}$ **9.** $\frac{31}{22}$ **10.** 40,920,000 people **11.** 21,120,000 people **12.** 9 students **13.** 11 students

14.

Class Intervals (scores)	Tally	Class Frequency (Number of Students)
40–49	|	1
50–59	|||	3
60–69	||||	4
70–79	~~||||~~	5
80–89	~~||||~~ |||	8
90–99	||||	4

15.

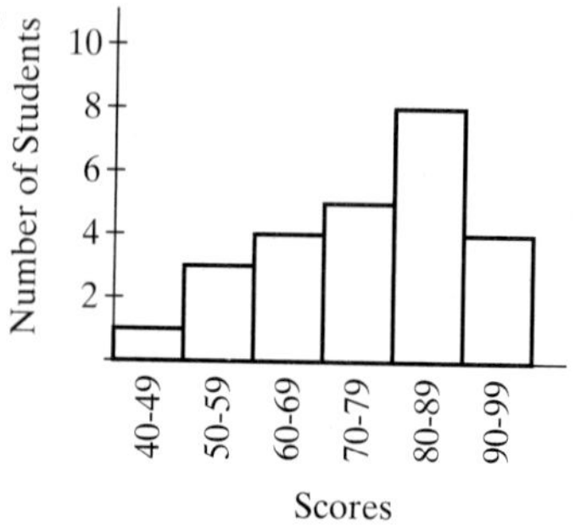

16. mean: 38.4; median: 42; no mode

17. mean: 12.625; median: 12.5; mode: 16 and 12 **18.** 3.07

19.

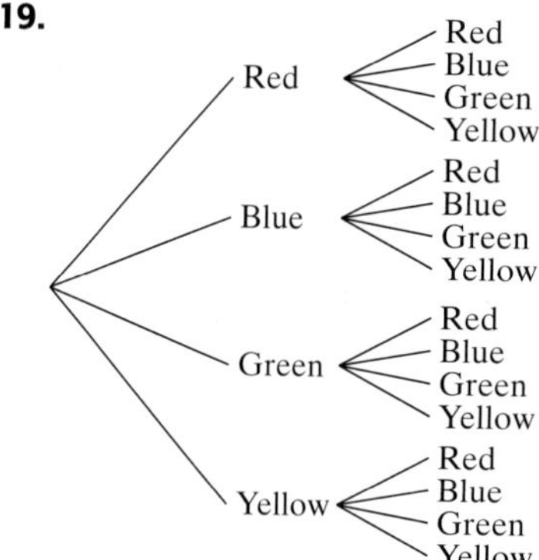

20. H: H, T; T: H, T

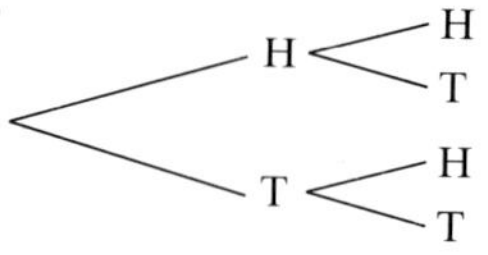

21. $\frac{1}{10}$ **22.** $\frac{1}{5}$

CUMULATIVE REVIEW

1. 47; Sec. 1.8, Ex. 12 **2.** −70 **3.** −12; Sec. 2.3, Ex. 11 **4.** 9 **5.** −3; Sec. 3.3, Ex. 5 **6.** −6 **7.** 2; Sec. 4.7, Ex. 5 **8.** −10 **9.** $7\frac{17}{24}$; Sec. 4.8, Ex. 11 **10.** $8\frac{3}{20}$ **11.** $5\frac{3}{5}$; Sec. 5.1, Ex. 8 **12.** $2\frac{4}{5}$ **13.** 3.432; Sec. 5.2, Ex. 5 **14.** 7.327 **15.** 0.0849; Sec. 5.3, Ex. 2 **16.** 0.0294 **17.** −0.052; Sec. 5.4, Ex. 3 **18.** 0.136 **19.** 4.09; Sec. 5.5, Ex. 7 **20.** 7.29 **21.** 0.25; Sec. 5.6, Ex. 1 **22.** 0.375 **23.** 0.7; Sec. 5.7, Ex. 5 **24.** 1.68 **25.** 5.657; Sec. 5.8, Ex. 8 **26.** 7.746 **27.** 0.046; Sec. 6.2, Ex. 4 **28.** 0.32 **29.** 0.0074; Sec. 6.2, Ex. 6 **30.** 0.027 **31.** 21; Sec. 6.3, Ex. 7 **32.** 14.4 **33.** 52; Sec. 6.4, Ex. 9 **34.** 38 **35.** 7.5%; Sec. 6.6, Ex. 2 **36.** 6.5% **37.** \$101.50; Sec. 6.7, Ex. 6 **38.** \$144.05 **39.** 80.5; Sec. 7.3, Ex. 4 **40.** 48.5 **41.** $\frac{1}{3}$; Sec. 7.4, Ex. 4 **42.** $\frac{1}{2}$

Chapter 8 GEOMETRY AND MEASUREMENT

CHAPTER 8 PRETEST

1. acute; 8.1B **2.** straight; 8.1B **3.** 88°; 8.1C **4.** $x = 55°$, $y = 125°$, $z = 125°$; 8.1D **5.** 110°; 8.1D **6.** $3\frac{2}{3}$ yd; 8.2A **7.** 62.5 km; 8.2D **8.** 60 in.; 8.3A **9.** 56.52 yd; 8.3B **10.** 6 sq cm; 8.4A **11.** 7 lb 12 oz; 8.5B **12.** 8 gal 3 qt; 8.6B **13.** 25,000 ml; 8.6C **14.** 47.8°C; 8.7B **15.** $\frac{1}{2}$; 8.8B **16.** 6; 8.8C **17.** 19.8 ft; 8.8D

EXERCISE SET 8.1

1. line; line yz or $\overleftrightarrow{yz}$, line zy or $\overleftrightarrow{zy}$ **3.** line segment; line segment LM or $\overline{LM}$, line segment ML or $\overline{ML}$ **5.** line segment; line segment PQ or $\overline{PQ}$, line segment QP or $\overline{QP}$ **7.** ray; ray UW or $\overrightarrow{UW}$ **9.** 15° **11.** 50° **13.** 65° **15.** 95° **17.** 90° **19.** 0°; 90° **21.** straight **23.** right **25.** obtuse **27.** right **29.** 73° **31.** 163° **33.** 42° **35.** 55° **37.** $\angle MNP$ and $\angle RNO$; $\angle PNQ$ and $\angle QNR$ **39.** $\angle SPT$ and $\angle TPQ$; $\angle SPR$ and $\angle RPQ$; $\angle SPT$ and $\angle SPR$; $\angle TPQ$ and $\angle QPR$ **41.** 54.8° **43.** 32° **45.** 75° **47.** $\angle x = 35°$; $\angle y = 145°$; $\angle z = 145°$ **49.** $\angle x = 77°$; $\angle y = 103°$; $\angle z = 77°$ **51.** $\angle x = 100°$; $\angle y = 80°$; $\angle z = 100°$ **53.** $\angle x = 134°$; $\angle y = 46°$; $\angle z = 134°$ **55.** $\frac{9}{8}$ or $1\frac{1}{8}$ **57.** $\frac{7}{32}$ **59.** $\frac{5}{6}$ **61.** $\frac{4}{3}$ or $1\frac{1}{3}$ **63.** $\angle a = 60°$; $\angle b = 50°$; $\angle c = 110°$; $\angle d = 70°$; $\angle e = 120°$ **65.** 45° **67.** No; answers may vary.

MENTAL MATH

1. 1 ft **3.** 2 ft **5.** 1 yd **7.** no **9.** yes **11.** no

EXERCISE SET 8.2

1. 5 ft **3.** 36 ft **5.** 8 mi **7.** $8\frac{1}{2}$ ft **9.** $3\frac{1}{3}$ yd **11.** 33,792 ft **13.** 13 yd 1 ft **15.** 3 ft 5 in. **17.** 1 mi 4720 ft **19.** 62 in. **21.** 17 ft **23.** 84 in. **25.** 12 ft 3 in. **27.** 22 yd 1 ft **29.** 8 ft 5 in. **31.** 5 ft 6 in. **33.** 3 ft 4 in. **35.** 50 yd 2 ft **37.** 10 ft 6 in. **39.** 13 ft 11 in. **41.** 15 ft 9 in. **43.** 3 ft 1 in. **45.** 86 ft 6 in. **47.** $105\frac{1}{3}$ yd **49.** 4000 cm **51.** 4 cm **53.** 0.3 km **55.** 1.4 m **57.** 15 m **59.** 83 mm **61.** 0.201 dm **63.** 40 mm **65.** 8.94 m **67.** 2.94 m or 2940 mm **69.** 1.29 cm or 12.9 mm **71.** 12.640 km or 12,640 m **73.** 54.9 m **75.** 1.55 km **77.** 9.12 m **79.** 26.7 mm **81.** 41.25 m or 4125 cm **83.** 3.35 m **85.** 6.009 km or 6009 m **87.** 15 tiles **89.** 21% **91.** 13% **93.** 25% **95.** Answers may vary. **97.** 6.575 m

EXERCISE SET 8.3

1. 64 ft **3.** 36 cm **5.** 21 in. **7.** 120 cm **9.** 48 ft **11.** 66 in. **13.** 21 ft **15.** 60 ft **17.** 346 yd **19.** 22 ft **21.** \$66 **23.** 36 in. **25.** 28 in. **27.** \$24.08 **29.** 96 m **31.** 66 ft **33.** 128 mi **35.** 17π cm; 53.38 cm **37.** 16π mi; 50.24 mi **39.** 26π m; 81.64 m **41.** $31\frac{3}{7}$ ft **43.** 12,560 ft **45.** 23 **47.** 1 **49.** 10 **51.** 216 **53. a.** 62.8 m; 125.6 m **b.** yes **55.** $(44 + 10\pi)$ m $\approx$ 75.4 m **57.** 6 ft **59.** 27.4 m

EXERCISE SET 8.4

1. 7 sq m **3.** $9\frac{3}{4}$ sq yd **5.** 15 sq yd **7.** 2.25π sq in. $\approx$ 7.065 sq in. **9.** 36.75 sq ft **11.** 28 sq m **13.** 22 sq yd **15.** $36\frac{3}{4}$ sq ft **17.** $22\frac{1}{2}$ sq in. **19.** 25 sq cm **21.** 86 sq mi **23.** 24 sq cm **25.** 36π sq in. $\approx 113\frac{1}{7}$ sq in. **27.** 72 cu in. **29.** 512 cu cm **31.** $12\frac{4}{7}$ cu yd **33.** $523\frac{17}{21}$ cu in. **35.** 75 cu cm **37.** $2\frac{10}{27}$ cu in. **39.** 8.4 cu ft **41.** 113,625 sq ft **43.** 168 sq ft **45.** 960 cu cm **47.** 9200 sq ft **49.** 381 sq ft **51.** $\frac{1372}{3}\pi$ cu in. or $\left(457\frac{1}{3}\right)\pi$ cu in. **53.** $7\frac{1}{2}$ cu ft **55.** $12\frac{4}{7}$ cu cm **57.** 36π cu in. $\approx$ 113.04 cu in. **59.** $10\frac{5}{6}$ cu in. **61.** 25 **63.** 9 **65.** 5 **67.** 20 **69.** perimeter **71.** area **73.** area **75.** perimeter **77.** 12-in. pizza **79.** $1\frac{1}{3}$ sq ft; 192 sq in. **81.** 298.5 sq m **83.** 2,583,283 cu m **85.** 2,583,669 cu m **87.** 26,696.5 cu ft **89.** No; answers may vary. **91.** 4605 ft

INTEGRATED REVIEW

1. 153°; 63° **2.** $\angle x = 75°$; $\angle y = 105°$; $\angle z = 75°$ **3.** $\angle x = 128°$; $\angle y = 52°$; $\angle z = 128°$ **4.** $\angle x = 52°$ **5.** 3 ft **6.** 2 mi **7.** $6\frac{2}{3}$ yd **8.** 18 ft **9.** 11,088 ft **10.** 38.4 in. **11.** 3000 cm **12.** 2.4 cm **13.** 2 m **14.** 1800 cm **15.** 72 mm **16.** 0.6 km **17.** perimeter = 20 m; area = 25 sq m **18.** perimeter = 12 ft; area = 6 sq ft **19.** circumference = 6π cm $\approx$ 18.84 cm; area = 9π sq cm $\approx$ 28.26 sq cm **20.** perimeter = 32 mi; area = 44 sq mi **21.** perimeter = 62 ft; area = 238 sq ft **22.** 64 cu in. **23.** 30.6 cu ft **24.** 400 cu cm **25.** $\frac{2048}{3}\pi$ cu mi $\approx 2145\frac{11}{21}$ cu mi

Mental Math

1. 1 lb **3.** 2000 lb **5.** 16 oz **7.** 1 ton **9.** no **11.** yes **13.** no

Exercise Set 8.5

1. 32 oz **3.** 10,000 lb **5.** 6 tons **7.** $3\frac{3}{4}$ lb **9.** $1\frac{3}{4}$ tons **11.** 260 oz **13.** 9800 lb **15.** 76 oz **17.** 1.5 tons **19.** 53 lb 10 oz **21.** 9 tons 390 lb **23.** 3 tons 175 lb **25.** 8 lb 11 oz **27.** 31 lb 2 oz **29.** 1 ton 700 lb **31.** 5 lb 8 oz **33.** 35 lb 14 oz **35.** 130 lb **37.** 211 lb **39.** 5 lb 1 oz **41.** 0.5 kg **43.** 4000 mg **45.** 25,000 g **47.** 0.048 g **49.** 0.0063 kg **51.** 15,140 mg **53.** 4010 g **55.** 13.5 mg **57.** 5.815 g or 5815 mg **59.** 1850 mg or 1.850 g **61.** 1360 g or 1.360 kg **63.** 13.52 kg **65.** 2.125 kg **67.** 8.064 kg **69.** 30 mg **71.** 250 mg **73.** 144 mg **75.** 6.12 kg **77.** 850 g or 0.85 kg **79.** 2.38 kg **81.** 0.25 **83.** 0.16 **85.** 0.875 **87.** Answers may vary.

Mental Math

1. 1 pt **3.** 1 gal **5.** 1 qt **7.** 1 c **9.** 2 c **11.** 4 qt **13.** no **15.** no

Exercise Set 8.6

1. 4 c **3.** 16 pt **5.** $2\frac{1}{2}$ gal **7.** 5 pt **9.** 8 c **11.** $3\frac{3}{4}$ qt **13.** 768 fl oz **15.** 9 c **17.** 22 pt **19.** 10 gal 1 qt **21.** 4 c 4 fl oz **23.** 1 gal 1 qt **25.** 2 gal 3 qt 1 pt **27.** 2 qt 1 c **29.** 17 gal **31.** 4 gal 3 qt **33.** 48 fl oz **35.** 2 qt **37.** yes **39.** 4.3 fl oz **41.** 9 qt **43.** 5000 ml **45.** 4.5 L **47.** 0.41 kl **49.** 0.064 L **51.** 160 L **53.** 3600 ml **55.** 0.00016 kl **57.** 22.5 L **59.** 4.5 L or 4500 ml **61.** 8410 ml or 8.410 L **63.** 10,600 ml or 10.6 L **65.** 3840 ml **67.** 162.4 L **69.** 1.59 L **71.** 18.954 L **73.** \$0.316 **75.** 944 ml **77.** $\frac{7}{10}$ **79.** $\frac{3}{100}$ **81.** $\frac{3}{500}$ **83.** Answers may vary. **85.** 1.5 cc **87.** 2.7 cc **89.** 54u or 0.54cc **91.** 86u or 0.86cc **93.** 256 fl drams

Mental Math

1. yes **3.** no **5.** no **7.** yes

Exercise Set 8.7

1. 19.55 fl oz **3.** 218.44 cm **5.** 40 oz **7.** 57.66 mi **9.** 3.77 gal **11.** 13.64 kg **13.** 3.94 in. **15.** 112.7 km per hr **17.** 0.008 oz **19.** yes **21.** 380 ml **23.** 90 mm **25.** 112.5 g **27.** 104 mph **29.** 26.24 ft **31.** 3 mi **33.** 8 oz **35.** A **37.** B **39.** C **41.** D **43.** D **45.** 5°C **47.** 40°C **49.** 140°F **51.** 239°F **53.** 16.7°C **55.** 61.2°C **57.** 197.6°F **59.** 61.3°F **61.** 50°C **63.** 80.6°F **65.** 21.1°C **67.** 37.9°C **69.** 244.4°F **71.** 260°C **73.** 462.2°C **75.** −29 **77.** $\frac{1}{9}$ **79.** −15 **81.** −24 **83.** 2.13 sq m **85.** 1.18 sq m **87.** 1.70 sq m **89.** 21.3 mg − 25.56 mg **91.** 800 sq m or 8606.72 sq ft **93.** 510,000,000°C

Exercise Set 8.8

1. congruent; SSS **3.** congruent; ASA **5.** $\frac{2}{1}$ **7.** $\frac{3}{2}$ **9.** 4.5 **11.** 6 **13.** 5 **15.** 13.5 **17.** 17.5 **19.** 8 **21.** 21.25 **23.** 500 ft **25.** 60 ft **27.** 14.4 ft **29.** 17.5 **31.** 81 **33.** 8.4 **35.** Answers may vary. **37.** 12 ft by 18 ft

Chapter 8 Review

1. right **2.** straight **3.** acute **4.** obtuse **5.** 65° **6.** 75° **7.** 108° **8.** 89° **9.** 58° **10.** 98° **11.** 90° **12.** 25° **13.** 133° and 47° **14.** 43° and 47°; 58° and 32° **15.** $\angle x = 100°$; $\angle y = 80°$, $\angle z = 80°$ **16.** $\angle x = 155°$; $\angle y = 155°$; $\angle z = 25°$ **17.** $\angle x = 53°$; $\angle y = 53°$; $\angle z = 127°$ **18.** $\angle x = 42°$; $\angle y = 42°$; $\angle z = 138°$ **19.** 9 ft **20.** 24 yd **21.** 13,200 ft **22.** 75 in. **23.** 17 yd 1 ft **24.** 3 ft 10 in. **25.** 4200 cm **26.** 820 mm **27.** 0.01218 m **28.** 0.00231 km **29.** 21 yd 1 ft **30.** 7 ft 5 in. **31.** 41 ft 3 in. **32.** 3 ft 8 in. **33.** 9.5 cm or 95 mm **34.** 5.26 m or 526 cm **35.** 9117 m or 9.117 km **36.** 1.1 m or 1100 mm **37.** 169 yd 2 ft **38.** 126 ft 8 in. **39.** 108.5 km **40.** 0.24 sq m **41.** 88 m **42.** 30 cm **43.** 36 m **44.** 90 ft **45.** 32 ft **46.** 440 ft **47.** 5.338 in. **48.** 31.4 yd **49.** 240 sq ft **50.** 140 sq m **51.** 600 sq cm **52.** 189 sq yd **53.** 49π sq ft $\approx$ 153.86 sq ft **54.** 4π sq in. $\approx$ 12.56 sq in. **55.** 119 sq in. **56.** 1248 sq cm **57.** 144 sq m **58.** 432 sq ft **59.** 130 sq ft **60.** $15\frac{5}{8}$ cu in. **61.** 84 cu ft **62.** $62{,}857\frac{1}{7}$ cu cm **63.** $346\frac{1}{2}$ cu in. **64.** $2\frac{2}{3}$ cu ft **65.** 307.72 cu in. **66.** $7\frac{1}{2}$ cu ft **67.** 0.5π cu ft **68.** 1260 cu ft

69. 28.728 cu ft **70.** 4.125 lb **71.** 4600 lb **72.** 3 lb 4 oz **73.** 4 tons 200 lb **74.** 1.4 g **75.** 40,000 g **76.** 21 dag **77.** 0.0003 dg **78.** 3 lb 9 oz **79.** 10 tons 800 lb **80.** 2 tons 750 lb **81.** 33 lb 8 oz **82.** 4.9 g or 4900 mg **83.** 9 kg or 9000 g **84.** 8.1 g or 8100 mg **85.** 50.4 kg **86.** 4 lb 4 oz **87.** 9 tons 1075 lb **88.** 7.85 kg **89.** 1.1625 kg **90.** 8 qt **91.** 5 c **92.** 27 qt **93.** 17 c **94.** 4 qt 1 pt **95.** 3 gal 3 qt **96.** 3800 ml **97.** 0.042 dl **98.** 1.4 kl **99.** 3060 cl **100.** 1 gal 1 qt **101.** 7 gal 1 qt **102.** 736 ml or 0.736 L **103.** 15.5 L or 15,500 ml **104.** 2 gal 3 qt **105.** 6 fl oz **106.** 10.88 L **107.** yes, 9.9 L **108.** 22.96 ft **109.** 10.55 m **110.** 4.62 gal **111.** 8.27 qt **112.** 425.26 g **113.** 10.44 kg **114.** 109 yd **115.** 180.4 lb **116.** 3.18 qt **117.** 2.36 in. **118.** 473° F **119.** 320° F **120.** 107.6° F **121.** 186.8° F **122.** 34° C **123.** 11° C **124.** 5.2° C **125.** 26.7° C **126.** 1.7° C **127.** 329° F **128.** 37.5 **129.** $13\frac{1}{3}$ **130.** 17.4 **131.** $6\frac{1}{2}$ **132.** 33 ft **133.** $x = \frac{5}{6}$ in., $y = 2\frac{1}{6}$ in. **134.** congruent: ASA **135.** not congruent

Chapter 8 Test

1. 12° **2.** 56° **3.** 50° **4.** $\angle x = 118°$; $\angle y = 62°$; $\angle z = 118°$ **5.** $\angle x = 73°$; $\angle y = 73°$; $\angle z = 73°$ **6.** 6.2 m **7.** 10 in. **8.** circumference $= 18\pi \approx 56.52$ in.; area $= 81\pi \approx 254.34$ sq in. **9.** perimeter $= 24.6$ yd; area $= 37.1$ sq yd **10.** perimeter $= 68$ in.; area $= 185$ sq in. **11.** $62\frac{6}{7}$ cu in. **12.** 30 cu ft **13.** 16 in. **14.** 18 cu ft **15.** 198.08 oz **16.** 23 ft 4 in. **17.** 10 qt **18.** 1.875 lb **19.** 5600 lb **20.** 2400 m **21.** 36 mm **22.** 0.43 g **23.** 830 ml **24.** 15.4 lb **25.** 21.59 cm **26.** 28.9° C **27.** 54.7° F **28.** 5.6 m **29.** 4 gal 3 qt **30.** 91.4 m **31.** 16 ft 6 in. or $16\frac{1}{2}$ ft **32.** 37,336 pt **33.** 0.2 oz **34.** 18.95 L **35.** 3.1 mi **36.** 7.5 **37.** 69 ft

Cumulative Review

1. 5; Sec. 3.4, Ex. 3 **2.** 6 **3. a.** $\frac{16}{625}$ **b.** $\frac{1}{16}$; Sec. 4.3, Ex. 9 **4. a.** $-\frac{1}{27}$ **b.** $\frac{9}{49}$ **5.** $8\frac{3}{10}$; Sec. 4.8, Ex. 13 **6.** $9\frac{11}{15}$ **7.** $20x - 10.9$; Sec. 5.2, Ex. 12 **8.** $1.2y + 1.8$ **9.** 6.35; Sec. 5.5, Ex. 8 **10.** 2.16 **11.** =; Sec. 5.6, Ex. 7 **12.** > **13.** −1.3; Sec. 5.7, Ex. 6 **14.** 30 **15.** 424 ft; Sec. 5.8, Ex. 12 **16.** 236 ft **17. a.** $\frac{5}{7}$ **b.** $\frac{7}{24}$; Sec. 6.1, Ex. 3 **18. a.** $\frac{1}{4}$ **b.** $\frac{4}{9}$ **19.** 22.4 cc; Sec. 6.1, Ex. 7 **20.** 7.5 cups **21.** 17%; Sec. 6.2, Ex. 1 **22.** 17% **23.** 200; Sec. 6.3, Ex. 10 **24.** 1200 **25.** 2.7; Sec. 6.4, Ex. 7 **26.** 12.6 **27.** 32%; Sec. 6.5, Ex. 5 **28.** 27% **29.** sales tax: $6.41, total price: $91.91; Sec. 6.6; Ex. 1 **30.** sales tax: $30; total price: $405 **31.** 57; Sec. 7.3, Ex. 3 **32.** 83 **33.** $\frac{1}{4}$; Sec. 7.4, Ex. 5 **34.** $\frac{2}{7}$ **35.** 42°; Sec. 8.1, Ex. 4 **36.** 43° **37.** 96 in.; Sec. 8.2, Ex. 1 **38.** 21 ft **39.** 4 tons 1650 lb; Sec. 8.5, Ex. 3 **40.** 18 lb 2 oz **41.** 45.6°C; Sec. 8.7, Ex. 6 **42.** 30°C

Chapter 9 Equations, Inequalities, and Problem Solving

Chapter 9 Pretest

1. $10x + 42$; 9.2B **2.** $-10y + 16$; 9.2B **3.** $x = 15$; 9.3A **4.** $b = 2$; 9.3A **5.** $m = -12$; 9.3A **6.** $y = -3$; 9.3A **7.** $x = -3$; 9.3A **8.** $x = 22.5$; 9.3B **9.** no solution; 9.3C **10.** 4; 9.4A **11.** 20, 22; 9.4A **12.** $A = 45$; 9.5A **13.** 9 ft; 9.5A **14.** $y = 8 - 2x$; 9.5B **15.** commutative property of multiplication; 9.2A **16.** associative property of addition; 9.2A **17.** identity property of addition; 9.2A **18.** $x \le 6$; 9.6B **19.** $y < -4$; 9.6C **20.** $x \ge 3$; 9.6D

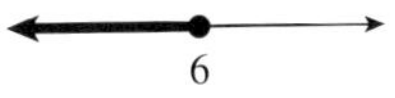

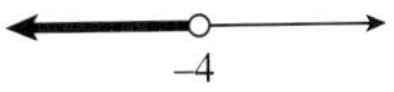

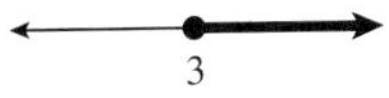

6 −4 3

Section 9.1

1. < **3.** > **5.** = **7.** < **9.** $32 < 212$ **11.** true **13.** false **15.** false **17.** true **19.** $30 \le 45$ **21.** $20 \le 25$ **23.** $6 > 0$ **25.** $-12 < -10$ **27.** $7 < 11$ **29.** $5 \ge 4$ **31.** $15 \ne -2$ **33.** 14,494; −282 **35.** −34,841 **37.** 475; −195 **39.** −4 −3 −2 −1 0 1 2 3 4 5 **41.** $-\frac{1}{4}$ $\frac{1}{2}$ −4 −3 −2 −1 0 1 2 3 4 **43.** −2.5 $-\frac{3}{2}$ $\frac{7}{4}$ 3.25 −4 −3 −2 −1 0 1 2 3 4

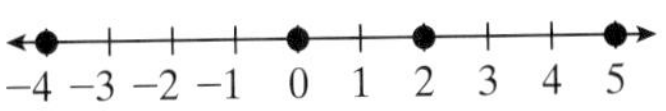

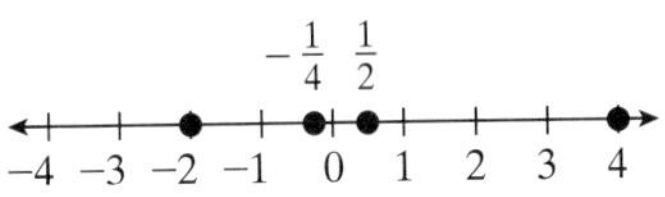

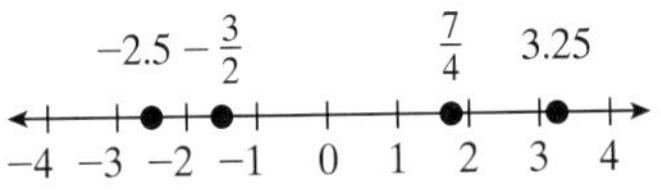

45. whole, integers, rational, real **47.** integers, rational, real **49.** natural, whole, integers, rational, real **51.** rational, real **53.** false **55.** true **57.** false **59.** false **61.** $>$ **63.** $=$ **65.** $<$ **67.** $<$ **69.** false **71.** true **73.** false **75.** true **77.** Blue Ridge Parkway **79.** $19.0 \geq 14.5$ **81.** $-0.04 > -26.7$ **83.** Sun **85.** Sun **87.** Answers may vary.

Section 9.2

1. $16 + x$ **3.** $y \cdot (-4)$ **5.** yx **7.** $13 + 2x$ **9.** $x \cdot (yz)$ **11.** $(2 + a) + b$ **13.** $4a \cdot (b)$ **15.** $a + (b + c)$ **17.** $17 + b$ **19.** 24y **21.** y **23.** $26 + a$ **25.** $-72x$ **27.** s **29.** Answers may vary. **31.** $4x + 4y$ **33.** $9x - 54$ **35.** $6x + 10$ **37.** $28x - 21$ **39.** $18 + 3x$ **41.** $-2y + 2z$ **43.** $-21y - 35$ **45.** $5x + 20m + 10$ **47.** $-4 + 8m - 4n$ **49.** $-5x - 2$ **51.** $-r + 3 + 7p$ **53.** $3x + 4$ **55.** $-x + 3y$ **57.** $6r + 8$ **59.** $-36x - 70$ **61.** $-16x - 25$ **63.** $4(1 + y)$ **65.** $11(x + y)$ **67.** $-1(5 + x)$ **69.** $30(a + b)$ **71.** -16 **73.** 8 **75.** -1.2 **77.** 2 **79.** $\frac{3}{2}$ **81.** $-\frac{6}{5}$ **83.** $\frac{6}{23}$ **85.** $-\frac{1}{2}$ **87.** commutative property of multiplication **89.** associative property of addition **91.** distributive property **93.** associative property of multiplication **95.** identity property of addition **97.** distributive property **99.** commutative and associative properties of multiplication **101.** 2 **103.** -23 **105.** -25 **107.** $-8; \frac{1}{8}$ **109.** $-x; \frac{1}{x}$ **111.** $2x; -2x$ **113. a.** commutative property of addition **b.** commutative property of addition **c.** associative property of addition **115.** Answers may vary. **117.** no **119.** yes

Calculator Explorations

1. solution **3.** not a solution **5.** solution

Section 9.3

1. -6 **3.** -3 **5.** 1 **7.** $\frac{9}{2}$ **9.** $\frac{3}{2}$ **11.** 0 **13.** 1 **15.** 4 **17.** -4 **19.** $\frac{19}{6}$ **21.** $\frac{14}{3}$ **23.** 2 **25.** -5 **27.** 10 **29.** 18 **31.** 3 **33.** 13 **35.** 50 **37.** 0.2 **39.** 1 **41.** $\frac{7}{3}$ **43.** -17 **45.** Answers may vary. **47.** all real numbers **49.** no solution **51.** no solution **53.** no solution **55.** Answers may vary. **57.** $(6x - 8)$ m **59.** $-8 - x$ **61.** $-3 + 2x$ **63.** $9(x + 20)$ **65.** 15.3 **67.** -0.2 **69.** $x = 4$ cm, $2x = 8$ cm

Integrated Review

1. whole, integers, rational, real **2.** natural, whole, integers, rational, real **3.** rational, real **4.** natural, whole, integers, rational, real **5.** integers, rational, real **6.** rational, real **7.** rational, real **8.** irrational, real **9.** $7d - 11$ **10.** $9z + 48$ **11.** 16 **12.** $-12x - 15$ **13.** 5 **14.** -1 **15.** -2 **16.** -2 **17.** $-\frac{5}{6}$ **18.** $\frac{1}{6}$ **19.** 1 **20.** 6 **21.** 4 **22.** 1 **23.** $\frac{9}{5}$ **24.** $-\frac{6}{5}$ **25.** all real numbers **26.** all real numbers **27.** 0 **28.** -1.6 **29.** $\frac{4}{19}$ **30.** $-\frac{5}{19}$ **31.** $\frac{7}{2}$ **32.** $-\frac{1}{4}$

Section 9.4

1. 1 **3.** -25 **5.** $-\frac{3}{4}$ **7.** -16 **9.** governor of Michigan = \$177,000; governor of Oregon = \$93,600 **11.** 1st piece: 5 in.; 2nd piece: 10 in.; 3rd piece: 25 in. **13.** 172 mi **15.** 1st angle: 37.5°; 2nd angle: 37.5°; 3rd angle: 105° **17.** cerulean: 7956 votes; blue: 11,322 votes **19.** smaller angle: 45°; larger angle: 135° **21.** shorter piece: 5 ft; longer piece: 12 ft **23.** diameter: 1 m; height: 13 m **25.** Sahara: 3,500,000 sq mi; Gobi: 500,000 sq mi **27.** Answers may vary. **29.** Texas and Florida **31.** Hawaii: \$37.9 million; Pennsylvania: \$23 million **33.** 2.2 million sq ft **35.** Notre Dame: 68; Purdue: 66 **37.** China: 59 medals; Australia: 58 medals; Germany: 57 medals **39.** 58°, 60°, 62° **41.** $\frac{1}{2}(x - 1) = 37$ **43.** $\frac{3(x + 2)}{5} = 0$ **45.** 34 **47.** 225π **49.** Answers may vary. **51.** 15 ft by 24 ft **53.** Answers may vary.

Section 9.5

1. $h = 3$ **3.** $h = 3$ **5.** $h = 20$ **7.** $c = 12$ **9.** $r = 2.5$ **11.** $T = 3$ **13.** $h = 15$ **15.** 131 ft **17.** 137.5 mi **19.** 50°C **21.** 96 piranhas **23.** 12,090 ft **25.** 6.25 hr **27.** 2 bags **29.** 515,509.5 cu in. **31.** one 16-in. pizza **33.** -109.3°F **35.** 500 sec or $8\frac{1}{3}$ min **37.** 449 cu in. **39.** 332.6°F **41.** 10.7 **43.** 44.3 sec **45.** 4.65 min

47. $h = \frac{f}{5g}$ **49.** $W = \frac{V}{LH}$ **51.** $y = 7 - 3x$ **53.** $R = \frac{A - P}{PT}$ **55.** $A = \frac{3V}{h}$ **57.** $a = P - b - c$

59. $h = \frac{S - 2\pi r^2}{2\pi r}$ **61.** 0.32 **63.** 2.00 or 2 **65.** 17% **67.** 720% **69.** $V = G(N - R)$

71. multiplies the volume by 8 **73.** $-40°$

Mental Math

1. $x > 2$ **3.** $x \geq 8$ **5.** -5 **7.** 4.1

Section 9.6

1. 
3. 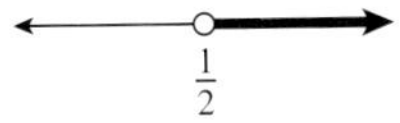
5. 
7. 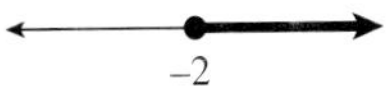

9. $\{x|x \geq -5\}$ 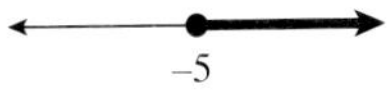
11. $\{y|y < 9\}$ 
13. $\{x|x > -3\}$ 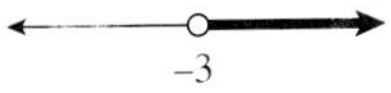
15. $\{x|x \leq 1\}$ 

17. $\{x|x < -3\}$ 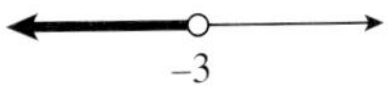
19. $\{x|x \geq -2\}$ 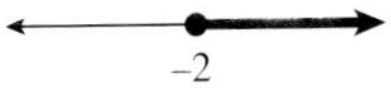
21. $\{x|x < 0\}$ 
23. $\left\{y|y \geq -\frac{8}{3}\right\}$ 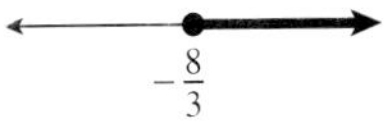

25. $\{y|y > 3\}$ 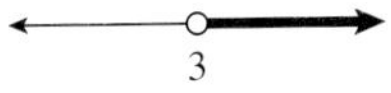
27. when multiplying or dividing by a negative number **29.** $\{x|x > -3\}$ **31.** $\left\{x|x \geq -\frac{2}{3}\right\}$ **33.** $\{x|x \leq -2\}$

35. $\{x|x \leq -8\}$ **37.** $\{x|x > 4\}$ **39.** $\{x|x > 3\}$ **41.** $\{x|x > -13\}$ **43.** $\{x|x \geq 0\}$ **45.** $x > -10$ **47.** 35 cm

49. 193 **51.** 81 **53.** 1 **55.** $\frac{49}{64}$ **57.** approx. 1000 **59.** 2000 and 2001 **61.** 2001 **63.** final exam ≥ 78.5

65. Answers may vary.

Chapter 9 Review

1. < **2.** > **3.** > **4.** > **5.** < **6.** > **7.** = **8.** = **9.** > **10.** < **11.** $4 \geq -3$ **12.** $6 \neq 5$

13. $0.03 < 0.3$ **14.** $400 > 155$ **15. a.** 1, 3 **b.** 0, 1, 3 **c.** $-6, 0, 1, 3$ **d.** $-6, 0, 1, 1\frac{1}{2}, 3, 9.62$ **e.** π

f. all numbers in set **16. a.** 2, 5 **b.** 2, 5 **c.** $-3, 2, 5$ **d.** $-3, -1.6, 2, 5, \frac{11}{2}, 15.1$ **e.** $\sqrt{5}, 2\pi$

f. all numbers in set **17.** Friday **18.** Wednesday **19.** commutative property of addition **20.** multiplicative identity property **21.** distributive property **22.** additive inverse property **23.** associative property of addition **24.** commutative property of multiplication **25.** distributive property **26.** associative property of multiplication **27.** multiplicative inverse property **28.** additive identity property **29.** -1

30. 3 **31.** 0 **32.** $-\frac{1}{5}$ **33.** -4 **34.** 2 **35.** -3 **36.** no solution **37.** no solution **38.** $\frac{3}{4}$ **39.** $-\frac{8}{9}$

40. 20 **41.** $-\frac{6}{23}$ **42.** $\frac{23}{7}$ **43.** $-\frac{2}{5}$ **44.** 102 **45.** 0.25 **46.** $3x + 3$ **47.** 6665.5 in. **48.** short piece: 4 ft; long piece: 8 ft **49.** Kellogg: 35 plants; Keebler: 18 plants **50.** $-39, -38, -37$ **51.** 3 **52.** -4 **53.** $w = 9$

54. $h = 4$ **55.** $m = \frac{y - b}{x}$ **56.** $s = \frac{r + 5}{vt}$ **57.** $x = \frac{2y - 7}{5}$ **58.** $y = \frac{2 + 3x}{6}$ **59.** $\pi = \frac{C}{D}$ **60.** $\pi = \frac{C}{2r}$

61. 15 m **62.** 40°C **63.** 
64. 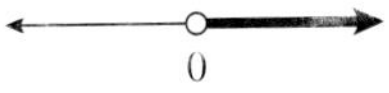
65. $\{x|x \leq 1\}$

66. $\{x|x > -5\}$ **67.** $\{x|x \leq 10\}$ **68.** $\{x|x < -4\}$ **69.** $\{x|x < -4\}$ **70.** $\{x|x \leq 4\}$ **71.** $\{y|y > 9\}$

72. $\{y|y \geq -15\}$ **73.** $\left\{x|x < \frac{7}{4}\right\}$ **74.** $\left\{x|x \leq \frac{19}{3}\right\}$ **75.** at least $2,500 **76.** score must be less than 83

Chapter 9 Test

1. $|-7| > 5$ **2.** $(9 + 5) \geq 4$ **3. a.** 1, 7 **b.** 0, 1, 7 **c.** $-5, -1, 0, 1, 7$ **d.** $-5, -1, \frac{1}{4}, 0, 1, 7, 11.6$ **e.** $\sqrt{7}, 3\pi$ **f.** all numbers in set **4.** associative property of addition **5.** commutative property of multiplication **6.** distributive property **7.** multiplicative inverse property **8.** 9 **9.** -3 **10.** $10y + 1$ **11.** $-2x + 10$ **12.** 8 **13.** no solution **14.** 0 **15.** 27 **16.** 3 **17.** 0.25 **18.** $\frac{25}{7}$ **19.** 21 **20.** 7 gal **21.** $x = 6$ **22.** $h = \frac{V}{\pi r^2}$ **23.** $y = \frac{3x - 10}{4}$ **24.** $\{x|x < -2\}$

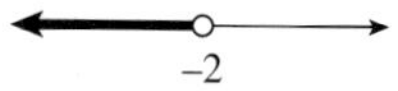

25. $\{x|x \leq -8\}$ **26.** $\{x|x \geq 11\}$ **27.** $\left\{x|x > \frac{2}{5}\right\}$ **28.** New York: 1077; Indiana: 427

Cumulative Review

1. a. 4 **b.** -5 **c.** -6; Sec. 2.1, Ex. 6 **2. a.** $<$ **b.** $=$ **c.** $>$ **d.** $<$ **e.** $>$ **3.** -19; Sec. 2.5, Ex. 10 **4.** $\frac{8}{3}$ **5.** -8; Sec. 2.2, Ex. 9 **6.** -19 **7.** 4; Sec. 2.2, Ex. 10 **8.** 8 **9.** -10; Sec. 2.2, Ex. 14 **10.** -0.3 **11.** $-\frac{8}{33}$; Sec. 4.5, Ex. 2 **12.** $-\frac{23}{40}$ **13.** $\frac{6 - x}{3}$; Sec. 4.5, Ex. 4 **14.** $\frac{10 + x}{2}$ **15.** $-\frac{8}{9}$ **16.** $\frac{11}{12}$ **17.** $-5\frac{3}{5}$; Sec. 4.8, Ex. 23 **18.** $6\frac{1}{4}$ **19.** $\frac{14}{15}$; Sec. 4.8, Ex. 24 **20.** $-1\frac{2}{25}$ **21.** 4.09; Sec. 5.5, Ex. 7 **22.** -2.4 **23.** 7; Sec. 5.8, Ex. 1 **24.** 8 **25.** $\frac{2}{5}$; Sec. 5.8, Ex. 6 **26.** $\frac{3}{10}$ **27.** 2; Sec. 5.7, Ex. 3 **28.** -5 **29.** true; Sec. 9.1, Ex. 3 **30.** true **31.** true; Sec. 9.1, Ex. 4 **32.** true **33.** false; Sec. 9.1, Ex. 5 **34.** true **35.** true; Sec. 9.1, Ex. 6 **36.** false **37.** $15 - 10z$; Sec. 9.2, Ex. 8 **38.** $-8x + 4$ **39.** $12x + 38$; Sec. 9.2, Ex. 12 **40.** $10x + 21$ **41.** 0; Sec. 9.3, Ex. 4 **42.** $\frac{30}{7}$ **43.** Republicans: 220; Democrats: 210; Sec. 9.6, Ex. 3 **44.** 1 **45.** 79.2 yr; Sec. 9.5, Ex. 1 **46.** $w = \frac{V}{lh}$ **47.**

; Sec. 9.6, Ex.2 **48.** $\{x|x > 10\}$; **49.** $\{x|x \geq 1\}$;Sec. 9.6, Ex. 8 **50.** $\{x|x > 4\}$;

Chapter 10 Exponents and Polynomials

Chapter 10 Pretest

1. $\frac{9}{16}$; 10.1A **2.** $8y^{13}$; 10.1B **3.** $\frac{b^{11}}{a^3}$; 10.1E **4.** 3; 10.1E **5.** -216; 10.2A **6.** $\frac{m^{16}}{n^{18}}$; 10.2B **7.** $4x^2 + 10x - 5$; 10.3D **8.** 8.14×10^{-7}; 10.2C **9.** 5; 10.3B **10.** 1; 10.3C **11.** $10x^2 + 1$; 10.4A **12.** $9y^2 - 5y - 3$; 10.4B **13.** $-a^2 - 16b^2$; 10.4D **14.** $-\frac{3}{8}n^9$; 10.5A **15.** $-6t^7 - 8t^5 + 16t^2$; 10.5B **16.** $10y^2 + 7y - 6$; 10.5C **17.** $49a^2 - 70a + 25$; 10.6B **18.** $16b^2 - 81$; 10.6C **19.** $4p^3 - 2p^2 + 5p$; 10.7A **20.** $5x + 2$; 10.7B

Mental Math

1. base: 3; exponent: 2 **3.** base: -3; exponent: 6 **5.** base: 4; exponent: 2 **7.** base: 5; exponent: 1; base: 3; exponent: 4 **9.** base: 5; exponent: 1; base: x; exponent: 2

Section 10.1

1. 49 **3.** -5 **5.** -16 **7.** 16 **9.** $\frac{1}{27}$ **11.** 112 **13.** Answers may vary. **15.** 4 **17.** 135 **19.** 150 **21.** $\frac{32}{5}$ **23.** x^7 **25.** $(-3)^{12}$ **27.** $15y^5$ **29.** $-24z^{20}$ **31.** $20x^5$ sq ft **33.** x^{36} **35.** p^7q^7 **37.** $8a^{15}$ **39.** $\frac{m^9}{n^9}$ **41.** $x^{10}y^{15}$ **43.** $\frac{4x^2z^2}{y^{10}}$ **45.** $64z^{10}$ sq decimeters **47.** $27y^{12}$ cu ft **49.** x^2 **51.** 4 **53.** p^6q^5 **55.** $\frac{y^3}{2}$ **57.** 1 **59.** -2

61. 2 **63.** Answers may vary. **65.** -25 **67.** $\frac{1}{64}$ **69.** z^8 **71.** $81x^2y^2$ **73.** 1 **75.** 40 **77.** b^6 **79.** a^9
81. $-16x^7$ **83.** $64a^3$ **85.** $36x^2y^2z^6$ **87.** $\frac{y^{15}}{8x^{12}}$ **89.** $3x$ **91.** $2x^2y$ **93.** -2 **95.** 5 **97.** -7 **99.** 343 cu m
101. volume **103.** x^{9a} **105.** a^{5b} **107.** x^{5a} **109.** \$1045.85

Calculator Explorations

1. 5.31 EE 3 **3.** 6.6 EE -9 **5.** 1.5×10^{13} **7.** 8.15×10^{19}

Mental Math

1. $\frac{5}{x^2}$ **3.** y^6 **5.** $4y^3$

Section 10.2

1. $\frac{1}{64}$ **3.** $\frac{7}{x^3}$ **5.** -64 **7.** $\frac{5}{6}$ **9.** p^3 **11.** $\frac{q^4}{p^5}$ **13.** $\frac{1}{x^3}$ **15.** z^3 **17.** $\frac{4}{3}$ **19.** $\frac{1}{9}$ **21.** $-p^4$ **23.** -2 **25.** x^4
27. p^4 **29.** m^{11} **31.** r^6 **33.** $\frac{1}{x^{15}y^9}$ **35.** $\frac{1}{x^4}$ **37.** $\frac{1}{a^2}$ **39.** $4k^3$ **41.** $3m$ **43.** $-\frac{4a^5}{b}$ **45.** $-\frac{6x}{7y^2}$ **47.** $\frac{a^{30}}{b^{12}}$
49. $\frac{1}{x^{10}y^6}$ **51.** $\frac{z^2}{4}$ **53.** $\frac{1}{32x^5}$ **55.** $\frac{49a^4}{b^6}$ **57.** $a^{24}b^8$ **59.** x^9y^{19} **61.** $-\frac{y^8}{8x^2}$ **63.** $\frac{25b^{33}}{a^{16}}$ **65.** $\frac{27}{z^3x^6}$ cu in.
67. 7.8×10^4 **69.** 1.67×10^{-6} **71.** 6.35×10^{-3} **73.** 1.16×10^6 **75.** 1.56×10^7 **77.** 1.36×10^4 **79.** 2.84×10^8 **81.** 0.0000000008673 **83.** 0.033 **85.** 20,320 **87.** 0.000000000000000000000397 **89.** 700,000,000 **91.** 0.000036 **93.** 0.0000000000000000028 **95.** 0.0000005 **97.** 200,000 **99.** 2.7×10^9 gal **101.** $-2x + 7$ **103.** $2y - 10$ **105.** $-x - 4$ **107.** -394.5 **109.** 1.3 sec **111.** a^m **113.** $27y^{6z}$ **115.** Answers may vary. **117.** Answers may vary.

Section 10.3

1. 1; $-3x$; 5 **3.** -5; 3.2; 1; -5 **5.** 1; binomial **7.** 3; none of these **9.** 4; binomial **11.** 1; binomial **13.** Answers may vary. **15.** Answers may vary. **17. a.** 6 **b.** 5 **19. a.** -2 **b.** 4 **21. a.** -15 **b.** -16 **23.** 184 ft **25.** 595.84 ft **27.** 212.06 million wireless subscribers **29.** $23x^2$ **31.** $12x^2 - y$ **33.** $7s$ **35.** $-1.1y^2 + 4.8$ **37.** $5x + 3 + 4x + 3 + 2x + 6 + 3x + 7x$; $21x + 12$ **39.** $4x^2 + 7x + x^2 + 5x$; $5x^2 + 12x$ **41.** 2, 1, 1, 0; 2 **43.** 4, 0, 4, 3; 4 **45.** $9ab - 11a$ **47.** $4x^2 - 7xy + 3y^2$ **49.** $-3xy^2 + 4$ **51.** $14y^3 - 19 - 16a^2b^2$ **53.** $10x + 19$ **55.** $-x + 5$ **57.** Answers may vary. **59.** $11.1x^2 - 7.97x + 10.76$

Section 10.4

1. $12x + 12$ **3.** $-3x^2 + 10$ **5.** $-3x^2 + 4$ **7.** $-y^2 - 3y - 1$ **9.** $8t^2 - 4$ **11.** $15a^3 + a^2 + 16$ **13.** $-x + 14$ **15.** $-2x + 9$ **17.** $2x^2 + 7x - 16$ **19.** $y^2 - 7$ **21.** $2x^2 + 11x$ **23.** $-2z^2 - 16z + 6$ **25.** $2u^5 - 10u^2 + 11u - 9$ **27.** $5x - 9$ **29.** $6y + 13$ **31.** $-2x^2 + 8x - 1$ **33.** $7x^2 + 14x + 18$ **35.** $3x - 3$ **37.** $7x^2 - 4x + 2$ **39.** $7x^2 - 2x + 2$ **41.** $4y^2 + 12y + 19$ **43.** $-15x + 7$ **45.** $-2a - b + 1$ **47.** $3x^2 + 5$ **49.** $6x^2 - 2xy + 19y^2$ **51.** $8r^2s + 16rs - 8 + 7r^2s^2$ **53.** $6x^2$ **55.** $-12x^8$ **57.** $200x^3y^2$ **59.** $(x^2 + 7x + 4)$ ft **61.** $(3y^2 + 4y + 11)$ m **63.** $-6.6x^2 - 1.8x - 4.9$ **65.** $-0.56x^2 + 10.32x + 35.72$
67. a. $2x$ **b.** x^2 **c.** $-2x$ **d.** x^2; Answers may vary.

Mental Math

1. x^8 **3.** y^5 **5.** x^{14}

Section 10.5

1. $24x^3$ **3.** $-12.4x^{12}$ **5.** x^4 **7.** $-\frac{2}{15}y^3$ **9.** $-24x^8$ **11.** $6x^2 + 15x$ **13.** $7x^3 + 14x^2 - 7x$ **15.** $-2a^2 - 8a$ **17.** $6x^3 - 9x^2 + 12x$ **19.** $3a^3 + 6a$ **21.** $-6a^4 + 4a^3 - 6a^2$ **23.** $6x^5y - 3x^4y^3 + 24x^2y^4$ **25.** $x^2 + 3x$ **27.** $x^2 + 7x + 12$
29. $a^2 + 5a - 14$ **31.** $x^2 + \frac{1}{3}x - \frac{2}{9}$ **33.** $12x^4 + 25x^2 + 7$ **35.** $12x^2 - 29x + 15$ **37.** $1 - 7a + 12a^2$
39. $4y^2 - 16y + 16$ **41.** $x^3 - 5x^2 + 13x - 14$ **43.** $x^4 + 5x^3 - 3x^2 - 11x + 20$ **45.** $10a^3 - 27a^2 + 26a - 12$ **47.** $49x^2y^2 - 14xy^2 + y^2$ **49.** $x^2 + 5x + 6$ **51.** $12x^2 - 64x - 11$ **53.** $2x^3 + 10x^2 + 11x - 3$

55. $x^4 - 2x^3 - 51x^2 + 4x + 63$ **57.** $25x^2$ **59.** $9y^6$ **61.** \$7000 **63.** \$500 **65.** Answers may vary. **67.** $(4x^2 - 25)$ sq yds **69.** $(6x^2 - 4x)$ sq in. **71. a.** $6x + 12$; Answers may vary. **b.** $9x^2 + 36x + 35$; Answers may vary. **73. a.** $a^2 - b^2$ **b.** $4x^2 - 9y^2$ **c.** $16x^2 - 49$ **d.** Answers may vary.

Section 10.6

1. $x^2 + 7x + 12$ **3.** $x^2 + 5x - 50$ **5.** $5x^2 + 4x - 12$ **7.** $4y^2 - 25y + 6$ **9.** $6x^2 + 13x - 5$ **11.** $6y^3 + 4y^2 + 42y + 28$ **13.** $x^2 + \frac{1}{3}x - \frac{2}{9}$ **15.** $8 - 26a + 15a^2$ **17.** $2x^2 + 9xy - 5y^2$ **19.** $x^2 + 4x + 4$ **21.** $4x^2 - 4x + 1$ **23.** $9a^2 - 30a + 25$ **25.** $x^4 + 10x^2 + 25$ **27.** $y^2 - \frac{4}{7}y + \frac{4}{49}$ **29.** $4a^2 - 12a + 9$ **31.** $25x^2 + 90x + 81$ **33.** $9x^2 - 42xy + 49y^2$ **35.** $16m^2 + 40mn + 25n^2$ **37.** Answers may vary. **39.** $a^2 - 49$ **41.** $x^2 - 36$ **43.** $9x^2 - 1$ **45.** $x^4 - 25$ **47.** $4y^4 - 1$ **49.** $16 - 49x^2$ **51.** $9x^2 - \frac{1}{4}$ **53.** $81x^2 - y^2$ **55.** $4m^2 - 25n^2$ **57.** $a^2 + 9a + 20$ **59.** $a^2 - 14a + 49$ **61.** $12a^2 - a - 1$ **63.** $x^2 - 4$ **65.** $9a^2 + 6a + 1$ **67.** $x^2 + 3xy - y^2$ **69.** $a^2 - \frac{1}{4}y^2$ **71.** $6b^2 - b - 35$ **73.** $x^4 - 100$ **75.** $16x^2 - 25$ **77.** $25x^2 - 60xy + 36y^2$ **79.** $4r^2 - 9s^2$ **81.** $\frac{5b^5}{7}$ **83.** $-\frac{2a^{10}}{b^5}$ **85.** $\frac{2y^8}{3}$ **87.** $(4x^2 + 4x + 1)$ sq ft **89.** $(24x^2 - 32x + 8)$ sq m

Integrated Review

1. $35x^5$ **2.** $32y^9$ **3.** -16 **4.** 16 **5.** $2x^2 - 9x - 5$ **6.** $3x^2 + 13x - 10$ **7.** $3x - 4$ **8.** $4x + 3$ **9.** $7x^6y^2$ **10.** $\frac{10b^6}{7}$ **11.** $144m^{14}n^{12}$ **12.** $64y^{27}z^{30}$ **13.** $16y^2 - 9$ **14.** $49x^2 - 1$ **15.** $\frac{y^{45}}{x^{63}}$ **16.** $\frac{x^{27}}{27}$ **17.** $2x^2 - 2x - 6$ **18.** $6x^2 + 13x - 11$ **19.** $2.5y^2 - 6y - 0.2$ **20.** $8.4x^2 - 6.8x - 12.3$ **21.** $x^2 + 8x + 16$ **22.** $y^2 - 18y + 81$ **23.** $2x + 8$ **24.** $2y - 18$ **25.** $7x^2 - 10xy + 4y^2$ **26.** $-a^2 - 3ab + 6b^2$ **27.** $x^3 + 2x^2 - 16x + 3$ **28.** $x^3 - 2x^2 - 5x - 2$ **29.** $6x^2 - x - 70$ **30.** $20x^2 + 21x - 5$ **31.** $2x^3 - 19x^2 + 44x - 7$ **32.** $5x^3 + 9x^2 - 17x + 3$

Mental Math

1. a^2 **3.** a^2 **5.** k^3

Section 10.7

1. $4x^2 + x + \frac{9}{5}$ **3.** $12x^3 + 3x$ **5.** $5p^2 + 6p$ **7.** $-\frac{3}{2x} + 3$ **9.** $-3x^2 + x - \frac{4}{x^3}$ **11.** $-1 + \frac{3}{2x} - \frac{7}{4x^4}$ **13.** $ab - b^2$ **15.** $x + 4xy - \frac{y}{2}$ **17.** $x + 1$ **19.** $2x + 3$ **21.** $2x + 1 + \frac{7}{x - 4}$ **23.** $4x + 9$ **25.** $3a^2 - 3a + 1 + \frac{2}{3a + 2}$ **27.** $2b^2 + b + 2 - \frac{12}{b + 4}$ **29.** $4x + 3 - \frac{2}{2x + 1}$ **31.** $2x^2 + 6x - 5 - \frac{2}{x - 2}$ **33.** $x^2 + 3x + 9$ **35.** $-3x + 6 - \frac{11}{x + 2}$ **37.** $2b - 1 - \frac{6}{2b - 1}$ **39.** 3 **41.** -4 **43.** $3x$ **45.** $9x$ **47.** $x^3 - x^2 + x$ **49.** $(3x^3 + x - 4)$ ft **51.** $(2x + 5)$ m **53.** Answers may vary.

Chapter 10 Review

1. base: 3; exponent: 2 **2.** base: -5; exponent: 4 **3.** base: 5; exponent: 4 **4.** base: x; exponent: 6 **5.** 512 **6.** 36 **7.** -36 **8.** -65 **9.** 1 **10.** 1 **11.** y^9 **12.** x^{14} **13.** $-6x^{11}$ **14.** $-20y^7$ **15.** x^8 **16.** y^{15} **17.** $81y^{24}$ **18.** $8x^9$ **19.** x^5 **20.** z^7 **21.** a^4b^3 **22.** x^3y^5 **23.** $\frac{4}{x^3y^4}$ **24.** $\frac{x^6y^6}{4}$ **25.** $40a^{19}$ **26.** $36x^3$ **27.** 3 **28.** 9 **29.** b **30.** c **31.** $\frac{1}{49}$ **32.** $-\frac{1}{49}$ **33.** $\frac{2}{x^4}$ **34.** $\frac{1}{16x^4}$ **35.** 125 **36.** $\frac{9}{4}$ **37.** $\frac{17}{16}$ **38.** $\frac{1}{42}$ **39.** x^8 **40.** z^8 **41.** r **42.** y^3 **43.** c^4 **44.** $\frac{x^3}{y^3}$ **45.** $\frac{1}{x^6y^{13}}$ **46.** $\frac{a^{10}}{b^{10}}$ **47.** 2.7×10^{-4} **48.** 8.868×10^{-1} **49.** 8.08×10^7 **50.** -8.68×10^5 **51.** 1.302079×10^9 **52.** 1.5×10^5 **53.** 867,000 **54.** 0.00386 **55.** 0.00086 **56.** 893,600 **57.** 1,431,280,000,000,000 cu km **58.** 0.0000000001 m **59.** 0.016 **60.** 400,000,000,000 **61.** 5 **62.** 2 **63.** 5 **64.** 6 **65.** 22; 78; 154.02; 400 **66.** $2a^2$ **67.** $-4y$ **68.** $15a^2 + 4a$ **69.** $22x^2 + 3x + 6$ **70.** $-6a^2b - 3b^2 - q^2$ **71.** cannot be combined **72.** $8x^2 + 3x + 6$ **73.** $2x^5 + 3x^4 + 4x^3 + 9x^2 + 7x + 6$ **74.** $-7y^2 - 1$

75. $-6m^7 - 3x^4 + 7m^6 - 4m^2$ **76.** $-x^2 - 6xy - 2y^2$ **77.** $-5x^2 + 5x + 1$ **78.** $-2x^2 - x + 20$ **79.** $6x + 30$ **80.** $9x - 63$ **81.** $8a + 28$ **82.** $54a - 27$ **83.** $-7x^3 - 35x$ **84.** $-32y^3 + 48y$ **85.** $-2x^3 + 18x^2 - 2x$ **86.** $-3a^3b - 3a^2b - 3ab^2$ **87.** $-6a^4 + 8a^2 - 2a$ **88.** $42b^4 - 28b^2 + 14b$ **89.** $2x^2 - 12x - 14$ **90.** $6x^2 - 11x - 10$ **91.** $4a^2 + 27a - 7$ **92.** $42a^2 + 11a - 3$ **93.** $x^4 + 7x^3 + 4x^2 + 23x - 35$ **94.** $x^6 + 2x^5 + x^2 + 3x + 2$ **95.** $x^4 + 4x^3 + 4x^2 - 16$ **96.** $x^6 + 8x^4 + 16x^2 - 16$ **97.** $x^3 + 21x^2 + 147x + 343$ **98.** $8x^3 - 60x^2 + 150x - 125$ **99.** $x^2 + 14x + 49$ **100.** $x^2 - 10x + 25$ **101.** $9x^2 - 42x + 49$ **102.** $16x^2 + 16x + 4$ **103.** $25x^2 - 90x + 81$ **104.** $25x^2 - 1$ **105.** $49x^2 - 16$ **106.** $a^2 - 4b^2$ **107.** $4x^2 - 36$ **108.** $16a^4 - 4b^2$ **109.** $(9x^2 - 6x + 1)$ sq m **110.** $(5x^7 - 3x - 2)$ sq mi **111.** $\frac{1}{7} + \frac{3}{x} + \frac{7}{x^2}$ **112.** $-a^2 + 3b - 4$ **113.** $a + 1 + \frac{6}{a-2}$ **114.** $4x + \frac{7}{x+5}$ **115.** $a^2 + 3a + 8 + \frac{22}{a-2}$ **116.** $3b^2 - 4b - \frac{1}{3b-2}$ **117.** $2x^3 - x^2 + 2 - \frac{1}{2x-1}$ **118.** $-x^2 - 16x - 117 - \frac{684}{x-6}$ **119.** $\left(5x - 1 + \frac{20}{x^2}\right)$ ft **120.** $\left(7a^3b^6 + a - 1\right)$ units

Chapter 10 Test

1. 32 **2.** 81 **3.** −81 **4.** $\frac{1}{64}$ **5.** $-15x^{11}$ **6.** y^5 **7.** $\frac{1}{r^5}$ **8.** $\frac{y^{14}}{x^2}$ **9.** $\frac{1}{6xy^8}$ **10.** 5.63×10^5 **11.** 8.63×10^{-5} **12.** 0.0015 **13.** 62,300 **14.** 0.036 **15.** 4 **16.** $-2x^2 + 12x + 11$ **17.** $16x^3 + 7x^2 - 3x - 13$ **18.** $-3x^3 + 5x^2 + 4x + 5$ **19.** $x^3 + 8x^2 + 3x - 5$ **20.** $3x^3 + 22x^2 + 41x + 14$ **21.** $6x^4 - 9x^3 + 21x^2$ **22.** $3x^2 + 16x - 35$ **23.** $16x^2 - 16x + 4$ **24.** $x^4 - 81b^2$ **25.** 1516 ft; 1372 ft; 940 ft; 220 ft **26.** $(4x^2 - 9)$ sq in. **27.** $\frac{x}{2y} + 3 - \frac{7}{8y}$ **28.** $x + 2$ **29.** $9x^2 - 6x + 4 - \frac{16}{3x+2}$

Cumulative Review

1. 64; Sec. 1.8, Ex. 5 **2.** 125 **3.** 32; Sec. 1.8, Ex. 7 **4.** 27 **5. a.** $7 + x$ **b.** $15 - x$ **c.** $2x$ **d.** $\frac{x}{5}$ **e.** $x - 2$; Sec. 1.9, Ex. 6 **6. a.** $x + 3$ **b.** $3x$ **c.** $2x$ **d.** $10 - x$ **e.** $5x + 7$ **7.** $2x - 9$; Sec 3.1, Ex. 11 **8.** $5x + 10$ **9.** $6x + 18$; Sec 3.1, Ex. 12 **10.** $-2y - 0.6z + 2$ **11.** −3; Sec. 3.2, Ex. 6 **12.** 140 **13.** −4; Sec. 3.4, Ex. 4 **14.** 19 **15.** 15; Sec. 4.7, Ex. 2 **16.** 24 **17.** $-\frac{2}{33}$; Sec. 4.7, Ex. 4 **18.** $-\frac{1}{25}$ **19.** $\frac{1}{8}$; Sec. 5.1, Ex. 9 **20.** $\frac{1}{4}$ **21.** $23\frac{1}{2}$; Sec. 5.1, Ex. 10 **22.** $10\frac{3}{4}$ **23.** $-105\frac{83}{1000}$; Sec. 5.1, Ex. 11 **24.** $-31\frac{7}{100}$ **25.** 1.88; Sec. 5.2, Ex. 10 **26.** −8.8 **27.** 1.012; Sec. 5.3, Ex. 12 **28.** −3.05 **29. a.** 11, 112 **b.** 0, 11, 112 **c.** −3, −2, 0, 11, 112 **d.** $-3, -2, 0, \frac{1}{4}, 11, 112$ **e.** $\sqrt{2}$ **f.** $-2, 0, \frac{1}{4}, 112, -3, 11, \sqrt{2}$; Sec. 9.1, Ex. 10 **30.** rational numbers, real numbers **31.** 4, Sec. 9.3, Ex. 5 **32.** 5 **33.** 10; Sec. 9.4, Ex. 1 **34.** 2 **35.** 40 ft; Sec. 9.5, Ex. 2 **36.** 32 ft **37.** $\{x|x \le 4\}$; Sec. 9.6, Ex. 6 **38.** $\{x|x \ge 10\}$ **39. a.** x^{11} **b.** $\frac{1}{16}$ **c.** $81y^{10}$; Sec. 10.1, Ex. 29 **40. a.** y^6 **b.** $\frac{8}{27}$ **c.** $64x^6$ **41.** $\frac{b^3}{27a^6}$; Sec. 10.2, Ex. 10 **42.** $\frac{y^2}{4x^6}$ **43.** $\frac{1}{25y^6}$; Sec. 10.2; Ex. 14 **44.** $\frac{1}{27x^{21}}$ **45.** $10x^3$; Sec. 10.3, Ex. 8 **46.** $8x^3$ **47.** $5x^2 - 3x - 3$; Sec 10.3, Ex. 9 **48.** $-5x^2 + 7x$ **49.** $7x^3 + 14x^2 + 35x$; Sec. 10.5, Ex. 4 **50.** $-2x^3 + 2x^2 + 2x$ **51.** $3x^3 - 4 + \frac{1}{x}$; Sec. 10.7; Ex. 2 **52.** $2x^6 - 6x + 1$

Chapter 11 Factoring Polynomials

Chapter 11 Pretest

1. $2x^2y(x - 3y)$; 11.1B **2.** $(x - 4)(y + 6)$; 11.1C **3.** $(a + 6)(a + 2)$; 11.2A **4.** prime; 11.2A **5.** $3x(x - 1)(x - 5)$; 11.2B **6.** $(2x - 3)(x + 4)$; 11.3A **7.** $7(2x + 5)(x + 2)$; 11.3B **8.** $(3b - 2)(8b - 3)$; 11.4A **9.** $(5y + 1)(3y + 7)$; 11.4A **10.** $(x + 12)^2$; 11.5B **11.** $(2x - 3y)^2$; 11.5B **12.** $(a - 7b)(a + 7b)$; 11.5C **13.** $(1 - 8t)(1 + 8t)$; 11.5C **14.** prime; 11.5C **15.** 18; 11.5A **16.** $x = 12, x = -5$; 11.6A **17.** $y = 0, y = 13$; 11.6A **18.** $m = 0, m = 4, m = -3$; 11.6B **19.** 8 in. × 15 in.; 11.7A **20.** −16 or 15; 11.7A

Mental Math

1. 2 **3.** 3 **5.** 7

Section 11.1

1. y^2 **3.** xy^2 **5.** 4 **7.** $4y^3$ **9.** $5x^2$ **11.** $3x^3$ **13.** $9x^2y$ **15.** $3(a + 2)$ **17.** $15(2x - 1)$ **19.** $x^2(x + 5)$ **21.** $2y(3y^3 - 1)$ **23.** $2x(16y - 9x)$ **25.** $4(x - 2y + 1)$ **27.** $3x(2x^2 - 3x + 4)$ **29.** $a^2b^2(a^5b^4 - a + b^3 - 1)$ **31.** $5xy(x^2 - 3x + 2)$ **33.** $4(2x^5 + 4x^4 - 5x^3 + 3)$ **35.** $\frac{1}{3}x(x^3 + 2x^2 - 4x^4 + 1)$ **37.** $(x + 2)(y + 3)$ **39.** $(x + 2)(8 - y)$ **41.** Answers may vary. **43.** $(x^2 + 5)(x + 2)$ **45.** $(x + 3)(5 + y)$ **47.** $(2x^2 + 5)(3x - 2)$ **49.** $(y - 4)(2 + x)$ **51.** $(2x + 1)(x^2 + 4)$ **53.** $(x - 2y)(4x - 3)$ **55.** Answers may vary. **57.** $x^2 + 7x + 10$ **59.** $b^2 - 3b - 4$ **61.** 2, 6 **63.** −1, −8 **65.** −2, 5 **67.** −8, 3 **69.** $2(3x^2 - 1)(2y - 7)$ **71.** $12x^3 - 2x; 2x(6x^2 - 1)$ **73.** $(n^3 - 6)$ units **75. a.** 2000 billion kw hr **b.** 2088 billion kw hr **c.** $-4(2x^2 - 15x - 500)$ or $4(-2x^2 + 15x + 500)$

Mental Math

1. +5 **3.** −3 **5.** +2

Section 11.2

1. $(x + 6)(x + 1)$ **3.** $(x - 9)(x - 1)$ **5.** $(x - 6)(x + 3)$ **7.** $(x + 10)(x - 7)$ **9.** prime **11.** $(x + 5y)(x + 3y)$ **13.** $(a^2 - 5)(a^2 + 3)$ **15.** $x^2 + 5x - 24$ **17.** Answers may vary. **19.** $2(z + 8)(z + 2)$ **21.** $2x(x - 5)(x - 4)$ **23.** $(x - 4y)(x + y)$ **25.** $(x + 12)(x + 3)$ **27.** $(x - 2)(x + 1)$ **29.** $(r - 12)(r - 4)$ **31.** $(x + 2y)(x - y)$ **33.** $3(x + 5)(x - 2)$ **35.** $3(x - 18)(x - 2)$ **37.** $(x - 24)(x + 6)$ **39.** prime **41.** $(x - 5)(x - 3)$ **43.** $6x(x + 4)(x + 5)$ **45.** $4y(x^2 + x - 3)$ **47.** $(x - 7)(x + 3)$ **49.** $(x + 5y)(x + 2y)$ **51.** $2(t + 8)(t + 4)$ **53.** $x(x - 6)(x + 4)$ **55.** $2t^3(t - 4)(t - 3)$ **57.** $5xy(x - 8y)(x + 3y)$ **59.** $2x^2 + 11x + 5$ **61.** $15y^2 - 17y + 4$ **63.** $9a^2 + 23a - 12$ **65.** $2x^2 + 28x + 66; 2(x + 3)(x + 11)$ **67.** $(x + 1)(y - 5)(y + 3)$ **69.** 3; 4 **71.** 8; 16 **73.** $(x^n + 2)(x^n + 3)$

Section 11.3

1. $x + 4$ **3.** $10x - 1$ **5.** $4x - 3$ **7.** $(2x + 3)(x + 5)$ **9.** $(y - 1)(8y - 9)$ **11.** $(2x + 1)(x - 5)$ **13.** $(4r - 1)(5r + 8)$ **15.** $(5x + 1)(2x + 3)$ **17.** $(3x - 2)(x + 1)$ **19.** $(3x - 5y)(2x - y)$ **21.** $(3x - 5)(5x + 3)$ **23.** $(x - 4)(x - 5)$ **25.** $(2x + 11)(x - 9)$ **27.** $(7t + 1)(t - 4)$ **29.** $(3a + b)(a + 3b)$ **31.** $(7x + 1)(7x - 2)$ **33.** $(6x - 7)(3x + 2)$ **35.** $x(3x + 2)(4x + 1)$ **37.** $3(7x + 5)(x - 3)$ **39.** $(3x + 4)(4x - 3)$ **41.** $2y^2(3x - 10)(x + 3)$ **43.** $(2x - 7)(2x + 3)$ **45.** $3(x^2 - 14x + 21)$ **47.** $(4x + 9)(2x - 3)$ **49.** $-1(x - 6)(x + 4)$ **51.** $x(4x + 3)(x - 3)$ **53.** $(4x - 9)(6x - 1)$ **55.** $b(8a - 3)(5a + 3)$ **57.** $(3x^2 + 2)(5x^2 + 3)$ **59.** $2y(3y + 5)(y - 3)$ **61.** $5x(2x - y)(x + 3y)$ **63.** $-1(2x - 5)(7x - 2)$ **65.** April and May **67.** decreased 0.1% **69.** $(y - 1)^2(4x^2 + 10x + 25)$ **71.** $-3xy^2(4x - 5)(x + 1)$ **73.** 5; 13 **75.** 4; 5

Section 11.4

1. $(x + 3)(x + 2)$ **3.** $(x - 4)(x + 7)$ **5.** $(y + 8)(y - 2)$ **7.** $(3x + 4)(x + 4)$ **9.** $(8x - 5)(x - 3)$ **11.** $(5x^2 - 3)(x^2 + 5)$ **13. a.** 9, 2 **b.** $9x + 2x$ **c.** $(2x + 3)(3x + 1)$ **15. a.** −20, −3 **b.** $-20x - 3x$ **c.** $(5x - 1)(3x - 4)$ **17.** $(3y + 2)(7y + 1)$ **19.** $(7x - 11)(x + 1)$ **21.** $(5x - 2)(2x - 1)$ **23.** $(2x - 5)(x - 1)$ **25.** $(2x + 3)(2x + 3)$ or $(2x + 3)^2$ **27.** $(2x + 3)(2x - 7)$ **29.** $(5x - 4)(2x - 3)$ **31.** $x(2x + 3)(x + 5)$ **33.** $2(8y - 9)(y - 1)$ **35.** $(2x - 3)(3x - 2)$ **37.** $3(3a + 2)(6a - 5)$ **39.** $a(4a + 1)(5a + 8)$ **41.** $3x(4x + 3)(x - 3)$ **43.** $x^2 - 4$ **45.** $y^2 + 8y + 16$ **47.** $81z^2 - 25$ **49.** $16x^2 - 24x + 9$ **51.** $(x^n + 2)(x^n + 3)$

Calculator Exploration

16, 14, 16; 16, 14, 16; 2.89, 0.89, 2.89; 171.61, 169.61, 171.61; 1, −1, 1

Mental Math

1. 1^2 **3.** 9^2 **5.** 3^2 **7.** $(3x)^2$ **9.** $(5a)^2$ **11.** $(6p^2)^2$

Section 11.5

1. yes **3.** no **5.** yes **7.** no **9.** no **11.** yes **13.** 8 **15.** $(x + 11)^2$ **17.** $(x - 8)^2$ **19.** $(4a - 3)^2$ **21.** $(x^2 + 2)^2$ **23.** $2(n - 7)^2$ **25.** $(4y + 5)^2$ **27.** $(xy - 5)^2$ **29.** $m(m + 9)^2$ **31.** prime **33.** $(3x - 4y)^2$ **35.** $(x + 7y)^2$ **37.** Answers may vary. **39.** $(x - 2)(x + 2)$ **41.** $(9 - p)(9 + p)$ **43.** $-1(2r - 1)(2r + 1)$ **45.** $(3x - 4)(3x + 4)$ **47.** prime **49.** $(-6 + x)(6 + x)$ **51.** $(m^2 + 1)(m + 1)(m - 1)$ **53.** $(x - 13y)(x + 13y)$ **55.** $2(3r - 2)(3r + 2)$ **57.** $x(3y - 2)(3y + 2)$ **59.** $25y^2(y - 2)(y + 2)$ **61.** $xy(x - 2y)(x + 2y)$ **63.** $9(5a - 3b)(5a + 3b)$ **65.** $3(2x - 3)(2x + 3)$ **67.** $(7a - 4)(7a + 4)$ **69.** $(13a - 7b)(13a + 7b)$ **71.** $(4 - ab)(4 + ab)$ **73.** $\left(y - \frac{1}{4}\right)\left(y + \frac{1}{4}\right)$ **75.** $\left(10 - \frac{2}{9}n\right)\left(10 + \frac{2}{9}n\right)$ **77.** $(5 - y)$ **79.** −5 **81.** 3

83. $-\frac{1}{2}$ **85.** 5 ft **87.** $(y-6-z)(y-6+z)$ **89.** $(m-3)(m+3)(n+8)$ **91.** $(x+1-6y)(x+1+6y)$
93. $(x^n+9)(x^n-9)$ **95.** perfect square trinomial **97. a.** 513 ft **b.** 273 ft **c.** 6 sec **d.** $(23-4t)(23+4t)$
99. a. 720 ft **b.** 384 ft **c.** 7 sec **d.** $16(7-t)(7+t)$

Integrated Review

1. $(x-3)(x+4)$ **2.** $(x-8)(x-2)$ **3.** $(x+2)(x-3)$ **4.** $(x+1)^2$ **5.** $(x-3)^2$ **6.** $(x+2)(x-1)$
7. $(x+3)(x-2)$ **8.** $(x+3)(x+4)$ **9.** $(x-5)(x-2)$ **10.** $(x-6)(x+5)$ **11.** $2(x-7)(x+7)$
12. $3(x-5)(x+5)$ **13.** $(x+3)(x+5)$ **14.** $(y-7)(3+x)$ **15.** $(x+8)(x-2)$ **16.** $(x-7)(x+4)$
17. $4x(x+7)(x-2)$ **18.** $6x(x-5)(x+4)$ **19.** $2(3x+4)(2x+3)$ **20.** $(2a-b)(4a+5b)$ **21.** $(2a-b)(2a+b)$
22. $(x-5y)(x+5y)$ **23.** $(4-3x)(7+2x)$ **24.** $(5-2x)(4+x)$ **25.** prime **26.** prime **27.** $(3y+5)(2y-3)$
28. $(4x-5)(x+1)$ **29.** $9x(2x^2-7x+1)$ **30.** $4a(3a^2-6a+1)$ **31.** $(4a-7)^2$ **32.** $(5p-7)^2$
33. $(7-x)(2+x)$ **34.** $(3+x)(1-x)$ **35.** $3x^2y(x+6)(x-4)$ **36.** $2xy(x+5y)(x-y)$ **37.** $3xy(4x^2+81)$
38. $2xy^2(3x^2+4)$ **39.** $2xy(1-6x)(1+6x)$ **40.** $2x(x-3)(x+3)$ **41.** $(x-2)(x+2)(x+6)$
42. $(x-2)(x-6)(x+6)$ **43.** $2a^2(3a+5)$ **44.** $2n(2n-3)$ **45.** $(x^2+4)(3x-1)$ **46.** $(x-2)(x^2+3)$
47. $6(x+2y)(x+y)$ **48.** $2(x+4y)(6x-y)$ **49.** $(5+x)(x+y)$ **50.** $(x-y)(7+y)$ **51.** $(7t-1)(2t-1)$
52. prime **53.** $(3x+5)(x-1)$ **54.** $(7x-2)(x+3)$ **55.** $(1-10a)(1+2a)$ **56.** $(1+5a)(1-12a)$
57. $(x-3)(x+3)(x-1)(x+1)$ **58.** $(x-3)(x+3)(x-2)(x+2)$ **59.** $(x-15)(x-8)$ **60.** $(y+16)(y+6)$
61. prime **62.** $(4a-7b)^2$ **63.** $(5p-7q)^2$ **64.** $(7x+3y)(x+3y)$ **65.** $-1(x-5)(x+6)$ **66.** $-1(x-2)(x-4)$
67. $(s+4)(3r-1)$ **68.** $(x-2)(x^2+3)$ **69.** $(4x-3)(x-2y)$ **70.** $(2x-y)(2x+7z)$ **71.** $(x+12y)(x-3y)$
72. $(3x-2y)(x+4y)$ **73.** $(x^2+2)(x+4)(x-4)$ **74.** $(x^2+3)(x+5)(x-5)$
75. Answers may vary. **76.** yes; $9(x^2+9y^2)$

Mental Math

1. 3, 7 **3.** −8, −6 **5.** −1, 3

Section 11.6

1. 2, −1 **3.** 6, 7 **5.** −9, −17 **7.** 0, −6 **9.** 0, 8 **11.** $-\frac{3}{2}, \frac{5}{4}$ **13.** $\frac{7}{2}, -\frac{2}{7}$ **15.** $\frac{1}{2}, -\frac{1}{3}$ **17.** −0.2, −1.5
19. $(x-6)(x+1)=0$ **21.** 9, 4 **23.** −4, 2 **25.** 0, 7 **27.** 0, −20 **29.** 4, −4 **31.** 8, −4 **33.** $\frac{7}{3}, -2$ **35.** $\frac{8}{3}, -9$
37. $0, \frac{1}{2}, -\frac{1}{2}$ **39.** $\frac{17}{2}$ **41.** $\frac{3}{4}$ **43.** $\frac{1}{2}, -\frac{1}{2}$ **45.** $-\frac{3}{2}, -\frac{1}{2}, 3$ **47.** −5, 3 **49.** $2, -\frac{4}{5}$ **51.** $-\frac{5}{6}, \frac{6}{5}$ **53.** $-\frac{4}{3}, 5$
55. −4, 3 **57.** 0, 8, 4 **59.** $x^2-12x+35=0$ **61.** $\frac{47}{45}$ **63.** $\frac{17}{60}$ **65.** $\frac{7}{10}$ **67.** didn't write equation in standard form
69. a. 300; 304; 276; 216; 124; 0; −156 **b.** 5 sec **c.** 304 ft **71.** $0, \frac{1}{2}$ **73.** 0, −15

Section 11.7

1. width $= x$; length $= x+4$ **3.** x and $x+2$ if x is an odd integer **5.** base $= x$; height $= 4x+1$ **7.** 11 units
9. 15 cm, 13 cm, 70 cm, 22 cm **11.** base = 16 mi; height = 6 mi **13.** 5 sec **15.** length = 5 cm; width = 6 cm
17. 54 diagonals **19.** 10 sides **21.** −12 or 11 **23.** slow boat: 8 mph; fast boat: 15 mph **25.** 13 and 7 **27.** 5 in.
29. 12 mm; 16 mm; 20 mm **31.** 10 km **33.** 36 ft **35.** 9.5 sec **37.** 20% **39.** length: 15 mi; width: 8 mi
41. 105 units **43.** 9600 thousand acres **45.** 9500 thousand acres **47.** end of 1998 **49.** Answers may vary.
51. width of pool: 29 m; length of pool: 35 m **53.** 70420: Abita Springs, LA; 70434: Covington, LA

Chapter 11 Review

1. $2x-5$ **2.** $2x^4+1-5x^3$ **3.** $5(m+6)$ **4.** $4x(5x^2+3x+6)$ **5.** $(2x+3)(3x-5)$ **6.** $(x+1)(5x-1)$
7. $(x-1)(3x+2)$ **8.** $(2x-1)(3x+5)$ **9.** $(a+3b)(3a+b)$ **10.** $(x+4)(x+2)$ **11.** $(x-8)(x-3)$ **12.** prime
13. $(x-6)(x+1)$ **14.** $(x+4)(x-2)$ **15.** $(x+6y)(x-2y)$ **16.** $(x+5y)(x+3y)$ **17.** $2(3-x)(12+x)$
18. $4(8+3x-x^2)$ **19.** $5y(y-6)(y-4)$ **20.** −48, 2 **21.** factor out the GCF, 3 **22.** $(2x+1)(x+6)$
23. $(2x+3)(2x-1)$ **24.** $(3x+4y)(2x-y)$ **25.** prime **26.** $(2x+3)(x-13)$ **27.** $(6x+5y)(3x-4y)$
28. $5y(2y-3)(y+4)$ **29.** $5x^2-9x-2$; $(5x+1)(x-2)$ **30.** $16x^2-28x+6$; $2(4x-1)(2x-3)$ **31.** yes **32.** no
33. no **34.** yes **35.** yes **36.** no **37.** yes **38.** no **39.** $(x-9)(x+9)$ **40.** $(x+6)^2$ **41.** $(2x-3)(2x+3)$
42. $(3t-5s)(3t+5s)$ **43.** prime **44.** $(n-9)^2$ **45.** $3(r+6)^2$ **46.** $(3y-7)^2$ **47.** $5m^6(m-1)(m+1)$
48. $(2x-7y)^2$ **49.** $3y(x+y)^2$ **50.** $(2x-1)(2x+1)(4x^2+1)$ **51.** −6, 2 **52.** $0, -1, \frac{2}{7}$ **53.** $-\frac{1}{5}, -3$ **54.** −7, −1
55. −4, 6 **56.** −5 **57.** 2, 8 **58.** $\frac{1}{3}$ **59.** $-\frac{2}{7}, \frac{3}{8}$ **60.** 0, 6 **61.** 5, −5 **62.** $x^2-9x+20=0$ **63.** c **64.** d

65. 9 units **66.** 8 units, 13 units, 16 units, 10 units **67.** width: 20 in.; length: 25 in. **68.** 36 yd **69.** 19 and 20
70. a. 17.5 sec and 10 sec; answers may vary. **b.** 27.5 sec **71.** 32 cm

Chapter 11 Test

1. $3x(3x - 1)$ **2.** $(x + 7)(x + 4)$ **3.** $(y + 11)^2$ **4.** $(4 - y)(a + 3)$ **5.** $(y - 12)(y + 4)$ **6.** $(3a - 7)(a + b)$
7. $(3x - 2)(x - 1)$ **8.** $5(6 - x)(6 + x)$ **9.** $3x(x - 5)(x - 2)$ **10.** $(6t + 5)(t - 1)$ **11.** $(y - 2)(y + 2)(x - 7)$
12. $x(1 - x)(1 + x)(1 + x^2)$ **13.** $(x + 12y)(x + 2y)$ **14.** $-7, 2$ **15.** $-7, 1$ **16.** $0, \frac{3}{2}, -\frac{4}{3}$ **17.** $0, 3, -3$ **18.** $-3, 5$
19. $-\frac{2}{3}, 1$ **20.** 17 ft **21.** 8 and 9 **22.** 8.25 sec **23.** hypotenuse: 25 cm; legs: 15 cm, 20 cm

Cumulative Review

1. -3; Sec. 2.5, Ex. 13 **2.** -14 **3.** -9; Sec. 2.5, Ex. 14 **4.** -72 **5.** $6y + 2$; Sec. 3.1, Ex. 2 **6.** $-4a - 1$
7. $-x + 5$; Sec. 3.1, Ex. 4 **8.** $7.3x - 6$ **9.** solution; Sec. 3.2, Ex. 1 **10.** solution **11.** -15; Sec. 3.2, Ex. 8 **12.** -2
13. 3; Sec. 3.3, Ex. 4 **14.** -6 **15.** $\frac{31}{2}$; Sec. 6.1, Ex. 5 **16.** -17 **17. a.** $9 \leq 11$ **b.** $8 > 1$ **c.** $3 \neq 4$; Sec. 9.1, Ex. 7
18. a. $5 \geq 1$; **b.** $2 \neq -4$ **19.** every real number; Sec. 9.3, Ex. 7 **20.** no solution **21.** $l = \frac{V}{wh}$; Sec. 9.5, Ex. 3
22. $x = \frac{2y + 5}{3}$ **23.** 5^{18}; Sec. 10.1, Ex. 14 **24.** 7^{18} **25.** y^{16}; Sec. 10.1; Ex. 15 **26.** x^{33} **27.** x^6; Sec. 10.2; Ex. 9
28. y^{22} **29.** $\frac{y^{18}}{z^{36}}$; Sec. 10.2, Ex. 11 **30.** $\frac{y^8}{x^2}$ **31.** $\frac{1}{x^{19}}$; Sec. 10.2, Ex. 13 **32.** y^{13} **33.** $4x$; Sec. 10.3, Ex. 6 **34.** $2y$
35. $13x^2 - 2$; Sec. 10.3, Ex. 7 **36.** $12y - 2y^2$ **37.** $4x^2 - 4xy + y^2$; Sec. 10.5, Ex. 8 **38.** $9x^2 + 6x + 1$
39. $t^2 + 4t + 4$. Sec. 10.6, Ex. 5 **40.** $x^2 - 8x + 16$ **41.** $x^4 - 14x^2y + 49y^2$; Sec. 10.6, Ex.8 **42.** $x^4 + 14x^2y + 49y^2$
43. $2xy - 4 + \frac{1}{2y}$; Sec. 10.7, Ex. 3 **44.** $4ab^2 - 1 + \frac{2}{a}$ **45.** $(x + 3)(5 + y)$; Sec. 11.1, Ex. 7 **46.** $(y - 2)(9 + x)$
47. $(x^2 + 2)(x^2 + 3)$; Sec. 11.2, Ex. 7 **48.** $(x^2 + 1)(x^2 - 5)$ **49.** $2(x - 2)(3x + 5)$; Sec. 11.4, Ex. 2 **50.** $5(x + 2)(2x + 1)$
51. 4 sec; Sec. 11.7, Ex. 1 **52.** $-12, 10$

Chapter 12 Rational Expressions

Chapter 12 Pretest

1. $x = -1, x = 10$; 12.1B **2.** $\frac{4}{x + 2}$; 12.1C **3.** $10(x + 2)(x + 3)$; 12.3B **4.** 3; 12.2A **5.** $\frac{5(x + 5)}{x^3(x - 5)}$; 12.2B
6. $\frac{1}{b - 11}$; 12.3A **7.** $\frac{7 - 4x}{x - 1}$; 12.4A **8.** $\frac{9}{x - 5}$; 12.4A **9.** $\frac{x^2 + 12}{(x + 4)(x - 4)(x - 3)}$; 12.4A **10.** $b = -7$; 12.5A
11. no solution; 5.5A **12.** $y = -1$; 12.5A **13.** $b = \frac{2A}{h}$; 12.5B **14.** $\frac{15n^6}{m^3}$; 12.7A **15.** $4a - 1$; 12.7A, B
16. $x = 5$; 12.6D **17.** 2 or 5; 12.6A **18.** $3\frac{1}{13}$ hr; 12.6B **19.** 250 mph; 12.6C

Mental Math

1. $x = 0$ **3.** $x = 0, x = 1$

Section 12.1

1. $\frac{7}{4}$ **3.** $-\frac{8}{3}$ **5.** $-\frac{11}{2}$ **7. a.** \$37.5 million **b.** \$85.7 million **c.** \$48.2 million **9.** $x = 0$ **11.** $x = -2$ **13.** $x = 4$
15. $x = -2$ **17.** none **19.** Answers may vary. **21.** $\frac{1}{4(x + 2)}$ **23.** $\frac{1}{x + 2}$ **25.** can't simplify **27.** 1 **29.** -1
31. -5 **33.** $\frac{1}{x - 9}$ **35.** $5x + 1$ **37.** $\frac{1}{x - 2}$ **39.** $x + 2$ **41.** $\frac{x + 5}{x - 5}$ **43.** $\frac{x + 2}{x + 4}$ **45.** $\frac{x + 2}{2}$ **47.** $\frac{11x}{6}$
49. -1 **51.** $\frac{x + 1}{x - 1}$ **53.** $\frac{m - 3}{m + 3}$ **55.** $\frac{3}{11}$ **57.** $\frac{50}{99}$ **59.** $\frac{4}{3}$ **61.** $\frac{117}{40}$ **63.** $x + y$ **65.** $\frac{5 - y}{2}$
67. Answers may vary. **69.** 24.9% **71.** 58.3%

Mental Math

1. $\frac{2x}{3y}$ **3.** $\frac{5y^2}{7x^2}$ **5.** $\frac{9}{5}$

Section 12.2

1. $\frac{21}{4y}$ **3.** x^4 **5.** $-\frac{b^2}{6}$ **7.** $\frac{x^2}{10}$ **9.** $\frac{1}{3}$ **11.** 1 **13.** $\frac{x+5}{x}$ **15.** $\frac{2}{9(x-5)}$ sq ft **17.** x^4 **19.** $\frac{12}{y^6}$ **21.** $x(x+4)$ **23.** $\frac{3(x+1)}{x^3(x-1)}$ **25.** m^2-n^2 **27.** $-\frac{x+2}{x-3}$ **29.** $\frac{x+2}{x-3}$ **31.** $\frac{5}{6}$ **33.** $\frac{3x}{8}$ **35.** $\frac{3}{2}$ **37.** $\frac{3x+4y}{2(x+2y)}$ **39.** $\frac{2(x+2)}{x-2}$ **41.** $-\frac{x+3}{4x}$ **43.** $\frac{(a+5)(a+3)}{(a+2)(a+1)}$ **45.** 1440 **47.** 1.93 cu yd **49.** 73 **51.** 411,755 sq yd **53.** 1119 ft per sec **55.** 1 **57.** $-\frac{10}{9}$ **59.** $-\frac{1}{5}$ **61.** $\frac{x}{2}$ **63.** $\frac{5a(2a+b)(3a-2b)}{b^2(a-b)(a+2b)}$ **65.** Answers may vary. **67.** 1834.86 euros

Mental Math

1. 1 **3.** $\frac{7x}{9}$ **5.** $\frac{1}{9}$ **7.** $\frac{17y}{5}$

Section 12.3

1. $\frac{a+9}{13}$ **3.** $\frac{3m}{n}$ **5.** 4 **7.** $\frac{y+10}{3+y}$ **9.** 3 **11.** $\frac{1}{a+5}$ **13.** $\frac{1}{x-6}$ **15.** $\frac{20}{x-2}$ m **17.** Answers may vary. **19.** $4x^3$ **21.** $8x(x+2)$ **23.** $(x+3)(x-2)$ **25.** $3(x+6)$ **27.** $6(x+1)^2$ **29.** $x-8$ or $8-x$ **31.** $(x-1)(x+4)(x+3)$ **33.** Answers may vary. **35.** $\frac{6x}{4x^2}$ **37.** $\frac{24b^2}{12ab^2}$ **39.** $\frac{18}{2(x+3)}$ **41.** $\frac{9ab+2b}{5b(a+2)}$ **43.** $\frac{x^2+x}{x(x+4)(x+2)(x+1)}$ **45.** $\frac{18y-2}{30x^2-60}$ **47.** $\frac{29}{21}$ **49.** $-\frac{5}{12}$ **51.** $\frac{7}{30}$ **53.** 3 packages hot dogs and 2 packages buns **55.** Answers may vary.

Section 12.4

1. $\frac{5}{x}$ **3.** $\frac{75a+6b^2}{5b}$ **5.** $\frac{6x+5}{2x^2}$ **7.** $\frac{11}{x+1}$ **9.** $\frac{3x-7}{(x-2)(x+2)}$ **11.** $\frac{35x-6}{4x(x-2)}$ **13.** $-\frac{2}{x-3}$ **15.** $-\frac{1}{x^2-1}$ **17.** $\frac{5+2x}{x}$ **19.** $\frac{6x-7}{x-2}$ **21.** $-\frac{y+4}{y+3}$ **23.** $\frac{-5x+14}{4x}$ or $-\frac{5x-14}{4x}$ **25.** 2 **27.** $3x^3-4$ **29.** $\frac{x+2}{(x+3)^2}$ **31.** $\frac{9b-4}{5b(b-1)}$ **33.** $\frac{2+m}{m}$ **35.** $\frac{10}{1-2x}$ **37.** $\frac{15x-1}{(x+1)^2(x-1)}$ **39.** $\frac{x^2-3x-2}{(x-1)^2(x+1)}$ **41.** $\frac{a+2}{2(a+3)}$ **43.** $\frac{y(2y+1)}{(2y+3)^2}$ **45.** $\frac{x-10}{2(x-2)}$ **47.** $\frac{-3-2y}{(y-2)(y-1)}$ **49.** $\frac{-5x+23}{(x-2)(x-3)}$ **51.** $\frac{2x^2-2x-46}{(x+1)(x-6)(x-5)}$ **53.** Answers may vary. **55.** $x=\frac{2}{3}$ **57.** $x=-\frac{1}{2}, x=1$ **59.** $x=-\frac{15}{2}$ **61.** $\frac{6x^2-5x-3}{x(x+1)(x-1)}$ **63.** $\frac{4x^2-15x+6}{(x-2)^2(x+2)(x-3)}$ **65.** $\frac{-2x^2+14x+55}{(x+2)(x+7)(x+3)}$ **67.** $\frac{2x-16}{(x-4)(x+4)}$ in. **69.** $\frac{P-G}{P}$ **71.** Answers may vary.

Mental Math

1. 10 **3.** 36

Section 12.5

1. 30 **3.** 0 **5.** −2 **7.** −5, 2 **9.** 5 **11.** 3 **13.** 1 **15.** −3 **17.** no solution **19.** 1 **21.** no solution **23.** 3, −4 **25.** 6, −4 **27.** 5 **29.** 0 **31.** 8, −2 **33.** −2 **35.** no solution **37.** 3 **39.** −11, 1 **41.** $I=\frac{E}{R}$ **43.** $Q=\frac{V}{T}$ **45.** $t=\frac{A-Bi}{i}$ **47.** $G=\frac{V}{N-R}$ **49.** $r=\frac{C}{2\pi}$ **51.** $x=\frac{3y}{3+y}$ **53.** $\frac{1}{x}$ **55.** $\frac{1}{x}+\frac{1}{2}$ **57.** $\frac{1}{3}$ **59.** 100°, 80° **61.** 22.5°, 67.5° **63.** $a=5$ **65.** No; multiplying both terms in the expression by 4 changes the value of the original expression.

Integrated Review

1. expression; $\frac{3 + 2x}{3x}$ **2.** expression; $\frac{18 + 5a}{6a}$ **3.** equation; 3 **4.** equation; 18 **5.** expression; $\frac{x - 1}{x(x + 1)}$ **6.** expression; $\frac{3(x + 1)}{x(x - 3)}$ **7.** equation; no solution **8.** equation; 1 **9.** expression; 10 **10.** expression; $\frac{z}{3(9z - 5)}$ **11.** expression; $\frac{5x + 7}{x - 3}$ **12.** expression; $\frac{7p + 5}{2p + 7}$ **13.** equation; 23 **14.** equation; 3 **15.** expression; $\frac{25a}{9(a - 2)}$ **16.** expression; $\frac{9}{4(x - 1)}$ **17.** expression; $\frac{3x^2 + 5x + 3}{(3x - 1)^2}$ **18.** expression; $\frac{2x^2 - 3x - 1}{(2x - 5)^2}$ **19.** expression; $\frac{4x - 37}{5x}$ **20.** equation; $-\frac{7}{3}$ **21.** equation; $\frac{8}{5}$ **22.** expression; $\frac{29x - 23}{3x}$ **23.** Answers may vary. **24.** Answers may vary.

Mental Math

1. c

Section 12.6

1. 2 **3.** −3 **5.** 5 **7.** 2 **9.** $2\frac{2}{9}$ hr **11.** $1\frac{1}{2}$ min **13.** \$108.00 **15.** 3 hr **17.** 20 hr **19.** 6 mph **21.** 1st portion speed: 10 mph; cooldown speed: 8 mph **23.** 30 mph **25.** 8 mph **27.** 63 mph **29.** $x = 6$ **31.** $x = 5$ **33.** $y = 21.25$ **35.** $y = 18$ ft **37.** $26\frac{2}{3}$ ft **39.** $\frac{1}{2}$ **41.** $\frac{3}{7}$ **43.** Castroneves: 219.5 mph; Junqueira: 215.9 mph **45.** Answers may vary.

Section 12.7

1. $\frac{2}{3}$ **3.** $\frac{2}{3}$ **5.** $-\frac{4x}{15}$ **7.** $\frac{4}{3}$ **9.** $\frac{27}{16}$ **11.** $\frac{m - n}{m + n}$ **13.** $\frac{2x(x - 5)}{7x^2 + 10}$ **15.** $\frac{1}{y - 1}$ **17.** $\frac{1}{6}$ **19.** $\frac{x + y}{x - y}$ **21.** $\frac{3}{7}$ **23.** $\frac{a}{x + b}$ **25.** $\frac{x + 8}{2 - x}$ or $-\frac{x + 8}{x - 2}$ **27.** $\frac{s^2 + r^2}{s^2 - r^2}$ **29.** Answers may vary. **31.** Steffi Graf **33.** Seles, Hinges, Sanchez-Vicario **35.** $\frac{13}{24}$ **37.** $\frac{R_1R_2}{R_2 + R_1}$ **39.** $\frac{2x}{2 - x}$ **41.** $\frac{1}{y^2 - 1}$

Chapter 12 Review

1. $x = 2, x = -2$ **2.** $x = \frac{5}{2}, x = -\frac{3}{2}$ **3.** $\frac{4}{3}$ **4.** $\frac{11}{12}$ **5.** $\frac{2}{x}$ **6.** $\frac{3}{x}$ **7.** $\frac{1}{x - 5}$ **8.** $\frac{1}{x + 1}$ **9.** $\frac{x(x - 2)}{x + 1}$ **10.** $\frac{5(x - 5)}{x - 3}$ **11.** $\frac{x - 3}{x - 5}$ **12.** $\frac{x}{x + 4}$ **13.** $\frac{x + a}{x - c}$ **14.** $\frac{x + 5}{x - 3}$ **15.** $\frac{3x^2}{y}$ **16.** $-\frac{9x^2}{8}$ **17.** $\frac{x - 3}{x + 2}$ **18.** $-\frac{2x(2x + 5)}{(x - 6)^2}$ **19.** $\frac{x + 3}{x - 4}$ **20.** $\frac{4x}{3y}$ **21.** $(x - 6)(x - 3)$ **22.** $\frac{2}{3}$ **23.** $\frac{1}{x - 3}$ **24.** $\frac{x}{x + 6}$ **25.** $\frac{1}{2}$ **26.** $\frac{3(x + 2)}{3x + y}$ **27.** $\frac{1}{x + 2}$ **28.** $\frac{1}{x - 3}$ **29.** $\frac{2x - 10}{3x^2}$ **30.** $\frac{2x + 1}{2x^2}$ **31.** $14x$ **32.** $(x - 8)(x + 8)(x + 3)$ **33.** $\frac{10x^2y}{14x^3y}$ **34.** $\frac{36y^2x}{16y^3x}$ **35.** $\frac{x^2 - 3x - 10}{(x + 2)(x - 5)(x + 9)}$ **36.** $\frac{3x^2 + 4x - 15}{(x + 2)^2(x + 3)}$ **37.** $\frac{4y - 30x^2}{5x^2y}$ **38.** $\frac{-2x + 10}{(x - 3)(x - 1)}$ **39.** $\frac{14x + 58}{(x + 3)(x + 7)}$ **40.** $\frac{-2x - 2}{x + 3}$ **41.** $\frac{5x + 5}{(x + 4)(x - 2)(x - 1)}$ **42.** $\frac{x - 4}{3x}$ **43.** $-\frac{x}{x - 1}$ **44.** $\frac{x^2 + 2x - 3}{(x + 2)^2}$ **45.** $\frac{x^2 + 2x + 4}{4x}; \frac{x + 2}{32}$ **46.** $\frac{29x}{12(x - 1)}; \frac{3xy}{5(x - 1)}$ **47.** 1 **48.** 30 **49.** 2 **50.** 3, −4 **51.** $-\frac{5}{2}$ **52.** no solution **53.** 1 **54.** 5 **55.** $\frac{9}{7}$ **56.** −6, 1 **57.** $b = \frac{4A}{5x^2}$ **58.** $y = \frac{560 - 8x}{7}$ **59.** 3 **60.** 2 **61.** fast car speed: 30 mph; slow car speed: 20 mph **62.** 20 mph **63.** $17\frac{1}{2}$ hr **64.** $8\frac{4}{7}$ days **65.** $x = 15$ **66.** $x = 6$ **67.** $x = 15$ **68.** $x = 60$ **69.** $-\frac{7}{18y}$ **70.** $\frac{2x}{x - 3}$ **71.** $\frac{6}{7}$ **72.** $\frac{2x^2 + 1}{x + 2}$ **73.** $\frac{3y - 1}{2y - 1}$ **74.** $-\frac{7 + 2x}{2x}$

CHAPTER 12 TEST

1. $x = -1, x = -3$ **2. a.** \$115 **b.** \$103 **3.** $\frac{3}{5}$ **4.** $\frac{1}{x-10}$ **5.** $\frac{1}{x+6}$ **6.** -1 **7.** $\frac{2m(m+2)}{m-2}$ **8.** $-\frac{1}{x+y}$ **9.** 15 **10.** $\frac{y-2}{4}$ **11.** $\frac{1-x}{2x+5}$ **12.** $\frac{3a-4}{(a-3)(a+2)}$ **13.** $\frac{3}{x-1}$ **14.** $\frac{2(x+5)}{x(y+5)}$ **15.** $\frac{x^2+2x+35}{(x+9)(x+2)(x-5)}$ **16.** $\frac{30}{11}$ **17.** -6 **18.** no solution **19.** $5, -2$ **20.** $\frac{xz}{2y}$ **21.** $\frac{5y^2-1}{y+2}$ **22.** 12 **23.** 1, 5 **24.** 30 mph **25.** $6\frac{2}{3}$ hr

CUMULATIVE REVIEW

1. 45%; Sec. 6.2, Ex. 16 **2.** 80% **3.** $66\frac{2}{3}\%$; Sec. 6.2, Ex. 17 **4.** $11\frac{1}{9}\%$ **5.** 150%; Sec. 6.2, Ex. 18 **6.** 375% **7.** commutative property of multiplication; Sec. 9.2, Ex. 15 **8.** commutative property of addition **9.** associative property of addition; Sec. 9.2, Ex. 16 **10.** associative property of multiplication **11.** shorter piece, 2 ft; longer piece, 8 ft; Sec. 9.4, Ex. 2 **12.** 16 ft, 16 ft, 13 ft **13.** $\frac{y-b}{m} = x$; Sec. 9.5, Ex. 4 **14.** $b = y - mx$ **15.** $x \leq -10$; 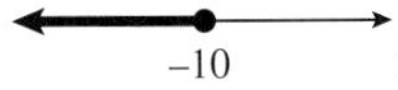 ; Sec. 9.6, Ex. 3 **16.** $\{x|x > -9\}$ **17.** x^3; Sec. 10.1, Ex. 21 **18.** x^8 **19.** $4^4 = 256$; Sec. 10.1, Ex. 22 **20.** 7^8 **21.** -27; Sec. 10.1, Ex. 23 **22.** 16 **23.** $2x^4y$; Sec. 10.1, Ex. 24 **24.** $13ab$ **25.** $\frac{2}{x^3}$; Sec. 10.2, Ex. 2 **26.** $\frac{9}{x^2}$ **27.** $\frac{1}{16}$; Sec. 10.2, Ex. 4 **28.** $-\frac{1}{27}$ **29.** $10x^4 + 30x$; Sec. 10.5, Ex. 5 **30.** $12y^3 - 6y$ **31.** $-15x^4 - 18x^3 + 3x^2$; Sec. 10.5, Ex. 6 **32.** $-35y^3 + 15y^2 - 5y$ **33.** $4x^2 - 4x + 6 + \frac{-11}{2x+3}$; Sec. 10.7, Ex. 6 **34.** $3x - 5 + \frac{9}{2x+1}$ **35.** $(x+3)(x+4)$; Sec. 11.2, Ex. 1 **36.** $(x+7)(x+10)$ **37.** $(5x+2y)^2$; Sec. 11.5, Ex. 5 **38.** $(6a-4b)^2$ **39.** $x = 11, x = -2$; Sec. 11.6, Ex. 4 **40.** $x = 3, x = -5$ **41.** $\frac{2}{5}$; Sec. 12.2, Ex. 2 **42.** $\frac{6x}{5}$ **43.** $3x - 5$; Sec. 12.3, Ex. 3 **44.** 4 **45.** $\frac{3}{x-2}$; Sec. 12.4, Ex. 2 **46.** $\frac{7x+6}{x^2-9}$ **47.** $t = 5$; Sec. 12.5, Ex. 2 **48.** $y = 2$ **49.** $2\frac{1}{10}$ hr; Sec. 12.6, Ex. 2 **50.** $7\frac{1}{5}$ hr **51.** $\frac{3}{z}$; Sec. 12.7, Ex. 3 **52.** $\frac{x-3}{9}$

Chapter 13 GRAPHING EQUATIONS AND INEQUALITIES

CHAPTER 13 PRETEST

1. 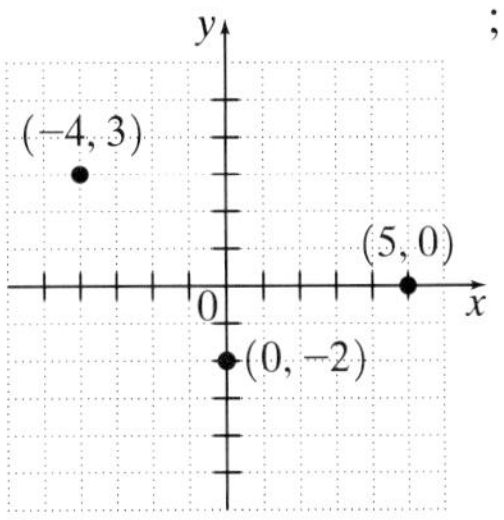

; 13.1A **2.** $(-2, -6)$; 13.1C **3.** 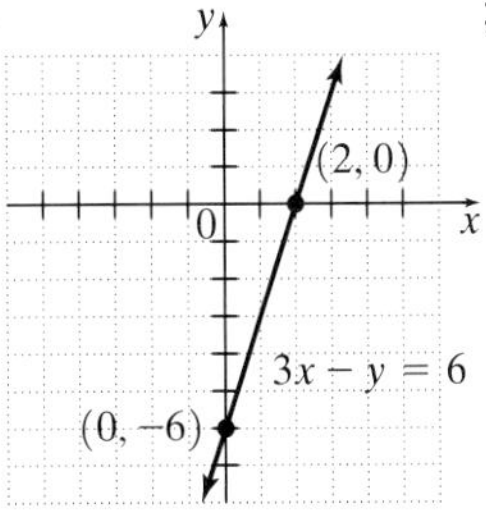

; 13.2A **4.** 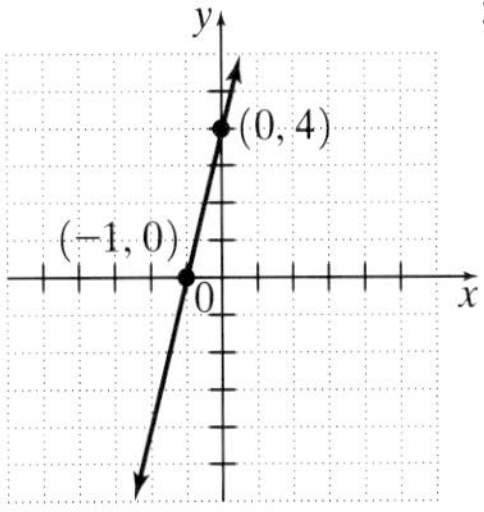

; 13.3B **5.** 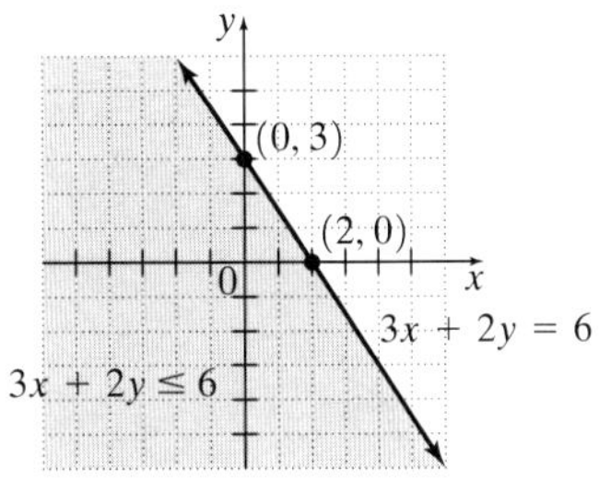

; 13.7B **6.** $-\frac{3}{10}$; 13.4A **7.** $\frac{4}{5}$; 13.4B **8.** undefined slope; 13.4C **9.** $x + 3y = -15$; 13.5C **10.** $x + 8y = 0$; 13.5D **11.** $2x - 7y = -98$; 13.5A **12.** domain: $\{-3, 0, 2, 7\}$; range: $\{-1, 6, 8\}$; 13.6A **13.** function; 13.6B **14.** not a function; 13.6B **15. a.** 11 **b.** 8 **c.** -22; 13.6D **16.** $y = \frac{40}{3}$; 13.8A

MENTAL MATH

1. Answers may vary; Examples (5, 5), (7, 3)

SECTION 13.1

1. ; (1, 5) is in quadrant I, $\left(-1, 4\frac{1}{2}\right)$ is in quadrant II, (−5, −2) is in quadrant III, (2, −4) is in quadrant IV, (−3, 0) lies on the x-axis, (0, −1) lies on the y-axis **3.** $a = b$ **5.** (0, 0) **7.** (3, 2) **9.** (−2, −2) **11.** (2, −1) **13.** (0, −3) **15.** (1, 3) **17.** (−3, −1)

19. a. (1995, 438), (1996, 438), (1997, 436), (1998, 435), (1999, 432), (2000, 434), (2001, 436)

b.

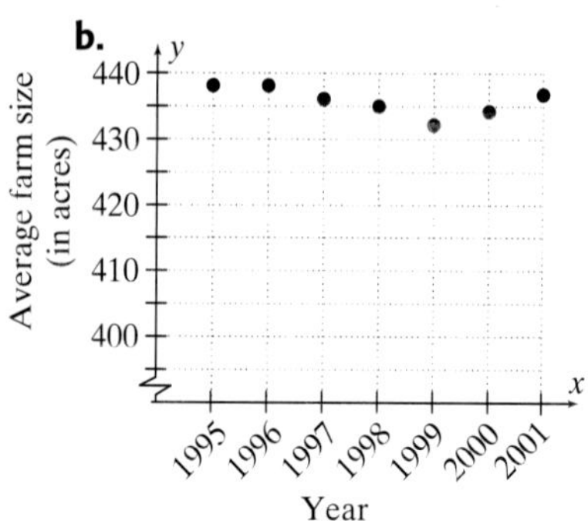

21. a. (2313, 2), (2085, 1), (2711, 21), (2869, 39), (2920, 42), (4038, 99), (1783, 0), (2493, 9)

b.

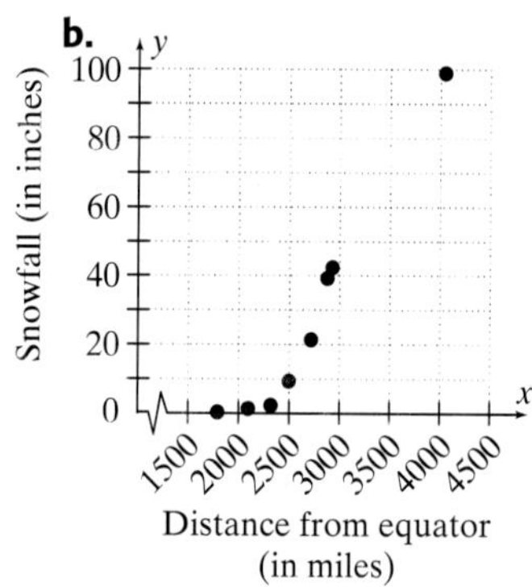

c. The farther from the equator, the more snowfall.

23. a. (0.50, 10), (0.75, 12), (1.00, 15), (1.25, 16), (1.50, 18), (1.50, 19), (1.75, 19), (2.00, 20)

b.

c. Answers may vary. **25.** (−4, −2); (4, 0) **27.** (0, 9); (3, 0) **29.** (11, −7); any x

31.

x	y
0	2
6	0
3	1

33.

x	y
0	−12
5	−2
3	−6

35.

x	y
0	$\frac{5}{7}$
$\frac{5}{2}$	0
−1	1

37.

x	y
3	0
3	−0.5
3	$\frac{1}{4}$

39.

x	y
0	0
−5	1
10	−2

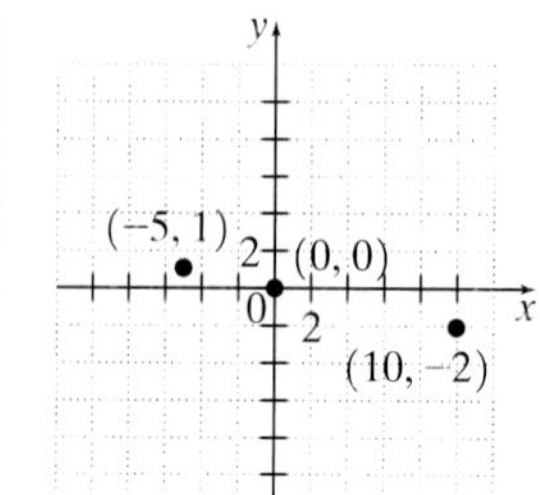

41. a. 13,000; 21,000; 29,000 **b.** 45 desks **43.** $y = 5 - x$ **45.** $y = \dfrac{5 - 2x}{4}$ **47.** $y = -2x$ **49.** 26 units **51.** \$21 million; \$23 million; \$24 million; \$25 million **53.** Answers may vary. **55. a.** 968.14; 955.48; 942.82 **b.** 1999

Graphing Calculator Explorations

1.

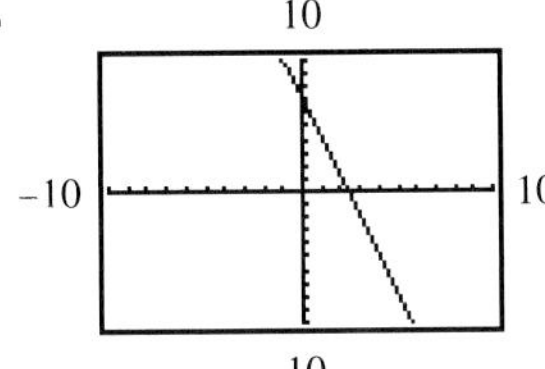

3.

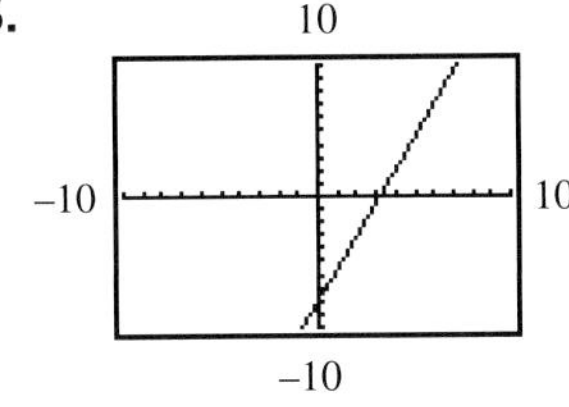

5.

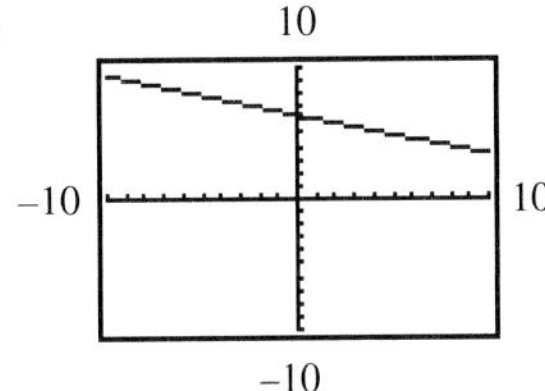

Section 13.2

1. $(6, 0)$; $(4, -2)$; $(5, -1)$

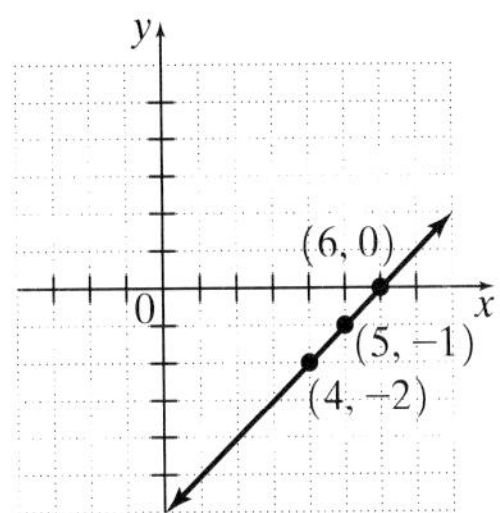

3. $(1, -4)$; $(0, 0)$; $(-1, 4)$

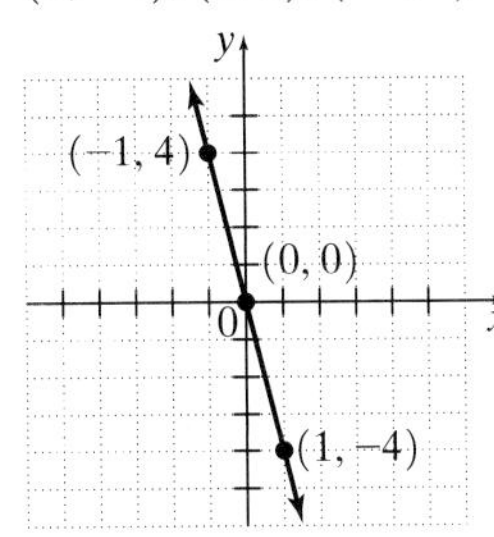

5. $(0, 0)$; $(6, 2)$; $(-3, -1)$

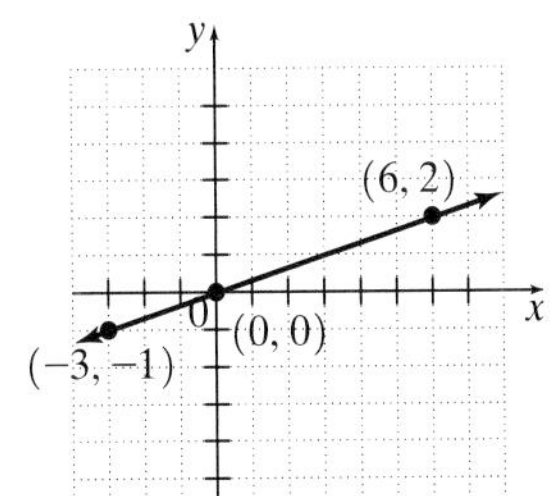

7. $(0, 3)$; $(1, -1)$; $(2, -5)$

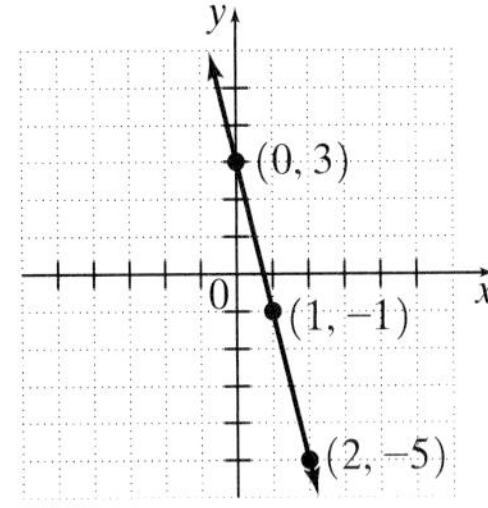

9.

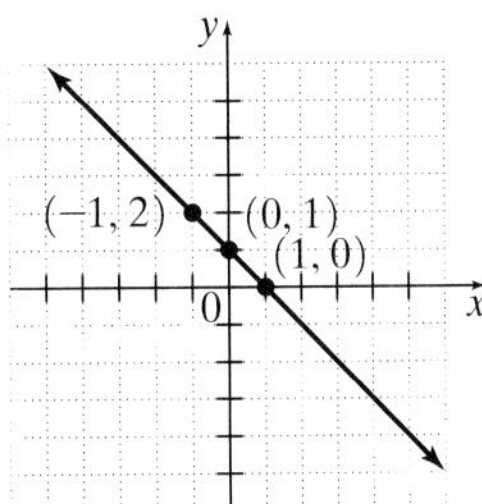

11.

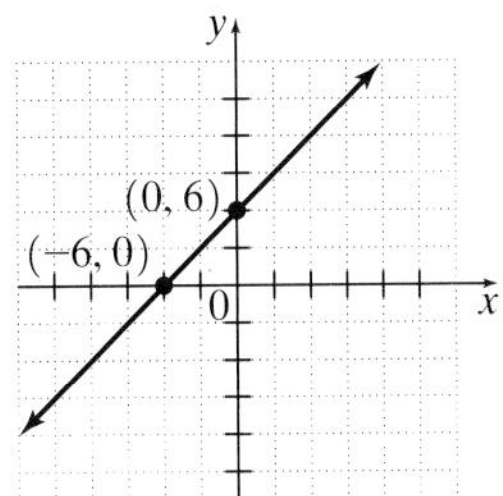

13.

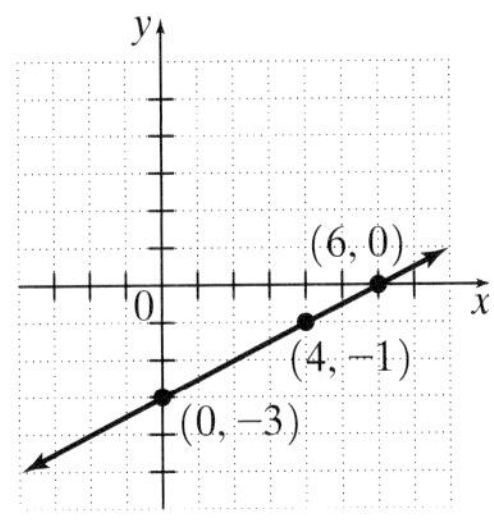

15.

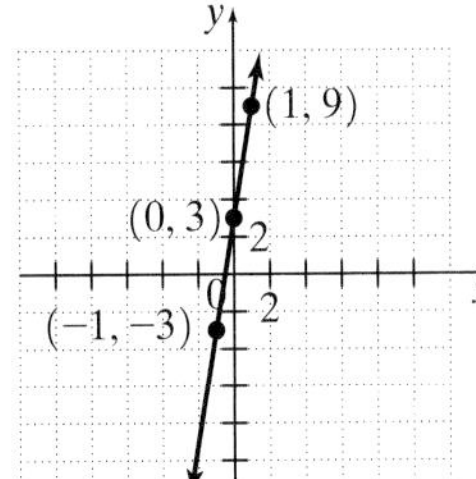

17.

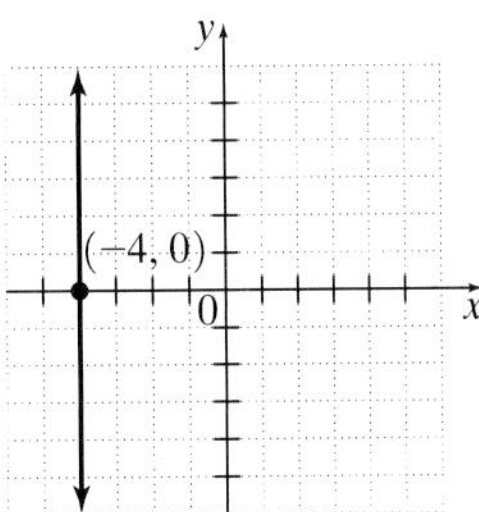

19.

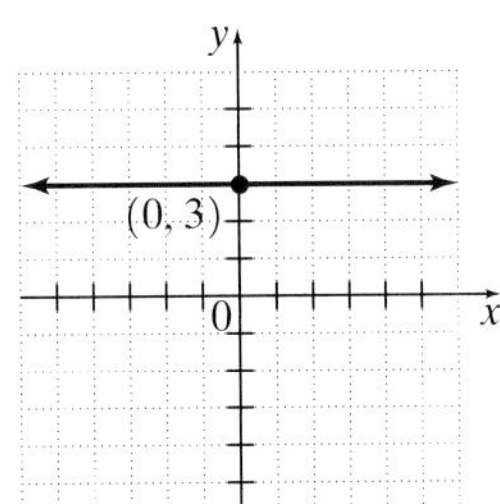

21.

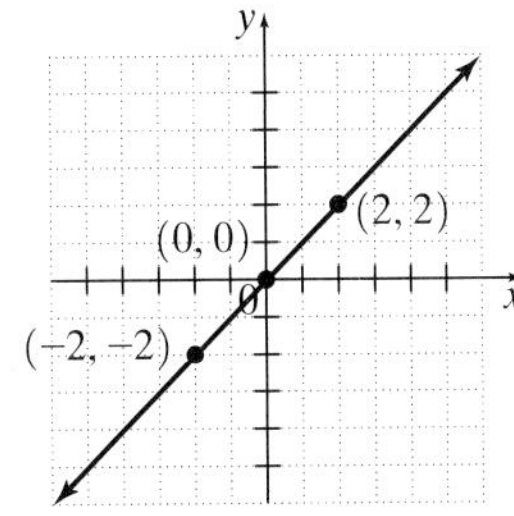

23.

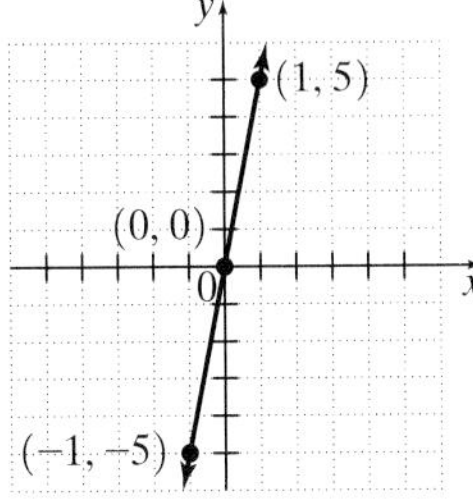

25.

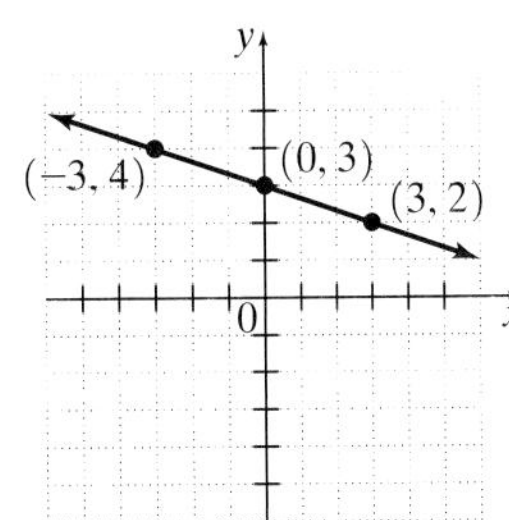

27.

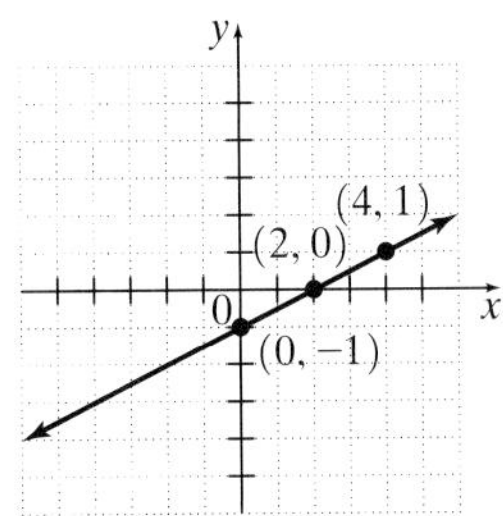

29.

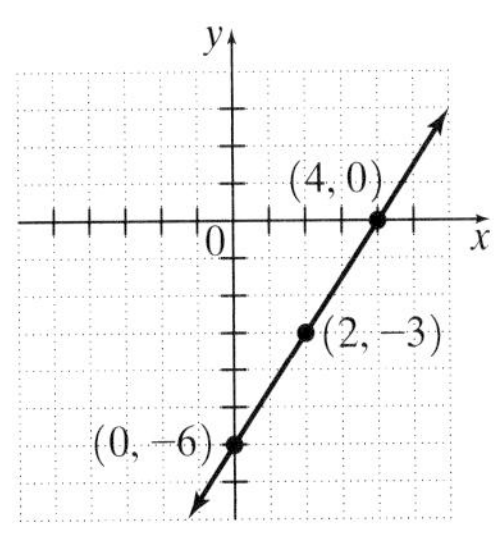

31.

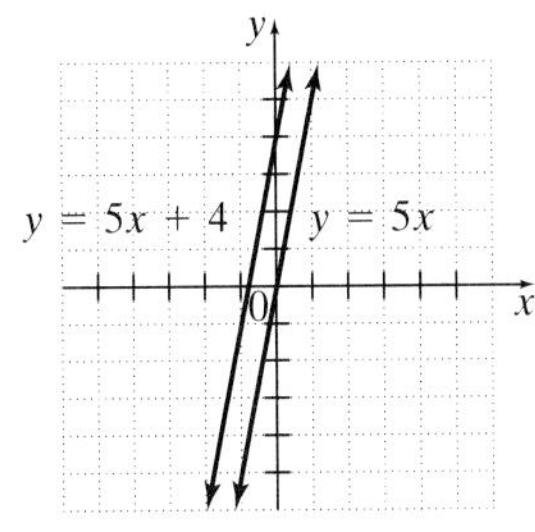

Answers may vary.

33.

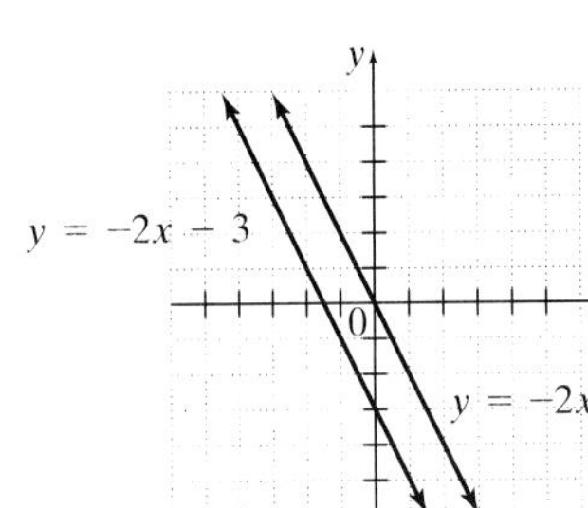

Answers may vary.

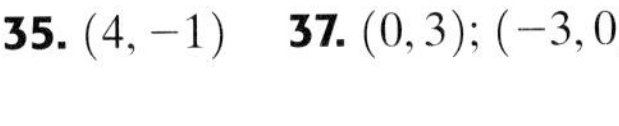
35. $(4, -1)$ **37.** $(0, 3)$; $(-3, 0)$ **39.** $(0, 0)$; $(0, 0)$

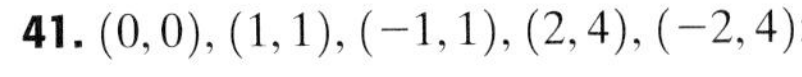
41. $(0, 0)$, $(1, 1)$, $(-1, 1)$, $(2, 4)$, $(-2, 4)$;

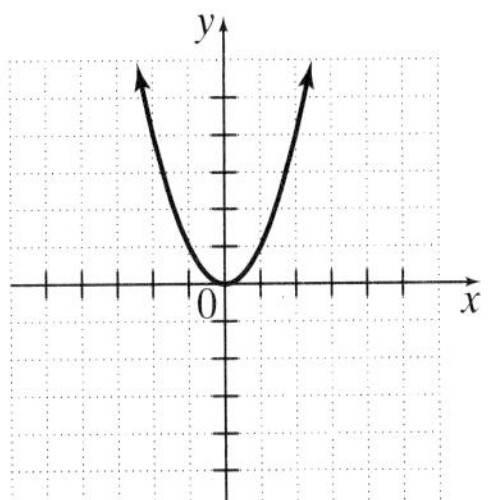

43. $x + y = 12$; 9 cm **45. a.**

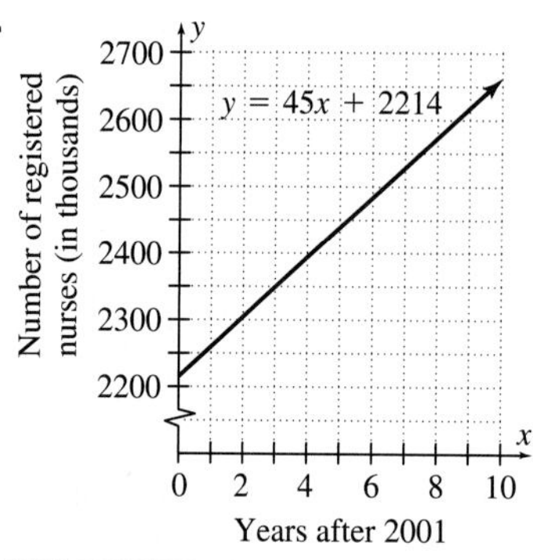

b. yes; answers may vary.

47. a.

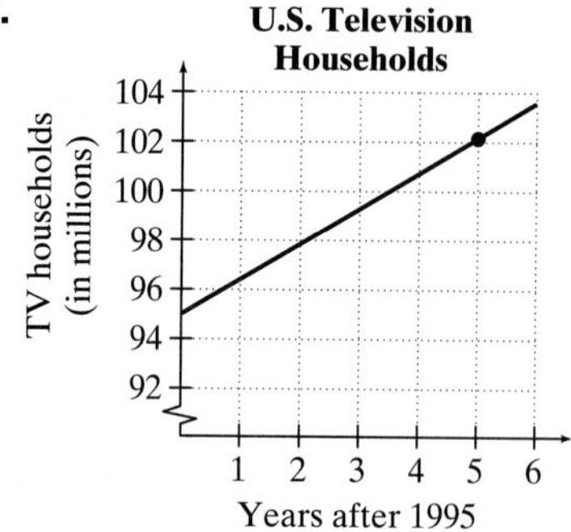

b. (5, 102.15)

c. In 2000, there were 102.15 million households in the United States with at least one television.

Graphing Calculator Explorations

1.

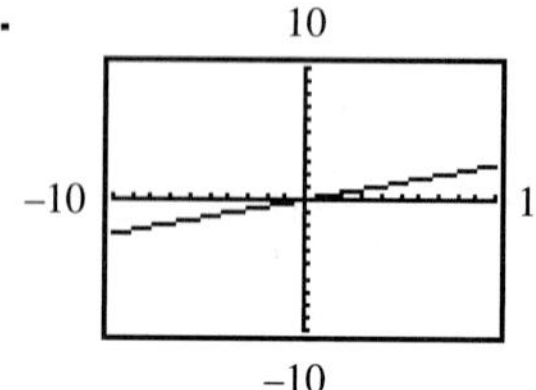

3.

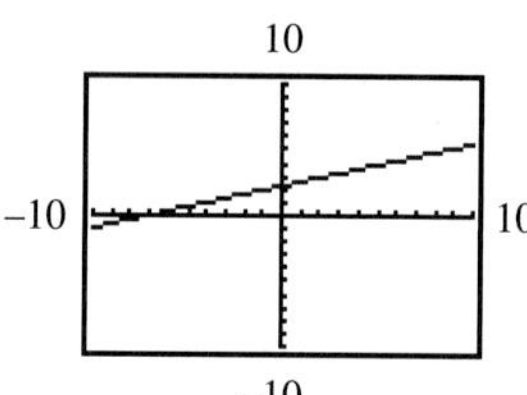

Mental Math

1. false **3.** true

Section 13.3

1. $(-1, 0); (0, 1)$ **3.** $(-2, 0); (1, 0); (3, 0); (0, 3)$ **5.** infinite **7.** 0

9.

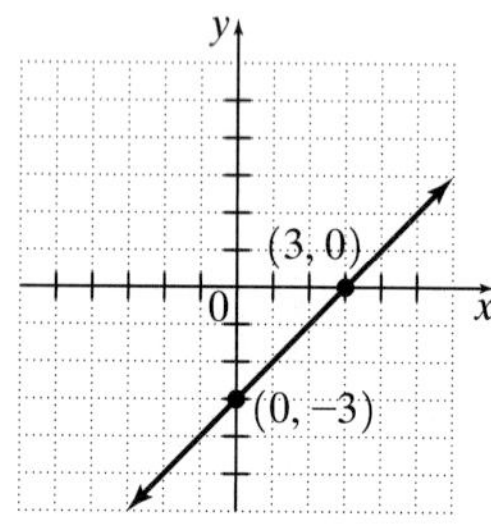

11.

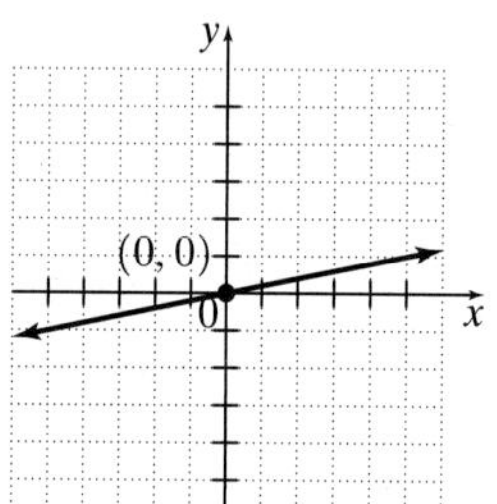

13.

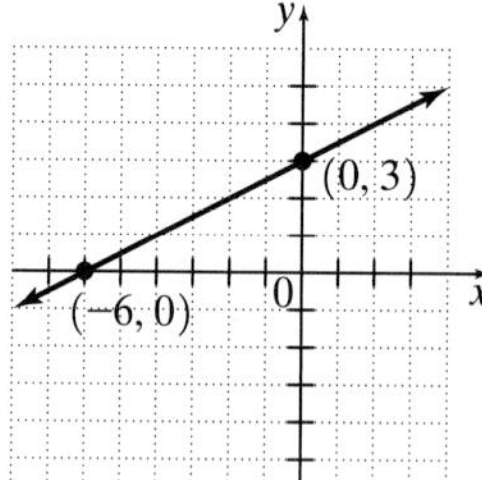

15.

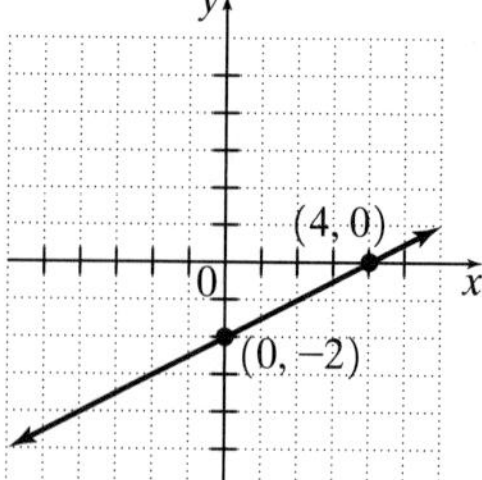

17.

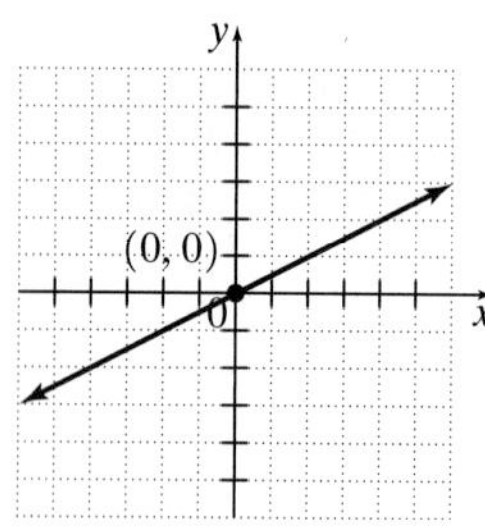

19.

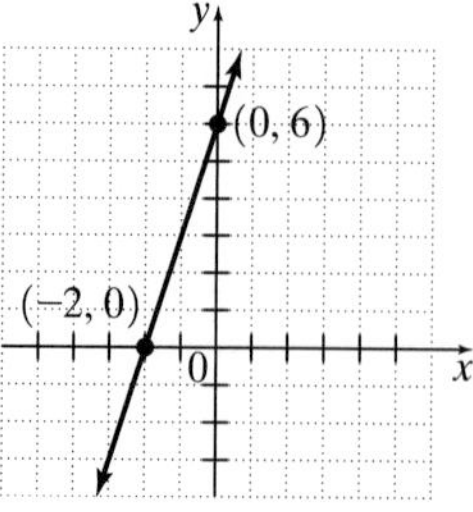

21.

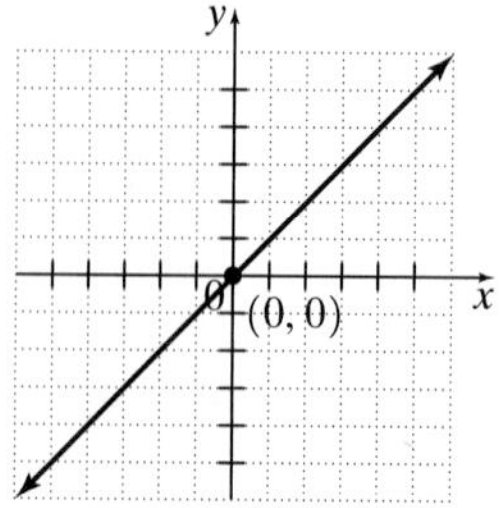

23.

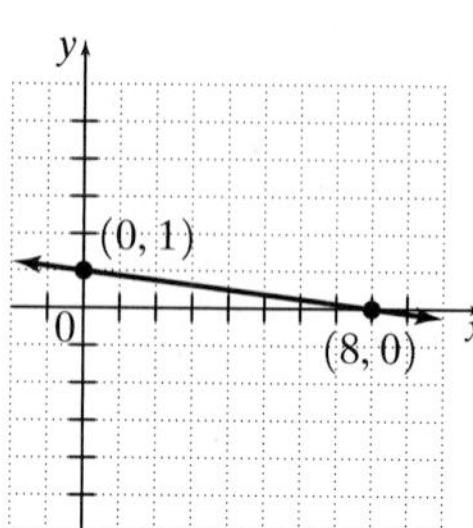

25.

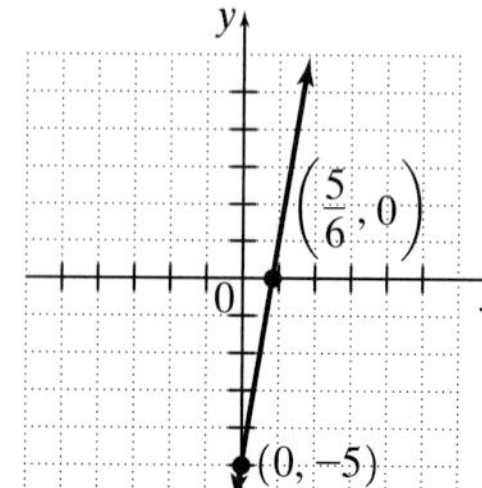

27.

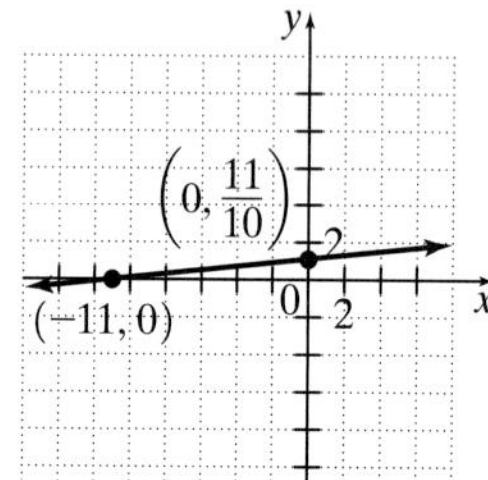

29.

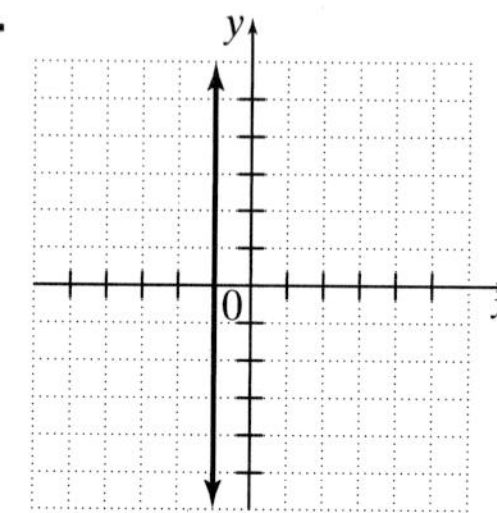

31.

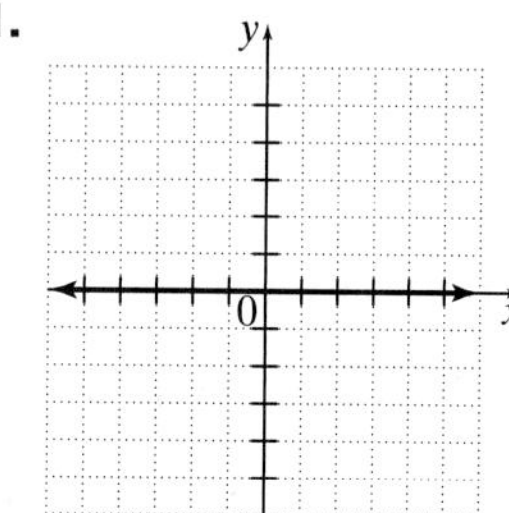

33. 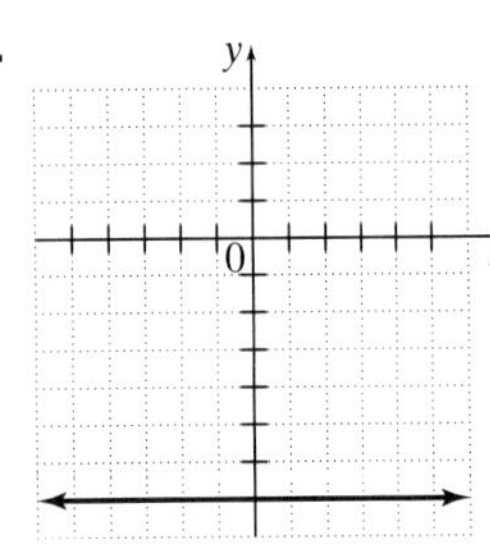

35. 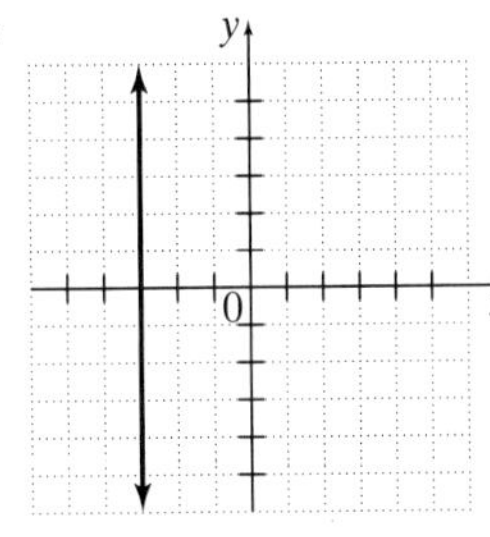

37. $\frac{3}{2}$ **39.** 6 **41.** $-\frac{6}{5}$ **43.** C **45.** A

47. Answers may vary. **49. a.** (0, 200); no chairs and 200 desks are manufactured **b.** (400, 0); 400 chairs and no desks are manufactured **c.** 300 chairs

51. 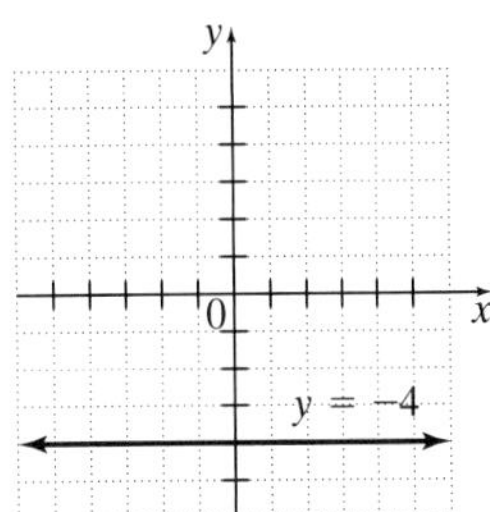

; $y = -4$ **53. a.** (0, 560.2) **b.** In 1996, the number of Disney Stores was about 560.2.

Calculator Explorations

1. 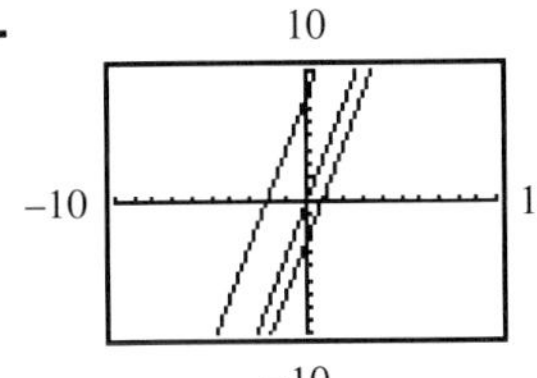

3. 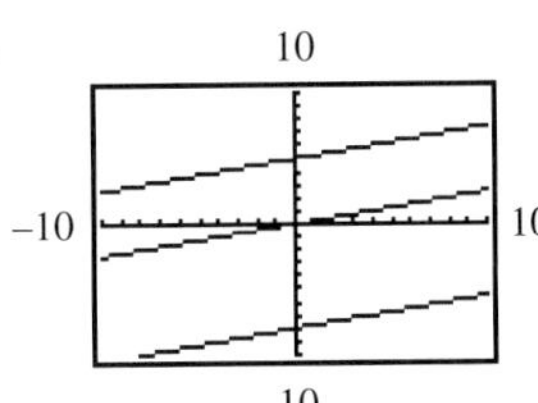

Mental Math

1. upward **3.** horizontal

Section 13.4

1. $-\frac{4}{3}$ **3.** $\frac{5}{2}$ **5.** $\frac{8}{7}$ **7.** −1 **9.** $-\frac{1}{4}$ **11.** $-\frac{2}{3}$ **13.** 0 **15.** line 1 **17.** line 2 **19.** 5 **21.** −2 **23.** $\frac{2}{3}$ **25.** $\frac{1}{2}$ **27.** undefined slope **29.** undefined slope **31.** 0 **33.** neither **35.** perpendicular **37.** parallel **39. a.** 1 **b.** −1 **41. a.** $\frac{9}{11}$ **b.** $-\frac{11}{9}$ **43.** $\frac{3}{5}$ **45.** 12.5% **47.** 40% **49.** 0.02 **51.** Every 1 year, there are/should be 15 million more Internet users. **53.** It costs $0.36 per 1 mile to own and operate a compact car. **55.** $y = 2x - 14$ **57.** $y = -6x - 11$ **59.** D **61.** B **63.** E **65.** 28.3 **67.** 1992; 27.6 **69.** from 1992 to 1993 **71.** $x = 6$ **73. a.** (1994, 782)(2001, 1132) **b.** 50 **c.** For the years 1994 through 2001, the price per acre of U.S. farmland rose $50 every year. **75. a.** (0, 1485) **b.** In 1998 there were 1485 million admissions to movie theaters in the U.S. and Canada. **c.** −30 **d.** For the years 1998 through 2000, the number of movie theater admissions has decreased at a rate of 30 million per year. **77.** The slope through (−3, 0) and (1, 1) is $\frac{1}{4}$. The slope through (−3, 0) and (−4, 4) is −4. The product of the slopes is −1 so the sides are perpendicular. **79.** −0.25 **81.** 0.875 **83.** The line becomes steeper.

Graphing Calculator Explorations

1. 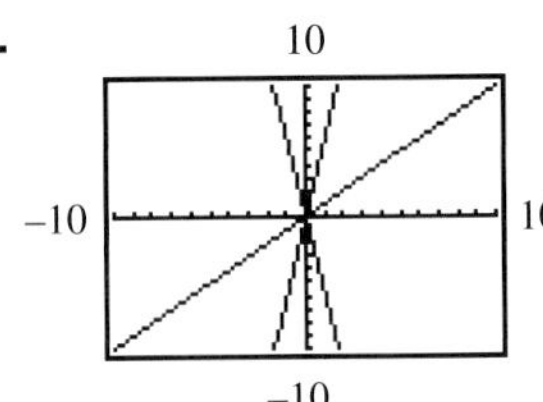

3. 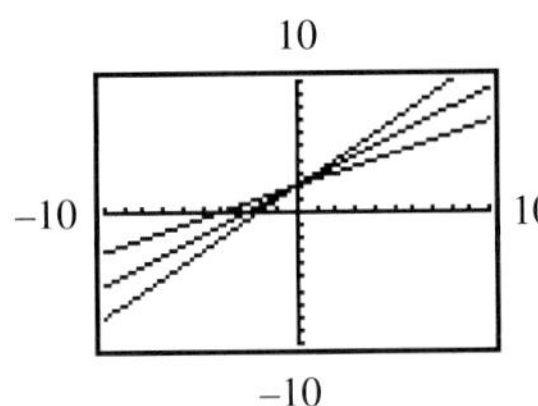

Mental Math

1. $m = 2$; $(0, -1)$ **3.** $m = 1$; $\left(0, \frac{1}{3}\right)$ **5.** $m = \frac{5}{7}$; $(0, -4)$ **7.** $m = 3$; answers may vary, Example $(4, 8)$

9. $m = -2$; answers may vary, Example $(10, -3)$ **11.** $m = \frac{2}{5}$; answers may vary, Example $(-1, 0)$

Section 13.5

1. $y = 5x + 3$ **3.** $y = \frac{2}{3}x$ **5.** $y = -\frac{1}{5}x + \frac{1}{9}$ **7.**

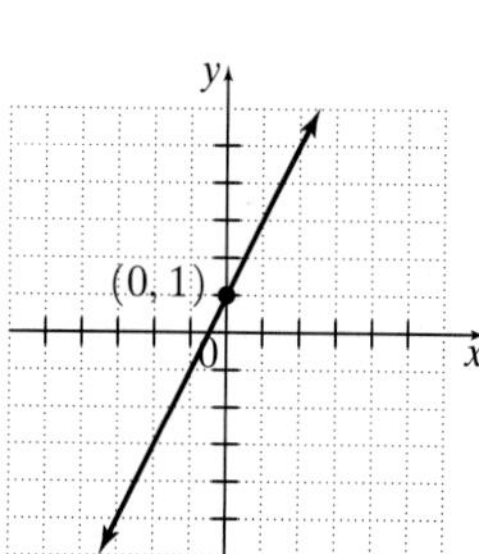

9.

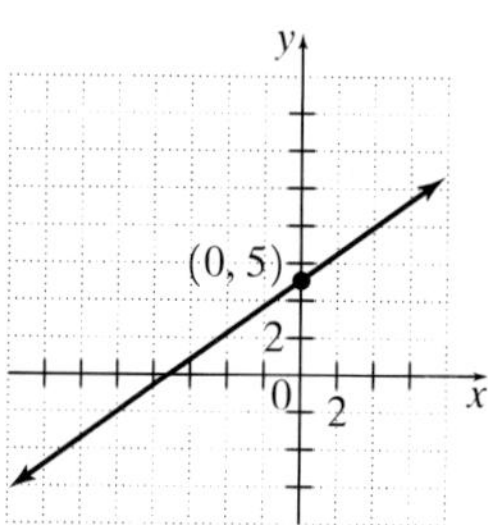

11.

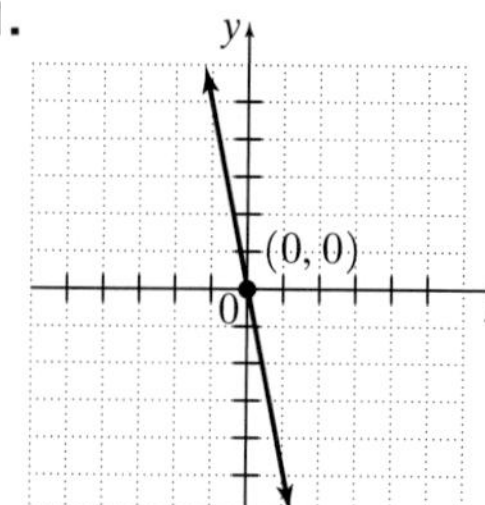

13.

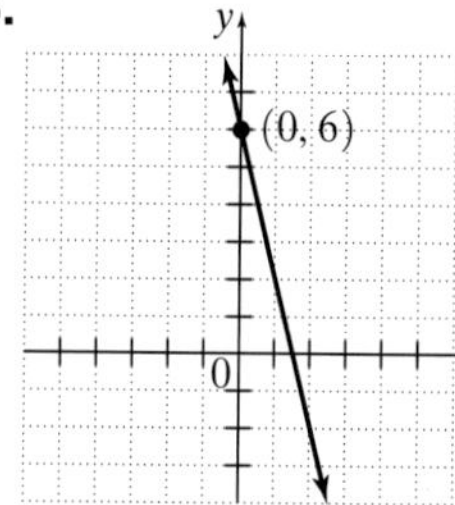

15.

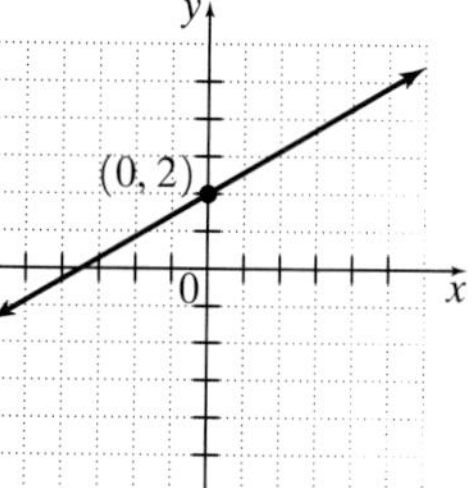

17. $-6x + y = -10$ **19.** $8x + y = -13$ **21.** $x - 2y = 17$ **23.** $x + 2y = -3$ **25.** $2x - y = 4$ **27.** $8x - y = -11$ **29.** $4x - 3y = -1$ **31.** $x + y = 17$ **33.** $x + y = 17$ **35.** $x + 8y = 0$ **37. a.** $s = 32t$ **b.** 128 ft/sec **39. a.** $y = 1174x + 5242$ **b.** 18,156 vehicles **41. a.** $y = 0.93x + 70.3$ **b.** 86.11 persons per sq mi **43. a.** $(0, 191)$, $(5, 260)$ **b.** $y = 13.8x + 191$ **c.** \$246.2 million **45. a.** $S = -1000p + 13{,}000$ **b.** 9500 Fun Noodles **47.** -1 **49.** 5 **51.** no **53.** yes **55.** B **57.** D **59.** $3x - y = -5$ **61. a.** $3x - y = -5$ **b.** $x + 3y = 5$

Integrated Review

1. 2 **2.** 0 **3.** $-\frac{2}{3}$ **4.** undefined **5.**

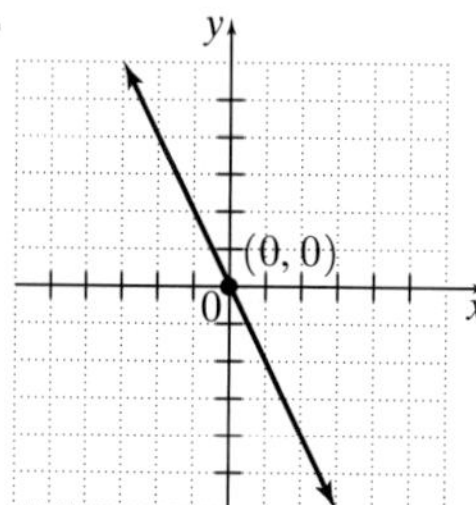

6.

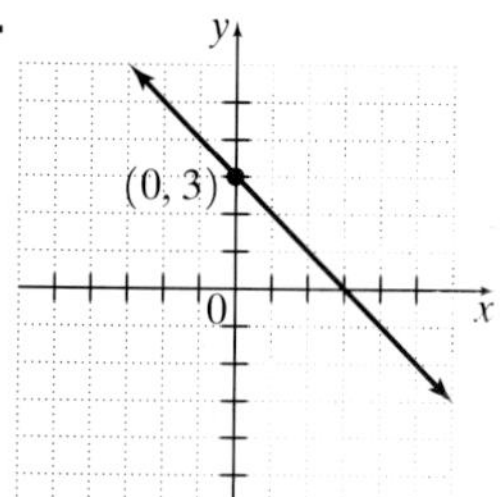

7.

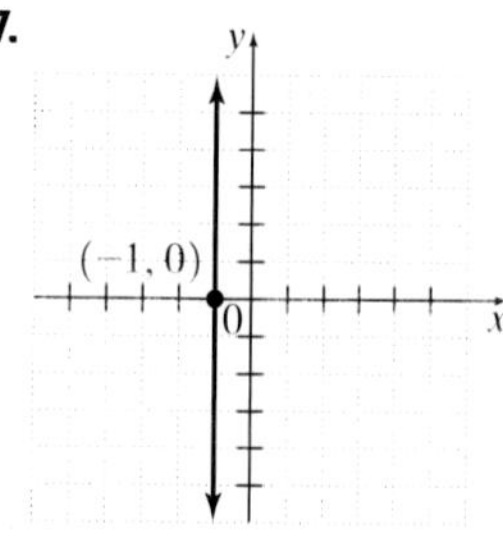

8.

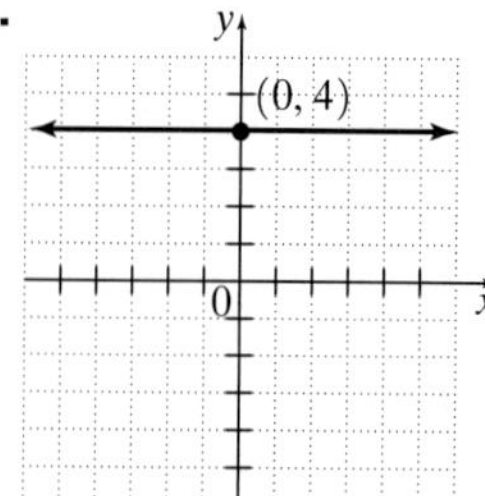

9.

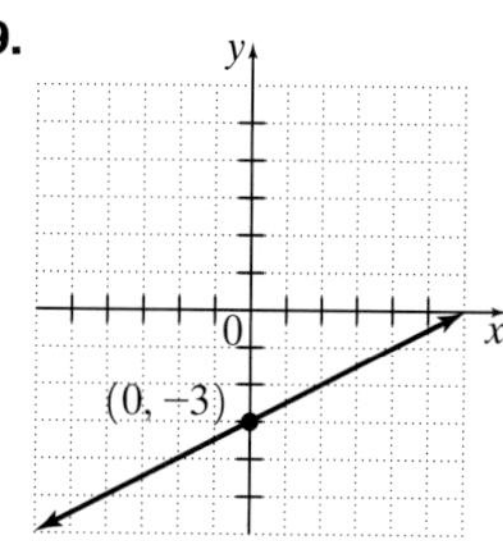

10.

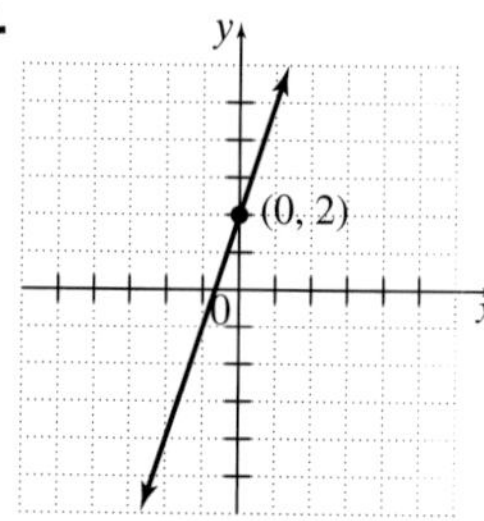

11. 3 **12.** -6 **13.** $-\frac{7}{2}$

14. 2 **15.** undefined **16.** 0 **17.** $y = 2x - \frac{1}{3}$ **18.** $4x + y = -1$ **19.** $-x + y = -2$ **20.** neither **21.** perpendicular **22. a.** $(1997, 11.6)$ $(2001, 15.4)$ **b.** 0.95 **c.** For the years 1997 through 2001, the number of grill units shipped increased at a rate of 0.95 million per year.

SECTION 13.6

1. domain: $\{-7, 0, 2, 10\}$; range: $\{-7, 0, 4, 10\}$ **3.** domain: $\{0, 1, 5\}$; range: $\{-2\}$ **5.** yes **7.** no **9.** no **11.** yes **13.** yes **15.** no **17.** 5:30 A.M. **19.** Answers may vary. **21.** 9:30 P.M. **23.** January 1 and December 1 **25.** yes; it passes the vertical line test **27.** \$4.75 per hour **29.** 1992 **31.** yes; answers may vary. **33.** -9; -5; 1 **35.** 6; 2; 11 **37.** -6; 0; 9 **39.** 2; 0; 3 **41.** 5; 0; -20 **43.** 5; 3; 35 **45.** $x < 1$ **47.** $x \geq -3$ **49.** $\frac{19}{2x}$ m **51. a.** 166.38 cm **b.** 148.25 cm **53. a.** Answers may vary. **b.** Answers may vary. **c.** Answers may vary. **55.** $f(x) = x + 7$

MENTAL MATH

1. yes **3.** yes **5.** yes **7.** no

SECTION 13.7

1. no; no **3.** yes; no **5.** no; yes

7.

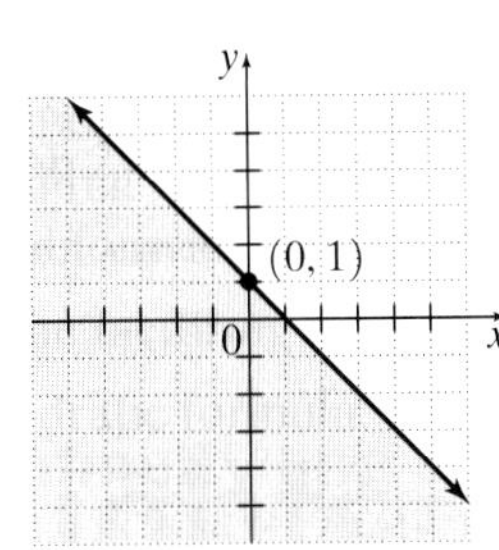

9.

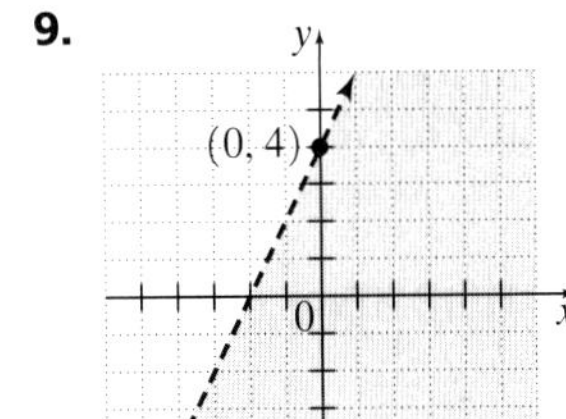

11.

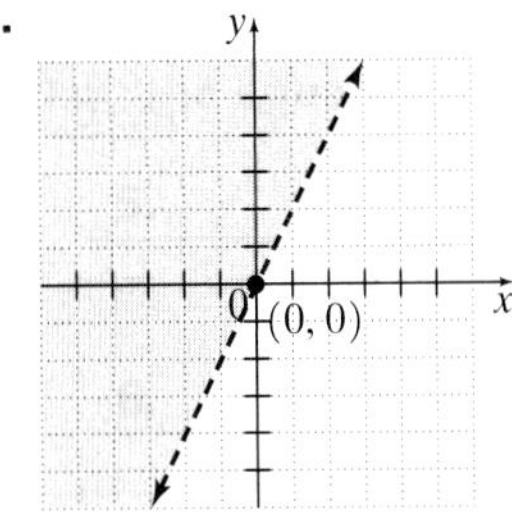

13.

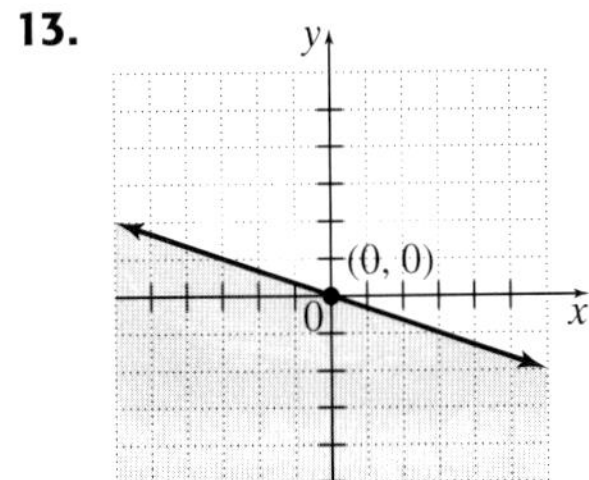

15.

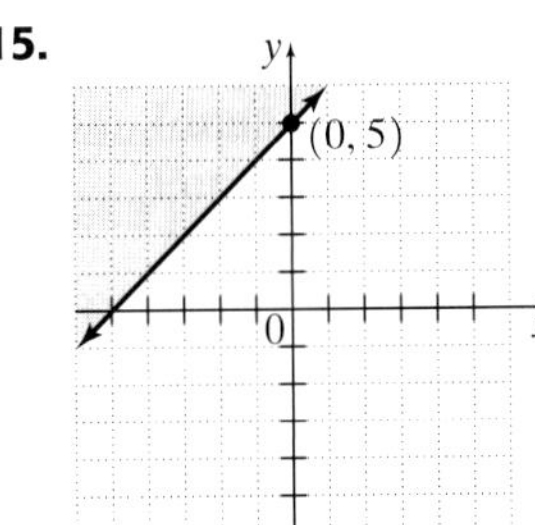

17.

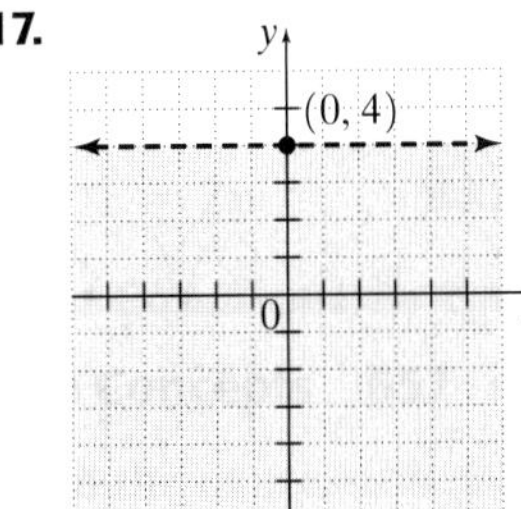

19.

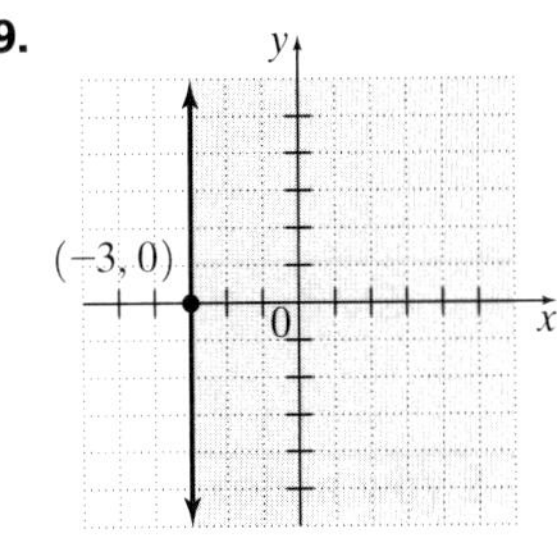

21.

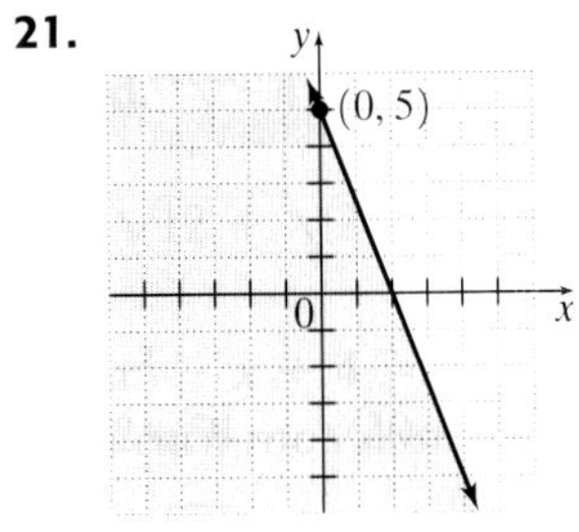

23.

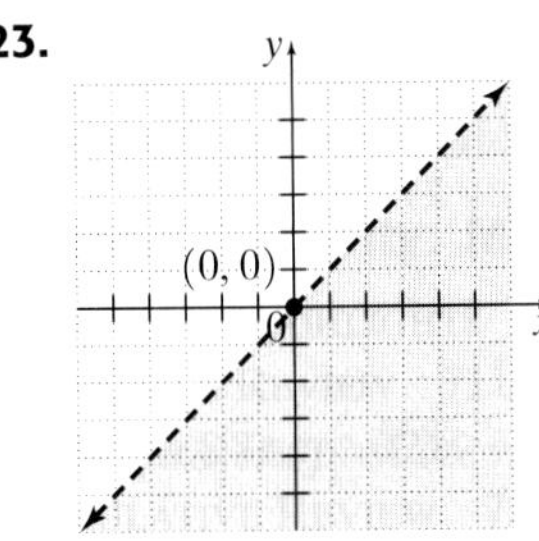

25.

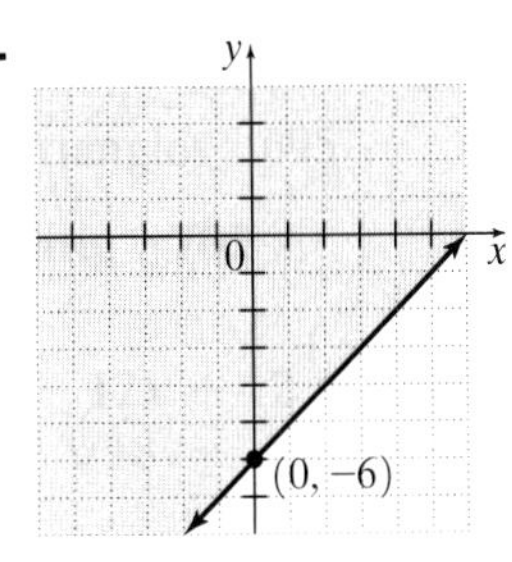

27.

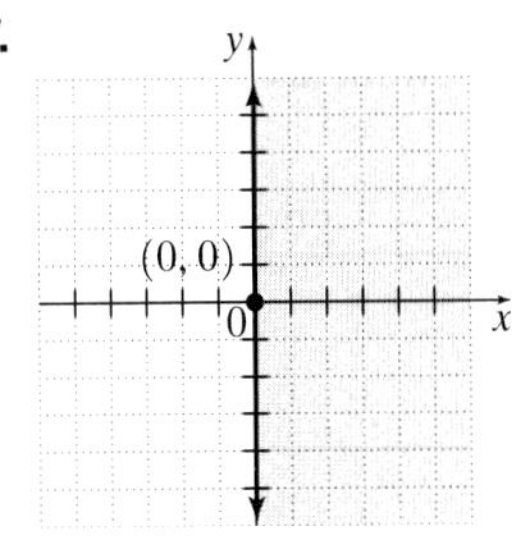

29.

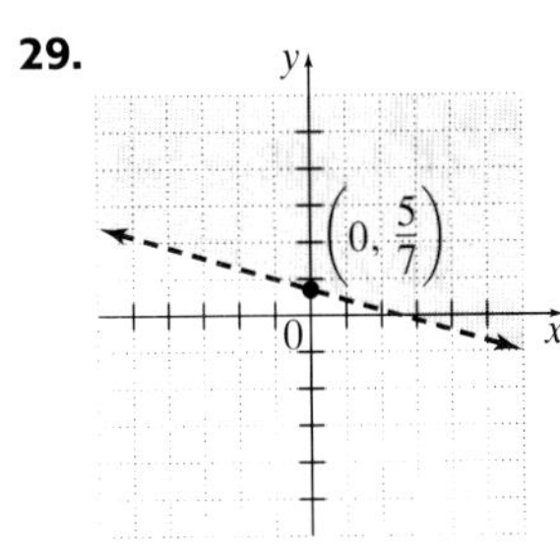

31.

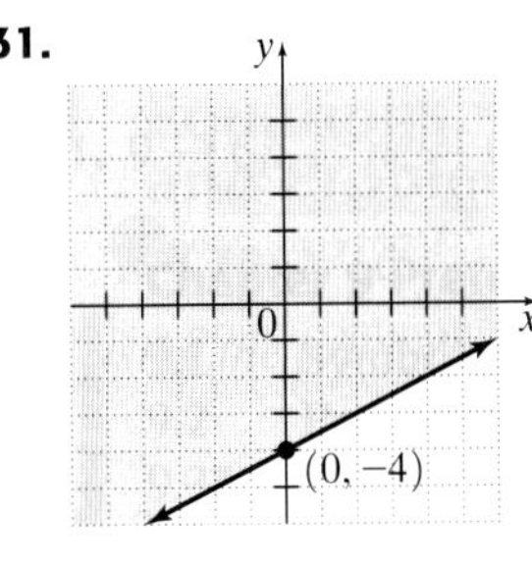

33. $(-2, 1)$ **35.** $(-3, -1)$ **37.** A **39.** B **41.** Answers may vary. **43.a.** $30x + 0.15y \leq 500$

b.

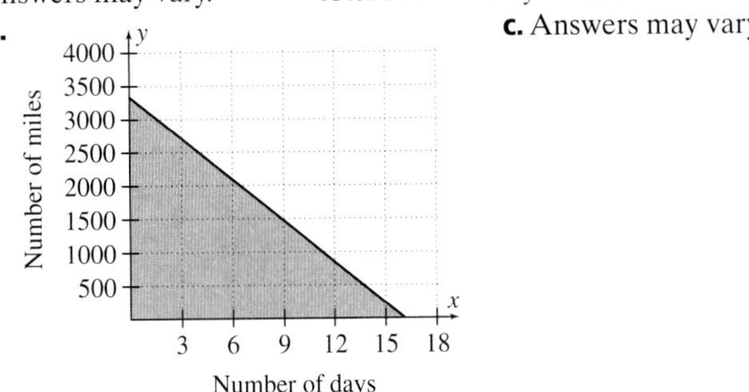

c. Answers may vary.

MENTAL MATH

1. direct **3.** inverse **5.** inverse **7.** direct

SECTION 13.8

1. $y = \frac{1}{2}x$ **3.** $y = 6x$ **5.** $y = 3x$ **7.** $y = \frac{2}{3}x$ **9.** $y = \frac{7}{x}$ **11.** $y = \frac{0.5}{x}$ **13.** $y = kx$ **15.** $h = \frac{k}{t}$ **17.** $z = kx^2$

19. $y = \frac{k}{z^3}$ **21.** $x = \frac{k}{\sqrt{y}}$ **23.** $y = 40$ **25.** $y = 3$ **27.** $z = 54$ **29.** $a = \frac{4}{9}$ **31.** \$62.50 **33.** \$6 **35.** $5\frac{1}{3}$ in. **37.** 179.1 lb **39.** 1600 ft **41.** $2y = 16$ **43.** $-4x = 0.5$ **45.** multiplied by 3 **47.** It is doubled.

Chapter 13 Review

1–6.

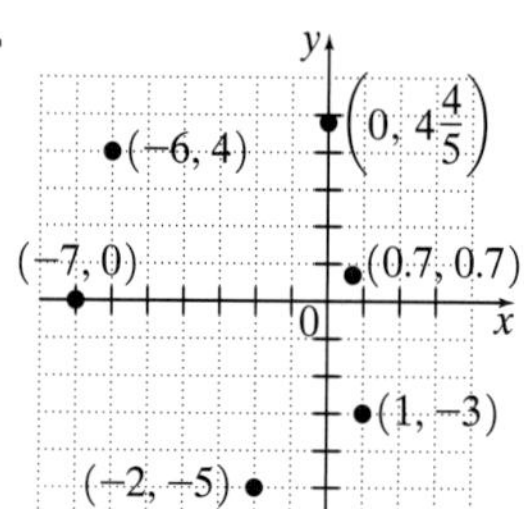

7. (7, 44) **8.** $\left(-\frac{13}{3}, -8\right)$

9.

x	y
−3	0
1	3
9	9

10.

x	y
7	5
−7	5
0	5

11.

x	y
0	0
10	5
−10	−5

12. a. 2005; 2500; 7000 **b.** 886 compact disc holders

13.

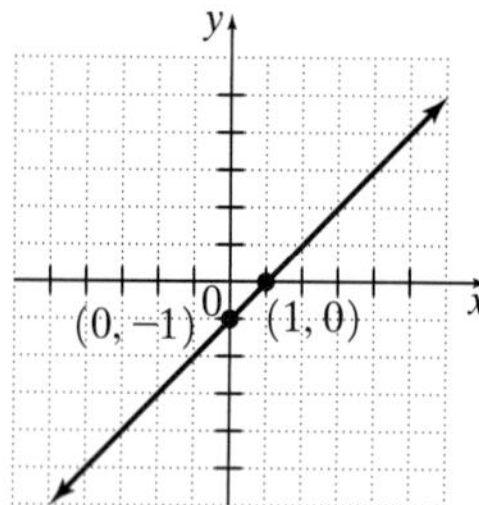

14.

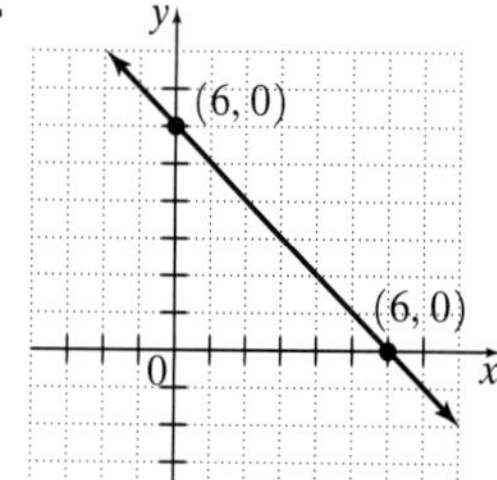

15.

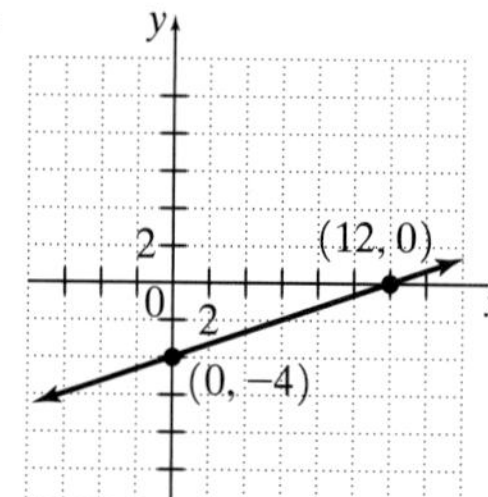

16.

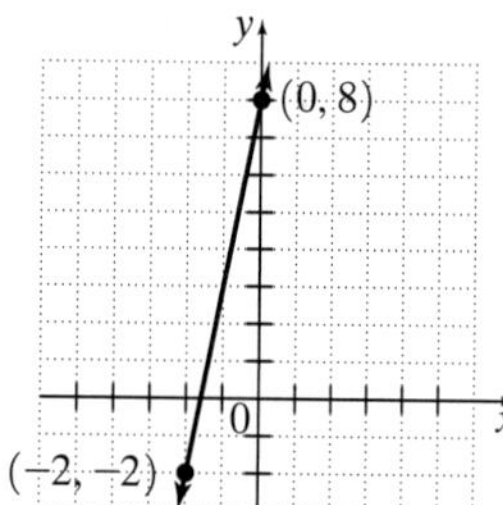

17.

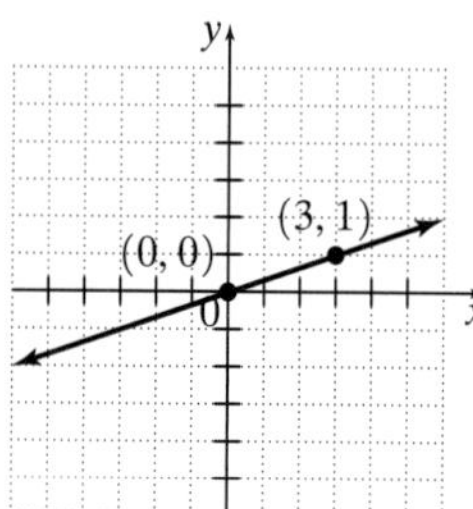

18.

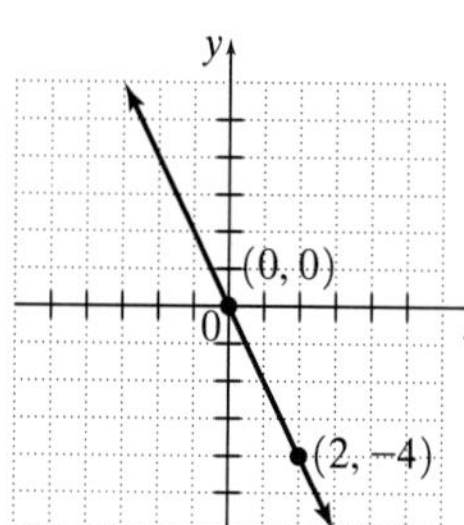

19. (4, 0); (0, −2) **20.** (−2, 0); (2, 0), (0, −2); (0, 2)

21.

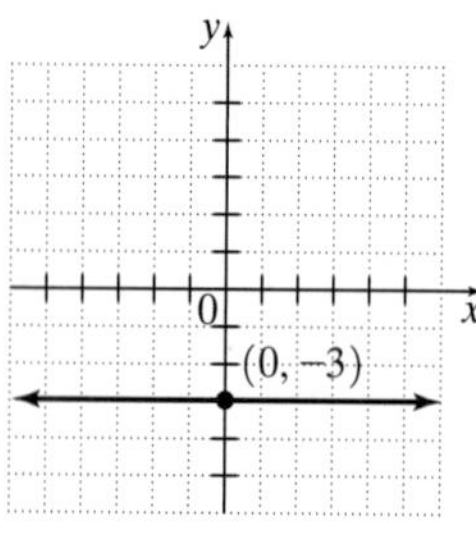

22.

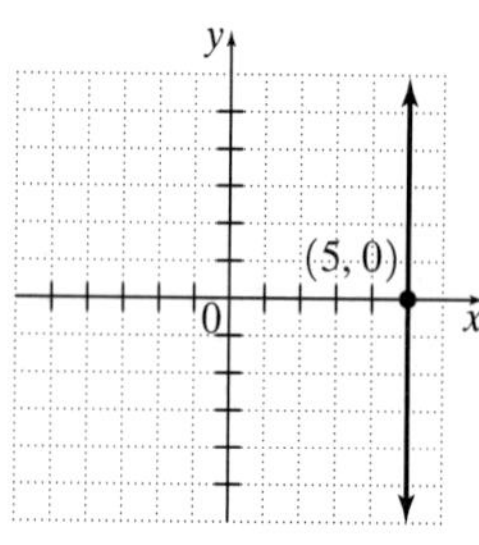

23. (0, −4); (12, 0) **24.** (0, 8); (−2, 0) **25.** $-\frac{3}{4}$ **26.** $\frac{1}{5}$

27. D **28.** B **29.** C **30.** A **31.** $\frac{3}{4}$ **32.** $\frac{5}{3}$ **33.** 4 **34.** −1 **35.** 3 **36.** $\frac{1}{2}$ **37.** 0 **38.** undefined **39.** perpendicular **40.** parallel **41.** neither **42.** Every 1 year, 0.96 million more persons have a bachelor's degree or higher. **43.** Every 1 year, 32.4 million more people go on vacations. **44.** −3; (0, 7) **45.** $\frac{1}{6}$; $\left(0, \frac{1}{6}\right)$ **46.** $y = -5x + \frac{1}{2}$ **47.** $y = \frac{2}{3}x + 6$ **48.** D **49.** C **50.** A **51.** B **52.** $-4x + y = -8$ **53.** $3x + y = -5$ **54.** $-3x + 5y = 17$

55. $x + 3y = 6$ **56.** $14x + y = 21$ **57.** $x + 2y = 8$ **58.** no **59.** yes **60.** yes **61.** yes **62.** no **63.** yes

64. a. 6 **b.** 10 **c.** 5  **65.**

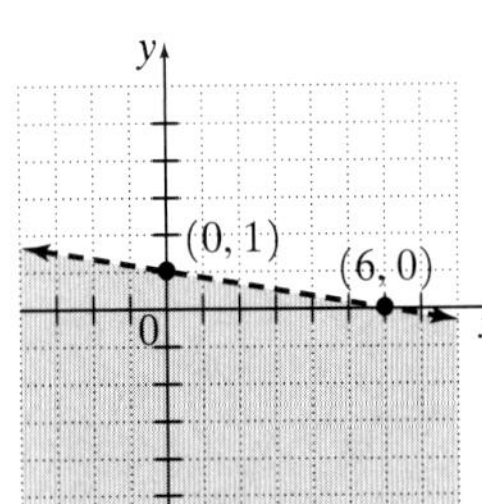

66.

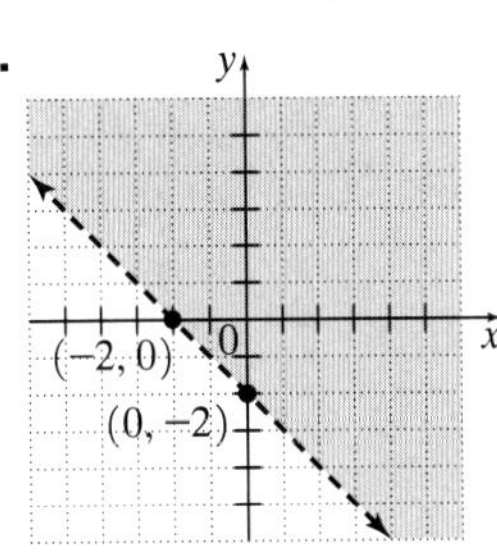

67.

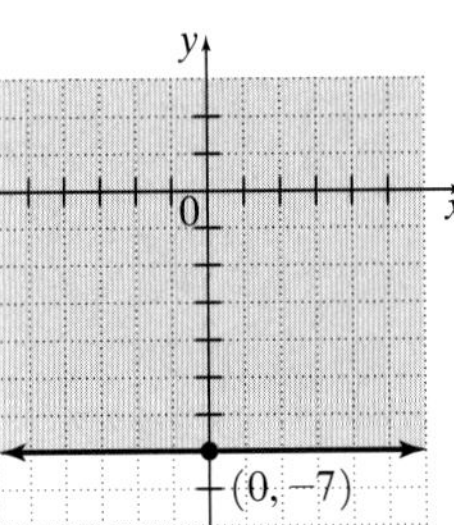

68.

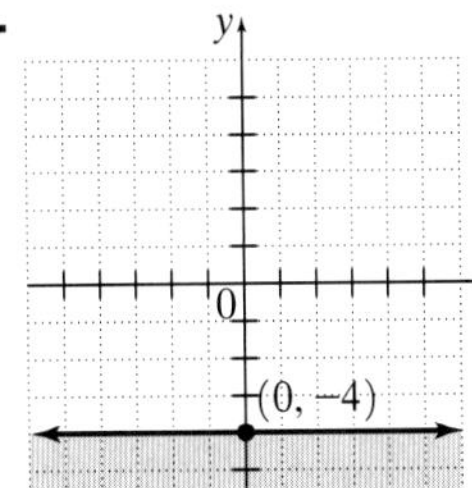

69.

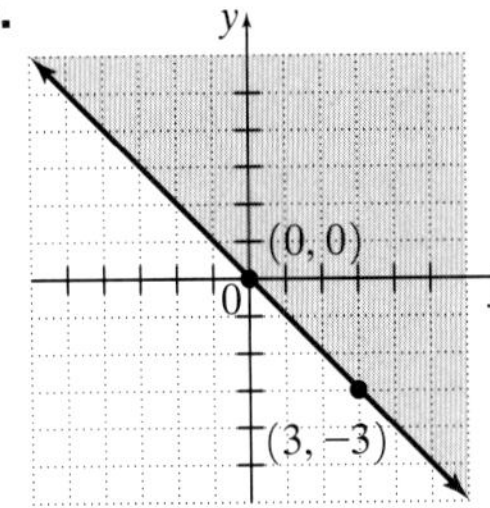

70.

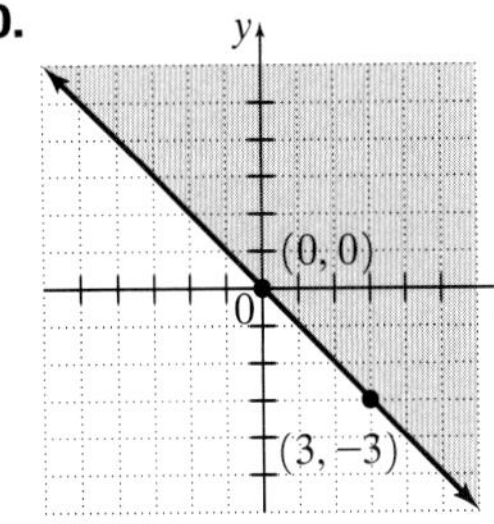

71. $y = 110$ **72.** $y = \frac{1}{2}$ **73.** $y = \frac{100}{27}$ **74.** $y = 700$ **75.** \$3960 **76.** $4\frac{4}{5}$ in.

CHAPTER 13 TEST

1. a. (1980, 38)
(1984, 47)
(1988, 51)
(1992, 54)
(1996, 59)
(2000, 55)

b.

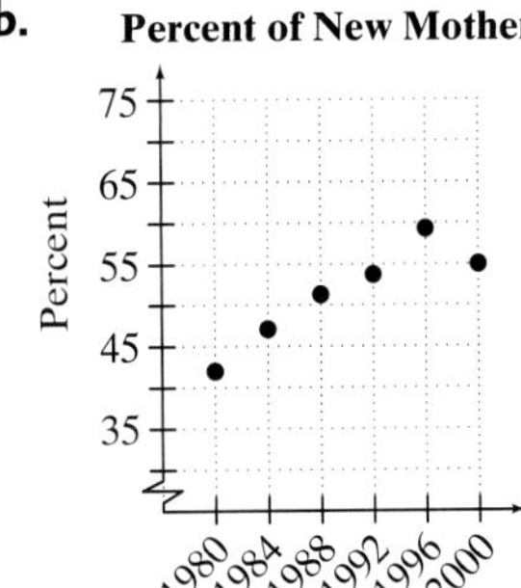

2.

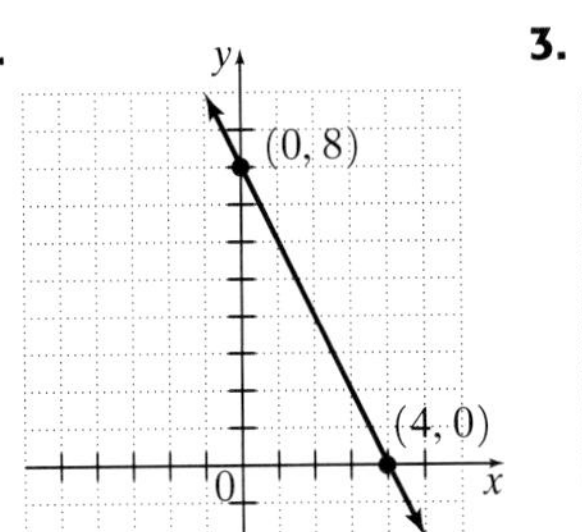

3.

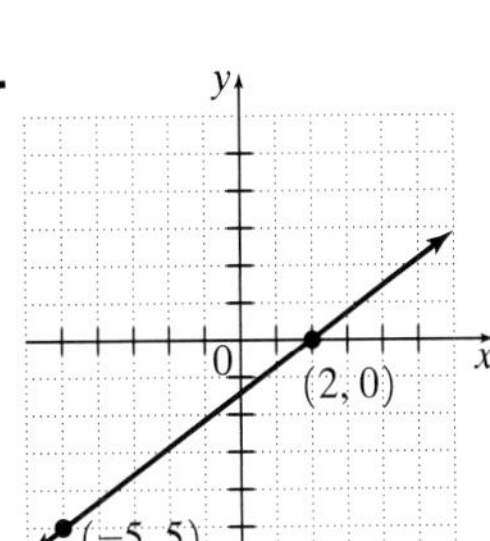

4.

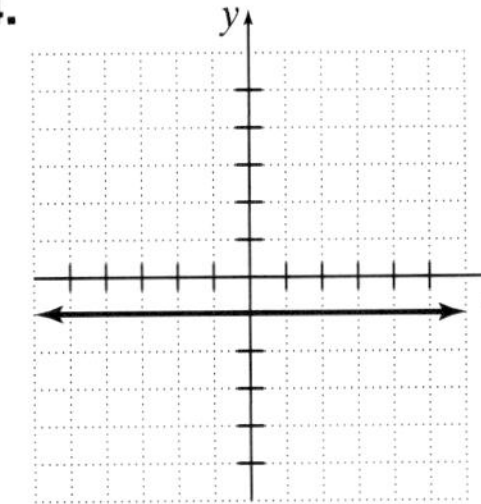

5.

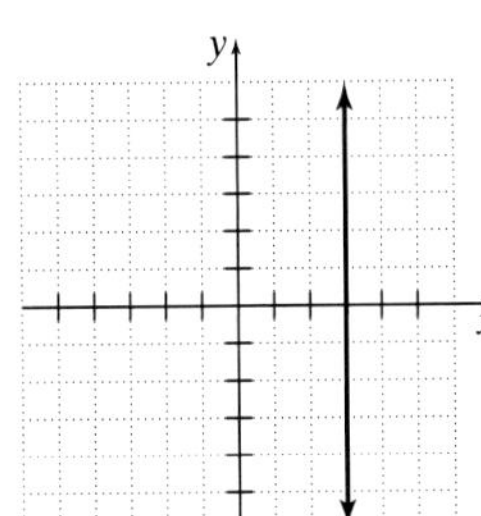

6.

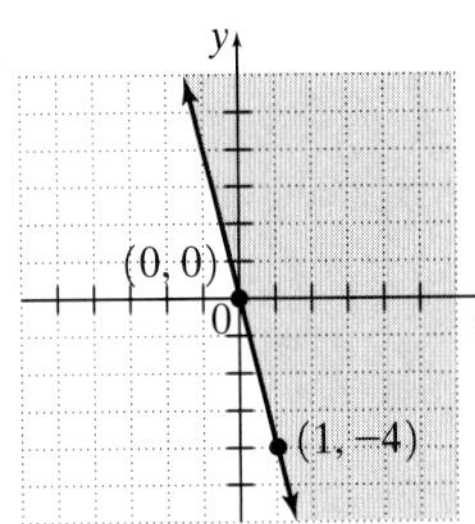

7.

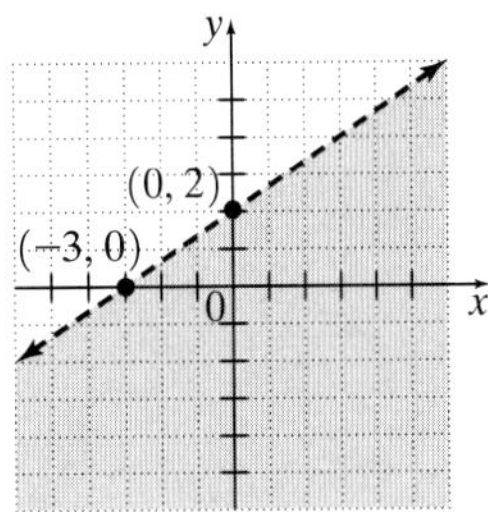

8. $\frac{2}{5}$ **9.** 0 **10.** −1 **11.** 3 **12.** undefined **13.** $m = \frac{7}{3}$ y-intercept: $-\frac{2}{3}$ **14.** neither **15.** $x + 4y = 10$ **16.** $7x + 6y = 0$ **17.** $8x + y = 11$ **18.** $x - 8y = -96$ **19.** yes **20.** no **21.** yes **22.** yes **23. a.** −8 **b.** −3.6 **c.** −4 **24. a.** 0 **b.** 0 **c.** 60 **25.** $x + 2y = 21$; $x = 5$ m **26.** Every 1 year, 105 million more movie tickets are sold. **27.** 28 **28.** $\frac{8}{9}$

CUMULATIVE REVIEW

1. 106,400 sq mi; Sec. 6, Ex. 6 **2.** 147 trees **3.** −8; Sec. 2.2, Ex. 15 **4.** −10 **5.** $\frac{32}{12x}$; Sec. 4.1, Ex. 21 **6.** $\frac{12}{8c}$ **7.** $3\frac{4}{7}$; Sec. 4.8, Ex. 16 **8.** $10\frac{3}{5}$ **9.** −2.6; Sec. 5.5, Ex. 9 **10.** −13.6 **11.** 3.14; Sec. 5.6, Ex. 5 **12.** 1.947 **13.** $\{x|x < -2\}$; Sec. 9.6, Ex. 5 14. $\{x|x < -3\}$ **15. a.** 2; trinomial **b.** 1; binomial **c.** 3; none of these; Sec. 10.3, Ex. 3 **16. a.** 1; binomial **b.** 2; trinomial **c.** 3; none of these **17.** $-4x^2 + 6x + 2$; Sec. 10.4, Ex. 2 **18.** $x^2 + 3x$ **19.** $9y^2 + 6y + 1$; Sec. 10.6, Ex. 4 **20.** $4x^2 - 20x + 25$ **21.** $3a(-3a^4 + 6a - 1)$; Sec. 11.1, Ex. 3 **22.** $x^3(2x^2 - 1)$ **23.** $(x - 2)(x + 6)$; Sec. 11.2, Ex. 3 **24.** $(x - 3)(x + 7)$ **25.** $(4x - 1)(2x - 5)$; Sec. 11.3, Ex. 2 **26.** $(5x + 2)(3x - 1)$ **27.** 4, 5; Sec. 11.6, Ex. 5 **28.** 7, 2 **29.** 1; Sec. 12.2, Ex. 7 **30.** $\frac{3x - 1}{10}$ **31.** $\frac{12ab^2}{27a^2b}$; Sec. 12.3, Ex. 9 **32.** $\frac{63x^3y}{99x^2y^2}$

33. $\frac{2m+1}{m+1}$; Sec. 12.4, Ex. 5 **34.** $\frac{1}{m+1}$ **35.** $-3, -2$; Sec. 12.5, Ex. 3 **36.** $-5, -2$ **37.** $\frac{x+1}{x+2y}$; Sec. 12.7, Ex. 5
38. $\frac{x+4}{x-4}$ **39. a.** $(0, 12)$ **b.** $(2, 6)$ **c.** $(-1, 15)$; Sec. 13.1, Ex. 3 **40. a.** $(0, -5)$ **b.** $(20, 0)$ **c.** $(12, -2)$

41.

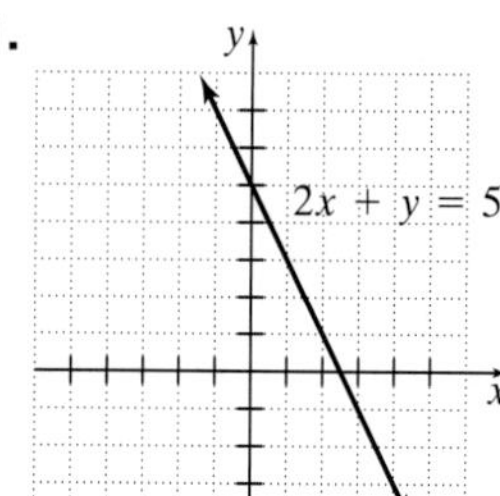

; Sec. 13.2, Ex. 1 **42.**

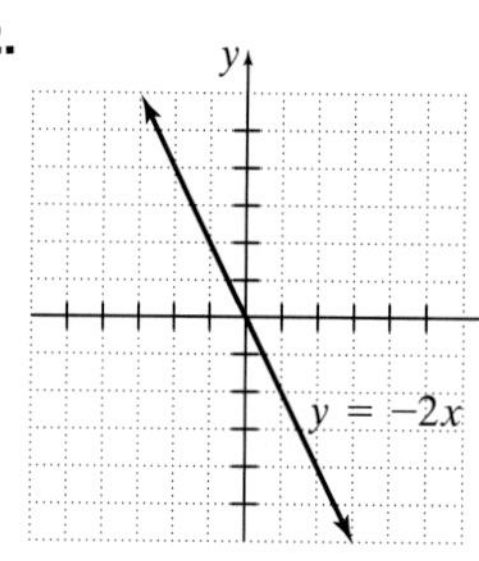

43. $\frac{2}{3}$; Sec. 13.4, Ex. 3 **44.** $-\frac{7}{4}$
45. $2x + y = 3$; Sec. 13.5, Ex.4
46. $5y + y = 3$ **47. a.** 1; $(2, 1)$ **b.** 1; $(-2, 1)$
c. -3; $(0, -3)$; Sec. 13.6, Ex. 6 **48. a.** 2; $(0, 2)$
b. 50; $(4, 50)$ **c.** 5; $(-1, 5)$

Chapter 14 Systems of Equations

Chapter 14 Pretest

1. $(6, -1)$; 14.1B

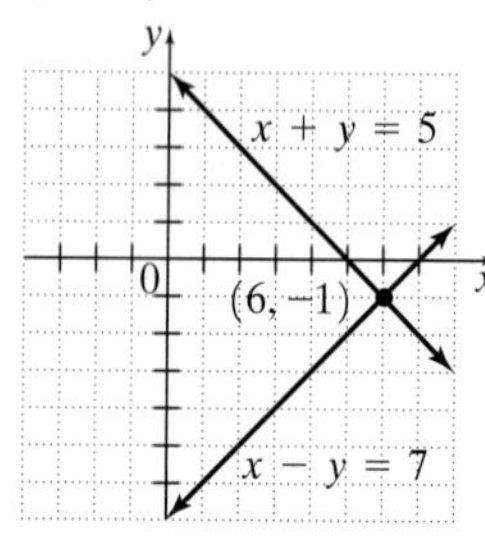

2. $(1, 4)$; 14.1B

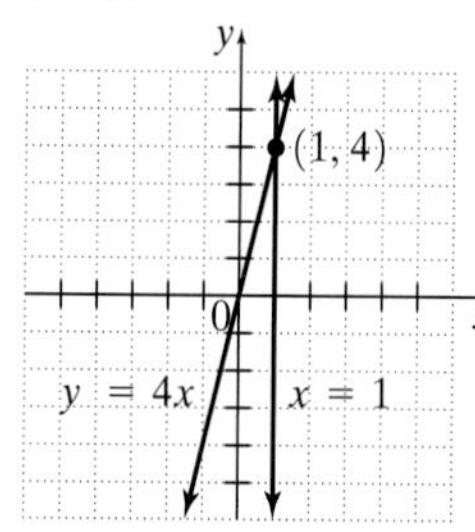

3. $(4, 2)$; 14.2A **4.** $(3, -2)$; 7.2A **5.** $(-1, -1)$; 14.2A
6. infinite number of solutions; 14.2A **7.** no solution; 14.2A
8. $\left(\frac{1}{4}, -\frac{5}{2}\right)$; 14.2A **9.** infinite number of solutions; 14.2A
10. $(-2, 1)$; 14.2A **11.** $(8, -5)$; 14.3A **12.** $(-2, -7)$; 14.3A
13. $(6, 7)$; 14.3A **14.** no solution; 14.3A
15. $(12, -12)$; 14.3A **16.** $\left(\frac{3}{8}, -\frac{5}{8}\right)$; 14.3A
17. infinite number of solutions; 14.3A **18.** $(0, -6)$; 14.3A
19. 81 and 16; 14.4A **20.** 32° and 58°; 14.4A

Calculator Explorations

1. $(0.37, 0.23)$ **3.** $(0.03, -1.89)$

Mental Math

1. 1 solution, $(-1, 3)$ **3.** infinite number of solutions **5.** no solution **7.** 1 solution, $(3, 2)$

Section 14.1

1. a. no **b.** yes **3. a.** yes **b.** no **5. a.** yes **b.** yes

7. $(3, 1)$

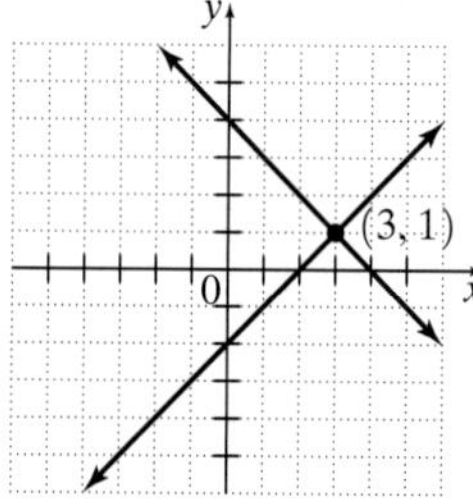

9. $(6, 0)$

11. $(-2, -4)$

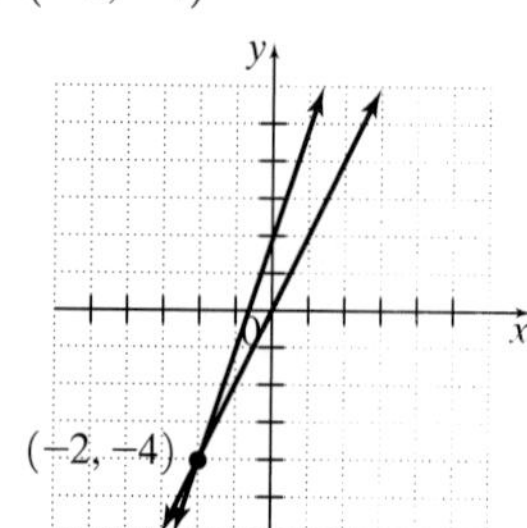

13. $(2, 3)$

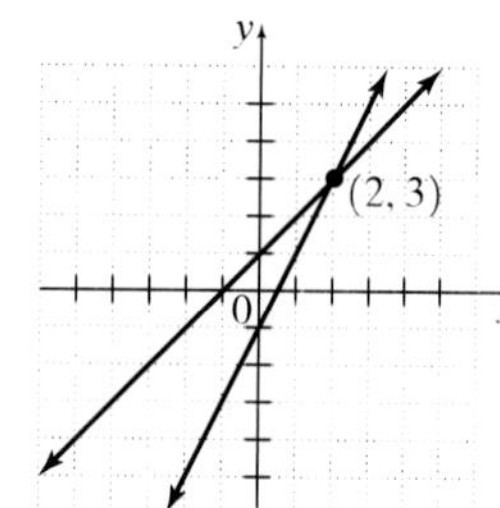

15. $(1, -2)$

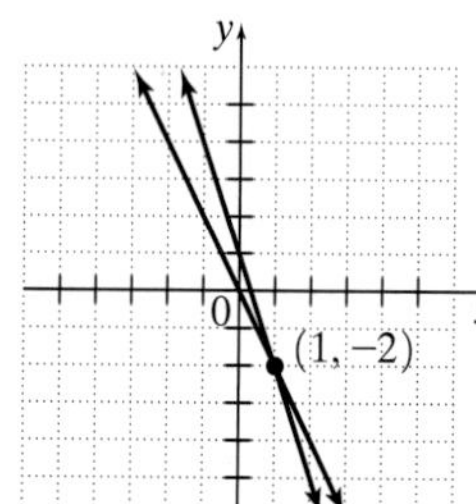

17. $(-2, 1)$

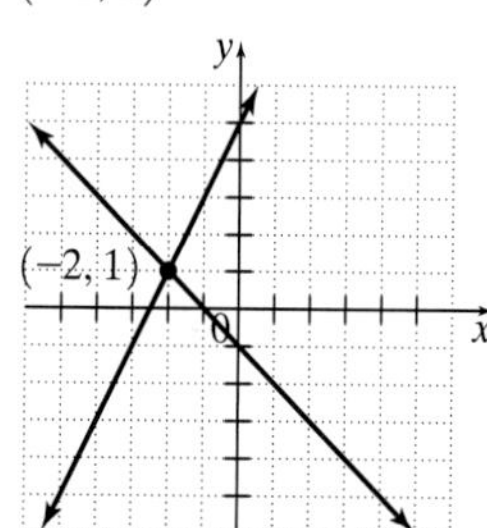

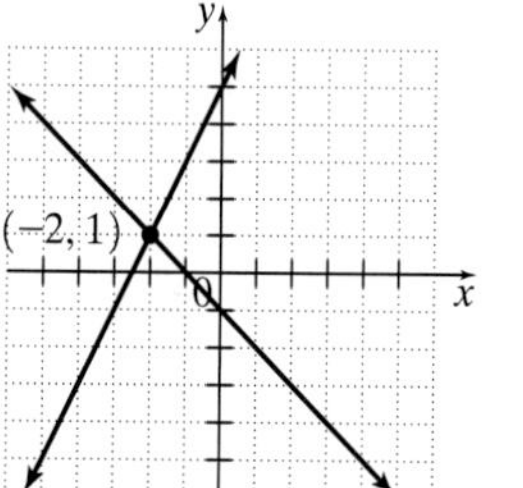

19. $(4, 2)$

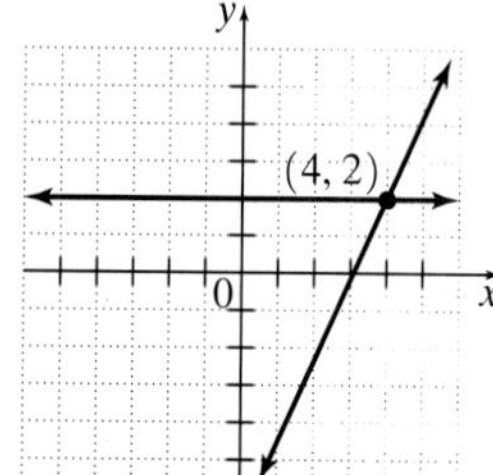

21. no solution

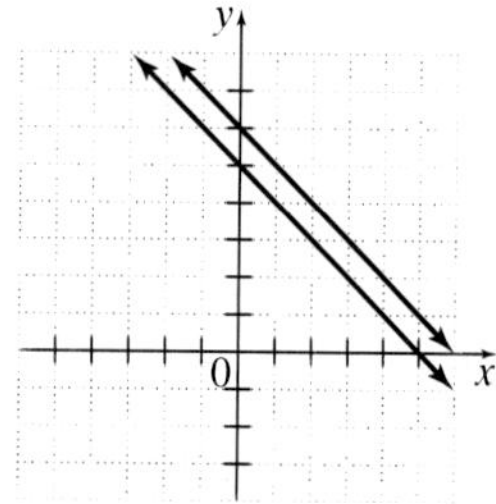

23. $(2, 0)$

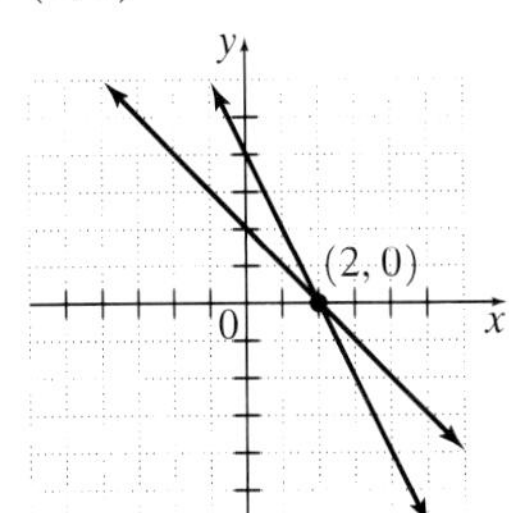

25. $(0, -1)$

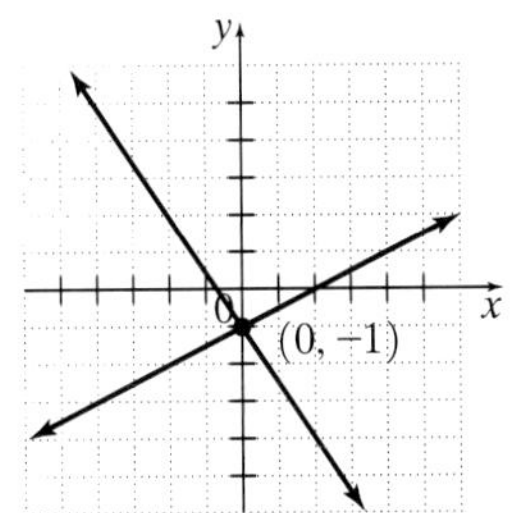

27. infinite number of solutions

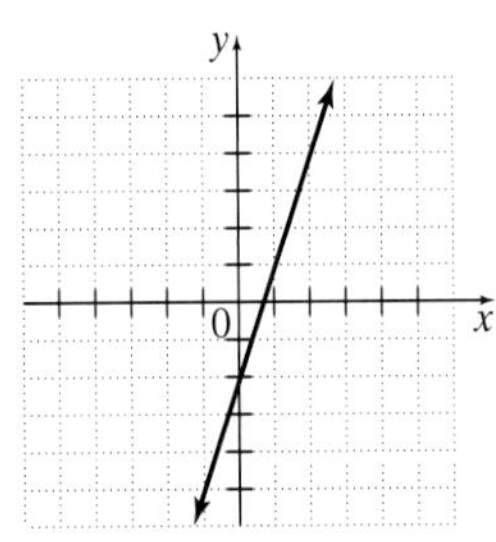

29. $(3, -1)$

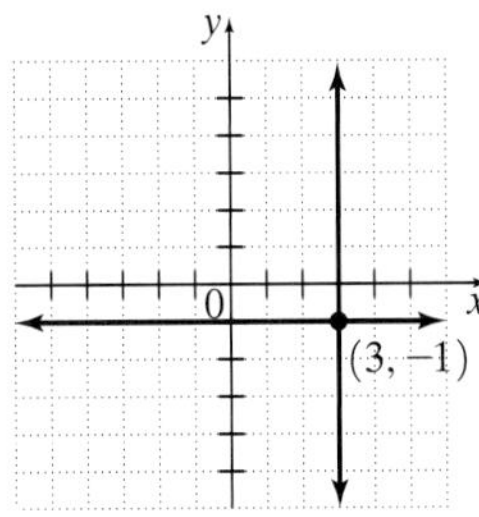

31. $(-5, -7)$

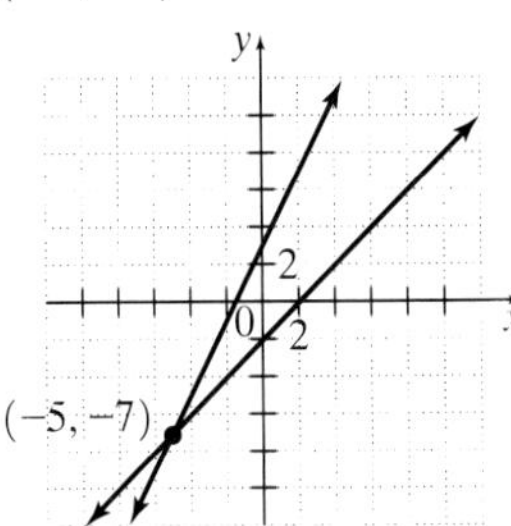

33. $(5, 4)$

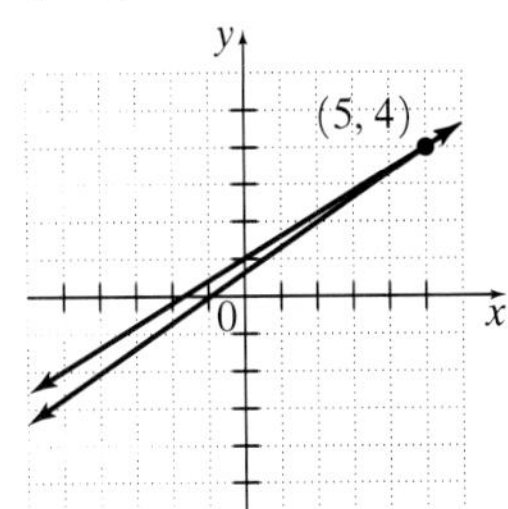

35. Answers may vary.

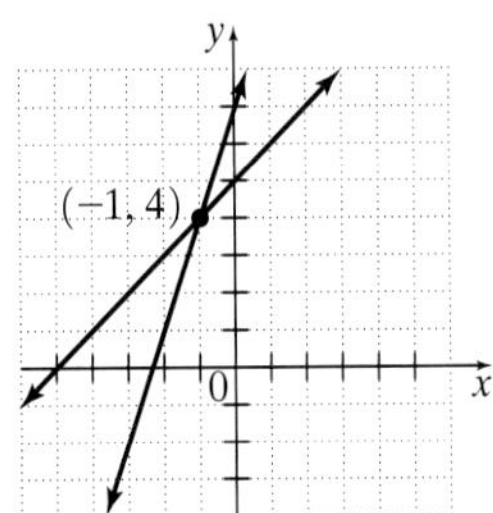

37. Answers may vary.

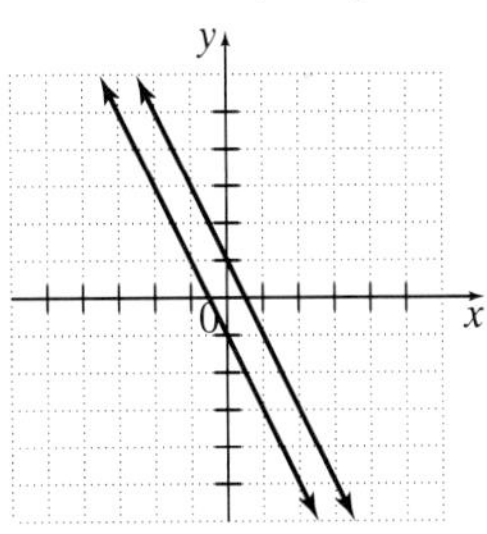

39. 1984, 1988 **41.** 1996 **43. a.** $(4, 9)$ **b.**

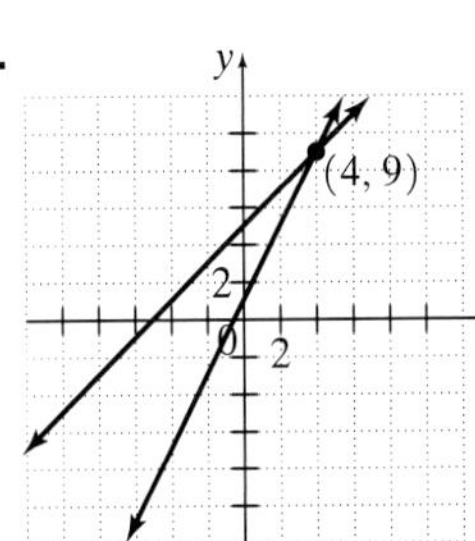

c. yes **45.** $x = -1$ **47.** $y = 3$ **49.** $z = -7$ **51.** Answers may vary. **53.** Answers may vary.

Section 14.2

1. $(2, 1)$ **3.** $(-3, 9)$ **5.** $(4, 2)$ **7.** $(10, 5)$ **9.** $(2, 7)$ **11.** $\left(-\frac{1}{5}, \frac{43}{5}\right)$ **13.** $(-2, 4)$ **15.** $(-2, -1)$ **17.** no solution **19.** $(3, -1)$ **21.** $(3, 5)$ **23.** $\left(\frac{2}{3}, -\frac{1}{3}\right)$ **25.** $(-1, -4)$ **27.** $(-6, 2)$ **29.** $(2, 1)$ **31.** no solution **33.** infinite number of solutions **35.** Answers may vary. **37.** $-6x - 4y = -12$ **39.** $-12x + 3y = 9$ **41.** $5n$ **43.** $-15b$ **45.** $(1, -3)$ **47.** $(-2.6, 1.3)$ **49.** $(3.28, 2.1)$ **51. a.** $(13, 18)$ **b.** In 1983, the percentage of U.S. households that used fuel oil equaled the percentage that used electricity. That percentage was 18%. **c.**

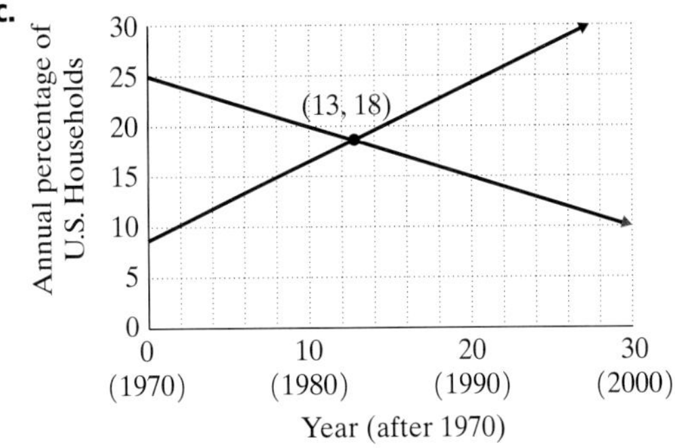

Section 14.3

1. $(1, 2)$ **3.** $(2, -3)$ **5.** $(5, -2)$ **7.** $(6, 0)$ **9.** $(-2, -5)$ **11.** $(-7, 5)$ **13.** $\left(\frac{12}{11}, -\frac{4}{11}\right)$ **15.** no solution **17.** no solution **19.** $\left(2, -\frac{1}{2}\right)$ **21.** $(6, -2)$ **23.** infinite number of solutions **25.** $(-2, 0)$ **27.** Answers may vary. **29.** $\left(\frac{3}{2}, 3\right)$ **31.** $(1, 6)$ **33.** infinite number of solutions **35.** $\left(-\frac{2}{3}, \frac{2}{5}\right)$ **37.** $(2, 4)$ **39.** $(-0.5, 2.5)$ **41.** $2x + 6 = x - 3$ **43.** $20 - 3x = 2$ **45.** $4(x + 6) = 2x$ **47. a.** $b = 15$ **b.** any real number except 15 **49.** $(-8.9, 10.6)$ **51.** b; answers may vary. **53. a.** $(13, 544)$ **b.** Answers may vary. **c.** 1994 to 2001

Integrated Review

1. $(2, 5)$ **2.** $(4, 2)$ **3.** $(5, -2)$ **4.** $(6, -14)$ **5.** $(-3, 2)$ **6.** $(-4, 3)$ **7.** $(0, 3)$ **8.** $(-2, 4)$ **9.** $(5, 7)$
10. $(-3, -23)$ **11.** $\left(\frac{1}{3}, 1\right)$ **12.** $\left(-\frac{1}{4}, 2\right)$ **13.** no solution **14.** infinite number of solutions **15.** $(0.5, 3.5)$
16. $(-0.75, 1.25)$ **17.** infinite number of solutions **18.** no solution **19.** Answers may vary. **20.** Answers may vary.

Mental Math

1. c **3.** b **5.** a

Section 14.4

1. $\begin{cases} x + y = 15 \\ x - y = 7 \end{cases}$ **3.** $\begin{cases} x + y = 6500 \\ x = y + 800 \end{cases}$ **5.** 33 and 50 **7.** 14 and -3 **9.** Jackson: 658 points; Holdsclaw: 516 points
11. child's ticket: \$18; adult's ticket: \$29 **13.** quarters: 53; nickels: 27 **15.** Ohio Art Co.: \$15; General Electric: \$50.77
17. labor: \$13.50 per hr; material: \$275 per ton **19.** still water: 6.5 mph; current: 2.5 mph **21.** still air: 455 mph; wind: 65 mph
23. $4\frac{1}{2}$ hr **25.** 12% solution: $7\frac{1}{2}$ oz; 4% solution: $4\frac{1}{2}$ oz **27.** \$4.95 beans: 113 lbs; \$2.65 beans: 87 lbs **29.** $60°, 30°$
31. $20°, 70°$ **33.** number sold at \$9.50: 23; number sold at \$7.50: 67 **35.** $2\frac{1}{4}$ mph and $2\frac{3}{4}$ mph **37.** 30%: 50 gal; 60%: 100 gal
39. length: 42 in.; width: 30 in. **41.** 16 **43.** $36x^2$ **45.** $100y^6$ **47.** width: 9 ft; length: 15 ft
49. Answers may vary.

Chapter 14 Review

1. a. no **b.** yes **2. a.** yes **b.** no **3. a.** no **b.** no **4. a.** no **b.** yes

5. $(3, 2)$

6. $(1, 2)$

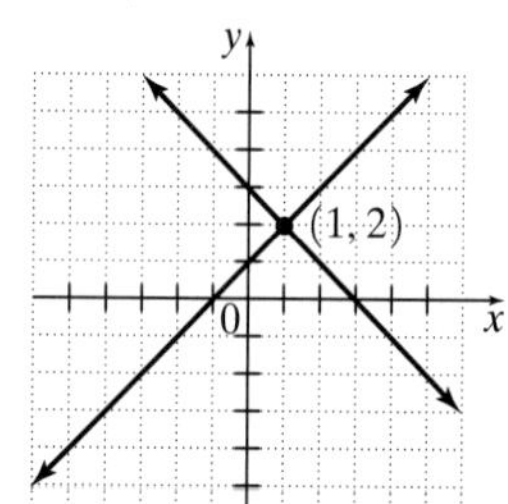

7. $(5, -1)$

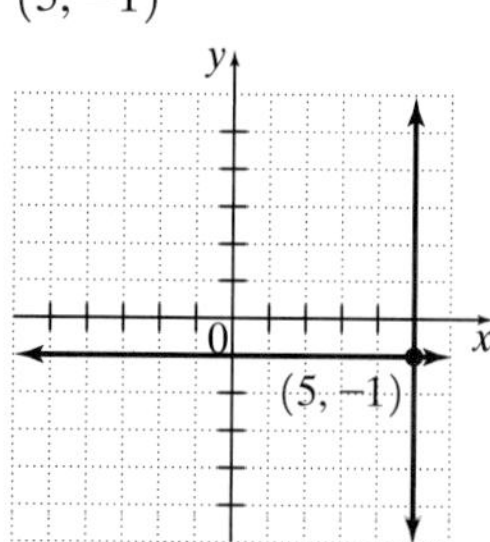

8. $(-3, 2)$

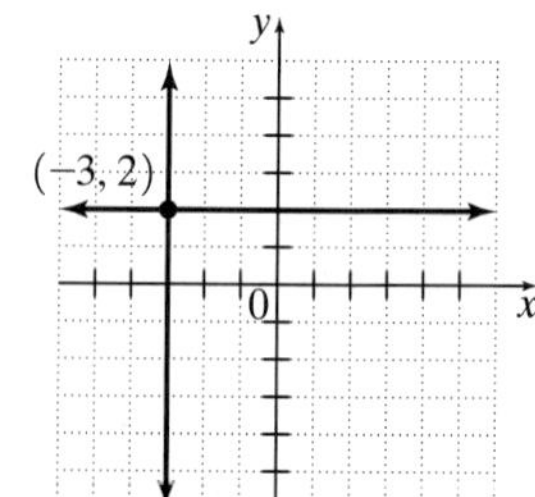

9. $(3, -1)$

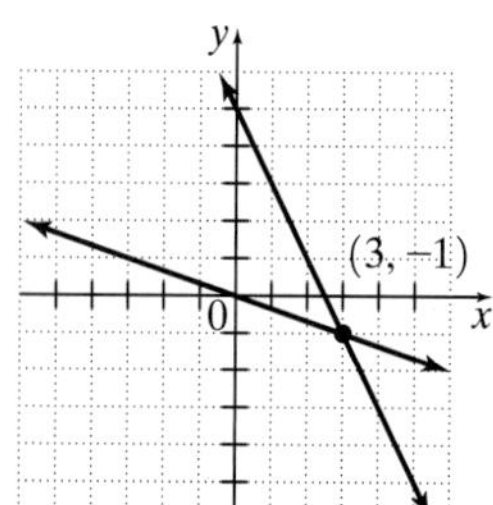

10. $(1, -5)$

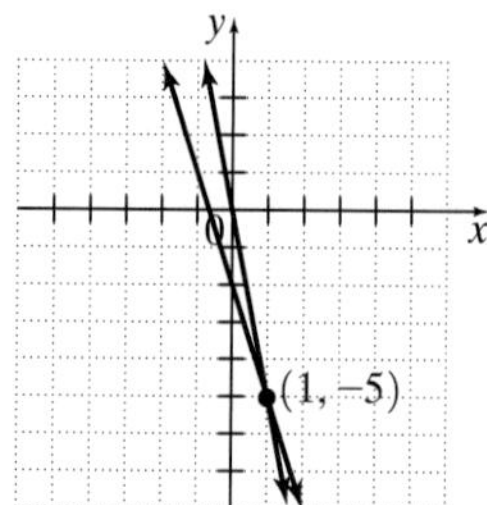

11. $(-3, -2)$

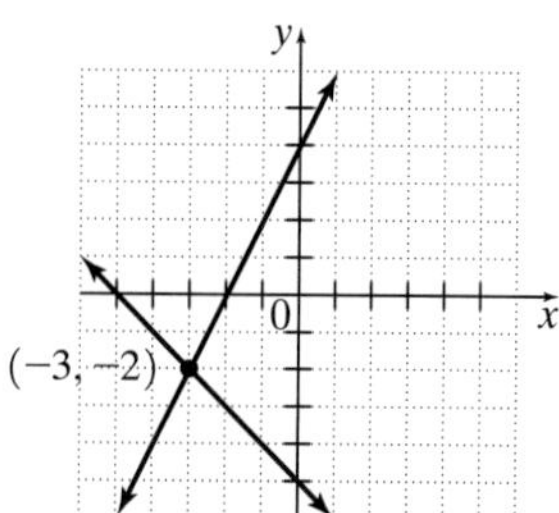

12. $\left(2\frac{1}{3}, -2\frac{2}{3}\right)$

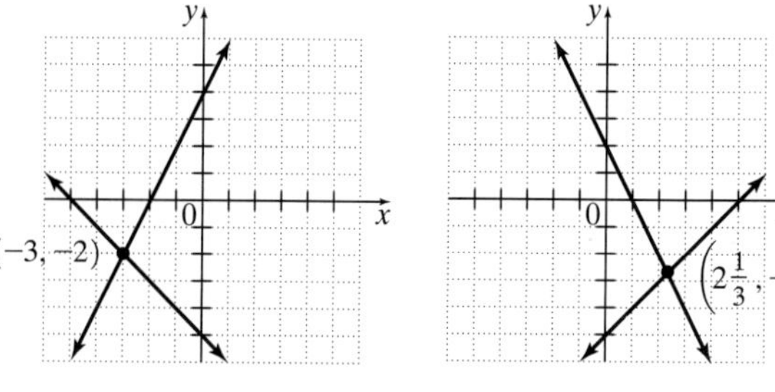

13. no solution

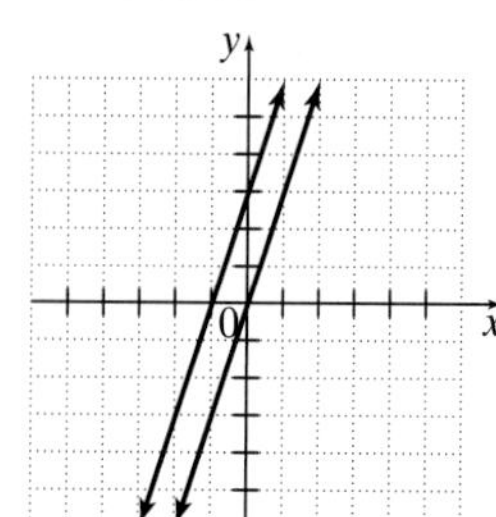

14. infinite number of solutions

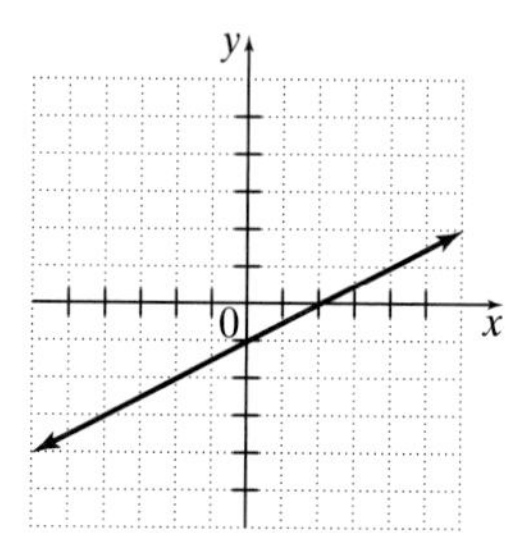

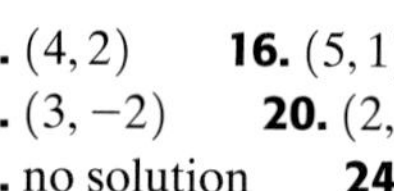

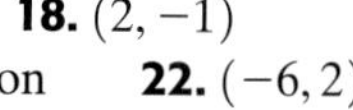

15. $(4, 2)$ **16.** $(5, 1)$ **17.** $(-1, 4)$ **18.** $(2, -1)$
19. $(3, -2)$ **20.** $(2, 5)$ **21.** no solution **22.** $(-6, 2)$
23. no solution **24.** no solution **25.** $(16, -2)$
26. $(11, -2)$ **27.** $(-6, 2)$ **28.** $(4, -1)$ **29.** $(3, 7)$
30. $(-2, 4)$ **31.** infinite number of solutions
32. no solution **33.** $(8, -6)$ **34.** $(10, -4)$
35. $\left(-\frac{3}{2}, \frac{15}{2}\right)$ **36.** infinite number of solutions
37. -6 and 22

38. orchestra: 255 seats; balcony: 105 seats **39.** current of river: 3.2 mph; speed in still water: 21.1 mph **40.** amount invested at 6%: \$6,180; amount invested at 10%: \$2,820 **41.** length: 1.85 ft; width: 1.15 ft **42.** 6% solution: $12\frac{1}{2}$ cc; 14% solution: $37\frac{1}{2}$ cc **43.** egg: 40¢; strip of bacon: 65¢ **44.** jogging: 0.86 hr; walking: 2.14 hr

Chapter 14 Test

1. $(-4, 2)$

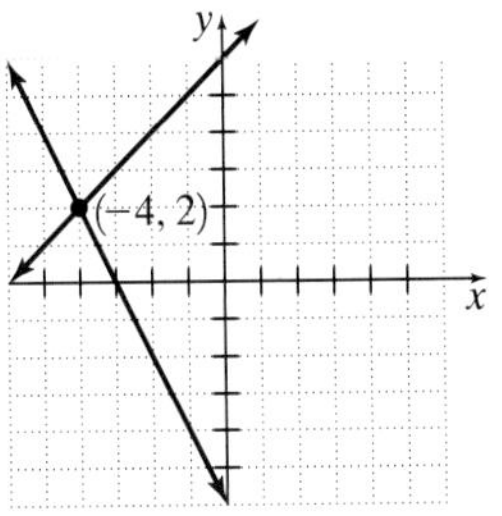

2. $(4, -5)$ **3.** $(-4, 1)$ **4.** $(20, 8)$ **5.** $(4, -2)$ **6.** $(-3, -7)$ **7.** infinite number of solutions **8.** $(7, 2)$ **9.** $\left(\frac{1}{2}, -2\right)$ **10.** $\left(9\frac{2}{5}, 9\frac{3}{5}\right)$ **11.** $(-5, 3)$ **12.** no solution **13.** $(5, -2)$ **14.** 78, 46 **15.** 120 cc **16.** \$1225 at 5%; \$2775 at 9% **17.** Texas: 226 thousand; Missouri: 110 thousand

Cumulative Review

1. -7; Sec. 2.3, Ex. 6 **2.** -3 **3.** 1; Sec. 2.3, Ex. 7 **4.** 5 **5.** 4; Sec. 3.2, Ex. 5 **6.** -2 **7.** $1.2 = 30\% \cdot x$; Sec. 6.3, Ex. 2 **8.** $9 = 45\% \cdot x$ **9.** 16%; Sec. 6.4, Ex.10 **10.** 25% **11.** 775 freshmen; Sec. 6.5, Ex. 2 **12.** 1450 apples **13.** no solution; Sec. 9.3, Ex. 6 14. $\frac{1}{2}$ **15.** $\left\{x \mid x > \frac{13}{7}\right\}$;

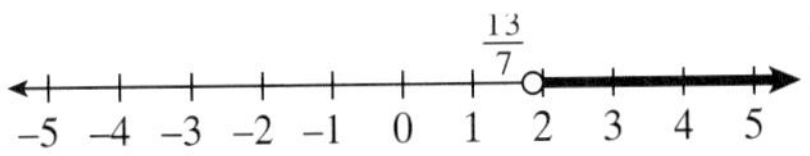

; Sec. 9.6, Ex. 7

16. $\{x \mid x \geq -2\}$;

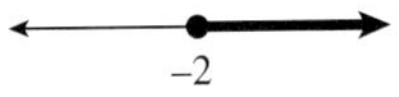

17. $\frac{m^7}{n^7}$, $n \neq 0$; Sec. 10.1, Ex. 19 **18.** $35x^7$ **19.** $\frac{16x^{16}}{81y^{20}}$, $y \neq 0$; Sec. 10.1, Ex. 20 **20.** $\frac{25x^4}{16y^6}$ **21.** $9x^2 - 6x - 1$; Sec. 10.4, Ex. 5 **22.** $8x + 4$ **23.** $2x + 4 - \frac{1}{3x - 1}$; Sec. 10.7, Ex. 5 **24.** $3x + 2 - \frac{2}{x - 1}$ **25.** $-\frac{1}{2}$, 4; Sec. 11.6, Ex. 6 **26.** 8, -3 **27.** 6, 8, 10; Sec. 11.7, Ex. 4 **28.** 11 **29.** 1; Sec. 12.3, Ex. 2 **30.** x **31.** $m = 0$; Sec. 13.4, Ex. 4 **32.** undefined **33.** $-x + 5y = 23$; Sec. 13.5, Ex. 5 **34.** $x + y = -1$ **35.** domain: $\{-1, 0, 3\}$; range: $\{-2, 0, 2, 3\}$; Sec. 13.6, Ex. 1 **36.** domain: $\{2\}$; range: $\{3, 0, -2, 4\}$; **37.** $\left(6, \frac{1}{2}\right)$; Sec. 14.2, Ex. 3 **38.** $(-3, 1)$ **39.** $\left(-\frac{15}{7}, -\frac{5}{7}\right)$; Sec. 14.3, Ex. 6 **40.** $(3, 12)$ **41.** 29 and 8; Sec. 14.4, Ex. 1 **42.** 42 and 33

Chapter 15 Roots and Radicals

Chapter 15 Pretest

1. -7; 15.1A **2.** $\frac{2}{5}$; 15.1A **3.** -4; 15.1B **4.** $2\sqrt{30}$; 15.2A **5.** $\frac{2\sqrt{6}}{y^3}$; 15.2C **6.** $2\sqrt[3]{14}$; 15.2D **7.** $-3\sqrt{15}$; 15.3A **8.** 0; 15.3B **9.** $\frac{7\sqrt{7}}{10}$; 15.3B **10.** $6\sqrt{3}$; 15.4A **11.** $2\sqrt{7} - \sqrt{10}$; 15.4A **12.** $y - 6\sqrt{y} + 9$; 15.4A **13.** $2x\sqrt{7}$; 15.4B **14.** $\frac{\sqrt{55}}{11}$; 15.4C **15.** $\frac{8\sqrt{2a}}{a}$; 15.4C **16.** $\frac{6 + 3\sqrt{x}}{4 - x}$; 15.4D **17.** $x = 49$; 15.5A **18.** $x = \frac{9}{4}$; 15.5B **19.** $4\sqrt{10}$ cm; 15.6A **20.** 2.52 in.; 15.6B

Calculator Explorations

1. 2.646 **3.** 3.317 **5.** 9.055 **7.** 3.420 **9.** 2.115 **11.** 1.783

Section 15.1

1. 4 **3.** 9 **5.** $\frac{1}{5}$ **7.** -10 **9.** not a real number **11.** -11 **13.** $\frac{3}{5}$ **15.** 30 **17.** 12 **19.** $\frac{1}{10}$ **21.** 5 **23.** -4 **25.** -2 **27.** $\frac{1}{2}$ **29.** -5 **31.** Answers may vary. **33.** 2 **35.** not a real number **37.** -5 **39.** 1 **41.** 2.646 **43.** 3.464 **45.** 6.083 **47.** 11.662 **49.** $\sqrt{2} \approx 1.41$; 126.90 ft **51.** z **53.** x^2 **55.** $3x^4$ **57.** $9x$ **59.** ab^2 **61.** $4a^3b^2$

63. $25 \cdot 2$ **65.** $16 \cdot 2$ or $4 \cdot 8$ **67.** $4 \cdot 7$ **69.** $9 \cdot 3$ **71.** 7 mi **73.** 3.1 in. **75.** 3 **77.** 1; 1.7; 2; 3
79. $(2, 0)$ **81.** $(-4, 0)$ **83.** 58 ft

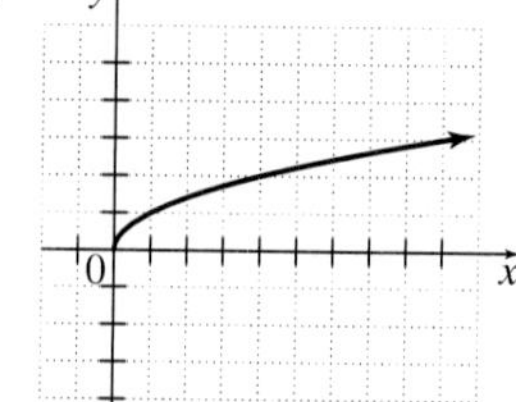

Mental Math

1. 6 **3.** x **5.** 0 **7.** $5x^2$

Section 15.2

1. $2\sqrt{5}$ **3.** $3\sqrt{2}$ **5.** $5\sqrt{2}$ **7.** $\sqrt{33}$ **9.** $2\sqrt{15}$ **11.** $6\sqrt{5}$ **13.** $2\sqrt{13}$
15. $\frac{2\sqrt{2}}{5}$ **17.** $\frac{3\sqrt{3}}{11}$ **19.** $\frac{3}{2}$ **21.** $\frac{5\sqrt{5}}{3}$

23. $\frac{\sqrt{11}}{6}$ **25.** $-\frac{\sqrt{3}}{4}$ **27.** $x^3\sqrt{x}$ **29.** $x^6\sqrt{x}$ **31.** $5x\sqrt{3}$ **33.** $4x^2\sqrt{6}$ **35.** $\frac{2\sqrt{3}}{y}$ **37.** $\frac{3\sqrt{x}}{y}$ **39.** $\frac{2\sqrt{22}}{x^2}$ **41.** $2\sqrt[3]{3}$
43. $5\sqrt[3]{2}$ **45.** $\frac{\sqrt[3]{5}}{4}$ **47.** $\frac{\sqrt[3]{7}}{2}$ **49.** $\frac{\sqrt[3]{15}}{4}$ **51.** $2\sqrt[3]{10}$ **53.** $14x$ **55.** $2x^2 - 7x - 15$ **57.** 0 **59.** $x^3y\sqrt{y}$

61. $x + 2$ **63.** $2\sqrt[3]{10}$ in. **65.** $2\sqrt{5}$ in. **67.** 5 in. **69.** \$1700 **71.** Answers may vary. **73.** $\frac{26}{15}$ sq m ≈ 1.7 sq m

Mental Math

1. $8\sqrt{2}$ **3.** $7\sqrt{x}$ **5.** $3\sqrt{7}$

Section 15.3

1. $-4\sqrt{3}$ **3.** $9\sqrt{6} - 5$ **5.** $\sqrt{5} + \sqrt{2}$ **7.** $6\sqrt{3}$ **9.** $-5\sqrt{2} - 6$ **11.** $7\sqrt{5}$ **13.** $2\sqrt{5}$ **15.** $6 - 3\sqrt{3}$
17. Answers may vary. **19.** $5\sqrt{3}$ **21.** $9\sqrt{5}$ **23.** $4\sqrt{6} + \sqrt{5}$ **25.** $x + \sqrt{x}$ **27.** 0 **29.** $x\sqrt{x}$ **31.** $9\sqrt{3}$
33. $5\sqrt{2} + 12$ **35.** $\frac{4\sqrt{5}}{9}$ **37.** $\frac{3\sqrt{3}}{8}$ **39.** $2\sqrt{5}$ **41.** $-\sqrt{35}$ **43.** $11\sqrt{x}$ **45.** $12x - 11\sqrt{x}$ **47.** $x\sqrt{3x} + 3x\sqrt{x}$

49. $8x\sqrt{2} + 2x$ **51.** $2x^2\sqrt{10} - x^2\sqrt{5}$ **53.** $x^2 + 12x + 36$ **55.** $4x^2 - 4x + 1$ **57.** $(4, 2)$ **59.** $8\sqrt{5}$ in. **61.** $\left(48 + \frac{9\sqrt{3}}{2}\right)$ sq ft

Mental Math

1. $\sqrt{6}$ **3.** $\sqrt{6}$ **5.** $\sqrt{10y}$

Section 15.4

1. 4 **3.** $5\sqrt{2}$ **5.** 6 **7.** $2x$ **9.** 20 **11.** $36x$ **13.** $3\sqrt{2xy}$ **15.** $4xy\sqrt{y}$ **17.** $\sqrt{10} + \sqrt{2}$ **19.** $2\sqrt{5} + 5\sqrt{2}$
21. $\sqrt{30} + \sqrt{42}$ **23.** -33 **25.** $\sqrt{6} - \sqrt{15} + \sqrt{10} - 5$ **27.** $16 - 11\sqrt{11}$ **29.** $x - 36$ **31.** $x - 14\sqrt{x} + 49$
33. $6y + 2\sqrt{6y} + 1$ **35.** 4 **37.** $\sqrt{7}$ **39.** $3\sqrt{2}$ **41.** $5y^2$ **43.** $5\sqrt{3}$ **45.** $2y\sqrt{6}$ **47.** $2xy\sqrt{3y}$ **49.** $\frac{\sqrt{15}}{5}$
51. $\frac{7\sqrt{2}}{2}$ **53.** $\frac{\sqrt{6y}}{6y}$ **55.** $\frac{\sqrt{10}}{6}$ **57.** $\frac{\sqrt{3x}}{x}$ **59.** $\frac{\sqrt{2}}{4}$ **61.** $\frac{\sqrt{30}}{15}$ **63.** $\frac{\sqrt{15}}{10}$ **65.** $\frac{3\sqrt{2x}}{2}$ **67.** $\frac{8y\sqrt{5}}{5}$ **69.** $\frac{\sqrt{3xy}}{6x}$
71. $3\sqrt{2} - 3$ **73.** $-8 - 4\sqrt{5}$ **75.** $5 + \sqrt{30} + \sqrt{6} + \sqrt{5}$ **77.** $\sqrt{6} + \sqrt{3} + \sqrt{2} + 1$ **79.** $\frac{10 - 5\sqrt{x}}{4 - x}$ **81.** $\frac{3\sqrt{x} + 12}{x - 16}$
83. $x = 44$ **85.** $z = 2$ **87.** $x = 3$ **89.** $130\sqrt{3}$ sq m **91.** Answers may vary. **93.** $\frac{2}{\sqrt{6} - \sqrt{2} - \sqrt{3} + 1}$

Integrated Review

1. 6 **2.** $4\sqrt{3}$ **3.** x^2 **4.** $y^3\sqrt{y}$ **5.** $4x$ **6.** $3x^5\sqrt{2x}$ **7.** 2 **8.** 3 **9.** -3 **10.** not a real number **11.** $\frac{\sqrt{11}}{3}$
12. $\frac{\sqrt[3]{7}}{4}$ **13.** -4 **14.** -5 **15.** $\frac{3}{7}$ **16.** $\frac{1}{8}$ **17.** a^4b **18.** x^5y^{10} **19.** $5m^3$ **20.** $3n^8$ **21.** $6\sqrt{7}$ **22.** $3\sqrt{2}$
23. cannot be simplified **24.** $\sqrt{x} + 3x$ **25.** $\sqrt{30}$ **26.** 3 **27.** 28 **28.** 45 **29.** $\sqrt{33} + \sqrt{3}$ **30.** $3\sqrt{2} - 2\sqrt{6}$
31. $4y$ **32.** $3x^2\sqrt{5x}$ **33.** $x - 3\sqrt{x} - 10$ **34.** $11 + 6\sqrt{2}$ **35.** 2 **36.** $\sqrt{3}$ **37.** $2x^2\sqrt{3}$ **38.** $ab^2\sqrt{15a}$ **39.** $\frac{\sqrt{6}}{6}$
40. $\frac{x\sqrt{5}}{10}$ **41.** $\frac{4\sqrt{6} - 4}{5}$ **42.** $\frac{\sqrt{2x} + 5\sqrt{2} + \sqrt{x} + 5}{x - 25}$

Section 15.5

1. 81 **3.** −1 **5.** 5 **7.** 49 **9.** no solution **11.** −2 **13.** $\frac{3}{2}$ **15.** 16 **17.** no solution **19.** 2 **21.** 4 **23.** 3 **25.** 2 **27.** 2 **29.** no solution **31.** 0, −3 **33.** −3 **35.** 9 **37.** 12 **39.** 3, 1 **41.** −1 **43.** 2 **45.** 16 **47.** 1 **49.** 1 **51.** $3x - 8 = 19; x = 9$ **53.** $2(2x + x) = 24$; length = 8 in. **55. a.** 3.2; 10; 31.6 **b.** no **57.** Answers may vary. **59.** 2.43 **61.** 0.48

Section 15.6

1. $\sqrt{13}$; 3.61 **3.** $3\sqrt{3}$; 5.20 **5.** 25 **7.** $\sqrt{22}$; 4.69 **9.** $3\sqrt{17}$; 12.37 **11.** $\sqrt{41}$; 6.40 **13.** $4\sqrt{2}$; 5.66 **15.** $3\sqrt{10}$; 9.49 **17.** 20.6 ft **19.** 11.7 ft **21.** 24 cu ft **23.** 54 mph **25.** 27 mph **27.** 59.1 km **29.** 3; −3 **31.** 10; −10 **33.** 8; −8 **35.** $x = 2\sqrt{10} - 4$ **37.** 201 mi **39.** Answers may vary. **41.** Answers may vary.

Chapter 15 Review

1. 9 **2.** −7 **3.** 3 **4.** 2 **5.** $-\frac{3}{8}$ **6.** $\frac{2}{3}$ **7.** 2 **8.** −2 **9.** c **10.** a, c **11.** x^6 **12.** x^4 **13.** $3y$ **14.** $5x^2$ **15.** $2\sqrt{10}$ **16.** $2\sqrt{6}$ **17.** $3\sqrt{6}$ **18.** $2\sqrt{22}$ **19.** $x^2\sqrt{x}$ **20.** $y^3\sqrt{y}$ **21.** $2x\sqrt{5}$ **22.** $5y^2\sqrt{2}$ **23.** $3\sqrt[3]{2}$ **24.** $2\sqrt[3]{11}$ **25.** $\frac{3\sqrt{2}}{5}$ **26.** $\frac{5\sqrt{3}}{8}$ **27.** $\frac{-5\sqrt{2}}{3}$ **28.** $\frac{-2\sqrt{3}}{7}$ **29.** $\frac{\sqrt{11}}{x}$ **30.** $\frac{\sqrt{7}}{y^2}$ **31.** $\frac{y^2\sqrt{y}}{10}$ **32.** $\frac{x\sqrt{x}}{9}$ **33.** $-3\sqrt{2}$ **34.** $-5\sqrt{3}$ **35.** $4\sqrt{5} + 4\sqrt{6}$ **36.** $-2\sqrt{7} + 2\sqrt{2}$ **37.** $5\sqrt{7} + 2\sqrt{14}$ **38.** $9\sqrt{3} - 4$ **39.** $\frac{\sqrt{5}}{6}$ **40.** $\frac{9\sqrt{11}}{20}$ **41.** $10 - x\sqrt{5}$ **42.** $2\sqrt{2x} - \sqrt{3x}$ **43.** $3\sqrt{2}$ **44.** $5\sqrt{3}$ **45.** $\sqrt{10} - \sqrt{14}$ **46.** $\sqrt{55} + \sqrt{15}$ **47.** $3\sqrt{2} - 5\sqrt{3} + 2\sqrt{6} - 10$ **48.** $2 - 2\sqrt{5}$ **49.** $x - 4\sqrt{x} + 4$ **50.** $y + 8\sqrt{y} + 16$ **51.** 3 **52.** 2 **53.** $2\sqrt{5}$ **54.** $4\sqrt{2}$ **55.** $x\sqrt{15x}$ **56.** $3x^2\sqrt{2}$ **57.** $\frac{\sqrt{22}}{11}$ **58.** $\frac{\sqrt{39}}{13}$ **59.** $\frac{\sqrt{30}}{6}$ **60.** $\frac{\sqrt{70}}{10}$ **61.** $\frac{\sqrt{5x}}{5x}$ **62.** $\frac{5\sqrt{3y}}{3y}$ **63.** $\frac{\sqrt{3x}}{x}$ **64.** $\frac{\sqrt{6y}}{y}$ **65.** $3\sqrt{5} + 6$ **66.** $8\sqrt{10} + 24$ **67.** $\frac{\sqrt{6} + \sqrt{2} + \sqrt{3} + 1}{2}$ **68.** $\sqrt{15} - 2\sqrt{3} - 2\sqrt{5} + 4$ **69.** $\frac{10\sqrt{x} - 50}{x - 25}$ **70.** $\frac{8\sqrt{x} + 8}{x - 1}$ **71.** 18 **72.** 13 **73.** 25 **74.** no solution **75.** 12 **76.** 5 **77.** 1 **78.** 9 **79.** $2\sqrt{14}$; 7.48 **80.** $\sqrt{117}$; 10.82 **81.** $4\sqrt{34}$ ft **82.** $5\sqrt{3}$ in. **83.** 2.4 in. **84.** 144π sq in.

Chapter 15 Test

1. 4 **2.** −5 **3.** 3 **4.** $\frac{3}{4}$ **5.** not a real number **6.** x^5 **7.** $3\sqrt{6}$ **8.** $2\sqrt{23}$ **9.** $y^3\sqrt{y}$ **10.** $2x^4\sqrt{6}$ **11.** 3 **12.** $2\sqrt[3]{2}$ **13.** $\frac{\sqrt{5}}{4}$ **14.** $\frac{y\sqrt{y}}{5}$ **15.** $-2\sqrt{13}$ **16.** $\sqrt{2} + 2\sqrt{3}$ **17.** $\frac{7\sqrt{3}}{10}$ **18.** $7\sqrt{2}$ **19.** $5 + 2\sqrt{35}$ **20.** $4x - 9$ **21.** $\sqrt{5}$ **22.** $2x\sqrt{5x}$ **23.** $\frac{\sqrt{6}}{3}$ **24.** $\frac{8\sqrt{5y}}{5y}$ **25.** $4\sqrt{6} - 8$ **26.** $\frac{3 + \sqrt{x}}{9 - x}$ **27.** 9 **28.** 5 **29.** 9 **30.** $4\sqrt{5}$ in. **31.** 2.19 m

Cumulative Review

1. 736.2; Sec. 5.1, Ex. 15 **2.** 328.2 **3.** 25.454; Sec. 5.2, Ex. 1 **4.** 17.052 **5.** no; Sec. 5.3, Ex. 13 **6.** yes **7.** $\frac{1}{6}$; Sec. 5.8, Ex. 5 **8.** $\frac{2}{5}$ **9.** 2; Sec. 9.3, Ex. 1 **10.** $\frac{2}{3}$ **11. a.** 102,000 **b.** 0.007358 **c.** 84,000,000 **d.** 0.00003007; Sec. 10.2, Ex. 17 **12. a.** 82,600 **b.** 0.099 **c.** 100,200 **d.** 0.008039 **13.** $6x^2 - 11x - 10$; Sec. 10.5, Ex. 7 **14.** $20x^2 + x - 1$ **15.** $(y + 2)(x + 3)$; Sec. 11.1, Ex. 8 **16.** $(4x^2 + 3)(4x - 7)$ **17.** $(3x + 2)(x + 3)$; Sec. 11.3, Ex. 1 **18.** $(9x + 4)(x - 1)$ **19. a.** $x = 3$ **b.** $x = 2, x = 1$ **c.** none; Sec. 12.1, Ex. 2 **20. a.** $x = 0$ **b.** none **c.** $x = 2, x = -2$ **21.** $\frac{x + 2}{x}$; Sec. 12.1, Ex. 5 **22.** $2(2x + y)$ **23. a.** 0 **b.** $\frac{15 + 14x}{50x^2}$; Sec. 12.4, Ex. 1 **24. a.** $-\frac{x}{10}$ **b.** $\frac{12 + 5a}{16a^2}$ **25.** $-\frac{17}{5}$; Sec. 12.5, Ex. 4 **26.** 2 **27.** Sec. 13.3, Ex. 7 **28.**

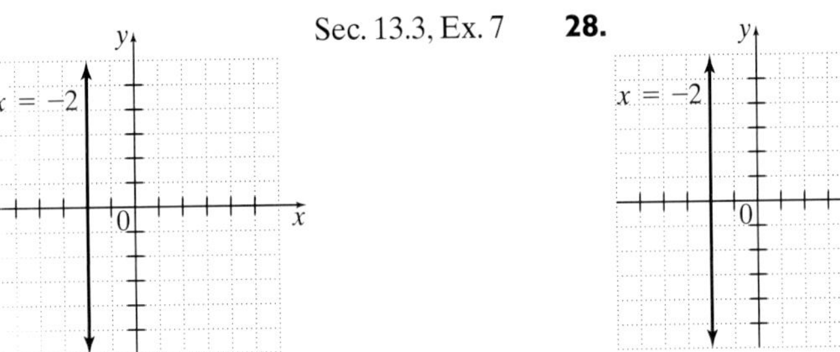

29. 1; Sec. 15.1, Ex. 6 **30.** 2 **31.** -3; Sec. 15.1 Ex.7 **32.** -2 **33.** $\frac{1}{5}$; Sec. 15.1, Ex. 8 **34.** $\frac{3}{4}$ **35.** $3\sqrt{6}$; Sec. 15.2, Ex. 1 **36.** $2\sqrt{10}$ **37.** $10\sqrt{2}$; Sec. 15.2, Ex. 3 **38.** $5\sqrt{5}$ **39.** $9\sqrt{3}$; Sec. 15.3, Ex. 6 **40.** $9\sqrt{3}$ **41.** $2x - 4\sqrt{x}$; Sec. 15.3, Ex. 8 **42.** $12x + 6\sqrt{x}$ **43.** $\frac{2\sqrt{7}}{7}$; Sec. 15.4, Ex. 9 **44.** $\frac{4\sqrt{5}}{5}$ **45.** $\frac{1}{2}$; Sec. 15.5, Ex. 3 **46.** 4

Chapter 16 Quadratic Equations

Chapter 16 Pretest

1. $a = 0, 6$; 16.1A **2.** $x = -\frac{1}{2}, 6$; 16.1A **3.** $b = \pm 12$; 16.1B **4.** $y = \frac{7 \pm 2\sqrt{6}}{2}$; 16.1B **5.** $x = 6, 8$; 16.2A **6.** $x = -\frac{1}{3}, 2$; 16.2B **7.** $x = -3, 9$; 16.3A **8.** $m = \frac{7 \pm \sqrt{145}}{8}$; 16.3A **9.** $x = \frac{-1 \pm \sqrt{73}}{4}$; 16.3A **10.** $x = \frac{-3 \pm 3\sqrt{2}}{5}$; 16.1B **11.** $x = -\frac{3}{4}, -\frac{3}{2}$; 16.1A **12.** $m = 3 \pm \sqrt{6}$; 16.2A **13.** $x = \frac{-4 \pm 3\sqrt{2}}{2}$; 16.3A **14.** $y = -7 \pm \sqrt{5}$; 16.1B

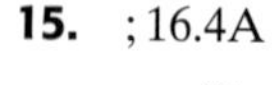 

15. ; 16.4A **16.** ; 16.4B **17.** ; 16.4B **18.** ; 16.4B

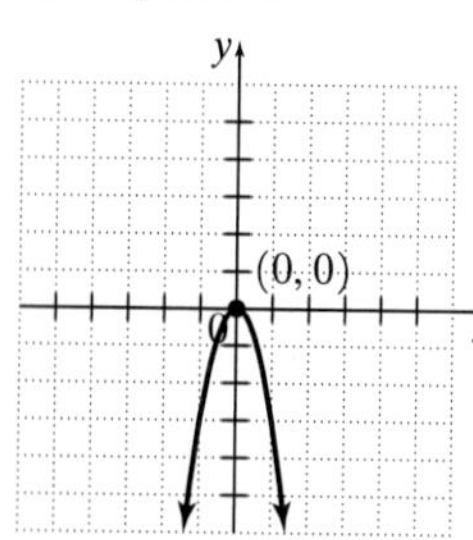

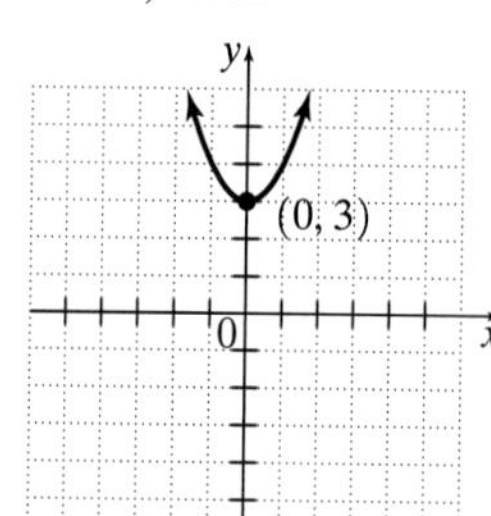

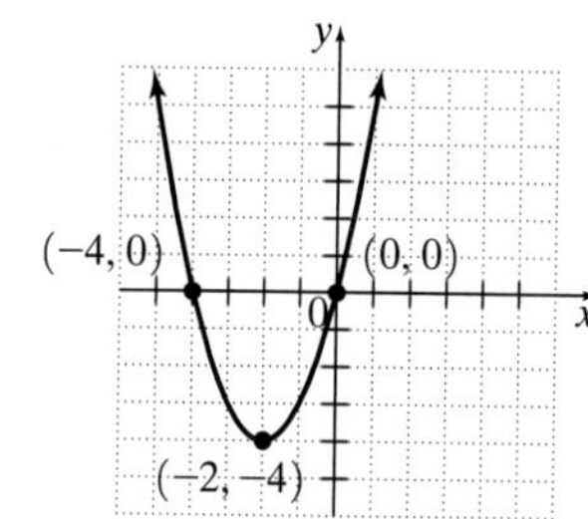

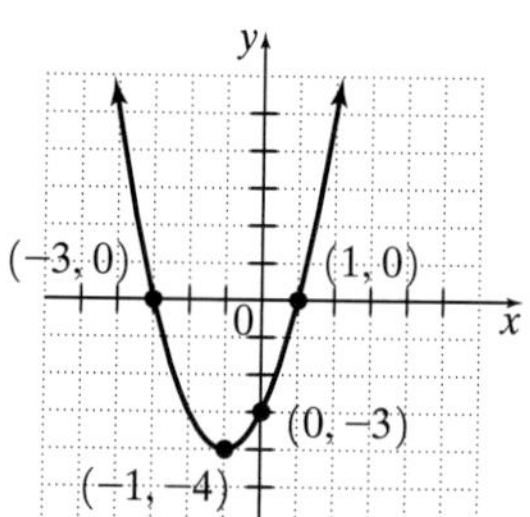

Section 16.1

1. ± 3 **3.** $-5, 3$ **5.** ± 4 **7.** ± 3 **9.** $-5, -2$ **11.** ± 8 **13.** $\pm\sqrt{21}$ **15.** $\pm\frac{1}{5}$ **17.** no real solution **19.** $\pm\frac{\sqrt{39}}{3}$ **21.** $\pm\frac{2\sqrt{7}}{7}$ **23.** $\pm\sqrt{2}$ **25.** Answers may vary. **27.** 12, -2 **29.** $-2 \pm \sqrt{7}$ **31.** 1, 0 **33.** $-2 \pm \sqrt{10}$ **35.** $\frac{8}{3}, -4$ **37.** no real solution **39.** $\frac{11 \pm 5\sqrt{2}}{2}$ **41.** $\frac{7 \pm 4\sqrt{2}}{3}$ **43.** 2.3 sec **45.** 72.7 sec **47.** $2\sqrt{5}$ in. $\approx$ 4.47 in. **49.** $\sqrt{3039}$ ft $\approx$ 55.13 ft **51.** $(x + 3)^2$ **53.** $(x - 2)^2$ **55.** 2, -6 **57.** $r = 6$ in. **59.** 5 sec **61.** -1.02, 3.76 **63.** 2004

Mental Math

1. 16 **3.** 100 **5.** 49

Section 16.2

1. $-6, -2$ **3.** $-1 \pm \sqrt{6}$ **5.** 0, 6 **7.** $\frac{-5 \pm \sqrt{53}}{2}$ **9.** $1 \pm \sqrt{2}$ **11.** $-1, -4$ **13.** $-3 \pm \sqrt{34}$ **15.** $\frac{3 \pm \sqrt{21}}{2}$ **17.** $-6, 3$ **19.** $-2, 4$ **21.** no real solution **23.** $\frac{3 \pm \sqrt{19}}{2}$ **25.** $-2 \pm \frac{\sqrt{6}}{2}$ **27.** $\frac{1}{2}, 1$ **29.** Answers may vary. **31.** $-\frac{1}{2}$ **33.** -1 **35.** $3 + 2\sqrt{5}$ **37.** $\frac{1 - 3\sqrt{2}}{2}$ **39.** $k = 8$ or $k = -8$ **41.** 2006 **43.** $-6, -2$ **45.** ≈ -0.68, 3.68

Mental Math

1. $a = 2, b = 5, c = 3$ **3.** $a = 10, b = -13, c = -2$ **5.** $a = 1, b = 0, c = -6$

Section 16.3

1. 2, 1 **3.** $\frac{-7 \pm \sqrt{37}}{6}$ **5.** $\pm\frac{2}{7}$ **7.** no real solution **9.** 10, −3 **11.** $\pm\sqrt{5}$ **13.** −3, 4 **15.** $-2 \pm \sqrt{7}$ **17.** $\frac{-9 \pm \sqrt{129}}{12}$ **19.** $\frac{4 \pm \sqrt{2}}{7}$ **21.** $3 \pm \sqrt{7}$ **23.** $\frac{3 \pm \sqrt{3}}{2}$ **25.** $-1, \frac{1}{3}$ **27.** $\frac{3 \pm \sqrt{13}}{4}$ **29.** $\frac{1}{5}, -\frac{3}{4}$ **31.** no real solution **33.** no real solution **35.** $1 \pm \sqrt{2}$ **37.** $-\frac{3}{4}, -\frac{1}{2}$ **39.** $\frac{7 \pm \sqrt{129}}{20}$ **41.** no real solution **43.** $\frac{1 \pm \sqrt{2}}{5}$

45.

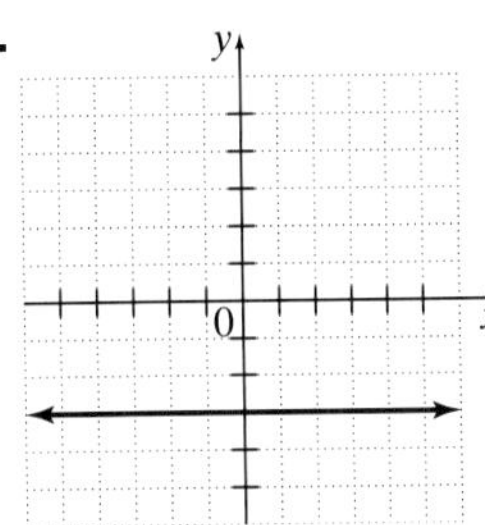

47.

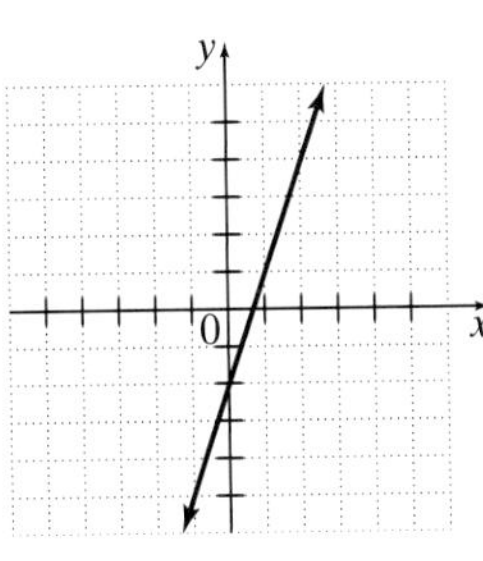

49. $\sqrt{51}$ m **51.** 3.9 ft by 8.9 ft **53.** $\frac{-3\sqrt{2} \pm \sqrt{38}}{2}$ **55.** Answers may vary. **57.** 3.9, −2.9 **59.** −0.9, 0.2 **61.** 8.1 sec **63.** 2005

Integrated Review

1. $\frac{1}{5}, 2$ **2.** $-3, \frac{2}{5}$ **3.** $1 \pm \sqrt{2}$ **4.** $3 \pm \sqrt{2}$ **5.** $\pm 2\sqrt{5}$ **6.** $\pm 6\sqrt{2}$ **7.** no real solution **8.** no real solution **9.** 2 **10.** 3 **11.** 3 **12.** $\frac{7}{2}$ **13.** ±2 **14.** ±3 **15.** 0, 1, 2 **16.** 0, −3, −4 **17.** −5, 0 **18.** $0, \frac{8}{3}$ **19.** $\frac{3 \pm \sqrt{7}}{5}$ **20.** $\frac{3 \pm \sqrt{5}}{2}$ **21.** $\frac{3}{2}, -1$ **22.** $\frac{2}{5}, -2$ **23.** $\frac{5 \pm \sqrt{105}}{20}$ **24.** $\frac{-1 \pm \sqrt{3}}{4}$ **25.** $5, \frac{7}{4}$ **26.** $\frac{7}{9}, 1$ **27.** $\frac{7 \pm 3\sqrt{2}}{5}$ **28.** $\frac{5 \pm 5\sqrt{3}}{4}$ **29.** $\frac{7 \pm \sqrt{193}}{6}$ **30.** $\frac{-7 \pm \sqrt{193}}{12}$ **31.** 11, −10 **32.** −8, 7 **33.** $-\frac{2}{3}, 4$ **34.** $2, -\frac{4}{5}$ **35.** 0.1, 0.5 **36.** 0.3, −0.2 **37.** $\frac{11 \pm \sqrt{41}}{20}$ **38.** $\frac{11 \pm \sqrt{41}}{40}$ **39.** $\frac{4 \pm \sqrt{10}}{2}$ **40.** $\frac{5 \pm \sqrt{185}}{4}$ **41.** Answers may vary.

Calculator Explorations

1. −0.41, 7.41 **3.** 0.91, 2.38 **5.** −0.39, 0.84

Section 16.4

1.

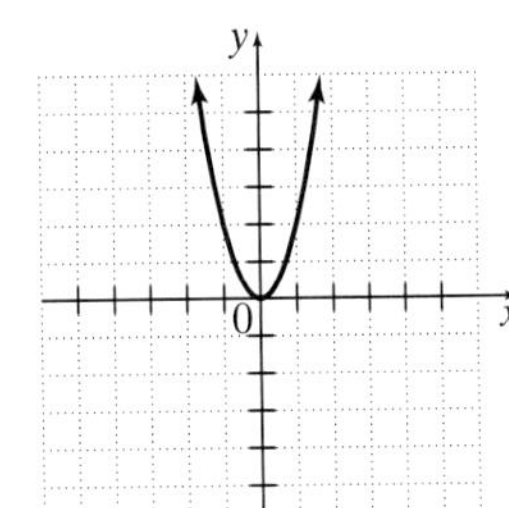

3.

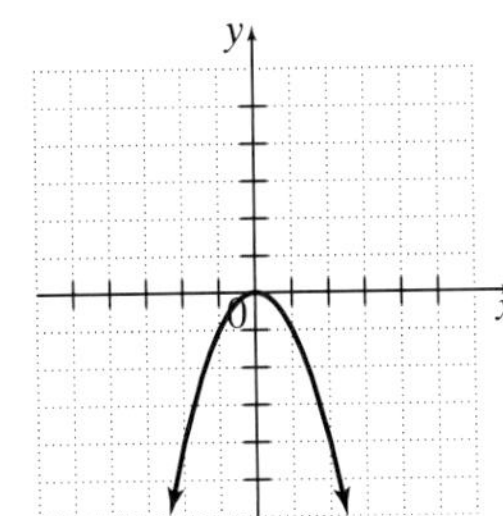

5.

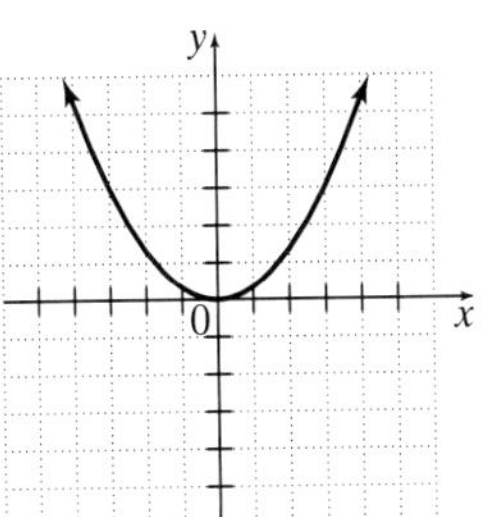

7

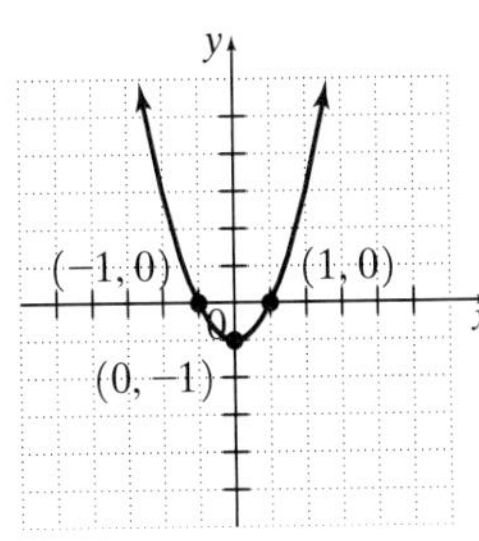

9.

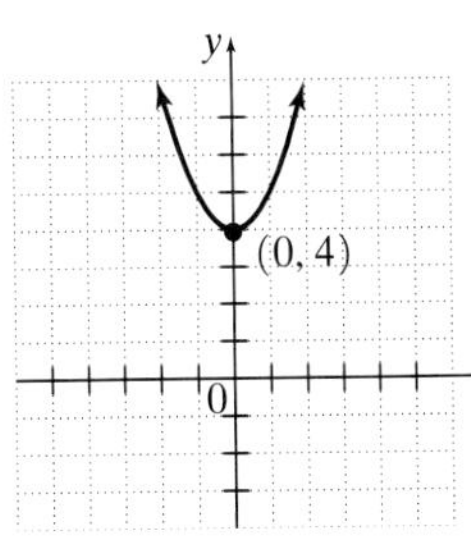

11.

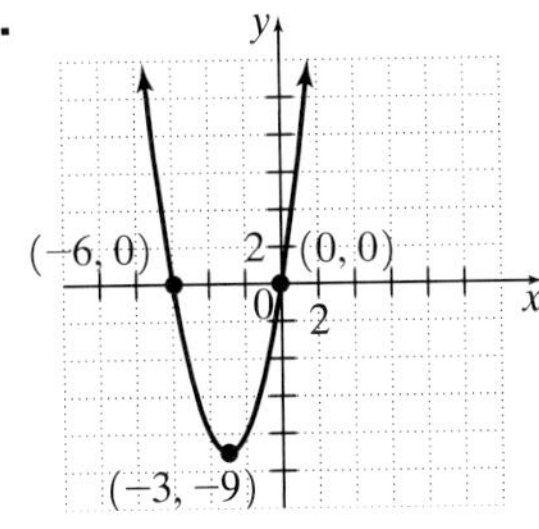

13.

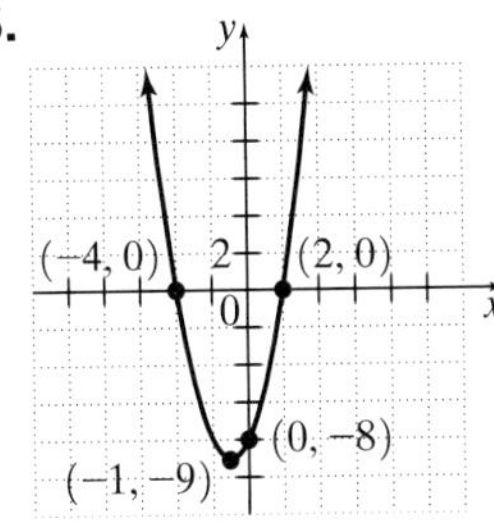

15.

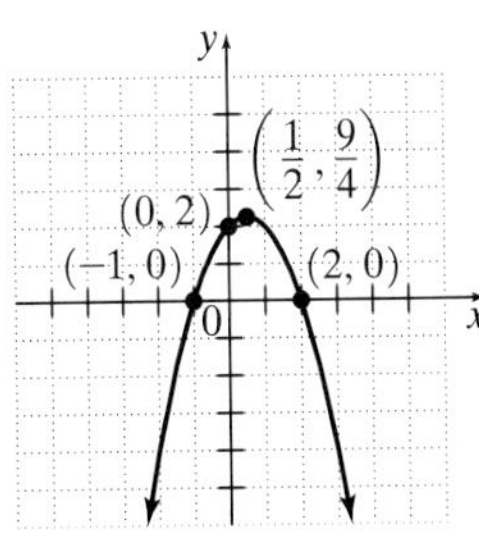

17.

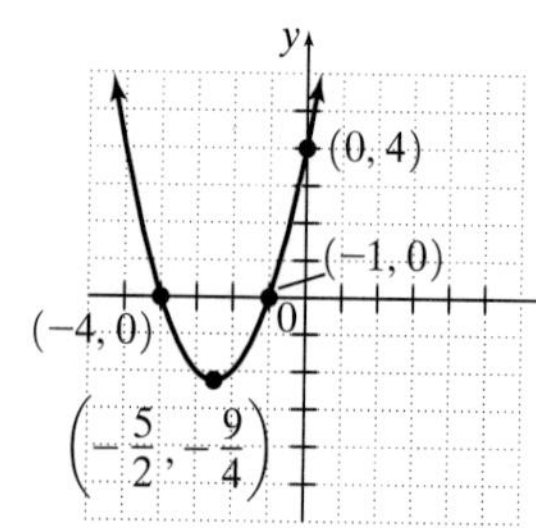

19.

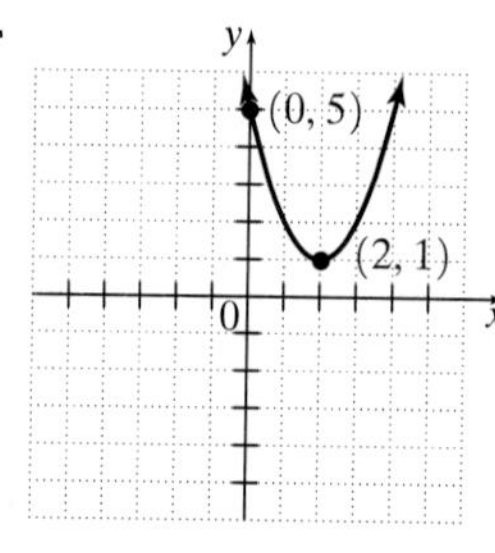

21.

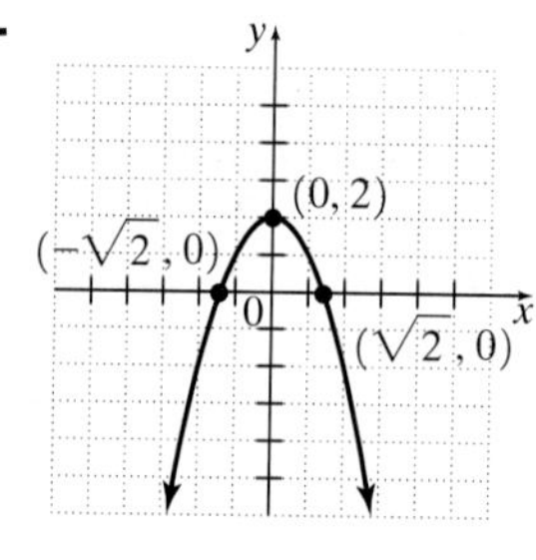

23.

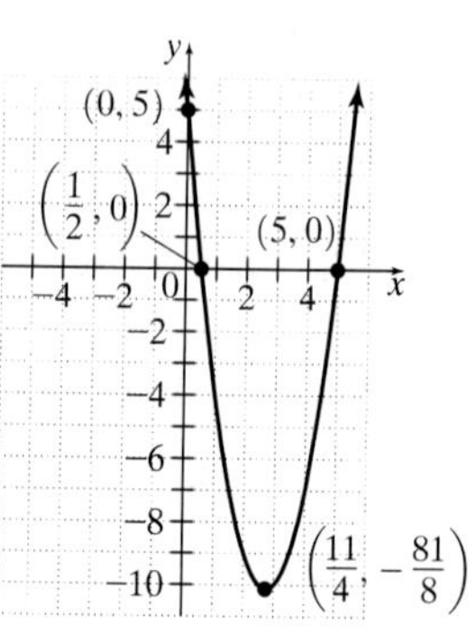

25.

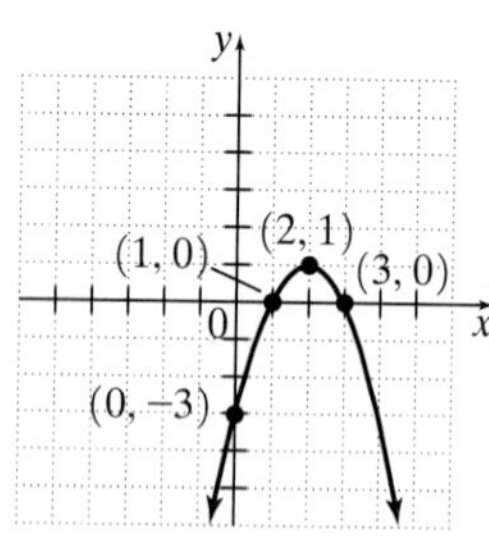

27. $\frac{5}{14}$ **29.** $\frac{x}{2}$ **31.** $\frac{2x^2}{x-1}$ **33.** $-4b$ **35. a.** 256 ft **b.** $t = 4$ sec **c.** $t = 8$ sec **37.** E **39.** C **41.** B

Chapter 16 Review

1. $4, -\frac{3}{5}$ **2.** $-7, -\frac{4}{3}$ **3.** $-\frac{1}{3}, 2$ **4.** $\frac{5}{7}, -1$ **5.** $0, 3, -3$ **6.** $-2, 2$ **7.** $6, -6$ **8.** $9, -9$ **9.** $\pm 5\sqrt{2}$ **10.** $\pm 3\sqrt{5}$ **11.** 4, 18 **12.** $7, -13$ **13.** $2, -3$ **14.** $1, -5$ **15.** 2.5 sec **16.** 40.6 sec **17.** $-2 \pm \sqrt{5}$ **18.** $4 \pm \sqrt{19}$ **19.** $3 \pm \sqrt{2}$ **20.** $-3 \pm \sqrt{2}$ **21.** $\frac{1}{2}, -1$ **22.** $\frac{-3 \pm \sqrt{13}}{2}$ **23.** $5 \pm 3\sqrt{2}$ **24.** $-2 \pm \sqrt{11}$ **25.** $\frac{1}{2}, -1$ **26.** $\frac{-3 \pm \sqrt{13}}{2}$ **27.** $-\frac{5}{3}$ **28.** $\frac{9}{4}$ **29.** $\frac{2}{5}, \frac{1}{3}$ **30.** $\frac{2}{3}, \frac{1}{5}$ **31.** no real solution **32.** no real solution **33.** 2004 **34.** 2012

35.

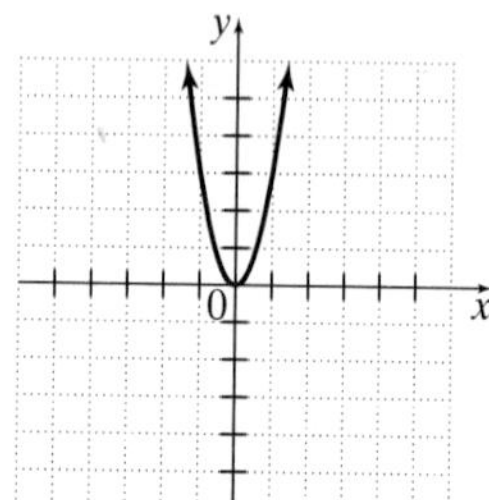

36.

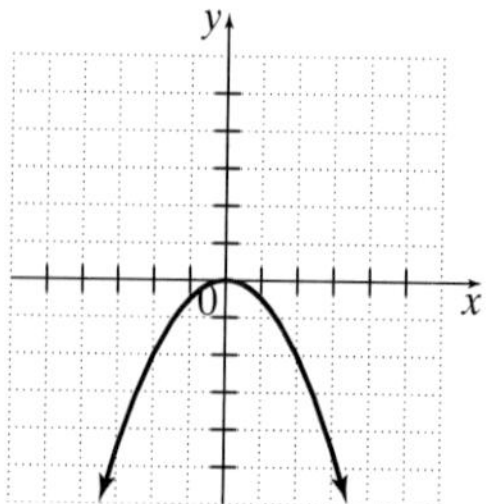

37.

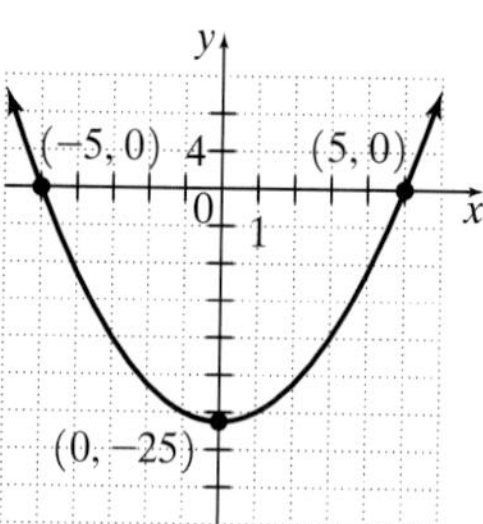

38.

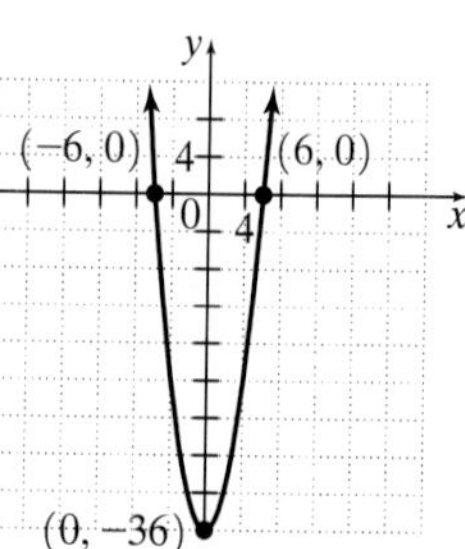

39.

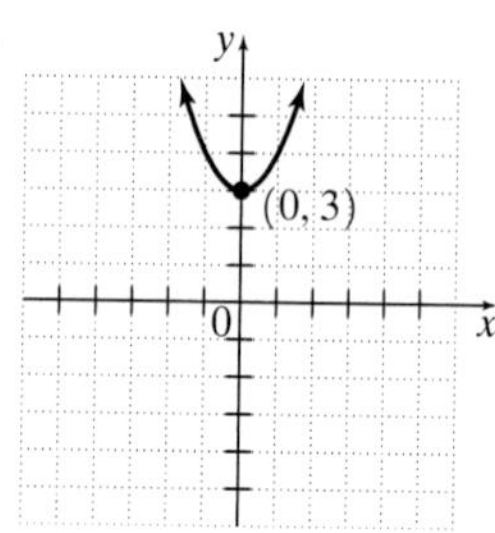

40.

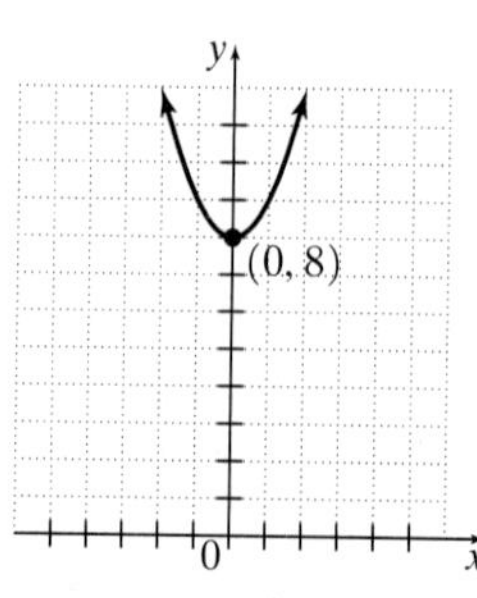

41.

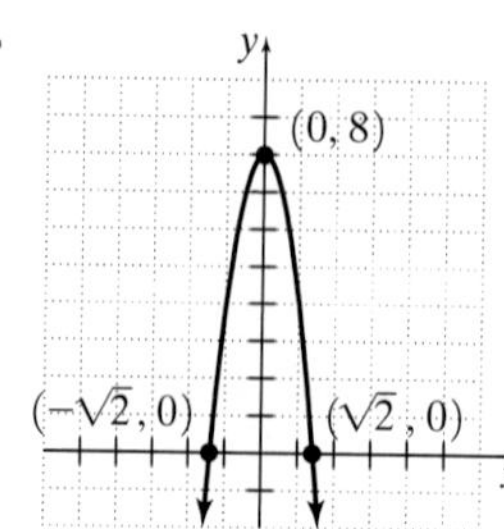

42.

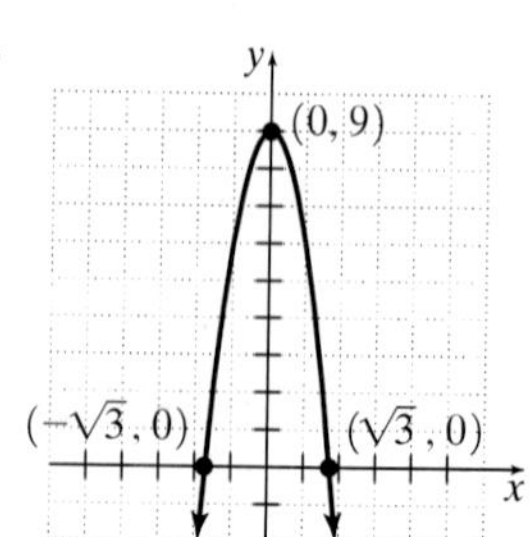

43.

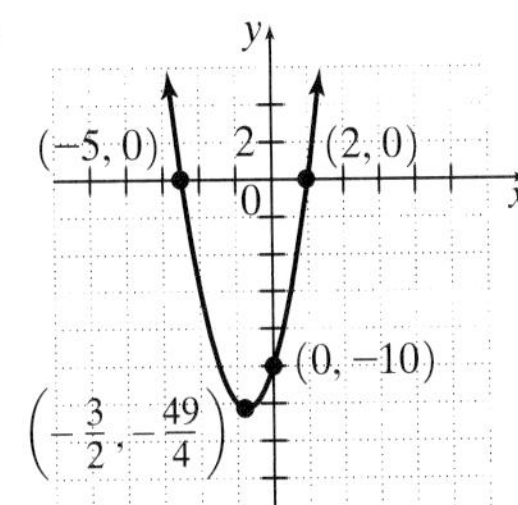

44.

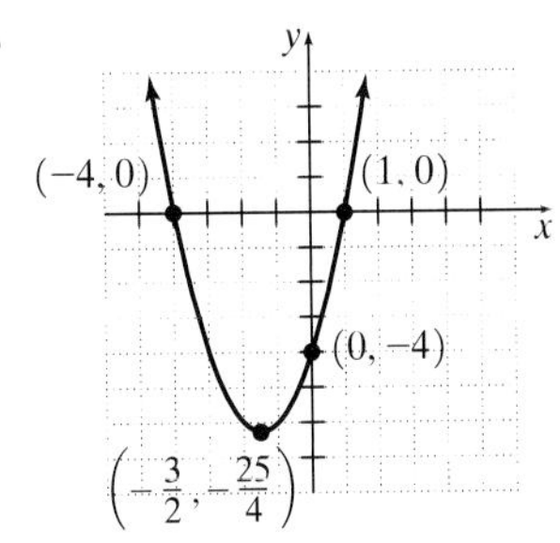

45.

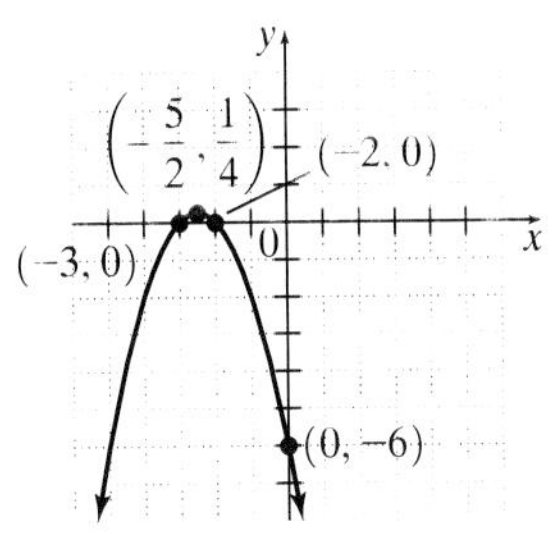

46.

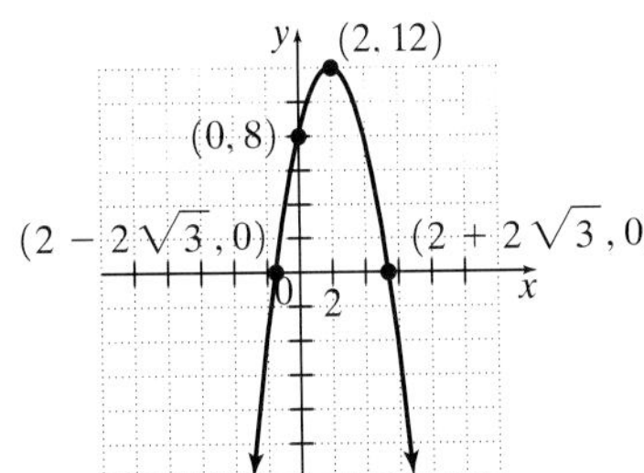

47.

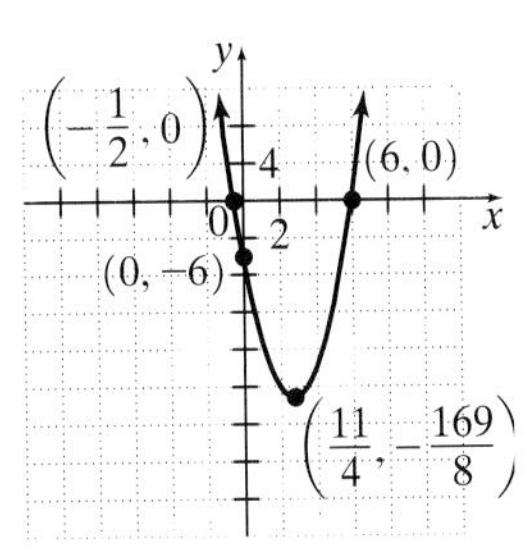

48.

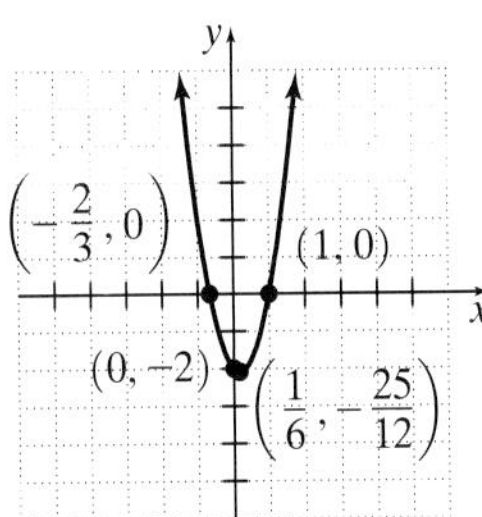

49. -2 **50.** $-\frac{3}{2}, 3$ **51.** no real solution **52.** $-2, 2$ **53.** A **54.** D **55.** B **56.** C

Chapter 16 Test

1. ± 4 **2.** $\frac{5 \pm 2\sqrt{2}}{3}$ **3.** $10, 16$ **4.** $\frac{-6 \pm 4\sqrt{3}}{3}$ **5.** $-2, 5$ **6.** $\frac{5 \pm \sqrt{37}}{6}$ **7.** $1, -\frac{4}{3}$ **8.** $-1, \frac{5}{3}$ **9.** $\frac{7 \pm \sqrt{73}}{6}$
10. $-7, 1$ **11.** $2, \frac{1}{3}$ **12.** $\frac{3 \pm \sqrt{7}}{2}$ **13.** $0, \frac{1}{3}, -\frac{1}{3}$

14.

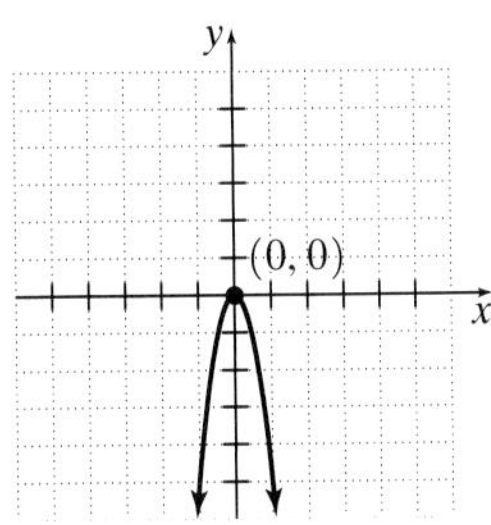

15.

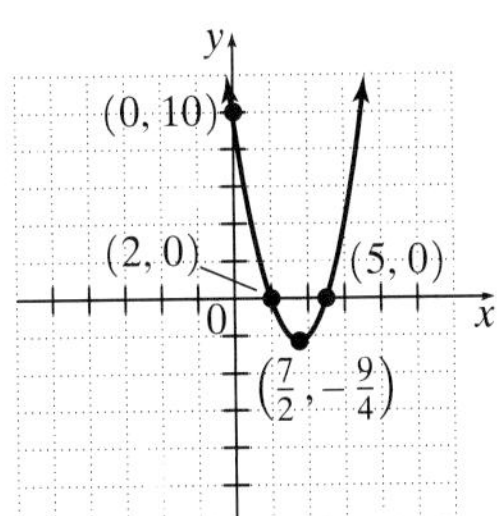

16. 2.7 sec

Cumulative Review

1. 0.0786; Sec. 5.4, Ex. 5 **2.** 0.818 **3.** 0.012; Sec. 5.4, Ex. 6 **4.** 0.0503 **5.** 71% **6.** 91% **7.** 28 in.; Sec. 8.3, Ex. 2
8. 25 ft **9.** 5.1 sq mi; Sec. 8.4, Ex. 2 **10.** 68 sq in. **11.** 3.21 L; Sec. 8.6, Ex. 7 **12.** 43.21 L **13.** $\frac{16}{3}$; Sec. 9.3, Ex. 2
14. $\frac{5}{6}$ **15.** 1; Sec. 10.1, Ex. 25 **16.** -1 **17.** $(r + 6)(r - 7)$; Sec. 11.2, Ex. 4 **18.** $(y - 7)(y + 10)$
19. $(2x - 3y)(5x + y)$; Sec. 11.3, Ex. 4 **20.** $(9x - y)(8x - 3y)$ **21.** $(2x - 1)(4x - 5)$; Sec. 11.4, Ex. 1 **22.** $(3x - 2)(5x + 2)$
23. a. $x(2x + 7)(2x - 7)$; Sec. 11.5, Ex. 16 **b.** $2(9x^2 + 1)(3x + 1)(3x - 1)$; Sec. 11.5, Ex. 17 **24. a.** $x(3x - 1)(3x + 1)$
b. $5(x^2 + 1)(x - 1)(x + 1)$ **25.** $\frac{1}{5}, -\frac{3}{2}, 6$; Sec. 11.6, Ex. 8 **26.** $-4, \frac{3}{4}, \frac{1}{10}$ **27.** $\frac{x + 7}{x - 5}$; Sec. 12.1, Ex. 4 **28.** $\frac{x - 5}{x + 7}$
29. -5; Sec. 12.6, Ex. 1 **30.** 4 **31. a.** -3 **b.** 0 **c.** -3; Sec. 13.1, Ex. 4 **32. a.** 7 **b.** $\frac{7}{2}$ **c.** -3 **33. a.** parallel
b. perpendicular **c.** neither; Sec. 13.4, Ex. 6 **34. a.** parallel **b.** perpendicular **c.** neither **35. a.** function
b. not a function; Sec. 13.6, Ex. 2 **36. a.** function **b.** not a function **37.** (4, 2); Sec. 14.2, Ex. 1 **38.** (1, 4)
39. 6; Sec. 15.1, Ex. 1 **40.** 9 **41.** $\frac{3}{10}$; Sec. 15.1, Ex. 4 **42.** $\frac{4}{5}$ **43.** $-1 + \sqrt{3}$; Sec. 15.4, Ex. 12 **44.** $7(\sqrt{5} + 2)$
45. 7, −1; Sec. 6.1, Ex. 5 **46.** −1, −7 **47.** $1 \pm \sqrt{5}$; Sec. 16.3, Ex.5 **48.** $\frac{5 \pm \sqrt{137}}{8}$

SUBJECT INDEX

H

I

T

U

V

W

PHOTO CREDITS

Chapter 1 CO © Phil Schermeister/CORBIS; **(p. 3)** Sepp Seitz/Woodfin Camp & Associates; **(p. 4)** Rachel Epstein/PhotoEdit Inc.; **(p. 17)** NASA/Johnson Space Center; **(p. 29)** Dwayne Newton/PhotoEdit Inc., Tom Roberts/Getty Images, Inc.–Hulton Archive Photos; **(p. 41)** Steve Shott/Dorling Kindersley Media Library, SuperStock; **(p. 42)** Dole Plantation, Hawaii, The Schiller Group, Ltd., Jerry L. Ferrara/Photo Researchers, Inc.; **(p. 50)** PhotoEdit; **(p. 52)** AP/Wide World Photos; **(p. 63)** Hyatt Corporation; **(p. 77)** Jonathan Ferrey/Allsport/Getty Images, Inc.–Allsport Photography; **(p. 87)** Jean-Claude LeJeune/Stock Boston

Chapter 2 CO www.comstock.com; **(p. 117)** John Elk/Bruce Coleman Inc.; © Derek Croucher/CORBIS; **(p. 124)** Voscar—The Maine Photographer; **(p. 132)** Mclancolia (engraving) By Albrecht Durer (1471-1528). Guildhall Library, Corporation of London, UK/Bridgeman Art Library, London/New York; **(p. 142)** Kathy Willens/AP/Wide World Photos; **(p. 145)** Jerry L. Ferrara/Photo Researchers, Inc., Lon C. Diehl/PhotoEdit Inc.

Chapter 3 CO © Owaki-Kulla/CORBIS; **(p. 176)** Collection of the The New York Public Library, Astor, Lenox and the Tilden Foundation/Art Resource, NY; **(p. 215)** Reuters America, Inc., Steve Morrell/AP/Wide World Photos, F. Rangel/The Image Works; **(p. 228)** Bob Krist/CORBIS BETTMANN

Chapter 4 CO © Larry Williams/CORBIS; **(p. 257)** Shuttle Mission Imagery/Getty Images, Inc., Kevin Horan/Stock Boston; **(p. 260)** American Red Cross; **(p. 266)** Paul L. Ruben; **(p. 270)** Tony Freeman/PhotoEdit Inc.; **(p. 271)** P. Lloyd/Weatherstock; **(p. 296)** © David Young-Wolff/PhotoEdit Inc.; **(p. 332)** Ray Massey/Getty Images, Inc.–Stone Allstock; **(p. 333)** © Gary W. Carter/CORBIS, ©Jeff Greenberg/PhotoEdit Inc.; **(p. 335)** Robert Harbison, Peter Welmann/Animals Animals/Earth Scenes

Chapter 5 CO © David Young Wolff/PhotoEdit Inc.; **(p. 362)** AP/Wide World Photos; **(p. 371)** CORBIS BETTMANN; **(p. 384)** William H. Johnson/Johnson's Photography; **(p. 397)** Angelo Hornak/CORBIS BETTMANN; **(p. 408)** Michael Newman/PhotoEdit Inc., Getty Images, Inc.; **(p. 426)** Jose Luis Pelaez, Inc./CORBIS BETTMANN, A. Ramey/PhotoEdit Inc.; **(p. 430)** Bill Bachmann/Stock Boston, Piet Mondrian, "Composition with Grid #1," 1918. Oil on canvas, 31 9/16 x 19 5/8 inc. (80.2 x 49.8 cm). The Museum of Fine Arts, Houston, Gift of Mr. and Mrs. Pierre Schlumberger © 2003 Mondrian/Holtzman Trust, c/o Beeldrecht/Artists Rights Society (ARS), New York

Chapter 6 CO Stephanie Maze/CORBIS BETTMANN; **(p. 462)** Michael Newman/PhotoEdit; **(p. 469)** C. Borland/Getty Images, Inc.; **(p. 477)** SIU/Photo Researchers, Inc., Getty Images, Inc.; **(p. 480)** Telegraph Colour Library/Getty Images, Inc.-Taxi; **(p. 501)** Getty Images, Inc.-Liaison; **(p 504)** Jeff Greenberg/Omni-Photo Communications, Inc.; **(p. 510)** AP/Wide World Photos; **(p. 511)** Michal Heron/Pearson Education/PH College, AP/Wide World Photos

Chapter 7 CO Keith Brofsky/Getty Images, Inc.

Chapter 8 CO Klaus Lahnstein/Getty Images, Inc.-Stone Allstock; **(p. 615)** AP/World Wide Photos, Will & Deni McIntyre/Photo Researchers, Inc.; **(p. 639)** Tom Bean/DRK Photo; **(p. 655)** Will & Deni McIntyre/Photo Researchers, Inc.; **(p. 656)** ©Bettmann/CORBIS; **(p. 683)** Philatelic Group; **(p. 685)** Bill Bachmann/ © Index Stock Imagery, Inc.; © Jon Feingersh/CORBIS, **(p. 686)** Philatelic Group

Chapter 9 CO Crayola, chevron and serpentine designs are registered trademarks, the rainbow/swash is a trademark of Binney & Smith used with permission. Photo courtesy of Binney & Smith; **(p. 733)** Photo Researchers, Inc.; **(p. 765)** AP/World Wide Photos; **(p. 769)** National Space Science Data Center; **(p. 770)** AP/World Wide Photos; **(p. 772)** Bill Bachmann/Stock Boston; **(p. 780)** Colin Braley Reuters/Getty Images, Inc. – Hulton Archive Photos; **(p. 781)** ©Norbert Wu/www.norbertwu.com , John Elk III/Stock Boston; **(p. 799)** Catherine Karnow/Woodfin Camp & Associates

Chapter 10 CO J.A. Kraulis/Masterfile Corporation; **(p. 826)** G. Brad Lewis/Innerspace Visions, Jim Pickerell/Stock Boston; **(p. 834)** Reuters/Getty Images, Inc.; **(p. 836)** NASA; **(p. 844)** Masterfile Corporation, Terrence Moore/Woodfin Camp & Associates; **(p. 862)** Getty Images, Inc.; **(p. 866)** CORBIS BETTMANN; **(p. 872)** Getty Images, Inc., National Space Science Data Center

Chapter 11 CO R. Ian Lloyd/Masterfile Corporation; **(p. 913)** R. Ian Lloyd/Masterfile Corporation, John Freeman/Getty Images, Inc.-Stone Allstock; **(p. 914)** Culver Pictures, Inc.; **(p. 935)** Stephen Dunn/Getty Images, Inc./Allsport Photography

Chapter 12 CO Getty Images, Inc.; **(p. 960)** AP/World Wide Photos; **(p. 965)** Matthew McVay/Stock Boston, Tebo Photography; **(p. 966)** AP/World Wide Photos; **(p. 969)** Gary Caskey Reuters/Getty Images, Inc.-Hulton Archive Photos, AP/World Wide Photos; **(p. 1005)** AP/World Wide Photos; **(p.1006)** AP/World Wide Photos

Chapter 13 CO Frank LaBua/Pearson Education/PH College, **(p. 1055)** AP/World Wide Photos, Superstock, Inc.,**(p. 1075)** Faraway Places; **(p. 1081)** Masterfile Corporation; **(p. 1086)** Woodfin Camp & Associates; **(p. 1092)** Woodfin Camp & Associates, PhotoEdit; **(p. 1093)** CORBIS/SABA Press Photos, Inc.

Chapter 14 CO Lambert/Getty Images, Inc.; **(p. 1166)** Lloyd Sutton/Masterfile Corporation; **(p. 1173)** MTV photograph used with permission by MTV: MUSIC TELEVISION. © 2001 MTV Networks. All Rights Reserved. MTV: Music Television, all related titles, characters and logos are trademarks owned by MTV Networks, a division of Viacom International, Inc.; **(p. 1178)** Photo of the Banquine act from the Cirque du Soleil show Quidam: © Cirque du Soleil, Inc. Photo by Al Seib. Costumes bby Dominique Lemieux; **(p. 1185)** AP/World Wide Photos

Chapter 15 CO CORBIS BETTMANN; **(p. 1217)** Photofest, Peter Poulides/Getty Images, Inc.-Stone Allstock

Chapter 16 CO Larry L. Miller/Photo Researchers, Inc.; **(p. 1266)** Photofest; **(p. 1268)** AP/World Wide Photos; **(p. 1269)** AP/World Wide Photos; **(p. 1270)** Tony Acevedo/National Astonomy and Ionosphere Center; **(p. 1296)** Lawrence Migdale/Photo Researchers, Inc.; **(p. 1300)** Ted Clutter/Photo Researchers, Inc.

READ THIS LICENSE CAREFULLY BEFORE OPENING THIS PACKAGE. BY OPENING THIS PACKAGE, YOU ARE AGREEING TO THE TERMS AND CONDITIONS OF THIS LICENSE. IF YOU DO NOT AGREE, DO NOT OPEN THE PACKAGE. PROMPTLY RETURN THE UNOPENED PACKAGE AND ALL ACCOMPANYING ITEMS TO THE PLACE YOU OBTAINED THEM. ***THESE TERMS APPLY TO ALL LICENSED SOFTWARE ON THE DISK EXCEPT THAT THE TERMS FOR USE OF ANY SHAREWARE OR FREEWARE ON THE DISKETTES ARE AS SET FORTH IN THE ELECTRONIC LICENSE LOCATED ON THE DISK:***

Single PC Site License

1. **GRANT OF LICENSE and OWNERSHIP:** The enclosed computer programs and any data ("Software") are licensed, not sold, to you by Pearson Education, Inc. publishing as Pearson Prentice Hall ("We" or the "Company") in consideration of your adoption of the accompanying Company textbooks and/or other materials, and your agreement to these terms. You own only the disk(s) but we and/or our licensors own the Software itself. This license allows instructors and students enrolled in the course using the Company textbook that accompanies this Software (the "Course") to use and display the enclosed copy of the Software on an unlimited number of computers, for academic use only, so long as you comply with the terms of this Agreement. You may make one copy for back up only. We reserve any rights not granted to you.

2. **USE RESTRICTIONS:** You may not sell or license copies of the Software or the Documentation to others. You may not transfer, distribute or make available the Software or the Documentation. You may not reverse engineer, disassemble, decompile, modify, adapt, translate or create derivative works based on the Software or the Documentation. You may be held legally responsible for any copying or copyright infringement that is caused by your failure to abide by the terms of these restrictions..

3. **TERMINATION:** This license is effective until terminated. This license will terminate automatically without notice from the Company if you fail to comply with any provisions or limitations of this license. Upon termination, you shall destroy the Documentation and all copies of the Software. All provisions of this Agreement as to limitation and disclaimer of warranties, limitation of liability, remedies or damages, and our ownership rights shall survive termination.

4. **DISCLAIMER OF WARRANTY: THE COMPANY AND ITS LICENSORS MAKE NO WARRANTIES ABOUT THE SOFTWARE, WHICH IS PROVIDED "AS-IS." IF THE DISK IS DEFECTIVE IN MATERIALS OR WORKMANSHIP, YOUR ONLY REMEDY IS TO RETURN IT TO THE COMPANY WITHIN 30 DAYS FOR REPLACEMENT UNLESS THE COMPANY DETERMINES IN GOOD FAITH THAT THE DISK HAS BEEN MISUSED OR IMPROPERLY INSTALLED, REPAIRED, ALTERED OR DAMAGED. THE COMPANY DISCLAIMS ALL WARRANTIES, EXPRESS OR IMPLIED, INCLUDING WITHOUT LIMITATION, THE IMPLIED WARRANTIES OF MERCHANTABILITY AND FITNESS FOR A PARTICULAR PURPOSE. THE COMPANY DOES NOT WARRANT, GUARANTEE OR MAKE ANY REPRESENTATION REGARDING THE ACCURACY, RELIABILITY, CURRENTNESS, USE, OR RESULTS OF USE, OF THE SOFTWARE.**

5. **LIMITATION OF REMEDIES AND DAMAGES: IN NO EVENT, SHALL THE COMPANY OR ITS EMPLOYEES, AGENTS, LICENSORS OR CONTRACTORS BE LIABLE FOR ANY INCIDENTAL, INDIRECT, SPECIAL OR CONSEQUENTIAL DAMAGES ARISING OUT OF OR IN CONNECTION WITH THIS LICENSE OR THE SOFTWARE, INCLUDING, WITHOUT LIMITATION, LOSS OF USE, LOSS OF DATA, LOSS OF INCOME OR PROFIT, OR OTHER LOSSES SUSTAINED AS A RESULT OF INJURY TO ANY PERSON, OR LOSS OF OR DAMAGE TO PROPERTY, OR CLAIMS OF THIRD PARTIES, EVEN IF THE COMPANY OR AN AUTHORIZED REPRESENTATIVE OF THE COMPANY HAS BEEN ADVISED OF THE POSSIBILITY OF SUCH DAMAGES.** SOME JURISDICTIONS DO NOT ALLOW THE LIMITATION OF DAMAGES IN CERTAIN CIRCUMSTANCES, SO THE ABOVE LIMITATIONS MAY NOT ALWAYS APPLY.

6. **GENERAL:** THIS AGREEMENT SHALL BE CONSTRUED IN ACCORDANCE WITH THE LAWS OF THE UNITED STATES OF AMERICA AND THE STATE OF NEW YORK, APPLICABLE TO CONTRACTS MADE IN NEW YORK, AND SHALL BENEFIT THE COMPANY, ITS AFFILIATES AND ASSIGNEES. This Agreement is the complete and exclusive statement of the agreement between you and the Company and supersedes all proposals, prior agreements, oral or written, and any other communications between you and the company or any of its representatives relating to the subject matter. If you are a U.S. Government user, this Software is licensed with "restricted rights" as set forth in subparagraphs (a)-(d) of the Commercial Computer-Restricted Rights clause at FAR 52.227-19 or in subparagraphs (c)(1)(ii) of the Rights in Technical Data and Computer Software clause at DFARS 252.227-7013, and similar clauses, as applicable.

Should you have any questions concerning this agreement or if you wish to contact the Company for any reason, please contact in writing: Customer Service Pearson Prentice Hall, 200 Old Tappan Road, Old Tappan NJ 07675.

Minimum System Requirements

Windows	Macintosh
Pentium II 300 MHz processor	Power PC G3 233 MHz or better
Windows 98 or later	Mac OS 9.x or 10.x
64 MB RAM	64 MB RAM
800 x 600 resolution	800 x 600 resolution
8x or faster CD-ROM drive	8x or faster CD-ROM drive
QuickTime 6.0 or later	QuickTime 6.0 or later